Climate Change and Cities
Second Assessment Report of the
Urban Climate Change Research Network

The Urban Climate Change Research Network's *Second Assessment Report on Climate Change in Cities* (ARC3.2) is the second in a series of global, science-based reports to examine climate risk, adaptation, and mitigation efforts in cities. The book explicitly seeks to explore the implications of changing climatic conditions on critical urban physical and social infrastructure sectors and intersectoral concerns. The ARC3.2 Report presents downscaled climate projections and catalogs urban disasters and risks, along with the effects on human health in cities. ARC3.2 gives concrete solutions for cities in regard to mitigation and adaptation; urban planning and urban design; equity and environmental justice; economics, finance, and the private sector; critical urban physical and social sectors such as energy, water, transportation, housing and informal settlements, and solid waste management; and governing carbon and climate in cities. Other key topics include ecosystems and biodiversity, and urban coastal zones. The primary purpose of ARC3.2 is to inform the development and implementation of effective urban climate change policies, leveraging ongoing and planned investments for populations in cities of developing, emerging, and developed countries.

This volume – like its predecessor – will be invaluable for a range of audiences involved with climate change and cities: Mayors, city officials, and policy-makers; urban planners; policy-makers charged with developing climate change mitigation and adaptation programs; and a broad spectrum of researchers and advanced students in the environmental sciences.

Cynthia Rosenzweig is a Senior Research Scientist at the NASA Goddard Institute for Space Studies, where she heads the Climate Impacts Group. She is Co-Chair of the New York City Panel on Climate Change (NPCC), a body of experts convened by the mayor to advise the city on adaptation for its critical infrastructure. She co-led the Metropolitan East Coast Regional Assessment of the U.S. National Assessment of the Potential Consequences of Climate Variability and Change, sponsored by the U.S. Global Change Research Program. She was a Coordinating Lead Author of Working Group II for the Fourth Assessment Report of the Intergovernmental Panel on Climate Change (IPCC). She is Co-Director of the Urban Climate Change Research Network (UCCRN), Co-Editor of the First and Second UCCRN Assessment Reports on Climate Change and Cities (the ARC3 series), and Co-Chair of the Urban Thematic Group for the United Nations UN Sustainable Development Solutions Network (SDSN) and the Campaign for an Urban Sustainability Development Goal (SDG). She serves as Chair of the Board of the New York City Climate Museum. She was named as one of "Nature's 10: Ten People Who Mattered in 2012" by the journal *Nature*, for her work preparing New York for climate extremes and change. A recipient of a Guggenheim Fellowship, she joins impact models with climate models to project future outcomes of both land-based and urban systems under altered climate conditions. She is a Professor at Barnard College and a Senior Research Scientist at The Earth Institute at Columbia University.

William Solecki is a Professor in the Department of Geography, Hunter College, City University of New York (CUNY). He has led or co-led numerous projects on the process of urban environmental change and transformation. As Director of the CUNY Institute for Sustainable Cities, he has worked extensively on connecting cutting-edge urban environmental science to everyday practice and action in cities. He most recently served as Co-Chair of the New York City Panel on Climate Change, as Co-Principal Investigator of the Integrated Assessment for Effective Climate Change Adaptation Strategies in New York State (ClimAID), and as Co-Leader of the Metropolitan East Coast Assessment of the U.S. National Assessment of the Potential Consequences of Climate Variability and Change. He is a Coordinating Lead Author of the IPCC Special Report on the Impacts of 1.5 Degree Warming and was a Lead Author of the IPCC Working Group II Fifth Assessment Report (AR5). He is also a member of International Geographical Union (IGU) Megacity Study Group and a member of the Scientific Steering Committee of the Urbanization and Global Environmental Change core project of the International Human Dimensions Programme (IHDP).

Patricia Romero-Lankao is an "interdisciplinary sociologist" by training. She has been a research scientist at the National Center for Atmospheric Research (NCAR) and is currently leading the "Urban Futures" initiative there. Her research explores the dynamics of urbanization and urban systems that shape urban emissions, vulnerabilities, and risk. She has also done research on why and how particular cities attempt to meet the challenges of reducing emissions while improving their response capacity (resilience) to environmental impacts. She was co-lead author of Working Group II of the Nobel prize–winning IPCC Fourth Assessment Report (AR4) and is convening author of IPCC: AR5, North American chapter. She has been a member of several scientific committees designing a research agenda on the interactions and feedbacks between urban development and the environment, including the carbon cycle, the climate system, and the water cycle (e.g., Global Carbon Project, Urbanization and Global Environmental Change and U.S. Carbon Cycle Science Program).

Shagun Mehrotra is Professor of Sustainable Development at The New School University, New York, and the founding Director of the Sustainable Development Solutions Center. He is a Lead Author of the IPCC Special Report on the Impacts of 1.5 Degree Warming. He serves on UNSDSN's Urban Thematic Group charged by the UN Secretary General as an external advisory group for the post-2015 development agenda. Over the past two decades, his research and advice has widely engaged governments and private sector in North America, Africa, Asia, and Latin America on climate change, infrastructure economics and finance, and poverty reduction in cities, particularly large slums. He has led more than fifty multidisciplinary teams to assess climate risk and craft response in global cities and a dozen teams on infrastructure reforms. Previously, he served on the staff of the World Bank, leading infrastructure reform of public utilities in Africa with a focus on expanding services to the urban poor. He has facilitated global strategic partnerships for the Sustainable Development Goals SDGs and Habitat III's New Urban Agenda. He has published extensively on solutions for global urbanization, including two recent books, the first ARC3 with Cambridge University Press and another on infrastructure economics with Oxford University Press. His work has featured in *Nature* and *Scientific American*, and he is a reviewer of policy-relevant research for *Science*. He has a PhD from Columbia University in Urban Planning and Infrastructure Economics.

Shobhakar Dhakal is an Associate Professor in the Energy Field of Study at the Asian Institute of Technology in Thailand. His areas of expertise are in urbanization, cities and climate change, and energy policies and modeling. He has been a visiting researcher at the National Institute for Environmental Studies, Japan, since 2012. He is a Coordinating Lead Author of Working Group III for the IPCC's Fifth Assessment Report AR5 of the IPCC. He serves as a member of the scientific steering committee of the Global Carbon Project, the premier scientific program under Future Earth. He was a guest research scholar at the International Institute for Applied System Analysis in Austria from 2010 to 2013. He has served as a lead author for the *Global Energy Assessment*, principal scientific reviewer for UNEP's Global Environmental Outlook-5, member of the Consensus Panel on Low-Carbon Cities of the Academy of Sciences of South Africa, member of the Cities Energy Modeling Group of the International Energy Agency, and an international expert to the Taskforce on Urban Development and Energy Efficiency of the China Council for International Cooperation on Environment and Development, among others. He is also one of the editors-in-chief of the journal *Carbon Management*.

Somayya Ali Ibrahim is the Associate Director of the Urban Climate Change Research Network (UCCRN). Based at the Earth Institute at Columbia University and the NASA Goddard Institute for Space Studies (NASA GISS), she develops and manages climate change projects and partnerships with large city groups, universities, development banks, and federal and UN agencies. She manages the UCCRN Secretariat in New York and its global network of projects and partnerships, including more than 800 urban scientists and practitioners and involving the establishment of the UCCRN Regional Hubs in Africa, Asia, Australia-Oceania, Europe, Latin America, and North America. She also manages the development and publication of the UCCRN Assessment Report on Climate Change and Cities (ARC3) series, an ongoing set of major global, interdisciplinary, science-based assessments on climate change and urban areas. She was one of ten individuals chosen to represent Columbia University at the UNFCCC 21st Conference of the Parties (COP21) in Paris in December 2015, and works with a leadership team striving to establish a Climate Museum in New York, one of the first of its kind in the world. She holds a B.Sc. (Hons.) in Management Sciences and an MBA from the University of Peshawar, and a Master's degree in Climate and Society from Columbia University.

Praise for the ARC3.2 Report

Anne Hidalgo, Mayor of Paris and Chair of C40

"ARC3.2 provides the critical knowledge base for city actions on climate change around the world."

Eduardo Paes, Former Mayor of Rio de Janeiro and Former Chair of C40

"The remarkable ARC3.2 will make a difference in developing effective and efficient climate change mitigation and adaptation policies in cities."

James Nxumalo, Former Mayor of Durban

"The full ARC3.2 report ... is the gold standard for science-based policymaking as we enter into the post-2015, climate change implementation era."

Joan Clos, Former Executive Secretary of UN-Habitat; Former Mayor of Barcelona

"... a great example of the benefit of interdisciplinary science-policy co-operation. ... ARC3.2 will help to ensure our future cities enable us to live more sustainably and to be more resilient."

Gino Van Begin, Secretary General of ICLEI-Local Governments for Sustainability

"The Climate Change in Cities report zooms in at the city level, providing us with a wealth of local climate data. And what these data tell us is that if we are to overcome the climate change challenge, we need more than ever the concerted efforts of all levels of government, multilateral institutions, civil society and the business sector."

Mark Watts, Executive Director for C40 Cities Climate Leadership Group

"With the international community now galvanized to put the world on a climate safe pathway, the evidence is stacking up that cities have a key role to play. The second edition of the ARC3 report from the Urban Climate Change Research Network provides a critical knowledge base for global cities as they respond to climate change challenges and seize the economic opportunities of low carbon, climate resilient development. Leading mayors, through network such as the C40, are learning from each other, exchanging ideas and thereby accelerating local action on the ground."

Senator Loren Legarda, Chair, Senate Committees on Foreign Relations, Finance, and Climate Change Global Champion for Resilience, United Nations Office for Disaster Risk Reduction (UNISDR)

"Urban areas will not stop from growing, but growth need not compromise the future. The climate crisis presents the opportunity to promote sustainable growth. Key science knowledge and practical insights are needed to allow our urban areas to meet the adaptation imperative to climate change. The ARC3.2 Report of the Urban Climate Change Research Network (UCCRN) provides vital inputs to this process."

Climate Change and Cities
Second Assessment Report of the Urban Climate Change Research Network

Edited by

Cynthia Rosenzweig

NASA Goddard Institute
for Space Studies
The Earth Institute, Columbia University

William Solecki

Hunter College,
City University of New York

Patricia Romero-Lankao

National Center for Atmospheric Research

Shagun Mehrotra

Milano School of
International Affairs, Management, and
Urban Policy, The New School

Shobhakar Dhakal

Asian Institute of Technology

Somayya Ali Ibrahim

The Earth Institute, Columbia University

NASA Goddard Institute
for Space Studies

URBAN CLIMATE CHANGE
RESEARCH NETWORK

CAMBRIDGE
UNIVERSITY PRESS

CAMBRIDGE
UNIVERSITY PRESS

University Printing House, Cambridge CB2 8BS, United Kingdom

One Liberty Plaza, 20th Floor, New York, NY 10006, USA

477 Williamstown Road, Port Melbourne, VIC 3207, Australia

314-321, 3rd Floor, Plot 3, Splendor Forum, Jasola District Centre, New Delhi - 110025, India

79 Anson Road, #06-04/06, Singapore 079906

Cambridge University Press is part of the University of Cambridge.

It furthers the University's mission by disseminating knowledge in the pursuit of
education, learning and research at the highest international levels of excellence.

www.cambridge.org
Information on this title: www.cambridge.org/9781316603338

Co-Editors: Cynthia Rosenzweig, William Solecki, Patricia Romero-Lankao, Shagun
Mehrotra, Shobhakar Dhakal, Somayya Ali Ibrahim

Project Manager: Somayya Ali Ibrahim

First published 2018

A catalogue record for this publication is available from the British Library

ISBN 978-1-316-60333-8 Paperback

Cover Photo:

*Rio de Janeiro, a city with 6.5 million residents (14.5 million in the Greater Rio de Janeiro area), is a frontrunner in climate change mitigation
and adaptation. Temperatures in Rio de Janeiro are projected to rise by 3.4°C, with sea level rise of 37cm–82cm, by the 2080s. Along with
developing a Climate Adaptation Plan, Rio de Janeiro is committed to reducing greenhouse gas emissions by 20% of 2005 levels by 2020.
(Photo: Somayya Ali Ibrahim)*

Contents

UCCRN ARC3.2 List of Boxes

Foreword – Anne Hidalgo, Mayor of Paris and Chair of C40

Taking action to fight climate change is increasingly a priority for cities around the world, where half the world's population now lives and where two-thirds of the inhabitants of our planet will be concentrated by 2050.

The best science has illuminated major risks facing our planet, and as mayors, we have heeded the call to tackle climate change head-on, both in terms of mitigation (reduction of greenhouse gases) and adaptation (development of resilience to climate stresses). Climate action is and must be incorporated into everyday urban planning and growth in a way that is sustainable and also financially viable. The Urban Climate Change Research Network (UCCRN) was established to meet the information needs of cities responding to climate change, and ARC3.2 provides the critical knowledge base for city actions on climate change around the world.

Because cities are both vulnerable to climate risks and are also sources of innovation for sustainable solutions, it is essential that they collaborate to find bold solutions based on cutting-edge research, such as that found in ARC3.2. The City of Paris was delighted to host the UCCRN European Hub and the launch of the *Second Assessment Report on Climate Change and Cities (ARC3.2)* at the Climate Summit for Local Leaders, hosted by the City of Paris at Paris City Hall and held during the UNFCCC Conference of the Parties 21 (COP21) in Paris at the end of 2015.

ARC3.2 focuses on key urban sectors (energy, transportation, water, and sanitation), as well as on human services in cities (health and housing). Chapters on urban ecology and coastal zones reveal important dimensions of urban climate change action. And in the chapter on equity and environmental justice, the volume highlights the need for cities to consider their most vulnerable citizens.

The City of Paris released its Climate and Energy Action Plan in 2007, updated it in 2012, and adopted a new Adaptation Roadmap in 2015 that presents comprehensive strategies for responding to the city's own climate change challenges.

The City of Paris is committed to the pathway of climate change solutions, sustainability, and transformation and looks forward to a continued partnership with UCCRN and the ARC3 series to achieve these goals.

Anne Hidalgo
Mayor of Paris
Chair of C40 Cities Climate Leadership Group

Foreword – Eduardo Paes, Former Mayor of Rio de Janeiro and Former Chair of C40

The coming years are critical to determining our future in regard to climate change. Scientists, leaders, and decision-makers are joining forces to balance growth with environmental protection and social justice.

The *Second Assessment Report on Climate Change and Cities (ARC3.2)*, developed by the Urban Climate Change Research Network (UCCRN), presents cutting-edge scientific information on climate change mitigation and adaptation in cities. It offers detailed information to support policy-makers in making better, more information-informed decisions about how climate change affects public health, local infrastructures, and the economy.

The past work of the Intergovernmental Panel on Climate Change (IPCC) has been decisive in changing the mindsets of world leaders, and the UCCRN *First Assessment Report on Climate Change and Cities (ARC3.1)* provoked similar transformations at the local level. The *Second Assessment Report* now highlights how poverty and biodiversity are intimately connected to the challenges of urban climate change.

The Report stresses the importance of addressing poverty and climate change together. There is no opposition between social development and environmental protection – we must do both. Climate change in cities affects the poorest and most vulnerable members of our societies. Transforming cities into successful low-carbon communities will only be possible if these changes are made in combination with social and environmental justice.

The ARC3.2 also emphasizes the importance of environmental preservation as a means of fostering urban resilience. Rio de Janeiro – as a coastal and tropical city – experiences heavy summer rains. Their impacts will likely grow more frequent and intense with climate change. For this reason, the protection of the city's biodiversity is vital, to avoid landslides and other adverse consequences of the increased rainfall.

Investing in quality green spaces is a means of strengthening resilience, while improving residents' quality of life.

Decision-makers and local leaders around the world need the support of the scientific community and the knowledge it provides; their work is complementary. That is why we have endorsed a new partnership with the Urban Climate Change Research Network to establish a Latin American Hub in Rio de Janeiro. Recently, we have been working with UCCRN on developing the Rio de Janeiro Resilience Plan. Our collaboration has resulted in studies on heat islands, the proliferation of dengue fever (and other vector-borne diseases), and other local development challenges, contributing to risk reduction for the city's residents and infrastructure.

Science works to understand the multiple dimensions of climate change hazards, imparting knowledge that is often lacking in cities. Developing cities represent the fastest-growing urban places in the world. The UCCRN Latin American Hub can identify and promote the resilience potential of cities in the region and reinforce our mitigation and adaptation policies. Lessons learned in Rio will benefit other cities in the region, just as studies elsewhere will help local policy-makers deal with their local challenges.

On the road from COP21, cities are central to supporting the ambitious commitments and to implementing the Paris agreements. Mayors from around the world have shown impressive leadership, but they need support to do more. The remarkable ARC3.2 will make a difference in developing effective and efficient climate change mitigation and adaptation policies in cities.

Eduardo Paes
Former Mayor of Rio de Janeiro
Former Chair of C40 Cities Climate Leadership Group

Foreword – James Nxumalo, Former Mayor of Durban

The future of our planet was shaped in 2015. In December, negotiating parties representing nations from around the globe met in Paris for the United Nations Framework Convention on Climate Change's 21st Conference of the Parties (COP21), and agreed on planet-saving measures to combat climate change. With a rapidly urbanizing globe, the role of cities and local governments is pivotal. Cities must be supported effectively because the challenge of climate change will be won and lost in urban areas. Cities offer twin transformative solutions, with the greatest opportunities for reducing greenhouse gas emissions through mitigation activities and localized climate risk reduction through urban adaptation. Given the uncertainty around climate change impacts at the local level, it is critical that the adaptive management decisions are informed by cutting-edge science and independent research.

The climate change challenge cuts across a broad range of disciplines. The development of research partnerships that connect scientists from different disciplines to work collaboratively to inform city-level management decisions is a high priority. The Urban Climate Change Research Network (UCCRN) is an excellent example of how urban-focused climate change research can bridge the divide between researchers and policy-makers. The *First Assessment Report on Climate Change and Cities* (ARC3.1) was published in 2011 and did exactly that. It provided a multidisciplinary, global assessment of climate risks, adaptation, mitigation, and policy mechanisms that is relevant to cities and based on sound scientific principles.

The Summary for City Leaders of the *Second UCCRN Assessment Report* (ARC3.2) was launched during COP21 in Paris in 2015 and will serve as a "call to action." The full ARC3.2 report, the *Second UCCRN Assessment Report on Climate Change and Cities*, is the gold standard for science-based policy-making as we enter into the post-2015, climate change implementation era. Developing partnerships to share knowledge products will be important so that all stakeholders – especially those in low-income countries – can benefit. In this respect, the development of a knowledge network, formed around UCCRN Hubs, offers a transformative solution. We are proud that, as of the launch in 2016, the city of Durban in South Africa will function as one of these UCCRN Hubs.

Furthermore, through an agreement to collaborate with UCCRN as its key knowledge partner, the Durban Adaptation Charter will enable scaling up the knowledge-to-policy process with its signatory base of more than a thousand cities. This empowering linkage from co-generated research to implementation partnership will support city leaders from low-income countries as they operationalize climate change mitigation and adaptation policies and action plans at transformative scales.

James Nxumalo
Former Mayor of Durban, eThekwini Municipality

Foreword – Joan Clos, Former Executive Secretary of UN-Habitat and Former Mayor of Barcelona

Cities and local governments are increasingly recognized as key actors in addressing climate challenges. They are strong sources of leadership that require enabling frameworks and a combination of global, national, and local measures in order to achieve the transformational change that is needed. It is important to provide decision-makers with the latest data at different levels of granularity, as well as with global platforms for exchange of information.

Since the Urban Climate Change Research Network (UCCRN) published its innovative *First Assessment Report on Climate Change and Cities* (ARC3.1) in 2011, we have seen a significant increase of attention to this issue, as well as considerable growth of the body of knowledge. This is exemplified by the recent publications of the Fifth Assessment Report of the Intergovernmental Panel on Climate Change (IPCC), which has dedicated two full chapters to the urban issue – one on mitigation and one on impacts, adaptation, and vulnerabilities. The report of the New Climate Economy has elaborated on the cost of inaction and the co-benefits of compact urban growth, connected infrastructure, and coordinated governance. This new edition, the *Second UCCRN Assessment Report on Climate Change and Cities* (ARC3.2), offers not only updated findings, but is expanded in scope and coverage. It includes an extensive database of case studies, which will allow for a continuous collection of key city data online and will enable users to compare cases and lessons learned across factors such as geography, sector, income levels, and size.

The nexus between cities and climate change is crucial for addressing the sustainable development challenges of the 21st century. More than half of the global population is already living in urban areas, and it is estimated that, by 2050, this figure will grow to more than two-thirds. Production and consumption is concentrated in urban areas, generating around 80% of gross domestic product (GDP) and more than 60% of all carbon dioxide, in addition to significant amounts of other greenhouse gas (GHG) emissions. Urban areas host most of the vulnerable populations as well as vital economic and social infrastructure. Hundreds of millions of people in urban areas across the world will be affected by rising sea levels, increased precipitation extremes, landslides, inland floods, more frequent and intense cyclones and storms, and periods of more extreme heat.

The year 2015 constituted a pivotal point for the global policy agenda on sustainable development and climate change with key conferences, such as the UN Summit for the Adoption of the Post-2015 Development Agenda, the Financing for Development Conference, the UN 21st World Conference on Disaster Risk Reduction, and the Conference of the Parties (COP21) of the United Nations Framework Convention on Climate Change (UNFCCC) in Paris. The urban issue is increasingly recognized as a key component in these global processes. Furthermore, the United Nations Conference on Housing and Sustainable Urban Development (Habitat III) took place in Quito in October 2016. Habitat III served to reinvigorate the global commitment to sustainable urbanization and to focus on the implementation of a "New Urban Agenda."

UN-Habitat is working with partners at various levels to integrate climate mitigation and adaptation concerns into policy and infrastructure planning processes, taking into account broader sustainability considerations and economic, environmental, and social co-benefits. It is doing so, for example, in the context of the Cities and Climate Change Initiative, which targets medium-sized cities in developing and least-developed countries, as well as through the Urban-LEDS project, which promotes Low-Emission Urban Development in Emerging Economies. Climate change is also among the cross-cutting issues mainstreamed throughout UN-Habitat as per its strategic plan. Complementing its intergovernmental activities, UN-Habitat is partnering with a broad range of stakeholders to advance the implementation of a number of initiatives launched at the Secretary-General's Climate Summit in September 2014, such as the Global Covenant of Mayors, the Cities Climate Financing Leadership Alliance, the Urban Electric Mobility Initiative, and the Resilient Cities Accelerator Initiative.

I am confident that this report will contribute significantly to the body of knowledge in this area and help guide decision-makers at the various levels in their quest for sustainable urbanization. It is a great example of the benefit of interdisciplinary science policy cooperation. Cities provide tremendous opportunities to mitigate climate change and increase resilience while also improving well-being and economic output. If well planned, equipped with the necessary capacity, and managed through the appropriate governance structures, cities can be places of innovation and efficiency. ARC3.2 will help to ensure that our future cities enable us to live more sustainably and be more resilient.

Joan Clos
Former Executive Secretary of UN-Habitat
Former Mayor of Barcelona

Foreword – Christiana Figueres, Former Executive Secretary, United Nations Framework Convention on Climate Change and Vice Chair of the Global Covenant of Mayors

With the majority of the world's population living in urban areas, action by cities holds great potential to curb emissions and build resilience to climate impacts. Cities are a powerful force in meeting the global challenge of climate change. This fact was underscored at the 2015 UN Climate Change Conference in Paris (COP21), where commitments to act by cities registered on the United Nations Framework Convention on Climate Change (UNFCCC).

The Paris Agreement is a transformative vision of growth shared by the 195 countries that adopted it. It is built on a foundation of national climate change action plans. For these plans to succeed, and for us to meet the climate change challenge, cities must align policy to national goals and the long-term goal of the Paris Agreement. I welcome the *Second UCCRN Assessment Report on Climate Change and Cities (ARC3.2)* because it will help cities choose a policy suite that accelerates their local transition to low-emission and highly resilient growth. With this report as a resource, I am confident that cities can be engaged in meeting our global goals while enhancing liveability in their communities.

Christiana Figueres
Former Executive Secretary of UNFCCC
Vice Chair, Global Covenant of Mayors for Climate & Energy

Preface

This volume is the Urban Climate Change Research Network's *Second Assessment Report on Climate Change and Cities* (ARC3.2). It contains the Summary for City Leaders, an introductory section, and four parts of the report, as well as the Case Study Docking Station Annex.

This report would not be possible without the great support of Aalborg University, the African Development Bank (AfDB), The Earth Institute at Columbia University, the Helmholtz Centre for Environmental Research (UFZ-Leipzig), the Inter-American Development Bank (IDB), the International Development Research Centre (IDRC), the Japan International Cooperation Agency (JICA), the NASA Goddard Institute for Space Studies (NASA GISS), Siemens, the Urbanization and Global Environmental Change Project (UGEC), the United Nations Environment Programme (UNEP), and the United Nations Human Settlement Programme (UN-Habitat).

We especially thank Balgis Osman-Elasha and Aymen Ali at AfDB; David Wilk at IDB; Mark Redwood, formerly at IDRC; Tomonori Sudo at JICA; Keith Alverson and Stuart Crane at UNEP; and Joan Clos, Rafael Tuts, Robert Kehew, Marcus Mayr, and Andrew Rudd at UN-Habitat. They are all exemplary international public servants committed to the development of effective ways for cities to confront climate change challenges and to leading in the implementation of solutions.

We are grateful for the substantive support provided to the Case Study Docking Station by Martin Lehmann at Aalborg University and the Joint European Master in Environmental Studies – Cities and Sustainability (JEMES CiSu), and to the Economics, Finance, and the Private Sector chapter by Reimund Schwarze at UFZ-Leipzig.

We thank Stefan Denig and Michael Stevns at Siemens for generously hosting the ARC3.2 Midterm Authors Workshop at The Crystal in London in September 2014.

We appreciate the sound advice provided by the members of the UCCRN ARC3.2 Steering Group: Keith Alverson, Martha Barata, Anthony G. Bigio, Richenda Connell, Richard Dawson, Stefan Denig, Shobhakar Dhakal, David Griggs, Alice Grimm, Saleemul Huq, Martin Lehmann, Yu Lizhong, Helena Molin Valdés, Claudia E. Natenzon, Catherine Neilson, Ademola Omojola, Rajendra Pachauri, Mark Redwood, Debra Roberts, Joyashree Roy, Patricia Romero-Lankao, Roberto Sanchez-Rodriguez, Joel Scheraga, Joel Towers, Rafael Tuts, David Wilk, and Carolina Zambrano-Barragan. We also recognize the contributions of JoAnn Carmin, a UCCRN Steering Group member whom we sadly lost during the development of this report.

We thank the leaders of the UNFCCC and the IPCC who have supported the need for ARC3 – in particular Christiana Figueres, former Executive Secretary of the UNFCCC, and Debra Roberts, Co-Chair of IPCC Working Group II.

We thank the city networks and our colleagues in each of them who strive to enable cities to fulfill their leadership potential for climate change mitigation and adaptation. At ICLEI, we commend the Secretary General, Gino Van Begin, and thank our close colleague Yunus Arikan for his tireless efforts; at C40, we thank Seth Schultz, Katie Vines, and Mandy Ikert; at Cities Alliance, we thank William Cobbett, Omar Siddique, and Julie Greenwalt; and at the Medellín Collaboration on Urban Resilience, we thank Tricia Holly Davis and Laura Kavanaugh. At the Global Environment Facility (GEF), we thank Naoki Ishii and Saliha Dobardzic; at GIZ, we thank Carmen Vogt; at the Nairobi Work Programme, we thank Xianfu Lu and Laureline Krichewsky-Simon. At the U.S. Environmental Protection Agency, we thank Anthony Socci for his long-term support of the UCCRN. At EDF France, we thank Claude Nahon, Carole Ory, and Marianne Najafi; at UNISDR, we thank Jerry Velasquez, Ebru Gencer, Abhilash Panda, and Julian Templeton.

We gratefully acknowledge the discussions and feedback during sessions with Mayors, their advisors, leaders of major institutions, urban policy-makers, and scholars during our international scoping events and thank the respondents of our Information Needs Assessment Survey in 2013–2014.

We extend special gratitude to the urban leaders who represent a diverse group of cities and who have commended UCCRN and ARC3.

This Assessment Report is the product of the work of the over 350 dedicated members of the UCCRN ARC3.2 writing team, representing more than 100 cities around the world. We express our sincere thanks to each of them for their sustained and sustaining contributions, and to their institutions for supporting their participation.

We especially thank and profoundly appreciate the work of Somayya Ali Ibrahim for her tremendous efforts as the UCCRN and ARC3 series Program Manager; without her, ARC3.2 could not have been completed in such a comprehensive manner.

We thank the founding Directors and partners of the UCCRN Regional Hubs: Chantal Pacteau and Luc Abbadie for the European Hub in Paris; Martha Barata and Emilio La Rovere for the Latin American Hub in Rio de Janeiro; Sean O'Donaghue and Mathieu Rouget for the African Hub in Durban; Kate Auty, Ken Doust, and Barbara Norman for the Australian-Oceania Hub in Melbourne, Sydney, and Canberra; Franco Montalto for the North American Hub in Philadelphia; Min Liu, Xiaotu Lei, and Ruishan Chen for the East Asian Hub in Shanghai; Patricia Iglecias and Oswaldo Lucon for the Center for Multilevel Governance in São Paulo; and Martin Lehmann for the Nordic Node in Aalborg. We also thank Emma Porio and Antonia Loyzaga for their efforts to establish a UCCRN Southeast Asian Hub in Manila and Muhammad Shah Alam for working to establish a UCCRN South Asian Hub in Dhaka.

We thank the city partners of the UCCRN Hubs and the people who make those partnerships happen. In particular, we thank Anne Girault, Aurelien Lechevallier, Nicolas de Labrusse, and Yann Francoise from Paris; Rodrigo Rosa, Bruno Neele, Pedro Junqueira, Laudemar Aguiar, Luciana Nery, and Camila Pontual from Rio de Janeiro; and Debra Roberts and Sean O'Donoghue from Durban.

We give special thanks to the many students at Columbia University, Hunter College of the City University of New York, and Milano School for International Affairs at The New School for their keen interest in the field of urban climate change, which has helped to move forward the ideas of this volume.

We acknowledge the exceptional commitment of the ARC3.2 research assistants and project interns: Ioana Blaj, Antonio Bontempi, Wim Debucquoy, Julia Eiferman, Jonah Garnick, Anna Gusman, Megan Helseth, Annel Hernandez, Jonathan Hilburg, Andrea Irazoque, Christina Langone, Ipsita Kumar, Carissa Lim, Brandon McNulty, Jovana Milić, Grant Pace, Victoria Ruiz Rincón, Samuel Schlecht, Stephen Solecki, Marta De Los Ríos White, and Megi Zhamo. We also thank Chris Barrett, Nicholas Belenko, Erin Friedman, Joseph Gilbride, Benjamin Marconi, Madeline McKenna, and Paul Racco. At the Columbia University Center for Climate Systems Research and the NASA Goddard Institute for Space Studies, we thank Daniel Bader, Manishka de Mel, Vivien Green, Shari Lifson, Danielle Manley, and Erik Mencos Contreras for their assistance and technical expertise.

We recognize with great esteem the Expert Reviewers of the ARC3.2 chapters, without whom the independent provision of sound science for climate change mitigation and adaptation in cities cannot proceed.

We especially thank Tom Bowman for his expertise and guidance in communicating climate change, for his important role in helping develop the Summary for City Leaders of the ARC3.2 volume, and for leading our Communications and Outreach Team, otherwise known as the "Troublemakers." We especially thank Ronaldo Barata for his long-term efforts in championing the UCCRN and ARC3.2.

It is a great honor that ARC3.2 is published by Cambridge University Press. We especially thank Matt Lloyd, Editorial Director for Science, Technology, and Medicine, Americas, and Mark Fox, Content Manager, Editorial/Production for Academic and Professional Books, for their expert partnership in the publication of this volume. We also thank Sathishkumar Rajendran and Allan Alphonse from Integra, for their support throughout the publication process.

Finally, we are deeply grateful to The Earth Institute at Columbia University, which hosts the UCCRN Secretariat; Jeffrey Sachs, its former Director; and Steven A. Cohen, in his role as Executive Director, who have enthusiastically supported UCCRN from its establishment.

Cynthia Rosenzweig, William Solecki, Patricia Romero-Lankao, Shagun Mehrotra, Shobhakar Dhakal, and Somayya Ali Ibrahim

Co-Editors
Second UCCRN Assessment Report on Climate Change and Cities (ARC3.2)

Climate Change and Cities

*Second Assessment Report of the Urban
Climate Change Research Network*

Summary for City Leaders

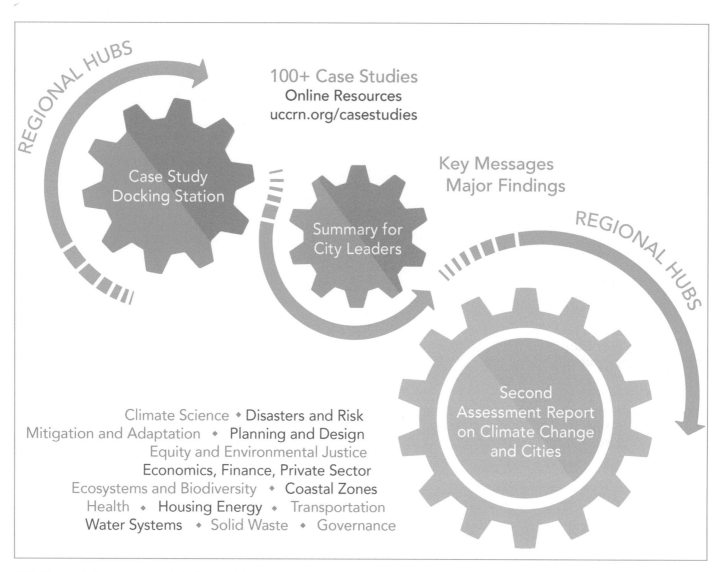

SCL Figure 1 *Components of the Second Assessment Report on Climate Change and Cities (ARC3.2) and their interactions.*

ARC3.2 Summary for City Leaders Authors

Cynthia Rosenzweig, William Solecki, Patricia Romero-Lankao, Shagun Mehrotra, Shobhakar Dhakal, Tom Bowman, and Somayya Ali Ibrahim

The *ARC3.2 Summary for City Leaders* was released at the UNFCCC COP21 Climate Summit for Local Leaders in Paris, December 2015

The *ARC3.2 Summary for City Leaders* should be cited as: Rosenzweig, C., W. Solecki, P. Romero-Lankao, S. Mehrotra, S. Dhakal, T. Bowman, and S. Ali Ibrahim. (2015). ARC3.2 Summary for City Leaders. In Rosenzweig, C., W. Solecki, P. Romero-Lankao, S. Mehrotra, S. Dhakal, and S. Ali Ibrahim (eds.), *Climate Change and Cities: Second Assessment Report of the Urban Climate Change Research Network*. Cambridge University Press, New York.

ARC3.2
SUMMARY FOR CITY LEADERS

This is the Summary for City Leaders of the Urban Climate Change Research Network (UCCRN) *Second Assessment Report on Climate Change and Cities* (ARC3.2) (see SCL Figure 1). UCCRN is dedicated to providing the information that city leaders – from government, the private sector, non-governmental organizations, and the community – need in order to assess current and future risks, make choices that enhance resilience to climate change and climate extremes, and take actions to reduce greenhouse gas emissions.

ARC3.2 presents a broad synthesis of the latest scientific research on climate change and cities.[1] Mitigation and adaptation climate actions of 100 cities are documented throughout the 16 chapters, as well as online through the ARC3.2 Case Study Docking Station (www.uccrn.org/casestudies). Pathways to Urban Transformation, Major Findings, and Key Messages are highlighted here in the ARC3.2 Summary for City Leaders. These sections lay out what cities need to do to achieve their potential as leaders of climate change solutions. UCCRN Regional Hubs in Europe, Latin America, Africa, Australia-Oceania, and Asia will share ARC3.2 findings with local city leaders and researchers.

The ARC3.2 Summary for City Leaders synthesizes Major Findings and Key Messages on urban climate science, disasters and risks, urban planning and urban design, mitigation and adaptation, equity and environmental justice, economics and finance, the private sector, urban ecosystems, urban coastal zones, public health, housing and informal settlements, energy, water, transportation, solid waste, and governance. This important information is based on climate trends and future projections for 100 cities around the world.

Climate Change and Cities

The international climate science research community has concluded that human activities are changing the Earth's climate in ways that increase risk to cities. This conclusion is based on many different types of evidence, including the Earth's climate history, observations of changes in the recent historical climate record, emerging new patterns of climate extremes, and global climate models. Cities and their citizens already have begun to experience the effects of climate change. Understanding and anticipating these changes will help cities prepare for a more sustainable future. This means making cities more resilient to climate-related disasters and managing long-term climate risks in ways that protect people and encourage prosperity. It also means improving cities' abilities to reduce greenhouse gas (GHG) emissions.

While projections for future climate change are most often defined globally, it is becoming increasingly important to assess how the changing climate will specifically impact cities. The risks are not the same everywhere. For example, sea level rise will affect the massive zones of urbanization clustered along the world's tidal coastlines, and most significantly those cities in places where the land is already subsiding. In response to the wide range of risks facing cities and the role that cities play as home to more than half of the world's population, urban leaders are joining forces with multiple groups, including city networks and climate scientists. They are assessing conditions within their cities in order to take science-based actions that increase resilience and reduce GHG emissions, thus limiting the rate of climate change and the magnitude of its impacts.

In September 2015, the United Nations endorsed the new Sustainable Development Goal 11, which is to "Make cities and human settlements inclusive, safe, resilient and sustainable." This new sustainability goal cannot be met without explicitly recognizing climate change as a key component. Likewise, effective responses to climate change cannot proceed without understanding the larger context of sustainability. As ARC3.2 demonstrates, actions taken to reduce GHG emissions and increase resilience can also enhance the quality of life and social equity.

---

1 Cities are defined here in the broad sense to be urban areas, including metropolitan and suburban regions.

Pathways to Urban Transformation

Hyderabad, India   Cairo, Egypt   Paris, France   Phnom Penh, Cambodia   New York, USA   Rio de Janeiro, Brazil

As is now widely recognized, cities can be the main implementers of climate adaptation, and mitigation, which is now understood as encompassing low emissions development and resilience. However, the critical question that ARC3.2 addresses is under what circumstances this advantage can be realized. Cities may not be able to address the challenges and fulfill their climate change leadership potential without transformation.

ARC3.2 synthesizes a large body of studies and city experiences and finds that transformation is essential in order for cities to excel in their role as climate change leaders. As cities mitigate the causes of climate change and adapt to new climate conditions, profound changes will be required in urban energy, transportation, water use, land use, ecosystems, growth patterns, consumption, and lifestyles. New systems for urban sustainability will need to emerge that encompass more cooperative and integrated urban–rural, peri-urban, and metropolitan regional linkages.

Five pathways to urban transformation emerge throughout ARC3.2 (see SCL Figure 17). These pathways provide a foundational framework for the successful development and implementation of climate action. Cities that are making progress in transformative climate change actions are following many or all of these pathways. The pathways can guide the way for the hundreds of cities – large and small; low-, middle-, and high-income – throughout the world to play a significant role in climate change action. Cities that do not follow these pathways may have greater difficulty realizing their potential as centers for climate change solutions. The pathways are:

Pathway 1: Actions that reduce greenhouse gas (GHG) emissions while increasing resilience are a win-win. Integrating mitigation and adaptation deserves the highest priority in urban planning, urban design, and urban architecture. A portfolio of approaches is available, including engineering solutions, ecosystem-based adaptation, municipal policies, and social programs. Taking the local context of each city into account is necessary in order to choose actions that result in the greatest benefits.

Pathway 2: Disaster risk reduction and climate change adaptation are the cornerstones of resilient cities. Integrating

these activities into urban development policies requires a new, systems-oriented, multi-timescale approach to risk assessments and planning that accounts for emerging conditions within specific, more vulnerable communities and sectors, as well as across entire metropolitan areas.

Pathway 3: Risk assessments and climate action plans co-generated with the full range of stakeholders and scientists are most effective. Processes that are inclusive, transparent, participatory, multisectoral, multijurisdictional, and interdisciplinary are the most robust because they enhance relevance, flexibility, and legitimacy.

Pathway 4: Needs of the most disadvantaged and vulnerable citizens should be addressed in climate change planning and action. The urban poor, the elderly, women, minorities, recent immigrants, and otherwise marginal populations most often face the greatest risks due to climate change. Fostering greater equity and justice within climate action increases a city's capacity to respond to climate change and improves human well-being, social capital, and related opportunities for sustainable social and economic development.

Pathway 5: Advancing city creditworthiness, developing robust city institutions, and participating in city networks enable climate action. Access to both municipal and outside financial resources is necessary in order to fund climate change solutions. Sound urban climate governance requires longer planning horizons, effective implementation mechanisms, and coordination. Connecting with national and international capacity-building networks helps to advance the strength and success of city-level climate planning and implementation.

A final word on urgency: Cities need to start immediately to develop and implement climate action. The world is entering into the greatest period of urbanization in human history, as well as a period of rapidly changing climate. Getting started now will help avoid locking in counterproductive, long-lived investments and infrastructure systems and ensure cities' potential for the transformation necessary to lead on climate change.

Climate Observations and Projections for 100 ARC3.2 Cities

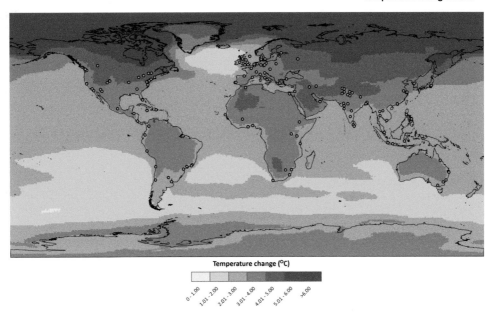

Temperature change 2080s

Temperature change (°C)

0 - 1.00 | 1.01 - 2.00 | 2.01 - 3.00 | 3.01 - 4.00 | 4.01 - 5.00 | 5.01 - 6.00 | >6.00

○ UCCRN Cities

SCL Figure 2 *Projected temperature change in the 2080s. Temperature change projection is mean of 35 global climate models (GCMs) and two representative concentration pathways (RCP4.5 and RCP8.5). Colors represent the mean change in mean annual temperature (2070–2099 average relative to 1971–2000 average). Dots represent ARC3.2 cities. ARC3.2 Cities include Case Study Docking Station cities, UCCRN Regional Hub cities, UCCRN project cities, and cities of ARC3.2 Chapter Authors.*

- Temperatures are already rising in cities around the world due to both climate change and the urban heat island effect. Mean annual temperatures in 39 ARC3.2 cities have increased at a rate of 0.12–0.45°C per decade over the 1961–2010 time period.[1]

- Mean annual temperatures in 153 ARC3.2 cities around the world are projected to increase by 0.7–1.6°C by the 2020s, 1.4–3.1°C by the 2050s, and 1.7–5.0°C by the 2080s (see SCL Figure 2).[2]

- Mean annual precipitation in 153 ARC3.2 cities around the world is projected to change by –7 to +10% by the 2020s, –9 to +14% by the 2050s, and –11 to +20% by the 2080s.

- Sea level in 71 ARC3.2 coastal cities is projected to rise 4–18 centimeters by the 2020s, 14–56 centimeters by the 2050s, and 22–118 centimeters by the 2080s.[3]

---

1 At the 99% significance level. Data are from the NASA GISS GISTEMP dataset.

2 Temperature and precipitation projections are based on 35 global climate models and two representative concentration pathways (RCP4.5 and RCP8.5). Timeslices are 30-year periods centered around the given decade (e.g., the 2050s is the period from 2040 to 2069). Projections are relative to the 1971–2000 base period. For each of the 153 cities, the low estimate (10th percentile) and high estimate (90th percentile) was calculated. The range of values presented is the average across all 153 cities.

3 Sea level rise projections are based on a four-component approach that includes both global and local factors. The model-based components are from 24 global climate models and 2 representative concentration pathways (RCP4.5 and RCP8.5). Timeslices are 10-year periods centered around the given decade (e.g., the 2080s is the period from 2080 to 2089). Projections are relative to the 2000–2004 base period. For each of the 71 cities, the low estimate (10th percentile) and high estimate (90th percentile) was calculated. The range of values presented is the average across all 71 cities.

4 Like all future projections, UCCRN climate projections have uncertainty embedded within them. Sources of uncertainty include data and modeling constraints, the random nature of some parts of the climate system, and limited understanding of some physical processes. In the ARC3.2 Report, the levels of uncertainty are characterized using state-of-the-art climate models, multiple scenarios of future greenhouse gas concentrations, and recent peer-reviewed literature. The projections are not true probabilities, and scenario planning methods should be used to manage the risks inherent in future climate.

What Cities Can Expect

Smog over Jakarta, Indonesia, November 2011. Photo: Somayya Ali Ibrahim

People and communities everywhere are reporting weather events and patterns that seem unfamiliar. Such changes will continue to unfold over the coming decades and – depending on which choices people make – possibly for centuries. But the various changes will not occur at the same rates in all cities of the world, nor will they all occur gradually or at consistent rates of change.

Climate scientists have concluded that although some of these changes will take place over many decades, even centuries, there is also a risk of crossing thresholds in the climate system that cause some rapid, irreversible changes to occur. One example would be melting of the Greenland and West Antarctic ice sheet, which would lead to very high and potentially rapid rates of sea level rise.

Major Findings

- Urbanization tends to be associated with elevated surface and air temperature, a condition referred to as the *urban heat island* (UHI). Urban centers and cities are often several degrees warmer than surrounding areas due to the presence of heat-absorbing materials, reduced evaporative cooling caused by lack of vegetation, and production of waste heat.

- Some climate extremes will be exacerbated under changing climate conditions. Extreme events in many cities include heat waves, droughts, heavy downpours, and coastal flooding; these are projected to increase in frequency and intensity.

- The warming climate, combined with the UHI effect, will exacerbate air pollution in cities.

- Cities around the world have always been affected by major, naturally occurring variations in climate conditions, includ-

ing the El Niño Southern Oscillation, the North Atlantic Oscillation, and the Pacific Decadal Oscillation. These oscillations occur over years or decades. How climate change will influence these recurring patterns in the future is not fully understood.

Key Messages

Human-caused climate change presents significant risks to cities beyond the familiar risks caused by natural variations in climate and seasonal weather patterns. Both types of risk require sustained attention from city governments in order to improve urban resilience. One of the foundations for effective adaptation planning is to co-develop plans with stakeholders and scientists who can provide urban-scale information about climate risks, both current risks and projections of future changes in extreme events.

Weather and climate forecasts of daily, weekly, and seasonal patterns and extreme events are already widely used on international, national, and regional scales. These forecasts demonstrate the value of climate science information that is communicated clearly and in a timely way. Climate change projections perform the same functions on longer timescales. These efforts now need to be carried out on the city scale.

Within cities, various neighborhoods experience different microclimates. Therefore, urban monitoring networks are needed to address the unique challenges facing various microclimates and the range of impacts of extreme climate effects at neighborhood scales. The observations collected through such urban monitoring networks can be used as a key component of a citywide climate indicators and monitoring system that enables decision-makers to understand the variety of climate risks across the city landscape.

Managing Disasters in a Changing Climate

SCL Figure 3 *Damaged homes in New York as a result of Hurricane Sandy, November 2012. Photo: Somayya Ali Ibrahim*

Globally, the impacts of climate-related disasters are increasing. These may be exacerbated in cities due to interactions of climate change with urban infrastructure systems, growing urban populations, and economic activities (see SCL Figure 3). Because the majority of the world's population is currently living in cities – and this share is projected to increase in the coming decades – cities need to focus more on climate-related disasters such as heat waves, floods, and droughts.

In a changing climate, a new decision-making framework is needed to fully manage emerging and increasing risks. This involves a paradigm shift away from impact assessments that focus on single climate hazards based on past events. The new paradigm requires integrated, system-based risk assessments that incorporate current and future hazards throughout entire metropolitan regions.

Major Findings

- The number and severity of weather and climate-related disasters is projected to increase in the next decades; because most of the world's population lives in urban areas, cities require specific attention to risk reduction and resilience building.

- The vulnerability of cities to climate-related disasters is shaped by the cultural, demographic, and economic characteristics of residents, local governments' institutional capacity, the built environment, the provision of ecosystem services, and human-induced stresses. The latter include resource exploitation and environmental degradation such as removal of natural storm buffers, pollution, overuse of water, and the UHI effect.

- Integrating climate change adaptation with disaster risk reduction involves overcoming a number of barriers: among others,

adding climate resilience to a city's development vision; understanding of the hazards, vulnerabilities, and attendant risks; closing gaps in coordination between various administrative and sectoral levels of management; and development of implementation and compliance strategies and financial capacity.

- Strategies for improving resilience and managing risks in cities include the integration of disaster risk reduction with climate change adaptation, urban and land-use planning and innovative urban design, financial instruments and public–private partnerships, management and enhancement of ecosystem services, building strong institutions and developing community capabilities, and resilient post-disaster recovery and rebuilding.

Key Messages

Disaster risk reduction and climate change adaptation are the cornerstones of making cities resilient to a changing climate. Integrating these activities with a city's development vision requires a new, systems-oriented approach to risk assessments and planning. Moreover, since past events cannot inform decision-makers about emerging and increasing climate risks, systems-based risk assessments must incorporate knowledge about current conditions and future projections across entire metropolitan regions.

A paradigm shift of this magnitude will require decision-makers and stakeholders to increase the capacity of communities and institutions to coordinate, strategize, and implement risk-reduction plans and disaster responses. This is why promoting multilevel, multisectoral, and multistakeholder integration is so important.

Integrating Mitigation and Adaptation as Win-Win Actions

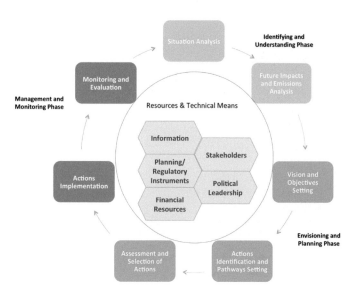

SCL Figure 4 *Main resources and technical means that can be used by cities in their planning cycle for integrating mitigation and adaptation.*

Urban planners and decision-makers need to integrate efforts to alleviate the causes of climate change (mitigation) and adjust to changing climatic conditions (adaptation). Actions that promote both goals provide win-win solutions. In some cases, however, decision-makers have to negotiate tradeoffs and minimize conflicts between competing objectives.

A better understanding of mitigation and adaptation synergies can reveal greater opportunities for urban areas. For example, strategies that reduce the UHI effect, improve air quality, increase resource efficiency in the built environment and energy systems, and enhance carbon storage related to land use and urban forestry are likely to contribute to GHG emissions reduction while improving a city's resilience. The selection of specific adaptation and mitigation measures should be made in the context of other sustainable development goals by taking current resources and the technical means of the city, plus needs of citizens, into account.

Major Findings

- Mitigation and adaptation policies have different goals and opportunities for implementation. However, many drivers of mitigation and adaptation are common, and solutions can be interrelated. Evidence shows that broad-scale, holistic analysis and proactive planning can strengthen synergies, improve cost-effectiveness, avoid conflicts, and help manage tradeoffs.

- Accurate diagnosis of climate risks and the vulnerabilities of urban populations and territory is essential. Likewise, cities need transparent and meaningful GHG emissions inventories and emission reduction pathways in order to prepare mitigation actions.

- Contextual conditions determine a city's challenges, as well as its capacity to integrate and implement adaptation and mitigation strategies. These include the environmental and physical setting, the capacities and organization of institutions and governance, economic and financial conditions, and sociocultural characteristics.

Integrated planning requires holistic, systems-based analysis that takes into account the quantitative and qualitative costs and benefits of integration compared to stand-alone adaptation and mitigation policies (see SCL Figure 4). Analysis should be explicitly framed within local priorities and provide the foundation for evidence-based decision support tools.

Key Messages

Integrating mitigation and adaptation can help avoid locking a city into counterproductive infrastructure and policies. Therefore, city governments should develop and implement climate action plans early in their administrative terms. These plans should be based on scientific evidence and should integrate mitigation and adaptation across multiple sectors and levels of governance. Plans should clarify short-, medium-, and long-term goals; implementation opportunities; budgets; and concrete measures for assessing progress.

Integrated city climate action plans should include a variety of mitigation actions (those involving energy, transport, waste management, water policies, and more) with adaptation actions (those involving infrastructure, natural resources, health, and consumption policies, among others) in synergistic ways. Because of the comprehensive scope, it is important to clarify the roles and responsibilities of key actors in planning and implementation. Interactions among the actors must be coordinated during each phase of the process.

Once priorities and goals have been identified, municipal governments should connect with federal legislation, national programs, and, in the case of low-income cities, international donors in order to match actions and foster helpful alliances and financial support.

Embedding Climate Change in Urban Planning and Urban Design

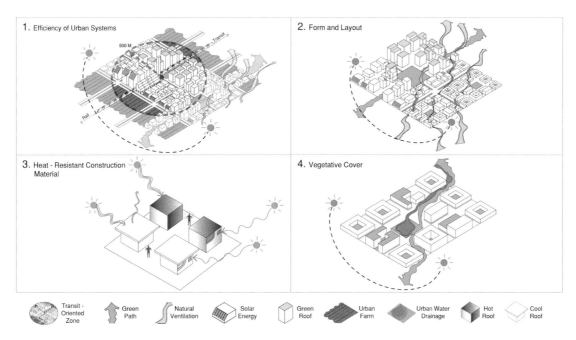

SCL Figure 5 *Main strategies used by urban planners and designers to facilitate integrated mitigation and adaptation in cities: (a) reducing waste heat and greenhouse gas emissions through energy efficiency, transit access, and walkability; (b) modifying the form and layout of buildings and urban districts; (c) use of heat-resistant construction materials and reflective surface coatings; and (d) increasing vegetative cover. Source: Jeffrey Raven, New York Institute of Technology, 2016*

Urban planning and urban design have a critical role to play in the global response to climate change. Actions that simultaneously reduce GHG emissions and build resilience to climate risks should be prioritized at all urban scales: metropolitan region, city, district/neighborhood, block, and building. This needs to be done in ways that are responsive to and appropriate for local conditions.

- Selecting construction materials and reflective coatings can improve building performance by managing heat exchange at the surface.

- Increasing the vegetative cover in a city can simultaneously lower outdoor temperatures – building cooling demand, runoff, and pollution – while sequestering carbon.

Major Findings

Urban planners and designers have a portfolio of climate change strategies that guide decisions on urban form and function (see SCL Figure 5).

- Urban waste heat and GHG emissions from infrastructure – including buildings, transportation, and industry – can be reduced through improvements in the efficiency of urban systems.

- Modifying the form and layout of buildings and urban districts can provide cooling and ventilation that reduce energy use and allow citizens to cope with higher temperatures and more intense runoff.

Key Messages

Climate change mitigation and adaptation strategies should form a core element in urban planning and urban design, taking into account local conditions. Decisions on urban form have long-term (>50 years) consequences and affect the city's capacity to reduce GHG emissions and respond to climate hazards. Investing in mitigation strategies that yield concurrent adaptive benefits should be prioritized.

Urban planning and urban design should incorporate long-range strategies for climate change that reach across physical scales, jurisdictions, and electoral time frames. These activities need to deliver a higher quality of life for urban citizens as the key performance outcome.

Equity, Environmental Justice, and Urban Climate Change

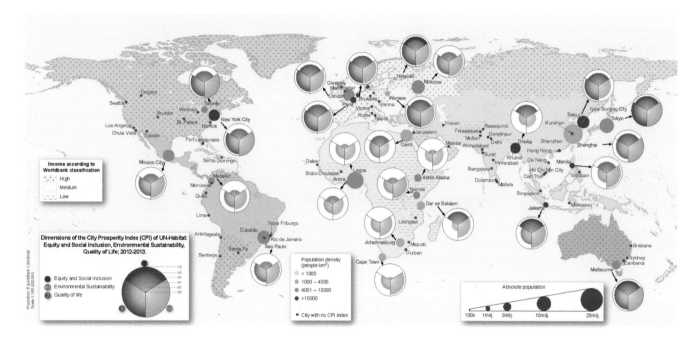

SCL Figure 6 *Case study cities included in this assessment, along with a number of dimensions of the City Prosperity Index (CPI) (where available) and including equity and social inclusion. The CPI is a multidimensional index developed by UN-Habitat (2013b) and comprising six dimensions with subdimensions (and indicators) that are measured for each city. Source: Adapted from Metz, 2000*

Cities are characterized by the large diversity of socioeconomic groups living in close proximity. Diversity is often accompanied by stratification based on class, caste, gender, profession, race, ethnicity, age, and ability. This gives rise to social categories that in turn affect the ability of individuals and various groups to endure climate stresses and minimize climate risks.

Differences between strata often lead to discrimination based on group membership. Poorer people and ethnic and racial minorities tend to live in more hazard-prone, vulnerable, and crowded parts of cities. These circumstances increase their susceptibility to the impacts of climate change and reduce their capacity to adapt and withstand extreme events.

Major Findings

- Differential vulnerability of urban residents to climate change is driven by four factors: (1) differing levels of physical exposure determined by the location of residential/occupational areas; (2) urban development processes that lead to risks, such as failure to provide access to critical infrastructure and services; (3) social characteristics that influence resources for adaptation; and (4) institutional and governance weaknesses

such as ineffective planning and absence of community engagement (see SCL Figure 6).

- For New York, London, Dar es Salaam, and Durban, risk levels increase dramatically for all key risks in the long-term, especially for the 4°C warming. But, at least for now, in the near-term, and mostly for the long-term with a 2°C temperature rise, a high level of adaptation can keep risks down. However, under a 4°C temperature rise, adaptation measures are likely to be ineffective not only in Dar es Salaam but also in cities with currently high levels of adaptation, such as New York and London.

- Climate change amplifies vulnerability and hampers adaptive capacity, especially for the poor, women, the elderly, children, and ethnic minorities. These people often lack power and access to resources, adequate urban services, and functioning infrastructure. Gender inequality is particularly pervasive in cities, contributing to differential consequences of climate changes.

- Frequently occurring climate events – such as droughts in many drought-prone areas – can, over time, undermine everyone's resource base and adaptive capacity, including better-off urban residents. As climate events become more frequent and intense, this can increase the scale and depth of urban poverty overall.

- Mobilizing resources to increase equity and environmental justice under changing climatic conditions requires (1) participation by impacted communities and the involvement of civil society, (2) nontraditional sources of finance, including partnerships with the private sector, and (3) adherence to the principle of transparency in spending, monitoring, and evaluation.

Key Messages

Urban climate policies should include equity and environmental justice as primary long-term goals. They foster human well-being, social capital, and sustainable social and economic development, all of which increase a city's capacity to respond to climate change. Access to land situated in nonvulnerable locations, security of tenure, and access to basic services and risk-reducing infrastructure are particularly important.

Cities need to promote and share a science-informed policy-making process that integrates multiple stakeholder interests and avoids inflexible, top-down solutions. This can be accomplished by participatory processes that incorporate community members' views about resilience objectives and feasibility.

Over time, climate change policies and programs need to be evaluated and adjusted in order to ensure that sustainability, resilience, and equity goals are achieved. Budgetary transparency, equitable resource allocation schemes, monitoring, and periodic evaluation are essential to ensure that funds reach target groups and result in equitable resilience outcomes.

Financing Climate Change Solutions in Cities

Since cities are the locus of large and rapid socioeconomic development around the world, economic factors will continue to shape urban responses to climate change. To exploit response opportunities, promote synergies between actions, and reduce conflicts, socioeconomic development must be integrated with climate change planning and policies.

Public-sector finance can facilitate action, and public resources can be used to generate investment by the private sector (see SCL Figure 7). But private-sector contributions to mitigation and adaptation should extend beyond financial investment. The private sector should also provide process and product innovation, capacity-building, and institutional leadership.

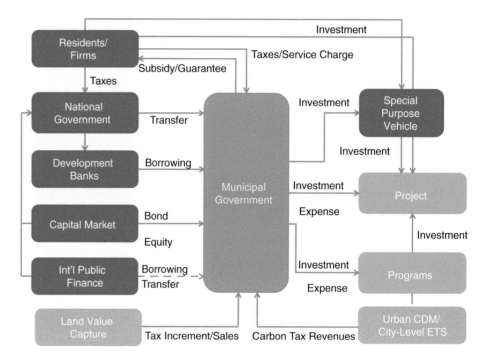

SCL Figure 7 *Opportunities of climate finance for municipalities.*

Major Findings

- Implementing climate change mitigation and adaptation actions in cities can help solve other city-level development challenges, such as major infrastructure deficits. Assessments show that meeting increasing demand will require a more than doubling of annual capital investment in physical infrastructure to more than US$20 trillion by 2025, mostly in emerging economies. Estimates of global economic costs from urban flooding due to climate change are approximately US$1 trillion a year.

- Cities cannot fund climate change responses on their own. Multiple funding sources are needed to deliver the large infrastructure financing that is essential to low-carbon development and climate risk management in cities. Estimates of the annual cost of climate change adaptation range between US$80 and US$100 billion, of which about 80% will be borne in urbanized areas.

- Public–private partnerships are necessary for effective action. Partnerships should be tailored to the local conditions in order to create institutional and market catalysts for participation.

- Regulatory frameworks should be integrated across city, regional, and national levels to provide incentives for the private sector to participate in making cities less carbon-intensive and more climate-resilient. The framework needs to incorporate mandates for local public action along with incentives for private participation and investment in reducing business contributions to emissions.

- Enhancing credit worthiness and building the financial capacity of cities is essential to tapping the full spectrum of resources and raising funds for climate action.

Key Messages

Financial policies must enable local governments to initiate actions that will minimize the costs of climate impacts. For example, the cost of inaction will be very high for cities located along coastlines and inland waterways due to rising sea levels and increasing risks of flooding.

Climate-related policies should also provide cities with local economic development benefits as cities shift to new infrastructure systems associated with low-carbon development.

Networks of cities play a crucial role in accelerating the diffusion of good ideas and best practices to other cities, both domestically and internationally. Therefore, cities that initiate actions that lead to domestic and international implementation of nationwide climate change programs should be rewarded.

Urban Ecology in a Changing Climate

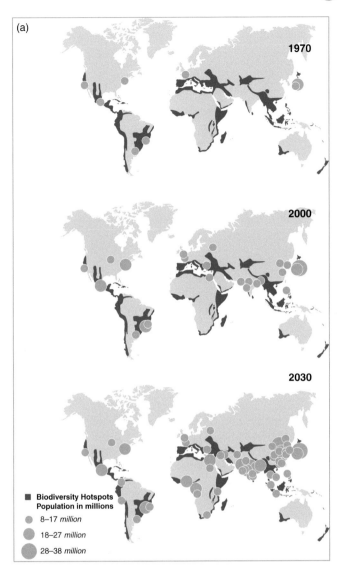

1970

2000

2030

■ Biodiversity Hotspots
Population in millions

- 8–17 *million*
- 18–27 *million*
- 28–38 *million*

SCL Figure 8a *Urban areas (green) with large populations in 1970, 2000, and 2030 (projected), as examples of urban expansion in global biodiversity hotspots (blue).*

SCL Figure 8b *Settlements on hillsides (bairros-cota); invaded areas belong to the Serra do Mar State Park. Such a settlement creates several hazards to its inhabitants, including landslides, floods, road accidents, and freshwater contamination.*

Almost all the impacts of climate change have direct or indirect consequences for urban ecosystems, biodiversity, and the critical ecosystem services they provide for human health and well-being in cities. These impacts are already occurring in urban ecosystems and their constituent living organisms.

Urban ecosystems and biodiversity have an important and expanding role in helping cities adapt to the changing climate. Harnessing urban biodiversity and ecosystems as adaptation and mitigation solutions will help achieve more resilient, sustainable, and livable outcomes.

Conserving, restoring, and expanding urban ecosystems under mounting climatic and nonclimatic urban development pressures will require improved urban and regional planning, policy, governance, and multisectoral cooperation.

Major Findings

- Urban biodiversity and ecosystems are already being affected by climate change.

- Urban ecosystems are rich in biodiversity and provide critical natural capital for climate adaptation and mitigation.

- Climate change and urbanization are likely to increase the vulnerability of biodiversity hotspots, urban species, and critical ecosystem services (see SCL Figure 8a and 8b).

- Investing in urban ecosystems and green infrastructure can provide cost-effective, nature-based solutions for adapting to climate change while also creating opportunities to increase social equity, green economies, and sustainable urban development.

- Investing in the quality and quantity of urban ecosystems and green infrastructure has multiple co-benefits, including improving quality of life, human health, and social well-being.

Key Messages

Cities should follow a long-term systems approach to ecosystem-based climate adaptation. Such an approach explicitly recognizes the role of critical urban and peri-urban ecosystem services and manages them to provide a sustained supply over time horizons of 20, 50, and 100 years. Ecosystem-based planning strengthens the linkages between urban, peri-urban, and rural ecosystems through planning and management at both urban and regional scales.

The economic benefits of urban biodiversity and ecosystem services should be quantified so that they can be integrated into climate-related urban planning and decision-making. These benefits should incorporate both monetary and non-monetary values of biodiversity and ecosystem services, such as improvements to public health and social equity.

Cities on the Coast: Sea Level Rise, Storms, and Flooding

Coastal cities have lived with extreme climate events since the onset of urbanization, but climatic change and rapid urban development are amplifying the challenge of managing risks. Some coastal cities are already experiencing losses during extreme events related to sea level rise. Meanwhile, urban expansion and changes and intensification in land use put growing pressure on sensitive coastal environments through pollution and habitat loss.

The concentration of people, infrastructure, economic activity, and ecology within the coastal zone merits specific consideration with regard to hazards exacerbated by a changing climate. Major coastal cities often locate valuable assets along the waterfront or within the 100-year flood zone, including port facilities, transport and utilities infrastructure, schools, hospitals, and other long-lived structures. These assets are potentially at risk for both short-term flooding and permanent inundation.

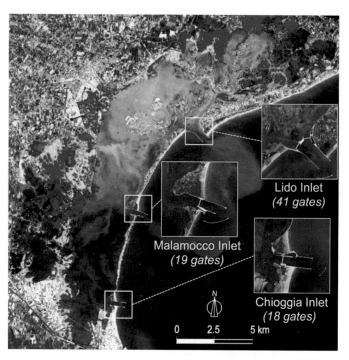

SCL Figure 9 *The MoSE project for the defense of the city of Venice from high tides. Yellow indicates marsh areas surviving at the beginning of the 21st century; red, marshes that have disappeared over the course of the 20th century. Source: Modified from Consorzio Venezia Nuova – Servizio Informativo*

Major Findings

- Coastal cities are already exposed to storm surges, erosion, and saltwater intrusion (see SCL Figure 9). Climate change and sea level rise will likely exacerbate these hazards.

- Around 1.4 billion people could live in the coastal zone, worldwide, by 2060. The population within the 100-year floodplain at risk to a 10–21 cm sea level rise could increase from around 286 million to 411 million people between 2030 and 2060. Three quarters of the exposed populations live in south and southeast Asia.

- Expansion of coastal cities is expected to continue over the 21st century. Although costs of coastal protection could reach US$12–71 billion by 2100, these expenses would be substantially less than taking no action.

- Climate-induced changes will affect marine ecosystems, aquifers used for urban water supplies, the built environment, transportation, and economic activities, particularly following extreme storm events. Critical infrastructure and precariously built housing in flood zones are vulnerable.

- Increasing shoreline protection can be accomplished by either building defensive structures or by adopting more natural solutions, such as preserving and restoring wetlands or building dunes. Modifying structures and lifestyles to "live with water" and maintain higher resiliency is a key adaptive measure.

Key Messages

Coastal cities must be keenly aware of the rates of local and global sea level rise and future sea level rise projections, as well as emerging science that might indicate more rapid rates (or potentially slower rates) of sea level rise.

An adaptive approach to coastal management will maintain flexibility to accommodate changing conditions over time. This involves implementing adaptation measures with co-benefits for the built environment, ecosystems, and human systems. An adaptive strategy requires monitoring changing conditions and refining measures as more up-to-date information becomes available.

Simple, less costly measures can be implemented in the short term while assessing future projects. Land-use planning for sustainable infrastructure development in low-lying coastal areas should be an important priority. Furthermore, cities need to consider transformative adaptation, such as large-scale relocation of people and infrastructure with accompanying restoration of coastal ecosystems.

Delivering integrated and adaptive responses will require robust coordination and cooperation on coastal management issues. This must be fostered among all levels of local, regional, and national governing agencies and include engagement with other stakeholders.

Managing Threats to Human Health

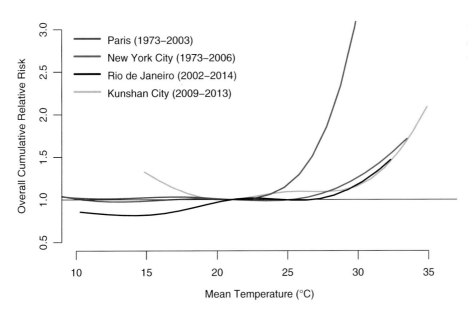

SCL Figure 10 *Overall cumulative heat-mortality relationships in Paris, New York, Rio de Janeiro, and Kunshan City (China).*

Climate change and extreme events are increasing risks of disease and injury in many cities. Urban health systems have an important role to play in preparing for these exacerbated risks. Climate risk information and early warning systems for adverse health outcomes are needed to enable interventions. An increasing number of cities are engaging with health adaptation planning, but the health departments of all cities need to be prepared.

Major Findings

- Storms, floods, heat extremes, and landslides are among the most important weather-related health hazards in cities (see SCL Figure 10). Climate change will increase the risks of morbidity and mortality in urban areas due to greater frequency of weather extremes. Children, the elderly, the sick, and the poor in urban areas are particularly vulnerable to extreme climate events.

- Some chronic health conditions (e.g., respiratory and heat-related illnesses) and infectious diseases will be exacerbated by climate change. These conditions and diseases are often prevalent in urban areas.

- The public's health in cities is highly sensitive to the ways in which climate extremes disrupt buildings, transportation, waste management, water supply and drainage systems, electricity, and fuel supplies. Making urban infrastructure more resilient will lead to better health outcomes, both during and following climate events.

- Health impacts in cities can be reduced by adopting "low-regret" adaptation strategies in the health system and throughout other sectors such as water resources, wastewater and sanitation, environmental protection, and urban planning.

- Actions aimed primarily at reducing GHG emissions in cities can also bring immediate local health benefits and reduced costs to the health system through a range of pathways, including reduced air pollution, improved access to green space, and opportunities for active transportation on foot or bicycle.

Key Messages

In the near term, improving basic public health and health care services; developing and implementing early warning systems; and training citizen groups in disaster preparedness, recovery, and resilience are effective adaptation measures.

The public health sector, municipal governments, and the climate change community should work together to integrate health as a key goal in the policies, plans, and programs of all city sectors.

Connections between climate change and health should be made clear to public health practitioners, city planners, policy-makers, and the general public.

Housing and Informal Settlements

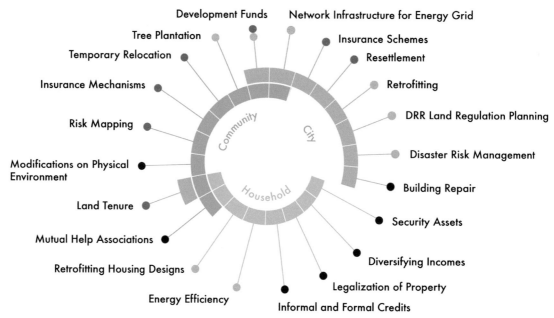

SCL Figure 11 *Overlapping coping, adaptation, and mitigation strategies at household, community, and citywide scales.*

● Coping Strategies
● Adaptation Strategies
● Mitigation Stategies

Addressing vulnerability and exposure in the urban housing sector can contribute to the well-being of residents. This is especially true in informal settlements, where extreme climate events present the greatest risks. Understanding the impacts of mitigation and adaptation strategies on the housing sector will help decision-makers make choices that improve quality of life and close development and equity gaps in cities (see SCL Figure 11).

regard to housing and land tenure are especially vulnerable to climate.

- Among informal settlements, successful adaptation depends on addressing needs for climate-related expertise, resources, and risk-reducing infrastructure.

Major Findings

- The effects of hazards, people's exposure, and their vulnerability collectively determine the types and levels of risk. Risks are associated with specific social and physical factors within each city. Mapping risks and developing early warning systems, preparedness plans, and pre-disaster recovery strategies – especially for informal settlements – can support decision-makers and stakeholders in reducing exposure and vulnerability.

- Developed countries account for the majority of the world's energy demand related to buildings. Incentives and other measures are enabling large-scale investments in mass retrofitting programs in higher income cities.

- Housing construction in low- and middle-income countries is focused on meeting demand for more than 500 million more people by 2050. Efficient, cost-effective, and adaptive building technologies can avoid locking in carbon-intensive and non-resilient options.

- Access to safe and secure land is a key measure for reducing risk in cities. Groups that are already disadvantaged in

Key Messages

City managers should work with the informal sector to improve safety in relation to climate extremes. Informal economic activities are often highly vulnerable to climate impacts, yet they are crucial to economies in low- and middle-income cities. Therefore, direct and indirect costs to the urban poor should be included in loss and damage assessments in order to accurately reflect the full range of impacts on the most vulnerable urban residents and the city as a whole.

Evidence of affordable insurance schemes in developing countries' low-income communities that fulfill adaptation goals is limited. Several implementation-related hurdles need to be addressed if insurance schemes are to be successful. These are excessive reliance on government and donor subsidies, lack of local distribution channels, poor financial literacy of communities, and overall limited demand.

Retrofits to housing that improve resilience create co-benefits such as more dignified housing, improvements to health, and better public spaces. Meanwhile, mitigating GHG emissions in the housing sector can create local jobs in production, operations, and maintenance, especially in low-income countries and informal settlements.

Energy Transformation in Cities

Demands on urban energy supply are projected to grow exponentially due to the growth trends in urbanization and the size of cities, industrialization, technological advancement, and wealth. Increasing energy requirements are associated with rising demands for vital services including electricity, water supply, transportation, buildings, communication, food, health, and parks and recreation.

With climate change, the urban energy sector is facing three major challenges. The first is to meet the rising demand for energy in rapidly urbanizing countries without locking into high-carbon-intensive fuels such as coal. The second is to build resilient urban energy systems that can withstand and recover from the impacts of increasingly extreme climate events. The third is to provide cities in low-income countries with modern energy systems while replacing traditional fuel sources such as biomass.

Major Findings

- Urbanization has clear links to energy consumption in low-income countries. Urban areas in high-income countries generally use less energy per capita than do nonurban areas due to the economies of scale associated with higher density.

- Current trends in global urbanization and energy consumption show increasing use of fossil fuels, including coal, particularly in rapidly urbanizing parts of the world.

- Key challenges facing the urban energy supply sector include reducing environmental impacts such as air pollution, the UHI effect, and GHG emissions; providing equal access to energy; and ensuring energy security and resilience in a changing climate.

- While numerous examples of energy-related mitigation policies exist across the globe, less attention has been given to adaptation policies. Research suggests that radical changes in the energy supply sector, customer behavior, and the built environment are needed to meet the key challenges.

- Scenario research that analyzes energy options requires more integrated assessment of the synergies and tradeoffs in meeting multiple goals: reducing GHG, increasing equity in energy access, and improving energy security.

Key Messages

In the coming decades, rapid population growth, urbanization, and climate change will impose intensifying stresses on existing and not-yet-built energy infrastructure. The rising demand for energy services – mobility, water and space heating, refrigeration, air conditioning, communications, lighting, and construction – in an era of enhanced climate variation poses significant challenges for all cities.

Depending on the type, intensity, duration, and predictability of climate impacts on natural, social, and built and technological systems, threats to the urban energy supply sector will vary from city to city. Local jurisdictions need to evaluate vulnerability and improve resilience to multiple climate impacts and extreme weather events.

Yet future low-carbon transitions may also differ from previous energy transitions because future transitions may be motivated more by changes in governance and environmental concerns than by the socioeconomic and behavioral demands of the past. Unfortunately, the governance of urban energy supply varies dramatically across nations and sometimes within nations, making universal recommendations for institutions and policies difficult, if not impossible. Given that energy-sector institutions and activities have varying boundaries and jurisdictions, there is a need for stakeholder engagement across the matrix of institutions to cope with future challenges in both the short and long term.

To achieve global GHG emission reductions through the modification of energy use at the urban scale, it is critical to develop an urban registry that has a typology of cities and indicators for both energy use and GHG emissions (see SCL Figure 12). This will help cities benchmark and compare their accomplishments and better understand the mitigation potential of cities worldwide.

Energy Transformation in Cities (continued)

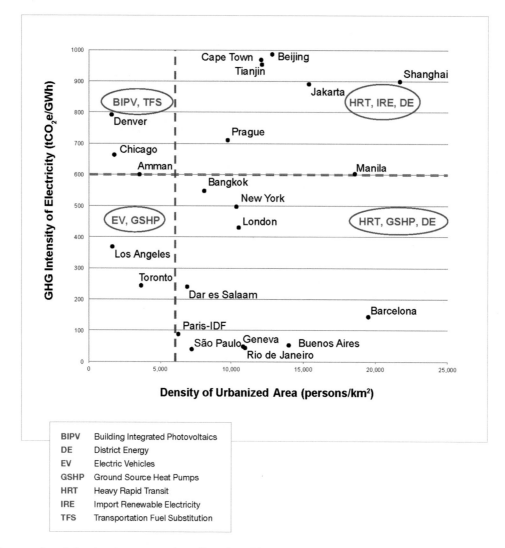

SCL Figure 12 *Low-carbon infrastructure strategies tailored to different cities based on both urban population density and average GHG intensity of the existing electricity supply. Both factors need to be taken into account in developing sustainable urban energy solutions. Source: Adapted from Kennedy et al., 2014*

Transport as Climate Challenge and Solution

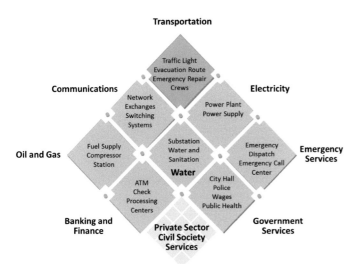

SCL Figure 13 *Urban transport's interconnectivity with other urban systems. Source: Adapted from Melillo et al., 2014*

Urban transport systems are major emitters of GHG and are essential to developing resilience to climate impacts. At the same time, cities need to move forward quickly to adopt a new paradigm that ensures access to clean, safe, and affordable mobility for all.

In middle-income countries, rising incomes are spurring demand for low-cost vehicles and, together with rapid and sprawling urbanization and segregated land use, are posing unprecedented challenges to sustainable development while contributing to climate change.

Expanded climate-related financing mechanisms such as the Green Climate Fund are being developed at national and international levels. Local policy-makers should prepare the institutional capacity and policy frameworks needed to access financing for low-carbon and resilient transport.

Major Findings

- Cities account for more than 70% of GHG emissions, with a significant proportion due to urban transport choices. The transport sector directly accounted for nearly 30% of total end-use, energy-related CO2 emissions. Of these, direct emissions from urban transport account for 40%.

- Urban transport emissions are growing at 2–3% annually. The majority of emissions from urban transport are from higher income countries. In contrast, 90% of the growth in emissions is from transport systems in lower income countries.

- Climate-related shocks to urban transportation have economy-wide impacts extending beyond disruptions to the movement of people and goods. The interdependencies between

transportation and other economic, social, and environmental sectors can lead to citywide impacts (see SCL Figure 13).

- Integrating climate risk reduction into transport planning and management is necessary in spatial planning and land-use regulations. Accounting for these vulnerabilities in transport decisions can ensure that residential and economic activities are concentrated in low-risk zones.

- Low-carbon transport systems yield co-benefits that can reduce implementation costs, yet policy-makers often need more than a good economic case to capture potential savings.

- Integrated low-carbon transport strategies – Avoid-Shift-Improve – involve avoiding travel through improved mixed land-use planning and other measures; shifting passengers to more efficient modes through provision of high-quality, high-capacity mass transit systems; and improving vehicle design and propulsion technologies to reduce fuel use.

- Designing and implementing risk-reduction solutions and mitigation strategies requires supportive policy and public–private investments. Key ingredients include employing market-based mechanisms; promoting information and communication technologies; building synergies across land-use and transport planning; and refining regulations to encourage mass transit and non-motorized modes.

Key Messages

Co-benefits such as improved public health, better air quality, reduced congestion, mass transit development, and sustainable infrastructure can make low-carbon transport more affordable and sustainable and can yield significant urban development advantages. For many transport policy-makers, co-benefits are primary entry points for reducing GHG emissions. At the same time, policy-makers should find innovative ways to price the externalities – the unattributed costs – of carbon-based fuels.

The interdependencies between transport and other urban sectors mean that disruptions to transport can have citywide impacts. To minimize disruptions due to these interdependencies, policy-makers should take a systems approach to risk management that explicitly addresses the interconnectedness among climate, transport, and other relevant urban sectors.

Low-carbon transport should also be socially inclusive because social equity can improve a city's resilience to climate change impacts. Automobile-focused urban transport systems fail to provide mobility for significant segments of urban populations. Women, the elderly, the poor, non-drivers, and disadvantaged people need urban transport systems that go beyond enabling mobility to fostering social mobility as well.

Sustaining Water Security

Water is both a resource and a hazard in climate change. As a resource, good-quality water is basic to the well-being of the ever-increasing number of people living in cities. Water is also critical for many economic activities, including peri-urban agriculture, food and beverage production, and industry. However, excess precipitation or drought can lead to hazards ranging from increased concentrations of pollutants – with negative health consequences, a lack of adequate water flow for sewerage, and flood-related damage to physical assets.

Projected deficits in the future of urban water supplies will likely have a major impact on both water availability and costs. Decisions taken now will have an important influence on future water supply for industry, domestic use, and agriculture.

Major Findings

- The impacts of climate change put additional pressure on existing urban water systems and can lead to negative impacts for human health and well-being, economies, and the environment (see SCL Figure 14). Such impacts include increased frequency of extreme weather events, leading to large volumes of storm water runoff, rising sea levels, and changes in surface water and groundwater.

- A lack of urban water security, particularly in lower income countries, is an ongoing challenge. Many cities struggle to deliver even basic services to their residents, especially those living in informal settlements. As cities grow, demand and competition

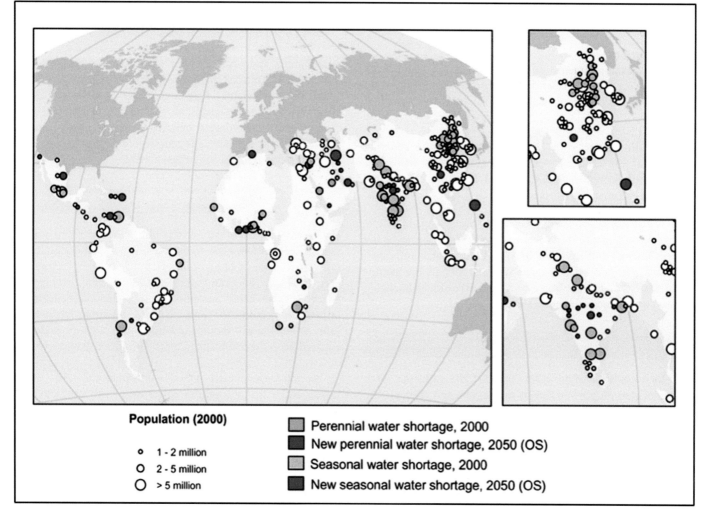

Population (2000)

- ○ 1 - 2 million
- ○ 2 - 5 million
- ○ > 5 million

- Perennial water shortage, 2000
- New perennial water shortage, 2050 (OS)
- Seasonal water shortage, 2000
- New seasonal water shortage, 2050 (OS)

SCL Figure 14 *Distribution of large cities (>1 million population in 2000) and their water shortage status in 2000 and 2050. Gray areas are outside the study area. Source: McDonald et al., 2011*

for limited water resources will increase, and climate changes are very likely to make these pressures worse in many urban areas.

- Water security challenges extend to peri-urban areas as well, where pressure on resources is acute and where there are often overlapping governance and administrative regimes.

- Governance systems have largely failed to adequately address the challenges that climate change poses to urban water security. Failure is often driven by a lack of coherent and responsive policy, limited technical capacity to plan for adaptation, limited resources to invest in projects, lack of coordination, and low levels of political will and public interest.

Key Messages

Adaptation strategies for urban water resources will be unique to each city since they depend heavily on local conditions.

Understanding the local context is essential to adapting water systems in ways that address both current and future climate risks.

Acting now can minimize negative impacts in the long term. Master planning should anticipate projected changes over a time frame of more than 50 years. Yet, in the context of an uncertain future, finance and investment should focus on low-regret options that promote both water security and economic development, and policies should be flexible and responsive to changes and new information that come to light over time.

Many different public and private stakeholders influence the management of water, wastewater, stormwater, and sanitation. For example, land-use decisions have long-lasting consequences for drainage, infrastructure planning, and energy costs related to water supply and treatment. Therefore, adapting to the changing climate requires effective governance as well as coordination and collaboration among a variety of stakeholders and communities.

Cities should capture co-benefits in water management whenever possible. Cities might benefit from low-carbon energy production and improved health with wastewater treatment. Investment strategies should include the application of life cycle analysis to water supply, treatment, and drainage; use of anaerobic reactors to improve the balance between energy conservation and wastewater treatment; elimination of high-energy options such as interbasin transfers of water wherever alternative sources are available; and recovering biogas produced by wastewater.

Managing and Utilizing Solid Waste

Municipal solid waste management is inextricably linked to increasing urbanization, development, and climate change. The municipal authority's ability to improve solid waste management also provides large opportunities to mitigate climate change and generate co-benefits such as improved public health and local environmental conservation.

Driven by urban population growth, rising rates of waste generation will severely strain existing municipal solid waste infrastructure in low- and middle-income countries. In most of these countries, the challenge is focused on effective waste collection and improving waste treatment systems to reduce GHG emissions. In contrast, high-income countries can improve waste recovery through reuse and recycling, and promote upstream interventions to prevent waste at the source (see SCL Figure 15).

Because stakeholder involvement, economic interventions, and institutional capacity are all important for enhancing solid waste management, integrated approaches involving multiple technical, environmental, social, and economic efforts will be necessary.

Major Findings

- Globally, solid waste generation was about 1.3 billion tons in 2010. Due to population growth and rising standards of living worldwide, waste generation is likely to increase significantly by 2100. A large majority of this increase will come from cities in low- and middle-income countries, where per capita waste generation is expected to grow.

- Up to 3–5% of global GHG emissions come from improper waste management. The majority of these emissions is methane – a gas with high greenhouse potential – that is produced in landfills. Landfills, therefore, present significant opportunities to reduce GHG emissions in high- and middle-income countries.

- Even though waste generation increases with affluence and urbanization, GHG emissions from municipal waste systems are lower in more affluent cities. In European and North American cities, GHG emissions from the waste sector account for 2–4% of total urban emissions. These shares are smaller than in African and South American cities, where emissions from the waste sector are 4–9% of the total urban emissions. This is because more affluent cities tend to have the necessary infrastructure to reduce methane emissions from municipal solid waste.

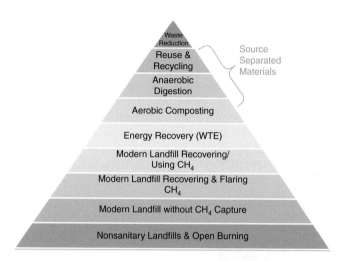

SCL Figure 15 *The hierarchy of sustainable solid waste management. Source: Kaufman and Themelis, 2010*

- In low- and middle-income countries, solid waste management represents 3–15% of city budgets, with 80–90% of the funds spent on waste collection. Even so, collection coverage ranges only from 25% to 75%. The primary means of waste disposal is open dumping, which severely compromises public health.

- Landfill gas-to-energy is an economical technique for reducing GHG emissions from the solid sector. This approach provides high potential to reduce emissions at a cost of less than US\$10 per tCO_2eq. However, gas-to-energy technology can be employed only at properly maintained landfills and managed dumpsites, and social aspects of deployment need to be considered.

Key Messages

Reducing GHG emissions in the waste sector can improve public health; improve quality of life; and reduce local pollution in the air, water, and land while providing livelihood opportunities to the urban poor. Cities should exploit the low-hanging fruit for achieving emissions reduction goals by using existing technologies to reduce methane emissions from landfills. In low- and middle-income countries, the best opportunities involve increasing the rates of waste collection, building and maintaining sanitary landfills, recovering materials and energy by increasing recycling rates, and adopting waste-to-energy technologies. Resource managers in all cities should consider options such as reduce, re-use, recycle, and energy recovery in the waste management hierarchy.

Urban Governance for a Changing Climate

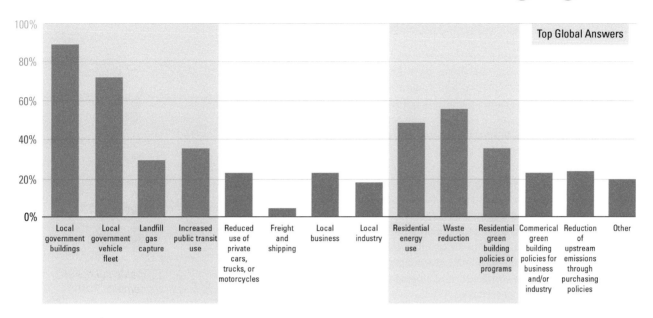

SCL Figure 16 *Mitigation interventions and uptake by cities, resulting in measurable emission reductions. Source: Aylett, 2014*

GHG emissions and climate risks in cities are not only local government concerns. They challenge a range of actors across jurisdictions to create coalitions for climate governance. Urban climate change governance occurs within a broader socioeconomic and political context, with actors and institutions at a multitude of scales shaping the effectiveness of urban-scale interventions. These interventions may be particularly powerful if they are integrated with co-benefits related to other development priorities, creating urban systems (both built and institutional) that are able to withstand, adapt to, and recover from climate-related hazards.

Collaborative, equitable, and informed decision-making is needed to enable transformative responses to climate change, as well as fundamental changes in energy and land-use regimes, growth ethos, production and consumption, lifestyles, and worldviews. Leadership, legal frameworks, public participation mechanisms, information sharing, and financial resources all work to shape the form and effectiveness of urban climate change governance.

Major Findings

- While jurisdiction over many dimensions of climate change adaptation and mitigation resides at the national level, along with the relevant technical and financial capacities, comprehensive national climate change policy is still lacking in most countries. Despite this deficiency, municipal, state, and provincial governmental and non-governmental actors are taking action to address climate change (see SCL Figure 16).

- Urban climate change governance consists not only of decisions made by government actors, but also by non-governmental and civil society actors in the city. Participatory processes that engage these interests around a common aim hold the greatest potential to create legitimate, effective response strategies.

- Governance challenges often contribute to gaps between the climate commitments that cities make and the effectiveness of their actions.

- Governance capacity to respond to climate change varies widely within and between low- and high-income cities, creating a profile of different needs and opportunities on a city-by-city basis.

- The challenge of coordinating across the governmental and non-governmental sectors, jurisdictions, and actors that is necessary for transformative urban climate change policies is often not met. Smaller scale, incremental actions controlled by local jurisdictions, single institutions, or private and community actors tend to dominate city-level actions.

- Scientific information is necessary for creating a strong foundation for effective urban climate change governance, but governance is needed to apply it. Scientific information needs to be co-generated if it is to be applied effectively and meet the needs and address the concerns of the range of urban stakeholders.

Urban Governance for a Changing Climate (continued)

Key Messages

While climate change mitigation and adaptation have become a pressing issue for cities, governance challenges have led to policy responses that are mostly incremental and fragmented. Many cities are integrating mitigation and adaptation, but fewer are embarking on the more transformative strategies required to trigger a fundamental change toward sustainable and climate-resilient urban development pathways.

The drivers, dynamics, and consequences of climate change cut across jurisdictional boundaries and require collaborative governance across governmental and non-governmental sectors, actors, administrative boundaries, and jurisdictions. Although there is no single governance solution to climate change, longer planning timescales; coordination and participation among multiple actors; and flexible, adaptive governance arrangements may lead to more effective urban climate governance.

Urban climate change governance should incorporate principles of justice in order that inequities in cities are not reproduced. Therefore, justice in urban climate change governance requires that vulnerable groups be represented in adaptation and mitigation planning processes, that priority framing and setting

The Rocinha Favela in Rio de Janeiro, October 2015. Photo: Somayya Ali Ibrahim

recognize the particular needs of vulnerable groups, and that actions taken to respond to climate change enhance the rights and assets of vulnerable groups.

UCCRN Regional Hubs

Building on a series of scoping sessions with stakeholders and members, UCCRN is transitioning from a report-focused organization to one that leads an ongoing, sustained global city-focused climate change knowledge assessment and solutions program, through the founding of the UCCRN Regional Hubs. The Regional Hubs operate at continental scale (e.g., Europe, Latin America, Africa, South and East Asia, Australia-Oceania) and will be linked to other relevant regional and global resource nodes.

The UCCRN Regional Hubs serve to promote enhanced opportunities for new urban climate change adaptation and mitigation knowledge and information transfer, both within and across cities, by engaging in a real-time monitoring and review process with cities through ongoing dialogue between scholars, experts, urban decision-makers, and stakeholders.

Since its launch, the Regional Hubs Program has made great strides. The first UCCRN Regional Hub was launched in Paris in July 2015, as the European Hub, shortly followed by the launch of the UCCRN Latin American Hub in Rio de Janeiro in October 2015, and the announcement of the UCCRN Australia-Oceania Hub, co-located in Canberra, Melbourne, and Sydney, at the COP21 conference in December 2015. The UCCRN African Hub was launched in Durban, South Africa in May 2016, followed by the UCCRN East Asian Hub in Shanghai in August 2016, and the UCCRN North American Hub in Philadelphia in November 2016.

In addition to the formal Hubs, a Nordic Node has also been established at Aalborg University to help coordinate Northern European urban climate change efforts. São Paulo State is the home of the UCCRN Center for Multilevel Governance, which has the overall objective of discussing implementation of climate policies at the subnational level and their jurisdictional circumstances. The UCCRN is exploring other potential Asian Hubs in Dhaka and Manila.

References

Aylett, A. (2014). *Progress and Challenges in the Urban Governance of Climate Change: Results of a Global Survey*. MIT Press.

Kaufman, S. M., and Themelis, N. J. (2010). Using a direct method to characterize and measure flows of municipal solid waste in the United States. *Journal of AI, and Waste Management Associations* **2009**(59), 1386–1390.

McDonald, R. I., Green, P., Balk, D., Fekete, B. M., Revenga, C., Todd, M., and Montgomery, M. (2011). Urban growth, climate change and freshwater availability. *Proceedings of the National Academy of Sciences* **108**, 6312–6317.

Melillo, J. M., Richmond, T., and Yohe, G. W. (eds.). (2014). *Climate Change Impacts in the United States: The Third National Climate Assessment*. U.S. Government Printing Office.

Metz, B. (2000). International equity in climate change policy. *Integrated Assessment* **1**, 111–126.

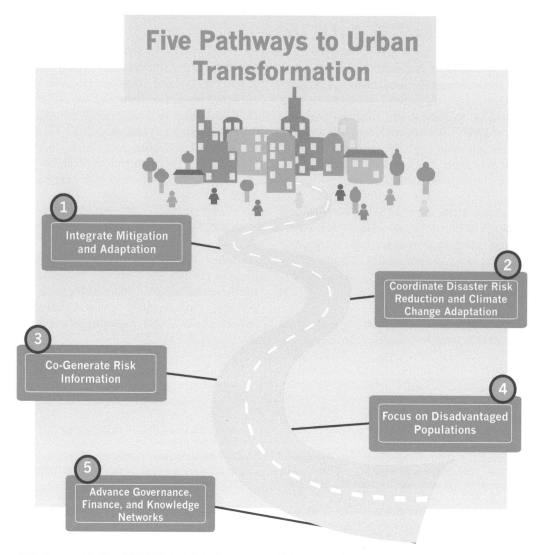

SCL Figure 17 *The UCCRN ARC3.2 Pathways to Urban Transformation*

Introduction

1

Pathways to Urban Transformation

Lead Authors

Cynthia Rosenzweig (New York), William Solecki (New Brunswick/New York), Patricia Romero-Lankao (Boulder/Mexico City), Shagun Mehrotra (New York/Indore), Shobhakar Dhakal (Bangkok/Kathmandu), Somayya Ali Ibrahim (New York/Peshawar)

This chapter should be cited as

Rosenzweig, C., Solecki, W., Romero-Lankao, P., Mehrotra, S., Dhakal, S., and Ali Ibrahim, S. (2018). Pathways to urban transformation. In Rosenzweig, C., W. Solecki, P. Romero-Lankao, S. Mehrotra, S. Dhakal, and S. Ali Ibrahim (eds.), *Climate Change and Cities: Second Assessment Report of the Urban Climate Change Research Network*. Cambridge University Press. New York. 3–26.

Pathways to Urban Transformation

The Five Pathways

Five pathways to urban transformation emerge throughout the *Second Urban Climate Change Research Network Assessment Report on Climate Change and Cities (ARC3.2)*. These pathways provide a foundational framework for the successful development and implementation of climate action in cities. Cities that are making progress in transformative climate change actions are following many or all of these pathways. The pathways can guide the way for hundreds of cities – large and small, low-, middle-, and high-income – throughout the world to play a significant role in climate change action. Cities that do not follow these pathways may have greater difficulty realizing their potential as centers for climate change solutions. The UCCRN ARC3.2 Pathways are:

- *Pathway 1 – Integrate Mitigation and Adaptation*: Actions that reduce greenhouse gas (GHG) emissions while increasing resilience are a win-win.

- *Pathway 2 – Coordinate Disaster Risk Reduction and Climate Change Adaptation*: Disaster risk reduction (DRR) and climate change adaptation (CCA) are the cornerstones of resilient cities.

- *Pathway 3 – Co-generate Risk Information*: Risk assessments and climate action plans co-generated with a full range of stakeholders and scientists are most effective.

- *Pathway 4 – Focus on Disadvantaged Populations*: Needs of disadvantaged and vulnerable citizens should be addressed in climate change planning and action.

- *Pathway 5 – Advance Governance, Finance, and Knowledge Networks*: Developing robust city institutions, advancing city creditworthiness, and participating in city research and action networks enable climate action.

1.1 Introduction

At the United Nations Framework Convention on Climate Change Conference of the Parties (COP21) held in Paris in December 2015, cities[1] were recognized as key actors in both mitigation and adaptation, which are now understood as encompassing low emissions development and resilience. The COP21 Paris Agreement, entered into force in November 2016, highlights the significant role that cities play in implementing national commitments: *"Agreeing to uphold and promote regional and international cooperation in order to mobilize stronger and more ambitious climate action by all Parties and non-Party stakeholders, including ... cities ..."* As is now widely acknowledged, cities can be the main implementers of climate resiliency, adaptation, and mitigation. The Second Urban Climate Change Research Network (UCCRN) Assessment Report on Climate Change and Cities (ARC3.2) addresses the critical question of under what circumstances this advantage can be realized. Cities will not be able to address the challenges and fulfill their climate change leadership potential without transformation.

ARC3.2 aims to provide the knowledge needed for cities to achieve transformation in order to fulfill their emerging role as prime actors in low emissions development and resilience. ARC3.2 synthesizes a large body of studies and city experiences and finds that transformation is essential if cities are to excel in their role as climate change leaders. As cities mitigate the causes of climate change and adapt to new climate conditions, profound changes will be required in urban energy, transportation, water resources, land use, ecosystems, growth patterns, consumption, and lifestyles. New systems for urban sustainability will need to emerge that encompass more cooperative and integrated urban-rural, peri-urban, and metropolitan regional linkages.

Cities are a prime source of greenhouse gas (GHG) emissions and thus collectively represent a significant opportunity to promote climate mitigation. In regard to resilience, climate change in cities encompasses a wide range of direct and indirect impacts, with more frequent extreme temperatures, exacerbated coastal and inland flooding, increases in vector-borne diseases, and heightened water shortages posing risks to infrastructure, resource availability, health, and ecosystems.

Although there is great potential for cities to respond to climate change with transformative solutions of global significance, early actions to date in cities have mostly been incremental. However, international urban climate change networks are gaining strength, city climate change programs are being funded by national governments and foundations, and individual cities are taking on responsibility for both reducing GHG emissions and building resilience. This leadership role of cities is likely to expand and deepen as the implementation phase of global climate action, initiated in Paris in 2015 and entered into force as international law in November 2016, gets under way.

Climate change impacts have widely varying consequences on cities as diverse as New York, Mexico City, Lagos, Shanghai, and Indore. As cities develop their own individual responses to increasing climate risks, a strong knowledge base of cutting-edge science and case studies of effective actions in other cities can contribute to effective and efficient decision-making. Furthermore, urban planning and decision-making at the city level needs to be complemented by policy-making and actions at state, regional, and national levels as well.

For ARC3.2, the Urban Climate Change Research Network (UCCRN) has engaged with urban decision-makers and communities of practice[2] to synthesize the necessary pathways to transformation – both the mechanisms by which urban areas respond to risks and the links between urban mitigation and adaptation. Only through transformation can cities rise to the dual challenge of protecting their vulnerable populations and economic activities from increasing climate risks while taking actions to reduce GHG emissions, the root cause of climate change.

1.2 Urbanization, Transformation, and Sustainable Development

We now live on an urban-dominated planet (see Box 1.1). More than half of the world's 7.3 billion people live in cities, and almost all of the projected population growth at least through the year 2050 is expected to take place as part of the urbanization process (UN Population Fund [UNPF], 2015). According to the UNPF, it is likely that two-thirds of the world's population will live in cities by 2050 and that urbanization will be especially dramatic throughout Asia and Africa and in smaller urban areas (Seto et al., 2011). Not only does the majority of the world's population live in cities, but also the large majority of the world's wealth-generating capacity (likely 90% or more) takes place in cities (Seto et al., 2012). As a result, the prospects for global sustainability will be determined primarily by what happens in cities. As centerpoints for human settlement and economic activity, cities have become the focus of attention with respect to GHG emissions and climate risk exposure and vulnerability, and thus leading actors in climate adaptation and mitigation.

Urbanization can be defined as a set of system-level processes through which population and human activities are concentrated at sufficient densities at which a variety of scalar factors become present that in turn can promote further agglomeration effects. The most obvious manifestation of the urbanization process is the conversion of non-urban land to urban land uses. Urbanization not

---

1 Cities are defined here in the broad sense to be urban areas, including metropolitan, suburban, and peri-urban regions.
2 City decision-makers and communities of practice encompass a broad range of stakeholders that includes municipal governments, civil society groups, local organizations, international agencies, and donors.

Box 1.1   Demographics and Climate Change

Deborah Balk

CUNY Institute for Demographic Research, Baruch College, New York

Daniel Schensul

United Nations Population Fund, New York

A few years ago, the world population turned for the first time from majority rural to majority urban. With high proportions of city-dwellers in the Americas, Europe, and industrial countries of Asia and Oceania, this trend is seen as irreversible. Countries vary widely in the proportion of the population living in cities. Most Asian countries are predominately rural today, even while being home to some of the world's largest urban areas (UN Dept. of Economic and Social Affairs, Population Division, 2014, 2015).

These trends have important implications for climate change. Low-lying coastal zones are more likely to be disproportionately urban (McGranahan et al., 2007) thus implying that urban residents, more so than rural ones, will experience the untoward effects of hazards associated with seaward climate-related change (e.g., increased frequency or severity of coastal flooding). Cities dwellers are different from the general population in other ways as well. Cities tend to have somewhat younger age structures than the rest of the population, in part because cities receive migrants from other cities, towns, and rural areas and because migrants themselves tend to be young (Montgomery et al., 2012).

The relationship between migration and climate change, including internal migration to cities, is commonly understood as flight from climate impacts or, in extreme instances, migration driven by existential threats, as in the case of some small-island developing states. Yet the history of natural disasters shows that most displacement is relatively short term and local (Tacoli, 2009). Environmental drivers have always been a component of mobility, but the full calculus of migration includes social and economic factors as well. Given rural to urban migration, often toward coastal cities, significant amounts of internal migration may be increasing people's geographic exposure to climate hazards while at the same time improving and diversifying livelihoods. The net effect of urbanization driven by rural-urban migration on climate resilience is therefore highly contingent on circumstances, climate threats, and the protection factors in place.

City size is an often overlooked characteristic of development, with much attention being paid to mega-cities. In the developing world, only 12% of urban population lives in cities of 10 million persons or more, whereas about one-quarter of urbanites live in relatively small cities with populations of 100,000–500,000. Small cities tend to grow faster than large cities (Balk et al., 2009). Yet, small cities with far fewer resources may find this faster growth particularly challenging. Decentralization and the pressures of local governance further complicate or hamper the task of climate adaptation, where many interventions will be local. Even in terms of mitigation, secondary and tertiary cities may be at the mercy of

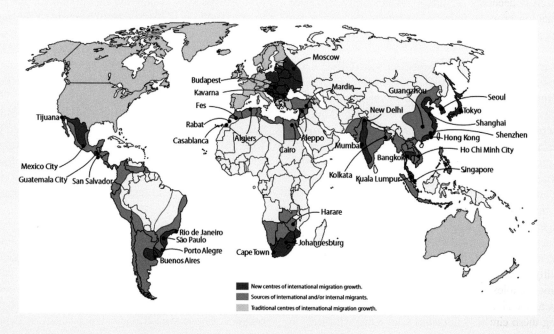

Box 1.1 Figure 1 *Global diversification of migration destinations.*

Source: IOM, 2015. Adapted from Skeldon, 2013

regulations designed for the biggest cities. City growth itself bears consideration: it may impact urban water availability well before the full effects of a warmer climate are realized in the second half of the 21st century (McDonald et al., 2011) (see Chapter 14, Urban Water Systems).

In terms of mitigation, the relationship between the demographic phenomenon of urbanization – or the increasing share of population that lives in urban areas – and GHG emissions is complex and is greatly dependent on a number of other factors. Urban areas are disproportionately the sources of emissions globally, and, in that sense, the trajectory of urbanization is correlated with the trajectory of emissions growth (O'Neill et al., 2010). However, much of this effect is driven by the concentration of wealth in cities; controlling for income, urban living is more energy efficient than rural living, meaning that the broad-scale shift of population from rural to urban provides the potential for a significant mitigation

benefit (Dodman, 2009a). Yet the link between urbanization and emissions also depends on income and wealth, including their distribution in society, as well as on technologies for energy use and urban form (Dodman, 2009b). Cities are also home to smaller households with more independent dwelling units overall; this is complicated because an increasing share of these households are elderly, which tends to lower emissions (O'Neill et al., 2010; Zagheni, 2011). Scenarios of population, economic factors, and technology – for instance the shared socioeconomic pathways (SSPs) (Hunter and O'Neill, 2014) – are useful in global and regional studies assessed by the Intergovernmental Panel on Climate Change (IPCC). Understanding trajectories of emissions and population change in cities is critical for identifying the best approaches to mitigation and adaptation.

For further coverage of this topic, see Box 6.4 in Chapter 6, Equity and Environmental Justice.

only transforms specific sites where cities are located and growing, it is a condition that also has created a web of global-scale resource supply, demand, and waste distribution chains resulting in impacts far beyond city borders. The metabolism of cities has impacts throughout the globe and is responsible for the transport into cities from nearby hinterlands and far-distant locations of primary resources such as energy and water, secondary resources including timber and building materials, and agricultural products.

Cities and their residents have the potential to play an important role in responding to climate change, but concerted transformative action is necessary to overcome the negative effects of the urbanization process. Given the clustering of economic activities, cities often become sites of increased per capita resource consumption in comparison to rural areas. Urban dwellers, particularly in low-income countries, may have relatively higher incomes than their rural counterparts so they tend to consume more. At the same time, density provides economies of scale and resource-use efficiencies so that more people are served with fewer inputs, albeit at higher aggregate consumption (Wenban-Smith, 2009). A case in point is GHG emissions: cities produce approximately 70% of CO_2 emissions (depending on measurement protocols) yet the per capita energy consumption of urban residents tends to be lower than that of rural residents in developed countries (UN-Habitat, 2011; Seto et al., 2014).

In regard to resilience, urbanization concentrates population, infrastructure, and economic activity thus potentially exacerbating vulnerability to extreme climate events. In addition, cities are defined by complex interdependent infrastructure systems and established social and financial networks. Understanding and integrating these circumstances into ongoing climate efforts presents a clear opportunity for enhanced resilience.

In ARC3.2, larger-scale mitigation and adaptation actions are presented within the context of transformation (see Box 1.2).

Transformations are defined as the conditions under which system-level changes take place when the integrated urban energy and risk-management regimes of a specific site, sector, or institution are fundamentally altered as one management regime is replaced by another regime. Transformation opportunities and contexts are explored explicitly in several chapters of ARC3.2 (see Chapter 3, Disasters and Risk; Chapter 4, Mitigation and Adaptation; Chapter 5, Urban Planning and Design; and Chapter 16, Governance and Policy).

1.2.1 Cities as Urban Social-Ecological Systems

In this volume, we understand cities to be complex social-ecological systems (SES), uniquely endowed with attributes and functions that enable them to be the first and leading responders to climate change challenges in both mitigation and adaptation (Redman et al., 2004). Urban system are dynamically interactive at multiple spatial or temporal scales (see Figure 1.1). They consist of social and ecological components (broadly defined) that have their own internal processes; at the same time, these processes interact across the entire urban system in a variety of ways to produce overall urban system forms and dynamics. Drivers external to the urban system are fundamentally important and affect the social and ecological components and processes with different strengths or intensity. This conceptual approach to studying urban SES is scale-independent and can therefore be applied at multiple spatial or temporal scales (see Chapter 8, Urban Ecosystems).

The role of technology in the structure, metabolism, and management of cities is profound. The operation and potential failure of the technological systems of cities have important implications for the resilience of urban areas. Climate extremes in urban contexts reveal the potential for catastrophic collapse resulting from large-scale disturbances and cascading system failures. At the same time, the integration of social, ecological,

Box 1.2   Urban Climate Change Transformation

David Simon

Royal Holloway, University of London

William Solecki

CUNY Institute for Sustainable Cities, Hunter-CUNY, New York

Transformation in the context of systems is found when a system, subsystem, or system components are no longer tenable and are replaced with a new system-level configuration. Expressed differently, this implies that the limits of resilience, adaptive capacity, and hence sustainability of the *status quo* are exceeded and incremental reforms are inadequate, with the result that systemic changes become inevitable and essential. Likewise, urban energy systems will undergo similar systemic change as forms of high-carbon development become untenable. The integration of larger-scale mitigation and adaptation actions are presented in the ARC3.2 within the context of transformation.

Transformations are defined as the conditions under which system-level changes (including value systems; regulatory, legislative, or bureaucratic regimes; financial institutions; and technological or biological systems) take place when the urban energy and risk-management regimes of a specific site, sector, or institution are fundamentally altered as one management regime is replaced by another new regime that integrates both mitigation and adaptation. Transformation opportunities and contexts are explored explicitly throughout this ARC3.2 volume.

Transformation opens new policy options once resilience and energy systems meet their limits. Transformation targets the root drivers of unmet sustainable development needs

where these constrain mitigation, adaptive capacity, and action (see examples in Marshall et al., 2012). Intentional transformation of one system or object may allow the maintenance of systems at other scales (e.g., relocation of households exposed to risk will be transformative for the households involved, for the places of origin, and for the destinations) and may require legislative change. At the same time, relocation may help maintain wider political and economic or social stability. Forced transformations may open greater scope for uncertainty in the behavior of surrounding systems. In comparison, transformation has been developed from a broad range of social science frames particularly focused on development theory and political ecology approaches (Welsh, 2014; Brown et al., 2013; Brown, 2014; Cote and Nightingale, 2011).

Transformations involve large, abrupt, and persistent changes in the structure and function of a physical or social system, such as changes in governance structures or policy objectives. They can open up the possibility of new rights being extended, of greater social and economic equality and greater political participation, and of sustainable development in response to a stressor or shock. Transformations provide abrupt redirecting to alternative development pathways for the system of interest. It may be that the system undergoing transformation is localized and discrete in sector terms, or it may be grand and all-encompassing. Thus, these shifts can be small and local or very widespread in their effects. A transformative state also can be one that is highly dynamic and potentially difficult to predict (Simon and Hayley, 2015; Solecki et al., 2016). As formulated in ARC3.2, urban climate change transformation integrates mitigation and resilience and leads to fundamental regime shift at both local and larger systems levels.

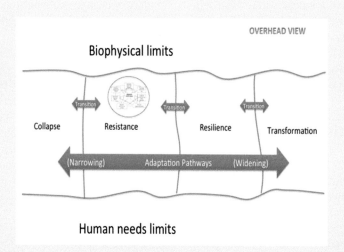

Box 1.2 Figure 1 *Adaptation activity sphere, transitions, and pathway. Time in this diagram is not left to right. Adaptation pathways can move from a lower state to a higher state (i.e., from left to right) or from a higher state to a lower state (i.e., right to left). Time is referenced from the current to moments or eras in a future time.*

Source: Solecki et al., 2017

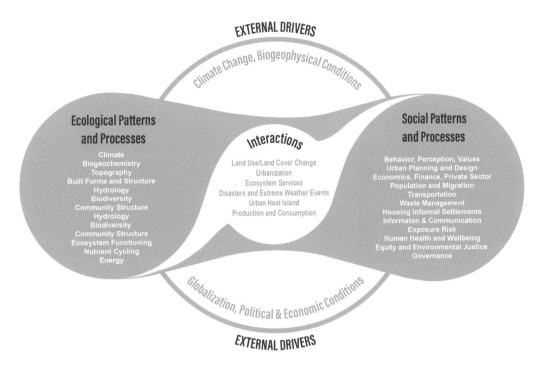

EXTERNAL DRIVERS

Climate Change, Biogeophysical Conditions

Ecological Patterns and Processes

Climate
Biogeochemistry
Topography
Built Forma and Structure
Hydrology
Biodiversity
Community Structure
Hydrology
Biodiversity
Community Structure
Ecosystem Functioning
Nutrient Cycling
Energy

Interactions

Land Use/Land Cover Change
Urbanization
Ecosystem Services
Disasters and Extreme Weather Events
Urban Heat Island
Production and Consumption

Social Patterns and Processes

Behavior, Perception, Values
Urban Planning and Design
Economics, Finance, Private Sector
Population and Migration
Transportation
Waste Management
Housing Informal Settlements
Informaton & Communication
Exposure Risk
Human Health and Wellbeing
Equity and Environmental Justice
Governance

Globalization, Political & Economic Conditions

EXTERNAL DRIVERS

Figure 1.1 *The urban social-ecological system.*

Source: Adapted from Redman et al., 2004

and technological systems in cities provides transformative avenues leading to urban climate adaptation and mitigation.

The use of an urban systems approach is valuable for assessing climate risks and impacts, as well as adaptation and mitigation opportunities and challenges. Systems can operate in a variety of ways including simple linear and complex non-linear interactions and responses. Urban system sectors typically involve relatively well understood, linearly structured engineering systems but are embedded in complex societal and ecological systems with non-linear structures. The systems approach provides a framework for understanding the role and significance of stresses on the operation of urban sectors, metrics of resilience, and early-warning signals of potential system crises and pending system tipping points.

Finally, planning and governance are key dimensions of cities as SES (see Chapter 5, Urban Planning and Design, and Chapter 16, Governance and Policy). Urban climate change governance is the set of formal and informal rules, rule-making systems, and actor networks at all levels (from local to global), both in and outside of government, that are established to steer cities toward mitigating and adapting to climate change (Biermann et al., 2009). Urban climate change governance occurs within the broad context of the SES, with actors and institutions at a multitude of scales shaping the effectiveness of interventions.

1.2.2 Disaster Risk Reduction and Climate Resilience

In laying out concepts related to disaster risk reduction (DRR), ARC3.2 moves from the earlier linear impacts-centric framing of climate hazards, vulnerability, adaptive capacity, resilience, and impacts to an ongoing process-based decision-centric framing that explicitly includes the roles of stakeholders and institutions, governance, capacity building, and exposure reduction (see Figure 1.2) (see Chapter 3, Disasters and Risk). This framing explicitly incorporates elements of DRR and climate change adaptation and highlights operational, management, and governance opportunities for defining connections between the two. Resiliency presents an effective vehicle for highlighting these linkages because it can be applied to both the post-disaster context as well as to the longer-term transformations associated with climate change adaptation.

Often, city governments are the first level of connection to address resilience and DRR in urban areas; however, they often lack technical, knowledge-based, and financial capacities. Regional and national governments need to be engaged through legislation that expands city mandates on DRR and climate change.

Currently, many initiatives are assisting city governments in addressing their needs for building urban resilience and reducing disaster risks. The newly implemented "Ten Essentials"[3] of the *Making Cities Resilient Campaign* of the United Nations Office

3 http://www.unisdr.org/files/26462_13.tenessentialschecklist.pdf

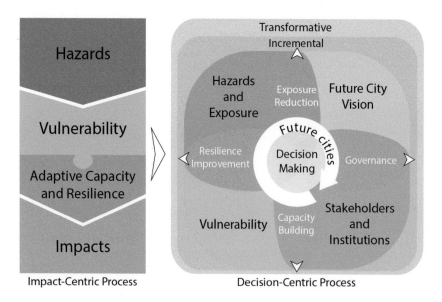

Figure 1.2 *Shift from impact-centric to decision-centric process for disaster risk reduction and climate resilience.*

Source: Xiaoming Wang and Ebru Gencer, 2014; adapted from Xiaoming Wang, 2014

for Disaster Risk Reduction (UNISDR) have provided city governments with practical tools and indicators to build their resilience (see Box 3.3). Currently, nearly 3,000 cities worldwide have joined the Making Cities Resilient Campaign and will be able to use these tools to advance their activities toward DRR and resilience building.

The United Nations Human Settlements Programme (UN-Habitat) City Resilience Profiling Programme (CRPP) is another initiative exploring and providing tools to measure resilience to multihazard impacts and is currently testing these tools in ten pilot cities (UN-Habitat, 2016). The Rockefeller Foundation's 100 Resilient Cities initiative is another program that focuses on building resilience through direct collaboration with city governments (Rockefeller Foundation, 2016) (see CAG 1.6).

1.2.3 Cities, Sustainability, and the Low-Carbon Urban Transition

Urban climate change transformation as presented in ARC3.2 requires the comprehensive integration of mitigation and adaptation. It brings profound changes in energy and land-use regimes, growth patterns, production and consumption, lifestyles, and worldviews (Denton et al., 2014). Some of these actions target the underlying drivers of GHG emissions and vulnerability, such as systems of production and consumption, and the social inequalities that give rise to the coexistence of sub-standard housing, illiteracy, and poverty alongside wealth-related consumptive patterns. As such, transformative climate change actions hold the potential to trigger a broader shift toward sustainable and resilient development pathways (Shaw et al., 2014; Burch et al., 2014).

Just as climate change responses in cities cannot proceed without understanding the larger context of sustainability,

sustainability goals cannot be met without explicit recognition of climate change and the role of cities. The year 2015 not only culminated in the Paris Agreement at the 21st Conference of the Parties for the United Nations Framework Convention on Climate Change (UNFCCC); it was also the year that the nations of the world adopted the Sustainable Development Goals (SDGs) (Sustainable Development Solutions Network [SDSN], 2013) (see Box 1.3). A worldwide campaign was successful in achieving a stand-alone urban sustainability goal, SDG11, to "Make cities and human settlements inclusive, safe, resilient and sustainable." SDG11 targets include:

> By 2020, substantially increase the number of cities and human settlements adopting and implementing integrated policies and plans towards inclusion, resource efficiency, mitigation and adaptation to climate change, resilience to disasters, and develop and implement, in line with the Sendai Framework for Disaster Risk Reduction 2015–2030, holistic disaster risk management at all levels (SDSN TG09, 2013).

Climate change and sustainability were thus explicitly intertwined in the major policy actions of 2015.

A critical aspect of the movement to sustainability and transformation is the transition to low-carbon cities (see Chapter 12, Urban Energy). The connection between urbanization and GHG emissions is complex (Bulkeley et al., 2012; Bulkeley et al., 2014; Sachs and Tubiana, 2014). Cities present opportunities for resource-use efficiency across a large population. At the same time, rural to urban migration is associated with increased energy demand because citizens in cities are often wealthier and better able to access energy-intensive technologies. Researchers and practitioners are actively attempting to define mechanisms to promote technologies and governance structures that enhance opportunities to promote lower energy demand and uses in cities, a multifaceted social-technological transition.

Box 1.3   The Sustainable Development Goals and Urban Areas

The dynamism of cities is a major sustainable development opportunity (Second Urban Sustainable Development Goal Campaign Consultation on Targets and Indicators: *Bangalore Outcome Document* 12–14 January 2015). By getting urban development right, cities can create jobs and offer better livelihoods, increase economic growth, improve social inclusion, promote the decoupling of living standards and economic growth from environmental resource use, protect local and regional ecosystems, reduce both urban and rural poverty, and drastically reduce pollution. Sound sustainable urban and regional development will accelerate progress toward achieving the sustainable development goals (SDGs).

The Urban Climate Change Research Network (UCCRN) partnered with many city groups on the Urban Sustainable Development Goal Campaign (SDSN, TG09, 2013).

The global Urban SDG Campaign[4] (urbansdg.org) supported by more than 400 cities, major urban networks, and institutions[5] played a major role in encouraging the inclusion of SDG 11: "*Make Cities and Human Settlements inclusive, safe, resilient and sustainable.*"

Recognizing this, the United Nations included a stand-alone urban SDG as part of the set of 17 SDGs, passed by the General Assembly in September 2015. Targets included in the urban goal include:

Target 11.1: By 2030, ensure access for all to adequate, safe, and affordable housing and basic services, and upgrade slums.
Target 11.2: By 2030, provide access to safe, affordable, accessible and sustainable transport systems for all, improving road safety, notably by expanding public transport, with special attention to the needs of those in vulnerable situations, women, children, persons with disabilities and older persons

Target 11.3: By 2030, enhance inclusive and sustainable urbanization and capacity for participatory, integrated and sustainable human settlement planning and management in all countries
Target 11.4: Strengthen efforts to protect and safeguard the world's cultural and natural heritage
Target 11.5: By 2030, significantly reduce the number of deaths and the number of people affected and substantially decrease the direct economic losses relative to global gross domestic product caused by disasters, including water-related disasters, with a focus on protecting the poor and people in vulnerable situations
Target 11.6: By 2030, reduce the adverse per capita environmental impact of cities, including by paying special attention to air quality and municipal and other waste management
Target 11.7: By 2030, provide universal access to safe, inclusive and accessible, green and public spaces, in particular for women and children, older persons and persons with disabilities
Target 11.a: Support positive economic, social and environmental links between urban, per-urban and rural areas by strengthening national and regional development planning
Target 11.b. By 2020, substantially increase the number of cities and human settlements adopting and implementing integrated policies and plans towards inclusion, resource efficiency, mitigation and adaptation to climate change, resilience to disasters, and develop and implement, in line with the Sendai Framework for Disaster Risk Reduction 2015-2030, holistic disaster risk management at all levels
Target 11.c: Support least developed countries, including through financial and technical assistance, in building sustainable and resilient buildings utilizing local materials

Source: www.sustainabledevelopment.un.org

1.3   The Urban Climate Change Research Network

To inform effective city-level action on climate change, UCCRN (www.uccrn.org) was established in May 2007 at a side event held during the C40 Large Cities Climate Summit in New York. Beginning with an initial group of about 100 researchers in 60 cities, UCCRN was created to provide knowledge to the C40 cities and other urban decision-makers to enhance climate science–based policy-making on low emissions development and resilience. To this end, UCCRN has begun a global process for an ongoing city-focused climate change knowledge assessment and solutions program targeted to cities of all geographies, sizes, and income levels. It aims to provide a knowledge base for cities just beginning to assess climate change challenges as well as to those

who are leading response policies and measures. UCCRN has grown to include over 800 members worldwide and is based at the Columbia University Earth Institute in New York, with Regional Hubs in Paris, Rio de Janeiro, Durban, Shanghai, Philadelphia, and an Australian-Oceania Hub co-located in Canberra, Melbourne, and Sydney (see Box 1.4).

1.3.1   Role of ARC3 Process

Research on climate change drivers, impacts, and solutions has proliferated in cities in recent years. The need exists to consolidate and assess the existing knowledge to make it relevant and accessible for all cities. One goal of the UCCRN Assessment Report on Climate Change and Cities (ARC3) series is to complement the Intergovernmental Panel on Climate Change (IPCC) work on human settlements as well as other urban assessments.

---

4   Initiated in September 2013.
5   Including UN-Habitat, UCLG, ICLEI, C40, SDSN, Communitas, WIEGO, and SDI.

Whereas in many ways, the ARC3 Reports may be conceived as an "International Panel on Climate Change for cities," there are significant differences with the IPCC. These include an interactive engagement with stakeholders throughout the assessment process, the inclusion of cutting-edge work that is developed by the cities themselves or by other expert groups outside of the peer-reviewed literature but is nonetheless internationally available, and a combined focus on mitigation and adaptation throughout the volume. ARC3.2 addresses mitigation and adaptation in each chapter, exploring synergies throughout.

ARC3.2 operates as an open process, focusing on stakeholder engagement through the co-creation of knowledge and an emphasis on accessible policy recommendations. It draws on both peer-reviewed and practitioner literature. Besides cities themselves, key stakeholders are global urban-focused multilateral institutions that are taking a growing interest in climate change and cities. The aim is for the ARC3 series to provide comprehensive knowledge assessments for its stakeholders' benefit, to increase communication and collaboration (see Figure 1.3).

What distinguishes the ARC3 process is the comprehensive global scale at which urban climate change issues are examined and its potential to become the key ongoing assessment process for climate change and urban areas. Rooted in urban areas throughout the world, the UCCRN network of authors is able to provide in-depth understanding of the needs of decision-makers in cities and deliver the information they require. ARC3 expertise spans a broad spectrum of urbanization, climate change,

mitigation, adaptation, resilience, and transformation, and many authors have real-world experience working with urban policy-makers to implement mitigation and adaptation measures "on the ground." Thus, the ARC3.2 report elaborates and expands on the coverage of both mitigation and adaptation provided by other institutional publications, providing practical information and lessons learned for urban policy-makers as they cope with the challenges that climate change poses to their cities.

1.3.2 First UCCRN Assessment Report on Climate Change and Cities (ARC3.1)

UCCRN's first major publication was *Climate Change and Cities: First Assessment Report of the Urban Climate Change Research Network (ARC3.1)* (ISBN-10: 1107004209), published by Cambridge University Press in 2011. The ARC3.1 represented a four-year effort by about 100 authors from dozens of cities around the world and was the first-ever global, interdisciplinary, cross-regional, science-based assessment to address climate risks, adaptation, mitigation, and policy mechanisms relevant to cities. The assessment articulated an urban climate risk framework, presented relevant climate science for cities, and derived policy implications for key urban sectors – water and sanitation, energy, transportation, public health – and system-wide issues such as land use and governance. To contextualize the knowledge, the ARC3.1 presented 46 Case Studies from a wide range of cities, providing lessons learned from mitigation and adaptation efforts in urban areas around the world.

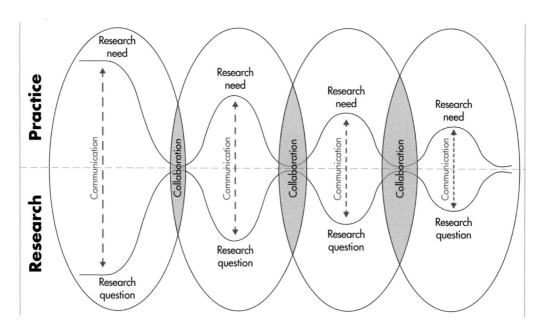

Figure 1.3 *The evolution of a collaborative partnership between practitioners and researchers. Each large oval represents a problem common to both the practitioner community and research community. The dashed line down the center of the figure is the conceptual boundary between research and practice. While the problem may be common, the motivations and ways of framing and addressing it are often distinct for each community. As practitioners and researchers communicate more, their mutual understanding of each other's professional language and culture grows, allowing those collaborative activities to become more complex and resulting in more integrated problem-solving. The net effect of the growth and evolution of these collaborative relationships is that the space shrinks between the research demand and the research supply, and the collaborative space grows. For greater detail, see Case Study 14.3 in Chapter 14, Urban Water Systems.*

Source: Ferguson et al., 2014

During the writing phase, the ARC3.1 was presented to and discussed with a broad group of urban stakeholders – governmental, private-sector, and civil society institutions – in many cities and at high-level urban policy forums. On publication, UCCRN disseminated the ARC3.1 to developing country practitioners and scholars in Asia, Latin America, and Africa, and at urban climate change conferences, meetings, and workshops held in cities throughout the world. The ARC3.1 has been used by urban officials and networks to shape city mitigation and adaptation plans and in undergraduate and graduate-level courses to train the next generation of urban professionals.

The report received strong endorsements from leading mayors and other officials with direct responsibility for cities, including the mayors of Mexico City, São Paulo, and Toronto, the Governor of Lagos State, and senators from Indonesia. These endorsements reinforce the value that decision-makers place on urban-specific climate knowledge. Many urban stakeholders expressed demand for a second and broader international climate change and cities assessment report, which is presented in this volume, *Climate Change and Cities: Second Assessment Report of the Urban Climate Change Research Network (ARC3.2)*.

1.3.3 The Second UCCRN Assessment Report on Climate Change and Cities (ARC3.2)

Exciting new directions in regard to cities and climate change have emerged since the publication of the ARC3.1 in 2011. In the past five years, cities are increasingly taking on the role of implementers of climate change mitigation and adaptation policies and programs. A large amount of material about climate change and cities has been published since work was completed on the ARC3.1 in late 2010. Assessing this body of literature and new reports from a broad range of sources, including peer-reviewed literature as well as city and private-sector expert reports, is a key task of the ARC3.2. Lessons from implementation and updated climate science are incorporated to provide cutting-edge information.

The ARC3.2 has become essential as urban actors and scholars continue to create knowledge in the rapidly evolving "space" of urban climate change solutions, city government officials, national government institutions, global city networks, and international organizations. New knowledge is being created at an unprecedented pace due to the increasing scale of action by policymakers and the growing interest of urban scholars. ARC3.2 contributes to the creation of this multidisciplinary space by capturing the aggregate implications of the new research and practice-based policy learning on climate change and cities.

ARC3 stakeholders include urban practitioners, civil society groups, scholars, and city leadership groups. Following the publication of ARC3.1, UCCRN held several scoping sessions at urban climate change conferences around the world, to obtain feedback on the first ARC3 report and solicit suggestions on additional topics of interest for the ARC3.2 report. Scoping sessions were held at the World Delta Summit (Jakarta, 11/2011),

American Association of Geographers (AAG) Annual Meeting (New York, 2/2012), the Third ICLEI Resilient Cities World Congress (Bonn, 5/2012), C40 Large Cities Climate Summit during the Rio+20 Conference (Rio de Janeiro, 6/2012), the Sixth World Urban Forum (Naples, 9/2012), the European Climate Change Adaptation Conference (Hamburg, 3/2013), and the Fourth ICLEI Resilient Cities World Congress (Bonn, 6/2013).

UCCRN also distributed widely an Information Needs Assessment Survey (2013–2014), to draw on user experiences with the ARC3.1 and to tailor ARC3.2 to provide the greatest benefit to city stakeholders and decision-makers. The UCCRN received 68 responses to the survey from 58 cities in 31 countries. Responses were received from individuals working in local government, the private sector, NGOs, academia, and intergovernmental organizations. They included civil servants, elected officials, decision-makers, researchers, and technical, political, financial, and policy experts. Survey respondents were asked, among other questions, how helpful they found the ARC3.1 report, which topics were most relevant to their work, which Case Studies were most relevant, how often they used scientific assessments in their work, which climate-related resources they relied upon most, and how likely they were to obtain climate information or data in their field of work. The results of this survey were presented at the Seventh World Urban Forum in Medellín in 2014.

Stakeholder consultations have proceeded throughout the ARC3.2 writing process, which began with the ARC3.2 Kick-off Workshop held at the Columbia University Earth Institute in September 2013, continued at the ARC3.2 Authors Meeting held in London in September 2014, and culminated at the Climate Summit for Local Leaders hosted at Paris City Hall during COP21 held in Paris in December 2015.

Based on these interactions, ARC3.2 is structured to communicate to a range of groups important for urban decision-making. These include national institutions with responsibility for urban development policies and finance and city leaders and their technical staff who inform the decisions of urban sectors like transport, energy, water, solid waste, and health. There are also key policy units and associated personnel involved in urban decision-making for climate change, including chief sustainability officers, urban planners, and design professionals. Civil society groups, including non-governmental organizations, often play a major role in climate change mitigation and adaptation programs, particularly in low-income countries. In addition, private-sector organizations often provide the technical expertise to implement climate change response plans in regard to both mitigation and adaptation strategies.

The ARC3.2 has undergone three rounds of a rigorous independent peer-review process (see Appendix C). Each chapter of the report has been peer-reviewed by scholars who are subject experts, city decision-makers, and representatives of institutions that organize programs for cities. There were 89 reviewers involved in the review process across three rounds of review. The chapters cite references from the peer-reviewed and internationally available literature.

The UCCRN presented the *ARC3.2 Summary for City Leaders* during the Climate Summit for Local Leaders during the COP21 in Paris. The full ARC3.2 was shared at the Habitat III Conference in Quito, Ecuador. Other launches will take place in major cities on all continents. For latest launch information and other news about the Network, visit www.uccrn.org.

UCCRN is active in cities throughout the world and will distribute ARC3.2 to city practitioners and urban scholars from developing countries via its Regional Hubs, with support from international donors (see Box 1.4).

Box 1.4   UCCRN Regional Hubs

The UCCRN Regional Hubs in Europe, Latin America, Australia, Africa, and Asia will disseminate the ARC3.2 in low-, middle-, and high-income cities across the globe and spur ongoing dialogue between stakeholders and scientists. The Hubs promote enhanced opportunities for urban climate change adaptation and mitigation knowledge and information transfer, both within and across cities, by engaging in ongoing dialogue between scholars, experts, urban decision-makers, and stakeholders (see Annex 1 for more details).

The first UCCRN Regional Hub was launched in Paris in July 2015 as the European Hub, in partnership with the Centre National de la Recherché Scientifique (CNRS), University

Pierre et Marie Curie (UPMC), and l'Atelier International du Grand Paris (AIGP).

The UCCRN Latin American Hub was launched in Rio de Janeiro in October 2015, with Instituto Oswaldo Cruz at FIOCRUZ, Universidade Federal do Rio de Janeiro, and the City of Rio de Janeiro. The UCCRN announced an Australian-Oceania Hub at COP21 in December 2015, co-located at the University of New South Wales in Sydney, the University of Melbourne, and the University of Canberra. The UCCRN African Hub was established in Durban, South Africa in May 2016, in partnership with the Durban Research Action Partnership (D'RAP), the University of KwaZulu-Natal, and eThekwini Municipality. The

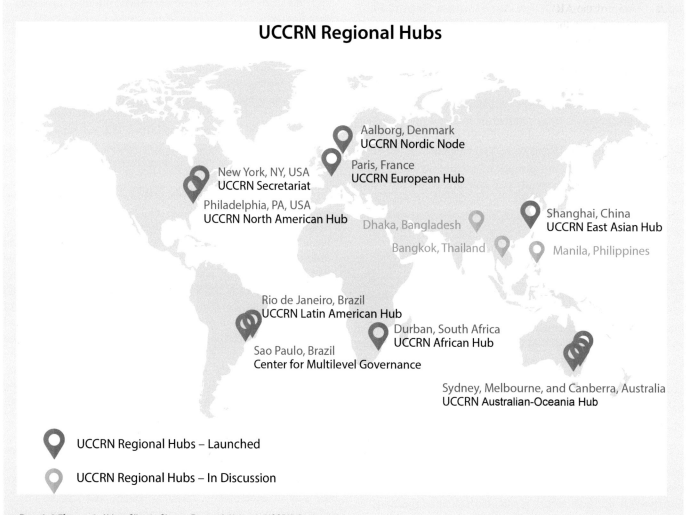

Box 1.4 Figure 1 *Urban Climate Change Research Network (UCCRN) Regional Hubs.*

UCCRN East Asian Hub was launched in Shanghai in August 2016, with East China Normal University and the Shanghai Meteorological Service, and the UCCRN North American Hub was launched at Drexel University in Philadelphia in November 2016, to strengthen a North American network of scholars and stakeholders dedicated to climate change and cities.

A Nordic Node has also been established at Aalborg University in Denmark, and São Paulo State, in partnership with the University of São Paulo, is the home of the UCCRN Center for Multilevel Governance. UCCRN is in discussion to launch a Southeast Asian Hub at the Ateneo de Manila University and the Manila Observatory, and others in Dhaka and Bangkok.

In addition to hosting and organizing region-specific, climate change and cities activities and knowledge sharing, Regional Hubs are also responsible for recruitment and outreach to local urban climate experts to expand the UCCRN network and access an increased diversity of knowledge; stakeholder engagement to connect climate change expertise with city leaders; production of locally focused research and down-scaled projections for the regions; fundraising to support research projects, coordination activities, staffing, and operational expenses; hosting regional and topical workshops for local scholars and stakeholders to facilitate the exchange of ideas around climate change and cities; promotion of the *Urban Climate Change Research Network Assessment Report on Climate Change and Cities (ARC3)* series of reports to targeted stakeholders and translation of reports and publications into regional languages; and liaising between the UCCRN Secretariat in New York and the region.

For more details on the UCCRN Regional Hubs, see Annex 1.

1.4   Structure of ARC3.2

The ARC3.2 is the primary product of UCCRN, with 16 chapters covering a range of topics. It builds on feedback from the readership of the ARC3.1 and a series of focus groups held at urban forums over the past five years that identified the key topics on which city decision-makers required information.

It is comprised of three main parts: the Summary for City Leaders (SCL); the ARC3.2; and the Case Study Docking Station (CSDS) (see Figure 1.4).

The purpose of the SCL is to make assessment results more accessible to urban decision-makers and other stakeholders and to explicitly address their needs for a focus on solutions (see Summary for City Leaders). The SCL is brief and much more concise than the primary volume. It organizes information differently than the main volume, highlights details that are of interest to the user community, and is written in language that is familiar to them.

The CSDS is a web-based, searchable database that serves to inform both research and practice on climate change and cities. Annex 3 of this volume describes the methods and procedures for development of the CSDS database, including a set of data protocols that enable comparisons across a range of social, bio-physical, and economic contexts. 117 Case Studies are included in the ARC3.2, some embedded within the chapters and the rest gathered in the CSDS Annex (see Figure 1.5 and Annex 5).

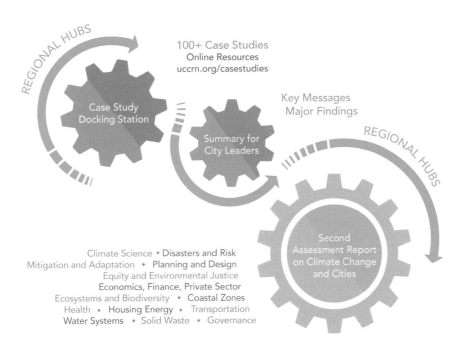

Figure 1.4 *Components of the Second UCCRN Assessment Report on Climate Change and Cities (ARC3.2) and their interactions.*

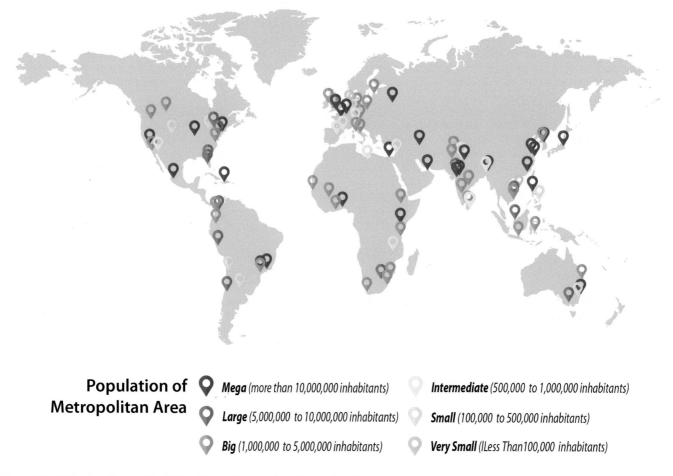

Population of Metropolitan Area

Mega (more than 10,000,000 inhabitants)

Large (5,000,000 to 10,000,000 inhabitants)

Big (1,000,000 to 5,000,000 inhabitants)

Intermediate (500,000 to 1,000,000 inhabitants)

Small (100,000 to 500,000 inhabitants)

Very Small (ILess Than100,000 inhabitants)

Figure 1.5 *ARC3.2 Case Study Docking Station cities and the population of their metropolitan areas.*

These Case Studies present empirical evidence on how cities are responding to climate change, across a diverse set of urban challenges and opportunities. The aims are to develop a mechanism by which to organize the Case Studies via a variety of metrics and sectoral and content elements and to engage a broad and diverse set of city examples for the ARC3.2.

Based on feedback from the stakeholder scoping sessions, the ARC3.2 presents the latest information on new topics that urban decision-makers see as crucial as they take on the challenges associated with adapting to the already-changing climate and mitigating GHG emissions, the root cause of climate change. The new topics covered in ARC3.2 include integration of climate mitigation and adaptation; urban planning and design; equity and environmental justice; economics, finance, and the private sector; urban ecosystems and biodiversity; coastal zones; housing and informal settlements; and urban solid waste. Other emerging topics covered are urban demographics (see Boxes 1.1 and 6.4); information and communications technology (see Box 7.2); and psychological, social, and behavioral challenges and opportunities for climate change decision-making (see Box 4.4). Topics covered in the ARC3.1 – including cities, disasters, and climate risk frameworks; urban climate science and modeling; urban energy; water and wastewater; transportation; human health; and governance – are updated in the ARC3.2 to reflect new research findings.

After the Summary for City Leaders, the main body of ARC3.2 is structured in four parts, along with introductory and concluding sections. The *Introduction* section describes Pathways to Urban Transformation and the role of UCCRN and ARC3 (Chapter 1), presents the latest research on Urban Climate Science (Chapter 2), and develops a new framework for the intersection of Climate Change, Disasters, and Risk (Chapter 3).

Part I presents the *Cross-Cutting Themes* of Integrating Mitigation and Adaptation: (Chapter 4); Urban Planning and Urban Design (Chapter 5); Equity and Environmental Justice (Chapter 6); and Economics, Finance, and the Private Sector (Chapter 7) (see Section 1.5).

Part II covers *Urban Ecosystems and Human Services*: Urban Ecosystems and Biodiversity (Chapter 8), Urban Areas in Coastal Zones (Chapter 9), Urban Health (Chapter 10), and Housing and Informal Settlements (Chapter 11).

Part III provides updates for the *Urban Infrastructure Systems* that were introduced in the ARC3.1, including Energy Transformation in Cities (Chapter 12) Urban Transportation (Chapter 13), and Urban Water Systems (Chapter 14), with the addition of a new chapter on Urban Solid Waste Management (Chapter 15).

Part IV, *Governance and Urban Futures*, describes the role of policy-making in effective climate change responses (Chapter 16).

The *Conclusions: Moving Forward* section presents the conclusions of the entire volume and the way forward for climate change and cities.

1.5 ARC3.2 Cross-Cutting Themes

Four cross-cutting themes are tracked throughout the volume: Integrating Mitigation and Adaptation (Chapter 4); Urban Planning and Design (Chapter 5); Equity and Environmental Justice (Chapter 6); and Economics, Finance, and the Private Sector (Chapter 7).

1.5.1 Integrating Mitigation and Adaptation

Urban planners and decision-makers need to integrate efforts to mitigate the causes of climate change (mitigation) and respond to changing climatic conditions (adaptation), for a global transition to a low-emissions economy and a resilient world (see Chapter 4). Actions that promote both goals provide win-win solutions. In some cases, however, decision-makers have to negotiate tradeoffs and minimize conflicts between competing objectives.

A better understanding of mitigation and adaptation synergies can reveal greater opportunities for urban areas. For example, strategies that reduce the urban heat island effect, improve air quality, increase resource efficiency in the built environment and energy systems, and enhance carbon storage related to land use and urban forestry are likely to contribute to GHG emissions reduction while improving a city's resilience. The selection of specific adaptation and mitigation measures should be made in the context of the current resources and technical means of the city, the needs of citizens, and the SDGs.

1.5.2 Urban Planning and Design

Urban planning and urban design have critical roles to play in the global response to climate change (see Chapter 5). Actions that simultaneously reduce GHG emissions and build resilience to climate risks should be prioritized at all urban scales – metropolitan region, city, district/neighborhood, block, and building. This needs to be done in ways that are responsive to and appropriate for city conditions.

Urban planning and urban design are emerging as important platforms for enabling more effective mitigation and adaptation responses to climate change challenges. Urban planning and design that integrates mitigation and adaptation can leverage the traditional influence and capabilities of practitioners and policy-makers, bringing together climate science, natural systems, and compact urban form to configure dynamic, desirable, resilient, sustainable, and healthy communities.

1.5.3 Equity and Environmental Justice

Cities are characterized by the large diversity of socioeconomic groups living in close proximity (see Chapter 6). Diversity is often accompanied by stratification based on class, caste, gender, profession, race, ethnicity, age, and ability. This gives rise to social categories that, in turn, affect the ability of individuals and groups to endure climate stresses and minimize climate risks. Such differences often lead to discrimination based on group membership. Poorer people and ethnic and racial minorities tend to live in more hazard-prone, vulnerable, and crowded parts of cities. These circumstances increase their susceptibility to the impacts of climate change and reduce their capacity to adapt to and withstand extreme events.

Countries and communities that have historically contributed least to global climate change might be impacted the most. The distribution of impacts between countries as well as within cities of documented natural disasters highlights the disproportionate impact of weather-related extreme events on the most vulnerable in society. This is illustrated by Bangkok's flood crisis in 2011, Typhoon Haiyan in the Philippines in 2013, the Chicago heat wave in 1992, and Hurricane Katrina in New Orleans in 2005.

Highly vulnerable people often share a set of similar and often interacting characteristics including minority status, race, ethnicity, gender, education, income, age, and health. These characteristics determine where people live, how sensitive they are, their assets, and how adaptive they are after one or several hazard occurrences. High exposure and sensitivity often coincide with low adaptive capacity, leading to higher risk and potentially serious impacts. These characteristics are exacerbated by existing poverty and inequality. They are further eroded over time through repeated coping and "risk accumulation processes" (the cumulative impact of relatively minor weather events is potentially significant to vulnerable groups), with indirect implications for chronic poverty. However, it is not just the most vulnerable who are impacted: regularly occurring events such as droughts and floods can also persistently act to undermine the resource base of better-off groups in society, ultimately leading to an increasing spread of those affected by conditions of poverty.

1.5.4 Economics, Finance, and the Private Sector

Because cities are the locus of large and rapid socioeconomic development around the world, economic factors will continue to shape urban responses to climate change (see Chapter 7). To exploit response opportunities, promote synergies between actions, and reduce conflicts, socioeconomic development must be integrated with climate change planning and policies.

Public sector finance can facilitate action, and public resources can be used to generate investment by the private sector. But private sector contributions to mitigation and adaptation should extend beyond financial investment. They should also provide process and product innovation, capacity building, and institutional leadership.

Cities are where socioeconomic change occurs in most nations. With rapidly burgeoning populations, the influence of cities will only grow in the 21st century. This, coupled with the increasing threat of climate change, puts cities at risk of major social and economic disruptions absent sound plans for climate change mitigation and adaptation. These plans must take economics, financing, and the private sector into account.

Any single source of finance, including international public funding, will be inadequate to deliver the infrastructure needs of financing low-carbon development and climate risk management at the city level unless investment is undertaken to earn returns in addition to reducing the impacts of climate change. Cities therefore must tap their full spectrum of sources to raise money for climate action. In most countries, national or federal ministries of environment currently handle climate change activities. However, it is not clear how national climate change budgets – if any – will finance city initiatives. Therefore, city-based approaches to economic decision-making and finance are crucial to meet climate challenges. Public funding is especially effective if it helps to overcome key problems of access to finance (e.g., builds creditworthiness), supports the development of capacities at the city level to utilize to diverse sources of funding and responsibly manage funds available at the city level.

The private sector has a strong motivation to make cities more climate resilient. For private efforts to be effective, however, the right regulatory framework has to be set, and the private and public sectors need to work in an integrated manner. A common vision, shared knowledge, and recognition of co-benefits are keys to successful partnerships.

1.6 UCCRN and Global City Networks

One aim of the ARC3 series of reports is to build the knowledge foundation for city networks, groups, and programs, such as the Cities Alliance Joint Work Programme (JWP) on Resilient Cities, C40 Cities Climate Leadership Group, the Durban Adaptation Charter (DAC), ICLEI—Local Governments for Sustainability, the International City/County Management Association (ICMA) CityLinks program, the Rockefeller Foundation's 100 Resilient Cities Program, UNISDR's Making Cities Resilient Campaign and the New Ten Essentials United Cities and Local Governments (UCLG), and UN-Habitat (see Chapter Annex 1.1, City Action Groups [CAGs]). ARC3 provides these and other groups with a comprehensive synthesis of current knowledge on the mechanisms by which urbanization and urban areas shape their own risks and the transformative links between urban mitigation and adaptation. One of the main findings of ARC3.2 is that cities should connect with national and international capacity-building networks because they can help to advance the strength and success of city-level climate planning and implementation.

The C40 Cities Climate Leadership Group, established in 2005, connects over 80 of the world's megacities, representing more than 550 million people and one quarter of the global economy (see CAG 1.1). Created and led by cities, C40 is focused on tackling climate change and driving urban actions that reduce GHG emissions and climate risks while increasing the health, well-being, and economic opportunities of urban citizens. UCCRN is an advisor to C40 on climate science and climate risks.

UCCRN is part of the Cities Alliance Joint Work Programme (JWP) on Resilient Cities, which was launched at the COP21 Cities and Regions Day in Paris in December 2015 (see CAG 1.2). The JWP convenes a diverse consortium of international organizations to enhance work at the global and local levels to build more resilient cities. The JWP also collaborates with initiatives such as the Medellín Collaboration on Urban Resilience (MCUR), a cooperative effort between major organizations to build urban resilience, and the Lima Paris Action Agenda (LPAA) Cities and Sub-nationals Working group.

The Durban Adaptation Charter, signed in 2011 at the UNFCCC 17th Conference of the Parties, is an evolving international network of more than 1,000 elected urban and local officials drawn together for the purpose of taking action to adapt to climate change and build local resilience (see CAG 1.3). UCCRN and the DAC have signed a Memorandum of Understanding to incorporate the ARC3 reports as a key climate science knowledge base for DAC communications and policies.

ICLEI–Local Governments for Sustainability is a leading network of more than 1,500 cities, towns, and regions committed to building a sustainable future (ICLEI, 2015) (see CAG 1.4). By helping the ICLEI Network to make their cities and regions sustainable, low-carbon, resilient, biodiverse, resource-efficient, and healthy with a green economy and smart infrastructure, ICLEI impacts more than 25% of the global urban population. ICLEI's mission is to build and serve a worldwide movement of city governments to achieve tangible improvements in global sustainability, with a specific focus on environmental conditions through cumulative local actions. UCCRN is a partner of ICLEI and launched the ARC3.2 Summary for City Leaders at the ICLEI Agora at COP21, during a joint session to promote the ICLEI-managed Transformative Actions Program (TAP).

UCCRN is a key science knowledge partner of the ICMA CityLinks Climate Adaptation Partnership Program, which allows cities to be paired with a Resource City to hone their technical and management skills and implement projects that will benefit their community (see CAG 1.5). UCCRN climate experts accompany Partner and Resource City delegates in a series of technical assistance exchange trips, to provide targeted climate science expertise and support.

The Rockefeller Foundation, as part of its 100 Resilient Cities program, is providing financial and logistical guidance for chief resilience officers in cities (see CAG 1.6). Selected cities also gain access to expertise from service providers and partners to develop and implement their own resilience strategies. They also become part of a network of cities that exchange best practices and lessons learned.

The UCLG represents and defends the interests of municipal governments on the world stage, regardless of the size of the communities they serve (Urban Cities and Local Governments [UCLG], 2015) (see CAG 1.7). Headquartered in Barcelona, the organization's stated mission is 'To be the united voice and world advocate of democratic local self-government, promoting its values, objectives and interests, through cooperation between municipal governments, and within the wider international community." The UCLG plays a major role in hosting and participating in international climate change negotiations on mitigation and adaptation in cities. They created a Local Government Climate Roadmap in 2007, organized UCLG World Council Meetings to address issues of climate change, and played a role in the 2013 World Mayor's Summit and the 2014 Climate Summit.

UN-Habitat, one of UCCRN's key partners, supports governments and local authorities, in line with the principle of subsidiarity,[6] to respond positively to the opportunities and challenges of urbanization (see CAG 1.8). Its Climate Change and Cities Initiative (CCCI) helps cities in developing and low-income countries address climate challenges by providing technical assistance and expertise. UN-Habitat utilizes the ARC series of reports as a guide for CCCI city decision-makers at various levels.

Other city and climate initiatives include UNISDR's Making Cities Resilient Campaign and the New Ten Essentials (see Box 3.3) and the World Bank's City Creditworthiness Initiative (see Box 7.1).

1.7 Pathways to Urban Transformation

As is now widely recognized, cities can be the main implementers of climate resiliency, adaptation, mitigation, and sustainable development. However, the critical question that ARC3.2 addresses is under what circumstances this advantage can be realized. Cities may not be able to address the challenges and fulfill their climate change leadership potential without transformation that integrates mitigation and adaptation in virtually every decision, plan, and implementation.

ARC3.2 synthesizes a large body of studies and city experiences and finds that transformation is essential in order for cities to excel in their role as climate change leaders. As cities mitigate the causes of climate change and adapt to new climate conditions, profound changes will be required in urban energy, transportation, water use, land use, ecosystems, growth patterns, consumption, and lifestyles. New systems for urban sustainability will need to emerge that encompass more cooperative and integrated urban-rural, peri-urban, and metropolitan regional linkages.

A set of five pathways to urban transformation emerges throughout ARC3.2 (see Figure 1.6). These pathways provide a foundational framework for the successful development and implementation of climate action. Cities that are making progress in transformative climate change actions are following many or all of these pathways. The five-fold pathway can guide the way for the hundreds of cities – large and small, low-, middle-, and high-income – throughout the world to play a significant role in climate change action. Cities that do not follow these pathways may have greater difficulty realizing their potential as centers for climate change solutions. The UCCRN ARC3.2 Pathways are:

Pathway 1 – Integrate Mitigation and Adaptation: *Actions that reduce GHG emissions while increasing resilience are a win-win.* Integrating mitigation and adaptation deserves the highest priority in urban planning, urban design, and urban architecture. A portfolio of approaches is available, including engineering solutions, ecosystem-based adaptation, energy policies, and social programs. Taking the local context of each city into account is necessary in order to choose integrated actions that result in the greatest benefits.

Pathway 2 – Coordinate Disaster Risk Reduction and Climate Change Adaptation: *Disaster Risk Reduction and climate change adaptation are the cornerstones of resilient cities.* Integrating these activities into urban development policies requires a new, systems-oriented, multi-timescale approach to risk assessments and planning that accounts for emerging conditions within specific, more vulnerable communities and sectors, as well as across entire metropolitan areas.

Pathway 3 – Co-generate Risk Information: *Risk assessments and climate action plans co-generated with the full range of stakeholders and scientists are most effective.* Processes

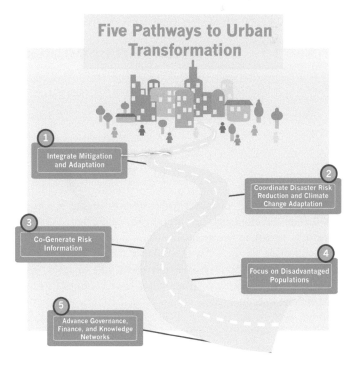

Figure 1.6 *The five ARC3.2 Pathways to Urban Transformation.*

---

6 **Subsidiarity** is defined as is an organizing principle that matters ought to be handled by the smallest, lowest or least centralized competent authority. Political decisions should be taken at a local level if possible, rather than by a central authority.

that are inclusive, transparent, participatory, multisectoral, multijurisdictional, and interdisciplinary are the most robust because they enhance relevance, flexibility, and legitimacy.

***Pathway 4 – Focus on Disadvantaged Populations*:** *Needs of the most disadvantaged and vulnerable citizens should be addressed in climate change planning and action.* The urban poor, the elderly, women, minority, recent immigrants and otherwise marginal populations most often face the greatest risks due to climate change. Fostering greater equity and justice within climate action increases a city's capacity to respond to climate change and improves human well-being, social capital, and related opportunities for sustainable social and economic development.

***Pathway 5 – Advance Governance, Finance, and Knowledge Networks*:** *Developing robust city institutions, advancing city creditworthiness, and participating in city research and action networks enable climate action.* Access to both municipal and outside financial resources is necessary to fund climate change solutions. Sound urban climate governance requires longer planning horizons and effective implementation mechanisms and coordination. Connecting with national and international capacity-building networks helps to advance the strength and success of city-level climate planning and implementation.

A final word on timing: Cities need to start immediately to develop and implement climate action. The world is entering into the greatest period of urbanization in human history, as well as a period of rapidly changing climate. Initiating planning and implementation now will help avoid locking in counterproductive long-lived investments and infrastructure systems and will ensure that cities achieve the transformation necessary to fulfill their leadership role on climate change and sustainable development.

Annex 1.1   City Action Groups (CAGs)

CAG 1.1   C40 Cities Climate Leadership Group

Kathryn Vines

C40 Climate Leadership Group

Today, a number of leading mayors are forging a path to low-carbon development and are already achieving economic growth by investing in sustainable city climate solutions. Since the publication of the *First Assessment Report of the Urban Climate Change Research Network* (ARC3.1) in 2011, C40 Cities Climate Leadership Group cities have continued to take bold actions to reduce GHG emissions and climate risks across multiple sectors. As of 2016, C40 cities have together taken more than 10,000 climate actions, nearly doubling the number of actions in just two years, according to C40's landmark report *Climate Action in Megacities 3.0* (CAM 3.0).

Officials from 98% of C40 cities report that climate change presents risks to their city. Because of this, cities understand the need to exchange best practices and prioritize management of climate risks to ensure their climate resilience. As part of its commitment to assist cities to reduce their climate risk, C40 has developed the City Climate Hazard Taxonomy. It aims to establish a clear and concise lexicon of the climate hazards that cities face today and to document how those hazards may change in the future – a language for cities to speak when discussing climate change adaptation.

Collaboration through Networks Results in Greater Climate Action

C40 research also shows that, by 2050, cities could cut annual greenhouse gas (GHG) emissions by 13 $GtCO_2e$ over what national policies are currently on track to achieve, the equivalent of cutting annual global coal use by more than half. By helping member cities create, share, and measure the impact of climate action, C40 accelerates results and helps to transmit the successful solutions around the world. C40 currently runs 16 networks across such initiative areas as Energy, Finance and Economic Development, Transportation, and Adaptation and Water. Every month, these networks bring member cities together to exchange ideas, solutions, and lessons learned and to collaborate on joint projects to combat common challenges.

These cities are actively advancing major climate actions that will reduce GHG emissions and reduce vulnerability to climate hazards. Data collected by C40 and its member cities show that these collaborations have led to significant results. As a result of taking part in C40 networks, 91 cities have taken 10,000 actions to combat climate change. These include the establishment of a climate change bureau in Ho Chi Minh City through the support of the Connecting Delta Cities Network and the City of Rotterdam and the launch of a Building Energy Challenge in London, working through the Private Building Efficiency Network, whereby 60 companies and 1,000 locations have committed to reduce energy usage. As signatories to the 2015 C40 Clean Bus Declaration of Intent developed by the Low Emission Vehicle Network, Buenos Aires and Rio de Janeiro have set new clean bus targets. Buenos Aires has committed to incorporate 100 bi-articulated buses with electric technology by 2020, and Rio has committed to convert 20% of its bus fleet to clean technologies by 2020.

Another key initiative area where C40 is accelerating local climate action is by helping cities gain access to finance and capital markets. A number of C40 cities do not have a credit rating or access to international capital and therefore cannot make the climate investments that simultaneously can drive economic growth and development and deliver significant health co-benefits. The C40 Creditworthiness program, with support and guidance from international experts, is helping cities break through these barriers and realize their full potential to build a low-carbon and climate-resilient future.

City Action Networks Beyond COP21

In December 2015, C40 mayors assembled in Paris for COP21 to demonstrate city leadership and innovation in the fight against climate change. Limiting global temperature rise to 2°C will require actions by countries, but cities are playing a crucial role in spearheading the way. Collectively, C40 cities have committed to reduce emissions by 1 gigaton by 2020, and they are on course to deliver even more while also taking necessary steps to boost climate resilience.

In June 2016, the Compact of Mayors and the Covenant of Mayors were merged into a new initiative, the Global Covenant of Mayors for Climate and Energy. Through this initiative, C40 and partners ICLEI–Local Governments for Sustainability, United Cities and Local Governments (UCLG), and the United Nations Special Envoy for Cities and Climate Change, with support from UN-Habitat, expect to see cities making even greater commitments. The Global Covenant of Mayors is the world's largest coalition of city leaders addressing climate change by pledging to reduce their GHG emissions, tracking their progress, and preparing for the impacts of climate change.

The Global Covenant of Mayors will drive more aggressive city climate actions and reaffirm existing targets while capturing the significance of these efforts through a consistent, transparent public reporting of cities' GHG data. Through the Covenant, cities are:
- Showing national governments the extent of actions that cities are already undertaking so that it might be incorporated into national-level strategies or further supported through more enabling policy environments and resourcing approaches
- Encouraging increased capital flows into cities to support local action
- Demonstrating the commitment of city governments to contribute positively to more ambitious, transparent, and credible national climate targets by voluntarily agreeing to meet standards similar to those followed by national governments
- Establishing a consistent and transparent accountability framework that can be used by national governments, private investors, or the public to ensure that cities can be held responsible for their commitments

The Global Covenant of Mayors has the potential to make cities' role as partners to nations truly evident – now and in the years to come.

CAG 1.2   Cities Alliance Joint Work Programme on Resilient Cities

Laura Kavanaugh

ICLEI–Local Governments for Sustainability, Bonn

Julie Greenwalt

Cities Alliance, Brussels

The Cities Alliance Joint Work Programme is a partnership supporting cities to become more resilient with a focus on slums, informality and the working urban poor.

Members of the Joint Work Program include C40 Cities Climate Leadership Group, the French Alliance for Cities and Territorial Development (PFVT), GIZ, Global Facility for Disaster Reduction and Recovery (GFDRR), ICLEI—Local Governments for Sustainability, the Inter-American Development Bank (IDB), 100 Resilient Cities, Slum Dwellers International (SDI), The Ecological Sequestration Trust (TEST), United Nations Environment Programme (UNEP), UN-Habitat, the United Nations Office for Disaster Risk Reduction (UNISDR), Women in Informal Employment: Globalizing and Organizing (WIEGO), the World Bank, and the World Resources Institute (WRI).

UCCRN is a Knowledge Partner of the JWP, along with Overseas Development Institute and the International Institute for Environment and Development.

Observers: AECOM, ARUP, The Organization for Economic Cooperation and Development (OECD), Rockefeller Foundation, Swiss State Secretariat for Economic Affairs (SECO).

Activities: Cities Alliance established a Joint Work Programme (JWP) that brings together members and partners to support city resilience with an emphasis on the challenges of slums, informality and the working urban poor. The partnership aims to (1) facilitate the flow of knowledge and resources to enhance city resilience, supporting the emerging Post-2015 framework, climate change and Habitat III processes and (2) promote local resilience strategies through inclusive, long-term, urban planning processes.

Launched at COP21 in December 2015, the JWP is a new type of partnership with a unique constellation of institutions: slum dweller networks, informal workers, city networks working on resilience and climate change, combined with development partners, foundations, and multilateral partners such as the World Bank. The JWP is also unusual in that it combines support for global knowledge, financing, tools, and dialogue, and connects them with on-the-ground technical assistance and implementation. While many other resilience organizations focus on one or more of these aspects, the Cities Alliance JWP bring them all together to support growth trajectories increasingly characterized by equity, inclusion, and environmental sustainability.

The JWP together with the Medellín Collaboration on Urban Resilience has supported the creation of a *Local Governments' Pocket Guide to Resilience* and the *online platform resilience tools.org* which aims to offer an overview of the global resources available for local governments to assess, measure, monitor, and improve city-level resilience.

During 2016, the JWP has supported the development of several joint initiatives among the members. One aims to connect local governments with funders and implementation partners to increase access to finance for transformative local resilience projects. Another will provide technical assistance for cities in Asia, Africa, and Latin America to set meaningful GHG emissions targets and develop energy management plans, incorporating informal settlements, in the context of broader climate change action plan. A third seeks to improve resilience at the household level by undertaking urban resilience audits of informal settlements and incorporating that information into city-wide resilience strategies. Through these and other activities, the JWP is leveraging global partnerships and international agreements to realize more sustainable, resilient, and inclusive cities and communities.

CAG 1.3   Durban Adaptation Charter

Sean O'Donoghue

eThekwini Municipality, Durban

Climate change is already affecting millions of people around the globe through extreme and unseasonable weather events. These impacts are likely to have a disproportionately greater impact upon nations from the Global South that have limited resources and infrastructure to adequately protect themselves and insufficient means to recover. Local governments, in particular, are most challenged because they are responsible for responding to climate impacts at ground level. It is at the local level where livelihoods are lost, water security and food security are impacted, and infrastructure is destroyed. Although local governments will suffer the full impacts of climate change, they are also most equipped to take rapid action now to prepare for and adapt to the impacts of climate change.

The Durban Adaptation Charter (DAC) commits local governments to local climate actions in their jurisdiction that will assist their communities to respond to and cope with climate change risks and thereby reduce vulnerability. By signing the DAC, they commit to *inter alia*:

1. Providing key information on all local government development planning;
2. Ensuring that adaptation strategies are aligned with mitigation strategies;
3. Promoting the use of adaptation that recognizes the needs of vulnerable communities and ensures sustainable local economic development;
4. Prioritizing the role of functioning ecosystems as core municipal green infrastructure;
5. Seeking innovative funding mechanisms.

The DAC was launched at the United Nations Framework Convention on Climate Change (UNFCCC) Conference of the Parties (COP17) held in the City of Durban (eThekwini Municipality), South Africa, in December 2011. The South African government, through the South African Local Government Association (SALGA), South African Cities Network (SACN), eThekwini Municipality, and the Department of Environmental Affairs partnered with ICLEI—Local Governments for Sustainability in hosting the *Durban Local Government Convention: Adapting to a Changing Climate – Towards COP17/ CMP7 and Beyond*.

The momentous signing of the DAC by 114 signatories representing 950 local government organizations from 27 countries builds on the recognition of local governments as important government stakeholders in the Cancun Agreement.

CAG 1.4   ICLEI–Local Governments for Sustainability

Yunus Arikan

ICLEI–Local Governments for Sustainability, Bonn

ICLEI–Local Governments for Sustainability is the leading global network of more than 1,500 cities, towns, and regions committed to building a sustainable future. By helping the ICLEI Network to make their cities and regions sustainable, low-carbon, resilient, biodiverse, resource-efficient, and healthy, with a green economy and smart infrastructure, ICLEI impacts more than 25% of the global urban population.

ICLEI Vision

ICLEI envisions a world of sustainable cities that confront the realities of urbanization, adapt to economic and demographic trends and prepare for the impacts of climate change and other urban challenges. This is why ICLEI unites local and regional governments in creating positive change through collective learning, exchange, and capacity building.

Assets and Achievements

For the past 25 years, ICLEI has maintained that local action is at the center of global change. Its multidisciplinary network continues to develop and apply practical strategies, tools and methodologies that bring about tangible local progress worldwide.

The growing ICLEI Network works collaboratively across the world, leveraging local assets to address pressing urban challenges.

Approach

The ICLEI Network takes an integrated approach to sustainable development, and our 10 Urban Agendas are an expression of our integrated approach. ICLEI forges strategic partnerships with business and financial institutions to strengthen its results and bring about global change with a coalition of able partners. ICLEI also works to ensure that strong policy environments support local action through our national and global advocacy advancements.

CAG 1.5   ICMA CityLinks Climate Adaptation Partnership Program

UCCRN has partnered with the International City/County Management Association (ICMA) CityLinks Program, via its Climate Adaptation Partnership Program (CAPP).

Through the CityLinks Climate Adaptation Partnership Program, local governments in developing countries can apply to become the partner city of a "resource city" that will aid in the development and implementation of a climate adaptation project, with technical assistance provided by municipal and climate experts. CityLinks pairs international partner cities with resource cities that face and have begun to address similar adaptation challenges. International partner city staff have the opportunity to further hone their technical and management skills and implement a project that will benefit their community. They observe solutions and best practices developed in the resource city.

To provide additional targeted technical support to the CityLinks partnerships, ICMA draws on expertise from the Urban Climate Change Research Network. The academic experts are selected UCCRN scholars that are based regionally or locally. They join the city representatives on technical exchanges to provide tailored climate science expertise and assistance, combining a science-based approach with practical application to assist the municipality in preparing for future climate impacts.

City-to-City partnerships last 9–12 months and include two technical assistance trips to the international partner city and one trip to the resource city. The technical exchange trips consist of site tours; visits to universities, environmental agencies, and organizations enacting climate projects in the area; meetings with key stakeholders; and working sessions. During the trips, a work plan is developed for a 9–12 month period that focuses on specific climate-related objectives, delineates the partnership goals, and defines concrete action items for the subsequent technical exchanges and periods in-between the exchange visits. Following the work planning trip, a team of two or three individuals from the international partner city travel to the resource community to gain hands-on experience and observe the implementation of technical solutions. At the conclusion of this trip, next steps are planned for a final return trip of the resource community back to the international partner city to complete the work plan objectives.

CityLinks teams work closely together over the period to develop strategic recommendations that are the basis of an implementation plan informed by local science data, city staff, community stakeholders, and leading global practices in urban adaptation.

The 2015 CAPP pairings included Portmore, Jamaica and Townsville, Australia; Shimla, India and Boulder, Colorado; La Ceiba, Honduras and Somerville, Massachusetts, USA; and Semarang, Indonesia and Gold Coast, Australia.

About CityLinks

The CityLinks model was designed by ICMA as a way to enable municipal officials in developing and decentralizing countries to draw on the resources of their U.S. counterparts to find sustainable solutions tailored to the real needs of their cities. It was formalized in collaboration with the U.S. Agency for International Development (USAID) in 1997 with the launch of a funded program, known at the time as Resource Cities. Based on the success of Resource Cities, USAID awarded ICMA a new

program with the CityLinks name in 2003 and a five-year City-to-City Partnerships cooperative agreement – now known as CityLinks – in 2011.

Source: http://icma.org/en/cl/about/what_is_citylinks

CAG 1.6 Rockefeller Foundation 100 Resilient Cities Program

Heather Grady

Rockefeller Philanthropy Advisors, New York

Several years ago, the Rockefeller Foundation decided to address the issues of climate change in ways that targeted a gap in foundation sector funding. It decided to focus on adaptation and go one step further – explore not just how to adapt to the impacts of climate change, but how to build resilience to climate change. In other words, the organization's members set out to find ways to help people, communities, and systems bounce back from and perhaps even thrive in the face of both long-term stresses and sudden shocks. In an innovative approach, they applied this to urban contexts at a time when few were yet talking about "urban resilience."

An initiative called Asian Cities Climate Change Resilience Network was the foundation for a more widespread effort launched in the Rockefeller Foundation's centenary year, a program called "100 Resilient Cities (100RC)." The 100RC is dedicated to helping cities around the world become more resilient to the physical, social, and economic challenges that are an increasing part of the 21st century.

With an interdisciplinary team that interacts with their growing network of cities, 100RC supports the adoption and incorporation of a view of resilience that is tackling what the cities themselves see as shocks and stresses – from earthquakes, fires and floods to high unemployment, overburdened transportation systems, and chronic food or water shortages. The 100RC is drawing on the expertise of municipal employees, academics, non-profit organizations, and businesses that are, in turn, working with their networks of colleagues to solve day-to-day problems and long-term challenges. With the world's burgeoning population increasing most rapidly by far in urban areas, 100RC is providing a laboratory for experimentation and learning

The 100 successful cities that were selected over three competitive rounds receive resources along four pathways. First, they receive financial and logistical guidance for establishing an innovative new position in city government, a Chief Resilience Officer who leads the city's resilience efforts. Second, if they do not have a resilience strategy yet, they can gain access to expertise to develop their own robust plan. Third, they gain access to service providers and partners from the private, public, and non-governmental organization sectors who can help them develop and implement resilience strategies. And, finally, they become part of a network of member cities exchanging learning and advice. This effort was launched and is currently funded by the Rockefeller Foundation and is managed as a sponsored project by Rockefeller Philanthropy Advisors (RPA), an independent 501(c)(3) nonprofit organization that provides governance and operational infrastructure to its sponsored projects.

CAG 1.7   United Cities and Local Governments

The United Cities and Local Governments (UCLG) represents and defends the interests of local governments on the world stage, regardless of the size of the communities they serve. Headquartered in Barcelona, the organization's stated mission is to be the united voice and world advocate of democratic local self-government, promoting its values, objectives, and interests through cooperation between local governments and within the wider international community.

UCLG's work program focuses on:
- Increasing the role and influence of local government and its representative organizations in global governance;
- Becoming the main source of support for democratic, effective, innovative local government;
- Ensuring an effective and democratic global organization.

The UCLG supports international cooperation between cities and their associations and facilitates programs, networks, and partnerships to build the capacities of local governments. The organization promotes the role of women in local decision-making and is a gateway to relevant information on local government across the world.

Source: http://www.uclg.org/en/organisation/about#sthash.SGvKEUnT.dpuf

CAG 1.8   UN-Habitat Cities and Climate Change Initiative (CCCI)

Marcus Mayr

Cities and Climate Change, UN-Habitat, Nairobi

UN-Habitat supports governments and local authorities, in concert with the principle of subsidiarity[7], to respond positively to the opportunities and challenges of urbanization. UN-Habitat provides advice and technical assistance on transforming cities and other human settlements into inclusive centers of vibrant economic growth, social progress, and environmental safety. Climate change and urbanization are two of the defining global trends of the 21st century. In response to this important linkage, UN-Habitat has brought together its multiple climate change activities under the flagship Cities and Climate Change Initiative (CCCI).

The CCCI is helping cities in developing and low-income countries to address climate challenges, with emphasis on a sound assessment of vulnerabilities and risks, urban planning, good governance, and practical initiatives for municipalities and their citizens. Launched in 2008 in just four cities, CCCI has expanded until, to date, it has assisted more than 45 cities in 23 countries. CCCI has been generously supported by the governments of Norway and Sweden and by the Cities Alliance.

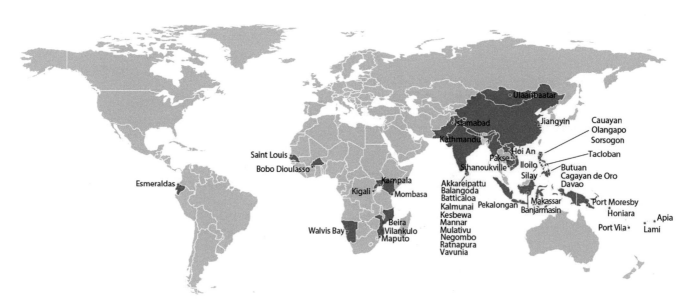

CAG 1.8 Figure 1 *Cities and countries in which the UN-Habitat Cities and Climate Change Initiative (CCCI) is active.*

---

7 **Subsidiarity** is a principle wherein political decisions and other matters are handled at a local level, by the smallest competent authority, rather than by a central authority.

UN-Habitat CCCI works at neighborhood and city-wide scales. The entry point for climate action varies from city to city according to its capacities, resources, vulnerabilities, and opportunities.

Neighborhood projects are effective ways to address climate change at a human scale. UN-Habitat helps local governments to become more responsive to climate change challenges and engage local partners to support climate projects. It develops and implements projects with local communities and businesses, including urban farming, ecosystem-based adaptation, and local energy and mobility solutions. CCCI has supported several pilot activities to demonstrate good practices that merited upscaling. In Bobo Dioulasso (Burkina Faso), Kathmandu (Nepal), and Kesbewa (Sri Lanka), CCCI supported different types of urban agriculture pilots, such as greenways in Bobo Dioulasso. In Kathmandu, the garden rooftop pilot is now being upscaled city-wide, based on an agreement between the City and Ministry of Urban Development. In Maputo, attention focused on environmental zoning in the ecologically fragile Costa da Sol neighborhood, where mangrove forests that served as a buffer to extreme weather events were under intense pressure from urban development. In coastal Saint Louis (Senegal) following a vulnerability assessment, UN-Habitat succeeded in mobilizing co-funding from the government of Japan to help relocate those low-income families that were most exposed to storm surge and extreme weather events.

Cities around the world are beginning to address climate change. Some partner cities have committed to carrying out greenhouse gas (GHG) inventories and to monitoring, reporting, and reducing GHG emissions within their constituencies, whether in established districts or via planned low-carbon city extensions. UN-Habitat brings cities together to learn, share among peers, and enhance the effectiveness of city-led climate actions and to contribute to global research and capacity building. A complementary set of tools has been developed to support cities in raising awareness of the impacts of climate change and undertaking mitigation and adaptation activities. Sorsogon is one of the cities in the Philippines that is most exposed to sea level rise and storm surge from tropical cyclones. An intensive participatory assessment of the city's vulnerability was conducted and used to update the city's statutory plans and to open policy discussion. The city climate strategy focuses on climate change-resilient housing and basic infrastructure, livelihood adaptation plans, environmental management, and disaster risk reduction (DRR), and it is currently being implemented step-by-step to increase urban resilience.

Metropolitan areas are being reshaped continuously as urbanization advances. For Kampala Capital City Authority (Uganda), a CCCI-led vulnerability assessment identified neighborhoods at risk, the urban development dynamics, and institutional framework, and examined opportunities for addressing climate change. Working with the city, CCCI then proposed an integrated approach to dealing with flood management, acknowledging the linkages among floods, development, and poverty. The integrated management strategy is beginning to inform the metropolitan government's flood policy and related investments.

National governments, in their efforts to shape market conditions and coordinate development efforts, increasingly recognize the need for climate change to be included in national urban polices. Through vertical integration, urban issues and local authorities as front-line actors are being included in national-level climate policies, mitigation strategies, and adaptation plans. UN-Habitat has helped several national governments to address urban issues in their climate change policies. Substantive inputs, generally provided by CCCI through regional offices, are backstopped by a policy note on "Addressing Urban Issues in National Climate Change Policies."

Global institutions and processes are slowly but surely recognizing the nexus of cities and climate change as a critical piece of the puzzle to address the sustainable development challenges of the 21st century. Shaping global frameworks and mechanisms that support nations to address climate change in a multilevel effort together with their cities is part of UN-Habitat's work, as well as providing a forum for discussing and shaping the New Urban Agenda at the third UN conference on housing and sustainable urban development (Habitat III) in 2016, in Quito, Ecuador. Another important milestone is the set of goals emerging from the Post 2015 process and the work undertaken to identify means to implement the Sustainable Development Goals (SDGs) in an urban world.

Chapter 1 Pathways to Urban Transformation

References

Balk, D., M. R. Montgomery, G. McGranahan, D. Kim, V. Mara, M. Todd, T. Buettner, and A. Dorelien (2009). Mapping urban settlements and the risks of climate change in Africa, Asia, and South America. In G. Martine, J. M. Guzman, G. McGranahan, D. Schensul, and C. Tacoli (eds.), *Population Dynamics and Climate Change* (pp. 80–103). United Nations Population Fund and International Institute for the Environment and Development.

Biermann, F., Betsill, M., Gupta, J., Kanie, N., Lebel, L., Liverman, D. D., Schroeder, H. H., and Siebenhüne, B. B. (2009). *Earth System Governance: People, Places, and the Planet: Science and Implementation Plan of the Earth System Governance Project*. Earth System Governance Report 1, IHDP Report 20. Bonn: The Earth System Governance Project.

Brown K. (2014). Global environmental change | a social turn for resilience? *Progress in Human Geography* **38**(1), 107–117.

Brown, K., O'Neill, S., and Fabricius, C. (2013). Social science understandings of transformation (pp. 100–106). World Social Science Report 2013: Changing Global Environments. ISSC, UNESCO.

Bulkeley, H., Broto, V. C., Hodson, M., and Marvin, S. (2012). *Cities and Low Carbon Transitions*. Routledge.

Bulkeley, H., Broto, V. C., Hodson, M., and Marvin, S. (2014). Low-carbon transitions and the reconfiguration of urban infrastructure. *Urban Studies Journal*, **51**(7), 1471–1486, Accessed January 12, 2015: http://usj.sagepub.com/content/51/7/1471.full.pdf+html

Burch, S., Shaw, A., Dale, A., and Robinson, J. (2014). Triggering transformative change: A development path approach to climate change response in communities. *Climate Policy* **14**(4), doi:10.1080/14693062.2014.876342.

Cote, M., and Nightingale, A. J. (2011). Resilience thinking meets social theory: Situating social change in socio-ecological systems (SES) research. Progress in Human Geography. doi:10.1177/0309132511425708.

Denton, F., Wilbanks, T. J., Abeysinghe, A. C., Burton, I., Gao, Q., Lemos, M. C., Masui, T., O'Brien, K. L., and Warner, K. (2014). Climate-resilient pathways: Adaptation, mitigation, and sustainable development. In C. B. Field, V. R. Barros, D. J. Dokken, et al. (eds.), *Climate Change 2014: Impacts, Adaptation, and Vulnerability. Part A: Global and Sectoral Aspects. Contribution of Working Group II to the Fifth Assessment Report of the Intergovernmental Panel on Climate Change* (pp. 1101–1131). Cambridge University Press.

Dodman, D. (2009a). Blaming cities for climate change: An analysis of urban greenhouse gas emissions inventories. *Urbanization and Environment* **21**(1), 185–201.

Dodman, D. (2009b). Urban form, greenhouse gas emissions and climate vulnerability. In J. M. Guzmán, G. Martine, G. McGranahan, D. Schensul, and C. Tacoli (eds.), *Population Dynamics and Climate Change* (pp. 64–79). UNFPA/IIED.

Ferguson, D. B., Rice, J., and Woodhouse, C. (2014). *Linking Environmental Research and Practice: Lessons from the Integration of Climate Science and Water Management in the Western United States.* Climate Assessment for the Southwest.

Grimm, N. B., Faeth, S. H., Redman, C. L., Wu, J., Bai, X., Briggs, J. and Golubiewski, N. E. (2008). Global change and the ecology of cities. *Science* **319**(5864), 756–760.

Hunter, L., and O'Neill, B. (2014). Enhancing engagement between the population, environment, and climate research communities: The shared socio-economic pathway process, *Population and Environment.* **35**, 231–242.

ICLEI (2015). *Who We Are.* Accessed September 16, 2015: http://www.iclei.org/iclei-global/who-is-iclei.html

International Organization for Migration (IOM) (2015). *World Migration Report 2015. Migrants and Cities: New Partnerships to Manage Mobility. International Organization for Migration.* Accessed January 12, 2016: http://publications.iom.int/system/files/wmr2015_en.pdf

Marshall, N. A., Park, S. E., Adger, W. N., Brown, K., and Howden, S. M. (2012). Transformational capacity and the influence of place and identity. *Environmental Research Letters* **7**(3), 1–9.

McDonald, R. I., Green, P. Balk, D., Fekete, B., Revenga, C., Todd, M., and Montgomery, M. (2011). Urban growth, climate change, and freshwater availability. *Proceedings of the National Academy of Sciences* **3**(21), 1–6.

McGranahan, G., Balk, D., and Anderson, B. (2007). The rising tide: Assessing the risks of climate change and human settlements in low elevation coastal zones. *Environment and Urbanization* **19**, 17–37.

Mehrotra, S., Rosenzweig, C., Solecki, W. D., Natenzon, C. E., Omojola, A., Folorunsho, R., and Gilbride, J. (2011). Cities, disasters and climate risk. In C. Rosenzweig, W. D. Solecki, S. A. Hammer, and S. Mehrotra (eds.), *Climate Change and Cities: First Assessment Report of the Urban Climate Change Research Network* (pp. 15–42). Cambridge University Press.

Montgomery, M. R. (2008, 8 February). The urban transformation of the developing world. *Science* **319**(5864), 761–764.

Montgomery, M. R., Balk, D., Liu, Z., and Kim, D. (2012, 22 October). Understanding City Growth in Asia's Developing Countries: The Role of Internal Migration, Working paper, Asia Development Bank.

O'Neill, B. C., Dalton, M., Fuchs, R., Jiang, L., Pachauri, S., and Zigova, K. (2010). Global demographic trends and future carbon emissions, *Proceedings of the National Academy of Sciences of the United States of America* **107**(41), 17521–17526.

Redman, C. L., Grove, J. M., and Kuby, L. H. (2004). Integrating social science into the Long-Term Ecological Research (LTER) network: Social dimensions of ecological change and ecological dimensions of social change. *Ecosystems* **7**(2), 161–171.

Sachs, J., and Tubiana, L. (2014). Pathways to deep decarbonization. Sustainable Development Solutions Network, Institute for Sustainable Development and International Relations, Accessed September 7, 2015: http://unsdsn.org/wp-content/uploads/2014/09/DDPP_Digit.pdf

Seto K. C., Fragkias, M., Güneralp, B., and Reilly, M. K. (2011). A meta-analysis of global urban land expansion. *PLoS ONE* **6**, e23777.

Seto, K. C., Reenberg, A., Boone, C. G., Fragkias, M., Haase, D., Langanke, T., Marcotullio, P., Munroe, D. K., Olah, B., and Simon, D. (2012). Urban land teleconnections and sustainability. *Proceedings of the National Academy of Sciences of the United States of America* **109**, 7687–7692. Accessed August 11, 2015: http://dx.doi.org/10.1073/pnas.1117622109

Seto K. C., Dhakal, S., Bigio, A., Blanco, H., Delgado, G. C., Dewar, D., Huang, L., Inaba, A., Kansal, A., Lwasa, S., McMahon, J. E., Müller, D. B., Murakami, J., Nagendra, H., and Ramaswami, A. (2014). Human settlements,

infrastructure and spatial planning. In Edenhofer, O., Pichs-Madruga, R., Sokona, Y., Farahani, E., Kadner, S., Seyboth, K., Adler, A., Baum, I., Brunner, S., Eickemeier, P., Kriemann, B., Savolainen, J., Schlömer, S., von Stechow, C., Zwickel, T., and Minx, J. C. (eds.), *Climate Change 2014: Mitigation of Climate Change. Contribution of Working Group III to the Fifth Assessment Report of the Intergovernmental Panel on Climate Change.* Cambridge University Press. Accessed July 12, 2015: http://www.ipcc.ch/pdf/assessment-report/ar5/ wg3/ipcc_wg3_ar5_chapter12.pdf

Shaw, A., Burch, S., Kristensen, F., Robinson, J., and Dale, A. (2014). Accelerating the sustainability transition: Exploring synergies between adaptation and mitigation in British Columbian communities. *Global Environmental Change.* Accessed March 13, 2015: http://dx.doi.org/10.1016/j.gloenvcha.2014.01.002

Simon, D. and Hayley, L. (2015). Sustainability challenges: Assessing climate adaptation in Africa. *Current Opinion in Environmental Sustainability* **13**, iv–viii.

Skeldon, R. (2013). *Global Migration: Demographic Aspects and Its Relevance for Development.* UN DESA Technical paper 2013/6. Accessed July 29, 2014: www.un.org/esa/population/migration/documents/EGM.Skeldon_17.12.2013.pdf

Solecki, W., Rosenzweig. C., Solecki, S., Patrick, L., Horton, R., and Dorsch, M. (2017). New York, United States of America – Case Study. In S Bartlett and D Satterthwaite (eds.), *Cities on a Finite Planet. Transformative Responses to Climate Change.* Routledge.

Solecki, W., Seto, K. C., and Marcotullio, P. (2013). It's time for an urbanization science. *Environment* **55**, 12–16.

Sustainable Development Solutions Network, Thematic Group on Sustainable Cities (SDSN, TG09). (2013). The urban opportunity: Enabling transformative and sustainable development. *Report to the High-Level Panel on Eminent Persons.* Accessed May 24, 2014: http://unsdsn.org/wp-content/uploads/2014/02/Final-052013-SDSN-TG09-The-Urban-Opportunity1.pdf

Sustainable Development Solutions Network (SDSN). (2013). *An Action Agenda for Sustainable Development: Report for the UN Secretary General,* The United Nations. Accessed February 25, 2014: http://unsdsn.org/wp-content/uploads/2013/06/140505-An-Action-Agenda-for-Sustainable-Development.pdf

United Nations, Department of Economic and Social Affairs, Population Division. (2012). *World Urbanization Prospects, the 2011 Revision. Final Report with Annex Tables.* United Nations.

United Nations, Department of Economic and Social Affairs, Population Division (2014). *World Population Prospects: The 2012 Revision, Methodology of the United Nations Population Estimates and Projections.* ESA/P/WP.235. The United Nations.

UN-Habitat. (2011). Urbanization and the challenge of climate change. In *Cities and Climate Change: Global Report on Human Settlement* (pp. 1–16). UN-Habitat.

United Nations, Department of Economics and Social Affairs, Population Division (2015). *World Population Prospects: The 2015 Revision.* New York: United Nations.

United Nations Office for Disaster Risk Reduction. (2012). *How to Make Cities More Resilient: A Handbook for Local Government Leaders.* United Nations. Accessed February 25, 2014: http://www.unisdr.org/files/26462_handbookfinalonlineversion.pdf

United Nations Population Fund (UNPF). (2007). *State of the World Population.* United Nations. Accessed September 7, 2015: www.unfpa.org/sites/default/files/pub-pdf/695_filename_sowp2007_eng.pdf

Urban Cities and Local Governments (UCLG). (2014). *About Us.* Accessed October 26, 2014: http://www.uclg.org/en/organisation/about

Wang, X., Khoo, Y. B., and Wang, C. H. (2014). Risk assessment and decision-making for residential housing adapting to increasing storm-tide inundation due to sea level rise in Australia. *Civil Engineering and Environmental Systems* **31**(2), 125–139.

Wenban-Smith, B. H. (2009). *Economies of scale, distribution costs and density effects in urban water supply: A spatial analysis of the role of infrastructure in urban agglomeration.* PhD thesis, The London School of Economics and Political Science (LSE).

Zagheni, E. (2011). The leverage of demographic dynamics on carbon dioxide emissions: Does age structure matter? *Demography* **48**, 371–399.

2

Urban Climate Science

Coordinating Lead Authors

Daniel A. Bader (New York), Reginald Blake (New York/Kingston), Alice Grimm (Curitiba)

Lead Authors

Rafiq Hamdi (Brussels/Oujda), Yeonjoo Kim (Seoul), Radley Horton (New York), Cynthia Rosenzweig (New York)

Contributing Authors

Keith Alverson (Nairobi), Stuart Gaffin (New York), Stuart Crane (Nairobi/London)

This chapter should be cited as

Bader, D. A., Blake, R., Grimm, A., Hamdi, R., Kim, Y., Horton, R., and Rosenzweig, C. (2018). Urban climate science. In Rosenzweig, C., W. Solecki, P. Romero-Lankao, S. Mehrotra, S. Dhakal, and S. Ali Ibrahim (eds.), *Climate Change and Cities: Second Assessment Report of the Urban Climate Change Research Network*. Cambridge University Press. New York. 27–60

What Cities Can Expect

People and communities around the globe are reporting weather events and patterns that seem unfamiliar. Such changes will continue to unfold over the coming decades and, depending on which choices people make, possibly for centuries. But the various changes will not occur at the same rates in all cities of the world, nor will they all occur gradually or at consistent rates of change.

Climate scientists have concluded that, whereas some of these changes will take place over many decades, even centuries, there is also a risk of crossing thresholds in the climate system that cause some rapid, irreversible changes to occur. One example would be melting of the Greenland and West Antarctic ice sheets, which would lead to very high and potentially rapid rates of sea level rise in many cities.

Major Findings

- Urbanization tends to be associated with elevated surface and air temperature, a condition referred to as the *urban heat island* (*UHI*). Urban centers and cities are often several degrees warmer than surrounding areas due to presence of heat-absorbing materials, reduced evaporative cooling caused by lack of vegetation, and production of waste heat.

- Mean annual temperatures in 153 ARC3.2 cities around the world are projected to increase by 0.7 to 1.6°C by the 2020s, 1.4 to 3.1°C by the 2050s, and 1.7 to 5.0°C by the 2080s. Mean annual precipitation in 153 ARC3.2 cities around the world is projected to change by –7 to +10% by the 2020s, –9 to +14% by the 2050s, and –11 to +20% by the 2080s. Sea level in 71 ARC3.2 coastal cities is projected to rise 4 to 18 cm by the 2020s; 14 to 56 cm by the 2050s, and 22 to 118 cm by the 2080s (see Annex 2, Climate Projections for ARC3.2 Cities).

- Some climate extremes will be exacerbated under changing climate conditions. Extreme events in many cities including heat waves, droughts, heavy downpours, and coastal flooding are projected to increase in frequency and intensity.

- The warming climate combined with the UHI effect will exacerbate air pollution in cities.

- Cities around the world have always been affected by major, naturally occurring variations in climate conditions including the El Niño Southern Oscillation, North Atlantic Oscillation, and the Pacific Decadal Oscillation. These oscillations occur over years or decades. How climate change will influence these recurring patterns in the future is not fully understood.

Key Messages

Human-caused climate change presents significant risks to cities beyond the familiar risks caused by natural variations in climate and seasonal weather patterns. Both types of risk require sustained attention from city governments in order to improve urban resilience. One of the foundations for effective adaptation planning is to co-develop plans with stakeholders and scientists who can provide urban-scale information about climate risks – both current risks and projections of future changes in extreme events.

Weather and climate forecasts of daily, weekly, and seasonal patterns and extreme events are already widely used at international, national, and regional scales. These forecasts demonstrate the value of climate science information that is communicated clearly and in a timely way. Climate change projections perform the same functions on longer timescales. These efforts now need to be carried out on the city scale.

Within cities, neighborhoods experience different microclimates. Therefore, urban monitoring networks are needed to address these unique challenges and the range of impacts from extreme climate effects at neighborhood scales. The observations collected through such urban monitoring networks can be used as a key component of citywide climate indicators and monitoring systems that enable decision-makers to understand and plan for a variety of climate risks across the city urban landscape.

Note: Temperature and precipitation projections are based on 35 global climate models and 2 representative concentration pathways (RCP4.5 and RCP8.5). Timeslices are 30-year periods centered around the given decade (e.g., the 2050s is the period from 2040 to 2069). Projections are relative to the 1971 to 2000 base period. For each of the 153 cities, the low estimate (10th percentile) and high estimate (90th percentile) was calculated. The range of values presented is the average across all 153 cities. Sea level rise projections are based on a 4-component approach that includes both global and local factors. The model-based components are from 24 global climate models and 2 representative concentration pathways (RCP4.5 and RCP8.5). Timeslices are 10-year periods centered around the given decade (e.g., the 2080s is the period from 2080 to 2089). Projections are relative to the 2000 to 2004 base period. For each of the 71 cities, the low estimate (10th percentile) and high estimate (90th percentile) was calculated. The range of values presented is the average across all 71 coastal ARC3.2 cities. ARC3.2 Cities include Case Study Docking Station cities, UCCRN Regional Hub cities, UCCRN project cities, and cities of ARC3.2 Chapter Authors.

Like all future projections, UCCRN climate projections have uncertainty embedded within them. Sources of uncertainty include data and modeling constraints, the random nature of some parts of the climate system, and limited understanding of some physical processes. In the ARC3.2 Report, the levels of uncertainty are characterized using state-of-the-art climate models, multiple scenarios of future greenhouse gas concentrations, and recent peer-reviewed literature. The projections are not true probabilities, and scenario-planning methods should be used to manage the risks inherent in future climate.

2.1 Introduction

Urban areas have special interactions with the climate system that produce heat islands, reduce air quality, and exacerbate runoff. This chapter presents information about these processes, observed climate trends, extreme events, and climate change projections for cities. It is essential that this information be developed and communicated in ways that contribute to science-based decisions made by city managers to enhance climate resilience.

Section 2.2 introduces the co-generation process that can be used to develop urban climate risk information. Section 2.3 provides an overview of the urban climate system, including a discussion of the urban heat island (UHI), air quality, and the role of urban monitoring networks. Section 2.4 describes the role of natural variability in influencing urban climate risk and how lessons from communicating seasonal climate forecast information may translate to future climate change action. Section 2.5 discusses observed climate trends in cities and the influence that urbanization may be having on these trends. Section 2.6 focuses on how climate change is projected to impact cities in the future and how climate science information, including projections, is being used for adaptation and resilience planning.

Throughout the chapter, city examples and focused case studies are provided to emphasize the key themes of urban climate science, which include urban vulnerability to extreme climate events and the need to advance the science (including climate science) of urban adaptation assessment and implementation.

2.2 Co-generation of Urban Climate Risk Information

The information about urban climate processes, observed climate trends, extreme events, and climate change presented in this chapter is tailored to support the decisions that city stakeholders make in planning for climate resilience. The development of climate risk information for cities is an interactive and iterative process between scientists and stakeholders. To improve policy efforts for enhancing resilience in urban settings, city managers need to identify key stakeholders, engage scientists in the process of risk analysis, work with specialized experts in the co-generation of climate information, and maintain lines of communication among the various groups. A growing body of literature supports the need for adaptation planning grounded in climate science (Moss et al., 2013; Lemos et al., 2013; Kerr, 2011).

The potential for and current use of scientific information in urban decision-making demonstrates the importance of improved communications. The diversity of urban systems and

urban climate science topics makes dissemination of information outside of the scientific community a daunting task; standardization of terminology and classification of phenomena can help to improve the dialogue (Oke, 2006). The numerous different foci of urban climate science research are exemplary of its large scope, and illustrate the large scope for further research on the use of scientific information in urban planning.

The complexity of cities presents unique challenges and opportunities for knowledge co-generation. Given the wide range of systems, operations, and perspectives that are characteristic of cities, an array of stakeholder representation has become a necessary complement to the knowledge co-generation process. Developing climate information with multi-sector considerations requires the bringing together of governmental and non-governmental organizations (NGOs), "knowledge providers" with scientific backgrounds, business leaders, planners, and utilities experts. By utilizing a framework that includes global climate scenarios, the exchange of local climate change data, and the illustration of socioeconomic and climate risk factors, such groups can communicate constructively and develop salient and usable local data. The use of remote sensing tools and the establishment of urban monitoring networks can enhance understanding of urban climate effects, helping to ensure that adaptation planning is grounded in science. Regular and iterative interactions among these stakeholder groups and scientists can produce climate risk information that is vital to a city's resilience planning and implementation (New York City Panel on Climate Change [NPCC], 2015) (see Figure 2.1).

Local climate risk information is an essential component of a city's comprehensive framework for responding to the risks of climate variability and change and to its implementation of adaptation strategies. This underlying climate science is critical for identifying vulnerabilities and for planning a response for urban resilience that integrates and is predicated on the ongoing development of scientific knowledge. Interactions between scientists and stakeholders, taking into consideration the many components of the urban climate system, lead to the co-generation of usable information that aids in the understanding of climate risk and in the planning for resilient cities. It is this collaboration between the climate science providers and the decision-makers that generates useful and practical climate science information that is needed to address the challenges that arise when climate change adaptation initiatives are implemented in urban areas.

2.3 Urban Climate Processes

A thorough understanding of the urban climate system is the starting point for the climate risk assessment process. Critical to this is the need for long-term, quality-controlled, observed climate data. In many cities, especially those in developing countries, the historical record is short and/or discontinuous or of uncertain quality. This makes trend analysis and climate change detection difficult. Without long-term historical records,

Figure 2.1 *Wetland restoration in New York, where co-generated climate science information is guiding the city's resiliency efforts.*

Source: New York City Department of City Planning

the role of climate variability cannot be adequately described, and climate change projections will not be supported by a strong historical baseline (Blake et al., 2011 and references therein). Even in places where a long-term record is available, there often exists scope to expand urban climate monitoring networks to better understand within-city variations and improve awareness of climate risks.

Understanding how the urban climate (e.g., temperature, precipitation, and winds) varies within cities has important implications for stakeholders when developing adaptation strategies. With a greater understanding of the microclimate environment, recognition of key vulnerabilities can advance, leading to targeted adaptation responses. All of the urban effects described here can impact city systems with varying magnitudes depending on the relative importance of climate hazards for a particular location.

This section describes the key components of the urban climate system and urban climate monitoring networks.

2.3.1 Urban Heat Island

Urbanization is often associated with elevated surface and air temperature, a condition referred to as the *urban heat island* (see Case Study 2.1 and Box 2.1). Urban centers and cities are often several degrees warmer than their surrounding areas (see Table 2.1). Due to the low albedo (reflectivity) of urban surfaces such as rooftops and asphalt roadways, the trapping of radiation within the urban canopy, differential heat storage, and greater surface roughness, cities "trap" heat (Oke, 1978). The reduction in evapotranspiration due to impervious surfaces also contributes to the UHI. In addition, the high density of urban environments often leads to intense anthropogenic heat releases within small

spatial scales – particularly from critical urban infrastructure systems such as transportation and energy – that can enhance the UHI by up to 1°C (Ohashi et al., 2007; Tremeac et al., 2012; Zhang et al., 2013a).

Urban sectors, such as energy and health, are readily influenced by the UHI effect. A study of a small city in western Greece found greater (lower) cooling (heating) demands in the summer (winter) in the urban center as compared to surrounding rural locations (Vardoulakis et al., 2013). A study of the heat island in Shanghai, China, found heightened heat-related mortality in urban regions, with worsening health effects from higher temperatures (Tan et al., 2010). The combined effects of increasing heat waves due to climate change and the UHI effect also pose serious health risks to urban populations (Li and Bou-Zeid, 2013).

In one study, three factors contributing to the UHI (urban geometry, impervious surfaces, and anthropogenic heat releases) were analyzed for their relative importance and combined effects. Results showed that, during the day, heat island intensity tends to be driven by impervious surfaces, whereas in the evening anthropogenic heat is the main factor (Ryu and Baik, 2012). Regional climate models (see Section 2.6) can be used to investigate the importance of these factors and their effect on heat islands (Giannaros et al., 2013; Chen et al., 2014). In addition, regional climate models also are used to understand how physical processes in the background climate and geography also contribute to UHI (Zhao et al., 2014).

A large body of research on the mechanisms of the UHI is focused on land use, land surface characteristics, and surface temperatures. An analysis of the heat island in Rotterdam found that the heat island is largest for parts of the city with limited

Box 2.1   NASA Monitoring of Urban Heat Islands

The National Aeronautics and Space Administration (NASA) missions and instruments play a critical role in the monitoring of urban microclimates, including the urban heat island (UHI) effect. Satellite remote sensing instruments (such as the Moderate Resolution Imaging Spectroradiometer [MODIS] and the Landsat satellite) are important tools used by researchers in cities around the globe to investigate how urbanization influences surface–air heat fluxes and how UHIs evolve with time. Images captured from the instruments help to identify the magnitude and spatial scale of microclimates, as well as to infer land surface types and changes that also impact the urban environment.

Using MODIS data, a study of the heat island in Milan identified two different heat island phenomena present in the city, one throughout the day (the surface UHI) and one present only during the evening (the canopy layer heat island) (Anniballe et al., 2014). In another study, Landsat thermal images were used to calculate the intensity and spatial extent of the heat island in Brno, Czech Republic (Dobrovolny, 2013). A study of

Bucharest (Cheval and Dumitrescu, 2009) used MODIS data to study the spatial extent and the intensity of the heat island of that city.

Analyses across multiple cities in North America (Imhoff et al., 2010) and Asia (Tran et al., 2006) are further examples of how NASA instruments are being utilized to better understand urban microclimates. For all thirty-eight cities analyzed in Imhoff et al. (2009), impervious surface area (measured by Landsat) was identified as the primary driver of urban heating, which explains 70% of the total variance in land surface temperature (measured by MODIS). In Tran et al. (2006), data from both Landsat and MODIS were used to explore results between UHIs and land surface type in Bangkok and Ho Chi Minh City, with the research showing that both vegetation and urban density influenced microclimates in those cities.

The images and analysis that are produced from tools such as MODIS and Landsat can contribute to urban planning and the development of adaptation strategies. Imhoff et al. (2010) found that in North American cities in forested biomes, the summertime heat island tends to be much stronger than that in winter. This finding has implications for energy demand because increased urbanization could require additional cooling in these regions. The effectiveness of climate change adaptation and mitigation strategies, such as increasing green spaces, parks, and vegetated areas, can also be evaluated through NASA remote sensing products. Such analysis has already been completed for Chicago (Mackey et al., 2012) and Tel Aviv (Rotem-Mindali et al., 2015), with results indicating that greening techniques have the ability to reduce the impacts of the heat island.

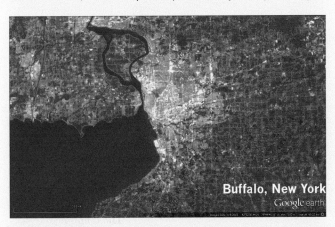

Box 2.1 Figure 1 *Urban heat island in Buffalo, New York measured by NASA satellites. Bright colors indicate higher temperatures, which are associated with more urbanized areas.*

Source: NASA

In addition to enabling a better understanding of the urban environment, the data collected from remote sensing monitoring systems can supplement (or serve as an alternative to) observed station-based data and be used to develop standardized metrics for cross-city comparison of heat islands and other urban phenomena.

vegetation and large amounts of impervious surface (and thereby a low albedo; Klok et al., 2012). Similar results were found in a study of Toronto, Canada (Rinner and Hussain, 2011) and Phoenix, Arizona, with a study finding that changes in the configuration of grass and impervious surfaces explained temperatures in industrial and commercial areas, whereas the proportion of land cover of grass and impervious surfaces alone best explained temperatures in residential areas (Connors et al., 2013). Land use and land cover changes and population shifts also have influences on UHIs (Zhang et al., 2013b).

In some locations, such as New York, within-city temperature variations, which can be attributed to the UHI or other factors such as wind direction and proximity to water (Rosenzweig et al., 2009), can be as large as the projected changes between the late 20th and the late 21st centuries (e.g., Horton et al., 2011; Rosenzweig and Solecki, 2010). The UHI intensities in Tokyo, Shanghai, and Delhi (ranging from 3 to 12°C) already exceed the mean temperature increases projected for these cities by the 2080s (1.5–2.5°C)[1] (Blake et al., 2011) (see Table 2.1).

---

1 For a description of the methods used for these climate projections, see Chapter 3 (Blake et al., 2011) of the *First UCCRN Assessment Report on Climate Change and Cities* (Rosenzweig et al., 2011).

Case Study 2.1   Urban Heat Island in Brussels

Rafiq Hamdi

Royal Meteorological Institute of Belgium, Brussels

Keywords	Urban heat island effect, urban growth, impervious surfaces, minimum temperature increase
Population (Metropolitan Region)	2,061,000 (UN, 2016)
Area (Metropolitan Region)	161.38 km² (Brussels Statistics, 2015)
Income per capita	US$46,010 (World Bank, 2017)
Climate zone	Cfb – Temperate, without dry season, warm summer (Peel et al., 2007)

The Brussels Capital Region (BCR) in Belgium has experienced a rapid transformation of agricultural land and natural vegetation to built areas (e.g., buildings, impermeable pavements) over the past century. There has been a rapid expansion of the city since the 1950s. The acceleration of urban growth is linked to widespread use of the car as a new mode of transport (Fricke and Wolf, 2002). The evolution of the fraction of impervious surfaces over the BCR since 1955 was studied by Vanhuysee et al. (2006). The results indicate a sharp increase in impervious surfaces areas, nearly a doubling from 26% in 1955 to 47% in 2006.

At the same time, a gradual increase in temperature has been observed at the national recording station of the Royal Meteorological Institute (RMI) of Belgium, located just south of the center of the capital in the Uccle suburb. This time series has a long history dating back to 1833 and has been homogenized within the EU IMPROVE project (Camuffo and Jones, 2002).[2]

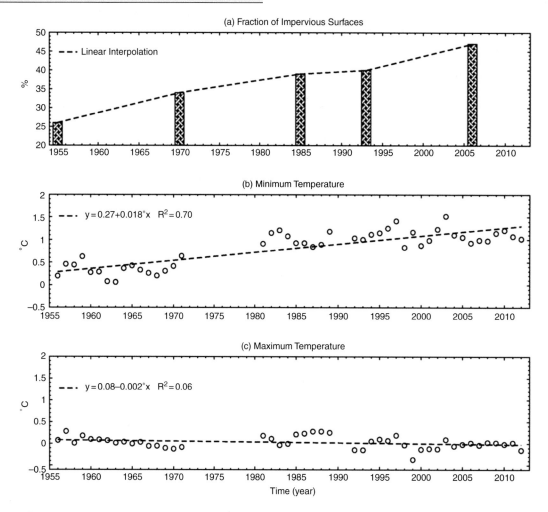

Case Study 2.1 Figure 1 *(a) Evolution of the average percent imperviousness of the Brussels Capital Region from 1955 to 2006; (b) and (c), annual mean urban heat island (UHI) effect on minimum and maximum, near-surface air temperature observed at Uccle, estimated as the difference with the rural climatological stations, with the linear trend, 1956–2012. R² is the coefficient of determination. The annual mean maximum temperature is substantially less affected by urbanization than the annual mean minimum temperature, which is consistent with previous studies in other cities (Landsberg, 1981; Kalnay and Cai, 2003; Hua et al., 2008). The increase of annual mean urban bias on minimum temperature may be attributed to: (1) higher thermal inertia, which, in combination with lower albedo of urban surfaces, delays the cooling of the cities at nights compared to rural areas (Hamdi and Schayes 2008) (2) limited evapotranspiration that prevents evaporative cooling of urban areas, and (3) the contribution of anthropogenic heat during night hours, which can also influence the long-term trend of near-surface air temperature.*

It is important to understand whether, and to what extent, estimates of warming trends at Uccle can be explained by the growth of the UHI of the city of Brussels due to urban sprawl. If observations of near-surface air temperatures in growing cities are used in the assessment of global warming trends, these trends may be overestimated.

Two stations belonging to the RMI climatological network, situated 13 kilometers away from the center of Brussels are used to assess the degree to which the Uccle near-surface temperature trends are amplified by urbanization.[3]

The UHI intensity is defined as the difference in near-surface air temperature between urban and rural stations. Estimates of urban bias (influences of urbanization on the observed trend) at the Uccle recording station on annual-mean minimum and maximum temperature calculated between 1956 and 2012 are plotted with the linear trends in Case Study 2.1 Figure 1. The UHI effect on minimum temperature is shown to be rising with increased urbanization, with a linear trend of 0.18°C (± 0.02°C) per decade. The coefficient of determination is $R2 = 0.70$, which indicates a strong dependence between the increase of urban bias on minimum temperature and the changes in the percentage of impervious surfaces.

Table 2.1 *Examples of cities with urban heat and urban precipitation islands*

City	Effect observed	Urban Heat Island intensity (°C)	Source	Urban Precipitation Island intensity (%)	Source
Athens	UHI	3.6	Kastoulis *et al.*, 1985	N/A	N/A
Cairo	UHI and UPI	1–4.5	Robaa, 2003	33–47	Robaa, 2003
Delhi	UHI	3.8–7.6	Mohan *et al.*, 2009	N/A	N/A
Houston	UHI and UPI	2–3	Streutker, 2003	22–25	Burian and Sheppard, 2005
Melbourne	UHI	−3.2–6	Morris *et al.*, 2000	N/A	N/A
Mexico City	UHI and UPI	3–7.8	Jauregui *et al.*, 1997	70	Jauregui, 1996
New York	UHI	1.5–8	Gaffin *et al.*, 2008 and Gedzelman *et al.*, 2003	N/A	N/A
São Paulo	UHI	5–10	Sobral, 2010	N/A	N/A
Shanghai	UHI and UPI	0.7–3	Jiong, 2004 and Hung *et al.*, 2006	5–9	Jiong, 2004
Tokyo	UHI	4–12	Hung *et al.*, 2006	N/A	N/A

2.3.2 Urban Precipitation, Moisture, and Wind Effects

Besides temperature, the dense urban environment can also have impacts on other climate variables (see Table 2.1). For example, there is evidence that precipitation can be enhanced downwind of highly urbanized areas (Burian and Shepherd, 2005; Shepherd, 2006; Han et al., 2014). One possible mechanism for this is that the buildings within a city provide a source of lift for air, which, combined with a destabilized environment due to the heat island, leads to cloud development and precipitation (Shepherd et al., 2010). Regional climate models (RCMs) (see Section 2.6) have been used to simulate precipitation patterns near urban centers and investigate the extent to which urbanization influences them. A study analyzing precipitation in Tokyo, Japan, found that precipitation in the city is enhanced by urbanization, with the regional climate model able to simulate the effects (Hiroyuki et al., 2013). RCM simulations of precipitation in the Baltimore-Washington metropolitan area show that urban surface characteristics, such as the presence of buildings, paved roads, and mass vehicle use, influence rainfall patterns (Li et al., 2013; Li and Bou-Zeid, 2013). Although these studies have supported the hypothesis of urbanization influencing local precipitation, other studies (looking at cities in Turkey) found no evidence that urbanization affects local precipitation patterns (Tayanc and Toros, 1997).

---

2 Corrections due to non-climatic factors such as changes in observation time, instrumentation, and relocation from the Gate of Schaerbeek in the city center to the Plateau of Uccle in 1890 were taken into account. More details about the homogenization of the time series of Uccle can be found in Demarée et al., 2002.

3 Without assurance of homogeneity, trend estimates are unreliable, and artifacts in long-term observations and rural/urban differences can be introduced and may bias the estimate of the UHI. For this reason, the assumption that the three rural sub-series could be linked to constitute a reference rural series was tested with respect to data homogeneity (see Hamdi and Van De Vyver, 2011 for full information on the testing methods).

There are also examples of greater incidence of extreme precipitation events over cities – one study noted 35% more heavy downpours over Houston, Texas, compared to adjacent rural areas, possibly due to enhanced convection due to the UHI (Burian and Shepard, 2005). The urban environment may also influence thunderstorm composition and structure. A study of storm progression across Indianapolis, Indiana, found that more than 60% of storms change structure over the city itself, compared to only 25% when passing over surrounding rural regions (Niyogi et al., 2010). This effect was more likely during the daytime. This study also used regional climate models to verify the influence urban areas have on the thunderstorms, with simulations unable to model the thunderstorms without the inclusion of the urbanized Indianapolis region. Whereas these two examples reveal that urban areas may be enhancing convective precipitation, other studies suggest that increased aerosol concentrations in urban areas can "interrupt" precipitation formation processes and thereby reduce heavy rainfall (Seifert et al., 2012).

An additional urban phenomenon, urban moisture excess (e.g., Holmer and Eliason, 1999; Kuttler et al., 2007) or the *urban moisture island* (Richards, 2005) refers to conditions where higher humidity values are observed in cities relative to more rural locations. The primary mechanisms for these differences are evaporation, condensation, advection, and anthropogenic emissions of water vapor. Thessaloniki, Greece, is one city where observations show that the urban center is moisture-rich compared to its surrounding, semi-rural areas; this condition is more prevalent at nighttime (Giannaros and Melas, 2012).

Cities can also experience faster or slower wind speeds compared to their adjacent suburbs and countryside. Although urban structures increase the roughness of the land surface and present a widespread impediment to wind, periods of strong convection in the urban heat island can overcome the friction effect and cause locally elevated wind speeds (Lee, 1979). The net result is that the urban boundary layer tends to weaken winds that are fast and strengthen winds that are slow (Childs and Raman, 2005). Street-scale studies have indicated that channeling is a prominent feature at the neighborhood level, increasing wind speeds in street canyons parallel to the prevailing winds and decreasing them in perpendicular ones (Dobre et al., 2005).

There is also evidence of interactions between the different microclimate effects. For example, in Melbourne, wind speed and cloud cover were linked to the strength of the UHI (Morris et al., 2001). Regional climate modeling of the UHI near Taipei, Taiwan, simulated how enhanced temperatures affect the location of precipitation and thunderstorms near the city (Lin et al., 2011). A study of streamflow near Ottawa, Canada, linked reduced severity of spring floods to the UHI effect (Adamowski and Prokoph, 2013). Placing these research results into the context of city decision-making, the findings from such studies could be of relevance to water managers in cities around the globe. The complexity and connections between the many urban climate effects reveals the need for continued advances in modeling and observational analysis.

2.3.3 Urban Climate Monitoring Networks

Using a network of weather monitoring stations and both satellite and ground-based remote sensing instruments, urban meteorology networks track multiple climate variables (e.g., temperature, precipitation, humidity, and wind speed) at high/fine spatial and temporal resolutions and can be used for a variety of applications, including urban micrometeorology research, and real-time tracking of extreme weather events (Muller et al., 2013a). Improvements and expansion of these systems can further the understanding of urban microclimates and improve high-resolution climate data availability in cities. Several cities across the globe have already established urban monitoring networks, including Helsinki (Wood et al., 2013), Hong Kong (Hung and Wo, 2012), and Tokyo (Takahashi et al., 2009) (see Table 2.2).

Table 2.2 *Examples of urban monitoring networks. Source: Adapted from Muller, 2013a.*

City	Monitoring network	Description of network	Source
Berlin	Berlin City Measurement Network	City-scale operational and research network	http://www.geo.fu-berlin.de/en/met/service/stadtmessnetz/index.html
Helsinki	The Helsinki Testbed	Open research network, advance knowledge of mesoscale meteorology	Dabberdt *et al.*, 2005; Koskinen et al., 2011; http://testbed.fmi.fi/
Hong Kong	Community Weather-Information Network (Co-WIN)	City-scale network to engage communities, promote education, and raise awareness of urban environmental issues, such as climate change, acid rain, and urban heat island	Hung and Wo, 2012; http://weather.ap.polyu.edu.hk/
London	London Air Quality Network (LAQN)	Mesoscale air quality monitoring with meteorological variables	http://www.londonair.org.uk/

Table 2.2 (*continued*)

City	Monitoring network	Description of network	Source
New York	NYC Mesonet	Integration of existing meteorological station data to provide real-time modeling	http://nycmetnet.ccny.cuny.edu/index.php
Taipei	Taipei Weather Inquiry-Based Learning Network (TWIN)	City-scale education/public information network	Chang *et al.*, 2010; http://www.aclass.com.tw/products.aspx?BookNo=weather_01
Tokyo	Metropolitan Environmental Temperature and Rainfall Observation System (METROS)	City-scale temperature and precipitation observing system	Mikami *et al.*, 2003; Takahashi *et al.*, 2009
Washington D.C.	DCNet	City-scale network for forecasting dispersion of hazardous materials	Hicks *et al.*, 2012

Knowledge of urban microclimates has been advanced through the use and development of city-scale climate monitoring networks. For example, data from the meteorological network in Atlanta show that the UHI can trigger convective rainfall activity in the city (Bornstein and Lin, 2000). Observations from the weather network in Paris have been used to validate computer model simulations of the city's urban temperature and humidity microclimates (Lemonsu and Masson, 2002). Understanding urban islands and microclimates within cities and their metropolitan areas allows for better identification of climate vulnerabilities and risks, thereby enhancing the initial steps of the adaptation process (Major and O'Grady, 2010).

Climate data are often lacking in many cities. These records serve as important tools for analyzing past and current climate risks and can help form the basis of future climate projections. Depending on the methodologies used for developing future climate projections (see Section 2.6), future modeled changes may be applied to observed climate data. Observed records are also of great importance for historical climate model validation. Although historical climate analysis can provide a basis for future planning, it is important to note that it is necessary for adaptation strategies to prepare for climate hazards beyond those already experienced (Milly et al., 2008).

Whereas increasing the number of meteorological stations is an important step for improving climate data in urban areas and advancing knowledge on urban climate science, there is also a need to harmonize collection practices, instrumentation, station location, and quality controls across cities to facilitate collaborative research and adaptation initiatives (Muller et al., 2013b). Novel techniques to increase the number of observations and improve data quality include Light Detection and Ranging (LIDAR), scintillometers[4], and low-cost sensors (Basara et al., 2011; Wood et al., 2013). There also may be

opportunities to crowdsource climate data through citizen science, social media, amateur weather stations and equipment, smart devices such as cell phones, and mobile platforms (e.g., sensors mounted on motor vehicles) (Overeem et al., 2013; Muller et al., 2015).

As meteorological stations and networks expand across cities, additional climate variables may be tracked and adaptation planning efforts can advance. With knowledge of the strategies that can be used to protect against a particular climate hazard, the implementation can now be targeted toward the most vulnerable communities. If particular neighborhoods are more prone to extreme heat events, based in part on observations from urban temperature monitoring stations (both maximum and minimum), cooling resources could be allocated there when heat waves are forecast. Urban precipitation monitoring networks can be used to identify portions of cities that are most at risk of flooding from intense precipitation events. Planning efforts to reduce urban flooding, such as the cleaning of storm drains in advance of heavy rainfall, could be concentrated in those vulnerable areas identified by the monitoring network (City of New York, 2013).

Data from meteorological networks in urban areas can also be used to evaluate the effectiveness of adaptation strategies. In Oberhausen, Germany, a network of meteorological observing stations was used to assess the potential of blue and green infrastructure to increase thermal comfort and reduce the thermal load of the city (Müller et al., 2013; Goldbach and Kuttler, 2013).

Urban monitoring is also done for mitigation as well as for adaptation and resilience purposes. For example, the Los Angeles Megacities Carbon Project tracks carbon dioxide, methane, and carbon monoxide with surface and satellite instruments (see Case Study 2.2).

---

4 A scintillometer is a scientific device used to measure small fluctuations of the refractive index of air caused by variations in temperature, humidity, and pressure. These instruments can measure the transfer of heat between the Earth's surface and atmosphere. Such measurements can be used to better understand the UHI effect.

Case Study 2.2   Los Angeles Megacities Carbon Project

Riley Duren

NASA Jet Propulsion Laboratory, California Institute of Technology, Pasadena

Kevin Gurney

Arizona State University, Tempe

Keywords	Climate mitigation, carbon, greenhouse gas, climate feedbacks
Population (Metropolitan Region)	13,340,068 (U.S. Census Bureau, 2015)
Area (Metropolitan Region)	12,557 km² (U.S. Census Bureau, 2010)
Income per capita	US$56,180 (World Bank, 2017)
Climate zone	Csa – Temperate, dry summer, hot summer (Peel et al., 2007)

Urbanization has concentrated more than 50% of the global population, approximately 70% of fossil fuel carbon dioxide (CO_2) emissions, and a significant amount of anthropogenic methane (CH_4) into a small fraction of the Earth's land surface. Were they a single nation, the 50 largest cities would have collectively ranked as the third largest emitter of fossil fuel CO_2 after China and the United States in 2010 (World Bank, 2010). Carbon emissions from cities and their power plants are projected to undergo rapid change over the next two decades with cases of growth and stabilization. Cities are emerging as "first responders" for climate mitigation – with an increasing number of voluntary carbon stabilization programs in the largest cities as well as active exploration of linked sub-national carbon emissions trading systems. The ability to accurately assess and project the carbon trajectories of cities depends on improved understanding of the urban carbon cycle including interactions between the atmosphere, built systems, land, ecosystems, and aquatic systems.

The following key questions motivate urban carbon cycle studies in general (Hutyra et al., 2014):

1. What are the urban anthropogenic carbon fluxes? How are these fluxes and associated carbon pools changing in time and space? How are they likely to change in the future?
2. What are the primary causes for discrepancies between research-grade and regulatory or "self-reported" emission inventories? Can we reconcile "top-down" and "bottom-up" approaches to quantifying fossil fuel CO_2 emissions?
3. Can we attribute fluxes to their underlying processes and resolve emissions in space and time?
4. How are these emissions manifested across cities, and what are their sensitivities to the many controlling factors in different urban environments: geography, topography, climate, ecosystem type, socioeconomics, and engineering/technological factors? Are there emerging urban typologies for carbon emissions?
5. How do we apply natural science information on urban carbon flows to support and assess climate change policy options and assess efficacy?

Answering these questions is challenging given the paucity of urban-scale carbon flux data – emissions estimates at the scale of cities are often nonexistent or highly uncertain (NRC, 2010). The Los Angeles Megacities Carbon Project (http://megacities.jpl.nasa.gov) is one of several pilot projects currently under way with the goal of increasing the availability, fidelity, and policy relevance of urban carbon flux data and improving scientific understanding of the controlling processes.

The Megacities Carbon Project began as a partnership between U.S. and French scientists in 2010 to establish measurement networks, data assimilation, and analysis in the megacities of Los Angeles (Kort et al., 2013) and Paris (Breon et al., 2014). The team is working with collaborators in São Paulo, London, and other megacities to coordinate research and data sharing. In addition to advancing the state of the art and promoting technical capacity building, these and other related efforts seek to contribute to the formation of a *urban carbon typology* with biogeophysical and socioeconomic characteristics that represent the diversity of processes controlling carbon fluxes for the world's cities. The Megacities Carbon Project also serves as a pathfinder for a potential global urban carbon monitoring system, one combining surface and satellite observations (Duren and Miller, 2012).

For the Los Angeles project, measurements of CO_2, CH_4, and carbon monoxide (CO) atmospheric mixing ratios are collected continuously from a network of 15 monitoring stations located in and around the basin (see Case Study 2.2 Figure 1). A mix of in-situ and remote-sensing methods are used to track carbon in the planetary boundary layer and total atmospheric column (Wunch et al., 2009; Kort et al., 2013). The surface measurements are traceable to international standards using calibration gas tanks at each site – offering a transparent, consistent, and traceable mechanism for intercomparison. In some locations, air samples are collected in flasks and returned to a central facility for analysis of radioisotopes to disentangle fossil fuel and biogenic sources. Satellites such the Japanese Space Agency's Greenhouse Gas Observing Satellite and NASA's Orbiting Carbon Observatory provide periodic total column CO_2 measurements for LA and other global megacities (Kort et al., 2012; Silva et al., 2013) that are validated with surface-based measurements in and around those cities. All of these atmospheric carbon measurements are combined with tracer transport models and meteorological data to generate "top-down" carbon flux estimates at the urban scale.

A high-resolution "bottom-up" fossil fuel CO_2 flux estimate has also been constructed for the LA megacity by researchers at Arizona State University and JPL using the "Hestia" methodology (Gurney et al., 2012). Hestia combines extensive data mining of public databases and a variety of model algorithms to estimate fossil fuel CO_2 fluxes at the building/street level every hour of the year. Among the data utilized are traffic volume measurements, census data, fuel statistics, tax assessor parcel data, hourly power plant emissions monitoring, and local air quality emissions reporting (e.g., Mendoza et al., 2013). The top-down and bottom-up datasets will be integrated with synthesis analysis, resulting in reduced uncertainty in carbon fluxes and improved understanding of the controlling processes. This information can help assess the impact of carbon management decisions including trends in emissions for specific municipalities, areas, and economic sectors. Improved understanding of the biogeophysical and socioeconomic processes that drive a city's carbon footprint and the linkages between them can help inform future projections and planning by urban stakeholders.

Case Study 2.2 Figure 1 *Conceptual illustration of the tiered observational system applied by the Megacities Carbon Project in Los Angeles is also representative of elements found in other urban carbon studies.*

Finally, a key goal of the Megacities Carbon Project is to deliver urban carbon data via a web-based portal for transparent data sharing. The objective is to provide decision-relevant carbon information to a broad range of stakeholders: urban planners, mayors, regional policy-makers, carbon markets, businesses, and members of the general public. The concept is that pilot efforts that demonstrate voluntary data sharing between cities could reduce barriers to broader visibility into carbon emissions globally, thus enabling greater collective confidence in climate mitigation efforts.

The Los Angeles Megacities Carbon Project is funded by the U.S. National Institute of Standards and Technology (NIST), National Aeronautics and Space Administration (NASA), and National Oceanographic and Atmospheric Administration (NOAA) as well as contributions from the California Air Resources Board and the Keck Institute for Space Studies (KISS).

2.3.4 Urban Air Quality

With the confluence of rapid urbanization, fast-growing populations due to mass migration from rural areas, and industrialization since the mid-20th century, observed urban air quality has been declining (Kura et al., 2013). Not only does this confluence result in increased air pollution, it also threatens sustainable urban living and has negative health impacts. Urban air pollution is linked to about 1 million premature deaths and 1 million prenatal deaths each year (Kura et al., 2013). In addition, it is costly to ameliorate. It is estimated that developed countries spend about 2% of gross domestic product (GDP) on urban pollution, whereas developing countries spend about 5% of GDP for urban pollution (Kura et al., 2013).

Urban air quality varies regionally. Using satellite observations, Lamsal et al. (2013) showed that regional differences in industrial development, per capita emissions, and geography were related to the population–pollution relationship. The study showed that, for the same population, a developed city might experience six times the pollution concentration of a developing city. A satellite-based multipollutant index illustrates these differences, with cities exhibiting higher levels of individual pollutant types based on region (Cooper et al., 2012). An additional study in which a multipollutant index for cities was developed found Dhaka, Beijing, Cairo, and Karachi as cities with the poorest air quality (Gurjar et al., 2008).

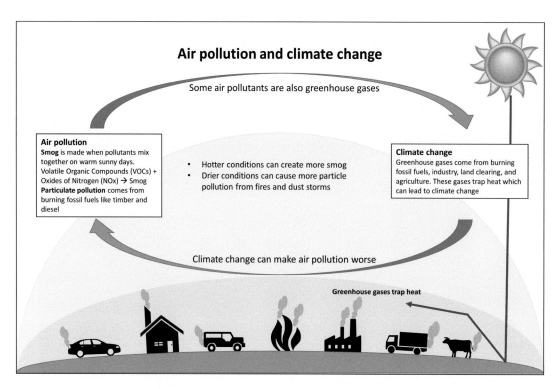

Figure 2.2 *The relationship between urban air pollution and climate change.*

Six main urban air pollutants are observed and measured: ground-level ozone, carbon monoxide, sulfur dioxide, particulate matter, lead, and nitrogen dioxide. Sources of these emissions include traffic, agriculture, fuel burning, natural sources (such as dust and salt), and industrial activities (see Figure 2.2). A recent study (Karagulian et al., 2015) found that for global particulate matter emissions, traffic is the largest source, followed by domestic fuel burning for cooking and heating. Looking specifically at emissions from vehicles, pollution concentrations can be directly correlated to daily rush hours and to the winter season. One study found that in Gothenberg, Germany, winter temperature inversions are associated with higher levels of traffic-related pollutants, including carbon monoxide, nitrous oxide, and nitrogen dioxide (Janhäll et al., 2006).

The spatially heterogeneous urban landscape with its inherently complex and highly variable emission sources makes both urban pollution measurement and modeling challenging endeavors. Zauli et al. (2004) reported that the spatial variability of pollution concentrations in cities often leads to mischaracterization of the urban environment and thereby complicates the study of the microclimate and its associated urban planetary boundary layer.

Climate change is projected to have impacts on the amounts of air pollutants and thus air quality in urban areas. Future changes are sensitive to both pollutant type and geographic region. The meteorological conditions (e.g., temperature inversions) that contribute to air quality in cities are also projected to change in the future (Jacob and Winner, 2009). Warming temperatures are linked to higher levels of ground-level ozone in many cities, and

ozone concentrations are projected to increase, particularly in the United States and Europe (Jacob and Winner, 2009; Katragkou et al., 2011). Future changes in particulate matter are less certain (Dawson et al., 2014).

Because they are prone to poor air quality, urban areas can invest in a variety of initiatives to have cleaner air. Pugh et al. (2012) showed that, along with controlling pollutant emissions and increasing dispersion (through changes in wind speed, wind direction, and atmospheric qualities), cities could improve air quality by increasing the deposition rates of pollutants. The study underscores the findings that green infrastructure (e.g., in urban street canyons) can be very effective deposition sites, reducing nitrogen dioxide (NO_2) by 40% and particulate matter by 60%. Along with other proven benefits to the urban center, particularly in the context of climate adaptation, vegetation can also be an efficient urban pollutant filter that helps to raise air quality at the street level in dense urban areas. For an example of the potential effectiveness of green infrastructure on reducing urban air pollution, a study of Toronto found that increasing the surface area for green roofs by 10–20% would greatly improve air quality in the city through pollutant removal (Currie and Bass, 2008).

To comprehensively study urban air quality in the context of climate change, an integrative strategy should include air quality measurements, which can be gathered through meteorological networks and satellites sensors, along with low-cost personal devices that can both collect and share real-time pollution data (Bertaccini et al., 2012). Emissions and pollutant data collected through urban observing networks can inform planning, thus allowing cities to be better able to issue air quality alerts. Such a system is already in place in Santiago de Chile (Gramsch et al., 2006). In order to track how clean and green technologies and other strategies within the climate adaptation framework are impacting urban pollution, monitoring ambient air quality, source emissions, and indoor air quality is required. This could lead to improved modeling of emissions and atmospheric dispersion (Kura et al., 2013).

2.4 Natural Climate Variability

Variations in the climate due to natural processes influence urban climate risk. The effects from human activities on the climate system are superimposed on the background natural climate variability. This section introduces the major modes of variability and identifies some examples of how they affect the climate in certain cities. The predictability of these modes, which in some cases allows for advanced preparedness, can serve as a learning tool for planning for future climate changes.

By analyzing natural climate variability, scientists and urban decision-makers can improve responses to current climate hazards and vulnerabilities and prepare for future changes.

2.4.1 Overview

The major modes of natural climate variability include the El Niño Southern Oscillation (ENSO), the North Atlantic Oscillation (NAO), and the Madden-Julian Oscillation (MJO). Additional modes include the Pacific Decadal Oscillation (PDO), Atlantic Multi-Decadal Oscillation (AMO), Indian Ocean Dipole (IOD), and the Pacific North American pattern (PNA) (see Table 2.3). To varying degrees, these modes represent coupled interactions or feedbacks between the atmosphere and ocean and can be thought of as preferred patterns in space and time. These modes are characterized by indices that measure their phase and/or strength. The influence of each mode varies depending on the season of the year. A given city's

Table 2.3 *Temporal and spatial scales of major modes of natural climate variability.*

Mode & Acronym	Temporal scale	Spatial scale
El Niño Southern Oscillation (ENSO)	Typically occurs every 3–7 years. El Niño and La Niña events (opposite phases of ENSO) generally last 9–12 months. They develop during March–June, reach peak intensity during December–April.	Equatorial eastern and central Pacific Ocean, with effects on circulation seen in tropics. Teleconnections[a] span the globe, depending on phase and season.
Madden-Julian Oscillation (MJO)	Cycles every 30–60 days Greatest level of activity is during the late fall, winter, and early spring. MJO activity is enhanced in connection with the warm sea surface temperatures during an El Niño. Activity peaks before El Nino events, then is absent throughout its duration. MJO activity strengthens during La Niña (Seiki et al., 2015) Strong year-to-year variability.	Tropical regions, primarily over the Indian and Pacific Oceans. Can also be present in the tropical Atlantic and over Africa. Teleconnections span the globe, depending on phase and season.
North Atlantic Oscillation (NAO)	Exhibits considerable interseasonal and interannual variability, and prolonged periods (several months) of each phase of the pattern are common The wintertime NAO also exhibits significant multi-decadal variability	Atlantic Ocean, with one center located over Greenland and the other near the Azores. Teleconnections in the eastern United States, western Europe, and the Mediterranean
Pacific Decadal Oscillation (PDO)	Shifts between warm and cool phases occurs every 20–30 years (Mantua et al., 1997). Phases identified by prolonged sea surface temperature anomalies[b] in the equatorial and North Pacific.	Pacific Ocean Teleconnections in the United States and South America
Atlantic Multidecadal Oscillation (AMO)	The warm and cool phases of AMO persist for approximately 20–40 years due to changes in the overturning circulation of water and heat in the North Atlantic.	North Atlantic Ocean Teleconnections in North America and the United States
Pacific/North American Pattern (PNA)	Low-frequency mode that varies on interannual-to-decadal time scales. Strongly influenced by the dynamical processes of the NAO, reacting with Rossby waves coming from Asia (Baxter and Nigam, 2013).	Pacific Ocean through North America Associated with strong fluctuations in the strength and location of the East Asian jet stream.
Indian Ocean Dipole (IOD)	Aperiodic	Indian Ocean Teleconnections in Australia and other countries that surround the Indian Ocean basin

[a] Teleconnections are linkages between weather and climate changes occurring in widely separated regions of the globe.
[b] An anomaly is the difference of (usually) temperature or precipitation for a given region over a specified period from the long-term average value for the same region.

Box 2.2   Modes of Natural Climate Variability and Cities

EL NIÑO SOUTHERN OSCILLATION

El Niño (La Niña) episodes are characterized respectively by the warming (cooling) of the tropical central and eastern Pacific surface. The main atmospheric manifestation, known as the Southern Oscillation, is a seesaw pattern of the global-scale tropical and subtropical surface pressure, which also involves changes in the trade winds and tropical circulation and precipitation (Rasmusson and Wallace, 1983).

Cities impacted by El Niño (La Niña) events include Singapore (Earnest et al., 2012), Caracas (Bouma and Dye, 1997), and Nairobi (see Case Study 2.3).

NORTH ATLANTIC OSCILLATION

The North Atlantic Oscillation (NAO) is a quasi-regular variation of atmospheric pressure between subtropical and high latitudes (Hurrell et al., 2003). It is the dominant mode of atmospheric circulation variability in the North Atlantic region. Its influence on climate extends over a much larger region, from North America to Europe, Asia, Africa, and even more remote regions. The NAO and its influence are stronger in northern hemisphere winter, but are present throughout the year. Although the NAO exhibits interannual variability, there has been a trend over approximately the past 30 years toward a more persistent positive state (Visbeck et al., 2001).

Examples of cities impacted by the North Atlantic Oscillation include Belgrade (Luković et al., 2015) and Oslo (Pozo-Vázquez, 2001). During the positive phase of the NAO, London and Paris generally experience warmer than average temperatures, while Athens is typically cooler than normal conditions (see Box 2.2 Figure 1).

MADDEN-JULIAN OSCILLATION

The Madden-Julian Oscillation (MJO) is the dominant mode of tropical intraseasonal climate variability (cycling every 30–60 days). It is a "pulse" of cloudiness and rainfall moving eastward in the equatorial region, in the Indian Ocean, and the western Pacific Ocean (Zhang, 2005), and it can excite atmospheric teleconnections that affect the climate and weather in many regions around the world.

Cities influenced by the MJO include Rio de Janeiro (see Case Study 2.3), Seattle (Bond and Vecchi, 2003), and Dakar (Conforth, 2013).

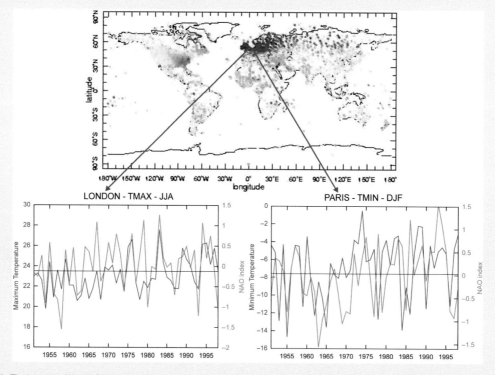

Box 2.2 Figure 1 *The impacts of the NAO on surface temperature. The upper panel shows the correlation between the NAO index and winter surface temperature (from http://www.ldeo.columbia.edu/res/pi/NAO/intro/correlations.html). Areas in yellow and red (blue and green) indicate regions where a positive (negative) NAO index is linked with above (below) normal temperatures. The two graphs plot temperature against the NAO index in London (for summer maximum temperature) and Paris (for winter minimum temperatures). In both cities, London during the summer and Paris during the winter, warmer than normal temperatures (red line) are associated with positive NAO values (green line).*

The cities provided here, as examples of those that are impacted by the major modes of natural variability, are a select group for each mode. A much larger number of cities across the globe, although not identified, may experience impacts of the different modes. The cities in Box 2.2 were chosen based on available literature documenting the relationships between their climate and natural variability.

Case Study 2.3 Rio de Janeiro Impacts of the Madden-Julian Oscillation

Alice Grimm

Federal University of Paraná, Curitiba

Keywords	Madden-Julian Oscillation, precipitation, rainstorm, disasters
Population (Metropolitan Region)	12,981,000 (UN, 2016)
Area (Metropolitan Region)	5,328.8 km² (IBGE, 2015)
Income per capita	US$8,840 (World Bank, 2017)
Climate zone	Am – Tropical monsoon (Peel et al., 2007)

The metropolitan region of Rio de Janeiro, Brazil, including Niteroi and other neighboring cities with a total population of more than 11.8 million, is vulnerable to rainstorm-related disasters. The characteristics of Rio de Janeiro's natural terrain, inadequate urban planning, and the city's sprawl over mountains and marshlands have made it prone to landslides and flooding. In the preceding century, one of the worst rainstorms happened in January 1966, when almost 250 millimeters of rain fell in less than 12 hours, causing loss of life and large economic damages (UEL, 2014; Corpo de Bombeiros, 2014). Another strong flood event happened in February 1988, when it rained over 430 millimeters over a 4-day period, killing 290 people and leaving close to 14,000 without homes (UEL, 2014; Corpo de Bombeiros, 2014; Costa, 2001; UNDRO, 1988). In the 21st century, the worst episode of floods and landslides in the metropolitan area occurred in April 2010, causing large-scale disruption to local livelihoods. Observations in the region near Rio de Janeiro show heavy rainfall events have been increasing over the past 50 years, creating a cause for concern and further analysis.

Rio's location within the South Atlantic Convergence Zone (SACZ)[5] makes the city prone to intense rainfall events during the southern hemisphere summer (December-February) rainy season, beginning in spring and ending in autumn. Intense precipitation events in Rio de Janeiro are mainly caused by frontal systems and occurrence of the SACZ (Dereczynski et al., 2009). The frequency and intensity with which these large-scale

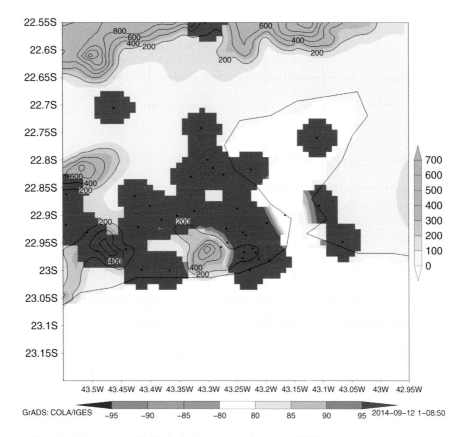

Case Study 2.3 Figure 1 *Meteorological stations used in Rio de Janeiro to assess the impact of MJO on the frequency of extreme events in summer (December, January, and February). Lines indicate topography (in meters), with the three massifs that surround the city (compact mountain groups). In the right half of the figure is the Bay of Guanabara and in the bottom is the Atlantic Ocean. The red color indicates the stations in which the number of extreme precipitation events significantly increases during Phase 1 of MJO in summer (level of confidence better than 95%).*

5 The SACZ is a local maximum in cloudiness, precipitation, and low-level convergence that extends across southeastern Brazil and the western South Atlantic.

(synoptic) phenomena occur is modulated by climate variability, including El Niño and La Niña episodes in the interannual time scale, and the Madden-Julian Oscillation (MJO) in the intraseasonal time scale (Grimm and Tedeschi, 2009; Tedeschi et al., 2015; Hirata and Grimm, 2015). Although the ENSO influence is less consistent, the MJO impact is very significant. A statistical assessment of the connection between Phase 1 of MJO, when enhanced equatorial rainfall associated with the event is located over the western Indian Ocean and extreme precipitation (defined as above the 90th percentile), confirmed with 95% confidence level that in all but one series of station precipitation data analyzed in the city of Rio de Janeiro the number of extreme summer rainfall events increased during Phase 1 of MJO (see Case Study 2.3 Figure 1).

The MJO is not the sole contributor to or necessarily the main influence on the intensity and frequency of extreme precipitation

events in Rio de Janeiro. As a city near the sea and surrounded by mountains, there are also local effects (such as land and sea breezes) producing heavy rainfall, and there have been very extreme events not associated with MJO. However, the relationship between the MJO and intense precipitation may allow for, in some cases, advance planning in order to reduce climate risks. The MJO's level of predictability contributes to Rio de Janeiro's precipitation-based planning; knowledge of the patterns of the MJO can prove to be useful in infrastructural and flood contingency efforts at the local scale. By using regional climate models to simulate the cyclical influence of the MJO, a more accurate projection of future interannual variability in Rio de Janeiro rainfall can be provided. With damaging floods in recent memory for this urban setting, precipitation prediction is especially important to inform policies that contribute to proactive adaptation and resilience efforts aimed at protection of local and precipitation-vulnerable people and their livelihoods.

climate variability maybe influenced by a combination of patterns, although one may have a dominant influence and they could act at different timescales (see Case Study 2.3). Box 2.2 presents examples of cities impacted by modes of natural variability.

The modes of natural climate variability can affect not only monthly, seasonal, and annual mean temperature or precipitation totals, but can also (and even more significantly) affect the frequency and intensity of extreme events that can produce natural disasters (Grimm and Tedeschi, 2009). It is important for urban stakeholders to be aware of climatic modes and how they may impact key urban sectors and infrastructure (Ning and Bradbury, 2015).

Understanding how climate change may influence natural climate variability is an active area of research. Although the confidence in future climate projections on how patterns of natural variability may change in the future is low (see Section 2.4.3), some modes (such as ENSO and MJO) have predictability on season-to-annual timescales. This allows for their evolution to be forecasted in advance. Advanced lead-time to climate risks on these shorter timescales can help decision-makers prepare ahead for possible impacts. The lessons learned through preparedness for extreme events associated with natural variability can potentially be applied to the planning of adaptation and resilience strategies for future climate extremes.

2.4.2 Natural Climate Variability and Urban Decision-Making

The availability and dissemination of seasonal forecast information, based in part on the modes of natural climate variability, have the potential to inform decision-making in many urban areas. Known teleconnections between a mode and climate variable for a given city could aid planning and management of urban sectors, including public health, water resources, and energy. For example, if water managers in a particular city are anticipating a drier-than normal-season

related to ENSO or the NAO, they might not release water from a dam even when faced with an immediate short-term risk of minor flooding so that water supplies remain adequate in the months to come. Droughts in Brazil, which can severely impact water supplies, have been linked to such patterns of natural climate variability (Marengo, 2004). Recognizing this relationship, utility managers in the river basin where the city of Rio de Janeiro's water supply comes from have developed new management practices (Formiga-Johnsson and Britto, 2009). Case Study 2.4 provides an illustration of how cities may prepare for climate extremes and health impacts associated with patterns of natural climate variability in part based on seasonal forecast information and how climate change may play a role (see Chapter 10, Urban Health).

Across Scandinavia, the NAO is linked to seasonal rainfall and therefore streamflow. This link in turn affects hydropower output (Cherry et al., 2005). In cities across the region (e.g., Copenhagen and Stockholm), this known relationship could allow for advanced planning in the energy sector, given that a seasonal forecast could be used to predict energy prices and estimate demand (see Chapter 12, Urban Energy).

Additionally, climate variability is often linked to water usage and supply (see Chapter 14, Urban Water Systems), which also presents opportunities for decision-making based on seasonal forecast information. In North America, Portland, Oregon demonstrates observed linkages between climate variability and water resources, illustrating the potential for forecasts to guide water resources management (Chang et al., 2014).

For an example in the public health sector, seasonal pollen in London has been linked to the phase of the NAO (Smith and Emberlin, 2006). Based on the state of this mode of natural variability, health officials can potentially have lead-time to preparing for high-impact events for allergy sufferers (see Chapter 10, Urban Health). Urban air quality in Hong Kong has also been linked to El Niño phases (Kim et al., 2013).

Case Study 2.4   Will Climate Change Induce Malaria in Nairobi?

Annie Ramsell

Climate Change Adaptation Unit, UNEP, Nairobi

Keywords	Climate science, malaria, El Niño, urban health, rainfall, science for cities, action plan
Population (Metropolitan Region)	6,547,547 (KNBS, 2010)
Area (Metropolitan Region)	30,389 km² (KNBS, 2010)
Income per capita	US$1,380 (World Bank, 2017)
Climate zone	Cwb – Temperate, dry winter, warm (Peel et al., 2007)

Kenya's capital city, Nairobi, is projected to be adversely affected by the impacts of climate change, with temperature and rainfall variability resulting in an array of cross-sectoral climate-related impacts (IPCC, 2013). The shifting climatic system of the South Indian Ocean Dipole (IOD) is believed to contribute to the increasing uncertainty of seasonal and interseasonal rainfall and spatial variability in Kenya (IPCC, 2013; Hashizume et al., 2012) and is already leading to a warmer climate in Nairobi. These changes in climate are thought to make highland areas in East Africa more suitable for malaria epidemics, potentially placing Nairobi as a future "at risk" city for malaria (Ermert et al., 2012). Improved knowledge of malaria threshold conditions and greater predictability of localized weather patterns would enable the Government of Nairobi to assess the threat of a malaria epidemic and determine the need for a response strategy.

The IOD is a coupled ocean–atmosphere phenomenon in the equatorial Indian Ocean defined as the difference between the sea surface temperature (SST) of the eastern and western parts of the Indian Ocean and is measured by the Dipole Mode Index (DMI). A negative IOD event is characterized by cooler than normal water in the tropical eastern part of the Indian Ocean and warmer than normal water in the western part of the Indian Ocean basin. Heavy and prolonged rains during the short rainy season (September, October, and November) in the East African highlands are most likely influenced by positive IOD events (Hashizume et al., 2012). Any prolonged short-rains season is therefore conducive to a seasonal malaria event since the short-rains season is followed by warm weather, enhancing the development of mosquitoes and malaria parasites. A shift in the IOD induced by climate change could therefore directly influence the weather in Nairobi, exposing the city to future malaria epidemics. Even if precipitation and the IOD remained unchanged, warmer temperatures thereafter due to climate change could also increase malaria risk.

According to the Intergovernmental Panel on Climate Change (IPCC) Fifth Assessment Report (AR5; Smith et al., 2014), highland areas in East Africa could experience increased malaria epidemics due to climate change. However, the complexity of malaria development and how it is affected by climate variations is a subject of uncertainty. Nonetheless, theories on malaria thresholds have been studied and observed (Cox et al., 1999), and thresholds

Case Study 2.4 Table 1 *Summary of Climate Variables for Kericho and Nairobi.*

Climate variable	Minimum malaria threshold	Kericho (2,182 m)	Nairobi (198 m)
Precipitation	Greater than 80 mm for 5 consecutive months	✓	X
Relative humidity	Greater than 60%	✓	✓
Mean temperature	Greater than 14.5°C	✓	✓

for malaria transmission have been established for temperature, humidity, and precipitation. Malaria transmission requires a relative humidity of minimum 60%, monthly average precipitation of 80 millimeters or more for 5 consecutive months (Cox et al., 1999), and for the malaria development cycle to be completed at monthly mean temperatures not below 14.5°C and 20°C for the parasites *Plasmodium vivax* and *P. falciparum*, respectively (Kim et al., 2012).

Malaria epidemics have long been common in the Kericho district in western Kenya, and malaria is now a seasonal occurrence within the district. High levels of malaria incidence emerge in Kericho after the long rains in March until the declining temperatures decrease the transmission in July. Changes in climate variability, increased resistance to antimalarial drugs, population movements, alterations in mosquito vectors (e.g., species), and impaired health services are all contributing factors to the re-emergence of malaria seasonality in Kericho (Shanks et al., 2005). Many studies have implied that Nairobi is malaria-free because of its location at an altitude of 1,798 meters and its resulting low mean temperature (Hay et al., 2002). However, Kericho, at 2,182 meters, is situated at a higher altitude than Nairobi and its mean monthly minimum temperatures are not significantly different. Because temperatures alone do not correlate with malaria transmission and minimum humidity and precipitation are also required (Cox et al., 1999), an assessment of data on precipitation, temperature, and humidity in Nairobi and Kericho from Kenya Meteorological Services (KMS) assessed the climatic differences related to malaria transmission between these two sites. These findings are summarized in Case Study 2.4 Table 1.

Prolonged short-rain seasons are recognized as triggers for increased malaria transmission in the East African Highlands region (Hashizume et al., 2012), and, as such, the primary focus on potential malaria introduction in Nairobi was assessing data on the short-rains season precipitation and timeline. Data on monthly precipitation over a 30-year period show that Nairobi does receive the minimum 80 millimeters precipitation during its short-rains season, but the season has not exceeded 5 consecutive months in current conditions.

In the comparison site of Kericho, the data revealed that although moderate malaria transmission occurs throughout the year, including the short-rains season, the highest levels of malaria transmission were actually found to occur during the long rains (Tonui et al., 2013). The study subsequently looked into whether malaria-suitable climate conditions occurred in Nairobi during any period of the year. The data revealed that Nairobi had precipitation of 80 millimeters or

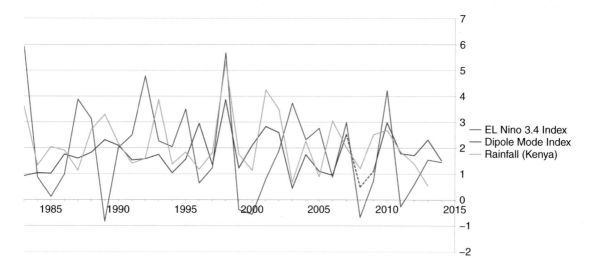

Case Study 2.4 Figure 1 *Average rainfall over Kenya from eight weather stations for the period of December, January, and February (DJF) annually from 1983 to 2013, against sea surface temperatures (SSTs) in the central Pacific Ocean region associated with the El Niño Southern Oscillation (ENSO 3.4) and the Indian Ocean Dipole (IOD) (represented by the Dipole Mode Index) for the same periods.*

more for only 4 consecutive months six times in the past 30 years. The observations show that the trigger point for malaria in Nairobi is increased precipitation, but no indication of increased precipitation for 5 consecutive months has been found in the past 30 years. Furthermore, extrapolating the precipitation trends to 2030 does not indicate that any malaria introduction will occur in Nairobi up to this future period.

This study shows that Nairobi is malaria–free because it does not meet the malaria threshold of more than 80 millimeters of precipitation over 5 consecutive months. Nevertheless, Nairobi was found to be very close to meeting all minimum thresholds. When correlating the precipitation data with El Niño events, the results show that 4 out of the 6 years in which Nairobi had 4 consecutive months of more than 80 millimeters precipitation are directly correlated to El Niño events. Nairobi is therefore only 1 month short from being climatically suitable for malaria during years with El Niño events. This indicates the need for better climate variability modeling in Nairobi so it will have the capacity to

project potential future outbreaks of malaria. This will make the city more resilient in preparing for effective preventative malaria response action.

Rainfall in East Africa is known to align closely with the ocean-atmospheric anomaly of the El Niño Southern Oscillation (ENSO) over the Pacific Ocean, reacting as well with the IOD in the Indian Ocean, as Case Study 2.4 Figure 1 demonstrates. The effect, known as a teleconnection, refers to climate anomalies that are related to one another over large distances. The 2014 IPCC AR5 indicates that future warming is highly likely to intensify interannual rainfall variability in East Africa, resulting in an increase in the number of extremely wet seasons (Smith et al., 2014). Monitoring and prediction of ENSO are therefore of large value for countries in the East Africa region.

In conclusion, improved climate modeling and monitoring of future El Niño events will enable the city government to plan for and reduce the potential impacts of future malaria epidemics in Nairobi.

It is important to recognize that despite the connections between natural variability and sectoral impacts, there are many other factors involved, and climate may not be the sole or best predictor of societal impacts. For example, in Phoenix, Arizona, although water usage is linked to climate variables (temperature and precipitation), using local observations does not necessarily aid managers because the water supplies are located far outside of the city (Balling and Gober, 2007). Furthermore, every occurrence of a particular mode of natural variability may not bring the expected impacts due to differences across the events themselves. Therefore, decision-making based on climate variability is constrained by uncertainty. However, these opportunities do illustrate that modes of natural variability do impact cities, and decision-making can take short-term seasonal information into account along with other factors. This is similar to the process of using future climate risk data in long-term planning.

Even if modes of variability and teleconnections remain the same, climate change could modify their impacts (see Section 2.4.3). For example, if climate change leads to a drier mean state in southern California, El Niño-driven precipitation events might be less likely to lead to saturated ground, reservoirs, and flooding. Alternatively, the same drier mean state could be associated with more fires and changes in vegetation cover that paradoxically increase flood risk once El Niño rains arrive.

The potential for and current use of scientific information in decision-making demonstrates the need for continuing interactions between scientists and stakeholders and also emphasizes the need for improved communication of scientific information. Research on the effectiveness of seasonal forecasts for planning has identified a set of constraints limiting the usefulness of the information and proposed ways to overcome them (Patt and Gwata, 2002). Scientists must incorporate the needs of the local users, clearly communicate uncertainties, and repeat the process regularly. These barriers and ways to overcome them were tested at rural sites in Zimbabwe; the outcomes may have application to many cities in the developing world (Patt and Gwata, 2002).

2.4.3 Modes of Natural Variability and Climate Change

Interactions between climate change and natural climate variability are complex and their predictability remains challenging. As decision-makers in cities become more familiar with using seasonal forecast information, clear expression of how these modes may change in the future will become increasingly important (Deser et al., 2012). As the climate changes globally, there may be changes in patterns of natural variability in terms of their strength, frequency, and the duration of the various modes. Even if the patterns of variability remain similar to what they are today, climate change may alter the background (or "mean state") thereby modifying teleconnection patterns that drive regional climate impacts. This could have implications for cities, especially those that use seasonal information for planning.

The Intergovernmental Panel on Climate Change (IPCC) Fifth Assessment Report (AR5) found that El Niño is projected to continue to remain the most prominent mode of natural variability and will continue to have global influences (Christensen et al., 2013). It remains uncertain as to whether a more El-Niño-like state due to warmer sea surface temperatures will become more dominant than La Niña with climate change (Collins et al., 2005). As for the frequency of El Niño events with climate change, Cai et al. (2014) project a doubling in occurrences of El Niño with climate warming, stemming from projected warming over the eastern equatorial Pacific Ocean. Other studies (e.g., Collins et al., 2010) suggest that it is not possible to project changes in El Niño frequency at this time.

The teleconnections associated with El Niño that may impact regional climate and thereby cities, may also change in the future. An eastward shift in the Pacific-North American (PNA) pattern is projected, and this shift may intensify rainfall anomalies along the west coast of North America and push El Niño-associated warming further east across the continent (Zhou et al., 2014). Cities potentially impacted by such changes include Seattle and Los Angeles. Precipitation at the regional scale associated with El Niño events is likely to intensify due to changes in available moisture (Christensen et al., 2013; Power et al., 2013).

The NAO is projected to become (on average) slightly more positive with climate change caused by increasing greenhouse gases (GHGs), with continuing large natural variations, as observed in the present climate (IPCC, 2013). A more positive NAO could be associated with wetter conditions in northern Europe, drier conditions in the Mediterranean, and fewer cold air outbreaks over eastern North America. Future projections of the NAO are highly sensitive to the climate models used because some are unable to simulate this pattern of observed variability in the present climate (Davini and Cagnazzo, 2014). Despite these difficulties, some climate models are able to simulate current and future teleconnections from the NAO with regard to temperature and precipitation in the United States (Ning and Bradley, 2015).

2.5 Observed Climate in Cities

Observations over the past century at the global and continental scales have shown that climate is already changing in response to increasing GHG concentrations. Global temperatures and sea levels have both been rising. Regionally, changes in both mean climate variables (e.g., seasonal and annual temperature and precipitation) and the frequency and intensity of extreme events (e.g., hot and cold days, days with intense precipitation, coastal storms, and floods) have been observed (IPCC, 2012; 2013). The climate in many cities is also changing, but attributing urban climate change to increasing GHG emissions is challenging because of the process of urbanization and its connection to the urban heat island effect (see Section 2.3.1) and the relatively small scale of analysis that increases variability. In many cases, changes observed in cities across the globe may serve as a starting point for identifying the climate risks that may increase in the future.

Attributing observed trends to natural and/or anthropogenic processes is a major focus of climate science study (Cramer et al., 2014). There is considerable scientific consensus that the observed warming trends in recent decades cannot be explained without anthropogenic forcing (IPCC, 2013). In contrast, recent changes in regional precipitation patterns may be more heavily driven by natural climate cycles (Hoerling et al., 2010). To better understand how the climate of urban environments is changing and how this relates to the process of urbanization, long-term climate data records are required, reinforcing the need for improved observations.

2.5.1 Urban Climate Trends

Across the globe, annual average temperature (combined land and ocean surface data) has increased by approximately 0.85°C since 1880 (IPCC, 2013). Global sea levels have also risen, with a rate of 1.2 mm/year for the 1901–1990 period and an increased rate of 3.0 mm/year for the 1993–2010 period (Hay et al., 2015). Although these trends are global, warming temperatures and rising sea levels have been observed in numerous cities (Blake et al., 2011). Documented climate trends also encompass high and low temperature extremes, heat waves, intense precipitation, and coastal flooding in many cities (Alexander et al., 2006; Mishra and Lettenmaier, 2011).

In one study of 217 urban areas across the globe, researchers found that these areas have experienced significant increases in the number of heat waves (defined as periods during which the daily maximum temperature stayed above the empirical 99th percentile consecutively for six or more days) in the past 40 years, whereas cold wave frequency (periods during which the daily maximum temperature stayed below the empirical 99th percentile consecutively for six or more days) has decreased (Mishra et al., 2015). This study also found that the number of hot days and hot nights (defined in as the 99th percentile of daily maximum and minimum temperatures, respectively) also increased in the majority of cities analyzed (see Case Study 2.5). A smaller number of cities showed an upward trend in extreme precipitation, and some cities showed significant downward extreme

---

Case Study 2.5   Climate Extreme Trends in Seoul

Yeonjoo Kim and Won-Tae Kwon

Yonsei University, Seoul

Keywords	Temperature extremes, coastal flooding, public health
Population (Metropolitan Region)	23,575,000 (Demographia, 2016)
Area (Metropolitan Region)	2,590 km² (Demographia, 2016)
Income per capita	US$27,600 (World Bank, 2017)
Climate zone	Dfa – Cold, without dry season, hot summer (Peel et al., 2007)

Seoul, South Korea has a temperate climate with four seasons and a relatively large temperature difference between the hottest days of summer and the coldest days of winter. It experiences hot, humid weather in the summer under the influence of the North Pacific high-pressure system and cold weather in the winter under the influence of the Siberian high-pressure system. The summer climate in Seoul is linked to the East Asian monsoon, a pattern of natural climate variability with connections to the Madden-Julian oscillation (Chi et al., 2014).

Temperatures from 1908 to 2013 were investigated with a focus on extremes. Averages of mean, minimum, and maximum temperatures all show increasing trends over the past 106 years. In particular, the minimum temperature during the winter increased at an average rate of 0.5°C/decade.

Temperature-related extremes – tropical days[6], tropical nights[7], and frost days[8] – were investigated. Results show that Seoul has been experiencing changes in extreme temperatures. In particular, the increase in the number of tropical nights is very significant, which can adversely affect human health and energy demand. Warmer nighttime temperatures may result in increased air conditioning usage, placing strain on energy supplies. Increasing nighttime temperatures may also impact vulnerable populations, such as the elderly, children, and those with pre-existing medical conditions.

Seoul experienced an average of 35.7 tropical days per year over the past century, with a slight decreasing trend of –0.6 days/decade, which is not statistically significant. For tropical nights, the average is 4.2 days per year, and there is a strong increasing trend with a rate of 0.9 days/decade (Case Study 2.5 Fig. 1). This trend reflects the fact that the minimum temperature during the summer has increased over the past 106 years at an average rate of 0.2°C/decade. Seoul experienced an average of 1.7 tropical nights for the earliest decadal period of the data record (1908–1957) and 6.2 days, a 600% increase, for the latest period analyzed (1958–2013). In 1994, the most frequent tropical nights (34 days), took place followed by 20 and 17 days in 2013 and 2012, respectively.

Seoul experienced an average of 107.9 frost days per year, with a decreasing trend of –4.5 days/decade (Case Study 2.5 Figure 2). For the 1908–1917 period, an average of 127.54 frost days was found, whereas for 2008–2013, an average of 95.5 days was observed. For the past 106 years, the minimum temperature during the winter showed marked increase of 0.5°C/decade, and such an increase was also present during the spring and fall. The largest drops in the annual frost days were found in March and November in Seoul. In March, an average of 22.3 days for the 1908–1917 period was found, compared to 11.8 days for the 2008–2013 period.

The recent observed trends are an example of how trends in extreme temperatures can influence urban decision-making. In Seoul, based on the climate data analysis, planning for an increased number of heat-related impacts in the health, energy, and other urban sectors may need to be considered.

Given anticipated increases in the frequency, duration, and intensity of heat waves due to climate change, Son et al. (2012) examined mortality from heat waves for 2000 through 2007. Heat waves are defined as two or more consecutive days with daily mean temperature at or above the 98th percentile for the warm season and mortality during heatwave days and non-heatwave days are compared. A significant increase in total mortality in Seoul was noted on heatwave days compared with non-heatwave days (8.4%; 95% confidence interval [CI]: 0.1%, 17.3%). As an additional analysis for the timing of the heatwave in the season, we compared mortality risk based on whether a

---

6 Tropical days are defined as days where daily maximum temperature exceeds 30°C.
7 Tropical nights are defined as days where the daily minimum temperature exceeds 25°C.
8 Frost days are defined as days where the daily minimum temperature falls below 0°C.

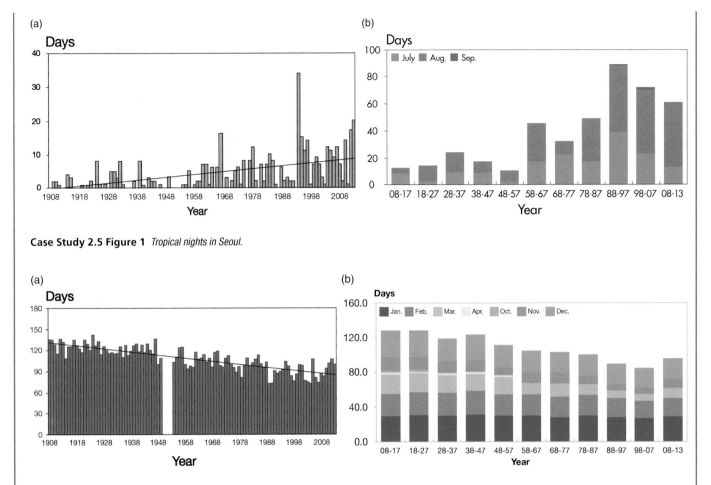

Case Study 2.5 Figure 1 *Tropical nights in Seoul.*

Case Study 2.5 Figure 2 *Frost days in Seoul.*

heatwave was the first in its summer in Seoul. The first heatwave of the summer had a larger estimated mortality effect than did later heatwaves.

Seoul has also been experiencing the urban heat island effect (UHI). Based on investigating near-surface air temperature data measured at automatic weather stations (AWSs) (Kim and Baik, 2005), the Seoul urban heat island deviates considerably from an idealized, concentric heat island pattern. This is attributed to the location of the main commercial and industrial sectors and the local topography. The temperature anomaly in southwestern and

southeastern Seoul is positive and its magnitude is large because these two regions have been built up for the two decades before 2001. It clearly shows increases in near-surface air temperature due to urbanization and industrialization. Furthermore, Lee and Baik (2010) performed statistical analysis of UHI intensity in Seoul, showing that the seasonal mean of UHI intensity is strongest in autumn and weakest in summer. The daily maximum UHI intensity is observed around midnight in all seasons except in winter, when the maximum occurrence frequency is around 08 LST, suggesting that anthropogenic heating contributes to the UHI in the cold season.

precipitation trends. Such events can have significant impacts on cities across multiple urban sectors. The results of Mishra et al. (2015) are similar to the observed changes found at the global scale in the IPCC Special Report on Extreme Events (IPCC, 2012).

2.5.2 Influences of Urbanization

How the UHI influences observed temperature trends globally (Parker, 2006; Hansen et al., 2010) and regionally (Arnfield, 2003; Peterson, 2003; Stone, 2007) is a focus of current research. While the UHI effect may have an influence on larger geographic scales (Stone, 2009), the IPCC

Fourth Assessment Report (Trenberth et al., 2007) states that the contribution of urbanization on global temperature trends is not significant, and any urban-related trend is an order of magnitude smaller than decadal and longer time-series trends. Similar results were presented in the IPCC Fifth Assessment Report (IPCC, 2013), which again found that the UHI effect has very limited influence on global temperature trends; however, the IPCC AR5 found that the influence on regional trends could be substantially larger (IPCC, 2013).

At the regional scale, some studies show that a percentage of the warming trends in cities can be linked to UHI effects, with estimates attributing nearly half of the observed decrease in diurnal temperatures range for stations in the United States to urbanization and land-use change (Kalnay and Cai, 2003). Hausfather et al. (2013) found that urbanization accounted for between 14% and 21% of the increase in minimum temperatures since 1895 and for 6% to 9% since 1960 for stations within the United States. In New York, it is estimated that approximately one-third of the warming experienced in New York from 1900 to 2002 can be attributed to the UHI (Gaffin et al., 2008). An analysis of five North American cities found that one metric to measure the effects of increasing urbanization on temperatures in cities is the day-to-day variability in maximum and minimum temperatures (Tam et al., 2015).

Land-use and the UHI have contributed to the warming trend in Glasgow, which presents an interesting case because the temperatures increased as the population of the city decreased (Emmanuel and Kruger, 2012). Despite the falling population, the UHI persists because much of the urban infrastructure and land use changes caused by past growth remain in place.

Similar results have been found for cities in east China (Yang et al., 2011), most notably in the Yangtze River Delta, where the highest level of urbanization has occurred. Densely populated metropolis cities in this region have warmed 0.285°C per decade more than rural areas, which see fewer impacts from urban-related warming. From 1981 to 2007, an increase in annual mean temperature over east China of 0.578°C per decade was observed. Specifically, within the city of Shanghai, this trend amounted to 0.961°C per decade. Through comparison with rural sites and reanalysis of temperature data, approximately one-quarter of the observed warming trend is attributable to urbanization and changes in land use.

Although urbanization may be contributing to warming trends within cities, some studies have shown that rates of observed warming do not significantly differ from the trends observed in more rural locations. Adjusting global temperature data to remove the impacts of urban effects revealed that for 42% of global stations, urban areas warmed at slower rates compared to the surrounding non-urban areas (Hansen et al., 2001). Although the raw data showed that large urban areas experienced 0.25°C warming in the 20th century, this urban effect is largely influenced by regional temperature variability, errors in measurement, and nonuniformity of the station data. In London and Vienna, where both cities observed temperature records display the influence of the UHI, rates of warming are comparable between the city and the surrounding rural locations (Jones et al., 2008).

2.6 Future Climate Projections

Future climate scenarios can be used to inform resiliency efforts in cities. This section discusses available climate risk information at the city scale and presents examples of where it has been used to implement adaptation strategies. It is important to develop a process by which local, national, and international researchers provide climate risk information to urban stakeholders and to ensure that the data are periodically updated through time. Regular updates are necessary to incorporate the latest climate science research available, including new models and methodologies, which are also described here.

2.6.1 Developing Climate Projections for Cities

Several steps can be followed to develop climate projections that can be used by cities as they plan and implement adaptation and mitigation strategies. A useful first step can be to scope out the main climate risk factors for cities; this is the subset of climate hazards that is of most consequence for a given city. This subset of hazards is selected on the basis of interactions between researchers and stakeholders and of expert judgment using quantitative and qualitative climate hazard information. Risk factors are generalized climate variables prioritized by consideration of their potential importance for a city. Quantitative (where possible) and qualitative statements of the likelihood of occurrence of these tailored climate risk factors, their potential impacts and their consequences are presented. Examples of climate risk factors include heat waves, coastal floods, intense precipitation events, and droughts, all of which can impact urban areas.

After initial scoping using climate risk factors, downscaled climate projections can be developed for cities. Global climate models (GCMs) and regional climate models (RCMs) are both tools that scientists use to project future climate at the localscale. GCM results are often statistically downscaled to provide information at higher spatial resolution. This is less common with RCMs, which are already downscaled dynamically. These techniques have advantages and disadvantages compared to each other (see Table 2.4 and Box 2.3), with the climate projections from both suitable to be incorporated into climate risk information for development of urban adaptation strategies, accompanied by a clear presentation of uncertainty (see Section 2.6.4) (Vaughan, 2016).

By incorporating the latest climate model outputs from the Coupled Model Intercomparison Project Phase 5[9] (CMIP5; Taylor et al., 2012), prepared for the IPCC Fifth Assessment Report (IPCC, 2013), projections can be provided for more detailed model-based probability distributions of climate variables. Using a combination of climate models and representation concentration pathways (RCPs; Moss et al., 2010) linked to possible future GHG concentrations produces a matrix of outputs for a given climate variable (see Box 2.3). This allows, for each time period, a model-based range of outcomes (i.e., a distribution that shows for any given threshold the number of climate models and RCP results that are at, above, or below the threshold) that can be used to inform risk-based decision-making. The presentation of such model-based outcomes to stakeholders should include clear statements about associated uncertainties, which include uncertainties in future GHG emissions and how the climate system will respond, natural climate variability, local processes not captured by the climate models, and the increase in uncertainty with spatial resolution (Willows et al., 2003).

---

9 CMIP provides a framework for standard protocols and comparison in global climate modeling, and the outputs are used in the IPCC assessments.

Table 2.4 *Comparison of Global Climate Models and Regional Climate Models for Downscaling for Urban Areas.*

	Global Climate Models (GCMs)	Regional Climate Models (RCMs)
Resolution (spatial and temporal)	GCMs have coarse horizontal resolution of approximately 30 to 65 km. Outputs range from 3-hourly to monthly; preferred for use on longer timescales.	RCM resolution is generally on the order of 25–50 km. Outputs range from hourly to monthly; preferred for use on shorter timescales.
Computational power	GCMs are less computationally intensive.	RCMs are more computationally intensive.
Downscaling considerations	GCMs depict relevant large-scale climate phenomena (e.g., synoptic weather patterns) that impact the local climate of interest.	Regional phenomena such as orography and land use are incorporated in order to determine local climate impacts.
Requirements	High-quality historical climate data are required for bias-correction and statistical downscaling approaches with GCMs.	GCM boundary conditions are used as input for temperature and wind factors; without high-quality inputs, RCMs are limited.
Urban applications	GCM outputs can be downscaled for resiliency planning at metropolitan region scales	RCMs are able to simulate components of the urban climate system. Projections from RCMs can be used for resiliency planning.
Examples	Coupled Model Intercomparison Project Phase 5 (CMIP5; Taylor et al., 2012),	North American Regional Climate Change Assessment Program (NARCCAP; Mearns et al., 2009)

Box 2.3   Climate Models and Representative Concentration Pathways

Global climate models (GCMs) are physics-based mathematical representations of the Earth's climate system over time that can be used to estimate the sensitivity of the climate system to changes in atmospheric concentrations of greenhouse gases (GHGs) and aerosols.

Regional climate models (RCMs) are similar to GCMs, with the primary difference being that regional models are run at a higher spatial resolution that allows for the use of more detailed physical parameterizations to simulate certain processes such as convective precipitation, sea breezes, or differences in elevation.

Representation concentration pathways (RCPs) represent the amount of radiative forcing (measured in watts per meter squared) caused by GHGs and other important agents such as aerosols over time. Each RCP is consistent with a trajectory of GHG emissions, aerosols, and land-use changes developed for the climate modeling community as a basis for long- and near-term climate modeling experiments. RCPs serve as inputs to global climate models, in order to project the effects of these climate drivers on future climate.

Projections for temperature, precipitation, and sea level for *Second Urban Climate Change Research Network Assessment Report on Climate Change and Cities (ARC3.2)* were generated from IPCC Coupled Model Intercomparison Project Phase 5 (CMIP5) GCM simulations based on two RCPs (Moss et al., 2010). The analysis uses two RCPs, RCP 4.5 and RCP 8.5, which represent relatively low and high GHG projections and radiative forcing, respectively.[10]

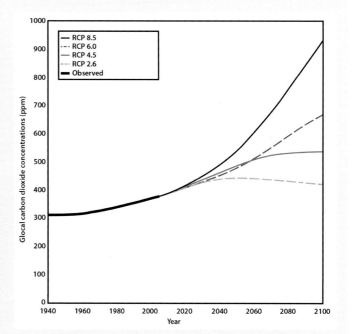

Box 2.3 Figure 1 *Observed CO_2 concentrations through 2005 and future CO_2 concentrations consistent with four representative concentration pathways (RCPs). Urban Climate Change Regional Network (UCCRN) climate projections are based on RCP 4.5 and RCP 8.5. Carbon dioxide and other greenhouse gas concentrations are driven by a range of factors, including carbon intensity of energy used, population and economic growth, and diffusion and adoption of new technologies, including energy efficiency and green energy.*

Source: NPCC, 2015

10 The Paris Agreement in December 2015 set the goal of limiting global temperatures from increasing more than 2°C above pre-industrial levels by the year 2100, with an additional aim for an even more ambitious stabilization target of 1.5°C. Therefore, inclusion of additional RCPs, such as RCP 2.6, may be of greater importance for future climate projections.

2.6.2 Temperature and Precipitation

Climate projections for temperature and precipitation developed for the *Second Urban Climate Change Research Network Assessment Report on Climate Change and Cities (ARC3.2)* are based on downscaled climate data from Coupled Model Intercomparison Project Phase 5 (CMIP5) multimodel data set (see Annex 2, Climate Projections for ARC3.2 Cities). Projections were obtained for 35 GCMs used in the IPCC AR5. Local projections for 153 ARC3.2 cities are based on GCM output from the single land-based model grid box centered over each of the cities and is used to develop city-specific climate change projections for temperature and precipitation.[11] Projections are based on two RCPs, RCP4.5 and RCP8.5. See Annex 2 in this volume for the full set of projections for the 153 ARC3.2 Cities.

Although it is not possible to predict the temperature or precipitation for a particular day, month, or year, GCMs are valuable tools for projecting the likely range of changes over multidecadal time periods. These projections, known as *timeslices*, are expressed relative to the 1971–2000 baseline period. The 30-year timeslices are centered on a given decade. For example, the 2050s timeslice refers to the period from 2040 to 2069.

The projections for temperature and precipitation are provided for a set of cities across five geographic regions (Africa,

Table 2.5 *Future climate projections for ARC3.2 Cities.*[12]

	2020s	2050s	2080s
Temperature	+ 0.7 to 1.6°C	+ 1.4 to 3.1°C	+ 1.7 to 5.0°C
Precipitation	–7 to +10%	–9 to +14%	–11 to +20%
Sea Level Rise	+ 4 to 18 cm	14 to 56 cm	+ 21 to 118 cm

Asia and Australia, Europe, Latin America, and North America) ARC3.2 Cities include Case Study Docking Station cities, UCCRN Regional Hub cities, UCCRN project cities, and cities of ARC3.2 Chapter Authors.

Temperatures in the ARC3.2 cities around the world are projected to increase by 0.7–1.6°C by the 2020s, 1.4–3.1°C t the 2050s, and 1.7–5.0°C by the 2080s (see Table 2.5).

Following from the UCCRN analysis and the IPCC AR5 (IPCC, 2013), the greatest increases in temperature by the 2080s are projected for cities in North America and northern Europe, with upward of between 6 and 7°C of warming by the end of the century under the high emissions scenario (see Figure 2.3). Cities located in and around these regions include Calgary, Helsinki, Warsaw, and Moscow. Non-equatorial regions of Africa are also projected to have large temperature increases, including Johannesburg.

Temperature change 2080s

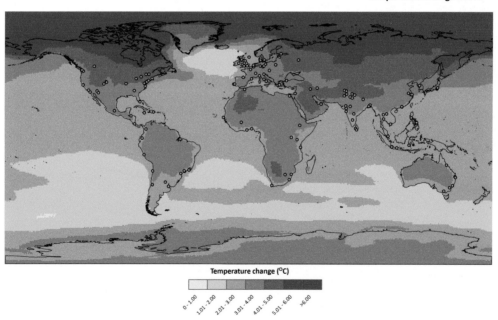

Figure 2.3 *Projected temperature change in the 2080s. Temperature change projection is mean of 35 global climate models (GCMs) and two representative concentration pathways (RCP4.5 and RCP8.5). Colors represent the mean change in mean annual temperature (2070–2099 average relative to 1971–2000 average). Dots represent ARC3.2 cities. ARC3.2 Cities include Case Study Docking Station cities, UCCRN Regional Hub cities, UCCRN project cities, and cities of ARC3.2 Chapter Authors.*

11 Because projected changes through time are generally similar between nearby gridboxes (Horton et al., 2011), a straightforward and easy-to-replicate single gridbox approach was used in this analysis.
12 Projections for temperature and precipitation are based on 35GCMs and 2 RCPs. For each of the 153 ARC3.2 cities, the low estimate (10th percentile) and high estimate (90th percentile) were calculated. The range of values presented here is the average for each percentile averaged across all ARC3.2 cities. Projections are relative to the 1971–2000 base period.
    Projections for sea level rise are based on a four-component approach that incorporates both local and global factors. For each of the 71 ARC3.2 coastal cities, the low estimate (10th percentile) and high estimate (90th percentile) were calculated. The range of values presented here is the average for each percentile averaged across all coastal ARC3.2 cities. The model-based components are from 24 GCMs and 2 RCPs. Projections are relative to the 2000-2004 base period.
    ARC3.2 Cities include Case Study Docking Station cities, UCCRN Regional Hub cities, UCCRN project cities, and cities of ARC3.2 Chapter Authors.

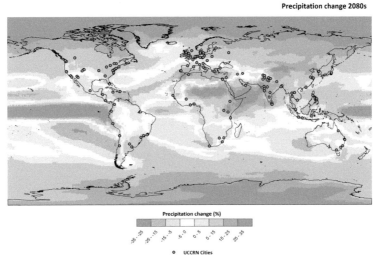

Precipitation change 2080s

Precipitation change (%)

○ UCCRN Cities

Figure 2.4 *Projected precipitation change in the 2080s. Precipitation change projection is mean of 35 global climate models (GCMs) and two representative concentration pathways (RCP4.5 and RCP8.5). Colors represent the percentage change in mean annual precipitation (2070–2099 average relative to 1971–2000 average). Dots represent ARC3.2 cities. ARC3.2 Cities include Case Study Docking Station cities, UCCRN Regional Hub cities, UCCRN project cities, and cities of ARC3.2 Chapter Authors.*

Similar rates of warming are projected for Asian cities such as Faisalabad, Tehran, and Delhi. Southern South America, South Asia, coastal Africa, and parts of Southeast Asia are also projected to see warming, however, at lower rates than these cities.

Precipitation in the ARC3.2 cities around the world is projected to change by −7 to +10% by the 2020s, −9 to +14% by the 2050s, and −11 to +20% by the 2080s (see Table 2.5). Projections for precipitation are characterized by greater uncertainty than those for temperature because GCMs have difficulty capturing the small-scale processes and climate variability that influence precipitation.

By the end of the century, increased precipitation is projected for near-equatorial areas in the Middle East, Africa, and portions of Asia including India and China (see Figure 2.4). Cities with projected precipitation increases include Bangalore, Colombo, and Dakar. Northern portions of North America and Europe are also projected to see increased precipitation.

Drier conditions are generally projected for northernmost Africa and southern Europe, including the cities of Rome, Naples, and Jerusalem. Some areas of Central and South America (Mexico City, Santiago, and Santo Domingo), Australia (Melbourne), and southern Africa (Cape Town) are also projected to see decreased precipitation by the end of the century under the high emissions scenario.

2.6.3 Sea Level Rise, Coastal Storms, and Flooding

Globally, many cities face faster local sea level rise than the global average due to subsidence caused by sediment compaction and groundwater withdrawal (Nicholls, 1995; Syvitski et al., 2009) (see Chapter 9, Coastal Zones). This is of significant concern because many of the world's largest cities are located along the coast (Hanson et al., 2011). The processes of subsidence and groundwater withdrawal accelerate local sea level rise and thereby increase the vulnerability of urban areas. A study analyzing 40 river deltas globally, where Bangkok, Cairo, Cartagena, and Shanghai, are located, found that anthropogenic activities (changes in land-water storage and sediments) are the dominant contributors to observed sea level rise in many of these locations (Ericson et al., 2006). Rates of sea level rise in Manila and Bangkok exceed the global average, with the relative rise attributed to groundwater extraction and subsidence (Cazenave and Le Cozannet, 2014).

Projections for sea level rise are developed using both local and regional components based on GCMs and expert assessment of scientific literature (see Annex 2, Climate Projections for ARC3.2 Cities). For the ARC3.2 sea level rise projections, the following four components are used: 1) changes in ocean height (local); 2) thermal expansion (global); 3) loss of ice from glaciers, ice caps, and land-based ice sheets (global); and 4) land water storage (global).[13] For each of these components of sea level rise, a set of distribution points are estimated and summed to give the total sea level rise projection (NPCC, 2015). These projections are expressed relative to the 2000–2004 baseline period. The timeslices are centered on a given decade. For example, the 2050s timeslice refers to the period from 2050 to 2059.

Sea level in 71 ARC3.2 coastal cities is projected to rise 4–18 centimeters by the 2020s; 14–56 centimeters by the 2050s, and 21–118 centimeters by the 2080s (see Table 2.5).[14] Local effects, such as groundwater extraction, are not captured in the methods used for sea level rise projections here. However, they should be included in comprehensive city assessments if the data are available.

13 Two additional local components that can be incorporated into city-specific sea level rise projections are vertical land motions and the gravitational, rotational, and elastic "fingerprints" of ice loss. For most cities, the additional change in sea level rise projected from these components is approximately an order of magnitude smaller over the 21st-century timescale than the total projections. Exceptions may include (1) some delta cities such as Jakarta experiencing rapid sinking associated, for example, with groundwater withdrawal and (2) cities near any glacier or ice sheet that experiences rapid 21st-century mass loss.
14 For each of the coastal cities, the low estimate (10th percentile) and high estimate (90th percentile) were calculated. The range of values presented here are the average for each percentile averaged across all 52 cities.

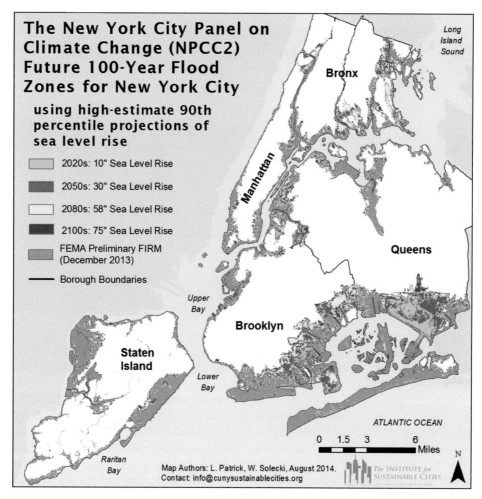

The New York City Panel on Climate Change (NPCC2) Future 100-Year Flood Zones for New York City

using high-estimate 90th percentile projections of sea level rise

- 2020s: 10" Sea Level Rise
- 2050s: 30" Sea Level Rise
- 2080s: 58" Sea Level Rise
- 2100s: 75" Sea Level Rise
- FEMA Preliminary FIRM (December 2013)
- Borough Boundaries

Map Authors: L. Patrick, W. Solecki, August 2014.
Contact: info@cunysustainablecities.org

The INSTITUTE for SUSTAINABLE CITIES

Figure 2.5 *Potential areas that could be impacted by the 100-year flood in the 2020s, 2050s, 2080s, and 2100 based on projections of the high-estimate 90th percentile New York City Panel on Climate Change (NPCC2) sea level rise scenario. Map developed using the static approach. Note: This map is subject to limitations in accuracy as a result of the quantitative models, datasets, and methodology used in its development. The map and data should not be used to assess actual coastal hazards, insurance requirements, or property values or be used in lieu of flood insurance rate maps (FIRMS) issued by the Federal Emergency Management Agency (FEMA). The flood areas delineated in no way represent precise flood boundaries but rather illustrate three distinct areas of interest: (1) areas currently subject to the 100-year flood that will continue to be subject to flooding in the future, (2) areas that do not currently flood but are expected to potentially experience the 100-year flood in the future, and (3) areas that do not currently flood and are unlikely to do so in the timeline of the climate scenarios used by the NPCC (end of the current century).*

Source: NPCC, 2015

Because sea level rise is not spatially uniform, cities will differ in how much sea level rise they will experience. For examples, some studies have found that the coastal Northeast United States, which includes large urban population centers such as New York, Boston, and Washington, D.C., have experienced, and may continue to experience, faster sea level rise than the global average due to a weakening of the Gulf Stream (Yin et al., 2009). Megacities are experiencing a triple threat of sea level rise associated with climate change, land sinking, and growing concentrations of people and assets near the coast (see Chapter 9, Coastal Zones).

In the future, the magnitude and frequency of coastal flooding may change due to changes in sea level and/or changes in coastal storms. Many coastal cities are not prepared for the coastal flood risks they face today, and the threat is exacerbated by sea level rise. A recent study found that cities in the United States saw more than 10-fold increases in nuisance flood events (floods that cause an inconvenience to the public, such as road closures or clogged storm drains) over the past 60 years (Sweet and Park 2014). For these locations, coastal flooding now occurs at times of high tide and in minor coastal flood events due to observed changes in sea level and local land processes.

How coastal storms like tropical cyclones and mid-latitude storms may change in the future is still difficult to project (IPCC, 2013), although some studies suggest that as the upper oceans warm, the strongest tropical cyclones may tend to become stronger, and mid-latitude storms may tend to move closer to the poles (IPCC, 2012; IPCC, 2013; Colle et al., 2013). Although the implications of coastal storm changes for individual cities are highly uncertain, it is very certain that sea level rise will lead to large changes in coastal flood frequency. For example, in New York, by the 2080s, the current 100-year flood (a flood with a 1% annual chance of occurrence) is projected to become an approximately once-in-eight-year event (NPCC, 2015) (see Figure 2.5 and Table 2.6).

Table 2.6 *Future flood recurrence intervals at the Battery, New York.*
Source: NPCC, 2015

	Low estimate (10th percentile)	Middle range (25th–75th percentile)	High estimate (90th percentile)
2020s			
Annual chance of today's 100-year flood (1%)	1.1%	1.1–1.4%	1.5%
2050s			
Annual chance of today's 100-year flood (1%)	1.4%	1.6–2.4%	3.6%
2080s			
Annual chance of today's 100-year flood (1%)	1.7%	2.0–5.4%	12.7%

The change in coastal flood frequency associated with a given amount of sea level rise is highly dependent on the baseline distribution of coastal flooding (Tibaldi et al., 2012; Kopp et al., 2014). For cities like Lagos and San Francisco that do not experience large storm surges, even small amounts of sea level rise can lead to large increases in the frequency of coastal flooding (Strauss et al., 2012). In contrast, coastal regions that historically experience large storm surges (such as much of the northern Gulf of Mexico coast and parts of the coastline along the northwest Pacific Ocean) are projected to see smaller increases in coastal flood frequency associated with the sea level rise effect. When storms do occur, sea level rise means that the area inundated expands dramatically for low-lying and flat areas; this is less of a concern for coastal cities with steeper coastlines, such as Lima (Weiss et al., 2011).

Given the increased vulnerability that coastal cities have to sea level rise and coastal flooding, employment and improvement of mapping techniques such as geographic information system (GIS), global positioning system (GPS), and Light Detection and Ranging (LIDAR) could help urban areas plan for the future. For example, a comprehensive geodetic-based mapping survey of Jakarta found that land subsidence in the city is strongly linked to ground water extraction (Abidin, 2001). Identification of coastal risks through these technologies could identify vulnerable neighborhoods, and, therefore, where resources could be targeted in future flooding events.

2.6.4 Use of Climate Projections

Future climate projections from both GCMs and RCMs are being used by cities across the globe in a variety of applications (see Annexes 4 and 5; ARC3.2 Case Study Docking Station, www.uccrn.org/casestudies). Projections can be used to inform citywide adaptation plans, as has been done in New York (Rosenzweig and Solecki, 2010), and London (LCCP, 2002). City sector-specific uses of climate projection examples include watershed modeling in Seoul (Chung et al., 2011), transportation planning in Stockholm (Gidhagen et al., 2012), and evaluating air pollution in Melbourne (Coutts et al., 2008). Outputs and projections from RCMs can be used to understand sector-specific climate impacts, such as how warming temperatures will impact public health in cities (Dessai, 2003; Bell et al., 2007; Knowlton et al., 2007; Casati et al., 2013).

Improving the process of the co-generation of climate risk information and interactions between scientists and stakeholders is under way (e.g., Lemos and Morehouse, 2005; Mauser et al., 2013), and these types of projections are already being used to inform adaptation strategies. Scientific expert panels have been formed in cities such as Quito, New York, and London for the purpose of providing the best available information for city governments to use in creating long-term municipal adaptation strategies to deal with the local impacts of climate change. Climate projections developed by these groups help to guide investments in flood management, infrastructure improvements, urban cooling techniques, and the development of emergency heat wave response plans (NPCC, 2015; Johnson and Breil, 2012; Zambrano-Barragán et al., 2011). Community involvement in the development of such climate actions plans is being solicited in cities like Quito and Durban. This information is being used to help implement building retrofit plans, form health network systems throughout the urban population, and improve emergency management during landslides and mudslides (Walsh et al., 2013; Johnson and Breil, 2012; eThekwini Municipality, 2011).

In places where infrastructure is not a viable option to prevent damage from extreme events, climate risk information is being used to identify vulnerable groups so that they can be moved out of harm's way. For example, in Ho Chi Minh City, stakeholder–scientist interactions that utilize flood impact models have led decision-makers to relocate low-income slum settlements out of the flood plain (Vietnam Climate Adaptation Partnership, 2013). These cases show that co-generated science is helping to prepare cities for the worst effects of climate change because the information is tailored to the needs of decision-makers at the local level, which is most relevant for city planning.

2.6.5 Advances in Urban Climate Science Research

In recent years, use of big data, new instruments, and falling costs of high-powered computing have enabled better understanding of extreme events and microclimatic variations within cities. Crowdsourcing, data mining (Muller et al., 2015), and the diffusion of small, inexpensive sensors that monitor climate-related variables are helping fill gaps in urban meteorological data that hinder understanding of these complex environments (Chen et al., 2012). Improved remote sensing technologies are being explored in Istanbul and Paris (Sismanidis et al., 2015), and Guangzhou (Guo et al., 2015). As these innovations emerge, they are also being used to facilitate the real-time monitoring of hazards, climate, and associated impacts such as air quality problems and health hazards in coastal waterways. Developing advanced climate change indicators and monitoring systems, such as in New York (Solecki et al., 2015), and building on existing networks to integrate new instruments and indicators as in Birmingham, UK (Chapman et al., 2015), will become essential for research and adaptation in the future.

Urban climate models (from the building to full-city scale) have diversified and become more sophisticated in the past decade. Recent model comparison efforts have identified common characteristics of the best models and opportunities for improvement, such as better measurement of key parameters (Best and Grimmond, 2014) and integration with urban climate monitoring (Chen et al., 2012). Other research priorities in urban climate modeling include refining the estimation of heat fluxes between urban surfaces and the surrounding air; improving the simulation of anthropogenic heat inputs from traffic and buildings; and integrating dynamics, aerosol, and hydrology components to advance urban precipitation modeling (Chen et al., 2012). Downscaling of global and regional climate model projections for urban areas continues to advance with improved statistical techniques (Lauwaet et al., 2015), while the models themselves better simulate city-relevant processes such as the urban canopy and ocean–atmospheric interactions in coastal cities (Rummukainen et al., 2015) and are developed for use in impacts research (Mearns et al., 2015).

2.7 Conclusions and Recommendations

Cities are uniquely vulnerable to climate hazards, as well as being centers of greenhouse gas emissions. From a climate perspective, the urban heat island increases the frequency, duration, and intensity of heat waves beyond that associated with climate change; coastal cities are also sinking due to urban activities at the same time that sea levels are rising along the majority of the world's coastlines. More research is needed on joint risks associated with climate change, such as (1) combined impacts of heat and humidity; (2) sequences of extreme events, such as a tropical cyclone followed by a heat wave; and (3) interactions between cold air outbreaks and poor air quality. Additionally and particularly for cities in developing countries, limited data availability and scientific capacity remain major challenges. More research is also needed on how climate information can inform the evaluation of adaptation and mitigation programs, particularly as these programs are being increasingly implemented in the world's innovative cities.

Chapter 2   Urban Climate Science

References

Abidin, H. Z., Djaja, R., Darmawan, D., Hadi, S., Akbar, A., Rajiyowiryono, H., Sudibyo, Y., Meilano, I., Kasuma, M. A., Kahar, J., and Subarya, C. (2001). Land subsidence of Jakarta (Indonesia) and its geodetic monitoring system *Natural Hazards* **23**(2–3), 365–387.

Adamowski, J., and Prokoph, A. (2013). Assessing the impacts of the urban heat island effect on streamflow patterns in Ottawa, Canada. *Journal of Hydrology* **496**, 225–237.

Alexander, L. V., Zhang, X., Peterson, T. C., Caesar, J., Gleason, B., Klein Tank, A. M. G., Haylock, M., Collins, D., Trewin, B., Rahimzadeh, F., Tagipour, A., Rupa Kumar, K., Revadekar, J., Griffiths, G., Vincent, L., Stephenson, D. B., Burn, J., Aguilar, E., Brunet, M., Taylor, M., New, M., Zhai, P., Rusticucci, M., and Vazquez-Aguirre, J. L. (2006). Global observed changes in daily climate extremes of temperature and precipitation. *Journal of Geophysical Research: Atmospheres* **111**(D5).

Anniballe, R., Bonafoni, S., and Pichierri, M. (2014). Spatial and temporal trends of the surface and air heat island over Milan using MODIS data. *Remote Sensing of Environment* **150**(0), 163–171.

Arnfield, A. J. (2003). Two decades of urban climate research: a review of turbulence, exchanges of energy and water, and the urban heat island. *International Journal of Climatology* **23**(1), 1–26.

Balling, R. C., and Gober, P. (2007). Climate variability and residential water use in the city of Phoenix, Arizona. *Journal of Applied Meteorology and Climatology* **46**(7), 1130–1137.

Basara, J. B., Illston, B. G., Fiebrich, C. A., Browder, P. D., Morgan, C. R., McCombs, A., Bostic, J. P., McPherson, R. A., Schroeder, A. J., and Crawford, K. C. (2011). The Oklahoma City micronet. *Meteorological Applications* **18**(3), 252–261.

Baxter, S., and Nigam, S. (2013). A subseasonal teleconnection analysis: PNA development and its relationship to the NAO. *Journal of Climate* **26**(18), 6733–6741.

Bell, M. L., Goldberg, R., Hogrefe, C., Kinney, P. L., Knowlton, K., Lynn, B., Rosenthal, J., Rosenzweig, C., and Patz, J. A. (2007). Climate change, ambient ozone, and health in 50 US cities. *Climatic Change* **82**(1), 61–76.

Bertaccini, P., Dukic, V., and Ignaccolo, R. (2012). Modeling the short-term effect of traffic and meteorology on air pollution in Turin with generalized additive models. *Advances in Meteorology* **2012**, 16.

Best, M. J., and Grimmond, C. S. B. (2014). Key conclusions of the first International Urban Land Surface Model Comparison Project. *Bulletin of the American Meteorological Society* **96**(5), 805–819.

Blake, R., Grimm, A., Ichinose, T., Horton, R., Gaffin, S., Jiong, S., Bader, D., and Cecil, D. W. (2011). Urban climate: Processes, trends, and projections. In Rosenzweig, C., Solecki, W. D., Hammer, S. A., and Mehrotra, S. (eds.), *Climate Change and Cities: First Assessment Report of the Urban Climate Change Research Network* (43–81). Cambridge University Press.

Bond, N. A., and Vecchi, G. A. (2003). The influence of the Madden–Julian Oscillation on precipitation in Oregon and Washington. *Weather and Forecasting* **18**(4), 600–613.

Bornstein, R., and Lin, Q. (2000). Urban heat islands and summertime convective thunderstorms in Atlanta: three case studies. *Atmospheric Environment* **34**(3), 507–516.

Bouma, M., and Dye, C. (1997). Cycles of malaria associated with El Niño in Venezuela. *Journal of the American Medical Association* **278**(21), 1772–1774.

Burian, S. J., and Shepherd, J. M. (2005). Effect of urbanization on the diurnal rainfall pattern in Houston. *Hydrological Processes* **19**(5), 1089–1103.

Cai, W., Borlace, S., Lengaigne, M., van Rensch, P., Collins, M., Vecchi, G., Timmermann, A., Santoso, A., McPhaden, M. J., Wu, L., England, M. H., Wang, G., Guilyardi, E., and Jin, F. -F. (2014). Increasing frequency of extreme El Nino events due to greenhouse warming. *Nature Climate Change* **4**(2), 111–116.

Casati, B., Yagouti, A., and Chaumont, D. (2013). Regional climate projections of extreme heat events in nine pilot Canadian communities for public health planning. *Journal of Applied Meteorology and Climatology* **52**(12), 2669–2698.

Cazenave, A., and Cozannet, G. L. (2014). Sea level rise and its coastal impacts. *Earth's Future* **2**(2), 15–34.

Chang, B., Wang, H. -Y., Peng, T. -Y., and Hsu, Y. -S. (2010). Development and evaluation of a city-wide wireless weather sensor network. *Journal of Educational Technology & Society* **13**(3), 270–280.

Chang, H., Praskievicz, S., and Parandvash, H. (2014). Sensitivity of urban water consumption to weather and climate variability at multiple temporal scales: The case of Portland, Oregon *International Journal of Geospatial and Environmental Research* **1**(1).

Chapman, L., Muller, C. L., Young, D. T., Warren, E. L., Grimmond, C. S. B., Cai, X. -M., and Ferranti, E. J. S. (2015). The Birmingham Urban Climate Laboratory: An open meteorological test bed and challenges of the smart city. *Bulletin of the American Meteorological Society* **96**(9), 1545–1560.

Chen, F., Bornstein, R., Grimmond, S., Li, J., Liang, X., Martilli, A., Miao, S., Voogt, J., and Wang, Y. (2012). Research priorities in observing and modeling urban weather and climate. *Bulletin of the American Meteorological Society* **93**(11), 1725–1728.

Chen, F., Yang, X., and Zhu, W. (2014). WRF simulations of urban heat island under hot-weather synoptic conditions: The case study of Hangzhou City, China *Atmospheric Research* **138** (2014), 364–377.

Cherry, J., Cullen, H., Visbeck, M., Small, A., and Uvo, C. (2005). Impacts of the North Atlantic Oscillation on Scandinavian hydropower production and energy markets. *Water Resources Management* **19**(6), 673–691.

Cheval, S., and Dumitrescu, A. (2009). The July urban heat island of Bucharest as derived from modis images. *Theoretical and Applied Climatology* **96**(1–2), 145–153.

Childs, P., and Raman, S. (2005). Observations and numerical simulations of urban heat island and sea breeze circulations over New York City. *Pure and Applied Geophysics* **162**(10), 1955–1980.

Christensen, J. H., Krishna Kumar, K., Aldrian, E., An, S. -I., Cavalcanti, I. F. A., Castro, M. D., Dong, W., Goswami, P., Hall, A., Kanyanga, J. K., Kitoh, A., Kossin, J., Lau, N. -C., Renwick, J., Stephenson, D. B., Xie, S. -P., and Zhou, T. (2013). Climate phenomena and their relevance for future regional climate change. In Stocker, T. F., Qin, D., Plattner, G. -K., Tignor, M., Allen, S. K., Boschung, J., Nauels, A., Xia, Y., Bex, V., and Midgley, P. M. (eds.), *Climate Change 2013: The Physical Science Basis. Contribution of Working Group I to the Fifth Assessment Report of the Intergovernmental Panel on Climate Change* (1217–1308). Cambridge University Press.

Chung, E. -S., Park, K., and Lee, K. S. (2011). The relative impacts of climate change and urbanization on the hydrological response of a Korean urban watershed. *Hydrological Processes* **25**(4), 544–560.

City of New York. (2013). *New York City Special Initiative on Rebuilding and Resiliency: A Stronger, More Resilient New York*. City of New York.

Colle, B. A., Zhang, Z., Lombardo, K. A., Chang, E., Liu, P., and Zhang, M. (2013). Historical evaluation and future prediction of eastern North American and western Atlantic extratropical cyclones in the CMIP5 models during the cool season. *Journal of Climate* **26**(18), 6882–6903.

Collins, M. (2005). El Niño- or La Niña-like climate change?. *Climate Dynamics* **24**(1), 89–104.

Collins, M., An, S. -I., Cai, W., Ganachaud, A., Guilyardi, E., Jin, F. -F., Jochum, M., Lengaigne, M., Power, S., Timmermann, A., Vecchi, G., and Wittenberg, A. (2010). The impact of global warming on the tropical Pacific Ocean and El Nino. *Nature Geoscience* **3**(6), 391–397.

Connors, J., Galletti, C., and Chow, W. L. (2013). Landscape configuration and urban heat island effects: Assessing the relationship between landscape characteristics and land surface temperature in Phoenix, Arizona *Landscape Ecology* **28**(2), 271–283.

Cooper, M. J., Martin, R. V., van Donkelaar, A., Lamsal, L., Brauer, M., and Brook, J. R. (2012). A satellite-based multi-pollutant index of global air quality *Environmental Science & Technology* **46**(16), 8523–8524.

Coutts, A. M., Beringer, J., and Tapper, N. J. (2008). Investigating the climatic impact of urban planning strategies through the use of regional climate modelling: A case study for Melbourne, Australia. *International Journal of Climatology* **28**(14), 1943–1957.

Cramer, W., Yohe, G. W., Auffhammer, M., Huggel, C., Molau, U., Dias, M. A. F. S., Solow, A., Stone, D. A., and Tibig, L. (2014). Detection and attribution of observed impacts. In Field, C. B., Barros, V. R., Dokken, D. J., Mach, K. J., Mastrandrea, M. D., Bilir, T. E., Chatterjee, M., Ebi, K. L., Estrada, Y. O., Genova, R. C., Girma, B., Kissel, E. S., Levy, A. N., MacCracken, S., Mastrandrea, P. R., and White, L. L. (eds.), *Climate Change 2014: Impacts, Adaptation, and Vulnerability. Part A: Global and Sectoral Aspects. Contribution of Working Group II to the Fifth Assessment Report of the Intergovernmental Panel of Climate Change* (979–1037). Cambridge University Press.

Currie, B. A., and Bass, B. (2008). Estimates of air pollution mitigation with green plants and green roofs using the UFORE model. *Urban Ecosystems* **11**(4), 409–422.

Dabberdt, W., Koistinen, J., Poutiainen, J., Saltikoff, E., and Turtiainen, H. (2005). Research: The Helsinki Mesoscale Testbed: An invitation to use a new 3-D observation network *Bulletin of the American Meteorological Society* **86**(7), 906–907.

Davini, P., and Cagnazzo, C. (2014). On the misinterpretation of the North Atlantic Oscillation in CMIP5 models. *Climate Dynamics* **43**(5), 1497–1511.

Dawson, J. P., Bryan, J. B., Darrell, A. W., and Christopher, P. W. (2014). Understanding the meteorological drivers of US particulate matter concentrations in a changing climate. *Bulletin of the American Meteorological Society* **95**(4), 521–532.

Deser, C., Knutti, R., Solomon, S., and Phillips, A. S. (2012). Communication of the role of natural variability in future North American climate. *Nature Climate Change* **2**(11), 775–779.

Dessai, S. (2003). Heat stress and mortality in Lisbon Part II. An assessment of the potential impacts of climate change. *International Journal of Biometeorology* **48**(1), 37–44.

Dobre, A., Arnold, S. J., Smalley, R. J., Boddy, J. W. D., Barlow, J. F., Tomlin, A. S., and Belcher, S. E. (2005). Flow field measurements in the proximity of an urban intersection in London, UK. *Atmospheric Environment* **39**(26), 4647–4657.

Dobrovolný, P. (2013). The surface urban heat island in the city of Brno (Czech Republic) derived from land surface temperatures and selected reasons for its spatial variability. *Theoretical and Applied Climatology* **112**(1–2), 89–98.

Earnest, A., Tan, S. B., and Wilder-Smith, A. (2012). Meteorological factors and El Niño southern oscillation are independently associated with dengue infections. *Epidemiology & Infection* **140**(07), 1244–1251.

Emmanuel, R., and Krüger, E. (2012). Urban heat island and its impact on climate change resilience in a shrinking city: The case of Glasgow, UK. *Building and Environment* **53**(0), 137–149.

Ericson, J. P., Vörösmarty, C. J., Dingman, S. L., Ward, L. G., and Meybeck, M. (2006). Effective sea-level rise and deltas: Causes of change and human dimension implications. *Global and Planetary Change* **50**(1–2), 63–82.

eThekwini Municipality. (2011). *Durbans Municipal Climate Protection Programme: Climate Change Adaptation Planning for a Resilient City* (36), Durban, South Africa, EThekwini Municipality Environmental Planning and Climate Protection Department.

Formiga-Johnsson, R., and Britto, A. (2009). *Climate Variability And Competing Demands For Urban Water Supply: Reducing Vulnerability Through River Basin Governance In Brazil*. Marseille: Fifth Urban Research Symposium 2009.

Gaffin, S. R., Rosenzweig, C., Khanbilvardi, R., Parshall, L., Mahani, S., Glickman, H., Goldberg, R., Blake, R., Slosberg, R. B., and Hillel, D. (2008). Variations in New York City's urban heat island strength over time and space *Theoretical and Applied Climatology* **94**(1–2), 1–11.

Gedzelman, S. D., Austin, S., Cermak, R., Stefano, N., Partridge, S., Quesenberry, S., and Robinson, D. A. (2003). Mesoscale aspects of the urban heat island around New York City. *Theoretical and Applied Climatology* **75**(1–2), 29–42.

Giannaros, T. M., and Melas, D. (2012). Study of the urban heat island in a coastal Mediterranean city: The case study of Thessaloniki, Greece. *Atmospheric Research* **118**(0), 103–120.

Gidhagen, L., Engardt, M., Lövenheim, B., and Johansson, C. (2012). Modeling effects of climate change on air quality and population exposure in urban planning scenarios. *Advances in Meteorology* **2012**, 12.

Goldbach, A., and Kuttler, W. (2013). Quantification of turbulent heat fluxes for adaptation strategies within urban planning. *International Journal of Climatology* **33**(1), 143–159.

Gramsch, E., Cereceda-Balic, F., Oyola, P., and von Baer, D. (2006). Examination of pollution trends in Santiago de Chile with cluster analysis of PM10 and Ozone data, *Atmospheric Environment* **40**(28), 5464–5475.

Grimm, A. M., and Tedeschi, R. G. (2009). ENSO and Extreme Rainfall Events in South America, *Journal of Climate* **22**(7), 1589–1609.

Guo, G., Wu, Z., Xiao, R., Chen, Y., Liu, X., and Zhang, X. (2015). Impacts of urban biophysical composition on land surface temperature in urban heat island clusters. *Landscape and Urban Planning* **135**, 1–10.

Gurjar, B. R., Butler, T. M., Lawrence, M. G., and Lelieveld, J. (2008). Evaluation of emissions and air quality in megacities. *Atmospheric Environment* **42**(7), 1593–1606.

Han, J. -Y., Baik, J. -J., and Lee, H. (2014). Urban impacts on precipitation, *Asia-Pacific Journal of Atmospheric Sciences* **50**(1), 17–30.

Hansen, J., Ruedy, R., Sato, M., Imhoff, M., Lawrence, W., Easterling, D., Peterson, T., and Karl, T. (2001). A closer look at United States and global surface temperature change. *Journal of Geophysical Research: Atmospheres* **106**(D20), 23947–23963.

Hansen, J., Ruedy, R., Sato, M., and Lo, K. (2010). Global surface temperature change. *Reviews of Geophysics* **48**(4), RG4004.

Hanson, S., Nicholls, R., Ranger, N., Hallegatte, S., Corfee-Morlot, J., Herweijer, C., and Chateau, J. (2011). A global ranking of port cities with high exposure to climate extremes. *Climatic Change* **104**(1), 89–111.

Hausfather, Z., Menne, M. J., Williams, C. N., Masters, T., Broberg, R., and Jones, D. (2013). Quantifying the effect of urbanization on US Historical Climatology Network temperature records. *Journal of Geophysical Research: Atmospheres* **118**(2), 481–494.

Hay, C. C., Morrow, E., Kopp, R. E., and Mitrovica, J. X. (2015). Probabilistic reanalysis of twentieth-century sea-level rise. *Nature* **517**(7535), 481–484.

Hicks, B. B., Callahan, W. J., Pendergrass, W. R., Dobosy, R. J., and Novakovskaia, E. (2012). Urban turbulence in space and in time. *Journal of Applied Meteorology and Climatology* **51**(2), 205–218.

Hiroyuki, K., Nawata, K., Suzuki-Parker, A., Takane, Y., and Furuhashi, N. (2013). Mechanism of precipitation increase with urbanization in Tokyo as revealed by ensemble climate simulations. *Journal of Applied Meteorology and Climatology* **53**(4), 824–839.

Hoerling, M., Eischeid, J., and Perlwitz, J. (2010). Regional precipitation trends: Distinguishing natural variability from anthropogenic forcing. *Journal of Climate* **23**(8), 2131–2145.

Holmer, B., and Eliasson, I. (1999). Urban–rural vapour pressure differences and their role in the development of urban heat islands. *International Journal of Climatology* **19**(9), 989–1009.

Horton, R. M., Gornitz, V., Bader, D. A., Ruane, A. C., Goldberg, R., and Rosenzweig, C. (2011). Climate hazard assessment for stakeholder adaptation planning in New York City. *Journal of Applied Meteorology and Climatology* **50**(11), 2247–2266.

Hung, T., D. Uchihama, S. Oci, and Y. Yasuoka (2006). Assessment with satellite data of the urban heat island effects in Asian mega cities. *International Journal of Applied Earth Observation and Geoinformation* **8**, 34–48.

Hung, T. K., and Wo, O. C. (2012). Development of a Community Weather Information Network (Co-WIN) in Hong Kong. *Weather* **67**(2), 48–50.

Hurrell, J. W., Kushnir, Y., Ottersen, G., and Visbeck, M. (2003). An overview of the North Atlantic oscillation. *Geophysical Monograph-American Geophysical Union* **134**, 1–36.

Imhoff, M. L., Zhang, P., Wolfe, R. E., and Bounoua, L. (2010). Remote sensing of the urban heat island effect across biomes in the continental USA. *Remote Sensing of Environment* **114**(3), 504–513.

Intergovernmental Panel on Climate Change (IPCC). (2013). *Climate Change 2013: The Physical Science Basis. Contribution of Working Group I to the Fifth Assessment Report of the Intergovernmental Panel on Climate Change.* Cambridge University Press.

Intergovernmental Panel on Climate Change (IPCC). (2012). Managing the Risks of Extreme Events and Disasters to Advance Climate Change Adaptation. A Special Report of Working Groups I and II of the Intergovernmental Panel on Climate Change. C. B. Field, V. R. Barros, T. F. Stocker et al. (eds.). Cambridge University Press. 582.

Jacob, D. J., and Winner, D. A. (2009). Effect of climate change on air quality. *Atmospheric Environment* **43**(1), 51–63.

Janhäll, S., Olofson, K. F. G., Andersson, P. U., Pettersson, J. B. C., and Hallquist, M. (2006). Evolution of the urban aerosol during winter temperature inversion episodes. *Atmospheric Environment* **40**(28), 5355–5366.

Jauregui, E. (1997). Heat island development in Mexico City. *Atmospheric Environment* **31**(22), 3821–3831.

Jauregui, E., and Romales, E. (1996). Urban effects on convective precipitation in Mexico city *Atmospheric Environment* **30**(20), 3383–3389.

Jiong, S. (2004). Shanghai's land use pattern, temperature, relative humidity, and precipitation. *Atlas of Shanghai Urban Geography*.

Johnson, K., and Breil, M. (2012). Conceptualizing urban adaptation to climate change: Findings from an applied adaptation assessment. In Carraro, C. (ed.), *Climate Change and Sustainable Development Series* (66), Venice, Italy, CMCC Research Paper.

Jones, P. D., Lister, D. H., and Li, Q. (2008). Urbanization effects in large-scale temperature records, with an emphasis on China. *Journal of Geophysical Research: Atmospheres* **113**(D16).

Kalnay, E., and Cai, M. (2003). Impact of urbanization and land-use change on climate. *Nature* **423**(6939), 528–531.

Karagulian, F., Belis, C. A., Dora, C. F. C., Prüss-Ustün, A. M., Bonjour, S., Adair-Rohani, H., and Amann, M. (2015). Contributions to cities' ambient particulate matter (PM): A systematic review of local source contributions at global level. *Atmospheric Environment* **120**, 475–483.

Katragkou, E., Zanis, P., Kioutsioukis, I., Tegoulias, I., Melas, D., Krüger, B. C., and Coppola, E. (2011). Future climate change impacts on summer surface ozone from regional climate-air quality simulations over Europe. *Journal of Geophysical Research: Atmospheres* **116**(D22).

Katsoulis, B. D., and Theoharatos, G. A. (1985). Indications of the urban heat island in Athens, Greece. *Journal of Climate and Applied Meteorology* **24**(12), 1296–1302.

Kerr, R. A. (2011). Time to adapt to a warming world, but where's the science?. *Science* **334**(6059), 1052–1053.

Kim, J. -S., Zhou, W., Cheung, H., and Chow, C. (2013). Variability and risk analysis of Hong Kong air quality based on Monsoon and El Niño conditions. *Advances in Atmospheric Sciences* **30**(2), 280–290.

Klok, L., Zwart, S., Verhagen, H., and Mauri, E. (2012). The surface heat island of Rotterdam and its relationship with urban surface characteristics *Resources, Conservation and Recycling* **64**, 23–29.

Knowlton, K., Lynn, B., Goldberg, R. A., Rosenzweig, C., Hogrefe, C., Rosenthal, J. K., and Kinney, P. L. (2007). Projecting heat-related mortality impacts under a changing climate in the New York City region. *American Journal of Public Health* **97**(11), 2028–2034.

Kopp, R. E., Horton, R. M., Little, C. M., Mitrovica, J. X., Oppenheimer, M., Rasmussen, D. J., Strauss, B. H., and Tebaldi, C. (2014). Probabilistic 21st and 22nd century sea-level projections at a global network of tide-gauge sites. *Earth's Future* **2**(8), 383–406.

Koskinen, J. T., Poutiainen, J., Schultz, D. M., Joffre, S., Koistinen, J., Saltikoff, E., Gregow, E., Turtiainen, H., Dabberdt, W. F., Damski, J., Eresmaa, N., Göke, S., Hyvärinen, O., Järvi, L., Karppinen, A., Kotro, J., Kuitunen, T., Kukkonen, J., Kulmala, M., Moisseev, D., Nurmi, P., Pohjola, H., Pylkkö, P., Vesala, T., and Viisanen, Y. (2010). The Helsinki testbed: A mesoscale measurement, research, and service platform. *Bulletin of the American Meteorological Society* **92**(3), 325–342.

Kura, B., Verma, S., Ajdari, E., and Iyer, A. (2013). Growing public health concerns from poor urban air quality: Strategies for sustainable urban living *Computational Water, Energy, and Environmental Engineering* **2**(02), 1.

Kuttler, W., Weber, S., Schonnefeld, J., and Hesselschwerdt, A. (2007). Urban/rural atmospheric water vapour pressure differences and urban moisture excess in Krefeld, Germany. *International Journal of Climatology* **27**(14), 2005–2015.

Lamsal, L. N., Martin, R. V., Parrish, D. D., and Krotkov, N. A. (2013). Scaling relationship for NO_2 pollution and urban population size: A satellite perspective. *Environmental Science & Technology* **47**(14), 7855–7861.

Lauwaet, D., Hooyberghs, H., Maiheu, B., Lefebvre, W., Driesen, G., Van Looy, S., and De Ridder, K. (2015). Detailed urban heat island projections for cities worldwide: Dynamical downscaling CMIP5 global climate models. *Climate* **3**(2), 391.

Lee, D. O. (1979). The influence of atmospheric stability and the urban heat island on urban-rural wind speed differences. *Atmospheric Environment (1967)* **13**(8), 1175–1180.

Lemonsu, A., and Masson, V. (2002). Simulation of a summer urban breeze over Paris. *Boundary-Layer Meteorology* **104**(3), 463–490.

Lemos, M. C., Agrawal, A., Eakin, H., Nelson, D. R., Engle, N. L., and Johns, O. (2013). Building adaptive capacity to climate change in less developed countries. In Asrar, G. R., and Hurrell, J. W. (eds.), *Climate Science for Serving Society* (437–457). Springer Netherlands.

Lemos, M. C., and Morehouse, B. J. (2005). The co-production of science and policy in integrated climate assessments. *Global Environmental Change* **15**(1), 57–68.

Li, D., and Bou-Zeid, E. (2013). Synergistic interactions between urban heat islands and heat waves: The impact in cities is larger than the sum of its parts. *Journal of Applied Meteorology and Climatology* **52**(9), 2051–2064.

Li, D., Bou-Zeid, E., Baeck, M. L., Jessup, S., and Smith, J. A. (2013), Modeling land surface processes and heavy rainfall in urban environments: Sensitivity to urban surface representations. *Journal of Hydrometeorology* **14**(4), 1098–1118.

Lin, C. -Y., Chen, W. -C., Chang, P. -L., and Sheng, Y. -F. (2010). Impact of the urban heat island effect on precipitation over a complex geographic environment in northern Taiwan *Journal of Applied Meteorology and Climatology* **50**(2), 339–353.

London Climate Change Partnership (LCCP). (2002). *London's Warming. The Impacts of Climate Change on London: Technical Report* (311). LCCP.

Luković, J., Blagojević, D., Kilibarda, M., and Bajat, B. (2015). Spatial pattern of North Atlantic oscillation impact on rainfall in Serbia. *Spatial Statistics* **14**(Part A), 39–52.

Mackey, C. W., Lee, X., and Smith, R. B. (2012). Remotely sensing the cooling effects of city scale efforts to reduce urban heat island. *Building and Environment* **49**(0), 348–358.

Major, D. C., and O'Grady, M. (2010). Adaptation Assessment Guidebook. In Rosenzweig, C., and Solecki, W. (eds.), *Climate Change Adaptation in New York City: Building a Risk Management Response* (229–292). Blackwell Publishing on behalf of the New York Academy of Sciences.

Mantua, N. J., Hare, S., Zhang, Y., Wallace, J. M., and Francis, R. C. (1997). A Pacific interdecadal climate oscillation with impacts on salmon production. *Bulletin of the American Meteorological Society* **78**(6), 1069–1079.

Marengo, A. J. (2004). Interdecadal variability and trends of rainfall across the Amazon basin *Theoretical and Applied Climatology* **78**(1), 79–96.

Mauser, W., Klepper, G., Rice, M., Schmalzbauer, B. S., Hackmann, H., Leemans, R., and Moore, H. (2013). Transdisciplinary global change research: The co-creation of knowledge for sustainability. *Current Opinion in Environmental Sustainability* **5**(3–4), 420–431.

Mearns, L. O., Lettenmaier, D. P., and McGinnis, S. (2015). Uses of results of regional climate model experiments for impacts and adaptation studies: The example of NARCCAP. *Current Climate Change Reports* **1**(1), 1–9.

Mikami, T., Ando, H., Morishima, W., Izumi, T., and Shioda, T. (2003). A new urban heat island monitoring system in Tokyo. In *Fifth International Conference on Urban Climate*, Lodz, Poland.

Milly, P. C. D., Betancourt, J., Falkenmark, M., Hirsch, R. M., Kundzewicz, Z. W., Lettenmaier, D. P., and Stouffer, R. J. (2008). Stationarity is dead: Whither water management? *Science* **319**(5863), 573–574.

Mishra, V., Ganguly, A. R., Nijssen, B., and Lettenmaier, D. P. (2015). Changes in observed climate extremes in global urban areas. *Environmental Research Letters* **10**(2), 024005.

Mishra, V., and Lettenmaier, D. P. (2011). Climatic trends in major US urban areas, 1950–2009. *Geophysical Research Letters* **38**(16).

Mohan, M., Y. Kikegawa, B. R. Gurjar, S. Bhati, A. Kandya, and K. Ogawa (2009). Assessment of Urban Heat Island Intensities over Delhi. *The Seventh International Conference on Urban Climate*. Yokohama, Japan.

Morris, C. J. G., and Simmonds, I. (2000). Associations between varying magnitudes of the urban heat island and the synoptic climatology in Melbourne, Australia. *International Journal of Climatology* **20**(15), 1931–1954.

Morris, C. J. G., Simmonds, I., and Plummer, N. (2001). Quantification of the influences of wind and cloud on the nocturnal urban heat island of a large city. *Journal of Applied Meteorology* **40**(2), 169–182.

Moss, R. H., Edmonds, J. A., Hibbard, K. A., Manning, M. R., Rose, S. K., van Vuuren, D. P., Carter, T. R., Emori, S., Kainuma, M., Kram, T., Meehl, G. A., Mitchell, J. F. B., Nakicenovic, N., Riahi, K., Smith, S. J., Stouffer, R. J., Thomson, A. M., Weyant, J. P., and Wilbanks, T. J. (2010). The next generation of scenarios for climate change research and assessment. *Nature* **463**(7282), 747–756.

Moss, R. H., Meehl, G. A., Lemos, M. C., Smith, J. B., Arnold, J. R., Arnott, J. C., Behar, D., Brasseur, G. P., Broomell, S. B., Busalacchi, A. J., Dessai, S., Ebi, K. L., Edmonds, J. A., Furlow, J., Goddard, L., Hartmann, H. C., Hurrell, J. W., Katzenberger, J. W., Liverman, D. M., Mote, P. W., Moser, S. C., Kumar, A., Pulwarty, R. S., Seyller, E. A., Turner, B. L., Washington, W. M., and Wilbanks, T. J. (2013). Hell and high water: Practice-relevant adaptation science. *Science* **342**(6159), 696–698.

Muller, C. L., Chapman, L., Grimmond, C. S. B., Young, D. T., and Cai, X. (2013a). Sensors and the city: A review of urban meteorological networks. *International Journal of Climatology* **33**(7), 1585–1600.

Muller, C. L., Chapman, L., Grimmond, C. S. B., Young, D. T., and Cai, X. (2013b). Toward a standardized metadata protocol for urban meteorological networks. *Bulletin of the American Meteorological Society* **94**(8), 1161–1185.

Muller, C. L., Chapman, L., Johnston, S., Kidd, C., Illingworth, S., Foody, G., Overeem, A., and Leigh, R. R. (2015). Crowdsourcing for climate and atmospheric sciences: current status and future potential. *International Journal of Climatology* **35**(11), 3185–3203.

Müller, N., Kuttler, W., and Barlag, A. -B. (2013). Counteracting urban climate change: adaptation measures and their effect on thermal comfort. *Theoretical and Applied Climatology* **115**(1), 243–257.

New York City Panel on Climate Change (NPCC). (2015). *Building the Knowledge Base for Climate Resiliency: New York City Panel on Climate Change 2015 Report*. Annals of the New York Academy of Sciences.

Nicholls, R. (1995). Coastal megacities and climate change. *GeoJournal* **37**(3), 369–379.

Ning, L., and Bradley, R. S. (2015). Winter climate extremes over the northeastern United States and Southeastern Canada and teleconnections with large-scale modes of climate variability. *Journal of Climate* **28**(6), 2475–2493.

Niyogi, D., Pyle, P., Lei, M., Arya, S. P., Kishtawal, C. M., Shepherd, M., Chen, F., and Wolfe, B. (2010). Urban modification of thunderstorms: An observational storm climatology and model case study for the Indianapolis urban region. *Journal of Applied Meteorology and Climatology* **50**(5), 1129–1144.

Ohashi, Y., Genchi, Y., Kondo, H., Kikegawa, Y., Yoshikado, H., and Hirano, Y. (2007). Influence of air-conditioning waste heat on air temperature in Tokyo during summer: Numerical experiments using an urban canopy model coupled with a building energy model. *Journal of Applied Meteorology and Climatology* **46**(1), 66–81.

Oke, T. R. (1978). *Boundary Layer Climate*. Routledge.

Oke, T. R. (2006). Towards better scientific communication in urban climate. *Theoretical and Applied Climatology* **84**(1–3), 179–190.

Overeem, A., Leijnse, H., and Uijlenhoet, R. (2013). Country-wide rainfall maps from cellular communication networks. *Proceedings of the National Academy of Sciences* **110**(8), 2741–2745.

Parker, D. E. (2006). A demonstration that large-scale warming is not urban. *Journal of Climate* **19**(12), 2882–2895.

Patt, A., and Gwata, C. (2002). Effective seasonal climate forecast applications: Examining constraints for subsistence farmers in Zimbabwe. *Global Environmental Change* **12**(3), 185–195.

Peterson, T. C. (2003). Assessment of urban versus rural *in situ* surface temperatures in the contiguous United States: No difference found. *Journal of Climate* **16**(18), 2941–2959.

Power, S., Delage, F., Chung, C., Kociuba, G., and Keay, K. (2013). Robust twenty-first-century projections of El Niño and related precipitation variability. *Nature* **502**(7472), 541–545.

Pozo-Vázquez, D., Esteban-Parra, J. M., Rodrigo, S. F., and Castro-Díez, Y. (2001). A study of NAO variability and its possible non-linear influences on European surface temperature. *Climate Dynamics* **17**(9), 701–715.

Pugh, T. A. M., MacKenzie, A. R., Whyatt, J. D., and Hewitt, C. N. (2012). Effectiveness of green infrastructure for improvement of air quality in urban street canyons. *Environmental Science & Technology* **46**(14), 7692–7699.

Rasmusson, E. M., and Wallace, J. M. (1983). Meteorological aspects of the El Niño/Southern Oscillation. *Science* **222**(4629), 1195–1202.

Richards, K. (2005). Urban and rural dewfall, surface moisture, and associated canopy-level air temperature and humidity measurements for Vancouver, Canada. *Boundary-Layer Meteorology* **114**(1), 143–163.

Rinner, C., and Hussain, M. (2011). Toronto's urban heat island—Exploring the relationship between land use and surface temperature *Remote Sensing* **3**(6), 1251–1265.

Robaa, S. M. (2003). Urban-suburban/rural differences over greater Cairo, Egypt. *Atmósfera* **16**(3), 157–171.

Rosenzweig, C. and Solecki, W. (eds.). (2010). Climate change adaptation in New York City: Building a risk management response. New York City Panel on Climate Change 2010 Report. *Annals of the New York Academy of Sciences* 1196, 354.

Rosenzweig, C., Solecki, W. D., Cox, J., Hodges, S., Parshall, L., Lynn, B., Goldberg, R., Gaffin, S., Slosberg, R. B., Savio, P., Watson, M., and Dunstan, F. (2009). Mitigating New York City's heat island: Integrating stakeholder perspectives and scientific evaluation. *Bulletin of the American Meteorological Society* **90**(9), 1297–1312.

Rotem-Mindali, O., Michael, Y., Helman, D., and Lensky, I. M. (2015). The role of local land-use on the urban heat island effect of Tel Aviv as assessed from satellite remote sensing. *Applied Geography* **56**(0), 145–153.

Rummukainen, M., Rockel, B., Bärring, L., Christensen, J. H., and Reckermann, M. (2015). Twenty-first-century challenges in regional climate modeling. *Bulletin of the American Meteorological Society* **96**(8), ES135–ES138.

Ryu, Y. -H., and Baik, J. -J. (2012). Quantitative analysis of factors contributing to urban heat island intensity. *Journal of Applied Meteorology and Climatology* **51**(5), 842–854.

Seifert, A., Köhler, C., and Beheng, K. D. (2012). Aerosol-cloud-precipitation effects over Germany as simulated by a convective-scale numerical weather prediction model. *Atmospheric Chemistry and Physics* **12**(2), 709–725. doi: 10.5194/acp-12-709-2012

Seiki, A., Nagura, M., Hasegawa, T., and Yoneyama, K. (2015). Seasonal onset of the Madden-Julian Oscillation and its relation to the southeastern Indian Ocean cooling. *Journal of the Meteorological Society of Japan. Ser. II* **93A**, 139–156.

Shepherd, J. M. (2006). Evidence of urban-induced precipitation variability in arid climate regimes. *Journal of Arid Environments* **67**(4), 607–628.

Shepherd, J. M., Carter, M., Manyin, M., Messen, D., and Burian, S. (2010). The impact of urbanization on current and future coastal precipitation: A case study for Houston. *Environment and Planning B: Planning and Design* **37**(2), 284–304.

Sismanidis, P., Keramitsoglou, I., and Kiranoudis, C. T. (2015). A satellite-based system for continuous monitoring of Surface Urban Heat Islands. *Urban Climate* **14**(Part 2), 141–153.

Smith, M., and Emberlin, J. (2006). A 30-day-ahead forecast model for grass pollen in north London, United Kingdom. *International Journal of Biometeorology* **50**(4), 233–242.

Sobral, H. R. (2005). Heat island in São Paulo, Brazil: Effects on health. *Critical Public Health* **15**(2), 147–156.

Solecki, W., Rosenzweig, C., Blake, R., de Sherbinin, A., Matte, T., Moshary, F., Rosenweig, B., Arend, M., Gaffin, S., Bou-Zeif, E., Rule, K., Sweeny, G., and Dessy, W. (2015). New York City Panel on Climate Change 2015 Report: Indicators and monitoring. *Annals of the New York Academy of Science* **1336**, 89–106.

Stone, B. (2007). Urban and rural temperature trends in proximity to large US cities: 1951–2000. *International Journal of Climatology* **27**(13), 1801–1807.

Stone, B. (2009). Land use as climate change mitigation. *Environmental Science & Technology* **43**(24), 9052–9056.

Strauss, B. H., Ziemlinski, R., Weiss, J. L., and Overpeck, J. T. (2012). Tidally adjusted estimates of topographic vulnerability to sea level rise and flooding for the contiguous United States. *Environmental Research Letters* **7**(1), 014033.

Streutker, D. R. (2003). Satellite-measured growth of the urban heat island of Houston, Texas. *Remote Sensing of Environment* **85**(3), 282–289.

Sweet, W. V., and Park, J. (2014). From the extreme to the mean: Acceleration and tipping points of coastal inundation from sea level rise. *Earth's Future* **2**(12), 579–600.

Syvitski, J. P. M., Kettner, A. J., Overeem, I., Hutton, E. W. H., Hannon, M. T., Brakenridge, G. R., Day, J., Vorosmarty, C., Saito, Y., Giosan, L., and Nicholls, R. J. (2009). Sinking deltas due to human activities. *Nature Geoscience* **2**(10), 681–686.

Takahashi, H., Mikami, T., and Takahashi, H. (2009). Influence of the urban heat island phenomenon in Tokyo in land and sea breezes. In *The Seventh International Conference on Urban Climate*, Yokohama, Japan.

Tam, B. Y., Gough, W. A., and Mohsin, T. (2015). The impact of urbanization and the urban heat island effect on day-to-day temperature variation. *Urban Climate* **12**(0), 1–10.

Tan, J., Zheng, Y., Tang, X., Guo, C., Li, L., Song, G., Zhen, X., Yuan, D., Kalkstein, A., Li, F., and Chen, H. (2010). The urban heat island and its impact on heat waves and human health in Shanghai. *International Journal of Biometeorology* **54**(1), 75–84.

Tayanc, M., and Toros, H. (1997). Urbanization effects on regional climate change in the case of four large cities of Turkey. *Climatic Change* **35**(4), 501–524.

Taylor, K. E., Stouffer, R. J., and Meehl, G. A. (2012). An overview of CMIP5 and the experiment design. *Bulletin of the American Meteorological Society* **93**(4), 485–498.

Tran, H., Uchihama, D., Ochi, S., and Yasuoka, Y. (2006). Assessment with satellite data of the urban heat island effects in Asian mega cities. *International Journal of Applied Earth Observation and Geoinformation* **8**(1), 34–48.

Tremeac, B., Bousquet, P., de Munck, C., Pigeon, G., Masson, V., Marchadier, C., Merchat, M., Poeuf, P., and Meunier, F. (2012). Influence of air conditioning management on heat island in Paris air street temperatures. *Applied Energy* **95**(0), 102–110.

Trenberth, K.E., Jones, P. D., Ambenje, P., Bojariu, R., Easterling, D., Klein Tank, A., Parker, D., Rahimzadeh, F., Renwick, J. A., Rusticucci, M., Soden, B., and Zhai, P. (2007). Observations: Surface and atmospheric climate change. In Solomon, S., Qin, D., Manning, M., Chen, Z., Marquis, M., Averyt, K. B., Tignor, M., and Miller, H. L. (eds.), *Climate Change 2007: The Physical Science Basis. Contribution of Working Group I to the Fourth Assessment Report of the Intergovernmental Panel on Climate Change.* Cambridge University Press.

Vanhuysse, S., Depireux, J., and Wolff, E. (2006). *Etude de l'évolution de l'imperméabilisation du sol en région de Bruxelles-Capitale.* Université Libre de Bruxelles, IGEAT.

Vardoulakis, E., Karamanis, D., Fotiadi, A., and Mihalakakou, G. (2013). The urban heat island effect in a small Mediterranean city of high summer temperatures and cooling energy demands. *Solar Energy* **94**, 128–144.

Vaughan, C. (2016). *An Institutional Analysis of the IPCC Task Group on Data and Scenario Support for Impacts and Climate Analysis (TGICA).* IPCC Task Group on Data and Scenario Support for Impacts and Climate Analysis, Geneva, Switzerland.

Vietnam Climate Adaptation PartnerShip. (2013). *Ho Chi Minh City Climate Adaptation Strategy.* Vietnam Climate Adaptation Partnership.

Visbeck, M. H., Hurrell, J. W., Polvani, L., and Cullen, H. M. (2001). The North Atlantic Oscillation: Past, present, and future. *Proceedings of the National Academy of Sciences* **98**(23), 12876–12877.

Walsh, C. L., Roberts, D., Dawson, R. J., Hall, J. W., Nickson, A., and Hounsome, R. (2013). Experiences of integrated assessment of climate impacts, adaptation and mitigation modelling in London and Durban. *Environment and Urbanization* **25**(2), 361–380.

Weiss, J. L., Overpeck, J. T., and Strauss, B. (2011). Implications of recent sea level rise science for low-elevation areas in coastal cities of the conterminous USA *Climatic Change* **105**(3), 635–645.

Willows, R., Reynard, N., Meadowcroft, I., and Connell, R. (2003). *Climate Adaptation: Risk, Uncertainty and Decision-making. UKCIP Technical Report, UK*. Climate Impacts Programme.

Wood, C. R., Järvi, L., Kouznetsov, R. D., Nordbo, A., Joffre, S., Drebs, A., Vihma, T., Hirsikko, A., Suomi, I., Fortelius, C., O'Connor, E., Moiseev, D., Haapanala, S., Moilanen, J., Kangas, M., Karppinen, A., Vesala, T., and Kukkonen, J. (2013). An overview of the urban boundary layer atmosphere network in Helsinki. *Bulletin of the American Meteorological Society* **94**(11), 1675–1690.

Yang, X., Hou, Y., and Chen, B. (2011). Observed surface warming induced by urbanization in east China. *Journal of Geophysical Research: Atmospheres* **116**(D14).

Yin, J., Schlesinger, M. E., and Stouffer, R. J. (2009). Model projections of rapid sea-level rise on the northeast coast of the United States. *Nature Geoscience* **2**(4), 262–266.

Zambrano-Barragán, C., Zevallos, O., Villacís, M., and Enríquez, D. (2011). Quito's climate change strategy: A response to climate change in the metropolitan district of Quito, Ecuador. In Otto-Zimmerman, K. (ed.), *Resilient Cities: Cities and Adaptation to Climate Change Proceedings of the Global Forum 2010* (515–529). Springer Science and Business Media.

Zauli, S., Scotto, F., Lauriola, P., Galassi, F., and Montanari, A. (2004). Urban air pollution monitoring and correlation properties between fixed-site stations. *Journal of the Air & Waste Management Association* **54**(10), 1236–1241.

Zhang, C. (2005). Madden-Julian Oscillation. *Reviews of Geophysics* **43**(2).

Zhang, G. J., Cai, M., and Hu, A. (2013a), Energy consumption and the unexplained winter warming over northern Asia and North America. *Nature Climate Change* **3**(5), 466–470.

Zhang, H., Qi, Z. -F., Ye, X. -Y., Cai, Y. -B., Ma, W. -C., and Chen, M. -N. (2013b). Analysis of land use/land cover change, population shift, and their effects on spatiotemporal patterns of urban heat islands in metropolitan Shanghai, China. *Applied Geography* **44**, 121–133.

Zhao, L., Lee, X., Smith, R. B., and Oleson, K. (2014). Strong contributions of local background climate to urban heat islands. *Nature* **511**(7508), 216–219.

Zhou, Z. -Q., Xie, S. -P., Zheng, X. -T., Liu, Q., and Wang, H. (2014). Global warming–induced changes in El Niño teleconnections over the North Pacific and North America. *Journal of Climate* **27**(24), 9050–9064.

Chapter 2   Case Study References

Case Study 2.1   Urban Heat Island in Brussels

Camuffo, D., and Jones, P. (2002). Improved understanding of past climatic variability from early daily European instrumental sources. In Camuffo, D., and Jones, P. (eds.), *Improved Understanding of Past Climatic Variability from Early Daily European Instrumental Sources* (1–4). Springer Netherlands.

Demarée, G. R., Lachaert, P. J., Verhoeve, T., and Thoen, E. (2002). The long-term daily Central Belgium Temperature (CBT) series (1767–1998) and early instrumental meteorological observations in Belgium. *Climatic Change* **53**(1–3), 269–293.

Fricke, R., and Wolff, E. (2002). The MURBANDY Project: Development of land use and network databases for the Brussels area (Belgium) using remote sensing and aerial photography. *International Journal of Applied Earth Observation and Geoinformation* **4**(1), 33–50.

Hamdi, R., and Schayes, G. (2008). Sensitivity study of the urban heat island intensity to urban characteristics. *International Journal of Climatology* **28**(7), 973–982.

Hamdi, R., and Van de Vyver, H. (2011). Estimating urban heat island effects on near-surface air temperature records of Uccle (Brussels, Belgium): An observational and modeling study. *Advances in Science and Research* **6**, 27–34.

Hua, L. J., Ma, Z. G., and Guo, W. D. (2008). The impact of urbanization on air temperature across China. *Theoretical and Applied Climatology* **93**(3–4), 179–194.

Kalnay, E., and Cai, M. (2003). Impact of urbanization and land-use change on climate. *Nature* **423**(6939), 528–531.

Landsberg, H. E. (1981). *The Urban Climate*. Academic Press.

Peel, M. C., Finlayson, B. L., and McMahon, T. A. (2007). Updated world map of the Köppen-Geiger climate classification. *Hydrology and Earth System Sciences Discussions* **4**(2), 462.

Vanhuysse, S., Depireux, J., and Wolff, E. (2006). *Etude de l'évolution de l'imperméabilisation du sol en région de Bruxelles-Capitale*. Université Libre de Bruxelles, IGEAT.

World Bank. (2017). 2016 GNI per capita, Atlas method (current US$). Accessed August 9, 2017: http://data.worldbank.org/indicator/NY .GNP.PCAP.CD

Case Study 2.2   Los Angeles Megacities Carbon Project

Breon, F. M., et al. (2014). An attempt at estimating Paris area CO2 emissions from atmospheric concentration measurements, *Atmos. Chem. Phys. Discuss.*, **14**, 9647–9703, 2014. Accessed January 21, 2015: www.atmos-chem-phys-discuss.net/14/9647/2014/ doi:10.5194/acpd-14–9647–2014

Duren, R., and C. E. Miller. (2012). Measuring the carbon emissions of megacities. *Nature Climate Change* **2** (8) 560–562. doi:10.1038/nclimate1629.

Gurney, K. R., Razlivanov, I., Song, Y. Zhou, Y., Benes, B., and Abdul-Massih, M. (2012). Quantification of fossil fuel CO_2 at the building/street scale for a large US city. *Environmental Science and Technology* **46** (21), 12194–12202. dx.doi.org/10.1021/es3011282.

Hutyra, L. R., Duren, R., Gurney, K. R., Grimm, N., Kort, E. A., Larson, E., and Shrestha, G. (2014). Urbanization and the carbon cycle: Current capabilities and research outlook from the natural sciences perspective. *Earth's Future* **2**(10), 473–495.

Kort, E. A., Angevine, W., Duren, R., and Miller, C. E. (2013). Surface observations for monitoring urban fossil fuel CO_2 emissions: minimum site location requirements for the Los Angeles megacity, *Journal of Geophysical Research* **118**, 1577–1584. doi: 10.1002/jgrd.50135.

Kort, E. A., Frankenberg, C., Miller, C. E., and Oda, T. (2012). Space-based observations of megacity carbon dioxide. *Geophysical Research Letters* **39**, L17806. doi:10.1029/2012GL052738

Mendoza, D., Gurney, K. R., Geethakumar, S., Chandrasekaran, V., Zhou, Y., and Razlivanov, I. (2013). US regional greenhouse gas emissions mitigation implications based on high-resolution on road CO_2 emissions estimation. *Energy Policy* **55**, 386–395.

NRC. (2010). *Verifying Greenhouse Gas Emissions: Methods to Support International Climate Agreements*. National Academies Press.

Peel, M. C., Finlayson, B. L., and McMahon, T. A. (2007). Updated world map of the Köppen-Geiger climate classification. *Hydrology and Earth System Sciences Discussions* **4**(2), 462.

Silva, S. J., Arellano, A. F., and Worden, H. M. (2013). Toward anthropogenic combustion emission constraints from space-based analysis of urban CO_2/CO sensitivity. *Geophysical Research Letters* **40**(18), 4971–4976.

U.S. Census Bureau (2010). 2010 Census Data. Accessed December 21, 2014: http://www.census.gov/topics/population/data.html

Wong, K. W., Fu, D., Pongetti, T. J., Newman, S., Kort, E. A., Duren, R., Hsu, Y.-K., Miller, C. E., Yung, Y. L., and Sander, S. P. (2015). Mapping CH_4 : CO_2 ratios in Los Angeles with CLARS-FTS from Mount Wilson, California. *Atmospheric Chemistry and Physics* **15**, 241–252. doi:10.5194/acp-15-241-2015.

World Bank. (2010). *Cities and Climate Change: An Urgent Agenda.* World Bank.

World Bank. (2017). 2016 GNI per capita, Atlas method (current US$). Accessed August 9, 2017: http://data.worldbank.org/indicator/NY.GNP.PCAP.CD

Wunch, D., Wennberg, P. O., Toon, G. C., Keppel-Aleks, G., and Yavin, Y. G. (2009). Emissions of greenhouse gases from a North American megacity. *Geophysical Research Letters* **36**, L15810. doi:10.1029/2009GL039825.

Case Study 2.3   Rio de Janeiro Impacts of the Madden-Julian Oscillation

Corpo de Bombeiros. (2014). *Histórico do Corpo de Bombeiros Militar do Estado do Rio de Janeiro.* Accessed September 7, 2015: http://www.cbmerj.rj.gov.br.

Costa, H., and Teuber, W. (2001). *Enchentes no Estado do Rio de Janeiro: Uma abordagem geral.* SEMADS.

Dereczynski, C. P., Oliveria, J. S., and Machado, C. O. (2009). Climatologia da precipita?? O no município do Rio de Janeiro. *Revista Brasileira de Meteorologia* **24**(1), 24–38.

Grimm, A. M., and Tedeschi, R. G. (2009). ENSO and extreme rainfall events in South America. *Journal of Climate* **22**(7), 1589–1609.

Peel, M. C., Finlayson, B. L., and McMahon, T. A. (2007). Updated world map of the Köppen-Geiger climate classification. *Hydrology and Earth System Sciences Discussions* **4**(2), 462.

Tedeschi, R. G., Grimm, A. M., and Cavalcanti, I. F. A. (2015). Influence of Central and East ENSO on extreme events of precipitation in South America during austral spring and summer. *International Journal of Climatology* **35**(8), 2045–2064.

Universidade Estadual de Londrina (UEL). (2014). Cidade alagada: Chuvas de verão, classe e estado no Rio de Janeiro 1966–1967.

United Nations Disaster Relief Organization (UNDRO). (1988). Brazil – Floods Feb 1988 UNDRO Information Reports 1–5.

World Bank. (2017). 2016 GNI per capita, Atlas method (current US$). Accessed August 9, 2017: http://data.worldbank.org/indicator/NY.GNP.PCAP.CD

Case Study 2.4   Will Climate Change Induce Malaria in Nairobi?

Cox, J., Craig, M., Le Sueur, D., and Sharp, B. (1999). *Mapping Malaria Risk in the Highlands of Africa.* MARA/HIMAL Technical Report (96).

Ermert, V., Fink, A. H., Morse, A. P., and Paeth, H. (2012). The impact of regional climate change on malaria risk due to greenhouse forcing and land-use changes in tropical Africa. *Environmental Health Perspectives* **120**(1), 77.

Hay, S. I., Cox, J., Rogers, D. J., Randolph, S. E., Stern, D. I., Shanks, G. D., Myers, M. F., and Snow, R. W. (2002). Climate change and the resurgence of malaria in the East African highlands. *Nature* **415**(6874), 905–909.

Hashizume, M., Chaves, L. F., and Minakawa, N. (2012). Indian Ocean Dipole drives malaria resurgence in East African highlands. *Scientific Reports* **2**, 269.

IPCC. (2013). *Climate Change 2013: The Physical Science Basis. Contribution of Working Group I to the Fifth Assessment Report of the Intergovernmental Panel on Climate Change.* Cambridge University Press.

Kim, Y. -M., Park, J. -W., and Cheong, H. -K. (2012). Estimated effect of climatic variables on the transmission of Plasmodium vivax malaria in the Republic of Korea. *Environmental Health Perspectives* **120**(9), 1314.

Peel, M. C., Finlayson, B. L., and McMahon, T. A. (2007). Updated world map of the Köppen-Geiger climate classification. *Hydrology and Earth System Sciences Discussions* **4**(2), 462.

Shanks, G. D., Hay, S. I., Omumbo, J. A., and Snow, R. W. (2005). Malaria in Kenya's western highlands. *Emerging Infectious Diseases* **11**(9), 1425–1432.

Smith, K. R., Woodward, A., Campbell-Lendrum, D., Chadee, D. D., Honda, Y., Liu, Q., Olwoch, J. M., Revich, B., and Sauerborn, R. (2014). Human health: Impacts, adaptation, and co-benefits. In Field, C. B., Barros, V. R., Dokken, D. J., Mach, K. J., Mastrandrea, M. D., Bilir, T. E., Chatterjee, M., Ebi, K. L., Estrada, Y. O., Genova, R. C., Girma, B., Kissel, E. S., Levy, A. N., MacCracken, S., Mastrandrea, P. R., and White, L. L. (eds.), *Climate Change 2014: Impacts, Adaptation, and Vulnerability. Part A: Global and Sectoral Aspects. Contribution of Working Group II to the Fifth Assessment Report of the Intergovernmental Panel of Climate Change* (709–754). Cambridge University Press.

Tonui, W. K. S., Otor, C. J., Kabiru, E. W., and Kiplagat, W. K. (2013). Patterns and trends of malaria morbidity in western highlands of Kenya. *International Journal of Education and Research* **1**(12), 1–8.

World Bank. (2017). 2016 GNI per capita, Atlas method (current US$). Accessed August 9, 2017: http://data.worldbank.org/indicator/NY.GNP.PCAP.CD

Case Study 2.5   Climate Extreme Trends in Seoul

Chi, Y., Zhang, F., Li, W., He, J., and Guan, Z. (2014). Correlation between the onset of the east Asian subtropical summer monsoon and the eastward propagation of the Madden–Julian Oscillation. *Journal of the Atmospheric Sciences* **72**(3), 1200–1214.

Demographia (2016). *Demographia World Urban Areas 2016.* Accessed August 9, 2017: http://www.demographia.com/db-worldua.pdf

Lee, S. H., and Baik, J. -J. (2010). Statistical and dynamical characteristics of the urban heat island intensity in Seoul. *Theoretical and Applied Climatology* doi: 10.1007/s00704-009-0247-1.

Peel, M. C., Finlayson, B. L., and McMahon, T. A. (2007). Updated world map of the Köppen-Geiger climate classification. *Hydrology and Earth System Sciences Discussions* **4**(2), 462.

Son, J. –Y., Lee, J. –T., Anderson, B., and Bell, M. L. (2012). The impact of heat waves on mortality in seven major cities in Korea. *Environmental Health Perspectives* **120**(4), 566–571.

World Bank. (2017). 2016 GNI per capita, Atlas method (current US$). Accessed August 9, 2017: http://data.worldbank.org/indicator/NY.GNP.PCAP.CD

3

Disasters and Risk in Cities

Coordinating Lead Authors

Ebru Gencer (New York/Istanbul), Regina Folorunsho (Lagos), Megan Linkin (New York)

Lead Authors

Xiaoming Wang (Melbourne), Claudia E. Natenzon (Buenos Aires), Shiraz Wajih (Gorakphur), Nivedita Mani (Gorakphur), Maricarmen Esquivel (Washington, D.C./San José), Somayya Ali Ibrahim (New York/Peshawar), Hori Tsuneki (Washington, D.C.), Ricardo Castro (Buenos Aires), Mattia Federico Leone (Naples), Dilnoor Panjwani (New York), Patricia Romero-Lankao (Boulder/Mexico City), William Solecki (New Brunswick/New York)

Contributing Authors

Brenda Lin (Melbourne), Abhilash Panda (Geneva)

This chapter should be cited as

Gencer, E., Folorunsho, R., Linkin, M., Wang, X., Natenzon, C. E., Wajih, S., Mani, N., Esquivel, M., Ali Ibrahim, S., Tsuneki, H., Castro, R., Leone, M., Panjwani, D., Romero-Lankao, P., and Solecki, W. (2018). Disasters and risk in cities. In Rosenzweig, C., W. Solecki, P. Romero-Lankao, S. Mehrotra, S. Dhakal, and S. Ali Ibrahim (eds.), *Climate Change and Cities: Second Assessment Report of the Urban Climate Change Research Network*. Cambridge University Press. New York. 61–98

Disasters and Risk in Cities

The effects of climate-related disasters are often exacerbated in cities due to interactions with urban infrastructure systems, growing urban populations, cultures, and economic activities. Because the majority of the world's population is currently living in cities – and with this share projected to increase in the coming decades – cities need to focus on improving responses to climate-related disasters such as heat waves, floods, and droughts.

In a changing climate, a new decision-making framework is needed in order to manage emerging and increasing risks. This involves a paradigm shift away from attention to single climate hazards based on past events. The new paradigm requires integrated, system-based risk assessments and interventions that address current and future hazards throughout entire metropolitan regions.

Major Findings

- The environmental baselines of cities have started to shift as climate change impacts take hold. More frequent climate and weather extreme events are being experienced in some urban areas. The frequency and severity of weather and climate-related disasters in urban areas are projected to increase in the coming decades.

- Cultural, demographic, and economic characteristics of urban residents, city governments, built environment, ecosystem services, and human-induced stresses, such as over-exploitation of resources and environmental degradation, define the vulnerability of cities to climate-related disasters. Environmental conditions resulting from unplanned urbanization including removal of natural storm buffers, air and water pollution, overuse of water, and the urban heat island effect exacerbate impacts of climate disasters.

- Given that more than half of the world's population lives in urban areas and that this percentage is expected to significantly increase in the next decades, cities must focus attention on disaster risk reduction and enhancing resilience, issues that most smaller cities have not addressed. Assuming that urban decision-makers have the necessary institutional capacity, their ability to ensure resilient futures could be redirected through strategic development initiatives such as effective risk management, adaptation, and urban planning systems.

- Integrating climate change adaptation with disaster risk reduction involves overcoming a number of barriers. The key barriers include lack of climate resilience in a city's development vision; limited understanding of the hazards, vulnerabilities, and resulting risks; lack of coordination between administrative and sectoral levels of city management; inadequate implementation and financial capacities; and poor connection between climate adaptation and risk management efforts and cities' development visions and strategies.

- Central strategies for improving resilience and managing risks in cities include the integration of disaster risk reduction with climate change adaptation; land-use planning and innovative urban design; financial instruments and public–private partnerships; management and enhancement of ecosystem services; strong institutions and communities; and effective pre- and post-disaster recovery and rebuilding.

Key Messages

Disaster risk reduction and climate change adaptation are the cornerstones of making cities resilient to a changing climate. Integrating these activities with a metropolitan region's development vision requires a new, systems-oriented approach to risk assessments and planning. Moreover, since past events can only partially inform decision-makers about emerging and increasing climate risks, risk assessments must incorporate knowledge about both current climate conditions and future projections.

A paradigm shift of this magnitude will require urban decision-makers and stakeholders to increase the institutional capacity of many communities and organizations to apply a systems lens to coordinating, strategizing and implementing risk-reduction, disaster response and recovery plans on a flexible and highly adaptive basis. As a result, the promotion of effective multilevel governance and multi-sectoral and multi-stakeholder integration is critically important (see Chapter 16, Governance and Policy). The demands for transformational adaptation will be significant and require high levels of governance capacity and financial resources.

3.1 Introduction

Climate change-driven extreme natural disasters and the severity of their impacts expose a need for an enhanced policy framework, particularly in urban areas, where the majority of the world's population lives. It is essential to understand the linkages between the impacts of climate change and disaster risks in urban areas and to address integrated strategies for disaster risk reduction (DRR), climate change adaptation (CCA), and resilience building. This chapter explores and assesses these issues.

First, it identifies the fundamental linkages between climate change, hazards, and risks in urban areas, and it explores hazard exposure and the current and potential impacts of climate change in urban areas. The next section focuses on urban vulnerability: socioeconomic, physical, institutional, and environmental vulnerabilities are assessed (see Chapter 6, Equity and Environmental Justice). A new risk assessment and management framework is presented that integrates CCA and DRR options for cities in making decisions that could lead to more effective system-level approaches to implementing strategies.

Disaster risk reduction mechanisms related to urban planning, financial instruments, building capacity, ecosystem services, and post-disaster recovery and reconstruction, are also assessed in this chapter. A particular focus is on integrated approaches to DRR, CCA, and greenhouse gas (GHG) mitigation and the benefits to using such integrated approaches.

A section on barriers, challenges, and opportunities serves as the conclusion for the chapter. The section discusses innovative actions and opportunities in urban areas for the integration of DRR and CCA and concludes by providing recommendations for decision-makers.

3.2 Climate Change and Disaster Risk

A connection exists between climate change and shifting patterns of risk in cities. In some cases, the linkage is direct because climate change is associated with more frequent and more intense extreme weather and climate events. In other contexts, the connection between the two is mediated by the pathways of urban development and local-scale environmental stresses and degradation. In the following sections, these multiple connection pathways are examined.

3.2.1 Climate Trends

Climate change is expected not only to affect the intensity and the frequency of extreme events, but also to amplify existing social and environmental risks and create novel risks for cities. These emerging conditions result from the interaction of climate hazards (including intense events such as heat weaves and long-term trends such as sea level rise) with the vulnerability and exposure of urban social ecological systems and populations, including their ability to cope, adapt, and transform. These changes have significant implications for cities and urban areas (see Table 3.1).

Table 3.1 *Projected impacts on urban areas of changes in extreme weather and climate events with expected likelihood statements. Source: Romero-Lankao, 2008; Revi et al., 2014*

Climate phenomena and their likelihood	Major projected impacts on urban areas
• Warmer and fewer cold days and nights • Warmer and more frequent hot days and nights *Very likely*	• Reduced energy demand for heating • Increased demand for cooling • Declining air quality • Reduced disruption to transport due to snow, ice • Effects on winter tourism
• Increases in frequency of heat waves *Very likely*	• Reduction in quality of life for people in warm areas without air conditioning • Impacts on elderly, very young, and poor
• Heavy precipitation events more frequent over most areas *Very likely*	• Disruption of settlements, commerce, transport, and societies due to flooding • Pressure on urban infrastructure • Loss of property
• Increases in areas affected by drought *Likely*	• Water shortages for households, industries, and services • Reduced hydropower generation potentials • Potential for population migration
• Increases in intense tropical storms *Likely*	• Damages by floods and high winds • Disruption of public water supply • Withdrawal of risk coverage in vulnerable areas by private insurers • Potential for population migration
• Increased incidence of extreme high sea level *Likely*	• Coastal flooding • Decreased freshwater availability due to saltwater intrusion • Potential for movement of population and infrastructure (also see tropical storms)

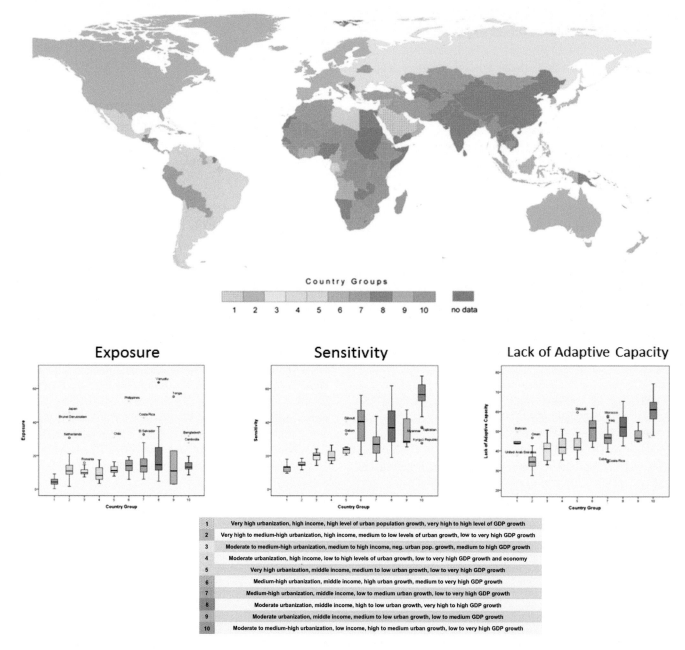

Figure 3.1 *Urbanization and vulnerability. Top: Global distribution of the countries in the ten country groups. Groups 1, 6, 8, and 10 display high urban and economic growth rates. The box plots illustrate the associations between the country groups (x-axes) and exposure, sensitivity, and lack of adaptive capacity. Outliers (i.e., countries with values greater than 1.5 interquartile ranges away from the first or the third quartile) are named.*

Source: Garschagen and Romero-Lankao, 2015

Recent assessment reports, such as the Intergovernmental Panel on Climate Change (IPCC) Fifth Assessment Working Group II Report, note that it is very likely that heat waves will occur more often and last longer and that extreme precipitation events will become more intense and frequent in many regions (IPCC, 2014a). The rise of the global mean sea level will increase the risks to coastal systems and low-lying areas (IPCC, 2014a). These impacts will have direct effects on cities. It is projected with *very high confidence*, "that climate change is to increase risks for people, assets, economies and ecosystems, including risks from heat stress, storms, extreme precipitation, inland and

coastal flooding, landslides, air pollution, drought, water scarcity, sea-level rise, and storm surges" in urban areas, especially for those "lacking essential infrastructure and services or living in exposed areas" (IPCC, 2014b).

3.2.2 Urbanization and Hazard Exposure

Urbanization and rapid or unplanned population growth lead to the concentration of population and assets in hazard-prone urban areas (Gencer 2013a). This exposure and the embedded conditions of socioeconomic, built-environment, spatial, and institutional vulnerabilities produce disaster risks when hazards occur (Gencer 2013a). Settlement patterns, urbanization, and changes in socioeconomic conditions have all influenced observed trends in exposure and vulnerability to climate extremes (Revi et al., 2014). These urban climate change risks, vulnerabilities, and impacts are increasing across the world in urban centers of all sizes, economic conditions, and site characteristics (Revi et al., 2014). However, certain regions of the world experience higher urbanization levels and more unplanned urbanization than others. These regions are not able to meet their cities' needs due to inadequate capacity, unstable governance structures, and substandard infrastructure, built-environment, and urban services. This leads to not only increased exposure, but also to urban vulnerabilities that will increase disaster risks. For instance, Latin America and the Caribbean (LAC) and Europe are currently among the most urbanized regions in the world; however, according to UN-DESA (2015), it is projected that within the next decades, African countries will face higher projected urbanization levels (see Table 3.1). Coupled with existing vulnerabilities in the region, African countries are expected to continue to experience high risk from disasters (Gencer, 2013b) (see Figure 3.2).

Scholarship on urbanization and vulnerability has focused mostly on global and national distributions of current and future exposure of urban areas to climate hazards (McGranahan et al., 2007; Balk et al., 2009). Other dimensions of urban vulnerability, such as sensitivity and capacity, have been insufficiently studied. Some scholars have explored the associations between levels and rates of urbanization of country groups with different levels of development and selected indicators of exposure, sensitivity, coping capacity, and adaptive capacity. Although country groups are at similar risks from exposure to floods, droughts, and other hazards, countries with rapid urbanization and economic growth, especially in Asia and Africa, face greater challenges with respect to higher sensitivity and lack of capacity (Garschagen and Romero-Lankao, 2015) (see Figure 3.1). These countries show significantly higher sensitivity and lower capacity than those with similar current income and urbanization levels but less dynamic urban growth.

Although the global distribution of urban risks is highly context-specific, dynamic, and uneven among and within urban areas and regions, absolute exposure to extreme events over the next few decades is projected to be concentrated in large cities and in countries with populations in low-lying coastal areas, as in many Asian nations (McGranahan et al., 2007). Indeed, as of 2010, there were 442 cities with populations of 1 million or more, and a large majority of them are located in low- and middle-income nations and in hazard-prone areas particularly in Africa, Asia, and Latin America, a trend that is expected to continue within the next decade (Gencer, 2013a).

The geographic location of cities makes them susceptible to certain climate hazards. Among the most challenging risks of climate change are the effects of sea level rise on coastal cities that may have critical infrastructure and large settlement areas in

Figure 3.2 *Metropolitan populations exposed to river flooding, according to Swiss Re's* Mind the Risk *(2013). Longer bars represent larger populations.*

Source: Swiss-Re (www.swissre.com/catnet)

Figure 3.3 *Damaged homes in Midland Beach, Staten Island as a result of Hurricane Sandy, October 2012.*

Photo: Somayya Ali Ibrahim

the most low-lying areas (see Chapter 9, Coastal Zones). Sixty-five percent of the world's urban population currently lives in coastal locations, and this percentage is expected to increase to 74% by 2025 (UN-Habitat, 2011; Gencer 2013a). Most megacities are either located on seacoasts or directly linked with rivers, thus increasing exposure in their hazard-prone areas (Gencer, 2013a). Coastal flooding, beach erosion and saltwater intrusion, river sedimentation, flooding, and landslides are some of the potential hazards than can affect coastal areas and cities built near rivers in all regions of the world (Wang et al., 2014; Stewart et al., 2014). Some cities, such as Santiago, Mexico City, Bogota, and Rio de Janeiro, are also located in geographical regions that are prone to landslides as a result of high climatic rainfall and rapidly changing terrain.

The New York Metropolitan Area offers an example of the importance of geographic location. In *A Stronger More Resilient New York*, the New York City Panel on Climate Change estimated that, sea level rise in the metropolitan region could reach 30 inches above present day by the 2050s (City of New York, 2013; NPCC, 2015). Swiss Re estimated the annual average loss from tropical cyclones in the city to increase from US$1.7 billion to US$4.4 billion by the 2050s due to climate change alone if resiliency measures are not implemented. A storm that causes an economic loss comparable to Hurricane Sandy is projected to increase in frequency, having a return period of 50 years by the 2050s (NPCC, 2015). Neighborhoods in New York vary in elevation and proximity to the coast, and therefore the vulnerability of the built environment changes across this urban landscape (NPCC, 2015). The Rockaway Peninsula in Queens and Midland Beach in Staten Island are two particularly exposed residential neighborhoods. Both felt the brunt of severe structural damage when Hurricane Sandy came ashore,

despite its having been downgraded to a tropical storm by the time it made landfall. Widespread beach erosion, flooding, and boardwalk damage occurred there, while homes and subway lines were inundated with floodwaters. Figure 3.3 shows the destructive power that the storm imposed on households in coastal neighborhoods.

Lower Manhattan was also critically impacted by the storm surge, affecting transportation hubs, electricity distribution, and businesses. At the time of the storm, climate risks were not fully incorporated into major infrastructure projects like the US$530 million construction of the South Ferry subway station from 2005 to 2009, which was not flood-proof despite being in a high-risk flood zone (Rosenzweig and Solecki, 2014). As a result, the subway station was one of the most critically impacted stations throughout the system, experiencing a 14.1-foot storm surge and severe flood damage (see Figure 3.4).

Likewise, storm surges from typhoons and their impacts are expected to worsen due to rising sea level in the Philippine Sea at an increasing rate of 12 millimeters annually. One of the strongest tropical cyclones on record, Typhoon Haiyan, made landfall in November 2013 and caused estimated economic damages between US$6.5 billion and US$14.5 billion. Only a small percentage of this damage (US$300 million–700 million) was covered by private insurance (AIR Worldwide, 2013). Typhoon Haiyan did not strike the most populated area of the Philippines; however, it is predicted that if the storm had struck Manila, losses would have been significantly higher.

In addition to coastal flooding and storms, cities are also expected to be affected by severe heat events. Extreme cold

Figure 3.4 *Aftermath of the flood damage to South Ferry station, October 2012.*

Photo: Somayya Ali Ibrahim

events could lead to increased use of energy and worsening air pollution conditions, whereas expected heat waves could worsen in cities with pronounced urban heat islands (UHIs) due to the heating up of the concrete buildings and paved areas[1] (see Chapter 2, Urban Climate Science, and Chapter 12, Urban Energy). Indeed, many cities in Europe were highly affected by heat waves in the past two decades, leading to deaths as well as to high monetary costs due to the impact on agricultural crops. The 2010 heat wave in Russia caused more than 50,000 deaths and resulted in US$400 billion in economic damages.[2] It also resulted in higher prices for specific food commodities such as wheat for a period of several months. A large proportion of the deaths occurred in Moscow and other larger cities in the region. According to the projections of the IPCC, heat waves and droughts are expected to continue to impact particularly the cities of Southern Europe through negative consequences

in agriculture and water supply and other sectors such as tourism and health and massive forest fires endangering the peripheries of urban areas (Gencer, 2014).

Changes in water availability and in the intensity and frequency of floods and droughts (see Table 3.1) will have profound consequences for cities in terms of both water resources and water management systems (see Chapter 14, Urban Water Systems). Santiago, Chile, illustrates this. The city is located within the Maipo Basin, a watershed supplied by Andean mountain glaciers and their associated snowmelt (Melo et al., 2010). The water provided by these glaciers feeds nearly the entire urban water supply, supplying 90% of potable drinking water, and the built environment in Santiago was constructed to catch this runoff for municipal water demands (Melo et al., 2010). The functioning of the region's economy depends on water being fed to this infrastructure network on a regular basis. However, climate change is predicted to increase temperatures as well as decrease precipitation in this region of the globe (Magrin et al., 2014). This poses several threats of floods, landslides, and droughts as a result of changes in the hydrological cycle that includes the glacial and snowmelt water supply to this catchment area. It is predicted that over the coming century, water availability fed by Andean glaciers will diminish by roughly 40%, while simultaneously the urban population grows from 6 million to more than 8 million by 2030 (Barton and Heinrichs, 2011; Ebert et al., 2010).

Climate and environmental hazards result not only from long-term anthropogenic GHG-driven climate change but are also driven by regional changes (e.g., in land use and water demands) induced by urbanization. Land use changes will in turn escalate the risks of short-term disaster events such as flooding and landslides in neighborhoods of Santiago due to increased exposure of residents to water runoff after rain events (Ebert et al., 2010). There has been growing evidence of these flood risks over the past two decades. For example, in May 2008, severe floods and subsequent landslides forced 13,000 Santiago residents from their homes. The rising population of these neighborhoods in proximity to rivers will continue to present new and challenging flood disaster risks over the coming decades (Ebert et al., 2010). While city planners will need to use the best available science about projected water supply levels as well as future flood risks in their long-term decision-making, they are only starting to incorporate climate into their policies and actions (Romero-Lankao et al., 2014) (see Chapter 16, Governance and Policy).

Urbanization and economic growth coupled with the impacts of climate change is likely to greatly increase the impacts of disasters in urban areas in the future. Climate change is expected not only to alter the intensity and the frequency of hazards, but also to increase the vulnerability of urban populations and places by affecting exposure patterns of settlements (Gencer 2013a). Additionally, economic growth and uncontrolled urbanization

---

1   Munich Re Group (2005). *Megacities – Megarisks: Trends and Challenges for Insurance and Risk Management*. Munich Re Group Knowledge Series. Accessed November 2, 2014: http:www.munichre.com (2006):25; in Gencer, 2013a.
2   According to EM-DAT Figures, http://www.emdat.be/country_profile/index.html

Figure 3.5 *Metropolitan populations exposed to storm surge, overlaid with the historical hurricane storm tracks from 1851 to 2012, according to Swiss Re's* Mind the Risk *(2013). Longer bars represent larger populations.*

Source: Swiss Re (www.swissre.com/catnet)

will add to the exposure and vulnerability of urban areas, which are inherently complex, thus requiring a new framework for decision-making for DRR, CCA, and resilience building.

3.3 Urban Risk and Vulnerability

Urban risk is not only determined by hazard and exposure but also by vulnerability, which is shaped by many factors including the cultural and economic characteristics of urban residents (see Chapter 6, Equity and Environmental Justice); level of technical and institutional capacity of city governments; built environment and infrastructure; quality of ecosystem services; and the threats from human-induced, interconnected stresses and actions such as resource overexploitation and environmental degradation of areas providing natural resources and services. This section analyzes these components and then presents a new decision-making framework that captures the complex and systemic nature of vulnerability in urban areas and recognizes the dynamics of urban risk and the influence of multiple stakeholders in urban systems.

Urban risk can be defined as the likelihood of occurrence of a hazard; the possibility of loss, injury, and other impacts; or the probability of the occurrence of an adverse event and the probable magnitude of its consequences. Risk scholarship has focused on how changes in an environmental hazard

or combination of hazards (e.g., temperature extremes, air pollution, and precipitation extremes) relate to such outcomes (risk proxies) as mortality, morbidity, and economic damage and on how sociodemographic, built environment, and institutional factors (e.g., age and gender, quality housing, and effective response systems) mediate the relationship between the urban hazard and risk (O'Brien et al., 2007; Romero-Lankao and Qin, 2011).

Studies on urban vulnerability tend to portray it as the degree to which a city, population, infrastructure, or economic sector (i.e., a system of concern) is susceptible to and unable to cope with the adverse effects of hazards or stresses, such as heat waves, storms, and political instability (Revi et al., 2014; Romero-Lankao et al., 2012). Urban vulnerability is a relational concept that captures a complex and dynamic reality. In addition to referring to the possibility that a system may be negatively affected by a hazard or stress, it is also a relative property defining both the sensitivity and the capacity to cope with that stressor. Therefore, vulnerability cannot be defined by the hazard alone, nor can it be represented strictly by internal properties of the system being stressed. Instead, it must be looked at as an interaction of these factors (Romero-Lankao and Qin, 2011).

The concepts, research questions, dimensions, and indicators of urban vulnerability and risk can be grouped into three lineages: *vulnerability as impact* (the most commonly applied approach), *inherent or contextual vulnerability*, and *urban resilience* (O'Brien et al., 2007; Romero-Lankao et al., 2012). Urban vulnerability is shaped by physical, demographic, economic, and environmental factors or processes that affect the differential susceptibility of urban households, neighborhoods, and communities – the focus of concern – to the impact of climate hazards.

3.3.1 Socioeconomic Inequality

Common to urban vulnerability research and policy intervention is the concern that differentiated capacities of urban populations to respond to heat waves, floods, and other hazards depend on differences in socioeconomic status (see Chapter 6, Equity and Environmental Justice). Urban population vulnerability springs from social inequality; in other words, from differential access to land property rights, education, income, employment, infrastructure, housing, and political power and from weak, ineffective, or lacking social security, planning, and early warning systems (Harlan et al., 2007; Gencer, 2008; Romero-Lankao et al., 2014).

The strong tie between vulnerability and social inequality in cities starts with the legacies of past decisions and policies around urban land use planning and access to sanitation, water, and other infrastructures and services. Particularly in developing countries, it includes some mechanisms of social exclusion such as formal and informal divisions of the ordered and spontaneous parts of a city (Romero-Lankao et al., 2014). In many informal settlements and peripheral municipalities insecure land titles add to the vulnerability of urban residents. These lead to the social fragility and difficult disaster recovery for these settlers, who can neither obtain government aid nor credit with their illegal titles (Gencer 2013a). Social exclusion, ethnic or immigrant status, poor education, and limited job opportunities add to the income poverty of these residents, limiting their mobility and ability to resettle and creating one of the biggest challenges for urban policy-making in the developing world.

These mechanisms of exclusion result in differentiated patterns of location, access to resources, rights, assets, infrastructures, and services, some of which define urban populations' resilience. Social inequality is not equal to social vulnerability, however, because vulnerability is a relational concept. At the microlevel, it is not only location in hazard-prone places or individual characteristics such as age, gender, and existing medical conditions that can make populations sensitive, but also the material and symbolic sources of assets, capital, or resources that can enhance population resilience, such as education, income, house quality, infrastructure and services, legal status, and social capital (e.g., participation in networks and family support) (Romero-Lankao et al., 2014). Of no less importance are city-wide disaster management and adaptation policies.

3.3.2 Physical Processes

The urban built environment is susceptible to climate extremes such as heat waves, floods, and other climate hazards as a result of physical location conditions and the performance of buildings and critical infrastructure. Implementation of up-to-date building standards and good urban planning actions can help manage or reduce disaster risks (Gencer, 2008; Solecki et al., 2011). However, many cities do not have standard design, building, and land-use regulations available; do not implement them due to a powerful construction and development sector; or corrupt development and building control practices.

In many instances, this phenomenon is observed in high-income residential construction that disregards planning decisions and the protection of ecosystems in order to secure locations in scenic areas such as close to water basins or in protected forest areas. Many examples of this practice are observed in gated complexes in Istanbul, on the fertile slopes of Mount Vesuvius in Naples, or in Mexico City (Gencer 2013a).

Box 3.1 Disasters and Social Vulnerability Index

Claudia Natenzon and Ricardo Castro

University of Buenos Aires

A disasters social vulnerability index (DSVI) was developed by the Natural Resources and Environment Research Program, University of Buenos Aires (Natenzon and González, 2010). It is a quantitative, statistical evaluation that allows researchers to identify different degrees of social vulnerability in administrative units. The usefulness of such an index is that it provides a primary evaluation of the heterogeneities in the geographic distribution of social vulnerability. The DSVI is composed of indicators that are grouped into three subsets: demography,

living standards, and economic capacity. The index allows one to distinguish the importance of the different aspects considered. The selected indicators thus aid in prioritizing interventions and are derived from publicly available data on social, demographic, and basic economic characteristics.

Such indicators are based on variables listed in the three dimensions (Vazquez-Brust et al., 2012): (1) demographic (e.g., dependent population and single-parent homes), (2) living standards (e.g., overcrowded housing, supply of drinking water, and sewage services), and (3) economic capacity (e.g., health, literacy/education, and work/employment status).

Figure 3.6 *Gated settlements in Istanbul's previously protected forest areas.*

Photo: Ebru Gencer, 2005

In other instances, corruption or shortcomings during construction increase risk from natural hazards. In Florida, investigations after the 1992 Hurricane Andrew found major shortcomings in construction techniques and code enforcement (Mileti, 1999). Mileti (1999) reported that in Southern Dade County, homes built after 1980 suffered more damages than pre-1980 constructions, including loss of roof materials that also led to damage to other buildings and cars (Gencer, 2008; Gencer, 2013a). According to Mileti, a review of the county's Board of Rules and Appeals found a number of instances in which changes were made under pressure from builders for construction cost savings such as allowing the use of staples instead of nails to install roofs.

Physical susceptibility to hazards that are heightened by climate change is most evident in the rapid expansion of urban areas and the creation of unplanned informal settlements (see Chapter 11, Housing and Informal Settlements). The increase of the urban poor and their exclusion from formal housing sectors result in growing informal settlements that create immense challenges in disaster risk management (DRM) for climate change (Gencer 2008; Gencer 2013a). Most informal settlements display physical vulnerabilities due to their location or construction practices because they are often located on land not deemed appropriate for habitation because of its steep terrain or geographical characteristics that make these areas prone to subsidence, landslides, or mudslides (UN-Habitat, 2003; Gencer 2013a). For instance, in the Caribbean nation of Belize, where the slum population is equal to nearly half of the urban population, the low-lying coastline accommodates approximately 45% of its total population in densely populated

urban areas such as Belize City (approximately 20.5% of total population), and these coastal centers represent some of the country's areas most vulnerable to storm events because they lie approximately 1–2 feet below sea level (WB and GFDRR, 2010; Gencer, 2013b).

The informal status of urban populations exerts a profound influence on both hazard exposure and capacity. Informality is a state of regulatory flux, where land ownership, land use and purpose, access to livelihood options, job security, and social security cannot be fixed or mapped according to any prearranged sets of laws, planning instruments, or regulations (Romero-Lankao et al., 2014). This condition leads to an ever-shifting relationship between the legal and the illegal, the legitimate and the illegitimate, the authorized and the unauthorized. Informality becomes the site of considerable power, where some forms of growth in risk-prone areas enjoy state sanctions while others are criminalized. For the latter, informal status becomes a systemic determinant of lack of access to assets and options for adaptation capacity. Conversely, the "regular," "legal," or "formal" status of a settlement gives security from eviction, becomes an incentive to invest in more structural adaptation actions (e.g., housing improvements to effectively respond to floods), is a requirement for infrastructure and service provision, and helps avoid the stigmatization that disempowers informal neighborhoods.

Many times, inadequate building materials accompany risk of physical exposure in squatter settlements because structures are often built with impermanent materials such as earthen floors,

Figure 3.7 *Use of non-resilient materials in hurricane-prone Jamaica.*

Photo: Ebru Gencer, 2012

mud-and-wattle walls, or straw roofs (UN-Habitat, 2003) (see Chapter 11, Housing and Informal Settlements). In many cases, these settlements lack municipal services and infrastructure, resulting in further disasters, such as waste disposal in riverbeds and ravines as well as the urbanization of watersheds and wetlands that leads to floods from heavy rainfall, as observed in the informal settlements of the Western tBalkans (Gencer, 2014). In addition, lack or inefficiency of public urban services and institutions – transportation networks, hospitals, fire and police stations – translates into lack of response capacities at times of disaster, thus further exacerbating vulnerability and disaster risk for these urban settlers (Gencer, 2008, Gencer 2013a).

Other factors that contribute to physical vulnerability include the type of terrain where structures are built, such as the soil quality, geomorphology, and surface and groundwater features of the landscape (Blanco et al., 2011). Additionally, the political and legal framework pertaining to land use further contributes to the risks that the built environment faces, including land-use planning practices, zoning codes, and legal rights to property ownership.

3.3.3 Institutional Processes

Institutional dynamics play a central role in urban vulnerability and risk. Vulnerability and risk may be amplified due to organizational constraints and exclusion of multiple stakeholders in urban governance (see Chapter 16, Governance and Policy). Several characteristics of ineffective or capacity-lacking institutions involved in DRM may include (Natenzon, 2005):
• Obstacles such as political distrust, lack of communication (e.g., unawareness of what other agencies are doing), and lack of coordination.
• Inability to scale up successful programs and projects that may be dismissed due to government changes
• Inadequate means or ways to communicate risks and disasters

to urban populations.
• Corruption in the policy-making and implementation process (see Chapter 16, Governance and Policy). Corruption affects the confidence of urban citizens in their institutions and destabilizes the society.

Fragmentation of policies and actions increases social vulnerability and generates a high degree of uncertainty that can affect the capacities of cities to respond to disasters. Poor institutions and policies are not adequate to confront long-term processes. Nevertheless, cultural norms and traditions of urban residents who experience increased vulnerability can play a positive role in risk reduction. Identification with place, and previous experience with local danger may invigorate vulnerable urban residents to establish local strategies such as applying traditional house-building techniques to reduce risk or establish neighborhood networks, such as voluntary organizations for risk communication, preparedness, or recovery. Recognition of these conditions can open the door to participatory governance, greater public involvement, and greater autonomous risk management.

3.3.4 Ecosystem Services and Environmental Processes

Urbanization is not only related to increased levels of GHG emissions, but also exerts pressure on ecosystems surrounding cities, increasing environmental vulnerability to hazards and climate change (see Chapter 8, Urban Ecosystems). Urbanization is a key driver of land fragmentation and loss of biodiversity (Grimm et al., 2008). Differences in land quality, land use, and functional characteristics determine urban vulnerability and risk in different and not yet fully understood ways. For instance, changes in vegetation cover are one of the factors influencing the risk of floods, rainfall-triggered landslides, and wildfires near or in urban centers (Braimoh and Onishi, 2007; Smyth and Royle,

---

Box 3.2 Management of Slope Stability in Communities

One particularly useful example for housing and shelter comes from an innovative methodology for assessing and reducing landslide risk in unplanned urban communities (see Chapter 11, Housing and Informal Settlements). Management of Slope Stability in Communities (MoSSaiC) is "based on identifying the localized physical causes of landslides (often related to inadequate drainage), designing appropriate engineering measures to address these causes (such as surface water drains), and constructing those measures to an adequate specification so that the root cause of the landslides is effectively addressed. This science- and engineering-based approach is embedded in community participation and the engagement of city government experts, policy-makers, and development agencies. It has been successfully applied in 12 communities in the Eastern Caribbean with funding from national governments, UNDP, USAID, and the World Bank." Improved, resilient housing is important because most housing is renewed at periods of 30 years or less (particularly in developing countries), and designs that address landslide hazards can make a substantial contribution.

2000). Changes in land use may cause changes in land surface physical characteristics (e.g., surface albedo) that have implications for water-related hazards such as droughts and floods because precipitation can be enhanced or reduced depending on climate regime, geographic location, and regional patterns of land, energy, and water use (Romero-Lankao et al., 2014).

Urban areas are key drivers of changes in carbon, water, and other biogeochemical cycles, but they are also vulnerable to extreme temperatures, air pollution, water degradation, and other hazard risks associated with these changes (Pataki et al., 2006). High levels of air pollution are known to increase the risks of negative health impacts on human populations, particularly when combined with adverse weather conditions (e.g., heat waves caused by climate change) (Bell et al., 2008). Health impacts from air quality and temperature changes become especially critical in rapidly growing low- and middle-income countries (Kan et al., 2008; Romero-Lankao et al., 2013; Revi et al., 2014). Climate hazards, such as floods and droughts, are another example. Increased vulnerability of urban water systems and their users can result from long-term processes such as poor construction, the inefficient operation of water infrastructure, and land use changes that increase impervious surfaces and are driven by urbanization (Romero-Lankao, 2010).

Ecosystem services of relevance to urban areas (e.g., flood protection) represent an important set of non-built assets and add value to the quality of life in urban areas. However, many environmental assets and ecosystem services are vulnerable to the changes associated with population growth, land use, and climate change. One of the greatest challenges for coastal cities is that of sea level rise and storm surge, which can cause inundation and damage to many important environmental assets such as wetlands and coastal recreational zones (see Figure 3.8).

Figure 3.8 *Coastal recreation zone in Victoria, Australia.*

Photo: Xiaoming Wang

Figure 3.9 *Buffer zone developed in the Central Coast, a peri-urban region in New South Wales, Australia.*

Photo: Xiaoming Wang

In Australia, Brisbane Water is one of the most heavily populated catchments in eastern New South Wales, and its environmental assets are heavily used by tourists and residents for water-based recreational activities such as boating and fishing. This area is also adjacent to a National Heritage Area of valuable wetlands that extend into Brisbane Water and provide scenic attraction, recreation, and environmental education, as well as economically valuable ecosystem goods and services (e.g., water filtration, storm surge protection, and fisheries) and tourism revenue of US$360 million per year (Harty and Cheng, 2003; Tourism Research Australia, 2007).

The coastal ecosystems in this region are already under stress from development and increasing fragmentation, and future sea level rise will have a large impact on the conservation and productive value of these coastal habitats (HCCREMS, 2010). The estuarine wetlands of mangrove and salt marsh communities are relatively rare and host a broad range of endangered and vulnerable plants and animals (Conacher Travers Pty. Ltd., 2001). They are also habitats in decline, with coastal inundation already affecting the persistence of present species (Mitchel and Adam, 1989; Mazumder et al., 2006). Lin and colleagues (2014) in a land use study of coastal communities show that a 1 meter inundation event has an annual occurrence probability of 66.7% at present, inundating 98% of the ecologically important habitat in a particularly threatened study area. This risk is expected to increase with future sea level rise and could potentially lead to the salinization and loss of habitat for freshwater or brackish ecosystems.

Protection of environmental assets presents a difficult set of adaptation decisions; the landward progression of salt marshes due to sea level rise is often restricted by the placement of roads or other hard structures that cause salt marshes to shrink or become more fragmented over time (Harty and Cheng, 2003). If present levels of urban development are to be maintained, less land will be apportioned for ecosystem habitat and migration, with present ecosystems competing for smaller areas of coastal land area. Management options to maintain ecosystems within the landscape (e.g., restoration of salt marshes or use of buffer zones) have been discussed, but can be labor intensive (Laegdsgaard, 2006) (see Figure 3.9). Therefore, the question of how and if ecosystem amenities and services should be maintained under future sea level rise scenarios becomes increasingly important.

3.4 Risk Assessment Framework and Decision-Making Process

Many concepts of DRM are based on a static comprehension of hazards and do not take into consideration the cumulative effects of climate change or the everyday risks to which urban communities are exposed (Gencer et al., 2013; Solecki et al., 2011). A new framework for disaster risk assessment and risk-based hazard management decision-making is based on a comprehensive concept encompassing the dynamics of urban development and the complexity of cities. This requires a paradigm change in urban DRM, making it a more comprehensive policy agenda that incorporates DRM and CCA agendas into a comprehensive policy and agenda for sustainable urban development (Gencer et al., 2013).

Such a paradigmatic shift is reflected in the *Special Report on Managing the Risks of Extreme Events and Disasters to Advance*

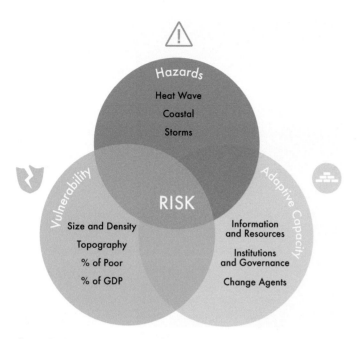

Figure 3.10 *Urban climate change vulnerability and risk assessment framework.*

Source: Mehrotra et al., 2011

Climate Change Adaptation that illustrated the significance of linking climate change adaptation with DRM and sustainable development (IPCC, 2012). The IPCC Working Group II highlighted the importance of analyzing vulnerability, risk, and adaptation as the most relevant subjects for understanding and managing climate change risks (IPCC, 2014a).

Climate change impacts, adaptation, and vulnerability assessments have been developed in different forms driven to address current and future challenges and uncertainties (Carter et al., 2007). In the Urban Climate Change Research Network (UCCRN)'s *First Urban Climate Change Research Network Assessment Report on Climate Change and Cities* (ARC3.1), an urban climate change vulnerability and risk assessment framework was introduced that was based on the interplay of hazard, vulnerability, and adaptive capacity to develop climate adaptation and disaster management (Mehrotra et al., 2011) (see Figure 3.10).

UCCRN's *Second Urban Climate Change Research Network Assessment Report on Climate Change and Cities* (ARC3.2) uses the framework of risk (R) expressed as a function of hazard (H), exposure (E), and vulnerability (V).[3] The IPCC Working Group II report (IPCC, 2014b) explains that risks from climate change impacts arise from the interactions among hazard (triggered by an event or trend related to climate change), vulnerability (susceptibility to harm), and exposure (people, assets, or ecosystems at risk).

A paradigmatic shift from single-impact hazards-focused risk assessment to system-based risk assessment is essential to identify current and future risks if we are to reduce disaster risks

and make cities resilient. An understanding of the city as a system involving multiple stakeholders and institutions, composed of urban vulnerabilities – as described in Chapter 1, Pathways to Urban Transformation, and Section 3.1 – and their horizontal and vertical integration, is essential in defining risks to current and future cities.

In the risk assessment framework, it is critical to make optimally targeted decisions in developing CCA and DRR across scales that can be spatial, temporal, and institutional. In this chapter, risk-based decision-making is shifting away from the impact-centric approach (ICA) to a decision-centric approach (DCA) as described in Figure 3.12. The decision-centric risk assessment framework includes:

- An H-E-V-based risk assessment that takes into account the hazards caused by climate change and variability to estimate risk distribution in terms of socioeconomic, environmental, institutional, and physical loss (or opportunity) faced by cities. Future city outlooks (e.g., population growth, demographic changes, land use change) and subsequent exposure and vulnerability to hazards are taken into account.

- Development of climate change actions and DRR options involving policies, planning, and local actions through both horizontal and vertical integration (at multiple scales of spatial, temporal, institutional aspects) to reduce exposure and/or vulnerability of socioeconomic, environmental, institutional, and physical aspects and eventually lead to the reduction of risks to cities.

- Understanding of the balance of avoided loss and additional benefit – with opportunity loss in adaptation and residual loss after adaptation – to optimize the selection of better options, which is also a dynamic, time-dependent, and iterative process.

The impact-centric approach to DRM and CCA decision-making for cities is considered a linear approach because it starts with an examination of the climate hazards to cities and then takes into account vulnerability to hazards by integrating the knowledge of existing city capacities. Disaster risk reduction and CCA strategies are then developed by enhancing resilience and adaptive capacity to reduce the impact. However, because decisions are based on hazard-driven impact assessment, DRR and CCA development could become disconnected from other parallel priorities (e.g., poverty reduction and sustainable livelihoods, environmental conservation, and economic development) and also from municipal and national needs. In addition, differing DRR and CCA strategies could be developed from different perspectives based on different interpretations of the impacts for different agendas, leading to ineffective results due to narrow views on DRM and CCA.

In contrast, the DCA leads to vision-driven urban DRM and CCA. Decisions aim to reduce hazard exposure in line with a future city vision, nurture resilience of cities for the potential impacts of hazards, build capacities to reduce the vulnerability of stakeholders, and establish governance that ensures

---

3   As explained by the equation: R = H × E × V, where all are time-dependent and spatially random variables and described by probability distribution functions.

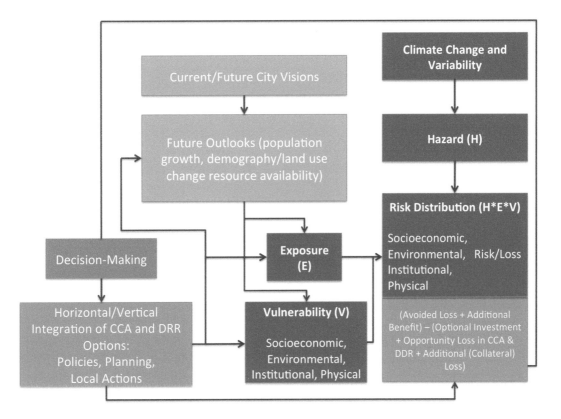

Figure 3.11 *Risk Assessment Framework, developed for UCCRN ARC3.2.*

Source: Xiaoming Wang and Ebru Gencer, 2014

resilience-building and capacity development to enable stakeholders' visions and goals. Ultimately, DCA aims to implement the visions and goals of future cities by minimizing risks of disaster and climate change impacts by taking transformative steps, in contrast to ICA that aims to reduce impacts in relation to specific disaster and climate change hazards.

3.4.1 Disaster Risk Reduction Policy Domains and the Movement toward Transformation

Opportunities for disaster risk management can contribute to the overarching aims of climate change adaptation as well as the larger goal of urban sustainable development. Much current DRR is focused on the enhancement of resiliency and flexibility to extreme events. Implicit or explicit in urban DRM are other risk management regimes including resistance (planning for stability), transformation (planning for fundamental change), and even collapse (no planning/failure to implement planning) (Solecki et al., 2017).

Many cities employ DRR policies focused on resistance. A resistance-focused DRR policy is oriented toward deploying risk management to achieve stability in underlying development. This may require major shifts and investment in nonprioritized or external elements so that resistance in one system may require collapse or transformation in another. For example, the construction of increasingly larger and more complex coastal defenses

to prevent any change in function, value, or appearance of coastal land may transform near-shore ecology and livelihoods or downstream hazards. Resistant systems may expend considerable resources on preventing change by attempting to manage the external environment. Resistance and resilience have similar qualities but different intentions.

A resilient system is able to adjust flexibly in the anticipation or experience of a hazard. A resilient system's functions and core aims are maintained with only slight adjustment, although these adjustments may be significant for subsystems or over time. In social systems, an example is adjustment to insurance regimes that allow continued habitation in places of risk through changes in payment rates. In contrast to resistant systems, resilient systems can anticipate, absorb, accommodate, or recover from the effects of a hazardous event in a timely and efficient manner through preservation, restoration, or improvement of the system's essential basic structures and functions. Essentially, the system responds by accepting loss and returning to its pre-shock/ stress state, which in turn may be perceived by dominant actors as the preferred state.

A transformative regime identifies the need for fundamental change in underlying development if unacceptable risk and future loss are to be avoided. This can include distributional as well as total loss concerns. Including transformation opens new policy options once resilience meets its limits (Pelling et al., 2015). Transformation can also target the root drivers of unmet

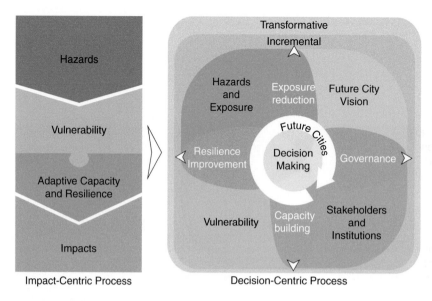

Figure 3.12 *Evolving impact-centric approach to decision-centric approach for urban vulnerability and risk assessment developed for UCCRN ARC3.2.*

Source: Xiaoming Wang and Ebru Gencer, 2014; adapted from Xiaoming Wang, 2014

sustainable development needs where these constrain adaptive capacity and action (Marshall et al., 2012). Intentional transformation of one system or object can allow the maintenance of systems at other scales (e.g., relocation of households exposed to risk will be transformative for households involved, for places of origin and destination, and may require legislative change). At the same time, relocation may help maintain resilience or resistance in wider political and economic or social systems.

City planners, risk managers, and stakeholders need to have transformation presented as an option within the range of policy domains (Solecki et al., 2017). Risk managers should address the extent to which the current policy conditions enable or fail to enable a pathway toward transformative sustainability and should consider the barriers and bridges of a transition to meaningful CCA and mitigation and enhanced sustainability. An overall premise of ARC3.2 is that fundamental transformational changes in risk management regimes are needed, given the gap between current urbanization trajectories, accelerating climate change, and the goals of sustainable development.

The ARC3.2 takes the position that transformation can occur within the contexts of multi-layered systems, formal and informal politics, and structural limits on the local action and agency of individuals and specific organizations. The requirements of urban sustainability and transformative adaptation have become linked to a series of preconditions and pathways through which people, communities, and places can move toward greater sustainability (Pelling et al., 2015; Folke et al., 2010). Although the concept of disaster resiliency holds significant theoretical and practical appeal in sustainability pathways, many argue that it does not fully explain or enable larger-scale changes within institutions and society needed to enhance the opportunities for genuine movement toward sustainability (Kates et al., 2012). This category of larger or more profound shifts is associated with the concept of transformation (see Chapter 1, Pathways

to Urban Transformation) (Bahadur and Tanner, 2014; Olsson et al., 2014). Resiliency and transformation approaches are increasingly being compared and illustrated via examples and future prospects (Pelling et al., 2015).

3.5 Urban Planning and Land-Use Tools

The Decision-Centric Approach is multidisciplinary in nature, recognizing the importance of links among climate change, hazards, and the wider environment (Lewis, 1999; Wisner et al., 2004; Tran and Shaw, 2007). Identification of a hazard and assessment of risks lay the basis of potential DCA strategies, which are shaped by social and cultural influences as well as by legal, institutional, and economic constraints (Gencer, 2008). City governments are usually the immediate responsible actors that undertake DRR and resilience-building activities in urban areas. There are several strategies that can help reduce disaster risks and increase resilience in urban areas. Some of the most used strategies are institutional organization and capacity development by city governments; urban planning and development actions; building codes and engineering practices, such as infrastructure retrofitting and investment; social capacity development; public awareness and education; financial capacity development, such as insurance and incentives; ecosystem services; early warning systems; and post-disaster recovery and planning. This section explores some of these planning strategies in detail.

Urban planning is a continuous process of foreseeing, anticipating, and preparing for the future. In order to manage such change in physical geographies, urban planning makes arrangements for future demands on the use of public and private land and seeks a balance among interests "to resolve conflicting demands on space" (ISOCARP, 2005; Gencer, 2008) (see Chapter 5, Urban Planning and Design). Urban planning

Box 3.3   Sendai Framework for Disaster Risk Reduction and the UNISDR Making Cities Resilient Campaign

Abhilash Panda

UNISDR, Geneva

In March 2015, representatives from 187 United Nations Member States adopted the first major agreement of the Post-2015 development agenda, the Sendai Framework for Disaster Risk Reduction (SFDRR). This builds on its predecessor, the Hyogo Framework for Action (HFA), and articulates seven targets and four priorities for action. (For further information, see UN, 2015.) Policy-makers will now need to reassess the significance of resilience-building in cities as they implement the SFDRR. To support the implementation of SFDRR and the more than 2,500 cities in the UNISDR Making Cities Resilient program, a group of city and expert partners in urban resilience, including the UCCRN, have now proposed a set of New Ten Essentials. The proposed New Ten Essentials build on the previous Ten Essentials, are linked with the priorities for action of the SFDRR, and represent a transition to implementation. The primary objective is to be operational, adaptive, and applicable to all and encourage cities to implement disaster risk reduction (DRR).

The New Ten Essentials offer a comprehensive approach to DRR because they undertake to cover the most important issues for cities to become more resilient (see Box 3.3 Figure 1): Essentials 1–3 cover governance, risk science, and financial issues; Essentials 4–8 cover the many dimensions of planning, capacity, and growth; and Essentials 9–10 cover post-event recovery and building back better.

The intention of the New Ten Essentials is to enable cities to establish a baseline measurement of their current level of disaster resilience, identify priorities for investment and action, and track their progress in improving their disaster resilience over time. The New Ten Essentials intend to guide cities toward optimal disaster resilience and challenge complacency. This demanding standard reminds cities that there is *always* more that could be done and to establish investment goals (of both time and effort) for achievement over a period of years.

"NEW TEN ESSENTIALS"
for making cities disaster resilient
critical and interdependent steps

Essential 1
Organize for disaster resilience

Essential 2
Identify, understand and use current and future risk scenarios

Essential 3
Strengthen financial capacity for resilience

Essential 4
Pursue resilient urban development and design

Essential 5
Safe guard natural buffers to enhance the protective functions offered by natural ecosystems

Essential 6
Strengthen institutional capacity for resilience

Essential 7
Understand and strengthen societal capacity for resilience

Essential 8
Increase infrastructure resilience

Essential 9
Ensure effective disaster response

Essential 10
Expedite recovery and build back better

Box 3.3 Figure 1 *The New Ten Essentials for Making Cities Resilient.*

can help to protect environmentally sensitive areas and reduce disaster vulnerability and risks by employing land-use zoning and implementing development actions. These efforts can increase resilience by upgrading and retrofitting poorly planned settlements, especially through a participatory process that can ensure implementation and sustainability (UPAG, 2015). Olhansky and Kartez (1998) in the Second Natural Hazard Assessment study in the United States identified land-use management tools to guide development in hazard-prone areas. These and other tools include (Olhansky and Kartez, 1998; Gencer, 2008; Rosenzweig et al., 2011) (see Chapter 5, Urban Planning and Design):

- *Building standards*, such as building codes, flood-proofing requirements, seismic design standards, and retrofit requirements for existing buildings;
- *Development regulations* including zoning and subdivision ordinances such as flood-zone regulations; setbacks from faults, steep slopes, and coastal erosion areas; and zoning standards for sensitive lands (e.g., wetlands, dunes, and hillsides);
- *Critical public facilities* policies that require location of these facilities outside of hazard areas in order to discourage development and reduce damages;
- *Land and property acquisition* in hazardous areas through public funds and use of these properties in minimally

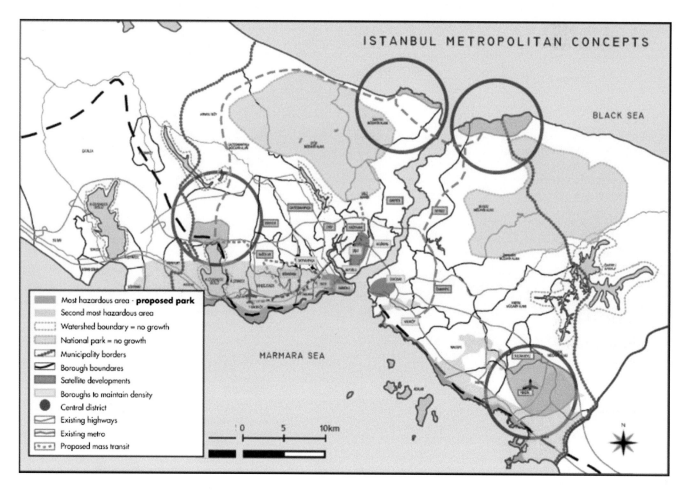

Figure 3.13 *Integrating disaster risk reduction into urban planning education in Istanbul.*

Source: S. Grava; K. Jacob, E. Gencer et al. Columbia University Urban Planning Studio: Disaster Resilient Istanbul, 2002

vulnerable ways and the acquisition of open space, recreational, or undeveloped lands for risk reduction, relocation of existing hazard area development, and acquisition of development rights;

- *Taxation and fiscal policies* to provide incentives for urban residents to reduce public costs in hazardous areas by applying regulations for safety or relocating and reducing population density in hazardous areas;
- *Information dissemination* to influence public behavior, especially of real estate customers by requiring hazard disclosure statements of real-estate sellers, providing public information by posting warning signs in high-hazard areas, and education of construction professionals.

Studies conducted in the United States examined the use of these tools by city governments and found that zoning ordinances and building standards are the most frequently used DRR tools by municipalities (Gencer, 2008; Gencer, 2013a).

In another study, the Institute for Business and Home Safety (IBHS) together with the American Planning Association (APA) and the American Institute of Certified Planners (AICP) surveyed nearly 1,500 municipal planners on Community Land-Use Evaluation for Natural Hazards (IBHS, 2002). The results of this study indicated the need for hazards planning, additional funding, support from elected officials, and technical assistance in addition to better mapping and data, states mandates for planning, and additional staff and legislative changes (Steinberg and Burby, 2002a).

However, in many developing countries, regulations, codes, standards, technical requirements, performance indicators, and best practices represent the capacity development needs of city governments and urban communities and challenges to implementation. Evaluation of the costs and benefits of specific DRR as well as climate change action alternatives can lead to the development of effective strategies aimed at exceeding minimum regulatory requirements. In addition, specific DRR/CCA requirements may be imposed by local regulations in response to site-specific hazards. Furthermore, when a change in urban land use is proposed, it must be determined whether this change triggers other risk conditions, thus requiring additional DRR and/or climate change action measures. For example,

Case Study 3.1   The Boulder Floods: A Study of Decision-Centric Resilience

Karen MacClune, Kanmani Venkateswaran

ISET-International, Boulder

Chris Allan

Pitcher Allan Associates, Boulder

Somayya Ali Ibrahim

Center for Climate Systems Research, Columbia University, New York

Keywords	Resilience, adaptation, disasters and risk, recovery, flooding
Population (Metropolitan Region)	319,372 (U.S. Census, 2015)
Area (Metropolitan Region)	1,880 km² (U.S. Census, 2010)
Income per capita	US$56,180 (World Bank, 2017)
Climate zone	BSk – Arid, steppe, cold (Peel et al., 2007)

On September 2013, Boulder County, Colorado received nearly a year's worth of rain in a week. Rain poured out of the Rocky Mountains onto the towns of the prairie. Creeks destroyed roads and bridges, tore out culverts and downed trees, flooded homes and businesses, and resulted in the evacuation of several towns. Boulder County and fourteen surrounding counties were declared federal disaster areas. Statewide, 1,852 homes and 203 commercial structures were destroyed and more than 18,000 people were evacuated (Colorado Division of Homeland Security and Emergency Management).[4]

Initial analysis estimates that the total cost of the disaster exceeded US$2 billion, including $430 million to rebuild roads and bridges and $760 million to repair public infrastructure. Much of the financial loss was borne by residents – only about 1% of homeowners in the state possess flood insurance, despite the fact that the City of Boulder is considered one of Colorado's riskiest areas (City of Boulder, 2014). The City of Boulder has fifteen major drainage ways, and approximately 13% of the city is located within the regulatory 100-year floodplain, including nearly 2,600 individual structures. Flash flood risk is exacerbated by the city's downtown location, positioned at the mouth of Boulder Creek canyon, and frequent droughts and forest fires that seal off the soil, reducing its ability to absorb water.

Yet in spite of the unprecedented scale of the event, only ten lives were lost, most core infrastructure was maintained, and the response and recovery have been strong, well-coordinated, and

Case Study 3.1 Figure 1 *Floodwaters destroy a canyon road connecting the City of Boulder and mountain communities, September 2013.*

Source: City of Boulder

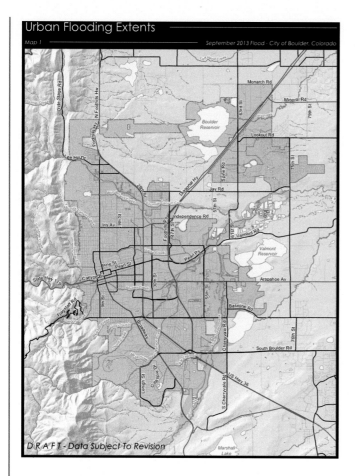

Case Study 3.1 Figure 2 *Urban flooding extent within the City of Boulder during the September 2013 flood.*

Source: City of Boulder

effective (Colorado Division of Homeland Security; City of Boulder). What made the area resilient to the devastation? Based on interviews, historical documents, and participant observation, this Case Study demonstrates that actions in three major categories increased resilience:
1. Physical systems
2. Human systems
3. Legal and cultural norms

PHYSICAL SYSTEMS

Plan for physical system failure: Virtually all physical systems will eventually fail; designing them to fail in nondamaging ways is critical to resilience. Open space or transportation greenways border many of the creeks in Boulder County, providing recreational opportunities and preserving ecosystem biodiversity. During floods, these trails and paths "failed" in their primary roles, taking on their planned secondary role for floodwater conveyance. This allowed space for creeks to overflow, entrain, and carry large debris and scour and deposit sediments with little impact to built infrastructure or people.

Additionally, the city has spent roughly US$45 million on flood risk reduction projects since 1999 (City of Boulder), knowing that eventually even the best-mitigated creek will overflow. The city has purchased and removed buildings in flood-prone areas and passed

building codes to prohibit new development in "high-hazard" flood zones. Bridges have been raised to accommodate deluges of water and debris. Automatic floodgates have been constructed around buildings that sit creek-side. This planning paid off during the floods, with floodwaters routed through Boulder City with relatively minimal impact. Nonetheless, Boulder was also lucky to receive significantly smaller flood peaks than many neighboring communities.

Prevent failure of critical physical systems through redundancy: Redundant systems, such as a backup diesel generator at the Boulder Water Treatment Plant, can prevent the loss of critical systems. However, main and backup systems must have different sources of vulnerability. Although having multiple roads providing access into the mountains appears redundant, in Boulder County, six of the seven roads between the plains and the mountain communities failed because they were at the bottoms of canyons next to creeks and were washed away. Furthermore, as a result of the loss of roads, it proved almost impossible to transport the diesel required to keep the backup generator at the Boulder Water Treatment Plant in operation. This highlights the need to consider the potential for cascading failures in assessing resilience and to assure that backup and redundant systems do not have common points of failure.

Build in diversity: Many of the physical systems that failed during the flood would benefit from a more diverse, distributed, multiple-small-solutions approach. For example, many homes outside the floodplain suffered substantial damage from sewage upwelling in basement drains. Although refurbishing the entire sewage system is prohibitively expensive, there are opportunities to strengthen system performance through lining sewer pipes and installing backflow devices in individual households. Over the longer term, moving to a distributed sewage treatment system or encouraging the development and adoption of composting toilets could dramatically improve resilience.

HUMAN SYSTEMS

Support individual capacity: Flood preparedness, response, and recovery were strongest where individuals had access to basic resources and were able to act with creativity to address the problems at hand. For example, a network of civilian ham radio operators became the backbone of the communications network for many mountain communities. Many of the operators received training following a previous wildfire in 2010. Similarly, the sewage and potable water systems in the City of Boulder were maintained primarily through the ingenuity and resourcefulness of staff that felt free to take needed action without fear of reprisal. Utilities personnel were out in the middle of the night building a concrete cradle around the main sewage pipe to the treatment plant where it had become exposed along the river; their unorthodox solution kept the pipe intact. Because the road to the water treatment plant was lost, diesel for the water treatment plant backup generator was loaded onto pickup trucks and the drivers told to find a way to the plant. Local mountain residents provided information and suggestions, and the drivers were able to create a route through fields and on abandoned four-wheel drive roads.

Develop networks: Strong collaboration among county nonprofits, the faith community, and city governments dramatically aided the initial response to the floods. Active community groups, originally developed for other purposes, rapidly came together to aid victims.

Maintain broad access to resources: Access to outside resources, such as private-sector stores in the Denver Metro Area, the national disaster clean-up industry, outside volunteer groups, strong financial institutions, insurance, and other sources of outside financing all sped response and recovery. In places where access has been limited, recovery has been slowed or halted. The city and county were still seeing new aid cases arrive sixteen months post-flood, for example from undocumented migrant communities where fear of reprisal initially prevented households from seeking aid.

Develop avenues for learning: Much of the resilience seen in flood response and recovery resulted from modifications made to improve upon less-resilient preparation and response to previous disasters. The economic downturn in 2008 left many organizations in the non-profit and faith communities scrambling for funds while they tried to respond to increased demand. The networking developed to address needs at that time strengthened the community as a whole and provided a strong foundation for rapid collaboration during recovery. A large wildfire in the County in 2010 provided useful lessons for emergency personnel, highlighting areas where better communications, early warning techniques, resident capacity for preparedness and response, and response and recovery coordination were needed. The progress made in all of these areas was readily apparent during the flood event.

The City of Boulder has made a point of learning, both from previous flooding events in the region, such as the Big Thompson Canyon flood of 1976 that killed 145 people in a neighboring county, as well as through activities like extensive flood-zone mapping. A culture of learning, at all levels of society, is critical to building and maintaining resilience. Where learning is inhibited, such as due to threat of litigation, it is critical to build in forums that foster cross-sectoral, multilevel learning from the disaster event, the response, and the recovery.

LEGAL AND CULTURAL NORMS

Adapt legal requirements to enable response and recovery: During and following the flood, Boulder County cities, towns, and the county as a whole modified many existing legal requirements around waste disposal, construction permits, and other flood-related issues to enable rapid response and reduce the financial burden on residents. However, laws about construction in floodplains continue to delay reconstruction in many places where creeks have moved. Where creeks moved, some residents are no longer in floodplains and other residents have become newly at risk. In both cases, the legal frameworks have not yet been updated to reflect new realities, and, until they are, government staff cannot approve building permits. This is further complicated by the challenges of understanding whether these

types of events are likely to become more frequent under climate change, in which case flood plain maps need to be modified to reflect new realities.

Promote a culture of collaborative self-help: Many citizens overestimated the assistance that government could provide in disasters. The Boulder County mountain communities have learned this lesson well. In response to the 2010 fires, the mountain communities established the fully volunteer Inter-Mountain Alliance. This group provided core communications capacity and self-rescue services for at least five days during and immediately following the flood when most of the mountain communities were inaccessible to rescue workers. Boulder County as a whole is now looking at ways to transfer these lessons to the plains communities and build a stronger culture of neighborhood collaboration and self-help.

A collection of Boulder County volunteers have established BoCo Strong, a county-wide resilience network whose first actions have been to launch a local Voluntary Organizations Active in Disasters group. The City of Boulder is similarly focusing on strengthening social networks and is in the process of hiring a city neighborhood coordinator tasked with outreach to and capacity-building of city neighborhoods and communities.

CONCLUSION

Modern society is increasingly dependent on complex, rapidly evolving systems for survival:
- Our food and water comes from distant sources that are beyond our control;
- Food, water, shelter, and livelihoods are often highly dependent on power and transportation systems, which can lead to a cascading failure of systems;
- In-person interactions increasingly involve transportation over miles, and in times of crisis we often do not know the people next door.

Climate change is likely to intensify rainfall, fire, and drought in Boulder County. This makes the resilience of core systems, of people and organizations, and of legal and cultural norms ever more important.

The City and County of Boulder are in the process of building back from the floods and, in doing so, are thinking about how to build back better. However, they also recognize that infrastructure alone is insufficient. A core element of the building back is developing stronger communities within the city and county across all sectors – individuals, aid organizations, government, private sector, faith groups – and educating those communities about the actions they can take to be more resilient.

development of the built environment and greater amounts of impervious surfaces can have dramatic impacts on the rate and volume of rainwater runoff, resulting in accelerated flood frequency.

3.5.1 Financial Instruments and Public–Private Partnerships

Financial capacity-building is another essential disaster risk reduction strategy in urban areas prone to natural hazards and

the impacts of climate change. Developed mostly through public–private partnerships, financial instruments such as insurance or tax incentives for retrofitting, relocation, and redevelopment practices are some of the ways that financial capacity for DRR can be developed (see Chapter 7, Economics, Finance, and the Private Sector).

As many researchers have pointed out, the poorest urban populations are usually hit hardest by disasters (see, for instance, Wisner et al., 2004; and see Chapter 6, Equity and

Box 3.4 Climate-Resilient Housing in Gorakhpur, India Selected as Lighthouse Activity 2013 by UNFCCC

Nivedita Mani and Shiraz A. Wajih

Gorakhpur Environmental Action Group

Community-Based Micro-Climate Resilience has helped the urban poor communities in Gorakhpur, India, to adapt to climate change by designing and building a new type of affordable flood-resilient house (see Chapter 11, Housing and Informal Settlements). Using locally available bricks and energy-saving techniques, this has proved to be environmental friendly, both in terms of optimization of resources and energy efficiency.

Mahewa ward of Gorakhpur is prone to flooding and waterlogging during the monsoon season. People living in this community are poor, marginalized, and particularly vulnerable to the impacts of climatic hazards, including floods, cyclones, changing rainfall patterns, and heat waves. Lack of affordability and technological knowledge resulting in inappropriate

construction of houses adds further vulnerability to their lives during disasters.

The Gorakhpur Environmental Action Group (GEAG) along with the technical support from Sustainable Environment and Ecological Development Society (SEEDS) India designed a low-cost model house to meet local needs that can be easily replicated throughout the community. This house features unique design elements that limit climate change impacts such as higher plinth levels to reduce the risk of waterlogging, walls constructed to moderate temperature, and earthquake-proofing. The house is resilient to climate and produces fewer carbon emissions. GEAG can help interested households in accessing bank loans for construction of this type of house. This initiative was awarded a 2013 "Lighthouse Activity" by the United Nations Framework Convention on Climate Change (UNFCCC)'s 19th session of the UN Conference of the Parties.

Environmental Justice, for more discussion). With no private insurance to rely on and no personal savings, the poor are often unable to recover economically, requiring often financial assistance from the government, which also bears the burdens of emergency services, debris clearing, and infrastructure repair. The inability of the poor to participate in local transactions feeds back onto the larger macro economy; reducing government revenue and delaying recovery further, resulting in a vicious cycle of nonrecovery (Schnarwiler and Tuerb, 2011).

Studies by the G20 and the Bank of International Settlements (BIS) detail the importance of sovereign disaster risk-financing strategies, particularly in low-income countries where many live in poverty. A study by the G20, *Improving the Assessment of Disaster Risk to Strengthen Financial Resilience* (WB, 2012), finds that the macroeconomic impacts of disasters in countries that are not adequately prepared to cope with them range from stunted economic growth due to decreased tax revenues, loss of employment, and increased poverty to health effects, such as impaired cognitive ability of children in the wake of natural disasters due to temporary malnutrition. The study from BIS, *New Evidence on the Macroeconomic Cost of Natural Catastrophes* (von Peter et al., 2012), draws similar conclusions on the long-term health and economic impacts of natural disasters and adds that societies with a mature insurance market recover more quickly after an event – in some instances even showing positive economic growth.

Insurance instruments designed by the private reinsurance industry to address the needs of governments and other public-sector entities can play a significant role in holistic disaster-risk financing schemes. These public–private insurance

partnerships transfer catastrophe risk from government entities to the private insurance market, provide governments with liquidity in the immediate wake of a natural disaster, and reduce the need for budget reallocation or tax increases to finance disaster recovery. These solutions also give government decision-makers an independent market-based estimate of climate and weather risk. By putting a price tag on unmitigated risk, a government can make more educated decisions in how to allocate its financial and human resources toward risk prevention and mitigation.

Numerous examples of successful public–private partnerships exist both in developing and developed economies. For instance, the government of Mexico uses a combination of traditional indemnity insurance and a catastrophe bond to protect itself from the impacts of tropical cyclones and floods.

The Caribbean Catastrophe Risk Insurance Facility (CCRIF SPC) provides the sixteen Caribbean community (CARICOM) countries with tropical cyclone protection. Since 2007, CCRIF has already paid out eight times, with a total of more than US$32 million, to a subset of its member countries. In 2014, the CCRIF announced plans for expansion into Central America to increase the financial resilience of these countries to tropical cyclones and floods.

In June 2014, the CCRIF launched excess rainfall insurance to protect its member countries against the financial impacts of heavy downpours and flooding. Heavy rains that occurred in St. Vincent and St. Lucia during Christmas 2013 caused more than US$100 million in damage and demonstrated the Caribbean's vulnerability to extreme rainfall produced by a trough of low pressure that moved through the region. Like its traditional natural catastrophe counterpart, the CCRIF XSR product has already demonstrated its benefit. Anguilla received a payout of US$493,000 after rains from Hurricane Gonzalo in October 2014 inundated the island nation. In November 2014, the government of Anguilla received its second payout, in the amount of US$560,000, while St. Kitts and Nevis received a payout of just over US$1 million.

Public–private partnerships have also been utilized by municipal entities. In 2013, after the New York Metropolitan Transit Authority (MTA) experienced a US$5 billion loss from Hurricane Sandy, the insurer for the MTA, First Mutual Transportation Assurance Company, sponsored a parametric, or index, catastrophe bond to protect the MTA against storm surge losses. The US$200 million MetroCat Re bond produces a 100% payout to the MTA if storm surge values exceed pre-defined triggers at various tidal gauges throughout the New York metropolitan area. This is the first storm surge-only catastrophe bond.

Natural hazards and climate change pose a real threat to urban economies and communities. The private sector can and should be an important partner to individuals, businesses, and governments at all levels by assuming some of the natural hazard risk and increasing the financial resiliency of governments, their populations, and economic producers in the aftermath of a natural catastrophe.

3.5.2 Ecosystem-Services Management

Ecosystems service management can mitigate disaster risks and build resilience in urban systems. Ecosystems sustain the livelihoods of communities and reduce their physical exposure to hazards (see Chapter 8, Urban Ecosystems)(CATALYST, 2013). Wetlands, forests, and coastal reefs serve as natural protective barriers against the impacts of storms, landslides, floods, and droughts. Ecosystems services include nutrient dispersal and cycling, seed dispersal, primary production, and provisioning services (MEA, 2005; de Groot et al., 2002). A wide variety of products and material and nonmaterial benefits can be experienced by urban residents and provide positive conditions for urban resilience. Key products and benefits include, for example, food, fuel, water, fodder, fiber, genetic resources, medicines, carbon sequestration and climate regulation, waste decomposition and detoxification, water and air purification, natural hazard mitigation, pest and disease control, erosion control, and cultural services (MEA, 2005; de Groot et at et al., 2002).

However, ecosystem services may and often do fall outside of the administrative boundaries of municipal governments. This implies the need for development approaches that consider a city within its larger landscape and administrative jurisdictions, even if some areas are governed by nonmunicipal entities. Rural–urban connections need to be strengthened for fully effective urban DRR and CCA to proceed.

Case Study 3.2 Adaptation to Flooding in the City of Santa Fe, Argentina: Lessons Learned

Claudia E. Natenzon

Universidad de Buenos Aires/FLACSO Argentina

Keywords	Floods, disasters, institutional adaptation, risk management, communication
Population (Metropolitan Region)	580,000 (Demographia, 2016)
Area (Metropolitan Region)	4,957 km² (Cardoso, 2011)
Income per capita	US$11,960 (World Bank, 2017)
Climate zone	Cfa – Temperate, without dry season, hot summer (Peel et al., 2007)

In 2011, the city of Santa Fe won the United Nations Sasakawa Award for Disaster Reduction for the progress made in effective risk communication to the public (Gobierno de la Ciudad de Santa Fe,

2011). It was the first city in Argentina to join the global campaign "Making Cities Resilient. My City Is Getting Ready."

The municipal government of Santa Fe has achieved this award by overcoming hard experiences of catastrophic flooding, particularly those that occurred in 2003 and 2007.

The city is located between the Salado River to the west, Setubal lagoon to the east, and the Santa Fe River (an arm of the Paraná River) and islands to the south and is bordered to the north by the municipalities of Recreo and Monte Vera. The city has been exposed to catastrophic flooding originating in both of the rivers that surround it and extreme precipitation.

In late April 2003, the Salado River overflowed, causing loss of life, a large number of evacuees, and very significant material damage:
- A third of the urban area was under water.
- More than 130,000 people, 35% of a total of nearly 370,000 inhabitants in 2001 according to the National Population Census (Gobierno de la Ciudad de Santa Fe, 2015) and 27,928 households were affected.
- There were 23 deaths officially recognized, and 129 deaths from indirect causes reported by human rights organizations.
- The economic losses were estimated to be US$1,025 million, equivalent to 12% of Santa Fe's Provincial[5] 2002 gross domestic

5 In Argentina, the Provinces are the federal political units, with the same status of the States in USA and Brazil.

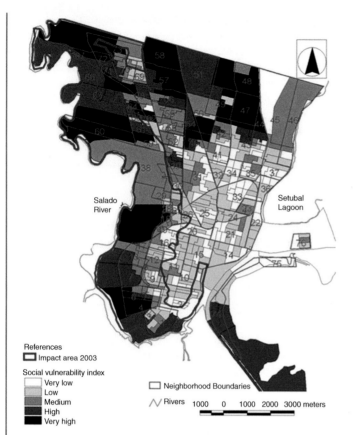

Case Study 3.2 Figure 1 *Social risk construction and the flood of 2003 in Santa Fe city. Geography Bachelor's Thesis, University of Buenos Aires.*

Source: Viand, 2009. "Before the Disaster"

product (GDP; CEPAL, 2003). These losses comprised direct damages of 35% and indirect of 65%. Private property and production sustained 91% of the losses; the public sector, 9%. By sector, losses were: 38% industry and commerce, 35% agriculture, 16% transportation, 8% housing, and 3% others. From the total impacted population, nearly 20% had no self-capability to respond, recover, or rebuild.

The floodwater entered the city from the northwest, in a 300-meter opening in an unfinished embankment defense. Water accumulation was reinforced by the inadequate aperture of the bridge (part of the Santa Fe–Rosario highway) located over the Salado River, which acted to retain the rising water and cause a backwater effect upstream. Within the city, the flood went south, covering the lowlands, and was retained because of the lack of finished levees and the inadequate or outright nonfunctional pumping stations.

The lack of reaction from authorities and protracted discussions with the contractor who built the highway over the embankment delayed the opening of channels to expel water, allowing it to reach the city

at two times the level of the river. Once flooding receded, response organizations were lacking, and many problems in reconstruction generated a widespread social crisis and distrust in state institutions.

Four years later, in March 2007, the city was flooded again. This time, the water came from prolonged convective rains over the city in unusual quantities, an extraordinary but predictable phenomenon. Runoff from these waters was hampered by the annual floods of the Paraná and Salado rivers. As a result of the flood, a million agricultural hectares of land were flooded. Many communities were disadvantaged, including the city of Santa Fe, which had 30,000 evacuees, and much of the urban infrastructure was affected.

According to the Argentine Red Cross (ARC) the region sustained

"electricity cuts (mostly preventative), damage to infrastructure making some areas inaccessible, suspension of school activity in some towns for weeks, and a significant loss of soybean and alfalfa crops. … At the early stages of the emergency, local discontent with the assistance being provided by local authorities grew, and local protests took place to demand improved assistance. Some warehouses and trucks containing humanitarian aid were looted by protesters. This created an atmosphere of insecurity and the police and naval authorities increased patrols and controls during the initial days of the emergency. At the beginning of the emergency, local supplies were less available and an increase in prices of basic food stuffs was registered" (DREF, 2008).

At the time of the second flooding, some defenses had been built, and the key plans to reduce the city's risk included (1) a system of clean drains, (2) reservoirs in good repair and of adequate depth to withstand heavy rains, and (3) a well-maintained pumping system to ensure operation. However, these planned actions were not realized. Technical reports prepared after the flood show that both reservoirs and sewage systems were not maintained and were insufficient and that 60% of the pumping stations (27 of a total of 45) were not operational at the beginning of the crisis. Nor had a warning system and a contingency plan been developed as promised after the flood of 2003.

This second great flood had a direct impact on the election for city mayor that year. The new city management for the 2007–2011 period put flood prevention on the political agenda as one of the central issues in the development plan of the city. Using the concept of disaster risk management (DRM), a communication program for the public was implemented, and this motivated the UN award. This Risk Communication Program has included workshops, conversations, and courses with more than sixty community organizations and forty-five neighborhood commissions; capacity-building for teachers; and training courses for journalists and social communicators. With the involvement of the Santa Fe Red Cross, some school plans for emergencies were made, and more than 5,000 people were invited to visit and inspect the city's drainage system. Finally, the Culture Department created "Agua Cuentos"("Water-tales"), performed for 3,000 students in twenty locations over the course of three years.

3.5.3 Building Community Institutions and Developing Capacities

Public participation in disaster risk management, including participatory planning and social capacity development, is essential to the effectiveness and implementation of disaster

risk reduction actions. Increasing the capacity and awareness of both governmental and nongovernmental stakeholders such as citizens, school children, media, elected representatives, and government officials helps in policy and programmatic changes, enforcement, prioritization, and ownership of resilience activities at the local level. It also helps in addressing areas of concern related to basic services, ecosystems, infrastructure, and land-use planning.

The experience of communities and their knowledge about local issues can be helpful in identifying problems and negotiating solutions. However, strategies focusing on collective groups rather than individual members of the community have been found to be most effective. This approach helps to improve decision-making in all sectors of the community, with the goal of producing effective disaster management interventions. To sustain disaster risk reduction, bottom-up approaches and multi-stakeholder engagement in the resilience-building process is a prerequisite for cities and especially critical in urban areas because they face resource challenges in terms of human, technical expertise, and budgetary allocations. This engagement can be achieved by building community institutions from neighborhood to city levels, developing their capacities as well as those of government officials, and raising awareness through the entire metropolitan region about disaster risk reduction and climate change action.

The city of Gorakhpur (Uttar Pradesh, India) has led the way with an experiment in community-led participation for urban climate resilience planning done by the local nongovernmental organization (NGO), Gorakhpur Environmental Action Group (GEAG) (see Box 3.4 and Figure 3.14). In

recent years, city planners and policy-makers have realized the limitations of a top-down approach to urban planning. Historical methods have failed to address specific community needs and overlooked the potential of mobilizing local resources and capacity to solve problems. Taking shortcomings of the past into account, city planners and policy-makers have placed increasing emphasis on a participatory approach to develop sustainable and long-lasting solutions. In Gorakhpur, an area prone to floods and serious waterlogging further worsened by top-down approaches to governance, the process of building urban climate change resilience was adopted with the participation of the community and other stakeholders. People's participation was at the core of every step. Tools and methodologies were developed to create an enabling environment for building participatory urban resilience.

Figure 3.14 showcases a bottom-up approach for building resilience to climate change that works at three levels: neighborhood, ward, and city levels. GEAG's experiences have shown that, in a city like Gorakhpur with poor basic infrastructure facilities and weak governance, addressing issues through community participation helps achieve better planning, stronger governance, and greater accountability of the government. In addition, the role of municipal governments in Gorakhpur has been crucial in the success of these initiatives for sustaining the activities and ensuring long-term sustainability

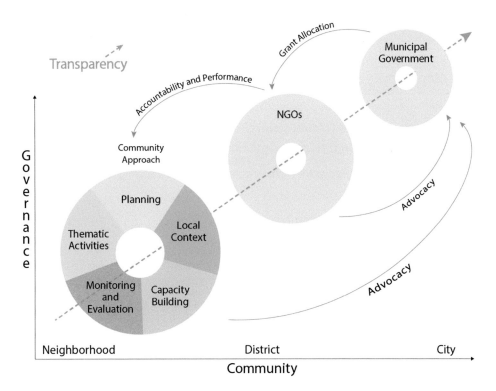

Figure 3.14 *Addressing systems, agents, and institutions through community participation.*

Source: Gorakhpur Environmental Action Group (GEAG), India

of the processes. The capacity-building programs aimed at government officials have enhanced their knowledge about climate-related issues and is playing a crucial role in enhancing the sustainability of community-based interventions in Gorakhpur.

3.5.4 Post-Disaster Recovery and Rebuilding

Post-disaster recovery, rebuilding, and reconstruction is as a tool for disaster risk reduction and resilience building, particularly if it is planned ahead of disasters. Quarantelli (1999) poses the question, "Is it enough to bring back the past, or is something new or different necessary?" Several researchers have demonstrated how disasters can open rare windows of opportunity for instituting long-term change and altering the course of resilience. From a practitioner perspective, any recovery activity following a disaster that fails to reduce the population's exposure to risks is merely sowing the seeds for future disasters (IRP, 2007). Consequently, DRR is becoming increasingly recognized as an integral component of successful disaster recovery policy and rebuilding programming.

The recovery stage provides opportunities for risk reduction and the chance to break the cycle of destruction due to disaster. For example, researchers emphasize that future vulnerability can be reduced and community resilience can be improved through incorporating DRR measures such as developing minimum building codes and land-use regulations (Berke et al., 1993; Reddy, 2000). Yasui's (2007) study of recovery in two communities after the Kobe earthquake demonstrates that development practices and capacity-building efforts employed during a recovery process reduced overall community vulnerability in the long run. Other researchers insist that a good recovery program begins with a serious commitment to incorporate DRR and preparedness strategies to reduce future damage (Comerio, 1998).

Recovery planning and rebuilding should be an integral component of making cities resilient and thought out in advance. However, due to a lack of priority and resources dedicated to recovery planning in advance of a crisis situation, it is still rarely considered until disaster strikes.

3.6 Integrated Approaches to DRR and CCA

Various strategies are used to reduce disaster risks and build resilience as well as to mitigate GHGs emissions and adapt to climate change in urban areas. There is frequently a disconnection between CCA and DRR research communities and a lack of collaborative integrated work on these areas (Solecki et al., 2011). This is due to differences of emphasis

Case Study 3.3 Preparedness, Response, and Reconstruction of Tacloban for Super Typhoon Haiyan in the Philippines

Xiaoming Wang

Commonwealth Scientific and Industrial Research Organization (CSIRO), Melbourne

Dennis G. de la Torre

Climate Change Commission, Manila

Perlyn M. Pulhin

The OML Centre, Manila

Keywords	Typhoon, storm surge, disaster response, recovery and reconstruction, housing
Population (Metropolitan Region)	242,000 (Philippine Statistics Authority, 2016)
Area (Metropolitan Region)	201.72 km² (Philippine Statistics Authority, 2014)
Income per capita	US$9,400 (World Bank, 2017)
Climate zone	Af – Tropical, rainforest (Peel et al., 2007)

The latest disaster caused by Super Typhoon Haiyan[6] (locally named Yolanda) in the Philippines led to around 6,000 fatalities and more than 27,000 injured with thousands missing. The impact on communities was significant, considering that more than 12 million were affected, which represented more than 2.5 million families. The scale of the impact was also shown in the direct loss up to US$9 billion as a result of damage to infrastructure and agriculture in addition to more than 1 million damaged houses.

Tacloban is a small city located in a typhoon-prone area and is vulnerable to storm surges. It is among the areas that sustained significant damage and losses as a result of the Super Typhoon Haiyan. A total of 1,012,790 houses were damaged with a cost estimated at PhP303,837.0 million among the total loss of about US$13 billion (The Philippine Government, 2013). There were 4.4 million people among 930,000 families displaced in affected areas, which were mostly occupied by informal settlements where the occupants' livelihoods were considerably disrupted. The service loss of critical infrastructure including power and water supplies further hindered the recovery of local livelihoods and businesses. Meanwhile, social services, including health services and schools, suffered significant disruption due to damages, as shown in Case Study 3.3 Figure 1, further hampering the recovery. Based on the impact of Super Typhoon Haiyan on Tacloban City, this Case Study reviews the preparedness, response, and reconstruction activities

6 'A super typhoon has wind speeds of at least 100 knots/120 miles per hour.

Case Study 3.3 Figure 1 *Damages in Tacloban caused by Super Typhoon Haiyan.*

in the area as well as relevant policies and planning for managing natural disaster risks.

Prior to the disaster, the Philippine government developed the Climate Change Act of 2009, creating the Climate Change Commission (CCC), the National Strategic Framework on Climate Change (NSFCC), and the National Climate Change Action Plan (NCCAP) to guide the mainstreaming of climate change adaptation (CCA) in policy and planning. It enacted the Philippine Disaster Risk Reduction and Management Act of 2010, creating the National Disaster Risk Reduction and Management Council (NDRRMC), correspondingly the NDRRM Framework and the NDRRM Plan, to guide disaster risk reduction (DRR) actions in the Philippines. It also established the People's Survival Fund as a long-term finance stream for climate change actions.

In reacting to the disaster, President Aquino appointed Senator Lacson as Presidential Assistant for Rehabilitation and Recovery, and the Department of Budget and Management issued a National Budget Memorandum including the creation of the Reconstruction Assistance for Yolanda (RAY) in the 2015 National Budget, calling for systematic integration of CCA/DRR. A total of 41 billion pesos was initially allotted for Yolanda rehabilitation, administered by NDRRMC and the Department of Social Welfare and Development. The Department of Environment and Natural Resources defined some areas in the city as "no-build zones" located 40 meters from shoreline, following the Philippine Water Code (the majority of residents, however, ignored these rules when rebuilding their homes). Finally, the Department of Labor and Employment established emergency employment programs with support from the International Labor Organization.

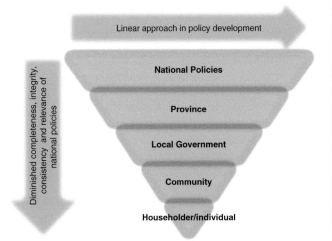

Case Study 3.3 Figure 2 *Gaps in development policy and actions for disaster risk reduction and climate adaptation.*

In hindsight, Super Typhoon Haiyan highlights critical gaps in managing disaster responses and recovery across tiers of government from national to city and local levels down to communities, households, and individuals, as shown in the Case Study 3.3 Figure 2, at both horizontal and vertical scales. The Philippines government has made significant efforts and great progress in developing national policies to mitigate the risks of natural disasters and climate change, but their relevance to implementation as well as their influence on practice diminished at the finer scale of provinces, municipal governments, communities, and householders, creating a significant barrier for the government to build a proper level of capacity for local communities and householders in response to natural disasters and climate change risks.

With a potential increase of extreme climate risks in the future as a result of climate change, policy development to enhance preparedness and prevention, especially of catastrophic events such as the Super Typhoon Haiyan, should be integrated into disaster management. It should also be reflected across all levels, as shown in Case Study 3.3 Figure 3, to harmonize DRR management and CCA policies into the legislative framework for sustainable development. The goal is to transform national policies into content aligned with the realities of provinces, city governments, communities, and households (and individuals, if necessary) while maintaining its consistency, completeness, and integrity with national policies in terms of visions and goals. Moreover, such finer-grained policies must supplement national policies at different levels, thus facilitating national capacity-building and DRR/CCA implementation across all levels. Finally, existing DRR/CCA

policies in the Philippines can be further improved based on the policy nexus concept.

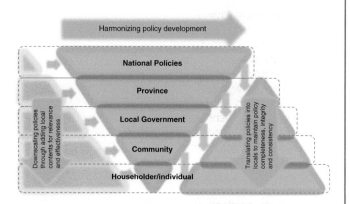

Case Study 3.3 Figure 3 *Implementing the policy nexus for climate adaptation and natural disaster risk reduction.*

between disaster risk and climate change research, with the former focused on the past and present and latter on the impacts of future risk (Thomalla et al., 2006; UNDP, 2004; Gencer, 2008).

Scholars and communities of practice have increasingly called for the need to apply a systems-based risk management and risk decision-making approach that takes into consideration both contexts (IPCC, 2012). Disaster risk reduction and CCA strategies are emerging worldwide as an important requisite for sustainable development. Disaster risk reduction and CCA policies and measures need to be implemented to build disaster-resilient societies and communities, with the aim of reducing risks while ensuring that development efforts do not exacerbate climate vulnerability.

Disaster risk reduction and CCA also share a common conceptual understanding of the components of vulnerability and the processes of building resilience that needs to consider the sensitivity and capacity as components of vulnerability, exposure, and magnitude and/or likelihood of the hazard. Both vulnerability and exposure are compounded by other societal and environmental trends, such as unplanned urbanization, ineffective governance structures, inequality, and environmental degradation. To reduce disaster and climate change risk, exposure needs to be minimized, sensitivity reduced, and capacity strengthened in ways that address both disaster and climate change risk simultaneously, in a dynamic process requiring continual effort across economic, social, cultural, environmental, institutional, and political domains. Thus, a multirisk analysis framework that accounts for the possible interactions among the threats, including cascading events, is needed, taking into account temporal (e.g., duration of the event and typical return period of extreme events) and spatial scales.

Performing quantitative multi-risk analysis presents many challenges. Disaster scenarios are often qualitative, related to

one reference event, and rarely account for the related uncertainties. Furthermore, the risks associated with different types of natural hazards (e.g., volcanic eruptions, landslides, earthquakes, and floods) are often estimated using different procedures so that the produced results are not comparable. The events themselves could be highly correlated, or one type of threat could be the result of another one (e.g., floods and debris flows could be triggered by an extreme storm event, so-called cascading effects). Key characteristics of the elements at risk, represented by vulnerability to specific threats, are not constant and change over time. In particular, exposure to one type of hazard might increase the vulnerability significantly to other types of hazards. The challenge is to find innovative, efficient ways to collect, organize, assess, and communicate to urban planners, designers, and decision-makers the risk and vulnerability data on hazards and impacts as well as on mitigation/adaptation and the criteria of alternative policy and development scenarios while also accounting for inherent spatial-temporal dynamics. This becomes particularly important in the context of climate change-induced increases in risk at different spatial and temporal scales.

The Urban Climate Change Resilience Framework (UCCRF) developed by Institute for Social and Environmental Transition (ISET) is a conceptual planning approach to building resilience to climate change (see Box 3.5). It is designed for practical application and has been developed from and tested in field situations. The Framework addresses the need for an approach that clarifies the sources of vulnerability and addresses the complexities of climate adaptation, yet is simple enough for local practitioners to apply in their own context.

Community-based, national, and regional projects have impediments to integrating DRR and CCA because of donor requirements, the sometimes conflicting goals of partner organizations, and underlying policy frameworks. Opportunities do

Box 3.5   Urban Climate Change Resilience Framework

The Urban Climate Change Resilience Framework (UCCRF) is structured to build a broad understanding of urban resilience by describing the characteristics of urban *systems*, the *agents* (people and organizations) that depend on and manage those systems, *institutions* (laws, policies, and cultural norms) that link systems and agents, and patterns of exposure to climate change (Tyler and Moench, 2012). The characteristics that make a system resilient include:

1. *Flexibility and diversity*. The ability to perform essential tasks under a wide range of conditions (e.g., multiple and geographically distributed water resources, both surface and underground)
2. *Redundancy and modularity*: A system that has spare capacity for contingency situations in order to accommodate extreme or surge pressures or demand (e.g., multiple routes in transportation, redundant cell phone transmission towers)
3. *Safe failure*: The ability to absorb sudden shocks, including those that exceed design thresholds (e.g., dikes that can be opened to fill flood retention zones outside city).

The agent is an essential subcomponent that includes individuals, households, and public and private-sector organizations. The *responsiveness* (capacity to organize and reorganize in an opportune time), *resourcefulness* (capacity to identify and anticipate problems; establish priorities, and

mobilize resources for action), and *capacity to learn* (from past experiences, avoid repeated failure, etc.) are important aspects of agents to be considered in resilience building.

The concept of institutions in social sciences refers to the rules or conventions that constrain human behavior and exchange in social and economic transactions (aspects like decision-making, rights and entitlements, learning and change, and information flows are examples of the characteristics of a resilient institution).

The framework operationalizes these concepts through structured and iterative shared learning approaches that allow local planners to define these factors in their own context in order to develop practical strategies for local action to build urban climate change resilience.

The UCCRF focuses on vulnerable populations in urban locations and their marginalized subsistence that often lacks secure access to services and depends on fragile urban systems. This makes them especially exposed to system failures in the wake of climate-related stress. In addition to this, the framework provides information on agents and institutions as enablers of resilient systems in a city, thus defining three pillars of resilience-building within a city system: (1) strengthening fragile systems, (2) strengthening social agents, and (3) strengthening institutions.

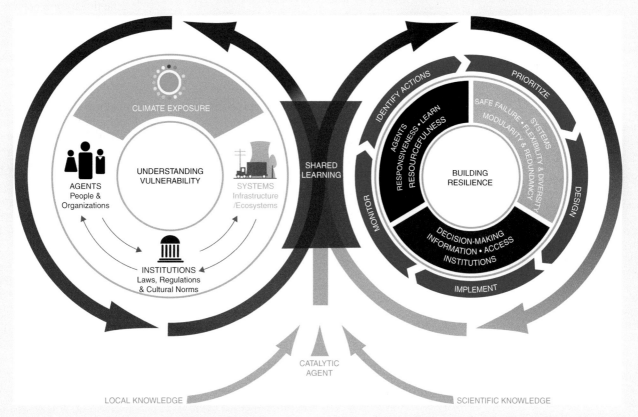

Box 3.5 Figure 1 *The Urban Climate Change Resilience Framework.*

Source: Tyler and Moench, 2012

exist, however, and leadership is needed to bring together these two often overlapping fields. Developing an awareness of both DRR- and CCA-related initiatives, organizations, and policy arrangements will reduce duplication of efforts and thus contribute to effectiveness. Not only will existing organizations be able to share relevant experiences and lessons learned (perhaps from CCA and DRR practitioners, or vice versa), they will also be aware of gaps and future needs. Opening the dialogue between these groups serves to initiate, develop, and maintain the good relationships that are crucial to creating a supportive institutional architecture. Furthermore, by addressing vulnerability at the local level in a holistic sense, the community will benefit by receiving a well-targeted program, and DRR and CCA will inherently be incorporated alongside risks such as health, nutrition, disease, and other issues. By sharing lessons learned and successful practices, as well as sharing practices that perhaps were not effective, organizations can better utilize the funds that are available to achieve advances in their resilience to disasters today and tomorrow.

3.7 Barriers, Challenges, and Opportunities

Cities in low-, middle-, and high-income countries are focused on a range of issues associated with improving disaster risk reduction and climate change adaptation. Economic losses from environmental hazards have increased exponentially within the past decades. There is an increasing awareness that sustainable development, along with international strategies and instruments aiming at poverty reduction and environmental protection, cannot be achieved without taking into account the risk of natural hazards and their impacts in a world of changing climate. In this section, we examine some of the barriers and challenges to DRR in the context of climate change (see Chapter 16, Governance and Policy). Some critical considerations are listed in Table 3.2.

Table 3.2 *Top five recommended actions for decision-makers.*

Capacity	Have a dedicated person on disaster risk reduction (DRR) and climate change adaptation (CCA) in each agency with a clear mandate and hold the agency accountable for demonstrating DRR and climate change integration into its development and planning agenda.
Resources	Work with financing and development agencies; develop public–private partnerships to develop resources for DRR and climate change actions; forge partnerships between neighboring city governments and international organizations to share technical and knowledge-based resources. Promote advocacy and policy dialogue at the national level as well as with key donors of development initiatives.

Table 3.2 (*continued*)

Information	Have protocols for gathering and sharing of information (with other agencies, with the public, etc.) and devote more resources to the development of science aimed at DRR/climate change. Work and develop partnerships with the scientific community, NGOs, and the private sector for the development of knowledge and innovative products.
Planning	Work on solving current gaps in land-use planning, environmental management, licensing for construction, etc.; incorporate the long-term.
Public Participation	Initiate bottom-up approaches and advocate the participation of urban communities and civil society organizations because public participation is key to effective implementation for DRR and climate change.

3.7.1 Coordination

An enabling governance framework is essential for disaster risk reduction and climate change adaptation. However, lack of coordination among different levels of government and across sectors, and even within the DRR and CCA sector itself, can hamper these efforts (see Chapter 16, Governance and Policy). Some of the barriers and challenges are:

At the national level: One of the main challenges of DRR and DRM has been to transition from a response-oriented framework to a proactive vulnerability and exposure-reduction framework. Disaster risk reduction includes identifying and understanding risk; reducing risk; preparing for and responding to emergencies when they arise; and recovering, reconstructing, and rebuilding in fair and sustainable ways. To achieve these objectives, DRR ideally needs to be embedded within each sector (e.g., transport, energy). It is not a task only for a disaster response agency. Both DRR and climate change action have great potential to help cities plan for the longer term, something that is essential, but generally tough to implement in practice because political terms and financial decisions generally have shorter-term horizons.

At the national level, responsibility for DRR and climate change action should be part of a ministry or agency that has sufficient authority to enable mainstreaming at the sectoral, territorial, and municipal levels and that has access to financial resources (UNISDR, 2011). In Latin America, Colombia made an effort in this direction with the creation in 1988 of the National System for Disaster Prevention and Response (*Sistema Nacional Para la Prevención y Atención de Desastres*, SNPAD) (CEPAL and BID, 2007). This body enables, at least at a legislative level, an integrated DRM approach that is also integrating issues of CCA. There is also an inter-sectoral commission on climate change.

> *Recommendation:* Pass national legislation that creates an integrated system of disaster risk reduction (DRR) and climate change adaptation (CCA) in one high-level agency with authority and resources. This is one way to integrate the topics and to give them decision-making relevance, thus providing a framework in which sectors and municipal governments can coordinate (PREDECAN, 2009).

At the regional level: Integrating DRR and CCA at city-relevant levels such as states and provinces is important mainly for two reasons: (1) it provides a link between the national and local, and (2) it allows for coordination between city governments. It is generally at the regional level that watershed and natural-resource management take place. However, regional bodies often do not exist or have a weak mandate and few financial resources.

Water, food, and other natural resources that cities depend on know no political boundaries. It is common, therefore, to find bodies at the regional level that manage resources such as water (e.g., the Mekong River Commission or the Autonomous Regional Corporations in Colombia, which are regional environmental agencies). Deficient regional planning can have a direct effect on city water resources and on urban flooding. It is important to have the right governance at the regional level to foster ecosystem services approaches and the valuation of natural resources.

At the municipal level: Because of the spatial nature of hazards and risk, the most immediate response and risk reduction measures happen at the municipal and city district levels. Small cities in developing countries may face resource challenges. Organizations such as ICLEI–Local Governments for Sustainability provide good platforms for city governments to exchange experiences. There are low-cost measures that can help reduce risk at the city level and that focus on being no-regret measures that have multiple co-benefits (e.g., employment creation, reduction of stagnant water, and focus on associated health hazards).

At the urban level, there can also be important gaps between policies and city government needs. In small countries like El Salvador, several services are the responsibilities of agencies at the national level (e.g., the Ministry of Public Works covers everything related to roads and water infrastructure). The intervention plan of such national agencies might not match city needs or interests. At the same time, another challenge is that there might not be enough human and financial resources at the local level, and there often exists needs for technical training and support.

> *Recommendation:* Create more spaces for dialogue and joint planning between different levels and sectors of government. This requires increasing the capacity of municipal and state governments so they have the necessary technical personnel. Another useful strategy to build the capacity and collaboration of city-relevant governments is to incentivize them to share resources and personnel when relevant and necessary.

3.7.2 Capacity

Risk assessments need communications strategies and decision criteria in order to be translated into useful risk management interventions. Lack of scientific information and data or issues with data sharing are often challenges for risk assessments (see Chapter 16, Governance and Policy), and there is often a lack of capacity to conduct or update assessments.

> *Recommendation:* More emphasis should be given to protocols for data collection and sharing, as well as for the sharing of technical equipment (e.g., meteorological stations) where needed. In addition, city governments need more professionals trained in risk and climate change at the urban level to conduct assessments and interpret them. They can weigh in when difficult choices have to be made.

The Open Data Institute (ODI) is a nonprofit organization that helps unlock supply, generates demand, and creates and disseminates knowledge to address local and global issues.[7] Using information gathered through the ODI, Resurgence,[8] an NGO promotes the use of open data, social media, and communication to foster risk reduction and resilience-building in city governments.

> *Recommendation*: Municipal governments can also benefit from increased capacity in the use of negotiation tools such as role-playing games (Mendler de Suarez et al., 2012). For capacity development in urban professionals who need to integrate DRR and CCA into urban planning, development, design, and construction fields, professional associations can develop continuing education programs. Educational programs that focus on DRR and CCA is needed for all ages.

3.7.3 Resources

One of the biggest challenges in disaster risk reduction and climate change adaptation is to show that it pays to invest *ex-ante* in risk reduction and adaptation, especially when there are limited resources and competing needs. One way to show the

---

7 For more information, see ODI's website, http://theodi.org
8 For more information, see Resurgence website, http://www.resurgence.io/

savings from early action is to monitor interventions and account for avoided losses. Another way is to improve the tools used to prioritize investments, such as cost-benefit analysis.

It can be challenging to find resources for DRR and CCA. Investments that can reduce risk and increase adaptive capacity (e.g., setbacks in vulnerable coastal areas) are often not prioritized because benefits may only show at a later stage and are thus heavily discounted. There is a need for improved methodologies to incorporate DRR and CCA criteria into public investment decision tools such as cost-benefit analysis (Vorhies, 2012). Measuring costs and benefits can be challenging, especially when dealing with environmental and social issues, and cost-benefit analysis is heavily dependent on choice of discount rate (which is used to calculate future benefits and costs in present value). There is also a need to differentiate between direct economic benefits and extended economic benefits.

With support from Germany's *Gesellschaft für Internationale Zusammenarbeit (GIZ)*, Peru has been working to incorporate DRR and CCA criteria into the formulation and approval processes for public investments (Public Investment and Climate Change Adaptation [IPACC]). The methodology analyzes the project costs with and without risk reduction measures and thus includes a measure of "avoided costs"[9] (GIZ, 2012). There are also new tools that aim to account for the benefits provided by ecosystem services (e.g., InVEST)[10]. Another way to create awareness in Ministries of Finance on the importance of investing in DRM and CCA, one widely used by multilateral development agencies such as the Inter-American Development Bank, is to use probabilistic risk assessments to show potential losses. This also provides a good baseline to support the provision of financial resources and the prioritizing of interventions. Highlighting the benefits of ecosystem services for risk reduction is also proving beneficial for green infrastructure schemes in coastal communities.[11]

Assessments made in countries such as the Philippines and Malaysia on current efforts to address disaster risks and climate change, focusing particularly on aspects that can help build linkages between DRM and CCA, revealed the following impediments (Senga, 2012):

- Inadequate provision of high-resolution meteorological data for detecting trends and validating models;
- Poor access to physical (e.g., hydrological) and socioeconomic datasets for assessing risk;
- Insufficient incorporation of implications of climate change in risk assessments;
- Analyses of potential climate change impacts stop short of identifying practical adaptation options;
- Gaps in awareness and understanding of risk and climate change projections;
- Relatively weak coordination mechanisms regarding CCA;
- Underdevelopment of a preventive DRR approach;
- Diversion of funds due to disaster emergency response.

3.7.4 Other Challenges

Limits of top-down approaches: Historical top-down approaches have sometimes failed to address specific local needs; therefore, bottom-up approaches – initiatives developed by urban communities and their participation in programs developed by municipal governments and other stakeholders – are a prerequisite for cities in building resilience. Inadequate government capacity and resources and the lack of reliable scientific data mean that the role of urban communities in disaster preparedness has become more crucial. The recent devastations by floods in Asian countries including Indonesia, Malaysia, Thailand, and India have raised serious questions in cities in regard to the extent to which development plans are meeting local needs, including addressing disaster risks. In this context, it is crucial to highlight a community's ability to reduce its own disaster risk as well as the areas where additional support are needed. Local initiatives at the grassroots level should be linked with appropriate top-down strategies and city government interventions and policies (Anderson and Woodrow, 1998; DFID, 2005; Fraser et al., 2006). This ensures the sustainability of the approaches adopted and enables access to outside knowledge that may assist in vulnerability reduction.

There is often a shortage of local municipality experience in incorporating climate change into DRM despite the desire to do so. Interviews showed that municipal governments in Costa Rica in general are concerned about the potentially hazardous impacts of climate change (Box 3.6) (Hori and Shaw, 2011). Even small municipalities in rural areas demonstrated their conceptual understanding of the importance of incorporating climate change impacts into DRM and local development planning, especially in the agriculture and tourism sectors important to their sustainable local economy. The municipalities receive information related to climate change from a variety of media and sources that include both international policy bodies (such as Conference of Parties or ICCP meeting information received via Internet or TV). However, these municipalities are uncertain about how to incorporate potential climate change impacts into local DRM planning in practice. This situation coincides with the many challenges that disaster risk re-education and CCA face, and thus local municipal governments often have no long-term solutions to their communities' problems nor have they developed their coping mechanisms and capacities.

Prabhakar et al. (2009) suggest that climate change mainstreaming into local development planning could be initiated with capacity-building by local DRM personnel and policy-makers. This concept corresponds to that of Perez (2008), who explains the importance of local key stakeholders' critical understanding of CCA. Vignola and colleagues (2009) assert that national policy-makers should empower local actors to facilitate adaptation processes. Thus, capacity-building and empowerment of local actors are considered key factors for establishing and improving DRM planning at the local level.

---

9   For more information see IPACC's webpage, http://www.ipacc.pe/.
10 See Natural Capital Project – Invest: http://www.naturalcapitalproject.org/InVEST.html
11 Building with nature for coastal resilience. Wetlands International, Deltares, The Nature Conservancy, Wageningen University. http://www.wetlands.org/NewsandEvents/
    NewsPressreleases/tabid/60/Default.aspx

Box 3.6 Incorporation of Potential Climate Change Impacts into Disaster Risk Management in Costa Rica

Tsuneki Hori

Inter-American Development Bank, Washington, D.C.

Rajib Shaw

Integrated Research on Disaster Risk Program, Kyoto University

Some municipalities in Costa Rica, especially those in the greater metropolitan area, are engaged in concrete planning actions related to the incorporation of climate change into disaster risk management (DRM). Travel agencies and the emergency committee in the Municipality of Tibas have recently formed a coordination mechanism to prepare for an increasing number of floods and protect tourists from future climate-related hazard events. The Santa Ana municipality is in the process of regulatory plan updates that will allow validation of land-use licenses for local tourism operations, including restaurants and hotels. The process of licensing incorporates reducing potential disaster risk related to recent flood damages. Many municipalities in Costa Rica do not sufficiently recognize the national policy planning instruments such as the National Tourism Development Plan 2010–2016 and the National Strategy on Climate Change (ENCC). Nonetheless, municipal governments do take small but necessary planning actions related to disaster risk reduction (DRR) in the context of climate change, independent from national policy priorities. In addition, the tourism sector is important for the local economies in Costa Rica, and municipalities in general are concerned about additional disaster impacts associated with climate change. This arena of local initiative constitutes an opportunity for improving DRR planning at the local level, independent of the national policy priorities.

Many Costa Rican municipalities engage in developing information related to longer-term disaster risk scenarios. For example, the municipality of Santa Ana is analyzing its future water demand, taking into account future real estate development for commercial infrastructure and residences. The risk analysis for this sector incorporates future flood estimation as one of its variables. The municipality does not have an in-depth capacity for this analysis, so national universities provide technical and analytical support. For sustainable groundwater use, the Municipality of Belen has developed a regulation that requires relevant information on potential impacts of climate-related hazards. The study was conducted in collaboration with the National Meteorological Institute (IMN).

These activities represent efficient local approaches to developing relevant information on climate-related impacts. First, municipalities conduct research often for their local needs, independent of national priorities or policies. Second, the municipalities do not always use advanced technology to develop information on climate-related risk scenarios generally used by the international science community; instead, they conduct research with the tools that they have. Third, when municipalities need additional technical support, they request support from technical institutions or universities. The essence of this approach may be applicable for many sectors and may be an opportunity to incorporate climate change into local DRR planning.

Source: Hori, T. and Shaw, R., 2011

Other development priorities: For the majority of cities, addressing the potential impacts of climate change is not their top priority (Bai, 2007). For example, in a study by Hori and Shaw (2011), certain municipalities of Costa Rica asserted that there are a variety of urgent local operational demands, including housing construction (the Municipality of Aguirre), permission for commercial construction (the Municipality of Desamparados), reduction of unemployment (the Municipality of Esparza), and reconstruction of infrastructure damage caused by recent disasters (the Municipality of Garabito). Moreover, most of the municipalities in this study have no or inadequate human resources that can be responsible for CCA. This situation is a serious constraint in the mainstreaming of hazards and the impacts of climate change into local development planning.

To mainstream the consideration of climate change impacts into each sector, coordinate policies with other priority issues, and incorporate these policies into local development planning processes, assigning a single point of coordination for CCA and DRR at the city government level may be a first step to binding local priorities and DRR or climate change initiatives. This would also address the lack of interdepartmental DRR co-ordination at the local level.

Urban development planning: It takes a long time for municipalities to develop local development plans, and this creates problems in implementation. One example is the Municipality of San Ramon in Costa Rica, which took more than eight years to develop its regulatory plan (Hori and Shaw, 2011). It is true that many of the municipalities in Costa Rica and in other developing countries lack technical knowledge, as well as the human and financial resources needed to produce local development plans. It is important to understand this reality and consider strategies for better scoping of local DRM planning on a longer time scale.

3.7.5 Recommendations for the Integration of DRR and CCA in Urban Areas

Dividing areas of action into sectors has often proved organizationally convenient in government and academia but can undermine a thorough understanding of the complexity of and interactions among the human and physical factors involved in the definition of a problem at urban scales. A more integrated approach would facilitate recognition of the complex relationships among diverse social, temporal, and spatial contexts in cities. Decision processes that employ participatory methods and decentralization within a supporting hierarchy of higher levels could enhance many urban DRM organizations currently faced with climate-related decisions. The following areas, some of which have been pursued by governments, civil society actors, and communities, are recommended or proposed to foster such integration among adaptation to climate change, DRR, and management in cities.

If progress is to be made in reducing the impacts of disaster and climate change on developing countries and particularly on the poor, an adequate information base for decision-making is needed, one that includes disaster risk exposure as well as the socioeconomic and environmental dimensions of vulnerability. Information needs to be provided in a format that meets the practical demands of the targeted stakeholders, with emphasis placed on a process of monitoring and updating if it is to reflect the dynamics of a changing climate and shifting parameters of municipal vulnerability. However, information is only useful if it is embedded in an enabling governance framework that allows for action aimed at reducing vulnerabilities at the local to national levels. The ability to coordinate expertise to galvanize action at all scales is crucial, including the ability to leverage financing for science. Furthermore, risk management cannot be viewed in isolation from other pressures of development but should instead be part of an integrative effort toward reducing vulnerability and promoting livelihoods.

Annex 3.1   Stakeholder Engagement

Ebru Gencer is the Founding Director of the Center for Urban Disaster Risk Reduction and Resilience (CUDRR+R) and the Co-Chair of the Urban Planning Advisory Group (UPAG) to the Special Representative of the Secretary-General of the United Nations for DRR. She has taken part in multiple DRR and resilience-building activities since the inception of the UCCRN ARC3.2 process including, among others, being part of the Expert Team that drafted the New Ten Essentials for Making Cities Resilient that can be used by city governments and other stakeholders; working with municipal governments for UNISDR/WMO's Building Resilience to Disasters in Western Balkans and Turkey project; and currently working on a CDKN/IDRC/FFLA-funded project; on Climate Resilient Development in Latin American cities.

Megan Linkin is specialized in public–private partnerships (PPPs) and works within the re-/insurance industry. She has had numerous meetings with public-sector entities covering the topics in this chapter.

Claudia Natenzon has been an expert of the Working Committee in Risk Management of the National Ministry of Science, Technology and Innovation in Argentina, advising this group in the formulation of a protocol to prevent, respond to, and rebuild urban areas after flash floods, as well as being an expert for the Third National Report to the Climate Change Convention of Argentina, diagnosing Argentina's vulnerability impacts, the identification of dangers coming from the climate change process (in collaboration with climatologists), and the evaluation of risk in this climate change context.

Mattia F. Leone has been involved, as part of the PLINIVS Study Centre of University of Naples Federico II, with

drafting agreements with the Campania Region and the National Department of Civil Protection, and for the drafting of preparatory guidelines for the implementation of the first regional regulations and local building codes for volcanic risk-prone areas. The specific topic of an integrated approach to geophysical and climate change-induced hazards has been developed in the context of an agreement between the Department of Architecture of University of Naples Federico II and the City of Poggiomarino as part of preliminary studies related to the development of the new Urban and Building Code.

Chapter 3   Disasters and Risk

References

AIR Worldwide. (2013). AIR estimates insured losses from super typhoon Haiyan between USD 300 million and USD 700 million. AIR Press Release. Accessed January 13, 2016: http://www.air-worldwide.com/Press-Releases/AIR-Estimates-Insured-Losses-from-Super-Typhoon-Haiyan-at-Between–USD-300-Million-and-USD-700-Million/

Anderson, M. B., and Woodrow, P. J. (1998). *Rising from the Ashes: Development Strategies in Times of Disaster.* Intermediate Technology Publications.

Bahadur, A., and Tanner, T. (2014). Transformational resilience thinking: Putting people, power and politics at the heart of urban climate resilience. *Environment & Urbanization* 26(1), 200–214.

Bai, X. (2007). Integrating global environmental concerns into urban management. *Journal of Industrial Ecology* 11(2), 15–29.

Balk, D., Montgomery, M. R., McGranahan, G., Kim, D., Mara, V., Todd, M., ... and Dorélien, A. (2009). Mapping urban settlements and the risks of climate change in Africa, Asia and South America. *Population Dynamics and Climate Change, 80.*

Barton, J., and Heinrichs, D. (2011). Santiago de Chile: Adaptation, water management, and the challenges for spatial planning. In C. Rosenzweig, W. D. Solecki, S. A. Hammer, and Mehrotra, S. (eds.), *Climate Change and Cities: First Assessment Report of the Urban Climate Change Research Network* (125–126). Cambridge University Press.

Bell, M. L., O'Neill, M. S., Ranjit, N., Borja-Aburto, V. H., Cifuentes, L. A., and Gouveia, N. C. (2008). Vulnerability to heat-related mortality in Latin America: A case-crossover study in São Paulo, Brazil, Santiago, Chile and Mexico City, Mexico. *International Journal of Epidemiology* 37(4), 796–804.

Berke, P. R., Kartez, J., and Wenger, D. (1993). Recovery after disaster: Achieving sustainable development, mitigation and equity. *Disasters* 17(2), 93–109.

Blanco, H., McCarney, P., Parnell, S., Schmidt, M., and Seto, K. C. (2011). The role of urban land in climate change. In Rosenzweig, C., Solecki, W. D., Hammer, S. A., Mehrotra, S. (eds.), *Climate Change and Cities: First Assessment Report of the Urban Climate Change Research Network* (217–248). Cambridge University Press.

Braimoh, A. K., and Onishi, T. (2007). Spatial determinants of urban land use change in Lagos, Nigeria. *Land Use Policy* 24(2), 502–515.

Capacity Development for Hazard Risk Reduction and Adaptation (CATALYST). (2013). *Before Disaster Strikes: Transformations in Practice and Policy.* CATALYST Best Practice paper on the Central America and the Caribbean region. Accessed August 20, 2015: www.catalyst-project.eu

Carter, T. R., Jones, R. N., Lu, X., Bhadwal, S., Conde, C. Mearns, L. O., O'Neill, B. C., Rounsevell, M. D. A., and M. B. Zurek. (2007). New assessment methods and the characterisation of future conditions. In Parry, M. L., Canziani, O. F., Palutikof, J. P., van der Linden, P. J., and Hanson, C. E. (eds.), *Climate Change 2007: Impacts, Adaptation and Vulnerability. Contribution of Working Group II to the Fourth Assessment Report of the Intergovernmental Panel on Climate Change* (133–171). Cambridge University Press.

City of New York. (2013). *PlaNYC: A Stronger, More Resilient New York.* Mayor's Office of the City of New York.

Comerio, M. C. (1998). *Disaster Hits Home: New Policy for Urban Housing Recover.* University of California Press.

Comisión Económica para América Latina y el Caribe (CEPAL), and Banco Inter-Americano de Desarrollo (BID), [Economic Commission for Latin America and the Caribbean (ECLAC) and the Inter-American Development Bank (IADB)]. (2007). Information for disaster risk management: Case study of five countries. Summary report. [Información para la gestión de riesgo de desastres. Estudios de caso de cinco países. Informe resumido]. Mexico City, Mexico.

Conacher Travers Pty. Ltd. (2001). *Gosford City Council Biodiversity Project: Winter–Spring 2000 Fauna Survey Component.* Local study, Gosford City Council.

de Groot, R., Wilson, M., and Boumans, R. (2002). A typology for the classification, description and valuation of ecosystem functions, goods and services. *Ecological Economics* **41**, 393–408.

Department for International Development (DFID). (2005). *Disaster Risk Reduction: A Development Concern.* DFID.

Ebert, A., Welz, J., Heinrichs, D., Krellenberg, K., and Hansjürgens, B. (2010). *Socio-environmental Change and Flood Risks: The Case of Santiago de Chile* (303–313). Erdkunde.

EM-DAT: The OFDA/CRED International Disaster Database. Brussels, Belgium: Université Catholique de Louvain, Center for Research on the Epidemiology of Disasters (CRED). Accessed September 14, 2015: http://www.em-dat.net

Folke, C., Carpenter, S. R., Walker, B. H., Scheffer, M., Chapin, T., and Rockstrom, J. (2010). Resilience thinking: Integrating resilience, adaptability and transformability. *Ecology and Society* **15**(4), 20.

Fraser, E. D. G., Dougill, A. J., Mabee, W. E., Reed, M., and McAlpine, P. (2006). Bottom up and top down analysis of participatory processes for sustainability indicator identification as a pathway to community empowerment and sustainable environmental management. *Journal of Environmental Management* **78** (2), 114–127.

Garschagen, M., and Romero-Lankao, P. (2015). Exploring the relationships between urbanization trends and climate change vulnerability. *Climatic Change.* **133**, 37–52. doi: 10.1007/s10584-013-081-6

Gencer, E. A. (2008). *Natural Disasters, Vulnerability, and Sustainable Development.* VDM Verlag.

Gencer, E. A. (2013a). *The Interplay between Urban Development, Vulnerability, and Risk Management: A Case Study of the Istanbul Metropolitan Area.* Vol. 7. Springer Science & Business Media. Springer Briefs in Environment, Security, Development and Peace. Heidelberg, New York, Dordrecht, London.

Gencer, E. A. (2013b). An Overview of Urban Vulnerability to Natural Disasters and Climate Change in Central America and the Caribbean Region. FEEM Nota di Lavoro 78.2013. Fondazione Eni Enrico Mattei (FEEM): Milan, Italy.

Gencer, E. A. (2014). *A Compendium of Disaster Risk Reduction Practices in Cities of the Western Balkans and Turkey.* United Nations Office for Disaster Risk Reduction (UNISDR) and World Meteorological Organization (WMO).

Gencer, E.A., Mysiak, J., and M. Breil, 2013. *Resilient City Characteristics and a Questionnaire to Assess Resiliency in Urban Areas.* Working paper. Centro Euro-Mediterraneo sui Cambiamenti Climatici: Venice, Italy. (unpublished).

Gibson, T. D., Pelling, M., Ghosh, A., Matyas, D., Siddiqi, A., Solecki, W., … and Du Plessis, R. (2016). Pathways for Transformation: Disaster Risk Management to Enhance Resilience to Extreme Events. *Journal of Extreme Events* **3**(01), 1671002.

GIZ. (2012). Disaster risk management and adaptation to climate change: Experience from German development cooperation. Deutsche Gesellschaft für Internationale Zusammenarbeit (GIZ) GmbH Sector Project 'Disaster Risk Management in Development Cooperation'. Bonn. Accessed July 17, 2015: file:///C:/Users/SAK/Downloads/giz2012-0275en-disaster-risk-management-climate-change.pdf

Grimm, N. B., Faeth, S. H., Golubiewski, N. E., Redman, C. L., Jianguo, W., Bai, X., and Briggs, J. M. (2008). Global change and the ecology of cities. *Science* **319**(5864), 756–760.

Harlan, S. L., Brazel, A. J., Darrel Jenerette, G., Jones, N. S., Larsen, L., Prashad, L., and Stefanov, W. L. (2007). In the shade of affluence: the inequitable distribution of the urban heat island. In *Equity and the Environment* (173–202). Emerald Group Publishing Limited.

Harty, C., and Cheng, D. (2003). Ecological assessment and strategies for the management of mangroves in Brisbane water—Gosford, New South Wales, Australia. *Landscape Urban Planning* **62**, 219–240.

Hori, T. and R. Shaw. (2011). Incorporation of potential climate change impacts into local disaster risk management in Costa Rica. *Journal of Risk, Hazards and Crisis in Public Policy (RHCPP)* **2**(4). Accessed November 27, 2014: http://onlinelibrary.wiley.com/doi/10.2202/1944-4079.1094/abstract.

Hunter and Central Coast Regional Environmental Management Strategy (HCCREMS). (2010). *Potential Impacts of Climate Change on the Hunter, Central and Lower North Coast of NSW.* Hunter Councils, NSW.

Intergovernmental Panel on Climate Change (IPCC). (2012). *Managing the Risks of Extreme Events and Disasters to Advance Climate Change Adaptation. A Special Report of Working Groups I and II of the Intergovernmental Panel on Climate Change.* Field, C.B., Barros, V., Stocker, T. F., Qin, D., Dokken, D. J., Ebi, K. L., Mastrandrea, M. D., Mach, K. J., Plattner, G. -K., Allen, S. K., Tignor, M., and Midgley P. M. (eds.). Cambridge University Press.

Intergovernmental Panel on Climate Change (IPCC). (2014a). *Climate Change 2014: Mitigation of Climate Change.* Contribution of Working Group III to the fifth assessment report of the Intergovernmental Panel on Climate Change, Annex I: Glossary. Cambridge University Press.

Intergovernmental Panel on Climate Change (IPCC). (2014b). *Climate Change 2014: Synthesis Report.* Cambridge University Press.

International Recovery Platform (IRP). (2007). *Learning from Disaster Recovery Guidance for Decision Makers.* UNISDR, Geneva.

International Society of City and Regional Planning (ISOCARP). (2005). *Four Decades of Knowledge Creation and Sharing.* ISOCARP, Madrid.

Institute for Business and Home Safety (IBHS). (2002). Are we planning safer communities? Results of a national survey of community planners and natural disasters. Accessed December 8, 2014: http://www.ibhs.org/publications

Kan, H., London, S. J., Chen, G., Zhang, Y., Song, G., Zhao, N., Jiang, L., and Chen, B. (2008). Season, sex, age, and education as modifiers of the effects of outdoor air pollution on daily mortality in Shanghai, China: The Public Health and Air Pollution in Asia (PAPA) study. *Environmental Health Perspectives* **116**(9), 1183.

Kates R. W., Travis, W. R., and Wilbanks, T. J. (2012). Transformational adaptation when incremental adaptations to climate change are insufficient. *PNAS* **109**(19), 7156–7161. doi 10.1073/pnas.1115521109

Laegdsgaard, P. (2006). Ecology, disturbance and restoration of coastal saltmarsh in Australia: A review. *Wetlands Ecological Management* **14**, 379–399.

Lewis. J. (1999). *Development in Disaster-prone Places: Studies of Vulnerability.* Intermediate Technology Publications.

Lin, B., Khoo, Y. B., Inman, M., Wang, C. H., Tapsuwan, S., and Wang, X. (2014). Assessing inundation damage and timing of adaptation: sea level rise and the complexities of land use in coastal communities. *Mitigation Adaptation Strategies Global Change* **19**, 551–568.

Magrin, G. O., Marengo, J. A., Boulanger, J. P., Buckeridge, M. S., Castellanos, E., Poveda, G., Scarano, F. R., and Vicuña, S. (2014). Central and South America. In Barros, V. R., Field, C. B., Dokken, D. J., Mastrandrea, M. D., Mach, K. J., Bilir, T. E., Chatterjee, M., Ebi, K. L., Estrada, Y. O., Genova, R. C., Girma, B., Kissel, E. S., Levy, A. N., MacCracken, S., Mastrandrea, P. R., and White, L. L. (eds.), *Climate Change 2014: Impacts, Adaptation, and Vulnerability. Part B: Regional Aspects.* Contribution of Working Group II to the Fifth Assessment Report of the Intergovernmental Panel on Climate Change (1499–1566). Cambridge University Press.

Mazumder, D., Saintilan, N., and Williams, R. J. (2006). Tropic relationships between itinerant fish and crab larvae in a temperate Australian saltmarsh. *Marsh Freshwater Research* **57**, 193–199.

McGranahan, G. (2007). *Urban Environments, Wealth and Health: Shifting Burdens and Possible Responses in Low and Middle-Income Nations.* Human Settlements Working Paper, Urban Environments No. 1, International Institute for Environment and Development (IIED), London.

Mehrotra, S., Rosenzweig, C., Solecki, W. D., Natenzon, C. E., Omojola, A., Folorunsho, R., and Gilbride, J. (2011). Cities, disasters and climate risks. In Rosenzweig, C., Solecki, W. D., Hammer, S. A., and Mehrotra, Mehrotra. (eds.), *Climate Change and Cities: First Assessment Report of Climate Change Research Network* (15–42). Cambridge University Press.

Melo, O., Vargas, X., Vicuna, S., Meza, F., and McPhee, J. (2010). Climate change economic impacts on supply of water for the M & I sector in the metropolitan region of Chile. In *2010 Watershed Management Conference: Innovations in Watershed Management Under Land Use and Climate Change*, August (23–27).

Mendler de Suarez, J., Suarez, P., Bachofen, C., Fortugno, N., Goentzel, J., Gonçalves, P., Grist, N., Macklin, C., Pfeifer, K., Schweizer, S., Van Aalst, M., and Virji, H. (2012). *Games for a New Climate: Experiencing the Complexity of Future Risks*. Pardee Center Task Force Report. Boston University.

Mileti, D. S. (1999). *Disasters by Design: A Reasssesment of Natural Hazards in the United States*. Joseph Henry Press.

Millenium Ecosystem Assessment (MEA). (2005). *Ecosystems and Human Well-being: Synthesis*. Island Press.

Mitchel, M. L., and Adam, P. (1989). The relationship between mangrove and saltmarsh communities in the Sydney region. *Wetlands (Australia)* **8**, 37–46.

Natenzon, C. E. (2005). Social vulnerability, disasters and climate change in Latin America. Thematic, theoretical and methodological approaches. In: *IIª Conferência Regional sobre Mudanças Globais: América do Sul*, San Pablo University – Brazil, November 7–10.

Natenzon, C. E., and González, S. G. (2010). Risk, social vulnerability and indicators development. Samples for Argentina [Riesgo vulnerabilidad social y construcción de indicadores. Aplicaciones para Argentina]. In *Argentina y Brasil posibilidades y obstáculos en el proceso de Integración Territorial* (195–217). Universidad de Buenos Aires.

NPCC. (2015). Building the knowledge base for climate resiliency: New York City Panel on Climate Change 2015 report Rosenzweig, C., and Solecki, W. (eds.). *Annals of the New York Academy of Sciences* **1336**, 1–149.

O'Brien, K., Eriksen, S., Nygaard, L. P., and Schjolden, A. (2007). Why different interpretations of vulnerability matter in climate change discourses. *Climate Policy* **7**(1), 73–88.

Olhansky, R. B., and Kartez, J. D. (1998). Managing land-use to build resilience. In *Cooperating with Nature: Confronting Natural Hazards with Land-Use Planning for Sustainable Communities* (167–201). Joseph Henry Press.

Olsson, P., Galaz, V., and Boonstra, W. J. (2014). Sustainability transformations: A resilience perspective. *Ecology and Society* **19**(4), 1.

Pataki, D. E., Alig, R. J., Fung, A. S., Golubiewski, N. E., Kennedy, C. A., McPherson, E. G., Nowak, D. J., Pouyat, R. V., and Romero Lankao, P. (2006). Urban ecosystems and the North American carbon cycle. *Global Change Biology* **12**(11), 2092–2102.

Pelling, M., O'Brien, K., and Matyas, D. (2015). Adaptation and transformation. *Climatic Change* **133**(1), 113–127.

Perez, R. T. (2008). A community-based flood risk management in the lower Pampanga River basin. *Journal of Environmental Science and Management* **11**(1), 55–63.

Prabhakar, S. V. R. K., Srinivasan, A., and Shaw, R. (2009). Climate change and local level disaster risk reduction planning: Need, opportunities and challenges. *Mitigation and Adaptation Strategies for Global Change* **14**, 7–33.

Proyecto Apoyo a la Prevención de Desastres en la Comunidad Andina (PREDECAN) [Support of Disaster Prevention in the Andean Community Project]. (2009). *Articulando Risk Management and Adaptation to Climate Change in Agriculture: General Guidelines for Sector Planning and Management* [Articulando la Gestión del Riesgo y la Adaptación al Cambio Climático en el sector agropecuario: lineamientos generales para la planificación y la gestión sectorial]. Andean Community, European Commission.

Quarantelli, E. L. (1999). *The Disaster Recovery Process: What We Know and Do Not Know From Research*. Disaster Research Center.

Reddy, S. D. (2000). Factors influencing the incorporation of hazard mitigation during recovery from disaster. *Natural Hazards* **22**(2), 185–201.

Revi, A., Satterthwaite, D. E., Aragón-Durand, F. Corfee-Morlot, J., Kiunsi, R. B. R., Pelling, M., Roberts, D. C., and Solecki, W. (2014), Urban areas. In Field, C. B., V. R. Barros, D. J. Dokken, K. J. Mach, M. D. Mastrandrea, T. E. Bilir, M. Chatterjee, K. L. Ebi, Y. O. Estrada, R. C. Genova, B. Girma, E. S. Kissel, A. N. Levy, S. MacCracken, P. R. Mastrandrea, and White, L. L. (eds.), *Climate Change 2014: Impacts, Adaptation, and Vulnerability. Part A: Global and Sectoral Aspects.* Contribution of Working Group II to the fifth assessment report of the Intergovernmental Panel on Climate Change. Cambridge University Press.

Romero-Lankao, P. (2008). *Urban Areas and Climate Change: Review of Current Issues and Trends*. Issues paper prepared for Cities and Climate Change: Global Report on Human Settlements 2011. Accessed September 4, 2015: http://www.ral.ucar.edu/staff/prlankao-staff.php

Romero-Lankao, P. (2010). Water in Mexico City: What will climate change bring to its history of water-related hazards and vulnerabilities? *Environment Urbanization* **22**(1), 157–178.

Romero-Lankao, P., Hughes, S., Qin, H., Hardoy, J., Rosas-Huerta, A., Borquez, R., and Lampis, A. (2014). Scale, urban risk and adaptation capacity in neighborhoods of Latin American cities. *Habitat International* **42**, 224–235.

Romero-Lankao, P., and Qin, H. (2011). Conceptualizing urban vulnerability to global climate and environmental change. *Current Opinion in Environmental Sustainability* **3**(3), 142–149.

Romero-Lankao, P., Qin, H., and Borbor-Cordova, M. (2013). Exploration of health risks related to air pollution and temperature in three Latin American cities. *Social Science & Medicine* **83**, 110–118.

Romero-Lankao, P., Qin, H., and Dickinson, K. (2012). Urban vulnerability to temperature-related hazards: *A meta-analysis and meta-knowledge approach. Global Environmental Change* **22**(3), 670–683.

Rosenzweig, C., and Solecki, W. (2014). Hurricane Sandy and adaptation pathways in New York: Lessons from a first-responder city. *Global Environmental Change* **28**(0),395–408.

Rosenzweig, C., Solecki, W., Hammer, S., and Mehrotra, S. (2010). Cities lead the way in climate-change action. *Nature* **467**, 909–911.

Rosenzweig, C., Solecki, W., Hammer, S., and Mehrotra, S. (eds.). (2011). *Climate Change and Cities: First Assessment Report of the Urban Climate Change Research Network*. Cambridge University Press.

Schnarwiler, R., and Tuerb, J. (2011). *Closing the Financial Gap*. Swiss Re. Accessed April 17, 2014: http://www.swissre.com/rethinking/crm/Closing_the_financial_gap.html

Senga, R. (2012). Natural or unnatural disasters: The relative vulnerabilities of South-East Asian megacities to climate change. In *Mega-Stress for Mega-Cities*. Accessed November 18, 2015: http://www.wwf.org.uk/research_centre/?3454/Mega-Stress-for-Mega-Cities

Smyth, C. G., and Royle, S. A. (2000). Urban landslide hazards: Incidence and causative factors in Niteroi, Rio de Janeiro State, Brazil. *Applied Geography*, **20**(2), 95–118.

Solecki, W., O'Brien, K., and Leichenko, R. (2011). Disaster risk reduction and climate change adaptation strategies: Convergence and synergies. *Current Opinion in Environmental Sustainability* **3**(3), 135–141.

Solecki W, Pelling M, Garschagen M. (2017). A framework for urban risk management regime shifts. Ecology & Society. 22(2): 38.

Steinberg, M., and Burby, R. (2002). Growing safe. *Planning* (April), 22–23.

Stewart, M. G., Wang, X., and Willgoose, G. R. (2014). Direct and indirect cost and benefit assessment of climate adaptation strategies for extreme wind events in Queensland. *Natural Hazards Review* **15**(4), 04014008. Accessed May 24, 2015: http://dx.doi.org/10.1061/(ASCE)NH.1527–6996.0000136

Swiss Re. (2013). *Natural catastrophes and man-made disasters in 2012: A year of extreme weather events in the U.S. Sigma 2013/2*. Author.

Thomalla, F., et al. (2006). Reducing hazard vulnerability: Towards a common approach between disaster risk reduction and climate adaptation. *Disasters* **30** (1), 39–48.

Tourism Research Australia. (2007). Tourism profiles for local government areas in regional Australia, New South Wales, City of Gosford. Accessed May 26, 2014: http://www.ret.gov.au/tourism/Documents/tra/local%20government%20area%20profiles/Gosford%20LGA.pdf

Tran, P., and Shaw, R. (2007). Towards an integrated approach of disaster and environmental management: A case study of Thua Thien Hue province in central Vietnam. *Environmental Hazards* **7**(4), 271–282.

Tyler, S., and Moench, M. (2012). A framework for urban climate resilience. *Climate and Development* **4**.4(2012), 311–326.

United Nations (UN). (2015). *Sendai Framework for Disaster Risk Reduction 2015–2030*. Third UN World Conference on Disaster Risk Reduction. Sendai, Japan. A/CONF.224/CRP.1. Accessed May 18, 2016: http://www.wcdrr.org/

United Nations Department of Economic and Social Affairs (UN-DESA). (2015). *World Urbanization Prospects*, United Nations.

United Nations Development Programme (UNDP). (2004). *Reducing Disaster Risk: A Challenge for Development*. Accessed August 10, 2015: http://www.undp.org/bcpr

United Nations Human Settlements Programme (UN-Habitat). (2003). The Challenge of Slums. Global Report on Human Settlements. Earthscan. London.

United Nations Human Settlements Programme (UN-Habitat). (2011). *Global Report on Human Settlements: Cities and Climate Change*. Earthscan.

United Nations International Strategy for Disaster Reduction (UNISDR). (2007). *Hyogo Framework for Action 2005–2015: Building the Resilience of Nations and Communities to Disasters*. United Nations.

United Nations International Strategy for Disaster Reduction (UNISDR). (2009). *2009 UNISDR Terminology on Disaster Risk Reduction*. United Nations.

United Nations International Strategy for Disaster Reduction (UNISDR). (2011). *2011 Global Assessment Report on Disaster Risk Reduction*. UNISDR.

UNISDR Urban Planning Advisory Group (UPAG). (2015). *Eight Frequently Asked Questions on DRR and Urban Planning*. Prepared for the Third UN World Conference on Disaster Risk Reduction, Sendai, Japan. (unpublished).

United States (U.S.) Census. (2010).

Vazquez-Brust, D., Plaza Ubeda, J. A., de Burgos Jiménez, J., and Natenzon, C. E. (eds.). (2012). *Business and Environmental Risks: Spatial Interactions between Environmental Hazards and Social Vulnerabilities in Ibero-America*. Springer.

Vignola, R., Locatelli, B., Martinez C., and Imbach, P. (2009). Ecosystem-based adaptation to climate change: What role for policy-makers, society and scientists? *Mitigation and Adaptation Strategies for Global Change* **14**, 691–696.

von Peter, G., von Dahlen, S., and Saxena, S. C. (2012). *Unmitigated Disasters? New Evidence on the Macroeconomic Cost of Natural Catastrophes*. Bank for International Settlements (BIS). Accessed September 3, 2015: http://www.bis.org/publ/work394.htm

Vorhies, F. (2012). *The Economics of Investing in Disaster Risk Reduction*. UNISDR Working Paper. UNISDR.

Wang, X., Khoo, Y. B., and Wang, C. H. (2014). Risk assessment and decision-making for residential housing adapting to increasing storm-tide inundation due to sea level rise in Australia. *Civil Engineering and Environmental Systems* **31**(2), 125–139.

Warner, K., et al. (2009). In search of shelter: Mapping the effects of climate change on human migration and displacement. CARE.

Wisner, B., et al. (2004). *At Risk: Natural Hazards, People's Vulnerability, and Disasters*. 2nd ed. Routledge.

World Bank (WB). 2012. *Improving the Assessment of Disaster Risks to Strengthen Financial Resilience*. World Bank. Accessed August 15, 2014: http://documents.worldbank.org/curated/en/2012/01/16499055/improving-assessment-disaster-risks-strengthen-financial-resilience

World Bank (WB) and the Global Facility for Disaster Reduction and Recovery (GFDRR). (2010). *Disaster Risk Management in Latin America and the Caribbean Region: GFDRR Country Notes*. World Bank.

Yasui, E. (2007). *Community Vulnerability and Capacity in Post-disaster Recovery: The Cases of Mano and Mikura Neighbourhoods in the Wake of the 1995 Kobe Earthquake*. University of British Columbia.

Chapter 3   Case Study References

Case Study 3.1   The Boulder Floods: A Study of Urban Resilience

City of Boulder Flood Management Program (2014).City of Boulder. Accessed February 8, 2015: https://www-static.bouldercolorado.gov/docs/flood-management-program-overview-1-201410060951.pdf

Peel, M. C., Finlayson, B. L., and McMahon, T. A. (2007). Updated world map of the Köppen-Geiger climate classification. *Hydrology and Earth System Sciences Discussions* **4**(2), 462.

World Bank. (2017). 2016 GNI per capita, Atlas method (current US$). Accessed August 9, 2017: http://data.worldbank.org/indicator/NY.GNP.PCAP.CD

Case Study 3.2   Adaptation to Flooding in the City of Santa Fe, Argentina: Lessons Learned

Cardoso, M. M. (2011). Impacto ambiental de los nuevos procesos urbanos en el área metropolitana de Santa Fe [Environmental impact of the new urban processes in the Santa Fe Metropolitan Area].*V Congreso Iberoamericano sobre Desarrollo y Ambiente de REDIBEC/V Jornadas de la Asociación Argentina Uruguaya de Economía Ecológica*. Santa Fe, Facultad de Ingeniería y Ciencias Hídricas, Universidad Nacional del Litoral; 12 a 14 de setiembre. Accessed September 14, 2014: http://fich.unl.edu.ar/CISDAV/upload/Ponencias_y_Posters/Eje08/Maria_Mercedes_Cardoso/V%20CISDA%20Trabajo%20completo%20Impacto.pdf

CEPAL – United Nations. (2003). *Evaluación del impacto de las inundaciones y el desbordamiento del río Salado en la provincia de Santa Fe, República de Argentina en 2003*. [Assessing the impact of flooding and overflow of the Salado River in the province of Santa Fe, Republic of Argentina in 2003]Buenos Aries, LC/BUE/R.254, Buenos Aries, CEPAL, LC/BUE/R.254, Restricted Distribution 89 p. Accessed October 14, 2014: http://repositorio.cepal.org/bitstream/handle/11362/28461/LCbueR254_es.pdf?sequence=1/

Demographia. (2016). *Demographia World Urban Areas 12th Annual Edition: 2016:04*. Accessed August 9, 2017: http://www.demographia.com/db-worldua.pdf

Disaster Relief Emergency Fund (DREF). (2008). *Argentina: Foods*. International Federation of Red Cross and Red Crescent Societies, DREF operation n° MDRAR002. GLIDE n°FL-2007–000044-ARG, 8 May 2008. Accessed October 14, 2014: http://www.ifrc.org/docs/appeals/rpts08/MDRAR002final.pdf

Gobierno de la Ciudad de Santa Fe. (2011). Premio Sasakawa de las Naciones Unidas Gestión de Riesgos (UN Sasakawa Award 2011). Accessed June 3, 2014: http://santafeciudad.gov.ar/blogs/gestionderiesgos/premio-sasakawa-de-las-naciones-unidas/

Gobierno de la Ciudad de Santa Fe. (2015). *Datos Generales [General Data]*. Accessed February 29, 2016: http://www.santafeciudad.gov.ar/ciudad/datos_generales/poblacion.html

INDEC. (2015). *Censo Nacional de Población 2010. In: Base de datos Redatam+Sp*. Accessed February 29, 2016: http://200.51.91.245/argbin/RpWebEngine.exe/PortalAction?&MODE=MAIN&BASE=CPV2010B&MAIN=WebServerMain.inl

Peel, M. C., Finlayson, B. L., and McMahon, T. A. (2007). Updated world map of the Köppen-Geiger climate classification. *Hydrology and Earth System Sciences Discussions* **4**(2), 462.

Viand, J. (2009). Before the disaster. Social risk construction and the flood of 2003 in Santa Fe city. Bachelor´s thesis, University of Buenos Aires.

World Bank. (2017). 2016 GNI per capita, Atlas method (current US$). Accessed August 9, 2017: http://data.worldbank.org/indicator/NY.GNP.PCAP.CD

Case Study 3.3 Preparedness, Response, and Reconstruction of Tacloban for Super Typhoon Haiyan in The Philippines

Philippine Statistics Authority. (2014). Leyte Province. Municipalities and cities. Accessed March 24, 2015: http://www.nscb.gov.ph/activestats/psgc/province.asp?regName=REGION+VIII+%28Eastern+Visayas%29&regCode=08&provCode=083700000&provName=LEYTE

Philippine Statistics Authority (2016). Population of Population of Region VIII – Eastern Visayas (Based on the 2015 Census of Population) Accessed June 28, 2017: http://psa.gov.ph/content/population-region-viii-eastern-visayas-based-2015-census-population

Philippine Government. (2013). Reconstruction Assistance on Yoland – Building Back Better.

World Bank. (2017). 2016 GNI per capita, Atlas method (current US$). Accessed August 9, 2017: http://data.worldbank.org/indicator/NY.GNP.PCAP.CD

Part I

Cross-Cutting Themes

4

Integrating Mitigation and Adaptation

Opportunities and Challenges

Coordinating Lead Authors

Stelios Grafakos (Rotterdam/Athens), Chantal Pacteau (Paris), Martha Delgado (Mexico City)

Lead Authors

Mia Landauer (Vienna/Espoo), Oswaldo Lucon (São Paulo), Patrick Driscoll (Copenhagen/Trondheim)

Contributing Authors

David Wilk (Washington, D.C.), Carolina Zambrano (Quito), Sean O'Donoghue (Durban), Debra Roberts (Durban)

This chapter should be cited as

Grafakos, S., Pacteau, C., Delgado, M., Landauer, M., Lucon, O., and Driscoll, P. (2018). Integrating mitigation and adaptation: Opportunities and challenges. In Rosenzweig, C., W. Solecki, P. Romero-Lankao, S. Mehrotra, S. Dhakal, and S. Ali Ibrahim (eds.), *Climate Change and Cities: Second Assessment Report of the Urban Climate Change Research Network*. Cambridge University Press. New York. 101–138

Integrating Mitigation and Adaptation: Opportunities and Challenges

Urban planners and decision-makers need to integrate efforts to mitigate the causes of climate change (mitigation) and adapt to changing climatic conditions (adaptation), for a global transition to a low-emissions economy and a resilient world. Actions that promote both goals provide win-win solutions. In some cases, however, decision-makers have to negotiate tradeoffs and minimize conflicts between competing objectives.

A better understanding of mitigation, adaptation, resilience and low-emissions development synergies can reveal greater opportunities for their integration in urban areas. For example, strategies that reduce the UHI effect, improve air quality, increase resource efficiency in the built environment and energy systems, and enhance carbon storage related to land use and urban forestry are likely to contribute to greenhouse gas (GHG) emissions reduction while improving a city's resilience. The selection of specific adaptation and mitigation measures should be made in the context of other Sustainable Development Goals by taking into account current resources and technical means of the city, plus the needs of citizens.

Major Findings

- Mitigation and adaptation policies have different goals and opportunities for implementation. However, many drivers of mitigation and adaptation are common, and solutions can be interrelated. Evidence shows that broad-scale, holistic analysis and proactive planning can strengthen synergies, improve cost-effectiveness, avoid conflicts, and help manage trade-offs.

- Diagnosis of climate risks and the vulnerabilities of urban populations and territory is essential. Likewise, cities need systematic GHG emissions inventories and emission reduction pathways in order to prepare mitigation actions.

- Contextual conditions determine a city's challenges, as well as its capacity to integrate and implement adaptation and mitigation strategies. These include the environmental and physical setting, the capacities and organization of institutions and governance, economic and financial conditions, and sociocultural characteristics.

- Integrated planning requires holistic, systems-based analysis that takes into account the quantitative and qualitative costs and benefits of integration compared to stand-alone adaptation and mitigation policies. Analysis should be explicitly framed within city priorities and provide the foundation for evidence-based decision support tools.

Key Messages

Integrating mitigation and adaptation can help avoid locking a city into counterproductive infrastructure and policies. Therefore, city governments should develop and implement climate action plans early in their administrative terms. These plans should be based on scientific evidence and should integrate mitigation and adaptation across multiple sectors and levels of governance. Plans should clarify short-, medium-, and long-term goals; implementation opportunities; budgets; and concrete measures for assessing progress.

Integrated city climate action plans should include a variety of mitigation actions (involving energy, transport, waste management, and water resources) and adaptation actions (involving infrastructure, natural resources, health, and consumption, among others) in synergistic ways. Because of the comprehensive scope, it is important to clarify the roles and responsibilities of key actors in planning and implementation. Interactions among the actors must be coordinated during each phase of the process.

Once priorities and goals have been identified, municipal governments should connect with federal legislation, national programs, and, in the case of low-income cities, with international donors in order to implement actions and foster helpful alliances and financial support.

4.1 Introduction

This chapter presents the challenges and opportunities related to the integration of adaptation and mitigation policies and practices in cities. The objective is to guide decision-makers, urban planners, and practitioners toward opportunities and challenges of integrated climate change planning. This is done by means of a literature review of mitigation and adaptation relationships and by presenting examples from selected cities to discover best practices with regard to integrated solutions. The chapter shows how urban planners and decision-makers can enhance synergies, negotiate tradeoffs, and minimize conflicts between adaptation and mitigation.

The focus of the chapter is threefold. First, we present relationships between adaptation and mitigation, which are now understood as encompassing low emissions development and resilience across different urban sectors, and we identify synergies and conflicts on the policy level. After introducing the integration of mitigation and adaptation (*Ad-Mit* and *Mit-Ad*, respectively – adaptation actions with mitigation effects, and, conversely, mitigation positively affecting adaptation), we show real-world examples of synergies and conflicts in selected city Case Studies. Finally, we discuss how a better understanding of the integration of mitigation and adaptation can provide opportunities for urban areas, but challenges as well.

Table 4.1 presents urban examples of potential adaptation and mitigation synergies, tradeoffs, and conflicts cutting across sectors. These illustrate the scales at which integration issues must be addressed. Learning from experience is an important element in understanding the necessary steps toward a successful process of implementation of climate change planning and management.

4.2 Climate Mitigation and Adaptation in the Urban Context

As of 2014, 54% of the world's population resided in urban areas, compared to 30% in 1950 and 66% projected for 2050 (UNDESA, 2014). Cities are major contributors to global greenhouse gas (GHG) emissions and, due to large population concentrations, also highly vulnerable to climate change impacts such as heat waves, floods, severe storms, and droughts (Sims and Dhakal, 2014; Fischedick et al., 2012; Lucon et al., 2014; Revi et al., 2014; Balaban and de Oliveira, 2013). Cities are at the forefront of climate policies (Rosenzweig et al., 2011), and the need for decision-makers and planners to respond to climate change is crucial for collaborative urban climate governance (Bulkeley, 2013) (see Box 4.1) (see Chapter 16, Governance and Policy).

Responses to climate change in cities consist of the design and implementation of policies and practices to reduce anthropogenic GHG emissions, known as *mitigation measures*, and responses to climate-related impacts and risks, known as *adaptation measures*. For mitigation planning, the primary goal is to reduce current and future direct and indirect GHG emissions, particularly from energy production, land use, waste, industry, the built environment infrastructure, and transportation. The primary goal of adaptation is to adjust the built, social, and ecological environment to minimize the negative impacts of both slow-onset and extreme events caused by climate change, such as sea-level rise, floods, droughts, storms, and heat waves.

The Intergovernmental Panel on Climate Change (IPCC) Fourth Assessment Report (AR4) chapters covering the relationships between adaptation and mitigation raised the importance of examining possibilities for integration of adaptation and mitigation policies and foundations for decision-making (Klein et al.,

Table 4.1 *Main differences between mitigation and adaptation. Source: Adapted from Dang et al., 2003*

	MITIGATION POLICY	ADAPTATION POLICY
Sectoral focus	All sectors that can reduce GHG emissions	Selected at-risk sectors
Geographical scale of effect	Global	Local, regional
Temporal scale of effect	Long term	Short, medium, and long term
Effectiveness	Reduction in global temperature rise commitment	Increases in climate resilience
Ancillary benefits (or co-benefits)	Multiple	Improved response to extreme events in current climate
Actor benefits	Through ancillary benefits	Almost fully through reduction of climate impact and ancillary benefits
Polluter pays	Yes	Not necessarily
Monitoring	Relatively easy (measuring the reduction of greenhouse gas emissions)	More difficult (measuring the reduction of climate risk)

Box 4.1   Cities' Commitment to Tackle Climate Change

By signing the Global Cities Covenant on Climate – known as the Mexico City Pact – in November 2010, mayors and municipal authority representatives demonstrated their voluntary commitments on the frontline of climate change response. Furthermore, climate finance for city governments is highlighted in the Nantes Declaration of Mayors and Subnational Leaders on Climate Change that was adopted in 2013. These city agreements and the culture of fostering city-to-city cooperation take a number of forms at different levels. The Global Covenant of Mayors for Climate & Energy (2016), the ICLEI–Local Governments for Sustainability, United Cities and Local Governments (UCLG), C40 Cities Climate Leadership Group, World Mayors Council on Climate Change, and World Association of the Major Metropolises are among the organizations that are convening city authorities to act against climate change. In October 2014, the Mayors Adapt Signature Ceremony in Brussels gathered more than 150 city and regional authority representatives committed to European initiatives on adaptation to climate change (European Climate Adaptation Platform, 2014). The leadership of cities in combatting climate change and advancing energy efficiency and renewable energy use is the subject of the Earth Hour City Challenge contest that the World Wildlife Fund (WWF) launched in 2013, awarding a city each year for its outstanding achievements (i.e., Vancouver in 2013, Cape Town in 2014, and Seoul in 2015). The Climate Summit for Local Leaders at Paris City Hall on December 4, 2015, brought to the fore the strength of the commitment made by city actors in the fight against climate change. The Paris Agreement secured at COP21 places the involvement of non-state actors on the cutting edge of research toward and implementation of climate solutions.

2007; Jones et al., 2007). The Fifth Assessment Report (AR5) of the IPCC addressed urban issues directly through chapters on adaptation in urban areas and mitigation in human settlements, infrastructure, and spatial planning but also indirectly through the subjects of integrated risk and uncertainty assessment of climate policies as well as sectoral chapters on topics such as buildings and transport (Revi et al., 2014; Seto and Dhakal, 2014; Kunreuther et al., 2014; Lucon et al., 2014; Sims et al., 2014).

The dichotomy between mitigation and adaptation is rooted in history (Pacteau and Joussaume, 2013). Mitigation has been considered a global-scale issue, whereas adaptation is seen as local. Furthermore, in the first years after the establishment of the United Nations Framework Convention on Climate Change (UNFCCC), mitigation issues had more importance politically, whereas adaptation is a newer issue to be dealt with. Yet, because the impacts of climate change are already being felt across the world, the role and associated responsibilities of both mitigation and of adaptation have been reconsidered. The need for future balance between adaptation and mitigation has led to a search for integrated climate policies across scales. The importance of bottom-up action is recognized along with increasing acceptance of the important role of cities, metropolises, and other subnational territories. Municipal authorities in cities and urban areas across the world have been driven to find ways to mitigate GHG emissions and seek innovative strategies to adapt to climate change based on individual capacities and networks (see Box 4.1).

4.3   Integrating Mitigation and Adaptation

A growing body of scientific evidence demonstrates the importance of implementing mitigation and adaptation in an integrative manner. This literature analyzes adaptation and mitigation relationships on a more conceptual level and provides in-depth empirical analyses and case studies of best practices in different cities, applying a range of qualitative and quantitative methods (Landauer et al., 2015). In the literature, the interrelationships are typically conceptualized as synergies and conflicts between the two climate policies or tradeoffs in cases where a balance is being sought (Klein et al., 2007). Based on the empirical evidence, mitigation policies such as promotion of energy-efficient technologies and actions for energy savings and efficiencies have traditionally been introduced by national governments and targeted toward specific sectors such as industry, power generation, transportation, and construction. Adaptation policies are newer on the agenda, particularly at the subnational level (de Oliveira, 2009). Ayers and Huq (2009) point out that the predominant focus on mitigation actions in vulnerable developing countries has hindered their engagement in adaptation due to lack of financial incentives.

Examining adaptation and mitigation in an integrated manner is considered particularly important at the city scale (McEvoy et al., 2006; Saavedra and Budd, 2009). This is so because the benefits of integration of the two policies can best be seen at the city level. Additionally, the integration of mitigation and adaptation has the potential to reduce the costs of emissions that influence urban climates, and adaptation helps cities prepare for both slow-onset and extreme events of climate change (Callaway, 2004). Especially in urban areas, integrated solutions can help avoid maladaptation and realize Sustainable Development Goals (Barnett and O'Neill, 2010; Döpp et al., 2010). The Case Studies in this chapter serve as an evidence base from cities in different geographical regions and contexts. However, it should be kept in mind that an action implemented in one place does not necessarily mean that it is suitable for another.

In terms of urban climate governance, the complex interactions of different actors, sectors, and scales make implementation of climate policies particularly challenging (see Chapter 16, Governance and Policy). Despite the complexity originating from the multiscale dynamics in urban areas, the integration of adaptation and mitigation strategies can succeed (Thornbush et al., 2013). City administrations have responsibility for both adaptation and mitigation, but climate policies at the city scale cannot be completely separated from their national and global contexts – or from the private sector (Hall et al., 2010; Swart and Raes, 2007). The costs and benefits of adaptation and mitigation and allocation of responsibilities to implement policies vary across urban sectors and levels of governance, which complicates planning and decision-making in cities (Piper and Wilson, 2009). Dual consideration explicitly takes into account the cross-sectoral and cross-scale nature of adaptation and mitigation.

4.3.1 Differences between Adaptation and Mitigation across Multiple Scales and Sectors

A dichotomy between adaptation and mitigation policies arises from a number of factors; these include differences in spatial, temporal, institutional, and administrative scales, as well as differences in research traditions and disciplines (Moser, 2012; Goklany, 2007; Swart and Raes, 2007; Wilbanks et al., 2007; Dymén and Langlais, 2013). The integration of adaptation and mitigation is often discussed in such a scale-related context. In

the case of mitigation, the main focus is global and national, whereas in the case of adaptation, it is local and territorial scales. Time frames also differ since mitigation is considered a long-term process, whereas adaptation often implies short-term actions ranging from seasonal to decadal (see Table 4.1). Adaptation also reduces pre-existing vulnerabilities to climate extremes that exist even without climate change.

In addition, the governance of adaptation and mitigation is placed at different jurisdictional and institutional scales, characterized by vertical or horizontal modes of governance, and different actors and their interactions (Kern et al., 2008) (see Chapter 16, Governance and Policy). Sometimes adaptation and mitigation are not separated in scale-based dichotomies but instead are considered on "continuous" scales: municipal to national, short-term to long-term, and local to global (Dymén and Langlais, 2013; Dantec and Delebarre, 2013). To identify the interrelationships between adaptation and mitigation, it is advantageous to have information on the scales at which the policy development is driven, how the policies transect one another,

Box 4.2 Linking Adaptation and Mitigation and the Prospects for Transformation

Notwithstanding the extent of the challenge, the importance of concerted and holistic action on climate and other forms of environmental change has recently been underscored in the global arena by the inclusion of such action as a specific target within the urban Sustainable Development Goal (Goal 11). Additional target 11.b states:

By 2030, substantially increase the number of cities and human settlements adopting and implementing integrated policies and plans towards inclusion, resource efficiency, mitigation and adaptation to climate change, resilience to disasters, and develop and implement, in line with the Sendai Framework for Disaster Risk Reduction 2015–2030, holistic disaster risk management at all levels.

This is intended to promote appropriate collaborative actions by national, regional, and city governments to promote urban sustainability.

Mitigation and adaptation interventions of all categories have particular cost and impact thresholds, from the proverbial low-hanging fruit yielding positive returns to expensive capital-intensive solutions that may provide only modest benefits that challenge their value. Incremental change and reform therefore have limitations. Partly in consequence, attention has recently been drawn increasingly to the actual or potential limitations of even integrated action on mitigation and adaptation where this is unlikely to be adequate to overcome manifestly unsustainable urbanism (Pelling, 2011, Pelling et al., 2012; Simon and Leck, 2015). The reasons for this could be many, including where the magnitude of forecast climate change will demand dramatic changes to the urban fabric, where biophysical or environmental constraints in arid and semi-arid regions will occur, and where obsolete built environments and infrastructure or highly polarized societies reflecting strongly unequal power relations present strong constraints to change.

Transformation was defined in Chapter 1 as *a fundamental change to the status quo and its underlying social-ecological relations or, in urban contexts, to the nature of the built environment and how it is used*. With respect to tackling the impacts of climate change, long-term unsustainability would arise where conventional mitigation and adaptation interventions, even as part of a holistic strategy, would prove inadequate in preventing inundation, desertification, or the persistence of widely uninhabitable conditions. Rising sea levels and the growing frequency and intensity of storms on the one hand and the changing frequency and intensity of rainfall in many regions on the other present different challenges to coastal and inland cities alike.

The profound upgrading of response and recovery strategies – as well as preventative measures – provoked in New Orleans by Hurricane Katrina and in metropolitan New York by Hurricane Sandy are instructive, but cities and societies with limited resources would be overstretched by such measures (Rosenzweig and Solecki, 2014; Solecki, 2015). There will be low-lying areas, for instance, that cannot be protected from frequent inundation, and steep-sloped neighborhoods may be rendered uninhabitable or their infrastructure unusable by floods and landslides.

This will necessitate profoundly difficult decisions about possible abandonment of such areas and organized resettlement of the inhabitants elsewhere. Although this provides opportunities to design or redesign substantial areas in accordance with new sustainability principles, with implications for the town or city as a whole in terms of overall sustainability, it will also require massive investment as well as more flexible and appropriate building, planning, and zoning regulations than are currently in place in most urban areas worldwide.

and where the policies are implemented (Landauer et al., 2015). Laukkonen et al. (2009) point out that development of tools and procedures is needed that can help actors at different scales find best practices for adaptation, mitigation, and their integration.

In urban planning practice, a few key scale-related differences exist between adaptation and mitigation. Both policies are driven by *institutional-scale* factors such as laws and regulations to support policy decisions and operating rules to govern climate change in cities. The institutional complexity makes implementation of adaptation actions more challenging compared to mitigation actions due to the great variety of sectors and actors involved (McEvoy et al., 2006).

In addition, the focus of the policy decisions and strategies differ: in the case of adaptation, they tend to be more city- and regional-level initiatives because the impacts of climate change depend on the likelihood of risk outcomes at a smaller scale, whereas for mitigation emissions, reduction should take place globally. In regard to *spatial scale*, the benefits of mitigation accrue globally, whereas the benefits of adaptation tend to aggregate at city and regional scales, encouraging policies ranging from the regional scale to even the building scale (Ayers and Huq, 2009; Balaban and de Oliveira, 2013).

In terms of the *temporal scale* – due to feed-forward delays in the carbon cycle in the atmosphere – benefits from mitigation measures are realized over longer time scales, while adaptation has more short-term effects by reducing vulnerability to immediate and near-term climate risks (McEvoy et al., 2006; IPCC, 2007; Ayers and Huq, 2009) (see Figure 4.1) (see Chapter 2, Urban Climate Science, and Chapter 3, Disasters and Risk).

Furthermore, mitigation costs are typically local – although benefits are mainly global (although reductions in energy costs can be also local) – whereas adaptation costs and benefits tend to be localized (Jones et al., 2007; Ayers and Huq, 2009). An exception is where the benefits of adaptation can also be seen globally through reduction of threats to natural systems (Goklany, 2007) (see Chapter 8, Urban Ecosystems). Moreover, mitigation co-benefits (or ancillary benefits) are often local, especially in cases where reduction of emissions leads to, for example, improvements in air quality, public health, or improved transportation systems (IPCC, 2007) (see Table 4.1). This is particularly relevant for cities since the search for an optimum planning and policy balance between mitigation and adaptation is to a large extent contingent upon capturing positive co-benefits and avoiding policy conflicts.

Mitigation and adaptation policy formation and implementation are conducted at different *jurisdictional scales*. Adaptation is the responsibility of mainly municipal-, provincial-, and national-level administrations, whereas national governments and supranational institutions are the legal governing institutions for mitigation actions (Ayers and Huq, 2009; Ford et al., 2011). However, in some countries, mitigation actions, laws, and policies have been adopted and implemented at the city level long before such measures or policies were adopted at the national level. These municipal mitigation actions and policies were commonly attributed to urban sectors – such as transportation, water management, and waste management – delivering urban development benefits simultaneously (see Case Study 4.6). Some authors suggest that the optimal combination of mitigation and adaptation depends on the magnitude of the climate impacts within each management jurisdiction (Saavedra and Budd, 2009; Jones et al., 2007).

Cities and municipal governments have different incentives, motivations, and dynamics for mitigating GHG emissions (as well as different beneficiaries of actions) than do national governments. For several reasons, city governments also apply GHG

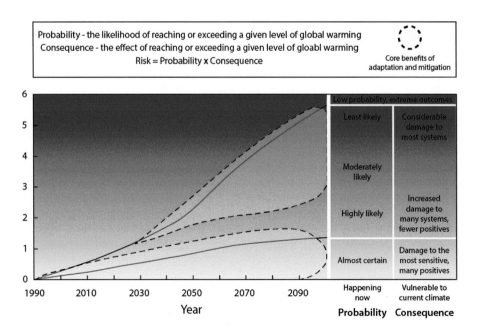

Figure 4.1 *Illustrative benefits and risks of climate policies, according to projected ranges of global warming.*

Source: Jones et al., 2004; Jones and Yohe, 2008

mitigation metrics that are different from national ones: lack of disaggregated data, leakages and spillover effects, and drivers for sectoral action (e.g., decarbonized transportation modes, energy-efficient appliances and equipment, "green" buildings). Very often municipal and regional actors work independently of the national governments and multilateral climate agreements since they are more influenced by the considerations of local civil society. However, in recent years – particularly after the COP16 in Cancun, 2010 – subnational governments have strived more intensely to be recognized as important players and to take an active role in the international climate change decision-making framework within the UNFCCC. This has resulted in the explicit negotiation of the role of cities in the Paris Agreement of COP21.

4.3.2 Adaptation and Mitigation Measures across Different Sectors

In urban areas, *adaptation* measures are implemented through urban planning and management sectors that focus mainly on zoning, building codes, water quality, flood protection, and surface runoff management (see Table 4.2). Adaptation also includes measures that increase the indoor climate comfort of buildings – such as heating, ventilation, and air conditioning (HVAC) and cool roofs (white and green) that address the UHI (McEvoy et al., 2006) (see Chapter 2, Urban Climate Science) – as well as green (vegetation, permeable surfaces) and blue (bodies of water) measures to increase climatic comfort, control flooding, and enhance urban biodiversity. In general, adaptation

Table 4.2 *Examples of synergies and conflicts between adaptation and mitigation*

Climate Policy	Practical measures by sectors	Examples of synergies between adaptation and mitigation	Examples of conflicts between adaptation and mitigation	Examples of sectors affected by implementation of measures	Source
	Building and Infrastructure				
Mitigation	Building orientation, height and spacing	Reduced need for conventional air-conditioning		Urban Planning, Health and Security, Energy	Barbhuiya et al. (2013)
Adaptation	Urban greening and green infrastructure practices	Carbon sequestration and reduction of heat stress, air pollution and flooding	High space demand	Urban Planning, Agriculture, Forest and Biodiversity (AFB), Water Management, Health and Security	Thornbush et al. (2013)
Adaptation	Ventilation and air-conditioning	Passive cooling combined with night ventilation	High energy demand	Energy, Health and Security	Gupta and Gregg (2013)
	Water Management				
Adaptation	Open storm water systems via urban wetlands		High space demand	AFB, Health and Security, Urban Planning	Laukkonen et al. (2009)
Adaptation	Water pumping to control flooding		High energy demand	Energy, Building and Infrastructure	Sugar et al. (2013)
Adaptation	Flood protection walls, dams, etc.		Emissions through material production and construction, biodiversity loss	Energy, AFB, Building and Infrastructure	Kenway et al. (2011)
Adaptation	Water saving	Reduction of energy use for water treatment/extraction		Energy, Infrastructure	Kenway et al. (2011)
	Urban Planning				
Mitigation	Urban densification		More built mass, less urban drainage, heat gains, storm water and flood risks, discomfort and health risks, more emissions from transportation, water pollution via poorly planned dense cities	Energy, Water Management, Health and Security, Transportation, AFB	Dymen and Langlais (2013); Hamin and Gurran (2009)

107

Table 4.2 (*continued*)

Climate Policy	Practical measures by sectors	Examples of synergies between adaptation and mitigation	Examples of conflicts between adaptation and mitigation	Examples of sectors affected by implementation of measures	Source
Energy					
Mitigation	Solar, wind, and wave energy	Reduction of risks of widespread power loss or peak power loads under storm events and temperature extremes		Building and Infrastructure, Health and Security	Hamin and Gurran (2009); Laukkonen et al. (2009); McEvoy et al. (2006); Sugar et al. (2013)
Transportation					
Mitigation	Multimodal and public transportation	Synergy if built along urban green corridors*		Energy, Health and Security	Sugar et al. (2013); Thornbush et al. (2013)

*Note: Urban green corridors are networks of green areas within the city and its surroundings. In addition to adaptation and mitigation functions (e.g., flood protection, carbon capture and storage, and surface temperature regulation), they provide many other benefits such as recreation and biodiversity protection.

measures are more difficult to retrofit over existing settlements than to implement in new areas.

Furthermore, adaptation of vulnerable sectors such as agriculture, forestry, and coastal zone management are often interlinked with urban decision-making. There is a growing body of literature highlighting the importance of urban and peri-urban linkages, particularly with regard to the agriculture and forestry sectors. Lwasa et al. (2014) emphasize the role of urban and peri-urban agriculture and forestry (UPAF) both in climate mitigation and adaptation. UPAF contributes to mitigation through the sequestration of carbon and by reducing the carbon emissions of food systems through the reduction of transport-related emissions for food consumption in cities. These UPAF contributions to adaptation come through the promotion of urban food security.

Mitigation measures cover efficiency, fuel decarbonization, and carbon recovery in sectors such as energy production, industry, buildings, infrastructure, transportation, waste management, and land use (see Table 4.2). In the energy sector, efficiency and decarbonization are important issues: consumption can be reduced for the same output, renewable sources can be substituted, and smart technologies can reduce emissions either directly or indirectly. Moreover, behavioral change can reduce demand and lifestyle-related impacts. In the transportation sector, low-carbon fuels, advanced technologies, efficient transport modes, and adequate planning can ensure efficiency and reduce dependency of fossil fuels. Urban infrastructure can be more efficient from a mitigation perspective, benefiting from densification of urban structure and multiple centers, as well as from public and non-motorized transportation. In the building sector, energy efficiency requirements and the resulting GHG mitigation goals can be achieved by means of several measures (design, materials, envelope, "greening," and albedo), where in many cases passive technologies are advantageous (McEvoy

et al., 2006; Gupta and Gregg, 2013; Barbhuyia et al., 2013; Lucon et al., 2014) (see Chapter 5, Urban Planning and Design). Municipal solid waste management measures include reuse and recycling programs, utilizing energy from methane in landfills, and waste-to-energy systems.

There are adaptation measures that also can be considered as mitigation (*Ad-Mit*) – for instance, white roofs and other reflective surfaces installed to primarily improve dwellers' thermal comfort but as a co-benefit mitigate global warming by reducing the solar energy (radiative forcing W/m^2 over a period) absorbed by the surface. Other measures can be considered as primarily for mitigation, with adaptation co-benefits (*Mit-Ad*) – for example, carbon sequestration by trees reducing the UHI effect; passive or zero-energy building designs that simultaneously save energy and improve comfort; and the use of improved fuel wood cook stoves reducing the pressure for deforestation in ecologically sensitive areas.

4.3.3 Synergies, Conflicts, and Tradeoffs between Adaptation and Mitigation across Urban Sectors and Scales

Climate change adaptation and mitigation measures are interrelated – in some cases positively (synergies), in others negatively (conflicts) – and sometimes decisions on implementation are based on difficult tradeoffs, thus necessitating choices between conflicting policy and planning goals (Klein et al., 2007). For the purposes of this chapter, a *synergy* is understood as an interaction between an adaptation and a mitigation plan, policy, strategy, or practical measure that produces an effect greater than the constituent components. A *conflict* is a plan, policy, strategy, or practical measure that counteracts or undermines one or more planning goals between adaptation and mitigation. Finally, a *tradeoff* is a situation that necessitates choosing (balancing) between one or more desirable, but sometimes conflicting, plans, policies, or strategies. Table 4.3 presents a list of real examples of cases where adaptation and mitigation integration happened (of which further details are provided in Case Studies 4.1–4.6).

Key issues are structured around four basic questions:
1. What types of interrelationships can be identified (synergies, conflicts, tradeoffs)?
2. Where do the interrelationships originate (drivers)?
3. Across which sectors do they typically cut (cross-sectoral interactions)?
4. Where can examples of integrated implementation be found (geographical location)?

Synergies between adaptation and mitigation can be found in the building sector (Barbhuiya et al., 2013; Gupta and Gregg, 2013; McEvoy et al., 2006) (see Chapter 5, Urban Planning and Design). In order to increase the indoor climate comfort of buildings while simultaneously reducing energy use, passive designs, changed behavioral measures, and more advanced technologies can be utilized for heating, cooling, and ventilation (Thornbush et al., 2013; Gupta and Gregg, 2013; Lucon et al., 2014). Adequate orientation and morphology of buildings and streets also target adaptation and mitigation in an integrated manner (Barbhuiya et al., 2013; Mills et al., 2010). House insulation and introduction of solar collectors for heating water increase energy efficiency and reduce related CO_2 emissions while at the same time increasing resilience to temperature changes (Barbhuiya et al., 2013). Furthermore, while ensuring energy efficiency of buildings by selection of materials and location of buildings, at the same time buildings should be resistant to heat waves, floods, and humidity.

Europe has faced exceptionally warm summers, such as the record years 2003 and 2010 have shown, and the probability of a summer experiencing mega heat waves is considered to increase by a factor of 5–10 within the next forty years (Barriopedro et al., 2011). Especially in Nordic countries such as Finland, people are not used to high temperatures, and sick and elderly people are highly vulnerable to heat waves (Hassi and Rytkönen, 2005) (see Chapter 10, Urban Health). Therefore, more frequent periods of heat will increase demand for cooling. District cooling is a prominent example where Ad-Mit synergy can be found: air-conditioning systems with high emissions can be replaced by district cooling using energy that would be otherwise wasted (Riipinen, 2013). The district heating and cooling system (DHCS) of the energy company Helen Oy in Helsinki represents state-of-the-art technology that contributes to both climate change mitigation and adaptation (see Figure 4.2). It has increased the energy efficiency of buildings significantly, improving air quality in Helsinki and simultaneously providing an energy-efficient adaptation tool to avoid conventional air-conditioning in summer time. District heating and cooling can also be a major resilience

Table 4.3 *Interrelationships between adaptation and mitigation: Examples of sectors and practice measures in selected cities*

Sectors and measures	Interrelationship type	Examples of benefits	Examples of challenges	Climate policy	City	Country	Case study number
Urban forestry: Reforestation	Synergy	Carbon sequestration, flood protection, biodiversity	Space requirements	Adaptation	Durban	South Africa	4.1
Water: Open storm water systems via urban wetlands	Synergy	Flood protection, carbon sequestration, biodiversity, recreation	Space requirements	Adaptation	Colombo	Sri Lanka	4.2
Urban structure and design: Compact urban design	Conflict/tradeoff	Less carbon emissions	Restriction of green structures to mitigate heat island effect	Mitigation	Jena	Germany	4.3
Implementation of measures across multiple sectors	Synergy	High level of stakeholder engagement: residents, businesses, and community representatives	Fiscal and jurisdictional challenges	Adaptation/ Mitigation	Chula Vista, California	United States	4.4
Implementation of measures across multiple sectors	Diverse types	Integrated climate action planning	Monitoring and evaluation in an integrated manner	Adaptation/ Mitigation	Quito	Ecuador	4.5
Implementation of measures across multiple sectors	Diverse types	Integrated climate action planning	Implementation responsibilities, economic feasibility of actions	Adaptation/ Mitigation	Mexico City	Mexico	4.6

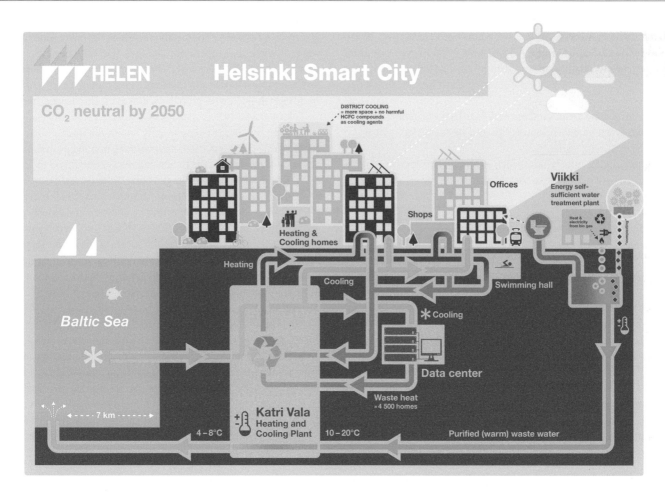

Figure 4.2 *Illustration of the district heating and cooling system in Helsinki.*

Source: Helen Oy; copyright Kirmo Kivelä

investment by reducing the risk and impact of power outages. Although the DHC system in Helsinki is still partially based on fossil fuels, energy savings by using combined heat and power production is equivalent to the consumption of 500,000 detached homes with conventional systems (Riipinen, 2013).

Synergies between adaptation and mitigation in the energy sector can also be found in decentralized renewable generation connected through smart grids (see Chapter 12, Urban Energy). Such options reduce GHG emissions and, at the same time, reduce risks of power shortages due to peak loads or supply disruptions under temperature extremes or storm-related power losses (Hamin and Gurran, 2009; Grafakos and Flamos, 2015). Furthermore, smart grids allow a large number of distributed energy generators to feed into the grid and thus improve system reliability in response to impacts of climate change on individual elements of energy production, transfer, and distribution. Small hydropower plants are based on renewable sources (hence mitigating GHG emissions) but require adequate design and operation in areas where scarce water supplies can reduce the adaptive capacity of ecosystems (Sugar et al., 2013). In the case of technical measures for adaptation such as water pumping and desalination, an option to minimize the conflict between mitigation and adaptation is to use renewable energy sources such as photovoltaic or wind power generating systems.

Urban greening is also a synergistic mitigation-adaptation measure (see Chapter 14, Urban Water Systems, and Chapter 8, Urban Ecosystems). The main benefits of urban greening are the capacity to absorb and store water, cool surrounding areas, improve biodiversity, and sequester carbon through wider subregional regeneration (Kithiia and Lyth, 2011; Newman, 2010; Piper and Wilson, 2009; Rankovic et al., 2012). An example of reforestation for carbon storage is presented from Durban, South Africa (see Case Study 4.1).

In Sri Lanka, a plan for recovering wetlands with replanted native trees aims to provide multiple benefits to the environment such as protecting biodiversity, providing flood protection for buildings and road infrastructure, increasing security of the population, and increasing carbon sequestration capacity (see Case Study 4.2).

Green roofs, roof gardens, and green walls for buildings help to mitigate climate change by providing carbon sinks, reducing albedo, regulating indoor temperatures while consuming less energy, improving water management, enhancing local biodiversity and landscapes, and even making urban agriculture possible (Williams et al., 2010; Lehman, 2015; Prochazka et al., 2015). These options, however, require adequate support and proper maintenance to avoid leakages and mold and to secure water

Case Study 4.1   Synergies, Conflicts, and Tradeoffs between Mitigation and Adaptation in Durban, South Africa

Sean O'Donoghue and Debra Roberts

eThekwini Municipality, Durban

Keywords	Renewable electricity, feed-in tariff, reverse auction, mitigation
Population (Metropolitan Region)	3,000,000 (eThekwini Municipality, 2015)
Area (Metropolitan Region)	2,297 km² (eThekwini Municipality, 2015)
Income per capita	US$12,860 (World Bank, 2017)
Climate zone	Cfa – Temperate, without dry season, hot summer (Peel et al., 2007)

DURBAN CASE STUDY

In the case of Durban, the synergies, conflicts, and tradeoffs between mitigation and adaptation action need to be understood within the context of a large, local development deficit. Durban has high levels of unemployment and poverty, high crime rates, substantial infrastructural backlogs, and high rates of HIV infection. These immediate needs compete with an issue like climate change for political attention and resources, so any climate protection action must have development co-benefits.

The city authority responsible for Durban is eThekwini Municipality. Unusually, this Municipality has prioritized adaptation in responding to the climate change challenge. This approach is supportive of the broader African agenda as relates to climate change. Hosting the United Nations Framework Convention on Climate Change (UNFCCC)'s 17th Conference of the Parties (COP17) negotiations in 2011 gave Durban an opportunity to advocate for climate change adaptation action at the city level, raise climate change awareness within both the Municipality and South Africa, develop partnerships and networks, and catalyze the implementation of adaptation and mitigation projects as part of the event greening program. This program aimed to reduce the ecological impact associated with hosting COP17. Core focus areas included carbon neutrality, resource and energy efficiency, ecological footprinting, and the production of an event greening set of guidelines with an awareness campaign focused on responsible accommodation and tourism.

EThekwini Municipality put out a public call for potential carbon offset options for COP17, but, of the five submissions received, none was regarded as suitable. The evaluation process used the UN Development Programme's Millennium Development Goals Carbon Safeguard Principles to assess the environmental and social sustainability of the proposed offset projects and the sustainability track record of the organizations involved. The negative outcome highlighted that carbon offsetting potential and sustainability are not necessarily synonymous and that the full range of benefits and disadvantages of any offset option must be carefully reviewed.

In the absence of suitable externally sourced offset options, the Municipality expanded the carbon sequestration work undertaken for the Durban 2010 FIFA™ World Cup through the initiation of a community-based reforestation project located adjacent to a local nature reserve. Over and above carbon sequestration, the project also helps improve the ecological health of the nearby river catchment, an important watershed with high levels of urban development and many poor communities, all of whom rely to some extent on the ecosystem services delivered within the catchment. Project implementation has occurred in partnership with local communities, nongovernmental organizations, the private sector, and provincial government and has employed 118 residents in tree planting (615,845 trees planted in 489 ha), invasive alien plant clearing (1.185 ha), ecosystem restoration, and fire protection, and as community facilitators supporting 495 "Treepreneurs" (i.e., indigent community members who source locally indigenous seeds and grow them into seedlings that are traded for critically needed supplies such as food, clothing, and building materials).

The project is also significant in that it resulted in the development of the Community-Ecosystem-Based Adaptation (CEBA) concept. This looks to eventually expand the original reforestation approach to embrace a more complete understanding of the link between communities and the ecosystems that underwrite their welfare and livelihoods by creating cleaner and greener neighborhoods that are less dependent on costly utilities and services (e.g., through water recycling and the use of renewable energy). The CEBA approach is now being implemented throughout the province of KwaZulu-Natal. There is a clear synergy among adaptation, mitigation, and the development needs of Durban's residents, and this is being used

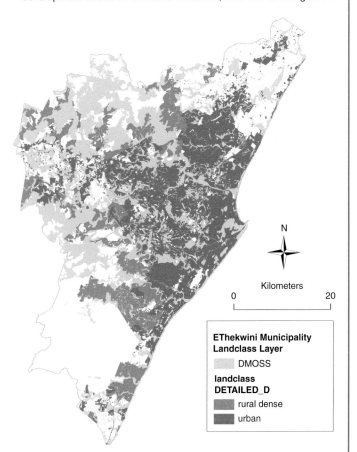

N

Kilometers
0   20

EThekwini Municipality Landclass Layer
 DMOSS

landclass DETAILED_D
 rural dense
 urban

Case Study 4.1 Figure 1 *EThekwini Municipality (29°51′31″ S; 31°01′19″ E) with urban/peri-urban areas and the Durban Metropolitan Open Space System (DMOSS).*

Source: Roberts and O'Donoghue, 2013

to define a new development paradigm, where natural infrastructure is used to generate multiple developmental benefits for local residents.

In Africa, the current rapid rate of urbanization means that there is an urgent need to ensure that an appropriate and integrated adaptation and mitigation framework is put in place to ensure that African cities take advantage of the opportunity to leapfrog the carbon-intensive and ecologically destructive development path of the past and that this framework engages in an appropriate way with the challenge of high levels of informality and still-evolving governance structures. This requires that the climate question is asked in a different way in Africa – that is, how does a low carbon development pathway create increased adaptive capacity, and how can that adaptive capacity be used to meet growing development needs in a more sustainable and resilient way? To enable this, innovative finance mechanisms that directly support city-level action are essential following COP21 in Paris, 2015.

Case Study 4.2   Pilot Application of Sustainability Benefits Assessment Methodology in Colombo Metropolitan Area, Sri Lanka

Stelios Grafakos, Somesh Sharma, and Alberto Gianoli

Institute for Housing and Urban Development Studies (IHS), Erasmus University, Rotterdam

Monali Ranade and Sarah Mills-Knap

The World Bank Group, Washington, D.C.

Keywords	Integrated assessment, sustainability benefits, adaptation–mitigation interrelationships, ecosystem based adaptation, floods
Population (Metropolitan Region)	2,195,000 (Demographia, 2016)
Area (Metropolitan Region)	3,684 km² (SLRCS, 2015)
Income per capita	US$11,970 (World Bank, 2017)
Climate zone	Af – Tropical, rainforest (Peel et al., 2007)

This case presents the green growth project Bedaggana Wetland Development in Colombo Metropolitan Area, Sri Lanka, and the pilot application of an integrated Sustainability Benefits Assessment (SBA) methodology developed by the Institute for Housing and Urban Development Studies (IHS) in cooperation with the World Bank for capturing and quantifying its sustainability benefits including those related to climate change mitigation and adaptation.

This project is part of the integrated improvement, management and maintenance of flood-detention areas around Parliament Lake, funded by The World Bank 2017. Part of the wetland will be replanted with native tree species. This will enhance the natural environment in the area and protect the rich biodiversity of the wetland, increase the flood-retention capacity, and facilitate climate change mitigation through carbon sequestration.

The SBA methodology is a combination of top-down and bottom-up approaches including geographical information system (GIS)-based scenario building (Fraser et al., 2006; Graymore et al., 2009). This helps in the process of identifying specific sustainability benefits of different types (i.e., social, environmental, and economic) including

mitigation and adaptation benefits that can be accrued at different levels (i.e., individual, local community, and global community) from various green growth projects. The methodology creates map-based scenarios that enable better identification and overall visualization of sustainability benefits. The *ex-ante* methodology consists of five steps, which are explained below as they apply to the Bedaggana Wetland Development project (see Case Study 4.2 Figure 1).

STEP 1: CREATE AN EXISTING SITUATION SCENARIO (BASELINE)

The first step of the SBA consists of creating the existing situation scenario (or baseline) that provides information on the current state of sustainability of the project area. A GIS is used to create and manage a multidimensional, multipurpose, and multitemporal database in a common frame of reference. A geospatial database was created for the entire Metro Colombo Area (see Case Study 4.2 Figure 2).

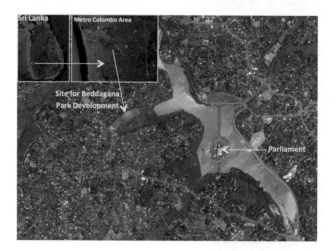

Case Study 4.2 Figure 1 *Location of Bedagana Wetland Development Project*

STEP 2: DEVELOP A "WITH-PROJECT" SCENARIO

The scenario "With-project" characterizes the incremental net changes brought about by the implementation of the project. It is based on a GIS map and provides a summary of the project's main objectives, activities, and expected outcomes in different sustainability benefit categories (see Case Study 4.2 Figure 3).

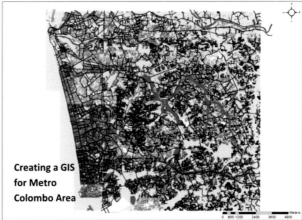

Case Study 4.2 Figure 2 *Geographical Information System (GIS) for Metro Colombo Area*

STEP 3: IDENTIFY OF PRIMARY PHYSICAL IMPACTS

At this stage, various expected sustainability benefits are associated with the actual physical changes (primary impacts) that the project will bring after implementation. The project scenario is compared with the existing situation scenario in terms of physical characteristics. In the case of the Bedaggana Biodiversity Park project, the analysis suggested that the development of the area will mainly bring the following three primary physical changes in the baseline:

- Increase of wetland area
- Increase in canopy cover
- Development of recreational area

As an illustration of one of the primary impacts, the map shows the development of a "pocket" of urban forest in the Biodiversity Park. The three changes in the baseline will lead to several sustainability benefits (see Case Study 4.2 Figure 4).

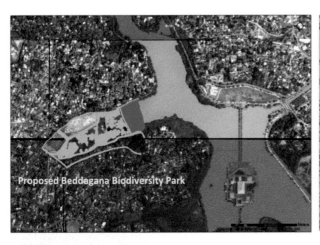

Case Study 4.2 Figure 3 *Mapping the Project Scenario*

Case Study 4.2 Figure 4 *Canopy cover in the Beddagana Biodiversity Park area*

STEP 4: MEASURE THE SUSTAINABILITY BENEFITS FOR MITIGATION AND ADAPTATION

At this stage, the benefits are categorized by type (social, environmental, economic) and level (individual, local community, global community) of benefit. The sustainability benefits expected from the Beddagana Biodiversity Park development project are:

- *Adaptation benefits* associated with increase in flood retention capacity. In order to analyze the effect of flood risk reduction in the area around Beddagana Biodiversity Park, a hydrological analysis for the area was carried out. When the building footprints and road infrastructure layers are overlaid, it can be estimated how many built assets and how much road length can benefit from flood prevention due to the implementation of the project.
- *Mitigation benefits* associated with increase in canopy cover. It was assumed that about 95 tons of carbon would be stored

per hectare of tree cover in the Bedaggana Biodiversity Park. In the existing situation scenario, total tree cover in the Beddagana wetland site is about 3 hectares, which will increase to 11 hectares after the trees are fully grown if planted according to the proposed design of the park. This will increase carbon storage capacity of the site from 285 tons per annum to around 1,045 tons per annum.

STEP 5: DOCUMENT ADAPTATION-MITIGATION BENEFITS IN THE SUSTAINABILITY BENEFITS MATRIX

Final step of the SBA methodology is to document all the sustainability benefits into a two-dimensional matrix, the Sustainability Benefits Matrix. The sustainability benefits are classified according to their level and type. Social benefits associated with the development of a recreational are also measured.

---

requirements (Laukkonen et al., 2009; Thornbush et al., 2013; Hodo-Abalo et al., 2012). Various attempts to quantify the costs and multiple (including adaptation and mitigation) benefits of green roofs have been made (see, e.g., Blackhurst et al., 2010; Bianchini and Hewage, 2012). There are also urban governments (e.g., Chicago, Montréal, and Portland, Oregon) that have commissioned or conducted studies to quantify the benefits of green roofs in their cities (City of Portland, 2008; City of Chicago, 2011; Gariepy, 2015). The Chicago Green Roof Initiative in 2011 resulted in about 5,470,000 square feet of green roofs that reduce the output of approximately 21 metric tons of carbon each year and absorb approximately 124 million gallons of storm water per year.

Tradeoffs between adaptation and mitigation often appear in situations where decisions have to be made on "hard" versus "soft" engineering and planning solutions, as well as in situations where the temporal scale of implementation sets limitations or uncertainties regarding planning horizons, availability of resources such as financing and staff, overall limits of authority, availability of expertise and data, and availability of physical space to implement integrated solutions (Jordan, 2009; Juhola et al., 2013; Dymén and Langlais, 2013).

Conflicts between adaptation and mitigation are often spatial in nature given that many of the adaptation measures (such as water management practices using urban forestry and urban greening) require significant land area in order to be effective. Poorly planned, such efforts may undermine urban densification efforts that are key to reducing transportation and energy demands (Dymén and Langlais, 2013; Viguié and Hallegatte, 2012; Hamin and Gurran, 2009). Expanding urban green space can increase emissions from transportation due to longer commuting needs – an example of adaptation that negatively affects mitigation.

Furthermore, some water sector adaptation measures potentially conflict with mitigation measures because they have high energy demand – such as desalination to tackle water scarcity and water pumping to reduce flooding (Cooley et al., 2012; Cook et al., 2012) (see Chapter 14, Urban Water Systems). Sometimes these measures can be implemented at the micro-scale level, with marginal effects on mitigation. In some cases, the conflict

can be minimized or eliminated by supplying renewable energy (e.g., based on photovoltaics) to meet the energy needs of water treatment and supply, as was the case in Amman, Jordan (Al-Karaghouli et al., 2011; Sugar et al., 2013). However, cases like the Chinese interbasin transfer project present significant energy requirements, as well as posing considerable risks of water shortages upstream (see Box 4.3).

The Chinese central government aims, through the Special Plan for Seawater Utilization, to produce 3 million tons (807 million gallons) of purified seawater a day by 2020, approximately four times the country's current capacity. At least 400 of the largest Chinese cities already suffer from water scarcity. According to Cooley et al. (2012), 12,000–18,000 kilowatt hours will be needed to desalinate a million gallons of seawater, whereas pumping groundwater to the surface requires less than 4,000 kilowatt hours per million gallons.

Additional examples of *mitigation-adaptation interrelationships* can be found in urban design and densification policies (see Chapter 5, Urban Planning and Design). On the one hand, urban densification maximizes agglomeration economies through more efficient resource use, waste management reduction, reductions in urban sprawl, and a lower reliance on motorized transport (Hickman et al., 2013). On the other hand, increased urban density may affect food belts, riparian areas, and wetlands that protect cities from floods. It can increase heat islands, for instance, by blocking free air flow, which may lead to pollution, discomfort, and health problems and to an increased need for conventional cooling (Hamin and Gurran, 2009; Laukkonen et al., 2009) (see Chapter 2, Urban Climate Science). Air conditioning that uses conventional fossil fuel electricity to provide cooling services conflicts with mitigation efforts (Dymén and Langlais, 2013; Sugar et al., 2013).

Furthermore, urban planning practices that support urban densification often result in the loss of permeable surfaces and tree cover, increased risk of flooding, and water pollution (Mees and Driessen, 2011). Densification of urban areas tends to reduce natural drainage possibilities, thereby making it more expensive and difficult to implement adaptation measures. Diminishing natural urban

Box 4.3   China's Eastern Route Project: A Challenge for South–North Water Diversion

China's Eastern Route Project (ERP) is a project aimed at supplying water to the northeastern part of the country by fossil-fuel based pumpings; Wang et al., 2006). A third of Beijing's annual demand is to be supplied by a new water-course pumping from the Danjiangkou Dam in the central province of Hubei to the capital. The more than US$62 billion ERP is part of a larger water diversion project: the second phase of which is known as the South-North Water Diversion Project. It is designed to supply the dry, urbanized, and farm-ing-intensive north of China. Initially conceived in 1952, the

ERP was commissioned in 1972 after a prolonged drought in Northern China; the first phase was commissioned in 2002. According to official sources, the 1,156-kilometer ERP will divert to the north 14.8 billion cubic meters of water from the Changjiang River flow of more than 600 billion cubic meters per year. In the first of thirteen engineering stages, fifty-one pumping stations with an installed capacity of 529 megawatts will be built. Waste water plants will also be built (Chinese Government, 2012).

drainage possibilities also affect the safety and security of urban dwellers if floods occur, especially if vulnerable urban dwellers cannot be relocated to flood-secure areas (cf. Sugar et al., 2013). Health impacts can also be expected if urban recreational possibilities are reduced (van Dillen et al., 2012). Biodiversity losses in urban areas can also result due to smaller and fewer green (and blue) spaces (Fontana et al., 2011). The use of biofuels by urban dwellers may have positive effects for mitigation due to decreased use of fossil fuels and consequently reduced carbon emissions, but this creates potential conflicts with broader-scale adaptation because biofuels production requires land; this competes with agricultural use and therefore affects food security (Smith et al., 2014).

Some of the conflicts stemming from adaptation measures (such as urban greening that requires considerable space) can be diminished and synergies enhanced by using multimodal and public transportation, hybrid vehicles and other cleaner

technologies, and planned transportation routes along green and blue areas (Saavedra and Budd, 2009; Hamin and Gurran, 2009). Adaptation measures can enhance mitigation when population resettlement is connected to the restoration of degraded areas. The São Paulo Case Study shows how adaptation measures (relocation and protection from flooding in landslide-prone areas) were combined with mitigation strategies (recovery of forests, storage of carbon in vegetation and soils) to address existing conflicts created by uncontrolled urbanization (see Case Study 4.B in Annex 5). Jakarta provides a similar successful example of slum upgrading and relocation of a vulnerable community to safe areas with energy-efficient homes connected to public transit and community-based electricity generation (Sugar et al., 2013).

Finally, there are interesting examples of integrated Mit-Ad/Ad-Mit strategies from middle-sized cities in Europe and the United States (see Case Studies 4.3 and 4.4).

Case Study 4.3   Jena, Germany Adaptation Strategy as an Essential Supplement to Climate Change Mitigation Efforts

Oliver Gebhardt

Helmholtz Centre for Environmental Research – UFZ, Leipzig

Keywords	Adaptation, mitigation, climate-proof urban planning, decision support, multicriteria analysis, heat stress, flood
Population (Metropolitan Region)	457,578 (Statistical Office State of Thuringia, 2015a)
Area (Metropolitan Region)	1,271 km² (Statistical Office State of Thuringia, 2015b)
Income per capita	US$49,530 (World Bank, 2017)
Climate zone	Cfb – Temperate, without dry season, warm summer (Peel et al., 2007)

Jena is a prosperous city of about 100,000 inhabitants located 250 kilometers southwest of Berlin, Germany in the hilly landscape of the Saale River valley. Since the late 19th century, owing to the activities of the entrepreneur Carl Zeiss, the city has become the center of the optical industry, which is known worldwide. The strong urban economy and large science and technology sector form the basis of the population's high standard of living. Due to its specific location, the city is exposed to various climate-related threats, especially heat stress, fluvial floods and pluvial downpours.

The city center is surrounded by steep shell limestone slopes, which operate as a thermal storage system, making Jena one of the warmest places in Central Germany. Based on current climate projections, an increase of heat stress events is expected. By the end of the century, the average maximum temperature in the summer will increase by 3°C (CMIP5, RCP 4.5) to 6°C (CMIP5, RCP 8.5). Accordingly, the number of hot days, i.e., days with a maximum temperature of at least 30°C, will rise from 11 in 1981–2010 to 35 (CMIP5, RCP 4.5), or 49°C (CMIP5, RCP 8.5).

Numerous tributaries flow from the surrounding plateau and discharge in the floodplain of the Saale River, which crosses the city

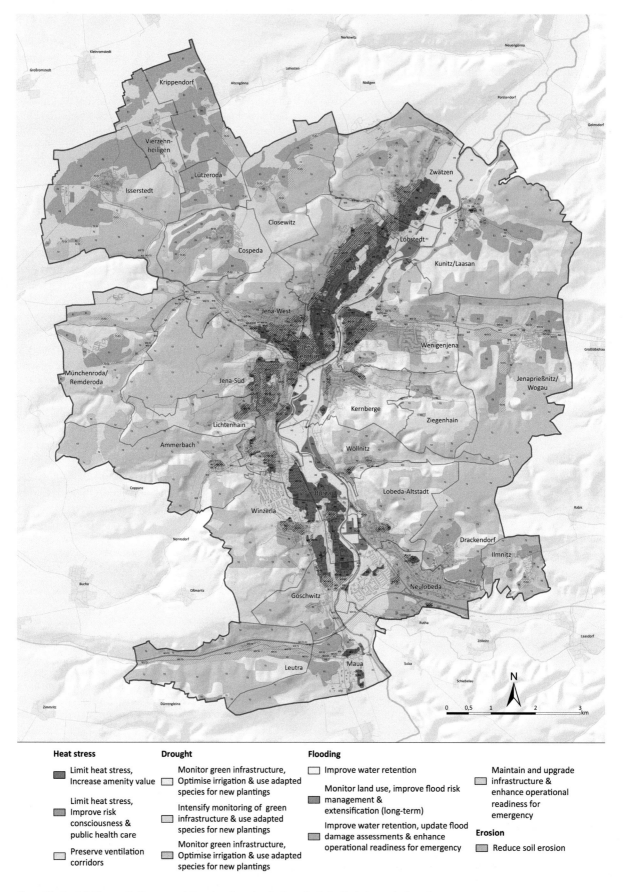

Heat stress

■ Limit heat stress, Increase amenity value

▨ Limit heat stress, Improve risk consciousness & public health care

□ Preserve ventilation corridors

Drought

□ Monitor green infrastructure, Optimise irrigation & use adapted species for new plantings

□ Intensify monitoring of green infrastructure & use adapted species for new plantings

□ Monitor green infrastructure, Optimise irrigation & use adapted species for new plantings

Flooding

□ Improve water retention

▨ Monitor land use, improve flood risk management & extensification (long-term)

▨ Improve water retention, update flood damage assessments & enhance operational readiness for emergency

□ Maintain and upgrade infrastructure & enhance operational readiness for emergency

Erosion

▨ Reduce soil erosion

Case Study 4.3 Figure 1 *Recommendations for urban planning in particularly affected areas in Jena, Germany.*

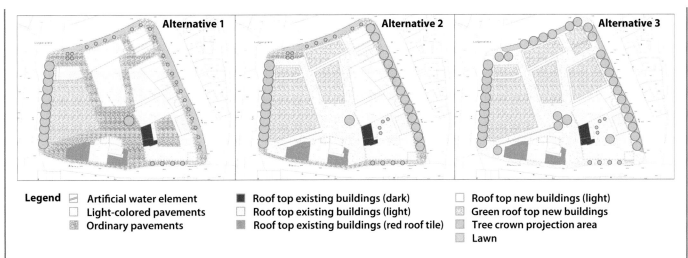

Legend
- ⊠ Artificial water element
- ☐ Light-colored pavements
- ▨ Ordinary pavements
- ■ Roof top existing buildings (dark)
- ☐ Roof top existing buildings (light)
- ▨ Roof top existing buildings (red roof tile)
- ☐ Roof top new buildings (light)
- ▨ Green roof top new buildings
- ☐ Tree crown projection area
- ☐ Lawn

Case Study 4.3 Figure 2 *Alternative project designs of an urban square in Jena used as a basis for an adaptation check.*

center and industrial areas. Heavy or long-lasting precipitation events repeatedly cause major floods. Experience from the recent past and modeling results from other German river basins suggest an increase of peak discharges especially for flood events occurring with a medium-to-high probability.

Given this situation, in 2005, urban planners and scientists started discussing how these risks might change over time and how related impacts could be managed. In 2009, the Department of Urban Development & City Planning (DUDCP) commissioned and financed a pilot study to analyze local climate change impacts, identify potential adaptation measures, and better understand the risk perceptions of relevant stakeholder groups. On the basis of the results of this study the decision was taken to develop a local climate change adaptation strategy *Jenaer KlimaAnpassungsStrategie (JenKAS)*. The development was initiated as well as steered by DUDCP and financially supported by the federal government of Germany. It involved experts from all relevant departments of the city administration and agencies of the federal state of Thuringia, interested stakeholder groups (e.g., associations and cooperatives), scientists, and politicians.

JenKAS was formally adopted by the City Council in May 2013 and consists of various elements. Its backbone is a handbook on climate-sensitive urban planning (City of Jena 2013), which includes information on current and future climatic conditions and their potential local impacts, information on legal aspects of climate change adaptation, economic assessments of adaptation options, and best practice examples of successful climate change adaptation in Jena and elsewhere. For each city district, impacts are described in detail, and related risks are visualized using a traffic-light labeling system. Recommendations for urban planning in particularly affected areas are presented (see Case Study 4.3 Figure 1).

The handbook is complemented by the decision support system *Jenaer Entscheidungsunterstützung für lokale Klimawandelanpassung (JELKA)*. This tool was developed to make climate risk information more accessible and to provide tailor-made recommendations for different target groups (e.g., suitable adaptation measures for a specific field of action or spatial unit).

In Jena, adaptation is understood not as a substitute but as an essential supplement to climate change mitigation efforts. Since the turn of the century, mitigation has been on the municipal political agenda as an important aspect of the city's sustainability goals. In 2004, the advisory board of the Local Agenda 21 started to develop an urban climate change mitigation concept, which was officially approved in

2007. The implementation of the concept is monitored, revised, and extended on a regular basis. The increasing demand for reducing carbon emissions triggered the development of an integrated mitigation strategy, which was completed and approved in 2015. On the basis of the manifold activities stimulated by the mitigation agenda, Jena was the first German city to be awarded the European Energy Award Gold. Until 2014, only two other German municipalities received this prestigious award. The achievements of the past years include such diverse activities as the use of 100% renewable energy in all public buildings, the development of a new urban cycling and public transport concept, numerous energy-saving activities, the introduction of car-sharing incentives, and the establishment of a façade greening award. The city's biannual energy action plan comprised more than fifty measures for the years 2014 and 2015.

Because there are not only synergies but also potential conflicts between adaptation and mitigation measures, special interest has been paid to take these contradictory effects into consideration when developing recommendations for improving climate resilience. Urban planners in Jena are guided by the urban design concept of the compact city. On the one hand, limiting outward urban expansion and promoting dense urban structures by efficiently using land resources improves energy efficiency, but, conversely, this density is likely to restrict the establishment of green structures to mitigate urban heat island effects. In Jena, this potential conflict was balanced by recommending that planners should retain the compact city as a guiding principle but also to preserve areas to allow ventilation of fresh and cold air to the inner-city residential and commercial areas. A map representing intranight airflows was made available to planners to inform them about these corridors.

The main focus for the implementation of JenKAS is on mainstreaming climate change adaptation into administrative decision-making. DUDCP promotes the consideration of adaptation-related aspects in these processes through various in-house activities, such as JELKA trainings. As a consequence of these efforts, a constantly growing number of land development plans refer to JenKAS when making recommendations or substantiating restrictions. It is expected that the results of ongoing and future research efforts (e.g., a highly awarded urban tree concept providing site-specific tree recommendations based on climatic, locational, aesthetic, and even historico-cultural considerations) will further promote this uptake. Beyond the actions directed at internal municipal processes, there are several activities addressing citizens and associations (e.g., a nature trail with display boards financed by local businesses that provide information about

117

important aspects of the changing urban climate as well as the city adaptation strategy).

One way of considering climate change in today's urban decision-making is to use adaptation checks when drafting plans for major (re)construction projects (see Case Study 4.3 Figure 2). Assisted by scientists from the Helmholtz Centre for Environmental Research – UFZ, probabilistic multicriteria analyses were conducted to facilitate the development of climate-proof detailed designs. It was intended that these drafts should not only suit current and future climate conditions but also take into account the manifold other factors (e.g., financial and aesthetic aspects) as well as stakeholder preferences affecting decision-making in urban planning. The results of these adaptation check rankings of the alternatives were calculated and rated their suitability from an adaptation perspective.

Due to the short period of time since the adoption of JenKAS, no systematic evaluation of its impacts has yet taken place, yet. However, several findings and recommendations for promoting urban climate change adaptation in Jena can still be presented:

- Potential conflicts of adaptation and mitigation efforts can be solved or at least limited by explicitly addressing these issues at an early stage of strategy and project development.
- Exchange between representatives of different administrative bodies, scientists, and consultants about adaptation activities at the various political scales and scientific progress in the field should be institutionalized and take place on a regular basis.
- Adaptation-related outreach activities of municipalities do not only raise awareness of the general public, but also improve civic and political support for adaptation action.
- Momentum created by the initial adoption of a local adaptation strategy should be maintained through projects that continuously update and expand the existing adaptation knowledge base.
- Trainings and hands-on workshops are essential to improve municipal staff's ability to use data and tools available for supporting adaptation.
- Public commitment of political decision-makers to support adaptation activities (e.g., the adoption of a local adaptation strategy by the city council) is pivotal for their success.

Case Study 4.4   Sustainable Win-Win: Decreasing Emissions and Vulnerabilities in Chula Vista, California

Oswaldo Lucon

São Paulo State Environment Secretariat, University of São Paulo

Keywords	Mitigation, adaptation, integrated plan
Population (Metropolitan Region)	265,757 (U.S. Census, 2015)
Area (Metropolitan Region)	128 km² (U.S. Census, 2010)
Income per capita	US$56,180 (World Bank, 2017)
Climate zone	BSk – Arid, steppe, cold (Peel et al., 2007)

The coastal city of Chula Vista (in Southern California, U.S., 32°37′40″N 117°2′53″W), although small (128 km², pop. 243,916 in 2010), is a global benchmark in terms of planning and implementing integrated climate change mitigation and adaptation strategies (United States Census, 2015). The impact of Chula Vista's climate policy is due to its replicability, resource efficiency, and focus. The relevance of Chula Vista resides in its comprehensiveness and level of implementation, driven by the vulnerabilities found in the Southern California Region, the level of awareness of the local community – a diverse population on a varied landscape, and a long history of progressive thinking on climate change (City of Chula Vista, 2011).

A Working Group comprising residents, businesses, and community representatives recommended eleven strategies to curb greenhouse gas (GHG) emissions and to adapt the community to key impacts within different sectors: energy and water supply, public health,

wildfires, ecosystem management, coastal infrastructure, and the local economy (City of Chula Vista, 2011).

In the 1990s, the city formed the Climate Change Working Group and created a baseline inventory of its GHG emissions. From this starting point, mitigation strategies covering transportation (such as a 100% clean fuel bus fleet), energy efficiency and solar retrofits (an appliance rebate program), and green buildings with smart growth strategies (a city-wide standard) were devised (Green, 2010).

The Working Group researched the best available data and engaged the community in the consensus-based decision-making process. A list of eleven strategies comprised mitigation-adaptation (Mit-Ad) and adaptation-mitigation (Ad-Mit) measures including (1) cool paving, (2) shade trees, (3) cool roofs, (4) local water supply and reuse, (5) storm water pollution prevention and reuse, (6) education and wildfires, (7) extreme heat plans, (8) open space management, (9) wetlands preservation, (10) sea level rise and land development codes, and (11) green economy (City of Chula Vista, 2011).

The process was conducted in three phases. The first was information gathering, including data collection via public forums. Twelve meetings featured presentations from Working Group members and regional experts on the different adaptation topics and discussed the current state of practices and predicted impacts to the San Diego region. Public notices were posted prior to meetings at various municipal locations. Additionally, more than sixty additional stakeholder groups and community members were invited for feedback. A newsletter helped to build public awareness about the climate planning process. In an open public forum, more than thirty community members shared their opinions and priorities for strategies. The group came up with 183 options based on the best available data, summarized into a planning matrix for each focus area (Green, 2010; City of Chula Vista, 2011).

Step two was evaluation. after risk levels were assigned to each identified vulnerability in consultation with researchers from the

Case Study 4.4 Figure 1 *Chula Vista, California.*

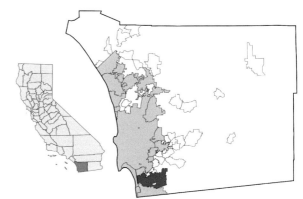

Case Study 4.4 Figure 2 *Chula Vista, California.*

Case Study 4.4 Table 1 *Chula Vista's Implementation stage compared to other cities in the San Diego Region. Source: Center for Sustainable Energy, 2017*

Local Climate Planning Efforts (2000–2015)

Jurisdiction	GHG Inventory	Climate Action Plan (CAP)	
		Adopted	Developing or Updating
Carlsbad	x	2015	
Chula Vista	x	2000, 2008	x
Coronado	x		
County of San Diego (unincorporated)	x	2012	x
Del Mar	x	2016	x
El Cajon	x		
Encinitas	x	2011	x
Escondido	x	2013	
Imperial Beach	x		x
La Mesa	x		x
Lemon Grove	x		x
National City	x	2011	
Oceanside	x		x
Port of San Diego	x	2013	
Poway	x		
San Diego	x	2005, 2015	
San Diego County Water Authority	x	2015	
San Marcos	x	2013	
Santee	x		
Solana Beach	x		x
Vista	x	2012	

University of California San Diego. Risk was defined as a product of the likelihood of the climate change impact occurring and the consequence of the impact. Each factor was scored from 1 to 5, and overall risk was categorized as "Low," "Medium," or "High." The Working Group also consulted the Resource Conservation Commission, the Environmental Health Coalition, and the San Diego Coastkeeper (City of Chula Vista, 2011).

Finally, for each vulnerability, a priority was assigned according to criteria such as jurisdiction, fiscal feasibility, and complementarity. These criteria relate to (1) a strategy falling within the city's jurisdiction, (2) a strategy being fiscally feasible (not relying on General Fund support for implementation), and (3) a strategy not duplicating or contradicting current climate mitigation measures, hence building on existing municipal efforts rather than creating new stand-alone policies or programs. No-regret actions or actions having multiple co-benefits were regarded as high priority. Specific implementation components were outlined, as well as critical steps, costs, and timelines. Implementation of all strategies were projected to cost approximately US$554,000 over the course of three years, plus US$337,000 annually for ongoing activities (City of Chula Vista, 2011).

Lessons learned included engaging stakeholders, stressing preparedness instead of resilience, lowering risks, avoiding analysis paralysis, focusing on areas where the city could actually have influence, and integrating action plans and programs.

4.3.4 Opportunities and Challenges

Opportunities and challenges for integrating mitigation and adaptation measures arise at all stages of planning, from initial assessments to implementation, monitoring, and evaluation. The goal is decision-making for integrated climate change management in cities. A city's capacity to undertake integrated actions for climate change mitigation and adaptation is determined by structural conditions that can either provide the necessary opportunities or, on the contrary, impede and hinder integrated climate change action. Resources and technical means are at cities' disposal to overcome these barriers and better manage climate change challenges in an integrated manner.

4.3.4.1 Structural Conditions

Structural conditions define the current context and boundaries of a city's operating system. Structural conditions are comprised of the environmental and physical setting, institutions and governance, economic and financial conditions, and sociocultural characteristics of a city (see Figure 4.3a). Structural conditions are difficult to change in the short run and often require coherent, continuous, and persistent action in order to influence them. Structural conditions to a large extent determine the level of a city's vulnerability and GHG emissions, but also its capacity to adapt to climate change impacts and reduce GHG emissions.

Environmental and physical setting: This refers to the main physical limits (e.g., all types of land uses, availability of freshwater), local conditions (e.g., urban traffic patterns and distribution, buildings characteristics, land-use zoning, hotspots such as UHIs), and infrastructure systems (the long-term, fixed nature of infrastructure creates path dependencies, thus diminishing or

enhancing a city's ability to adapt and reduce GHG emissions in the short term). For example, coastal cities are threatened by sea level rise and storm surges and therefore require investments for flood defense measures.

Institutions and governance setting: This entails existing policies and institutions including current plans, standards, and regulatory frameworks that could determine opportunities or constraints for integrated climate actions as well as interactions between different levels of jurisdictions. For example, urban areas may find it hard to adopt an integrated climate change action plan for the entire metropolitan region because of inter-jurisdictional conflicts or regulatory systems that contradict each other (UN-Habitat, 2015). Overlaps between different policy instruments and ineffective coordination of programs within and between municipal departments as well as among multiple levels of government (national, provincial, and municipal) can limit integrated approaches to climate change planning and management (Burch, 2010; Moser, 2012).

Economic development and municipal finances: These are also important structural conditions that determine a city's capacity to adapt and mitigate. Cities with advanced economic development and diversification can have high adaptive and mitigation capacity and thus the ability to develop and implement efficient low-carbon technologies that also increase climate resilience (Bizikova et al., 2008). Economic development and wealth enhance adaptation and mitigation capacity (Bergquist et al., 2012). The feasibility of implementing instruments for climate change mitigation is highly dependent on a city's financial and governance capability (Seto and Dhakal, 2014). In addition, long-term sustainable growth requires long-term budgetary equilibrium (Georgeson et al., 2016).

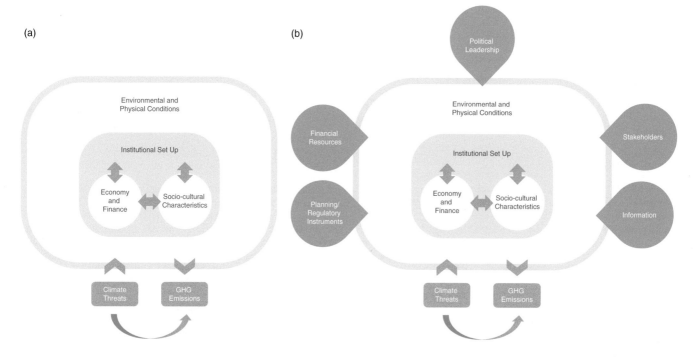

(a) (b)

Figure 4.3a *Structural conditions that determine a city's capacity to adapt and mitigate.*

Figure 4.3b *Resources and technical means for cities to overcome barriers for integrated climate change response due to specific structural conditions.*

Sociocultural characteristics: These include cultural values and worldviews that play an important role in how climate-related risks are perceived among individuals and organizations, and how policies and practices to respond to climate impacts and risks could or should be implemented. They may influence the acceptance of low-carbon and risk-reduction solutions or lead to a misperception of impacts and their causes, consequently affecting preferences for responsibility and behavioral patterns (Shove, 2010; Greenham et al., 2012). There is also a challenge to change older perspectives that view mitigation and adaptation as conflicting alternatives rather than complementary ones, a view that often leads to the perception that the implementation of adaptation policies would imply abandonment of mitigation policies (Moser, 2012). Cultural values and worldviews may also affect perceptions of equity and justice (Creutzig et al., 2014) (see Chapter 6, Equity and Environmental Justice).

4.3.4.2 Resources and Technical Means

Certain resources and technical means can be used to overcome barriers to integrated responses that might exist or barriers that can appear due to deficits of specific structural conditions or other constraints (see Figure 4.3b). Key means and resources are stakeholder engagement and participation in the planning and decision-making process and information in all dimensions and forms (such as awareness-raising campaigns and education). In addition, financial resources and mechanisms at all stages of policy development, project initiation, and implementation, along with planning and regulatory instruments, are parts of the capacity needed in cities. Political leadership is a vital factor that can often drive climate policy and determine its successful implementation (Lesueur et al., 2015; Burch, 2010; Johnson and Breil, 2012) (see Chapter 16, Governance and Policy)

A wide range of urban actors (government, practitioners, public and private companies, the scientific community, and stakeholders from civil society such as boundary organizations) are needed for effective planning and implementation and broad outreach during the preparation and execution of policies and actions. Transparency helps to build mutual trust, avoiding unequal distribution of information, and combatting corruption or other types of harmful influences from certain political pressure groups (Gavin, 2010). Evidence suggests the increasing role of partnerships (public–private and private–private) and nongovernmental actors in areas traditionally governed by municipal agencies (Broto and Bulkeley, 2013).

A valuable resource linked with stakeholder participation is the availability and provision of information (including decision support systems). Enhancing social awareness of and preparedness for climate change (notably climate-induced risks and disasters) is a major goal for communicating information necessary for climate action. Collecting and accessing data for climate change planning purposes – such as vulnerability assessments and GHG emissions inventories, assessment of climate change actions, and monitoring and evaluation – are critical activities that enable planning for climate change. Other aspects include capacity-building (e.g., different types of training), technology transfer (from local to international), networking, and best practices exchange (Greenham et al., 2012).

Financial resources and mechanisms are essential (see Chapter 7, Economics, Finance, and the Private Sector). Mitigation and adaptation projects require mobilizing a combination of financial resources from federal, state, and municipal governments; development banks; private investments; multilateral and bilateral funding; concessional lending; and existing and new climate funds such as the Global Environmental Facility (GEF), Green Climate Fund (GCF), Climate Investment Fund (CIF), and others. On the other hand, there are many examples of successful bottom-up community-based approaches, particularly with regard to resilience-related projects (Smith et al., 2014). Chapter 7 provides a detailed discussion of possible financial resources and mechanisms available for cities to develop climate change actions and plans to address both adaptation and mitigation. Public–private partnerships and private-sector engagement are crucial means for financing the implementation of climate change measures particularly related to capital-intensive, energy-efficient, and climate-resilient infrastructure. Chapter 7 analyzes the opportunities and important issues to be taken into account when establishing public–private partnerships and private-sector involvement in financing climate change actions.

Implementation of climate change mitigation and adaptation actions entails the use of different planning systems, policy instruments, and steering mechanisms. Climate change (mitigation and/or adaptation) actions can be mainstreamed into existing sectoral plans and policies, whereas existing plans and actions in different sectors can incorporate climate change objectives. Actions can be implemented so that urban and infrastructure plans (e.g., for land use, transport, water and sanitation) contain climate considerations. The provision of services (e.g., water, transport) can incorporate low-carbon and climate-proof regulations and specifications. In addition, a special climate change unit can be created within the municipal structure to be held responsible for climate policy (within an existing unit or as a separate unit), or climate considerations can be mainstreamed into a range of municipal units. Examples of the creation of climate change units can be found in Copenhagen, Mexico City, and Durban to name a few. The city may then adopt a Climate Change Action Plan that sets GHG emissions and vulnerability reduction targets. Furthermore, better results can be achieved when there are policies and actions integrated with neighboring cities, harmonized with provincial, national, and international policies (European Commission, 2011; UN-Habitat, 2015) (see Chapter 16, Governance and Policy).

Political brinkmanship also poses a significant challenge to the success of integrated climate policies, leading in many cases to ineffective micromanagement and communication greenwashing. Hence, strong political leadership is required for the adoption of an ambitious integrated climate change program that brings both climate and local benefits (Burch, 2010; Johnson and Breil, 2012; Moser, 2012). Successful experiences include the role of the city mayor as champion of the agenda or a strong city council advocating for climate actions. The cases of Quito and Mexico City are examples of how effective strong political leadership can be in planning and implementing integrated climate change policies (see Case Studies 4.5 and 4.6).

121

Case Study 4.5 Integrating Mitigation and Adaptation in Climate Action Planning in Quito, Ecuador

Carolina Zambrano

Avina Foundation, Ecuador

Diego Enriquez

Municipality of the Metropolitan District of Quito, Ecuador

Keywords	Integration, multiple benefits, stakeholder engagement, impact assessment, adaptation and mitigation
Population (Metropolitan Region)	2,365,000 (Demographia, 2016)
Area (Metropolitan Region)	4,230 km² (STHV, 2010)
Income per capita	US$5,820 (World Bank, 2017)
Climate zone	Cfb – Temperate, without dry season, warm summer (Peel et al., 2007)

The Metropolitan District of Quito – Ecuador's capital, nested within the Andes Mountains – is highly vulnerable to the effects of climate change. Between 1891 and 1999, temperature rose by 1.2–1.4°C, which, combined with an increase in the frequency and intensity of landslides, floods, and forest fires, has increased risk for the city

(Municipality of the Metropolitan District of Quito [MDMQ], 2012a; Zambrano-Barragán et al., 2011). In recent years, extreme weather events have affected infrastructure, human settlements, agriculture, and forests, while the loss of glaciers and highland ecosystems, known as *páramos*, threatens food security and water and hydropower supplies (MDMQ, 2012b; Rockefeller Foundation, 2013).

In response to current and potential impacts, and with the aim of reducing the city's carbon footprint, the Municipality of the Metropolitan District of Quito adopted a Climate Change Strategy in 2009, followed by an Action Plan in 2012. Although the Strategy made a clear distinction between adaptation and mitigation policies (MDMQ, 2009), the Action Plan addresses adaptation and mitigation in an integrated way. The actions are to be implemented during a five-year period and focus on ten strategic sectors: water resources, disaster risk management, sustainable transportation, agriculture, land-use planning, energy, waste management, industries, health, and ecosystems (MDMQ, 2012b).

Political leadership was the main driver for integrating adaptation and mitigation. In a context of limited financial and human resources, municipal leaders prioritized measures that had multiple benefits (including resilience and sustainable development). This decision not only responded to technical analyses carried out by the Municipality, but was also supported by a diverse group of stakeholders that participated throughout the Plan's design. One of the success factors of the stakeholder engagement process was to allow them to define the criteria of prioritization for the final set of measures that were included in the Action Plan; among the most supported criteria were precisely the capacity to deliver multiple benefits, as well as cost-effectiveness.

Case Study 4.5 Figure 1 *View of Quito*

Integration also offered officials from the Environment Secretariat – the institution that coordinates climate change management and the Action Plan process – the opportunity to mainstream climate change and generate political support across sectors and institutions. During the design process, the lead team analyzed the adaptation and mitigation potential of actions that were already planned by different sectors and promoted the recognition of their climate benefits in the Action Plan and across society. Other institutions saw the added value of their interventions, felt empowered, and supported the adoption of Quito's first Climate Action instrument.

While the planning process helped foster commitment from different stakeholders, integrating adaptation and mitigation has also proved to be a challenge. Actions with multiple benefits were more easily identified in sectors like land use, ecosystems, water, agriculture, and sustainable building, whereas health, waste management, and transportation were more related to either adaptation or mitigation. Moreover, as implementation moves forward, measuring the Plan's impact in a holistic manner is also challenging, In light of the fact that impacts related to adaptation and mitigation vary in temporal, spatial, and institutional scales, it is hard to measure and demonstrate performance *ex-post* (Klein et al., 2007). Communicating impact and defining targets for vulnerability reduction and adaptation are demanding tasks for urban policy-makers, making it a challenge to provide a balanced discourse between adaptation and mitigation.

The new city administration, which took office in 2014, has given continuity to the Climate Action Plan and its integrated vision for adaptation and mitigation. Policies for both an increase in resilience and a reduction of the carbon footprint were included in the new Development and Land Use Plan 2015–2025, and strategic actions with multiple benefits in sectors like sustainable building, water

resources management, and land-use planning are being strengthened (MDMQ, 2015).

Moving forward, in early 2016, the Municipality performed an impact assessment and evaluation of the 2012–2016 Climate Action Plan. Based on performance results, and prioritizing on-the-ground implementation, it updated this planning instrument and its GHG inventory before the convening of Habitat III, the United Nations Conference on Housing and Sustainable Urban Development.[1] As the host city, Quito aimed to contribute to the debate on the New Urban Agenda by providing concrete examples of resilience and mitigation policies and actions.

Recognizing the need for further appropriation of climate action by municipal agencies and civil society in Quito and Ecuador in the following years, city officials expect to prioritize replicable, tangible, and visible initiatives. Since communicating the impact of adaptation initiatives in the city remains one of the major challenges, indirect, sectorial indicators are being developed and used to measure success. As an example, the city measured its water footprint and is using it as the base for the creation of a reduction and compensation mechanism by the private sector, the municipality, and other civil society stakeholders.

Although progress has been made in areas like institutional frameworks, information and knowledge, and collaboration for climate action in Quito, their scale and scope are still insufficient. In addition to win-win actions, transformational changes in Quito are increasingly required in sectors such as energy, water, mobility, and disaster risk management, particularly in terms of adaptation so that human rights are upheld. Efforts to promote equity and modify power relations in the city are key for vulnerability reduction and should remain at the core of urban climate action in Ecuador and Latin America.

4.4 Approaches to Integrated Response to Climate Change in Cities

When cities approach adaptation and mitigation activities, they tend to follow, to a large extent, a general planning cycle process. This general planning process for climate change can be found in numerous policy documents and scientific articles (Bizikova et al., 2008; ICLEI and UN-Habitat, 2009; Moser and Ekstrom, 2010; UN-Habitat, 2014, 2015). Based on a review of municipal climate change action plans of cities, planning for climate change in cities usually focuses either on climate adaptation or climate mitigation depending on the local context and city priorities, whereas there are few examples of integrated approaches. Ignoring one of the two agendas might create conflicts between mitigation and adaptation objectives or miss potential but important synergies. If conflicts can be avoided and synergies enhanced by identifying their drivers at the early stage of the planning process, adaptation and mitigation can be successfully integrated into urban planning and implemented in tandem in practice (see Case Study 4.6). Therefore, based on current practices and empirical evidence, we introduce five possible pathways for cities to make decisions and plan for climate change response:

1. *Stand–alone single approach*: Cities develop stand-alone municipal climate change (either adaptation or mitigation) plans, without considering any possible interrelationships (synergies and conflicts) between adaptation and mitigation objectives. This is the general case in many cities, which have largely focused on mitigation only.
2. *Stand-alone parallel/combined approach*: Cities develop stand–alone municipal climate change plans both for adaptation and mitigation in parallel without considering interrelationships between them (e.g., New York, London, Danang).
3. *Adaptation driven with mitigation co-benefits*: Cities develop municipal climate adaptation plans considering mitigation co-benefits and tradeoffs (e.g., Durban, Quito, Vancouver).
4. *Mitigation driven with adaptation co-benefits*: Cities conduct climate mitigation action plans considering adaptation co-benefits and tradeoffs (e.g., Paris, Buenos Aires).
5. *Integrated approach*: Cities develop municipal climate change action plans that incorporate both mitigation and adaptation objectives taking interrelationships into consideration (e.g., Mexico City, Wellington).

---

1 Habitat III, the United Nations Conference on Housing and Sustainable Urban Development, will take place in Quito, Ecuador, in October 2016. It will focus on the implementation of a New Urban Agenda, building on the Habitat Agenda of Istanbul in 1996.

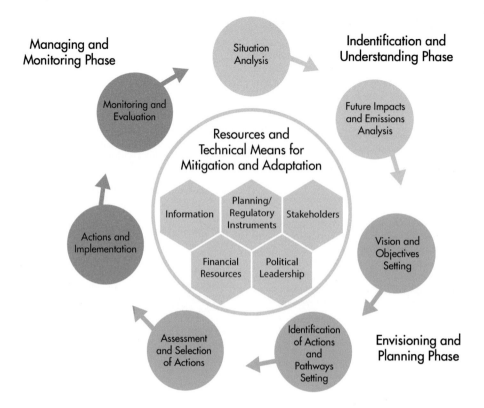

Figure 4.4 *Planning for low emissions development and resilience in cities.*

Source: Adapted from Bizikova et al., 2011 and Moser and Ekstrom, 2010

Next, we present an approach for integrating climate change mitigation and adaptation for metropolitan regimes. We identify, at each stage of the planning process, the challenges and opportunities of integrating climate mitigation and adaptation that are related either to structural conditions or to the availability of technical means and resources (see Table 4.4).

4.4.1 Phase 1: Identifying and Understanding

4.4.1.1 Step 1: Situation Analysis

A starting point in the planning process is the *situation analysis* of the city, considering the current baseline data for multiple variables. This includes the availability and development of datasets for a range of socioeconomic, environmental, climate, and land-use variables. Vulnerability maps (using GIS) and vulnerability indicators at appropriate spatial scales allow for the identification of current climate risk, taking into account vulnerability factors such as exposure, sensitivity, and current level of adaptive capacity. Climate impacts should be differentiated according to their temporal scale: (1) extreme events (immediate and short term) such as floods, heat waves, landslides, and storm surges and (2) long-term (annual/decadal) climate threats such as variations in average temperature or other slow-onset events such as sea level rise.

Simultaneously, sufficiently disaggregated city-level (and, whenever possible, metropolitan-level) GHG inventories are the starting point for urban mitigation measures, characterization of emissions (direct and whenever possible also indirect, e.g., from power generation), adequate sectoral breakdown (e.g., transport, buildings, industries, waste management, land-use change), and activity data that drive emissions (e.g., energy production by different means, passenger kilometers traveled, production outputs, floor space of different commercial facilities, household characteristics, waste generation/recycled/treated), both for community and municipal government. Figure 4.4 addresses planning for these aspects.

4.4.1.2 Step 2: Future Impacts and Emissions Analysis

Climate projections (downscaling based on global and regional climate models) of variables such as temperature, precipitation, and sea level rise are needed to be able to understand the likely range of climate change impacts in cities, address uncertainties, and develop future scenarios (see Chapter 2, Urban Climate Science). Assessment of future climate impacts and future emissions also requires consideration of the current and projected future growth of multiple urban sectors such as infrastructure, transport, energy, buildings, and an estimation of the probabilities of risk outcomes and damages throughout the metropolitan region; and the level of carbon emissions in the various sectors.

Uncertainty of future climate impacts at the city level is one of the main challenges that municipal governments must address. It is beneficial to co-generate the climate modeling risk information with a research center and city agency so that it can be updated and easily used by decision-makers (see Chapter 2, Urban Climate Science). One of the main challenges is the large time-frame discrepancy between climate change projections (e.g., 20 or even 50 to 100 years) and political decision-making (usually about 4 years). However, even in cases of large uncertainty in climate projections or lack of political support for climate change, measures with local benefits can be generated to improve the overall resilience and sustainability of a city to current extreme events, and by extension to future risks (Johnson and Breil, 2012).

Conducting assessments and collecting data for vulnerability assessments, GHG emissions inventories, and modeling of future emissions and impacts scenarios require significant technical capacity and cost. Therefore, lack of financial resources might lead to tradeoffs between mitigation and adaptation objectives for the same limited municipal budget. However, cities have been developing significant experiences in conducting both vulnerability assessments and GHG emission inventories in recent years. Vulnerability assessment often relies on the engagement of different stakeholders to identify the level of risk and the adaptive capacity of different communities, whereas engineers or experts usually conduct a GHG emissions inventory. Conducting vulnerability assessments, GHG emissions inventories, and scenario analysis is by definition intersectoral and participatory. Hence, coordination and collaboration among different city departments and jurisdictions are critical during this phase.

4.4.2 Phase 2: Envisioning and Planning Phase

4.4.2.1 Step 3: Vision, Objectives Setting, and Institutional Set-Up

This step requires creating a vision for the future of the city by identifying development objectives and priorities in the context of current trends and future climate impacts and emissions. This step also helps in the identification of climate change adaptation and mitigation actions in order to meet overall long-term priorities. A clear list of existing city development objectives from other plans (such as the master plan, environmental plan, land use plan, transport plan) and strategies should be used as part of the process. While developing the vision of the city with regard to climate change issues, identifying and understanding different stakeholders' priorities and objectives are also necessary. Furthermore, it requires analysis and understanding of the overall city-level objectives vis-à-vis climate change planning priorities (see Case Study 4.6).

Inclusion of multiple stakeholders at this step is essential for harmonizing different priorities and making the climate change action plan workable, as examples from Mexico City and Quito have shown. One example is Mexico City's Climate Action Plan (2008–2012) that included important environmental synergies with objectives established earlier in the city's Green Plan

(2007), but also synergies between mitigation and adaptation actions.

Setting up an institutional and regulatory framework with regard to coordination and collaboration among multiple city departments and jurisdictions is required in order to incorporate the objectives of multiple departments in planning and implementation. The cross-sectoral and cross-jurisdictional relationships might create conflicts in addressing both mitigation and adaptation. Agreeing on specific goals can be challenging, particularly in regard to the differences in spatial and temporal scales of adaptation and mitigation benefits (Moser, 2012).

4.4.2.2 Step 4: Actions Identification and Pathway Setting

During this step, cities should identify possible adaptation and mitigation actions that would meet the vision and achieve the identified objectives. Climate change adaptation and mitigation actions can be policies, projects, programs, and practices that can be undertaken to reduce a city's vulnerability and GHG emissions and develop its capacity to adapt and mitigate. It is very common to identify different portfolios (combinations) of measures in the form of distinct strategies or to explore possible alternative pathways for meeting cities' climate-resilience and low-carbon development objectives (Klein et al., 2007).

Different stakeholders and city departments can identify a variety of actions, and therefore their inclusion is essential. Existing plans that already lay down sectoral actions could help the process of identifying climate change actions (e.g., as done in Quito) (see Case Study 4.5).

4.4.2.3 Step 5: Assessment and Selection of Actions

Technical expertise is necessary to assess and prioritize mitigation and adaptation options while taking into account costs and multiple co-benefits. This is highly desirable but not always possible due to lack of capacity, constrained resources, and the complexity of assessment processes. Based on the type of assessment and information needed, city governments select assessment and prioritization methodologies.

Mitigation and adaptation actions (or portfolios of actions) are assessed against multiple objectives and criteria, whereas tradeoffs between different objectives can be also identified and assessed. These aspects are very relevant when cities assess both mitigation and adaptation actions in an integrated manner. Several assessment methods have been applied in cities to conduct or support urban climate change action plans. Assessment and prioritization methods can be classified as (1) economics-based approaches, including cost-benefit analysis and cost-effectiveness analysis); (2) integrated approaches such as multiple criteria analysis and integrated modeling; and (3) sectoral approaches (Cartwright et al., 2013; Haque et al., 2012;

Case Study 4.6   Climate Action Program in Mexico City 2008–2012

Martha Delgado

Fundación Pensar. Planeta, Política, Persona
Secretariat of the Global Cities Covenant on Climate, Mexico City
Harvard University, Cambridge, MA

Keywords	Climate adaptation plan, mitigation, transportation, water management, housing
Population (Metropolitan Region)	21,157,000 (UN, 2016)
Area (Metropolitan Region)	2,072 km² (Demographia, 2016)
Income per capita	US$17,740 (World Bank, 2017)
Climate zone	(Cfb) – Temperate, without dry season, warm summer (Peel et al., 2007)

The Mexico City's Climate Action Program 2008–2012 (PACCM) established two main goals: (1) reduction of 7 million tons of CO_2 equivalent (ton CO_2-eq) and (2) development of a plan for adaptation to climate change (Secretaria del Medio Ambiente, 2012). Although the program included both mitigation and adaptation actions, they were specifically designed to perform their primary objective either to reduce greenhouse gas (GHG) emissions or to achieve greater urban resilience to climate change. However, we can analyze some lessons learned regarding the interaction of both types of activities:

1. Implementation of PACCM led to the creation of a specific department in the government to lead and coordinate the implementation of both mitigation and adaptation actions: the Directorate of Climate Change and CDM Projects. This office was an important base that enabled the city's government to develop a deep and comprehensive dialogue among all stakeholders that carried out the program. During the program implementation, actions were clearly separated between mitigation and adaptation; however, having both issues integrated within this department allowed us to identify co-benefits very easily, to start new activities and goals, and prioritize and analyze these measures together as a new way of planning.

2. Some PACCM actions that contributed to both mitigation and adaptation are a program to reduce water consumption by 10% in the central government, the improvement of energy-efficient water pumping equipment, a wastewater treatment plant replacement program, and the development of water networks and pipe rehabilitation to reduce leakages. The actions on water management referred to the Climate Action Program in 2012 and achieved a reduction of GHG emissions of 4,670 tons CO_2eq. PACCM actions regarding transportation are the bike-sharing program ECOBICI, the renewal of the public transportation vehicle fleet, and the introduction of electric taxis (Villagran, 2012). Further actions are tax incentives for green roofs, conservation decrees for 33 urban ravines, an urban

reforestation program, and a Certification Program on Sustainable Buildings. The actions taken in the context of the sustainable housing programs, the comprehensive environmental improvement project, and social development in housing units were all carried out by the Institute of Housing and the Federal District Social Attorney's Office. They achieved a reduction of GHG emissions of 30,527 ton CO_2-eq by 2012.

Regarding adaptation, a total of 12 strategies were developed. Examples are early warning systems geared toward identifying risks and threats to the Mexico City population, micro-basin management of urban ravines, assistance to people who are vulnerable to extreme climate events such as heavy rains or intense heat or cold waves, epidemiological monitoring, vitalization of native crops to maintain the biodiversity and resilience of agro-systems, and monitoring of forest fires during the dry season by remote sensing detection.

3. PACCM originated the creation and signing of the Global Cities Covenant on Climate in 2010 in Mexico City. This Pact is an international instrument through which currently more than 340 mayors around the world have pledged to take climate action. The pact includes commitments to execute both adaptation and mitigation actions, and the signatories are required to report annually on both types of measures.

4. Mexico City's Climate Action Program was the first plan of its kind in Mexico and Latin America. It was published even before the Mexican federal government had a national plan. That represented an enormous challenge for planning and implementation.

Some of the challenges faced were:

Planning stage: The need to involve experts, officials, and citizens and to standardize and evaluate their proposals was quite challenging. The plan had to be ambitious but feasible at the same time. Many proposed measures were not economically feasible or had not been implemented anywhere before and could not be assessed. Other measures were simple, but it was difficult to consider them at the same level as other more complex ones. Finally, there were cross-sectoral measures (e.g., energy and transport, energy and water) that made it hard to determine which area of government should be responsible for implementing them.

Implementation stage: There was no allocated budget for some of the selected measures and therefore they were not implemented. At the same time, there were other programs that were not initially included in the Plan, but subsequently were devised and implemented successfully.

Monitoring and evaluation stage: One challenge was to involve managers of other departments in achieving the goals of the plan. Most officials realized that the government's goals were of common interest, but in daily life they lost interest or did not prioritize those measures because they had other important things to do. Political leadership shown by the Mayor drove the process of information provision, planning, and implementation of some measures. Meetings were organized as an opportunity to take high-level decisions to fulfill the plan.

Scrieciu et al., 2014; Grafakos et al., 2016b; Walsh et al., 2013; Charoenkit and Kumar, 2014). According to Johnson and Breil (2012) who conducted a study of seven major cities, only a limited number of cases have quantified the costs and benefits of individual projects.

This step requires the collection and development of multiple types of data and information regarding the likely impacts of different climate mitigation and adaptation actions. It further requires technical expertise and capacity that many cities lack. However, in the past few years, a large number of tools and

Table 4.4 *Technical means and resources in different phases of planning for climate change (CC) in cities*

Technical means and resources	Identifying and understanding phase		Envisioning and planning phase			Management and monitoring phase	
	Situation analysis	Future impacts and emissions analysis	Vision and objectives setting	CC actions identification and pathways setting	CC actions assessment and selection	CC actions implementation	Monitoring and evaluation
Information	x	x			x		x
Stakeholders	x		x	x	x	x	
Planning and regulatory instruments		x		x		x	x
Financial resources	x	x			x	x	
Political leadership			x	x		x	

resources have become available for cities to conduct prioritization of climate change adaptation or mitigation actions, although few have been developed in an integrated manner to address both. See, for instance, the CLIMACT Prio tool developed by the Institute for Housing and Urban Development Studies, the Climate Filter of the Emerging and Sustainable Cities Initiative by the Inter-American Development Bank, the Urban Integrated Assessment Facility by the Tyndall Centre for Climate Change Research, and the Integrated Assessment tool by the eThekwini Municipality (Walsh et al., 2013).

Building technical capacity can be costly; hence, the lack of financial resources could be a conflict for cities regarding the assessment of different mitigation and adaptation actions. Addressing both mitigation and adaptation objectives could increase the complexity of the assessment process, leading to even higher demand for resources.

4.4.3 Phase 3: Management and Monitoring Phase

4.4.3.1 Step 6: Implementation

Implementing both adaptation and mitigation actions requires the involvement of a range of institutions and departments. Actual implementation of different climate change actions (particularly structural ones) can be financially challenging. Mainstreaming climate actions into existing plans (e.g., sectoral plans) can help to ensure proper implementation and accountability. According to Johnson and Breil (2012), the administrative level of institutional actors involved in urban adaptation planning determines the range, scope, and capacity to trigger implementation. Furthermore, strong political leadership and support for the climate agenda are essential for effective action. Integrating climate actions at the sectoral level within existing policies and plans is an effective way to ensure funding and

implementation. Therefore, planning frameworks, stakeholders, and financial resources are important means for effective implementation of actions.

4.4.3.2 Step 7: Monitoring and Evaluation

Monitoring and evaluation systems track and analyze results before, during, and after implementation, enabling improvements and modifications through feedback processes. In this stage, the level of achievement of the climate change adaptation and mitigation objectives is measured through information and data collection for monitoring and evaluation. However, there are major differences in measuring adaptation and mitigation outcomes and impacts (different metrics, time scales, and uncertainties) that should be considered. Monitoring of actions is a challenging task in integrated climate change policy (Grafakos et al., 2016a).

Table 4.4 summarizes relevant resources and technical means that cities can use in different phases of integrated planning for climate change.

4.5 Future Research and Recommendations

Given that climate change planning is a rapidly evolving in cities and that bottom-up actions are going to be an important cornerstone of the future climate regime, there are key knowledge gaps that need to be addressed. The most pressing areas for further research can be grouped under the following headings: (1) integrated assessment methods and decision-support tools; (2) holistic, intersectoral, and nexus studies; (3) longitudinal studies; and (4) basic terminology and the need for structured taxonomies.

4.5.1 Integrated Assessment Methods and Decision-Support Tools

The need to develop integrated assessment methods and frameworks that capture both adaptation and mitigation aspects and user-friendly decision-support tools that address these multiple aspects at different levels of governance is manifested worldwide (see Annex 3, Case Study Docking Station, and Annex 5, Case Study Annex). It is essential that the development of decision-support tools should incorporate the needs of the users and allow broad participation of multiple stakeholders while integrating knowledge and information from different disciplines and agendas. Furthermore, studies assessing the costs and benefits of integrated implementation approaches in comparison to the option of implementing adaptation and mitigation separately will provide useful insights to researchers, policy-makers, and planners. Climate change policy assessments should compare different portfolios of options instead of individual ones and explore their robustness across different plausible future scenarios and outcomes (Scrieciu et al., 2014).

4.5.2 Holistic, Intersectoral, and Nexus Studies

There is a need to better understand how urban system works in an integrated manner from a climate change and sustainability point of view (Sattherthwaite, 2007, Jones et al., 2014) (see Chapter 1, Pathways to Urban Transformation). Most studies to date have been either sectoral or specific to mitigation or adaptation, failing to treat the city as a system to be optimized for better response to climate (Leseaur et al., 2015). Systemic aspects and their climate change potentials in cities are key to improving understanding (Seto and Dhakal, 2014). Therefore, holistic studies and interdisciplinary research frameworks such as urban metabolism studies that explore material flows and mass balances of water, energy, food, and waste provide the knowledge necessary to design and plan climate adaptation and mitigation in cities. Understanding and quantifying the complex relationships at the energy, water, and food nexus in urban areas would help to elucidate the complex interrelationships of climate adaptation and mitigation policies across different urban systems and sectors.

4.5.3 Need for Longitudinal Behavioral Studies

Given the large range of possible factors that contribute to why cities do or do not approach mitigation and adaptation in an integrated manner, it is surprising how few truly global (or even national) studies have been conducted to tease out what is specific to the local context and what is universal. In the field of integrated climate change planning, there is a clear need for larger longitudinal studies that move beyond individual or a small number of case studies to develop a more nuanced understanding of the relative impacts and effects of enablers and barriers to integrated planning approaches (see Box 4.4). It is still an open question to determine the main independent and dependent variables that condition cities' responses to mitigation and adaptation (Dupuis and Biesbroek, 2013). Much of the existing body of research is highly theoretical, necessitating more empirical fieldwork.

4.5.4 Understanding Basic Terminology and the Need for Structured Taxonomies

Integrating mitigation and adaptation requires an extra effort to understand the research and/or policy object (i.e., what exactly is being addressed). The terminology employed frequently varies between individuals because there is very little synthesized knowledge to date, and methods and theoretical frameworks vary widely. There is also little in the way of agreed-upon taxonomies of causal linkages between the various drivers and responses to climate change. Basic questions, such as how to separate adaptation in general from climate-related adaptation; what constitutes effective or ineffective mitigation and adaptation plans, programs, and policies; and how to best merge mitigation and adaptation strategies, have not been resolved. One possible remedy may be to draw on the classificatory sciences to develop more structured taxonomies of key terms and definitions (e.g., what is a barrier, what is a driver, what is meant by synergy, and so on). Possessing a common understanding of basic terminology may help to clarify what exactly is the object of study.

4.5.5 Recommendations for Policy-Makers

In this chapter, we presented the complexity of mitigation and adaptation interrelationships as well as the opportunities and challenges of integrating the two policies and mainstreaming climate actions in urban planning and decision-making. We provide the following recommendations for urban policy-makers:

- *Diagnose key risks and vulnerabilities*: Cities must have an accurate diagnosis of the current and future climate-related risks to and vulnerabilities of their population and territory. Likewise, cities must have sound emissions inventories and emission scenarios to evaluate mitigation potential. The use of scientific tools and approaches are therefore essential, in particular to strengthen the legitimacy for politically sensitive measures or major investments.

- *Start planning and executing programs early in the administration*: City governments are often short-lived. In many countries, municipalities have a three- or four-year period to realize their programs. Therefore, policy-makers should start climate action planning early in the administration term and ensure that enough legal and budgetary mechanisms are available for policy implementation in the medium term.

- *Evaluate and take advantage of resources and technical means at city's disposal*: Economic analysis is necessary to support decisions regarding the most cost-effective measures to reduce GHG emissions and to adapt to climate change impacts. These kinds of studies provide useful insights as well as transparency for decision-making. However, often they fall short on addressing other important aspects, impacts, and co-benefits that cannot be monetized, such as reductions in air pollution and traffic congestion and the amelioration of ecosystem services; therefore, they should be complemented by interdisciplinary studies and participatory approaches such as Multi-Criteria Decision Analysis.[2]

- *Consider adaptation and mitigation in an integrated manner in climate action plans*: Specific goals, budgets, and concrete measures should be identified and tracked over time. An integrated climate action plan should include a range of mitigation actions in different sectors such as energy, transportation, waste management, and water management, as well as adaptation actions in sectors such as infrastructure, natural resources, and health sectors along with ways in which these actions can create cross-sectoral synergies. Define precisely the key stakeholders and their responsibilities and roles in each phase of the policy implementation process. The action plans should involve the community, the private sector, and universities and the scientific community in the planning stages, and plans must be communicated to citizens to ensure support, transparency, and opportunities for participation. The plan should establish a monitoring system to evaluate the implementation process and to consider the legislative, fiscal, and economic settings required for success.
- *Identify actions and achieve alliances*: These need to mutually benefit city and national climate policies and include financial support and multilateral or international aid and assistance for sustainable development, particularly in cities in low-income countries. It is important that cities coordinate policies and financial efforts with federal and subnational governments, and with neighboring municipal governments, to make sure that the actions to be financed are not contradictory with actions that the province or nation has already undertaken.

4.6 Conclusions

There is a broad range of initiatives and actions demonstrating that adaptation and mitigation are inextricably linked, especially in cities. While the nature of such linkages is clear in many cases, it is yet unclear in other instances. These linkages appear in the form of positive (i.e., synergistic), negative (i.e., conflicting), and "balanced" (i.e., with tradeoffs) interrelationships between the two policy objectives. A clear identification of these interrelationships in policies and actions and their extent is important and should take into account the multiscale dynamics between adaptation and mitigation.

For climate mitigation, key actions are efficiency, decarbonization, improving carbon sinks, systemic intervention such as reducing consumption patterns and urban spatial planning, and local co-benefits. For climate adaptation, the key actions are assessing and reducing risks at the city level, prioritizing options, and allowing for adequate capacities (institutional, financial, and

behavioral) to be built, always envisioning a resilience-based perspective for the urban environment.

The structural conditions of cities (i.e., the environmental and physical setting, institutions and governance, economic and financial conditions, and sociocultural characteristics) determine the current context and boundaries of operating systems regarding climate change adaptation and mitigation and how technical means and resources can be used to create opportunities and overcome barriers for integrating them. A holistic consideration and quantification of the costs and multiple benefits of integrating adaptation and mitigation policies in comparison to stand-alone policies is necessary within the framework of municipal priorities, supported by user-friendly, evidence-based decision-support tools.

Recognized municipal climate policies and actions that address both mitigation and adaptation are frequently incidental, often as a co-benefit of the other. When synergies occur, they are usually highlighted as a means of promoting win-win strategies. However, in many instances, the costs of adaptation or mitigation are allocated unevenly, are not well calculated (in terms of magnitude and likelihood), or are simply ignored.

Given differences in the priorities and needs of cities, any attempt at mitigation and adaptation integration should be embedded in the local context. Adaptation and mitigation in low- and middle-income countries should reflect the urban context of sustainable development, where social and economic aspects often have higher priority than environmental objectives. There are very large differences in adaptive capacity between urban areas in different parts of the world. Carbon-intensive and climate-vulnerable infrastructure lock-ins should be avoided as much as possible, and development should aim at measures and investments that allow leapfrogging to more sustainable pathways.

Making a city resilient to climate change requires integrative approaches that account for multiple goals: adaptation, mitigation, and sustainable development. It means avoiding one-size-fits-all solutions. It also means adequately negotiating tradeoffs and avoiding conflicts among initiatives. Making a city resilient to climate change and sustainable also means ensuring reliable and fair provision of services as well as climate change responses.

Annex 4.1 Stakeholder Engagement

This chapter was prepared in collaboration with a multidisciplinary team of scientists and urban policy-makers and planners from cities across the world. In addition to the scientific writing team (Stelios Grafakos, Rotterdam; Chantal Pacteau, Paris; Mia Landauer, Vienna/Espoo; Oswaldo Lucon, São Paulo), the chapter benefited from valuable inputs from policy and planning

---

2 A consideration of multiple, usually conflicting, criteria in the decision-making process.

Box 4.4   Psychological, Social, and Behavioral Challenges and Opportunities for Climate Change Decision-Making

Diana Reckien

University of Twente, Enschede

David Maleki

Inter-American Development Bank, Washington, D.C.

Somayya Ali Ibrahim

Center for Climate Systems Research (CCSR), Columbia University, New York

Sabine Marx

Center for Research on Environmental Decisions (CRED), Columbia University, New York

Here, we seek to shed light on aspects of behavior by focusing on psychological and social factors that are particularly relevant for decision-making in the context of climate change in cities (World Bank, 2015; Gifford et al., 2011; Kollmuss and Agyeman, 2002).

CLIMATE CHANGE AS PSYCHOLOGICAL, SOCIAL, AND BEHAVIORAL CHALLENGE

Because climate change is a complex phenomenon and long-term in nature, acting on climate change becomes a decision made under uncertainty, one that conveys perceived risks of action as well as inaction. Risk perceptions differ between individuals and are shaped by a person's mental model as well as his or her worldview and core values. A mental model represents a person's thought process of how something works. It is based on incomplete facts, past experiences, and intuition, which are all employed to process new information. People's worldviews relate to social and political power structures (CRED, 2009). For example, if someone believes in a hierarchical world order, they will trust that leaders will make decisions for them and that they should not question the government's way of approaching climate change actions. Values serve as personal guiding principles for people and relate, for example, to the importance of security, financial prudence, and environmentalism. However, even someone who does not value the environment inherently can lend himself to engaging in climate change actions if it is framed as, for example, a national security issue for people with this core value. Using a person's or decision-maker's mental model, worldview and core values can help in communicating and planning for climate change risks more effectively.

Moreover, because climate change can be difficult to understand and even harder to feel given its long-term nature, it is often believed to be a challenge for the future rather than of the present. Here, a number of psychological concepts help explain these dynamics, such as *psychological distance*, the *availability heuristic*, the *recency effect*, and the *affect heuristic*.

Psychological distance applies when the consequences of an action are only felt after a time delay. This results in a perceived decrease in the importance of that consequence (Mischel et al., 1969; Read, 2001). This could be, for example, a factor explaining hesitancy to implement energy-efficiency measures even if they would pay off over the medium- or long-term.

The *availability heuristic* implies that people make likelihood estimates based on the ease with which they can retrieve or generate examples for a phenomenon (Tversky and Kahneman, 1973). This is based on the assumption that the future will be similar to the present or to former experience (Sunstein, 2006). For example, when asked to judge whether the probability of a blizzard is greater for November or January, people will try to recall storms that they remember occurring in either November or January (e.g., by using "on Thanksgiving weekend" or "after New Year's Eve" as mental aids). This rule of thumb works relatively well under static conditions but is often misleading in dynamic environments like a world characterized by a changing climate.

Moreover, not all easily recalled events are similarly likely to occur again. Some easily recalled events may have taken place more recently (*recency effect*) or be associated with strong emotions (*affect heuristic*) because one is personally affected or due to the strong media coverage that particularly high-intensity, low-probability events receive. In such cases, people will remember the event strongly and overestimate its likelihood (Hertwig et al., 2004). Very rare events, however, are commonly underestimated because they have normally not occurred in the recent past (Weber et al., 2004).

These and other psychological effects do not only apply to the perception of climate change, but may also stand in the way of finding solutions to deal with it. In the following sections, we look at five different sectors where a focus on psychological factors could make a difference in acting on climate change.

ERRONEOUS RISK PERCEPTION IN THE RUN-UP TO HURRICANE SANDY

Baker et al. (2012) report the findings of a real-time survey of the risk perceptions of residents as Hurricane Sandy approached the U.S. Mid-Atlantic coast in October 2012. While almost 90% of residents in threatened locations took some preparatory action, this was ultimately insufficient for the sheer magnitude of the storm, which was the second most costly natural disaster to ever affect the United States, after Hurricane Katrina in 2005 (National Oceanic and Atmospheric Administration [NOAA], 2013). Most preparations were temporary, short-term measures, such as stocking up on extra supplies, rather than longer-term or more costly actions like installing storm shutters, developing an evacuation plan, or purchasing flood insurance (see, e.g., Rosenzweig and Solecki, 2014).

A key factor in this ineffective short-term action was the mistaken belief that the area was under a hurricane watch, rather

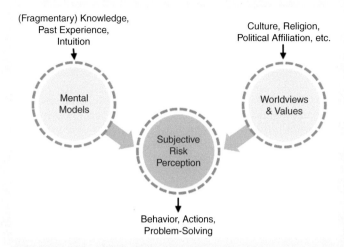

Box 4.4 Figure 1 *Psychological and social factors that shape people's risk perception.*

Source: CRED, 2009

than a hurricane warning[3]. The day before Sandy hit the coast, only 36% of surveyed residents believed they were under a hurricane warning, and this number rose to more than 50% only on the day of landfall. Many residents remained unaware of the increase in the intensity of the storm throughout the night. Moreover, most coastal residents misconstrued the primary threat of Sandy as coming from wind forces, rather than flooding and storm surge. This may have been affected by how hurricanes are defined – by wind speed – and how most televised weather alerts direct public attention to the track of the storm and particularly its eye. Moreover, much of the information on storms is usually gathered from a single synthesized source that is seen as authoritative, such as television, rather than in a more disaggregated manner from peers or sources such as the Internet. The Internet carries far more detailed information about a storm than that conveyed by television broadcast (Meyer et al., 2013; Morss and Hayden, 2010).

There was also the possibility that the residents' concern may have been dampened by their experiences with Hurricane Irene the year before – 76% of the households surveyed indicated that they had some experience living through hurricanes, although only a minority reported suffering any damage from Irene. Such prior experience can affect the propensity to take action for the next incidence (Orlove, 2011). People had stronger intentions to take protective action if they had less experience with hurricanes (Meyer et al., 2013). Prior storm experience thus can have a suppressing effect.

HOW HOMEOWNERS FAIL TO PROTECT THEMSELVES AGAINST FLOODS THROUGH DISASTER RISK INSURANCE

Insurance is an important mechanism to help homeowners deal with the financial impacts of extreme weather events. However, research has demonstrated that individuals often refrain from purchasing such protection even if reasonably priced. For example, in the United States, many homeowners in flood-prone areas have not purchased flood insurance despite government subsidies. This may be influenced by a difference in the perception of flood risk by residents and the actual risk as calculated by insurers based on hydrological data and models.

One explanation for such behavior is that individuals systematically underestimate the likelihood of rare events, thus not deeming it necessary to protect themselves against them (Browne et al., 2012; Kunreuther et al., 2013). A second reason is the tendency of individuals to buy insurance as a reaction to a natural disaster, even if the likelihood of occurrence does not change or, for some hazards, decreases after an incidence (Kahneman, 2011). Third, if homeowners have not collected on their insurance policy, many do not renew it, considering it a bad investment. The absence of flooding over several years is mistakenly understood as an indication that flood risk has decreased or that the policy was never justified in the first place (Kunreuther et al., 2013).

To deal with these challenges, it is important to develop transparent insurance products and to provide homeowners in flood-prone areas with easy-to-understand information on the likelihood and impact of disasters. Products linking insurance to the property rather than the owners is also recommended because it helps prevent incorrect flood risk perception.

CURBING ENERGY CONSUMPTION BY APPEALING TO SOCIAL NORMS

The right kind of information can also aid in curbing energy consumption. Studies have shown that residents achieve measurable reductions in electricity consumption when they are informed about how much electricity they consume in comparison to others, particularly peers (Allcott, 2011). The best-known large-scale field experiment in this regard was conducted by the energy company OPOWER and included 600,000 households in the United States. As part of this program, residential customers received energy reports that summarized their electricity consumption and compared it to that of neighboring homes with similar characteristics in terms of size and heating technology. Presumably perceiving their neighbors' consumption as a desirable social norm to follow customers after the intervention reduced average electricity consumption by 2% (Allcott, 2011).

PROMOTING WASTE SEPARATION IN HOUSEHOLDS THROUGH MEDIA CAMPAIGNS AND HABIT DEVELOPMENT

Effective waste management is another important determinant of urban sustainability. Most urban waste is brought outside of the cities of origin causing transportation, land

---

3   A warning means that hurricane conditions are expected, whereas a watch means that conditions are possible (NOAA, 2015). During a hurricane warning, residents are expected to complete their storm preparations and immediately leave the threatened area if directed to by local officials. During a hurricane watch, people are advised to prepare their homes and review plans for evacuation in case a hurricane or tropical storm warning is issued. People are also advised to listen closely to instructions from local officials.

consumption, and greenhouse gas (GHG) emissions. The impact of waste processing could be reduced substantially if organic waste were separated from all-purpose waste and kept in the city of origin for use in urban parks and gardens. In developed countries like Sweden, about half of the garbage (45–55%) is biodegradable material such as food and garden waste (Aberg et al., 1996), which is less suited for incineration due to its high water content. Nevertheless, studies show that neither a pro-environmental attitude nor an increase in information provision alone causes people to separate waste (Chan, 1998; Aberg et al., 1996; Gifford et al., 2011). For example, in a survey in Hong Kong, 98% of the respondents agreed that individuals have the responsibility to protect the environment, whereas only 30% of them separated waste (Chan, 1998). Other studies show that a combination of a pro-environmental attitude, influential social norms, and the person's perception of sufficient influence and control over one's actions are jointly important in generating pro-environmental behavior (Chan, 1998; Fishbein and Ajzan, 1975; Tucker and Speirs, 2003).

However, as with other environmental behaviors, there is a difference between the determinants of initiation and persistence (i.e., starting to separate waste and keeping on with it). Initiation seems to be strongly influenced by the mass media, whereas durability is connected to comfort and daily routines. For the latter, it is important that people perceive separating waste as little a change to former routines as possible. If larger routine changes are necessary, little-by-little changes are preferable. The maintenance of emptying containers is also important because it relates to comfort. People separated less waste when the timing of emptying containers and weather circumstances decreased the rate of composting (and thus increased odors and flies). Moreover, surveys in Hong Kong showed that doorway or curbside recycling was more acceptable than bring systems, and binary waste sorting was more popular than multisorting (Aberg et al., 1996).

City governments can play a key role in increasing environmental awareness for pollution and waste reduction by publicly promoting green behaviors (Chan, 1998). It has been shown that it is particularly beneficial to communicate waste separation behavior as being responsible, rewarding, sensible, and good for maintaining a good place to live. This, together with emphasizing a "feel good factor" when participating in recycling schemes, is believed to be personally rewarding and may ensure greater participation levels (Tonglet et al., 2004). Recycling campaigns should focus on reinforcing the positive attitudes of recyclers and potentially change the negative attitudes of non-recyclers (Emery et al., 2003; Evison and Read, 2001). Research has also shown that economic incentives can be successful in increasing domestic waste separation (Yau, 2010).

PRICING INSTRUMENTS AND CONSUMER BEHAVIOR IN THE PUBLIC TRANSPORTATION SECTOR

Pricing instruments are often an efficient mechanism to reduce demand or modify consumer behavior in the public transportation sector (Eliasson and Mattsson, 2006; Nakakura and Kockelman, 2002; Rotaris and Danielis, 2014). A combination of incentives and disincentives may encourage changes in commuters' travel behaviors, shifting such behaviors from car travel to other modes such as public transport, walking, and cycling (Miyoshi and Rietveld, 2015).

Disincentives to car travel may include fixed tolls, congestion pricing, fuel taxes, parking charges, subsidies, and Pay-as-You-Drive (PAYD) programs. Both fixed tolls and congestion pricing are used in Singapore, London, and Stockholm, whereas Bogota and Chicago use only congestion pricing. New York, Sheffield, and Edinburgh collect parking charges. The congestion pricing program in London – the first in a major European city – started in 2003 and has since then significantly reduced traffic congestion, improved bus and taxi service, and generated substantial revenues. Public acceptance has grown, and there is now support to expand the program to other parts of London and to other cities in the United Kingdom (Mehrotra et al., 2011; Lorenzoni et al., 2007).

In many other cities, such as Los Angeles, San Francisco, Mumbai, and Delhi, transit fares are subsidized. This provides another incentive for car commuters to switch to public transport. PAYD is yet another mechanism to reduce vehicular miles traveled, using insurance premiums based on per-mile charges instead of driving records and other traditional risk factors. This provides motorists an opportunity to lower their insurance costs by driving less, which can benefit the environment (Ferreira and Minikel, 2010). PAYD insurance was introduced in California in 2009 to help reduce vehicle miles traveled and associated GHG emissions.

Bicycling is being encouraged as an alternative urban transport mode in many cities with the aim of reducing automobile dependency and associated GHG emissions. Strategies take the form of bicycle rental stations, being used especially in European cities, and the provision of bike lanes. Although many cities in developing country have a high share of non-motorized transport, this could change with rising incomes. Safety, convenience, and the possibility of improving daily commutes are again important psychological and social parameters. Another noteworthy, low-carbon alternative to car transport is the adoption of bus rapid transit systems (BRTs), as seen in cities such as Mexico City, Delhi, Curitiba, and Istanbul. BRT systems often include BRT-only lanes traveling along established major transport routes and may replace more chaotic, informal transit alternatives that are used in mixed-use traffic lanes (EMBARQ, 2013). BRT systems may offer significantly reduced travel times for passengers (e.g., the Istanbul Metrobüs system, which connects the European and Asian sides of the city, saves an average passenger 52 minutes per day and reduces CO_2 emissions by 167 tons per day; Alpkokin and Ergun, 2012). A well-run BRT system with a high frequency of service, such as the Guangzhou BRT system's 350 buses per hour, may reduce passengers' waiting times by a significant amount (Guangzhou Transport Research Institute, 2012). The Guangzhou BRT system was planned together with a bike sharing system along the same corridor, thus offering travelers an even more environmentally friendly option (EMBARQ,

2013). Reduced travel time, safety, comfort, and the potential for increased routine physical activity all serve as incentives to change individuals' behavior.

HEALTH RISK PERCEPTION IN THE CONTEXT OF HEAT WAVES

Heat waves are slow-onset extreme climatic events expected to increase with climate change. In many countries, heat waves put more lives at risk than do rapid-onset hazards like hurricanes, floods, and landslides (e.g., in the United States; Klinenberg, 2002). A number of psychological factors come into play.

Klinenberg (2002) studied the disastrous Chicago heat wave of 1992 that caused the death of 739 people within a week. He concluded that the risk from this hazard had been underestimated due to factors such as the low visibility of heat damage (often only reported as number of excess deaths) and victims as compared with the structural damages that other hazards cause and the subsequent lack of visual and other reporting materials such as pictures (lack of signal value) or tangible experiential reports. The social, economic, and institutional situations of residents also play a role. People living alone, not leaving home daily, lacking access to transportation, being sick or bedridden, not having social contacts nearby, and, of course, not having an air conditioner were found to be most at risk. Older men were at twice the risk of older women due to fewer social contacts, and black communities were more at risk than Hispanic communities.

The case of Chicago shows that heat wave risk has to be taken seriously, particularly in large, dense, and socially differentiated cities. Because communities and residents differ in their coping strategies for heat waves and other risks, community-targeted communication strategies are key. These include sending warnings and press releases through effective media, opening cooling centers and providing free bus transportation to them, addressing residents in risk categories (e.g., by phoning elders or sending police officers/city workers to do door-to-door check-ups on seniors who live alone), or increasing social contact and social embeddedness in communities (Klinenberg, 2002).

IMPROVING COMMUNICATION FOR CLIMATE CHANGE ACTION

In many decision-making processes, perceptions matter more than facts. How we feel about a risk (i.e., our subjective perceptions of risk) influences what we pay attention to in complicated situations and how we approach and solve problems. This can explain the gap between what experts perceive as risk and what the public perceives as risk. Based on utilizing a person's mental model, worldview, and core values, a number of additional aspects can be summarized as aiding climate change communication (CRED, 2009; CRED and ecoAmerica, 2014):

- Climate change communication is best framed in a relevant and relatable way, one speaking to people's worldviews and values, connected to a personal emotional relation, concentrated on positive aspects and potential gains, and formulated in a way that ends with a doable action (i.e., what a person willing to act can do now).
- Communication may generally try to highlight the impacts of climate change that are already being experienced in the present or are likely to occur in the very near future. This will create an urgency to act now (Gifford, 2011). It is also advisable to focus on facts that are assumed to be certain rather than those that are uncertain – for example, the fact that hurricanes with storm surges are highly likely to make landfall in the mid-Atlantic region.
- Moreover, people tend to think that it will be easier for them to act tomorrow or pay in the future – not now – as demonstrated by research on retirement savings (Benartzi, 2012). Applied to climate preparedness or energy conservation programs, participation may be greater if communicators ask people to commit now with the actual or more cost-intensive action being implemented later (e.g., by committing to weatherizing their home in the following year).
- Finally, in any communication, recommendations and/or advice emphasizing local variation can be taken into account to tailor and contextualize information, draw on past experience, and thereby achieve personal relevance with people. It should be kept in mind that different locations and contexts may need different approaches.

practitioners: Patrick Driscoll from Smart Cities Catalyst consulting company, Copenhagen; Martha Delgado, the General Director of the Global Cities Covenant on Climate Secretariat, Mexico City; and David Wilk from the Inter-American Development Bank, Washington D.C. In addition, inputs and advice for the case examples were provided by Sean O'Donoghue, Manager of the Climate Protection Branch in eThekwini Municipality, Durban, South Africa; and Carolina Zambrano, National Representative for Avina Foundation (Fundación Avina), Quito and stakeholders from the energy company Helen Oy from Helsinki.

The chapter writing team organized a special session and roundtable discussion "Integrating Adaptation and Mitigation Strategies in Urban Areas: The UCCRN Assessment Report on Climate Change" (chaired by Stelios Grafakos) at the second biennial European Climate Change Adaptation Conference in Copenhagen (May 12–14, 2015). The purpose of the session was to increase interactions among stakeholders and scientists and engage them in discussions of the experiences of cities from different geographical regions. The session provided insights into key principles when dealing with climate change planning in cities from a mitigation and adaptation perspective and used best practices and lessons learned from different cities. Approximately thirty practitioners, urban policy-makers and planners, researchers, scientists, and private-sector participants attended the session.

Chapter 4 Integrating Mitigation and Adaptation: Opportunities and Challenges

References

Aberg, H., Dahlman, S., Shanahan, H., and Säljö, R. (1996). Towards sound environmental behaviour: Exploring household participation in waste management. *Journal of Consumer Policy* 19, 45–67.

Al-Karaghouli, A. A., and Kazemerski, L. L. (2011). Renewable energy opportunities in water desalination. In Schorr, M. (ed.), *Desalination, Trends and Technologies* (149–185), InTech. Accessed November 2, 2014: http://cdn.intechopen.com/pdfs/13758/InTech-Renewable_energy_opportunities_in_water_desalination.pdf

Allcott, H. 2011. Social Norms and Energy Conservation. *Journal of Public Economics* 95(9), 1082–1095. doi: DOI 10.1016/j.jpubeco.2011.03.003

Alpkokin, P., and Ergun, M. (2012). Istanbul Metrobüs: First intercontinental bus rapid transit. *Journal of Transport Geography* 24(0), 58–66.

Ayers, J. M., and Huq, S. (2009). The value of linking mitigation and adaptation: A case study of Bangladesh. *Environmental Management* 43(5), 753–764.

Baker, E. J., Broad, K., Czajkowski, J., Meyer, R., and Orlove, B. (2012). *Risk Perceptions and Preparedness among Mid-Atlantic Coastal Residents in Advance of Hurricane Sandy*. Working paper #2012–18, Wharton School, University of Pennsylvania.

Balaban, O., and de Oliveira, J. A. (2013). Understanding the links between urban regeneration and climate-friendly urban development: Lessons from two case studies in Japan. *Local Environment* 19(8), 868–890.

Barbhuiya, S., Barbhuiya, S., and Nikraz, H. (2013). Adaptation to the future climate: A low carbon building design challenge. *Procedia Engineering* 51, 194–199.

Barnett J., and O'Neill, S. (2010). Maladaptation. *Global Environmental Change* 20, 211–213.

Barriopedro, D., Fischer, E. M., Luterbacher, J., Trigo, R. M., and García-Herrera, R. (2011). The hot summer of 2010: Redrawing the temperature record map of Europe. *Science* 332(6026), 220–224.

Benartzi, S. (2012). *Save More Tomorrow: Practical Behavioral Finance Solutions to Improve 401(k) Plans*. Penguin.

Bergquist, P. (2012). *Urban Climate Change Adaptation: Case Studies in Ann Arbor and Grand Rapids, Michigan*. Doctoral dissertation, University of Michigan.

Bianchini, F., and Hewage, K. (2012). Probabilistic social cost-benefit analysis for green roofs: A lifecycle approach. *Building and Environment* 58, 152–162.

Bizikova, L., Burch, S., Robinson, J., Shaw, A., and Sheppard, S. (2011). Utilizing participatory scenario-based approaches to design proactive responses to climate change in the face of uncertainties. In Gramelsberger, G., and Feichter, J. (eds.), *Climate Change and Policy* (171–190). Springer.

Bizikova, L., Neale, T., and Burton, I. (2008). *Canadian Communities' Guidebook for Adaptation to Climate Change. Including an Approach to Generate Mitigation Co-Benefits in the Context of Sustainable Development*. First edition. Environment Canada and University of British Columbia.

Blackhurst, M., Hendrickson, C., and Mathews, S. (2010). Cost effectiveness of green roofs. *Journal of Architectural Engineering* 16 (4), 136–143

Broto, V. C., and Bulkeley, H. (2013). A survey of urban climate change experiments in 100 cities. *Global Environmental Change* 23(1), 92–102.

Browne, M. J., Knoller, and C., Richter, A. (2012). *Behavioral Bias, Market Intermediaries and the Demand for Bicycle and Flood Insurance*. Working paper 10, Munich Risk and Insurance Center.

Bulkeley, H. (2013). *Cities and Climate Change*. Routledge.

Burch, S. (2010). Transforming barriers into enablers of action on climate change: Insights from three municipal case studies in British Columbia, Canada. *Global Environmental Change* 20(2), 287–297.

Callaway, J. M. (2004). *Assessing and linking the benefits and costs of adapting to climate variability and climate change. The benefits of climate change policy*. Organisation for Economic Co-operation and Development (OECD).

Cartwright, A., Blignaut, J., De Wit, M., Goldberg, K., Mander, M., O'Donoghue, S., and Roberts, D. (2013). *Economics of climate change adaptation at the local scale under conditions of uncertainty and resource constraints: The case of Durban*, South Africa. Environment and Urbanization.

Center for Sustainable Energy. (2017). Equinox Project. Accessed April 8, 2017: http://energycenter.org/equinox/dashboard/climate-action-plan-progress.

Center for Research on Environmental Decisions (CRED). (2009). *The Psychology of Climate Change Communication: A Guide for Scientists, Journalists, Educators, Political Aides, and the Interested Public*. New York: Columbia University. Accessed March 27, 2015: http://guide.cred.columbia.edu/pdfs/CREDguide_full-res.pdf

Center for Research on Environmental Decisions (CRED) and ecoAmerica. (2014). *Connecting on Climate: A Guide to Effective Climate Change Communication*. Accessed July 10, 2015: http://ecoamerica.org/wp-content/uploads/2014/12/ecoAmerica-CRED-2014-Connecting-on-Climate.pdf.

Chan, K. (1998). Mass communication and pro-environmental behaviour: waste recycling in Hong Kong. *Journal of Environmental Management* 52(4), 317–325.

Charoenkit, S., and Kumar, S. (2014). Environmental sustainability assessment tools for low carbon and climate resilient low income housing settlements. *Renewable and Sustainable Energy Reviews* 38, 509–525.

Chinese Government. (2012). South-to-north water diversion. Office of the South-to-North Water Diversion Project Commission of the State Council. Accessed June 26, 2014: http://www.nsbd.gov.cn/zx/english/erp.htm

City of Chicago. (2011). Chicago green roof initiative. Accessed April 13, 2014: http://www.artic.edu/webspaces/greeninitiatives/greenroofs/main.htm and http://www.gisclimat.fr/sites/default/files/MBerkshire.pdf

City of Portland. (2008). Cost benefit evaluation of eco-roofs. Accessed March 5, 2015: http://www.portlandoregon.gov/bes/article/261053

Cook, S., Hall, M., and Gregory, A. (2012). Energy use in the provision and consumption of urban water in Australia: An update. CSIRO Water for a Healthy Country Flagship.

Cooley, H., Wilkinson, R., Heberger, M., Allen, L., Gleick, P. H., and Nuding, A. (2012). Implications of future water supply sources for energy demands. Alexandria, VA: WaterReuse Foundation.

Creutzig, F., Hedahl, M., Rydge, J., and Szulecki, K. (2014). Challenging the European climate debate: Can universal climate justice and economics be reconciled with particularistic politics? *Global Policy* 5(s1), 6–14.

Dang, H. H., Michaelowa, A., and Tuan, D. D. (2003). Synergy of adaptation and mitigation strategies in the context of sustainable development: The case of Vietnam. *Climate Policy* 3(sup1), S81–S96.

Dantec, R., and Delebarre, M. (2013). Local governments in the run-up to Paris Climate 2015: From local stakeholders to global facilitators. Directorate-General of Global Affairs, Development and Partnerships 2013.

de Oliveira, J. A. P. (2009). The implementation of climate change related policies at the subnational level: An analysis of three countries. *Habitat International* 33(3), 253–259.

Döpp, S., Hooimeijer, F., and Maas, N. (2010). *Urban Climate Framework: A System Approach Towards Climate Proof Cities*. TNO, Built Environment and Geosciences.

Dupuis, J., and Biesbroek, R. (2013). Comparing apples and oranges: The dependent variable problem in comparing and evaluating climate change adaptation policies. *Global Environmental Change* 23(6), 1476–1487.

Dymén, C., and Langlais, R. (2013). Adapting to climate change in Swedish planning practice. *Journal of Planning Education and Research*, 33(1),108–119.

Eliasson, J., and Mattsson, L. G. (2006). Equity effects of congestion pricing: Quantitative methodology and a case study for Stockholm. *Transportation Research Part A: Policy and Practice* 40(7), 602–620.

EMBARQ. (2013). Social, environmental and economic impacts of bus rapid transit systems. Accessed December 18, 2014: http://www.wrirosscities.org/sites/default/files/Social-Environmental-Economic-Impacts-BRT-Bus-Rapid-Transit-EMBARQ.pdf

Emery, A. D., Griffiths, A. L., and Williams, K. P. (2003). An in-depth study of the effects of socio-economic conditions on household waste recycling practices. *Waste Manade Res* 21(3), 180–190

European Climate Adaptation Platform (Climate-ADAPT). (2014). Accessed December 8, 2015: http://mayors-adapt.eu/events2014/

European Commission. (2011). Cities of tomorrow – Challenges, visions, ways forward. European Union Regional Policy. Accessed June 28, 2015: http://ec.europa.eu/regional_policy/sources/docgener/studies/pdf/citiesoftomorrow/citiesoftomorrow_final.pdf

Ferreira, J., and Minikel, E. (2010). Pay-As-You-Drive auto insurance in Massachusetts: A risk assessment and report on consumer, industry and environmental benefits. Accessed March 27, 2014: http://web.mit.edu/jf/www/payd/PAYD_CLF_Study_Nov2010.pdf

Fischedick, M., Schaeffer, R., Adedoyin, A., Akai, M., Bruckner, T., Clarke, L. and Wright, R. (2012). Mitigation potential and costs. In Edenhofer, O., et al. (eds.), *IPCC Special Report on Renewable Energy Sources and Climate Change Mitigation* (791–864), Cambridge University Press.

Fishbein, M., and Ajzen I. (1975). *Belief, Attitude, Intention, and Behavior: An Introduction to Theory and Research*. Addison-Wesley.

Fontana, S., Sattler, T., Bontadina, F., and Moretti, M. (2011). How to manage the urban green to improve bird diversity and community structure. *Landscape and Urban Planning* 101(3), 278–285.

Ford, J. D., Berrang-Ford, L., and Paterson, J. (2011). A systematic review of observed climate change adaptation in developed nations: A letter. *Climatic Change* 106 (2), 327–336.

Fundación Pensar (2013). Global Cities Covenant on Climate Second Annual Report 2012. Mexico City: Fundación Pensar. Planeta, Política, Persona (163 p). Accessed December 8, 2015: http://mexicocitypact.org/pactomexicocity/downloads/texto-original/Segundo-Reporte-Anual-2012_1.pdf

Gariepy, B. (2015). La conquête des toits dans une perspective de développement durable. In Prochazka, A., Breux, S., Séguin Griffith, C., and Boyer-Mercier, P. (eds.), *Toit urbain: Les défis énergétique et écosystémique d'un nouveau territoire* (305–328), PUL.

Gavin, N. T. (2010). Pressure group direct action on climate change: The role of the media and the web in Britain—a case study. *The British Journal of Politics & International Relations* 12(3), 459–475.

Georgeson, L., Maslin, M., Poessinouw, M., and Howard, S. (2016). Adaptation responses to climate change differ between global megacities. *Nature Climate Change*. doi: 10.1038/nclimate2944. Accessed March 16, 2017: http://www.nature.com/nclimate/journal/vaop/ncurrent/full/nclimate2944.html

Gifford, R. (2011). The dragons of inaction: Psychological barriers that limit climate change mitigation. *American Psychologist* 66, 290–302.

Gifford, R., Kormos, C., and McIntyre, A. (2011). Behavioral dimensions of climate change: Drivers, responses, barriers, and interventions. *WIREs Climate Change* 2, 801–827. doi: 10.1002/wcc.143

Goklany, I. M. (2007). Integrated strategies to reduce vulnerability and advance adaptation, mitigation, and sustainable development. *Mitigation and Adaptation Strategies for Global Change* 12(5), 755–786.

Grafakos, S., and Flamos, A. (2015). Assessing low-carbon energy technologies against sustainability and resilience criteria: Results of a European experts' survey. *International Journal of Sustainable Energy*, 1–15. Accessed January 11, 2016: http://dx.doi.org/10.1080/14786451.2015.1047371

Grafakos, S., Gianoli, A., and Tsatsou, A. (2016a). Towards the development of an integrated sustainability and resilience benefits assessment framework of urban green growth interventions. *Sustainability* 8(5), 461; doi: 10.3390/su8050461.

Grafakos, S., Gianoli, A., Olivotto, V., and Kazsmarski, C. (2016b). Measuring the immeasurable: Towards an integrated evaluation framework of climate change adaptation projects. In Madu, C. N., and Kuei, C. (eds.), *Handbook of Disaster Risk Reduction and Management*, World Scientific Publishing, London (forthcoming).

Green, J. (2010). How to create a climate change mitigation and adaptation plan. *American Society of Landscape Architects*. Accessed March 25, 2014: http://dirt.asla.org/2010/09/27/how-to-create-a-climate-change-mitigation-and-adaptation-plan/

Greenham, T., Ryan-Collins, J., Werner, R., and Jackson, A. (2012). *Where does money come from? A guide to the UK monetary and banking system*. New Economics Foundation.

Guangzhou Transport Research Institute. (2012). Guangzhou BRT win UN lighthouse award. Accessed August 13, 2014: http://www.gztri.com/chengguo-do.asp?id=219

Gupta, R., and Gregg, M. (2013). Preventing the overheating of English suburban homes in a warming climate. *Building Research & Information* 41(3), 281–300.

Hall, J. W., Dawson, R. J., Barr, S. L., Batty, M., Bristow, A. L., Carney, S., and Zanni, A. M. (2010). City-scale integrated assessment of climate impacts, adaptation and mitigation. In Bose, R. K. (ed.), *Energy Efficient Cities: Assessment Tools and Benchmarking Practices* (43–64). World Bank.

Hamin, E. M., and Gurran, N. (2009). Urban form and climate change: Balancing adaptation and mitigation in the U.S. and Australia. *Habitat International* 33(3), 238–245.

Haque, A. N., Grafakos, S., and Huijsman, M. (2012). Participatory integrated assessment of flood protection measures for climate adaptation in Dhaka. *Environment and Urbanization* 24(1), 197–213.

Hassi, J., and Rytkönen, M. (2005). *Climate Warming and Health Adaptation in Finland*. FINADAPT Working Paper 7. Finnish Environment Institute.

Hertwig, R., Barron, G., Weber E. U., and Erev, I. (2004). Decisions from experience and the effect of rare events in risky choice. *Psychological Science* 15(8), 534–539.

Hickman, R., Hall, P., and Banister, D. (2013). Planning more for sustainable mobility. *Journal of Transport Geography* 33, 210–219.

Hodo-Abalo, S., Banna, M., and Zeghmati, B. (2012). Performance analysis of a planted roof as a passive cooling technique in hot-humid tropics. *Renewable Energy* 39(1), 140–148.

ICLEI and UN-Habitat. (2009). *Sustainable Urban Energy Planning: A Handbook for Cities and Towns for Developing Countries*. Author.

Intergovernmental Panel on Climate Change (IPCC). (2007). *Climate Change 2007: Impacts, Adaptation and Vulnerability: Contribution of Working Group II to the Fourth Assessment Report of the Intergovernmental Panel on Climate Change*. Volume 4. Cambridge University Press.

Jiang, Y. (2009). China's water scarcity. *Journal of Environmental Management* 90 (11), 3185–3196.

Johnson, K., and Breil, M. (2012). Conceptualizing urban adaptation to climate change-findings from an applied adaptation assessment framework. Accessed February 17, 2015: http://eucities-adapt.eu/cms/assets/Uploads/UserFiles/32/Johnson%20and%20Breil%202012_38.pdf

Jones, R. N. (2004). Managing climate change risks. In Agrawal, S., and Corfee-Morlot, J. (eds.), *The Benefits of Climate Change Policies: Analytical and Framework Issues* (249–298), OECD.

Jones, R. N., Dettmann, P. Park, G., Rogers, M., and White, T. (2007). The relationship between adaptation and mitigation in managing climate change risks: A regional response from North Central Victoria, Australia. *Mitigation and Adaptation Strategies for Global Change* 12, 685–712.

Jones, R. N., Patwardhan, A., Cohen, S. J., Dessai, S., Lammel, A., Lempert, R. J., and von Storch, H. (2014). Foundations for decision making. In *Climate Change 2014: Impacts, Adaptation, and Vulnerability. Part A: Global and Sectoral Aspects. Contribution of Working Group II to the Fifth Assessment Report of the Intergovernmental Panel on Climate Change* (195–228), Cambridge University Press. Accessed February 2, 2015: http://www.ipcc.ch/pdf/assessment-report/ar5/wg2/WGIIAR5-Chap2_FINAL.pdf

Jones, R., and Yohe, G. (2008). Applying risk analytic techniques to the integrated assessment of climate policy benefits. *Integrated Assessment* 8(1), 123–149.

Jordan, F. (2009). Urban responses to climate change. *Regional Development Dialogue* 30(2), 60–75.

Juhola, S., Driscoll, P., de Suarez, J. M., and Suarez, P. (2013). Social strategy games in communicating trade-offs between mitigation and adaptation in cities. *Urban Climate* 4, 102–116.

Kahneman, D. (2011). *Thinking, Fast and Slow*. Allen Lane.

Kenway, S. J., Lant, P., and Priestley, A. (2011). Quantifying the links between water and energy in cities. *Journal of Water and Climate Change* 1, 247–259.

Kern, K., Alber, G., Energy, S., and Policy, C. (2008). Governing climate change in cities: Modes of urban climate governance in multi-level systems. In *Competitive Cities and Climate Change* (171), OECD.

Kithiia, J., and Lyth, A. (2011). Urban wildscapes and green spaces in Mombasa and their potential contribution to climate change adaptation and mitigation. *Environment and Urbanization* 23(1), 251–265.

Klein R. J. T., Saleemul, H., Fatima, D., Downing, T. E., Richels, R. G., Robinson, J. B., and Toth F. L. (2007). Inter-relationships between adaptation and mitigation. In Parry, M. L., Canziani, O. F., Palutikof, J. P., van der Linden, P. J., and Hanson, C. E. (eds.), *Climate Change 2007: Impacts, Adaptation and Vulnerability. Contribution of Working Group II to the Fourth Assessment Report of the Intergovernmental Panel on Climate Change* (745–778), Cambridge University Press.

Klinenberg, E. (2002). *Heat Wave: A Social Autopsy of Disaster in Chicago*. University of Chicago Press. Accessed February 17, 2014: http://www.press.uchicago.edu/Misc/Chicago/443213in.html

Kollmuss A, and Agyeman J. (2002). Mind the gap: Why do people act environmentally and what are the barriers to pro-environmental behavior? *Environmental Education Research* **8**(3), 239–260.

Kunreuther, H. C., Pauly, M., and McMorow, S. (2013). *Insurance & Behavioral Economics. Improving Decisions in the Most Misunderstood Industry*. Cambridge University Press.

Kunreuther, H., Gupta, S., Bosetti, V., Cooke, R., Duong, M. H., Held, H., and Weber, E. (2014). Integrated risk and uncertainty assessment of climate change response policies. In Edenhofer, O., Pichs-Madruga, R., Sokona, Y., Farahani, E., Kadner, S., Seyboth, K., Adler, A., Baum, I., Brunner, S., Eickemeier, P., Kriemann, B., Savolainen, J., Schlömer, S., von Stechow, C., Zwickel, T., and Minx, J. C. (eds.), *Climate Change 2014: Mitigation of Climate Change. Contribution of Working Group III to the Fifth Assessment Report of the Intergovernmental Panel on Climate Change* (151–206), Cambridge University Press.

Landauer, M., Juhola, S., and Söderholm, M. (2015). Inter-relationships between adaptation and mitigation: A systematic literature review. *Climatic Change* **31**(4), 505–517.

Laukkonen, J., Blanco, P. K., Lenhart, J., Keiner, M., Cavric, B., and Kinuthia-Njenga, C. (2009). Combining climate change adaptation and mitigation measures at the local level. *Habitat International* **33**(3), 287–292.

Lehman, S. (2015). Green districts: Increasing walkability, reducing carbon and generating energy. In Prochazka, A., Breux, S., Séguin Griffith, C., and Boyer-Mercier, P. (eds.), *Toit urbain: Les défis énergétique et écosystémique d'un nouveau territoire*, PUL, pp. 39–64.

Lesueur, A., Rosenzweig, C., Pacteau, C., Dépoues, V., Abbadie, L., Bordier, C., and Ali Ibrahim, S. (2015). Scientific brief on the implications of local and regional jurisdictions for mitigation and adaptation to climate change. LPAA Focus on Cities & Regions Climate Action. Accessed February 11, 2016: http://uccrn.org/files/2016/01/COP21-LPAA_UCCRN_I4CE.pdf

Lorenzoni, I., Nicholson-Cole, S., and Whitmarsh, L. (2007). Barriers perceived to engaging with climate change among the UK public and their policy implications. *Global Environmental Change* **17**, 445–459.

Lucon O., Ürge-Vorsatz, D., Zain Ahmed, A., Akbari, H., Bertoldi, P., Cabeza, L. F., Eyre, N., Gadgil, A., Harvey, L. D. D., Jiang, Y., Liphoto, E., Mirasgedis, S. Murakami, S., Parikh, J., Pyke, C., and Vilariño, M. V. (2014). Buildings. In Edenhofer, O., Pichs-Madruga, R., Sokona, Y., Farahani, E., Kadner, S., Seyboth, K., Adler, A., Baum, I., Brunner, S., Eickemeier, P., Kriemann, B., Savolainen, J., Schlömer, S., von Stechow, C., Zwickel, T., and Minx, J. C. (eds.), *Climate Change 2014: Mitigation of Climate Change. Contribution of Working Group III to the Fifth Assessment Report of the Intergovernmental Panel on Climate Change* (671–738), Cambridge University Press. Accessed January 21, 2016: http://www.ipcc.ch/report/ar5/wg3/

Lwasa, S., Mugagga, F., Wahab, B., Simon, D., Connors, J., and Griffith, C. (2014). Urban and peri-urban agriculture and forestry: Transcending poverty alleviation to climate change mitigation and adaptation. *Urban Climate* **7**, 92–106. doi: http://dx.doi.org/10.1016/j.uclim.2013.10.007

McEvoy, D., Lindley, S., and Handley, J. (2006). Adaptation and mitigation in urban areas: Synergies and conflicts. *Proceedings of the ICE-Municipal Engineer* **159**(4),185–191. Accessed March 3, 2017: http://mayors-adapt.eu/

Mees, H. L. P., and Driessen, P. P. (2011). Adaptation to climate change in urban areas: Climate-greening London, Rotterdam, and Toronto. *Climate Law*, **2**(2), 251–280.

Mehrotra, S., Lefevre, B., Zimmerman, R., Gerçek, H., Jacob, K., and Srinivasan, S. (2011). Climate change and urban transportation systems. In Rosenzweig, C, Solecki, W. D., Hammer, S. A., Mehrotra, S. (eds.), *Climate Change and Cities: First Assessment Report of the Urban Climate Change Research Network.* (145–177), Cambridge University Press.

Meyer, R., Broad, K., Orlove, B., and Petrovic, N. (2013). Dynamic simulation as an approach to understanding hurricane risk response: Insights from the Stormview Lab. *Risk Analysis*, **33**, 1532–1552.

Mills, G., Cleugh, H., Emmanuel, R., Endlicher, W., Erell, E., McGranahan, G., and Steemer, K. (2010). Climate information for improved planning and management of mega cities (needs perspective). *Procedia Environmental Sciences* **1**, 228–246.

Mischel, W., Grusec, J., and Masters, J. C. (1969). effects of expected delay time on the subjective value of rewards and punishments. *Journal of Personality and Social Psychology* **11**, 363–373.

Miyoshi, C., and Rietveld, P. (2015). Measuring the equity effects of a carbon charge on car commuters: A case study of Manchester Airport. *Transportation Research Part D: Transport and Environment* **35**, 23–39.

Morss, R. E., and Hayden M. H. (2010). Storm surge and "certain death": Interviews with Texas coastal residents following Hurricane Ike. *Weather, Climate, and Society* **2**(3),174–189.

Moser, S. C. (2012). Adaptation, mitigation, and their disharmonious discontents: An essay. *Climatic Change* **111**(2), 165–175.

Moser, S. C., and Ekstrom, J. A. (2010). A framework to diagnose barriers to climate change adaptation. *Proceedings of the National Academy of Sciences* **107**(51), 22026–22031.

Nakamura, K., and Kockelman, K. M. (2002). Congestion pricing and roadspace rationing: An application to the San Francisco Bay Bridge corridor. *Transportation Research Part A: Policy and Practice* **36**(5), 403–417.

National Oceanic and Atmospheric Administration (NOAA). (2013). *Service Assessment: Hurricane/Post-Tropical Cyclone Sandy, October 22–29, 2012* (**10**), U. S. Department of Commerce. Accessed February 20, 2014: http://www.nws.noaa.gov/os/assessments/pdfs/Sandy13.pdf

National Oceanic and Atmospheric Administration (NOAA). (2015). What is the difference between a hurricane watch and a warning? Accessed March 12, 2016: http://oceanservice.noaa.gov/facts/watch-warning.html

Newman, P. (2010). Green urbanism and its application to Singapore. *Environment and Urbanization Asia* **1**(2), 149–170.

Orlove, B. (2011). Waiting for Hurricane Irene in New York. *Weather, Climate and Society* **3**(3), 145–147.

Pacteau, C., and Joussaume, S. (2013). *Adaptation au changement climatique*. In Euzen, A., Eymard, L. and Gaill, F. (eds.), *Le développement durable à découvert*. CNRS Editions.

Pelling, M (2011). *Adaptation to climate change: From resilience to transformation*. Routledge.

Pelling, M., Manuel-Navarette, D., and Redclift, M. (eds.). (2012). *Climate Change and the Crisis of Capitalism: A Chance to Reclaim Self, Society and Nature*. Routledge.

Piper, J., and Wilson, E. (2009). Built environment. In Berry P. (ed.), *Biodiversity in the Balance – Mitigation and Adaptation Conflicts and Synergies*. Sofia: Pensoft (147–170).

Prochazka, A., Breux, S., Séguin Griffith, C., and Boyer-Mercier, P. (eds.). (2015). *Toit urbain: Les défis énergétique et écosystémique d'un nouveau territoire*, PUL.

Rankovic, A., Pacteau, C., and Abbadie, L. (2012). *Trames vertes: Le point de vue de l'écologie fonctionnelle, Hors-série Adaptation aux changements climatiques et trames vertes: Quels enjeux pour la ville ?* Vertigo.

Read, D. (2001). Is Time-discounting hyperbolic or sub-additive? *Journal of Risk and Uncertainty* **23**, 5–32.

Revi, A., Satterthwaite, D. E., Aragón-Durand, F., Corfee-Morlot, J., Kiunsi, R. B. R., Pelling, M., Roberts, D. C., and Solecki, W. (2014). Urban areas. In *Climate Change 2014: Impacts, Adaptation, and Vulnerability. Part A: Global and Sectoral Aspects. Contribution of Working Group II to the Fifth Assessment Report of the Intergovernmental Panel on Climate Change* (535–612), Cambridge University Press.

Riipinen, M. (2013). District heating and cooling in Helsinki. International Energy Agency CHP/DHC Collaborative & Clean Energy. Ministerial CHP/DHC Working Group Joint Workshop, 12–13 February 2013, Paris. Accessed January 21, 2016: http://www.iea.org/media/workshops/2013/chp/markoriipinen.pdf

Rosenzweig, C., Solecki, W. D., Hammer, S. A., and Mehrotra, S. (eds.). (2011). *Climate Change and Cities: First Assessment Report of the Urban Climate Change Research Network*. Cambridge University Press.

Rosenzweig, C., and W. Solecki (2014). Hurricane Sandy and adaptation pathways in New York: Lessons from a first-responder city. Glob. Environ. Change, 28, 395–408, doi:10.1016/j.gloenvcha.2014.05.003.

Rotaris, L., and Danielis, R. (2014). The impact of transportation demand management policies on commuting to college facilities: A case study at the University of Trieste, Italy. *Transportation Research Part A: Policy and Practice* **67**, 127–140.

Saavedra, C., and Budd, W. (2009). Climate change and environmental planning: Working to build community resilience and adaptive capacity in Washington State, USA. *Habitat International* **33**(3), 246–252.

Satterthwaite, D. (2007). *Adapting to climate change in urban areas: The possibilities and constraints in low-and middle-income nations.* Volume 1. International Institute for Environment and Development (IIED).

Scrieciu, S. Ş., Belton, V., Chalabi, Z., Mechler, R., and Puig, D. (2014). Advancing methodological thinking and practice for development-compatible climate policy planning. *Mitigation and Adaptation Strategies for Global Change* **19**(3), 261–288.

Seto, K. C., and Dhakal, S. (2014). *Human settlements, infrastructure, and spatial planning. In* Edenhofer, O., Pichs-Madruga, R., Sokona, Y., Farahani, E., Kadner, S., Seyboth, K., Adler, A., Baum, I., Brunner, S., Eickemeier, P., Kriemann, B., Savolainen, J., Schlömer, S., von Stechow, C., Zwickel, T., and Minx, J. C. (eds.), Climate Change 2014: Mitigation of Climate Change. *Contribution of Working Group III to the Fifth Assessment Report of the Intergovernmental Panel on Climate Change* (923–1000), Cambridge University Press.

Shove, E. (2010). Beyond the ABC: Climate change policy and theories of social change. *Environment and Planning A* **42** (6), 1273–1285.

Simon, D. (2013). Climate and environmental change and the potential for greening African cities. *Local Economy* **28** (2), 203–217. doi: 10.1177/0269094212463674. Accessed October 23, 2015: http://lec .sagepub.com/content/28/2/203.full.pdf±html

Simon, D. (2014). New evidence and thinking on urban environmental change challenges. *International Development Planning Review* **36**(2), v–xi. doi: http://dx.doi.org/10.3828/idpr.2014.9

Simon, D., and Leck, H. (2015). Understanding climate adaptation and transformation challenges in African cities. *Current Opinion in Environmental Sustainability* **13**, 109–116. doi: http://dx.doi .org/10.1016/j.cosust.2015.03.003

Sims, R., Schaeffer, R., Creutzig, F., Cruz-Núñez, X., D'Agosto, M., Dimitriu, D., Figueroa Meza, M. J., Fulton, L., Kobayashi, S., Lah, O., McKinnon, A., Newman, P., Ouyang, M., Schauer, J. J., Sperling, D., and Tiwari, G. (2014). Transport. In Edenhofer, O., Pichs-Madruga, R., Sokona, Y., Farahani, E., Kadner, S., Seyboth, K., Adler, A., Baum, I., Brunner, S., Eickemeier, P., Kriemann, B., Savolainen, J., Schlömer, S., von Stechow, C., Zwickel, T., and Minx, J. C. (eds.), *Climate Change 2014: Mitigation of Climate Change. Contribution of Working Group III to the Fifth Assessment Report of the Intergovernmental Panel on Climate Change* (599–670), Cambridge University Press. Accessed June 21, 2015: http://www .ipcc.ch/pdf/assessment-report/ar5/wg3/ipcc_wg3_ar5_chapter8.pdf

Smith, B., Brown, D., and Dodman, D. (2014). *Reconfiguring Urban Adaptation Finance.* Working paper. IIED. Accessed April 30, 2015: http://pubs.iied.org/10651IIED ISBN 978–1-84369–993–4

Solecki, W. (2015). Hurricane Sandy in New York, extreme climate events and the urbanization of climate change: Perspectives in the context of sub-Saharan African cities. *Current Opinion in Environmental Sustainability* **13**, 88–94. doi: http://dx.doi.org/10.1016/j.cosust.2015.02.007

Sugar, L., Kennedy, C., and Hoornweg, D. (2013). Synergies between climate change adaptation and mitigation in development: Case studies of Amman, Jakarta, and Dar es Salaam. *International Journal of Climate Change Strategies and Management* **5**(1), 95–111.

Sunstein, C. (2006). *Infotopia: How Many Minds Produce Knowledge.* Oxford: Oxford University Press.

Swart, R., and Raes, F. (2007). Making integration of adaptation and mitigation work: Mainstreaming into sustainable development policies? *Climate Policy* **7** (4), 288–303.

Thornbush, M., Golubchikov, O., and Bouzarovski, S. (2013). Sustainable cities targeted by combined mitigation–adaptation efforts for future-proofing. *Sustainable Cities and Society* **9**, 1–9.

Tonglet, M., Phillips, P. S., and Read, A. D. (2004). Using the theory of planned behavior to investigate the determinants of recycling behavior: A case study from Brixworth, UK. *Resources, Conservation and Recycling* **41**(2004), 191–214.

Tucker, P., and Speirs, D. (2003). Attitudes and behavioural change in household waste management behaviours. *Journal of Environmental Planning and Management* **46**(2), 289–307.

Tversky, A., and Kahneman, D. (1973). Availability: A heuristic for judging frequency and probability. *Cognitive Psychology* **5**(2), 207–232.

UNDESA. (2014). *World Urbanization Prospects: The 2014 Revision, Highlights* (ST/ESA/SER.A/352). United Nations, Department of Economic and Social Affairs, Population Division. Accessed July 30, 2015: http://esa.un.org/unpd/wup/Highlights/WUP2014-Highlights.pdf

UN-Habitat. (2014). *Planning For Climate Change: A Strategic–Values Based Approach for Urban Planners.* Author.

UN-Habitat. (2015). *Guiding Principles for City Climate Action Planning.* Available at http://e-lib.iclei.org/wp-content/uploads/2016/02/Guiding-Principles-for-City-Climate-Action-Planning.pdf.

Van Dillen, S. M., de Vries, S., Groenewegen, P. P., and Spreeuwenberg, P. (2012). Greenspace in urban neighbourhoods and residents' health: Adding quality to quantity. *Journal of Epidemiology and Community Health* **66**(6), e8.

Viguié, V., and Hallegatte, S. (2012). Trade-offs and synergies in urban climate policies. *Nature Climate Change* **2**(5), 334–337.

Walsh, C. L., Dawson, R. J., Hall, J. W., Barr, S. L., Batty, M., Bristow, A. L., and Zanni, A. M. (2013). Experiences of integrated assessment of climate impacts, adaptation and mitigation modelling in London and Durban. *Environment and Urbanization* **25**(2), 1–20.

Wang, C., Wang, Y., and Wang, P. (2006) Water quality modeling and pollution control for the Eastern route of South to North water transfer project in China. *Journal of Hydrodynamics, Ser. B.* 18(3):253–261.

Weber, E. U., Shafir, S., and Blais, A. R. (2004). Predicting risk sensitivity in humand and lower animals: Risk as variance or coefficient of variation. *Psychology Review* **111**(2), 430–445.

Wilbanks, T. J., Leiby, P., Perlack, R., Ensminger, J. T., and Wright, S. B. (2007). Toward an integrated analysis of mitigation and adaptation: Some preliminary findings. *Mitigation and Adaptation Strategies for Global Change* **12**(5), 713–725.

Williams, N. S. G., Rayner, J. P., and Raynor, K. J. (2010). Green roofs for a wide brown land: Opportunities and barriers for rooftop greening in Australia. *Urban Forestry and Urban Greening* **9**, 245–251.

World Bank. (2017). *World Development Report 2015: Mind, Society, and Behavior*, World Bank. doi: 10.1596/978–1-4648–0342–0

Yau, Y. (2010). Domestic waste recycling, *collective action and economic incentive: The case in Hong Kong, Waste Management* **30**(2010), 2440–2447.

Chapter 4 Case Study References

Case Study 4.1 Synergies, Conflicts, and Tradeoffs between Mitigation and Adaptation in Durban, South Africa

eThekwini Municipality. (2015). eThekwini Municipality Integrated Development Plan. 5 Year Plan: 2012/13 to 2016/17. Annual review 2015/2016, Durban, South Africa.

Peel, M. C., Finlayson, B. L., and McMahon, T. A. (2007). Updated world map of the Köppen-Geiger climate classification. *Hydrology and Earth System Sciences Discussions* 4(2), 462.

Roberts, D., and O'Donoghue, S. (2013). Urban environmental challenges and climate change action in Durban, South Africa. *Environment and Urbanization* 25(1), 299–319.

United Nations Development Programme. (2014). Human Development Index (HDI). Accessed August 9, 2017: http://hdr.undp.org/sites/ default/files/hdr14_statisticaltables.xls

World Bank. (2017). 2016 GNI per capita, Atlas method (current US$). Accessed August 9, 2017: http://data.worldbank.org/indicator/NY.GNP .PCAP.CD.

Case Study 4.2 Pilot Application of Sustainability Benefits Assessment Methodology in Colombo Metropolitan Area, Sri Lanka

Demographia. (2016). *Demographia World Urban Areas 12th Annual Edition: 2016:04.* Accessed August 9, 2017: http://www.demographia.com/db-worldua.pdf

Department of Census and Statistics, Sri Lanka (DCSSL). (2012). Population of Municipal Councils and Urban Councils by Sex Census, 2012. Accessed July 13, 2014: http://www.statistics.gov.lk/Abstract2014/CHAP2/2.4.pdf

Fraser, E., Dougill, A., Mabee, W., Reed, M., and McAlpine, P. (2006). Bottom up and top down: Analysis of participatory processes for sustainability indicator identification as a pathway to community empowerment and sustainable environmental management. *Journal of Environmental Management* **78**, 114–127.

Graymore, M., Wallis, A., and Richards, A. (2009). An index of regional sustainability: A GIS-based multiple criteria analysis decision support system for progressing sustainability. *Ecological Complexity* **6**, 453–462.

Peel, M. C., Finlayson, B. L., and McMahon, T. A. (2007). Updated world map of the Köppen-Geiger climate classification. *Hydrology and Earth System Sciences Discussions* **4**(2), 462.

Sri Lanka Red Cross Society (SLRCS). 2015. *Districts and local authorities.* Accessed April 11, 2016: http://www.redcross.lk/sri-lanka-country-profile/districts-and-local-authorities/

World Bank. (2017). GNI per capita, Atlas method (current US$). Accessed: http://databank.worldbank.org/data/download/GNIPC.pdf

Case Study 4.3 Jena, Germany Adaptation Strategy as an Essential Supplement to Climate Change Mitigation Efforts

City of Jena. (2013). Handbook on climate-sensitive urban planning, Jenaer KlimaAnpassungStrategie - JenKAS. Accessed August 5, 2014: http://www.jenkas.de/index.php/ergebnisse/handbuch

Peel, M. C., Finlayson, B. L., and McMahon, T. A. (2007). Updated world map of the Köppen-Geiger climate classification. *Hydrology and Earth System Sciences Discussions* **4**(2), 462.

Statistical Office State of Thuringia. (2015a). Population, including non-German, by gender. Accessed February 1, 2016: http://www.statistik.thueringen.de/startseite.asp

Statistical Office State of Thuringia. (2015b). Land use. Accessed February 1, 2016: http://www.statistik.thueringen.de/startseite.asp

World Bank. (2017). 2016 GNI per capita, Atlas method (current US$). Accessed August 9, 2017: http://data.worldbank.org/indicator/NY.GNP.PCAP.CD

Case Study 4.4 Sustainable Win-Win: Decreasing Emissions and Vulnerabilities in Chula Vista, California

City of Chula Vista. (2011). Climate adaptation strategies implementation plans, May 2011. Accessed August 5, 2014: http://38.106.5.202/home/showdocument?id=5443

Equinox Center. (2013). Dashboard 2013 in depth: Action on climate change. Accessed September 29, 2014: http://www.equinoxcenter.org/sustainability-blog/Dashboard-2013-In-Depth-Action-on-Climate-Change.html

Green, J. (2010). How to create a climate change mitigation and adaptation plan. American Society of Landscape Architects. Accessed August 5, 2014: http://dirt.asla.org/2010/09/27/how-to-create-a-climate-change-mitigation-and-adaptation-plan/

Peel, M. C., Finlayson, B. L., and McMahon, T. A. (2007). Updated world map of the Köppen-Geiger climate classification. *Hydrology and Earth System Sciences Discussions* **4**(2), 462.

U.S. Census Bureau. (2010). Decennial census, summary file 1. Accessed August 5, 2014: http://www.census.gov/population/metro/files/CBSA%20Report%20Chapter%203%20Data.xls

World Bank. (2017). 2016 GNI per capita, Atlas method (current US$). Accessed August 9, 2017: http://data.worldbank.org/indicator/NY.GNP.PCAP.CD

Case Study 4.5 Integrating Mitigation and Adaptation in Climate Action Planning in Quito, Ecuador

Demographia. (2016). *Demographia World Urban Areas 12th Annual Edition: 2016:04.* Accessed August 9, 2017: http://www.demographia.com/db-worldua.pdf

Klein, R. J. T., Huq, S., Denton, F., Downing, T. E., Richels, R. G., Robinson, J. B., and Toth, F. L. (2007). Inter-relationships between adaptation and mitigation. In Parry, M. L., Canziani, O. F., Palutikof, J. P., van der Linden, P. J., and Hanson, C. E. (eds.), *Climate Change 2007: Impacts, Adaptation and Vulnerability. Contribution of Working Group II to the Fourth Assessment Report of the Intergovernmental Panel on Climate Change* (745–777), Cambridge University Press.

Municipality of the Metropolitan District of Quito (MDMQ). (2009). Quito's climate change strategy.

Municipality of the Metropolitan District of Quito (MDMQ). (2012a). Quito's environmental agenda 2011–2016.

Municipality of the Metropolitan District of Quito (MDMQ). (2012b). Plan de acción climático de Quito 2012–2016.

Municipality of the Metropolitan District of Quito. (2015). Ordenanza Metropolitana No. 0041: Plan Metropolitano de Desarrollo y Ordenamiento Territorial 2015–2025.

Peel, M. C., Finlayson, B. L., and McMahon, T. A. (2007). Updated world map of the Köppen-Geiger climate classification. *Hydrology and Earth System Sciences Discussions* **4**(2), 462.

Rockefeller Foundation. (2013). 100 Resilient Cities: Quito, Ecuador. Accessed February 4, 2014: http://100resilientcities.rockefellerfoundation.org/cities/entry/quitos-resilience-challenge

Secretario Metropolitano De Territorio, Habitat Y Vivienda (STHV). (2010). Poblacion e indicadores del distrito metropolitano de Quito. Accessed October 23, 2014: http://sthv.quito.gob.ec/images/indicadores/parroquia/Demografia.htm

World Bank. (2017). 2016 GNI per capita, Atlas method (current US$). Accessed August 9, 2017: http://data.worldbank.org/indicator/NY.GNP.PCAP.CD

Zambrano-Barragán, C., Zevallos, O., Villacís, M., and Enríquez, D. (2011). Quito's climate change strategy: A response to climate change in the metropolitan district of Quito, Ecuador. *Local Sustainability 1* **1**(Part 6), 515–552.

Case Study 4.6 Climate Action Program in Mexico City 2008–2012

Demographia. (2016). *Demographia World Urban Areas 12th Annual Edition: 2016:04.* Accessed August 9, 2017: http://www.demographia.com/db-worldua.pdf

INEGI. (2010a). Censo general de población y vivienda 2010. Accessed October 23, 2014: http://www3.inegi.org.mx/sistemas/mexicocifras/default.aspx?e=09

INEGI. (2010b). Censo General de Población y Vivienda 2010. Accessed October 23, 2014: http://www.inegi.org.mx/est/contenidos/espanol/sistemas/cezm11/estatal/default.htm

Peel, M. C., Finlayson, B. L., and McMahon, T. A. (2007). Updated world map of the Köppen-Geiger climate classification. *Hydrology and Earth System Sciences Discussions* **4**(2), 462.

Secretaria Del Medio Ambiente. (2012). *Informe Final del Programa de Acción Climática 2008–2012.* Gobierno del Distrito Federal.

SEDESOL. (2007). Delimitación de las zonas metropolitanas en México 2005. Accessed September 7, 2014: http://www.conapo.gob.mx/es/CONAPO/Zonas_metropolitanas_2000

United Nations Development Programme. (2014). Human Development Index (HDI). Accessed April 30, 2015: http://hdr.undp.org/sites/default/files/hdr14_statisticaltables.xls

Villagran, L. (2012). Mexico City powers up electric taxi service: MEXICO CITY – Is there an emissions-free future for Mexico City taxis? *ZDNet 6 June 2012.* Accessed April 11, 2015: http://www.zdnet.com/article/mexico-city-powers-up-electric-taxi-service/

World Bank. (2017). 2016 GNI per capita, Atlas method (current US$). Accessed August 9, 2017: http://data.worldbank.org/indicator/NY.GNP.PCAP.CD

5

Urban Planning and Urban Design

Coordinating Lead Author

Jeffrey Raven (New York)

Lead Authors

Brian Stone (Atlanta), Gerald Mills (Dublin), Joel Towers (New York), Lutz Katzschner (Kassel), Mattia Federico Leone (Naples), Pascaline Gaborit (Brussels), Matei Georgescu (Tempe), Maryam Hariri (New York)

Contributing Authors

James Lee (Shanghai/Boston), Jeffrey LeJava (White Plains), Ayyoob Sharifi (Tsukuba/Paveh), Cristina Visconti (Naples), Andrew Rudd (Nairobi/New York)

This chapter should be cited as

Raven, J., Stone, B., Mills, G., Towers, J., Katzschner, L., Leone, M., Gaborit, P., Georgescu, M., and Hariri, M. (2018). Urban planning and design. In Rosenzweig, C., W. Solecki, P. Romero-Lankao, S. Mehrotra, S. Dhakal, and S. Ali Ibrahim (eds.), *Climate Change and Cities: Second Assessment Report of the Urban Climate Change Research Network*. Cambridge University Press. New York. 139–172

Embedding Climate Change in Urban Planning and Urban Design

Urban planning and urban design have a critical role to play in the global response to climate change. Actions that simultaneously reduce greenhouse gas (GHG) emissions and build resilience to climate risks should be prioritized at all urban scales – metropolitan region, city, district/neighborhood, block, and building. This needs to be done in ways that are responsive to and appropriate for local conditions.

Major Findings

Urban planners and urban designers have a portfolio of climate change strategies that guide decisions on urban form and function:

- Urban waste heat and GHG emissions from infrastructure – including buildings, transportation, and industry – can be reduced through improvements in the efficiency of urban systems.

- Modifying the form and layout of buildings and urban districts can provide cooling and ventilation that reduces energy use and allow citizens to cope with higher temperatures and more intense runoff.

- Selecting low heat capacity construction materials and reflective coatings can improve building performance by managing heat exchange at the surface.

- Increasing the vegetative cover in a city can simultaneously lower outdoor temperatures, building cooling demand, runoff, and pollution, while sequestering carbon.

Key Messages

Integrated climate change mitigation and adaptation strategies should form a core element in urban planning and urban design, taking into account local conditions. This is because decisions on urban form have long-term (>50 years) consequences and thus strongly affect a city's capacity to reduce GHG emissions and to respond to climate hazards over time. Investing in mitigation strategies that yield concurrent adaptation benefits should be prioritized in order to achieve the transformations necessary to respond effectively to climate change.

Consideration needs to be given to how regional decisions may affect neighborhoods or individual parcels and vice versa, and tools are needed that assess conditions in the urban environment at city block and/or neighborhood scales.

There is a growing consensus around integrating urban planning and urban design, climate science, and policy to bring about desirable microclimates within compact, pedestrian-friendly built environments that address both mitigation and adaptation.

Urban planning and urban design should incorporate long-range mitigation and adaptation strategies for climate change that reach across physical scales, jurisdictions, and electoral timeframes. These activities need to deliver a high quality of life for urban citizens as the key performance outcome, as well as climate change benefits.

5.1 Introduction

Key concepts, challenges, and pathways for adaptation and mitigation of climate change through recent advances in the planning and design of cities are reviewed in this chapter. Section 5.2 presents the concept of integrated mitigation and adaptation as a framework and introduces the factors of urban-scale form and function as influences on urban climate. Section 5.3 explains how urban microclimates are embedded in zones of human occupation and links metropolitan-scale urbanization with heat and storm-water impacts. Section 5.4 focuses on planning and design innovations that can be applied to achieve integrated mitigation and adaptation. Section 5.5 describes a process for implementing climate-responsive urban planning and urban design. Section 5.6 identifies key climate-resilient urban planning and urban design stakeholders and a set of value propositions to engage a broader constituency. Section 5.7 describes the challenges in cross-sector linkages between the scientific, design, and policy-making communities. Section 5.8 identifies knowledge gaps and future research opportunities, and Section 5.9 presents conclusions and recommendations for practitioners and policy-makers. Case Studies are distributed throughout the chapter to illustrate on-the-ground, effective implementations of the planning and design strategies presented.

5.2 Framework for Sustainable and Resilient Cities

Urban planning and urban design encompasses multiple disciplines, providing critical input to inform systems, management, and governance for sustainability and resilience to climate change (see Box 5.1). They configure spatial outcomes that yield consequences for and constitute responses to climate change (see Figure 5.1). The spatial form of a city – from the scale of the metropolitan region to the neighborhood block – strongly predetermines per capita greenhouse gas (GHG) emissions. With each 10% reduction in urban sprawl, per capita emissions are reduced by 6% (Laidley, 2015). Although compact urban form generally contributes positively to mitigation, it can paradoxically exacerbate local climate effects, requiring creative forms of adaptation. Research in this area is expanding, and, as a result, planning and design strategies are increasingly providing win-win solutions for compact urban morphology.

However, not all existing urban areas are compact (see Figure 5.2). Low-density areas continue to contribute disproportionately to emissions because of the excess mobility required by long distances, few alternatives to the private car, and scant possibilities for shared building envelopes. Whether such patterns are the result of planning or a lack thereof, it is

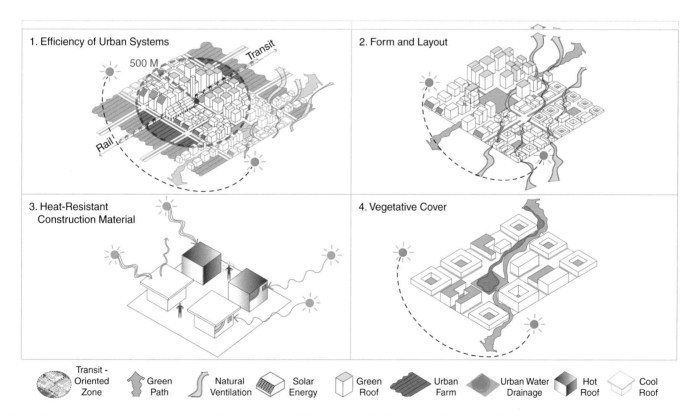

Figure 5.1 *Strategies used by urban planners and urban designers to facilitate integrated mitigation and adaptation in cities: (1) reducing waste heat and greenhouse gas emissions through energy efficiency, transit access, and walkability; (2) modifying form and layout of buildings and urban districts; (3) use of heat-resistant construction materials and reflective surface coatings; and (4) increasing vegetative cover.*

Source: Jeffrey Raven, 2016

Box 5.1   Key Definitions for Urban Planning and Urban Design

Urban planning: A field of practice that helps city leaders to transform a sustainable development vision into reality using space as a key resource for development and engaging a wide variety stakeholders in the process. It generally takes place at the scale of the city or metropolitan region whose overall spatial pattern it sets. Good urban planning formulates medium- and long-term objectives that reconcile a collective vision with the rational organization of the resources needed to achieve it. It makes the most of municipal budgets by informing infrastructure and services investments and balancing demands for growth with the need to protect the environment. And it ideally distributes economic development within a given urban area to reach wider social objectives (UN-Habitat 2013).

Urban design: Urban design involves the arrangement and design of buildings, public spaces, transport systems, services, and amenities. Urban design is the process of giving

form, shape, and character to groups of buildings, to whole neighborhoods, and to a city. It is a framework that orders the elements into a network of streets, squares, and blocks. Urban design blends architecture, landscape architecture, and city planning together to make urban areas functional and attractive.

Urban design is about making connections between people and places, movement and urban form, nature and the built fabric. Urban design draws together the many strands of place-making, environmental stewardship, social equity, and economic viability into the creation of places with distinct beauty and identity. Urban design is derived from but transcends planning, transportation policy, architectural design, development economics, engineering, and landscape. It draws together create a vision for an urban area and then deploys the resources and skills needed to bring the vision to life (urbandesign.org).

---

clear that the planning and design disciplines will increasingly need to prioritize the retrofitting of these areas for greater land-use efficiency (UN-Habitat, 2012).

A high proportion of urban areas that will need to minimize GHG emissions and adapt to climate change have not yet been built. Beyond aiming for appropriate levels of compactness, new urban development can and must be strategic about location (avoiding, e.g., areas particularly vulnerable to heat, flooding, or landslides).

5.2.1 Integrated Mitigation and Adaptation

Urban planning and urban design can be critical platforms for integrated mitigation and adaptation responses to the challenges of climate change. They have the opportunity to expand on the traditional influence and capabilities of practitioners and policy-makers and integrate climate science, natural systems, and urban form – particularly compact urban form – to configure dynamic, desirable, and healthy communities.

Traditionally, urban planning and urban design have focused on settlement patterns, optimized land use, maximized proximity, community engagement, place-making, quality of life, and urban vitality. Their focus is increasingly expanding to include principles such as resilience, comfort, resource efficiency, and ecosystem services (see Chapter 8, Urban Ecosystems). Applying these principles to urban policy helps to identify and strengthen prescriptive measures and performance standards and broaden urban performance indicators.

If future cities are to be sustainable and resilient, they must develop the physical and institutional capacities to respond to constant change and uncertainty. This will require strategies for

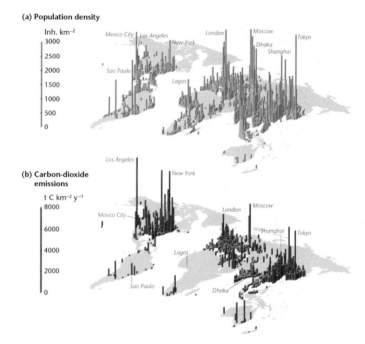

Figure 5.2 *Each bar represents an entire metropolitan area (i.e., the city and the continuous urban footprint surrounding it), including often much lower-density suburbs.*

Source: A. L. Brenkert, Oak Ridge National Laboratory. Maps created by Andreas Christen, UBC

long-term commitments across multiple electoral cycles and often among many political jurisdictions that constitute functional metropolitan areas (see Chapter 16, Governance and Policy).

Global climate risk is accumulated in urban areas because people, private and public assets, and economic activities become more concentrated in cities (Mehrotra et al., 2011; Revi et al., 2014). Recognition of the growing vulnerability of urban populations to climate-related health threats requires that the climate management activities of municipal governments be broadened (see Chapter 10, Urban Health).

One means of doing so is to prioritize investments in mitigation strategies that yield concurrent adaptive benefits over those

Figure 5.3 *"Green and blue fingers" in Thanh Hoa City, Vietnam, planned for 2020: Contiguous green corridors and canal circulation networks aligned with prevailing summer breezes, punctuated by stormwater retention bodies as urban design amenities.*

Source: Jeffrey Raven, Louis Berger Group, 2008

that do not (see Chapter 4, Mitigation and Adaptation). At present, nonintegrated mitigation and adaptation is most commonly pursued, with the majority of mitigation funds directed to energy projects that produce no secondary benefits for local populations in the form of heat management and enhanced flood protection or reduced damage to private property and public infrastructure. For example, mitigation strategies involving the substitution of a lower carbon-intensive fuel, such as natural gas, for a higher carbon-intensive fuel, such as coal, are an effective means of lowering CO_2 emissions yet provide few benefits related to climate adaptation.

5.2.2 Form and Function

Forward-thinking cities are beginning to exploit the positive potential of built and natural systems – including green infrastructure, urban ventilation, and solar orientation – to "future-proof" the built environment in response to changing conditions (see Figure 5.3 and Box 5.2). These passive urban design strategies "lock in" long-term resilience and sustainability, protecting

---

Box 5.2 Urban Form and Function

The physical character of cities can be described by three aspects of form: land cover, urban materials, and morphology. On the other hand, the flow of materials through a city describes its metabolism, the character of which is regulated by its functions (see, e.g., Decker et al., 2000).

Surface cover (Form): The replacement of natural land covers by impermeable materials limits the infiltration of precipitation into the substrate, increases runoff, and decreases evapotranspiration.

Construction materials and surface coating (Form): Common urban materials such as concrete have high conductivity and heat capacity values that can store heat efficiently. Also, many urban materials are dark colored and reflect poorly (e.g., asphalt).

Morphology (Form): The configuration and orientation of the built environment, from regional settlement patterns to buildings, create a corrugated surface that slows and redirects near-surface airflow and traps radiation.

Urban activities (Function): Cities concentrate material, water, and energy use that must be acquired from a much larger area. Some is used to build the city (changing its form), but most is employed to sustain its economy and society. Once used, the wastes and emissions are deposited into the wider environment, degrading soil, water, and air quality, and increasing heat through the exacerbated greenhouse effect.

the city from future decisions that could undermine its adaptability. They also remove the risk of relying on bolted-on, applied technologies that may require expensive maintenance or become obsolete in a short time. These form-based, contextually specific urban planning and urban design strategies are the ultimate guarantors of successful life cycle costs, payback, and liveability.

Integrated mitigation and adaptation in cities can assume many forms across spatial scales, urban systems, and physical networks (see Figure 5.4); a wide range of strategies adopted in service of urban sustainability already advances this objective. Enhanced urban transit, for example, has the effect of reducing both carbon emissions from single-occupant vehicle use and waste heat emissions that contribute to the urban heat island (UHI) effect. Investments in pedestrian and cycling corridors, particularly when integrated with parks and other green spaces in cities, can reduce carbon emissions, enhance carbon sequestration, and, perhaps most effectively, cool cities through evapotranspiration and shading. Sustainability strategies across urban systems can contribute to climate management goals under the umbrella of integrated mitigation and adaptation (see Figure 5.5).

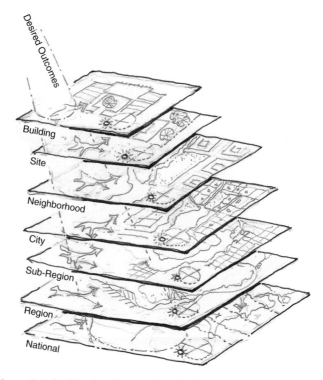

Figure 5.4 *Spatial scales relevant to urban planning and urban design for climate change mitigation and adaptation.*

Source: Jeffrey Raven, 2008

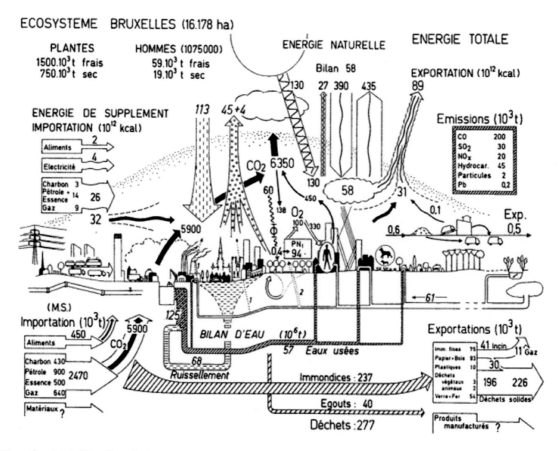

Figure 5.5 *L' Ecosystème Urbain (Urban Ecosystem).*

Source: Duvigneaud, P. and Denayer-de Smet, S., 1975

5.3 Climate in Cities

Urban areas occupy a small percentage (perhaps less than 3%) of the planet's land area, but this area is intensively modified (Miller and Small, 2003; Schneider et al., 2009). The landscape changes that accompany urbanization modify climate across a spectrum of scales, from the micro-scale (e.g., street), city-scale, and regional scales (see Chapter 2, Urban Climate Science).

The magnitude of the modification is evaluated by comparing the urban climate with its background climate, which is taken to be the "natural" climate. Because each city has a unique geographical region (latitude and topography primarily), the natural climate is assessed in the same region but over a non-urban surface (Lowry, 1977). One of the challenges posed by global climate change is that the background climate is itself changing and that cities contribute significantly to this change through the emission of GHGs.

The most profound changes occur in the layer of air below roof height. Here, access to sunlight is restricted, wind is slowed and diverted, and energy exchanges between buildings are the norm. The spatial heterogeneity of the urban landscape creates a myriad of microclimates associated with individual buildings and their relative disposition, streets, and parks (Errel et al., 2012). This is also the layer of intense human occupation, where building heating and cooling demand is met, emissions of waste heat and pollution from traffic are concentrated, and humans are exposed to a great variety of indoor and outdoor urban climates.

The climate effect of cities extends well beyond the urbanized area. As air flows over the urban surface, a boundary layer forms that deepens with distance from the upwind edge. This envelope may be 1–2 kilometers thick by mid-afternoon and is distinguishable as a warm and turbulent atmosphere that is enriched with contaminants, including GHGs. The extent of urban influence depends on the character of the city (e.g., its area, built density, and the intensity of its emissions) and on the background climate, which regulates the spread and dilution of the urban envelope. As a result, cities contribute significantly to regional and global air pollution (Guttikunda et al., 2003; Monks et al., 2009). Moreover, meeting urban energy demand accounts for up to three-quarters of CO_2 emissions from global energy use and thus represents a significant driver of global climate change (IPCC, 2014) (see Chapter 12, Urban Energy).

The magnitude of the urban climate effect is linked to both the form and function of cities (Box 5.2). The former refers to aspects of the physical character of cities, including the extent of paving and the density of buildings. The latter describes the nature of urban occupancy including the energy used in buildings, transport, and industry. *Integrated mitigation and adaptation* strategies focus on managing urban form and function together to moderate and respond to climate changes at urban, regional, and global scales.

5.3.1 Urban Climate Zones

The changes that accompany urbanization have profound impacts on the local environment and are clearly seen in aspects of climate and hydrology (Hough, 1989) (see Chapter 2 Urban Climate Science). The magnitude of these urban effects depends on both the form and functions of individual cities. However, cities are highly heterogeneous landscapes, and impacts vary across the urbanized area as well. Detailed mapping of urban layout, including aspects of form (e.g., impervious land cover) and of function (e.g., commercial land use), provides a basis for examining climate at a local scale.

For example, Stewart and Oke (2012) have developed a simple scheme that classifies urban neighborhoods mainly by form into local climate zones (LCZ) (see Figure 5.6). Each LCZ is characterized by typical building heights, street widths, vegetative cover, and paved area. Not surprisingly, the most intense local climate impacts are found where building density is greatest, streets are narrowest, and there is little vegetation (e.g., compact high-rises or dense slums). In many of these areas, the population is highly vulnerable due to poverty or age (see Chapter 6, Equity and Environmental Justice). Cities comprise many LCZ types that occupy varying proportions of the urbanized landscape. This

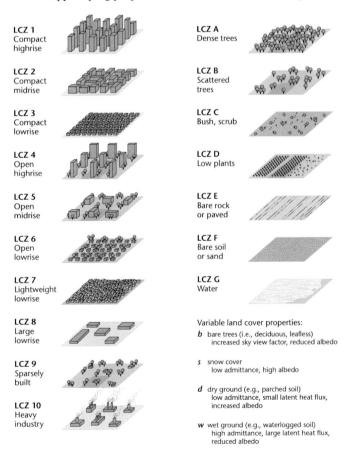

Figure 5.6 *Local climate zone type. Admittance, or thermal admittance, is a measure of a material's ability to absorb heat from, and release it to, a space over time. Albedo is the proportion of the incident light or radiation that is reflected by a surface back into space.*

Source: Stewart and Oke, 2012

chapter describes win-win form/function strategies to mitigate local climate impacts in compact districts.

5.3.2 Urbanization as Amplifier of Global Climate Change

Global climate change is modifying the background climate within which cities are situated, altering the frequency and intensity of extreme weather experienced (see Chapter 2, Urban Climate Science). The most recent Intergovernmental Panel on Climate Change (IPCC) assessment (IPCC, 2014a) concludes that global climate change has already resulted in warming both days and nights over most land areas and will cause more frequent hot days and nights in the future.

One of the most widely recognized climate impacts of urbanization is the UHI effect (e.g., Arnfield, 2003; Roth, 2007) (see Chapter 2, Urban Climate Science). The magnitude of the UHI is measured as the difference in air and surface temperatures between the city and proximate rural areas; these differences increase from the edge of the city to the center, where it is usually at a maximum. It is strongest during calm and clear weather but exhibits different impacts on surface and air temperatures. When measured as differences in air temperature between urban and non-urban surfaces, the UHI is strongest at night (due to heat retention), whereas differences in surface temperatures are largest during daytime (due to solar absorption).

Both types of UHI show a clear correlation with the amount of impervious surface cover and building density, whereas parks and green areas appear as cooler spots. The maximum value of the UHI as measured by air temperatures is likely to be between 2°C and 10°C, depending on the size and built density of the city, with largest values occurring in densely built and impervious neighborhoods (Oke, 1981). The magnitude of the surface temperature UHI depends greatly on the material characteristics of the surface, especially its albedo (i.e., reflectivity) and moisture status (see, e.g., Doulos et al., 2004).

In urban areas, the UHI adds to current warming trends due to global climate change contributes to poor air quality, increases energy demand for cooling, and elevates the incidence of heat stress (Akbari, et al., 2001; Grimmond, 2007; Oleson et al., 2015).

Once built, many aspects of the urban form are difficult to change (overall layout and morphology especially), so immediate emphasis must focus on altering aspects of surface cover and construction materials in the short term. At the same time, the role of urban planning to shape the potential doubling of cities' total physical footprint within the next 15 years must not be ignored because it represents a significant opportunity for mitigating future climate change at the global scale. New urban development – particularly since much future urban development will occur in warmer climate zones – can lower its emissions drastically by pursuing compact development that employs mixed-use zoning and public transit (Zhao et al., 2017; Resch et al., 2016).

Projections of climate change show that there will be distinct urban impacts. The locations of cities tend to be at low elevation, close to coasts and in river valleys/basins, which exposes urban areas to hazards such as high winds and flooding (McGranahan et al., 2007; Miller and Small, 2003). The concentration of population and infrastructure in cities makes them especially vulnerable to the impacts of natural hazards. Land-use and land-cover strategies designed to regulate these urban effects (such as urban greening to mitigate urban flooding and heating) can complement global climate change adaptation strategies that emphasize resilience.

Urbanization also has a dramatic impact on local hydrological processes and water quality (see, e.g., Brabec et al., 2002; Paul and Meyer, 2001). Impervious surface cover reduces the rate of infiltration to the underlying soil, thus limiting storage. While sewers and channelized rivers improve the hydraulic efficiency of drainage networks, the net effect is to increase the risk of flooding by increasing the volume and intensity of runoff (Kravčík et al., 2007; Konrad, 2003). In addition, the water that washes off impermeable urban surfaces during rain events adds warm and polluted water to river courses, further degrading water quality.

Moderating the magnitude of these urban effects in cities requires altering aspects of urban form, especially surface cover and materials. Vegetation in particular has an important role to play as a versatile tool that can cool surfaces through shading and cool air via evaporation (Shashua-Bar et al., 2010). Green areas can also play a key role in water management by delaying urban runoff and using soils as a filter to improve water quality. Where the landscape is densely built, green roofs can both insulate buildings, moderate air temperature above the urban surface, and slow urban runoff (Mentens et al., 2006). Similarly, changing surface albedo by applying surface coatings or replacing impervious surfaces with permeable materials can moderate urban effects (Gaffin et al., 2012; Santamouris, 2014). Altering urban morphology once in place is a more challenging prospect. However, where change is possible, design goals are to ensure access to the sun and provide shade, protection from wind, or ventilation by breezes (Bottema, 1999; Knowles, 2003; Emmanuel et al., 2007; Chen et al., 2010). Urban areas not yet built, particularly those in the developing world, have the advantage of being able to design along these parameters in advance of construction.

The role of urban design is critical because the urban climate impact is a product of both its physical character and the background climate. The best solution in a city where the climate is cool and wet will not be the same for a city in an arid and warm climate. Similarly, where cities are already substantially built, the opportunities for change will differ for each neighborhood. Nevertheless, managing the outdoor climate can have multiple benefits including reduced demand for indoor cooling/heating and increased use of outdoor spaces for health and improved air quality (Akbari et al., 2001) (see Case Study 5.4).

In cities, emissions of GHGs arise mainly from buildings (residential and commercial), transportation, and industries, but the proportions vary based on the character of the urban economy and the source of energy (Kennedy et al., 2009) (see Chapter 12, Urban Energy). The contributions of buildings and transport have received the most attention because each is amenable to management at the urban scale using a variety of measures, including building energy codes, public transit systems, and land-use management (ARUP, 2014). Much of the current evidence indicates that densely occupied cities are more efficient in their use of energy (and generate less waste heat as a consequence) (Resch et al., 2016). The evidence is especially strong for transport energy, which is largely based on cities where there are mass transit systems (Newman and Kenworthy, 1989). Increasing urban population density through policies that co-manage land-use and transport networks is an important strategy for reducing urban GHG emissions (Dulal et al., 2011).

Good urban planning and urban design are critical to achieving climate change objectives at city, regional, and global scales. Compact and densely occupied cities do not have to feature impermeable, densely built, and high-rise neighborhoods associated with unwanted urban effects. In a study of urban spatial structure and the occurrence of heat wave days across more than 50 large U.S. cities, for example, Stone et al. (2010) found the annual frequency of extreme heat events to be rising more slowly in compact cities than in sprawling cities. A wealth of studies find the enhancement of vegetation and surface reflectivity in dense urban environments to measurably reduce urban temperatures at the urban and regional scale (see Taha et al., 1999; USEPA, 2008; Gaffin et al. 2012; Stone et al., 2014). Further, tall buildings in cities impede direct sunlight from reaching the ground (see Figure 5.7).

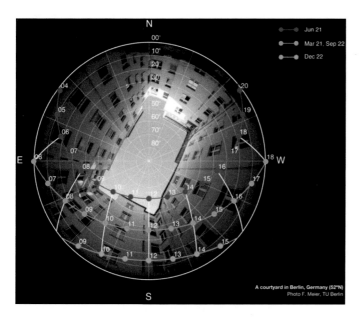

Figure 5.7 *The sky view from street level in Berlin, Germany. The hemispheric image shows the extent to which the sky is obscured by the surrounding buildings. The path of the sun at different times of the year is plotted to show the loss of direct sunlight at street level.*

Source: F. Meier, TU Berlin

5.4 Innovations

In this section of the chapter, we explore the implementation of strategies that utilize the four urban climate factors – urban function, form, construction materials, and surface cover – to achieve integrated mitigation and adaptation (see Figure 5.1).

5.4.1 Transportation, Energy, and Density

Since the middle of the 20th century, built environments the world over have tended to increase outward from central cities, consuming great swaths of previously undeveloped land while reinvestment in city centers falters. The infrastructure network needed to maintain this sprawling development pattern, particularly roads, has resulted in development that is land and infrastructure inefficient. It has also led to increased reliance on motor vehicles to get from one place to another. This reliance on motor vehicles has consequently led to a significant increase in vehicle miles (or kilometers) traveled (VMT), a concomitant increase in GHG emissions, and an amplification of the UHI effect through increased imperviousness, reduced green cover, and enhanced waste heat emissions (Stone et al., 2010). By developing in a denser, more compact form that mixes land use and supports mass transit use, cities may begin to reverse these trends (see Figure 5.8) (see Chapter 13, Urban Transportation).

If cities are to reduce VMT, then they must change their sprawling development pattern into one that relies on compact development. This focuses on regional accessibility through multiple transportation modes, including walking and bicycling, and clustered land-use patterns incentivized with vehicle distance traveled–based fees. At the core of this strategy is transit-oriented development (TOD) (Zheng and Peeta, 2015). TOD is compact, pedestrian-friendly development that incorporates housing, retail, and commercial growth within walking distance of public transportation, including commuter rail, light rail, ferry, and bus terminals (see Figure 5.8). It has become an essential and sustainable economic development strategy that responds to changing demographics and the need to reduce GHG emissions and health-related impacts.

Changing the built form from conventional suburban sprawl to compact, walkable, mixed-use and transit-oriented neighborhoods reduces travel distances (VMT). The meta-study *Effects of the Built Environment on Transportation: Energy Use, Greenhouse Gas Emissions, and Other Factors*, prepared by the National Renewable Energy Laboratory and Cambridge Systematics, Inc. (March 2013), notes that residents of compact, walkable neighborhoods have about 20–40% fewer VMTs per capita, on average, than residents of less-dense neighborhoods. Other studies have found a doubling of residential densities in U.S. cities to be associated with a 5–30% reduction in VMT (Gomez-Ibanez et al., 2009; Stone et al., 2010).

1. Efficiency of Urban Systems

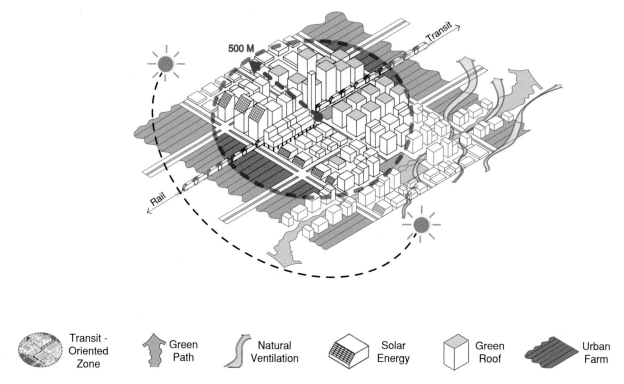

Figure 5.8 *Efficiency of urban systems.*

Source: Jeffrey Raven, 2016

Not only does compact, walkable TOD lower VMT, but it also requires less energy (see Chapter 12, Urban Energy). As Nolon (2012) reports, residential and commercial buildings used an extraordinary amount of electricity and energy in the past generation. In 2008, U.S. residential and commercial buildings consumed 29.29 quadrillion BTUs, which represented 73.2% of all electricity produced in the United States (Nolon, 2012). By 2035, the U.S. Department of Energy estimates that residential and commercial buildings will use 76.5% of the total electricity in the United States (Nolon, 2012). This energy consumption is also highly inefficient due to the systems used to produce and transmit it.

Two-thirds of the energy used to produce electricity in the United States is vented as waste heat that escapes into the atmosphere during generation and contributes to UHI formation (Nolon, 2012). Additionally, up to 15–20% of the net energy produced at these plants is then lost during electricity transmission (Nolon, 2012). By increasing the density of the built environment and reducing the distances that both electricity and people must travel, energy efficiency is notably increased in compact, transit-centered development (see Figure 5.8). As discussed in Jonathan Rose Companies (2011), a single-family home located in a compact, transit-oriented neighborhood uses 38% less energy than the same size home in a conventional suburban development (149 million BTU/year versus 240 million BTU/year).

Because compact development reduces VMTs and is more energy efficient, it also lessens GHG and waste heat emissions (see Figure 5.9). In a 2010 report to Congress, the U.S. Department of Transportation concluded that land-use strategies relying on compact, walkable, TOD could reduce U.S. GHG emissions by 28–84 million metric tons carbon dioxide equivalent (CO_2-eq) by the year 2030. Benefits would grow over time to possibly double that amount annually in 2050 (U.S. DOE, 2010).

Integrated mitigation and adaptation in urban planning and urban design can be successfully implemented through the configuration of low-carbon compact settlements configured for local microclimates. The mitigation perspective focuses on compact TOD prototypes reconfigured as low-carbon ecodistricts (see Box 5.3). The integrated mitigation and adaptation paradigm also addresses stormwater runoff and the UHI in high-density zones through material composition, urban morphology, and ecosystem services. This paradigm effectively "locks-in" long-term resilience.

The efficient use and recycling of energy and resources is a cornerstone of a resilient city and should be integral to the concept of integrated mitigation and adaptation, along with other climate-management strategies. This suggests the need for two levels of integrated mitigation and adaptation: passive and active. Passive Integrated Mitigation and Adaptation

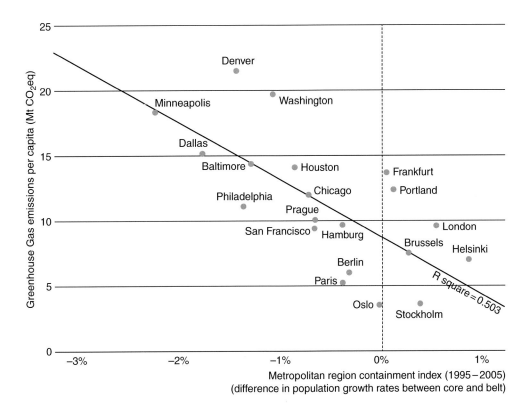

Figure 5.9 *Metropolitan region containment index (1995–2005).*

Source: Philipp Rode, 2012

Box 5.3   Multisectoral Synergies for Transit-Oriented, Low-Carbon Districts

Multisectoral approaches that integrate land use, mass transit, green buildings, and green districts to promote healthy, climate-resilient cities can be described as transit-synergized development (TSD). TSD leverages the greater scale, density, and economic value of transit-oriented development (TOD; nominal 1 km² urban districts around transit nodes) to create compact, vibrant, mixed-use communities that increase urban efficiency and reduce transport-related energy use, congestion, pollution, and greenhouse gas (GHG) emissions. As a co-benefit, the more compact development at the heart of TSD reduces pressures on interstitial spaces between transit corridors that can provide critical green infrastructure for managing climate impacts within dense urban zones. In China, promoting mass transit could generate up to 4 QBTUs (4.2 Exajoules) in energy savings per year (McKinsey Global Institute, March 2009). A number of cities in Canada and the United States are beginning to retrofit their urban fabric through TOD, and others in Latin America – most notably Curitiba, Brazil – have successfully used bus rapid transit (BRT) as a centerpiece of wider urban revitalization (Lindau et al., 2010).

TSD is a "node and network" model of sustainable urbanism. At each transit node, TSD combines passive and active green building design (high-performance envelope and mechanical, electrical, and plumbing [MEP] systems) with passive and active green district design (integrated urban design and advanced district infrastructure). District infrastructure provides a platform for the reuse and recycling of energy and resources among the buildings within the district. It is also

a platform for innovation and "'forward integration'" of new technologies, important attributes of a robust, resilient, and adaptive community (Lee, 2012).

At the network level, the transit nodes collectively provide diversity, redundancy, and synergy, effectively transforming the transit network into a framework for a robust, resilient, and adaptive city (Walker and Salt, 2006).

Box Figure 5.3 Figure 1 *Transit-synergized development concept.*

Source: iContinuum Group

149

(PIMA) includes climate-responsive designs such as green cover, reflective ground surface, natural ventilation, and solar orientation. PIMA represents good design practice and should be the basic design strategy for all buildings and urban areas. In high-density urban districts, however, Active Integrated Mitigation and Adaptation (AIMA) may be required. AIMA deploys advanced building systems and district infrastructure such as integrated building energy management, renewable energy, energy storage, district energy systems, water recycling, and on-site wastewater treatment that actively reduce energy and climate impacts.

District-scale AIMA infrastructure can be inherently more cost-effective to "upgrade" than individual building systems and can therefore better "climate-proof" the built environment against changing conditions. An example of this is Singapore's Marina Bay District Cooling System, which is envisioned as an "energy platform that enables forward integration of new energy technologies" (Tey Peng Kee, Managing Director, Singapore District Cooling Pte Ltd.; Interview, 2012).

5.4.2 Climate-Resilient Urban Form

Urban morphology is defined as the three-dimensional form and layout of the built environment and settlement pattern. From regional, urban, and district scales to finer-grained street grids that promote walkability and social cohesion, climate-resilient

planning and design strategies include configuration of urban morphology influenced by solar design, urban ventilation, and enhanced vegetation (see Figure 5.10). There are almost infinite combinations of different climate contexts, urban geometries, climate variables, and design objectives. A starting point in any project is to assess the micro- and macroclimatic characteristics of the site, an exercise that will indicate appropriate bioclimatic design strategies (Brophy et al., 2000). As the climate heats up, compact communities offer attractive alternatives to suburban sprawl by featuring comfortable, healthy microclimates with comparable natural amenities.

Wind velocities in cities are generally lower than those in the surrounding countryside due to the obstruction to air flow caused by buildings. In dense, compact communities, natural ventilation is challenging during warm months, often leading to increased cooling demand. This is partly because natural ventilation systems require very little energy but may need more space to accommodate low-resistance air paths (Thomas, 2003). Built-up areas with tall buildings may lead to complex air movement through a combination of wind channeling and resistance, often resulting in wind turbulence in some areas and concentrated pollution where there are wind shadows (Brophy et al., 2000). In general, denser developments result in a greater reduction in wind speeds but proportionally increased turbulence. Compact developments have less heat loss because there is generally less surface area for the volume enclosed due to shared wall space (Thomas, 2003).

2. Form and Layout

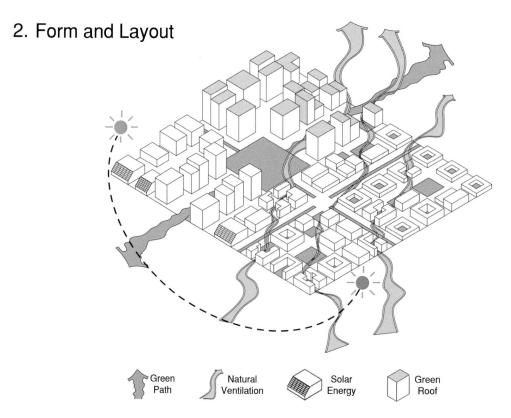

Green Path   Natural Ventilation   Solar Energy   Green Roof

Figure 5.10 *Urban form and layout.*

Source: Jeffrey Raven, 2016

3. Heat-Resistant Construction Material

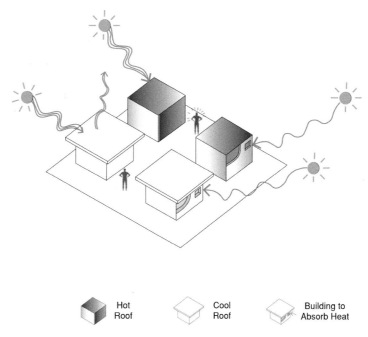

Hot Roof   Cool Roof   Building to Absorb Heat

Figure 5.11 *Surface reflectivity.*

Source: Jeffrey Raven, 2016

Street-level ventilation for warm and humid climates is key, using approaches that do not necessarily require changing morphology. A simulation exercise of the likely urban warming effects of the planned urban growth trajectories in the warm, humid city of Colombo, Sri Lanka, indicates that there are significant differences in the likely warming rates between different urban growth trajectories (Emmanuel et al., 2007). A moderate increase in built cover (at an LCZ class of compact midrise) appears to lead to the least amount of warming. At the neighborhood scale, streets oriented to the prevailing wind directions with staggered building arrangements together with street trees appear to offer the best possibility to deal with urban warming in warm humid cities. The combined approach could eliminate the warming effect due to the heat island phenomenon. The Hong Kong example (see Case Study 5.3) illustrates how existing high-rise districts can be retrofitted to exploit passive urban ventilation.

Exploiting prevailing breezes is a key factor in implementing district-wide passive cooling strategies (see Figure 5.10). Wind affects temperature, rates of evaporative cooling, and plant transpiration and is thus an important factor at a microclimatic level (Brophy et al., 2000). Urban morphology is responsible for varying the "porosity" of the city and the extent of airflow through it, and it is a lynchpin for using passive cooling to reduce energy loads in the built environment (Smith et al., 2008). Wind flow across evapotranspiring surfaces and water bodies provide cooling benefits. The morphology and surface roughness of the built environment has significant impacts on the effectiveness of urban ventilation.

Passive methods to increase comfort and reduce energy loads through solar design include orienting street and public space layout to reduce solar gain during hot months, shading through the configuration of adjacent vegetation, orienting neighborhood configurations to the sun's path to maximize daylight in ground floor living rooms, placing tall buildings to the north edges of a neighborhood to preserve solar potential for photovoltaic arrays, varying building heights and breaks in the building line to reduce shadowing and increase solar access during cold months, and maximizing use of cool surfaces and reflective roofs in hot climates. Figures 5.10, 5.11, and the Masdar example (Case Study 5.4) illustrate these approaches.

5.4.3 Construction Materials

Increasing the surface reflectivity or *albedo* of urban materials is a well-established urban heat management strategy. Due to the darkly hued paving and roofing materials distributed throughout cities, a larger quantity of solar energy is often absorbed in cities than in adjacent rural areas with higher surface reflectivity, thus contributing to a lower albedo (see Chapter 2, Urban Climate Science). Unable to compensate for an enhanced absorption of solar energy through an increase in evapotranspiration, a larger percentage of this absorbed energy is returned to the atmosphere as sensible heat and longwave radiation, raising temperatures (see Figure 5.11).

Recognition of the potential to measurably cool cities through the application of highly reflective coatings to roofing surfaces

Case Study 5.1 Green Infrastructure as a Climate Change Adaptation Option for Overheating in Glasgow, UK

Rohinton Emmanuel

Glasgow Caledonian University, Glasgow

Keywords	Urban overheating, green infrastructure, green area ratio method, planning and design
Population (Metropolitan Region)[a]	606,340 (National Records of Scotland, 2016)
Area (Metropolitan Region)[b]	3,345.97 km² (Office for National Statistics, 2012)
Income per capita	US$42,390 (World Bank, 2017)
Climate zone	Cfb – Temperate, without dry season, warm summer (Peel et al., 2007)

[a]Counting the following Local Authority areas: East Dunbartonshire; East Renfrewshire; Glasgow City; Inverclyde; North Lanarkshire; Renfrewshire; South Lanarkshire; West Dunbartonshire

[b]Counting the following Local Authority areas: East Dunbartonshire; East Renfrewshire; Glasgow City; Inverclyde; North Lanarkshire; Renfrewshire; South Lanarkshire; West Dunbartonshire

From its medieval ecclesiastical origins, Glasgow (originally *Glaschu* – 'dear green place') expanded into a major port in the 18th century, and, with the advent of the Industrial Revolution, added a massive industrial base to its already well-developed built fabric. However, the success of its industrial base could not withstand the pressures of globalization, and, by the early 20th century, the city had begun to lose population. This decline appears to have been arrested in recent years. The long history of growth, decline, and regrowth provides Glasgow a historic opportunity to recreate its "green" past.

Emmanuel and Kruger (2012) showed that even when urban growth had subsided, Glasgow's local warming that results from urban morphology (increased built cover, lack of vegetation, pollution, anthropogenic heat generation) continues to generate local heat islands. Such heat islands are of the same order of magnitude as

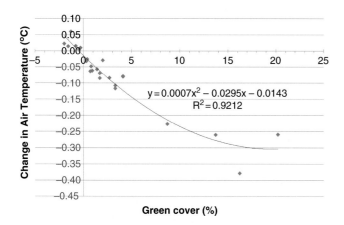

Case Study 5.1 Figure 1 *Summer daytime temperature.*

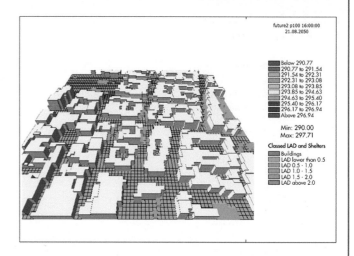

Case Study 5.1 Figure 2 *Extra green cover needed in reduction against green cover in Glasgow City Center to mitigate overheating.*

the predicted warming due to climate change by 2050. And the microscale variations are strongly related to local land cover/land-use patterns.

MITIGATING URBAN OVERHEATING

An option analysis exploring the role of green infrastructure (landscape strategies) in and around the Glasgow (Glasgow Clyde Valley, GCV) Region revealed the following (Emmanuel and Loconsole, 2015):

1. Green infrastructure could play a significant role in mitigating the urban overheating expected under a warming climate in the GCV Region.
2. A green cover increase of approximately 20% over the present level could eliminate a third to a half of the expected extra urban heat island (UHI) effect in 2050.
3. This level of increase in green cover could also lead to local reductions in surface temperature by up to 2°C.
4. More than half of street users would consider a 20% increase in green cover in the city center to be thermally acceptable, even under a warm 2050 scenario.

ACHIEVING GREEN COVER

Not all green areas contribute equally to local cooling, nor are they equal in their other environmental and sustainability benefits. Recognizing this, planners have begun to develop weighting systems that capture the relative environmental performance of different types of green cover. The most widely used among these is the Green Area Ratio (GAR) method (Keeley, 2011). GAR is currently implemented in Berlin and has been adapted in Malmo (Sweden), several cities in South Korea, and Seattle (USA). Elements of GAR include:

Impermeable surfaces (i.e., surfaces that do not allow the infiltration of water)

Includes roof surfaces, concrete, asphalt and pavers set upon impermeable surfaces or with sealed joints). = *0.0*

Impermeable surfaces from which all storm water is infiltrated on property

Includes surfaces that are disconnected from the sewer system. Collected water is instead allowed to infiltrate on site in

Case Study 5.1 Table 1 *Alternative approaches to increasing green cover by 20% in Glasgow City Center.*

Scenario	Permeable vegetated area (m²)	Street trees (Nos.)	Intensive roof gardens (m²)	Extensive roof gardens (m²)	Green façades
1. A single large park	1,056				
2. Street trees only		528			
3. 50% of additional greenery in street trees, balance intensive roof gardens		264	755		
4. 50% of additional greenery in street trees, balance extensive roof gardens		264		1,056	
5. Mix of intensive (50%) and extensive (50%) roof gardens			755	1,056	
6. 50% of all "sun facing (i.e., South & West) façade covered by green façades					1,268

a swale or rain garden. Guidelines for preventing groundwater and soil contamination must be followed. = *0.2*

Nonvegetated, semi-permeable surfaces
Includes cover types that allow water infiltration, but do not support plant growth. Example include brick, pavers and crushed stone. = *0.3*

Vegetated, semi-permeable surfaces
Includes cover types that allow water infiltration and integrate vegetation such as grass. Examples include wide-set pavers with grass joints, grass pavers, and gravel-reinforced grassy areas. = *0.5*

Green façades
Includes vines or climbing plants growing (often from ground) on training structures such as trellises that are attached to a building. The façade's area is measured as the vertical area the selected species could cover after 10 years of growth up to a height of 10 meters; window areas are subtracted from the calculation. = *0.5*

Extensive green roofs
Includes green roofs with substrate/soil depths of less than 80 centimeters. However, Berlin excludes green roofs constructed on high-rise buildings. = *0.5*

Intensive green roofs and areas underlain by shallow subterranean structures
Includes green roofs with substrate/soil depths of greater than 80 centimeters. This category includes subterranean garages. = *0.7*

Vegetated areas
Any area that allows unobstructed infiltration of water without evaluation of the quality or type of vegetation present. Examples range from lawns to gardens and naturalistic wooded areas. = *1.0*

A system of target setting is initially required that takes into account the severity of the environmental risk faced by a particular urban neighborhood. Once the target green cover is determined, the above-indicated weighting is used to develop alternate green infrastructure scenarios.

Case Study 5.1 Table 1 shows alternate approaches to a 20% increase in green cover in Glasgow city center. These employ an urban park, street trees, roof gardens, façade greening, or combinations of these.

and streets has led to the development of new product lines known as "cool" roofing and paving materials. For roofing surfaces concealed from ground view, such as atop a flat industrial building, very high-albedo, cool material coatings can be applied to reflect away a substantial percentage of incoming solar radiation. Industry analyses of these materials have found that the surface temperature of roofing materials can be reduced by as much as 50°F/10°C during periods of intense solar gain (Gaffin et al., 2012).

To explore the extent to which cool materials could reduce temperatures not only within the treated buildings themselves but also throughout the ambient urban environment, scientists at the Lawrence Berkeley Labs and the Columbia Center for Climate Systems Research have modeled extensive albedo enhancement strategies (Rosenzweig et al., 2014). Measured on a scale of

0 to 1, average surface albedos in U.S. cities tend to range from 0.10 to 0.20, much lower than the albedos of 0.6 to 0.8 associated with cool roofing and paving materials. In densely settled districts such as Manhattan, the potential to raise average albedos is great, but all cities can enhance their reflectivity through the use of higher albedo materials in routine resurfacing over time. Finding optimal values of reflectivity adjustment, rather than an all-out pursuit of maximum attainable values over large swaths of surface areas, will limit potential hydroclimatic trade-offs while still attaining temperature reduction goals (Georgescu et al., 2014; Jacobson and Ten Hoeve, 2012).

An important advantage of albedo enhancement over other urban climate management strategies is its relatively low cost. Cool roofing treatments can be applied to low-sloping roofs for a cost premium of between US$0.05 to US$0.10 per square foot, raising the cost of a 1,000 square foot roofing project by as little as US$100. Balanced against this low initial cost are annual energy savings estimated by the U.S. Environmental Protection Agency (EPA) to be about U.S. $0.50 per square foot, an estimate accounting for potentially greater winter heating costs (U.S. EPA, 2008). Also advantageous is the immediacy of beneficial returns from cool materials strategies, especially in semi-arid areas where the use of water for vegetation is not sustainable. In contrast to tree planting and other vegetative programs through which maximum cooling benefits are not realized until plants reach maturity, high-albedo coatings yield maximum benefits upon installation, with benefits diminishing somewhat thereafter with weathering, and as roofs become soiled.

5.4.4 Green and Blue Infrastructure

The interaction of green and blue components in the urban environment links together integrated mitigation and adaptation strategies at different scales – from buildings and open spaces design to landscape design and metropolitan region planning – and can yield many co-benefits (see Figure 5.12) (see Chapter 8, Urban Ecosystems). A comprehensive climate-based design supports developing and maintaining a network of green and blue infrastructure integrated with the built environment to conserve ecosystem functions and provide associated benefits to human populations (STAR Communities, 2014). Urban planning and urban design strategies focusing on green infrastructure and

---

Case Study 5.2 Adapting to Summer Overheating in Light Construction with Phase-Change Materials in Melbourne, Australia

Jun Han[1,2], Xiaoming Wang[1], Dong Chen[1]

[1]*Commonwealth Scientific and Industrial Research Organization (CSIRO), Melbourne*
[2]*Heriot-Watt University, Dubai Campus*

Keywords	Thermal comfort, residential buildings, passive cooling, phase change materials, mitigation and adaptation, planning and design
Population (Metropolitan Region)	4,258,000 (UN, 2016)
Area (Metropolitan Region)	9,999.5 km² (Australian Bureau of Statistics, 2013)
Income per capita	US$54,420 (World Bank, 2017)
Climate zone	Cfb – Warm temperate, fully humid, warm summer (Peel et al., 2007)

Melbourne, ranked one of the most livable cities around the world since 2011, is the capital city in the state of Victoria and the second most populous city in Australia (2006 Census QuickStats). It has a population of 3.99 million living in the greater metropolis (Australian Bureau of Statistics, 2013).

As a result of recent rapid urbanization, the city has undergone an outward expansion. The recent construction boom in both the Central Business District and nearby suburbs has led to a significant change in land use. This may imply that more buildings will be constructed in the near future and, consequently, more greenhouse gas (GHG) emissions from building operations. Change in land use such as replacement of green space by construction and increasing concrete or paved roads can be anticipated if appropriate urban planning for climate adaptation is lacking.

Existing studies have recognized the challenges facing current urbanization posed by the urban heat island (UHI) effect and global

warming. Exacerbated thermal conditions in the urban built environment and increasing human health issues can be expected without proper intervention. In this regard, we now face challenges not only in designing low-energy buildings to reduce GHG emissions for mitigating global warming but also to meet thermal comfort requirements without sacrificing indoor environment quality (IEQ) to actively adapt to climate change.

However, modern construction methods introduce more lightweight buildings. These methods employ offsite, prefabrication strategies to reduce construction time. Consequently, there is a potential risk of overheating and deteriorated thermal comfort conditions with lightweight construction products, which are likely to be exacerbated in a warming climate. Occupant thermal comfort in lightweight buildings therefore is receiving increasing attention among architects and designers.

In this Case Study, phase-change materials (PCMs) are used as a heat sink to absorb heat from the sun during the day and to reduce rapid room temperature rise due to added thermal stability. To examine the effectiveness of a lightweight building using PCMs, a one-dimensional numerical model was developed and solved by an enthalpy[1] method with an explicit scheme. The performance of PCMs for cooling a lightweight building with a brick veneer residential wall during the hot summer of 2009 in Melbourne was predicted numerically. The study reveals that application of PCMs in lightweight buildings could achieve better thermal comfort and energy savings in summer.

Bio-based PCM, applied instead of conventional wall insulation is made with a mix of soy-based chemicals that change from liquid to solid and vice versa at specific melting or solidification temperatures. The advantage of using PCMs is increased thermal storage capacity. It is estimated that about 30% of heating and cooling costs would be reduced.

Case Study 5.2 Figure 1 depicts the geometric configuration of the thermally enhanced brick veneer wall with PCMs and its original configuration without PCMs before modification. The thermally enhanced PCM wall is composed of 110-millimeter brickwork, 40-millimeter air gap, 20-millimeter PCMs, and 10-millimeter plasterboard.

---

1 *Enthalpy* is the thermodynamic quantity equivalent to the total heat content of a system. It is equal to the internal energy of the system plus the product of pressure and volume.

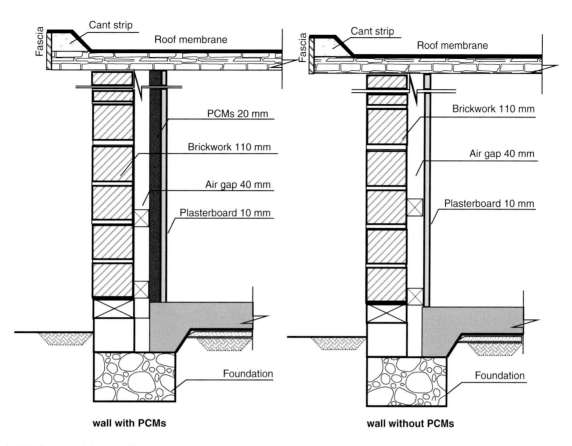

Case Study 5.2 Figure 1 *Schematic of lightweight brick veneer wall with integrated phase-change materials (left), and without (right).*

To understand the occupancy comfort level, the peak wall temperatures were considered when evaluating the performance of the PCM wall. The interior surface temperatures for the west-facing walls are compared for the PCM brick veneer wall and for the reference wall without PCM, as shown in Case Study 5.2 Figure 2. It was found that the interior surface temperature of the PCM wall in the day is lower than the conventional wall due to the presence of PCM heat storage during the daytime. The maximum peak temperature of the conventional brick veneer wall reached almost 48°C, whereas the maximum of the PCM wall was around 32°C. It is generally believed that lower surface temperatures result in greater occupant thermal comfort and energy savings in summer. A significant peak cooling load reduction therefore would be expected.

The successful testing of PCMs in lightweight building materials in the weather conditions of Melbourne demonstrated the effectiveness and validity of both mitigation and adaptation strategies. Combined with other sustainable energy technologies as an integrated approach for climate change mitigation and adaptation, PCMs could be useful for other cities with similar conditions.

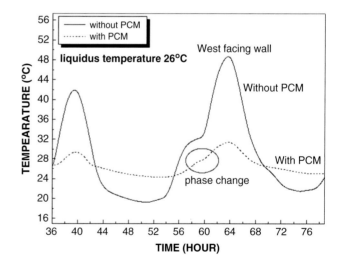

Case Study 5.2 Figure 2 *Surface temperature profiles of two west-facing walls with and without phase-change materials.*

sustainable water management help restore interactions between built and ecological environments. This is necessary to improve the resilience of urban systems, reduce the vulnerability of socio-economic systems, and preserve biodiversity (UNEP, 2010).

Integration of water management with urban planning and urban design represents an effective opportunity for climate change adaptation (UNEP, 2014) (see Figure 5.13). This has been demonstrated by the emerging *Water Sensitive Urban Design* (*WSUD*) approach (Ciria, 2013; ARUP, 2011; Flörke et al., 2011; Hoyer et al., 2011; BMT WBM, 2009). All the elements of the water cycle and their interconnections are considered to achieve together an outcome that sustains a healthy natural environment while addressing societal needs and reducing climate-related risks (Ciria, 2013). The implementation of integrated water cycle management as adaptive design strategy should be based on a

155

4. Vegetative Cover

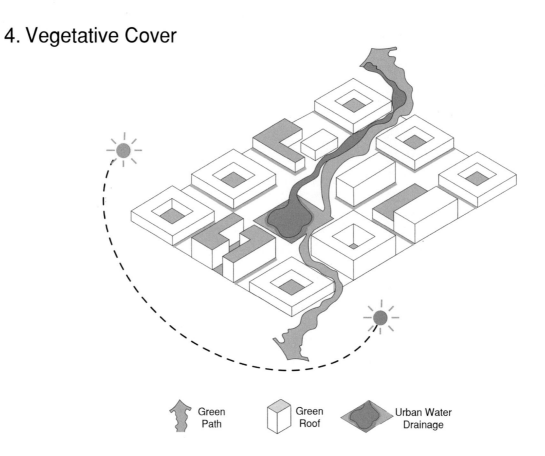

Green Path

Green Roof

Urban Water Drainage

Figure 5.12 *Surface cover.*

Source: Jeffrey Raven, 2016

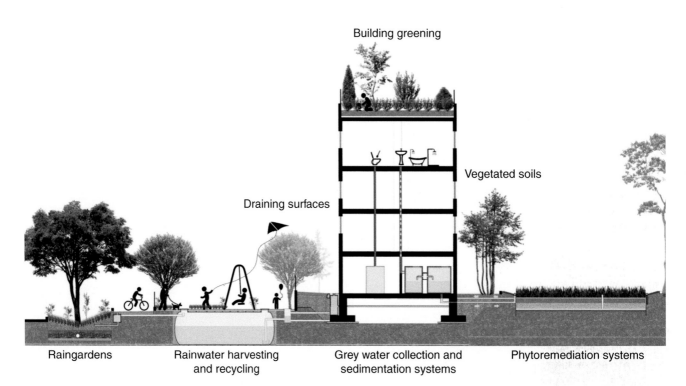

Building greening

Vegetated soils

Draining surfaces

Raingardens

Rainwater harvesting and recycling

Grey water collection and sedimentation systems

Phytoremediation systems

Figure 5.13 *Green and blue infrastructure design: building/open space scale, Naples, Italy.*

Source: Cristina Visconti and Mattia Leone

Case Study 5.3   Application of Urban Climatic Map to Urban Planning of High-Density Cities: An Experience from Hong Kong

Edward Ng and Chao Ren

Chinese University of Hong Kong

Lutz Katzschner

University of Kassel

Keywords	Heat island effect, high density, urban climatic map, ventilation corridors, planning and design
Population (Metropolitan Region)	7,310,000 (GovHK, 2016)
Area (Metropolitan Region)	1,105.7 km² (GovHK, 2016)
Income per capita	US$60,530 (World Bank, 2017)
Climate zone	Cwa – Monsoon-influenced humid subtropical, hot summer (Peel et al., 2007)

Hong Kong is located on China's south coast and situated in a subtropical climate region with hot and humid summers. As a high-density city with a population of 7.3 million living on 25 square kilometers of land, Hong Kong has a hilly topography and 40% of the territory is classified as country-park, where development is prohibited; hence only about 25% is built-up. Due to limited land area and increasing land prices, taller and bulkier buildings with higher building plot ratios, very limited open space, large podium structures, and high building-height-to-street ratios have been built. These tall and wall-like buildings in the urban areas block the incoming wind and sea breezes. This leads to a worsening of urban air ventilation and exacerbates the city's urban heat island (UHI) intensity. The number of very hot days (maximum air temperature greater than 33°C) and very hot nights (maximum air temperature greater than 28°C) has increased, whereas the mean wind speeds recorded in urban areas over the past 10 years have decreased. This intensifies uncomfortable urban living, heat stress, and related health problems and increases energy consumption.

The Hong Kong Observatory has conducted studies that note that Hong Kong's urban temperature has been increasing over the decades (Leung et al., 2004). Good urban air ventilation is an effective adaptation measure for the UHI effect and rising temperatures under climate change. However, Hong Kong's urban wind environment is deteriorating due to intensive urban development that increases the surface roughness and blocks the free flow of air, leading to weaker urban air ventilation and higher urban thermal heat stress. Higher air temperatures and a higher occurrence and longer duration of heat waves will have a severe impact on urban living; therefore, there is a need to plan and design the city to optimize urban climatic conditions and urban air ventilation based on a better understanding of the UHI phenomenon and the urban climate to reduce the impact of urban climate and climate change.

The Planning Department of the Hong Kong SAR Government produces the Hong Kong Urban Climatic Map System (PlanD, 2012) to provide an evidence-based tool for planning and decision making.

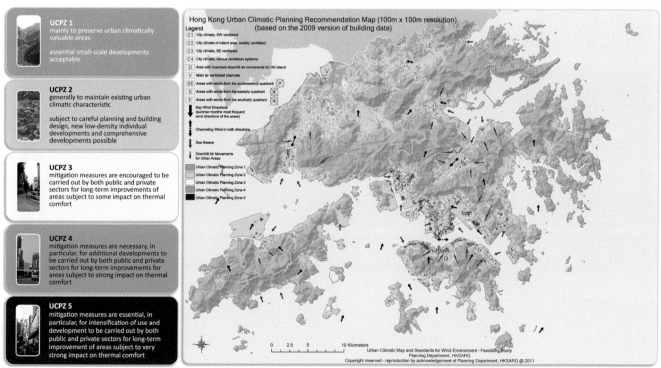

Case Study 5.3 Figure 1 *The Urban Climatic (Planning Recommendation) Map of Hong Kong.*

Source: Hong Kong Planning Department, HKSAR Government

The Urban Climatic (Planning Recommendation) Map classifies Hong Kong's urban and rural areas into five planning recommendation zones. General planning advice is given for each zone. Detailed advice is contained in the map's accompanying notes.

Based on a scientific understanding of the Hong Kong Urban Climatic Maps, future planning scenarios may be tested and effective adaptation measures (including advice on building density, site coverage, building height, building permeability, and greening) may be developed (PlanD, 2012). Prescriptive guidelines and performance-based methodologies in the Hong Kong Urban Climatic Map System provide further quantification. With a better understanding of urban climate, planners can balance various planning needs and requirements when making their final decisions.

Based on an understanding of the Urban Climatic Maps, the following planning and design measures should thus be taken into account in project planning and in the formulation of development parameters. They could help improve the urban climate and reduce the impact of climate change:

The UC-ReMap provides a strategic and comprehensive urban climatic planning framework and information platform for Hong Kong that can be also applied to other high-density cities. It helps to clarify and identify appropriate planning and design measures for the formulation of planning guidelines on matters related to urban climate and climate change, and it provides a strategic urban planning and development process for future development (e.g., maximizing the adaptation opportunities within urban climate planning zones (UCPZs) 3, 4, and 5) and accommodating comprehensive new development areas in UCPZ 2 with prudent planning and building design measures (PlanD, 2012). It also provides an urban climatic planning framework for reviewing outline zoning plans and formulating suitable planning parameters.

Case Study 5.3 Table 1 *Planning and design measures to be taken into account in project planning.*

Planning parameters	Recommendations
Building volume	Site plot ratio of 5 or less. Higher plot area must be adapted using other planning parameters.
Building permeability	25–33% of the project site's frontal elevation. Lower permeability must be adapted using other planning parameters.
Building site coverage	70% of the site area. Higher site coverage must be adapted using other planning parameters.
Air paths and breezeways	Open spaces must be linked with landscaped pedestrianized streets from one end of the city to the other end in the direction of the prevailing wind.
Building heights	Vary building heights so that there is a mixture of building heights in the area with an average aggregated differential of 50%.
Greenery	20–30% of tree planting preferably at grade, or essentially in a position less than 20 m from the ground level. Trees with large canopy and a leaf area index of more than 6 are preferred.

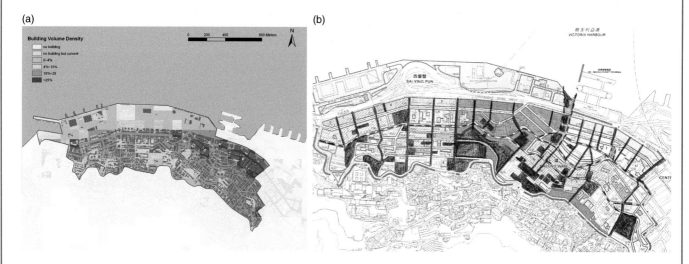

Case Study 5.3 Figure 2 *Building volume density study of the area (left); open spaces (blue) and air paths (red lines) suggested for the area (right), Hong Kong.*

Case Study 5.4   An Emerging Clean-Technology City: Masdar, Abu Dhabi, United Arab Emirates

Gerard Evenden, David Nelson,
Irene Gallou

Foster + Partners, London

Keywords	Carbon-free technologies, nearly-zero energy buildings, microclimate comfort, planning and design
Population (Metropolitan Region)	1,179,000 (UN, 2016)
Area (Metropolitan Region)	803 km² (Demographia, 2016)
Income per capita	US$72,850 (World Bank, 2017)
Climate zone	Bwh – Arid, desert, hot (Peel et al., 2007)

Masdar City is an emerging clean-technology cluster located in what aims to be one of the world's most sustainable urban developments powered by renewable energy. The project continues to be a work in progress. Located about 17 kilometers from downtown Abu Dhabi, the area is intended to host companies, researchers, and academics from across the globe, creating an international hub focused on renewable energy and clean technologies. The master plan is designed to be highly flexible, to benefit from emergent technologies, and to respond to lessons learned during the implementation of the initial phases. Expansion has been anticipated from the outset, allowing for growth while avoiding the sprawl that besets so many cities (Bullis, 2009; Manghnani and Bajaj, 2014).

The aim of a new development settlement characterized by a comfortable living environment in such an extreme desert climate required the implementation of adaptive design strategies to effectively respond to issues related to scarcity of precipitation, seasonal drought, high temperatures, and wide daily temperature range. The carbon-free new development sets new standards for climate change mitigation in arid countries through the adoption of nearly zero energy standards and building-integrated energy production from renewable sources.

The design concept explores the adoption of sustainable technologies and planning principles of traditional Arab settlements combined with contemporary city spatial-functional needs and state-of-the-art technological solutions to develop a carbon-neutral and zero-waste community despite the extreme climatic conditions. The quest for a mixed-use, low-rise, and high-density development, entirely car-free, with a combination of personal and public transit systems and pedestrian areas, is achieved through an extensive use of traditional solutions such as narrow streets and optimal orientation; shaded windows; exterior walls and walkways to control solar radiation; thick-walled buildings to maximize thermal mass and reduce energy consumption; courtyards and wind towers for natural ventilation; and

Case Study 5.4 Figure 1 *Masdar, Carbon-Neutral Development Case Study: Abu Dhabi, UAE.*

Source: Foster + Partners

Case Study 5.4 Figure 2 *Masdar wind tower.*

Source: Foster + Partners

(a)

(b)

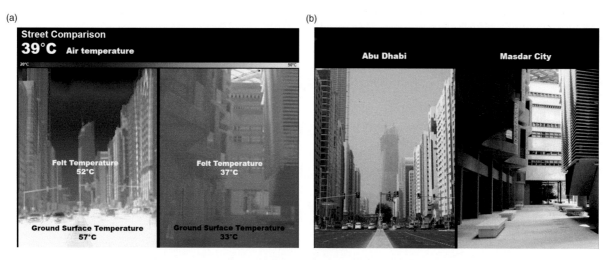

Case Study 5.4 Figure 3 *Thermal imaging comparing streetscapes in Abu Dhabi and Masdar City.*

Source: Foster + Partners

vegetation design with optimized water management to improve the local micro-climatic conditions of open spaces.

The northeast-southwest orientation of the city makes best use of the cooling night breezes and lessens the effect of hot daytime winds. Green parks separate built-up areas, not only to capture and direct cool breezes into the heart of the city, but also to reduce solar gain and provide cool pleasant oases throughout the city. The intelligent design of residential and commercial spaces, based on building standards currently set by internationally recognized organizations, reduces demand for artificial lighting and air conditioning. Such standards, adapted to the local climatic context, contributed to the

development of Abu Dhabi's "Estidama" rating system for sustainable building (Abu Dhabi Estidama Program, 2008).

Carefully planned landscape and water features lower ambient temperatures while enhancing the quality of the street. The elimination of cars and trucks at street level not only makes the air cleaner for pedestrians, but also allows buildings to be closer together, providing more shade but allowing maximum natural light. The placement of residential, recreational, civic, leisure, retail, commercial, and light industrial areas across the master plan, along with the public transportation networks, ensures that the city is pedestrian friendly and a pleasant and convenient place in which to live and work.

(a)

12:00 23rd July 2011

(b)

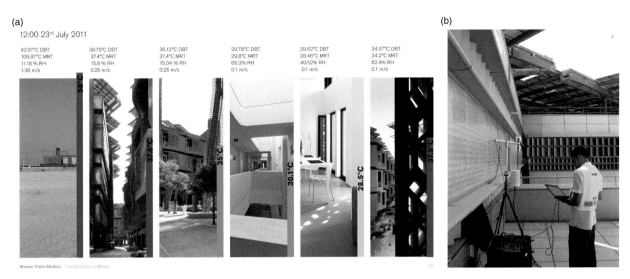

42.97°C DBT
106.97°C MRT
11.18 % RH
1.36 m/s

38.75°C DBT
37.4°C MRT
15.8 % RH
0.25 m/s

36.12°C DBT
37.4°C MRT
15.04 % RH
0.25 m/s

29.78°C DBT
29.8°C MRT
65.3% RH
0.1 m/s

29.62°C DBT
28.46°C MRT
49.52% RH
0.1 m/s

34.57°C DBT
34.2°C MRT
62.4% RH
0.1 m/s

Case Study 5.4 Figure 4 *Masdar field studies in the desert.*

Source: Foster + Partners

Wind towers, which can be found on both sides of the Arabic Gulf, are a traditional form of enhancing thermal comfort within the courtyard houses of the region. Masdar is exploring the possibility of using the principle to passively ventilate undercroft spaces and the public realm, thereby reducing the demand for mechanical ventilation systems.

Urban devices from local vernacular architecture – such as colonnades, wind towers, green canopies, and fountains – can bring the felt temperature down by 20 degrees compared to open desert. Cumulatively, all of these design principles have the effect of prolonging the moderate season in the city.

The land surrounding the city will contain wind and photovoltaic farms, as well as research fields and plantations, allowing the community to be entirely energy self-sufficient (Foster+Partners, 2009).

The design process has been based on studies and simulations of energy and thermal comfort, solar and wind analysis, material heat gain, and thermal imaging. Field measurement studies (Case Study 5.4 Figures 3 and 4) have been conducted to assess the microclimatic performance of spaces in Masdar City. Infrared thermal imaging was used in addition to hand-held equipment to track variations in temperature in urban spaces and then compared them to similar urban spaces in central Abu Dhabi and the desert. The comparisons of images show the superior performance of Masdar City due to the shade provided by built form, correct use of materials, natural ventilation, and evaporative cooling strategies.

dual definition of water as both resource and hazard (see Chapter 9, Coastal Zones). In the framework of urban design and planning, managing water as a resource addresses environmental quality and microclimate conditions of urban spaces, availability of water, rebalancing of ecosystem exchange, and the hydrological cycle in buildings and open spaces (ARUP, 2011). When considering water as a hazard, design should focus on the control of water discharge through runoff management and infiltration measures able to achieve wastewater retention and employing a decentralized sewage system.

Best practices of adaptation-driven urban policies worldwide provide significant examples of how the paradigm shift toward water-sensitive and water-resilient cities allows for the implementation of an integrated approach that combines risk prevention with a regeneration of urban fabric driven by adaptive design solutions (Kazmierczak and Carter, 2010). The Sydney Water Sensitive Urban Design (WSUD) Program and the post-Sandy "Rebuild by Design" initiatives in New York demonstrate such practices.

Recirculation of water on site is among the main concepts of a water-sensitive approach to urban design and planning.

It represents a key priority for enhancing water resilience and requires an integrated set of complementary measures, including decentralization of water discharge, harvesting and recycling, draining, and vegetated surfaces, thus improving urban microclimate and flood prevention.

Evaporative cooling processes, fostered by the development of green spaces in cities, allow for sustainable management of the water cycle and a reduction of the UHI effect. Building greening measures (sustainable roofs and vegetative facades) reduce the amount of water flowing into sewage systems, mitigate temperature extremes, provide thermal insulation, and increase biodiversity in urban areas (Ciria, 2007b; Schimdt et al., 2009; Nolde et al., 2007, Steffan et al., 2010; UNEP, 2012). Green streetscapes, including the use of permeable paving, provide shade and reduce thermal radiation.

Greening and permeable paving, as elements of storm-water management, have the potential to retain water that is then evaporated while delaying and reducing runoff (Scholz and Grabowiecki, 2007). Such a design approach for urban open spaces is strengthened by the integration of sustainable drainage systems (SuDS). This is a set of measures aimed at retaining and infiltrating storm water (bio-swales, rain gardens, retention basins, bio-lakes, wetlands, rainwater harvesting systems). This allows for the control of water discharge and reduces flood risk (Ciria, 2007a, 2010; Charlesworth, 2010; Poleto, 2012) (see Chapter 14, Urban Ecosystems).

The location and form of green infrastructure should be determined in relation to the built environment and aligned in relation to natural systems, including water bodies, solar impacts, and prevailing winds. A network of local microclimates can comprise small green spaces, planted courtyards, shaded areas, and "urban forests" to moderate temperature, as demonstrated in the Manchester, United Kingdom, Case Study. Vegetation should be sited to maximize the absorption rate of solar radiation. Localized water bodies can moderate temperature extremes through their high thermal storage capacity and through evaporative cooling.

5.5 Steps to Implementation

A planning and design approach to urban climate intervention should follow a four-phase approach: climate analysis mapping, public space evaluation, planning and design intervention, and post-intervention evaluation (see Figure 5.14).

5.5.1 Climate Analysis and Mapping

Considering climate in urban planning and urban design, the first step is to understand large-scale climatic conditions and individual inner-city local climates, including their reciprocal interactions (see Figure 5.14). Considerations include:

- Regional occurrence and frequency of air masses exchange (ventilation) and their frequencies;
- Seasonal occurrence of the thermal and air quality effects of urban climate (stress areas, insolation rates, shading conditions);
- Regional presentation and evaluation of the impact area and stress areas; and
- Energy optimization of location based on urban climate analysis with regard to areas with heat load, cooler air areas, and building density.

One also has to address sectoral planning (see Table 5.1).

Climate analyses and maps provide a critical first step in identifying urban zones subject to the greatest impacts associated with rising temperatures, increasing precipitation, and extreme weather events (see Figure 5.14). A climate analysis map may be developed in consecutive steps on the basis of spatial reference data. The spatial resolution is tailored to the planning level (see Table 5.1). Commonly employed climate analysis maps include urban heat hotspot and flood zone maps routinely employed in urban planning applications. Geographical Information System (GIS) layers include topographical information, buildings, roughness and greenery needed to create an urban climate map (see Figure 5.14).

5.5.2 Evaluation of Public Space

Urban climate is an essential part of urban planning evaluation. Urban climate maps are increasingly used in the planning process for urban development as well as for open space design. The public should be involved at all stages through the

Table 5.1 *Meteorological scales for planning. Source: Lutz Katzschner*

Instruments and Plans		Scale, Spatial Resolution of Maps	Climate Analysis Components (Air Quality and Human Biometeorology)
Regional Planning	Regional Land-Use Plan	1:50,000 to 1:100,000 ≥100 m	Meso-scale climate Comprehensive pollution control maps Thermal stress areas (overhead areas) Ventilation lanes Cool air production areas Planning recommendation map
Urban Planning Land-Use	Preliminary Urban Land Planning: Land-use plan	1:5000 to 1:25,000, 25 m to 100 m	Meso-scale climate Area-related ambient air quality maps Air exchange Thermal stress areas (overheated areas) Planning recommendation map
	Mandatory Urban Land Planning: Local development plan Planning permission and procedure	≤ (1:1000), 2m to 10 m	Micro-scale climate Local ambient quality calculations for "most severely affected areas" Neighborhood considerations Air exchange Human bio-meteorological suitability tests for "highly relevant areas" Planning recommendation map

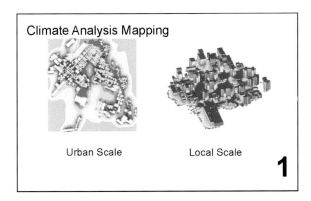

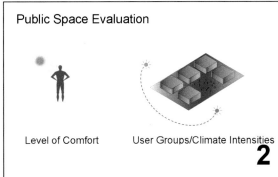

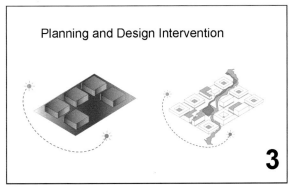

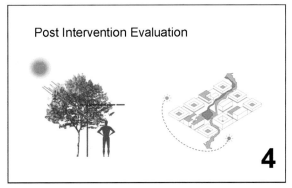

Figure 5.14 *Urban climate planning and design process.*

Source: Jeffrey Raven, 2016

use of interactive geographical information systems and/or surveys that help citizens foresee potential land-use changes (see Figure 5.14). This is enabled by climatic evaluation through spatial and temporal quantitative descriptions and specifications. At the regional level, areas worth protecting by virtue of their climatic functions, e.g., areas of heat load, fresh air supply, and ventilation pathways, are identified as key targets for planning measures.

5.5.3 Planning and Design Interventions

The task of planning and designing interventions relevant to urban climatology is to improve thermal conditions and air quality (see Figure 5.14):

*   Reduction of UHIs (heat islands being an indication of thermal comfort/discomfort) through open space planning
*   Optimization of urban ventilation via air exchange and wind corridors
*   Prevention of stagnating air in stationary temperature inversion conditions by eliminating barriers to air exchange
*   Maintenance and promotion of fresh air or cool air generation areas to further air exchange and improve air quality

Regional specifications may include sustaining cool air or fresh air generation areas (slopes) or ventilation lanes and taking into consideration building orientation, building height, and density of development. Such specifications may be implemented according to building codes in urban land-use planning.

In addition, mandatory regulations for areas designated as open spaces due to urban climate analyses are possible in the zoning plan. Regulations at the regional level should also be reviewed. Climatic concerns are thus considered in regional as well as city planning.

5.5.4 Post-Intervention Evaluation

Field measurement studies (see Case Study 5.4) should be conducted to assess the microclimatic performance of the urban design and planning intervention (see Figure 5.14). Infrared thermal imaging and/or population surveys can be undertaken to assess temperature variations compared to conditions prior to intervention. Climate-resilient strategies can have the effect of prolonging moderate temperatures with associated benefits to public health and energy savings.

5.6 Stakeholders and Public Engagement

Building resilient urban environments through integrated mitigation and adaptation approaches requires a strong framework for ongoing engagement with a broad range of stakeholders – households, regional communities, and local and national tiers of government, as well as academia, private businesses, health care providers, and civil society organizations – that make up the urban landscape.

Figure 5.15 *Urban design intervention – future climate scenario.*

Source: Urban Climate Lab, Graduate Program in Urban & Regional Design, New York Institute of Technology with Klimaat Consulting, 2014

Public engagement is as much about discourse as it is about design. From an urban planning and urban design perspective, community engagement and public participation are essential for operationalizing any policy, program, or intervention and offer many tangible and intangible benefits. Despite the potential obstacles for developing and carrying out a successful public outreach process, the absence of a robust stakeholder engagement process has many higher costs and ramifications. For instance, it can result in the development of locally

inappropriate solutions, increased conflict and tension between community groups, absence of a shared vision for the future, "rebound effect" (whereby appropriate solutions are not appropriately used due to lack of community awareness), and, ultimately, lack of preparedness and resiliency. Alternatively, structured conversations can facilitate participation, knowledge exchanges, shared decision-making, and ultimately, resiliency actions. While the process can often be messy and contentious because each group is pursuing different goals and interests, a robust public engagement process with genuine stakeholder participation and partnerships can ensure the sustained, inclusive, and meaningful transformation of regulations, built environments, and society (Kloprogge and Van Der Sluijs, 2006).

Community-led and place-based initiatives recognize and leverage local knowledge and expertise. Top-down or expert-driven outreach processes that lack genuine grassroots organizing and leave little opportunity for community-led initiatives can act as a barrier to an effective public outreach campaign. Civic engagement is often driven in an expert-dominated and exclusive manner (Velazquez et al., 2005). Participatory actions on the ground can be undermined or even neutralized by governing institutions' lack of willingness to change (Warburton and Yoshimura, 2005). These attitudes may lead to local resistance to implementing identified solutions. More functionally, simple, clear, and comprehensive resources must be readily available to the public in order to increase knowledge and awareness.

Even if decision-makers are actively seeking to develop genuine engagement and partnerships with stakeholders, an abundance of confusing and contradictory resources can serve as a barrier to robust engagement. Stakeholders and inhabitants must be provided tailored resources with clear and easy steps on how best to contribute to developing integrated mitigation and adaptation solutions for their communities. This not only helps ensure that integrated mitigation and adaptation solutions are appropriate for the local context (i.e., a good "fit" for the community and/or city), but also serves to empower communities and strengthen the relationships among government, the private sector, and citizens.

Public engagement is a means of ensuring the diverse needs of communities, particularly those of the most vulnerable groups, including populations with disabilities or chronic health conditions; seniors and children; those socially isolated, historically underrepresented, or otherwise marginalized; and people living below the poverty line. It is vital that these populations are integrated into the decision-making process and solutions (see Chapter 6, Equity and Environmental Justice).

Genuine and sustained stakeholder participation (as opposed to simply assessing opinions or asking for rubber-stamp approval on already-made plans) can draw out disagreements early, provide opportunities to work through different scenarios, and move plans toward a shared vision, thereby actually saving time, money, and political will. In order for stakeholders to be more committed, decision-makers and experts need to identify joint solutions that break down institutional and disciplinary silos. Urban governance can facilitate a robust civic participation process that creates mechanisms for systematic learning and capacity-building in communities as well as a transparent and open system in which responsibilities and accountabilities are clearly defined at the local level (see Chapter 6, Urban Governance). The dynamic and variable conditions that climate change introduces call for a robust stakeholder engagement process to help ensure integrated mitigation and adaptation responses are not simply implemented as one-off, discrete protection measures, but rather are incrementally adjusted and become part of the mechanisms for systematic learning, engagement, and transformation in a community. Community engagement offers decision-makers an opportunity to prototype a wider range of innovative solutions by creatively brainstorming, testing, and iterating. Despite focus by decision-makers on formal planned responses, most adaptation responses are actually carried out informally by individuals, households, and organizations. Therefore, communities that have been provided with information and resources and that possess increased knowledge about complex climate challenges and the integrated mitigation and adaptation solutions needed to adequately address them will more likely be willing to adopt resilient practices and policies (Yohe and Leichenko, 2010).

Although physical interventions often provide protection from only a single hazard or risk, communities that are integrated into the mitigation and adaptation planning process increase their capacity to prepare for, withstand, and recover from a wider range of climate-related disasters (not just a single hazard) as well as everyday challenges that span health, income, and equity considerations. Given that most cities have pre-existing vulnerabilities and that the potential for institutional and/or systems failures is ever-present, iterative, flexible, and redundant responses are needed to build ongoing capacity for adaptation. In other words, a robust stakeholder outreach process can yield benefits beyond simply helping to implement policies and instead offer opportunities for improved quality of life, public health, and equity.

The role of local authorities in the public engagement process is evolving as integrated mitigation and adaptation and sustainable urban development goals become a more significant priority for urban residents (Kloprogge and Van Der Sluijs, 2006). Local authorities must help facilitate and negotiate competing interests between urban challenges (sprawl, fragmentation of spaces, complexity of scales), social necessities (health, education, employment, culture, access to basic services), and environmental concerns (GHG emissions, ecosystems protection, resource management, and conservation) (Broto, 2017).

Local decision-makers can use a range of urban planning and urban design tools to overcome the obstacles to engaging the public. There are different levels of local participation (Donzelot, 2009). The lowest level is the simple distribution of information to inhabitants, whereas the highest level is direct engagement of communities to share decision-making and prioritize the results of consultations (Donzelot, 2009). For example, *charrettes*, an intense, multiday design exercise with the community, are commonly used in urban planning and design projects to engage the community to help create a plan for a particular site, area, or neighborhood.

The private sector is also at the forefront of eco-innovation systems to develop more sustainable cities (see Chapter 7, Economics, Finance, and the Private Sector). Examples include smarter solutions for energy efficiency and energy provision, development of renewable energy like geothermal systems, braking energy recovery from light rail systems, energy smart grids, and solutions to improve intermodal transport.

Although public engagement is a critical and necessary step for scaling integrated mitigation and adaptation solutions, there are real challenges that can inhibit ongoing engagement and, ultimately, action. As Bai et al. (2010) suggest, there is frequently an inherent temporal ("not in my term"), spatial ("not in my patch"), and institutional ("not my business")-scale mismatch between urban decision-making and global environmental concerns. Also, the involvement of stakeholders requires a general level of trust and cooperation (Tilly, 2005). A complication lies in the fact that decision-makers must often act as referees to solve potential conflicts and facilitate negotiations. However, to create consensus, promote cooperation, and move toward an equitable environmental management process, an intimate understanding of the motivations and drivers of the different stakeholder groups is needed (Schaltegger, 2003). Lack of transparency between the public and the private sectors can be a key obstacle to reaching consensus among different stakeholders. Stakeholders need to know who is accountable for the local decisions on sustainability and climate solutions and how they can contribute to the decision-making process.

5.7 Sector Linkages

There is a growing consensus around integrating urban planning and urban design, climate science, and policy to bring about desirable microclimates within compact, pedestrian-friendly built environments. However, there remains much work to do to bridge the gaps in tools, methods, and language between the scientific, design, and policy-making communities. Conveying a compelling investment/payback narrative also remains a challenge for setting priorities with stakeholder groups. *Ad hoc*, disconnected approaches fail to exploit synergies between professional practitioners, and departments within government administrations are often insufficiently coordinated to capitalize on cross-disciplinary actions. Silos of expertise are difficult to harness over the long-term due to different departmental missions. A central challenge remains the poor interdisciplinary connections between the various policy experts, technical specialists, and urban planners/urban designers. The absence of objective evaluation methodologies in the practice of climate-resilient urban design illustrates this divergence, as does the lack of a common interdisciplinary methodology for addressing various spatial scales (Odeleye et al., 2008). For example, urban climatology research produces sophisticated but theoretical results that resist easy integration with empirical, design-oriented findings of urban design (Ali-Toudert et al., 2005).

Although experts in sustainable buildings and ecological footprints are familiar with sustainability metrics to measure progress, the absence of objective evaluation of urban morphology across spatial scales is a challenge in the urban design profession. In a recent survey on sustainability methods and indicators in three built-environment professions in the United Kingdom (planning, urban design, and architecture), sustainability guidance documents had not led to a range of formal, systemic, or

morphological sustainability indicators in the urban design sector – instead continuing to emphasize "place-making" and the quality/vitality of the public realm (Odeleye et al., 2008).

Although there has been extensive research on the UHI effect, the relationship between urban geometry and thermal comfort is by far less well understood and the numbers of studies are very few (Ali-Toudert et al., 2005). Traditional urban design tools must be refined and expanded to serve climate-resilient urban design.

From parcel and neighborhood scale to municipal and regional scale, a discontinuity of policy between scales challenges the public's understanding of holistic urban form. Consideration needs to be given to how regional decisions may affect neighborhoods or individual parcels and vice versa, and few tools have been developed to assess conditions in the urban environment at city block or neighborhood scale (Brophy et al., 2000; Miller et al., 2008).

For urban areas, the key is a coordinated response that addresses issues simultaneously rather than individually. For example, an important concern underlying natural ventilation is air quality, which means that transport management and building microclimate need to be linked. Whereas one of the most effective passive ventilation strategies is to introduce cooler night air to the perimeters of buildings, excessive pollution from nearby vehicles and industry often undermines this strategy in compact communities. This anthropogenic pollution is exacerbated by lack of space for air movement, resulting in insufficient ground-level air movement to disperse pollutants (Odeleye, 2008). The totality of the environment – including noise, activity, climate, and pollution – affects human health and well-being.

For compact development, proximity is the principal goal. Proximity requires integrating infrastructure, housing, and sustainable development into land-use planning to reduce the carbon footprint through compact development patterns. At the urban scale, a comprehensive approach to transportation from the perspective of sustainable development requires a holistic view of planning (see Chapter 13, Urban Transportation). The challenge remains to establish national frameworks and policies for integrated mitigation and adaptation to address sustainable development that encourages all sectors to coordinate and integrate their activities (Hall, 2006) (see Chapter 16, Governance and Policy).

5.8 Knowledge Gaps and Future Research

Urban areas alter their regional climates by adjusting the overlying airshed. A substantial number of observational studies across the world have illustrated the prevalence of warmer and drier conditions within cities, degraded air quality regimes, and altered hydrological patterns resulting from impacts on precipitation and changes in drainage associated with increased impervious cover. Advances in the physical understanding of the urban climate system together with progress in computing technologies has enabled the development and refinement of complex process-based models that characterize urban areas and their interaction with the overlying atmosphere in a mathematical framework (Chen et al., 2011). Such process-based models have considerable planning and design utility and are increasingly applied to examine the impact of urban expansion and of commonly proposed urban adaptation strategies to a long-term globally changing climate (Georgescu et al., 2014).

To support utility for the planning and design process of cities, such process-based models require two important improvements. The spatial extent and morphology of urban areas remain simplistic in contemporary modeling approaches (Chen et al., 2011). The nature of this representation is assumed to vary by urban land use and land cover, which is conditional on the density of urban structures. However, implied in this assumption is that a diversity of key morphological characteristics (e.g., sky view factor) within a particular urban cover (e.g., high-density residential) is nonexistent, an important condition that presents limitations for place-based planning and design decisions. Therefore, the realistically heterogeneous representation of cities within current modeling frameworks remains an important but as-yet unrealized objective. How effective landscape configuration can be as an adaptive strategy has only recently become a ripe area of research in climate modeling (Connors et al., 2013; Rosenzweig et al., 2014).

Waste heat resulting from energy use within cities (primarily from building heating, ventilation, and air conditioning [HVAC] systems) is also only crudely accounted for in current representations (Sailor, 2011). There does not yet exist a database of waste heat profiles for modeling applications for a diverse set of cities, and although efforts are under way to develop spatially explicit and time-varying heating profiles, such datasets remain absent for most urban areas (Chow et al., 2014). In regard to energy use, suburbs in the United States account for roughly 50% of the total domestic household carbon footprint due to longer commutes (Jones and Kammen 2014).

Such concerns highlight the importance of future cooperation between urban climatologists, planners, urban designers and architects. A key asset of process-based urban climate modeling frameworks is their ability to offer insights by examining the adaptive capacity of "what-if" growth scenarios and growth management strategies to inform the planning and design process prior to incipient stages of development.

Other major knowledge gaps concern the GHG emissions of different cities and the association between GHG emissions and different urban forms. The IPCC AR5 highlights the importance of the next several decades for influencing low-carbon urbanization. It is essential to develop urban GHG emission inventories and experiment with alternative urbanization patterns that facilitate low-carbon urban development. This is crucially related to medium-sized cities in developing counties such as China and India, which are expected to accommodate the majority of expected urban population growth.

The impacts of medium-sized cities on the climate at urban, regional, and global scales is a topic of considerable debate, but their comparatively small size poses a conundrum for researchers: how do we acquire and incorporate the relevant information into global understanding? Much research has focused on mapping these urban centers using demographic and administrative information often supplemented by remote sensing. However, these data provide little information on the internal makeup of cities, which is crucially important for understanding their GHG emissions and vulnerabilities. The absence of such information inhibits international comparisons, knowledge transfer, and effective integrated mitigation and adaptation.

5.9 Conclusions

This chapter has endorsed the concept of integrated mitigation and adaptation: climate management activities designed to reduce global GHG emissions while producing regional benefits related to urban heat, flooding, and other extremes.

Cities shaped by integrated mitigation and adaptation principles can reduce energy consumption in the built environment, strengthen community adaptability to climate change, and enhance the quality of the public realm. Through energy-efficient planning and urban design, compact morphology can work synergistically with high-performance construction and landscape configuration to create interconnected, protective, and attractive microclimates. The long-term benefits are also significant, ranging from economic savings through lowered energy consumption to the improved ability of communities to thrive despite climate-related impacts (Raven, 2011). A community's capacity to cope with adversity, adapt to future challenges, and transform in anticipation of future crises yields greater social resilience with particularly positive benefits for poor and marginalized populations (Keck and Sakdapolrak, 2013).

Annex 5.1 Stakeholder Engagement

The contributors to the ARC3.2 chapter on Urban Planning and Urban Design engaged stakeholder groups and experts throughout the chapter production process, with specific forums in Asia, Europe, and the United States. The International Conference on Urban Climate (ICUC9), held in Toulouse, France, in July 2015, is an international forum for global urban climatologists. Chapter Coordinating Lead Authors (CLAs), Lead Authors, Contributing Authors, and Case Study Authors participated in the conference, including Gerald Mills, Lutz Katzschner, Matei Georgescu, and Jeffrey Raven. The Chapter Key Findings and Major Messages were presented by the chapter CLA at the launch of the Urban Climate Change Research Network (UCCRN) European Hub, in partnership with the Centre National de la Recherche Scientifique, the Pierre and Marie Curie University, and l'Atelier International du Grand Paris launched in Paris, July 2015.

At the ARC3.2 Midterm Authors Workshop in London, in September 2014, which was attended by key chapter stakeholders, the Coordinating Lead Author and Lead Authors configured the scientific basis for the chapter's Major Messages and Key Findings. The chapter CLA presented Major Messages and Key Findings at Beijing University for Civil Engineering and Architecture (BUCEA) and Tongji University in Shanghai, in October 2014.

The 4th China–Europa Forum in Paris was held in December 2014, focusing on the theme "Facing Climate Change: Rethinking Our Global Development Model," in preparation for the UN FCCC Conference of the Parties (COP21) in Paris 2015. It brought together more than 300 participants through two plenary sessions and three round tables as well as twelve thematic workshops. As part of the conference, at the Project EAST workshop with lead author Pascaline Gaborit in Brussels, the Chapter CLA presented ARC3.2 key scientific findings and other research findings. At the closing plenary held on December 5 at the Town Hall of the 4th Arrondissement of Paris, ARC3.2 CLA Raven presented the ARC3.2 draft Major Messages. This included the most relevant points in the ARC3.2 chapter that will guide urban decision-makers.

At the New York City Department of City Planning, the ARC3 draft chapter Key Findings and Major Messages were presented to experts in the city government. Co-presenters included ARC3.2 lead author Gerald Mills, CLA Jeffrey Raven, and the President of the American Institute of Architects (AIANY). The ARC3.2-based presentation provided an operational framework, Case Studies, and a policy framework for NYC municipal government. The chapter CLA and editor introduced draft chapter Key Findings in discussions during the National Science Foundation Research Coordination Network project. The chapter CLA presented chapter Key Findings to climate and health experts at the Center for Disease Control in Atlanta, July 2014. The chapter CLA presented chapter Key Findings at the American Institute of Architects Dialogues on the Edge of Practice, February 2015, and presented ARC3.2 chapter Key Findings and Major Messages at the Center for Architecture in NYC, April 2015. Lead author Brian Stone and CLA Raven presented climate-resilient urban planning and urban design research related to the ARC3.2 chapter at the New York conference Extreme Heat: Hot Cities, November 2015.

The chapter's Key Findings and Major Messages was presented by the CLA at the New York Institute of Technology-Peking University Sustainable Megacities conference in Beijing, China in October 2015. A Critical Climate Change Debrief: COP21 Paris Conference was held at the Center for Architecture in New York, January 2016, where chapter CLA Jeffrey Raven presented how ARC3.2 strategies were tested in a Paris district during COP21, in collaboration with lead authors Gerald Mills and Mattia Leone.

The ARC3.2 Urban Planning and Urban Design chapter abstract was presented at the Design Solutions for Climate Change in Urban Areas conference in Naples, Italy, in July 2016. Jeffrey Raven (CLA), and Lead Author Mattia Leone each led sessions during the conference. Joining them in presenting ARC3.2's integrated cross-disciplinary framework was Chantal Pacteau who is Coordinating Lead Author of the ARC3.2 Mitigation and Adaptation chapter and Co-Director of the UCCRN European Hub. Jeffrey Raven (CLA) presented ARC3.2 urban planning and urban design research-action at the National Science Foundation Workshop, International Conference on Sustainable Infrastructure (ICSI), Shenzhen, China, in October 2016.

5 Urban Planning and Urban Design

References

Akbari, H., Pomerantz, M., and Taha, H. (2001). Cool surfaces and shade trees to reduce energy use and improve air quality in urban areas. *Solar Energy*, **70**(3), 295–310.

Ali-Toudert, F., and Mayer, H. (2005). Effects of street design on outdoor thermal comfort. Meteorological Institute, University of Freiburg.

Alley, R., Kwok, K., Lam, D., Lau, W., Watts, M., and Whyte, F. (2011). Water resilience for cities. ARUP. http://publications.arup.com/publications/u/urban_life_water_resilience_for_cities

American Planning Association (APA). (2015). What is planning?. American Planning Association. Accessed March 20, 2016: https://www.planning.org/aboutplanning/whatisplanning.htm

Arnfield, A. J. (2003). Two decades of urban climate research: A review of turbulence, exchanges of energy and water, and the urban heat island. *International Journal of Climatology* **23**(1), 1–26.

ARUP. (2014). C40 Climate action in megacities: A quantitative study of efforts to reduce GHG emissions and improve urban resilience to climate change in C40 cities. Accessed July 30, 2015: C40.org.

Australian Sustainable Built Environment Council (ASBEC). (2013). What is urban design? Accessed April 11, 2014: http://www.urbandesign.org.au/whatis/index.aspx

BMT WBM Pty Ltd. (2009). *Evaluating options for water sensitive urban design: A national guide*. Joint Steering Committee for Water Sensitive Cities (JSCWSC), Australia.

Bottema, M. (1999). Towards rules of thumb for wind comfort and air quality. *Atmospheric Environment* **33**(24),4009–4017.

Brophy, V., O'Dowd, C., Bannon, R., Goulding, J., and Lewis, J. O. (2000). Sustainable urban design. Energy Research Group, University College Dublin.

Broto, V.C. (2017). Urban Governance and the Politics of Climate Change. World Development. 93:1–15.

Charlesworth, S. (2010). A review of the adaptation and mitigation of Global Climate Change using Sustainable Drainage in cities. *Journal of Water and Climate Change* **1**(3), 165–180

Chen, F., Kusaka, H., Bornstein, R., Ching, J., Grimmond, C. S. B., Grossman-Clarke, S., and Zhang, C. (2011). The integrated WRF/urban modelling system: Development, evaluation, and applications to urban environmental problems. *International Journal of Climatology* **31**(2), 273–288.

Chen, L., Ng, E., An, X. P., Ren, C., He, J., Lee, M. Wang, U., and He, J. (2010). Sky view factor analysis of street canyons and its implications for intra-urban air temperature differentials in high-rise, high-density urban areas of Hong Kong: A GIS-based simulation approach. *International Journal of Climatology* **32**(1), 121–136. doi: 10.1002/joc. 2243 .(EdNg)

Chow, W. T., Salamanca, F., Georgescu, M., Mahalov, A., Milne, J. M., and Ruddell, B. L. (2014). A multi-method and multi-scale approach for estimating city-wide anthropogenic heat fluxes. *Atmospheric Environment* **99**, 64–76.

Connors, J. P., Galletti, C. S., and Chow, W. T. (2013). Landscape configuration and urban heat island effects: Assessing the relationship between landscape characteristics and land surface temperature in Phoenix, Arizona. *Landscape Ecology* **28**, 271–283. doi:10.1007/s10980-012-9833-1

Decker, E., Smith, B., and Rowland, F. (2000). Energy and material flow through the urban ecosystem. *Annual Review of Energy and the Environment* **25**, 685–740.

Dickie, S., Ions, L., McKay, G., and Shaffer, P. (2010). *Planning for SuDS – Making it Happen*. Ciria.

Donzelot, J. (2009). *La ville à trois vitesses*, Éditions de la Villette, 2009. Éditions de la rue d'Ulm, 2009 and presentation in Sénart 2009 Energy-Cités Info and newsletters.

Doulos, L., Santamouris, M., and Livada, I. (2004). Passive cooling of outdoor urban spaces. The role of materials. *Solar Energy* **77**(2), 231–249.

Duvigneaud, P., and Denayer-de Smet, S. (eds.). (1975). L'Ecosystème Urbain—Application à l'Agglomération bruxelloise. Publication de Colloque International organisé par L'Agglomération de Bruxelles 14 et 15 septembre 1974, Bruxelles.

Early P., Gedge D., and Wilson S. (2007b). *Building Greener: Guidance on the Use of Green Roofs, Green Walls and Complementary Features on Buildings*. Ciria.

Erell, E., Pearlmutter, D., and Williamson, T. (2012). *Urban Microclimate: Designing the Spaces Between Buildings*. Routledge.

Flörke, M., Wimmer, F., Laaser, C., Vidaurre, F., Tröltzsch, J., Dworak, T., Stein, U., Marinova, N., Jaspers, F., Ludwig, F., Swart, R., Long, H., Giupponi, G., Bosello, F., and Mysiak, J. (2011). *Final Report for the Project Climate Adaptation – Modeling Water Scenarios and Sectoral Impacts*. CESR – Center for Environmental Systems Research.

Gaffin, S.R., M. Imhoff, C. Rosenzweig, R. Khanbilvardi, A. Pasqualini, A.Y.Y. Kong, D. Grillo, A. Freed, D. Hillel, and E. Hartung, 2012: Bright is the new black – Multi-year performance of high-albedo roofs in an urban climate. Environ. Res. Lett., 7, 014029, doi:10.1088/1748-9326/7/1/014029.

Georgescu, M., Morefield, P. E., Bierwagen, B. G., and Weaver, C. P. (2014). Urban adaptation can roll back warming of emerging megapolitan regions. *Proceedings of the National Academy of Sciences* **111**(8), 2909–2914.

Gomez-Ibanez, D. J., Boarnet, M. G., Brake, D. R., Cervero, R. B., Cotugno, A., Downs, A., Hanson, S., Kockelman, K. M., Mokhtarian, P. L., Pendall, R. J., Santini, D. J., and Southworth, F. (2009). *Driving and the built environment: The effects of compact development on motorized travel, energy use, and CO2 emissions*. Oak Ridge National Laboratory (ORNL).

Grimmond, S. (2007). Urbanization and global environmental change: Local effects of urban warming. *Geography Journal* **173**, 83–88.

Guttikunda, S. K., Carmichael, G. R., Calori, G., Eck, C., and Woo, J. H. (2003). The contribution of megacities to regional sulfur pollution in Asia. *Atmospheric Environment* **37**(1), 11–22.

Hall, R. (2006). *Understanding and Applying the Concept of Sustainable Development to Transportation Planning and Decision-Making in the U.S.* Doctoral dissertation, MIT.

Hoyer, J., Dickhaut, W., Kronawitter, L., and Weber, B. (2011). *Water sensitive urban design*. Jovis Verlag GmbH.

Hough, M. (1989). *City Form and Natural Process: Toward a New Urban Vernacular*. Routledge.

International Panel on Climate Change (IPCC). (2014a). *Climate Change 2014: Synthesis Report. Contribution of Working Groups I, II and III to the Fifth Assessment Report of the Intergovernmental Panel on Climate Change [Core Writing Team*, Pachauri, R. K., and Meyer, L. A. (eds.)]. IPCC.

Jacobson, M. Z., and Ten Hoeve, J. E. (2012). Effects of urban surfaces and white roofs on global and regional climate. *Journal of Climate* **25**(3), 1028–1044.

Jonathan Rose Companies. (2011). Location efficiency and housing type: Boiling it down to BTUs. Accessed December 8, 2014: https://www.epa.gov/sites/production/files/2014-03/documents/location_efficiency_btu.pdf

Jones, C., and Kammen, D. M. (2014). Spatial distribution of U.S. household carbon footprints reveals suburbanization undermines greenhouse gas benefits of urban population density. *Environmental Science and Technology* **48**(2), 895–902.

Kazmierczak, A., and Carter, J. (2010). Adaptation to climate change using green and blue infrastructure: A database of case studies. University of Manchester, Interreg IVC Green and blue space adaptation for urban areas and eco-towns (GRaBS). Accessed December 20, 2015: https://orca.cf.ac.uk/64906/1/Database_Final_no_hyperlinks.pdf

Keck, M., and Sakdapolrak, P. (2013). What is social resilience? Lessons learned and ways forward. *Erdkunde* **67**(1), 5–19. Accessed August 20, 2014: http://www.academia.edu/3110553/What_is_Social_Resilience_Lessons_Learned_and_Ways_Forward

Kennedy, C., Steinberger, J., Gasson, B., Hansen, Y., Hillman, T., Havranek, M., Pataki, D., Phdungsilp, A., Ramaswami, A., and Mendez, G. V. (2009). Greenhouse gas emissions from global cities. *Environmental Science & Technology* **43**(19), 7297–7302.

Kloprogge, P. and Van Der Sluijs, J.P. (2006). The inclusion of stakeholder knowledge and perspectives in integrated assessment of climate change. Climatic Change. 75:359–389.

Knowles, R. (2003). The solar envelope: Its meaning for energy and buildings. *Energy and Buildings* **35**, 15–25.

Konrad, C. P. (2003). Effects of urban development on floods. U.S. Geological Survey Fact Sheet 076–03.

Kravčík, M., Pokorný, J., Kohutiar, J., Kováč, M., and Tóth, E. (2007). *Water for the Recovery of the Climate – A New Water Paradigm*. Municipality of Tory.

Laidley, T. (2015). Measuring sprawl: A new index, recent trends, and future research. *Urban Affairs Review* 1078087414568812. Accessed April 6, 2016: http://uar.sagepub.com/citmgr?gca=spuar%3B1078087414568812v1

Lee, J. (2012). *Transit Synergized Development – Framework for a Smart, Low-Carbon, Eco-City*. Accessed August 20, 2014: http://www.academia.edu/6286908/Transit_Synergized_Development_Framework_for_a_Smart_Low-Carbon_Eco-City_Executive_Summary_

Li, W., Chen, S., Chen, G., Sha, W., Luo, C., Feng, Y., and Wang, B. (2011). Urbanization signatures in strong versus weak precipitation over the Pearl River Delta metropolitan regions of China. *Environmental Research Letters* **6**(3), 034020.

Lindau, L.A., Hidalgo, D., and Facchini, D. (2010). Curitiba, the Cradle of Bus Rapid Transit. Build Environment (1978). 35(3):274–282.

Lowry, W. P. (1977). Empirical estimation of urban effects on climate: A problem analysis. *Journal of Applied Meteorology* **16**(2), 129–135.

Luhmann N. (2006). La Confiance un mécanisme de réduction de la complexité sociale., 2–111.

McGranahan, G., et al. (2007). The rising tide: Assessing the risks of climate change and human settlements in low elevation coastal zones. *Environment and Urbanisation* **19**, 17–37.

Mehrotra, S., Lefevre, B., Zimmerman, R., Gercek, H., Jacob, K., and Srinivasan, S. (2011). Climate change and urban transportation systems. In Rosenzweig, C., Solecki, W. D., Hammer, S. A., Mehrotra, S. (eds.), *Climate Change and Cities: First Assessment Report of the Urban Climate Change Research Network* (145–177). Cambridge University Press.

Mentens, J., Raes, D., and Hemy, M. (2006). Green roofs as a tool for solving the rainwater runoff problem in the urbanized 21st century? *Landscape and Urban Planning* 77, 217–226.

Miller, N., Cavens, D., Condon, P., Kellett, N., and Carbonell, A. (2008). Policy, urban form and tools for measuring and managing greenhouse gas emissions: The North American problem. Climate change and urban design: The third annual congress of the Council for European Urbanism, Oslo.

Miller, R. B., and Small, C. (2003). Cities from space: Potential applications of remote sensing in urban environmental research and policy. *Environmental Science & Policy* **6**(2), 129–137.

Mills, G. (2005). Urban form, function and climate. Dept. of Geography, University of California, Davis. Accessed August 30, 2014: epa.gov/heatisland/resources/pdf/GMills4.pdf

Mills, G. (2004). The urban canopy layer heat island. IAUC Teaching Resources. Accessed June 18, 2015: www.epa.gov/heatislands/resources/.../HeatIslandTeachingResource.pdf

Monks, P. S., Granier, C., Fuzzi, S., Stohl, A., Williams, M. L., Akimoto, H., Amann, M., Baklanov, A., Baltensperger, U., Bey, I., Blake, N., Blake, R. S., Carslaw, K., Cooper, O. R., Dentener, F., Fowler, D., Fragkou, E., Frost, G. J., Generoso, S., Ginoux, P., Grewe, V., Guenther, A., Hansson, H. C., Henne, S., Hjorth, J., Hofzumahaus, A., Huntrieser, H., Isaksen, I. S. A., Jenkin, M. E., Kaiser, J., Kanakidou, M., Klimont, Z., Kulmala, M., Laj, P., Lawrence, M. G., Lee, J. D., Liousse, C., Maione, M., McFiggans, G., Metzger, A., Mieville, A., Moussiopoulos, N., Orlando, J. J., O'Dowd, C. D., Palmer, P. I., Parrish, D. D., Petzold, A., Platt, U., Poeschl, U., Prevot, A. S. H., Reeves, C. E., Reimann, S., Rudich, Y., Sellegri, K., Steinbrecher, R., Simpson, D., ten Brink, H., Theloke, J., van der Werf, G. R., Vautard, R., Vestreng, V., Vlachokostas, C., and von Glasow, R. (2009). Atmospheric composition change – global and regional air quality. Atmospheric *Environment* 43, 5268–5350. doi:10.1016/j.atmosenv.2009.08.021

Morgan, C., Bevington, C., Levin, D., Robinson, P., Davis, P., Abbott, J., and Simkins, P. (2013). *Water Sensitive Urban Design in the UK – Ideas for Built Environment Practitioners*. Ciria.

Newman, P. G., and Kenworthy, J. R. (1989). *Cities and Automobile Dependence: An International Sourcebook*. Gower Technical.

Nolde, E., Vansbotter, B., Rüden, H., König, K.W. (2007). *Innovative water concepts. Service water utilization in buildings*. Berlin Senate Department for Urban Development. Accessed January 17, 2015: http://www.stadtentwicklung.berlin.de/internationales_eu/stadtplanung/download/betriebswasser_englisch_2007.pdf

Nolon, J. R. (2012). Land use for energy conservation and sustainable development: A new path toward climate change mitigation. *Journal of Land Use & Environmental Law* 27. Accessed June 21, 2014: http://digitalcommons.pace.edu/lawfaculty/793/

Odeleye, D., and Maguire, M. (2008). Walking the talk? Climate change and UK spatial design policy. Climate change and urban design: The Third Annual Congress of the Council for European Urbanism, Oslo.

Oke, Tim R. (1981). Canyon geometry and the nocturnal urban heat island: Comparison of scale model and field observations. *Journal of Climatology* 1 (3), 237–254.

Oleson, K. W., Monaghan, A., Wilhelmi, O., Barlage, M., Brunsell, N., Feddema, J., Hu, L., Steinhoff, D. F. (2015). Interactions between urbanization, heat stress, and climate change. Climatic Change. 129:525–541.

Ong, B. L. (2003). Green plot ratio: An ecological measure for architecture and urban planning. *Landscape and Urban Planning* 63(4), 197–211.

Paul, M. J., and Meyer, J. L. (2001). Streams in the urban landscape. *Annual Review of Ecological Systems* 32, 333–365.

Poleto, C., and Tassi, R. (2012). Sustainable urban drainage systems. In Javaid, M. S. (ed.), *Drainage Systems* (55–72). InTech.

Raven, J. (2011). Cooling the public realm: Climate-resilient urban design · resilient cities. In Otto-Zimmermann, K. (ed.), *Cities and Adaptation to Climate Change: Local Sustainability* (Vol. 1, 451–463), Springer.

Resch, E., Bohne, R.A., Kvamsdal, T., and Lohne, J. (2016). Impact of urban density and building height on energy use in cities. Energy Procedia. 96:800–814.

Revi, A., Satterthwaite, D. E., Aragón-Durand, F., Corfee-Morlot, J., Kiunsi, R. B. R., Pelling, M., Roberts, D. C., and Solecki, W. (2014). Urban areas. In Field, C. B., Barros, V. R., Dokken, D. J., Mach, K. J., Mastrandrea, M. D., Bilir, T. E., Chatterjee, M., Ebi, K. L., Estrada, Y. O., Genova, R. C., Girma, B., Kissel, E. S., Levy, A. N., MacCracken, S., Mastrandrea, P. R., and White, L. L. (eds.), *Climate Change 2014: Impacts, Adaptation, and Vulnerability. Part A*: Global and Sectoral Aspects. *Contribution of Working Group II to the Fifth Assessment Report of the Intergovernmental Panel on Climate Change* (535–612). Cambridge University Press. Accessed October 20, 2015: http://www.ipcc.ch/pdf/assessment-report/ar5/wg2/WGIIAR5-Chap8_FINAL.pdf

Rosenzweig, C., Horton, R. M. Bader, D. A. Brown, M. E. DeYoung, R. Dominguez, O. Fellows, M. Friedl, L. Graham, W. Hall, C. Higuchi, S. Iraci, L. Jedlovec, G. Kaye, J. Loewenstein, M. Mace, T. Milesi, C. Patzert, W. Stackhouse, P. W., and Toufectis, K. (2014). Enhancing climate resilience at NASA centers: A collaboration between science and stewardship. *Bulletin of the American Meteorological Society* 95(9), 1351–1363. doi:10.1175/BAMS-D-12–00169.1

Rosenzweig, C., Solecki, W., Pope, G., Chopping, M., Goldberg, R., and Polissar, A. (2004). Urban heat island and climate change: An assessment of interacting and possible adaptations in the Camden, New Jersey region. New Jersey Department of Environmental Protection.

Roth, M. (2007). Review of urban climate research in (sub) tropical regions. *International Journal of Climatology* 27(14), 1859–1873.

Sailor, D. J. (2011). A review of methods for estimating anthropogenic heat and moisture emissions in the urban environment. *International Journal of Climatology* 31(2), 189–199.

Santamouris, M. (2014). Cooling the cities – A review of reflective and green roof mitigation technologies to fight heat island and improve comfort in urban environments. *Solar Energy* 103, 682–703.

Schaltegger, S., Burit, R., and Petersen, H. (2003). *An Introduction to Corporate Environmental Management- Striving for Sustainability*. Greenleaf Publishing.

Schmidt M. (2009). Rainwater harvesting for mitigating local and global warming. In *Proceedings Fifth Urban Research Symposium: "Cities and Climate Change: Responding to an Urgent Agenda*. World Bank.

Schneider, A., Friedl, M. A., and Potere, D. (2009). A new map of global urban extent from MODIS satellite data. *Environmental Research Letters* 4(4), 044003.

Scholz, M., and Grabowiecki, P. (2007). Review of permeable pavement systems. *Building and Environment* 42(11), 3830–3836.

Schuler, M. (2009). The Masdar Development – showcase with global effect. *Urban Futures 2030*. Visionen künftigen Städtebaus und urbaner Lebensweisen, Band 5 der Reihe Ökologie, Herausgegeben von der Heinrich-Böll-Stiftung. Accessed March 27, 2014: http://www.boell.de/downloads/Urban-Future-i.pdf

Shashua-Bar, L., Potchter, O., Bitan, A., Boltansky, D., and Yaakov, Y. (2010). Microclimate modelling of street tree species effects within the varied urban morphology in the Mediterranean city of Tel Aviv, Israel. *International Journal of Climatology* 30(1), 44–57.

Smith, C., and Levermore, G. (2008). Designing urban spaces and buildings to improve sustainability and quality of life in a warmer world. *Energy Policy*. doi:10.1016/j.enpol.2008.09.011

Steffan C., Schmidt M., Köhler M., Hübner I., Reichmann B. (2010). *Rain water management concepts, greening buildings, cooling buildings planning, construction, operation and maintenance guidelines*. Berlin Senate Department for Urban Development. Accessed June 12, 2014: http://www.stadtentwicklung.berlin.de/bauen/oekologisches_bauen/download/SenStadt_Regenwasser_engl_bfrei_final.pdf

Stewart, I. D., and Oke, T. R. (2012). Local climate zones for urban temperature studies. *Bulletin of the American Meteorological Society* 93(12),1879–1900.

Stone, B., Hess, J., and Frumkin, H. (2010). Urban form and extreme heat events: Are sprawling cities more vulnerable to climate change? *Environmental Health Perspectives* 118, 1425–1428.

Stone, B., Vargo, J., Liu, P., Habeeb, D., DeLucia, A., Trail, M., Hu, Y., Russel, A. (2014). Avoided heat-related mortality through climate adaptation strategies in three U.S. cities. *Plos One* 9, e100852.

Taha, H., Konopacki, S., and Gabersek, S. (1999). Impacts of large scale surface modifications on meteorological conditions and energy use: A 10-region modeling study. *Theoretical and Applied Climatology* 62, 175–185.

Thomas, R., and Ritchie, A. (eds.). (2003). *Sustainable Urban Design: An Environmental Approach*. Spon Press.

Tilly C. (2005). *Trust and Rule*. Cambridge Studies in Comparative Politics. Cambridge University Press

United Nations Environment Programme (UNEP). (2010). *Green Economy Developing Countries Success Stories*. UNEP.

United Nations Environment Programme (UNEP). (2012). *Using ecosystems to address climate change: Ecosystem based adaptation Regional seas program Report*. UNEP.

United Nations Environment Programme (UNEP). (2014). *Green Infrastructure Guide for Water Management: Ecosystem-based management approaches for water-related infrastructure projects*. UNEP.

UN-Habitat. (2012). *Urban Patterns for a Green Economy: Leveraging Density*. UN-Habitat.

UN-Habitat. (2013). *Urban Planning for City Leaders*. UN-Habitat.

U.S. Department of Energy. (2013). *Effects of the Built Environment on Transportation: Energy Use, Greenhouse Gas Emissions, and Other Factors*. U.S. DOE.

U.S. Environmental Protection Agency (EPA). (2008). *Reducing Urban Heat Islands: Compendium of Strategies*. U.S. EPA.

Velazquez, J., Yashiro, M., Yoshimura, S., and Ono, I. (2005). *Innovative Communities: People-centred Approaches to Environmental Management in the Asia-Pacific Region*. United Nations University Press.

Walker, B., and Salt, D. (2006). *Resilience Thinking*. Island Press.

Warburton, D., and Yoshimura, S. (2005). Local to global connections. In Velasquez, J., Yashiro, M., Yoshimura, S., and Ono, I. (eds.), *Innovative Communities: People-centred Approaches to Environmental Management in the Asia-Pacific Region*, United Nations University Press.

Woods-Ballard, B., Kellagher, R., Martin, P., Jefferies, C., Bray, R., and Shaffer, P. (2007). *The SuDS Manual*. Ciria.

Yohe, G. and Leichenko, R. (2010), Chapter 2: Adopting a risk-based approach. Annals of the New York Academy of Sciences, 1196: 29–40. doi:10.1111/j.1749-6632.2009.05310.x

Zhao, S.X., Guo, N.S., Li, C.L.K., Smith, C. (2017). Megacities, the World's Largest Cities Unleashed: Major Trends and Dynamics in Contemporary Global Urban Development. World Development. 98:257–289.

Zheng, H. and Peeta, S. (2015). Network design for personal rapid transit under transit-oriented development. Transportation Research Part C. 55:351–362.

Chapter 5   Case Study References

Case Study 5.1   Green Infrastructure as a Climate Change Adaptation Option for Overheating in Glasgow, UK

Emmanuel, R., and Krüger, E. (2012). Urban heat islands and its impact on climate change resilience in a shrinking city: The case of Glasgow, UK. *Building and Environment* **53**, 137–149. Accessed June 21, 2014: http://dx.doi.org/10.1016/j.buildenv.2012.01.020

Emmanuel R., and Loconsole, A. (2015). Green infrastructure as an adaptation approach to tackle urban overheating in the Glasgow Clyde Valley Region. *Landscape and Urban Planning* **138**, 71–86. Accessed April 4, 2016: http://dx.doi.org/10.1016/j.buildenv.2012.01.020

Glasgow City Council. (2010). Climate change strategy & action plan. Accessed June 21, 2014: https://www.glasgow.gov.uk/CHttpHandler.ashx?id=7609andp=0

Keeley, M. (2011). The green area ratio: An urban site sustainability metric. *Journal of Environmental Planning and Management* **54**, 937–958, http://dx.doi.org/10.1080/09640568.2010.547681

Office for National Statistics. (2012). 2011 Census: Population and household estimates for the United Kingdom Table 2. Accessed October 12, 2015: http://www.ons.gov.uk/ons/rel/census/2011-census/population-and-household-estimates-for-the-united-kingdom/rft-table-2-census-2011.xls

Peel, M. C., Finlayson, B. L., and McMahon, T. A. (2007). Updated world map of the Köppen-Geiger climate classification. *Hydrology and Earth System Sciences Discussions* **4**(2), 462. Accessed July 31, 2015: http://www.hydrol-earth-syst-sci-discuss.net/4/439/2007/hessd-4-439-2007.pdf

United Nations Development Programme. (2014). Human Development Index (HDI). Accessed November 4, 2015: http://hdr.undp.org/sites/default/files/hdr14_statisticaltables.xls

World Bank. (2017). 2016 GNI per capita, Atlas method (current US$). Accessed August 9, 2017: http://data.worldbank.org/indicator/NY.GNP.PCAP.CD

Case Study 5.2   Adapting to Summer Overheating in Light Construction with Phase-Change Materials in Melbourne, Australia

Australian Bureau of Statistics. (2013). 2011 Census Community Profiles, Greater Capital City Statistical Areas, Greater Melbourne. Accessed January 16, 2015: http://www.censusdata.abs.gov.au/census_services/getproduct/census/2011/communityprofile/2GMEL?opendocument&navpos=230

Peel, M. C., Finlayson, B. L., and McMahon, T. A. (2007). Updated world map of the Köppen-Geiger climate classification. *Hydrology and Earth System Sciences Discussions* **4**(2), 462. Accessed July 31, 2015: http://www.hydrol-earth-syst-sci-discuss.net/4/439/2007/hessd-4-439-2007.pdf

World Bank. (2017). 2016 GNI per capita, Atlas method (current US$). Accessed August 9, 2017: http://data.worldbank.org/indicator/NY.GNP.PCAP.CD

Case Study 5.3   Application of Urban Climatic Map to Urban Planning of High-Density Cities: An Experience from Hong Kong

Census SAR. (2011). Census and Statistics Department, Government of Hong Kong Special Administrative Region (SAR). Accessed June 12, 2014: http://www.census2011.gov.hk/en/main-table/A202.html

Census SAR. (2013). Information Services Department, Census and Statistics Department, Hong Kong Special Administrative Region Government. Accessed March 27, 2014: http://www.gov.hk/en/about/abouthk/factsheets/docs/population.pdf

GovHK. (2015). Hong Kong – the facts. Accessed: http://www.gov.hk/en/about/abouthk/facts.htm

Leung, Y. K., Yeung, K. H., Ginn, E. W. L., and Leung, W. M. (2004). Climate change in Hong Kong. Technical Note No. **107**, Hong Kong Observatory.

Peel, M. C. Finlayson, B. L. , and McMahon, T. A. (2007). Updated world map of the Köppen-Geiger climate classification. *Hydrology and Earth System Sciences Discussions* **4**(2), 462.

PlanD. (2012). Final Report, Urban Climatic Map and Standards for Wind Environment – Feasibility Study. Planning Department, Hong Kong Government. Accessed January 16, 2015: http://www.pland.gov.hk/pland_en/p_study/prog_s/ucmapweb/ucmap_project/content/reports/final_report.pdf

World Bank. (2017). 2016 GNI per capita, Atlas method (current US$). Accessed August 9, 2017: http://data.worldbank.org/indicator/NY.GNP.PCAP.CD

Case Study 5.4   An Emerging Clean-Technology City: Masdar, Abu Dhabi, United Arab Emirates

Bullis, K. (2009). A zero-emissions city in the desert. *MIT Technology Review* 56–63.

Demographia. (2016). *Demographia World Urban Areas 12th Annual Edition: 2016:04*. Accessed August 9, 2017: http://www.demographia.com/db-worldua.pdf

Clarion Associates. (2008). Abu Dhabi estidama program: Interim estidama community guidelines assessment system for commercial, residential, and institutional development. Prepared for the Emirate of Abu Dhabi Urban Planning Council.

Foster & Partners. (2009). Masdar development. Accessed February 25, 2014: http://www.fosterandpartners.com/projects/masdar-development

Manghnani, N., and Bajaj, K. (2014). Masdar City: A model of urban environmental sustainability. *International Journal of Engineering Research and Applications* **4**(10) 38–42.

Peel, M. C., Finlayson, B. L., and McMahon, T. A. 2007. Updated world map of the Köppen-Geiger climate classification. *Hydrology and Earth System Sciences Discussions*, **4** (2), 462.

SCAD. (2011). Statistic Centre of Abu Dhabi, United Emirates, Accessed February 25, 2015: http://www.scad.ae/en/pages/default.aspx

United Nations, Department of Economic and Social Affairs, Population Division (2016). The World's Cities in 2016 – Data Booklet (ST/ESA/SER.A/392).

World Bank. (2017). 2016 GNI per capita, Atlas method (current US$). Accessed August 9, 2017: http://data.worldbank.org/indicator/NY.GNP.PCAP.CD

6

Equity, Environmental Justice, and Urban Climate Change

Coordinating Lead Authors
Diana Reckien (Enschede/Berlin), Shuaib Lwasa (Kampala)

Lead Authors
David Satterthwaite (London), Darryn McEvoy (Melbourne), Felix Creutzig (Berlin), Mark Montgomery (Cambridge, MA/New York), Daniel Schensul (New York), Deborah Balk (New York), Iqbal Alam Khan (Toronto/Dhaka)

Contributing Authors
Blanca Fernandez (Berlin), Donald Brown (London), Juan Camilo Osorio (Cambridge, MA/New York), Marcela Tovar-Restrepo (New York), Alex de Sherbinin (New York), Wim Feringa (Enschede), Alice Sverdlik (London/Berkeley), Emma Porio (Manila), Abhishek Nair (Enschede), Sabrina McCormick (Washington, D.C.), Eddie Bautista (New York)

This chapter should be cited as
Reckien, D., Lwasa, S., Satterthwaite, D., McEvoy, D., Creutzig, F., Montgomery, M., Schensul, D., Balk, D., and Khan, I. (2018). Equity, environmental justice, and urban climate change. In Rosenzweig, C., W. Solecki, P. Romero-Lankao, S. Mehrotra, S. Dhakal, and S. Ali Ibrahim (eds.), *Climate Change and Cities: Second Assessment Report of the Urban Climate Change Research Network*. Cambridge University Press. New York. 173–224

Equity, Environmental Justice, and Urban Climate Change

Cities are characterized by a great diversity of socioeconomic groups living in close proximity. Diversity is often accompanied by stratification based on class, caste, gender, profession, race, ethnicity, age, and ability. This gives rise to social categories that, in turn, affect the ability of individuals and various groups to endure climate stresses and minimize climate risks.

Differences between strata often lead to discrimination based on group membership. Poorer people and ethnic and racial minorities tend to live in more hazard-prone, vulnerable, and crowded parts of cities. These circumstances increase their susceptibility to the impacts of climate change and reduce their capacity to adapt to and withstand extreme events.

Major Findings

- Differential vulnerability of urban residents to climate change is driven by four factors: (1) differing levels of physical exposure determined by the location of residential/occupational areas; (2) urban development processes that lead to risks, such as failure to provide access to critical infrastructure and services; (3) social characteristics that influence resources for adaptation; and (4) institutional and governance weaknesses such as ineffective planning and absence of community engagement.

- For New York, London, Dar es Salaam, and Durban, risk levels increase dramatically for all key risks in the long-term, especially for the 4°C warming. But, at least for now, in the near-term, and mostly for the long-term with a 2°C temperature rise, a high level of adaptation can keep risks down. However, under a 4°C temperature rise, adaptation measures are likely to be ineffective not only in Dar es Salaam but also in cities with currently high levels of adaptation, such as New York and London.

- Climate change amplifies vulnerability and hampers adaptive capacity, especially for the poor, women, the elderly, children, and ethnic minorities. These people often lack power and access to resources, adequate urban services, and functioning infrastructure. Gender inequality is particularly pervasive, cutting across a number of documented inequities such as income, disability, and literacy, which in turn contribute to differential consequences of climate changes.

- Frequently occurring climate events, such as droughts in many drought-prone areas, can, over time, undermine everyone's resource base and adaptive capacity, including better-off urban residents. As climate events become more frequent and intense, this can increase the scale and depth of urban poverty overall.

- Mobilizing resources to increase equity and environmental justice under changing climatic conditions requires the participation of impacted communities and the involvement of civil society; nontraditional sources of finance, including partnerships with the private sector; and adherence to principles of transparency in spending, monitoring, and evaluation.

Key Messages

Urban climate policies should include equity and environmental justice as primary long-term goals. Equity fosters human well-being, social capital, and sustainable social and economic urban development, all of which increase a city's capacity to respond to climate change. Access to land situated in nonvulnerable locations, security of tenure, and access to basic services and risk-reducing infrastructure are particularly important.

Cities need to promote and share a science-informed policy-making process that integrates multiple stakeholder interests to avoid inflexible, top-down solutions. This can be accomplished by participatory processes that incorporate community members' views about resilience objectives and feasibility.

Over time, climate change policies and programs need to be evaluated and adjusted to ensure that resilience and equity goals are reached. Periodic monitoring and evaluation using fair indicators and progress measurements, budgetary transparency, and equitable resource allocation schemes are essential to ensure that funds reach target groups and result in equitable resilience outcomes.

6.1 Introduction

This chapter focuses on equity and environmental justice aspects of climate change in and across cities. Within cities, climate change equity and environmental justice aspects are important due to distinctive urban characteristics related to socioeconomic and ethnic diversity as well as to high density (UN-Habitat, 2010). Diversity is often accompanied by stratification – a permanent social process – based on, for example, class, caste, gender, profession, race, ethnicity, age, and ability. This gives rise to social categories within urban populations that affect the capacities of individuals and groups to endure climate stresses and to adapt to or minimize climate risks (Dodman, 2009; Marino and Ribot, 2012). Differences between strata potentially lead to discrimination and unfair treatment of individuals due to group membership, thus raising equity issues (Dovidio et al., 2010). For example, poorer people and ethnic minorities tend to live in more vulnerable locations and crowded parts of cities, increasing their susceptibility to climate change impacts and lowering their adaptive capacities to withstand risks.

Between cities, climate change equity and environmental justice issues arise for cities that have traditionally emitted greenhouse gases (GHGs) at low rates per capita but experience higher than average climate change impacts, such as many cities in low-income countries. For example, the 23 mega-disasters (those with more than 10,000 fatalities, excluding epidemics) that occurred between January 1975 and October 2008 mainly affected low-income countries, causing 78% of mortality from only 0.26% of the total number of events (UNISDR, 2009). Cities that are affected by higher-than-average climate change impacts show particularly high population growth rates – especially those in low-income countries of Asia and Africa (Wheeler, 2011), where nearly 90% of the increase in urban population between now and 2050 is expected to take place (UNDESA, 2014).

These trends put ever-increasing numbers of people at risk from climate change and will potentially amplify equity and environmental justice issues because the growth of cities in Asia and Africa is likely to correspond with an increase in slum populations. Currently, 62% of the population in Africa and about 30% in Asia live in slums (UN-Habitat, 2010). Although, the proliferation of densely populated informal settlements – often fueled by rural–urban migration – is also a sign of growth and anticipated development and therefore of the "success" of a city, rapid urban growth poses a major challenge for city managers. It is often associated with unplanned and unregulated settlements in risk-prone areas (UN-Habitat, 2013a; Revi et al., 2014).

A changing climate will also act to further amplify equity issues in cities of high-income countries, as shown by extreme heat waves in Central Europe in 2003, causing between 22,000 and 35,000 deaths of mainly elderly people (Schär and Jendritzky, 2004); in Chicago in 1992, claiming 739 excess deaths in mainly lower-income neighborhoods (Klinenberg,

2003); and in Melbourne in 2009, with 374 heat-related deaths (Victorian Government Department of Human Services, 2009).

6.1.1 Objectives

In this chapter, we assess how climate change risk, impacts, adaptation, and mitigation actions in cities relate to equity. For each of these dimensions, the chapter provides a synthesis of the various factors that affect equity issues under current climate conditions and discusses how further climate change might potentially increase or decrease equity in the future. Equity issues affecting urban centers in low-, middle-, and high-income countries are considered. We stress the particularities of urban contexts – including socioeconomic characteristics, location, and potential for policy interventions – and the wide spectrum of capacity to respond.

Section 6.1 provides an introduction and an overview of the chapter. It provides explanations of key terminology and introduces the relation of equity to climate change in cities. Section 6.2 discusses the nexus of climate change impacts and adaptation, and equity, with a focus on common impacts and risk factors as well as on hazard-specific risk factors such as high temperatures, heavy precipitation, and sea level rise. Section 6.3 highlights equity concerns in regard to mitigation, including aspects of transportation, land use, energy, and waste management. In Section 6.4, we draw attention to lessons from the implementation of climate change policies and practices in cities. Section 6.5 evaluates a number of frameworks for assessing equity and presents an Urban Equity Impact Assessments (EquIA-urban) Guide for policy-makers. Section 6.6 summarizes knowledge and research gaps. The chapter concludes with policy recommendations for city leaders and national-level decision-makers.

6.1.2 Definitions, Principles, and Domains

Promoting equity is an implicit (and sometimes explicit) goal of many local and regional climate initiatives (McDermott and Schreckenberg, 2009). However, it is often unclear which aspects of equity (see Box 6.1) are being referred to: equity in the distribution of costs and benefits or in privileges and burdens; between individuals such as women and men or between households within communities; between urban districts, local groups, and national stakeholders or generations of urban residents. These aspects refer to one dimension of equity concerns – outcome-based aspects – whereas, three scholarly dimensions are distinguished:

1. Outcome-based, distributive, or consequential equity;
2. Process-oriented or procedural equity (Metz, 2000; McDermott et al., 2011);
3. Contextual equity (McDermott and Schreckenberg, 2009).

Outcome-based equity relates to the consequences of a policy, action, or developmental trend, which is acknowledged to be important for both low-and high- income countries. Procedural equity refers to impartiality and fairness in the process of delivering and administering justice (Shukla, 1999). This is more often discussed in relation to urban climate initiatives in low-income

Box 6.1 Definitions of Terms: Fairness, Justice, Equity, and Equality

The terms "fairness," "justice," and "equity" are often used interchangeably in development discussions (Metz, 2000; Kallbekken et al., 2014). Some distinctions has been suggested by Rawls (1971), who saw justice and fairness as being distributive concepts and equity as the normative criterion for judging this distribution (Fahmi et al., 2014; Dankelman et al., 2008; WEDO and UNFPA, 2009). Soltau (2009) uses the term "fairness" for the more general concept of distributional norms and the term "equity" for a particular subset of these norms (Kallbekken et al., 2014). However, what exactly qualifies as "fair" or "just" depends on personal or cultural judgment (Rawls, 1971, 1993; Barry, 1995; Linnerooth-Bayer, 2009; McDermott et al., 2011). It is therefore important to consider who decides what is equitable, fair, and just.

Fairness practices are thought to enhance the quality of social life, and a good society has been defined as one in which norms of fairness play a significant role in guiding human behavior and government policies (Kallbekken et al., 2014; Moser, 2011). The fact that some fairness principles are frequently invoked and rarely disputed indicates that they have some "normative clout" (Kallbekken et al., 2014; Tovar-Restrepo, 2010; Schildberg, 2014). This means that some constraints on the actions of self-interest are generally agreed to be favorable for all in the long run.

Justice can refer to either social or environmental justice. The social justice movement seeks to establish fair distributions of wealth, opportunity, and privileges by means of fair treatment, proportional distribution, and the meaningful involvement of all people in social decision-making. The goals of the environmental justice movement are healthy environments and protection from environmental hazards for all people, regardless of race, nationality, origin, or income (EPA, 2011). Environmental justice interacts with environmental risk, exposure, impacts, sensitivity, and adaptive capacity. In that respect, social justice and environmental justice are inextricably linked. However, growing environmental concerns have arguably detracted from efforts to increase social justice in the political arena (Agyeman et al., 2003; Khosla and Masaud, 2010).

Distinguishing equity and equality is more straight-forward: *equality* refers to an equal treatment of equal cases or a "state or quality of correspondence in quantity, degree, value, rank, or ability" (Random House, 2014), for example with respect to status, rights, or opportunities. In urban areas, crucial components of equality include the right to adequate housing and security of tenure; and affordability, accessibility, location, culture, and availability of services, infrastructure, and facilities. *Equity* refers to the impartial treatment of cases that may differ in important respects (Kallbekken et al., 2014) (see Box 6.1. Table 1).

To simplify the discussion in this chapter, we mostly use the term "equity," with the aim of exploring the issues of fairness, justice, equity, and equality arising from climate change in cities.

Box 6.1 Table 1 *Broadly accepted fairness principles. Source: Kallbekken et al., 2014*

Fairness principle	Summary definition	Explanation
Equality	Equal treatment of equal cases	Relevant differences not important, e.g., all urban residents shall have equal right of security of tenure
	Proportional treatment of similar cases	Relevant differences important, e.g., different, but fair and equitable payment for jobs requiring different skills
Equity	Exceptional treatment of dissimilar cases	Relevant differences very large and/or cases dissimilar, e.g., special treatment for entities with no moral responsibility for damage and/or very low problem-solving capacity

countries (Bulkeley et al., 2013). Contextual equity links the first two dimensions by taking into account pre-existing political, economic, and social conditions.

In order to operationalize equity, McDermott et al. (2011) relate these dimensions to three parameters: target, goal, and process (see Figure 6.1). Operationalization is further based on principles and indicators, of which a large number have been proposed (Metz, 2000; Klinsky and Dowlatabadi, 2009; Cazorla and Toman, 2000) (see Table 6.1).

Support for equity principles and operational indicators differ between low-income and high-income countries (Shukla, 1999; Kallbekken et al., 2014) and potentially on subnational and local levels. Among delegates to the climate change negotiations of the United Nations Framework Convention on Climate Change (UNFCCC) the "polluter pays" principle had most support, at least in a short-term perspective (i.e., ≤20 years). This was followed by "the exemption of the poorest" and "ability to pay." An "egalitarian" principle (equal mitigation pledges) was not supported by many delegates and even more objected to the "sovereignty" principle (i.e., the full right and power of a state to govern itself and decide on its mitigation pledges) (Kallbekken et al., 2014; Lange et al., 2010). Cazorla and Toman (2000) conclude that efforts to find a "magic" solution to equity disputes are likely to be in vain, and the question of which international climate policies will be equitable over the long term will require a great deal of additional time and effort to resolve. However,

Table 6.1 *Commonly applied equity domains in international climate change mitigation efforts Kallbekken et al., 2014*

Focus on	Object to be allocated (distributed) on basis of ...	
	Costs (obligations)	**Benefits (rights)**
Causes of the problem	Moral responsibility ("guilt" in having caused the problem)	Previous contributions (to providing the benefits under consideration) (not frequently claimed, at least in UNFCCC negotiations)
Consequences of the solution	Capabilities (capacity to contribute to problem solving)	Need for (or right to) the outcome to be achieved, i.e., goods or services of a policy

UNFCCC: United Nations Framework Convention on Climate Change

after COP21 in Paris and its agreement, the establishment of Nationally Determined Contributions (NDCs) offers another approach to the problem of who should pay and which amount. Now, each nation presents its voluntary commitments to reduce the GHG emissions, which brings a sense of ownership to the countries and also makes compliance enforcement easier.

6.1.3 Equity and Climate Change

Equity, equality, and environmental justice issues first entered the debate on climate change when it was recognized that countries that historically contributed least to global warming might be impacted the most by climate change in the future (UNEP, 2014b, UNDP, 2004; Revi et al., 2014), although, due to interdependencies, will likely also increase impacts in high-income countries (Nabangchang et al., 2015). Consequently, initial discussions revolved around mitigation responsibility and pledges. However, Metz (2000) stresses that the equity discussions around climate change should not only consider mitigation, but also explicitly take account of impacts and adaptation. This is of particular importance for urban areas because it is at local and regional scales where differential impacts and adaptation needs will unfold. Discussion on justice in adaptation has recently gained momentum (Sovacool et al., 2015; Shi et al.,

2016). And because the impacts of climate change and risks are reduced by adaptation and mitigation, all three dimensions play a role (see Figure 6.2).

Impacts of climate change differ among people and groups because of interacting socioeconomic conditions based on income; assets; and discrimination related to minority status, race or ethnicity, sex and gender, age, poor health, and impaired mobility. These characteristics influence where people live and how severely they are affected. However, it is not just the most susceptible who are impacted. Regularly occurring events can also gradually undermine the resource base of more resilient groups in society, ultimately leading to increases in the scale and depth of urban poverty (Tyler and Moench, 2012; Tompkins et al., 2013).

High exposure and sensitivity to climate impacts often coincide with low adaptive capacity (The World Bank, 2010). In this respect, it is important to recognize the current vulnerability of many cities in low- and middle-income nations and the limited capacity of their governments and inhabitants to adapt to a changing climate (Revi et al., 2014). Differential vulnerability can often be attributed to deficiencies with respect to the quality and location of infrastructure and housing, availability of social services and facilities, opportunities and access to education, and effectiveness of planning systems, as well as lack of resources and low levels of community and individual adaptive capacity (Sen, 1999; Taylor, 2013; Revi et al., 2014). Adaptive capacity is also eroded over time through repeated coping and "risk accumulation processes" (Satterthwaite et al., 2007; Rodin, 2014), with knock-on effects for chronic poverty (UNISDR, 2009).

Mitigation issues are also a concern of contemporary urban planning because an approximate 70% of the global total CO_2 emissions are from urban areas (depending on measurement protocols), principally from cities in middle- and high-income nations (Seto et al., 2014). Some cities, with encouragement from city networks such as the C40 Cities Climate Leadership Group and ICLEI—Local Governments for Sustainability, have shown farsighted leadership in setting targets (Reckien et al., 2014) and devising and implementing plans to reduce GHG emissions, but it is important to evaluate such commitments with respect to the distribution of the benefits and burdens.

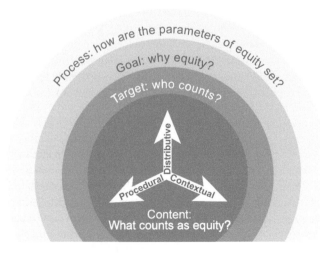

Figure 6.1 *The equity framework.*

Source: McDermott et al., 2011

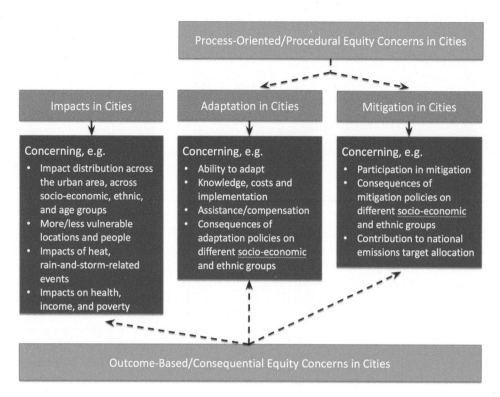

Figure 6.2 *Conceptualization of equity in climate policy.*

Source: Adapted from Metz, 2000

6.2 Equity, Urban Impacts, and Adaptation

The extent to which climate change or extreme weather events pose risks to urban residents and lead to immediate or long-lasting impacts depends on a combination of several factors (Adelekan, 2010; Fuchs, 2010). Physical exposure, as determined by the location of a community, is one factor, whereas urban development processes or so-called inherent or "constructed risk" is another (Eiser et al., 2012; McBean, 2012). Risk is also associated with the social, economic, and demographic characteristics of a population (Barrios et al., 2006), leading to vulnerability that couples with institutional, power, and governance aspects (Bulkeley et al., 2009; UN-Habitat, 2008a). Most of these factors are interrelated. They play out in low-, middle-, and high-income nations and in large, medium and small cities as a result of historical as well as contemporary urban development processes (Adelekan, 2012; Awuor et al., 2008; Fuchs, 2010).

6.2.1 Exposure, Social Context, Demographics, Governance, and Power

There is growing evidence that impacts of both gradual climate change and extreme weather events (such as tropical storms, heat waves, and excessive precipitation) disproportionately affect people with low incomes and low social status, for example in Indian and South American cities (Reckien et al., 2013; Reckien, 2014; Hardoy and Pandiella, 2009), and especially women (see Box 6.2). However, it is important to note that

the most vulnerable are not the only ones impacted; regularly occurring events such as droughts and floods also undermine the resource base of better-off groups in society (Tyler and Moench, 2012; Tompkins et al., 2013). People may also be susceptible to multiple risks (e.g., infants, young children, and older age groups with impaired mobility).

Geographic and locational factors also have a major influence on climate risks. In cities in low- and many middle-income nations, such as Lagos (Adelekan, 2010), Cairo, Alexandria (Hereher, 2010), Rio de Janeiro (de Sherbinin and Hogan, 2011), Dhaka (Khan et al., 2011), and other Asian cities (Fuchs, 2010), residents with low social status and low incomes characteristically inhabit areas more exposed to climate risks, such as low-elevation coastal zones and flood plains. Housing located in high-risk urban areas is often constructed illegally and without adhering to building codes. These areas typically have high population densities and poor-quality buildings (UNISDR, 2009), leaving residents, predominantly from the lower social classes, exposed to climate risks and potentially severe impacts.

Figure 6.3 shows an indicator for (1) Equity and Social Inclusion, as measured by UN-Habitat (2013b) as part of its City Prosperity Index (CPI). It subsumes the subdimensions (and indicators) of economic equity (urban Gini coefficient and poverty rate), social inclusion (slum households, youth unemployment), and gender inclusion (equitable secondary school enrollment); (2) Environmental Sustainability with the subdimensions of air quality (PM2.5 concentration), waste management (wastewater treatment), and energy (share of renewable energy;

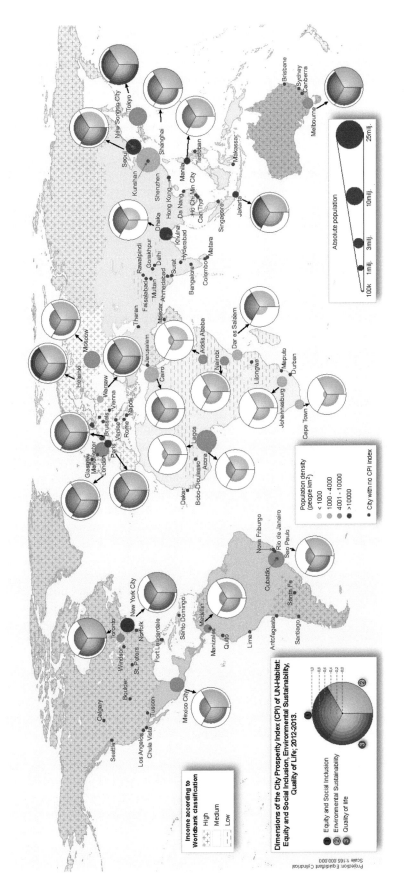

Figure 6.3 *Case study cities included in this assessment and, where available, the three dimensions of the City Prosperity Index (CPI) (2012/2013) relevant for this chapter, namely (1) Equity and Social Inclusion, (2) Environmental Sustainability, and (3) Quality of Life. The CPI is a multidimensional index, developed by UN-Habitat (2013b) and comprises six dimensions with sub-dimensions and respective indicators that are measured for cities. Indicators are standardized using internationally observed benchmarks. For further explanation see text.*

CO$_2$ emissions); and (3) Quality of Life, by aggregating the subdimensions of health (life expectancy at birth; under-five mortality rate), education (literacy rate; mean years of schooling), and safety and security (homicide rate). The other three dimensions of the CPI (i.e., Productivity, Infrastructure, and Governance and Legislation), are not shown (not relevant or no data).

Across the case studies represented in Figure 6.3 we see that equity and social inclusion is generally low in cities of central and southern Africa and in cities of South America. In African cities, low equity and social inclusion go along with relatively low quality of life, meaning poor health, education and safety. This is likely to be accentuated by climate change impacts, although there is low- to medium-level evidence and a low level of confidence. Dar es Salaam is an exception to the African pattern, showing relatively high equity but very low quality of life. In comparison, South American cities are able to sustain a medium quality of life despite low levels of equity, but the interactions with climate risks in these cities is uncertain.

Equity and Social Inclusion in the North American cities is higher than in the South American cities but lower than in the European cities on average. An exception for Europe marks the city of Moscow, showing a relatively low level of equity and social inclusion. Cities in Asia show a mixed picture regarding equity and social inclusion. In the cities of high-income nations, cities such as Tokyo and Melbourne, equity is high. In cities of middle-income nations, cities such as Manila, Jakarta, and Dhaka, equity is somewhat lower. In these cities (as well as in Dar es Salaam), the equity index is higher than the index for quality of life. This constellation generally shows that large parts of the urban population live in low-quality conditions.

We acknowledge that these interpretations are based on a very small sample and advise caution about any generalizations across the case study cities (based on data from UN-Habitat State of the World Cities 2012/2013 – Prosperity of Cities).

Evidence also points to the role of legal, governance, and investment practices in shaping or constructing urban risks. The disproportionate impact of extreme weather events on low-income urban populations is strongly associated with the lack of risk-reducing infrastructure (including piped water, sanitation, drainage and solid waste collection, all-weather roads and paths) and services (including health care and emergency services) (Revi et al., 2014). This disproportionate impact on low-income urban neighborhoods is often underpinned by a lack of capacity within urban governments or their voluntary refusal to address the large infrastructure and service deficits, particularly in low- and middle- but also in high-income countries. Both Hurricane Katrina in New Orleans and Hurricane Sandy in New York disproportionately impacted social groups with lower incomes and social status, particularly ethnic minorities and women (David and Enarson, 2012; Elliott and Pais, 2006; Brodie et al., 2006; Blake et al., 2013). However, risks are even higher for many residents in most cities of low- and middle-income countries and affect a greater proportion of the urban population.

Government refusal to address existing vulnerabilities leads to the social construction of risk (Eiser et al., 2012; Singh and Fazel, 2010) or "development-accumulated" risk (Satterthwaite, 2013; McBean, 2012). Some recent local initiatives have made commitments to incorporate equity and environmental justice in plans to build resilience to climate change, for example New York, 2015 "OneNYC Plan: The Plan for a Strong and Just City" (City of New York, 2015). In most cities, however, low- and medium-income neighborhoods continue to be located in hazard-prone areas as a result of historical development trajectories. Infrastructure in these neighborhoods is less resilient to climate shocks. There are usually deficiencies in risk-reducing and disaster-response infrastructure, and less attention is paid to lessons from previous disasters (Singh and Fazel, 2010) (see Chapter 3, Disasters and Risk). This has resulted in the accumulation of risk over time, documented by increasing disaster losses in cities from mega-debris flows, floods, earthquakes, tsunamis, and tropical storms in the past two decades (Allen, 2006; Annez et al., 2010; Rao, 2013).

6.2.2 Equity and Climate Hazards

6.2.2.1 Heat

Heat-related impacts are among the main hazards associated with climate change in cities. Two dynamics converge: (1) the global increase in average temperature and (2) the urban heat island (UHI) effect (i.e., the temperature gradient between dense built-up environments and rural areas around them) (see Chapter 2, Urban Climate Science). These dynamics can be beneficial when reducing the mortality and morbidity risks of cold temperatures, but often evolve into particular risk situations during periods of excessive heat or heat waves (White-Newsome et al., 2009). Heat waves are prolonged periods of heat crossing either an absolute or relative threshold above a long-term temperature average that differs by city; typical relative thresholds are two or three standard deviations above mean temperatures (Tong et al., 2010) (see Chapter 2, Urban Climate Science). Heat waves pose a major climate-related risk because more fatalities occur from heat waves than from other climate hazards such as floods and hurricanes (Satterthwaite et al., 2007; Klinenberg, 2003).

Heat waves in cities can cause increased morbidity and mortality rates as a result of direct heat stress and other indirect effects (Kinney, 2012) (see Chapter 10 Urban Health). Direct heat stress is particularly harmful when night-time temperatures are high, which prevents the human body from rest, repose, and regeneration (Amengual et al., 2014). Indirect effects on health arise through the interaction of heat and other environmental factors, particularly air and water pollution (Petkova et al., 2013; Petkova et al., 2014). For example, air pollutants and heat can cause higher ozone concentrations (Neidell and Kinney, 2010; Sheffield et al., 2011), reducing lung function and irritating the respiratory system. As a result, heat waves can cause heart attacks and aggravate asthma, bronchitis, and other cardiopulmonary diseases, leading to premature death. Effects of ozone and direct heat stress are additive (Kosatsky, 2005).

Box 6.2 Climate Change and Gender Inequality in Cities

Marcela Tovar-Restrepo

Women's Environment and Development Organization (WEDO), New York

Climate change exacerbates existing social and economic inequalities among diverse social groups. However, gender inequality is more pervasive than any other form of inequality (Brady, 2009) since it cuts across other forms of exclusion and inequity (Brady and Kall, 2008). Gender differences intersect with other identity markers such as income, ethnicity, race, religion, ability/disability, age, literacy, and geographical location. All these factors lead to differential exposure to risk, preparedness, and coping capacities to recover from climate change impacts (Chen et al., 2005; UNIFEM, 2008). Women from low- and middle-income countries living in poverty are typically more vulnerable to climate change impacts than men because of the discrimination they face with respect to wealth and capital goods, health, access to technologies, education, services and information, and opportunities to generate financial and productive assets. Due to differentiated gender roles, climate change extremes also increase the number of underpaid and nonpaid hours of care-work that women have to devote to their domestic and community spheres (Tronto, 1993; Fahmi et al., 2014). All these further challenge women to adapt to climate change and recover from impacts in cities.

To target climate change causes and consequences in a sustainable and equitable way, urban policies, plans, and projects need to be formulated to reduce *ex-ante* vulnerabilities of individuals and social groups, taking into account gender roles and women's needs. Finally, engaging at the household level and involving women in leadership roles in community organizing processes and political representation will help to develop effective coping strategies to mitigate and adapt to climate change (Alber, 2011; WEDO et al., 2013).

Box 6.2 Table 1 *The impacts of climate change events on women and girls in cities, as compared to men. Sources: GGCA, 2009; Dankelman, 2010; Tovar-Restrepo, 2010; Levy, 2013*

Climate Change Event	Impact on Cities	Impact on Women and Girls
Heat waves; flooding; land slides	**Reduced or no access to potable water, drainage, and sanitation infrastructure**	• Women are more likely to be affected from heat stress than men (see Section 6.2.1). • Women face loss of income from their home-based activities and often water-based economic activities like cleaning, washing clothes, or cooking food products, particularly in informal settlements. • Women have to spend more hours fetching water from water trucks or tanks. • Reduction in food supply or increase in food prices may cause malnutrition or low calorie intake, especially in older women and young girls because of gendered diet hierarchies. • Women overwhelmingly take care of children, old, and sick family and community members who tend to suffer diarrheal, respiratory, and other health problems, placing themselves at the risk of infections. • Immigrant women who do not speak the dominant language have less access and understanding of risks and preparedness information; they may have less education and less contact with the public sphere.
Sea level rise; hurricanes; cyclones; heavy rains	**Damage or loss of shelter, urban infrastructure, or services such as electricity, transport facilities, roads, and community public spaces**	• Women are present in greater numbers in the urban informal economic sector. The loss of small productive assets such as sewing machines from extreme weather especially impacts their home-based businesses. • Women are more vulnerable to losing their jobs given that they need to devote more time to nonpaid care-work. • Women may be more severely impacted by damages to the public transport infrastructure because they make more daily trips than men and are greater public transport users.
General	**Lack of gender-specific facilities and policies during recovery**	• Because women have less access to secure land tenure, they may be less eligible for financial credits or subsidies in climate change recovery stages. Displacement or relocation plans usually do not take into account differentiated gender needs and roles. For example, land-use planning, public space, and transportation facilities are especially central to women since they need to have access to community services and child-care facilities. In post-disaster camps and temporary accommodations, women often face serious risks of sexual harassment and violence.

Heat-related risk might be expected to impact all citizens equally because both heat and ozone affect the overall urban environment. However, heat-related risk is stratified across the population and is linked to both "intrinsic" person-specific characteristics and "extrinsic" socioeconomic factors. Intrinsic factors include physiological attributes such as age, sex, disabilities, and medical status. Extrinsic factors refer to social, environmental, and location-specific characteristics such as socioeconomic status, gender, and living and working conditions.

A meta-analysis of eighteen recent studies that allowed for consistent quantification concluded that, among intrinsic factors, age is the most determinant risk factor contributing to excess (above normal) heat-related mortality (with reported relative risk ratios [RRR][1] ranging from 1.3 to 3.7) (Gouveia et al., 2003; Pirard et al., 2005; Simón et al., 2005; Michelozzi et al., 2005; Canoüï-Poitrine et al., 2005; Garssen et al., 2005; Baccini et al., 2008; Johnson et al., 2005; Yang et al., 2013; Tran et al., 2013). Only one study shows higher RRR for children in comparison to the overall population (Gouveia et al., 2003).

Females have a relatively higher risk of heat-related mortality than males (RRR = 1.0–1.4 and possibly higher) (Pirard et al., 2005; Michelozzi et al., 2005; Canoüï-Poitrine et al., 2005; Yang et al., 2013; Nogueira et al., 2005). Women may be more heat intolerant than men due to potential physiological and thermo-regulatory differences (Druyan et al., 2012; Racine et al., 2012). However, women may also typically experience more exposure to heat than men due to the time spent in interior spaces undertaking labor such as cooking in houses without adequate air flow or air-conditioning (Jabeen, 2014).

In terms of medical status, vulnerability to heat waves is higher in people who are less mobile and confined to bed, the latter with an odds ratio (OR)[2] of 3–9 (Semenza et al., 1996; Vandentorren et al., 2006). People suffering from cardiovascular diseases are also at relatively higher risk (OR, 4.05 for the overall population and up to 34.1 for the elderly) (Vandentorren et al., 2006; Baccini et al., 2008; Tran et al., 2013; Nitschke et al., 2013).

Extrinsic, socioeconomic factors for heat-related effects reported in the literature are mainly related to location-specific characteristics. Although data on relative risk linked to socioeconomic levels are not systematically reported, the overall trend indicates that lower socioeconomic status (using a deprivation index based on education, occupation, unemployment, number of household members, overcrowding, and household ownership data) and lower education levels increase relative vulnerability to heat stress (Begum and Sen, 2005; Michelozzi et al., 2005; Harlan et al., 2006; Yang et al., 2013). Loughnan et al. (2013) note that heat disproportionately impacts socioeconomically disadvantaged households because of their residence in areas with less access to urban green infrastructure and their reduced ability to fund, maintain, and develop private green space. The existence of open spaces and water (such as pools) are risk-reducing environments because they cool their immediate surroundings (Nogueira et al., 2005; Harlan et al., 2006).

Comparing risks across urban regions, people living in inner cities are generally more at risk than those living in the suburbs (Reid et al., 2009; Harlan et al., 2013). Specific climatic regions modify the relative risks across urban or regional areas (Michelozzi et al., 2005; Baccini et al., 2008). A study on the effects of the 2003 heat wave in Europe reported that death rates increased by 42% in London compared to an average of 16% over the whole of England and Wales (Johnson et al., 2005). This corresponds to an RRR of up to 2.6 for excess mortality in London, possibly because the UHI accentuates climatic heat stress.

Working and living conditions also influence vulnerability to heat waves (White-Newsome et al., 2012), especially due to the heat stress experienced by members of low-income households working from home. Tran et al. (2013) reported an OR of 1.86 in Ahmedabad, India (see also White-Newsome et al., 2012). People living under the roof or on the upper floor face similar high risk (RRR of 4.7 and 5.4, respectively) (Semenza et al., 1996; Vandentorren et al., 2006; Canoüï-Poitrine et al., 2005). Additionally, studies suggest loneliness (among the elderly and particularly unmarried men) and related behaviors as risk factors (Canoüï-Poitrine et al., 2005; Klinenberg, 2003). Considering the additive nature of intrinsic and extrinsic factors, indoor heat stress is reported to be particularly prevalent among women, children, and elderly people living in inadequate housing (White-Newsome et al., 2012).

Measures to assist populations with higher risks of heat-related mortality can address both intrinsic and extrinsic factors (see Chapter 10, Urban Health). At the intrinsic level, individuals at risk could receive special health care and broader social support (i.e., neighbors or community group members may check in on elderly living alone during periods of excessive heat). Air conditioning – although a common way of dealing with heat – is not a sustainable measure. Most current systems consume electricity and, if generated from fossil-fuels, contribute to GHG emissions. Moreover, purchase and running costs may prohibit use by many poorer residents. To add to this, extreme heat can often produce electricity black-outs or brown-outs, resulting in the unavailability of air conditioning altogether. Urban planning and public health managers may therefore provide more effective support to poorer households with measures at the extrinsic level. This includes traditional and new adaptive architecture in areas with high risk and technology that cools rather than traps heat (e.g., through albedo modifications, greening, and landscape planning) (see Chapter 5, Urban Planning and Design).

---

1 When two groups were compared (e.g., the elderly with the overall population; or women and men), relative risk ratios (RRR) are calculated from the difference of mortality risk between the two groups, given as ratio of the percentages of excess of mortality.

2 Odds ratios (OR) are calculated as the risk of death among subjects as compared with those without the characteristic in question.

6.2.2.2 Rain

Precipitation-related hazards present a range of significant risks to human well-being, such as those connected to inland flooding, landslides, and drought. Inland flooding can occur on a massive scale – as in Pakistan in 2010 (Atta-ur-Rahman and Khan, 2013), Australia in 2011 (Coumou and Rahmstorf, 2012), and Thailand in 2011 (Komori et al., 2012) – but localized floods can also cause substantial damage and threaten health, lives, and livelihoods. In many cities, informal settlements have arisen on flood plains that experience regular flooding or on steep slopes where heavy precipitation regularly triggers dangerous landslides (Dodman, 2013; Carcellar et al., 2011; Moser and Stein, 2011; Hardoy and Pandiella, 2009; Douglas et al., 2008b; UNISDR, 2009, 2011). However, insufficient precipitation and rainfall that is mistimed relative to the agricultural growing season can also severely impact urban residents since these events can cause water shortages, crop failures, and food price increases, with negative consequences for low-income populations.

6.2.2.3 Inland Flooding

Urban flood risk in low- and middle-income countries stems from a number of factors: impermeable surfaces that prevent water from being absorbed and instead cause rapid run-off, the general scarcity of parks and other green spaces to absorb such flows, inadequate drainage systems that are often clogged by waste and quickly overloaded with water, and the ill-advised development of housing on marshlands and other natural buffers (Jha et al., 2012; Revi et al., 2014).

The urban poor are often more exposed than other city dwellers to these environmental hazards because the housing they can afford tends to be located in environmentally riskier areas (such as floodplains and slopes) and of poorer quality, and because municipal governments overseeing such neighborhoods often fail to establish and maintain proper drainage and waste collection and disposal. According to the Asian Development Bank (ADB) (2010), 40% of urban dwellers in Asia can be classified as living in substandard housing or slums, which are often found along a city's rivers and canals – areas that tend to be publicly owned and thereby typically less problematic to settle on than private land (Taylor, 2013). Living close to urban waterways is in many instances a consequence of the pressure for land in fast-growing cities and can be attributed to a lack of tenure security for the urban poor and new migrants, in turn leading to population displacement and the disruption of livelihoods and social support networks when flooding reoccurs (Hardoy and Pandiella, 2009). Other indirect effects are related to poor-quality housing and unsanitary conditions (Haines et al., 2013). For example, when flooding occurs, hazardous materials frequently contaminate open waters and wells, elevating the risks of water-borne, respiratory, and skin diseases (Ahern et al., 2005; Kovats and Akhtar, 2008; Akanda and Hossain, 2012; Khan et al., 2011) (see Chapter 10, Urban Health).

6.2.2.4 Landslides

Landslide risks have not received as much attention as flooding and coastal hazards. Cepeda et al. (2010) note that landslides are usually not separate from other natural hazard triggers, such as extreme precipitation, earthquakes, or floods in the natural disaster databases. This contributes to reducing the awareness and concern of both authorities and the general public about landslide risk. Yet in many cities, landslides present significant threats to human well-being.

Rainfall-triggered landslides are the product of a combination of geo-hydrological and locational factors. Geo-hydrological factors refer to duration and intensity of precipitation. Locational factors include slope, rock strength, rock susceptibility to fracturing, soil moisture, and vegetation cover, as identified in risk analyses carried out in Indonesia (Cepeda et al., 2010) and El Salvador with comparisons to Nepal and Sri Lanka (NGI, 2012). Both these studies drew on DESINVENTAR[3] data on the occurrence of rainfall-induced landslides and compared them to population density and socioeconomic composition. However, when holding geo-hydrological exposure constant, the studies reached different conclusions.

In El Salvador, better-off municipalities experienced greater landslide mortality than did less well-off ones. By contrast, the Indonesian analysis found the expected negative association between the Human Development Index (HDI) and landslide mortality, net of physical exposure. These examples – based on similar methods – underscore the need for caution in making generalizations about the nature of the poverty–hazard vulnerability relationship with regard to landslides.

The need for caution also extends to policy measures driven by equity concerns. Well-intentioned infrastructure measures and public policies to protect or upgrade settlements may actually increase risk in these or adjacent neighborhoods. For example, in Medellín, infrastructure improvements are criticized as detrimental to environmentally protective infrastructure such as parks on hill slopes that reduce the risk of landslides and flooding (Drummond et al., 2012; Guerrero, 2011). In Rio de Janeiro, the paving of walkways in *favelas* as part of slum upgrading has increased runoff to the low-lying areas (de Sherbinin and Hogan, 2011). In Rio, several studies that have focused on modeling landslides converge on the nexus between slope instability, rainfall intensity, and soil hydrology as determinants of landslides (Moreiras, 2005). This is true whether the slopes are disturbed or not (Caine, 1980). The convergence of geophysical factors with locational factors (in turn related to patterns of deprivation) distributes landslide risk inequitably. Historical conflicts and control of some areas by organized crime gangs notwithstanding, public policies and low infrastructure investments raise equity issues in these cities and poorer neighborhoods.

6.2.2.5 Drought

It is seldom appreciated that many cities in low- and middle-income countries are located in dryland ecosystems, where precipitation is low but can also be erratic and unpredictable. McGranahan et al. (2005) estimated that about 45% of the population living in drylands is urban. Safriel et al. (2005) estimated that drylands ecosystems cover 41% of the Earth's surface while providing a home to some 2 billion people. Low- and middle-income countries account for about 72% of the land area and some 87–93% of the population of the drylands.

Drought can have many effects in urban areas, including increases in water shortages, electricity shortages (where hydropower is a source), water-related diseases (through use of contaminated water), and food price increases due to reduced supplies (Revi et al., 2014). An estimated 150 million people currently live in cities with perennial water shortages, defined as less than 100 liters per person per day of sustainable surface and groundwater flow within their urban extent. This may increase to up to 1 billion people by 2050 (McDonald et al., 2011).

Among the people estimated to live with perennial water shortages in the future, women are likely to be disproportionally represented because they often belong to the poorest of the poor. Many of the described impacts, particularly increased food prices and food insecurity in cities, will disproportionally impact women because they reduce food intake compared with other family members if food is scare and/or expensive (GGCA, 2009).

Urban risks of drought and water shortages are likely to be accompanied by intense, damaging rainfall events. As Safriel et al. (2005) note, in dryland ecosystems the expected annual rainfall typically occurs in a limited number of intensive, highly erosive storms. For example, in the Sahel, Descroix et al. (2012) describe a severe flood that affected the middle Niger River valley in 2010, inundating 3.1 square kilometers of the city of Niamey, where an estimated 5,000 people lost their homes. The trigger was damaging intense rainfall, which is infrequent but not unusual in the Sahel dryland areas, while the damaging impact was mainly attributed to land-use changes in the region. Another example refers to São Paulo, where downpours caused severe flooding in the midst of a serious drought in February 2015. In just one hour, São Paulo experienced 96 millimeters of rainfall while at the same time undergoing the most severe drought in the past eighty years (Fox, 2015). Ironically, while leaving the city under water and causing substantial traffic disruptions, the amount of rainfall was not enough to replenish the city's reservoirs further away from the urban area, and water rationing had to continue.

If droughts occur in combination with other extreme events, they increase the knock-on effects of disasters and cause cumulative impacts because cities may still be struggling to recover from the previous event (Kates et al., 2012). Intensive flooding as a result of heavy rainfall during droughts is one example; the risk of fires for heat-sensitive ecosystems in water-stressed cities is another (Ziervogel et al., 2010). Cape Town is an example where droughts increase the risk of fires in the surrounding heat-sensitive ecosystems (Jenerette and Larsen, 2006; Vairavamoorthy et al., 2008).

Moreover, uncertainties about long-term drought risk from climate change may have an influence on current access to water in cities, as observed in Mexico City (Romero Lankao, 2010). This raises equity issues because the poor are reported to spend from 4.2% to 4.7% of their income on water, whereas rich people pay 0.4% to 0.5% and consume more than twice as much water (Olmstead et al., 2007; Ruijs et al., 2008); though percentages differ across countries. Literature suggests the need for progressive block prices, full-cost prices, and income-dependent price systems and not egalitarian pricing systems (Olmstead et al., 2007). However, increasing block rates (prevalent in Europe) may have negative impacts on the welfare of low-income groups in the long run when structured on a volume basis rather than per-capita (Bithas, 2008). Similarly, availability of treated water (and thus of water treatment plants) is also unequal and positively correlated with income levels (Awad, 2012).

6.2.2.6 Storm Surge and Coastal Flooding

Storm-related hazards (tropical and extratropical cyclones and storm surges) are often connected to precipitation-related hazards and constitute major risks to urban populations. In coastal regions, global warming–induced sea level rise, combined in places with subsidence of coastal land and increasing storm intensity, put large coastal populations at risk from storm surges (see Chapter 9, Coastal Zones). Recent examples of coastal flood disasters include the flooding caused by Hurricane Katrina in New Orleans in 2005, Cyclone Nargis in southern Myanmar in 2008, Hurricane Sandy in New York in 2012, and Super Typhoon Haiyan in the Philippines in 2013 (Temmerman et al., 2013). Hurricane Katrina reached up to 10 meters (Fritz et al., 2007) and Hurricane Sandy to almost 4 meters above normal tide levels (Blake et al., 2013; McGranahan et al., 2007; IPCC, 2007).

Urban dwellers are more likely than rural villagers to be exposed to the risks of cyclones and storm surges because urbanites are more likely to live on or near the coast: cities and towns account for nearly two of every three residents of coastal areas worldwide (McGranahan et al., 2005; McGranahan et al., 2007). In Asia, 18% of the population lives in the low-elevation coastal zone – the highest percentage across all world regions – and 12% of the urban land is at low elevation and near the coast (McGranahan et al., 2007). Moreover, many of Asia's largest cities are located in coastal areas that are cyclone-prone, such as Mumbai, or Karachi (World Bank, 2008; Kovats and Akhtar, 2008). Flooding and storm surges also threaten coastal African cities, such as Port Harcourt and Lagos in Nigeria (de Sherbinin et al., 2014; Güneralp et al., 2015) (see Box 6.3). Similar vulnerabilities affect Mombasa and various cities in Latin America (Douglas et al., 2008a; Awuor et al., 2008; Hardoy and Pandiella; 2009; Revi et al., 2014).

Vulnerable people and households are more likely to be affected during extreme weather events, such as hurricanes and storm surge flooding (Hartman and Squires, 2006; Cutter et al., 2006), partly because these groups live disproportionately in low-lying areas and flood plains in many world regions, as documented from various cities in Africa, Asia, and Latin America (Hardoy et al., 2001; Balk et al., 2009). Vulnerable people and households, such as women with low incomes or ethnic minorities, also possess substantially fewer resources to cope if damage occurs. For example, Hurricane Katrina disproportionately affected African-American residents and elderly people (Hartman and Squires, 2006; Cutter et al., 2006; Curtis et

al., 2007). Case studies suggest that mortality rates of women and men vary significantly for both climatologic and other natural disasters. During the cyclone in Bangladesh in 1991, death rates were 71 per 1,000 for women and 15 per 1,000 for men. In the 2004 Tsunami in Amapara, Sri Lanka, fatalities were 3,972 and 2,124 among women and men, respectively (David and Enarson, 2012).

In addition to human fatalities and infrastructure damages, a number of health risks are associated with coastal (and inland) flooding, such as cholera, cryptosporidiosis, typhoid fever, diarrheal diseases, and leptospirosis (Kovats and Akhtar, 2008) (see

Box 6.3 Mapping Exposure to Coastal Stressors, Social Vulnerability, and Population Growth in West Africa

The African Resilience to Climate Change project mapped the exposure of coastal systems and vulnerable populations to projected sea-level rise along the West African coast (de Sherbinin et al., 2014).

Results show areas of high social vulnerability and high exposure, particularly in the Niger Delta, Lagos, Cotonou (Benin), and Abidjan. These same areas are projected to see high levels of population growth. Projections to 2050 in the near coastal zone (0–5 m elevation) suggest a more than three-fold increase in population (from 15 to 57 million people), with most of that concentrated in Nigeria. For the 0–20 meter low elevation coastal zone, the projected increase is from 33

million to 115 million people, most of them urban residents. Climate change and associated sea level rise will put substantially more socially vulnerable people at risk along the West African coast in the future.

A composite social vulnerability index (SVI) was constructed from indicators representing population density, population growth (2000–2010), subnational poverty and extreme poverty, maternal education levels, market accessibility, and conflict events (de Sherbinin et al., 2014). Population projections to 2050 used Shared Socioeconomic Pathway 4, representing a socially divided world with high levels of rural-to-urban migration (Riahi et al., 2017).

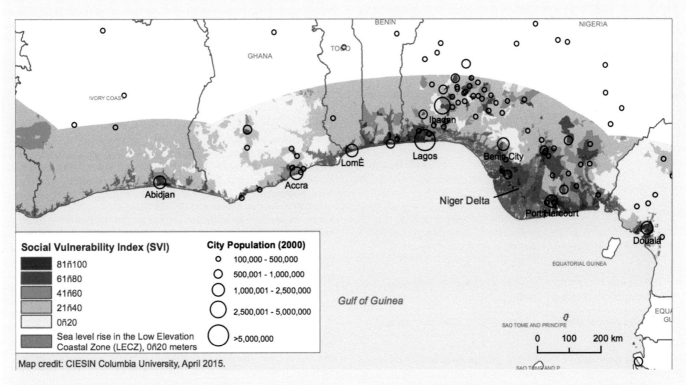

Box 6.3 Figure 1 *Social vulnerability index, low-elevation coastal zones (<20 m above sea level) and location of large urban areas along the coast of West Africa.*

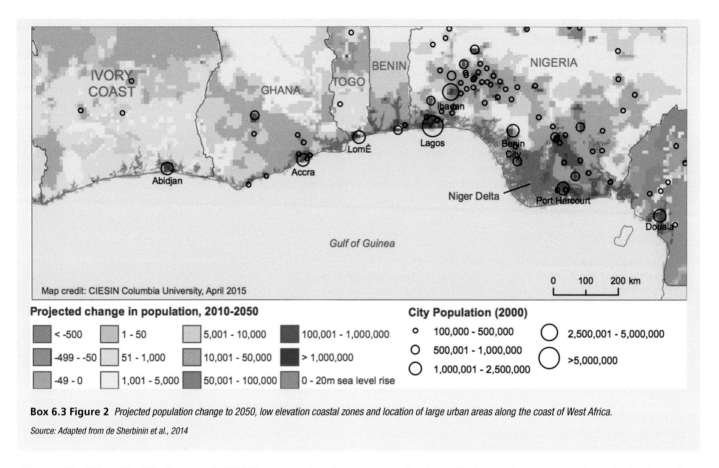

Box 6.3 Figure 2 *Projected population change to 2050, low elevation coastal zones and location of large urban areas along the coast of West Africa.*

Source: Adapted from de Sherbinin et al., 2014

Chapter 10, Urban Health). Lau et al. (2010) suggest that the combination of climate change, flooding, population growth, and urbanization will almost certainly lead to an escalation of leptospirosis, with high risks for urban slums, low-lying areas, and small island states. Storms are also expected to lead to water contamination with chemicals, heavy metals, and other hazardous substances for populations living near industrial areas on the coast (see Case Study 6.1). Moreover, Smith et al. (2014b) describe a newly emerging literature on the mental health consequences of flooding and other extreme events (see also Kinney et al., 2015).

Case Study 6.1   Building Climate Justice in New York: NYC-EJA's Waterfront Justice Project, The Sandy Regional Assembly, and the People's Climate March

Juan Camilo Osorio
Massachusetts Institute of Technology, Cambridge, MA

Eddie Bautista
New York City Environmental Justice Alliance (NYC-EJA)

Keywords	Environmental justice, climate justice, storm surge, Waterfront Justice Project, Sandy Regional Assembly, People's Climate March
Population (Metropolitan Region)	20,153,634 (U.S. Census Bureau 2016)
Area (Metropolitan Region)	17,319 km² (U.S. Census Bureau, 2010)
Income per capita	US$53,380 (World Bank, 2017)
Climate zone	Dfa – Cold, without dry season, hot summer (Peel et al., 2007)

Coastal storms, storm surge, and extreme winds all pose risks to urban coastal populations in most low-, middle-, and high-income countries in the future. For example, climate change projections suggest that, in the North Atlantic, the most intense hurricanes will increase in frequency by the 2050s (NPCC2, 2013), whereas the number of hurricanes might remain relatively unchanged. The combination of extreme weather events and climate change is creating new challenges for environmental justice for many communities living along industrial waterfronts.

Industrial waterfront communities have historically been the site of clusters of polluting industry and infrastructure. For instance in New York, most manufacturing zoning districts are located along the waterfront and have been linked with the inequitable distribution of noxious matter in these communities (Maantay 2002; Fahmi et al., 2014). However, research addressing the impacts of climate change, storm surge, and extreme winds on these waterfronts and the risks to low-income communities and communities of color living in close proximity to them is lacking. Therefore, environmental justice communities in New York are working to (1) conduct research to the threats in their communities; (2) identify proactive policies and programs to promote climate resilience that reflect local priorities; and (3) build climate-resilient industrial waterfronts while preserving jobs.

In 2010, the New York City Environmental Justice Alliance (NYC-EJA) launched the Waterfront Justice Project, a research and advocacy campaign to promote climate-resilient industrial waterfront communities. NYC-EJA's research has focused on six Significant Maritime and Industrial Areas (SMIAs), where industrial uses and public/private infrastructure are clustered and that are located in areas at risk of storm surge and high winds as projected by the New York State Office of Emergency Management. The majority of those vulnerable areas are in Brooklyn (NYC-DOHMH, 2010), including some SMIAs located within the 100-year floodplain of the Federal Emergency Management Agency (FEMA) and therefore at particular risk of rising sea levels. By 2050, sea levels are projected to rise by at least 2.5 feet (NYC-SIRR, 2013; NPCC2, 2013).

Before the launch of NYC-EJA's Waterfront Justice Project, the City of New York had not considered the risk of toxic exposures associated with clusters of heavy industrial uses in such vulnerable locations. After five years of research and advocacy work by NYC-EJA, the NYC Panel on Climate Change has incorporated potential hazardous exposures that can occur in the event of severe weather as an important threat affecting industrial waterfront communities that are vulnerable to climate change impacts (Kinney et al., 2015).

Low-income residents and people of color in the communities who live and work in and around the SMIAs are especially vulnerable to the potential release of contaminants in the event of strong winds, flooding, and storm surges, which are projected to increase in severity and frequency. According to the 2010 U.S. Census, approximately 622,600 New Yorkers lived in census tracts that fall within a half mile of the SMIAs and are vulnerable to storm surge. Of that number,

approximately 430,000 are people of color (U.S. Census Bureau, 2010b). In addition, these areas present some of the highest levels of uninsured populations, implying limited access to health care in the event of toxic exposures (NYC-DOHMH, 2010). NYC-EJA has successfully advocated for changes in the City of New York's updated Waterfront Revitalization Program (i.e., NYC's official coastal zone management plan) and initiated conversations with local policy-makers, government agencies, residents, and businesses. The Waterfront Justice Project is increasing awareness of hazardous substances in the context of climate change impacts to industrial waterfront neighborhoods in New York (Bautista et al., 2014). NYC-EJA's Waterfront Justice Project shows how affected communities are leading the call to integrate climate adaptation and pollution prevention into planning and development strategies in industrial waterfront communities.

After Hurricane Sandy, NYC-EJA extended this work to participate in the recovery planning process by co-convening the Sandy Regional Assembly, a coalition of community, environmental justice, labor, and civic groups from NYC, Long Island, and New Jersey. The Sandy Regional Assembly participated in the recovery process by advocating for green infrastructure and climate adaptation projects in low-income communities and communities of color, commenting on government reports, and promoting community-driven resiliency planning across the New York-New Jersey region.

In that mission, NYC-EJA was a co-coordinator of the massive People's Climate March in September 2014 (see Case Study 6.1 Figure 1) – at the time, the largest climate march in history – with an estimated 400,000 participants, 1,500 organizational sponsors, and more than 2,000 solidarity marches and rallies across the globe. NYC-EJA helped build this diverse mobilization of labor unions, environmental justice organizations, social justice, community-based

Case Study 6.1 Figure 1 *People's Climate March in New York, September 21, 2014.*

Photo: Climate Action Network International

organizations, faith-based organizations, and environmentalists (Bautista et al., 2015).

In April 2016, NYC-EJA released *The NYC Climate Justice Agenda: Strengthening the Mayor's OneNYC Plan* (NYC-EJA, 2016). This was the first comprehensive analysis of Mayor de Blasio's *OneNYC:* *The Plan for a Strong and Just City* (City of New York, 2015) from an environmental/climate justice perspective. The Mayor's Office publicly welcomed this partnership and has been exploring opportunities for further collaboration to continue advocating for the implementation of NYC-EJA's efforts to address environmental justice issues in New York.

6.2.3 Equity in Relation to Urban Climate Change Adaptation

Equity and environmental justice issues related to climate change in cities include inequalities in the capacity to cope and adapt, which in particular affects low-income groups (Dodman, 2013; Hardoy and Pandiella, 2009). This section focuses on inequality in relation to urban risks arising from the failure to adapt, inadequate adaptation, or maladaptation to climate change.

Differentials in the scale and nature of risks among settlements relate to the extent of infrastructure and services provision (see, e.g., Krishna et al., 2014, for a discussion on infrastructure and service provision among different informal settlements in Bengaluru). Differentials in risk arising from inadequate or no infrastructure and services can emerge in relation to age, sex, and health status (Bartlett, 2008) but can also be socially constructed, as in the case of discrimination (e.g., by gender; Dankelman et al., 2008). An analysis of the impacts of floods in Lagos in 2011 revealed differences in vulnerability among low-income women created by the intersection of gender relations and gender roles in household structure, occupation, and access to health care (Ajibade et al., 2013). Differentials in risk also arise from the lack of voice for particular groups (e.g., informal settlers) and the lack of accountability to them by government agencies (Bulkeley et al., 2014; Adger, 2013). It is thus relevant to consider the extent to which adaptation measures acknowledge these differentials, identify the groups most at risk, and take action to reduce them.

Within almost all cities in high-income countries, development has greatly reduced risk from extreme weather. There is universal provision for piped treated water, adequate drainage, and implemented building standards for structural safety. Cities in high-income countries have citywide sewer systems and storm drains with the capacity to cope with extreme precipitation as well as all-weather roads, health care systems, and emergency services with little or no "inequality" in their provision as these serve everyone (see Figure 6.3). These services are not provided as a response to climate change and therefore not "adaptation" *per se*, but they promote resiliency to climate change impacts, and strengthen the institutions and financial systems that make resiliency possible (Satterthwaite, 2013). In well-governed cities, there may be high levels of inequality in income, assets, and the quality of housing and urban services between favorable and unfavorable locations, but far less inequality in benefits from risk-reducing infrastructure and services and thus in exposure to risks from climate change. This includes groups generally considered vulnerable because the universal provision of infrastructure and services reduces or removes exposure to risks. This, however, is not to claim that all inequalities in risk are addressed, as work on environmental justice has shown (Schlosberg and Collins, 2014).

The disproportionate impacts of extreme weather events on low-income populations in urban centers in low- and most middle-income countries are strongly associated with the lack of risk-reducing infrastructure (piped water, sanitation, effective drains, all-weather roads and paths) and risk-reducing services (including health care and emergency services) (Revi et al., 2014; Dodman and Satterthwaite, 2009). The lack of risk-reducing infrastructure is often underpinned by a lack of capacity within urban governments to address infrastructure and service deficits (Parry et al., 2009). In low-income and many middle-income nations, most urban authorities have very small budgets and even less investment capacity (UCLG, 2014). Housing development on dangerous sites – especially in flood plains, alongside rivers, or on steep slopes (Hardoy et al., 2001; Hardoy and Pandiella, 2009; Dodman, 2013) – is exacerbated by inappropriate building regulations and land-use/zoning practices that restrict the supply of affordable housing plots (Aylett, 2010; Lwasa and Kinuthia-Njenga, 2012; Lwasa, 2012). Undefined property rights and land tenure also contribute, as documented in cities like Nairobi, Dar es Salaam, Dhaka, Dakar, Maputo, Manila, and Kolkata (Dodman, 2013; Hardoy and Pandiella, 2009; Jenkins, 2000; Owens, 2010; Rao, 2013; Roy, 2009).

The Intergovernmental Panel on Climate Change (IPCC) Fifth Assessment Report (AR5) highlights the wide variety of urban areas' adaptive capacity (Revi et al., 2014). At one extreme, there are a billion people living in urban centers with very little capacity to adapt to climate change and with large deficits in risk-reducing infrastructure and services. Another 1.5 billion live in urban centers with limited capacities and significant infrastructure and service deficits, whereas a very small proportion of the world's urban population live in urban centers with universal provision for risk-reducing infrastructure and services and active climate change adaptation policies.

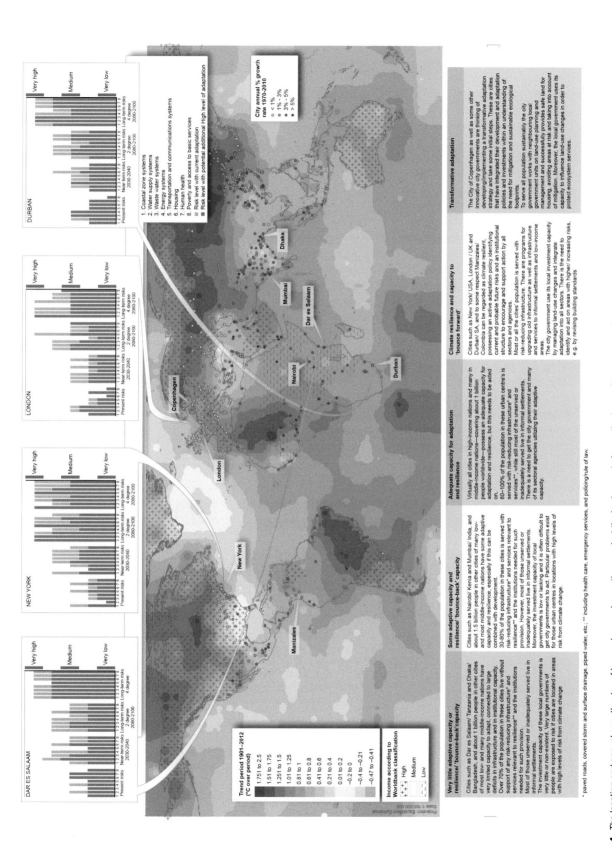

Figure 6.4 *The location and annual growth rates of large urban agglomerations against the background of recently observed temperature change. Insets (above) show current and future climate change risks in selected urban areas and (below) describe the spectrum of adaptive capacity in urban centers. (1) The map shows large urban agglomerations in 2010 across the globe and their population growth rates (1970–2010) against observed climate change (trend period 1901–2012). Many cities with the highest population growth rates are located in areas of moderate recent temperature increase, but many of them are also in low- and middle-income nations with low adaptive capacity. (2) The bar charts above the map show key climate risks to urban systems and services with and without adaptation for Dar es Salaam, New York, London, and Durban. Risk levels are identified based on an assessment of the literature and expert judgments by the Intergovernmental Panel on Climate Change, AR5, WGII, and Chapter 8 authors (Revi et al., 2014), ranging from very low to very high. For the near-term era of committed climate change (2030–2040), projected levels of global mean temperature increase do not diverge substantially across emission scenarios. For the longer-term era of climate options (2080–2100), risk levels are presented for global mean temperature increases of 2°C and 4°C above pre-industrial levels. For each time frame, risk levels are estimated for a continuation of current adaptation (pink bars) and for a hypothetical highly adapted state (red bars). Please note that the climate risks should be compared across time for each city individually; cross-city comparisons are difficult. (3) Risk levels of the selected cities can be compared to the large spectrum in adaptive capacity of urban centers to adapt to climate change as highlighted in the bottom table.*

Source: Adapted by authors with data from Revi et al., 2014

Figure 6.4 shows the risk levels for a range of key sectors for New York, London, Dar es Salaam, and Durban and their potential to adapt to current levels of risk and those anticipated in the near-term (2030–2040) and long-term (2080–2100) future, using scenarios of 2°C and 4°C warming above pre-industrial levels. Perhaps not surprisingly, risk levels increase dramatically for all key risks in the long-term, especially for the 4°C warming. But, at least for now, in the near-term, and mostly for the long-term with a 2°C temperature rise, a high level of adaptation can keep risks down. For example, in the case of Dar es Salaam (see Case Study 9.6) adaptation can potentially be effective in protecting energy and transport systems, safeguarding human health, and maintaining progress on poverty reduction. Currently, however, only a small proportion of the city's population has piped water supply to their home, sewers, covered drains, and solid waste collection; the city lacks the capacity to address these issues and implement other urgently needed adaptation measures. Dar es Salaam is an example with a large gap between adaptation needs and adaptive capacity (Kiunsi, 2013). However, under a 4°C temperature rise, adaptation measures are likely to be ineffective not only in Dar es Salaam but also in cities with currently high levels of adaptation, such as New York and London.

Changes in land-use planning and regulatory frameworks (for buildings, infrastructure, and zoning) are an important part of adaptation to climate change, as are fiscal incentives and infrastructure investments that respond to current and projected future climate risks (see Chapter 7, Economics, Finance, and the Private Sector). Land-use planning and management should play critical roles in ensuring there is sufficient land for housing that avoids dangerous sites and in protecting ecological services and systems. There is also growing awareness of the need for gender-sensitive adaptation processes and intersectional analyses in order to develop inclusive, contextually specific interventions and policies (Alston, 2013; Sultana, 2013; Kaijser and Kronsell, 2014). Adaptation practices also need to intersect with mitigation concerns (see Chapter 4, Mitigation and Adaptation). If these aspects, are covered, the IPCC AR5 speaks of transformative adaptation (see Figure 6.4; Revi et al., 2014; and Chapter 1, Pathways to Urban Transformation).

It is not only inadequate government capacity that underpins lack of attention to climate change adaptation but also deliberate choices by city or national governments (Bulkeley et al., 2014). Thailand's flood crisis in 2011 is an example of how policy interventions translate into a redistribution of risks. When city officials in Bangkok sought to protect the city center by diverting floodwaters to other areas, it heightened the disproportionate impact on those suburbs and communities outside the defenses (Nabangchang et al., 2015). The refusal to address risks to poor and politically underrepresented groups in urban areas is often related to the low priority that national governments and international agencies have given to such equity issues.

A concern for equity in climate change adaptation also means a concern for avoiding maladaptation in policies, public investments, and responses to climate risks. For example, the choices made in the management of floodwaters in and around Bangkok could be considered maladaptive practice because it protected

Case Study 6.2   Citizen-led Mapping of Urban Metabolism in Cairo

Heba Allah Essam E. Khalil

Cairo University

Dave Ron

Ecocity Builders, Oakland

Keywords	Participatory mapping, heat waves, urban metabolism, research justice, environmental justice
Population (Metropolitan Region)	14,629,360 (Cairo Governorate, 2014)
Area (Metropolitan Region)	4,692.7 km² (Cairo Governorate, 2014)
Income per capita	US$11,110 (World Bank, 2017)
Climate zone	BWh – Arid, desert, hot (Peel et al., 2007)

Cairo, the cultural and economic capital of Egypt, is the largest city in the Middle East and Africa. With a population of approximately 20 million people and three governorates, the continued expansion of Cairo has created multiple stressors on the environment, quality of life, and existing infrastructure.

Traffic congestion by the nearly 5 million cars on its roads, coupled with unregistered smelters and other industries, has resulted in Cairo having one of the highest pollution rates of any city second only to Delhi. This will be further exacerbated by the effects of global climate change, which are projected to include sea level rise into Egypt's fertile Delta region and freshwater scarcity from desertification.

Informal areas, which house approximately 60% of the city's population, are anticipated to be highly vulnerable to increasing heat waves. Such areas also tend to be the most densely populated in a city that averages 89,000 people per square kilometer (Khalil, 2010). The degree of urban consolidation can pose challenges for mitigation and adaptation measures in response to climate change.

The Eco-Citizen World Map Project (EWMP) comprises three distinct yet interwoven components: the Partnership, the Platform, and the Pedagogy. The Partnership is led by U.S. nongovernmental organization (NGO) Eco-City Builders and joined by ESRI, the Association of American Geographers, Eye on Earth (a partnership of UNEP, Abu Dhabi Global Environmental Data Initiative), Cairo University, Mundiapolis University, University of California at Berkeley, local NGOs, and community partners.

The Platform of EWMP provides the incentive and understanding for communities to crowdsource urban data and holistically assess the condition of their neighborhoods. This promotes more democratic and grassroots leadership in proposing and planning interventions that directly enhance the sustainability and equitability of cities. Geographic information systems (GIS) and urban metabolism information systems (UMIS) are the two primary methods employed for organizing and displaying data through the Platform. UMIS describes a system, along with all of its components, to account for and analyze resource flows as they move from the natural (i.e., a source) through the built (i.e., a city) back to the natural environment (i.e., a sink). Sankey diagrams are a means to represent this, whereby the width of arrows in a linear flow is proportional to their quantity (Google Developers, 2015). The Platform displays data in visually accessible ways that communities can customize and interact with directly – specifically, spatially dynamic online maps with multiple dataset layers and Sankey diagrams.

The Pedagogy of EWMP is defined through a research justice framework, breaking down existing structural barriers between researcher and researched, and it includes a training-of-trainers methodology to support capacity-building among students and citizens. These trainers then engage in knowledge-transfer activities with citizens, thus facilitating bottom-up data collection, analysis, and publication for the Platform.

The Project's piloting in Cairo, established in early 2014, has been led in part by El-Balad, a local community-based organization (CBO) in the neighborhood of Imbaba. As one of the most densely populated informal areas in Egypt, Imbaba is among the oldest districts to host rural immigrants. Originally, Imbaba was agricultural land, subdivided illegally and built by local inhabitants. The area has since been consolidated and is a preferred destination for low- to middle-income households given its proximity to other Cairene districts and low rental costs. The Project involves El-Balad mobilizing citizens through its existing networks and students at Cairo University being trained on adapting and applying public participation techniques, GIS, and UMIS by academic faculty and the EWMP Partnership.

The students participate in labs, lectures, and group activities that touch on these topics. A community roundtable invites elders, youth, and leadership from the selected study area to a presentation on the EWMP. The event includes a discussion that refines the purpose, scope, preferred decision-making models (e.g., consensus-based), study area boundaries, data-reporting standards, and ownership of research outcomes. At the Imbaba roundtable, the community prioritized concerns around access and quality of freshwater supply. The roundtable is also an opportunity to propose realistic targets, relevant indicators, and forms of participation for data collection, input, and analytical processes. Examples of specific activities can include training on GPS devices, co-facilitation methods, computer data entry, digitizing, and geocoding techniques.

An intensive session is held as a two-day training-of-trainers event where the CBO and students facilitate workshops and initiate the citizen-led collection of data (Case Study 6.2 Figure 1). Teams are arranged according to different collection methods: namely, environmental assessments (e.g., air and water quality tests), quality-of-life questionnaires, and parcel audits of resource management such as water demand. At the designated workstations set up within the study area, data are digitized, samples are tested against established baselines, and results are analyzed.

Data for Imbaba published through the Platform have brought attention to water quality, access, and management issues. In terms of the former, quality testing has revealed an increase in the percentage

Case Study 6.2 Figure 1 *Intensive training presentation in neighborhood of Imbaba, Cairo.*

Case Study 6.2 Figure 2 *Awareness flyers to reduce water consumption.*

of coliform bacteria as a result of broken piping infrastructure that mixes potable water and sanitary streams, as well as an increase in copper traces due to old piping infrastructure and minimal storage maintenance.

In terms of management, parcel audits have shown poor water flow from the main supply lines and the disproportionate use of water demand according to building archetypes. This presents citizens with visual guides to suggest areas for conservation (e.g., minimizing use in cooking), efficiency (e.g., low-flow shower heads), cascades (e.g., gray water use for rooftop gardens), and advocacy for municipal upgrades of upstream infrastructure (e.g., retrofitting crumbling concrete plumbing with more enduring materials to minimize water

loss) (see Case Study 6.2 Figure 2). A series of printed materials and awareness events organized by El-Balad and the students disseminated the pilot study results and ensured transparent flow of data to the local community.

As an initial intervention, participants formed an advocacy group to mobilize funds for household interventions and to push for governmental financing of upgrades to the area's water network.

The EWMP helps ensure equitable and grassroots interventions to reduce Imbaba's vulnerability to climate change. Technical capacity gained through the process has enabled the group to approach other informal areas including Gezirat Al-Dahab and Warak, Giza.

the wealthy and placed an increasing burden on the more vulnerable in society (see also McEvoy et al., 2014, for examples in Vietnam and Bangladesh). Other maladaptive practices relate to constraining land supplies, forcibly resettling people in areas far from their employment, or evicting people with no compensation and pushing up land and housing costs. These practices are likely to increase the proportion of the population living in vulnerable conditions and informal settlements. Maladaptation leads to further impoverishment of vulnerable groups, often in the name of "development," when low-income households are displaced by the expansion of roads and highways and other measures to reduce infrastructure deficits. Forced evictions constitute gross intrusions on human rights because they indirectly and directly violate the full spectrum of civil, cultural, economic, political, and social rights.

Those who live in settlements on sites that are dangerous and lack risk-reducing infrastructure and services often take

measures to reduce risks to their household, homes, and assets. They often work at a community level to collectively address risks – especially in informal settlements or urban centers where there is low government interest or adaptive capacity. Revi et al. (2014) provide a review of these responses, such as maintaining local water sources, toilets, and washing facilities and constructing and improving drainage systems. Such measures can make an important contribution toward risk reduction, especially if they are community-wide responses, but they cannot provide a larger network of infrastructure on which local adaptation measures depend (e.g., water and waste-water treatment plants, and water, sewer, and drainage mains). However, an important lesson of these experiences is that the adaptive capacity and resilience of communities can be sustainably increased by providing appropriate support for community-based initiatives. Support should preferably be provided in the form of economic incentives and opportunities for low- and middle-income households.

6.3 Equity and Climate Change Mitigation

Climate change will affect urban residents not only by climate change-related risks, potential impacts, and adaptation, but also by mitigation actions, policy, and planning. Urban mitigation actions are currently planned and undertaken mostly in cities in high-income countries because these have historically emitted the bulk of GHG emissions, still emit the largest share of urban emissions on a per capita basis, and more often have the means to act. Research on mitigation actions in cities in low- and middle-income countries is currently limited.

6.3.1 Spatial Planning

With respect to spatial planning, cities in low- and most middle-income countries face challenges very different from those in high-income countries (see Chapter 5, Urban Planning and Design). In low- and middle-income countries, spatial planning policies are often outpaced by rapid population growth, whereas city budgets are usually inadequate to meet the ensuing need for expanded or adjusted infrastructure and service provision (UCLG, 2014). Additionally, spatial planning controls are often weak and, where they exist, there is limited enforcement (Bartlett et al., 2009). Particularly in low-income countries, most city governments have not yet begun to plan for climate change. There are only a few cities where climate change mitigation is currently being integrated into city plans (see Hardoy and Ruete [2013] for Rosario; and Roberts [2010], Roberts et al., [2012], and Roberts and O'Donoghue [2013] for Durban).

In cities of high-income countries, mitigation strategies often build on anti-sprawl policies, which attempt to reverse low-density development and encourage urban revitalization. The main principles of this "Smart Growth" planning framework are denser development, walkable neighborhoods, enhancement of mixed land use, and the conservation of open spaces (Wilson et al., 2008). A key objective is to reduce transport and related GHG emissions (see Chapter 13, Urban Transportation). However, these principles are not only praised for reducing traffic, but also criticized for pushing up housing prices and the subsequent displacement of low-income residents unable to afford the higher rents (Addison et al., 2013; Burton, 2000; Cox, 2008; Ewing et al., 2014; Smyth, 1996; Wendell, 2011). In particular, low-carbon building and construction standards, driven by a narrow technocratic vision of climate change mitigation, may undermine socially progressive housing policy and result in housing designs that are unaffordable for low- and medium-income households (Bradshaw et al., 2005; Golubchikov and Deda, 2012). This process is referred to as "environmental gentrification" (Checker, 2011; Curran and Hamilton, 2012; Jennings et al., 2012; Johnson-Gaither, 2011; Todes, 2012). Densification has also been linked to unequal distribution of domestic living space (Burton, 2000) and increases in residents' exposure to air pollutants, particularly if compactness is not accompanied by improved public transportation (Schindler and Caruso, 2014).

Compact city policies may also unequally curtail access to public facilities as well as open and green space thereby exacerbating already existing inequalities (Burton, 2000; Dempsey et al., 2012; Newman, 1972). Access to urban green space often negatively correlates with the racial/ethnic and socioeconomic characteristics of a neighborhood (Dai, 2011; Joassart-Marcelli, 2010; Joassart-Marcelli et al., 2011). In the United States, low-income households in Hispanic, African-American, and Asian communities have less green space within their neighborhoods than city averages (Byrne et al., 2009; Dai, 2011; Landry and Chakraborty, 2009; Sister et al., 2010; Wolch et al., 2014). Similarly, in South Africa, communities of black, low-income residents have sharply reduced access to public green space, as compared with predominantly white residential areas (McConnachie and Shackleton, 2010). Urban mitigation policies aiming at densification need to explicitly focus on an equitable increase in urban density and the distribution of opportunities and benefits of improved environments.

There is no consensus on how fiscal anti-sprawl policies shift burdens on residents with different income levels (Burton, 2000; Sharpe, 1982; Smyth, 1996). However, development taxes imposed to cover infrastructure-related costs seem to imply a lower burden on low-income groups than other instruments such as zoning and growth management policies that mandate which areas can be developed and under which conditions (Bento et al., 2011, 2006; Brueckner, 1997).

6.3.2 Accessibility and Transport Policies

Infrastructure investments aiming at improving public and private transportation services are a frequent and successful means to address climate change mitigation in urban areas (see Chapter 13, Urban Transportation). However, if implemented incautiously investments can potentially impact lower-income households by way of changes of housing affordability, costs of transportation, and accessibility. This is also the case for some policies whose main goal is not mitigation – for instance, policies to restrict private car use or improve public transport to lessen congestion or lower air pollution.

One strategy with the potential to mitigate climate change in cities is transit-oriented development (TOD), i.e., the improvement of access to public transportation (see Chapter 13, Urban Transportation). In addition to its mitigation potential, TOD has positive socioeconomic effects for residents because it brings a larger area of land into the employment catchment area. For example, in Medellin, the implementation of a cable car connected isolated low-income communities with the urban center and contributed to the enhancement of economic conditions and the reduction of violence (Brand and Dávila, 2011; Cerdá et al., 2012). However, studies have also shown that TOD can have negative effects on low-income groups because it influences the level of housing affordability (Deng and Nelson, 2010, 2011; Zhang and Wang, 2013; Smith and Gihring, 2006). In an open housing market, better access to transportation usually leads to

Case Study 6.3   Growth Control, Climate Risk Management, and Urban Equity: The Social Pitfalls of the Green Belt in Medellín

Isabelle Anguelovski

Institute for Environmental Science and Technology, Autonomous University of Barcelona, Barcelona

Keywords	Adaptation, resilience, disaster risk management, growth containment, socio-spatial justice, greenbelt, green infrastructure, resettlement, intense precipitation, heavy downpours, flood, landslides
Population (Metropolitan Region)	3,731,000 (Alcaldía de Medellín, 2015)
Area (Metropolitan Region)	1,152 km² (Alcaldía de Medellín, 2015)
Income per capita	US$13,910 (World Bank, 2017)
Climate zone	Am – Tropical, monsoon (Peel et al., 2007)

For the past fifteen years, the city of Medellín has been actively working on rebranding itself, changing its image of violent crime and drug trafficking to a more welcoming and safer place, better in tune with its environment. Thanks to many social urbanism and "urban acupuncture" projects targeting the urban poor and to the construction of new infrastructure, Medellín received the award for "Most Innovative City of the World" by the Urban Land Institute in 2013.

Announced in 2012, the future Metropolitan Green Belt (73 km, US$249 million) is part of this process of urban reinvention and urban resilience. Conceived as a   planning strategy to consolidate the metropolitan territory in a balanced and equitable way, the greenbelt is meant to restrain unregulated growth[4] and sprawl in the hillsides around the city, protect water basins and forests key to the region's biodiversity and to climate control, and, most importantly, reduce risks of landslides during extreme weather events (Agudelo Patino, 2013). Risks of climate disasters exist in Medellín because of a prolonged rainy season with a high prevalence of torrential downfalls and an increase in frequency of extreme rainfall events. Experts predict that extended dry periods will make the unstable soil along the steep slopes even more treacherous, while intense rainfall will increase the incidence of landslides. Today, 180,000 households are located on hillsides and ravines that are at risk of mudslides and other climate-related disasters.

Case Study 6.3 Figure 1 *The Medellín Green Belt.*

Photo: Municipality of Medellín

---

4   From 1955 to 2013, the population of Medellín almost tripled from 500,000 to 3 million residents, a growth driven by industrialization and internal displacement from armed conflicts.

The greenbelt is based on three levels of intervention and will affect 230,000 residents (see Case Study 6.3 Figure 1) who live above the 1,800 meter altitude set by the project, not all of which are considered "high-risk." The first level is a Protection Zone (in green) that is the "Green Belt" itself with natural habitat preservation, ecological restoration of hillsides, recuperation of rural corridors, natural and community tourism, carbon sinks protection, and rural habitat improvement. The second level is a Transition Zone (in yellow), close to the greenbelt, with the highest concentration of residents living in conditions that often lack basic amenities and that have expanded beyond the city limits. This Transition Zone will receive new metropolitan parks, farming projects, education gardens, bike paths, and risk mitigation measures (see Case Study 6.3 Figures 2 and 3). The third zone is a Consolidation Zone (in orange), meant to "re-conquer the Valley," with the creation of longitudinal parks, the construction of high-rises for new residents, structural intervention and habitat improvement projects, land entitlement, and a network of public services. At the heart of the Green Belt, the municipality is also planning a Clean Mobility Corridor (Alcaldía de Medellín, 2013). This project has strong potential to protect the city against more extreme and frequent weather events and to regulate uncontrolled growth on the slopes.

However, the construction of the Green Belt is also raising planning controversies related to sociospatial equity. First, because the municipality is planning to relocate thousands of households living on unstable terrain from the Zone of Transition to the Zone of Consolidation, community concerns have emerged (see Case Study 6.3 Figure 4). Most low-income residents do not want to be relocated and would prefer alternative solutions to risk management (interviews in 2013 and 2015). For instance, in Comuna 8, where the municipality is planning to relocate 6,600 households (but possibly up to 39,200 households), residents are opposed to the idea of being evicted from the houses they built during the armed conflict and being moved into city-built tower blocks far away from their original settlement. Controversies over relocation also highlight the political nature of risk assessments and maps since discrepancies exist between different sources of assessment – the Risk Zone map of the city, the Geological Aptitude Map (a map of geological risks associated with different land types), and resident-produced estimates of the number of households in "non-recoverable risk areas."

Second, decisions over which communities have to move away from the greenbelt highlight spatial inequity in urban development and relocation. Higher-income neighborhoods (El Poblado, Cedro

Case Study 6.3 Figures 2 and 3 *Renderings of the Medellín Green Belt.*

Source: Global Site Plans

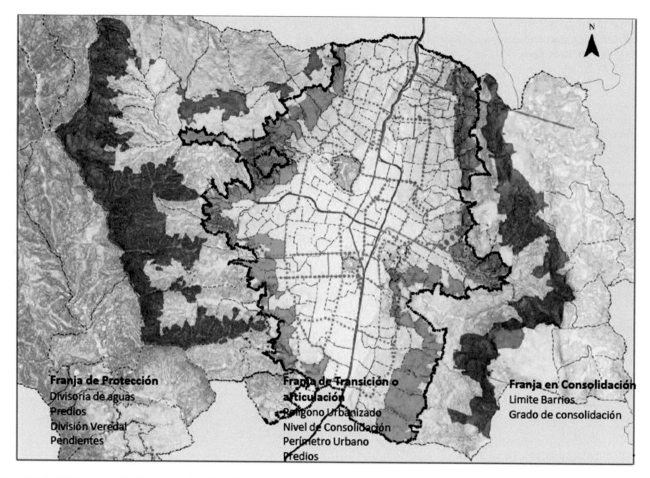

Case Study 6.3 Figure 4 *The three zones of the Medellín Green Belt.*

Source: Urban Solutions Platform

Verde, Alto de las Palmas) seem to be able to further expand toward the top of the hills in South East Medellin even though they have expanded beyond the city border. Gated communities such as Alto de Escobero, which is next to important reserves of native forest, are continuing to grow without a mention of their need for resettlement (Arango, 2012). In contrast, the municipality is planning to move residents from lower-income neighborhoods (Comuna 8).

Third, although the future system of longitudinal connectivity and mobility – to be built along with the Green Belt – linking new mobility projects such as bike paths, hiking trails, and a monorail system to the city's current urban transit system will create new environmental benefits for the city's residents, it could bring new forms of environmental privilege. Low-income residents fear that this monorail will attract tourists and wealthier residents benefiting from easier connections to new recreational areas and parks such as Park Arví (Interviews 2013 and 2015). Additionally, the Metrocable and other forms of public transport would stop before the furthest extent of these communities, and there is no plan to improve access to the city for the most vulnerable residents. By the same token, low-income residents would lose access to green space around the Pan de Azúcar Mountain on which they rely for their livelihoods and sources of fresh food. Even though the Pan de Azúcar is farmed by many residents, public officials see it as

an important ecological resource to preserve and transform into a recreational area.

Last, low-income residents from communities impacted by the Green Belt are concerned about a lack of meaningful engagement with affected vulnerable communities and a lack of recognition of their land-use and planning experience. In Comuna 8, using the three pillars of social and ecological function of property, direct participation of citizens in decision-making, and equitable distribution of costs and benefits of urbanization from the 1991 Constitution, residents have prepared a community development plan asking for the municipality to articulate the Green Belt project with their Declaration of Needs and Wants, including integral barrio-upgrading projects, food security and urban agriculture, risk management with the construction of proper sewage systems and retention walls, and housing and transportation improvements. Yet, to date, the dialogue between residents and the municipality has not produced tangible results and solutions.

In sum, the Green Belt project reveals that, although green infrastructure offers much resilience and climate adaptation opportunity for cities affected by climate impacts, such efforts may produce accelerated cultural, economic, and physical displacement for the most vulnerable residents while overlooking needs for social cohesion, community recognition, and livelihood protection.

an increase in property values and rents in zones of improved infrastructure services and particularly transportation nodal points. As a result, low-income groups may be forced to migrate to other locations with limited access to transportation but more affordable rents and housing prices (Boarnet, 2007; Deng and Nelson, 2011; Munoz-Raskin, 2010; Zhang and Wang, 2013).

Another aspect of concern with regard to social inequality is the usually high up-front costs required to access discount fares, such as annual or monthly transit passes. Low-income households may be obliged to purchase weekly or daily passes, which can be up to three times more expensive than longer term passes (Nuworsoo et al., 2009; Schweitzer, 2011). Moreover, an increase in the costs of transfers or the removal of unlimited-use passes mostly affects lower-income riders, youth, and minorities because these groups make more trips and transfer more frequently than others (Cheng et al., 2013; Nuworsoo et al., 2009).

Gender-based violence, harassment, and crime in public transport are also sources of concern and should be addressed through gender-sensitive transportation-based mitigation actions. For instance, Montreal, Bogotá, Malmö, Vienna, and Berlin are noteworthy pioneer cities that have mainstreamed gender in their mobility plans. These cities have sought to improve the accessibility, safety, and comfort of public transport and prioritize women's perspectives (Maffii et al., 2014; Levy, 2013; Clarke, 2012).

Transport policies aiming to improve private transportation infrastructure often show contradictory results with regard to their effects on equity mainly due to the range of policy options available (whether these include tax revenue recycling and/or other additional benefits[5]) and the use of different methodologies in defining inequality (Schweitzer, 2011). For example, fuel taxes and vehicle registration fees, although highly dependent on instrument design and the associated revenue recycling measures, generally show very small effects across income groups (Dill et al., 1999; Walls and Hanson, 1999; Fullerton et al., 1980; Bento et al., 2005). However, registration fees that are based on pollutant emission rates typically affect low-income drivers more than those based on distance traveled, because low-income residents drive vehicles that pollute more per mile than do those owned by wealthier groups of the society.

Cordon charges (location- or time-based charging tolls on a road network or upon entering a defined zone) include two main types of pricing strategies: cordon tolls and area charges. The key difference is that the first one tolls travelers per crossing whereas the second one tolls users for a license (i.e., one-day period) to enter or travel inside an area. On average, area-based schemes tend to perform better than crossing schemes in terms of equity effects because, with the same boundary, they affect a higher volume of demands in the network compared to cordon tolls (Maruyama and Sumalee, 2007). However, very specific time-based cordon tolls for peak hours are also progressive. For example, in Stockholm, the relative burden change amounted to 0.35% for low-income groups if revenues were allocated to public transit or tolls were low enough (Eliasson and Mattsson, 2006). Area-based schemes showed a higher burden of more than 2% for low-income motorists (in the case of Paris), but this changes to 0.5% when all commuters are factored in (Bureau and Glachant, 2008). Overall, cordon charges can produce beneficial outcomes to low-income households by reducing traffic congestion and improving air quality when considering social and environmental benefits and/or revenue recycling (Creutzig and He, 2009; Schweitzer, 2011; Creutzig et al., 2012).

Speed limit schemes can cause burden changes from 5% to 17%, depending on the integration (or not) of commuting patterns in the policy design (Schweitzer, 2011; Wang, 2013). Finally, in terms of environmental inequity, charges based on road usage seems more effective than those based on low emission zones (LEZs)[6] because the latter might cause an increase in emissions in the surrounding areas – with potential effects on lower-income communities. Moreover, the quality benefits sought by LEZs may also occur in a do-nothing scenario due to fleet renewal processes (Carslaw and Beevers, 2002; Mitchell, 2005).

Overall, both urban geography and individual characteristics of policies determine who benefits from spatial planning and transport mitigation strategies (Santos and Rojey, 2004). Mitigation strategies may only exacerbate inequalities if the design and implementation does not take into account differential impacts on low-income and other vulnerable groups. Distributional impacts are city-specific and require a case-specific evaluation on multiple scales, such as the travel behavior of different communities.

More generally, there is the issue of where responsibility for decarbonization in transport systems is located and the worry that neo-liberal mechanisms will exacerbate inequalities in travel patterns (Schwanen et al., 2011). Hence, policy-makers should try to anticipate negative effects and offset them by adopting additional measures or modifying policy designs. Dulal et al. (2011) note that, in the context of urban growth, ensuring that all neighborhoods in an urban settlement are equally served by efficient public transport routes can be effective not only in reducing transport volumes and private vehicle use, but also in lessening distributional and welfare inequality of GHG emissions reduction (Grazi and van den Bergh, 2008).

6.3.3 Waste Management and Renewable Energy

Sector-specific policies for urban waste management have not yet been explicitly analyzed in terms of equity performance, although some studies offer insights on specific issues, for example waste-pickers. Waste-picking constitutes the major reuse and recycling business in urban centers in many low- and

---

5 Revenue "recycling": Use of tax revenues to lower other taxes or to finance explicit public investments. Additional benefits include environmental benefits (i.e., lowering pollution) and social benefits (i.e., lowering time congestion).
6 Low emission zones (LEZs): An air quality management tool in which most polluting vehicles are barred from a specific area.

middle-income countries, thereby helping to avoid substantial GHG emissions (King and Gutberlet, 2013). One study in Ribeirão Pires, Brazil showed that the informal/cooperative recycling sector was capable of achieving GHG emissions reductions similar to those in formal solid waste management, recycling, and landfill gas capture (King and Gutberlet, 2013). Although being sustainable, inclusive, and integrated, it is a survival strategy that sometimes faces strong opposition from authorities (Hunt, 1996 Hayami et al., 2006; Chen et al., 2013). When improved waste collection and management becomes a public priority, pickers are often displaced and become unrecognized (Rouse and Ali, 2001; Ahmed and Ali, 2004; Scheinberg and Anschütz, 2006; Wilson et al., 2006; Medina, 2008; Betancourt, 2010), regardless of their environmental contribution and the subsequent social impacts (Huysman, 1994; Baud et al., 2001; Moreno-Sánchez and Maldonado, 2006). However, some progressive cities have devised contractual arrangements for waste-pickers (Fergutz et al., 2011; Kareem and Lwasa, 2011; Vergara and Tchobanoglous, 2012; Campos and Zapata, 2013). Pro-poor recycling strategies like those in Maputo and Bangalore have strengthened waste-pickers' cooperatives through improvements in infrastructure, governance, and skills, thereby obtaining benefits in a number of sustainability dimensions (Storey et al., 2013).

Renewable energy schemes may also lead to unequal burden shifts because low-income households often contribute a larger fraction of their income to such schemes than do higher-income households. This reflects a fundamental bias of incentive-based mechanisms that leave high-income households following their old consumption practices (and paying a bit more) while leaving low-income households scrambling to adjust (Earl and Wakeley, 2009; Perry et al., 2013).

6.4 Innovations and Lessons from Implementation

A growing number of city governments have made innovations in climate change adaptation and mitigation. These include many cases where cities have taken action despite the lack of supporting national policies and international funding (Reckien et al., 2014; Bulkeley and Castán Broto, 2013), including in low- and middle-income nations (Bicknell et al., 2009; Anguelovski and Carmin, 2011; Bulkeley and Castán Broto, 2013; Castán Broto and Bulkeley, 2013; Carmin and Dodman, 2013). Recently, more attention has also been given to adaptation, but few policies explicitly address the risks of low-income groups, especially the billion urban residents in informal settlements (Satterthwaite and Mitlin, 2014).

Policy programs tend to distribute benefits and costs along existing social class lines and thereby often overlook the exclusion of low-income people and other economic, social, or ethnically underrepresented groups – who often live in informal settlements – from the environmental policy process (Pelling,

1998; de Sherbinin et al., 2007). This is evident in housing, infrastructure, and disaster risk reduction programs. The exclusion leads to an underrepresentation of the needs of these groups in formal environmental policies and a failure to recognize the value of "autonomous adaptation" initiatives (Huq et al., 2007; Bartlett et al., 2009). Thus, these groups often receive little support from formal governance processes (Roy et al., 2012).

Urban community actors – often elite groups and/or commercial enterprises based in the community – may fill the vacuum of formal governance systems by establishing their own informal rights for the settlers and by acting as intermediaries between the settlers and wider urban institutions (Khan, 2000). Informal governance structures can transform slums into areas of innovation, developing the social capital and cohesiveness required for adaptation to climate change (Pelling, 1998; Khan, 2000). For example, in the Dharavi section of Mumbai, inventers and innovators at the household level have transformed the slum through waste economies, installation of infrastructure, negotiations for land rights, and the mobilization of people in the wider community (Appadurai, 2000; 2001). Such measures help secure collective survival and enable people to adapt to climate change through social bonding and the sharing of costs and benefits. However, informality can also make it more difficult to manage uncertain or unpredictable hazards created by climate extremes.

Moreover, informal governance may also be conducive to crime and give rise to poorly integrated social structures (Galea and Vlahov, 2005; Roy et al., 2012). Poor and underrepresented groups often become the victims of injustice created by informal institutions – in a way similar to those produced by formal governments – and their structural (re)arrangements. In cities such as Mumbai, Nairobi, Lagos, Delhi, Manila, and Dhaka, access to water, sanitation, other services, and infrastructure is often controlled by commercial enterprises, elites, or other influential community actors (Khan, 2000; Akanda and Hossain, 2012; Cullis et al., 2011; Roy et al., 2012). These informal governance regimes are a response to lack of formal provisions of services to informal settlements often based on discrimination due to race, caste, class, and gender (Cullis et al., 2011; Akanda and Hossain, 2012). It is the goal of many environmental and social justice institutions and policy programs to break this exploitative cycle (Franzen and Vogl, 2013; Deneulin, 2014; Rakodi, 2014).

There is an increasingly greater recognition of the importance of city governments in both adaptation and mitigation and of the need for city governments to align their agendas for development, poverty reduction, and disaster risk reduction with adaptation and mitigation (see Chapter 1, Pathways to Urban Transformation). A review of disaster risk reduction actions of more than fifty cities within the United Nations International Strategy for Disaster Reduction (UNISDR) "Making My City Resilient" campaign shows how resilience to disasters is being conceived and addressed by city governments, especially with regard to changes in their institutional frameworks and engagement with communities (Roy et al., 2012; Johnson and Blackburn, 2014) (see Chapter 3, Disasters and Risk). This commitment is also expressed in efforts to mobilize finance, undertake multi-hazard risk assessments, upgrade informal settlements, adjust urban planning procedures, and implement building codes. Many cities report paying particular attention to vulnerable groups and encourage them to actively participate in risk reduction decision-making, policy-making, planning, and implementation (Johnson and Blackburn, 2014).

These initiatives demonstrate the potential overlap between building resilience to climate change and poverty reduction. Cities in low- and middle-income nations that have taken action to upgrade informal settlements and expand provision of infrastructure and services have thereby helped to reduce the risk differentials between neighborhoods and residents. Thus, these measures can also be labeled as climate change adaptation because they help build resilience to climate change impacts.

However, adaptation investments may also prioritize the protection of the formal city infrastructure and ignore informal settlements (Roy et al., 2012). City governments in low-income (but sometimes also in high-income) countries often bulldoze informal settlements to make room for infrastructure "improvements" that serve central districts and middle- and upper-income groups (Macharia, 1992; Collins and Shester, 2011). Another example of urban activities that have gone against the needs of the most vulnerable is the filling up of natural water bodies to make way for construction or their conversion to economic uses. This can have consequences for traditional drainage systems and lead to an increase of waterlogging and the frequency or severity of floods (Tanner et al., 2009). Research in Bangladesh and Vietnam has investigated climate impacts and adaptation options in Satkhira and Hue, respectively (McEvoy et al., 2014). In both cases, local human intervention is undoubtedly having an influence on flooding and waterlogging. Important equity issues are raised in instances that have downstream consequences for the most vulnerable in society. However, the opposite can also be observed (i.e., the poor may encroach on traditional water bodies through landfill, narrowing river beds, and causing water areas to shrink, thereby contributing to increased flooding risk that may affect wider urban areas).

Most cities are affected by local political constraints and powerful vested interests that may oppose equity-sensitive adaptation and mitigation, especially if this restricts the land available for development or imposes measures and standards that may limit profits. Cross-municipal and cross-departmental action in urban agglomerations also faces governance challenges. For example, the Asian Cities Climate Change Resilience Network (ACCCRN) has faced challenges in sharing learning among different interest groups in politicized urban environments. However, the engagement and support of all relevant sectoral departments are needed for implementation (Roberts, 2010; Roberts et al., 2012). Several cities have therefore created a climate change focal point to help coordinate climate action across

government departments or agencies (Roberts, 2008; Roberts, 2010; Anguelovski and Carmin, 2011; Hunt and Watkiss, 2011; Brown et al., 2012). However, locating it in the environment department may not ensure sufficient attention because environment departments are typically weak, with limited budgets and influence, as in Durban (Roberts, 2008; Roberts, 2010), Boston (City of Boston, 2011), and Sydney (Measham et al., 2011). In contrast, New York's climate change adaptation agenda is guided by a Climate Change Adaptation Taskforce anchored within the Mayor's Long-Term Planning and Sustainability Office (Rosenzweig et al., 2010; Solecki, 2012). The Taskforce includes various city and state agencies and private companies involved in critical infrastructure, and it is advised by the New York City Panel on Climate Change.

Providing a space for discussion of vulnerability and resilience in each city's particular context is essential (Roy et al., 2012; Reed et al., 2013). Budgetary transparency and metrics to measure progress on adaptation and mitigation can also help to institutionalize changes in planning and policy practice (OECD, 2012). There still is very limited documentation of the design and implementation of climate change adaptation initiatives and its monitoring, particularly in cities in low-income and many lower-middle-income countries, but household and community-based adaptation, both its importance and limitations, has been a main focus (Moser and Satterthwaite, 2008; Carcellar et al., 2011; UN-Habitat, 2011; Dodman and Mitlin, 2013; Wamsler and Brink, 2014). A number of capacity-building initiatives have been developed, such as the Urban Climate Change Research Network (Rosenzweig et al., 2011) to address these shortcomings. However, additional targeted support is needed to secure the engagement of smaller cities in climate change efforts (Reckien et al., 2014) and to include guidance and training for city staff (Moser, 2006; Carmin et al., 2013; Tavares and Santos, 2013), particularly with respect to the full range of equity concerns that are characteristic of cities.

City and municipal governments need support from multi-level governance frameworks through which provincial and national governments enable and support city and municipal action (Corfee-Morlot et al., 2009; Revi et al., 2014) (see Chapter 16, Governance and Policy). Some national governments have developed new laws, funds, and regulatory frameworks to channel such support; many of these are focused on disaster risk reduction (Hardoy and Pandiella, 2009; IFRC, 2010; Carcellar et al., 2011; Kehew et al., 2012) and on increasing the resilience of the most vulnerable groups. However, in some countries, current policies, especially at the federal level, have major negative consequences for low-income and ethnic communities. Urban equity policies therefore cannot fully address the "root of the problems" and will struggle to achieve justice if the institutions that are creating policy – higher-level, national and international policy environments – fail to engage with equity and environmental justice efforts. WE ACT, a local environmental justice organization in the United States, specifically tries to address environmental injustices that are caused by national (climate change) policies.[7]

Case Study 6.4   Individual, Communal, and Institutional Responses to Climate Change by Low-Income Households in Khulna, Bangladesh

Anika Nasra Haque

Department of Geography, University of Cambridge

David Dodman

International Institute for Environment and Development, London

Md. Mohataz Hossain

University of Nottingham

Keywords	Poverty, flood, resilience, household adaptation, community-based adaptation, institutional adaptation, environmental justice
Population (Metropolitan Region)	759,618 (BBS, 2013)
Area (Metropolitan Region)	72.6 km² (BBS, 2013)
Income per capita	US$1,330 (World Bank, 2017)
Climate zone	Aw – Tropical savannah (Peel et al., 2007)

Khulna is the third largest metropolitan city in Bangladesh, located in the coastal region in the southwest of the country (see Case Study 6.4 Figure 1). Climate-related hazards have long been experienced in and around Khulna, for instance, floods, storms, limited availability of fresh water, waterlogging, and heat waves. This study examines the underlying drivers of vulnerability as they affect extremely low-income residents of the city. In addition, it examines the potential for actions taken at the household and community levels in urban areas to go beyond offering short-term "coping" solutions in response to specific events and instead result in more transformational changes that address the underlying drivers of vulnerability.

Other than the conventional "socioeconomic" and "biophysical" vulnerability, this research identifies a third type of vulnerability in the study area, which can be termed "legal vulnerability" and is derived from the tenure insecurity of the low-income urban residents. Tenure insecurity is a major factor dissuading low-income households from investing in their housing to make it more adaptive to climatic shocks because they live in constant fear of eviction. A wide range of specific adaptation-related activities can be identified as responding to these vulnerabilities, and these can be grouped into three main categories: individual, communal, and institutional. The study examines the extent to which institutional actions meant to address these underlying vulnerabilities are merely coping or whether they create the conditions in which individuals and households can strengthen their own long-term resilience. Similarly, it examines the extent to which individual and communal responses are coping or whether they have the potential to generate broader political change that strengthens the position of marginalized groups in the city.

ADAPTATION PRACTICES: HOUSEHOLD, COMMUNAL, AND INSTITUTIONAL

Low-income residents in Khulna are already taking a wide range of actions to respond to climate-related hazards (see Case Study

6.4 Figure 2). These are largely spontaneous or "impact minimizing" rather than planned or "preventative," often because residents lack the means to make more substantial changes. They are also ingenious and varied, particularly recognizing the severe technical, locational, and economic constraints under which these households operate.

Most of these actions involve making modifications to the physical dwelling and its immediate surroundings to deal with different types of threats (see Figure 2A). Many of these deal with hazards related to heavy rainfall and flooding. Polythene sheets or the covers from cement bags are placed on the roof and in wall openings as protection from heavy rain. Plinths are elevated or houses are built on stilts to avoid waterlogging. If flooding does take place, household goods are placed on shelves near the ceiling, and furniture is lifted off the floor using bricks or wood. In addition, a wide range of approaches using locally available materials, such as *golpata* (Nypa leaves), are used to repair damaged houses. Residents also use community kitchens during disasters to reduce costs. As part of climate-proofing or recovery from extreme weather events, communal activities involve building or repairing common services (i.e., tube wells, drains, toilets, elevated pathways, and small retaining walls at the edge of water bodies to prevent land erosion).

A considerable number of institutions operate in these areas, both public (City Corporation) and private (nongovernmental organizations [NGOs] like Save the Children, Bangladesh Rural Advancement Committee [BRAC], Water Aid). The City Corporation is mostly engaged with post-disaster relief. The NGOs work more generally on community development as well as on providing emergency services during disasters.

TOWARD TRANSFORMATIONAL CHANGE THAT REDUCES VULNERABILITY

None of the existing responses addresses the underlying social and political marginalization of the communities, which is perhaps the single most important feature contributing to their vulnerability to climate variability and change. While all the inhabitants of the city are entitled to basic service provision, the City Corporation in Khulna fails to acknowledge the existence of many informal settlements. This means that they are excluded from the City Corporation's provision of basic services. Improved provision of basic services and infrastructure would be a considerable contribution to vulnerability reduction by strengthening the adaptive capacity of individuals and households and by reducing exposure to flooding and waterlogging. In turn, such public investment could increase the motivation of residents to invest in improvements.

Another significant underlying factor contributing to vulnerability is the position of these low-income and informal settlements, most of which are in locations categorized as low-lying or agricultural rather than residential because areas designated for residential use are unaffordable to the very poor. Some of these are in areas that are exposed to particular hazards. Although responses to informality have gradually recognized the advantages of *in situ* upgrading (as opposed to relocating), the changing risk context as a consequence of climate change may result in the acceptance of the need to move some residents. However, this requires strong relationships of trust between organized community groups and local authorities.

One possible response to risk is to use the skills and knowledge that already exist in these communities. These skills and knowledge are

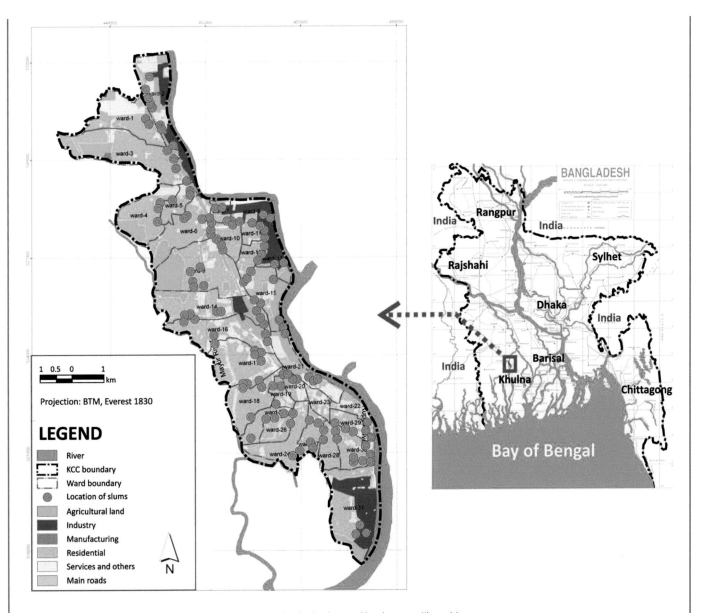

Case Study 6.4 Figure 1 *Location map of the city of Khulna (showing land use and low-income settlements).*

more likely to be used by residents if they are more confident that their efforts will not be lost through forced evictions. Their knowledge can be supplemented through awareness-raising and training workshops that can also help to prevent maladaptation. For example, although elevating the level of paths and walkways may reduce flooding in one location, if this is done without adequate consideration for broader drainage patterns it can worsen the issue elsewhere.

Another response that strengthens the adaptive capacity of low-income communities is to support the development of climate-resilient livelihood strategies. People living in low-income areas of Khulna have already been identified as undertaking different ways of earning money, and these have often been supported by NGOs – for example, through providing financial services (assets or capital) to develop small businesses such as sewing, handicraft production, and retail outlets. Households with savings have greater coping ability during a crisis (Dodman et al., 2010), and the city corporation and NGOs are implementing different types of saving schemes. NGOs have also been seeking to strengthen women's access to and control over assets and resources, thereby boosting the ability to make decisions, participate in the process of city governance,

and thereby strengthen their resilience. However, efforts also need to be made to improve information-sharing about climate-related risks. Because the majority of the households surveyed do not have televisions, radios, or mobile phones, they lack access to climate change information from media sources and are therefore unable to take actions to reduce the consequences of particular climate-related events, in addition to being uninformed of longer-term climate trends. Although the City Corporation provides public service announcements using loudspeakers, this is only delivered to river-bank settlements.

All of these issues will require greater engagement and accountability from institutional actors. The Khulna Development Authority is responsible for enforcing regulations (which are also monitored by the Khulna City Corporation) to ensure that landlords meet the minimum standards of housing infrastructure, while the presence of active NGOs can also reduce the likelihood of forced evictions. These responses can provide an incentive to low-income residents to invest in their housing, make physical adjustments to their shelter, and sometimes even improve the settlement to better adapt to climatic variability and change.

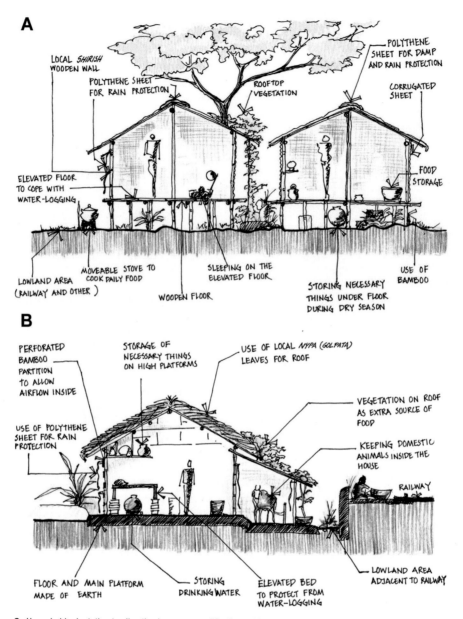

A

LOCAL *SHIRISH* WOODEN WALL

POLYTHENE SHEET FOR RAIN PROTECTION

ROOFTOP VEGETATION

POLYTHENE SHEET FOR DAMP AND RAIN PROTECTION

CORRUGATED SHEET

ELEVATED FLOOR TO COPE WITH WATER-LOGGING

FOOD STORAGE

LOWLAND AREA (RAILWAY AND OTHER)

MOVEABLE STOVE TO COOK DAILY FOOD

WOODEN FLOOR

SLEEPING ON THE ELEVATED FLOOR

STORING NECESSARY THINGS UNDER FLOOR DURING DRY SEASON

USE OF BAMBOO

B

PERFORATED BAMBOO PARTITION TO ALLOW AIRFLOW INSIDE

STORAGE OF NECESSARY THINGS ON HIGH PLATFORMS

USE OF LOCAL *NYPA (GOLPATA)* LEAVES FOR ROOF

USE OF POLYTHENE SHEET FOR RAIN PROTECTION

VEGETATION ON ROOF AS EXTRA SOURCE OF FOOD

KEEPING DOMESTIC ANIMALS INSIDE THE HOUSE

RAILWAY

FLOOR AND MAIN PLATFORM MADE OF EARTH

STORING DRINKING WATER

ELEVATED BED TO PROTECT FROM WATER-LOGGING

LOWLAND AREA ADJACENT TO RAILWAY

Case Study 6.4 Figure 2 *Household adaptation to climatic changes – modifications to household structure to reduce impacts of climatic events.*

The experiences of these residents of Khulna indicate that many activities are already being taken that respond to particular threats. However, while these short-term practices may respond to urgent needs, they will be insufficient in the longer term. Long-term meaningful resilience is not possible without institutional support to households and communities – and this in turn will not happen unless households and communities have effective ways of influencing the processes of urban governance. Equally, this needs to be supported by national policies that grant responsibility, autonomy, and resources to local authorities to address local and urban development concerns. National, urban, institutional, communal, household, and individual adaptation are therefore all required to achieve meaningful and enduring resilience.

More effective international, national, and municipal financing mechanisms that are able to channel resources toward local projects, particularly those that help the poorest and most vulnerable groups to adapt to climate change, are also needed (Smith et al., 2014a) (see Chapter 7, Economics, Finance, and the Private Sector). In this respect, the municipal government is a duty bearer for financial support. For example, the Kuyasa Fund in Cape Town, South Africa, provides microfinance lending for housing, targeting the most vulnerable groups, particularly women. The funds are frequently used to finish external house walls to increase space and improve thermal efficiency (Houston, 2010). This fund could, in the future, be redesigned to encourage additional retrofitting, ideally combined with energy-saving measures (Mills, 2007; UN-Habitat, 2008b; Alber, 2011).

Overall, to secure fairness and sustainability in climate change actions, the combined further development of the formal institutional, regulatory, financial, economic, and social frameworks is paramount (Satterthwaite, 2013). Women in particular should be recognized as important agents of change; they create solutions

Case Study 6.5   Public-Private-People Partnerships for Climate Compatible Development in Maputo, Mozambique

Vanesa Castán Broto

Bartlett Development Planning Unit, University College London

Charlotte Allen

Independent Consultant, Los Angeles

Keywords	Flood, participatory methodologies, partnerships, climate compatible development, environmental justice
Population (Metropolitan Region)	2,655,000 (Demographia, 2016)
Area (Metropolitan Region)	414 km² (Demographia, 2016)
Income per capita	US$480 (World Bank, 2017)
Climate zone	Aw – Tropical Savannah (Peel et al., 2007)

Mozambique is highly vulnerable to natural disasters, particularly those of hydro-meteorological origin such as floods, drought, and cyclones. In Maputo, the capital city with about 1.2 million inhabitants, the main hazards associated with climate change are likely to be temperature increase, extreme precipitation events, and sea level rise leading to increased flood risk in low-lying areas (Maputo Municipal Council et al., 2012).

Climate change in Maputo must be understood in the context of vulnerability linked to poverty and deficient infrastructure (Castán Broto et al., 2013). Approximately 54% of residents live below the poverty line of US$1.50 per day, and 70% live in dense, unplanned neighborhoods with deficient water supplies, sanitation, and drainage (Maputo Municipal Council et al., 2012). Many of these neighborhoods occupy low-lying areas.

The climate project Public-Private-People Partnerships for Climate Compatible Development (4PCCD) ran from 2011 to 2013 and aimed to empower people living in Maputo's poor neighborhoods, to design and implement activities to adapt to climate change.[8] Among others, the project worked in the neighborhood of "Chamanculo-C," where the municipality was preparing an upgrading program.

4PCCD was funded by the Climate Development Knowledge Network and conceived collaboratively by the Mozambican government's Environment Fund (FUNAB) and a group of UK academics led by the Bartlett Development Planning Unit, University College London. Although it was initially interested in climate change mitigation, recurrent flooding in neighborhoods such as Chamanculo necessarily oriented the project toward adaptation (Case Study 6.5 Figure 1).

Case Study 6.5 Figure 1 *Flooding in Chamanculo-C Neighborhood, Maputo.*

8 http://www.bartlett.ucl.ac.uk/dpu/4pccd

Case Study 6.5 Figure 2 *Residents group meeting, Chamanculo-C Neighborhood, Maputo.*

The project involved three stages: (1) a review of studies on climate change in Maputo to characterize key impacts and identify vulnerabilities (January to December 2012); (2) the implementation of a participatory methodology in a specific neighborhood in order to share climate change information, identify potential impacts, and develop potential solutions (mainly from January to June 2013, but continuing into 2014); and (3) the presentation of the community's proposals to a wide range of municipal and national institutions to establish partnerships for implementation (May to July 2013) in order to make explicit the knowledge and priorities of the communities.

The project used participatory action plan development (PAPD), a consensus-building tool that seeks to identify and solve environmental problems with community support and input through different participatory techniques and principles, such as the recognition of the wide range of stakeholders and of their diverse interests and their full engagement (Castán Broto et al., 2015). Using this methodology, the driver of action was the community, represented by the Climate Planning Committee (CPC).

In Chamanculo-C, five groups identified at the start of the PAPD exercise by the community (elderly, young people, traders, employees, and housewives) analyzed the causes and impacts of flooding and potential future impacts and solutions (Case Study 6.5 Figure 2). Existing impacts, including loss of access, damage to property, and vector-borne diseases, are keenly felt by residents, and vulnerability to climate change impacts became a focal point for discussion. The groups elected the CPC which, with facilitation, compiled the community's proposals into a Community Plan for Climate Change Adaptation and presented it at a multi-stakeholder workshop to seek support. Proposals included improving drainage and waste

management through community organization, development of a recycling center, repairing water supply networks, and community-led environmental education.

Actor-mapping was used to understand the key players who were delivering climate change interventions in Maputo and who might become partners in the project. This led to the involvement of other key stakeholders, including municipal departments, Eduardo Mondlane University, UN-Habitat, neighborhood leaders, and civil society organizations working in Chamanculo-C. Following presentation of the Community Plan, the project enrolled other actors suggested by the CPC: local businesses, the Mozambican recycling association, the water utility, and the Ministry for Environmental Coordination.

IMPACT AND SCALE

The project worked directly with a small community of one administrative block *(quarteirão),* containing 82 households with 570 members. However, the CPC also worked with the leaders of neighboring blocks, and the Community Plan included proposals relevant to the whole of Chamanculo-C, with a population of about 26,000. Given its experimental character, this project could not have been implemented at a larger scale without a prior proof of concept – which 4PCCD has now provided.

The PAPD process strengthened community organization and representation through the establishment of the CPC, whose expertise and legitimacy have been acknowledged by stakeholders and policy-makers. Now, some months after the initial project, the residents are mobilizing external funding and moving toward implementation of their proposals.

The project team is positive about the long-term viability of the outcomes. The CPC put forward feasible proposals, including community organization to clean and maintain the drainage channel, a community waste separation and composting center, and community-led environmental education. Key institutions including FUNAB and Maputo Municipality have expressed commitment to their implementation. There is no evidence of policy impact yet, but, following the project, the municipality has embarked on deeper climate change planning processes.

LESSONS LEARNED

4PCCD demonstrated that, in Maputo, local communities are capable of organizing themselves for collective action; engaging with climate information, uncertainties, and future scenarios when these relate to their daily experiences; and developing and presenting sensible proposals that directly tackle climate change vulnerabilities. Communities have a grounded understanding of climate change and can do a lot with limited resources by drawing on their own human capital. The project also demonstrates that government institutions and businesses have much to gain from listening to local perspectives.

In Maputo, the participatory process has been a means to build and share an understanding of the challenges for communities faced by climate change. Longer timeframes are nonetheless required to show whether the community's ideas are practicable.

---

to environmental problems and alleviate GHG emissions in their daily activities. For example, the ninety women heads of households belonging to the women's group Guardianas de la Ladera (Guardians of the Hillside) in Manizales, Colombia, carried out traditionally male work in order to preserve their houses and the environment on unstable city hillsides (UNDP, 2009). Another project called Girls in Risk Reduction Leadership in Ikageng, a township of Potchefstroom, South Africa, aims to reduce the social vulnerability of underrepresented adolescent girls using practical capacity-building initiatives. Girls were trained by experts in areas such as personal and public health, fire safety, counseling, and disaster planning. Girls can help design risk reduction plans for the community to improve resilience (UNISDR et al., 2009).

6.5 Equity Impact Assessments in Cities

To ensure that formal and informal adaptation and mitigation policies and programs do not have detrimental effects on urban equity, recent international reviews propose carrying out structural equity impact assessments (Kallbekken et al., 2014). The aim is to support the equitable and sustainable implementation of environmental and climate change policies, similar to the environmental impact assessments that are applied ahead of major infrastructural investments in many countries.

Four assessment approaches are currently discussed in the literature: (1) the Equity Reference Framework (ERF) (CAN Equity Group, 2014), (2) an open indicator approach, (3) the template of indicators approach (Kallbekken et al., 2014), and (4) the equity framework of McDermott et al. (2011):

1. The ERF was suggested by the Climate Action Network (CAN Equity Group, 2014) as a way to operationalize common but differentiated responsibilities in international climate negotiations. In this respect, its usefulness has been questioned (Kallbekken et al., 2014). A key concern is that local equity may be undermined or existing inequalities exacerbated when nationally and globally aggregated or uniform indicators are used without considering local circumstances (McDermott et al., 2011).

Another point of contention is the differential treatment of adaptation and mitigation; the latter is substitutable, whereas adaptation is assumed to be largely place-specific. UN Framework Convention on Climate Change (UNFCCC) Conferences of the Parties (COP17, COP18, and COP19) have long associated equity issues with both mitigation and adaptation, even though equity concerns expressed at the international level mostly relate to different allocation approaches for climate change mitigation commitments. However, equity issues themselves are generally seen as not place-specific (Rawls, 1971).

2. An "open indicator" approach allows for all conceivable indicators to be included. Difficulties with indicator-based approaches, such as the ERF, highlight geographic variation in costs of action, impacts, and capabilities as well as related divergent views on good principles and the indicators to be used (Kallbekken et al., 2014). With an open indicator approach, however, each national government might use those indicator(s) most suitable to its own interests in order to minimize ambition and display its stance in a more favorable light (Kallbekken et al., 2014). This could undermine the implementation of measures. Some scholars therefore call for region-specific equity indicators (e.g., McDermott et al., 2011).

3. A "bounded flexibility" approach offers a spectrum of commitments within a "template" of agreed indicators (Kallbekken et al., 2014). The idea is to develop a finite, official list of indicators, agreed upon by an expert process, and allow each entity to decide which of these indicators to monitor.

4. The equity framework approach of McDermott et al. (2011) allows analyzing how equity is addressed in policies and assessing its baseline status. This approach is based on analysis of the local and regional policy arena and on identifying particular social contexts, norms, and values. The framework targets four parameters of equity that should be considered in the planning or assessment of a policy or project: its content, target, goals, and process. McDermott et al. (2011) suggest that policy analysts start by analyzing equity goals. Goals may be to maximize or improve equity, to "do no harm," or simply not to worsen

the current situation. When setting targets, it is important to determine who counts and at what spatial, temporal, and social scales. At this stage, crucial decisions have to be made, and a more inclusive assessment is generally preferable, especially the participation of vulnerable groups. It is assumed that the equity framework helps in analyzing tradeoffs between sectors or impacts on different target groups and in ordering or weighting rights that can conflict with each other, such as growth versus sustainability, rights of indigenous people versus migrants (or majorities), and current versus future responsibilities and duties (McDermott et al., 2011).

We use the ERF as starting point and develop it further to reflect more fully those concerns related to equity in urban areas. We see the need to provide one framework for mitigation (see Figure 6.5a) and another for adaptation (see Figure 6.5b) in order to increase specificity, although adaptation and mitigation efforts in urban areas should be conceptualized jointly to avoid mal-adaptation or mal-mitigation. The mitigation framework for urban areas recognizes the importance of the Paris Agreement on the total global efforts needed to keep global warming to 2°C above pre-industrial levels (with the ambition to limit warming to 1.5°C), but also the key contributions required from municipal governments and the extent to which achieving this global goal depends on appropriate support for city and municipal governments from national governments and international agencies. The adaptation framework (see Figure 6.5b) recognizes urban centers and their urban governments as its starting point (not national

governments and international agencies), acknowledging what they can do directly and how they can support contributions from citizens, civil society, the private sector, and other stakeholders.

Additionally, we present the Urban Equity Impact Assessment (EquIA-urban) tool based on McDermott et al. (2011), for use by urban decision-makers and community groups that are implementing climate actions and want to review the likely effects in light of potential equity issues. This assessment offers hands-on, step-by-step guidance when implementing climate change policy measures in urban areas (see Table 6.2).

6.6 Knowledge Gaps, Future Research, and Awareness-Raising

To improve assessment of the influence of climate change on equity in cities, research is needed on all dimensions of climate change (i.e., impacts and risks, adaptation and mitigation).

6.6.1 Data Requirements and Research Needs

- To date, most urban climate change-related impacts and risks across gender, age/generation, ethnicity, social classes, and demographic groups have been qualitatively described but are not well quantified, thus impeding structural comparisons. Shortcomings also exist with respect to adaptation and adaptive capacity.

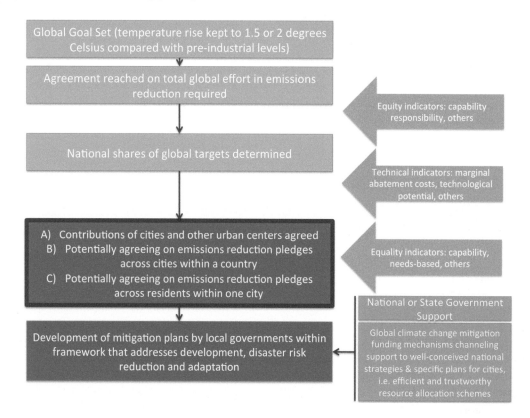

Figure 6.5a *Equity framework for addressing mitigation in urban areas.*

Source: Based on CAN Equity Group, 2014. Modified for urban areas.

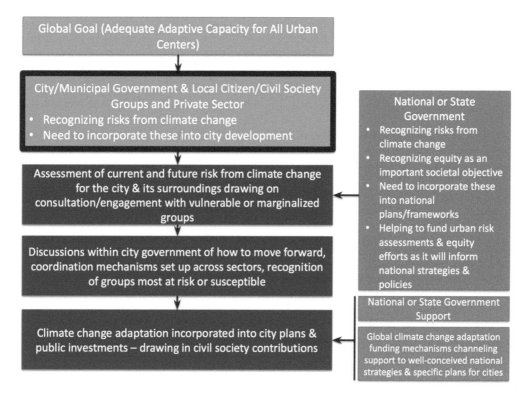

Figure 6.5b *Equity framework for addressing adaptation and adaptive capacity in urban areas.*

Source: Based on CAN Equity Group, 2014. Modified for urban areas.

Table 6.2 *Urban Equity Impact Assessment (EquIA-urban) tool: A step-by-step guide. Source: Adapted from McDermott et al., 2011*

The Urban Equity Impact Assessment (EquIA-urban) tool: A step-by-step guide
This guide summarizes important questions that need to be considered when defining the dimensions and parameters of climate change policies in order to understand their likely impact on urban equity. Following this guide and discussing these questions will minimize the risk that a policy measure will have unintended, inequitable consequences following its implementation.

1) **Process: How are the parameters of urban equity set?**
   What is the decision-making process in framing the initiative?
   How is it established and at what scale of decision-making?
   Who is included/excluded in the decision-making process?
   Who defines the goals, targets, and content of the initiative?

2) **Goal: Why equity? What is the explicit/implicit goal?**
   Is the goal to maximize urban equity, to improve equity, or to do no harm? Or are equity impacts not under consideration?

3) **Target: Who counts as a subject of equity?**
   At which scale(s) is equity considered: individual, household, community, city, value chain, or regional? Consider multiple scales to prevent negative side-effects for others.
   How are the needs of current and future urban generations taken into account?
   How are the needs of nonhuman species or urban ecosystems taken into account?

4) **Content: What counts as a matter of urban equity?**
   4a) Distributive equity
      Has the distribution of benefits, costs, and risks across urban residents been considered?
      What is the intended basis for the distribution of benefits: e.g., equal shares, net social welfare, merit, needs?
      What is the intended cost-benefit distribution, and what will its impacts be?
   4b) Procedural/participatory equity
      Who is participating in decision-making, and who is left out?
      Which underrepresented groups are recognized? Can they voice their interests and be heard?
   4c) Contextual equity (incorporating capabilities, access, power)
      Do decisions reflect the interests of the under-represented groups?
      Do under-represented individuals have access to the resources (e.g., land, capital) necessary to secure benefits of the initiative?
      What new capabilities are being developed (e.g., economic opportunities)?
      What local institutions provide "safety nets"?
      Are the causes of inequity identified? Are they addressed?

Adaptation and adaptive capacity are difficult to assess for a variety of reasons, including lack of accepted definitions and the abstract nature of the qualities of flexibility and inventiveness. Regarding mitigation, research needs relate to calculating fair assessments of pledges for cities, neighborhoods, and social classes.

• In general, the spatial, social, and temporal resolution of statistical data has to be improved (Martine and Schensul, 2013; Guzman and Miguel, 2009; Montgomery et al., 2003).

The resolution of most available data – including climate records, disaster incidence, disaster losses, and climate projections – is generally much coarser than a city's spatial extent. And yet even finer spatial disaggregation is needed if we are to understand how the risks and vulnerabilities of any given city are likely to vary across its heterogeneous districts and neighborhoods. Similarly, more information is needed on the spatial extent of cities and their expansion or contraction. It would be useful to have data for separate risk areas, such as coastal regions. There have been important advances drawing on satellite imagery (World Bank, 2015; Angel et al., 2011), but much work remains to be done on methodology, image-processing algorithms, and validation before a globally consistent, spacio-temporal view of urban areas can be created (Small, 2005).

• Demographic data on cities will also need to improve (Montgomery et al., 2003).

Most demographic information is currently based on national samples, which are inadequate to portray urban realities and hide the scale and depth of inequality within urban populations (with the exception of a few estimates of demographic characteristics that are available for very large cities). Moreover, national data and censuses are usually only undertaken every ten years, which is insufficiently frequent to capture more rapid urban developments. Providing census data in fine geographic detail on the neighborhood level could be a worthwhile starting point.

• Studies need to examine the correlates, root causes, and cascading impacts of climate risk in cities, and urban risk typologies (i.e., confounding, reinforcing, and dampening factors of risk and their interrelationships).

Research to date includes instructive general accounts of the consequences of extreme events that are vividly illustrated by compelling case studies. But in its present form, the evidence base for impact comparison across hazards is grossly inadequate, particularly for local and smaller-scale events. For example, causes of death and health risks are well recorded in high- but not in many low- and middle-income countries. The most important international disasters database, DISINVENTAR (EM-DAT, 2010), does not record the precise locations of events within a country, nor is it able to go beyond tallying the numbers of people killed or affected to provide demographic information.

If combined with spatially disaggregated socioeconomic information from population censuses or related sources such as the World Bank's Small-Area Poverty Mapping project (Elbers et al., 2003, 2005), the DESINVENTAR data could shed light on the occurrence and impact of disasters at a community level (Marulanda et al., 2010). Shortcomings also exist with respect to research on financial consequences. The costs of rebuilding as well as the costs of adaptation or prevention are policy concerns that need to be addressed in research.

• Micro-level longitudinal and retrospective sample surveys can enable reconstruction of before-and-after portraits of affected individuals and neighborhoods (Fernanda Rosales, 2014).

Linking GIS mapping of impacts or potential risk indicators with estimations of social vulnerability is another promising way forward to assess relations between climate-related impacts and their socioeconomic effects (Cutter and Finch, 2008).

• There is also the need for more qualitative social research.

This should include improving the gender perspective, which relates to both the disproportional impacts of climate change on women in cities and the contributions of women as key agents of change for mitigation and adaptation. With respect to social processes, research should explore the differentials based on gender, age, ethnicity, social class, and demography with respect to access, use and control of resources, level of assets, and disaster preparedness. Progress is also needed in integrating attitudes, perception, and abilities in the quantification of climate risks and vulnerabilities because these are related to adaptation and mitigation willingness for action. One interesting approach is *fuzzy cognitive mapping*, which aids in assessing the relative impact of weather events among individuals and groups producing socially feasible adaptation options (Reckien et al., 2011, 2013, 2014).

• Research needs to investigate enablers of change toward more resilient and equitable urban environments.

The research community has often identified key risk factors but not the enablers of change toward more resilient and equitable urban environments. Because the processes that exacerbate inequity, vulnerability, and risk may not match the processes that promote equity and resilience, there is substantial need to investigate factors that can foster positive change, such as involving women's organizations in climate change planning processes.

• Overall, there is a need for conceptual frameworks, methodologies, and tools for measuring and monitoring climate change-related equity aspects in cities.

The EquIA-urban tool presented in Table 6.2 provides an important entry point, but needs to be further developed to allow for operational, comparable assessments.

Table 6.3 *Concrete policy recommendations for climate change and policy domains covered in this chapter*

Climate change domain	Policy domains	Potentially useful recommendations	Responsibility
All	General equity concerns	• Provide risk-reducing infrastructure and services across all urban neighborhoods • Account for "development-accumulated" risk and take into account lessons from previous disasters; avoid creating "new risk" • Create awareness among low-income populations about likely impacts of climate change, including differential impacts on women, children, the aged, immobile residents, and ethnic groups • Disseminate the principles of minimal safety standards in low-income housing • Ensure safety standards in legal housing • Build capacity of community-based organizations to address local risks through local infrastructure solutions and services	City authorities, civil society organizations, neighborhood organizations and/or committees
Impacts and risks	Policy recommendations *Heat Stress*	• Establish neighborhood watch groups to support lonely elderly people during heat waves • Inform residents of the differential vulnerabilities of women to heat related to physiological characteristics as well as to work environments • Provide sustainable public green space (plus management) in low-income urban areas • Promote urban forestry and establishment of urban nature reserves • Look after open water bodies in urban areas, particularly to protect them from littering and encroachment	Neighborhood and housing organizations, planning offices, landscaping professionals and contractors
Impacts and risks	Policy recommendations *Inland Flooding*	• Establish safe destination areas for newly arrived migrants • Adopt sustainable urban drainage design principles in addition to stabilization of riverbanks to prevent flooding of residential areas • Increase drainage capacities and systems, particularly in low-income communities • Improve the functioning of upstream and downstream drainage systems across entire urban area	City authorities, planning departments, engineering departments, developers
Impacts and risks	Policy recommendations *Landslides*	• Develop environmental protective infrastructure on hill slopes (i.e., plant trees and other soil-fixing greenery) • Implement sustainable-upgrading measures like paving of streets and walkways on hill slopes; include mandatory improvement and reinforcement of draining systems in order to channel runoff during and after heavy rainfall events	City authorities, planning departments, engineering departments, developers
Impacts and risks	Policy recommendations *Drought*	• Diversify urban water supply in cities in dryland ecosystems • Diversify the electricity supply in dryland regions • Plan for an increase in intensive rainfall during periods of droughts (i.e., through infrastructure solutions that cater to both extremes – water shortage and water excess at the same time) • Provide additional support to low-income households and women in times of high food prices and food insecurity	City authorities, community organizations
Impacts and risks	Policy recommendations *Storm Surge and Coastal Flooding*	• Keep flood plains and coastal zones free of new developments in order to avoid constructing new risks • Establish coastal buffer zones (e.g., beaches, marshes, dunes, and mangroves) • When redeveloping coastal zones, integrate measures to prevent pollution from industrial harbor activities • Inform low-income households and other people at risk of health risks and the transmission pathways of diseases likely to occur after floods and storm surges • After an event, target support to low-income communities, ethnic minorities, the elderly, and other groups with low coping capacity • Establish disaster relief aid, such as bus lines and neighborhood watch programs • Inform particularly vulnerable citizens about community relief programs	City authorities, planning departments, engineering departments, developers

Table 6.3 (*continued*)

Climate change domain	Policy domains	Potentially useful recommendations	Responsibility
Adaptation	Policy recommendations *Adaptation*	• City governments are advised to provide the network infrastructure and institutional services required to build community resilience, particularly in low-income neighborhoods • City governments are advised to take measures to counteract policy decisions by national governments that increase the vulnerability of the poor, women, and other underrepresented groups • Foster individual and community self-help, where possible, linked to new economic opportunities for low- and middle-income residents • Establish disaster response agencies and strengthen the emergency services; take measures to build their capacity	City authorities, planning departments, developers
Mitigation	Policy recommendations *Spatial Planning*	• Focus on mixed land use and urban density • Be aware of and attempt to lower possible negative side effects of compact city spatial planning models on low-income neighborhoods and ethnic communities (e.g., by using social policy to cap accommodation prices and rents for households in need) • Secure the continuation of social housing policies even for housing built to (expensive) low-carbon housing and construction standards • Combine densification programs with measures to improve public transport • Find innovative ways to provide recreational areas in compact neighborhoods (e.g., through green roofs and including recreational facilities in multistory structures), so people don't have to drive to recreation areas • Secure availability and access to green space in low-income and ethnic communities • Finance densification with development taxes imposed on new development to cover infrastructure-related costs, to reduce the relative burden for low-income groups	City authorities, planning departments, developers
Mitigation	Policy recommendations *Accessibility and Transport Policy*	• All new urban areas shall be serviced by efficient public transport routes • Improve the accessibility, safety, and comfort of public transportation • Prioritize women's perspectives in public transport schemes (e.g., allow request stops during the night) • Reduce out-of-pocket fees for public transport schemes (e.g., by allowing monthly and annual passes to be paid for in multiple installments) • Provide unlimited-use passes for public transport (not multiride schemes based on number of transfers, which disproportionally affect the poor and women) • For private transport, create high occupancy toll lanes	City authorities, planning departments, developers

6.6.2 Awareness-Raising

Meeting information needs refers not only to generating the required data but also to ensuring that these data are communicated or made accessible, and understandable to the people who "need to know." Thus, actions to spread awareness and transmit know-how are a necessary (but not sufficient) intervention to promote equity and climate resilience in cities. The forgoing discussions show that environmental justice policies have evolved in some cities while in many other cities existing policies have inequitable impacts on the urban poor and other disadvantaged groups. There are cases of cities with long-standing municipal laws and policies that obscure the realities of equity issues, such as those revealed by flood disasters in Bangkok and New Orleans.

Raising awareness of the importance of equity and equality issues in relation to climate change in cities will be key for sustainable urban futures. Working with environmental justice groups can make an important contribution to increasing awareness, bringing equity issues to the forefront and promoting the inclusion of equity into city plans and policies. Particular attention should be paid to the need for equity between women and men because gender inequality is often overlooked in climate impacts, adaptation, and mitigation studies; disaster-relief programs; and policies. Another concern relates to particularly hard-to-reach populations, which are often underrepresented and unheard. (Sampson et al., 2013).

In summary, the following actions can help raise awareness of equity in cities:

1. Document and communicate climate change impacts and adaptation capacities as well as information on access, use, and control of resources, assets, and preparedness in relation to gender, age/generation, ethnicity, social class, and demographic groups.
2. Document and communicate the correlates, root causes, and cascading impacts of climate risk in cities and their fine-grained spatial distribution.
3. Develop capacity-building tools for urban stakeholders and authorities to mainstream gender, intersectionality, and other equity issues into their plans and actions.
4. Identify ongoing community processes and women's initiatives that should be acknowledged and incorporated into municipal action plans.
5. Ensure gender-balanced decision-making processes and integration of other underrepresented groups into policy-making spheres. Integration should transcend numerical representation to encompass issues of social empowerment and political influence, keeping in mind the need to reach out to particularly hard-to-reach populations.

Despite the existing challenges to the integration of equity in city policies related to current climate change risks, impacts, and adaptive capacities, the need to respond to newly emerging risks will be even more difficult to communicate. A multi-level approach to raising awareness is needed to enable communities and actors of both public and private sectors to understand the potential magnitudes and impacts of risks that have yet to become apparent and the need to account for uncertainties in long-term climate change plans and policies.

6.7 Policy Recommendations

Projected ongoing climate change and the likely occurrence of more climate extreme events, in combination with continuing rural–urban migration across the world, make addressing inequity and inequalities a major challenge for urban policy-makers. In this final section, we propose a series of policy recommendations based in part on the results of a stakeholder survey (see Annex Box 6.1).

Urban policy should integrate equity and environmental justice as a primary long-term goal because equity fosters human well-being, social capital, and sustainable social and economic urban development. There is an urgent need to complement attention to short-term needs with measures to address climate risks projected in the medium to long term. The goal is to avoid short-term responses that may ultimately prove to be "maladaptive," in particular by placing an increasing burden on the most vulnerable in society.

Climate policy should show greater sensitivity toward the vulnerability of the urban poor, women, and the elderly, as well as to the discrimination faced by women, ethnic minorities, and other underrepresented groups who lack access to adequate urban services and infrastructure. In framing and implementing climate policy, it is essential to continually take into account

gender outcomes and other equity issues. Particular attention must be paid to the economic and social consequences of climate change policy in urban areas.

Access to land, security of tenure, basic services, and risk-reducing infrastructure for all urban residents, and particularly for vulnerable and newly established urban communities, is crucial for reducing inequalities and the equitable distribution of risk-reducing benefits. Growing urban centers require increased investment in services and infrastructure, and this needs to be targeted at vulnerable groups and neighborhoods.

Urban policy should incorporate community views regarding what is resilient and feasible by promoting participatory approaches and avoiding top-down, inflexible solutions. Policy-makers need to learn to appreciate existing cultural knowledge and values and to integrate these into the planning processes. Engaging local residents at the beginning of the process and partnering with existing grassroots organizations are key; this can create important educational opportunities and develop the trust and consensus necessary for moving from conceptualization to implementation of adaptation actions.

There is need for cities to promote a shared science–policy interface for improved integration of multiple (sometimes conflicting) stakeholder interests in the policy-making process. Shared science–policy interfaces enable better communication among scientific institutions, think-tanks, and application-oriented stakeholders and policy-makers while facilitating the identification of common interests and shared understandings. At a local level, this work can usually be led by a "taskforce" or "focal point" prominently located in a large and financially strong municipal department (i.e., close to the Mayor's office). Close coordination between urban and national representatives is indispensable to avoid national policy frameworks that impede equity and environmental justice efforts at the urban level.

Long-term financial mobilization and institutional support for addressing both climate change and equity concerns are essential to ensure that adaptation and mitigation interventions in urban areas are equitable and inclusive. Combining traditional and nontraditional sources of finance, including community contributions, is crucial for equitable risk reduction. Poorer communities need stronger multilevel governance arrangements supported by more effective international, national, and municipal financing mechanisms that are able to channel resources toward local projects that help the poorest and most vulnerable groups to adapt to climate change.

Fair resource allocation schemes, budgetary transparency, and progress measurement are essential to make sure that funds reach their target groups. Metrics to measure progress on adaptation and mitigation should also monitor their effects on equity, equality, and environmental justice in urban areas. Urban statistical databases need to be developed and improved (see World Council on City Data, www.dataforcities.org). Cities should build up their own urban statistical data-gathering and monitoring

Box 6.4   Demographic Composition and Change

Deborah Balk

CUNY Institute for Demographic Research, Baruch College, New York

Daniel Schensul

United Nations Population Fund

Demography, the study of human population, provides a view into understanding inequalities of populations at risk of the consequences of climate change and inherent inequalities in the relationship between population change and emissions, with important resulting implications for both adaptation and mitigation. Populations vary in their size, structure, and distribution of the population and subgroups thereof. They change over time and space, with urbanization currently and over the coming decades projected to be one of the most significant population changes.

As seen in Box 6.4 Table 1, the world today is home to 7.3 billion persons. By the end of the century, the United Nations estimates that the world population total will exceed 11 billion, with more than 4 billion residents each in Africa and Asia. While the population of Asia will grow and then decline in the coming century, the largest rate of growth will occur in Africa. According to the UN, the world population currently continues to grow, although more slowly than in the recent past. Ten years ago, world population was growing by 1.24% per year. Today, it is growing by 1.18% per year, or approximately an additional 83 million people annually.

As with any type of projection (such as that shown in Box 6.4 Table 1, for the medium-variant), there is a degree of uncertainty surrounding these population projections. The medium-variant projection assumes a decline of fertility for countries where large families are still prevalent as well as a

Box 6.4 Table 1 *Population of the world and major areas, 2015, 2030, 2050, and 2100, according to the medium-variant projections. Source: UN, 2015*

Major area	Population (millions)			
	2015	**2030**	**2050**	**2100**
World	7,349	8,501	9,725	11,213
Africa	1,186	1,679	2,478	4,387
Asia	4,393	4,923	5,267	4,889
Europe	738	734	707	646
Latin America and Caribbean	634	721	784	721
Northern America	358	396	433	500
Oceania	39	47	57	71

slight increase of fertility in several countries with fewer than two children per woman on average. Survival prospects are also projected to improve (i.e., mortality rates are expected to decline) in all countries.

Uncertainty tends to increase the further out in time one projects. The UN Population Division uses statistical methods to make statements about the degree of uncertainty around this medium-variant projection. Box 6.4 Figure 1 shows that one can say with a 95% degree of confidence that global population will be between 8.4 and 8.6 billion in 2030 and between 9.5 and 13.3 billion in 2100. In other words, global population is virtually certain to rise in the short- to medium-term future. Later in the century, global population is likely to continue to rise, but there is roughly a 23% chance that it could stabilize or begin to fall before 2100 (UN, 2015).

The world of the future will, demographically speaking, be older and more urban. The world of today has twice as many children under age 15 than persons over age 60, but by mid-century the number of those 60 and older will have doubled, and there will be parity between those under age 15 and those over 60. Some regions, notably Europe, already have around one-quarter of their population aged 60 or older. A few years ago, the world population turned for the first time from majority rural to majority urban. With high proportions of city-dwellers in the Americas, Europe, and industrial countries of Asia and Oceania, this trend is seen as irreversible.

Migrants, while usually better off than those who cannot migrate, tend to be worse off than the native population in the destination location (Greenwood, 1997; Montgomery et al., 2012). In addition to being younger than the native population, they tend to be less educated. Migrants to cities, as well as the urban poor, may also have less access to affordable and safe housing, potable water and sanitation, social services (health care and education), and employment, although systematic evidence finds that migrants are not uniformly disadvantaged (NRC, 2003). This unequal access, where it exists, places migrants and the poor at greater risk for the consequences of climate change or other natural disasters. As fertility declines, it is likely that migration will play an increasingly important role in the character of cities: thus the importance of improving our understanding of migration and migrant well-being, as well as that of the urban poor more generally.

Cities are densely populated relative to the surrounding areas. Population density is not necessarily thought of as a dimension of inequality, in some part because density allows for the provision of municipal services and goods that reduce inequality. However, cities are not uniformly dense. Poor persons tend to live in more crowded parts of cities, and crowding may exacerbate adverse climate conditions (e.g., excess heat or flooding), foster disease transmission, limit escape routes where public space is insufficient, and cut people off

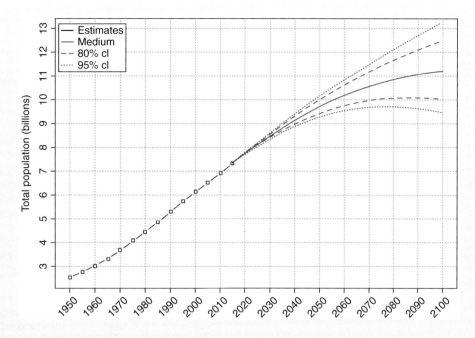

Box 6.4 Figure 1 *Population of the world: estimates, 1950–2015, medium-variant projection and 80% and 95% confidence intervals, 2015–2100.*

Source: UN, 2015

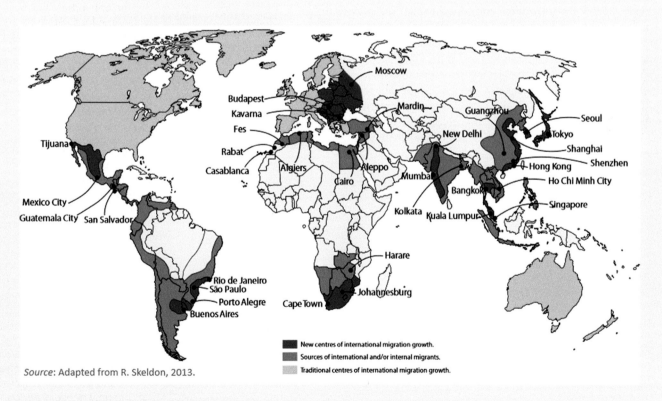

Source: Adapted from R. Skeldon, 2013.

Box 6.4 Figure 2 *Global diversification of migration destinations.*

Source: IOM, 2015. Adapted from Skeldon, 2013

from access to critical services and emergency response. In a case study of Vietnam, it was found that the urban poor were much more likely to live in the low-lying areas of cities in low-lying coastal zones rather than in higher elevation portions of those cities or in higher-elevation cities altogether (Balk et al., 2009).

To better understand the change that cities will undergo in the coming century, demographic data on cities will need to improve. Demographic data are typically collected through population censuses, surveys, and vital registration systems. While the reporting unit varies, typically, demographic estimates from censuses are tallied for administrative units such as counties rather than cities *per se*, except for the very largest ones. Surveys that produce demographic estimates are usually representative at the national level or, at best, the first-order subnational level, such as states or provinces, but rarely cities, with the occasional exception of the largest or capital cities. Surveys usually allow for tallying population into urban and rural strata, but that is not sufficient to understand different demographic profiles by city type and characteristics. Census data from national statistical offices, even when reported for small enumerator areas, are never reported for specialized land areas such as coastal regions; thus, understanding the lives of coastal city dwellers requires additional access to fine-resolution or micro-data records that are often impossible to acquire. In order to better meet the needs of city and regional planners and for national policy-makers to understand the demographic shifts of the future, particularly in the context of climate and other environmental change, greater attention to the spatial location of census data and sampling frames for survey data are essential (Martine and Schensul, 2013).

For more coverage on this topic, see Box 1.1

offices and lobby national agencies to collect city-specific data on equity and equality markers. Focusing on these issues may also contribute to institutionalizing progressive changes in planning and policy practice.

Policy needs to be continuously re-evaluated, reassessed, and readjusted in order to make cities sustainably equitable. Because social categorization and stratification are always evolving, equity and environmental justice are not an outcome achieved once and for all, but one that is negotiated in an ongoing process. The EquIA-urban tool is a step-by-step guide supporting such a process. Achieving an equitable response by cities to the multiple challenges of climate change will require participation from all social groups and ongoing commitment to both climate change and equity objectives by policy-makers and civil society. More detailed recommendations for particular climate change events and policy domains are summarized in Table 6.3

Annex 6.1   Stakeholder Engagement

Every urban stakeholder is an important source of knowledge of how equity in cities might be affected by climate change and a potential user of the final product of this assessment. As co-producers of knowledge, stakeholders were involved in the production of this chapter as authors, advisors, and reviewers. Lead stakeholders submitted contributions and examples of their work as Case Studies (see Boxes 6.2 and 6.3). They were also involved in the chapter's new piece of research, conducting a questionnaire survey to elicit the experience of UN-Habitat partner organizations with equity and environmental justice issues in cities. The results of the questionnaire survey contributed to recommendations for policy (see Annex Box 6.1). Lead stakeholders were the New York City Environmental Justice Alliance (NYC-EJA) and the Women's Environment and Development

---

Annex Box 6.1   Policy Recommendations from the Ground, Elicited through Stakeholders of UN-Habitat

As part of the chapter's research, stakeholders of UN-Habitat were contacted and asked about their experience of equity and environmental justice issues in their work. The questionnaire also asked respondents to suggest indicators that should be monitored after the implementation of climate change policies in order to ensure that there are no negative side effects on equity dimensions. The following text summarizes the responses:

Who is disproportionately affected?
- Low-income groups and people living in low-standard housing, as well as middle- and high-income residents
- People dependent on natural systems and climate-dependent livelihoods (fisheries, agriculture, tourism) are particularly affected, as are women, children, the elderly, and people with disabilities and reduced rights (migrants, minorities, socially and ethnically discriminated groups)

What are the reasons for vulnerability?
- Residency in vulnerable locations, such as riverine areas, urban fringes, and areas affected by inadequate service provision and deficits in urban planning (e.g., lack of protective infrastructure and/or deficient infrastructure, failure to enforce building standards)
- Inadequate urban governance and absence of commitment by locally elected representatives to tackle environmental issues

- Deficiencies with respect to empowerment, entitlement, management, literacy, and social capital among the vulnerable; a general lack of assets, wealth, and insurance cover were also mentioned

Which factors should be monitored to ensure no detrimental effects on equity from climate change policies?
- Achievement of fundamental aims of climate change policy, such as reduced impacts or increased resilience
- Proportion of population living in disaster-prone locations, substandard housing, below the poverty line, and with reduced access to basic urban services
- Unintended side effects of climate change policies on equity and effects on regional and local water supplies, as well as effects on women and the young

Important factors believed to increase equity in urban areas under conditions of climate change:
- Risk-reducing infrastructure and services (e.g., public roads with drainage systems) should be provided across the entire metropolitan region
- Construction of stronger houses, prevention of housing in riverine areas, control of soil erosion and surface runoff,

and promotion of greenbelts and urban forestry (as windbreaks and natural protection)
- Permanent financing mechanisms for climate change policies and actions in cities through harnessing of local and international finance mechanisms and partnerships; prioritization of credits to finance small-scale enterprises and those based on the use of natural resources
- Institutional arrangements (e.g., support provided by community-based and non-governmental organizations (CBOs and NGOs)
- Policies that focus on increasing economic opportunities for residents (e.g., as seen in the green and smart cities initiatives)
- Training and education to raise awareness of climate change risks and impacts and build the capacities of communities to take appropriate action; education of citizens about the role of cities and city governments in responding to climate change
- Mobilization of community leaders to encourage commitment of elected representatives and to provide examples of municipal leadership
- Resilience and integration of low-carbon development into broader sustainable urban development initiatives

Organization (WEDO). Contributing stakeholders included the Municipality of Bobo-Dioulasso (Burkina Faso), Ministry of Environment and Fishery Resources (Burkina Faso), and UN-Habitat (Main Office and Regional Office for Asia and the Pacific [ROAP]).

Chapter 6   Equity and Environmental Justice

References

Asian Development Bank (ADB). (2010). *Access to Justice for the Urban poor – Towards Inclusive Cities*. Asian Development Bank (ADB).

Addison, C., Zhang, S., and Coomes, B. (2013). Smart growth and housing affordability: A review of regulatory mechanisms and planning practices. *Journal of Planning Literature* **28**, 215–257.

Adelekan, I. O. (2010). Vulnerability of poor urban coastal communities to flooding in Lagos, Nigeria. *Environment and Urbanization* **22**, 433–450.

Adelekan, I. O. (2012). Vulnerability to wind hazards in the traditional city of Ibadan, Nigeria. *Environment and Urbanization* **24**, 597–617.

Adger, N. W. (2013). Emerging dimensions of fair process for adaptation decision-making. In Palutikof, J., Boulter, S. L., Ash, A. J., Smith, M. S., Parry, M., Waschka, M., and Guitart, D. (eds.), *Climate Adaptation Futures* (69–74). John Wiley & Sons.

Agyeman, J., Bullard, R., and Evans, B. (eds.). (2003). *Just Sustainabilities: Development in an Unequal World*. MIT Press.

Ahern, M., Kovats, R. S., Wilkinson, P., Few, R., and Matthies, F. (2005). Global health impacts of floods: Epidemiologic evidence. *Epidemiologic Reviews* **27**, 36–46.

Ahmed, S. A., and Ali, M. (2004). Partnerships for solid waste management in developing countries: Linking theories to realities. *Habitat International* **28**, 467–479.

Ajibade, I., Mcbean, G., and Bezner-Kerr, R. (2013). Urban flooding in Lagos, Nigeria: Patterns of vulnerability and resilience among women. *Global Environmental Change* **23**, 1714–1725.

Akanda, A. S., and Hossain, F. (2012). The climate-water-health nexus in emerging megacities. *Eos, Transactions American Geophysical Union* **93**, 353–354.

Alber, G. (2011). Gender, cities and climate change. *Thematic report prepared for Cities and Climate Change: Global Report on Human Settlements 2011*. UN-Habitat.

Allen, K. M. (2006). Community-based disaster preparedness and climate adaptation: Local capacity-building in the Philippines. *Disasters* **30**, 81–101.

Alston, M. (2013). Women and adaptation. *Wiley Interdisciplinary Reviews: Climate Change* **4**, 351–358.

Amengual, A., Homar, V., Romero, R., Brooks, H., Ramis, C., and Alonso, S. (2014). Projections of heat waves with high impact on human health in Europe. *Global and Planetary Change* **3**, 71–84.

Angel, S., Parent, J., Civco, D. L., Blei, A., and Potere, D. (2011). The dimensions of global urban expansion: Estimates and projections for all countries, 2000–2050. *Progress in Planning* **75**, 53–107.

Anguelovski, I., and Carmin, J. (2011). Something borrowed, everything new: Innovation and institutionalization in urban climate governance. *Current Opinion in Environmental Sustainability* **3**, 169–175.

Annez, P., Buckley, R., and Kalarickal, J. (2010). African urbanization as flight? Some policy implications of geography. *Urban Forum* **21**, 221–234.

Appadurai, A. (2000). Grassroots globalization and the research imagination. *Public Culture* **12**, 1–19.

Appadurai, A. (2001). Deep democracy: Urban governmentality and the horizon of politics. *Environment and Urbanization* **13**, 23–43.

Atta-ur-Rahman, and Khan, A. N. (2013). Analysis of 2010-flood causes, nature and magnitude in the Khyber Pakhtunkhwa, Pakistan. *Natural Hazards* **66**, 887–904.

Awad, I. M. (2012). Using econometric analysis of willingness-to-pay to investigate economic efficiency and equity of domestic water services in the West Bank. *The Journal of Socio-Economics* **41**, 485–494.

Awuor, C. B., Orindi, V. A., and Adwera, A. O. (2008). Climate change and coastal cities: The case of Mombasa, Kenya. *Environment and Urbanization* **20**, 231–242.

Ayletta, A. (2010). Participatory planning, justice, and climate change in Durban, South Africa. *Environment and Planning A* **42**, 99–115.

Baccini, M., Biggeri, A., Accetta, G., Kosatsky, T., Katsouyanni, K., Analitis, A., Anderson, H. R., Bisanti, L., D'ippoliti, D., Danova, J., Forsberg, B., Medina, S., Paldy, A., Rabczenko, D., Schindler, C., and Michelozzi, P. (2008). Heat effects on mortality in 15 European cities. *Epidemiology* **19**, 536–543.

Balk, D., Montgomery, M. R., McGranahan, G., Kim, D., Mara, V., Todd, M., Buettner, T., and Dorelien, A. (2009). Mapping urban settlements and the risks of climate change in Africa, Asia, and South America. In Martine, G., Guzman, J. M., Mcgranahan, G., Schensul, D. and Tacoli, C. (eds.), *Population Dynamics and Climate Change* (80–103). United Nations Population Fund and International Institute for the Environment and Development.

Barrios, S., Bertinelli, L., and Strobl, E. (2006). Climatic change and rural–urban migration: The case of sub-Saharan Africa. *Journal of Urban Economics* **60**, 357–371.

Barry, B. (1995). *Justice as Impartiality*. Oxford University Press.

Bartlett, S. (2008). Climate change and urban children: Impacts and implications for adaptation in low- and middle-income countries. *Urbanization and Environment* **20**, 501–519.

Bartlett, S., Dodman, D., Hardoy, J., Satterthwaite, D., and Tacoli, C. Social aspects of climate change in urban areas in low-and middle-income nations. In Hoornweg, D., Freire, M., Lee, M. J., Bhada-Tata, P., and Yuen, B. (eds.), *Fifth Urban Research Symposium, Cities and Climate Change: Responding to an Urgent Agenda* (670–726). World Bank.

Baud, I., Grafakos, S., Hordijk, M., and Post, J. (2001). Quality of life and alliances in solid waste management. *Cities* **18**, 3–12.

Begum, S., and Sen, B. (2005). Pulling rickshaws in the city of Dhaka: A way out of poverty? *Environment and Urbanization* **17**, 11–25.

Bento, A. M., Franco, S. F., and Kaffine, D. (2006). The efficiency and distributional impacts of alternative anti-sprawl policies. *Journal of Urban Economics* **59**, 121–141.

Bento, A. M., Franco, S. F., and Kaffine, D. (2011). Is there a double-dividend from anti-sprawl policies? *Journal of Environmental Economics and Management* **61**, 135–152.

Bento, A. M., Goulder, L. H., Henry, E., Jacobsen, M. R., and Haefen, R. H. V. (2005). Distributional and efficiency impacts of gasoline taxes: An econometrically based multi-market study. *American Economic Review* **95**, 282–287.

Betancourt, A. A. (2010). *Waste Pickers in Bogotá: From Informal Practice to Policy*. Master's thesis, Massachusetts Institute of Technology.

Bicknell, J., Dodman, D., and Satterthwaite, D. (eds.). (2009). *Adapting Cities To Climate Change: Understanding and Addressing the Development Challenges*. Earthscan.

Bithas, K. (2008). The sustainable residential water use: Sustainability, efficiency and social equity. The European experience. *Ecological Economics* **68**, 221–229.

Blake, E. S., Kimberlain, T. B., Berg, R. J., Cangialosi, J. P., and Beven, J. L.,II. (2013). Tropical cyclone report: Hurricane Sandy (22–29 October 2012). National Hurricane Center.

Boarnet, M. G. (2007). *Conducting Impact Evaluations in Urban Transport: Doing Impact Evaluation*. World Bank.

Bradshaw, W., Connelly, E. F., Cook, M. F., Goldstein, J., and Pauly, J. (2005). *The Costs & Benefits of Green Affordable Housing*. New Ecology, The Green CDCs Initiative.

Brady, D. (2009). *Rich Democracies, Poor People – How Politics Explain Poverty*. Oxford University Press.

Brady, D., and Kall, D. (2008). Nearly universal, but somewhat distinct: The feminization of poverty in affluent Western democracies, 1969–2000. *Social Science Research* **37**, 976–1007.

Brand, P., and Davila, J. D. (2011). Mobility innovation at the urban margins: Medellin Metrocables. *City: A*, **15**, 647–661.

Brodie, M., Weltzien, E., Altman, D., Blendon, R. J., and Benson, J. M. (2006). Experiences of Hurricane Katrina evacuees in Houston shelters: Implications for future planning. *American Journal of Public Health* **96**, 1402–1408.

Brown, A., Dayal, A., and Rumbaitis Del Rio, C. (2012). From practice to theory: Emerging lessons from Asia for building urban climate change resilience. *Environment and Urbanization* **24**, 531–556.

Brueckner, J. K. (1997). Infrastructure financing and urban development: The economics of impact fees. *Journal of Public Economics* **66**, 383–407.

Bulkeley, H., Carmin, J., Castán Broto, V., Edwards, G. A. S., and Fuller, S. (2013). Climate justice and global cities: Mapping the emerging discourses. *Global Environmental Change* **23**, 914–925.

Bulkeley, H., and Castán Broto, V. (2013). Government by experiment? Global cities and the governing of climate change. *Transactions of the Institute of British Geographers* **38**, 361–375.

Bulkeley, H., Edwards, G. A. S., and Fuller, S. (2014). Contesting climate justice in the city: Examining politics and practice in urban climate change experiments. *Global Environmental Change* **25**, 31–40.

Bulkeley, H., Schroeder, H., Janda, K., Zhao, J., Armstrong, A., Chu, S. Y., and Ghosh, S. (2011). The role of institutions, governance, and urban planning for mitigation and adaptation. In Hoornweg, G., Freire M., Lee, M.J., Bhada-Tata, P., and Yuen, B. (eds.), *Cities and Climate Change: Responding to an Urgent Agenda* (125–159). World Bank.

Bureau, B,. and Glachant, M. (2008). Distributional effects of road pricing: Assessment of nine scenarios for Paris. *Transportation Research Part A: Policy and Practice* **42**, 994–1007.

Burton, E. (2000). The compact city: Just or just compact? A preliminary analysis. *Urban Studies* **37**, 1969–2006.

Byrne, J., Wolch, J., and Zhang, J. (2009). Planning for environmental justice in an urban national park. *Journal of Environmental Planning and Management* **52**, 365–392.

Caine, N. (1980). The rainfall intensity: Duration control of shallow landslides and debris flows. *Geografiska Annaler. Series A, Physical Geography* **62**, 23–27.

Campos, M. J. Z., and Zapata, P. (2013). Switching Managua on! Connecting informal settlements to the formal city through household waste collection. *Environment and Urbanization* **25**, 225–242.

Can Equity Group. (2014). *Equity Reference Framework at the UNFCCC Process*. A CAN discussion paper. In Climate Action Network (CAN).

Canoüï-Poitrine, F., Cadot, E., and Spira, A. (2005). Excess deaths during the August 2003 heat wave in Paris, France. *Revue d'Epidemiologie et de Sante Publique* **54**, 127–135.

Carcellar, N., Co, J. C. R., and Hipolito, Z. O. (2011). Addressing disaster risk reduction through community-rooted interventions in the Philippines: Experience of the Homeless People's Federation of the Philippines. *Environment and Urbanization* **23**, 365–381.

Carmin, J., and Dodman, D. (2013). Engaging science and managing scientific uncertainty in urban climate adaptation planning. In Moser, S. C., and Boykoff, M. T. (eds.), *Towards Successful Adaptation to Climate Change – Linking Science and Policy in a Rapidly Changing World* (220–234). Routledge.

Carmin, J., Dodman, D., and Chu, E. (2013). Urban climate adaptation and leadership: From conceptual to practical understanding. *OECD Regional Development working papers*. OECD.

Carslaw, D. C., and Beevers, S. D. (2002). The efficacy of low emission zones in central London as a means of reducing nitrogen dioxide concentrations. *Transportation Research Part D: Transport and Environment* **7**, 49–64.

Castán Broto, V., and Bulkeley, H. (2013). A survey of urban climate change experiments in 100 cities. *Global Environmental Change* **23**, 92–102.

Cazorla, M. and Toman, M. (2000). *International equity and climate change policy*. Climate Issue Brief No. 27. Washington, D.C.: Resources for the Future 1–22.

Cepeda, J., Smebye, H., Vangelsten, B., Nadim, F., and Muslim, D. (2010). Landslide risk in Indonesia. *Global assessment report on disaster risk reduction*. ISDR.

Cerdá, M., Morenoff, J. D., Hansen, B. B., Tessari Hicks, K. J., Duque, L. F., Restrepo, A., and Diez-Roux, A. V. (2012). Reducing violence by transforming neighborhoods: A natural experiment in Medellín, Colombia. *American Journal of Epidemiology* **175**, 1045–1053.

Checker, M. (2011). Wiped out by the "Greenwave": Environmental gentrification and the paradoxical politics of urban sustainability. *City and Society* **23**, 210–229.

Chen, M., Jhabvala, R., Kanbur, R., and Richards, C. (2013). *Membership Based Organizations of the Poor.* Routledge.

Chen, M., Vanek, J., Lund, F., Heintz, J., Jhabvala, R., and Bonner, C. (2005). *Progress of the World's Women 2005: Women, Work & Poverty.* UNIFEM (United Nations Development Fund for Women).

Cheng, L., Bi, X., Chen, X., and Li, L. (2013). Travel behavior of the urban low-income in China: Case study of Huzhou City. *Procedia – Social and Behavioral Sciences* **96**, 231–242.

City of Boston. (2011). A climate of progress: City of Boston climate action plan update 2011. *Update report.* City of Boston.

City of New York. (2015). *One New York: The plan for a strong and just city.* City of New York. Accessed February 29, 2016: http://www.nyc.gov/html/onenyc/downloads/pdf/publications/OneNYC.pdf

Clarke, M. (2012). *Making Transport Work for Women and Men: Challenges and Opportunities in the Middle East and North Africa (MENA) Region – Lessons from Case Studies. Working paper.* World Bank.

Collins, W. J., and Shester, K. L. (2011). *Slum Clearance and Urban Renewal in the United States. NBER Working paper series 1748.* National Bureau of Economic Research (NBER).

Corfee-Morlot, J., Kamal-Chaoui, L., Donovan, M. G., Cochran, I., Robert, A., and Teasdale, P. -J. (2009). *Cities, Climate Change and Multilevel Governance. OECD Environment working paper No. 14.* OECD.

Coumou, D., and Rahmstorf, S. (2012). A decade of weather extremes. *Nature Climate Change* **2**, 491–496.

Cox, W. (2008). *How Smart Growth Exacerbated the International Financial Crisis.* The Heritage Foundation.

Creutzig, F., and He, D. (2009). Climate change mitigation and co-benefits of feasible transport demand policies in Beijing. *Transportation Research Part D: Transport and Environment* **14**, 120–131.

Creutzig, F., Rainer, M., and Julia, R. (2012). Decarbonizing urban transport in European cities: Four cases show possibly high co-benefits. *Environmental Research Letters* **7**, 04(4042).

Cullis, J., Strzepek, K., Tadross, M., Sami, K., Havenga, B., Gildenhuys, B., and Smith, J. (2011). Incorporating climate change into water resources planning for the town of Polokwane, South Africa. *Climatic Change* **108**, 437–456.

Curran, W., and Hamilton, T. (2012). Just green enough: Contesting environmental gentrification in Greenpoint, Brooklyn. *Local Environment: The International Journal of Justice and Sustainability* **17**, 1027–1042.

Curtis, A., Mills, J. W., and Leitner, M. (2007). Katrina and vulnerability: The geography of stress. *Journal of Health Care for the Poor and Underserved* **18**, 315–330.

Cutter, S. L., Emrich, C. T., Mitchell, J. T., Boruff, B. J., Gall, M., Schmidtlein, M. C., Burton, C. G., and Melton, G. (2006). The long road home: Race, class, and recovery from Hurricane Katrina. *Environment: Science and Policy for Sustainable Development* **48**, 8–20.

Cutter, S. L., and Finch, C. (2008). Temporal and spatial changes in social vulnerability to natural hazards. *Proceeding of the National Academy of Science* **105**, 2301–2306.

Dai, D. (2011). Racial/ethnic and socioeconomic disparities in urban green space accessibility: Where to intervene? *Landscape and Urban Planning* **102**, 234–244.

Dankelman, I. (2010). Introduction: Exploring gender, environment and climate change. In Dankelman, I. (ed.), *Climate Change and Gender: An Introduction.* Earthscan.

Dankelman, I., Alam, K., Ahmed, W. B., Gueye, Y. D., Fatema, N., and Mensah-Kutin, R. (2008). Gender, climate change and human security lessons from Bangladesh, Ghana and Senegal. In The Women's Environment and Development Organization (WEDO) with ABANTU for Development in Ghana, Action Aid Bangladesh and ENDA in Senegal.

David, E., and Enarson, E. (eds.). (2012). *The Women of Katrina – How Gender, Race, and Class Matter in an American Disaster.* Vanderbilt University Press.

De Sherbinin, A., Chai-Onn, T., Jaiteh, M., Mara, V., Pistolesi, L., and Schnarr, E. (2014). Mapping the exposure of socioeconomic and natural systems of West Africa to coastal climate stressors. *Report for the USAID African and Latin American Resilience to Climate Change (ARCC) project.* United States Agency for International Development (USAID).

De Sherbinin, A., and Hogan, D. (2011). Box 3.1 Climate Proofing Rio de Janeiro, Brazil. In Rosenzweig, C., Solecki, W., Mehrotra, S., and Hammer, S. A. (eds.), *Climate Change and Cities: First Assessment Report of the Urban Climate Change Research Network* (50). Cambridge University Press.

De Sherbinin, A., Schiller, A., and Pulsipher, A. (2007). The vulnerability of global cities to climate hazards. *Environment and Urbanization* **19**, 39–64.

Dempsey, N., Brown, C., and Bramley, G. (2012). The key to sustainable urban development in UK cities? The influence of density on social sustainability. *Progress in Planning* **77**, 89–141.

Deneulin, S. (2014). Creating more just cites: The right to the city and capability approach combined. *Bath papers in international development and well-being (BPIDW).* Centre for Development Studies, University of Bath.

Deng, T., and Nelson, J. D. (2010). The impact of bus rapid transit on land development: A case study of Beijing, China. *International Scholarly and Scientific Research & Innovation* **4**, 949–959.

Deng, T., and Nelson, J. D. (2011). Recent developments in bus rapid transit: A review of the literature. *Transport Reviews* **31**, 69–96.

Descroix, L., Genthon, P., Amogu, O., Rajot, J. -L., Sighomnou, D., and Vauclin, M. (2012). Change in Sahelian Rivers hydrograph: The case of recent red floods of the Niger River in the Niamey region. *Global and Planetary Change* **98–99**, 18–30.

Dill, J., Goldman, T., and Wachs, M. (1999). California vehicle license fees: Incidence and equity. *Journal of Transportation and Statistics* 133–147.

Dodman, D. (2009). Blaming cities for climate change? An analysis of urban greenhouse gas emissions inventories. *Environment and Urbanization* **21**, 185–201.

Dodman, D. (2013). The challenge of adaptation that meets the needs of low-income urban dwellers. In Palutikof, J., Boulter, S. L., Ash, A. J., Smith, M. S., Parry, M., Waschka, M., and Guitart, D. (eds.), *Climate Adaptation Futures* (227–234). John Wiley & Sons.

Dodman, D., and Mitlin, D. (2013). Challenges for community-based adaptation: Discovering the potential for transformation. *Journal of International Development* **25**, 640–659.

Dodman, D., and Satterthwaite, D. (2009). The costs of adapting infrastructure to climate change. In Parry, M., Arnell, N., Berry, P., Dodman, D., Fankhauser, S., Hope, C., Kovats, S., Nicholls, R., Satterthwaite, D., Tiffin, R., and Wheeler, T. (eds.), *Assessing the Costs of Adaptation to Climate Change: A Review of the UNFCCC and Other Recent Estimates.* IIED and Grantham Institute.

Douglas, I., Alam, K., Maghenda, M., Mcdonnell, Y., Mclean, L., and Campbell, J. (2008a). Unjust waters: Climate change, flooding and the urban poor in Africa. *Environment and Urbanization* **20**, 187–205.

Dovidio, J. F., Hewstone, M., Glick, P., and Esses, V. M. (2010). Prejudice, stereotyping and discrimination: Theoretical and empirical overview. In Dovidio, J. F., Hewstone, M., Glick, P., and Esses, V. M. (eds.), *The SAGE Handbook of Prejudice, Stereotyping and Discrimination* (3–28). SAGE Publications Ltd.

Drummond, H., Dizgun, J., and Keeling, D. J. (2012). Medellín: A city reborn? *Focus on Geography* **55**, 146–154.

Druyan, A., Makranz, C., Moran, D., Yanovich, R., Epstein, Y., and Heled, Y. (2012). Heat tolerance in women: Reconsidering the criteria. *Aviation, Space, and Environmental Medicine* **83**, 58–60.

Dulal, H. B., Brodnig, G., and Onoriose, C. G. (2011). Climate change mitigation in the transport sector through urban planning: A review. *Habitat International* **35**, 494–500.

Earl, P. E., and Wakeley, T. (2009). Price-based versus standards-based approaches to reducing car addiction and other environmentally destructive activities. In Holt, R. (ed.), *Post Keynesian and Ecological Economics* (158–177). Edward Elgar Publishing.

Eiser, J. R., Bostrom, A., Burton, I., Johnston, D. M., Mcclure, J., Paton, D., Van Der Pligt, J., and White, M. P. (2012). Risk interpretation and action: A conceptual framework for responses to natural hazards. *International Journal of Disaster Risk Reduction* **1**, 5–16.

Elbers, C., Lanjouw, J. O., and Lanjouw, P. (2003). Micro-level estimation of poverty and inequality. *Econometrica* **71**, 355–364.

Elbers, C., Lanjouw, J. O., and Lanjouw, P. (2005). Imputed welfare estimates in regression analysis. *Journal of Economic Geography* **5**, 101–118.

Eliasson, J., and Mattsson, L. -G. (2006). Equity effects of congestion pricing. Quantitative methodology and a case study for Stockholm. *Transportation Research Part A: Policy and Practice* **40**, 602–620.

Elliott, J. R., and Pais, J. (2006). Race, class, and Hurricane Katrina: Social differences in human responses to disaster. *Social Science Research* **35**, 295–321.

Em-Dat. (2010). The OFDA/CRED International Disaster Database: On-line database maintained by the Université Catholique de Louvain. Accessed October 6, 2014: www.emdat.be

Ewing, R., Richardson, H. W., Bartholomew, K., Nelson, A. C., and Bae, C. -H. C. (2014). *Compactness vs. Sprawl Revisited: Converging Views. CESIFO working paper No. 4571*. Ludwig-Maximilian-University.

Fahmi, F. Z., Hudalah, D., Rahayu, P., and Woltjer, J. (2014). Extended urbanization in small and medium-sized cities: The case of Cirebon, Indonesia. *Habitat International*, **42**, 1–10.

Farley, K. A., Tague, C., and Grant, G. E. (2011). Vulnerability of water supply from the Oregon Cascades to changing climate: Linking science to users and policy. *Global Environmental Change* **21**, 110–122.

Fergutz, O., Dias, S., and Mitlin, D. (2011). Developing urban waste management in Brazil with waste picker organizations. *Environment and Urbanization* **23**, 597–608.

Fernanda Rosales, M. (2014). *Impact of Early Life Shocks on Human Capital Formation: Evidence from El Nino Floods in Ecuador. Working paper; Job Market Paper*. Department of Economics, University of Chicago.

Fox, E. (2015). *Drought-stricken São Paulo hit by floods; Torrential rains leave Brazil's largest city under water after months of severe drought*. Aljazeera. Accessed March 6, 2016: http://www.aljazeera.com/news/2015/02/drought-stricken-sao-paulo-hit-floods-15022610111(9755).html

Franzen, A., and Vogl, D. (2013). Time preferences and environmental concern. *International Journal of Sociology* **43**, 39–62.

Fritz, H. M., Blount, C., Sokoloski, R., Singleton, J., Fuggle, A., Mcadoo, B. G., Moore, A., Grass, C., and Tate, B. (2007). Hurricane Katrina storm surge distribution and field observations on the Mississippi Barrier Islands. *Estuarine, Coastal and Shelf Science* **74**, 12–20.

Fuchs, R. J. (2010). *Cities at risk: Asia's coastal cities in an age of climate change. Analysis from the East-West Center: Asia Pacific Issues*. East-West Center.

Fullerton, D., Devarajan, S., and Musgrave, R. A. (1980). Estimating the distribution of tax burdens: A comparison of different approaches. *Journal of Public Economics* **13**, 155–182.

Galea, S., and Vlahov, D. (2005). Urban health: Evidence, challenges, and directions. *Annual Review Public Health* **26**, 341–365.

Garssen, J., Harmsen, C., and De Beer, J. (2005). The effect of the summer 2003 heat wave on mortality in the Netherlands. *Eurosurveillance* **10**, 165–167.

GGCA. (2009). *Training Manual on Gender and Climate Change*. International Union for Conservation of Nature (IUCN), United Nations Development Programme (UNDP), as well as Gender and Water Alliance, ENERGIA, International Network on Gender and Sustainable Energy, United Nations Educational, Scientific and Cultural Organization (UNESCO), Food and Agriculture Organization (FAO), Women's Environment and Development Organization (WEDO) as part of the Global Gender and Climate Alliance (GGCA).

Golubchikov, O., and Deda, P. (2012). Governance, technology, and equity: An integrated policy framework for energy efficient housing. *Energy Policy* **41**, 733–741.

Gouveia, N., Hajat, S., and Armstrong, B. (2003). Socioeconomic differentials in the temperature-mortality relationship in São Paulo, Brazil. *International Journal of Epidemiology* **32**, 390–397.

Grazi, F., and Van Den Bergh, J. C. J. M. (2008). Spatial organization, transport, and climate change: Comparing instruments of spatial planning and policy. *Ecological Economics* **67**, 630–639.

Greenwood, M. J. (1997). Internal migration in developed countries. In Rosenzweig, M., and Stark, O. (eds.), *Handbook of Population and Family Economics* (647–720). Elsevier Science.

Guerrero, A. (2011). *Rebuilding trust in government via service delivery: The case of Medellin, Colombia. Companion technical note*. World Bank.

Güneralp, B., Güneralp, İ., and Liu, Y. (2015). Changing global patterns of urban exposure to flood and drought hazards. *Global Environmental Change* **31**, 217–225.

Guzman, G. M., and Miguel, J. (2009). Population dynamics and climate change: Recasting the policy agenda. In Palosuo, E. (ed.), *Rethinking Development in a Carbon-Constrained World – Development Cooperation and Climate Change* (71–85). Ministry for Foreign Affairs of Finland.

Haines, A., Bruce, N., Cairncross, S., Davies, M., Greenland, K., Hiscox, A., Lindsay, S., Lindsay, T., Satterthwaite, D., and Wilkinson, P. (2013). Promoting health and advancing development through improved housing in low-income settings. *Journal of Urban Health* **90**, 810–831.

Hardoy, J., and Pandiella, G. (2009). Urban poverty and vulnerability to climate change in Latin America. *Environment and Urbanization* **21**, 203–224.

Hardoy, J., and Ruete, R. (2013). Incorporating climate change adaptation into planning for a liveable city in Rosario, Argentina. *Environment and Urbanization* **25**, 339–360.

Hardoy, J. E., Mitlin, D., and Satterthwaite, D. (2001). *Environmental problems in an urbanizing world: Finding solutions for cities in Africa, Asia, and Latin America*. Earthscan.

Harlan, S. L., Brazel, A. J., Prashad, L., Stefanov, W. L., and Larsen, L. (2006). Neighborhood microclimates and vulnerability to heat stress. *Social Science and Medicine* **63**, 2847–(2863).

Harlan, S. L., Declet-Barreto, J. H., Stefanov, W. L., and Petitti, D. B. (2013). Neighborhood effects on heat deaths: Social and environmental predictors of vulnerability in Maricopa county, Arizona. *Environmental Health Perspectives* **121**, 197–204.

Hartman, C. W., and Squires, G. D. (eds.). (2006). *There Is No Such Thing as a Natural Disaster: Race, Class, and Hurricane Katrina*. Routledge.

Hayami, Y., Dikshit, A. K., and Mishra, S. N. (2006). Waste pickers and collectors in Delhi: Poverty and environment in an urban informal sector. *Journal of Development Studies* **42**, 41–69.

Hereher, M. E. (2010). Vulnerability of the Nile Delta to sea level rise: An assessment using remote sensing. *Geomatics, Natural Hazards and Risk* **1**, 315–321.

Herrfahrdt-Pähle, E. (2010). South African water governance between administrative and hydrological boundaries. *Climate and Development* **2**, 111–127.

Houston, A. (2010). *Housing Support Services for Housing Microfinance Lending in East and Southern Africa: A Case Study of The Kuyasa Fund*. FinMark Trust and Rooftops Canada.

Hunt, A., and Watkiss, P. (2011). Climate change impacts and adaptation in cities: A review of the literature. *Climatic Change* **104**, 13–49.

Hunt, C. (1996). Child waste pickers in India: The occupation and its health risks. *Environment and Urbanization* **8**, 111–118.

Huq, S., Kovats, S., Reid, H., and Satterthwaite, D. (2007). Editorial: Reducing risks to cities from disasters and climate change. *Environment and Urbanization* **19**, 3–15.

Huysman, M. (1994). Waste picking as a survival strategy for women in Indian cities. *Environment & Urbanization* **6**, 155–174.

Intergovernmental Panel on Climate Change (IPCC). (2007). Summary for policymakers. In Solomon, S., Qin, D., Manning, M., Chen, Z., Marquis, M., Averyt, K. B., Tignor, M., and Miller, H. L. (eds.), *Climate Change 2007: The Physical Science Basis. Contribution of Working Group I to the Fourth Assessment Report of the Intergovernmental Panel on Climate Change*. Cambridge University Press.

International Organization for Migration (IOM). (2015). *World Migration Report 2015. Migrants and Cities: New Partnerships to Manage Mobility*. International Organization for Migration. Accessed February 20, 2016: http://publications.iom.int/system/files/wmr2015_en.pdf

Jabeen, H. (2014). Adapting the built environment: The role of gender in shaping vulnerability and resilience to climate extremes in Dhaka. *Environment and Urbanization* **26**, 147–165.

Jenerette, D. G., and Larsen, L. (2006). A global perspective on changing sustainable urban water supplies. *Global and Planetary Change* **50**, 202–211.

Jenkins, P. (2000). Urban management, urban poverty and urban governance: Planning and land management in Maputo. *Environment and Urbanization* **12**, 137–152.

Jennings, V., Gaither, C. J., and Gragg, R. S. (2012). Promoting environmental justice through urban green space access: A synopsis. *Environmental Justice* **5**, 1–7.

Jha, A. K., Bloch, R., and Lamond, J. (2012). *Cities and Flooding: A Guide to Integrated Urban Flood Risk Management for the 21st Century.* World Bank.

Joassart-Marcelli, P. (2010). Leveling the playing field? Urban disparities in funding for local parks and recreation in the Los Angeles region. *Environment and Planning A* **42**, 1174–1192.

Joassart-Marcelli, P., Wolch, J., and Salim, Z. (2011). Building the healthy city: The role of nonprofits in creating active urban parks. *Urban Geography* **32**, 682–711.

Johnson-Gaither, C. (2011). Latino park access: Examining environmental equity in a "New Destination" county in the South. *Journal of Park and Recreation Administration* **29**, 37–52.

Johnson, C., and Blackburn, S. (2014). Advocacy for urban resilience: UNISDR's making cities resilient campaign. *Environment and Urbanization* **26**, 29–52.

Johnson, H., Kovats, R., Mcgregor, G., Stedman, J., Gibbs, M., and Walton, H. (2005). The impact of the 2003 heat wave on daily mortality in England and Wales and the use of rapid weekly mortality estimates. *Eurosurveillance* **10**, 168–171.

Kaijser, A., and Kronsell, A. (2014). Climate change through the lens of intersectionality. *Environmental Politics* **23**, 417–433.

Kallbekken, S., Sælen, H., and Underdal, A. (2014). *Equity and Spectrum of Mitigation Commitments in the 2015 Agreement.* TemaNord.

Kareem, B., and Lwasa, S. (2011). From dependency to interdependencies: The emergence of a socially rooted but commercial waste sector in Kampala City, Uganda. *African Journal of Environmental Science and Technology* **5**, 136–142.

Kates, R. W., Travis, W. R., and Wilbanks, T. J. (2012). Transformational adaptation when incremental adaptations to climate change are insufficient. *Proceedings of the National Academy of Sciences of the United States of America* **109**, 7156–7161.

Kehew, R., Kolisa, M., Rollo, C., Callejas, A., and Alber, G. (2012). Urban climate governance in the Philippines, Mexico and South Africa: National- and state-level laws and policies. In Otto-Zimmermann, K. (ed.), *Resilient Cities 2, Local Sustainability* 2 (305–315). Springer Netherlands.

Khan, A. E., Ireson, A., Kovats, S., Mojumder, S. K., Khusru, A., Rahman, A., and Vineis, P. (2011). Drinking water salinity and maternal health in coastal Bangladesh: Implications of climate change. *Environmental Health Perspectives* **119**, 1328–1332.

Khan, I. A. (2000). *Struggle for Survival: Networks and Relationships in a Bangladesh Slum.* University of Bath.

Khosla, P., and Masaud, A. (2010). *Cities, Climate Change and Gender: A Brief Overview.* Earthscan.

King, M. F., and Gutberlet, J. (2013). Contribution of cooperative sector recycling to greenhouse gas emissions reduction: A case study of Ribeião Pires, Brazil. *Waste Management* **33**, 2771–2780.

Kinney, P. L. (2012). Health: A new measure of health effects. *Nature Climate Change* **2**, 233–234.

Kinney, P. L., Matte, T., Knowlton, K., Madrigano, J., Petkova, E., Weinberger, K., Quinn, A., Arend, M., and Pullen, J. (2015). New York City Panel on Climate Change 2015: Public health impacts and resiliency. *Annals of the New York Academy of Sciences* **1336**, 67–88.

Kiunsi, R. (2013). The constraints on climate change adaptation in a city with a large development deficit: The case of Dar es Salaam. *Environment and Urbanization* **25**, 321–337.

Klinenberg, E. (2003). *Heat Wave: A Social Autopsy of Disaster in Chicago.* University of Chicago Press.

Klinsky, S., and Dowlatabadi, H. (2009). Conceptualizations of justice in climate policy. *Climate Policy* **9**, 88–108.

Komori, D., Nakamura, S., Kiguchi, M., Nishijima, A., Yamazaki, D., Suzuki, S., Kawasaki, A., Oki, K., and Oki, T. (2012). Characteristics of the 2011 Chao Phraya River flood in Central Thailand. *Hydrological Research Letters* **6**, 41–46.

Kosatsky, T. (2005). The 2003 European heat waves. *Eurosurveillance* **10**, 148–149.

Kovats, S., and Akhtar, R. (2008). Climate, climate change and human health in Asian cities. *Environment and Urbanization* **20**, 165–175.

Krishna, A., Sriram, M. S., and Prakash, P. (2014). Slum types and adaptation strategies: Identifying policy-relevant differences in Bangalore. *Environment and Urbanization* **26**, 1–18.

Landry, S. M., and Chakraborty, J. (2009). Street trees and equity: Evaluating the spatial distribution of an urban amenity. *Environment and Planning A* **41**, 2651–2670.

Lange, A., Löschel, A., Vogt, C., and Ziegler, A. (2010). On the self-interested use of equity in international climate negotiations. *European Economic Review* **54**, 359–375.

Lau, C. L., Smythe, L. D., Craig, S. B., and Weinstein, P. (2010). Climate change, flooding, urbanisation and leptospirosis: Fuelling the fire? *Transactions of the Royal Society of Tropical Medicine and Hygiene* **104**, 631–638.

Levy, C. (2013). Travel choice reframed: "Deep distribution" and gender in urban transport. *Environment and Urbanization* **25**, 47–63.

Linnerooth-Bayer, J. (2009). Climate change and multiple views of fairness. In Toth, F. L. (ed.), *Fair Weather: Equity Concerns in Climate Change* (44–64). Earthscan.

Loughnan, M. E., Carroll, M., and Tapper, N. (2013). Learning from our older people: Pilot study findings on responding to heat. *Australasian Journal on Ageing* 33 (4), 271–277.

Lwasa, S. (2012). Planning innovation for better urban communities in sub-Saharan Africa: The education challenge and potential responses. *Town and Regional Planning* **60**, 38–48.

Lwasa, S., and Kinuthia-Njenga, C. (2012). Reappraising urban planning and urban sustainability in East Africa. In Polyzos, S. (ed.), *Urban Development.* InTech. doi: 10.5772/35133. Accessed November 1, 2014: http://www.intechopen.com/books/urban-development/reappraising-urban-planning-and-urban-sustainability-in-east-africa.

Macharia, K. (1992). Slum clearance and the informal economy in Nairobi. *The Journal of Modern African Studies* **30**, 221–236.

Maffii, S., Malgieri, P., and Bartolo, C. D. (2014). Gender equality and mobility: Mind the gap! CIVITAS WIKI policy analyses. Accessed November 26, 2015: http://www.civitas.eu/sites/default/files/civ_pol-an2_m_web.pdf

Marino, E., and Ribot, J. (2012). Special Issue Introduction: Adding insult to injury: Climate change and the inequities of climate intervention. *Global Environmental Change* **22**, 323–328.

Martine, G., and Schensul, D. (eds.). (2013). *The Demography of Adaptation to Climate Change.* UNFPA, IIED and El Colegio de Mexico.

Marulanda, M. C., Cardona, O. D., and Barbat, A. H. (2010). Revealing the socioeconomic impact of small disasters in Colombia using the DesInventar database. *Disasters* **34**, 552–570.

Maruyama, T., and Sumalee, A. (2007). Efficiency and equity comparison of cordon- and area-based road pricing schemes using a trip-chain equilibrium model. *Transportation Research Part A: Policy and Practice* **41**, 655–671.

McBean, G. A. (2012). Integrating disaster risk reduction towards sustainable development. *Current Opinion in Environmental Sustainability* **4**, 122–127.

McConnachie, M. M., and Shackleton, C. M. (2010). Public green space inequality in small towns in South Africa. *Habitat International* **34**, 244–248.

McDermott, M., Mahanty, S., and Schreckenberg, K. (2011). *Defining Equity: A Framework for Evaluating Equity in the Context of Ecosystem Services.* Working paper. REDD-net and Ecosystem Services for Poverty Alleviation (ESPA).

McDermott, M. H., and Schreckenberg, K. (2009). Equity in community forestry: Insights from North and South. *International Forestry Review* **11**, 157–170.

McDonald, R. I., Green, P., Balk, D., Fekete, B. M., Revenga, C., Todd, M., and Montgomery, M. (2011). Urban growth, climate change, and freshwater availability. *Proceedings of the National Academy of Sciences of the United States of America* **108**, 6312–6317.

McEvoy, D., Ahmed, I., Trundle, A., Sang, L. T., Diem, N. N., Suu, L. T. T., Quoc, T. B., Mallick, F. H., Rahman, R., Rahman, A., Mukherjee, N., and Nishat, A. (2014). In support of urban adaptation: A participatory assessment process for secondary cities in Vietnam and Bangladesh. *Climate and Development* **6**, 205–215.

McGranahan, G., Balk, D., and Anderson, B. (2007). The rising tide: Assessing the risks of climate change and human settlements in low elevation coastal zones. *Environment and Urbanization* **19**, 17–37.

McGranahan, G., Marcotullio, P., Bai, X., Balk, D., Braga, T., Douglas, I., Elmqvist, T., Rees, W., Satterthwaite, D., Songsore, J., and Zlotnik, H. (2005). Urban systems. In Hassan, R., Scholes, R., and Ash, N. (eds.), *Millennium Ecosystems Assessment: Ecosystems and Human Wellbeing: Current State and Trends* (18–19). Island Press.

Mearns, R., and Norton, A., eds. (2010). *The Social Dimensions of Climate Change – Equity and Vulnerability in a Warming World.* World Bank.

Measham, T. G., Preston, B. L., Smith, T. F., Brooke, C., Gorddard, R., Withycombe, G., and Morrison, C. (2011). Adapting to climate change through local municipal planning: Barriers and challenges. *Mitigation and Adaptation Strategies for Global Change* **16**, 889–909.

Medina, M. (2008). The informal recycling sector in developing countries: Organizing waste pickers to enhance their impact. *Gridlines.* PPIAF, World Bank.

Metz, B. (2000). International equity in climate change policy. *Integrated Assessment* **1**, 111–126.

Michelozzi, P., De'donato, F., Bisanti, L., Russo, A., Cadum, E., Demaria, M., D'ovidio, M., Costa, G., and Perucci, C. (2005). The impact of the summer 2003 heat waves on mortality in four Italian cities. *Eurosurveillance* **10**, 161–165.

Mills, S. (2007). The Kuyasa Fund: Housing microcredit in South Africa. *Environment and Urbanization* **19**, 457–469.

Mitchell, G. (2005). Forecasting environmental equity: Air quality responses to road user charging in Leeds, UK. *Journal of Environmental Management* **77**, 212–226.

Montgomery, M. R., Stren, R., Cohen, B., and Reed, H. E. (eds.). (2003). *Cities Transformed: Demographic Change and Its Implications in the Developing World.* National Academies Press.

Montgomery, M. R., Balk, D., Liu, Z., and Kim, D. (2012, 22 October). *Understanding City Growth in Asia's Developing Countries: The Role of Internal Migration.* Working paper. Asia Development Bank.

Moreiras, S. M. (2005). Landslide susceptibility zonation in the Rio Mendoza Valley, Argentina. *Geomorphology* **66**, 345–357.

Moreno-Sánchez, R. D. P., and Maldonado, J. H. (2006). Surviving from garbage: The role of informal waste-pickers in a dynamic model of solid-waste management in developing countries. *Environment and Development Economics* 11 (3), 371–391.

Moser, C. (2011). A conceptual and operational framework for pro-poor asset adaptation to urban climate change. In Hoornweg, D., Freire, M., Lee, M. J., Bhada-Tata, P., and Yuen, B. (eds.), *Cities and Climate Change: Responding to an Urgent Agenda* (225–253). World Bank.

Moser, C., and Satterthwaite, D. (2008). *Towards Pro-Poor Adaptation to Climate Change in the Urban Centres of Low- and Middle-Income Countries. Human settlements discussion paper series.* International Institute for Environment and Development (IIED).

Moser, C., and Stein, A. (2011). Implementing urban participatory climate change adaptation appraisals: A methodological guideline. *Environment and Urbanization* **23**, 463–485.

Moser, S. C. (2006). Talk of the city: Engaging urbanites on climate change. *Environmental Research Letters* **1**, 1–10.

Munoz-Raskin, R. (2010). Walking accessibility to bus rapid transit: Does it affect property values? The case of Bogotá, Colombia. *Transport Policy* **17**, 72–84.

Nabangchang, O., Allaire, M., Leangcharoen, P., Jarungrattanapong, R., and Whittington, D. (2015). Economic costs incurred by households in the 2011 greater Bangkok flood. *American Geophysical Union: Water Resources Research* **51**, 58–77.

National Research Council (NRC). (2003). *Cities Transformed: Demographic Change and Its Implications in the Developing World,* M. R. Montgomery, et al. (eds.). National Academies Press.

Neidell, M., and Kinney, P. L. (2010). Estimates of the association between ozone and asthma hospitalizations that account for behavioral responses to air quality information. *Environmental Science & Policy* **13**, 97–103.

Newman, O. (1972). *Defensible Space: Crime Prevention through Urban Design.* Macmillan.

New York City Environmental Justice Alliance. (2016). The NYC Climate Justice Agenda: Strengthening the Mayor's OneNYC Plan. Accessed March 31, 2017: http://nyc-eja.org/public/publications/NYC_Climate-JusticeAgenda.pdf

NGI. (2012). *Landslide Hazard and Risk Assessment in El Salvador.* Background paper prepared for the Global Assessment Report on disaster risk reduction 2013. UNISDR global assessment report 2013 – GAR13. United Nations.

Nitschke, M., Hansen, A., Bi, P., Pisaniello, D., Newbury, J., Kitson, A., Tucker, G., Avery, J., and Dal Grande, E. (2013). Risk factors, health effects and Behaviour in older people during extreme heat: A survey in South Australia. *International Journal of Environmental Research and Public Health* **10**, 6721–(6733).

Nogueira, P. J., Falcão, J. M., Contreiras, M. T., Paixão, E., Brandão, J., and Batista, I. (2005). Mortality in Portugal associated with the heat wave of August 2003: Early estimation of effect, using a rapid method. *Eurosurveillance* **10**, 150–153.

NPCC2 (2013). Climate risk information 2013: Observations, climate change projections, and maps. In Rosenzweig, C., and Solecki, W. (eds.). NPCC2.

Nuworsoo, C., Golub, A., and Deakin, E. (2009). Analyzing equity impacts of transit fare changes: Case study of Alameda–Contra Costa Transit, California. *Evaluation and Program Planning* **32**, 360–368.

NYC-DOHMH. (2010). *Community health survey data.* New York City Department of Health and Mental Hygiene (NYC-DOHMH).

NYC-SIRR. (2013). *A stronger more resilient New York.* New York City Special Initiative for Rebuilding and Resiliency (NYC-SIRR).

OECD. (2012). *Greening Development: Enhancing Capacity for Environmental Management and Governance.* OECD.

Olmstead, S. M., Michael Hanemann, W., and Stavins, R. N. (2007). Water demand under alternative price structures. *Journal of Environmental Economics and Management* **54**, 181–198.

Owens, G. R. (2010). Post-colonial migration: Virtual culture, urban farming and new peri-urban growth in Dar es Salaam, Tanzania, 1975–2000. *Africa* **80**, 249–274.

Parry, M., Arnell, N., Berry, P., Dodman, D., Fankhauser, S., Hope, C., Kovats, S., Nicholls, R., Satterthwaite, D., Tiffin, R., and Wheeler, T. (2009). *Assessing the Costs of Adaptation to Climate Change. A review of the UNFCCC and other recent estimates.* International Institute for Environment and Development (IIED) and Grantham Institute for Climate Change.

Pelling, M. (1998). Participation, social capital and vulnerability to urban flooding in Guyana. *Journal of International Development* **10**, 469–486.

Perry, N., Rosewarne, S., and White, G. (2013). Clean energy policy: Taxing carbon and the illusion of the equity objective. *Ecological Economics* **90**, 104–113.

Petkova, E. P., Gasparrini, A., and Kinney, P. L. (2014). Heat and mortality in New York City since the beginning of the 20th century. *Epidemiology* **25**, 554–560.

Petkova, E. P., Horton, R. M., Bader, D. A., and Kinney, P. L. (2013). Projected heat-related mortality in the U.S. urban Northeast. *International Journal of Environmental Research and Public Health* **10**, 6734–(6747).

Pirard, P., Vandentorren, S., Pascal, S., Laaidi, K., Le Tertre, A., Cassadou, S., and Ledrans, M. (2005). Summary of the mortality impact assessment of the 2003 heat wave in France. *Eurosurveillance* **10**, 153–156.

Racine, M., Tousignant-Laflamme, Y., Kloda, L. A., Dion, D., Dupuis, G., and Choinire, M. (2012). A systematic literature review of 10 years of research on sex/gender and experimental pain perception – Part 1: Are there really differences between women and men? *Pain* **153**, 602–618.

Rakodi, C. (2014). Religion and social life in African cities. In Parnell, S. and Pieterse, E. (eds.), *Africa's Urban Revolution* (82–109). Zed Books Ltd.

Random House. (2014). *Random House Dictionary.com Unabridged.* Random House. Accessed March 11, 2015: http://dictionary.reference.com/browse/equality

Rao, P. (2013). Building climate resilience in coastal ecosystems in India: Cases and trends in adaptation practices. In Leal Filho, W. (ed.), *Climate Change and Disaster Risk Management* (335–349). Springer.

Rawls, J. (1971). *A Theory of Justice*. Harvard University Press.

Rawls, J. (1993). *Political Liberalism*. Columbia University Press.

Reckien, D. (2014). Weather extremes and street life in India: Implications of fuzzy cognitive mapping as a new tool for semi-quantitative impact assessment and ranking of adaptation measures. *Global Environmental Change* **26**, 1–13.

Reckien, D., Flacke, J., Dawson, R. J., Heidrich, O., Olazabal, M., Foley, A., Hamann, J. J. P., Orru, H., Salvia, M., De Gregorio Hurtado, S., Geneletti, D., and Pietrapertosa, F. (2014). Climate change response in Europe: What's the reality? Analysis of adaptation and mitigation plans from 200 urban areas in 11 countries. *Climatic Change* **122**, 331–340.

Reckien, D., Wildenberg, M., and Bachhofer, M. (2013). Subjective realities of climate change: How mental maps of impacts deliver socially sensible adaptation options. *Sustainability Science* **8**, 159–172.

Reckien, D., Wildenberg, M., and Deb, K. (2011). Understanding potential climate change impacts and adaptation options in Indian megacities. In Otto-Zimmermann, K. (ed.), *Resilient Cities: Cities and Adaptation to Climate Change: Proceedings of the Global Forum 2010* (111–121). Springer.

Reed, S. O., Friend, R., Toan, V. C., Thinphanga, P., Sutarto, R., and Singh, D. (2013). "Shared learning" for building urban climate resilience – experiences from Asian cities. *Environment and Urbanization* **25**, 393–412.

Reid, C. E., O'neill, M. S., Gronlund, C. J., Brines, S. J., Brown, D. G., Diez-Roux, A. V., and Schwartz, J. (2009). Mapping community determinants of heat vulnerability. *Environmental Health Perspectives* **117**, 1730–1736.

Revi, A., Satterthwaite, D., Aragón-Durand, F., Corfee-Morlot, J., Kiunsi, R. B. R., Pelling, M., Roberts, D. C., and Solecki, W. (2014). Urban areas. In Field, C. B., Barros, V. R., Dokken, D. J., Mach, K. J., Mastrandrea, M. D., Bilir, T. E., Chatterjee, M., Ebi, K. L., Estrada, V., Genova, R. C., Girma, B., Kissel, E. S., Levy, A. N., Maccracken, S., Mastrandrea, P. R., and White, L. L. (eds.), *Climate Change 2014: Impacts, Adaptation, and Vulnerability. Part A: Global and Sectoral Aspects. Contribution of Working Group II to the Fifth Assessment Report of the Intergovernmental Panel on Climate Change* (535–612). Cambridge University Press.

Riahi, K., Van Vuuren, D.P., Kriegler, E., Edmonds, J., O'neill, B.C., Fujimori, S., Bauer, N., Calvin, K., Dellink, R., Fricko, O., and Lutz, W. (2017). The shared socioeconomic pathways and their energy, land use, and greenhouse gas emissions implications: an overview. *Global Environmental Change*, 42, 153–168.

Roberts, D. (2008). Thinking globally, acting locally – institutionalizing climate change at the local government level in Durban, South Africa. *Environment and Urbanization* **20**, 521–537.

Roberts, D. (2010). Prioritizing climate change adaptation and local level resilience in Durban, South Africa. *Environment and Urbanization* **22**, 397–413.

Roberts, D., Boon, R., Diederichs, N., Douwes, E., Govender, N., Mcinnes, A., Mclean, C., O'donoghue, S., and Spires, M. (2012). Exploring ecosystem-based adaptation in Durban, South Africa: "Learning-by-doing" at the local government coal face. *Environment and Urbanization* **24**, 167–195.

Roberts, D., and O'Donoghue, S. (2013). Urban environmental challenges and climate change action in Durban, South Africa. *Environment and Urbanization* **25**, 299–319.

Rodin, J. (2014). *The Resilience Dividend – Being Strong in a World Where Things Go Wrong*. PublicAffairs.

Romero Lankao, P. (2010). Water in Mexico City: What will climate change bring to its history of water-related hazards and vulnerabilities? *Environment and Urbanization* **22**, 157–178.

Rosenzweig, C., Solecki, W., Hammer, S. A., and Mehrotra, S. (2010). Cities lead the way in climate-change action. *Nature* **467**, 909–911.

Rosenzweig, C., Solecki, W. D., Hammer, S. A., and Mehrotra, S. (eds.). (2011). *Climate Change and Cities – First Assessment Report of the Urban Climate Change Research Network*. Cambridge University Press.

Rouse, J., and Ali, M. (2001). *Waste Pickers in Dhaka: Using the sustainable livelihoods approach–Key findings and field notes*. WEDC, Loughborough University.

Roy, M. (2009). Planning for sustainable urbanisation in fast growing cities: Mitigation and adaptation issues addressed in Dhaka, Bangladesh. *Habitat International* **33**, 276–286.

Roy, M., Jahan, F., and Hulme, D. (2012). *Community and Institutional Responses to the Challenges Facing Poor Urban People in Khulna, Bangladesh in an Era of Climate Change. BWPI working paper*. Brooks World Poverty Institute, University of Manchester.

Ruijs, A., Zimmermann, A., and Van Den Berg, M. (2008). Demand and distributional effects of water pricing policies. *Ecological Economics* **66**, 506–516.

Safriel, U., Adeel, Z., Niemeijer, D., Puigdefabregas, J., White, R., Lal, R., Winslow, M., Ziedler, J., Prince, S., Archer, E., and King, C. (2005). Dryland systems. In Hassan, R., Scholes, R., and Ash, N. (eds.), *Millenium Ecosystems Assessment: Ecosystems and Human Well-being: Current State and Trends* (623–662). Island Press.

Sampson, N. R., Gronlund, C. J., Buxton, M. A., Catalano, L., White-Newsome, J. L., Conlon, K. C., O'neill, M. S., Mccormick, S., and Parker, E. A. (2013). Staying cool in a changing climate: Reaching vulnerable populations during heat events. *Global Environmental Change* **23**, 475–484.

Santos, G., and Rojey, L. (2004). Distributional impacts of road pricing: The truth behind the myth. *Transportation* **31**, 21–42.

Satterthwaite, D. (2013). The political underpinnings of cities' accumulated resilience to climate change. *Environment and Urbanization* **25**, 381–391.

Satterthwaite, D., Huq, S., Reid, H., Pelling, M., and Romero Lankao, P. (2007). *Adapting to Climate Change in Urban Areas: The Possibilities and Constraints in Low- and Middle-Income Nations. Human settlements discussion paper series*. IIED.

Satterthwaite, D., and Mitlin, D. (2014). *Reducing Urban Poverty in the Global South*. Routledge.

Schär, C., and Jendritzky, G. (2004). Hot news from summer 2003. *Nature* **432**, 559–560.

Scheinberg, A., and Anschütz, J. (2006). Slim pickin's: Supporting waste pickers in the ecological modernization of urban waste management systems. *International Journal of Technology Management and Sustainable Development* **5**, 257–270.

Schildberg, C. (ed.). (2014). *A Caring and Sustainable Economy*. Friedrich-Ebert-Stiftung (FES).

Schindler, M., and Caruso, G. (2014). Urban compactness and the trade-off between air pollution emission and exposure: Lessons from a spatially explicit theoretical model. *Computers, Environment and Urban Systems* **45**, 13–23.

Schlosberg, D., and Collins, L. B. (2014). From environmental to climate justice: Climate change and the discourse of environmental justice. *Wiley Interdisciplinary Reviews: Climate Change* **5**, 359–374.

Schwanen, T., Banister, D., and Anable, J. (2011). Scientific research about climate change mitigation in transport: A critical review. *Transportation Research Part A: Policy and Practice* **45**, 993–(1006).

Schweitzer, L. (2011). The empirical research on the social equity of gas taxes, emissions fees, and congestion charges. *Special Report: Equity of Evolving Transportation Finance Mechanisms*. Transportation Research Board.

Semenza, J. C., Rubin, C. H., Falter, K. H., Selanikio, J. D., Flanders, W. D., Howe, H. L., and Wilhelm, J. L. (1996). Heat-related deaths during the July 1995 heat wave in Chicago. *New England Journal of Medicine* **335**, 84–90.

Sen, A. (1999). *Development as Freedom*. Alfred A. Knopf.

Seto, K. C., Bigio, A., Bento, A., Cervero, R., Torres Martinez, J., Christensen, P., C, S. K., Dhakal, S., Bigio, A., Blanco, H., Delgado, G. C., Dewar, D., Huang, L., Inaba, A., Kansal, A., Lwasa, S., McMahon, J. E., Müller, D. B., Murakami, J., Nagendra, H., and Ramaswami, A. (2014). Human settlements, infrastructure, and spatial planning. In Edenhofer, O., Pichs-Madruga, R., Sokona, Y., Farahani, E., Kadner, S., Seyboth, K., Adler, A., Baum, I., Brunner, S., Eickemeier, P., Kriemann, B., Savolainen, J., Schlömer, S., Von Stechow, C., Zwickel, T., and Minx, J. C. (eds.), *Climate Change 2014: Mitigation of Climate*

Change. Contribution of Working Group III to the Fifth Assessment Report of the Intergovernmental Panel on Climate Change (923–1000). Cambridge University Press.

Sharpe, R. (1982). Energy efficiency use patterns and equity of various urban land. *Urban Ecology* **7**, 1–18.

Sheffield, P. E., Knowlton, K., Carr, J. L. and Kinney, P. L. (2011). Modeling of regional climate change effects on ground-level ozone and childhood asthma. *American Journal of Preventive Medicine* **41**, 251–257.

Shi, L., Chu, E., Anguelovski, I., Aylett, A., Debats, J., Goh, K., Schenk, T., Seto, K. C., Dodman, D., Roberts, D., Roberts, J. T., and Vandeveer, S. D. (2016). Roadmap towards justice in urban climate adaptation research. *Nature Climate Change* **6**, 131–137.

Shukla, P. R. (1999). Justice, equity and efficiency in climate change: A developing country perspective (134–144). In Ference, T. (ed.), *Fairness Concerns in Climate Change*. Earthscan.

Simón, F., Lopez-Abente, G., Ballester, E., and Martínez, F. (2005). Mortality in Spain during the heat waves of summer 2003. *Eurosurveill* **10**, 156–160.

Singh, J. P., and Fazel, S. (2010). Forensic risk assessment: A meta review. *Criminal Justice and Behavior* **37**, 965–988.

Sister, C., Wolch, J., and Wilson, J. (2010). Got green? Addressing environmental justice in park provision. *GeoJournal* **75**, 229–248.

Skeldon, R. (2013). *Global Migration: Demographic Aspects and Its Relevance for Development*. UN DESA Technical paper 2013/6. Accessed September 28, 2014: www.un.org/esa/population/migration/documents/EGM.Skeldon_17.12.2013.pdf

Small, C. (2005). A global analysis of urban reflectance. *International Journal of Remote Sensing* **26**, 661–681.

Smith, K. R., Swisher, J., and Ahuja, D. R. (1993). Who pays (to solve the problem and how much)? InHayes, P., and Smith, K. R. (eds.), *The Global Greenhouse Regime: Who Pays?* (70–98). London: Earthscan.

Smith, B., Brown, D., and Dodman, D. (2014a). *Reconfiguring Urban Adaptation Finance. IIED working paper*. IIED.

Smith, J. J., and Gihring, T. A. (2006). Financing transit systems through value capture: An annotated bibliography. *American Journal of Economics and Sociology* **65**, 751–786.

Smith, K. R., Woodward, A., Campbell-Lendrum, D., Chadee, D. D., Honda, Y., Liu, Q., Olwoch, J. M., Rewich, B., and Sauerborn, R. (2014b). Human health: Impacts, adaptation, and co-benefits. In Field, C. B., Barros, V. R., Dokken, D. J., Mach, K. J., Mastrandrea, M. D., Bilir, T. E., Chatterjee, M., Ebi, K. L., Estrada, Y. O., Genova, R. C., Girma, B., Kissel, E. S., Levy, A. N., Maccracken, S., Mastrandrea, P. R., and White, L. L. (eds.). *Climate Change 2014: Impacts, Adaptation, and Vulnerability. Part A: Global and Sectoral Aspects. Contribution of Working Group II to the Fifth Assessment Report of the Intergovernmental Panel on Climate Change* (709–754). Cambridge University Press.

Smyth, H. (1996). Running the Gauntlet: A compact city within a doughnut of decay. In Jenks, M., Burton, E., and William, K. (eds.), *The Compact City: A Sustainable Urban Form?* Spon.

Solecki, W. (2012). Urban environmental challenges and climate change action in New York City. *Environment and Urbanization* **24**, 557–573.

Soltau, F. (2009). *Fairness in International Climate Change Law and Policy*. Cambridge University Press.

Sovacool, B. K., Linner, B. -O., and Goodsite, M. E. (2015). The political economy of climate adaptation. *Nature Climate Change* **5**, 616–618.

Storey, D., Santucci, L., Aleluia, J., and Varghese, T. (2013). Decentralized and integrated resource recovery centers in developing countries: Lessons learnt from Asia-Pacific. *International Solid Waste Association (ISWA) Congress*. United Nations Economic and Social Commission for Asia and the Pacific (ESCAP).

Sultana, F. (2013). Gendering climate change: Geographical insights. *The Professional Geographer* **66**, 1–10.

Tanner, T., Mitchell, T., Polack, E., Guenther, B., Tanner, T., Mitchell, T., Polack, E., and Guenther, B. (2009). *Urban Governance for Adaptation: Assessing Climate Change Resilience in Ten Asian Cities. IDS working paper.* Institute of Development Studies.

Tavares, A. O., and Santos, P. P. D. (2013). Re-scaling risk governance using local appraisal and community involvement. *Journal of Risk Research* **17**, 923–949.

Taylor, J. (2013). *When Non-climate Urban Policies Contribute to Building Urban Resilience to Climate Change: Lessons Learned from Indonesian Cities. Asian cities climate resilience working paper series.* IIED.

Temmerman, S., Meire, P., Bouma, T. J., Herman, P. M. J., Ysebaert, T., and De Vriend, H. J. (2013). Ecosystem-based coastal defence in the face of global change. *Nature* **504**, 79–83.

Todes, A. (2012). Urban growth and strategic spatial planning in Johannesburg, South Africa. *Cities* **29**, 158–165.

Tompkins, E. L., Mensah, A., King, L., Long, T. K., Lawson, E. T., Hutton, C., Hoang, V. A., Gordon, C., Fish, M., Dyer, J., and Bood, N. (2013). *An investigation of the evidence of benefits from climate compatible development.* Sustainability Research Institute.

Tong, S., Wang, X. Y., and Barnett, A. G. (2010). Assessment of heat-related health impacts in Brisbane, Australia: Comparison of different heat-wave definitions. *PLoS ONE* 5.

Tovar-Restrepo, M. (2010). Climate change and indigenous women in Columbia. In Dankelman, I. (ed.), *Gender and Climate Change: An Introduction* (145–152). Earthscan.

Tran, K. V., Azhar, G. S., Nair, R., Knowlton, K., Jaiswal, A., Sheffield, P., Mavalankar, D., and Hess, J. (2013). A cross-sectional, randomized cluster sample survey of household vulnerability to extreme heat among slum dwellers in Ahmedabad, India. *International Journal of Environmental Research and Public Health* **10**, 2515–2543.

Tronto, J. (1993). *Moral boundaries: A political argument for an ethic of care.* Routledge.

Tyler, S., and Moench, M. (2012). A framework for urban climate resilience. *Climate and Development* **4**, 311–326.

UCLG (ed.). (2014). *Basic Services for All in an Urbanizing World.* Routledge.

United Nations, Department of Economics and Social Affairs (UNDESA), Population Division (2014). *World Urbanization Prospects: The 2014 Revision, Highlights.* United Nations.

United Nations, Department of Economics and Social Affairs (UNDESA), Population Division (2015). *World Population Prospects: The 2015 Revision.* United Nations.

UNDP. (2004). *Reducing Disaster Risk: A Challenge for Development.* United Nations.

UNDP. (2009). *Resource Guide on Gender and Climate Change.* UNDP.

UNEP. (2014b). *Global Risk Data Platform.* United Nations Environment Program (UNEP)/GRID.

UN-Habitat. (2008a). *The State of African Cities: A Framework for Addressing Challenges in Africa.* UN-Habitat.

UN-Habitat. (2008b). *State of the World's Cities 2008/2009 – Harmonious Cities.* UN-Habitat.

UN-Habitat. (2010). *State of the World's Cities 2010/2011 – Bridging the Urban Divide.* UN-Habitat.

UN-Habitat. (2011).*Global Assessment Report: Cities and Climate Change.* UN-Habitat.

UN-Habitat. (2013a). *Planning and Design for Sustainable Urban Mobility. Global Report on Human Settlements.* UN-Habitat.

UN-Habitat. (2013b). *State of the World's Cities 2012/2013 – Prosperity of Cities.* United Nations Human Settlements Programme.

UNIFEM. (2008). *Who Answers to Women? Gender and Accountability: Progress of the World's Women 2008/2009.* UNIFEM.

UNISDR. (2009). *Global Assessment Report on Disaster Risk Reduction 2009: Risk and Poverty in a Changing Climate.* United Nations.

UNISDR. (2011). *Global Assessment Report on Disaster Risk Reduction: Revealing Risk, Redefining Development.* United Nations.

U.S. Census Bureau. (2010). Decennial census, summary file 1. Accessed July 2, 2014: http://www.census.gov

U.S. Environmental Protection Agency (EPA). (2011). Environmental justice: Frequently asked questions. Accessed August 14, 2014: http://www.epa.gov/earth1r6/6dra/oejta/ej/ejfaq.htm#What_is_Environmental

Vairavamoorthy, K., Gorantiwar, S. D., and Pathirana, A. (2008). Managing urban water supplies in developing countries – Climate change and water scarcity scenarios. *Physics and Chemistry of the Earth, Parts A/B/C* **33**, 330–339.

Vandentorren, S., Bretin, P., Zeghnoun, A., Mandereau-Bruno, L., Croisier, A., Cochet, C., Ribéron, J., Siberan, I., Declercq, B., and Ledrans, M. (2006). August 2003 heat wave in France: Risk factors for death of elderly people living at home. *European Journal of Public Health* **16**, 583–591.

Vergara, S. E., and Tchobanoglous, G. (2012). Municipal solid waste and the environment: A global perspective. *Annual Review of Environment and Resources* **37**, 277–309.

Victorian Government Department of Human Services. (2009). *January 2009 Heatwave in Victoria: An Assessment of Health Impacts.* State Government of Victoria.

Walls, M., and Hanson, J. (1999). Distributional aspects of an environmental tax shift: The case of motor vehicle. *National Tax Journal* **52**, 53–65.

Wamsler, C., and Brink, E. (2014). Moving beyond short-term coping and adaptation. *Environment and Urbanization* **26**, 86–111.

Wang, S. (2013). Efficiency and equity of speed limits in transportation networks. *Transportation Research Part C: Emerging Technologies* **32**, 61–75.

WEDO, IUCN, and GGCA. (2013). *Linking Data and Actions – Connections between IPCC AR5 Data, Gender Differentiated Data and Climate Change Actions.* WEDO, IUCN, GGCA.

WEDO and UNFPA. (2009). *Climate Change Connections.* WEDO, UNFPA.

Wendell, C. (2011). *The Housing Crash and Smart Growth. Policy report.* National Center for Policy Analysis.

Wheeler, D. (2011). *Quantifying Vulnerability to Climate Change: Implications for Adaptation Assistance. Working paper.* Center for Global Development.

White-Newsome, J., O'neill, M. S., Gronlund, C., Sunbury, T. M., Brines, S. J., Parker, E., Brown, D. G., Rood, R. B., and Rivera, Z. (2009). Climate change, heat waves, and environmental justice: Advancing knowledge and action. *Environmental Justice* **2**, 197–205.

White-Newsome, J. L., Sánchez, B. N., Jolliet, O., Zhang, Z., Parker, E. A., Timothy Dvonch, J., and O'Neill, M. S. (2012). Climate change and health: Indoor heat exposure in vulnerable populations. *Environmental Research* **112**, 20–27.

Wilson, D. C., Velis, C., and Cheeseman, C. (2006). Role of informal sector recycling in waste management in developing countries. *Habitat International* **30**, 797–808.

Wilson, S., Hutson, M., and Mujahid, M. (2008). How planning and zoning contribute to inequitable development, neighborhood health, and environmental injustice. *Environmental Justice* **1**, 211–216.

Wolch, J. R., Byrne, J., and Newell, J. P. (2014). Urban green space, public health, and environmental justice: The challenge of making cities 'just green enough'. *Landscape and Urban Planning* **125**, 234–244.

World Bank. (2008). *Climate-Resilient Cities: 2008 Primer.* World Bank.

World Bank (2015). *East Asia's Changing Urban Landscape – Measuring a Decade of Spatial Growth. Urban Development Series.* World Bank.

Yang, J., Liu, H. Z., Ou, C. Q., Lin, G. Z., Ding, Y., Zhou, Q., Shen, J. C. and Chen, P. Y. (2013). Impact of heat wave in 2005 on mortality in Guangzhou, China. *Biomedical and Environmental Sciences: BES,* **26**, 647–654.

Zhang, M., and Wang, L. (2013). The impacts of mass transit on land development in China: The case of Beijing. *Research in Transportation Economics* **40**, 124–133.

Ziervogel, G., Shale, M., and Du, M. (2010). Climate change adaptation in a developing country context: The case of urban water supply in Cape Town. *Climate and Development* **2**, 94–110.

Chapter 6   Case Study References

Case Study 6.1   Building Climate Justice in New York: NYC-EJA's Waterfront Justice Project, The Sandy Regional Assembly and the People's Climate March

Bautista, E., Hanhardt, E., Osorio, J. C., and Dwyer, N. (2014). New York City Environmental Justice Alliance (NYC-EJA) Waterfront Justice Project. *Local Environment: The International Journal of Justice and Sustainability* 1–19.

Bautista, E., Osorio, J. C., and Dwyer, N. (2015). Building climate justice and reducing industrial waterfront vulnerability. *Social Research: An International Quarterly* 82 (3), 821–838.

Fahmi, F. Z., Hudalah, D., Rahayu, P., and Woltjer, J. (2014). Extended urbanization in small and medium-sized cities: The case of Cirebon, Indonesia. *Habitat International* **42**, 1–10.

Kinney, P. L., Matte, T., Knowlton, K., Madrigano, J., Petkova, E., Weinberger, K., Quinn, A., Arend, M., and Pullen, J. (2015). Public health impacts and resiliency. *Annals of the New York Academy of Sciences* **1336**, 67–88.

Maantay, J. A. (2002). Industrial zoning changes in New York City and environmental justice: A case study in "expulsive" zoning. *Projections: The Planning Journal of Massachusetts Institute of Technology (MIT) 63–108. (Special issue: Planning for Environmental Justice).*

New York City Department of Health and Mental Hygiene (NYC-DOHMH). (2010). Community health survey data. Accessed March 27, 2015: http://www.nyc.gov/html/doh/downloads/pdf/epi/nyc_commhealth_atlas09.pdf

New York City Special Initiative for Rebuilding and Resiliency (SIRR). (2013). A stronger more resilient New York. Accessed March 16, 2015: http://www.nyc.gov/html/sirr/html/report/report.shtml

New York City Panel on Climate Change (NPCC). (2013). Climate risk information 2013 Observations, climate change projections, and maps. City of New York Special Initiative on Rebuilding and Resiliency. Accessed May 22, 2015: http://www.nyc.gov/html/planyc2030/downloads/pdf/npcc_climate_risk_information_2013_report.pdf

Peel, M. C., Finlayson, B. L., and McMahon, T. A. (2007). Updated world map of the Köppen-Geiger climate classification. *Hydrology and Earth System Sciences Discussions* 4(2), 462.

Sandy Regional Assembly. (2013). Sandy regional assembly recovery agenda. Accessed June 18, 2015: http://www.nyc-eja.org/public/publications/SandyRegionalAssembly_SIRRAnalysis.pdf/

U.S. Census Bureau. (2010). Decennial census, summary file 1. Accessed May 3, 2014: http://www.census.gov/population/metro/files/CBSA%20Report%20Chapter%203%20Data.xls

U.S. Census Bureau. (2016). Annual Estimates of the Resident Population: April 1, 2010 to July 1, 2016. Accessed March 27, 2017: https://factfinder.census.gov/faces/tableservices/jsf/pages/productview.xhtml?pid=PEP_2016_PEPANNRES&prodType=table

World Bank. (2017). 2016 GNI per capita, Atlas method (current US$). Accessed August 9, 2017: http://data.worldbank.org/indicator/NY.GNP.PCAP.CD

Case Study 6.2   Citizen-led Mapping of Urban Metabolism in Cairo

Brookings Institute. (2012). Global MetroMonitor. Accessed May 3, 2014: http://brook.gs/1IqM6Ns

Cairo Governorate. (2014). Cairo portal. Accessed October 27, 2015: www.cairo.gov.eg.

Central Agency for Public Mobilization and Statistics. (2015). Population estimates 2015. Accessed March 11, 2016: www.capmas.gov.eg

Global Construction Review. (2014). Arabtec signs $40bn Egyptian housing contract. Accessed May 28, 2015: http://bit.ly/1CmU0FF

Giza Governorate. (2015). Giza governorate website. Accessed March 30, 2016: www.giza.gov.eg

Global Construction Review. (2014). Arabtec signs $40bn Egyptian housing contract. Accessed May 25, 2015: http://bit.ly/1CmU0FF

Khalil, H. (2010). New urbanism, smart growth and informal areas: A quest for sustainability. In Lehmann, S., AlWaer, H., and Al-Qawasmi, J. (eds.), *Sustainable Architecture & Urban Development* (137–156). CSAAR.

Peel, M. C., Finlayson, B. L., and McMahon, T. A. (2007). Updated world map of the Köppen-Geiger climate classification. *Hydrology and Earth System Sciences Discussions* 4 (2), 462.

Qaliobia Governorate. (2015). Qaliobia gOVERNORATE website. Accessed February 2, 2016: www.qaliobia.gov.eg

Sankey diagrams. (2014). Accessed April 17, 2016: http://bit.ly/1BU1qSm

World Bank. (2014). Country at a glance: Egypt. Accessed May 25, 2015: http://bit.ly/1CZXdNj

World Bank. (2017). 2016 GNI per capita, Atlas method (current US$). Accessed August 9, 2017: http://data.worldbank.org/indicator/NY.GNP .PCAP.CD

Case Study 6.3 Growth Control, Climate Risk Management, and Urban Equity: The Social Pitfalls of the Green Belt in Medellín

Agudelo Patiño, L. C. (2013). Formulación del Cinturón Verde Metropolitano del Valle de Aburrá. Área Metropolitana del Valle de Aburrá and Universidad Nacional, Medellín.

Alcaldía de Medellin. (2013). *Presentación Cinturón Verde Metropolitano.* Accessed February 6, 2015: http://www.medellincomo vamos.org/download/presentacion-cinturon-verde-metropolitano-parte-4-2013/

Arango, S. (2012). Radiografía al Cinturón Verde Metropolitano. La Ciudad Verde. Accessed February 16, 2015: http://tinyurl.com/kfbhvuq

Peel, M. C., Finlayson, B. L., and McMahon, T. A. (2007). Updated world map of the Köppen-Geiger climate classification. *Hydrology and Earth System Sciences Discussions* **4** (2), 462.

Urban Land Institute (2013). Medellín voted City of the Year. Accessed April 6, 2014: http://uli.org/urban-land-magazine/medellin-named-most-innovative-city/

World Bank. (2017). 2016 GNI per capita, Atlas method (current US$). Accessed August 9, 2017: http://data.worldbank.org/indicator/NY .GNP.PCAP.CD

Case Study 6.4 Individual, Communal and Institutional Responses to Climate Change by Low-Income Households in Khulna, Bangladesh

Bangladesh Bureau of Statistics (BBS). (2013). Population census 2011 (Dhaka and Khulna). Ministry of Planning, Government of the People's Republic of Bangladesh.

Dodman, D., Mitlin, D., and Rayos, C. J. (2010). Victims to victors, disasters to opportunities: Community-driven responses to climate change in the *Philippines. International Development Planning Review* **32**(1), 1–26.

Peel, M. C., Finlayson, B. L., and McMahon, T. A. (2007). Updated world map of the Köppen-Geiger climate classification. *Hydrology and Earth System Sciences Discussions*, **4**(2), 462.

World Bank. (2017). 2016 GNI per capita, Atlas method (current US$). Accessed August 9, 2017: http://data.worldbank.org/indicator/NY.GNP .PCAP.CD

Case Study 6.5 Public-Private-People Partnerships for Climate Compatible Development (4PCCD) in Maputo, Mozambique

Castán Broto, V., Oballa, B., and Junior, P. (2013). Governing climate change for a just city: Challenges and lessons from Maputo, Mozambique. *Local Environment: The International Journal of Justice and Sustainability* **18**, 678–704.

Castán Broto, V., Boyd, E., and Ensor, J. (2015). Participatory urban planning for climate change adaptation in coastal cities: Lessons from a pilot experience in Maputo, Mozambique. *Current Opinion in Environmental Sustainability* **13**, 11–18.

Demographia. (2016). *Demographia World Urban Areas 12th Annual Edition: 2016:04.* Accessed August 9, 2017: http://www.demographia .com/db-worldua.pdf

Instituto Nacional da Estatística. (2007). Recenseamento Geral Da População E Habitação (2007). Accessed September 20, 2014: http://www .ine.gov.mz/estatisticas/estatisticas-demograficas-e-indicadores-sociais/populacao/projeccoes-da-populacao

Peel, M. C., Finlayson, B. L., and McMahon, T. A. (2007). Updated world map of the Köppen-Geiger climate classification. *Hydrology and Earth System Sciences Discussions* **4**(2), 462.

Maputo Municipal Council, UN-Habitat and Agriconsulting. (2012). *Availação detalhada dos impactos resultantes dos eventos das mudanças climáticas no Município de Maputo.* UN-Habitat.

World Bank. (2017). 2016 GNI per capita, Atlas method (current US$). Accessed August 9, 2017: http://data.worldbank.org/indicator/NY .GNP.PCAP.CD

7

Economics, Finance, and the Private Sector

Coordinating Lead Authors

Reimund Schwarze (Leipzig), Peter B. Meyer (New Hope), Anil Markandya (Bilbao/Bath)

Lead Authors

Shailly Kedia (New Delhi), David Maleki (Washington, D.C.), María Victoria Román de Lara (Bilbao), Tomonori Sudo (Tokyo), Swenja Surminski (London)

Contributing Authors

Nancy Anderson (New York), Marta Olazabal (Bilbao), Stelios Grafakos (Rotterdam/Athens), Saliha Dobardzic (Washington, D.C.)

This chapter should be cited as

Schwarze, R., Meyer, P. B., Markandya, A., Kedia, S., Maleki, D., Román de Lara, M. V., Sudo, T., and Surminski, S. (2018). Economics, finance, and the private sector. In Rosenzweig, C., W. Solecki, P. Romero-Lankao, S. Mehrotra, S. Dhakal, and S. Ali Ibrahim (eds.), *Climate Change and Cities: Second Assessment Report of the Urban Climate Change Research Network*. Cambridge University Press. New York. 225–254

Financing Climate Change Solutions in Cities

Since cities are the locus of large and rapid socioeconomic development around the world, economic factors will continue to shape urban responses to climate change. To exploit response opportunities, promote synergies among actions, and reduce conflicts, socioeconomic development must be integrated with climate change planning and policies.

Public-sector finance can facilitate action, and public resources can be used to generate investment by the private sector. But private-sector contributions to mitigation and adaptation should extend beyond financial investment. The private sector should also provide process and product innovation, capacity building, and institutional leadership.

Major Findings

- Implementing climate change mitigation and adaptation actions in cities can help solve other city-level development challenges, such as major infrastructure deficits. Assessments show that meeting increasing demand will require more than a doubling of annual capital investment in physical infrastructure to more than US$20 trillion by 2025, mostly in emerging economies. Estimates of global economic costs from urban flooding due to climate change are approximately US$1 trillion a year.

- Cities cannot fund climate change responses on their own. Multiple funding sources are needed to deliver the large infrastructure financing that is essential to low-carbon development and climate risk management in cities. Estimates of the annual cost of climate change adaptation range between US$80 billion and US$100 billion, of which about 80% will be borne in urbanized areas.

- Public–private partnerships are necessary for effective action. Partnerships should be tailored to the local conditions in order to create institutional and market catalysts for participation.

- Regulatory frameworks should be integrated across city, regional, national, and international levels to provide incentives for the private sector to participate in making cities less carbon-intensive and more climate-resilient. The frameworks need to incorporate mandates for local public action along with incentives for private participation and investment in reducing business contributions to emissions.

- Enhancing credit worthiness and building the financial capacity of cities are essential to tapping the full spectrum of resources and raising funds for climate action.

Key Messages

Financial policies must enable local governments to initiate actions that will minimize the costs of climate impacts. For example, the cost of inaction will be very high for cities located along coastlines and inland waterways due to rising sea levels and increasing risks of flooding.

Climate-related policies should also provide cities with local economic benefits as cities shift to new infrastructure systems associated with low-carbon development.

Networks of cities play a crucial role in accelerating the diffusion of good ideas and best practices to other cities, both domestically and internationally. Cities that initiate actions that lead to domestic and international implementation of nationwide climate change programs should be rewarded.

7.1 Introduction

This chapter provides an overview of the economic factors shaping urban responses to climate change. In four parts, it moves from a description of how climate change shapes urban economic options to an examination of the ways in which public-sector finance can facilitate action, including the use of public monies to generate private-sector investment. It then examines the many roles the private sector can play and ends with conclusions and policy recommendations.

7.1.1 Economic Challenges and Opportunities for Cities

Cities are where the biggest and fastest socioeconomic changes take place. Global climate risk is accumulated in urban areas because people, private and public assets, and economic activities become more concentrated there (Mehrotra et al., 2009; Revi et al., 2014). With rapidly expanding populations, the influence of cities will only grow in the 21st century. This dynamic, coupled with the increasing threat of climate change, puts cities at risk of major social and economic disruption in the absence of sound plans for climate change mitigation and adaptation.

Mitigation and adaptation are closely related tasks at the city level in terms of synergies and conflicts. They can be complementary (as with energy efficiency measures) or in conflict (as when climate adaptation relies on air conditioning). At the same time, measures to improve carbon sequestration in urban green spaces must be coherently connected with the goals of biodiversity and urban development. To exploit synergies and reduce conflicts, we must integrate socioeconomic development with climate change policies. The Urban Climate Change Research Network (UCCRN) conceives this as a joint effort: one without the other would be insufficient. More detailed analysis of the interrelationships (synergies and conflicts) of mitigation and adaptation can be found in Chapter 4 (Mitigation and Adaptation) of this volume.

In most countries, national or federal Ministries of Environment are recognized as responsible for climate change, whereas in reality such activities are far more distributed across multiple levels of public-sector jurisdictions (see Chapter 16, Governance and Policy). It is not clear how national-level climate change budgets – if any – will finance city initiatives. Thus, we emphasize the importance of citywide approaches to economic decision-making and finance to meet insurgent climate challenges. We recommend pursuit of policies that:

1. Enable local governments to initiate climate action. For cities historically located near water (such as ports and/or along rivers for waterpower), the costs of inaction will be very high in the face of rising seas and inland flooding, with the global economic costs from flooding in cities due to climate change projected to amount to US$1 trillion a year by mid-century (World Bank Group, 2013a).

2. Provide cities with local economic development benefits as economies shift to new realities associated with weaning themselves from reliance on fossil fuel.

3. Reward cities that initiate actions that can lead to domestic implementation of nationwide climate change programs.

4. Empower cities to stimulate and accelerate the diffusion of good ideas and best practices to other cities internationally and thus to lead by example.

7.1.2 The Costs of Inaction

Climate change and extreme weather patterns in cities across the globe have already demonstrated the risk of major socioeconomic disruptions in urban areas. Because of their spatially integrated infrastructure, city economic sectors are strongly interdependent and interlinked, sharing many potential consequences from climate change although individual neighborhoods may be less tightly linked. While effective adaptation methods, notably diversification of the city economy away from dependence on one sector, can reduce vulnerability, such moves are risky and add policy uncertainty. Uncertainty about the extent of change and costs from inaction as well as the inability to reliably calculate costs and benefits cannot be a barrier to action: cities should act on their current knowledge and evolve their approaches as new evidence and scenarios emerge (Rosenzweig and Solecki, 2014).

Hallegatte et al. (2008) elaborate the importance of "localizing" understanding of benefits and costs associated with climate change action plans by tapping into regional business, lawmaker, and stakeholder knowledge and experience. They argue for downscaling present knowledge of global and regional climate and socioeconomic scenarios to the municipal or even neighborhood level in order to inform local discourse about mitigation and adaptation plans. Cost-benefit analysis at the city-level must also include a vision of how different local communities will develop over the course of time as demographic, economic, and technological changes occur (UN Framework Convention on Climate Change [UNFCCC], 2011). A city's ripple effects on its surroundings must also be included in the analysis because cities are not isolated economic systems, but part of an interrelated spatial network.

Cost-benefit analysis (CBA), however, can never be more than a partial contributor to decision-making at the city level because many systemic effects cannot be quantified and monetized. Consider lives lost due to climate disasters: economists' measures are fraught with ethical, cultural, and social differences across any city population (Hallegatte, 2006). Valuation of the future poses another barrier to decision-making using CBA; discounting involves incorporating the fact that future costs and benefits are less important to current decision-makers than are those impacts that occur more immediately. The problem is that even massive disruptions in the distant future can be rendered insignificant in current terms in CBA with even a relatively low discount rate. Multi-criteria analysis (MCA), which considers noneconomic impacts and those not easily aggregated, is often used to complement CBA in order to incorporate aspects that cannot be quantified or monetized and, at the same time, to enhance stakeholders' participation (Bell et al., 2003).

7.1.3 Economic Development Benefits

City economic development and efforts to improve energy efficiency in transportation or buildings, generate renewable energy, and adapt to imminent threats from climate change recognize a range of local economic benefits beyond simple financial returns. Unlike financial incentive programs and infrastructure investments that depend on returns from the attraction or retention of successful businesses, many climate change investments, in addition to reducing carbon emissions, provide direct short- and medium-term secondary local economic benefits such as potential expanded sales and employment gains associated with lower business energy costs, multiplier effects of recirculating more local income through reduced utility bills, and the benefits of more stable and predictable energy supplies and costs (Meyer et al., 2013a). Furthermore, evidence shows that more compact urban growth combined with mixed-use development and efficient public transport systems can not only increase economic productivity and generate other benefits, but also have a substantial impact on reducing carbon emissions (Floater et al., 2014a) (see Chapter 5, Urban Planning and Design). These benefits include:

Job creation: Direct jobs in new construction and more permanent positions are associated with the continued operation of any new economic activity involving operation and maintenance of new transit systems, energy management for buildings, or distributed energy systems. Additional indirect jobs are created as the result of the new local spending generated from the payroll associated with the direct jobs. This growth further fuels urban and regional economic gains if it is not offset by loss of fossil fuel industry jobs (van den Berge, 2010).

Energy cost savings: Reducing a city's traditional energy consumption through efficiency gains or renewable power generation can expand the local economy simply by saving money for local utility customers. Any changes that increase the ability of a local economy to provide for its own needs keeps money within the local economy longer, thus multiplying the direct impact of increased payrolls or energy-cost savings and raising local well-being.

Higher energy cost certainty: The cost of consumed power is determined in part by fuel cost – and climate-friendly solar, wind, and geothermal fuel supplies are free. Once a renewable energy infrastructure is in place, there is lower risk of rising energy-supply costs due to an increase in fuel or carbon prices. This increased certainty, independent of any cost savings, reduces overall climate change risks and facilitates further response by eliminating the costs of hedging fluctuating fossil fuel prices and utility bills (Grafakos and Flamos, 2015).

Higher electricity supply certainty: Distributed energy and local microgrids made possible by renewable energy provide both climate change adaptation and mitigation returns. Renewables replace fossil fuels and reduce emissions, but the microgrids can also provide electricity service when the grid is down due to weather damage or other system problems, a major climate change adaptive response (see Chapter 12, Urban Energy).

Improved local business competitiveness: Local businesses benefit from climate change adaptation and mitigation investments because these enable them to outperform market competitors, to expand the range of goods and services they offer, or to broaden the markets they serve. Savings linked to lower energy consumption or greater power certainty thus can help to further expand the local economy.

Improved property values and higher tax revenues: Given limited funds available for occupancy of premises, lower power bills make higher-cost mortgages or rentals more affordable for households and businesses alike. Lower mortgage costs may have the same effect, so urban areas may take advantage of national or subnational policies, such as the availability of lower interest rates on government-insured energy efficiency mortgages in the United States, to reduce housing costs by investing in energy efficiency (Meyer et al., 2013a). Buyers and renters with more money to spend may drive up property valuations (Pivo, 2014; Fuerst et al., 2013). Those higher prices may be important to local revenues if real estate value is taxed. If business expansions and associated growth in payrolls are made possible by lower operating costs associated with responses to climate change, they may contribute to government revenues via sales or value-added taxation and/or payrolls and profits, if those gains are subject to taxation.

Sectoral clustering: Enterprises select locations based on the availability of markets, of resource supplies that are expensive to move, and/or of specialized technological knowledge and worker skills. Promoting climate change response investments may enable development efforts to stimulate new sectoral clusters presenting opportunities to a city's economy. An example of climate technology clustering can be found in Copenhagen (Floater et al., 2014a).

Marketing and reputation: Success often breeds success, and special features of a winning economic development effort can provide the basis for a city's marketing campaign that complements and strengthens local efforts to provide new economic opportunities. Cities and local authorities pursuing new investments in real estate or infrastructure that qualify as "zero net carbon" or promoting other "green" or "climate proof" characteristics often are doing so for the reputational gains, not just the environmental benefits. That may explain why many of these efforts are located in the economic development, rather than the environment or sustainability offices in their local governments. These green investments are intended to promote the attraction and retention of skilled workers and businesses.

Quality of life: Actions taken to address climate change can also benefit urban areas through such outcomes as reduced health costs from air pollution, lower construction costs from more compact urban development, increased social inclusion from higher-quality housing and better public transport links, amenity opportunities and reduction of heat island effect from urban greening, and improvements in environmental, social, and economic equity (see Chapter 6, Equity and Environmental Justice; Chapter 10, Urban Health; and Chapter 11, Housing and Informal Settlements).

7.1.4 Leading Domestic Implementation

Examples of cities more actively engaged in climate policy than their own national governments are quite common (Bulkeley, 2010). This is unsurprising because cities often bear the brunt of concentrated effects of extreme climate events. Such events are less significant at the national than the local scale, so cities may take the lead in climate change action, providing great "bottom-up" learning. There is evidence of proactive cities demonstrating (1) planning and policy implementation for adaptation, risk management and disaster response; and (2) target setting of climate goals (e.g., greenhouse gas [GHG] emissions reduction targets and other mitigation efforts) (Erickson and Tempest, 2014).

Rio de Janeiro has put in place a cross-sectoral Low-Carbon City Development Program (LCCDP), which is an ISO-compliant environmental management system helping the city government to plan, implement, monitor, and account for low-carbon investments and climate change mitigation actions across all sectors in the city over time. The LCCDP Assessment Protocol is aligned with ISO 14064–2, ISO 14001, and the World Resources Institute and World Business Council for Sustainable Development Greenhouse Gas Protocol. Although flexibility has been at the core of the Protocol design and its application – in order to meet the needs of cities globally in developing city-wide, low-emission strategies, regardless of their scale – the requirements of the LCCDP Assessment Protocol provide a concrete and tangible way for cities to operationalize what might otherwise remain vaguely described and summarized as low-carbon development efforts (Scholz and Sugar, 2012; Scholz et al., 2014).

Portland, Oregon, established a locally generated green building program to enhance competitiveness. The city pursued a strategy of incentivizing nationally prominent energy-efficient buildings while developing a local green economic sector that attracted specialized companies and labor from around the country (Allen and Potiowsky, 2008). While Denmark and Sweden are energetically pursuing climate policies at the national level, Copenhagen and Stockholm are pursuing even more stringent policies than their national governments. The 600 local governments in the United States that have established climate change plans have enjoyed far greater success in effective implementation of actions to mitigate carbon emissions than cities without such plans (Millard-Ball, 2012).

Evidence suggests that aspirations of local governments to achieve significant reductions of carbon emissions and reduction

of climate change risk often lead them to outperform their countries. According to a recent Stockholm Environment Institute study, city actions could decrease global GHG emissions by 7 Gigatons of carbon dioxide equivalents ($GtCO_2e$) below what national actions are currently on track to achieve in 2030, and by 13 $GtCO_2e$ in 2050 (Erickson and Tempest, 2014).

7.1.5 Accelerating Diffusion of Innovations

If the economic development benefits of climate actions are realized, climate-active cities will gain relative to those that do not undertake comparable efforts. Their successes are likely to have a demonstration effect on cities that have not yet addressed climate change (Rosenbloom, 2008). In turn, this will create interurban competition for economic activity and populations that will drive more city climate change efforts. The first generation of global impacts arising from the efforts of a small number of cities will be multiplied as other cities adopt their practices or adapt them to their situations. In some places, legal constraints or unique environmental conditions may constrain local authority participation, but, in an increasingly globalized economy linked by the flow of goods, services, and now by electronic information, rapid diffusion of knowledge about the successes of initial initiatives should be expected.

The issue, then, becomes how to stimulate and accelerate that diffusion. Assuring financial capacity to act is an essential first step.

Cities could potentially be key allies for disseminating and diffusing a wide range of disruptive innovations that provide jobs for the poor, support green investments, and buck the trend toward building massive, poorly performing infrastructure. Examples of actions that could also address both mitigation and adaptation include distributed solar power, rooftop gardens, urban farming, green spaces through recycled water, cashless transactions, and innovative mobility solutions, as well as recapturing public space for greening and reuse of streets formerly dedicated to transport, such as for street art, street cuisine, and theater (see Chapter 5, Urban Planning and Design).

7.2 Role of Funding and Finance

Adequate financial resources are essential to undertake climate change mitigation and adaptation activities, particularly for those cities in low-income countries that also need to reduce poverty and ensure economic and social development (see Table 7.1)

Access to such financial resources will become increasingly significant as cities take on more responsibilities to mitigate GHG emissions and adapt to ongoing changes. This is to be expected because the impacts of climate change are felt at the local level, and city governments typically are the first responders (Rosenzweig and Solecki, 2014); those impacts will become more prevalent. At the same time, existing development demands and day-to-day tasks have already strained the financial capacity of many cities, while external funding for specific climate change programs is scarce.

Table 7.1 *Costs for climate change actions and green bonds subsidies. Source: OECD, Organization for Economic Co-operation and Development*

Mitigation	Scale	Timeframe	Cost (US$)	Source
Investment in capital infrastructure	Global	–	$25 trillion	World Bank Group (2013)
Investment gap without climate change	Global	Annual	$1 trillion	World Bank Group
Potential savings in terms of energy costs	Global	–	$950 million	Climate Major Group
Money used for climate-related projects and programs	Cities	2012	$359 billion	Climate Policy Initiative
Impacts and Adaptation	**Scale**	**Timeframe**	**Cost (US$)**	**Source**
Costs from flooding due to climate change per year (impacts)	Global	2025	$25 trillion	World Bank Group (2013)
Response to anticipated 2°C temperature rise	Global	–	$14 trillion	World Economic Forum
Annual cost of climate change adaptation	Global	Annual	$80–100 billion	–
Average annual infrastructure investment needed for 2°C temperature rise	Cites	Annual	$5.7 trillion	World Economic Forum
Green Bonds	**Scale**	**Timeframe**	**Cost (US$)**	**Source**
Issuance of green bonds by State of Massachusetts	Regional	2013	$100 million	Climate Bond Initiative
HSBC estimate for green bond issuance by end of 2014	Cities	2015	$40 billion	Climate Bond Initiative
HSBC estimate for green bond issuance for 2015	Cities	2016	$100 billion	Soffiatti (2012); Merk et al. (2012)
Funds used by UK Green Investment Bank for modern green infrastructure	Cities	–	$5.9 billion	UKgov.greeninvestmentbank.com
Issuance of green bonds Gothenburg	Cities	2013	$79 million	Climate Bond Initiative
Issuance of green bonds Johannesburg	Cities	2013	$136 million	Climate Bond Initiative
Other	**Scale**	**Timeframe**	**Cost (US$)**	**Source**
Pension fund assets at the end of 2012 in OECD countries	Global	2012	$78 trillion	OECD (2013)
Sovereign wealth funds held by national governments	Global	2014	$7.2 trillion	SWFI (2015)
Money raised by São Paulo using Land Value Capture	Cities	–	$1.2 billion	Soffiatti (2012); Merk et al. (2012)
World Bank loan for Mexico Urban Transport	Cities	2010	$200 million	World Bank Group (2010a)
Clean Technology loan for Mexico Urban Transport	Cities	2010	$200 million	World Bank Group (2010a)
Profits from "Rail plus Property" model in Hong Kong	Cities	2013	$940 million	World Bank Group (2010a)
Investment by Curitiba for conversion of highway to BRT corridor	Cities	–	$600 million	Soffiatti (2012); Merk et al. (2012)

Municipalities also face legal and structural financial difficulties due to national regulations on local governance and fiscal management, as well as limited financial absorptive capacity. Particularly for adaptation, Revi et al. (2014) state that there is limited current commitment to provide finance from different levels of government and international agencies. We discuss the challenges that confront municipalities attempting to fund climate change activities, available funding sources, and how to increase the access of cities to those resources.

7.2.1 Infrastructure Financing Needs in Cities Related to Climate Change

The World Bank recognizes the need for additional partnerships to make more climate change funding available for local governments (World Bank Group, 2013). Partnerships with the private sector and key stakeholders are essential in successful adaptation and mitigation processes at the city level. The World Economic Forum (WEF) summarized estimates of necessary infrastructure investment calculated by several institutions such as the International Energy Agency, Food & Agricultural Organization, Organization for Economic Co-operation and Development, and the United Nations Environmental Program (World Economic Forum [WEF], 2013). Under a business-as-usual scenario (i.e., without taking climate change into account), the WEF arrived at an investment gap of US$100 trillion. Responding to an anticipated 2°C temperature rise will add only $14 trillion, or 14% to the total gap. The biggest investment challenges, therefore, appear to exist independent of climate change. Mobilizing the funding for infrastructure that will be required through 2030 is a daunting task, but, it may be made easier by the threat of climate change and the role that climate-resilient public infrastructure can play in reducing the risks involved and in catalyzing private investments.

National infrastructure gaps are much larger than those faced by any one city or even the totality of global conurbations. But, due to their density and total populations, cities are the places most likely to suffer the adverse consequences of inadequate infrastructure. This suggests that cities should make sure that funding dedicated to closing the gap is deployed in a manner that takes climate change considerations into account. Thus, it is crucial that mitigation and adaptation not be thought of as activities separate from urban infrastructure development. Rather, climate change considerations must be mainstreamed into such investments, especially due to their long-term nature and the need for climate-resilient and low-carbon infrastructure development paths.

7.2.2 Challenges for Cities in Financing Climate Change Activities

Lack of reliable financing often goes back to capacity and regulatory barriers that limit cities' access to finance for climate change activities (Beltran, 2012). Project development and management capacity may be limited by the absence of specialist staffing and limited availability of geographical information systems (GIS) and risk exposure mapping, GHG inventories, and/or acceptance of the measurement, reporting, and validation (MRV) protocols required for adaptive program planning. Those capacities also may be limited by constraints on local authorities' powers to act independently, requirements for bidding and procurement that limit access to specialists, and/or barriers to subnational public bodies gaining access to international financing and its associated expertise (New Climate Economy [NCE], 2014).

The biggest barrier by far, however, in accessing capital for urban infrastructure is the perceived lack in private capital markets of creditworthiness on the part of city governments. An analysis by Lall (2013) of the 500 largest cities in developing countries shows that only about 4% are creditworthy in international financial markets and only 20% are creditworthy in local markets. In relation to this, the World Bank notes that US$1 invested in raising creditworthiness can leverage more than US$100 in private-sector financing for smart infrastructure (World Bank Group, 2013b). Creditworthiness of the city

government may depend on the (1) types and bankability of the projects executed by the city municipality; (2) fiscal stability and governance, including transparency; and (3) national financial regulations and institutions. In particular, fiscal stability depends on the effectiveness of the local tax and service charge collection system of the city government. The capacity to manage revenues and expenditures in the local fiscal budget is a key for municipalities to strengthen their creditworthiness – and thus their ability to leverage external funds. In this context, the World Bank and its partners have designed a City Creditworthiness Program to help city financial officers conduct thorough reviews of their municipal revenue management systems and take actions to qualify for a rating (see Box 7.1).

For cities to take effective action on climate change, their fiscal management and project development and implementation capacities must be upgraded. Specific climate change–related tools such as GHG inventories, vulnerability assessments, and action plans can strengthen cities' capacity to develop and execute climate change projects and leverage the required resources.

7.2.3 Finance Opportunities for Cities

In an important tracking effort, the Climate Policy Initiative (2013) estimated that US$359 billion flowed into climate-related projects and programs in 2012. Comparing this to the average annual US$5.7 trillion infrastructure investment needed to achieve the 2°C stabilization target (WEF, 2013), it is clear that public funding mechanisms will be inadequate – even with stepped-up contributions to the Green Climate Fund, a global platform that invests in low-emission and climate-resilient development. Cities therefore must tap into a full spectrum of opportunities to raise money for climate action.

Figure 7.1 shows how municipal governments could raise climate finances and how this could be invested in programs and projects, although both the inflow and outflow of a municipality's finances will vary depending on its level of fiscal autonomy.

7.2.3.1 Domestic Public Finance

Domestic public finance is a key source of finance for climate change activities at the subnational level. For municipalities, there are four sources:

- Local taxes and service charges
- Transfers from the federal or state governments
- Borrowing from domestic financial institutions
- Bond and equity finance from domestic capital markets

7.2.3.1.1 Local Tax, Service Charges, and Transfers from the National Government

Indigenous revenues from local taxes and service charges are a limited but stable source of finance for cities. Although most countries collect taxes through national systems, local governments

**Box 7.1 The World Bank City Creditworthiness Initiative: Innovation to Improve Cities'
Access to Funding for Low-Carbon, Resilient Infrastructure**

Julie Podevin

The World Bank Group, Washington, D.C.

Cities are challenged to deliver basic services to their populations, and, with the pressure from accelerated growth, urbanization, and climate change, the matter of financing infrastructure services takes on additional urgency. City budgets alone are often unable to meet these growing demands, and subnationals' weak creditworthiness is a major constraint to raising other financing. A World Bank analysis of the 500 largest cities in developing countries shows that only a fraction of them are deemed creditworthy: Approximately 20% have access to local market financing, and a mere 4% can access financing in international markets. Helping cities access private financing is a smart investment. Internal estimates from the World Bank indicate that every dollar invested in the creditworthiness of a developing country city has the potential to mobilize more than US$100 in private-sector financing for low-carbon and climate-resilient infrastructure.

In this context, the World Bank launched the City Creditworthiness Initiative in 2013, in partnership with the Public Private Infrastructure Advisory Facility (PPIAF), the Korean Green Growth Trust Fund, the Rockefeller Foundation, and UN-Habitat.[1] The Initiative is designed to systematically identify and reach reform-minded cities with customized technical assistance to assist them in accessing long-term financing for green growth and climate-smart infrastructure. Design of the Initiative was informed by a prior engagement the World Bank and PPIAF had with the Metropolitan Municipality of Lima, Peru, which led to the city achieving investment grade ratings and raising capital that helped finance the city's bus-rapid transit (BRT) system.

Among other things, the City Creditworthiness Initiative assists city financial officers in conducting thorough reviews of their municipal revenue management systems, in understanding how rating agencies and potential investors assess credit quality, and in taking the first steps to qualify for a rating while recognizing that achieving an investment grade will likely take several years of effort. Improving credit standing is important even where private capital lending is not yet possible because the factors that contribute to creditworthiness can be broadly interpreted to stand for good governance and administration.

The City Creditworthiness Initiative comprises different elements/stages designed to achieve impact with efficient use of limited resources (as summarized in Figure 1).

Engagement typically starts with the delivery of an "Academy," a five-day interactive workshop dealing with the full range of factors affecting cities' financial management performance, including issues determined by the national enabling environment and options for financing, including the use of special-purpose vehicles and public–private partnerships.

During the Academy, participants complete self-assessments that lead to customized draft action plans to improve their overall management and facilitate their ability to plan, finance, build, and operate infrastructure projects. More targeted support is provided during the post-Academy phase, when cities finalize and approve these action plans.

Technical assistance provided through the Initiative may encompass everything from improving national legal and regulatory frameworks for local government finance, promoting the use of data in decision-making and policy formulation, and improving revenue collection and management systems and procedures, to reforming local capital planning and budgeting processes. Some financial advisory support will also be provided for obtaining private capital investment for selected adaptation and/or mitigation projects. Knowledge management and sharing is a strong focus of the Initiative. An online information repository is being developed to provide

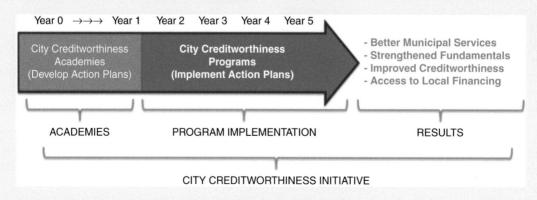

Box 7.1 Figure 1 *Elements and stages of the World Bank City Creditworthiness Initiative.*

1 Additional partners support specific components of the Initiative such as the "Creditworthiness Academies" or technical assistance programs. For example, IFC, USAID, Findeter, Fitch Ratings, the Municipal Institute of Learning (MILE), C40, and Agusto were all key partners for specific Academies.

opportunities for asking questions and sharing experiences related to creditworthiness.

MAIN CHARACTERISTICS AND HIGHLIGHTS

Scalability/replicability: The Initiative aims to assist 300 local governments from as many as 50–60 developing countries directly. It is expected that as many as 3,000 subnational entities will benefit through knowledge sharing and dissemination and take action to improve their creditworthiness gaps.

Flexibility and low cost of first engagement: It is estimated that Academies cost less than US$2,000 per city, making it a relatively low-cost form of first engagement. The core academy curriculum addresses all creditworthiness factors and is adapted to each host country and to the characteristics of invited participants.[2] The materials are adapted for work with local governments of various sizes, from small districts in low-income countries to megacities in middle-income countries. As of April 2015, four Academies have been delivered: in Nairobi (for East Africa), Seoul (for East and Southeast Asia), Arusha (for Tanzania), and Bogotá (for Colombia). These academies covered 23 countries and included 83 cities and 258 participants. Four additional Academies occurred in May 2015: three organized by the World Bank and covering Uganda, Rwanda, Jordan, and West Bank Gaza. The fourth, organized by the C40 network of cities is for 10 global megacities and was held in Jordan.

Practicality and results: Local stakeholder engagement is key. The self-assessment process throughout the initial engagement invites the participants themselves to prioritize critical challenges and potential actions. To date, more than 150 cities have joined the Initiative, completed self-assessments producing diagnostics, and developed basic action plans as a result. Identification and implementation of post-academy support is ongoing and includes:

- US$1 million raised for technical assistance programs for 34 municipalities in Tanzania, already in full implementation
- Similar funding levels being secured in the short-term for Colombia, Uganda, and Rwanda
- Ongoing support to the Kampala Capital City Authority, in Uganda, with more than 80% in enhanced revenues and the first credit rating delivered to a local government in the country
- Support to Dakar, Senegal, with own-source revenues enhancement activities and transaction support for a bond issuance

The City Creditworthiness Initiative's long-term objective is to facilitate municipalities' creditworthiness to make them more attractive to private investors and help them access markets to get finance flowing for low-carbon planning. The Initiative complements other development goals, and, along the way, municipalities can reap benefits from short- and medium-term achievements.

often have legal authority to collect residual taxes on their own. In practice, there is a tension between national and local taxes and between different localities that impose lower taxes to attract industrial and commercial investment, thus limiting the municipal revenue that can be achieved. Another relatively limited source of revenue for cities is the collection of service fees. Cities provide public services such as public transport, waste collection/disposal, and drinking water supply. These services generate stable but price-sensitive sources of municipal revenue that add to local budgets but are often earmarked for specific use.

Thus, local governments have only limited opportunities to raise discretionary revenues. Typically, they are financially reliant on national governments. For example, whereas the highly decentralized national government in Indonesia has limited influence over urban policy, it can provide a financial incentive through the Specific Allocation Fund (DAK-EE) to encourage urban investment to reduce air pollution, increase adaptation, improve basic services, or otherwise contribute to green growth (Indonesia Ministry of Finance, 2014). In addition to transferring part of their budgets to the local government, in the case of large infrastructure projects (e.g., mass transit systems), national governments often take responsibility for undertaking urban infrastructure development. Even though the national government can manage larger public investments in more climate-resilient infrastructure than can municipalities, the national government may not necessarily invest.

In some instances, despite limited resources, national governments compensate local authorities for the positive environmental spillovers of their spending. Examples include Brazil's tax-based Payment for Ecosystem Services and Sweden's Climate Investment Program (KLIMP) (Revi et al., 2014).

Local tax revenue, service fees, and allocations from national governments can form core stable finance sources for cities. The scale and stability of these sources of revenue are the key factor in municipal creditworthiness. Municipalities need to build their capacity to manage these core revenues and expenditures in order to strengthen their creditworthiness, thereby helping them to manage their funds effectively and attract private investment.

7.2.3.1.2 *National and Regional Development Banks*

Some countries have their own national development banks. In general, they are established as publicly owned entities and national governments are major shareholders, but they collect funds from the market and/or savings and deposits. National development banks play an important role

---

2 Host country partners assist in identifying the core set of local governments to be invited.

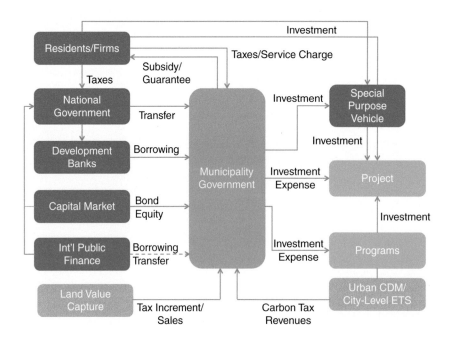

Figure 7.1 *Climate finance opportunities for municipalities.*

as agents for governments to provide long-term finance in line with their policies (IDFC, 2013). City governments and municipalities may be able to apply to these domestic public banks for resources under special lines with favorable lender conditions in the field of climate change mitigation and adaptation (KfW, 2013).

One emerging case of a regional development bank is the New York Green Bank that has been established as a public–private partnership (PPP). It is a state-sponsored specialized financial entity designed to address gaps in clean energy financing and to transform those markets as part of an integrated strategic state-wide energy plan (greenbank.ny.gov). Another example is the UK Green Investment Bank, founded in 2012 by the government of the United Kingdom with US$5.9 billion to leverage private funds for modern green infrastructure.[3]

7.2.3.1.3 Local Government Bonds

Depending on the powers granted by higher levels of government, some cities may be able to issue local government bonds in domestic and/or international markets. The bond issuer can earmark the use of proceeds as well as designate the source of funds for repayment, so the tool can be focused on financing climate change actions. Some bonds earmarked for environmental or climate purposes are being called "green bonds" or "climate bonds." In 2013, Massachusetts initiated this form of bonds, authorizing the municipal issuance of US$100 million in green bonds. Gothenburg, Germany, and Johannesburg, South Africa, followed, issuing green bonds for SEK 500 million (US$79 million) and ZAR1.45 billion (US$136 million), respectively. The green bond market is growing rapidly, and HSBC estimated that green bond

issuance would reach US$40 billion by the end of 2014, rising to US$100 billion in 2015 (Climate Bond Initiative, 2014).

Subnational entities are often perceived to be high-risk borrowers, which increases their borrowing costs. By obtaining formal public credit ratings, creditworthy subnationals can increase their lenders pool, raise cheaper funds, and borrow without sovereign guarantees. *Shadow*, or *confidential*, ratings allow subnationals to identify the issues that need to be addressed to improve their creditworthiness before obtaining a formal rating. The World Bank Group's PPIAF's Subnational Technical Assistance (SNTA) program can assist subnational entities to prepare for and obtain credit ratings.[4] The SNTA program can also provide technical assistance to improve a subnational's creditworthiness and address weaknesses highlighted by a rating assessment. Although the possibility of bond issuance as well as borrowing from financial institutions depends on national and local regulations for fiscal management of municipalities, municipalities should explore this possibility for direct access to climate finance.

7.2.3.2 International Public Finance

Bilateral and multilateral donors have increasingly focused on providing financing specifically for climate action, and corresponding programs and funds have been established that support activities in urban areas.

International public funds dedicated to climate change include:
• Multilateral Development Banks
• Global Environment Facility
• Climate Investment Funds

---

3   UKgov.greeninvestmentbank.com
4   (http://www.ppiaf.org/page/sub-national-technical-assistance).

- Green Climate Fund
- Sustainable Development Goals Fund (SDG-F)
- Least-Developed Countries Fund
- Special Climate Change Fund
- Millennium Development Goal Achievement Fund
- Adaptation Fund
- Global Facility for Disaster Reduction and Recovery (GFDRR)
- Bilateral sources (national donor funding)

Most international donors and funds channel their resources through national governments of the recipient country (World Bank Group, 2011). The Mexico Urban Transport Transformation Program, for example, is co-financed by a US$200 million World Bank loan and by another US$200 million loan from the Clean Technology Fund. Although these resources will benefit Mexican municipalities that reduce GHG emissions in the urban transportation sectors, participating cities access them through the Banco Nacional de Obras rather than receiving them directly from the donor institutions (World Bank Group, 2010). Multilateral grant finance benefiting cities is usually distributed through or at least in close cooperation with national governments.

There are several reasons for the central role of national governments in the distribution of multilateral funding to sub-national actors. Most importantly, the activities of international donors usually follow agreements negotiated with national governments, for example bilateral contracts or the World Bank's Country Assistance Strategies. Furthermore, internationally funded projects need to be planned and implemented in a manner consistent with national development plans, which are likely disclosed to funders before monies are awarded. For lending operations and guarantee instruments, the role of national governments is even more important because a sovereign guarantee is usually required for these modalities to be used. If cities want to benefit from multilateral funding, they will therefore usually have to negotiate access with their national government.

Nevertheless, under certain conditions, donors can deal directly with city governments. In the case of the Inter-American Development Bank, loans and guarantees can be provided to municipalities through the Bank's private-sector facility without requiring a sovereign guarantee. Multilateral sources providing direct access for cities also include the Adaptation Fund and the Millennium Development Goal Achievement Fund (World Bank Group, 2011). Nevertheless, consistency with country strategies and the non-objection of the national government are still required. Therefore, even though cities could access multilateral finance directly, these provisions are rarely used. This is unfortunate since the incentive and capacity to leverage external funds may be greater at the local than the national level.

The limits and barriers to accessing international donor funding imply that most of the funding needed to tackle the challenges of climate change in cities will have to be mobilized by local governments themselves, in the best case with support from their national governments. As such, it is crucial that cities mainstream climate change considerations into their sectoral infrastructure

activities. This is especially true because responding to climate change can be achieved by closing the current development gap, although special attention to low-carbon and resilience transformation will be required (World Bank Group, 2011).

7.2.3.3 Private Sources of Finance

7.2.3.3.1 Private Investment in City Infrastructure

In some instances, private finance takes the form of actual ownership of, rather than lending for, needed infrastructure. That is, the private sector may construct or purchase public assets and operate them, thus freeing up public-sector resources for other investments as the private firms manage and deliver public services. Private capital has funded water and sewage systems, transportation systems, telecommunications, and other needed infrastructure for decades (Smith, 1999; Harris, 2003). Privatization has its costs in terms of some decline in the level of public control over prices and quality of services, but its benefits may include making funds available for public efforts that no private investor would fund itself, such as population relocations and other climate adaptation measures for the most vulnerable urban populations with minimal ability to pay for such protection.

Over the long term, the need for private investments to return profits to their owners means that public services or facilities owned by private firms may generate lower public benefits than those owned publically, especially after the debts incurred in their construction are paid for (Harris, 2003; Kessides, 2004). But provision of needed infrastructure through private ownership may be one way for cities to overcome the problems of their limited creditworthiness when pursuing loans, even from favorably inclined development banks. As climate risks become more severe and the need for adaptation measures more acute, the public benefits to be gained from freeing capital for adaptation investments that would not otherwise be available in urban areas with low credit status may warrant support for private ownership of assets that traditionally have been held in public hands. On the other hand, such developments may pose problems for public provision due to concerns over allocation of associated risks (Ng and Loosemore, 2007).

7.2.3.3.2 Pension Funds

By far the largest individual investment pools in the world include public and private pension or superannuation funds. Pension fund assets at the end of 2012 in countries of the Organization for Economic Co-operation and Development (OECD) alone totaled more than US$78 trillion (Organization for Economic Cooperation and Development [OECD], 2013b). Globally, sovereign-wealth funds held by national governments, some of which provide pensions to their citizens with these public assets, amounted to US$7.2 trillion in 2014 (Sovereign Wealth Fund Institute [SWFI], 2015). With the responsibility of investing to provide funds for retirees, such funds can be characterized as "patient capital." That is, they do not need to show immediate returns to impatient individual investors, but rather need to have earned returns by the time their members reach retirement age; that is, typically decades in the future. At the same time, they have stringent fiduciary responsibilities and must concern themselves with capital preservation. Thus, they can only accept innovative opportunities if their risk of loss is minimized. They may, however, be modernizing their investment criteria (OECD, 1998, 2008).

To the extent that climate change becomes more acute over time, any investments that successfully respond to those changes, whether providing adaptation and/or mitigation services, will increase in value over time. If the services are sufficiently in demand today that their provision involves a break-even investment, sector-level losses are unlikely and future profits could be exceptionally high. City climate action programs that can meet this threshold condition, therefore, may be able to access the large amounts of capital controlled by pension funds.

7.2.3.4 Market Mechanisms and Other Innovative Finance Sources

7.2.3.4.1 Clean Development Mechanism

The Clean Development Mechanism (CDM) is one of the "flexibility mechanisms" defined under the Kyoto Protocol.[5] Its objective is to assist developing countries in achieving sustainable development and mitigating GHG emissions that cause climate change. In addition, CDM aims to assist industrialized countries in achieving compliance with their quantified emissions limitations. Despite its great success, with more than 7,500 CDM projects registered within many countries and sectors, some important emission sources, sectors, and countries are still underrepresented within the CDM, especially in some of the least developed countries of Africa and Asia (Spalding-Fecher et al., 2012). A relatively high number of approved methodologies are applicable in the urban context, and several successes demonstrate that CDM activities are possible in cities.

Urban areas have the highest populations, which leads to increased demand for energy resources and high levels of GHG emissions. Implementing sustainability and emission mitigation measures in cities has great potential to be replicated in other cities and countries and may lead to positive co-benefits. Although the CDM instrument, given its evolving nature, has limitations for wider application in mitigating carbon emissions in cities, mitigation measures in cities that are initiated by city councils or municipalities should cover more than one sector/technology. The United Nations Environment Program (UNEP) and Gwangju City (2012) demonstrate that the CDM has evolved by introducing the concept of a Program of Activities (PoA) that allows for the combination of an unlimited number of emission mitigation activities under a single umbrella using different methodologies.

The World Bank's citywide approach proposal follows the basic principles of a CDM PoA that is based on a multisector

approach (World Bank, 2010). Under such an approach, the coordinating entity (e.g., a municipality) would have the flexibility to combine relevant technology options across different sectors given its financial and development abilities. This PoA concept refers to relevant CDM methodologies for quantification of the emissions reductions. Note, however, that the demand for Kyoto credits remains low, possibly due to the uncertain future of CDM under the Post Kyoto Paris Agreement framework. According to the World Bank (2014), there is a growing feeling in the CDM market that demand is saturated. With little prospect of a significant recovery, the biggest players have begun to leave the market, along with their skills and expertise.

Other innovative approaches include setting aside the funds saved from increased energy efficiency investments for adaptation or further mitigating efficiency efforts (Meyer et al., 2013b; Revi et al., 2014) or revolving loan pools that replenish themselves as projects mature and that might be funded through revenue streams from CDM projects (Puppim de Oliveira, 2009).

7.2.3.4.2 City-Level Emissions Trading Systems

A city-level GHG emissions trading system (ETS) is a mitigation approach to encourage municipalities and their private sectors to foster low-carbon project financing. The Governor of Tokyo, Shintaro Ishihara, submitted a bill to the Tokyo Metropolitan Assembly in June 2008 that introduced mandatory targets for reductions in overall GHG emissions for large-scale emitters as part of an emissions trading program. The Tokyo Metropolitan Assembly passed the bill, thus introducing Japan's first cap-and-trade emissions trading program, which took effect in fiscal 2010. Since then, the Tokyo Metropolitan Government (TMG) developed a cap-and-trade program that many advanced nations and regions are also moving to implement. TMG's program is the first one to be implemented in Japan and Asia (Tokyo Metropolitan Government, 2010) (see Case Study 7.3).

The government of China announced its plan to develop seven official ETS pilot programs (Beijing, Shanghai, Tianjin, Chongqing, Guangdong, Hubei, and Shenzhen) in 2011. This plan began phase-in during 2013. By April 2014, six of the seven pilot schemes started trading. Carbon markets are now officially open for business in China. The total 2013 allocations of these six pilots combined amount to 1,115 MtCO2e, making China the second largest carbon market in the world, after the European Union (EU) ETS. The Guangdong ETS, the largest of the Chinese ETS pilots, itself covered 388 $MtCO_2e$ in 2013, equivalent to of France's GHG emissions in 2012 (World Bank, 2014).

7.2.3.4.3 Land Value Capture

Cities usually own substantial assets that could be managed to facilitate climate change mitigation and adaptation. One approach, already frequently used for financing transportation

---

5 www.cdm.unfccc.int

systems, is land value capture. Land value capture, like local taxes, user charges, and licenses are local public sources of finance. As Smith and Gihring (2006) describe, transit infrastructure can be at least partly funded by capturing the increase in the land value of properties close to transit stations.

There are different variations of land value capture: development impact fees, tax incremental financing, public land leasing and development right sales, land readjustment programs, connection fees, joint developments, and cost-benefit-sharing (NCE, 2014). For example, in Hong Kong, the government's "Rail plus Property" model captures the uplift in property values along new transit routes, ensuring efficient urban form while delivering US$940 million in profits in 2009 for the 76% government-owned MTR Corporation (Rode et al., 2013). São Paulo has raised more than US$1.2 billion in 6 years using related instruments, and Curitiba is funding the conversion of a highway into a BRT corridor, complemented by higher-density, mixed-use spaces and green areas – an investment of US$600 million (Soffiatti, 2012; Merk et al., 2012).

Certain climate change–related activities provide immediate co-benefits that similarly could help with project finance and leverage appropriated or granted project-specific funds. For example, providing public green spaces would increase surface permeability and improve air quality, but it also would increase the value of the surrounding properties and thus the real estate taxes of the municipality. Also, vulnerability reduction could raise land and property values and generate savings by reducing insurance premiums. Where value capture is possible, the challenge is likely to be mobilizing upfront financing rather than medium- or long-term cost recovery.

Certain city assets can also be used directly for climate action. A prime example is public space designed to help in reducing risk. Green improvements to publicly–owned lands could reduce the need for more expensive gray infrastructure, another form of leveraging. Interventions like the creation of green corridors, green roofs, and urban gardens to separate sidewalks from vehicle traffic or the creation or redesign of parks as a network that facilitates runoff and retention of water, thus create mitigating inundations (GeoAdaptive, 2013) (see Chapter 5, Urban Planning and Design). At the same time, such interventions can reduce GHG emissions and provide additional co-benefits such as reducing the urban heat island effect. The effect of heat waves can be mitigated by simple measures like painting roofs in light colors that reflect rather than absorb solar radiation. These are only some of the many climate action measures that cities can take without having to resort to expensive infrastructure. Mainstreaming such innovations into city operations across sectors is therefore of utmost importance.

7.3 The Role of the Private Sector

As concentrations of people, cities produce agglomeration effects and attract business enterprises (NCE, 2014). Hence, the potential role of the private sector in urban climate mitigation and resilience is important and multifaceted. While private-sector engagement has become a buzz-word for policy-makers and climate experts, its role is poorly understood (Surminski, 2013; Averchenkova et al., 2016). This section aims to present modalities of private-sector engagement in urban climate mitigation and adaptation by looking at the challenges, barriers, and opportunities such engagement presents. The scope of the private sector here is restricted to businesses and does not include community, nongovernmental organization (NGO)-led, and household interventions.

7.3.1 Modalities of Private Sector Engagement

Policy-makers have growing expectations of how the private sector may be involved with meeting public policy goals for climate change, although it is still somewhat unclear what action qualifies as private-sector adaptation or if and how companies consider their efforts in the context of future climate trends rather than current risks (Averchenkova et al., 2016; Isaho and Surminski, 2015). A review of climate change strategies published by different cities (especially Tokyo, Delhi, Mexico City, New York, São Paulo, Dhaka, Calcutta, and Karachi) and published case studies (e.g., Copenhagen) (Floater et al., 2014a) show the different roles that the private sector may play. Key lessons learned from these cases are:

- The private sector may be involved in partnerships with city governments for the provision of technologies, the construction and operation of infrastructure, and the provision of insurance, to cite a few emergent fields. "Cleantech" clustering is a promising strategy for building partnerships and networks among private entities, the research community, and public institutions, which facilitates the commercialization of new products and services (applicable to almost all cities).
- Companies are a key vehicle for implementing climate change strategies because they have to comply with regulations for energy saving and GHG emissions reduction, such as standards, rating systems, or cap-and-trade schemes. They also have to comply with requirements of environment performance disclosure. This ensures their motivation in engaging in climate mitigation and adaptation activities (applicable to all cities).
- Low-carbon and resilient urban development provides business opportunities for innovative products and services in myriad sectors, including transport, waste, energy, civil construction, urban planning, food, insurance, knowledge management, and R&D. Local and national policies can create business opportunities for local firms to apply their technologies (many cities, but outstanding examples are Mexico City and São Paulo).
- Companies may receive support for financing the implementation of low-carbon technologies, for example, through collateralized bond obligations (CBO) or support instruments for renewable energy deployment, and for exploiting new business opportunities (e.g., Tokyo and Mexico City).
- Companies can increase efficiency and productivity and provide funding for climate-related solutions if correctly incentivized (e.g., São Paulo).

Case Study 7.1   London Climate Change Partnership: Public- and Private-Sector Collaboration

Swenja Surminski, Hayley Leck, and Jillian Eldridge

Grantham Research Institute on Climate Change and the Environment (GRI), London School of Economics and Political Science

Keywords	Resilient city, multisectoral partnership, policy, flood, heat waves, economics and finance
Population (Metropolitan Region)	14,031,830 (Eurostat, 2015a)
Area (Metropolitan Region)	12,091 km² (Eurostat, 2015b)
Income per capita	US$42,390 (World Bank, 2017)
Climate zone	Cfb – Temperate, without dry season, warm summer (Peel et al., 2007)

Mounting cross-cutting climate risks cannot be addressed successfully at any single institutional or spatial scale or by any one category of actor. Multisectoral partnerships (MSPS) are increasingly central to the new wave of climate governance. They hold the potential for innovative solutions but also raise considerable challenges in terms of power relations, accountability, equity, and effectiveness. Partnerships are often underpinned by complex multiscale governance arrangements that need to be better understood. One example of a long-established effort to bring together public- and private-sector players within an urban context is the London Climate Change Partnership (LCCP).

Launched in 2001 by the then Mayor, Ken Livingstone, the LCCP supports climate risk reduction and climate change adaptation across London. As a large city with complex cross-boundary environmental risks, London´s collaborative management of climate risks across spatial, political, and organizational boundaries is critical. Such risks cannot be dealt with solely by just one category of actor, and LCCP's approach has focused on harnessing the understanding and expertise of local, national, and London-specific organizations and representatives, including a range of public and private groups. This strategy has facilitated the delivery of advice, research, and understanding of how London can become a climate-resilient city. The LCCP has a long-term outlook with a range of actions to prepare London for extreme weather events and future impacts of climate change (see Table 1).

Coordination and facilitation of the LCCP is government-led, with funding from the Environment Programme budget from the Greater London Authority (GLA), the city government for London. There are over 20 members consisting of experts in the fields of environment, finance, health and social care, development, housing, government, utility, communications, transport, and retail sectors. The partnership's work is structured around several projects that involve research on specific climate risks as well as resilience actions (see Table 2). These include additional projects for climate resilience, such as Drain London; a cross-boundary strategy to develop surface water management plans for London and its boroughs (city government).

THE PARTNERSHIP APPROACH: WHAT CAN WE LEARN?

The LCCP provides an important example of how an urban partnership approach can address climate risks. Although the LCCP has proved effective in many ways, the partnership also faces multiple challenges including financial constraints and complex funding arrangements, political barriers, and divergent perspectives and expectations amongst partners, as well as difficulties in assessing impacts of partnership activities and functions. This is particularly relevant in the context of the broader scope of adaptation. Our survey of LCCP members has revealed that the dominant focus of past and current work is on knowledge-sharing and information dissemination, whereas the implementation

Case Study 7.1 Table 1 *An overview of the LCCP including key actions.*

Characteristic	Detail
Members	Twenty-four members representing a range of public-, private-, and community-sector organizations.
Spatial scope	Greater London
Sector	Public-based organization, but members are public, private, or voluntary
Key actions	• Collecting and sharing information about expected climate change impacts on London and possible adaptation options • Raising awareness of organizations and individuals of the impacts of climate change • Facilitating and encouraging adaptation in London • Informing policy with local evidence • Monitoring London's preparedness for climate change • Seeking opportunities to improve resilience to climate change
Established	Established in 2001 by the Government Office for London and run by the Greater London Authority
Wider scope	LCCP members work with the London Resilience team, and LCCP is part of Climate UK, a UK-based community interest group.

Case Study 7.1 Table 2 *Key past and present projects delivered under the LCCP.*

Project	Involved lead partners
Adaptation Economy	Greater London Authority
Observing London	The Met Office, Greater London Authority, Reading University, Lloyd's of London
Retrofitting London	Sustainable Homes, Greater London Authority, Thames Water and the Environment Agency
Resilient Business	Greater London Authority
Overheating Thresholds for Londoners	Environment Agency and Greater London Authority
Joint Strategic Needs Assessment Guidance	Greater London Authority and London Boroughs
Capturing Adaptation Research for London	UK Climate Impacts Programme (UKCIP), Environment Agency
Retrofitting social housing: Barking and Dagenham	Sustainable Homes, London Borough of Barking and Dagenham, Mayor of London, Sprunt, United House, Environment Agency
London Health and Social Care Climate Action Plan	London Climate Change Partnership

of adaptation measures rests predominantly with the individual partnership members and other stakeholders. This Case Study forms part of a large European Union-funded research project ENHANCE. ENHANCE aims to develop and analyze new ways to enhance society's resilience to catastrophic natural hazard impacts and develop supportive multisectoral partnerships and to understand the relationships between partners in delivering climate mitigation and resilience.

- Companies can help in advocacy and generate mass awareness, especially when they are subject to disclosure requirements about their GHG emissions and climate risks (many cities).
- Private business can collaborate with tertiary education and research institutions in the search of innovation excellence in the area of mitigation (e.g., Copenhagen).
- Scientific information is key to inform business management decisions to ensure business resilience within climate change mitigation and adaptation strategies (e.g., New York).
- Local entrepreneurship and foreign private capital are drivers of urban transformation, growth, and competitiveness, but financial and information barriers to attracting finance remain in the clean technology sector. These barriers can be overcome through coordination between national and local-level policies (e.g., Copenhagen).

These examples demonstrate that private-sector engagement can occur in many shapes and forms. These can be categorized according to four modalities: business continuity, business opportunities, business finance, and risk transfer (Khattri et al., 2010). Each type of engagement is considered next, highlighting its relevance for the broader capacity of cities to respond to climate change.

7.3.1.1 Business Continuity

Climate change is already having negative impacts on many types of businesses, including operational disruptions, increased costs of maintenance and materials, and higher insurance premiums. Catastrophic events are drawing the attention of companies to such climate-related risks, which are expected to increase and become more severe in the future. Consequently, businesses are interested in participating in both mitigation and adaptation for their own private benefits (NRT/NBS, 2012). Business continuity is a mode of private-sector engagement that derives from safe-guarding business interests by climate-proofing supply chains and operations (Khattri et al., 2010).

Reduction of GHG emissions by private-sector entities is likely to be driven by the benefits of corporate social responsibility (CSR). Leadership in this area can enhance reputation, which is of high value to companies. For example, the relationship between stock performance and disclosure of climate change strategies is positive where companies face pressure about climate change issues. As awareness of global warming increases, such investments could become more attractive (Zieger et al., 2011).

While adaptation actions are also influenced by CSR considerations, they can have in addition a direct effect on the company and improve its performance under more volatile conditions. An effective adaptation strategy can result in competitive advantage, but, at the same time, it can be based on cooperation between companies (WRI, 2012). Certain business enterprises are also interested in contributing to the adaptation of the cities where they undertake their activity because these provide the infrastructure, human capital, and potential markets that companies require. Resiliency of urban infrastructure, communication, and transport systems helps to minimize climate-related risks. Well-designed cities are also less exposed to risks and thus to insurance price increases, delays or shortages of key inputs, and other often-fluctuating expenses.

Climate-related risks to a business's competitiveness and profits are of both a direct and indirect nature. In the city context, four types of risks are becoming prominent: hazard-related, financial, operational, and strategic (see Table 7.2).

Some companies are starting to consider these risks in their business or risk management plans: they conduct vulnerability assessments, utilize climate-specific risk models or other instruments to inform their decisions, and/or rely on insurance (Crawford and Seidel, 2013). Preparing for the effects of climate change will become increasingly important as businesses seek to maintain their current operations and competitive advantage. To name some examples, AT&T, Monsanto, Coca-Cola, and Munich Re have identified potential threats of climate change for their activities and are working to minimize risks through appropriate adaptation measures. Company-specific efforts and public-sector efforts need to be integrated.

Consideration of climate risks may affect decisions about the location of new and existing facilities. Several companies are already trying to minimize exposure to severe weather conditions by taking into account projected threats when selecting locations (Castán Broto and Bulkeley, 2013; Crawford and Seidel, 2013; Sussman and Freed, 2008). Thus, municipalities able to guarantee the security of investments by promoting urban resilience will be more able to attract and retain companies (Revi et al., 2014). In this way, response to climate change becomes a source of "competitive differentiation" both for cities and companies (Carbon Disclosure Project [CDP], 2012).

In cases of cities with limited capacity to manage climate risks, the private sector may add to municipal risk management capability, knowledge, skills, and resources. By forming partnerships with insurance and engineering companies, for example, cities can gain access both to information about risks and expertise in resilience planning (public–private partnerships (PPPs) are described in more detail in Section 7.3.1.3). Companies and cities both benefit by joining efforts to generate information about future climatic and socioeconomic trends at the urban level (Carmin et al., 2013). For example, in the design of the Chicago Climate Action Plan, the city authorities collaborated with private companies to finance science-based assessments of alternative policies and plans (Rasker, 2012).

Table 7.2 *Typology of climate risks faced by the private sector (Crawford and Seidel, 2013), based on a review of documentation from companies listed in the Standard and Poor's (S&P) Global 100 Index.*

Risk type	Short description	Relevance by sectors (% of companies in a sector)
Hazard risks	Inability to do business due to damage to facilities, communications or transport systems[a] Increasing maintenance costs Supply chain interruptions	Banking & Financial Services (20%) Consumer Goods, Healthcare, Materials (10%) ICT & Services, Manufacturing & Industrials (5%)
Financial risks	Increased operational cost due to higher costs for key supplies, backup power or other commodity price shocks[a] Increased capital cost due to plant or equipment upgrades, higher insurance and business loans prices[a]	Banking & Financial Services (60%) ICT & Services, Consumer Goods (50%) Healthcare (50%) Materials (40%) Manufacturing & Industrials (10%) Consumer Goods (30%) Banking & Financial Services (25%) Materials (10%) ICT & Services (5%)
Operational risks	Reduction/disruption in production capacity due to power outage, shortage of key input, changing resource availability and quality[a] Reputational risk: customer obligations not met due to supply interruption	Healthcare (90%) Materials (75%) ICT & Services (>60%) Consumer Goods, Manufacturing & Industrials (>50%) Banking & Financial Services (40%)
Strategic risks	Reduced demand for goods/services due to shifting market preferences or ability to pay[a] First-mover advantage for meeting new market demands	Banking & Financial Services (25%) Consumer Goods, Healthcare, ICT & Services (12%)

[a] *Top five current or expected impacts from climate change within the risk types. Figures indicate percentage of companies in a sector that have identified the different risks.*

Case Study 7.2   Public Enabling of Private Real Estate in New York

Jesse M. Keenan

Harvard University, Cambridge, MA

Keywords	Adaptive capacity, real estate, private sector, economics, coastal, storm surge
Population (Metropolitan Region)	20,153,634 (U.S. Census Bureau, 2016)
Area (Metropolitan Region)	17,319 km² (U.S. Census Bureau, 2010)
Income per capita	US$56,180 (World Bank, 2017)
Climate zone	Dfa – Cold, without dry season, hot summer (Peel et al., 2007)

John Jacob Astor, the first millionaire in America, earned his wealth almost entirely through the speculation of real estate in New York (NYC). As in Hong Kong and London, the real estate sector in NYC today is a significant component of the economy of the city. The real estate industry in NYC accounts for US$106 billion in annual economic output, which equals approximately 13% of the Gross City Product (GCP) (AKRF, 2014). At just over 519,000 jobs, the real estate industry makes up an estimated 11% of the city's employment and contributes US$15.4 billion in annual taxes to the city or 38% of total municipal tax revenues (AKRF, 2014). The initial land use of the 18th and 19th centuries relating to the commerce of the sea dictated the continuous expansion of a working waterfront, one often created through tenuous infill development through the leveling of the city's topography. As industrial uses now give way to the waterfront as an amenity for residential populations, there is a resurgence in the city's relationship with the waterfront. Yet this re-engagement of the waterfront through increased real estate and infrastructure development is not without risk (see generally, Metropolitan Waterfront Alliance, 2015).

In the past decade alone, the city has been subject to several tropical storms that led to floods of varying intensity, including the devastation wrought by Hurricane Sandy in September 2012. The New York City Panel on Climate Change (NPCC) estimates that not only it is more likely than not that there will be an increase in the number and strength of such intense storms, but also that more frequent precipitation downpour events and inundation from sea level rise are likely to pose unprecedented risks to the city over the next several generations (2015). Based on recently revised flood insurance rate maps (FIRMS), the city comptroller estimates that US$129 billion dollars of real estate is at risk within the 100-year floodplain (City of New York, 2014). Such flood events, and other similar events such as power outages from heat waves, pose significant risks to the commercial operations of the city even for those properties that are not directly affected by on-premises casualty losses because business continuity insurance is less than accessible in terms of cost and geographic availability (RAND, 2013).

TECHNICAL ENABLING OF DESIGN AND CONSTRUCTION PRACTICES

Following Hurricane Sandy, Mayor Bloomberg impaneled the NYC Building Resiliency Task Force (BRT) composed of public-sector regulators and private-sector actors in real estate, insurance, engineering, design, and various construction trades. With adaptation being defined not only by specific interventions, but also by a capacity to implement those decisions (Adger et al., 2005), the BRT was designed with a capacity to provide a continuous dialogue by and between the public and private sectors so as to advance building codes that enable technologies, designs, and materials that serve to reduce known environmental risks – and, hence, promote the resiliency of buildings (New York BRT, 2013). The primary impetus of the BRT was to incorporate elements of the International Building Code, as well as best practices from high-risk states such as Florida, into the NYC building code. Each proposed element was subject to peer review evaluated by a cost-benefit analysis benchmarked against anticipated risk reduction, as well as by qualitative political, legal, and market considerations for practical implementation.

The technical recommendations include everything from allowing properties to capture excess flash flood water to requiring backflow valves to prevent sewage backflow in buildings located in special flood hazard areas. Beyond water-related elements, additional recommendations included breaking down barriers found in existing regulations to allow buildings to accommodate power outages with co-generation, solar, and natural gas autonomous power generation facilities. Beyond autonomous systems, passive building systems for potable water and lighting were also advanced. Aside from the technical recommendations, consideration was given to advancing preapproved emergency inspectors and recovery agreements, as well as accommodations in reducing liability for supporting the reconstruction work of Good Samaritans following disasters. Of the thirty-two recommendations, fifteen were officially implemented as a matter of local law.

STRATEGIC ENABLING OF REAL ESTATE OWNERS AND INVESTORS

The banking, private equity, insurance, and many other service-based sectors have advanced adaptive capacities through corporate disclosures and a broader effort of bringing transparency to markets where vulnerabilities to climate change may be overlooked or underappreciated. Unlike most markets, real estate is highly localized, and, as such, ongoing risk assessment and reduction are timely and potentially costly endeavors. However, the Department of City Planning, through the Vision 2020 Comprehensive Waterfront Plan (2011), and the New York Economic Development Corporation, through the Waterfront Vision and Enhancement Strategy (WAVES) (2014), have taken significant steps in comprehensively evaluating risks and promoting experiments and pilot projects that reduce those risks while setting new standards for performance.

Experiments have included sponsoring multiple competitions for marine construction, resilient technologies, and ecologically sensitive landscape designs, which are integrated with stormwater management programs. The implications of these efforts, specifically the pilot projects, have been to set a benchmark for the private sector with regard to (1) estimating more accurate construction costs, (2) setting risk performance thresholds, and (3) providing a roadmap for environmental regulation, which has historically resulted in unpredictable outcomes.

Interviews with stakeholders and regulators have suggested that this final component of environmental regulation is perhaps the single greatest barrier in implementing innovative techniques and technologies that are flexible in adapting to changing and evolving risks. This friction highlights the multiple layers of regulation that are often beyond the control of the City. However, it is anticipated that, by building a coalition of public- and private-sector actors, there will be greater political leverage in advancing experimentation. By bringing measures of certainty and predictability to the development of infrastructure and real estate, the private real estate sector's adaptive capacity is arguably more robust by virtue of these collective public-sector actions and strategies.

7.3.1.2 Business Opportunities

Climate change is also a source of business opportunities. The risks can create opportunities in the form of increased demand for new and existing products/services, potential for winning over new customers, technology development, reduced operational costs, increased production capacity and investment, higher staff retention rates, and good publicity (Crawford and Seidel, 2013; Schroeder et al., 2013). These are elaborated in Table 7.3, which compiles examples of business opportunities for high-potential sectors.

Table 7.3 *High-potential sectors and examples of business opportunities. Source: Adapted from Khattri et al. 2010*

Sectors	Examples
Banking, financial services, micro-insurance, and micro-finance	Financing infrastructure resilience and affordable housing solutions Catastrophe-linked securities to transfer risks of extreme weather events Financing for farmers affected by weather risks Weather-related insurance for crops and forests Weather derivatives for electricity utilities
Health care and preventive care	Products to prevent water-related illnesses Treatments and vaccines for climate-related diseases Mosquito control Eye-care products for treatment of sun exposure, allergies, glaucoma, and infections
Urban infrastructure	Waste management and sanitation Water management and desalination Low-energy buildings, green infrastructure and retrofitting Off-grid energy and renewable energy Urban public transport, bikeways and railways Drainage and roadway construction/elevation
Materials	Insulating foams for temperature regulation Chemicals to bond roof tiles in hurricane areas Stronger building materials Systems to protect dikes from wave impacts
Manufacturing and industrials	Electric and hybrid vehicles New lighting solutions Cleaner coal technologies Biodegradable products Flood/drought-resistant seeds
Livelihood promotion	Mobile auction system enabling farmers to improve their livelihoods SMS broadcast to distribute messages on weather information Restoration of ecosystems providing income-generation opportunities (including revenues from carbon market through carbon sequestration)
Information and communication technology	Mobile technology to manage water footprint Digital solutions to dematerialize processes and services Smart grid solutions
Services and consulting	Risk assessments and management frameworks Energy efficiency and management solutions (including energy service company [ESCOs]) Environmental analysis and training
Waste and pollution control and recycling	Mobile and stationary air pollution source controls Water conservation, wastewater treatment and reuse technologies Pulp and paper, aluminum, and electronic recycling

Cities are major markets for new products and services. As intersections of transport, industries, supply nets, buildings, and infrastructure, they have a large potential for the implementation of new technologies to generate energy savings (see Box 7.2). The Climate Group (2008) provides estimates of the potential savings in terms of energy costs (US$950 million) and emissions (15% by 2020 globally) associated with the implementation of new technologies in cities. The technologies involved are LED/other energy-efficient lighting, solar electricity generation, low-energy buildings, compressed natural gas (CNG), hybrid vehicles, and smart grids (U.S. Conference of Mayors, 2014). Dissemination of new lighting solutions in cities could result in significant energy savings (60–70% in offices, schools, hospitals, etc.) according to Verhaar (2009). In addition to generating energy savings, smart grids enhance the competitiveness of megacities, improve the security of supply, and create green jobs (Evans, 2009). Consulting and engineering companies can specialize in providing climate-specific solutions in the design of urban form and in providing frameworks for decision-making under uncertainty (Bellamy and Patwardhan, 2009; Galal, 2009).

Box 7.2   Information and Communications Technology and Climate Change

Peter Adams

Acclimatise, New York

Michaël Houle

MIT Climate CoLab, Cambridge, MA/ICLEI Canada, Ottawa

INTRODUCTION

The information and communications technology (ICT) sector is a critical part of the global economy and plays a significant role in the operations and economies of cities. ICT is also a key part of the equation as cities address climate change, where there are three primary links:

1. *ICT and mitigation*: As a growing source of GHG emissions, largely through high energy consumption, ICT providers face increasing regulation to reduce emissions through increased efficiency.
2. *ICT and climate risk*: With complex, sprawling networks of infrastructure and technology, ICT providers also face direct and indirect physical risks because they are exposed to changing weather and climate conditions that can stress thresholds and operating parameters.
3. *ICT and emergency management*: Functioning, resilient communication is vital for coordinating disaster response during extreme weather events associated with climate change.

Despite the multiple climate issues facing ICT planners and providers, there is limited research in this field to date. This section highlights overarching issues facing cities, ICT providers, and emergency services toward addressing challenges at the intersection of ICT and climate change.

ICT AND MITIGATION

The ICT sector's total emissions rose from 0.53 to 0.91 $GtCO_2e$ between 2002 and 2011 and are expected to rise to 1.27 $GtCO_2e$ by 2020 (GeSI, 2012). This represents 2.3% of global emissions in a business-as-usual growth emission scenario (GeSI, 2012). Even though the ICT sector's potential to abate the annual GHG emissions of all sectors is seven times higher than the ICT sector's own direct emissions (GeSI, 2012; Jacob et al., 2011). Many of the most transformative economic trends (e.g., social media, big data) are dependent on cloud computing for the storage, transfer, and processing of digital information. Operating large-scale data centers accounts for approximately 1.3% of worldwide electricity consumption. This percentage is expected to grow as high as 8% by 2020 (Deng et al., 2014).

However, the sector also offers technologies that support climate mitigation, such as intelligent building systems and intelligent transportation systems. These play a critical role in helping other sectors tackle climate change.

Data Centers' Sources of Energy

The carbon footprint of a data center is influenced by the type of energy used ($GtCO_2e$/kWh ratio) and its energy efficiency level. Building location influences a data center's carbon footprint owing to market differences in the mix of primary energy used for generating electricity (Shehabi et al., 2011). Locating data centers where renewable energy markets are strong and reliable could contribute to the reduction of their carbon emissions. However, powering cloud data centers with renewable energies is challenging due partly to the fact that global users require cloud services throughout the day and night. The intermittent nature of renewable energy consequently represents a challenge for data center operators that require consistent sources of power (Deng et al., 2014).

Data Centers' Energy Efficiency Level

The most common measure of energy efficiency is *power usage effectiveness*, which is the ratio of overall power drawn by the data center facility to the power delivered to IT hardware (Accenture, 2010). Heating, ventilating, and air conditioning (HVAC) is the dominant component of non-IT energy consumption (Shehabi et al., 2011). Enterprises often cool internal systems by routing outside air – a method called *free cooling* – into the data center, thereby reducing the energy demand associated with its operation. There has been a recent trend to deploy data centers in cooler climates. For example, Facebook revealed plans to locate one of their data centers near the Arctic Circle in Sweden (Stevenson, 2014). Data centers are also beginning to use "dry coolers (closed loops) ... in combination with tech equipment capable of operating at higher temperatures based upon revised guidelines" (Adams et al., 2014).

Enterprises with local data center and cloud providers will have to find the appropriate balance between cost minimization, renewable and fossil fuel energy, and energy efficiency if they are to prevent the ICT abatement potential from decreasing.

ICT AND CLIMATE RISK

ICT networks rely on distributed physical infrastructure that can be at risk to a wide range of climate impacts. Climate change presents an increasing intensity and frequency of extreme events that can physically damage this infrastructure or sensitive equipment, resulting in service disruptions. Slower, incremental changes in climate also can lead to damage. As climatic conditions move outside the designed operational capacity and tolerance thresholds of ICT systems, this will result in heightened wear and tear that reduces the lifetime of assets or increases the need for maintenance (Adams et al., 2014). The ICT sector is also dependent on other sectors, such as water and energy, to function – sectors that are themselves at high risk from climate change. All of these risks can raise costs, reduce return on investments, and reduce quality of service.

There is also the threat of severe ripple effects to all other sectors that rely on ICT to facilitate their communications and operations, as well as the coordination of emergency responses associated with extreme weather (see the following section). A compounding risk factor is the trend toward companies sharing infrastructure or elements of the infrastructure (e.g., underground cables, transmission towers) to save on capital costs (Ricardo AEA, 2014). Business efficiency in this case might negatively affect ICT's redundancy level and thus its resilience.

Impacts from extreme weather are increasingly prevalent in the sector and will likely increase in severity. Reducing ICT's vulnerability to natural hazards can be achieved through different means, including relocating critical infrastructures and elevating the equipment off the ground. Local and remote data centers used by emergency response agencies for gathering critical information must also be designed to be disaster-resilient. Similarly, the deployment of smart grids and local power generation can help prevent localized disruptions in services caused by external stress from grid failures or power outages. Despite these challenges and opportunities to build resilience, there is still little awareness of climate risk in the ICT sector, and thus there has been limited action in the private or public sector to address these risks (Horrocks et al., 2010; Adams et al., 2014).

ICT AND EMERGENCY MANAGEMENT

High performance in disasters requires the ability to access critical information and to expand coordination between emergency response agencies. The proliferation of new technologies in urban areas generates a reconfiguration of the communication environment and offers new insights into public and private sectors for better disaster planning, management, and response. For example, ICT provides the backbone of advanced traffic management systems and intelligent transportation systems. The data collected through and information generated by these systems are critical for the emergency response (e.g., the management of evacuation and detour routes, movement and distribution of relief supplies). Other tools such as disaster management information systems – supported by ICT – are essential to interorganizational communications and to the creation of a common operating protocol among emergency response agencies (Houle, 2015).

The importance of information calls for new ways of ensuring data gathering and information sharing during a crisis. In emergency management, "an inability to communicate is a threat to national and human security and puts business value at risk" (Adams et al., 2014). The enhancement of the ICT sector's resilience must therefore include risk reduction measures to ensure lines of communication are maintained in the event of a disaster. One way of doing that is through a cloud-supported crisis response and management process (Buscher et al., 2014). Cloud disaster response can ensure rapid deployment of information services and resource management capabilities in the event that local infrastructures are damaged (Grolinger et al., 2013). However, these new advances may simultaneously represent a reduction of the responding agencies' ability to operate independently of other elements of the emergency system (Ricardo AEA, 2014).

CONCLUSION

Cities have a growing impetus to assess and address how the ICT systems they rely upon are tied to climate change through mitigation, adaptation, and emergency management. Cities must not only address these challenges in the context of municipal ICT systems, but must also engage with major telecom companies to develop robust and resilient responses.

The participation of the private sector in climate change "experiments" provides evidence of business opportunities in activities that promote sustainability and urban resilience. Coined by Castán Broto and Bulkeley (2013), an "'experiment" is understood as an intervention to test innovative solutions in the context of climate change uncertainty. According to the review of case studies by Castán Broto and Bulkeley (2013), there are about 170 cases where the private sector has been involved in climate change experiments as a partner and almost 100 cases where the private sector was the leading actor. Their research shows that the private sector has already participated in transformative initiatives, but, in many cases, the public sector has taken the leading role. In the majority of the cases, climate experiments were led by a local government and consisted of greening of infrastructure and consumer services provided by different authorities or of supporting initiatives led by other actors through information and resource provision and partnerships. When experiments were led by private companies, governments were partners in only 15% of the cases.

Many of these opportunities are still underdeveloped and require market-making support by the public sector (Khattri et al., 2010). For instance, municipalities can take the lead in the introduction of information and communication technologies (ITC) through the "dematerialization" of public services, by expanding the infrastructures of broadband, by implementing advanced technologies in their operations, and more. Such leadership translates into job creation, sources of tax revenue, and competitiveness for cities. The city platform called Forum Virium Helsinki is an example of cooperation between city actors to develop ICT-based services (http://www.forumvirium.fi/en) such as mobile phone services (that facilitate urban travel and living) and the opening up of public data (that has enabled companies and citizens to create new services) (Anttiroiko et al., 2014). Floater et al. (2014a) describe a number of actions municipalities can take to facilitate the exploitation of new business opportunities, such as scaling the market for energy efficiency, facilitating access to finance of the clean tech sector, and exploring new models of PPPs for delivering goods and services.

Other opportunities arise from innovative business models that create compelling proposals for consumers. This is the case of car-sharing/-pooling, for which the market has increased by more than 20% annually (Cohen, 2013). In addition, governments and private companies can work together on education programs about the benefits of smart technology (Osborne Clarke, 2015). According to North and Nurse (2014), an effective way in which local authorities can support transformation in the private sector is by promoting the diffusion of "war stories" from those entrepreneurs whose personal experience has led them to change their practices. Change-adverse business people might be convinced of the benefits of transformation when they become aware of the advantages of concrete experiences of other peers (e.g., cost savings, attraction of better human capital, reputation, among others). To facilitate the communication of these stories among small and medium enterprise (SME) owners is especially important, given that SMEs represent 60% of industrial GHG emissions. Today, green-tech innovation is concentrated in a few dynamic clusters close to cities. According to Kamal-Chaoui and Roberts (2009), between 2004 and 2006, 73% of green patents in the renewable energy sector in the OECD were produced in urban regions. The authors provide some examples like the Lahti Clean Tech Cluster in Finland or the London Hydrogen Partnership. By playing a facilitating and enabling role, cities can promote the creation of green clusters that will attract and retain innovative companies and propel competitiveness. The vision of becoming a global example of an eco-energy city guided the design of a successful policy mix (i.e., promotion of a renewable energy cluster, stimulation of local demand, and attraction of investment) in Dezhou, China. The Development Plan of this city is a good example of how to draw on local strengths to build a renewable industry cluster from scratch (Yong, 2013).

7.3.1.3 Business Finance

The lack of access to funding has been usually identified as one of the most critical limiting factors for climate actions in cities (Kernaghan and da Silva, 2014). The private sector is a source of finance, especially when there are opportunities to improve business performance, continuity, or opportunities to participate in profitable partnerships (Schuttenbelt, 2013). Finance organizations and companies across the world have shown their interest in the transformation of the city landscape, provided that the risk–reward proposition is attractive (Osborne Clarke, 2015).

Private investment can be also be motivated by noneconomic factors, such as reputational benefits or networking opportunities. In this regard, for example, the Bangka Botanical Garden in Pangkalpinang (Indonesia) was launched as a Corporate Social Responsibility project (Hardjosoesilo, 2011b).

Innovative governance structures such as partnerships between the private and public sector can stimulate private sector co-financing (Tompkins and Eakin, 2012). Partnerships are generally developed to meet the common interests between two or more parties. In the case of cities, PPPs are described as coalitions between local authorities that lack public funding and private entities and are commonly used to deliver critical infrastructure, housing affordability, and urban regeneration (Harman et al., 2015; Johannessen et al., 2013). The role of the municipality in PPPs consists of facilitating project development by removing barriers, while the private sector assumes part of the risk, provides funding, and manages the project. As a type of externalization, the proliferation of PPPs can be linked with budget constraints and the expiration of subsidy schemes (Bacheva-McGrath et al., 2008). PPPs are usually associated with innovation, spillover benefits, cost-efficiency, increased productivity, and political independence (IFC, 2011; Johannessen et al., 2013). Some promising PPPs are privately financed schemes in which the private sector expects a profit from the investment. However, PPPs are also criticized because the interests of the private entity might go against public interests (Harman et al., 2015).

In the case of water, sanitation, and hygiene (WASH) systems, Johannessen et al. (2014) argue that PPPs are a good solution to address the potential lack of public funding and thus to possibly manage resources more effectively. They also argue that to build long-term resilience, it is important to integrate long-term risks of climate change into business management and to build partnerships through a series of strategies for investments. These include re-examining the profitability of existing WASH investments taking into account ecosystem services, acknowledgment of the needs of adequate land area for water resources, creation of an institutional culture for private-sector investments, development of a better understanding of the customer base (especially focused on poor, more-vulnerable communities), creation of support for a new segment of private entrepreneurs, development of

Case Study 7.3   Raising Awareness of Negotiating, and Implementing Tokyo's Cap-and-Trade System

Magali Dreyfus

National Center for Scientific Research (CNRS), Paris

Keywords	Cap-and-trade system, mitigation
Population (Metropolitan Region)	37,750,000 (Demographia, 2016)
Area (Metropolitan Region)	8,547 km² (Demographia, 2016)
Income per capita	US$42,870 (World Bank, 2017)
Climate zone	Cfa – Temperate, without dry season, hot summer (Peel et al., 2007)

Tokyo's Emissions Trading System (ETS) is the first city-level cap-and-trade system applied to greenhouse gases (GHG) emissions worldwide. Its overall target is to reduce CO_2 emissions by 25% below 2000 levels by 2020.

The ETS applies to the heaviest emitters of the industrial and commercial sectors: Industrial factories, office buildings, administrative institutions, and commercial buildings. Most of Tokyo's skyscrapers fall under the scheme. Around 1,400 large-scale facilities (1,100 business facilities and 300 industrial facilities), which account for approximately 20% of Tokyo's GHG emissions, have been selected on the basis of their energy consumption. The baseline is a total consumption of fuels, heating, and electricity of at least 1,500 kiloliters per year (crude oil equivalent) (Tokyo Metropolitan Government [TMG], 2010).

These facilities are awarded a limited number of allowances (i.e., "cap") that determine the total quantity of GHG emissions that they are authorized to release for a given time period. Facility tenants have the responsibility to control their CO_2 emissions and to adopt mitigation measures. Units that emit less than the credits they have can "trade" their unused allowances to other participants who exceed their cap (Lee and Colopinto, 2010).

Allowances are *grandfathered*, which means awarded free of charge. Their number is fixed on the basis of past emissions. Thanks to the "Tokyo CO_2 Emissions Reduction Program" launched in 2000, which included a voluntary emissions reduction plan with a mandatory reporting scheme for targeted facilities, data on emissions have been collected since that time. On this basis, the ETS base year is calculated as a function of the average emissions of the facilities over the 3 years between 2002 and 2007.

Allowances are allocated at the beginning of each compliance period. Their number is calculated as follows (first period):

Base year emissions – required reduction or "compliance factor" (6% for industrial buildings or 8% for rest of the buildings)] × compliance period (5 years) (Lee and Colopinto, 2010).

They are two compliance periods: the first one, from 2010 to 2014, foresees a reduction of 6% of GHG emissions for the 5-year annual average; the second one, from 2015 to 2019, aims at a reduction of 17% of emissions for the 5-year annual average. Monitoring and reporting are required on a yearly basis.

To support the efficient functioning of the scheme, a system of alternative credit offsets is also established. It consists of small and medium-sized installation credits within the Tokyo area and renewable energy certificates in the whole country. This second mechanism is favored by TMG. It allows companies to get credits on the basis of renewable energy certificates received by the company thanks to installations located outside of Tokyo and using renewable energy. Yet this is limited to up to one-third of the company's obligations (IETA, 2014).

The ETS started functioning in April 2010, and the first performance evaluation was made in 2015. As of 2014, only a few allowances have been traded because about 93% of facilities have largely met their targets for the first compliance period (IETA, 2014). In 2016, the Tokyo Metropolitan Government reported that the program had achieved a 25% reduction in emissions in its first 5 years. For participants who do not meet their targets, the system foresees financial fines and moral "shaming," where names of defaulting institutions are released. For now, the ETS covers only one GHG, namely CO_2. However the ETS plans to be extended to other gases in the future (TMG, 2010).

The ETS is an innovative tool in several ways. It is the first city-level carbon trading scheme that closely connects a global-level challenge with local action. It is also the only mandatory ETS with an absolute volume cap in Japan, and it is the only local ETS focusing on commercial activities and energy end-users.

The success of the scheme lies in the close relationship between the facilities concerned and the TMG. Through a series of meetings, they defined together the targets and potential actions. The mandatory reporting of data to TMG has been a key step in building up a sound dataset upon which the cap-and-trade mechanism could be further refined. It allows data collection, familiarity, and capacity-building for the stakeholders involved in the program. The evolutionary nature of the scheme, from voluntary activities to binding regulations, is also a relevant feature of the Tokyo experience.

The Tokyo process could inspire other cities of the world. Many cities today have emission targets, yet they lack concrete regulatory and mandatory measures to achieve their goals. The process that led to the adoption and success of Tokyo's cap-and-trade program could be an inspiring example for other cities.

micro-insurance mechanisms in dialogue with the most-vulnerable groups, and, linked to this, the creation of (micro)financial opportunities.

In the area of climate-related projects, the Chicago Climate Action Plan is a good example of an effective PPP for the design and implementation of adaptation policies at city level. Local business participated actively in the monitoring of policies and problem-solving and co-financed the downscaling of climate models, the assessment of the costs of inaction, and the implementation of the plan (Rasker, 2012). The London Climate Change Partnership, running since 2001, exemplifies a multisectoral partnership that crosses spatial, political, and organizational boundaries to collaboratively tackle the complex goal of enhancing city resiliency (see Case Study 7.1) (Surminski and Leck, 2016).

Another example of an innovative partnership enabling a successful climate policy can also be found in California: the Community Energy Partnership is a collaboration between seven cities and a utility, funded via a consumption fee, that promotes energy efficiency in households and companies (by means of energy audits, improvement of installations, etc.). The Association of Bay Area Governments' Energy Watch, funded by a utility, enhances the energy efficiency of the local government. In addition to associations with the private sector, the key element for the success of these initiatives is the generation of cost savings and economic development benefits. Using PPPs to facilitate the implementation of pilot projects of new technologies in urban areas has been demonstrated to work well. That is the case of Rotterdam with carbon capture and sequestration technologies (Rotterdam Climate Initiative [RCI], 2011).

Although PPPs have been demonstrated to work well in a number of cases, there is evidence from other fields, however, suggesting that PPP have not always delivered the expected advantages. In particular, the costs associated with these schemes have been often underestimated and used to evade restrictions on public budgets. Here, we list a series of recommendations about how to guarantee beneficial PPP projects (Bacheva-McGrath et al., 2008; Bunning, 2014):

- Use realistic predictions to calculate affordability.
- Establish restrictions, like annual limits on the total amount of payments that the authorities can commit.
- Contrast the convenience of PPPs with alternatives like public procurement, community-based initiatives, joint-ventures, cooperatives, and the like.
- Foster transparency and accountability in order to limit opportunities for corruption and inflation of projects.
- Specify conditions for the termination of contracts, penalties for poor performance, and benefit-sharing schemes for refinancing benefits.

An assessment of 165 empirical ex-post studies examining policies on low-carbon technologies suggests that there exists tradeoffs between efficiency and effectiveness, resulting in government-led policy instruments being more effective and private-led interventions being more efficient (Auld et al., 2014).

7.3.1.4 Risk Transfer

With increasing climate change impacts, the role of insurance as a tool for risk-sharing and transfer to address uncertainties is receiving growing attention (Mills, 2009a; Surminski et al., 2016). Some insurance sector initiatives (e.g., Climate Wise., ClimateWise and UNEPFI's Insurance Working Group), and industry organizations (e.g., the Chartered Insurance Institute) have actively engaged in policy debates, assessed climate

impacts and opportunities, and initiated adaptation activities (Mills, 2009b; Surminski, 2014). Very recently, the Global Innovation Lab for Climate Finance announced its Energy Savings Insurance scheme. This is a pilot initiative consisting of insuring the value of savings generated by energy-efficiency projects. The initiative has received the support of the Inter-American Development Bank and the Danish Energy Agency.

While some authors note that insurance is not a silver bullet, it can help in driving response to climate change (Ranger et al., 2011; Mechler et al., 2014). In the urban context, insurance has three functions:

- *Compensate losses and fund recovery efforts*: Risk transfer is more cost-effective for increasing resilience than ex-post disaster aid (Ranger et al., 2011). Recent examples where insurance provision for climate risks has been taken into account in an urban context are Mumbai (Ranger et al., 2011) and New York (Aerts et al., 2011). In the case of Mumbai, Ranger et al. (2011) estimate that indirect losses could be halved if insurance penetration rate would achieve 100%.
- *Reduce the financial risk of investments*: Insurance could reduce barriers to private investment in climate actions (Surminski, 2013). Positive urban examples are New York and Rotterdam, where flood insurance schemes help control the vulnerability to flood risks and also reduce the barriers to potential private investments in the waterfront and port areas (Aerts and Botzen, 2011).
- *Incentivize risk management activities*: Purchasing an insurance risk transfer product can influence the behavior of those at risk (Surminski and Oramas-Dorta, 2013). If not correctly structured, it can provide disincentives, but otherwise, it puts a price tag on risks, signaling the need to address underlying risks (Kunreuther, 1996; Botzen et al., 2009; Shilling et al., 1989; Treby et al., 2006).

One particular type of risk-transfer instrument that could be suitable for climate risks in cities is the *catastrophe bond*. This is a financial instrument developed by insurers or governments to pass extreme risks on to private investors who are willing to assume them in exchange for high interest rates. For example, Allianz recently issued a flood bond for London. However, it must be noted that this instrument does not reduce risks if the proceeds are not used in risk reduction measures. In addition, catastrophe bonds might not be an appropriate instrument to protect against climatic risks in that they are narrowly designed for specific events in specific locations, tending to protect private investor interests (Keogh et al., 2011; Brugmann, 2012).

7.3.2 Challenges and Enablers of Private Sector Engagement

Key barriers for the private sector when implementing adaptive strategies are the lack of understanding of the uncertainties, poorly perceived risks, and limited knowledge and expertise. According to Crawford and Seidel (2013), many companies often lack in-house knowledge or expertise about extreme weather and climate change. Because of this, it is important to engage with suppliers of key knowledge inputs to improve event response planning and capacity building.

Regarding mitigation, the main barrier is the lack of appropriate incentives. According to the Global Commission on the Economy and Climate, despite the high and multidimensional costs of the business–as-usual urban development model (i.e., urban sprawl), market and governance failures causing current problems continue in many cases unaddressed (Floater et al., 2014b; Rode et al., 2013). The NCE Report (2014) presents the case of the Netherlands as an example of how costly fuel subsidies increase the use of cars and the case of the Multilateral Development Bank's financial support as an example of how funding can be directed toward a model of development incompatible with resilience and sustainability. Low-carbon technology and resilient infrastructure also encounter barriers to attracting private investment at scale; one reason is the lack of appropriate information for investors, companies, and public authorities (Floater et al., 2014a). This could be addressed by complying with some of the recommendations about how to guarantee beneficial PPPs (see Section 7.3.1.3).

Enablers of private-sector engagement can be classified in three groups:

- *Demand-side enablers* such as financial products to enhance capacity to pay; demand generation by awareness-building programs; information-disclosure mandates; voluntary labeling initiatives; community buy-in with demonstration effects; public pilot and demonstration projects; city, regional, and national government procurement; and city-led initiatives of cross-border collaboration to create strong regional markets for low-carbon and other green products and services;
- *Financial enablers* can be public or private; for example, local/international seed capital for technical assistance by banks; high city credit ratings; project liquidity; monetization of avoided losses; innovative financial products to reduce risk including value-capture instruments, insurance, and reinsurance; catastrophe bonds; social impact bonds; securitization and structured finance; public subsidies and support programs; and the green bond market;
- *Supply side enablers* such as forging partnerships with market aggregators; PPP models; microfranchisors and technology platforms; associations with R&D organizations; public interventions to address externalities, coordinate policies and actors, and generate and disseminate information on the scale of opportunities and risks; and those types of enabling actions that capture opportunities for comparative advantage in global markets for green products and services.

Local authorities can play an important role as facilitators of all three groups by providing regulatory and fiscal environments that encourage the reduction of risks and the transition to a low-carbon economy (Cleverley, 2009). They can:

- *Provide incentives*: Encouraging or requiring the implementation of risk-management practices; guaranteeing the stability of policy interventions across levels and areas to correct externalities; setting standards for efficiency with consistent metrics for monitoring and verification; promoting behavioral changes toward sustainability goals, in some cases through market-based instruments or through providing education and information, including environmental labeling and support to R&D; reducing risks by setting long-term targets and supporting pilot projects; building capacity by promoting community engagement, civic movements, economic networks, partnerships, and clusters; and establishing enabling conditions in markets (e.g., Payment for Ecosystem Services) that deliver urban adaptation;
- *Redirect support*: Transferring incentives from industries flourishing under the business-as-usual development model to low-carbon, resilient businesses. Among the sectors offering positive outcomes in development and climate action are the renewable energy sector, electric cars, energy-efficient devices, affordable housing, green infrastructure, ICTs, and biodiversity conservation (da Schio, 2013);
- *Mainstream adaptation*: Implementing appropriate risk-reduction strategies; improving the efficiency of public-sector data centers, buildings, and operations; and sponsoring virtual infrastructure (i.e., cloud computing); policies and planning must incorporate the adaptation dimension to ensure that investments and actions by the private sector are protected (Kazmierczak and Carter, 2010; Mees and Driessen, 2011).

7.4 Areas for Further Research and Considerations for Policy

7.4.1 Further Research

Empirical evidence suggests that successful climate action at the local level benefits local inhabitants. However, this effect is rarely well-documented in the public domain, and local outcomes could be better communicated within a co-benefit framework of city-level climate policy. At the same time, we find that important spillover benefits from local climate action are already accruing at the national level and to other cities, as exemplified by diverse demonstration and learning effects. Both, in turn, justify public policy support for networking the activities of cities in this field and global research infrastructures, such as UCCRN, to document and communicate these effects.

There is an order of magnitude gap between mitigation and adaptation needs and available funding at the urban level. This gap, however, turns out to be more of a problem of access to private and public funds than a problem of availability of investment capital. Programs to improve access to finance the sustainability of cities (e.g., the World Bank Group) are under way but must be closely monitored with regard to identifying the key factors of success and failure, given the multiple barriers assessed in this chapter. The better documented those programs are, the more subsequent efforts may attract the masses of private-sector capital that have yet to consider climate change investments as key elements of their portfolios.

Given the similar efforts of funding programs for capacity-building and technology innovation, we recommend a special assessment report on "Sustainability Finance for Cities" in 2020 as a UCCRN Special Report. The Case Study Docking Station (CSDS) of UCCRN could be a promising source of knowledge for this report if structured for that need through the collection of more detailed cost, benefit, and non-monetized impact data on the cases profiled.

7.4.2 Policy Considerations

Private-sector agents respond to incentives. Accordingly, policy design is crucial. Policy measures for urban transformation will affect private-sector actors and can help reduce unavoidable resistance to change. But climate action will not necessarily generate losers if industries are able to incorporate into their plans the urgent need for an alternative development model and innovate accordingly. The private sector as a source of finance, capacity, and innovation constitutes a key partner for cities in the search for transformation. Cities and businesses can build a mutually beneficial relationship based on their shared objectives and complementary resources and capabilities.

Companies can better minimize climate risks and benefit from new business opportunities in those cities where the public sector is committed to playing a leading role in urban transformation. At the same time, municipalities can neither achieve the required transformation of unsustainable practices nor prepare themselves for the impacts of climate change without coordination with the private sector. Many cities have the legal power and responsibility to implement urban development and risk reduction, but they often lack the capacity and resources to tackle ambitious transformative projects. Thus, collaboration with the private sector may give cities access to finance and expertise on risk management, resource optimization, and technological solutions, thus facilitating at the same time the creation of jobs as well as addressing climate change challenges.

Annex 7.1 Stakeholder Engagement

This chapter benefited from important contributions and feedback from a variety of parties at conferences and meetings. The launch meeting included inputs from Richenda Connell of Acclimatise (London, UK), Stefan Denig and Michael Stevns from the London office of SIEMENS, and Mussa Natty of the Kinondoni Municipal Council, Dar es Salaam, Tanzania. Inputs during the process of chapter development were then collected from a webinar on "Show Me the Money" conducted for the U.S. Department of Energy in December 2013; the National Conference on Science and the Environment conference on Building Climate Solutions in Washington, D.C., January 2014; the ICLEI–Local Governments for Sustainability – Resilient Cities Congress in Bonn, May 2014; a workshop

on cross-cutting chapters held after that ICLEI meeting; the CITY FUTURES III conference in Paris June 2014; the 11th Symposium of International Urban Planning and Environment Association in La Plata, Argentina, September 2014; the London Climate Change Partnership, with a survey of members on the role of the private sector in managing urban risks in November 2014; the Climate Change Risk Roundtables discussion on the role of insurance in February 2015; a panel debate on the role of insurance for climate risks at the Third UN World Conference on Disaster Risk Reduction in March 2015; and a workshop on the chapter in Bonn, June 2015, that incorporated views from GEF, World Bank Group, and the Rockefeller Foundation.

Acknowledgments

Maria Victoria Román de Larais is funded by Strategic Challenges in International Climate and Energy Policy (CICEP), one of three centers for social science–based research on environmentally friendly energy established by the Research Council of Norway in 2011 (http://www.cicep.uio.no/).

Marta Olazabal is funded by the Spanish Ministry of Economy and Competitiveness (MINECO).

Chapter 7 Economics, Finance, and the Private Sector

References

Accenture Microsoft Report. (2010). Cloud computing and sustainability: The environmental benefits of moving to the cloud. Accessed November 12, 2015: http://www.accenture.com/sitecollectiondocuments/pdf/accenture_sustainability_cloud_computing_theenvironmentalbenefitsofmovingtothecloud.pdf

Adams, P., Steeves, J., Ashe, F., Firth, J., and Rabb, B. (2014). *Climate risk study for telecommunications and data center services: Report prepared for the general services administration.* Accessed March 19, 2016: https://sftool.gov/Content/attachments/GSA%20Climate%20Risks%20Study%20for%20Telecommunications%20and%20Data%20Center%20Services%20-%20FINAL%20October%202014.pdf

Aerts, J. C. J. H., and Wouter Botzen, W. J. (2011). Flood-resilient waterfront development in New York City: Bridging flood insurance, building codes, and flood zoning. *Annals of the New York Academy of Sciences* 1227, 1–82. doi:10.1111/j.1749–6632.2011.06074.x

Allen, J. H., and Potiowsky, T. (2008, November). Portland's green building cluster. *Economic Trends and Impacts. Economic Development Quarterly* 22, 303–315.

Anttiroiko, A. V., Valkama, P., and Bailey, S. J. (2014). Smart cities in the new service economy: Building platforms for smart services. *AI Society* 29, 323–334. doi:10.1007/s00146–013–0464–0

Averchenkova, A., Crick, F., Kocornik-Mina, A., Leck, H., and Surminski, S. (2016). Multinational and large national corporations and climate adaptation: Are we asking the right questions?: A review of current knowledge and a new research perspective. *Wiley Interdisciplinary Reviews: Climate Change* 7(4), 517–536.

Auld, G., Mallett, A., Burlica, B., Nolan-Poupart, F., and Slater, R. (2014). Evaluating the effects of policy innovations: Lessons from a systematic review of policies promoting low carbon technology. *Global Environmental Change* 29, 444–458.

Bacheva-McGrath, F., Cisarova, E., Eger, A., Gallop, P., Kalmar, Z., and Ponomareva, V. (2008). Never mind the balance sheet. The dangers posed by public-private partnerships in central and eastern Europe. CEE Bank Watch Network. Accessed September 9, 2014: http://bankwatch.org/documents/never_mind_the_balance_sheet.pdf

Bell, M. L., Hobbs, B. F., and Ellis, H. (2003). The use of multi-criteria decision-making methods in the integrated assessment of climate change: Implications for IA practitioners. *Socio-Economic Planning Sciences* **37**(4), 289–316.

Bellamy, W., and Patwardhan, A. (2009). CH2M Hill today. Presentation by the Vice President and Technology Fellow of CH2M Hill. In *A Dialogue on Cities and Climate Change Workshop*, organized by the World Bank. Accessed September 9, 2014: http://siteresources.worldbank.org/INTUWM/Resources/340232–1254279511547/Bellamy.pdf

Beltran, P. T. (2012). International financing options for city climate change interventions – An introductory guide. Accessed August 14, 2015: http://www.cdia.asia/wp-content/uploads/International-Financing-Options-for-City-Climate-Change-Interventions1.pdf

Botzen, W. J. W., Aerts, J. C. J. H., and van den Bergh, J. C. J. M. (2009). *Willingness of homeowners to mitigate climate risk through insurance. Ecological Economics* **68**(8–9), 2265–2277.

Brugmann, J. (2012). Financing the resilient city. *Environment and Urbanization* **24**(1), 215–232.

Bulkeley, H. (2010). Cities and the governing of climate change. *Annual Review of Environment and Resources* **35**, 229–255.

Bunning, J. (2014). Governance for regenerative and decarbonised eco-city regions. *Renew. Energy, Renewable Energy for Sustainable Development and Decarbonisation* **67**, 73–79. doi:10.1016/j.renene.2013.11.041

Buescher, M., Easton, C., Kuhnet, M., Wietfield, C., Ahlén, M., Pottebaum, J. And van Veelen, B. (2014). Cloud ethics for disaster response. In S.R. Hiltz, M.S. Pfaff, L. Plotnick and P.C. Shih (eds.), *Proceedings 11th International Conference on Information Systems for Crisis Response and Management.* (284–288). Pennsylvania State University Press, ICSRAM, 284–288.

Carbon Disclosure Project (CDP). (2012). *Insights into Climate Change Adaptation by UK Companies*. Author. Accessed March 27, 2015: https://www.cdproject.net/CDPResults/insights-into-climate-change-adaptation-by-uk-companies.pdf

Carmin, J., Dodman, D., and Chu, E. (2013). *Urban Climate Adaptation and Leadership*. OECD regional development working papers. Organization for Economic Co-operation and Development.

Castán Broto, V., and Bulkeley, H. (2013). A survey of urban climate change experiments in 100 cities. *Global Environmental Change* **23**, 92–102. doi:10.1016/j.gloenvcha.2012.07.005

Cleverley, M. (2009). The private sector, cities and climate change. Presentation by Mark Cleverley from IBM Corporation. In *A Dialogue on Cities and Climate Change Workshop*, organized by the World Bank. Accessed January 11, 2016: http://siteresources.worldbank.org/INTUWM/Resources/340232–1254279511547/Cleverley.pdf

Climate Bond Initiative. (2014). Bonds and climate change – The state of the market in 2014. Accessed February 25, 2016: http://www.climate-bonds.net/files/files/CB-HSBC-15July2014-A4-final.pdf

The Climate Group. (2008). SMART 2020: Enabling the low carbon economy in the information age. Accessed March 11, 2014: http://www.smart2020.org/_assets/files/02_smart2020Report.pdf

Climate Policy Initiative (CPI). (2013). The global landscape of climate finance 2013. Accessed January 31, 2015: http://climatepolicyinitiative.org/wp-content/uploads/2013/10/The-Global-Landscape-of-Climate-Finance-2013.pdf

Cohen, A. (2013). Innovative mobility carsharing outlook: Carsharing market overview, analysis, and trends – Summer 2013. Accessed January 25, 2015: http://www.innovativemobility.org/publications/Carsharing_Innovative_Mobility_Industry_Outlook.shtml

Crawford, M., and Seidel, S. (2013). Weathering the storm: Building business resilience to climate change. Accessed November 6, 2014: http://www.c2es.org/docUploads/business-resilience-report-07–2013-final.pdf

Da Schio, N. (2013). *Stimulating Renewable Energy through Public and Private Procurement*. City in focus: Austin, U.S. ICLEI Local Governments for Sustainability/IRENA, International Renewable Energy Agency.

Deng, W., Fangming, L., Hai, J. and Li, D. (2014). Harnessing renewable energy in cloud datacenters: Opportunities and challenges. *IEEE Network* **28**(1), 48–55.

Erickson, P., and Tempest, K. (2014). Advancing climate ambition: Cities as partners in global climate action. A report to the UN Secretary-General from the UN Secretary General's Special Envoy for Cities and Climate Change, in partnership with the C40 Cities Climate Leadership. Accessed April 28, 2015: www.C40.org/research

Evans, P. (2009). Smart grid. Presentation of the Director of Global Strategy and Planning of General Electric Energy. GeSI – Global e-Sustainability Initiative. In *A Dialogue on Cities and Climate Change Workshop*, organized by the World Bank. Accessed September 24, 2014: http://siteresources.worldbank.org/INTUWM/Resources/340232–1254279511547/Evans.pdf

Floater, G., Rode, P., Robert, A., Kennedy, C., Hoornweg, D., Slavcheva, R., and Godfrey, N. (2014b). *Cities and the New Climate Economy: The Transformative Role of Global Urban Growth*. New Climate Economy Cities Paper 01. LSE Cities. London School of Economics and Political Science.

Floater, P., Rode, D., Zenghelis, M., Ulterino, D., Smith, K., Baker, K. and Heeckt, C. (2014a). *Copenhagen: Green Economy Leader Report*. LSE Cities. Economics of Green Cities Programme, LSE Cities. London:London School of Economics and Political Science.

Fuerst, F., McAllister, P., Nanda, A., and Wyatt, W. (2013). *Final Project Report: An Investigation of the Effect of EPC Ratings on House Prices*. Department of Energy & Climate Change.

Galal, H. (2009). A collaborative approaches to cities' development. Presentation by the Global Leader for Cities and Local Government in Price Waterhouse Coopers. In *A Dialogue on Cities and Climate Change Workshop*, organized by the World Bank. Accessed September 4, 2014: http://siteresources.worldbank.org/INTUWM/Resources/340232–1254279511547/Galal.pdf

GeoAdaptive. (2013). *Cambio Climático y Espacio Público. Adaptaciones Estratégicas al Cambio Climático a través de los Espacios Públicos en Colombia*. Informe Final.

GeSI. (2012). SMARTer2020: The role of ICT in driving a sustainable future. Accessed August 18, 2014: http://gesi.org/assets/js/lib/tinymce/jscripts/tiny_mce/plugins/ajaxfilemanager/uploaded/SMARTer%202020%20-%20The%20Role%20of%20ICT%20in%20Driving%20a%20Sustainable%20Future%20-%20December%202012._1.pdf

Grafakos, S., and Flamos, A. (2015). *Assessing low carbon energy technologies against Sustainability and Resilience criteria: Results of a European experts survey. International Journal of Sustainable Energy* **0**(0), 1–15.

Grolinger, K., Capretz, M., Mezghani, E., and Exposito, E. (2013). *Knowledge as a Service Framework for Disaster Data Management*. Paper presented at the Enabling Technologies: Infrastructure for Collaborative Enterprises (WETICE), 2013 IEEE 22nd International Workshop.

Hallegate, S. (2006). *A Cost-Benefit Analysis of the New Orleans Flood Protection System*. AEI-Brookings Joint Center.

Hallegatte, S., Henriet, F., and Corfee-Morlot, J. (2008). *The Economics of Climate Change Impacts and Policy Benefits at City Scale: A Conceptual Framework*. OECD Environment Working Paper 4, ENV/WKP (2008)3. OECD.

Hardjosoesilo, A. (2011b). Pangkalpinang, Indonesia. Public private partnership: From mining to re-claiming spoilt lands. ICLEI case studies #127. ICLEI Local Governments for Sustainability

Harman, B. P., Taylor, B. M., and Lane, M. B. (2015). Urban partnerships and climate adaptation: Challenges and opportunities. *Current Opinion in Environmental Sustainability* **12**, 74–79. doi:10.1016/j.cosust.2014.11.001

Harris, C. (2003). *Private Participation in Infrastructure in Developing Countries: Trends, Impacts and Policy Lessons*. World Bank.

Horrocks, L., Beckford, J., and Hodgson, N. (2010). *Adapting the ICT sector to the impacts of climate change – Final report, Defra contract number RMP5604*. Defra. Accessed October 8, 2015: https://www.gov.uk/government/uploads/system/uploads/attachment_data/file/183486/infrastructure-aea-full.pdf

Horton, et al. (2011). *Developing coastal adaptation to climate change in the New York City infrastructure-shed: Process, approach, tools, and strategies. Climatic Change* **106**(1), 93–127.

Houle, M. (2015). *The Potential of Information and Communications Technology for the Enhancement of Organizational Resilience during Hurricane Evacuation –New York City: A Case Study*. Unpublished Master's thesis, University of Montreal, Institute of Urban Planning.

IDFS. (2013). Mapping of Green Finance Delivered by IDFC Members in 2012, Accessed April 29, 2014: https://www.idfc.org/Downloads/Publications/01_green_finance_mappings/IDFC_Green_Finance_Mapping_Report_2013_11-01-13.pdf

IFC. (2011). Climate change PPPs. *Handshake*. Issue #2. International Finance Corp. World Bank Group. Accessed December 19, 2014: http://www.ifc.org/wps/wcm/connect/1deb920048e7c234be3bfe3e-a49e3f47/Handshake_ClimateChange.pdf?MOD=AJPERES

Indonesia Ministry of Finance. (2014). Urbanization in Indonesia. Presentation by Mochamad Bara Ampera, Head of Transportation, Energy & Other Sectors, Centre for Climate Change Financing and Multilateral Policies, Ministry of Finance at the Towards Green Growth in Southeast Asia Regional Workshop, organized by the OECD and the Indonesia Ministry of Finance, June 12–13, 2014.

Isoaho, K., and Surminski, S. (2015). *Does It Matter What You Call It? Reflections on How Companies Voluntarily Disclose Their Adaptation Activities*. Centre for Climate Change Economics and Policy Working Paper No. 236 Grantham Research Institute on Climate Change and the Environment Working Paper No. 210. Accessed January 27, 2016: http://www.lse.ac.uk/GranthamInstitute/wp-content/uploads/2015/09/Working-Paper-210-Isoaho-and-Surminski.pdf

Jacob, K., et al. (2011). Telecommunications. In Rosenzweig, C., Solecki, W., DeGaetano, A., O'Grady, M., Hassol, S., and Grabhorn, P. (eds.), *Responding to Climate Change in New York State: The ClimAID Integrated Assessment for Effective Climate Change Adaptation in New York State* (363–396). New York State Energy Research and Development Authority (NYSERDA).

Johannessen, A., Rosemarin, A., Gerger Swartling, A., Han, G., Vulturius, G., and Stenström, T. A. (2013). *Linking Investment Decisions with Disaster Risk Reduction in Water Sanitation and Hygiene (WASH): The Role of the Public and Private Sectors, Potentials for Partnership and Social Learning*. Stockholm Environment Institute.

Kamal-Chaoui, L., and Robert, A. (eds.). (2009). *Competitive Cities and Climate Change*. OECD.

Kazmierczak, A., and Carter, J. (2010). Adaptation to climate change using green and blue infrastructure. A database of case studies. The University of Manchester. Accessed November 15, 2014: http://orca.cf.ac.uk/64906/1/Database_Final_no_hyperlinks.pdf

Keogh, B., Westbrook, J., and Suess, O. (2011). *The trouble with catastrophe bonds. Bloomberg Businessweek Magazine* (April 25, 2011). Accessed August 17, 2014: www.businessweek.com/magazine/content/11_18/b4226055260651.htm

Kernaghan, S., and da Silva, J. (2014). Initiating and sustaining action: Experiences building resilience to climate change in Asian cities. *Urban Climate* 7, 47–63. doi:10.1016/j.uclim.2013.10.008

Kessides, L. N. (2004). *Reforming Infrastructure: Privatization, Regulation, and Competition*. International Bank for Reconstruction and Development.

KfW. (2013). Energy-efficient Urban Redevelopment Programme: Tapping the potential of neighbourhoods. Commissioned by the Federal Ministry for the Environment, Nature Conservation, Building and Nuclear Safety. Accessed April 3, 2014: http://www.energetische-stadtsanierung.info/energy-efficient-urban-redevelopment/overview/?changelang=2

Khattri, A., Parameshwar, D., and Pellech, S. (2010). *Opportunities for Private Sector Engagement in Urban Climate Change Resilience Building*. Intellecap.

Kunreuther, H. (1996). Mitigating disaster losses through insurance. *Journal of Risk and Uncertainty* 12, 171–187.

Lall, S. V., Lozano-Gracia, N., Agarwal, O. P., Dowall, D., Klien, M., and Wang, H. G. (2013). *Planning, Connecting, and Financing Cities-Now: Priorities for City Leaders*. World Bank. Accessed March 12, 2015: http://documents.worldbank.org/curated/en/2013/01/17197253/planning-connecting-financing-cities-now-priorities-city-leaders

Mechler, R., Bouwer, R. M., Linnerooth-Bayer, J., Hochrainer-Stigler, S., Aerts, J. C. J. H., Surminski, S., and Williges, K. (2014). Managing unnatural disaster risk from climate extremes. *Nature Climate Change* **4**, 235–237.

Mees, H. P., and Driessen, P. P. (2011). Adaptation to climate change in urban areas: Climate-greening London, Rotterdam, and Toronto. *Climate Law* **2**(2), 251–280.

Mehrotra, S., Natenzon, C. E., Omojola, A., Folorunsho, R., Gilbride J., and Rosenzweig, C. (2009). *Framework for city climate risk assessment*. World Bank. Accessed June 8, 2015: http://siteresources.worldbank.org/INTURBANDEVELOPMENT/Resources/336387–1256566800920/6505269–1268260567624/Rosenzweig.pdf

Merk, O., Saussier, S., Staropoli, C., Slack, E., and Kim, J. H. (2012). *Financing Green Urban Infrastructure*. OECD regional development working papers 2012/10. Organization for Economic Co-operation and Development. Accessed September 17, 2015: http://dx.doi.org/10.1787/20737009

Meyer, P. B., Yount, K. R., Barnes, G., and Weiss, J. (2013a). *Spurring Local Economic Development with Clean Energy Investments: Lessons from the Field*. U.S. Department of Energy.

Meyer, P. B., Yount, K. R., Barnes, G., and Weiss, J. (2013b). *Financing Clean Energy Programs: A Guide*. Draft report submitted to the U.S. Department of Energy by the Center for Climate Initiatives, Washington, D.C. Available from the author (PBM@TheEPSystemsGroup.com).

Millard-Ball, A. (2012). Do city climate plans reduce emissions? *Journal of Urban Economics* **71**(3), 289–311.

Mills, X. (2009a). From risk to opportunity: Insurer responses to climate change (2009). A Ceres report. Accessed June 8, 2015: www.ceres.org

Mills, X. (2009b). A global review of insurance industry responses to climate change. *The Geneva Papers* **34**, 323–359. doi:10.1057/gpp.2009.14

New Climate Economy (NCE). (2014). Better growth, better climate. *The New Climate Economy Report*.

Ng, A., and Loosemore, M. (2007). Risk allocation in the private provision of public infrastructure. *International Journal of Project Management* **XXV**(1), 66–76.

NRT/NBS. (2012). Managing the business risks and opportunities of a changing climate. A primer for executives on adaptation to climate change. National Round Table on the Environmental and the Economy. Network for Business Sustainability. Accessed September 17, 2015: http://nbs.net/wp-content/uploads/Adaptation-to-Climate-Change-Primer.pdf

Organization for Economic Cooperation and Development (OECD). (1998). *Institutional Investors in the New Financial Landscape*. OECD.

Organization for Economic Cooperation and Development (OECD). (2008). *Upgrading the Investment Policy Framework in Public Pension Funds*. OECD.

Organization for Economic Cooperation and Development (OECD). (2010). *Cities and climate change*. OECD. Accessed March 12, 2015: http://dx.doi.org/10.1787/9789264091375-en.

Organization for Economic Cooperation and Development (OECD). (2013a). OECD DAC statistics climate-related aid. OECD. Accessed March 12, 2015: http://www.oecd.org/dac/stats/Climate%20change-related%20Aid%20Flyer%20-%20November%202013.pdf

Organization for Economic Cooperation and Development (OECD). (2013b). *Pension Markets in Focus, 2013*. OECD.

Osborne, C. (2015). Smart cities in Europe. Enabling innovation. Accessed March 20, 2016: http://www.cleanenergypipeline.com/Resources/CE/ResearchReports/Smart%20cities%20in%20Europe.pdf

Pivo, G. (2014). The effect of sustainability features on mortgage default prediction and risk in multifamily rental housing. *Journal of Sustainable Real Estate* **5**(1), 149–170.

Puppim de Oliveira, J. A. (2009). The implementation of climate change related policies at the subnational level: An analysis of three countries. *Habitat International* **33**(3), 253–259.

Ranger, N., Hallegatte, S., Bhattacharya, S., Bachu, M., Priya, S., Dhore, K., and Herweijer, C. (2011). An assessment of the potential impact of climate change on flood risk in Mumbai. *Climatic Change* **104**(1), 139–167.

Rasker, R. (2012). Implementing climate change adaptation. Lessons learned from ten examples. Headwaters Economics. Accessed June 8, 2015:

http://headwaterseconomics.org/wphw/wp-content/uploads/Climate_Adaptation_Lessons_Learned.pdf

Revi, A., Satterthwaite, D. E., Aragón-Durand, F., Corfee-Morlot, J., Kiunsi, R. B. R., Pelling, M., Roberts, D. C., and Solecki, W. (2014). Urban areas. In Field, C. B., Barros, V. R. Dokken, D. J., Mach, K. J., Mastrandrea, M. D., Bilir, T. E., Chatterjee, M., Ebi, K. L., Estrada, Y. O., Genova, R. C., Girma, B., Kissel, E. S., Levy, A. N., MacCracken, S., Mastrandrea, P. R., and White, L. L. (eds.), *Climate Change 2014: Impacts, Adaptation, and Vulnerability. Part A: Global and Sectoral Aspects. Contribution of Working Group II to the Fifth Assessment Report of the Intergovernmental Panel on Climate Change* (535–612). Cambridge University Press.

Ricardo AEA. (2014). Climate Change Agreement for data centers – a recognition of the sector's contribution. Accessed September 17, 2015: http://www.ricardo-Horrocks et al.com/cms/climate-change- agreement-for-data-centers-a-recognition-of-the-sector-s-contribution-2/#.U9t7kONdV1E

Rode, P., Floater, G., Kandt, J., Baker, K., Montero, M., Heeckt, C., Smith, D., and Delfs, M. (2013). *Going Green: How Cities Are Leading the Next Economy*. LSE Cities.

Rosenbloom, J. L. (2008). *Quantitative Economic History: The Good of Counting*. Routledge Explorations in Economic History.

Rosenzweig, C., and Solecki, W. (2014). Hurricane Sandy and adaptation pathways in New York: Lessons from a first-responder city. *Global Environmental Change* 28(2014), 395–408.

Rotterdam Climate Initiative (RCI). (2011). CO2 capture and storage in Rotterdam. A network approach. RCI. Accessed April 13, 2014: http://www.rotterdamclimateinitiative.nl/documents/CO2%20network%20approch.pdf

Scholz, S. M., Rescalvo, M., Sugar, L., Lasa, M., D'Silva, N., Barrios, R., and Mata, J. (2014). *The Low Carbon City Development Program (LCCDP) Guidebook: A Systems Approach to Low Carbon Development in Cities*. World Bank Group.

Scholz, S. M., and Sugar, L. (2012). *The Rio de Janeiro Low Carbon City Development Program: A Business Model for Green and Climate-Friendly Growth in Cities. Directions in Urban Development*. World Bank.

Schroeder, H., Burch, S., and Rayner, S. (2013). Novel multisector networks and entrepreneurship in urban climate governance. *Environment and Planning C: Government and Policy* 31(5),761–768.

Schuttenbelt, P. (2013). Financing climate change resilient urban infrastructure. Presentation of Paul Schuttenbelt – South Asia Coordinator, Cities Development Initiative for Asia (CDIA), January 2013. Accessed January 4, 2015: http://www.teriin.org/events/Paul_Schuttenbelt.pdf

Shehabi, A., et al. (2011). Data center design and location: Consequences for electricity use and greenhouse-gas emissions. *Building and Environment* 46(5), 990–998.

Shilling, J. D., Sirmans, C. F., and Benjamin, J. D. (1989). Flood insurance, wealth redistribution, and urban property values. *Journal of Urban Economics* 26(1), 43–53.

Smith, A. (1999). *Privatized Infrastructure: The Role of Government*. Thomas Telford

Smith, J. J., and Gihring, T. (2006). Financing transit systems through value capture. An annotated bibliography. *American Journal of Economics and Sociology* 65(3).

Soffiatti, R. V. F. (2012). *A contribuição de melhoria como instrumento de recuperação da Mais-Valia Fundiária Urbana: Estudo de caso Eixo Urbano "Linha Verde."* Pontifícia Universidade Católica Do Paraná.

Sovereign Wealth Fund Institute (SWFI). (2015). Sovereign Wealth Fund rankings. Accessed March 3, 2016: http://www.swfinstitute.org/fund-rankings/

Spalding-Fecher, R., Narayan Achanta, A., Erickson, P., Haites, E., Lazarus, M., Pahuja, N., Pandey, N., Seres, S., and Tewari, R. (2012). Assessing the impact of the Clean Development Mechanism. CDM policy dialogue. Accessed January 4, 2015: http://www.cdmpolicydialogue.org/research/1030_impact.pdf

Stevenson, R. (2014). An inconvenient (data center) truth. Data Center Knowledge. Accessed September 12, 2015: http://www.datacenter-knowledge.com/archives/2014/03/13/inconvenient-data-center-truth/

Surminski, S. (2013). Private-sector adaptation to climate risk. *Nature Climate Change* 3, 943–945.

Surminski, S. (2014). The role of insurance in reducing direct risk – The case of flood insurance. *International Review of Environmental and Resource Economics* 7(3–4) 241–278.

Surminski, S., Bouwer, L., and Linnerooth-Bayer, J. (2016). *How insurance can support climate resilience. Nature Climate Change*, in press.

Surminski, S., and, Oramas–Dorta, D. (2015). *Flood insurance schemes and climate adaptation in developing countries. International Journal of Disaster Risk Reduction* 7, 154–164.

Sussman, F. G., and Freed, J. R. (2008). *Adapting to Climate Change: A Business Approach*. Pew Center on Global Climate Change Arlington.

Tokyo Metropolitan Government. (2010). Tokyo cap-and-trade program: Japan's first mandatory emissions trading scheme. Tokyo Metro Gotvernment. Accessed February 18, 2014: http://www.kankyo.metro.tokyo.jp/en/attachement/Tokyo-cap_and_trade_program-march_2010_TMG.pdf

Tokyo Metropolititan Government Bureau of Environment. (2016). Tokyo Cap-and-Trade Program achieves 25% reduction after 5th year. Accessed February 23, 2017: http://www.kankyo.metro.tokyo.jp/en/files/3c08a5ad895b5130cb1d17ff5a1c9fa4.pdf

Tompkins, E. L., and Eakin, H. (2012). Managing private and public adaptation to climate change. *Global Environmental Change* 22, 3–11. doi:10.1016/j.gloenvcha.2011.09.010

Treby, E. J., Clark, M. J., and Priest, S. J. (2006). Confronting flood risk: Implications for insurance and risk transfer. *Journal of Environmental Management* 81(4) 351–359.

United Nations Environment Programme (UNEP) and Gwangju City. (2012). Cities and carbon finance: A feasibility study on an Urban CDM. United Nations Environment Programme, Division of Technology, Industry and Economics. Accessed March 3, 2015: http://www.unep.org/urban_environment/PDFs/UNEP_UrbanCDMreport.pdf

United Nations Framework Convention on Climate Change (UNFCCC). The Nairobi Work Programme on Impacts, Vulnerability, and Adaptation to Climate Change. (2011). Assessing the costs and benefits of adaptation options: An overview of approaches. Accessed July 29, 2014: http://unfccc.int/resource/docs/publications/pub_nwp_costs_benefits_adaptation.pdf

U.S. Conference of Mayors. (2014). Climate mitigation and adaptation actions in America's cities. A 282-city survey. Mayors Climate Protection Center. The United States Conference of Mayors. Accessed February 20, 2015: http://usmayors.org/pressreleases/uploads/2014/0422-report-climatesurvey.pdf

van den Berge, J. (2010). Employment opportunities from climate change mitigation policies in the Netherlands. *International Journal of Labour Research* 2(2), 211–231.

Verhaar, H. (2009). The benefits of energy efficient lighting in tackling climate change in our cities and buildings. A triple win for people environment and economy. PPT presentation. Accessed June 2, 2014: http://siteresources.worldbank.org/INTUWM/Resources/340232-1254279511547/Verhaar.pdf.

World Bank. (2010). *A city-wide approach to carbon finance. Carbon partnership facility innovation series*. World Bank. Accessed March 9, 2015: http://www.citiesalliance.org/sites/citiesalliance.org/files/A_city-wide_approach_to_carbon_finance.pdf

World Bank. (2014). *State and trends of carbon pricing*. World Bank. Accessed June 2, 2015: http://www-wds.worldbank.org/external/default/WDSContentServer/WDSP/IB/2014/05/27/000456286_20140527095323/Rendered/PDF/882840AR0Carbo040Box385232B00OUO090.pdf

World Bank Group. (2010a). Climate finance in the urban context. Accessed April 1, 2014: http://wbi.worldbank.org/wbi/Data/wbi/wbicms/files/drupal-acquia/wbi/578590revised0101Public10DCFIB0141A.pdf

World Bank Group. (2011). Guide to climate change adaptation in cities. Accessed November 20, 2015: http://iaibr3.iai.int/twiki/pub/ForoEditorial2012/WebHome/Urban_Handbook_Final.pdf

World Bank Group. (2013a). Building resilience. Accessed April 26, 2014: http://www.worldbank.org/content/dam/Worldbank/document/SDN/Full_Report_Building_Resilience_Integrating_Climate_Disaster_Risk_Development.pdf

World Bank Group. (2013b). Planning and Financing Low-Carbon, Livable Cities. Accessed March 25, 2015: http://www.worldbank.org/en/news/feature/2013/09/25/planning-financing-low-carbon-cities

World Economic Forum (WEF). (2013). *The Green Investment Report – The Ways and Means to Unlock Private Finance for Green Growth. A report of the Green Growth Action Alliance. Author.*

WRI. (2012). Why businesses must focus climate change mitigation and adaptation. Accessed March 6, 2015: http://www.wri.org/blog/2012/11/why-businesses-must-focus-climate-change-mitigation-and-adaptation

Yong, W. (2013). Green economic development with renewable energy industries. City in focus: Dezhou,China. ICLEI/IRENA. Accessed February 22, 2014: http://www.iclei.org/fileadmin/PUBLICATIONS/Case_Studies/1_Dezhou_-_ICLEI-IRENA_2012.pdf

Ziegler, A., Busch, T., and Hoffmann, V. H. (2011). Disclosed corporate responses to climate change and stock performance: An international empirical analysis. *Energy Economics* 33(2011), 1283–1294.

Chapter 7   Case Study References

Case Study 7.1   London Climate Change Partnership: Public and Private Sector Collaboration

Eurostat. (2015a). Population by sex and age groups on 1 January. European Commission. Accessed March 21, 2016: http://appsso.eurostat.ec.europa.eu/nui/show.do?dataset=met_pjanaggr3&lang=en

Eurostat. (2015a). Population by sex and age groups on 1 January. European Comission. Accessed March 21, 2016: http://appsso.eurostat.ec.europa.eu/nui/show.do?dataset=met_pjanaggr3&lang=en

Eurostat. (2015b). Area of the regions [met_d3area]. European Commission. Accessed March 21, 2016: http://appsso.eurostat.ec.europa.eu/nui/show.do?dataset=met_d3area&lang=en

LCCP. (2013). London Climate Change Partnership. Accessed March 6, 2015: http://climatelondon.org.uk/lccp/

ONS. (2015). Mid-year population estimates from NOMIS. Accessed March 20, 2016: https://www.nomisweb.co.uk/query/construct/submit.asp?menuopt=201&subcomp

Peel, M. C., Finlayson, B. L., and McMahon, T. A. (2007). Updated world map of the Köppen-Geiger climate classification. *Hydrology and Earth System Sciences Discussions* 4(2), 462.

Surminski, S., and Leck, H. (2016). *You Never Adapt Alone – The Role of MultiSectoral Partnerships in Addressing Urban Climate Risks.* Centre for Climate Change Economics and Policy Working Paper No. 262 Grantham Research Institute on Climate Change and the Environment Working Paper No. 232 London School of Economics. Accessed February 23, 2017: http://www.lse.ac.uk/GranthamInstitute/wp-content/uploads/2016/03/Working-Paper-232-Surminski-and-Leck.pdf

World Bank. (2017). 2016 GNI per capita, Atlas method (current US$). Accessed: August 9, 2017: http://data.worldbank.org/indicator/NY.GNP.PCAP.CD

Case Study 7.2   Public Enabling of Private Real Estate in New York

Adger, W. N., Arnell, N. W., and Thompkins, E. (2005). Successful adaptation to climate change across scales. *Global Environmental Change* 15, 77–86. doi:10.1016/j.gloenvcha.2004.12.005

AKRF, Inc. (2014). *The Invisible Engine: The Economic Impact of New York City's Real Estate Industry.* Report prepared for the Real Estate Board of New York. Accessed July 31, 2015: http://www.rebny.com/content/rebny/en/newsroom/research/policay_reports/Economic_Impact_NYC_Real_Estate_Industry.html

City of New York. (2011). *Vision 2020: New York City comprehensive waterfront plan.* Department of City Planning. Accessed February 22, 2014: http://www.nyc.gov/html/dcp/html/cwp/index.shtml

City of New York. (2014). *On the frontlines: $129 billion in property at risk from flood waters.* Office of the Comptroller. Accessed January 16, 2015: http://comptroller.nyc.gov/wp-content/uploads/documents/Policy_Brief_1014.pdf

Horton, R., Bader, D., Kushnir, Y., Little, C., Blake, R., and Rosenzweig, C. (2015). New York City Panel on Climate Change 2015 report: Climate observations and projections. *Annals of the New York Academy of Sciences* 1336(1), 18–35. Accessed February 28, 2016: http://onlinelibrary.wiley.com/doi/10.1111/nyas.12586/abstract

Metropolitan Waterfront Alliance. (2015). Waterfront edge design guidelines. Accessed March 30, 2016: http://www.waterfrontalliance.org/WEDG

New York City Building Resiliency Task Force. (2013). *Building Resiliency Task Force: Report to Mayor Michael Bloomberg and Speaker Christine C. Quinn.* U.S. Green Building Council. Accessed July 25, 2015: http://urbangreencouncil.org/content/projects/building-resiliency-task-force

New York City Economic Development Corporation. (2014). *Waterfront vision and enhancement strategy.* Accessed June 11, 2015: http://www.nycedc.com/project/waterfront-vision-and-enhancement-strategy

Peel, M. C., Finlayson, B. L., and McMahon, T. A. (2007). Updated world map of the Köppen-Geiger climate classification. *Hydrology and Earth System Sciences Discussions* 4(2) 462.

RAND. (2013). *Flood insurance in New York City following Hurricane Sandy.* New York City Mayor's Office of Long-Term Planning and Sustainability. Accessed June 29, 2014: www.rand.org/content/dam/rand/pubs/research_reports/RR300/RR328/RAND_RR328.pdf

U.S. Census Bureau. (2010). Decennial census, summary file 1. Accessed August 26, 2015: http://www.census.gov/population/metro/files/CBSA%20Report%20Chapter%203%20Data.xls

U.S Census Bureau. (2016). Table 1. Annual Estimates of the Resident Population: April 1, 2010 to July 1, 2016 – Metropolitan Statistical Area; – 2016 Population Estimates. U.S. Census Bureau. Accessed March 28, 2017: https://factfinder.census.gov/faces/tableservices/jsf/pages/productview.xhtml?pid=PEP_2016_PEPANNRES&prod-Type=table

World Bank. (2017). 2016 GNI per capita, Atlas method (current US$). Accessed: August 9, 2017: http://data.worldbank.org/indicator/NY.GNP.PCAP.CD

Case Study 7.3   Raising Awareness of Negotiating, and Implementing Tokyo's Cap-and-Trade System

Demographia World Urban. (2015). Demographia world urban 2015. Accessed August 9, 2017: http://www.demographia.com/db-worldua.pdf

IETA. (2014). Tokyo – The world's carbon markets: A case study guide to emissions trading. Accessed January 24, 2015: http://www.ieta.org/assets/EDFCaseStudyMarch2014/tokyo%20ets%20case%20study%20march%202014.pdf

Lee, M., and Colopinto, K. (2010). *Tokyo's Emissions Trading System: A Case Study.* World Bank, Urban Development and Local Government Unit.

Peel, M. C., Finlayson, B. L., and McMahon, T. L. (2007). Updated world map of the Köppen-Geiger climate classification. *Hydrology and Earth System Sciences Discussions* 4(2) 462.

Tokyo Metropolitan Government (TMG). (2010). *Tokyo cap-and-trade program: Japan's first mandatory emissions trading scheme.* Accessed July 8, 2014: http://www.kankyo.metro.tokyo.jp/en/climate/cap_and_trade.html

World Bank. (2017). 2016 GNI per capita, Atlas method (current US$). Accessed: August 9, 2017: http://data.worldbank.org/indicator/NY.GNP.PCAP.CD

Part II

Urban Ecosystems and Human Services

8

Urban Ecosystems and Biodiversity

Coordinating Lead Authors

Timon McPhearson (New York), Madhav Karki (Kathmandu)

Lead Authors

Cecilia Herzog (Rio de Janeiro), Helen Santiago Fink (Vienna/D.C.), Luc Abbadie (Paris), Peleg Kremer (New York), Christopher M. Clark (Washington D.C.), Matthew I. Palmer (New York), Katia Perini (Genoa)

Contributing Authors

Marielle Dubbeling (Leusden)

This chapter should be cited as

McPhearson, T., Karki, M., Herzog, C., Santiago Fink, H., Abbadie, L., Kremer, P., Clark, C. M., Palmer, M. I., and Perini, K. (2018). Urban ecosystems and biodiversity. In Rosenzweig, C., W. Solecki, P. Romero-Lankao, S. Mehrotra, S. Dhakal, and S. Ali Ibrahim (eds.), *Climate Change and Cities: Second Assessment Report of the Urban Climate Change Research Network*. Cambridge University Press. New York. 257–318

Urban Ecology in a Changing Climate

Almost all of the impacts of climate change have direct or indirect consequences for urban ecosystems, biodiversity, and the critical ecosystem services they provide for human health and well-being in cities. These impacts are already occurring in urban ecosystems and their constituent living organisms.

Urban ecosystems and biodiversity have an important and expanding role in helping cities adapt to and mitigate the impacts of changing climate. Harnessing urban biodiversity and ecosystems as adaptation and mitigation solutions will help achieve more resilient, sustainable, and livable outcomes for cities and urban regions.

Conserving, restoring, and expanding urban ecosystems under mounting climatic and non-climatic urban development pressures will require improved urban and regional planning, policy, governance, and multisectoral cooperation.

Major Findings

• Urban biodiversity and ecosystems are already being affected by climate change.

• Urban ecosystems are rich in biodiversity and provide critical natural capital for climate change adaptation and mitigation.

• Climate change and urbanization are likely to increase the vulnerability of biodiversity hotspots, urban species, and critical ecosystem services.

• Urban ecosystems and green infrastructure can provide cost-effective, nature-based solutions for adapting to climate change while also creating opportunities to increase social equity, green economies, and sustainable urban development.

• Investing in the quality and quantity of urban ecosystems and green infrastructure has multiple co-benefits, including improving quality of life, human health, and social well-being.

Key Messages

Cities should take a long-term, system-based approach to climate adaptation and mitigation. Nature-based approaches to address climate change in cities explicitly recognize the critical role of urban and peri-urban ecosystem services (UES) that require thoughtful management in order to ensure sustainable supply of environmental goods and services to residents who need them over the next 20, 50, and 100 years. Ecosystem-based planning can strengthen the linkages between urban, peri-urban, and rural ecosystems through participatory planning and management for nature-based solutions at both city and regional scales.

The economic benefits of urban biodiversity and ecosystem services should be quantified so that they can be integrated into climate-related urban resilience and sustainability planning and decision-making. These benefits should incorporate both monetary and non-monetary values of biodiversity and ecosystem services, including how they relate to physical and mental health and social equity in access to services.

8.1 Introduction

Climate change is already affecting cities and urbanized regions around the world impacting human populations and the built environment, as well as urban ecosystems and their associated biota (While and Whitehead 2013). Almost all of the impacts of climate change have direct or indirect consequences for urban ecosystems,[1] biodiversity,[2] and the critical ecosystem services[3] they provide for human health and well-being in cities (e.g., urban heat island [UHI] reduction) (The Economics of Environmentalism and Biodiversity [TEEB], 2011; Elmqvist et al., 2013). Increasing knowledge of the benefits of urban ecosystems for the livelihoods of urban residents suggests an important and expanding role for urban ecosystems and biodiversity in adaptation to local effects of climate change. However, conserving, restoring, and expanding urban ecosystems to enhance climate resilience and other co-benefits under mounting climatic and non-climatic stresses of growing urbanization and development processes will require improved urban and regional planning, policy and governance, and multisectoral cooperation to protect and manage urban ecosystems and biodiversity (Elmqvist et al., 2013; Solecki and Marcotullio, 2013; McPhearson et al., 2014).

In this chapter, we review key concepts, challenges, and ecosystem-based pathways for adaptation and mitigation of climate change in cities. This leads to and supports concepts, strategies, and tools of ecosystem-based adaptation, disaster risk reduction, and green infrastructure planning. Section 8.1 reviews the relationships among urban ecosystems, biodiversity, and ecosystem services as critical resources for climate adaptation and, to some extent, mitigation. Sections 8.2 and 8.3 discuss current and future climate-related challenges including hazards, risks, and vulnerabilities for urban biodiversity and ecosystem services. Section 8.4 discusses examples of how ecosystems can provide adaptive capacity and be used innovatively to reduce effects of climate change in urban systems, whereas Section 8.5 presents ecosystem-based adaptation as an effective entry point for nature-based solutions to building climate resilience in cities. Section 8.6 discusses the economic cost-effectiveness of ecosystem-based adaptation, with particular emphasis on investing in green infrastructure. Section 8.7 discusses how urban ecosystems intersect with urban planning and design (see Chapter 5, Urban Planning and Design), the importance of engaging with diverse stakeholders, and how ecosystem-based planning and management can help address issues of social equity and environmental justice while yielding multiple socioeconomic benefits. Section 8.8 discusses important planning, governance, and management tools (see Chapter 16, Governance and Policy). Sections 8.9 and 8.10 present the need for better linking science with policy, in particular for building urban climate resilience. Section 8.11 identifies remaining knowledge gaps and suggests avenues for future research. Section 8.12 provides a summary of recommendations for cities to harness urban biodiversity and ecosystems as nature-based solutions to adapt to the effects of and mitigate climate change that will help achieve more sustainable, resilient, and livable cities. Case Studies are provided throughout the chapter to illustrate effective, on-the-ground implementation of many of the ecosystem-based adaptation and mitigation strategies and approaches reviewed.

8.1.1 A Systems Approach to Ecology in, of, and for Cities

Cities and urban areas are complex systems with social, ecological, economic, and technical/built components interacting dynamically in space and time (Grimm et al., 2000, 2008; Pickett et al., 2001; McPhearson et al., 2016a). The complex nature of urban systems[4] can make it challenging to predict how ecosystems will respond to climate change in cities (Batty 2008; Bettencourt and West, 2010; McPhearson et al., 2016b). This complexity is driven by many intersecting feedbacks affecting ecosystems, including climate, biogeochemistry, nutrient cycling, hydrology, population growth, urbanization and development, human perceptions and behavior, and more (Bardsley and Hugo, 2010; Pandey and Bardsley, 2013; Alberti, 2015).

In urban ecology, cities and urbanized areas are understood to be complex human-dominated ecosystems (Pickett et al., 1997, 2001; Niemelä et al., 2011). These systems interrelate dynamically with the social, ecological, economic, and technological/built infrastructure of the city (Grimm et al., 2000; McDonnell and Hahs, 2013; McPhearson et al., 2016a) (see Figure 8.1). Patterns and processes of urban systems in this view emerge from the interactions and feedbacks between components and systems in cities, emphasizing the need to consider multiple sources of social-ecological patterns and processes to understand reciprocal interactions between climate change and urban ecosystems (see Figure 8.1).

Urban social-ecological systems (SES) consist of social and ecological components (broadly defined) that have their own internal patterns and processes, but these patterns and processes interact across the system in a number of ways to produce overall urban system dynamics, behavior, and emergent phenomena. Drivers external to the urban system are fundamentally important but can affect social and ecological components and processes within the urban system with different strengths

---

1 Urban ecosystems include all vegetation, soil, and water-covered areas that may be found in urban and peri-urban areas at multiple spatial scales (parcel, neighborhood, municipal city, metropolitan region), including parks, cemeteries, lawns and gardens, green roofs, urban allotments, urban forests, single trees, bare soil, abandoned or vacant land, agricultural land, wetlands, streams, rivers, lakes, and ponds (Gómez-Baggethun et al., 2013).

2 "Biological diversity" means the variability among living organisms from all sources including, inter alia, terrestrial, marine, and other aquatic ecosystems and the ecological complexes of which they are part; this includes diversity within species, between species, and of ecosystems.

3 Ecosystem services are the benefits that people obtain directly or indirectly from ecosystem functions, such as protection from storm surges and heat waves, air quality regulation, and food, fiber, and fresh water (MA, 2005; TEEB, 2010; Gomez-Baggethun et al., 2013).

4 Urban systems are defined here as those areas where the built infrastructure covers a large proportion of the land surface or those in which people live at high densities (Pickett et al., 2001).

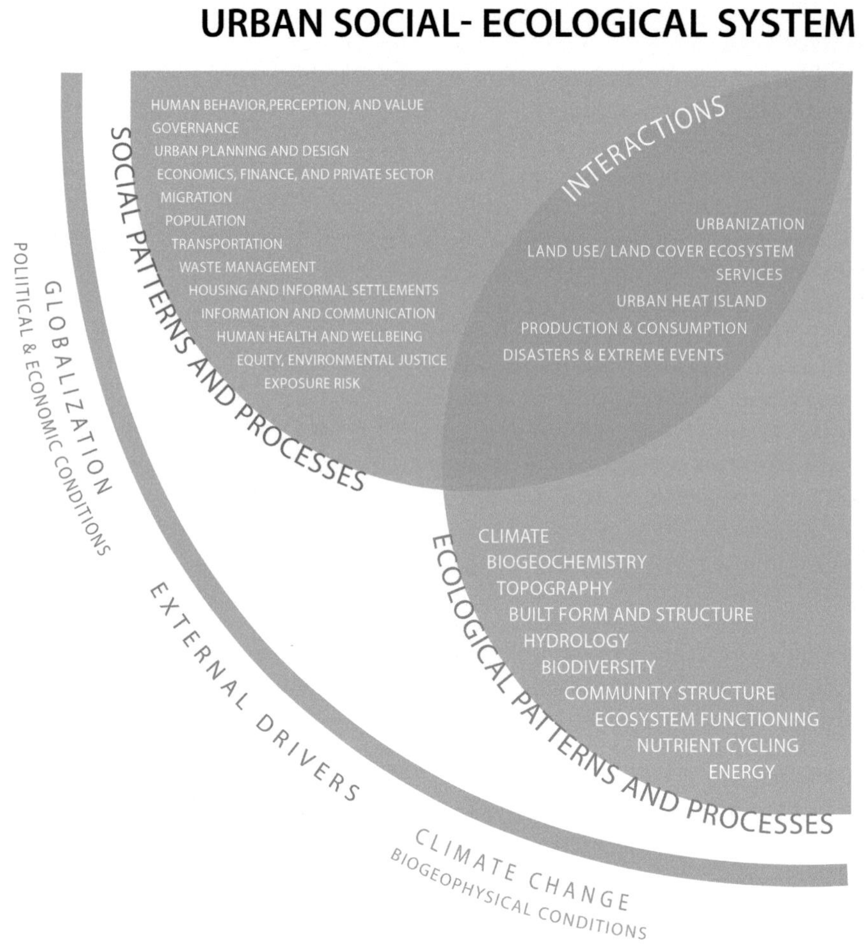

URBAN SOCIAL- ECOLOGICAL SYSTEM

Figure 8.1 *Urban systems are complex and dynamically interactive and can be conceptualized and studied as social-ecological systems (SES) at multiple spatial or temporal scales.*

or intensity. This conceptual approach to studying urban SES is scale-independent and can therefore be applied at multiple spatial or temporal scales in urban areas. The urban ecosystem approach has developed rapidly in the past two decades, incorporating methods and approaches from the social sciences, biophysical sciences, urban planning, and design to provide insight for developing and managing urban areas to meet the needs of expanding populations in a changing climate (McDonnell, 2011; McPhearson et al., 2016a). We focus here on biodiversity and ecosystem functions and services provided by natural systems within urban and peri-urban areas.

Studies of the ecology *in* the city as well as ecology *of* the city (Grimm et al., 2000; Pickett et al., 2001) are both domains of urban ecology, a science increasingly focused on applying sustainability and resilience science *for* cities (Childers et al., 2014, 2015; McPhearson et al., 2016a). Defining clear boundaries for ecosystems *in* the city is challenging due to the fact that species and many of the relevant fluxes and interactions necessary to understand the functioning of urban ecosystems extend beyond the city boundaries defined by political borders (Solecki and Marcotullio, 2013). For example, nutrients, water, species, and humans all move across political boundaries, emphasizing the importance

of regional planning and management. Thus, the relevant scope of urban ecosystem analysis reaches far beyond the municipal boundary. It comprises not only the ecological areas within cities, but also the peri-urban areas and linkages to nearby rural areas that are directly affected by the energy and material flows from the urban core, including city water catchments, peri-urban forests, and nearby cultivated fields (Grimm et al., 2000; Pickett et al., 2001; La Rosa and Privitera, 2013). Urban ecosystems therefore include all vegetation, soil, and water-covered areas that may be found in urban and peri-urban areas at multiple spatial scales (parcel, neighborhood, municipality, metropolitan region), including parks, cemeteries, lawns and gardens, green roofs, urban allotments, urban forests, single trees, bare soil, abandoned or vacant land, agricultural land, wetlands, streams, rivers, lakes, and ponds (Gómez-Baggethun et al., 2013).

The social and biophysical context of urban areas influences resilience to climate change and other social-ecological challenges (Marcotullio and Solecki, 2013; Solecki and Marcotullio, 2013). For example, the bio-geophysical context of the city or urban area may determine how ecosystems in cities respond to climate change, extreme events, and urbanization (Schewenius et al., 2014). Urbanization and suburbanization in urban areas often reduce both species richness (i.e., the number of species) and evenness (i.e., the distribution of species) for most biotic communities (Paul et al., 2001; McKinney, 2002). Changes in species richness and evenness have been found to affect the stability of ecosystems and their ability to deliver needed services for mitigating and adapting to climate change (Grimm, 2008; Cardinale et al., 2012). Additionally, many of the changes taking place in urban areas have analogues to those driven by climate change (e.g., elevated CO_2, higher temperatures, changes in precipitation), thus making urban systems useful models for examining the interaction of social and biophysical patterns and processes in changing climates (Grimm et al., 2008; Collins et al., 2000). Therefore, urban ecological approaches to improving climate adaptation and mitigation should employ a systems approach characterized by interdisciplinary, multiscalar studies and a focus on interactions and feedbacks to further develop an ecology *of* and *for* cities (see Figure 8.1) (Grimm et al., 2000; Pickett et al., 2001; McDonald et al., 2013; Childers et al., 2014, 2015; McPhearson et al., 2016). Green infrastructure and ecosystem-based adaptation are important components of nature-based solutions for climate mitigation and adaptation.

8.1.2 Urban Green Infrastructure

Many cities have already made significant progress employing urban ecological resources for climate change adaptation and mitigation as part of urban infrastructure design, planning, and development (Frischenbruder and Pellegrino, 2006). Green infrastructure is becoming a widely utilized nature-based solution for climate change adaptation and mitigation in cities (Florgard, 2007). We consider green infrastructure as a network of natural and semi-natural areas, features, and green spaces in rural and urban, terrestrial, coastal, and marine areas, which together enhance ecosystem health and climate change resilience, contribute to biodiversity, and benefit human populations through the maintenance and enhancement of ecosystem services (Pauleit et al., 2011; Kopperoinen et al., 2014). Green infrastructure is often also examined as a specific management tool for combining engineered and ecological systems (e.g., bioswales) in place of engineered non-ecological systems (e.g., concrete sewer drains) to provide ecosystem services such as cooling, stormwater management, UHI reduction, carbon storage, flood protection, and recreation (Novotny et al., 2010).

Urban green infrastructure is a key strategy for mitigating and adapting to the effects of climate change. For example, the UHI effect can be reduced by several degrees through enhanced transpiration and the shading provided by street trees, green roofs, and parks (Onishi et al., 2010; Petralli et al., 2006; Rosenzweig et al., 2006; Susca et al., 2011; Taha, 1997). Vegetation also decreases energy use for heating and air conditioning (McPherson et al., 1997; Akbari et al., 2001, UNEP, 2011). Akbari et al. (2001) estimated that about 20% of the national cooling demand in the United States can be avoided through a large-scale implementation of heat-island mitigation measures such as urban green infrastructure, particularly urban forestry. Vegetation also contributes to a city's mitigation efforts by capturing CO_2

through photosynthesis and absorbing atmospheric pollutants through dry deposition on leaves and branches, and uptake by stomata (Fowler, 2002; Ottelé et al., 2010). Green roofs and vegetated areas, including trees, increase rainwater infiltration and reduce peak flood discharge and associated water pollution while also providing mental and physical health benefits such as providing spaces for recreation and relaxation and decreasing the level of citizen stress (Dunnett and Kingsbury, 2008; Scholz-Barth, 2001; Czemiel Berndtsson, 2010; Carson et al., 2013) (see Figure 8.2).

New York, for example, launched the Green Infrastructure Plan in 2010 designed to invest in new and restored green infrastructure for stormwater management instead of traditional gray infrastructure. This included committing US$1.5 billion for green infrastructure development over the next 20 years (NYC Environmental Protection, 2010) (see Case Study 8.2). Similarly, the city of Taizhou, China, located on the southeast coast of Zhejiang Province with 5.5 million inhabitants, developed a zoning plan that utilized green infrastructure to adapt urban growth to deal with potential impacts of climate change including preventing stormwater related floods and maintaining food production areas (Yu and Li, 2006). The Taizhou plan incorporated ecological areas at multiple scales (local to regional) to maintain critical natural processes and flows including hydrology and biodiversity while simultaneously protecting cultural heritage sites and recreation areas (Yu and Li, 2006; Ahern, 2007; Gotelli et al., 2013;). These and other relevant Case Studies described in this chapter demonstrate the importance and cost-effective benefits of incorporating urban ecosystems explicitly into urban design, management, planning, and policy for mitigating and adapting to the effects of climate change.

8.1.3 Urban Biodiversity and Ecosystem Services

Nature in cities plays a crucial role as the ecological basis for human–nature interactions and the production of Urban Ecosystem Services (UES) (see Box 8.1 and Figure 8.3) (Kowarik 2005; Bolund and Hunhammar, 1999; Gómez-Baggethun et al., 2013; TEEB, 2011; Kremer et al., 2016a). Biodiversity and ecosystems in cities are increasingly linked to human health and well-being, livability, and the quality of urban life (McGranahan et al., 2005; Gómez-Baggethun et al., 2013; McPhearson et al., 2013). For example, urban trees can remove harmful air pollution, provide shade during heat waves, absorb and store carbon, and create spaces for contemplation, aesthetic and spiritual enjoyment, and social cohesion (see Table 8.1 and Figure 8.2) (TEEB, 2011; McPhearson 2011; Gómez-Baggethun et al., 2013; Andersson et al., 2014; Nowak et al., 2013).

Biodiversity is the fundamental basis for the generation of ecosystem services (see Figure 8.3) (Elmqvist et al., 2013; Gomez-Baggethun et al., 2013). There are many ecosystem services that cannot be imported and must be supplied locally within urban ecosystems (McPhearson et al., 2013b, 2014; Andersson et al., 2015a). For example, utilizing urban parks, green walls and roofs, and street trees to adapt to and mitigate impacts of

Case Study 8.1   Coastal Natural Protected Areas in Mediterranean Spain: The Ebro Delta and Empordà Wetlands

Sandra Fatorić and Ricard Morén-Alegret

Department of Geography, Autonomous University of Barcelona

Christos Zografos

Institute of Environmental Science and Technology (ICTA), Autonomous University of Barcelona

Keywords	Sea level rise, vulnerability, ecosystems-based adaptation, coastal natural protected area
Population (Study Region)	Ebro Delta: 48,031 Empordà wetlands: 43,354 (IDESCAT, 2015)
Area (Study Region)	Ebro Delta: 299.4 km² Empordà wetlands: 123 km² (IDESCAT, 2015)
Income per capita	US$27,520 (World Bank, 2017)
Climate zone	Csa – Temperate, dry summer, hot summer (Peel et al., 2007)

Case Study 8.1 Figure 1 *Locations of the three municipalities in the Ebro Delta and coastal natural protected area.*

Climate change is an increasingly significant global problem with potentially far-reaching consequences for coastal human communities, livelihoods, and ecosystems in the Mediterranean region. Seven economically, socially, and environmentally dynamic urban towns across coastal natural protected areas in Mediterranean Spain, the Ebro Delta (see Case Study 8.1 Figure 1), and Empordà wetlands have been particularly vulnerable to three aspects of climate change: (1) air and sea temperature rise (2) sea level rise, and (3) decreased river flows (see Case Study 8.1 Table 1). In addition, intensification of coastal erosion, flooding, saltwater intrusion, and deficits in river sediment supply have been affecting natural habitats and livelihoods in these areas (Barnolas and Llasat, 2007; Candela et al., 2007; CIIRC, 2010; Day et al., 2006; Guillén and Palanques 1992; Jiménez et al., 1997; Martín-Vide et al., 2012; Sánchez-Arcilla et al., 2008).

This Case Study is based on studies that identified a local dimension of climate change adaptation relevant for maintaining a wide range of livelihoods while facing current and future climate change (Fatorić, 2010, 2014). These studies build on the work of Smit and Wandel (2006) who highlighted that adaptation is an outcome of the interaction of environmental, social, cultural, political, and economic forces. Analytically, adaptation is conceptualized in this paper as a set of technical options to respond to specific risks (Nelson et al., 2007), where the need for local stakeholder involvement has been increasingly acknowledged (Bormann et al., 2012; Cote et al., 2014; Eriksen et al., 2011; Mozumder et al., 2011). Different stakeholders may hold different knowledge, opinions, and understandings of the local context of adaptation. Thus, which specific sources of knowledge are recognized and used in decision-making process is crucial for determining which interests, development paths, and solutions are prioritized (Eriksen et al., 2011).

Local, regional, and national stakeholders belonging to various economic sectors (e.g., employees of local tourist information centers,

farmers, peasants, engineers); public administrations (e.g., governmental officers), environmental organizations and research centers (e.g., members of environmental groups, scientists), and social organizations (e.g., members of social and ethnic organizations) linked to the Ebro Delta and Empordà wetlands were selected to participate in the studies.

The results showed that adaptation appears to have taken place in the Ebro Delta and Empordà wetlands during the past few decades, but mainly through unsustainable measures (e.g., artificial or hard structures).

More than half of interviewed stakeholders reported that they favor "natural" adaptation measures such as (1) building and/or restoring coastal sand dunes and (2) raising ground level. Approximately one-quarter were in favor of "artificial" adaptation measures such as (1) seawalls, groins, and breakwaters; (2) flood and underwater gates; (3) beach nourishment; and (4) rainwater harvesting. The remaining stakeholders considered combining both types of measures. Stakeholders were also asked to consider coastal relocation as an adaptation option.

With respect to natural adaptation measures, building sand dunes parallel to the shoreline where none exists and/or restoring and stabilizing the existing ones was perceived as the optimal adaptation measure in both protected coastal areas. This option was often considered as the cheapest one for both study areas, and it is compatible with environmental sustainability actions. Moreover, building and restoring dunes can increase socioecological resilience in both areas and produce benefits in the absence of climate change effects.

Regarding raising ground level, the other natural adaptation measure, interviewees expressed little support for elevating ground level by a few centimeters. This might be due to weak technical and urban design skills among most stakeholders.

Artificial measures, on the other hand, did not have such unified support among stakeholders. For instance, dykes, seawalls, and breakwaters generated different opinions. About one-third of stakeholders were against "artificialization" mainly due to the current ecological value of both areas. Stakeholders perceived these measures as too costly to build and maintain. A small number of stakeholders were willing to preserve an already attractive landscape for economic activities (especially tourism) by implementing artificial measures.

Flood and saltwater intrusion gates were suggested and discussed, but gates may be not suitable measures because they entail significant investments. Beach nourishment was perceived by a minority of stakeholders as a suitable measure that

is aesthetically pleasing and that sometimes can be implemented with a reasonable budget.

The studies also revealed that rainfall capture and storage in those parts of Mediterranean Spain where precipitation is likely to decrease and become more variable has not yet been prioritized.

Regarding coastal relocation, it was interesting to note a difference between the two study areas: according to population data (IDESCAT, 2015), the rate of registered foreign immigrants is higher in Empordà wetlands than in the Ebro Delta, and, interestingly, among the interviewed stakeholders in Empordà wetlands there was more willingness for relocation elsewhere as an "adaptation" measure than in the Ebro Delta. In this sense, it emerged that place

Case Study 8.1 Table 1 *Comparison of socioeconomic, environmental, and climatic characteristics of the Ebro Delta and Empordà wetlands. Source: ACA, 2014; BirdLife International, 2014a and 2014b; CREAF, 2013; Ebre Observatory, 2015; IDESCAT, 2015; Meteo Estartit, 2015; Ninyerola et al., 2004; SMC, 2015; Wetlands International, 1992 and 1995*

	Ebro Delta	Emporda wetlands
Municipalities	Amposta, Deltebre, Sant Carles de la Rapita	Castello d'Empuries, Escala, Roses, Sant Pere Pescador
Physical territory	Coastal lagoons, marshlands, beaches, dunes, saltpans	Coastal lagoons, inland freshwater ponds, marshlands, beaches, dunes
Protection	Ramsar Convention on Wetlands (1986), Natura 2000, Special Protection Area for Birds	Ramsar Convention on Wetlands (1992), Natura 2000, Special Protection Area for Birds
Protected area	11,530 ha	10,830 ha
River(s)	Ebro	Muga, Fluvia
River regulation	Mequinega, Flix, Riba-Roja dams	Boadella dam
Vegetation	Arthrocnemetum fruticosi, Crucianelletum maritimae, Scirpetum maritimi-littoralis, Agropyretum mediterraneum	Salix alba, Fraxinus angustifolia, Rosa sempervivens, Arum iotalicum, Aristo lochia rotunda, Typha latifolia
Fauna	Anas strepera, Phoenicopterus roseus, Botaurus stellaris, Ardea purpurea, Larus audouinii	Coracias garrulus, Lanius minor, Buteo buteo, Falco subbuteo, Bos taurus domestica
Climate	Mediterranean	Mediterranean
Air temperature (1990–2014)	Increase by 0.37°C/decade (Ebre Observatory)	Increase by 0.19°C/decade (Sant Pere Pescador)
Precipitation (1990–2014)	Slight increase (Ebre Observatory)	Decrease (Sant Pere Pescador)
River flow (1990–2012)	Decrease by 14 m³/s/decade (Ebro)	Decrease by 3.5 m³/s/decade (Fluvia)
Sea level (1990–2014)	Increase by 3.9 cm/decade (Estartit)	Increase by 3.9 cm/decade (Estartit)
Sea temperature (1990–2014)	Increase by 0.18°C (sea surface), 0.17°C (20 m), 0.28°C (50 m), 0.13°C (80 m) per decade (Estartit)	Increase by 0.18°C (sea surface), 0.17°C (20 m), 0.28°C (50 m), 0.13°C (80m) per decade (Estartit)
Tourism	Ecotourism, birdwatching	Campsites, second homes, hotels, ecotourism, birdwatching, marina
Agriculture	Rice, citrus fruits	Fruits trees, vines, olives
Fishery	14% of Catalonia's fish production	11% of Catalonia's fish production
N° population (2014)	48,031	43,354

attachment (and previous migration experience) among local residents is relevant when considering relocation as an adaptation measure.

One of the lessons that can be drawn from this Case Study is the need to gather and integrate local understanding, perception, and knowledge with scientific knowledge in order to develop successful response to climate change, empower local decision-making, and preserve current ecosystems, livelihoods, and communities. Encouraging local communities and policy-makers to undertake short- and medium-term thinking and to develop adaptation planning with more desirable sustainable outcomes should be a priority in the Ebro Delta and Empordà wetlands. Stakeholders can help to

raise awareness in order to implement adaptation measures based on technical solutions that would reduce the vulnerability of natural and socioeconomic systems and take advantage of any potential opportunities and benefits (Fatorić and Chelleri, 2012; Fatorić and Morén-Alegret, 2013; Fatorić et al., 2014).

Another lesson that emerged from the research is that the optimal adaptation measure according to the stakeholders in both coastal protected areas is building and/or restoring coastal dunes, which is likely to be the most efficient and least expensive protection against various climate change effects (see Case Study 8.1 Figure 2). This highlights the need for dune conservation and maintenance as climate change reinforces the value of its protective capacity.

Case Study 8.1 Figure 2 *Dunes as "natural" adaptation measure in Empordà wetlands.*

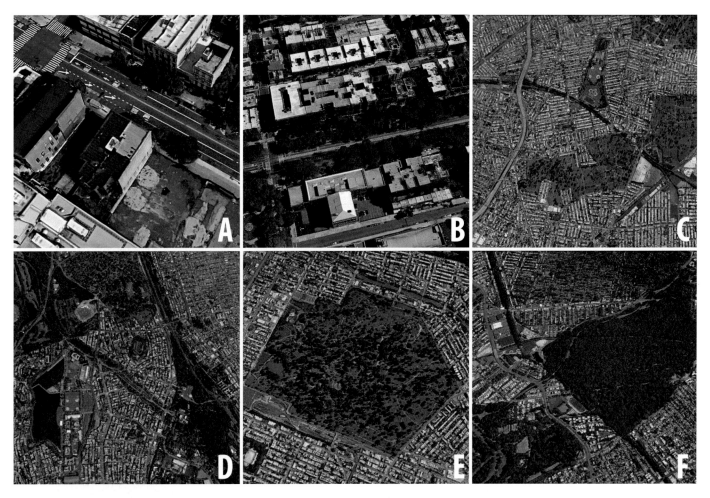

Figure 8.2 *Urban trees and other types of green infrastructure, shown here from New York, can provide important climate adaptation and mitigation in cities. The quantity and quality of benefits (e.g., carbon storage, urban heat island mitigation, stormwater absorption) depend on the urban context and configuration of trees in an urban landscape. A. Individual scattered street trees at 125th Street and Madison Avenue, Harlem. B Dense street trees at Eastern Parkway and Classon Avenue, Brooklyn. C. Street trees as a corridor connecting small- and medium-sized urban green spaces, Elmhurst, Juniper Valley Parks, and Lutheran All Faiths Cemetery, Queens. D. Dense urban street trees connecting large urban parks: Bronx Park and Van Cortlandt Park. E. Disconnected urban green space with scattered trees, Green-Wood Cemetery, Brooklyn. F. Large urban forest, Forest Park, Queens.*

climate change such as urban heat must occur locally (Gill et al., 2007; Pataki et al., 2011). Urban ecosystems are therefore especially important in delivering climate-related ecosystem services with direct impact on human health, well-being, and security (Novotny et al., 2010; Elmqvist et al., 2013; McPhearson et al., 2015). Additionally, investing in urban ecosystems for climate adaptation and mitigation can create multiple co-benefits by simultaneously generating other ecosystem services important to human health and well-being in cities (see Figure 8.4).

8.2 Challenges for Maintaining Urban Biodiversity and Ecosystem Services

Biodiversity protection and adaptive urban ecosystem management, planning, and restoration are critical to maintain a resilient supply of climate-relevant UES in the face of global environmental change (McPhearson et al., 2014a). Globally, urban land cover is projected to increase by 1.2 million square kilometers by 2030, nearly tripling the urban area in 2000; this could result in considerable loss of habitats in key biodiversity

hotspots, including the Guinean forests of West Africa, the tropical Andes, the Western Ghats of India, and Sri Lanka (Seto et al., 2012). Mediterranean habitat types are particularly affected by urban growth because they support a large concentration of cities as well as many habitat-restricted endemic species – species that occur nowhere else in the world (Elmqvist 2013). Although urban land area globally comprises a small fraction of total land area, the impacts of urbanized land on biodiversity, ecosystem services, and other environmental impacts are wide-reaching (McPhearson et al., 2013c; Schewenius et al., 2014).

For example, expansion of urban development into the world's remaining hotspots (see Figure 8.5) for species and genetic diversity has implications for both urban and global biodiversity. These changes have downstream impacts on local ecosystem service provisioning that can feed back to influence urban climate and regional climate change. The direct and indirect effects of land-use changes outside of cities, which can include damming of rivers, water diversions, and agricultural practices, can also have effects on the capacities of ecosystems inside cities to function and produce services (Schewenius et al., 2014; Seto,

Case Study 8.2   New York's Staten Island Bluebelt

Jack Ahern

Department of Landscape Architecture and Regional Planning, UMass Amherst

Robert Brauman

NYC Department of Environmental Protection, New York

Keywords	Urban stormwater management, urban biodiversity, blue-green network, ecosystem-based adaptation
Population (Metropolitan Region)	20,153,634 (U.S. Census Bureau, 2016)
Area (Metropolitan Region)	17,319 km² (U.S. Census Bureau, 2010)
Income per capita	US$56,180 (World Bank, 2017)
Climate zone	Dfa – Continental, fully humid, hot summer (Peel et al., 2007)

SUMMARY

New York faces growing impacts from climate change, an with increasing frequency of extreme weather events such as the 2012 Hurricane Sandy. The city's Staten Island Bluebelt stormwater management practice is one of the best cases of an integrated ecosystem based adaptation (EbA) and disaster risk reduction (DRR) response wherein traditional water bodies and depressions are managed to accommodate and slow flood water. Native vegetation sites are developed by expanding, buffering, and linking with existing parks and conservation areas to form an ecological network to deliver multiple ecosystem services.

CASE DESCRIPTION

The Staten Island Bluebelt is a system of created wetlands developed since the 1990s to provide alternative, ecosystem-based stormwater management services in a rapidly developing borough of New York (NYC). The Bluebelt has become a model for providing multiple ecosystem services including stormwater management, water quality improvement, wildlife habitat provisioning, environmental education, and increased property values.

CLIMATE CHANGE ISSUES

NYC has been facing growing hazards and risks of climate change such as sea-level rise, storm surge, rising temperatures, and other related issues. Hurricane Sandy was an example of an extreme climate event. Mean annual temperature and precipitation in NYC increased 4.4°F (2.44°C) and 7.7 inches (19.55 cm), respectively, from 1900 to 2011, and sea level (at the Battery) has risen 1.1 feet (33.5 cm) since 1900. Climate models projections predict that, by 2050, the temperature will rise by 6°F (3.33°C) and precipitation by 15% (New York City Panel on Climate Change [NPCC2], 2013). NYC

is now working with academia, civil society, and others to make the city's infrastructure and population more resilient and its infrastructure development sustainable. The high water table, poor drainage, and extensive wetlands of Staten Island challenge the development of a conventional stormwater drainage system. Cities across the world can learn from this good practice of stormwater management.

ADAPTATION STRATEGY

In 1990, the NYC Department of Environmental Protection conceptualized the Bluebelt program and began constructing stormwater best management practices (BMP) along stream and wetland corridors to attenuate routine storm flow and improve water quality and flood flow (Ryan, 2006). The Bluebelt concept had two principal goals: (1) to provide basic stormwater infrastructure and (2) to preserve the last remaining wetlands in Staten Island. Since 1995, more than 50 sites have been developed under the Bluebelt program, all of which were justified by a cost-benefit analysis comparing Bluebelt development costs with those of a conventional piped stormwater storage system. The cost-benefit analysis indicated a direct saving of US$30 million (http://cooper.edu/isd/news/waterwatch/statenisland).

The Bluebelt's principal function is to slow, store, treat, and attenuate stormwater in created wetlands and stormwater BMP in a self-regulating native ecosystem. Bluebelt facilities are designed as a "treatment train" of BMPs starting with a constructed "micropool" or fore-bay that receives stormwater from a trunk outlet. The stormwater flow then passes to an extended detention wetland where water is attenuated through contact with native wetland plants and soils. Native wetland plants sequester nutrients and add oxygen to wetland soils, facilitating the bacterial breakdown of nitrogen and phosphorous. Field stones are installed in culvert bottoms to reduce stream velocity and provide fish habitat.

Bluebelt design practices emphasize native plant species and communities including rare and near-extinct plants. Wetland plants are sourced from local nurseries or rescued from local development sites using custom excavating buckets to enable transplanting the full soil profile along with the wetland trees and shrubs. Bioengineering techniques including fascines, mats, and rolls are used to restore and stabilize slopes and stream banks with native wetland tree and shrub species. Bluebelts are constructed to intentionally include habitat "niches" with brush piles, downed trees, and boulder piles. Removed trees with roots attached are placed in the bottoms of Bluebelt ponds to create diverse microhabitats for fish and amphibians. Dead trees are left standing to provide habitat for cavity-nesting birds (Brauman et al., 2009).

Bluebelts are carefully designed to fit and complement their community context. For example, dams and bridges are built from fieldstone to evoke the character of the region's many historical bridges and dams. Bluebelt sites are selected to expand, buffer, and link up with existing parks and conservation areas, forming an ecological network to deliver multiple ecosystem services. Bluebelt trails are designed to link adjacent parks and provide direct community access for recreation. The Adopt-a-Bluebelt program has been successful in engaging community residents and environmental groups with basic maintenance tasks.

Water quality and flow monitoring by the U.S. Environmental Protection Agency found nutrient-removal rates exceeded the national standards for pollutant removal. Wildlife monitoring by the

Case Study 8.2 Figure 1 *Extended detention weir, Conference House Park, Staten Island Bluebelt.*

Photo: Jack Ahern

Case Study 8.2 Figure 2 *An aesthetic bridge connecting different ecosystems.*

Audubon Society has found a large number of breeding birds in the Bluebelt, including green herons, wood thrushes, and great-crested flycatchers. Fish passage provided by fish ladders support migratory breeds, such as the American eel, that go upstream to spawn. Mosquitoes are controlled through the Bluebelt's constant through-flow of water that minimizes their breeding grounds as well as the support that BMP provides to populations of beneficial insects that feed on mosquitoes.

LESSONS LEARNED

The Bluebelt is a good example of a "green infrastructure" – a hybrid engineered and natural system designed to provide a suite of specific urban infrastructure and ecosystem services. It represents an example of an efficient system because of its innovations and collateral ecosystem services, including wildlife habitat provision, community recreation and education and increased property values.

Motivated by the success of this case, the Bluebelt concept is being exported to other NYC boroughs under the City's multiple plans, including the High Performance Infrastructure and new stormwater management plans and the NYC sustainability plans "PlaNYC 2030" and "OneNYC." Bluebelts are also being considered to address ongoing combined sewer overflow (CSO) problems in other boroughs under the Jamaica Bay Watershed Plan. However, in other NYC boroughs, land use is more intensive and there are few existing wetlands and large areas of undeveloped land. In these boroughs, blue belts will be built on public lands, including highway verges and parks.

CONCLUSION

The Bluebelt is an effective adaptation response to effects of climate change on an urban environment. Staten Island was directly in the path of the 2012 Hurricane Sandy, and the Bluebelt demonstrated its resilience and adaptability. Although the storm surge and intense precipitation from Sandy exceeded the treatment capacity of the Bluebelt, it returned to a functional condition soon after the storm passed. The Bluebelt has saved the city more than US$80 million in comparison with a conventional stormwater drainage system (Mayor's Office, 2012).

Box 8.1   Urban Ecosystem Services

Urban ecosystem services (UES) refer to those ecosystem functions that are used, enjoyed, or consumed by humans in urban areas and can range from material goods (such as water, raw materials, and medicinal plants) to various non-market services (such as climate regulation, water purification, carbon sequestration, and flood control) (Gómez-Baggethun et al., 2013). The Millennium Ecosystem Assessment (MA) classified ecosystem services into four different categories: (1) provisioning services, (2) supporting services, (3) regulating services, and (4) cultural services (Convention on Biological Diversity [CBD], 2009), which have been modified and updated by The Economics of Ecosystems and Biodiversity (TEEB, 2010) project and applied to the urban context (TEEB, 2011) (see Figure 8.2). Provisioning services include the material products obtained from ecosystems, including food, fiber, freshwater and genetic resources. Regulating services include water purification, climate regulation, flood control and mitigation, soil retention and landslide prevention, pollination, and pest and disease control. Cultural services are the nonmaterial benefits from ecosystems including recreation, aesthetic experience, spiritual enrichment, and cognitive development, as well as their role in supporting knowledge systems, social relations, and aesthetic values (Andersson et al., 2015b; Chan et al., 2011). Finally, supporting or habitat services are those that are necessary for the production of all other ecosystem services including provisioning of habitat for species, primary production, nutrient cycling, and maintenance of genetic pools and evolutionary processes (Gómez-Baggethun et al., 2013).

Table 8.1 *Key abiotic, biotic, and cultural functions of green urban infrastructure. Source: Adapted from Ahern, 2007*

Abiotic	Biotic	Cultural
Surface-groundwater interactions	Habitat for generalist species	Direct experience of natural ecosystems
Soil development process	Habitat for specialist species	Physical recreation
Maintenance of hydrological regime(s)	Species movement routes and corridors	Experience and interpretation of cultural history
Accommodation of disturbance regime(s)	Maintenance of disturbance and successional regimes	Provisions of sense of solitude and inspiration
Buffering of nutrient cycling	Biomass production	Opportunities for healthy social interactions
Sequestration of carbon	Provision of genetic reserves	Stimulus of artistic/abstract expression(s)
Modification and buffering of climatic extremes	Support of flora-fauna interactions	Environmental education

2013; Ignatieva et al., 2011). Moreover, the ability of species to move within and among urban landscapes is considered a key issue of biodiversity adaptation to climate change, one that suggests the need for cities to improve habitat connectivity and use green corridors for healthy, functioning urban ecosystems.

Cities and urban regions often have a perhaps surprisingly high level of biodiversity, including both native species and non-native species from around the world (Müller et al., 2010; Aronson et al., 2014). Urban species can therefore be an important component of regional and global biodiversity. Cities are often concentrated along coastlines, major rivers, and islands, which are also areas of high species richness and endemism, with many cities existing in close proximity to protected areas (see Figure 8.6) (Güneralp et al., 2013; McDonald et al., 2013). However, because expanding urban areas encompass an increasingly larger percentage of global biodiversity hotspots, it is all the more critical to safeguard urbanized biodiversity hotspots and promote ecological conservation in urban, peri-urban, and nearby rural areas.

Ecosystems are highly fragmented in urban areas, which can alter the genetic diversity and long-term survival of sensitive species. To ensure viable urban populations, urban planners and designers need to understand species' needs for habitat quality and connectivity among suitable habitat patches. For example, the connectivity of the habitat network within the urban area can play a major role for ground-dwelling animal movement, as for the European hedgehog in Zurich (Braaker et al., 2014). Understanding and planning for greater habitat connectivity through the use of green corridors is a key tool for city planners to design appropriate management and conservation strategies of urban biodiversity and to improve the resilience of species to climate change. Furthermore, it is important to understand how the impacts of climate change in cities will create risks and affect the vulnerability of urban ecosystems. The ability of ecosystems to sustain levels of biodiversity at or above the thresholds necessary for maintaining ecosystem integrity is critical to sustainable delivery of ecosystem services important for meeting urban sustainability and resilience goals (Andersson et al., 2015a).

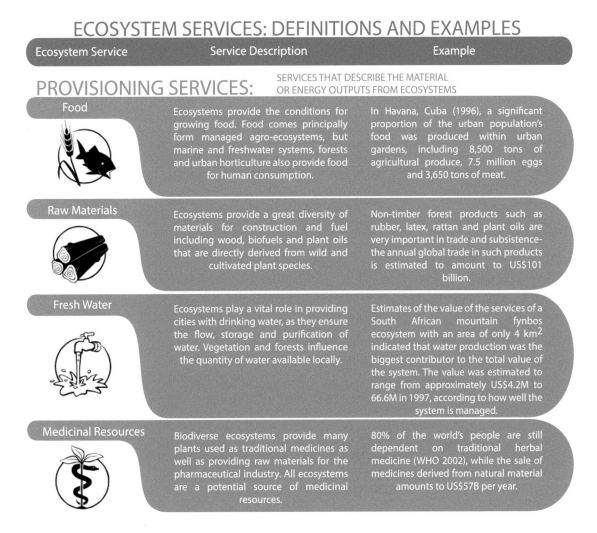

ECOSYSTEM SERVICES: DEFINITIONS AND EXAMPLES

Ecosystem Service	Service Description	Example
PROVISIONING SERVICES:	SERVICES THAT DESCRIBE THE MATERIAL OR ENERGY OUTPUTS FROM ECOSYSTEMS	
Food	Ecosystems provide the conditions for growing food. Food comes principally form managed agro-ecosystems, but marine and freshwater systems, forests and urban horticulture also provide food for human consumption.	In Havana, Cuba (1996), a significant proportion of the urban population's food was produced within urban gardens, including 8,500 tons of agricultural produce, 7.5 million eggs and 3,650 tons of meat.
Raw Materials	Ecosystems provide a great diversity of materials for construction and fuel including wood, biofuels and plant oils that are directly derived from wild and cultivated plant species.	Non-timber forest products such as rubber, latex, rattan and plant oils are very important in trade and subsistence- the annual global trade in such products is estimated to amount to US$101 billion.
Fresh Water	Ecosystems play a vital role in providing cities with drinking water, as they ensure the flow, storage and purification of water. Vegetation and forests influence the quantity of water available locally.	Estimates of the value of the services of a South African mountain fynbos ecosystem with an area of only 4 km^2 indicated that water production was the biggest contributor to the total value of the system. The value was estimated to range from approximately US$4.2M to 66.6M in 1997, according to how well the system is managed.
Medicinal Resources	Biodiverse ecosystems provide many plants used as traditional medicines as well as providing raw materials for the pharmaceutical industry. All ecosystems are a potential source of medicinal resources.	80% of the world's people are still dependent on traditional herbal medicine (WHO 2002), while the sale of medicines derived from natural material amounts to US$57B per year.

Figure 8.3 *Investing in urban ecosystems for climate adaptation and mitigation can create multiple co-benefits by simultaneously generating other ecosystem services important to human health and well-being in cities. Ecosystem services can be divided into four categories: provisioning services, regulating services, habitat or supporting services, and cultural services, with examples of each. For references in this table, see TEEB Manual for Cities (2011).*

Source: Modified and adapted from TEEB Manual for Cities (2011). For references in this Figure, see TEEB Manual for Cities

REGULATING SERVICES: SERVICES THAT ECOSYSTEMS PROVIDE BY REGULATING THE QUALITY OF AIR AND SOIL OR PROVIDING FLOOD AND DISEASE CONTROL, ETC

Local climate and air quality regulation

Trees and green space lower the temperature in cities whilst forests influence rainfall and water availability both locally and regionally. Trees or other plants also play an important role in regulating air quality by removing pollutants from the atmosphere.

In Cascine Park in Florence, Italy, the urban park forest was shown to have retained its pollutant removal capability of about 72.4 kg per hectare per year (reducing by only 3.4 kg/ha to 69.0 kg/ha after 19 years, despite some losses due to cutting and extreme climate events). Harmful pollutants removed included O3, CO, SO2. NO2, and particulate pollutants as well as CO2.

Carbon sequestration and storage

Ecosystems regulate the global climate by storing greenhouse gases. As trees and plants grow, they remove carbon dioxide from the atmosphere and effectively lock it away in their tissues; thus acting as carbon stores.

Urban trees too, are important in carbon sequestration: in the United States, their annual gross carbon sequestration amounts to 22.8 million tons of carbon per year (as calculated in 2002). This is equivalent to the entire USA population's emissions in five days. This sequestration service is valued at US$460 million per year, and US$14,300 million in total.

Moderation of extreme events

Ecosystems and living organisms create buffers against natural disasters, thereby preventing or reducing damage from extreme weather events or natural hazards including floods, storms, tsunamis, avalanches and landslides. For example, plants stabilize slopes, while coral reefs and mangroves help protect coastlines from storm damage.

In the case of the Californian Napa City, USA, the Napa river basin was restored to its natural capacity by means of creating mudflats, marshes and wetlands around the city. This has effectively controlled flooding to such an extent that a significant amount of money, property, and human lives could be saved.

Waste-water treatment

Ecosystems such as wetlands filter effluents. Through the biological activity of microorganisms in the soil, most waste is broken down. Thereby pathogens (disease causing microbes) are eliminated, and the level of nutrients and pollution is reduced.

In Louisiana, USA, it was found that wetlands could function as alternatives to conventional wastewater treatment, at an estimated cost saving of between US$785 to 34,700 per hectare of wetland (in 1995).

Erosion prevention and maintenance of soil fertility

Soil erosion is a key factor in the process of land degradation, desertification and hydroelectric capacity. Vegetation cover provides a vital regulating service by preventing soil erosion. Soil fertility is essential for plant growth and agriculture and well-functioning ecosystems supply soil with nutrients required to support plant growth.

A study estimated that the total required investment to slow erosion to acceptable rates in the USA would amount to US$8.4 billion, yet the damage caused by erosion amounted to US$44 billion per year. This translates into a US$5.24 saving for every US$1 invested.

Pollination

Insects and wind pollinate plants which is essential for the development of fruits, vegetables and seeds. Animal pollination is an ecosystem service mainly provided by insects but also by some birds and bats.

Some 87 out of the 115 leading global food crops depend upon animal pollination including important cash crops such as cocoa and coffee.

Biological control

Ecosystems are important for regulating pests and vector borne diseases that attack plants, animals and people. Ecosystems regulate pests and diseases through the activities of predators and parasites. Birds, bats, flies, wasps, frogs and fungi all act as natural controls.

Water hyacinth was brought under control in southern Benin using three natural enemies of that plant. Whereas the biological control project cost only US$2.09 million in present value, its accumulated value is estimated to amount to US$260 million in present value (assuming the benefits stay constant over the following 20 years), representing a very favourable 124:1 benefit cost ratio.

Figure 8.3 *(continued)*

HABITAT OR REGULATING SERVICES:

THESE SERVICES UNDERPIN ALMOST ALL OTHER SERVICES. ECOSYSTEMS PROVIDE LIVING SPACES FOR PLANTS OR ANIMALS: THEY ALSO MAINTAIN A DIVERSITY OF PLANTS AND ANIMALS.

Habitats for species

Habitats provide everything that an individual plant or animal needs to survive: food, water, and shelter. Each ecosystem provides different habitats that can be essential for a species' lifecycle. Migratory species including birds, fish, mammals and insects all depend upon different ecosystems during their movements.

In a March 2010 article, IUCN reports that habitat loss is the single biggest threat to European butterflies, and may lead to the extinction of several species. Habitat loss was said to occur most often as a result of changes in agricultural practice, climate change, forest fires, and expansion of tourism.

Maintenance of genetic diversity

Genetic diversity (the variety of genes between, and within, species populations) distinguishes different breeds or races from each other, providing the basis for locally well-adapted cultivars and a gene pool for developing commercial crops and livestock. Some habitats have an exceptionally high number of species which makes them more genetically diverse than others and are known as 'biodiversity hotspots'.

In the Philippines, an initiative to conserve local varieties of rice aided in the development of rice strains that are better adapted to local conditions - giving greater yield, a quality seed supply, and decreasing dependence on plant breeders - at a much lower cost than that of formal plant breeding.

CULTURAL SERVICES:

THESE INCLUDE THE NON-MATERIAL BENEFITS PEOPLE OBTAIN FROM CONTACT WITH ECOSYSTEMS. THEY INCLUDE AESTHETIC, SPIRITUAL AND PSYCHOLOGICAL BENEFITS.

Recreation and mental and physical health

Walking and playing sports in green space is a good form of physical exercise and helps people to relax. The role that green space plays in maintaining mental and physical health is increasingly becoming recognized, despite difficulties of measurement.

A review article examined the monetary value of ecosystem services related to urban green space, based on 10 studies, including 9 cities from China and 1 from the USA. It reported that on average, 'Recreation and Amenity' and 'Health effects' contributed a value of US$5.882 and US$17.548 per hectare per year respectively to the total average of US$29.475 per hectare per year provided by the seven identified ecosystem services in the various studies.

Tourism

Ecosystems and biodiversity play an important role for many kinds of tourism which in turn provides considerable economic benefits and is a vital source of income for many countries.
In 2008 global earnings from tourism summed up to US$944 billion. Cultural and eco-tourism can also educate people about the importance of biological diversity.

Based on the amounts of money people spent on travel and local expenditure in order to visit Coral reefs in Hawaii, it was estimated that the value associated with these reefs amounted to US$97 million per year. This implies that reef tourism resulted in significant income generation for individuals, companies, and countries.

Aesthetic appreciation and inspiration for culture, art and design

Language, knowledge and the natural environment have been intimately related throughout human history. Biodiversity, ecosystems and natural landscapes have been the source of inspiration for much of our art, culture and increasingly for science.

Prehistoric rock art of southern Africa, Australia, and Europe, and other examples like them throughout the world, present evidence of how nature has inspired art and culture since very early in human history. Contemporary culture, art and design are similarly inspired by nature.

Spiritual experience and sense of place

In many parts of the world natural features such as specific forests, caves or mountains are considered sacred or have a religious meaning. Nature is a common element of all major religions and traditional knowledge, and associated customs are important for creating a sense of belonging.

In the example of the Maronite church of Lebanon, the church committed to protecting a hill in their possession, comprising rare remainders of intact Mediterranean forest, independent of scientific and legal arguments, because this was in line with Maronite culture, theology and religion.

Figure 8.3 *(continued)*

Figure 8.4 *Investing in urban ecosystems and green infrastructure can provide multiple co-benefits. This shows a cultural co-benefit of urban and peri-urban trees through tapping sugar maple (Acer saccharum) trees in Pound Ridge, New York. Maple sugar tapping represents a seasonally occurring peri-urban and urban food production ecosystem service that has long-standing cultural traditions in many northern countries.*

Photo: Timon McPhearson. Adapted from Andersson et al., 2015b

8.2.1 Current Effects of Climate Change on Urban Biodiversity and Ecosystems

All urban ecosystems will experience the effects of climate change. Additionally, many cities are located in geographic areas that are especially vulnerable to both existing and projected climate hazards, such as coastal flooding, landslides, and extreme events. Climate change is impacting a broad spectrum of urban ecosystem functions, biodiversity, and ecosystem services (Rosenzweig et al., 2011; Solecki and Marcotullio, 2013; UN-Habitat, 2011). Urban ecosystems are already under general stress from development, pollution, and direct human use (Elmqvist et al., 2013), and climate change variability poses additional challenges for urban species and ecosystems. For example, in a comprehensive review of the potential impacts of climate change on urban biodiversity in London, Wilby and Perry (2006) highlight the importance of four threats to biodiversity in the city: competition from non-native species, pressure on saltmarsh habitats from rising sea levels, drought effects on wetlands, and changing phenology of multiple species as earlier springs occur more frequently (Hunt and Watkiss, 2011).

The UHI effect in cities can change the reproductive and population dynamics of animals. Insect life cycles and migration

---

Case Study 8.3 The Serra do Mar Project, Baixada Santista Metropolitan Region (BSMR), São Paulo State

Oswaldo Lucon

São Paulo State Environment Secretariat

Keywords	Resettlement, biodiversity protection, climate resilience, urban ecology, floods, landslides, ecosystems
Population (Metropolitan Region)	1,664,136 (IBGE, 2015)
Area (Metropolitan Region)	2,405 km² (IBGE, 2015)
Income per capita	US$14,810 (World Bank, 2017)
Climate zone	Af – Tropical rainforest (Peel et al., 2007)

A partnership of the Inter-American Development Bank (IDB) and the São Paulo State Government, the Serra do Mar and Mosaics System Recovery Program has been recognized as an international standard for resettling communities in disaster-prone, ecologically sensitive areas. *Mosaics* are sets of protected areas located nearby or juxtaposed to each other. Their main purpose is to promote integrated and participatory management of their components, respecting different categories of management and conservation objectives. Mitigation strategies comprise halting deforestation, reforestation, and wastewater treatment. Adaptation strategies are based on the resettlement of populations living in landslide- and flood-prone areas. The Program started in the city of Cubatão and part of the

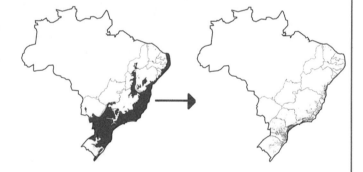

Case Study 8.3 Figure 1 *Brazilian Atlantic Forest, before (left) and currently (right).*

Baixada Santista Metropolitan Region (BSMR) of 2,405 square kilometers (IBGE, 2015). Topography varies from cliffs (700 m) to plains (average 3 m above sea level). The Atlantic Forest is a UNESCO Biosphere Reserve, one of the planet's biologically richest regions and also one of the most endangered. Overexploitation and biome devastation have resulted in only 7% of the Brazilian Atlantic Forest being preserved in fragments of more than 100 acres (see Case Study 8.3 Figure1).

In the Southeastern State of São Paulo, the Atlantic Forest is concentrated in the *Serra do Mar*, squeezed between the coastal BSMR (nine cities, 1.6 million people) and the São Paulo Metropolitan Region (19 cities and 20 million people). The 133,000 square kilometers of the Serra do Mar State Park cover twenty-four municipalities in the state. Additionally, three mosaics (Paranapiacaba, Jureia-Itatins, and Jacupiranga) allow for buffer zones between urban and native preserved areas. Despite having been reduced and highly fragmented, the Atlantic Forest is habitat to more than 20,000 plant species – a wealth of diversity greater than that found in North America (17,000 species) and Europe (12,500 species). Out of the native plant species, 8,000 are endemic; that is, native species that only exist in Brazil (IAD and São Paulo State Government, 2009). Degradation of the forest had its origins in the construction of roads. In the highly industrialized city of Cubatão, settlements on hillsides (*bairros-cota*) invaded areas belonging to the Serra do Mar State Park. Illegal occupations harmed not only the Park, but also created several hazards to its inhabitants: landslides, floods, road accidents, and freshwater contamination (see Case Study 8.3 Figure 2).

In the first stage, the São Paulo Government contributed 65% of the US$470 million budget, and the IDB allocated the remaining 35%. Geotechnical studies mapped and classified risk areas.

Potential damage to dwellings and their residents was estimated, considering their positions and distances to critical slopes plus the degree of building vulnerability (construction pattern and level of urban consolidation). A joint analysis of these criteria established a mapping of risk sectors, with hotspots defined (IDB and São Paulo State Government, 2013; see Case Study 8.3 Figure 3). In 2007, new settlements were halted ("frozen") through supervision of the Military Police, with protective measures for the Park that included preventing deforestation, fires, and the capture of wild animals and extraction of plant species, and monitoring of the various sectors of the *bairros-cota* to prevent their expansion (IDB and São Paulo State Government, 2013).

A total of 7,388 irregular households were identified, with around 7,760 families and 7,843 buildings. The resettlement program was

Case Study 8.3 Figure 2 *Settlements on hillsides (*bairros-cota*) invaded areas belonging to the Serra do Mar State Park, creating several hazards to its inhabitants, including landslides, floods, road accidents, and freshwater contamination.*

Case Study 8.3 Figure 3 *Designated risk areas.*

followed by an environmental education program, an enrollment process, sealing buildings, commissioning basic housing and urbanization projects, obtaining environmental licenses, conducting public hearings, and negotiating with the IDB for co-financing. Benefits include improved living conditions for around 3.2 million people in the surrounding area, an increase to 60,000 visitors per year in the Park, improved biodiversity, improved water quality, strengthened management and protection of conservation units (an additional 20,000 hectares of Atlantic Forest; recovery of 1,240 hectares of State Park), and lowered disaster risk, plus more sustainable sources of income.

More than 5,000 families living in at-risk or protected areas have been resettled and assisted with housing and upgraded infrastructure. Living in new structured communities, they have also benefited from professional training programs for construction professionals, gardeners, and nurserymen to work on the reforestation of the reclaimed areas. The second phase of the program aims at assisting approximately 25,000 families with resettlement or infrastructural upgrading. Building improvements included two or three types of houses with diversified typologies, accessibility for the disabled, preservation of significant green areas, and improved urban infrastructure. Family assistance combines social, cultural, economic, and environmental aspects. Resettlement has brought innovations

that enabled families to feel sufficiently engaged before and after moving from their homes, including the choice of one of fifteen housing options. Housing units were not donated, and leaving a house where one had lived for a long time is not an easy decision, even if it means moving to better conditions. Therefore, for families who live in rural or peri-urban areas, other methods have been developed.

To anchor all actions, synergy among institutions has proved decisive. In 2009, the state joined the United Nations Environment Program (UNEP)'s Sustainable Social Housing (Sushi) initiative for building sustainable social housing for low-income populations. A pilot neighborhood (Residencial Rubens Lara) in Cubatão City has been developed and today is recognized by the UNEP as a replicable model for other countries. In 2012, the Serra do Mar Social and Environmental Recovery Program earned the Greenvana GreenBest award, the highest distinction conferred in Brazil for environmental initiatives. The Serra do Mar Program went beyond the limits of the City of Cubatão. It now covers the whole of the Atlantic Forest of São Paulo, extending throughout the Park (north and south of the state) to the Jureia-Itatins territory and the Units for Marine Conservation. The extended program is called Serra do Mar and the Atlantic Forest Mosaics System Social and Environmental Recovery Program (CDHU, 2012; São Paulo, 2013).

---

patterns have been well-documented, with changes in the life cycle of certain insects having already occurred in response to urban warming (Parmesan and Yohe 2003; Parmesan, 2006). Butterfly species in Ohio, for example, appear to have shifted when they fly in response to urban warming. Some native butterfly species appear to be at risk due to the shortening of their flight periods (Kingsolver et al., 2013). In Raleigh, North Carolina, the abundance of the gloomy scale butterfly (*Melanaspis tenebricosa*) increases with increases in impervious surfaces that create warmer forest temperatures and therefore drive increased reproduction rates, thus contributing to greater population growth for this urban forest pest (Dale and Franck 2014). This suggests that urban trees could face greater herbivory in the future as a consequence of the increased fitness of some herbivorous arthropods under warming scenarios. However, more research is needed to generalize these results to other urban areas.

Although other ecological and socioeconomic factors are affecting vegetation in urban areas, many of the non-native invasive species colonizing cities originate in warmer areas and are benefiting from changing climate conditions (Sukopp and Wurzel, 2003). In mountain regions, climate is already causing changes in vegetation structure and diversity (Theurillat and Guisan 2001; ICIMOD, 2009). The response of trees to extreme climatic events may be species-specific. For example, in Dresden, a study of oak trees showed that *Quercus petrea* and *Q. rubra* are better adapted to warm and dry conditions than are *Acer platanoides* and *A. pseudoplatanus* (Gillner et al., 2014).

Trees have been perhaps better studied than other taxonomic groups in urban areas. Urban trees experience multiple forms of stress including heat stress, low air humidity, and soil drought. Rapid climate change can have a significant impact on the distribution and biology of trees. In Philadelphia, climate change

is influencing the biology of urban tree pathogens and pests. Results from a recent study indicate that the future climate in Philadelphia will become less optimal for multiple tree species since major pests and diseases are likely to become more problematic (Yang, 2009).

Comparing urban and rural species has yielded a useful understanding of urban biodiversity responses to changing climate. Woodall et al. (2010) compared tree species compositions in northern urban areas to tree compositions in forestland areas. They found that some tree species native to eastern U.S. forests of southern latitudes have been planted or are present in northern urban forests, indicating the tolerance of southern species in northern urban ecosystems. Although urbanization and climate change can both profoundly alter biological systems, scientists often analyze their effects separately. Recent studies are beginning to look at these impacts on organisms simultaneously to better understand how multiple simultaneous stressors might affect species, but more research is needed (Kingsolver et al., 2013).

8.2.2 Projecting Impacts of Climate Change on Urban Biodiversity and Ecosystems

Projecting impacts of climate change on the distribution of species is complex, with many factors to consider including dispersal ability, species interactions, and evolutionary changes (Pearson and Dawson 2003; Gilman et al., 2010; Urban et al., 2012). Still, future climate change in cities, when combined with additional urban stressors such as short-lived climate pollutants, land use change, and direct human impacts is expected to pose difficult challenges for urban species and ecosystems. Maintaining adequate levels of biodiversity and managing urban ecosystems to ensure a resilient supply of critical ecosystem services that are necessary for expanding urban populations may

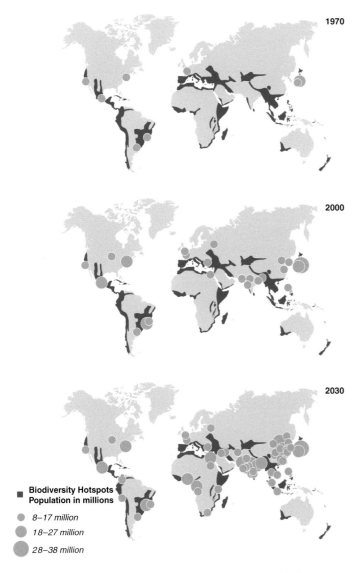

1970

2000

2030

Biodiversity Hotspots
Population in millions

○ *8–17 million*

● *18–27 million*

● *28–38 million*

Figure 8.5 *As urban areas expand into some of the most sensitive biodiversity areas globally, urban biodiversity conservation and ecological planning will have an increasing impact on global biodiversity and ecosystems and the services they provide to both urban and rural residents. Urban areas with large populations in 1970, 2000, and 2030 (projected) are shown in green as examples of urban expansion in global biodiversity hotspots (in blue). The cities shown have a projected population of more than 8 million in 2030, according to UN World Urbanization Prospects 2014.*

become increasingly challenging in the future as climate change intensifies its effects on cities. Which ecosystems will be most affected in the near and longer term may be signaled by current species' responses to climate change (Parmesan, 2006; Gillner et al., 2014). The risks and vulnerabilities associated with climate change in urban ecosystems are likely to vary with temporal and spatial scale and nature of change (e.g., chronic vs. acute), although in general they are expected to increase over the next several decades (Solecki and Marcotullio, 2013).

We present here new regionally downscaled climate projections using the Intergovernmental Panel on Climate Change (IPCC) Fifth Assessment Report (AR5) scenarios for forty global cities spanning small, large, and megacities in multiple contexts including coastal and inland cities in the Global North

and Global South in the 2020s, 2050s, and 2080s (see Figure 8.7a, 8b) (see Chapter 2, Urban Climate Science, and ARC3.2 Annex 2, Climate Projections for ARC3.2 Cities). Projections for both temperature and precipitation show wide variation in cities around the world, with temperature generally increasing and precipitation both increasing and decreasing depending on location. Effects on ecosystems will vary considerably from city to city, and therefore it is not possible to suggest general management or planning approaches. Instead, decision-makers in cities and urban areas will need to take into account locally relevant climate projections combined with data on sensitive species or ecosystems to develop plans and adaptive management strategies to safeguard urban ecosystems and the benefits they provide for climate adaptation and mitigation (as well critical co-benefits for human well-being). These downscaled climate projections suggest that urban planning, policy, and management must pay close attention to decisions and actions involving urban ecosystems that may be directly impacted by uncertain climate futures. Climate change in cities is already having significant impacts on urban biodiversity and ecosystems. Further impacts of rising temperatures and increasing or decreasing precipitation suggest increasing ecological impacts over time, with concurrent affects on the ability of urban ecosystems to provide nature-based solutions for building climate resilience in cities.

8.3 Climate Change Hazards, Risks, and Vulnerabilities for Urban Ecosystems

When combined with socioeconomic changes, there are multidimensional vulnerabilities affecting biodiversity and urban ecosystems. Heat stress, inland and coastal flooding, droughts, cyclones, fire, and extreme rainfall pose risks to urban ecosystems, populations, and economies (Revi et al., 2014). Massive land conversion from natural ecosystems to a built environment exposes urban landscapes to loss of biodiversity, flash floods, droughts, and pollution while urban sprawl and poor urban design further threaten urban biodiversity (Munaung et al., 2013; Revi et al., 2014). Recent studies demonstrate how climate change is reinforcing urban ecosystem vulnerability through unsustainable development, agricultural land conversion, and degradation of ecosystem services that affect the ability of ecosystems to meet urban climate adaptation and mitigation goals (Satterthwaite et al., 2007; UN-Habitat, 2009, 2011).

8.3.1 Climate Hazards and Risks

Urban climate hazards are defined as the climate-induced stressors or drivers that affect urban ecosystems. Examples include elevated temperature, changes in precipitation patterns, sea level rise, and the build-up of short-lived climate pollutants such as black carbon (see Figure 8.7), as well as changes in the frequency and intensity of extreme events such as storm surge, flash floods, heat and cold waves, and wild fires (UNEP, 2011). The cascading effects of climate change can have both direct and indirect effects on biodiversity and ecosystems. Climate change

275

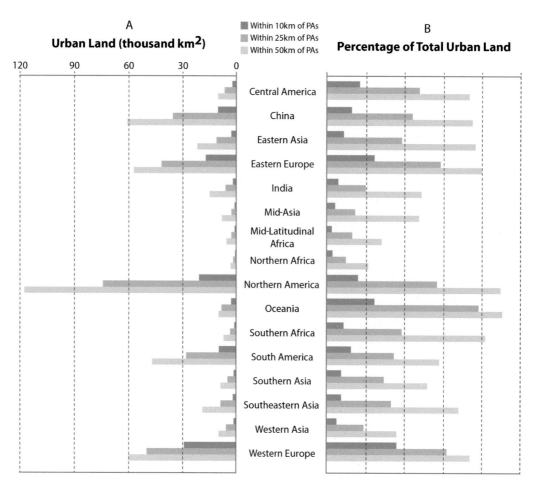

Figure 8.6 *Urban areas are expanding into protected areas in all parts of the world. Figure shows (a) urban extent and (b) percentage of total urban extent within a distance of, from top to bottom, 10, 25, and 50 kilometers of protected areas (PAs, e.g., national parks) by geographic region circa 2000.*

Source: Adapted from McDonald et al., 2013; Güneralp and Seto, 2013

Case Study 8.4   Ecosystem-Based Climate Change Adaptation in the City of Cape Town

Pippin M. L. Anderson

Department of Environmental and Geographical Science, University of Cape Town, Cape Town

Keywords	Ecosystem-based adaptation, disaster risk reduction. flood water, wetlands, stormwater management
Population (Metropolitan Region)	3,740,025 (City of Cape Town, 2015)
Area (Metropolitan Region)	2,461 km² (City of Cape Town, 2015)
Income per capita	US$12,860 (World Bank, 2017)
Climate zone	Csb – Temperate, dry summer, warm summer (Peel et al., 2007)

SUMMARY

Cape Town is adapting to growing urban climate change vulnerability and impacts. The city, with its rich biodiversity and unique ecosystems, historically used hard engineering measures to reduce growing flood and storm surge risks. However, in recent years, the role of ecosystem services is being recognized and included in urban climate change adaptation plans. Recent initiatives by the city administration to identify and spatially map urban ecosystem services (UES), in particular in relation to the bionet (network of green open spaces) map, to establish critical connectivity corridors suggest a good start in mainstreaming climate change in urban development planning and environment conservation.

CASE DESCRIPTION

Cape Town, with an area of 2,460 square kilometers and a population of approximately 3.7 million has close to 38% low-income households, indicating high poverty incidence. The city's population also has a high disease burden due to the high prevalence of HIV and tuberculosis. More than 58% of the adult population has below a

Case Study 8.4 Figure 1 *Iconic Table Mountain of Cape Town viewed from the Durbanville Conservation Area.*

Photo: Pippin Anderson

high school education, and 16.9% of the population is unemployed. The city is characterized by urban sprawl and rapidly expanding informal poor settlements on the lowland areas that are known as the Cape Flats. The increasing demand for housing continues to place a burden on city authorities and on remnant urban biodiversity.

Cape Town is located in the Cape Floristic Region, the smallest and most diverse floral kingdom on earth: the region hosts almost 9,000 plant species on 90,000 square kilometers – some 44% of the flora of the subcontinent on a mere 4% of its land area. Of approximately 3,350 indigenous plant species within the metropolitan boundary, 190 are endemic to the city that also hosts 19 of 440 National Vegetation Types (Cilliers and Siebert, 2012). The process of urbanization has significantly contributed to the erosion of local biodiversity, putting further stress on eleven nationally recognized critically endangered vegetation types in the city. The City is host to 83 mammal, 364 bird, 60 reptile, 27 amphibian, and 8 freshwater fish species. The lowlands historically hosted the greatest vegetation-type and floral diversity, and the majority of this has been lost to urban settlements. Some 450 of these indigenous plant species are listed as threatened or near-threatened, and 13 species are known to be extinct.

Remnant natural ecosystems are highly fragmented, with little connectivity. Fire is used as a management tool in a burning rotation of 10–15 years, which poses a management challenge in an urban setting, threatening both property and life. Introduced invasive plant species suppress indigenous biodiversity and yield high fuel loads that, under a rising temperature regime, lead to hotter and more dangerous fires.

CLIMATE CHANGE VULNERABILITY AND IMPACTS

Climate change is occurring faster in South Africa than the global average. Mean annual temperatures have increased faster than the global average during the past 50 years. Extreme rainfall and drought events have also increased in frequency (Ziervogel et al., 2014). Urban areas are particularly vulnerable due to stormwater

surge, flooding, uncontrolled fire, and coastal erosion. The Cape Town region is likely to face significant climate change risks with predicted increases in temperature in all seasons, reductions in rainfall, greater evaporation, more intense and frequent winds, and greater coastal erosion and storm surge with changes in the frequency and intensity of extreme weather events. Increased rainfall intensity will exacerbate flooding, especially in high water table areas on the Cape Flats. Flooding is exacerbated due to the canalized nature of rivers where natural vegetation buffers have been removed.

Cape Town is a water-scarce area. Current climate change predictions suggest increased rainfall variability with associated future increases in periods of drought and water shortages. Climate change predictions suggest hotter, more frequent, and runaway fires. Cape Town, with its 307 kilometers of coastline, is at threat from climatic hazards such as sea level rise and increased storm surge.

ADAPTATION STRATEGY

The City has adopted an integrated water resource management (IWRM) approach that includes demand-side water management. Acknowledging the role of invasive plant species in reducing water availability, the government public works program seeks to train and employ unskilled and unemployed labor to clear invasive vegetation, producing positive outcomes in biodiversity, social benefits, and water yield.

Adaptation measures to increased flood risk include both engineering and ecological solutions that includes the creation of retention ponds and resilient infrastructure, regular drain cleaning, better disaster warning systems, the decanalizing of rivers, and the restoration of riparian vegetation to vulnerable areas. However, engineering solutions get less attention mostly due to high costs and flood disaster-relief funding structures.

Fire, an ecologically necessary measure to promote indigenous flora, is being used more judiciously. The intensity and season of firing

Case Study 8.4 Figure 2 *Controlled burning of urban vegetation.*

are being regulated to have positive implications for biological processes of recruitment and regeneration. Fire regimes during periods of drought, higher wind speeds, and generally greater climate variability are being used strictly for assured biodiversity and employment generation. Government public works – "Working for Water" and "Working on Fire" programs – are used to reduce large fuel loads and minimize runaway fires. These programs train firefighters in ecological fire management using higher public safety protocols. In general, these programs, set up to address various environmental and social issues, have proved to be an important vehicle for generating adaptive capacity and change in the face of threats posed by climate change.

DRIVERS

The City Administration has taken a number of measures to adapt to climatic changes and mitigate threats. Historical measures, such as sea embankments to protect infrastructure, are now recognized as extremely expensive to maintain and as sometimes ineffective. Acknowledging the high costs of these engineering measures, the City is employing more ecosystem-based approaches including the protection and restoration of extensive wetlands sites that can

absorb large volumes of water and dissipate wave energy (ICLEI, 2012). Efforts are on to restore dune vegetation and to open paths to improve sand supply to these mobile systems that have frequently become cut off due to hard engineering solutions employed in the past. These ecosystem-based adaptation (EbA) measures are providing green employment and thus contributing to the City's poverty reduction goal.

IMPACT AND LESSONS LEARNED

These multipronged adaptation approaches have drawn involvement from multiple stakeholders and worked to create better impacts and synergy. The City is trying to secure the establishment of a "bionet" – a network of green open spaces – that would serve to improve biodiversity areas by allowing for greater flexibility and opportunity for species conservation, provide vegetated areas for water infiltration, and reduce flooding and storm surge impacts. The role of ecosystem services will become critical in the face of climate change. The initiatives by City government to identify and maintain ecosystem services and a biodiversity corridor suggest a good start in mainstreaming climate change in urban development planning and environment conservation.

also has significant economic and human impacts that can extend from infrastructure and built environment sectors to natural ecosystems (Frumkin et al., 2008; Keim, 2008; Hallegatte et al., 2010; Ranger et al., 2010; Solecki and Marcotullio, 2013). For example, in cities with diminishing precipitation, the vegetated cover of green roofs may face drought risks. Increased exposure due to rising populations and growth of human settlements in flood- and landslide-prone areas exacerbate climatic hazards as well as socioeconomic risks, thus emphasizing the sensitive interactions among climate urban ecosystems, and communities.

8.3.1.1 Thermal Hazards

Changing temperature regimes (see Figure 8.7a) can have both direct and indirect effects on the organisms that live in

cities and the ecosystem services that they provide. At the individual and species population level, many physiological processes such as photosynthesis, respiration, growth, and flowering of plants are affected by changing temperature. Elevated temperature can affect growth and reproductive rates either positively or negatively for plants (Hatfield, 2011), while also inducing a range of landscape-level impacts on biogeochemical cycles and watershed hydrology (Suddick et al., 2012). Warming conditions in New York, for example, have led to changes in tissue chemistry in tree seedlings relative to cooler, non-urban settings, resulting in more rapid shoot growth but reduced root mass (Searle et al., 2012). Higher temperatures can also lead to increased physiological stress on wildlife, affecting their behavior and reproduction (Marzluf, 2001) (see also Section 8.2.1).

Table 8.2 *Effects of urban climate and environment on urban agriculture. Source: Adapted from Wortman and Lovell, 2013*

Drivers of plant production	Compared to rural areas	Observed positive effects	Observed negative effects	Resulting impact on urban crop yield	Expected future dynamics of drivers
Length of growing season	7 to 8 days longer	Potential of double cropping systems		Higher	Increase
Time to flowering	Earlier		Risk of asynchrony between timing of flowers and pollinator presence	Lower	Increase
CO_2 concentration	Higher	Increased photosynthesis rate in many vegetable crops (C_3 plants)		Higher	Increase
Temperature	Higher	Increased photosynthesis rate up to threshold	Decreased photosynthesis rate (in case of extreme temperature), increased irrigation water demand	Higher, lower	Increase
Wind speed	Lower	Reduced plant mechanical damages	Increased leaf gas exchanges	Higher	?
Vapor pressure deficit	Higher (less air humidity)		Greater plant transpiration, moisture stress, reduced photosynthesis rate, reduced rainwater infiltration in soil, lower soil moisture	Lower	?
Ground-level ozone concentration	Higher (sometimes lower)		Decreased photosynthesis rate, lower root-to-shoot ratio, premature leaf senescence	Lower-higher	Increase
NO_2 concentration	Higher	More efficient nitrogen nutrition	Delayed flowering, accelerated plant senescence	Higher-lower	Increase
Soil water infiltration	Lower		Higher moisture stress	Lower	Increase-decrease

8.3.1.2 Drought Hazards

More intensive increases or decreases in precipitation can lead to significant water-related urban hazards including drought and severe water shortages (IPCC, 2013). Reductions in precipitation can be exacerbated by warming temperatures, which increase water losses to evapotranspiration driven by hydrological alterations from surface-water diversion and groundwater extraction (Pataki et al., 2011). Increased frequency and duration of droughts exacerbated by warming can also increase evapotranspiration (Leipprand and Gerten, 2006). Increased evapotranspiration reduces water availability and groundwater resources, often leading to increased salinization and water stress affecting both the quality and quantity of water for plants, with negative consequences on floral and faunal biodiversity and productivity (Alberti and Marzluf, 2004). For example, projected drought conditions in Manchester, England, are likely to reduce the cooling services provided by grasslands, which may increase the local UHI and wild fires (Gill et al., 2013). Current drought in California is affecting drinking water supplies and is also having dramatic effects on peri-urban agriculture; this has led to historic water conservation measures to deal with drought stress. Drought affects both street trees and urban parklands and will likely have cascading effects on herbivores, soil fauna, and other components of urban biodiversity, as well as effects on urban residents through decreased water availability affecting livability (Wilby and Perry 2006; Gillner et al., 2014).

8.3.1.3 Flood Hazards

More frequent and increased precipitation (see Figure 8.7b) can lead to significant urban flood hazards. Flash floods, in addition to damaging critical infrastructure and directly impacting the lives of urban dwellers, also are harmful to urban water supplies and drainage systems and can have lasting negative impacts on ecosystems (IPCC, 2013). Increasing extreme precipitation events in combination with land-cover changes and increased frequency of tropical cyclones and subsequent altered water flow in urban watersheds is likely to result in an increased incidence of flooding in many cities (Depietri et al., 2012; IPCC, 2013). Flood hazards include the short-term impacts of the force of moving water (e.g., flash floods), inundation, and drowning, which cause longer-term impacts resulting from sediment movement (erosion and deposition), soil processes, and the distribution of pathogens that precipitate negative public health impacts (ICIMOD, 2012; Teegavaerapu and UNESCO, 2012; Wisner et al., 2003; Walker et al., 2008). For cities along rivers and coastlines, rising sea levels and increasing storm surges will increase urban flooding as well (Mosely 2014). Coastal flooding due to sea level rise can

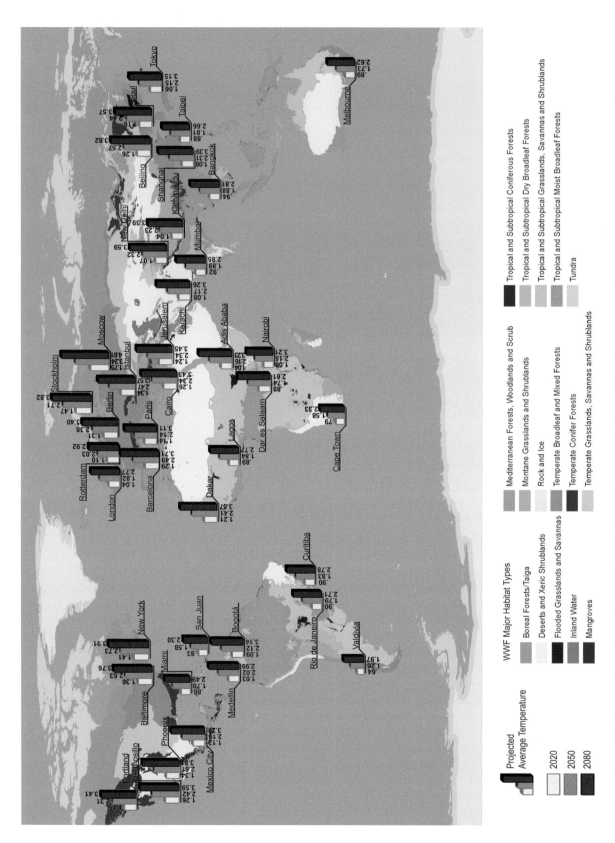

Figure 8.7 a) *Here we show projected average temperature and precipitation for forty global cities in 2020, 2050, and 2080 (see Chapter 2, Urban Climate Science, and ARC3.2 Annex 2, Climate Projections for ARC3.2 Cities) and major habitat types. Cities represent a range of small, large, and megacities in the Global North and Global South, including inland and coastal cities. Temperature (8.7a) (in °C) and precipitation (8.7b) (% change) projections are based on thirty-three Global Climate Models and two Representative Concentration Pathways (RCP4.5 and RCP8.5) downscaled from regional to city spatial extent. Changes are relative to the 1971–2000 base period. The time slices are the 30-year periods on which the projections are centered (e.g., the 2020s is the period from 2010 to 2039).*

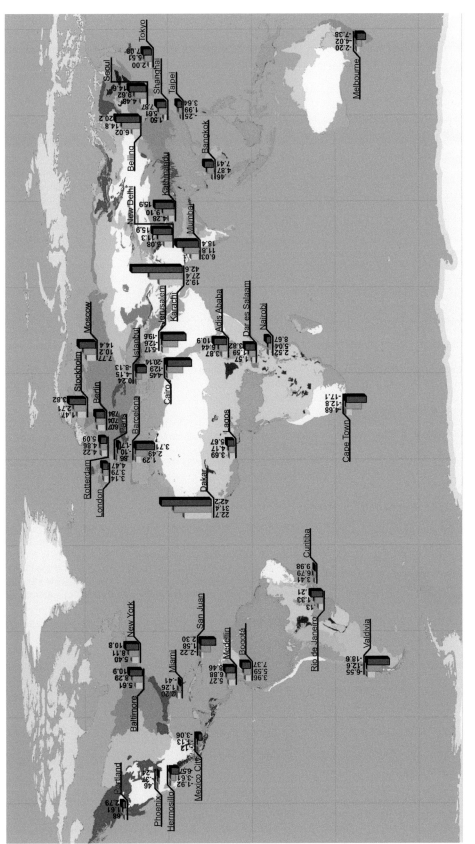

Figure 8.7 b) *(continued)*

lead to increased salinization and reduced groundwater recharge (Chan et al., 2011; IPCC, 2013), which can decrease habitat quality for biodiversity.

Climate change in cities will lead to increased precipitation in some places and decreased precipitation in others (see Figure 8.7a). In cities projected to receive increased precipitation,

increased discharge into surface waters will have ecosystem consequences. For example, urban development affected the ability of watersheds in Baltimore, Maryland, to retain nitrogen, and urban watersheds showed increased sensitivity to climate variation (Kaushal et al., 2008). Loss of this urban ecosystem function in Baltimore (nitrogen retention) led to increased nitrogen downstream, with negative consequences for the ecology

---

Case Study 8.5   Jerusalem Gazelle Valley Park Conservation Program

Naomi Tsur

Former Deputy Mayor, Jerusalem Green Fund

Helene Roumani

Jerusalem LAB Coordinator, Local Action for Biodiversity

Keywords	Biodiversity, urban planning, development, Jerusalem mountain gazelle, ecosystems
Population	839,000 (CIA World Factbook, 2015)
Area	125 km² (www.worldcat.org, 2017)
Income per capita	US$36,190 (World Bank, 2017)
Climate zone	Csa – Temperate, dry summer, hot summer (Peel et al., 2007)

CASE DESCRIPTION

The historic city of Jerusalem is also a well-known place for its rich natural heritage of Biblical flora and fauna that has developed as an integral part of the city landscape. The city is a significant habitat for half a billion birds since it lies on one of the most important global bird migration routes following the course of the Great Rift Valley. The credit for creating this rich ecosystem and wealth of biodiversity within a city lying in a water-scarce area goes to the Gazelle Valley Park Conservation Program (Krasny, 2015).

The major issue of climate change faced by the city is shortage of water and the threat of desertification. Temperatures in the Middle East region are not rising as fast as in other parts of the world, but the region is already experiencing weather extremes and the process of desertification is on the rise. Although rainfall has increased, so has evaporation. The impact of climate change on the region's natural flora and fauna is still mild mainly because of their historical adaptive capacity to withstand moisture stress and high temperatures. However, future predictions are that, due to extreme heat and water stress, plants and animals will have difficulty surviving.

CLIMATE CHANGE ADAPTATION STRATEGY

The City of Jerusalem has assumed the responsibility for improving and maintaining its unique desert and hilly ecosystems to preserve its floral and faunal biodiversity in the face of increasing climate change stresses. In 2009, Jerusalem joined the ICLEI–Local Governments for Sustainability/Local Action for Biodiversity (ICLEI/ LAB) Network to further pursue sustainable development measures.

In the context of Jerusalem's LAB Legacy project for the International Decade of Biodiversity, Jerusalem has established the Gazelle Valley Conservation Program to protect and restore one of the city's unique biodiversity areas and to plan the development of a park for both wildlife preservation and local recreation at the site. The area has recently been designated as an urban nature park – a local model of sustainable development.

The Gazelle Valley is situated on a sixty-acre undeveloped tract of land in southwest Jerusalem, between two residential neighborhoods and closed in by major roadways. After being used for agricultural purposes during the 1960s and 1970s, the land, a rich wildlife habitat, was left as open space while the surrounding urban area continued to develop. The mountain gazelle (*Gazella gazella*), an indigenous species particularly prevalent in this part of the Jerusalem hills, has been roaming the valley and sustaining itself on its natural resources since ancient times. It is also the site of ancient terraces with orchards that still bear fruit.

In the late 1990s, a residential plan was established for the Gazelle Valley, threatening to destroy the gazelle habitat and remove a vital open space in the city. The Jerusalem branch of the Society for the Protection of Nature (SPNI) opposed the development plan, citing that it was a reversal of established urban planning principles. Local residents and activists joined SPNI to launch a campaign to save the Valley.

The Gazelle Valley Citizen Action Committee was thus formed. Understanding the need for a comprehensive plan, the Committee, together with SPNI, commissioned an alternative plan focusing on conservation and restoration of the site's unique biodiversity. After 10 years of rigorous grassroots opposition, the city decided to withdraw the residential plan and designate the Gazelle Valley a natural heritage

Case Study 8.5 Figure 1 *Mountain gazelles, examples of revived urban biodiversity urban biodiversity.*

site. In addition, the conservation plan was approved by the Local Planning Committee in 2009, marking the first time that local authorities approved a development plan initiated by residents. This civil society initiative for environmental protection in Jerusalem was also a significant victory for the environmental movement in the region.

CLIMATE CHANGE ADAPTATION

The development of the Gazelle Valley Park in Jerusalem plays an important role in the city's promotion of climate change adaptation. Water conservation is a significant aspect of the park design. Apart from the need to regulate the drainage basin, water features prominently in the plan as a vital natural resource for sustaining the local biodiversity. In addition, regulation of existing water systems is being planned to enhance the beauty of the site and serve to attract visitors. The plan includes a series of runoff collection pools that will have the capacity to store 20,000 cubic meters of rainwater and seepage. In addition, a runoff filtration system is also planned for sedimentation of solids in water entering the park. To control seasonal flood zones, the valley's natural irrigation system will be rehabilitated, facilitating the restoration of the site's ancient agricultural terraces. In order to

prevent erosion and control channel flow, two gravel-lined streams will be dug in alignment with the local topography. Proper rainwater management will not only create a buffer zone between the conservation area and the adjacent recreational area, it will also help mitigate climate change effects and the effects of increased urbanization around the Valley.

IMPACT AND LESSON LEARNED

The park is expected to serve local residents and visitors with a public activity core (differentiated from the animal habitat), including pedestrian and bike paths, gazelle observation points, a bird watching route, agricultural gardens, and an educational visitor center. The Gazelle Valley Conservation Program in Jerusalem demonstrates that through proper planning practices, conservation efforts in an urban setting can facilitate both climate change adaptation and promote efficient ecosystem management. In the case of Jerusalem, it is anticipated that this effort will also produce an effective interface between biodiversity and human activity. The city government is taking the lead in mobilizing stakeholders in steering this green adaptation project.

Case Study 8.5 Figure 2 *Gazelle Valley Park, Jerusalem City.*

and economy of the Chesapeake Bay. On the other hand, freshwater wetlands with reduced hydrologic inputs could become even further water-limited, with negative effects on both ecosystem services and biological diversity (World Bank, 2015). The modification of climate within and around cities combined with increasing drought stress from decreased precipitation illustrates how climate change will affect many urban ecosystems worldwide (UN-Habitat, 2011).

8.3.1.4 *Hazards Related to Shifting Species Distributions*

Species movement in response to climatic regime shifts has already been well-documented (Parmesan and Yohe, 2003; Parmesan 2006; ICIMOD, 2009; Porter et al., 2013). Organisms

at the edge of their distributions may decline as temperature or other climate conditions shift outside their physiological tolerances. For example, plant and animal species may shift northward seeking cooler temperatures following climate shifts, meaning that cities in the tropical and subtropical belts may lose species faster (Gonzalez et al., 2010; Grimm et al., 2013). Changing climate may also affect the introduction of new species in urban ecosystems by reducing noninvasive and native species while favoring weedy and urban-adapted species (Kendle and Forbes 1997; Booth et al., 2003; Heutte et al. 2003). Warming cities (see Figure 8.7a) may find new problems with invasive species and pests that had formerly been limited by cold conditions. The most adaptive species in an era of changing urban climate are likely to include more weeds, pests, and invasive species, such as the introduced Burmese python in Florida (IPCC, 2014). Species that are highly specialized and heat sensitive may be threatened with local extinction driven also by an inability to move to new areas as urban development expands.

The distribution of pathogens is also likely to shift with changing climate, with consequences for both resident organisms and the ecosystem functions they provide. For example, climate change is likely to influence the distribution of the mosquito *Aedes aegypti*, the primary urban vector of dengue and yellow fever viruses (Eisen et al., 2014) (see Chapter 10, Urban Health). Ultimately, changes in species distributions are expected to modify the ecological interaction networks in cities and have the potential to promote invasive species, which could accelerate the loss of urban biodiversity (Nobis et al., 2009; Kendal et al., 2012).

It is important to recognize that none of these hazards and risks operates in isolation. For example, changes in CO_2 concentrations may or may not amplify the impacts of changes in precipitation, temperature, or other climate hazards on urban vegetation (Zavaleta et al., 2003), suggesting the need for further research to better understand critical feedbacks in the urban system. Thus, integrating all of the ecosystem processes and recognizing that there are critical feedbacks among ecological, built, and social components of urban systems will yield a more thorough understanding of climate risks to urban biodiversity and ecosystems.

8.3.2 Urban Ecosystem Vulnerability

Vulnerability may be considered a lack of resilience or a reduction in adaptive capacity (Tyler and Moench, 2012). However, the complexity of urban ecosystems is characterized by vulnerability along multiple dimensions. Urban ecosystems share many of the same types of climatic vulnerabilities as non-urban ecosystems. However, urban ecosystems are also exposed to a number of unique stressors and therefore they experience greater exposure to hazards such as a high concentration of pollution, the inherent role of non-climate stressors, and the UHI phenomenon (Farrell et al., 2015). The extent of human conversion of the landscape and anthropogenic inputs of materials, energy, and organisms are all greater in cities, which can affect climate vulnerability in

a variety of ways (Fitzpatrick et al., 2008, Loarie et al., 2008). Rapid urban growth and local landscape dynamics[5] contribute to national, regional, and global-scale climate change driven by elevated rate of greenhouse gas (GHG) emissions, radiative forcing[6] of non-greenhouse gases, and alteration of rainfall patterns by short-lived climate pollutants (e.g., black carbon, tropospheric ozone, and methane) (Cerveny and Balling, 1998; Pielke et al., 2002; Parmesan, 2006; UNEP, 2011).

In the urban ecosystem context, exposure to multiple stressors is a real concern, particularly in developing countries where socioeconomic and political drivers along with climate variability play important roles (Leichenko and O'Brien, 2008). However, there are very few studies that have assessed the multidimensional nature of urban ecosystem vulnerability important for planning appropriate adaptation measures (IPCC 2007, Williams et al., 2008). Assessing the vulnerability of urban ecosystems to climate change is critical to include as part of urban planning, policy, and design processes that intend to ensure sustainable delivery of ecosystem services into the future (McPhearson et al., 2015a) (see Chapter 5, Urban Planning and Design).

Vulnerability of urban ecosystems can be assessed at multiple levels within the urban system, including for the individual organism (e.g., physiological health and reproductive success of humans, plants, and other biota), populations, and for larger landscapes (e.g., land use and land cover, biogeochemical cycling) (UNEP-WCMC, 2009; Vignola et al., 2009; Kalusmeyer et al., 2011; Violin et al., 2011). Most studies to date examine vulnerability at the species level or, in some cases, landscape level. Williams et al. (2008) and Glick et al. (2011) have developed species-level vulnerability assessments in which they define "species vulnerability as a function of climate change–related impacts and the adaptive capacity of the species." However, given the strong connections among urban, peri-urban, and rural landscapes, it is important to assess combined and connected cumulative effects of exposure and sensitivity to climate change. Kalusmeyer et al. (2011) argue that assessing vulnerability at landscape level is cost effective and a more useful tool for decision-makers than, for instance, than, for example, vulnerability assessments focused on single species (see Box 8.2).

Box 8.2 Ecosystem Vulnerability

Klausmeyer et al. (2011) used a vulnerability assessment tool for climate change impacts on biodiversity using landscape-scale indicators in California. This method allows biodiversity managers to focus analysis on the species likely to be most vulnerable and to decide on the best adaptive strategies to reduce vulnerability to climate change. Based on results, the authors recommended that state biodiversity managers focus on minimizing current threats to biodiversity (9% area), reducing constraints to adaptation (28%), reducing exposure to climatic changes (24%), and implementing all three (9%). In 12% of the high-vulnerability areas, current conservation goals have to change; in remaining areas, no additional actions are required. This tool can also help to identify adaptation measures focused on endangered species only.

---

5 Landscape dynamics is a concept of landscape equilibrium highlighting the spatial and temporal scaling of disturbance regimes and their influence on equilibrium/non-equilibrium dynamics in a particular landscape (Pielke et al., 2002).

6 "Radiative forcing is a measure of the influence a factor has in altering the balance of incoming and outgoing energy in the Earth-atmosphere system, and is an index of the importance of the factor as a potential climate change mechanism" (IPCC, 2001).

8.4 Adaptive Capacity and Urban Ecosystem Resilience

The adaptive capacity of species in urban landscapes is a function of ecology, physiology, and genetic diversity (Kalusmeyer et al., 2011; Williams et al., 2008). The adaptive capacity of an urban ecosystem is also the degree to which system dynamics can be modified to reduce risk. Traditionally, adaptive capacity focused on human actors and institutions, but, in the context of urban biodiversity and ecosystems, nonhuman actors, behavior, species interactions, and human–ecological interventions are also important. For example, human-induced adaptive capacity could include planting species that are more tolerant of higher temperatures and droughts. Nonhuman-derived adaptive capacity could include natural processes that change ecosystem components rapidly for organisms like insects populations persisting despite changing climate. Adaptation measures such as introducing green infrastructure (e.g., urban green spaces, constructed wetlands, agricultural land in outlying flood-prone areas) can reduce thermal loads and flood hazards and improve water and air quality for vulnerable biota (Depietri et al., 2012) (see also Case Study 8.2, Staten Island Bluebelt, and Case Study 8.5, Jerusalem Gazelle Park). In addition, cities are dependent on urban and peri-urban ecosystems for food production, water provision, and air-quality regulation, meaning that the adaptive capacity of a specific urban area depends at least partially on local-to-regional considerations (Tyler and Moench, 2012).

Resilience to climate change is a growing priority among urban decision-makers. Improving resilience will require transformations in social, ecological, and built infrastructure components of urban systems (Tyler and Moench, 2012; Ernstson et al., 2010) (see Chapter 1, Pathways to Urban Transformation). Urban ecosystems are important components when building urban resilience through their ability to absorb climate-induced shocks and ameliorate the worst effects of extreme climate events (McPhearson et al., 2015a). However, disturbances of sufficient magnitude or duration, such as prolonged drought, can push biodiversity and ecological relationships beyond safe thresholds for reliable production of ecosystem services and may require new approaches to land-use planning and adaptation that focus on building ecological resilience (Folke et al., 2004).

Cities are increasingly seeking to enhance adaptive capacity of urban ecosystems through, for example, green infrastructure, including urban agriculture, landscape conservation, green roofs, green walls, and other green and open spaces that conserve ecosystem values and functions (Kremer et al., 2016a). Building urban parks and other green spaces and adding vegetation strips to densely built neighborhoods can help reduce thermal hazards, manage stormwater, and enhance health benefits, thus enhancing climate change resilience (e.g., in Rio de Janeiro). From a climate and resource-efficiency perspective, the spatial configuration of green spaces is particularly important to mitigate the UHI effect and to conserve water and energy use. Cities with a combination of a high percentage of green areas, high edge

density (distribution of the green space), and high patch density (number of patches per unit area) can more effectively respond to climate extremes such as heat waves and heavy precipitation (European Environment Agency [EEA], 2015; Maimaitiyiming et al., 2014). This suggests that policy and urban planning should ideally prioritize connected green corridors of critical mass rather than a multitude of fragmented green spaces; nevertheless, the total percentage of green space independently is likely most impactful for climate resiliency and, in practice, is often more feasible to create.

Many cities are vulnerable to the hazards associated with climate change as a function of their location (UN-Habitat, 2011). For example, cities are disproportionately distributed along coasts and major rivers, which increases their vulnerability to floods and storm surges. Urban ecosystem managers planning species- and landscape-level adaptation often have multiple goals such as protecting land, restoring habitat, encouraging compatible lands uses, and reducing fragmentation (Heller and Zavaleta, 2009). Building resilient urban ecosystems therefore needs flexible, modulating, and safe-to-fail approaches that can adapt to uncertainty and extreme climate events such as typhoons and hurricanes (e.g., Hurricane Sandy in New York, 2012) (Tyler and Moench, 2012). Also, greater coordination and networks among governance structures that manage local ecosystems and urban biodiversity, including cemeteries, golf courses, urban parks, and neighborhood gardens, would strengthen ecosystem functioning as well as the associated and essential social-ecological engagement (Ernstson et al., 2010).

8.4.1 Interactions between Social and Ecological Infrastructure

The vulnerability of urban ecosystems and biodiversity is intrinsically linked to human activities that drive urban system dynamics. The urban population, with its resource consumption and waste-generation activities, the built infrastructure system (buildings, transportation infrastructure, utilities), and the direct and indirect modifications to the landscape (e.g., changes in vegetation, water courses and storage, microclimate) all create a distinct set of vulnerabilities for the systems and biota embedded in cities (Alberti, 2015). These vulnerabilities are manifested at multiple spatial scales. At the very local level, the altered microclimate and hydrology of a city street will affect the ecosystem services generated by local trees, wildlife, and microbes. Within larger ecosystems embedded in cities, such as remnant forests and urban agriculture and wetlands, the direct effects of human activities and infrastructure need to consider both local and landscape-level management to reduce hazards exposure and risks simultaneously at multiple levels.

8.4.2 Adaptive Management of Vulnerable Urban Biota

The multidimensional nature of urban vulnerability impacts urban biodiversity components including the diversity of plants, animals, and microbes within city boundaries. These groups are all influenced by environmental changes associated with both urbanization and human management. City managers should support both biological communities that have persisted since before urban development (e.g., remnant forest patches, indigenous wildlife that have adapted to urban conditions) and novel communities that depend on human inputs (e.g., pests, deliberately or accidentally introduced species) (Aronson et al., 2014). For example, cities create novel ecosystems and habitats outside their natural biome, such as warm subway tunnels in cold regions, lakes and ponds in arid areas, and dry soils

in humid areas that contribute to increased biodiversity levels often observed in urban areas compared to surrounding ecosystems (McKinney, 2002).

Urban biodiversity vulnerability can be mediated by direct and indirect human management of habitats. The response of indigenous species in remnant ecosystems is affected by regional climate shifts, local ecological dynamics, and the local impact of the city itself (e.g., augmented warming, altered water resources, direct human impact). These urban influences can be moderated by direct human management that reduces their exposure and

Case Study 8.6   Medellín City: Transforming for Life

Leonor Echeverri

Administrative Department of Planning, Medellín City Council

Keywords	Urban development; transportation; adaptive urban planning; resilient infrastructure, ecosystems
Population (Metropolitan Region)	3,731,000 (Alcaldía de Medellín, 2015)
Area (Metropolitan Region)	1,152 km² (Alcaldía de Medellín, 2015)
Income per capita	US$13,910 (World Bank, 2017)
Climate zone	Am – Monsoon-influenced humid subtropical (Peel et al., 2007)

Medellín is an inspiring example of sustainable and innovative urban development with good governance, community participation, and business partnerships. City leaders can take some credit for transforming the city into a vibrant, socially cohesive, and more environmentally resilient city through initiating adaptive and flexible urban planning strategies with effective implementation. The positive impact of mass transportation, green spaces, and equitable benefit sharing resulted from citizen participation, stakeholder involvement, and government support for urban development. The effective use of social networks and good communication by city leaders sustained community support.

CASE DESCRIPTION

Medellín, located in mountainous Aburrá Valley, is a Colombian city with a history of sustainable urban development processes. Many of the poor communities living on the mountainous slopes were challenged for safety and access to essential city services. City leadership has since provided public safety, security, easy mobility, access, amenities, and opportunities. Medellín also developed affordable mass transport systems – the world's first cable car system – the Metrocable – and also a Metro to address both access and pollution problems. Today, Medellín is famous for its social cohesion, business-friendly environment, people and environment-centric city governments and a high quality of life. How Medellín transformed itself from a city with high socioeconomic challenges to one described as "a great inspiration to other cities facing similar issues" can be attributed to Medellín's bold, visionary leadership, which

encompassed diverse stakeholders to deliver a series of small-scale but high-impact, innovative green urban projects (Eveland, 2012).

SOCIOECONOMIC AND ENVIRONMENTAL ISSUES

By the 1970s, Medellín demographically had grown by almost twenty-fold – from around 60,000 in 1905 to more than 1 million – to become the second-largest city of Colombia. A large number of poor were living in precarious socioeconomic and ecological conditions, suffered exclusion, and were struggling with a high cost of living. Medellín entered a cycle of decline in its economic base, which led to a consolidation of a segregated, unequal, and conflict-ridden society and degrading ecosystem. By the early 1980s, Medellín faced a host of social and economic upheavals that led to government failure. In response, the city unleashed social mobilization processes that constituted the genesis for a collective construction of a new vision of urban development, one that led to political and strategic processes that began major change and development pathways in the city. The community responded with significant efforts toward collective dialogue in which a broad cultural and pedagogical process laid the groundwork for civic and citizen-led projects. This included environmentally sensitive urban development planning and program implementation.

ADAPTIVE CHANGE PROCESS

Affordable mobility played an important role in achieving equitable connectivity between urban and rural sections in Medellín (Moreno et al., 2013) As well as setting a process of forming neighborhoods in response to functional interests and a population demanding specific interventions, expanding the city's services to include green spaces throughout the city improved its greening index, making the city more climate resilient (Green City Index [GCI], 2010).

Milestones were conceived during the 1980s and 1990s as the Strategic Plan of Medellín and the Metropolitan Area 2015. This generated a broad and pluralistic project of continuity and consistency in a society that was in crisis. A participatory process was developed for sustainable development that became a foundation for environmentally friendly policies and practices.

In 1995, the Metro became operational – a point of origin for the Integrated Mass Transit System that linked physical, institutional, virtual, sustainable, and environmental modes of mass transit with efficiency and effectiveness. The Metro system serves the current as well as future transportation needs of all inhabitants. This has helped Medellín to minimize its ecological footprint and protect biodiversity and ecosystem services in spaces freed-up by the Metro.

A joint exercise between government and development agencies has been proposed to assess the vulnerability dynamics of the territory, along with implementing sustainable alternatives to mitigate climate impacts in both urban and rural areas of the Valle de Aburrá. With a holistic adaptation approach, criteria are enforced to ensure the security of both the people and the ecosystem within the city's territory. The Integrated Transport System of the Aburrá Valley (SITVA), the Inventory of Greenhouse Gases, the Environmental Classrooms Program of Integrated Solid Waste Management, Linear Parks and Ecological Corridors, Best Practices of Sustainable Consumption and Production, the More Forests to Medellín project, and Integral Water Management among others have positively contributed to the improvement of indicators of an adaptive urban system. Linear parks in particular help the city protect itself from storms and increased pollution.

IMPACTS AND LESSONS LEARNED

The city's Green Belt encourages conditions and opportunities for integral human resources development in the transition zone between urban and rural regions. The Green Belt is important as a way to regulate city expansion into sensitive ecosystems and has helped to conserve and protect natural habitat.

The Medellín River Park is another urban renewal project, connecting the city with efficient mobility, public space, and environmental interventions. Engineering and urbanism work hand-in-hand so that the city's rivers can form the structural axis of civic life.

This ongoing process of transformation has shown that it is possible to build a community-driven, environmentally friendly project in a city. The city's development plan is based on a territorial focus on its urban–rural areas and contains a systemic view to overcome inequities. This has inspired bottom-up planning processes and public–private partnerships to find innovative alternatives.

Medellín's Home for Life initiative recognizes that a participative society and good governance are combined in an institution that seeks equity as a result of political and social rationality. Here, the urban development goes beyond different forms of land use and integrates a combined human–environment urban ecosystem framework. The lessons learned in these efforts will prove useful in confronting the daunting challenges of adapting to climate change.

CONCLUSION

The main driver of Medellín's transformation has been city government's efforts to be inclusive, fair, participatory, and environmentally sound in urban development governance. These approaches transformed Medellín into a model of sustainable urban livability and earned it the 2014 Lee Kuan Yew World City Prize Special Mention. Medellín aspires to continue advancing as an innovative and intelligent city, and hopes to facilitate the exchange of experiences and the advancement of collective knowledge among cities and their inhabitants. To promote sound green design and appropriate policies embracing multidimensional development, building resilient rules, regulations, capacity, and citizen's participation have been the key factors.

Case Study 8.6 Figure 1 *Parque de Los Pies Descalzos (Barefoot Park)*

Photo: Municipality of Medellín and Development Plan 2012–2015: Medellín, a home for life

sensitivity, for example by eradicating pest organisms or creating conservation programs for rare or endangered and endemic species. The Gazelle Valley Park Conservation Program is an example of how climate change adaptation can be combined with biodiversity conservation through ecosystem management (see Case Study 8.5).

Another example is how integrated urban water management can reduce the vulnerability of urban ecosystems and biodiversity (see Case Studies 9.5 and 14.B, Rotterdam). Management of water resources for drinking and sanitation, as well as the hazards associated with water (flooding, landslides, etc.), can alter water flow and storage for the benefit of urban plants. Management of urban hydrological systems through improved greening can decrease the vulnerability of urban ecosystems. For example, during drought periods, a small share of water resources may be reserved as environmental flow for use by plants and animals, thus allowing ecological systems such as forests, wetlands, and streams to survive and maintain adaptive capacity. While drought may affect an entire region, urban ecosystems where water resources are well managed can reduce the impact of such climate-driven water stress, but only provided that urban ecosystem management activities are part of a larger system-level urban resilience plan.

8.5 Ecosystem-Based Adaptation and Nature-Based Solutions

In the urban context, healthy ecosystems can replace or complement often expensive "hard" or engineered infrastructure (e.g., sea-walls, dykes or embankments for river control, and shelters). EbA[7] and similar nature-based solutions have been widely recognized as "soft," safe-to-fail, and often less expensive approaches to climate resilience that values and uses ecological services for adaptation (Huq et al., 2013; CBD, 2009). EbA approaches can generate numerous co-benefits and indirect ecological and social benefits to both non-urban and urban communities in ways that support urban transitions to sustainable, livable communities (UNFCCC Secretariat, 2011, 2013; UNEP, 2012; Huq et al., 2013; Zandersen et al., 2014). The contribution of green infrastructure to EbA in the form of urban parks, avenue plantation, and urban forestry can also provide small levels of GHG mitigation by storing carbon in soils and vegetation. Multiple co-benefits are also expected from the integration of climate change adaptation and biodiversity conservation measures (ICLEI-Africa, 2013; UNEP, 2009). For example, although not a direct goal of climate adaptation, city green spaces have been shown to have important societal co-benefits including but not limited to lower crime rates, reduced level of stress, enhanced cognitive capacities, and improved public health (Troy et al., 2012; Demuzere et al., 2014). The Singapore Case Study (Case Study 8.7) illustrates how EbA can be integrated with disaster risk reduction strategies.

Urban ecosystem services may play an increasingly vital role as cities grow in population size and contain larger senior age cohorts than any other settled topography. The 2003 heat wave in Europe resulting in more than 70,000 deaths can be seen as an early warning of more severe climatic conditions to come, with climate change being viewed as a new public health threat (Petkova et al., 2014). In the United States, heat is the greatest weather-related cause of human death because increasing temperatures above 90°F (32°C) aggravate air pollution and ozone levels and result in greater health risks, including respiratory illness (e.g., asthma) and heart attacks, particularly

---

Box 8.3   Urban and Peri-Urban Agriculture and (Agro) Forestry for Climate Change Adaptation and Mitigation

Marielle Dubbeling

RUAF Foundation, Leusden

The IPCC's AR5 (IPCC, 2013) projects that, due to climate change, there is likely to be a loss of food production and productive arable lands in many regions. Cities with a heavy reliance on food imports and the urban poor will be significantly affected. Adaptation options and local responses mentioned in the report include support for urban and peri-urban agriculture, green roofs, local markets, enhanced social (food) safety nets, and the development of alternative food sources, including inland aquaculture (University of Cambridge and ICLEI, 2014).

Urban and peri-urban agriculture have long been recognized as a potential food security and income strategy (Zezza and Tasciotti, 2010; De Zeeuw et al., 2011; FAO, 2014; Porter et al., 2014). However, its potential contribution as a climate change adaptation and, to a lesser extent, mitigation strategy has only been more recently studied and acknowledged (Lwasa, 2013; Dubbeling, 2014: Lwasa and Dubbeling, 2015). Because a clear framework and tools for monitoring the contributions of urban and peri-urban agriculture and forestry (UPAF) to climate change mitigation and adaptation was not available until recently, the potential to integrate UPAF into city climate change plans was limited.

A recent (2012–2014) collaboration between UN-Habitat and the Resource Centers on Urban Agriculture and Food Security (RUAF) Foundation-International aimed to respond to these gaps by (1) enhancing the awareness of local authorities and

---

7   Ecosystem-based adaptation is an approach to planning and implementing climate change adaptation by considering ecosystem services and their uses for human well-being (MEA, 2005).

Box 8.3 Figure 1 *Multifunctional design of urban greenways in Bobo Dioulasso (Burkina Faso).*

Source: F. Skarp

other stakeholders involved in urban climate change programs, land use, agriculture, and green spaces regarding the potentials (and limitations) of UPAF for climate change adaptation and mitigation; and (2) assisting interested cities and other local actors to integrate urban agriculture into local climate change and land-use policies and strategies and to initiate pilot actions that "showcase" replicable urban agriculture models. At the same time, RUAF Foundation and the Climate Development Knowledge Network (CDKN) developed and tested a monitoring framework in an attempt to quantify the impacts of UPAF on climate change adaptation and mitigation in participating cities.

One of the partner cities, Bobo-Dioulasso, in Burkina Faso, is characterized by increasing urbanization resulting from industrial and economic growth, as well as from the return of migrants following an internal crisis in Ivory Coast. The city is built up in housing blocks with square open urban lots between them. The municipality is trying to preserve these lots (called greenways) for greening (agroforestry) and multifunctional (recreation and urban agriculture production) activities as part of their parks and gardens program and climate change adaptation strategy.

Satellite images and remote sensing data were used to quantify the effect of land uses on land-surface temperatures (LSTs). A comparison of 1991–2013 data showed that LST differences between urban and peri-urban areas increased approximately 6% a year. The study also showed that mean LSTs over a 10-year period were consistently cooler (0.3°C) in the three specific green infrastructure areas analyzed than in adjacent urbanized areas (Di Leo et al., 2015). This may have important effects on human well-being.

In addition, the greenways will contribute to increased infiltration and retention of stormwater, with a reduced runoff coefficient of 4%, thus possibly reducing flood risks in periods of intensive rainfall. Monitoring has also shown that from a first harvest cycle (August–October 2013) from open fields can contribute to at least 6% of the monthly food expenditures of the households involved in the project (RUAF Foundation, 2014).

In the same way, these urban agriculture greenways contribute to a more permanent availability of home-produced food for these households. Such increased diversification of food and income sources helps to increase the resilience of poor households, which are generally vulnerable to increases in food prices (Dubbeling, 2014). Furthermore, preservation of green infrastructure is highly relevant because municipalities in Africa, as elsewhere, regularly encourage infill developments and higher housing densities that lead to the reduction or loss of green spaces and gardens.

In the period 2013–2014, the municipality of Bobo-Dioulasso decided to (1) install and institutionalize a municipal committee for the future management of the greenways; (2) draft and adopt a technical statute for the greenways promoting their productive and multifunctional use; and (3) adopt a set of specifications applicable to the exploitation of the greenways. The draft legal texts were submitted to and adopted by the Environment and Local Development Commission of the Municipality in January 2014. On March 26, 2014, the proposal to install the municipal committee was unanimously adopted by the municipal council. A provision of €20,000 was made in the 2013–2014 municipal budget to cover the functioning and activities of the Greenway Management

Box 8.3 Figure 2 *Rehabilitated paddy areas with vegetables growing on raised bunds in Kesbewa, Sri Lanka.*

Photo: Janathakshan

Committee and to support maintenance of the existing productive greenways as well as their replication on other lots (RUAF Foundation, 2014).

In comparison to Bobo-Dioulasso, where increasing urban temperatures and urban heat islands are the main predicted climate change impacts, the city of Kesbewa, Sri Lanka, has to deal not only with increasing temperatures but also with more intense rainfall and regular flooding (see Box 8.3 Figure 2). Kesbewa is a medium-sized, rapidly expanding city located 25 kilometers from the capital Colombo. Kesbewa city used to be characterized by a large presence of agricultural and rice-producing fields, the latter in lower-lying areas and flood zones. Much of the agricultural activity has been abandoned due to rice production from the north of the country being more economically viable and due to sale of land for urbanization. The rapid filling and conversion of the paddy lands to residential and commercial areas has significantly altered natural water flows and drainage. Coupled with an increase in average rainfall and heavy rainfall events, this has resulted in recurrent flooding and related damage to infrastructure, utility supply, and the urban economy in some parts of Kesbewa (CDKN, 2014).

It was for this reason that the Ministry of Agriculture, Western Province, with support of UN-Habitat, RUAF, and the local non-governmental organization Janathakshan, decided to implement a pilot project on rehabilitating abandoned paddy lands by promoting the production of traditional varieties of salt-resistant paddy rice that fetch good market prices combined with the growing of vegetables on raised beds. By re-establishing the flood regulation and ecosystem services of these areas, this strategy will not only contribute to reducing flood risk but

also to increasing urban food production and income generation for farming households. Support for urban agriculture as a flood risk or stormwater management strategy was also taken up in cities like Bangkok after the 2012 flooding (Boossabong, 2014) and in New York (Cohen and Wijsman, 2014).

Support for this program and its expansion is being institutionalized in policy uptake at different levels. At the local level, the preservation of agricultural areas and flood zones is included in the Kesbewa Urban Development Plan. At the provincial level, urban agriculture is now considered a climate change adaptation strategy for the province. The current Climate Change Adaptation Plan 2015–2018 of the Western Province of Sri Lanka (Ministry of Agriculture, 2014) now specifically includes action lines regarding the expansion of urban and peri-urban agriculture and agroforestry, the management of paddy lands as a flood risk reduction strategy, and the reduction of food miles by promoting localized production. And, at the national level, prescribed land use for low-lying urban and peri-urban rice fields now allows for the new production model as part of the revised "Paddy Act" (RUAF Foundation, 2014).

More localized food production may also have positive impacts on reducing energy use and emissions related to food transport (cold) storage, and packaging. In both Kesbewa and Rosario, Argentina, urban consumption patterns and food flows were analyzed and scenarios developed to calculate the potential impacts of increased local food production. Assuming similar production systems would be applied in both distant rural (current production locations) and peri-urban areas for production of the main consumed

Box 8.3 Figure 3 *Horticulture production in Rosario's greenbelt.*

vegetables in Rosario (potato, squash, beans, and lettuce), CO_2 emissions related to such food consumption would be reduced by about 95%. Analysis of production capacity in Rosario's peri-urban areas shows that such local production is feasible.

In response, the Municipality of Rosario has already zoned – in its urban development plan – an additional 4,000 hectares of its remaining greenbelt for vegetable production and has, in collaboration with the Province of Sante Fe, started a pilot program to support horticulture producers using ecological production techniques and opportunities for direct marketing. A first agreement was signed with the Association of Hotels and Restaurants for this purpose (Dubbeling, 2015).

The amount of food that can actually be produced in urban and peri-urban areas was more recently the subject of study in Almere, the Netherlands, and in Toronto, Canada. A 2012 scenario study done in Almere found that 20% of total food demand (in terms of potatoes, vegetables, fruits, milk, and eggs) projected for a future population of 350,000 inhabitants can be produced locally, within a radius of 20 kilometers of the city. More than 50% of the needed area is devoted to animal production (grass and fodder). By replacing 20% of the food basket with local production in Almere while at the same time promoting fossil fuel reduction in production, processing, and cooling through the use of renewable energy sources, energy savings (363 TJ) would add up to the equivalent of the energy use of 11,000 Dutch households. Savings in GHG emissions (27.1 Kt CO_2 equivalent) would equal carbon sequestration in about 1,360 hectares of forest or the emissions of 2,000 Dutch households. The largest savings are due to reduction in transport, replacing fossil fuel use with renewable energy sources (solar, wind energy); use of excess heat from greenhouses; and replacing conventional with organic – or ecological – production (Jansma et al., 2012).

among urban dwellers. High GHG emissions compound temperature levels, leading to forecasts of even higher U.S. summer temperatures and health concerns in coming years (Kenward et al., 2014). As climate change becomes increasingly viewed in the context of public health, cities that incorporate green infrastructure in urban planning and the built environment, and that safeguard local biodiversity, will optimize their urban ecosystems services for temperature mitigation, thus strengthening climate resiliency as well as improving quality of life (Santiago Fink, 2016).

8.6 Urban Agriculture and Forestry

With increasing urbanization, climate change, and growing urban demand for food, cities need to address the triple challenge of climate change mitigation and adaptation, as well as the provision of basic services, including food, to vulnerable residents. Barthel et al. (2013, 2015) suggest that urban agriculture production in its many forms has been supporting urban resilience throughout the history of urban development. The examples here show that urban and peri-urban agriculture and forestry may provide helpful strategies to address this triple challenge (see Box 8.3). The future upscaling of these interventions will need new urban design concepts and the development of local and provincial climate change action and city development plans that recognize urban agriculture as an accepted, permitted, and encouraged land use. The involvement of the subnational (e.g., provincial) government is key to addressing agriculture and land-use planning at larger scales (outside municipal boundaries), facilitating access to financing, and developing the regional policies that must accompany city-level strategies (Dubbeling, 2014).

8.7 Ecosystem-Based Mitigation Strategies

Urban areas are likely to face the most adverse impacts of climate change due to high concentration of people, resources, and infrastructure (Revi et al., 2014). Climate change mitigation is therefore required to reduce the sources and enhance the sinks of GHGs, especially carbon. Combining green infrastructure and EbA may increase urban CO_2 sinks (Rogner et al., 2007), although estimates for different kinds of green infrastructure remain contested (Pataki et al., 2011). Urban land-use changes have significant impact on GHG emissions and carbon sequestration, as well as on albedo, which plays an important role in radiative forcing. New and updated urban plans warrant the inclusion of both climate change resilience measures as well as long-term mitigation strategies that need to be supported by metrics and decision-support tools that demonstrate GHG reductions; land use and transportation as well as green infrastructure indicators are needed (Condon et al., 2009).

Integrated urban planning that incorporates a multidisciplinary perspective to target schemes that also support increased use of green infrastructure, forest restoration, and other EbA approaches can help advance sustainable urban development while reinforcing climate mitigation and enhancing the quality and quantity of UES (RUAF, 2014; Ecologic Institute, 2011; Georgescu et al., 2014). For example, incorporating green infrastructure in urban design, especially in warmer climates, can potentially reduce the use of air conditioning, cause significant energy savings, and therefore indirectly reduce GHG emissions (Alexandri and Jones, 2008; Georgescu et al., 2014).

8.8 Cross-Cutting Themes

8.8.1 Urban Planning and Design

Designing, planning, and managing complex urban systems for climate resilience and human health and well-being require ecosystems to be resilient to effects of climate change and be able to sustainably and reliably provide critical ecosystem services over time (McPhearson et al., 2014b). Urban planning and design are key processes that determine the quantity, quality, and accessibility of urban residents to UES (see Chapter 5, Urban Planning and

Design). Urban development often replaces natural elements with built and impervious surfaces, which can degrade and eliminate ecosystems, natural processes and flows (e.g., water and nutrients cycles), and biodiversity (Alberti, 2008; Colding, 2011; Novotny et al., 2010). Impervious surfaces also exacerbate climate-related problems such as the UHI effect, flooding, and other stormwater management concerns. To counter these trends, ecosystem-based approaches in urban planning and design practices are emerging. Ecosystem-based approaches can include urban green infrastructure in ways that enhance ecosystem services and restore native biodiversity. In a growing number of cities, local communities and city planners are collaborating to create new green spaces and improve existing ones using GIS and other holistic spatial planning tools and technologies (Pickett and Cadenasso 2008).

Over the past few decades, "ecocities" and "green cities" theories began to emphasize the importance of ecosystems within cities and in linked rural areas as a way to provide important ecosystem services to city residents (Yang, 2013). Innovative urban planning theories such as Ecological Design (Rottle and Yocom, 2011), New Urbanism, Sustainable Urbanism (Farr, 2008), Ecological Urbanism (Mostafavi and Doherty, 2010), Agricultural Urbanism (De La Salle and Holland, 2010), Landscape Urbanism (Waldheim, 2007), Green Urbanism (Beatley, 2000), Biophilic Urbanism (Beatley, 2009), Ecocities (Register, 2006), and Ecopolises (Ignatieva et al., 2011) emphasize ecological restoration and connected multifunctional green infrastructure in dense, compact cities. These new approaches in urban planning are beginning to prioritize walkable and mixed land uses, emphasizing designs that cater to the needs of people and other living things (Register, 2006). In this way, urbanizing areas can start to facilitate climate mitigation and adaptation as co-benefits with efforts to reduce waste and consumption (Register, 2006).

Sustainable urban design seeks to maximize the quality of the built environment and minimize impacts on the natural environment, transforming impervious areas into high-performance landscapes (McLennan, 2004). Inter- and transdisciplinary, collaborative and strategic urban planning and design, based on restoration and reconnection of green areas at different scales, can offer numerous benefits (Breuste et al., 2008; Colding, 2011;

Box 8.4 Key Messages: Cities Biodiversity Outlook

1. Rich biodiversity can exist in cities.
2. Biodiversity and ecosystem services are critical natural capital.
3. Maintaining functioning urban ecosystems can significantly enhance human health and well-being.
4. Urban ecosystem services and biodiversity can help contribute to climate change mitigation and adaptation.
5. Ecosystem services must be integrated into urban policy and planning.

Source: Adapted from Secretariat of the Convention on Biological Diversity, 2012

Novotny et al., 2010; McDonald and Marcotullio, 2011; Pauleit et al., 2011; Ignatieva et al., 2011; Ahren, 2013). For example, urban planning and design that promotes habitat connectivity through linkages or clustering of landscapes, parks, and green infrastructure can increase the provision of multiple ecosystem services such as recreation, stormwater management, and biodiversity preservation (Colding, 2011). More recent approaches to urban green infrastructure design also acknowledge ecosystem disservices (see Box 8.5), the need to account for disservices as well as tradeoffs and synergies in biodiversity, and different ecosystem services (Von Döhren and Haase, 2015; Gomez-Baggethun et al., 2013; Kronenberg, 2015).

In both urban and non-urban contexts, climate change is associated with the increased frequency and intensity of extreme events and accelerated loss of urban biodiversity (Thomas et al., 2004). Adapting to urban climate change in the face of an uncertain magnitude of risk, vulnerability, and impacts means that urban planners should have both short- and long-term adaptation options for which a constant flow of information and knowledge is critical. Ongoing assessment of the state of urban ecosystems and ecosystem services across multiples scales and functions can support the planning and design of interconnected urban social-ecological systems (Kremer et al., 2016a).

8.8.2 Equity, Environmental Justice, and Urban Ecosystem Services

Human and nonhuman vulnerabilities are intimately intertwined at the urban scale, and the most vulnerable (including both human and nonhuman) species lack the power and capacity to respond to climate change impacts (Steele et al., 2015) (see Chapter 6, Equity and Environmental Justice). From an environmental justice[8] perspective, the quantity, quality, and accessibility of urban ecosystems and their services is unevenly distributed across urban populations, with the poor and minorities often disproportionally affected by environmental hazards and ecosystem disservices and lack of access to essential ecosystem services (Pham et al., 2012; McPhearson et al., 2013a).

For example, the location, structure, and quality of urban parks present a long-term environmental justice challenge. Access to parks provides ecosystem services benefits such as recreation, physical activity, public health, aesthetic value, education, and sense of place. Historically, it has been demonstrated that the urban poor were often forced to leave their homes to create space for the creation of urban parks (Taylor, 2011). More recent research shows that the health and well-being of minorities and low-income populations are affected by the lack of access to high-quality, large, urban parks (Boone et al., 2009; Loukaitou-sideris and Stieglitz, 2002; Miyake et al., 2010) and other kinds of green spaces – such as urban vacant lots – that produce social and ecological benefits (McPhearson et al., 2013). A recent study in Bogotá, Colombia (Escobedo et al., 2015) identified marked

Box 8.5   Ecosystem Disservices

Although urban planning and design is increasingly embracing urban ecosystems as cost-effective design solutions, it has yet to deal with the emerging knowledge of how ecosystems can also create negative impacts on human well-being, known as "ecosystem disservices." Although green infrastructure provides a wide range of services, they also generate disservices (Gomez-Baggethun et al., 2013; McPhearson, 2014; Döhren and Haase, 2015) that are important to take into account in urban planning and management.

For example, green roofs have value in improving the quality of runoff by reducing pollutant release (Dunnett and Kingsbury, 2008), but some studies have noted the negative effect of the roofing materials used on the quality of runoff water due to chemicals or metal compounds (Bianchini and Hewage, 2012) and also on air quality by the emission of volatile organic compounds and nitrogen oxides (Kaye, 2004).

Similarly, urban trees can produce pollen that negatively affects allergy sufferers and may affect asthma rates in cities. Working closely with local ecological experts when developing green infrastructure will be important to understand the inherent tradeoffs to maximize ecosystem services and minimize disservices associated with particular species or species assemblages (McPhearson et al., 2014; Döhren and Haase, 2015).

inequalities in ecosystem services provision by urban trees. The poorest socioeconomic stratum had the lowest tree and crown size, whereas the wealthiest stratum had the largest tree attributes.

Minorities and the poor are also more likely to use urban biodiversity directly as a source of livelihood and thus are more impacted by the effect of climate change and pollution on natural resources such as fisheries and urban agriculture, especially in low- and middle-income countries (Corburn, 2005; National Environmental Justice Advisory Council, 2002). It will be important to consider the spatial distribution of environmental justice in planning and decision-making on policies related to ecosystem services. For example, the location of new green infrastructure can improve environmental justice by locating natural spaces and elements in proximity to otherwise underserved populations. The opposite may be true if new green infrastructure is located at the expense of such populations, where, for example, gentrification processes together with new green space development increase the cost of housing and force low-income residents to relocate (Wolch et al., 2014). Addressing environmental justice issues requires participatory planning and community-based strategies to address the structural changes that may be required (e.g., by improving the access of marginalized groups to green spaces and providing them with opportunities for recreation, urban agriculture, flood protection, urban heat reduction, and other ecosystem services without forcing the displacement of affected groups).

8   Environmental justice is a normative concept and social movement concerned with the spatial distribution of environmental goods and ills (Ernstson, 2013), as well as with the social structure and institutional context in which environmental decisions are made (Cole and Foster, 2001).

Case Study 8.7 Singapore's Ecosystem-Based Adaptation

Lena Chan, Geoffrey W. H. Davison

National Biodiversity Centre, National Parks Board, Singapore

Keywords	Flood risks, ecosystem-based adaptation, resilience, greeneries
Population (Metropolitan Region)	5,607,300 (Department of Statistics Singapore, 2016)
Area (Metropolitan Region)	719.2 km² (Department of Statistics Singapore, 2016)
Income per capita	US$51,880 (World Bank, 2017)
Climate zone	Af – Tropical rainforest (Peel et al., 2007)

SUMMARY

Singapore has taken an integrated and interdisciplinary approach to urban biodiversity conservation and restoration of ecosystems by adopting both biological and engineering approaches to climate change adaptation and mitigation. Multidimensional strategies are planned and implemented to address multiple climate stressors such as temperature and sea level rise and increased water-induced hazards. Restoring terrestrial and marine biodiversity through both *in-situ* and *ex-situ* conservation work and building green infrastructure such as urban parks, wetlands, and roadside avenues have increased urban greeneries and carbon sequestration and reduced flood disaster risks. These integrated ecosystem-based adaptation and disaster risk reduction measures have made a resilient Singapore.

CASE DESCRIPTION

Singapore has taken a holistic approach to addressing climate change vulnerability and impact to its urban ecosystems. It carried out two national climate change studies incorporating vulnerability assessments that investigated physical and meteorological parameters by using statistical and/or dynamical downscaling to better understand the implications of latest Intergovernmental Panel on Climate Change (IPCC) Fifth Assessment Report (AR5) climate change projections at regional and local levels[9] (National Climate Change Secretariat [NCCS], 2012). It was followed by studying a range of downstream impacts that fed into adaptation plans based on a risk assessment exercise done across all government agency levels (NCCS, 2012).

The aim of adaptation and mitigation plans includes reducing emissions across sectors (NEA, 2013), building capabilities to adapt to the impact of climate change, and harnessing green growth opportunities, as well as forging partnerships on climate change actions. The approach assesses Singapore's physical vulnerabilities to climate change based on a resilience framework (RF) to guide measures against potential climate change impacts. The RF ensures that appropriate adaptation measures are identified and implemented by adopting a cyclical approach to risk appraisal and adaptation planning. The cycle is shown in Case Study 8.7 Figure 1.

CLIMATE CHANGE IMPACTS ON BIODIVERSITY

While assessing risks and planning adaptation measures, an understanding of biological and environmental assets was gained through risk identification and quantification. Biodiversity assets are understood through continuing surveys, such as site- or habitat-specific studies including the Terrestrial Sites Survey (2002–2003), Natural Areas Survey (2005–2007), and Comprehensive Marine Biodiversity Survey (2011–2015) to update information on the flora and fauna of Singapore. Regular and *ad-hoc* assessments of biodiversity and environmental assets are undertaken as part of long-term adaptation planning.

VULNERABILITY AND IMPACTS ASSESSMENT AND ADAPTATION PLANNING

The first vulnerability assessment looked at plant groups particularly vulnerable to climate change, such as figs (as keystone species for vertebrates), *dipterocarp* trees (whose bi-annual mass flowering events are keyed to the intensity and frequency of El Niño events), bryophytes (group susceptible to drought), and the effects on planted roadside trees.

Challenges have been encountered in the administrative definition and categorization of natural assets (e.g., whether each tree, each species, or each population in different areas is to be considered a separate asset) and in suggesting biological thresholds or tipping points that might be related to the various climate change parameters (rainfall, sea level, sea surface temperature, wind) in a way that facilitates risk assessment.

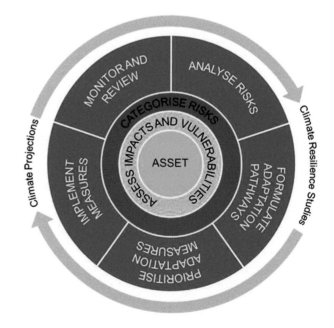

Case Study 8.7 Figure 1 *Cyclical approach to adaptation planning.*

---

9   The Strategy, developed with public and private-sector consultation, is available at https://www.nccs.gov.sg/sites/nccs/files/NCCS-2012.pdf

Case Study 8.7 Figure 2 *A well-planned urban ecosystem.*

Past fragmentation of Singapore's forests makes them vulnerable to future long-term changes such as increased likelihood or duration of drought and higher average temperatures. Wetlands are exposed to rainfall changes, sea level rise, or water quality changes related to warming and changes in precipitation. Sea level rise will be a challenge for mangroves, which cannot retreat inland because of competing land uses. Corals, which require sunlight, might not be able to grow upward quickly enough to keep pace with rising sea levels. In addition, a 1–2°C rise in sea water temperature will lead to coral bleaching. The strategies adopted to build up the resilience of these taxonomic groups are to conserve as broad a spectrum of species as possible and to safeguard known sources of propagules.

ECOSYSTEM-BASED ADAPTATION

To help Singapore's biodiversity withstand the potential impacts of climate change, National Parks is working with other agencies and the community to safeguard existing species, increase connectivity of various green areas across the island, and enhance the resilience of ecosystems. This includes measures to restore forest and mangrove areas through planting and through minimization of other pressures (e.g., by removing alien invasive competitors and controlling ship wakes on mangrove coasts). Singapore has a very high proportion of planted roadside trees; efforts are made to diversify plant species used, intensify planting, create more complex 3D layering, and increase connectivity between green areas.

To keep the city green, tree management and maintenance is being intensified and enhanced. National Parks manages approximately 350 parks and 3,500 kilometers of roadside greenery islandwide as part of the effort to lower ambient temperatures. Parks and greenery are not viewed as merely the passive victims of climate change, but

Case Study 8.7 Figure 3 *Innovations in urban greenery development.*

as tools for adaptation and mitigation. In addition, National Parks continues to support research that investigates the responses of coral reef communities to climate change triggers and promotes strategies that increase biodiversity resilience.

BUILDING A RESILIENT WATER SYSTEM

Water resource management is a key priority for Singapore. An increase in weather variability may bring more frequent or more severe cycles of floods and droughts threatening the reliability of the city's water supply. To ensure a sustainable water supply for Singapore's population and industry, Singapore has built a robust and diversified water supply through four "national taps": namely, local catchment water, imported water, treated recycled water (NEWater), and desalinated water. In particular, NEWater and desalinated water are not dependent on rainfall and are thus more resilient sources in times of dry weather. Regarding flood water management, efforts are made to enhance resilience against coastal erosion and inundation associated with rising sea levels coupled with short-lived, extreme meteorological events (MEWR, 2014). A risk map study was done to better identify the specific coastal areas at risk of inundation and the potential associated damage. The results will help develop long-term coastal protection strategies.

Recognizing the impact of greater weather uncertainties as well as the constraints to drainage planning posed by increasing urbanization, Singapore has revamped its drainage management approach to strengthen its flood resilience. The strategy is to optimize the management of stormwater using a holistic source-pathway-receptor approach that looks at catchment-wide ecosystem-based solutions to achieve higher drainage and flood protection standards. It covers the entire drainage system, addressing not just the pathway over which the rainwater travels. A new provision has been added to its surface-water drainage regulations, which requires developers/owners of land size 0.2 hectares or more to implement measures to slow surface runoff and reduce the peak flow of stormwater into the public drainage system by implementing on-site detention measures such as green roofs, rain gardens, and detention tanks.

CONCLUSION

Comprehensive and ecosystem-based adaptation strategies used by Singapore to enhance urban resilience are broad and interdisciplinary. These are approached from a multi-agency and multidisciplinary perspective. Additional efforts are continuously entrained in coordination with agencies and development partners under a common framework on risk, adaptation, and mitigation.

Pascual et al. (2010) argue that the institution of payment for ecosystem services (PES) is another policy area where distributive justice has critical importance. Although PES theory commonly disregards distributed justice questions, actual programs are often required to take such issues into consideration for legitimacy and stakeholder buy-in. Depending on the fairness criterion used (e.g., equal distribution, need, compensation), the outcome of PES programs are determined by an equity–efficiency interdependency analysis (Pascual et al., 2010). By including a fairness criterion of some kind, programs can offer a mechanism to more systematically include equity and justice issues in management and planning for UES (Salzman et al., 2014).

Climate change effects in coastal cities expose the complexities and challenges of developing policy to address issues of distributive justice. For example, in New York, Hurricane Sandy in 2012 devastated many coastal communities. Federal, state, and city programs determined the redevelopment path of such communities with some areas purchased for the purpose of creating new protective natural coastal buffer zones (NYS, 2013). This effort aimed at improving the adaptive capacity of the entire city to future extreme weather events. Some low-income urban residents were unable to rebuild their houses and had to relocate (Sandy Redevelopment Oversight Group, 2014). In other cases, newly required building elevations and other building reinforcement policies may mean that the individual's or community's ability to pay determines whether a family is able to rebuild its residence or instead has to relocate (Consolo et al., 2013). Planning decisions can, be complex, such as determining how best to serve residents in low-income countries where informal urban settlements are located in flood-prone areas, thus emphasizing the need to consider the broader complexity of the social, ecological, and economic linkages of the urban system.

8.8.3 Economics of Ecosystem-Based Adaptation and Green Infrastructure

Green infrastructure and other types of urban ecosystems in urban areas generate monetary and nonmonetary value through the provision of ecosystem services (Gomez-Baggethun et al., 2013; Kremer et al., 2016b) (see Chapter 7, Economics, Finance, and the Private Sector). A major advantage of green infrastructure and EbA strategies is that they offer some of the most cost-effective adaptation options available to cities (TEEB, 2011). Around the world, evidence is mounting that effective planning, design, and management of nature in urban areas can provide multiple benefits and cost-effective solutions where traditional "gray" infrastructure solutions alone have been prohibitively costly. Linking green and gray infrastructure can provide cities both cost-effectiveness and improved function.

For example, management of stormwater runoff through green infrastructure is becoming increasingly popular among cities due to the cost savings it provides by reducing the need for new gray infrastructure to reduce local flooding and sewage overflows in combined sewage systems (see Case Studies 9.4 or 14.B, Rotterdam). Green infrastructure methods such as green streets, tree plantings, and rain barrel installations are estimated to be three to six times more effective for stormwater management than further expanding gray infrastructure (Foster et al., 2011). A U.S. Environmental Protection Agency report analyzing thirteen case studies from cities such as New York, Philadelphia, Portland (Oregon), and Seattle found that although each municipality or entity used different cost and benefit matrices, in most cases green infrastructure was found to cost less than gray alternatives and to provide multiple benefits (NYC, 2010). Portland's Cornerstone project to disconnect downspouts resulted in the removal of approximately 1.5 billion gallons of runoff from the city's combined sewer system (Foster et al., 2011), and, in Philadelphia, more than 100 green acres were constructed and 3,000 rain barrels distributed to support increased stormwater absorption. A life cycle analysis of Low Impact Development (LID) in a New York neighborhood found a strategy that included permeable pavement and street trees to be cost effective even though it only considered energy saving in downstream treatment plants; this has mirrored similar studies conducted in other cities.

Other important examples of ecosystem services include flood risk reduction by extending time lag between floods and storm runoff and temperature regulation, ground water recharge, and air purification. Rezoning areas for green infrastructure or restricted development are cost-effective ways to address flood risks (Foster et al., 2011). Kousky et al. (2013) evaluate avoided flood damages against the cost of preventing development on flood-sensitive lots in Wisconsin and New York. Their findings highlight the importance of the spatially specific characteristics of the lot as a way to create a cost-effective flood protection plan.

UHI research shows that the loss of urban vegetation increases the energy costs of cooling (McPherson et al., 1997).

Case Study 8.8   Seattle's Thornton Creek Water Quality Channel

Nate Cormier

SvR Design Company, Seattle

Keywords	Stormwater treatment, green infrastructure, public space, ecosystem-based adaptation
Population (Metropolitan Region)	3,613,621 (U.S. Census Bureau, 2010)
Area (Metropolitan Region)	15,209 km² (U.S. Census, Bureau, 2010)
Income per capita	US$56,180 (World Bank, 2017)
Climate zone	Csb – Temperate, dry summer, warm summer (Peel et al., 2007)

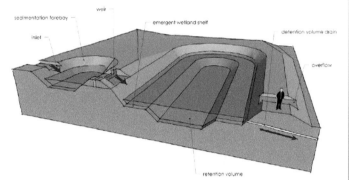

Case Study 8.8 Figure 1 *An innovative natural drainage design.*

The Thornton Creek Water Quality Channel, located at the headwaters of the South Branch of Thornton Creek, Washington, is a multipurpose water management project providing multiple environmental and social benefits to the urban population of Seattle. This facility addresses the problem of both heavy sedimentation and polluted water flow into the natural creek in the hilly catchments of Seattle. The integrated water treatment and management plant captures runoff from the human-populated upstream watershed areas and treats it before it flows into Thornton Creek and Lake Washington. The environmentally sound water cleaning facility occupies minimal space but provides multiple spatial and environmental benefits to the local community. It has also led to the development of a new neighborhood that is emerging as a growing urban center of the city. The facility can be termed as classic example of urban green infrastructure.

INNOVATIVE AND RESILIENT DESIGN

The project uses natural drainage system revival technology simulating the natural process of water flow to clean polluted and silted water and allow the cleaned water to flow through natural percolation and seepage systems year-round. The environmentally friendly design (Case Study 8.8 Figure 1) has developed natural landscaping and public pathways giving easy access to citizens to different public facilities and private buildings located throughout the area.

This model project offers the last-available opportunity to improve the quality of stormwater runoff before it reaches the creek. The channel design diverts stormwater from the drainage pipe under the site to a series of surface swales landscaped with special soils and native plants. These ponds interrupt runoff speed, allowing water to seep into the soil and removing pollutants in the process. The channel regulates the water flow both during wet and dry weather, allowing for continuous cleaning of stormwater.

The community-driven project turned into a collective action effort that met the broad objectives of major stakeholders and fulfilled their common goals. The design has allowed development of diverse types of residential buildings, job-creating private-sector enterprises, retail shops, and rest and recreation places while preserving a natural environment. This is in contrast to what existed before – a gray and brown parking lot. The provision of public open space has been used to raise environmental awareness thus providing long-term benefits, albeit of intangible nature. The facility has attracted significant private-sector investment in terms of the residential and commercial complex. The modest US$14.7 million that it cost to build the Thornton Creek facility is believed to have generated more than US$200 million in the form private-sector–led investment in the city, thus catalyzing the Northgate neighborhood as a vibrant urban center of Seattle (Benfield, 2011).

ADAPTATION STRATEGY

Carved out of a former mall parking lot, the Thornton Creek Water Quality Channel provides public open space for Seattle's Northgate neighborhood while treating urban stormwater runoff from 680 acres of North Seattle. This project grew out of grassroots efforts to transform the piped Thornton Creek that ran under the parking lot to a natural water catchment system. Political leaders overcame a number of barriers that stood between developers and environmentalists by establishing a broad-based Northgate Stakeholder Group to find a way to integrate private development, public open space, and a major stormwater facility. What resulted through these collective efforts is an adaptive and resilient urban ecosystem management project providing multiple climate change adaptation and social benefits.

Opened in 2009, this catalytic natural space provides pedestrian connectivity among a major transit hub, community services, housing, and retail outlets. There is a continuous expression of water flowing, pooling, and cascading in the channel. During and after storms, the full capability of the broad channel bottom is engaged for water-quality treatment. Overlooks and bridges allow users to enjoy the channel habitats and wildlife. Seat walls, benches, and interpretive artwork contribute to an inviting environment where visitors can linger and learn in a high-performance landscape (see Case Study 8.8 Figure 2).

The project has resulted in:
- A successful community process that balances public and private goals in support of environmentally compatible development and socioeconomic sustainability developed in a highly contested urban space

- The ability to catalyze more than US$200 million in investment in adjacent private residential and commercial development, generating jobs and economic opportunities
- An illustration of how to transform a former mall parking lot, a common "grayfield" in many American communities, into an aesthetically and environmentally productive urban landscape.
- Water-quality treatment for runoff from 680 acres within a beautiful setting where visitors can learn about natural systems and the restoration of a historic creek.
- Increases in open space in the Northgate Urban Center by 50% to provide an oasis of native vegetation for neighbors and wildlife, thus promoting urban biodiversity.

The key lessons learned are that (1) multistakeholder processes and community-driven initiatives lead to change in developing urban resilience, and (2) both bottom-up and top-down processes are necessary, provided the city government recognizes and internalizes both in urban ecosystem-based adaptation planning and implementation.

Case Study 8.8 Figure 2 *Thornton Creek in Seattle: An example of human-developed biodiversity and ecosystem.*

Significant savings can accrue due to the reduction of power generation through the implementation of green infrastructure (ACCCRN, 2015; Rosenzweig et al., 2009). For example, one study in Los Angeles showed that increasing pavement reflectivity by 10–35% could produce a 0.8°C decrease in UHI temperature and an estimated savings of US$90 million per year from lower energy use and reduced ozone levels (Foster et al., 2011).

8.8.4 Payment for and Valuation of Ecosystem Services

Many cities are developing programs for valuation of and payment for ecosystem services. However, PES programs are not often decided based on proper valuation and often fail to address the issue of social equity; in some cases, they exacerbate poverty and equity by raising prices or introducing a fee on previously low-priced or free services (Pascual et al., 2009).

Common valuation methods include preference-based approaches and biophysical approaches (Sukhdev et al., 2010). *Preference-based approaches* include all monetary and nonmonetary societal value settings, and *biophysical approaches* include assessments that are grounded in the processes, flows, and structures of the ecosystem (Sukhdev et al., 2010; Gomez-Baggethun et al., 2013). An important characteristic of urban green infrastructure is that it generates multiple benefits and different types of values (Kremer et al., 2016b). One of the challenges in the evaluation of the cost-effectiveness of green infrastructure is accounting for societal and cultural benefits and values that are not easily quantifiable in monetary terms. For example, a study of flood protection strategies in the Netherlands found engineering methods to be most cost-effective when not considering nonmonetary benefits, but when those were included (e.g., social and cultural values), green infrastructure became more competitive. Such integration is at the forefront of current UES research (Haase et al., 2014).

8.8.5 Economic Valuation Tools

Because of a growing effort to support the integration of green infrastructure into the urban landscape, software tools are becoming increasingly available to urban planners and decision-makers for the evaluation of certain ecosystem services and benefits. For example, i-Tree[10] is a suite of software tools built by the U.S. Department of Agriculture Forest Service that allows the quantification of ecosystem services benefits from urban trees; the Green Values Calculator[11] is a tool for comparing performance, costs, and benefits of green infrastructure practices; and InVEST[12] is a suite of software models for the assessment of ecosystem services values and tradeoffs (Nowak et al., 2013). Such tools enable the valuation of UES and support the integration of green infrastructure into urban planning. However, major gaps remain in the capacity to value urban green infrastructure and the ecosystem services it provides, including public participation in the valuation process, the integration of monetary and nonmonetary values through multicriteria analysis and other methods, scale- and thresholds-dependent values, and bridging supply and demand for the purpose of valuation (Gómez-Baggethun et al., 2013; Haase et al., 2014). Additionally, costs of EbA and nature-based solutions for climate mitigation and adaptation often have to be estimated, especially with respect to future costs, since adaptation is a long-term process. In most cases, obtaining reliable cost data will continue to be a challenge requiring several sources of evidence ranging from project case studies to national-level assessments.

8.8.6 Combining Adaptation and Mitigation in Climate Resilience Strategies

Although adaptation is necessary to minimize the unavoidable impacts of climate-induced risks and hazards, mitigation is needed to reduce urban GHG emissions and their impacts in the short- and long-term. An integrated strategy that combines all types of adaptation and resilience building measures together with mitigation strategies will have the highest level of co-benefits for human well-being (Satterthwaite et al., 2008; Karki et al., 2011) (see Chapter 4, Mitigation and Adaptation). Risks and vulnerabilities are shaped by local environmental conditions,

site characteristics, natural resource availabilities, and environmental hazards (IPCC, 2007; Satterthwaite et al., 2008). Urban adaptation aims at reducing vulnerabilities and enhancing the resiliency of systems, agents, and institutions, and it needs to be planned by taking a holistic view of the broader urban landscape since urban areas depend on surrounding peri-urban and rural areas for ecosystem services (Tyler and Moench, 2012).

Strategies for urban ecosystem adaptation and mitigation need to recognize that climate change may undermine the ability of contiguous urban and peri-urban social ecological systems to provide critical ecosystem services (Satterthwaite et al., 2008). Therefore, urban adaptation and mitigation planning should ensure the sustained flow of provisioning (e.g., food, water) and regulating (e.g., clean air) ecosystem services to urban communities (Locatelli et al., 2010; McPhearson et al., 2015). In many parts of the world, the relationships among urban ecosystems, adaptation, mitigation, and livelihoods are changing in fundamental ways as urban economic systems diversify across the urban–peri-urban spectrum, thus creating mixed or interlinked economic and environmental systems. Understanding these changes and their implications on the vulnerability of urban populations and ecosystems is essential to developing integrated adaptation and mitigation strategies for cities.

For example, Singapore has taken steps to restore its biodiversity and enhance UES (see Case Study 8.7, Singapore). The city has increased green cover from 35.7% to 46.5% in 20 years and also has set aside approximately 10% of its total land for green infrastructure (Lye, 2010) to provide increased climate change mitigation in the city in addition to improving ecosystem services that support adaptation. Similarly Seattle, Edmonton, Stockholm, Copenhagen, and many other cities have restored or created new urban ecosystems that ensure a more sustainable flow of ecosystem goods and services to the city-dwellers now and in the future (Zandersen et al., 2014).

8.8.7 Biodiversity Governance for Human Well-Being in Cities

In many parts of the world, the relationship between urban ecosystems and overall urban development is changing in fundamental ways (Tzoulas et al., 2007). Understanding these changes and their implications is necessary for holistic sustainable urban development. Specifically, over the coming decades, two interacting forces will influence urban economic and ecological systems especially in developing countries: (1) intensifying processes of technological and economic globalization that are already increasing pressures on urban/peri-urban ecosystems through shifting patterns of dependency; and (2) multiple environmental stress at all levels – from local to regional – mainly due to the impacts of climate change. These changes will likely undermine the ability of complex urban ecosystems to provide

---

10 https://www.itreetools.org
11 http://greenvalues.cnt.org/national/calculator.php
12 http://www.naturalcapitalproject.org/InVEST.html

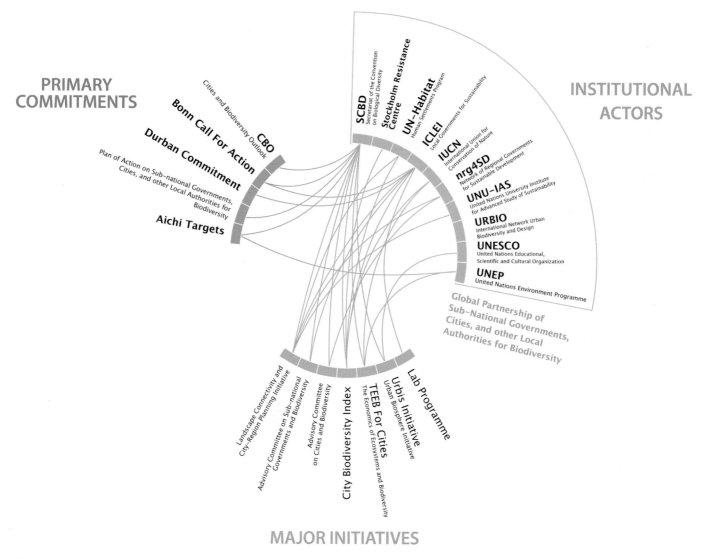

PRIMARY COMMITMENTS

INSTITUTIONAL ACTORS

SCBD — Secretariat of the Convention on Biological Diversity
Stockholm Resistance Centre
UN-Habitat — Human Settlements Program
ICLEI — Local Governments for Sustainability
IUCN — International Union for Conservation of Nature
nrg4SD — Network of Regional Governments for Sustainable Development
UNU-IAS — United Nations University Institute for Advanced Study of Sustainability
URBIO — International Network Urban Biodiversity and Design
UNESCO — United Nations Educational, Scientific and Cultural Organization
UNEP — United Nations Environment Programme

CBO — Cities and Biodiversity Outlook
Bonn Call For Action
Durban Commitment — Plan of Action on Sub-national Governments, Cities, and other Local Authorities for Biodiversity
Aichi Targets

Global Partnership of Sub-National Governments, Cities, and other Local Authorities for Biodiversity

Landscape Connectivity and City-Region Planning Initiative
Advisory Committee on Sub-national Governments and Biodiversity
Advisory Committee on Cities and Biodiversity
City Biodiversity Index
TEEB For Cities — The Economics of Ecosystems and Biodiversity
Urbis Initiative — Urban Biosphere Initiative
Lab Programme

MAJOR INITIATIVES

Figure 8.8 *Overview of global governance arrangements for urban biodiversity and ecosystem services.*

Source: UN-Habitat, 2012

critical services – water, energy, food, clean air, and healthy and livable habitats – to their population, thus underlining the critical importance of urban planning, policy, and governance to safeguard urban and peri-urban biodiversity and ecosystem services (see Chapter 16, Governance and Policy).

The decisions and deliberations of the UN Convention on Biological Diversity (CBD, 2009), as well as many others (ICLEI-Africa, 2013; UNEP, 2009; IUCN, 2009), have created an emerging global effort to enhance urban ecosystem governance structures by capturing the nexus between urban biodiversity and climate change. These efforts have called for biodiversity vulnerability assessments and research on links between biodiversity loss and urbanization (Wilkinson et al., 2010, 2013; Schewenius et al., 2014). Figure 8.8 illustrates the current international governance landscape for urban biodiversity and ecosystem services, although it is constantly evolving. Progress remains at a formative stage as stakeholders and urban actors struggle to fully understand their respective roles and establish coordination mechanisms to exploit the latent potential

of biodiversity and ecosystem services in cities for climate change adaptation and mitigation.

Increasing a city's capacity to meet growing challenges can depend on the development of a holistic governance approach in which the city is understood as a dynamically interacting social-ecological system (Frantzeskaki and Tilie, 2014) (see Chapter 1, Pathways to Urban Transformation and Chapter 16, Governance and Policy). Increased linkages among strategies, projects, and actors (Meyer et al., 2012), including the active involvement of local citizens, is important for identifying needs, challenges, and design policies and efficiently implementing them (Ward et al., 2013; Wardekker et al., 2010). Creating strong links between formal governance and informal, on-the-ground participants and managers is crucial to forming holistic governance with greater potential for successful urban ecosystem management outcomes. However, informal participation and management is seldom translated into formal governance in urban settings (Colding, 2013). The Thornton Creek Water Quality initiative of Seattle is an example that shows how a local community-driven project supported by city government is generating multiple benefits and synergies (see Case Study 8.8).

Governing ecosystem processes requires coordination across different levels of policy, legislation, and jurisdictional authorities. Urban ecosystems and biodiversity benefits often transcend administrative boundaries, thus necessitating collaboration among national, regional, and local-level agencies (Ernstson et al., 2010; McPhearson et al., 2014). The multiscalar and multi-sectoral relationships that impact urban biodiversity and ecosystem services create urban governance and policy challenges because decisions by one branch and one level of fragmented urban and national government structures can create long-term implications for the entire urban ecosystem landscape (Asikainen and Jokinen, 2009; Ernstson et al., 2010; Borgström et al., 2006). Apart from this scale mismatch issue, there is also a functional mismatch between ecosystems and the institutions managing them (Cumming et al., 2006), because different decision-makers are operating within and beyond the city and urban landscape boundaries. However, if different units of city and peri-urban governments worked in tandem, a number of synergies in the governance of urban biodiversity and ecosystem services is possible (Raudsepp-Hearne et al., 2010). This requires functioning and dynamic science–policy linkages at regional scales, challenging the current structure of governance frameworks, practices, and institutions.

Despite new research initiatives and science–policy platforms, significant challenges remain in equitably managing biodiversity and ecosystem services in urban and peri-urban areas for the mutual benefit of humans and other species (Schewenius et al., 2014). Rapid urbanization is occurring in places that face some of the most severe challenges to public health and urban biodiversity conservation. Additionally, these same urban systems are where systems of formal government and planning tend to be weak and the capacity to influence policy inadequate (Wilkinson et al., 2013).

Effective city governance will play a key role in determining the future of biodiversity across the world, not least because cities are rapidly expanding into the world's biodiversity hotspots (see Figure 8.5). Significant urban ecosystem policy changes will need to accompany or even precede effective governance practices in order to direct future urban growth so that biodiversity and the ecosystems services it provides are safeguarded (Seto et al., 2012; Wilkinson et al., 2013). Ecosystem protection in cities will rely on increasing efforts by parks and natural area managers to focus on those management outcomes that seek to maximize ecosystem functioning for services – in many places an abrupt shift from existing or past management goals. Supporting a diversity of governance systems, from official regulations to informal governance systems (e.g., local governance of urban allotment gardens) can provide a multilayered protection system and strengthen support through multiple stakeholders (Schewenius et al., 2014). Additionally, flexible policies and regulations will be needed to accommodate unanticipated climate changes and ecological responses.

8.9 Science–Policy Linkages

Adaptive and resilient urban ecosystem conservation requires policies that are based on synthesized and relevant knowledge systems, local evidence, and multilevel, multidisciplinary, and multistakeholder consultations and inputs (IUCN, 2009; Van Kerkhoff and Lebel, 2006) (see Chapter 16, Governance and Policy). Three important challenges inhibit the use of climate and conservation science in managing and planning urban ecosystems: (1) research capacity and activities are often scattered so that policy-influencing efforts are uneven in distribution and quality across sectors and regions (Haase et al., 2014); (2) data and information availability and usability are limited due to knowledge gaps and scale-appropriate specificity (e.g., in Sweden, local planners and decision-makers found it difficult to implement national biodiversity strategies because they were too general and abstract); and (3) policy-makers and other decision-makers often have limited capacity to access, interpret, and act on research information on technical subjects such as biodiversity and climate change, particularly where results are complex and reflect inherent uncertainties at multiple scales (Amin, 2007). Urban policy-makers therefore face significant challenges when seeking to increase the resilience of communities and the built environment to the effects of climate change (Frantzeskaki et al., 2016).

An integrated and coordinated urban planning, design, and implementation policy that considers biodiversity and ecosystem services should address multiple co-benefits from human health improvement, climate change adaptation, biodiversity conservation, and disaster management (see Chapter 5, Urban Planning and Design). For example, increasing green space, tree cover, and water bodies in urban areas, in addition to moderating UHI effects, will also sequester carbon, control pollution such as aerosol dusts, regulate hydrological processes, and influence regional climate (Hamstead et al., 2015; Larondelle et al., 2014). Policy actions that take advantage of the complex concepts of multifunctionality, synergies, and tradeoffs require science- and evidence-based policy processes that integrate ecosystem-based approaches in governance areas such as disaster management, community actions, and linking adaptation with sustainable development goals and practices (Elmqvist, 2013). Such processes involve key urban actors (including politicians) and civil societies in urban conservation activities that can translate research-based information and knowledge into use by policy-makers and other decision-makers (Elmqvist et al., 2013; Mincy et al., 2013; OECD/CDRF, 2009).

Co-production of knowledge, where the users of the knowledge are involved from the beginning in the research and review process, is another key component of successful science-based policy-making. The Asian Cities Climate Change Research Network (ACCRN) has concluded that urban process needs to be based on multiple-stakeholder engagement and iterative shared-learning dialogue that can bring a broad range of perspectives to city managers. Urban policies are rarely completely objective or neutral. Shared learning processes and co-production of knowledge can help raise awareness and empower stakeholders with new and consolidated

Box 8.6   WWF's Earth Hour City Challenge

The World Wildlife Fund (WWF) supports a vision of the world where people and nature thrive. In an increasingly urbanizing world, achieving this vision means working together with cities to make them livable and sustainable. WWF is the world's largest conservation organization, working not only on wildlife protection but also on food, oceans, forests, water, and climate change. WWF is bringing together its network of experts on renewable energy, public engagement, nature-based adaptation, and many other disciplines to address the issues cities are facing in the 21st century, in particular the threat of climate change and its associated hazards.

WWF's signature program for cities is the Earth Hour City Challenge (EHCC). EHCC was created in 2011 to mobilize action and support from cities in the global transition toward a sustainable future. It has since grown to encompass cities in twenty countries around the globe. Last year, 166 participating cities reported their climate data, commitments, and a total of 2,287 mitigation actions on the carbon Climate Registry (cCR) for review by an esteemed jury of experts. The jury, comprising high-level representatives from key city networks, development banks, institutions, universities, and enterprises, evaluate the participating cities' goals and strategies. Every year, one city from each participating country is awarded the title National Earth Hour Capital. From among these inspiring finalists, the jury then selects one Global Earth Hour Capital. WWF offices in twenty countries support cities on EHCC communications and low-carbon project implementation.

One key objective of the EHCC has been to gather a critical mass of city reporting on their climate commitments;

and climate actions in order to raise the awareness of decision-makers involved in global climate negotiations and increase aspirations and actions at the national level.

The We Love Cities campaign profiles finalists and spurs interaction between cities and their citizens through social media. Public engagement and raising awareness around the positive stories on local climate action are key components of the program. We Love Cities invites citizens from around the world to express their love through votes, tweets, and Instagram pictures and by submitting suggestions on how their cities can be more sustainable. These suggested improvements are shared among all the participating cities. More than 300,000 people who truly love their cities and want to see them become more sustainable have engaged in this campaign.

WWF works closely with the ICLEI–Local Governments for Sustainability to run the EHCC as well as many country-level programs that extend technical and communications support to cities around the world. In addition to ICLEI, WWF is partnering with other leaders to address climate change including C40 Cities Climate Leadership Group, Compact of Mayors, Rockefeller 100 Resilient Cities, U.S. Agency for International Development.

Technical guidance and original research from WWF are also available to support cities including the Green Recovery and Reconstruction Training Toolkit; Green Flood Risk Management Guidelines; Urban Solutions for a Living Planet; Measuring Up 2015; Financing the Transition: Sustainable Infrastructure in Cities; and Reinventing the City.

---

knowledge (Sutcliffe and Court, 2005; Institute for Social and Environmental Transition [ISET], 2010).

Strategies to effectively link science with policy and action need to (1) involve key actors (local residents, planners, designers, managers, policy-makers, NGOs) in the process of identifying problems and the actions they can take (i.e., shared learning and knowledge coproduction), (2) produce grounded evidence where action can be used to respond to ecosystem changes that are relevant to members of communities and key sectoral decision-makers, (3) effectively communicate evidence to an array of end users so that they understand and can act on it (translation of research results into use depends critically on how they are communicated via direct experience, accessible products, and, for academic and policy global audiences, peer-reviewed articles), and (4) design research outputs to respond to the types of information different types of actors need and can relate to (e.g., cost-benefit analyses and regulatory regimes for government and multilateral investors, new business opportunities for the private sector, equity concerns for the community groups, and examples of tangible solutions to common climate vulnerabilities that individuals and households face). These approaches can help to build incremental science–policy linkages that support efforts to transition cities toward sustainability and resilience.

8.10   Knowledge Gaps and Areas for Further Research

Sustainable generation and management of urban biodiversity and ecosystem services in the face of the challenges posed by climate change, population growth, poverty, and environmental degradation requires adaptive human and institutional capacity that can enhance resilience, human well-being, and conservation (TEEB, 2010; Elmqvist et al., 2013; RUAF, 2014). However, a common problem that urban policy-makers and city managers face when dealing with climate change is bridging significant knowledge gaps. This is especially challenging in the context of climate change effects on cities and urban areas (Elmqvist et al., 2013). The release of the IPCC AR5 report made headway in bridging the knowledge gap at the global level (IPCC, 2014), but for many urban areas, especially in developing countries, data and knowledge gaps remain a problem at both local and regional levels. Enhancing urban biodiversity and ecosystem services while tackling climate change and a host of social issues in cities requires a continuous flow of knowledge-based solutions (see ARC3.2 Annex 1, UCCRN Regional Hubs). Missing empirical evidence and practical ecological knowledge on urban biodiversity and ecosystems management often prevents city managers from recognizing the value of ecosystems for the development of more climate-resilient urban systems.

Additionally, significant knowledge gaps remain in understanding the current status of biodiversity in cities. Despite growing databases and new global analyses of urban biodiversity and ecosystem services (Elmqvist et al., 2013; Gomez-Baggethun et al., 2013; Aronson et al., 2014), most cities, especially in low- and middle-income countries do not have adequate data on the status and extent of biodiversity and urban ecosystem resources. Leveraging UES for climate resilience is hampered by this lack of data, with multiple global and local agencies and institutions calling for national, regional, and local biodiversity and ecosystem assessments.

Producing tools and guidelines on how to effectively manage and govern urban ecosystems so that critical services are available to local populations remains an area in need of additional research and practice-based expertise (Schewenius et al., 2014). Benefits of biodiversity and ecosystems are not equally distributed in urban areas. Often, poor communities or housing for them (e.g., Cape Flats) are blamed for biodiversity loss and habitat fragmentation in spite of their low per capita impact or having been pushed to the most marginal and fragile sites (Ernstson et al., 2010a). Improving equitable distribution and access to ecosystem services, whether it is for shade relief from urban heat waves or protection from climate-driven extreme events such as flooding in coastal cities, depends on increasing equality and reducing mismatches between ecosystem services supply and social demand for these services (McPhearson et al., 2014; Salzman et al., 2014).

Although cities and urbanized regions depend on biodiversity in ecosystems to sustain human health and well-being (TEEB, 2011), this relationship is not well understood for all ecosystem services, and the connection between biodiversity and human livelihoods has yet to be widely incorporated in urban policy and planning (Hansen et al., 2015; McPhearson et al., 2014; McPhearson et al., 2016). We also still know little about how biodiversity, ecosystem function, and ecosystem services are related in urban environments. Empirical and theoretical research on the relationships among biodiversity (including native and non-native species), ecosystem function, and ecosystem services is critical for developing design standards for climate-resilient green infrastructure.

8.11 Recommendations for Policy-Makers

The growing impacts of climate change and climate variability on interconnected human–environmental urban systems are increasing the vulnerability of both human and ecosystems in cities. Cities are particularly at risk. Ecosystems in urban contexts underpin the security of public health, water, food, industrial activities, biodiversity conservation, energy, and transport, as well as recreation and tourism sectors. Effective management of urban ecosystems using multisector and multiscale approaches

will be key in the pursuit of climate-resilient, sustainable urban development.

Adaptive management of ecosystems at landscape or watershed scales involving all stakeholders across municipal boundaries is critical to safeguarding ecological resources for climate adaptation and mitigation. Investing in green infrastructure and EbA is particularly relevant for cities and urbanized regions because they can integrate climate change adaptation and disaster risk reduction, providing cost-effective nature-based solutions for addressing climate change in cities (UNEP, 2012; Munroe et al., 2010). Investment in green infrastructure and EbA can generate multiple co-benefits for human well-being by mainstreaming climate and environmental considerations across urban systems and encouraging the sustainable management of ecological resources to improve the resiliency of inhabitants, built environments, and urban infrastructure. These approaches have the potential to mainstream environmental and climate change information into urban planning, decision-making in urban design, and management and implementation processes. Research in urban systems is making clear the cost-effective, widely beneficial impacts of investing in biodiversity and urban ecosystems for climate adaptation. We suggest the following policy-relevant recommendations:

1. Invest in ecosystem-based adaptation and green infrastructure planning as critical components of climate adaptation strategies and urban development, as well as for improved health, disaster risk reduction, and sustainable development.
2. Incorporate the monetary and nonmonetary values of biodiversity and ecosystem services into cost-benefit analyses for climate adaptation and urban development and develop innovative means of financing (e.g., public–private partnerships) for urban ecosystem and biodiversity protection, restoration, and enhancement.
3. Utilize a systems approach to ecosystem-based climate adaptation, explicitly recognizing the social-ecological relationships that co-produce ecosystem services and drive ecological dynamics in urban systems.
4. Plan and manage for a sustained supply of critical urban and peri-urban ecosystem services over longer-term time horizons (20, 50, 100 years).
5. Strengthen urban–peri-urban–rural linkages through integrated and multidisciplinary urban and regional ecosystem planning and management and involve local communities and diverse stakeholders to reduce the vulnerability of urban poor and minorities.
6. Launch collaborative, cross-boundary, and co-designed urban biodiversity and ecosystem research and advocacy programs to inform policies and planning and further develop nature-based solutions toward more resilient, livable, and sustainable urban futures.

8.12 Conclusions

Urban areas all over the globe, especially in developing countries, are growing rapidly in both population and area and are putting pressure on urban biodiversity and ecosystems to support livability, sustainability, and climate resilience. Climate change and its impacts on cities amplify the effects of urban stressors for ecosystems. Urban biodiversity and ecosystems will need to be safeguarded and enhanced to support climate mitigation and adaptation efforts and deliver critical, nature-based co-benefits for human well-being in cities. Urban ecosystems can help offset the worst impacts of climate change, including reducing the impact of extreme events by regulating hydrology, moderating local temperature, and providing critical ecosystem services. City leaders need to recognize the interdependence of the city with peri-urban and rural surroundings and continue to broaden their planning horizon to regional levels to account for the fact that species, ecosystems, and people cross municipal boundaries and so must planning, management, and governance.

Urban and peri-urban ecosystems provide critical natural capital for climate change adaptation in cities and urban regions. Ecological spaces in cities, including all forms of green infrastructure, provide important ecosystem services such as UHI reduction, coastal flood protection, and stormwater management. Urban ecosystems are already and will continue to be affected by climate change. Cities should utilize, protect, and restore these ecosystems when seeking to improve urban resilience to the effects of climate change. City planners, managers, and decision-makers can utilize nature-based solutions to design and implement climate adaptation and mitigation strategies in combination with more traditional built infrastructure solutions. Investing in natural capital is a cost-effective strategy that also generates multiple co-benefits that enhance human well-being. In this way, urban ecosystems simultaneously provide means for improving urban resilience, livability, equity, and sustainability.

Building climate-resilient urban communities entails a socio-ecological framework as opposed to socio-technological approaches (Berkes and Folke, 1998) that can reconnect cities to the biosphere (Andersson et al., 2014). Investing in urban ecosystems for climate adaptation and mitigation makes good sense because it is cost-effective and provides numerous co-benefits that can improve equity and livability in cities. Mounting evidence of the benefits of urban ecosystems as nature-based solutions calls for strengthening climate resiliency by investing in good governance, flexible institutions, and collaborative programs. We find through this review that urban biodiversity and ecosystem services are critical to the development of climate-resilient cities.

Annex 8.1   Stakeholder Engagement

To better gauge stakeholder understanding of urban ecosystems and biodiversity for climate change adaptation and mitigation, a two-pronged approach was used to reach a diverse array of groups. Chapter authors met informally with stakeholders at workshops and meetings in Berlin, New York, Rotterdam, Stockholm, and Paris, engaging with a multidisciplinary, global group of actors who contributed broad perspectives on urban climate change and development issues. Despite these engagements being held in the United States and Europe, the stakeholders involved were geographically, gender, and ethnically diverse, capturing views of managers, designers, citizens, planners, and policy-makers and other decision-makers. Additionally, an electronic survey was conducted to gather the views of a wider community for more formalized engagement with stakeholders (see Annex 8.2). The goal of the survey was to better understand how a broad range of stakeholders perceives urban ecosystems, their value, and their role in reducing climate change impacts and improving the resilience of cities.

Annex 8.2   Urban Ecosystem and Biodiversity Stakeholder Engagement Survey

Authored by Helen Santiago Fink, Quynn Nguyen, and Chapter Lead Authors

The stakeholder survey period was October 20–November 21, 2014, during which sixty-two responses were collected and then analyzed.

This basic survey instrument (Exhibit 8.A) was designed as part of the research for this chapter to solicit information from two key stakeholders groups (see Exhibit 8.B for Stakeholder List): (1) Urban professionals/practitioners or those entities involved in shaping or influencing the physical urban space, including planners, architects, engineers, political/regulatory decision-makers, real estate and construction industry professionals, environmental NGOs, and others; and (2) Urban end-users, which includes everyone who uses and benefits from the urban environment, from the general public to households; visitors; business enterprises; social, service, and learning facilities; and many others. The survey was developed by a subgroup of the chapter authors and reviewed by external reviewers. The twenty survey questions were structured in four parts: (1) Profile of the anonymous responder, (2) Role/value of urban ecosystem and biodiversity, (3) Relationship of ecosystems services to climate change, and (4) Socioeconomic and policy measures to support urban ecosystems services.

The survey received responses from many regions of the world, including Africa, South America, Asia, and the United States and Europe (with the largest representation). In respect to responder profiles, statistics indicated that 90% were urban dwellers; 40% were government employees; and 60% held a master's degree. Awareness of the term "urban ecosystem services" (UES) was indicated by 59% of the responders, with 28% never having heard of the term before this survey and 13% somewhat aware. The role of UES was seen by 80% of responders as valuable for aesthetics, recreation, health, pollution control, and climate mitigation and adaptation; however, climate change recorded the lowest (9%) among them all. Rural areas and wealthy populations were seen to benefit more, despite 57% of all responders acknowledging the benefits of UES for all groups. When asked to rate the value of UES among sixteen potential attributes, air quality (80%) and a healthy life (78%) followed by water quality (66%) received the highest responses. Physical,

Annex Box 8.1   Stakeholder Survey

Chapter authors pursued a two-pronged approach to engage a multidisciplinary group of stakeholders to contribute broader perspectives on the chapter's themes: (1) Consultations at relevant international conferences and workshops and (2) A detailed online survey. Although the survey had limitations in sample size ($n = 62$), it offers insights into the views of an international audience and suggests key points for broader stakeholder engagement. Key findings included:

- Stakeholders are strongly in support of protecting and enhancing UES for climate action, human well-being, and general quality of life. Despite 27% of survey respondents reporting that they had not heard of the term "urban ecosystem services," survey results suggest wide public support for investing in urban ecosystems for climate adaptation and mitigation.
- The role of urban ecosystems services was seen by 80% of survey responders as valuable for a multitude of issues including aesthetics, recreation, health, and pollution

control; however, climate change was among the lowest (9%) reported benefit.
- Stakeholders' "strong concern for climate change" and high agreement on "associated benefits of ecosystem services" calls attention to the value of increasing awareness to better communicate the multiple benefits of urban ecosystems to society including as ecosystem-based climate adaptation and mitigation.
- Survey results indicated a favorable "willingness to pay" for ecosystem services that provide climate adaptation and mitigation.
- Research is needed to fully understand how to positively encourage "human/personal attachment to the natural environment" (acknowledged by 89% of survey responders). The goal of the stronger attachment is to promote environmental stewardship and the development of stronger policy actions and fiscal instruments to advance climate decision-making and investment for natural capital and nature-based solutions.

---

psychological, and spiritual well-being, as well as recreation/leisure, were rated high by approximately 53% of responders. Climatic benefits such as carbon sequestration, temperature reduction, and lessened impacts of extreme weather events (e.g., landslide prevention) received 50%. UES ratings were 21% for cultural and sport activities, and education received 48% with pollination of crops being an outlier with only one response.

Because of the small sample size of sixty-two responders, the chapter authors are cognizant of the limitations to generalizing the findings of the survey's results. It is unrealistic to correlate the results to the wider population. Nevertheless, the survey highlights some key points for further investigation on the developing role of UES in the climate change agenda. Engagement with practitioners and decision-makers (e.g., city managers, administrators, policy-makers) at multiple levels, including with active end-users of ecological infrastructure (e.g., urban naturalists, conservationists, researchers, non-profits, NGOs, governments, social institutions, museums, community groups, and citizenry) could benefit from increased social-learning models promoting environmental education as an opportunity for increased stakeholder engagement. Communicating the critically important role that the natural environment and biodiversity play in both climate adaptation and mitigation, as well as in their nexus, is a cornerstone to elevating the climate and sustainable cities dialogue and practical action in urban areas.

In a number of separate questions, the relationship between UES and climate change was highly correlated, with 82% acknowledging a connection (see Annex Figure 8.A); similarly, 72% of responders indicated being "very concerned" about climate change. UES were perceived as important to help or be "able to protect" health (93%), water and sewage (82%), and property

values (80%), while both food supply and employment recorded a lower response (65%). There was general willingness to support UES for climate change action through economic and financial measures by around 65% of respondents. Combinations of fiscal instruments were favored by almost half of responders, yet when viewed individually, specific measures such as a carbon tax (46%), general government budget (51%), and penalty for polluters (43%) rated among the highest (see Annex Figure 8.B). The level of support was strongly recorded on a personal basis, with 68% of responders willing to volunteer and participate in a planning process for urban ecosystems for environmental protection. Regulatory encumbrances on land use were also overwhelmingly supported (62%) in order to provide more green space in cities, including restrictions on responders' private property.

Overall, the survey suggests strong support for protecting and enhancing UES for climate action, with strong co-benefits for human well-being. The result is encouraging, given the fact that 27% of responders had not heard of the term "urban ecosystem services" before and thus indicating potentially wider multistakeholder support. The high rating for "strong concern for climate change" and a majority agreeing on reaping "associated benefits of ecosystem services" calls attention to the need for increased awareness in building efforts to educate the public on the multiplicity of benefits provided by urban ecosystems to society. The survey result indicating a general "willingness to pay" for UES to contribute to both climate mitigation and adaptation, and this suggests increasing opportunities to incorporate EbA and green infrastructure development (among other measures) into local and national urban policies and practices. Research is needed to understand how to positively exploit the strong response of a "human/personal attachment to the natural environment" (acknowledged by 89% of survey responders) toward the

(a)

(b)

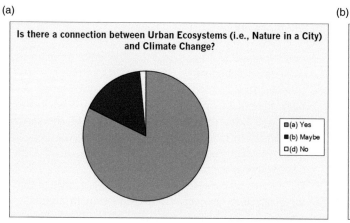

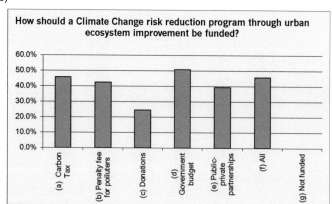

Annex Figures 8.A and 8.B *Results of stakeholder engagement survey: A. Demonstrates broad understanding of relationship between urban nature and climate change; B. Shows possible climate change risk reduction programs and how they are prioritized among stakeholder respondents. Highest are government budget, carbon tax, and penalty fee for polluters.*

development of stronger policy actions, integrated planning, and fiscal instruments to advance climate change decision-making and investment in building climate-resilient cities.

Exhibit 8.A Urban Ecosystem and Biodiversity Stakeholder Engagement Survey

1. Where do you live? (Choose one only)
   (a) Urban area (city) Name and Country: _____
   (b) Non-urban areas (suburban and rural areas) – Name and Country: _____

2. What is your employment? (Choose one only)
   (a) Government employee
   (b) Private sector employee
   (c) Self-employed
   (d) Civil society/NGO/nonprofit
   (e) Development partner (donor) agencies
   (f) Others

3. Which best describes your household status?
   (a) Single
   (b) Married with no children
   (c) Married with children
   (d) Other

4. What is your highest level of education?
   (a) Post/doctoral degree
   (b) Master degree
   (c) Bachelor degree
   (d) High school degree (12 years of education)
   (e) Less than high school degree

5. What in your view is the role of Ecosystems (or Nature such as trees, parks, gardens, animals, lakes, rivers, wetlands, green spaces) in a city?
   (a) Aesthetic value
   (b) Recreation
   (c) Pollution control
   (d) Climate change adaptation and mitigation
   (e) Others
   (f) All

6. In your view which section or who in society benefits the most from Ecosystem(s) in your community/city?
   (a) Rich class
   (b) Middle class
   (c) Poor class
   (d) Other
   (e) Everyone

7. Which sector or community benefits most in the world from Ecosystems?
   (a) Urban populations
   (b) Rural populations
   (c) Global population
   (d) Governments
   (e) Businesses

8. Select the importance of each of the benefits of Ecosystems/Nature in a city (rating from 1 to 5 [highest]):

(a)	Recreation and leisure	1	2	3	4	5
(b)	Sports activities	1	2	3	4	5
(c)	Cultural activities	1	2	3	4	5
(d)	Healthy life	1	2	3	4	5
(e)	Water quality	1	2	3	4	5
(f)	Air quality	1	2	3	4	5
(g)	Physical well-being	1	2	3	4	5
(h)	Psychological well-being	1	2	3	4	5
(i)	Spiritual well-being	1	2	3	4	5

(j)	Urban temperature	1	2	3	4	5
(k)	Flood prevention	1	2	3	4	5
(l)	Landslide prevention	1	2	3	4	5
(m)	Food production	1	2	3	4	5
(n)	Carbon sequestration	1	2	3	4	5
(o)	Pollination of crops	1	2	3	4	5
(p)	Others _____	1	2	3	4	5

9. Have you heard about *Urban Ecosystem Services* before this Survey?
(a) Yes
(b) Somewhat
(c) No

10. How concerned are you about Climate Change?
(a) Very concerned
(b) Somewhat concerned
(c) Not concerned

11. Is there a connection between Urban Ecosystems (Nature in a city) and Climate Change?
(a) Yes
(b) Maybe
(c) No

12. Can Urban Ecosystems HELP with or PROTECT the following?
12.1 PEOPLE

(a)	Health	Yes	Somewhat	No
(b)	Injuries/risks (on personal level)	Yes	Somewhat	No
(c)	Security (on community/city level)	Yes	Somewhat	No
(d)	Food supply	Yes	Somewhat	No
(e)	Water supply	Yes	Somewhat	No

12.2 ASSETS

(a)	Property values	Yes	Somewhat	No
(b)	Business losses	Yes	Somewhat	No
(c)	Foreign/domestic Investment	Yes	Somewhat	No

12.3 INFRASTRUCTURE

(a)	Roads	Yes	Somewhat	No
(b)	Electric power	Yes	Somewhat	No
(c)	Water and sewage	Yes	Somewhat	No
(d)	Public transportation	Yes	Somewhat	No
(e)	Employment/jobs	Yes	Somewhat	No

13. Can expanding urban green areas in the city help prevent global warming and/or reduce greenhouse gas emissions (GHGs)?
(a) Yes
(b) Maybe
(c) No

14. Can Urban Ecosystems in the city offer a better quality of life?
(a) Yes
(b) Maybe
(c) No

15. Would you be willing to pay for preserving and/or expanding Urban Ecosystems to help with Climate Change risks?
(a) Yes
(b) Somewhat
(c) No

16. How should a Climate Change risk reduction program through urban ecosystem improvement be funded?
(a) Carbon tax
(b) Penalty fee for pollution/carbon emissions
(c) Donations
(d) Government budget
(e) Public–private partnerships

17. Would you be willing to participate in the process of Urban Ecosystem planning and environmental protection/conservation on a personal/volunteer level?
(a) Yes
(b) Maybe
(c) No

18. Would you be willing to have more regulations/laws that would require more green spaces in public and private areas?
(a) Yes
(b) Maybe
(c) No

19. Would be you be willing to have a restriction(s) on your property/land in order to have a greener city/community?
(a) Yes
(b) Maybe
(c) No

20. Do you feel a personal attachment to the natural environment?
(a) Yes
(b) No
(c) Somewhat

Exhibit 8.B   Urban Ecosystems and Biodiversity Stakeholders List

Chapter 8   Urban Ecosystems and Biodiversity

Institutional (representatives of):
Public:

1. *City (Executive)*
   a. Climate Change – Adaptation/Mitigation
   b. Urban Planning
   c. Environment
   d. Housing
   e. Health
   f. Education
   g. Transportation
   h. Urban Conservation (Infrastructure)
   i. Parks and Gardens (Green Areas)

2. *Metropolitan/State (Executive)*
   a. Climate Change – Adaptation/Mitigation
   b. Environment
   c. Economy
   d. Transportation
   e. Health
   f. Education
   g. Housing
   h. Infrastructure

3. *City and State Councils (Lawmakers)*
   Representatives of related areas (as above)

Private Economic Sectors Associations
a. Real Estate
b. Infrastructure
c. Industry
d. Commerce
e. Transportation
f. Tourism

Private Civil Sociesty Associations
a. Residents
b. Arts and Culture, Education, Social and Political Groups, and Environmental NGOs
c. Professionals Urban Planners, Architects, Landscape Architects, Engineers, Foresters, Agronomists, etc.
d. Informal active organized groups

Individuals (users)
a. Parks and trails
b. Public transportation
c. Pedestrians in busy streets
d. Private transportation (car drivers)

References

Agencia Catalana de l'Aigua (ACA). (2014) Accessed April 17, 2015: http://acaweb. gencat.cat/aca/appmanager/aca/aca?_nfpb=true&_pageLabel =P51200158891423828204014

Ahern, J. (2007). Green infrastructure for cities: The spatial dimension. In Novotny, W, and Brown, P. (eds.), *Cities of the Future: Towards integrated landscape and water management* (267–283). IWA Publishing.

Ahern, J. (2013). Urban landscape sustainability and resilience: the promise and challenges of integrating ecology with urban planning and design. *Landscape Ecology* **28**(6): 1203–1212. Accessed April 25, 2015: https://scholar-google- com.proxy.wexler.hunter.cuny.edu/scholar?oi=bibs&clus ter=2360019842460228842&btnI=1&hl=en

Akbari, H., Pomerantz, M., and Taha, H. (2001). Cool surfaces and shade trees to reduce energy use and improve air quality in urban areas. *Solar Energy* **70**(3), 295–310.

Alberti, M. (2008). *Advances in Urban Ecology: Integrating Humans and Ecological Processes in Urban Ecosystems*. Springer Science + Business Media.

Alberti, M. (2015). Eco-evolutionary dynamics in an urbanizing planet. *TREE*, in press. Accessed January 8, 2016: http://dx.doi.org/10.1016/j.tree.(2014).11.007

Alberti, M., and Marzluff, J. M. (2004). Ecological resilience in urban eco-systems: Linking urban patterns to human and ecological functions. *Urban Ecosystems* **7**, 241–265.

Alexandri, E., and Jones, P. (2008). Temperature decreases in an urban can-yon due to green walls and green roofs in diverse climates. *Building and Environment*, **43**, 480–493. Accessed March 2, 2014: http://www.sciencedirect.com/science/article/pii/S0360132306003957

Amin, A. (2007). Re-thinking the urban social. *City* **11**(1), 100–114.

Andersson, E., Barthel, S., Borgström, S., Colding, J., Elmqvist, T., Folke, C., and Gren, Å. (2014). Reconnecting cities to the biosphere: Stewardship of green infrastructure and urban ecosystem services. *Ambio* 43, 445–453. Accessed February 27, 2015: http://dx.doi.org/10.1007/ s13280-014-0506-y.

Andersson, E., McPherson, T., Kremer, P., Gomez-Baggethun, E., Haase, D., Tuvendal, M., and Wurster, D. (2015a). Scale and Context Dependence of Ecosystem Service Providing Units. *Ecosystem Services* (Special Issue). Accessed March 25, 2016: http://dx.doi.org/10.1016/j.ecoser.2014.08.001

Andersson, E., Tengö, M., McPherson, T., and Kremer, P. (2015b). Cultural ecosystem services as a platform for working towards urban sustain-ability. *Ecosystem Services* (Special Issue). Accessed February 6, 2016: http://dx.doi.org/10.1016/j.ecoser.2014.08.002

Aronson, M. F. J., La Sort, F. A., Nilon, C. H., Katti, M., Goddard, M. A., Lepczyk, C. A. et al. (2014). A global analysis of the impacts of urban-ization on bird and plant diversity reveals key anthropogenic drivers. *Proceedings of the Royal Society B* **281**(2013), 3330.

Asian Cities Climate Change Resilience Network. (2015). Urban Climate Change Resilience in Action: Lessons from Projects in 10 ACCCRN Cities. ACCCRN & The Rockefeller Foundation. Accessed March 30, 2016: http://acccrn.net/sites/default/files/publication/attach/ACCCRN_ProjectsInsightsPaper.pdf

Asikainen, E., and Jokinen, A. (2009). Future natures in the making: Implementing biodiversity in sub-urban land use planning. *Planning Theory and Practice* **10**(3), 351–368.

Bardsley, D. K., and Hugo, G. J. (2010). Migration and climate change: Examining thresholds of change to guide effective adaptation decision-making. *Population and Environment* **32**, 238–262. doi:10.1007/s11111-010-0126-9

Barthel, S., and Isendahl, C. (2013). Urban gardens, agriculture, and water management: Sources of resilience for long-term food security in cit-ies. *Ecological Economics* **86**(2), 224–234.

Barthel, S., Parker, J., and Ernstson, H. (2015). Food and green space in cities: A resilience lens on gardens and urban environmental movements. *Urban Studies* **52**(7), 1321–1338.

Batty, M. (2008). The size, scale, and shape of cities. *Science* **319**, 769–771. doi:10.1126/science.1151419

Beatley, T. (2000). *Green Urbanism*. Island Press.

Beatley, T. (2009). *Biophilic Cities*. Island Press.

Berkes, F., and Folke, C. (1998). *Linking Social and Ecological Systems: Management Practices and Social Mechanisms for Building Resilience*. Cambridge University Press.

Bettencourt, L., and West, G. (2010). A unified theory of urban living. *Nature*, **467**, 912–913. doi:10.1038/467912a.

Bianchini, F., and Hewage, K. (2012). Probabilistic social cost-benefit analysis for green roofs: A lifecycle approach. *Building and Environment* **58**, 152–162. doi:10.1016/j.buildenv.(2012).07.005

Bolund, P., and Hunhammar, S. (1999). Ecosystem services in urban areas. *Ecological Economics* **29**, 293–301. doi:10.1016/S0921-8009(99)00013-0

Boone, C. G., Buckley, G. L., Grove, J. M., and Sister, C. (2009). Parks and people: An environmental justice inquiry in Baltimore, Maryland. *Annals of the Association of American Geographers* **99**(July 2008), 767–787.

Booth, D.B., Haugerud, R.A., and Troost, K.G. 2003. Geology, watershed, and Puget lowland rivers. In Montgomery, D.R., Bolton, S., Booth, D.B. and Wall, L. (Eds.) *Restoration of Puget Sound Rivers*. The University of Washington Press. Seattle, WA. pp. 14–45.

Boossabong, P. (2014). Coping with flooding in Bangkok. *Urban Agriculture Magazine* 27.

Borgström, S. T., Elmqvist, T., Angelstam, P., and Alfsen-Norodom, C. (2006). Scale mismatches in management of urban landscapes. *Ecology And Society* **11**(2), 16. doi:10.1097/MCC.0b013e32807f2aa5

Braaker, S., Moretti, M., Boesch, R., Ghazoul, J., Obrist, M. K., and Bontadina, F. (2014). Assessing habitat connectivity for ground-dwelling animals in an urban environment. *Ecological Applications* **24**, 1583–1595.

Breuste, J., Niemelä, J., and Snep, R. P. H. (2008). Applying landscape ecological principles in urban environments. *Landscape Ecology* **23**(10), 1139–1142. doi:10.1007/s10980-008-9273-0

Cardinale, B. J., Duffy, J. E., Gonzalez, A., Hooper, D. U., Perrings, C., Venail, P., and Naeem, S. (2012). Biodiversity loss and its impact on humanity. *Nature* **486**(7401), 59–67. doi:10.1038/nature11148

Carson, T. B., Marasco, D. E., Culligan, P. J., and McGillis, W. R. (2013). Hydrological performance of extensive green roofs in New York City: Observations and multi-year modelling of three full-scale systems. *Environmental Research Letters* **8**(02), 4036. doi:10.1088/1748-9326/8/2/024036

CDKN. (2014). Inside stories on climate compatible development. Climate Development Knowledge Network (CDKN) Accessed December 21, 2015: http://cdkn.org/wp-content/uploads/2014/05/SriLanka_Inside_Story_final_web-res.pdf

Cerveny, R. S., and Balling, R. C., Jr. (1998). Weekly cycles of air pollutants, precipitation and tropical cyclones in the coastal NW Atlantic region. *Nature* **394**, 561–563. doi:10.1038/29043

Chan, K. M., Satterfield, T., and Goldstein, J. (2011). Rethinking ecosystem services to better address and navigate cultural values. *Ecological Economics* **74**, 8–18. doi:10.1016/j.ecolecon.2011.11.011

Childers, D.L., Cadenasso, M.L., Grove, J.M., Marshall, V., MacGrath, B., and Pickett, S. 2015. An ecology for cities: A transformational nexus of design and ecology to advance climate change resilience and urban sustainability. *Sustainability* 7:3774–3791.

Childers, D. L., Pickett, S. T. A., Grove, J. M., Ogden, L., and Whitmer, A. (2014). Advancing urban sustainability theory and action: Challenges and opportunities. *Landscape & Urban Planning* **125**, 320–328.

Cohen, N., and Wijsman, K. (2014). Urban agriculture as green infrastructure: The case of New York City. *Urban Agriculture Magazine* 27.

Colding, J. (2011). The role of ecosystem services in contemporary urban planning. In Niemela, J. (ed.), *Urban Ecology*. Oxford University Press, 228–237.

Colding, J. (2013). Local assessment of Stockholm: Revisiting the Stockholm Urban Assessment. In Elmqvist, T., Fragkias, M., Goodness, J., Güneralp, B., Marcotullio, P. J., McDonald, R. I., Parnell, S., Schewenius, M., et al. (eds.), *Urbanization, Biodiversity and Ecosystem Services – Challenges and Opportunities* (313–335). Springer Netherlands. doi:10.1007/978-94-007-7088-1

Cole, L. W., and Foster, S. R. (2001). *From the Ground Up: Environmental Racism and the Rise of the Environmental Justice Movement* (**244**). NYU Press.

Collins, J. P., Kinzig, A., Grimm, N. B., Fagan, W. F., Hope, D., Wu, J., and Borer, E. T. (2000). A new urban ecology. *American Scientist* **88**, 416–425.

Condon, P. M., Cavens, D., and Miller, N. (2009). Urban planning tools for climate change mitigation. *Policy focus report/Code PF021*. Accessed May 28, 2015: www.lincolninst.edu

Consolo, T., Faux, M., Becker, M., Hamstead, Z., Kremer, P., Malone, E., and Segal, P. (2013). Equitable re-development of land acquired by New York City and New York State in Sandy-affected areas. Planner Network. Accessed March 13, 2014: http://596acres.org/en/news/2013/12/19/what-our-new-mayor-should-do-redevelopment-in-sandy-affected-communities/

Convention on Biological Diversity (CBD). (2009). Connecting biodiversity and climate change mitigation and adaptation: Report of the second ad hoc Technical Expert Group on Biodiversity and Climate Change. Technical series No. 41. Secretariat of the Convention on Biological Diversity, Montreal.

Corburn, J. (2005). *Street Science*. MIT Press.

Cumming, G. S., Cumming, D. H. M., and Redman, C. L. (2006). Scale mismatches in social-ecological systems: Causes, consequences, and solutions. *Ecology and Society* **11**(1), 20.

Czemiel Berndtsson, J. (2010). Green roof performance towards management of runoff water quantity and quality: A review. *Ecological Engineering* **36**, 351–360. doi:10.1016/j.ecoleng.(2009).12.014

Dale, A. G., and Frank, S. D. (2014). Urban warming trumps natural enemy regulation of herbivorous pests. *Ecological Applications* **24**, 1596–1607.

De la Salle, J., and Holland, M. (2010). *Agricultural Urbanism: Handbook for Building Sustainable Food Systems in 21st Century Cities*. Green Frigate Books.

Demuzere, M., Orru, K., Heidrich, O., Olazabal, E., Geneletti, D., Orru, H., Bhave, A. G., and Faehnle, M. (2014). Mitigating and adapting to climate change: Multi-functional and multi-scale assessment of green urban infrastructure. *Journal of Environmental Management* **146**, 107–115.

Depietri, Y., Renaud, F. G., and Kallis, G. (2012). Heat waves and floods in urban areas: A policy-oriented review of ecosystem services. *Sustainability Science* 7, 95–107. doi:10.1007/s11625-011-0142-4

De Zeeuw, H., van Veenhuizen R., and Dubbeling M. (2011). The role of urban agriculture in building resilient cities in developing countries. *Journal of Agricultural Science* **149**(S1), 153–163.

Di Leo, N., Escobedo, F. J., and Dubbeling, M. (2015). The role of urban green infrastructure in mitigating land surface temperature in Bobo-Dioulasso, Burkina Faso. *Environment Development and Sustainability*. doi:10.1007/s10668-015-9653-y.

Dubbeling, M. (2014). Integrating urban agriculture and forestry into climate change action plans: Lessons from Sri Lanka. Accessed November 26, 2015: http://www.ruaf.org/publications/integrating-urban-agriculture-and-forestry-climate-change-action-plans-lessons-sri-0

Dubbeling, M. (2015). Integrating urban agriculture and forestry into climate change action plans: Lessons from Western Province, Sri Lanka and Rosario, Argentina. RUAF Foundation.

Dunnett, N., and Kingsbury, N. (2008). *Planting Green Roofs and Living Walls*. Timber Press.

Ebre Observatory. (2014). Accessed March 11, 2015: http://www.obsebre.es/en/

Ecologic Institute. (2011). Analysis of socio-economic driving forces in built-up area expansion in Xiamen, China. In *Issues in Ecological Research and Application: 2011 Edition*. Scholarly Editions.

Eisen, L., Monaghan, A. J., Lozano-Fuentes, S., Steinhoff, D. F., Hayden, M. H., and Bieringer, P. E. (2014). The impact of temperature on the bionomics of *Aedes* (*Stegomyia*) *aegypti*, with special reference to the cool geographic range margins. *Journal of Medical Entomology* **51**(3), 496–516. doi:http://dx.doi.org/10.1603/ME13214

Elmqvist, T., Redman, C. E., Barthel, S., and Costanza, R. (2013). History of urbanization and the missing ecology. In Elmqvist, T., Fragkias, M., Goodness, J., Güneralp, B., Marcotullio, P. J., McDonald, R. I., and Wilkinson, C. (eds.), *Urbanization, Biodiversity and Ecosystem Services: Challenges and Opportunities* (13–30). Springer. doi:10.1007/978-94-007-7088-1_2

Ernstson, H. (2013). The social production of ecosystem services: A framework for studying environmental justice and ecological complexity in urbanized landscapes. *Landscape and Urban Planning* **109**(1), 7–17. doi:10.1016/j.landurbplan.(2012).10.005

Ernstson, H., Barthel, S., Andersson, E., and Borgström, S. T. (2010). Scale-crossing brokers and network governance of urban ecosystem services: The case of Stockholm. *Ecology and Society* **15**(4), 28. Accessed October 24, 2015: http://www.ecologyandsociety.org/vol15/iss4/art28/main.html

Escobedo, F. J., Clerici, N., Staudhammer, C. L., and Corzo, G. T. (2015). Socio-ecological dynamics and inequality in Bogotá, Colombia's public urban forests and their ecosystem services. *Urban Forestry & Urban Greening* **14**(4), 1040–1053.

European Environmental Agency. (2015). Exploring nature-based solutions – The role of green infrastructure in mitigating the impacts of weather- and climate change-related natural hazards. EEA Technical Report No. 12.

FAO. (2014). Growing greener cities in Latin America and the Caribbean. A FAO report on urban and peri-urban agriculture in the region. Rome, Food and Agriculture Organization of the United Nations.

Farr, D. (2008). *Sustainable Urbanism: Urban Design With Nature*. John Wiley & Sons.

Farrell, C., Szota, C., and Arndt, S. K. (2015). Urban plantings: "Living laboratories" for climate change response. *Trends in Plant Science* **20**(10), 597–599.

Fitzpatrick, M. C., Gove, A. D., Sanders, N. J., and Dunn, R. R. (2008). Climate change, plant migration, and range collapse in a global biodiversity hotspot: The Banksia (Proteaceae) of Western Australia. *Global Change Biology* 14, 1337–1352.

Florgård, C. (2007). Treatment measures for original natural vegetation preserved in the urban green infrastructure at Jarvafaltet, Stockholm. In Stewart, M., Ignatieva, M., Bowring, J., Egoz, S., and Melnichuk, I. (eds.), *Globalisation of Landscape Architecture: Issues for Education and Practice* (100–102). St. Petersburg's State Polytechnic University Publishing House.

Folke, C., Carpenter, S., Walker, B., Scheffer, M., Elmqvist, T., Gunderson, L., and Holling, C. S. (2004). Regime shifts, resilience, and biodiversity in ecosystem management. *Annual Review of Ecology, Evolution, and Systematics* 35, 557–581.

Folke, C. (2010). How resilient are ecosystems to global environmental change? *Sustainability Science* **5**, 151–154. doi:10.1007/s11625-010-0109-x

Fook, L. L., and Gang, C. (eds.). (2010). *Towards a Liveable and Sustainable Urban Environment: Eco-Cities in East Asia*. World Scientific Publishing.

Foster, J., A. Lowe, and S. Winkelman. (2011). *The Value of Green Infrastructure for Urban Climate Adaptation*. Washington, D.C.: Center for Clean Air Policy. Accessed May 5, 2015: http://mi.mi.gov/documents/dnr/TheValue_347538_7.pdf.

Fowler, D. (2002). Pollutants deposition and uptake by vegetation. In Bell, J. N. B., and Treshow, M. (eds.), *Air Pollution and Plant Life.*, second edition(43). Wiley.

Frantzeskaki, N., and Tilie, N. (2014). The dynamics of urban ecosystem governance in Rotterdam, The Netherlands. *Ambio* **43**(4), 542–555. Accessed August 12, 2015: http://doi.org/10.1007/s13280-014-0512-0

Frantzeskaki, N., Kabisch, N., and McPhearson, T. (2016). Advancing urban environmental governance: Understanding theories, practices and processes shaping urban sustainability and resilience. *Environmental Science and Policy* (Special Issue) **62**, 1–6. Accessed May 18, 2017: http://dx.doi.org/10.1016/j.envsci.2016.05.008

Frischenbruder, M., and Pellegrino, P. (2006). Using greenways to reclaim nature in Brazilian cities. *Landscape and Urban Planning* **76**, 67–78.

Frumkin, H., Hess, J., Luber, G., Malilay, J., and McGeehin, M. (2008). Climate change: The public health response. *American Journal of Public Health* **98**(3), 435–445. doi:10.2105/AJPH.2007.119362

Georgescu, M., Morefield, P. E., Bierwagen, B. G., and Weaver, C. P. (2014). Urban adaptation can roll back warming of emerging megapolitan regions. *Proceedings of the National Academy of Science U.S.A.* **111**, 2909–2914. doi:10.1073/pnas.1322280111.

Gill, S. E., Handley, J. F., Ennos, A. R., and Pauleit, S. (2007). Adapting cities for climate change: The role of the green infrastructure. *Built Environment* **33**(1), 115–133.

Gill, S. E., Rahman, M. A., Handley, J. F., and Ennos, A. R. (2013). Modelling water stress to urban amenity grass in Manchester UK under climate change and its potential impacts in reducing urban cooling. *Urban Forestry & Urban Greening* **12**, 350–358. Accessed June 4, 2014: http://dx.doi.org/10.1016/j.ufug.2013.03.005

Gilman, Sarah E., Urban, M. C., Tewksbury, J., Gilchrist, G. W., and Holt, R. D. (2010). A framework for community interactions under climate change. *Trends in Ecology and Evolution* **25**, 325–331. Accessed April 25, 2014: doi:10.1016/j.tree.2010.03.002

Gillner, S., Bräuning, A., and Roloff, A. (2014). Dendrochronological analysis of urban trees: climatic response and impact of drought on frequently used tree species. *Trees* 28, 1079. doi:10.1007/s00468-014-1019-9

Girot, P., Ehrhart, C., and Oglethorpe, J. (2009). *Integrating Community and Ecosystem-Based Approaches in Climate Change Adaptation Responses*. Ecosystem and Livelihoods Adaptation Network (ELAN).

Glick, P., Stein, B. A., and Edelson, N. A. (eds.). (2011). *Scanning the Conservation Horizon: A Guide to Climate Change Vulnerability Assessment*. National Wildlife Federation.

Gómez-Baggethun, E., Gren, Å., Barton, N. D., Langemeyer, J., McPhearson, T., O'Farrell., P., and Kremer, P. (2013). Urban ecosystem services. In Elmqvist, T., Fragkias, M., Goodness, J., Güneralp, B., Marcotullio, P. J., McDonald, R. I., and Wilkinson, C. (eds.), *Urbanization, Biodiversity and Ecosystem Services: Challenges and Opportunities* (175–251). Springer Netherlands. doi:10.1007/978-94-007-7088-1

Gonzalez, P., R.P. Nielson, R.P., Lenihan, J.M., and Drapek, R.J. 2010. Global patterns in the vulnerability of the ecosystems to vegetation shifts due to climate change. *Global Ecology and Biogeography* **19**:755–768.

Gotelli, N. J., and Chao, A. (2013). Measuring and estimating species richness, species diversity, and biotic similarity from sampling data. In Levin, S. A. (ed.), *Encyclopedia of Biodiversity*, second edition, volume **5**(195–211). Academic Press. Accessed November 27, 2015: http://www.elsevier.com/locate/permissionusematerial

Grimm, N. B., Chapin, F. S., Bierwagen, B., Gonzalez, P. Groffman, P. M., Luo, Y., Melton, F., Nadelhoffer, K., Pairis, A., Raymond, P. A., Schimel, J., and Williamson, C. E. (2013). The impacts of climate change on ecosystem structure and function. *Frontiers in Ecology and the Environment* **11**, 474–482.

Grimm, N. B., Foster, D., Groffman, P., Grove, J. M., Hopkinson, C. S., Nadelhoffer, K. J., Peters, D. P. (2008). The changing landscape: Ecosystem responses to urbanization and pollution across climatic and societal gradients. *Frontiers in Ecology and the Environment* **6**(5), 264–272. doi:10.1890/070147

Grimm, N.B., Grove, J.M., Pickett, S.T.A., and Redman, C.A. (2000). Integrated approaches to long-term studies of urban ecological systems. *BioScience* **50**(7), 571. doi:10.1641/0006-3568(2000)050[0571:IATLTO]2.0.CO;2

Güneralp, B., Mcdonald, R. I., Fragkias, M., Goodness, J., Marcotullio, P. J., and Seto, K. C. (2013). Urbanization, biodiversity and ecosystem services: Challenges and opportunities. In Elmqvist, Fragkias, M. Goodness, J., Güneralp, B., Marcotullio, P. J., McDonald, R. I., and Wilkinson, C. (eds.), *Urbanization, Biodiversity and Ecosystem Services: Challenges and Opportunities* (437–452). Springer Netherlands. doi:10.1007/978-94-007-7088-1

Güneralp, B., and Seto, K. C. (2013). Futures of global urban expansion: Uncertainties and implications for biodiversity conservation. *Environmental Research Letters* **8**(1), 014025. doi:10.1088/1748-9326/8/1/014025

Haase, D., Haase, A., Kabisch, N., Kabisch, S., and Rink, D. (2012). Actors and factors in land-use simulation: *The challenge of urban shrinkage. Environmental Modelling and Software* **35**, 92–103. doi:10.1016/j.envsoft.2012.02.012

Haase, D., Larondelle, N., Andersson, E., Artmann, M., Borgström, S., Breuste, J., Gomez-Baggethun, E., Gren, Å., Hamstead, Z., Hansen, R., Kabisch, N., Kremer, P., Langemeyer, J., Rall, E. L., McPhearson, T., Pauleit, S., Qureshi, S., Schwarz, N., Voigt, A., Wurster, D., and Elmqvist, T. (2014). A quantitative review of urban ecosystem services assessments: Concepts, models, and implementation. *AMBIO* (Special Issue) **43**, 413–433.doi:10.1007/s13280-014-0504-0

Hallegatte, S., Ranger, N., Mestre, O., Dumas, P., Corfee-Morlot, J., Herweijer, C., and Wood, R. M. (2010). Assessing climate change impacts, sea level rise and storm surge risk in port cities: A case study on Copenhagen. *Climatic Change* **104**(1), 113–137. doi:10.1007/s10584-010-9978-3

Hatfield, J. L., Boote, K. J., Kimball, B. A., Ziska, L. H., Izaurralde, R. C., Ort, D., Thomson, A. M., and Wolfe, D. W. (2011). Climate impacts on agriculture: Implications for crop production. *Agronomy Journal* **103**(2011), 351–370.

Hamstead, Z. A., Larondelle, N., Kremer, P., McPhearson, T., and Haase, D. (2015). Classification of the heterogeneous structure of urban landscapes (STURLA) as an indicator of landscape function applied to surface temperature in New York City. *Ecological Indicators* (Special Issue). Accessed April 12, 2016: http://dx.doi.org/10.1016/j.ecolind.2015.10.014

Hansen, R., Frantzeskaki, N., McPhearson, T., Rall, E., Kabisch, N., Kaczorowska, A.,Kain, J. -H., Artmann, M., and Pauleit, S. (2015). The uptake of the ecosystem services concept in planning discourses of European and American cities. *Ecosystem Services* (Special Issue). doi:10.1016/j.ecoser.2014.11.013

Heller, N. E., and Zavaleta, E. S. (2009). Biodiversity management in the face of climate change: A review of 22 years of recommendations. *Biological Conservation* **142**, 14–32.

Heutte, R., Bella, E., Snyder, J., and Shephard, M. 2003. Invasive plants and exotic weeds of Southeast Alaska. *USDA Forest Service*. Anchorage.

Hunt, A., and Watkiss, P. (2011). Climate change impacts and adaptation in cities: A review of the literature. *Climatic Change* **104**, 13–49.

Huq, N., Renaud, F., and Sebesvari, Z. (2013). *Ecosystem Based Adaptation (EBA) to Climate Change – Integrating Actions to Sustainable Adaptation.* United Nations University, Institute for Environment and Human Security (UNU-EHS), UN Campus, Bonn, Germany.

ICLEI Africa Secretariat 2013 *Annual Report.* (2013). ICLEI Local Governments for Sustainability.

ICIMOD. (2009). *Local Responses to Too Much and Too Little Water in the Greater Himalayan Region.* ICIMOD. Accessed September 15, 2014: http://lib.icimod.org/record/8021

Ignatieva, M., Stewart, G.H., and Meurk, C. (2011). Planning and design of ecological networks in urban areas. *Landscape Ecol Eng.* 7:17–25.

Intergovernmental Panel on Climate Change (IPCC). (2001). *Climate Change 2001: The Scientific Basis. A Report of Working Group I of the Intergovernmental Panel on Climate Change.* Cambridge University Press.

Intergovernmental Panel on Climate Change (IPCC). (2007a). *Climate Change 2007: Synthesis Report.* Cambridge University Press.

Intergovernmental Panel on Climate Change (IPCC). (2013). Summary for policymakers. In Stocker, T. F., Qin, D., Plattner, G._K., Tignor, M., Allen, S. K., Boschung, J., and Midgley, P. M. (eds.), *Climate Change 2013: The Physical Science Basis. Contribution of Working Group I to the Fifth Assessment Report of the Intergovernmental Panel on Climate Change.* Cambridge University Press.

Intergovernmental Panel on Climate Change (IPCC). (2014). *Climate Change 2014: Impacts, Adaptation, and Vulnerability. IPCC Working Group II Contribution to AR5.* Accessed August 13, 2015: http://ipcc-wg2.gov/AR5/

Institute for Social and Environmental Transition (ISET). (2010). *The Shared Learning Dialogue: Building Stakeholder Capacity and Engagement for Resilience Action.* Resilience in Concept and Practice Series Working Paper No. 1. ISET.

IUCN. (2009). Ecosystem-based Adaptation (EBA): Policy Briefing. *Fifth session of the UNFCCC Ad Hoc Working Group on Long-Term Cooperative Action under the Convention (AWG-LCA),* March 29–April 8, 2009, Bonn.

Jansma, J. E., Sukkel, W., Stilma, E. S. C., vanOost, A. C. J., and Visser, A. J. (2012).The impact of local food production on food miles, fossil energy use and greenhouse gas (GHG) emission: The case of the Dutch city of Almere. In Viljoen, A., and Wiskerke, J. S. C. (eds.), *Sustainable Food Planning: Evolving Theory and Practice* (307–321). Wageningen Academic Publishers.

Karki, M.B., Shrestha, A.B., and Winiger, M. (2011). Enhancing Knowledge Management and Adaptation Capacity for Integrated Management of Water Resources in the Indus River Basin. *Mountain Research and Development.* 31(3):242–251.

Kaushal, S. S., Groffman, P. M., Band, L. E., Shields, C. A., Morgan, R. P., Palmer, M. A., Belt, K. T., Swan, C. M., Findlay, S. E. G., and Fisher, G. T. (2008). Interaction between urbanization and climate variability amplifies watershed nitrate export in Maryland. *Environmental Science and Technology* **42**, 5872–5878.

Kaye, J. P., Burke, I. C., Mosier, A. R., and Guerschman, J. P. (2004). Methane and nitrous oxide fluxes from urban soils to the atmosphere. *Ecological Applications* **14**, 975–981.

Keim, M. E. (2008). Building human resilience: The role of public health preparedness and response as an adaptation to climate change. *American Journal of Preventive Medicine* **35**(5), 508–16. doi:10.1016/j.amepre.2008.08.022

Kendal, D., Williams, N. S. G., and Williams, K J. H. (2012). A cultivated environment: Exploring the global distribution of plants in gardens, parks and streetscapes. *Urban Ecosystems* **15**, 637–652. doi:10.1007/s11252-011-0215-2

Kendle T. and Forbes S. 1997. Urban nature conservation. *E. & F.N. Spon.* London.

Kenward, A., et al. (2014). *Summer in the City: Hot and Getting Hotter.* Climate Central. Accessed November 27, 2015: http://www.climatecentral.org/

Kingsolver, J. G., Diamond, S. E., and Buckley, L. B. (2013). Heat stress and the fitness consequences of climate change for terrestrial ectotherms. Functional Ecology. **27**(6):1415–1423. https://scholar.google.com/scholar?oi=bibs&cluster=10819040795748572617&btnI=1&hl=en

Klausmeyer, K. R., Shaw, R., MacKenzie, J. B., and Cameron, D. R. (2011). Landscape-scale indicators of biodiversity's vulnerability to climate change. *Ecosphere* **2**(8), art88. Accessed May 5, 2014: http://dx.doi.org/10.1890/ES11-0(0044).1

Kopperoinen, L., Itkonen, P., and Niemelä, J. (2014). Using expert knowledge in combining green infrastructure and ecosystem services in land use planning: An insight into a new place-based methodology. *Landscape Ecology* **29**, 1361–1375. doi:10.1007/s10980-014-0014-2

Kousky, C., Olmstead, S., Walls, M., and Macauley, M. (2013). Strategically placing green infrastructure: Cost-effective land conservation in the floodplain. *Environmental Science & Technology* **47**(8), 3563–3570.

Kowarik, I. (2005) Wild Urban Woodlands: Towards a Conceptual Framework. *Wild Urban Woodlands.* 1–32.

Kremer, P., Hamstead, Z., Haase, D., McPhearson, T., Frantzeskaki, N., Andersson, E., Kabisch, N., Larondelle, N., Lorance Rall, E., Avlonitis, G., Bertram, C., Baró, F., Gómez-Baggethun, E., Gren, A., Hansen, R., Kaczorowska, A., Kain, J. -H., Kronenberg, J., Langemeyer, J., Muehlmann, P., Pauleit, S., Rehdanz, K., Schewenius, M., van Ham, C., Voigt, A., Wurster, D., and Elmqvist, T. (2016a). Key insights for the future of urban ecosystem services research. *Ecology and Society* **21**(2), 29. Accessed January 12, 2017: http://dx.doi.org/10.5751/ES-08445-210229

Kremer, P., Hamstead, Z. A., and McPhearson, T. (2016b). The value of urban ecosystem services: A spatially explicit multicriteria analysis of landscape scale valuation scenarios in NYC. *Environmental Science and Policy* (Special Issue). Accessed February 15, 2017: http://dx.doi.org/10.1016/j.envsci.2016.04.012

Kronenberg, J. (2015). Why not to green a city? Institutional barriers to preserving urban ecosystem services. Ecosystem Services. **12**:218–227. https://scholar-google-com.proxy.wexler.hunter.cuny.edu/scholar?oi=bibs&cluster=3678534514957264970&btnI=1&hl=en

Larondelle, N, Hamstead, Z. A., Kremer, P., Haase, D., and McPhearson, T. (2014). Applying a novel urban structure classification to compare the relationships of urban structure and surface temperature in Berlin and New York City. *Applied Geography* 53, 427–437. Accessed February 7, 2015: http://dx.doi.org/10.1016/j.apgeog.2014.07.004

Laros, M., Birch, S., and Clover, J. (2013). *Ecosystem-Based Approaches to Building Resilience in Urban Areas: Towards a Framework for Decision-Making Criteria*. Background paper. ICLEI-Africa.

La Rosa, D., and Privitera, R. (2013). Characterization of non-urbanized areas for land-use planning of agricultural and green infrastructure in urban contexts. *Landscape and Urban Planning* 109(1), 94–106. doi:10.1016/j.landurbplan.(2012).05.012

Leichenko, R. M., and O'Brien, K. L. (2008). *Environmental Change and Globalization: Double Exposures*. Oxford University Press.

Leipprand, A., and Gerten, D. (2006). Global effects of doubled atmospheric CO2 content on evapotranspiration, soil moisture and runoff under potential natural vegetation. *Hydrological Sciences Journal* 51, 171–185.

Loarie, S. R., Carter, B. E., Hayhoe, K., McMahon, S., Moe, R., Knight, C. A., and Ackerly, D. D. (2008). Climate change and the future of California's endemic flora. *PLoS ONE* 3, e2502.

Locatelli, B., Brockhaus, M., Buck, A., and Thompson, I. (2010). Forests and adaptation to climate change: Challenges and opportunities. In Mery, G., et al. (eds.), *Forest and Society: Responding to Global Drivers of Change*. IUFRO World Series.

Loukaitou-sideris, A., and Stieglitz, O. (2002). Children in Los Angeles parks. *Town Planning Review* 73, 4(April), 467–488.

Lye, L.H. (2010) The Judiciary and Environmental Governance in Singapore. *Journal of Court Innovation*. 3:133.

Lwasa S., and Dubbeling, M. (2015). Urban agriculture and climate change. In de Zeeuw, H., and Dreschel, P. (eds.), *Cities and Agriculture: Developing Resilient Urban Food Systems*. Earthscan-Routledge.

Lwasa, S., Mugagga, F., Wahab, B., Simon, D., Connors, J., and Griffith, C. (2013). Urban and peri-urban agriculture and forestry: Transcending poverty alleviation to climate change mitigation and adaptation. *Urban Climate*. doi:10.1016/j.uclim.(2013).10.007

Maimaitiyiming, M., et al. (2014). Effects of green space spatial pattern on land surface temperature: Implications for sustainable urban planning and climate change adaptation. *Journal of Photogrammetry and Remote Sensing* 89, 59–66. Accessed December 27, 2015: http://dx.doi.org/

Marcotullio, P. J., and Solecki, W. D. (2013). Sustainability and cities – Meeting the grand challenge for the twenty-first century. In Sygna, L., O'Brien, K., and Wolf, J. (eds.), *A Changing Environment for Human Security: Transformative Approaches to Research, Policy and Action* (496). Routledge.

Marzluff, J. M. (2001). Worldwide urbanization and its affects on birds. In Marzluff, J. M., Bowman, R., and Donnelly, R. (eds.), *Avian Conservation and Ecology in an Urbanizing World* (19–47). Kluwer Academic Publishers.

McDonald, R., and Marcotullio, P. (2011). Global effects of urbanization on ecosystem services. In Niemelä, J. (ed.), *Urban Ecology* (193–205). Oxford University Press.

McDonald, R. I., Marcotullio, P. J., and Güneralp, B. (2013). Urbanization and global trends in biodiversity and ecosystem services. In Elmqvist, T., Fragkias, M., Goodness, J., Güneralp, B., Marcotullio, P. J., McDonald, R. I., and Wilkinson, C. (eds.), *Urbanization, Biodiversity and Ecosystem Services: Challenges and Opportunities* (31–52). Springer Netherlands. doi:10.1007/978-94-007-7088-1_3

McDonnell, M. J. (2011). The history of urban ecology. In Niemelä J., Breuste, J. H., Guntenspergen, G., McIntyre, N. E., Elmqvist, T., and James, P. (eds.), *Urban Ecology: Patterns, Processes, and Applications*. Oxford University Press.

McDonnell, M. J., and Hahs, A. (2013). The future of urban biodiversity research: Moving beyond the "low-hanging fruit." *Urban Ecosystems* 16, 397–409. doi:10.1007/s11252-013-0315-2

McGranahan, G., Marcotullio, P. J., Bai, X., Balk, D., Braga, T., Douglas, I., and Zlotnik, H. (2005). Urban systems. In Hassan, R., Scholes, R.,

and Ash, N. (eds.), *Current State and Trends: Findings of the Condition and Trends Working Group. Ecosystems and Human Well-being* (795–825). Island Press.

McKinney, M. L. (2002). Urbanization, Biodiversity and Conservation. *BioScience* 52(10), 883–890.

McLennan, J. F. (2004). *The Philosophy of Sustainable Design: The Future of Architecture*. Ecotone Publishing. Accessed February 22, 2014: http://books.google.com/books?id=-Qjadh_0IeMC&pgis=1

McPhearson, T. (2011). Toward a sustainable New York City: Greening through urban forest restoration. In Slavin, M. (ed.), *The Triple Bottom Line: Sustainability Principles, Practice, and Perspective in America's Cities* (181–204). Island Press.

McPhearson, T., Auch, R., and Alberti, M. (2013c). Urbanization trends, biodiversity patterns, and ecosystem services in North America. In Elmqvist, T., Fragkias, M., Goodness, J., Güneralp, B., Marcotullio, P. J., McDonald, R. I., Parnell, S., Schewenius, M., Sendstad, M., SEto, K. C., and Wilkinson, C. (eds.), *Cities and Biodiversity Outlook: Urbanization, Biodiversity and Ecosystem Services: Challenges and Opportunities* (279–286). Springer Netherlands. doi:10.1007/978-94-007-7088-1_14

McPhearson, T., Haase, D., Kabisch, N., Gren, Å. (2016). Advancing understanding of the complex nature of urban systems. *Ecological Indicators* 70, 566–573. Accessed April 18, 2017: http://dx.doi.org/10.1016/j.ecolind.2016.03.054

McPhearson, T., Hamstead, Z., and Kremer, P. (2014). Urban ecosystem services for resilience planning and management in New York City. *AMBIO* 43, 502–515.doi:10.1007/s13280-014-0509-8

McPhearson, T., Kremer, P., and Hamstead, Z. (2013). Mapping ecosystem services in New York City: Applying a social-ecological approach in urban vacant land. *Ecosystem Services* (2013), 11–26. Accessed October 29, 2015: http://dx.doi.org/10.1016/j.ecoser.2013.06.005

McPhearson, T., Andersson, E., Elmqvist, T., and Frantzeskaki, N. (2015). Resilience of and through urban ecosystem services. Ecosystem Services. 12:152–156. http://www.sciencedirect.com/science/article/pii/S2212041614000837

McPhearson, T., Pickett, S. T. A., Grimm, N., Alberti, M., Elmqvist, T., Niemelä, J., Weber, C., Haase, D., Breuste, J., and Qureshi, S. (2016). Advancing urban ecology toward a science of cities. *BioScience* 66(3), 198–212. doi:10.1093/biosci/biw002

McPherson, E. G., Nowak, D., Heisler, G., Grimmond, S., Souch, C., Grant, R., and Rowntree, R. (1997). Quantifying urban forest structure, function, and value: The Chicago Urban Forest Climate Project. *Urban Ecosystems* 1(1), 49–61. doi:10.1023/A:1014350822458

Meteo Estartit. (2015). Accessed January 5, 2016: http://www.historique-meteo.net/europe/espagne/estartit/2015/09/

Meyer, H., Nillesen, A. L., and Zonneveld, W. (2012). Rotterdam: A city and a mainport on the edge of a delta. *European Planning Studies* 20, 71–94. doi:10.1080/09654313.2011.638498.

Millennium Ecosystem Assessment. (2005). *Ecosystems and Human Well-Being: Millennium Ecosystem Assessment Scenarios for the Future of Ecosystem Services*. Island Press.

Mincy, S. K., Hutten, M., Fischer, B. C., Evans, T. P., Stewart, S. I., and Vogt, J. M. (2013). Structuring institutional analysis for urban ecosystems: A key to sustainable urban forest management. *Urban Ecosystems* 16(3), 553–571. doi:10.1007/s11252-013-0286-3

Ministry of Agriculture, Agrarian Development, Minor Irrigation, Industries. Environment, Arts and Cultural Affairs, Western Province. (2014). Climate change adaptation action plan (2015–2018). Western Province of Sri Lanka. Accessed September 18, 2015: http://www.ruaf.org/sites/default/files/Western%20Province%20Climate%20Change%20Action%20Plan_2015-2018.pdf

Miyake, K. K., Maroko, A. R., Grady, K. L., Maantay, J. A., and Arno, P. S. (2010). Not just a walk in the park: Methodological improvements for determining environmental justice implications of park access in New York City for the promotion of physical activity. *Cities and the Environment* 3(1), 1–17. Accessed August 14, 2014: http://www.pubmedcentral.nih.gov/articlerender.fcgi?artid=3160641&tool=pmcentrez&rendertype=abstract

Mosely, S. (2014). Coastal cities and environmental change. *Environment and History* **20**, 517–533. doi:10.3197/096734014X14091313617280

Mostafavi, M., and Doherty, G. (2010). *Ecological Urbanism*. Lars Muller Publishers.

Müller, N. (2010). Most frequently occurring vascular plants and the role of non-native species in urban areas – A comparison of selected cities of the old and new worlds. In N. Müller, P. Werner, and J. G. Kelcey (Eds.), Urban biodiversity and design (pp. 227–242). Hoboken: Wiley-Blackwell.

Munang, R., Thiaw, I., Alverson, K., Mumba, M., Liu, J. and Rivington, M. (2013). Climate change and Ecosystem-based Adaptation: a new pragmatic approach to buffering climate change impacts. *Current Opinion Environmental Sustainability*. 5:1–5.

Munroe, R., Roe, D., Doswald, N.,Spencer, T., Möller, I., Vira, B., Reid, H., Kontoleon, A., Giuliani, A., Castelli I., and Stephens, J. (2010). *Review of the Evidence Base for Ecosystem-based Approaches for Adaptation to Climate Change*. International Institute for Environment and Development (IIED).

National Environmental Justice Advisory Council. (2002). Fish consumption and environmental justice. Accessed January 4, 2015: http://www.epa.gov/compliance/ej/resources/publications/nejac/fish-consump-report_1102.pdf

New York State (NYS). (2013). New York State Community Development Block Grant Disaster Recovery (CDBG-DR) Program. Accessed April 12, 2016: https://stormrecovery.ny.gov/housing/buyout-acquisition-programs

Niemelä, J., Breuste, J. H., Guntenspergen, G., McIntyre, N. E., Elmqvist, T., and James, P. (2011). *Urban Ecology: Patterns, Processes, and Applications*. Oxford University Press.

Nobis, M. P., Jaeger, J. A. G., and Zimmerman, N. E. (2009). Neophyte species richness at the landscape scale under urban sprawl and climate warming. *Diversity and Distributions* **15**, 928–939. doi:10.1111/j.1472-4642.2009.00610.x

Nowak, D. J., Hirabayashi, S., Bodine, A., and Hoehn, R. (2013). Modeled PM2.5 removal by trees in ten U.S. cities and associated health effects. *Environmental Pollution* **178**, 395–402. doi:10.1016/j.envpol.2013.03.050

Novotny, V., Ahern, J., and Brown, P. (2010). *Water Centric Sustainable Communities: Planning, Retrofitting, and Building the Next Urban Environment*. Wiley.

Observatora de l'Ebre. (2014). Accessed December 7, 2015: http://www.obsebre.es/en/

OECD/CDRF. (2009). *Trends in Urbanisation and Urban Policies in OECD Countries: What Lessons for China?* OECD/CDRF.

Onishi, A., Cao, X., Ito, T., Shi, F., and Imura, H. (2010). Evaluating the potential for urban heat-island mitigation by greening parking lots. *Urban Forestry and Urban Greening* **9**(4), 323–332. doi:10.1016/j.ufug.(2010).06.002

Ottelé M., Van Bohemen H., Fraaij A. L. A. (2010). Quantifying the deposition of particulate matter on climber vegetation on living walls. *Ecological Engineering* **36**, 154–162.

Pandey, R., and Bardsley, D. K. (2013). Human ecological implications of climate change in the Himalaya: Pilot studies of adaptation in agro-ecosystems within two villages from Middle Hills and Tarai, Nepal. In *Impacts World 2013, International Conference on Climate Change Effects*, Potsdam, May 27–30.

Parmesan, C. (2006). Ecological and evolutionary responses to recent climate change. *Annual Review of Ecology Evolution and Systematics* **37**, 637–669.

Parmesan, C., and Yohe, G. (2003). A globally coherent fingerprint of climate change impacts across natural systems. *Nature* **421**(6918), 37–42.

Pascual, U., Muradian, R., Rodríguez, L. C., and Duraiappah, A. (2010). Exploring the links between equity and efficiency in payments for environmental services: A conceptual approach. *Ecological Economics* **69**(6), 1237–1244. doi:10.1016/j.ecolecon.2009.11.004

Pataki, D. E., Boone, C. G., Hogue, T. S., Jenerette, G. D., McFadden, J. P., and Pincetl, S. (2011). Ecohydrology bearings—invited commentary: Socio-ecohydrology and the urban water challenge. *Ecohydrology* **4**, 341–347. doi: 10.1002/eco.209

Paul, M. J., and Meyer, J. L. (2001). Streams in the urban landscape. *Annual Review of Ecological Systems* **32**, 333.

Pauleit, S., Liu, L., Ahern, J., and Kazmierczak, A. (2011). Multifunctional green infrastructure planning to promote ecological services in the city. In Niemelä, J. (ed.), *Urban Ecology* (272–286). Oxford University Press.

Pearson, R. G., and Dawson, T. P. (2003). Predicting the impacts of climate change on the distribution of species: Are bioclimate envelope models useful ? *Global Ecology & Biogeography* **12**, 361–371.

Petkova, E. P., Bader, D. A., Anderson, G. B., Horton, R. M., Knowlton, K., and Kinney, P. L. (2014). Heat-related mortality in a warming climate: Projections for 12 U.S. cities. *International Journal of Environmental Research and Public Health* **11**(11), 11371–11383.

Petralli, M., Prokopp, A., Morabito, M., Bartolini, G., Torrigiani, T., and Orlandini, S. (2006). Ruolo delle aree verdi nella mitigazione dell'isola di calore urbana: Uno studio nella città di Firenze. *Rivista Italiana Agrometeorologia* **1**, 51–58.

Pham, T. T. H., Apparicio, P., Séguin, A. M., Landry, S., and Gagnon, M. (2012). Spatial distribution of vegetation in Montreal: An uneven distribution or environmental inequity? *Landscape and Urban Planning* **107**(3), 214–224. Accessed October 15, 2014: http://dx.doi.org/10.1016/j.landurbplan.(2012).06.002

Pickett, S., Burch W. R. Jr., Dalton, S. D. and Foresman, R. W. 1997. Integrated urban ecosystem research. *Urban Ecosyst.* **1**:183–84.

Pickett, S., Cadenasso, M., Grove, J., Groffman, P., Band, L., Boone, C., Burch, W., Grimmond, S., Hom, J., Jenkins, J., Law, N., Nilon, C., Pouyat, R., Szlavecz, K., Warren, P., and Wilson, M. (2008). Beyond urban legends: An emerging framework of urban ecology, as illustrated by the Baltimore ecosystem study. *BioScience* **58**(2), 139–150.

Pickett, S. T. A., Cadenasso, M. L., Grove, J. M., Nilon, C. H., Pouyat, R. V, Zipperer, W. C., and Costanza, R. (2001). Urban ecological systems: Linking terrestrial ecological, physical, and socioeconomic components of metropolitan areas. *Annual Review of Ecology and Systematics*, **32**, 127–57.

Pielke, R. A., Marland, G., Betts, R. A., Chase, T. N., Eastman, J. L., Niles, J. O., and Running, S. W. (2002). The influence of land-use change and landscape dynamics on the climate system: Relevance to climate-change policy beyond the radiative effect of greenhouse gases. *Philosophical Transactions. Series A, Mathematical, Physical, and Engineering Sciences* **360**(1797), 1705–1719. doi:10.1098/rsta.2002.1027

Porter, E. M., et al. (2013). Interactive effects of anthropogenic nitrogen enrichment and climate change on terrestrial and aquatic biodiversity. *Biogeochemistry* **114**(1–3), 93–120.

Porter, J. R., Dyball, R., Dumaresq, D., Deutsch, L., and Matsuda, H. (2014). Feeding capitals: Urban food security and self-provisioning in Canberra, Copenhagen and Tokyo. *Global Food Security* **3**(1), 1–7.

Pötz, H., and Bleuze, P. (2012). *Urban Green-Blue Grids for Sustainable and Dynamic Cities*. Coop for Life.

Ranger, N., Hallegatte, S., Bhattacharya, S., Bachu, M., Priya, S., Dhore, K., and Corfee-Morlot, J. (2010). An assessment of the potential impact of climate change on flood risk in Mumbai. *Climatic Change* **104**(1), 139–167. doi:10.1007/s10584-010-9979-2

Rasul, G. (2012). Contribution of Himalayan ecosystems to water, energy and food security in South Asia: A nexus approach. ICIMOD. Accessed August 1, 2014: http://lib.icimod.org/record/1898

Raudsepp-Hearne, C., Peterson, G. D., and Bennett, E. M. (2010). Ecosystem service bundles for analyzing tradeoffs in diverse landscapes. *PNAS* **107**, 5242–5247. doi:10.1073/pnas.0907284107.

Register, R. (2006). *Ecocities: Rebuilding Cities in Balance with Nature*. New Society Press.

Revi, A., Satterthwaite, D. E., Aragón-Durand, F., Corfee-Morlot, J., Kiunsi, B. R., Pelling Roberts, D. C., and Solecki, W. (2014). Urban areas. In *Climate Change 2014: Impacts, Adaptation and Vulnerability. IPCC Working Group II Contribution to AR5* (535–612). Cambridge University Press.

Rogner, H. -H., Zhou, D., Bradley., R., Crabbé, P., Edenhofer, O. (Australia), B. H., and Yamaguchi, M. (2007). Introduction. In Metz, B., Davidson, O. R., Bosch, P. R., Dave, R., and Meyer, L. A. (eds.), *Climate Change 2007: Mitigation. Contribution of Working Group III to the Fourth*

Assessment Report of the Intergovernmental Panel on Climate Change. Cambridge University Press.

Rosenzweig, C., Solecki, W. D., Cox, J., Hodges, S., Parshall, L., Lynn, B., and Dunstan, F. (2009). Mitigating New York City's heat island: Integrating stakeholder perspectives and scientific evaluation. *Bulletin of the American Meteorological Society* **90**(9), 1297–1312. doi:10.1175/2009BAMS2308.1

Rosenzweig, C., Solecki, W., DeGaetano, A., O'Grady, M., Hassol, S., and Grabhorn, P. (2011). *Responding to Climate Change in New York State: The ClimAID Integrated Assessment for Effective Climate Change Adaptation: Synthesis Report.* New York State Energy Research and Development Authority.

Rosenzweig, C., Solecki, W. D., and Slosberg, R. B. (2006). *Mitigating New York City's Heat Island with Urban Forestry, Living Roofs, and Light Surfaces.* New York City Regional Heat Island Initiative.

Rottle, N., and Yocom, K. (2011). *Basics Landscape Architecture 02: Ecological Design.* AVA Publishing.

RUAF Foundation. (2014 March). Promoting urban agriculture as a climate change adaptation and disaster risk reduction strategy. *Urban Agriculture* 27.

Salzman, J., Arnold, C. A., Garcia, R., Hirokawa, K. H., Jowers, K., LeJava, J., and Olander, L. P. (2014). The most important current research questions in urban ecosystem services. *Duke Environmental Law & Policy Forum* (1–47). Accessed June 20, 2015: http://scholarship.law.duke.edu/faculty_scholarship/3350

Sandy Redevelopment Oversight Group. (2014). Equitable re-development of land acquired by New York City and New York State in Sandy-affected areas. Accessed May 27, 2015: http://596acres.org/en/news/2013/12/19/what-our-new-mayor-should-do-redevelopment-in-sandy-affected-communities/

Santiago Fink, H. (2016). Human-nature for climate action: Nature-based solutions for urban sustainability, *Sustainability* **8**(3), 254. doi:10.3390/su8030254

Satterthwaite, D., Huq, S., Pelling, M., and Reid, H. (2008). *Adapting to Climate Change in Urban Areas: The Possibilities and Constraints in Low- and Middle-Income Nations. Human Settlements Discussion Paper Series; Theme: Climate Change and Cities.* IIED.

Satterthwaite, D., Huq, S., Pelling, M., Reid, H., and Romero, P. L. (2007 and 2008). *Adapting to Climate Change in Urban Areas: The Possibilities and Constraints in Low- and Middle-Income Nations. IIED working paper.* IIED.

Schewenius, M., McPhearson, T., and Elmqvist, T. (2014). Opportunities for increasing resilience and sustainability of urban social–ecological systems: Insights from the URBES and the Cities Biodiversity Outlook Projects. *AMBIO*(Special Issue) **43**, 434–444.doi:10.1007/s13280-014-0505-z

Scholz-Barth, K. (2001). Green roofs: Stormwater management from the top down. *Environmental Design and Construction.* January/February 2001, Posted on edcmag.com on 1/15/2001. Accessed December 17, 2014: http://www.greenroofs.com/pdfs/archives-katrin.pdf

Searle, S. Y., Turnbull, M. H., Boelman, N. T., Schuster, W. S. R., Yakir, D., and Griffin, K. L. (2012). Urban environment of New York City promotes growth in northern red oak seedlings. *Tree Physiology* **32**, 389–400. doi:10.1093/treephys/tps027

Secretariat of the Convention on Biological Diversity. (2012). *Cities and Biodiversity Outlook* (64 pages). Montreal.

Seto, K. C., Parnell, S., and Elmqvist, T. (2013). A global outlook on urbanization. In Elmqvist, T., Fragkias, M., Goodness, J., Guʻneralp, B., Marcotullio, P. J., Mcdonald, R. I., Parnell, S., Schewenius, M., Sendstad, M., Seto, K. C., and Wilkinson, C. (eds.), *Urbanization, Biodiversity and Ecosystem Services: Challenges and Opportunities* (1–12). Springer. doi:10.1007/978-94-007-7088-1.

Seto, K. C., Reenberg, A., Boone, C. G., Fragkias, M., Haase, D., Langanke, T., Marcotullio, P., Munroe, D. K., Olah, B., Simon, D. (2012). Urban land teleconnections and sustainability. *PNAS* **109**, 7687–7692. doi:10.1073/pnas.1117622109.

SMC Coastal Modeling System. (2015). Accessed February 9, 2016: http://smc.ihcantabria.es/SMC25/en/

Solecki, W., and Marcotullio, P. J. (2013). Urbanization, biodiversity and ecosystem services: Challenges and opportunities. In Elmqvist, T., Fragkias, M., Goodness, J., Güneralp, B., Marcotullio, P. J., McDonald, R. I., and Wilkinson, C. (eds.), *Urbanization, Biodiversity and Ecosystem Services: Challenges and Opportunities* (485–504). Springer Netherlands. doi:10.1007/978-94-007-7088-1

Steele, W., Mata, L., and Fünfgeld, H. (2015). Urban climate justice: Creating sustainable pathways for humans and other species. *Current Opinion in Environmental Sustainability* **14**, 121–126.

Susca, T., Gaffin, S. R., and Dell'Osso, G. R. (2011). Positive effects of vegetation: Urban heat island and green roofs. *Environmental Pollution* **159**, 2119–2126. doi:10.1016/j.envpol.2011.03.007

Suddick, E. C., and Davidson, E. A. (eds.) (2012). The role of nitrogen in climate change and the impacts of nitrogen-climate interactions on terrestrial and aquatic ecosystems, agriculture, and human health in the United States: A technical report submitted to the U.S. National Climate Assessment. Falmouth, MA: North American Nitrogen Center of the International Nitrogen Initiative (NANC-INI), Woods Hole Research Center.

Sukhdev, P., Wittmer, H., Schröter-Schlaack, C., Nesshöver, C., Bishop, J., Ten, P., and Simmons, B. (2010). The Economics of Ecosystems and Biodiversity: Mainstreaming the economics of nature. A Synthesis of the approach, conclusions and recommendations of TEEB. Accessed April 12, 2016: http://www.urbanhabitats.org/v01n01/climatechange_full.html

Sukopp, H., and Wurzel, A. (2003). The effects of climate change on the vegetation of Central European cities. *Urban Habitats* **1**(1), 66–86.

Sutcliffe, S., and Court, J. (2005). *Evidence-Based Policymaking: What Is It? How Does It Work? What Relevance for Developing Countries?* Overseas Development Institute (ODI).

Taha, H. (1997). Urban climates and heat islands: Albedo, evapotranspiration, and anthropogenic heat. *Energy and Buildings* **25**(2), 99–103. Accessed September 30, 2015: http://www.sciencedirect.com/science/article/pii/S0378778896009991

Taylor, D. E. (2011). The evolution of environmental justice activism, research, and scholarship. *Environmental Practice* 13(December), 280–301.

The Economics of Environmentalism and Biodiversity (TEEB). (2010). *The Economics of Ecosystems and Biodiversity – Ecological and Economic Foundations*, ed. P. Kumar. Earthscan.

The Economics of Environmentalism and Biodiversity (TEEB). (2011). TEEB Manual for cities: Ecosystem services in urban management. In UNEP and the European Union (ed.), *Ecosystem Services in Urban Management.* Accessed October 23, 2014: http://www.teebweb.org/Portals/25/Documents/TEEB_Manual_for_Cities_Ecosystem_Services_for_Urban_managment__FINAL_(2011).pdf

Theurillat, J. -P., and Guisan, A. (2001). Potential impact of climate change on vegetation in the European Alps: A review. *Climatic Change* **50**(1–2), 77–109.

Thomas, C. D., Cameron, A., Green, R. E., Bakkenes, M., Beaumont, L. J., Collingham, Y. C., and Williams, S. E. (2004). Extinction risk from climate change. *Nature* **427**(6970), 145–148. doi:10.1038/nature02121

Troy, A. J., Grove, M., and O'Neil-Dunne, J. (2012). The relationship between tree canopy and crime rates across an urban–rural gradient in the great Baltimore region. *Landscaping and Urban Planning* 106, 62–270.

Tyler, S., and Moench, M. (2012). A framework for urban climate resilience. *Climate and Development* **4**(4), 311–326. Accessed April 10, 2016: http://dx.doi.org/10.1080/17565529.2012.745389

Tzoulas, K., Korpela, K., Venn, S., Yli-Pelkonen, V., Kaźmierczak, A., Niemela, J., and James, P. (2007). Promoting ecosystem and human health in urban areas using Green Infrastructure: A literature review. *Landscape and Urban Planning* **81**(3), 167–178. doi:10.1016/j.landurbplan.2007.02.001

UNEP. (2009). *The Role of Ecosystem in Climate Change Adaptation and Disaster Risk Reduction; Copenhagen Discussion Series.* UNEP.

UNEP. (2011a). Supporting countries in meeting the climate challenge: UNEP's priorities for catalysing a green economy. Accessed May 27, 2014: www.unep.org/pdf/flagship_unep.pdf

UNEP. (2011b). Towards a green economy: Pathways to sustainable development and poverty eradication – A synthesis for policy makers. Accessed October 13, 2015: www.unep.org/greeneconomy

UNEP. (2012). UNEP Regional Seas Releases Report on Ecosystem Based Adaptation. Accessed August 28, 2014: http://sdg.iisd.org/news/unep-regional-seas-releases-report-on-ecosystem-based-adaptation/?rdr=climate-l.iisd.org

UNFCCC Secretariat. (2011). Ecosystem-based approaches to adaptation: Compilation of information. Thirty-fifth session of the Subsidiary Body for Scientific and Technological Advice. Accessed August 13, 2015: http://unfccc.int/resource/docs/2011/sbsta/eng/inf08.pdf

UN-Habitat. (2009). *Annual Report.*

UN-Habitat. (2011). *Cities and Climate Change: Global Report on Human Settlements (2011).* Earthscan.

University of Cambridge and ICLEI. (2014). *Climate change: Implications for cities. Key findings from the Intergovernmental Panel on Climate Change Fifth Assessment Report.* Accessed December 16, 2015: http://www.cisl.cam.ac.uk/Resources/Climate-and-Energy/Climate-Change-Implications-for-Cities

Urban, M. C., De Meester, L., Vellend, M., Stoks, R., and Vanoverbeke, J. (2012). A crucial step toward realism: responses to climate change from an evolving metacommunity perspective. *Evolutionary Applications* 5:154–167.

Van Kerkhoff, L., and Lebel, L. (2006). Linking knowledge and action for sustainable development. *Annual Review of Environment and Resources* 31(1),445–477.

Vignola, R., Locatelli, B., Martinez, C., and Imbach, P. (2009). Ecosystem-based adaptation to climate change: What role for policy-makers, society and scientists? *Mitigation and Adaptation Strategies for Global Change* 14, 691–696.

Violin, C. R., Cada, P., Sudduth, E. B., Hassett, B. A., Penrose, D. L., and Bernhardt, E. S. (2011). Effects of urbanization and urban stream restoration on the physical and biological structure of stream ecosystems. *Ecological Applications* 21(6), 1932–1949. doi:10.1890/10-1551.1

Von Döhren, P., and Haase, D. (2015). Ecosystem disservices research: A review of the state of the art with a focus on cities. *Ecological Indicators* 52(2015),490–497.

Waldheim, C. (2007). Precedents for a North American landscape urbanism. In D. Almy and M. Benedikt (eds.), *Center 14 – On Landscape Urbanism* (292–303). Center for American Architecture and Design.

Walker, M., Wilcox, B., and Wong, M. (2008). Waterborne zoonoses and changes in hydrologic response due to watershed development. In Fares, A., and ElKadi, A. (eds.), *Coastal Watershed Management* (349–367). WIT Press.

Ward, P. J., Pauw, W. P., van Buuren, M. W., and Marfai, M. A. (2013). Governance of flood risk management in a time of climate change: The cases of Jakarta and Rotterdam. *Environmental Politics* 22, 518–536. doi:10.1080/09644016.2012.683155.

Wardekker, J. A., de Jong, A., Knoop, J. M., and van der Sluijs, J. P. (2010). Operationalising a resilience approach to adapting an urban delta to uncertain climate changes. *Technological Forecasting and Social Change* 77, 987–998.

While, A., and Whitehead, M. (2013). Cities: *Urbanisation and climate change. Urban Studies* 50, 1325–1331. doi:10.1177/0042098013480963

Wilby, R. L., and Perry, G. L. W. (2006). Climate change, biodiversity and the urban environment: A critical review based on London, UK. *Progress in Physical Geography* 30(1), 73–98. doi:10.1191/0309133306pp470ra

Wilkinson, C., Porter, L., and Colding, J. (2010). Metropolitan Planning and Resilience Thinking: A Practitioner's Perspective. *Critical Planning.* 17:2–20.

Wilkinson, C., Sendstad, M., Parnell, S., and Schewenius, M. (2013). Urban governance of biodiversity and ecosystem services. In Elmqvist, T., Fragkias, M., Goodness, J., Güneralp, B., Marcotullio, P. J., McDonald, R. I., Parnell, S., Schewenius, M., et al. (eds.), *Urbanization, Biodiversity and Ecosystem Services – Challenges and Opportunities* (539–587). Springer Netherlands. doi:10.1007/978-94-007-7088-1.

Williams, S. E., Shoo, L. P., Isaac, J. L., Hoffmann, A. A., and Langham, G. (2008). Towards an integrated framework for assessing the vulnerability of species to climate change. *PLoS Biology* 6, 2621–2626.

Wisner, B., Blaikie, P., Cannon, T., and Davis, I. (2003). *At Risk: Natural hazards, people's vulnerability and disasters*, second edition. Routledge Taylor & Francis Group.

Wolch, J. R., Byrne, J., and Newell, J. P. (2014). Urban green space, public health, and environmental justice: The challenge of making cities "just green enough." *Landscape and Urban Planning* 125(1), 234–244.

Woodall, C. W., Nowak, D. J., Likens, G. C., and Westfall, J. A. (2010). Assessing the potential for urban trees to facilitate forest tree migration in the eastern United States. USDA. Accessed November 8, 2015: http://www.nrs.fs.fed.us/pubs/jrnl/2010/nrs_2010_woodall_004.pdf

Wortman, E., and Lovell, S. T. (2013). Environmental challenges threatening the growth of urban agriculture in the United States. *Journal of Environmental Quality.* doi:10.2134/jeq2013.01.0031

Yang, J. (2009). Assessing the impact of climate change on urban tree species selection: A case study in Philadelphia. *Journal of Forestry* 107(7), 364–372.

Yang, Z. (2013). *Eco-Cities: A Planning Guide.* Taylor and Frances Group.

Yu, K., and Padua, M. (2005). *The Art of Survival: Recovering Landscape Architecture.* The Images Publishing Group Pty Ltd.

Yu, K., and Li, D. (2006). The growth pattern of Taizhou City based on ecological infrastructure. In Yu, K., and Padua, M., *The Art of Survival (64–80).* The Images Publishing Group.

Zandersen, M., Jensen, A., Termansen, M., Buchholtz, G., Munter, B., Bruun, H. G., and Andersen, A. H. (2014). Ecosystem based approaches to Climate Adaptation – Urban Prospects and Barriers. Aarhus University, DCE – Danish Centre for Environment and Energy, 94 Scientific Report from DCE – Danish Centre for Environment and Energy No. 83. Accessed December 7, 2015: http://dce2.au.dk/pub/SR83.pdf

Zavaleta, E. S., Shaw, M. R., Chiariello, N. R., Thomas, B. D., Cleland, E. E., Field, C. B., and Mooney, H. A. (2003) Grassland responses to three years of elevated temperature, CO2, precipitation, and N deposition. *Ecological Monographs* 73, 585–604.

Zezza, A., and Tasciotti, L. (2010). Urban agriculture, poverty and food security: Empirical evidence from a sample of developing countries. *Food Policy* 35, 265–273.

Ziervogel, G., New, M., van Garderen, E.A., Midgley, G., Taylor, A., Hamann, R., Stuart-Hill, S., Myer, J., and Warburton, M. 2014. Climate change impacts and adaptation in South Africa. *WIREs Clim Change* 5:605–620.

Chapter 8   Case Study References

Case Study 8.1   Coastal Natural Protected Areas in Mediterranean Spain: The Ebro Delta and Emporda Wetlands

Agencia Catalana de l'Aigua(ACA). (2014) Accessed April 17, 2015: http://acaweb.gencat.cat/aca/appmanager/aca/aca?_nfpb=true&_pageLabel=P5120015889142382820401

Barnolas, M., and Llasat, M. C. (2007). A flood geodatabase and its climatological applications: The case of Catalonia for the last century. *Natural Hazards and Earth System Sciences* 7, 271–281.

BirdLife International. (2014a). Important bird areas factsheet: Ebro Delta. Accessed November 8, 2015: http://www.birdlife.org/datazone/sitefactsheet.php?id=1811

BirdLife International. (2014b). Important bird areas factsheet: Ampurdán marshes. Accessed December 7, 2015: http://www.birdlife.org/datazone/sitefactsheet.php?id=1808

Bormann, H., Ahlhorn, F., and Klenke, T. (2012). Adaptation of water management to regional climate change in a coastal region – Hydrological change vs. community perception and strategies. *Journal of Hydrology* 454–455, 64–75.

Candela, L., Fabregat, S., Josa, A., Suriol, J., Vigués, N., and Mas, J. (2007). Assessment of soil and groundwater impacts by treated urban wastewater reuse. A case study: Application in a golf course (Girona, Spain). *Science of the Total Environment* 374, 26–35.

CIIRC. (2010). *Estat de la zona costanera a Catalunya.* CIIRC.

Cote, M., Hurley, B., and Pratt, J. (2014). National Adaptation Planning: A Compendium of Stakeholder Workshops for USAID. Accessed June 26, 2015: http://pdf.usaid.gov/pdf_docs/PA00K2TB.pdf

CREAF. (2013). Mapa de Coberturas del Suelo de Catalunya. Accessed September 30, 2014: http://www.creaf.uab.es/mcsc/esp/index.htm

Day, J. W., Maltby, E., and Ibañez, C. (2006). River basin management and delta sustainability: A commentary on the Ebro delta and the Spanish national hydrological plan. *Ecological Engineering* **26**, 85–99.

Eriksen, S., Aldunce, P., Bahinipati, C. S., D'Almeida Martins, R., Molefe, J. I., Nhemachena, C., O'Brien, K., Olorunfemi, F., Park, J., Sygna, L., and Ulsrud, K. (2011). When not every response to climate change is a good one: Identifying principles for sustainable adaptation. *Climate and Development* **3**, 7–20.

Fatorić, S. (2010). *Vulnerability to the Effects of Climate Change and Adaptation: The Case of Ebro Delta*. Master's thesis, Autonomous University of Barcelona.

Fatorić, S. (2014). Vulnerability and Adaptation to Climate Change in the Mediterranean region: Climate out of balance in Aiguamolls de l'Empordà? Doctoral thesis, Autonomous University of Barcelona.

Fatorić, S., and Chelleri, L. (2012). Vulnerability to the effects of climate change and adaptation: The case of Spanish Ebro Delta. *Ocean and Coastal Management* **60**, 1–10.

Fatorić, S., and Morén-Alegret, R. (2013). Integrating local knowledge and perception for assessing vulnerability to climate change in economically dynamic coastal areas: The case of natural protected area Aiguamolls de l'Empordà, Spain. *Ocean and Coastal Management* **85**(A), 90–102.

Fatorić, S., Morén-Alegret, R., and Kasimis, C. (2014). Exploring climate change effects in Euro-Mediterranean protected coastal wetlands: The cases of Aiguamolls de l'Empordà, Spain and Kotychi-Strofylia, Greece. *International Journal of Sustainable Development & World Ecology* **21**(4), 346–360.

Guillén, J., and Palanques, A. (1992). Sediment dynamics and hydrodynamics in the lower course of a river regulated by dams: The Ebro river. *Sedimentology* **39**, 567–579.

IDESCAT. (2015). The municipality in figures. Accessed February 23, 2016: http://www.idescat.cat/emex/?lang=en

Jiménez, J. A., Sánchez-Arcilla, A., Valdemoro, H. I., Gracia, V., and Nieto, F. (1997). Processes reshaping the Ebro delta. *Marine Geology* **144**, 59–79.

Martín-Vide, J. P., Rodríguez-Máñez, E., Ferrer-Boix, C., Núñez-González, F., and Maruny-Vilalta, D. (2012). Estudio de la dinámica morfológica del río Fluvià. *Alcances y métodos frente a la escasez de datos. Tecnología y Ciencias del Agua* **III**(3), 115–133.

Meteo Estartit. (2015). Accessed March 18, 2016: http://www.historique-meteo.net/europe/espagne/estartit/2015/09/

Mozumder, P., Flugman, E., and Randhir, T. (2011). Adaptation behaviour in the face of global climate change: Survey responses from experts and decision makers serving the Florida Keys. *Ocean and Coastal Management* **54**(1), 37–44.

Nelson, D. R., Adger, N., and Brown, K. (2007). Adaptation to environmental change: Contributions of a resilience framework. *The Annual Review of Environment and Resources* **32**, 395–419.

Ninyerola, M., Pons, X., and Roure, J. M. (2004). Digital climatic atlas of Catalonia. Accessed July 24, 2014: http://www.opengis.uab.cat/acdc/english/en_cartografia.htm

Observatora de l'Ebre. (2014). Accessed December 7, 2015: http://www.obserbre.es/en

Peel, M. C., Finlayson, B. L., and McMahon, T. A. (2007). Updated world map of the Köppen-Geiger climate classification. *Hydrology and Earth System Sciences Discussions* **4**(2), 462.

Sánchez-Arcilla, A., Jiménez, J. A., Valdemoro, H. I., and Gracia, V. (2008). Implications of climatic change on Spanish Mediterranean low-lying coasts: The Ebro delta case. *Journal of Coastal Research* **24**(2), 306–316.

SMC Coastal Modeling System. (2015). Accessed February 10, 2016: http://smc.ihcantabria.es/SMC25/en

Smit, B., and Wandel, J. (2006). Adaptation, adaptive capacity and vulnerability. *Global Environmental Change* **16**, 282–292.

World Bank. (2017). 2016 GNI per capita, Atlas method (current US$). Accessed August 9, 2017: http://data.worldbank.org/indicator/NY.GNP.PCAP.CD

Wetlands International. (1992). Delta del Ebro. Accessed March 11, 2015: http://sites.wetlands.org/reports/ris/3ES019en_FORMER_

Wetlands International. (1995). Aiguamolls de l'Empordà. Accessed March 19, 2015: http://sites.wetlands.org/reports/ris/3ES018en_FORMER_(1995).pdf

Case Study 8.2 New York's Staten Island Bluebelt

Agencia Catalana de l'Aigua (ACA). (2014) Accessed April 17, 2015: http://acaweb.gencat.cat/aca/appmanager/aca/aca?_nfpb=true&pageLabel=P5120015889142382804014

Brauman, R. J., Gumb, D. F., and Duerkes, C. (2009). Designing for wildlife in the Blue Belt. *Clearwaters Winter* 2009, 41–43.

Peel, M. C., Finlayson, B. L., and McMahon, T. A. (2007). Updated world map of the Köppen-Geiger climate classification. *Hydrology and Earth System Sciences Discussions* **4**(2), 462.

Mayor's Office, New York City. (2012). PlaNYC: A Stronger more resilient New York. Accessed August 16, 2015: http://www.nyc.gov/html/planyc2030/html/home/home.shtml

New York City Department of Design and Construction, and the Design Trust for Public Space. (2005). High performance infrastructure guidelines. Accessed November 29, 2015: http://www.nyc.gov/html/ddc/downloads/pdf/hpig.pdf

New York City Panel on Climate Change (NPCC). (2013). *New York City Panel on Climate Change Climate Risk Information 2013; Observations.* Climate change projections, and maps. City of New York.

Ryan, G. C. (2006). *The return of the natives. The Urban Audubon* XXV11(4),1,4.

U.S. Census Bureau. (2010). Decennial census, summary file 1. Accessed November 17, 2014: http://www.census.gov/population/metro/files/CBSA%20Report%20Chapter%203%20Data.xls

U.S Census Bureau. (2016). Table 1. Annual Estimates of the Resident Population: April 1, 2010 to July 1, 2016 – Metropolitan Statistical Area; – 2016 Population Estimates. U.S. Census Bureau. Retrieved March 28, 2017. Accessed August 9, 2017: https://factfinder.census.gov/faces/tableservices/jsf/pages/productview.xhtml?pid=PEP_2016_PEPANNRES&prodType=table

World Bank. (2017). 2016 GNI per capita, Atlas method (current US$). Accessed August 9, 2017: http://data.worldbank.org/indicator/NY.GNP.PCAP.CD

Case Study 8.3 The Serra do Mar Project, Baixada Santista Metropolitan Region (BSMR), São Paulo State

CDHU. (2012). Novos Bairros. Accessed March 9, 2014: http://www.agem.sp.gov.br/pdf/projeto_serra_do_mar_novos_bairros.pdf

IAD, São Paulo. (2009). *Programa de Recuperação Socioambiental da Serra do Mar e do Sistema de Mosaicos da Mata Atlântica (Br-L1241). Estratégia Ambiental e Social do Programa. 05 de Outubro de (2009). Banco Interamericano de Desarrollo, Governo do Estado de São Paulo Brasil.* Accessed July 1, 2015: fflorestal.sp.gov.br/files/2012/02/easabidaserraadoamar.doc

IBGE. (2015). *Instituto Brasileiro de Geografia e Estatística.* Accessed August 5, 2014: http://www.ibge.gov.br/home/estatistica/populacao/censo2010/sinopse/sinopse_tab_rm_zip.shtm

IDB and São Paulo State Government. (2013). *Serra do Mar and the Atlantic Forest Mosaics System – A Social and Environmental Recovery Project. 1st Ed. Inter-American Development Bank (IDB), São Paulo State Government. São Paulo.* Accessed May 8, 2014: http://www.ambiente.sp.gov.br/serradomar/files/2014/03/INGL%C3%8AS_CARTILHA_Serra-do-Mar_Internet.pdf

Peel, M. C., Finlayson, B. L., and McMahon, T. A. (2007). Updated world map of the Köppen-Geiger climate classification. *Hydrology and Earth System Sciences Discussions* **4**(2), 462.

São Paulo. (2013). Programa Serra do Mar e Mosaicos da Mata Atlântica (in Portuguese). Accessed May 5, 2014: http://www.agem.sp.gov.br/pdf/projeto_serra_do_mar_apresentacao.pdf

World Bank. (2017). 2016 GNI per capita, Atlas method (current US$). Accessed August 9, 2017: http://data.worldbank.org/indicator/NY.GNP.PCAP.CD

Case Study 8.4 Ecosystem-Based Climate Change Adaptation in the City of Cape Town

Cilliers, S. S., and Siebert, S. J. (2012). Urban ecology in Cape Town: South African comparisons and reflections. *Ecology and Society* **17**(3), 33. Accessed November 5, 2014: http://dx.doi.org/10.5751/ES-05146–170333

City of Cape Town. (2015). City statistics. Accessed March 28, 2016: https://www.capetown.gov.za/en/stats/Pages/CityStatistic.aspx

Laros, M. (2012). Cape Town, South Africa: An urban biodiversity network. ICLEI Case Study Series (part of the Local Sustainability). Kelleher, S., and Simpson, R., eds., ICLEI World Secretariat. Accessed August 15, 2014: http://www.iclei.org/fileadmin/PUBLICATIONS/Case_Studies/ICLEI_cs_138_Cape_Town.pdf

Peel, M. C., Finlayson, B. L., and McMahon, T. A. (2007). Updated world map of the Köppen-Geiger climate classification. *Hydrology and Earth System Sciences Discussions* **4**(2), 462.

World Bank. (2017). 2016 GNI per capita, Atlas method (current US$). Accessed August 9, 2017: http://data.worldbank.org/indicator/NY.GNP.PCAP.CD

Case Study 8.5 Jerusalem Gazelle Valley Park Conservation Program

CIA World Factbook. (2015). Major Urban Areas – Population. Central Intelligence Agency. Accessed: August 4, 2017. https://www.cia.gov/library/publications/the-world-factbook/fields/2219.html

Krasny, M. E. (2015). *Civic Ecology: Adaptation and Transformation from the Ground Up (Urban and Industrial Environments)*. MIT Press.

Peel, M. C., Finlayson, B. L., and McMahon, T. A. (2007). Updated world map of the Köppen-Geiger climate classification. *Hydrology and Earth System Sciences Discussions* **4**(2), 462.

World Bank. (2017). 2016 GNI per capita, Atlas method (current US$). Accessed August 9, 2017: http://data.worldbank.org/indicator/NY.GNP.PCAP.CD

Case Study 8.6 Medellín City: Transforming for Life

Eveland, Jennifer. (2012). Medellín Transformed: From Murder Capital to Model City. Urban Redevelopment Authority. Accesed September 23, 2015: http://www.Leekuanyewworldcityprize.com.sg/Features_Medellin_Transformed.htm

Green City Index (GCI). (2010). Latin American Green City Index (2010). *Assessing the Environmental Performance of Latin America's Major Cities*. Siemens AG Corporate Communications and Government Affairs, Munich.

Case Study 8.7 Singapore's Ecosystem-Based Adaptation

Moreno, E. L. et al. (2013). *State of the World's Cities 2012/2013*: Prosperity of Cities. UN-Habitat, New York.

Peel, M. C., Finlayson, B. L., and McMahon, T. A. (2007). Updated world map of the Köppen-Geiger climate classification. *Hydrology and Earth System Sciences Discussions* **4**(2), 462.

World Bank. (2017). 2016 GNI per capita, Atlas method (current US$). Accessed August 9, 2017: http://data.worldbank.org/indicator/NY.GNP.PCAP.CD

Department of Statistics Singapore. (2015). Singapore in Figures 2015. Accessed January 18, 2016: http://www.singstat.gov.sg/docs/default-source/default-document-library/publications/publications_and_papers/reference/sif(2015).pdf

Department of Statistics Singapore. (2016) Yearbook of Statistics Singapore 2016. Accessed February 19, 2017: http://www.singstat.gov.sg/docs/default-source/default-document-library/publications/publications_and_papers/reference/yearbook_2016/yos2016.pdf

Peel, M. C., Finlayson, B. L., and McMahon, T. A. (2007). Updated world map of the Köppen-Geiger climate classification. *Hydrology and Earth System Sciences Discussions* **4**(2), 462.

MEWR. (2014). Ministry of the Environment and Water Resources, The Singapore Green Plan 2012. Accessed July 9, 2015: http://www.uncsd(2012).org/content/documents/The%20Singapore%20Green%20Plan%20(2012).pdf

National Climate Change Secretariat Prime Minister's Office Republic of Singapore (NCCS). (2012). Climate change & Singapore: Challenges, opportunities, partnerships. Accessed July 24, 2014: https://www.nccs.gov.sg/sites/nccs/files/NCCS-(2012).pdf

NEA. (2013). Climate Change, 2 September 2013. Accessed December 3, 2014: http://www.nea.gov.sg/energy-waste/climate-change

World Bank. (2017). 2016 GNI per capita, Atlas method (current US$). Accessed August 9, 2017: http://data.worldbank.org/indicator/NY.GNP.PCAP.CD

Case Study 8.8 Seattle's Thornton Creek Water Quality Channel

Kaid Benfield's Blog. (2011). Outstanding urbanism, transit, and state-of-the-art green infrastructure, beautifully mixed In: SWITCHBOARD, Natural Resources Defense Council (NRDC) staff blog. Accessed October 2, 2014: http://switchboard.nrdc.org/blogs/kbenfield/outstanding_urbanism_and_state.html

Peel, M. C., Finlayson, B. L., and McMahon, T. A. (2007). Updated world map of the Köppen-Geiger climate classification. *Hydrology and Earth System Sciences Discussions* **4**(2), 462.

Seattle Public Utilities. (2015). Thornton Creek water quality channel. Accessed March 14, 2016: http://www.seattle.gov/util/MyServices/DrainageSewer/Projects/ThorntonCreekWaterQualityChannel/index.html

U.S. Census Bureau. (2010). Decennial census, summary file 1. Accessed September 23, 2014: http://www.census.gov/population/metro/files/CBSA%20Report%20Chapter%203%20Data.xls

World Bank. (2017). 2016 GNI per capita, Atlas method (current US$). Accessed August 9, 2017: http://data.worldbank.org/indicator/NY.GNP.PCAP.CD

9

Urban Areas in Coastal Zones

Coordinating Lead Authors

Richard J. Dawson (Newcastle upon Tyne), M. Shah Alam Khan (Dhaka), Vivien Gornitz (New York)

Lead Authors

Maria Fernanda Lemos (Rio de Janeiro), Larry Atkinson (Norfolk, VA), Julie Pullen (Hoboken), Juan Camilo Osorio (Cambridge, MA/New York)

Contributing Authors

Lindsay Usher (Norfolk, VA)

This chapter should be cited as

Dawson, R.J., Khan, M.S.A., Gornitz, V., Lemos, M. F., Atkinson, L., Pullen, J., and Osorio, J. C. (2018). Urban Areas in Coastal Zones. In Rosenzweig, C., W. Solecki, P. Romero-Lankao, S. Mehrotra, S. Dhakal, and S. Ali Ibrahim (eds.), *Climate Change and Cities: Second Assessment Report of the Urban Climate Change Research Network*. Cambridge University Press. New York. 319–362

Cities on the Coast: Sea-Level Rise, Storms, and Flooding

Coastal cities have been subjected to extreme weather events since the onset of urbanization. Climatic change, in particular sea level rise, coupled with rapid urban development are amplifying the challenge of managing risks to coastal cities. Moreover, urban expansion and changes and intensification in land use further pressure sensitive coastal environments through pollution and habitat loss.

The concentration of people, infrastructure, economic activity, and ecology within the coastal zone merits specific consideration of hazards exacerbated by a changing climate. Major coastal cities often locate valuable assets along the waterfront or within the 100-year flood zone, including port facilities, transport and utilities infrastructure, schools, hospitals, and other long-lived structures. These assets are potentially at risk for both short-term flooding and permanent inundation.

Major Findings

- Coastal cities are already exposed to storm surges, erosion, and saltwater intrusion. Climate change and sea level rise will likely exacerbate these hazards.

- Around 1.4 billion people could live in the coastal zone, worldwide, by 2060. The population within the 100-year floodplain at risk to a 10–21 cm sea level rise could increase from around 286 million to 411 million people between 2030 and 2060. Three quarters of the exposed populations live in south and southeast Asia.

- Expansion of coastal cities is expected to continue over the 21st century. Although costs of coastal protection could reach US\$12–71 billion by 2100, these expenses would be substantially less than taking no action.

- Climate-induced changes will affect marine ecosystems, aquifers used for urban water supplies, the built environment, transportation, and economic activities, particularly following extreme storm events. Critical infrastructure and precariously built housing in flood zones are vulnerable.

- Increasing shoreline protection can be accomplished by either building defensive structures or by adopting more natural solutions, such as preserving and restoring wetlands or building dunes. Modifying structures and lifestyles to "live with water" and maintain higher resiliency are key adaptive measures.

Key Messages

Coastal cities must become keenly aware of the rates of local and global sea level rise and future sea level rise projections, as well as emerging science that might indicate more rapid (or potentially slower) rates of sea level rise.

An adaptive approach to coastal management will maintain flexibility to accommodate changing conditions over time. This involves implementing adaptation measures with co-benefits for the built environment, ecosystems, and human systems. An adaptive strategy requires monitoring changing conditions and refining measures as more up-to-date information becomes available.

Simple, less costly measures can be implemented in the short term while assessing future projects. Land-use planning for sustainable infrastructure development in low-lying coastal areas should be an important priority. Furthermore, cities need to consider transformative adaptation, such as large-scale relocation of people and infrastructure with accompanying restoration of coastal ecosystems.

Delivering integrated and adaptive responses will require robust coordination and cooperation on coastal management issues. This must be fostered among all levels of local, regional, and national governing agencies and include engagement with other stakeholders.

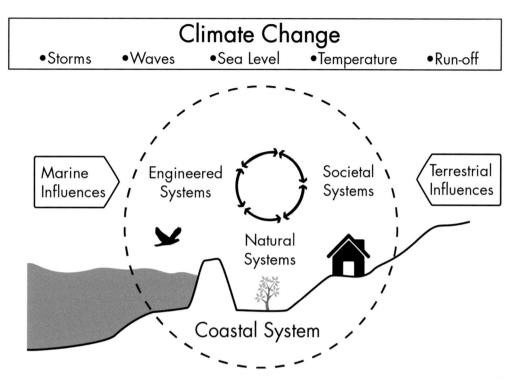

Figure 9.1 *Climate change and the coastal system.*

Source: Adapted from Nicholls et al., 2007

9.1 Introduction

Coastal zones around the world represent a wide diversity of features and a high variability in their ecosystems and environment, resources, and socioeconomic activities (see Figure 9.1). Moreover, these zones are extremely dynamic because of the continually changing balance of energy conveyed by tides, currents and waves, river runoff, sediment deposition, and erosion. Other shoreline changes have been induced by artificial structures intended to prevent or reduce erosion. This poses particular challenges for coastal cities, which have experienced some of the worst losses during extreme climatic events (Hall et al., 2005; Jongman et al., 2012; Hallegatte et al., 2013), while urban expansion, intensification of and changes in land use place growing pressure on sensitive coastal environments through pollution and land loss (Arthurton and Korateng, 2006). There is limited understanding of the interaction between urban and coastal systems (Pelling and Blackburn, 2014), yet it is significant from a climate perspective because of the concentration of people, infrastructure, and ecology within the coastal zone. Coastal cities therefore merit specific consideration of how they will be affected by and respond to hazards exacerbated by a changing climate.

The low-elevation coastal zone (LECZ), the area often considered to be "coastal" and the definition used here, is described by McGranahan et al. (2007) as the area below 10 meters in elevation that is hydrologically connected to the sea. The LECZ is home to approximately 10% of the world's total population and 13% of its urban population (McGranahan et al., 2007). Small and Nicholls (2003) estimated that in 1990 about 23% of the

world's population and 5 million square kilometers of land area were in the LECZ. Proximity to the coast is a driver of urbanization, with coastal cities thought to provide more open access to trade, investment, resources, and tourism than those in the hinterland (Henderson and Wang, 2007). Many of the world's largest, most populous cities are coastal: of the 26 megacities in the world in 2011, 16 were coastal and vital to shipping, fisheries, and international commerce (UN-DESA, 2012). Coastal populations have rapidly increased over the past 100 years, particularly in low- and middle-income economies (UN-Habitat, 2008).

Coastal settlements are uniquely exposed to climate hazards such as sea level rise, storm surges, shoreline erosion, and saltwater intrusion. Major coastal cities often locate valuable assets along the waterfront or within the 100-year flood zone, including port facilities, transport and utilities infrastructure, schools, hospitals, and other long-lived structures, potentially at risk to both short-term and permanent flooding. Analysis of the 136 world's largest port cities (population exceeding 1 million in 2005) shows that the value of assets at risk exceeded US$3.0 trillion or 5% of Gross World Product in 2005 (Hanson et al., 2011). Globally, expansion of coastal cities is expected to continue over the 21st century (Nicholls, 2004; O'Neill et al., 2014), with some analyses suggesting that more than half the global population could live in cities in the coastal zone by the middle of the 21st century (Aerts et al., 2013) and annual coastal flood losses could reach US$71 billion by 2100 (Hinkel et al., 2014).

Urban centers built on low-lying deltas are especially vulnerable. Major deltaic cities include Khulna, Shanghai, Guangzhou, Ho Chi Minh City, Bangkok, and Rotterdam. Many are affected by anomalously high rates of relative sea level rise often exacerbated by land subsidence caused by groundwater overdraft, sediment compaction, long-term geologic subsidence, enlarging of coastal inlets, dredging of ports and waterways, and upstream trapping of sediments in reservoirs (Syvitski et al., 2009). Highly fertile deltaic soils of places like the Chao Phraya delta, near Bangkok, Thailand (with a relative sea level rise of +13 to 150 mm/yr; Syvitski et al., 2009) or the Yangtze delta, near Shanghai, China (+3 to 28 mm/yr; Syvitski et al., 2009) can be ideal locations for growing essential food crops such as rice (Akinro et al., 2008; Malm and Esmailian, 2013). Crops are threatened not only by increased frequency of coastal flooding, but by increasing saltwater intrusion and soil salinization as saltwater encroaches further up estuaries and rivers and contaminates water (Bear, 1999; Mazi et al., 2013; Le Dang et al., 2014).

Moreover, urban coastal ecosystems perform multiple beneficial ecological services (see Chapter 8, Urban Ecosystems). Coastal wetlands, for example, help dampen wave action and protect against storm surges. They also provide important habitat for birds, fish, and other nearshore marine life, and they offer recreational opportunities. However, many urban salt marshes or mangroves have been deteriorating for decades due to land clearing, draining, water pollution, and coastal protection measures (Torio et al., 2013; Nitto et al., 2014).

Coastal cities have lived with weather extremes for centuries, but climatic change and rapid urban development are amplifying the challenge of managing coastal risks. This chapter reviews the main climatic hazards, key vulnerabilities, and adaptation options, focusing on issues most relevant to coastal cities. The few activities that reduce greenhouse gas (GHG) emission specific to coastal cities are briefly considered before exploring a number of cross-cutting issues relevant to policy-makers.

9.2 Climate Risks

Climate risks in coastal cities are shaped by geological, oceanographic, and environmental factors as well as by socioeconomic factors (see Figure 9.1). Rising sea levels will lead to more frequent coastal flooding even without other changes in storm behavior or further urban development (e.g., Tebaldi et al., 2012) (see Chapter 2, Urban Climate Science). This section reviews coastal hazards that are influenced by climate drivers and how the nature of the built, social, and environmental characteristics of a city and its environs can shape its vulnerability to these hazards.

9.2.1 Coastal Hazards and Climate Drivers

Findings from the Intergovernmental Panel on Climate Change (IPCC)'s Fifth Assessment Report (AR5) can be summarized as (Wong et al., 2014):
- Coastal systems are particularly sensitive to three key drivers related to climate change: sea level, ocean temperature and ocean acidity.

- Coastal systems and low-lying areas will increasingly experience adverse impacts such as submergence, coastal floodings and coastal erosion due to relative sea level rise.
- Acidification and warming of coastal waters will continue with significant negative consequences for coastal ecosystems.
- The population and assets exposed to coastal risks as well as human pressures on coastal ecosystems will increase significantly in the coming decades due to population growth, economic development, and urbanization.

9.2.1.1 Sea Level Change

Global sea level rose 1.7+/−0.2 mm/yr between 1900 and 2010 (Wong et al., 2014; IPCC, 2013). Since 1993, both satellites and tide gauges register a sea level rise of ~3.4±0.4 mm/yr (Church et al., 2013; Nerem et al., 2010; also http://www.sealevel.colorado.edu) High-resolution proxy and modern sea level data suggest that, relative to the last few millennia, the rate of sea level rise has accelerated since the late 19th century (Kemp et al., 2011; Engelhart and Horton, 2012; Gehrels and Woodworth, 2013), and there are indications of a more rapid acceleration in the past few decades (Masters et al., 2012; IPCC, 2013).

Local sea level changes may differ considerably from the global mean due to vertical land motions (e.g., neotectonics, glacial isostatic adjustments, subsurface fluid withdrawal [water, oil, gas], ocean dynamic processes, and gravitational changes resulting from recent ice mass loss and terrestrial water storage). Observed relative sea level changes range from an extreme drop of −17.59 mm/yr (1944–2014) at Skagway, Alaska (glacial rebound, including that from recent [19th–20th centuries] melting of nearby glaciers at Glacier Bay), and rise of 9.03 mm/yr (1947–2014) at Grand Isle, Louisiana (oil and gas extraction; sediment compaction; NOAA Tides and Currents, 2016). Anomalously high rates of sea level rise in cities like Manila; Norfolk, Virginia; and Bangkok are probably caused by excess groundwater extraction (Raucoules et al., 2013; Eggleston and Pope, 2013).

Changes in ocean water density affect different regions to varying degrees. For example, a weaker Atlantic Meridional Ocean Circulation (AMOC) due to freshening of the North Atlantic Ocean from increased ice loss from the Greenland Ice Sheet (and glaciers) could chill northwestern Europe, at the same time increasing thermal expansion of sea level along the northeast coast of North America and raising regional sea level there by an additional 0.4 to 0.55 meters by 2100 (Hu et al., 2011).

Changing ocean dynamics including Gulf Stream slowing are being linked to anomalous sea level rise off the East Coast of the United States (Yin and Goddard, 2013; Ezer, 2013), emphasizing that local and global sea level changes will have a significant impact on coastal cities. In the future, the highly populated coastal northeastern United States (including major cities such as Boston, New York, Baltimore, and Washington, D.C.) is expected to become a "hotspot" of enhanced regional sea level rise largely owing to these ocean circulation changes (Yin et al., 2010; Carson et al., 2016).

9.2.1.2 Coastal Storms: Storm Surge, Waves, and Winds

Current and future flooding and storm damage frequency depend on various factors including changes in sea level, surge, and storm intensity, duration, and wave height. These also significantly affect coastal erosion. The frequency of tropical cyclones, their intensity, and the number of countries impacted has remained stable since the 1970s (Peduzzi et al., 2012), although the intensity of the strongest North Atlantic cyclones has increased (Elsner et al., 2008). Whether the total number of tropical cyclones will increase as a result of climate change is uncertain, but the number of more intense cyclones may grow (Bender et al., 2010; Knutson et al., 2010). Within the past three decades, maximum intensities of tropical cyclones have shifted away from the equator (Kossin et al., 2014). A continuation of this trend would increase flood hazards for nontropical coastal communities, which hitherto were less exposed to such damaging storms.

Evidence of changes in mean and extreme winds is weak, although analysis suggests that wave heights have been increasing in the Northeast Atlantic between 1958 and 2002 and in the Southern Ocean between 1985 and 2008 (Wong et al., 2014). Increases in coastal populations and economic development along with rising tropical cyclone intensities and sea level will magnify cyclone risks and damages (Mendelsohn et al., 2012; Hallegatte et al., 2013; Estrada et al., 2015).

Extratropical cyclones show no consistent large-scale changes in behavior during the past half-century, and, projected changes in extra-tropical storm activity remain uncertain. Rising sea level has led to increased coastal flooding, as for example in New York and elsewhere along the U.S. East Coast (Grinsted et al., 2012; Talke et al., 2014). Heavy precipitation from extratropical cyclones adds to the impacts of coastal flooding because of the large synoptic scale of the storms and their duration over several tidal cycles, often leading to overflow of stormwater drainage systems.

Case Study 9.1   Norfolk, Virginia: A City Dealing with Increased Flooding

Larry Atkinson and Tal Ezer

Old Dominion University, Norfolk, VA

Keywords	Sea level rise, flooding, adaptation
Population (Metropolitan Region)	246,393 (U.S. Census Bureau, 2015)
Area (Metropolitan Region)	140.2 km² (U.S. Census Bureau, 2010)
Income per capita	US$56,180 (World Bank, 2017)
Climate zone	Cfa – Temperate, without dry season, hot summer (Peel et al., 2007)

Norfolk, Virginia was settled more than 400 years ago by European immigrants. The deep protected harbor and many creeks provided a perfect location to establish trade routes and access to the interior. The city has a population of approximately 250,000 in a larger region (Hampton Roads) of about 1.5 million people in many cities including the largest city in the state: Virginia Beach. Norfolk is home to the largest navy base in the world and many other defense-related activities. Nearly one-half the region's economy is defense related. Those same harbors and creeks abut very low land with highest elevations of less than 10 meters. Over time, with filling of waterfront areas and small creeks combined with sea level rise, the city is experiencing increased flooding (Ezer and Atkinson, 2014; Sweet and Parks, 2014). This has become especially noticeable since the 1970s, as Case Study 9.1 Figure 1 (third panel) shows.

The high local sea level rise rate in Norfolk (~4.5 mm/y over the past 80 years and ~6 mm/y over past 10 years) is two to three times faster than the global rate and is accelerating (Boon, 2012; Ezer and Corlett, 2012; Ezer, 2013). This is the result of three main factors: (1) global sea level rise due to thermal expansion and land ice

melt, (2) local land subsidence due to post-glacial rebound and local underground water extraction (Boon et al., 2010), and (3) potential climate change-related slowdown of the Gulf Stream (Ezer et al., 2013). Norfolk is situated in the mid-Atlantic region that has been declared a "hotspot for accelerated sea level" (Sallenger et al., 2012) and a "hotspot for accelerated flooding" (Ezer and Atkinson, 2014). The alleged slow-down of the Gulf Stream may be caused by inter-decadal variability (e.g., Watson et al., 2016; Yin and Goddard, 2013). But is still considered likely if future ice melting of Greenland continues or intensifies.

Norfolk faces two types of coastal flooding threats. One is often called minor or nuisance flooding (Sweet and Park, 2014). This occurs during high tides without any local storm (some offshore wind or a weakening Gulf Stream can cause this to happen), when many streets near the water start to flood (see Case Study 9.1 Figure 2). The hours that this occurs have already increased in the past due to sea level rise (see Case Study 9.1 Figure 1, third panel). This is expected to dramatically increase in the future (see Case Study 9.1 Figure 3), such that certain streets will be under water and not usable for extensive parts of the year.

The second is significant (major) flooding that is related to the effect of tropical and extra-tropical storms, winter storms, and Nor'easters. While the impact of climate change on the frequency and intensity of tropical storms and hurricanes is complex, the frequency of and damage from major storm surges has already increased in recent years because of sea level rise (see Case Study 9.1 Figure 1, first panel). Storms that in the past caused only minor or moderate flooding are now causing major floods and more damage to property because of the additional sea level rise on top of storm surge.

ADAPTATION TO RISING SEAS AND STORM SURGES

In 2011, it became clear that Norfolk was experiencing more flooding from both nuisance flooding and from passing hurricanes and winter storms. The Mayor of Norfolk at the time recognized this and directed the city to formally plan for increased flooding.

Norfolk is using a three-pronged approach (Norfolk, 2013):

Prepare – Nature can be unpredictable. Norfolk's residents and its government must address that unpredictability with thoughtful preparation. Community preparation resources include community response teams, detailed transportation and evacuation strategies, sound planning practices and use of a National Incident Management System to vet preparedness concepts and principles.

Mitigate (i.e., reduce risks) – Like other coastal cities, Norfolk is vulnerable to the increased severity of storms and flooding caused by relative sea level rise. Immediate and long-term solutions range from simple landscaping techniques that allow adequate storm water drainage to complex engineering projects designed to reroute and deflect water.

Communicate – Communication is a critical link in implementing Norfolk's flooding strategy. To prepare for future events as well as to cope during an event, direct and timely communication between the government and citizens is fundamental. Norfolk actively seeks input from residents. The City uses a wide range of communication tools to achieve these goals.

In 2014, Norfolk was designated one of 100 Resilient Cities by the Rockefeller Foundation (Norfolk, 2014).

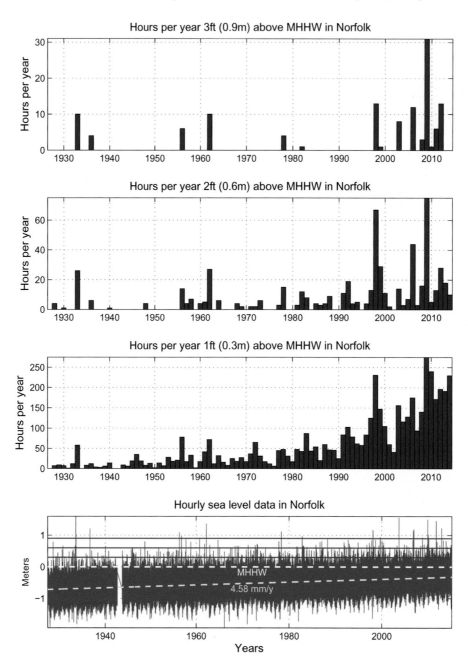

Case Study 9.1 Figure 1 *Hours per year that street flooding occurs in parts of downtown Norfolk, Virginia for (from top to bottom) major floods, moderate floods, and minor floods. The hourly water level data are also shown. MHHW = mean higher high water.*

Case Study 9.1 Figure 2 *Example of minor street flooding that is occurring more often in Norfolk, Virginia because of sea level rise.*

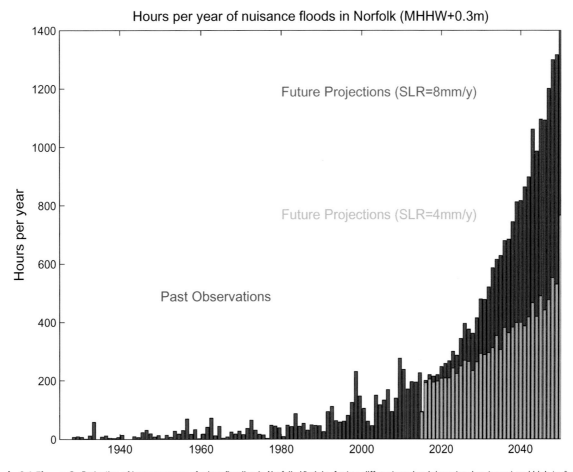

Case Study 9.1 Figure 3 *Projection of hours per year of minor flooding in Norfolk, Virginia, for two different sea level rise rates: low (green) and high (red), relative to past flooding (blue). MHHW = mean higher high water.*

CREATION OF INTERGOVERNMENTAL PILOT PROJECT

As the flooding threat was increasingly recognized, a plan was developed to bring together the many federal agencies (primarily the Department of Defense) with the local cities to work together toward an adaptation strategy (Steinhilber et al., 2015):

Initiated in June 2014, the Hampton Roads Sea Level Rise Preparedness and Resilience Intergovernmental Pilot Project (Intergovernmental Pilot Project or IPP) convened at Old Dominion University (ODU) is an effort to use the knowledge skills and expertise of all regional stakeholders. The goal

is to create a framework or template for intergovernmental strategic planning that can be used outside the region and to implement that integrated strategy in Hampton Roads, Virginia. This is creating an effective and efficient method for planning holistically for sea level rise and recurrent flooding. Old Dominion University acts as the convener of the IPP and supports it with expert faculty, research facilities, and access to partnerships within academia (e.g., the Virginia Institute of Marine Science (VIMS) and William and Mary's Virginia Coastal Policy Center).

This pilot project is planned to be the initial phase of a regional, state, and federally organized approach to coastal resilience.

Meanwhile, a statistically significant trend in the frequency of high surge events has been observed, generally associated with landfalling Atlantic tropical cyclones (Grinsted et al., 2012). Extreme surge events tend to occur in warmer years. Projected 21st-century warming patterns furthermore suggest that the frequency of storm surges similar in magnitude to those associated with Hurricane Katrina could increase by a factor of 2 to 7 for each 1°C rise in temperature (Grinsted et al., 2013). A higher surge generates more potential energy and therefore poses a greater risk to life and property. The surge-related threat will be exacerbated by rising sea level. A recent study that combines the effects of tropical cyclone intensity with a flood index based on duration and excessive high water during hurricane season along the eastern U.S. finds major increases in flood risk (Little et al., 2015). In the conservative RCP2.6 emissions scenario, the flood index increases by 4–25 times by 2080–2099 relative to 1986–2005, soaring upward by a factor of 35–350 in the high RCP8.5 scenario. This study, however, omits non-oceanographic components of sea level change, such as ice mass loss and gravitational or glacial isostatic effects, which may increase local sea levels even further.

The frequency of "nuisance flooding" (less-extreme tidally related coastal flood events) has been increasing in areas of rising sea level, including much of the United States (Sweet and Mara, 2015; Sweet et al., 2014). After removing the sea level rise component caused by anthropogenic-induced temperature increase, Strauss et al. (2016) conclude that two-thirds of U.S. "nuisance flood" days since 1950 can be attributed to climate change. Human-induced sea level rise has increased the number of flood days by more than 80% between 1955–1984 and 1985–2014.

Coastal populations face increasing risks to the combined effects of sea level rise and coastal flooding. The number of people within the 100-year flood plain exposed to flooding for a 10–21 cm rise in global sea level could increase from around 286 million to 411 million between 2030 and 2060 (Neumann et al., 2015). Of these, 75% live in south and southeast Asia, with a substantial growth expected in Africa (e.g., in the Nile and Niger Deltas). Economic losses and adaptation costs will also rise. One study estimates that average flood losses for the 136 largest coastal cities may increase from US$6 billion in 2005 to $52 billion by 2050, due to socioeconomic changes alone (Hallegatte et al., 2013). Adding a 20–40 cm sea level rise plus subsidence

would increase these costs to $60–63 billion by 2050, even with adaptations to present flood probabilities. Another study finds that 0.2–4.6% of the world's population could face annual floods under a sea level rise of 25–123 cm by 2100 (Hinkel et al., 2014). Costs of coastal protection (e.g., dikes) could range between US$12 to 71 billion by 2100, but would be much lower than costs of no protection.

9.2.1.3 Other Drivers of Urban Coastal Risks

Freshwater Inflows

Heavy rainfall accompanying coastal storms can cause severe inland flooding and also compound flood risks associated with high surge levels (Wahl et al., 2015). Precipitation has generally increased over Northern Hemisphere mid-latitude land areas, particularly since the 1950s (IPCC, 2013). In many parts of the world, the frequency, intensity, and/or total precipitation have increased over the past century (Jenkins et al., 2008; Jones et al., 2013; Colle et al., 2015;). A high storm surge in conjunction with heavy precipitation can lead to excess runoff and/or increased river discharge. The number of compound flood events has increased significantly during the past century for many major U.S. cities, exacerbated by long-term sea level rise (Wahl et al., 2015). More frequent, heavy rainfall will also increase flood risk for cities along tidal rivers (e.g., London, Rotterdam, New Orleans, Bangkok, Shanghai, Haiphong), which face flooding from both high surges and overflowing rivers. Moreover, low-lying cities increasingly rely on energy to pump water from their drainage systems that are below high water levels. Conversely, anthropogenic activities such as the construction of upstream reservoirs can reduce the freshwater flows and lead to fluctuations in acidity and salinity and associated impacts (Haque, 2006; Das et al., 2012; Gao et al., 2012).

Salinization/Salt Water Intrusion

An important consequence of sea level rise is saltwater intrusion upstream and into coastal aquifers, potentially jeopardizing urban drinking water supplies and contaminating agricultural soils (see Case Study 9.2). This is a slow process, which can be accelerated by human-induced activities such as excessive extraction from aquifers (Ferguson and Gleeson, 2012).

Case Study 9.2 Vulnerabilities and Adaptive Practices in Khulna, Bangladesh

M. Shah Alam Khan

Bangladesh University of Engineering and Technology, Dhaka

Keywords	Vulnerability, planned and autonomous adaptation, salinity, drainage, coastal
Population (Metropolitan Region)	1,013,000 (UN, 2016)
Area (Metropolitan Region)	72.6 km² (BBS, 2013)
Income per capita	US$1,330 (World Bank, 2017)
Climate zone	Aw – Tropical savannah (Peel et al., 2007)

Khulna, the third-largest metropolitan city in Bangladesh, is situated in the southwestern Gangetic tidal floodplains. The city is one of the five largest river ports in the country and a gateway to the Sundarbans – the world's largest mangrove forest and a Ramsar site. This central urban corridor of the southwest coastal region of Bangladesh is important for its proximity to the second-largest national seaport at Mongla (Murtaza, 2001). Apart from agriculture, shrimp export processing activities are major contributors to the national economy (Aqua-Sheltech Consortium, 2002). Globally, Khulna is one of the 15 most climate-change vulnerable cities (IIED, 2009) and the eighth most vulnerable city in terms of projected flood loss and gross domestic product (GDP) losses (Hallegatte et al., 2013).

VULNERABILITIES

The geographic setting of Khulna makes it sensitive to climate change. The city area is generally flat, with an average ground elevation of 2.5 meters from mean sea level and elevations ranging between 0.45 and 5.4 meters (ADB, 2011). This low elevation causes tidal flooding and drainage congestion during high tides in the adjacent Rupsha and Bhairab Rivers (see Case Study 9.2 Figure 1). Increasing frequency of severe cyclones and storm surges, rising tidal water levels, saline water incursion, changing rainfall patterns, and rising temperature are the principal climate change–induced stresses that have major implications for the citizens' lives and livelihoods (Khan et al., 2013).

The rate of temperature rise in Khulna is higher than that observed or projected elsewhere in the country. The number of extremely cold nights is decreasing and the heat index is increasing. Sunshine duration shows a decreasing trend, and humidity shows an increasing trend. Rainfall is increasing in terms of both magnitude and number of rainy days. However, the annual maximum rainfall and the number of days with high-intensity rainfall have remained almost the same. The monsoon is apparently strengthening toward the end of the rainy season. The annual maximum tidal high water level is increasing, and

Case Study 9.2 Figure 1 *Low-lying areas in Khulna, Bangladesh, vulnerable to tidal flooding.*

the annual minimum low water level is decreasing at a rate of 7–18 millimeters and 4–8 millimeters per year, respectively. Variation in water salinity, tidal water levels, and river discharges during different time periods indicates that human interventions through upstream water diversion and coastal polder construction have also contributed significantly to the hydro-morphological changes in the region (Mondal et al., 2013).

ADAPTIVE RESPONSES

Various adaptive responses, both planned and autonomous, are observed in Khulna. The planned responses at the city level include infrastructure development for drainage, water supply augmentation, and flood management whereas the autonomous responses by communities and households involve improving water and food security and creating alternative livelihood opportunities (Khan et al., 2013). To reduce dependency on saline groundwater, Khulna Water Supply and Sewerage Authority (KWASA) has planned to construct a treatment plant and increase the size of the impounding reservoir

(ADB, 2011). Such initiatives would still exclude marginalized communities from the city supply coverage. Further interventions are needed to improve institutional capacities (finance, manpower, skill, treatment, cost recovery, etc.) and community capacities (rainwater harvesting, wastage reduction awareness, etc.) to cope with the vulnerabilities. A drainage master plan, yet to be implemented by Khulna City Corporation (KCC), aims to reduce waterlogging in the Khulna Aqua-Sheltech Consortium (2002).

The vulnerable communities usually devise a range of autonomous adaptive practices to reduce their vulnerabilities. Various forms of rainwater harvesting are seen that provide drinking water for domestic uses. This option, however, is feasible for only a few months. Construction and maintenance of most of the systems require special skill and support. Communities often install hand-dug tube wells for drinking water supplies, whereas pond water is used for other domestic uses. New forms of collective actions to cope with adverse conditions are also seen in Khulna. In waterlogged areas, communities construct temporary foot-bridges made of brick and bamboo. Temporary drains are also excavated to relieve drainage congestion.

Groundwater Flooding

In some locations, including Miami and much of south Florida (see Case Study 9.3), sea level rise can increase the risk of groundwater flooding. A rise in sea level will simultaneously raise water tables, saturate the soil, expand wetlands, and increase flooding during heavy rainfalls (Rotzoll and Fletcher, 2013). However, abstraction from aquifers lowers water tables, which can reduce the likelihood of flooding. Thus, groundwater withdrawals can potentially mitigate the effects of a rising water table that is driven by sea level, although this may instead increase the likelihood of saltwater intrusion.

Shoreline Erosion

High waves and/or water levels during intense storms lead to beach erosion and shoreline retreat. This occurs along the California coast during a strong El Niño event, such as during the winter of 2015–2016 (e.g., see NOAA, 2015; Southall, 2016). Sea level rise will generally increase erosion rates (Bird, 2008; IPCC, 2014). Around developed areas, this can lead to disruption of sediment movements and coastal "squeeze," i.e., loss of land and environmental degradation.

Acidification

Increased CO_2 has been linked to increased ocean acidification, which is considered to be one of the largest threats to marine organisms and ecosystems (Billé et al., 2013). Coastal zone waters are much more sensitive to these changes than open ocean, and acidity may increase from a pH of 8.16 in 1850 to 7.83 by the end of the 21st century (Lerman et al., 2011), albeit with considerable spatial variability. For example, Cai et al. (2011) project a decline in the Northern Gulf of Mexico of pH 0.74 over the same timeframe. A lower ocean pH decreases calcification and inhibits coral growth. Corals are furthermore under stress due to ocean warming, which results in bleaching and increased mortality (Wong et al., 2014). Ocean acidification

will also adversely affect most other marine organisms including fish, crustaceans, molluscs, and the calcified single-celled marine organisms (e.g., foraminifera and coccoliths) that play an important role in regulating the ocean's inorganic carbon cycle. This would thereby detrimentally impact various economic sectors (e.g., fisheries, aquaculture, tourism) and consequently the coastal cities that depend upon these activities (Cooley and Doney, 2009; Narita et al., 2011; Burke et al., 2011).

Sea Surface Temperature

Increased sea surface temperature can increase the likelihood of a number of climate-sensitive diseases such as cholera and parasitic diseases (Chou et al., 2010; Cash et al., 2013; Baker-Austin, 2013). Limited evidence has associated harmful algal blooms with sea surface temperature variability (Glibert et al., 2014).

9.2.2 Vulnerabilities

Disasters are not caused solely by climatic events but by their interactions with urban systems (IPCC, 2012; Birkmann et al., 2010). The scale of a disaster, and risk more generally, is in part a function of the nature and magnitude of the hazard, but also the vulnerability of the urban system, including infrastructure, transportation networks, residences, and communications, which is locally specific to each city. Understanding these vulnerabilities is crucial for effective adaptation and disaster reduction (see Adger, 2006; Sherbinin et al., 2013).

9.2.2.1 Vulnerability in Natural Systems

Ecosystems

Coastal ecosystems have a richer flora and fauna than many other natural environments and consequently provide a number of useful ecosystem services (Jones et al., 2011). These include supporting wildlife habitats, food-web support for fish and shell-fish, flood buffers, shoreline erosion control, recreation, and waste assimilation (Hopkinson et al., 2008).

Ecosystems in the coastal zone are susceptible to climate hazards through a number of mechanisms. Sea level rise and other climate-sensitive drivers are expected to increase the rate of erosion and salinity intrusion in many parts of the world, but anthropogenic activity is also an important driver of degradation for coastal ecosystems and biodiversity reduction (Moser et al., 2012). These present major threats to coastal ecosystems and are already resulting in widespread global losses (Nicholls et al., 2008b; Nicholls et al., 2012).

Losses in ecosystem services can lead to other significant knock-on impacts. Removal or disturbance of coastal vegetation, for example, makes the coastal zone more vulnerable to erosion and sea level rise (IPCC, 2014b). Wetlands (salt marshes, mangrove forests, seagrass meadows, etc.), as well as dunes and beaches, provide important habitats and also mechanisms for dissipating wave energy during storms. Under natural conditions these systems are able to accommodate rising sea levels by retreating landward, but human settlements and infrastructure often confine coastal ecosystems leading to squeeze and ultimately total loss (IPCC, 2012). Ocean warming and acidification can affect animals and algae, particularly coral reefs through bleaching and reduction in calcium carbonate production.

Aquifers

Aquifers are susceptible to saltwater intrusion, furthered by sea level rise and flooding by waves and storm surges (Nicholls and Cazenave, 2010). They are also affected by reduced recharge from increases in temperature and evaporation and from changes in precipitation patterns. The risk to aquifers is often increased by unsustainable groundwater extraction rates and anthropogenic pollution (Ferguson and Gleeson, 2012). The risk related to water supply reduction is very high all over the world (Moser et al., 2012). Urban areas on small islands are often particularly dependent on aquifers, so in such places the risk is particularly acute.

Land

Land losses are expected to increase in the next decades due to submergence, coastal erosion, and reduced sediments from rivers flowing into deltas (often resulting from human activities to control rivers and reduce coastal erosion elsewhere). It is not only territory loss and its direct consequences to the economy, but the increased cost in protection measures is also expected to be significant (Hanson, 2011; Giosan, 2014). In addition to environmental drivers, conflict and economic, technological, and political factors can all influence population movements and lead to long-term shifts in the location and viability of coastal cities (McLeman, 2011).

These trends are most evident in the Asian and African mega-deltas that are seeing uneven spatial economic development

as populations move into the largest cities (Seto, 2011). Nearly half a billion people live on or near the world's deltas, many in major Asian cities such as Shanghai, Dhaka, and Bangkok. Deltas are the landform most vulnerable to sea level rise owing to low elevation, low topographic gradient, and erodible sediments (Schiermeier, 2014; Schmidt, 2015). Construction of dams, levees, and other floodplain engineering structures have starved deltas of sediments, adding to the natural land subsidence and risk of inundation (Syvitski et al., 2009). Deltaic environments, such as lagoons, wetlands, and dunes, harbor hotspots of biodiversity including thousands of species of plants and land and aquatic wildlife that would be threatened by inundation.

9.2.2.2 Vulnerability of Socioeconomic and Engineered Systems

Growing populations and urban development add to the increased vulnerability posed by natural coastal hazards. Much of this new growth will occur in flood-prone, low-lying coastal locations. Many of Asia's largest port cities are already and will be increasingly at risk (Fuchs, 2010) (see Table 9.1). By 2070, roughly half of the world's population exposed to coastal flooding will be concentrated in just ten megacities, all but one of which are in Asia (Hanson et al., 2011) (see also Table 9.1). The concentration of commerce, industry, tourism, and trade makes coastal cities vulnerable in distinctive ways.

Built Environment

Buildings and their cities are directly impacted by land degradation and losses. Resilience in these cities is related to the quality and distribution of infrastructure, transportation, and urban facilities, as well as the availability and quality of natural resources.

Urban planning, often shaped by national policy and governance, is central to avoiding exposure (Hanson et al., 2011; Walsh et al., 2013). Buildings are often ill-adapted to accommodate and resist coastal risks such as flooding and storms. Informal settlements, often located in the most risk-prone urban areas and vulnerable to landslide-susceptible slopes, flooding, river erosion, and the like continue to grow, particularly in developing nations. Reduced adaptive capacity of the occupants is compounded by building conditions such as deficiency in materials, structural safety, accessibility, overcrowded homes, and the lack of adequate urban infrastructure (Satterthwaite et al., 2007; Handmer et al., 2012).

Economic Infrastructure

Urban infrastructure systems (e.g., transport, water, energy, communications, wastewater and solid waste management) are often readily disrupted by extreme events (Handmer et al., 2012). Interruptions to the energy supply can lead to a cascading impact on other infrastructures that require power to operate. Damages to transportation systems can propagate more widely by stopping flows of people, goods, and services, with economic

Table 9.1 *ARC3.2 coastal cities: Current population, observed rate of relative sea level rise, and projected relative sea level rise for the 2050s and 2080s.*

Note: *ARC3.2 Cities include Case Study Docking Station cities, UCCRN Regional Hub cities, UCCRN project cities, and cities of ARC3.2 Chapter Authors.*

City[a]	Population	Observed RSLR, mm/yr[b]	Projected RSLR, 2050s[c]	Projected RSLR, 2080s[d]
Abu Dhabi	1,765,000		+15 to 60 cm	+23 to 124 cm
Accra	4,010,050		+17 to 58 cm	+23 to 119 cm
Antofagasta	296,900	−0.79	+13 to 55 cm	+20 to 116 cm
Antwerp	1,744,860		+18 to 70 cm	+25 to 140 cm
Athens	3,475,000		+15 to 55 cm	+22 to 116 cm
Bangkok[efgh]	15,645,000	1.07	+15 to 57 cm	+21 to 122 cm
Boston[h]	4,794,447	2.81	+16 to 70 cm	+23 to 141 cm
Brisbane	2,065,000	0.58	+15 to 58 cm	+23 to 124 cm
Buenos Aires	15,355,000	1.52	+14 to 54 cm	+19 to 115 cm
Can Tho	1,200,000		+14 to 58 cm	+21 to 122 cm
Cape Town	3,740,000	1.94[i]	+16 to 57 cm	+22 to 118 cm
Chula Vista	3,095,310		+14 to 56 cm	+21 to 118 cm
Colombo	5,000,000		+14 to 59 cm	+22 to 122 cm
Copenhagen	1,290,000	0.55	+16 to 70 cm	+25 to 140 cm
Cubatão	1,664,140		+14 to 56 cm	+22 to 118 cm
Dakar	3,137,200		+15 to 56 cm	+22 to 118 cm
Dar es Salaam	4,364,540		+15 to 60 cm	+22 to 122 cm
Dhaka[f]	14,171,600		+14 to 57 cm	+21 to 122 cm
Dublin	1,205,000	2.14	+15 to 65 cm	+24 to 131 cm
Durban	3,440,000		+14 to 58 cm	+20 to 118 cm
Fort Lauderdale	5,564,640		+15 to 62 cm	+21 to 123 cm
Genoa	615,000	1.18	+15 to 56 cm	+22 to 118 cm
Gold Coast	620,000	1.63	+15 to 58 cm	+23 to 124 cm
Gothenburg	590,000	−1.67; 0.34	+17 to 70 cm	+24 to 140 cm
Hayama	32,835		+13 to 65 cm	+20 to 129 cm
Helsinki	1,090,620	−2.31	+18 to 75 cm	+26 to 148 cm
Hong Kong[g]	7,187,000	3.25	+15 to 60 cm	+22 to 123 cm
Istanbul	13,755,000		+15 to 55 cm	+22 to 116 cm
Jakarta[fh]	9,610,000		+14 to 58 cm	+21 to 122 cm
Khulna[f]	759,620		+14 to 57 cm	+21 to 122 cm
Kingston	585,000		+14 to 57 cm	+20 to 119 cm
Kunshan	1,970,000		+13 to 59 cm	+21 to 122 cm
La Ceiba	204,140		+15 to 59 cm	+21 to 122 cm
Lagos[f]	21,000,000		+16 to 58 cm	+23 to 120 cm
Lima	9,752,000		+14 to 55 cm	+21 to 117 cm
London	14,031,830		+17 to 70 cm	+24 to 139 cm
Los Angeles	12,828,840	0.96	+14 to 56 cm	+21 to 118 cm
Makassar	1,484,000		+15 to 58 cm	+22 to 122 cm

Table 9.1 (*continued*)

City[a]	Population	Observed RSLR, mm/yr[b]	Projected RSLR, 2050s[c]	Projected RSLR, 2080s[d]
Manila	11,856,000		+14 to 60 cm	+21 to 124 cm
Maputo	2,615,000		+15 to 58 cm	+22 to 120 cm
Masdar	921,000		+15 to 60 cm	+23 to 124 cm
Matara	761,370		+14 to 59 cm	+22 to 122 cm
Ebro Delta (Mediterranean Spain)			+15 to 57 cm	+22 to 119 cm
Melbourne	4,246,350		+12 to 56 cm	+18 to 116 cm
Miami[f g h]	5,564,640	2.39	+15 to 62 cm	+21 to 123 cm
Middelfart	37,980		+16 to 70 cm	+25 to 140 cm
Mumbai[e f g h]	20,748,400	0.81	+15 to 58 cm	+23 to 122 cm
Naples	4,996,100	2.33	+15 to 56 cm	+22 to 116 cm
New Songdo City	2,879,780		+13 to 60 cm	+20 to 124 cm
New York[f g h]	18,897,100	2.84	+16 to 71 cm	+23 to 139 cm
Norfolk	1,671,700	4.61	+15 to 67 cm	+23 to 129 cm
Portmore	685,000		+14 to 57 cm	+20 to 119 cm
Rio de Janeiro	11,835,710		+15 to 56 cm	+21 to 118 cm
Rotterdam[g]	1,175,480		+18 to 70 cm	+25 to 140 cm
San Diego	3,215,000	2.15	+14 to 56 cm	+21 to 118 cm
San Francisco	6,455,000	1.94	+14 to 56 cm	+21 to 118 cm
Santo Domingo	3,265,050		+14 to 56 cm	+20 to 117 cm
Seattle	3,439,810	2.03	+14 to 56 cm	+21 to 118 cm
Semarang	1,670,000		+14 to 58 cm	+21 to 122 cm
Shanghai[f g]	24,750,000		+13 to 59 cm	+21 to 122 cm
Shenzhen[h]	12,775,000	3.25	+15 to 60 cm	+22 to 123 cm
Singapore	5,469,700	3.2	+15 to 58 cm	+21 to 122 cm
Surat[h]	6,081,320		+15 to 58 cm	+23 to 122 cm
Sydney	4,391,670	0.60; 0.96	+16 to 64 cm	+24 to 128 cm
Tacloban	221,170		+14 to 61 cm	+21 to 125 cm
Tangerang Seltan	1,443,400		+14 to 57 cm	+21 to 122 cm
Tokyo[f g]	37,833,000	1.88	+13 to 65 cm	+20 to 129 cm
Townsville	170,000	1.85	+17 to 70 cm	+25 to 141 cm
Trondheim	863,130	−2.12, −1.11	+15 to 56 cm	+22 to 117 cm
Venice	340,000	2.47	+14 to 56 cm	+21 to 118 cm
Victoria	1,765,000	1.37	+15 to 60 cm	+23 to 124 cm

[a] *Cities and population data from ARC3.2 Case Study Docking Station (CSDS), unless otherwise indicated.*

[b] *Historic relative (local) sea level rise trends, in mm/yr (uncorrected for local land movement). Data sources: U.S. and some international stations: NOAA Tide and Currents (Last accessed August 7, 2017). Other: http://www.psmsl.org/products/trends/txt (last updated May 15, 2017).*

[c] *Sea level rise projections for the 2050s. Numbers represent the 10th and 90th percentiles of the model-based frequency distribution, in cm. Methodology based on Horton et al., 2015. (Glacial isostatic adjustment (GIA) and glacial "fingerprint" components are not included, which may increase/decrease the projected sea level rise for the selected city).*

[d] *The same for the 2080s.*

[e] *Population data from: Demographia World Urban Areas. 13th Annual Edition: 2017(04); United States Census Bureau.*

[f] *Among the top twenty cities ranked in terms of population exposed to a sea level rise of 0.5 m and 1 in 100-year flood event by the 2070s (after Hanson et al., 2011).*

[g] *Among the top twenty cities ranked in terms of assets exposed to a sea level rise of 0.5 m and 1 in 100-year flood event by the 2070s.*

[h] *Among the top twenty cities ranked by average losses (in US$ million) for a sea level rise of 20 cm by 2050, with adaptation (after Hallegatte et al., 2013).*

[i] *Simon's Bay, near Cape Town, South Africa.*

Note: Additional information on sea level rise impacts on many of these cities can be found in Surging Seas: Risk Zone Maps*: Interactive maps of the area flooded by high water levels (tides, surges, sea level rise) in increments of 1 meter (or 1 foot) for the selected city. Mapping Choices: Interactive maps of the area inundated by sea level rise for temperature increases of 2°C and 4°C, in 2050 and 2100, for the selected city. Source: http://sealevel.climatecentral.org. See also Strauss et al., 2015. Carbon choices determine U.S. cities committed to futures below sea level. Proceedings of the National Academy of Sciences 112(44):13508–13513. The Climate Central website also provides a Risk Finder for U.S. cities that lists basic socioeconomic data, vulnerable infrastructure, and other relevant information.*

consequences (Wilbanks and Fernandez, 2014). This can result in the total closure of tunnels and jeopardize population safety because disaster contingency plans are usually dependent on safe evacuation routes (Dawson et al., 2011a). Disruptions in even a small area can become more significant if they involve an important transport interchange, such as a sea- or airport. Many of these are situated on low-lying reclaimed land. Drainage and sewer systems are sensitive to sea level rise and flooding, which can lead to discharge pollution in nonseparated systems. Depending on the degree of sea level rise or storm surge, drainage can be temporarily or permanently blocked, requiring structural adaptation or energy-intensive pumping. Furthermore, the salinization of sewer systems as consequence of various hazards can lead to a change of chemistry and failure of the system (Bjerklie et al., 2012).

Tourism

Along with ports, military instillations, and fisheries, tourism is a major component of the economy in many urban coastal areas. Out of fifteen top tourism destination countries, twelve of them have coastlines (UNEP, 2009). One of the primary climate change concerns for the tourism industry is the impact on tourist flows and seasonality for many destinations (Perry, 2006; Amelung et al., 2007; Bigano et al., 2008). Changing temperatures will mean that destinations have longer shoulder seasons and could change the time of year when more tourists visit (Perry, 2006; Amelung et al., 2007). Studies in Australia and Europe have used the Tourism Climate Index (TCI) and climate projections to determine the future suitability of destinations for certain activities and changes in tourist numbers due to rising temperatures: some popular destinations will become too hot for tourists and others will become more popular because of more agreeable temperatures for activities such as sunbathing (Hein et al., 2009; Coombes and Jones, 2010; Amelung and Nicholls, 2014). These changing weather patterns may increase domestic trips for many tourists (Rossello-Nadal, 2014), which could mean increased pressure for coastal urban areas. This will mean that host communities will need to adapt to the changing climate, increasing or decreasing visitor numbers, and changes in activity types.

However, according to multiple case studies, many tourism stakeholders (primarily business owners) did not see adapting to climate change as an immediate priority. While stakeholders were optimistic about their capacity to adapt to climate change, inadequate technical, human resource, and financial capacities were found to be barriers to adaptation. Tourism stakeholders in the case studies viewed government as the entity responsible for climate change adaptation (Scott et al., 2012).

Health, Education, and Wealth

Inequality and poverty are probably the most important factors of vulnerability due to the precarious conditions of life (in urban environment, housing, income, health, etc.) that are usually imposed by them. Those lead to an unequal and unfair distribution of risks (UN-Habitat, 2011) that can perversely maintain a very high condition of vulnerability in the poorest urban population. This vulnerability is defined by the sensitivity of building and infrastructure conditions and low adaptive capacity due to lack of financing resources, education, and environmental and personal health, in addition to fragile governance conditions.

Toxic Exposures

Research conducted by the New York City Environmental Justice Alliance (NYC-EJA) has shown that coastal communities in close proximity to former or current industrial uses are especially vulnerable to the potential release of contaminants in the event of extreme weather events (Handmer et al., 2012). Vulnerable residents (e.g., low-income, elderly, children) living/working in industrial waterfront neighborhoods are particularly exposed to the potential release of contaminants in the event of severe weather (Bautista et al., 2015a). These releases can come from current operating sites where toxic materials are used or stored, or from sites contaminated from former industrial processes and/or waste storage practices.

9.2.2.3 Assessment of Vulnerability

To assess the growing vulnerability of coastal populations, settlements, and ecosystems to natural and anthropogenic hazards, various global to regional-scale databases have been developed. Among these is the DIVA Coastal Database for impact and vulnerability analysis to sea level rise (Vafeidis et al., 2008) that integrates information on geography, landforms, topography, bathymetry, tidal ranges, wetlands, surge levels, administrative units, uplift/subsidence, land use, gross domestic product (GDP) per capita, storm surges, and waves. In another approach, a Coastal City Flood Vulnerability Index (CCFVI) has been calculated using three elements: (1) A hydro-geological component (sea level rise, river discharge, soil subsidence, cyclones, storm surge), (2) A socioeconomic component (exposed populations, vulnerable groups), and (3) A politico-administrative component (institutional organizations, flood risk maps, flood protection measures) (Balica et al., 2012) (see Table 9.2).

9.3 Adaptation

9.3.1 Adaptation Strategies

Many strategies to manage risks in coastal cities are present (Aerts et al., 2012). Some of these can be implemented readily and rapidly, whereas others may require large investment or long-term planning and implementation. Actions to reduce exposure to natural hazards and consequent vulnerability include moving people and infrastructure, building "hard" engineering protection, and also adopting "soft" solutions, such as planting and protection of mangroves and other natural vegetation (e.g., Möller et al., 2014). These actions directly influence the sensitivity of the system or the adaptive capacity to build resilience (Allenby and Fink, 2005; Adger et al., 2005).

Case Study 9.3   Coastal Hazard and Action Plans in Miami

Vivien Gornitz

Center for Climate Systems Research, Columbia University, New York

Keywords	Sea level rise, flooding, salt water intrusion, coastal, community based adaptation
Population (Metropolitan Region)	2,693,117 (U.S. Census Bureau, 2010)
Area (Metropolitan Region)	4,915 km² (U.S. Census Bureau, 2010)
Income per capita	US$56,180 (World Bank, 2017)
Climate zone	Aw – Tropical savannah (Peel et al., 2007)

Southeast Florida enjoys a subtropical climate with relatively dry, sunny winters and hot, wet summers. The hurricane season extends from late June through November. Along with rising temperatures, south Florida can expect increased variability in rainfall – more intense downpours during the summer wet season, also longer droughts (Heimlich et al., 2009). Regional sea level rise lies close to the 20th-century global mean: approximately 2 mm/yr (NOAA, 2016). South Florida is geologically stable with no appreciable land subsidence. The City of Miami, in Miami-Dade County, is especially vulnerable to the effects of sea level rise due to its low topography, large urban population, highly vulnerable water supplies, and its important climate-dependent tourist industry.

Sea level rise projections for southeast Florida are 8–18 centimeters by 2030; 23–61 centimeters by 2060; and 48–145 centimeters by 2100 (Southeast Florida Regional Compact Climate Change, 2011). Large percentages of the urban portions of Miami-Dade could be inundated by sea level rise; for example a 30 centimeter rise would inundate 18.2% of the county; 60 centimeter, 28.2%; and 90 centimeter, 33.6%. More intense hurricanes (e.g., see Knutson et al., 2010) coupled with sea level rise will lead to higher surge levels and stronger waves and winds, causing increased coastal damage.

Major coastal hazards include coastal and inland flooding, storm surge inundation, wind and wave damage from tropical cyclones (including hurricanes), beach erosion, and saltwater intrusion into the Biscayne Aquifer, which is the city's major water resource.

Water management and the threat of salinization are major concerns in Miami-Dade County. Unique geological features include the low average elevation of under 10 meters; a highly porous and permeable limestone bedrock; the shallow, near-surface Biscayne aquifer; and infiltration of saltwater into existing aquifers during heavy storms or at times of low rainfall. Farther north, due to ongoing saltwater intrusion, the city of Hallandale Beach, Florida, plans to relocate its water well field inland, to the west (City of Hallandale Beach, 2013).

The Southeast Florida Regional Climate Action Plan (2012; 2014) recommends regionally coordinated water management plans to address stormwater use and disposal, future water resources, wastewater treatments, and water conservation. The Action Plan urges a unified effort among government, businesses, and consumers in order to implement near- and long-term strategies to lessen adverse climate change effects on water supplies while diversifying existing water supplies. Water conservation and development of less vulnerable, new water resources are given high priority. Implementation may require changes in policy and regulatory frameworks, additional funding for infrastructure, development of alternative water supplies, and public education.

The Action Plan further advocates development of regional inventories of water management infrastructure and additional requirements to prepare for anticipated climate change. It also recommends development of improved saltwater intrusion models and refined inland and tidal inundation maps for increased sea level. The Plan also advises the consideration of climate change impacts in site selection and design of new water and wastewater facilities, as well as in the relocation of existing wellfields and other water management infrastructure (Southeast Florida Regional Compact Climate Change, 2014). Improvements in efficiency of flood/control/drainage infrastructure, recycling of stormwater, reclamation of wastewater, protection of ecosystems, and expansion of green infrastructure are also urged, as well as greater coordination with other federal agencies, such as the U.S. Army Corps of Engineers, U.S. Geological Survey, and NOAA. The ultimate goal is a regional Integrated Water Management Plan.

Miami-Dade County, as an active member of the Southeast Florida Regional Climate Change Compact, has adopted the Unified Sea Level Rise Projection (see above), initiated a county-wide vulnerability assessment, and increased regional collaboration. Near-term goals are development and implementation of a sustainability plan that also includes community stakeholder and general public inputs (Hefty, 2011; Miami-Dade County, 2015).

9.3.1.1   Shoreline Protection: Hard Solutions

Shoreline protection derives from engineering structures or enhanced natural features designed to withstand current and anticipated shoreline retreat, storm surge, and sea level rise. Protection includes "armoring" the shoreline with "hard" defenses and "soft" defenses that mimic natural processes (e.g., NRC, 1995; Titus and Craghan, 2009; Gornitz, 2013). Shoreline armoring is typically applied to defend important assets. Hard structures include seawalls, bulkheads, boulder ramparts (revetments, riprap), groins, jetties, and breakwaters (see Figure 9.2). The first three types of structures strengthen the existing shoreline by preventing slumping or erosion of soft, poorly consolidated sediments. While resisting flooding from average storm surges and wave heights, they can still be overtopped by extreme events. Seawalls and revetments may, however, intensify basal erosion. This can be reduced by careful placement of rubble. Groins and jetties project outward and trap sand, widening the beach (but often intensify erosion downdrift). Breakwaters shelter a harbor or beach from extreme wave action but, if poorly designed, can also induce erosion.

Other structures such as dikes, tidal gates, and storm surge barriers protect against extreme floods or permanent

Table 9.2 *Adaptation strategies for coastal cities. Source: Adapted from Dawson et al., 2011b*

INTERVENTION	EFFECT OF ACTION	POTENTIAL MODIFICATION OF CLIMATE RISKS
River and Coastal engineering measures	Hard engineering structures (e.g., river diversions, levees, dikes, breakwaters, seawalls, riprap, barrages) reduce the probability of flooding by providing greater protection against higher water levels, increasing capability of excess water removal or storage. Soft engineering measures (e.g., beach nourishment, vegetation management) reduce the vulnerability of defenses through dissipation of wave energy.	The effectiveness of flood defenses may be assessed by calculating changes in the probability of flooding upon implementation.
Rural runoff reduction and storage	Reduce flood severity from altered runoff properties through changing the infiltration, storage, and conveyancing properties of catchments and floodplains.	Decreases the probability of flooding.
Urban runoff reduction and storage	Reduce the probability of flooding using a combination of storage, infiltration, routing, and drainage capacity management.	Decreases the probability of flooding.
Incident management	Improved flood forecasting and early warning systems provide information to flood risk managers, local authorities, and emergency services to give the public sufficient time to take effective mitigative actions before actual flooding occurs. Proactive pre-incident activities ensure adequate public, emergency services, and other key stakeholder preparation to undertake appropriate actions.	Most flood incident measures act to change the depth–damage relationship of floods (if followed by appropriate action by the public) and increase public safety and reduced health impacts of flooding. However, some flood-fighting actions (e.g., reinforcing failing defenses) can reduce the probability of flooding, and their success is tied to timely responses to specific flood events.
Flood-proofing	Reduce flood damage	Flood-proofing measures change the depth–damage relationship for the properties in which they are implemented. These could be retrofitted to old properties or designed into new builds.
Land-use planning	Limit construction of buildings and infrastructure in the flood plain, hence decrease vulnerability.	Appropriate land-use planning measures lessen overall potential damages through time by reducing floodplain development.
Building codes	Reduced flood damage. Improved flood-proofing measures in new buildings will be more reliable than those in retrofitted properties. For example, raising buildings on stilts.	Flood-proofing measures change the depth–damage relationship for newly built properties in which they are implemented.
Risk spreading (e.g., insurance)	Redistribution of the cost of damage across the population and through time	As well as redistributing risk, insurance is a potent means of communicating flood risk through an economic signal in order to reduce overall future damages by discouraging development in high risk-areas.
Health and social measures	Reduced social, health, and associated economic impacts of flooding	Health, social measures could be incorporated if an appropriate health/social or secondary economic impacts damage function were available.

inundation. While dikes hold back the sea, low-lying terrain behind the dike may need to be pumped dry. Tide gates open and close with the tides, allowing water to drain out at low tide. Storm surge barriers close only during extreme surges or tides, permitting water flows and shipping to continue at other times (e.g., Gilbert, 1986; London's Thames barrier; Environment Agency, 2012); the MoSE system in Venice (Venice Water Authority, Consorzio Venezia Nuova, 2012; private communication); and the Maeslant Barrier, protecting Rotterdam (Delta Committee, 2008; Deltawerken online). However, most present coastal defenses designed for current sea level and storminess will eventually need to be retrofitted, for example by raising the height of seawalls, dikes, tidal gates, and the like or by reinforcing them to resist stronger and higher waves. New

structures should be built to withstand anticipated higher water levels (i.e., relative sea level rise plus storm surge, high tides, and waves).

A rising sea level will require periodic strengthening and raising of hard defenses. The timing and extent of work would depend on the rate of sea level rise, which also varies spatially, as discussed earlier. In London, for example, the defenses will likely be reinforced within the next 25–60 years, with the option to build a new barrier after that (Environment Agency, 2012). The Netherlands is already planning to upgrade its sea defense system, taking an integrated approach that includes "building with nature" and allowing "room for the river" (see later discussion and the

Box 9.1   Middle East and North Africa Coastal Cities: Findings from a World Bank Study

Anthony G. Bigio

George Washington University, Washington, D.C.

In 2011, the World Bank completed a 2-year study on the vulnerability and adaptation to climate change and natural hazards in the cities of Alexandria, Casablanca, and Tunis in the Middle East and North Africa (MENA) region. It also covered the major urban development project under way in the Bouregreg Valley between the Moroccan cities of Rabat and Salé. The study focused on a relatively short time horizon (2010–2030), considered the most significant for current-day national and local authorities, because it coincided with the timeframe of infrastructure and urban master plans under preparation.

Its major objectives were to provide national and local authorities with an integrated assessment of urban risks facing these major cities, as well as with detailed adaptation and resilience action plans focused on options for priority and no-regret measures to be implemented soon. The study's objectives

were achieved through the active participation of and consultations with scientific, administrative, and civil society organizations in the three countries and at the regional scale.

Under the guidance and supervision of a World Bank team, a consortium of scientific and technical consultants conducted a number of crucial inquiries into the present and projected future levels of urban risk. Climate downscaling was carried out for each location for temperature and precipitations. Probabilistic risk assessment was conducted for seismic and tsunamic activities. Land subsidence was measured via satellite earth observation and spectral interferometry. Digital elevation models of the urban terrain and hydrological models were updated to simulate sea level rise and flooding risks. Economic valuations of the likely cumulative damages and losses due to natural hazards and climatic impacts were carried out.

The *hazards* examined were seismic and tsunamic activities, flooding, land subsidence, coastal erosion, storm surges, and sea-level rise, as well as urban water scarcity in the context of a changing climate. The urban *exposure* was based

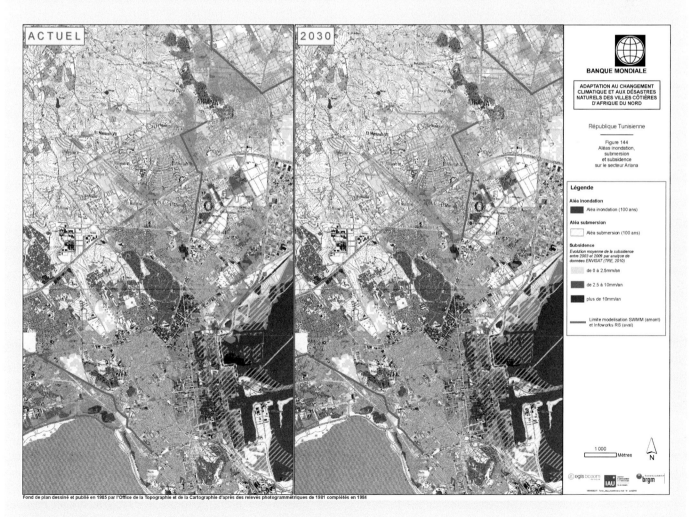

Box 9.1 Figure 1 *Central Tunis, showing current flooding and land subsidence risks.*

on the 2010 urban footprints and characteristics of the built environment and key infrastructure, as well as on the urban projected 2030 footprints resulting from demographic and spatial growth projections. The *vulnerability* analysis took into account the characteristics of the urban fabrics as well as the resilience and adaptation capacity of the urban systems and responsible institutions as of 2010.

Finally, *urban risk* was calculated as the combination of these parameters. Because the same methodology had been applied to the four urban sites at the two distinct points in time, varying degrees of urban risks were assessed for each of the hazards examined, in each urban location, for the current and forecasted scenarios. Overall, the assessments showed a medium to high *current* urban risk in the three cities, with Tunis ranking the highest, followed by Alexandria, and then by Casablanca. *Future* urban risk by 2030 increases to high to very high, with flooding and coastal erosion expected to increase considerably in all locations.

The net present value (NPV) of the likely cumulative damages and losses that each of the cities faces for the 2010–2030 period is more than US$1 billion. The percentage of such NPV due to climate change impacts was, however, considered very low at the outset of the period and estimated to reach only about 20% by 2030, given the relevance of natural hazards such as seismic activities and land subsidence. This was a sobering finding of the study because it indicated the need for these cities to urgently address current urban risks stemming from natural hazards as a precondition to planning for and implementing further climate adaptation measures. A backlog of unaddressed urban risks dwarfs in the short-term the likely impacts of climate change. Because climatic conditions are expected to worsen by mid-century and beyond, the percentage of cumulative damages and losses due to climate change is expected to increase proportionally.

The adaptation and resilience action plans developed for each of the three cities took into account the specific risks for each of the hazards reviewed and offered implementable responses organized in three areas: institutional and governance responses, urban planning measures, and green and gray infrastructure. The costs of the proposed actions were compared to the value of the potential damages and losses to be avoided through their implementation. The most

cost-effective measures turned out to be those related to the better functioning of the institutions charged with risk prevention, early warning systems, hydro-met services, and the management of climate forecasting and climate adaptation planning. These, however, appeared also as the most difficult ones to implement due to inherent resistance to organizational reform.

Urban planning responses consisted of strategies to prevent urban growth in areas at high risk and to incorporate the detailed knowledge of impacts in the retrofitting of the current built environment and in the planning and design of future urban expansions. Such actions are highly relevant in cities whose populations are expected to grow in the two-decades period from a minimum of 33% in the case of Tunis to a maximum of 65% in the case of Alexandria, with Casablanca at 55%. Such growth and the even faster paced expansion of their urban footprints will expose significantly more population and assets to natural hazards and climatic risks.

Finally, infrastructure responses proposed for the three cities included the provision of additional coastal defenses; beach nourishment; stabilization of buildings in areas subject to land subsidence; the expansion and improvement of drainage systems, including the use of eco-system services and on-site water absorption and retention systems; the "hardening" of critical infrastructure, such as power-stations, water-treatment plants, key roadways, and port systems; and improved water usage management and waste-water reuse.

The study has had a significant impact on the national and local institutions in the three countries involved in terms of focusing on the "clear and present dangers" their cities are facing. Their representatives were able to exchange views and set the stage for regional collaboration via the good offices of the Center for Mediterranean Integration in Marseilles (CMIM), which fosters innovation and exchanges across the Mediterranean region. The study's reports were widely disseminated across MENA, resonated in the regional media, and have contributed to improving the general understanding of urban risks and climate change. However, due primarily to the political upheaval that has swept across the region following the 2011 Arab Spring, no implementation of the proposed action plans has yet occurred.

Rotterdam Case Study; Delta Committee, 2008; Kabat et al., 2009; Katsman et al., 2011). Likewise, coastal polders in Bangladesh are being raised to provide additional protection against the rising sea level.

Analysis of the costs associated with coastal defense systems to protect against flooding in the context of three IPCC RCP scenarios (2100 with respect to 1985–2005), and under two adaptation strategies, shows that without adaptation, 0.2–4.6% of the world's population could be flooded each year in 2100, with a sea level rise of 25–123 centimeters, costing 0.3–9.3% of GDP. While coastal protection and maintenance costs are high, amounting to an estimated US$12–71 billion per year in 2100,

these costs are much smaller than damages arising from doing nothing (Hinkel et al., 2014).

9.3.1.2 Shoreline Protection: Soft Solutions

"Soft" defenses have become a preferred means of shore protection in many places because of the negative impacts of hard stabilization on beaches (NRC, 1995; Bird, 1996; Duarte et al., 2013; Arkema et al., 2013). These include beach nourishment and rehabilitation of dunes and coastal wetlands. Stable beaches and saltmarshes not only provide recreational opportunities and important habitat for a wide variety of wildlife including fish, shellfish, waterfowl, and small amphibians, reptiles, and

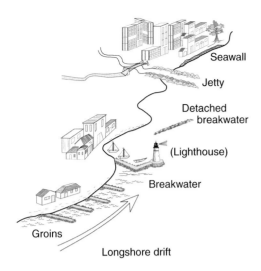

Figure 9.2 *Hard structures for shoreline protection.*

Source: Gornitz, 2013

mammals, but also protect the hinterlands against storm surges and high waves (e.g., Nordenson et al., 2010; Arkema et al., 2013; IGCI, 2015).

"Ecological engineering" is emerging as a no- or low-regret approach to coastal zone management, recognizing that shorelines are vulnerable to multiple sources of change including climate change, urbanization, and development (Cheong et al., 2013). One such approach replaces traditional engineering with ecological solutions. For example, a successfully restored oyster reef, when fully grown, acts as a shoreline protection system by damping incoming wave energy and reducing erosion. The new reef furthermore removes algae and suspended organic matter, thereby curbing turbidity, and creates suitable habitat for fish, crabs, and other marine wildlife (Cheong et al., 2013). This approach is closely related to that of *living shorelines,* which are an example of coastal management that maintains or simulates natural processes (Titus and Craghan, 2009). Strategic

Case Study 9.4   Venice: Human-Natural System Responses to Environmental Change

Marco Marani

Nicholas School of the Environment and Pratt School of Engineering, Duke University, Durham, NC
Department of Civil, Environmental, and Architectural Engineering, University of Padova

Sonia Silvestri

Nicholas School of the Environment and Pratt School of Engineering, Duke University, Durham, NC

Keywords	Sea-level rise, subsidence, erosion, coastal floods, ecosystem-based adaptation, storm surge barriers, salt marsh restoration
Population (Metropolitan Region)	863,133 (ISTAT, 2015)
Area (Metropolitan Region)	2,462 km² (Comune di Venezia, 2015)
Income per capita	US$31,590 (World Bank, 2017)
Climate zone	Cfa – Temperate, without dry season, hot summer (Peel et al., 2007)

ABSTRACT

The artistic and historical patrimony of Venice is part of humanity's global heritage. Its preservation involves issues related to global environmental change, environmental protection and sustainability, economic development, and cultural heritage preservation. The city and its lagoon have been transformed by human interventions over more than a millennium and are now a powerful symbol of the coexistence

of the natural and the built environments, of the tension between sustainable and unsustainable uses of natural resources. To preserve the lagoon, the Venetian Republic diverted three major rivers directly out to the sea over the course of about three centuries (1330–1664) (D'Alpaos, 2011). These massive works did invert the increasing sedimentation trend of the lagoon but caused the current state of diffuse erosion. In more recent years, industrial development left its marks on the lagoon water quality and has contributed to the city's subsidence, while increased tourists fluxes and cruise ship navigation are exerting an ever-growing pressure on the city and its environment as a whole. Overall, ancient and more recent human interventions have greatly increased the vulnerability of the city and of its lagoon to climatic changes, in particular sea level rise acceleration.

The protection of the city from increasingly more intense and frequent high tides, the MoSE system (Modulo Sperimentale Elettromeccanico, i.e., Experimental Electromechanic Module) is a complex engineering project that has required a complex decision-making process involving a large number of local and global stakeholders and a difficult equilibrium between a large number of often-conflicting uses of environmental resources. The project's far-reaching environmental consequences have perhaps not yet been fully understood, but will certainly be felt for several generations to come.

VENICE AS A NATURAL–HUMAN SYSTEM

The very origin of the city of Venice lies in the lagoon surrounding it – a natural defense, a means of transportation, and a source of food. Naturally destined to silt up due to fluvial sediment inputs, the lagoon sediment balance has been deeply changed by major river diversions carried out by the Venetian Republic in the 14th and 15th centuries. These diversions have transformed the lagoon into a sediment-starved environment, thus setting up the stage for the modern conservation issues. The sediment balance was definitively compromised by the completion, in the early 20th century, of three jetties to maintain navigable depths at the inlets connecting the lagoon and the sea. The jetties created a circulation asymmetry between flood and ebb tidal phases that causes sediments to be ejected far into the Adriatic Sea during tidal ebb, such that they cannot be transported back into the lagoon during the tidal flood (D'Alpaos, 2011). The current sediment loss is between 500,000 and 700,000 cubic meters per year, corresponding to a layer of about 0.9–1.3 millimeters per year, if uniformly distributed throughout the lagoon.

The preindustrial rate of sea level rise in the Adriatic Sea in the 20th century amounted to about 1.2 millimeters per year. However, groundwater withdrawals, mainly between 1950 and 1971 (when industrial artesian wells were closed), have induced additional loss of about 9 centimeters over 30 years (Carbognin et al., 2010). Relative sea level rise has been approximately 25 centimeters since the start of the 20th century (Carbognin et al., 2010).

The historical and, particularly, the recent relative sea level changes have had significant implications for flooding in the city. Flooding starts at a water level of about 80 centimeters above mean sea level (AMSL) – the lowest pavement level in St. Mark's Square – whereas significant flooding is considered to occur above 110 centimeters AMSL. The frequency with which this flood threshold is exceeded has already significantly increased since the start of the 20th century, from one event every 2.2 years to four events per year. Such frequency is expected to further increase, due to projected sea level rise and subsidence, to between 20 and 250 flooding events per year (Carbognin et al., 2010). This circumstance poses a serious threat to the conservation of the city of Venice and has spurred the development of "hard" protection measures in the form of a tidal-storm surge barrier, the MoSE project.

ENVIRONMENTAL MANAGEMENT

Defending the City from High Tides: The MOSE Project

After a disastrous flooding event in November 1966 that lasted for two days with sustained water levels above 2 meters AMSL, a governmental panel of experts suggested the construction of gates to close the lagoon from the sea as the only effective solution to preserve the city. The design, developed several years later by a designated consortium of private companies, is known as MoSE and is based on a system of seventy-eight planar gates that can be raised to close the three inlets of the lagoon. The project has been highly controversial, particularly because of its potential associated environmental impacts. The realization of the MoSE system may also have major impacts on the viability of the Venice harbor, currently the second most important commercial harbor in the Adriatic (after Trieste).

Managing the Lagoon Environment

Human interventions on the lagoon have greatly changed the lagoon sediment balance and induced net erosion throughout its extent. In particular, the lack of sediment input has had a primary role in determining marsh lateral erosion and elevation loss, such that more than 50% of the marsh area has been lost in the past century (see Case Study 9.4 Figure 1) (Marani et al., 2007).

Mitigation strategies to offset this erosive trend have been put in place. Material obtained from dredging navigable channels has been used (when levels of pollution allowed it) to restore and build salt marshes for about 30 years (more than 20 million cubic meters of sediments have been reused to build about 11 square kilometers of salt marshes). This effort has partially compensated for the loss of ecosystem services, but has also spurred controversy over the methods of restoration and the location of the reconstructed marshes.

In 2010, a panel appointed by the Venice Water Authority to develop a Master Plan for managing the lagoon and its environment made the following recommendations: (1) Construction of artificial salt marshes along the main channels to limit the transport of eroded sediment to the sea; (2) Protection, using green engineering methods, of marsh margins

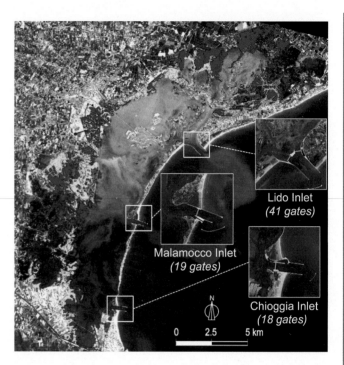

Case Study 9.4 Figure 1 *The MoSE project for the defense of the City of Venice from high tides (see insets). In yellow, marsh areas surviving at the beginning of the 21st century. In red, marshes that have disappeared over the course of the 20th century.*

Source: Modified from Consorzio Venezia Nuova – Servizio Informativo; www.salve.it; accessed in 2010

to stop the lateral erosion due to boat and wind waves; (3) Experimental artificial tidal flats to boost sediment availability and sustain marsh growth; and (4) Protection and partial reintroduction of water and sediment from the Brenta River toward the restoration of a neutral sediment balance. The Master Plan has not been acted upon as of early 2015.

OUTLOOK

Considering the recent IPCC scenarios for sea level rise, the further increase in tidal/storm surge flooding of Venice and the associated potential damage can only be addressed through hard measures regulating flows between the lagoon and the sea. Soft measures are unlikely to significantly control high tides in the city, essentially because the historical – and current – states of the system are not compatible with current environmental forcings. The current erosional trend experienced by the lagoon and its landforms is irreversible through sustainable solutions. Balancing the elevation deficit of the lagoon (i.e., stopping net erosion and balancing the elevation loss associated with sea level rise and subsidence) requires about 2–10×10^6 m^3/yr of sediments. It has been estimated that the reintroduction of the Brenta River into the lagoon would only yield about 70×10^5 m^3/yr. Hence, the delivery to the lagoon, through "natural" and sustainable means, of an amount of sediment sufficient to offset the effects of climatic changes is hardly possible. The only alternative is the transport and subsequent distribution of extraneous sediment through mechanical means. Whether this is a viable and desirable solution remains to be determined. Finally, an alternative method to balance the effects of sea level rise would be raising the city and its lagoon by means of water injections into the deep subsurface. This solution has been theoretically evaluated and experimented on a small scale (Comerlati et al., 2004), but its feasibility in the context of a fragile urban fabric, such as that of the city of Venice, is far from established. Economic and technical feasibility also remain to be determined.

Figure 9.3 *Widening and extending the dune ridges seaward and replanting with native vegetation along the Delfland coast, The Netherlands, 2013.*

Source: Vivien Gornitz

emplacement of plants, stone, sand, and other materials trap sediments and reduce wave energy, thereby cutting beach erosion and wetlands losses. These measures also protect the coast against future sea level rise. The restored broad beach and dune ridges, replanted with native vegetation between the Hague and Hoek van Hollan along the Delfland coast, has created a whole new nature district (see Figure 9.3). The Dutch also employ ecological engineering methods in protecting inhabited areas, such as in the coastal town of Scheveningen. There, multifamily dwellings stand behind a raised levee, fronted by a broad sand beach that is widened by beach nourishment (see Case Study 9.5).

Because of the protection provided by tidal wetlands, their integrity should be preserved as much as possible. Saltmarshes generally keep pace with current sea level rise, except for rapid subsidence (e.g., Louisiana and parts of the Chesapeake Bay), low sediment supply, or altered natural biogeochemical cycles (e.g., Jamaica Bay, New York) (Hartig et al., 2002; Cahoon et al., 2009; Kirwan and Megonigal, 2013). Nevertheless, most marshes would likely submerge when exposed to rapid sea level rise, except under macrotidal regimes or in regions of high sedimentation rates (Kirwan et al., 2010). Saltmarsh restoration involves adding sediment and replanting marsh grasses, sedges, or rushes. Marsh replanting, often with native vegetation, can be combined with engineered structures or submerged stone sills to reinforce the marsh. Coastal uplands should be preserved as buffer zones for future landward saltmarsh migration.

In tropical regions, intact mangroves, which shield the hinterland from severe storm surges, should remain protected from development, and cleared areas should be reforested (McIvor et al., 2012; Van Lavieren et al., 2012). Increasing soil salinity affects the Ganges-Brahmaputra delta of Bangladesh, exposed to both river and coastal flooding. Bangladesh's Char Development

and Settlement Project Phase III aims to provide protection for the coastal regions of the country, including the Sundarbans – world's largest mangrove forest – from saltwater intrusion and flooding by building embankments, sluice gates, drainage channels, protective tree belts, and cyclone shelters, as well as by improving local economic opportunities (Heering et al., 2010). *Char* is newly accreted land in the delta formed by sediments deposited by rivers. Although the Bangladesh Forest Department imposes a 20-year period for replanting and growth of mangroves on the char, new forests are often cut and settled before then. Replanting of mangroves on newly accreted land reduces erosion, stabilizes the soil, and shields against cyclone damage. Adaptation measures include planting of more salt-tolerant crops and conversion to shrimp farming. However, most of the low-lying urban coastal areas like Khulna remain vulnerable to flooding and drainage congestion due to rising high tide levels (see Case Study 9.2).

A broad beach with an elevated dune ridge provides the first line of defense against high seas. Dunes are stabilized by planting grasses and installing fences to trap sand. Beach nourishment also serves as an important means of shoreline protection. Sand from offshore or inland deposits is added to replace erosional losses and to widen and raise the beach. Because of historical erosion, sand needs periodic replacement along the U.S. East and Gulf Coasts. For example, the East Coast from New York to Key West, Florida, has undergone numerous beach nourishment projects since the 1920s. Approximately US$1.3 billion (in 1996 dollars) has been spent on beach stabilization along the East Coast between 1960 and 1996, with a cumulative cost in the United States of US$2.4 billion since the 1920s (Valverde, 1999).

Sea level rise will likely worsen erosion, creating shorter sand replacement cycles and increased beach nourishment costs. For example, by the second half of this century, beaches in the New York metropolitan region will need as much as a 26% additional volume of sand for a 60 centimeter to 1 meter rise in sea level (Gornitz, 2002). Beach nourishment will ultimately become economically and physically nonviable for significant increases in sea level.

9.3.1.3 Accommodation

Coastal populations will need to accommodate the anticipated upward climb of sea level and consequent increase in storm floods, as well as other anthropogenic and climate-induced stressors. Various planned and autonomous adaptation measures are possible. For example, building codes can help strengthen structures to make them more storm-resilient. Providing adequate space for water reduces flooding risks. In flood-prone areas, buildings can be raised above the current (or projected) 100-year flood zone, constructed on stilts or pilings, or have ground floors used for nonresidential purposes, such as business, parks, or recreation. One approach is to create "green infrastructure" – planting trees and grass along sidewalks or expanding parks to improve drainage and enhance water infiltration into the ground (e.g., Aerts et al., 2009; Enarrson, 2011). Another approach is to design neighborhoods around floating buildings and houseboats (e.g., Rotterdam, Seattle, Sausolito, Bangkok; Dircke et al., 2012). Innovative multiuse flood defenses (such as dikes) combine surge protection with housing, parking, parks, and commercial activities (see later discussion; Aerts et al., 2009; Stalenberg, 2012). Underground garages provide storage space for excess water at times of high river or ocean levels (Rotterdam Climate Proof, 2010). Multipurpose examples from Dordrecht, the Netherlands; Hamburg, Germany; and Tokyo, Japan, can serve as models for other coastal cities (Stalenberg, 2012). Preserving community freshwater ponds, harvesting rainwater, floating vegetable gardening, and other community-based practices are commonly seen in coastal Bangladesh.

9.3.1.4 Retreat

Defensive measures, and even accommodation, may provide increased protection against sea level rise and storm surges in the short term, but may offer a false sense of security about changes in the long run, thus serving to encourage further development in inherently high-risk areas. It may ultimately become impossible to defend even the most heavily developed shorelines, particularly on barrier islands lined with high-rise condos outside of major urban centers (Jin et al., 2013). Beyond a certain point, repeated rebuilding after storms, massive protection, or even raising land may become too expensive or ineffective, necessitating a retreat option. As resources available for coastal protection become increasingly stretched, densely populated coastal megacities will probably continue to be defended, whereas smaller settlements are already being considered for realignment. For example, defenses protecting the cliff top village of Happisburgh on the east coast of the United Kingdom are being allowed to fail in order to restore natural coastal processes that will protect downdrift floodplains (Dawson et al., 2009).

Long-term, prudent, coastal management may avert some of the adverse consequences long before a rising sea renders many coastal areas uninhabitable. For example, appropriate land use would limit housing density and building size in flood-prone areas, replacing them instead with natural "buffer zones" of parks, restored wetlands, or recreational facilities. A number of options exist to limit development in flood-prone areas and to plan for eventual retreat. Governments may acquire title to the property at risk through eminent domain, or the property can be donated voluntarily to a conservation organization. Other measures include the creation of erosion setbacks and easements that establish buffer zones for coastal wetlands or beaches. *Buyout programs* reimburse shorefront landowners for abandoning property in high-risk zones. In such cases, the public bears the cost. *Setbacks* restrict shore construction based on erosion or elevation thresholds. In the United States, regulations vary by state. In North Carolina, single-family houses and small structures must be set back thirty times the historic average annual erosion rate or a minimum of 60 feet (18.2 m). For larger structures such as condominiums, the setback is sixty times the historic erosion rate. In Florida, Coastal Construction Control Lines (CCCLs),

based on the 100-year storm surge zone, are intended to protect upland property and control coastal erosion. Construction is limited seaward of CCCL (Florida DEP, 2006). In addition, no new construction is permitted seaward of the projected 30-year erosion line (based on historic trends), with some exceptions, such as building as far landward as possible or behind the barrier dunes. However, numerous loopholes exist in the regulations that do not exclude all forms of development. Setbacks based on historic erosion rates presuppose a future continuation of those rates – unlikely, given anticipated sea level rise. Significantly, accurate historic erosion data may not always exist.

Implementation remains a challenge: setbacks have faced challenges in courts from landowners who claim that they constitutes a "taking" (i.e., government seizure of private property without due compensation) when a property cannot be developed or construction is limited by the setback. In response, some U.S. states, such as Maine, Massachusetts, Rhode Island, South Carolina, and Texas have therefore adopted variants of *rolling easements* (Titus and Craghan, 2009), which allow landowners to keep and develop their property, but prevent shoreline armoring. In *conservation easements*, a conservation organization such as the Nature Conservancy buys the right to prevent further development, but permits the landowner to still remain on the land. Since easements are voluntary, adjacent land can still be armored, but greater erosion downdrift may cancel any conservation benefits. However, easements may not be practical for already heavily developed shorelines. Although designed for present conditions, many coastal management programs can be strengthened by anticipating increased future erosion rates and wider flood zones.

The most extreme option is managed relocation. Extensive use of coastal structures and beach nourishment is expected to become infeasible under future sea level rise (Pilkey and Young, 2009), and transition policies should include coastal retreat and ending financial incentives to rebuild after storms. Individual structures can be moved landward, as has been done most impressively for the historic Cape Hatteras Lighthouse, North Carolina, in 1999 (Bodzin, 1999; National Park Service, 2016). More frequently, houses are moved some tens of meters landward within a given shorefront property, or other buildings threatened by imminent collapse due to coastal erosion are torn down.

Some countries are beginning to take landward relocation seriously. Great Britain now views "managed realignment" as a long-term planning tool, especially for estuaries and relatively undeveloped land (De la Vega-Leinert and Nicholls, 2008). The UK government recognizes that, in the long term, it will become uneconomic to defend many small communities along eroding coasts. "Realignment" involves moving existing coastal defenses inland. However, because most people tend to be unaware of coastal hazard and risk until after damages occur, setbacks and managed realignment are generally highly unpopular and politically contentious. Greater stakeholder involvement from the onset may, however, encourage increased public acceptance (De la Vega-Leinert and Nicholls, 2008).

9.3.2 Adaptive Capacity

A diversity of adaptation measures and practices, ranging from hard and soft solutions to accommodation and retreat, is seen in urban areas around the world. Although these measures are mostly needs-based, the emerging diversity also results from the variation in adaptive capacity of communities and nations.

In coastal urban areas, seen as socio-ecological systems, adaptive capacity refers to the ability of individuals or communities to utilize their resources and capitals to resist and adapt to the present and future hazards (Brooks and Adger, 2004). This capacity varies across and within societies, conditioned by the degree to which the geophysical, biological, and socioeconomic systems are susceptible to the adverse impacts of climate change (Adger et al., 2006; Füssel and Klein, 2006) while being dependent on several factors such as economic wealth, technology, information and skills, infrastructure, institutions, and equity (Smit et al., 2001). In many low-income coastal areas, economic conditions are important for not only affordability of adaptation measures but also for social vulnerability caused by poverty and inequality. In general, adaptation choices are largely influenced by access to technology and information that enhances risk recognition and creates alternative choices for strategies and tools, whereas the effectiveness and appropriateness of adaptation actions depend on the equitable distribution of infrastructure and services. The adaptive capacity of communities is enhanced by increased stability, integrity, and reliability of institutions and governance, a working relationship among local communities and actors (Brooks and Adger, 2004; Smit and Pilifosova, 2003; Smit et al., 2001), high social capital, strong social networks, and community empowerment (Adger, 2003).

Coastal cities often face situations where adaptive capacity and financial capital are enhanced at the cost of increasing climatic risks. The oil exploration and petroleum industry in Hammerfest on the northern coast of Norway triggered economic growth since 2002, which also enhanced the local adaptive capacity by improving institutional capacity, elevating employment and education, and increasing city government investment in schools and urban facilities. However, these transformations were accompanied by a huge increase in GHG emissions due to the industrial activities (Angell and Stokke, 2014).

The importance of information and skills in raising adaptive capacity has been highlighted in Jakarta, the capital of Indonesia, on the northeast coast of Java Island. An environmental vulnerability assessment carried out by Yoo et al. (2014) show that the greatest difficulty in implementing adaptation policies in the city is the local people's lack of awareness and the lack of professionals and experts. Environmental awareness, environmental policy foundation, regional GDP, and infrastructure are the key indicators for adaptive capacity that have positive influences on the city's vulnerability. The less vulnerable region of the city has flood warning system, environmental education programs, and support program for low-income communities after environmental disasters.

Considering the rising risk of coastal flooding and storms, coastal cities will need to invest not only in the spread of information but also in building the autonomy and commitment of local communities for adaptation. Many contingency plans around the world are playing important roles in increasing local adaptive capacity. Rio de Janeiro, a coastal city of around 6.5 million inhabitants, is built over wetlands and mountains on which prevail the poorest population and precarious buildings highly susceptible to landslides. A contingency plan of the city includes preparing local community leaders, mobilizing voluntary groups, and acting through simulation, awareness, and educational programs in primary and secondary schools that function as strategic points for shelter during emergency situations. The whole process fosters awareness and knowledge of risks and the dissemination of information and contributes to community empowerment (Prefeitura da Cidade do Rio de Janeiro, 2013).

9.4 Mitigation Strategies in the Coastal Zone

Mitigation here refers to the reduction in anthropogenic GHG emissions that are a major driver of anthropogenic climate change (as opposed to mitigation of flooding or wave impact energy, which is sometimes used within coastal management terminology). Several mitigation strategies unique to the coastal zone are considered. Although these mitigation options are not necessarily implemented within a coastal city, they usually require supporting urban infrastructure for their maintenance and operation.

9.4.1 Coastal Energy Infrastructure

The coast can provide excellent renewable energy resources and is convenient for siting more traditional power plants due to the availability of cooling water that is not exposed to the risk of low flows, as can be the case for power plans sited next to rivers (Byers et al., 2014). Brown et al. (2014) analyzed nonrenewable coastal energy infrastructure in Europe and identified 158 major oil, gas, liquid natural gas, and tanker terminals (40% on the North Sea coast), and 71 (37% of the European total) operating nuclear reactors on the coast and noted that the United Kingdom has three times more coastal energy facilities than any other European country. Ensuring the security of this infrastructure is important for reducing risks, but, as efforts to mitigate GHG emissions advance, the nature of coastal energy infrastructure is changing. For example, as alternative energy sources such as wind, solar, or tidal power become more prevalent, the need to locate nuclear or conventional power plants at the shore will diminish.

Tidal and wave energy can provide non-fossil fuel energy sources to assist global mitigation. Marine renewables could theoretically provide fifteen times global energy demand, but technical constraints mean it could deliver 1–10% of this (Resch et al., 20108. Potential for marine renewable varies around the world (Arinaga and Cheung, 2012; Moriarty and Honnery, 2012) according to tidal range and wave climate.

Tidal energy technologies include barrages and turbines. The former are damlike structures that capture tidal energy as water flows into and out of a bay or river. Turbines installed next to or within the barrage turn generators that produce electricity. However, tidal barrages may impact estuary ecology and sediment movement (Wolf et al., 2009; Kadiri et al., 2012). Tidal turbines, similar to wind turbines, operate underwater, driven by strong tidal currents.

Currently, some of the world's large tidal power stations include Sihwa Lake Tidal Power Station, South Korea (254 MW) and La Rance Barrage, Brittany, France (240 MW), with

smaller plants in Jiangxia, China; the Kislaya Guba (Barents Sea, Russia); and the Annapolis Royal Generating Station (Bay of Fundy, Canada). In 2013, the Scottish government approved installation of Europe's largest tidal turbine in the Pentland Firth between Orkney and the Scottish mainland. The Pentland Firth holds tremendous potential for tidal energy, but its remote location and harsh environmental conditions pose significant challenges for tidal development. Engineers estimate that a potential output of 1.9 gigawatts could eventually satisfy 43% of Scotland's domestic energy needs (Adcock et al., 2013). In 2012, permission was granted to install up to thirty tidal power turbines in the East River, New York, to harness the extremely

Case Study 9.5   Adaptation Benefits and Costs of Residential Buildings in Greater Brisbane

Chi-Hsiang Wang, Xiaoming Wang, and Yong Bing Khoo

Commonwealth Scientific and Industrial Research Organisation (CSIRO), Melbourne

Keywords	Wetlands, flood control, flooding, resilience, adaptation, ecosystem-based adaptation, coastal
Population (Metropolitan Region)	2,274,560 (Australian Bureau of Statistics, 2014)
Area (Metropolitan Region)	15,826 km² (Australian Bureau of Statistics, 2011)
Income per capita	US$54,420 (World Bank, 2017)
Climate zone	Cfa – Temperate, without dry season, hot summer (Peel et al., 2007)

The Greater Brisbane area, Queensland, with a population of 2.1 million, is one of the fastest growing areas in Australia. Since 2010, it has been inundated by storms and rainfalls brought by tropical cyclones in December 2010–January 2011, and again in January 2013, inflicting insured losses of more than AU$2 billion (US$1.6 billion) to this area and more than AU$3.5 billion (US$2.81 billion) to Queensland. This urban area will remain subject to great risks from coastal inundation as the Intergovernmental Panel on Climate Change (IPCC)'s Fifth Assessment Report (AR5) affirms continuing sea level rise (Rogelj et al., 2012). Furthermore, tropical cyclones under a future climate may become more intense (Knutson et al., 2010), although by how much is not yet clear.

In view of the challenges ahead, the Queensland Government in 2012 adopted the IPCC A1FI projected sea level rise (Meehl et al., 2007) in its coastal development plan and assumed that future tropical cyclone intensity will increase by 10% (DERM, 2012). With these assumptions and the population growth projected by the Australian Bureau of Statistics (ABS) up to 2056 (ABS, 2008), we applied cost-benefit analysis to investigate raising building floors for the adaptation of coastal residential buildings to future coastal inundation events. For houses designed according to the building code, significant damage costs can occur only under extreme

events; therefore, this Case Study investigates the consequence due to annual extreme storm tides.

NEW BUILDINGS ADAPTED VERSUS ALL BUILDINGS ADAPTED

For the inundation analysis, an assortment of data from Australian Bureau of Meteorology, Australian National Tidal Centre, Light Detection and Ranging (LiDAR), Shuttle Radar Topography Mission, Geoscience Australia, and Queensland Government were collected. Ground and building elevations were obtained with average grid spacing of 2 meters from survey data collected by the LiDAR technology. Given a storm-tide height, the areas that are hydrologically connected to the ocean and lower than the water level are projected to be inundated. The number of future residential buildings is assumed to grow at the same rate of population growth projected by the ABS (2008). The spatial distributions of new residential buildings are determined by Monte Carlo simulation under the assumption that they are distributed in accordance with the current distributions of their respective mesh blocks, and the building floor elevations above the Australian Height Datum are determined by the elevation recorded by LiDAR. For damage cost estimation, a discount rate of 4% per annum is assumed to convert future cost to the 2011 value.

The Case Study shows that the increase of future damage costs due to sea level rise is moderate compared to the increase due to building stock growth. This is a result of projected strong growth in the future population of the region. Interestingly, simultaneous consideration of sea level rise and building stock growth results in higher damage costs than the sum of damage costs caused separately by the two. This phenomenon is consistent with the type II extreme value distributions, which exhibit a heavy right tail (i.e., the damage cost is a convex function of the logarithm of average recurrence interval) (Beirlant et al., 2004).

We compare the adaptation options when only new buildings are adapted and when both old and new buildings are adapted. New buildings are regarded as those constructed after 2011. Buildings are adapted by lifting their floors to 1.57 meters above the Australian Height Datum, roughly corresponding to the current 100-year storm-tide level. The adaptation extent is the geographical area inundated by the current 100-year storm-tide events. We found that the benefit of lifting old buildings grows only marginally up to 2056, whereas the bulk of benefit is attributed to the lifting of new buildings. Considering the fact that the modest gain in benefit of lifting old buildings could

Case Study 9.5 Figure 1 *The Brisbane River.*

Case Study 9.5 Figure 2 *Buildings developed at waterfront with a high exposure to inundation by extreme storm tides.*

be easily offset by the social cost involved and the high building retrofit cost, lifting new buildings alone appears to be a more cost-effective adaptation policy.

COST-EFFECTIVE ADAPTATION

The larger the adaptation extent, the lower the expected future damage cost. Because of limited resources, however, an infinitely high level of adaptation is untenable. In practical situations, the extent of adaptation is constrained by competing policies and the planning

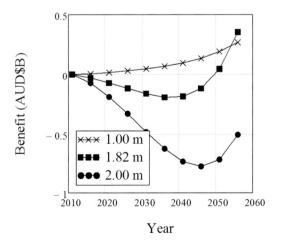

Case Study 9.5 Figure 3 *Benefit for adaptation extents determined by 1.00, 1.82, and 2.00 meter storm tides.*

time horizon (i.e., the future point in time that the decision-maker sets for achieving the planning objective). A properly chosen adaptation plan can avoid the problem of "underadaptation" that leaves some potential benefits unrealized or that of "overadaptation" that consumes too many resources to the detriment of other policy objectives.

We demonstrate three alternative adaptation extents assuming a planning time horizon of 2056 (the same time horizon for population growth projected by the ABS), as shown in Figure 3. If an adaptation extent determined by a 1.00 meter storm tide is chosen, then it may have the advantage of having an almost always positive net benefit up to the planning time but with modest positive net benefit. For an adaptation extent determined by a 1.82 meter storm tide, it results in negative net benefit during a period of time, but it achieves a higher benefit value than the 1.00 meter storm tide inundation extent. An overaggressive option (e.g., adaptation extent inundated by a 2.00 meter storm tide) requires more resources and may never reach a positive benefit value up to the planning horizon, an example of overadaptation.

Nevertheless, even with limited, suboptimal adaptation, immediate net benefits could be achieved by focusing adaptation on coastal housing with comparatively lower floor height. Similarly, overadaptation in the next decade that lifts more coastal housing to future storm tides gives longer term net benefits, although it incurs higher initial costs. However, too much overadaptation may be undesirable because it incurs unreasonably high initial costs and the benefits of it could only be reaped over a very long time. Therefore, the extent of adaptation should be decided in accordance with the planning time horizon, available resources, and balanced consideration of other policy objectives to avoid either under- or overadaptation.

powerful currents in the tidal strait (Verdant Power, 2014). When fully operational, the turbines will generate 1,050 kilowatts of energy, enough to serve approximately 9,500 local residents (U.S. Department of Energy, 2012).

Marine current power harnesses the kinetic energy of marine currents (e.g., the Gulf Stream) that are generated from a combination of temperature, salinity, wind, bathymetry, and the rotation of the earth. The predictability and potential scale of the resource makes this a potentially significant but not yet widely harnessed source of renewable energy. Wave power harnesses potential energy in the form of water displaced from the mean sea level that has been generated through the transformation of kinetic energy from the wind. However, wave energy varies over short time periods and seasonally making this a less predictable energy source (U.S. Department of Energy, 2010).

Ocean thermal energy conversion exploits gradients in temperature between deep and shallow water to drive a heat engine (Liu et al., 2014). Huge energy resources have been estimated globally, more than any other marine renewable, but currently implementation is limited (Lewis et al., 2011). However, because it can provide a continuous and stable energy source, it is useful for base-load power supply. Similarly useful for base load power supply are salinity gradients, caused by the mixing of freshwater and seawater, which releases energy that could be captured (Scramesto et al., 2009)

9.4.2 Carbon Sequestration

Carbon sequestration is the process of capturing and storing CO_2 emissions. This may involve removing CO_2 from the atmosphere or directly capturing the emissions at their source.

9.4.2.1 *Coastal Habitat Restoration*

Vegetated coast habitats represent a major terrestrial carbon sink and therefore hold great potential to mitigate climate change (Duarte et al., 2013). Consisting of both submerged (i.e., seagrass, microalgae) and periodically emerged vegetation (i.e., mangroves, saltmarshes), this habitat is highly endangered, having lost some 25–50% of its area within the past half century, largely due to anthropogenic intervention (Duarte, 2013). Coastal vegetation not only sequesters carbon through its high productivity but also traps sediment, elevates the sea floor through salt marsh accretion and upward growth, and lessens wave energy. Thus, it plays an important dual role in acting as both a CO_2 sink and a natural coastal defense system. Therefore, restoration of degraded coastal wetlands and preservation of existing habitat (e.g., using approaches described in Section 9.3.1.2), offers a safe and cost-effective mitigation strategy.

9.4.2.2 *Carbon Capture and Storage*

A number of engineering techniques can be used to sequester carbon offshore. These include deep oceanic injection, or converting CO_2 into stable carbonates (such as chalk). These are not usually directly relevant to coastal, cities but some cities are situated above or provide supporting infrastructure for injection into geological strata. This involves capture, liquefaction, transport, and injection of industrial CO_2 into deep strata including coal seams, old oil wells, stable rock strata, or saline aquifers (Lal, 2008).

9.5 Cross-Cutting Theme Linkages

9.5.1 Governance

Multilevel governance strategies that involve international to local public, private, and nonprofit stakeholders are central to addressing issues such as climate change that extend beyond local administrative boundaries into the social, environmental, and economic systems with which they connect. Climate change hazards pose governance challenges for coastal cities due to the uncertainties associated with the extension of its impact across the territory, thus increasing the complexity of decisions to prioritize adaptation measures. For example, hazards can require population resettlement in retreat adaptation strategies, which can lead to considerable social and economic impacts and thereby demand the empowerment and engagement of all actors involved.

To respond to climate change, governance depends deeply on government's political and institutional capacity to carry out and/or enforce mitigation and adaptation actions. For example, in Mexico City "downsizing and retrenchment of the state, liberalization, decentralization and deregulation" are limiting institutional capacity to adapt (Romero-Lankao, 2007). In many countries the lack of government reliability due to corruption, but also lack of continuity and long-term policies, can be a great barrier for governance, limiting its capacity to respond to all previous challenges. Moreover, in developing countries, climate change adaptation can be delayed by fragile government structures and deficient basic infrastructure (Bulkeley et al., 2011). In the Dar es Salaam Case Study, a port city of 4 million in sub-Saharan Africa, investment requirements to address lack of infrastructure and informal housing and economies, as well as to achieve mobility, solid waste management, and sanitation, deprioritize the need for local-level climate change mitigation and adaptation issues as a public urgency (Kiunsi, 2013).

Coastal cities must build capacity and mobilize the necessary political and financial resources to deliver measures to protect the population and adapt their infrastructure (Bulkeley et al., 2011). Civil society and NGOs play a determinant role in prioritizing climate change adaptation (Bulkeley et al., 2011; Broto and Bulkeley, 2013). This emphasizes that climate change responses cannot be exclusively the responsibility of the public sector (Broto, Obala, and Junior, 2013). As Broto and Bulkeley (2013) have shown, in Asia's economic context, for example, private actors have prevailed in leading urban infrastructure investments seeking climate adaptation.

Case Study 9.6   Adapting to Climate Change in Coastal Dar es Salaam

Liana Ricci and Silvia Macchi

Department of Civil, Building, and Environmental Engineering, Sapienza University of Rome

Keywords	Adaptation; mainstreaming; peri-urban; vulnerability; seawater intrusion
Population (Metropolitan Region)	4,364,541 (National Bureau of Statistics, 2013)
Area (Metropolitan Region)	1,393 km² (National Bureau of Statistics, 2013)
Income per capita	US$820 (World Bank, 2017)
Climate zone	Aw – Tropical savannah (Peel et al., 2007)

Dar es Salaam is located in the eastern part of Tanzania. It is bounded by the Indian Ocean on the east and the Coast Region on the west. The main physical features are the coastal plains, composed of limestone; the alluvial plains, with a series of valleys; and the upland plateau, 100–200 meters in altitude. It is the third fastest-growing city in Africa, the largest one in Tanzania, and the administrative and economic hub of the country. It has the status of a city-region with both a regional administration and four local government authorities (LGAs), the City Council, and the three Municipal Councils of Ilala, Kinondoni, and Temeke.

About 80% of residents live in spontaneous low-density settlements where livelihoods are based on skillful and dynamic combinations of urban and rural activities and resources. Preserving natural resources is fundamental to residents' survival and capacity to adapt to climate change.

RELEVANCE OF THE ACTION TO CLIMATE CHANGE ADAPTATION

The Adapting to Climate Change in coastal Dar es Salaam[1] project is aimed at improving the effectiveness of municipalities supporting coastal peri-urban dwellers who depend on natural resources in their efforts to adapt to climate change (Ricci, 2014). The project was co-funded by the European Union and implemented between 2010 and 2014. It focused on communities in the coastal plain where groundwater salinization, due to the combined effects of urban expansion and climate change, seriously affects local communities because they depend heavily on boreholes for accessing water for domestic and productive purposes (see ACC Dar Papers accessible from the project website; ACC, n.d.).

The project provided Dar's municipalities with enhanced methodologies for mainstreaming climate change adaptation into their Urban Development and Environment Management plans and programs and increased their understanding of adaptation practices. The interplay between climatic and nonclimatic stressors (i.e., developing methodologies for assessing groundwater salinization and urban sprawl) was explicitly recognized. The action aimed to improve the capacity to integrate climate change concerns at the local level, thus contributing to the implementation of the National Adaptation Programme of Action.[2]

Three sets of activities were implemented: (1) improve understanding of climate change adaptation; (2) develop methodologies for designing adaptation initiatives; and (3) build the local government authorities' capacity to understand climate change issues specific to peri-urban livelihood in the coastal plain and identify effective measures for supporting coastal peri-urban inhabitants in their efforts to adapt to climate change.

ACTION AND POLICY DRIVERS

Sapienza University of Rome designed and coordinated the project with Ardhi University of Dar es Salaam as a partner and the Dar es Salaan City Council as an associate. The three municipalities were the main stakeholders. The coastal peri-urban residents were involved in the household survey, borehole monitoring, and participatory scenario building (see Case Study 9.6 Figure 1). It ensured that their concerns over environmental change, coping experiences, and future expectations played a major role in the definition of adaptation objectives. Other relevant stakeholders took part in the activities (e.g., the Wami Ruvu Basin Authority was involved in capacity-building to share competences and responsibilities and strengthen the relationships with the municipalities). Nongovernmental organizations (NGOs) (Haki Ardhi, Forum Climate Change, Environmental Engineering and Pollution Control Organization [EEPCO]) also took part in the capacity-building program and community-based organizations (CBOs) (Kigamboni Community Centre and Club Wazo) were key players in participatory activities (i.e., community-based scenario exercises combining a backcasting approach and forum theatre) (Macchi and Ricci, 2016; Macchi and Tiepolo, 2014). The whole process of knowledge and methodology development, together with the continual interactions and sharing of the results with local decision-makers, was crucial to improving capacity in adaptation mainstreaming and bridging the gap between knowledge and action (Rugai and Kassenga, 2014).

IMPACT AND SCALE

The actions focused on climate change impacts on peri-urban areas within the coastal plain, slow and incremental environmental changes, and risks to the sustainability of rural–urban livelihoods, rather than on extreme weather events and disasters. Addressing the salinization of the coastal aquifer in connection with seawater intrusion and urban sprawl, the project carried out a detailed analysis of the current situation and future scenarios (Faldi and Rossi, 2014; Congedo and Munafò, 2014). This led to the formulation of amendments for the current Urban Development and Environmental Management plans and programs. These amendments were formulated using an ad-hoc methodology to support adaptation efforts through the water conservation and secured access to water resources for domestic and productive uses (see Case Study 9.4 Figure 2) (Shemdoe et al., forthcoming).

LESSONS LEARNED

By strengthening the relationship between knowledge institutions and local government authorities (LGAs), and by involving multiple

1   www.planning4adaptation.eu
2   See http://unfccc.int/national_reports/napa/items/2719.php.

Case Study 9.6 Figure 1 *Community-based scenario exercise combining backcasting approach and forum theatre techniques.*

Photo: Laura Fantini

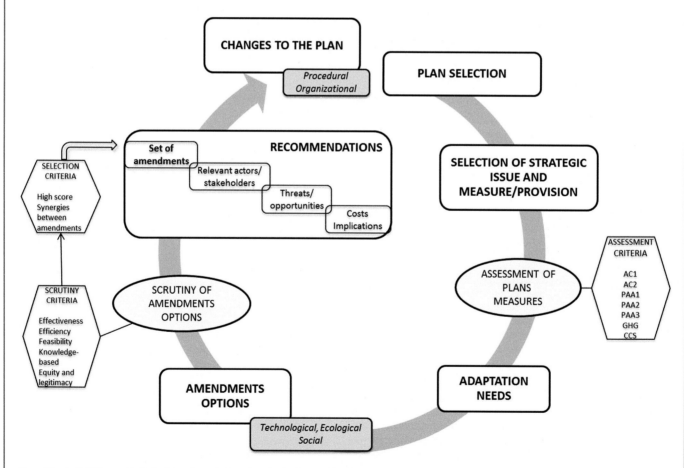

Case Study 9.6 Figure 2 *Methodology for mainstreaming climate change adaptation into Urban Development and Environmental Management plans and programs.*

Source: Liana Ricci

actors and stakeholders, the territorial approach[3] proved to be highly suitable for tackling adaptation challenges. The importance of the local dimension in determining the efficacy of adaptation mainstreaming in Urban Development and Environmental Management was confirmed. However, given the poor implementation of many urban and environmental plans, the viability of climate change mainstreaming should be carefully evaluated (Macchi and Ricci, 2014). Moreover, an in-depth analysis of autonomous adaptation strategies and practices is necessary to understand vulnerabilities, their structural causes, and the diverse responses that are emerging. Autonomous adaptation includes strategies to cope with current and future climate-related environmental stresses and changes in coastal peri-urban areas (IPCC, 2014).

At the same time, future uncertainties require the consideration of multiple adaptation strategies to be revised as new challenges and opportunities arise.

The relationships between access to natural resources and climate change vulnerability needs to be deeply investigated and particular attention should be paid to the intersection between uneven access to resources, environmental processes, and human activities affecting the quality and availability of resources such as water. The collaboration among knowledge and research institutions, LGAs, and stakeholders was crucial for integrating different types of knowledge and competences in decision-making to support effective adaptation actions and reduce vulnerability.

Stakeholders in coastal areas can be very heterogeneous, from working waterfronts including small local fishing or industrial businesses to large utility companies, residential and commercial neighborhoods, and stewards of recreational open spaces and ecological sanctuaries. To engage everyone in the resiliency-building process, governance requires transparency in city government actions, inclusive planning, accessibility in the distribution of technical information, and ample opportunities for community oversight (Adger et al., 2006).

9.5.2 Co-Benefits, Barriers, and Bridges

Adaptation efforts in coastal cities provide potential co-benefits for climate change mitigation and local economic development. Although these reflect some of the distinct features of their location in the coastal zone (e.g., marine trade networks, coastal habitats), they are not fundamentally different from other cities although possibly more acute. For example, coastal megacities will experience more intensified conflicts related to land-use due to "squeeze," whereas coastal erosion, salt water intrusion, loss of habitats for fish and wildlife, and deteriorating marine environment aggravate natural disasters.

A significant challenge for coastal cities will be to develop and grow in a more climate-sensitive manner. It is cheaper and easier to protect a small compact area from rising sea level than a large sprawl, whether by erecting dikes, elevating the urban area, or rebuilding inland. Retreat is less expensive with a dense area because there is less infrastructure (e.g., roads and sewers) that needs to be rebuilt. Furthermore, higher densities can reduce energy use and lower infrastructure costs.

9.5.3 Urban Planning and Design

The challenge to urban planners and designers in coastal cities is to find positive solutions to adapt, resist, or mitigate coastal climate risks while being sensitive to the coastal environment and adding value to the development. A growing body of

innovative architectural and engineering design work is emerging that embraces these challenges within urban design and planning (Building Futures, 2010). Central to these are use of new building design approaches, multifunctional flood defense infrastructures that benefit local people and businesses, and planning systems that enable integrated and long-term local strategies.

Possible secondary functions of flood defenses include housing, transport, shipping, agriculture, habitat, and amenity (Anvarifar, 2013). For example, a flood defense with a road on top is multifunctional. Governance for these structures must be considered because they require maintenance and inspection to ensure safety. It must be noted that there is potential to create conflicts of interest with multiple stakeholders invested in the structure (Voorendt, 2015).

Low-lying coastal or riverine cities often develop around a series of canals (the "Venetian solution"). Venice is best-known, but Amsterdam, Bruges, Copenhagen, Suzhou, and Bangkok also boast extensive canal networks. Although many of Bangkok's canals have recently been filled and paved over, Rotterdam plans to construct additional moats and canals to store excess discharge at times of high water. The expanded canal system would additionally create a supplementary transportation corridor. Innovative harbor redevelopment proposals include formation of an artificial island archipelago and reefs using dredge material or reconfiguring the shoreline by lowering the seaward gradient, rehabilitating or expanding remnant wetlands, creating new parks, and building a "feathered edge" of piers and slips to lessen storm surge impacts (Nordenson et al., 2010). Some of these inventive ideas have already been implemented or are under consideration. For example, with only a few cruise ships still docking directly at Manhattan piers, New York has built an extensive network of parks and pedestrian and bicycle pathways along the waterfront. Floating buildings (like large houseboats) represent yet a different mode of living with a higher water level. A floating neighborhood of approximately 100 houses has been constructed in Ijberg, eastern Amsterdam, and more are planned (Finney, 2013). Other buildings sit on a

---

3   The territorial approach for adaptation is a place-based approach in which local authorities interact with a range of other local and multilevel actors (e.g., universities) to make the most of existing political and institutional resources. LGAs also cooperate with national-level institutions to implement National Adaptation Programs of Action (NAPA). Moreover, climate change adaptation mainstreaming of LGAs not only improves the national climate change agenda, but also contributes to the effective use of resources and enhances the potential to attract other resources.

Case Study 9.7   Rotterdam: Commitment for a Climate-Proof City

Vivien Gornitz

Center for Climate Systems Research, Columbia University, New York

Keywords	Sea level rise, river flooding, climate resilience, coastal
Population (Metropolitan Region)	1,175,477 (Statistics Netherlands, 2014)
Area (Metropolitan Region)	782,43 km² (Statistics Netherlands, 2015)
Income per capita	US$46,310 (World Bank, 2017)
Climate zone	Cfb – Warm temperate, fully humid, warm summer (Peel et al., 2007)

Like many other major port cities, Rotterdam faces increasing vulnerability to sea level rise. The Delta Committee (2008) anticipates a sea level rise of 0.65 to 1.3 meters for the Netherlands by 2100, as compared with 0.4 to 1.05 meter after Katsman et al. (2011).

Rotterdam has embarked on an ambitious plan to "climate-proof" the city, anticipating up to 1.3 meters higher sea level and greater winter rainfall (causing additional river flooding) by 2100. The Delta Committee (2008) has therefore proposed to tighten current flood risk standards by creating a "closable yet open" Rhine estuary through new movable flood barriers that would divert water into safer directions depending on hydraulic conditions. Other options include introducing new interconnections within the existing canal and river network or even building entirely new waterways. Added coastal hazard protection from storm surges and sea level rise including reinforcing water defenses – the dikes and the neighborhoods beyond the dikes – as well as undertaking appropriate land-use planning, constructing water-adapted buildings, and allowing water to drain or to temporarily retain excess rainwater. Large areas of unembanked land reserved for future urban expansion will need to be raised by 1–1.5 meters to meet the revised base flood level (the minimum threshold level for new buildings and infrastructure) requirement of 3.9 meters above the Amsterdam Ordnance Datum (Rijcken, 2010; Van Veelen, 2010).

The Floating Pavilion, installed in June 2010, previewed other floating districts planned for the Stadshavens section of Rotterdam. It represents a prototype for future floating districts (Aboutaleb, 2009). Floating homes already exist in various places, such as the sampans in Hong Kong harbor and houseboats along the shores of Lake Union, Seattle, Washington, and Sausolito, California. Former barges, now comfortable homes, line the canals of Amsterdam and Rotterdam. Rotterdam even boasts a floating three-star hotel in historic Wijnhaven. Rotterdam foresees entire future floating residential districts, complete with office complexes and parks, and will expand

Case Study 9.7 Figure 1 *Multipurpose dike in Scheveningen, a coastal resort near the Hague, the Netherlands.*

Photo: Vivien Gornitz

its network of moats and canals. Impermeable concrete or asphalt associated with urbanization prevents rainwater from infiltrating into the soil and recharging aquifers and causes street and basement flooding after heavy rains. In Rotterdam, excess rainwater is temporarily stored beneath municipal parking or in water plazas that also serve as parks or playgrounds when dry. "Green roofs" also helps curb excess runoff (Aerts, 2009).

Following the disastrous North Sea flood of 1953, which left 300,000 people homeless and caused 1,800 deaths, the Netherlands began construction of the Delta Works (a series of dams, surge barriers, dikes, and sluices along the Rhine-Meuse-Scheldt delta), which closed off some of the estuarine outlets to the sea. Approximately 40% of the country lies below sea level, and an even greater proportion is vulnerable to coastal or river flooding. The lowest areas are highly populated and encompass the bulk of the nation's economic activity. Acceptable risk levels of dike failure were set at 1 in 10,000 years for North and South Holland; other high risk areas are set at 1 in 4,000 years, and South Holland river flooding at 1 in 1,250 years. A movable barrier, the Maeslant Barrier completed in 1997, now closes off the New Waterway (Nieuwe Waterweg) whenever a high storm surge threatens the city of Rotterdam and surrounding dikes. These impressive engineering structures have been designed for a sea level rise of only 20 to 50 centimeter per

century (Delta Committee, 2008), far short of the projected rise. Sea level rise will lead to more frequent barrier closures: three closures per year for an 85 centimeter rise and an average of seven closures per year for a 1.3 meter rise. The Delta Committee (2008) therefore recommended that existing or new storm surge barriers should withstand a worst-case scenario of regional sea level rise of 0.65–1.3 meters by 2100, and up to 2–4 meters by 2200 (using the SRES A1F1 emissions scenario) (Nakicenovic et al., 2000). Present flood protection standards in the diked areas must be raised tenfold by 2050. Dikes, such as a massive new one 1 kilometer long and 12 meters high under construction in the coastal resort of Scheveningen, near the Hague, will serve multiple purposes (see Case Study 9.5 Figure 1). It will be overtopped by a sweeping new boulevard with broad stairways to ensure ready access to the beach, which is also being widened to dampen the power of the waves and buffer against floods.

Rotterdam, like many other delta or tidal river cities, faces inland flooding as well as marine incursions. Dutch winters are expected to grow wetter while summers become drier. Therefore, water management plans will defend against both inland and coastal flooding and also ensure an adequate water supply during dry spells. New developments beyond the dikes will be designed not to interfere with river discharge or lake water levels.

series of artificial islands on the Ijsselmeer, accessible by bridges and walkways. To deal with extreme hazards, tsunami-resilient building designs are starting to emerge using principles of performance-based design – rather than try to resist the force of the tsunami, the lower floors can be opened up (given sufficient warning) or designed so the walls easily "blow-through" while the upper floors and main supports remain intact (Designs Northwest, 2015).

9.5.4 Equity and Environmental Justice

Climate change poses risks to urban coastal populations, but the uneven spatial distribution of these risks has not been fully understood (Bautista et al., 2015b). This constitutes a critical gap for communities currently challenged by poverty and/or environmental injustice (and their cumulative impacts), vulnerabilities that can be exacerbated by severe weather and climate change (Milligan et al., 2009; Nicholson-Cole and O'Riordan, 2009). Equity and environmental justice issues are therefore critical factors to assessing local vulnerability to floods and other climate change impacts in urban areas (Maantay and Maroko, 2009; Bautista et al., 2015a). Coastal cities, particularly those in mid- and low-income countries, concentrate the most vulnerable communities in high-risk areas – informal settlements and substandard housing mainly located in flood-prone areas or in landslide susceptible hills. These vulnerable populations experience a cumulative risk effect that adds to the underlying adversities posed by the lack of employment, poor health conditions, and deficient urban services and infrastructure, among others (Satterthwaite, 2013). This is the case in cities like Chittagong, the biggest port city on the coast of Bangladesh, where landslides have caused 400 deaths since 1997 (Ahammad, 2011), or in Rio de Janeiro, the biggest Brazilian coastal city. Rio is

highly threatened by summer storms and had experienced 373 deaths from climate-related disasters from 1966 to 2010, when the city was hit by a huge storm (Prefeitura da Cidade do Rio de Janeiro, 2015). In 2015, the municipal government launched its new resilience plan, Rio Resiliente, that prioritizes reducing the vulnerability of thousands of households under risk of potential landslides (mainly in the slums) by adding to previous plans that were specifically focused on disaster prevention (Prefeitura da Cidade do Rio de Janeiro, 2015).

The challenge posed then is to guarantee that adaptation and mitigation measures do not reinforce inequalities (Olsson et al., 2014). This creates an opportunity for environmental justice advocates, community-based planners, and other local stakeholders to lead efforts in identifying strategies to build climate resiliency and to address the needs of vulnerable populations (Bautista et al., 2015a). The New York City Environmental Justice Alliance's (NYC-EJA) approach to building community resilience illustrates the potential for innovative, locally driven resiliency planning. This approach "focuses on building healthy, resilient communities while advocating for long-term climate adaptation, mitigation, and resiliency measures in order to ensure that vulnerable communities are stronger and healthier before disaster hits" (Bautista et al., 2015b).

In the aftermath of Hurricane Sandy, there have been extensive conversations regarding opportunities to reduce the vulnerability of the New York region to flooding and storm surge, but government reports had almost exclusively focused on the built environment. NYC-EJA co-convened and facilitated the Sandy Regional Assembly (SRA), an association of grassroots stakeholders from communities vulnerable to severe weather events, in order to address this gap. As a result of this process, the SRA issued a Recovery Agenda, including comprehensive

349

Figure 9.4 *Post Hurricane Sandy community forum in New York.*

Photo: UPROSE

rebuilding/resiliency recommendations and capital projects – some of which were incorporated in government plans (Sandy Regional Assembly, 2013). NYC-EJA has also partnered with one of its member organizations, The Point CDC (www.thepoint .org), to create the South Bronx Community Resiliency Agenda and engage local communities in creating a comprehensive climate resiliency agenda to strengthen the physical and social resiliency of five of the most vulnerable communities in NYC – and in the United States (Bautista et al., 2015b; NYC-EJA, 2015). The Climate Justice & Community Resiliency Center that is being developed by the community-based planning organization UPROSE in Sunset Park, Brooklyn, is another example where an environmental justice community is leveraging on-the-ground knowledge and grassroots organizing with advice from urban planners and public health scientists to create strategies for a climate-resilient industrial waterfront community (www .uprose.org/).

Addressing existing vulnerabilities to build community resiliency constitutes a central pivot point for the governance of coastal areas (Paavola et al., 2006). Over the past decade, environmental justice advocates have been striving for the recognition of principles of equity and environmental justice in climate change discussions. In part as the result of these efforts, in 2015, the de Blasio Administration published its sustainability blueprint, "One New York, The Plan for a Strong and Just City," articulating strategies to address equity as a cross-cutting priority in the city's long-term economic growth, sustainability and resiliency agendas (City of New York, 2015). Nonetheless, to guarantee that these efforts reach those who need them most, such government initiatives require the direct continued participation of all sectors of civil society in the decision-making process – both during the planning and design phases as well as throughout its implementation (Scholsberg, 2014; Bulkeley, 2014).

9.6 Conclusions, Policy Recommendations, and Areas for Future Research

Climate change, and sea level rise in particular, will likely exacerbate natural hazards to which coastal cities are uniquely exposed (e.g., storm surges, beach erosion, and saltwater intrusion). Urban centers built on low-lying deltas will be especially vulnerable. Many coastal cities face additional risks from river and groundwater flooding, loss of protection from offshore (coral) reefs, and increased wave damage at the shore. These climate-induced changes will, in turn, affect marine ecosystems, aquifers used for urban water supplies, the built environment, and disruption to transportation and economic activities, particularly following extreme storm events. The intensity of the risk will vary from place to place, depending on the extent of local changes in sea level rise, ocean warming, precipitation, or river runoff. In addition to natural hazards, cities face vulnerabilities in exposure of critical infrastructure in flood zones, precariously built housing in poor neighborhoods, and sharp contrasts in income distribution that affect adaptive capacity and personal health.

Various strategies exist to manage climate-induced risks affecting coastal cities. These range from increasing shoreline protection, both by building defensive structures or by adopting "soft," more natural solutions, such as dune building or wetlands preservation and restoration. Other adaptive strategies include accommodating structures and lifestyles to a more aquatic presence. It may ultimately become impossible to defend further development in extremely high-risk areas. However, a long-term integrated approach to coastal management and inclusive governance is essential to adapt to climate change impacts and manage cities in the coastal zone in the third millennium (Nicholls et al., 2015).

Case Study 9.8   Preparing for Sea-Level Rise, Coastal Storms, and Other Climate Change–Induced Hazards in New York

Vivien Gornitz and David C. Major

Center for Climate Systems Research, Columbia University, New York

Keywords	Sea level rise, coastal flooding, structural adaptations
Population (Metropolitan Region)	20,153,634 (U.S. Census Bureau, 2016)
Area (Metropolitan Region)	17,319 km²
	(U.S. Census Bureau, 2010)
Income per capita	US$56,180 (World Bank, 2017)
Climate zone	Dfa – Cold, without dry season, hot summer (Peel et al., 2007)

New York, with a population of 8.4 million (Metropolitan Statistical Area population 20.1 million [2016]), is the largest city in the United States and is a major center of global finance and commerce. Significant economic assets and a population vulnerable to coastal flooding line its 837 kilometers of shoreline. The city has long led climate change adaptation efforts (NPCC, 2010; Rosenzweig et al., 2011), and new projects have been initiated since Hurricane Sandy in October 2012 (SIRR, 2013; New York City Office of Emergency Management, 2014). The city's population residing within the 100-year flood zone[4] exceeds that of other vulnerable U.S. coastal cities, including Houston, New Orleans, and Miami (SIRR, 2013). The New York-Newark-New Jersey area ranks among the world's ten port cities most vulnerable to coastal flooding (by assets, not population), not counting future coastal protection (Hanson et al., 2011). This case study highlights major conclusions about accelerated sea level rise and increased coastal flooding and some of New York's ongoing and planned responses.

Underscoring New York's vulnerability to intense coastal storms, Hurricane Sandy, in October 2012, caused forty-four deaths, hospitals and nursing home evacuations, inundation of some subway and other tunnels and 17% of the city's land area, power outages affecting nearly 2 million people, and major transportation disruptions. An estimated 70,000–90,000 buildings were flooded and/or damaged. Storm damages totaled nearly US$20 billion (New York, PlaNYC, 2014).

Following Sandy, New York developed more comprehensive plans to reduce present and future climate-related risks, drawing upon the scientific expertise of city and regional agencies, universities, the private sector, and the New York City Panel on Climate Change (NPCC), an advisory group inaugurated by Mayor Michael Bloomberg in 2008 from academia, government, and the private sector (SIRR, 2013; NPCC, 2013, 2015; New York City PlaNYC, 2014). Implementation of resiliency measures has already begun. This case study highlights

major NPCC conclusions regarding accelerated sea level rise and increased coastal flooding, with some of New York City's ongoing and planned responses.

CLIMATE, SEA LEVEL RISE, AND COASTAL STORMS

New York experiences a hot-summer continental climate (Köppen-Geiger, Dfa), with a mean annual temperature of 12.2°C between 1971 and 2000, climbing 0.2°C per decade between 1900 and 2013 (NPCC, 2015). Rainfall ranges between 109 and 127 centimeters per year, rising by 2.0 centimeters per decade between 1900 and 2013. Variability has increased noticeably over the past 40 years (NPCC, 2015). In the future, the NPCC finds that mid-range temperatures could increase by 2.3–3.2°C by the 2050s, 2.95–4.9°C by the 2080s, and 3.3–5.8°C by 2100 (NPCC, 2015)[5].

Twentieth-century regional sea level rise ranged between 2.2 and 3.8 centimeters per decade in the New York metropolitan area[6] (NOAA, 2016), as compared to the global average rise of 1.7 centimeter per decade (IPCC, 2013). The NPCC (2015) finds that mid-range sea level at the Battery (southern tip of lower Manhattan) will increase 28–53 centimeters by the 2050s and 46–99 centimeters by the 2080s, relative to 2000–2004. High-end estimates reach 76 centimeters by the 2050s, 147 centimeters by the 2080s, and 190 centimeters by 2100.

Past winter cyclones and hurricanes have flooded parts of the city. Hurricane Sandy (an extra-tropical-tropical storm in NYC) generated the highest recorded water level (3.38 m) at the Battery (the southern tip of Manhattan) that has been recorded in nearly 200 years due to strong easterly winds, surge amplification,[7] maximum storm surge at high tide and full moon, plus historic sea level rise (0.44 m since 1856 [NOAA, 2016]).

Using the just noted sea level rise projections, updated Federal Emergency Management Agency (FEMA, 2013) flood return curves, and assuming unchanged storm characteristics, flood heights for the 100-year storm (excluding waves) would rise from the present 3.4 meters in the 2000s to 3.9–4.5 meters by the 2080s (mid-range). The annual likelihood of such a flood would increase from 1 to 2.0–5.4% by the 2080s, with a high estimate of 12.7%. The area potentially at risk to flooding would consequently expand in the future.

VULNERABILITY TO SEA LEVEL RISE AND COASTAL STORMS

Major critical New York assets, including port facilities, major transportation routes, oil tanks and refineries, power stations, and wastewater treatment plants lie along the waterfront or within the 100-year flood zone. Coastal wetlands in Jamaica Bay that buffer storm surges and waves and provide important wildlife habitat, recreation, and water pollution filtration have deteriorated within the past few decades (Hartig et al., 2002), although a restoration program is under way (New York City Audubon). Wetland restoration to

---

4   The area flooded by a storm having an estimated 1% probability of occurrence per year.
5   Future temperature and precipitation projections derive from a suite of 35 global climate models (Coupled Model Intercomparison Project Phase 5, CMIP5) and two GHG emission pathways (RCP 4.5 and RCP 8.5; IPCC, 2013b), which produce a 70-member model-based distribution for risk-based decision-making. The 25th to 75th percentile of the model-based distribution represents the mid-range; the 90th percentile represents a high-end estimate. Projected temperatures and precipitation are averaged over selected 30-year time intervals; 10-year intervals for sea level rise, of 24 GCMs are used.
6   Of this, 0.86–1.1 centimeters per decade is due to glacial isostatic adjustments from collapse of the peripheral bulge south of the last ice sheet, following its retreat (NOAA, 2016; PSML, n.d.).
7   Landward funneling of storm surge due to near right-angle geometry of New Jersey and Long Island shorelines.

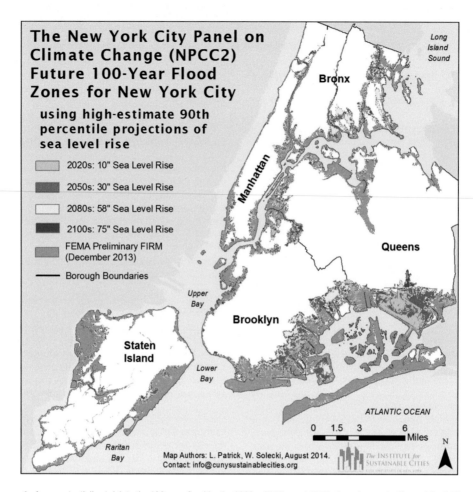

The New York City Panel on Climate Change (NPCC2) Future 100-Year Flood Zones for New York City

using high-estimate 90th percentile projections of sea level rise

- 2020s: 10" Sea Level Rise
- 2050s: 30" Sea Level Rise
- 2080s: 58" Sea Level Rise
- 2100s: 75" Sea Level Rise
- FEMA Preliminary FIRM (December 2013)
- Borough Boundaries

Map Authors: L. Patrick, W. Solecki, August 2014. Contact: info@cunysustainablecities.org

The INSTITUTE *for* SUSTAINABLE CITIES

Case Study 9.8 Figure 1 *Areas potentially at risk to the 100-year flood by the 2020s, 2050s, and 2080s based on projections of the 90th percentile sea level rise.*

Source: NPCC , 2015

preserve habitat and buffer storm surges in Jamaica Bay is ongoing in partnership with the New York City Department of Parks, New York State, and the U.S. Army Corps of Engineers. Continual erosion of the city's sandy beaches requires periodic nourishment with dredged sand by the U.S. Army Corps of Engineers. Although diverse income groups share a similar exposure to coastal flood hazards in New York and surroundings, the poor, aged, and disabled urban populations are more vulnerable and less able to cope with natural disasters (Buonaiuto et al., 2011). Following Hurricane Sandy, the City's Office of Emergency Management (OEM), Center for Economic Opportunity (CEO), and other city agencies worked closely with grassroots community groups, such as the New York City Environmental Justice Alliance and its member organizations, to improve local preparedness and response capacity in extreme weather events (e.g., New York City Office of Emergency Management, 2014; SIRR, 2013).

ADAPTATION TO RISING SEAS AND STORM SURGES

New York intends to strengthen its resiliency to climate change risks (SIRR, 2013; New York City, PlaNYC, 2014). Ongoing and proposed initiatives call for improved coastal flood mapping, continued collaboration with the NPCC to update and refine local climate projections, and strengthening of coastal defenses using a variety of approaches tailored to specific neighborhood needs. These include raising bulkheads and seawalls; enhancing beach nourishment;

building local levees and storm surge barriers; implementing stricter building codes to reduce flood risk; increasing protection of critical infrastructure, including port facilities, utilities, telecommunications, sewer and drainage systems, and transportation arteries; and restoring wetlands and beach dunes. The city also recommends several smaller, strategically placed local storm surge barriers (SIRR, 2013). Many of these adaptations are included in "The Big U," an imaginative proposal for a protective system around the southern part of Manhattan stretching from West 57th Street south to the Battery and up to East 42nd Street (Inexhibit, n.d.). An important element of protection for part of lower Manhattan that is now moving forward is the East Side Coastal Resilience Project, which includes flood barriers and other measures to protect the area from East 23rd Street south to Montgomery Street (New York City, PlaNYC, 2015).

CONCLUSION

Global warming in New York is anticipated to increase temperatures, produce more frequent and prolonged heat waves, create more intense downpours, raise sea level, and lead to more damaging coastal flooding. In response, the city has embarked on a comprehensive long-term program to improve its resiliency to climate change, working closely with the NPCC, FEMA, and other state and national agencies. New York City's proposed strategies for enhancing coastal urban resiliency stand as a model for other urban coastal centers to prepare for climate change.

9.6.1 Policy Recommendations

9.6.1.1 Adaptive Management of Coastal Cities

An adaptive management framework for urban coastal adaptation involves assessing climate change risks, implementing various adaptive measures, monitoring outcomes periodically, and refining these successively as more up-to-date climate change data become available (e.g., Environment Agency, 2012; NPCC, 2010, 2015). An adaptive approach maintains flexibility to accommodate to changing conditions over time. While simple, less costly measures can be implemented in the short-term, the door is kept open for major projects needed for long-term protection as needed. Such an approach is already being applied in London (e.g., the Thames Estuary 2100 Plan; Environment Agency, 2012), in New York (New York City Office of Emergency Management, 2014), and across the Netherlands (Delta Commission, 2013).

9.6.1.2 Integrated Coastal City Management

Coastal cities need to develop and adopt integrated management strategies that plan adaptation to climate risks, encompassing where possible co-benefits for the built environment, ecosystems, and human systems. Appropriate land-use planning for sustainable infrastructure development in low-lying coastal areas should become an important priority.

9.6.1.3 Inclusive Governance and Learning from Change

Delivering adaptive and integrated responses will require greater coordination and cooperation on coastal management issues. This must be fostered among all levels of local, regional, and national governing agencies and be participative, engaging with other stakeholders. Learning from experience, especially dealing with change in coastal cities, is challenging. We understand how to defend a coast much better than how we can live in cities on a dynamic coast. Action research should be undertaken on this issue, and the lessons must be captured and incorporated into coastal policy formulation.

9.6.1.4 Integrated Decision-Support Tools and Skills

Central to delivering sustainable and climate-sensitive coastal cities will be the monitoring of critical climate indicators and the development of approaches for integrated assessment of coastal cities to support decision-making. The necessary cross-disciplinary approaches, including the research to support them, require education programs for early career researchers and practitioners to facilitate future development and the implementation of integrated assessment and decision-making in practice (Nicholls et al., 2015).

9.6.2 Knowledge Gaps

- Significant uncertainties remain in understanding climate change processes, but in particular for upper-end scenarios such as ice sheet collapse.

- Information is lacking on the response and long-term behavior of deltas and other complex, dynamic, coastal environments.
- Understanding of interactions between coastal and urban processes is limited: these include atmospheric flows that influence air pollution and human health and biogeochemical processes that can extend the influence of coastal cities into the atmosphere and thousands of kilometers into the ocean.
- There is a need for more systematic integrated assessment to address a limited understanding of and to provide quantitative evidence-based understanding of tradeoffs and co-benefits.
- The costs of coastal climate management choices are poorly understood, thus hindering robust evaluation of the benefits and costs of coastal management.
- Investment decisions and implementation of successful transitions in coastal city management have both technical and social dimensions, and analysis capable of delivering this is needed to develop adaptation pathways under uncertain futures.

Annex 9.1 Stakeholder Engagement

Coastal cities face particular challenges but are no less diverse than other cities. The international authorship of this chapter was therefore augmented with a stakeholder engagement program that involved three main mechanisms: (1) Chapter authors exploited their own networks and attendance at international workshops and conferences in the United Kingdom, France, the United States, Bangladesh, India, Netherlands, China, and Brazil to engage with a multidisciplinary, global group of stakeholders who contributed a broader and holistic perspective to the challenges of coastal cities; (2) Chapter authors worked closely with local actors and stakeholders (e.g., the NYC-EJA, New York City Mayor's Office, Khulna Mayor's Office, and local communities) in the development of detailed case studies; and (3) More formally, the chapter was reviewed by international experts from academia, industry, and international agencies.

Chapter 9 Urban Areas in Coastal Zones

References

Aboutaleb, A. (2009). Rising water level approach offers opportunities. *Change Magazine: Water and Climate* **5** (1), 4–5.

ADB. (2011). *Adapting to Climate Change: Strengthening the Climate Resilience of the Water Sector Infrastructure in Khulna, Bangladesh, Mandaluyong City.* The Asian Development Bank.

Adcock, T. A., Draper, S., Houlsby, G. T., Borthwick, A. G., and Serhadlıoğlu, S. (2013). The available power from tidal stream turbines in the Pentland Firth. Proceedings of the Royal Society A: Mathematical, *Physical and Engineering Science* **469**(2157), 20130072.

Adger, N. (2003). Social aspects of adaptive capacity. In Smith, J., Klein, R., and Huq, S. (eds.), *Climate Change: Adaptive Capacity and Development* (29–50). Imperial College Press.

Adger, N. (2006). Vulnerability. *Global Environmental Change* **16**(3), 268–281.

Adger, W. N., Hughes, T. P., Folke, C., Carpenter, S. R., and Rockström, J. (2005). Social-ecological resilience to coastal disasters. *Science* **309**(5737), 1036–1039.

Adger, W. N., Paavola, J., and Huq, S., eds. (2006). Toward justice in adaptation to climate change. In Adger, W. N., Paavola, J., and Huq, S. (eds.), *Fairness in Adaptation to Climate Change* (1–19). MIT Press.

Aerts, J., Botzen, W., Bowman, M. J., Ward, P. J., and Dircke, P., eds. (2013). *Climate Adaptation and Flood Risk in Coastal Cities*. Earthscan, Taylor and Francis

Aerts, J., Major, D. C., Bowman, M. J., Dircke, P., and Marfai, M. A. 2009. *Connecting Delta Cities: Coastal Cities, Flood Risk Management and Adaptation to Climate Change*. Vrije Universiteit Press.

Ahammad, R. (2011). Constraints of pro-poor climate change adaptation in Chittagong city. *Environment and Urbanization* 23(2), 503–515.

Akinro, A. O., Opeyemi, D. A., and Ologunagba, I. B. (2008). Climate change and environmental degradation in the Niger Delta region of Nigeria: Its vulnerability, impacts and possible mitigations. *Research Journal of Applied Sciences* 3(3), 167–173.

Allenby, B., and Fink, J. (2005). Toward inherently secure and resilient societies. *Science*, 309, 1034–1036.

Amelung, B., and Nicholls, S. (2014). Implications of climate change for tourism in Australia. *Tourism Management* 41, 228–244.

Amelung, B., Nicholls, S., and Viner, D. (2007). Implications of global climate change for tourism flows and seasonality. *Journal of Travel Research* 45, 285–296.

Angell, E., and Stokke, K. B. (2014). Vulnerability and adaptive capacity in Hammerfest, Norway. *Ocean and Coastal Management* 94, 56–65.

Anvarifar, F., Zevenbergen, C., and Thissen, W. (2013). An exploration of multifunctional flood defences with an emphasis on flexibility. In *Proceedings of the International Conference on Flood Resilience: Experiences in Asia and Europe*, held in Exeter, United Kingdom, September 5–7, 2013.

Aqua-Sheltech Consortium (2002). *Draft Structure Plan, Master Plan and Detailed Area Plan (2001–2020) for Khulna City, Volume I: Urban Strategy'*. Khulna Development Authority, Ministry of Housing and Public Works, Government of the People's Republic of Bangladesh.

Arinaga, R. A., and Cheung, K. F. (2012). Atlas of global wave energy from 10 years of reanalysis and hindcast data. *Renewable Energy* 39(1), 49–64.

Arkema, K. K., Guannel, G., Verutes, G., Wood, S. A., Guerry, A., Ruckelshaus, M., Kareiva, P., Lacayo, M., and Silver, J. M. (2013). Coastal habitats shield people and property from sea-level rise and storms. *Nature Climate Change* 3, 913–918. doi:10.1038/NCLIMATE1944.

Arthurton, R. and Korateng, K. (2006). Coastal and marine environments. In *Africa Environment Outlook 2* (ch. 5). UNEP, Nairobi, Kenya.

Baker-Austin, C., Trinanes, J. A., Taylor, N. G. H., Hartnell, R., Siitonen, A., and Martinez-Urtaza, J. (2013). Emerging Vibrio risk at high latitudes in response to ocean warming. *Nature Climate Change* 3(1), 73–77.

Ballica, S. F., Wright, N. G., and van der Meulen, F. (2012). A flood vulnerability index for coastal cities and its use in assessing climate change impacts. *Natural Hazards* 64, 73–105.

Bautista, E., Hanhardt, E., Osorio, J. C., and Dwyer, N. (2015a). New York City Environmental Justice Alliance (NYC-EJA) Waterfront Justice Project. *Local Environment: The International Journal of Justice and Sustainability,* 20(6), 664–682, DOI: 10.1080/13549839.2014.949644.

Bautista, E., Hanhardt, E., Osorio, J. C., and Dwyer, N. (2015b). The New York City Environmental Justice Alliance's Efforts to Build Climate Justice and Reduce Industrial Waterfront Vulnerability. *Social Research: An International Quarterly.* 82(3), 821–838.

Bear, J. (ed.) (1999). *Seawater Intrusion in Coastal Aquifers*. Springer.

Bender, M. A., Knutson, T. R., Tuleya, R. E., Sirutis, J. J., Vecchi, G. A., Garner, S. T., and Held, I. M. (2010). Modeled impact of anthropogenic warming on the frequency of intense Atlantic hurricanes. *Science* 327, 454–458.

Bigano, A., Bosello, F., and Roson, R. (2008). Economy-wide impacts of climate change: A joint analysis for sea level rise and tourism. *Mitigation and Adaptation Strategies for Global Change* 13, 765–791.

Billé, R., Kelly, R., Biastoch, A., Harrould-Kolieb, E., Herr, D., Joos, F., … and Gattuso, J. P. (2013). Taking action against ocean acidification: A review of management and policy options. *Environmental Management* 52(4), 761–779.

Bird, E. C. F. (1996). *Beach Management* (281). John Wiley and Sons.

Bird, E. (2008). *Coastal Geomorphology: An Introduction* (ch. 14). John Wiley and Sons.

Birkmann, J., Garschagen, M., Kraas, F., and Quang, N. (2010) Adaptive urban governance: New challenges for the second generation of urban adaptation strategies to climate change. *Sustainability Science* 5(2), 185–206.

Bjerklie, D. M., Mullaney, J. R., Stone, J. R., Skinner, B. J., and Ramlow, M. A. (2012). *Preliminary Investigation of the Effects of Sea-level Rise on Groundwater Levels in New Haven, Connecticut.* USGS OFR 2012–1025. U.S. Geological Survey.

Bodzin, A. M. (1999). *Carolina Coastal Science: Relocating a Lighthouse—Cape Hatteras Lighthouse Timeline*. North Carolina State University (NCSU). Accessed August 11, 2015: http://www.ncsu.edu/coast/chl/timeline.html

Brooks, N., and Adger, W. N. (2004). Assessing and enhancing adaptive capacity, Technical Paper 7. In Lim, B., and Spanger-Siegfried, E. (eds.), *Adaptation Policy Frameworks for Climate Change: Developing Strategies, Policies and Measures* (165–182). United Nations Development Programme and Cambridge University Press.

Broto, V., and Bulkeley, H. (2013). A survey of urban climate change experiments in 100 cities. *Global Environmental Change* 23, 92–102.

Broto, V., Obala, B., and Junior, P. (2013). Governing climate change for a just city: Challenges and lessons from Maputo. *Mozambique. Local Environment* 18(6), 678–704.

Brown, S., Hanson, S., and Nicholls, R. J. (2014). Implications of sea-level rise and extreme events around Europe: A review of coastal energy infrastructure. *Climatic Change* 122(1–2), 81–95.

Building Futures and Institution of Civil Engineers. (2010). Facing up to rising sea-levels: Retreat? Defend? Attack? Accessed October 19, 2015: http://www.buildingfutures.org.uk/projects/building-futures/facing-up

Bulkeley, H., Edwards, G., and Fuller, S (2014). Contesting climate justice in the city: Examining politics and practice in urban climate change experiments. *Global Environmental Change* 25, 31–40.

Bulkeley, H., Schroeder, H., Janda, K., Zhao, J., Armstrong, A., Chu, S., and Ghosh, S. (2011). The role of institutions, governance, and urban planning. In Hoornweg, D., Freire, M., and Lee, M. J. (eds.), *Cities and Climate Change: Responding to an Urgent Agenda*. World Bank Publications.

Burke, L. M., Reytar, K., Spalding, M., and Perry, A. (2011). *Reefs at Risk Revisited*. World Resources Institute, p. 114.

Byers, E. A., Hall, J. W., and Amezaga, J. M. (2014). Electricity generation and cooling water use: UK pathways to 2050. *Global Environmental Change* 25, 16–30.

Cahoon, D. R., Reed, D. J., Kolker, A. S., Brinson, M. M., Stevenson, J. C., Riggs, S., Christian, R., Reyes, E., Voss, C., and Kunz, D. (2009). Coastal wetland sustainability. *Coastal sensitivity to sea-level rise: a focus on the mid-Atlantic region. US Climate Change Science Program Synthesis and Assessment Product 4.1.* 57–72. U.S. Government Printing Office.

Cai, W., Hu, X., Huang, W., Murrell, M. C., Lehrter, J. C., Lohrenz, S. E., Chou, W., Zhai, W., Hollibaugh, J. T., Wang, Y., Zhao, P., Guo, X., Gundersen, K., Dai, M., and Gong, G. (2011). Acidification of subsurface coastal waters enhanced by eutrophication. *Nature Geoscience* 4, 766–770.

Carson, M., Köhl, A., Stammer, D., Slangen, A. B. A., Katsman, C. A., van de Wal, R. S. W., Church, J., and White, N. (2016). Coastal sea level changes, observed and projected during the 20th and 21st century. *Climatic Change* 134, 269–281.

Cash, B. A., Rodó, X., Ballester, J., Bouma, M. J., Baeza, A., Dhiman, R., and Pascual, M. (2013). Malaria epidemics and the influence of the tropical South Atlantic on the Indian monsoon. *Nature Climate Change* 3(5), 502–507.

Cheong, S. -M., Silliman, B., Wong, P. P., van Wesenbeeck, B., Kim, C. -K., and Guannel, G. (2013). Coastal adaptation with ecological engineering. *Nature Climate Change* 3, 787–791.

Chou, W. -C., Wu, J. -L., Wang, Y. -C., Huang, H., Sung, F. -C., and Chuang, C. -Y. (2010). Modeling the impact of climate variability on diarrhea-associated diseases in Taiwan (1996–2007). *Science of The Total Environment* 409(1), 43–51.

City of Hallandale Beach. (2013). *City of Hallandale Beach-Water Supply Strategy*. White Paper, June 14, 2013. Accessed June 13, 2015: http://www.hallandalebeachfl.gov/files/2013-06-17-HBCRA/Item_10A/Water_Supply_White_Paper-June_17_2013_Commission_Workshop, pdf/

City of New York. (2015). One New York, The Plan for a Strong and Just City. Accessed February 20, 2016: http://www1.nyc.gov/html/onenyc/index.html

Colle, B. A., Booth, J. F., and Chang, E. K. M., 2015. A review of historical and future changes of extratropical cyclones and associated impacts along the U.S. East Coast. *Current Climate Change Reports* **1**, 125–143.

Cooley, S. R., and Doney, S. C. (2009). Anticipating ocean acidification's economic consequences for commercial fisheries. *Environmental Research Letters* **4**, 024007. doi:10.1088/1748–9326/4/2/024007

Coombes, E. G., and Jones, A. P. (2010). Assessing the impact of climate change on visitor behaviour and habitat use at the coast: A UK case study. *Global Environmental Change* **20**, 303–313.

Das, A., Justic, D., Inoue, M., Hoda, A., Huang, H., and Park, D. (2012). Impacts of Mississippi River diversions on salinity gradients in a deltaic Louisiana estuary: Ecological and management implications. *Estuarine Coastal and Shelf Science*, **111**, 17–26.

Dawson, R. J., Ball, T., Werritty, J., Werritty, A., Hall, J. W., and Roche, N. (2011b). Assessing the effectiveness of non-structural flood management measures in the Thames Estuary under conditions of socio-economic and environmental change, *Global Environmental Change* **21**(2), 628–646. doi:10.1016/j.gloenvcha.2011.01.013

Dawson, R. J., Dickson, M. E., Nicholls, R. J., Hall, J. W., Walkden, M. J. A., Stansby, P., Mokrech, M., Richards, J., Zhou, J., Milligan, J., Jordan, A., Pearson, S., Rees, J., Bates, P., Koukoulas, S., and Watkinson, A. (2009). Integrated analysis of risks of coastal flooding and cliff erosion under scenarios of long term change, *Climatic Change* **95**, 249–288. doi: 10.1007/s10584–008–9532–8

Dawson, R. J., Peppe, R., and Wang, M. (2011a). An agent based model for risk-based flood incident management, *Natural Hazards* **59**(1), 167–189. doi: 10.1007/s11069–011–9745–4

De la Vega-Leinert, A. C., and Nicholls, R. J. (2008). Potential implications of sea-level rise for Great Britain. *Journal of Coastal Research* **24**(2), 342–357.

Delta Commissie (Committee). (2008). Working together with water: A living land builds for its future. Accessed April 29, 2014: http://www.deltacommissie.com/doc/deltareport_full.pdf

Delta Commission. (2013). *Delta Programme 2014*, The Ministry for Infrastructure and Environment and The Ministry of Economic Affairs.

Deltawerken online. (2014). Accessed October 29, 2015: http://www.deltawerken.com/Deltaworks/23.html

Designs Northwest. (2015). Accessed February 7, 2016: http://www.designsnw.com/portfolio/#fancyboxID-5

Dircke, P., Molenaar, A., and Aerts, J. (2012). Climate adaptation and flood management in the City of Rotterdam. In Aerts, J., Botzen, W., Bowman, M., Ward, P. J., and Dircke, P. (eds.), *Climate Adaptation and Flood Risk in Coastal Cities*. Earthscan.

Duarte, C. M., Losada, I. J., Hendriks, I. E., Marzarrasa, I., and Marba, N., 2013. The role of coastal plant communities for climate change mitigation and adaptation. *Nature Climate Change* **3**, 961–968.

Eggleston, J., and Pope, J. (2013) Land subsidence and relative sea-level rise in the southern Chesapeake Bay region: U.S. Geological Survey Circular 1392. Accessed February 18, 2014: http://dx.doi.org/10.3133/cir1392.

Elsner, J. B., Kossin, J. P., and Jagger, T. H. (2008). The increasing intensity of the strongest tropical cyclones. *Nature* **455**(7209), 92–95.

Enarrson, L. (2011). Box 8.10. Climate change adaptation in Stockholm, Sweden. In Blanco, H., McCarney, P., Parnell, S., Schmidt, M., and Seto, K. C. (eds.), The role of urban land in climate change. In Rosenzweig, C., Solecki, W. D., Hammer, S. A. and Mehrotra, S. (eds.), *Climate Change and Cities: First Assessment Report of the Urban Climate Change Research Network* (217–248). Cambridge University Press.

Engelhart, S. E., and Horton, B. P. (2012). Holocene sea level database for the Atlantic coast of the United States. *Quaternary Science Reviews* **54**, 12–25.

Environment Agency (2012). *Thames Estuary 2100: Managing Flood Risk Through London and the Thames Estuary: TE2100 Plan*. Strategic Environmental Assessment, Environmental Report Summary, London, UK. Accessed June 30, 2014: http://www.environment-agency.gov.uk

Estrada, F., Botzen, W. J., and Tol, R. S. J. (2015). Economic losses from U.S. hurricanes consistent with an influence from climate change. *Nature Geoscience* **8**, 880–884.

Ezer T., 2013. Sea level rise, spatially uneven and temporally unsteady: Why the U.S. East Coast, the global tide gauge record, and the global altimeter data show different trends. *Geophysical Research Letters* **40**, 1–6. doi:10.1002/2013GL057952

Ferguson, G., and Gleeson, T. (2012). Vulnerability of coastal aquifers to groundwater use and climate change. *Nature Climate Change* **2**, 342–345.

Finney, C. (2013). A future afloat. *Geographical Magazine*. Accessed October 3, 2015: http://www.geographical.co.uk/Magazine/Floating_houses_-_Jul_13.html

Florida Department of Environmental Protection (2006). The homeowner's guide to the Coastal Construction Control Line Program (Section 161.053), Florida Statutes. Accessed August 12, 2015: http://www.dep.state.fl.us/beaches/publications/pdf/propowner.pdf

Fuchs, R. J. (2010). Cities at risk: Asia's coastal cities in an age of climate change. AsiaPacific Issues, Analysis from the East-West Center No. 96. Accessed October 13, 2014: http://scholarspace.manoa.hawaii.edu

Füssel, H. M., and Klein, R. J. (2006). Climate change vulnerability assessments: an evolution of conceptual thinking. *Climatic Change*, **75**(3), 301–329.

Gao, Y., Cornwell, J. C., Stoecker, D. K., and Owens, M. S. (2012). Effects of cyanobacterial-driven pH increases on sediment nutrient fluxes and coupled nitrification-denitrification in a shallow fresh water estuary. *Biogeosciences Discussions* **9**(1), 1161–1198.

Gehrels, W. R., and Woodworth, P. L. (2013). When did modern rates of sea-level rise start? *Global and Planetary Change* **100**, 263–277.

Giosan, L. (2014). Protect the world's deltas. *Nature* **516**, 31–33.

Glibert, P. M., Icarus Allen, J., Artioli, Y., Beusen, A., Bouwman, L., Harle, J., Holmes, R., and Holt, J. (2014). Vulnerability of coastal ecosystems to changes in harmful algal bloom distribution in response to climate change: Projections based on model analysis. *Global Change Biology*. doi:10.1111/gcb.12662

Gilbert, S. (1986). *The Thames Barrier*. Thomas Telford Ltd.

Gornitz, V. (2013). Coping with the rising waters. In Gornitz, V (ed.), *Rising Seas: Past, Present, Future* (211–247). Columbia University Press.

Grinsted, A., Moore, J. C., and Jevrejeva, S. (2012). Homogeneous record of Atlantic hurricane surge threat since 1923. *Proceedings of the National Academy of Sciences* **109**(48), 19601–19605.

Grinsted, A., Moore, J. C., and Jevrejeva, S. (2013). Projected Atlantic hurricane surge threat from rising temperatures. *Proceedings of the National Academy of Sciences* **110**(14), 5369–5373.

Hall, J. W., Sayers, P. B., and Dawson, R. J. (2005), National-scale assessment of current and future flood risk in England and Wales, *Natural Hazards* **36**(1–2), 147–164.

Hallegatte, S., Green, C., Nicholls, R. J., and Corfee-Morlot, J. (2013). Future flood losses in major coastal cities. *Nature Climate Change* **3**(9), 802–806.

Handmer, J., Honda, Y., Kundzewicz, Z. W., Arnell, N., Benito, G., Hatfield, J., Mohamed, I. F., Peduzzi, P., Wu, S., Sherstyukov, B., Takahashi, K., and Yan, Z. (2012). Changes in impacts of climate extremes: Human systems and ecosystems. In *Managing the Risks of Extreme Events and Disasters to Advance Climate Change Adaptation. A Special Report of Working Groups I and II of the Intergovernmental Panel on Climate Change (IPCC)* (231–290). Intergovernmental Panel on Climate Change. Cambridge University Press.

Hanson, S., Nicholls, R., Ranger, N., Hallegatte, S., Corfee-Morlot, J., Herweijer, C., and Chateau, J. (2011). A global ranking of port cities with high exposure to climate extremes. *Climatic Change* **104**, 89–111.

Haque, S. A. (2006). Salinity problems and crop production in coastal regions of Bangladesh. *Pakistan Journal of Botany* **38**(5), 1359–1365.

Hartig, E. K., Gornitz, V., Kolker, A., Mushacke, F., and Fallon, D. (2002). Anthropogenic and climate-change impacts on salt marsh morphology in Jamaica Bay, New York City. *Wetlands* **22**, 71–89.

Heering, H., Firoz, R., and Khan, Z. H. (2010). Climate change adaptation measures in the coastal zone of Bangladesh. In *Deltas in Times of Climate Change*, Rotterdam, Sept. 29–Oct. 1, 2010, PDD1.5–09 (Abstr.).

Hefty, N. L. (2011). Miami-Dade County: A case study on mitigation and adaptation. Climate information for managing risks symposium. Accessed June 28, 2015: www.conference.ifas.ufl.edu/cimr/Presentations/Wednesday/pm/0315(2)_Hefty.pdf

Heimlich, B. N., Bloetscher, F., Meeroff, D. E., and Murley, J. (2009). *Southeast Florida's Resilient Water Resources: Adaptation to Sea Level Rise and Other Impacts of Climate Change*. Center for Urban and Other Environmental Solutions at Florida Atlantic University, 11/15/09. Accessed December 11, 2015: www.fau.edu/SE_Florida_Resilient_Water_Resources.pdf

Hein, L., Metzger, M. J., and Moreno, A. (2009). Potential impacts of climate change on tourism; A case study for Spain. *Current Opinion in Environmental Sustainability* 1, 170–178.

Henderson, J. V. and Wang, H. G (2007). Urbanization and city growth: The role of institutions *Regional Science and Urban Economics* 37(3), 283–313.

Hinkel, J., Lincke, D., Vafeidis, A. T., Perrette, M., Nicholls, R. J., Tol, R. S. J. et al. (2014). Coastal flood damage and adaptation costs under 21st century sea-level rise. *Proceedings of the National Academy of Sciences* 111(9), 3292–3297.

Hopkinson, C. S., Lugo, A. E., Alber, M., Covich, A. P., and Van Bloem, S. J. (2008) Forecasting effects of sea-level rise and windstorms on coastal and inland ecosystems. *Frontiers in Ecology and Environment* 6, 255–263.

Horton, R., Bader, D.A., Kushner, Y., Little, C., Blake, R. and Rosenzweig, C. (2015). *New York City Panel on Climate Change 2015 Report: Climate observations and projections.* Ann. New York Acad. Sci., 1336, 18-35, doi:10.1111/nyas.12586.

Hu, A., Meehl, G. A., Han, W., and Yin, J. (2011). Effect of the potential melting of the Greenland Ice Sheet on the Meridional Overturning Circulation and global climate in the future. *Deep-Sea Research II* 58, 1914–1926.

Intergovernmental Panel on Climate Change (IPCC). (2012). *Managing the Risks of Extreme Events and Disasters to Advance Climate Change Adaptation. A Special Report of Working Groups I and II of the Intergovernmental Panel on Climate Change.* Field, C. B., Barros, V., Stocker, T. F., Qin, D., Dokken, D. J., Ebi, K. L., Mastrandrea, M. D., Mach, K. J., Plattner, G. -K., Allen, S. K., Tignor, M., and Midgley, P. M. (eds.). Cambridge University Press.

Intergovernmental Panel on Climate Change (IPCC); Church, J. A., Clark, P. U., Cazenave, A., Gregory, J. M., Jevrejeva, S., Levermann, A., Merrifield, M. A., Milne, G. A., Nerem, R. S., Nunn, P. D., Payne, A. J., Pfefer, W. T., Stammer, D., and Unnikrishnan, A. S. (2013). Sea level change. In Stocker, T. F., Qin, D., Plattner, G. -K., Tignor, M., Allen, S. K., Boschung, J., Nauels, A., Xia, Y., Bex, V., and Midgley, P. M. (eds.), *Climate Change 2013: The Physical Science Basis. Contribution of Working Group I to the Fifth Assessment Report of the Intergovernmental Panel on Climate Change.* Cambridge University Press.

Intergovernmental Panel on Climate Change (IPCC). (2014a). Summary for policymakers. In Field, C. B., Barros, V. R., Dokken, D. J., Mach, K. J., Mastrandrea, M. D., Bilir, T. E., Chatterjee, M., Ebi, K. L., Estrada, Y. O., Genova, R. C., Girma, B., Kissel, E. S., Levy, A. N., MacCracken, S., Mastrandrea, P. R., and White, L. L. (eds.), *Climate Change 2014: Impacts, Adaptation, and Vulnerability. Part A: Global and Sectoral Aspects. Contribution of Working Group II to the Fifth Assessment Report of the Intergovernmental Panel on Climate Change* (1–32). Cambridge University Press.

Intergovernmental Panel on Climate Change (IPCC). (2014b). *Climate Change 2014: Impacts, Adaptation, and Vulnerability. Part A: Global and Sectoral Aspects. Contribution of Working Group II to the Fifth Assessment Report of the Intergovernmental Panel on Climate Change.* Cambridge University Press.

International Global Change Institute (IGCI). (2015). Accessed March 7, 2016: http://igci.org.nz/dunes/solutions/solution-1/

Jenkins, G. J., Perry, M. C., and Prior, M. J. (2008). *The Climate of the United Kingdom and Recent Trends.* Met Office Hadley Centre, Exeter, UK.

Jin, D., Ashton, A. D., and Hoagland, P. (2013). Optimal responses to shoreline changes: An integrated economic and geological model with application to curved coasts. *Natural Resource Modeling* 26(4), 572–604.

Jones, L., Angus, S., Cooper, A., Doody, P., Everard, M., Garbutt, A., Gilchrist, P., Hansom, J., Nicholls, R., Pye, K., Ravenscroft, N., Rees, S., Rhind, P., and Whitehouse, A. (2011). Chapter 11: Coastal margins. In *The UK National Ecosystem Assessment Technical Report. UK National Ecosystem Assessment*. UNEP-WCMC, Cambridge.

Jones, M. R., Fowler, H. J., Kilsby, C. G., and Blenkinsop, S. (2013). An assessment of changes in seasonal and annual extreme rainfall in the UK between 1961 and 2009. *International Journal of Climatology* 33, 1178–1194.

Jongman, B., Ward, P. J., and Aerts, J. C. (2012). Global exposure to river and coastal flooding: Long term trends and changes. *Global Environmental Change* 22(4), 823–835.

Kabat, P., Fresco, L. O., l Stive, M. J. F., Veerman, C. P., van Alphen, J. S. L., Parmat, B. W. A. H., Hazeleger, W., and Katsman, C. A. (2009). Dutch coasts in transition. *Nature Geoscience* 2, 450–452.

Kadiri, M., Ahmadian, R., Bockelmann-Evans, B., Rauen, W., and Falconer, R. (2012). A review of the potential water quality impacts of tidal renewable energy systems. *Renewable and Sustainable Energy Reviews* 16(1), 329–341.

Katsman, C. A., Sterl, A., Beersma, J. J., van den Brink, H. W., Church, J. A., Hazeleger, H., Kopp, R. E., Kroon, D., Kwadijk, J., Lammersen, R., Lowe, J., Oppenheimer, M., Plag, H. -P., Ridley, J., van Storch, H., Vaughn, D. G., Vellinga, P., Vermeersen, L. L. A., van de Wal, R. S. W., and Weisse, R. (2011). Exploring high-end scenarios for local sea level rise to develop flood protection strategies for a low-lying delta—the Netherlands as an example. *Climatic Change* 109, 617–645.

Kemp, A. C., Horton, B. P., Donnelly, J. P., Mann, M. E., Vermeer, M., and Rahmstorf, S. (2011). Climate related sea-level variations over the past two millennia. *Proceedings of the National Academy of Sciences* 108(27), 11017–11022.

Khan, M. S. A., Mondal, M. S., Kumar, U., Rahman, R., Huq, H., and Dutta, D. K. (2013). Climate change, salinity intrusion and water insecurity in peri-urban Khulna, Bangladesh. In Prakash, A., and Singh, S. (eds.), *Water Security in Peri-urban South Asia: Adapting to Climate Change and Urbanisation* (9–30). SaciWATERs and IDRC.

Kirwan, M. L., Guntenspergen, G. R., D'Alpaos, A. D., Morris, J. T., Mudd, S. M., and Temmermann, S. (2010). Limits on the adaptability of coastal marches to rising sea level. *Geophysical Research Letters* 37, L23401. doi:10.1012/2010GL045489

Kirwan, M. L., and Megonigal, J. P. (2013). Tidal wetland stability in the face of human impacts and sea-level rise. *Nature* 504, 53–60.

Kiunsi, R. (2013). The constraints on climate change adaptation in a city with a large development deficit: The case of Dar es Salaam. *Environment and Urbanization* 25(2), 321–337.

Knutson, T. R., McBride, J.L., Chan, J., Emanuel, K., Holland, G., Landsea, C., Held, I., Kossin, J.P., Srivastava, A.K., and Sugi, M. (2010). Tropical cyclones and climate change. *Nature Geoscience* 3, 157–163.

Kossin, J. P., Emanuel, K. A., and Vecchi, G. A. (2014). The poleward migration of the location of tropical cyclone maximum intensity. *Nature* 509, 349–352.

Lal, R. (2008). Carbon sequestration. *Philosophical Transactions of the Royal Society B: Biological Sciences* 363(1492), 815–830.

Le Dang, H., Li, E., Nuberg, I., and Bruwer, J. (2014). Understanding farmers' adaptation intention to climate change: A structural equation modelling study in the Mekong Delta, Vietnam. *Environmental Science and Policy* 41, 11–22.

Lerman, A., Guidry, M., Andersson, A. J., and Mackenzie, F. T. (2011). Coastal ocean last glacial maximum to 2100 CO2-carbonic acid-carbonate system: A modeling approach. *Aquatic Geochemistry* 17, 749–773.

Lewis, A., Estefen, S., Huckerby, J., Musial, W., Pontes, T., and Torres-Martinez, J. (2011). Ocean energy. In Edenhofer, O., Pichs-Madruga, R., Sokona, Y., Seyboth, K., Matschoss, P., Kadner, S., Zwickel, T., Eickemeier, P., Hansen, G., Schlömer, S., and von Stechow, C. (eds.), *IPCC Special Report on Renewable Energy Sources and Climate Change Mitigation*. Cambridge University Press.

Lichter, M., Vafeidis, A. T., Nicholls, R. J., and Kaiser, G. (2010). Exploring data-related uncertainties in analyses of land area and population in the Low-Elevation Coastal Zone(LECZ). *Journal of Coastal Research* **27**(4), 757–768.

Little, C. M., Horton, R. M., Kopp, R. E., Oppenheimer, M., Vecchi, G. A., and Villarini, G. (2015). Joint projections of U.S. East Coast sea level and storm surge. *Nature Climate Change* **5**, 1114–1120.

Liu, L. (2014). Feasibility of large-scale power plants based on thermoelectric effects, *New Journal of Physics* **16**(12) (123019).

Maantay, J., and Maroko, A. (2009) Mapping urban risk: Flood hazards, race, and environmental justice in New York. *Applied Geography* **29**(1), 111–124.

Malm, A., and Esmailian, S. (2013) Ways in and out of vulnerability to climate change: Abandoning the Mubarak Project in the northern Nile Delta, Egypt. *Antipode* **45**(2), 474–492.

Masters, D., Nerem, R. S., Choe, C., Leuliette, E., Beckley, B., White, N., and Ablain, M. (2012). Comparison of global mean sea level time series from TOPEX/Poseidon, Jason-1, and Jason-2. *Marine Geodesy* **35**(suppl. 1), 20–41.

Mazi, K., Koussis, A. D., and Destouni, G. (2013). Tipping points for seawater intrusion in coastal aquifers under rising sea level. *Environmental Research Letters* **8**(1), 014001.

McGranahan, G., Balk, D., and Anderson, B. (2007). The rising tide: Assessing the risks of climate change and human settlements in low elevation coastal zones. *Environment and Urbanization* **19** (1), 17–37.

McIvor, A. L., Möller, I., Spencer, T., and Spalding. M. (2012). *Reduction of Wind and Swell Waves by Mangroves*. Natural Coastal Protection Series: Report 1. Cambridge Coastal Research Unit Working Paper 40. The Nature Conservancy and Wetlands International. Accessed July 28, 2015: http://www.naturalcoastalprotection.org/documents/reduction-of-wind-and-swell-waves-by-mangroves

McLeman, R. A. (2011). Settlement abandonment in the context of global environmental change. *Global Environmental Change* **21**, S108–S120.

Mendelsohn, R., Emanuel, K., Chonabayashi, S., and Bakkensen, L. (2012). The impact of climate change on global tropical cyclone damage. *Nature Climate Change* **2**, 205–209.

Milligan, J., O'Riordan, T., Nicholson-Cole, S. A., and Watkinson, A. R. (2009). Nature conservation for sustainable shorelines: Lessons from seeking to involve the public. *Land Use Policy* **26**(2), 203–213.

Möller, I., Kudella, M., Rupprecht, F., Spencer, T., Paul, M., van Wesenbeeck, B., Wolters, G., Jensen, K., Bouma, T. J., Miranda-Lange, M., and Schimmels, S. (2014). Wave attenuation over coastal salt marshes under storm surge conditions. *Nature Geoscience* **7** (Oct. 2014). doi:10.1038/NGEO2251

Mondal, M. S., Jalal, M. R., Khan, M. S. A., Kumar, U., Rahman, R., and Huq, H. (2013, March). Hydro-meteorological trends in southwest coastal Bangladesh: Perspectives of climate change and human interventions. *American Journal of Climate Change, Scientific Research* 62–70.

Moriarty, P., and Honnery, D. (2012). What is the global potential for renewable energy?. *Renewable and Sustainable Energy Reviews* **16**(1), 244–252.

Moser, S., Williams, S., and Boesch, D. (2012). Wicked challenges at Land's End: Managing coastal vulnerability under climate change. *Annual Review of Environment and Resources* **37**, 51–78.

Murtaza, G. (2001). Environmental problems in Khulna City, Bangladesh: A spatio-household level study. *Global Built Environment Review* **1**(2), 32–37.

Nakicenovic, N., and Swart, R. (eds.). (2000, July). *Special Report on Emissions Scenarios*. Cambridge University Press.

Narita, D., Rehdanz, K., and Tol, R. S. J. (2011). Economic costs of ocean acidification: A look into the impacts on global shellfish production. *Climate Change.* doi:10.1007/s10584–011–0383–3

National Park Service (NPS). U.S. Dept. of Interior (2016). Moving the Cape Hatteras Lighthouse. Accessed April 6, 2017: http://nps.gov/caha/historyculture/movingthelighthouse.htm

National Research Council (NCR). (1995). *Beach Nourishment and Protection*. National Academies Press. Washington, D.C.

Neumann, B., Vafeidis, A.T., Zimmermann, J., and Nicholls, R.J. 2015. Future coastal population growth and coastal flooding – a global assessment. *PLoS ONE* 10(3):e0118571;doi:10.1371/journal.pone.0118571.

New York City Environmental Justice Alliance (NYC-EJA). (2015). Project Description for the South Bronx Community Resiliency Agenda. Accessed April 9, 2016: http://www.nyc-eja.org/

New York City Office of Emergency Management. (2014). NYC Hazard Mitigation Plan. Accessed August 10, 2015: http://www.nyc.gov/html/oem/html/planning_response/planning_hazard_mitigation_2014.shtml

New York City Panel on Climate Change (NPCC). (2010). *Climate change adaptation in New York City: Building a risk management response*, C. Rosenzweig and W. Solecki, eds. *Annals of the New York Academy of Sciences*. Accessed December 2, 2015: http://www.nyas.org

New York City Panel on Climate Change (NPCC). (2013). *Climate Risk Information 2013: Observations, Climate Change Projections, and Maps*, C. Rosenzweig and W. Solecki, eds. Prepared for use by the City of New York Special Initiative on Rebuilding and Resiliency. Accessed July 9, 2015: www.nyc.gov/planyc, www.nyc.gov/resiliency

New York City Panel on Climate Change (NPCC). (2015). *Building the knowledge base for climate resiliency*, C. Rosenzweig and W. Solecki, eds. *Annals of the New York Academy of Sciences* **1336**.

Nicholls, R. J. (2004). Coastal flooding and wetland loss in the 21st century: Changes under the SRES climate and socio-economic scenarios. *Global Environment Change* **14**, 69–86.

Nicholls, R. J., Wong, P. P., Burkett, V. R., Codignotto, J. O., Hay, J. E., McLean, R. F., Ragoonaden, S., and Woodroffe, C. D. (2007). Coastal systems and low-lying areas. In Parry, M. L., Canziani, O. F., Palutikof, J. P., van der Linden, P. J., and Hanson, C. E. (eds.), *Climate Change 2007: Impacts, Adaptation and Vulnerability. Contribution of Working Group II to the Fourth Assessment Report of the Intergovernmental Panel on Climate Change* (315–356). Cambridge University Press.

Nicholls, R. J., Wong, P. P., Burkett, V. R., Woodroffe, C. D., and Hay, J. E. (2008). Climate change and coastal vulnerability assessment: Scenarios for integrated assessment. *Sustainability Science* **3**(1), 89–102.

Nicholls, R. J., Woodroffe, C. D., Burkett, V., Hay, J., Wong, P. P., and Nurse, L. (2012). Scenarios for coastal vulnerability assessment. In Wolanski, E. and McLusky, D. (eds.), *Treatise on Estuarine and Coastal Science*: Volume 12: Ecological Economics of Estuaries and Coasts (289–303). Elsevier.

Nicholls, R. J., Dawson, R. J., and Day, S. (2015). *Broad Scale Coastal Simulation: New Techniques to Understand and Manage Shorelines in the Third Millennium*, Springer.

Nicholson-Cole, S. A., and O'Riordan, T. (2009). Adaptive governance for a changing coastline: Science, policy and publics in search of a sustainable future. In Adger, W. N., Lorenzoni, I., and O'Brien, K. L. (eds.), *Adapting to Climate Change: Thresholds, Values, Governance*. Cambridge University Press.

Nitto, D., Neukermans, G., Koedam, N., Defever, H., Pattyn, F., Kairo, J. G., and Dahdouh-Guebas, F. (2014). Mangroves facing climate change: Landward migration potential in response to projected scenarios of sea level rise. *Biogeosciences* **11**(3), 857–871.

NOAA. (2015). Coastal flooding in California. Accessed March 19, 2016: http://oceanservice.noaa.gov/news/dec15/california-flooding.html

NOAA Tides and Currents. Accessed May 15, 2017: (2016). http://www.tidesandcurrents/noaa.gov.

Nordenson, G., Seavitt, C., and Yarinsky, A. (2010). *On the Water/Palisade Bay*. Hatje Cantz Verlag and the Museum of Modern Art.

Norfolk Virginia. (2013). Coastal resilience strategy. Accessed July 9, 2015: http://www.norfolk.gov/DocumentCenter/View/16292

Olsson, L., Opondo, L., Tschakert, P., Agrawal, A., Eriksen, S., Ma, S., Perch, L., and Zakieldeen, S. (2014). Livelihoods and poverty. In Field, C. B., V. R. Barros, D. J. Dokken, K. J. Mach, M. D. Mastrandrea, T. E. Bilir, M. Chatterjee, K. L. Ebi, Y. O. Estrada, R. C. Genova, B. Girma, E. S. Kissel, A. N. Levy, S. MacCracken, P. R. Mastrandrea, and L. L. White (eds.), *Climate Change 2014: Impacts, Adaptation, and Vulnerability. Part A: Global and Sectoral Aspects. Contribution of Working Group II to the Fifth Assessment Report of the Intergovernmental Panel on Climate Change* (793–832). Cambridge University Press.

O'Neill, B. C., Kriegler, E., Riahi, K., Ebi, K. L., Hallegatte, S., Carter, T. R.,Mathur, R. and van Vuuren, D. P. (2014). A new scenario framework for climate change research: The concept of shared socioeconomic pathways. *Climatic Change* 122(3), 387–400.

Paavola, J., Adger, W. N., and Huq, S. (2006). Multifaceted Justice in Adaptation to Climate Change. In Adger, W. N., Paavola, J., and Huq, S. (eds.), *Fairness in Adaptation to Climate Change* (263–277). MIT Press.

Peduzzi, P. et al. (2012). Global trends in tropical cyclone risk. *Nature Climate Change* 2, 289–294.

Pelling, M., and Blackburn, S. (eds.). (2014). *Megacities and the Coast: Risk, Resilience and Transformation.* Routledge.

Perry, A. (2006). Will predicted climate change compromise the sustainability of Mediterranean tourism? *Journal of Sustainable Tourism* 14(4), 367–375.

Pilkey, O., and Young, R. (2009). *The Rising Sea.* Island Press.

Prefeitura da Cidade do Rio de Janeiro. (2013). Plano de Contingência. Accessed July 22, 2014: http://www.rio.rj.gov.br/web/defesacivil/plano-de-contingencia

Prefeitura da Cidade do Rio de Janeiro. (2015). Rio Resiliente: Diagnóstico e Áreas de Foco. Accessed March 10, 2016: http://www.centrodeoperacoes.rio.gov.br/

Raucoules, D., Le Cozannet, G., Wöppelmann, G., de Michele, M., Gravelle, M., Daag, A., and Marcos, M. (2013). High nonlinear urban ground motion in Manila (Philippines) from 1993 to 2010 observed by DInSAR: Implications for sea-level measurement. *Remote Sensing of Environment* 139, 386–397.

Resch, G., Held, A., Faber, T., Panzer, C., Toro, F., and Haas, R. (2008). Potentials and prospects for renewable energies at global scale. *Energy Policy* 36(11), 4048–4056.

Rijcken, T. (2010). Rhine Estuary closable but open—An integrated systems approach to floodproofing the Rhine and Meuse estuaries in the 21st century. In *Rotterdam's Climate Adaptation Research Summaries 2010* (10–11). Accessed May 8, 2014: www.rotterdamclimateinitiative.nl

Romero-Lankao, P. (2007). How do local governments in Mexico City manage global warming? *Local Environment* 12, 519–535.

Rossello-Nadal, J. (2014). How to evaluate the effects of climate change on tourism. *Tourism Management* 42, 334–340.

Rotterdam Climate Proof. (2010). *Rotterdam's Climate Adaptation—Research Summaries 2010.* Accessed October 5, 2014: http://www.rotterdamclimateinitiative.nl

Rotzoll, K., and Fletcher, C. H. (2013). Assessment of groundwater inundation as a consequence of sea-level rise. *Nature Climate Change* 3(5), 477–481.

Rowe, M. (2013). On shaky ground. *Geographical Magazine.* Accessed June 9, 2014: http://www.geographical.co.uk/Magazine/Asias_inking_cities_-_May_13.html/

Sandy Regional Assembly. (2013). Sandy Regional Assembly recovery agenda. Accessed September 18, 2014: http://www.nyc-eja.org/public/publications/SandyRegionalAssembly_SIRRAnalysis.pdf/

Satterthwaite, D. (2013). The political underpinnings of cities' accumulated resilience to climate change. *Environment and Urbanization* 25(2), 381–391.

Satterthwaite, D., Huq, S., Pelling, M., Reid, H., and Romero-Lankao, P. (2007). *Adapting to Climate Change in Urban Areas: The Possibilities and Constraints in Low- and Middle-Income Nations.* Human Settlements Working Paper Series, Climate Change and Cities No.1. International Institute for Environment.

Schiermeier, Q. (2014). Holding back the tide. *Nature* 508, 164–166.

Schlosberg, D., and Collins, L. (2014). From environmental to climate justice: Climate change and the discourse of environmental justice. *WIREs Climate Change* 5(3), 359–374.

Schmidt, C. (2015). Alarm over a sinking delta. *Science* 348, 845–846.

Scott, D., Hall, C. M., and Gossling, S. (2012). Government, industry and destination adaptation to climate change. In Scott, D., Hall, C. M., and Gossling, S. (eds.), *Tourism and Climate Change: Impacts, Adaptation and Mitigation* (265–297). Routledge.

Scråmestø, O. S.,Skilhagen, S. -E., and Nielsen, W. K. (2009). Power production based on osmotic pressure. In Waterpower XVI Conference, Spokane, Washington, July 27–30, 2009.

Seto, K. C. (2011). Exploring the dynamics of migration to mega-delta cities in Asia and Africa: Contemporary drivers and future scenarios. *Global Environmental Change* 21, S94-S107.

Sherbinin, A., Schiller, A., and Pulsipher, A. (2013). The vulnerability of global cities to climate hazards. *Environment and Urbanization* 19(1), 39–64.

SIRR and PlaNYC (2013, June). *A Stronger, More Resilient New York.* City of New York, Mayor Michael R. Bloomberg.

Small, C., and Nicholls, R. J. (2003). A global analysis of human settlement in coastal zones. *Journal of Coastal Research*, 584–599.

Smit, B., Pilifosova, O., Burton, I., Challenger, B., Huq, S., Klein, R. J. T., Yohe, G., Adger, N., Downing, T., Harvey, E., Kane, S., Parry, M., Skinner, M., Smith, J., and Wandel, J. (2001). Adaptation to climate change in the context of sustainable development and equity. In McCarthy, J. J., Canziani, O. F., Leary, N. A., Dokken, D. J., and White, K. S. (eds.), *Climate Change 2001: Impacts, Adaptation and Vulnerability. Contribution of Working Group II to the Third Assessment Report of the Intergovernmental Panel on Climate Change.* Cambridge University Press.

Smit, B. and Pilifosova, O. (2003). From adaptation to adaptive capacity and vulnerability reduction. In Smith, J., Klein, R., and Huq, S. (eds.), *Climate Change: Adaptive Capacity and Development.* Imperial College Press.

Southall, A. (2016). El Niño Puts Pacifica Cliff Apartments at risk. *New York Times.* Accessed April 3, 2016: http://www.nytimes.com/2016/01/27/us/el-nino-storms-put-pacifica-cliff-apartments-at-risk.html?_r=0

Southeast Florida Regional Compact Climate Change. (2011). *A Unified Sea Level Rise Projection for Southeast Florida.* Accessed June 7, 2015: www.broward.org/NaturalResources/ClimateChange/Documents/SE_FL_Sea_Level_Rise_White_Paper_April_2011ADAFINAL.pdf

Southeast Florida Regional Compact Climate Change. (2012). *A Region Responds to a Changing Climate: Southeast Regional Compact Counties Regional Climate Action Plan,* October 2012. Accessed October 22, 2014: http://www.southeastfloridaclimatecompact.org/wp-content/uploads/2014/09/regional-climate-action-plan-final-adn-compliant.pdf

Southeast Florida Regional Compact Climate Change. (2014). *Integrating Climate Change and Water Supply Planning in Southeast Florida, June 14, 2014.* Accessed March 11, 2015: http://southeastfloridaclimatecompact.files.wordpress.com/2014/06/rcap-igd-water-supply-final-v-3.pdf

Stalenberg, B. (2012). Innovative flood defences in highly urbanized water cities. In Aerts, J., Botzen, W., Bowman, M. J., Ward, P. J. and Dircke, P. (eds.), *Climate Adaptation and Flood Risk in Coastal Cities* (145–164). Earthscan.

Strauss, B. H., Kopp, R. E., Sweet, W. V., and Bittermann, K. (2016). *Unnatural Coastal Floods: Sea level Rise and the Human Fingerprint on U.S. Floods Since 1950* (1–16). Climate Central Research Report.

Sweet, W. V., and Marra, J. J. (2015). 2014 State of nuisance tidal flooding. Accessed March 19, 2016: http://www.noaa.gov/pub/data/cmb/special-reports/sweet-marra-nuisance-flooding-2015.pdf

Sweet, W., Park, J., Marra, J., Zervas, C., and Gill, S. (2014). *Sea Level Rise and Nuisance Flood Frequency Changes around the United States.* NOAA Technical Report NOS CO-OPS 073.

Syvitski, J. P. M., Kettner, A. J., Overeem, I., Hutton, E. W. H., Hannon, M. T., Brakenridge, G. R., Day, J., Vörörosmarty, Saito, Y., Giosan, L., and Nicholls, R. J. (2009). Sinking deltas due to human activities. *Nature Geoscience* 2, 681–686. doi:10.1038/NGE0629

Talke, S., Orton, P., and Jay, D. (2014). Increasing storm tides at New York City, 1844–2013. *Geophysical Letters* **41**. doi:10.1002/2014GL059574

Tebaldi, C., Strauss, B. H., and Zervas, C. E. (2012). Modelling sea level rise impacts on storm surges along U.S. coasts. *Environmental Research Letters* **7**, 014032.

Titus, J. G., and Craghan, M. (2009). Shore protection and retreat. In Titus, J. G. et al. (eds.), *Coastal Sensitivity to Sea Level Rise: A Focus on the Mid-Atlantic Region. U.S. Climate Change Science Program Synthesis and Assessment Product 4.1*. (87–103). U.S. Government Printing Office.

Torio, D. D., and Chmura, G. L. (2013). Assessing coastal squeeze of tidal wetlands. *Journal of Coastal Research* **29**(5), 1049–1061.

UN-DESA (2012). *World Urbanization Prospects: The 2011 Revision*. United Nations, Department of Economic and Social Affairs, Population Division.

UN-Habitat. (2008). *State of the World's Cities: Harmonious Cities*. Earthscan.

UN-Habitat. (2011). *Global Report on Human Settlements 2011: Cities and Climate Change*. Earthscan.

United Nations Environment Programme (UNEP). (2009). *Sustainable Coastal Tourism: An Integrated Planning and Management Approach*. UNEP.

U.S. Department of Energy (DOE). (2010). *Energy Effi ciency and Renewable Energy Marine and Hydrokinetic Database*. US DOE. Accessed September 23, 2014: www.eere.energy.gov/windandhydro/hydrokinetic/default.aspx.

U.S. Department of Energy (DOE). (2012). Turbines off NYC East River will provide power to 9, 500 residents. Accessed February 8, 2015: http://energy.gov/articles/turbines-nyc-east-river-will-provide-power-9500-residents/

Vafeidis, A. T., Nicholls, R. J., McFadden, L., Tol, R. S., Hinkel, J., Spencer, T., Grashoff, P. S., Boot, G. and Klein, R. J., 2008. A new global coastal database for impact and vulnerability analysis to sea-level rise. Journal of Coastal Research, pp.917–924.

Valverde, H. R., Trembanis, A. C., and Pilkey, O. H. (1999). Summary of beach nourishment episodes on the U.S. East Coast Barrier Islands. *Journal of Coastal Research* **15**, 1100–1118.

Van Lavieren, H., Spalding, M., Alongi, D., Kainuma, M., Clüsener-Godt, M., and Adeel, Z. (2012). Securing the future of mangroves. A policy brief. UNU-INWEH, UNESCO-MAB with ISME, ITTO, FAO, UNEP-WCMC, and TNC (319). Earthscan.

van Veelen, P. (2010). Adaptive building is an important component of flood-proofing strategy. In *Rotterdam's Climate Adaptation Research Summaries 2010* (12–15). Accessed May 2, 2015: www.rotterdamclimateinitiative.nl

Verdant Power. (2014). The RITE Project, New York, NY (East River-East Channel). Accessed May 8, 2015: http://verdantpower.com/rite-project.html/

Voorendt, M. G. (2015). *Examples of Multifunctional Flood Defences*. Working Paper, TU Delft. Accessed April 6, 2016: http://www.citg.tudelft.nl/ileadmin/Faculteit/CiTG/Over_de_faculteit/Afdelingen/Afdeling_Waterbouwkunde/Hydraulic_Engineering/Research_Groups/Hydraulic_Structures_and_Flood_Risk/Members/Voorendt_MZ/doc/Examples_of_multifunctional_flood_defences.pdf

Wahl, T., Jain, S., Bender, S., Meyers, S. D., and Luther, M. E. (2015). Increasing risk of compound flooding from storm surge and rainfall for major cities. *Nature Climate Change* **5**, 1093–1097.

Walsh, C. L., Roberts, D., Dawson, R. J., Hall, J. W., Nickson, A., and Hounsome, R. (2013). Experiences of integrated assessment modelling in London and Durban, *Environment and Urbanisation* **25**(2), 257–376.

Wilbanks, T. J., and Fernandez, S. (2014). *Climate Change and Infrastructure, Urban Systems, and Vulnerabilities: Technical Report for the US Department of Energy in Support of the National Climate Assessment*. Island Press.

Wolf, J., Walkington, I. A., Holt, J., and Burrows, R. (2009). Environmental impacts of tidal power schemes. *Proceedings of the ICE-Maritime Engineering* **162**(4), 165–177.

Wong, P. P., Losada, I. J.,Gattuso, J. -P., Hinkel, J.,Khattabi, A. McInnes, K. L., Saito, Y., and Sallenger, A. (2014). Coastal systems and low-lying areas. In Field, C. B., Barros, V. R., Dokken, D. J., Mach, K. J., Mastrandrea, M. D., Bilir, T. E., Chatterjee, M., Ebi, K. L., Estrada, Y. O., Genova, R. C., Girma, B., Kissel, E. S., Levy, A. N., MacCracken, S., Mastrandrea, P. R., and White, L. L. (eds.), *Climate Change 2014: Impacts, Adaptation, and Vulnerability. Part A: Global and Sectoral Aspects. Contribution of Working Group II to the Fifth Assessment Report of the Intergovernmental Panel on Climate Change*. Cambridge University Press.

Yin, J. and Goddard, P.B, Oceanic control of sea level rise patterns along the East Coast of the United States, 2013. *Geophysical Research Letters* 40, 5514–5520, doi:10.1002/2013GL057992.

Yin, J., Griffies, S. M., and Stouffer, R. J. (2010). Spatial variability of sea level rise in twenty-first projections. *Journal of Climate* **23**, 4585–4608.

Yoo, G., Kim, A., and Hadi, S. (2014). A methodology to assess environmental vulnerability in a coastal city: Application to Jakarta, Indonesia. *Ocean and Coastal Management* **102**, 169–177.

Chapter 9 Case Study References

Case Study 9.1 Norfolk, Virginia: A City Dealing with Increased Flooding Now

Boon, J. D. (2012). Evidence of sea level acceleration at U.S., and Canadian tide stations, *Atlantic coast, North America. Journal of Coastal Research* **28**(6), 1437–1445. doi:10.2112/JCOASTRES-D-12–00102.1

Boon, J. D., Brubaker, J. M., and Forrest, D. R. (2010). *Chesapeake Bay Land Subsidence and Sea Level Change: An Evaluation of Past and Present Trends and Future Outlook*. Applied Marine Science and Ocean Engineering Report No. 425. Virginia Institute of Marine Science, Gloucester Point, VA.

Ezer, T. (2013). Sea level rise, spatially uneven and temporally unsteady: Why the U.S. East Coast, the global tide gauge record and the global altimeter data show different trends. *Geophysical Research Letters* **40**(20), 5439–5444. doi:10.1002/2013GL057952

Ezer, T., and Atkinson, L. P. (2014), Accelerated flooding along the U.S. East Coast: On the impact of sea level rise, tides, storms, the Gulf Stream and the North Atlantic Oscillations. *Earth's Future* **2**(8), 362–382. doi:10.1002/2014EF000252

Ezer, T., Atkinson, L. P. Corlett, W. B., and Blanco, J. L. (2013). Gulf Stream's induced sea level rise and variability along the U.S. mid-Atlantic coast, *Journal of Geophysical Research* **118**(2), 685–697. doi:10.1002/jgrc.20091

Ezer, T., and Corlett, W. B. (2012). Is sea level rise accelerating in the Chesapeake Bay? A demonstration of a novel new approach for analyzing sea level data. *Geophysical Research Letters* **39**(19), L19605. doi:10.1029/2012GL053435

Norfolk. (2013). Coastal resilience strategy. Page 2. Accessed November 21, 2015: http://www.norfolk.gov/DocumentCenter/View/16292

Norfolk. (2014). Norfolk's Resilience Challenge. Accessed February 11, 2015: http://www.100resilientcities.org/cities/entry/norfolks-resilience-challenge#/-_/

Peel, M. C., Finlayson, B. L., and McMahon, T. A. (2007). Updated world map of the Köppen-Geiger climate classification. *Hydrology and Earth System Sciences Discussions* **4**(2), 462.

Sallenger, A. H., Doran, K. S., and Howd, P. (2012) Hotspot of accelerated sea-level rise on the Atlantic coast of North America. *Nature Climate Change* 2, 884–888. doi:10.1038/NCILMATE1597

Steinhilber, E. E., Considine, C., and Whitelaw, J. (2015). Hampton Roads Sea Level Rise Preparedness and Resilience Intergovernmental Pilot Project. Accessed February 26, 2015: http://digitalcommons.odu.edu/hripp_reports/1/

Sweet, W. V., and Park, J. (2014). From the extreme to the mean: Acceleration and tipping points of coastal inundation from sea level rise. *Earth's Future* **2**(12), 579.

U.S. Census Bureau. (2010). Decennial census, summary file 1. Accessed February 11, 2015: http://www.census.gov/population/metro/files/CBSA%20Report%20Chapter%203%20Data.xls

World Bank. (2017). 2016 GNI per capita, Atlas method (current US$). Accessed August 9, 2017: http://data.worldbank.org/indicator/NY.GNP.PCAP.CD

Case Study 9.2   Vulnerabilities and Adaptive Practices in Khulna, Bangladesh

ADB. (2011). *Adapting to Climate Change: Strengthening the Climate Resilience of the Water Sector Infrastructure in Khulna, Bangladesh.* The Asian Development Bank.

Aqua-Sheltech Consortium. (2002). *Draft Structure Plan, Master Plan and Detailed Area Plan (2001–2020) for Khulna City, Volume I: Urban Strategy.* Khulna Development Authority, Ministry of Housing and Public Works, Government of the People's Republic of Bangladesh.

Hallegatte, S., Green, C., Nicholls, R. J., and Corfee-Morlot, J. (2013) *Future flood losses in major coastal cities. Nature Climate Change* **3**, 802–806.

IIED. (2009). *Climate Change and the Urban Poor: Risk and Resilience in 15 of the World's Most Vulnerable Cities.* International Institute for Environment and Development.

Khan, M. S. A., Mondal, M. S., Kumar, U., Rahman, R., Huq, H., and Dutta, D. K. (2013), Climate change, salinity intrusion and water insecurity in peri-urban Khulna, Bangladesh. In Prakash, A., and Singh, S. (eds.), *Water Security in Peri-urban South Asia: Adapting to Climate Change and Urbanisation* (9–30). SaciWATERs and IDRC.

Mondal, M. S., Jalal, M. R., Khan, M. S. A., Kumar, U., Rahman, R., and Huq, H. (2013). Hydro-meteorological trends in southwest coastal Bangladesh: Perspectives of climate change and human interventions. *American Journal of Climate Change, Scientific Research* March, 62–70.

Murtaza, G. (2001). Environmental problems in Khulna City, *Bangladesh: A spatio-household level study. Global Built Environment Review* **1**(2), 32–37.

Peel, M. C., Finlayson, B. L., and McMahon, T. A. (2007). Updated world map of the Köppen-Geiger climate classification. *Hydrology and Earth System Sciences Discussions* **4**(2), 462.

World Bank. (2017). 2016 GNI per capita, Atlas method (current US$). Accessed August 9, 2017: http://data.worldbank.org/indicator/NY.GNP.PCAP.CD

Case Study 9.3   Coastal Hazard and Action Plans in Miami

City of Hallandale Beach. (2013). *City of Hallandale Beach-Water Supply Strategy.* White paper, June 14, 2013. Accessed December 5, 2015: http://www.hallandalebeachfl.gov/files/2013–06–17-HBCRA/Item_10A/Water_Supply_White_Paper-June_17_2013_Commission_Workshop, pdf/

Hefty, N. L. (2011). Miami-Dade County: A case study on mitigation and adaptation. Climate Information for Managing Risks Symposium. Accessed September 25, 2015: http://conference.ifas.ufl.edu/cimr/Presentations/Wednesday/pm/0315%20(2)%20%20Hefty.pdf

Heimlich, B. N., Bloetscher, F., Meeroff, D. E., and Murley, J. (2009). Southeast Florida's resilient water resources: Adaptation to sea level rise and other impacts of climate change. Center for Urban and Other Environmental Solutions at Florida Atlantic University, 11/15/09. Accessed November 21, 2015: http://www.ces.fau.edu/files/projects/climate_change/SE_Florida_Resilient_Water_Resources.pdf

Knutson, T. R., McBride, J.L, Chan, J., Emanuel, K., Holland, G., Landsea, C., Held, I., Kossin, J.P, Srivastava, A.K, and Sugi, M. (2010). Tropical cyclones and climate change. *Nature Geoscience* **3**, 157–163.

Miami-Dade County (2015). *Miami Dade Green.* Accessed January 3, 2016: http://www. miamidade.gov/green/

NOAA Tides and Currents (2016). Accessed February 12, 2017: http://www.tidesandcurrents.noaa.gov/

Peel, M. C., Finlayson, B. L., and McMahon, T. A. (2007). Updated world map of the Köppen-Geiger climate classification. *Hydrology and Earth System Sciences Discussions* **4**(2), 462.

Southeast Florida Regional Compact Climate Change (2011). *A Unified Sea Level Rise Projection for Southeast Florida.* Accessed November 27, 2014: https://www. broward. org/NaturalResources/ClimateChange/Documents/SE%20FL%20Sea%20Level%20Rise%20White%20Paper%20April%202011%20ADA%20FINAL. pdf

Southeast Florida Regional Compact Climate Change Compact Counties. (2012). *A Region Responds to a Changing Climate: Southeast Regional Compact Counties Regional Climate Action Plan.* Accessed December 18, 2015: http://www.southeastfloridaclimatecompact.org/wp-content/uploads/2014/09/regional-climate-action-plan-final-ada-compliant.pdf

Southeast Florida Regional Compact Climate Change (2014). *Integrating Climate Change and Water Supply Planning in Southeast Florida, June 14 2014.* Accessed July 4, 2015: http://southeastfloridaclimatecompact.files.wordpress.com/2014/06/rcap-igd-water-supply-final-v-3.pdf.

U.S. Census Bureau. (2000). *Decennial census, summary file 1.* Accessed December 1, 2015: http://www. census. gov/population/metro/files/CBSA%20Report%20Chapter%203%20Data. xls

World Bank. (2017). 2016 GNI per capita, Atlas method (current US$). Accessed August 9, 2017: http://data.worldbank.org/indicator/NY.GNP.PCAP.CD

Case Study 9.4   Venice: Human-Natural System Responses to Environmental Change

Carbognin, L., Teatini, P., Tomasin, A., and Tosi, L. (2010). Global change and relative sea level rise at Venice: What impact in term of flooding, *Climate Dynamics* **35**, 1039–1047. doi:10.1007/s00382-009-0617-5

Comerlati A., Ferronato, M., Gambolati, G., Putti, M., and Teatini P. (2004). Saving Venice by seawater. *Journal of Geophysical Research* **109**, F03006. doi:10.1029/2004JF000119

Citta di Venezia. (2015). *Ecografico e Territorio- Superfici amministrative.* Accessed April 8, 2016: http://www.comune.venezia.it/flex/cm/pages/ServeBLOB.php/L/IT/IDPagina/18700

D'Alpaos, L. (2011). *Fatti e Misfatti di Idraulica Lagunare.* Istituto Veneto di Scienze Lettere ed Arti, Venezia.

ISTAT. (2015). *Istituto Nazionale di Statistica.* Accessed January 6, 2016: http://www.istat.it/en/

Peel, M. C., Finlayson, B. L., and McMahon, T. A. (2007). Updated world map of the Köppen-Geiger climate classification. *Hydrology and Earth System Sciences Discussions* **4**(2), 462.

Marani, M., D'Alpaos, A., Lanzoni, S., Carniello, L., and Rinaldo, A. (2007). Biologically-controlled multiple equilibria of tidal landforms and the fate of the Venice lagoon. *Geophysical Research Letters* **34**, L11402. doi:10.1029/2007GL030178

World Bank. (2015). 2016 GNI per capita, Atlas method (current US$). Accessed August 9, 2017: http://data.worldbank.org/indicator/NY.GNP.PCAP.CD

Case Study 9.5   Adaptation Benefits and Costs of Residential Buildings in Greater Brisbane

ABS (Australian Bureau of Statistics). (2008). *Population Projections, Australia, 2006 to 2101.* Cat. no. 3222.0. ABS.

ABS (Australian Bureau of Statistics). (2011a). *Australian Statistical Geography Standard (ASGS): Volume 1 – Main Structure and Greater Capital City Statistical Areas, July 2011.* ABS.

ABS (Australian Bureau of Statistics). (2011b). *Basic Community Profile.* Canberra: ABS.

Beirlant, J., Goegebeur, Y., Segers, J., and Teugels, J. (2004). *Statistics of Extremes: Theory and Applications.* John Wiley & Sons.

DERM. (2012). *Queensland Coastal Plan.* Department of Environment and Resource Management, State of Queensland, Australia.

Knutson, T. R., McBride, J. L., Chan, J., et al. (2010) Tropical cyclones and climate change. *Nature Geoscience* 3, 157–163.

Meehl, G. A., Stocker, T. F., Collins, W. D., Friedlingstein, P., Gaye, A. T., Gregory, J. M., Kitho, A., Knutti, R., Murphy, J. M., Noda, A., Raper, S. C. B., Watterson, I. G., Weaver, A. J., and Zhao, Z. -C. (2007). Global climate projection. In Solomon, S., Qin, D., Manning, M., Chen, Z., Marquis, M., Averyt, K. B., Tignor, M., and Miller, H. L. (eds.), *The Physical Science Basis. Contribution of Working Group I to the Fourth Assessment Report of the Intergovernmental Panel on Climate Change.* Cambridge University Press.

Peel, M. C., Finlayson, B. L., and McMahon, T. A. (2007). Updated world map of the Köppen-Geiger climate classification. *Hydrology and Earth System Sciences Discussions* 4(2), 462.

Rogelj, J., Meinshausen, M., and Knutti, R. (2012) Global warming under old and new scenarios using IPCC climate sensitivity range estimates. *Nature Climate Change*, 2,248–253.

World Bank. (2017). 2016 GNI per capita, Atlas method (current US$). Accessed August 9, 2017: http://data.worldbank.org/indicator/NY.GNP.PCAP.CD.

Case Study 9.6   Adapting to Climate Change in Coastal Dar es Salaam

Adapting to Climate Change in Coastal Dar es Salaam (ACC). (n.d.) Dar Papers. Accessed October 29, 2015: http://www.planning4adaptation.eu/041_Papers.aspx

Congedo, L., and Munafò, M. (2014). Urban sprawl as a factor of vulnerability to climate change: Monitoring land cover change in Dar es Salaam. In Macchi, S., and Tiepolo, M. (eds.), *Climate Change Vulnerability in Southern African Cities: Building Knowledge for Adaptation.* Springer. doi 2014: 39-56.10.1007/978-3-319-00672-7_3. Accessed January 19, 2015: http://link.springer.com/chapter/10.1007%2F978-3-319-00672-7_5

Faldi, G. and Rossi, M. (2014). Climate change effects on seawater intrusion in coastal Dar es Salaam: Developing exposure scenarios for vulnerability assessment. In Macchi, S., and Tiepolo, M. (eds.), *Climate Change Vulnerability in Southern African Cities: Building Knowledge for Adaptation.* Springer. doi: 2014: 39-56.10.1007/978-3-319-00672-7_3. Accessed March 21, 2015: http://link.springer.com/chapter/10.1007%2F978-3-319-00672-7_4#

Intergovernmental Panel on Climate Change (IPCC), Agard, J., E. L. F.Schipper, J. Birkmann, M. Campos, C. Dubeux, Y.Nojiri, L. Olsson, B.Osman-Elasha, M. Pelling, M. J. Prather, M. G. Rivera-Ferre, O. C.Ruppel, A. Sallenger, K. R. Smith, A. L. St. Clair, K. J. Mach, M. D. Mastrandrea and T. E. Bilir (eds.). (2014). Annex II: Glossary. In Barros, V. R., Field, C. B., Dokken, D. J., Mastrandrea, M. D., Mach, K. J., Bilir, T. E., Chatterjee, M., Ebi, K. L., Estrada, Y. O., Genova, R. C., Girma, B., Kissel, E. S., Levy, A. N., MacCracken, S., Mastrandrea, P. R., and White, L. L. (eds.), *Climate Change 2014: Impacts, Adaptation, and Vulnerability. Part B: Regional Aspects. Contribution of Working Group II to the Fifth Assessment Report of the Intergovernmental Panel on Climate Change* (1757–1776). Cambridge University Press.

Macchi, S., and Ricci, L. (2014). Mainstreaming adaptation into urban development and environmental management planning: A literature review and lesson from Tanzania. In Macchi, S., and Tiepolo, M. (eds.), *Climate Change Vulnerability in Southern African Cities: Building Knowledge for Adaptation.* Springer. doi: 2014:

39–56.10.1007/978–3–319–00672–7_3. Accessed February 26, 2015: http://link.springer.com/chapter/10.1007/978–3–319–00672–7_7

Macchi S., and Tiepolo M. (2014). *Climate Change Vulnerability In Southern African Cities. Building Knowledge for Adaptation.* Springer. Accessed July 4, 2015: http://www.springer.com/series/11741?detailsPage=titles

National Bureau of Statistics NBS. (2013). Accessed December 12, 2014: http://www.nbs.go.tz/takwimu/references/Tanzania_in_figures2012.pdf

Peel, M. C., Finlayson, B. L., and McMahon, T. A. (2007). Updated world map of the Köppen-Geiger climate classification. *Hydrology and Earth System Sciences Discussions* 4(2),462.

Ricci, L. (2014). Linking adaptive capacity and peri-urban features. In Macchi, S., and Tiepolo, M. (eds.), *Climate Change Vulnerability in Southern African Cities: Building Knowledge for Adaptation.* Springer. doi:2014: 39–56.10.1007/978–3–319–00672–7. Accessed May 17, 2015: http://link.springer.com/chapter/10.1007%2F978–3–319–00672–7_6

Rugai, D., and Kassenga, G. R. (2014). Climate change impacts and institutional response capacity in Dar es Salaam, Tanzania. In Macchi, S., and Tiepolo, M. (eds.), *Climate Change Vulnerability in Southern African Cities: Building Knowledge for Adaptation.* Springer. doi:2014: 39–56.10.1007/978–3–319–00672–7_3. Accessed March 31, 2015: http://link.springer.com/chapter/10.1007/978–3–319–00672–7_3#

Shemdoe, R., Kassenga, G., Ricci, L., Norero, C., Macchi, S., and Sappa G. (forthcoming). Mainstreaming adaptation into existing urban development and environmental management plans: Guidelines and an application to four plans and programs in Dar es Salaam. In Macchi, S., and Tiepolo, M. (eds.), *Climate Change Vulnerability in Southern African Cities: Building Knowledge for Adaptation.* Springer.

United Nations Development Programme. (2014). Human Development Index (HDI). Accessed May 5, 2015: http://hdr.undp.org/sites/default/files/hdr14_statisticaltables.xls

United Republic of Tanzania (URT). (2012). National Climate Change Strategy. VicePresident's Office (Division of Environment). Accessed January 24, 2015: http://climatechange.go.tz/

World Bank. (2017). 2016 GNI per capita, Atlas method (current US$). Accessed August 9, 2017: http://data.worldbank.org/indicator/NY.GNP.PCAP.CD

Case Study 9.7   Rotterdam: Commitment for a Climate-Proof City

Aboutaleb, A. (2009). Rising water level approach offers opportunities. *Change Magazine: Water and Climate* 5(1), 4–5.

Aerts, J., Major, D. C., Bowman, M. J., Dircke, P., and Marfai, M. A. (2009). *Connecting Delta Cities: Coastal Cities, Flood Risk Management and Adaptation to Climate Change.* Vrije Universiteit Press.

Delta Commissie (Committee). (2008). Working together with water: A living land builds for its future. Accessed March 27, 2016: http://www.deltacommissie.com/doc/deltareport_full.pdf

Nakicenovic, N., and Swart, R., eds. (2000, July). *Special Report on Emissions Scenarios.* Cambridge University Press.

Netherlands Statistics. (2014). Population dynamics Rotterdam SG, Rotterdam 2012. Accessed September 30, 2015: http://statline.cbs.nl/Statweb/publication/?DM=SLENandPA=37259engandD1=0–1, 3, 8–9, 14, 16, 21–22, 24andD2=0andD3=70, 92, 933andD4=0, 10, 20, 30, 40 (l-1)-landLA=ENandHDR=TandSTB=G1, G2, G3andVW=T

Netherlands Statistics. (2015). Land Use Rotterdam SG, Rotterdam 2012. Accessed January 26, 2016: http://statline.cbs.nl/Statweb/publication/?DM=SLENandPA=70262engandD1=0, 2, 6, 16, 20, 25, 28, 31, 41andD2=71, 94, 600andD3=3–6andLA=ENandHDR=TandSTB=G1, G2andVW=T

Peel, M. C., Finlayson, B. L., and McMahon, T. A. (2007). Updated world map of the Köppen-Geiger climate classification. *Hydrology and Earth System Sciences Discussions* 4(2), 462.

Rijcken, T. (2010). Rhine estuary closable but open—An integrated systems approach to floodproofing the Rhine and Meuse estuaries in the 21st century. In *Rotterdam's Climate Adaptation Research Summaries 2010* (10–11). Accessed October 29, 2015: www.rotterdamclimateinitiative.nl

van Veelen, P. (2010). Adaptive building is an important component of flood-proofing strategy. In *Rotterdam's Climate Adaptation Research Summaries 2010* (12–15_. Accessed October 8, 2015: www.rotterdamclimateinitiative.nl

World Bank. (2017). 2016 GNI per capita, Atlas method (current US$). Accessed August 9, 2017: http://data.worldbank.org/indicator/NY.GNP.PCAP.CD

Case Study 9.8   Preparing for Sea-Level Rise, Coastal Storms, and Other Climate Change–Induced Hazards

Buonaiuto, F., Patrick, L., Gornitz, V., Hartig, E., Leichenko, R., Stediger, J., Tanski, J., Vancura, P., and Waldman, J. (2011). Coastal zones. In Rosenzweig, C., Solecki, W., DeGaetano, A., O'Grady, M., Hassol, S., and Grabhorn, P. (eds.), *Responding to Climate Change in New York State: The ClimAID Integrated Assessment for Effective Climate Change Adaptation in New York State.* Annals of the New York Academy of Sciences.

Federal Emergency Management Agency (FEMA). (2013). *Flood Insurance Study: City of New York, New York, Bronx County, Richmond County, New York County, Queens County, Kings County.* FEMA.

Hanson, S., Nicholls, R., Ranger, N., Hallegate, S., Corfee-Morlot, J., Herweijer, J.,Chateau, J. (2011). A global ranking of port cities with high exposures to climate extremes. *Climatic Change* **104**, 89–111.

Hartig, E. K., Gornitz, V., Kolker, A., Mushacke, F., and Fallon, D. (2002). Anthropogenic and climate-change impacts on salt marsh morphology in Jamaica Bay, New York City. *Wetlands* **22**, 71–89.

Inexhibit (n.d.). BIG U winning project for Rebuild by Design in New York. Accessed May 17, 2015: http://www.inexhibit.com/marker/big-u-winning-design-new-york/

NOAA Tides and Currents. (2016). Accessed February 10, 2017: http://www.tidesandcurrents/noaa.gov

New York City Audubon. Jamaica Bay Project. Accessed May 1, 2015: http://www.nycaudubon.org/jamaica-bay-project

New York City Office of Emergency Management. (2014). NYC Hazard Mitigation Plan. Accessed August 31, 2015: http://www.nyc.gov/html/oem/html/planning_response/planning_hazard_mitigation_2014.shtml

New York City, PlaNYC. (2014). Progress Report (2014). Sustainability and Resiliency. Accessed September 2, 2015: http://www.nyc.gov/html/planyc2030/downloads/pdf/140422_PlaNYCP-Report_FINAL_Web.pdf

New York City, PlaNYC. (2015). A stronger, more resilient New York: East Side Coastal Resilience Project. Accessed April 6, 2016: http://www.nyc.gov/html/planyc/downloads/pdf/150319_ESCR_FINAL.pdf

New York City Panel on Climate Change (NPCC). (2010). New York City Panel on Climate Change 2010: Building a risk management response. Rosenzweig, C., and Solecki, W. (eds.). Annals of the New York Academy of Sciences 1196, 354.

New York City Panel on Climate Change (NPCC). (2013). New York City Panel on Climate Change Climate Risk Information 2013. *Climate Risk Information: Observations, Climate Change Projections, and Maps.*

New York City Panel on Climate Change (NPCC). (2015). New York City Panel on Climate Change 2015: Building the knowledge base for climate resilience, Rosenzweig, C., and Solecki, W. (eds.), *Annals of the New York Academy of Sciences* **1336**, 150.

Peel, M. C., Finlayson, B. L., and McMahon, T. A. (2007). Updated world map of the Köppen-Geiger climate classification. *Hydrology and Earth System Sciences Discussions* **4**(2), 462.

Permanent Service for Mean Sea Level (PSML). (n.d.). Peltier GIA data sets. Accessed September 2, 2015: http://www.psmsl.org/train_and_info/geo_signals/gia/peltier/index.php

Rosenzweig, C., Solecki, W., Hammer, S., and Mehrotra, S., eds. (2011). *Climate Change and Cities: First Assessment Report of the Urban Climate Change Research Network.* Cambridge University Press.

SIRR, PlaNYC, June (2013). *A Stronger, More Resilient New York.* The City of New York, Mayor Michael R. Bloomberg.

U.S. Census Bureau. (2010). Decennial census, summary file 1. Accessed December 1, 2015: http://www.census.gov/population/metro/files/CBSA%20Report%20Chapter%203%20Data.xls

U.S Census Bureau. (2016). Table 1. Annual Estimates of the Resident Population: April 1, 2010 to July 1, 2016 – Metropolitan Statistical Area; – 2016 Population Estimates. U.S. Census Bureau. Retrieved March 28, 2017.

World Bank. (2017). 2016 GNI per capita, Atlas method (current US$). Accessed August 9, 2017: http://data.worldbank.org/indicator/NY.GNP.PCAP.CD

10

Urban Health

Coordinating Lead Authors
Martha M. L. Barata (Rio de Janeiro), Patrick L. Kinney (New York), Keith Dear (Canberra/Kunshan)

Lead Authors
Eva Ligeti (Toronto), Kristie L. Ebi (Seattle), Jeremy Hess (Atlanta), Thea Dickinson (Toronto), Ashlinn K. Quinn (New York), Martin Obermaier (Rio de Janeiro), Denise Silva Sousa (Rio de Janeiro), Darby Jack (New York)

Contributing Authors
Livia Marinho (Rio de Janeiro), Felipe Vommaro (Rio de Janeiro), Kai Chen (Nanjing), Claudine Dereczynski (Rio de Janeiro), Mariana Carvalho (Rio de Janeiro), Diana Pinheiro Marinho (Rio de Janeiro)

This chapter should be cited as
Barata, M. M. L., P. L. Kinney, K. Dear, E. Ligeti, K. L. Ebi, J. Hess, T. Dickinson, A. K. Quinn, M. Obermaier, D. Silva Sousa, D. Jack (2018). Urban Health. In Rosenzweig, C., W. Solecki, P. Romero-Lankao, S. Mehrotra, S. Dhakal, and S. Ali Ibrahim (eds.), *Climate Change and Cities: Second Assessment Report of the Urban Climate Change Research Network*. Cambridge University Press. New York. 363–398

Managing Threats to Human Health

Climate change and extreme events are increasing risks of disease and injury in cities. Urban health systems have a significant role to play in preparing for these exacerbated risks. Climate risk information and early warning systems for adverse health outcomes are needed to enable interventions. An increasing number of cities are engaging with health adaptation planning, but health departments of all cities need to be prepared.

Major Findings

- Storms, floods, heat extremes, and landslides are among the most important weather-related health hazards in cities. Climate change will increase the risks of morbidity and mortality in urban areas due to greater frequency of weather extremes. Children, the elderly, the sick, and the poor in urban areas (i.e., those with proportionally low incomes in comparison to the local residents) are particularly vulnerable to extreme climate events.

- Some chronic health conditions (e.g., respiratory and heat-related illnesses) and infectious diseases will be exacerbated by climate change. These conditions and diseases are often prevalent in urban areas.

- The public's health in cities is highly sensitive to the ways in which climate extremes disrupt buildings, transportation, waste management, water supply and drainage systems, electricity, and fuel supplies. Making urban infrastructure more resilient will lead to better health outcomes, both during and following climate events.

- Health impacts in cities can be reduced by adopting "low-regret" adaptation strategies in the health system and throughout other sectors, such as water resources, wastewater and sanitation, environmental protection, and urban planning.

- Actions aimed primarily at reducing greenhouse gas emissions in cities can also bring immediate local health benefits and reduced costs to the health system through a range of pathways, including reduced air pollution, improved access to green space, and opportunities for active transportation on foot or bicycle.

Key Messages

In the near term, improving basic public health and health care services, developing and implementing early warning systems; and training citizens' groups in disaster preparedness, recovery, and resilience are effective adaptation measures.

The public health sector, municipal governments, and the climate change community should work together to integrate health as a key goal in the policies, plans, and programs of all city sectors.

Connections between climate change and health should be made clear to public health practitioners, city planners, policy-makers, and to the general public. Collaborative efforts, focused for example on the identification of vulnerable residents and resources, have been recognized as effective for enhancing community resiliency during extreme events.

10.1 Introduction

Cities are complex, and the actions of different urban sectors influence their populations' health (see Figure 10.1). Climate change is expected to affect the frequency and severity of existing diseases and may also threaten progress toward reducing the burden of climate-related disease and injury (Smith et al., 2014). Urban health systems should be prepared for potentially enhanced disease risks related to climate change (see Box 10.1).

The effectiveness of programs and measures to address any health burden attributable to climate change depends upon a variety of factors, including the current burden of disease, the effectiveness of current interventions, and projections for where, when, and how climate change is expected to affect the health burden. Equally relevant is the feasibility of implementing additional programs in light of different stakeholders' engagement with the problem and the surrounding social, economic, and political setting.

In this context, the use of scientific knowledge can be vital to the protection of urban citizens' health and well-being. For example, health sector specialists can use climate information effectively in epidemic early warning systems to help select which interventions to initiate (Ghebreyesus et al., 2009). The challenge for the scientific community is to generate and communicate this knowledge in a way that can usefully inform policy choices based on the realities of the urban environment. Improved data and knowledge will be critical to reduce citizens' vulnerability to climate hazards, as well as existing inequities and social injustice in cities, and to assess the health co-benefits of strategies to mitigate greenhouse gas (GHG) emissions.

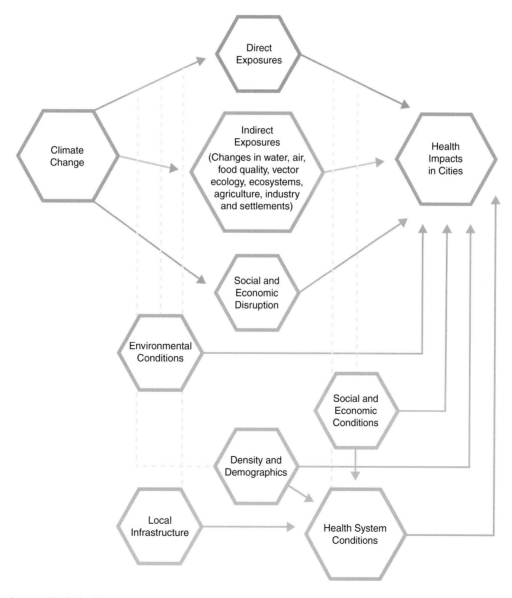

Figure 10.1 *Climate change and health in cities.*

Source: Adapted from IPCC, 2007

Box 10.1 Health and Health Systems in Cities

This chapter uses the definition applied by the World Health Organization (WHO) for *health* and *health systems* where *health* "is a state of complete physical, mental and social well-being and not merely the absence of disease or infirmity" (World Health Organization [WHO], 1946). This definition strengthens the idea of health as the ability to adapt and to self-manage (Rydin et al., 2012). Within this notion of *health*, Rydin includes Amartya Sen's idea of *justice,* entailing the ability to live a life one has reason to value. In cities, *health* is associated with social, economic, and environmental determinants (Barata et al., 2011), including climate change. There is a large and continually expanding body of research presenting that the way in which cities are planned and managed can make a substantial difference to the health

of their residents (Rydin et al., 2012). The WHO writes that *health systems* include all the organizations, institutions, and resources that are devoted to produce actions principally aimed at improving, maintaining, or restoring health (WHO, 2005). Therefore, a public health system should advocate for healthy policies, plans, and projects for all urban sectors. Long-term projections of global health outcomes now explicitly include factors such as unsafe water, food, and residence; poor sanitation; urban air pollution; and indoor air pollution – all of which are aggravated by climate change (see Section 10.3). Thus, a health impact assessment of urban polices, plans, and projects is critical when planning and managing cities as well as when implementing city adaptation and mitigation strategies related to climate change hazard.

Urban Adaptation and Mitigation Strategies

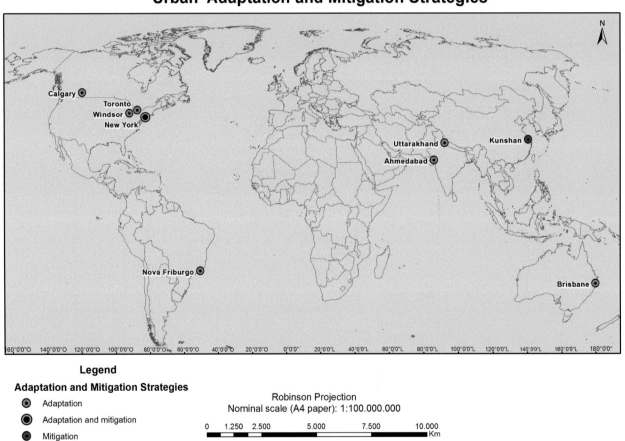

Figure 10.2 *Examples of ARC3.2 Case Study Docking Station cities that have implemented mitigation and adaptation strategies with co-benefits for their health systems.*

Since publication of the First Urban Climate Change Research Network Assessment Report on Climate Change and Cities (ARC3.1) chapter on climate change and human health, important progress has been made. Reviews of climate and urban climate change adaptation strategies in key cities (Carmin et al., 2012a; Castán Broto and Bulkeley, 2013; Revi et al., 2014) find that an increasing number of cities are engaging with health

adaptation planning. These reports also provide a number of research and Case Studies of the potential health co-benefits of mitigation and adaptation activities and include examples drawn from the Global South. Meanwhile, the Fifth Assessment Report (AR5) developed by the Intergovernmental Panel on Climate Change (Intergovernmental Panel Climate Change [IPCC], 2014) provides an updated report of the international scientific

Table 10.1 *Climate change–related drivers and benefits from implemented adaptation and mitigation strategies.*

Drivers	Country	City	Health endpoint	Adaptation and Mitigation Strategies	Benefits
	Brazil	Nova Friburgo	Increase in the incidence of leptospirosis and dengue fever	Syndromic approach	Prepare the health sector to react to extreme events
Heavy rainfall	Canada	Calgary	Mental health illness	Improvement in infrastructure	Reduce city infrastructure and operational vulnerabilities
		Toronto	Loss of well-being and foodborne illnesses	Investment in order to reduce the impact of future flood events	Improve wastewater and storm water collection systems and increase public education efforts
	India	Uttarakhand (13 districts)	Deaths and loss of well-being	Improvement in infrastructure	Optimal utilization of water resources
	Canada	Windsor	Deaths	Heat alert and response plan	Create a strong health system that can protect the most vulnerable populations
	Australia	Brisbane	Deaths and years-of-life-lost	Urban green infrastructure	Reduce local ambient temperature
Heat wave	India	Ahmedabad	Dehydration; acute heat illnesses; cardiovascular diseases; kidney diseases; respiratory diseases	Heat action plan and extreme heat early warning system	Alert the population on heat risks and prevention measures and forecast extreme weather temperatures
	USA	New York	Deaths and respiratory illness	Urban green infrastructure (Million Trees NYC Project)	Reduce atmospheric CO_2; reduce local ambient temperature; improve air quality and reduce storm water runoff
Reduction of CO_2	China	Kunshan	Growing health system demand	Building of three new healthcare facilities using green technology	Energy and water efficiency; disaster-resilient health system with self-contained tertiary health care facilities

community's most current knowledge of the human health impacts of climate change. This includes emerging human health risks in cities and applied responses to prevent and mitigate them (Mustelin et al., 2013).

In this chapter, we summarize the scientific evidence concerning climate-related human health risks in urban areas and discuss responses for reducing these risks in light of climate change. In Section 10.2, the process of stakeholder engagement is discussed. The extent to which the health sector and other relevant stakeholders confront the challenges of climate change and promote the well-being of all citizens is considered. In Section 10.3, an overview of the current knowledge of climate change hazards, health impacts, and urban health vulnerabilities is presented. Section 10.4 turns to adaptation strategies to protect city residents from some of the health impacts and risks posed by climate change. In Section 10.5 we discuss mitigation strategies that, while aimed at reducing GHG emissions,

also provide so-called co-benefits to human health. Section 10.6 presents barriers and bridges to mitigation and adaptation strategies. Section 10.7 highlights knowledge gaps and recommended areas for further research, and in Section 10.8 some policy recommendations are presented as a conclusion of the assessment presented in this chapter.

10.2 Public Engagement

Public engagement (Greenwood, 2007) is essential to all stages of climate change assessment, from the initial planning stage through the implementation, monitoring, response, and evaluation phases of successful urban risk management response to climate change (URMRCC). Box 10.2 presents the process, advantages, barriers, and challenges of the implementation of a stakeholder engagement strategy.

Box 10.2 Stakeholder Engagement Process, Steps, Barriers, and Challenges

Stakeholder engagement provides a wealth of benefits to planning and response processes. First, this engagement fosters innovation by harnessing collective potential to identify key issues. Stakeholders can empower those outside the decision-making process and ensure that community concerns are taken into account. They can also help increase credibility, identify vulnerable populations, mobilize resources and networks, aid in fact-checking, and provide different social perspectives for planning. Stakeholders' long-term contributions include improved community relations, lasting capability for action, and conflict mitigation. Finally, fostering stakeholder engagement makes room for the development of alternative solutions, which can reduce response times, mitigate risks, and stimulate learning (IPCC, 2007; Admassie et al., 2008; Littell, 2010).

PROCESS AND STEPS FOR STAKEHOLDER ENGAGEMENT

The literature presents many frameworks for engaging stakeholders. According to Kema Inc. (2012), the process of stakeholder engagement begins with establishing a team, identifying goals and an audience, and determining a timeline and resources. It concludes with developing outreach materials and implementing an engagement strategy. Similarly, a five-step stakeholder engagement process is used by the Canadian government, starting by identifying stakeholders and proceeding toward understanding the reasons for stakeholder engagement, planning the engagement process, commencing the dialogue, and, last, maintaining the dialogue and delivering on commitments (Hohnen, 2007; Schmeer, K. 1999).

Pragmatically speaking, the level at which climate and health activities are likely to be pursued – and the capacity to pursue

these activities at that level – are both important factors to consider in adaptation planning. It is important to house adaptation activities within the appropriate level of department or organization to ensure that both willingness and capacity are maximized. Similarly, endorsement by senior management is central, but the ongoing engagement of mid-level stakeholders is also important for the result of the adaptation process.

LESSONS, BARRIERS, AND CHALLENGES TO STAKEHOLDER ENGAGEMENT

Stakeholder engagement can be time-consuming and costly. To go beyond research to action, it is important that stakeholders take ownership of the adaptation process. Additionally, the provision of appropriate staff time and funds is important for successful stakeholder involvement. Finally, the process should not be focused on technical modeling issues and reports written in technical jargon (adapted from CAP, 2007).

Several barriers exist to stakeholder engagement, including time commitments and impractical timescales (e.g., when the time requested to deliver on goals is inadequate). When working in international settings, gender, language, and cultural barriers (e.g., males and females working together) can also pose challenges to successful stakeholder engagement. Trust must be cultivated, especially in community settings where stakeholders from different interest groups (e.g., the private sector, government, nongovernmental organizations [NGOs], citizen's groups, and academia) may be involved. Stakeholders previously engaged in other similar settings may have already worked to meet challenges and may be reluctant to engage due to fears of redundancy, and power struggles between groups and community members challenge stakeholder engagement when visions are not shared (Weible, 2007; Stott and Walton, 2013).

O'Haire et al. (2011) provide a helpful list of factors that can transform these barriers into successful endeavors, beginning with:

- Engaging stakeholders early in the process to gauge their ability to participate (identifying time constraints and the type and amount of information each group or member can provide)
- Ensuring that the engagement is consistent and ongoing (building trust with the stakeholders)
- Preparing and providing material for the stakeholders to ensure an equal level of knowledge (providing additional information if needed can decrease issues later in the engagement process)

- Allowing each stakeholder (individual or group) equal participation in the process to reduce any unnecessary power dynamics that may decrease engagement effectiveness
- Reinforcing understanding of stakeholders' duties so that *scope-creep*, in which duties extend beyond those originally identified, does not occur.

Facilitators should ensure proper time management as much as possible, be neutral and encourage participation from all stakeholders, and be familiar not only with the health issue but with the cultural and community setting that they are working.

Stakeholders provide diverse perspectives and data to the decision-making process. Examples of possible health impacts related to climate change and the potential relevant stakeholders who are able to provide information and support for reducing those impacts to populations in cities, are listed here:

- *Vector-borne and infectious diseases*: Epidemiologists (to identify changes in infection rates), hospitals (to respond to public health emergencies and treat patients), and social workers and local community members and groups (to help identify vulnerable populations and respond to environmental health needs)
- *Heat-related illnesses (including stroke, respiratory and cardiovascular distress)*: Local municipalities' decision-makers (to develop and implement heat-health warning and response policies), the media (to alert the public to extreme heat events and locations of cooling centers), and independent power producers and utilities (who provide electricity for cooling and maintain infrastructure)
- *Water quality and water-borne diseases:* Emergency preparedness organizations (first responders for flood events), municipal planning departments (to upgrade sewer and drainage systems), and water management departments (to detect changes in water quality)
- *Air quality, asthma, allergies*: Meteorology services, air quality managers, public health and/or medical schools, NGOs and research scientists (to conduct research on air quality and health impacts), private sectors (who may contribute to GHG emissions but may also produce valuable products, including medications for respiratory distress)

Engaging stakeholders and improving governance in light of the health risks of climate change in cities requires understanding how the health system is structured. As the WHO writes, a health system involves the organizations, institutions, resources, and people whose main purpose is to improve health. Bowen et al. (2012) present an integrated view of the institutions and sectors relevant to the health risks of climate change (see Figure 10.3).

The WHO is responsible for providing global and sectoral leadership. The WHO shapes the global health research

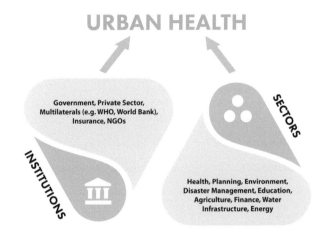

Figure 10.3 *Integrated view of the institutions and sectors relevant to the health risks of climate change.*

Source: Adapted from Bowen et al., 2012

agenda, sets norms and standards, articulates evidence-based policy options, provides technical support to countries, and monitors and assesses health trends (information available on their website). WHO headquarters takes the lead in coordinating climate change and health work across United Nations agencies. Regional and country offices work directly with national ministries of health, providing technical support and guidance.

In most countries, at the national level, the ministry of health is the primary contact for efforts to manage the health risks of climate change. The ministry of health often coordinates and collaborates with a national climate change team and serves as a resource for regional and local adaptation efforts. Such cooperative efforts need to be enhanced, particularly in low- and middle-income countries. For example, a policy analysis of Fiji's National Climate Change Policy, National Climate Change Adaptation Strategy, and Public Health Act found that the health risks of climate change were considered only to a minor extent and that supporting documents in sectors such as water and agriculture did not consider health risks (Morrow and Bowen, 2014).

In Fiji, the incidence of dengue, waterborne diseases, and under-nutrition will likely be affected by climate change, which means that these risks should be addressed with sectorial and climate change–specific policies. International and academic collaboration may be also useful, as can be seen in Case Study 10.1.

Although some cities are responsible for their own health system's functions, others are embedded within a multilayered

hierarchy of responsibilities directed by the ministry of health. Capacity to monitor and respond to climate change health threats at a municipal level is likely to vary alongside capacities in other areas and may be lower than at the national level (Paterson et al., 2012; Bierbaum et al., 2013). Managing the health risks of climate change is complex; it often requires a cross-sectorial collaboration (see Case Study 10.1), and it should be considered when designing a stakeholder engagement strategy.

Case Study 10.1 Participative Development of a Heat-Health Action Plan in Ahmedabad, India

Martin Obermaier

Energy Planning Program, Federal University of Rio de Janeiro

Keywords	Heat waves, adaptation action plan, early warning system, India
Population (Metropolitan Region)	6,352,254 (Census Bureau of India, 2011)
Area (Metropolitan Region)	1,866 km² (Ahmedabad Urban Development Authority, 2016)
Income per capita	US$1,680 (World Bank, 2017)
Climate zone	BSh – Arid, steppe, hot arid (Peel et al., 2007)

The city of Ahmedabad is an important economic and industrial hub in western India, located in the center-north of the State of Gujarat along the banks of Sabarmati River. The city's urban population has increased from 4.5 million (2001) to 6.3 million (2011) in just 10 years (Government of India, 2011; Census Bureau of India, 2011), making it one of the fastest growing urban regions within India and worldwide (Kotkin, 2010).

The climate in Ahmedabad on the Köppen climate scale is hot semi-arid (BSh) with three distinct seasons: winter, summer, and monsoon. Temperatures usually get extremely hot before the monsoon arrives: average monthly high temperatures (1971–2000) are 39.7°C (April), 41.6°C (May), and 38.7°C (June), with record high temperatures climbing as high as 47–48°C.

Heat waves pose a threat for the public health system already today – not only in Ahmedabad, but also in many other Indian cities. Heat waves cause fatalities and are related to dehydration, acute heat illnesses (heat exhaustion or heat stroke), cardiovascular diseases, kidney diseases, and respiratory diseases (Knowlton et al., 2014). In Ahmedabad, this particularly affects poor slum dwellers (25.8% of the city's population in 2006) (Tran et al., 2013) where adaptation options such as adequate drinking water supply or air condition are frequently missing and poor environmental conditions (e.g., water quality, air pollution) and work conditions (especially for outdoor workers) already exacerbate pressures for human health (Knowlton et al., 2014).

Climate change is expected to increase the frequency and intensity of prolonged periods of high temperatures (Smith et al., 2014), with urban heat islands likely exacerbating vulnerability to extreme heat caused by global warming (Smith et al., 2014).

In this context, a team of American and Indian research institutions, the Indian Institute of Public Health, and the Ahmedabad Municipal Corporation joined forces following a severe heat wave in 2010 to start the Ahmedabad Heat and Climate Study Group (see Knowlton et al., 2014). The group's task focused on the design and implementation of two distinct but integrated instruments: a Heat Action Plan (HAP), and an Extreme Heat Early Warning System that informs the HAP (Knowlton et al., 2014; Ahmedabad Municipal Corporation [AMC], 2013, 2015).

The HAP integrates awareness-building (alerting the population on heat risks and prevention measures) through different media, capacity-building programs for health professionals, and a simple warning system. Classification of heat emergencies according to their degree of severity triggers a different set of responsibilities and actions for each institutional actor participating in the HAP. The Extreme Heat Early Warning System comprises a probabilistic weather forecasting system for extreme temperatures and the probability that specific extreme temperature thresholds will in fact be exceeded. Heat waves are projected with a seven-day advance lead time in order to allow Ahmedabad officials time to plan and coordinate public health and interagency responses.

The design and implementation process of both HAP and the Extreme Heat Early Warning System included considerable stakeholder feedback: for example, the project's needs assessment – focused on the characterization of susceptible populations, evaluation of earlier heat wave impacts, assessment of health sector capacity to deal with heat wave conditions, and the like – organized comprehensive focus group meetings and roundtables with experts such as medical professionals, meteorologists, and health agency personnel, but also included community leaders and media experts. Site visits to hospitals and informal settlements to understand community needs and to gain trust also among non-expert participants complemented this broad approach (Knowlton et al., 2014).

This open and participatory approach has been conducted at all steps of HAP and Extreme Heat Early Warning System development, from the initial project design phase to data collection to intervention development and project implementation. In fact, maintaining communications open and coordinated at all stages of the project, including the possibility to discuss working methodologies with experts and non-experts, has strongly contributed to establishing trusted relationships between the project team and partners while in turn also increasing overall acceptance of the outcomes and buy-in from relevant stakeholders (Knowlton et al., 2014).

Both the HAP and Extreme Heat Early Warning System are currently implemented and in use (AMC, 2013, 2015), and evaluation of effectiveness and efficiency of the project is currently ongoing. Future plans for Ahmedabad include the integration of additional

Heat Alert

Dos and Don'ts During a Heat Wave:

•Drink water and other liquids (no soft drinks)
•Stay out of the sun
•Find a place to cool down
•Wear light clothing
•Check in with friends and family

Symptoms to Watch For:

•Lack of sweating despite the heat
•Heavy sweating and weakness
•Muscle weakness or cramps
•Headache and nausea
•Heat Rash or cramps
•Red, hot and dry skin
•Nausea and vomiting

People at Greater Risk: Children, the Elderly, and Pregnant Women

Case Study 10.1 Figure 1 *Alerting the population on heat risks and prevention measures.*

Source: Adapted from Knowlton et al., 2014

components into the project approach (e.g., promoting green infra-structure for cooling microclimate or developing dedicated heat avoidance training programs for children and their mothers), but also studying existing air pollution problems in Ahmedabad as a possibly confounding factor for heat-related mortality and transferring exist-ing measures to other cities across Gujarat State.

Box 10.3   Stakeholder Engagement for an Effective Heat Wave Early Warning and Response System

Establishing an effective heat wave early warning and response system involves, at minimum, representatives of public health, meteorology, hospital, and media groups engaged in planning and outreach activities.

White-Newsome et al. (2014) surveyed the steps for estab-lishing an effective heat wave early warning and response system. They interviewed leaders from government and nongovernmental organizations representing the public health, general social services, emergency management, meteorology, and environmental planning sectors in four U.S. cities chosen for their diverse demographics, climates, and climate adaptation strategies (Detroit, Michigan; New York, New York; Philadelphia, Pennsylvania; and Phoenix, Arizona). The interviews identified activities that could reduce the harmful effects of high ambient tempera-tures and described the obstacles faced in implementing them. Cooling centers, heat wave early warning systems, programs to distribute fans and/or air conditioning, and

outreach programs were common across the cities. The local context (including local political will, availability of resources, and the maturity of local community organiza-tions) was an important consideration both for the success of efforts to identify and reach the most vulnerable popu-lations and for optimizing the use of health statistics. The primary obstacles faced were financial constraints, promot-ing the effective use of cooling centers, and communication issues; addressing these issues requires a multisectoral approach.

Considering that stakeholder engagement should include citizens' groups, individual citizens need to know how to recognize symptoms of heat exposure and learn basic ways to stay cool as they access other available community-level heat-health resources. Educating citizens to understand how to protect and increase their resilience must be a key com-ponent of overall community response to health challenges related to climate change (see Case Study 10.1).

10.3   Climate Hazards, Vulnerabilities, and Opportunities

Urban areas are vulnerable to the health impacts of climate change due to their high population density, concentration of vulnerable populations, higher temperatures compared to sur-rounding areas, and, in many cases, exposure to coastal storms. Increase in climate-related temperatures, amplified intensity of storms and related flooding, storm surge in coastal and low-lying urban areas, and changes in vector-borne and infectious diseases are all examples of climate-related health impacts that have the potential to affect public health in urban areas. In addition, the health of urban residents is closely linked to key infrastructure, such as public transportation, sewerage, waste management, the efficiency of rainwater drainage systems, and the utility sup-ply. When climate-related extreme events threaten critical urban infrastructure, reverberations can be felt in the health of urban populations (see Figure 10.4).

10.3.1 Storms and Flooding

Urban populations in coastal and low-lying regions are often vulnerable to the impacts of extreme storms and flooding. The number of people living near coasts and within areas likely to be impacted by storm-related flooding events is large (NCA 2014, ch. 2). Health impacts of extreme storms depend on interactions between the storm's particular characteristics and the characteristics of the affected communities and may include direct effects (e.g., death and injury) and indirect, long-term effects (e.g., infrastructure damage, contamination of water and soil, and changes in vector-borne diseases, respiratory health, and mental health) (Lane et al., 2013). Flooding can induce the creation of breeding sites for vectors and can lead to bacterial contamination of water sources, resulting in outbreaks of infectious disease. Storms and flooding can also mobilize chemical toxins from industrial or contaminated sites (Ruckart et al., 2008). Elevated indoor moisture and mold levels associated with flooding of interior spaces have been identified as risk factors for cough, wheeze, and childhood asthma (Jaakkola et al., 2005). Mental health impacts may be among the most common and long-lasting impacts of extreme storms; to date, however, they have received relatively little study (Berry et al., 2010). Evacuation and disruptions in access to communications and health care delivery are all correlates of extreme storms that can adversely affect health. The stress of evacuation, property damage, economic loss, and household disruption, as well as knowledge of trauma and deaths, are among the triggers of adverse mental health impacts that have been identified among storm-affected populations (Weisler et al., 2006).

10.3.2 Temperature Extremes

Heat has a direct impact on all-cause mortality, with peaks in mortality occurring on the same day or shortly after exposure to heat. In urban areas, residential environments may pose significant risk for fatal heat exposure during heat waves; for example, more than 80% of heat stroke deaths in New York have been attributed to exposure at home (Wheeler et al., 2013). Demographic groups including children, the elderly, and people with pre-existing health impairments may be particularly vulnerable to such stresses (Romero-Lankao et al., 2013).

While mortality represents the most severe health outcome of extreme heat, increases in emergency room visits and in hospital admissions for heat-sensitive diseases have also been observed during heat episodes (Knowlton et al., 2009; Lin et al., 2009). A limited amount of evidence suggests that exposure to elevated temperatures may also have an impact on adverse birth outcomes such as preterm birth (Basu et al., 2010).

For both urban mortality and nonfatal temperature-related illnesses (e.g., heat stroke, heat exhaustion, hyperthermia), temperature threshold levels are city-specific and depend on latitude and normal average temperatures (Harlan and Ruddell, 2011). In their study of twelve urban agglomerations in low- and middle-income countries, McMichael et al. (2008) found that thresholds for heat-related morbidity ranged from 16°C to 31°C, with higher threshold levels in cities with warmer climates. Figure 10.5 presents the overall cumulative heat–mortality relationships in four cities in different regions of the world. There

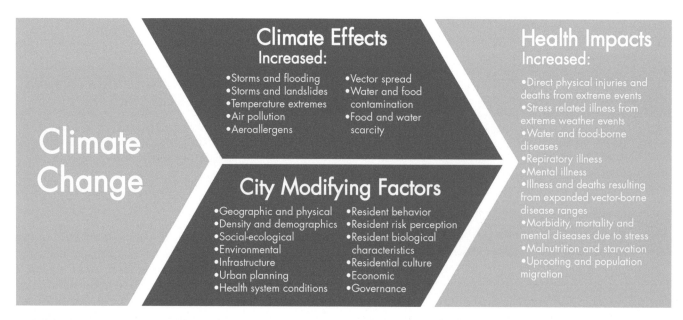

Figure 10.4 *How climate change affects citizens' health.*

Source: Rozenzweig et al., 2011

Case Study 10.2   Health and Social Cost of Disaster: Nova Friburgo, State of Rio de Janeiro, Brazil

Diana Pinheiro Marinho

ENSP/Fiocruz, Rio de Janeiro

Livia Marinho and Martha M. L. Barata

IOC/Fiocruz, Rio de Janeiro

Keywords	Rainfall, extreme event, health, health care costs, syndromic approach
Population (Metropolitan Region)	184,786 (Instituto Brasileiro de Geografia e Estatística [IBGE], 2015)
Area (Metropolitan Region)	933,415 km² (Instituto Brasileiro de Geografia e Estatística [IBGE], 2015)
Income per capita	US$8,840 (World Bank, 2017)
Climate zone	Cfb – Warm temperate, fully humid, warm summer (Peel et al., 2007)

In the Mountain Region of the state of Rio de Janeiro in Brazil, urban sprawl, urban population density, and deficits in basic sanitation services increase the severity of the environmental and human impacts of extreme events. Additionally, among the 14 municipalities located in this region, Nova Friburgo, Petrópolis, and Teresópolis have the highest vulnerability due to their geomorphology and high population density. These cities were vulnerable and not well prepared to avoid damages caused by precipitation.

In January 2011, heavy rains fell on the Mountain Region, causing one of the most severe natural disasters ever recorded in Brazil. The United Nations ranked it as the eighth largest landslide to have occurred worldwide in the past 100 years. More than 900 deaths occurred, and nearly 35,000 people were left homeless or displaced in the affected municipalities of the Mountain Region.

Because of the disaster, 3,000 landslides were registered in Nova Friburgo, with resulting damages to water, energy, transport, telecommunications, and health services. In this city, 429 people died and 3,220 were left homeless (Pereira et al., 2014). The disaster also triggered a significant increase in the incidence of leptospirosis and dengue fever in the city (see Case Study 10.2 Figures 1 and 2).

A partial economic assessment was conducted to evaluate the social cost of the leptospirosis and dengue fever cases attributed to this natural disaster in Nova Friburgo. It considered both direct (health care costs) and indirect costs (loss for society, considering loss of school and work days, as well as lost productivity). The study used restricted-access secondary data provided by the Municipal Health Surveillance and Primary Care Health Sector. Income data included the 2011 prevailing national and state minimum wages and the estimated average income of the local population.

The study also evaluated avoided costs achieved via a syndromic surveillance system adopted by the local Service of Epidemiological Surveillance just after the disaster. It consisted of the administration of specific medication for disease treatment even if there was no diagnostic confirmation, a procedure in which the presence of three major characteristic symptoms of the disease was required. This measure aimed to prevent worsening of patients' condition before their diagnoses were confirmed and avoid increased social costs caused by the disease. The necessary information for calculating the avoided cost of treatment was obtained from charts used for syndromic surveillance; namely, the medications administered, duration of treatment, and dates of procedures. The cost of the syndromic approach comprised drug expenses as well as the cost of the health care team performing the procedure. As reported by the Epidemiological Surveillance Service of the municipality, the State of Rio de Janeiro Health Department professionals performed the first procedures and interventions at the calamity site. After this team left, the city staff continued the job.

The total social cost of leptospirosis cases attributed to the 2011 disaster ranged between US$22,000 and US$66,000. The adopted empirical therapy (syndromic approach) represented a total avoided cost of US$14,800, in addition to a reduction in lethality (i.e., thirty-one deaths were avoided among confirmed cases of the disease, and no deaths resulted from the leptospirosis cases attributed to the natural disaster) (Instituto Oswaldo Cruz-Fiocruz, 2013).

The disaster also triggered a significant increase in the incidence of dengue fever in Nova Friburgo. Moreover, there was an increase in sites for mosquito breeding, which facilitated the proliferation of dengue fever in this city, where its incidence had previously been low. The social cost of the disease was evaluated using the same methodology applied for the cases of leptospirosis raging between US$29,000 and US$219,000 (Instituto Oswaldo Cruz-Fiocruz, 2013).

The increase in these diseases was associated with extensive changes in the city's environment (e.g., sanitation and urban cleaning programs were interrupted in the post-event period).

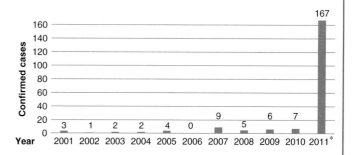

Case Study 10.2 Figure 1 *Confirmed cases of leptospirosis in Nova Friburgo, Rio de Janeiro State between 2001 and 2011.*

Source: SINAN

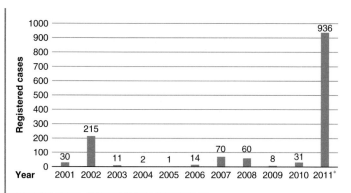

Case Study 10.2 Figure 2 *Registered cases of dengue in Nova Friburgo, Rio de Janeiro State between 2001 and 2011.*

Source: SINAN

In fact, there has been a significant post-disaster long-term rise in leptospirosis and dengue fever incidence in the municipality. This study illustrates the potential for increased cases of a vector-borne infectious disease following a natural disaster. A disease outcome such as this brings significant costs to the health sector and to society. Underestimation of these costs and underappreciation of these risks are clearly problems to be confronted in urban planning for climate change adaptation. Estimation of the social cost of natural disaster–related disease outcomes can help encourage the adoption of ongoing preventive measures that can protect urban residents from these types of adverse health outcomes.

Source: Archive of authors, 2012. Produced from the database of the Brazilian Institute of Geography and Statistics (IBGE); Pereira et al. (2014b; 2014c)

is consensus that exposure–response functions for health effects due to heat exposure do not follow a strict linear trend but instead have U-, V-, or J-shapes (McMichael et al., 2008; Romero-Lankao et al., 2013). In addition to heat extremes, cold temperatures can also be associated with ill-health, although the causal

pathways remain unclear in many cases (Smith et al. 2014). In a meta-analysis of 224 city studies, Romero-Lankao et al. (2012) found that deaths increase as temperatures fall below or rise above certain threshold values. The thresholds are noteworthy since they may be assumed as measures of heat or cold tolerance,

Case Study 10.3   Windsor Heat Alert and Response Plan: Reaching Vulnerable Populations

Thea Dickinson

University of Toronto

Keywords	Heat, public health, response plan
Population (Metropolitan Region)	210,891 (Government of Canada, 2015)
Area (Metropolitan Region)	146.32 km² (Government of Canada, 2015)
Income per capita	US$43,660 (World Bank, 2017)
Climate zone	Dfa – Cold, without dry season, hot summer (Peel et al., 2007)

Cities are built to accommodate high-density populations, favoring impervious concrete surfaces and low-albedo material over natural forests and sprawling ecosystems. These planning designs create, absorb, and retain heat (urban heat island) increasing temperatures by up to 12°C over surrounding rural areas (Oke, 1997). In August 2003, the temperature in France went 12°C above the mean maximum temperature for nine successive days (Fouillet, 2006) resulting in more than 70,000 deaths across Europe (Robine et al., 2008). Following this extreme heat event, many cities moved to proactively develop heat-related policies that would (1) create stronger public health systems and (2) implement adaptation measures to decrease residents' vulnerability to these life-threatening heat events. To further complicate matters, climate change is projected to cause an

increase in the frequency and intensity of extreme heat events and having the potential to triple heat-related deaths (IPCC, 2014).

Following the release of Environment Canada's climate projections for Southern Ontario, the City of Windsor and the County of Essex enacted a three-tier Heat Alert and Response Plan (City of Windsor, 2013). The plan goes beyond Public Health Unit involvement to include nongovernmental organizations (NGOs), community groups, decision-makers, first responders, hospitals, pharmacist, and utilities to help identify and reach the most vulnerable populations before and during extreme heat events.

When a heat event occurs,[1] the Windsor-Essex Heat Alert and Response Plan triggers Public Health Units, the Red Cross, the City of Windsor and County Municipalities, NGOs, and neighboring cities to coordinate, prepare, notify, report, and exchange information and resources relating to the heat event. Each group has defined roles that include:

- Monitoring Environment Canada's Humidex forecast, April through October
- Preparing first responders with heat-health messaging
- Assisting with the dissemination of information regarding extreme heat events
- Notifying media and community partners of change in alert level
- Providing a list of public facilities with air conditioning (and generators) with normal hours of operation that the general public can access if required
- Coordinating with Windsor Essex County Health Unit to develop heat-health messaging
- Providing emergency medical response to the public during extreme heat events
- Providing volunteers, including member of the First Aid Service (FAS) team to provide support

1   In Canada, a heat wave is defined as three or more consecutive days in which the maximum temperature is greater than or equal to 32°C.

- Exchanging information with the Windsor Essex County Health Unit prior to and during extreme heat events
- Coordinating with Detroit to develop heat-health messaging
- Reporting to the Medical Officer of Health the prevalence of heat-related illnesses

Case Study 10.3 Table 1 provides a brief illustration of each level of the system.

Prior to the heat event, the plan identifies vulnerable populations and individuals who will be contacted (during the alert) by phone or visited in their communities by volunteer groups and NGOs. A Community Emergency Management Coordinator (CEMC) will contact the local utilities to ensure that disaster plans are in place in case of a power outage. Pharmacists are asked to complete a report at the end of the day that will help identify any changes in access during the heat event. A municipal website (staycoolwindsoressex.com) is maintained and updated regularly before and during the heat alert along with a call center to help residents easily access information including cooling center locations and signs and symptoms relating to heat-related illnesses. These actions are pre-empted by a public awareness campaign that is held prior to the start of each warm season to remind and educate residents of the potential impacts of heat to their health and well-being.

All these measures are vitally important to create a strong public health system that protects the most vulnerable populations with the aim of having them understand the risks associated with extreme heat events and to aid them in finding assistance prior to them being at risk. These

Case Study 10.3 Table 1 *City of Windsor and the County of Essex Three-Tier Heat Alert and Response Plan. Source: Adapted from http://www .staycoolchatham-kent.com*

Monitoring	The Windsor-Essex County Health Unit is monitoring the Humidex forecast for extreme heat events. The Humidex is a combination of temperature (°C) and humidity (%) to reflect perceived temperature
Level 1	One or more days reaching Humidex 40
Level 2	Four or more days reaching Humidex 40
	One or more days reaching Humidex 45
	Four or more nights above Humidex 28
Level 3	A heat emergency is issued in response to a severe or prolonged emergency, such as power outages or water shortages

proactive measures allow cities to (1) understand who is vulnerable (2) identify where these vulnerable populations live (3) determine the cities hot spots, and (4) develop and enact plans with community partners. A final component of the plan is to research adaptation strategies in order to build a more heat-resilient community that will reduce a changing climate's impact on their community.

---

or as the "comfort range" that changes depending on such factors as city location and age and that is expected to be affected by climate change (Romero-Lankao et al., 2012). Vardoulakis et al. (2014) developed a comparative assessment of the effects of climate change on heat- and cold-related mortality in cities of United Kingdom and Australia. According to their analysis, cold-related mortality currently accounts for more deaths than heat-related mortality in cities in the United Kingdom (around 61 and 3 deaths per 100,000 population per year) and from Australia (around 33 and 2 deaths per 100,000 population per year), respectively (Vardoulakis et al., 2014). Others have argued that high death rates in winter are not caused by cold temperatures but rather by other factors that vary across seasons (Kinney et al., 2015).

10.3.3 Vector-Borne Diseases

Many diseases are spread by vector organisms such as ticks and mosquitoes. Dengue and malaria are important mosquito-borne diseases contributing significantly to the burden of death and disease in urban areas in the developing world (Murray et al., 2012; Bhatt et al., 2013). Vector-borne disease incidence is influenced by temperature and precipitation, which can affect the range, prevalence, and reproductive cycle of disease vectors, among other determinants (Kelly-Hope et al., 2009). For example, trends in malaria in east Africa have been associated with warming trends observed there over multiple decades (Omumbo et al., 2011). There is evidence that the Lyme disease vector,

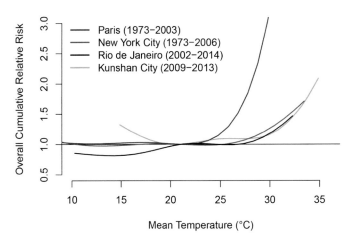

Figure 10.5 *Overall cumulative heat-mortality relationships in Paris, New York, Rio de Janeiro, and Kunshan City, China. Relative risks were calculated using a distributed-lag nonlinear model with a natural cubic spline with 4 degree of freedom for temperature and the lag, a maximum of 6 lag days, and 21°C as a reference temperature. Warm season (May–September) data for Paris (1973–2003), New York (1973–2006), and Kunshan City (2009–2013); November–March data for Rio de Janeiro (2002–2014).*

Source: Kai Chen and Patrick L. Kinney, with data from Surveillance Secretariat in Health from the Municipal Secretary of Rio de Janeiro

the tick species *Ixodes scapularis*, has expanded its range from the United States northward into Canada over the past several decades in part due to warming temperatures (Ogden et al., 2008; Ogden et al., 2010).

Dengue fever transmitters, the *Aedes aegypti* mosquito, can be easily found in regions of tropical and subtropical climate, having an ideal temperature of transmission of between 23°C and 27°C, although temperatures from 18°C can also trigger its transmission. In contrast, temperatures below 5°C and above 36°C can impair and hinder the vectors' survival (Colón-González et al., 2013).

Climate change may lead to changes in the seasonal cycle and spatial distribution of some vector-borne diseases, although it is important to note that climate is only one of many drivers of vector-borne disease distribution (see Case Study 10.2). While climate change might lead to an expansion in the range of important disease vectors, it is also possible that optimal temperature conditions for certain vector species will be exceeded

Case Study 10.4   Health Impacts of Extreme Temperatures in Brisbane, Australia

Cunrui Huang

Sun Yat-sen University, Guangdong

Xiaoming Wang

Commonwealth Scientific and Industrial Research Organisation (CSIRO), Melbourne

Keywords	Temperature, public health, years-of-life-lost
Population (Metropolitan Region)	2,065,996 (Australian Bureau of Statistics Census, 2011a)
Area (Metropolitan Region)	15,826 km² (Australian Bureau of Statistics Census, 2011b)
Income per capita	US$54,420 (World Bank, 2017)
Climate zone	Cfa – Temperate, without dry season, hot summer (Peel et al., 2007)

Brisbane is the third largest city in Australia, located on the Brisbane River, with its eastern suburbs by the shores of Moreton Bay. The greater Brisbane region is on the coastal plain east of the Great Dividing Range. The urban area is partially elevated by two large hills reaching up to 300 meters (980 ft), Mount Coot-tha and Mount Gravatt in the south. The city area covers 1,327 square kilometers, and the population was 992,176 in 2006. Brisbane is situated near the coast and has a subtropical climate. Summers are hot and wet, and winters are mild and dry.

In Australia, exposure to extreme temperatures has become a great public health concern over the past decade, largely because of numerous studies linking daily temperatures with daily mortality. Most previous studies were designed to examine temperature-related excess deaths or mortality risks. However, in previous analyses, deaths occurring on the same day were combined and differences in ages were ignored. Although some analyses stratified by several age groups, they treated all deaths as equally important within age groups.

To guide policy decision and resource allocation, it is desirable to know the actual burden of temperature-related mortality. Years of life lost is an indicator of premature mortality that accounts for the age at which deaths occurred by giving greater weight to deaths at younger ages. In comparison with mortality risk, years-of-life-lost is a more informative measurement for quantifying premature mortality. In this study, we conducted a time-series analysis to estimate years of life lost associated with season and temperature in Brisbane, Australia.

We also projected future temperature-related years-of-life-lost under different climate change scenarios (Huang et al., 2012).

Daily mortality data on nonexternal causes from 1996 to 2004 were requested from the Office of Economic and Statistical Research of the Queensland Treasury. All deaths were residents of Brisbane city. These data included date of death, sex, and age. Years-of-life-lost were estimated by matching each death by age and sex to the Australian national life tables. The daily total years-of-life-lost were made by summing the years-of-life-lost for all deaths on the same day. Separate sums were made for men and women. The daily total years-of-life-lost contains information on the number of deaths and the characteristic of the deaths.

The Australia Bureau of Meteorology provided daily weather data. We used daily values of maximum temperature, minimum temperature, and relative humidity from the Archerfield station located near the city center. Maximum and minimum temperatures were the highest and lowest hourly measurements each day in degrees Celsius, with the mean temperature as their average. When data were missing for the Archerfield station (<2%), data from Brisbane Airport were used.

A regression model was used to estimate the association between daily mean temperature and years-of-life-lost, with adjustments for trend, season, day of the week, and daily humidity. To examine the nonlinear and delayed effects of temperature, we used a distributed lag nonlinear model. We plotted the mean and 95% confidence intervals for the years-of-life-lost against month and temperature.

The relative risk of mortality associated with changes in seasonality and temperature were estimated. We used the same independent variables as the years-of-life-lost model but with a dependent variable of the daily number of deaths, which we assumed followed a Poisson distribution. These results were intended to show the difference between a standard analysis of mortality and the analysis of years-of-life-lost.

According to the Intergovernmental Panel on Climate Change (IPCC), evidence for future temperature changes in variability is sparse, and the patterns of changes in extremes are shown in accordance with a general warming. Therefore, we assumed that climate change will cause increasing average temperatures but no change in variability. We simulated future daily temperatures by adding 1–4°C to the observed daily temperature data from 1996 to 2003. We used these 8 years as our baseline to reduce the influence of any unusual temperatures from any one year and because these data are centered on the year 2000. The increases of 1–4°C aimed to simulate daily temperatures in 2046–2053, centered on the year 2050.

The projected temperature-related years-of-life-lost in 2050 were calculated. The future health impacts were based on the nonlinear relationships between temperature and years-of-life-lost, separately for men and women. We assumed no human physiological

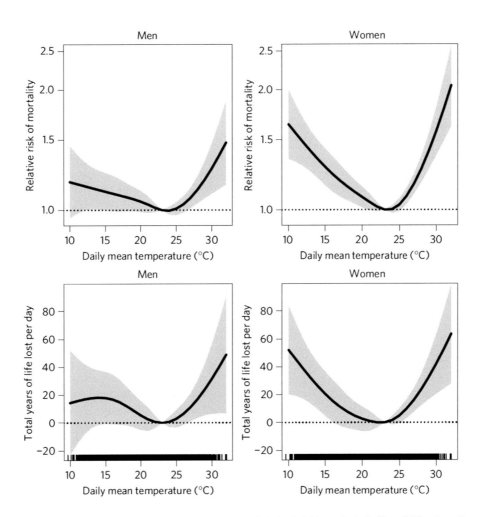

Case Study 10.4 Figure 1 *The effects of temperature on mortality risk and years-of-life-lost in Brisbane, Australia. The solid line shows the mean; gray areas show the 95% confidence intervals. The relative risk of mortality is shown in a log scale. The fine lines show the observed daily mean temperatures. The reference temperature is 23°C.*

acclimatization to higher temperatures. We also assumed that the population size, age structure, and life expectancy will remain constant, allowing the differences in years-of-life-lost to be those due to climate change. Estimates of the burden of heat-related mortality due to climate change can be somewhat offset by reductions in cold-related mortality. We therefore divided our estimates into heat- and cold-related years-of-life-lost.

We show that, in Brisbane, daily mortality has a strong seasonal pattern, with the highest numbers in winter (June–August) and lowest in summer (December–February). When using daily mortality counts, we observed a significant effect of season on mortality even after

adjusting for daily temperature and humidity. However, there was no seasonal pattern in years-of-life-lost for both men and women. The association between temperature and years-of-life-lost is U-shaped, with increased years-of-life-lost for cold and hot temperatures. We also found comparable results for mortality risks, particularly in terms of the general U-shape and turning points (Case Study 10.4 Figure 1). Assuming a temperature increase more than 2°C but without any level of adaptation, large increases in heat-related years-of-life-lost are projected, and the increased heat-related years-of-life-lost will not be offset by decreased cold-related years-of-life-lost. This study highlights that public health adaptation to climate change is necessary.

and thus potentially reduce the risk of infection, particularly under high warming scenarios (Smith et al., 2014).

10.3.4 Water- and Food-Borne Diseases

Humans can be exposed to water- and food-borne pathogens through a variety of routes, including via the ingestion of polluted drinking water, consumption of contaminated food, inhalation of aerosols containing bacteria, and by direct contact with

recreational or floodwaters. A number of pathogens that cause water- and food-borne illnesses in humans are sensitive to climate parameters, including increased temperature, changing precipitation patterns, extreme precipitation events, and associated changes in seasonal patterns in the hydrological cycle. While specific relationships vary by pathogen, increased temperatures appear to increase the incidence of common North American diarrheal diseases such as campylobacteriosis and salmonellosis (Curriero et al., 2001; European Centre for Disease Prevention and

Control, 2011; Semenza et al., 2012). Floods enhance the potential for runoff to carry sediment and chemical pollutants to water supplies (CCSP, 2008), and waterborne illnesses from exposures to pathogens and chemical residues from pesticide runoff (especially from freshly treated properties) in recreational waters have been shown to increase in the hours after extreme rainfall events (Patz et al., 2008). Risk of water-borne illness is greater among the poor, infants, elderly, pregnant women, and immune-compromised individuals (Rose et al., 2001; USGCRP, 2008).

Food and water scarcities impact human health. According to Sena (2014), drought is often a hidden risk and a potential silent public health disaster. It is difficult to identify precisely when this hidden health risk begins and ends because drought's impact on a population emerges gradually and depends on biological and social vulnerability. In the case of Brazil, half of all natural disasters are drought-related, and the effects of drought are largely concentrated in the semiarid Northeastern region of the country. Using available census data for 1991, 2000, and 2010, Sena (2014) found that the 1,133 municipalities most affected by drought (out of 5,565 municipalities in Brazil) were located in the Northeast region. Notably, health and well-being indicators for this region were worse than in the rest of the country. Current interactions between low social and environmental conditions will continue to be aggravated by climate change.

10.3.5 Air Pollution

It has been estimated that one in seven global deaths are related to air pollution, and a large fraction of these deaths result from exposures to fine particle components that also affect the climate (Lim et al., 2012. Short-lived climate-active pollutants such as ozone, methane, and black carbon contribute significantly to health burdens. Significant opportunities exist for improving urban health while also contributing to climate mitigation.

In addition, climate change has the potential to increase morbidity and mortality from respiratory and cardiovascular causes through its effects on air pollution. Emissions, transport, dilution, chemical transformation, and eventual deposition of air pollutants all can be influenced by meteorological variables such as temperature, humidity, wind speed and direction, and mixing height (Kinney, 2008). Ozone and particulate matter (e.g., particles with aerodynamic diameter <2.5 μm [$PM_{2.5}$] and <10 μm [PM_{10}]) are pollutants of particular concern due to their association with adverse health effects and their difficulty to control in the urban environment. Although air pollution emission trends for volatile organic compounds (VOCs) and nitrogen oxides will likely play a dominant role in future ozone levels, climate change will make it harder to achieve air quality goals (Jacob and Winner, 2009). Thus, we will need more aggressive reductions in VOCs and nitrogen oxides (NOx) to achieve ozone concentration targets.

Ground-level ozone is produced on hot, sunny days from a combination of NOx and VOCs, many of the sources of which also emit particles. NOx are emitted primarily by fossil fuel combustion. VOCs are emitted by fuel use as well as by vegetation. VOC levels are also high during surfacing and resurfacing of roads. This includes the traffic marking lines that are refreshed during warmer months when smog and ozone levels are highest. Ozone production is dependent on temperature and the presence of sunlight, with higher temperatures and still, cloudless days leading to increased production. Thus, ground-level ozone concentrations have the potential to increase in some regions in response to climate change (Ebi and McGregor, 2008; Tsai et al., 2008; Cheng et al., 2010). Although it has been noted that climate change could decrease ozone in remote areas with low levels of NOx, this is unlikely to be the case in urban environments, where traffic emissions are high. Exposure to ozone is associated with decreased lung function, increased premature mortality, increased cardiopulmonary mortality, increased hospital admissions, and increased emergency room visits (Dennekamp and Carey, 2010; Kampa and Castanas, 2008; Kinney, 2008; Smith et al., 2009).

$PM_{2.5}$ is a complex mixture of solid and liquid particles, including primary particles directly emitted from sources and secondary particles that form via atmospheric reactions of precursor gases. $PM_{2.5}$ is emitted in large quantities by combustion of fuels by motor vehicles, furnaces, power plants, wildfires, and, in arid regions, wind-blown dust. Because of their small size, $PM_{2.5}$ particles have relatively long atmospheric residence times (on the order of days) and may be carried long distances from their source regions. Research on health effects in urban areas has demonstrated associations between both short- and long-term average ambient $PM_{2.5}$ concentrations and a variety of adverse health outcomes including premature deaths related to heart and lung diseases (Schwartz, 1994; Samet et al., 2000; Pope et al., 2002).

Forest fire is a source of particle emissions and can lead to increased cardiac and respiratory disease incidence, as well as direct mortality (Rittmaster et al., 2006; Ebi et al., 2008; Jacobson et al., 2014). Studies of North American forest fires have demonstrated a link between climate change and the increased frequency and intensity of fires (Westerling, 2006). Although few urban areas are forested, the pollution emitted from wildfires can travel long distances and can affect respiratory health in urban centers. Simulation studies have predicted increases in ozone concentrations downwind of fire sites of between 10 parts per billion (ppb) and 50 ppb over eight-hour measurement periods (Pfister et al., 2008; Mueller and Mallard, 2011). Forest fires also exhibit potential synergistic effects on health when they coincide with heat waves, as occurred in Russia in 2010 (De Sario et al., 2013).

Changes in climate, such as temperature, humidity, and precipitation, also have the potential to alter the concentration and allergenicity of aeroallergens (pollen from trees, grasses, and weeds; and mold), as well as the length of their seasons (Sheffield et al., 2011; Ziska et al., 2011). Exposure to pollen has been associated with a range of allergic outcomes, including exacerbations of allergic rhinitis and asthma (Cakmak et al., 2002; Delfino, 2002; Villeneuve et al., 2006). Temperature and precipitation in the months prior to the pollen season affect production of many types of tree and grass pollens (Reiss and

Kostic, 1976; Minero et al., 1998; Lo and Levetin, 2007; U.S. Environmental Protection Agency [EPA], 2008).

In urban areas, the indoor environment can be a major source of exposure to airborne chemicals and aeroallergens. Climate adaptation measures, such as increased air conditioner use, and mitigation measures, such as building weatherization/energy efficiency measures, can have impacts on indoor air quality and human health. Bacteria and mold can grow in air conditioning systems. Air-tight buildings have fewer air changes per hour to rid the air of chemical contaminants from building materials, paints, sealants, cleaning agents, and more. Air conditioning tends to diminish indoor penetration of outdoor ozone and pollens, while tighter building envelopes can prevent the water intrusion and damage that contribute to mold growth but also reduce air exchange between indoors and outdoors, thus potentially increasing exposures to pollutants generated indoors, such as second-hand cigarette smoke, NO_2 from gas stoves, and indoor allergens. Strict air quality standards applied to emissions from the road transport sector can reduce fine PM concentrations with considerable gains in life expectancy (Harlan and Ruddell, 2011).

10.3.6 Climate Risks and the Urban Poor

Exposure to climate risks and health inequities are closely related. Poverty and poor health status are highly correlated, and growing economic inequality – whether as a result of climate change or other socioeconomic stressors such as rapid urbanization – will likely magnify the gulf between the health status of the wealthy and the poor (Satterthwaite et al., 2008; Smith et al., 2014; Walters and Gaillard, 2014). Poor people in cities commonly have limited access to healthy housing and high-quality health care (Satterthwaite et al., 2008), and essential public health services are themselves vulnerable to being adversely affected by climate variability and climate change (Frumkin and McMichael, 2008). Health outcomes for the urban poor often occur through indirect pathways: for example, rising food prices due to agricultural productivity losses are likely to disproportionately affect the urban poor who have no means of producing their own food (Porter et al., 2014).

Research on recovery from the Hurricane Katrina disaster, for instance, found that psychological distress, post-traumatic stress, and perceived stress were particularly elevated among low-income parents, mainly non-Hispanic black single mothers (Sastry and Gregory, 2013; Fussell and Lowe, 2014). Main drivers for this outcome were "indirect" correlates of the disaster, including household breakup, negative impacts on children, and living in severely affected dwellings and communities (Sastry and Gregory, 2013).

Poor people, particularly in low-income countries, are often highly reliant on informal health services, such as those provided by traditional healers, shopkeepers, and in the home. Reliance on these services may be beneficial or detrimental to health outcomes. Based on a comprehensive literature review, Sudhinaraset et al. (2013) found that informal health services are often the first

choice for care in developing countries. This review outlined many health care problems potentially affecting the informal health sector, including inadequate drug provision, poor adherence to national clinical guidelines, and gaps in knowledge and provider practice; problems, however, were also found in the formal health sectors of these countries. Conversely, informal health networks may bolster social capital and thus aid post-disaster recovery in the case of climate-related disasters in urban areas (Aldrich, 2012). In New Orleans, for example, close-knit family relationships benefited a community that was devastated by Hurricane Katrina (Chamlee-Wright and Storr, 2011).

Last, it is worth noting that individual climate impacts do not always disproportionately affect the poor. The heavy 2011 landslides in Nova Friburgo (State of Rio de Janeiro) led to 177 confirmed leptospirosis cases and 937 confirmed dengue fever cases, and poor people were not found to be more susceptible than other social groups (Pereira et al., 2014b, 2014c) (see Case Study 10.2). The harmful effects of "ubiquitous" exposures such as dengue-carrying mosquitos and air pollution are also likely to be shared across socioeconomic strata. Using climate, mortality, socioeconomic, and pollution data for the cities of Bogotá, Mexico City, and Santiago, Romero-Lankao et al. (2013a) found that air pollution levels and socioeconomic vulnerabilities do not clearly correlate, affecting the poor and the wealthy alike.

10.3.7 Food Insecurity and Displaced Populations

Within urban areas there is often an enormous range in the level of access to nutritious foods, leading to unhealthy extremes of both under- and overnutrition. Climate change along with rapid urbanization may exacerbate food insecurity in cities that experience existing disparities. Poor urban consumers are extremely sensitive to price variations caused by climatic impacts on food production and/or distribution because they do often not produce their own food (Porter et al., 2014). Families may be forced to either limit the quantity or quality of food consumed, with strong potential impacts for human health. Children may be particularly affected by such impacts and suffer from wasting or stunting. In Box 10.4 we present examples of strategies of urban agriculture that are being implemented in Latin American Cities in order to feed poor families.

Urban areas are often the destination for people displaced from their rural livelihoods when drought leads to crop failures, with substantial potential for social disruption.

10.3.8 Role of Critical Urban Infrastructure in Supporting Population Health Following Extreme Climate Events

Many health hazards associated with extreme climate events in urban areas operate through disruptions in critical infrastructure. Lack of electricity can make it difficult or impossible to control the interior climate, refrigerate food, physically move about in high-rise buildings, pump water to upper floors, and operate medical support equipment (Beatty et al., 2006). These infrastructure disruptions can lead to a wide range of adverse health effects depending on the age, health status, and economic resources of residents in the affected households. For example, exposure to ambient heat or cold in the absence of climate control may lead to heat- or cold-related illness or exacerbate underlying chronic conditions. Carbon monoxide poisoning from backup generators or cooking equipment used improperly is another potential risk. Increases in overall mortality rates have been observed after widespread power outages (Anderson and Bell, 2012).

Storms can adversely affect power and gas delivery, transportation, communications and health care infrastructure, and food distribution systems, all of which are critical to maintaining the public's health in densely populated urban centers (Lane et al., 2013). Morbidity and mortality effects of heat, meanwhile, may be especially severe if a power outage (blackout) occurs during an extreme heat event. While a lack of air conditioning in homes increases the risk of heat-related death (O'Neill et al., 2005; Wheeler, 2013), air conditioning also contributes to higher power demand during heat waves, which increases the risk of power disruptions and blackouts. When blackouts occur, exposure to heat increases, with a corresponding increase in health risks. Blackouts can also increase risk of carbon monoxide poisoning from improper use of generators and cooking equipment. During August 2003, the largest blackout in U.S. history occurred in the Northeast. Although this particular blackout did not coincide with a heat wave, it occurred during warm weather and resulted in approximately 90 excess deaths and an increase in respiratory hospitalizations (Lin et al., 2011; Anderson and Bell, 2012).

Box 10.4   Contributions of Urban and Peri-urban Agriculture to Adaptation for Food Security, Nutrition and Health

Cristina Tirado and Sharona Sokolow

Institute of Environment and Sustainability, UCLA, Los Angeles

Latin America (LA) is characterized by high levels of urbanization, health disparities, and the highest inequity index in the world (World Bank, 2016). The gap between the rich and poor is widening within some LA cities in terms of income, access to health services, adequate housing, sanitation, food security, and nutrition. Climatic conditions have contributed to the increasing rural–urban migration trends occurring over the past 50 years in LA (IOM, 2014). High rates of urbanization in LA force the conversion of peripheral land into informal settlements for newly arriving migrants. These settlements are located in areas that lack infrastructure such as water and sanitation and adequate food supplies.

The exponential increase in urban populations, coupled with widening income disparities in LA cities, has eroded the ability of urban infrastructures to offer adequate food access to their populations, which affects nutrition and health. LA cities have the challenge of addressing the double burden of malnutrition; millions of people are undernourished, and there is an increasing trend in the number of obese who suffer from lifestyle-related chronic illnesses. The Americas is the region with highest obesity rates in the world (PAHO, 2014).

There are many potential contributions of urban and peri-urban agriculture and forestry (UPAF) to climate mitigation and adaptation for food security, nutrition, health, and well-being in LA cities. UPAF can play a strong role in enhancing food security and nutrition for the urban poor, greening the city, improving air quality, lowering temperatures, stimulating water storage and drought resistance, reusing urban organic wastes, and reducing the urban energy footprint (Prain and Dubbeling, 2010). Several cities in LA have already promoted or engaged in these type of initiatives.

For example, in Lima, Peru, a city-level policy adopted in 2012 supports urban agriculture. It is included in the municipal development plans, which set up special municipal structures and budget allocation to urban agriculture (FAO, 2014). In Lima, urban agriculture is widely practiced at different scales and for multiple reasons. Commercial horticulture and livestock production take over in the peri-urban and transition areas. In the intra-urban areas producers grow mostly vegetables and small livestock. Many of the low-income households on the hillsides also raise chickens, guinea pigs, and pigs, mostly for local sales and for home consumption. Beyond the contribution to food security, there are psychological benefits of having one's own food production (Prain and Dubbeling, 2010). Women producers in Lima have the perception that by growing their own food they were caring for the environment as well as for their own and their families' health by eating fresh and non-contaminated foods. They also consider the physical activity in itself healthy and contributing to their sense of well-being and relaxation. In this sense, they feel that urban agriculture is enhancing their quality of life (Prain and Dubbeling, 2010).

In Managua, Nicaragua, the success of urban agriculture is focused on water availability. Because the rainy season occurs strictly between May and November, the remainder of the year sees little precipitation. This discrepancy challenged the community in developing their backyard urban gardens. Within Managua, Los Laureles Sur is a community with high rates of both food insecurity and malnutrition, with an average of fruit and vegetable consumption of 60 g per capita per day (15% of recommended consumption levels by FAO and WHO); thus, it was targeted as an important place to integrate urban agriculture. In combination with a training program, rooftop rain collection systems were installed in homes to capture and store water for use during the dry months; this program was extremely successful to improve nutrition through healthier diets (FAO, 2014). This rainwater collection system helped boost urban agriculture in the city to improve average vegetable consumption by 60% (FAO, 2014).

In Mexico City, one specific success in urban agriculture was accomplished in conjunction with government and private organizations to create green rooftops across Mexico's urban areas. These green rooftop spaces support hydroponic gardens where the growth of succulents help reduce the impacts of air pollution in a city suffering daily from poor air quality (FAO, 2014). The program has been quite successful: more than 12,300 square meters of city rooftops have been converted to green urban gardens. Additionally, within the heart of the city, the Huerto Romita, a 56 square meter gardening center, was created to promote and provide an area for community urban vegetable production. The Huerto Romita center also functions to teach garden maintenance techniques and helps to install gardens in schools and throughout the community (FAO, 2012), contributing to food and nutrition security.

In Rio de Janeiro, urban agriculture pilot programs have provided blueprints for working inside fragile communities, for reutilizing abandoned lands, for remediating degraded space, for increasing food and nutrition security, and for generating income (Rekow, 2015). Being able to cultivate or play in a garden is one of the most profound and therapeutic experiences available when living in a dense urban favela. The ability to congregate in an aestheticized, productive public space and pick produce to take home is testimony to the enormous difference that food security programs are making in the daily lives of at least some residents living under pacification in Rio de Janeiro's inner-city slums (Rekow, 2015).

The everyday basic infrastructure deficiencies of informal settlements such as slums, meanwhile, can exacerbate the health impacts of disasters in these highly populated areas. Outbreaks of water-borne infectious diseases are one health concern of particular importance in slum areas, where crowding and poor sanitation are prevalent (OMS, 2012; Freitas et al., 2014).

10.4 Adaptation Strategies to Protect Health in Cities

The question of how to facilitate urban adaptation to climate change health impacts quickly leads into issues of understanding what the likely impacts are and how they might best be managed in specific urban settings. In turn, the answers to these questions feed into the planning processes and management strategies that can help cities increase their resilience to threats from climate change.

In addition, understanding the multidimensional aspects that confer population vulnerability to climate extremes is essential for effective and long-term sustainable adaptation because it provides elements for the formulation of adaptation strategies for protection in relation to the expected impacts (Confalonieri et al., 2014).

One important consideration throughout is that cities are complex systems. They have many distinct components with a wide range of potential interactions, and these components self-organize over time and exhibit behavioral patterns that emerge from these diverse and complex interactions (Batty, 2008). This complexity influences what health impacts are likely to emerge, how they might be detected, and how they might best be managed. Recognizing and exploring this complexity can also help clarify what management strategies may be most effective and may highlight knowledge gaps that can be filled through modeling, targeted learning, and other strategies.

Case Study 10.5 Kedarnath: Flood and Humanitarian Loss in Uttarakhand Districts in India

Thea Dickinson

University of Toronto

Keywords	Flood, monsoon, rainfall, extreme event, health
Population (Metropolitan Region)	10,116,752 (Census Bureau of India, 2011)
Area (Metropolitan Region)	53,484 km² (Census Bureau of India, 2011)
Income per capita	US$1,680 (World Bank, 2017)
Climate zone	Csa – Hot-summer Mediterranean (Peel et al., 2007)

The second largest humanitarian disaster in 2013 occurred in thirteen cities in Uttarakhand, North India, with the town of Kedarnath being the most impacted by the flood. It illustrates how many developing countries are not yet well prepared for the interconnected and cascading impacts of climate change. During June 2013, tens of thousands of people were on a pilgrimage when the monsoon rains began. Kedarnath lies on the southern slope of the Himalayas, and the land traversed by the pilgrims is almost entirely mountainous, with more than 60% covered in forest. The rains that fell in June 2013 were stronger than in the 80 years prior, surprising pilgrims and tourists by arriving earlier in the season than expected. The rains fell for 50 hours and produced more than 500 millimeters of total precipitation in the cities of Uttarakhand (Munich Re, 2014; Dube et al., 2013; Dube et al., 2014)); this was 374% above normal monsoon precipitation of 65.9 millimeters (Anumeha et al., 2013). This triggered a catastrophic chain of events. Snowmelt, including from the retreating Chorabari Glacier, added to the heavy precipitation, causing glacial lake outbursts and flooding the Mandakini river. Water levels in the

Case Study 10.5, Figure 1 *An early start to the 2013 monsoon rains led to severe flooding in Uttarakhand, India.*

Source: UNCS, GAUL, Natural Earth

cities increased by upwards of 7 meters, leading to catastrophic flooding, erosion, and landslides.

Landslides and flood waters trapped 75,000 pilgrims in Kedarnath. A reported 5,748 lives were lost and hundreds of others were sickened with gastroenteritis from water-borne disease due to contaminated spring water and groundwater contamination in broken pipelines.

In the past few decades, the government has increased development in areas along the popular Himalayan pilgrimage routes.

Natural forest cover has been reduced in order to build roads and dams, thus decreasing the stabilization of mountains. Infrastructure, plans and policies, and early warning detection systems to prevent such calamities are either lacking or underdeveloped. These factors led to increased vulnerability to extreme events. The resulting 7 meter floodwaters caused not only losses to property, schools, health centers, hydroelectric power stations, city infrastructure, cash crops, and fisheries, but also claimed the lives of 6,054 locals and tourists. A total of 504,473 people were injured and/or displaced. Economically, the overall losses were estimated at $1.5 billion with only $600 million covered in insured losses.

In the wake of the June floods, the government announced a new water policy: the Uttarakhand Water Management and Regulatory Act, which came into effect on April 4, 2014. Because flood events are expected to increase with a changing climate, this Act is intended to improve the management of future risks from water-related gastrointestinal illnesses.

The Act:
* Provides for the establishment of a Water Management Regulatory Authority to ensure judicious and equitable management of water resources in the state as well as its proper allocation and optimal utilization.
* Creates a five-member authority with the powers of a civil court and the mandate to carry out developments in the state in an eco-friendly and sustainable manner.
* Devises a new water policy to manage rivers.
* Fixes rates for water use for industrial, drinking, power, agricultural, and other purposes, fixes and taxes on land benefited by flood protection and drainage works.
* Deters construction on surface recharge areas and improves tourism plans (because flooding deterred tourism) (Indian Express, 2013)

As in Calgary (see Case Study 10.A in Annex 5), the many layers of vulnerability magnified the impacts of this event: a shifting monsoon season, lack of government intervention, and substantial changes to land use. All these factors converged to create catastrophic – and perhaps somewhat preventable – loss of life. The new Act provides a step in the right direction, but it needs to be implemented along with other no-regret policies and actions that consider climate change while providing co-benefits to the region's residents and protecting tourists when they visit.

10.4.1 Understanding Health Risks and Possible Responses

Cities are developed with an unspoken assumption of a stationary climate – an assumption made invalid by climate change. As a result, most cities are at baseline not well adapted to the threats that climate change is likely to pose. Increasing resilience in urban areas means focusing on the strategies, policies, and measures needed to address the (1) considerable existing adaptation deficit and (2) modifications needed to current and newly developed policies and measures to manage these projected health risks. Doing so requires consideration of how health risks could change over temporal and spatial scales and of the other sectors whose activities could potentially enhance or harm population health. Effective and efficient adaptation requires a multisectoral and multigovernance response (Bowen et al., 2012).

Urban health risks of climate change relate to the natural and built environment, to the population's exposure to climate change hazards, and also to the ability of individuals and institutions to prepare for, cope with, respond to, and recover from the exposure (IPCC, 2012; NRC, 2012). Therefore, careful investments in infrastructure and environmental and health care systems must be applied (Barata et al., 2011). Adaptation measures tend to focus on one of these factors, with the particular concerns depending on the health outcome of concern. For example, an approach focused on preventing *exposure* to storm surges and associated flooding may be effective for reducing the resulting adverse health effects in some coastal urban populations (Solecki et al., 2010). Some other health impacts, such as outbreaks of climate-sensitive vector-borne and zoonotic diseases, may necessitate a multisectoral approach to disease surveillance and control, as was seen with an outbreak of West Nile virus in Texas: the outbreak was associated with weather patterns, geographic characteristics, and historic hotspots for vector activity (Chung et al., 2013). Even when a narrow emphasis may be efficient and expedient, it is important to note that this has the potential to compromise risk management when there are assumptions of full functionality in other domains. For example, an approach focusing on air conditioning as a preventive measure for reducing heat-related morbidity and mortality is vulnerable to failure when the underlying infrastructure – the electrical grid – weakens or fails during a summer heat wave (Anderson and Bell, 2012).

In addition to understanding the likely health effects of climate change in a given urban environment, it is important to understand how health care services are delivered and to consider other major factors that affect population health status. Among them are the cumulative health risks to vulnerable urban populations from multiple climate change-exacerbated exposure; one example is the long-term diminishment of resilience among U.S. Gulf Coast communities after repeated storms, flooding, displacement, and storm-related pollution.

10.4.2 Planning a Response to Health Risks

The processes of designing, implementing, monitoring, and evaluating health adaptation options are similar to those used in

other sectors. Several health adaptation assessment guidelines (e.g., WHO, 2011; ECDC, 2011; Health Canada, 2011) all share similar features. Although these are not focused on urban areas, the process is the same. Figure 10.6 is an example of the process for conducting a vulnerability and adaptation assessment developed by the WHO.

One novel strategy may facilitate planning processes at the urban level. Carlsson-Kanyama and colleagues employed a novel back-casting approach to develop adaptation case studies in Swedish cities and identified several potential barriers to adaptation at the municipal level (Carlsson-Kanyama et al., 2013). They found strong interdependencies related to water supply and quality, energy, the built environment, and services and care for elders. Importantly, their research supported the contention that effective adaptation is contingent on strong intersectorial communication, a prominent theme in climate change adaptation. Their methods may be useful for other regions looking to strengthen municipal climate change adaptation efforts specifically.

Much of the current focus of health adaptation policies and measures (in urban environments and otherwise) is on increasing health protection in a changing climate, such as increasing resilience to the current pattern of extreme weather and climate events (Lesnikowski et al., 2013). The attention of most adaptation efforts has been on strengthening existing policies and

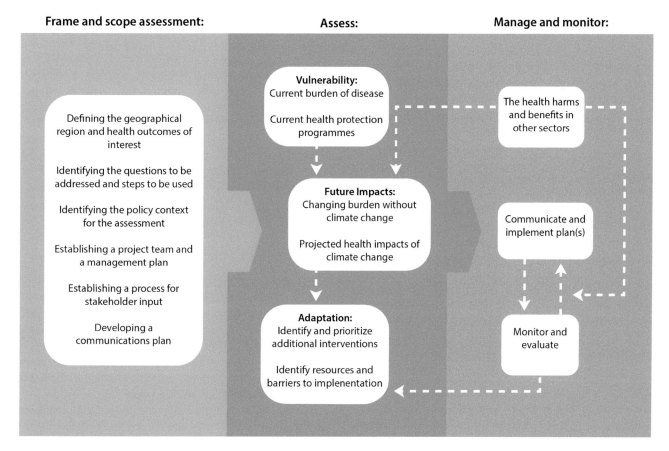

Figure 10.6 *Process for conducting a vulnerability and adaptation assessment in cities.*

Source: WHO, 2011

measures to address current weather patterns (e.g., early warning systems; ensuring vector-borne disease control programs incorporate information on recent changes in temperature and/or precipitation). However, these efforts may not be sufficient as climate change progresses (Kates et al., 2012).

A major impediment to planning a response to the health risks of climate change is a shortage of evidence that can be used to determine the effectiveness of particular adaptation strategies in the face of climate change. That is, we lack a sense of public health's coping range to address the hazards associated with climate change. A systematic review of public health interventions related to vector-borne diseases, water-borne diseases, and heat stress assessed the strength of evidence for addressing climate change risks (Bouzid et al., 2013). The interventions evaluated included environmental interventions to control vectors, household and community water treatment, greening cities, and community advice. Overall, the quality of evidence for the positive impacts of environmental interventions was low, owing both to poor study design and to high heterogeneity among study results. Furthermore, for some extreme weather and climate events (floods and droughts) and for food safety there were no satisfactory systematic reviews of public health interventions. These are clearly priority areas for additional research and evaluation. In terms of the evaluated interventions, in some cases, there may merely be insufficient evidence to make a determination of an intervention's effectiveness. The greater concern, however, is that the evidence available suggests a general lack of effectiveness of existing adaptation strategies. This could become a significant public health concern if hazard intensity increases as expected.

A strategy that has met with some success in other sectors is adaptive management, a structured, iterative process of decision-making in the face of imperfect information (Ebi, 2011; Hess et al., 2012). Adaptive management recognizes the uncertainties associated with projecting future outcomes and considers a range of possible future outcomes when formulating interventions. It also explicitly incorporates models of complex systems to support decisions and requires regular updating of models to support institutional learning and iterative decision-making, both of which can facilitate effective management of complex systems. In an adaptive management approach, interventions are intended to be flexible, taking into account various stakeholder objectives and preferences, and must be subject to adjustment in an iterative, social learning process. Identifying best practices from experiences with using adaptive management approaches would help urban areas as they augment current policies and measures to address the health risks of climate change.

10.4.3 Management of Health Risk and Adaptation Strategies

Management of climate change adaptation strategies includes implementation, monitoring, and evaluation (Moser and Ekstrom,

2010). As noted, much adaptation activity to date has been located at the national level and has been in the planning stage, although implementation is now under way in several areas.

A survey of European countries on health adaptation identified measures being taken to strengthen health systems to manage the health risks of climate change (Wolf et al., 2014). In this survey, representatives of the working group on health in climate change in the WHO European Member States were interviewed to address the question "How far are we in implementing climate change and health action in the WHO European Region?" Twenty-two Member States provided answers to a comprehensive questionnaire that focused on eight thematic areas, with the countries scored on their performance in each: the maximum score was reached by 16 countries in governance; 17 in conducting vulnerability and adaptation assessments; four in developing adaptation strategies and action plans; four in reducing GHG emissions; seven in strengthening health systems; six in raising awareness and building capacity; five in greening health services; and six in sharing best practices. These results indicate considerable possibilities for improvement.

Adaptation should fit within country and city development goals to increase the likelihood of success and of securing the necessary human and financial resources. Panic and Ford (2013) reviewed national-level adaptation planning for infectious disease risks in a changing climate in 14 countries of the Organization for Economic Cooperation and Development (OECD) (Australia, Belgium, Canada, Chile, France, Ireland, Luxembourg, Mexico, New Zealand, Slovenia, Spain, Switzerland, the United Kingdom, and the United States). Although the adaptation plans varied widely, in general, adaptations were mainstreamed into existing public health programs, indicating that the plans fit into health sector planning. However, the plans did not prioritize multidisciplinary approaches that indicate that opportunities to ensure adaptations implemented in sectors such as water and energy also promote population health. Further limitations were negligible consideration of the needs of vulnerable population groups, limited emphasis on local risks, and inadequate attention to implementation logistics, such as available funding and timelines for evaluation.

As a welcome departure from the largely nationally focused planning that has traditionally been exercised to date, three multicountry health adaptation projects covering fourteen low- and middle-income countries are completed or nearing completion. They are:

- Barbados, Bhutan, China, Fiji, Jordan, Kenya, and Uzbekistan[2] in a UNDP/WHO GEF project "Piloting Climate Change Adaptation to Protect Human Health"
- China,[3] Jordan,[4] and the Philippines[5] in the health components of the Millennium Development Goals (MDG) Achievement Fund

---

2 http://www.who.int/globalchange/projects/adaptation/en/
3 http://www.mdgfund.org/program/chinaclimatechangepartnershipframework
4 http://www.mdgfund.org/program/adaptationclimatechangesustainjordan%E2%80%99smdgachievements
5 http://www.mdgfund.org/program/strengtheningphilippines%E2%80%99institutionalcapacityadaptclimatechange

- Albania, Kazakhstan, Kyrgyzstan, Macedonia, Russia, Tajikistan, and Uzbekistan[6] in the WHO EURO project "Protecting Health from Climate Change: A Seven-Country Initiative" funded by the International Climate Initiative of the German Federal Ministry for the Environment, Nature Conservation, and Nuclear Safety

Although few of the specific activities within these projects focused on urban areas, they considered issues that are of high relevance to urban areas, such as access to safe water and the possible geographic spread of vector-borne diseases. The lessons learned and best practices from these projects will provide valuable guidance for health adaptation in urban areas.

Progress is also being made in many regions concerning data exchange and cross-national disease surveillance. For example, the European Centre for Disease Control and Prevention (ECDC) is promoting integrating surveillance for infectious diseases through its European Environment and Epidemiology (E3) Network (Semenza et al., 2013). The E3 portal (http://E3geoportal.ecdc.europa.eu/) supports data exchanges and sustained collaborations between European Union Member States, researchers, and other interested collaborators to, among other aims, provide technical support for the reporting, monitoring, analysis, and mapping data, as well as to enhance the analytical capacity of existing resources in Europe. Such a resource will be helpful to urban areas in Europe and will provide best practices and lessons learned to urban areas in other regions as they augment surveillance programs to include the risks of climate change.

Monitoring and evaluation are important aspects of management and strengths in public health generally, although, as noted, public health does not always invest in developing a strong evidence base as much as it might want, and there has been relatively little use of cluster-randomized trials and other available tools to evaluate potential adaptation activities. Public health has extensive experience with disease surveillance, however, and this experience can be leveraged to facilitate climate change adaptation (Frumkin et al., 2008). Surveillance activities are key to establishing trends in climate-sensitive diseases that might serve as indicators of shifts in population health as a result of climate change (McMichael, 2001). Evidence-based public health is an increasingly important tool for assessing the effectiveness of interventions and for facilitating the dissemination of effective practices (Knowlton et al., 2014).

Vulnerability mapping is an efficient tool for planning the implementation of transitional adaptation measures in cities (Frumkin et al., 2008) as well as monitoring and evaluating its results. Examples are presented in the Case Study Docking Station Annex (see Case Study 2.B and Case Study 3.B in Annex 5).

Romero-Lankao et al. (2012) applied a meta-analysis of 54 papers looking at urban vulnerability to temperature-related hazards, covering 222 urban areas globally. They found that the vast majority of papers focus on the epidemiological linkages between temperature and mortality, with very few attempting to understand the structural mechanisms determining differences in vulnerability to temperature-related hazards within and across communities and cities (Romero-Lankao et al., 2012). In the United States, Reid et al. (2009) mapped a set of proposed community determinants of vulnerability to heat. The factors explaining most of the difference, among a large group of potential variables, were (1) a combination of social and environmental factors (2) social isolation (3) prevalence of air conditioning, and (4) the proportion of elderly and diabetics in the population. The map Reid et al. (2009) produced helps the public health sector identify areas likely to contain residents at greater risk of the adverse health consequences of heat.

10.4.4 Applying Urban Adaptation Strategies to Protect Human Health

Because each city and its populations are unique and the dynamics of climate change are complex, the process of health adaptation planning must be tailored and flexible to make necessary adjustments, especially considering the uncertainty in a long-term perspective. Furthermore, because resources are limited in developing countries, planning for these important and growing urban aspects may require new and innovative approaches. City authorities are willing and able, but they need timely information on long-term trends, potential tipping points, and possibilities for surprise. Integration of mitigation and adaptation measures into broader development goals may be particularly relevant in this respect (Rosenzweig et al., 2010).

Complex thinking stresses that the development of a plan that anticipates all future climate-related hazards and their associated impacts on human well-being represents a challenge. Instead, incremental efforts need to be tried and tested in different cities of the world. Implementing and evaluating diverse projects will increase understanding of how best to improve urban health outcomes in specific contexts (Rydin et al., 2012). Nevertheless, a recent global study of 468 member communities of the ICLEI—Local Governments for Sustainability network found that health concerns rank relatively low on a list of perceived climate impacts (Carmin et al., 2012b).

Smith et al. (2014) suggest that the most effective adaptation strategies for health in the near-term are measures that improve basic public health and health care services, employ technology to promote the adoption of effective policies, and aid in implementing early warning systems.

---

6 http://www.euro.who.int/__data/assets/pdf_file/0019/215524/PROTECTING-HEALTH-FROM-CLIMATE-CHANGE-A-seven-country-initiative.pdf

In Latin America and the Caribbean, countries that have been impacted by extreme events have instituted building safety codes in line with the specific threats characteristic of their regions (Centro Regional De Información Sobre Desastres [CRID], 2009). Local health institutions are also proactively conducting vulnerability studies. For example, Chile and Mexico have both used the Pan American Health Organization's (PAHO, 2009) hospital security index: low-cost, rapid means of evaluating whether a given health institution could cope in a disaster. In addition to this, there is also an initiative called SMART hospitals, which aims at making hospitals resilient to disasters, but which also adds a mitigation component of measuring and reducing their GHG emissions. Pereira et al., 2014a presents the experiences of Latin American countries regarding the preparation and adaptation of the health sector to climate changes.

Relevant adaptation strategies for the protection of citizens' health in urban areas must also include the strengthening of all oversight programs. Good governance should include efforts to integrate a concern for public health into all programs, plans, and projects on the municipal level (Smith et al., 2014). For example, the City of Toronto identified the interdependencies of infrastructure, services, and priority populations as a key challenge to developing and adopting effective and equitable climate adaptation actions (City of Toronto, 2014). Through a cross-corporate working group, and by reaching out to individual city divisions, Toronto Public Health (TPH) encourages health and equity to be considered in a range of municipal climate adaptation and mitigation policy and program decisions. Developing strong partnerships and identifying health co-benefits of climate adaptation and mitigation actions led to outcomes such as a pilot project to reduce vulnerability to heat in multiresidential settings, use of heat vulnerability maps developed by TPH to inform decisions about where to do new tree-planting in the City (Toronto Public Health [TPH], 2011), and support for TPH's ongoing role in supporting sustainable transportation systems (e.g., TPH, 2014a, 2014b).

Other sectors, including food and water distribution, ecosystem services, infrastructure, energy and transportation, and land-use management, also play important roles in determining the urban risk of disease and injury resulting from climate change (Smith et al., 2014).

Implementing basic measures that improve citizens' health quality, such as the provision of clean water and sanitation as well as the reduction of flooding and poverty in cities, is relevant to reduce diseases related to climate change. Projecting climate-sensitive disease will likely play an important part in adaptation strategies, such as efforts to do seasonal forecasts of dengue vector incidence, which could help detect possible incidences months in advance of their actual occurrence.

Two Case Studies in the ARC3.2 Case Study Annex (Annex 5) present the consequences of heavy rainfall and floods on human health in urban centers. They are Case Study 10.A, "Economic Cost and Mental Health Impact of Climate Change in Calgary, Canada" and Case Study 10.C, "City of Toronto Flood: A Tale of Flooding and Preparedness." Case Study 10.5, "Kedarnath: Flood and Humanitarian Loss in Uttarakhand Districts in India," also provides another example. A comparative analysis of those three cases shows that in Uttarakhand districts, 6,054 people died and 504,473 people were evacuated because of the flooding caused by heavy rainfall. Similar events occurred in the cities of Calgary and Toronto in Canada, where preventive adaptation strategies were well implemented. In particular, 50,000 people in 26 communities were evacuated. Mental health illness was the reported consequence in Calgary, and the loss of well-being and potential food-borne disease was the health consequence in Toronto. In addition, property damage and economic loss caused by the heavy rainfall and flood were reported.

10.5 Mitigation Strategies and Co-Benefits

As we saw earlier, human health is at risk from climate change in many ways. Mitigating climate change by reducing GHG emissions will therefore deliver health gains, indirectly and in the long term, by containing these increased risks (Haines et al., 2009). However, there are more immediate health gains that can result in the short term through actions aimed at mitigation – these are the so-called "health co-benefits" of action on climate change. We consider these first, then move to the reverse: the beneficial effects on the environment of steps taken primarily to improve health.

10.5.1 Health Co-Benefits of Mitigation

Wherever a policy aimed at climate change mitigation also directly benefits health, the reverse is also true: if the same policy is adopted primarily for its health benefits, it will also mitigate climate change. An important example is the promotion of low-GHG diets. Useful reductions in GHG emissions can be achieved by reducing consumption of red meat, especially beef. This is for two reasons: first, the operation of the beef industry is itself very energy-intensive in water use, fertilizer, and transport; second, the digestive process in ruminants releases methane, the second most important GHG. Reducing beef consumption in favor of other foods is therefore an available mitigation strategy, but it also brings well-known health benefits through reductions in obesity, cardiovascular disease, and colon cancer (Friel et al., 2009). These health benefits can be a co-benefit of policies and actions aimed primarily at mitigation. This double gain can also be looked at the other way: reducing meat consumption for health reasons, whether through individual choice or through public health policy initiatives, will bring mitigation co-benefits through reducing emissions.

Similarly, implementing active transport strategies can reduce both GHG emissions and heart disease (Woodcock

Case Study 10.6 Kunshan Eastern Health Care Center, Kunshan City, China

Keith Dear

Duke Kunshan University, Kunshan

Keywords	Green technology, energy efficiency, water conservation.
Population (Metropolitan Region)	1,970 million (Demographia, 2016)
Area (Metropolitan Region)	557 km² (Demographia, 2016)
Income per capita	US$8,260 (World Bank 2017)
Climate zone	Cfa – Warm temperate, fully humid, hot summer (Peel et al., 2007)

The city of Kunshan lies 37 kilometers west of Shanghai in the Yangtze River Delta area of China. With a population approaching 2 million, including migrant workers, and a gross domestic product (GDP) of US$32 billion (2010), it is the single richest county in China and aspires to Western standards of education, industry, and health care.

To provide health care services for its growing population, Kunshan is building three new health care facilities. A new hospital in the east of the city will provide Western medical services. A second new hospital to the west will deliver traditional Chinese medicine, whereas the third center will house the Kunshan Center for Disease Control and Prevention, Health Authority, Red Cross blood banks, health promotion centers, and maternal and child health services.

Kunshan is investing in green technology as part of this next stage of its economic development. The Eastern Health Care Center will include a new 1,200-bed hospital, designed by an international consortium from Scandinavia. Its modular construction, materials, and air system are designed to reduce energy consumption both during construction and in operation, thus contributing to climate change mitigation through reduced greenhouse gas emissions.

Efficient heating will be achieved by recycling heat from shower water, supplemented by solar panels. Passive cooling will be achieved by circulating water to nearby lakes.

The hospital has also been designed to be water-efficient, using 70% less water than hospitals currently operating in China. As one strategy to achieve this, the hospital will use a centralized vacuum system for flushing toilets, similar to aircraft systems. The system will use 1.2 liters per flush instead of the 6 liters that is the U.S. Environmental Protection Agency's current efficiency baseline, an 80% reduction. Note that the LEED-2009 guideline is only for a 20% reduction (U.S. Green Building Council, 2010). This will reduce the dependence of the hospital on the city's water supply, increasing its resilience to extreme events, especially drought and flood. It also reduces the need for power-demanding ventilation, reduces hazardous waste storage, and reduces the risk of infection.

As one of the largest public buildings in Kunshan, this new hospital complex will set a standard in environmentally sensitive design and advanced green technology for the city, helping Kunshan follow a sustainable development path.

Case Study 10.6, Figure 1 *Kunshan Eastern Health Care Center.*

et al., 2009; Silva et al., 2012; Patz, 2014). Reducing domestic and industrial coal-burning in and around cities will reduce dangerous black carbon pollution in the air: but suspended carbon particles also contribute to global warming, so reducing particle emissions also mitigates climate change (Smith et al., 2009; Humphreys, 2014; Fleck, 2014). Policies motivated by either of these two gains will bring benefits in the other.

In considering mitigation strategies that can aid health, we should not overlook the health system itself. Hospitals are important consumers of energy, so gains in energy efficiency in hospitals can benefit the environment through energy conservation and also benefit the health of populations through cost savings to the health system (U.S. Energy Information Administration, 2007). As part of the America's Climate Action Plan, the U.S. Department of Health and Human Services has created a "Sustainable and Climate Resilient Health Care Facilities Initiative." This initiative includes a guide to enhancing resilience in the health sector.[7] In China, the Kunshan Hospital case study is another example (see Case Study 10.6).

These considerations strengthen the case for longer term climate change mitigation by pointing to additional current-day benefits to society, health co-benefits that can arise from mitigation policies. These health benefits would apply even in the absence of climate change, providing "no-regret" strategies that can be defended as being of value regardless of what view is taken of future climate change.

Finally, the health benefits may be delivered explicitly through climate change mitigation policies if public money is diverted from climate-damaging fuel subsidies and instead put toward enhancing national health systems (Yates, 2014).

10.5.2 Co-Benefits of Adaptation

Mitigating climate change by reducing GHG emissions will not be sufficient to eliminate the impact of climate change. Faced with the inevitability of climate change, cities and individuals must adapt to protect their health and livelihoods. Here again, the principle of co-benefits applies: policies and individual choices aimed at protecting health can also have co-benefits in the sense of mitigating climate change.

For example, an important step that cities can take to reduce the adverse health impact of climate change is to reduce the intensity of their urban heat islands (UHI). Cities are well known to be significantly warmer than the surrounding countryside due to the heat capacity of their buildings and roads and the relative lack of cooling provided by vegetation and open water. Green city policies, such as tree-planting and the provision of open water within cities, will therefore go some way toward reducing inner-city temperatures, and this is an important option for adaptation to climate change

(see Case Study 10.B in Annex 5). White roofs and light-colored roads also help (Rosenzweig et al., 2006). However, such initiatives will also yield benefits for climate change mitigation in several ways: (a) reduced temperatures in summer will mean reduced use of air conditioning, with reduced energy consumption and hence reduced GHG emissions; (b) the improved inner-city amenity will encourage active transport, reducing fuel use.

A second type of adaptive action is to increase the health of the population generally, thereby increasing its overall resilience to environmental threats like heat waves. After smoking, the greatest threat to urban health in both developed and developing countries is obesity. Actions to reduce levels of overweight include reduced calorie intake (derived from meat and other sources) and increased exercise, with active transport (walking/bicycling) forming an important component of everyday movement. Both these approaches will also reduce GHG emissions and thus have mitigation co-benefits in addition to their primary, direct health benefits.

10.5.3 Unintended Consequences

It is not all good news. Mitigation actions may have unintended, detrimental side effects on health, such as energy price increases that, while intended to reduce emissions, also make it harder to heat and cool homes. Actions aimed at health adaptation can be expensive in GHG emissions, such as increased use of air-conditioning: clearly protective of human health in cities at risk of heat waves, air conditioning is a massive consumer of electricity, leading to increased GHG emissions and accelerated climate change.

The case for a local climate change adaptation plan is less compelling if it merely moves the problem elsewhere. It might appear that electric transport, both individual (e-bikes) and mass (trains), delivers benefits both in terms of reduced GHG emissions and improved urban air quality. But if the electricity is generated by burning fossil fuels at remote power stations, it is important to weigh the benefit of this policy against the cost of both the GHGs and of the air pollution generated by the power stations.

10.6 Barriers and Bridges to Mitigation and Adaptation

Although mitigation and adaptation activities are beneficial to citizens' health, there are also barriers to their implementation. These barriers include:

- *Limited institutional capacity*: Unclear/unhelpful administrative boundaries, lack of sufficient knowledge, concentrated expertise (often in marginalized environmental departments), and lack of financial and human resources (Bulkeley, 2010)
- *Lack of local government resources* and very limited capacities to invest (e.g., when the majority of local revenues go toward recurrent expenditures and/or debt repayment)

---

7 http://toolkit.climate.gov/image/662

- *Difficulty and/or reluctance to persuade the electorate of the importance of tackling climate change,* when faced with more immediate, local concerns that are more likely to win votes
- *Local governments that are unrepresentative of and unaccountable* to precisely those sections of their populations that are most vulnerable to climate change: those living in informal settlements and working within the informal economy, whom they regard as "the problem" (Satterthwaite in Bulkeley, 2010)
- *Uncertainty related to the health risks of climate change* (e.g., as concerns changing disease distributions and newly emergent diseases, see Section 10.3)
- *Difficulties related to stakeholder engagement* (see Section 10.2), even for planning and implementing early warning systems in cities

Despite these barriers, on a positive note, there are many "bridge options" that encourage the implementation of adaptation and mitigation strategies. These refer to "no-regret" interventions that generate net social benefits under all future scenarios of climate change (Heltberg et al., 2009). In fact, there are many examples of interventions adopted by different sectors that reduce the risk of climate change to health. Some of them are presented here:

- *Community-based programs designed for disaster risk management*: Interventions in low-income urban settings, small-scale loans, hygiene education, local control and maintenance of water supplies, and neighborhood waste management strategies (Dodman et al., 2010; Obermaier, 2013)
- *Improved flooding-prevention infrastructure* (see Case Studies 3.1, Boulder; 4.3, Jena; 11.B, Dakar; and Chapter 9, Coastal Zones)
- *Revegetation of cities to reduce ambient air temperature and improve air quality*, as presented in Case Study 10.B in Annex 5

The assessment of the health co-benefits of adaptation and/or mitigation strategies can be regarded as a bridge to implementation, but it is important to note that health dis-benefits may also result from climate change adaptation and mitigation strategies. The term "health dis-benefits" refers to interventions adopted by different sectors that increase the adverse impacts of climate change on health. As an example, urban wetlands designed primarily for flood control may promote mosquito breeding (Medlock and Vaux et al., 2011). A full consideration of the health consequences of any adaptation or mitigation strategy is the only way to determine whether a particular adaptation will be beneficial or detrimental to health.

10.7 Knowledge Gaps and Recommended Areas for Further Research

Research and practice that crosses disciplinary boundaries is vital for supporting evidence-based policies and programs to effectively and efficiently address the health risks of climate variability and change in the context of multistressor environments. The recent survey by the WHO regional office for Europe, which has been working to promote health adaptation since the late 1990s, indicated areas where further adaptation research is needed, particularly in developing adaptation and mitigation strategies and action plans; designing and implementing integrated climate, environment, and health surveillance; and sharing best practices (Wolf et al., 2014).

A particular challenge with health adaptation is that local and regional contextual factors are often key determinants of the effectiveness of policies and measures. Therefore, it is important to determine which factors determining the success of particular interventions were broadly applicable and thus can be transferred to other regions and which are unique to a location (such as strong commitment of an individual policy-maker to health adaptation). Stakeholder engagement is crucial to building successful adaptation projects.

Adaptation should focus not just on shorter term outputs to address climate variability, but also on establishing processes to address longer term climate change. To do so, research is needed on projecting the health risks of climate change and on effective approaches to iteratively managing risks that will evolve as the climate continues to change and as development proceeds. The literature remains limited on projections of the magnitude and pattern of possible future health risks of climate change and the efficacy of adaptation in reducing those risks, which means adaptation projects have a limited basis for putting shorter term adaptation into longer term perspectives. Projecting the extent to which alterations in weather patterns may affect future health burdens requires moving beyond simple models based on exposure–response relationships and projected temperature/precipitation change to models that incorporate a range of plausible environmental and socioeconomic futures (Ebi and Rocklov, 2014). Finer temporal and spatial scale models are sorely needed to inform decision-making.

Effectively managing the health risks of climate variability and change requires interventions to explicitly consider risks changing over spatial and temporal scales, with high degrees of uncertainty as to the magnitude, rate, and pattern of changes in a particular location at a particular time. This includes risks from a changing climate as well as from changes in other factors that determine the distribution and incidence of climate-sensitive health outcomes. Here, the issue of cities as incompletely understood complex systems is particularly relevant. In the future, cities are likely to experience climatic risks differently than other settings, and risks may be compounded by migration and other specifically urban factors.

Better understanding the cumulative health effects from multiple climate change–related impacts is also relevant when considering urban adaptation strategies considering the long term.

A critical short-term knowledge gap concerns efficient approaches to capacity-building in the health risks of climate change for the full range of actors: from public health and health care professionals, to the general public, to decision- and policy-makers within the health sector and across ministries. This includes facilitating and developing methods, tools, and guidance documents to support countries as they implement adaptation programs and activities. In its broadest sense, an effective capacity-building initiative would not only develop relevant skills and expertise, but also build the evidence base and enhance collaboration across sectors outside public health.

Another knowledge gap is the need for indicators. Defining indicators for the health risks of climate change is an emerging field (English et al., 2009). An agreed minimum set of indicators, similar to those defined for measuring meteorological and climatological variables, along with means of verification, is needed to help establish baselines and for measuring the degree of success of health adaptation activities. This set could then inform indicators chosen within adaptation projects. Having a common set across projects would help future evaluations.

Research is needed on how to most effectively address adaptation and mitigation jointly to reduce the magnitude of climate change to which health systems will need to adapt later in the century. This is not to divert attention from adaptation, but to note that some activities could beneficially incorporate mitigation at low cost. For example, an organization concerned with adapting health care facilities in urban areas to changes in the magnitude and frequency of extreme weather and climate events could encourage applying for supplementary funding to green the health care sector as well (see Case Study 10.6).

Considering that the urban health system should be prepared for the potentially enhanced disease risks related to climate change, it is relevant to improve research considering current and emerging health threats. Key questions include: How can the populations living in tropical regions be affected by urban heat waves? How can we implement adaptation options for the poorest people in cities?

Some cities currently are experiencing socioenvironmental pressure, like the increase of slums, inequalities, and scarcity of water among others. Increases in research that incorporates climate change pressure on the health system and possible "no-regret" adaptation strategy are necessary. In Case Study 10.1, Box 10.3, and Case Study 10.B, we have examples of adaptation strategies applied in cities. Should they be replicated in other cities around the world?

The increase in urbanization and incomes imply new pressures, including increased demands for food, increased waste, and, if unchecked, increased contamination. Understanding the implications of these issues for urban health systems that are also facing climate change deserves more research.

10.8 Conclusions and Recommendations for Policy

In cities, *health* is associated with social, economic, and environmental determinants (Barata et al., 2011), including climate change. Urban planners usually make decisions in one sphere (e.g., transportation) without complete information on the impacts of these decisions on other spheres (e.g., human health, energy, and natural systems). Therefore, a public health system should advocate for healthy policies, plans, and projects for all urban sectors, including the assessment of the risk of climate change to the health of urban residents.

The potential impact of climate change in the population of different cities varies according to their vulnerability. Therefore, it is relevant to:
- Consistently apply health impact assessment, which can provide a quantitative evaluation of the potential health impacts of scenarios, policies, plans, and projects tailored to the city
- Incorporate climate change projections into city standards, policies, and codes
- Alter existing urban conditions that predispose a population to increased vulnerability during extreme climatic events;
- Apply health impact assessments to proposed adaptation and mitigation strategies intended for reducing climate change health risks in order to adopt:
  - GHG mitigation strategies that result in co-benefits for health of the population and the health system
  - Climate change adaptation strategies that reduce health impacts while reducing GHG emissions
  - "No-regret" adaptation and mitigation strategies that will be beneficial to the population under all future climate change scenarios
- Map intra-urban differences in vulnerability to the impacts of climate change, especially in large cities in the developing world that are highly heterogeneous in terms of equity, infrastructure, housing, population density, and health care
- Implement an early warning systems for extreme events such as storms/floods and heat waves; these can have a positive impact on prevention of morbidity and mortality in urban settings

Preparing for the health consequences of climate-related events necessitates a multisectoral approach. Consequently, it is relevant that:
- Connections between climate change and health in cities are made clear to public health practitioners, city planners, policy makers, and the general public
- Urban practitioners take immediate responsibility for integrating climate change projections in all areas of urban planning
- Adaptation strategies focus on activities that eliminate health disparities, improve neighborhood conditions, and protect those who will be most impacted by climate change
- Successful methodologies for climate change adaptation strategies designed and implemented in cities count on the

collaboration of researchers and consultants in urban management so that they can be tailored to other cities
- International networks be established, maintained, and supported to provide forums for data exchange, disease surveillance, intervention evaluation, and ongoing research into the health impacts of climate change in urban areas

Annex 10.1 Stakeholder Engagement

Stakeholder engagement is critical for achieving success in managing health risk due to climate change in cities. In Figure 10.3, we present an integrated view of institutions and sectors that are relevant in managing those risks in order to protect urban populations.

We aimed to develop a chapter that is both scientifically robust and helpful to authorities and professionals involved in managing health climate risk in cities. To better respond to this challenge, we decided to invite as reviewers for our chapter experts in climate change and health from the WHO, the National Resource Defense Council, and the Health Research Institution, as well as outstanding city health managers and other city health advisors. The participation of stakeholders as chapter reviewers was relevant for the content and shape of this chapter. Protecting the health of the world's urban population requires the involvement of stakeholders specialized in diverse disciplines such as health, planning, engineering, meteorology, ecology, epidemiology, and others. Finally, we hope stakeholders, city governments, the business and academic communities, consultants, or citizens interested in managing health risks due to climate change in cities will find this chapter useful.

Chapter 10 Urban Health

References

Admassie, A., Adenew, B., and Tadege, A. (2008). Perceptions of stakeholders on climate change and adaptation strategies in Ethiopia. *IFPRI Research Brief* **15**(6).

Aldrich, D. P. (2012). *Building Resilience: Social Capital in Post-Disaster Recovery*. University of Chicago Press.

Anderson, G. B., and Bell, M. L. (2012). Lights out: Impact of the August 2003 power outage on mortality in New York, NY. *Epidemiology* **23**(2), 189.

Barata, M., Ligeti, E., de Simone, G., Dickinson, T., Jack, D., Penney, J., Rahman, M., and Zimmerman, R. (2011). Climate change and human health in cities. In Rosenzweig, C., Solecki, S. A. Hammer and Mehrotra, S. (eds.), *Climate Change and Cities: First Assessment Report of the Urban Climate Change Research Network* (183–217). Cambridge University Press.

Basu, R., Malig, B., and Ostro, B. (2010). High ambient temperature and the risk of preterm delivery. *American Journal of Epidemiology* **172**(10), 1108–1117. Accessed December 9, 2015: http://doi.org/10.1093/aje/kwq170

Batty, M. (2008). *Cities as complex systems: Scaling, interactions, networks, dynamics and urban morphologies*. In *The Encyclopedia of Complexity & System Science*. Springer.

Beatty, M. E., Phelps, S., Rohner, C., and Weisfuse, I. (2006). Blackout of 2003: Public health effects and emergency response. *Public Health Reports* **121**(1), 36–44.

Berry, H. L., Bowen, K., and Kjellstrom, T. (2010). Climate change and mental health: A causal pathways framework. *International Journal of Public Health* **55**(2), 123–132.

Bhatt, S., Gething, P. W., Brady, O. J., Messina, J. P., Farlow, A. W., Moyes, C. L., Drake, J. M., Brownstein, J. S., Hoen, A. G., Sankoh, O., Myers, M. F., George, D. B., Jaenisch, T., Wint, G. R. W., Simmons, C. P., Scott, T. W., Farrar, J. J., and Hay, S. I. (2013). The global distribution and burden of dengue. *Nature* **496**(7446), 504–507

Bierbaum, R., Smith, J. B., Lee, A., Blair, M., Carter, L., Chapin III, F. F., Fleming, P., Rufio, S., Stults, M., McNeely, S., Wasley, E., and Verduzco, L. (2013). A comprehensive review of climate adaptation in the United States: More than before, but less than needed. *Mitigation and Adaptation Strategies for Global Change* **18**(3), 361–406.

Bouzid, M., Hooper, L., and Hunter, P. R. (2013). The effectiveness of public health interventions to reduce the health impact of climate change: A systematic review of systematic reviews. *PLoS One* **8**(4). Accessed November 8, 2014: http://www.readcube.com/articles/10.1371/journal.pone.0062041

Bowen, K. J., Friel, S., Ebi, K., Butler, C. D., Miller, F., and McMichael, A. J. (2012). Governing for a healthy population: Towards an understanding of how decision-making will determine our global health in a changing climate. *International Journal of Environmental Research and Public Health* **9**(1), 55–72.

Bulkeley, H. (2010). Cities and the governing of climate change. *Annual Review of Environmental Resources* **35**, 229–253.

Cakmak, S., Dales, R. E., Burnett, R. T., Judek, S., Coates, F., and Brook, J. R. (2002). Effect of airborne allergens on emergency visits by children for conjunctivitis and rhinitis. *Lancet* **359**(9310), 947–948. Accessed May 23, 2014: http://doi.org/10.1016/S0140-6736(02)08045-5

CAP. (2007). *Cities Preparing for Climate Change A Study of Six Urban Regions*. Clean Air Partnership.

Carlsson-Kanyama, A., Carlsen, H., and Dreborg, K.H (2013). Barriers in municipal climate change adaptation: Results from case studies using backcasting. *Futures* **49**, 9–21.

Carmin, J., Anguelovski, I., and Roberts, D. (2012). Urban climate adaptation in the Global South: Planning in an emerging policy domain. *Journal of Planning, Education, and Research* **32**, 18–32.

Carmin, J., Nadkarni, N., and Rhie, C. (2012). *Progress and Challenges in Urban Climate Adaptation Planning: Results of a Global Survey*. MIT Press.

Castán Broto, V., and Bulkeley, H. (2013). A survey of urban climate change experiments in 100 cities. *Global Environmental Change* **23**, 92–102.

CCSP. (2008). *Weather and climate extremes in a changing climate: Regions of focus: North America, Hawaii, Caribbean and U.S. Pacific Islands*. In Karl, T. R., Meehl, G. A., Miller, C. D., Hassol, S. J., Waple, A. M., and Murray, W. L. (eds.), *Synthesis and Assessment Product 3.3, Report by the United States Climate Change Science Program (CCSP) and the Subcommittee on Global Change Research*. Department of Commerce, NOAA's National Climatic Data Center.

Centro Regional De Información Sobre Desastres (CRID). (2009). *Casos de estudio*. Accessed June 3, 2014: http://www.crid.or.cr/CD/CD_hospitales_Seguros/casos-estudio.html

Chamlee-Wright, E., and Storr, V. H. (2011). Social capital as collective narratives and post-disaster community recovery. *Sociology Review* **59**, 266–282. doi:10.1111/j.1467-954X.2011.02008.x

Cheng, D. S., Campbell, M., Li, Q., Li, G., Auld, H., Day, H., Pengelly, D., Gingrich, S., Klaassen, J., MacIver, D., Comer, N., Mao, Y., Thompson, W., and Lin, H. (2010). Differential and combined impacts of extreme temperatures and air pollution on human mortality in south-central Canada. Part II: Future estimates. *Air Quality and Atmospheric Health* 2008(**1**), 223–235.

Chung, W. M., Buseman, C. M., Joyner, S. N., Hughes, S. M., Fomby, T. B., Luby, J. P., and Haley, R. W. (2013). The 2012 West Nile encephalitis epidemic in Dallas, Texas. *JAMA* **310**(3), 297–307.

City of Toronto. (2014). Resilient city: Preparing for a changing climate. Accessed January 14, 2015: http://www.toronto.ca/legdocs/mmis/2014/pe/bgrd/backgroundfile-70623.pdf

Colón-González, F. J., Fezzi, C., Lake, I. R., and Hunter, P. R. (2013). The effects of weather and climate change on dengue. *PLoS Neglected Tropical Diseases* 7(11), e2503. doi:10.1371/journal.pntd.0002503

Confalonieri, U., Barata, M. M. L., and Marinho, D. (2014). Vulnerabilidade Climática no Brasil. In Chang, M. et al. (eds.), *Metodologias de Estudos de Vulnerabilidade à Mudança do Clima*. Coletâneas Mudanças Globais, Ed. Interciências. Volume 5. IVIG/COPPE-UFRJ.

Curriero, F. C., Patz, J. A., Rose, J. B., and Lele, S. (2001). The association between extreme precipitation and waterborne disease outbreaks in the United States, 1948–1994. *American Journal of Public Health* 91(8), 1194–1199.

De Sario, M., Katsouyanni, K., and Michelozzi, P. (2013). Climate change, extreme weather events, air pollution and respiratory health in Europe. *European Respiratory Journal* 42, 826–843.

Delfino, R. J. (2002). Epidemiologic evidence for asthma and exposure to air toxics: Linkages between occupational, indoor, and community air pollution research. *Environmental Health Perspectives* 110(Suppl 4), 573–589.

Dennekamp, M., and Carey, M. (2010). Air quality and chronic disease: Why action on climate change is also good for health. *New South Wales Public Health Bulletin* 21(5–6), 115–121. Accessed December 1, 2014: http://doi.org/10.1071/NB10026

Dodman, D., Mitlin, D., and. Rayos, C. J. (2010). Victims to victors, disasters to opportunities: Community driven response to climate change in the Philippines. *International Development Planning Review* 32(1), 1–26.

Ebi, K. (2011). Climate change and health risks: Assessing and responding to them through "adaptive management." *Health Affairs (Millwood)* 30(5), 924–930.

Ebi, K. L., Grambsch, A. E., Sussman, F. G., and Wilbanks, T. J. (2008). Effects of global change on human health. In *Analyses of the Effects of Global Change on Human Health and Welfare and Human Systems*. UNT Digital Library. Accessed November 30, 2014: http://digital.library.unt.edu/ark:/67531/metadc12033/citation/

Ebi, K. L., and McGregor, G. (2008). Climate change, tropospheric ozone and particulate matter, and health impacts. *Environmental Health Perspectives* 116(11), 1449–1455. Accessed June 8, 2015: http://doi.org/10.1289/ehp.11463

Ebi, K. L., and Rocklov, J. (2014). Climate change and health modelling: Horses for courses. *Global Health Action* 7, 24154. Accessed September 28, 2015: http://dx.doi.org/10.3402/gha.v7.24154

European Centre for Disease Prevention and Control (ECDC). (2011). *Climate Change and Communicable Diseases in the EU Member States: A Handbook for National Vulnerability, Impact and Adaptation Assessments*. ECDC.

English, P. B., Sinclair, A. H., Ross, Z., Anderson, H., Boothe, V., Davis, C., Ebi, K., Kagey, B., Malecki, K., Schultz, R., and Simms, E. (2009). Environmental health indicators of climate change for the United States: Findings from the State Environmental Health Indicator Collaborative. *Environmental Health Perspective* 117, 1673–1681.

FAO. (2011). Global food losses and food waste. Accessed December 1, 2014: http://www.fao.org/fileadmin/user_upload/sustainability/pdf/Global_Food_Losses_and_Food_Waste.pdf

FAO. (2014). *Growing Greener Cities in Latin America and the Caribbean*. A FAO Report on Urban and Peri-urban Agriculture in the Region.

Fleck, F. (2014). Bringing air pollution into the climate change equation. *Bulletin of the World Health Organization* 92, 553–554.

Freitas, C. M., Silva, M. E., and Osorio-de-Castro, C. G. S. (2014). A redução dos riscos de desastres naturais como desafio para a saúde coletiva. *Ciência e Saúde Coletiva (Impresso)* V. 19, 3628–3628

Friel, S., Dangour, A. D., Garnett, T., Lock, K., Chalabi, Z., Roberts, I., and McMichael, A. J. (2009). Health and Climate Change 4 Public health benefits of strategies to reduce greenhouse-gas emissions: Food and agriculture. *Nutrition*. doi:10.1016/S0140-6736(09)61753-0

Frumkin, H., Hess, J., Luber, G., Malilay, J., and McGeehin, M. (2008). Climate change: The public health response. *American Journal of Public Health* 98(3), 435–445.

Frumkin, H., McMichael, A. J., and Hess, J. J. (2008). Climate change and the health of the public. *American Journal of Preventive Medicine* 35(5), 401–402. Accessed August 26, 2014: http://doi.org/10.1016/j.amepre.2008.08.031

Fussell, E., and Lowe, S. R. (2014). The impact of housing displacement on the mental health of low-income parents after Hurricane Katrina. *Social Science and Medicine* 113, 137–144. doi:10.1016/j.socscimed.2014.05.025

Ghebreyesus, T. A., Tadesse, Z., Jima, D., Bekele, E., Mihretie, A., Yihdego, Y. Y., Dinku, T., Connor, S. J., and Rogers, D. P. (2009). Using climate information in the health sector. Field Actions Science Reports, Vol. 2. Accessed July 19, 2014: http://factsreports.revues.org/178

Greenwood, M. (2007). Stakeholder engagement: Beyond the myth of corporate responsibility. *Journal of Business Ethics* 74, 315–327.

Haines, A., McMichael, A. J., Smith, K. R., Roberts, I., Woodcock, J., Markandya, A., and Wilkinson, P. (2009). Public health benefits of strategies to reduce greenhouse-gas emissions: Overview and implications for policy makers. *The Lancet* 374(9707), 2104–2114. doi:10.1016/S0140-6736(09)61759-1.

Harlan, S. L., and Ruddell, D. M. (2011). Climate change and health in cities: Impacts of heat and air pollution and potential co-benefits from mitigation and adaptation. *Current Opinions in Environmental Sustainability* 3, 126–134.

Health Canada (2011). *Adapting to Extreme Heat Events: Guidelines of Assessing Health Vulnerability*. Health Canada. Accessed November 18, 2015: www.healthcanada.gc.ca

Heltberg, R., Siegel, P. B., and Jorgensen, S. L. (2009). Addressing human vulnerability to climate change: Toward a 'no-regrets' approach. *Global Environmental Change* 19, 89–99.

Hess, J. J., McDowell, J. Z., and Luber, G. (2012). Integrating climate change adaptation into public health practice: Using adaptive management to increase adaptive capacity and build resilience. *Environmental Health Perspectives* 120, 171–179.

Hohnen, P. (2007). *Corporate Social Responsibility: An Implementation Guide for Business*. International Institute for Sustainable Development, Canada. Accessed January 9, 2016: http://www.iisd.org/pdf/2007/csr_guide.pdf

Humphreys, G. (2014). Reframing climate change as a health issue. *Bulletin of the World Health Organization* 92, 551–552.

Intergovernmental Panel on Climate Change (IPCC). (2007). *Climate Change 2007: 2.3.2 Stakeholder involvement in Impacts, Adaptation and Vulnerability. Contribution of Working Group II to the Fourth Assessment Report of the Intergovernmental Panel on Climate Change*, Parry, M. L., Canziani, O. F., Palutikof, J. P., van, P. J.der Linden and Hanson, C. E. (eds.). Cambridge University Press.

Intergovernmental Panel on Climate Change (IPCC). (2012). *Managing the Risks of Extreme Events and Disasters to Advance Climate Change Adaptation. A Special Report of Working Groups I and II of the Intergovernmental Panel on Climate Change*, Field, C. B., Barros, V., Stocker, T. F., Qin, D., Dokken, D. J., Ebi, K. L., Mastrandrea, M. D., Mach, K. J., Plattner, G. -K. Allen, S. K., Tignor, M., and Midgley, P. M. (eds.). Cambridge University Press.

Intergovernmental Panel on Climate Change (IPCC). (2014). *Climate Change 2014: Impacts, Adaptation, and Vulnerability, Summaries, Frequently Asked Questions, and Cross-Chapter Boxes. A Contribution of Working Group II to the Fifth Assessment Report of the Intergovernmental Panel on Climate Change*, Field, C.B., Barros, V. R., Dokken, D. J., Mach, J. K., Mastrandrea, M. D., Bilir, T. E., Chatterjee, M., Ebi, K. L., Estrada, Y. O., Genova, R. C., Girma, B., Kissel, E. S., Levy, A. N., MacCracken, S., Mastrandrea, P. R., and White, L. L. (eds.). World Meteorological Organization (in Arabic, Chinese, English, French, Russian, and Spanish). Accessed January 21, 2016: http://ipcc-wg2.gov/AR5/images/uploads/WGIIAR5-IntegrationBrochure_FINAL.pdf

Jaakkola, J. J. K., Hwang, B. F., and Jaakkola, N. (2005). Home dampness and molds, parental atopy, and asthma in childhood: A six-year population-based cohort study. *Environmental Health Perspectives* 113(3), 357–361.

Jacob, D. J., and Winner, D. A. (2009). Effect of climate change on air quality. *Atmospheric Environment* 43(1), 51–63. Accessed March 2, 2016: http://doi.org/10.1016/j.atmosenv.2008.09.051

Jacobson, L. da S. V., Hacon, S. de S., de, H. A., Castro, Ignotti, E., Artaxo, P., Saldiva, P. H. N., and de Leon, A. C. M. P. (2014). Acute effects of particulate matter and black carbon from seasonal fires on peak expiratory flow of schoolchildren in the Brazilian Amazon. *PloS One* **9** (8), e104177. Accessed February 9, 2015: http://doi.org/10.1371/journal.pone.0104177

Kampa, M., and Castanas, E. (2008). Human health effects of air pollution. *Environmental Pollution (Barking, Essex: 1987)* **151**(2), 362–367. Accessed November 30, 2014: http://doi.org/10.1016/j.envpol.2007.06.012

Kates, R. W., Travis, W. R., et al. (2012). Transformational adaptation when incremental adaptations to climate change are insufficient. *Proceedings of the National Academy of Sciences* **109**(19), 7156–7161.

Kelly-Hope, L. A., Hemingway, J., and McKenzie, F. E. (2009). Environmental factors associated with the malaria vectors *Anopheles gambiae* and *Anopheles funestus* in Kenya. *Malaria Journal* **8**(268), 1–8.

KEMA, Inc. (2012). *A guide to developing a climate action plan using regionally integrated climate action plan suite.* City/County Association of Governments of San Mateo County, California, USA. Accessed October 9, 2014: http://www.smcenergywatch.com/sites/default/files/RICAPS_Users%20Guide_v1.pdf

Kinney, P. L. (2008). Climate Change, Air Quality, and Human Health. *American Journal of Preventive Medicine* **35**(5), 459–467. Accessed October 18, 2014: http://doi.org/10.1016/j.amepre.2008.08.025

Kinney, P. L., Schwartz, J., Pascal, M., Petkova, E., Le Tertre, A., Medina, S., and Vautard, R. Winter seasons mortality: Will climate warming bring benefits? *Environmental Research Letters* 10(2015), 064016. doi:10.1088/1748-9326/10/6/064016.

Knowlton, K., Kulkarni, S. P., Azhar, G. S., Mavalankar, D., Jaiswal, A., Connolly, M.,Nori, A. -Sarma, Rajiva, A., Dutta, P., Deol, B., Sanchez, L., Khosla, R., Webster, P. J., Toma, V. E., Sheffield, P., Hess, J. J., the Ahmedabad Heat and Climate Study Group (2014). Development and Implementation of South Asia's First Heat-Health Action Plan in Ahmedabad (Gujarat, India). *International Journal of Environmental Research and Public Health* **11**, 3473–3492. doi:10.3390/ijerph110403473

Knowlton, K., Rotkin, M. Ellman, King, G., Margolis, H. G., Smith, D., Solomon, G.,English, P. (2009). The 2006 California heat wave: Impacts on hospitalizations and emergency department visits. *Environmental Health Perspectives* **117**(1), 61–67. Accessed February 9, 2015: http://doi.org/10.1289/ehp.11594

Lane, K., Charles -Guzman, K., Wheeler, K., Abid, Z., Graber, N., and Matte, T. (2013). Health effects of coastal storms and flooding in urban areas: A review and vulnerability assessment. *Journal of Environmental and Public Health.* Article ID 913064.

Lesnikowski, A. C., Ford, J. D., Berrang-Ford, L., Barrera, M., Berrz, P., Henderson, J., and Hezmann, S. J. (2013). National-level factors affecting planned, public adaptation to health impacts of climate change. *Global Environmental Change* **23**(5), 1153–1163

Lim, S. S., Vos, T., Flaxman, A. D., Danaei, G., Shibuya, K., Adair-Rohani, H., and Andrews, K. G. (2012). A comparative risk assessment of burden of disease and injury attributable to 67 risk factors and risk factor clusters in 21 regions, 1990–2010: A systematic analysis for the Global Burden of Disease Study 2010. *The Lancet* **380**(9859), 2224–2260. Accessed November 30, 2014: http://doi.org/10.1016/S0140-6736(12)61766-8

Lin, S., Fletcher, B. A., Luo, M., Chinery, R., and Hwang, S. -A. (2011). Health impact in New York City during the Northeastern blackout of 2003. *Public Health Reports* (Washington, D.C.: 1974) **126**(3), 384–393.

Lin, S., Luo, M., Walker, R. J., Liu, X.,Hwang, S. -A., and Chinery, R. (2009). Extreme high temperatures and hospital admissions for respiratory and cardiovascular diseases. Epidemiology **20**(5), 738–746. Accessed October 9, 2014: http://doi.org/10.1097/EDE.0b013e3181ad5522

Littell, J. (2010). Engaging stakeholders in climate change adaptation. CAKE. Accessed October 18, 2014: http://www.cakex.org/virtual-library/engaging-stakeholders-climate-change-adaptation

Lo, E., and Levetin, E. (2007). Influence of meteorological conditions on early spring pollen in the Tulsa atmosphere from 1987–2006. *Journal of Allergy and Clinical Immunology* **119** (1),S101. Accessed June 15, 2015: http://doi.org/10.1016/j.jaci.2006.11.612

McMichael, A. J. (2001). Global environmental change as "risk factor": Can epidemiology cope? *American Journal of Public Health* **91**(8), 1172–1174.

McMichael, A. J., Wilkinson, P., Kovats, R. S., Pattenden, S., Hajat, S., Armstrong, B., and Nikiforov, B. (2008). International study of temperature, heat and urban mortality: The "ISOTHURM" project. *International Journal of Epidemiology* **37**(5), 1121–1131. Accessed May 14, 2014: http://doi.org/10.1093/ije/dyn086

Medlock, J. M., and Vaux, A. G. C. (2011). Assessing the possible implications of wetland expansion and management on mosquitoes in Britain. *European Mosquito Bulletin* 29, 38–65.

Minero, F. J. G., Candau, P., Morales, J., and Tomas, C. (1998). Forecasting olive crop production based on ten consecutive years of monitoring airborne pollen in Andalusia (southern Spain). *Agriculture Ecosystems & Environment* **69**(3), 201–215. http://doi.org/10.1016/S0167-8809(98)00105-4

Morrow, G., and Bowen, K. (2014). Accounting for health in climate change policies: A case study in Fiji. *Global Health Action* 7, 23550. Accessed June 25, 2015: http://dx.doi.org/10.3402/gha.v7.23550

Moser, S. C., and Ekstrom, J. A. (2010). A framework to diagnose barriers to climate change adaptation. *Proceedings of the National Academy of Sciences* **107**(51), 22026–22031.

Mueller, S. F., and Mallard, J. W. (2011). Contributions of natural emissions to ozone and PM2.5 as simulated by the Community Multiscale Air Quality (CMAQ) model. *Environmental Science & Technology* **45**(11), 4817–4823. Accessed December 20, 2014: http://doi.org/10.1021/es103645 m

Murray, C. J. L., Rosenfeld, L. C., Lim, S. S., Andrews, K. G., Foreman, K. J., Haring, D., Fullman, N., Naghavi, M., Lozano, R., and Lopez, A. D. (2012). Global malaria mortality between 1980 and 2010: A systematic analysis. *The Lancet* **379**, 413–431

Mustelin, J., Kuruppu, N., Kramer, A. M., Daron, J., Bruin, K., and Noriega, A. G. (2013). Climate adaptation research for the next generation. *Climate and Development* **5**(3), 189–193.

National Research Council (NRC). (2012). *Climate and Social Stress: Implications for Security Analysis. Committee on Assessing the Impacts of Climate Change on Social and Political Stresses,* J. D. Steinbruner, P. C. Stern and J. L. Husbands (eds.), Board on Environmental Change and Society, Division of Behavioral and Social Sciences and Education (253). The National Academies Press.

Obermaier, M. (2013). *City-Level Climate Change Adaptation Strategies: The Case of Quito, Ecuador.* ELLA Case Study Brief. Practical Action Peru, Lima.

Ogden, N. H., St-Onge, L., Barker, I. K., Brazeau, S.,Bigras, M. -Poulin, Charron, D. F., and Thompson, R. A. (2008). Risk maps for range expansion of the Lyme disease vector, Ixodes scapularis, in Canada now and with climate change. *International Journal of Health Geographics* 7, 24. Accessed January 23, 2016: http://doi.org/10.1186/1476-072X-7-24

Ogden, N. H., Bouchard, C., Kurtenbach, K., Margos, G., Lindsay, L. R., Trudel, L., and Milord, F. (2010). Active and passive surveillance and phylogenetic analysis of Borrelia burgdorferi elucidate the process of Lyme disease risk emergence in Canada. *Environmental Health Perspectives* **118**(7), 909–914. Accessed April 28, 2014: http://doi.org/10.1289/ehp.0901766

O'Haire, C., McPheeters, M., Nakamoto, E. K., LaBrant, L., Most, C., Lee, K., Graham, E., Cottrell, E., and Guise, J-M. (2011). Methods for engaging stakeholders to identify and prioritize future research needs. Methods Future Research Needs Report No. 4. AHRQ Publication No. 11-EHC044-EF. Agency for Healthcare Research and Quality. June 2011. Accessed January 27, 2016: www.effectivehealthcare.ahrq.gov/reports/final.cfm.

OMS. (2012). *Atlas of Health and Climate.* World Health Organization.

Omumbo, J., Lyon, B., Waweru, S., Connor, S., and Thomson, M. (2011). Raised temperatures over the Kericho tea estates: Revisiting the climate in the East African highlands malaria debate. *Malaria Journal* **10**(12).

O'Neill, M. S., Zanobetti, A., and Schwartz, J. (2005). Disparities by race in heat-related mortality in four U.S. cities: The role of air conditioning prevalence. *Journal of Urban Health: Bulletin of the New York Academy of Medicine* **82**(2), 191–197. Accessed December 11, 2014: http://doi.org/10.1093/jurban/jti043

Pan American Health Organization (PAHO). (2014). Plan of action for the prevention of obesity in children and adolescents. Pan American Health Organization and World Bank. Accessed August 10, 2015: http://www.paho.org/hq/index.php?option=com_docman&task=doc_view&Itemid=270&gid=28890&lang=en

Pan American Health Organization (PAHO). (2009). What is the Hospital Safety Index?. Accessed July 1, 2014: http://www1.paho.org/english/dd/ped/SafeHospitalsChecklist.htm

Panic, M., and Ford, J. D. (2013). A review of national-level adaptation planning with regards to the risks posed by climate change on infectious diseases in 14 OECD nations. *International Journal of Environmental Research and Public Health* **10**, 7083–7109.

Paterson, J. A., and Ford, J. D., et al. (2012). Adaptation to climate change in the Ontario public health sector. *BMC Public Health* **12**(1), 452.

Patz, J. A., Frumkin, H., Holloway, T., Vimont, D. J., and Haines, A. (2014). Climate change challenges and opportunities for global health. *JAMA* **312**(15), 1565–1580. doi:10.1001/jama.2014.13186

Patz, J. A., Vavrus, S. J., Uejio, C. K., and McLellan, S. L. (2008). Climate change and waterborne disease risk in the Great Lakes region of the U.S. *American Journal of Preventive Medicine* **35**(5), 451–458. Accessed April 11, 2014: http://doi.org/10.1016/j.amepre.2008.08.026

Pereira, C. A. R., and Barata, M. M. L. (2014a). Organização dos serviços urbanos de saúde frente à mudança do clima e ao risco de desastres na América Latina. *Saude em Debate* **38**, 624–634.

Pereira, C. A. R., Barata, M. M. L., Hoelz, M. P. C., Medeiros, V. N. L. O., Marincola, F. C. V., Neto, C. C., Marinho, D. P., Oliveira, T. V. S., Trigo, A. M., and Medeiros, K. (2014b). Economic evaluation of cases of dengue fever attributed to the disaster of 2011 in Nova Friburgo, Brasil. *Ciênc. saúde Coletiva* **19**(9), 3693–3704.

Pereira, C. A. R., Barata, M. M. L., and Trigo, A. M. (2014c). Social cost of leptospirosis cases attributed to the 2011 disaster striking Nova Friburgo, Brazil. *International Journal of Environmental Research and Public Health* **11**, 4140–4157.

Pfister, G. G., Wiedinmyer, C., and Emmons, L. K. (2008). Impacts of the fall 2007 California wildfires on surface ozone: Integrating local observations with global model simulations. *Geophysical Research Letters* **35**(19), L19814. Accessed October 17, 2014: http://doi.org/10.1029/2008GL034747

Prain, G., and Dubbeling, M. (2010). *Case Studies of the Cities of Accra, Nairobi, Lima, and Bangalore Undertaken.* RUAF Foundation.

Prain, G., and Dubbeling, M. (2010). Case Studies of the Cities of Accra, Nairobi, Lima, and Bangalore Undertaken. RUAF Foundation.

Pope, C. A., 3rd, Burnett, R. T., Thun, M. J., Calle, E. E., Krewski, D., Ito, K., and Thurston, G. D. (2002). Lung cancer, cardiopulmonary mortality, and long-term exposure to fine particulate air pollution. *Journal of the American Medical Association* **287**(9), 1132–1141.

Porter, J. R., Xie, L., Challinor, A. J., Cochrane, K., Howden, S. M., Iqbal, M. M., Lobell, D. B., and Travasso, M. I. (2014). Food security and food production systems. In Field, C. B., Barros, V. R., Dokken, D. J., Mach, K. J., Mastrandrea, M. D., Bilir, T. E., Chatterjee, M., Ebi, K. L., Estrada, Y. O., Genova, R. C., Girma, B., Kissel, E. S., Levy, A. N., MacCracken, S., Mastrandrea, P. R., and White, L. L. (eds.), *Climate Change 2014: Impacts, Adaptation, and Vulnerability. Part A: Global and Sectoral Aspects. Contribution of Working Group II to the Fifth Assessment Report of the Intergovernmental Panel of Climate Change* (485–533). Cambridge University Press.

Reid, C. E., O'Neill, M. S., Gronlund, C. J., Brines, S.J, Brown, D.G, Diey-Roux, A.V, and Schwarty, J. (2009). Mapping community determinants of heat vulnerability. *Environmental Health Perspectives* **117**, 1730–1736.

Reiss, N. M., and Kostic, S. R. (1976). Pollen season severity and meteorologic parameters in central New Jersey. *Journal of Allergy and Clinical Immunology* **57**(6), 609–614.

Rekow, L. (2015). Fighting insecurity: Experiments in urban agriculture in the favelas of Rio de Janeiro. *Field Actions Science Reports*, 8.

Revi, A., Satterthwaite, D. E., Aragón-Durand, F., Corfee-Morlot, J., Kiunsi, R. B. R., Pelling, M., Roberts, D. C., and Solecki, W. (2014).

Urban areas. In Field, C. B., Barros, V. R., Dokken, D. J., Mach, K. J., Mastrandrea, M. D., Bilir, T. E., Chatterjee, M., Ebi, K. L., Estrada, Y. O., Genova, R. C., Girma, B., Kissel, E. S., Levy, A. N., MacCracken, S., Mastrandrea, P. R., and White, L. L. (eds.), *Climate Change 2014: Impacts, Adaptation, and Vulnerability. Part A: Global and Sectoral Aspects. Contribution of Working Group II to the Fifth Assessment Report of the Intergovernmental Panel on Climate Change* (535–612). Cambridge University Press.

Rittmaster, R., Adamowicz, W. L., Amiro, B., and Pelletier, R. T. (2006). Economic analysis of health effects from forest fires. *Canadian Journal of Forest Research* **36**(4), 868–877. Accessed September 9, 2014: http://doi.org/10.1139/X05-293

Romero-Lankao, P., Hughes, S.,Rosas, A. -Huerta, Borquez, R., and Gnatz, D. M. (2013a). Institutional capacity for climate change responses: An examination of construction and pathways in Mexico City and Santiago. *Environmental Planning and Government Policy* **31**, 785–805.

Romero-Lankao, P., Qin, H., and Borbor-Cordova, M. (2013). Exploration of health risks related to air pollution and temperature in three Latin American cities. *Social Science and Medicine* **83**, 110–118.

Romero-Lankao, P., Qin, H., and Dickinson, K. (2012). Urban vulnerability to temperature-related hazards: A meta-analysis and meta-knowledge approach. *Global Environmental Change* **12**, 670–683.

Rose, J. B., Epstein, P. R., Lipp, E. K., Sherman, B. H., Bernard, S. M., and Patz, J. A. (2001). Climate variability and change in the United States: Potential impacts on water- and foodborne diseases caused by microbiologic agents. *Environmental Health Perspectives* **109**(Suppl 2), 211–221.

Rosenzweig, C., Solecki, W., Hammer, S. A., and Mehrotra, S. (2010). Cities lead the way in climate-change action. *Nature* **467**, 909–911.

Rosenzweig, C., and Solecki, W. (2010). Introduction to climate change adaptation in New York City: Building a risk management response. *Annals of the New York Academy of Sciences* **1196**(1), 13–17.

Rosenzweig, C., Solecki, W. D., and Slosberg, R. B. (2006). *Mitigating New York City's Heat Island with Urban Forestry, Living Roofs, and Light Surfaces.* New York City Regional Heat Island Initiative, Final Report 06–06. New York State Energy Research and Development Authority.

Ruckart, P. Z., Orr, M. F., Lanier, K., and Koehler, A. (2008). Hazardous substances releases associated with Hurricanes Katrina and Rita in industrial settings, Louisiana and Texas. *Journal of Hazardous Materials* **159**(1), 53–57.

Rydin, Y., Bleahu, A., Davies, M., Dávila, J. D., Friel, S., De, G.Grandis, N., Groce, Hallal, P. C., Hamilton, I., Howden-Chapman, P., Lai, K-M., Lim, C., Martins, J., Osrin, D., Ridley, I., Scott, I., Taylor, M., Wilkinson, P., and Wilson, J. (2012). Shaping cities for health: Complexity and the planning of urban environments in the 21st century. *The Lancet* **379**, 2079–2108.

Samet, J.M, Zeger, S.L., Dominici, F., Curriero, F., Coursac, I., Dockery, D. W., Schwarty, J., and Yanobetti, A. (2000). The National Morbidity, Mortality, and Air Pollution Study Part II: Morbidity and mortality from air pollution in the United States. Accessed August 6, 2015: http://www.cabq.gov/airquality/documents/pdf/samet2.pdf

Sastry, N., and Gregory, J. (2013). The effect of Hurricane Katrina on the prevalence of health impairments and disability among adults in New Orleans: Differences by age, race, and sex. *Social Science and Medicine* **80**, 121–129.

Satterthwaite, D., Huq, S., Reid, H., Pelling, M., and Romero-Lankao, P. (2008). *Adapting to Climate Change in Urban Areas: The Possibilities and Constraints in Low- and Middle-Income Nations.* IIED.

Schmeer, K. (1999). *Guidelines for Conducting a Stakeholder Analysis.* Partnerships for Health Reform, Abt Associates Inc for the World Health Organization.

Schwartz, J. (1994). Air pollution and daily mortality: A review and meta-analysis. *Environmental Research* **4**, 36–52

Semenza, J. C., Herbst, S., Rechenburg, A., Suk, J. E., Höser, C., Schreiber, C., and Kistemann, T. (2012). Climate change impact assessment of food- and waterborne diseases. *Critical Reviews in Environmental Science and Technology* **42**(8), 857–890. Accessed December 11, 2014: http://doi.org/10.1080/10643389.2010.534706

Semenza, J. C., Sudre, B., Oni, T., Suk, J. E., and Giesecke, J. (2013). Linking environmental drivers to infectious diseases: The European Environment and Epidemiology Network. *PLoS Neglected Tropical Diseases* **7** (7), e2323.

Sena, A. L., Barcellos, C. L., Freitas, C., and Corvalan, C. (2014). Managing the health impacts of drought in Brazil. *International Journal of Environmental Research and Public Health* **11**, 10737–10751.

Sheffield, P. E., Knowlton, K., Carr, J. L., and Kinney, P. L. (2011). Modeling of regional climate change effects on ground-level ozone and childhood asthma. *American Journal of Preventive Medicine* **41**(3), 251–257; quiz A3. Accessed February 27, 2016: http://doi.org/10.1016/j.amepre.2011.04.017

Silva, C. B. P., Saldiva, P. H. N., Lourenço, L. F. A., Silva, F. R., and Miraglia, S. G. K. (2012). Evaluation of the air quality benefits of the subway system in São Paulo, Brazil. *Journal of Environmental Management* **101**, 191–196

Smith, K. R., Jerrett, M., Anderson, H. R., Burnett, R. T., Stone, V., Derwent, R., and Thurston, G. (2009). Health and Climate Change 5 Public health benefits of strategies to reduce greenhouse-gas emissions: Health implications of short-lived greenhouse pollutants. *The Lancet* **374**(9707), 2091–2103. doi:10.1016/S0140-6736(09)61716-5

Smith, K. R., Woodward, A., Campbell-Lendrum, D., Chadee, D. D., Honda, Y., Liu, Q., Olwoch, J. M., Revich, B., and Sauerborn, R. (2014). Human health: Impacts, adaptation, and co-benefits. In Field, C. B., Barros, V. R., Dokken, D. J., Mach, K. J., Mastrandrea, M. D., Bilir, T. E., Chatterjee, M., Ebi, K. L., Estrada, Y. O., Genova, R. C., Girma, B., Kissel, E. S., Levy, A. N., MacCracken, S., Mastrandrea, P. R., and White, L. L. (eds.), *Climate Change 2014: Impacts, Adaptation, and Vulnerability. Part A: Global and Sectoral Aspects. Contribution of Working Group II to the Fifth Assessment Report of the Intergovernmental Panel on Climate Change* (709–754). Cambridge University Press.

Solecki, W., Patrick, L., Brady, M., Grady, K., and Maroko, A. (2010). Climate protection levels: Incorporating climate change into design and performance standards. New York City Panel on Climate Change. *Annals of the New York Academy of Sciences* **1196**(1), 293–352.

Stott, P., and Walton, P. (2013). Attribution of climate-related events: Understanding stakeholder needs. *Weather* **68**(10), 274–279.

Sudhinaraset, M., Ingram, M., Lofthouse, H. K., and Montagu, D. (2013). What is the role of informal healthcare providers in developing countries? A systematic review. *PLoS ONE* **8**, e54978.

Toronto Public Health (TPH). (2011). Protecting vulnerable people from the health impacts of extreme heat. Accessed August 13, 2015: http://app.toronto.ca/tmmis/viewAgendaItemHistory.do?item=2011.HL6.3

Toronto Public Health (TPH). (2014a). Path to healthier air. Accessed November 14, 2015: http://app.toronto.ca/tmmis/viewAgendaItemHistory.do?item=2014.HL30.1

Toronto Public Health (TPH). (2014b). Active city: Designing for health. Accessed November 9, 2015: http://app.toronto.ca/tmmis/viewAgendaItemHistory.do?item=2014.HL31.1

Tsai, D. -H., Wang, J. -L., Wang, C. -H., and Chan, C. -C. (2008). A study of ground-level ozone pollution, ozone precursors and subtropical meteorological conditions in central Taiwan. *Journal of Environmental Monitoring: JEM* **10**(1), 109–118. Accessed September 7, 2014: http://doi.org/10.1039/b714479b

U.S. Energy Information Administration. (2007). Energy characteristics and energy consumed in large hospital buildings in the United States in 2007. Accessed June 12, 2014: http://www.eia.gov/consumption/commercial/reports/2007/large-hospital.cfm

U.S. Environmental Protection Agency (EPA) National Center for Environmental Assessment, I. O. (2008). A review of the impact of climate variability and change on aeroallergens and their associated effects (final report). Accessed July 23, 2015: http://cfpub.epa.gov/ncea/cfm/recordisplay.cfm?deid=190306

U.S. Global Change Research Program. (2014). National Climate Assessment. Accessed September 22, 2015: http://nca2014.globalchange.gov/node/1961

USGCRP. (2008). *Analyses of the Effects of Global Change on Human Health and Welfare and Human Systems (SAP 4.6)*. Washington, D.C.: U.S. Environmental Protection Agency.

Vardoulakis, S., Dear, K., Hajat, S., Heaviside, C., Eggen, B., and McMichael, A. J. (2014). Comparative assessment of the effects of climate change on heat and cold related mortality in cities of United Kingdom and Australia. *Environmental Health Perspectives* **122**(12), 1285–1292.

Villeneuve, P. J., Doiron, M. -S., Stieb, D., Dales, R., Burnett, R. T., and Dugandzic, R. (2006). Is outdoor air pollution associated with physician visits for allergic rhinitis among the elderly in Toronto, Canada? *Allergy* **61**(6), 750–758. Accessed March 16, 2014: http://doi.org/10.1111/j.1398-9995.2006.01070.x

Walters, V., and Gaillard, J. C. (2014). Disaster risk at the margins: Homelessness, vulnerability and hazards. *Habitat International* **44**, 211–219.

Weible, C. M. (2007). An advocacy coalition framework approach to stakeholder analysis: Understanding the political context of California Marine Protected Area Policy. *Journal of Public Administration Research and Theory* **17**, 95–117.

Weisler, R. H., Barbee, J. G., and Townsend, M. H. (2006). Mental health and recovery in the Gulf Coast after Hurricanes Katrina and Rita. *JAMA* **296**(5), 585–588.

Westerling, A. L. (2006). Warming and earlier spring increase western U.S. forest wildfire activity. *Science* **313**(5789), 940–943. Accessed January 6, 2015: http://doi.org/10.1126/science.1128834

Wheeler, K., Lane, K., Walters, S., and Matte, T. (2013). Heat illness and deaths – New York City, 2000–2011. (Cover story). *MMWR: Morbidity & Mortality Weekly Report* **62**(31), 617–621.

White-Newsome, J. L., McCormick, A., Sampson, N., Buxton, M. A., O'Neill, M. S., Gronlund, C. J., Catalano, L., Conlon, K., and Parker, E. A. (2014). Strategies to reduce the harmful effects of extreme heat events: A four-city study. *International Journal of Environmental Research and Public Health* **11**, 1960–1988.

Wolf, T., Sanchez Martinez, G., Cheong, H. -K., Williams, E., and Menne, B. (2014). Protecting health from climate change in the WHO European region. *International Journal of Environmental Research and Public Health* **11**, 6265–6280.

Woodcock, J., Edwards, P., Tonne, C., Armstrong, B. G., Ashiru, O., Banister, D., Beevers, S., Chalabi, Y., Chowdhury, Y., Cohen, A., Franco, O.H, Haines, A., Hickman, R., Lindsay, G., Mittal, I., Mohan, D., Tiwari, G., Woodward, A., and Roberts, I. (2009). Public health benefits of strategies to reduce greenhouse-gas emissions: Urban land transport. *The Lancet* **374**(9705), 1930–1943. doi:10.1016/S0140-736(09)61714-1.

World Bank. (2016). Gini index (World Bank estimate), Development Research Group. Accessed March 16, 2017: http://data.worldbank.org/indicator/SI.POV.GINI

World Health Organization (WHO). (1946). *Preamble to the Constitution of the World Health Organization as adopted by the International Health Conference*, New York, 19–22 June 1946; signed on 22 July 1946 by the representatives of 61 States (Official Records of the World Health Organization, no. 2, p. 100) and entered into force on 7 April 1948. Accessed November 11, 2015: http://www.who.int/about/definition/en/print.html

World Health Organization (WHO). (2005). *Strengthened health systems save more lives: An insight into WHO's European health systems' strategy*. The World Health Organization Regional Office for Europe, Denmark. Accessed July 3, 2014: http://www.euro.who.int/__data/assets/pdf_file/0011/78914/healthsys_savelives.pdf

World Health Organization (WHO). (2011). *Protecting Health from Climate Change: Vulnerability And Adaptation Assessment*. World Health Organization, 72.

Yates, R. (2014). Recycling fuel subsidies as health subsidies. *Bulletin of the World Health Organization* **92**, 547–547A.

Ziska, L., Knowlton, K., Rogers, C., Dalan, D., Tierney, N., Elder, M. A., and Frenz, D. (2011). Recent warming by latitude associated with increased length of ragweed pollen season in central North America. *Proceedings of the National Academy of Sciences of the United States of America* **108**(10), 4248–4251. Accessed August 1, 2014: http://doi.org/10.1073/pnas.1014107108.

Chapter 10   Case Study References

Case Study 10.1   Participative Development of a Heat-Health Action Plan in Ahmedabad, India

Ahmedabad Municipal Corporation (AMC). (2013). Ahmedabad Heat Action Plan. Guide to Extreme Heat Planning in Ahmedabad, India. Accessed August 30, 2015: http://www.egovamc.com/downloads/healthcare/healthpdf/heat_action_plan.pdf

Ahmedabad Municipal Corporation (AMC). (2015). Ahmedabad Heat Action Plan. Guide to Extreme Heat Planning in Ahmedabad, India. Accessed December 27, 2014: http://www.ndma.gov.in/images/pdf/HAP2015.pdf

Ahmedabad Urban Development Authority (2016). Accessed January 29, 2017: http://www.auda.org.in/Content/about-us-42

Census Bureau of India (2011). Population Census 2011. Accessed February 4, 2016: http://www.census2011.co.in/

Government of India. (2011). Census of India. Registrar General & Census Commissioner, India, New Delhi. Accessed July 28, 2015: http://censusindia.gov.in/

Knowlton, K., Kulkarni, S. P., Azhar, G. S., Mavalankar, D., Jaiswal, A., Connolly, M., Nori-Sarma, A., Rajiva, A., Dutta, P., Deol, B., Sanchez, L., Khosla, R., Webster, P. J., Toma, V. E., Sheffield, P., and Hess, J. J., the Ahmedabad Heat and Climate Study Group (2014). Development and implementation of south Asia's first heat-health action plan in Ahmedabad (Gujarat, India). *International Journal of Environmental Research and Public Health* 11, 3473–3492. doi:10.3390/ijerph110403473.

Kotkin, J. (2010). The world's fastest growing cities. Forbes. Accessed October 28, 2015: http://www.forbes.com/2010/10/07/cities-china-chicago-opinions-columnists-joel-kotkin.html.

Peel, M. C., Finlayson, B. L., and McMahon, T. A. (2007). Updated world map of the Köppen-Geiger climate classification. *Hydrology and Earth System Sciences Discussions* **4**(2), 462.

Smith, K. R., Woodward, A., Campbell-Lendrum, D., Chadee, D. D., Honda, Y., Liu, Q., Olwoch, J. M., Revich, B., and Sauerborn, R. (2014). Human health: Impacts, adaptation, and co-benefits. In Field, C. B., Barros, V. R., Dokken, D. J., Mach, K. J., Mastrandrea, M. D., Bilir, T. E., Chatterjee, M., Ebi, K. L., Estrada, Y. O., Genova, R. C., Girma, B., Kissel, E. S., Levy, A. N., MacCracken, S., Mastrandrea, P. R., and White, L. L. (eds.), *Climate Change 2014: Impacts, Adaptation, and Vulnerability. Part A: Global and Sectoral Aspects. Contribution of Working Group II to the Fifth Assessment Report of the Intergovernmental Panel of Climate Change* (709–754). Cambridge University Press.

Tran, K. V., Azhar, G. S., Nair, R., Knowlton, K., Jaiswal, A., Sheffield, P., Mavalankar, D., and Hess, J. (2013). A cross-sectional, randomized cluster sample survey of household vulnerability to extreme heat among slum dwellers in Ahmedabad, India. *International Journal of Environmental Research and Public Health* 10, 2515–2543. doi:10.3390/ijerph10062515.

World Bank. (2017). 2016 GNI per capita, Atlas method (current US$). Accessed August 9, 2017: http://data.worldbank.org/indicator/NY.GNP.PCAP.CD

Case Study 10.2   Health and Social Cost of Disaster: Nova Friburgo, State of Rio de Janeiro, Brazil

Instituto Brasileiro de Geografia e Estatística (IBGE). (2015). Population Census. *Cidades/Rio de Janeiro/Nova Friburgo*. Accessed January 25, 2016: http://cidades.ibge.gov.br/

Instituto Oswaldo Cruz-Fiocruz. (2013). *Social Cost Estimate of Hydrological Disaster in Nova Friburgo (RJ) – Health Sector*. IOC/Fiocruz,.

Peel, M. C., Finlayson, B. L., and McMahon, T. A. (2007). Updated world map of the Köppen-Geiger climate classification. *Hydrology and Earth System Sciences Discussions* **4**(2), 462.

Pereira, C. A. R., Barata, M. M. L., and Trigo, A. M. (2014). Social cost of leptospirosis cases attributed to the 2011 disaster striking Nova Friburgo, Brazil. *International Journal of Environmental Research and Public Health* **11**, 4140–4157.

World Bank. (2017). 2016 GNI per capita, Atlas method (current US$). Accessed August 9, 2017: http://data.worldbank.org/indicator/NY.GNP.PCAP.CD

Case Study 10.3   Windsor Heat Alert and Response Plan: Reaching Vulnerable Populations

City of Windsor. (2013). Stay Cool Windsor-Essex Heat Alert and Response Plan. Accessed December 23, 2014: www.windsorfire.com/wp-content/uploads/2013/08/Heat-Alert-Response-Plan.pdf

Fouillet, A., Rey, G., Laurent, F., Pavillon, G., Bellec, S., Guihenneuc-Jouyaux, C., Clavel, J., Jougla, E., and Hémon, D. (2006). Excess mortality related to the August 2003 heat wave in France. *International Archives of Occupational and Environmental Health* **80**(1), 16–24.

Government of Canada. (2015). Population and dwelling counts, for Canada and census subdivisions (municipalities), 2011 and 2006 censuses. Accessed February 3, 2016: http://www12.statcan.gc.ca/census-recensement/2011/dp-pd/hlt-fst/pd-pl/Table-Tableau.cfm?LANG=Eng&T=301&S=3&O=D

IPCC. (2014). Summary for policymakers. In Field, C. B., V. R. Barros, D. J. Dokken, K. J.

Mach, M., Mastrandrea, D., Bilir, T. E., Chatterjee, M., Ebi, K. L., Estrada, Y. O., Genova, R. C., Girma, B., Kissel, E. S., Levy, A. N., MacCracken, S., Mastrandrea, P. R., and White, L. L. (eds.), *Climate Change 2014. Impacts, Adaptation, and Vulnerability. Part A: Global and Sectoral Aspects. Contribution of Working Group II to the Fifth Assessment Report of the Intergovernmental Panel on Climate Change* (1–32). Cambridge University Press.

Oke, T. R. (1997). Urban climates and global change. In Perry A., and Thompson R. (eds.), *Applied Climatology: Principles and Practices*. Routledge.

Peel, M. C., Finlayson, B. L., and McMahon, T. A. (2007). Updated world map of the Köppen-Geiger climate classification. *Hydrology and Earth System Sciences Discussions* **4**(2), 462.

Robine, J., Cheung, S., Le Roy, S., Michel, J., and Herrmann, F. (2008). Death toll exceeded 70,000 in Europe during the summer of 2003. *Comptes Rendus – Biologies* **331**(2), 171–178

World Bank. (2017). 2016 GNI per capita, Atlas method (current US$). Accessed August 9, 2017: http://data.worldbank.org/indicator/NY.GNP.PCAP.CD

Case Study 10.4   Health Impacts of Extreme Temperatures in Brisbane, Australia

Australian Bureau of Statistics (ABS). (2011a). *Census of Population and Housing*. Australian Bureau of Statistics, Canberra.

Australian Bureau of Statistics (ABS). (2011b.) *Queensland Australian Statistical Geography Standard (ASGS) Edition 2011*. Australian Bureau of Statistics, Canberra.

Huang, C., Barnett, A. G., Wang, X., and Tong, S. (2012). The impact of temperature on years of life lost in Brisbane, Australia. *Nature Climate Change* **2**(4), 265–270.

Peel, M. C., Finlayson, B. L., and McMahon, T. A. (2007). Updated world map of the Köppen-Geiger climate classification. *Hydrology and Earth System Sciences Discussions* **4**(2), 462.

World Bank. (2017). 2016 GNI per capita, Atlas method (current US$). Accessed August 9, 2017: http://data.worldbank.org/indicator/NY .GNP.PCAP.CD

Case Study 10.5 Kedarnath: Flood and Humanitarian Loss in Uttarakhand Districts in India

Census Bureau of India (2011). Population Census 2011. Accessed February 24, 2015: http://www.census2011.co.in/

Dube, A., Ashrit, R., Ashish, A., Sharma, K., Iyengar, G.R., Rajagopal, E.N., and Basu, S. (2013). Performance of NCMRWF Forecast models in predicting the Uttarakhand Heavy Rainfall Event during 17–18 June 2013. National Centre for Medium Range Weather Forecasting, Ministry of Earth Sciences. Accessed February 3, 2016: http://www.ncmrwf.gov.in/KEDARNATH_REPORT_ FINAL.pdf

Dube, A., Ashrit, R., Ashish, A., Sharma, K., Iyengar, G. R., Rajagopal, E. N., and Basu, S. (2014). Forecasting the heavy rainfall during Himalayan flooding-June 2013. *Weather and Climate Extremes* 4, 22–34.

Indian Express. (2013). Uttarakhand yet to implement water management act. Accessed June 16, 2014: http://archive.indianexpress.com/news/ uttarakhand-yet-to-implement-water-management-act/1147808/

Munich Re (2014). *Natural catastrophes* 2013: Analyses, assessments, positions. Topics Geo, 2013. Accessed June 9, 2015: http://www.munichre .com/site/corporate/get/documents_E1043212252/mr/assetpool .shared/Documents/5_Touch/_Publications/302-08121_en.pdf

Peel, M. C., Finlayson, B. L., and McMahon, T. A. (2007). Updated world map of the Köppen-Geiger climate classification. *Hydrology and Earth System Sciences Discussions* **4**(2), 462.

World Bank. (2017). 2016 GNI per capita, Atlas method (current US$). Accessed August 9, 2017: http://data.worldbank.org/indicator/NY.GNP .PCAP.CD

Case Study 10.6 Kunshan Eastern Health Care Center, Kunshan City, China

Hairong, W. (2016). An inspiring visit. *Beijing Review* 26, June 30. Accessed February 24, 2017: http://usa.bjreview.com/Arts/201606/t20160628_ 800060747.html

Peel, M. C., Finlayson, B. L., and McMahon, T. A. (2007). Updated world map of the Köppen-Geiger climate classification. *Hydrology and Earth System Sciences Discussions* **4**(2), 462.

U.S. Green Building Council. (2010, updated 2014). LEED 2009 for Healthcare. USGBC, Washington, D.C. Accessed July 23, 2015: http://www.usgbc.org/ sites/default/files/LEED%202009%20RS_HC_4–2014_cover.pdf

World Bank. (2017). 2016 GNI per capita, Atlas method (current US$). Accessed August 9, 2017: http://data.worldbank.org/indicator/NY.GNP .PCAP.CD

11

Housing and Informal Settlements

Coordinating Lead Authors

Nathalie Jean-Baptiste (Leipzig/Dar es Salaam), Veronica Olivotto (Rotterdam)

Lead Authors

Emma Porio (Manila), Wilbard Kombe (Dar es Salaam), Antonia Yulo-Loyzaga (Manila)

Contributing Authors

Ebru Gencer (New York/Istanbul), Mattia Leone (Naples), Oswaldo Lucon (São Paulo), Mussa Natty (Dar es Salaam)

This chapter should be cited as

Jean-Baptiste, N., Olivotto, V., Porio, E., Kombe, W., and Yulo-Loyzaga, A. (2018). Housing and informal settlements. In Rosenzweig, C., W. Solecki, P. Romero-Lankao, S. Mehrotra, S. Dhakal, and S. Ali Ibrahim (eds.), *Climate Change and Cities: Second Assessment Report of the Urban Climate Change Research Network*. Cambridge University Press. New York. 399–440

Housing and Informal Settlements

Addressing vulnerability and exposure in the urban housing sector is critical in informal settlements where extreme climate events present multiple risks for millions of people. Understanding the impacts of mitigation and adaptation strategies on the housing sector will help decision-makers make choices that improve quality of life and close development and equity gaps in cities.

Major Findings

- The effects of climate hazards, people's exposure and vulnerability to them collectively determine the types and levels of risk faced by cities. Risks are associated with specific socio economic, political and physical factors exacerbating climate risks in cities. Adaptation actions like mapping of risks, developing early warning systems (EWS), preparedness plans, and preventive risk strategies – especially in informal settlements – can support decision-makers and stakeholders in reducing exposure and vulnerability in the housing sector.

- Developed countries account for the majority of the world's energy demand related to buildings. Incentives and other measures are enabling large-scale investments in mass-retrofitting programs in higher income cities.

- Housing construction in low- and middle-income countries is focused on meeting demand for more than 500 million more people by 2050. Cost-effective and adaptive building technologies can avoid locking-in carbon-intensive and non-resilient options.

- Access to safe and secure land is a key measure for reducing risk in cities. Groups that are already disadvantaged with regard to housing and land tenure are especially vulnerable to climate change.

- In informal settlements, successful adaptation depends upon addressing needs for climate related expertise and resources at different government levels as well as risk-reducing physical infrastructure and social structures

Key Messages

City managers should work with the informal sector to improve safety in relation to climate extremes. Informal economic activities are often highly vulnerable to climate impacts, yet they are crucial to economies in low- and middle-income cities. Therefore, direct and indirect costs to the urban poor should be included in loss and damage assessments in order to accurately reflect the full range of impacts on the most vulnerable and the city as a whole.

Evidence of affordable insurance schemes in developing countries' low-income communities that fulfill adaptation goals is limited. Several implementation-related hurdles that need to be addressed if insurance schemes are to be successful. These are excessive reliance on government and donor subsidies, lack of local distribution channels, poor financial literacy of communities, and overall limited demand.

Retrofits to housing that improve resilience create co-benefits, such as more dignified housing, improvements to health, and better public spaces. Meanwhile, mitigating greenhouse gas (GHG) emissions in the housing sector can create local jobs in production, operations, and maintenance – especially in low-income countries and informal settlements.

11.1 Introduction

Cities across the globe are facing the reality of climate change. Hazards such as floods, storms, and slow-onset sea level rise are affecting human, economic, and environmental assets. Knowledge of the impacts of mitigation and adaptation strategies employed in the housing sector is critical to inform policies and decisions to achieve combined benefits on improving quality of life and closing the development and equity gap in cities worldwide.

While cities contribute to global emissions at various degrees, the impacts of climate change are local and differ across social groups. In developed countries, economic activity, zoning, mobility, and design and construction standards define source contributions to national and global greenhouse gas inventories and their mitigation potential. However, where informal settlements are the dominant urban form, adaptation to climate change is considered necessary to reduce growing socioecological vulnerability (Satterthwaite et al., 2007). In fact, combined strategies for climate change adaptation and mitigation are critical for responding to climate-related challenges in a more cost-effective way, ultimately improving quality of life both globally and locally. Reducing GHG emissions, especially at municipal levels such as waste, transport, and building sectors, appears to be the low hanging fruit of urban GHG mitigation, with cities placing considerable emphasis on energy efficiency because it allows for multiple agendas to be amalgamated (financial savings, energy security, air pollution, fuel poverty, and climate change). Measures that may have larger impacts on GHG reduction, such as low-carbon and renewable energy infrastructure systems, remain the exception (UN-Habitat, 2011). Moreover, actively addressing vulnerability and exposure in sectors such as urban housing, may reduce potential for further decline in the well-being of informal settlement populations. Adopting integrated urban strategies may contribute to closing the development equity gap in cities around the world by reducing the potential occurrence of climate-related shocks and ensuring a level of well-being through those socioecological interventions that address the persistent exposure and vulnerability of the poor (Laukkonen et al., 2009).

This chapter highlights the specific impacts of climate change on the housing sector and responses in the areas of adaptation and mitigation. We address how extreme events have transformed housing and, particularly, what the drivers and impacts of climate-related events are and how they have shaped vulnerability in informal settlements. To do this, coupled socio-environmental research perspectives were reviewed for the latest scientific input of the past decade, and several workshops were conducted to address the question of housing in connection with growing demands at the intersections of formal and informal systems and pre-conditions for adaptation to climate change (see Annex 11.2 Table 1).

While considering a broad spectrum in processes and practices relevant to the housing sector, the focused is geared toward housing and informal settlements in middle- and low-income countries because of their greater exposure and vulnerability. They account for about 80% of the world's urban population and are likely to house most of the world's growth in the next 10–20 years (Satterthwaite et al., 2013). Such growth does not come without consequences in a globally and climatically changing environment (Jean-Baptiste et al., 2013; UN-Habitat, 2014). The lack of coping and adaptive capacities in urban areas of low- and middle-income nations increase their risk to floods, storm surge, and rise in sea level (Pelling and Blackburn, 2013). Coastal urbanization trends offer increased economic opportunities for both formal and informal economies. Such livelihood prospects will draw the poor toward greater exposure and vulnerability to climate-associated hazards (Porio, 2014).

11.2 Urbanization, Climate Risk, Housing, and Informal Settlements

In 2012, the Intergovernmental Panel on Climate Change (IPCC) reported that populations in Asia, Africa, and small island nations would be the most exposed to floods and tropical cyclones. Demographic changes and urbanization were among the major reasons for exposure (IPCC, 2012). Moreover, coastal urbanization in these regions continues to drive the increases in number of people exposed to the impacts of climate change (UNDESA, 2014).

Today, the majority of the world's population lives in urban areas. Most were originally established along sources of food and water, with many growing cities in Asia and Africa being founded as a result of the property ownership laws and planning practices of their long colonial histories (Huq et al., 2007).

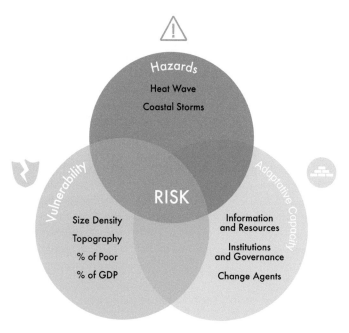

Figure 11.1 *Urban climate change vulnerability and risk assessment framework.*
Source: Mehrotra et al., 2011

While 70–80% of the population of North America, Europe, and Latin America lives in cities today, cities in Asia and Africa will soon be the main source of urban growth worldwide. Moreover, the pace of urbanization in Nigeria, China, and India will soon surpass that of the developed world. Notably, Asia will be home to half of the world's urban population by 2050 (UNDESA, 2014).

The United Nations International Strategy for Disaster Reduction (UNISDR) defines disaster risk as the result of the compounding effect of hazards, exposure, and vulnerability. The pre-existing physical exposure to environmental hazards, urban densities, and social marginalization of informal settlements renders them at the highest risk to the impacts of climate change (UNISDR, 2013). In a globally warmer world, increases in temperature have amplified the likelihood of hazards such as droughts and extreme rainfall events, flooding, rainfall-induced landslides, and rise in sea level. While threatening the safety of communities and the sustainability of cities, these hazards especially endanger those whom urban governance fails, namely, those without access to adequate housing, basic utilities and infrastructure, and social services (Loyzaga et al., 2014).

Disaster risk, which is associated with climate change, is further defined by the IPCC as, "a product of the hazards influenced by climate change and the vulnerabilities of the human systems and ecological systems" (IPCC, 2014). Although the type and intensity of hazards may differ in each geographic region and city, it is the exposure and vulnerability in urban socioecological systems that largely determine the probability, nature, and magnitude of risk. Poverty and housing are the two critical dimensions of vulnerability that drive risk. This is most clearly seen in low-elevation coastal zones where macro- and micro-economic drivers influence not only coastal settlements, but also sources of water, food, and livelihood for both the formal and informal communities (Pelling et al., 2013). The risks of coastal and riverine flooding are further enhanced by human activities with subsidence due to groundwater extraction. While the geomorphology of coastal cities may influence ground movement, the combination of subsidence, tides, and sea level rise further increases the risk of flooding from storm surge and waves in coastal cities of Africa and Asia (World Bank, 2013a).

Today, city governance systems struggle not only to cope with the effects of GHGs from industrial use, rapid land-use change, and land-cover loss, but also from overwhelming population growth. The Urban Climate Change Risk Framework presents risk as the intersection of climate hazards, vulnerability, and adaptive capacity (Mehrotra et al., 2011) (see Figure 11.1).

The World Risk Report of 2014 discusses urban risk and vulnerability in light of susceptibility, coping, and adaptive capacity. While susceptibility may be a characteristic of exposed communities, coping and adaptive capacity depend primarily on the responsiveness of government institutions to current and long-term human needs (Bündnis Entwicklung Hilft, 2014). In cities where 30–70% of the population is without access to clean water or basic health services, this failure in governance is at the root of urban risk to the impacts of climate change (Lavell [2002], cited in Dodman and Satterthwaite 2008; Porio, 2015; Pauleit et al., 2015).

Informal settlements include slums[1], which in addition to the definitions by UN-Habitat (2013) and United Nations (2014), are known as dilapidated "inner-urban settlements with substandard living conditions which are, unlike informal or marginal settlements in peri-urban or newly urbanized areas, originally understood as an emergency accommodation in dilapidated parts of an existing city. Overall housing standards and the infrastructure conditions in slums are correspondingly poor" (Bündnis Entwicklung Hilft, 2014).

The UN Department of Economic and Social Affairs (UNDESA) estimates nearly one billion slum dwellers today, with the potential for growth to as many as three billion by 2050 (UNDESA, 2013). In Asia and Africa, they are estimated to constitute from 30% to as much as 70% of urban populations. Slum dwellers often live in hazardous areas that are generally undesirable or too expensive to develop commercially. Table 11.1 presents the possible impacts of climate change. It integrates the table adapted by Dodman and Satterthwaite with a listing of possible impact areas on slums based on the five deprivations and other characteristics of poor housing as defined by UN-Habitat (2003, 2006/2007). The analysis presents hazards, exposure, and vulnerability as a basis for designing specific and integrated interventions for the informal sector.

Therefore, risk reduction in informal settlements requires a dynamic contextualization in the specific intersections between the social and physical geographies of each city. However, localized decision-making must also be grounded on global political and economic forces, which directly influence human development. Furthermore, this may only be achieved with a shift in conceptualizing disasters as *systems rather than events*. This approach brings both human exposure and vulnerability to the center of the debate on risk. In doing so, this allows for innovative and strategic approaches to be included into policies and practice. Integrated decision-support systems could then incorporate economic, social, health, and environmental considerations into the informal sector. Last, this approach would effectively mainstream their needs into national and local development programs, which should cascade both horizontally and vertically.

---

1 The term "slum" – widely debated since the publication of the global assessment of slums in 2003 – is generally described as a heterogeneous type of settlements with a certain degree of deprivation (UN-Habitat, 2003; 2006/2007). Commonly, slums are considered to lack one or more of these five basic housing conditions: (1) durable housing of a permanent nature, (2) sufficient living space, (3) safe access to water, (4) access to adequate sanitation, (5) security of tenure. While we recognize the negative charge and/or connotation associated with the term "slum," we also realize the futility of attempting to avoid the term in this chapter. The reality is that the term is still widely used and generally conveys deteriorated, overcrowded housing with lack of services, increased diseases, and poverty. Further differentiation between "slums," "informal settlements," and "inadequate housing" is offered in Annex 11.2 Table 1 in an attempt to distinguish the terminology most relevant to this chapter.

Table 11.1 *Climate change impacts on urban areas.*

Change in climate	Possible impact on urban areas	Possible impacts areas on informal settlements based on 5 deprivations and other characteristics of poor housing
Changes in means		
Temperature	Increased energy demands for heating/cooling Worsening of air quality Exaggerated by urban heat islands	Disease, morbidity, and mortality Health and life expectancy Access to water and sanitation
Precipitation	Increased risk of flooding Increased risk of landslides Distress migration from rural areas Interruption of food supply networks	Life expectancy Access to water and sanitation Access to adequate living space Durability of dwellings Security of tenure
Sea-level rise	Coastal flooding Reduced income from agriculture and tourism Salinization of water sources	Durability of dwellings Access to water and sanitation Security of tenure Labor and employment
Changes in extremes		
Extreme rainfall	More intense flooding Higher risk of landslides Disruption to livelihoods and city economies Damage to homes and businesses	Durability of dwellings Disease, morbidity, and mortality Life expectancy Access to adequate living space Security of tenure Labor and employment
Drought	Water shortages Higher food prices Disruption of hydro-electricity Distress migration from rural areas	Access to water and sanitation Security of tenure Disease, morbidity, and mortality
Heat- or cold-waves	Short-term increase in energy demands for heating/cooling	Life expectancy Disease, morbidity, and mortality
Abrupt climate change	Possible significant impacts from rapid and extreme sea-level rise Possible significant impacts from rapid and extreme temperature change	Access to water and sanitation Durability of dwellings Labor and employment
Changes in exposure		
Population movements	Movements from stressed rural habitats	Security of tenure Access to adequate living space
Biological changes	Extended vector habitats	Access to water and sanitation Disease, morbidity, and mortality

Critical interventions toward risk reduction for the informal sector lie in adaptive governance that would enable access to social services, incomes, and material resources to immediately enhance coping capacities. Over the long term, it is necessary to design and implement institutional changes at both macro-economic and city levels to reduce historical socioeconomic inequities (Bündnis Entwicklung Hilft, 2014). Although climate change affects both the formal and informal housing sector, it is only by re-conceptualizing risk to specifically include those marginalized by informality that transformative change toward urban resilience can happen. This chapter highlights marginality and informality as both drivers of vulnerability and transformation toward building urban resilience.

11.3 Informality, Informal Settlements, and Vulnerabilities

11.3.1 Understanding informality in the context of development

Cities worldwide have been successful in attracting more than 80% of global economic activities, partly owing to the high concentration of people and goods (World Bank, 2013b). Rapid urbanization often has positive outcomes on human development. In fact, highly urbanized countries also tend to have high incomes, more stable economies, and stronger and better performing institutions (UN-Habitat, 2006/2007). However, the role of cities, especially their productivity and functionality as engines of economic growth, by and large depends on the quality of their spatial structure, level of basic infrastructure, services, and, not least, on how they are managed (World Bank, 2013b). Thus, cities with poor access to basic services such as potable water, sanitation, power, accessibility, and stormwater control are often congested and have poor housing and insecure property rights or tenure. Such cities are unlikely to attract much investment to invigorate economic growth and improve the well-being of inhabitants. Generally, the capacity of poorly serviced and managed cities to maximize the benefits of urbanization is low because such cities tend to have narrow windows of opportunities to harness their potential as engines of economic growth (World Bank, 2013a).

Most cities in the low-income countries (Africa, Asia, and Latin America) are far from being managed efficiently and effectively. In part, this is manifested in the dominance of informal/unplanned settlements, which have persisted in existing urban plans, laws, and regulations. Currently, informal settlements accommodate between 20% and 80% of the urban population in developing cities. Of the world's 1 billion population living in informal settlements, 61% are in Asia (ADB, 2013) and between 30% and 85% of sub-Saharan Africa's urban population are living in informal housing. In Dar es Salaam, about 80% of the built housing area comprises informal settlements (Kombe and Kreibich, 2006).

The proliferation of informal settlements poses growing challenges to the stability of the social and political fabric of cities because informal settlements lack basic infrastructure services and facilities and have a high concentration of poverty. For instance, Lagos, one of the fastest growing cities in Africa, has 70% of the population living in informal settlements without direct access to piped water or a managed sewerage system. Nairobi, the economic powerhouse of East Africa, has one of the largest informal areas in the world, Kibera, where more than 400, 000 people live without basic services (UN-Habitat, 2003). In most cities of developing countries, one also notes widespread informality in terms of income and employment generation, economic activities that are unregulated or do not follow predetermined urban development rules, laws, procedures, and other regulations that are sanctioned by the state or formal institutions (Dodman et al., 2013).

Literature shows that contemporary economic processes of globalization and neo-liberal urban policy responses and their effects on urban labor markets are key drivers of the informal sector income- and employment-generation activities (Watson, 2009). Recent economic trends have led to exploding informality in low-income cities (Watson, 2009).

The informal economy accounts for 60% of the active labor force and provides 90% of new jobs in Africa. Meanwhile, in Asia, the informal sector accounts for about 25–75% of economic activity (Jamil, 2013). The bulk of the inhabitants of

Table 11.2 *Summary matrix of climate-related threats and their areas of impact*

Threat	Directly Impacted Areas
Increased storm activity in coastal areas	Low-lying settlements Settlements on steep slopes Settlements in ravines
Sea level rise	Low-lying settlements
Increased temperature	All urban residents
Increased rainfall	Low-lying settlements and settlements in or near flood plains
Increased air pollution	All urban residents, though possibly more severe for those near polluting industry and transportation

informal settlements draw their livelihoods from the informal income and employment sectors. Despite varying perceptions and political positions, this sector is expanding rapidly. Some view the informal economy as a symbol of developmental backwardness, whereas others see it as a positive and dynamic sector that enables many to gain access to sources of living in urban areas (Dewar, 2005). Most importantly, in most cities of developing countries, there is an increasing trend toward informalization (Dodman et al., 2013). Informality is also the main force driving and facilitating the poor to secure a living space and thus enhancing their integration and inclusiveness, albeit in precarious forms of employment (Kombe and Kreibich, 2006).

Although informality in terms of categories of income, employment generation, mode of settlement development, or land servicing are unregulated by the state or informal regulatory structures, it is acknowledged that informal social institutions and networks play a critical role in organization and support, particularly in informal settlements (Simone, 2004). They help alleviate vulnerability and other undesirable challenges associated with informality. For instance, informal social networks, norms, and structures play an important role in holding economies and communities together despite daunting economic, political, and environmental challenges (Meagher, 2007). In some countries, especially in Sub-Saharan Africa, social institutions and regulatory networks have been regulating spatial structure in informal settlements and facilitating land markets; they also provide security of tenure. The noble roles being played by social regulators have made a difference in many cases, making informal settlements not only attractive to the poor, but also to the affluent (Kombe and Kreibich, 2006). In other words, informality is gradually becoming a domain for both poor and middle-income households in many developing countries (Roy and AlSayyad, 2005).

Residents in low-income informal settlements are disproportionately more vulnerable due to:
- *Greater exposure to hazards due to geophysical location*: Many settlements consolidate on unstable slopes, low-elevation coastal zones (LECZ), and in proximity to rivers, making them susceptible to harmful events

- *Land availability*: The lack of access to affordable land for housing influences low-income households to settle outside formal housing markets, typically on unregulated land controlled by informal land dealers
- *Underinvested infrastructure*: Resulting in poor sanitation, lack of drainage system, and limited waste management and water supply services
- *Housing quality*: Poor construction materials used in settlements where there is a lack of economic and financial capital to invest in better structure render settlers more

susceptible to hazards (UNESCAP and UNISDR, 2012; UNESCAP, 2013)
- *Limited relief support* in the event of a disaster (Feiden, 2011)

11.3.2 Informal Settlements

Geographic conditions as well as diverse social groups influence the morphology of informal settlements. Box 11.1 illustrates different types of informal settlements across Africa, Asia, and Latin America and demonstrates that although these types of

---

Box 11.1   Typologies of Informal Settlements

Informal settlements are diverse and take different forms; thus, addressing risk reduction, adaptation, and mitigation may be complex. While these settlements worldwide share some characteristics and challenges, they cannot be regarded as homogeneous because every one of them has emerged under particular geographic, ethnic, economic, and historical situations that have determine its composition, typology, construction materials used, and residents. Thus, the vulnerabilities faced by informal settlements in the periphery of Mexico City or Samar will be completely different from those faced in Abidan or Dar es Salaam (Box 11.1 Figures 1, 2, 3, and 4).

Box 11.1 Figure 1 *Informal settlements on the outskirts of Mexico City.*

Box 11.1 Figure 3 *Informal settlements in Dar es Salaam.*

Box 11.1 Figure 2 *Middle-income housing in Informal settlement, Ibadan, Nigeria.*

Box 11.1 Figure 4 *Informal settlements in Samar, Manila.*

settlements share some common characteristics, they cannot be considered homogeneous.

Increasingly, informal housing and land markets have become a symbol that ascertains citizens' demands for or realization of their rights to the city. Informality is, therefore, the process and force generating and sustaining informal settlements. In most low-income countries, informality (i.e., forms of settlements and housing developments and modes of income generation) has become a *modus operandi* of urban expansion (Kombe and Kreibich, 2006; Castillo, 2000; Roy, 2009; Watson, 2009). It is the major force producing and transforming the socioeconomic life and spatial character of cities in most of the developing countries. On the other hand, it is also the major urban development challenge that cities are confronted with in the 21st century. Primarily, this is because informality generates settlements that are spatially unstructured and socially and economically marginalized, particularly in terms of access to basic services and facilities. Owing to the geographical location of informal settlements on fragile ecosystems, the poor socioeconomic and environmental conditions of the inhabitants, climate change–induced hazards will have a devastating impact (Vedeld et al., 2015). Climate change will accentuate the already stressful living and working conditions in informal settlements that are predominantly occupied by the urban poor in fragile ecosystems.

Although the formal/informal binary debates remain inconclusive, urban settlers often weave between the two modes and strategies of urban life as they deem appropriate. The report "State of African Cities" acknowledges that "inadequate urban governance policies" and "low urban institutional capacities," as well as limited options for the poor to access urban land, contribute to informality and proliferation of informal settlements (UN-Habitat, 2010). In this respect, informality and informal settlements seems closely interlinked and mutually reinforcing. Hernando De Soto (2001) perceives informality through the prism of capital, noting that property occupants of informal settlements in many cities of the developing countries and ex-communist states are unable to capitalize their property and therefore opt for extra-legal business transaction modes. In other words, because the legal property rights system does not provide for poor people's needs and expectations, they have little option but to opt out of the formal system (De Soto, 2001).

Therefore, the central issue regarding informality – especially informal settlements – is how to enable the informal land development sector to perform better and thereby improve the quality of living and working environments created by informal sector actors. First, this implies accepting informality as a mode of urbanization (Duminy, 2011); second, it means supporting and working with those parceling and constructing buildings in informal settlements rather than working against them.

11.3.3 Source and Drivers of Vulnerability of Informal Settlements

At the global level, the main sources of vulnerability are population growth, rapid urbanization, and poverty (Adger, 2006). Urban growth has an impact on housing quality and safety, especially when government institutions and the private sector are unable to cope with the need for adequate and affordable urban housing. As presented in Section 11.2, rapid urbanization, especially in low- and middle-income economies, is often associated with growing inequalities, marginalization, and exclusion of the urban poor from formal service delivery systems, all factors that contribute to climate change–related vulnerabilities (Satterthwaite et al., 2007; McGranahan et al., 2007).

Climate change is set to worsen (directly and/or indirectly) the already existing vulnerability of informal urban housing – the most disaster-prone of all urban areas – where livelihood opportunities are generated by surrounding economic activity (Porio, 2014; Jamil, 2013). Rapid urbanization (McGranahan et al., 2007) and poor spatial planning has increased the concentration of people in LECZs (i.e., less than 10 m above sea level) at risk of sea level rise and extreme weather events. Accommodating human activity in naturally unstable environments such as LECZs requires driving resources from beyond urban, national, and even continental boundaries (Seto et al., 2010) and introducing measures that secure natural coastal protection (e.g., mangrove deforestation, excavations, diversions, soil sealing). This, in turn, increases the risks for people and economic activities that reside in LECZ. Informal settlements in LECZ areas of Asia and Africa are increasingly vulnerable to disaster risks associated with climate change (e.g., incidences of flooding) (UN-Habitat, 2008). This happens as a combination of factors: (1) the decrease of natural costal protection from factors explained earlier; (2) the fact that informal subdivisions, whether on low-lying areas or on hills and steep slopes, depend on the decisions of informal land dealers and follow the logics of economic profit, which often prevents any intervention with a perspective to reduce risk or promote adaptation (Castillo, 2013); and (3) the morphology and road structure is not compliant with formal urban planning, a factor that may increase risks deriving from excess rainfall on steep slopes or emergency services access for evacuation.

Key drivers of vulnerability in informal settlements are also connected to the partial or total lack of access to resources (natural, financial, social, political, economic, and human) that allow people to prevent potential risks and rebuild after disasters (Moser and Satterthwaite, 2008; Lopez-Marrero and Yarnal, 2010; Romero Lankao et al., 2014; Islam et al., 2014). In addition, the partial or total lack of basic services such as provision of potable water and electricity, solid waste collection, and safe disposal of wastewater increase the sensitivity of informal dwellers to external stressors. For

Case Study 11.1 Water-Related Vulnerabilities to Climate Change and Poor Housing Conditions in Lima

Liliana Miranda Sara

Cities for Life Forum, Lima

Keywords	Rainfall, flooding, vulnerability, water, community based adaptation
Population (Metropolitan Region)	9,752,000 (Instituto Nacional de Estadística e Informática [INEI], 2015)
Area (Metropolitan Region)	2,819.3 km² (Metropolitan Planning Institute [IMP], 2013)
Income per capita	US$5,950 (World Bank, 2017)
Climate zone	BWh – Arid, desert, hot (Peel et al., 2007)

Between 1940 and 2007, the urban population in Peru increased 9.5 times, whereas the rural one grew only 1.6 times. Peru is one of the ten countries most vulnerable to climate change in the world and one of the most affected by the El Niño System Oscillation (ENSO).

CLIMATE-RELATED HAZARDS AND DRIVERS OF VULNERABILITY

At the national level, experts agree that the country as it currently exists will experience a series of climatic events with extreme peaks forming various regional scenarios ranging from prolonged drought to heavy rains (MML, 2014). According to the Intergovernmental Panel on Climate Change (IPCC), temperatures will rise by at least 2°C and sea level by 1 centimeter (0.4 inches) per year.

The main effects of El Niño are caused by increases in seawater temperature. This raises evaporation levels and causes extreme rainfall, resulting in overflowing rivers and large landslides that seriously affect roads and neighborhoods located on the hills, which are usually inhabited by the urban poor. This negatively impacts the national gross domestic product (GDP) (the Economy Ministry has indicated that, due to climate change, the GDP could decrease by around 6% in future years).

PHYSICAL AND SOCIOECONOMIC VULNERABILITIES

Lima is located on three coastal valleys formed by the mouths of rivers descending from the Andean highlands. The city has more than 9 million inhabitants and is growing by more than 80,000 per year (Miranda and Baud, 2014). The lack of rain in the city (about 9 mm per year) requires SEDAPAL – the water-supply company – to transfer water from the upper basin of the Mantaro River on the other side of the Andean mountain. However, this region faces growing water stress due to melting glaciers and overexploitation of groundwater.

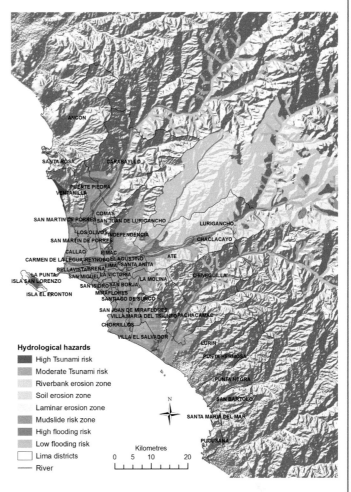

Case Study 11.1 Figure 1 *Map of hydrological hazards in Lima.*

Source: Miranda et al., 2014b, chance2sustain.

The social distribution of water in Lima is uneven. Inhabitants of the richest areas consume around 460 liters per person per day, while those in poor areas consume less than 50 liters per person per day (SEDAPAL, n.d.; 2007; Instituto Nacional de Estadística e Informática [INEI], 2007).[2] Those who have no connection to the water network pay 10 times more than those who have a connection, and their consumption is below 25 liters per person per day (Miranda, 2014a).

Usually, these people live in precarious constructions settled on slopes of 20 degrees or greater and at high risk of landslides; on less compact sandy soils; or on the banks and beds of seasonal rivers that overflow during the monsoon. The "Barrio Mío" Program (*My Neighborhood*) of the Municipality of Lima estimates that there are about 600,000 homes built in areas at high risk of disaster during heavy rains (Escudero, 2014).

---

2 Analyzed by Miranda Liliana and Karin Pffeffer under chance2sustain.

In 2025, Lima will have more than 12 million inhabitants. The urbanization will be by densification rather than by expansion, including the peripheries settled by the most vulnerable. In central districts, modern low-density buildings will be built, and infrastructures and services will be improved. However, in the suburbs, buildings of high density (3.8 persons per room) will continue to grow through self-built construction, thus consolidating urbanization in high-risk disaster areas (Miranda, 2014b).

According to these trends, the peripheral areas surrounding the center of Lima are "slumifying," becoming overcrowding, vulnerable, polluted, or insecure, while the center has the best residential appeal, equipment, infrastructure, environmental quality, urban green spaces, security, and relatively low population density (albeit higher constructive density). So far, nothing indicates that these trends will change.

EXISTING POLICIES AND APPROACHES ADOPTED TO RESPOND TO IDENTIFIED HAZARDS AND ADAPTATION STRATEGIES

A former president of SEDAPAL said that if each person would save 10 liters of water daily, especially those who consume more, water stress would decrease to the point where it would not be necessary to build new dams in the short term to provide water for everyone in the city. However, this has not occurred, and the water company continues to make large investments to transfer water from the other side of the Andes and to build desalination plants.

During 2011–2014, the Municipality of Lima developed a strategy to adapt the city to climate change based on a reduction of water demand in rich districts and vulnerability in the poorest sectors of the periphery. Thus, while the Program Barrio Mío developed a poverty map of the city, SEDAPAL gave priority attention to projects in underserved areas.

In 2014, the Barrio Mío program articulated usually fragmented Metropolitan Municipality of Lima (MML) investments that prioritized neighborhoods from twenty district municipalities and invested around US$115 million to build pedestrian walkways, stairs (60 km), and retaining walls (50 km) in peripheral neighborhoods. With the participation of leaders and neighbors, Urban Integrated Plans have been developed to decide on new infrastructure and services. Barrio Mío has trained 36,430 people in risk management through civil defense, creating 315 community committees and groups for the purpose. Barrio Mío has restored five public spaces and has planted 42,600 trees to help stabilize hills and improve environmental quality, the landscape, and livability. It has also trained another 11,000 people from 17 district municipalities through the "Adopt a Tree Program" (A1A) (Escudero, 2014). All of its work has been done via community organization participation (from prioritization, design, and civil defense groups, as well as community work) integrating the work of citizens and neighborhood organizations with the work of different entities and companies from Lima Municipality, as well as District Municipalities and INDECI (the Civil Defense Institute of Peru).

Despite being an urban intervention program that successfully developed a strategy of comprehensive urban development and achieved excellent results during its short implementation time, the Barrio Mío program was deactivated in January 2015.

SUMMARY AND KEY MESSAGES

This document summarizes the initial findings of a study on the hydroclimatic vulnerabilities of metropolitan cities in the context of climate change, informal developments, and the risk perception of vulnerable populations.

In the future, Lima will experience more extreme weather events. This will exacerbate gaps in the distribution and supply of water, especially among the poor and vulnerable. Given the complexity of the problem, decentralization and intersectoral integration are essential to delineate adaptation strategies to climate change. A combination of infrastructure projects and small-scale local interventions are needed to better define priorities and decide on investments that will reach the most vulnerable.

instance, the inefficiency of solid waste collection in Dar es Salaam (58% of household waste collected) causes drainage channel blockages in many hazard-prone settlements, therefore increasing the likelihood of flooding. An infrastructure backlog characterizes many informal settlements globally and plays a key role in the ability of exposed communities to overcome risk and disasters (see Section 11.4). At the institutional level, informal settlements are considered more vulnerable due to weak governance (including both capacities and resources availability), which in turn manifests in planning deficits and poor technical knowledge of disaster risk and vulnerability (UN-Habitat, 2010; Rojas, 2014).

11.4 Climate Change Threats and Impact of Extreme Weather Events on Housing

Climate threats impact settlements differently, depending on the nature of the hazard and the geographic and sociodemographic characteristics of the area. The 2014 World Risk Index reported that "much of urban growth takes places in highly exposed coastal and delta regions, particularly in developing and emerging economies" (Bündnis Entwicklung Hilft, 2014) and that, in the 616 cities assessed, river flooding posed a threat to more than 379 million residents, the majority of these living in low-income countries. The housing sector is particularly vulnerable in cities most affected by climate-related hazards and their impacts.

In the United States, the storm surge from Hurricane Sandy covered 16.6% of the land in New York (Koepnick and Weselcouch, 2013) and damaged or destroyed 305,000 housing units and 72,000 buildings in New Jersey (The Furman Center and the Moelis Institute for Affordable Housing Policy, 2013). Estimated repair and response costs for the housing sector was US$4.921 billion in New Jersey, representing the second highest amount after the business sector, and US$9.672 billion in New York and surrounding counties, representing the second highest amount after that put forward for schools, utilities, and individual assistance (U.S. Department of Commerce, 2013). In Italy, a recent report (ANCE/CRESME, 2012) highlights the extent of the socioeconomic factors connected to the vulnerability of

Italian territory to natural hazards, especially hydro-geological risk (82% of the territory and 5.7 million people living in risk-prone areas). From 1944 to 2012, the total cost of damage caused by landslides and floods was estimated at more than €242.5 billion (at 2011 prices), about €3.5 billion per year; 75% of this amount is associated with earthquakes and 25% with hydrological hazards.

Notwithstanding high damage costs incurred by the housing sector, cities in high-income countries generally benefit from more readily available resources, more expertise, and stronger governance systems. In contrast, the impacts in low- and middle-income countries are expected to be the greatest because current levels of investments are far from adequate, leading to high vulnerability to climatic impacts. This is termed the "adaptation deficit" (Burton [2004], in Parry et al., 2009) which is of course also a "development deficit". A 2009 study concluded that removing the housing and infrastructure deficit in low- and middle-income countries would cost around US$6.3 trillion, resulting in the need to invest US$315 billion per year over 20 years. But to adapt this upgraded infrastructure to specifically meet the added risks brought by climate change will cost an additional US$16–63 billion per year (Parry et al., 2009).

There are also huge differences in the way cities are resilient to weather extremes. There is a twenty-fold difference in mortality between the Philippines and Japan when hit by a typhoon of the same intensity. This is the result of differences in the resilience of the building stock and in the effectiveness of government arrangements to develop resilience. Similarly, in Ibadan (Nigeria), a very small increase in the intensity of windstorms caused heavy damage to buildings in the city center (Adelekan, 2010). An event of similar intensity in a high-income city would be unlikely to cause such damage. Therefore, infrastructure and housing that was once resilient under one climatic regime may not be so in another. Hence, strengthening, protecting, and adapting the assets and abilities of households and communities is far more important in low- and middle-income nations rather than in high-income ones (Satterthwaite, 2013).

The compilation of disaster risk and impact information is essential to inform sustainable development and disaster risk reduction. International loss data sources like EMDAT collect global and regional data that is homogenous and comparable across countries, and these data are used by practitioners when validating national data. Instead, national-level data can inform decision-makers about risk trends and support disaster loss accounting, forensics, and risk modeling. Most datasets, however, account for few physical impact indicators (e.g., houses damaged or destroyed) and are very poor at collecting economic loss value (UNISDR, 2015).

These characteristics are also present in disaster bulletins and most academic papers. Table 11.3 presents a list of intensive disasters that occurred in emerging and developing countries from the late 1990s until now. The list is by no means comprehensive and intends to illustrate the fact that damages and losses in the housing sector are substantial and are, to a large extent, reported in terms of aggregated direct costs and physical damage. Few detailed studies attempt to calculate disaggregated costs (direct, indirect) before, during, and after disasters at the household level (Danh, 2014). These can include damage to housing, including total or partial destruction of housing units and in-house components of electricity and water or supply-sanitation systems, household goods and equipment, and products of home-based micro-enterprises. Whereas, in terms of loss, these can include cost of demolition and rubble removal, temporary loss of rental income, cost of temporary housing of homeless people (met by government, international assistance, and/or the private sector), and the cost of associated transport to/from temporary camps and places of work. What is also clear is that information related to damages and losses in informal settlements is less systematically reported due to the illegal and therefore unregistered status of dwellers and their lack of access to insurance (whether formal or informal).

More significantly, assessing the impacts of such events in terms of value of property damaged or destroyed can be misleading in an informal context; extensive devastation in terms of deaths, injuries, and loss of property may have low economic impact because of the low value assigned to the damaged or destroyed housing. Instead, the meaning and importance of losses and damages to the homes, livelihoods, and services of the urban poor and informal settlers is more complex.

Poor households wrestle with many kinds of risks, as well as with the desire to meet unfulfilled needs and wants, and this leads to tradeoffs between immediate expenditures on household maintenance against investments to recover lost resources or anticipate future risks. This makes it difficult to replace savings or productive assets once those have been spent on coping (Pelling, 2011). To illustrate, for breadwinners whose families' daily subsistence depend on their earnings (e.g., those in petty services, hawkers), failure to report to work due to flooding could mean food scarcity and hunger for the entire family or no money to pay for water or fuel. Protracted and destructive rounds of coping can lead to household collapse as more fundamental aspects of short-term (health) and long-term social capital fail to materialize. Repeated losses and damages for the poor and those in the informal sector, then, have tremendous implications to their survival, and, considering the importance of the informal sector for many economies, also to the way cities function (Jamil, 2013).

Disaster impacts also lead to the difficult task of rebuilding and rehousing people where further costs and vulnerability may occur but are often left unaccounted. Poorer households in formal and especially in informal settlements often fall short of receiving necessary aid to start the recovery process or lack the means (e.g., insurance) to reconstruct near their sources of livelihood (Porio, 2004; Masozera et al., 2006). Lack of land

Table 11.3 *Selected examples of disaster impacts in emerging and developing countries with emphasis on the housing sector and the urban poor from 1998 until 2014.*

Year	Event	Location (City, Province or Region)	Direct damages of disasters in developing and emergent countries with a focus on housing in low-income areas or slums (where available damages to properties either in housing units or in equivalent value are displayed)
2014	Heavy downpours and flooding	Abidjan (Côte d'Ivoire)	Four shanty towns, home to some 25,000 people are completely destroyed, while 50,000 other informal settlement residents are to be relocated. (1)
2013	Super Typhoon Haiyan	Tacloban City (Leyte Province); Metro Cebu (Cebu); Surigao (Mindanao); townships in Bohol, Panay, Eastern Samar Provinces (Philippines)	11.3 million people affected, 273,375 houses damaged, 153,098 destroyed, and 1.9 million people displaced. (2)
2012	Hurricane Sandy	New York New Jersey (USA)	The surge from Sandy reached nearly 76,000 buildings affecting more than 300,000 housing units (nearly 9% of the total housing units in the city). (3) Estimated repair and response costs for the housing sector are US$4,921 million in New Jersey and US$9,672 million in New York and surrounding counties. (4) Due to their extremely low income, low-income renters in New York are particularly vulnerable of being unable to find affordable new housing. (5)
2012	Tropical Cyclone Evan	Fiji (Samoa)	Damaged almost 1,000 houses and destroyed almost 700, both of legal dwellers and informal settlers. Assessment of the 41 informal settlements showed that 177 households were affected. At the peak of the emergency, approximately 14,000 people were in more than 242 evacuation centers. (6,7)
2012	Typhoon Bopha	Eastern Mindanao Region (The Philippines)	Over 6.2 million people affected and over 230,000 houses destroyed. (8)
2011	Flood	Ibadan (Nigeria)	Over US$1.92 million in direct damage (30 million Nigerian Naira); those living in flood-prone areas were mostly affected. (9)
2011	Heavy rainfall and flooding	Eastern KwaZulu-Natal province (South Africa)	At least 13,000 homes damaged by floods across the country, including thousands of shacks washed away, according to the National Disaster Management Center. Squatter communities known as "informal settlements" have been most affected. The government has put the flood damage at US$211 million, but this is an early estimate and expected to rise. (10)
2011	Floods	Bangkok metropolitan region (Thailand)	300,000 homes damaged in the greater Bangkok metropolitan region alone, 700,000 total residential units impacted, total economic losses to households estimated at THB84.0 billion (US$2.7 billion). (11)
2011	Floods	Sindh and Balochistan Provinces (Pakistan)	Damage was estimated at US$2.5 billion with the housing and agricultural subsectors again being the most affected. (12)
2011–2008	Flooding during four major events	São Paulo and Rio de Janeiro and other cities in South-East Brazil	Losses of more than US$4 billion in the housing sector (of total US$9 billion in damages including many other sectors). (13)
2010–2009	Floods	Benin, Burkina Faso, Ghana, Niger, Senegal among many others in West Africa	Benin, US$131 million in damages (most affected sectors: housing, infrastructure, education, agriculture). Togo, US$23 million in damages (most affected sectors: housing, infrastructure, education). Niger (29,252 homes destroyed). Senegal US$56 million in damages (most affected sectors housing, infrastructure, transport, education). (14)
2009	Tropical Storm Ketsana	Metropolitan Manila, Bulacan, Laguna, Rizal, Southern Luzon Provinces (Philippines)	More than 2.2 million people were reported directly affected by the typhoon and approximately 736,000 people were displaced. An estimated worth of damages to property and infrastructure reached 2 billion pesos (US$43.5 billion) and left more than a million Filipinos homeless. (15,16)
2009	Typhoon Ondoy and Pepeng	Metropolitan Manila (Philippines)	US$730.3 million in damages to housing. 700,000 people displaced from shelters. The storms severely disrupted livelihoods in informal areas, with about 170 million workdays – equivalent to about 664,000 1-year jobs – lost due to their impacts. (17,18)

Table 11.3 (continued)

Year	Event	Location (City, Province or Region)	Direct damages of disasters in developing and emergent countries with a focus on housing in low-income areas or slums (where available damages to properties either in housing units or in equivalent value are displayed)
2008	Flash flood after Typhoon Frank	Iloilo city (Philippines)	4,139 families were left homeless, the majority of affected are urban poor settlements on riverbanks and foreshore; 22 dead, 174 injured, 1,445 totally damaged structures (statistics does not include affected families from adjacent towns that were also hit by the typhoon). (19)
2008	Tropical Cyclone Nargis	Ayeyarwady and Yangon Divisions (Myanmar)	790,000 houses were damaged or destroyed. (20)
2006	Super typhoon Reming	Mt. Mayon volcano region, Bicol (Philippines)	655 deaths, 2,437 injured, 445 missing, damages (infrastructure and agriculture) estimated 608 billion PhP (around US$14 billion). (21)
2005	Hurricane Katrina	New Orleans (USA)	134,000 housing units – 70% of all occupied units – suffered damage from the hurricane and subsequent flood. More than 400,000 residents displaced. (22,23)
2005	Floods	Mumbai (India)	US$240–250 million estimated total loss to marginalized populations in most affected neighborhoods. (24)
2004	Debris flow	Jimani (Dominican Republic)	3,000 people displaced and at least 870 homes destroyed, mainly in informal settlements. (25)
1999	Debris flow	Vargas State (Venezuela)	Estimated 60,000–80,000 houses destroyed or severely damaged with large impacts on informal settlements. (25)
1999	Hurricane Georges	Santo Domingo and surrounding provinces (Dominican Republic)	60% of homes damaged and 25% destroyed, thus displacing 2,500 people. Direct damage US$1,337 million. (26)
1998	1998 Flood	Dhaka (Bangladesh)	30% of units destroyed (2/3 owned by lower-middle classes and poorest). (27)
1998	Hurricane Mitch	Honduras (Nicaragua) El Salvador (Guatemala)	124,068 houses destroyed or damaged. Total direct damage US$3,078 million. (28)

(1) IRIN, 2014; (2) The Christian Science Monitor, 2013; Loyzaga et al., 2014; (3) The Furman Center and the Moelis Institute for Affordable Housing Policy, 2013; (4) U.S. Department of Commerce, 2013; (5) Koepnick and Weselcouch, 2013; (6) ReliefWeb, 2012; (7) Shelter Cluster, 2014; (8) UNCHOA, 2013; (9) Agbola et al., 2012; (10) Global Post, 2011; (11) Impact Forecasting LLC and Aon Benfield, 2011; (12) UNISDR and UNESCAP, 2012; (13) The World Bank, 2012; (14) Wielinga, D. and Dingel, C., 2011; (15) The International Federation of Red Cross and Red Crescent Societies, 2010; (16) Abon et al., 2011; (17) Baker, J. L., 2012; (18) Government of the Republic of the Philippines; The World Bank, 2009; (19) Shack/Slum Dwellers International, 2008; (20) Oxfam International, unknown; (21) Orense and Ikeda, 2007; (22) Geaghan, 2011; (23) The Data Center Research, 2014; (24) Hallegatte, 2010; (25) Doberstein, 2013; (26) U.S. Agency for International Development, 1998; U.S. Department of Homeland Security, 1998; (27) Alam and Rabbani, 2007; (28) Charvériat, 2000.

tenure is also an impediment to accessing government-driven housing reconstruction. Often, these types of schemes are targeted at people who can prove their title to land or buildings, thus discriminating against informal settlers, particularly tenants (Payne, 2004). For instance, the cash-for-shelter program implemented after the Pisco earthquake in Peru was abandoned after at least three-quarters of the intended beneficiaries could not prove their property ownership. With lack of documentation, it is difficult to prove evidence of a claim to these sites (Ferradas et al., 2011). In these situations, the ability to leverage social networks and political connections becomes crucial for reclaiming land, whereas people with poor social networks risk falling into further vulnerability.

Under market and political pressure, post-disaster housing processes in low- and middle-income countries often follow one-size-fits-all approaches despite geographical and cultural differences across countries. Little attention is given to the visual and physical aspects of buildings and there is poor consideration of people's lifestyles (e.g., selling crafts or farming) and complex social contexts (Esther [2011] in Tran et al., 2013; Schildermann and Lyons, 2011; Boano, 2009; Ruwanpura, 2008; Williams, 2008; Norton and Chantry, 2008). For example, in Andhra Pradesh home owners abandoned their reinforced concrete and "cyclone-resistant houses" because they began cracking during winter and were too hot during summer. As a result, many households rebuilt traditional *kutcha* huts as living space or made adjustments to their concrete house to improve thermal comfort. This scheme, subsidized by local nongovernmental organizations (NGOs) and the state government, also generated significant debt on low-income families (Bosher, 2011).

Case Study 11.2   Vulnerability and Climate-Related Risks on Land, Housing, and Informal Settlements in Dar es Salaam

Nathalie Jean-Baptiste

Helmholtz Centre for Environmental Research – UFZ, Leipzig

Wilbard Kombe

Institute of Human Settlements Studies (IHSS), Ardhi University, Dar es Salaam

Mussa Natty

Kinondoni Municipal Council, Dar es Salaam

Keywords	Heat, droughts, seal level rise, resilient development, vulnerability, urban infrastructure, housing
Population (Metropolitan Region)	5,166,570 (National Bureau of Statistics [NBS], 2015)
Area (Metropolitan Region)	1,393 km² (NBS, 2013)
Income per capita	US$900 (World Bank, 2017)
Climate zone	Aw – Tropical savannah (Peel et al., 2007)

Dar es Salaam, located on the Tanzanian Eastern coastline bordering the Indian Ocean, is one of the ten largest cities on the African continent in terms of population and by far the largest city in Tanzania. The population increased from 270,000 in 1967 and already reached 4.5 million in 2012 (United Republic of Tanzania [URT], 2012). It accounts for around one-third of the urban population in Tanzania and it continues to grow at annual rates of between 4% and 8% due to rural-urban migration and high fertility rates (Hussein, 2013; World Bank, 2002; START, 2011). The metropolitan area, including eight offshore islands, stretches over approximately 1,350 square kilometers. Dar es Salaam is the commercial, industrial, and administrative center of Tanzania; some 40% of the manufacturing industry is located there (Hussein, 2013).

Dar es Salaam lies in the Inter Tropical Convergence Zone (ITCZ) and has two rainy seasons: the long rainy season lasts from mid-March to the end of May (*Masika*), the short one takes place from mid-October until the end of December (*Vuli*). Between May and September, the rains are influenced by the southeast monsoon winds, and the northeast monsoon winds influence the area from October to March. In addition, El Niño Southern Oscillation (ENSO) and tropical cyclones influence the rainfall distribution across the region (START, 2011).

In the city of Dar es Salaam, the mean annual maximum temperature is 30.8°C and the mean annual minimum temperature is 21.3°C (START, 2011). Extreme rainfalls are experienced between March and May and can reach up to 50 millimeters. Recorded floods triggered by strong El Niño influences were first reported in 1983. Since then and until 2014, more than ten severe flood incidences were recorded. The latest floods in April 2014 and May 2015 resulted not only in material and agricultural losses but also caused dozens of human losses.

HAZARDS

Temperature: Mean annual temperatures are expected to further increase in Tanzania. Projections suggest temperatures to rise between 2°C in the northeastern parts and 4°C in the central and western parts until 2075. Increases in mean annual temperatures are expected to result in intensified and frequent droughts in different regions of the country, also impacting its capital city Dar es Salaam. These result in food shortages, food insecurity, water scarcity, and power shortages (URT, 2011).

Precipitation: Rainfall patterns decreased over the past decades. Rainfall patterns within the two rainy seasons also shifted in timing. While *Masika* tends to start earlier than usual, the *Vuli* rains "become almost negligible" in terms of timing (URT, 2011). However, other scenarios project an increase of rainfall during *Vuli* of up to 6% by 2100 (START, 2011). Hence, there is uncertainty regarding climate change impacts in terms of precipitation.

Sea level rise: Sea level rise is a major concern for the Tanzanian coastline and for Dar es Salaam in particular because currently some 8% of the city lies below sea level, threatening at least 143.000 people (START, 2011). Impacts of a sea level rise include "loss of land, accelerated coastal erosion, loss of coastal and marine ecosystems, saline water intrusion in freshwater bodies, inundation of low-lying coastal areas and reduced freshwater flows" (URT, 2011: 118). In terms of monetary losses, an expected sea level rise of 0.5 meters could add up to US$48–82 million in losses for Dar es Salaam only. These figures exclude associated losses and challenges for inhabitants dependent on fishery and tourism in the region.

Drought: Tanzania in general and Dar es Salaam in particular are subject to droughts. In 2006, the country suffered the consequences of a drought event that particularly affected access to clean water and food, causing diseases and malnutrition. Furthermore, periods of drought affect the availability of electricity due to its dependence on water (e.g., hydropower) (START, 2011).

VULNERABILITY

Multidimensional aspects of vulnerability in Dar es Salaam are related to the urban morphology and existing infrastructure of the city. Unplanned and informal settlements play a major role in this regard. These settlements such as the Msasani bonde la Mpunga, Msimbazi Valley, Jangwani, Tandale, Suna, Kunduchi and Bahari Beaches, the Ocean Road beach area, and the Temeke River Kizinga areas, as well as the city center, are often prone to flooding (Pauleit et al., 2015; Jean-Baptiste et al., 2013; John et al., 2014). Vulnerability in terms of housing is high, with more than fifty-five unplanned settlements accommodating 70% of the city's population (URT, 2011; START, 2011). These settlements show limited access to health services, solid waste management, sanitation, and clean water. This situation could worsen with a changing climate in the region. This would imply further cases of waterborne diseases such as cholera, malaria, lymphatic filariasis, and diarrhea.

The different forms of income generation in Dar es Salaam are related to informal modes of economic activity and micro-enterprises. Statistics from 2000 reveal 42% of the population was employed, 43% were self-employed, and 38% remained in poverty (START, 2011).

Drainage, particularly during flood events, is a mayor challenge. The city possesses 1,100 kilometers of ditches as well as 600 kilometers of piped stormwater drainage, all lacking maintenance. Water supply is also of concern in terms of quality, distribution, and availability.

INSTITUTIONAL ADAPTIVE CAPACITY

Today, urban planning in Dar es Salaam is still regulated by the city's master plan of 1979. Thirteen years later (1992), a subdivision plan, prepared by the City Council and the Ministry of Lands and Human Settlements allowed housing units to be constructed in areas previously marked as hazardous in 1979. To counterbalance this development, the National Land Policy of 1995 focused on the need of proper land management. In 2000, the National Human Settlements Development Policy was set up to explicitly restrict ongoing construction in hazardous areas. Despite these policies, constructions continued to rise in flood-risk areas.

Among concrete programs and projects that address urban planning and development as well as risks in Dar es Salaam are:

1. The Strategic Urban Development Plan (SUDP) of 1992, which dealt with urban management in several flood-prone areas. Despite enormous resources used, SUDP was, however, not adopted by the government, primarily due to conceptual weakness and apprehension among bureaucrats. Preparation of a new master plan is in progress.
2. The Community Infrastructural Upgrading Program (CIUP), which aimed not only at upgrading urban infrastructure in informal settlements but also at building community capacities to participate in planning and maintenance of their infrastructure.

CIUP was supported by the World Bank and UNDP and was implemented in two phases from 2004 until 2011.
3. The African Urban Risk Analysis Network (AURAN), set up in 2004, which focused on the reduction of disaster risks in urban areas. AURAN regroups NGOs, community-based organizations, and African universities and has been responsible for several projects, including three case studies in Dar es Salaam on health risks and disaster prevention.

Initiatives regarding formalization of properties are ongoing in Dar es Salaam. These include provisions of land or property licenses as well as rights of occupancy. This will ultimately allow the provision of basic infrastructure services including stormwater drainage, water supply, waste collection, and tenure rights (Kyessi and Kyessi, 2007).

Adaptation to climate change at a national level through the National Adaptation Program of Action (NAPA) remains weak due to financial constraints. In Dar es Salaam, concrete plans include tree planting along beaches, roadsides, and open spaces with a particular focus on managing exposed coastal areas (START, 2011). The Kinondoni Integrated Coastal Area Management Project (KICAMP) formulated a comprehensive plan to protect land, mangroves, and water resources in coastal areas. This has led to banning the excavation of sands in critical areas such as Kunduchi Beach and Bahari Beach (Baker, 2012)

Despite the uncertainty in projected rainfall patterns in Dar es Salaam, the intensity of heavy rainfall should be expected to persist, which increases the risk of flash floods throughout the city. While several projects on adaptation and mitigation focus on expected impacts for agricultural production, forestry, and wildlife conservation, there is a need to address current and future socio-economic consequences for coastal urban dwellers and consider multilevel actions that can render benefits in both short and long terms.

11.5 Options and Processes for Adaptation in the Housing Sector

11.5.1 Addressing Disaster Risk and Adaptation in Informal Settlements

Most climate change adaptation is ex-ante (anticipatory) and top-down, lending itself to large-scale, technological solutions (Tanner and Mitchell, 2008). This approach largely ignores the social determinants of vulnerability (Earnson et al., 2007; Prowse and Scott 2008; Cutter, 2009), resulting in a range of more inductive community-based approaches to adaptation that build on the existing risk-coping strategies[3] of individuals and communities (Reid and Huq, 2007; Dodman and Mitlin, 2011; Robledo et al., 2012; Wamsler and Lawson, 2011). The distinction between coping and adapting is also becoming increasingly

accepted (Wamsler and Brink, 2014; Haque and Dodman, 2014: Pelling, 2011). Typically, "coping" stands for short-term and "survival today" responses that individuals, households, and governments take with the assumption that actions taken during previous events can serve as a guide for similar events (Wisner et al., 2004). A significant structural change like migration or decreasing dependence from a certain livelihood activity (e.g., shifting from farming to livestock rearing, innovative housing designs, and insurance schemes) would be an expression of adaptation.

Coping strategies can be both preventive of risk or help post-disaster recovery. Cases from Mombasa, Kenia, and Esteli, Nicaragua (Moser et al., 2010); Dhaka and Khulna, Bangladesh (Jabeen et al., 2010; Haque and Dodman, 2014); San Salvador (Wamsler and Lawson, 2011); and Bonde la Mpunga and Magomeni Suna in Dar es Salaam (John et al., 2014) reveal a

---

3 "Risk coping strategies" is one way of using the term "coping." Wamsler and Lawson (2011) noted that many nuances of the term exist, such as "local coping capacity," "local coping strategies," and " local adaptive capacity" without being properly defined.

large diversity of both structural and non-structural measures at both household and community levels (see Table 11.4). Non-structural measures involve the use of households or community finances as a preventive or recovery measure. Examples of preventive measures at the household level are the pursuit of property legalization to increase house value. Various forms of mutual help with the expectation of obtaining direct or indirect compensation in times of need can be undertaken with the expectation of prolonged periods of drought or for post-disaster recovery. Recovery measures can be donations, formal and informal credit, remittances from relatives and family living abroad, income diversification, and riskier strategies such as borrowing money from the informal sector. Structural measures relate to preventing loss or damage to housing structures and interiors or preventing loss or damage by temporarily reinforcing neighborhood structures. In some cases, we see the combination of recovery actions, such as the use of locally sourced materials to repair damaged houses, with actions that have mitigation co-benefits (e.g., allow vegetation to grow on rooftops to increase strength and natural cooling during heat waves).

While authors recognize the diversity and importance of such strategies, they are very often insufficient to keep pace with the growing impacts of climate change and their related uncertainty and magnitude. Wamsler and Lawson (2011) recognized the following limitations of using non-structural coping strategies by residents of informal settlements: (1) households may default on their obligations toward relatives and neighbors; (2) different income levels in informal settlements may lead better-off households to opt out of mutual arrangements; (3) generally, informal settlers have little to sell; (4) informal settlers need to compete economically, which often privileges specialization rather than diversification; and (5) because they are forced to make tradeoffs between household maintenance, recovery, and anticipated future risks, they can easily experience continuous vulnerability even after a disaster is over.

Compared with strategies undertaken by individuals and households, community-level strategies are characterized by a high degree of self-organization and the ability to network with other partners, to document what they achieve, and to lobby governments to increase the scale and scope of their actions (Satterthwaite, 2013; Satterthwaite, 2011; Cruz-Mudimu, 2013; Archer, 2012; Rayos and Christopher, 2010).

11.5.2 Saving Schemes

Saving schemes take various forms mostly designed to encourage savings through small but regular deposits or deductions from salaries for various purposes. In the context of adaptation, saving schemes commonly serve as coping strategies that allow households to overcome the impact of an event. There are limits, however, to the amount of savings that organized community

groups can collect, which do not normally extend beyond small loans for housing improvements. To achieve a lasting impact, communities and civil society organizations have started looking toward larger schemes established within a district, a city, and even at a national level.

Community development funds (CDFs) are up-scaled funds of larger programs from the Asian Coalition for Community Action (ACCA) of the Asian Coalition for Housing Rights (ACHR). Also, institutions such as Slum/Shack Dwellers International (SDI)[4] play a role in creating responsive relationships between local government and residents living in informal settlements. For instance, members of the Namibian Shack Dwellers Federation were included in Namibia's larger housing program, which aims to build 185,000 dwellings by 2030. The Federation's loan fund is expected to receive a portion (approximately US$5 million of local government funds to address the needs of those living in informal settlements. The Federation has so far supported 5,591 households and constructed 3,403 houses (Mitlin, 2014).

CDFs can bolster the credibility of smaller savings groups and attract funding from larger international agencies. For instance, the Homeless People Federation Philippines Inc. (HPFPI) obtained funding from the GIZ, the World Bank/Cities Alliance (WB/CA), and the Asian Development Bank/Japan Fund for Poverty Reduction (ADB/JFPR). The group also worked with network partners, such as the Slum/Shack Dwellers International (SDI), the Asian Coalition for Housing Rights (ACHR), and other institutions like the Latin American, Asian, and African Social Housing Service (SELAVIP), Homeless International (HI), and the International Institute for Environment and Development (IIED). One of the community-scale projects supported by the SDI in 2014 focuses on housing improvement and community/settlement upgrades in response to recent flooding events in Lilongwe, Malawi. This project involves repairing houses damaged by recent heavy rain and storm events affecting the southern region of Malawi, including buildings as well as social and technical infrastructure. The project has had an impact on government policy, allowing urban poor to increase their position in annual plot allocations and develop their own resource mobilization skills. Based on community saving schemes, the federation has developed expertise in composting toilets and the use of adobe and compressed earth block for constructions. Challenges include the fact that the Malawi Homeless People's Federation was unable to implement the revolving fund process at the start of the project. In consequence, the labor charge for builders was not repaid.

This type of community finance helps provide affordable long-terms assets to the poor. The ACCA program in particular has boosted hundreds of saving schemes across Asia. While the ACCA works as an aggregator of financial capital at the local

---

4 SDI is a transnational network of community-based organizations operating 115 projects through the Urban Poor Fund International (UPFI, 2014) in thirty-three countries of Africa, Asia, and Latin America. This network is organized in different cities as federations and has been known to mobilize vulnerable groups, develop strategies for community upgrading, set up public amenities, and advocate for alternatives to eviction (Dodman et al., 2010).

Table 11.4 *Summary of preventative and impact minimizing coping strategies.*

Preventive and recovery coping strategies	
Nonstructural measures (socioeconomic)	
HOUSEHOLD LEVEL	**COMMUNITY LEVEL**
Formal and Informal credits (e.g., micro-finance, borrowing from informal sector lenders at higher interest rates)	Informal financial services (e.g., illegally accessing formal insurance mechanisms)
Informal donations and formal monetary compensations	Community emergency funds (through community savings)
Diversifying income (in the form of multiple jobs, subletting spare rooms to other tenants, domestication of animals to secure alternative food sources and income)	Creating/leveraging linkages with government and (mostly) non-governmental institutions
The selling price of acquired assets	Mutual Help Associations
Pursue legalization of property to enhance selling price of dwelling if funds are needed	
Various forms of mutual help (pre-/post-disaster actions in the hope of obtaining direct or indirect compensation, e.g., money or object lending, taking care of neighbors' children)	
Remittances from abroad (risk sharing)	
Structural measures (physical/built environment)	
HOUSEHOLD LEVEL	**COMMUNITY LEVEL**
Modifications within/around the house: Building up stores of food and sealable assets (increase the height of furniture and plinths, arrange higher storage facilities, build retaining walls around homes, improve housing materials, use of wood planks flooring to prevent waterlogging, create perforated bamboo partitions between floor and roof to enhance ventilation, spread wood and ash on slippery mud floors, reinforce foundations, build steps with tires). Use of locally sourced materials to repair damaged houses in a cost-effective way (local varieties of leaves for roofs and woods for walls) or corrugated sheets and allow vegetation to grow on it to increase strength and cooling during heat waves; use mud from rivers to restore mud plinths	*Upgrading of neighborhood's physical environment*: Use sandbags, clean up drainage pipes, dig drainage tunnels, construct elevated bamboo pathways during waterlogging; mobilize senior inhabitants reach out to local authorities regarding needed repairs

level, CDFs instead represent the institutionalization of community-scale savings at a city scale (Archer, 2012).

While these schemes come closest to housing finance, they generally require some subsidy for land or infrastructure development. Although responsibility for repayments is spread across individuals to communities, thus increasing the business risk for lenders, community finance helps build community engagement toward asset maintenance and social cohesion (e.g., by allowing individuals to rely on neighbors rather than costly money lenders in the event of emergencies). This kind of finance also tries to fill the gap of market failures in the supply of housing finance. In fact, the ACCA operates in countries where mortgage finance is unavailable at a cost that exceeds the cost of government borrowing, with the exception of Vietnam. Community finance also differs from micro-finance for its collectivist approach, government investment, and long-term support aspects.

11.5.3 Relocation

Human mobility was identified as one of the effects of climate change in the first IPCC report published in 1990 (Houghton et al., 1990), Subsequently, although a substantial body of work has been developed on various forms of human mobility from the development, humanitarian, and disaster risk perspectives, those coming from the climate change spectrum will not only have to navigate through existing terminologies, but will also need to venture through relatively new territory with regard to relocation. Ferris defines relocation as "the physical movement of people instigated, supervised, and carried by State authorities (whether national and local)" (Ferris, 2014: 8). Relocation practices that are temporary are known as *evacuation*, whereas more permanent efforts to counteract slow unset effects of climate change imply resettling in another area and usually require planning. The idea of planning relocation comes to the fore when the associated implications for living with climate change means that certain places will be inhabitable. This is the case for Small Island Developing States (SIDS); communities of the Cateret Islands, Papua New Guinea; Vunidogoloa and Narikoso Villages in Fiji; and Funafuti Atoll, Tuvalu.

Relocation is a multisectoral issue that needs to be addressed in various parts of local government. Relocation touches not only aspects of urban planning and infrastructure, but also concerns

Box 11.2 Community Finance through Community Development Funds and the Asia Coalition for Community Action Program

The Asian Coalition for Community Action (ACCA) is a 3-year program that has set out to transform development options for Asia's urban poor by supporting a process of community-led change in 150 cities in fifteen Asian countries. The program began in November 2008. By 2014, it had expanded to 165 cities in nineteen countries (Asian Coalition [ACHR], 2014).

The program consists of a series of coordinated, comprehensive activities initiated by a community. The program implements four functions: it provides a small grant to the community for small-scale infrastructure (up to US$15,000 per city); it then provides a loan to the community for larger housing project (up to US$40,000); it helps the community design its own housing and infrastructure by providing architectural and planning assistance; and it helps the community acquire formal land title by negotiating land purchase, a land grant, or a long-term lease from the owners (World Bank, 2013a; Archer, 2012). To access funds, communities are required to organize into a community development fund (CDF) by linking together their savings. The amount of savings each member contributes varies. For instance, one community may decide to invest 40% of its savings in the CDF and keep the rest as contingency funds in case of emergencies. The injection of money in the fund puts the city CDF in evidence and encourages contributions from other city actors and formal financial sector organizations, potentially

including climate funds. Committees comprised of community representatives manage CDFs while local authority representatives act as chairs in an advisory role (Archer, 2012). A regional revolving fund provides 2-year loans for up to US$50,000 to country groups at 4% annual interest, and it is currently experimenting with lending and repayment in local currency, which lessens the burden of fluctuating exchange rates (ACHR, 2014). So far, six cities in Nepal, Cambodia, Sri Lanka, and the Philippines have used eight loans.

Basic services are a key area of activity for small projects with improvements in water, sanitation, drainage, solid waste, and electrical services. The projects also often include embankments and roads and bridges to connect settlements to city grids (Boonyabancha and Mitlin, 2012). The rationale of small projects is to make collectively undertaken physical upgrading visible to local authorities, thus bringing the community together and strengthening its negotiating powers toward authorities. As of November 2014, the total amount of small projects approved in 3 years was 1,424, although those actually implemented amount to 2,139 in 2,021 communities spanning 207 cities.

The total budget approved was US$2,859,100 coming from a variety of sources, including 24% from ACCA, 17% from communities, 53% from government, and 6% from other sources, including donor agencies (ACHR, 2014).

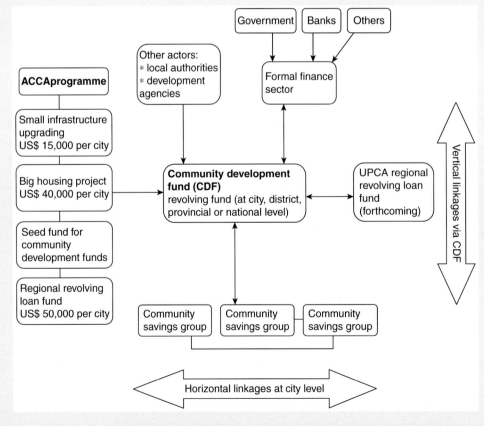

Box 11.2 Figure 1 *The Asian Coalition for Community Action (ACCA) program as a linking mechanism between different funding sources.*

(a)

(b)

Box 11.2 Figures 2a and 2b *Before and after shots of the paved alleyway that links 150 poor households in Block 3, Ward 5 in the city of Ben Tre, Vietnam.*

EXAMPLE OF SMALL PROJECTS IN VIETNAM

Figures 2a and b depict the before and after of the paved alleyway that links 150 poor households in Block 3, Ward 5 in the city of Ben Tre. Like all Vietnamese ACCA cities, the communities in Ben Tre use ACCA small project funds as loans (at 4–6% annual interest) through their CDF, rather than as grants, so the money can revolve and help more communities. And, like most other small projects in Vietnam, this community used the ACCA loan (US$3,369) to leverage a much bigger amount from community members (US$3,190) and from their Ward Office (US$5,199) to replace a muddy and perpetually flooded walkway with a paved road they built themselves (World Bank, 2013a).

ACCA's intent is not to simply address individual households' needs but rather to catalyze change through a people-driven approach to housing interventions. ACCA's large projects focus largely on housing. Funds for big projects serve two purposes: to finance the project and to seed (or add to) a city-based CDF. The project loans are channeled through this fund, and repayments are made back to it. As of November 2014, 51 housing projects were completed, with 41 underway. Twelve projects were unable to begin due to difficulties related to land. Only 19% of the big projects (28 projects) involved the relocation of whole communities, whereas more than 47% (69 projects) had been able to upgrade or reconstruct in-situ. Twenty-two percent of the projects (32 projects) provided loans for housing improvements to households in scattered locations, and 12% (17 projects) created new communities of previously scattered squatters on new land (ACHR, 2014). As of 2014, the big projects budget amounted to 5% contributions from ACCA, 12% from community, an impressive 80% from government, and 3% from other sources.

EXAMPLE OF PROJECTS IN THE PHILIPPINES

In a context where relocating poor communities to remote resettlement sites is still the norm, the housing project being built by the Binina Homeowners Association is an important example of "in-barangay" relocation. These seventy-six squatter families collectively bought a small piece of private land (1,260 square meters) in the same barangay for US$71,820, partly with their savings and partly with loans from the Community Mortgage Programme. The US$40,000 from ACCA is being used to seed the new citywide revolving loan fund, with the first batch of housing loans going to the families at Binina to construct double-unit rowhouses, with one loft unit up and one down (ACHR, 2014).

Community networks in several countries are using ACCA support for disaster relief and rehabilitation. By the end of October 2014, a total of thirty community-driven disaster rehabilitation projects had been approved in eleven countries: Cambodia (one project), Nepal (one project), Myanmar (three projects), Philippines (eight projects), Vietnam (four projects), Lao PDR (one project), Sri Lanka (one project), Thailand (one project), Pakistan (one project), and Japan (one project). This represents an almost equal outcome compared to 2012, where the same number of projects were implemented in ten countries (ACHR, 2014) It can be argued that many of the small and big projects also contribute to reinforce disaster prevention and recovery, since access to reliable basic services, better housing and resettlement policies, good drainage, and paved roads can all contribute to climate adaptation.

An independent review of selected ACCA program locations (World Bank, 2013a: 2) concluded that CDFs "offer a reliable alternative in the provision of housing and infrastructure for low-income communities. This alternative does not aim to replace already existing institutional and financial structures, but through its incremental process, promotes deep transformational and systemic changes." This is underscored by policy revisions being undertaken in several countries as a result of the ACCA program. Examples include new national housing policy based on community-driven or partnership-based upgrading projects in Cambodia, political support in four Indonesian cities for infrastructure upgrading, in-situ upgrading of riverside settlements and free land for housing, and first cases of informal settlement upgrading and long-term land leases in Lao PDR" (World Bank, 2013a).

One of the greatest challenges for maintaining CDFs is the ability to retain flexibility in the management of the schemes (which is one of the main reasons communities choose this option) without accumulating bureaucratic rigidities or recreating local elite powers dynamics (Archer, 2012).

human rights. For instance, existing tribes in Washington State in the United States are working with government officials and universities to develop an adaptation scheme in line with their traditions, preserving indigenous land rights and avoiding forced relocation. Those engaged in discussion with policy-makers on mobility argue that the question of relocation must be placed under the umbrella of a national plan rather than disaster risk and climate change alone. The cost of relocation at a community scale is extremely high (e.g., US$400 million for Kivalina, Alaska). Many communities have to bear these expenses themselves, which is unfeasible for residents of low-income countries. This highlights the need for adequate climate financing, in particular, the need for high-income countries to meet their commitment to mobilize US$100 billion in climate financing by 2020 (OECD, 2015).

It is frequently argued that there are ethical and social limits related to relocation among low-income households and communities (UNHCR, 2014). Several scholars have argued over the loss of traditional and cultural identity through relocation (Cernea, 2009; Ferris, 2012; Modi, 2009). The extent to which those affected by sudden or recurrent events are entitled to draw on resources greatly depends on the capacity of the surrounding system, which in turn may still exacerbate vulnerability. In Fiji, the relocation site chosen by the government was considered unattractive to fishermen. In Magomeni Suna, the relocation site located 30 kilometers away from Dar es Salaam's business district was highly criticized by resettlers, who were faced with mobility issues (e.g., distance to and from their place of work) and livelihood safety issues (e.g., loss of clientele from petty traders), as well as loss of social networks and capital. Experiences with relocation largely depend on whether affected communities are given a choice. *Voluntary relocation* versus *forced relocation* depends on the extent to which communities are driving the process and whether existing frameworks and guidelines are put in place to support them.

Because relocation may occur at various scales, successful practices will depend on actions taken at various levels. On a global scale, there are opportunities for more international engagement, policy, and approaches on relocation:

1. *National Adaptation Program of Action (NAPA)*: If relocation is included in such a framework, developing countries that usually lack the financial resources will have a way to access funds for relocation. Such references have already been made to NAPA from small islands such as Solomon Islands, Kiribati, and Tuvalu (McDowell, 2013).
2. *Adaptation frameworks* (Cancun agreements) emphasize the need for countries to take measures concerning relocation.
3. *Loss and damage* (Decision3 CP/18) that calls for understanding the impact of climate change on patterns of migration and human mobility. One limitation is that the focus is largely on impacts of extreme weather events and less on slow-onset events. Additionally, it is estimated that mechanisms to respond to slower changes could take 15–20 years or longer to deliver adequate finances whereas solutions on the ground require more urgent actions.

4. *Peninsula principles*: A set of international principles put forward by a group of scholars under the coordination of NGO Displacement Solutions to establish the rights of people displaced by climate change. The set of eighteen principles aim to provide a "comprehensive normative framework" based on existing international law and human rights values. The principles address inland migration issues and put forward protection and assistance that are consistent with international guiding principles such as the UN internal displacement chart. The Peninsula principle considers relocation as a last resort and uses a participatory approach to deal with displacement actions at community levels.

11.5.4 Insurance Mechanisms

Weather-related insurance claims have increased fifteen-fold over the past 30 years. According to many climate models, one of the primary and immediate risks of climate change is a likely acceleration of that trend, with more frequent severe weather events (Zurich Financial Services Group, 2009). In events such as flooding and other natural disasters that may cause housing damage or destruction, the insurance industry can act as a bridge between public and private sectors by addressing risk awareness through assessment and mapping, physical recovery through policies and regulations, and financial preparedness. The demand and supply challenges to provide property-related insurance in relation to climate hazards are significant. They include, on the supply side, the issue of insurability (i.e., that climate uncertainty may lead to an inability to quantify hazards accurately and the risk of excessive amounts of claims) and factors that push premiums higher and threaten affordability. On the demand side, populations need to be aware of risks, should be willing to insure, and should be able to afford the premiums (Lamond and Penning-Roswell, 2014).

The latter is particularly difficult in low-income settlements and virtually nonexistent in informal settlements. As explained in the initial paragraphs of Section 11.1, in times of natural events and disasters, the poor and those living in informal settlements largely rely on informal risk strategies and their social capital and trusted networks for assistance and recovery support (Kabisch et al., 2015).

The micro-finance industry may offer some windows of opportunity to lower income settlers, although adjustments to lending practices may be required for the sector to fulfill adaptation goals. Agrawala and Maëlis (2010) found that micro-finance in Bangladesh (the world's largest micro-finance industry) and Nepal worked as a good option for the small and short-term transactions involved in adaptation because there are strong links between the existing activities funded through micro-finance and what is required for adaptation. Property-related micro-insurance, however, is infrequent with only few examples mentioned. For instance, the Integrated Development Foundation (IDF) in Bangladesh provides long-term loans (eight-year loans with flexible repayment times) to disaster-affected clients in order to build safer houses with more resilient materials and specific housing loans to support the construction of weather-resistant houses with locally sourced materials. In general, evidence of affordable insurance schemes for low-income communities that fulfill adaptation goals at the same time is very poor.

The combination of both risk financing and risk reduction for developing countries is important for spreading losses spatially and temporally (Linnerooth-Bayer, 2011) and can influence behavior in terms of reducing moral hazard context (when insurance can lead to risky behavior) or as an incentive (where insurance triggers risk reduction investments or the undertaking of prevention measures). However, evidence of the existence of flood-related schemes showing a link between risk reduction and risk transfer, especially in least developed countries, is very poor and not based on empirical observations (Surminski and Oramas-Dorta, 2014). Recently, the Munich Climate Insurance Initiative (MCII) proposed a closer link between risk reduction and insurance through

incentives, along with considerations for risk reduction activities to be prerequisites for participation in climate risk insurance.

Evidence of the implementation and effectiveness of measures is, however, very limited, with scholarly debate mainly focusing on insurance in developed countries. In 2014, the European insurance industry was the largest in the world (35%), followed by North America (30%) and Asia (28%), whereas Latin America, Oceania and Africa accounted for 6% (3% each) of global insurance premiums (CEA, 2011 in Surminski and Oramas-Dorta, 2014). There are also wide differences in insurance schemes for agriculture and flooding. A review of 123 risk-transfer initiatives, part of the International Labor Organization's Micro-insurance Compendium in middle- and lower-income countries, highlighted that very few nonagriculture-related schemes are currently operational or running as pilot schemes, and those that exist mainly focus on indemnity (what is paid by the insurance after loss). The highest concentration of flood insurance schemes is in Latin America and the Caribbean, whereas Sub-Saharan Africa has no recorded scheme (Surminski and Oramas-Dorta, 2014).

Case Study 11.3   Sheltering from a Gathering Storm: Urban Temperature Resilience in Pakistan

Fawad Khan, Sharmeen Malik, and Atta Rehman

Institute for Social and Environmental Transition (ISET-Pakistan), Islamabad

Keywords	Resilience, urban heat island effect. temperature minimums, cooling, shades
Population (Metropolitan Region)	Rawalpindi-Islamabad: 2,590,000 Multan: 1,950,000 Faisalabad: 3,675,000 (Demographia, 2016)
Area (Metropolitan Region)	Rawalpindi-Islamabad: 427 km² Multan: 207 km² Faisalabad: 181 km² (Demographia, 2016)
Income per capita	US$5,580 (World Bank, 2017)
Climate zone	Rawalpindi: Cfa – Temperate, without dry season, hot summer Multan: BWh – Arid, desert, hot Faisalabad: BWh – Arid, desert, hot (Peel et al., 2007)

Changes in temperature are the most accurately predicted consequence of global climate change. Summer temperatures in Pakistan have increased by an estimated 3°C from 1961 and 2007. The impacts of temperature changes and the return on investment in climate-resilient shelter in Rawalpindi, Faisalabad, and Multan were studied. Temperature increases will make cities unaffordable

for the poor. Shelter design modifications (i.e., passive cooling) to reduce heat impacts are likely to have positive economic returns at the household level; however, these measures need to be supplemented by measures to improve the provision of municipal services.

Pakistan is experiencing a considerable increase in temperatures. Heat wave events are a major cause of weather-related morbidity and mortality in Pakistan. Rawalpindi (at the foot of the Hindu Kush mountain range), Faisalabad (in the central plains), and Multan (in the hot desert) represent a range of conditions in Pakistan and rank among the top five most populous cities in the province of Punjab (see Case Study 11.3 Figure 1) and some of the fastest urbanizing centers in the world. In all of these locations, increases in temperature are a central concern. Summer peak temperatures of 50°C in Multan have been officially recorded, with unofficial sources reporting temperatures that exceed 53°C. Even Rawalpindi, at the base of the mountains, has recorded temperatures in excess of 46°C.

Climate extremes may often cause damages linked to structural and owned assets – however, in the case of heat in Pakistan, heat does not affect the housing structure but causes direct harm to the people who live in these shelters. Shelter design can alleviate or exacerbate the extent of heat stress. Newly migrated (in the past 20 years) low-income households residing in formal and informal settlements in Rawalpindi, Faisalabad, and Multan are often well-established, with brick and reinforced concrete houses. Brick houses in these cities have a high thermal mass and little to no ventilation, which make them vulnerable to current and future heat extremes and the associated health impacts (Vandentorren et al., 2006). The thermal performance of concrete, the most common roof material in Pakistan, makes it an unsuitable construction material for heat-resilient housing. Concrete roofs elevate nighttime temperature within houses by approximately 3°C. Households in Multan are already spending half of their income on measures to reduce heat stress and health care costs arising from heat exposure. Expected temperature rise due

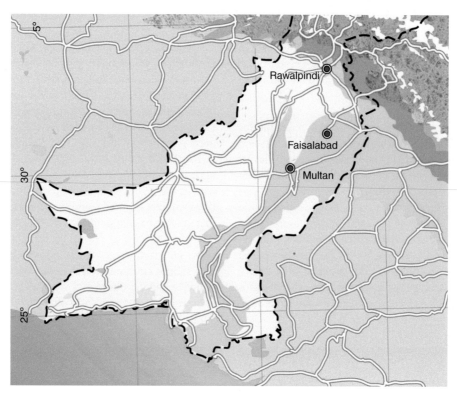

Case Study 11.3 Figure 1 *Vulnerability study sites in Pakistan – Faisalabad, Multan, and Rawalpindi.*

to climate change would make these costs higher than the income of these households in all locations except Rawalpindi, which will exhibit conditions similar or worse than present-day Multan according to current climate change projections.

IMPACT OF HEAT WITH RISING TEMPERATURES

Temperatures within settlements currently reach levels that have a substantial impact on human health and productivity, and these levels are likely to increase substantially over time due to climate change (see Case Study 11.3 Figure 2). Temperatures in the northern cities of Pakistan are projected to increase markedly over the coming decades. Temperature maxima during the hot season are projected to increase by 2–2.5°C, whereas temperature minima are expected to increase by 2.5–3°C. Projections show that heat-related impacts on people and infrastructure will start earlier in the year, be more intense throughout the summer, and last longer in the fall.

While the impacts of rising daily maximum temperatures can be somewhat addressed through various strategies such as providing shade, it is particularly the increase in consecutive nights with very high minimum temperatures and humidity (measured through the heat index) that are of concern (Zahid and Rasul, 2008; Mustafa, 2011). Currently, heat stress increases significantly when nighttime temperatures exceed approximately 28°C, with even more significant physiological impacts as minima approach 37°C for more than three consecutive days. Nighttime temperatures exceeding 37°C are expected to more than double in the coming three to four decades (Cheema et al., 2010). The heat-related economic burden increases disproportionately with every degree of temperature increase (see Case Study 11.3 Figure 2). In addition, the projected loss of overnight respite will affect how well the human body can recover from daytime heat exposure and thus affect health and productivity. With

few resources, poor infrastructure, and inconsistent electrical supply, the poor will struggle to adapt as the need for active cooling increases. These impacts will affect women in particular due to cultural norms that have them more likely to stay inside their homes and due to infrastructural failures that prevent those homes from being cross-ventilated.

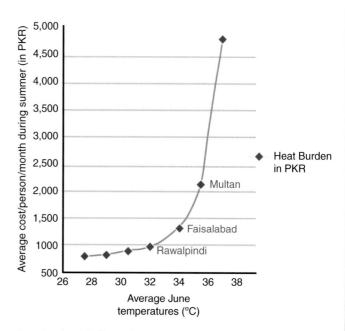

Case Study 11.3 Figure 2 *Heat-related expenditures in different temperature scenarios.*

TEMPERATURE RISK REDUCTION OPTIONS

Shelter design modifications (i.e., passive and active cooling) to reduce the impact of heat are likely to have positive economic returns at the household level under current conditions in cities like Multan and for most cities with projected climate change. Cost-effective techniques for reducing the impact of peak daily temperatures on in-house conditions primarily involve increasing shading and reflectivity and improving insulation. Improvements in ventilation would also improve conditions greatly but would require concurrent improvements in overall sanitation within neighborhoods. Importantly, no interventions currently available can address the impact of extended high ambient temperature minima. Insulation, shading, and similar interventions can ensure that temperatures inside houses remain well below daytime peak temperatures. They cannot, however, reduce temperatures below nighttime minima, particularly when nighttime minima remain high for extended periods.

IMPLEMENTATION

More resources need to go into material and shelter in climate change adaptation design research for temperature resilience. Data and knowledge-sharing can help to predict and plan for these risks. Knowledge generation and dissemination at the local level may be the best place to start. This primarily involves training and awareness-building among local masons and contractors as a means to integrate climate-resilient principles in housing construction.

There are, however, fundamental challenges to creating and implementing change. With lack of financing and insurance mechanisms in the housing sector, which are a reflection of the inability to enforce property rights, the state and national levels of government cannot be expected to play a significant role in either incentivizing or enforcing climate-resilient designs. However, provision of services such as electricity, water, and sanitation could greatly alleviate the burden of heat stress in the medium term. Unplanned settlements, when assimilated into municipal boundaries, lack basic services such as water and sanitation, as well as open areas. These deficiencies already impact the health of millions of people, and the situation will only be exacerbated by the expected climate change.

In Pakistan, the most common cause for families to fall into poverty is repeated health shocks (Heltberg and Lund, 2008). Improvement in health services now may generate the resources needed by poor households to deal with heat stress better in the future. Conversely, relief from temperature stress may help households to better deal with health problems. However, in the longer term, more radical adaptation strategies may be needed to cope with large-scale heat stress.

Source: Adapted from Khan, F., Malik, S., & Rehman, A., 2014. Sheltering from a gathering storm: Temperature resilience in Pakistan. ISET-International.

Some challenges for flood- and physical risk reduction insurances are lack or unreliability of risk data (modeling and exposure information), obsoleteness of networks and data infrastructure, and unreliable government asset databases. Specifically for insurance, implementation is hindered by excessive reliance on government and donor subsidies, lack of local distribution channels, poor financial literacy of communities, and overall limited demand (ClimateWise, 2011; Surminski and Oramas-Dorta, 2014).

11.5.5 Early Warning Systems and Risk Mapping

The purpose of early warning systems (EWS) is to warn the population of potentially fatal phenomenon that might cause potential causalities or structural damage to housing or infrastructure. EWS are recognized as contributing to disaster reduction, but are also an effective adaptation measure because they strengthen the database needed for assessing climate-related vulnerabilities. Designing, implementing, and maintaining EWS can require enormous amounts of resources. EWS demand scientific, managerial, technical, and social components integrated with communication systems. In this respect, they require governments that mainstream disaster risk into policy processes and design and maintain the system. EWS demand local community participation to ensure that risk is adequately communicated.

Sophisticated and more expensive EWS tend to appear in larger cities, driven by the high concentration of people and assets, more complex logistics, and coordination. For example, Makati City in the Philippines set up Command, Control, and Communication Centre (C3) under the Makati City Disaster Reduction and Management Council to act as city liaison between and among national government agencies, NGOs, and other city stakeholders in times of disasters and to issue warnings to communities (UNESCAP and UNISDR, 2012). Similar efforts to decentralize disaster risk management are ongoing in Sri Lanka, where, since the 2004 tsunami, localized disaster risk management centers were established, national risk-level profiles issued, and a vulnerability atlas prepared based on information and participation of locally affected communities. Preparedness has also been strengthened. For example, repeated evacuation exercises are shortening the response time during alert trials and the first mass alert warning system for Sri Lanka, the Disaster and Emergency Warning Network, now issues public alerts through cell broadcasts (Weeresinghe, 2013).

By decentralizing disaster risk reduction resources from a national to local level, municipalities (e.g., Chacao, Venezuela) have found ways to introduce less complex disaster risk management systems, including wireless technology. Other examples include monitoring equipment that feeds into real-time slope stability measures on a city scale (e.g., Ancona, Italy). The presence of a strong local leadership also helps to reinforce decentralization efforts. For instance, the Mayor of San Francisco City in the Philippines, who received several awards for his efforts in building disaster resilience, has been engaging children in programs for risk assessments, community drills, and simulation exercises. These have also helped to increase disaster resilience and reduce social vulnerability by including early school leavers in training exercises, rescue simulations, and emergency response (UNESCP and UNISDR, 2012).

Box 11.3 Principles of Climate Compatible *in Situ* Development in Informal Settlements

The African Center for Cities (ACC) and the Climate and Development Knowledge Network (CDKN) developed a vision for adaptive African cities. This vision comes from an attempt to embed the concept of "climate-compatible development" (CCD; Mitchell and Maxwell, 2010) within the growth and development of African cities. In a nutshell, CCD addresses the overlaps among development, adaptation/resilience, and mitigation strategies. When developing the eight principles, the ACC and CDKN take into account that adaptation overlaps naturally with development because vulnerability and poverty reduction strategies contribute to increasing adaptive capacity. From a development perspective, climate change will surely affect production possibilities and prices. Consequently, investing in high-emission solutions to achieve energy security is likely to commit African cities to higher fuel costs in a world where fossil fuel may be constrained and oil prices uncertain. Therefore, although climate mitigation agendas are not and should not be a priority for low-income countries, it makes sense to look for mitigation benefits when designing development strategies. According to ACC and CDKN, CCD interventions in informal settlements need to be designed to:

1. Achieve tangible and rapid results in improving people's safety and quality of life that incrementally generates a driving force for larger scale, longer term transformative change – working toward a *hierarchy of improvements*. For example, start with local, off-grid, safe, affordable, renewable energy technologies to generate energy for local cooking and lighting, but with a view to getting the network infrastructure in place for local generators to sell excess energy into the citywide grid.

2. *Demonstrably reduce climate vulnerabilities based on careful assessment and tracking, while looking for interventions and innovations that provide both climate adaptation and mitigation benefits (i.e., emissions reductions) where possible.* This might require new expertise and project partners in addition to those who would otherwise be included in a traditional development project that doesn't explicitly factor in current and future changes to the climate.

3. *Include affordability as a key criterion in the design of technologies and service-delivery models, not simply for installation/construction but also for ongoing maintenance and repairs,* to ensure widespread access and financial sustainability. In this respect, smart, innovative design and low-tech options that meet the adaptation needs of the urban poor should be given priority. For example, look at the feasibility of distributed, locally administered savings cooperatives to finance the maintenance of neighborhood bio-gas digesters to reuse waste, reduce methane emissions, and produce a local source of energy rather than relying solely on government support programs.

4. *Match the selection of technologies and servicing models with local skills to deliver, install, and maintain thus strengthening existing livelihood portfolios rather than creating new competing markets.* This might require initial "up-skilling" and training of trainers, seeking to avoid ongoing reliance on outside expertise (linked to the affordability principle just described) and to create local employment opportunities. For example, look at registering and improving the capability of existing informal food vendors to refrigerate and store fresh food under conditions of increasing heat and humidity while improving education about nutrition and health rather than removing informal stalls and forcing people to travel further to large retail chains to access food.

5. Push for *softer forms of regulation that support informal practices of entrepreneurship, social innovation, and private service provision* in slums while protecting consumers and employees by enforcing basic standards and limiting negative impacts on human health and the environment: for example, extending food safety standards to accommodate street food and having health inspectors visit street food vendors in informal settlements to discuss methods for increasing hygiene and to set a date for a return visit to measure improvements before facing a fine. This may become increasingly important under changing climate conditions as heavier rainfall events lead to more contaminants in water and higher temperatures encourage pathogens.

6. Work toward *enhancing security of land tenure*, fostering a sense of stability and a shared future. This can help shift the perception of (previously) informal settlements from being temporary and marginal in need of removal to that of being a legitimate, integral, and valuable part of the city as a whole – as places for investment in and servicing of permanent, higher quality infrastructures that are more robust against a range of climatic conditions including heavy rains, strong winds, hotter temperatures, and the like (e.g., insulated ceilings, paved footpaths, and vegetated parks).

7. *Take a reflexive learning approach* to factor in complexities, contingencies, and uncertainties, allowing for adjustments within the project cycle. Informal settlements are highly dynamic settings and are poorly understood (i.e., minimal plans, maps, census data). Similarly, climate change is a new and emerging field of knowledge, especially on the local scale, so many local climate dynamics and feedback loops are still unclear, especially in under-researched cities. However, we know enough about both to recognize an imperative to act. So we need to act and learn iteratively, with clear goals in mind but with the flexibility to adjust our approach as we progress (i.e., building adaptive capacity), documenting and sharing new knowledge as it is produced.

Source: Taylor, A., and Peter, C. (2014)

The incorporation of climate-related considerations into the disaster profile of cities also requires new capacities and measures. For instance, city leaders in Surat, India, understood that the likelihood of flooding overflow would only increase and, with support from the Asian Cities Climate Change Resilience Network (ACCCRN), are improving their EWS by turning it into an end-to-end system that will improve communication among many stakeholders, thus reducing flood warning announcement time (Brown et al., 2012). Similarly, person-to-person and person-to-city horizontal knowledge transfer can improve disaster risk reduction by spreading knowledge of facts learned from experience or observations and thereby contribute to EWS (i.e., knowledge of storm routes, wind patterns, cloud formations, animal behavior) (ADPC, 2009).

The essential components of EWS, risk mapping, and communication present a series of challenges in informal settlements. Lack of general infrastructure, census data, and cadastre data makes it difficult to identify and map risks. On the one hand, government-led efforts to generate usable data are often viewed suspiciously by settlers as they seek to remain undetected by official authorities. On the other hand, collecting household data is complicated by the expectation that slumlords, tenants, and subtenants have regarding upgrading or relocation, which may result in practices that cloud data

collection (e.g., slumlords registering all houses to themselves rather than their tenants, families overstating their numbers to receive more plots). Building on efforts that communities already undertake is one way to deal with these complex situations. For instance, since 2000, the Homeless People's Federation of the Philippines (HPFP) began many enumeration initiatives that led to more reliable data on structural, tenure, and sanitarian information of flood-prone settlements thereby easing the identification of disaster reduction project beneficiaries (Satterthwaite, 2011).

Communities alone cannot address conditions that create disasters in the first place. It is coordinated efforts by civil society's organizations, organized communities, and government support that can reduce the underlying causes of vulnerability. A case in point is the city of Manizales, Colombia, that since the early 1990s integrated disaster risk into city planning through a collaborative process involving the municipality's agencies, universities, and citizens. The disaster risk reduction program achieved several goals by mixing community-led risk mapping, a risk management index, taxation mechanisms, voluntary insurance premium, school education, and women-led slope monitoring (Warner et al., 2007; Hardoy and Velasquez Barrero, 2014). Undoubtedly, there is a need for institutional environments that recognize and support the contribution of community-level organizations to

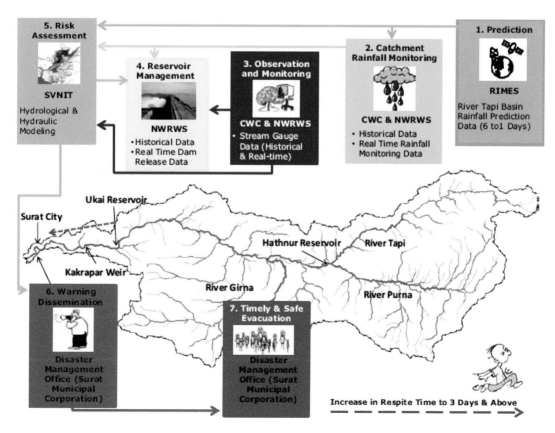

Figure 11.2 *End-to-end early warning system in Surat, India.*

Source: Brown et al., 2012

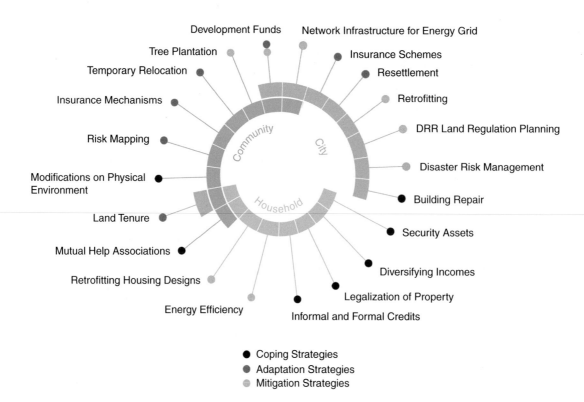

Figure 11.3 *Overlapping coping, adaptation and mitigation at household, community and city-wide scales.*

reduce disaster risk and a financial environment that allows them to address broader issues than only climate change (Pelling, 2011; Satterthwaite, 2011). Fig. 11.3, shows that many adaptation and mitigation actions at household, community and city scale are all associated with each other and should engage with each other to achieve meaningful and enduring resilience.

11.6 Options and Processes for Mitigation of Climate Change

Evidence of links between housing and climate change is commonly described in terms of global trends in housing emissions. These are mainly CO_2 emissions from various combustion sources across cities based on different economies, geographical context (regional climate variation, hot/cold weather), and building types (commercial, residential). The International Energy Agency estimated residential areas were responsible for 18% of direct CO_2 emissions in 2008 (WHO, 2011). The level of energy consumption per unit of space is lower in middle- and low-income households than in high-income ones. However, as cities continue to grow and more households gain access to electricity for cooking, lighting, and cooling, the impact of the household energy sector in developing countries will be relevant to address building-related emissions. Hence, the housing sector has the potential for reducing GHG emissions provided cost-effective and adequate technologies exist.

Mitigation solutions related to this sector can be implemented in a shorter period of time compared to other sectors (e.g., transport, agriculture, forestry, waste) (WHO, 2011). Current global trend equations need to consider issues of housing density, growth of informal settlements, and urban infrastructure services for the poor. While energy consumption in buildings varies widely across developed and developing cities, it is important to consider a broader range of factors beyond economic development that influence housing emission; these include built character of cities (forms of urban sprawls and density), quality of buildings, urban infrastructure, climate variation, policies, and environmental behavior.

Significant pressure on cities – and hence energy usage and related emissions – is occurring by the increased access for billions of people in developing countries to adequate housing, electricity, and improved cooking facilities. Substantial new construction taking place in developing countries represents both a significant risk and an opportunity from a mitigation perspective. By mid-century, energy consumption may triple when compared to levels in 2010, or it may decline if adequate land-use planning, best cost-effective practices, and more efficient technologies are applied and broadly diffused. In any case, it is clear that ambitious and immediate measures are urgent due to the very long life spans of buildings and other urban infrastructure. To avoid locking-in carbon-intensive options for several decades, a shift to electricity and modern fuels needs to be accompanied by energy-saving solutions (technological, architectural), as well as renewable sources, adequate management, and sustainable lifestyles.

Modern knowledge and techniques can be used to improve vernacular designs. Principles of low-energy design often provide comfortable conditions, thereby reducing the pressure to install energy-intensive cooling equipment such as air conditioners. These principles are embedded in vernacular designs throughout the world and have evolved over centuries in the absence of active energy systems (Lucon et al., 2014).

Furthermore, addressing fuel poverty requires ensuring a certain level of consumption of domestic energy services (especially heating and cooking) in an affordable way and preventing impacts such as indoor air pollution. A cleaner fuel used in efficient devices (e.g., improved cooking stoves) in well-built households (avoiding leak heat outside) is a way of providing better services while saving costs. Treated pumped water, lighting, and basic refrigeration are needs that can also be supplied efficiently, bringing about co-benefits such as improved public health.

Effective GHG reduction strategies are to be implemented both in developed and developing countries, taking into account scenarios for the next 20–30 years both in terms of changing climate and socioeconomic development. The major role played by the construction sector within this challenge comes from the big share of life cycle emissions (from extraction of raw materials to building construction and management) attributed to buildings (about 40%). This is coupled with the relative cost-effectiveness of emission reductions through increased energy efficiency of about US\$35/tCO$_2$, compared to US\$10/tCO$_2$ and US\$20/tCO$_2$ in the transport and power sector, respectively (Metz et al., 2007).

Existing building stock conditions and projected building-sector growth represent key factors to assess mitigation measures to be implemented (see Table 11.5). A major distinction among developing and developed countries is represented by the focus given to new construction or building retrofitting in view of potential value of the market and its "green share" (see Figure 11.4). This is intended as an opportunity to achieve significant results in terms of energy efficiency and environmental quality, given that developed countries currently account for the majority of the world's existing building-related energy demand and CO$_2$ emissions.

In developing countries, the housing sector will show a high share of new construction due to the need to provide shelter for more than 500 million people (and provide access to electricity for an estimated 1.5 billion people) by 2050. In this context, the green building paradigm for new construction also entails significant economic benefits compared to a "business as usual" approach and calls for green retrofitting over residential buildings' life spans.

Developed countries, by contrast, should primarily implement measures to meet the ambitious goals on CO$_2$ emissions reduction set up at the international level (e.g., G8 countries' target of −80% by 2050 or European Union (EU) targets of −20% by 2020). Given the limited share of new construction, the priority is to implement measures and incentives enabling large-scale investments in mass retrofitting programs. Indeed, as outlined by the European Climate Foundation, "it is virtually impossible to achieve an 80% GHG reduction across the economy without a 95 to 100% de-carbonized power sector" (European Climate Foundation, 2010, Volume 1), thus requiring a significant transformation in the way energy is produced and distributed (shifting from fossil fuels to renewable energy and carbon capture and storage).

11.6.1 Housing Design

In developed countries, the transition toward a financially sustainable "green retrofitting" paradigm appears to be increasingly feasible thanks to the availability of technologies and relatively low-cost processes (see Table 11.6). Depending on climate zones, different residential retrofitting approaches

Table 11.5 *Comparison of key indicators connected to CO$_2$ emissions variation for different countries, 2050 projections (derived by chapter authors from literature and expert knowledge).*

	Economic growth	Population growth	Urbanization	CO2 emissions
Western Europe	=/+	=/−	=	=/+
Russia	=/+	=/−	=/+	+
Japan	=/+	=/−	=	=/+
USA	=/+	+	=/+	++
India	++	=/+	++	+++
China	++	=	+++	+++
South-East Asia	++	=/+	++	+++

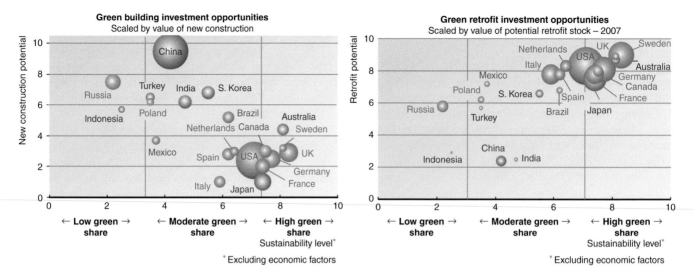

Figure 11.4 *Expected value of potential new construction and retrofitting markets for representative countries compared with the achievable share of "green" construction.*

have been experimented in the past decades, especially in the United States and the European Union, addressing the energy performance upgrade of building envelopes and technical systems. This is aimed at reducing energy demand and maximizing energy production from renewable sources. Although, on the one hand, in colder climates space heating on average represents 60% of residential energy consumption, followed by domestic hot water (DHW) at 18% (UNEP SBCI, 2007), on the other hand, regions in the developed south show the highest consumption in hot seasons. This is mainly due to widespread use of conventional air conditioning systems that also entails a significant increase in the urban heat island (UHI) phenomena and the consequent increase in energy demand and CO_2 emissions.

Housing retrofitting in developed countries is based on the application of either "passive" or "active" technologies. In the case of passive technologies, bioclimatic concepts (such as orientation, thermal insulation and inertia, natural ventilation, and daylight) are exploited to balance the local climatic conditions and reduce thermal exchange with the external environment, thus reaching comfort conditions with a limited need of heat, ventilation, and air conditioning (HVAC) and artificial lighting systems. The "active" approach is based on the efficient and combined use of new technologies for clean energy production (building integrated photovoltaic, micro-wind turbines, advanced glazing, domestic appliances, etc.) and automated building management systems.

More recently, different standards and operational scenarios have been tested with the aim of balancing desired energy performance levels with cost-effective solutions. A number of studies suggest that energy retrofitting in the housing sector should address a target value of 50–60% energy demand reduction to achieve 2050 climate goals, thus requiring a

"whole-house" approach (cf., among others, IEA, 2013; Boermans et al., 2012).

At the same time, national and international energy policy goals require residential mass retrofitting programs to be implemented, addressing the majority of existing homes in developed countries. As an example, the 2020 EU target of 20% energy saving would require an "average of 40% savings in half of the existing housing stock if all sectors and all end-uses within each sector were to contribute equally" (Neme et al., 2011).

To this aim, following the success of the Retrofit for the Future program, the United Kingdom launched ambitious housing retrofitting programs in which the Technology Strategy Board (TSB) funded private owners to achieve 80% carbon savings (TSB, 2013). The target for 2015 includes attic/loft insulation for 10 million houses (about 50% of UK single-family housing stock) and wall-cavity insulation for 7.5 million houses (plus solid-wall insulation for 2.3 million houses by 2022). The next phase is linked to the "scaling up retrofit" program, aimed at the market uptake of housing retrofitting through the funding of consortia to develop commercial design and technical solutions as assessed in relation to the unit cost and the achievable benefit in terms of kWh/m^2 reduction.

In France, mass retrofitting of residential stock started in 2007 with the Grenelle Environnement, aimed at mitigating climate change impacts and managing energy demand. Retrofit interventions concern both historical buildings and social housing complexes. In particular, the effort of social housing agencies that manage about 5 million houses in France aims to renovate 800,000 residential units with energy consumption of above 230 kWh per square meter by 2020 – about 70,000 per year – with an average energy saving of 40% and the aim of reaching a Factor 4 reduction in CO_2 emissions by 2050.

Table 11.6 *Summary of the major opportunities for green housing in different countries.*

	Building retrofitting	New construction
Developed Countries	**Key focus** Housing stock under efficiency levels compared to current regulations (e.g., EU) Housing stock durability increase (e.g., Japan) Appliance efficiency (e.g., United States) Obsolescence of multifamily buildings (e.g., Europe)	**Secondary focus** High potential to meet green standards (such as zero-carbon, zero-waste, and low-impact materials) thanks to advanced building regulations (e.g., United States and Japan).
Developing Countries	**Secondary focus** Informal housing lacking of energy efficiency standards (e.g., Brazil) Obsolescence of multifamily buildings (e.g., China, Brazil, and Russia) Single homes lacking of minimum liveability requirements such as basic electricity, better cooking fuels and durability (e.g., India)	**Key focus** Huge housing shortage and opportunity for green buildings through public and private financed new housing (e.g., India, China, Brazil, Russia, and other emerging economies)

In the United States, residential buildings account for about 60% of all cost-effective energy efficiency potential in 2020 within the building sector, of which two-thirds are associated with the improvement of building envelopes and HVAC system efficiency (McKinsey, 2009). Single states have defined specific targets, such as retrofitting 100% of existing housing stock by 2030 in Maine or the Prescriptive Whole House Retrofit Program in California, in which a performance-based approach has been established to support market penetration of residential retrofitting by defining a progressive logic for the measures to be implemented by private owners (1) air sealing (2) insulation (3) HVAC systems (4) DHW systems, and (5) renewables.

In the southern regions of Europe (e.g., Italy, Spain, or Greece), while the regulatory field has taken steps forward, significant gaps still exist in reaching the minimum acceptable level of sustainability for interventions. Gaps include, among others, the effectiveness of public programs and are mainly due to the low technical and operational levels of local administrations, enterprises, and professionals compared to northern EU regions (Dioudonnat et al., 2014).

In addition to the specific approaches of different countries, a shared vision emerges from mid- to long-term retrofitting strategies in developed countries. In the European Union and the United States, after the initial phase of best-practices, the need for widespread actions on the built-up area required the definition of governance strategies aimed at improving the market uptake of whole-house retrofits through complementary regulations for public and private initiatives and the fostering of technological innovation to promote cost-effective techniques and a performance-based approach to deep-retrofit design.

On the pathway to 2050, it is expected that a major focus will concern a customized "additive" design approach and the role of contractors in the retrofitting processes. In fact, even

though public incentives have proved to give a significant boost to private action, the financial effort required to deliver, within a single-step housing refurbishment, adequate measures to address the 2050 goals still represents an obstacle in achieving mass retrofitting using a "deep" and "whole-house" approach. The role of designers and contractors is to support a gradual improvement of housing energy efficiency through a customized sequence of actions and by raising households' awareness of the expected benefits of each measure to be implemented on both building envelope and technical systems.

Such a systemic approach to housing retrofitting should also include enhancing households' awareness of the energy retrofitting potential in a long-term cost-benefit perspective by highlighting a suitable sequence of retrofitting actions to be implemented over time and the need for integrated actions that allow cost-benefit maximization (e.g., air tightness and insulation; HVAC + DHW generators), as well as the potential co-benefits (e.g., in terms of aesthetics, air quality, comfort, durability).

Achieving mass-scale implementation of deep residential efficiency retrofits will require a multipronged strategy that is focused on both driving demand and ensuring adequate technical and market capacity to deliver quality work. To this aim, as shown in Figure 11.5, the delivery strategy should be based on the active engagement of private-sector product and service providers, financing institutions, government authorities, community organizations, and the market, eventually including performance-based obligations on one or more entities in the market (Neme et al., 2011).

The example given by the main programs promoted in Europe and the United States highlights the adjustment of international directives to local contexts by allowing access to public funding. Moreover, many opportunities arise from the potential transfer of regulatory frameworks, construction process models, and design principles from high-income to emerging economies in view of the definition of a competitive economic framework for urban regeneration and building retrofitting as a key priority for climate change mitigation.

11.6.2 Improving Housing Energy Consumption

Following the principle of climate-compatible development (CCD) (Mitchell and Maxwell, 2010), mitigation measures in the housing sector should achieve GHGs reduction benefits while simultaneously improving the existing standard of living of individuals, households, and communities in urban areas. This should be in line with ensuring minimum standards for household energy services to support a decent standard of living. This is also supported by foundations like the Gold Standard, which encourages sustainable development impact assessments of mitigation measures based on criteria such as environmental, economic, and social benefits in relation to existing baselines, as well as governance and capacity building. A recent study by the foundation (Gold Standard Foundation, 2014) in both urban poor and middle-income communities in New Delhi assessed the feasibility of a series of mitigation measures in the housing sector based on available national government support programs and subsidies and on existing evidence of performance. This assessment was key to rule out options that would be unsuccessful in the specific context of Delhi's poor and middle-income communities. For instance, according to the Energy Environment Partnership of Southern and East Africa, the use of renewable biomass briquettes for cooking was ruled out due to operational challenges such as inconsistent supply of raw feedstock, supply chains for biomass briquette production and sale, and inconsistency in the quality of the finished product. Similar challenges are indeed also found in East African countries where briquette markets are particularly

vibrant (Energy and Environment Partnership/Southern and East Africa, 2012).

Suggested options in Delhi ranged from replacing inefficient, kerosene-based cook stoves used by poorer communities with improved stoves (e.g., that significantly reduce the need for wood fuel or do without it entirely) or with solar cookers; replacing inefficient kerosene lamps or incandescent light bulbs with either solar lamps or CFL/LED bulbs; shifting from liquid propane or electric geysers to solar water heaters; moving from electric fans for space cooling to either solar-powered fans or efficient cooling appliances; using composting instead of open landfilling (as an option, compost can not only be marketed in Delhi's agricultural belt, but it can also reduce waste dumping that exacerbates flooding). Table 11.7 presents some of the preferred mitigation options for Delhi's communities with the addition of adaptation benefits (where they exist).

These climate-friendly housing policies and measures can crucially contribute to closing what the World Health Organization (WHO) calls the "health quality gap." The emergence of low-energy appliances and access to solar electricity can reduce the health impacts of, for instance, fuel-based lighting or conventional cooling measures such as air conditioning systems because those exposed to the noise and heat island effect may be among those who can least afford air conditioning (WHO, 2011).

Mitigation measures in low-income countries and informal settlements should seek local production, operation, and maintenance, thus stimulating local job creation. According to the International Energy Agency (2013), an estimated 952 TWH of electricity will be required annually under a

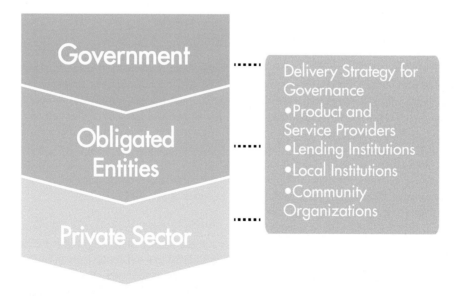

Figure 11.5 *Governance strategy for mass retrofitting performance-based delivery.*

universal access scenario. Measures that are locally administered through, for instance, a social cooperative model, are also more likely to ensure affordability of upfront and maintenance costs of devices distributed for electricity, water, and gas connection. The experience of the iShack Project is instrumental in this respect (see Box 11.4). The brand name is used to develop and implement a sustainable enterprise model for delivering ecological services and utilities (like solar electricity) to people living in shacks on an affordable pay-as-you-go (PAYG) basis in the informal settlement of Enkanini, Cape Town.

Accessing carbon finance for urban mitigation actions is not easy, particularly in peripheral municipalities. It depends on institutional arrangements requiring expertise in climate change science, political awareness of low-carbon technologies, and competency in international climate finance. A favorable regulatory environment, financial credibility, and information awareness at a local level are also crucial (UNDP, 2011; Corfee-Morlot et al., 2012). Moreover, urban mitigation projects represent less than 10% of the "compliance market" (the Clean Development Mechanism [CDM] and Joint Implementation) of all projects and are concentrated in sectors such as water, waste, and energy efficiency and distribution networks (Clapp et al., 2010; World Bank, 2010). This investment gap was also acknowledged in a statement by the Cities Climate Finance Leadership Alliance during New York's Climate Summit in 2014. The Cities Development Initiative for Asia (CDIA) suggested that to deal

Table 11.7 *Sustainable development assessment with adaptation benefits.*

Activity	Mitigation measure	Sustainable development dimension	Criteria and impact (compared to baseline)
Energy (cooking, heating)	Improved cook stove; solar cooker; solar water heater	Environment	Improvement of indoor air quality
		Social	Fuel cost saving; time saving in fuel collection
		Economic and technology development	Skilled/unskilled employment provided to local people for operation and maintenance/social cooperatives; know-how training
		Adaptation	Reliable solar source during blackouts due to disasters
Lighting	Solar lighting	Environment	Improved air quality
		Social	Fuel cost saving, access to renewable energy and lighting, increase safety
		Economic and technology development	Skilled/unskilled employment provided to local people/social cooperatives for operation and maintenance; know-how training;
		Adaptation	Reliable light source during blackouts due to disasters
Solid Waste Management	Composting	Environment	Improved solid waste management practice
		Social	Improved social cohesion
		Economic and technology development	Skilled/unskilled employment provided to local people/social cooperatives for operation and maintenance; income earning opportunities
		Adaptation	Improved drainage clogging
Water & Sanitation	Zero energy water filters; solar water disinfection	Environment	Indoor air quality improvement
		Economic and technology development	Access to clean drinking water
			Skilled/unskilled employment provided to local people/social cooperatives for operation and maintenance
		Adaptation	In connection with rainwater harvesting a reliable drinking water access during disasters

Box 11.4   Solar-Powered Shacks in Informal Settlements: Toward a Financially Viable Social Enterprise Model in Enkanini (Western Cape Town, South Africa)

Enkanini is a large informal settlement 1.5 kilometers northwest of the municipality of Stellenbosch (Western Cape), where about 8,000 people live. The settlement is predominantly made up of young job seekers who left the Eastern Cape to find employment in the city. Enkanini is strategically positioned next to transport linkages and employment opportunities in nearby Stellenbosch. The settlement began in 2006 on municipal-owned land and fast outgrew the limited water supply and bulk toilet facilities set up by the municipality in 2009. By 2012, the initially provided numbers of toilets and tap services were largely insufficient, forcing people to share services and creating congestions. The areas farthest from the center of the settlement do not have access to toilets nor tap points. The municipality maintains that connecting the settlement to the electricity grid is impossible due to the steep slopes and low cost recovery. Also the land is not yet zoned for residential use, so the residents are still some years away from attaining secure tenure and formal, household-denominated services. This means that residents spend a significant amount of their income on heating, cooking, and lighting using paraffin, coal, gas, wood, and candles (Smith, 2012).

In 2011, a group of Stellenbosch post-graduates from the Sustainability Institute (SI) at Lynedoch began a research project whose aim was to co-produce knowledge with a variety of groups in order to seek and implement alternative logics of infrastructure upgrading in the settlements. One of the first results of this process was an 18-month transdisciplinary project that resulted in a concept dwelling called the iShack, where the "i" stands for "improved" shack. Following this exploratory academic process, the iShack has now become a the brand name (iShack Project) to develop

and implement a sustainable enterprise model for delivering ecological services and utilities (like solar electricity) to people living in shacks on an affordable pay-for-use basis. While many articles have already reported on the broader ecological design principles that can be used to improve a shack-type dwelling (Smith, 2012; Earthworks Magazine, 2012; Hope Project, 2012; www.sustainabilityinstitute.net/programmes/ishack), the focus of the iShack Project is now in the process of establishing a viable and sustainable enterprise model to deliver ecologically designed products and services in an informal settlement, starting with the provision of solar electricity. One of the key steps has been to secure a contract for service delivery with the local municipality. The municipality's contribution will be up to 25% of the running costs to help ensure the sustainable delivery of the basic solar electricity service without the municipality having to invest large, centralized electricity infrastructures prior to formalization of tenure. An important part of the business model is to leverage the energy service for other social and economic multipliers for the community, such as skills development and job creation.

Most of the technical installation and maintenance work is done by local residents who have been recruited and trained by the project. Eventually, a core team of iShack Agents will run the service as micro-franchisees and will be incentivized financially to ensure a high level of service maintenance to their clients. By mid-2015, 12 Enkanini residents had been trained not only in operating, maintaining, and repairing the solar systems but also in client services and marketing. Depending on their level of training, they earn between R3500 and R5800 per month (1 Rand = US$0.083). People willing to subscribe contribute a down

Box 11.4 Figures 1 and 2 *Visiting students from neighboring Kayamandi are educated about solar electricity and the iShack project. On the right, solar-fitted homes in Enkanini.*

Photos: iShack Project

payment of R150 if they bring along another neighbor; otherwise, the fee is R320. This payment will grant access to the first month of electricity and TV service. After installation, iShack clients pay a set monthly fee (currently R150) that goes toward the maintenance costs of the service. The pricing is set factoring in the municipal subsidy and is subject to changes based on more accurate information about true running costs – which depend on factors such as the life span of batteries. People who cannot afford the standard monthly fee have been included in a pay-as-you-go (PAYG) pilot within the project to test the viability of more flexible pricing structures. Currently the PAYG pilot has approximately 50 clients while the standard contract counts about 600 users. Around 50–70 new households sign up monthly. The system is also modular and able to integrate more complexity following the income and expenditure patterns of households (e.g., fridge upgrade option). Some of the challenges to this model include keeping the PAYG viable and accessible by creating smarter platforms and dynamic pricing based on daily usage levels across the community, retaining clients in highly transient places like informal settlements, and ensuring high payment compliance. By December 2014, the project had installed 700 solar systems in 700 shacks with roll-out commencing in October 2013 (Personal Communication with Damian Conway, Manager of the iShack Project and director of the Sustainability Institute Innovation Lab June 2015).

Preliminary evaluations on a limited number of users reveal that households with access to solar electricity spend far less money on other unhealthy and unsafe fuel sources such as paraffin and candles. Access to solar electricity is also more reliable than candles and, in some cases, even of conventional electricity (which, in South Africa is currently subject to nationwide load-shedding due to serious capacity constraints). Children can stay up longer to finish homework, and, in some cases, mothers feel safer when their kids are at home watching TV rather than outside wandering alone (Personal Communication with Damian Conway, Manager of the iShack Project and director of the Sustainability Institute Innovation Lab June 2015).

The project is a case in point for the added value that collaboration among universities, municipalities, grassroots organizations, corporations, and citizens can bring to find solutions for complex urban development challenges. The project brings together Stellenbosch University's Tsama Hub, the Sustainability Institute Innovation Lab, Stellenbosch Municipality, Enkanini residents, and Specialized Solar Systems (the corporation supplying the solar systems). All training material is also translated into "how to" videos that can be easily used in other informal settlements in South Africa. The iShack Project is now using its acquired experience to develop a franchise model in order to support other organizations in adaptively replicating the social enterprise model in other contexts around the world.

with this shortage municipalities should request technical assistance and create partnerships with NGOs, the private sector, and other non-state actors to access climate finance in the first place (Beltran, 2012).

Municipalities are learning from Cape Town, where CDMs have attracted financing for a housing energy-efficiency measurement system. City leaders have made use of their findings and momentum to attract bilateral donor money and decide for themselves how to make use of it in relation to green energy and efficiency. Through the intervention of policy-makers from the Environment and Resource Management department, the Mamre Ceiling Retrofit project was framed as an issue of improving the city's housing stock by focusing on providing insulated ceilings to nearly half of the low-quality houses built under the post-apartheid Reconstruction and Development Program (RDP). The project employed locally trained workers and mobilized NGOs such as ICLEI — Local Governments for Sustainability Africa. Findings from the project reveal that although at least 31% of the people involved in the project still struggle to meet the current costs of energy, there has been a significant decrease in the use of unsafe heating sources (such as wood fires), which has also reduced the circulation of hazardous toxins and reducing instances of tuberculosis in households. Although the overall amount of energy used in the household

was not reduced, insulated ceilings allowed for homes to stay warmer and route energy from heating to power other appliances or to be used as savings to support other livelihood activities (Bulkeley et al., 2015). By aligning mitigation measures to existing policy imperatives, the project was able to harness climate change discourse and finance to generate direct economic impact and longer term social considerations, such as the right to dignified housing and household health improvements. Therefore, to be successful, interventions should seek to deploy multiple instruments (not just finance, but capacity development, technical assistance, network building) and multiple stakeholders from different ministries and geographical or sectoral foci.

For scaling up and replicating interventions, the Organization for Economic Cooperation and Development (OECD) also emphasizes the importance of at least three types of information: first, the project itself and its environmental performance; second, the project's financial structure and the strategies used to overcome economic or financial barriers; and third, information about the broader context within which the intervention is situated to ensure that changes required for replication in other areas can be accounted for (Kato et al., 2014). For instance, the initiative to install solar hot water (SHW) in social housing by the Housing and Urban Development Company (HUD.C.) in

São Paulo, shows how the success of a pilot of fifty SHWs in the peripheries of the state expended to 35,000 installations and is gaining traction in the metropolitan centers. This, however, required metabolic adjustments to higher urban densities and land values, emphasizing that low-carbon experimentation is not only about rolling out new technologies, but also about how these technologies "interact over time with pre-existing technological arrangements as well as final users" (Bulkeley et al., 2015).

11.7 Conclusion

Past climatic events have had severe consequences on the human, economic, and environmental capital of cities. Extreme events experienced in the past decade have resulted in both financial and social losses. Since Hurricane Katrina in 2005, the world has seen heavy flooding across Africa and Asia in 2009, 2011, and 2013; severe damage caused by Hurricane Sandy in North America and the Caribbean islands in 2012, as well as the destruction from Super Typhoon Haiyan in the Philippines; and high economic losses in Europe due to damaging flooding events in 2013 and 2014.

Variable weather patterns and booming processes of urbanization will make the whole issue of housing even more relevant in the future. The housing sector will need to address long-term, complex, and nonlinear socioenvironmental processes as cities transform toward becoming more stable, stronger, and better performing urban ecosystems. Yet cities don't perform equally. Low-income states face greater development challenges and less opportunity to harness existing resources for growth. This is evident, as the pressure coming from the housing sector and the proliferation of informal settlements remains a widespread settlement structure. Approximately 1 billion people experience lack of basic infrastructure, insufficient living spaces, low-quality housing, and insecure land tenure.

What options and processes for adaptation and mitigation exist, given that the geography and economy of cities determine the nature and dynamics of disaster risk? While savings schemes and CDFs continue to scale up to larger financial programs among low-income communities, the insurance industry is gaining momentum by addressing risk awareness through assessments, policy recommendations, and physical recovery in higher income cities. EWS and risk mapping are being recognized as effective adaptation measures because they strengthen the information base needed to address vulnerability and also serve as great tools for mainstreaming climate change data and disaster risk across governmental sectors. Such measures can be tailored to reflect the degree of complexity in logistics and coordination of actions in large, medium, or small cities. In this vein, the decentralization of risk management procedures is key.

In addition to adapting to the different effects of climate change, mitigation measures should also be considered, particularly in environments where substantial new construction is taking place (Asia and Africa). As cities continue to grow, so will the housing sector and the impact of building-related emissions. This calls for cost-effective best practices and efficient technologies to avoid locking-in carbon-intensive options for several decades. Buildings have long life spans. Therefore, energy-saving solutions should consider, on the one hand, housing design options relative to the building's envelope and retrofitting and, on the other hand, the improvement of housing energy consumption. The latter may well contribute to closing the "health quality gap" between economies stipulated by the WHO. For example, replacing inefficient kerosene-fired cook stoves offers co-benefits such as better indoor air quality, improved living standards, and reduced use of wood-based fuel.

The dichotomy of communities living in poverty in informal settlements is that their challenges are demanding on many fronts and their options are limited. Adaptation and mitigation technologies (passive or active) must be cost-effective and ultimately integrate measures that can lead the poor toward a more resilient path. If climate change data as well as disaster risk and impact information are to inform sustainable development, these need to take into account self-organization modus operandi and the complex manner in which the informal sector operates. These also need to consider the different scales at which adaptation and mitigation measures are most effective.

Annex 11.1 Stakeholder Engagement

Overall, five international workshops were organized from February until July 2014 in Africa, Asia, and Europe. They aimed at gaining a better understanding on existing limitations that have resulted in increasing impacts in the housing sector and particularly in informal settlements.

The first workshop on climate change adaptation among marginal communities in Asian cities took place in Chiang Mai, Thailand, in February 2014. The workshop focused on identifying the drivers of vulnerability in environmentally degraded environments. The adaptation options identified included housing and livelihood-based support from local governments and NGOs, and international rehabilitation. The second workshop was conducted in Dar es Salaam, Tanzania, in March 2014. The workshop aimed at sharing the specific impacts and innovative responses toward climate adaptation at the intersection of formal and informal settlement processes. The discussions were sensitive to ongoing processes in informal settlements such as relocation, housing finance, and investment, as well as urban planning. The third workshop took place in Bangkok, Thailand, in June 2014 within the frame of the Asian ministerial meetings on disasters. The exchanges focused in particular on building the capacities of local governments as well as stakeholders from informal settlements on participatory and gender-sensitive risk

Annex 11.1 Table 1 *Stakeholder engagement strategy.*

Stakeholder/ organization	Description
Participation and contribution in international conferences	
ICLEI-LOCS – Local Climate Solution for Africa 2013 Congress (Dar es Salaam, November 11, 2012–January 12, 2013; http://locs4africa.iclei.org)	UFZ chaired Session A3 on "Research Driving Innovation: Strengthening the Science-Policy Interface in African Local Governments," providing a platform for discussing how to bridge the knowledge–action and science–policy divides on climate change resilience at the local government level. The session touched on the latest research addressing climate change issues such as "Prediction of Climate Change Impacts in Tanzania Using Mathematical Models," "The Role of Green Infrastructure for Adapting African Cities to Climate Change," and "Social Vulnerability and Building Structures." These initial discussions served as a good starting point to investigate the state of the art of urban planning implications, which are also reflected in informal settlements and buildings, particularly in the Global South
IAPS Conference on "Transitions to Sustainable Societies: Designing Research and Policy for Changing Lifestyles and Communities" (Timisoara, Romania, June 23– 27, 2014)	UFZ is organizing a symposium within the conference in the Frame of IAPS's Housing Network. The symposium, titled "Housing in a Changing Environment: Assessing Vulnerability and Promoting Resilience," focuses on specific impacts and innovative responses toward climate adaptation in housing particularly at the intersections of formal and informal systems. It will put forward the following themes: 1. Asian perspective on housing and informal settlements and the impact of climate change 2. Quality and ownership of housing after flooding in Europe 3. Vulnerability and resilience of urban built structures and lifelines in African cities Results from these exchanges will be synthesized into a Symposium Protocol that will feed into this chapter.
Organization of Thematic Workshops	
Thematic workshop on climate change adaptation in Asian Cities (Asia-Pacific Sociological Conference, Chiang Mai, Thailand, February 14–15, 2014)	The workshop focused on identifying the drivers of vulnerability among vulnerable, marginal communities in Asian cities as well as marginal communities' adaptive capacities, given their environmentally degraded environments. Adaptation included strengthening their housing and livelihood bases, and support from local governments, NGOs, and international rehabilitation agencies.
Thematic workshop on adaptation strategies and housing and informal settlements (Tanzania, March 2014)	The workshop focused on past deficiencies that have resulted in the proliferation of informal settlements: Land-use planning and regularization (2) legal systems, and (3) land markets and tenure. Stakeholders were asked to pinpoint the challenges faced as they attempt to adapt to climate change at local levels with regard to the roles and capacities of local governments.
Thematic workshop on climate change and disasters (Asian Ministerial Meetings on Disasters, Bangkok, Thailand, June 25–27, 2014)	The workshop focused on assessing climate change impacts and building adaptive capacities of both local governments and informal settlements. The workshop included recommendations for investing in community resilience-building programs (e.g., training in participatory risk assessment, gender-sensitive risk assessment and planning, and community-based disaster risk reduction and management.
Thematic workshop on the cross-cutting theme of economics and finance (Germany, June 2014)	The workshop focused on the linkages among informal settings, co-benefits in adaptation and mitigation, and economics and finance. Chapter authors as well as practitioners were invited to discuss how the topics interlinked and what strategies can be implemented to bridge gaps between scientists and stakeholders.
Thematic workshop on public–private partnerships in building urban resilience, organized by Manila Observatory in partnership with Zuellig Foundation and the Romulo Foundation, July 8–9, 2014	This workshop aimed at identifying investments in building urban resilience through public–private partnership. In particular, the session chaired by Antonia Yulo-Loyzaga outlined the challenges in bridging investments across public–private sectors, especially urban communities and civil society organizations.
Interviews with Key Stakeholders	
Stakeholder interviews allow stakeholders to comment on, evaluate, and feed into elaborated drafts.	Key organizations identified are: • Slum Dwellers International (SDI) • Asian Coalition for Housing Rights (ACHR) • Local government authorities • Union of Cities and Local Government (UCLG) • Inter-American bank officials • IIED and SEI scholars • International Housing Coalition • C40 Regional Director for Africa, Hastings Chikoko • Sustainability Institute, Stellenbosch University, Cape Town

assessments and planning. In parallel, in June 2014, a fourth workshop was conducted in Bonn, Germany, following the ICLEI Annual Conference on Resilient Cities. Discussions centered on the linkages among informal settings, co-benefits in adaptation and mitigation, and economics and finance. Finally, the Manila Observatory organized a thematic workshop on public–private partnerships in building urban resilience in July 2014 in the Philippines. Part of the discussions was devoted to the challenges in bridging investments across different sectors, especially the urban communities and civil society organizations. Annex 11.2 Table 1 summarizes the activities conducted to elicit ideas and feedback from a range of stakeholder for this chapter on housing and informal settlements.

Annex 11.2   Glossary of Terms

Annex 11.2 Table 1 *Glossary of terms.*

TERM	DEFINITION	ORGANIZATION	SOURCE
Informal settlements	1. Areas where groups of housing units have been constructed on land that the occupants have no legal claim to, or occupy illegally; 2. unplanned settlements and areas where housing is not in compliance with current planning and building regulations (unauthorized housing).	United Nations Statistics Division Glossary of Environment Statistics	United Nations (2014), *Environment Glossary*, Accessed September 22, 2015: http://unstats.un.org/unsd/environmentgl/gesform.asp?getitem=665
Slums	Areas of older housing that are deteriorating in the sense of their being underserviced, overcrowded, and dilapidated	United Nations Statistics Division Glossary of Environment Statistics	United Nations (2014), *Environment Glossary*, Accessed September 22, 2015: http://unstats.un.org/unsd/environmentgl/gesform.asp?getitem=1046
	Slums are operationally defined by inadequate access to safe water, inadequate access to sanitation and other infrastructure, overcrowding, insecure residential status, and poor structural quality of housing	UN-Habitat (2003)	UN-Habitat (2003), in UN-Habitat (2010), The Human Settlements Financing Tools and Best Practices Series, Informal Settlements and Finance in Dar es Salaam, Tanzania, Nairobi, UN-Habitat
Inadequate housing	Housing is not adequate if: 1. Its occupants do not have a degree of tenure security 2. Its occupants do not have availability of services, materials, facilities, and infrastructure 3. If its cost compromises the occupants' enjoyment of other human rights 4. If it doesn't guarantee physical safety or adequate space 5. If the specific needs of disadvantaged and marginalized groups are not taken into account 6. If its location is cut off from services and infrastructure or is located in dangerous areas 7. If it doesn't take into account the expression of cultural identity	Office of the United Nations High Commissioner for Human Rights/ UN-Habitat	Office of the United Nations High Commissioner for Human Rights/ UN-Habitat, The Right to Adequate Housing, Fact Sheet No. 21/rev. 1. Accessed September 22, 2015: http://www.ohchr.org/Documents/Publications/FS21_rev_1_Housing_en.pdf
Adequate housing	For housing to be adequate it must, at minimum, meet the following criteria: Security of tenure; availability of services, materials, facilities, and infrastructure; affordability; habitability; accessibility; location; and cultural adequacy	Office of the United Nations High Commissioner for Human Rights/ UN-Habitat	Office of the United Nations High Commissioner for Human Rights/ UN-Habitat, The Right to Adequate Housing, Fact Sheet No. 21/rev. 1. Accessed September 22, 2015: http://www.ohchr.org/Documents/Publications/FS21_rev_1_Housing_en.pdf
Squatter settlements	Areas of housing units that have been constructed or erected on land to which the occupants do not have a legal claim. See also *informal settlements*.	United Nations Statistics Division Glossary of Environment Statistics	United Nations (2014), *Environment Glossary*, Accessed September 22, 2015: http://unstats.un.org/unsd/environmentgl/gesform.asp?getitem=1077
Marginal settlements	Housing units that, lacking basic amenities, are not considered fit for human habitation.	Glossary of Environment Statistics, Studies in Methods, Series F, No. 67, United Nations, New York, 1997.	Glossary of Environment Statistics, Studies in Methods, in OECD Glossary of Statistical terms. Accessed September 22, 2015: http://stats.oecd.org/glossary/detail.asp?ID=1592

Chapter 11 Housing and Informal Settlements

References

ADB. (2013). Annual Report 2012. Advancing regional Cooperation and Development in Asia and the Pacific. Accessed March 15, 2014: https://www.adb.org/sites/default/files/institutional-document/33806/adb-annual-report-2012.pdf

Abon, C. C., David, C. P. C., and Pellejera, N. E. B. (2011). Reconstructing the tropical storm Ketsana flood event in Marikina River, Philippines. *Hydrology and Earth System Sciences, 15*, 1283–1289.

Adelekan, O. I. (2010). Vulnerability of urban poor coastal communities to flooding in Lagos, Nigeria. *Environment and Urbanization, Institute for Environment and Development (IIED)* **22**(2), 312–333.

Adger, W. N. (2006). Vulnerability. *Global Environmental Change* **16**(3), 268–281.

ADPC. (2009). *Using Risk Assessments to Reduce Landslide Risk.* (September), 1–8. Accessed January 6, 2015: http://www.adpc.net/Igo/category/ID225/doc/2013-y17Gx3-ADPC-Safer_Cities_26.pdf

Agbola, B. S., Ajayi, A., Taiwo, O. J., and Wahab, B. W. (2012). The August 2011 flood in Ibadan, Nigeria: Anthropogenic causes and consequences. *International Journal of Disaster Risk Science* **3**(4), 207–217.

Alam, M., and Rabbani, M. D. G. (2007). Vulnerabilities and responses to climate change for Dhaka. *Environment & Urbanization* 19 Accessed November 11, 2015: http://eau.sagepub.com/content/19/1/81.

Agrawala, S., and Carraro, M. (2010). *Assessing the Role of Microfinance. Fostering Adaptation to Climate Change.* OECD Environmental Working Paper, 15, 2010. OECD Publishing.

ANCE/CRESME. (2012). *Primo Rapporto ANCE/CRESME.Lo stato del territorioitaliano 2012. Insediamento e rischiosismico e idrogeologico.* CRESME.

Archer, D. (2012). Finance as the key to unlocking community potential: Savings, funds and the ACCA programme. *Environment and Urbanization* **24**(2), 423–440. Accessed July 3, 2014: http://eau.sagepub.com/cgi/doi/10.1177/0956247812449235

Asian Coalition for Housing Rights (ACHR). (2014). ACCA Fifth Year Report. Accessed February 27, 2015: http://www.achr.net/upload/files/1%20ACCA%20How%20it%20works%201-4.pdf

Baker, J. L. ed. (2012). *Climate Change, Disaster Risk and the Urban Poor. Cities Building Resilience for a Changing World.* World Bank.

Beltran, P. T. (2012). International financing options for city climate change interventions: An introductory guide. Manila. Accessed August 1, 2014: http://cdia.asia/wp-content/uploads/2014/09/international-financing-options-for-city-climate-change-interventions.pdf

Boano, C. (2009). Housing anxiety and multiple geographies in post-tsunami Sri Lanka. *In Disasters* **33**(4), 762–785.

Boermans T., Bettgenhäuser K., Offermann M., and Schimschar S. (2012). *Renovation Tracks for Europe up to 2050: Building Renovation in Europe, What Are the Choices? Ecofys.*

Boonyabancha, S., and Mitlin, D. (2012). Urban poverty reduction: Learning by doing in Asia. *Environment & Urbanization* **24**(2), 403–421. doi:10.1177/0956247812455770

Bosher, L. (2011). Household and governmental perceptions of risk: Implications for the appropriateness of housing provision in South India. *Housing Studies* **26**(2), 241–257. doi:10.1080/02673037.2011.548507

Brown, A., Dayal, A., and Rumbaitisdel Rio, C. (2012). From practice to theory: Emergency lessons from Asia for building urban climate change resilience. *Environment and Development* **24**(2), 531–556.

Bulkeley, H., Castan Broto, V., and Luque-Ayala, A. (2015). Solar hot water and housing systems in São Paulo, Brazil. In Bulkeley, H.A., Broto, V.C., and Edwards, G. A. (eds.), *An Urban Politics of Climate Change: Experimentation and the Governing of Socio-Technical Transitions* (157–177). Routledge.

Bulkeley, H., Castan Broto, V., and Silver, J. (2015). Financing climate change experiments across Cape Town's housing infrastructure. In

Bulkeley, H. A., Broto, V. C., and Edwards, G. A. (eds.), *An Urban Politics of Climate Change: Experimentation and the Governing of Socio-Technical Transitions* (119–137). Routledge.

Bündnis Entwicklung Hilft. (2014). *World Risk Report 2014-Focus: The City as Risk Area.* Bündnis Entwicklung Hilft (Alliance Development Works), and United Nations University – Institute for Environment and Human Security (UNU-EHS).

Castillo, J. M. (2000). *The Urbanization of the Informal: Spatial Transformation in the Urban Fringe of Mexico City.* Doctoral thesis, Havard Design School, Ann Arbor, Michigan.

Castillo, M. M. (2013). Urban patterns and disaster risk: The informal city on the hills. In Kim, Y. O., Park, H. T., and Seo, K. W. (eds.), *Proceedings of the 2013 International Space Syntax Symposium* (vol. 31). Seoul, Korea.

CEA. (2014). European Insurance – key facts. Brussels 2014. Accessed March 23, 2015: http://www.insuranceeurope.eu/uploads/Modules/Publications/european-insurance–key-facts-2014.pdf.

Cernea, M. M. (2009). The benefit-sharing principle in resettlement. In R. Modi (ed.), *Beyond Relocation. The Imperative of Sustainable Resettlement* (7–62). Sage.

Charvériat, C. (2000). Natural disasters in Latin America and the Caribbean: An overview of risk. Inter-American Development Bank. Accessed December 23, 2014: http://www.iadb.org/res/publications/pubfiles/pubwp-434.pdf

Christian Science Monitor. (2013). By the numbers: Typhoon Haiyan's human toll, property damage, and aid pledges. Accessed September 24, 2014: http://www.csmonitor.com/World/Global-News/2013/1115/By-the-numbers-Typhoon-Haiyan-s-human-toll-property-damage-and-aid-pledges.

Clapp C., Leseur, A., Sartor, O.,Briner, G. and MCorfee-Morlot, J. (2010). *Cities and Carbon Market Finance: Taking Stock of Cities' Experience with Clean Development Mechanism (CDM) and Joint Implementation (JI).* OECD Environmental Working Paper No. 29. OECD.

Climate Summit. (2014). The Cities Climate Finance Leadership Alliance statement of action. Accessed February 3, 2016: http://www.un.org/climatechange/summit/wp-content/uploads/sites/2/2014/09/CITIES-Cities-Climate-Finance-Leadership-Alliance.pdf

Climate Wise. (2011). ClimateWise Compendium of disaster the developing world. London risk transfer initiatives in London. Accessed June 12, 2014: http://www.climatewise.org.uk/climatewise-compendium.

Corfee-Morlot, J., Marchal, V., and Dahou, K. (2012). *Towards a Green Investment Policy Framework: The Case of Low-Carbon, Climate Resilient Infrastructure.* OECD.

Cutter, S., Emrich, C. T., Webb, J. J., and Morath, D. (2009). *Social Vulnerability to Climate the Literature Social Vulnerability to Climate Literature,* Final Report to Oxfam America. Columbia: Hazards & Vulnerability Research Institute, University of South Carolina.

Danh, V. T. (2014). *Household Economic Losses of Urban Flooding: Case Study of Can Tho City, Vietnam.* Asian Cities Climate Resilience Working Paper Series 12. IIED. Accessed February 12, 2015: http://pubs.iied.org/10715IIED.html

The Data Center Research. (2014). Facts for Features: Katrina Impact. Accessed December 18, 2015: http://www.datacenterresearch.org/data-resources/katrina/facts-for-impact/

De Soto, H. (2001). *The Mystery of Capital: Why Capitalism Triumphs in the West and Fails Everywhere Else.* Black Swan.

Dewar, D. (2005). A conceptual framework of the formulation of an informal trader policy for South African towns and cities.*Urban Forum* **16**(1), 1–16.

Dioudonnat, M., Marti, X., Strangis, G., Sciacca, G., and Séon, A. (2014). *Responding to Challenges Regarding Energy Efficiency and Renewable Energy in Mediterranean Buildings.* Working paper from ELIH MED Final Conference "Enhancing Energy Efficiency in Low Income Housing (LIH) in the Mediterranean Area," November 28, 2014, Bruxelles.

Doberstein, B. (2013). Towards guidelines for post-disaster vulnerability reduction in informal settlements. *Disasters* **37**(1), 28–47. Accessed November 9, 2015: http://www.academia.edu/6461977/Towards_guidelines_for_post-disaster_vulnerability_reduction_in_informal_settlements.

Dodman, D., and Mitlin, D. (2011). Challenges for community-based adaptation: Discovering the potential for transformation. *Journal of International Development* **25**(5), 640–659.

Dodman, D., and Satterthwaite, D. (2008). Institutional capacity, climate change adaptation and the urban poor. *IDS Bulletin* **39**(4), 67–74.

Dodman, D., Mitlin, D., and Co, J. R. (2010). Victims to victors, disasters to opportunities: Community-driven responses to climate change in the Philippines. *International Development Planning Review* **32**(1), 1–26.

Dodman, D., McGranahan, G., and Dalal-Clayton, B. (2014). *Integrating the environment in urban planning and management: Key principles and approaches for cities in the 21st Century.* United Nations Environment Programme.

Duminy, J. (2011). *Literature Survey: Informality and Planning, African Centre for Cities.* University of Cape Town Press.

Earnson, E. (2007). Identifying and addressing social vulnerabilities. In Waugh, W. L, and Tierney, K. L. (eds.), *Emergency Management: Principles and Practice for Local Government* (366). ICMA Press.

Earthworks Magazine (April/May 2012). Accessed February 4, 2015: http://earthworksmagazine.co.za/projects/ishack-more-than-just-shelter/)

Energy and Environment Partnership/Southern and East Africa (EEP & SEA). (2012). Analysing briquettes markets in Tanzania, Kenya. Uganda Report, Gauteng, Republic of South Africa.

European Climate Foundation (ECF). (2010). *Roadmap 2050. A practical guide to a prosperous, low-carbon Europe.* ECF.

Esther, C. (2011). *Home, sustainable home. Making cities work.* Australia: Paul Noonan. In Tran, A. T., Tran, P., and Tuan, T. H. (2012). *Review of Housing Vulnerability Implications for Climate Resilient Houses.* from Sheltering From a Gathering Storm No. 1. Institute for Social and Environmental Transition-International.

Feiden, P. (2011). *Adapting to Climate Change: Cities and the Urban Poor.* International Housing Coalition.

Ferradas, P., Schilderman, T., Vilela, A., and Mariscal, J. (2011). *Participative Reconstruction and Risk Reduction in Ica.* WRC Innovation Competition Entry. The World Bank.

Ferris, E. (2014). *Planned Relocations, Disasters and Climate Change: Consolidating Good Practices and Preparing for the Future* (18). United Nations High Commissioner for Refugees (UNHCR), Brookings, Georgetown University.

Ferris, E. (2012). *Protection and Planned Relocations in the Context of Climate Change. Legal and Protection Policy Research Series. Division of International Protection.* United Nations High Commissioner for Refugees (UNHCR).

The Furman Center and the Moelis Institute for Affordable Housing Policy. (2013). Sandy's Effects on Housing in New York City: Fact Brief.

Geaghan., K. (2011). Forced to move: An analysis of Hurricane Katrina movers. Accessed December 18, 2015: https://www.census.gov/hhes/www/hlthins/publications/HK_Movers-FINAL.pdf.

Global Post. (2011). South Africa: Floods kill 120 and destroy crops. Accessed November 9, 2015: http://www.globalpost.com/dispatch/south-africa/110127/south-africa-floods-natural-disaster.

Gold Standard Foundation. (2014). *Financing Cities of the Future: Tools to Scale-Up Clean Urban Development.* Gold Standard Foundation.

Government of the Republic of the Philippines and The World Bank. (2009). Philippines. Typhoons Ondoy and Pepeng: Post-disaster needs assessment. Accessed July 29, 2015: http://www.pdrf.org/pdf/POPJAVolume1.pdf.

Hallegatte, S., Ranger, N., Bhattacharya, S., Bachu, M., Priya, S., Dhore, K., and Patwardhan, A. (2010). *Flood Risks, Climate Change Impacts and Adaptation Benefits in Mumbai: An Initial Assessment of Socio-Economic Consequences of Present and Climate Change Induced Flood Risks and of Possible Adaptation Options.* OECD Publishing.

Hardoy, J., and Velasquez Barrero, S. L. (2014). Rethinking Bio-Manizales: Addressing climate change adaptation in Manizales Colombia. *Environment & Urbanization* **26**(1), 53–68. doi:10.1177/0956247813518687

Haque, A., Dodman, D., and Hossain, M. (2014) Individual, communal and institutional responses to climate change by low-income households in Khulna, Bangladesh. *Environment and Urbanization April* **26**(1), 112–129.

Hope Project (2012). iShack: A bright idea. Accessed March 8, 2015: http://thehopeproject.co.za/hope/blog/Lists/Posts/Post.aspx?ID=54

Houghton, J.T., Jenkins, G.J., and Ephraums, J.J. (eds.) (1990). Climate Change: The IPCC Scientific Assessment. Report prepared for Intergovernmental Panel on Climate Change by Working Group Cambridge University Press, Cambridge, Great Britain, New York, NY, USA and Melbourne, Australia 410 pp.

Huq, S., Kovats, S., Reid, H., and Satterthwaite, D. (2007). Editorial: Reducing risks to cities from disasters and climate change. *Environment and Urbanization* **19**(1), 3–15.

Impact Forecasting LLC and Aon Benfield. (2011). 2011 Thailand floods. Event recap report. Impact forecasting. Accessed October 14, 2015: http://thoughtleadership.aonbenfield.com/Documents/20120314_impact_forecasting_thailand_flood_event_recap.pdf.

Intergovernmental Panel on Climate Change (IPCC). (2012). *Managing the Risks of Extreme Events and Disasters to Advance Climate Change Adaptation: A Special Report of Working Groups I and II of the Intergovernmental Panel on Climate Change.* Cambridge University Press.

Intergovernmental Panel on Climate Change (IPCC). (2014). *Climate Change 2014: Synthesis Report. Contribution of Working Groups I, II and III to the Fifth Assessment Report of the Intergovernmental Panel on Climate Change*, Core Writing Team, Pachauri, R. K., and Meyer, L. A. (eds.). IPCC.

International Energy Agency (IEA). (2013). *Transition to Sustainable Buildings.Strategies and Opportunities to 2050.* IEA Publications.

International Federation of Red Cross and Red Crescent Societies. (2010). World Disaster Report: Focus on urban risk. Accessed December 21, 2015: http://www.ifrc.org/Global/Publications/disasters/WDR/WDR2010-full.pdf.

IRIN. (2014). Ivoirian floods highlight disaster preparedness shortcomings. Accessed January 8, 2015: http://www.irinnews.org/report/100304/ivoirian-floods-highlight-disaster-preparedness-shortcomings.

iShack Project iShack brochure. (n.d.) Accessed January 10, 2016: http://www.ishackproject.co.za/sites/default/files/iShack_Brochure.pdf

Islam, M. M., Sallu, S., Hubacek, K., and Paavola, J. (2014). Limits and barriers to adaptation to climate variability and change in Bangladeshi coastal fishing communities. *Marine Policy* **43**, 208–216.

Jabeen, H., Johnson, C., and Allen, A. (2010). Built-in resilience: Learning from grassroots coping strategies for climate variability. *Environment and Urbanization* **22**(2), 415–431. doi: 10.1177/0956247810379937

Jamil, S. (2013). *Connecting the Dots: The Urban Informal Sector and Climate Vulnerabilities in Southeast Asia's Mega-Cities, NTS Alert no. AL13–01.* RSIS Centre for Non-Traditional Security (NTS) Studies for NTS-Asia.

Jean-Baptiste, N., Kabisch, S., and Kuhlicke, C. (2013). Urban vulnerability assessment in flood-prone areas in West and East Africa. In Rauch, S., Morrison, G. M., Schleicher, N., and Norra, S. (eds.), *Urban Environment* (203–218). Springer.

John, R., Jean-Baptiste, N., and Kabisch S. (2014). Vulnerability assessment of urban populations in Africa: The case of Dar es Salaam, Tanzania. In Edgerton, E., Romice, O., and Thwaites, K. (eds.), *Bridging the Boundaries. Human Experience in the Natural and Built Environment and Implications for Research, Policy, and Practice.* Advances in people and environment studies, vol. **5 (233–244)**. Hogrefe Publishing.

Kabisch, S., Jean-Baptiste N., John R., and Kombe W (2015). Assessing social vulnerability of households and communities in flood prone urban areas. In Pauleit, S., Coly, A., Fohlmeister, S., Gasparini, P., Jorgensen, G., Kabisch, S., Kombe, W., Lindley, S., Simonis, I, and Kumelachew, Y. (eds.), *Urban Vulnerability and Climate Change in Africa: A Multidisciplinary Approach* (ch. 6). Future City series. Springer. Accessed September 10, 2016: http://www.springer.com/life+sciences/ecology/book/978-3-319-03984-8

Kato, T, Ellis, J., Pauw., P., and Caruso, R. (2014). Scaling-up and Replicating Effective Climate Finance Interventions. Climate Change Expert Group, OECD. Accessed September 12, 2015: http://www.oecd.org/env/cc/Scaling_up_CCXGsentout_May2014_REV.pdf

Koepnick, B., and Weselcouch, M. (2013). Sandy's Effects on Housing in New York City, The Furman Center for Real Estate and Urban Policy.

Accessed November 22, 2014: http://furmancenter.org/files/publications/SandysEffectsOnHousingInNYC.pdf

Kombe, W., and Kreibich, V. (2006). *Governance of Informal Urbanisation in Tanzania*. Mkukina Nyota Publishers.

Lamond, J., and Penning-Roswell, E. (2014). The robustness of flood insurance regimes given changing risk resulting from climate change. *Climate Risk Management* **2**(2014), 1–10.

Laukkonen, J., Blanco, P. K., Lenhart, J., Keiner, M., Cavric, B., and Kinuthia-Njenga, C. (2009). Combining climate change adaptation and mitigation measures at the local level. *Habitat International* **33**(3), 287–292.

Linnerooth-Bayer, J., Mechler, R., and Hochrainer, S. (2011). Insurance against losses from natural disasters in developing countries: Evidence, gaps and the way forward. *Journal of IntegrDisaster Risk Management* **(1)**, 1–23.

López-Marrero, T., and Yarnal, B. (2010). Putting adaptive capacity into the context of people's lives: A case study of two flood-prone communities in Puerto Rico, *Natural Hazards* **52**(2), 277–297.

Loyzaga, A. et al. (2014). *Assessing the Vulnerability of Metro Manila to Climate Change and Disasters.* Paper presented at the 6th Asian Ministerial Conference on Disaster Risk Reduction, Bangkok, Thailand, June 25–29, 2014.

Lucon O., Ürge-Vorsatz, D., Zain Ahmed, A., Akbari, H., Bertoldi, P., Cabeza, L. F., Eyre, N., Gadgil, A., Harvey, L. D. D., Jiang, Y., Liphoto, E., Mirasgedis, S., Murakami, S., Parikh, J., Pyke, C., and Vilariño, M. V. (2014). Buildings. In Edenhofer, O., Pichs-Madruga, R., Sokona, Y., Farahani, E., Kadner, S., Seyboth, K., Adler, A., Baum, I., Brunner, S., Eickemeier, P., Kriemann, B., Savolainen, J., Schlömer, S., von Stechow, C., Zwickel, T., and Minx J. C. (eds.), *Climate Change 2014: Mitigation of Climate Change. Contribution of Working Group III to the Fifth Assessment Report of the Intergovernmental Panel on Climate Change* (671–738). Cambridge University Press.

Masozera, M., Bailey, M., and Kerchner, C. (2006). Distribution of impacts of natural disasters across income groups: A case study of New Orleans. *Ecological Economics* (**63**), 299–306. Accessed April 7, 2015: www.elsevier.com/locate/ecolecon

McGranahan, G., Balk, D., and Anderson, B. (2007). The rising tide: Assessing the risks of climate change and human settlements in low elevation coastal zones. *Environment and Urbanisation* **19**(1), 17–37.

McKinsey. (2009). *Unlocking Energy Efficiency in the U.S. Economy*. Technical Report, McKinsey & Company.

Meagher, K. (2007). Manufacturing disorder: Liberalization, informal enterprise and economic ungovernance in African small firm clusters, *Development and Change* **38**(3), 473–503.

Metz, B., Davidson, O. R., Bosch, P. R., Dave, R., and Meyer, L. A. (eds.). (2007). *Contribution of Working Group III to the Fourth Assessment Report of the Intergovernmental Panel on Climate Change, 2007*. Cambridge University Press.

Mitchell, T., and Maxwell, S. (2010). Defining climate compatible development. CDKN ODI Policy Brief November 2010/A. Accessed June 6, 2014: http://r4d.dfid.gov.uk/pdf/outputs/cdkn/cdkn-ccd-digi-master-19nov.pdf

Modi, R. 2009. *Beyond Relocation. The Imperative of Sustainable Resettlement*. Sage.

Moser, C., and Satterthwaite, D. (2008). *Towards pro-poor adaptation to climate change in urban centers of low- and middle-income countries*. International Institute for Environment and Development (IIED), 16–34. Accessed September 4, 2014: http://pubs.iied.org/pdfs/10564IIED.pdf.

Moser, C. O. N., Norton, A., Stein, A., and Georgieva, S. (2010). *A Pro-Poor Adaptation to Climate Change in Urban Centres: Case Study of Vulnerability and Resilience in Kenya and Nicaragua*, The World Bank.

Mitlin, D. (2014). Namibia shows how to support low-cost housing. International Institute for Environment & Development Blogs. Accessed February 18, 2016: http://www.iied.org/namibia-shows-how-support-low-cost-housing

Neme, C., Gottstein, M., and Hamilton, B. (2011). *Residential Efficiency Retrofits: A Roadmap for the Future*. Regulatory Assistance Project (RAP).

Norton, J., and Chantry, G. (2008). Vaccinate your home against the storm: Reducing vulnerability in Vietnam. *Open House International* **33**, 26–31.

OECD. (2015). Climate finance in 2013–14 and the USD 100 billion goal, a report by the Organisation for Economic Co-operation and Development (OECD) in collaboration with Climate Policy Initiative (CPI). Accessed January 15, 2016: http://www.oecd.org/environment/cc/OECD-CPI-Climate-Finance-Report.htm

Orense, R. P., and Ikeda, M. (2007). Damage caused by typhoon-induced lahar flows from Mayon volcano, Philippines. *Soils and Foundations* 47 (6), 1123–1132. Accessed June 10, 2015: https://www.jstage.jst.go.jp/article/sandf/47/6/47_6_1123/_pdf.

Oxfam International. (n.d.). Myanmar cyclone. Accessed September 15, 2015: http://www.oxfam.org/en/emergencies/myanmar-cyclone.

Pauleit, S., Coly, A., Fohlmeister, S., Gasparini, P., Jørgensen, G., Kabisch, S., Kombe, W., Lindley, S., Simonis, I., and Yeshitela, K. (2015). *Urban Vulnerability and Climate Change in Africa – A Multidisciplinary Approach*. Springer.

Parry, M., Arnell, N., Berry, P., Dodman,D., Fankhauser, S., Hope, C., Kovats, S., Nicholls, R., Satterthwaite, D., Tiffin, R., and Wheeler, T. (2009). *Assessing the Costs of Adaptation to Climate Change: A Review of the UNFCCC and Other Recent Estimates*. International Institute for Environment and Development and Grantham Institute for Climate Change.

Payne, G. (2004). Land tenure and property rights: An introduction. *Habitat International* **28**(2), 167–179.

Pelling, M. (2011). *Adaptation to Climate Change: From Resilience to Transformation* (204).Routledge.

Pelling, M., and Blackburn, S. (2013). *Megacities and the Coast: Risk, Resilience and Transformation*. Routledge.

Porio, E. (2014). Climate change vulnerability and adaptation in metro Manila: Challenging governance and human security needs of urban poor communities. *Asian Journal of Social Science* **42**, 75–102.

Porio, E. (2015). *Sustainable Development Goals and Quality of Life Targets: Insights from Metro Manila in Current Sociology*. Sage.

Porio, E., with Crisol, C. (2004). Property rights and security of tenure among the poor in metro Manila. *Habitat International* **28**(2), 203–219.

Prowse, M., and Scott, L. (2008). Assets and adaptation: An emerging debate. DS Bulletin Vol. 39 (4). Accessed July 29, 2015: http://www.odi.org/sites/odi.org.uk/files/odi-assets/publications-opinion-files/3431.pdf

Reid, H., and Huq, S. (2007). *Community-Based Adaptation: A Vital Approach to the Threat Climate Change Poses to the Poor*. International Institute for Environment and Development (IIED). Briefing Paper. IIED.

ReliefWeb. (2012). Tropical Cyclone Evan - Dec 2012. Accessed November 24, 2014: http://reliefweb.int/disaster/tc-2012-000201-wsm.

Robledo, C., Clot, N., Hammmill, A., and Riché, B. (2012). The role of forest ecosystems in community-based coping strategies to climate hazards: Three examples from rural areas in Africa. *Forest Policy and Economics* 24, 20–28.

Rojas, R., and Ward, P. J. (2014). Increasing stress on disaster risk finance due to large floods. *Nature Climate Change*. Accessed June 15, 2015: http://dx.doi.org/10.1038/nclimate2124.

Romero-Lankao, P., Hughes, S., Qin, H., Hardoy, J., Rosas-Huerta, A., Borquez, R., and Lampis, A. (2014). Scale, urban risk and adaptation capacity in neighbourhoods of Latin American cities. *Habitat International* (**42**), 224–235.

Roy, A. (2009). Why India cannot plan its cities: Informality, insurgence and the idiom of urbanisation. *Planning Theory* **8**(1), 76–87.

Roy, A., and Al Sayyad, N. (eds.). (2005). *Urban Informality: Transnational Perspective from the Middle East, South Asia and Latin America*. Lexington Books.

Ruwanpura, N. K. (2008). Temporality of disasters: The politics of women's livelihoods "after" the 2004 tsunami in Sri Lanka. *Singapore Journal of Tropical Geography* 29(3), 325–340.

Satterthwaite, D. (2011). What role for low-income communities in urban areas in disaster risk reduction. Global Assessment Report on Disaster Risk Reduction (GAR) (1–48). Accessed May 30, 2014: http://www.preventionweb.net/english/hyogo/gar/2011/en/home/download.html.

437

Satterthwaite, D. (2013). Eight points on financing climate change adaptation in urban areas. Accessed April 16, 2014: www.iied .org/8-points-financing-climate-change-Adaptation.

Satterthwaite, D., Huq, S. Pelling, M., Reid., H., and Romero-Lankao, P. (2007). *Adapting to Climate Change in Urban Areas: The Possibilities and Constraints in Low and Middle Income Nations.* Climate Change and Cities Series, Discussion Paper No. 1. International Institute for Environment and Development (IIED), 107.

Schilderman, T., and Lyons, M. (2011). Resilient dwellings or resilient people? Towards people-centred reconstruction. *Environmental Hazards* **10**(3–4), 218–231. Accessed March 12, 2016: http://www .tandfonline.com/doi/abs/10.1080/17477891.2011.598497

Seto, K. C., Sánchez-Rodríguez, R., and Fragkias, M. (2010).The new geography of contemporary urbanization and the environment. *Annual Review of Environment and Resources* **35**(1), 167–194.

Shelter Cluster. (2014). Emergency shelter assessment tropical cyclone Evan Fiji - fact sheet, informal settlements. Accessed November 22, 2015: http://www.docstoc.com/docs/171417467/Emergency-Shelter-Assessment-Tropical-Cyclone-Evan-Fiji---Fact-Sheet_-Informal-Settlements.

Shack/Slum Dwellers International. (2008). Typhoon and flashflood hit Iloilo city, 48, 495 families affected. Accessed August 13, 2014: http:// www.sdinet.org/media/upload/countries/documents/hpfpi_typhoon_ frank_doc_26june08_.pdf.

Simone, A. (2004). *For the City Yet to Come, Changing African Life in Four Cities.* Duke University Press.

Smith, L. (2012). The challenge of informal settlements. In Swilling, M., Sebitosi, B., and Loots, R. (eds.), *Sustainable Stellenbosch Opening Dialogues.* African Sun Media.

Surminski, S., and Oramas-Dorta, D. (2014). Flood insurance schemes and climate adaptation in developing countries. *International Journal of Disaster Risk Reduction* **7**, 154–164.

Tanner, T., and Mitchell, T. (2008). Introduction: Building the case for pro-poor adaptation. *IDS Bulletin* **39**(4), 1–5.

Taylor, A., and Peter, C. (2014). *Strengthening Climate Resilience in African cities: A Framework for Working with Informality.* African Centre for Cities and the Climate and Development Knowledge Network.

Technology Strategy Board (TSB) (2013). *Annual Reports and Accounts, 2012–2013.* House of Commons of England. Accessed January 5, 2015: https://www.gov.uk/government/uploads/system/uploads/ attachment_data/file/246469/0567.pdf

Tran, A. T., Tran, P., and Tuan, T. H. (2013). *Review of Housing Vulnerability: Implications for Climate Resistant Houses.* ISET Discussion Papers Series: Sheltering from a Gathering Storm. ISET.

UNDESA. (2013). World Economic Survey 2013: Sustainable development challenges. Accessed August 13, 2014: http://sustainabledevelopment .un.org/content/documents/2843WESS2013.pdf

UNDESA. (2014). World urbanization prospects: The 2014 Revision Population Division Highlights. Accessed March 1, 2016: http://esa.un .org/unpd/wup/Highlights/WUP2014-Highlights.pdf

UNDP. (2011). Readiness for climate finance: A framework for understanding what it means to be ready to use climate finance. Accessed June 1, 2014: http://www.undp.org/content/dam/undp/ library/Environment%20and%20Energy/Climate%20Strategies/ Readiness%20for%20Climate%20Finance_12April2012.pdf

UNEP SBCI. (2007). *Buildings and Climate Change: Status, Challenges, and Opportunities.* United Nations Environment Programme, Sustainable Buildings and Construction Initiative.

UN-Habitat. (2003). *The Challenge of Slums, Global Report on Human Settlements.* UN-Habitat.

UN-Habitat. (2006/2007). *State of the World's Cities: The Millenium Development Goals and Urban Sustainability: 30 years of Shaping the Habitat Agenda.* EarthScan.

UN-Habitat. (2010). *The State of African Cities 2010, Governance, Inequality and Urban Land Markets.* UNEP.

UN-Habitat (2011). *Cities and Climate Change: Global Report on Human Settlements 2011.* Earthscan.

UN-Habitat. (2014). World Habitat Day 2014: "Voices from slums" background paper. Accessed February 18, 2015: http://unhabitat .org/wp-content/uploads/2014/07/WHD-2014-Background-Paper.pdf.

United Nations Office for Disaster Risk Reduction (UNISDR). (2012). Towards a post-2015 framework for disaster risk reduction. Accessed May 24, 2014: http://www.beyond2015.org/sites/default/files/ Towards%20a%20Post-2015%20Framework%20for%20Disaster%20 Risk%20Reduction.pdf.

United Nations Office for Disaster Risk Reduction (UNISDR). (2013). *From Shared Risk to Shared Value –The Business Case for Disaster Risk Reduction. Global Assessment Report on Disaster Risk Reduction.* UNISDR.

United Nations Office for Disaster Risk Reduction (UNISDR). (2015). *Global Assessment Report on Disaster Risk Reduction (GAR).* UNISDR.

United Nations Office for Disaster Risk Reduction (UNISDR) and the United Nations Economic and Social Commission for Asia and the Pacific (UNESCAP). (2012). *Reducing Vulnerability and Exposure to Disasters: The Asia-Pacific Disaster Report. Bangkok – Thailand.* UNISDR.

United Nations Office for the Coordination of Humanitarian Affairs (UNCHOA). (2013). Philippines: Snapshot on Mindanao. Accessed December 21, 2015: http://reliefweb.int/sites/reliefweb.int/ files/resources/Mindanao%2BSnapshot%2BJune2013%2Bfinal.pdf.

UPFI. (2014). Household EcoSan toilet construction in Lilongwe. Accessed November 9, 2014: http://upfi.info

U.S. Department of Commerce. (2013). Economic impact of Hurricane Sandy: Potential economic activity lost and gained in New Jersey and New York. Economics and Statistics Administration. Accessed February 16, 2016: http://www.esa.doc.gov/sites/default/files/reports/ documents/sandyfinal101713.pdf.

Vedeld, T., Kombe, W., Kweka-Msale, C., and Hellevik, S. (2015). Multilevel governance and coproduction in urbanflood-risk management: The case of Dar es Salaam. Inderberg, T. H., Eriksen, S., O'Brien, K., and Sygna, L. *Climate Change Adaptation and Development. Transforming Paradigms and Practices* (117–139). Routledge.

Wamsler, C., and Brink, E. (2014). Moving beyond short term coping and adaptation. In *Environment & Urbanization* **26**(1), 86–111. doi:10.1177/0956247813516061

Wamsler, C., and Lawson, N. (2011). The role of formal and informal insurance mechanisms for reducing urban disaster risk: A south–north comparison. *Housing Studies* **26**(2), 197–223. Accessed August 10, 2014: http://www.tandfonline.com/doi/abs/10.1080/02673037.2011.542087

Warner, K., Bouwer, L., and Ammann, W. (2007). Financial services and disaster risk finance: Examples from the community level. *Environmental Hazards* **7**(1), 32–39. Accessed June 16, 2014: http:// www.tandfonline.com/doi/abs/10.1016/j.envhaz.2007.04.006

Watson, V. (2009). The planned city sweeps the poor away: Urban planning and 21st century urbanisation. *Progress in Planning* **72**, 151–193.

Weeresinghe, S. (2013). Achieving disaster resilience through the Sri Lankan early warning system: Good practices of disaster risk reduction and management. Second Annual ANDROID Conference. Limassol, Cyprus, 1–8.

Wielinga, D., and Dingel, C. (2011). *Disaster Risk Management in 2011: Investing Smart for Results. West Africa Floods 2010.* Respond to prevent presentation, Global Facility for Disaster Reduction and Recovery, The World Bank.

Williams, S. (2008). Rethinking the nature of disaster: From failed instruments of learning to a post-social understanding. *Social Forces* **87**(2), 1115–1138.

Wisner, B., Blaikie, P., Cannon, T., and Davis, I. (2004). *At Risk: Natural Hazards, People's Vulnerability and Disasters* (2nd edition). Routledge.

World Bank. (2010). *A City –Wide Approach to Carbon Finance.* Carbon Partnership Facility Innovation Series. World Bank.

World Bank. (2012). Brazil: A plan to manage disasters could save money for development. Accessed August 17, 2015: http://www .worldbank.org/en/news/feature/2012/11/19/Brazil-natural-disaster-management-costs-development.

World Bank. (2013a). *Harnessing Urbanisation to End Poverty and Boost Prosperity in Africa; An Action Agenda for Transformation.* World Bank.

World Bank. (2013b) *Planning, Connection and Financing Cities-Now.* World Bank.

World Health Organization (WHO). (2011). *Health in the Green Economy: Health Co-benefits of Climate Change Mitigation – Housing Sector*. WHO.

Zurich Financial Services Group. (2009). The climate risk challenge: The role of insurance in pricing climate-related risks. Accessed October 31, 2014: http://www.zurich.com/NR/rdonlyres/E2B5B53E-11DB-47AF-91E4-01ED6A2BD.C.A3/0/ClimateRiskChallenge.pdf.

Chapter 11 Case Study References

Case Study 11.1 Water-Related Vulnerabilities to Climate Change and Poor Housing Conditions in Lima

Escudero, L. (2014). Programa de gestión de riesgos en las laderas de lima: Barrio mío, Internal report. Metropolitan Municipality of Lima.

International Energy Agency (IEA). (2013). *Transition to Sustainable Buildings. Strategies and Opportunities to 2050*. IEA.

Instituto Metropolitano de Planificación (IMP). (2013). MML Concerted Development Regional Plan. Metropolitan Municipality of Lima.

Instituto Nacional de Estadística e Informática (INEI). (2007). Population and housing census. Metropolitan Municipality of Lima.

Instituto Nacional de Estadística e Informática (INEI). (2015). Lima cuenta con … La Republica Perú. Accessed January 7, 2016: http://lare-publica.pe/17-01-2015/inei-lima-tiene-9-millones-752-mil-habitantes

Mehrotra, S., Rosenzweig, C., Solecki, W. D., Natenzon, C. E., Omojola, A., Folorunsho, R., and Gilbride, J. (2011). Cities, disasters and climate risk. In Rosenzweig, C., Solecki, W. D., Hammer, S. A., and Mehrotra, S. (eds.), *Climate Change and Cities: First Assessment Report of the Urban Climate Change Research Network* (15–42). Cambridge University Press.

Miranda Sara, L., and Baud, I. S. A. (2014). Knowledge building in adaptation management: Concertation processes in transforming Lima water and climate change governance. *Environment and Urbanization* **26**(2), 505–524.

Miranda Sara, L. (2014a). ¿Acceso al agua para todos, justo y ecoeficiente, en Lima?. *Revista Actualidad Gubernamental. Sección Informes Especiales* **72**(23), 1.

Miranda Sara, L. (2014b). ¿Viviendas y barrios sostenibles en el contexto del Cambio climático en el Perú? *Revista Actualidad Gubernamental. Sección Informes Especiales*. **68**(23), 1.

Miranda Sara, L., Hordijk, M. A., and Khan, S. (2014a). Actors' capacities to address water vulnerabilities in metropolitan cities facing climate change: Exploring actor network configurations, discourse coalitions, power relations and scenario building processes as social constructions of knowledge for multi-scalar water governance. WP4 Thematic Report.

Miranda Sara, L., Takano, G., and Escalante, C. (2014b). Metropolitan Lima and the sustainability challenge growing cities. Growing Economies City Report Metropolitan Lima and Callao. Accessed December 21, 2015: http://www.chance2sustain.eu/fileadmin/Website/Dokumente/Dokumente/Publications/D4.3_City_Report__Peru_.pdf

Municipality of Lima (MML). (2014). Estrategia de Adaptación y Acciones de Mitigación de la Provincia de Lima al Cambio Climático. with support from Foundation Avina and Foro Ciudades Para La Vida.

Municipality of Lima (MML) (2014). Lima Climate Change Strategy Metropolitan Municipality of Lima.

Peel, M. C., Finlayson, B. L., and McMahon, T. A. (2007). Updated world map of the Köppen-Geiger climate classification. *Hydrology and Earth System Sciences Discussions* **4**(2), 462.

SEDAPAL. (2007). Commercial water consumption per block. commercial data base. SEDAPAL.

SEDAPAL. (n.d.) Plan Maestro Optimizado 2009–2030. Accessed March 19, 2016: http://www.sunass.gob.pe/websunass/index.php/eps/planes-maestros-optimizados-pmo/cat_view/419-regulacion-tarifaria/211-planes-maestros-optimizados/212-planes-maestros-optimizados/320-lima?start=10

World Bank. (2017). 2016 GNI per capita, Atlas method (current US$). Accessed August 9, 2017: http://data.worldbank.org/indicator/NY.GNP.PCAP.CD

Case Study 11.2 Vulnerability and Climate–Related Risks on Land, Housing, and Informal Settlements in Dar es Salaam

Baker, J. L. (ed.). (2012). *Climate Change, Disaster Risk, and the Urban Poor: Cities Building Resilience for a Changing World*. World Bank. Accessed February 4, 2015: https://openknowledge.worldbank.org/handle/10986/6018

Hussein, M. S. (2013). *Assessing the Impacts of Flooding in Urban Informal Settlements. The Case of Kigogo, Kinondoni Municipality*. Ardhi University.

Jean-Baptiste, N., Kabisch, S., and Kuhlicke, C. (2013). Urban vulnerability assessment in flood-prone areas in west and east Africa. In Rauch, S., Morrison, G. M., Schleicher, N., and Norra, S. (eds.), *Urban Environment* (203–218). Springer.

John, R., Jean-Baptiste, N., and Kabisch, S. (2014). Vulnerability assessment of urban populations in Africa: The case of Dar Es Salaam, Tanzania. In Edgerton, E., Romice, O. and Thwaites, K. (eds.), *Bridging the Boundaries: Human Experience in the Natual and Built Environmenta and Implications for Research, Policy and Practice* (vol. **5**, 234–244). Hogrefe Publishing.

Kyessi, S. A., and Kyessi A. G. (2007). *Regularization and Formalization of Informal Settlements in Tanzania: Opportunities and Challenges in the Case of Dar es Salaam City*. Paper presented at the FIG Working Week, Hong Kong SAR, China, May 13–17, 2007.

National Bureau of Statistics (NBS). (2013). Accessed November 12, 2015: http://www.nbs.go.tz/takwimu/references/Tanzania_in_figures2012.pdf

Pauleit, S., Coly, A., Fohlmeister, S., Gasparini, P., Jørgensen, G., Kabisch, S., Kombe, W., Lindley, S., Simonis, I., and Yeshitela, K. (2015). *Urban Vulnerability and Climate Change in Africa: A Multidisciplinary Approach*. Springer International.

Peel, M. C., Finlayson, B. L., and McMahon, T. A. (2007). Updated world map of the Köppen-Geiger climate classification. *Hydrology and Earth System Sciences Discussions* **4**(2), 462.

Romero-Lankao, P., Hughes, S., Qin, H., Hardoy, J., Rosas-Huerta, A., Borquez, R., and Lampis, A. (2014). Scale, urban risk and adaptation capacity in neighborhoods of Latin American cities. *Habitat International* 42, 224–235. doi: http://dx.doi.org/10.1016/j.habitatint.2013.12.008

Ruwanpura, N. K. (2008). Temporality of disasters: The politics of women's livelihoods "after" the 2004 tsunami in Sri Lanka. *Singapore Journal of Tropical Geography* **29**(3), 325–340.

START. (2011). *Urban Poverty and Climate Change in Dar es Salaam, Tanzania: A Case Study*. Final Report. START.

Tran, A. T., Tran, P., and Tuan, T. H. (2013). *Review of Housing Vulnerability: Implications for Climate Resistant Houses*. ISET Discussion Papers Series: Sheltering from a Gathering Storm. ISET.

United Nations Office for Disaster Risk Reduction (UNISDR) and the United Nations Economic and Social Commission for Asia and the Pacific (UNESCAP). (2012). *Reducing Vulnerability and Exposure to Disasters: The Asia-Pacific Disaster Report. Bangkok – Thailand*. UNISDR.

United Nations Office for Disaster Risk Reduction (UNISDR). (2015). *Global Assessment Report on Disaster Risk Reduction (GAR)*. UNISDR.

United Republic of Tanzania Vice Presidents Office (URT). (2011). *The Dar es Salaam City Environment Outlook 2011*. URT.

U.S. Agency for International Development. (1998). Caribbean, Dominican Republic, Haiti - Hurricane Georges Fact Sheet #6. Accessed April 19, 2015: http://reliefweb.int/report/antigua-and-barbuda/caribbean-dominican-republic-haiti-hurricane-georges-fact-sheet-6.

World Bank (2002). *Upgrading Low Income Urban Settlements. Country Assessment Report, Tanzania*. World Bank.

World Bank. (2017). 2016 GNI per capita, Atlas method (current US$). Accessed August 9, 2017: http://data.worldbank.org/indicator/NY.GNP .PCAP.CD

World Health Organization (WHO). (2011). Health in the Green Economy: Health Co-benefits of Climate Change Mitigation – Housing Sector. WHO.

Case Study 11.3 Sheltering from a Gathering Storm: Urban Temperature Resilience in Pakistan

Cheema, S. B., Rasul, G., Ali, G., and Kazmi, D. H. (2010). A comparison of minimum temperature trends with model projections. *Pakistan Journal of Meteorology* **8**(15), 39–52.

Demographia. (2016). *Demographia World Urban Areas 12th Annual Edition: 2016:04*. Accessed August 9, 2017: http://www.demographia .com/db-worldua.pdf

Heltberg, R., and Lund, N. (2008). Shocks, coping, and outcomes for Pakistan's poor: Health risks predominate. *Journal of Development Studies* **45**(6), 889–910. Accessed May 27, 2014: http://ssrn.com/abstract=1150408

Mustafa, Z. (2011). Climate change and its impact with special focus in Pakistan. *Pakistan Engineering Congress Symposiums* **33**, 99–117.

Peel, M. C., Finlayson, B. L., and McMahon, T. A. (2007). Updated world map of the Köppen-Geiger climate classification. *Hydrology and Earth System Sciences Discussions* **4**(2), 462.

Vandentorren, S., Bretin, P., Zeghnoun, A., Mandereau-Bruno, L., Croisier, A., Cochet, C., and Ledrans, M. (2006). August 2003 heat wave in France: Risk factors for death of elderly people living at home. *European Journal of Public Health* **16**(6), 583–591. doi: 10.1093/ eurpub/ckl063

World Bank. (2017). 2016 GNI per capita, Atlas method (current US$). Accessed August 9, 2017: http://data.worldbank.org/indicator/NY.GNP .PCAP.CD

Zahid, M., and Rasul, G. (2008). Rise in summer heat index over Pakistan. *Pakistan Journal of Meteorology* **6**(12), 85–96.

Part III

Urban Infrastructure Systems

12

Energy Transformation in Cities

Coordinating Lead Authors

Peter J. Marcotullio (New York), Andrea Sarzynski (Newark, DE), Joshua Sperling (Denver)

Lead Authors

Abel Chávez (Gunnison), Hossein Estiri (Seattle), Minal Pathak (Ahmedabad), Rae Zimmerman (New York)

Contributing Authors

Jonah Garnick (New York), Christopher Kennedy (Victoria), Edward J. Linky (New York), Claude Nahon (Paris)

This chapter should be cited as:

Marcotullio, P. J., Sarzynski, A. Sperling, J., Chavez, A., Estiri, H., Pathak, M., and Zimmerman, R. (2018). Energy transformation in cities. In Rosenzweig, C., W. Solecki, P. Romero-Lankao, S. Mehrotra, S. Dhakal, and S. Ali Ibrahim (eds.), *Climate Change and Cities: Second Assessment Report of the Urban Climate Change Research Network*. Cambridge University Press. New York. 443–490

Major Findings

- Urbanization has clear links to energy consumption in low-income countries. Urban areas in high-income countries generally use less energy per capita than non-urban areas due to economies of scale associated with higher density.

- Current trends in global urbanization and energy consumption show increasing use of fossil fuels, including coal, particularly in rapidly urbanizing parts of the world.

- Key challenges for the urban energy supply sector include reducing environmental impacts, such as air pollution, the urban heat island effect, and greenhouse gas (GHG) emissions; providing equal access to energy; and ensuring energy security and resilience in a changing climate.

- While numerous examples of energy-related mitigation policies exist across the globe, less attention has been given to adaptation policies. Research suggests that radical changes in the energy supply sector, customer behavior, and the built environment are needed to meet the key challenges.

- Scenario research that analyzes energy options requires more integrated assessment of the synergies and tradeoffs in meeting multiple goals: reducing GHGs, increasing equity in energy access, and improving energy security.

Key Messages

In the coming decades, rapid population growth, urbanization, and climate change will impose intensifying stresses on existing and not-yet-built energy infrastructure. The rising demand for energy services (e.g., mobility, water and space heating, refrigeration, air conditioning, communications, lighting, and construction) in an era of enhanced climate variability poses significant challenges for all cities.

Depending on the type, intensity, duration, and predictability of climate impacts on natural, social, and built and technological systems, threats to the urban energy supply sector will vary from city to city. Local jurisdictions need to evaluate vulnerability and improve resilience to multiple climate impacts and extreme weather events.

Yet future low-carbon transitions may also differ from previous energy transitions because future transitions may be motivated more by changes in governance and environmental concerns than by the socioeconomic and behavioral demands of the past. Unfortunately, the governance of urban energy supply varies dramatically across nations and sometimes within nations, making universal recommendations for institutions and policies difficult, if not impossible. Given that energy sector institutions and activities have varying boundaries and jurisdictions, there is a need for stakeholder engagement across the matrix of institutions to cope with future challenges in both the short and long term.

In order to achieve global GHG emission reductions through the modification of energy use at the urban scale, it is critical to develop an urban registry that contains a typology of cities and indicators for both energy use and GHG emissions. This will help cities benchmark and compare their accomplishments and better understand the mitigation potential of cities worldwide.

12.1 Introduction

Energy has enabled human development (International Energy Agency [IEA], 2010). The supply of energy to cities has been at the center of human progress since before the Roman Empire (Keirstead and Shah, 2013) when settlements required proximate sources of water, food, and fuel for cooking, warmth, and light. Modern urban life and human activities require ever greater amounts of energy. Given the tremendous projected growth in urbanization, wealth, industrialization, technological advancement, and the associated demands for vital services including electricity, water supply, transportation, buildings, communication, food, health, and parks and recreation, demands on energy supply will grow into the foreseeable future.

Meeting increasing energy demands related to urbanization given current climate change projections amplifies the challenges of the urban energy supply sector. Contemporary urban energy use is fueled largely from fossil sources, creating greenhouse gas (GHG) emissions. The energy supply sector is already one of the largest sources of GHG emissions (see, e.g., U.S. Environmental Protection Agency [U.S. EPA], 2015 for U.S. shares of emissions), and, if current urbanization trends continue, urban energy use will increase more than threefold from 2005 to 2050 (Creutzig et al., 2014). Urban energy system components are directly and indirectly vulnerable to climate change impacts. Coastal cities, for example, often have power plants located at low elevations. Moreover, with increasing urbanization, providing secure, clean, modern energy to all urban citizens is an increasingly important planning goal. Therefore, three key energy challenges motivate the focus of this chapter: (1) mitigating GHGs: over the past 20 years, the increasing demand for energy in rapidly urbanizing countries has largely been met with the burning of coal (World Coal Association, 2012), a powerful GHG producing fossil fuel. (2) Building resilient urban energy systems: extreme weather and climate risks are making cities more vulnerable to loss of electric power and damage to energy infrastructure (Evans and Fox-Penner, 2014). And (3) achieving just cities via equitable modern energy access: in 2010, approximately 179 million urban residents globally do not have access to electricity, and 447 million do not have access to modern, clean cooking fuels (World Bank, 2015).

Throughout the chapter, we focus our assessment on trends in the growth and complexity of the urban energy supply sector and the related challenges, opportunities, and barriers to moving toward low-carbon, resilient, and just energy supply systems. We use examples demonstrating reductions in environmental impacts including transitioning toward alternative technologies, fuels and changing behaviors, energy systems designed for new climatic conditions, and opportunities to increase energy access and reliability. We organize information around three key questions:

1. *What are the current states, patterns, and trends for the urban energy supply sector?*

2. *What are the mitigation, adaptation, and development challenges of the urban energy supply sector?*
3. *What are the opportunities, limits, and barriers for transforming the urban energy supply sector to reduce environmental impact and increase access and resiliency?*

To present answers to these questions, the chapter is laid out in five sections: (1) an overview of the urban energy supply sector and a framework by which the central questions can be addressed; (2) a review of trends, conditions, and drivers of urban energy supply focusing on infrastructure, energy resources, governance, and policy; (3) an evaluation of three key challenges to these systems: environmental impact, system resilience, and energy access; (4) a review of previous energy transitions and future scenarios research; and (5) options for low-carbon, resilient, and just energy supply systems and how and why they are being implemented. Throughout this chapter, Case Studies are shared to illustrate how cities with different histories, geographies, institutions, and policies address current and perceived future challenges.

12.2 Overview

Urban energy systems comprise physical systems that include infrastructures and technologies, natural systems from which humans draw raw materials and services, and socioinstitutional systems involved in social relations, governance, and the management of energy services. Figure 12.1 presents a framework to explore and understand the urban energy supply sector within urban energy systems and the challenges and opportunities for transitioning to low-carbon, resilient, and just cities. The core of the urban energy supply sector is the physical infrastructure that converts primary energy resources like coal or gas or sunlight to usable energy such as electricity or heat and then delivers that energy to end-users such as businesses and residences (see Section 12.3.1 for a variety of services; e.g., heating, cooking, transportation). The design, operation, and management of the urban energy supply sector thus depend on available energy resources (see Section 12.3.2), the policy and governance context (see Section 12.3.3), and the underlying drivers of consumer energy demand (Section 12.3.4). The operation of the urban energy supply sector shapes mitigation, adaptation, and sustainable development challenges (see Section 12.4). Likewise, all components of the system face limits, barriers, and opportunities for improvement (see Section 12.5).

12.3 Trends and Conditions

12.3.1 Urban Energy Supply Infrastructure

Urban energy supply infrastructure refers to the engineered systems that provide energy to more than half of the world's population by bringing primary energy resources such as coal (often from around the world) into the city region, converting primary

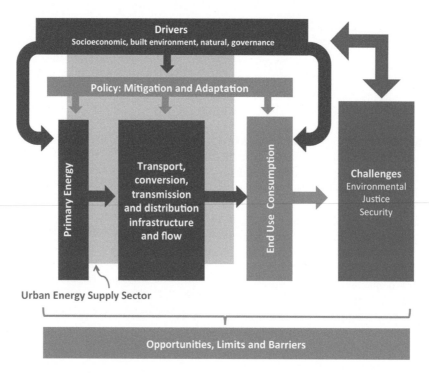

Figure 12.1 *Framework for understanding the urban energy supply system.*

resources into modern energy such as electricity, and transmitting and distributing this energy within and between urban areas (Bruckner et al., 2014). Supply networks can range from localized renewable energy generation to systems that span thousands of kilometers, linking mining and refining activities of solid, gas, and liquid fuels to energy conversion facilities and large manufacturing plants (Schock et al., 2012).

Urban energy supply infrastructure systems can be centralized, distributed (Hammer et al., 2011), or both. Centralized electric generation takes advantage of economies of scale offered by large power plants and the concentration of population and human activities. Large power plants can be fueled by different primary energy sources, including coal, natural gas, biomass, solid waste, or nuclear fuels, and can be located far from urban centers. Overhead wires and underground cables called *circuits* or *grids* connect electric generation technology with users. Rapidly advancing intelligent electronic technologies that enable broader consumer involvement in defining and controlling electricity needs are helping to integrate systems into "smart grids."

Some renewable energy systems, including large wind farms, geothermal power plants, or concentrating solar power facilities, can be large-scale, allowing them to fit relatively easy into a centralized generation and distribution model. Large-scale renewable energy remains a challenge because the natural variability of supply must be balanced with demand. Small amounts of non-hydro renewable energy can be easily accommodated.

Centralized thermal power systems are common in cities with extreme temperatures in winter or summer months.

These district heating and cooling systems (DHC) produce steam or hot and cold water centrally (often also producing electricity, known as combined heat and power or CHP) and then distribute this energy via a network of underground pipes. Examples of such systems are found in cities such as Copenhagen, Seoul, Austin, Goteborg, New York, and Paris (Hammer et al., 2011).

Distributed forms of electricity generation (DG) and heat distribution include smaller power production units located at or near the point of energy use, as in buildings. DG electrical systems link directly to the building's electric wiring system and therefore have lower transmission loss, reduced system vulnerability to service disruptions, and can more easily incorporate energy generated from renewable sources or technologies such as CHP than centralized generation (Lovins et al., 2002). DG systems are typically smaller than centralized systems, with capacities of 10 megawatts or smaller. These systems include energy storage (batteries) and sometimes connect through microgrids for greater reliability. DG systems can also be connected to larger centralized systems.

12.3.2 Energy Resources

Table 12.1 presents the levels of use, available resources, and percent change over the past 20 years for the major energy carriers. The table reveals that fossil fuels (oil, gas, and coal) provide the majority of world's energy, and the greatest recent increases in growth remain in these sources (IEA, 2013). During this period, coal (and peat) consumption has increased the most (68%), followed by natural gas (a "cleaner" fossil fuel).

Table 12.1 *Global energy resources, 2010. Source: Schock et al., 2012; Rogner et al., 2012; Bruckner et al., 2014; World Energy Council, 2013*

Source	World Energy Supply Percent	World Energy Supply EJ	Reserves at current production rate (years)	Percent Growth 1993–2011 (%)
Oil	34.1	170	40	25
Gas	22.4	114	60	62
Coal (proven) and Peat	28.4	151	132	68
Nuclear	2	10	100	13
Hydro	2.3	12		21
Geothermal, Solar, Wind, etc	0.6	3		n/a
Combustible renewables and waste	10.2	53		23
Total	100.0	513	332	

As consumption rises, energy debates emphasize the potential for overuse (Deffeyes, 2001). The "peak debate," as it has come to be called, has moved from a focus on conventional oil (Aleklett et al., 2010; Hubbert, 1981; Hughes and Rudolph, 2011) to include coal, gas, and uranium (Dittmar, 2013; Maggio and Cacciola, 2012; Heinberg and Fridley, 2010). Nevertheless, the primary concern for this chapter is not energy resource availability, but rather environmental impacts, disparities in energy access, and vulnerability and resilience to climate hazards (see Section 12.4).

12.3.3 Governance and Policy

Social, economic, and institutional mechanisms all shape demand for energy and help to oversee its supply and distribution within society. These mechanisms vary substantially from city to city depending on characteristics of the city's governance processes, characteristics of the energy system, and other local contextual factors including geography, culture, and history (Morlet and Keirstead, 2013; Jaglin, 2014). The governance of urban energy systems varies in many ways and therefore requires localized knowledge and perspectives.

The governance of urban energy systems is particularly complex given the "public good" nature of energy and the negative externalities associated with certain forms of energy generation (Florini and Sovacool, 2009; Morlet and Keirstead, 2013). Historically, urban energy systems were limited to cooking and heating fuels and supplied through private, decentralized markets that brought fuels from the hinterland to the city (Rutter and Keirstead, 2012). Throughout the Industrial Revolution, energy demand exploded, and energy systems were increasingly centralized and publicly supported. By the mid-20th century, many countries had nationalized electricity networks and distribution systems for fuels such as natural gas (Coase, 1950), shifting energy governance to the national scale. In some locations, aspects of energy and environmental governance are international and multilevel (i.e., Europe; Marks et al., 1996).

International institutions such as the International Energy Agency and the Asian Development Bank emerged to provide energy financing and policy support at the national and global scales (Florini and Sovacool, 2009). The International Atomic Energy Agency and the Nuclear Energy Agency of the Organization for Economic Cooperation and Development (OECD) emerged to govern the safe development of nuclear power. Today, approximately 80% of global nuclear capacity is in OECD countries (IEA, 2014).

In some countries such as Thailand, energy systems remain publicly owned and centrally operated (Kunchornrat and Phdungsilp, 2012), and energy decisions remain in the hands of national authorities. Therefore, efforts to green the energy supply and improve its resilience depend on priorities of the national government. Similarly, some countries such as South Africa may allow municipal utilities to distribute power within cities, but generation decisions remain with the central utility, leaving municipalities like Cape Town vulnerable to unreliable state-owned systems (Jaglin, 2014).

In other countries, urban energy systems are regionally coordinated, adding an additional layer of subnational governance between the market structuring functions at the national level and utility operations at the local level. In England, nine regional general-purpose governments coordinate renewable energy development among local energy providers in pursuit of national goals, although each region sets its own targets and strategies (Smith, 2007). Smooth coordination among multiple government bodies cannot be presumed, especially when urban priorities conflict with higher level goals (Jaglin, 2014). When municipal priorities diverged from their regional or national counterparts, cities like Hanover turned to transnational networks including ICLEI—Local Governments for Sustainability's Cities for Climate Protection to help provide guidance and support for local initiatives (Emelianoff, 2014).

In yet other countries, municipal authorities have some control over energy supply where interest and institutional capacity allows. In Los Angeles, the city's Department of Water and Power has begun to transition from using almost entirely coal-fired and nuclear power plants to a more diverse supply arrangement including purchasing from natural gas plants and wind and solar farms (Monstadt and Wolff, 2015). In Hanover, Germany, the municipal authority disinvested its share of local nuclear power and shifted its energy purchasing toward coal but also combined heat and power (CHP), wind, small hydro, solar, and biomass (Emelianoff, 2014). Likewise, in Vaxjo, Sweden, the municipal authority shifted entirely to biomass for its thermal

447

power generation, pushing to become a "fossil-fuel-free city" (Emelianoff, 2014).

Whereas municipal decisions are often within the context of infrastructure demands or higher level governmental choices, it is true that regional-to-local governments often have a suite of policy tools afforded to them for achieving sustainability and resilience across the energy system. The suite of policy tools can generally be articulated in three categories: regulatory, market-based, and voluntary type of instruments. Such a well-rounded set of options has proved to be an effective approach for subnational energy governance in aggregate, even though these three categories are very different in scope and outcomes.

Regulatory approaches are often associated with top-down and command-and-control actions, and they often yield high participation rates due to higher costs of noncompliance. Federally set regulations (e.g., 1990 U.S. Clean Air Act for Acid Rain) help set a standard for subnational governments (U.S. Environmental Protection Agency, 2014). For example, the United States' public utility commissions (PUC) have set grid-wide renewable portfolio standards (RPS) that are already showing respectable carbon intensity gains (National Renewable Energy Laboratory [NREL], 2015) at the city scale (e.g., with the City of Aspen Colorado achieving 100% renewable energy; Aspen, 2015). Like the United States, the European Union (EU) also has aggressive RPS regulatory targets that are reaching community-level implementation (European Union, 2015).

Market-based approaches are centered on their incentive-based regulatory scheme to drive participation and catalyze energy development. Pollution "caps" are examples of market-based instruments often used by federal governments, and, even though they are far less obvious at community levels, they do present potentials for innovation at local government scales. Exemplary cases include China's increased electricity prices and India's cap-and-trade, both for energy-intensive producers (C2ES, 2014).

Voluntary policies often yield substantially lower participation than other approaches and are mostly focused on driving consumer behavior change. Actions such as behavioral feedbacks, green energy purchases, or weatherization upgrades result in minimal energy system advances (Ramaswami et al., 2012).

Over the past several decades, many countries have experimented with privatization and deregulation of energy markets, shifting the balance of power back toward the private sector and investor-owned utilities (Al-Sunaidy and Green, 2006; Wallston et al., 2004). Moves to privatize were made in response to severe fiscal challenges facing the public sector and the rise of neo-liberal ideology (Monstadt, 2007). Concurrently, many governments shifted their energy governance from governing by authority (i.e., regulation) toward enabling of voluntary private actions (Bulkeley and Kern, 2006). Such shifts reduced public control over infrastructure investments and environmental outcomes. In Berlin, voluntary agreements with private utilities

were unable to achieve intended energy efficiency and solar energy targets, even though investments increased in these areas (Monstadt, 2007).

Concern over energy security has renewed interest in decentralized and distributed modes of energy generation, such as CHP and DHC at city and small/community scales. Cities have set distributed energy targets, such as 25% by 2025 in London, and several UK municipalities have adopted the Merton Rule requiring new buildings to supply 10% of their energy using on-site renewables (Lo, 2014). UK municipalities have experimented with various forms for governing DHC, including a municipal-owned DHC company in Woking, a nonprofit-led DHC in Aberdeen, and a public–private partnership for DHC in Birmingham (Hawkey et al., 2013). Decentralized generation and provision of energy services (heating and lighting) to small communities have often been provided by individuals and small and medium enterprises termed "ecopreneurs" (Schaper, 2002; Monstadt, 2007). Community-scaled renewables show promise for promoting community energy resilience by civil society actors without requiring much or any government involvement (Aylett, 2013; Frantzeskaki et al., 2013).

12.3.4 Drivers of Urban Energy Demand

Much scholarship has investigated the drivers of urban energy demand and associated impacts, from the household to city to region and global scales (Sattherwaite, 2009; Grubler et al., 2012a; Blanco et al., 2014; Marcotullio et al., 2014). These drivers can be categorized according to four sets of characteristics: socioeconomic, behavioral, geography and natural conditions, and built environment, as shown in Table 12.2.

A recent analysis of 225 cities worldwide found that affluence and fuel prices were the two most reliable predictors of total urban energy demand at end-use (Creutzig et al., 2015). Additionally, those cities could be appropriately classified as one of eight types, using indicators of the four characteristics identified earlier (gross domestic product [GDP] per capita for socioeconomic, fuel prices for behavioral, heating degree days for natural conditions, and population density for built environment). The drivers affected each city type differentially (see Section 12.4.1), indicating the importance of understanding the local context when selecting climate and energy policies for individual cities (see Section 12.5).

12.4 Mitigation, Adaptation, and Sustainable Development Challenges

Three key challenges emerge from the design and operation of the urban energy supply sector: (1) significant environmental impacts requiring mitigation efforts, (2) system vulnerabilities to climate change and associated impacts requiring adaptation efforts, and (3) disparities in access to modern energy requiring sustainable development efforts.

Table 12.2 *Drivers of urban energy consumption and greenhouse gas emissions. Source: Marcotullio et al., 2014*

DRIVER	DEPENDENT VARIABLE	SOURCE
Socioeconomic	Population (+), income (+) [*income also has a + association with housing size, automobile use, heating, and industrial fuel use], urbanization (+), regional production (+), high density of energy intensive industries, service, and industrial-sector economic base (+) versus recreation-based economy (~); population ageing (–); institutional maturity and know-how on emission regulations (–); governance arrangements (–/+); race/ethnicity (+) [** race/ethnicity also influences housing characteristics]	Ciccone and Hall, 1996; Schock et al., 2012; Poumanyyong and Kaneko, 2010; Schulz, 2010b; Satterthwaite, 2009; Weisz and Steinberger, 2010; Kahn, 2009; Hoornweg et al., 2011; Dhakal, 2009; Marcotullio et al., 2012; O'Neill et al., 2012; Estiri, 2014, 2015
Behavioral	Increasing energy prices (–), social norms and values (–/+); psychological factors [attitude, personal norm, awareness of consequences] (–/+); energy reporting (–); lifestyle-related choice including housing type (+), commuting distances (+), goods/services consumption (+), social contact (–)	Martinsen et al., 2007; Thøgersen and Olander, 2002; Schultz et al., 2007; Abrahamse and Steg, 2009; Allcott, 2011; Baiocchi et al., 2010; Heinonen et al., 2013
Geography and Natural Conditions	Weather and climate (–/+); geographic location of an urban area [e.g., coastal, mountainous, desert, by river or sea] (~); proximity to types of ecosystems (~) [e.g., temperate or tropical forests]	Pauchari, 2004; Pauchari and Jiang, 2008; Neumayer, 2002; Estiri et al., 2013; Kennedy et al., 2011
Built Environment	Technologies (–/+); physical infrastructure and related materials (–/+); design (–/+), building/infrastructure age (+); building type (–/+) and size (+); land-use mix (–); urban form; high population and employment densities (–); high connectivity street patterns; destination accessibility to jobs and services	Seto et al., 2014; Grubler et al., 2012; Chester et al., 2014; Norman et al., 2006; Estiri, 2014, 2015

Notes: Symbols in parentheses denote the direction of the relationship between each particular factor and urban energy use. The plus sign indicates a positive relationship (increases energy use), negative sign indicates a negative relationship (decreasing vulnerability), ~ (unknown).

12.4.1 Environmental Impacts

The urban energy supply sector generates local and global environmental impacts, including varying levels of GHG emissions, air pollution, urban heat island (UHI) effects, and habitat and ecosystem disturbances. As the world continues to urbanize, the environmental impacts of the energy supply sector may grow by virtue of the magnitude of overall energy use (not per capita) without substantial effort toward mitigation (see Section 12.5). This chapter focuses on GHG emissions to illustrate environmental impacts most closely related to climate change, but we direct readers to other sources for additional environmental impacts (see, e.g., Apte et al., 2012, 2015; Kryzanowski et al., 2014) and for interactions with GHG emissions in forcing climate change (e.g., Pandey et al., 2006, for urban particulates).

Global-Scale Impacts: GHG, Energy Patterns, and Trends

In 2010, the energy supply sector accounted for 49% of all energy-related GHG emissions (JRC/PBL, 2013) and 35% of all anthropogenic GHG emissions, up 13% from 1970, making it the largest sectoral contributor to global emissions. According to the Emissions Database for Global Atmospheric Research (EDGAR), global energy supply sector GHG emissions increased by 35.7% from 2000 to 2010 and grew on average nearly 1% per year faster than global anthropogenic GHG emissions (Bruckner et al., 2014). Asia, Europe, and the United States contributed the most to global annual energy GHGs, with Asia's contribution increasing rapidly in recent years.

Few studies estimate the relative urban and rural shares of global GHG emissions (Dhakal, 2010; Seto et al., 2014) due to the lack of comparable urban-scale energy and GHG emissions data. Moreover, debates remain as to the best way to inventory GHG emissions at the local level (see Box 12.2), and inconsistencies often exist among city inventories. Comparability and standard accounting protocols (e.g., the Global Protocol for Community-Scale GHG Emissions [GPC] now used by more than 100 cities globally; (ICLEI, C40 and WRI, 2012)[1] are needed for benchmarking, scalability, and guiding adaptive mechanisms. Box 12.1 discusses the challenges to acquiring data for these different types of accounting procedures (production and consumption). Given the lack of research infrastructure and data at the urban level, it remains difficult to rigorously quantify emissions, and this is particularly true in developing world cities. Box 12.2 gives a perspective of different tools and protocols used in cities around the world.

Nevertheless, studies suggest that cities are responsible for more than two-thirds of global energy use (IEA, 2008; Grubler et al., 2012a) and for approximately 70–75% of global energy-related CO_2 emissions (IEA, 2008; Grubler et al., 2012a; Marcotullio et al., 2013). When accounting for more than CO_2 emissions, however, studies suggest that the global urban contribution drops to less than 50% of total (Marcotullio et al., 2013), arguably owing to the largely non-urban contributions of methane worldwide (Satterthwaite, 2008).

1   See http://www.ghgprotocol.org/city-accounting

Box 12.1 Table 1 *Randomly selected cities from Annex 1 and non-Annex 1 economies for coupled footprints. Source: Abel Chávez*

Annex 1 Cities	New York	Toronto	London	Berlin	Paris	
	Barcelona	Madrid	Tokyo	Sydney	Melbourne	
Non-Annex 1 Cities	Mexico City	São Paulo	Rio de Janeiro	Bogotá	Cape Town	Durban
	Johannesburg	Beijing	Shanghai	Delhi	Manila	Jakarta

As the world's production and consumption of goods and services continues to rise, communities, both urban and rural, are at the forefront of the innovations required to provide these needs. If communities are to develop or maintain economies that help fill global niches, it is necessary that they plan, design, and construct infrastructures that are economically feasible, environmentally benign, and socially accessible to all residents. However, in order to plan, design, and construct the necessary infrastructures, it is imperative that communities use sound material and energy flow data to identify current conditions and trends. Therefore, measuring, benchmarking, and tracking community-scale material and energy flows becomes increasingly critical.

As noted in Box 12.2, material and energy flows can come in two broad types: production and consumption. For example, production footprints account for flows associated with all in-boundary activities and trans-boundary flows of key infrastructures, whereas consumption footprints account for all in- and trans-boundary flows associated only with local household consumption. The two approaches often yield different "footprint" estimates for any one community (see Chávez and Ramaswami, 2013), yet, despite this, measuring, benchmarking, and tracking the two – coupled and side-by-side – is rarely done. Thus, the chapter authors sought to compile coupled footprints for a host of cities that included an examination of the suite of mitigation strategies afforded to cities.

The authors randomly selected a sample of 10–12 cities from Annex 1 and non-Annex 1 economies each, for a total of 20–24 cities (see Box 2.1 Table 1). The principal approach was to apply primary and publicly available data to compute the coupled footprints. The data included *production* – inventories/footprints or climate action plans; and *consumption* – household and consumer expenditure surveys. The exercise revealed, however, large and fragmented data gaps in global cities. Of the Annex 1 cities, the team located primary production and consumption data for 80% and 60% of the cities, respectively. Meanwhile, of the non-Annex 1 cities, primary production and consumption data was located for 67% and 42% of the cities, respectively. Thus, far more data investigation and development is necessary to be able to embark on the vital coupled footprint analysis. However, data needs do not end here!

Upon examining trends in community development and population growth, the need for robust production and consumption data is magnified. Although for good reason much research is devoted to megacities, smaller cities and rural communities present substantial opportunities for low-carbon, adaptive, and just development. It is noted that 51% of today's urban population live in small cities of less than 500,000 people, and that by 2030 more than 40% of the urban population will be in even smaller cities of less than 300,000 (Chávez, 2016). Thus, as communities develop, measuring, benchmarking, and tracking community-scale material and energy flows are imperative toward creating the necessary infrastructures. Crucially, action is needed to generate and compile both production- and consumption-related data at the urban scale for communities of a variety of sizes and locations, but particularly for those in the developing world.

Urban Impacts: Energy Consumption

Many studies estimate urban energy demands and associated environmental impacts for selected cities worldwide (for reviews, Grubler et al., 2012a; Seto et al., 2014; Creutzig et al., 2015). The estimates vary extensively for reasons outlined earlier. Nevertheless, to illustrate the underlying variability, Table 12.3 provides the average per capita urban energy use from selected cities based on the Economist Intelligence Unit (EIU) data from 2002 to 2011. The table illustrates significant differences in energy use estimates among cities, recognizing, however, that databases are not always compatible for comparative purposes. Within one region such as Asia, differences appear to be large (nearly 200 GJ per capita in Guangzhou to 5.7 GJ per capita in Calcutta). As expected, European and North American cities fall toward the upper end of the energy use spectrum and also reflect large differences, ranging from nearly 150 GJ per capita in Dublin and Atlanta to 36 GJ per capita in Istanbul and about 10 GJ per capita in Cleveland. High energy demand in some cities results from the use of older and inefficient fossil fuel–based energy technologies, heavy industry, and high automobile use. Notably, energy use in all cities in

Box 12.2   Managing Greenhouse Gases Emissions in Cities: The Role of Inventories and Mitigation Actions Planning

Flavia Carloni, Vivien Green, Tomas Bredariol, Carolina Burle S. Dubeux, and Emilio Lèbre La Rovere

Federal University of Rio de Janeiro

Andrea Sarzynski

University of Delaware, Newark

Abel Chávez

Western State Colorado University, Gunnison

Apart from the differences in the institutional and political structures of cities, the majority of them may be seen as a planning unit for mitigation management purposes (Carloni, 2012). With both city specificities and similarities in mind, it is possible to profit from individual experiences that can be exchanged and define a common framework to improve mitigation actions.

Some cities' government officials think that the efforts to reduce greenhouse gas (GHG) emissions could jeopardize their economic and social agendas (Dubeaux, 2007). However, some initiatives show the opposite, combining development and climate action that results in a win-win situation.

In regard to the methodologies to be applied, there are differences between accounting methods for emissions of a city (the city's GHG inventory): (1) comparison of the absolute emissions in 1 year with respect to another and (2) monitoring the mitigation effects of particular actions through the comparison of emissions from a baseline scenario with the absence of action (see Box 12.2 Figure 1) (Carloni, 2012).

In addition to the choice regarding the methods, they must follow a step-by-step methodology to ensure both a good analysis during the process and also the possibility of future comparison between different inventories. The U.S. Environmental Protection Agency (U.S. EPA, 2015) presents on its website a very simple and direct guide to conduct a GHG inventory: (1) set the boundaries, either physical, operational, or governmental; (2) define the scope, considering which emission sources should be included in the report and also which gases are going to be investigated; (3) choose the quantification approach by considering data availability and the purpose of the inventory, then adopt either a top-down, bottom-up, or hybrid approach; (4) set the baseline by determining the benchmark year; (5) engage stakeholders by bringing them into the process in the very beginning with the intention of collecting more data and information and helping construct a public acceptance; and (6) consider certification; a third-party review and certification of the methods and data is highly advisable to assure the high quality, consistency, and transparency of the report.

There are two possible approaches to monitoring: compare two or more emissions inventory data (total values or sectorial and subsectorial) or use scenario-building techniques to assess the mitigation outcomes of specific policies, projects, and measures.

It is important to acknowledge the complementarity of approaches; both must be analyzed together and in parallel. Challenges faced when applying these approaches to cities include:

- Whether and how to account for indirect emissions (leakages[2])

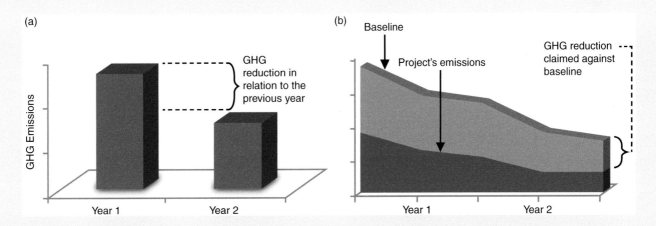

(a)   GHG reduction in relation to the previous year

GHG Emissions

Year 1   Year 2

(b)   Baseline   Project's emissions   GHG reduction claimed against baseline

Year 1   Year 2

Box 12.2 Figure 1 *Quantification of greenhouse gas emission reduction based on the analysis of an inventory (a) and based on the analyses of a mitigation action (b).*

Source: Adapted from Carloni, 2012

2   Carbon leakage is defined as the increase in emissions outside a region as a direct result of the policy to cap emissions in this region. Carbon leakage means that the domestic climate mitigation policy is less effective and more costly in containing emission levels, a legitimate concern for policy-makers. For example, if one city decides to create a carbon tax for a specific industry activity, companies may migrate to another city instead of improving their processes to mitigate emissions.

- How to ensure additionality[3] and reduction of GHG emissions by investment in projects or through the purchase of credits generated[4]
- Whether and how to demonstrate cause-and-effect relationships between a given action and a given emission reduction (Carloni, 2012)

GHG INVENTORIES IN CITIES

Different methodologies and protocols for GHG inventories have been developed: the Intergovernmental Panel on Climate Change (IPCC) Guidelines for National Greenhouse Gas Inventories (Intergovernmental Panel on Climate Control [IPCC], 2006), which is frequently adapted for cities; the

Box 12.2 Table 1 *Tools used by cities to conduct GHG emission inventories. Source: Based on data from Carloni, 2012*

Tool	Characteristics	Use
CO_2 Gronbolianz/ EMSIG	EMSIG (Emission Simulation in Gemeinden/Emission Simulation in Communities) was developed by Austria's energy agency. CO_2 Grobillanz is a simpler version. The tools comes with data from Austria regarding emission factors, goods consumption, and economic activities-related emissions. Both use geographical frontiers as boundaries.	Communities in Austria
ECO_2 Region	Supports the calculation of public authority and/or territory GHG emissions. The framework is mostly compatible with the IPCC (2006) methodology. It is also possible to include emissions of local pollutants such as particulate matter. Average emission factors for some countries are included.	Cities in Germany, Switzerland, and Italy
GRIP	The Greenhouse Gas Regional Inventory Protocol was developed by the University of Manchester and the United Kingdom's environmental agency. Initially, it was designed for metropolitan areas, but it has been used for smaller cities, too. The methodology used follows the IPCC Guide (2006), allowing greater comparability between cities. A tool for scenario construction is also available.	Cities in the United Kingdom as well as in some other cities in Europe and the United States
Bilan Carbone Collectivités – Territoires	This tool was developed based on work from the France Environmental Agency. It supports accountability for all gases included in the Kyoto Protocol, as well as chlorofluorocarbon (CFC) and water vapor emitted by airplanes. French cities' emission factors are available.	Municipalities in France
CO_2 – Beregner	A result of the work of the Denmark environmental agency in cooperation with a private consulting group, this instrument consider cities as a geographical entity – even though it may be adapted to account solely for the local authority. Only CO_2, CH_4, and N_2O are supported, but the reporting framework follows the IPCC (2006) guidelines. It requires a great range of data, enabling complex inventories. Furthermore, the tool comes with a guide with thirty-seven possible mitigation actions, and their impacts may be calculated.	Cities in Denmark
Project 2°	This project is a cooperative effort between the Clinton Climate Initiative, ICLEI and the Microsoft Corporation. It is based on HEAT, a tool developed by ICLEI. Therefore, the resulting inventories are consistent with IEAP. All six Kyoto Protocol gases are supported, and the methodology used is in accordance with the IPCC Guide (2006). One may account emissions by the territory or the governmental authority. Additionally, emission separation in scopes (1, 2, 3) is possible.	C40 (a network of the world's megacities committed to addressing climate change)
CACPS	The Clean Air and Climate Protection Software was developed by ICLEI and follows the IEAP model. This software supports the accountability of traditional air pollutants as well as GHG. It also assists in the elaboration of emission reduction strategies through the evaluation of policies and action plans.	Mostly by cities in United States, but also elsewhere
GPC	Provides a framework for accounting and reporting citywide GHG emissions. The tool was finalized after a pilot test in 2013 and global public comments in 2012 and 2014. It replaces all previous draft versions of the GPC and supersedes the International Local Government Greenhouse Gas Emissions Analysis Protocol (community section) published by ICLEI in 2009 and the International Standard for Determining Greenhouse Gas Emissions for Cities published by the World Bank, United Nations Environment Programme (UNEP), and UN-Habitat in 2010. Several programs and initiatives have adopted the GPC, including the Compact of Mayors, Carbon Climate Registry, and CDP, among others.	To date, more than 100 cities across the globe have used the GPC (current and previous versions)

---

3 Additionality is the requirement that GHG emissions after implementation of a clean development mechanism (CDM) project activity are lower than would have occurred in the most plausible alternative scenario to the implementation of the CDM project activity.

4 Carbon credit is a commercial unity that represents a ton of CO_2 or CO_2e removed from the atmosphere; it can be used to offset damaging carbon emissions that are or have been generated. The purchase is` usually a way to get this credit from different companies or countries.

International Local Government GHG Emissions Analysis Protocol (IEAP) (ICLEI, 2009), which was a first attempt to provide a program adjusted for local inventories; and the International Standard for Determining Greenhouse Gas Emissions for Cities (UNEP, UN-HABITAT, World Bank, 2010).

Despite the progress made in the past years, there are still some questions regarding GHG accounting at the local level (e.g., How to draw the boundaries? What to measure? How to measure?).

Nikolas Bader and Raimund Bleischwitz (2009) reviewed six tools that have been used in Europe: Project 2 Degrees (developed by ICLEI, Microsoft, and the Clinton Climate Foundation; in English; used by some C40 cities); GRIP (developed by University of Manchester, UK; in English; used by several European regions); CO_2 Grobbilanz (developed by Austria's energy agency; in German only); Eco2Regio (developed by Ecospeed; in German, French, and Italian; used by several Climate Alliance cities); Bilan Carbone (developed by French energy agency; in French); and the CO_2 Calculator (developed by Danish National Environmental Research Institute; in Dutch). They explain that the six tools vary according to the GHGs included (CO_2 vs. other GHGs), the global warming potentials (GWP), the scope of measurement (direct vs. indirect), the definitions of sectors, how emissions were quantified (top-down vs. bottom-up), how closely the tool follows the IPCC guidelines, and the usability of the tool (e.g., simplicity of use, available languages). More recently, economic functions that result in energy and GHG flows in and across communities have been articulated as production (purely territorial), consumption, or hybrid (encompassing trans-boundary flows). In practice, the economic sector composition for communities

has been shown to follow this line of understanding, resulting in distinct planning, policy, and development pathways (Chavez and Ramaswami, 2013).

We present here some of the tools used by cities to elaborate GHG inventories (Box 12.2 Table 1):

Even with the availability of these tools, cities are still in a long way from a common framework. There are significant differences in boundary setting and reporting sectors. In addition, a number of communities do not elaborate GHG inventories regularly, thus hindering comparison in time and action planning for emissions reduction (Neves and Dopico, 2013).

At COP20, in 2014, the first widely endorsed standard for cities to measure and report their GHG emissions was launched. The Global Protocol for Community-Scale Greenhouse Gas Emission Inventories (GPC)[5] uses a robust and clear framework to establish credible emissions accounting and reporting practices, thereby helping cities develop an emissions baseline, set mitigation goals, create more targeted climate action plans, and track progress over time.

Regarding other initiatives, in 2012, in a joint initiative by the Bonn Center for Local Climate Action and Reporting (carbon*n*) and the carbon*n* Cites Climate Registry (cCCR) a platform was created to be an open space where cities could report their GHG emissions reduction and climate adaptation targets, accomplishments, and actions. Also, the Carbon Disclosure Project, another global initiative, is achieving popularity in different fields including cities, private companies, shareholders, customers and governments with its own methodology of self-report. However, it is stressed that this initiative is not a GHG inventory methodology, but, in fact, a way to report accomplishments.

African and Latin America falls below the lowest European or North American city (except for Cleveland), with African cities ranging from 18 GJ per capita in Tunis to 0.8 GJ per capita in Maputo, and Latin American cities ranging from 13 GJ per capita in Buenos Aires to 3.3 GJ per capita in Lima (Economist Intelligence Unit, 2012).

Case Studies provide additional detail on the variable sources of energy, extent of modern energy coverage, and rates of growth in demand (see Box 12.3). As demonstrated in Quito, Seattle, and Delhi, the sources of energy are diverse. Sometimes they reflect the resources of the country and the larger continent (i.e., Quito's reliance on hydro reflects the high use of hydropower throughout Latin America) and sometimes they do not (i.e., Seattle's reliance on hydro diverges with the more typical fossil fuel–powered energy in the rest of the United States).

At the metropolitan level, as distinct from the urban scale, Kennedy et al. (2015) examined energy use in twenty-seven of the world's megacities (i.e., metropolitan areas with more than 10 million residents) in 2011, revealing a similarly wide range

worldwide and within regions. Total megacity energy use varied from 2.8 TJ in New York (population 22 million) to just 0.68 TJ in Kolkata (population 14 million). On a per capita basis, Moscow led the other megacities with approximately 147 GJ per capita, followed by New York and Los Angeles (127 and 104 GJ per capita, respectively). By contrast, Mumbai and Kolkata both consumed the least energy per person (8.58 and 4.88 GJ per capita, respectively). While comparable data were not available for seven megacities, Kennedy et al. (2015) note the extraordinary growth in total energy use in Moscow, Karachi, and Los Angeles from 2001 to 2011 (1039%, 637%, and 350%, respectively). Of the studied megacities, total energy use declined only in London and Paris from 2001 to 2011 (–12% and –3% respectively), illustrating the difficulty of achieving reductions at the (mega)city scale.

Across the twenty-seven megacities assessed, electricity comprised 23.6% of the 23.4 EJ used in 2011 (Kennedy et al., 2015). Electricity comprised the majority of the total energy used in 2011 in Shenzhen and Kolkata megacities (70% and 65%, respectively), while comprising only a small share in Mexico City, Moscow, Dhaka, and Tehran (8–12%) and a tiny share in Lagos (<1%).

---

5 http://www.ghgprotocol.org/city-accounting

Table 12.3 *Selected urban area energy consumption estimates, 2011 (average gigajoules [GJ] per capita). Source: Rae Zimmerman, with data from Economist Intelligence Unit (EIU) reports (EIU, 2011a, 2011b, 2011c, 2011d, 2011e, 2011f)*

Region	No. of Cities	Ave. GJ/capita	Maximum Users	Minimum Users
North America (Economist Intelligence Unit [EIU], 2011f)	27	52.2	Atlanta (152.4) Orlando (117.7) Sacramento (98.9)	Cleveland (10.3) Minneapolis (23.3) San Francisco (24.5)
Latin America[a] (EIU, 2011e)	17	7.2	Buenos Aires (13.0) Monterrey (12.9) Puerto Alegre (11.8)	Bogotá (3.3) Lima (3.3) Quito (4.2)
Europe[b] (EIU, 2011c)	30	80.9	Dublin (156.5) Ljubljana (105.9) Stockholm (104.9)	Istanbul (36.2) Belgrade (41.1) Warsaw (49.8)
Africa (EIU, 2011a)	15	6.4	Tunis (18.1) Cape Town (13.9) Durban (11.3)	Lagos (0.8) Maputo (0.8) Luanda (1.0)
Asia[a] (EIU, 2011b)	22	66.4	Guangzhou (197.0) Shanghai (169.7) Beijing (124.7)	Calcutta (5.7) Bangalore (9.5) Mumbai (14.2)

[a] Energy usage was computed from numerical values for GDP per capita and Gigajoules per capita; otherwise, they are the average of country values as given directly in the sources.

[b] The selection of cities in Germany for Europe are based on Economist Intelligence Unit (2011c), although there is a German report as well that includes other cities in Germany (EIU, 2011d).

Urban Impacts: GHG Emissions

The environmental impact from urban energy use in terms of GHG emissions requires assessment of both total energy consumption as well as life cycle assessment of the types of fuels and technologies used for energy generation (Heath and Mann, 2012). Cities with similarly sized energy consumption can have dramatically different carbon emissions depending on the carbon intensity of the fuel source (i.e., gCO_2/MJ) (Brown et al., 2008). Cities powered predominantly by hydropower (São Paulo, Rio de Janeiro) or nuclear electricity (Paris) have low carbon intensity and thus low total GHG emissions compared to cities powered predominantly by fossil fuels, and, among fossil fuels, coal-fired electricity (Kolkata, Shenzhen, Guangzhou) has a much higher carbon intensity than does gas-powered electricity (Moscow, Istanbul, Delhi). Comparisons of GHG emissions across cities are challenging not only due to variations in energy use and carbon intensity, as well as electrification rates, but also based on economic typologies (e.g., net-producer, net-consumer, and trade-balanced cities) (Chavez, 2012; Seto et al., 2014).

Table 12.4 highlights selected GHG emissions for cities using data from the EIU reports. The table reveals high variability across cities around the world and the need for better standardization of data. Within Asian cities, for example, there is variation between city GHG emissions levels (8–9 tCO_2 per capita in Beijing, Guangzhou, and Shanghai compared with 0.5–1.1 tCO_2 per capita in Bangalore, Mumbai and Delhi). In Latin American and African cities, urban GHG emissions are remarkably lower. Notably, however, the Latin American and African city data reported by EIU are only carbon emissions from electricity use

and thus underestimate total GHG emissions from all energy production. In places with low electrification rates like Lagos, the noted GHG emissions exclude nearly all emissions associated with energy use in the city.

Energy use and GHG emissions reported by the EIU are closely associated among Asian and among African cities, moderately associated in European and Latin American cities, and not associated in North American cities. The high association among Asian cities appears to reflect the carbon intensity of their electricity sources and the city's development status, with higher income Chinese cities having larger energy use and carbon emissions while lower income Indian cities have lower energy use and carbon emissions. The lack of association among North American cities reflects the wide variability in carbon intensity among all relatively high-income cities. Cleveland in particular illustrates the dichotomy: its energy use is quite low compared to other U.S. cities (10.3 GJ/capita) but its carbon emissions are very high (29.1 tCO_2/capita), reflecting its reliance on carbon-intensive fuels. The generally lower carbon emissions in Canadian cities compared to American cities also reflects source fuel availability and policy choices of Canadian cities to limit environmental impact (see also Section 12.5).

12.4.2 Urban Energy Supply Sector Vulnerabilities to Climate Change

The latest IPCC report suggests that climate change–related vulnerabilities are increasing across the world's urban centers (Revi et al., 2014). That is, cities are increasingly predisposed to be adversely affected by climate impacts (Eakin and Lynd Luers, 2006,

Box 12.3 Energy Supply in Quito, Seattle, and Delhi

Daniel Carrion

Mailman School of Public Health, Columbia University, New York

Hossein Estiri

University of Washington, Seattle

Harvard Medical School, Cambridge, MA

Joshua Sperling

National Renewable Energy Laboratory, Denver

The electrical grid is vast in Quito, reaching 98% of the population. Empresa Electrica de Quito (EEQ), the region's energy provider, planned a grid expansion to 99.5% of the population by 2016 (EEQ, 2012). Overall energy consumption is 3,066.4 GWh per year (United Nations Development Program [UNEP], 2011). Hydroelectric and diesel combustion are the two principal sources of energy production (EEQ, 2012). Only one diesel plant currently operates, with an electrical capacity of 31.2 MW. Five hydroelectric plants carry the remaining load, with one more forthcoming (EEQ, 2014). It is a timely expansion of the electrical grid because population growth and increasing reliance on electrical infrastructure are evident.

Two utility companies supply energy for the City of Seattle: Seattle City Light (SCL) and Puget Sound Energy (PSE). Seattle City Light, a department of the City of Seattle, is one of the largest municipally owned utilities in the United States and is the primary electricity supplier for the city, servicing 340,000 residential and 40,000 commercial customers, with a service area of 340 square kilometers. In general, most of the net electricity generation (more than 45%) in the state of Washington is from hydropower and other non-nuclear clean energy sources. According to the U.S. Energy Information Administration, in 2013, the state of Washington was the leading producer of electricity from hydropower in the United States, where 29% of the net hydroelectricity is generated. As a result, energy sector emissions are lower in Washington than in most other states. Seattle has the lowest electric rates of all twenty-five largest cities in the United States. In 2013, 88.9% of SCL's electricity was generated from hydropower and 8.4% from other clean resources (i.e., nuclear, wind, and landfill gas), while only 1.7% comes from coal.

In 2011, the Central Electric Authority of India projected that Delhi's power requirements would nearly double over a 5-year period (2009–2014) from an average requirement of 4,500 MW to 8,700 MW and therefore began planning ahead. As of April 2013, the North Capital Territory (NCT) of Delhi was estimated to have installed electricity generation capacity of 7,163 MW, with central, state, and private sectors constituting 75%, 23%, and 2% of total capacity, respectively, and with renewable power (including small hydro) representing 10% of the mix. At the state level, total system power capacity reached 18,007 MW by 2012, with roughly 70% of power from coal, 8% from natural gas, 19% hydro, and 3% nuclear.

Between 2005 and 2013, peak electricity demand in the NCT of Delhi grew at a compound annual growth rate of 7%, and peak demand deficit in the state has increased from 2% to 5% over that same period, often resulting in daily power cuts. According to the Delhi Statistical Handbook, the number of Delhi electricity consumers increased from 2,565,000 in 2002–2003 to 4,301,000 in 2011–2012. This included a total of nearly 3,465,000 domestic consumers. Only a year later, the National Sample Survey (administered in July–December 2012) showed that 99% of urban households in Delhi now had electricity access (*note*: this may be an overestimate due to the survey missing harder-to-reach informal/slum households lacking reliable access).

IPCC, 2014b). These impacts are important for the energy sector (see Figure 12.2a), and some countries including the Philippines are conducting vulnerability assessments for the energy supply sector and identifying and pursuing climate-proofing programs to upgrade standards for energy systems and facilities (Petilla, 2014).

Climate-related vulnerability is comprised of three elements: exposure to a hazard, sensitivity to that exposure, and capacity to adapt to the hazard (Fussel and Klein, 2006). The *hazard* or shock is measured in terms of its size, intensity, and duration. Weather- and climate-related hazards could include an increased number or intensity of tropical cyclones, sea level rise, duration and intensity of heat waves, and droughts. *Exposure* refers to the inventory of people, property, or other valued items in areas where hazards may occur (United Nations Office for Disaster Risk Reduction [UNISDR], 2009; Cardona et al., 2012). The *sensitivity* includes the degree to which the agent or system is affected by a hazard (Olmos, 2001). Portions of the population (i.e., the elderly, young and poor) are typically more sensitive to the effects of climate change, such as extreme temperature events. *Adaptive capacity* is defined as the ability of an agent or system to prepare for stresses and changes in advance of the shock or to adjust in the response to the effects of the shock (Smit and Wandel, 2006).

Prior assessments have noted that climate change presents vulnerabilities to urban energy supply in at least three different ways: to primary energy feedstocks, to power generation, and to transmission and distribution networks (Hammer et al., 2011).

Feedstocks: In developing countries, areas heavily dependent on different types of biomass as primary energy feedstocks may be vulnerable if climate change affects the availability of the material. Biomass is sensitive to changing temperature levels if plants reach the threshold of their biological heat tolerance or if storms or drought reduce plant or tree growth levels (Williamson et al., 2009).

Table 12.4 *Selected Urban Area Carbon Dioxide Emissions (Average tCO$_2$/capita)^a. Source: Rae Zimmerman, with data from Economist Intelligence Unit (EIU) reports (EIU, 2011a, 2011b, 2011c, 2011d, 2011e, 2011f)*

Region	No. of Cities	Average annual tCO^2/capita^b	Maximum Emitters (tCO$_2$/capita)	Minimum Emitters (tCO$_2$/capita)
North America (EIU, 2011f)	27	14.51	Cleveland (29.1) St. Louis (27.1) Houston (25.8)	Vancouver (4.2) Ottawa (6.9) Toronto (7.6)
Latin America (EIU, 2011e)	17	0.20^c	Monterrey (0.72) Buenos Aires (0.53) Santiago (0.46)	São Paulo (0) Brasilia (0.01) Curitiba (0.07)
Europe (EIU, 2011c)	30	5.21	Dublin (9.79) Prague (8.05) Lisbon (7.47)	Oslo (2.19) Istanbul (3.25) Ljubljana (3.41)
Africa (EIU, 2011a)	15	0.98^c	Cape Town (4.10) Durban (3.50) Pretoria (3.05)	Maputo (0.000) Luanda (.003) Addis Ababa (.016)
Asia (EIU, 2011b)	22	4.62	Shanghai (9.4) Guangzhou (9.2) Beijing (8.2)	Bengaluru (0.5) Delhi (1.1) Mumbai (1.0)

Notes:

^{a} Units and the base used can differ across regions; therefore, comparisons among cities should only be drawn within regions not between them.

^{b} Units for carbon dioxide are in tons (t) for total energy emissions from all sources; in some cases, other units were given and converted to tons (see notes for Latin America and Africa).

Figures are based on data generally from 2005 to 2009 with some data from 2002, and the dates vary for different cities. For example, the Economist Intelligence Unit (2011f: 16) notes that U.S. city data is from 2002, whereas Canada city data is from 2008.

^{c} For Latin America and Africa, carbon dioxide emission figures are for electricity use only. They were originally in terms of kilograms per capita and converted to tons for comparability for the average and city figures listed.

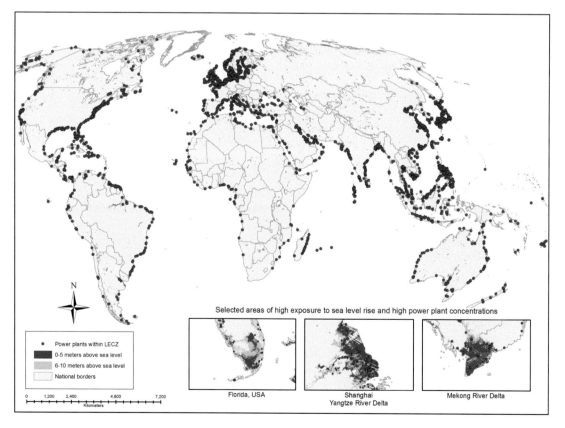

Figure 12.2a *Map of power plants around the world within 10 meters of sea level, 2009.*

Source: This map was constructed using the USGS Global Multi-resolution Terrain Elevation Data 2010 (GMTED2010) digital elevation model at 15 arc seconds resolution using mean elevation per cell and power plant data point file from the Carbon Monitoring for Action (CARMA v.3, 2012). Elevation was extracted between 0 and 10 meters above sea level and identified those power plants within the zone

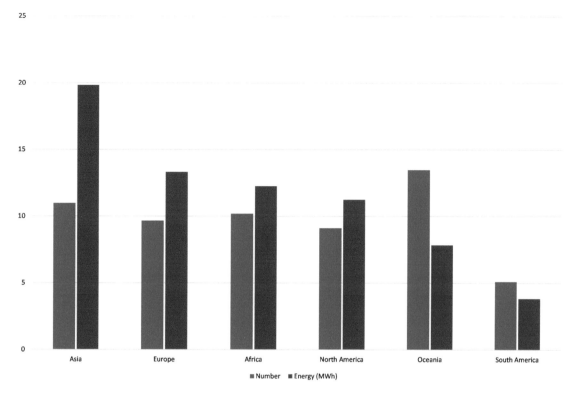

Figure 12.2b *Percent of number and power generation of thermal power plants within 10 meters of sea level, by, region, 2009.*

Source: Global Multi-resolution Terrain Elevation Data 2010 (GMTED2010), The Carbon Monitoring for Action (CARMA v.3, 2012)

Oil and gas drilling operations and refineries are also subject to exposure to extreme events, including flooding and high winds, and are vulnerable to thawing permafrost that can cause damage to infrastructure as well as decreasing water availability given the volumes of water required for enhanced oil recovery, hydraulic fracturing, and refining (Bull et al., 2007; U.S. Department of Energy, 2014). Closure of these facilities and fuel terminals in the Gulf of Mexico during and after Hurricane Katrina were linked to fuel price increases across the United States.

Energy generation: Sea level rise may also expose populations and energy and other infrastructure to risk (see, e.g., McGranahan et al., 2007; Nicholls et al., 2008; Hanson, 2011). The low elevation coastal zone (LECZ), defined as contiguous land areas along the coast that are within 10 meters of sea level, may be exposed to flooding risks, including, for example, storm surge, high tides, and extreme precipitation events, all of which are compounded by increases in sea level over the coming decades. As the map in Figure 12.2a demonstrates, across the world. more than 6,700 power generation plants that provided almost 15% of power generation in 2009 are within this zone. Figure 12.2b suggests that Asia has the highest percent of plant generation within the LECZ; almost 20% of Asian power generation. South America has the lowest share; approximately 4% of Latin American power generation.

Both total electricity demand and peak electricity demand will increase with rising temperatures, although peak demand will increase at a much faster rate (Smith and Tirpak, 1989; Baxter and Calandri, 1992; ICF, 1995; Franco and Sanstad, 2008). Increasing temperatures are associated with higher air conditioning (U.S. Department of Energy, 2013). Research for Boston, Massachusetts, suggests per capita energy demand will be at least 20% higher in 2030 compared to the 1960–2000 average (Kirshen et al., 2008). In developing cities, air conditioning increases rapidly with income. Under modest assumptions about income growth, all warm areas around the globe will reach near universal saturation of air conditioners (Davis and Gertler, 2015). Moreover, heavy use of further air conditioning raises nighttime temperatures by as much as 1°C (Salamanca et al., 2014).

Future high temperatures may affect large urban populations and hence energy demand. As Figure 12.3a and 12.3b suggest, a significant number of cities with large populations may experience extremely warm summers by the end of the century. These estimates suggest that if trends remain unchanged, by 2050, more than 9% of the world's urban population will experience average summer temperatures of more than 35°C. By 2100, the share will increase to approximately 17% of the global urban population and include approximately 1.5 billion people. Of this number, more than 99% are predicted to live in Africa and Asia.

A city's power generation capacity is sized to meet the highest summertime peak demand. By contrast, when peak demand growth outpaces total demand growth, spare capacity is in short supply, increasing the risk of blackouts and brownouts (Miller

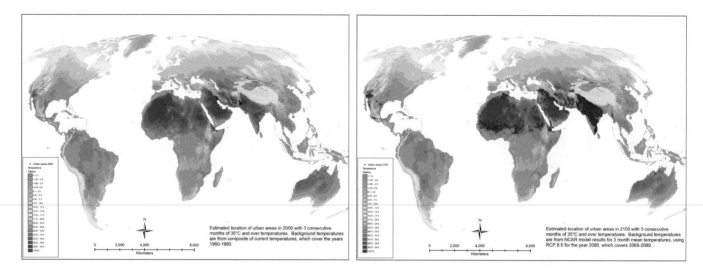

Figure 12.3a *Maps of estimates for cities under extreme summer temperatures (35°C and higher) in 2000 (left) and 2100 (right). The maps were produced by identifying the highest three consecutive mean monthly temperatures for each cell, averaged from 32 global climate models for 2080 (covering 2070–2099) and a composite map for the current temperature (1960–1990). For 2000, the background colors are the averages for the three highest consecutive monthly means for each cell for that year. The 2100 background values are presented using only the NCAR climate model (ccsm4) for Representative Concentration Pathway (RCP) 8.5. Grid cells are 2.5 arc minutes, approximately 5 km². The 2100 map demonstrates the concentration of cities in Africa and Asia that may experience extreme summer temperatures under the RCP8.5 scenario.*

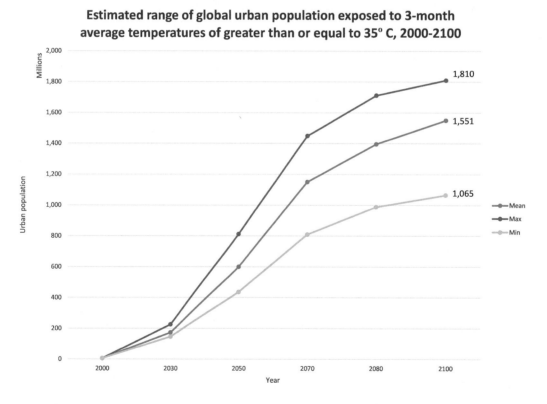

Figure 12.3b *Chart of estimated urban population under different summer temperature conditions, 2000–2100. The figure presents the growth in numbers of global urban residents over time for cities experiencing three consecutive months of 35°C and higher temperatures, with mean, minimum, and maximum estimates of population.*

et al., 2008). In California, a summer 2006 heat wave led to blackouts across the state because sustained high nighttime temperatures prevented the transformers from cooling down before demand increased again the next morning (Miller et al., 2008; Vine, 2012). In developing countries, during 2005, hot and cold

weather explained 13% of the variability in energy productivity (Farrell and Remes, 2009).

Drought can affect power generation due to the cooling needs of plants. Three potential impacts include (1) reduction of stream flow, (2) increase in temperatures downstream of power plants beyond allowable limits due to waste heat exhaust, and (3) increase of water temperature into plants (ICF, 1995; U.S. Department of Energy, 2013). In some situations, power plants may be asked to scale back operations. Moreover, drought is associated with wildfire and increased temperatures that damage energy supply infrastructure (CDP, C40, and AECOM, 2014).

Power generation facilities reliant on renewable resources may be affected by climate change. Hydroelectric facilities fed by glacial and snow-melt have historically benefited from the ability of glaciers to regulate and maintain the water levels of rivers and streams throughout the summer. With increasing temperatures, snow levels are decreasing and glaciers are shrinking, thus jeopardizing the amount of hydroelectric production available to serve many urban areas (Markoff and Cullen, 2008; Madnani, 2009). The densely populated Mediterranean region may face a 20–50% decrease in hydropower potential by the 2070s, although changing precipitation patterns are expected to increase hydropower production by roughly 15–30% in northern and eastern Europe over the same period (Lehner et al., 2001). The elevation at which precipitation occurs is key because retention dams serve different functions based on their elevation and have different water release rules that could affect the availability of power at different times of the year (Aspen Environmental Group and M Cubed, 2005; Franco, 2005; Vine, 2012).

Alternatively, as mentioned earlier, power generation facilities are also affected by too much rainfall. Geoje, South Korea, reported that frequent and intense rainfall reduced operational hours at the company shipyard resulting in delayed deliveries. Typhoon Maemi, in 2003, caused approximately US$20 million in damages (CDP, C40 and AECOM, 2014).

Climate-induced changes may be important in limiting solar and wind production in certain areas. More cloudy days may result in a decline in solar radiation in the United States by 20% (Pan et al., 2004) but only a 2% decline in solar radiation in Norway (Fidje and Martinsen, 2006). Wind patterns may also change. Research finds no clear signal in the Baltic Sea (Fenger, 2007), although onshore wind speeds in the United Kingdom and Ireland are expected to decrease in summer and increase in winter (Harrison et al., 2008). In the United States, wind speeds may decline from 1% to 15% (Breslow and Sailor, 2002).

Energy transmission and distribution networks: Typical electrical transmission losses range from 6% to 15% of net electricity produced (Lovins et al., 2002; International Electrotechnical Commission [IEC], 2007). The effect on aboveground lines is moderated by cooler ambient air, while wires below the ground are cooled by moisture in the soil. As temperatures increase, the cooling capacity of the ambient air and soil declines, conductivity declines, and lines may begin to fail (Hewer, 2006; Mansanet-Bataller et al., 2008; U.S. Department of Energy, 2013). Extreme events take a toll on transmission, including intense precipitation, flooding, storm surge, and high winds (McKinley, 2008;

U.S. Department of Energy, 2013). For example, heavy snows in central and southern China in 2008 blocked rail networks and highways used for delivering coal to power plants, forcing seventeen of China's thirty-one provinces to ration power and affecting hundreds of millions of people in cities across the country (French, 2008).

Increased size and intensity of hurricanes can affect oil and gas transmission. Hurricane Katrina caused the shutdown of major pipelines from the Gulf region resulting in a full disruption of the supply of gas, crude oil, and refined products to other U.S. regions (Hibbard, 2006). On the other hand, Larsen et al. (2008) note that Arctic transport routes and energy infrastructure for moving oil and gas across Alaska are located across areas at high risk of permafrost thaw as temperature rise. System stresses from this impact not only slow supply to cities, but also increase also energy prices.

The result of extreme event impacts on the energy supply sector appears in the form of outages. Understanding recovery patterns and trends is critical to identifying adaptation measures. Simonoff, Restrepo, and Zimmerman (2007) analyzed U.S. electric power outages from a variety of causes and found not only an increasing trend over the years but, since the early 2000s, increasing duration of the outages as well. The ability of electric power systems to recover varies across different types of facilities and circumstances. Zimmerman (2014) analyzed the recovery of electric power in New York citywide and by borough and found high levels of restoration within a couple of weeks after super storm Sandy.

Critical interdependencies with other infrastructure: as mentioned in Hammer et al. (2011), energy systems provide the "life blood" to cities. Energy supply is a critical resilience priority: if energy systems fail, results pose additional stresses on the ability to provide potable water supply, food, transportation, sanitation, communications, health care, and so on. Energy supply disruptions can lead to cascading failures across the economy, government, communities, and multiple other infrastructure sectors. This is not to mention the disproportional impacts of power outages and lack of access to modern energy services on the elderly and poor, especially during periods of extreme heat or other hazard events, as demonstrated by the recent heat waves in India and Pakistan and Hurricane Sandy in New York (see also Section 12.4.2).

Since the urban energy supply sector provides multiple benefits to society and enables improved standards of living (Pasternak, 2000), reducing vulnerabilities can help to avoid cascading failures ranging from reduced hours and services for hospitals (Hess et al., 2011; Schwartz et al., 2011), disruptions of critical human activities (cooking, boiling water, space heating, cooling), reduced transport that enables access to livelihoods, breaks in social networks, and reduction in industrial production and communication (McMichael et al., 1994; Saatkamp et al., 2000; Wilkinson et al., 2007).

Box 12.4 A Framework to Evaluate Climate Vulnerability of Urban Energy Systems

Hossein Estiri

University of Washington, Seattle

Harvard Medical School, Cambridge, MA

This box proposes a framework for evaluating the vulnerability of the urban energy supply sector to climate change impacts. Climate impacts can be categorized into two types: gradual and spontaneous. Gradual impacts occur slowly over time (e.g., incremental changes in temperature and precipitation), and, depending on severity of the effects, local jurisdictions should have enough time to plan for them. These effects can be considered as causing low vulnerability for the energy supply sector unless they are of high severity. In contrast, spontaneous impacts are often extreme and hard to predict (e.g., hurricanes, floods, heat waves, and prolonged droughts). Any spontaneous climate impact can raise the urban energy supply sector's vulnerability. In addition, climate impacts on urban energy systems can be characterized as having both direct and indirect effects (Box 12.4 Figure 1). This framework can be used to evaluate the vulnerability of components of the energy supply sector and the entire system as a whole.

Gradual impacts:
- Gradual damages to energy production and transmission equipment and structures that decrease efficiency and increase distribution losses
- Increases in energy demand for cooling
- Decreases in capacity to generate hydropower resulting in limited ability for cooling thermal plants as a result of changes in snow and rain dynamics

- Availability of biomass for energy generation due to adverse impacts of climate change on agricultural yields

Spontaneous impacts:
- Increases in frequency of extreme events such as hurricanes and floods can directly damage energy production, transmission, and distribution infrastructures, such as nuclear plants, offshore oil drilling platforms, and energy distribution systems and power lines – high vulnerability
- Sea-level rise can also pose a major direct threat to energy supply facilities such as coastal power plants – high vulnerability
- Extreme climate events can influence the urban energy supply sector. For instance:
  - Droughts and floods can produce shifting paradigms in availability of water resources used for cooling thermal plants – medium vulnerability
  - Heat waves can pose significant shocks to the urban energy supply sector by producing temporal peaks in energy demand – high vulnerability

Climate change is a complex phenomenon and so is predicting its impacts. In most cases, the urban energy supply sector is highly vulnerable to spontaneous climate impacts and less so to climate change's gradual effects. Highest vulnerabilities happen when both demand and supply of energy change in reverse directions (i.e., simultaneous increase in demand and decrease in supply). It is crucial for local and regional jurisdictions around the world to systematically evaluate possible impacts of changes in local, regional, and global climate on their energy systems and on other systems dependent on these energy systems, and adopt suitable adaptation strategies to improve their resilience.

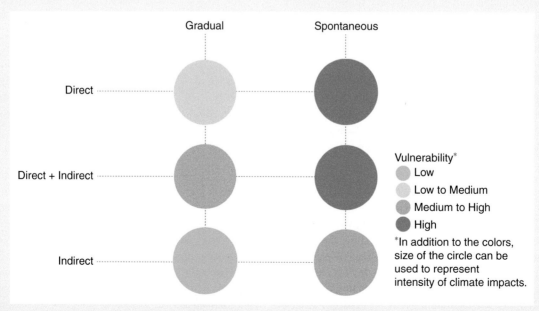

Box 12.4 Figure 1 *A framework to evaluate vulnerability of the urban energy supply sector.*

12.4.3 Urban Energy Supply Access Challenges

Energy access is critical for human development (Modi et al., 2005). Better provision of electricity and clean fuels improves health, literacy, and primary school completion rates. Similarly, better access to electricity lowers costs for businesses and increases trade and investment, driving economic growth and helping to reduce poverty.

Worldwide in 2010, however, more than 179 million urban residents lacked access to electricity and nearly 477 million urban residents lacked access to non-solid fuels (see Table 12.5). The urban population share without electricity has dropped slightly from around 5.7% to 5.1% over the past 20 years, but the large urbanization that has occurred has pushed up the absolute numbers. From 1990 to 2010, the numbers of urban population *with* access electricity jumped from 2.1 to almost 3.4 billion. At the same time, the number without electricity jumped from 127 to 179 million. The greatest gains in access over the past 20 years were in Asia, which also experienced the most intense urbanization. For example, from 1990 to 2010, the percentage without access in South Asia dropped from 13.9% to 6.9%.

Figure 12.4 demonstrates the gains made in populations with electricity access, with a focus on urban households in selected countries from 1990 to 2015 and using data from the USAID Demographic and Health Survey program. For the purposes of comparison, and from just these 13 randomly selected countries – including Indonesia, Peru, Philippines, Ghana, Bangladesh, Senegal, Rwanda, Zambia, Madagascar, Burkina Faso, Uganda, Tanzania, and Malawi -total gains (from 1990 to 2015) for urban residents now having electricity exceeds the

Table 12.5 *Urban population without access to energy, 1990–2010 (millions). Source: World Bank, 2015; UN, 2014*

Region	Electricity			Non-solid fuels
	1990	2000	2010	2010
Developing	127	132	179	442
Least developed	44	68	85	177
Developed	2.0	1.1	0.0	5.0
Africa	56	84	112	191
Asia	63	39	54	223
Europe	2.0	1.1	0.0	5.0
Latin American and Caribbean	8	8	12	28
North America	0.0	0.0	0.0	0.0
Oceania	0.6	0.7	0.8	0.2
Total	129	134	179	447

Note: Calculations were based upon the national percentage of urban population with access to electricity and non-solid fuels (World Bank) and total national urban population (UN).

total urban population of the largest 50 cities in the United States in 2014 (~48.4 million; U.S. Census Bureau). Figure 12.4 also shows that in 1990, the majority of African nations remained well below 50% coverage in terms of urban households with electricity access. Positive trends are recognizable for all selected countries across Asia, Latin America, and Africa, with

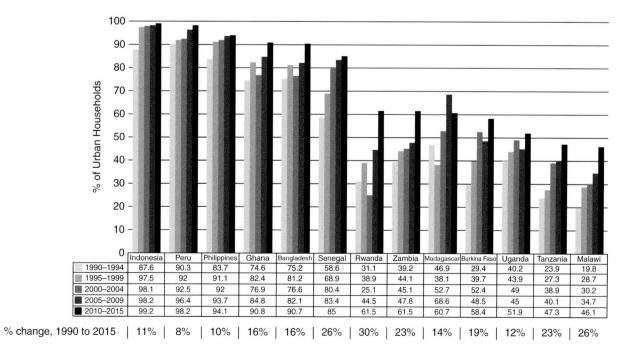

	Indonesia	Peru	Philippines	Ghana	Bangladesh	Senegal	Rwanda	Zambia	Madagascar	Burkina Faso	Uganda	Tanzania	Malawi
1990–1994	87.6	90.3	83.7	74.6	75.2	58.6	31.1	39.2	46.9	29.4	40.2	23.9	19.8
1995–1999	97.5	92	91.1	82.4	81.2	68.9	38.9	44.1	38.1	43.9	27.3	28.7
2000–2004	98.1	92.5	92	76.9	76.6	80.4	25.1	45.1	52.7	52.4	49	38.9	30.2
2005–2009	98.2	96.4	93.7	84.8	82.1	83.4	44.5	47.8	68.6	48.5	45	40.1	34.7
2010–2015	99.2	98.2	94.1	90.8	90.7	85	61.5	61.5	60.7	58.4	51.9	47.3	46.1

% change, 1990 to 2015 | 11% | 8% | 10% | 16% | 16% | 26% | 30% | 23% | 14% | 19% | 12% | 23% | 26%

Figure 12.4 *Increases in electricity access for urban households of select countries of the Global South: 1990 to 2015. Approximately 53.5M more urban residents in these countries may have access to electricity. Reliable 24-hour access is likely lower.*

Source: USAID, 2016; the Demographic and Health Survey Program. Data accessed at: http://beta.statcompiler.com/

almost a doubling in urban coverage in the countries of Rwanda, Tanzania, and Malawi (from 1990 to 2015).

At the same time, the most significant problems with energy access still remains in Sub-Saharan Africa, where in 2010, approximately 36.8% of urban residents lacked access to electricity and nearly 64% of urban residents lacked access to non-solid fuels (World Bank, 2015). The combined power generation capacity of the forty-eight countries of Sub-Saharan Africa is 68 gigawatts (GW), no more than that of Spain. Currently, the installed capacity per capita in Sub-Saharan Africa (excluding South Africa) is a little more than one-third of South Asia's (the two regions were equal in 1980) and about one-tenth of that of Latin America.

Moreover, electricity coverage in Sub-Saharan Africa is skewed to more affluent households. Among the poorest 40% of the population, coverage of electricity services is well below 10%. Conversely, the vast majority of households with coverage belong to the more affluent 40% of the population (Eberhard et al., 2011).

In recent years, external factors have exacerbated the already precarious power situation in Sub-Saharan Africa. Drought has seriously reduced the power available to hydro-dependent countries in western and eastern Africa. Countries with significant hydropower installations in affected catchments – Burundi, Ghana, Kenya, Madagascar, Rwanda, Tanzania, and Uganda – have switched to expensive and highly polluting diesel power (Eberhard et al., 2011).

Access to clean (non-solid) fuels is lowest among poor households in urban areas, especially in slums or informal settlements, and also in peri-urban areas. The types of energy sources used by the urban poor vary with very-low-income households relying exclusively on fuel wood, charcoal, animal dung, and waste materials, and with slightly better-off households using coal, kerosene, and some electricity. Only a small proportion of urban households in low-income nations use electricity or liquid propane gas (LPG) for cooking (Karekezi et al., 2012). Since many informal fuels do not have organized markets, limited information exists on supply chains for biomass and informal fuels in cities (Satterthwaite and Sverdlik, 2013). Sources such as animal dung may be available in the immediate vicinity; however, fuel wood supply can extend to the hinterland, especially for peri-urban households. In Malawi, nearly 60% of charcoal wood for the four major cities came from surrounding areas, including protected areas, forest reserves, and national parks in 2007 (Zulu, 2010). If current trends continue, the number of people in Sub-Saharan African relying on traditional biomass for cooking will increase sharply over the next two decades (Brew-Hammond, 2010). Important to note is that energy justice issues are not only an issue in developing-country cities. Even in the United States, survey results of residents in the city of Detroit show "almost 27 percent of low-income households have fallen behind on utility payments and an additional seven percent have experienced a utility shut-off" (Hernandez, 2015).

Considering that traditional sources will remain in the urban energy mix in the near future, near-term efforts are needed to formalize the sustainable production, promotion, and distribution

processes of informal fuels and to transform cook stoves and available fuel infrastructure choices (Karekezi et al., 2008; Karekezi et al., 2012). In the mid-term, the supply of clean and affordable energy in cities should be prioritized and embedded in local and national development plans. This effort will require investments in commercialization of clean energy complemented with strategies to improve affordability through financial reforms including targeted subsidies, micro-financing to spread upfront costs, and other innovative mechanisms.

Despite or rather because of current trends, energy justice for the world's urban population is high on the international agenda. The new Sustainable Development Goals (7 and 11) elevate energy justice and access to basic services to a high priority. Specifically, Target 7.1 calls for ensuring universal access to affordable, reliable, and modern energy services, whereas Target 11.1 calls for ensuring access for all to adequate, safe, and affordable housing and basic services and to upgrading slums, both by 2030 (UN, 2015).

While there were 17 goals, 169 targets, and more than 200 indicators as of January 2016, and a headline goal on climate, there was no headline goal on air pollution. With many areas undergoing transitions in access to electricity and industrialization, rapid growth in air pollution has emerged as a major challenge affecting urban populations today. From coal briquettes to oil-fired heating, energy and air pollution in cities is arguably more important than GHG emissions, at least in the near term. While climate impacts (e.g., heat and sea level rise) will have direct impacts on infrastructure, people, and energy systems, current challenges in the context of air pollution, especially in cities not meeting air quality standards (e.g., Los Angeles., Beijing, Mexico City) are critical.

12.5 Opportunities, Limitations, and Barriers to Achieving Adaptation, Mitigation, and Sustainable Development Goals for the Urban Energy Supply Sector

12.5.1 Energy Transitions and the Urban Energy Supply Sector

Transitions are breaks or inflections in long-term trends involving complex sociotechnical systems (National Research Academy, 1999). Energy supply transitions are crucial to development. For example, in countries such as China and India, sharp increases in the human development index are being achieved via relatively small increases in energy use (Sperling and Ramaswami, 2013; Steinberger and Roberts, 2009).

Historical transitions in energy supply systems reflect significant shifts in the role of different primary fuels, such as the transition from wood and water power to coal in the 19th century, or innovations in conversion technologies such as electrification in the late 19th century (Verbong and Geels, 2007; Smith, 2010; Grin et al., 2012; Jiusto, 2009). These transitions were due to a combination of the increased efficiency of the new carrier (higher energy intensity), the convenience of its use due to increased supply, the expansion of technologies that facilitated the use of the energy in a variety of forms, and its lower price (Grubler, 2004, 2012; Smil, 2010; Fouquet and Pearson, 2012), and they unfolded over long periods of time (40–70 years) (Fouquet, 2010; Grubler, 2012).

Contemporary energy transitions may not be driven by the same factors as those in the past (Grubler et al., 2012b). Recent urban energy supply transitions are fundamentally different from those of the past. When measured against the developed world experience, cities in the developing world demonstrate consistent and clear patterns of divergence. In developing world cities, the staged transitions between primary fuels (biomass to coal to liquid fossil fuels to natural gas) are occurring sooner during development (at lower levels of economic income), changing faster, and emerging simultaneously as opposed to sequentially (Marcotullio et al., 2005). Given these changes, how long future transitions might take is not well understood.

12.5.2 Energy Scenarios and the Urban Energy Supply Sector

Scenario efforts seek to predict future global energy demands, trajectories, uncertainties, and alternative futures. Examples include the World Energy Outlook (IEA, 2012), the Global Energy Assessment (GEA) (2012), the IPCC's latest concentration pathways (2014a), and the World Energy Council (WEC, 2013). All recent scenarios predict increased demand for energy in the future with the majority of the increase from non-OECD countries.

As the world urbanizes, energy systems increasingly exist to supply energy to urban residents and businesses. Global urban energy demand (end-use) in 2005 was estimated at 240 EJ, serving 3.2 billion urban residents (Creutzig et al., 2015). If current trends continue, including a doubling of the urban population and a dramatic increase in economic development worldwide, global urban energy use may increase more than threefold to 730 EJ in 2050 (Creutzig et al., 2015). Policy and planning interventions, including increasing fuel prices and population density, may enable reduction of that global urban energy demand by 180 EJ, resulting in only 540 EJ by 2050 (Creutzig et al., 2015). Nearly all the potential reductions from this "urban mitigation wedge" are expected in Asia, Africa, and the Middle East, with only minimal reductions likely in OECD nation cities given their expected slow growth.

In addition, more than half of the land area expected to be urban in 2030 remains to be built (Seto et al., 2012), and this has important implications for energy supply, especially at current rates of declining densities among developing-country cities where a doubling of urban population over the next 30 years may require a tripling of built-up areas (Angel, 2012).

As a result, all energy scenarios suggest that how the world's cities develop will be critical to achieving global sustainability (WEC, 2010; Calderon et al., 2014; Seto et al., 2014). In fact, cities have always been at the center of social, technological, and environmental change, and, by 2020, the world's rural population

will begin decreasing, thus making all subsequent future population changes occur in metropolitan areas (UN, 2014). Importantly, the technological, sociopolitical, cultural, and ecological drivers of energy transitions will increasingly be urban-related.

Whether cities will aid in the just and resilient development of energy resources is an open question. According to the GEA (2012), energy resources pose no inherent limitation to meeting the rapidly growing global energy demand as long as adequate upstream investment is forthcoming in exploration, production technology, and capacity for renewable technologies (Rogner et al., 2012). The problem is not the amount but the uneven distribution of resources, resource use, and related environmental impacts.

Energy scenarios differ as to whether mitigation, adaptation, and sustainable development goals can be simultaneously achieved, with most scenarios not yet focused on adaptation. The GEA (2012) states that achieving the objectives of providing almost universal access to affordable clean cooking fuel and electricity for the poor, limiting air pollution and health damages from energy use, improving energy security throughout the world, and limiting climate change are simultaneously possible using a variety of different pathways. Yet the IEA (2012) argues that about a billion people will remain without electricity in 2030 (using their middle range scenario) and that the United States is the only country that may achieve energy security by 2030, given its wealth, technological level, and energy resources. Many plausible futures suggest that business as usual will put the world on a course of high warming by the end of the century, with extreme and potentially irreversible impacts (Calderon et al., 2014). These impacts require new types of energy scenarios addressing not only mitigation, but also adaptation concerns.

12.5.3 Climate Mitigation and the Urban Energy Supply Sector

Urban areas play a key role in global climate stabilization as demand centers for power generation. Four categories of mitigation actions, as described in Chester et al. (2014), are available to governments, households, the private sector, civil society, and communities, and these have shown potential to position the urban energy supply sector on low-carbon trajectories, often with co-benefits for public health, economic development, and other city goals:

1. *Planning actions*: The IPCC describes the planning of low-carbon development patterns as: (1) high population and employment densities that are co-located, (2) compact urban form, (3) mixed land uses, (4) high connectivity street patterns, and (5) destination accessibility to jobs and services
2. *Policy actions*: Regulatory, market-based, and voluntary policy instruments make up the range of policy approaches to mitigation; efforts can include renewable portfolio standards, new pricing structures, smart growth policies, new codes for design and operation of infrastructures, and distribution of home energy meters (Davis and Weible, 2011)
3. *Technology actions*: Such as fuel switching, energy-efficiency upgrades, changing electricity generation mix to rely on lower to no-carbon sources (Jacobson and Delucchi, 2011),

increasing cost-effectiveness of renewables, electric vehicles, and improving storage

4. *Behavior change actions*: Such as residential and commercial building occupant behavior; active transport choices recycling; and shifting diets, lifestyles, purchasing habits, and values (Semenza et al., 2008)

Local plans aimed at energy and its impacts occur in a number of different forms: climate action plans (CAPs), energy plans, and sustainability plans. The existence and coverage of local plans has been sporadic (for a review, see Zimmerman, 2012). For example, in the United States, climate plans exist in 31 states; of those, only 19 had both state and local plans and 5 had local plans only (U.S. EPA, 2011; Zimmerman and Faris, 2011). Although the coverage of CAPs has been controversial (Zimmerman, 2012), many localities, such as Beijing, have adopted energy use and emission reduction policies without a local climate plan (Zhao, 2010). Aznar et al. (2015: 3) point out that "While the CAPs are a good indication of cities' planned actions and goals, they do not fully capture what actions cities actually implement." They also recognize the need to study CAPs more fully to ascertain what they are accomplishing (Aznar et al., 2015). However, they do conclude that the existence of climate, energy, and sustainability plans does spur action (Aznar et al., 2015).

Integrating actions in the development of the urban energy supply sector can reap great benefits. For example, in Shanghai, an international team of experts are designing a distributed low-carbon energy system that matches specific local challenges, using technologically mature and economically viable solutions (see Box 12.5).

While the Shanghai case demonstrates a few types of mitigation actions, all four types of mitigation actions are present in New York's efforts (Table 12.6), where plans, policies, and technologies for urban energy supply sector components (production, transformation, transmission, and end use) are reducing GHG emissions. Given that electricity systems cannot be viewed as isolated from social and institutional systems, behavior change efforts in government incentive programs or utility pricing schemes are also described.

Although these examples demonstrate the wide range of mitigation responses for locations like Shanghai and New York and options that are commercially available now, no single "silver bullet" solution to achieve low-carbon development exists. Instead, cities and energy suppliers must consider multiple approaches for achieving mitigation goals (see Section 12.5.5.4). Listed here are illustrative (not comprehensive) mitigation actions being adopted by cities (Aylett, 2014; NREL, 2015).

Small-scale actions for urban residents and businesses for mitigation in energy supply sector:
- Use less carbon-based energy
- Reuse and recycle goods
- Live close to work
- Walk, bike, carpool instead of driving solo
- Take public transportation
- Consume less meat
- Design buildings to use less energy
- Undertake weatherization and energy audits
- Upgrade heating/air conditioning
- Purchase energy-efficient appliances
- Install/increase use of renewables

Box 12.5   Shanghai, Lingang: An Innovative Local Energy Concept to cut CO_2 Emissions by Half

Marianne Najafi

EDF France, Paris

The Lingang District is one of Shanghai's nine satellite towns, located 70 kilometers southeast of the city's center. The initial development of this greenfield site will be started on a 1 square kilometer section, then extended to a larger district of 42 square kilometers. In order to obtain innovative green urban design proposals for this district, real estate developers launched an International Competition of Conceptual Urban Design in July 2014. The Sino-French team made up of the Urban Planning and Design Institute Tongji of Shanghai; EDF, a French electricity company; and l'AUC, a French architectural firm, won first place in the city's urban conceptual design competition in November 2014.

By taking into account energy at the earliest stages of urban planning and by using systemic approaches, EDF, along with Tongji, and l'AUC were able to put forward a transformative new low-carbon and energy-efficient urban solution. The proposed solutions needed to address two specific local energy challenges: CO_2 emissions reduction and energy system optimization. Three dimensions structured the proposal:

1. *Reducing energy demand* (electricity, heat, cold): The cooperative urban design focused on low-energy solutions, leveraging mixed-space use and higher urban density. Moreover, while following Chinese "green building" standards, new technologies to reduce thermal losses and promote heat recovery were also incorporated into the proposal, which greatly improved global energy efficiency in building heating and cooling.

2. *Proposing the most appropriate local energy mix*: The proposal reduced the use of coal production while harnessing the potential of local renewable energy sources.

3. *Designing the local energy system*: The team's energy system design spatially coupled heating and cooling networks with specifically sized energy centers to optimize efficiency and allow for the solution's easy scaling to the 42 square kilometer Comprehensive Zone during the next phase of construction.

The result of these designs is an estimated 50% reduction in CO_2 emissions, 50% reduction of electricity peak load, and 10% reduction in consumer energy bills.

Table 12.6 *Summary of four types of mitigation actions from New York. Source: Chester et al., 2014*

Mitigation Strategies in Energy Sector:	Planning	Governance/Policy	Technology	Behavior Change
Key Learning, Knowledge, and Policy Networks: Carbonn Climate Registry ICLEI C40	Rezoning for denser development and affordable transit-oriented housing; planning for large-scale wind turbines in federal offshore waters 20 miles outside NYC (anticipated output of 350–700 MW), and renewable biogas from anaerobic digesters at wastewater treatment plants	Local Law 84 (requiring annual large buildings energy use benchmarking) Local Law 85 (meeting energy codes for major renovations), Local Law 87 (requiring energy audits and retro-commissioning once every 10 years), Local Law 88 (requiring lighting upgrades and submeters by 2025) for large buildings	Reducing fossil fuel combustion via construction of new transmission lines providing 1,000 MW of new hydropower; Solar PV growth from 1 MW (2007) to nearly 20 MW (in mid-2013) with SunEdison developing its largest solar PV system of 10 MW and more soon to be added at Freshkills	Pricing, demand-side management; options for customer-sited clean and distributed energy resources (NY State Public Service Commission)

Larger-scale actions (directly or indirectly) achieving mitigation in the energy supply sector include:

- Switch to higher mix of renewables
- Use cleaner fossil fuels (e.g., coal to natural gas)
- Increase use of large carbon-neutral technologies, carbon capture and sequestration (CCS) or carbon capture utilization and storage (CCUS) (see below), and shifts to nuclear power (Zwaan, 2013)
- Increase production efficiencies and minimize transmission losses
- Enforce environmental labeling and use policies (e.g., use of EnergyStar equipment in the United States)
- Target financing (e.g., tax carbon usage according to its social impact)
- Plan urban areas to reduce spatial inefficiencies
- Undertake energy audits of existing buildings and upgrade/ retrofit (e.g., to higher energy efficiency office lighting, energy management systems to control heating/cooling in buildings)
- Increase purchase of and city targets for fuel-efficient vehicle fleets, hybrid electric vehicles, low-carbon transportation fuels, charging stations for electric vehicles
- Increase biking/walking trails, expand dedicated bike lanes on streets and bike parking facilities, improve public transport, install showers/changing facilities for employees, create car free zones, use congestion charging and other travel demand management strategies (Note: such transportation strategies may increase in relevance for the energy supply sector; as for the U.S. electric power industry, projections indicate "a 400% growth in annual sales of plug-in electric vehicles by 2023 may substantially increase electricity usage and peak demand in high adoption areas." (U.S. Department of Energy, 2014)

Whereas the actions listed here show a wide variety of options, which set of policies might work in specific locations is not well understood. Research is needed to demonstrate the range of local contextual factors that are key to usable science for implementation (Dilling and Lemos, 2011), large participation and adoption (Ramaswami et al., 2012), and increased effectiveness of programs (Stern et al., 1985).

12.5.3.1 Reduction in Energy Consumption and Emissions

One of the most direct strategies for reducing energy consumption and emissions is demand side management (DSM). Energy DSM programs can cut across energy and other sectors, such as buildings, transportation networks, and other infrastructure installations for heating, ventilation, and air conditioning (HVAC) and lighting. Alternatively, technological actions have also been very popular, including, for example, light-emitting diode (LED) installations in buildings and on streets. Such advances have served to reduce energy consumption in North American cities, including Calgary and Denver (EIU, 2011f: 41). Beijing estimates its "Green Lighting Programme" will save the city 39 MW of electricity per year (Zhao, 2010). Toronto has used cold water to provide air conditioning, estimating a savings of 61 MW per year (EIU, 2011f: 21). Planning of residential building retrofits combined with behavioral education resulted in a decrease in electricity and gas use in the Department for Communities and Local Government (DCLG) "low-carbon communities" program in the United Kingdom, although the wide range of demonstrated impacts illustrate the importance of broader behavioral and contextual factors in determining energy demand (Gupta et al., 2014).

Los Angeles has deployed integrated environmental, land-use, and development planning strategies to reduce overall energy consumption and has pursued infrastructure revitalization that reduces energy demand (EIU, 2011f). Similar "community energy management" strategies that combine land-use planning with community-based energy technology/infrastructure investments like DHC were estimated to result in an average energy consumption reduction of 15–30% and associated CO_2 emissions reduction of 30–45% from 1995 to 2010 in four communities in British Columbia (Jaccard et al., 1997). Rio de Janeiro has also sought to reduce energy consumption and associated emissions through a variety of infrastructural changes including modernization of its electricity network (see Case Study 12.1).

Globally, mitigation efforts explore the use of pricing schemes such as carbon taxes and cap-and-trade markets that discourage energy use and associated carbon emissions through

Case Study 12.1   Urban GHG Mitigation in Rio de Janerio

Andrea Nuñez

Catholic University of Honduras, San Pedro Sula

Keywords	Energy distribution, informal settlements, drought
Population (Metropolitan Region)	11,835,708 (IBGE, 2015)
Area (Metropolitan Region)	5,328.8 km² (IBGE, 2015)
Income per capita	US$8,840 (World Bank, 2017)
Climate zone	Am – Tropical monsoon (Peel et al., 2007)

Rio de Janeiro represents one of the dynamic axes of the Brazilian southeast region whose economy is the second most prosperous of the country. As the energy matrix of Brazil is already rather clean, depending mostly on renewable resources, one of the key energy challenges has to do with improving efficiency in energy distribution (ENEL Foundation Research Project, 2014). At the same time, and despite the progress of the past decade, recent demonstrations and upheaval signaled that Rio still faces severe social challenges. For example, despite ongoing state programs, roughly 20% of the city's population lives in informal settlements known as *favelas* or slums, with very limited access to public services. Electricity distribution is often informally accessed. Electricity infrastructure is generally available; however, until very recently, the nontechnical losses in some favelas (largely related with electricity theft) reached 95% of the total electricity in the grid (Light, 2013).

There are manifold distribution challenges in large cities such as Rio, and wasteful consumption and nontechnical losses (e.g., energy theft) are high, imposing large costs for utilities, governments, and ratepayers (Coelho, 2010). Moreover, distribution grids require modernization to cope with societal expectations and new regulatory frameworks on smart metering, hourly tariffs, distributed generation, and the like. Apart from the costs, local government and regulators have been actively paving the way toward smart-grid solutions (e.g., through new regulations and RandD funding), but there is a long way to go (ENEL Foundation Research Project, 2014).

In the current municipal administration, climate change measures are coordinated by Rio de Janeiro's Municipal Secretariat of the Environment (Secretaria Municipal de Meio Ambiente [SMAC]) through its Climate Change Office (CCO), along with the Mayor's Office (EIA, 2013). This management involves transversality with different areas of municipal administration and partnerships with academic institutions through shared actions and innovative activities in several sectors, such as solid waste management, transport, urban planning, energy, and civil defense, among others. The goal is to achieve sustainability, mitigation of GHG emissions, and adaptation to climate change impacts (Cities, 2014). Management actions are based on the development of a regulatory framework to enable feasible actions. the CCO also develops links with institutions of excellence in the public and private sectors and with civil society organizations.

The main piece of the Regulatory Framework is Law n. 5.248/2011 that establishes the Climate Policy of the City and sets measurable, reportable, and verifiable reduction targets for GHG emissions for 2012 (8%), 2016 (16%), and 2020 (20%) based on emissions recorded in Greenhouse Gas Inventory of Rio de Janeiro City, published in 2011 (City of Rio de Janeiro, 2011). The Law also establishes city adaptation policies to face climate change effects (Cities, 2014).

Rio's GHG Inventory was developed by the Centre for Integrated Studies on Climate Change and the Environment (Centro Clima/ COPPE) and Federal University of Rio de Janeiro (UFRJ), with alternative scenarios due to emissions mitigation actions in different sectors and across the city as a whole. It presents the emissions resulting from transportation and from residential and commercial buildings, public buildings, and refineries, and tracks land use and forestry; residential, industrial, and commercial wastewater; industrial processes; and solid waste (COPPE, 2011). Alternative scenarios consider projects and actions incorporated into the municipality's planning. Law n. 5.248/2011 establishes the elaboration, updating, and publication of the GHG Municipal Inventory every 4 years, but it does not determine specific targets for emissions that are the responsibility of the municipal administration. Most of the reduction in emissions will occur as a result of governmental actions, mainly infrastructural changes, such as new bus rapid transit (BRT) systems, subway expansion, and modernization of public lighting.

Despite the clean electricity matrix, new generation sources and energy efficiency policies are increasingly important issues in Brazil (Geller et al., 2004) not only because of the source of the energy but also because of the size of the facility and its impacts. For example, building new dams faces strong social and political opposition (notably in Amazonia), and dry seasons bring hydro-generation into jeopardy. Recently, one policy-supported solution is to use mini- and micro- (hydro) generation (e.g., in rivers). This issue is not only a matter of finding new sources, but is also search for better energy efficiency and innovation in electricity distribution. This search is increasingly supported by governments and the regulator Aneel (Agência Nacional de Energia Elétrica/National Agency for Electrical Energy), for example, through appliance-substitution programs for low-income households, educational initiatives, tight regulatory standards for energy losses, and technological innovation on many fronts (e.g., Aneel, 2008).

Aneel reports that:

1. 0.5% of the operational profits of distribution companies should be applied to actions and programs to fight energy inefficiency and wasteful consumption (with a special emphasis on low-income households)
2. 0.2% of the operational profits of distribution companies should be applied to research and development (including not only "pure" research and prototypes but also market applications such as smart grid/smart city projects)
3. Only a maximum amount of about 30% (e.g., for Light S.A.) of nontechnical energy losses (e.g., theft) can be transmitted to the distribution tariff, pushing the development of innovative solutions (i.e., different types of peak–off peak hourly tariff, micro- and mini-generation, and smart metering) to tackle the issue

Improving the quality of energy distribution (i.e., reliability, cost reduction, and efficiency) and reducing energy losses are the key drivers for the country's recent investment in the development of smart grids (CGEE, 2012).

SMART GRID PROGRAM OF LIGHT S.A.

To reduce losses and improve operational efficiency, Light Sociedade Anónima (Light S.A.), one of Brazil's main electricity distribution companies, which supplies the City of Rio, has recently formalized a smart grid program following-up on the company's past efforts to implement remote metering solutions in its concession area. The program comprises the development, prototyping, and early application of a portfolio of different technologies and solutions, including grid automation and smart metering technologies, charging stations for electric vehicles, distributed energy generation, and demand-side management. Light S.A. cooperates with other companies in the energy industry (CEMIG, AXXIOM) and other national and international technology providers and knowledge institutes (e.g., Lactec, CPqD, CAS, GE).

Currently, the Brazilian Agency for Industrial Development (Agência Brasileira De Desenvolvimento Industrial [ABDI]) is studying the development of a tailor-made industrial policy for the smart-grid field, and these regulatory efforts, first pilots, and R&D count already on strong financial support from the Brazilian Science and Technology Policy. Another very important initiative has been the Inova Energia program, a joint initiative of Aneel, the Funding Authority for Studies and Projects (Finep), and the Brazilian Development Bank (BNDES), which supports the development of different smart city pilots in the country, including in Rio de Janeiro (where important distribution companies and technology players are located). Both Ampla – another energy company that supplies part of the metropolitan region and Rio state, but not the city – and Light S.A. have active smart grid programs benefitting from government funding and close cooperation with national and international technology partners.

Current national government policies regarding energy use go from energy loss prevention initiatives that unfolded in Rio in the past decade to current activities that aim to reduce emissions through local government operations (Case Study 12.1 Table 1).

Curbing nontechnical electricity loss in informal settlements is an issue for many cities around the world. Rio de Janeiro is an example that shows that good technical solutions are required, and the implementation of technology has necessarily to go hand-in-hand with societal embedding efforts: building utility–community relationships and instituting behavioral and cultural change.

Case Study 12.1 Table 1 *Energy Emissions Inventory for Rio de Janeiro. Source: Carbon Disclosure Program CDP (Cities, 2014)*

Emissions reduction activity	Projected emissions reduction over lifetime (metric tons CO2e)	Action
Outdoor Lighting > LED / CFL / other luminaire technologies	640,000	Estimated value for 2020.
Other:	100,000	Project Morar Carioca provides urbanization for slums, and entails full reurbanization and waste management, public lighting, water, drainage, garbage collection, slope contention and public equipment. This project "My House, My Life," is part of a federal housing project that provides homes to those previously living in high-risk areas of slums. Estimated value for 2020.
Energy Supply > Transmission and distribution loss reduction	11,400,000	Fugitive emissions. Estimated value for 2020.
Transport > Improve rail, metro, and tram infrastructure, services and operations	529,700,000	Establishment of four new BRT systems: TransOeste (150,000 riders / day), TransCarioca, 1st phase (380,000 riders/day), TransCarioca, 2nd phase (150,000 riders/day), TransOlímpica (100,000 riders/day), TransBrasil (900,000/day). the BRT System is being implemented in Copacabana. Subway expansion (230,000 riders/day). Acquisition of new subway trains (+550,000/day) Expansion of cycle lanes, 300 km. Estimated value for 2020.

higher energy prices. For example, the Tokyo Metropolitan Government, in 2010, developed the world's first cap-and-trade program at a city level targeting energy-related CO_2 as a market-based approach to mitigation. Other cities developing systems for carbon emissions trading and offsets include Chicago, London, Sydney, and Tianjin (Broto and Bulkeley, 2013). The European Union's Emission Trading Scheme provides a model and important lessons for other emissions trading efforts, the most important of which is to establish a reliable, quantitative emissions baseline and tracking and verification system (Brown et al., 2008).

An important mitigation strategy for energy-related emissions is to lower the carbon intensity of the urban electricity supply, with efforts under way in London and Milan (Croci et al., 2010), throughout the Netherlands (Rotmans et al., 2001), in Beijing (Zhao, 2010), and now across all of China (Hsu, 2015). Kennedy, Ibrahim, and Hoornweg (2014) identify two key thresholds by which cities will be more effective in pursuing strategies for low-carbon development (Box 12.6).

Another remediation strategy is the deployment of carbon capture and sequestration (CCS) or carbon capture utilization and storage (CCUS). These processes captures CO_2 emissions from sources like coal-fired power plants, and store, reuse, and remove the emissions from the atmosphere. Storage is typically provided in geologic formations including oil and gas reservoirs, unmineable coal seams, and deep saline reservoirs. Today CCS and nuclear generation are the only large-scale technologies that are believed to significantly reduce the emissions from fossil fuels. CCS is, however, still at the pilot stage in places like Rotterdam and Shanghai, and its future is uncertain, mainly because of the high costs (WEC, 2013). Lower cost carbon sequestration options include urban greening and reforestation, practiced throughout the world in cities like Bogota, Quito, São Paulo, Cairo, Lagos, Johannesburg, London, Madrid, and Hong Kong (Broto and Bulkeley, 2013). Some urban greening efforts can be quite expensive, however, and thus should be given careful consideration as part of a city's mitigation plan (Kovacs et al., 2013).

12.5.3.2 Use of Renewables

To achieve a warming level of below 2°C by 2050, the recent IPCC (2014a) reports suggest that future scenario trajectories needed to reduce GHG emissions by 40–70% relative to 2010 levels are defined by a global share of low-carbon electricity supply comprising renewable energy, nuclear, natural gas, and some use of carbon capture and sequestration. Currently, in most countries, however, renewables account for less than 10% of the energy supply and usually less than 5%. The use of renewables by cities varies sharply around the world. For example, about half of the seventeen Latin American cities identified by the EIU draw more than 80% of their electricity supply from renewables (EIU, 2011e). In 2015, renewables excluding large hydro accounted for (the first time) a majority of new electricity-generating capacity (UNEP, 2016). We list here examples of a variety of renewable energy programs in cities around the world.

Solar energy: Solar energy is growing as a renewable source. A number of African cities are installing solar water heaters (EIU, 2011a). In the United States, Environment America (2015) identified fifty-seven cities installing solar photovoltaic systems ranging from less than 1 to 132 MW of cumulative capacity. Cities with solar installations exceeding 90 MW at the end of 2013 were Los Angeles, San Diego, Phoenix, San Jose, and Honolulu. Boston's Solar Boston initiative supports the adoption of solar power through permitting, financing, technology development, and implementation that increased capacity in 2010 to 3.1 MW with a 2015 projected increase to 25 MW (EIU, 2011f). Chicago built the United States' largest solar power plant in an urban area, estimated to provide 10 MW of power and save 14,000 tons of GHG emissions annually (EIU, 2011f). Minneapolis built the largest Midwest solar array, located at the top of its convention center and saving 540 metric tons of CO_2 annually (EIU, 2011f).

Hydropower: A number of cities especially in Latin America and parts of Africa use hydropower as a main source of energy. Hydropower supplies 100% of São Paulo's power (EIU, 2011e). The Economist Intelligence Unit (2011a: 16) estimates the use of hydropower at 69% in seven Sub-Sahara African cities (excluding South Africa). Seattle also produces the majority (89%) of its electricity from hydropower (see Box 12.2 and Case Study 12.2), although climate change impacts in the region may force diversification of its electricity sources to accommodate less hydropower generation in summer, estimated at 12–15% by 2040s and 17–21% by the 2080s (Hamlet et al., 2010).

Wind: Wind energy is considered the fastest growing renewable energy source, although its share of energy is still low. Several African cities are planning wind energy facilities; for example, Cape Town will be installing its country's first commercial wind plant (EIU, 2011a). Beijing is increasing its wind generating capacity at its Guanting Wind Farm and other nearby farms to more than 115 MW capacity, expecting to reduce CO_2e emissions by nearly 125 kton per year (Zhao, 2010).

Waste: Many areas, including Buenos Aires, Denver, Phoenix, Birmingham (UK), Dhaka, Cape Town, Hong Kong, Mumbai, Mexico City, and Ho Chi Minh City are capturing landfill gas to produce electricity or heat, serving to reduce the GHG emissions from the waste sector and reduce the demand and associated emissions from other energy fuels (Broto and Bulkeley, 2013). Beijing is recovering methane from chicken manure, reducing CO_2 emissions by approximately 88 kton per year (Zhao, 2010). Similarly, New York is experimenting with producing renewable gas from wastewater (see Case Study 12.3).

Renewable energy targets: Many countries have adopted renewable energy targets that will influence the cities located within those countries. The targets vary from percentage share of the renewable fuels category to fuel-specific target quantities. From 2005 to 2015, the number of countries adopting such targets increased from 43 to 164 countries with 12 more in non-OECD countries (International Renewable Energy Agency, 2015). Subnational governments are also adopting renewable targets, as with the 90% renewable electricity target by 2020 adopted by Australia's capital city, Canberra (see Case Study 12.4). To date, more than 15 ICLEI — Local Governments for Sustainability cities and regions have committed to using 100% renewable energy between 2020 and 2050. A national renewable energy standard is still being debated in the United States, as well as a clean power plan rule by 2030 to limit carbon emissions from power plants

12.5.3.3 Emission Reduction Targets

Many cities have now set specific GHG emissions reduction targets that could be achieved using a variety of mitigation strategies. C40, ARUP, and others (2014) estimated cumulative savings for 228 cities given the targets of 2.8, 6.1, and 13.0 Gt CO_2-equivalent by 2020, 2030, and 2050, respectively. Annual reductions estimates were 454 Mt CO_2e per year, 402 Mt CO_2e

Box 12.6   Low-Carbon Infrastructure Strategies for Cities: Carbon Intensity and Population Density Is Key

Christopher Kennedy

University of Victoria, British Columbia

Daniel Hoornweg

University of Ontario Institute of Technology, Toronto

Sustainable Development Network, World Bank, Washington, D.C.

The urban development and technological strategies that cities can pursue to reduce their greenhouse gas (GHG) emissions or grow with low-carbon trajectories differ depending on urban form, environment/climate, technological, economy (e.g., energy pricing), governance, and sociodemographic factors. Among the most important determining characteristics of city GHG emissions are the carbon intensity of electricity supply and the population density of the urbanized area (see Box 12.6 Figure 1). Cities where the carbon intensity of power grids is below approximately 600 $tCO_2e/$GWh can broadly pursue electrification and alternative space-heating strategies for low-carbon development (e.g., adopt electric vehicles or use heat pumps in the place of natural gas furnaces). Above about 600 $tCO_2e/$GWh, electrification becomes self-defeating; overall emissions are increased due to the high carbon content of the electricity supply. Other strategies for district energy systems or substantial low-carbon public transportation systems are broadly only economically viable at medium to high urban densities of more than about 6,000 persons per square kilometer. Based on local conditions and aspirations, strategies and key leverage points for low-carbon development will differ for cities such as Denver, Toronto, Rio, and Beijing.

In cities expected to undergo rapid growth (e.g., Dar es Salaam) and with commensurate needs for increased electricity supply, the future expected GHG intensity of electricity supply needs to be considered. In many fast-growing cities potential future hydroelectric opportunities may be exhausted, and the carbon intensity of new supply could vary markedly (e.g., likely mix of coal, natural gas, nuclear, renewables).

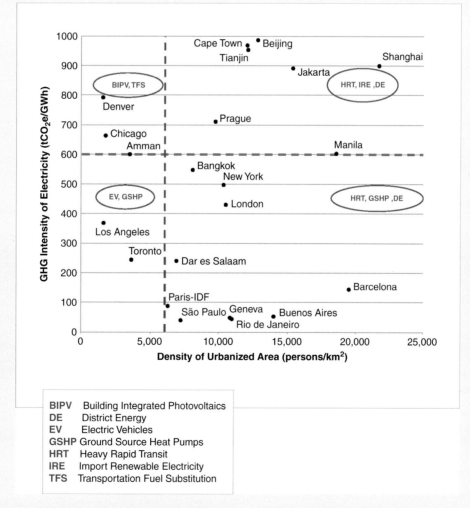

BIPV	Building Integrated Photovoltaics
DE	District Energy
EV	Electric Vehicles
GSHP	Ground Source Heat Pumps
HRT	Heavy Rapid Transit
IRE	Import Renewable Electricity
TFS	Transportation Fuel Substitution

Figure 1 *Examples of low-carbon infrastructure strategies tailored to different cities. Prioritization according to both urban population density and the average greenhouse gas intensity of existing electricity supply. Both factors need to be taken into account in developing sustainable urban energy solutions.*

Source: Adapted from Kennedy et al., 2014

Case Study 12.2 Climate Change and the Energy Supply System in Seattle

Hossein Estiri

University of Washington, Seattle

Keywords	Energy supply, adaptation, mitigation, hydro power
Population (Metropolitan Region)	3,043,878 (U.S. Census Bureau, 2010)
Area (Metropolitan Region)	15,208 km² (U.S. Census Bureau, 2010)
Income per capita	US$56,180 (World Bank, 2017)
Climate zone	Csb – Temperature, dry summer, warm summer (Peel et al., 2007)

Situated on a narrow isthmus between Puget Sound and Lake Washington, Seattle is the seat of King County and the largest city in the state of Washington. With an estimated population of 652,405 residents (and 3.61 million in the metropolitan area) in 2013, Seattle is one of the fastest growing major cities in the United States and home to some of world's most recognized technology companies, including the Boeing Commercial Airplane Group, Microsoft, and Amazon. Seattle has a milder and wetter climate than many other parts of the world, with less extreme variations in temperature and more cloudy days.

Seattle is a leader in climate action in the United States and globally, with several incentive programs and plans dating back to 2000. In 2006, Seattle was one of the first cities in the United States to adopt a climate action plan (CAP; City of Seattle, 2013) The city is also among the nine major cities in the United States that have passed building energy benchmarking and disclosure policies.

Climate change is already taken seriously by city officials in Seattle because its impacts on the Puget Sound area are being observed. Some of the most important climate impacts on Seattle's energy supply system are more variable and generally reduced mountain snow pack, earlier and faster spring melt of mountain snow pack, and reduced summer river levels for hydro power. The City of Seattle (and Seattle City Light [SCL] as one of it departments) is taking action to reduce its impact on climate change and reduce its adverse effects on the city. Seattle's climate mitigation and adaptation strategies focused on its energy supply system. These are either part of the Seattle CAP (with a holistic approach to the entire metropolitan region) or independently developed by SCL's active climate change program.

ENERGY SUPPLY AND CLIMATE MITIGATION IN SEATTLE

According to a report by the American Council for Energy-Efficient Economy (ACEEE) in May 2015, Seattle is the fifth most energy-efficient city in the United States (Ribeiro et al., 2015). Seattle's climate mitigation strategies for its energy supply systems focus on promoting energy conservation and clean energy resources.[6] In 2005, SCL was the first large electric utility in the United States to become carbon neutral. SCL's conservation program is among the longest-running

in the country. Since 1977, SCL's has been taking actions to reduce demand for fossil fuels that contribute to climate change.

To improve energy efficiency and reduce energy demand in buildings, the City of Seattle's CAP (2013) has envisioned that, by 2030, information from the Energy Benchmarking reports will be publicly accessible and disclosing home energy use or a home energy rating at the point of sale for single-family homes will be required.

By 2030, CAP also has envisioned that Seattle buildings will be using a portfolio of renewable and low- or no-carbon energy sources, and that these clean energy sources will be provided either by SCL's maintained carbon-neutral electricity or by neighborhood district energy systems that use renewable and waste heat sources, such as district energy, solar energy, and geothermal energy.

ENERGY SUPPLY AND CLIMATE ADAPTATION IN SEATTLE

Seattle's energy supply system is highly reliant on the local climate, particularly the timing, type, and amount of precipitation. In 2009, SCL contracted with the University of Washington's Climate Impact Group to study climate change effects on regional climate, streamflow, and stream temperature in order to support SCL's assessment of impacts of projected climate change on operations at its hydroelectric projects and on future electricity load in its service territory.

SCL has incorporated the results of its continued research on climate impacts into its Integrated Resource Plan (IRP) (Seattle City Light, 2014) and its updates. According to the SCL's 2014 IRP update, SCL hydropower generation is threatened by changes in snowpack and glaciers due to long-term climate change. River flows and generation are expected to gradually increase during the winter and decline in the summer due to the overall warming. Winter is currently the peak season for electricity use in the Puget Sound area, but climate impacts can change this seasonality. SCL is working with the National Park Service and the University of Washington to inventory and forecast future flows from the glaciers and snowpack.

SLC has identified its main concerns related to climate impacts. Due to warmer temperatures and more frequent heat waves, higher energy demand in summer and de-rating of the overhead lines, which can lead to reduced transmission capacity, are plausible. These impacts can specially create high risk for vulnerable populations. Due to the expected increase in winter rain, lower snowpack, and loss of glacier runoff, hydroelectric generation is expected to increase in winter, but decrease in the summer, when energy demand is likely to increase. Also, frequency of summer water conflicts and spilling for flood control are expected to upsurge. Due to the risk of more frequent wildfires, landslides, and floods, sea-level rise, higher frequency of transmission and distribution outages, and equipment damage (or reduced life expectancy of equipment) are likely to occur.

Some of SCL's adaptation strategies to prepare for climate impacts and reduce its adverse effects are:
- Developing a utility-wide adaptation plan
- Leveraging present tools to plan for hydro-climatic variability and prepare for high winds and storms
- Establishing a rate stabilization fund
- Relicensing with the Federal Energy Regulatory Commission (FERC)
- Upgrading with new equipment for landslides and lightning
- Assessing fire risk and preparing for greater fire frequencies

---

6 See Box 12.3 for details on Seattle's energy supply.

Case Study 12.3   Renewable Gas Demonstration Projects in New York

Annel Hernandez

New York City Environmental Justice Alliance (NYC-EJA)

Keywords	Renewable energy, wastewater management, organic waste management, infrastructure, Newtown Creek
Population (Metropolitan Region)	20,153,634 (US Census Bureau, 2016)
Area (Metropolitan Region)	17,319 km² (US Census Bureau, 2010)
Income per capita	US$56,180 (World Bank, 2017)
Climate zone	Dfa – Continental, fully humid, hot summer (Peel et al., 2007)

The development of the Newtown Creek Renewable Natural Gas Project is an example of how sustainable waste management strategies can produce renewable energy. Wastewater treatment operations maintain safe and clean waterways in urban areas to address public health concerns, create more livable spaces for city residents, and protect marine ecosystems. Wastewater treatment plants (WWTP) processes also create byproducts that include sludge, biosolids, solid waste, and methane. In New York, thirteen of the fourteen Department of Environmental Protection (DEP) WWTPs utilize the methane byproduct of operations to power boilers and other plant equipment.

The largest of these plants, Newtown Creek Wastewater Treatment Plant, has the daily capacity to treat 330 million gallons of wastewater and presently produces 500 million cubic feet of biogas annually, or on average 1.37 million cubic feet of biogas daily. The biogas, otherwise known as renewable gas or biomethane, is produced in anaerobic digesters where the organic sludge removed from treated water is heated to 95°F for approximately 15–20 days. During this process, microorganisms convert organic matter into biogas, which is about 50–60% methane and 40–50% carbon dioxide. Currently, the Newtown Creek WWTP applies 40% of this excess biogas to powering the plant operations, with the remaining amount of biogas flared into the atmosphere.

New York City's DEP and the international utility company National Grid have struck a long-deliberated deal to conduct the Newtown Creek Renewable Natural Gas Project. Under the agreement, both entities have entered into a 20-year contract that states National Grid will fund the design, construction, operation, and maintenance of the new demonstration. Moreover, DEP will provide the biogas free of charge for the first 5 years after renewable gas operations commence. After the 5-year mark, any operational surplus will be divided equally between both organizations. One of the reasons this project is viable is the proximity of both operations – the Newtown Creek WWTP facility is located just a few city blocks from National Grid's Greenpoint Energy Center. Furthermore, the existing pipeline infrastructure running alongside the Newtown Creek WWTP streamlined the project because there was no additional private or public land needed to develop the project. AECOM, an international infrastructure firm with previous experience working with the Newtown Creek WWTP, was contracted by National Grid to partner with Ennead Architects to develop the project. The project encountered various challenges and setbacks along the way due to the complex regulatory context in the state. This project is setting a new precedent for public–private partnerships between energy utilities and water utilities. The project began operations in the autumn of 2016.

The Newtown Creek Renewable Natural Gas Project is expected to reduce GHG emissions by 90,000 metric tons each year, which is equivalent to removing 19,000 cars from the city's crowded roads or planting 2 million additional trees with 10 years worth of growth and its associated carbon uptake. Furthermore, the project is estimated to produce enough renewable gas to heat 5,200 homes in the city, with more energy potential in the future. The project strives to ensure the 100% of the biogas is utilized in efficient ways (New York City Department of Environmental Protect [DEP], 2013).

Many consumers have the perception that renewable gas is of inferior quality to the more accepted natural gas found in fossil fuel reserves, frequently alongside oil operations. Although renewable gas has a slightly different composition than natural gas it is of the same quality because both are predominantly methane and are derived from the decay of organic matter. Once the anaerobic digestion process is complete, the renewable gas enters the upgrading and cleanup process before being injected into the gas distribution pipelines. First, the renewable gas enters the compression phase, followed by the gas-drying phase that extracts any remaining water (H_2O). Then the renewable gas enters the cleaning and conditioning phase where the methane (CH_4) is separated from the remaining CO_2 through a process known as *pressure swing adsorption* (PSA). The remaining CO_2 or tail gas is then flared into the atmosphere. The now pipeline-quality renewable gas in odorized with the distinct gas scent, as is all natural gas for safety concerns, before being injected into the distribution system. Additionally, to maintain and monitor the quality of the renewable gas, National Grid will conduct analytical chromatography, sample gas, and install meters (National Grid, 2014).

The Newtown Creek Renewable Natural Gas Project also provides an alternative to the growing costs of local solid waste management. In recent years, the New York City Department of Sanitation (DSNY) has been expanding the collection of organic waste and investing in modern processing. The city hopes to lower costs by diverting organic waste from landfills and reducing the cost of exporting the total solid waste streams. Organic food waste management is a major issue because it accounts for approximately 25–30% of New York's entire waste stream. Previous co-digestion studies show that organic food waste coupled with organic wastewater streams increase the rates of methane production and decrease the costs of solid waste management. To support the demonstration project, Waste Management of New York (WMNY), in partnership with DEP and DSNY, opened a specifically designated organics collection facility that is not geared toward composting efforts. Instead, organic food scraps are converted into a liquefied feedstock or *engineered bioslurry* at the Varick I transfer station, also situated along Newtown Creek, and sent to the WWTP. WMNY utilizes patented technology called the Centralized Organic Recycling equipment (CORe) process, which will potentially handle about 250 tons of organic waste per day.

The Newtown Creek WWTP has the long-term potential daily capacity to process 500 tons of organic waste, with short-term potential estimated at a daily capacity 250 tons within the next few years.

Case Study 12.3 Figure 1 *The Newton Creek Wastewater Treatment Plant. Source: National Grid*

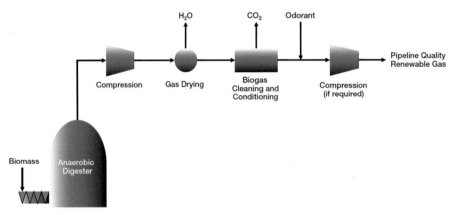

Case Study 12.3 Figure 2 *A diagram illustrating the process in which waste is broken down and converted into useable gas. Source: National Grid*

An organic waste diversion strategy of 153,000 tons annually would account for approximately 54,500 metric tons of GHG reductions (Waste Management of New York, 2014). In the United States, there have been various local success stories of anaerobic digestion and co-digestion applications (U.S. EPA, 2014).

Once the Newtown Creek Renewable Natural Gas Project starts operations, the policy-making sphere and the energy industry will closely monitor progress and performance. This project presents renewable gas production as a feasible, replicable, and scalable strategy for cities internationally.

per year, and 430 Mt CO$_2$e per year in 2020, 2030, and 2050, respectively (see Table 12.7). They provide specific emission reduction targets for 28 cities ranging from 20% to 100% over 10- to 60-year periods using baselines from 1990 to 2014 (C40 Cities and ARUP, 2014). Many cities also are covered by the emission reduction targets set by their states in subnational initiatives, including the Regional Greenhouse Gas Initiative in the northeastern United States, the Western Climate Initiative in the western United States and Canadian provinces, and the Midwestern Regional Greenhouse Gas Reduction Accord in the Midwestern United States.

12.5.3.4 Measuring Effectiveness

Despite pledges to cut GHG emissions, few cities have demonstrated quantifiable reductions to date that can be verified with publicly available information (Pierce, 2015). Furthermore, many of the cities with demonstrated reductions do not face the urbanization and development pressures seen in developing-world megacities or the smaller rapidly industrializing cities. These latter types of cities are expected to dominate future urbanization and GHG mitigation opportunities.

Case Study 12.4   The Benefits of Large-Scale Renewable Electricity Investment in Canberra

Cameron Knight

Environment and Planning Directorate, ACT Government, Canberra

Barbara Norman

*ACT Climate Change Council,
Canberra Urban and Regional Futures, University of Canberra*

Keywords	Renewable electricity, feed-in tariff, reverse auction, mitigation, energy
Population (Metropolitan Region)	387,069 (ACT Government, 2015)
Area (Metropolitan Region)	2,358 km² (Australian Bureau of Statistics [ABS], 2007)
Income per capita	US$60,070 (World Bank, 2015)
Climate zone	Cfb – Temperate, without dry season, warm summer (Peel et al., 2007)

In 2010, the Australian Capital Territory (ACT) government established a series of emissions reduction targets in legislation, specifically designed to meet targets of:
- 40% less than 1990 emissions by 2020
- 80% less than 1990 emissions by 2050
- Zero net greenhouse gas (GHG) emissions by 2050

Close engagement with local stakeholders and academic experts has been important to the ACT Government's pursuit of its emissions reduction targets. The ACT Climate Change Council was also established under the Act. The Council comprises a range of specialist and community interests to provide expert advice to the Minister for the Environment and Climate Change on climate change policy and implementation strategies (ACTCCC, 2015).

ESTABLISHING A 100% RENEWABLE ELECTRICITY TARGET

Electricity is by far Canberra's greatest source of GHG emissions. Electricity in Australia is generated predominantly from fossil fuels, particularly coal, from which 74% of the national electricity market's electricity is sourced (Garnaut, 2008; AEMO, 2014).

The high emissions intensity of Australia's electricity means that it is the source of 61% of ACT's emissions. Meeting the ACT's ambitious emission reduction targets requires a significant cut to the emissions intensity of ACT electricity. In 2012, a 90% renewable energy target became the central policy of the ACT's climate change strategy and action plan, AP2. In May 2016, this target was increased to 100% by 2020 (ACT Government 2016).

DELIVERING LARGE-SCALE RENEWABLE ELECTRICITY INVESTMENT AT THE LOWEST COST

The ACT Government estimated that 640 MW of new large-scale renewable energy investments would be required to achieve the 100% renewable energy target.

The ACT Government's first major investment in large-scale renewable electricity was the 20 MW Royalla Solar Farm, created by Fotowatio Renewable Ventures in August 2014. At the time of completion, it was the largest solar farm constructed in Australia. Feed-in tariff entitlements were also awarded to Zhenfa Australia and OneSun Capital for a 13 MW solar farm at Mugga Lane and a 7 MW solar farm at Williamsdale, respectively (ACT Government 2015b).

The ACT's investment in large-scale electricity generation has been achieved through an innovative feed-in tariff (FiT) reverse auction process. Renewable electricity project proponents are required to put forward bids against a set of criteria, including price. The winners of the auction process become eligible for a FiT at a fixed price. Because a "contract for difference" approach is applied, and because the price is fixed and not subject to inflation, the subsidy costs to Canberra will decrease as the value of wholesale electricity prices rise over time (Buckman et al., 2014).

On March 12, 2014, the Minister for the Environment, Simon Corbell MLA, announced a 200 MW Wind Auction to be conducted by a competitive reverse auction process. This was the first of three auctions offering feed-in tariff entitlements up to a total capacity of 600 MW, with a final announcement of successful tenders made in August 2016. The successful proponents were:
- 19.4 MW Coonooer Bridge wind farm being developed by Windlab situated near Bendigo, Victoria;
- 309 MW Hornsdale wind farm, stages 1, 2, and 3, being developed by developed by French renewable energy company Neoen International SAS in partnership with Australian company Megawatt Capital Investments, near Port Augusta, South Australia;
- 80.5 MW Ararat wind farm being developed by RES Australia near Ararat, Victoria;
- 100 MW Sapphire Wind Farm 1 developed by CWP Renewables in northern New South Wales; and
- 91 MW Crookwell 2 Wind Farm developed by Union Fenosa Wind Australia, 15 kilometers southeast of Crookwell NSW.

The successful proponents of this process are outlined in Case Study 12.4 Figure 1. In relative terms, this is the biggest step change reduction in greenhouse gas emissions of any Australian jurisdiction (ACT Government, 2015c).

Through competitive processes and an innovative FiT structure, the ACT government has been able to deliver this step change to renewables at an average cost of around AU$1.79 (US$1.44) per household per week. This is part of the estimated AU$4.67 (US$3.75) per week electricity price impact required to achieve the 90% renewables electricity target (ACT Government, 2015c). This demonstrates that moving to high levels of renewable electricity is both achievable and affordable.

ECONOMIC BENEFITS OF RENEWABLE ELECTRICITY INVESTMENT

ACT's investment in renewable electricity is stimulating investment in strategic priority areas of the local economy – building local infrastructure, intellectual property, and knowledge and skills of international significance – while creating opportunities for exports and sustainable job creation. The economic benefits flowing from successful reverse auction projects total more than AU$500 million (US$400.92 million).

LOCATION OF CANBERRA'S WIND AND SOLAR FARMS WITHIN THE NATIONAL ELECTRICITY MARKET

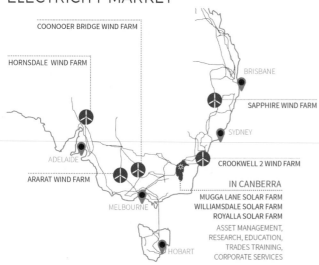

Case Study 12.4 Figure 1 *Wind farms funded under the ACT Government's reverse wind auction.*

Source: ACT Government, 2015b

The ACT Renewable Energy Local Investment Framework (ACT Government, 2014b) is designed to enhance opportunities for job creation resulting from the ACT's investments in large-scale renewable projects. All developers participating in the wind auction were required to demonstrate best-practice community engagement processes for their projects and contributions to local industry development.

The three successful wind auction proponents will deliver a range of benefits for the ACT through a AU$50 million (US$40.09 million) economic stimulus package, including the establishment of new operations centers, research and development partnerships with local universities, a new national trades training center, an innovation fund for small Canberra renewables businesses, and a AU$7 million (US$5.61 million) investment in new courses at the Canberra Institute of Technology (CIT) and the Australian National University (ANU).

As a result of investment by wind auction proponent, Neoen CIT will be developing its new Renewable Energy Skills Centre of Excellence to target national and international students looking for hands-on learning in renewable energy asset development and management.

The ACT government is also targeting skilled professionals around the country, transitioning from work in decommissioned coal and gas generation assets across Australia's national electricity market. Supported by WindLab and Renewable Energy Systems, the ANU is expanding its renewables programs and will establish Australia's first master's degree course in wind energy development, complementing the existing master's degree in energy change program. This is reinforcing the international reputation of the ANU and its Energy Change Institute as leaders in education and applied research in the energy field and in creating new opportunities for business–research collaborations.

Important for the ACT is the continued growth of renewable electricity investment in the region. A key partner in this is the South East Region of Renewable Energy Excellence (SERREE), which is working to develop a vibrant cluster of renewable energy businesses in Canberra and the surrounding region. A requirement of the wind auction was for proponents to invite tenders from and to contract with local businesses in the asset development and operational stages of the wind farms. SERREE will provide an important vehicle for this investment, supporting local jobs and creating international exposure for small business in the ACT and its surrounds.

All wind farms will be run from new management and operations headquarters in Canberra. In the short term, it is expected that these operations hubs will directly employ eleven highly skilled, full-time personnel, with employee numbers expected to grow substantially over time as new wind farms in Australia and overseas are developed and managed from these facilities. Local small businesses and start-ups will be supported through a new AU$1.2 million (US$0.96 million) Renewable Energy Innovation Fund supported by Neoen.

A big local winner out of the wind auction process is the Canberra-based company, WindLab. As a result of the wind auction, WindLab projects that investment in salaries and related costs are expected to grow to in excess of AU$240 million (US$192.49 million) over the 20-year FiT period.

To offer pathways for young people, WindLab and RES have partnered to deliver a Renewables in Schools program, introducing Canberra high school students – both government and nongovernmental – to the world of renewable energy. The program will outline to students opportunities to contribute to the growth of this exciting field through further tertiary education, research, or trades training.

CONCLUSION

Canberra exemplifies the increasingly important role of cities in responding to rising global GHG emissions. Its goal of supplying 100% of its electricity from renewable sources by 2020 has minimal impacts on local energy prices and has already generated significant benefits for the local economy.

The Territory's wind auctions have received significant industry attention, attracting both domestic and international proposals. Competition was intense, both in terms of FiT price and contributions to the Local Investment Framework.

Under the Renewable Energy Local Investment Framework, the government has recognized that renewable energy industries provide a strategic growth opportunity for the Canberra economy. The Framework sets out a vision of Canberra as an internationally recognized center for renewable electricity innovation and investment, and the city is well on the way to achieving that goal. The local investment benefits achieved through the wind auction demonstrate a concerted effort by the ACT government to develop renewable electricity as a strategic opportunity for Canberra. It also reflects a recognition by industry of Canberra as a good place to invest – a high-skills economy well placed in the global renewable energy revolution.

Table 12.7 *Urban carbon emission reduction targets. Computed from C40, ARUP and others, 2015.*

Number of cities	Percent of total	Target Year	Target: Percent of cumulative emission reductions
144	63	2020	12
27	12	2012–2030	13
57	25	2050	15

Efforts to maximize GHG mitigation potential need further research, especially into how multiple actions that transcend political/institutional boundaries are being implemented, as well as which mitigation options are cost-effective, scalable, align with existing local priorities or conditions, have high participation rates and leadership, and/or result in undesirable "rebound" effects. To assist with some of these types of analyses, efforts are needed to calculate the expected (or potential) and monitor the actual effectiveness of mitigation strategies over time. Equation 1 illustrates how effectiveness can be quantitatively assessed as a product of the baseline emissions from the city, the anticipated savings from the activity change, and the participation rate in the activity:

$$Mitigated\ Amount = Baseline \times Anticipated\ Savings \times Participation\ Rate$$

(Equation 1)

Future research that measures and "ground-truths" GHG reduction effectiveness via randomized case-control trials can generate information for decision-makers to better understand actual emission reductions and participation rates in GHG mitigation action programs that are voluntary or market- or regulatory-based (see Section 12.3.3).

Singapore, and specifically the Housing and Development Board of Singapore (HDB), serves as one model for using quantitative scenario-based urban information modeling comparing climate mitigation potential of various urban planning strategies to inform decision-making (see Case Study 12.5).

Many cities have begun to use quantitative analyses and tools to compare the effectiveness of various mitigation actions, and here we highlight two examples from China:
- *Changing commercial district in Shanghai*: Identified 58 actions for reducing energy use and emissions, including building retrofits, greening the energy supply, improving building codes, and clean transport. Each action was assessed according to its cost to implement and energy and GHG savings potential. The energy supply improvements from purchasing renewable energy, on-site distributed generation, and "phasing out" transformers together were estimated to reduce GHG emissions by 57 kton CO_2e during 2011–2015,

compared to a total estimated reduction potential from all 58 actions of 177 kilotons CO_2e, although improving building energy efficiency had a larger overall potential for emissions mitigation in the district (World Bank, 2013).
- *Xiamen City, China*: Found fuel switching toward lower carbon options had by far the most energy and emissions reduction potential in 2007–2020 as compared with other reduction strategies such as improving industrial energy efficiency, improving large public building efficiency, expanding public transit, or investing in renewable energy sources (Lin et al., 2010).

Results from these studies indicate that the relative effectiveness of different reductions strategies depends on the characteristics of the studied community and the drivers reviewed earlier (Section 12.3.4), and thus it is difficult to generate "best practices" that apply in all communities. Nevertheless, interest in identifying "no- and low-regrets" policies is high (Ostertag, 2012; Ruester et al., 2013). No- or low-regrets policies are those in which the benefits to society from energy or emissions reductions (and other goals such as job creation or reducing conventional air pollution) outweigh the implementation costs, regardless of the severity of future climate change impacts, including some energy-efficiency and demand-side management (Prasad et al., 2008; Ebinger and Vergara, 2011).

12.5.3.5 Institutional Barriers and Ways of Overcoming Them

Reducing the environmental impact of the urban energy supply sector requires substantial institutional capacity to identify and implement appropriate mitigation strategies. Twelve types of institutional barriers to effective environmental management are uncoordinated institutional framework; limited community engagement, empowerment, and participation; limits of regulatory framework; insufficient resources (capital and human); unclear, fragmented roles and responsibilities; poor organizational commitment; lack of information, knowledge, and understanding in applying integrated adaptive forms of management; poor communication; no long-term vision, strategy; technocratic path dependencies; little or no monitoring and evaluation; and lack of political and public will (Brown and Farrelly, 2009). Insufficient resources is relevant to all cities but especially smaller cities not having the same capacity and resources of the megacities of the world: less training to assess and mitigate urban energy-related GHG emissions, less financing, fewer knowledge networks, and often a higher need to prioritize investments in economic development relative to environmental conservation.

Furthermore, higher population growth and spatial expansion in small and medium-sized cities is often accompanied by fewer planning resources and weaker capacities to ensure provisions of public services and infrastructure, leaving GHG mitigation to be lower on the priority list unless it aligns with other local development priorities (see Case Study 12.6).

Case Study 12.5   The City of Singapore's 3D Energy Planning Tool as a Means to Reduce CO$_2$ Emissions Effectively

Marianne Najafi

EDF France, Paris

Keywords	Emissions, urban energy, technology, 3D modeling, EDF
Population (Metropolitan Region)	5,607,300 (Department of Statistics Singapore, 2016)
Area (Metropolitan Region)	719.2 km² (Department of Statistics Singapore, 2016)
Income per capita	US$72,711 (Department of Statistics Singapore, 2017)
Climate zone	Af – Tropical rainforest (Peel et al., 2007)

The Housing and Development Board of Singapore (HDB), Singapore's biggest public housing provider, uses a ground-breaking urban modeling tool to compare various urban planning strategies and select the most appropriate one for achieving the city's goals. This analysis allows the city to harness clean technology and, by doing so, reduce its CO$_2$ emissions.

EDF, a major electricity company, has developed the IT tool to facilitate a system's approach to energy and urban systems and their interactions at the very early stage of the planning process. Based on this systemic approach and expert advice, energy systems and CO$_2$ emissions can be optimized during the planning phase by using effective levers of action such as urban morphology, density, mixed land use (commercial and residential uses), renewable energy potential, efficient cooling and heating networks, and optimization of building consumption and emissions related to mobility.

The tool simultaneously maps three energy system dimensions:
- *Energy demand and its evolution*: Energy demand in buildings and energy efficiency actions, mobility and electric mobility development, public lighting
- *Local energy supply*: Distributed energy production and local renewable potentials
- *Electric and thermal networks*: Enablers of the integration of renewables and giving flexibility to the energy systems thanks to demand response and energy storage

The systemic approach of the tool facilitates collaboration between the Singaporean authorities and other industrial partners, incorporating location maps and 3D representations of buildings as well as graphs and tables of consumption data. The planning approach, in combination with use of the tool, is stakeholder inclusive and provides understanding of co-benefits or the positive externalities related to enhancing quality of life and air quality.

THE FIRST RESULTS

This IT tool was initially developed for the new residential district of Yuhua, Jurong East, in Western Singapore, in 2014. Results suggest a potential reduction of energy consumption by half in 2030 as compared to 2010 and a potential for photovoltaic (PV) renewable energy use of approximately 20% to 30% of energy consumption. In one plausible scenario, GHG emissions could be cut by more than half with simulations showing approximately 21,000 tons of GHG avoided over 15 years of PV operation. Measures include energy efficiency in building air conditioning systems, integration of solar panels, green roof development, water and domestic waste management, and improved urban mobility.

THE LESSONS LEARNED FROM THE DEPLOYMENT OF THE TOOL

Adapting to the Local Specificities

Each city or territory is highly specific and has its own characteristics in terms of natural resources, history, culture, and local energy production. Urban energy reduction solutions therefore must vary. However, for all cities, the global approach is the same: first look at energy demand and energy efficiency, then evaluate the potential for energy resources and specifically renewables, and finally design efficient, reliable, affordable energy supply networks. The approach must be cognizant of the challenges of balancing different interests, and it must focus on the ongoing operation and maintenance of facilities to meet initial design goals

Thinking Long-Term

Analyzing energy needs and energy resources over a 20- to 30-year horizon provides the context for good energy decision-making, in addition to determining early on positive and negative impacts, costs and benefits, and both negative and positive externalities (local energy production, sustainable mobility). While there are many uncertainties, comparing different strategies and impacts in the long run helps decision-makers to think of both today's cost and longer term benefits.

Acceptability and Stakeholder Engagement

Energy projects can only succeed if they rely on strong political will across scales capable of mobilizing the various stakeholders around a shared vision and trajectory. Given stretched municipal budgets, greater stakeholder engagement and public–private partnership are increasingly attractive.

12.5.4   Climate Adaptation and the Urban Energy Supply Sector

Vulnerability and risk assessment of energy systems are critical to inform adaptation strategies that improve the resilience of cities and their inhabitants to energy system stresses. Appropriate adaptation strategies depend on the specific climate hazards and vulnerabilities facing each city, as well as on the adaptive capacity of its residents.

The capacity of urban residents to adapt to climate vulnerabilities is embodied in both the physical infrastructure within the urban environment as well as the urban socioeconomic and political processes and structures. The adaptive capacity of residents, for example, is a function of the quality of provision and coverage of infrastructure and services, investment capacity, and land-use management (Revi et al., 2014). Urban adaptive capacity during extreme events depends on: (1) proper and simultaneous functioning of lifeline systems including transportation, water, communications, and power; (2) the robustness of critical facilities for public health, public safety, and education; and (3) preparedness programs and response and relief capabilities (Wenzel et al., 2007).

In the short term, parts of energy supply systems have been designed to cope with climate-related risks. For example, substation sites in San Diego are graded to divert waters away from facilities and to prevent erosion (ICLEI, 2012). Facilities including oil and gas drilling operations; thermal power plants; and hydro, wind, solar, and biomass generation can be better designed or managed on-site to withstand climate hazards such as higher winds, storm surge, or drought (Ebinger and Vergera, 2011). In the long-term, relocation of distribution lines and generation facilities will be required (Wilbanks et al., 2007) as well as increasing levels of redundancy, flexibility, and reliance on distributed generation systems that allow for avoiding certain design or service deficiencies involved with citywide or regional distribution grids, blackouts, and other types of service disruptions (Lovins et al., 2002). New facility siting decisions, water-efficient energy generation systems, community-based renewables, back-up diesel generators, early warning systems,

and hazard preparedness plans are other forms of adaptations (Tyler and Moench, 2012).

Table 12.8 identifies examples of energy system adaptation strategies that include reducing exposures and sensitivity and improving adaptive capacities via planning, policy, technology, and behavior change approaches. As with mitigation, these four categories of urban adaptation approaches are pathways to more resilient energy systems, defined by characteristics of having spare capacity, flexibility, limited or "safe" failure, rapid rebound, and planning/policy processes that catalyze constant learning. Importantly, these strategies will vary in relevance by city, region, and local contextual factors such as weather. In the future, integrated approaches toward low-carbon, climate-resilient, and just energy systems will require planning, policy, technology, and behavior change suited to local contexts. Meanwhile, city exchange of ideas, knowledge, and resources toward these goals – sharing what has worked and what has not worked – is increasing (e.g., ICLEI, C40, 100RC) although further mapping efforts can help to catalyze co-benefits integration (see Section 12.5.7), scaling, and replication.

A recent survey of 350 global cities (Aylett, 2014) identified that, of those cities with local government-operated electrical utilities, only 15% are focusing on adaptation planning. Therefore, we provide an illustrative list of options that can be adopted in cities. These options are gathered from a review of several reports and studies (World Bank, 2011; Royal Academy of Engineering, 2011).

Case Study 12.6   Managing Polluting and Inadequate Infrastructure Systems and Multiple Environmental Health Risks in Delhi

Joshua Sperling

National Renewable Energy Laboratory, Denver

Keywords	Energy supply, heat wave, GHG emissions reduction
Population (Metropolitan Region)	21,753,486 (IndiaStat, 2015)
Area (Metropolitan Region)	1,483 km² (Delhi Government, 2015)
Income per capita	US$1,680 (World Bank, 2017)
Climate zone	BSh – Arid, steppe, hot (Peel et al., 2007)

Current energy infrastructure conditions in Delhi are poor, with unscheduled power cuts, 8% still using solid fuels for cooking, many

lacking access to reliable/affordable electricity, and average pollutant concentrates up to four times higher than national outdoor air quality standards. Actions adopted by the Delhi government underline the importance of managing energy infrastructure systems given multiple environmental health risks that can be driven by urbanization, air pollution, and climate-related extreme weather (e.g., the rolling blackouts and more than 2,000 deaths in the North India heat wave early in the summer of 2016).

The local government has proactively planned for a number of activities contributing to improved management of energy systems, including conversion of coal-based to gas-based power plants, use of clean natural gas (CNG) for transportation, and reductions in supply losses. Stand-by loss reduction (Prakash, 2014) can have significant impacts, especially as these power losses make up 25% of the total Delhi electricity produced.[7] On the demand side, efficiency standards for appliances and lighting that make up the bulk of Delhi's residential energy demand have also been a focus, as well as Delhi's Transportation Department vision aiming to implement a comprehensive multimodal system of approximately 500 kilometers of metro rail, bus priority lanes, and use of CNG across the entire

---

7   See Box 12.3 for details on Delhi's energy supply.

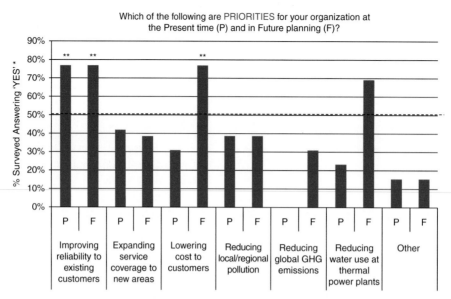

Which of the following are PRIORITIES for your organization at the Present time (P) and in Future planning (F)?

Footnotes:
* Computed based on "yes/no" responses from 12 of 14 EO's; remaining two will be captured (n = 12).
**,Ranked in the Top-3 by >50% of respondents who provided a ranking in addition to a 'yes/no' (n = 7).

Case Study 12.6 Figure 1 *Priorities of electricity infrastructure operators.*

Source: Cohen, 2014

bus fleet. However, the twin goals of reduced emissions and risks to populations in Delhi remain critical, and challenges remain including poorly maintained transmission lines, extreme heat, and overloaded grids (in general) due to the struggle to meet rapidly rising demand. Plans for heat wave–proof transmission and distribution systems are fundamental to reliable electricity and the heat action plans under development should consider use of cooling centers in public transit stations during extreme heat events. New policies under consideration for net metering, reducing supply losses, and achieving air pollution and GHG mitigation co-benefits are critical and will require long-term planning efforts.

In the short term, and despite these laudable efforts,– Delhi was ranked by the World Health Organization (WHO) in 2014 as the worst city globally among 1,600 cities worldwide in terms of particulate ($PM_{2.5}$) air pollution concentrations, and many inhabitants continue to lack basic infrastructure so that a focus on reducing current health burdens due to civil infrastructure (e.g., energy, water, transport) or infrastructure-related environmental factors (e.g., air and water quality, extreme weather events) could be a significant opportunity and strong motivator for low-carbon development, especially with 19% of classified deaths in Delhi (in 2008) potentially related to such factors. In fact, infrastructure intervention at present could reduce mortality by about 4% even through one action to reduce PM_{10} levels in the city (Sperling, 2014). Specifically in the context of electricity infrastructure operators (EOs), surveys by Cohen (2014) indicate that among seven identified priorities,

improving reliability, expanding service, and reducing pollution were highest present priorities, with reliability, lowering costs, and reducing water use at thermal power plants highest priority for future planning; note that reducing GHG emissions was not considered a priority for any of the surveyed EOs at present (see Case Study 12.6 Figure 1). The same survey also found that the factors contributing most to current power outages are insufficient generating capacity, heat waves/drought, and fuel supply disruptions to thermal power stations.

These examples, challenges, and opportunities illuminate important questions going forward:
1. What is the potential for energy and emissions mitigation for Delhi's energy system given such EO priorities?
2. Can mitigation actions in Delhi have pollution and health risk reduction co-benefits?
3. How and why are growing cities such as Delhi introducing technology, planning, policy, and behavioral change approaches to both mitigate and adapt to climate change?
4. What new studies, actions, and programs are needed to improve understanding of and develop solutions for the urban energy supply sector in Delhi?
5. What will future demands look like, and which risks to Delhi energy systems and local populations are highest priority, especially under rapid growth conditions and a changing climate where increased frequency and intensity of extreme events impact energy systems?

Small-scale adaptation actions of energy supply systems:
• Move to distributed generating capacity and systems
• Create public cooling centers, emergency shelters, health facilities with on-site and back-up power supplies to provide safe places to go during heat waves, wildfires, floods, and other extreme events

• Create on-site renewable energy generation for community-based resilience centers that help ensure on-site communications, alternative water treatment/sewer capacity, and on-site food and medicine refrigeration capabilities
• Ensure redundant power systems for operation of critical infrastructures, government buildings, health/disease

Table 12.8 *Summary of four types of adaptation actions. Source: Adapted from Chester et al., 2014*

Adaptation Strategies in Energy Sector	Planning	Policy	Technological	Behavior Change
Key Learning, Knowledge, and Policy Networks: 100 Resilient Cities, Asian Cities Climate Change Resilience Network, ICLEI, C40, UCCRN, Mayor's Adaptation Forum, and others	Heat-Health Action Plans (Knowlton et al., 2014) focused on maintaining power to critical facilities/vulnerable groups; cooling centers in public transit stations during extreme heat	Renewables and energy efficiency incentive programs; early warning systems such as heat/flooding alerts to mobile phones and other forms of media/communications (TV, radio, short online videos, door-to-door)	Expanding capacity and reducing stand-by losses (Prakash, 2014) to avoid overloaded energy systems in the longer term due to rising temperatures and increasing energy demand (Archer et al., 2014)	Use of home heating and cooling systems (Stern, 2016); infographics, dashboards, and real-time energy meters; community-based social marketing techniques; social norming (Schultz et al., 2007; Goldstein et al., 2008)

surveillance monitoring systems, and surge health care services
- Redesign/shift renewable power site locations or change operations to minimize hazards
- Improve severe weather early warning systems and prediction capabilities so utility operators can better prepare for and manage extreme events
- Use smart meters and grids to play a part in managing variability in demand and supply

Larger scale actions (directly or indirectly) for adaptation in the energy supply sector:
- Create detailed risk assessments for energy assets and facilities to examine the likely conditions they will be exposed to (e.g., floods, storms, drought)
- Combine the use of centralized and distributed energy supply systems
- Apply efficiency measures and prioritization of critical infrastructures during supply shortages
- Use strategic siting or relocation of electric power generation plants (e.g., due to sea level rise)
- Institute river basin management to protect hydropower potential
- Track needs for greater generating capacity during times of peak demand
- Make changes in regulation to allow for electricity operator cooperation/coordination
- Determine dependencies of energy infrastructure (e.g., on water infrastructure for cooling; ICT infrastructure for control, management, and communications; and transport infrastructure for the supply of fuel for power generation and the distribution of oil and gas, as well as to enable access for energy infrastructure operators and maintenance staff)

As with mitigation, context is critical to the adoption of any of the listed measures, and few studies illustrate the effectiveness of different adaptation strategies. The effectiveness of adaptation may be estimated quantitatively by the number of buildings that are removed from risk, although more often such estimates depend on context, sequencing and interaction of adaptation actions, and behavioral responses that are difficult to anticipate

(Adger et al., 2005). Some effort has been made in estimating the effects of urban greening strategies, which may provide both mitigation and adaptation benefits. For instance, urban trees in Beijing in 2002 were estimated to reduce average air temperatures by 1.6°C, store 0.2 million tons of CO_2, and reduce summertime electricity demand from coal-fired power plants, subsequently reducing emissions (Yang et al., 2005). Even relatively small investments in "green" infrastructure within cities through planning and design (such as urban parks of approximately 1 hectare with widely spaced shade trees and good water sources) can effectively reduce the urban heat island and reduce cooling energy demand in urban neighborhoods (Müller et al., 2014). Likewise, adding 10% more green space in high-density urban environments through green roofs was estimated to adequately maintain urban temperatures at baseline levels in the Greater Manchester region despite projected increased temperatures through 2080 from climate change (Gill et al., 2007). These examples illustrate the need for increased attention to mitigation and adaptation synergies and tradeoffs.

12.5.5 Mitigation and Adaptation Co-Benefits and Interactions

There are both benefits and risks posed by urban energy infrastructure systems (Sperling and Ramaswami, 2013). Complementing the four categories described of mitigation and adaptation options, a key focus area for cities in terms of policy development has been regulation, economic/fiscal instruments/investments, and capacity-building focused on several key areas of the urban energy supply sector including efficiency; greater use of clean fuels and processes; access, reliability, and energy security; and resilience.

With these multiple key areas for decision-making, measuring both synergistic and antagonistic electricity supply pathways toward low-carbon, resilient, and just cities is of great importance. As just one example of energy for just cities in the context of public health, a perhaps antagonistic pathway could be defined as the need to provide clean drinking water systems placing new energy demands on cities by requiring new water treatment plant facilities, resulting in increases in GHG emissions, meanwhile reducing waterborne diseases. As such, it is worth exploring synergistic pathways that reduce GHG emissions, improve urban air quality, and generate social justice co- benefits (such as in Beijing leading up to the Olympics; Zhao, 2010) and antagonistic pathways that may increase health benefits while worsening GHG emissions or lead to "maladaptations" (Noble et al., 2014). Similarly, efforts to build long-term resilience may reduce GHG emissions yet do little for current health or pressing development issues.

New scenario tools are needed, and a few are under development that estimate quantitative impacts for health and carbon emissions while also mapping/monitoring the vulnerability and resilience of the energy supply sector. At a minimum, actual performance assessment is needed. Equally important are the multiple qualitative tools such as the "action impact matrix" that allows comparison of various policy options against co-benefits criteria, including poverty alleviation, biodiversity, air quality, and water scarcity (Munasinghe and Swart, 2000) and the "adaptation matrix" that illustrates co-dependencies among urban systems including energy, transport, health, and water (Kirshen et al., 2007).

While significant complexity exists, interdisciplinary and systems-based efforts are still needed that help identify synergies, co-benefits, interactions, and tradeoffs that have yet to be evaluated. Some mitigation and adaptation co-benefit assessments have been conducted and have proved enlightening for decision-making in select cities (e.g., Harlan and Rudell, 2011; Ruth, 2010). Urban infrastructure sectors are described as a key area where opportunities for synergies are greater (Wilbanks et al., 2007) (see Table 12.9). Data availability and a lack of

Table 12.9 *Energy supply examples of co-benefits (top row) and tradeoffs (bottom row).*

Mitigation	Adaptation
Reduce emissions by expanding use of renewable sources	Reduce vulnerability to widespread power grid outages by encouraging distributed generation from multiple renewable sources.
Reduce emissions by improving efficiency of energy and water delivery systems	Reduce potential for grid overload and failure by decreasing demand.
Energy shifts to low-carbon natural gas or nuclear electricity production	Water constraints during periods of drought exacerbated by new low-carbon energy supply and generation systems

standard assessment methods remain key challenges to increasing the comparability of efforts and effectiveness.

A recent survey of more than 412 local and regional governments by the ICLEI (2015) reports that the majority of mitigation and adaptation actions are focused on policy, action planning, and infrastructure investments. The top five ranked co-benefits for local climate mitigation and adaptation actions, taken together, include improving air quality and urban livelihoods, boosting the urban economy, protecting urban ecosystems, and safeguarding urban health.

Quito, Ecuador, has sought to integrate its mitigation and adaptation efforts with its overall urban development strategy and thus pursue co-benefits, especially from improvements in its urban energy system (see Case Study 12.7).

Case Study 12.7   Energy and Climate Change in Quito

Daniel Carrion

Mailman School of Public Health, Columbia University, New York

Keywords	Mitigation, energy efficiency, methane capture, renewable energy, adaptation
Population (Metropolitan Region)	2,239,191 (Secretario Metropolitano de Territorio, Habitat y Vivienda [STHV], 2010)
Area (Metropolitan Region)	4,230 km² (STHV, 2010)
Income per capita	US$5,820 (World Bank, 2017)
Climate zone	Cfb– Temperate, without dry season, warm summer (Peel et al., 2007)

Quito is a leader among cities worldwide, evidenced by its involvement with horizontal government leadership initiatives such as C40 and ICLEI. The country as a whole has twenty-five registered Clean Development Mechanisms, many in Quito (UNEP, 2011). In 2009, the city introduced a robust and coherent framework to address climate change via targeted mitigation and adaptation strategies (Municipio del Distrito Metropolitano de Quito [MDMQ], 2009). This plan is known as the Estrategia Quiteña al Cambio Climática (EQCC) or Quito's Climate Change Strategy.

GREENHOUSE GAS INVENTORY

A 2011 report found that Quito was responsible for 4.5% of Ecuador's total greenhouse gas (GHG) emissions (MDMQ, 2011). CO_2 was the most abundantly emitted of the three GHGs measured in 2011. Total GHG emissions by sector were: 57% energy, 7% agriculture, 18% waste, and 18% in biomass use and land-use change (MDMQ, 2011). These classifications follow Intergovernmental Panel on Climate Change (IPCC) guidelines (IPCC, 2006). Total emissions are estimated at 2.55 tons CO_2 equivalent per person per year (MDMQ, 2014).[8]

---

8   See Box 12.3 for details on Quito's energy supply.

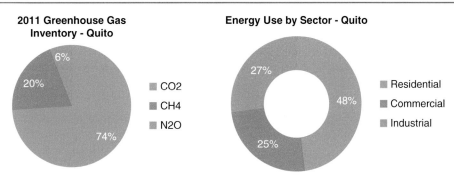

Total = 6,180,064 tons of GHG in CO$_2$ equivalent

Case Study 12.7 Figures 1 and 2 *The 2011 Greenhouse Gas Inventory for Quito and Quito's energy use by sector.*

MITIGATION EFFORTS

GHG mitigation projects are well under way in Quito. Ecuador is estimated to contribute less than 1% of the world's GHG emissions annually (MDMQ, 2014). Quito only comprises a fraction of that 1%. Despite this, the city demonstrates a strong commitment to sustainable development and leadership. Quito upholds initiatives pertaining to citizen education, transparency, and awareness, as outlined in the EQCC. The city has countless documents, presentations, reports, educational materials, curricula, and more ready for download on its website, further indicating this commitment. The EQCC seeks community support, but also encourages adaptation and mitigation starting in the home.

Energy efficiency and alternative energy initiatives: Quito has begun to incentivize both. The city's public housing and government buildings have begun implementation of mixed energy improvements, including photovoltaics and energy efficiency upgrades (MDMQ, 2010). These efforts may also be used to encourage private residential and commercial adoption, through incentivizing the use of clean energies in new construction. Key urban infrastructure has been targeted for implementation of photovoltaics, namely at bus stops (MDMQ, 2010). Photovoltaics may be one of Quito's best alternative energy options if done correctly because its high altitude and moderate temperatures may improve solar panel efficiency (Kawajiri et al., 2011). The Empresa Electrica de Quito (EEQ) is currently monitoring wind characterization throughout the region to determine the viability of wind power (EEQ, 2014b).

In 2011, Quito's waste sector had an annual output of 1,100,155 tons CO$_2$-equivalent GHGs. None of those emissions consisted of CO$_2$. Instead, almost 97% of those emissions were of methane. Despite some recent objections, methane is regarded as an efficient and clean combustible (Brandt et al., 2014). The City of Quito has decided to capture methane from waste (i.e., landfills) for energy. This technique demonstrates Quito's innovative methods to mitigate and promote development (MDMQ, 2010).

The municipality has created an admirable goal of 9 square meters of green space per city resident called La Red Verde Urbana (Green Urban Network). Furthermore, the construction or implementation of green roofs on new or existing infrastructures is being incentivized (MDMQ, 2010, 2012).

Green space can yield substantial energy-related paybacks. Research has found that green space in cities can counter traditional urban heat island effects caused by dark, impervious surfaces (Vasilakopoulou et al., 2014). This would attenuate the need for cooling during summer months. Maintenance of porous surfaces

allows for natural drainage of water, reducing the need for manmade drainage and energy-intensive wastewater treatment systems (Berndtsson, 2010). The Green Urban Network demonstrates that urban planning strategies offer highly effective tools against climate change.

Another urban planning technique with powerful mitigation potential is the concept of urban containment (Seto et al., 2014). Within Quito's 2011–2016 Climate Plans is a strategy for the collaboration of architects, regional planners, contractors, and real estate professionals to identify vacant or underutilized areas for (re)development within the city. Doing so offers planned densification throughout the city (MDMQ, 2012). This type of planning can reduce energy use and ecological footprint while avoiding unnecessary deforestation (Seto et al., 2014).

ADAPTATION STRATEGIES

Quito has already witnessed a 1.2–1.4°C increase in temperature from 1891 to 1999. In addition, changes in precipitation patterns have also been observed and more are expected (Zambrano-Barragan et al., 2010). In response, the city is involved in creating a metropolitan atmospheric network, mitigating risk and vulnerability to extreme weather events, reconfiguring the urban landscape, and diversifying its energy portfolio.

The Metropolitan Atmospheric Network of Quito (REMMAQ) collects and generates data on ozone, glaciers, carbon flows, and other climate-relevant factors that would inform short- and long-term decision-making. Such information would be essential to operationalizing any form of climate-resilient energy planning, especially in the context of disaster-prone areas (MDMQ, 2012).

Quito has a sizeable population living in informal settlements and disaster-vulnerable sections of the city, often at its outskirts. A 2011 UNEP report estimates that 53% of all settlements were informal and 443 neighborhoods were illegal (UNEP, 2011). The high electrical coverage of the population coupled with a large number of precarious houses implies an electrical grid under equally vulnerable conditions. The municipality has created programs to identify, prioritize, and manage high-risk areas. Threats can be addressed via engineering in some cases, while in others families must be relocated. An estimated US$10 million has been used to relocate 1,500 families between 2011 and 2012 (MDMQ, 2010). Those families may be best served if redirected toward the core of the city as part of the urban containment efforts.

In Quito, renewables are recognized simultaneously as mitigation and adaptation via "energy diversification" (MDMQ, 2009). The city's energy supply comes largely from hydroelectric, which may be vulnerable to the predicted precipitation changes and decreased glacier runoff (Zambrano-Barragan et al., 2010). Notable proportions of energy do still come from a diesel combustion plant (EEQ, 2012). The city, and country, is planning to replace diesel with natural gas (Ludeña and Wilk, 2013; Ministerio del Ambiente de Ecuador [MAE], 2000; MDMQ, 2012). Plans even include promotion of natural gas-powered vehicles alongside electric and hybrid (MAE, 2000). These adaptations, while not entirely shock-resistant, will offer some stability against the specter of climate phenomena.

12.6 Conclusions

This chapter reviews the urban energy supply sector in the context of the major challenges and opportunities for climate change mitigation, adaptation, and sustainable development.

The main challenges to the urban energy supply sector include environmental impacts and energy access and the vulnerabilities of energy systems to weather- and climate-related events. Given trends in urbanization, energy consumption, inequality, and climate change, these challenges will only increase if nothing is done to improve access, resilience, and lower impact.

The review cites both significant barriers and incentives to changing urban supply systems to meet these challenges. While previous transitions may not provide a good roadmap of what might happen in the future because the nature of change is significantly different now than in the past, there remain key indicators that help to provide insight. The length of the transition cycle has probably not changed so we can expect the process to take several decades to occur. This may be even more true today than in the past because in the current era the shift to low-carbon fuels will largely be driven by governance and natural influences because socioeconomically and behaviorally both the price and the convenience of non-hydro renewables is not more attractive than current fossil fuel options.

Moreover, given the diversity of urban conditions and demands, the review suggests that there will not be one single solution to energy supply challenges, but rather there will be numerous solutions to fit unique local needs. This does not mean that localities cannot learn from one another's successes and failure, but rather that the search for a one-size-fits-all set of policies may not be fruitful.

The diversity of urban energy supply solutions is already demonstrated in the large and growing number of efforts emerging across cities globally to mitigate the environmental impact of the energy supply sector and adapt it to climate impacts. This chapter identified a few of these efforts from Rio de Janeiro to Seattle, to Canberra, to Delhi. These policies have generated much interest, but also demonstrate that there remains much to do to transition away from the current trajectory of increasing GHG emissions, vulnerability, and energy inequity. Radical changes in institutions may be required, and, in some instances, cities are responding.

Finally, future scenarios all predict greater urbanization and energy consumption. While some future scenario research points to the possibility of lower carbon cities, but where many urban residents remain without access and few have energy security, there is also research that suggests it is possible to overcome current barriers and meet all three key challenges. We conclude that while there is much to be done, with greater participation, better technologies, political will, and institutions that support low-carbon, equity, and resilience, there is no reason why the urban future cannot support globally sustainable development and access to resilient, clean, modern energy supplies for all.

Chapter 12 Urban Energy

References

Abrahamse, W., and Steg, L. (2009). How do socio-demographic and psychological factors related to households' direct and indirect energy use and savings? *Journal of Economic Psychology,* **30** (5), 711–720.

Adger, W. N., Arnell, N. W., and Tompkins, E. L. (2005). Successful adaptation to climate change across scales. *Global Environmental Change* **15** 77–86.

Al-Sunaidy, A., and Green, R. (2006). Electricity deregulation in OECD (Organization for Economic Cooperation and Development) countries. *Electricity Market Reform and Deregulation* **31** 769–787.

Aleklett, K., Hook, M., Jakobsson, K., Lardelli, M., Snowden, S., and Soberbergh, B. (2010). The peak of the oil age – analyzing the world oil production reference scenario in World Energy Outlook 2008. *Energy Policy* **38**, 1398–1414.

Allcott, H. (2011). Social norms and energy conservation *Journal of Public Economics* **95**(9–10), 1082–1095.

Apte, J. S., Bombrun, E., Marshall, J. D., and Nazaroff, W. W. (2012). Global intra-urban intake fractions for air pollutants from vehicles and other distributed sources *Environmental Science, and Technology* **46**, 3415–3423.

Apte, J. S., Marshall, J. D., Cohen, A. J., and Brauer, M. (2015). Addressing global mortality from $PM_{2.5}$ *Environmental Science & Technology*. doi: 10.1021/acs.est.5b01236

Angel, S. (2012). *Planet of Cities*. Lincoln Institute of Land Policy.

Archer, D., Almansi, F., DiGregorio, M.m Roberts, D., Sharma, D., and Syam, D. (2014). Moving towards inclusive urban adaptation: Approaches to integrating community-based adaptation to climate change at city and national scale. *Climate and Development* **6**, 345–356.

ARUP, and C40 Cities. (2014). Climate action in megacities. C40 cities baseline and opportunities. Volume 2.0. Accessed May 17, 2015: www.arup.com/Projects/C40_Cities_Climate_Action_in_Megacities_report.aspx.

Aspen Environmental Group and M Cubed. (2005). Potential Changes in Hydropower Production from global climate change in California and western United States. Prepared in support of the 2005 Integrated Energy Policy Report Proceeding (Docket # 04-IEPR-01 G). California Energy Commission.

Aylett, A. (2013). Networked urban climate governance: Neighborhood-scale residential solar energy systems and the example of Solarize Portland. *Environment and Planning C: Government and Policy* **31**, 858–875.

Aylett, A. (2014). *Urban Climate Governance Report*. MIT.

Aznar, A., Day, M., Doris, E., Mathur, S., and Donohoo-Vallett, P. (2015). *City-Level Energy Decision Making: Data Use in Energy Planning, Implementation, and Evaluation in U.S. Cities*. National Renewable Energy Laboratory (NREL). Accessed January 8, 2016: http://www.nrel.gov/docs/fy15osti/64128.pdf

Bader, N., and Bleischwitz, R. (2009). Measuring urban greenhouse gas emissions: The challenge of comparability. *SAPIENS* **2**, 1–15.

Baiocchi, G., Minx, J., and Hubacek, K. (2010). The impact of social factors and consumer behavior on carbon dioxide emissions in the United Kingdom. *Journal of Industrial Ecology* 14, 50–72.

Baxter, L. W., and Calandri, K. (1992). Global warming and electricity demand: A study of California. *Energy Policy* 20, 233–244.

Blanco, G., Gerlagh, R., Suh, S., Barrett, J., de Coninck, H. C., Morejon, C. F. D., Mathur, R., Nakicenovic, N., Ahenkorah, A. O., Pan, J., Pathak, H., Rice, J., Richels, R., Smith, S. J., Stern, D. I., Toth, F. L., and Zhou, P. (2014). Drivers, trends and mitigation. In Edenhofer, O., Pichs-Madruga, R., Sokona, Y., Farahani, E., Kadner, S., Seyboth, K., Adler, A., Baum, I., Brunner, S., Eickemeier, P., Kriemann, B., Savolainen, J., Schlömer, S., von Stechow, C., Zwickel, T., and Minx, J. C. (eds.), *Climate Change 2014: Mitigation of Climate Change. Contribution of Working Group III to the Fifth Assessment Report of the Intergovernmental Panel on Climate Change* (351–411). Cambridge University Press.

Breslow, P. B., and Sailor, D. J. (2002). Vulnerability of wind power resources to climate change in the continental United States. *Renewable Energy* 27, 585–598.

Brew-Hammond, A. (2010). Energy access in Africa: Challenges ahead. *Energy Policy* 38, 2291–2301.

Broto, V. C., and Bulkeley, H. (2013). A survey of urban climate change experiments in 100 cities. *Global Environmental Change* 23, 92–102.

Brown, M. A., Southworth, F., and Sarzynski, A. (2008). *Shrinking the Carbon Footprint of Metropolitan America*. The Brookings Institution.

Brown, R. R., and Farrelly, M. A. (2009). Delivering sustainable urban water management: A review of the hurdles we face. *Water Science and Technology* 59, 839–846.

Bruckner, T., Bashmakov, I. A., Mulugetta, Y., Chum, H., De la Vega Navarro, A., Edmonds, J., Faaij, A., Fungtammasan, B., Garg, A., Hertwich, E., Honnery, D., Infield, D., Kainuma, M.,Khennas, S. Kim, S., Bashir Nimir, B., Riahi, K., Strachan, N., Wiser, R., and Zhang, X. (2014). Energy systems. In Edenhofer, O., Pichs-Madruga, R., Sokona, Y., Farahani, E., Kadner, S., Seyboth, K., Adler, A., Baum, I., Brunner, S., Eickemeier, P., Kriemann, B., Savolainen, J., Schlömer, S., von Stechow, C., Zwickel, T., and Minx, J. C. (eds.), *Climate Change 2014: Mitigation of Climate Change. Contribution of Working Group III to the Fifth Assessment Report of the Intergovernmental Panel on Climate Change* (511–597). Cambridge University Press.

Bulkeley, H., and Kern, K. (2006). Local government and the governing of climate change in Germany and the UK. *Urban Studies* 43, 2237–2259.

Bull, S. R., Bilello, D. E., Ekmann, J., Sale, J. M., and Schmalzer, D. K. (2007). Effects of climate change on energy production and distribution in the United States. In *Effects of Climate Change on Energy Production and Use in the United States.* U.S. Climate Change Science Program.

C2ES. (2014). *Market-Based Climate Mitigation Policies in Emerging Economies*. Center for Climate and Energy Solutions.

C40 Cities, and ARUP. (2014). Working together: Global aggregation of city climate commitments. Accessed January 11, 2016: http://c40-production-images.s3.amazonaws.com/researches/images/24_Working_Together_Global_Aggregation.original.pdf

Calderon, F., Stern, N., Bonde, I., Burrow, S., Yuan, C., Clark, H., Diogo, L., Doctoroff, D. L., Gopalakrishan, S., Gurria, A., Holliday, C., Polman, P., Indrawati, S. M., Koch-Weser, C., Lagos, R., Lies, M. M., Manuel, T., Nakao, T., Paes, E., Parker, A., Shafik, N., Stoltenberg, J., van der Hoeven, M., and Zhu, L. (2014). *The New Climate Economy*. World Resource Institute.

Cardona, O. -D., Aalst, M. K. V., Birkmann, J., Fordham, M., McGregor, G., Perez, R., Pulwarty, R. S., Schipper, E. L. F., and Sinh, B. T. (2012). Determinants of risk: Exposure and vulnerability. In Field, C. B., Barros, V., Stocker, T. F., Qin, D., Dokken, D. J., Ebi, K. L., Mastrandrea, M. D., Mach, K. J.,Plattner, G. -K., Allen, S. K., Tignor, M., and Midgley, P. M. (eds.), *Managing the Risks of Extreme Events and Disasters to Advance Climate Change Adaptation* (65–108). Cambridge University Press.

Carlonil, F. B. B. A. (2012). *Gestão do inventário e do monitoramento de emissões de gases de efeito estufa em cidades: O caso do Rio de Janeiro*. UFRJ/COPPE.

CDP, C40, and AECOM. (2014). Protecting our capital: How climate adaptation in cities creates a resilient place for business. London, UK. Accessed December 5, 2015: http://www.c40.org/researches/protecting-our-capital

Chavez, A. (2012). Comparing city-scale greenhouse gas (GHG) emissions accounting methods: Implementation, approximations, and policy relevance. Doctoral dissertation, University of Colorado.

Chávez, A. (2016). Community metabolism footprints: Scaling communities towards global sustainability. In preparation.

Chavez, A., and Ramaswami, A. (2013). Articulating a trans-boundary infrastructure supply chain greenhouse gas emission footprint for cities: Mathematical relationships and policy relevance. *Energy Policy* 54, 376–384.

Chester, M. V., Sperling, J., Stokes, E., Allenby, B., Kockelman, K., Kennedy, C., Baker, L. A., Keirstead, J., and Hendrickson, C. T. (2014). Positioning infrastructure and technologies for low-carbon urbanization. *Earth's Future* 2, 533–547.

Ciccone, A., and Hall, R. E. (1996). Productivity and the density of economic activity. *American Economic Review,* 86 (1), 54–70.

City of Aspen. (2015). Renewable energy – city of Aspen. Aspen, Colorado, USA. Accessed February 2, 2016: http://www.aspenpitkin.com/Living-in-the-Valley/Green-Initiatives/Renewable-Energy/.

City of Seattle. (2013). Seattle climate action plan. Accessed May 27, 2014: http://www.seattle.gov/Documents/Departments/OSE/2013_CAP_20130612.pdf

Coase, R. H. (1950). The nationalization of electricity supply in Great Britain. *Land Economics* 26, 1–16.

Creutzig, F., Baiocchi, G., Bierkandt, R.,Pichler, P. -P., and Seto, K. C. (2015). Global typology of urban energy use and potentials for an urbanization mitigation wedge. *Proceedings of the National Academy of Science* 112, 6283–6288.

Croci, E., Melandri, S., and Molteni, T. (2010). A Comparative Analysis of Global City Policies in Climate Change Mitigation: London, New York, Milan, Mexico City and Bangkok. Bocconi IEFE, Working Paper Series No. 32. Center for Research on Energy and Environmental Economics and Policy at Bocconi University.

Davis, L. W., and Gertler, P. J. (2015). Contribution of air conditioning adoption to future energy use under global warming. *Proceedings of the National Academy of Science* 119, 5962–5967.

Davis, M., and Weible, C. M. (2011). Linking social actors and social theories: Toward improved GHG mitigation strategies. *Carbon Management* 2, 483–491.

Deffeyes, K. S. (2001). *Hubbert's Peak: The Impending World Oil Shortage*. Princeton University Press.

Dhakal, S. (2009). Urban energy use and carbon emissions from cities in China and policy implications. *Energy Policy* 37, 4208–4219.

Dhakal, S. (2010). GHG emission from urbanization and opportunities for urban carbon mitigation. *Current Opinion in Environmental Sustainability* 2, 277–283.

Dilling, L., and Lemos, M. (2011). Creating usable science: Opportunities and constraints for climate knowledge use and their implications for science policy. *Global Environmental Change* 21, 680–689.

Dittmar, M. (2013). The end of cheap uranium. *Science of the Total Environment* 461–462, 792–798.

Dubeux, C. B. S. (2007). Mitigação de Emissões de Gases de Efeito Estufa por Municípios Brasileiros: Metodologias para Elaboração de Inventários Setoriais e Cenários de Emissões como Instrumentos de Planejamento. Tese de D.Sc, Planejamento Energético, COPPE, Universidade Federal do Rio de Janeiro, Rio de Janeiro.

Eakin, H., and Lynd Luers, A. (2006). Assessing the vulnerabililty of social-environmental systems. *Annual Review of Environment and Resources* 31, 365–294.

Eberhard, A., Rosnes, O., Shkaratan, M., and Vennemo, H. (2011). *Africa's Power Infrastructure: Investment, Integration, Efficiency*. World Bank.

Ebinger, J., and Vergera, W. (2011). Climate impacts on energy systems: Key issues for energy sector adaptation (No. 60051). Accessed April 5, 2014: http://documents.worldbank.org/curated/en/2011/01/13888226/climate-impacts-energy-systems-key-issues-energy-sector-adaptation

Economist Intelligence Unit. (2011a). *African Green City Index*. Siemens AG.

Economist Intelligence Unit. (2011b). *Asian Cities Green City Index*. Siemens AG.

Economist Intelligence Unit. (2011c). *European Green City Index*. Siemens AG.

Economist Intelligence Unit. (2011d). *German Green City Index*. Siemens AG.

Economist Intelligence Unit. (2011e). *Latin American Green City Index*. Siemens AG.

Economist Intelligence Unit (EIU). (2011f). *North American Green City Index*. Siemens AG.

Economist Intelligence Unit (EIU) (2012). *The Green City Index: A Summary of the Green City Index Research Series*. Siemens AG.

EEQ. (2012). Plan Estrategico 2012–2015.

EEQ. (2014). La Generación de electricidad en la EEQ: Gerencia de Generación y Subtransmisión.

Emelianoff, C. (2014). Local energy transition and multilevel climate governance: The contrasted experiences of two pioneer cities (Hanover, Germany, and Växjö, Sweden). *Urban Studies* 51, 1378–1393.

Environment America Research and Policy Center. (2015). Shining cities, at the forefront of America's solar energy revolution. Accessed March 7, 2016: http://www.environmentamerica.org/sites/environment/files/reports/EA_ShiningCities2015_scrn.pdf

EPA- United States Environmental Protection Agency. (2015). Developing a greenhouse gas inventory. Accessed March 29, 2016: http://www.epa.gov/statelocalclimate/state/activities/ghg-inventory.html

Estiri, H. (2014). Building and household x-factors and energy consumption at the residential sector. *Energy Economics* 43, 178–184. doi: 10.1016/j.eneco.2014.02.013

Estiri, H. (2015). A structural equation model of energy consumption in the United States: Untangling the complexity of per-capita residential energy use. *Energy Research & Social Science* 6, 109–120. doi:10.1016/j.erss.2015.01.002.

Estiri, H., Gabriel, R., Howard, E., and Wang, L. (2013). Different regions, differences in energy consumption: Do regions account for the variability in household energy consumption? Working paper 134. University of Washington, Seattle. Accessed May 27, 2014: http://www.csss.washington.edu/Papers/wp134.pdf

European Union. (2015). Renewable energy: The promotion of electricity from renewable energy sources. Accessed March 16, 2016: http://eur-lex.europa.eu/legal-content/EN/TXT?uri=URISERV:l27035

Evans, P. C., and Fox-Penner, P. (2014). Resilient and sustainable infrastructure for urban energy systems. *Solutions* 5, 48–54.

Farrell, D., and Remes, J. (2009). Promoting energy efficiency in the developing world. *The McKinsey Quarterly* 14.

Fenger, J. (2007). Impacts of climate change on renewable energy sources: Their role in the nordic energy system. Nordic Council of Ministers.

Fidje, A., and Martinsen, T. (2006). Effects of climate change on the utilization of solar cells in Nordic region. In *European Conference on Impacts of Climate Change on Renewable Energy Resources*. Reykjavik, Iceland.

Florini, A., and Sovacool, B. K. (2009). Who governs energy? The challenges facing global energy governance. *Energy Policy* 37, 5239–5248.

Fouquet, R. (2010). The slow search for solutions: Lessons from historical energy transitions by sector and service. *Energy Policy* 38, 6586–6596.

Fouquet, R., and Pearson, P. J. G. (2012). Past and prospective energy transitions: Insight from history. *Energy Policy* 50, 1–7.

Franco, G. (2005). Climate change impacts and adaptation in California. Prepared in support of the 2005 Integrated Energy Policy Report Proceeding (Docket # 04-IEPR-01E). California Energy Commission.

Franco, G., and Sanstad, A. H. (2008). Climate change and electricity demand in California. *Climatic Change* 87, S139–S151.

Frantzeskaki, N., Avelino, F., and Loorbach, D. (2013). Outliers or frontrunners? Exploring the (self-) governance of community-owned sustainable energy in Scotland and the Netherlands. In Michalena, E., and Hills, J. M. (eds.), *Renewable Energy Governance* (101–116). Springer.

French, H. (2008). Severe snow storms batter China. *New York Times*, January 28.

Fussel, H. -M., and Klein, R. J. T. (2006). Climate change vulnerability assessments: An evolution of conceptual thinking. *Climatic Change* 75, 301–329.

Garnaut, R. (2008). *The Garnaut Climate Change Review*. Cambridge University Press. Accessed April 5, 2014: http://www.garnautreview.org.au/2008-review.html

GEA. (2012). *Global Energy Assessment – Toward a Sustainable Future*. Cambridge University Press/International Institute for Applied Systems Analysis.

Gill, S. E., Handley, J. F., Ennos, A. R., and Pauleit, S. (2007). Adapting cities for climate change: The role of the green infrastructure. *Built Environment* 33, 115–133.

Goldstein, N. J., Cialdini, R. B., and Griskevicius, V. (2008). A room with a viewpoint: Using social norms to motivate environmental conservation in hotels. *Journal of Consumer Research* 35(3), 472–482.

GPC. (2014). *Global Protocol for Community-Scale Greenhouse Gas Emission Inventories: An Accounting and Reporting Standard for Cities*. World Resources Institute, C40 Cities, ICLEI.

Grin, J., Rotmans, J., and Schot, J. (2012). *Transitions to Sustainable Development: New Directions in the Study of Long Term Transformative Change*. Routledge Studies in Sustainability Transitions. Taylor and Francis.

Grubler, A. (2004). Transitions in energy use. In *Encyclopedia of Energy, Volume 6* (163–177). Elsevier.

Grubler, A. (2012). Energy transitions research: Insights and cautionary tales. *Energy Policy* 50, 8–16.

Grubler, A., Bai, X., Buettner, T., Dhakal, S., Fisk, D. J., Ichinose, T., Keirstead, J. E., Sammer, G., Satterthwaite, D., Schulz, N. B., Shah, N., Steinberger. J., and Weisz, H. (2012a). Chapter 18 - Urban Energy Systems. In *Global Energy Assessment - Toward a Sustainable Future*, Cambridge University Press, Cambridge, UK and New York, NY, USA and the International Institute for Applied Systems Analysis, Laxenburg, Austria, pp. 1307–1400.

Grubler, A., Johansson, T. B., Mundaca, L., Nakicenovic, N., Pachauri, S., Riahi, K.,Rogner, H. -H., and Strupeit, L. (2012b). Chapter 1 - Energy Primer. In *Global Energy Assessment - Toward a Sustainable Future*, Cambridge University Press, UK and New York, NY, USA and the International Institute for Applied Systems Analysis, Laxenburg, Austria, pp. 99–150.

Gupta, R., Barnfield, L., and Hipwood, T. (2014). Impacts of community-led energy retrofitting of owner-occupied dwellings. *Building Research & Information* 42, 446–461.

Hanson, S., Nicholls, R., Ranger, N., Hallegatte, S., Corfee-Morlot, J., Herwijer, C., and Chateau, J. (2011). A global ranking of port cities with high exposure to climate extremes. *Climatic Change* 104(1), 89–111.

Hamlet, A.,Lee, S. -Y., Mickelson, K. B., and Elsner, M. (2010). Effects of projected climate change on energy supply and demand in the Pacific Northwest and Washington State. *Climatic Change* 102, 103–128.

Hammer, S. A., Keirstead, J., Dhakal, S., Mitchell, J., Colley, M., Connell, R., Gonzalez, R., Herve-Mignucci, M., Parshall, L., Schulz, N., and Hyams, M. (2011). Climate change and urban energy systems. In Rosenzweig, C., Solecki, W. D., Hammer, S. A., and Mehrotra, S. (eds.), *Climate Change and Cities: First Assessment Report of the Urban Climate Change Research Network*. Cambridge University Press.

Harlan, S. L., and Ruddell, D. M. (2011). Climate change and health in cities: Impacts of heat and air pollution and potential co-benefits from mitigation and adaptation. *Current Opinion in Environmental Sustainability* 3, 126–134.

Harrison, G. P., Cradden, L. C., and Chick, J. P. (2008). Preliminary assessment of climate change impacts on the UK onshore wind energy resource. *Energy Sources, Part A: Recovery, Utilization and Environmental Effects* 30, 1286–1299.

Hawkey, D., Webb, J., and Winskel, M. (2013). Organisation and governance of urban energy systems: District heating and cooling in the

UK. *Special Issue: Advancing Sustainable Urban Transformation* **50**, 22–31.

Heath, G. A., and Mann, M. K. (2012). Background and reflections on the life cycle assessment harmonization project. *Journal of Industrial Ecology* **16**, 28–211.

Heinberg, R., and Fridley, D. (2010). The end of cheap coal. *Nature* **468**, 367–369.

Heinonen, J., Jalas, M., Juntunen, J.K, Ala-Mantila, S., and Junnila, S. (2013). Situated lifestyles: I. How lifestyles change along with the level of urbanization and what the greenhouse gas implications are – a study of Finland. *Environmental Research Letters* 8. doi:10.1088/1748-9326/8/2/025003

Hernández, D. (2015). Sacrifice along the energy continuum: A call for energy justice. *Environmental Justice* **8**, 151–156.

Hess, J. J., Bednarz, D., Bae, J. Y., and Pierce, J. (2011). Petroleum and health care: Evaluating and managing health care's vulnerability to petroleum supply shifts. *American Journal of Public Health* **101**, 1568–1579.

Hewer, F. (2006). Climate Change and Energy Management. UK Met Office.

Hibbard, P. J. (2006). *U.S. Energy Infrastructure Vulnerability: Lessons from the Gulf Coast Hurricanes.* Analysis Group.

Hoornweg, D., Sugar, L., Lorena, C., and Gomez, T. (2011). Cities and greenhouse gas emissions: Moving forward. *Environment and Urbanization* **23**, 207–227.

Hsu, A. (2015). Five key takeaways from China's climate pledge. Yale environmental performance index blog. Accessed February 25, 2016: http://epi.yale.edu/the-metric/five-key-takeaways-chinas-new-climate-pledge

Hubbert, M. K. (1981). The world's evolving energy system. *American Journal of Physics* **49**, 1007–1029.

Hughes, L., and Rudolph, J. (2011). Future world oil production: Growth, plateau, or peak? *Current Opinion in Environmental Sustainability* **3**, 225–235.

ICF. (1995). Potential Effects of climate change on electric utilities. Prepared for Central Research Institute of Electric Power Industry (CRIEPI) and Electric Power Research Institute (EPRI).

ICLEI. (2009). International Local Government GHG Emissions Analysis Protocol (IEAP) Version 1.0. Accessed April 5, 2014: http://carbonn.org/fileadmin/user_upload/carbonn/Standards/IEAP_October2010_color.pdf

ICLEI. (2012). U.S. community protocol for accounting and reporting of greenhouse gas emissions. Version 1.0: ICLEI — Local Governments for Sustainability USA.

ICLEI. (2015). Carbon Climate Registry 2014–2015 Digest. Accessed January 10, 2016: http://carbonn.org/fileadmin/user_upload/cCCR/ccr-digest-2014-2015/ccr-digest-2014–2015-online-final.pdf

ICLEI, C40, and World Resources Institute. (2012). Global Protocol for Community-Scale Greenhouse Gas Emission Inventories: An accounting and reporting standard for cities. Accessed December 16, 2014: http://www.ghgprotocol.org/city-accounting

Intergovernmental Panel on Climate Change (IPCC). (2006). *2006 IPCC Guidelines for National Greenhouse Gas Inventories*. IGES.

Intergovernmental Panel on Climate Change (IPCC). (2014a). Climate Change 2014, Synthesis Report, Summary for Policymakers. IPCC.

Intergovernmental Panel on Climate Change (IPCC). (2014b). Summary for Policymakers. In Field, C. B., Barros, V. R., Dokken, D. J., Mach, K. J., Mastrandrea, M. D., Bilir, T. E., Chatterjee, M., Ebi, K. L., Estrada, Y. O., Genova, R. C., Girma, B., Kissel, E. S., Levy, A. N., MacCracken, S., Mastrandrea, P. R., and White, L. L. (eds.), *Climate Change 2014: Impacts, Adaptation, and Vulnerability. Part A: Global and Sectoral Aspects. Contribution of Working Group II to the Fifth Assessment Report of the Intergovernmental Panel on Climate Change* (1–34). Cambridge University Press.

International Energy Agency (IEA). (2008). *World Energy Outlook*. IEA.

International Energy Agency (IEA). (2010). *World Energy Outlook*. IEA.

International Energy Agency (IEA). (2012). *World Energy Outlook*. IEA.

International Energy Agency (IEA). (2013). *World Energy Outlook*. IEA.

International Energy Agency (IEA). (2014). *World Energy Outlook*. IEA.

International Electrotechnical Commission (IEC). (2007). *Efficient Electrical Energy Transmission and Distribution*. IEC.

International Institute for Applied Systems Analysis (IIASA). (2012). *Global Energy Assessment – Toward a Sustainable Future*. Cambridge University Press,.

International Renewable Energy Agency. (2015). *Renewable Energy Target Setting*. IRENA.

Jaccard, M., Failing, L., and Berry, T. (1997). From equipment to infrastructure: Community energy management and greenhouse gas emission reduction. *Energy Policy* **25**, 1065–1074.

Jacobson, M. Z., and Delucchi, M. A. (2011). Providing all global energy with wind, water, and solar power, Part I: Technologies, energy resources, quantities and areas of infrastructure, and materials. *Energy Policy* **39**, 1154–1169.

Jaglin, S. (2014). Urban energy policies and the governance of multilevel issues in Cape Town. *Urban Studies* **51**, 1394–1414.

Jiusto, S. (2009). Energy transformations and geographic research. In Castree, N., Demeritt, D., Liverman, D., and Rhoads, B. (eds.), *A Companion to Environmental Geography* (533–551). Wiley-Blackwell.

JRC/PBL.(2013).Emission Database for Global Atmospheric Research (EDGAR). Release Version 4.2 FT2010. Accessed July 3, 2015: http://edgar.jrc.ec.europa.eu

Kahn, M. E. (2009). Urban growth and climate change. *Annual Review of Resource Economics* 1, 333–349.

Karekezi, S., Kimani, J., and Onguru, O. (2008). Energy access among the urban poor in Kenya. *Energy for Sustainable Development* **12**, 38–48.

Karekezi, S., McDade, S., Boardman, B., and Kimani, J. (2012). Energy, poverty and development. In *Global Energy Assessment – Toward a Sustainable Future* (153–190). Cambridge University Press.

Keirstead, J., and Shah, N. (2013). *Urban Energy Systems: An Integrated Approach*. Routledge.

Kennedy, C., Ramaswami, A. Carney, S., and Dhakal, S. (2011). Greenhouse gas emission baselines for global cities and metropolitan regions. In Hoornweg, D., Freire, M., Lee, M. J., Bhada-Tata, P., and Yuen, B. (eds.), *Cities and Climate Change: Responding to an Urgent Agenda* (15–54). World Bank.

Kennedy, C. A., Ibrahim, N., and Hoornweg, D. (2014). Low-carbon infrastructure strategies for cities. *Nature Climate Change* **4**, 343–346.

Kennedy, C. A., Stewart, I., Facchini, A., Cersosimo, I., Mele, R., Chen, B., Uda, M., Kansal, A., Chiu, A.,Kim, K. -G., Dubeux, C., Lebre La Rovere, E., Cunha, B., Pincetl, S., Keirstead, J., Barles, S., Pusaka, S., Gunawan, J., Adegbile, M., Nazariha, M., Hoque, S., Marcotullio, P. J., Otharán, F. G., Genena, T., Ibrahim, N., Farooqui, R., Cervantes, G., and Duran Sahin, A. (2015). Energy and material flows of megacities. *Proceedings of the National Academy of Science* **112**, 5985–5990.

Kirshen, P., Ruth, M., and Anderson, W. (2007). Interdependencies of urban climate change impacts and adaptation strategies: A case study of Metropolitan Boston, USA. *Climatic Change* **86**, 105–122.

Knowlton, K., Kulkarni, S., Azhar, G.S, Mavalankar, D., Jaiswal, A., Connolly, M., Nori-Sarma, A., Rajiva, A., Dutta, P., and Deol, B. (2014). Development and implementation of South Asia's first heat-health action plan in Ahmedabad (Gujarat, India). *International Journal of Environmental Research and Public Health* 11, 3473–3492.

Kovacs, K. F., Haight, R. G. Jung, S., Locke, D. H., and O'Neil-Dunne, J. (2013). The marginal cost of carbon abatement from planting street trees in New York City. *Ecological Economics* **95**, 1–10.

Krzyzanowski, M., Apte, J. S., Bonjour, S. P., Brauer, M., Cohen, A. J., and Prüss-Üstün, A. (2014). Air pollution in the megacities. *Current Environmental Health Reports* **1**, 185–191.

Kunchornrat, J., and Phdungsilp, A. (2012). Multi-level governance of low-carbon energy systems in Thailand. *Energies* **5**, 531–544.

Larijani, K. M. (2009). Climate Change Effects on High-Elevation Hydropower System in California. Doctoral dissertation, University of California, Davis.

Larsen, P. H., Goldsmith, S., Smith, O., Wilson, M. L., Strzepek, K., Chinowsky, P., and Saylor, B. (2008). Estimating future costs for Alaska public infrastructure at risk from climate change. *Global Environmental Change* **18**, 442–457.

Lehner, B., Czisch, G., and Vassolo, S. (2001). *Europe's hydropower potential today and in the future*. University of Kassel, Center for Environmental Systems Research.

Lin, J., Cao, B., Cui, S., Wang, W., and Bai, X. (2010). Evaluating the effectiveness of urban energy conservation and GHG mitigation measures: The case of Xiamen city, China. *Energy Policy* **38**, 5123–5132.

Lo, K. (2014). Urban carbon governance and the transition toward low-carbon urbanism: Review of a global phenomenon. *Carbon Management* **5**, 269–283.

Lovins, A. B., Datta, E. K. Feiler, T., Rábago, K. R., Swisher, J. N., Lehmann, A., and Wicker, K. (2002). *Small Is Profitable: The Hidden Economic Benefits of Making Electrical Resources the Right Size*. Rocky Mountain Institute.

Maggio, G., and Cacciola, G. (2012). When will oil, natural gas, and coal peak? *Fuel* **98**, 111–123.

Mansanet-Bataller, M., Herve-Mignucci, M., and Leseur, A. (2008). Energy Infrastructures in France: Climate Change Vulnerabilities and Adaptation Possibilities. Mission Climate Working Paper, No. 2008–1. Caisse des Depots.

Marcotullio, P. J., Hughes, S., Sarzynski, A., Pincetl, S., Sanchez-Pena, L., Romero-Lankao, P., Runfola, D., and Seto, K. C. (2014). Urbanization and the carbon cycle: Contributions from social science. *Earth's Future* **2**, 496–514.

Marcotullio, P. J., Sarzynski, A., Albrecht, J., and Schulz, N. (2012). The geography of urban greenhouse gas emissions in Asia: A regional analysis. *Global Environmental Change* **22**, 944–958.

Marcotullio, P. J., Sarzynski, A., Albrecht, J., Schulz, N., and Garcia, J. (2013). The geography of global urban greenhouse gas emissions: An exploratory analysis. *Climatic Change* **121**, 621–634.

Marcotullio, P. J., Williams, E. W., and Marshall, J. D. (2005). Faster, sooner, and more simultaneously: How recent transportation CO2 emission trends in developing countries differ from historic trends in the United States of America. *Journal of Environment and Development* **14**, 125–148.

Markoff, M. S., and Cullen, A. C. (2008). Impact of climate change on Pacific Northwest hydropower. *Climatic Change* **87**, 451–469.

Marks, G., Hooghe, L., and Blank, K. (1996). European integration from the 1980s: State-centric v. multi-level governance. *Journal of Common Market Studies* **34**, 341–378.

Martinsen, D., Krey, V., and Markewitz, P. (2007). Implications of high energy prices for energy system and emissions—The response from an energy model for Germany. *Energy Policy* **35**, 4504–4515.

McGranahan, G., Balk, D., and Anderson, B. (2007). The rising tide: Assessing the risks of climate change and human settlements in low elevation coastal zones. *Environment and Urbanization* **19**(1), 17–37.

McKinley, J. C. (2008). Crews from 31 states in Texas to restore power. *New York Times* A18.

McMichael, A., Woodward, M., and Leeuwen, R. (1994). The impact of energy use in industrialised countries upon global population health. *Medicine & Global Survival* **1**, 23–32.

Miller, N., Hayhoe, K., Jin, J., and Auffhammer, M. (2008). Climate extreme heat and electricity demand in California. *Journal of Applied Meteorology and Climatology* **47**, 1834–1844.

Modi, V., McDade, S., Lallement, D., and Saghir, J. (2005). *Energy Servcies for the Millennium Development Goals*. World Bank.

Monstadt, J. (2007). Urban governance and the transition of energy systems: Institutional change and shifting energy and climate policies in Berlin. *International Journal of Urban and Regional Research* **31**, 326–343.

Monstadt, J., and Wolff, A. (2015). Energy transition or incremental change? Green policy agendas and the adaptability of the urban energy regime in Los Angeles. *Energy Policy* **78**, 213–224.

Morlet, C., and Keirstead, J. (2013). A comparative analysis of urban energy governance in four European cities. *Energy Policy* **61**, 852–863.

Müller, N., Kuttler, W., and Barlag, A. -B. (2014). Counteracting urban climate change: Adaptation measures and their effect on thermal comfort. *Theoretical and Applied Climatology* **115**, 243–257.

Munasinghe, M., and Swart, R. (2000). Climate change and its linkages with development, equity, and sustainability. In *Proceedings of the IPCC Expert Meeting Held in Colombo, Sri Lanka*. LIFE/RIVM/World Bank.

National Renewable Energy Laboratory (NREL). (2015). State & local governments: Renewable portfolio standards. Accessed January 5, 2016: http://www.nrel.gov/tech_deployment/state_local_governments/basics_portfolio_standards.html

National Research Academy. (1999). *Our Common Journey, A Transition Toward Sustainability*. National Academy Press.

Neumayer, E. (2002). Can natural factors explain any cross-country differences in carbon dioxide emissions? *Energy Policy*, **30**, 7–12.

Neves, C. G., and Dopico, Y. B. C. (2013). Análise de Metodologias de Produção de Inventários de Gases de Efeito Estufa de Cidades – Rio de Janeiro: UFRJ/ Escola Politécnica.

Nicholls, R. J., Hanson, S., Herweijer, C., Patmore, N., Hallegatte, S., Corfee-Morlot, and J.,Muir-Wood, R. (2008). *Ranking Port Cities with High Exposure and Vulnerability to Climate Extremes: Exposure Estimates*. Accessed July 3, 2015: http://www.aia.org/aiaucmp/groups/aia/documents/pdf/aias076737.pdf

Noble, I. R., Huq, S., Anokhin, Y. A., Carmin, J., Goudou, D., Lansigan, F. P., Osman-Elasha, B., and Villamizar, A. (2014). Adaptation needs and options. In Field, C. B., Barros, V. R., Dokken, D. J., Mach, K. J., Mastrandrea, M. D., Bilir, T. E., Chatterjee, M., Ebi, K. L., Estrada, Y. O., Genova, R. C., Girma, B., Kissel, E. S., Levy, A. N., MacCracken, S., Mastrandrea, P. R., and White, L. L. (eds.), *Climate Change 2014: Impacts, Adaptation, and Vulnerability, Contribution of Working Group II to the Fifth Assessment Report of the Intergovernmental Panel on Climate Change* (833–868). Cambridge University Press.

O'Neill, B.C., Ren, X., Jiang, L., and Dalton, M. (2012). The effect of urbanization on energy use in India and China in the iPETS model. *Energy Economics* 34(Supplement 3), S339–S345.

Olmos, S. (2001). Vulnerability and adaptation to climate change: Concepts, issues, assessment methods. Climate Change Knowledge Network. Accessed November 13, 2015: www.cckn.net

Ostertag, K. (2012). *No-regret Potentials in Energy Conservation: An Analysis of Their Relevance, Size and Determinants*. Springer Science & Business Media.

Pachauri, S. (2004). An analysis of cross-sectional variations in total household energy requirements in India using micro survey data. *Energy Policy*,**32**(15), 1723–1735.

Pachauri, S., and Jiang, L. (2008). The household energy transition in India and China. *Energy Policy* **36**(11), 4022–2035.

Pan, Z., Segal, M., Arritt, R. W., and Takle, E. S. (2004). On the potential change in solar radiation over the U.S. due to increases of atmospheric greenhouse gases. *Renewable Energy* **29**, 1923–1928.

Pandey, K. D., Wheeler, D., Ostro, B., Deichmann, U., Hamilton, K., and Bolt, K. (2006). Ambient particulate matter concentrations in residential and pollution hotspot areas of world cities: New estimates based on the Global Model of Ambient Particulates (GMAPS). Working paper. World Bank.

Pasternak, A. (2000). Global energy futures and human development: A framework for analysis. Lawrence Livermore National Laboratory, U.S. Department of Energy.

Petilla, C. J. L. (2014). *Philippine Energy Plan 2012–2030*. Department of Energy. Republic of the Philippines. Accessed November 7, 2015: https://www.doe.gov.ph/pep/philippine-energy-plan-2012-2030

Pierce, N. (2015). 19 cities report documented reductions in greenhouse-gas emissions. *Citiscope*. Accessed April 1, 2016: http://citiscope.org/story/2015/19-cities-report-documented-reductions-greenhouse-gas-emissions

Prasad, N., Ranghieri, F., Shah, F., Trohanis, Z., Kessler, E., and Sinha, R. (2008). *Climate Resilient Cities: A Primer on Reducing Vulnerabilities to Climate Change Impacts and Strengthening Disaster Risk Management in East Asian Cities*. The World Bank.

Poumanyvong, P., and Kaneko, S. (2010). Does urbanization lead to less energy use and lower CO_2 emissions? A cross-country analysis. *Ecological Economics* **70**(2), 434–444.

Ramaswami, A., Bernard, M., Chavez, A., Hillman, T., Whitaker, M., Thomas, G., and Marshall, M. (2012). Quantifying carbon mitigation wedges in U.S. cities: Near-term strategy analysis and critical review. *Environmental Science & Technology* **46**, 3629–3642.

Revi, A., Satterthwaite, D., Aragon-Durand, F., Corfee-Morlot, J., Kiunsi, R. B. R., Pelling, M., Roberts, D., Solecki, W., da Silva, J., Dodman, D., Maskrey, A., Gajjar, S. P., and Tuts, R. (2014). Urban areas. In

Field, C. B., Barros, V. R., Dokken, D. J., Mach, K. J., Mastrandrea, M. D., Bilir, T. E., Chatterjee, M., Ebi, K. L., Estrada, Y. O., Genova, R. C., Girma, B., Kissel, E. S., Levy, A. N., MacCracken, S., Mastrandrea, P. R., and White, L. L. (eds.),*Climate Change 2014: Impacts, Adaptation and Vulnerability* (535–612). Cambridge University Press.

Rogner, H. -H., Aguilera, R. F., Archer, C. L., Bertani, R., Bhattacharya, S. C., Dusseault, M. B., Gagnon, L., Haberl, H., Hoogwijk, M., Johnson, A., Rogner, M. L., Wagner, H., and Yakushev, V. (2012). Chapter 7 – Energy Resources and Potentials. In *Global Energy Assessment – Toward a Sustainable Future*, Cambridge University Press, Cambridge, UK and New York, NY, USA and the International Institute for Applied Systems Analysis, Laxenburg, Austria, pp. 423–512.

Rotmans, J., Kemp, R., and van Asselt, M. (2001). More evolution than revolution: Transition management in public policy. *Foresight* **3**, 15–31.

Royal Academy of Engineering. (2011). *Infrastructure, Engineering, and Climate Change Adaptation: Ensuring Services in an Uncertain Future*. Published by RAE on behalf of Engineering the Future.

Ruester, S., Schwenen, S., Finger, M., and Glachant, J. -M. (2013). A strategic energy technology policy towards 2050: No-regret strategies for European technology push. *International Journal of Energy Technology and Policy* **9**, 160–174.

Ruth, M. (2010). Economic and social benefits of climate information: Assessing the cost of inaction. *Procedia Environmental Sciences* 1, 387–394.

Rutter, P., and Keirstead, J. (2012). A brief history and the possible future of urban energy systems. *Energy Policy* **50**, 72–80.

Saatkamp, B., Masera, O., and Kammen, D. (2000). Energy and health transitions in development: Fuel use, stove technology, and morbidity in Jarácuaro, México. *Energy for Sustainable Development* **4**, 7–16.

Salamanca, F., Georgescu, M., Mahalov, A., Moustaoui, M., and Wang, M. (2014). Anthropogenic heating of the urban environment due to air conditioning. *Journal of Geophysical Research: Atmospheres* **119**, 5949–5965.

Satterthwaite, D. (2008). Cities' contribution to global warming: Notes on the allocation of greenhouse gas emissions. *Environment and Urbanization* **20**, 539–549.

Satterthwaite, D. (2009). The implications of population growth and urbanization for climate change. *Environment and Urbanization* **21**, 545–567.

Satterthwaite, D., and Sverdlik, A. (2013). Energy access and housing for low-income groups in urban areas. In Grubler, A., and Fisk, D. (eds.), *Energizing Sustainable cities* (73–94). Routledge.

Schaper, M. (2002). The essence of ecopreneurship. *Greener Management International* **38**, 26–30.

Schock, R. N., Skimms, R., Bull, S., Larsen, H., Likhachev, V., Nagano, K., Nilsson, H., Vuori, S., Yeager, K., and Zhou, L. (2012). Chapter 15 – Energy Supply Systems. In *Global Energy Assessment – Toward a Sustainable Future*, Cambridge University Press, Cambridge, UK and New York, NY, USA and the International Institute for Applied Systems Analysis, Laxenburg, Austria, pp. 1131–1172.

Schultz, P. W., Nolan, J. M., Cialdini, R. B., Goldstein, N. J., and Grisevicius, V. (2007). The constructive, destructive, and reconstructive power of social norms. *Psychological Science* **18**(5), 429–434.

Schulz, N. (2010a). Delving into the carbon footprint of Singapore: Comparing direct and indirect greenhouse gas emissions of a small and open economic system. *Energy Policy* **38**(9), 4848–4855.

Schulz, N. (2010b). Urban energy consumption database and estimations of urban energy intensities.

Working paper. International Institute for Applied Systems Analysis, Laxenburg, Austria. Accessed March 26, 2017: http://www.iiasa.ac.at/web/home/research/Flagship-Projects/Global-Energy-Assessment/KM18_City_energy_DB.pdf

Schwartz, B., Parker, C., Hess, J. J., and Frumkin, H. (2011). Public health and medicine in an age of energy scarcity: The case of petroleum. *American Journal of Public Health* **101**, 1560–1567.

Semenza, J., Hall, D., Wilson, D., Bontempo, B., Sailor, D., and George, L. (2008). Public perception of climate change: Voluntary mitigation and barriers to behavior change. *American Journal of Preventive Medicine* **35**, 479–487.

Seto, K. C., Dhakal, S., Bigio, A., Blanco, H., Delgado, G. C., Dewar, D., Huang, L., Inaba, A., Kansal, A., Lwasa, S., McMahon, J., Mueller, D., Murakami, J., Nagendra, H., and Ramaswami, A. (2014). Human settlements, infrastructure and spatial planning. In Edenhofer, O., Pichs-Madruga, R., Sokona, Y., Farahani, E., Kadner, S., Seyboth, K., Adler, A., Baum, I., Brunner, S., Eickemeier, P., Kriemann, B., Savolainen, J., Schlömer, S., von Stechow, C., Zwickel, T., and Minx, J. C. (eds.), *Climate Change 2014: Mitigation of Climate Change, 5th Assessment Report* (923–1000). Cambridge University Press.

Seto, K. C., Guneralp, B., and Hutyra, L. (2012). Global forecasts of urban expansion to 2030 and direct impacts on biodiversity and carbon pools. *Proceedings of the National Academy of Sciences of the United States of America* **109**, 552–563.

Simonoff, J. S., Restrepo, C. E., and Zimmerman, R. (2007). Risk management and risk analysis-based decision tools for attacks on electric power. *Risk Analysis* **27**, 547–570.

Smil, V. (2010). *Energy Transitions: History, Requirements, Prospects*. Praeger.

Smit, B., and Wandel, J. (2006). Adaptation, adaptive capacity and vulnerability. *Global Environmental Change* **16**, 282–292.

Smith, A. (2007). Emerging in between: The multi-level governance of renewable energy in the English regions. *Energy Policy* **35**, 6266–6280.

Smith, J. B., and Tirpak, D. (1989). The Potential Effects of Global Climate Change on the United States. U.S. Environmental Protection Agency *(EPA-230–05–89–050)*.

Smith, K. R. (2010). What's cooking? A brief update. *Energy for Sustainable Development* **14**, 251–252.

Sperling, J. (2014). Exploring the nexus of infrastructures, environment and health in Indian cities: Integrating multiple infrastructures and social factors with health risks. Doctoral dissertation, University of Colorado, Denver.

Sperling, J., and Ramaswami, A. (2013). Exploring health outcomes as a motivator for low-carbon city development: Implications for infrastructure interventions in Asian cities. *Habitat International* **37**, 113–123.

Steinberger, J., and Roberts, J. (2009). *Across a Moving Threshold: Energy, Carbon and the Efficiency of Meeting Global Human Development Needs*. Institute of Social Ecology.

Stern, P. C., et al. (2016). Opportunities and insights for reducing fossil fuel consumption by households and organizations. *Nature Energy* **1**, 16043.

Stern, P., Aronson, E., Darley, J., Hill, D., Hirst, E., Kempton, W., and Wilbanks, T. (1985). The effectiveness of incentives for residential energy conservation. *Evaluation Review* **10**, 147–176.

Thøgersen, J., and Olander, F. (2002). Human values and the emergence of a sustainable consumption pattern: A panel study. *Journal of Economic Psychology* **23**(5), 605–630.

Tyler, S., and Moench, M. (2012). A framework for urban climate resilience. *Climate and Development* **4**, 311–326.

UN. (2014). *World Urbanization Prospects, 2014 Revisions*. Department of Economic and Social Affairs, United Nations.

UN. (2015). Open working group proposal for sustainable development goals. Accessed February 14, 2016: http://undocs.org/A/68/970

United Nations Environment Program (UNEP). (2011). ECCO Metropolitan District of Quito. Accessed April 9, 2015: http://www.unep.org/

United Nations Environment Program (UNEP). (2016). Global trends in renewable energy investment 2016. Frankfurt School of Finance and Management. Accessed January 25, 2017: http://fs-unep-centre.org/sites/default/files/publications/globaltrendsinrenewableenergyinvestment2016lowres_0.pdf

UNEP, UN-HABITAT, and World Bank. (2010). International Standard for Determining Greenhouse Gas Emissions for Cities (Version 2.1). Accessed October 28, 2014: http://siteresources.worldbank.org/INTUWM/Resources/340232-1205330656272/4768406-1291309208465/Annexes.pdf

United Nations Office for Disaster Risk Reduction (UNISDR). (2009). Terminology on disaster risk reduction. UNISDR.

U.S. Department of Energy. (2013). U.S. energy sector vulnerabilities to climate change and extreme weather. USDOE.

U.S. Department of Energy. (2014). Evaluating electric vehicle charging impacts and customer charging behaviors – experiences from six smart grid investment grant projects. USDOE.

U.S. Environmental Protection Agency. (2011). State climate and energy program. Accessed October 28, 2014: http://www.epa.gov/statelocalclimate/

U.S. Environmental Protection Agency. (2014). Guide to the Clean Air Act. Accessed April 27, 2015: http://www.epa.gov/airquality/peg_caa/acidrain.html.

U.S. Environmental Protection Agency. (2015). Inventory of U.S. greenhouse gas emissions and sinks: 1990–2013. U.S. EPA, Accessed April 12, 2016: https://www3.epa.gov/climatechange/Downloads/ghgemissions/U.S.-GHG-Inventory-2015-Main-Text.pdf.

Verbong, G., and Geels, F. (2007). The ongoing energy transition: Lessons from a socio-technical, multi-level analysis of the Dutch electricity system (1960–2004). *Energy Policy* **35**, 1025–1037.

Vine, E. (2012). Adaptation of California's electricity sector to climate change. *Climatic Change* **111**, 75–99.

Wallston, S., Clarke, G., Haggarty, L., Keneshiro, R., Noll, R., Shirley, M., and Xu, L. C. (2004). New tools for studying network industry reforms in developing countries: the telecommunications and electricity regulation database. Policy Research Working Paper No. 3286. World Bank.

Weisz, H., and Steinberger, J. K. (2010). Reducing energy and material flows in cities. *Current Opinion in Environmental Sustainability* **2**(3), 185–192.

Wenzel, F., Bendimerad, F., and Sinha, R. (2007). Megacities – megarisks. *Natural Hazards* **42**, 481–491.

Wilbanks, T. J., Bhatt, V., Bilello, D. E., Bull, S. R., Ekmann, J., Horak, W. D., Huang, Y. J., Levine, M. D., Sale, M. J., Schmalzer, D. K., and Scott, M. J. (2007). Effects of climate change on energy production and use in the United States. Report for the U.S. Climate Change Science Program and the Subcommittee on Global Change Research by the U.S. Department of Energy; Office of Biological and Environmental Research.

Wilkinson, P., Smith, K. R., Joffe, M., and Haines, A. (2007). Energy and health 1 – A global perspective on energy: Health effects and injustices. *The Lancet* **370**, 965–978.

Williamson, L. E., Connor, H., and Moezzi, M. (2009). *Climate-proofing Energy Systems.* HELIO International.

World Bank. (2011). Guide to climate change adaptation in cities. Accessed October 10, 2014: http://siteresources.worldbank.org/INTURBANDEVELOPMENT/Resources/336387–1318995974398/Guide-ClimChangeAdaptCities.pdf

World Bank. (2013). Applying abatement cost curve methodology for low-carbon strategy in Changning District, Shanghai (No. 84068 v1). World Bank.

World Bank. (2015). World Bank indicators: Climate change, CO_2 emissions (kt). Accessed January 31, 2016: http://data.worldbank.org/indicator/EN.ATM.CO2E.KT/countries?display=default

World Coal Association. (2012). *Coal Matters: Coal in the Global Energy Supply.* World Coal Association

World Energy Council (WEC). (2010). *Energy and Urban Innovation.* WEC.

World Energy Council (WEC). (2013). *World Energy Resources 2013 Survey.* WEC.

Yang, J., McBride, J., Zhou, J., and Sun, Z. (2005). The urban forest in Beijing and its role in air pollution reduction. *Urban Forestry & Urban Greening* **3**, 65–78.

Zhao, J. (2010). Climate change mitigation in Beijing, China. Accessed December 13, 2014: http://unhabitat.org/wp-content/uploads/2012/06/GRHS2011CaseStudyChapter05Beijing.pdf

Zimmerman, R. (2012). *Transport, the Environment and Security. Making the Connection.* Edward Elgar Publishing, Ltd.

Zimmerman, R. (2014). Planning restoration of vital infrastructure services following Hurricane Sandy: Lessons learned for energy and transportation. *Journal of Extreme Events* 1. Accessed September 25, 2015: http://www.worldscientific.com/doi/pdf/10.1142/S2345737614500043

Zimmerman, R., and Faris, C. (2011). Climate change mitigation and adaptation in North American cities. *Current Opinion in Environmental Sustainability* **3**, 181–187.

Zulu, L. C. (2010). The forbidden fuel: Charcoal, urban woodfuel demand and supply dynamics, community forest management and woodfuel policy in Malawi. *Energy Policy* **38**, 3717–3730.

Zwaan, B. (2013). The role of nuclear power in mitigating emissions from electricity generation. *Energy Strategy Reviews* **1**(4), 296–301.

Chapter 12 Case Study References

Case Study 12.1 Urban GHG Mitigation in Rio de Janeiro

Aneel (National Electric Energy Agency). (2008). *Program for R&D and Electric Energy Innovation: Manual and Guidelines.* Aneel.

Cities, C. (2014). CDP Cities 2014 Information Request. Accessed November 6, 2015: https://www.cdp.net/sites/2014/76/31176/CDP%20Cities%202014/Pages/DisclosureView.aspx

CGEE (Centre for Management and Strategic Research). (2012). Smart grids: National context. CGEE.

City of Rio de Janeiro (2011). *Greenhouse Gas Inventory and Emissions Scenario of Rio de Janeiro City – Brazil.* Accessed March 23, 2014: http://www.rio.rj.gov.br/dlstatic/10112/1712030/DLFE-23(7703).pdf/Inventarioapresentacao.pdf

CDP Cities. (2014). Information Request Prefeitura do Rio de Janeiro

Coelho, J. S. (2010). *Regulation for NonTechnical Losses.* Rio de Janeiro: ANEEL National Electric Energy Agency.

COPPE, U. (2011). *Greenhouse Gas Inventory and Emissions Scenario of Rio de Janeiro City.* Rio de Janeiro: Official Gazette of Rio de Janeiro.

ENEL Foundation Research Project (2014). Energy Transitions in Cities. Lifestyle, experimentation and change. Fifth case study. Aleteia Communication.

Geller, H., Shaeffer, R., Szklo, A., and Tolmasquim, M. (2004). Geller, H., Schaeffer, R., Szklo, A., and Tolmasquim, M. (2004). Policies for advancing energy efficiency and renewable energy use in Brazil. *Energy Policy,* **32**(12), 1437–1450.

IBGE. (2015). Instituto Brasileiro de Geografia e Estatistica. Table 3.2. Accessed January 19, 2016: http://www.ibge.gov.br/home/estatistica/populacao/censo2010/sinopse/sinopse_tab_rm_zip.shtm.

Light. (2013). Presentation, Smart Cities & Smart Grids in the electric sector, International Seminar Portugal-Brazil, October 25, 2013, GESEL, UFRJ – Rio de Janeiro Federal University.

Peel, M. C., Finlayson, B. L., and McMahon, T. A. (2007). Updated world map of the Köppen-Geiger climate classification. *Hydrology and Earth System Sciences Discussions* **4**(2), 462.

World Bank. (2017). 2016 GNI per capita, Atlas method (current US$). Accessed August 9, 2017: http://data.worldbank.org/indicator/NY.GNP.PCAP.CD

Case Study 12.2 Climate Change and the Energy Supply System in Seattle

City of Seattle. (2013). *Seattle Climate Action Plan.* Seattle. Accessed May 29, 2014: http://www.seattle.gov/Documents/Departments/OSE/2013_CAP_2013 (0612).pdf

Peel, M. C., Finlayson, B. L., and McMahon, T. A. (2007). Updated world map of the Köppen-Geiger climate classification. *Hydrology and Earth System Sciences Discussions* **4**(2), 462.

Ribeiro D., Hewitt, V., Mackres, E., Cluett, R., Ross, L. M., Vaidyanathan, S., and Zerbonne S. (2015). The 2015 city energy efficiency scorecard. Accessed January 19, 2016: http://aceee.org/research-report/u1502

Seattle City Light. (2014). Seattle City Light integrated resource plan update and progress report. Accessed May 14, 2014: http://www.seattle.gov/light/news/issues/irp/docs/SeattleCityLight2014_IRPUpdateandProgressReport.pdf

United Nations Development Programme. (2014). Human Development Index (HDI). Accessed June 11, 2015: http://hdr.undp.org/sites/default/files/hdr14_statisticaltables.xls

U.S. Census Bureau. (2010). Decennial census, summary file 1. Accessed February 7, 2014: http://www.census.gov/population/metro/files/CBSA%20Report%20Chapter%203%20Data.xls

World Bank. (2017). 2016 GNI per capita, Atlas method (current US$). Accessed August 9, 2017: http://data.worldbank.org/indicator/NY.GNP.PCAP.CD

Case Study 12.3 Renewable Gas Demonstration Projects in New York

AECOM. (2014). Awarded contract for Newtown Creek renewable natural gas project. Accessed February 2, 2015: http://www.aecom.com/News/Press±Releases/_carousel/

Environmental Protection Agency. (2014). The benefits of anaerobic digestion of food waste at wastewater treatment facilities. Accessed October 29, 2015: http://www.epa.gov/region9/organics/ad/Why-Anaerobic-Digestion.pdf

Environmental Protection Agency. (2014). Waste, resource conservation, food waste: Anaerobic digestion. Accessed November 19, 2015: http://www.epa.gov/foodrecovery/fd-anaerobic.htm

Gardiner, B. (2014). Biogas, a low-tech fuel with a big payoff. *New York Times*.

National Grid. (2014). Renewable gas: A vision for a sustainable gas network. Accessed July 8, 2015: http://www.nationalgridus.com/non_html/ng_renewable_wp.pdf

National Grid. (2014). Role of renewable natural gas in closing the carbon cycle. Accessed March 5, 2015: http://energy.columbia.edu/files/2014/02/3-Cavanagh-Role-of-RNG-in-Closing-Carbon-Cycle.pdf

National Grid. (2014). News Release: National Grid in partnership with NYC EPA begins design and construction of the Newtown Creek renewable natural gas project. Accessed February 20, 2015: http://www.nationalgridus.com/aboutus/a3-1_news2.asp?document=8765

New York City Department of Environmental Protection. (2013). City announces innovative new partnerships that will reduce the amount of organic waste sent to landfills, produce a reliable source of clean energy and improve air quality. Accessed May 7, 2014: http://www.nyc.gov/html/dep/html/press_releases/13–121pr.shtml–.VInZxN6DTww

Peel, M. C., Finlayson, B. L., and McMahon, T. A. (2007). Updated world map of the Köppen-Geiger climate classification. *Hydrology and Earth System Sciences Discussions* **4**(2), 462.

Rulkens, W. H. (2009). Opportunities to improve energy recovery from sewage sludge. *Water* 21, 24.

U.S. Census Bureau. (2010). Decennial census, summary file 1. Accessed November 15, 2014: http://www.census.gov/population/metro/files/CBSA%20Report%20Chapter%203%20Data.xls

U.S Census Bureau. (2016). Table 1. Annual Estimates of the Resident Population: April 1, 2010 to July 1, 2016 – Metropolitan Statistical Area; – 2016 Population Estimates. U.S. Census Bureau. Retrieved March 28, 2017. Accessed February 20, 2017: https://factfinder.census.gov/faces/tableservices/jsf/pages/productview.xhtml?pid=PEP_2016_PEPANNRES&prodType=table

Waste Management of New York. (2014). WM Varick I CORe – Newtown Creek Co-Digestion Project. Converting NYC food waste into clean renewable energy source. Accessed July 1, 2015: http://www.wm.com/NYCMA/WMCORe%20varick-factsheet%20073114.pdf

World Bank. (2017). 2016 GNI per capita, Atlas method (current US$). Accessed August 9, 2017: http://data.worldbank.org/indicator/NY.GNP.PCAP.CD

Case Study 12.4 The Benefits of Large-Scale Renewable Electricity Investment in Canberra

ACTCCC. (2015). ACT Climate Change Council. Accessed February 27, 2016: http://www.environment.act.gov.au/cc/climate%5Fchange%5Fcouncil

ACT Government. (2012). AP2: A new climate change strategy and action plan for the Australian Capital Territory. Accessed December 18, 2015: http://www.environment.act.gov.au/__data/assets/pdf_file/0006/581136/AP2_Sept12_PRINT_NO_CROPS_SML.pdf

ACT Government. (2014a). Adapting to a changing climate: Directions for the ACT. Accessed April 16, 2015: http://www.environment.act.gov.au/__data/assets/pdf_file/0008/597653/Adaptation-Framework-Directions.pdf

ACT Government. (2014b). Renewable energy local investment framework. Accessed March 29, 2015: http://www.environment.act.gov.au/__data/assets/pdf_file/0003/581700/Renewable-Energy-Local-Investment-Framework-v2.pdf

ACT Government. (2015a). Australian demographic statistics – September quarter 2014. Accessed February 27, 2016: http://apps.treasury.act.gov.au/__data/assets/pdf_file/0008/644813/ERP.pdf

ACT Government. (2015b). Review of AP2. Accessed March 30, 2016: http://www.environment.act.gov.au/cc/what-government-is-doing/emissions-and-mitigation

ACT Government. (2015c). How wind will power Canberra homes. Accessed January 24, 2016: http://www.environment.act.gov.au/energy/wind_Fpower

ACT Government (2016) 100% Renewable Energy Target. Accessed January 12, 2017: http://www.environment.act.gov.au/energy/cleaner-energy.

AEMO. (2014). AEMO annual report 2014. Accessed September 28, 2015: http://www.aemo.com.au/About-AEMO/Corporate-Publications/AEMO-Annual-Report

Australian Bureau of Statistics (ABS). (2015). Unincorporated ACT. Accessed April 14, 2016: http://stat.abs.gov.au/itt/r.jsp?RegionSummary&region=89399&dataset=ABS_NRP9_LGA&geoconcept=REGION&maplayeridLGA2012&measure=MEASURE&datasetASGS=ABS_NRP9_ASGS&datasetLGA=ABS_NRP9_LGA&regionLGA=REGION&regionASGS=REGION

Australian Bureau of Statistics (ABS). (2007). 1307.8 Australian capital territory in focus. Accessed July 30, 2014: http://www.abs.gov.au/AUSSTATS/abs@.nsf/Lookup/(1307).8Main±Features12007?OpenDocument

Buckman, G., Sibley, J., and Bourne, R. (2014). The large-scale solar feed-in tariff reverse auction in the Australian Capital Territory, Australia. *Energy Policy* **72**, 14–22.

Department of Foreign Affairs and Trade. (2015). Australian Capital Territory economic indicators. Accessed January 1, 2016: http://www.dfat.gov.au/trade/resources/Documents/act.pdf

Garnaut, R. (2008). The Garnaut climate change review. Accessed July 6, 2014: http://www.garnautreview.org.au/2008-review.html

Independent Competition and Regulatory Commission. (2014). ACT greenhouse gas inventory report 2011–12. Accessed September 28, 2015: http://www.environment.act.gov.au/__data/assets/pdf_file/0007/644326/ACT-GHG-Inventory-Report-2011–12.pdf

Peel, M. C., Finlayson, B. L., and McMahon, T. A. (2007). Updated world map of the Köppen-Geiger climate classification. *Hydrology and Earth System Sciences Discussions* **4**(2), 462.

United Nations Development Programme. (2014). Human Development Index (HDI). Accessed August 8, 2015: http://hdr.undp.org/sites/default/files/hdr14%5Fstatisticaltables.xls

World Bank. (2017). 2016 GNI per capita, Atlas method (current US$). Accessed August 9, 2017: http://data.worldbank.org/indicator/NY.GNP.PCAP.CD

Case Study 12.5 The City of Singapore's 3D Energy Planning Tool as a Means to Reduce CO$_2$ Emissions Effectively

Department of Statistics Singapore. (2015). Singapore in figures 2015. Accessed March 16, 2016: http://www.singstat.gov.sg/docs/default-source/default-document-library/publications/publications_and_papers/reference/sif2015.pdf

Peel, M. C., Finlayson, B. L., and McMahon, T. A. (2007). Updated world map of the Köppen-Geiger climate classification. *Hydrology and Earth System Sciences Discussions* **4**(2), 462.

World Bank. (2017). 2016 GNI per capita, Atlas method (current US$). Accessed August 9, 2017: http://data.worldbank.org/indicator/NY .GNP.PCAP.CD

Case Study 12.6   Managing Energy Systems for Reducing Emissions and Risks in Delhi

Census of India. (2011). The Registrar General & Census Commissioner, India. Accessed August 2, 2014: http://www.censusindia.gov .in/2011census/PCA/PCA_Highlights/PCA_Data_highlight.html

Cohen, E. (2014). The water footprint of urban energy systems: concepts, methods and applications for assessing electricity supply risk factors. Master's thesis, University of Colorado, Denver.

IndiaStat. (2015). Datanet India Pvt. Accessed March 16, 2016: http://www .indiastat.com/default.aspx

Peel, M. C., Finlayson, B. L., and McMahon, T. A. (2007). Updated world map of the Köppen-Geiger climate classification. *Hydrology and Earth System Sciences Discussions* 4(2), 462.

Prakash, S. (2014). Energy conservation through standby power reduction. *Middle-East Journal of Scientific Research* 19(7), 990–994. Accessed August 8, 2015: www.idosi.org/mejsr/mejsr19(7)14/19.pdf

Sperling, J. B., and Ramaswami, A. (2012). Exploring health outcomes as a motivator for low-carbon city development: Implications for infrastructure interventions in Asian cities. *Habitat International.* doi: 10.1016/j.habitatint.2011.12.013

World Bank. (2017). 2016 GNI per capita, Atlas method (current US$). Accessed August 9, 2017: http://data.worldbank.org/indicator/NY .GNP.PCAP.CD

Case Study 12.7   Energy and Climate Change in Quito

Berndtsson, J. C. (2010). Green roof performance towards management of runoff water quantity and quality: A review. *Ecological Engineering* 36(4), 351–360.

Brandt, A. R., Heath, G. A., Kort, E. A., O'Sullivan, F., Pétron, G., Jordaan, S. M., and Harriss, R. (2014). Methane leaks from North American natural gas systems. *Science* 343(6172), 733–735.

Constante, S. (2014). Ecuador prepara el terreno para eliminar el subsidio de gas. El Pais. Accessed February 2, 2015: http://internacional.elpais .com/internacional/2014/08/20/actualidad/1408569837_695217 .html

EEQ (Empresa Electrica de Quito). (2012). Plan estrategico 2012–2015. Pichincha, Ecuador.

EEQ. (2014a). Programa de eficiencia energetica para cocción pr inducción y calentamiento de agua con electricidad. Pichincha, Ecuador.

EEQ (2014b). La Generación de electricidad en la EEQ: Gerencia de generación y subtransmisión. Pichincha, Ecuador.

Kawajiri, K., Oozeki, T., and Genchi, Y. (2011). Effect of temperature on PV potential in the world. *Environmental Science & Technology* 45(20), 9030–9035.

Ludeña, C. E., and Wilk, D. (2013). Ecuador: Mitigación y adaptación al cambio climático. Inter-American Development Bank.

Ministerio del Ambiente de Ecuador (MAE). (2000). National communication: Republic of Ecuador to the United Framework Convention on Climate Change. Accessed July 8, 2014: http://www.ambiente.gob.ec/

Municipio del Distrito Metropolitano de Quito (MDMQ), Secretaria de Ambiente. (2009). Estrategia Quitena al cambio climatico. Accessed March 27, 2015: http://www.quitoambiente.gob.ec/

Municipio del Distrito Metropolitano de Quito (MDMQ), Secretaria de Ambiente. (2010). 10 Acciones de Quito frente al cambio climático. Accessed June 9, 2015: http://www.quitoambiente.gob.ec/

Municipio del Distrito Metropolitano de Quito (MDMQ), Secretaria de Ambiente. (2011). Inetario de emisiones de gases del efecto de invernadero en el districto metropolitano de Quito. Año 2017. Accessed April 21, 2014: http://www.quitoambiente.gob.ec/

Municipio del Distrito Metropolitano de Quito (MDMQ), Secretaria de Ambiente. (2014). Inetario de emisiones de gases del efecto de invernadero en el districto metropolitano de Quito. Año 2011. Accessed January 4, 2015: http://www.quitoambiente.gob.ec/

Municipio del Distrito Metropolitano de Quito (MDMQ), Secretaria de Ambiente. (2012). Plan de acción climático de Quito 2012–2016. Accessed March 27, 2015: http://www.quitoambiente.gob.ec

Peel, M. C., Finlayson, B. L., and McMahon, T. A. (2007). Updated world map of the Köppen-Geiger climate classification. *Hydrology and Earth System Sciences Discussions* 4(2), 462.

Secretario Metropolitano de Territorio, Habitat y Vivienda (STHV). (2010). *Poblacion e indicadores del distrito metropolitano de Quito.* Accessed September 13, 2014: http://sthv.quito.gob.ec/images/indicadores/parroquia/Demografia.htm

Seto K. C., Dhakal, S., Bigio, A., Blanco, H., Delgado, G. C., Dewar, D., Huang, L., Inaba, A., Kansal, A., Lwasa, S., McMahon, J. E., Müller, D. B., Murakami, J., Nagendra, H., and Ramaswami, A. (2014). Human settlements, infrastructure and spatial planning. In Edenhofer, O., Pichs-Madruga, R., Sokona, Y., Farahani, E., Kadner, S., Seyboth, K., Adler, A., Baum, I., Brunner, S., Eickemeier, P., Kriemann, B., Savolainen, J., Schlömer, S., von Stechow, C., Zwickel, T., and Minx, J. C. (eds.), *Climate Change 2014: Mitigation of Climate Change. Contribution of Working Group III to the Fifth Assessment Report of the Intergovernmental Panel on Climate Change* (932–1000). Cambridge University Press.

UNEP. (2011). ECCO metropolitan district of Quito. Accessed May 22, 2014: http://www.unep.org/

United Nations Framework Convention on Climate Change (UNFCCC). (2015). Clean development mechanism. Accessed January 23, 2016: http://cdm.unfccc.int/index.html

Vasilakopoulou, K., Kolokotsa, D., and Santamouris, M. (2014). Cities for smart environmental and energy futures: Urban heat island mitigation techniques for sustainable cities. In *Cities for Smart Environmental and Energy Futures* (215–233). Springer.

World Bank. (2017). 2016 GNI per capita, Atlas method (current US$). Accessed August 9, 2017: http://data.worldbank.org/indicator/NY .GNP.PCAP.CD

Zambrano-Barragan, C. (2010). Quito's climate change strategy: Policies for planned adaptation and reducing vulnerability. Accessed April 26, 2014: http://www.quitoambiente.gob.ec/

13

Urban Transportation

Coordinating Lead Authors

Shagun Mehrotra (New York/Indore), Eric Zusman (Hayama)

Lead Authors

Jitendra N. Bajpai (New York/Mumbai), Lina Fedirko (New York/San Francisco), Klaus Jacob (New York), Michael Replogle (Washington D.C./New York), Matthew Woundy (Boston), Susan Yoon (New York)

Contributing Authors

Ken Doust (Lismore) and John Black (Sydney)

This chapter should be cited as

Mehrotra, S., Zusman, E., Bajpai, J. N., Fedirko, L., Jacob, K., Replogle, M., Woundy, M., and Yoon, S. (2018). Urban transportation. In Rosenzweig, C., W. Solecki, P. Romero-Lankao, S. Mehrotra, S. Dhakal, and S. Ali Ibrahim (eds.), Climate Change and Cities: Second Assessment Report of the Urban Climate Change Research Network. Cambridge University Press. New York. 491–518

Transport as Climate Challenge and Solution

Urban transport systems are major emitters of greenhouse gases (GHGs) and are essential to developing resilience to climate impacts. At the same time, cities need to move forward quickly to adopt a new paradigm that ensures access to clean, safe, and affordable mobility for all.

In middle-income countries, rising incomes are spurring demand for low-cost vehicles, and, together with rapid and sprawling urbanization and segregated land use, this is posing unprecedented challenges to sustainable development while contributing to climate change.

Expanded climate-related financing mechanisms are being developed at national and international levels, such as the Green Climate Fund. Local policy-makers should prepare the institutional capacity and policy frameworks needed to access financing for low-carbon and resilient transport.

Major Findings

- Cities account for approximately 70% of CO_2 emissions (depending on measurement protocols), with a significant proportion due to urban transport choices. The transport sector directly accounted for nearly 30% of total end-use energy-related CO_2 emissions. Of these, direct emissions from urban transport account for 40%.

- Urban transport emissions are growing at 2–3% annually. The majority of emissions from urban transport are from higher-income countries. In contrast, 90% of the growth in emissions is from transport systems in lower-income countries.

- Climate-related shocks to urban transportation have economy-wide impacts, beyond disruptions to the movement of people and goods. The interdependencies between transportation and other economic, social, and environmental sectors can lead to citywide impacts.

- Integrating climate risk reduction into transport planning and management is necessary in spatial planning and land-use regulations. Accounting for these vulnerabilities in transport decisions can ensure that residential and economic activities are concentrated in low-risk zones.

- Low-carbon transport systems yield co-benefits that can reduce implementation costs, yet policy-makers often need more than a good economic case to capture potential savings.

- Integrated low-carbon transport strategies – Avoid-Shift-Improve – involve avoiding travel through improved mixed land-use planning and other measures; shifting passengers to more efficient modes through provision of high-quality, high-capacity mass transit systems; and improving vehicle design and propulsion technologies to reduce fuel use.

- Designing and implementing risk-reduction solutions and mitigation strategies require supportive policy and public–private investments. Key ingredients include employing market-based mechanisms; promoting information and communication technologies; building synergies across land-use and transport planning; and refining regulations to encourage mass transit and non-motorized modes.

Key Messages

Co-benefits such as improved public health, better air quality, reduced congestion, mass transit development, and sustainable infrastructure can make low-carbon transport more affordable and can yield significant urban development advantages. For many transport policy-makers, co-benefits are primary entry points for reducing GHG emissions. Moreover, policy-makers should find innovative ways to price the externalities – the unattributed costs – of carbon-based fuels.

The interdependencies between transport and other urban sectors mean that disruptions to transport can have citywide impacts. To minimize disruptions due to such interdependencies, policy-makers should take a systems approach to risk management that explicitly addresses the interconnectedness of climate, transport, and other relevant urban sectors.

Low-carbon transport should also be socially inclusive because social equity can improve a city's resilience to climate change impacts. Automobile-focused urban transport systems fail to provide mobility for significant segments of urban populations. Women, the elderly, the poor, non-drivers, and disadvantaged people need urban transport systems that go beyond enabling mobility to fostering social mobility as well.

13.1   Introduction

Over US$2 trillion dollars are spent on infrastructure – transport, energy, and water – globally each year, half in developing countries (International Monetary Fund [IMF], 2014). Cities account for approximately 70% of CO_2 emissions (depending on measurement protocols) with a significant proportion due to urban transport choices (Seto et al., 2014). The transport sector directly accounted for 28% of total end-use energy-related carbon dioxide (CO_2) emissions (Sims et al., 2014). Direct urban transport emissions accounts for 40% of these emissions (see Figure 13.1). The share of urban transport emissions becomes even greater with the inclusion of indirect transport emissions. For example, emissions resulting from the production of fuels and building of transit systems, energy consumed in non-travel functions of urban transportation departments such as offices and maintenance yards, and, counterintuitively, adaptation measures such as construction of transit-oriented development.

Urban transport emissions are growing at 2–3% annually (International Energy Agency [IEA], 2014). The majority of the stock of GHG emissions is from developed-country urban transport services. In contrast, 90% of the flow – growth in emissions – is expected to come from developing-country transit services (IEA, 2014). In developing countries, rising incomes that spur demand for low-cost vehicles accompanied by rapid and sprawling urbanization and segregated land use pose an unprecedented challenge to sustainable development. They are also a fast-growing cause of climate change. In response, billions of dollars are being invested in developing low-carbon transit systems – from metro rail systems and light rail networks

in Latin America to bus rapid transit (BRTs) systems in Asia. However, there is a knowledge gap when it comes to understanding the sustainability of these interventions, particularly as they pertain to urban climate change impacts; technological, institutional, and commercial viability; and benefits to the poor.

Urban transportation systems consist of passenger and freight subsectors. By modal choice, transportation services are classified into land-, water-, and air-based systems. Ownership and management of transit systems are split between public and private sectors and combinations of the two. More than 70% of transport emissions are from road-based systems (Sims et al., 2014), therefore, in this chapter we focus on climate change risks, mitigation, adaptation, and policy mechanisms for intra-urban road and rail systems for passengers and freight.

Our objective is to assess the state-of-the-knowledge on urban transport and climate change. The assessment is structured around four subcomponents: urban transport and climate change risks, adaptation solutions, mitigation choices, and policy mechanisms nested within enabling environments that can operationalize action. The assessment has engaged leading transport scholars and practitioners to evaluate existing research as well as to identify tacit knowledge. Case Studies of New York, Lagos, Johannesburg, and Delhi are integrated into the chapter to contextualize solutions. The chapter also builds on and extends themes in the first urban transport assessment (Mehrotra et al., 2011) and has iteratively informed and benefited from the endeavors of the chapter authors' engagement and leadership role in the United Nations Sustainable Development Goals process on issues of urbanization and transport as well as the 21st United Nations Framework Convention on Climate Change (UNFCCC) Climate Summit in Paris (COP21).

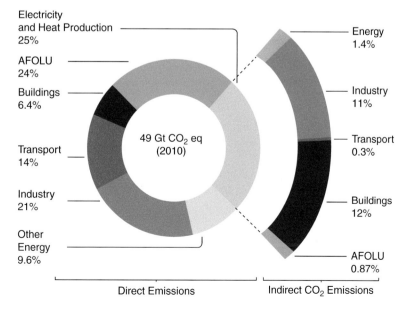

Greenhouse Gas Emission by Economic Sectors

Electricity and Heat Production 25%
AFOLU 24%
Buildings 6.4%
Transport 14%
Industry 21%
Other Energy 9.6%

49 Gt CO_2 eq (2010)

Energy 1.4%
Industry 11%
Transport 0.3%
Buildings 12%
AFOLU 0.87%

Direct Emissions   Indirect CO_2 Emissions

Figure 13.1 *Share of urban transport's greenhouse gas emissions among economic sectors.*

Source: Adapted from IPCC, 2014

13.2   Climate Risk for Urban Transport

To comprehensively assess climate risk to urban transport, we utilize the urban climate change risk assessment framework (Mehrotra et al., 2009) that combines three vectors: hazards, vulnerabilities, and response capacity (see Figure 13.2). Global climate risk is accumulated in urban areas and increases as people, private and public assets, and economic activities become more concentrated in cities (Mehrotra et al., 2009, 2011; Revi et al., 2014). Urban transport systems that enable movement of people, goods, and services within and between cities are at risk, particularly during extreme events such as heat waves, hurricanes, floods. These events can disrupt transport systems and interdependent urban services during extreme weather conditions with impacts lasting through the relief and recovery phase. Transport systems utilized for emergency response to rescue vulnerable populations (e.g., hospitalized patients) can be temporarily impaired due to system disruptions ranging from lack of fuel supply to inundated roads and tunnels (New York [NYC], 2013).

The first vector in the urban climate risk framework is climate hazards, which captures external forcings due to observed and projected change in mean and extreme precipitation, temperature, and sea levels. Hazards impacting urban transport systems include intense precipitation, increasing temperature, rising sea levels, and growing frequency and intensity of coastal storms.

More frequent and intense precipitation increases the probability and extent of threats to underground systems such as tunnels for rail and road. The structural strength of bridge footings may be compromised due to corrosion from saltwater intrusion and soil erosion. Storms with high-velocity winds impact static and mobile transport systems differently. Static systems such as roads, bridges, and railway networks may be inundated due to

Figure 13.2 *Urban climate risk framework.*

Source: Adapted from Mehrotra et al., 2009

storm surges or create additional stress on critical infrastructure and cause structural damage. Mobile assets and vehicles, such as cars and buses or train coaches, may be damaged due to flying debris. Heat waves can compromise the design strength of roads and rail structures due to damage to physical infrastructure. For example, roads can experience cracking, rutting, and peeling of asphalt or buckling of concrete and steel structures. Furthermore, road, rail, and water-based system connections can compound the impacts of climate hazards felt in one transport system and carry over to another. For a summary of climate change impacts on urban transport, see Table 13.1. Hazards are context-related, varying for land-based systems at local and national scales, and

Table 13.1 *Climate hazards and urban transportation impacts. Source: Adapted from Hodges, 2011*

Climate Hazard	Likely impacts (Impacts differ by vulnerabilities and response capacities)
Precipitation: Intense rainfall and sudden snowstorms	Damage and incapacitation of transit networks due to flooding of tracks, roadways, tunnels, maintenance facilities Blockage of roadways and drainage systems Landslides as a consequence of precipitation Scour on bridge as a result of precipitation
Temperature: Heat waves	Buckling of railroad tracks Damage to paved surfaces Passenger and worker health and safety
Sea level rise and land subsidence	Flooding of roadways, tracks, maintenance facilities, and tunnels Corrosion resulting from exposure to seawater
Coastal storms: Typhoons, cyclones, and hurricanes	Coastal inundation from storm surge Rainfall of sudden or intense nature Damage from debris and high winds Demand for evacuation services

therefore require context-appropriate assessments (PlaNYC, 2013; Transport Research Board [TRB], 2014).

Gradual change in climate-induced hazards receives less attention than more immediate changes in relevant research. Long-term gradual changes in temperature and precipitation patterns stress and test the design strength of urban transit systems. Climate hazards have indirect or secondary impacts as well. Heat waves may impose health hazards on passengers or damage freight. Likewise, extreme heat and precipitation events limit the time of day and duration of safe work hours for experts to restore transit system disruptions (Hodges, 2011; U.S. Global Change Research Program, 2014).

The second set of vectors are internal vulnerabilities due to the physical, social, and economic conditions of the city and its transportation system. These can be measured through the size and density of the population, topography, proportion of poor population, concentration of economic activity in cities, and the condition of transportation systems. Aging and obsolete transportation assets such as bridges, roads, and railways make urban transport infrastructure in the developed world vulnerable (American Society of Civil Engineers [ASCE], 2013). The absence of adequate and reliable urban transport systems and the widespread reliance of poor populations on walking and use of bicycles in developing-country cities increase the vulnerabilities of the poor during heat waves or floods (Mehrotra, 2012a; UN-Habitat, 2013; Revi et al., 2014). Each city's modal mix is different and combines the preferences of travelers, land-use-induced trip origins and destinations, and supply of transportation alternatives, ranging from individual modes such as cars and bicycles to mass transit systems of trains and buses. The particular combination of these urban mobility choices along with existing socioeconomic, topographical, and other urban conditions in a city shape the vulnerability of the urban transit systems and its users (European Commission, 2013).

The third set of vectors is response capacity. Response capacity measures refer to a city's urban transport–related institutional attributes and its actors. These capacities determine the degree of its capability to respond to potential climate change impacts. Variables that can determine the extent of a city's ability to adapt its transport systems include the structure and capacity of institutions, presence of adaptation and mitigation programs, and motivation of change agents. The greater the ability (institutional structure, caliber, resources, information, analysis) and willingness of actors to proactively respond to climate change, the higher the capacity to act, and *ceteris paribus,* the lower the impact. New York's transit system offers an example of reducing risk through preventive measures (NYC, 2013). Key actors for such response include city governments, their urban transport-related constituent departments, the private sector, civil society, nongovernmental organizations (NGOs), and academics (Mehrotra, 2009).

In sum, urban transport risks are shaped by the hazard exposure, existing system vulnerabilities, and response capacity, both ex-ante and ex-post. Extreme events convert risks to costs to the urban transport infrastructure. Hurricane Sandy demonstrated that even a well-managed city like New York was vulnerable to climate risks despite a robust early warning system (Kaufman et al., 2012). In resource-constrained cities in the developing world, impacts of severe climate events are more disruptive. Road networks of major cities like Mumbai frequently flood during monsoons and lack adequate access for emergency and evacuation efforts, while settlements of the urban poor are placed at greater risk by a lack of access to public transit or emergency planning (Revi et al., 2014). A storm of the magnitude of Hurricane Sandy, which brought storm surges of more than 9 feet to New York (Blake et al., 2013), pose many-fold higher risks for cities that lack preparedness.

While there is limited research estimating costs of climate change on urban transport infrastructure (Koetse and Rietveld, 2009), extreme weather events offer some insights. Hurricanes Katrina and Sandy (Jacob et al., 2011) in the United States or typhoon Haiyan in the Philippines (Kostro et al., 2013) cost billions of dollars in direct and indirect transit system damages and disruption to the economy (Hodges, 2011; Koetse, 2009). In New York, the Metropolitan Transportation Authority reported US$5.8 billion

Table 13.2 *Transportation sector interdependencies. Adapted from URS, 2010*

Transportation Depends on Energy	Energy Dependence on Transport
Transport sector is reliant on electricity for signage and lighting	Operator access to power stations during severe weather
Electrified rail network	Access to local distribution facilities
Flooding of oil refineries and depots disrupt supply	Raw material supply depends
Electric cars reliant on recharging stations	Road, rail, and ports disruption curtails fuel supply
Transportation Depends on Water	**Water Dependence on Transport**
Drain design and maintenance impact road and rail flooding	Wastewater sludge requires transport to off-site disposal
Water supply for rail and road vehicle cleaning and refill provisions	Road access to treatment works, pumping stations
	Overland transport of potable water during floods or service disruption
Transportation Depends on ICT	**Intermodal Dependencies**
ICT enabled transport road signage or rail signaling	Road failure impact on rail
Disruption of telecommunication and Internet services curtail communication for transport	Airports and seaports dependent on road and rail
	Rail failure leads to diversion of traffic onto roads

in recovery projects after landfall by Hurricane Sandy, while damage to additional transportation infrastructure stock in New York totaled US$2.5 billion (Blake et al., 2013). Hurricanes Katrina and Rita in 2005 cost approximately US$1.1 billion in infrastructure damages in New Orleans (Grenzeback and Lukmann, 2007). Flooding affects noncoastal urban transport systems as well (U.S. DOT, 2014). In Chicago, the damage to flooded transport systems and associated delays amounted to US$48 million in 1996 (Changnon, 1999). In developing-country cities, the cost of damages to land-based transport systems, including buses, cars, and rail, underrepresent the total cost to firms and households because poor households and small firms rely on several alternate modes like bicycles, carts, walking, and other non-motorized modes.

The preceding sets of risks are significant by themselves, but urban transport systems are interconnected. Thus, failure in one system produces failures in transport and other systems. Road and rail networks are increasingly dependent on the energy, water, and information and communications technology sectors and are interlinked to other modes of transportation (URS, 2010). Transportation infrastructure risks compound risks to households and firms (Revi et al., 2014). Table 13.2 illustrates interdependencies between transport and other sectors.

Cascading failures of transportation systems impede disaster response. City transportation planners increasingly need to prepare transportation networks for pre- and post-disaster response (e.g., the evacuation of residents prior to storms, floods, and landslides, and the delivery of food, water, medical supplies). Not all cities are adequately prepared for extreme climate events (Revi et al., 2014). To illustrate, the impacts of Typhoon Haiyan were made worse because the storm destroyed roads that link secondary cities to better equipped primary urban centers (Kostro et al., 2013). Power failure in parts of New York during Hurricane Sandy disrupted traffic control systems and pump operations, making it difficult to remove water from subway facilities and to fuel cars at gas stations (PlanNYC, 2013).

13.3 Adaptation Options and Strategies for Climate-Resilient Transport

13.3.1 Types of Adaptation

Adaptation solutions can offer the twin benefits of reducing impacts of climate change on urban transport as well as downstream consequences (see Figure 13.3). Adaptation measures to reduce impacts of extreme climate events as well as gradual change in climate trends can be unpacked by scales and scopes of impact into short, medium, and long-term *tactics* (quick-fix), *projects* (tasks), *programs* (systems), and *policies* (institutional arrangements and processes). Some recommended measures include tactics such as relocating facilities and vehicles to avoid damage during storms; projects such as retrofitting facilities to protect existing infrastructure, altering train speeds and schedules, and upgrading stormwater drainage systems; programs such

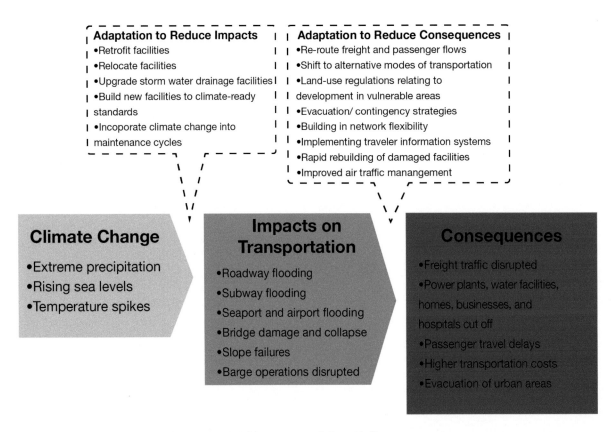

Figure 13.3 *Adaptation to reduce impacts and consequences: policies, program, projects, and tactics.*

Source: Adapted from Melillo et al., 2014

as building new facilities to climate-resilient standards; and policies that incorporate climate projections into maintenance cycles (Mehrotra et al., 2011; Zimmerman, 2012; Melillo et al., 2014).

Adaptation measures to reduce the consequences of climate change impacts on urban transport in the short-term range from the tactics of rerouting passenger flows and substituting modes of transport, to projects such as building in network flexibility and rebuilding damaged facilities. In the medium term, programs such as creating traveler information systems and policies for regulating land use in vulnerable areas and developing evacuation contingency strategies should be considered (Mehrotra et al., 2011; Melillo et al., 2014). Long-term adaptation measures for risk reduction include recalibrating design standards that incorporate climate change projections and considering the service life of transportation systems. In some cases, short- and medium-term policies are more effective for addressing flooding and sea level rise over a 30- to 40-year period. Long-term policies may focus on projected temperature variations over a 40- to 100-year period (Meyer, 2008). Table 13.3 summarizes the climate change hazards, impacts, and associated adaptation solutions. Temporal parsing of policies is desirable, as is constant monitoring and recalibration of policies (Mehrotra, 2012a) that respond to changing climate risk information.

13.3.2 Urban Planning, Transport Engineering, and Design Solutions

To mainstream adaptation measures in the planning, design, and implementation of transport projects, cities need new policies, standards, and codes that address transport system vulnerabilities. With long-term change in environmental factors that are likely to expose urban transport infrastructure to intense and frequent floods, strong winds, and rising temperature, cities should consider reducing risks through the integration of high-probability scenarios in the design of transport systems and associated citywide interdependent systems (Meyer, 2008). These practices and procedures are equally relevant for new capital-intensive investments as well as for routine retrofits for rail systems, roadways, bridges, vehicles, and other interdependent urban systems that have life expectancies of 50–100 years.

For many cities, urban transport adaptation requires a comprehensive reassessment of design practices and procedures. These include climate change risk assessment and the identification and selection of the appropriate mix of policies, programs, projects, and tactical responses parsed over the short and long term that incorporate new research, recalibrate maintenance and operations procedures, and asset management (Hodges, 2011; Filosa, 2015).

Examples of climate change integration into transport design standards and adaptation initiatives can be found predominantly at the national level. Canadian, British, Danish, and New Zealand departments of transport have developed adaptation action plans and set goals for resiliency (Filosa, 2015). Canadian public infrastructure programs have re-engineered construction protocols to reduce system vulnerabilities. The Danish road department has incorporated Intergovernmental Panel on Climate Change (IPCC)-based precipitation projections to recalibrate drainage standards and road regulations. Korean, British, and New Zealand transport agencies have altered design standards, improved drainage structures, and retrofitted bridges in accordance with precipitation projections and recalibrated flood levels (Filosa, 2015). New York's Metropolitan Transportation Authority offers a rare example of urban transport-specific comprehensive adaptation planning and implementation (see Case Study 13.1).

Table 13.3 *Climate change hazards, impacts, and adaptation solutions*

Hazard (Impact)	Transport module	Adaptation Solution
Temperature (Speed)	Roads, rail, transit vehicles, private vehicles	Milling out ruts (tactic); laying of more heat resistant materials such as asphalt for roads (project) Heat-tolerant metals for rail and rail connections (project)
Floods (Congestion, delays, accidents)	Drainage system; roads, rails, subways; transit vehicles, private vehicles	Deploy early warning systems – remote sensing technology to detect flood levels (program) Dikes and barriers (project) Enhancing the performance of drainage systems and contingency rechanneling of water (program) Enhance use of pumps (project) Elevation of structures (project)
Storms (Lower visibility, delays, accidents, cancelled trips, closure or reduction in capacity of routes)	Roads, bridges, rails, airports, subways, transit and private vehicles	Arrangement to park vehicles in secure garages (tactic) Enhancing emergency evacuation planning (program) Constructing protective barriers (project) Increasing clearance for bridges (project) Anchoring vulnerable structures (tactic) Preparedness of equipment and material for repairs (tactic, management)

Note: Tactics (quick-fix), projects (tasks), programs (systems), and policies (institutions and processes).

Source: Mehrotra et al. (2011, 2015)

Developing-country cities face constraints when integrating adaptation into urban planning. Dar es Salaam, a city of 4 million, lacks an adaptation plan (Kiunsi, 2013). Symptomatic of many developing-country cities (UN-Habitat, 2013; Revi et al., 2014), a large proportion of Dar es Salaam's population lacks access to basic services: networked water supply and sanitation, storm water drains, and solid waste collection. More than half the population lives in informal settlements located on low-lying flood-prone embankments. Underdevelopment and deprivation is further exacerbated by the lack of fiscal and managerial capacity to address climate impacts. The primary adaptation challenge for cities with large slums and deep development deficits is to prioritize the integration of climate change adaptation into ongoing and planned urban transport and city-wide development processes. Integration is imperative because for many cities the incremental cost of adaptation is negligible, whereas the cost of neglecting adaptation is high. In Malaysia, for example, the cost of landslide prevention in high-risk road segments was one-fifth of post-landslide road repair (URS, 2010). The New Zealand Transport Agency Bridge Manual points to significant economic benefits of ex-ante retrofitting bridges and culverts at the design stage instead of ex-post retrofitting after construction. Likewise, the New Zealand and Korean long-term transport plans integrate climate data into asset management systems to analyze hazards and integrate

adaptation measures into routine asset management procedures (Filosa, 2015).

Most urban transport adaptation literature focuses on supply-side planning, engineering, and design solutions. However, a significant space for adaptation solutions involves land-use planning. In some cases, land-use planning provides benefits for adaptation and mitigation. Mehrotra et al. (2011) discuss the mutually interdependent relationship between land use and transport systems and associated GHG reductions. Land use, zoning, and building codes along with other urban development policies determine the distribution of population density and mix of activities associated with urban land use. These non-transport urban development policies shape travel demand. The existing research and policy literature, however, lack a discussion of adaptation potential of such land- and real estate-derived policy instruments. Likewise, adaptation potential for emerging flexible and information technology-enabled flood response systems that can respond to real-time changes in travel demand or the elevation of structures (Meyer, 2008) should be explored, as should low-cost and low-technology solutions, particularly for resource-constrained cities.

Urban transport adaptation planning and investments should incorporate intermodal and intersectoral interdependencies (see Figure 13.4) to prevent cascading failures (Melillo et al., 2014).

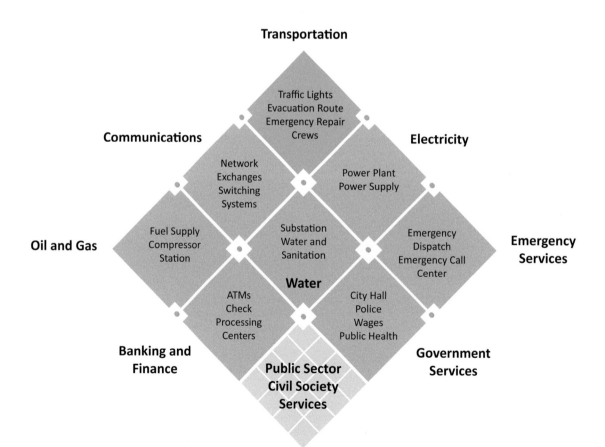

Figure 13.4 *Urban transport's interconnectivity with urban systems.*

Source: Adapted from Melillo et al., 2014

Case Study 13.1 New York Metropolitan Transportation Authority: Climate Adaptation Solutions Leading up to Hurricane Sandy

Susan Yoon

Sustainable Development Solutions Center, New York

Keywords	Transportation, adaptation, infrastructure, climate risk
Population (Metropolitan Region)	20,153,634 (U.S. Census Bureau, 2016)
Area (Metropolitan Region)	17,319 km² (U.S. Census Bureau, 2010)
Income per capita	US$56,180 (World Bank, 2017)
Climate zone	Dfa – Cold, without dry season, hot summer (Peel et al., 2007)

On October 29, 2012, Hurricane Sandy imposed approximately US$5 billion of losses on the New York State Metropolitan Transportation Authority's (MTA's) infrastructure (Blake et al., 2013; MTA, 2013) (see Case Study 13.1 Figure 1). These damages occurred despite a well-prepared adaptation planning process. In the decade leading up to Hurricane Sandy, the MTA had undergone a review of its emergency response systems, had upgraded its pumping and sewage infrastructure, and had commissioned storm and climate adaptation committees. In the days leading up to the storm, the MTA implemented its hurricane plan, activated its Incident Command Center and Situation Room, and responded to the immediate impact of the storm via direct response and in collaboration with federal, state, and local agencies.

Case Study 13.1 Figure 1 *Map of Hurricane Sandy inundation area and MTA New York Transit routes.*

Source: Spatiality, FEMA Modeling Task Force (MOTF)

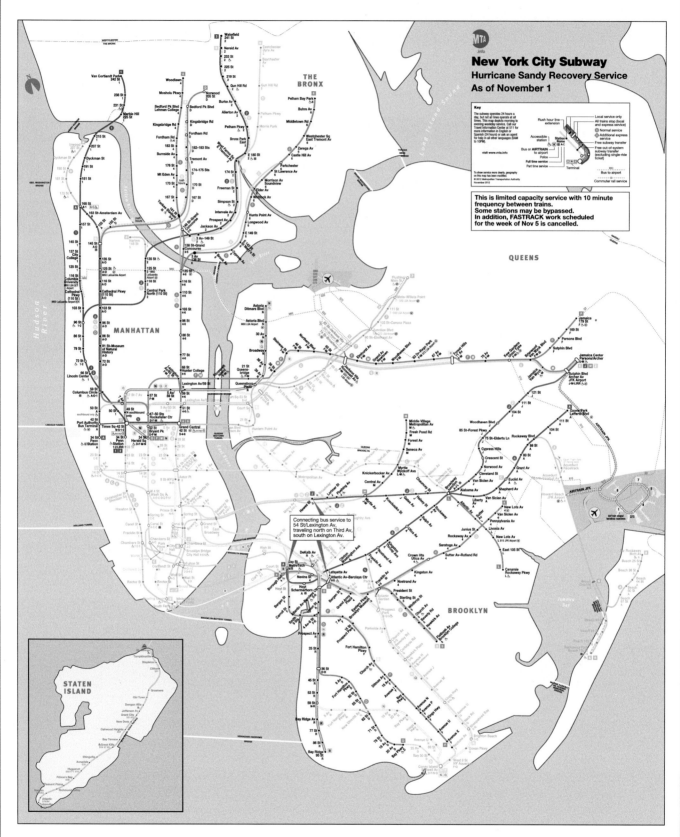

Case Study 13.1 Figure 2 *Example of subway service updates posted on November 1, 2012, on the MTA website and distributed to all major news outlets. Subway routes represented in brighter colors remained functional; the lighter colors represent parts of the network that were out of service.*

Source: Metropolitan Transportation Authority (used with permission)

The MTA is the largest public transportation provider in the Western Hemisphere, with infrastructure valued at more than US$1 trillion dollars. The Authority serves a 13,000 square kilometer metropolitan region with an average weekday ridership of 8.7 million (MTA, 2014).

Prior to 2012, the MTA had begun preparing for extreme climate events – prompted by two intense precipitation events in 2004 and 2007. In response to the 2004 storm, which delivered 2 inches of rain per hour and resulted in 1,156 subway train cancellations, the MTA authorized an Inspector General investigation and formed an MTA Board Task Force (MTA, 2006; MTA, 2007; Kaufman, 2012). The 2007 storm, which delivered 3.5 inches of rain in one morning, prompted the MTA to form an official Storm Task Force and issue the August 8, 2007 Storm Report (MTA, 2007). Additionally, in the fall of 2007, the MTA convened the Blue Ribbon Commission on Sustainability with a dedicated Climate Adaptation Working Group tasked with developing near-term strategies as well as long-range climate adaptation recommendations (Jacob et al., 2008; MTA, 2009).Findings from these committees included inadequacies in stormwater drainage systems, shortcomings in internal organizational communication, and the need for increased capacity to provide public service alerts. The findings led to an analysis of flood-prone stations, upgrades to valves for pumping water and preventing sewage backflow, installation of Doppler radar access throughout the network, embankment stabilization, elevation of 30 subway station exits, retrofits of 1,500 ventilation grates, and the recalibration of the MTA's hurricane plan. The experience and lessons from these prior climate events led the MTA to initiate its first-ever transit service suspension prior to Hurricane Irene in 2011 (Kaufman et al., 2012).

In 2012, three days prior to Hurricane Sandy's landfall, the MTA activated its hurricane plan. One day before the storm, the MTA began suspending service system-wide on subways, buses, and the commuter rail systems for the second time in its history. The Authority implemented hurricane and flood preparedness measures across the network. These actions included the pre-deployment of mobile pumping units, the covering of ventilation grates and subway entrances, the relocation of vehicles and rolling stock to higher ground, and the removal of signal and control systems from tunnels expected to be flooded. Removal of equipment prevented saltwater damage, allowed for a swifter resumption of subway service, and decreased the duration of bridge and tunnel closures. The adaptation response policy and procedures detailed in the hurricane plan outlined specific responsibilities for MTA officials and strategies for deployment of personnel and equipment. The plan also included communication protocols between internal agencies and between MTA officials and local, state, and federal agencies, as well as with news and media agencies (Coyle, 2014; MTA, 2013).

Adaptation policy measures and procedures reduced damage to the urban transit systems, enhanced passenger and personnel safety, and minimized the duration of service disruption, allowing the MTA to reinstate service shortly after the storm. By November 3, 2012, 80% of the 108-year-old subway system was operational (see Case Study 13.1 Figure 2). The rest of the MTA system was fully operational by November 16, 2012 (Kaufman et al., 2012). Despite having these adaptation protocols in place, major portions of the networks, including nine under-river subway tubes, two vehicular tunnels, and critical transit hubs such as South Ferry Station, were inundated with saltwater and suffered severe damage. The Montague Tube, connecting the Boroughs of Manhattan and Brooklyn, did not reopen until September 2014, and the process of restoring the network to its pre-Sandy condition will take years (MTA, n.d.). For additional information on the MTA's recovery, see Zimmerman (2014).

Furthermore, cost-benefit assessments of adaptation alternatives are best incorporating both impacts and consequences. The accounting costs of damage to transportation networks as well as long-term social, environmental, and economic impacts exacerbated by sectoral interdependencies should also be factored into analyses (Koetse, 2009).

Urban transport adaptation efforts in New York illustrate the value of multistakeholder engagement in addressing interdependencies inherent in transport systems. New York's Metropolitan Transportation Authority works across its transport departments and agencies that are responsible for planning, regulation, enforcement, operations, and maintenance to determine interdependencies of regional transport operations and investments. The adoption of a multistakeholder, multilevel engagement process helped strengthen coordination and dialogue between sectors and agencies – for example, between those tasked with environmental assessments and flood and climate forecasting (Jacob et al., 2011). The emphasis on engagement helped to formulate urban development and transit plans and policies, and review and screen investment incorporating climate risk assessments, monitoring climate indicators, evaluating program performance, and developing an effective urban transport adaptation system.

13.4 Low-Carbon Transport

Although adaptation can make transport systems resilient to climate change, low-carbon transport can mitigate climate impacts. In 2010, transport accounted for 22% of energy-related CO_2 emissions and 20% of global energy use. Urban transport accounts for approximately 40% of these totals (IEA, 2012). Even with advances in technology and fuel, energy consumption from urban transport is expected to double by 2050 (IEA, 2013). Globally, motor vehicle emissions could increase threefold between 2010 and 2050 (UN-Habitat, 2013). Developing countries will be the fastest-growing source of these emissions. Rising incomes and urbanization has already pushed up travel demand, motorization, and emissions in China and India (Pucher et al., 2007). In China specifically, transport is the fastest-growing emission source (Baeumler et al., 2012).

The preceding estimates are based on top-down methods that multiply national fuel use by emission factors to reach emissions estimates. Increasingly, urban transport emissions are estimated from bottom-up models that consider urban transport activity levels and mode structures. Bottom-up modeling shows that Mexico

City has significant emission reduction potential from new technologies and public transport (Chavez-Baeza and Sheinbaum-Pardo, 2014). Models in the four cities of Barcelona, Spain; Malmö, Sweden; Sofia, Bulgaria; and Freiburg, Germany, demonstrate reductions of up to 80% from 2010 to 2040 as well as cleaner air, reduced noise pollution, and other co-benefits (Creutzig et al., 2012). Meanwhile, models of CO_2 emissions in Osaka highlight the interactions between built environments and lifecycle stages (Waygood et al., 2014).

13.4.1 Emissions Drivers

Several interconnected drivers contribute to emission increases. Drivers can be classified as operating on the supply and demand sides. These categories parallel the top-down and bottom-up estimation techniques. The supply side concentrates on the fuel quality and emissions intensities. The demand side focuses on vehicle activity and modal structure (Schipper et al., 2000). Four drivers that work chiefly on the demand side are particularly relevant to urban transport and are thus emphasized here.

The first driver is urbanization. Approximately 54% of the global population lives in urban areas and will grow to 66% by 2050 (United Nations Department of Economic and Social Affairs [UN DESA], 2014). Globally, 2.5 billion people will be added to cities by 2050, with nearly 90% of that growth concentrated in developing areas of Asia and Africa (UN DESA, 2014). While Africa's population is predominantly rural, its urban population is greater than the United States' population and rapidly growing. Asia is home to the fastest-growing cities. Asia's urban population has grown sevenfold, from 232 million to 2.13 billion people between 1950 and 2015 (UN-Habitat/ESCAP, 2015). In Latin America,

three-quarters of the population lives in cities (UN DESA, 2014). Urbanization characteristics – land use mix, population density, and income levels – and associated transportation modal choices will determine urban transit emissions (see Figure 13.5) and its mitigation opportunities. Much of the urban growth will be in cities with populations below 1 million (UN DESA, 2014). This could cause a further spike in emissions as research shows that smaller cities tend to use energy less efficiently (Seto et al., 2014).

The second core driver of emissions is motorization. Motorized modes accounted for about 47% of urban trips globally in 2005; this figure could grow to 54% without significant policy changes (Pourbaix, 2012). Private vehicle ownership in China is projected to rise from 40 to 310 per 1,000 people between 2010 and 2035. Similar trends are anticipated for India. Other rapidly emerging economies are witnessing growth of 20% annually in vehicles (IEA, 2012). Figure 13.6 illustrates motorization in the developing world over the next three to four decades.

The third emissions driver is infrastructure investment. China's urban roads have doubled between 1990 and 2003, while Kenya invested more than US$530 million in 143 kilometers of urban road construction and repair between 2008 and 2012 (UN-Habitat, 2013). Increased connectivity will be essential for development, but short-sighted infrastructure planning can create path dependencies that lock in energy-intensive development patterns for decades (Sims et al., 2014). Path dependencies not only apply to roads and bridges but also to institutional and individual behaviors; to exemplify the connection to behavior, research illustrates that investments in infrastructure to ease traffic and mobility can boost vehicle kilometers traveled (VKT). For instance, studies show a 10–20%

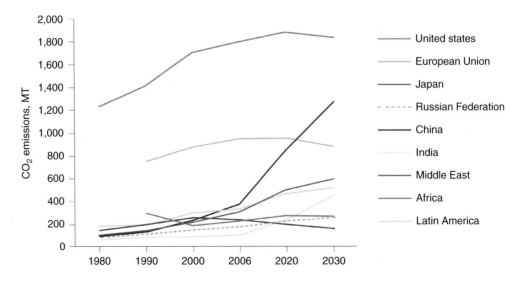

Figure 13.5 *Actual and projected global CO$_2$ emissions from the transport sector, 1980–2030.*

Source: Fulton and Cazzola, 2008; in Suzuki et al., 2013

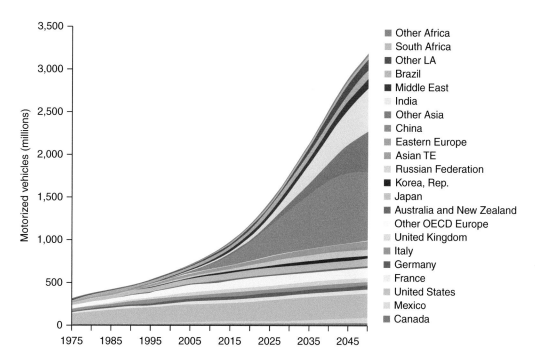

Figure 13.6 *Current and projected growth rates in private vehicle ownership in select countries and regions, 1975–2050.*

Source: Fulton and Cazzola, 2008; in Suzuki et al., 2013

jump in VKT is associated with a doubling of infrastructure capacity (Seto et al., 2014).

The fourth emissions driver is land-use change. Many cities have grown out rather than up. Reports show that between 1995 and 2005 the vast majority of the developed world's seventy-eight largest cities experienced faster growth in suburbs than in urban cores (UN-Habitat, 2013). Outward expansion causes increased trip lengths, motor vehicle dependencies, and emission increases (Kennedy et al., 2009; Rickwood et al., 2011; Seto et al., 2014). Projections showing that between 2005 and 2025 private motorized transport will increase by 80% reinforce the need for farsighted land use and infrastructure planning to curb emissions (UN-Habitat, 2013).

13.4.2 Mitigation Options: Avoid-Shift-Improve (ASI)

Civil society, industry, and governments are increasingly cooperating in promoting an approach known as *Avoid-Shift-Improve (ASI)* to mitigate emissions (Dalkmann and Brannigan, 2007). ASI organizes mitigation options into actions that (1) avoid unnecessary travel through smart urban planning, compact cities, and transit-oriented development (TOD); (2) shift travel to non-motorized or high-occupancy modes; and (3) improve efficiencies with cleaner vehicles and fuels (Dalkmann and Brannigan, 2007).

Arguably the greatest reductions in cities involve urban and spatial planning that avoid unnecessary travel. The scatterplot in Figure 13.7 illustrates the advantage of "avoid" by showing that greater density is often associated with lower emissions.

The benefits of density are also suggested in studies that show urban density above 35 person per hectare reduces automobile dependency (Newman and Kenworthy, 2006). Findings from the United States underscore that when mixed land-use is combined with transit-oriented development policies, residential densities can double and VKT can fall by 25% in the long run (EPA, 2014).

Some cities have attempted to increase densities, lower emissions, and achieve other development goals. To make an aging population more mobile, Toyama, Japan, successfully introduced a compact city design with dense networks of residences and businesses linked by monorail (Takami and Hatoyama, 2008). Others cities have emphasized not just density but *articulated density*, referring to pockets strategically concentrated across metropolitan areas (Suzuki et al., 2013). Yet other cities have found urban connectivity – determined by block size, number of intersections in road systems, and space for pedestrians – to lower VKT and emissions (Seto et al., 2014).

The "shift" options involve increasing mode shares of higher occupancy and more efficient vehicles. This often involves different forms of mass transit trips that can cut VKT per person (see Figure 13.8a). This frequently means improving "public transport that has been the backbone of development" (Wright, 2004).

BRTs are an increasingly common "shift" option (see Figure 13.8b). BRTs are bus systems that run on a segregated lane. They currently operate or are being built in more than 300

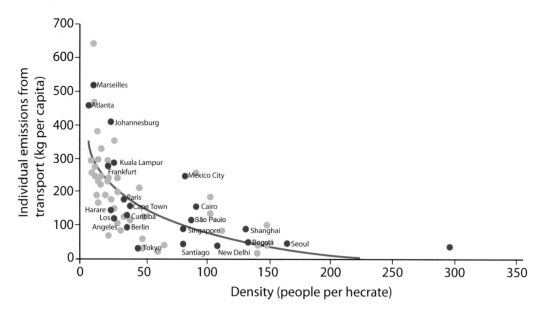

Figure 13.7 *Urban densities and per capita emissions from transport are much lower in denser cities.*

Source: World Bank, 2009. Based on 1995 data

cities (UN-Habitat, 2013; Carrigan et al., 2013). Most BRTs are located in Latin America (64%) and Asia (27%), but some are being built in Africa (UN-Habitat, 2013; Carrigan et al., 2013; see Case Study 13.2). The spread of BRTs is attributable to costs that are one-third to one-fifth of light rail and one-tenth of metro or subway systems. The BRTs also bring a range of benefits. Cities with established systems, such as Curitiba and Bogota, have seen reduced automobile trips, fuel use, and emissions (UN-Habitat, 2013; Suzuki et al., 2013; Turner et al., 2012).

In larger megacities, subways that can move higher occupancies (up to 30,000 passengers per hour) are becoming a more common "shift" option (De Jong et al., 2010). (UN-Habitat, 2013). Currently, 187 cities have metro systems – up from 40 in 1970 (Metrobits, 2012). China is investing heavily in subways; eighteen cities are currently constructing metros and light rail and another twenty-two are at various points in planning stages (UN-Habitat, 2013).

Non-motorized transport (NMT), the lowest emission "shift" option, accounts for half of all trips in many cities (SLoCat, 2014a). Among NMT options, cycling has considerable promise to improve mobility and keep emissions lower than private vehicles (European Cyclists' Federation, 2011). Despite these benefits, many urban areas have seen the prevalence of bicycles decline. India witnessed bicycle use fall from a 30% to an 11% mode share between 1994 and 2008 (UN-Habitat, 2013). Walking is another important NMT "shift" option. Walking currently accounts for between 40% and 60% of trips in several Asian cities (Institute for Transportation and Development Policy [ITDP], 2014). Retaining these percentages requires pedestrian-friendly infrastructure (UN-Habitat, 2013).

The third set of *ASI* options involves improving vehicle technologies and fuels. These improvements often involve leapfrogging or skipping stages of technological development. Leapfrogging through continuing fuel efficiency improvements could reduce 50% of fuel use per kilometer in new conventional vehicles by 2030 (Fulton, 2013). Improved technologies in larger trucks and buses could consume 20–30% less fuel using

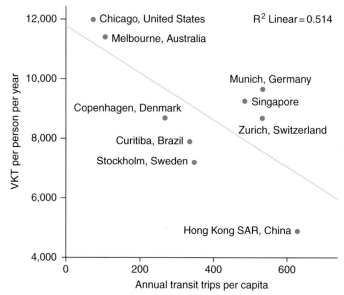

Note: VKT = vehicle kilometers traveled.

Figure 13.8a *Annual vehicle kilometers traveled and transit trips for selected global cities.*

Source: Suzuki et al., 2013 Based on data from UITP, 2006

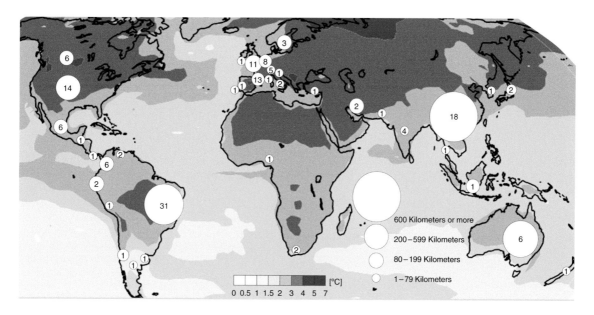

Figure 13.8b *Global temperature projections and distribution of bus rapid transit systems.*

Source: Sustainable Development Solutions Center analysis with data from IPCC, 2014; UN-Habitat, 2013; brtdata.org, and Hidalgo, 2012

hybrid electric or hydraulic hybrid drive trains (Chandler et al., 2006; AEA Consulting, 2011).

Electric vehicles reduce emissions by converting fuel to energy more efficiently and cutting tailpipe emissions. Studies show electric motors convert 59–62% of electricity drawn from the grid into energy at the wheel, whereas gasoline engines convert 17–21% (U.S. Department of Energy [DOE], n.d.). A global launch of electric vehicles could cost US\$500 billion but save US\$2 trillion through 2025. The UN set a goal for electric vehicles to make up 30% of urban motorized travel by 2030 to capture these savings (UN, 2014). Improvement in battery performance and reductions in battery prices make this goal feasible (U.S. DOE, n.d.). The impacts of electric vehicles nevertheless require changes to the overall energy system, or emissions will shift from vehicles to power plants.

Clean fuels, another "improve" option, reduce GHG emissions. Brazil's flex-fuel and biodiesel programs are two notable initiatives promoting these clean fuel alternatives (Ribeiro and de Abreu, 2008). However, clearing carbon-rich lands for biofuel production can create CO_2 emissions (Fargione et al., 2008). Reductions in energy prices from biofuels can increase energy usage and emissions (Fargione et al., 2008). Additional research is needed on emissions across the lifecycle of biofuels.

13.4.3 Co-Benefits of Low-Carbon Transport

Reducing transport emissions can yield additional benefits beyond mitigating GHGs or co-benefits (see Table 13.4).

Capturing co-benefits can make low-carbon transport more affordable and sustainable. The highest value co-benefits tend to be those that improve public health from reduced air pollution (see Chapter 10, Urban Health). To illustrate, bringing emissions levels within World Health Organization (WHO) guidelines in twenty-five European cities could add 6 months of life expectancy (for 30-year-olds) in half the cities and save €31 billion (Pascal et al., 2012). In Asia, such savings project to be even greater due to poorer air quality (Chen and Whalley, 2010). In Taipei, Taiwan, a 9–14% reduction in carbon monoxide (CO) and particulate matter (PM) from public rail could deliver welfare gains of US\$264 million annually. Importantly, the air quality benefits often accrue to poor residents who are more exposed to pollution because they live closer to highways and traffic. Co-benefits therefore can also address inequities.

Another co-benefit is reducing congestion and travel times. Time lost in traffic costs an estimated 2% of gross domestic product (GDP) in Europe's cities and 2–5% in Asia's cities (UN-Habitat, 2013). In Dakar, Senegal, and Abidjan, Cote d'Ivoire, the costs of traffic congestion amount to 5% of GDP (UN-Habitat, 2013). In American cities, congestion created 5.5 billion extra hours of travel, required the purchase of an extra 2.9 billion gallons of fuel, and cost drivers US\$121 billion dollars in 2011 (Schrank et al., 2012). In Western Europe, reduced noise pollution could save 1 million lost life years from sleep deprivation, cognitive impairments in children, and other defects (WHO, 2011). Yet another co-benefit of some avoid-and-shift options is improved traffic safety (UN-Habitat, 2013).

Case Study 13.2   Bus Rapid Transit in Lagos and Johannesburg: Establishing Formal Public Transit in Sub-Saharan Africa

Matthew Woundy

Sustainable Development Solutions Center, Boston

Keywords	BRT, Africa, mobility, emissions reductions
Population (Metropolitan Region)	Lagos: 21 million (Lagos State, 2013) Johannesburg: 7.2 million (Statssa, 2015)
Area (Metropolitan Region)	Lagos: 3,569 km² (Lagos State, 2013) Johannesburg: 6,009 km² (Statssa, 2015)
Income per capita	Nigeria: US$2,450 South Africa: US$5,480 (World Bank, 2017)
Climate zone	Lagos: Aw – Tropical savannah Johannesburg: BSk – Arid, steppe, cold (Peel et al., 2007)

Formal public transit remains virtually nonexistent in Sub-Saharan Africa (UN-Habitat, 2013), although this is beginning to change with recent and planned implementation of Bus Rapid Transit (BRT) systems in the region. The pioneers of this movement toward formal mass transit are the cities of Lagos and Johannesburg, the latter of which is credited with launching the first BRT system in the region only because the (slightly) older system is Lagos is described as BRT "lite." This distinction exists because the Lagos system was designed to capture most of the desirable traits of BRT while requiring less capital investment, making the project feasible for the city. The result is a high-quality system at US$2.75 million per kilometer, or half the cost of the usual BRT system (UN-Habitat, 2013). This increase in affordability is balanced by slightly slower transit times and lower capacities due to the lack of separated busways and other capital-intensive infrastructure changes, but the net gains for passengers remain substantial. In both Lagos and Johannesburg, informal taxis and minibuses dominated the public transport sector, usually exacting high fares and lacking in efficiency. BRT has improved economic, environmental, safety, and other outcomes for residents in both cities.

In Lagos, the BRT "lite" corridor opened in 2008 with a planned capacity of 60,000 passengers per day, a number that had reached 220,000 per day in 2010 (UN-Habitat, 2013). The World Bank provided technical advice and US$100 million in financing, with Lagos state providing US$35 million (World Bank, n.d.). To complete the financing and operation of the system a public–private partnership was established where the Lagos transit authority (LAMATA) provided facilities and terminals while private operators bought and operated the buses (World Bank, n.d.) The result for passengers was a 30% decrease in average fares, a 40% decrease in travel time, and a 35% decrease in waiting times (UN-Habitat, 2013). The buses are responsible for 25% of all trips along the BRT corridor while representing just 4% of total vehicle traffic (World Bank, 2015b). As a result, CO_2 emissions have been cut by 13% and overall GHG emissions by 20% according to LAMATA (World Bank, n.d.). The BRT corridor has also created direct employment opportunities for 1,000 people and indirect employment for 500,000 (UN-Habitat, 2013).

Launched in 2009, the Rea Vaya BRT in Johannesburg was the first full BRT system in Africa (Allen, 2011), and cost US$5.5 million per kilometer to construct (UN-Habitat, 2013). Rea Vaya was born of a transit plan by the city of Johannesburg with a goal of having 85% of residents within 1 kilometer of a BRT trunk corridor or a feeder route (City of Johannesburg, 2013). The awarding of the FIFA World Cup to South Africa in 2010 also provided motivation to get formal public transit systems up and running in host cities (Carrigan et al., 2013; UN-Habitat, 2013). Funding was achieved chiefly through fiscal transfers from national to city government in the form of a transport infrastructure grant. In 2011, the first phase of Rea Vaya was completed at a length of 122 kilometers and carried 434,000 passengers per day (UN-Habitat, 2013). Rea Vaya passengers reduced their travel times by 10–20% based on surveys, and a study by EMBARQ (Carrigan et al., 2013) shows that travel time savings and avoided road fatalities were the chief benefits of the BRT system, accounting for 30% and 28% of total benefits, respectively. Johannesburg launched the second phase of Rea Vaya (Phase 1b) in 2013, and it has registered both phases with the Voluntary Carbon Standard. The expected emissions reductions from the two phases are 400,000 metric tons of CO_2 over a 10-year period (Swiss Association for Quality and Management Systems, 2011).

Both cities are currently expanding their BRT systems, with Lagos planning on adding 13.5 kilometers to the existing corridor and Johannesburg constructing a third phase to service the northern parts of the city (C40 Cities Climate Leadership Group, 2013; Carrigan et al., 2013). Stakeholder engagement was key for both cities – in Johannesburg, the emphasis was on including existing taxi operators in the development and ownership of Rea Vaya to lower resistance and to lessen the impact on livelihoods of those in the transit sector (Allen, 2011). To make up lost revenues as taxis were taken off the road, taxi operators were offered ownership stakes in the Rea Vaya Company as compensation, and taxi drivers were retrained and absorbed into Rea Vaya as bus drivers (Allen, 2011).

In Lagos, community engagement was a key factor in building demand for BRT because there was a history of poor transport system delivery in the city and an overall lack of experience with formal public transit (UN-Habitat, 2013). LAMATA also worked to build a partnership with the National Union of Road Transport Workers, forming a cooperative to manage the BRT and building knowledge on the benefits of formal mass transit (World Bank, n.d.). The success of Lagos and Johannesburg in providing clean, efficient, and affordable public transit while working with the community and the dominant informal sector demonstrates a path forward to the rest of the continent to enhance mobility for urban residents while also improving the economic, environmental, and safety outcomes of transit.

Table 13.4 *Potential co-benefits from the transport sector beyond GHG mitigation. Source: Wright, 2009*

Economic co-benefits	Environmental co-benefits	Social co-benefits
Congestion reduction	Reduction of air pollutants (particulate matter, sulfur oxides, nitrogen oxides, carbon monoxide, volatile organic compounds)	Health improvements
Consumer spending savings		Crime reduction/security enhancement
Employment creation	Noise reduction	Gender equity promotion
Small and medium sized enterprise development	Solid waste reduction	Universal access for physically disabled
Traffic accident reduction	Water contaminant reduction	Convenience and comfort
Technology transfer		Community sociability
Reduced dependence on imports/ energy security		Reduction in community severance
Economic productivity/efficiency improvements		

13.5 Policy Frameworks and Enabling Environments

A well-designed low-carbon transport strategy will typically involve combining the *Avoid* and *Shift* demand-side with *Improve* supply-side options. Combining demand- and supply-side solutions is important to ensure net benefits and avoid unintended increase in emissions from a single intervention. Studies show a 2% increase in driving can offset 10% fuel consumption reductions from efficiency improvements (Greene et al., 1999). Potential synergies between and within the avoid-shift-improve categories is another reason for combining options (Schipper et al., 2000). The greatest benefits come from synergies between reductions in unnecessary travel and travel powered by advanced technologies and clean fuels. For example, this could involve equipping new buses with more efficient engines (Zusman et al., 2012).

Combining options requires support from multiple government and nongovernmental actors at different decision-making levels. It also necessitates well-trained staff and institutions capable of aligning stakeholder interests. Studies on applying multilevel, multistakeholder governance to transport suggest that significant improvements in institutional design are often needed to implement low-carbon strategies in the United Kingdom (Marsden and Rye, 2010) and Indonesia (Jaeger et al., 2015). In response to local conditions, mitigation policies and measures have varied across cities, and cities such as Delhi have shown that the actors and change agents that drive policy changes vary as well.

In 1995, a World Bank study found that poor air quality posed health hazards for households, causing an estimated 1 death every 70 minutes in Delhi, and branded Delhi as one of the most polluted cities in the world (Brandon and Hommann, 1995). The report generated public outrage in the city, and the Centre

for Science and Environment (CSE, 1998) started a campaign demanding clean air. The pollution emitted from poorly managed public transport was identified as one of the main causes of poor air quality.

The campaign galvanized citizen support through involvement of professional associations, media, academies, and other stakeholders and involved bringing the message directly to the attention of the national political leadership. In response, the Supreme Court issued a judgment in 1998 requiring the government of Delhi to stem air pollution by introducing compressed natural gas (CNG)-operated public transport and to augment the supply of mass transit within a prescribed timeframe of 3 years, as well as requiring adoption of stringent emission standards within 5 years. Additionally, to address resistance from automobile firms, the Supreme Court issued stringent directives including appointing a CNG czar to ensure compliance of new regulations; instituted large penalties for defaulters, including state and federal agencies; and increased funding of research and development. Both supply and demand for CNG and safety regulations were addressed through institutional mechanisms.

In about five years (1998–2002), all public transport in Delhi was converted to CNG-operated retrofitted buses. Furthermore, this effort triggered several projects to increase the supply of efficient and clean public transport systems like the Delhi metro, thus contributing to broader climate change mitigation efforts within the city. However, the key lesson from this case was that change agents are diverse – researchers, civil society, and the supreme court – and require creative and persistent efforts as well as the willingness to learn by doing (Mehrotra et al., 2009; with data from Centre for Science and Environment and C40 Cities). Building off past efforts, for the first 15 days of January, in 2016, the Government of Delhi introduced restrictions on car users based on car license plates to curb air pollution. Cars with even-numbered license plates were allowed to operate on even dates while those with odd license plates were able to operate on odd dates.

Financing is also a key element of an effective enabling environment (see Chapter 7, Economics, Finance, and the Private Sector). Geography, age, and types of infrastructure influence the cost of building, upgrading, and operating a transport system. Policy-makers will need to tailor financing arrangements to different contexts. For most cities, however, public financing from various taxes and fees, government

borrowing, bond financing, and public–private partnerships will constitute the majority of funding. Multilateral and regional development banks may play the role of intermediaries, providing grants and low-interest financing for projects in developing countries.

At the same time, cities can adopt reform synergies and unlock financing as part of an enabling environment, on the one hand, and redirect ongoing and planned investments toward mitigation (see Case Study 13.3) and adaptation (see Case Study 13.1), on the other (Mahendra et al., 2013). The next section profiles four

Case Study 13.3   London's Crossrail: Integrating Climate Change Mitigation in Construction and Operations

Danielle L. Petretta

Graduate School of Architecture, Planning and Preservation, Columbia University, New York

Keywords	Infrastructure, transportation, mass transit, mitigation
Population (Metropolitan Region)	14,031,830 (Eurostat, 2015a)
Area (Metropolitan Region)	12,091 km² (Eurostat, 2015b)
Income per capita	US$42,390 (World Bank, 2017)
Climate zone	Cfb – Temperate, without dry season, warm summer (Peel et al., 2007)

Large-scale mass transit projects, while consuming energy in both construction and operations, can still achieve net emission savings over time. By taking a proactive, holistic approach that seeks to minimize GHG emissions during all phases of construction and operation, contributions to climate change can be reduced and even prevented (Omega Centre, 2012). Since being approved by the UK

Parliament in 2008, Crossrail Ltd. (CRL) has worked to realize savings by institutionalizing mitigation strategies, accountability mechanisms, and an overall sustainable governance structure. According to the UK Department of Transport (2007), "rail's biggest contribution to tackling global warming comes from increasing its capacity, so that it can accommodate demand growth." Crossrail is a product of this strategy.

PROJECT SPECIFICATIONS

London's Crossrail is a new east–west rail line that, for the first time, will directly link London's major financial and commercial centers of Heathrow, the West End, the City of London, and Canary Wharf. The line will connect Maidenhead and Redding in the west to Shenfield and Abbey Wood (see Case Study 13.3 Figure 1) to the east with thirty-seven stations and 110 kilometers of track, including 21 kilometers of new twinbore tunnels dug deep under the City. At £14.9 billion (US$19.4b), Crossrail is the largest infrastructure project under construction in Europe and the largest project in the United Kingdom in more than 50 years. Crossrail is fully funded and set to start operations in 2018.

Deemed essential to London's global competitiveness, Crossrail will increase rail capacity by 10% and east-west capacity by 40%, greatly alleviating severe overcrowding and already at-peak capacity routes (Mayor of London, 2014). Poised to meet growing

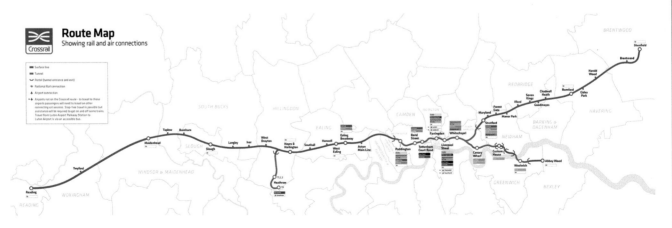

Case Study 13.3 Figure 1 *Crossrail Route Map.*

Source: Crossrail LTD

travel demand,[1] the line is designed to be highly efficient, capable of handling 72,000 passengers per hour and 200 million riders per year (Crossrail, 2014). Crossrail promises to cut travel times, increase access, and improve connectivity throughout Greater London and beyond. The UK economy is expected to benefit by "at least £42 billion (US$54.7b) over the next 60 years" (deSilva and Paris, 2014) through job creation and higher productivity and property value.[2] The project is expected to generate "added value" of residential and commercial property by as much as £5.5 billion (US$7.16b) between 2012 and 2022. Development along its corridor is expected to support and accelerate an additional 57,000 new homes and 3.25 million square meters of office floor space all within 1 kilometer of the new stations (GVA, 2012). London's economy alone could benefit by £1.24 billion (US$1.61b) annually (Roukouni and Medda, 2012).

CRL will bring an additional 1.5 million people within a 45-minute manageable commute to employment centers, including thousands in some of London's most deprived areas (HM Treasury, 2007). According to the Crossrail, "97% of Crossrail related contracts are being delivered by UK-based companies … with about 60% going to small and medium businesses" (Crossrail, 2014).

OVERSIGHT AND ACCOUNTABILITY

Aligning with national and local climate mitigation policies CRL developed a performance-based sustainability strategy that addresses seven sustainability themes.[3] Fifteen Key Sustainability Initiatives (KSI) translate the themes into actions with senior management assigned to meet objectives. Initiatives associated with the theme

to "Address Climate Change and Energy," include minimization of resource usage, minimization of energy use, optimization of logistics supply chain, minimization of environmental impacts, and reuse of excavation materials (de Silva and Paris, 2014). Indicators for each initiative are monitored and reviewed by in-house managers, further aiding adjustments and decision-making.

CONSTRUCTION

GHG Emissions: To meet initiatives, CRL first measured its construction carbon footprint and then sought to shrink it by 8%. It is on track to meet this target and reduce emissions by 57,000 tons of carbon.[4] Actions include fitting 73% of machinery with diesel particle filters and utilizing cement-free concrete (Avidan, 2015; Crossrail, 2014). Once construction is complete, the payback period will likely be 9–13 years, after which CRL expects annual net savings of 70,000 to 225,000 tons of CO_2, largely due to the displacement of car journeys and replacement of diesel trains on the existing network (Crossrail, 2014).

Landfill Diversion: About 99% of excavated tunnel spoil will be reused or recycled according to CRL. In Essex, 3 million tons have already been delivered and are being used to build a new, 670-hectare wetland habitat (Smale, 2015). The Wallasea Island Wild Coast Project, developed in partnership with the Royal Society (n.d.) for the Protection of Birds, will not only help to restore coastline, encourage biodiversity, and provide a buffer to rising sea levels, but will also divert waste from landfills, significantly reducing GHG emissions further, particularly methane (see Case Study 13.3 Figures 2 and 3).

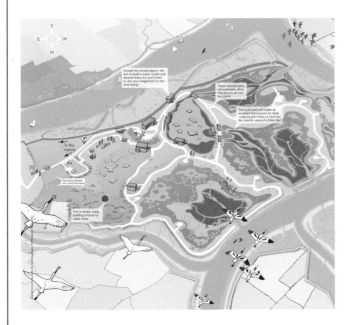

Case Study 13.3 Figures 2 and 3 *Wallasea Island Wild Coast Project.*

Sources: Royal Society for the Protection of Birds and Geographical Association (UK)

1 By 2050, London's population is expected to grow by 37% and its workforce by 29%. Tourism is expected to grow by 40% by 2022. TfL predicts "public transport trips could increase by 50–60% over the next decades" (Mayor of London, 2014).
2 Benefits from transport generally include travel time savings, increased speed of travel, reduced crowding, reduced waiting times, seating availability, improved safety, etc. (Gomez-ibanez, 2008). Estimates of Crossrail's impact on "wider benefits," include moves to more productive jobs, higher economic productivity, agglomeration benefits, increased labor force participation, and leveling of imperfect competition (Crossrail, 2011).
3 For instance, CRL's efforts to reduce carbon emissions is in line with recommendations laid out in the UK's 2013 Infrastructure Carbon Review, which if followed could reduce emissions from infrastructure by 24 million tons by 2050, thus contributing to the UK's climate change commitments. CRL also aligns with the sustainable strategies put forth by the Department for Transport, Transport for London, and the Mayor of London.
4 CRL estimates total CO_2 emissions from construction at 1.7 million tones.

OPERATIONS

Rolling stock and procurement requirements: CRL requires all contractors to meet stringent energy-efficient standards for its new rolling stock and applies sustainable standards throughout its entire supply chain. Bombardier's winning £1 billion (US$1.3b) bid to supply and maintain CRL's rolling stock promises to deliver lightweight trains of only 350 tons, thereby reducing the typical amount of energy needed to move and stop trains (Crossrail, 2014).

Station design: All stations will incorporate energy-efficient materials and products from LED lighting and presence detectors to regenerative braking on trains whereby energy used to slow and stop the train is captured and converted or stored for later use. The stations are also designed to reduce energy use. Utilizing a humpback design, trains arrive to the station on a slight incline and depart on a slight decline, thus capitalizing on the force of gravity (Smith, 2015).

Large-scale urban transport projects can contribute to climate change through their energy consumption during construction and operations. Nevertheless, sustainable outcomes are possible, particularly when calculating offsets gained through reduced energy use, energy efficiency, and mode shifts away from other higher GHG-emitting forms of transport that mass transit provides. As this case shows, it is even more likely that mega urban transit projects have the potential to provide net environmental benefits when applying and adhering to a holistic focus maintained throughout the life of the project. By combining stringent procurement requirements, sustainable supply-chain management, diverting waste from landfills, and using energy-efficient products and materials, CRL is working to mitigate its impact on climate.

---

key enabling reforms: managing land use, congestion pricing, taxes and auctions, and land value capture. These reforms are an illustrative example, not an exhaustive list, of the kinds of changes cities can introduce to expand the fiscal space for the transport sector.

13.5.1 Managing Land Use

Several enabling reforms can promote mixed land-use and transit-oriented development (TOD). In the United States, state governments provide tax credits to promote a form of mixed-land-use planning known as "smart growth" (Bolen et al., 2002) In Ahmadabad, India, policy-makers impose surcharges on land-owners for the right to increase densities (measured in floor area ratios [FAR]) and reinvest revenues in affordable housing near the BRT (UN-Habitat, 2013). In Curitiba, Brazil, officials offer developers the ability to increase density in exchange for open spaces to promote mixed-use development (Seto et al., 2014). Many cities nonetheless adopt poorly designed zoning and codes and fail to coordinate public transport and land-use rules. In consequence, they underuse urban space, face overcrowding, and confront extraordinarily high land prices in city centers (Bertaud, 2004, 2015).

13.5.2 Congestion Pricing

Several cities – Singapore, Stockholm, and London – have adopted congestion pricing to encourage environmentally friendly travel, mode shifts, and compact development (Seik, 1998; Givoni, 2012; Börjesson, 2012). Discontent from drivers, impacts on the poor, infringements on privacy from travel tracking, and a shortage of alternative modes have nonetheless often led to political backlash for congestion pricing. Resistance to the congestion pricing initiative in New York is a high-profile example of such contestation (Schaller, 2010). To avoid similar resistance, many cities have adopted less contentious parking restraints and moderate pricing policies. These alternatives to congestion pricing programs may be less effective as drivers carpool or change parking rather than forgo driving altogether (Marsden, 2006). In other cases, web-based real-time information on bus routes has helped efficient trip planning, thereby supplementing congestion pricing (MTA, 2014; Chicago Transit Authority [CTA], n.d.).

13.5.3 Taxes and Auctions

Taxing the carbon content of fuel can be more efficient than directly charging for vehicle use. A fuel tax can affect fuel and vehicle choice. Norway not only imposes fuel taxes but also an annual tax on vehicle ownership and registration (Naess and Smith, 2009). Other parts of the world employ different methods to capture externalities. Singapore auctions a rationed number of "Certificates of Entitlement" to own a vehicle (Seik, 1998). Cities such as Athens, Seoul, San Paulo, Lagos, and Mexico City limit the use of vehicles on days of the week based on license plate numbers (Mehrotra et al., 2011; World Bank, 2009). Well-designed vehicle restraint schemes discouraged peak driving by 14% in São Paulo (Viegas, 2001). Distance-based charging and pay-as-you-drive (PAYD, an insurance premium per mile based on driving record) have been recommended to affect driving behavior, consolidate loading, and improve efficiencies for the freight industry (Litman, 2012). However, a related challenge is posed by the extraordinarily high global subsidy on carbon-based fuels. The International Monetary Fund (IMF) estimates post-tax energy subsidies to be US$5.3 trillion dollars or 6.5% of global GDP in 2015 (Coady et al., 2015).

13.5.4 Land Value Capture

Value capture can supplement traditional revenue for transit infrastructure in dense and congested cities while promoting mixed-use development and facilitating development of affordable housing (Murakami, 2012; Cervero, 2009; Ardila-Gomez et al., 2016). Value capture can be achieved through a wide variety of techniques, including land sale or lease, joint development between the government and the private sector, sale of air rights, land readjustment (see Figure 13.9), or urban redevelopment schemes (see Chapter 5, Urban Planning and Design). Tokyo has enjoyed considerable success with its land value capture system. Tokyo finances a third of land redevelopment for new transit, grants land readjustment ability under the Housing Highway Integration Law, and allows for land owners and developers to capture value through the Urban Redevelopment Law (Suzuki et al., 2015) (see Figure 13.10). Cities such as Los Angeles have also employed land value capture to derive funds from neighboring commercial business owners, funding 10% of its Red Line metro-rail.

13.6 Transport Solutions: Leveraging Sustainable Development Goals, Paris Agreement, and the New Urban Agenda

The discussion on sustainable transport has evolved from defining and discussing the need for sustainable transport to determining the best strategies to implement sustainable transport. With this evolution, the dialogue is less one that is specific to the transport community and more one that needs to be addressed by the wider development community, particularly cities.

The year 2015 marked a critical juncture for sustainable transport. The UN General Assembly finalized and approved a set of 17 Sustainable Development Goals (SDGs) that are intended to inspire and guide the global development community. SDG 11, "make cities inclusive, safe, resilient and sustainable," includes a specific target on transportation. The target states that "by 2030, provide access to safe, affordable, accessible, and sustainable transport systems for all, improving road safety, notably by expanding public transport, with special attention to the needs of those in vulnerable situations, women, children, persons with disabilities and older persons. Additionally, transport-related targets are included in eight of the 17 SDGs. This suggests that it is likely to receive even more attention as an enabler of sustainable low-carbon development between 2015 and 2030 (UN-Habitat, UNEP, SLoCaT, 2015). This chapter's authors, as members of organizations like the UN Sustainable Development Solutions Network (SDSN), an external advisory group appointed by the UN Secretary General, have continued to advocate for a Post-2015 urban development agenda that offers cities as a space for transformative solutions, including in the New Urban Agenda, as part of the UN General Assembly-mandated Habitat III process – a multiyear engagement that led up to the United Nations Conference on Housing and Sustainable Urban Development, which took place in Quito, Ecuador in October 2016 (Mehrotra, 2012b; Revi et al., 2013).

In addition, a global climate change agreement for the post-2020 period was adopted in December 2015 in Paris. The Partnership on Sustainable Low-Carbon Transport (SLoCaT)

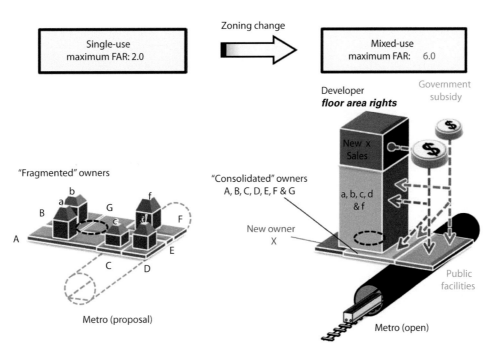

Source: Adapted from Ministry of Land, Infrastructure, Transport, and Tourism 2013.
Note: FAR = floor area ratio.

Figure 13.9 *Financing urban transport through land readjustment.*

Source: Suzuki et al., 2015

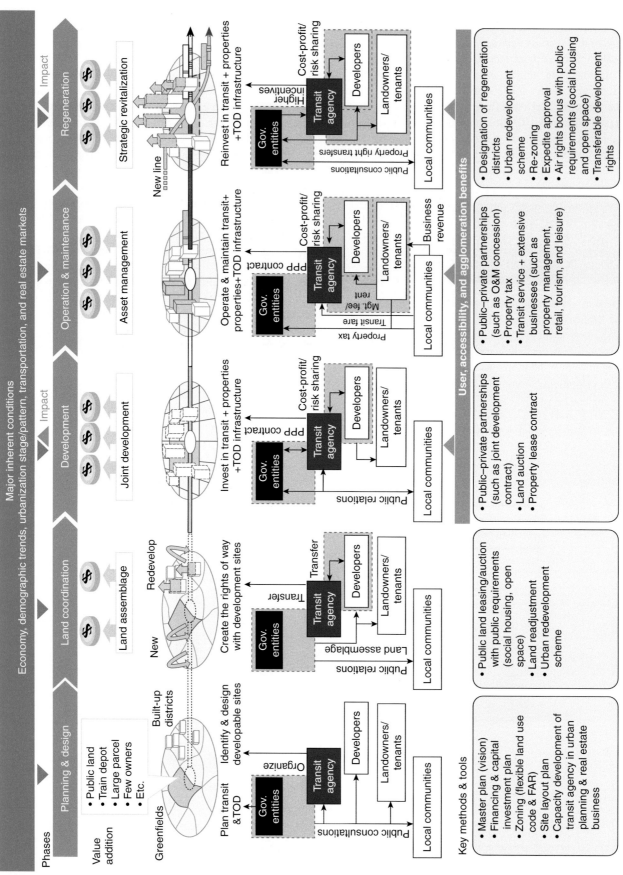

Figure 13.10 *Land-value capture implementation process.*

Note: FAR=floor area ratio; O&M=operation and maintenance; TOD=transit-oriented development.

Source: Suzuki et al., 2015

Results Framework on Sustainable Transport describes the potential contribution of land transport to SDGs by bringing together three targets on improving rural, urban, and national access and connectivity and three targets on reducing road fatalities and serious injuries, air pollution, and GHGs. Sustainable transport has been mainstreamed across seven of the seventeen SDGs adopted by the UN General Assembly. Making sustainable transport an integral part of global policies on sustainable development and climate change needs to go hand-in-hand with the development of effective international funding and financing frameworks for sustainable transport. Although the public sector will have to take the lead in increasing national and local funding and defining long-term strategic planning and investment frameworks, it is now widely acknowledged that greater private-sector involvement is required to provide both financing and technical expertise. Private-sector and international climate finance also have important roles to play in redirecting transport funding from car-centric to more sustainable transport models. Transport systems that utilize low-carbon and mass transit modes, which are designed to reduce the impacts of climate change, cater to urban mobility, and contribute toward creating reasonable densities with mixed land-use development, can transform cities toward sustainability (see Chapter 7, Economics, Finance, and the Private Sector).

These global decisions could make a significant difference in helping national and subnational governments scale up already effective urban transport policy, infrastructure, service development, and mobility management for climate mitigation and adaptation. If sustainable transport is not well-integrated into a global sustainable development and climate policy framework, many cities and companies will continue to press forward with effective actions. But many will not. Delays in helping the world's cities move forward quickly to adopt a new paradigm that ensures access for all to clean, safe, and affordable mobility and climate-resilient communities should be avoided.

Annex 13.1   Stakeholder Engagement

Urban transport and climate change research and practice are interdisciplinary and necessitate multistakeholder engagement. The combined experience of chapter authors represents over a century and a half of engagement on sustainable urban transport spanning scholarly research and high-level policy-making in developed and developing countries. During the 3-year preparatory process for the chapter, the research, policy engagement, and advocacy work have been shaped by, and have shaped, urban transport and climate change processes and associated state-of-the-knowledge. The community of engagement has spanned young policy-makers, graduate students, leading private-sector executives, high-level policy-makers, multilateral development banks, UN institutions, leading think-tanks and foundations, governments, city leaders, and an even wider spectrum of city stakeholders engaged in the Post 2015 development agenda, Paris COP 21 process, and Habitat III preparatory process.

Michael Replogle led a large-scale consultative process on mainstreaming urban transport into the SDGs, transitioning from a research and advocacy role to leading New York City's Department of Transportation as Deputy Commissioner for Policy. Eric Zusman, based in Japan, led high-level policy dialogues and research at Asia's premier environmental think-tank, the Institute for Global Environmental Studies, exploring co-benefits of climate policies in developing Asia's transport sector. Shagun Mehrotra, building on two decades of his research, policy engagement, and teaching experience, in 2012 was instrumental in the conceptualization and development of the community of practice (urban thematic group) and strategy for the dedicated urban SDG for the UN Secretary General's external advisory group of high-level experts. Through this consultative engagement and the efforts of leading urban city networks and stakeholders, a dedicated Urban SDG was adopted by the UN General Assembly, with a high priority accorded to sustainable and equitable urban transport solutions. Klaus Jacob, leading scholar and reflective practitioner with four decades of experience, has been instrumental during the course of this knowledge assessment in shaping the adaptation plan for one of the world's oldest and largest urban transit authority – New York City's Metropolitan Transportation Authority (MTA). He has been widely sought by the global urban transport and climate change community for consultations. Jitendra N. Bajpai, brought several decades of high-level urban transport policy-making and international development advisory and lending experience to inform the assessment as well as expand the community of stakeholders ranging from graduate students to leading policy forums. Susan Yoon has taken the knowledge assessed in this chapter into her practice of policy-making at New York's MTA. Mathew Woundy has engaged in leadership training of private- and public-sector leaders and education of business executives. Lina Fedirko has engaged with creative private-sector firms, bringing a unique blend of policy and design. All these author engagements, both individually and collectively, have brought a deep and diverse stakeholder perspective, have tested and enriched the knowledge assessment, and have made it relevant to a wide range of intellectual and applied contexts.

Global Public Engagement on Sustainable Transport

A growing number of local and national governments are adopting an Avoid-Shift-Improve paradigm for transport policy and investment. This emphasizes: (1) avoiding unnecessary or low-value travel through smarter urban planning, transport pricing, sound logistics, and telecommunications; (2) shifting travel to lower polluting modes like public transport, walking, cycling, and rail freight; and (3) improving the remaining transport with cleaner vehicles and fuels and more efficient network operations. Civil society, industry, and governments are increasingly cooperating in promoting this new paradigm and moving away from the failed approach of *predict and provide*, which sought to simply build more roads to accommodate growing traffic demand.

Partnerships between the public and private sector and civil society play a growing role in advancing sustainable transport. The Partnership on Sustainable Low-Carbon Transport (SLoCaT) was formed in 2009 and now brings together more than 90 civil society groups, UN agencies, and multilateral development banks, as well as industry and research organizations (SLoCaT, 2014c; SLoCaT, 2015a). SLoCaT and its members helped pull together an unprecedented Rio+20 voluntary commitment by the eight largest multilateral development banks of US$175 billion over the coming decade for more sustainable transport (African Development Bank, 2012), with annual reporting on progress (Replogle and Huizenga, 2015). The transport sector responded to the call for bold action on climate change by Secretary General Ban Ki-moon at his Climate Summit in September 2014 by announcing five major transport commitments (SLoCaT, 2015b). These actions on transport will result in scaling up urban public transport and doubling global mode share, greater use of more efficient passenger and freight rail transport, accelerated introduction of urban electric vehicles, more fuel efficient passenger vehicles, and renewed action plans on green freight. Collectively, these actions can reduce the carbon footprint of at least half of all passenger and freight trips made by 2025. The International Energy Agency (2013) has estimated that these actions, if implemented at a global scale, can result in cumulative savings of US$70 trillion by 2050, due to reduced investment needs for vehicles, fuel, and transport infrastructure.

It is clear that the discussion on sustainable transport has evolved from defining and discussing the need for sustainable transport to determining the best strategies to implement sustainable transport. With this evolution, the discussion is decreasingly one that is specific to the transport community and increasingly one that needs to be addressed by the wider development community. That shift should lead to a more stable climate and sustainable cities.

Chapter 13   Urban Transportation

References

AEA Consulting. (2011). Reduction and testing of greenhouse gas (GHG) emissions from heavy duty vehicles—Lot 1: Strategy. European Commission—DG Climate Action. Accessed September 13, 2014: http://ec.europa.eu/clima/policies/transport/vehicles/docs/ec_hdv_ghg_strategy_en.pdf

Allen, H. (2011). Africa's first full rapid bus system: The Rea Vaya Bus System in Johannesburg, Republic of South Africa. Accessed April 26, 2014: http://unhabitat.org/wp-content/uploads/2013/06/GRHS.2013.Case.Study.Johannesburg.South.Africa.pdf

American Society of Civil Engineers (ASCE). (2013). Report card for America's infrastructure. American Society of Civil Engineers. Accessed October 1, 2014: http://www.infrastructurereportcard.org/wp-content/uploads/2013ReportCardforAmericasInfrastructure.pdf

Ardila-Gomez, A., Ortegón, A., and Rubiano, L. C. (2016). Sustainable Urban Transport Financing from the Sidewalk to the Subway: Capital, Operations, and Maintenance Financing. Washington, D.C.: World Bank. World Bank. Accessed April 27, 2017: https://openknowledge.worldbank.org/handle/10986/23521

Baeumler, A., Ijjasz-Vasquez, E., and Mehndiratta, S. (2012). Sustainable low-carbon city development in China. World Bank. Accessed October 8, 2014: http://siteresources.worldbank.org/EXTNEWSCHINESE/Resources/3196537-1202098669693/4635541-1335945747603/low_carbon_city_full_en.pdf

Bertaud, A. (2004). The perfect storm: The four factors restricting the construction of new floor space in Mumbai. Accessed July 9, 2014: http://alain-bertaud.com/AB_Files/AB_Mumbai_FSI_FAR_conundrum.pdf

Bertaud, A. (2015). The Spatial Distribution of Land Prices and Densities: The Models Developed by Economists. Working paper number 23. Marron Institute of Urban Management. Accessed January 10, 2016: http://marroninstitute.nyu.edu/uploads/content/Bertaud_-_The_Spatial_Distribution_of_Land_Prices_and_Densities.pdf

Börjesson, M., Eliasson, J., Hugosson, M. B., and Brundell-Freij, K. (2012). The Stockholm congestion charges—5 years on. Effects, acceptability and lessons learnt. Transport Policy 20, 1–12.

Bolen, E., Brown, K., Kierman, D., and Konshnik, K. (2002). Smart growth: State by state. Accessed March 22, 2014: http://gov.uchastings.edu/public-law/docs/smartgrowth.pdf

Brandon, C., and Hommann, K. (1995). The Cost of Inaction: Valuing the Economy-Wide Cost of Environmental Degradation in India. The World Bank.

Carrigan, A., Duduta, N., King, R., Raifman, M., and Velasquez, J. (2013). Social, environmental, and economic impacts of BRT systems: Bus rapid transit case studies from around the world. EMBARQ. Accessed September 7, 2015: http://www.embarq.org/sites/default/files/Social-Environmental-Economic-Impacts-BRT-Bus-Rapid-Transit-EMBARQ.pdf

Centre for Science and Environment. (1998) Leap frog factor Delhi. Accessed August 1, 2014: http://cseindia.org/challenge_balance/readings/LeapfrogFactor_Delhistory.pdf Accessed August 1, 2014: http://www.cseindia.org/taxonomy/term/1/menu

Cervero, R., and Murakami, J. (2009). Rail + property development in Hong Kong: Experiences and extensions. Urban Studies 46(10), 2019–2043.

Chandler, K., Eberts, E., and Eudy, L. (2006). New York City transit hybrid and CNG transit buses: Interim evaluation results. Accessed February 16, 2014: www.afdc.energy.gov/pdfs/38843.pdf

Changnon, S. (1999). Record flood-producing rainstorms of 17–18 July 1996 in the Chicago metropolitan area. Part III: Impacts and responses to the flash flooding. Journal of Applied Meteorology 38, 273–280.

Chavez-Baeza, C., and Sheinbaum-Pardo, C. (2014). Sustainable passenger road transport scenarios to reduce fuel consumption, air pollutants and GHG (greenhouse gas) emissions in the Mexico City Metropolitan Area. Energy 66 (2014): 624–634.

Chen, Y., and Whalley, A. (2010). Green machines: The effects of urban mass transit on air quality. Accessed May 21, 2015: http://www.psrc.org/assets/10670/Green_Machines.pdf

Chicago Transit Authority (n.d.). CTA Bus Tracker. Accessed February 10, 2017: http://www.ctabustracker.com/bustime/home.jsp

City of Johannesburg. (2013). Strategic integrated transport plan framework for the City of Joburg. Accessed February 12, 2015: http://www.joburg.org.za/images/stories/2013/May/CoJ percent20SITPF percent20Draft percent2013 percent20May percent202013.pdf

Coady, D., Parry, I., Sears, L., and Shang, B. (2015). How Large Are Global Energy Subsidies? International Monetary Fund, Fiscal Affairs Department working paper WP/15/105. International Monetary Fund. Accessed January 8, 2016: http://www.imf.org/external/pubs/ft/wp/2015/wp15105.pdf http://www.c40cities.org/

Creutzig, F., Mühlhoff, R., and Römer, J. (2012). Decarbonizing urban transport in European cities: Four cases show possibly high co-benefits. Environmental Research Letters 7(4), 044042.

Dalkmann, H., and Brannigan, C. (2007). Transport and Climate Change: Module 5e, Sustainable Transportation Sourcebook: A Sourcebook for Policy-Makers in Developing Countries. GIZ. Accessed September 30, 2015: http://www.sutp.org/en-dn-th5

De Jong, M., Ma, Y., Mu, R., Stead, D., and Xi, B. (2010). Introducing public–private partnerships for metropolitan subways in China: What is the evidence? Journal of Transport Geography 18(2), 301–313.

Diaz, R., and Bongardt, D. (2013). Financing sustainable urban transport. International review of national urban transport policies and programmes. Deutsche Gesellschaft für Internationale Zusammenarbeit and EMBARQ. Accessed September 30, 2015: http://sustainabletransport .org/?wpdmact=processanddid=MzMuaG90bGluaw==

EMBARQ Brazil. (2013). Panorama of BRT and bus corridors in the world. Accessed January 12, 2015: http://www.brtdata.org/ – /info/about

European Commission (EC). (2013). Adapting infrastructure to climate change. Accessed April 8, 2015: http://ec.europa.eu/clima/policies/ adaptation/what/docs/swd_2013_137_en.pdf

European Cyclists' Federation. (2011). Quantifying CO2 savings of cycling. Accessed September 21, 2014: http://www.ecf.com/wp-content/uploads/ ECF_BROCHURE_EN_planche.pdf

Fargione, J., Hawthorne, P., Hill, J., Polasky, S., and Tilman, D. (2008). Land clearing and the biofuel carbon debt. *Science* **319**(5867), 1235–1238.

Filosa, G. A. (2015). *International Practices on Climate Adaptation in Transportation: Findings from a Virtual Review*. U.S. Department of Transportation, John A. Volpe National Transportation Systems Center. Accessed March 31, 2016: http://www.fhwa.dot.gov/environment/climate_change/ adaptation/publications_and_tools/international_practices/fhwahep 15011.pdf

Fulton, L. (2013). *How Vehicle Fuel Economy Improvements Can Save $2 Trillion and Help Fund a Long-Term Transition to Plug-In Vehicles*. Working paper 9. Accessed September 21, 2014: http://www .fiafoundation.org/media/44075/wp9-fuel-economy-improvements.pdf

Fulton, L., and Cazzola, P. (2008). Transport, energy, and CO$_2$ in Asia: Where are we going and how do we change it? Accessed November 10, 2014: http://www.slideshare.net/EMBARQNetwork/transport-energy-and- co2-in-asia-where-are-we-going-and-how-do-we-change-it-presentation

Greene, D. L., Kahn, J. R., and Gibson, R. C. (1999). Fuel economy rebound effect for U.S. household vehicles. *The Energy Journal* **20**, 1–31.

Grenzeback, L., and Lukmann, A. (2007). Case study of the transportation sector's response to and recovery from hurricanes Katrina and Rita. Accessed December 7, 2014: http://onlinepubs.trb.org/onlinepubs/sr/ sr290GrenzenbackLukmann.pdf

Givoni, M. (2012). Re-assessing the results of the London congestion charging scheme. *Urban Studies* **49**(5), 1089–1105.

Hidalgo, D. (2012). BRT around the world: Update 2012 and future evolution. Accessed July 16, 2014: http://www.slideshare.net/EMBARQ Network/brt-status-webinar-113012-by-dario-hidalgo

Hodges, T. (2011). Flooded bus barns and buckled rails: Public transportation and climate change adaptation. Accessed October 10, 2015: http://www.fta .dot.gov/documents/FTA_0001_-_Flooded_Bus_Barns_and_Buckled_ Rails.pdf

Institute for Transportation and Development Policy (ITDP). (2014). Cycling and walking. Accessed July 11, 2015: https://www.itdp.org/ what-we-do/cycling-and-walking/

Intergovernmental Panel on Climate Change (IPCC). (2014). Summary for policymakers. *In* Edenhofer, O., Pichs-Madruga, R., Sokona, Y., Farahani, E., Kadner, S., Seyboth, K., Adler, A., Baum, I., Brunner, S., Eickemeier, P., Kriemann, B., Savolainen, J., Schlömer, S., von Stechow, C., Zwickel, T., and Minx, J. C. (eds.), *Climate Change 2014, Mitigation of Climate Change. Contribution of Working Group III to the Fifth Assessment Report of the Intergovernmental Panel on Climate Change* Cambridge University Press. Accessed January 27, 2015: http://www.ipcc.ch/pdf/assessment-report/ar5/wg3/ipcc_wg3_ar5_ summary-for-policymakers.pdf

International Energy Agency (IEA). (2012). World energy outlook 2012. Accessed December 30, 2014: http://www.worldenergyoutlook.org/ publications/weo-2012/

International Energy Agency (IEA). (2013). Policy pathways: A tale of renewed cities. Accessed March 12, 2014: http://www.iea.org/publications/freepublications/publication/Renewed_Cities_WEB.pdf

International Energy Agency (IEA). (2014). CO$_2$ emissions from fuel combustion: Highlights, 2014 edition. Accessed January 14, 2015: http:// www.iea.org/publications/freepublications/publication/CO2EmissionsFromFuelCombustionHighlights2014.pdf

International Monetary Fund (IMF). (2014). Is it time for an infrastructure push? The macroeconomic effects of public investment. In *World Economic Outlook 2014: Legacies, Clouds, Uncertainties*. Accessed March 11, 2015: http://www.imf.org/external/pubs/ft/weo/2014/02/

Jacob, K., Deodatis, G., Atlas, J., Whitcomb, M., Lopeman, M., Markogiannaki, O., Kennett, Z., Morla, A., Leichenko, R., and Vancura, P. (2011). Responding to climate change in New York State (ClimAID). In Transportation. In Rosenzweig C., Solecki W., DeGaetano A., O'Grady M., Hassol S., and Grabhorn P. (eds.), *Responding to Climate Change in New York State*. New York State Energy Research and Development Authority. Accessed December 11, 2014: file:///C:/Users/SY/Desktop/ temp/ClimAID-Transportation.pdf

Jacob, K., Rosenzweig, C., Horton, R., Major, D., and Gornitz, V. (2008). MTA adaptation to climate change: A categorical imperative. Accessed March 11, 2015: http://web.mta.info/sustainability/pdf/Jacob_et%20 al_MTA_Adaptation_Final_0309.pdf

Jaeger, A., Nugroho, S. B., Zusman, E., Nakano, R., and Daggy, R. (2015). Governing sustainable low-carbon transport in Indonesia: An assessment of provincial transport plans. *Natural Resources Forum* **39**(1), 27–40.

Kaufman, S., Quing, C., Levenson, N., and Hansen, M. (2012). Transportation during and after Hurricane Sandy. Rudin Center for Transportation, NYU Wagner Graduate School of Public Services, Nov. 2012. Accessed June 1, 2014: http://wagner.nyu.edu/files/rudincenter/sandytransportation.pdf

Kennedy, C., Steinberger, J., Gasson, B., Hansen, Y., Hillman, T., Havránek, M., Pataki D., Phdungsilp, A., Ramaswami, A., and Mendez, G. A. (2009). Greenhouse gas emissions from global cities. *Environmental Science & Technology* **43**, 7297–7302.

Kiunsi, R. (2013). The constraints on climate change adaptation in a city with a large development deficit: The case of Dar es Salaam. *Environment and Urbanization* **25**(2), 321–337.

Koetse, M., and Rietveld, P. (2009). The impact of climate change and weather on transport: An overview of empirical findings. *Transportation Research Part D: Transport and Environment* **14**(3), 205–221.

Kostro, S., Bertman, L., and Pickens, B. (2013). Super typhoon Haiyan: With so many still suffering, why keep our eyes on recovery? Accessed May 12, 2015: http://csis.org/publication/super-typhoon-haiyan-so-many- still-suffering-why-keep-our-eyes-recovery

Litman, T. (2012). Distance-based vehicle insurance as a TDM strategy. Accessed August 15, 2014: http://www.islandnet.com/~litman/dbvi.pdf

Mahendra, A., Dalkmann H., and Raifman, M. (2013). Financing needs for sustainable transport systems for the 21st century. Accessed November 17, 2015: http://www.wricities.org/sites/default/files/Financing-Needs-for-Sustainable-Transport-Systems-21st-Century.pdf

Marsden, G. (2006). The evidence base for parking policies – a review. *Transport Policy* **13**(6),447–457.

Marsden, G., and Rye, T. (2010). The governance of transport and climate change. *Journal of Transport Geography* **18**, 669–678.

Mehrotra, S. (2012a). *Reinventing Infrastructure Economics: Theory and Empirics*. Doctoral dissertation, Academic Commons, Columbia University.

Mehrotra, S. (2012b). Cities: Smart, healthy, and productive—solutions for a global urban future. Making a case for an urban sustainable development goal. United Nations' Sustainable Development Solutions Networks (UNSDSN). Accessed August 26, 2015: http://urbansdg.org/wp-content/ uploads/2014/02/nov_2012_sustainable_cities_presentation.pdf

Mehrotra, S., Lefevre, B., Zimmerman, R., Gercek, H., Jacob, K., and Srinivasan, S. (2011). Climate change and urban transportation systems. In Rosenzweig, C., Solecki, W. D., Hammer, S. A., and Mehrotra, S. (eds.), *Climate Change and Cities: First Assessment Report of the Urban Climate Change Research Network* (145–177), Cambridge University Press.

Mehrotra, S., Natenzon, C. E., Omojola, A., Folorunsho, R., Gilbride J., and Rosenweig, C. (2009). *Framework for City Climate Risk Assessment*. World Bank. Accessed July 7, 2015: http://siteresources.worldbank .org/INTURBANDEVELOPMENT/Resources/336387– 1256566800920/6505269–1268260567624/Rosenzweig.pdf

Melillo, J. M., Richmond, T., and Yohe, G. W. (eds.). (2014). *Climate Change Impacts in the United States: The Third National Climate Assessment*. U.S. Government Printing Office.

Metrobits. (2012). World metro database. Accessed May 12, 2015: http:// mic-ro.com/metro/table.html

Metropolitan Transportation Authority. (2009). Greening mass transit and metro regions: The final report of the Blue Ribbon Commission on Sustainability and the MTA. Accessed December 10, 2015: http://web.mta.info/sustainability/pdf/SustRptFinal.pdf

Metropolitan Transportation Authority. (2013). Timeline of the storm. Accessed December 12, 2014: http://web.mta.info/sandy/timeline.htm

Metropolitan Transportation Authority. (2014). MTA bus time. Accessed September 14, 2015: http://bustime.mta.info/

Meyer, M. (2008). Design standards for U.S. transport infrastructure – The implications of climate change. Accessed June 1, 2014: http://onlinepubs.trb.org/onlinepubs/sr/sr290Meyer.pdf.

Murakami, J. (2012). Transit value capture: New town co-development models and land market updates in Tokyo and Hong Kong. In Ingram, G. K., and Hong, Y. -H. (eds.), *Value Capture and Land Policies* (285–320). Lincoln Institute of Land Policy.

Naess, E., and Smith, T. (2009). Environmentally related taxes in Norway: Totals and divided by industry. Accessed February 18, 2014: http://www.cbd.int/financial/fiscalenviron/norway-envtaxes.pdf

Newman P., and Kenworthy J. (2006). Urban design to reduce auto dependency. *Opolis* 2(1). Accessed August 30, 2014: http://www.naturaledgeproject.net/documents/newmankenworthyurbandesign.pdf

New York (NYC). (2013). NYC Hurricane Sandy after action: Report and recommendations to Mayor Michael R. Bloomberg. Accessed September 3, 2014: http://www.nyc.gov/html/recovery/downloads/pf/sandy_aar_5.2.13.pdf

Pascal, M., Corso, M., Chanel, O., Declercq, C., Badaloni, C., Cesaroni, G., and Medina, S. (2012). Assessing the public health impacts of urban air pollution in 25 European cities: Results of the Aphekom project. *Science of the Total Environment* 449, 390–400.

PlaNYC. (2013). A stronger, more resilient New York City. Accessed May 28, 2014: http://www.nyc.gov/html/sirr/html/report/report.shtml

Pourbaix, J. (2012). Towards a smart future for cities. *Journeys* May, 7–13. Accessed December 22, 2015: http://www.lta.gov.sg/ltaacademy/doc/J12 May-p07Jerome_Towards a Smart Future for Cities.pdf

Pucher, J., Peng, Z., Mittal, N., Zhu, Y., and Korattyswaroopam, N. (2007). Urban transport trends and policies in China and India: Impacts of rapid economic growth. *Transport Reviews* 27(4), 379–410. Accessed November 21, 2015: http://www.tandfonline.com/doi/abs/10.1080/01441640601089988#.VOzZ2nzF-1U

Replogle, M., and Huizenga, C. (2015). Two years in—How are the world's multilateral development banks doing in delivering on their $175 billion pledge for more sustainable transport? Accessed February 23, 2016: https://www.itdp.org/two-years-in-how-are-the-worlds-multilateral-development-banks-doing-in-delivering-on-their-175-billion-pledge-for-more-sustainable-transport/

Revi, A., Rosenzweig, C., Mehrotra, S., Solecki, W., et al. (2013). *The Urban Opportunity: Enabling Transformative and Sustainable Development.* Background paper for the United Nations secretary-general's high-level panel of eminent persons on the post- 2015 development agenda. Accessed September 30, 2015: https://sustainabledevelopment.un.org/content/documents/2579Final-052013-SDSN-TG09-The-Urban-Opportunity.pdf

Revi, A., Satterthwaite, D. E., Aragón-Durand, F., Corfee-Morlot, J., Kiunsi, R. B. R., Pelling, M., Roberts, D. C., and Solecki, W. (2014). Urban areas. In Field, C. B., Barros, V. R., Dokken, D. J., Mach, K. J., Mastrandrea, M. D., Bilir, T. E., Chatterjee, M., Ebi, K. L., Estrada, Y. O., Genova, R. C., Girma, B., Kissel, E. S., Levy, A. N., MacCracken, S., Mastrandrea, P. R., and White, L. L. (eds.), *Climate Change 2014: Impacts, Adaptation, and Vulnerability. Part A: Global and Sectoral Aspects. Contribution of Working Group II to the Fifth Assessment Report of the Intergovernmental Panel on Climate Change* (535–612). Cambridge University Press. Accessed April 10, 2015: http://www.ipcc.ch/pdf/assessment-report/ar5/wg2/WGIIAR5-Chap8_FINAL.pdf

Ribeiro, S., and de Abreu, A. (2008). Brazilian transport initiatives with GHG reductions as a co-benefit. *Climate Policy* 8(2), 220–240.

Rickwood P., Glazebrook, G., and Searle, G. (2011). Urban structure and energy – A review. *Urban Policy and Research* 26, 57–81. doi: 10.1080/08111140701629886, ISSN: 0811–1146

Schaller, B. (2010). New York City's congestion pricing experience and implications for road pricing acceptance in the United States. *Transport Policy* 17, 266–273.

Schipper, L., Marie-Lilliu, C., and Gorham, R. (2000). *Flexing the Link between Transport and Greenhouse Gas Emissions: A Path for the World Bank.* International Energy Agency. Accessed November 28, 2014: http://www.ocs.polito.it/biblioteca/mobilita/FlexingLink1.pdf

Schrank, D., Eisele, B., and Lomax, T. (2012). Texas A&M Transportation Institute 2012 urban mobility report. Accessed October 10, 2014: http://d2dtl5nnlpfr0r.cloudfront.net/tti.tamu.edu/documents/mobility-report-2012.pdf

Seik, F. (1998). A unique demand management instrument in urban transport: The vehicle quota system in Singapore. *Cities* 15(1), 27–39.

Seto K. C., Dhakal, S., Bigio, A., Blanco, H., Delgado, G. C., Dewar, D., Huang, L., Inaba, A., Kansal, A., Lwasa, S., McMahon, J. E., Müller, D. B., Murakami, J., Nagendra, H., and Ramaswami, A. (2014). Human settlements, infrastructure and spatial planning. In Edenhofer, O., Pichs-Madruga, R., Sokona, Y., Farahani, E., Kadner, S., Seyboth, K., Adler, A., Baum, I., Brunner, S., Eickemeier, P., Kriemann, B., Savolainen, J., Schlömer, S., von Stechow, C., Zwickel, T., and Minx, J. C. (eds.), *Climate Change 2014: Mitigation of Climate Change. Contribution of Working Group III to the Fifth Assessment Report of the Intergovernmental Panel on Climate Change.* Cambridge University Press. Accessed August 2, 2015: http://www.ipcc.ch/pdf/assessment-report/ar5/wg3/ipcc_wg3_ar5_chapter12.pdf

Sims R., Schaeffer, R., Creutzig, F., Cruz-Núñez, X., D'Agosto, M., Dimitriu, D., Figueroa Meza, M. J., Fulton, L., Kobayashi, S., Lah, O., McKinnon, A., Newman, P., Ouyang, M., Schauer, J. J., Sperling, D., and Tiwari, G. (2014). Transport. In Edenhofer, O., Pichs-Madruga, R., Sokona, Y., Farahani, E., Kadner, S., Seyboth, K., Adler, A., Baum, I., Brunner, S., Eickemeier, P., Kriemann, B., Savolainen, J., Schlömer, S., von Stechow, C., Zwickel, T., and Minx, J. C. (eds.), *Climate Change 2014: Mitigation of Climate Change. Contribution of Working Group III to the Fifth Assessment Report of the Intergovernmental Panel on Climate Change*, Cambridge University Press. Accessed November 8, 2015: http://www.ipcc.ch/pdf/assessment-report/ar5/wg3/ipcc_wg3_ar5_chapter8.pdf

SLoCaT. (2014a). Non-motorized transport. Accessed August 22, 2015: http://www.slocat.net/content-stream/183

SLoCaT. (2015a). SLoCaT partnership Accessed April 1, 2016: http://www.slocat.net/slocatpartnership

SLoCaT. (2015b). United Nations secretary general's climate summit. Accessed February 28, 2016: http://www.slocat.net/climatesummit

Suzuki, H., Cervero, R., and Iuchi, K. (2013). *Transforming Cities with Transit: Transit and Land-Use Integration for Sustainable Urban Development.* World Bank. Accessed October 10, 2014: https://openknowledge.worldbank.org/handle/10986/12233

Suzuki, H., Murakami, J., Hong, Y., and Tamayose, B. (2015). *Financing Transit-Oriented Development with Land Values: Adapting Land Value Capture in Developing Countries.* Urban Development Series. World Bank. Accessed February 21, 2016: http://www-wds.worldbank.org/external/default/WDSContentServer/WDSP/IB/2014/11/04/000333037_20141104220722/Rendered/PDF/922500WP0Box380REPORT0COMING0IN0DEC.pdf

Swiss Association for Quality and Management Systems (SQS). (2011). VCS Validation report BRT Rea Vaya phase 1A and 1B, South Africa. Accessed September 17, 2014: https://mer.markit.com/br-reg/PublicReport.action?getDocumentById=trueanddocument_id=100000000006631

Takami, K., and Hatoyama, K. (2008). Sustainable regeneration of a car-dependent city: The case of Toyama toward a compact city. In Kidokoro, T., Harata, N., Leksono, P. S., Jessen, J., Motte, A., and Seitzer, E. P. (eds.), *Sustainable City Regions* (183–200). Springer.

Transport Research Board (TRB). (2014). Strategic issues facing transportation, Volume 2: Climate change, extreme weather events, and the highway system. Transport Research Board, National Cooperative Highway Research Program Report Number 750. Accessed February 26, 2015: http://onlinepubs.trb.org/onlinepubs/nchrp/nchrp_rpt_750v2.pdf

Turner, M., Kooshian, C., and Winkelman, S. (2012). Case study: Bus rapid transit (BRT) development and expansion. Accessed September 17, 2014: http://www.ccap.org/docs/resources/1080/Colombia-case percent20study-final.pdf

UN-Habitat/ESCAP. (2015). The state of Asian and Pacific Cities 2015: Urban transformations shifting from quantity to quality. Accessed January 31, 2016: http://unhabitat.org/books/the-state-of-asian-and-pacific-cities-2015/

UN-Habitat, UNEP, SLoCaT (with contributions from IEA, Economic Commission for Latin America and the Caribbean (ECLAC), United Nations Conference on Trade and Development (UNCTAD), European Bank for Reconstruction and Development (EBRD), The International Road Assessment Programme (iRAP), International Maritime Organization (IMO), and The International Council on Clean Transportation (ICCT)). (2015). Analysis of the transport relevance of each of the 17 SDGs. Accessed January 3, 2016: https://sustainabledevelopment.un.org/content/documents/8656Analysis%20of%20transport%20relevance%20of%20SDGs.pdf.

United Nations (UN). (2014). Transport action plan: Urban electric mobility initiative. United Nations Climate Summit. Accessed July 27, 2015: http://www.un.org/climatechange/summit/wp-content/uploads/sites/2/2014/09/TRANSPORT-Action-Plan-UEMI.pdf

United Nations Department of Economic and Social Affairs (UN DESA). (2014). World urbanization prospects. Revision. Accessed September 20, 2015: http://esa.un.org/Unpd/Wup/

United Nations Human Settlements Program (UN-Habitat). (2013). Global report on human settlements: Planning and design for sustainable urban mobility. Accessed September 17, 2014: http://unhabitat.org/planning-and-design-for-sustainable-urban-mobility-global-report-on-human-settlements-2013

URS. (2010). Adapting energy, transport and water infrastructure for long-term impact of climate change. Accessed September 15, 2014: https://www.gov.uk/government/uploads/system/uploads/attachment_data/file/183472/infrastructure-full-report.pdf

U.S. Department of Energy (U.S. DOE). (n.d.). Energy efficiency & renewable energy: All-electric vehicles. Accessed March 23, 2017: https://www.fueleconomy.gov/feg/evtech.shtml

U.S. Department of Transportation (U.S. DOT). (2014). *Ensuring Transportation Infrastructure and System Resilience.* U.S. Department of Transportation Climate Adaptation Plan.

U.S. Environmental Protection Agency (EPA). (2014). Inventory of U.S. greenhouse gas emissions and sinks: 1990–2012. Accessed October 19, 2015: http://www.epa.gov/climatechange/Downloads/ghgemissions/U.S.-zGHG-Inventory-2014-Main-Text.pdf

U.S. Global Change Research Program. (2014). 2014 National climate assessment. Accessed August 19, 2015: http://nca2014.globalchange.gov/downloads

Viegas, J. (2001). Making urban road pricing acceptable and effective: Searching for quality and equity in urban mobility. *Transport Policy* 8(4), 289–294.

Waygood, E. O. D., Sun, Y., and Susilo, Y. O. (2014). Transportation carbon dioxide emissions by built environment and family lifecycle: Case study of the Osaka metropolitan area. *Transportation Research Part D: Transport and Environment* 31, 176–188.

World Bank. (2009). *World Development Report 2009: Reshaping Economic Geography.* World Bank. Accessed September 17, 2014: https://openknowledge.worldbank.org/handle/10986/5991

World Bank. (n.d.). Lagos bus rapid transit (BRT): Fighting congestion and climate change. Accessed February 3, 2017: http://siteresources.worldbank.org/INTAFRICA/Resources/FINAL_STORY_green-growth-lagos.pdf

World Health Organization (WHO). (2011). Burden of disease from environmental noise: Quantification of healthy years lost in Europe. Accessed September 15, 2014: http://www.euro.who.int/__data/assets/pdf_file/0008/136466/e94888.pdf

Wright, L. (2004). The limits of technology: Achieving transport efficiency in developing nations. Accessed October 3, 2014: http://discovery-dev.ucl.ac.uk/108/1/Lloyd_Wright,_Bonn,_Germany,_Transport_and_climate_change.pdf

Wright, L. (2009). *Win-win Solutions to Climate Change and Transport.* United Nations Commission on Regional Development (UNCRD).

Zimmerman, R. (2012). *Transport, the Environment and Security, Making the Connection.* Edward Elgar Publishing, Ltd.

Zusman, E., Srinivasan, A., and Dhakal, S. (2012). *Low carbon transport in Asia: Strategies for optimizing co-benefits.* Taylor & Francis.

Chapter 13   Case Study References

Case Study 13.1   New York Metropolitan Transportation Authority's Adaptation Solutions

Blake, S., Kimberlain, B., Berg, J., Cangialosi P., and Beven, L. (2013). Tropical cyclone report, Hurricane Sandy (AL182012). Accessed April 25, 2014: http://www.nhc.noaa.gov/data/tcr/AL182012_Sandy.pdf

Coyle, D. C. (2014). MTA Headquarters, personal communication, August 4, 2014.

Jacob, K., Gornitz, V., Horton, R., Major, D., and Rosenzweig, C. (2008). MTA adaptation to climate change: A categorical imperative. Accessed March 20, 2014: http://web.mta.info/sustainability/pdf/Jacob_et%20al_MTA_Adaptation_Final_0309.pdf

Kaufman, S., Quing, C., Levenson, N., and Hansen, M. (2012). Rudin Center for Transportation. NYU Wagner Graduate School of Public Services. Accessed May 2, 2014: http://wagner.nyu.edu/files/rudincenter/sandytransportation.pdf

Metropolitan Transportation Authority, Office of the Inspector General. (2006). Subway flooding during heavy rainstorms: Prevention and emergency response. MTA/OIG #2005-64, February 2006.

Metropolitan Transportation Authority. (2007). August 8, 2007 storm report. Accessed August 19, 2015: http://web.mta.info/mta/pdf/storm_report_2007.pdf

Metropolitan Transportation Authority. (2009). Greening mass transit and metro regions: The final report of the blue ribbon commission on sustainability and the MTA. Accessed May 27, 2015: http://web.mta.info/sustainability/pdf/SustRptFinal.pdf

Metropolitan Transportation Authority. (2013). Capital program 20 year needs letter. Accessed December 9, 2014: http://web.mta.info/mta/news/books/docs/Capital-Program-20-Year-Needs_Letter-from-Chairman-Prendergast.pdf

Metropolitan Transportation Authority. (2014). MTA Capital Program 2015–2019. Accessed August 5, 2015: http://web.mta.info/capital/pdf/Board_2015-2019_Capital_Program.pdf

Metropolitan Transportation Authority. (n.d.). Timeline of the storm. Accessed March 17, 2017: http://web.mta.info/sandy/timeline.htm

Peel, M. C., Finlayson, B. L., and McMahon, T. A. (2007). Updated world map of the Köppen-Geiger climate classification. *Hydrology and Earth System Sciences Discussions* 4(2), 462.

U.S. Census Bureau. (2010). Decennial census, summary file 1. Accessed October 9, 2015: http://www.census.gov/population/metro/files/CBSA%20Report%20Chapter%203%20Data.xls

U.S Census Bureau. (2016). Table 1. Annual Estimates of the Resident Population: April 1, 2010 to July 1, 2016 – Metropolitan Statistical Area; – 2016 Population Estimates. U.S. Census Bureau. Retrieved March 28, 2017. Accessed January 14, 2017: https://factfinder.census.gov/faces/tableservices/jsf/pages/productview.xhtml?pid=PEP_2016_PEPANNRES&prodType=table

World Bank. (2017). 2016 GNI per capita, Atlas method (current US$). Accessed August 9, 2017: http://data.worldbank.org/indicator/NY.GNP.PCAP.CD

Zimmerman, R. (2014), Planning restoration of vital infrastructure services following Hurricane Sandy: Lessons learned for energy and transportation, *Journal of Extreme Events* 1 (1). doi:10.3141/2532-08

Case Study 13.2   BRT in Lagos and Johannesburg: Establishing Formal Public Transit in Sub-Saharan Africa

Allen, H. (2011). Africa's first full rapid bus system: The Rea Vaya bus system in Johannesburg, Republic of South Africa. In *Global Report on Human Settlements 2013.* UN-HABITAT. Accessed September 2, 2015: http://unhabitat.org/wpcontent/uploads/2013/06/GRHS.2013.Case_.Study_.Johannesburg.South_.Africa.pdf

C40 Cities Climate Leadership Group (C40). (2013). Lagos Bus Rapid Transit system garners praise from World Bank leader. Accessed January 13, 2015: http://c40.org/c40blog/lagos-bus-rapid-transitsystemgarners-praise-from-world-bank-leader

Carrigan, A., King, R., Velasquez, J. M., Raifman, M., and Duduta, N. (2013). Social, environmental, and economic impacts of BRT systems: Bus rapid transit case studies from around the world. EMBARQ: A program of the World Resources Institute. Accessed July 5, 2015: http://www.embarq.org/sites/default/files/SocialEnvironmental-Economic-Impacts-BRT-Bus-Rapid-Transit-EMBARQ.pdf

Lagos State. (2013). Digest of statistics 2013. Accessed July 28, 2014: http://www.lagosstate.gov.ng/2013_Digest%20_of_Statistics.pdf

Peel, M. C., Finlayson, B. L., and McMahon, T. A. (2007). Updated world map of the Köppen-Geiger climate classification. *Hydrology and Earth System Sciences Discussions* **4**(2), 462.

Statssa. (2015). Statistics South Africa. Accessed April 1, 2016: http://www.statssa.gov.za/?page_id=1021andid=city-of-johannesburg-municipality

United Nations Human Settlements Program (UN-Habitat). (2013). *Global Report on Human Settlements 2013: Planning and Design for Sustainable Urban Mobility.* Author.

World Bank. (2015). Lagos bus rapid transit (BRT): Fighting congestion and climate change. Africa on the move: The seeds of green growth: A World Bank series. Accessed April 12, 2016: http://siteresources.worldbank.org/INTAFRICA/Resources/FINAL_STORY_green-growth-lagos.pdf

World Bank. (2017). 2016 GNI per capita, Atlas method (current US$). Accessed August 9, 2017: http://data.worldbank.org/indicator/NY.GNP.PCAP.CD

Case Study 13.3   London's Crossrail: Integrating Climate Change Mitigation in Construction and Operations

Avidan, A. A. (2015). Reducing greenhouse gas emissions on Crossrail. Bechtel Corporation. Accessed January 15, 2016: https://www.youtube.com/watch?v=OlgbBQJp1LI

BBC. (2012). Wallasea Island nature reserve project construction begins. Accessed July 5, 2015: http://www.bbc.com/news/science-environment-19598532

Crossrail. (n. d.). Environmental policy and objectives. Accessed January 17, 2017: http://www.crossrail.co.uk/sustainability/environment/objectives

Crossrail. (2011). Crossrail business case update: Summary report: July 2010. Accessed July 28, 2014: http://webarchive.nationalarchives.gov.uk/20111005174015/http:/assets.dft.gov.uk/publications/crossrail-business-case-update/crossrail-business-case-update-summary-report-july-2011.pdf

Crossrail. (2014). Crossrail sustainability report 2014. Accessed April 20, 2015: http://www.crossrail.co.uk/sustainability/

De Silva, M., and Paris, R. (2014). Building Crossrail – A holistic approach to sustainability. International Conference on Sustainable Infrastructure 2014, American Society of Civil Engineers, Long Beach, CA.

European Environment Agency. (2009). Diverting waste from landfill. Effectiveness of waste-management policies in the European Union. EEA Report No. 7/2009. Accessed October 26, 2014: https://www.scribd.com/document/18188424/EEA-Report-Diverting-waste-from-landfill-Effectiveness-of-waste-management-policies-in-the-European-Union-2009

Eurostat. (2015a). Population by sex and age groups on 1 January. European Comission. Accessed April 5, 2016: http://appsso.eurostat.ec.europa.eu/nui/show.do?dataset=met_pjanaggr3andlang=en

Eurostat. (2015b). Area of the regions [met_d3area]. European Commission. Accessed April 13, 2016: http://appsso.eurostat.ec.europa.eu/nui/show.do? dataset=met_d3areaandlang=en

Geographical Association (UK). (2015). Wallasea Island Case Study: Gallery.

Gomez-Ibanez, J. (2008). Crossrail (A): The business case. Kennedy School of Government Case Program. CR14–08–1898.0.

GVA. (2012). Crossrail property impact study. Accessed October 26, 2014: https://www.gva.co.uk/regeneration/crossrail-property-impact-study.

Her Majesty's Treasury. (2010). Business rate supplements: Guidance for local authorities. Accessed June 29, 2014: https://www.gov.uk/government/uploads/system/uploads/attachment_data/file/8306/business_rate_supplements_localauthority_guidance.pdf

Her Majesty's Treasury. (2007). *Business Rate Supplements: A White Paper.* October 2007. Accessed August 11, 2014: https://www.gov.uk/government/publications/business-rate-supplements-a-white-paper

Leuven. (2014). Leuven in numbers. Accessed January 24, 2015: http://www.leuven.be/bestuur/leuven-in-cijfers

Mayor of London. (2014). Long term infrastructure investment plan: Progress report. Accessed June 22, 2015: https://www.london.gov.uk/file/16744/download?token=5St1w4GC

Omega Centre. (2012). *Mega Projects: Executive Summary-Lessons for Decision-Makers: An Analysis of Selected International Large-Scale Transport Infrastructure Projects.* Bartlett School of Planning, University College London. Accessed August 11, 2014: http://www.omegacentre.bartlett.ucl.ac.uk/publications/reports/mega-project-executive-summary/

Peel, M. C., Finlayson, B. L., and McMahon, T. A. (2007). Updated world map of the Köppen-Geiger climate classification. *Hydrology and Earth System Sciences Discussions* **4**(2), 462.

Roukouni, A., and Medda, F. (2012). Evaluation of value capture mechanisms as a funding source for urban transport: The case of London's Crossrail. *Social and Behavioral Sciences* **48**, 2394–2404.

Royal Society for the Protection of Birds. (n.d.) Wallasea Island wild coast project. Accessed January 12, 2017: http://www.rspb.org.uk/whatwedo/campaigningfornature/casework/details.aspx?id=tcm:9-235089

Smale, K. (2015). The last shipment of excavated material donated by Crossrail has arrived at the nature reserves at Wallasea Island in Essex. *New Civil Engineer.* Accessed April 14, 2016: https://www.newcivilengineer.com/the-gallery-and-video-excavated-material-from-crossrail-creates-nature-reserve/8681853.article

Smith, K. (2015). Crossrail's sustainability strategy bears fruit. *Rail Journal.* Accessed March 6, 2016: http://www.railjournal.com/index.php/commuter-rail/crossrails-sustainability-strategy-bears-fruit.html

UK Department for Transport. (2007). *Delivering a Sustainable Railway.* White paper CM 7176. Accessed August 23, 2014: https://www.gov.uk/government/publications/delivering-a-sustainable-railway-whitepaper-cm-7176

World Bank. (2017). 2016 GNI per capita, Atlas method (current US$). Accessed August 9, 2017: http://data.worldbank.org/indicator/NY.GNP.PCAP.CD

14

Urban Water Systems

Coordinating Lead Authors

Sebastian Vicuña (Santiago), Mark Redwood (Ottawa)

Lead Authors

Michael Dettinger (San Diego), Adalberto Noyola (Mexico City)

Contributing Authors

Daniel B. Ferguson (Tucson), Leonor P. Guereca (Mexico City), Christopher M. Clark (Washington, D.C.), Nicole Lulham (Ottawa), Priyanka Jamwal (Bangalore), Anja Wejs (Aalborg), Upmanu Lall (New York), Liqa Raschid (Colombo), Ademola Omojola (Lagos), David C. Major (New York)

This chapter should be cited as

Vicuña, S., Redwood, M., Dettinger, M., and Noyola, A. (2018). Urban water systems. In Rosenzweig, C., W. Solecki, P. Romero-Lankao, S. Mehrotra, S. Dhakal, and S. Ali Ibrahim (eds.), *Climate Change and Cities: Second Assessment Report of the Urban Climate Change Research Network*. Cambridge University Press. New York. 519–552

Sustaining Water Security

In regard to climate change, water is both a resource and a hazard. As a resource, good-quality water is basic to the well-being of the ever-increasing number of people living in cities. Water is also critical for many economic activities, including peri-urban agriculture, food and beverage production, and industry. However, excess precipitation or drought can lead to hazards ranging from increased concentrations of pollutants (with negative health consequences), a lack of adequate water flow for sewerage, and flood-related damage to physical assets.

Projected deficits in the future of urban water supplies will likely have a major impact on both water availability and costs. Decisions taken now will have an important influence on future water supply for industry, domestic use, and agriculture.

Major Findings

- The impacts of climate change put additional pressure on existing urban water systems (UWS) and can lead to negative impacts for human health and well-being, economies, and the environment. Such impacts include increased frequency of extreme weather events leading to large volumes of stormwater runoff, rising sea levels, and changes in surface water and groundwater.

- A lack of urban water security, particularly in lower income countries, is an ongoing challenge. Many cities struggle to deliver even basic services to their residents, especially those living in informal settlements. As cities grow, demand and competition for limited water resources will increase, and climate changes are very likely to make these pressures worse in many urban areas.

- Water security challenges extend to peri-urban areas as well, where pressure on resources is acute and where there are often overlapping governance and administrative regimes.

- Governance systems have largely failed to adequately address the challenges that climate change poses to urban water security. Failure is often driven by a lack of coherent and responsive policy, limited technical capacity to plan for adaptation, limited resources to invest in projects, lack of coordination, and low levels of political will and public interest.

Key Messages

Adaptation strategies for urban water resources will be unique to each city since they depend heavily on local conditions. Understanding the local context is essential to adapting water systems in ways that address both current and future climate risks.

Acting now can minimize negative impacts in the long term. Master planning should anticipate projected changes over a time frame of more than 50 years. Yet, in the context of an uncertain future, finance and investment should focus on low-regret options that promote both water security and economic development, and policies should be flexible and responsive to changes and new information that come to light over time.

Many different public and private stakeholders influence the management of water, wastewater, stormwater, and sanitation. For example, land-use decisions have long-lasting consequences for drainage, infrastructure planning, and energy costs related to water supply and treatment. Therefore, adapting to the changing climate requires effective governance and coordination and collaboration among a variety of stakeholders and communities.

Cities should capture co-benefits in water management whenever possible. Cities might benefit from low-carbon energy production and improved health with wastewater treatment. Investment strategies should include the application of life cycle analysis to water supply, treatment, and drainage; use of anaerobic reactors to improve the balance between energy conservation and wastewater treatment; elimination of high-energy options, such as interbasin transfers of water, wherever alternative sources are available; and the recovery of biogas produced by wastewater.

14.1   Introduction

According to the Intergovernmental Panel on Climate Change (IPCC) (2013), much of the impact of climate change will be felt in the water sector. Given the critical role of water resources for health, well-being, and economic activities, the impact on both people and economies has the potential to disrupt development significantly. The impacts of climate change on the urban water sector can be categorized in two ways: water as a resource and water as a hazard. As a resource, the availability of good-quality water is the basis for the well-being of the ever-increasing number of people living inside cities. It is also critical for many economic activities in and around cities, including peri-urban agriculture, food and beverage industries, and other industrial activities. Meanwhile, excess precipitation or drought can lead to hazards ranging from an increased concentration of pollutants (with negative health consequences), lack of adequate water flow for sewerage, and flood-related damage to physical assets. Huge expected deficits in urban water supply will likely have a major impact on the future availability and cost of water. Decisions taken now by cities will have an important influence on future water supply for industry, domestic use, and agriculture.

In 2011, the First Assessment Report of the Urban Climate Change Research (UCCRN) Network (ARC3.1) published a chapter on "Climate Change, Water and Wastewater" (Major et al., 2011). The chapter offered one of the earlier overviews of climate change impacts on cities, rightly arguing that, as of 2011, the IPCC Assessment Reports and the wider research community had not paid specific attention to cities, water, and climate change. The chapter was an overview of the range of challenges faced by cities in the context of environmental change. Since then, three major trends and milestones have emerged that the present chapter reflects: (1) the growing body of research, particularly in developing countries, on urbanizing watersheds; (2) the increasing numbers of available case studies in smaller and medium-sized cities; and (3) the launch of the IPCC's Working Group II Fifth Assessment Report (AR5) in 2014, with a substantive section on climate change and cities, which concludes that water scarcity is already affecting cities and that reducing the basic service deficit is essential for longer term resilience.

This chapter updates the range of issues faced by cities in the context of climate change initially presented in the first ARC3 report (Rosenzweig et al., 2011). In particular, it focuses on water security – essentially the sustainable availability of water for different uses and the avoidance of water-related disasters – and how expected climate change will complicate the ability of water resource managers to secure water for future urban uses. It also emphasizes improvements in frameworks for adaptation in the water sector, which, to date, have been challenging to communicate and difficult to implement. The chapter starts off by defining water security and outlining the main components of a "water secure" city (Section 14.1.2). This section includes an overview of stormwater and drainage, an issue only touched on in ARC3.1. The range of stakeholders that must be engaged to support adaptive water management is then presented (Section 14.2), followed by an analysis of the main climate risks faced by urban water systems (UWS) (Section 14.3). The subsequent Section (14.4) integrates the IPCC perspective on adaptation in cities, with specificity added by the authors of this chapter through examples of available adaptation options for improved water management. The main argument made is that cities at different scales of development face different climate impacts and will need to determine their own context-specific adaptation pathways because there is no common solution available to all cities. Finally, while the chapter emphasizes adaptation, a section on mitigation (Section 14.5) discusses how to integrate water management with greenhouse gas (GHG) reductions. Throughout the chapter, case studies from a range of cities, both developed and developing, help to connect the "real-world" challenge with the concepts presented.

14.1.1   Water Security as a Framing Concept

Achieving water security in the context of a changing climate remains a central challenge for cities. UN Water defines water security as:

> The capacity of a population to safeguard sustainable access to adequate quantities of acceptable quality water for sustaining livelihoods, human well-being, and socio-economic development, for ensuring protection against water-borne pollution and water-related disasters, and for preserving ecosystems in a climate of peace and political stability. (UN Water, 2013: vi)

In the urban context, water security emphasizes the importance of mediating conflicts between competing options and sequencing of infrastructure investments, acknowledging the complex interplay among industry, human health, and well-being (Braga, 2001; World Bank, 2014).

Considerable progress has been made in recent years in securing water for cities. For example, between 1990 and 2012, a remarkable 2 billion people across Latin America, India, and China were able to improve their access to water and sanitation, largely due to economic growth and improved service delivery. Despite this, 700 million people worldwide remain underserved in regard to water, with half of this number in Sub-Saharan Africa (WHO-UNICEF, 2014). One typical problem is the lack of access to water services (e.g., supply, wastewater collection, and sanitation), which hinders the productive development of cities. In African cities, 40% of the population is still not able to access a safe source of drinking water (UN-Habitat, 2014). Consequently, informal systems often proliferate, particularly in developing countries, as a response to water needs. In the case of water quality (both supply and treatment), lack of access to services and low water quality threaten the health of urban water users and ecosystems located downstream of cities (WHO-UNICEF, 2014).

When referring to UWS, this chapter refers to the infrastructure that cities rely on for managing the availability of good-quality water for different users. Key determinants of a well-functioning system include water for health and well-being, economic productivity, recreational and cultural benefits, and environmental services, as well as a system that is designed with respect to the available operational capacity (Hellestrom et al., 2000). The range of physical infrastructure is broad and involves large capital costs for construction, as well as for operations and maintenance. Typically, this includes water collection and storage facilities at source sites; transport via aqueducts (canals, tunnels, and/or pipelines); water treatment facilities; water treatment, storage, and distribution systems; wastewater collection systems and treatment; and urban drainage works to manage surface runoff.

Water systems have historically evolved sequentially as specific needs were identified and funding was obtained. So, for example, as cities learned and developed better systems for managing wastewater, the management of stormwater also improved. Generally, "mature" cities (i.e., those with a long history and a consolidated urban form) have a reduced rate of urban expansion compared to the rapid growth scenario commonly observed in emerging and less developed countries. In mature cities, there is often established water infrastructure, although service delivery may be inadequate. In the scenario of rapid growth, the pace of expansion often exceeds the ability of municipal authorities to provide basic services. This results in an increased burden on formal and informal water systems in those areas. However, the lack of pre-existing infrastructure presents an opportunity to address challenges not previously considered, such as climate change.

Stormwater management is now acknowledged as a critical component of "climate smart" urban infrastructure (Gill et al., 2007). Many cities are seeing trends of increased annual precipitation (see Chapter 2, Urban Climate Science), as well as heavy rains falling in short time frames during storm events, which lead to water runoff and flooding (see Section 14.3.5). As a result, stormwater management has become a key concern for many cities. In mature cities, this is managed in two forms: (1) combined sewer systems and (2) municipal separate storm sewer systems (MS4). Combined sewer systems are designed to collect rainwater runoff, domestic sewage, and industrial wastewater in the same pipe and deliver it to treatment plants. During heavy precipitation events, the capacity of combined sewers can be exceeded, leading to discharges of untreated wastewater directly to nearby water bodies. In MS4 systems, stormwater and wastewater are managed separately. Therefore, during intense precipitation events, stormwater can be discharged, but not sewage.

Developing cities often lack formal stormwater management systems, with water draining naturally along hydrologic gradients. Each system has its own challenges with respect to maintaining water security. For instance, MS4 systems are better for managing sewage discharge during large storm events than are combined sewers or informal systems. However, with climate variability and longer term change, existing network capacity will almost certainly be exceeded at times, with threats posed by runoff and pollution. Moreover, combined sewer systems are more common in older cities, where infrastructure is already

built and adaptation options are more limited since wholesale retrofitting is very expensive. Areas with no infrastructure do not suffer from this limitation, although the significant upfront costs and a possible lack of human and financial capital to build these systems may impede the improvement of water security.

14.2 Stakeholders of Urban Water Systems

The challenge for planners responsible for adapting urban infrastructure to address future changes in climate is compounded by the range of actors whose engagement is important for an initiative to succeed. A variety of stakeholders are involved in the management of water, wastewater, and sanitation systems at the city level (Manila Water, 2007, 2013) (see Case Study 14.4). Notably, utilities (both public and privately owned) are often responsible for water supply and wastewater management. Municipal governments, with the exception of some countries in Africa where governance remains a challenging issue, are often responsible for the implementation of service delivery and the contracting of such services within a framework provided by national governments. Moreover, elected officials in cities often have different priorities than a city's technical staff. For example, the motivation to be re-elected may drive attention toward populist topics as opposed to the provision of basic services. Community organizations and nongovernmental organizations (NGOs) often play a key role in water security and, as such, need to be considered and consulted in infrastructure planning, particularly when slum neighborhoods are being retrofitted (Satterthwaite and Mitlin, 2013).

Industries are often heavy users of water and rely on secure supply for business continuity. In an increasing number of urban watersheds in upper middle-income countries such as India, Brazil, and Chile, heavy users in industries such as food and beverage processing, mining, and agriculture require a significant quantity of water to maintain operations. For example, the India Infrastructure Report of 2011 estimates that industrial water use in India is about 13% of the country's total freshwater withdrawal and that water demand for industrial uses and energy production will grow at a rate of 4.2% per year, rising from 67 billion cubic meters in 1999 to 228 billion cubic meters by 2025 (Aggarwal and Kumar, 2011). For industry, failure to adapt to changing water conditions due to climate change could lead to unsustainable business practices and reduced profits. Finally, individual consumers are now increasingly recognized as "clients," whose purchasing power can influence water demand considerably, particularly in contexts where effective pricing mechanisms and demand management are deployed (Cornish et al., 2004). Consumers may include, for instance, property owners who have potential to influence local hydrologic conditions (i.e., drainage and runoff) through the design of their private land.

In many developing countries, the range of actors in UWS is even more complex. For example, services are often supplied by informal providers, including delivery of water by truck. These informal systems can be efficient, although they often provide services at an exorbitant cost (Cornish et al., 2004). Equally important is the fact that informal providers are not regulated, nor are they generally willing to engage in formal water management since this would imply potentially losing access to their market. Another problematic approach leading to weak water governance is conflict between national ministries of water and state or city providers. Not only do political affiliations often differ between jurisdictions, there are also considerable overlaps in terms of administrative responsibilities. Thailand, as an extreme example, has more than 50 laws and more than 30 state organizations under 7 ministries working on water management (Nikomborirak and Ruenthip, 2013).

The challenge of managing conflicting stakeholder needs for water often arises during disasters (see Chapter 3, Disasters and Risk). Since 2010, there have been some high-profile failures that suggest a deep need for reform of federal-, state-, and local-level water management systems. For example, a postmortem of the 2011 flood disaster in Thailand noted that the primary problem was governance failure and poor decision-making as the flood was occurring, which resulted in nearly U.S. $47 billion in damages and a commensurate impact on employment (Haraguchi and Lall, 2015). There were many challenges, including water supply canals being overrun and failing sanitation systems, which led to enhanced health risks for people living in Bangkok. Similar failures in administrative planning and response for hydro-meteorological disasters were observed during the Indus floods in Pakistan in 2010 (affecting nearly 20% of the country), the Uttarakand floods in northern India in 2013, and flooding relating to Typhoon Haiyan in the Philippines in 2013. Each flood was the result of an extreme weather event that affected cities located either directly in the path of the storm (e.g., Tacloban in the Philippines) or downstream (e.g., in the case of Thailand, the Indus River, and India). As Section 14.4 elaborates, better governance is a precondition for effective climate-related disaster management.

Finally, there is growing acknowledgment that key private-sector actors (including companies, banks, and investors) have a much larger role to play in supporting climate change adaptation. In fact, there is currently a "missing middle" – a large gap between available financing and the number of viable project ideas – which is a key reason for the current limited investment into adaptation being made by the private sector (Pegels and Pauw, 2013).

14.3 Climate Risks to Urban Water Systems

Current projections for global climate change in the 21st century identify a number of risks that are expected to be particularly challenging for UWS in terms of managing water supply, distribution, waste, and stormwater runoff[1]. These include:
- increasing temperatures (with attendant changes in evaporative demands, availability, and quality)

- changing precipitation regimes
- changing extreme weather regimes
- sea level rise and storm surges
- changing surface-water and groundwater availability and conditions.

14.3.1 Warming Temperatures

The WG1 contribution to the AR5 concludes that, in the next 20 years, global temperatures are likely to increase by at least +0.5°C, but are very unlikely to increase beyond +1.5°C. Regional differences are expected, however, such that warming will proceed more rapidly over land, particularly in continental interiors, and at high latitudes, followed by the tropics and subtropical lands (Kirtman et al., 2014). Unless significant reductions in GHG emissions occur by the end of century, the global temperature will likely have increased by between +1.5°C and +4.8°C (depending on emissions levels), with high-latitude warming continuing to be more rapid than the global mean (Collins et al., 2014).

Warmer temperatures result in larger demands for water in many cities (Schleich and Hillenbrand, 2009), particularly for household consumption and thermoelectric cooling. The extent of this temperature sensitivity, however, depends considerably upon climate, land uses, and energy dependency within cities (Zhou et al., 2000; Ruth et al., 2007; O'Hara and Georgakakos, 2008; House-Peters and Chang, 2011; Almutaz et al., 2012; Breyer, 2014; Donkor et al., 2014; Stoker and Rothfeder, 2014) (see Chapter 2, Urban Climate Science). For example, climate change may have a larger impact in cities that are reliant on older, less water-efficient, and coal-based thermoelectric plants than in cities that rely on newer, more water-efficient natural gas combined-cycle thermoelectric plants (Scanlon et al., 2013 a, 2013b).

Half or more of residential water is used outside of the home, mostly for irrigation (Hof and Wolf, 2014; Mini et al., 2014). This does, however, vary dramatically within and among cities, depending on the density of urban development and on climatic conditions (e.g., Lwasa et al., 2014). Extreme heat waves, which are projected to become more common and severe in the future, typically lead to disproportionately elevated demands for water (Zhou et al., 2001; Breyer, 2014). Such water demands are likely to be among the most affected by increasing temperatures (see Chapter 2, Urban Climate Science).

Warmer air temperatures and more extreme temperature ranges may physically damage UWS structures (Institution of Mechanical Engineers [IME], 2014), especially in high latitudes (Boyle et al., 2013), including areas with thawing permafrost or where warming results in soil desiccation (Vardon, 2014). Engineering materials and structures are most vulnerable to climate change in extreme cases of wet and dry conditions, high and low humidity, and solar radiation (Valdez et al., 2010).

---

1 A comprehensive list of potential impacts of climate change on UWS, from warming temperatures as well as from flooding and sea level rise, is available at: http://water.wuk1 .emsystem.co.uk/home/policy/publications/archive/industry-guidance/asset-management-planning. Accessed December 29, 2014.

The overall effects of warming on the chemistry, biology, and contamination level in water supply and wastewater systems will likely be unique to each location. Warmer water temperatures can change the solubility and transport of contaminants as well as promote algal and other biological outbreaks (including invasive species) in water supply and wastewater/sanitation systems (Whitehead et al., 2009; Cisneros et al., 2014). However, warmer temperatures are generally conducive to improved biological reactions in water and wastewater treatment processes (Tchobanoglous et al., 2003; Whitehead et al., 2009). Some seasons or situations (e.g., extreme weather or runoff) may be more impactful on wastewater treatment than long-term average warming. For example, Plosz et al. (2009) showed in a Norwegian case study that changes in temperature during the winter may impact treatment processes more than changes in annual mean temperatures while also taking into account snowmelt (Langeveld et al., 2013).

Warmer temperatures can affect water use for industry and power generation that are either located in or serve urban systems (e.g., Koch and Vogele, 2009; Rebetez et al., 2009; Linnerud et al., 2011; Golombeck et al., 2012). For instance, high temperatures can decrease the efficiency of cooling for thermoelectric power generation, which is the largest or second largest (behind irrigation) use of water in many developed countries. As a result, decreasing efficiency can have a dramatic effect on electricity generation in urban areas (Kenny et al., 2009, Dell et al., 2014).

14.3.2 Changing Precipitation Regimes

IPCC's AR5 (2014) concludes that, at the largest of geographical scales, precipitation will very likely increase at high latitudes and, more likely than not, decline in the subtropics. The general rule that "the wet will get wetter, the dry will get drier, and the variable will get more variable" is projected to hold as warming progresses (Kirtman et al., 2014; Polade et al., 2014). However, most of these trends are small relative to the large natural variability in precipitation at regional scales, at least in the near term. Over continents, the mean projected change in precipitation in the 21st century is less than about 20% of historical totals and is in fact much less in many regions. There are some exceptions where precipitation is expected to rise by as much as 40–50%, including in Eurasia (reaching from Scandinavia to northern China) and around the Horn of Africa (Collins et al., 2014) (see Chapter 2, Urban Climate Science).

Mediterranean climates worldwide are projected to experience significant declines in precipitation (Polade et al., 2014; Seager et al., 2014). Drought, as measured by various indices, will become a normal state in many mid-latitude areas (including the Middle East, Central America, and Brazil) by the end of the century, unless GHG emissions are significantly reduced (Sillmann et al., 2013; Seneviratne et al., 2012; Collins et al., 2014; Polade et al., 2014). In Asia, increases in precipitation are very likely to be experienced at higher latitudes by mid-century, and for southern and eastern Asia by the end of the century (Hijioka et al., 2014).

In addition to changes in the quantity of precipitation, seasonal timing and the form of precipitation (e.g., snow vs. rain, heavy vs. light) are expected to change in response to atmospheric warming. In general, because of higher temperatures, precipitation is expected to fall increasingly as rain rather than snow (Berghuijs et al., 2014). Also, in some regions, the contrast between wet and dry seasons will increase (IPCC, 2013). Such changes in the timing and form of precipitation impacts the balance between the management of water supply and flood risk, with more runoff entering cities and water storage structures during winter seasons. Consequently, this results in more runoff taking the form of floods rather than steadier, more reliable, and manageable flows (e.g., Vicuña et al., 2013).

The nature of UWS vulnerability to climate change depends intrinsically on how precipitation is expected to change, and, unfortunately, vulnerabilities exist at both ends of the spectrum (Revi et al., 2014, Rosenzweig et al., 2010). In regions where precipitation increases, UWS can be threatened by insufficient conveyance systems leading to urban flooding and to combined sewer overflows that can contaminate water supply. This could also be the case in regions where precipitation decreases, since the overall frequency and magnitude of large storms, as well as dry spells, could increase (Polade et al., 2014). This could exacerbate the vulnerability associated with reductions in water supply that comes with a precipitation decrease.

UWS depends on water availability (McDonald et al., 2011; Cisneros et al., 2014; Revi et al., 2014), which is dependent on the nature, reliability, and diversity of sources that a city draws from. Changes in precipitation affect urban water demand (especially for exterior uses) (Ruth et al., 2007; Schleich and Hillenbrand, 2009; House-Peters and Chang, 2011), as well as water availability. For instance, drought often results in increased groundwater withdrawals (Konikow, 2013; Villholth et al., 2013), which have a much slower recovery rate than surface water sources. Even deep groundwater supplies can be vulnerable to precipitation changes over the long term (Cisneros et al., 2014; Georgakakos et al., 2014), which is further compounded by poor aquifer management (e.g., Suárez et al., 2014). This is an important concern because many cities depend on groundwater. It has been estimated that, in 2000, some 1.7 billion people were living in areas with threatened groundwater supply (Gleeson et al., 2012). The more overdrawn or generally threatened such supplies are, the more likely and quickly climate change impacts will become evident (Taylor et al., 2012). In terms of impacts on UWS, those systems that rely on local surface water supply may be most immediately at risk to changing precipitation regimes, followed by systems fed by more geographically diverse water sources, and, last, by systems that depend on shallow and then deep aquifer systems (O'Hara and Georgakakos, 2008; Crosbie et al., 2010; Newcomer et al., 2013).

14.3.3 Extreme Events

The WG1 contribution to the AR5 continues to support the long-standing expectation that heavy precipitation events will increase globally in both the near and long term (e.g., Min et al., 2011; Seneviratne et al., 2012). However, significant regional variation is expected, and extreme storms are not well represented in many climate models. Kirtman et al. (2014) conclude that the frequency and intensity of heavy precipitation will likely increase over many land areas, but may be masked by natural variability and other anthropogenic influences (e.g., deforestation) in the near term. Extreme precipitation events (e.g., with return periods >20 years) are projected to increase in frequency by 10–20% by the end of the 21st century for most mid-latitude land areas and even more so over wet tropical regions (Kharin et al., 2013).

More extreme precipitation could result in changes in frequency, extent, timing, and rapidity of stormwater runoff. This could cause flooding in many urban settings, especially given the impervious surfaces of most cities. Furthermore, this could pose added risks to public health and safety, property, and infrastructure (including UWS). Water quality could be affected by these extreme runoff events due to the increased concentration and build-up of contaminants during dry or low-flow conditions that are then released into the water supply with increased water flow (Langeveld et al., 2013). Many UWS infrastructural hubs are located in low-lying areas (e.g., California) that are more susceptible to flood damage (Porter et al., 2011). UWS infrastructure can also be at risk from wind damage and other non-precipitation storm effects, especially in the case where such infrastructure is either aging or exposed.

14.3.4   Sea Level Rise and Storm Surges

Current projections of likely global sea level rise (SLR) by the end of 21st century range from 0.5 to 1.2 meters, depending on whether aggressive climate mitigation measures are implemented in the coming decades and on the fates of polar ice caps (Horton et al., 2014, Church et al., 2014) (see Chapter 2, Urban Climate Science). SLR will vary around the globe from city to city, reflecting local to regional differences in plate tectonics, land subsidence (natural and anthropogenic), and long-term circulation and salinity variations across ocean basins (Church et al., 2014). For example, ensemble means suggest SLR will be higher in regions such as the central coast of Asia and the northeast coast of the United States, compared with lower increases off the southwest coast of South America (Church et al., 2014) (see Chapter 9, Coastal Zones).

In most settings, long-term sea level trends will have most immediate impacts by elevating the baseline upon which shorter term extreme sea level fluctuations (e.g., inter-annual fluctuations, wind-driven waves, storm surges, and even astronomical tides) will be superimposed (Lowe et al., 2010; Hunter, 2011; Obeysekera and Park, 2012). This may greatly increase the frequency and inland reach of otherwise "normal" high-water stands. (More details on sea level trends and impacts can be found in Chapter 9, Coastal Zones.)

Large portions of the world's urban populations (i.e., 13 of the 20 most populous cities, Hanson et al., 2011) and economies

(>70–80% of world trade) (Hanson and Nicholls 2012) are located in coastal settings. In many such cities, wastewater and sanitation systems have important hubs (e.g., treatment plants and outfalls) located at or very near sea level to take advantage of the gravity-feed and marine-outfall options (e.g., Jacob et al., 2007; Aerts et al., 2013). These hubs and systems will be among the infrastructure that is most immediately at risk by SLR and/or increased storm surge conditions (Revi et al., 2014).

Some important water supply conveyance systems also have links that are near to sea level and thus are at risk to structural disruptions or water-quality impacts from SLR and storm surges. On a large scale, California's Delta is a key example (Hanak and Lund, 2012), and Hurricane Sandy uncovered others in the eastern United States (Manuel, 2013). Coastal cities that depend on local groundwater sources (e.g., coastal aquifers) for water supply will, in many cases, face risks of increased seawater intrusion into freshwater aquifers (Revi et al., 2014, Cisneros et al., 2014). However, Ferguson and Gleeson (2012) concluded that the direct impact of groundwater extraction in the United States has been and will be much more significant than the impact of SLR by the end of the 21st century. The IPCC WGII concluded that human-induced extractions from coastal aquifers will continue to be the main driver for aquifer salinization during the next century, with changing precipitation, increased storm frequency, and SLR further exacerbating these problems (Wong et al., 2014).

Similarly, when assessing impacts of SLR, it is important to take into account impacts of non-climate drivers such as subsidence. For example, Higgins et al. (2013) present research that land subsidence rates exceed local and global average SLR by nearly 2 orders of magnitude in the Yellow River Delta in China. Similar findings have been presented from the Nile River, where subsidence caused in part by the Aswan Dam is exacerbating SLR because sediment trapped in upstream reservoirs is reducing aggradation in deltas (Syvitski et al., 2009).

14.3.5   Changing Water Availability

Climatic pressures will interact at different spatial scales to have a synergistic impact on water availability, which depends not only on the amount of water at different sources, but also on water quality, infrastructural integrity, arrangements among competing users, and strength of institutions, as discussed in the next section. Figure 14.1 presents a conceptual depiction of these possible connections, some of which are further described later.

In response to warming, evaporation is projected to increase over most land surfaces globally, except primarily southern Africa and Australia, where declines in soil moisture availability are sufficient to reduce overall evaporation (Collins et al., 2014). Net water deficits (i.e., evapotranspiration minus precipitation) are projected to increase over most subtropical to mid-latitude lands, and decline over the higher latitudes, with precipitation increases compensating for increases in evaporation caused by warming temperatures (Kirtman et al., 2014). Projections of net deficits in the tropics are mixed. In most regions where net deficits

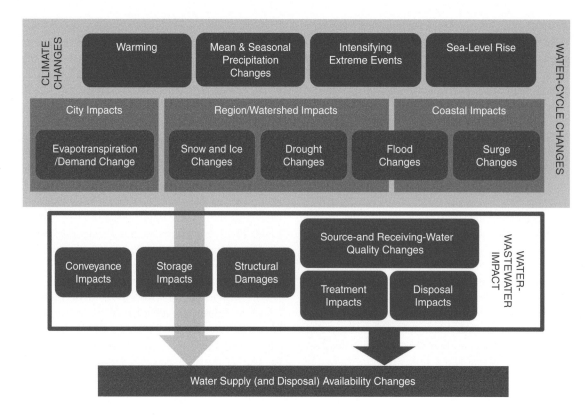

Figure 14.1 *Climate drivers impacting water availability.*

increase, runoff and recharge may be expected to decline, such that water availability is likely to suffer. Even in regions where net deficits decline somewhat, the amount of runoff and recharge derived from each unit of precipitation will likely decline due to enhanced evapotranspiration (e.g., Das et al., 2011; Georgakakos et al., 2014).

Warming-induced declines in snowpack, glaciers, and less seasonally persistent snowpack (e.g., earlier snowmelt) are expected to change the timing of water availability for 70% of major rivers and for water supplies around the world that depend on mountain-based seasonal snows as their source (Vuviroli and Weingartner, 2008). This alters the natural storage of water from cooler seasons with low water demands to warm seasons when demands are commonly highest. This shift in the seasonal timing of water availability is expected to challenge water management systems in many parts of the world (Barnett et al., 2005; Oberts, 2007; Kenney et al., 2008; Wiley and Palmer, 2008; Meza et al., 2014; Buytaert and De Bièvre, 2012).

Another potential impact on water supply has to do with climate change affecting erosion and turbidity levels, thus inhibiting water extraction from natural sources. For example, Mukundan et al. (2013) showed the negative effects of future climate on soil erosion and sediment yield, which are affecting a watershed that supplies water to New York. Recent high-turbidity events have occurred in the city of Santiago, Chile, that could be associated with the rising snow line, which leads to more silt in runoff,

combined with intense precipitation events. Similarly, changes in water discharge could also alter critical water conditions, such as temperature, that could in turn affect water availability (van Vliet et al., 2013).

Together, these processes and connections will likely affect water availability in ways not previously experienced (Cisneros et al., 2014). McDonald et al. (2011) illustrate (see Figure 14.2) a projected set of current and future water availability challenges in large cities around the developing world. They found that many cities will have less water available under several of the climate change and land-use change scenarios that they studied. In India, for instance, an analysis of twelve major basins found deficits in the range of 38 billion cubic meters, the result of both climate and anthropogenic drivers (Gosain et al., 2006). These challenges in turn have the potential to threaten water supply, demand, and quality in many urban settings around the world, along with wastewater, stormwater, and sanitation systems.

Cities very often draw from water sources in areas located much further away from their water supply (McDonald et al., 2014). Therefore, urban water supply is very much dependent on climatic changes in surrounding areas in addition to climate pressures on supplies located within cities. Different water uses and users in urban settings have different water supply and wastewater/sanitation requirements. As Figure 14.1 illustrates, climatic risks for UWS are a complex set of considerations that are not

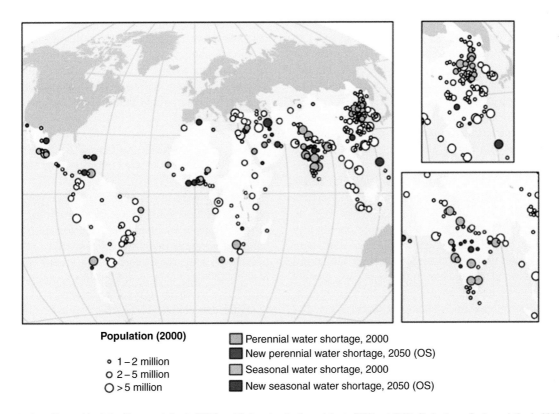

Population (2000)

- ∘ 1–2 million
- ⊙ 2–5 million
- ○ >5 million

- ▣ Perennial water shortage, 2000
- ■ New perennial water shortage, 2050 (OS)
- ▢ Seasonal water shortage, 2000
- ■ New seasonal water shortage, 2050 (OS)

Figure 14.2 *Distribution of large cities (>1 million population in 2000) and their water shortage status in 2000 and 2050. Circle sizes reflect population in 2000; colors indicate statuses. Gray areas are outside the study area.*

Source: McDonald et al., 2011

necessarily related to each other. This creates a deep uncertainty in determining risks to UWS. Increasingly, the most appropriate response strategy is, to paraphrase Kennel (2009), "mitigate globally, assess regionally, and adapt and prepare locally."

14.4 Adaptation Strategies for Urban Water Systems

Since the ARC3.1 report was published in 2011, there has been a shift in the adaptation research community from analysis of impacts to an implementation/practitioner approach.[2] Adaptation is now widely acknowledged more as an iterative process than as an end in and of itself (Hinkel and Bisaro, 2016). Several key questions underpin the fundamental challenges of adaptation and are represented in Figure 13.3.

14.4.1 Identifying Adaptation Needs: Why Does the Urban Water Sector Need to Adapt?

Adaptation should either preserve or ideally improve water security within a city, without putting in peril water availability and quality for other uses within the basin. Understanding

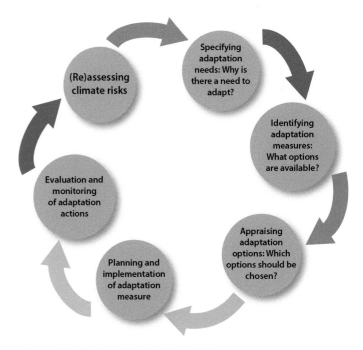

Figure 14.3 *Planning and implementing adaptation measures.*

Source: Adapted from PROVIA, 2013

---

2 For example, the emphasis on solutions is captured by the report *PROVIA Guidance on Assessing Vulnerability, Impacts and Adaptation to Climate Change* (PROVIA, 2013).

climate impacts is critical, but this must be complemented by a parallel process of identifying adaptation needs in order to prevent, moderate, or adjust to these impacts. This "vulnerability-led approach" starts with diagnosing existing problems and then identifying additional climate risks that can be addressed in parallel with existing problems, such as poverty and exposure to risks (Lafontaine, 2012). This assessment is critical because, in most cases, the barriers for implementing climate change adaptation are not due to a lack of knowledge about the future impacts of climate change, but rather the result of cognitive and institutional barriers to taking action toward long-term future change (Adger et al., 2009; Moser and Ekstrom, 2010). Barriers include the challenge of assessing future needs over a 30- to 50-year time horizon, short political windows of opportunity (often 4–5 years) in which to act, the immediacy of short-term challenges (i.e., health problems caused by lack of sanitation) that trump longer term plans, and making decisions in a context of uncertainty. In this regard, there are significant capacity differences between developed and developing world cities. For example, institutional flexibility, clear regulatory frameworks,

enforcement capacity, and water price-setting are all elements that strengthen an institution's ability to respond effectively to climate challenges. In developing countries, where poverty and weak institutions play a significant role in shaping the magnitude of risks and opportunities, capacity analysis could be even more important than impact assessments (Hinkel and Bisaro, 2016).

14.4.2 How to Address Climate Risks? Adaptation Options for the Urban Water Sector

Since the publication of the ARC3.1 report in 2011, a variety of adaptation options have been promoted that are not only sensible ways of adapting to climate change, but are also sound options for sustainable development. Table 14.1 provides a list of adaptation measures, categorized according to the two components of water security (i.e., water as a resource or as a hazard). This section presents several strategies that are increasingly recognized as fundamental parts of a sensible adaptation strategy for urban water security.

Table 14.1 *Urban water system adaptation measures.*

Objective (Parameter of water security)	Primary Climate Risk(s)	Adaptation Strategy	Specific Options
Water as a resource: Sustainable access to adequate quantities of acceptable quality water for sustaining livelihoods, health, well-being and economic development.	• Precipitation reduction, glacier retreat, land erosion, or sea level rise leads to a reduction in water availability and/or worsening of water quality. • Ecosystems are threatened by excess stress on water resources caused by both climate factors and poor management.	(1) Ensure adequate quantities to sustain livelihoods and ecosystems.	Reuse of wastewater Groundwater use/recharge Distribution efficiency improvements Groundwater Transfer from other sectors Desalination Green infrastructure Reservoirs/Increase storage capacity Point source separation in new construction
		(2) Reconsider "adequate" and identify different water needs (both quantity and quality) for different uses.	Demand management through tariffs (user fees) or other demand management options Cultural changes Standards Restrictions Incentives
Water as a hazard: Ensuring protection against water-borne pollution and water-related disasters.	• A reduction in water flows could lead to an increased concentration in pollutants.	(3) Ensure that there is adequate quantity and flow to dilute pollution.	Similar Options as in Adaptation Strategy (1) Restrictions and flow control
		(4) Reduce vulnerability to pollution of marginal communities.	Water quality standards Water treatment
	• Increase in precipitation intensity, storms, storm surges could increase threat associated with floods.	(5) Reduce the exposure of people and infrastructure to floods/related disasters.	Riparian buffering Increase in percolation Fluvial flood protection Green infrastructure Coastal set-back lines Land use regulations Relocation
		(6) Reduce vulnerability to flooding of marginal communities.	Adaptive planning Housing improvements/modified building codes

Case Study 14.1   Climate Adaptation through Sustainable Urban Water Development in Can Tho City, Vietnam

Minh Nguyen, Steven Cook, Magnus Moglia, Luis Neumann, and Xiaoming Wang

Commonwealth Scientific and Industrial Research Organisation (CSIRO), Melbourne

Nguyen Hieu Trung

College of Environment and Natural Resources, Can Tho University

Keywords	Sustainable development, climate adaptation, integrated urban water management, strategic planning, pilot demonstration
Population (Metropolitan Region)	1,200,000 (CanTho Portal, 2015)
Area (Metropolitan Region)	1,400 km² (Statistical Office of Can Tho City, 2009)
Income per capita	US$2,050 (World Bank, 2017)
Climate zone	Aw – Tropical, Savannah (Peel et al., 2007)

The Climate Adaptation through Sustainable Urban Development Project was a research initiative aimed at bringing sustainable principles into practice as an effective means of adapting to climate change. The project demonstrated a sustainable development framework to improve the planning of urban water services to enhance the resilience to climate change of local communities and the government of Can Tho City of Vietnam, as a case study. The project was undertaken in collaboration by CSIRO Australia, Can Tho University, and Can Tho Climate Change Coordination Office.

Can Tho is the central city of the Mekong Delta of Vietnam, with an approximate area of 1,400 square kilometers and population of 1.2 million (see Case Study 14.1 Figure 1). The region has very low-lying and flat terrain, with a dense network of waterways. Waterways are central to people's livelihoods and underpin the local economy, which is still based on agriculture and aquaculture. However, there is a rapid transition of the city toward regional and industrial services. The city has very mixed land uses, where old urban, new urban, peri-urban, industrial, and rural areas co-exist even within urbanized districts.

The sustainability of the water systems in Can Tho City, including both its physical infrastructure and its ecology, is under pressure from rapid urbanization and industrialization. Results from the project's large survey of 1,200 households and a comprehensive

Case Study 14.1 Figure 1 *Can Tho City, Vietnam.*

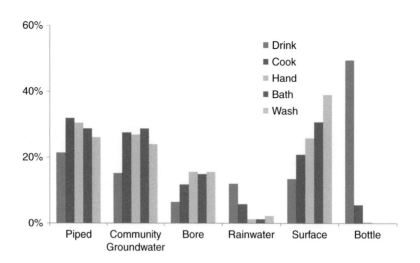

Case Study 14.1 Figure 2 *A result from the 1,200 household survey: Mixed uses of water for different purposes in peri-urban areas.*

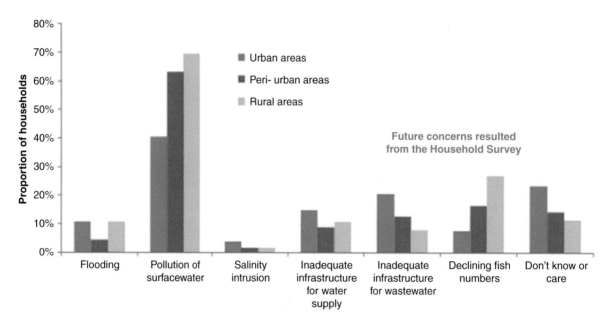

Case Study 14.1 Figure 3 *A result from the 1,200 household survey: Future concerns regarding water systems and environment.*

sector review (Neumann et al., 2011, 2013) indicated that water service provisions were greatly fragmented among urban and rural areas. Many peri-urban households draw water from multiple sources and use it for different purposes (see Case Study 14.1 Figure 2), presenting adverse implications for their health. Inadequate physical infrastructure is the main issue, resulting in limited access to clean water and sanitation, frequent urban floods, and increasing pollution in waterways, which was the utmost public concern (see Case Study 14.1 Figure 3). These issues are exacerbated by the impacts of a changing climate, such as more prolonged and frequent drought, heat waves, and inundations, which have already been experienced in recent decades.

INTEGRATION OF STRATEGIC PLANNING AND DEMONSTRATION

To deal with such a complex system of water and sanitation services and the environment, the research team applied the integrated urban water management (IUWM) principles (Maheepala et al., 2010). This provides state-of-the-art integrated assessment methodologies and participatory processes to assist in the strategic planning of urban water systems that are sustainable for specific conditions. IUWM aims to plan, design and manage the overall water cycle, including water supply, stormwater, and wastewater in a coordinated manner. This helps to minimize the impacts on the environment, maximize contributions to economic

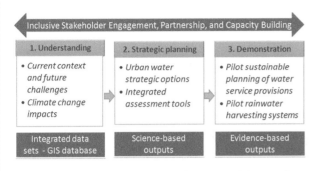

Case Study 14.1 Figure 4 *Summary of the case study's approach with three focus areas and results.*

development, engender overall community well-being, and enhance the resilience capacity of local communities to future challenges, including climate change. The case study includes three focus areas, as summarized in Case Study 14.1 Figure 4 (Nguyen et al., 2012).[3]

A set of urban water strategic adaptation options was developed for the city water systems. Through an inclusive participatory engagement process, the stakeholders were positioned to play the main role in identifying the strategies, developing implementation pathways, and assessing factors influencing their likelihood of success (see

Case Study 14.1 Figure 5). The strategies identified ranged widely from social behavior changes and household measures to infrastructure planning and covered all water supply, demand, and sewerage aspects. This process challenged stakeholders' preconceived notions, allowed them to develop integrated systems thinking, and highlighted the responsibilities of individuals or agencies in implementation (Moglia et al., 2013).

Pilot demonstrations of the identified strategic options were conducted to provide the stakeholders local evidence-based examples of sustainable development practices. An integrated conceptual design and planning exercise was undertaken to assess the sustainability of different integrated water supply and wastewater servicing options in a peri-urban area of 150 houses in Chau Van Liem ward, O Mon District, using a state-of-the-art sustainability assessment approach (Sharma et al., 2010). The purpose was to build the capacity of the local research partners and planning agencies to identify suitable and sustainable integrated water service options that consider cost-effective solutions, environmental impacts, service levels, community expectations, and suitable institutional arrangements to manage the system (Nguyen et al., 2012). The project also included another demonstration with two pilot systems of rainwater harvesting to evaluate rainwater quality and the use of rainfall to augment water supply with simple filtering techniques (see Case Study 14.1 Figure 6). A *Rainwater Harvesting Guidebook* was published to assist the public and agencies in appropriately harvesting, filtering, and

(a)

(b)

Case Study 14.1 Figure 5 *Structured workshops providing opportunities for local stakeholders to contribute, lead, and develop ownership of the outcomes.*

---

3   The first focus area of the study (Figure 4) focused on understanding the local context through a number of activities, mainly for data collection, which can be divided into two approaches:

(1) Top-down activities include

   a. A participatory workshop where stakeholders played the key role in defining key issues, identifying hotspots, and identifying data types and possible sources for the issues related to water systems and environment. In this way, the project set up an understanding and common language and obtained some commitment of providing data from the stakeholders.

   b. Collection of the identified data (and beyond) from governmental agencies (e.g., Can Tho DONRE, 2008). This was done/coordinated by the local research partners.

   c. CSIRO conducted a comprehensive water sector literature review that sought to understand the institutional context, identify data sources and data gaps, and crystallize the critical dilemmas facing the city (Neumann et al., 2011).

(2) Bottom-up activities include

   a. A survey of 1,200 households on issues of water access, water quality, groundwater, and flooding (Neumann et al., 2013)

   b. Two similar surveys of a few hundred of households targeted at specific areas or "hot-spots" for demonstration of planning and design

   c. Other informal surveys, meetings, and interviews with local people, managers from different levels of government agencies, and experts, including researchers and academics.

Case Study 14.1 Figure 6 *Demonstration of rainwater harvesting systems for safe water supply: One pilot system on Can Tho University campus for research and testing water quality, and one at a peri-urban household in Le Binh ward, Cai Rang District, for reality check.*

maintaining rainwater fit for purpose to ensure safe domestic uses (Trung et al., 2004a).

LESSONS LEARNED

The project found that the water system in Can Tho City is highly fragmented and that a "one size fits all" solution would not be adequate to address all the issues for different areas of the city. However, the framework developed in this Case Study can be applied to other locations that are facing the same fragmentation problem, given the following two important considerations are met.

First, there is a need for strong inclusive engagement with local partners and stakeholders, not only through workshops and interviews, but also through delivering tangible outputs with pilot demonstrations that show directly relevant benefits to the stakeholders. This ensures that the partners and stakeholders have a buy-in and an ownership in the outputs, and it also ensures that the developed solutions are locally suitable and acceptable. This project developed a strong collaborative relationship between the research partners and key relevant stakeholders. Many tangible outputs, including a City Water Atlas (Mapbook) and Web-GIS (Trung et al., 2013, 2014b) served effectively as a collaborative platform among stakeholders

in planning and management. The *Rainwater Harvesting Guidebook* (Trung et al., 2014a) and the Synthesis Report with recommendations to the city (Nguyen et al., 2012) have been adopted by various city agencies and nongovernmental organizations. As a result, the research was underpinned by strong partnerships and the active involvement of local research partners and governmental agencies; it received the 2012 Can Tho City Award for its contribution to the City's Adaptation Plan.

Second, there is a need for capacity-building for the local institutions, to allow them not only to effectively participate, but also to maintain any knowledge and technology developed in the project. Local capacity-building also helps improve institutional capacity for planning and maintaining the water systems. This project had a strong focus on capacity-building, which included consistent hands-on training during the project for researchers from Can Tho University. They have become active trainers for many training workshops for local stakeholders on the project outcomes. This "train the trainers" program using a practical hands-on approach (Nguyen et al., 2012) resulted in the leading role of Can Tho University in delivering many final outputs of the project and bringing the knowledge to students, the next generation of experts for the city and the region.

14.4.2.1 Water Supply and Access

Water supply options work both within and beyond city limits. Water storage infrastructure, water transfers, and new sources of water supply (e.g., desalinization) are all examples of options that are implemented beyond city limits (Vicuña et al., 2014a). Available options are very much dependent on the relative position of cities within a river basin, as discussed in Case Study 14.2, which presents a basin-level approach for addressing urban water supply in Santiago, Los Angeles, and Bangalore. The relative position of a city within a basin determines the type of water supply options available, as well as water-pollution issues, with consideration for other users located either up- or downstream of the city.

14.4.2.2 Water Demand Management

A core principle for adapting and securing urban water supply to future climate change is to ensure that demand is managed well. Water demand management aims to reduce water use and limit the need for new sources of water supply without affecting well-being or economic productivity. Common tools used to manage demand include introducing pricing structures where excessive water users pay higher rates, wastewater reuse, low-flow household plumbing and appliances, drip irrigation, and social marketing to inform consumers about the environmental impacts of their daily practices (Breyer and Chang, 2014). Critical to ensuring adequate quantities of water for different purposes is understanding end-use consumption patterns in order

Case Study 14.2   Using a Basin-Level Approach to Address Climate Change Adaptation of Urban Water Supply: The Case of Santiago, Los Angeles, and Bangalore

Sebastian Vicuña

Centre on Global Change, Pontifical Catholic University of Chile, Santiago

Priyanka Jamwal

Ashoka Trust for Research in Ecology and the Environment, Bangalore

Keywords	River basins, adaptation, water supply, demand management
Population (Metropolitan Region)	Santiago: 7,007,600 (Instituto Nacional de Estadistica [INE], 2012) Los Angeles: 12,828,837 (U.S. Census Bureau, 2010) Bangalore: 9,621,551 (Census Population Data, 2011)
Area (Metropolitan Region)	Santiago: 15,403.2 km² (INE 2012) Los Angeles: 12,557.43 km² (U.S. Census, 2010) Bangalore: 2,196 km² (Census Population Data, 2011
Income per capita	Chile: US$23,270 United States: US$56,180 India: US$1,680 (World Bank, 2017)
Climate zone	Santiago: Csb – Temperate, dry summer, warm summer Los Angeles: Csa – Temperate, dry summer, hot summer Bangalore: Aw – Tropical savannah (Peel et al., 2007)

Cities have a number of different adaptation options at their disposal for addressing shortages in water supply linked to climate change or climate variability. At the city level, these options can be considered through two lenses: (1) by considering the city as an isolated system with limited inputs and (2) by taking into account water supply options that fall within the city boundaries. For the former, measures could include reducing total water consumption, either through improved efficiency around water use or improving the distribution system (see Case Study 14.2 Table 1). For the latter, they may include groundwater extraction or wastewater reuse.

An additional influence on which options are available to secure supply in a changing climate depends largely on the position of a city in the river basin, for example, whether the city is located near the headwater or the outlet of the basin. In a situation where a city is facing a reduction in surface water availability, a potential adaptation measure available to both "headwater cities" and "outlet cities" could be constructing a reservoir to be drawn on in times of need. Other measures adopted by these cities would differ due to their relative position in the basin. For instance, a "headwater city" may have only one option: reduce water use by neighboring users located upstream (assuming that the costs are high to pump water upland and over great lengths). An "outlet city" located in this same basin, however, has additional options including:

- Improving water use efficiency in the agriculture sector, including for irrigated areas located in the lower portions of the basin
- Increasing water flow via an interbasin transfer or by drawing from a tributary river
- Sourcing water from the ocean through the use of a desalinization plant

To highlight this point further, the cities of Santiago, Chile; Los Angeles, United States; and Bangalore, India are compared here.

Case Study 14.2 Table 1 *Comparison of Santiago, Los Angeles, and Bangalore.*

Variables	Cities		
	Santiago, Chile	**Los Angeles, USA**	**Bangalore, India**
Area (metropolitan region) km²	15,403.2 (INE, 2012)	12,557.43 (U.S. Census Bureau, 2010)	2,196 (Census Population Data, 2011
Population (metropolitan region)	7,007,600 (INE, 2012)	12,828,837 (U.S. Census Bureau, 2010)	9,621,551 (Census Population Data, 2011)
Density (city, metro area) (inhabitants/km²)	8,464	2,729	11,905
Latitude and longitude	33.45° S, 70.67° W	34.05° N, 118.25° W	12.97° N, 77.56° E
Climate zone (Köppen-Geiger Climate Zones)	Temperate, dry summer, warm summer (Csb)	Temperate, dry summer, hot summer (Csa)	Tropical savannah (Aw)
Human Development Index	[0.52–1]	>0.9	0.753 in 2001 (for Bangalore Urban Dist.)
Water supply options available (1)	A; S	A; S; R; ST; T	A; S
Water management strategies (2)	T in summer months; A	C; A	Informal water supply system (Tanker Market), A

Case Study 14.2 Table 1 (*continued*)

Variables	Cities		
	Santiago, Chile	**Los Angeles, USA**	**Bangalore, India**
Relative location	City located at Andes Mountains foothills	City located in Southern California coast	City located on Deccan plateau
Shares water with (3)	A	A	A
Users upstream (3)	H	H, A, RE, C	A
Users downstream (3)	A, RE, C	None	A
Potential climate change impacts (4)	W, D, T, F	W, D, T, F, SI	W, F
References	Meza et al. (2014)	http://www.ladpw.org/wmd/irwmp/	Lele et al. (2014)

(1) A, Aquifer; S, Surface; R, Reuse; D, Desalination; ST, Storm runoff capture; T, Transfers

(2) T, Tarif controls; C, Conservation measures; L, Lawn irrigation restrictions; A, Awareness campaigns

(3) H, Hydropower; A, Agriculture; RE, Recreation/Environment; C, Cities

(4) W, Reduction in water supply; D, Increase in water demand; T, Increase in high turbidity events; SI, Salt water intrusion/Storm surge; F, Increase flooding

SANTIAGO DE CHILE

Santiago is the largest city in Chile, home to nearly 7 million people and producing nearly 40% of the nation's total gross domestic product (GDP). It is located in a semi-arid, Mediterranean climate at the foothills of the Andes Mountains. The city of Santiago relies on the Maipo River, which runs from the Andes Mountains to the Pacific Ocean, for 80% of its water needs. The remainder is derived from groundwater extraction. To address seasonal and interannual variability, the city operates a 200 million cubic meter reservoir located in the Andes mountains. In addition to pressure from growth in population and industry (Puertas et al., 2014), Santiago faces potential impacts from climate change, including a projected increase in temperature (+1.5 to +3.5 C by the end of the century) and reduction in precipitation (–10 to –40% by the end of the century), all of which is expected to further hinder water availability (Meza et al., 2014). Depending on the global circulation model (GCM) projection, the greenhouse gas (GHG) emission scenario, and the time horizon being considered, the Maipo River may face a reduction in total discharge of 10–40%, with river runoff peaking 1–4 weeks earlier than it does currently (Meza et al., 2014). Similar runoff projections are expected in other snowmelt-dominated basins in central Chile (Vicuña et al., 2010). A series of adaptation options were studied by Bonelli et al. (2014) in order to address this situation and to begin planning appropriately. One of the options considered is a reduction of water distribution losses that are currently near 30% of surface water extractions. When applying a basin perspective to urban water supply, the main option available is to increase the portion of water rights held by the city in relation to the agriculture sector. According to Bonelli et al. (2014), the share of water rights should increase from the current 24% to at least 40% by 2050 in order to cope with climate change impacts and population growth.

LOS ANGELES

The Greater Los Angeles region (including Los Angeles, Orange, San Bernardino, Riverside, and Ventura Counties) is located on the coast of southern California. This region is home to more than 14 million people and supports more than US$700 billion in economic activity (2010). Similar to Santiago, the climate in the Greater LA region is Mediterranean and semi-arid. However, the region has a more diverse portfolio in terms of water supply, in part due to its location at the outlet of a basin or system of basins. According to the Integrated Regional Water Management Plan for the Greater Los Angeles County (GLAC) – which covers the Greater Los Angeles region with the exception of Riverside County[4] – 57% of the region's water is imported from three different regions: (1) the Sacramento River in Northern California through the State Water Plan system of aqueducts, (2) the Colorado River through the Colorado River Aqueduct, and (3) the Mono Basin and Owens Valley through the Los Angeles Aqueduct. The remainder of the region's water supply is sourced from groundwater extraction (35%) and recycled water (1.5%). Demand management strategies, such as water use efficiency and conservation, account for 3% savings in water extraction (GLAC IRWM, 2013, Plan Update). According to the GLAC Plan Update, climate change could affect imported and local water supply due to a progressive reduction in average annual runoff. Also, there may be a missed opportunity to capture rainwater from more intense storms. A series of water management strategies is listed in the GLAC Plan Update for dealing with shortages in water supply, which include water desalination, groundwater management, combined use of surface water and groundwater, water storage, improved water conservation and efficiency of urban water use,[5] water recycling, and water transfers from different regions in the state.

---

4 http://www.ladpw.org/wmd/irwmp/

5 The State of California faced an extreme drought that required a declaration of a State of Emergency by Governor Brown in January 17, 2014. Enforceable water conservation measures were one of the key strategies. Governor Brown sought to reduce water usage by all Californians by 20% to confront this emergency. Accessed August 17, 2015: http://www.waterboards.ca.gov/board_decisions/adopted_orders/resolutions/2014/rs2014_0038_regs.pdf http://www.sacbee.com/2014/07/29/6591112/new-statewide-water-waste-prohibitions.html

BANGALORE

The city of Bangalore, located in southern India and home to nearly 10 million people, has more than doubled its population over the past 20 years. It has also seen rapid growth of investment in business, with companies such as Boeing, Samsung, Tata, Toyota, and a number of IT firms opening their doors in the city. These industries, including agriculture in the city's periphery, rely on water from the Arkavathy River basin, where the city is located. Researchers at the Ashoka Trust for Research in Ecology and the Environment, recipients of the government of Canada fast-start climate financing, have developed a land-use map for the basin and are investigating the impacts of rapid environmental change on water availability. The situation is dire: one of two major reservoirs has dried up entirely, and the other operates at one-fifth of its capacity due to extreme water shortage. The city instead imports water from more than 100 kilometers away, at great cost, while rural areas continue to pull from a depleting groundwater aquifer. Groundwater levels have dropped dramatically over the past 20 years, with bore well depths at 800–1,000 feet (down from at 200–300 feet 20 years ago), and the rate of replenishment is falling far short. The research suggests that while climate change may compound the problem, unregulated groundwater extraction, industrial pollution, and changes in land use are likely the greatest threats to water quality and availability. Preliminary recommendations include regulating groundwater extraction, increasing water reuse, distributing river water more evenly among users, and encouraging users to diversify their water sources.

The portfolio of adaptation options for addressing water supply that are available to a city located at the headwaters of a basin (e.g., Santiago), compared with one located at a halfway position (e.g., Bangalore) or at its outlet (e.g., Los Angeles) are different, as depicted in Case Study 14.1 Figure 1. There are other issues that also become evident when applying a basin perspective to adaptation for securing urban water supply. When the efficiency of water use inside a city increases, or if there are transfers from relatively inefficient agricultural users to urban users, then the downstream flow of water can be reduced, thus affecting supply. In a simple theoretical example, Vicuña and Meza (2012) show that with a 20% reduction in water availability, the cascade of adaptation measures (mostly improvements in irrigation efficiency for neighboring and downstream irrigation districts) used by a city located in the headwaters of a basin would result in more than 80% reduced flow at the basin's outlet. Users located near the outlet of the basin would also be affected disproportionally through the implementation of these types of measures. This is an issue that is also relevant in the case of Bangalore, India, for example (Lele et al., 2013). In sum, additional efficiency upstream may paradoxically create a more challenging situation downstream.

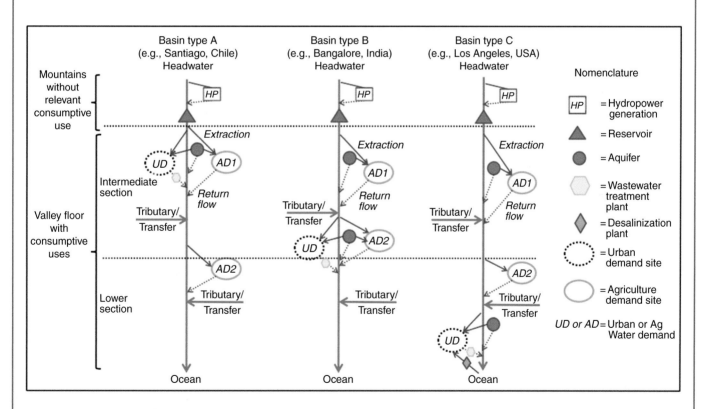

Case Study 14.2 Figure 1 *Schematics representing three possible urban water supply systems using a basin perspective.*

Source: Adopted from Vicuña et al., 2014b

to determine how much water will be required in the future. Demand management options are among the most politically viable actions and can help ensure that consumers are able to adapt to different water availability scenarios.

Key to the adoption of demand management is adequate governance, finance, and enforcement for ensuring compliance. For example, safe wastewater use is an increasingly popular way of matching water quality with water use (i.e., drinking water need not be used for agriculture). Safe use of wastewater, however, requires an operational capacity that is not always present in many developing countries (Drechsel et al., 2009). A graduated stepwise approach to implementing demand management needs to be adopted with a realistic assessment of the capacity to implement.

14.4.2.3 Storm Runoff Management

Increases in precipitation or storm surge intensities can increase the risk associated with urban flooding due to systems' inability to quickly absorb or redirect excess water or manage river overflow. A series of strategies can be implemented to diminish the physical causes of such risks. For example, peri-urban wetlands slow excess floodwaters, treat polluted waters, and provide valuable habitats for animals and plants (Kadlec and Wallace, 2008). The preservation of wetlands in and around cities is a key "ecosystem service" that is currently at risk from climate change and human pressures. As described in the study by Day et al. (2007), Hurricanes Katrina and Rita unveiled the vulnerability of coastal communities, and the authors described how human activities that caused the deterioration of the Mississippi Deltaic Plain served to exacerbate this vulnerability (see Chapter 8, Urban Ecosystems). Another key strategy is the revision of design criteria for urban drainage infrastructure that takes into account climate change impacts (Mailhot and Duchesne, 2010). Finally, in the case that the physical causes of risk cannot be avoided, strategies such as risk-based urban planning could be implemented (see Chapter 5, Urban Planning and Design).

14.4.3 Planning and Implementing Adaptation Options

The implementation of adaptation actions requires a deliberate investment in and planning of options that reduce exposure to future climate risks (Agarwala and Fankenhauser, 2008). There is no one-size-fits-all solution. Adaptation is necessarily local, and it is an iterative process that requires continual assessment and renewal of decisions as new climate information becomes available (Resurreccion et al., 2008). Although local in nature, the assessment of adaptation options should also consider multiple scales. For instance, the case study of the city of Naples in Italy (see Case Study 5.A in Annex 5) provides a good example of how different scales are connected in a storm runoff/flooding adaptation strategy, including components at the household, neighborhood, city, and basin levels.

14.4.3.1 Dealing with Uncertainty, the Importance of No-Regrets, and Avoiding Maladaptation

Once potential adaptation measures are identified, an appraisal is done to assess those that are best to implement given the context. Key in this appraisal is to take into account current climate/risk/vulnerability conditions and to assess uncertainties relating to future climate and non-climate drivers. In some cases, it may be impossible to perform a formal probabilistic assessment of future scenarios (Kiparsky et al., 2012). Risk-based decisions for engineering solutions using quantitative tools (such as decision matrices and probabilistic risk assessments) could be used, as suggested by Rosner et al. (2014). Recognizing the difficulties associated with uncertainty in climate change projections, experts conducting adaptation options appraisals have shifted their focus from a future climate (GCM) impact-based approach to more of a vulnerability-stakeholder driven process. Robust decision-making (RDM) or adaptation pathways are some of the recent frameworks that have been applied to water case studies, some of which are focused on urban areas. RDM provides an iterative decision framework for identifying robust strategies that fit a wide range of future scenarios. On the other hand, adaptation pathways provide an analytical approach for exploring and sequencing a set of possible actions based on alternative external developments over time. See specific examples of the application of these methods in Brown et al. (2012), Lempert and Groves (2010), and Haasnoot et al. (2013).

In most cases, however, the complexity of adaptation and the inability to precisely connect climate scenarios with impacts has led decision-makers to focus on options that offer a positive development pathway under a range of climate scenarios.[6] Such options for the water sector are generally considered as standard good practice for environmental stewardship and/or conservation. For example, of the options proposed by several of the authors and of those that are used in practice in many cities (e.g., leakage control, demand management, wastewater reuse, and restrictive land-use planning, see Table 14.1), most are already considered common-sense applications for smart water management (see Hallegatte, 2009). At its core, adaptation is local; it is about processes that improve decision-making, have embedded flexibility, and present decisiveness with imperfect information over long-term time horizons (Tyler and Moench, 2012; Ziervogel et al., 2010; Hallegatte, 2009).

The issue of maladaptation is a considerable risk to cities trying to meet urban water supply needs and must be avoided. Maladaptation refers to a situation in which an action, even if well intended, has other negative consequences. For instance, desalination, while addressing immediate water needs, can be considered maladaptive since it requires an enormous amount of energy and therefore GHG emissions (Hallegatte, 2009). Similarly, interbasin transfer of water (e.g., from the Cauvery River located more than 100 kilometers away to supply water to Bangalore) is not only costly, but also energy intensive. Such investments are often capitalized through debt but may not have adequate management to ensure that operation and maintenance costs are covered in the long term. The results are long-term financial liabilities that are difficult to manage. Such large-scale interbasin transfers are popular engineering projects but are frequently promoted before lower cost demand management solutions, which can be politically challenging

---

6 Sometimes called "low-regret" or "no-regret" options.

(i.e., raising prices to cover water supply and treatment costs). The implementation of some of these projects could also be related to a second form of maladaptation in which a given investment can address short-term climate variability but have negative long-term consequences. For example, investments in irrigation infrastructure can increase cultivated acreage in the short term but stress water basin capacity limits (e.g., Vicuña et al., 2014b). One useful way to implement the best adaptation options and avoid maladaptation is to sustain a close relation between researchers and decision-makers (see Case Study 14.3).

14.4.3.2 Financing Adaptation

Adaptation of the water sector in cities is expected to cost tens of billions of dollars, although specific estimates are required in order to improve planning and responses (Stern, 2006). Yet, due to the lack of reliability in available data, it is extremely difficult to accurately forecast the cost of adaptation. Estimates of the investment required to adapt the urban water

supply and sanitation sector vary. Early cost estimates from the IPCC (2007) suggested that an additional investment of US$9–11 billion annually would be required to accommodate water infrastructure to climate change impacts, not including the costs for extending services to unserved areas (IPCC, 2007). This amount was considered too low by others (Parry, 2009). Such variance stems from the lack of adequate data with which to specify costs and benefits of different adaptation options, as well as the large range of possible futures on the basis of existing climate scenarios. For the urban water sector, most of the benefit of adaptation comes in terms of avoided damage costs associated with extreme weather. This interests insurers to the extent that they have invested in economic assessments of damage in different sectors, including the water sector (see, e.g., Swiss Re, 2014).

Understanding the costs and benefits of adaptation is a key concern for implementation. Some adaptation actions are low cost but require behavioral change, whereas others are high in capital and operation and maintenance expenditures

---

Case Study 14.3   Denver, Seattle, Tucson: How Can Climate Research Be Useful for Urban Water Utility Operations?

Daniel B. Ferguson and Connie Woodhouse

University of Arizona, Tucson

Jennifer Rice

University of Georgia, Athens

Keywords	Water management, utilities, multistakeholder, climate information
Population (Metropolitan Region)	Seattle: 3,733,580 Tucson: 1,010,025 Denver: 2,814,330 (U.S. Census Bureau, 2015)
Area (Metropolitan Region)	Seattle: 15,209 km² Tucson: 23,794 km² Denver: 21,616 km² (U.S. Census Bureau, 2010)
Income per capita	US$56,180 (World Bank, 2017)
Climate zone	Seattle: Csb – Temperate, dry summer, warm summer Tucson: BSh – Arid, steppe, hot Denver: Dfa – Cold, without dry season, hot summer (Peel et al., 2007)

Municipal water utilities in Denver, Colorado; Seattle, Washington; and Tucson, Arizona, each interacted with climate researchers.

These experiences yielded a series of insights about how to foster productive collaboration between water management practitioners and climate researchers. The cities in which this research was conducted represent a range of water resource management settings for middle-sized cities (population of around 600,000) in the United States. Annual precipitation ranges from 300 (Tucson) to 940 millimeters (Seattle), resulting in notably different supply contexts. Seattle relies on two major watersheds for municipal water; Denver primarily utilizes four watersheds; Tucson – in the Sonoran Desert and without access to perennial surface water – has historically depended on groundwater. In recent years, however, supply has been bolstered by the Central Arizona Project, a 540 kilometer-long canal that brings water from the Colorado River to parts of Arizona that don't have access to perennial surface water.

These different contexts shape the climate information each utility needs to support its decision-making, but the processes for how information flows back and forth across the science/practice divide are relatively consistent. A central lesson from this work is the need to shrink the conceptual distance between the way in which climate researchers and water resource practitioners formulate climate-related problems for water resources, as well as potential solutions, in order to grow collaborative space for more integrated work.

This research yielded a series of ten heuristics or rules of thumb (Ferguson et al., 2014) meant to guide researchers and water resource practitioners who are seeking to collaborate. Three fundamental concepts underpin these rules of thumb. First, a key indicator of successful collaboration is persistence. Most of the heuristics point toward the need for sustained commitment to the common goal of solving a complex problem despite the challenges that arise from researcher–practitioner collaborations. Next, processes that encourage purposeful, thoughtful, and iterative interaction across the science–practice divide are the cornerstone of successful collaborations. A stand-alone workshop to disseminate research results or to gather stakeholder input may be a useful tool, but it is insufficient for building lasting collaborations to solve complex environmental problems. This highlights the critical importance of interdisciplinary research operating across social and natural sciences, as well as with practitioners, in order to bridge the sometimes disparate approaches to problem-solving.

Case Study 14.3 Figure 1 is an idealized diagram of the evolution of a collaborative partnership between resource management professionals and scientists. Each large oval represents a problem common to both the practitioner community and research community. The dashed line down the center of the figure is the conceptual boundary between research and practice. While the problem may be common (e.g., better understanding of long-term streamflow variability in a basin), the motivations and ways of framing and addressing it are often distinct for each community. The left side of the figure illustrates early efforts to collaborate, where communication between the two communities may

be infrequent and unfocused, as suggested by the dashed vertical lines. These problems become more commonly defined as more communication and tangible collaborations take place (e.g., moving from left to right across the figure). In this idealized scenario, the series of activities represented by the "collaboration" ovals may initially involve tasks focused on relationship-building and improving communication, like co-convening a workshop to discuss the particular problem (e.g., on the left side of the figure).

As practitioners and researchers communicate more, their mutual understanding of each other's professional language and culture grows, allowing those collaborative activities to become more complex and resulting in more integrated problem-solving. The net effect of the growth and evolution of these collaborative relationships is that the space shrinks between the research demand and the research supply, and the collaborative space grows. It is important to note that while the space between research needs and research questions shrinks, it never disappears. Even in fully collaborative, long-term relationships between researchers and practitioners, these are distinct communities with different motivations and mandates.

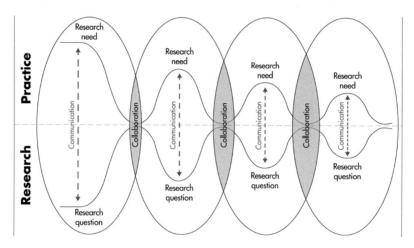

Case Study 14.3 Figure 1 *Shrinking the distance between research supply and demand by growing collaborative space.*

Source: Ferguson et al., 2014

(Agarwala and Fankenhauser, 2008). In India, an analysis of different adaptation measures for the water sector of the nation found that a US$24 billion investment in a mixture of rainwater harvesting, water metering, desalination, drainage, and channel improvements would greatly help reduce the deficit between supply and demand. Desalination, which some argue to be maladaptive due to the high energy costs and demand, would account for an expected reduction of 50% of the deficit alone (Markandya and Mishra, 2010; Hallegatte, 2009). A system of wastewater reuse based on existing infrastructure costing US$110 million per year would help meet around 6% of the expected deficit.

The coverage of available information on the economics of adaptation for cities and, in particular, the water sector is sparse compared with agriculture (Agarwala and

Fankenhauser, 2008). The literature, however, points to some important elements for consideration in developing a city-based adaptation strategy. For example, Hallegatte (2009) assesses how uncertainty is a natural fact of climate science, and decisions are required without perfect information. Tyler and Moench (2012) have developed "A Framework for Urban Climate Resilience," which profiles the preconditions for successful adaptation to climate change and touches on the water sector. Tested in ten Asian cities, the research concluded that social learning – a process whereby local planners engage with a range of stakeholders to identify the preferred strategy to adapt – is likely to lead to the strongest solutions and help secure stronger adaptive capacity.

Despite a range of initiatives to finance adaptation in the water sector, access to finance, particularly in resource constrained–countries (i.e., much of Africa) is a significant challenge (Tippmann et al., 2013). If adaptation is not connected with existing financial priorities, it is likely not to receive adequate funding. Therefore, mainstreaming adaptation into ministries of water and urban development is beneficial. Financing adaptation will likely mean increased water bills for different users, which is politically delicate. Private finance in the water sector, with riskier investments underwritten by development banks and public finance, and with incentives to provide adaptive solutions, are promising options; however, there is large variance among countries in the degree to which this is being done (Biagini and Miller, 2013). An option could be the development of "adaptation export credits" in countries that export materials and technologies used in infrastructure for conserving water. In most instances, the urban water sector benefits from having well-established monitoring regimes in place that help manage water supply and demand. This monitoring should be extended to longer time horizons, incorporating relevant climate data that can then provide feedback to inform decision-making.

14.4.3.3 Good Governance: The Precondition for Adaptation

For water management to be "adaptive," a significant investment in the quality of water resource governance is required (Pahl-Wostl, 2007; see Chapter 16, Governance and Policy). Adaptation requires adequate political and administrative will, but action is hampered by the lack of good governance in many instances. Urban resilience and the ability to implement deliberate adaptation is based on existing political and administrative structures that are accumulated over time and through city development (Satterthwaite, 2013). Adaptation also requires policy regimes that are flexible and that can adjust as new evidence becomes available (Swanson and Bhadwal, 2009). This point is underpinned by a study of 100 urban climate change projects that have pointed to experimentation and iterative social learning as key determinants in the success of such initiatives (Castan Broto and Bulkeley, 2013). Yet, in many cities, a lack of accountability and the inability to provide even basic services suggests that adaptation will be a key challenge (Lockwood, 2013). Research on how to adapt Cape

---

Case Study 14.4   Operationalizing Urban Climate Resilience in Water and Sanitation Systems in Manila

Julian Doczi

Water Policy, Overseas Development Institute, London

Mark Mulingbayan

Manila Water Company, Inc., Quezon City

Keywords	Water and sanitation, adaptation, infrastructure, business, floods
Population (Metropolitan Region)	11,855,975 (Philippine Statistics Authority, 2015)
Area (Metropolitan Region)	620 km² (Philippine Statistics Authority, 2015)
Income per capita	US$3,580 (World Bank, 2017)
Climate zone	Aw – Tropical savannah (Peel et al., 2007)

Manila is home to more than 16 million people within 17 cities/municipalities. The densely populated metropolis sits on a low-lying isthmus and includes many unplanned slums. It is among those areas of Southeast Asia most vulnerable to climate change because of its high population density, high risk from climatic hazards, and only moderate adaptive capacity (Yusuf and Francisco, 2009).

The metropolis has made significant progress toward more climate-resilient water supply and sanitation services since 1997. Prior to 1997, a public utility managed these services and was inefficient, overstaffed, and overburdened with debt. This had negative impacts on the quality and availability of these services for the population.

To expand and improve these services, the Philippine government privatized them in 1997. It split the metropolis into "east" and "west" zones and awarded each to a different business consortium via build-operate-transfer contracts. These contracts are a form of project financing in which a private entity receives a concession from the public sector to finance, design, construct, and operate a facility or service, with the public sector retaining asset ownership. The Manila Water Company, Inc. won the east zone and Maynilad Water Services, Inc. won the west zone. This Case Study focuses on Manila Water in the east zone.

PROGRESS TOWARD RESILIENT SERVICES

Manila Water has significantly improved water supply and sanitation service levels and its climate resilience since 1997, distinguishing itself as among the most proactive Philippine companies on climate change. Case Study 14.4 Table 1 displays data on these service improvements.

These service improvements strengthened the company's financial position, providing opportunities to pilot additional adaptation measures. Manila Water was among the first Philippine companies to develop a climate policy in 2007, updating it in 2013 (Manila Water, 2007, 2013). The policy includes activities on energy and fuel efficiency, vulnerability assessment, climate-resilient assets, disaster risk reduction and management, water source protection and development, and multistakeholder partnerships.

The company continuously recalibrates its disaster planning based on actual experience with the climate impacts that affect the country. This includes experience outside Manila when it provided water-related humanitarian assistance (e.g., mobile treatment facilities) after major climatic calamities, including Typhoons Bopha (in 2012) and Haiyan (in 2013). Manila Water was an initial convenor of the Philippine Disaster Recovery Foundation, a coalition of

(a)

(b)

Case Study 14.4 Figure 1 *The award-winning and flood-resilient Olandes (left) and Poblacion (right) sewage treatment plants of Manila Water.*

businesses that share lessons and coordinate their responses with the government.

Manila Water regularly assesses the vulnerability of its infrastructure and operations to climate hazards. It revised asset standards to retrofit existing facilities and construct new ones to be climate-re-silient. Case Study 14.4 Figure 1 depicts two of its facilities that won international awards for their innovative, flood-resistant designs. Medium-term plans include establishing interoperability protocols among the metropolis's lifeline utilities (power, communications, and fuel supply) and strengthening the resilience of its supply chain, which is a major dependency for restoring disaster-affected operations.

PROGRESS TOWARD A RESILIENT POPULATION

There is also evidence that Manila Water's improved services pos-itively affected the individual resilience of their customers. Doczi (2012) performed an impact evaluation of Manila Water's services in the east zone compared to Maynilad's in the west zone from 1997 to 2007.

The study assessed the comparative impact of these water supply and sanitation services on their population's health, wealth, and education using regression analysis on national survey data. It found a positive impact on the wealth and education of that portion of the population who reported receiving Manila Water's water supply services com-pared to those of Maynilad, particularly for poorer groups who reported

receiving their water supply in the form of a public standpipe in their community. However, these services provided little health benefit, par-ticularly for standpipes. This may be because Manila Water focused on improving water supply during this 10-year period, making less prog-ress on sanitation until recently. The study also argues that the greater health detriment from standpipes suggests that these are more easily contaminated. This needs to be balanced with the finding that these standpipes may boost wealth and education outcomes – and therefore individual climate resilience – as compared to a lack of service.

Doczi's study supports Manila Water's service impact but suffered important shortcomings. It used a national dataset with significant errors and missing data that affected the analysis. It also assumed that respondents received their services from either Manila Water or Maynilad. However, Cheng (2014) suggests that informal providers still serve significant portions of these zones. Nonetheless, Doczi's study remains unique in its assessment of Manila Water's services at the level of impacts rather than outputs.

LESSONS AND CHALLENGES

At least four factors drove Manila Water's progress, and these offer lessons to other cities: favorable initial conditions, proactive cor-porate culture, performance-based management systems, and strong branding to build public trust (Rivera, 2014; Luz and Paladio-Melosantos, 2012; Wu and Malaluan, 2008). The company received a smaller share of the public utility's debt in 1997 than did Maynilad, suffering less as a result during the Asian Financial Crisis of

Case Study 14.4 Table 1 *Comparative service improvements in Manila's East Zone from 1996 to 2012. Source: Rivera, 2014*

City	Population (Million)	24/7 Water Availability (% of Network)	Water Coverage (% of Population)	Non-Revenue Water[a] (% of Production)	Staff per 1,000 Connections
Manila East Zone (1996)	~2	26	67	63	9.8
Manila East Zone (2012)	>6	99	89	11	1.4

[a] Nonrevenue water is water that leaks or is stolen from supply networks between the production source and end users

1997–1999. The company was also better able to motivate the staff it inherited from the public utility through decentralized decision-making, target-based systems to promote personal accountability, and institutionalized values of integrity and customer centricity. These in turn promoted climate-resilient service improvements and helped to create a strong brand.

Challenges remain, including Manila Water's enabling environment in terms of its relationship with its public regulator. A public agency regulates both Manila Water and Maynilad, and their relationship dynamics have occasionally hindered progress. The agency ensures that the companies deliver on their service obligations while keeping water prices affordable. This involves reviewing the companies' infrastructure investment plans, for which expenses should be recovered through the water tariff. In times when the desire to temper tariff increases prevails, Manila Water's investments may be deferred or disallowed by the agency, which places at risk the company's commitments on adaptation, water availability, or environmental

compliance. The agency is not immune to political pressure, illustrating the importance of creating regulatory environments that incentivize – rather than hinder – progress.

In conclusion, Manila Water succeeded in rapidly absorbing and revitalizing the staff and services of an inefficient and indebted public utility. This drove progress on service improvements that positively affected their customers' resilience and provided fiscal space to pilot dedicated adaptation initiatives. Challenges remain, including the company's regulatory environment. The less successful case of Maynilad prior to 2007 also highlights that progress is not guaranteed even within a similar context and that public–private relationships must be carefully considered. Nonetheless, Manila Water's case offers learning about how business can – in certain circumstances – drive progress toward delivering resilient urban water supply and sanitation services.

Town's water supply infrastructure, for example, also points to the value of seeing adaptation more as a process than an outcome if it is to ensure the sustainability of water supply solutions (Ziervogel et al., 2010). An additional challenge is the rapid rise of secondary cities, many of which have only rudimentary systems of governance and, in some cases, zero climate change knowledge or adaptive capacity. Table 14.2 documents how existing water resource

management and infrastructure relates to the ability of a city to adapt to a changing climate; basic governance is a precondition for adaptation.

Other barriers exist that impede effective decision-making regarding the implementation of adaptation strategies. For example, the uncertainty of climate scenarios and the temporal scale

Table 14.2 *Adaptive capacity and urban water systems. Source: Adapted from IPCC Working Group II, Chapter 8 (2014) with additions from authors*

Level of Adaptive Capacity	Very little adaptive capacity or resilience/ 'bounce-back' capacity	Medium adaptive capacity or resilience/ 'bounce-back' capacity	Adequate capacity for adaptation and resilience/ 'bounce-back' but needs to be acted on
LEVEL OF WATER SECURITY	**LOW**	**MODERATE**	**HIGH**
Basic service delivery (water and sanitation)	0–30% of the urban center's population served; most of those unserved or inadequately served living in informal settlements. Women and children heavily burdened. Unequal access.	30–80% of the urban center's population served; most of those unserved or inadequately served living in informal settlements	80–100% of the urban center's population served; most of those unserved or inadequately served living in informal settlements
Service financing	Major challenge – unclear rules lead to limited investment; weak institutions limit cost recovery; donor dependency; private sector largely inactive	Some donor dependency; cost recovery evolving; PPPs; limited ability to leverage funds from capital markets without loan guarantees	Cost recovery adequate (few problems with bill payments); ability to secure funds from capital markets for investment in infrastructure; private sector active; regulations enforced
Institutional set up	Informal institutions predominant; lack of visibility of city authorities in some settlements; land tenure unclear; overlapping policy/mandates/unclear legal frameworks; traditional systems of governance; significant resource limitations; informal settlements predominant	Some formal institutions are present; may have active state intervention in service delivery; mix of traditional and modern governance; service provision in transition; mix of formal and informal settlements; legal framework for services evolving	Clear rules, roles, and responsibilities; resources available to support service delivery; adaptive policy framework with built-in flexibility; reasonable level of accountability in governance
Examples	Dar es Salaam; Khulna; Ouagadougou; Dhaka; Kinshasa	Mumbai, Nairobi, Colombo, Dakar, Accra	Most developed country cities; Santiago de Chile; Rio de Janeiro; Istanbul

of such scenarios are often discordant with short- or immediate-term decision-making needs for water and sanitation. Short time horizons among urban authorities also impede long-term decision-making that is required for climate change adaptation. Economic incentives are also skewed to short-term returns without adequately considering the long-term costs and benefits of adaptation. Moreover, the benefits of adaptation are extremely difficult to assess using conventional economic methods that exacerbate uncertainty (Chambwera et al., 2012). An additional barrier to adaptation is unequal power dynamics between different water users and a lack of ability to mediate interests in the context of uncertainty. An effort led by the International Institute for Environment and Development (IIED) pioneered a stakeholder-based cost-benefit analysis methodology as one tool that can help to manage negotiations between different water users. A central element of the approach is using climate information to consider the differentiated impacts of solutions on different stakeholder groups. For example, a subsidized investment in a drip irrigation system in Morocco was found to have more economic benefits for medium- and large-scale farmers than for small-scale farmers. Instead of aggregating impacts, a stakeholder-based cost-benefit analysis helps identify "winners and losers" and can improve compensation regimes and the allocation of benefits (Chambwera et al., 2012).

These challenges are compounded by underlying governance problems that impede effective stakeholder engagement and basic governance, let alone longer climate change adaptation decision-making. Examples of successful city-level multistakeholder engagement exist, such as in Cape Town, where researchers from the University of Cape Town have capitalized on a strong relationship with the city to work in several informal settlements on adaptation planning (Wadell, 2014). The main challenge for service provision to areas affected by SLR and coastal flooding is a failure of trust between these settlements and the city. As brokers, university staff have helped to rekindle some of this trust. The city has also shown leadership and established a set-back line, across which development cannot occur, in order to limit wasted investment in water and sanitation infrastructure that will be damaged through future storm surges and flood events. Case Study 14.5 explores the role that citizens could have in climate change adaptation.

Case Study 14.5   New Citizen Roles in Climate Change Adaptation: The Efforts of the Middle-Sized Danish City of Middelfart

Anja Wejs

Aalborg University

Keywords	Cloudbursts, urban-based adaptation, sustainable urban drainage system (SUDS), citizen involvement, floods
Population (Metropolitan Region)	37,981 (Statistics Denmark, 2015b)
Area (Metropolitan Region)	298.79 km² (Statistics Denmark, 2015a)
Income per capita	US$56,730 (World Bank, 2017)
Climate zone	Dfb – Cold, without dry season, warm summer (Peel et al., 2007)

A preliminary study of the city of Middelfart shows that, in the absence of climate change adaptation, current problems with flooding from heavy rain will worsen as the climate is expected to change over the next 50 years. Middelfart Wastewater Utility has over the past years invested tens of millions DKR to increase the capacity of the sewer system in parts of the city. However, now the utility together with Middelfart Municipality (Middelfart Municipality/Water Utility, n.d.) has decided to change the strategy to create solutions in synergy with the development of urban space to retain rainwater locally and increase urban livability. This involves changing the current roles of citizens from passive service receivers to actively handling rainwater on their private ground.

This is in line with a growing body of urban water professionals who are focused on transitioning to more resilient urban water management

due to the impacts of climate change (Brown et al., 2009). The experienced challenges with especially heavy rain and increased incidents of flooding demand more resilient cities and water management systems. Flexible and differentiated systems include both centralized pipe solutions below and above surface – solutions such as sustainable urban drainage systems (SuDS) to relieve the system in situations of heavy rain. The SuDS also correspond with the growing view on green and blue urban structures as desirable to create a livable city.

This transition demands new roles and responsibilities in relation to the design, operation, and maintenance of the water management system. Achieving this transition requires that the municipality and the wastewater utility work together and find new ways of engaging citizens in designing, developing, and, to a certain extent, maintaining the water management system. A very tangible example is that, in order to retain more rainwater locally instead of leading it through the sewer system, the individual plot owners have to be part of the process.

Citizens may further be regarded as a valuable knowledge resource when it comes to their local area and neighborhood and as a creative resource in urban development and the making of "livable cities" and innovative partnerships. Even though citizens do not have knowledge of water systems or urban planning, they are experts in their local neighborhood and have a unique knowledge of, for example, where there are local issues with flooding and how the local area serves different recreational purposes. This is knowledge that can be used (e.g., in risk assessments and in designing new solutions that fit local needs).

The city of Middelfart, Denmark, is working with green and blue structures, aboveground SuDS, and new citizens roles through two different projects:

• the transformation of private gardens into "rain gardens:"

- the development of a resilient urban area called Kongebro, with the aim to be Denmark's most beautiful climate change adaptation project

THE RAIN GARDENS

Over the course of 18 months, the municipality together with the Danish garden society have worked with the citizens through workshops and individual garden visits. The project is realized in collaboration with the Municipality of Middelfart, Middelfart Wastewater utility, and citizens (garden owners). The process has strengthened the already good neighborhood and also made it possible to combine some SuDS elements across garden boundaries.

As a pilot project, seven very different gardens have been established to investigate and show the potential of private gardens in climate change adaptation. Common to the seven gardens is that 95% of the roof and surface water is percolated into the garden, and the remaining 5% is surface discharged to a small stream in the woods.

This is a more aesthetic and more environmentally friendly option than separate sewer systems for both the utility and the garden owner. Furthermore, it has taught the professionals as well as the citizens to think of rainwater as a resource. Also, the process has inspired citizens in the area to redesign the gardens for both recreative and functional purposes.

DENMARK'S MOST BEAUTIFUL CLIMATE CHANGE ADAPTATION PROJECT

This project includes the implementation of an urban design and the subsequent implementation of a climate change adaptation project in the Kongebro area (KlimaByen, 2015; Middelfart Kommune, 2014). Citizens have provided input during the pre-study phase of the area

about the local qualities such as viewpoints, nature elements, and meeting spots, as well as about the problems they wanted to address in the urban development (e.g., flooding, road safety, accessibility). Citizens were invited to an introductory workshop held by the project group. The project group involves representatives from Middelfart Municipality and Middelfart Wastewater Utility. They provide input to an architectural competition program for the development of the area, and the dialogue group will then contribute to further develop the winning projects in collaboration with architects and engineers. Furthermore, a dialogue group of citizens has been established in which citizens and businesses are volunteering to take active part in the process.

In addition, two sixth-grade groups of school children who live in the area of Kongebro have contributed to the project by taking place in a "Walk and Talk Workshop." At the workshop, the children were first introduced to basic principles of local rain water management and the idea of using rain water for recreational purposes in the urban environment. Then the children went for a walk in the neighborhood and mapped the path they follow from the school to the sporting facility in the Kongebro area. Along the way, the children were asked to reflect on ways in which rainwater and green elements in the city can be used to increase the livability of the city. The school and its pupils are also involved in the development of a new playground that is designed to use rainwater as an element in children's' play, and the project is used to support teaching about climate change. Finally, residents of the area and other citizens have been given the opportunity get information and contribute local perspectives that will be incorporated into the program for an architectural competition. Later in the process, on-site workshops – so called "End-of-the-Road Workshops" – will be launched. Here, the residents are invited to talk to engineers and landscape architects about the implications of the project for their gardens and their local part of the Kongebro area. Along with these activities, citizens are informed about project development through newsletters and updates on the project website and through social and local media.

14.5 Mitigation Strategies for Urban Water Systems

Many cities are faced with significant needs in terms of rehabilitating and modernizing their UWS because many of their components have reached the end of their useful life (i.e., 40–50 years). This lack of adequate infrastructure is an opportunity to build treatment technologies that take into account an efficient use of energy.

UWS produce harmful GHGs in two ways. First, water systems are generally energy intensive, especially at the stage of pumping stations and wastewater treatment, where there are the most opportunities to reduce CO_2 emissions. For example, Friedrich et al. (2009) carried out a life cycle assessment of the urban water cycle for the Durban municipality in South Africa and found that, among the individual processes involved in the provision of water and wastewater, treatment of wastewater with activated sludge technology produced the highest emissions (41% of the total from the UWS) due to its high energy

consumption; the distribution of potable water came in second place (18%). A second source of emissions comes from off-gassing from the wastewater itself (e.g., methane emissions from biological waste).

New infrastructure investments into water systems can be optimized by identifying potential adaptation and mitigation benefits. There is a growing range of options available to reduce emissions associated with water and wastewater management. For example, biogas recovery for electricity production is a mitigation measure that may be applied in conventional municipal wastewater treatment plants (i.e., activated sludge–sewage that is aerated and broken down by micro-organisms). This approach not only reduces the emission of methane, but also helps substitute the draw-down of electricity from the grid. The amount of recovery can be as much as 100% of the total electric consumption in large treatment plants when combined heat and power equipment and other energy-efficiency measures are considered (McCarty et al., 2011; Nowak et al., 2011). San Antonio, for example, installed biogas recovery in a large wastewater treatment plant that generates US$200,000 in revenue per year, helping to cover operation and maintenance costs (Casey, 2010). In the case of developing countries with warm climate conditions, anaerobic technologies are increasingly applied to direct sewage treatment through the use of up-flow anaerobic sludge blanket (UASB) reactors, an effective method to treat wastewater in tanks and remove organic pollutants (see Figure 14.4).

A recent survey in Latin America (Noyola et al., 2012) sampled 2,734 wastewater treatment plants in six countries to assess GHG emissions for different categories of treatment. The survey identified three major treatment technologies for municipal wastewater: (1) activated sludge was the most significant based on treatment capacity (nearly 60% of total wastewater flow is treated by activated sludge), (2) stabilization ponds were the most common based on the number of facilities, and (3) UASB reactors. Each system has both advantages and disadvantages. In general, activated sludge is a compact but energy-intensive technology. Stabilization ponds, on the other hand, have low operating costs. However, in pond systems, methane is released into the atmosphere without burning, which is a major drawback for this kind of treatment systems. In contrast, the UASB reactor has also low energy needs and includes an arrangement for biogas capture and burning for electricity generation. An additional

advantage is that, as a compact technology, UASB reactors require a very small footprint, which is an important advantage for urban areas. A drawback of anaerobic sewage treatment is that 20–30% of the methane produced in the process leaves the system as dissolved gas in the effluent, which, depending on post-treatment, can lead to increased emissions (Noyola et al., 1988; Souza et al., 2011). Also, biogas – a useful source of energy – produced in small treatment plants cannot be used for energy purposes due to the limited amount of organic matter in conventional sewage (400–600 mg/L of chemical oxygen demand, COD). In these cases, biogas is best burned to reduce the warming potential of these emissions (Noyola et al., 2012). As Box 14.1 Table 1 illustrates, systems that allow for biogas recovery and burning for electricity generation have the greatest benefit in terms of emissions and energy efficiency.

To provide a solid evidence base for decisions on what type of wastewater treatment for which situation, Noyola et al. (2013) developed a life cycle analysis of municipal wastewater treatment technologies (e.g., stabilization ponds, activated sludge, and UASB reactors) in Latin America. Focusing on GHG emission potential, the results showed that, in descending order of impact, stabilization ponds have the highest impact due to the venting of methane. Extended aeration, a variant of activated sludge, ranks second due to the indirect CO_2 emissions generated by electricity use in the aeration tank. The contribution of conventional activated sludge to emissions comes from the anaerobic sludge digesters where biogas is produced. This in turn can be an asset for reducing electricity consumption from the grid and diminishing GHG emissions from the technology. Finally, UASB reactors showed the lowest impact due to efficient methane capture and burning. In addition, UASB reactors have a better overall environmental performance due to low energy requirements and the limited amount of excess sludge produced.

The future adoption of anaerobic treatment technologies in developing countries would reduce GHG emissions, accomplishing at the same time lower capital investments and reduced operational costs when compared to conventional full aerobic options. This technological option will therefore result in more sustainable systems, thus representing an attractive policy measure for developing countries. Box 14.1 presents an assessment of emission reduction options for municipal wastewater treatment technologies in Mexico.

14.6 Conclusions

In cities, successful adaptation to climate change will rely on good planning and governance, appropriate sources of finance for investment, social equity in provision, ensuring that the right capacity exists for implementation, and operational infrastructure that is resilient to rapidly changing conditions. With the increasing number of people living in cities and informal urban settlements, particularly in developing countries, water security is a growing concern under changing climate conditions. This chapter has profiled the threat posed by climate change to UWS

Figure 14.4 *A pilot UASB reactor installed in Belo Horizonte, Brazil.*

Source: Sustainable Sanitation Alliance, 2013

and, ultimately, water security in and around cities. Planners attempt to address these challenges by not only managing water as a resource, but also by avoiding the risks associated with extremes in water flows (e.g., flooding, scarcity, and pollution). Climate change will, in most cases, exacerbate the situation through changing precipitation patterns, extreme events, and SLR. A key challenge is the uncertainty embedded in future climate scenarios – an uncertainty that raises questions about when thresholds are crossed. For example, when poor water management and climate change leads to the unsustainable exploitation of water in a basin and thus leads to direct impacts on people and economies.

Many solutions exist that can be brought to the municipal scale. The use of alternative sources of water (i.e., safe use of wastewater), embedded flexibility to alter policy as new information comes to light, and finding the political will to establish proper cost recovery in order to ensure the sustainability of investments in water and sanitation systems are just some of the solutions presented in this chapter. Ensuring a balance between water and energy benefits is also critical. Importantly for decision makers, most solutions are not new options and have been a part of classic good practice in urban water management for many years. In other words, these are not necessarily radical ideas.

Mitigation options for reducing emissions were described in this chapter, several of which stand out as important areas for

---

Box 14.1 Municipal Wastewater Treatment as a Greenhouse Gas Emission Mitigation Strategy in Mexico

Adalberto Noyola and Leonor P. Guareca

National Autonomous University of Mexico, Mexico City

As with many developing countries and emerging economies, Mexico requires important investment in urban wastewater treatment infrastructure to increase its current coverage. At present, only 45% of collected municipal sewage is treated, while the difference is discharged either untreated or partially treated into bodies of water used for irrigation, resulting in higher risks to public health and the environment. To tackle this problem, investments are being made in new treatment facilities. The proper selection of treatment technologies is an important opportunity for contributing to Mexico's ambitious national climate change strategy goal of a 30% reduction in greenhouse gas (GHG) emissions by 2020 and 50% by 2050 (SEMARNAT, 2013).

To support decision-making for more sustainable water treatment systems, five technology options were analyzed, aiming at a 100% treatment coverage by 2030. A "business as usual" scenario was considered as a baseline (i.e., maintaining the present rate of building new infrastructure and using the present distribution of wastewater treatment technologies). Five improvement scenarios were considered based on currently employed technologies, as well as using a combined approach

of upflow anaerobic sludge blanket (UASB) reactors followed by aerobic processes for post-treatment (see Box 14.1 Table 1).

Results show that GHG emissions from sewage treatment in Mexico could be reduced by up to 34% relative to the baseline "business as usual" scenario by using the combined anaerobic-aerobic processes. This technology has a 95% methane burning efficiency and also provides opportunities for electricity co-generation in facilities with treatment capacity above 500 L/s. If electricity production through biogas recovery is not considered, then reduction of emissions is limited to 14%. The research also shows that the adoption of anaerobic reactors, where technically feasible, can reduce operating costs by around 40% compared to conventional aerobic wastewater treatment systems.

The study estimated emissions based on the Guidelines for Greenhouse Gas Inventories proposed by the IPCC (2006). Emissions can vary according to on-site characteristics, so there is a need to determine emission factors for each country or region, taking into account specific conditions at least for the most representative wastewater treatment technologies. As a result, more precise emission inventories from wastewater treatment could be calculated, allowing the identification of more effective mitigation strategies in developing countries.

Box 14.1 Table 1 *GHG emissions from municipal wastewater treatment in Mexico under different technological scenarios, including expected future scenarios.*

Scenario	2010	2015	2020	2025	2030	Percentage reduction (%)
			Gg CO$_2$ eq.			
Baseline (BL)	13,334	12,367	12,302	12,140	11,867	–
WA	13,334	12,388	12,086	11,687	11,176	6
A	13,334	12,321	11,878	11,339	10,688	10
An+A	13,334	11,639	11,270	10,805	10,227	14
ZME	13,334	11,251	10,630	9,912	9,083	23
BE	13,334	10,793	9,931	8,926	7,809	34

WA, Water agenda (BL at 100% coverage); A, Aerobic-only processes; An + A, Anaerobic followed by aerobic post-treatment; ZME, Zero methane emission; BE, Biogas to energy

investment: (1) ensuring the application of life cycle analysis to decision-making for water supply, treatment, and drainage; (2) the use of anaerobic reactors to improve the balance between energy conservation and wastewater treatment; and (3) eliminating high-energy options such as interbasin transfers of water wherever there are alternatives available. It is imperative that available options be carefully considered within local contexts to identify those that are most appropriate and cost-effective. It is also critical to assess options in the context of available local capacity. A major drawback of many infrastructure investments in the past has been the disregard for the operation and maintenance requirements and costs that can lead to failed systems.

A central criterion for all policy-making is finding ways to maximize public good in the most cost-effective manner. The strongest overarching conclusion we draw from this review of climate change in the water sector is the importance of building our collective capacity to act now in adapting UWSs and make them more resilient to climate-related risk. This will help reduce longer term costs (e.g., health, infrastructure, and industry) associated with reduced water security in the future.

Annex 14.1   Stakeholder Engagement

Stakeholder engagement in this chapter consisted of one formal event and a number of informal meetings at which ideas on urban adaptation in the water sector was the core topic. The formal side event at the ICLEI Resilient Cities meeting in Bonn, 2014, brought together ten practitioners and researchers to review the draft chapter and provide feedback. Informal meetings on the sidelines of larger events took place on numerous occasions, including the Istanbul Water Forum in 2014 and COP20 in Lima, 2014, where input was solicited. Each event provided additional substance for the chapter. Authors also had copy reviewed by colleagues to verify accuracy and substance. Generally, the stakeholder engagement either profiled work and case studies that the authors were unaware of or challenged assumptions within the text and added value by providing more specificity to arguments.

Chapter 14   Urban Water Systems

References

Adger, W. N., Dessai, S., Goulden, M., Hulme, M., Lorenzoni, I., Nelson, D. R., Naess, L. O., Wolf, J., and Wreford, A. (2009). Are there social limits to adaptation to climate change?. *Climatic Change* **93**(3–4), 335–354.

Aerts, C. J. H., Botzen, W. J. W., de Moel, H., and Bowman, M. (2013). Cost estimates for flood resilience and protection strategies in New York City: *Annals of the New York Academy of Sciences* **1294**, 1–104.

Agrawala, S., and Fankhauser, S. (eds.). (2008). *Economic Aspects of Adaptation to Climate Change Costs, Benefits and Policy Instruments: Costs, Benefits and Policy Instruments*. OECD.

Aggarwal, S. C., and Kumar S. (2011). *Industrial Water Demand in India: Challenges and Implications for Water Pricing in The 2011 India Infrastructure Report*. IFDI.

Almutaz, I., Ajbar, A., Khalid, Y., and Ali, E. (2012). A probabilistic forecast of water demand for a tourist and desalination dependent city—Case of Mecca, Saudi Arabia. *Desalination* **294**, 53–59.

Barnett, T. P., Adam, J. C., and Lettenmaier, D. P. (2005). Potential impacts of a warming climate on water availability in snow-dominated regions: *Nature* **438**, 303–309.

Berghuijs, W. R., Woods, R. A., and Hrachowitz, M. (2014). A precipitation shift from snow towards rain leads to a decrease in streamflow. *Nature Climate Change* **4**(7), 583–586.

Biagini, B., and Miller, A. (2013). Engaging the private sector in adaptation to climate change in developing countries: Importance, status, and challenges. *Climate and Development* **5**(3), 242–252.

Boyle, J., Cunningham, M., and Dekens, J. (2013). Climate change adaptation and Canadian infrastructure—A review of the literature: International Institute for Sustainable Development report. Accessed September 17, 2014: http://www.iisd.org/pdf/2013/adaptation_can_infrastructure.pdf

Braga, B. P. F. (2001). Integrated urban water resources management: A challenge into the 21st century. *International Journal of Water Resources Development* **17**(4), 581–599.

Breyer, B., and Chang, H. (2014). Urban water consumption and weather variation in the Portland, Oregon metropolitan area. *Urban Climate*. Accessed August 17, 2015: http://dx.doi.org/10.1016/j.uclim.2014.05.001

Breyer, E. Y. (2014). *Household Water Demand and Land Use Context—A Multilevel Approach*. Master's thesis (paper 1670), Portland State University. Accessed May 16, 2015: http://pdxscholar.library.pdx.edu/open_access_etds/1670/

Brown, C., Ghile, Y., Laverty, M., and Li, K. (2012). Decision scaling: Linking bottom-up vulnerability analysis with climate projections in the water sector. *Water Resources Research* **48**(9), W09537.

Buytaert, W., and De Bièvre, B. (2012). Water for cities: The impact of climate change and demographic growth in the tropical Andes. *Water Resources Research* **48** (8).

Casey, T. (2010). San Antonio gets the scoop on first commercial biogas from municipal sewage. Accessed August 15, 2014: http://cleantechnica.com/2010/10/24/san-antonio-gets-the-scoop-on-first-commercial-biogas-from-municipal-sewage/

Castán Broto, V., and Bulkeley, H. (2013). A survey of urban climate change experiments in 100 cities. *Global Environmental Change* **23**(1), 92–102.

Chambwera, M., Baulcomb, C., Lunduka, R., de Bresser, L., Chaudhury, A., Wright, H., Loga, D., and Dhakal, A. (2012). *Stakeholder-Focused Cost Benefit Analysis in the water Sector: A Guidance Report*. International Institute for Environment and Development (IIED).

Church, J. A., Clark, P. U., Cazenave, A., Gregory, J. M., Jevrejeva, S., Levermann, A., Merrifield, M. A., Milne, G. A., Nerem, R. S., Nunn, P. D., Payne, A. J., Pfeffer, W. T., Stammer, D., and Unnikrishnan, A. S. (2014). Sea level change. In Stocker, T. F., Quin, D., Plattner, G. K., Tignor, M. M. B., Allen, S. K., Boschung, J., Nauels, A., Xia, Y., Bex, V., and Midgley, P. M. (eds.), *Climate Change 2013, The Physical Science Basis—Working Group I Contribution to the Fifth Assessment Report of the Intergovernmental Panel on Climate Change* (1137–1216). Cambridge University Press.

Cisneros, B. E. J., Oki, T., Arnell, N. W., Benito, G., Cogley, J. G., Doll, P., Jiang, T., and Mwakalila, S. S. (2014). Freshwater Resources. In Field, C. B., Barros, V. R., Dokken, D. J., Mach, K. J., Mastrandrea, M. D., Bilir, T. E., Chatterjee, M., Ebi, K. L., Estrada, Y. O., Genova, R. C., Girma, B., Kissel, E. S., Levy, A. N., MacCracken, S., Mastrandrea, P. R., and White, L. L. (eds.), *Climate Change 2013, Impacts, Adaptation, and Vulnerability—Working Group II Contribution to the Fifth Assessment Report of the Intergovernmental Panel on Climate Change* (229–269). Cambridge University Press.

Collins, M., Knutti, R., Arblaster, J., Dufresne, J. -L., Fichelet, T., Friedlingstein, P., Gao, X., Gutowski, W. J., Johns, T., Krinner, G., Shongwe, M., Tebaldi, C., Weaver, A. J., and Wehner, M. (2014). Long-term climate change: Projections, commitments and irreversibility. In Stocker, T. F., Quin, D., Plattner, G. K., Tignor, M. M. B., Allen, S. K.,

Boschung, J., Nauels, A., Xia, Y., Bex, V., and Midgley, P. M. (eds.), *Climate Change 2013, The Physical Science Basis: Working Group I Contribution to the Fifth Assessment Report of the Intergovernmental Panel on Climate Change* (1028–1136). Cambridge University Press.

Cornish, G., Bosworth, B., Perry, C. J., and Burke, J. J. (2004). *Water Charging in Irrigated Agriculture: An Analysis of International Experience.* Volume **28**. FAO Water Reports. United Nations.

Crosbie, R. S., McCallum, J. L., Walker, G. R., and Chiew, F. H. S. (2010). Modelling climate-change impacts on groundwater recharge in the Murray-Darling Basin, Australia. *Hydrogeology Journal* **18**, 1639–1656.

Das, T., Pierce, D. W., Cayan, D. R., Vano, J. A., and Lettenmaier, D. P. (2011). The importance of warm season warming to western U.S. streamflow changes. *Geophysical Research Letters* **38**(L23403) 5.

Day, J. W., Boesch, D. F., Clairain, E. J., Kemp, G. P., Laska, S. B., Mitsch, W. J., and Whigham, D. F. (2007). Restoration of the Mississippi Delta: Lessons from hurricanes Katrina and Rita. *Science* **315**(5819), 1679–1684.

Dell, J., Tierney, S., Franco, G., Newell, R. G., Richels, R., Weyant, J., and Wilbanks, T. J. (2014). Energy supply and use. In Melillo, J. M., Richmond, T. C., and Yohe, G. W. (eds.), *Climate Change Impacts in the United States: The Third National Climate Assessment (113–129).* U.S. Global Change Research Program. doi: 10.7930/J0BG2KWD

Donkor, E. A., Mazzuchi, T. A., Soyer, R., and Roberson, J. A. (2014). Urban water demand forecasting—Review of methods and models. *Journal of Water Resources Planning and Management* **140**, 146–159.

Drechsel, P., Scott, C. A., Raschid-Sally, L., Redwood, M., and Bahri, A. (2009). *Wastewater Irrigation and Health.* Earthscan.

Ferguson, G., and Gleeson T. (2012). Vulnerability of coastal aquifers to groundwater use and climate change. *Nature Climate Change* **2**, 342–345.

Friedrich, E., Pillay S., and Buckley C. A. (2009). Environmental life cycle assessments for water treatment processes – A South African case study of an urban water cycle. *Water SA* **35**(1), 73–84.

Georgakakos, A., Fleming, P., Dettinger, M., Peters-Lidard, C., Richmond, T. C., Reckhow, K., White, K., and Yates, D. (2014). Water resources. In Melillo, J. M., Richmond, T. C., and Yohe, G. W. (eds.), *Climate Change Impacts in the United States: The Third National Climate Assessment (69–112).* U.S. Global Change Research Program.

Gill, S. E., Handley, J. F., Ennos, A. R., and Pauleit, S. (2007). Adapting cities for climate change: The role of the green infrastructure. *Built Environment* **33**(1), 115–133.

GLAC IRWM. (2013). Plan update. Accessed August 15, 2014: http://www.ladpw.org/wmd/irwmp/

Gleeson T., Wada, Y., Bierkens, M. F. P., and van Beek, L. P. H. (2012). Water balance of global aquifers revealed by groundwater footprint. *Nature* **488**, 197–200.

Golombek, R., Kittelsen, S. A. C., and Haddeland, I. (2012). Climate change–Impacts on electricity markets in Western Europe. *Climatic Change* **113**, 357–370.

Gosain, A. K., Rao, S., and Basuray, D. (2006). Climate change impact assessment on hydrology of Indian river basins. *Current Science* **90**(3), 346–353.

Haasnoot, M., Kwakkel, J. H., Walker, W. E., ter Maat, J. (2013). Dynamic adaptive policy pathways: A method for crafting robust decisions for a deeply uncertain world. *Global Environmental Change* **23**, 485–498.

Hallegatte, S. (2009). Strategies to adapt to an uncertain climate change. *Global Environmental Change* **19**(2), 240–247.

Hanak, E., and Lund, J. R. (2012). Adapting California's water management to climate change. *Climatic Change* **111**, 17–44.

Hanson. S., and Nicholls, R. J. (2012). Extreme flood events and port cities through the twenty-first century. In Asariotis, R., and Benemara, H. (eds.), *Maritime Transport and the Climate Change Challenge.* Earthscan/Routledge.

Hanson, S., Nicholls, R. J., Ranger, N., Hallegatte, S., Corfee-Morlot, J., Herweijer, C., and Chateau, J. (2011). A global ranking of port cities with high exposure to climate extremes. *Climatic Change* **104**, 89–111.

Haraguchi, M., and Lall, U. (2015, December). Flood risks and impacts: A case study of Thailand's floods in 2011 and research questions for supply chain decision making. *International Journal of Disaster Risk Reduction* **14**(3), 256–272.

Hellestrom, D., Jeppsson, U., and Karrman, E. (2000). A framework for systems analysis of sustainable urban water management. *Environmental Impact Assessment Review* **20**, 311–321.

Higgins, S., Overeem, I., Tanaka, A., and Syvitski, J. P. (2013). Land subsidence at aquaculture facilities in the Yellow River delta, China. *Geophysical Research Letters* **40**(15), 3898–3902.

Hijioka, Y., Lin, E., Pereira, J. J., Corlett, R. T., Cui, X., Insarov, G. E., Lasco, R. D., Lindgren, E., and Surjan, A. (2014). Asia. In Field, C. B., Barros, V. R., Dokken, D. J., Mach, K. J., Mastrandrea, M. D., Bilir, T. E., Chatterjee, M., Ebi, K. L., Estrada, Y. O., Genova, R. C., Girma, B., Kissel, E. S., Levy, A. N., MacCracken, S., Mastrandrea, P. R., and White, L. L. (eds.), *Climate Change 2014: Impacts, Adaptation, and Vulnerability. Part B: Regional Aspects. Contribution of Working Group II to the Fifth Assessment Report of the Intergovernmental Panel on Climate Change* (1327–1370). Cambridge University Press.

Hinkel, J., and Bisaro, A. (2016). Methodological choices in solution-oriented adaptation research: A diagnostic framework. *Regional Environmental Change* **16**(1), 7–20.

Hof, A., and Wolf, N. (2014). Estimating potential outdoor water consumption in private urban landscapes by coupling high-resolution image analysis, irrigation water needs and evaporation estimation in Spain. *Landscape and Urban Planning* **123**, 61–72.

Horton, B. P., Rahmstorf, S., Engelhart, S. E., and Kemp, A. C. (2014). Expert assessment of sea—level rise by AD2100 and AD2300. *Quaternary Science Reviews* **84**, 1–6.

House-Peters, L. A., and Chang, H. (2011). Urban water demand modeling—Review of concepts, methods and organizing principles. *Water Resources Research* **47**(W05401), 15.

Hunter, J. (2011). A simple technique for estimating an allowance for uncertain sea-level rise. *Climatic Change.* doi: 10.1007/s10584-011-0332-1

Institution of Mechanical Engineers (IME). (2014). Climate change—Adapting to the future? IME report. Accessed May 16, 2015: http://www.imeche.org/knowledge/themes/environment/climate-change/adaptation/adaptation-report

Intergovernmental Panel on Climate Change (IPCC). (2007). *Climate Change 2007: Synthesis Report. Contribution of Working Groups I, II and III to the Fourth Assessment Report of the Intergovernmental Panel on Climate Change* [Core Writing Team, Pachauri, R.K and Reisinger, A. (eds.)]. IPCC.

Intergovernmental Panel on Climate Change (IPCC). (2013). Summary for policymakers. In Stocker, T. F., Qin, D., Plattner, G. -K., Tignor, M., Allen, S. K., Boschung, J., Nauels, A., Xia, Y., Bex, V., and Midgley, P. M. (eds.), *Climate Change 2013: The Physical Science Basis. Contribution of Working Group I to the Fifth Assessment Report of the Intergovernmental Panel on Climate Change.* Cambridge University Press.

Intergovernmental Panel on Climate Change (IPCC). (2014). Climate Change 2014: Impacts, Adaptation, and Vulnerability. *Part A: Global and Sectoral Aspects. Contribution of Working Group II to the Fifth Assessment Report of the Intergovernmental Panel on Climate Change,* Field, C. B., Barros, V. R., Dokken, D. J., Mach, K. J., Mastrandrea, M. D., Bilir, T. E., Chatterjee, M., Ebi, K. L., Estrada, Y. O., Genova, R. C., Girma, B., Kissel, E. S., Levy, A. N., MacCracken, S., Mastrandrea, P. R., and White, L. L. (eds.). Cambridge University Press.

Intergovernmental Panel on Climate Control. (2006). *2006 Guidelines for National Greenhouse Gas Inventories. Vol. 5: Waste.* Eggleston H. S., Buendia L., Miwa K., Ngara T., and Tanabe, K. (eds.). IGES.

Jacob, K. H., Gornitz, V., and Rosenzweig, C. (2007). Vulnerability of the New York City metropolitan area to coastal hazards, including sea-level rise–Inferences for urban coastal risk management and adaptation policies. In McFadden L., Nicholls R. J., and Penning-Roswell E. (eds.), *Managing Coastal Vulnerability* (61–88). Elsevier.

Kadlec, R. H., and Wallace, S. (2008). *Treatment Wetlands*. CRC Press.

Kennel, C. F. (2009). Climate change—Think globally, assess regionally, act locally. *Issues in Science and Technology* **25**, 46–52.

Kenney, D., Klein, R., Goemans, C., Alvord, C., and Shapiro, J. (2008). *The Impact of Earlier Spring Snowmelt on Water Rights and Administration: A Preliminary Overview of Issues and Circumstances in the Western States*. Western Water Assessment.

Kenny, J. F., Barber, N. L., Hutson, S. S., Linsey, K. S., Lovelace, J. K., and Maupin, M. A. (2009). Estimated use of water in the United States in 2005. *U.S. Geological Survey Circular* **1344**, 52. Accessed June 5, 2014: http://pubs.usgs.gov/circ/1344/

Kharin, V. V., Zwiers, F. W., Zhang, X., and Wehner, M. (2013). Changes in temperature and precipitation extremes in the CMIP5 ensemble. *Climatic Change* **119**, 345–357.

Kiparsky, M., Milman, A., and Vicuña, S. (2012). Climate and water: Knowledge of impacts to action on adaptation. *Annual Review of Environmental Resources* **37**, 163–194.

Kirtman, B., Power, S. B., Adedoyin, A. J., Boer, G. J., Bojariu, R., Camilloni, I., Doblas-Reyes, F., Fiore, A. M. M., Kimoto, M., Meehl, G., Prather, M., Sarr, A., Schar, C., Sutton, R., van Oldnborough, G. J., Vecchi, G., and Wang, H. J. (2014). Near-term climate change—Projections and predictability. In Stocker, T. F., Quin, D., Plattner, G. K., Tignor, M. M. B., Allen, S. K., Boschung, J., Nauels, A., Xia, Y., Bex, V., and Midgley, P. M. (eds.), *Climate Change 2013, The Physical Science Basis—Working Group I Contribution to the Fifth Assessment Report of the Intergovernmental Panel on Climate Change* (953–1028). Cambridge University Press.

Koch, H., and Vogele, S. (2009). Dynamic modeling of water demand, water availability and adaption strategies for power plants to global change. *Ecological Economics* **68**, 2031–2039.

Konikow, L. F. (2013). *Groundwater Depletion in the United States (1900–2008)*. U.S. Geological Survey Scientific Investigations Report 2013−5079.

Lafontaine, A, Oladipo Adejuwon, J., Dearden, P., and Quesne, G. (2012). *Final Evaluation of the IDRC/DFID Climate Change Adaptation in Africa Program*. IDRC.

Langeveld, J. G., Schilperoort, R. P. S., and Weijers, S. R. (2013). Climate change and urban wastewater infrastructure—There is more to explore. *Journal of Hydrology* **476**, 112–119.

Lempert, R., and Groves, D. (2010). Identifying and evaluating robust adaptive policy responses to climate change for water management agencies in the American west. *Technological Forecasting & Social Change* **77** (6), 960–974.

Linnerud, K., Mideksa, T. K., and Eskeland, G. S. (2011). The impact of climate change on nuclear power supply. *Energy Journal* **32**(1), 149–168.

Lockwood, M. (2013). What can climate-adaptation policy in sub-Saharan Africa learn from research on governance and politics? *Development Policy Review* **31**(6), 647–676.

Lowe, J. A., Woodworth, P. L., Knutson, T., McDonald, R., McInness, K. L., Woth, K., von Storch, H., Wolf, J., Swail, V., Bernier, N. B., Gulev, S., Horsburgh, K. J., Unnikrishnan, A. S., Hunter, J. R., and Weisse, R. (2010). Past and future changes in extreme sea levels and waves. In Church, J. A., Woodworth, P. L., Aarup, T., and Wilson, W. S. (eds.). *Understanding Sea-Level Rise and Variability*. Wiley-Blackwell. doi: 10.1002/9781444323276.ch11

Lwasa, S., Mugagga, F., Wahab, B., Simon, D., Connors, J., and Griffith, C. (2014). Urban and peri-urban agriculture and forestry—Transcending poverty alleviation to climate change mitigation and adaptation. *Urban Climate* **7**, 92–106.

Mailhot, A., and Duchesne, S. (2010). Design criteria of urban drainage infrastructures under climate change. *Journal of Water Resources Planning and Management* **136**, 201–208.

Major, D. C., Omojola A., Dettinger, M., Hanson, R. T., and Sanchez-Rodriguez, R. (2011). Climate change, water, and wastewater in cities. In Rosenzweig, C., Solecki, W. D., Hammer, and S. A., Mehrotra, S. (eds.), *Climate Change and Cities: First Assessment Report of the Urban Climate Change Research Network* (113–143). Cambridge University Press.

Manuel, J. (2013). The long road to recovery-Environmental health impacts of Hurricane Sandy. *Environmental Health Perspectives* **121**, A152–A159.

Markandya, A., and Mishra, A. (2010). *Costing Adaptation: Preparing for Climate Change in India*. TERI.

McCarty P. L., Bae, J., and Kim, J. (2011). Domestic wastewater treatment as a net energy producer – can this be achieved? *Environmental Science and Technology* **45**, 7100–7106.

McDonald, R. I., Green, P., Balk, D., Fekete, B. M., Revenga, C., Todd, M., and Montgomery, M. (2011). Urban growth, climate change and fresh-water availability. *Proceedings of the National Academy of Sciences* **108**, 6312–6317.

McDonald, R. I., Weber, K., Padowski, J., Flörke, M., Schneider, C., Green, P. A., Gleeson, T., Eckman, S., Lehner, B., Balk, D., Boucher, T., Grill, G., and Montgomery, M. (2014). Water on an urban planet—Urbanization and the reach of urban water infrastructure. *Global Environmental Change* **27**, 95–105.

Meza, F. J., Vicuña, S., Jelinek, M., Bustos, E., and Bonelli, S. (2014). Assessing water demands and coverage sensitivity to climate change in the urban and rural sectors in central Chile. *Journal of Water and Climate Change* **5**(2), 192–203.

Middelfart Municipality/Water Utility (n.d.). Project homepage. Accessed February 5, 2017: www.klima-byen.dk

Min, S., Zhang, X., Zwiers, F., and Hegerl, G. (2011). Human contributions to more-intense precipitation extremes. *Nature* **470**, 378–381.

Mini, C., Hogue, T. S., and Pincetl, S. (2014). *Estimation of residential water use in Los Angeles: Landscape and Urban Planning* **127**, 124–135.

Moser, S. C., and Ekstrom, J. A. (2010). A framework to diagnose barriers to climate change adaptation. *Proceedings of the National Academy of Sciences* **107**(51), 22026–22031.

Mukundan, R., Pradhanang, S. M., Schneiderman, E. M., Pierson, D. C., Anandhi, A., Zion, M. S., Matonse, A. H., Lounsbury, D. G., and Steenhuis, T. S. (2013). Suspended sediment source areas and future climate impact on soil erosion and sediment yield in a New York City water supply watershed, USA. *Geomorphology* **183**, 110–119.

Newcomer, M. E., Gurdak, J. J., Sklar, L. S., and Nanus, L. (2013). Urban recharge beneath low impact development and effects of climate variability and change. *Water Resources Research* **50**, 1716–1734.

Nikomborirak, D., and Ruenthip, K. (2013). *History of Water Resource and Flood Management in Thailand*. TDRI.

Nowak O., Keil, S., and Fimml, C. (2011). Examples of energy self-sufficient municipal nutrient removal plants. *Water Science and Technology* **64**(1), 1–6.

Noyola, A., Capdeville, B., and Roques, H. (1988). Anaerobic treatment of domestic sewage with a rotating-stationary fixed-film reactor. *Water Researh* **22**(12), 1585–1592.

Noyola, A., Morgan-Sagastume, J. M., and Güereca, L. P. (2013). *Selección de tecnologías para el tratamiento de aguas residuales municipales. Guía de apoyo para ciudades pequeñas y medianas*. Instituto de Ingeniería, Universidad Nacional Autónoma de México. Accessed June 29, 2015: http://proyectos2.iingen.unam.mx/LACClimateChange/LibroTratamiento.html

Noyola A., Padilla-Rivera, A., Morgan-Sagastume, J. M., Güereca, L. P., and Hernández-Padilla, F. (2012). Typology of wastewater treatment technologies in Latin America. *CLEAN – Soil, Air, Water* **40**(9), 926–932.

Oberts, G. L. (2007). Influence of snowmelt dynamics on stormwater runoff quality. *Watershed Protection Techniques* **1**, 55–61.

Obeysekera, J., and Park, J. (2012). Scenario-based projection of extreme sea levels. *Journal of Coastal Research* **29**, 1–7.

O'Hara, J. K., and Georgakakos, K. P. (2008). Quantifying the urban water supply impacts of climate change. *Water Resources Management* **22**, 1477–1497.

Pahl-Wostl, C. (2007). Transitions towards adaptive management of water facing climate and global change. *Water Resources Management* **21**(1), 49–62.

Parry, M. L. (2009). *Assessing the Costs of Adaptation to Climate Change: A Review of the UNFCCC and Other Recent Estimates*. IIED.

Pegels, P., and Pauw, A. (2013). Private sector engagement in climate change adaptation in least developed countries: An exploration. *Climate and Development* 5(4), 257–267.

Plósz, B. G., Liltved, H., and Ratnaweera, H. (2009). Climate change impacts on activated sludge wastewater treatment: A case study from Norway. *Water Science and Technology* 60, 533–541.

Polade, S. D., Pierce, D. W., Cayan, D. R., Gershunov, A., and Dettinger, M. D. (2014). The key role of dry days in changing regional climate and precipitation regimes. *Nature Scientific Reports* 4(4364), 8.

Porter, K., Wein, A., Alpers, C., Baez, A., Barnard, P., Carter, J., Corsi, A., Costner, J., Cox, D., Das, T., Dettinger, M., Done, J., Eadie, C., Eymann, M., Ferris, J., Gunturi, P., Hughes, M., Jarrett, R., Johnson, L., Dam Le-Griffin, H., Mitchell, D., Morman, S., Neiman, P., Olsen, A., Perry, S., Plumlee, G., Ralph, M., Reynolds, D., Rose, A., Schaefer, K., Serakos, J., Siembieda, W., Stock, J., Strong, D., Sue Wing, I., Tang, A., Thomas, P., Topping, K., and Wills, C. (2011). *Overview of the ARkStorm Scenario*. U.S. Geological Survey open-file report 2010–1312.

PROVIA. (2013). PROVIA guidance on assessing vulnerability, impacts and adaptation to climate change. Consultation document. United Nations Environment Programme.

Rebetez, M., Dupont, O., and Girond, M. (2009). An analysis of the July 2006 heat wave extent in Europe compared to the record year of 2003. *Theoretical and Applied Climatology* 95, 1–7.

Resurreccion, B. P., Sajor, E., and Fajber, E. (2008). *Climate Adaptation in Asia: Knowledge Gaps and Research Issues in South East Asia*. ISET, IDRC, and DFID.

Revi, A., Satterthwaite, D., Aragon-Durand, F., Corfee-Morlot, J., Kiunsi, R. B. R., Pelling, M., Roberts, D., and Solecki, W. (2014). Urban areas. In Field, C. B., Barros, V. R., Dokken, D. J., Mach, K. J., Mastrandrea, M. D., Bilir, T. E., Chatterjee, M., Ebi, K. L., Estrada, Y. O., Genova, R. C., Girma, B., Kissel, E. S., Levy, A. N., MacCracken, S., Mastrandrea, P. R., and White, L. L. (eds.), *Climate Change 2013, Impacts, Adaptation, and Vulnerability—Working Group II Contribution to the Fifth Assessment Report of the Intergovernmental Panel on Climate Change*. Cambridge University Press.

Rosenzweig, C., and Solecki, W. D. (eds.). (2010). Climate change adaptation in New York City–Building a risk management response. *Annals of the New York Academy of Sciences* 1196, 5–6.

Rosenzweig, C., Solecki, W., Hammer, S., and Mehrotra, S. (eds.). (2011). *Climate Change and Cities First Assessment Report of the Urban Climate Change Research Network*. Cambridge University Press.

Rosner, A., Vogel, R. M., and Kirshen, P. H. (2014). A risk-based approach to flood management decisions in a nonstationary world. *Water Resources Research* 50. doi: 10.1002/2013WR014561

Ruth, M., Bernier, C., Jollands, N., and Golubiewski, N. (2007). Adaptation of urban water supply infrastructure to impacts from climate and socioeconomic changes—The case of Hamilton. *New Zealand: Water Resources Management* 21, 1031–1045.

Satterthwaite, D. (2013). The political underpinnings of cities' accumulated resilience to climate change. *Environment and Urbanization* 25(2), 1–11. doi:10.1177/0956247813500902

Satterthwaite, D., and Mitlin, D. (Eds.). (2013). *Empowering Squatter Citizens: Local Government, Civil Society and Urban Poverty Reduction*. Routledge.

Scanlon, B. R., Duncan, I., and Reedy, R. C. (2013b). Drought and the water-energy nexus in Texas. *Environmental Research Letters* 8(4), 045033.

Scanlon, B. R., Reedy, R. C., Duncan, I., Mullican, W. F., and Young, M. (2013a). Controls on water use for thermoelectric generation: Case study Texas, U.S. *Environmental Science & Technology* 47(19), 11326–11334.

Schleich, J., and Hillenbrand, T. (2009). Determinants of residential water demand. *Ecological Economics* 68,1756–1769.

Seager, R., Liu, H., Henderson, N., Simpson, I., Kelley, C., Shaw, T., Kushnir, Y., and Tin, M. (2014). Causes of increasing aridification of the Mediterranean region in response to rising greenhouse gases. *Journal of Climate* 27, 4655–4676.

SEMARNAT. (2013). *National Climate Change Strategy, 10-20-40 Vision*. Ministry of the Environment and Natural Resources (Secretaría de Medio Ambiente y Recursos Naturales), Mexico.

Seneviratne, S. I., et al. (2012). Changes in climate extremes and their impacts on the natural physical environment. In Field, C., et al. (eds.), *Managing the Risks of Extreme Events and Disasters to Advance Climate Change Adaptation—A Special Report if Working Groups I and II of the Intergovernmental Panel on Climate Change* (109–230). Cambridge University Press.

Sillmann, J., Kharin, V. V., Zwiers, F. W., Zhang, X., and Bronaugh, D. (2013). Climate extremes indices in the CMIP5 multimodel ensemble—Part 2: Future climate projections. *Journal of Geophysical Research* 118, 2473–2493.

Souza, C. L., Chernicharo, C. A. L., and Aquino, S. F. (2011). Quantification of dissolved methane in UASB reactors treating domestic wastewater under different operating conditions. *Water Science & Technology* 64(11), 2259–2264.

Stern, N. H. (2006). *Stern Review: The Economics of Climate Change* (Volume 30). HM Treasury.

Stoker, P., and Rothfeder, R. (2014). Drivers of urban water use. *Sustainable Cities and Society* 12, 1–8.

Suárez, F., Muñoz, J. F., Fernández, B., Dorsaz, J. M., Hunter, C. K., Karavitis, C. A., and Gironás, J. (2014). Integrated water resource management and energy requirements for water supply in the Copiapó River basin, Chile. *Water* 6(9), 2590–2613.

Swanson, D., and Bhadwal, S. (eds.). (2009). *Creating Adaptive Policies: A Guide for Policymaking in an Uncertain World*. IDRC.

Swiss Re. (2014). *Mind the Risk: A Global Ranking of Cities under Threat from Natural Disasters*. Swiss Reinsurance Company (Swiss Re).

Syvitski, J. P. M., Kettner, A. J., Overeem, I., Hutton, E. W. H., Hannon, M. T., Brakenridge, G. R., Day, J., Vörösmarty, C., Saito, Y., Giosan, L., and Nicholls, R. J. (2009). Sinking deltas due to human activities. *Nature Geoscience* 2, 681–686.

Taylor, R. G., Scanlon, B., Döll, P., Rodell, M., van Beek, R., Wada, Y., Longuevergne, L., Leblanc, M., Famiglietti, J. S., Edmunds, M., Konikow, L., Green, T. R., Chen, J., Taniguchi, M., Bierkens, M. F. P., MacDonald, A., Fan, Y., Maxwell, R. M., Yechieli, Y., Gurdak, J. J., Allen, D. M., Shamsudduha, M., Hiscock, K., Yeh, P. J. -F., Holman, I., and Treidel, H. (2012). Ground water and climate change. *Nature Climate Change* 3, 322–329.

Tchobanoglous, G., Burton, F. L., and Stensel, H. D. (eds.). (2003). *Wastewater Engineering: Treatment and Reuse*. McGraw-Hill Education.

Tippmann, R., Agoumi, A., Perroy, L., Doria, M., Henders, S., and Goldmann, R. (2013). *Assessing Barriers and Solutions to Financing Adaptation Projects in Africa*. IDRC.

Tyler, S., and Moench, M. (2012). A framework for urban climate resilience. *Climate and Development* 4(4), 311–326.

UN-Habitat. (2014). Water and sanitation. Accessed March 24, 2015: http://unhabitat.org/urban-themes/water-and-sanitation-2/

UN Water. (2013). Water security & the global water agenda. Accessed November 16, 2014: http://www.unwater.org/downloads/watersecurity_analyticalbrief.pdf

Valdez, B., Schorr, M., Quintero, M., Garcia, R., and Rosas, N. (2010). Effect of climate change on durability of engineering materials in hydraulic infrastructure—An overview. *Corrosion Engineering, Science and Technology* 45, 34–41.

van Vliet, M. T. H., Franssen, W. H. P., Yearsley, J. R., Ludwig, F., Haddeland, I., Lettenmaier, D. P., and Kabat, P. (2013). Global river discharge and water temperature under climate change. *Global Environmental Change* 23, 450–464.

Vardon P. J. (2014). Climatic influence on geotechnical infrastructure: A review. *Environmental Geotechnics*. doi: 10.1680/envgeo.13.00055

Vicuña, S., Alvarez, P., Melo, O., Dale, L., and Meza, F. (2014a). Irrigation infrastructure development in the Limarí Basin in Central Chile: Implications for adaptation to climate variability and climate change. *Water International* 39(5), 620–634.

Vicuña, S., Bonelli, S., Bustos, E., and Uson, T. (2014b). *Beyond City Limits: Using a Basin Perspective to Assess Urban Adaptation to*

Climate Change: The Case of the City of Santiago in Chile. Presented at Resilient Cities 2014 Congress.

Vicuña, S., Gironas, J., Meza, F., Cruzat, M., Jelinek, M., Bustos, E., Poblete, D., and Bambach, N. (2013). Exploring possible connections between hydrologic extreme events and climate change in central south Chile. *Hydrological Sciences Journal* 58(8),1598–1619.

Villholth, K. G., Tottrup, C., Stendel, M., and Maherry, A. (2013). Integrated mapping of drought risk in the Southern Africa Development Community (SADC) region. *Hydrogeology Journal* 21, 863–885.

Vuviroli, D., and Weingartner, R. (2008). Water towers—A global view of the hydrological importance of mountains: In E. Wiegandt (ed.), *Mountains-Sources of Water, Sources of Knowledge* (15–20). Springer.

Wadell, J. (2014). *Multi-Actor Flood Governance in Cape Town's Informal Settlements. Unpacking the Barriers to Collaborative Governance.* Presented at Resilient Cities 2014 Congress, Bonn.

Whitehead, P. G., Wilby, R. L., Battarbee, R. W., Kernan, M., and Wade, A. J. (2009). A review of the potential impacts of climate change on surface water quality. *Hydrological Sciences Journal* 54, 101–123.

Wiley, M. W., and Palmer, R. N. (2008). Estimating the impacts and uncertainty of climate change on a municipal water supply system. *Journal of Water Resources Planning and Management* 134, 239–246.

Wong, P. P., Losada, I. J., Gattuso, J. -P., Hinkle, J.,Khattabi, A., McInnes, K., Saito, Y., and Sallenger, A. (2014). Coastal systems and low-lying areas. In Field, C. B., Barros, V. R., Dokken, D. J., Mach, K. J., Mastrandrea, M. D., Bilir, T. E., Chatterjee, M., Ebi, K. L., Estrada, Y. O., Genova, R. C., Girma, B., Kissel, E. S., Levy, A. N., MacCracken, S., Mastrandrea, P. R., and White, L. L. (eds.), *Climate Change 2013, Impacts, Adaptation, and Vulnerability—Working Group II Contribution to the Fifth Assessment Report of the Intergovernmental Panel on Climate Change* (361–409). Cambridge University Press.

World Bank. (2014). Integrated urban water management. Accessed March 12, 2016: http://water.worldbank.org/iuwm

WHO-UNICEF. (2014). *Progress on Drinking Water and Sanitation.* WHO-UNICEF.

Zhou, S. L., McMahon, T. A., Walton A., and Lewis, J. (2000). Forecasting daily urban water demand—A case study of Melbourne. *Journal of Hydrology* 236,153–164.

Zhou, S., McMahon, T., and Wang, Q. (2001). Frequency analysis of water consumption for metropolitan area of Melbourne. *Journal of Hydrology* 247, 72–84.

Ziervogel, G., Shale, M., and Du, M. (2010). Climate change adaptation in a developing country context: The case of urban water supply in Cape Town. *Climate and Development* 2, 94–110.

Chapter 14   Case Study References

Case Study 14.1   Climate Adaptation through Sustainable Urban Water Development in Can Tho City, Vietnam

Can Tho DONRE. (2008). *Report on Current Situation of Environment of Can Tho City*. Department of Natural Resources and Environment, Can Tho City.

Cantho Portal. (2015). Cantho City. Accessed February 6, 2016: http://cantho.gov.vn/

Maheepala, S., Blackmore, J., Diaper, C., Moglia, M., Sharma, A., and Kenway, S. (2010). *Manual for Adopting Integrated Urban Water Management for Planning, Water*. Research Foundation.

Moglia, M., Nguyen, M. N., Neumann, L. E., Cook, S., and Trung, N. H. (2013). Integrated assessment of water management strategies: The case of Can Tho City, Vietnam. The 20th International Congress on Modelling and Simulation MODSIM 2013, MSSANZ, Adelaide, November 2013.

Neumann, L., Nguyen, M., Moglia, M., Cook, S., and Lipkin, F. (2011). Urban water systems in Can Tho, Vietnam: Understanding the current context for climate change adaptation. CSIRO Technical Report EP115086, CSIRO Land and Water, Climate Adaptation Flagship, Highett, VIC, Australia.

Neumann, L. E., Moglia, M., Cook, S., Nguyen, M. N., Sharma, A. K., Trung, N. H., and Nguyen, B. V. (2013). Water use, sanitation and health in a fragmented urban water system: Case study and household survey. *Urban Water Journal* 11(3), 198–210. doi: 10.1080/1573062X.2013.768685

Nguyen, M., Cook, S., Moglia, M., Neumann, L. E., Trung, N. H. (2012). *Planning for Sustainable Urban Water Systems in Adapting to a Changing Climate: A Case Study in Can Tho City, Vietnam.* Commonwealth Scientific and Industrial Research Organisation (CSIRO Australia).

Peel, M. C., Finlayson, B. L., and McMahon, T. A. (2007). Updated world map of the Köppen-Geiger climate classification. *Hydrology and Earth System Sciences Discussions* 4(2), 462.

Sharma, A. K., Tjandraatmadja, G., Grant, A. L., Grant, T., and Pamminger, F. (2010). Sustainable sewerage servicing options for peri-urban areas with failing septic systems. *Water Science & Technology* 62(3), 570–585.

Statistical Office of Can Tho City. (2009). Statistical Yearbook Can Tho City. Accessed November 16, 2014: http://www.gso.gov.vn/default_en.aspx?tabid=509&idmid=1&itemid=2656

Trung., N. H., Nguyen, M. N., Tri, L. Q., Tuan, D. D. A., Thinh, L. V., Doan, T. C., Loc, N. H., Neumann, L., Cook, S., and Moglia, M. (2013). *A Mapbook of Water Systems and Environment of Can Tho City.* Can Tho University Press.

Trung, N. H., Tuan, D. D. A., Hoang, N. X., Tri, L. Q., and Nguyen, M. N. (2014a). *Rainwater Harvesting Guidebook for the Mekong Delta.* Agriculture Publishing House.

Trung, N. H., Tuu, N. T., Doan, T. C., Thinh, L. C., Tuan, D. D. A., and Nguyen, M. (2014b). Application of GIS to support urban water management in adapting to a changing climate: A case study in Can Tho City, Vietnam. International Symposium on Geoinformatics for Spatial Infrastructure Development in Earth and Allied Sciences, Da Nang, Vietnam, December 2014.

World Bank. (2017). 2016 GNI per capita, Atlas method (current US$). Accessed August 9, 2017: http://data.worldbank.org/indicator/NY.GNP.PCAP.CD

Case Study 14.2   Using a Basin-Level Approach to Address Climate Change Adaptation of Urban Water Supply: The Case of Santiago, Los Angeles, and Bangalore

Bonelli, S., Vicuña, S., Meza, F. J., Gironas, J., and Barton, J. 2014, Incorporating climate change adaptation strategies in urban water supply planning: The case of central Chile. *Journal of Water and Climate Change* 5(3), 357–376.

Greater Los Angeles County Region. (2013). Integrated regional water management 2013 plan update. Accessed April 26, 2014: https://dpw.lacounty.gov/wmd/irwmp/

Instituto Nacional de Estadistica (INE). (2012). Compendio Estadistico. Estadisticas Demograficas. Accessed March 12, 2016: http://www.ine.cl/canales/menu/publicaciones/compendio_estadistico/pdf/2012/estadisticas_demograficas_2012.pdf

Lele, S. V., Jamwal, S. P., Thomas, B. K., Eswar, M., and Zuhail, T. (2013). *Water Management in Arkavathy Basin: A Situation Analysis.* Environment and development discussion paper No. 1. Ashoka Trust for Research in Ecology and the Environment.

Meza, F. J., Vicuña, S., Jelinek, M., Bustos, E., and Bonelli, S. (2014). Assessing water demands and coverage sensitivity to climate change in the urban and rural sectors in central Chile. *Journal of Water and Climate Change* 5(2), 192–203.

Peel, M. C., Finlayson, B. L., and McMahon, T. A. (2007). Updated world map of the Köppen-Geiger climate classification. *Hydrology and Earth System Sciences Discussions* **4**(2), 462.

Puertas, O. L., Henríquez, C., and Meza, F. J. (2014). Assessing spatial dynamics of urban growth using an integrated land use model. Application in Santiago Metropolitan Area, 2010–2045. *Land Use Policy* **38**, 415–425.

Vicuña, S., Garreaud, R., and McPhee, J. (2010). Climate change impacts on the hydrology of a snowmelt driven basin in semiarid Chile. *Climatic Change* **105**(3–4), 469–488.

Vicuña, S., and Meza, F. (2012). Los nuevos desafíos para los recursos hídricos en Chile en el marco del cambio global', Centro de Políticas Públicas UC. Temas de la Agenda Pública. Año 7 No. 55. Santiago, Chile. Accessed August 3, 2014: http://politicaspublicas.uc.cl/publicaciones/ver_publicacion/112

World Bank. (2017). 2016 GNI per capita, Atlas method (current US$). Accessed August 9, 2017: http://data.worldbank.org/indicator/NY.GNP.PCAP.CDf

Case Study 14.3 Denver, Seattle, Tucson: How Can Climate Research Be Useful for Urban Water Utility Operations?

Ferguson, D. B., Rice, J., and Woodhouse, C. (2014). *Linking Environmental Research and Practice: Lessons from the Integration of Climate Science and Water Management in the Western United States.* Climate Assessment for the Southwest.

Peel, M. C., Finlayson, B. L., and McMahon, T. A. (2007). Updated world map of the Köppen-Geiger climate classification. *Hydrology and Earth System Sciences Discussions* **4**(2), 462.

United Nations Development Programme (UNDP). (2014). Human Development Index (HDI). Accessed January 20, 2015: http://hdr.undp.org/sites/default/files/hdr14_statisticaltables.xls

U.S. Census Bureau. (2010). Decennial census, summary file 1. Accessed March 12, 2016: http://www.census.gov/population/metro/files/CBSA%20Report%20Chapter%203%20Data.xls

World Bank. (2017). 2016 GNI per capita, Atlas method (current US$). Accessed August 9, 2017: http://data.worldbank.org/indicator/NY.GNP.PCAP.CD

Case Study 14.4 Operationalizing Urban Climate Resilience in Water and Sanitation Systems in Manila

Cheng, D. (2014). The persistence of informality: Small-scale water providers in Manila's post-privatisation era. *Water Alternatives* **7**(1), 54–71.

Doczi, J. (2012). *Quantifying the Impact of Water and Sanitation Service with a Climate Resilience Perspective: A Case Study in Metro Manila, Philippines.* Master's dissertation, University of East Anglia.

Luz, J. M., and Paladio-Melosantos, M. L. (2012). Manila, Philippines. In Chiplunkar, A., Seetharam, K., and Tan, C. K. (eds.), *Good Practices in Urban Water Management: Decoding Good Practices for a Successful Future.* Asian Development Bank and National University of Singapore.

Manila Water. (2007). *Climate Change Policy – 2007.* Manila Water.

Manila Water. (2013). *Climate Change Policy – 2013.* Manila Water.

Peel, M. C., Finlayson, B. L., and McMahon, T. A. (2007). Updated world map of the Köppen-Geiger climate classification. *Hydrology and Earth System Sciences Discussions* **4**(2), 462.

Philippine Statistics Authority. (2015). Quickstat. On national capital region – August 2015. Accessed January 9, 2016: https://psa.gov.ph/sites/default/files/attachments/ird/quickstat/Quickstat_ncr_0.xls

Rivera, V. C., Jr. (2014). *Tap Secrets: The Manila Water Story.* Asian Development Bank and Manila Water Company, Inc.

World Bank. (2017). 2016 GNI per capita, Atlas method (current US$). Accessed August 9, 2017: http://data.worldbank.org/indicator/NY.GNP.PCAP.CD

Wu, X., and Malaluan, N. A. (2008). A tale of two concessionaires: A natural experiment of water privatisation in Metro Manila. *Urban Studies* **45**(1), 207–229.

Yusuf, A. A., and Francisco, H. (2009). *Climate Change Vulnerability Mapping for Southeast Asia.* Economy and Environment Program for Southeast Asia.

Case Study 14.5 New Citizen Roles in Climate Change Adaptation: The Efforts of the Middle-Sized Danish City Middelfart

Brown, R. R., Keath, N., and Wong, T. H. F. (2009). Urban water management in cities: Historical, current and future regimes. *Water Science & Technology* **59**(5), 847–855.

KlimaByen. (2015). The Climate City project – The most beautiful climate adaptation project in Denmark. Accessed February 12, 2016: http://www.klima-byen.dk/the%20climate%20city

Middelfart Kommune. (2014). Forslag til klimatilpasningsplan (Municipal climate change adaptation plan) p. 22. Accessed January 20, 2014: https://www.middelfart.dk/~/media/Files/Planer/Kommuneplan/Klimatilpasningsplan%20HANDLEPLAN%20samlet.ashx

Peel, M. C., Finlayson, B. L., and McMahon, T. A. (2007). Updated world map of the Köppen-Geiger climate classification. *Hydrology and Earth System Sciences Discussions* **4**(2), 462.

Statistics Denmark. (2015a). Area by region. Accessed April 13, 2016: http://www.statbank.dk/statbank5a/default.asp?w=1680

Statistics Denmark. (2015b). Official population statistics for Q3–2015. Accessed February 12, 2016: http://www.statbank.dk/statbank5a/default.asp?w=1680

Teknologisk Institut. (2015). Water in urban areas project. Accessed February 4, 2016: http://vandibyer.dk/english/

World Bank. (2017). 2016 GNI per capita, Atlas method (current US$). Accessed August 9, 2017: http://data.worldbank.org/indicator/NY.GNP.PCAP.CD

15

Urban Solid Waste Management

Coordinating Lead Authors

Martin Oteng-Ababio (Accra)

Lead Authors

Ranjith Annepu (Hyderabad/New York/Abu Dhabi), A. C. 'Thanos' Bourtsalas (New York/London), Rotchana Intharathirat (Udon Thani), Sasima Charoenkit (Phitsanulok)

Contributing Authors

Nicole Kennard (Atlanta)

This chapter should be cited as

Oteng-Ababio, M., Annepu, R., Bourtsalas, A., Intharathirat, R., and Charoenkit, S. (2018). Urban solid waste management. In Rosenzweig, C., W. Solecki, P. Romero-Lankao, S. Mehrotra, S. Dhakal, and S. Ali Ibrahim (eds.), *Climate Change and Cities: Second Assessment Report of the Urban Climate Change Research Network*. Cambridge University Press. New York. 553–582

Managing and Utilizing Solid Waste

Municipal solid waste (MSW) management is inextricably linked to increasing urbanization, development, and climate change. The municipal authority's ability to improve solid waste management also provides large opportunities to mitigate climate change and generate co-benefits, such as improved public health and local environmental conservation.

Driven by urban population growth, rising rates of waste generation will severely strain existing MSW infrastructure in low- and middle-income countries. In most of these countries, the challenge is focused on effective waste collection and improving waste treatment systems to reduce greenhouse gas (GHG) emissions. In contrast, high-income countries can improve waste recovery through reuse and recycling and promote upstream interventions to prevent waste at the source.

Because stakeholder involvement, economic interventions, and institutional capacity are all important for enhancing the solid waste management, integrated approaches involving multiple technical, environmental, social, and economic efforts will be necessary.

Major Findings

- Globally, solid waste generation was about 1.3 billion tons in 2010. Due to population growth and rising standards of living worldwide, waste generation is likely to increase significantly by 2100. A large majority of this increase will come from cities in low- and middle-income countries, where per capita waste generation is expected to grow.

- Up to 3–5% of global GHG emissions come from improper waste management. The majority of these emissions are methane – a gas with high greenhouse potential – that is produced in landfills. Landfills, therefore, present significant opportunities to reduce GHG emissions in high- and middle-income countries.

- Even though waste generation increases with affluence and urbanization, GHG emissions from municipal waste systems are lower in more affluent cities. In European and North American cities, GHG emissions from the waste sector account for 2–4% of the total urban emissions. These shares are smaller than in African and South American cities, where emissions from the waste sector are 4–9% of the total urban emissions. This is because more affluent cities tend to have the necessary infrastructure to reduce methane emissions from MSW.

- In low- and middle-income countries, solid waste management represents 3–15% of city budgets, with 80–90% of the funds spent on waste collection. Even so, collection coverage ranges from only 25% to 75%. The primary means of waste disposal is open dumping, which severely compromises public health.

- Landfill gas-to-energy is an economical technique for reducing GHG emissions from the solid sector. This approach provides high potential to reduce emissions at a cost of less than US$10 per tCO_2-eq. However, gas-to-energy technology can be employed only at properly maintained landfills and managed dumpsites, and social aspects of deployment need to be considered.

Key Messages

Reducing GHG emissions in the waste sector can improve public health; improve quality of life; and reduce local pollution in the air, water, and land while providing livelihood opportunities to the urban poor. Cities should exploit the low-hanging fruit for achieving emissions reduction goals by using existing technologies to reduce methane emissions from landfills. In low- and middle-income countries, the best opportunities involve increasing the rates of waste collection, building and maintaining sanitary landfills, recovering materials and energy by increasing recycling rates, and adopting waste-to-energy (WTE) technologies. Resource managers in all cities should consider options such as reduce, re-use, recycle, and energy recovery in the waste management hierarchy

15.1 Introduction

Municipal solid waste (MSW) management is inextricably linked to urbanization, development, and climate change (Oteng-Ababio, 2014). Currently, over half of the global population and a significant portion of the human livelihood activities that impact global climate change are concentrated in cities (Rayner and Malone, 1997; Kates et al., 1998; O'Meara, 1999). Estimated urban population in 2050 – 6 billion – will be equal to the world's entire population in 2000 (UN-Habitat, 2014). This provides considerable opportunities for city authorities to shape appropriate policies over land-use planning and play a more important role in transportation issues and energy consumption, all of which have implications for greenhouse gas (GHG) emissions (Collier, 1997; Rayner and Malone, 1997; Agyeman et al., 1998; DeAngelo and Harvey, 1998; Kates et al., 1998; Bulkeley, 2000). The authority's ability to improve MSW management also provides opportunities to mitigate climate change and generate co-benefits such as improved public health and local environmental conservation.

Globally, rates of waste generation have been increasing. It is forecasted that the volume of MSW will double from the current waste generation rate of 1.3 billion tons per year in 2012 to 2.2 billion tons per year by 2025 (World Bank, 2012). The highest rate of waste generation is projected for the Asia-Pacific region,

particularly in China, as shown in Figure 15.1, which presents the projected MSW generation globally in 2012 and 2025. These growing waste generation rates in developing countries experiencing increasing affluence have been phenomenal (UN-Habitat, 2011). Even though waste generation increases with affluence and urbanization (Barker et al., 2007), GHG emissions from MSW are lower in more affluent cities. In European and North American cities, GHG emissions from the waste sector account for 2.29–4.32% of the total urban GHG emissions. These shares are smaller than those of cities in Africa and South America, which have a higher share of GHG emissions from the waste sector, about 4.48–9.36% of their total urban GHG emissions (Marcotullio et al., 2014). This is because more affluent cities tend to have the appropriate infrastructure to reduce methane (CH_4) emissions from MSW that contribute to global anthropogenic GHG emissions (UNFCCC, 2005). Driven by urban population growth, increasing waste generation rates will severely strain existing MSW infrastructure in the urban areas of low- and middle-income countries.

While urbanization is a challenge, it creates a high concentration of people and services, which provides an opportunity to deliver efficient MSW services. In most developing countries, the challenge relates more to effective waste collection and better waste treatment systems to reduce GHG emissions from the waste sector. In contrast, developed countries have to contend with improved waste recovery through reuse and recycling as well as upstream

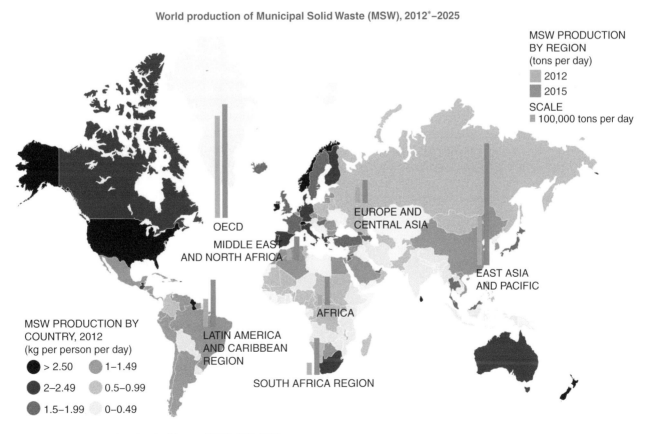

Figure 15.1 *World production of municipal solid waste (MSW), 2012–2025.*

Source: World Bank, 2012

interventions to prevent waste at source. MSW technologies alone are not sufficient to handle ever-growing waste problems. Stakeholder involvement, economic interventions, and institutional capacity are all important for enhancing the management of MSW. Therefore, multiple efforts involving with technical, environmental, social, and economic aspects must be considered when dealing with the complex task of MSW, thus indicating the need for an integrated approach to MSW management (Marshall and Farahbakhsh, 2013).

This chapter examines the concept of integrated solid waste management (ISWM). It demonstrates the current practices found in both developed and developing countries and highlights the challenges in achieving effective MSW management. The chapter examines ways through which proper, well-planned, and efficient SWM systems can mitigate climate change. The Case Studies suggest a number of ways in which city governments can address GHG emissions and also highlights several obstacles for local decision-makers. It also emphasizes the impacts of MSW on climate change emissions and climate change impacts related to discarded materials. Section 15.2 provides the background of the ISWM concept by focusing the definition of MSW and the development from the waste hierarchy concept to ISWM. Section 15.3 presents an overview of current MSW practices in developed and developing countries and their challenges in delivering better MSW management. Section 15.4 describes GHG emissions from MSW management practices, followed by the impacts of MSW on climate change in Section 15.5. Last, Section 15.6 focuses on the carbon market as a financial opportunity for GHG mitigation from waste.

15.2 Sustainable Solid Waste Management

15.2.1 MSW: Definition, Quantity, and Composition

The definition of MSW can be highly varied among countries. Usually, MSW refers to solid waste generated from community activities (e.g., residential, commercial, and business establishments). While construction waste and hazardous wastes are excluded as MSW in European countries, they are considered as MSW in most developing countries (Karak et al., 2012). Box 15.1 shows the comprehensive list of sources of MSW throughout the world.

Despite the inclusion of construction and hazardous wastes in MSW in some developing countries, the amount of MSW generated from developing countries is 648 million tons per year (World Bank, 2012). Notably, this amount generated by more than 170 countries is nearly as same as waste generation in developed countries comprising only 20 countries. The average values of waste generation per capita of developing countries are therefore relatively low, with a range of 0.45–1.1 kilograms per day, in comparison to those countries in the Organization for Economic Cooperation and Development (OECD) with an average value of 2.2 kilogram per person per day (World Bank, 2012).

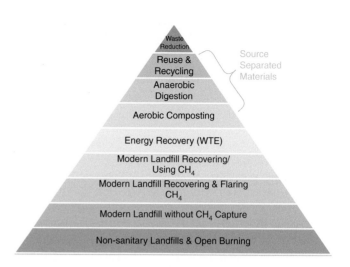

Figure 15.2 *The hierarchy of sustainable solid waste management.*
Source: Kaufman and Themelis, 2010

Common types of MSW are biodegradable material (e.g., food, garden waste), recyclable waste (e.g., paper, glass, metal, plastics), and other (e.g., textiles, leather). Based on the World Bank's data (2012), most low- and upper-income countries have a higher proportion of biodegradable waste, accounting for about 40–80% of the total volume of MSW. On the other hand, high-income countries have different waste composition. They have a higher share of paper and glass with a much lower portion of biodegradable waste, which represents about 30–40%.

15.2.2 Solid Waste Management Hierarchy

An important roadmap to ensure sustainable SWM is to address the concept of the sustainable SWM hierarchy, which is well recognized throughout the world (Kaufman and Themelis, 2010) (see Figure 15.2). Technically, all waste management strategies must aim primarily to prevent the generation of waste and to reduce its harmfulness. Where this is not possible, waste materials should be reused, recycled or recovered, or used as a source of energy. As a final resort, waste should be disposed of safely (e.g., in sanitary landfills or monitored dumpsites).

Box 15.1   The Definition of Municipal Solid Waste

Municipal solid waste (MSW), which is commonly called garbage, trash, or refuse refers to waste generated from the following activities:
- Residential (single and multifamily dwellings)
- Commercial (offices, stores, hotels, restaurants)
- Institutional (schools, prisons, hospitals, airports)
- Industrial (manufacturing, fabrication, etc., when the municipality is responsible for their collection)
- Nonrecycled construction and demolition debris
- Municipal services (street cleaning, landscaping)

As a concept, the historical antecedent of the waste hierarchy is traceable to the 1970s, when some environmental movements raised concerns about the then wholesale waste disposal-based approaches (Gertsakis and Lewis, 2003). The hierarchy is based on the "4Rs": reduce, reuse, recycle, and resource recovery (see Box 15.2), which collectively ensure waste reduction. As far as possible, disposal of materials in landfills should be considered only if none of the 4Rs is applicable and is therefore ranked lowest in priority.

15.2.3 Integrated Solid Waste Management (ISWM)

The concept of ISWM has been developed as a comprehensive approach that considers multidimensional aspects of SWM management in the integrated manner (McDougall et al., 2001). Both technical and nontechnical aspects of SWM must be incorporated because they are interdependent (UNEP, 2005). The aim of ISWM is to achieve a sustainable solution balancing environmental effectiveness, social acceptability, and economic affordability (McDougall et al., 2001; van de Klundert and Angchutz, 2001). As shown in Figure 15.3, ISWM requires stakeholder involvement and the consideration of six main aspects – environmental, political/legal, institutional, sociocultural, financial, and technical – for making decisions on waste systems that consist of the methods that will be used to sort, collect, transport, treat and dispose, reduce, reuse, recycle, and recover from waste. Unlike the priority order of the waste hierarchy, ISWM proposes a flexible framework for waste treatment systems. Rather than prioritizing reduction, recycling, and reuse of waste over treatment or

Box 15.2 Defining the 4Rs with the Waste Hierarchy

Reduce: This refers to waste avoidance and materials management (i.e., avoiding or reducing primary/virgin materials for manufacturing and preserving natural resources). This requires reducing financial and environmental resources in the collection, transport, treatment, and disposal of waste. For example, wastage can be minimized through reduced packaging, improved design, and use of durable materials.

Reuse: This refers to the practice of using materials over and over again for the same purpose for which they were intended. Reusing waste may require collection but relatively little or no processing.

Recycle: This refers to any activity that involves the collection, sorting, and processing of used or unused items that would otherwise be considered as waste into raw material that is then remanufactured into new products.

Resource recovery: This encompasses recycling, reprocessing, and energy recovery consistent with the most efficient use of the waste material. Resource recovery includes converting organic matter into useable products (such as compost and digestate) or energy recovery in the form of electricity and/or heat.

Disposal: If none of the above options is possible, then waste should be disposed of in a controlled manner. This includes using a sanitary landfill or pretreating the waste in other ways to prevent harmful impacts on public health or the environment.

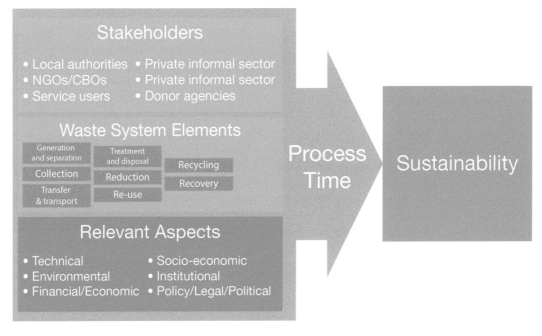

Figure 15.3 *Integrated solid waste management (ISWM) framework.*

Source: van de klundert and Angchutz, 2001

disposal, ISWM focuses on the use of a range of different methods to optimize resource conservation and limit final disposal (UNEP, 2005; Nordone et al., 1999). The combination of appropriate treatment methods, such as recycling, incineration, and landfilling, is necessary for MSW management because no single solution is sufficient for MSW disposal problems (Hoornweg and Bhada-Tata, 2012, in Menikpura et al., 2013).

15.3   MSW Management Practices Worldwide

MSW practices can be divided into four main activities:

- *Sorting and collection*: Waste sorting is the process of separating MSW into different types. Waste sorting can occur before or after the waste is collected. The collection process involves collecting waste from households, from community and street bins, or from bulk generators into larger containers or vehicles. It extends to activities such as driving between stops, idling, loading, and on-vehicle compaction of waste.
- *Recycling*: After waste sorting, recyclables are reprocessed into products.
- *Transfer and transportation*: This process involves the delivery of collected waste to transfer stations or treatment facilities.
- *Treatment and disposal*: Waste treatment is the process of disposing of waste after collection. Waste can be buried at landfills or burned through the incineration process. Non-recyclable waste items can be converted into compost or energy as various forms of useable heat, electricity, or fuel.

These four activities are used to analyze the current MSW practices as described next.

15.3.1   MSW Management Practices in Developed Countries

As recently as the 1970s, the practices in most developed countries were similar to the present situation in most transitional/developing countries with low waste collection rates and improper disposal of waste. The stimulus for change was damage to public health due to improper practices inside cities, at disposal sites, or in surface water or groundwater, which attracted several political and media commentaries (UN-Habitat, 2010). At this point, it became necessary to phase out open dumps and develop and operate state-of-the-art sanitary landfills. Another strong influence was public opposition (not-in-my-backyard or NIMBY) to new waste management projects, based at least in part on bad experiences with previously uncontrolled sites. Over the past few decades, high-income countries gradually overcame NIMBY by making project planning and implementation transparent, engaging communities, and implementing safe management practices.

15.3.1.1   *Waste Sorting and Collection*

In developed countries, the public is often trained to segregate their waste. Households usually carry out waste sorting at the source. Collected waste is segregated further at material sorting facilities, and the materials are recovered through recycling. For source sorting, waste is put into different containers, such as bags, bins, or rack sacks. The number of waste streams separated varies depending on the waste collection policy of each municipality. Usually, glass and paper are sorted at the source, and biodegradable waste is sorted and collected separately in most municipalities (Xevgenos et al., 2015).

Developed countries have developed technologically advanced SWM systems with waste collection coverage of greater than 90%. Collection systems that have a high degree of mechanical handling of the waste are usually employed in high-income countries; these include:

- *Full-service schemes* (door-to-door) and curbside collection services in neighborhoods where collection trucks provide door-to-door or curbside services
- *Drop-off systems* (or communal container collection); this system requires individual households to bring their waste to containers placed for each community or to drop-off centers
- *Private cars delivering waste to collection points* or driving a personal car with a load of waste to a facility and back to the house
- *Pneumatic pipe systems* used by multifamily neighborhoods; these can transport waste from a building's garbage chute using a vacuum created by electric fans and vents to a central collection point. Waste from the central collection points is then collected by trucks (Rypdal and Winiwarter, 2001; Monni et al., 2006).

Dual collection and highly mechanized vehicles like compactor trucks are commonly used for collecting waste. The former is efficient for collecting recyclables and the latter has the capacity to carry a large amount of dry waste (UNEP, 2005).

In addition, railways and barges are some of the most commonly used modes of waste transportation, although communities that want to use barges or railways should have access to existing port and rail infrastructure, respectively (Mogensen and Holbech, 2007; Saxena, 2009).

15.3.1.2   *Recycling*

Recyclable waste items, which are sorted at the source or at recycling centers, are reprocessed into products. There are three types of reprocessing operation: upcycling, recycling, and downcycling. *Upcycling* is the reprocessing of waste materials into products that have higher value than the original. *Recycling* processes produce products that have the same value with the original or can be used for the same purpose, whereas *downcycling* converts waste materials into products that have less value than the original and serve for lower application. On average, 22% of the total MSW was recycled in high-income countries (World Bank, 2012), and an ambitious target of 50% of municipal waste to be recycled by 2020 is set for European countries (European Union [EU], 2010).

15.3.1.3 Waste Transfer and Transport

Collected waste is delivered to transfer stations, recycling centers, treatment facilities, or disposal sites through optimized routes.

15.3.1.4 Treatment and Disposal

Two technologies are widely applied in developed countries for MSW treatment:

- *Thermal technologies* refer to technologies that operate at high temperatures to produce heat or electricity as a primary byproduct. Thermal technologies, such as gasification and pyrolysis, are the advanced form of incineration and are suitable for processing dry waste with low moisture content.
- *Biological technologies* require lower temperatures than thermal technologies for the operation. Examples of these technologies are anaerobic digestion, composting, biodiesel, and catalytic cracking. They are considered appropriate treatment systems for biodegradable waste. Byproducts of these technologies include electricity, biogas, compost, and chemicals.

Approximately, 21% and 11% of MSW was incinerated and composted in high-income countries (World Bank, 2012), respectively. Apart from energy and other byproducts, solid residuals are created during waste treatment operations. These solid residuals, which represent about 42.5% of MSW, are then disposed of at sanitary landfills using a system to capture landfill gases for energy recovery or flaring.

A critical component that keeps such highly advanced SWM systems running in developed countries is the government's ability to implement existing policies and regulations, which requires significant human and financial resources (World Bank, 2012). Table 15.1 presents a summary of instruments adopted in developed countries to promote sustainable waste management and efficient resource use.

15.3.2 MSW Practices in Developing Countries

15.3.2.1 Waste Sorting and Collection

Unlike high-income countries where there is public awareness of waste sorting, sorting activities at the household level in low- and middle-income countries are still limited. Therefore, the MSW generally consists of mixed waste containing food and other types of waste. Waste sorting is usually conducted by poor families to earn extra income from selling recyclable materials. Despite the high amount of municipal budget spent for waste collection, about 80–90% of the total MSW budget, the efficiency of MSW collection is still very low in many countries, particularly those in Sub-Saharan Africa, which has collection rates ranging from 17.7% to 55% (World Bank, 2012). Due to low waste collection efficiency, dumping waste on the roadside is a common practice (APO, 2007). Although the use of covered and compactor trucks for collecting waste is increasing, transporting MSW by inefficient and open vehicles is a common practice in urban areas (APO, 2007).

15.3.2.2 Recycling

In contrast to high-income countries where recyclables are collected through curbside or drop-off systems, informal sectors, like the waste pickers sector, play a significant but largely unrecognized role in handling such activities (Troschinetz and Mihelcic, 2009). Waste pickers collect recyclable waste materials from collection and disposal points, and mobile purchasers carry out house-to-house services (APO, 2007). Informal recycling contribution varies in different cities. It is estimated that the recycling rate of MSW collected by waste pickers is in the range of 3–8% of waste transported to disposal sites in Indonesia (Sasaki and Araki, 2014) and 15% and 20% of the waste generated in India and Vietnam, respectively (Chintan, 2009; APO, 2007). Figure 15.4 shows several waste pickers in Accra, Ghana, sorting and baling recyclable materials prior to taking them to their respective recycling companies. Baling reduces the volume of waste to be collected and optimizes transport. The participating shops save significantly on their monthly waste collection bills.

15.3.2.3 Waste Transfer and Transport

In developing cities, street networks characterized by improper planning, small size, and poor condition are a major barrier to enhancing the efficiency of waste collection.

15.3.2.4 Waste Treatment and Disposal

Open dumping and landfilling (see Figure 15.5) are the most common methods of MSW disposal in developing countries, mainly because they are cheap when social and environmental impacts are not considered (Renou et al., 2008; Ali et al., 2014). Together, these two methods account for about 70–90% of the total MSW (World Bank, 2012). Compared to other treatment methods, open dumping and landfilling pose the highest risk to environmental and human health, causing deterioration of soil and water quality, air pollution, and the spread of disease by insects and rodents. The use of incineration is still very limited due to high investment cost and inappropriate waste composition dominated by inerts and biodegradable waste. Composting has long been promoted as a mean of treating biodegradable waste and generating extra income for communities; however, several problems arise for composting practices at household and municipal levels. Most households lack motivation for separating their food waste, and composting plants operated by municipalities often face technical problems due to lack of expertise and the use of mixed MSW, which produces poor-quality compost (Furedy, 2004). The application of composting practices is therefore still limited to small-scale or pilot projects.

The overview of MSW management practices in both developed and developing countries is given in Table 15.2. It clearly shows several problems in the handling, collection, transfer and transport, treatment, and disposal of solid waste in developing countries. This implies an urgent need to shift from poor management of MSW to a more effective MSW management system

Table 15.1 *Economic and regulatory instruments employed in developed countries to achieve sustainable management of MSW. Source: Xevgenos et al., 2015*

TYPE OF INSTRUMENT	OBJECTIVE	REQUIREMENT	ISSUES
Economic instruments			
Extended Producer Responsibility (EPR)	To extend the responsibility of post-consumer waste to the manufacturers of the goods by requiring them to redesign products using fewer materials and with increased recycling potential. Different waste streams are included in EPR schemes such as packaging waste, batteries, end-of-life-vehicles (ELVs), oils, waste electronic and electronic devices (WEEEs).	The operational responsibilities for a producer responsibility scheme are: 1. Physical responsibility for collecting the products 2. Financial responsibility for paying fees to support collecting and recycling activities. The fees are country-specific and can be weight-based or/and material-specific.	Requirement for clear distinctions in the allocation of the operational responsibilities between the local authorities and the industry in collecting wastes.
Deposit-Refund	To increase and capturing the used packaging (i.e., mainly beverage bottles/cans) for recycling.	Customers are required to pay for a deposit on top of the product's price, when they buy a product. This deposit will be reimbursed to them partially or fully when the product is returned to a trader or a specialized treatment facility.	Decrease in the use of reusable packaging leads to the higher prices of products contained in reusable containers and the preference of users for separate collection in view of convenience.
Landfill/Incineration Tax	To internalize external costs of landfilling, to provide incentives for diverting waste from landfills.	A charge levied by a public authority to the individual households on the disposal of waste and is usually calculated based on the amount of waste disposed (weight-based).	Lack of a direct incentive to citizens for reducing their waste because the tax is not based on the amount of waste generated by each household. Consideration of optimized charges as low charges may not provide sufficient incentive against landfilling while high charges may lead to illegal disposal.
Pay-As-You-Throw (PAYT)	To internalize external costs of MSW disposal based on the amount of waste disposed by each household. This is aimed to create an incentive for households to recycle more and to generate less waste.	Residents are charged for the collection of municipal solid waste – ordinary household trash – based on the amount of waste they generate.	High charge rate can lead to illegal waste dumping. Suitable rate for different waste stream. It is suggested that a PAYT scheme should charge: a) The highest fee for residual waste b) A lower fee for biowaste c) Zero fee for kitchen waste d) A low or zero fee for dry recyclables
Regulatory Instruments			
Bans and Restrictions (e.g., a landfill ban)	To reduce dependency on landfills and to shift waste management up the waste hierarchy.	A variety of bans and restrictions are found in different countries such as. Landfill bans on unsorted/untreated waste or residual waste. Restriction on separated waste collection. A ban on the use of plastic bags in restaurants, large supermarkets, and all retail stores.	Bans and restrictions on residual waste have low potential for material recovery and usually result in increased incineration rates. Although bans and restrictions are powerful means, they do not create revenues.
Mandatory Source Separation	To increase recycling rate and enhance the efficiency of MSW management.	The municipalities or households must comply with the requirement to separate waste before disposal. Those who do not comply with the requirement have to pay a fine.	Requires a consideration of the number of waste streams needed to be separated.

Case Study 15.1 The Solid Waste Management Challenge of a Rapidly Developing Economy City: The Case of Rio de Janeiro

A. C. (Thanos) Bourtsalas

Earth and Environmental Engineering, Columbia University, New York

Keywords	Municipal waste management, poor source separation, formalized informal sector
Population (Metropolitan Region)	11,835,708 (IBGE, 2015)
Area (Metropolitan Region)	5,328.8 km² (IBGE, 2015)
Income per capita	US$8,840 (World Bank, 2017)
Climate zone	Am –Tropical monsoon (Peel et al., 2007)

Case Study 15.1 Figure 1 *Rio de Janeiro landfill.*

Like many municipalities in Brazil, waste collection and disposal services in Rio de Janeiro are coordinated by the municipality of Rio (COMLURB), and collection fees are incorporated into Rio's property taxes (Monteiro, 2013; IBGE, 2015). Generally, collection systems in low-income communities are inadequate, mainly because these areas are not easily accessed with traditional collection vehicles. COMLURB employs approximately 20,100 employees and has a reported annual budget of US$500 million (IBGE, 2015). Municipal authorities often spend between 20% and 30% of their budgets on cleaning and waste disposal, with around 70% related to transportation costs (Brookings Institution, 2015).

Rio de Janeiro citizens produce approximately 4,500 tons of MSW per day (IBGE, 2015), exhibiting lower calorific value (~8 MJ/kg), compared to the EU average (~10 MJ/kg) and the U.S. average (~11 MJ/kg). Reported recycling rates are at about 98%. However, recycling is mainly supported by the informal sector, where waste pickers manually sort materials to be recovered from waste bins and recycling facilities (IBGE, 2015). However, the public perception of scavengers is often negative, leading to the social isolation of these groups, and the importance of these groups for a sustainable waste management is not recognized. In addition, these groups are normally collecting recyclable waste without protective equipment, thus creating a significant problem for public health (World Bank, 2013).

Four transfer stations and two material recovery facilities are operating in Rio, employing approximately 160 people from the low-income communities (World Bank, 2013).

Rio recently closed its primary disposal site (i.e., the public Gramacho landfill) and began sending the bulk of its waste to a new sanitary landfill (Seropedica) located nearly 70 kilometers outside the city. The new sanitary landfill, which has an expected lifetime of 30 years, is operated and owned by a private consortium. However, the Seropedica landfill in Rio de Janeiro is being built over an underground water reservoir and concerns about public health are emerging. At the Gramacho landfill, a new biogas purification plant will deliver 10,000 cubic meters of high-grade gas per day to one of the country's main refinery complexes through a 5,500 meter pipeline (WorldinTwelve, 2015).

Approximately 50% of household waste is organic material. The city currently operates a 200 ton-per-day composting plant that uses

feedstock mainly from wholesale food markets and organic waste that has been separated at the material recovery facilities. The municipality of Rio diverts other waste streams, such as tree prunings from city parks and streets, to compost.

The new Brazilian policy on Solid Waste Management (PNRS), law 12.305 took effect on August 2, 2010, after 20 years of discussions in Congress. PNRS is a promising step forward; however, it is linked to many different lobbies and has resulted in a misleading legislation document. The main bottlenecks can be summarized as follows:

- It does not adopt or propose the waste management hierarchy to be followed by the involved parties.
- It defines landfilling as an "environmentally acceptable" solution, juxtaposing the widely accepted waste management hierarchy.
- It contains only fifteen objectives; therefore, a more comprehensive approach is needed. The involved bodies should take into consideration the Waste Directives implemented by the European Commission, such as the Waste Framework Directive and the zero waste approach toward a circular economy.

Rio participates in the Climate and Clean Air Coalition (CCAC), which is a voluntary partnership uniting governments, intergovernmental and nongovernmental organizations, and representatives of civil society and the private sector in a global effort to address ways of mitigating emissions of short-lived climate pollutants, including methane and black carbon, as a collective challenge. The CCAC Municipal Solid Waste Initiative is working with government officials, sanitation engineers, private entrepreneurs, and other stakeholders in Rio to help build capacity to improve waste management.

Policies, financing, and creative programs are providing the backdrop for urban revitalization and sustainable growth in Rio de Janeiro. For example, the Morar Carioca program is improving housing and services in informal settlements, while BikeRio is creating a cycling culture around cleaner and more accessible transport, and Bolsa Verde do Rio de Janeiro lays the groundwork for innovative emissions, effluent, and ecosystem services markets. Across this city, agencies from all levels of government, nongovernmental organizations, academia, and the private sector are working together to approach long-standing and emerging challenges with fresh ideas and clear commitment; however, lots of effort is still required (WorldinTwelve, 2015).

Case Study 15.2   The Challenge of Developing Cities: The Case of Addis Ababa

Martin Oteng-Abanio

University of Ghana, Accra

Keywords	Municipal waste management, no-source separation, open dumping, informal recycling systems
Population (Metropolitan Region)	32,220,000 (Government of Ethiopia, 2012)
Area (Metropolitan Region)	540 km² (Government of Ethiopia, 2012)
Income per capita	US$590 (World Bank, 2017)
Climate zone	Cfb – Temperate, without dry season, warm summer (Peel et al., 2007)

In Addis Ababa, on average, 0.4 kilograms per capita per day of municipal solid waste (MSW) is produced, whereas more than 200,000 tons was collected in 2013 (Future Mega Cities, 2015). The city's MSW composition exhibited high organic content (~60% of total MSW), and the recyclables fraction was about 15% in 2013. The high amounts of organic waste as well as paper and cardboard that go for disposal can result in large amounts of greenhouse gases (GHG) being emitted, and here the largest reduction potential can be found. The city is divided into 549 zones, representing 800–1,000 households. In each zone, one micro- and small-scale enterprise is responsible for the collection of MSW, employing 5,815 operators (Government of Ethiopia, 2012). The financing for solid waste management is volume-based at a rate of 30 birr per cubic meter (approx. US$1.50). Services charges are fixed with respect to water consumption, taking into account the ability and willingness of the residents to pay. MSW collection is regular and covers 80% of the city's needs. Sorting of waste takes place at various levels in the waste management process. The municipality increased the collection rate from 60% to 80%, and the sources of waste generated can be summarized as follows:

- 76% households
- 18% institutions, commercial, factories, hotels
- 6% street sweepings

The first level of source separation is at the household level, where plastic, glass, and bottles are considered as valuable materials and typically sorted out for reuse. There are independent collectors supporting the informal sector and active at the second stage of source separation, such as street boys, private-sector enterprises, and scavengers at municipal landfill and the korales, collecting metal, wood, tires, electronic products and appliances, old shoes, and plastic, which are further used by local plastic companies, shoe manufacturers, and metal factories. The municipality's role in recycling is absent and mainly focuses on collection, storage, transportation, and disposal of solid waste.

The highest level in the transportation system is represented by the municipality. The role of the private sector in transportation of solid waste is limited. There is currently one open dumpsite with a surface area of 0.25 square kilometers (0.1 square mile) ("Rappi" or "Koshe") where all collected waste is disposed of; it is located 13 kilometers southwest of the city center; it opened almost 50 years ago (Regassa, 2011). The major problems associated with the disposal site are that it is reaching its capacity and is surrounded by housing areas and institutions, creating a nuisance and health hazard for people living nearby. There is no daily cover with soil, leachate containment or treatment, rainwater drain-off, or odor or vector control. The present method of disposal is open dumping: hauling the wastes by truck, spreading and leveling by bulldozer, and compacting by compactor or bulldozer. Environmental sanitation activities and campaigns should start operating in Addis Ababa, and the installation of bins and the development of effective collection systems, transfer stations, and new sanitary landfills are important. An emphasis on recycling and reusing should be implemented.

The collection of information regarding the actors involved in the waste management system and how the material and resource flow through a megacity is a great challenge in any large urban center in a developing country because of the complexity of the system.

In Addis Ababa, there is neither legislation advocating sustainable waste management nor a plan for integrated waste management. There is a need to incorporate the informal recycling sector within the formal sector and, through educational programs, to advance the role of waste pickers and scavengers in the society. In addition, the municipality of Addis Ababa should create robust secondary markets to aid waste pickers in their role of advancing waste management in the city. This could be a great step forward; however, a lot still needs to be done by the citizens and Addis Ababa authorities to achieve sustainable waste management; an achievement that cannot be done overnight.

It is in this light that the attempt by the Addis Ababa City Administration, UNDP MDG Carbon, and UNDP Ethiopia Country Office to work together to support the development of the Repi Landfill Gas Clean Development Mechanism (CDM) Project under the UN Framework Convention on Climate Change (UNFCCC) was commendable (see UNDP, n.d.). Conceptually, the CDM project is based on the capture and destruction of the harmful greenhouse gas (GHG) methane produced by decomposing organic matter at the landfill site. The action of capturing and flaring methane has been made possible through revenue from the sale of certified emissions reductions (CERs). When successfully implemented and operated, the project was to generate

Case Study 15.2 Figure 1 *Addis Ababa landfill.*

a combination of economic, social, and environmental benefits. First, carbon credits were to help to make economically viable a project that would not otherwise happen by bringing additional revenues to the City of Addis Ababa. Second, social benefits would arise from green jobs, which would have been created for scavengers who currently live on the landfill. Third, the project would have delivered important environmental benefits through reducing GHGs that would otherwise be emitted into the atmosphere and contribute to global warming.

The project failed, however, due to financial and other administrative challenges (Bond et al., 2012). Furthermore, the difficulty of the task was compounded by a lack of a robust database (e.g., quantities and composition of waste generated) and by the fact that a large part of the waste and resources is managed and recovered informally or at the interface between the informal and formal sectors. This requires multiple perspectives to understand the problems associated with waste management in a megacity and a transdisciplinary approach for the collection of data and information.

using innovative and integrated approaches. The major challenge is the lack of adequate administrative and financial resources to support such transformation, the absence of effective and comprehensive legislative frameworks, inadequate enforcement mechanisms, the use of improper treatment technologies, and the lack of stakeholder involvement (Guerrero et al., 2013). Consequently, great efforts should be made to develop suitable financing mechanisms for enabling effective MSW management, the intervention of proper treatment technologies for the waste characteristics and local contexts of developing countries, the development of proper standards and laws enforcing waste separation/minimization, capacity-building of the local authorities in MSW management, and cooperation between citizens and local authorities in the planning and implementation of management activities to achieve appropriate and effective waste management practices.

15.4 GHG Mitigation Potential of Sustainable Waste Management

MSW management activities like collection, transportation, treatment, and disposal generate GHG emissions. The majority of GHGs are emitted during the disposal phase in sanitary landfills and dumpsites. Comparatively, GHG emissions from other activities like collection, transportation, and treatment are low. In principle, all these activities entail the movement of waste from the generation point to other facilities, which involves the use of different sources of energy and fuels, thus potentially resulting in GHG emissions. Other sources of GHG emissions involve compaction of waste and maintenance of waste collection and transport equipment including bins, containers, and vehicles, as well as construction of infrastructure and facilities. The following subsections highlight some of these sources.

Globally, it is expected that waste generation per capita will increase by approximately 30% from current levels, while total MSW generation will increase almost threefold (Hoornweg et al., 2012). With increasing waste generation also comes an increasing amount of biodegradable organic waste, which in turn leads to increased GHG emissions due to anaerobic decomposition in landfills and dumpsites. Waste prevention seems to be a promising approach to minimize the amount of waste. Reducing waste

through product design and reusing materials and through concepts like circular economy hold enormous potential for indirect reduction of GHG emissions through the conservation of raw materials, improved energy and resource efficiency, and fossil fuel avoidance (Saxena, 2009). With improved material management that uses a combination of reduced packaging, reduced use of non-packaging paper products (e.g., magazines, newspapers, and textbooks), and extended life of personal computers in U.S. industry, high amounts of GHG emissions reduction, up to 255 MMTCO$_2$e per year can be achieved (Environmental Protection Agency [EPA], 2009).

Waste prevention and reduction can also mitigate GHG emissions through:
- Substituting virgin raw material and reducing GHG emissions from virgin raw material procurement and manufacturing (i.e., avoiding baseline emissions attributable to current production)
- Forest carbon sequestration, in the case of paper products (also treated as negative emissions)
- Zero waste management GHG emissions (EPA, 2009)

15.4.1 GHG Emissions, Waste Sorting, and Collection

Considering the high amount of mixed wastes disposed of in developing countries, high amounts of GHG emissions are generated from the degradation process of biodegradable waste. Source separation of organics from other waste streams therefore provides great potential for reducing GHG emissions from landfill sites. The study of MSW practices in China indicates the possibility to reduce about 23% of GHG emissions through source-separated collection compared with the existing practice using a mixed waste collection system (Dong et al., 2013).

Collection systems involve both mechanical and manual handling of waste. While collection systems with a higher degree of manual handling reduce GHG emissions, they might have other drawbacks that also need to be considered. In estimating the GHG emissions associated with waste collection, only the energy used when operating the collection trucks is considered. Table 15.3 presents diesel consumption per ton of waste collected from different waste generation sources.

(a)

(b)

Figure 15.4a and 15.4b *A typical dumping site in Accra, Ghana.*

Photos: Ranjith Annepu, www.wastewise.be

Box 15.3   Informal Recyclers in Accra

Informal waste pickers handle large quantities of waste that would otherwise have to be collected and disposed of by Accra's authorities. By doing so, the informal recycling sector saves the city 20% or more of its municipal solid waste (MSW) budget, which by implication means that the poor are subsidizing the rest of the city. The city has a major opportunity to build on the existing recycling systems to increase its existing recycling rates further and to protect and develop people's livelihoods while reducing the costs of managing residual wastes.

Box 15.3 Figure 1 *E-waste scavengers in Accra burning wires to harvest copper.*

15.4.2   GHG Emissions and Transportation of Waste

GHG emissions from the transportation of waste also depend on the density of the material transported and the degree of compaction it was subjected to. Modern materials like plastic, paper, and cardboard have low density but are more compactable than are metals or organic and inorganic materials that have a higher density. Studies (Spielmann et al., 2004; Securities and Exchange Commission [SEC], 2006; Environmental Design of Industrial Products [EDIP], 2004) show that fuel consumption is higher for materials with low density when assessed per ton of material transported.

Table 15.2 *Municipal solid waste management practices worldwide. Source: The World Bank, 2012*

Activity	Developing country		Developed country
	Low-income country	**Middle-income country**	**High-income country**
Source reduction	Not organized, but reuse and low per capita waste generation rates are common.	Some discussion of source reduction, but rarely incorporated into any organized program.	Organized educational programs are beginning to emphasize source reduction and reuse of materials.
Collection	Sporadic and inefficient. Service is limited to high-visibility areas, the wealthy, and businesses willing to pay.	Improved service and increased collection from residential areas. Larger vehicle fleet and more mechanization.	Collection rate >90 percent. Compactor trucks and highly mechanized vehicles are common.
Recycling	Most recycling is through the informal sector and waste picking.	Informal sector still involved; some high technology sorting and processing facilities. Materials are often imported for recycling.	Recyclable material collection services and high-technology sorting and processing facilities. Increasing attention toward long-term markets.
Composting	Rarely undertaken formally even though the waste stream has a high percentage of organic material.	Large composting plants are generally unsuccessful; some small-scale composting projects are more sustainable.	Becoming more popular at both backyard and large-scale facilities. Waste stream has a smaller portion of compostable than in low- and middle-income countries.
Incineration	Not common or successful because of high capital and operation costs, high moisture content in the waste, and high percentage of inert material.	Some incinerators are used, but experiencing financial and operational difficulties; not as common as in high-income countries.	Prevalent in areas with high land costs. Most incinerators have some form of environmental controls and some type of energy recovery system.
Landfilling	Low-technology sites, usually characterized by open dumping of wastes.	Some controlled and sanitary landfills with some environmental controls. Open dumping is still common.	Sanitary landfills with a combination of liners, leak detection, leachate collection, and treatment systems.
Costs	Collection costs represent 80–90% of the municipal solid waste management budget. Waste fees are regulated by some local governments, but the fee collection system is very inefficient.	Collection costs represent 50–80% of the municipal solid waste management budget. Waste fees are regulated by some local and national governments, more innovation in fee collection.	Collection costs can represent <10% of the budget. Large budget allocations to intermediate waste treatment facilities. Upfront community participation reduces costs and increases options available to waste planners (e.g., recycling and composting).

More obvious factors that influence GHG emissions from transportation are the distance between the waste generation source and final disposal site and the size of the waste container. The bigger the size of the container, the less the GHG emissions rate per tons of waste transported per kilometer. An extreme case of GHG emissions from waste transportation is small cars and motor carts transporting small amounts of waste over the road, but such modes are used around the world where collection and transportation services are inadequate or expensive (Larsen et al., 2009).

All MSW management activities consume energy, either through the use of electricity[1] (e.g., to power pneumatic collection systems, balers, trains), diesel fuel (e.g., for trucks, trains), petrol (e.g., for private vehicles), bunker oil (e.g., barges, coasters, container ships), or natural gas (e.g., for forklifts and trucks) (Gertsakis and Lewis, 2003). Fruergaard et al. (2009) summarized the volume of GHG emissions (expressed

as kg/CO_2-eq) generated from different sources (see Tables 15.4 and 15.5).

15.4.3 GHG Emissions and Recycling

The GHG emission benefits from recycling are quite substantial as compared to other methods of waste management (see Table 15.6). Recycling can potentially reduce emissions because less waste is brought to the landfill and less virgin resources are extracted, hence the energy required for extraction and processing of primary resources is reduced. A comparative study of treatment practices in the Netherlands shows that high-quality recycling saves 2.3 $MtCO_2$ per year, which is higher than that achieved from improved efficiency incineration systems, which could reduce only 0.7 $MtCO_2$ per year (Corsten et al., 2013). Table 15.6 demonstrates the potential GHG emission reduction from recycling activities. In 2002, Canada recycled 4.3 million

---

1 Electricity is considered a mixed energy supply since in some countries nuclear, hydro, solar, and wind power may, together with fossil energy sources, contribute to the national grid mix

Table 15.3 *GHG emissions (kg CO$_{2\text{-eq}}$) from the use of fuel per ton of waste collected. Source: Larsen et al., 2009; StatBank Denmark, 2008; Mogensen and Holbech, 2007*

Collection	GHG emissions (kg CO$_{2\text{-eq.}}$/ton)
Full service/curbside collection:	
Residual waste, city center	9.3–9.9
Residual waste, apartment blocks	5.0–5.4
Residual waste, single-family houses	10.2–11.5
Residual waste, rural areas	19.5–32.3
Paper waste, apartment blocks	6.8–11.2
Paper waste, single-family houses	12.7–21.1
Drop-off containers:	
Glass waste, 0.7–2.5 m^3	11.5–15.7
Paper waste, 0.7–2.5 m^3	15.2–15.7
Private car:	
5–10 km one way carrying 15 kg	100–300
5–10 km one way carrying 100 kg	16–45
Pneumatic systems:	
Stationary systems	17.5–77.1
Mobile systems	43.0–45.6

Case Study 15.3 Integrated Community-Based Waste Management toward a Low-Carbon Eco-City in Tangerang Selatan, Indonesia

O. C. Dewi and H. H. Al-Rasyid

Institute for Economic and Social Development (BEST), Tangerang Selatan

R. Salam

Regional Environmental Agency (BLHD), Tangerang Selatan

D. Priyandana

Department of City Planning Building and Settlement (DTKBP), Tangerang Selatan

M. T. Rohmadi

Department of Cleanliness, Parks and Cemetery (DKPP), Tangerang Selatan

Keywords	Integrated waste management, material recovery, voluntary emission reduction, community-based actions
Population (Metropolitan Region)	1,290,322 (Badan Pusat Statistik, 2017)
Area (Metropolitan Region)	147.19 km^2 (Badan Pusat Statistik, 2017)
Income per capita	US$11,220 (World Bank, 2017)
Climate zone	Af – Tropical, rainforest (Peel et al., 2007)

Five years after administratively separating from the Tangerang Regency, the city of Tangerang Selatan has shown very fast development. The city is located in Banten Province and is part of the Greater Jakarta Metropolitan area, and about 30% of the city's inhabitants work in Jakarta. The share of residential areas reached about 52% of the total city area in 2013 with 3–4% annual population growth (DTKBP, 2011, 2014).

Facing the challenges of limited land availability and growing gastronomic tourism forces the city to confront problems in the waste sector. The city urgently needs innovative strategies to cope with these challenges through integrated regions that can support sustainable development, especially in the waste sector, in order to move toward a low-carbon eco-city.

Tangerang Selatan's landfill, the Cipeucang Landfill, was re-established by the city's authorities in 2011. Currently, this 2.5 hectare landfill has reached more than 70% of its maximum capacity and another 1.5 hectares was developed in 2015 (DKPP, 2014). This critical condition forced the city to change its waste management strategy toward waste reduction at the source level, mainly through integrated community-based waste management.

THE WASTE SECTOR AND A LOW-CARBON ECO-CITY

Waste management in Tangerang Selatan follows a holistic approach. It applies at multiple levels from the source to the landfill, with the goal of minimizing waste at the landfill, maximizing the utilization of recyclables, and avoiding greenhouse gas (GHG) emissions from organic waste, in parallel with the government's *"Indonesia Bersih Sampah 2020"* or "Indonesia Clean Waste 2020" program.

The city generates 700–1000 tons of waste per day (DKPP, 2014), which contains 51% organics, 35% non-organics, and 15% residue (DKPP, 2014). About 114 tons of mixed waste is brought to the landfill per day. The rest is treated at Recycle-Banks (RB), Material Recovery Facilities (MRF), by the private sector and through illegal burning and dumping (DKPP, 2014).

The Department of Cleanliness, Parks and Cemeteries (DKPP) serves 41% of the residents (DKPP, 2014). At the source, the city authority relies on community-based waste management. The city plans to implement 54 MRFs (one per subdistrict) by 2016, along with 572 Recycle Banks (RB) (one per neighborhood association). By the end of 2016, the city expects to reduce by 20% incoming waste to the landfill through this integrated waste management program. Also, communities are expected to play active roles in utilizing the waste, following Reduce, Reuse, and Recycle (3R) principles such as organic and non-organic waste separation (see Case Study 15.3 Figure 1).

An RB or *Bank Sampah*, uses the concept of reutilizing recyclable waste for useful products (bags, pencil cases, book covers, etc.). Recycle Banks were introduced by the Indonesian Ministry of Environment. The operations of RBs are driven by housewives or community initiatives (see Case Study 15.3 Figure 2). Many RB activities are not limited to handicrafts but can also include agricultural activities, such as the planting of herbs and houseplants. Currently, the city has 145 RBs, which serve around 8,500 inhabitants and have

handled around 700–900 tons of non-organic waste since they were first opened.

The concept of the Material Recovery Facilities (MRF) or *Tempat Pengelolaan Sampah* 3R was introduced by the nongovernmental organization (NGO) Bremen Overseas Research and Development Association (BORDA) and partners (Bina Ekonomi Sosial Terpadu, Lembaga Pengembangan Teknologi Pedesaan, and Bali Fokus) in 2004, and adopted by the Indonesian Ministry of Public Works in 2007. In Tangerang Selatan, implementation of MRFs started in 2010. The MRF concept emphasizes waste management at the local level through community-based organizations (CBO) to deal with organic and non-organic waste (see Case Study 15.3 Figure 3). Currently, 51 MRFs run in the city and serve around 125,000 inhabitants.

Within the last four years, the implementation of RBs and MRFs has encouraged about 10% of residents to be actively and voluntarily involved in waste reduction at the source level. This contributes to emission reductions from the waste sector in two ways: preventing emissions generation at the landfill and preventing emissions generation at source. This number is expected to increase in the coming years and contribute significantly to load reduction at the landfill and to reducing the burden of the Department of Cleanliness, Parks and Cemeteries.

FROM VISION TO MISSION: CLIMATE CHANGE MITIGATION

The city authority developed a study of climate change mitigation in Tangerang Selatan. This study reported potential GHG generation from different city sectors, including industries, transportation, energy, and waste. The results are used as guideline for further policies, strategies, and action plans in climate change mitigations.

The city has enforced previous acts for environmental protection, such as Local Regulation No. 13/2013 on Environmental Treatment that regulates the environment in Tangerang Selatan (Perda No. 13/2013 tentang Pengolahan Lingkungan Hidup yang Mengatur Pengelolaan Lingkungan, Badan Lingkungan Hidup Daerah; BLHD, 2013) and Local Regulation No. 3/2013 on solid waste management (PERDA No. 3/2013 tentang Pengelolaan Sampah, Dinas Kebersihan Pertamanan dan Pemakaman; DKPP, 2013). For the waste sector, the city authority obliges housing developers to connect to one of the MRFs in every new housing settlement.

Case Study 15.3 Figure 1 *Waste separation by the community at the Material Recovery Facility.*

Source: BORDA, 2014

Case Study 15.3 Figure 2 *Recycle Bank activities, which focus on recyclables.*

Source DKPP, 2014

Case Study 15.3 Figure 3 *Material Recovery Facility, with the composting by the community.*

Source: DKPP, 2014

Case Study 15.3 Figure 4 *Monitoring of the compost temperature at the Material Recovery Facility.*

Source: BEST, 2014

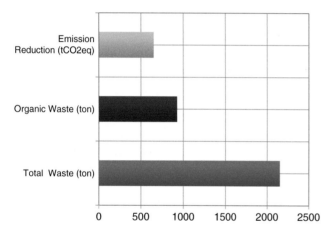

Case Study 15.3 Figure 5 *KIPRAH Voluntary Greenhouse Gas Emissions Reduction Program (VER) Numbers and Figures in Tangerang Selatan, June 2013–October 2014.*

Source: BORDA (2014)

KIPRAH VER PROGRAM: LOCAL ACTIONS FOR GLOBAL IMPACT

One of the most successful of the city's programs is the Voluntary Emission Reductions (VER) trading scheme under the KIPRAH (Kita-Pro-Sampah or We-Pro-Waste) Program, initiated by BORDA Indonesia and partners Lembaga Pengembangan Teknologi Pedesaan (LPTP) and Bina Ekonomi Sosial Terpadu (BEST) in 2006. The program is the first GHG emissions reduction community-based composting project worldwide, registered with the Gold Standard Foundation (www.goldstandard.org) in 2014. Carbon credits obtained as a result of verified emission reductions can be sold on the voluntary carbon market, thus contributing directly to international climate mitigation efforts.

KIPRAH VER promotes aerobic composting using an innovative bamboo aerator method, with a standard Measurement Reporting and Verification (MRV) method at the community level. Aerobic composting of organic waste avoids methane emissions that would result from burying the organic waste. One of the monitoring standards, measuring the compost temperature, is shown in Case Study 15.3 Figure 4. Currently nine MRFs in Tangerang Selatan are part of the VER program with significant amounts of GHG emission reduction in the Gold Standard verification process (see Case Study 15.3 Figure 5) under the partner area of BEST Tangerang Selatan.

LESSONS LEARNED

The goodwill and commitment of city government leaders, communities, and NGOs who have the vision to move forward and improve environmental quality and public health have motivated all sectors to be actively involved within the government's waste management programs. The city is committed to reaching its target of 20% reduced waste going into the landfill using RBs and MRFs. Tangerang Selatan also has the potential to become the Indonesian city with the highest GHG emission reductions from the community-based waste sector.

List of Abbreviations (terms in English).

BAPPEDA	*Badan Perencanaan dan Pembangunan Daerah* or Planning and Regional Development Agency
BEST	Bina Ekonomi – Sosial Terpadu (NGO) or Institute for Economic and Social Development
BLHD	*Badan Lingkungan Hidup Daerah* or Regional Environmental Agency
BORDA	Bremen Overseas Research and Development Association (NGO)
DKPP	*Dinas Kebersihan Pertamanan dan Pemakaman* or Department of Cleanliness, Parks and Cemetery
DTKBP	*Dinas Tata Kota Bangunan dan Pemukiman* or Department of City Planning Building and Settlement
LPTP	Lembaga Pengembangan Teknologi Pedesaan (NGO) or Foundation for the Development of Rural Technology

ACKNOWLEDGMENTS

The authors would like to thank BORDA Indonesia, the City Mayor of Tangerang Selatan, BEST Tangerang Selatan, DTKBP, BLHD, DKPP, and LPTP for their cooperation during the working group of the Case Study Tangerang Selatan, Indonesia.

Table 15.4 *GHG emissions (kg CO$_{2\text{-eq}}$) from use of fuel for collection, transfer and transport of 1 ton of waste, calculated in the six different examples. Source: Fruergaard et al., 2009*

Example	Collection	Transfer	Transport	Total
1. Residual waste with 20 km of transport by truck	43–46	–	4–11	47–57
2. Residual waste with 20 km of transport by truck	5.0–5.5	–	4–11	9–17
3. Residual waste with 150 km of transport by truck	10–11	0.05–4.5	14–28	24–44
4. Recyclable paper with 2,000 km of transport by truck	6–11	0.05–4.5	182–380	189–396
5. Recyclable paper with 3,000 km of transport by diesel train	6–11	0.05–4.5	6–174	13–190
6. Recyclable materials with 10,000 km of transport by ship	100–300	0.10–8.9	29–59	129–368

Table 15.5 *GHG emissions (kg CO$_{2\text{-eq}}$) from provision and combustion of fuel used in waste collection and transport. Source: Fruergaard et al., 2009*

Type of process/emission	Emission factor range	Unit
Provision of diesel oil	0.4–0.5	kg CO$_{2\text{-eq.}}$/lt
Provision of gasoline	0.7	kg CO$_{2\text{-eq.}}$/lt
Provision of fuel oil (heavy)	0.4–0.6	kg CO$_{2\text{-eq.}}$/lt
Provision of fuel oil (light)	0.4–0.5	kg CO$_{2\text{-eq.}}$/lt
Provision of natural gas	0.2–0.3	kg CO$_{2\text{-eq.}}$/Nm^3
Combustion of diesel oil	2.7	kg CO$_{2\text{-eq.}}$/lt
Combustion of gasoline	2.3	kg CO$_{2\text{-eq.}}$/lt
Combustion of fuel oil (heavy)	2.9	kg CO$_{2\text{-eq.}}$/lt
Combustion of fuel oil (light)	2.7	kg CO$_{2\text{-eq.}}$/lt
Combustion of natural gas	2.2	kg CO$_{2\text{-eq.}}$/Nm^3
Provision of electricity	0.1–0.9	kg CO$_{2\text{-eq.}}$/kWh
Provision of heat (EU-25)	0.075	kg CO$_{2\text{-eq.}}$/MJ

tons of materials, avoiding 12 million tons of GHG and saving 6.3 million G J of energy (0.4 million barrels of oil).

15.4.4 GHG Emissions and Waste Treatment and Disposal Practices

15.4.4.1 Anaerobic Digestion

GHG emissions from anaerobic digestion facilities are generally limited to system leaks from gas engines used to generate power from biogas, fugitive emissions, and CO$_2$ from combustion methane, and during system maintenance. There are also possible traces of methane emitted during maturation of the solid organic output. Anaerobic digestion requires energy input but is generally self-sustaining and can make several contributions to climate change mitigation.

First, digesters capture biogas or landfill gas that would have been emitted anyway because of the nature of organic waste management at the facility where the digester is in operation. Second, the displacement of fossil fuel–based energy that occurs when biogas is used to produce heat or electricity is an important contribution. Finally, GHG emissions are also reduced when the nutrient-rich digester created from anaerobic digestion is used to displace fossil fuel–based fertilizers used in crop production. This digestate can make a natural fertilizer that is produced with renewable energy as opposed to fossil fuels (Bogner et al., 2007; Hoorweg and Bhada-Tata, 2012; Annepu, 2013).

Anaerobic digestion can be well-suited to source-separated food wastes, particularly in developing countries where MSW contains 50% or more of food wastes, once the technological challenge the method imposes is surmounted. A critical impediment to its adoption in the developing world is the cost of separate collection and the initial capital investment, which is more than US$500 per ton of installed annual capacity (Arsova, 2010). This is true to the extent that, even in rich countries, it is not adopted on a large scale since the energy yield is around 0.2 MWhe per ton of organic waste compared with waste-to-energy (WTE) high-efficiency plants that can reach 0.8 MWhe per ton if mixed waste is used (no need to collect separately).

15.4.4.2 Aerobic Composting

Aerobic composting refers to the degradation of organic waste by micro-organisms in a controlled environment and in the presence of oxygen to produce a stable product – compost. The process, which is ineffective for the management of MSW high in plastics, metals, and glass content, can directly emit varying levels of gases including nitrous oxide, depending on how the closed system is managed (Mohee and Bundhoo, 2015). A review of several studies show that MSW composting

Table 15.6 *GHG Emissions reductions and energy saving of benefits of recycling (Canada). Source: Recycling Council of Ontario, 2002*

Recyclables	Recycled (tons)	GHG Savings (tons of CO$_{2-eq}$)	Energy savings (GJ)	Equivalent of barrels of oil saved	Equivalent value of oil saved (based on $ 62/ barrel)
Newsprint	8,000,043	1,224,066	5,160,277	793,889	49,221,107
Cardboard & Boxboard	705,856	2,498,730	6,013,893	925,214	57,363,288
Mixed paper	1,519,958	6,657,416	24,030,536	3,697,006	229,214,343
Glass	339,132	40,696	569,742	87,653	5,434,460
Ferrous metals	808,596	970,315	10,196,396	1,568,676	97,257,927
Copper	5,369	22,067	385,011	59,232	3,672,413
Aluminum	51,737	336,808	4,519,744	695,345	43,111,407
PET- plastic	97,450	354,718	8,313,460	1,278,994	79,297,614
HDPE - plastic	54,816	125,528	3,531,231	543,266	33,682,514
Total	4,382,957	12,230,344	62,720,290	9,649,275	598,255,072

emits 0.12–9 kilograms methane per ton of treated waste and 0–0.43 kilogram N$_2$O-N per ton of treated waste (Sánchez et al., 2015).

Composting is suited as a waste management technology in developing countries that have a high portion of biodegradable waste, but to date composting is mostly practiced in developed countries. In 2010, the fraction of MSW composted in Austria was more than 30%, whereas in Belgium and the Netherlands, it was greater than 20% (European Environment Agency [EEA], 2013). Most composting processes tend to be unsuccessful in developing countries due to the composting of commingled instead of segregated MSW, resulting in poor-quality compost. Composting output, which can be used as a substitute for the primary production of fertilizers, provides environmental benefits, yet it is beset with problems of quality and market for the products (UNEP, 2006).

15.4.4.3 Waste-to-Energy

There are more than 800 WTE power plants worldwide producing electricity and district heating by combusting waste. In Switzerland, Japan, France, Germany, Sweden, and Denmark, more than 50% of the waste that is not recycled is sent to WTE industries thereby reducing the amount of waste disposed of in landfills to as little as 4% of the total waste generated (Be Waste Wise, 2013). Incinerators that do not generate energy are net energy users and contribute to GHG emissions. In that respect, incineration without energy recovery is not recommended (UNEP, 2010). Advanced thermal treatment technologies, such as gasification and pyrolysis, may emit fewer GHG emissions compared to mass-burn incineration, and even negative GHG emissions if the energy produced by these

technologies is taken into account. This is clearly shown in the assessment of global warming potential from different treatment technologies in Aalborg. The shift to incineration technology with energy recovery significantly reduced about −400 kgCO$_2$e per ton of waste, compared to the use of incineration without energy recovery, which emitted 251.5 per ton of waste (Habib et al., 2013).

15.4.4.4 Landfill Gas-to-Energy

Methane generated in landfills may be flared, which reduces emissions into the atmosphere. If captured, methane can be burned to produce energy, thereby offsetting emissions from fossil fuel consumption (EPA, 2006). These landfill sites with flaring and electricity generation emit much less GHGs than those without gas collection. A study of direct GHG emissions from South African landfill sites show that about 40–75 kgCO$_2$e per ton of waste can be saved by disposing of MSW in landfill sites with flaring or energy recovery instead of general landfill sites (Friedrich and Trois, 2013).

Landfill gas-to-energy (LFGTE) is the most economical method to reduce GHG emissions from MSW when compared to all other treatment and disposal alternatives (see Table 15.7). LFGTE provides the highest potential to reduce GHG emissions at a cost of less than US$10 per tCO$_{2-eq}$. This potential rests mainly in non-OECD countries where financing waste management can provide many other co-benefits.

15.4.4.5 Landfilling

The organic content in waste sent to landfill (e.g., food, biomass, paper) naturally decomposes under anaerobic conditions.

Table 15.7 *Economic reduction potential of methane emissions from landfill waste by level of marginal costs for total GHG emission reduction assessed for the year 2030. Source: OECD, 2012*

CH4 reduction (Tg of CO$_{2-eq}$)		USD/t CO$_{2-eq}$				
Category	Region	0	10	20	50	100
Anaerobic digestion	OECD	0	0	1	5	5
	EIT	0	0	0	20	24
	Non-OECD	0	0	30	68	95
	Global	0	0	31	94	124
Composting	OECD	0	0	0	0	3
	EIT	0	0	0	6	19
	Non-OECD	0	0	0	58	81
	Global	0	0	0	64	102
Mechanical Biological Treatment	OECD	0	0	0	0	0
	EIT	0	0	0	0	0
	Non-OECD	0	0	0	0	19
	Global	0	0	0	0	19
LFG recovery- energy	OECD	27	43	41	23	22
	EIT	56	29	15	0	0
	Non-OECD	328	368	306	138	43
	Global	411	440	362	162	65
LFG recovery- flaring	OECD	0	6	1	0	0
	EIT	0	17	0	0	0
	Non-OECD	0	12	0	0	0
	Global	0	34	1	0	0
Waste incineration with energy recovery[a]	OECD	124	222	237	266	266
	EIT	0	101	156	156	140
	Non-OECD	0	0	166	515	653
	Global	124	323	558	936	1,059
Total	OECD	151	270	280	295	296
	EIT	56	147	171	182	182
	Non-OECD	328	380	501	779	890
	Global	535	797	953	1,255	1,369

[a] Combustion of waste also causes fossil CO$_{2-eq}$ emissions, which have been taken into account in the calculations, but this table only presents emissions savings from landfills. However, these emissions are typically overcompensated by the corresponding savings when waste-based energy replaces fossil fuels in the energy systems.

The decay, usually initiated by bacteria and microbes, can lead to the production and release of GHGs such as methane, carbon dioxide, and some trace gases that are environmentally unfriendly. Indeed, such emissions can persist for half a decade and more after waste has been disposal of (UNEP, 2006). The situation in most developing countries is worrying because most landfills do not include high-quality liners, leak detection leachate collection systems, or adequate gas collection and treatment systems (Hoorweg and Bhada-Tata, 2012). For biodegradable waste, landfills are the largest emitters of GHG compared to other treatment systems. As presented in Figure 15.5, the landfill option emitted nearly 1,200 kilograms of CO_2 for 1 ton of food waste in the European Union in 2008 while composting emitted negligible amounts of GHG. Further decreases in GHG emissions from treating biodegradable waste can be achieved from incineration, home composing, and anaerobic digestion.

15.5 Impacts of SWM on Climate Change

Generally, post-consumer waste is a small contributor to global GHG emissions, estimated at approximately 3–5% of total anthropogenic emissions or less than 50% with total emissions of approximately 1,300 $MtCo_2eq$ in 2005 (Bogner et al., 2008; UNEP, 2010) (see Table 15.8). The actual magnitude of these emissions in current terms is difficult to determine due to poor data on global waste generation, composition, and management as well as inaccuracies in emission models. The OECD nations, however, have an installed WTE capacity of more than 200 million tons of MSW and also 200 million tons of sanitary landfilling that either uses or flares an estimated 59% of the methane emitted. Developing countries, on the other hand, dispose of an estimated 900 million tons of MSW in nonsanitary landfills and waste dumps.

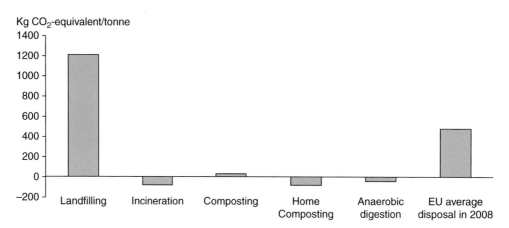

Figure 15.5 *Net emissions for different treatment options for biodegradable waste.*

Source: EEA, 2011

Table 15.8 *Trends for GHG emissions from waste using (a) 1996 and (b) 2006 IPCC inventory guidelines, extrapolations, and projections (MtCO$_2$-eq, rounded).*
Source: IPCC, 2007

Source	1990	1995	2000	2005	2010	2015	2020	2030	2050
Landfill methane (CH$_4$)[a]	760	770	730	750	760	790	820		
Landfill CH$_4$[b]	340	440	450	520	640	800	1000	1500	2900
Landfill CH$_4$ (average of[a] **and**[b]**)**	550	585	590	635	700	795	910		
Incineration CO$_2$[b]	40	40	50	50	60	60	60	70	80
Total GHG emissions	1120	1205	1250	1345	1460	1585	1740		

Notes: Emissions estimates and projections as follows:

[a] Based on reported emissions from national inventories and national communications, and (for nonreporting countries) on 1996 inventory guidelines and extrapolation (EPA, 2006).

[b] Based on 2006 inventory guidelines and BAU projection (Monni et al., 2006).

Total includes landfill CH$_4$ (average), wastewater CH$_4$, wastewater N$_2$O, and incineration CO$_2$.

15.5.1 Formal and Informal Recycling and Climate Change

Most developing countries face increasing challenges when it comes to waste recycling. While formal recycling programs appear to be the most plausible option, their applicability and practicality are complicated by a number of drawbacks such as technology, cost, and institutional inadequacies, among others (Potter et al., 2008). As a result, the most popular option is the use of informal and rudimentary approaches, mechanism, and practices where reusable and recyclable material are gathered at the individual, family, and household levels by poor scavengers who make a good business from their activities even if they are overly exploited by middlemen and well-organized pickers and unions/associations (Samson, 2009).

In terms of its livelihood generation potential, Potter et al. (2008) indicate that as many as 20,000 people live and work on municipal dumps in Kolkata, India, whereas Mexico City has some 15,000 such workers. Further research by Chaturvedi (2010) indicates that women and socially marginalized groups numbering up to 1.5 million people in India are engaged in waste picking; there are 18,000 *recicladores* in Bogotá, Colombia; 15,000 *clasificadores* in Montevideo, Uruguay; and 9,000 *cartoneros* in Buenos Aires, Argentina (Schamber et al., 2007). Table 15.9 presents the livelihood potentials of both formal and informal recycling in Asian cities.

The significance of the activities of informal recyclers and their operations lies not only in the reduction of waste to dumpsites – with its attendant environmental and health challenges – but also in their contribution in reducing GHG emissions. Tables 15.10, 15.11, and 15.12 present waste recovery rates and the carbon footprints of the formal and informal recycling sectors in seven cities.

Naturally, the activities of these formal and informal recycling sectors not only improve public health and sanitation but also guarantee environmental sustainability by way of reduced GHG emission. Additionally, the informal subsidy of SWM necessarily saves scare capital needed by city authorities for other pressing development issues. A recent UN report has regretted how some city and municipal authorities in developing countries continue to exploit waste gatherers who collect between 50% and 100% of MSW at no cost (UN-Habitat, 2010).

GHG emissions from the informal recycling sector are extremely low compared to formal collection systems. Informal recyclers use comparatively less motorized transportation and use numerous transfer points to collect and store recyclables for transportation. Some recommendable modes of waste collection used by informal recyclers are bicycles, tricycles, and other three-wheeled trolleys to collect waste from households. Some informal recyclers set up shop at the end of every street and are known in the locality so that the public can approach them whenever they have items that need to be salvaged. Some methods, like using non-motorized transportation for collection, can be emulated by formal systems, or the informal recyclers who perform such duties regularly can be integrated into formal systems. While dry recyclable items can be stored to optimize

Table 15.9 *Informal and formal livelihoods in six cities. Source: Scheinberg et al., 2010*

City/Indicator	Cairo	Cluj	Lima	Lusaka	Pune	Quezon
Livelihoods in informal waste sector (persons)	33,000	3,226	17,643	480	8,850	10,105
Livelihoods in the formal waste sector (persons)	8,834	330	13,777	800	4,545	5,591
Ratio of persons working in the informal sector to those in the formal sector	3.7	9.8	1.3	0.6	1.9	1.8
Average informal workers' earnings (€/year)	2.721	345[a]/2.070	1.767	586	1.199	1.667

[a] Represents actual earnings from about 50 days of labor per year of €345 multiplied by 6 for purposes of comparison with other cities.

Table 15.10 *Waste recovery rates in seven cities by sector. Source: Scheinberg et al., 2010*

Waste recovery rate in seven cities (CWG-GIZ/Scheinberg et al., 2010).	Belo Horizonte (Brazil)	Canete (Peru)	Delhi (India)	Dhaka (Bangladesh)	Managua (Nicaragua)	Moshi (Tanzania)	Quezon City (Philippines)
Recovered by formal sector (%)	0.10%	1%	7%	0%	3%	0%	8%
Recovered by informal sector (%)	6.90%	11%	27%	18%	15%	18%	31%
Total recovered all sectors (tons)	145,134	1,412	841,070	210,240	78,840	11,169	287,972

transportation, it is not suitable for wet or mixed waste, which is generally collected by the formal municipal collection systems. Wet or mixed waste cannot be stored for long due to decomposing materials and the risk of disease.

Existing and functioning informal recycling systems in developing countries have to be integrated into formal systems to

reduce GHG emissions from SWM. Additionally, diversifying the livelihoods of waste collectors and waste recyclers is sine qua non to the economic and social empowerment of women, children, and other marginalized groups who may be engaged in that business. This can prove significantly crucial in the global poverty alleviation agenda (Scheinberg et al., 2010).

15.5.2 Landfills and Climate Change Mitigation

There are two major strategies to reduce landfill methane emissions: implementation of standards that require or encourage its recovery and a reduction in the quantity of biodegradable waste that is landfilled (Price, 2001). In some instances, methane reduction efforts are complicated by countries that wish to trade their recovery standard for economic gains. This is particularly true in the case of the United Kingdom where the Non-fossil Fuel Obligation, which was meant to generate electricity per a certain standard, instead led to a compromise in the 1980s and 1990s. Also, periodic tax credits in the United States have provided an economic incentive for landfill gas utilization.

It is thought that landfill methane recovery across the developing world will likely increase in coming decades primarily because of improved and/or controlled waste disposal/management practices. And, with the emergence of the CDM that

Table 15.11 *Comparison of material recovery by formal and informal sector in six different cities. Source: Scheinberg et al., 2010*

City	Formal sector		Informal sector	
	Tons	% of total	Tons	Percent of total
Cairo	433,200	13	979,400	30
Cluj	8,900	5	14,600	8
Lima	9,400	0.3	529,400	19
Lusaka	12,000	4	5,400	2
Pune	–	0	117,900	22
Quezon City	15,600	2	141,800	23

Table 15.12 *Comparison of carbon footprint by formal and informal sector in cities. Source: Scheinberg et al., 2010*

City	Formal sector		Informal sector	
	GHG (tons CO_{2-eq})	Total net cost (benefit) of GHG emissions (€/year)	GHG (tons CO_{2-eq})	Total net cost (benefit) of GHG emissions (€/year)
Cairo	1,689,200	16,244,800	−28,900	−277,500
Cluj	103,600	1,295,300	−38,200	−478,000
Lima	448,500	4,313,400	−496,700	−4,776,800
Lusaka	25,800	247,700	−57,700	554,600
Pune	210,600	2,025,000	−295,000	−2,837,200
Quezon City	472,800	4,546,700	−249,200	−2,397,000

Box 15.4   Income-Generating Potential of Waste Pickers

The amount of income earned by waste pickers varies almost in tandem with the country's minimum wages as well as with the type of work that men and women do. In most instances, up to 91% of those engaged in the informal activities overtly or covertly depend on incomes from scavenging, as in the case of Cairo, Egypt (Scheinberg et al., 2010). In Belgrade, waste pickers may earn an average amount of US$100 per month (Simpson-Hebert et al., 2005) as compared to a poverty line of US$105 or 80 euros in that country.

By contrast, waste pickers in Cambodia could go home with a paltry US$1 a day (International Labor Office [ILO/IPEC], 2004). Relatedly, in Santa Cruz, Bolivia, about 59% of waste pickers earn below the minimum wage, while Brazilian and Mexican waste pickers earn more than the minimum wage. Crivellari et al. (2008) indicate that about 34% of waste pickers in Brazil earn about 1.01–1.50 times the minimum wage, whereas men earn more than women in all age groups (Crivellari et al., 2008).

champions the course of development through environmentally friendly practices such as carbon sink and sequestration, the future could not be any brighter (Sceinberg et al., 2010).

Due to many countries facing challenges on the basic way forward to maximize recycling and materials recovered, the selection of truly efficient and sustainable waste management strategies is paramount. To achieve appreciable GHG emissions mitigation, the elimination of open dumping sites is an absolute priority (see Table 15.14).

15.5.3 Climate Change Adaptation and SWM

Scholarly literature on the impacts of climate change on SWM is limited. However, a number of studies have been carried out in recent years by the development community showing that climate change can significantly impact SWM services both directly and indirectly (Bebb and Kersey, 2003; USAID, 2012). It can directly affect SWM through the impacts to the waste management infrastructure and, indirectly, through the

changes that would occur to the surrounding environment. For example, elevated temperatures and changes in hydrology could increase odor, litter, and decomposition rate, and may necessitate more frequent waste collection and better landfill management (to prevent leachate, landfill degradation). Similarly, extreme climate events (e.g., flooding, rainfall, erosion, sea level rise, storm surge) could affect the critical infrastructure (transport means, buildings, machinery) necessary for waste collection, transfer, disposal, and recycling (Bebb and Kersey, 2003). Table 15.13 shows potential ways in which the impacts of climate change affect waste management. These are just examples; the impacts would differ from city to city depending on the extent of impact, location, current practices of waste management, and prevailing infrastructure. Therefore, accessing the risks of climate change to waste management processes and sites at the early stage is very helpful.

Table 15.14 further shows the vulnerability of various waste management technologies and practices along with adaptation and mitigation implications and other sustainability dimensions (IPCC, 2006).

Table 15.13 *Climate change impact to solid waste management sector. Source: USAID, 2012*

	Collection	Processing	Disposal
Temperature change	Increased odor pest activity requiring more frequent waste collection Overheating of collection vehicles requiring additional cooling capacity. including to extend engine life	Overheating of sorting equipment	Altered decomposition rates Increased maintenance and construction costs due to thawing permafrost Increased risk of fire at disposal sites
	Greater exposure of workers to flies, which are a major cause of infectious diseases (flies breed more quickly in warm temperatures and are attracted to organic waste)		
Precipitation change	Flooding of collection routes and landfill access roads, making them inaccessible Increased stress on collection vehicles and workers from waterlogged waste	Increased need for enclosed or covered sorting facilities	Increased flooding in/around sites Increased leachate that needs to be collected and treated Potential risk of fire if conditions become too dry and hot
Sea level rise	Narrowed collection routes Potentially increased waste in a concentrated area as people crowd into higher elevations within an urban area	Damage to low-lying processing facilities Increased need for sorting and recycling to minimize waste storage needs	Deterioration of impermeable lining Water infiltration of pit leading to possible overflow of waste
	Permanent inundation of collection, processing, and disposal infrastructure		
Storm surge	Temporary flooding of and diminished access to roadways, rails, and ports for waste collection, sorting, and disposal Closure of facilities due to infrastructure damage		
Extreme wind	Dispersal of waste from collection sites, collection vehicles, processing sites, and landfills Reduced access to collection and landfill access routes due to damage and debris		

Table 15.14 *Summary of adaptation, mitigation, and sustainable development issues for the waste sector. Source: IPCC, 2006*

Technologies and practices	Vulnerability to Climate change	Adaptation implications and strategies to minimize emissions	Sustainable development dimensions			Comments
			Social	Economic	Environmental	
Recycling, reuse and waste minimization	Indirect low vulnerability or no vulnerability	Minimal implication	Usually positive Negative for waste scavenging without public health or safety controls	Positive Job creation	Positive Negative for waste scavenging from open dumpsites with air and water pollution	Indirect benefits for reducing GHG emissions from waste Reduces use of energy and raw materials. Requires implementation of health and safety provisions for workers
Thermal processes including incineration, industrial co-combustion, and more advanced processes for waste-to-energy (e.g., fluidized bed technology with advanced flue gas cleaning)	Low vulnerability	Minimal implications Requires source control and emission controls to prevent emissions of heavy metals, acids gases, dioxins and other air toxics	Positive Odor reduction (non-CH_4 gases)	Positive Job creation Energy recovery potential	Positive Negative for improperly designed or managed facilities without air pollution controls	Reduces GHG emissions relative to landfilling Costly, but can provide significant mitigating potential for the waste sector, especially in the short term Replaces fossil fuels
Aerobic biological treatment (composting) **Also a component of mechanical-biological treatment (MBT)**	Indirect low vulnerability or positive effects: Higher temperatures increase rates of biological processes (Q_{10})	Minimal implications or positive effects Produces CO_2 (biomass) and compost Reduces volume, stabilizes organic C, and destroys pathogens	Positive Odor reduction (non-CH_4 gases)	Positive Job creation Use of compost products	Positive Negative for improperly designed or managed facilities with, odors, air and water pollution	Reduces GHG emissions Can produce useful secondary materials (compost) provided there is quality control on material inputs and operations Can emit N_2O and CH_4 under reduced aeration or anaerobic conditions
Anaerobic biological treatment (anaerobic digestion) **Also a component of mechanical-biological treatment (MBT)**	Indirect low vulnerability or positive effects: Higher temperatures increase rates of biological processes	Minimal implications Produces CH_4, CO_2, and biosolids under highly controlled conditions Biosolids require management	Positive Odor reduction (non-CH_4 gases)	Positive Job creation Energy recovery potential Use of residual biosolids	Positive Negative for improperly designed or managed facilities with, odors, air and water pollution	Reduces GHG emissions CH_4 in biogas can replace fossil fuels for process heat or electrical generation Can emit minor quantities of CH_4 during start-ups, shutdowns and malfunctions
Sanitary landfilling with landfill gas recovery and utilization	Indirect low vulnerability or positive effects: Higher temperatures increase rates of microbial methane oxidation rates in cover materials	Minimal implications May be regulatory mandates or economic incentives Replaces fossil fuels for process heat or electrical generation	Positive Odor reduction (non-CH_4 gases)	Positive Job creation Energy recovery potential	Positive Negative for improperly managed sites with air and water pollution	Primary control on landfill CH_4 emissions with >1,200 commercial projects Important local source of renewable energy: replaces fossil fuels Landfill gas projects comprise 12% of annual registered CERs under CDM. Oxidation of CH_4 and NMVOCs in cover soils is a smaller secondary control on emissions

CDM, Clean Development Mechanism; CER, certified emissions reductions; CH_4, methane; CO_2, carbon dioxide; GHG, greenhouse gas; N_2O, nitrous oxide; NMVOCs, non-methane volatile organic compounds

15.6 Carbon Market and Finance for GHG Mitigation from Waste

For an effective integrated solid waste management program, behavioral, technological, and management elements are essential, which necessitates new and innovative policies, better institutional coordination, and effective financial arrangements. Some of the policy strategies, which are also linked to financial mechanisms, are shown in Table 15.1. Different modes of financing for waste management are possible, however, in the context of climate change mitigation, and many studies in the scholarly as well as the development community have already shown that this sector is a cost-effective and "low-hanging fruit" in the entire portfolio of climate change mitigation options. Therefore, in climate change–related projects and financing systems, the waste sector has attracted many projects (see Figure 15.6). In the global architecture of carbon markets and financing, the CDM, a flexible mechanism of the Kyoto Protocol, is prominent SWM project type. By the end of 2012, issued carbon credits (CERs) from 407 landfill gas projects under CDM amounted to 71 million.[2] Potdar et al. (2015) reported a total of 350 SWM CDM projects globally (by May 2015), of which 102 CDM projects (12.8 mn tCO_2eq) were in China followed by 45 projects in Brazil (10.6 mn tCO_2), and 28 projects in Mexico (3 mn tCO_2e).

Composting, anaerobic digestion, WTE combustion, landfill gas capture, and flaring are all approved for CDM credits. Even though studies show that thermal processes potentially and efficiently exploit the energy value of post-consumer waste, the high capital investment of WTE plants invariably restricts its application in many less endowed countries (Bogner et al., 2007). It

must be added that, at this time when the international market is uncertain, it is continual traction at the regional and subnational levels that shows some promise for the future.

15.7 Conclusion

Consistently increasing generation of MSW as a result of an increasing urban population and a rising standard of living has resulted in increasing amounts of biodegradable organic carbon and, by extension, GHG emissions. Implementing integrated waste management will push both the private and public sectors to rework the management process to decrease CO_2 emissions from the energy used for solid waste transport as well as to reduce methane and other non-CO_2 GHGs from landfills. This chapter has described many options for reducing GHG from the waste sector. There is a dire need to facilitate a shift from "waste management" to "resource efficiency," a paradigm shift that captures the entire value chain, thus merging the concept of sustainability and its subcomponents (e.g., the hierarchy) into programs that are effective across multiple sectors, disciplines, communities, and professions. In the case of cities in developed countries, continuous attempts are being made to divert waste from landfills to some advanced recycling facilities but the costs of environmental protection at treatment and disposal sites have also increased. Developing countries may lack access to such advanced technologies. However, technologies must be sustainable in the long term, and there have been many examples of advanced, but unsustainable, technologies for managing MSW that have been implemented in developing countries.

A number of economic, regulatory, and information-based policy instruments are available to implement these options to

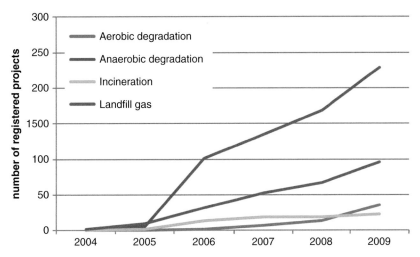

Figure 15.6 *Progression of number of waste-related projects with the status "registered" under the UN Framework Convention on Climate Change (UNFCCC) clean development mechanism between 2004 and 2009 according to the technology used (assessment of 400 out of 456 published projects).*

Source: Seibel et al., 2013

2 http://www.cdmpipeline.org/cdm-projects-type.htm

municipal government. Some of these instruments are extended producer responsibility (EPR), deposit-refund systems, landfill/incineration taxes, pay-as-you-throw (PAYT) fees, bans and restrictions (e.g., a landfill ban), mandatory source separation, and labeling and product information disclosure. To institute any of these measures, municipal authorities cannot work in isolation; they must involve all stakeholders in planning, implementing, and monitoring the changes. Successful cities (Vienna) have demonstrated a range of good practices – consultation, communication, and involvement with users; participatory and inclusive planning; inclusivity in siting facilities; and institutionalized inclusivity – that constitute the solid waste "platform." A strong and transparent institutional framework is an essential proxy indicator of good governance in solid waste.

One of the major hindrances, one that jeopardize improvements in waste management especially in developing countries, is lack of capital. It has been estimated that MSW management consumes between 3% and 15% of the total recurrent municipal budget, or between 0.1% and 0.7% of the per capita GDP. Recent advances in the carbon market

and in carbon financing have provided opportunities to the waste sector, but the process has slowed in the past 2 years as stakeholders wait on the outcome of the UNFCCC's COP21 in December 2015. The prospect remains high, however, for further development of carbon markets and finance instruments.

It goes without saying that to achieve appreciable GHG emissions mitigation, the selection of truly efficient and sustainable waste management strategies is paramount. Clear budgets and lines of accountability are essential. Above all, it needs the political will to see waste management as a key component in the infrastructure of modern life. And people have to be prepared to pay for proper waste management systems. Until the benefits of good waste management are recognized, good systems will not be developed, people will not pay, and waste will continue to be uncollected or dumped. We face huge challenges in many cities, but the basic way forward is to maximize the recycling of materials, which has been achieved in many countries through the informal sector, and to maximize WTE or landfilling with energy recovery of residual waste. The elimination of waste dumps is an absolute priority.

Case Study 15.4   Sustainable Waste Management: The Successful Example of Vienna

A. C. (Thanos) Bourtsalas

Earth and Environmental Engineering, Columbia University, New York

Keywords	Municipal waste management, effective, waste avoidance strategies
Population (Metropolitan Region)	1,791,298 (Comet, 2015)
Area (Metropolitan Region)	995,73 km² (Wien.at, 2015)
Income per capita	US$45,230 (World Bank, 2017)
Climate zone	Dfb – Continental, without dry season, warm summer (Peel et al., 2007)

Vienna holds the international top position regarding separate collection and sustainable waste management. This was achieved not only through Vienna's long-term planning of waste management and waste avoidance strategies, but also through environmental awareness training for children and adults (City of Vienna, 2015). In 2013, 1 million tons of municipal solid waste (MSW) was produced, 35% of which was source separated, a method that started in early 1980s and, by early 1990s, effectively covered the entire city. In the city of Vienna, there are 200,000 containers for recyclables, 19 waste

collection centers, and 112 mobile and stationary collection points for hazardous waste (Thon, 2015). In addition to the waste recycling and utilization industry, measures for waste avoidance also influence collection: thus waste collection centers also accept functioning used appliances and similar items, which are then sold at nominal prices at a bazaar organized by the municipality. Vienna's toy collection campaign with its own specially designed containers, introduced in 2006, is another example of a collection scheme developed to prolong the useful life of products.

Additionally, the City of Vienna complies with the principles of short distances and autonomous disposal, thus making a valuable contribution to environmental protection. Biogenic waste is fully treated on Vienna's municipal territory, and the Viennese population benefits from all results of waste processing, such as high-quality compost, electricity, and district heating. Moreover, the waste-to-energy residues are processed, the metals extracted have significant value, and the mineral fraction is extensively used as secondary recycled aggregate, predominantly as a sub-base and capping material in numerous civil engineering applications. Financing for the collection and treatment of all municipal waste is based on the residual waste fraction in order to create an incentive for separate waste collection (Comet, 2015; Thon, 2015). Thus, property owners are charged a quarterly waste management fee calculated from the volume of the residual waste containers installed on their properties and the frequency of bin emptying. The more material collected separately, the smaller the container volume that needs to be installed and the lower the cost (City of Vienna, 2015).

The minimum container capacity for residual waste is 120 liters; for hygienic reasons, every residual waste container must be emptied at least once a week. The collection and treatment of packaging material, used electrical appliances, and batteries is financed via manufacturers and importers according to the principle of manufacturer responsibility. Vienna has used three distinct initiatives to help reduce consumer waste and two additional initiatives for business wastes (see European Commission, 2012):

- *Web Flea Market*: An Internet-based exchange platform for consumer goods, construction tools and materials, and gardening equipment
- *Repair and Service Center (RUSZ):* Twenty-three local small repair shops provide affordable repair services for electrical household appliances and break down appliances for material recycling
- *Promotion of lifestyle change*: Encouraging spending on services and culture instead of material goods
- The city targets business waste through the following two measures:
- *The ÖkoBusinessPlan (EcoBusiness Plan)*: Targets small and medium-sized enterprises (SMEs). Launched in 1998 by Vienna metropolitan authorities, the initiative provides subsidized cleaner production and eco-efficiency consulting services to Vienna businesses.
- *OekoKauf Wien (EcoPurchasing Vienna)*: The City of Vienna spends around €5 billion on goods annually. This initiative has developed guidelines for ecologically sound purchasing methods.

The RUSZ centers repair approximately 400 tons of appliances annually, and the Internet Flea Market sells 450 tons of used appliances. The Vienna authorities calculate that about 11,000 tons of waste is saved through the RUSZ centers, while the flea market saves around 1,000 tons of waste annually. Since 1998, the ÖkoBusinessPlan has advised more than 600 businesses and helped to save an estimated €34 million, with more than 100,000 tonnes of waste prevented catalogues for the green public procurement of some 60 product groups.

In recent years, numerous waste-related measures implemented in Vienna have contributed to a reduction in climate-relevant emissions. Vienna's waste management system generates 130,000 tons of CO_2 credits. While waste treatment in 2010 triggered the generation of 420,000 tons of CO_2 equivalents, the emission volume avoided totaled 550,000 tons. This was made possible by the generation of district heat from residual waste incineration, the fermentation of kitchen scraps at Vienna's biogas plant, and waste separation and recycling activities, as well as the use of compost in organic farming. Projections for 2020 from the city of Vienna show a further decrease of CO_2-eq by approximately 650,000 tons. For the City of Vienna and all municipal actors concerned with waste management, active climate protection will remain a central task.

Annex 15.1 Stakeholder Engagement

Conscious efforts were made to engage some of the key stakeholders responsible for adapting infrastructure in respect of planning, implementation, monitoring, and evaluation of solid waste prevention, collection, and disposal initiatives that can impact on future climate. The challenge related to the variety of formal and informal actors in the management of solid waste at different spaces and scales, including the various ministerial, regional, and local authorities' advisory panels, and community- and faith-based organizations. To reach this multiplicity of stakeholders, we relied on in-depth interviews, focus group discussions, participatory events (forums), and interactive workshops. We also made use of knowledge exchange groups and social media engagements (including online discussions groups/forums) as well as academic publications (e.g., reports, policy briefs). Apart from highlighting the challenges posed by the multiplicity of actors in the industry, our engagements also revealed the catalytic and supportive role played by international organizations notably the World Bank, the World Health Organization, and the UN Development Program in bridging the gap between long- and short-term solid waste infrastructure provision.

Chapter 15 Urban Solid Waste

References

Agyeman, J., Evans, B., and R. W. Kates (1998). Greenhouse gases special: Thinking locally in science, *practice and policy. Local Environment* **3**(3), 382–383.

Ali, S. M., Pervaiz, A., Afzal, B., Hamid, N., and Yasmin, A. (2014). Open dumping of municipal solid waste and its hazardous impacts on soil and vegetation diversity at waste dumping sites of Islamabad city. *Journal of King Saud University – Science* 26(1), 59–65. doi:10.1016/j.jksus.2013.08.003

Annepu, R. K. (2013). *Observations from India's Waste Crisis*. Waste-to-Energy Research and Technology Council. Columbia University.

APO. (2007). *Solid Waste Management: Issues and Challenges in Asia*. The Asian Productivity Organization.

Arsova, L. (2010). *Anaerobic Digestion of Food Wastes*. Master's thesis. Columbia University. Available on Google.

Barker, T., Pollitt, H., and Summerton, S. (2007). *The Macroeconomic Effect of Unilateral Environmental Tax Reforms in Europe, 1995–2012*. Paper presented at the Eighth Global Conference on Environmental Taxation Innovation, Technology and Employment: Impacts of Environmental Fiscal Reforms and Other Market-Based Instruments, October 18–20, 2007. Munich, Germany.

Be Waste Wise. (2013). Solid waste management and climate change: Be waste wise. Accessed August 3, 2014: http://wastewise.be/2013/08/solid-waste-management-and-climate-change/

Bebb J., and Kersey, J. (2003). *Potential Impacts of Climate Change on Waste Management*. Report of the R&D Project X1–042. Entec UK Limited, Environmental Agency, Bristol, United Kingdom. Accessed November 5, 2014: https://www.gov.uk/government/uploads/system/uploads/attachment_data/file/290358/sx1–042-tr-e-e.pdf

Bogner, J., Abdelrafie, M.-Ahmed, Diaz, C., Faaij, A., Gao, Q., Hashimoto, S., Mareckova, K., Pipatti, R., and Zhang, T. Waste management. In B. Metz, O. R. Davidson, P. R. Bosch, R. Dave, and L. A. Meyer (eds.), *Climate Change 2007: Mitigation. Contribution of Working Group III to the Fourth Assessment Report of the Intergovernmental Panel on Climate Change*. Cambridge University Press.

Bulkeley, H. (2000). Down to earth: Local government and greenhouse policy in Australia, *Australian Geographer*, **31**(3), 289–308.

Chaturvedi, B. (2010). Mainstreaming Waste Pickers and the Informal Recycling Sector in the Municipal Solid Waste, Handling and Management Rules. A Discussion Paper. Women in Informal Employment: Globalizing and Organizing (WIEGO).

Chintan (2009). *Cooling Agents: An Analysis of Climate Change Mitigation by the Informal Recycling Sector in India*. Report prepared in association with The Advocacy Project, Washington, D.C.

Collier, U. (1997). Local authorities and climate protection in the EU: Putting subsidiarity into practice, *Local Environment* **2**(1), 39–57.

Corsten, M., Worrell, E., Rouw, M., and van Duin, A. (2013). The potential contribution of sustainable waste management to energy use and greenhouse gas emission reduction in the Netherlands. *Resources, Conservation and Recycling* 77, 13–21. doi:10.1016/j.resconrec.2013.04.002

Crivellari, H. M. T., Dias, S. M., and de S. Pena, A. (2008). *Informaçãoe trabalho: Uma leitura sobre os catadores de material reciclávela partir das bases públicas de dados.* Catadores na cena urbana.

DeAngelo, B., and Harvey, L. D. (1998). The jurisdictional framework for municipal action to reduce greenhouse gas emissions: Case studies from Canada, *USA and Germany, Local Environment* 3(2), 111–136.

Dong, J., Ni, M., Chi, Y., Zou, D., and Fu, C. (2013). Life cycle and economic assessment of source-separated MSW collection with regard to greenhouse gas emissions: A case study in China. *Environmental Science and Pollution Research International* 20(8), 5512–5524. doi:10.1007/s11356-013-1569-1

Environmental Design of Industrial Products (EDIP). (2004). *Lifecycle Assessment Database.* Developed by the Danish Environmental Protection Agency in 1996, 2nd update. EDIP, Copenhagen, Denmark.

EPA (2006). *Global anthropogenic non-CO₂ greenhouse gas emissions: 1990–2020.* Office of Atmospheric Programs, Climate Change Division.

European Environment Agency (EEA). (2013). *Managing Municipal Solid Waste – A Review Of Achievements In 32 European Countries.* European Environment Agency EEA Report no. 2/2013.

European Union (EU). (2010). Being wise with waste: The EU's approach to waste management. Accessed June 16, 2015: http://ec.europa.eu/environment/waste/pdf/WASTE%20BROCHURE.pdf

Environmental Protection Agency (EPA). (2009). National Waste Report for 2007. EPA.

Friedrich, E., and Trois, C. (2013). GHG emission factors developed for the collection, transport and landfilling of municipal waste in South African municipalities. *Waste Management* 33(4), 1013–1026. doi:10.1016/j.wasman.2012.12.011

Fruergaard, T., Astrup, T., and Ekvall, T. (2009). Energy use and recovery in waste management and implications for accounting of greenhouse gases and global warming contributions. *Waste Management and Research* 27.

Furedy, C. (2004). Urban organic solid waste: Reuse practices and issues for solid waste management in developing countries. In I. Baud, J. Post, and C. Furedy (Eds.), *Solid Waste Management and Recycling: Actors, Partnerships, and Policies in Hyderabad, India and Nairobi, Kenya (197–212).* Kluwer Academic Publishers.

Gertsakis, J., and Lewis, H. (2003). *Sustainability, and the waste management hierarchy. A discussion paper on the waste management hierarchy and its relationship to sustainability.* Ecorecycle Victoria.

Guerrero, L. A., Maas, G., and Hogland, W. (2013). Solid waste management challenges for cities in developing countries. *Waste Management* 33(1), 220–232. doi:10.1016/j.wasman.2012.09.008

Habib, K., Schmidt, J. H., and Christensen, P. (2013). A historical perspective of global warming potential from municipal solid waste management. *Waste Management* 33(9), 1926–1933. doi:10.1016/j.wasman.2013.04.016

Hoornweg, D., Bhada-Tata, P., and Kennedy, C. (2015). Peak waste: When is it likely to occur? *Journal of Industrial Ecology* 19 (1), 117–128.

International Labor Office (ILO)/International Programme on the Elimination of Child Labor (IPEC). (2004). *Addressing the Exploitation of Children in Scavenging (Waste Picking): A Thematic Evaluation of Action on Child Labour.* International Labor Organization

Intergovernmental Panel on Climate Change (IPCC). (2006). Solid waste disposal. In *Guidelines for National Greenhouse Gas Inventories, Volume 5: Waste* (chapter 3). National Greenhouse Gas Inventories Programme.

Intergovernmental Panel on Climate Change (IPCC). (2007). Summary for Policymakers. In Solomon, S., D. Qin, M. Manning, Z. Chen, M. Marquis, K. B. Averyt, Tignor, M., and H. L. Miller (eds.), *Climate Change 2007: The Physical Science Basis. Contribution of Working Group 1 to the Fourth Assessment Report of the Intergovernmental Panel on Climate Change.* Cambridge University Press

Karak, T., Bhagat, R. M., and Bhattacharyya, P. (2012). Municipal solid waste generation, composition, and management: The world scenario. *Critical Reviews in Environmental Science and Technology* 42(15), 1509–1630. doi:10.1080/10643389.2011.569871.

Kates, R. W., Mayfield, M. W., Torrie, R. D., and Witcher, B. (1998). Methods for estimating greenhouse gases from local places. *Local Environment* 3(3), 279–297.

Kaufman, S. M., and Themelis, N. J. (2010). Using a direct method to characterize and measure flows of municipal solid waste in the United States. *Journal of AI, and Waste Management Associations* 2009(59), 1386–1390.

Larsen, A. W., Vrgoc, M., Christensen, T. H., and Lieberknecht, P. (2009). Diesel consumption in waste collection and transport and its environmental significance. *Waste Management, and Research* 27, 652–659.

Marcotullio, P. J., Sarzynski, A., Albrecht, J., and Schulz, N. (2014). A top-down regional assessment of urban greenhouse gas emissions in Europe. *Ambio* 43(7), 957–968. doi:10.1007/s13280-013-0467-6

Marshall, R. E., and Farahbakhsh, K. (2013). Systems approaches to integrated solid waste management in developing countries. *Waste Management* 33(4), 988–1003. doi:10.1016/j.wasman.2012.12.023

McDougall, F., White, P. R., Franke, M., and Hindle, P., 2001. *Integrated Solid Waste Management: A Lifecycle Inventory,* second ed. Blackwell Science.

Menikpura, S. N. M., Sang-Arun, J., and Bengtsson, M. (2013). Integrated solid waste management: An approach for enhancing climate co-benefits through resource recovery. *Journal of Cleaner Production* 58, 34–42. doi:10.1016/j.jclepro.2013.03.012

Mogensen, B., and Holbech, A. (2007). Miljøøkonomisk vurdering af indsamling af dagrenovation og papir 2005 (Environmental Economic Assessment of Collection of Household and Paper Waste 2005, in Danish). PricewaterhouseCoopers and R98, Copenhagen, Denmark.

Mohee, R., and Bundhoo, M. A. Z. (2015). A comparative analysis of solid waste management in developed and developing countries. In R. Mohee,and T. Simelane, T. (eds.), *Future Directions of Municipal Solid Waste Management in Africa.* Africa Institute of South Africa.

Monni, S., Pipatti, R., Lehtila, A., Savolainen, I., and Syri, S. (2006). *Global Climate Change Mitigation Scenarios for Solid Waste Management.* VTT Publications, No. **603**, 51. ESPOO Technical Research Centre of Finland.

Nordone, A. J., White, P. R., McDougall, F., Parker, G., Carmendia, A., and Franke, M. (1999). Integrated waste management. In Smith, C. S. R., Cheeseman, C., and Blakey, N. (eds.), Waste management and minimization. *Encyclopedia of Life Support Systems* (ELOSS) and United Nations Education, Scientific and Cultural Organization (UNESCO).

O'Meara, M. (1999). *Reinventing Cities for People and the Planet.* Worldwatch Institute.

OECD. (2012). Greenhouse gas emissions and the potential for mitigation from materials management within OECD countries. Accessed July 12, 2014: http://www.oecd.org/env/waste/50034735.pdf

Oteng-Ababio M. (2014). Rethinking waste as a resource: Insights from a low-income community in Accra, Ghana City, Territory and Architecture. *Accra Ghana* 1, 10.

Potdar, A., Singh, A., Unnnikrishnan, S., Naik, N., Naik, M., and Nimkar, I. (2015). Innovation in solid waste management through Clean Development Mechanism in India and other countries. *Process Safety and Environmental Protection,* in press. doi:10.1016/j.psep.2015.07.009.

Potter, B., Binns, T., Elliott, J., and Smith, D. (2008). *Geographies of Development: An Introduction to Development Studies.* Pearson/Prentice Hall.

Price, J. L. (2001). The landfill directive and the challenge ahead: Demands and pressures on the householder. *Resources, Conservation and Recycling* 32, 333–348.

Rayner, S., and Malone, E. L. (1997). Zen and the art of climate maintenance. *Nature* **390**, 332–334.

Renou, S., Givaudan, J. G., Poulain, S., Dirassouyan, F., and Moulin, P. (2008). Landfill leachate treatment: Review and opportunity.

Journal of Hazardous Materials **150**(3), 468–493. doi:10.1016/j.jhazmat.2007.09.077

Rypdal, K., and Winiwarter, W. (2001). Uncertainties in greenhouse gas emission inventories, evaluation, comparability and implications. *Environmental Science and Policy* **4**, 107–116.

Samsom, M. (2009).*Refusing to be Cast Aside: Waste Pickers Organizing Around the World*. WIEGO.

Sánchez, A., Artola, A., Font, X., Gea, T., Barrena, R., Gabriel, D., Sánchez-Monedero, M. A., Roig, A., Cayuela M. L., and Mondini, C. (2015). Greenhouse gas emissions from organic waste composting. *Environmental Chemistry Letters* **13**(3), 223–238. doi:10.1007/s10311–015–0507–5

Sasaki, S., and Araki, T. (2014). Estimating the possible range of recycling rates achieved by dump waste pickers. The case of Bantar Gebang in Indonesia. *Waste Management Research* **32**, 474–481. doi:10.1177/0734242X14535651.

Saxena, K. (2009). *Greenhouse Gas Emissions: Estimation and Reduction*. Asian Productivity Organization.

Schamber, P., Suarez, F., and Vades, E. (Eds.). (2007). *Recicloscopio: Miradas sobre recuperadores Urban de residuos de Americ Latina*. Ediciones de La Unla.

Scheinberg, A., Simpson, M. H., and Gupt, Y. (2010). Economic Aspects of the Informal Sector in Solid Waste, Final Report and Annexes. GIZ (German International Co-operation), the CWG (Collaborative Working Group on Solid Waste Management in Low- and Middle-income Countries, and the German Ministry of Foreign Affairs, Eschborn, Germany. Accessed June 16, 2015: www.GIZ.de.

Securities and Exchange Commission (SEC). (2006). *Commision Staff Working Document, Annex to the Communication on the Promotion of the Inland Waterways Transport, COM 20066 final*. Commission European Communities, Brussels, Belgium.

Siebel, M., Rotter, V., Nabenda, A., et al. (2013). *Osterr Wasser-und Abfallw*. **65**: 42, doi: 10.1007/s00506-012-0052-4. Accessed July 12, 2014: http://link.springer.com/article/10.1007/s00506-012-0052-4

Simpson-Hebert, M., Mitrovic, A., Zajic, G., and Petrovic, M. (2005). *A Paper Life: Belgrade's Roma in the Underworld of Waste Scavenging and Recycling*. WEDC, Loughborough University.

Spielmann, M., Kagi, T., Stradler, P., and Tietje, O. (2004). *Life Cycle Inventories of Transport Services*. ECOINVENT Report No. 14. Swiss Centre for Life Cycle Inventories, Dubendorf, Switzerland

Statistics Denmark. (2008). StatBank Denmark. Accessed February 23, 2014: http://www.statbank.dk/statbank5a/default.asp?w=1280

Troschinetz, A. M., and Mihelcic, J. R. (2009). Sustainable recycling of municipal solid waste in developing countries. *Waste Management* **29**(2), 915–923. doi:10.1016/j.wasman.2008.04.016

UNEP. (2005). Solid waste management. Accessed December 9, 2015: http://www.unep.org/ietc/Portals/136/SWM-Vol1-Part1-Chapters1to3.pdf.

UNEP. (2006). CD4CDM, capacity development for CDM. UNEP Risø Centre on Energy, Climate and Sustainable Development. Accessed December 21, 2015: http://www.cd4cdm.org

UNEP. (2010). *Waste and Climate Change: Global Trends and Strategy Framework*. UNEP.

UNFCC. (2005). Caring for climate 2005: A guide to the climate change convention and the Kyoto Protocol. Accessed December 4, 2014: http://unfccc.int/resource/docs/publications/caring2005_en.pdf.

UN-Habitat. (2010). *Collection of Municipal Solid Waste in Developing Countries*. United Nations Centre for Human Settlements.

UN-Habitat. (2011). Cities and climate change: Global report on human settlements. Accessed October 16, 2015: http://unhabitat.org/books/cities-and-climate-change-global-report-on-human-settlements-2011-abridged/

UN-Habitat. (2014). World urbanization prospects. Accessed June 27, 2015: http://esa.un.org/unpd/wup/Highlights/WUP2014-Highlights.pdf

USAID. (2012). Solid waste management: Addressing climate change impacts on infrastructure. Accessed October 20, 2015: https://www.usaid.gov/sites/default/files/documents/1865/Infrastructure_Solid WasteManagement.pdf

van de Klundert, A., and Anschutz, J. (2001). Integrated sustainable waste management – the concept: Tools for decision-makers. In A. Schein-

berg (ed.), *Experiences from the Urban Waste Expertise Programme* (1995–2001). Urban Waste Expertise Programme, Netherlands.

World Bank. (2012). What a waste: A global review of solid waste management. Urban Development Series Knowledge Papers. Accessed October 20, 2015: http://siteresources.worldbank.org/INTURBANDEVELOPMENT/Resources/336387–1334852610766/What_a_Waste2012_Final.pdf

Xevgenos, D., Papadaskalopoulou, C., Panaretou, V., Moustakas, K., and Malamis, D. (2015). Success stories for recycling of MSW at municipal level: A review. *Waste and Biomass Valorization* **32** (1), 58–73.

Chapter 15   Case Study References

Case Study 15.1   The Challenge of a Rapidly Developing Economy: The Case of Rio de Janeiro

Brookings Institution. (2015). Global Cities Initiative: Rio de Janeiro metropolitan area profile. Accessed March 31, 2016: http://www.brookings.edu/

IBGE. (2015). Instituto Brasileiro de Geografia e Estatistica. Accessed March 26, 2016: http://www.ibge.gov.br/home/estatistica/populacao/censo2010/sinopse/sinopse_tab_rm_zip.shtm

Monteiro, J. (2013). Sustainable waste management in Rio de Janeiro. CCAC, Vancouver. Accessed September 5, 2015: https://www.sustainabledevelopment.un.org

Peel, M. C., Finlayson, B. L., and McMahon, T. A. (2007). Updated world map of the Köppen-Geiger climate classification. *Hydrology and Earth System Sciences Discussions* **4**(2), 462.

World Bank. (2013). Inequality and economic development in Brazil: The World Bank Country Study. Accessed December 21, 2015: https://www.openknowledge.worldbank.org/bitstream/handle/10986/14913/301140PAPER0Inequality0Brazil.pdf

World Bank. (2017). 2016 GNI per capita, Atlas method (current US$). Accessed August 9, 2017: http://data.worldbank.org/indicator/NY.GNP.PCAP.CD

WorldinTwelve. (2015). Why Rio de Janeiro?. Accessed April 8, 2016: http://www.worldintwelve.com/en/cities/riodejaneiro

Case Study 15.2   The Challenge of Developing Cities: The Case of Addis Ababa

Bond, P., Sharife, K., Allen F., Amisi B., Brunner K, Castel-Branco, R., Dorsey D., Gambirazzio, G., Hathaway, T., Nel, A., and Nham, W. (2012). *The CDM Cannot Deliver the Money to Africa*. Why the Clean Development Mechanism Won't Save the Planet from Climate Change, and How African Civil Society Is Resisting. Report No. 2. Environmental Justice Organisations, Liabilities and Trade (EJOLT).

Future Mega Cities. (2015). Solid waste management in Addis Ababa (Ethiopia): Model based strategic planning. Accessed March 30, 2016: http://future-megacities.org/fileadmin/documents/forschungsergebnisse/aktuell/ADI-AB1.pdf

Government of Ethiopia. (2012). Ethiopian government portal. Accessed September 5, 2015: www.ethiopia.gov.et/stateaddisababa

Peel, M. C., Finlayson, B. L., and McMahon, T. A. (2007). Updated world map of the Köppen-Geiger climate classification. *Hydrology and Earth System Sciences Discussions* **4**(2), 462.

Regassa, N., Sundaraa, R. D., and Seboka, B. B. (2011). Challenges and opportunities in municipal solid waste management: The case of Addis Ababa City, Central Ethiopia *Journal Human Ecology* **33**(3), 179–190

UNDP (n.d.). Addis Ababa to improve solid waste management practices. Accessed January 20, 2017: http://www.undp.org/content/dam/undp/library/Environment%20and%20Energy/MDG%20Carbon%20Facility/MDG_Carbon_Ethiopia.pdf

World Bank. (2014). Economic update II: Laying the foundation for achieving middle income status. Accessed August 6, 2015: https://www.openknowledge.worldbank.org

World Bank. (2017). 2016 GNI per capita, Atlas method (current US$). Accessed August 9, 2017: http://data.worldbank.org/indicator/NY.GNP.PCAP.CD

Case Study 15.3 Integrated Community-Based Waste Management toward a Low-Carbon Eco-City in Tangerang Selatan, Indonesia

Badan Pusat Statistik. (2017). Sensus Penduduk 2010. Accessed August 27, 2015: http://sp2010.bps.go.id/

BLHD. (2013). Perda No. 13/2013 tentang Pengolahan Lingkungan Hidup yang Mengatur Pengelolaan Lingkungan, Badan Lingkungan Hidup Daerah [Local Regulation No. 13/2013 about Environmental Treatment regulating the environment in Tangerang Selatan], Badan Pengelolaan Lingkungan Hidup Daerah.

BORDA (2014). Decentralized Solid Waste Management, DESWAM. BORDA Bremen. Accessed July 16, 2015: http://www.borda-net.org/fileadmin/borda-net/Knowledge/Education/Bildungsbroschuere_B2_DESWAM(en)_web.pdf

BPS Kota Tangerang Selatan. (2013a). Luas Wilayah Menurut Kecamatan Tahun 2013. Accessed June 1, 2014: http://tangselkota.bps.go.id/webbeta/frontend/linkTableDinamis/view/id/5

BPS Kota Tangerang Selatan (2013b). Jumlah Penduduk Kota Tangerang Selatan Menurut Jenis Kelamin tahun 2013. Accessed April 7, 2014: http://tangselkota.bps.go.id/webbeta/frontend/linkTableDinamis/view/id/2DKPP 2010

DKPP. (2013). PERDA No. 3/2013 tentang Pengelolaan Sampah [Local Regulation No. 3/2013 about Solid Waste Management]. Dinas Kebersihan Pertamanan dan Pemakaman.

DKPP. (2014). Mon-Ev Kinerja Pengelolaan Sampah tangerang Selatan (on-going) [Mon-Ev Waste Management Performance in Tangerang Selatan]. Dinas Kebersihan Pertamanan dan Pemakaman.

DTKBP. (2011). Perda Kota Tangerang Selatan No. 315 Tahun 2011 tentang RTRW Kota Tangerang Selatan [Regional Regulation Tangerang Selatan City no. 315, Year 2011 about Master Plan Tangerang Selatan City]. Dinas Tata Kota Bangunan dan Pemukiman.

DTKBP (2014). laporan antara: Penyusunan dokumen RP3KP bidang perumahan dan permukiman kota tangerang selatan (Interim Report: Framing the Documents RP3KP, Board Housing and Settlement Tangerang Selatan City). Dinas Tata Kota Bangunan dan Pemukiman.

Peel, M. C.,. Finlayson, B. L., and McMahon, T. A. (2007). Updated world map of the Köppen-Geiger climate classification. *Hydrology and Earth System Sciences Discussions* **4**(2), 462.

World Bank. (2017). 2016 GNI per capita, Atlas method (current US$). Accessed August 9, 2017: http://data.worldbank.org/indicator/NY.GNP.PCAP.CD

Case Study 15.4 Sustainable Waste Management: The Successful Example of Vienna

City of Vienna. (2015). Vienna in Figures 2015. Vienna City Administration; Vienna, Austria. Accessed April 8, 2016: https://www.wien.gv.at

Comet. (2015). Vienna region. Accessed March 6, 2016: http://www.oeaw.ac.at/isr/comet/documents/Final_Results/COMET_deliv_no11_leafletVienna_final.pdf

Economic Affairs, Labour and Statistics, Vienna City Administration. (2015). Vienna in figures. Accessed January 10, 2016: https://www.wien.gv.at/english/administration/statistics/

European Commission (DG Environment). (2012). Waste prevention – Handbook: Guidelines on waste prevention programmes. Accessed April 7, 2014: http://www.ec.europa.eu

Peel, M. C., Finlayson, B. L., and McMahon, T. A. (2007). Updated world map of the Köppen-Geiger climate classification. *Hydrology and Earth System Sciences Discussions* **4**(2), 462.

Thon, J. (2015). Waste management in Vienna: Municipal Department 48 – Waste management, street cleaning and vehicle fleet. City of Vienna, Austria. Accessed October 5, 2016: https://www.wien.gv.at

World Bank. (2017). 2016 GNI per capita, Atlas method (current US$). Accessed August 9, 2017: http://data.worldbank.org/indicator/NY.GNP.PCAP.CD

Part IV

Governance and Urban Futures

16

Governance and Policy

Coordinating Lead Authors

Patricia Romero-Lankao (Boulder/Mexico City), Sarah Burch (Waterloo), Sara Hughes (Toronto)

Lead Authors

Kate Auty (Canberra), Alex Aylett (Montreal), Kerstin Krellenberg (Leipzig), Ryoko Nakano (Hayama), David Simon (Gothenburg/Cape Town/London), Gina Ziervogel (Cape Town)

Contributing Authors

Anja Wejs (Aalborg)

This chapter is dedicated to Alex Aylett, a bright and kind co-author. His engagement and passion will always be remembered.

This chapter should be cited as

Romero-Lankao, P., Burch, S., Hughes, S., Auty, K., Aylett, A., Krellenberg, K., Nakano, R., Simon, D., and Ziervogel, G. (2018). Governance and policy. In Rosenzweig, C., W. Solecki, P. Romero-Lankao, S. Mehrotra, S. Dhakal, and S. Ali Ibrahim (eds.), *Climate Change and Cities: Second Assessment Report of the Urban Climate Change Research Network*. Cambridge University Press. New York 585–606

Urban Governance for a Changing Climate

Greenhouse gas (GHG) emissions and climate risks in cities are not only municipal government concerns. They challenge a range of actors across jurisdictions to create coalitions for climate governance. Urban climate change governance occurs within a broader socioeconomic and political context, such as the landmark international negotiations during the 21st Conference of the Parties (COP21) to the United Nations Framework Convention on Climate change in Paris (2015). As such, actors and institutions at a multitude of scales shape the effectiveness of urban-scale interventions. These interventions may be particularly powerful if they are integrated with co-benefits related to other development priorities (such as health, biodiversity, and poverty reduction), thus creating urban systems (both built and institutional) that are able to withstand, adapt to, and recover from climate-related hazards.

Collaborative, equitable, and informed decision-making is needed to enable transformative responses to climate change, as well as fundamental changes in energy and land-use regimes, growth ethos, production and consumption, lifestyles, and worldviews. Leadership, legal frameworks, public participation mechanisms, information sharing, and financial resources all work to shape the form and effectiveness of urban climate change governance.

Major Findings

- While jurisdiction over many dimensions of climate change adaptation and mitigation resides at the national level, along with the relevant technical and financial capacities, comprehensive national climate change policy is still lacking in most countries. Despite this deficiency, municipal, state, and provincial governmental and non-governmental actors are taking action to address climate change.

- Urban climate change governance consists not only of decisions made by government actors, but also by non-governmental and civil society actors in the city. Participatory processes that engage these interests around a common aim hold the greatest potential to create legitimate, effective response strategies.

- Governance challenges, such as disconnects between electoral cycles and climate change planning horizons, as well as inconsistencies or contradictions among climate policies at different levels of government, often contribute to gaps between the climate commitments that cities make and the effectiveness of their actions.

- Governance capacity to respond to climate change, including human resources, financial resources, legal frameworks, and legitimate institutions, varies widely within and between low- and high-income cities, creating a profile of different needs and opportunities on a city-by-city basis.

- The challenge of coordinating across the governmental and non-governmental sectors, jurisdictions, and actors that is necessary for transformative urban climate change policies is often not met. Smaller scale, incremental actions controlled by local jurisdictions, single institutions, or private and community actors tend to dominate city-level actions.

- Scientific information is necessary for creating a strong foundation for effective urban climate change governance, but governance is needed to apply it. Scientific information needs to be co-generated in order for it to be applied effectively and meet the needs and address the concerns of the range of urban stakeholders.

Key Messages

While climate change mitigation and adaptation have become pressing issues for cities, governance challenges have led to policy responses that are mostly incremental and fragmented. Many cities are integrating mitigation and adaptation, but fewer are embarking on the more transformative strategies required to trigger a fundamental change toward sustainable and climate-resilient urban development pathways.

The drivers, dynamics, and consequences of climate change cut across jurisdictional boundaries and require collaborative governance across governmental and non-governmental sectors, actors, administrative boundaries, and jurisdictions. Although there is no single governance solution to climate change, longer planning timescales; coordination and participation among multiple actors; and flexible, adaptive governance arrangements may lead to more effective urban climate governance.

Urban climate change governance should incorporate principles of justice in order that inequities in cities are not reproduced. Therefore, justice in urban climate change governance requires that vulnerable groups are represented in adaptation and mitigation planning processes; priority framing and setting recognize the particular needs of vulnerable groups; and actions taken to respond to climate change enhance the rights and assets of vulnerable groups.

16.1 Governance and Institutional Capacity for Mitigation and Adaptation

Cities account for approximately 70% of CO_2 emissions (depending on measurement protocols) yet, in many cases, only a small fraction of the emissions produced within a city are under the direct control of municipal governments (Seto et al., 2014). Other jurisdictions and actors, such as national governments, the private sector, and individuals often control a significant portion of GHG emissions. At the same time, urban populations, economic activities, and infrastructures (see Section III of this volume) are vulnerable to a suite of negative impacts that climate change might aggravate (e.g., mortality from heat waves and damages from floods). Furthermore, carbon and climate are cross-scale issues. As such, urban areas are affected by actions beyond their boundaries, and urban emissions, risks, and actions create effects far outside of the demarcations of city limits. Therefore, GHG emissions and risks in cities are not only municipal governmental concerns; they challenge a range of actors across sectors to create coalitions for climate governance in order to mitigate emissions and adapt to climate risks (Aylett, 2013).

In this context, we define urban climate change governance as the set of formal and informal rules, rule-making systems, and actor networks at all levels (from local to global), both in and outside of government, established to steer cities toward mitigating and adapting to climate change (Biermann et al., 2009). Thus, urban climate change governance occurs within a broader socioeconomic and political context, with actors and institutions at a multitude of scales shaping the effectiveness of urban-scale interventions.

Although mitigation and adaptation are emerging as some of the most pressing issues faced by urban areas, both the study and practice of climate governance have historically addressed them separately, with mitigation referring to measures aimed at reducing resource-use impacts and adaptation referring to actions aimed at managing these impacts, before or after they are experienced (Field et al., 2014). However, in reality, these goals interact in potentially synergistic or conflicting ways (see Section 16.3.2). The notions of *response* and *response capacity* are ways of defining the institutional capacity to govern carbon and climate in cities (Tompkins and Adger, 2005). A *response* is any action taken by governmental and non-governmental actors to manage environmental change, either in anticipation of known change or after change has happened. *Responses* are fashioned through power, though consensus, compromise, or coercion, often by actors who frame mitigation and adaptation in the context of other environmental concerns (e.g., energy and disaster risk management), development pressures and goals (e.g., economic growth and human well-being), and in pursuit of a range of often conflicting values and priorities (Romero-Lankao et al., 2013). *Response capacity* relates to a pool of resources and assets that governmental and non-governmental actors may draw on for climate change mitigation and adaptation while attending to other development needs (Yohe and Tol, 2002; Burch and Robinson, 2007; Romero-Lankao et al., 2013).

This chapter will explore some of the dimensions of the capacity to develop governance solutions for carbon and climate change in cities including:

1. Different types of mitigation and adaptation actions developed and implemented in urban areas
2. Actors and networks at multiple levels
3. The nature, opportunities, barriers, and limits that multilevel governance poses to local climate policy
4. Gaps between the policy discourse and the challenges that local climate action needs to address under real-world conditions

Given that GHG emissions continue to increase quickly in many jurisdictions (Field et al., 2014), there is growing recognition that the scale of the challenge is greater than can be addressed by the modest and isolated responses that are most common in cities around the globe. Incremental reform may prove inadequate, requiring instead a transformative approach that fundamentally alters elements of the system such as energy and land-use regimes and their underlying power relations, worldviews, market structures, and governance systems (Park et al., 2012). Concerted action by governments, disruptive innovation in the private sector, and pressure from civil society may be required to trigger such transformative processes. However, the question of what exactly needs to be transformed, and why, how, and in whose interest remains open (O'Brien, 2012). Moreover, the factors driving or triggering the necessary fundamental transformations remain uncertain. Therefore, rather than prescribing the most appropriate scope of responses needed to achieve transformational policies, in this chapter we build on previous work identifying the challenges cities face (Park et al., 2012; Moser and Ekstrom, 2010) to evaluate the scope and scale of existing response actions to climate change.

This chapter also seeks to outline a governance framework that can be applied to urban areas (Romero-Lankao et al., 2013). Its interconnected elements include the issue of concern (i.e., mitigation of GHG emissions and adaptation to climate risk[1]); the response actions with their guiding goals and targets (e.g., to reduce fossil fuel use and to avoid or lessen climate change impacts (Parris and Kates, 2003); the governmental and non-governmental actors; the broader social and environmental context in which they operate; and the limits, barriers, and options[2] these actors face during the different phases of the decision-making process, from problem definition to implementation and evaluation (Park et al., 2012).

Urban responses for mitigating and adapting to climate change, as identified in previous research and practice, range

---

1 Following (Field et al., 2012), we define risk as the possibility of loss, injury, and other climate-related impacts on things we value; as the outcome of exposure to hazards and the capacity to perceive and respond to these hazards. See Chapter 2 of this report.

2 While barriers have been described as being "mutable, subjective, and socially constructed" (Adger et al., 2009: 338), limits, which are more rigid and fixed (Rothman et al., 2013).

from short- to long-term and vary widely in their effectiveness and outcomes. Some examples are shown in Figure 16.1 and include the following domains (see Sections 16.2.1 and 16.2.2):

1. *Understanding of the problem*: Through an inventory of GHG emissions, which provides a baseline against which mitigation targets can be assessed; assessments of the climate risks urban populations may face under a changing climate; and assessments of the drivers of both emissions and risks

2. *Incremental responses*: For example, mitigation actions focused on municipal government buildings and vehicle fleets or adaptation actions that build on ongoing disaster risk management.

3. *Broader scope, longer term responses* seeking to alter urban form, institutions, and behavior. These include:

   a. Actions and infrastructural investments that: (i) reduce vehicle kilometers traveled, promote mixed-use development, improve destination accessibility, and reduce distance to transit by concentrating development and thus reducing transport energy use (Hamin and Gurran, 2009); (ii) discourage growth in risk-prone areas and protect or restore the ecosystem functions and services such as infiltration, flood, surge protection, and temperature regulation. These may influence not only GHG emissions, but may also shape the vulnerability of individuals, populations, and sectors to climate hazards;

   b. Actions that build capacity by enhancing the assets and options afforded to individuals from diverse socioeconomic groups to use low-carbon energy sources and to adapt to the impacts of climate change;

   c. Actions that reduce hazard exposure including risk mitigation (e.g., through engineered protection systems such as dikes and barriers)

4. *Transformative responses* that contribute to profound changes in energy and land-use regimes, growth ethos, production and consumption, lifestyles, and worldviews (Field et al., 2014). Some of these actions target the underlying drivers of GHG emissions and vulnerability, such as systems of production and consumption, and the social

inequalities that give rise to the coexistence of substandard housing, illiteracy, and poverty alongside wealth-related consumptive practices that are at the heart of our climate challenge. As such, transformative actions hold the potential to trigger a broader shift toward sustainable and resilient development pathways (Shaw et al., 2014; Burch et al., 2014).

Responses are initiated and shaped by state, community, and private-sector *actors,* defined here as individuals, communities, organizations, and networks that participate in decision-making related to urban mitigation and adaptation (Biermann et al., 2009). Such actors are involved both in defining the issue of concern and in seeking solutions. As such, they hold varied and often conflicting interests and visions about the best course of climate change action. For example, whereas some urban actors might consider nuclear energy to be a safe and proven alternative to fossil fuels, or hard infrastructures a feasible option to provide fresh water and sanitation and protect coastal cities from sea level rise, other actors might see these as poor choices (Romero-Lankao and Gnatz, 2013) (see Section 16.4.3). These differences result in competing discursive and material constructions of *response action*s and the potential for resulting fragmented or conflicting policies (Pelling and Manuel-Navarrete, 2011).

Furthermore, scholarship on urban climate governance has been concerned with the gap between the rhetoric and reality of carbon and climate action (Betsill and Bulkeley, 2007; Burch and Robinson, 2007) and, for the reasons just outlined, on the mechanisms by which multilevel governance arrangements shape responses. This chapter explores key factors or drivers shaping the scope and effectiveness of responses, which vary with context. These factors include multilevel actors and interactions (Section 16.3.2), mechanisms in place for actor engagement and participation (Section 16.4.3), legal frameworks (Section 16.4.1), generation and transmission of different ways of knowing (Section 16.4.2), financial resources, decision-making power (Section 16.3.3), and leadership (Section 16.4.4).

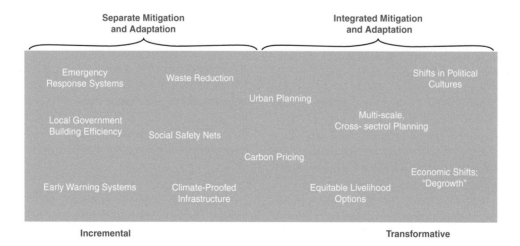

Figure 16.1 *A sample of urban climate change governance strategies, ranging from incremental responses that address adaptation and mitigation in isolation to potentially transformative strategies that integrate mitigation and adaptation. Relative location along the y-axis is not significant.*

Taken together, these dimensions reveal the complexity and dynamism inherent in urban carbon and climate governance. There is no single aspect of a governance system that can guarantee effective action to address climate change. Rather, it is the combination, synergy, and tradeoffs of several elements that may facilitate, or hinder, inclusive and actionable decision-making. Compelling examples of action, novel approaches to stakeholder engagement, and deeper insights into the drivers of effective governance have emerged in the communities of research and practice. This chapter weaves together case-based analysis with an assessment of established and evolving literature to identify the key challenges and opportunities of governing climate change in urban spaces (see Case Study 16.1).

16.2 Urban Climate Governance: A Brief Overview of Approaches to Mitigation and Adaptation

The initial recognition of the importance of cities to climate change governance came in the mid-1990s in the form of emissions reduction programs developed by urban actors in Europe and North America (Betsill and Bulkeley, 2007) and in frontrunner cities of middle-income countries (e.g., Cape Town and Mexico City) (Holgate, 2007; Romero-Lankao, 2007). Adaptation is a more recent addition to urban climate change governance, and first came into focus in low-income countries where a long tradition in disaster management exists and the impacts of climate change were more obvious before being taken up by middle- and high-income countries (Satterthwaite et al., 2007; Betsill and Bulkeley, 2007; Krellenberg and Hansjürgens, 2014). It has become apparent that the prospects for effective adaptation and sustainable development depend on accomplishing substantial mitigation. This poses the need for transformative change in existing technological, economic, social, and institutional systems (Field et al., 2014). Sections 16.2.1 and 16.2.2 outline the different urban mitigation and adaptation governance strategies identified to date, as well as their scope (see Figure 16.1). Sections 16.2.3 and 16.2.4 reflect on the ways in which these options relate to each other, with other key local priorities, and with what transformative approaches would entail.

16.2.1 Adaptation Responses

Five categories of adaptation responses available to governments, households, the private sector, and communities have been shown to hold the potential to address risk by reducing hazard exposure and vulnerability and by enhancing the capacities of urban actors. These categories are not necessarily mutually exclusive and can be pursued simultaneously.

1. *Institutional and behavioral actions* focus on changing the procedures, incentives, or actions, and often work through existing urban competencies and hybrid actor arrangements in sectors such as planning, health, and water (Fisher, 2013). Examples include disaster risk reduction (DRR), early warning systems, climate-sensitive management protocols, and disease surveillance in the health sector, or demand-side management in the water sector (McDonald et al., 2011). These are often initiated by municipal governments and bolstered by national or regional governments (Ziervogel and Parnell, 2014; Vasi, 2007), business, or civil-society actors (Brown et al., 2012).

2. *Technological and infrastructural actions* seek to discourage growth in risk-prone areas and protect urban infrastructure systems through changes to design, operational, and maintenance practices. Early responses to long-term climatic variability in transport, water, sanitation, telecommunications, and green infrastructure are often less costly than deferring action (Revi et al., 2014). Proactive planning helps to address the inherent uncertainty in many of these interventions, as do multistakeholder planning processes. The Thames Estuary 2100 project, which has drawn upon a wide range of actors to assess infrastructure for managing changing tidal risk, is an example of this strategy (Ranger et al., 2013). Some technological interventions may be driven exclusively by governments, such as investments in major flood defenses or siting decisions for infrastructure investments in the transportation, water, and sanitary services sectors (Brown et al., 2012). They may also result from community-based adaptive strategies, such as elevating houses in informal settlements or building floodwater drainage and sewer systems (Jabeen et al., 2010). Yet many community responses are not sufficient on their own to significantly reduce the vulnerabilities of urban populations. State interventions are therefore needed to foster adaptive capacity, for instance, by working collaboratively with different non-governmental actors so that municipal infrastructure investments are designed to enable more effective responses on the part of civil-society actors (Romero-Lankao et al., 2014).

3. *Economic and regulatory instruments* seek to create an enabling environment for autonomous action on the part of governmental, private and civil society actors and to support broader development goals. A recent survey of 350 ICLEI – Local Governments for Sustainability member cities (The Urban Climate Change Governance Survey, or UCGS) shows that most cities have been unable to effectively link their adaptation policies to their other local development goals (Aylett, 2014). Identifying such synergies is particularly important for gathering broad support in low- and middle-income countries (Huq et al., 2007). Prospects for progressing and mainstreaming climate change agendas, therefore, depend on being able to demonstrate how these are not in conflict with development priorities, as often claimed, but instead are essential and complementary to them (Simon, 2011).

4. *Urban planning*, or the policy process through which strategic decisions about a city's future are made, is a key instrument for anticipating climate change impacts and fostering early action, yet it does not always achieve these aims on the ground (Carmin et al., 2012; Bracken, 2014). For instance, some authorities might be concerned with avoiding growth in risk-prone areas or incorporating climate hazards into planning. These priorities, however, often compete for regulatory space within a policy agenda that is already coping with a very wide range of economic and capitalistic drivers of development (Romero-Lankao et al., 2013a). Furthermore, the level

of authority and autonomy of urban planning can vary significantly between cities. Even so, in some cities, a shift in the urban planning tradition is occurring away from reactive and toward more proactive approaches (Hansen et al., In press).

5. *Funding programs* from public and private sectors are fundamental. By strategically allocating funding (whose scale and sources vary widely and depends in part on how much local authorities can tax residents, property owners, and business), urban governments can effectively respond to risks. Still, according to a survey of 468 cities conducted by Carmin et al. (2012), 60% of city governments are not receiving any financial support to undertake adaptation. Furthermore, many ongoing activities may not be explicitly called adaptation, but may be considered to reduce vulnerability and enhance adaptive capacity. Subsequent research by Aylett (2014) identified a lack of financial resources (and associated lack of staff time and institutional resources) as the dominant challenges that cities report as affecting their climate change planning and implementation (for both mitigation and adaptation) (see Figure 16.2).

Although these responses can expand the capacity of urban actors and areas to cope with a changing climate, barriers and limits to adaptation exist. Barriers to action include short-term planning horizons, uncertainty of climate change impacts, and other socially constructed obstacles (Burch, 2010; Moser and Ekstrom, 2010). Limits occur when actions to avoid intolerable risks are not possible or not currently available (Rothman et al., 2013; Field et al., 2014). These limits suggest that transformational change may be a requirement for sustainable urban development in a changing climate.

16.2.2 Mitigation Responses

Mitigation responses are being undertaken by city governments through auto-regulation (a government addressing its own emissions), mandatory regulations, economic incentives, and facilitation (Kern et al., 2008; Castán Broto and Bulkeley, 2013). As with adaptation, many mitigation actions require city governments to work with non-governmental actors and other scales of government (see Section 16.3). Here, we explore how these options induce changes in energy and land use by affecting urban form and accessibility, consumption, living and housing type, infrastructures, and the carbon content of energy.

1. *Auto-regulation,* or voluntary self-regulation. These are the most common actions implemented by local authorities. A survey of 350 cities conducted by Aylett (2014) found that these have focused on municipal government buildings (89%) and vehicle fleets (72%), on waste reduction (55%), and public transit use (36%) The UCGS found that auto-regulation is the most common way in which city governments have achieved measurable emissions reductions (Aylett, 2014). The most common areas where emissions reductions have been made are municipal government buildings (89%) and vehicle fleets (72%) and waste reduction (55%) (see Figure 16.3).

2. *Mandatory regulations,* the most effective but least pursued by city governments, are enacted to reduce GHG emissions when urban authorities have legal jurisdiction over such sectors as energy, transport, land use, and waste. Authorities can introduce codes and ordinances for new building constructions as well as retrofits that enhance building energy efficiency and environmental performance (e.g., Toronto's green roof bylaw [in 2009], or Vancouver's Neighbourhood Energy Connectivity Standards and Green Homes Program [in 2014]; (Mehdi et al., 2006; Lutsey and Sperling, 2008). Other actions being used in middle- and low-income countries include vehicle emissions standards, fuel standards, appliance efficiency labeling, and renewable electricity portfolio standards.

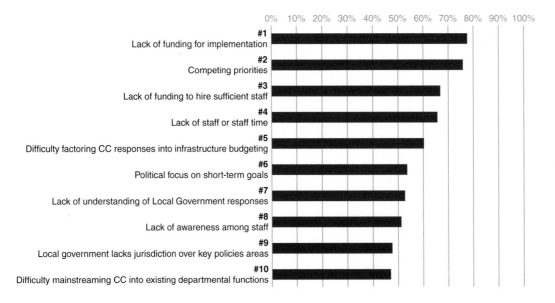

Figure 16.2 *Top ten challenges reported by cities in the Urban Climate Change Governance Survey, ranked according to the percentage of cities reporting that these issues were significant or major challenges to their climate change planning and implementation work.*

Source: Aylett, 2014

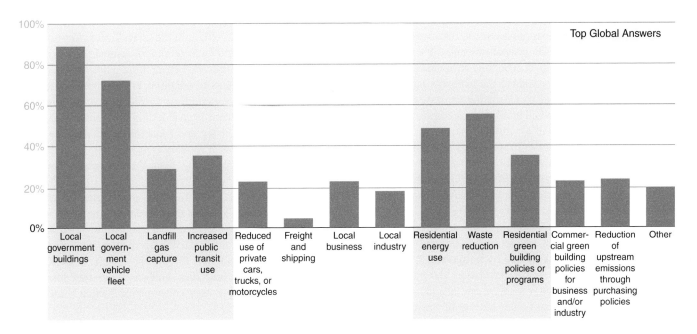

Figure 16.3 *Where cities have made measurable emission reductions. Cities can pursue a wide range of activities to reduce their greenhouse gas emissions. This figure shows the percentage of cities globally that report making measurable emissions reductions across fourteen different areas.*

Source: Aylett, 2014

3. *Economic incentives.* Municipal governments (when they have legal jurisdiction) offer grants, subsidies, tax credits, and other economic and financial incentives to facilitate the adoption of sustainable technologies, build efficiency retrofits, and create small-scale renewable energy systems (e.g., the Toronto Atmospheric Fund) (Zimmerman and Faris, 2011; Fisher, 2013). Market-based tools such as carbon taxes and cap-and-trade systems, while not typically initiated at the urban scale, directly affect urban actors and emissions.

4. *Facilitating measures* mostly focus on new services and infrastructure development. For example, cities can pursue climate change mitigation through public–private partnerships, through educational campaigns to citizens, or through guidance to the private sector. Incentives can also help to foster initial innovative actions. As illustrated by Portland's Clean Energy Works and Solarize Portland programs, the synergies associated with reducing local GHG emissions can also serve as additional motivations for non-governmental actors to support climate-related initiatives (Aylett, 2013; Burch et al., 2013; Castán Broto and Bulkeley, 2013).

Despite the rising prominence of mitigation in the policy arena and the emergence of ambitious targets, mitigation actions remain fragmented, focused on auto-regulation, and the realities of achieving reductions in GHG emissions are often more challenging than anticipated. The following sections will explore some of the factors why this is so.

16.2.3 Different Metrics Used to Evaluate Mitigation and Adaptation

Evaluating and increasing the effectiveness of both mitigation and adaptation responses is progressively becoming

a focus for researchers and practitioners. For mitigation, the impact of a policy can be measured by assessing the total tonnes of CO_2 equivalent reduced. These reductions are measured against a baseline inventory, which many cities are conducting globally using a variety of methodologies proposed and promoted by groups such as ICLEI, UN-Habitat, the World Bank, and the World Resources Institute. The multiplicity of approaches has meant that comparisons between cities and across time are problematic. A push to harmonize the different reporting techniques led to the creation of a Global Protocol for Community-Scale Greenhouse Gas Emissions reporting (GPC) in 2012, which has since been the basis for revised emissions accounting procedures such as ICLEI's Harmonized Emissions Analysis Tool (HEAT+). An additional challenge is presented by the gap between urban areas' pledges to achieve GHG emissions reduction targets and the real mitigation potential of their actions, which are often limited based on city government control and jurisdiction. As a handful of high-income country studies illustrate, many urban mitigation actions fall short of the emissions reduction targets needed to avoid a 2°C increase in global mean temperature (Reckien et al., 2014). The global mitigation impact of urban responses is, hence, unclear (Hutyra et al., 2014).

Although for adaptation no universally accepted assessment metric exists, adaptation policies are generally effective if they reduce negative impacts of climate change or enhance the underlying adaptiveness or resilience of populations, infrastructures, and other systems at risk. This complicates matters, not least because vehicles such as the Kyoto Protocol's Adaptation Fund encounter challenges when trying to compare the adaptive effect of projects in order to allocate funds efficiently (Stadelmann et al., 2011). Of the cities surveyed by Carmin et al. (2012), 65% expect impacts on stormwater management, 39% loss of natural systems, 35% droughts, 34% coastal erosion, 30% urban heat island effects, and 29% loss of economic revenue. These problems can be addressed by reducing the exposure and sensitivity of people and assets and by enhancing capacity to perceive and respond to these hazards, but adaptation responses still lag behind mitigation (Carmin et al., 2012; Revi, 2014). Following up on Carmin's work, Aylett's (2014) survey of ICLEI member cities found that 73% of respondents were planning for both mitigation and adaptation and are treating the issues in an integrated way that takes into consideration the synergies and conflicts between planning in the two areas (see Figure 16.4).

This research shows that adaptation has established itself in a policy space formerly dominated by mitigation. It also highlights the importance of better understanding the tradeoffs and synergies that exist between mitigation and adaptation, as well as between climate action and other local policy priorities. The broader sustainability implications of climate change action are increasingly being considered, leading to calls for more holistic and even transformative planning in cities. These issues are addressed in greater detail in Section 16.2.5.

16.2.4 Barriers Resulting from Different Timeframes at Which Mitigation and Adaptation Operate

One challenge for mitigation and adaptation policies is the lag between the timing of investments in implementation and the point at which investments yield financial, environmental, or social returns. Minor efficiency measures may recoup their costs quickly, but the associated GHG reductions are limited. Large-scale investments in infrastructure and urban form, such as smart grid technologies or public transit, take decades to recoup their high upfront costs and to realize their mitigation potential. This

is similar for adaptation measures that increase the resilience of urban infrastructure. Although the overall costs are justified by projections of avoided damage and associated recovery costs, it can take more than a decade before those gains are realized (and these gains may not just be adaptive, but may also reduce GHGs in the long term). For example, a robust strategy to protect New York's electrical system from flooding and wind damage while also increasing efficiency and reducing GHG emissions would require an investment of roughly US$3 billion. It would take 15 years for the financial returns from these investments (in terms of avoided damage and increased efficiency) to hit the break-even point (Arup RPA, 2014). In the short term, it is cheaper to pay for repairs than to make the investments needed to increase resilience. This is exacerbated by the short time horizon of local electoral cycles that can push officials to favor immediate returns and lower costs (Bulkeley, 2010). Nevertheless, there are also low-cost adaptation activities, such as those that foster green spaces or immediate energy conservation, which show positive short-term benefits.

Financial returns are not the only domain in which the timeframes of action and outcomes are likely to be misaligned. In a world of short-term political and media cycles, it can be challenging to capture and maintain the interest of elected officials and the public to sustain climate change action and build governance structures.

16.2.5 Overlaps, Synergies, and Conflicts among Adaptation, Mitigation, and Urban Development

In an effort to increase the direct local benefits and political attractiveness of climate change response strategies, much work has gone into identifying the synergies and co-benefits among adaptation, mitigation, and broader development priorities (Beg et al.,2002; UN-Habitat, 2011; Shaw et al., 2014).

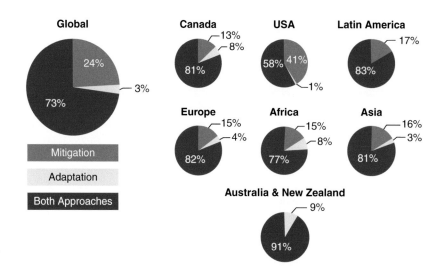

Figure 16.4 *Percentage of respondents in the 2014 MIT-ICLEI Urban Climate Change Governance Survey who reported that their climate change work was focused on adaptation, mitigation, or both.*

Source: Aylett, 2014

Mitigation measures can reduce local air pollution and improve respiratory health or produce financial savings through energy efficiency (while reducing energy associated emissions), for example. Likewise, adaptive measures to address flooding can also benefit local populations by improving local water quality, or providing sanitary infrastructure (Bai, 2007; Gore, 2010). But despite these documented examples of effective synergies in the research and best-practices literature, most cities have yet to effectively link their emissions mitigation work to achieving other local development priorities. The Aylett survey (2014) shows that across a broad list of possible social, economic, environmental, and infrastructure priorities, cities consistently report that mitigation measures have made little to no contribution; the key exceptions to this are actions directly linked to other environmental goals (e.g., increasing access to basic services) (see Figure 16.5). Reasons for this can include inertia behind out-of-date planning practices, lack of political will or of expertise in identifying and exploiting synergies, and the relatively nascent state of climate change policy.

In addition to interacting with other development priorities, mitigation and adaptation measures can also overlap and reinforce each other. Smart grid technologies, for example, can both decrease local GHG emissions and create a robust local energy system that is better able to withstand the impacts of extreme weather events. Green areas can also play a crucial role because they can serve as carbon sinks, provide flood mitigation services, and reduce the head island effect (Müller and Höfer, 2014; Locke et al., 2014).

These synergies are reframed and further developed in the emerging discourses around "smart" and "resilient" cities (Lööf

et al., 2012; Stumpp, 2013). Through technologically enabled interventions into urban space, smart urban systems (ranging from traffic control, energy, and water management) hold the promise of creating infrastructure that is both low-carbon and more adapted to the potential impacts of a changing climate. Thus, the resilient city is also a social effort. Technological solutions are only one component of enhanced resilience. Effective synergies among adaptation, mitigation, and other local development priorities also function as part of the connective tissue that underlies urban resilience. Co-benefits help enable multiple urban actors to collaborate in order to create urban systems (both built and institutional) that are able to withstand, adapt to, and recover from climate-related hazards. These co-benefits already exist in many projects around the world but have not been well captured by research.

Identifying synergies, or what are now often called co-benefits, and building them into the design of climate change actions provides technical and financial benefits by allowing actors to realize multiple objectives simultaneously. Capitalizing on synergies or co-benefits also removes climate change policies from a narrowly "environmental" category and anchors them to other local priorities, particularly those faced by cities in many low- and middle-income countries, as explained earlier. This has powerful political benefits because it helps climate policies to move through the complex political economy of municipal decision-making and the multiple priorities of city governments. It also facilitates the mainstreaming of climate policies and programs into existing planning and decision-making mechanisms rather than having to deal with them as exceptions or a separate budget category in competition for scarce funds (Simon, 2011).

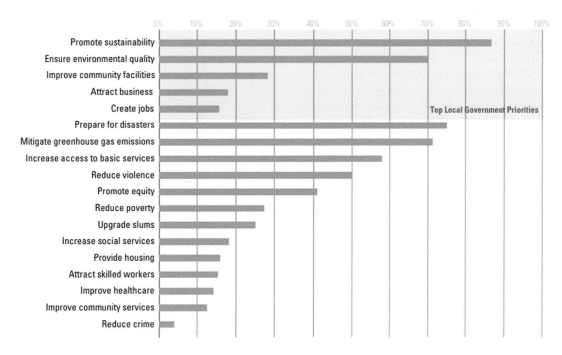

Figure 16.5 *Contribution of mitigation to other development priorities. This figure shows the percentage of cities that report that their climate change mitigation work has contributed significantly to other local development priorities. Priorities highlighted in blue were the top overall development priorities reported by cities.*

Source: Aylett, 2014

However, not all climate change actions will find obvious synergies with local development goals or other climate change-related policies. Diesel-powered emergency electrical generators may increase resilience to climatic disruptions to regional electrical grids, but at the cost of increased emissions. Reducing GHG emissions in locations dependent on carbon-intensive industries may require difficult decisions and tradeoffs (Simon, 2012). Local economic development priorities based on growth and increased consumption will in most cases also result in increased energy use and GHG emissions. Similarly, adaptation measures may not affect all urban residents equally, so it is particularly important to develop context-specific measures to reduce vulnerability (Hughes, 2013; Krellenberg and Welz, 2016). In some cases, for example, when a development is established in floodplains, it may be necessary to relocate populations, businesses, and infrastructure. While the long-term benefits of doing so may justify these decisions, attention and compensation to those who bear the costs of these types of tradeoffs is critical. The strategic identification and operationalization of co-benefits can be a powerful tool to driving forward the design, implementation, and mainstreaming of local responses to climate change. But there are limits to what can be accomplished in this fashion. Strategically using synergies and co-benefits to build coalitions of support that are robust enough to help address inevitable tradeoffs is an important component of successful local climate governance.

16.2.6 The Challenge of Consistency and Coherence

Neither adaptation nor mitigation is a discrete area for policy-making or action. They overlap, and multiple other local development priorities implicate a varied network of public and private actors and have costs and impacts that cross multiple geographical and temporal scales. Maintaining coherence and consistency in this reality is a critical challenge. Integrating mitigation and adaptation planning into broader spatial and development planning processes is one way to help ensure that policies and actions do not work at cross-purposes. Another key is to pay strategic attention to the relationships between various levels of both public and private organizations. In the next section, we will focus on this multilevel governance approach.

16.3 Multilevel Governance, Actors, and Interactions

16.3.1 Categories of Actors

As introduced in Section 16.1, climate change is a socially and biophysically pervasive phenomenon, challenging actors at different scales to come together and create multilevel governance coalitions. These actors have both consistent and inconsistent values, goals, and priorities (Adger, Lorenzoni, and O'Brien, 2009). None of these groups is homogeneous, while shifting alliances and varying levels of power create challenges

for coherent and legitimate urban climate change governance. The mix of actors engaged in responding to climate change more broadly is also changing: what was once the domain of formal state-to-state negotiations has become the contested territory of an evolving array of governmental and non-governmental actors. Networks of actors play multiple roles in urban climate change governance: as providers of resources, facilitators of interactions with other cities that face similar challenges, and shapers of the climate change discourse more broadly (Betsill and Bulkeley, 2007). Many cities are independently taking action even in the absence of national climate change policy frameworks. Some actors, such as the private sector, may operate independently to address climate change within their own domain or form partnerships to achieve a common goal. Suboptimal outcomes may result, however, if these autonomous actions are not integrated with other development and environmental activities.

In some cases, national governments have helped to stimulate local action. For example, in 2008, the government of India developed the National Climate Change Action Plan. The city of Delhi subsequently adopted its own plan, the Climate Change Action Plan of 2009 (the Delhi Plan) to pursue an agenda of being a climate change leader in India. The Delhi Plan determines targets, objectives, strategies, and responsible departments and blends mitigation and adaptation objectives. Each objective is aligned with a national priority, and the city strategically bundled climate change and development issues to achieve broader outcomes (Aggarwal, 2013). Some risks and pressures, such as population growth and water security challenges, were described as "emergent opportunities" even as the exposure to extreme coastal flooding acted as a more routine hazard. However, the local political and administrative environments present barriers – particularly the short termism of election cycles and the multiple overlapping jurisdictions governing Delhi that inhibit the development of institutional capacity (Aggarwal, 2013). Delhi's climate action plan foregrounds the ability of cities to respond to national-level initiatives, but success rests on improved coordination of institutional action. Emerging and significant criticisms of the Delhi Plan include its relatively weak adaptation agenda and its lack of focus on poverty and justice inherent in addressing climate change in the city (Aggarwal, 2013; Hughes, 2013).

However, in countries as diverse as Canada, the United States, Mexico, and South Africa, innovative climate change governance has been largely led by state and municipal governments, often in spite of the lack of comprehensive or ambitious climate change policy at the national level (Rabe, 2007; Macdonald, 2009; Romero-Lankao et al., 2013; Roberts and O'Donoghue, 2013; Ziervogel and Parnell, 2014). Despite their ambitions, these urban actors are constrained in their capacity to influence national climate policy, and many face barriers not only related to influence, resources, and institutional culture but also from the competing priorities with which urban decision-makers grapple (Gore, 2010; Burch, 2010; Measham et al., 2012).

One example of a city that has successfully encouraged private-sector involvement is Berlin. The city of Berlin has reduced emissions from 1,300 public buildings with twenty-five Energy Savings Partnerships since 1996. An audit in 2011 reported €60 million in private investment in energy efficiency measures and the reduction of 600,000 tonnes of CO_2. To make the projects viable, the annual (overarching) energy bill must be at least €250,000. As a result, Berlin set up a system allowing building owners to pool buildings into one project tender. Since unprofitable buildings are also integrated, "pooling" leads to a profitable cross-calculation. The Berlin public–private financing arrangements have been exported to develop projects in Slovenia, Bulgaria, Latvia, Poland, Mexico, and Estonia. Project partners are required to innovate to overcome financing barriers.

This complex landscape creates both opportunities and barriers. While more creative, locally relevant, and equitable strategies may be uncovered if engagement is undertaken in a meaningful way, conflicting jurisdictions or vested interests frequently present obstacles to effective governance. This suggests the need for a variety of actions that include adaptive management, deep and ongoing public participation, private-sector innovation, grassroots initiatives, contestation (e.g., Climate March 2014), and holistic planning that consider the synergies and tradeoffs among various policy priorities.

16.3.2 Governing Complex Actor Interactions

Although much of climate change governance for the past two decades has centered on the nation-state level, other actors (including cities) are now taking on significant responsibilities. This leads to a variety of interactions: transnational networks of cities, bilateral partnerships between cities, coalitions between state- and non-state actors (such as private companies and municipal governments), and independent actions by non-state actors. Three elements increasingly characterize the governance of these complex actor interactions: extensive vertical and horizontal interplay between actors, cross-sector partnerships, and networks at a variety of scales.

Actors, and the institutions of which they are a part, do not operate in isolation but rather in the context of a complex web of interactions. The issue of interplay brings to light the ways in which institutional arrangements at varying (vertical interplay) and similar (horizontal interplay) levels of organization are interdependent (Young, 2002; Urwin and Jordan, 2008). This creates challenges: the actors participating in the governance of climate change in urban spaces often have very different mandates, consider different time scales, and utilize different expertise or ways of knowing. For instance, in Cape Town, South Africa, both the provincial and city governments pursue climate change mitigation. Even so, differences in ruling parties and politics mean few mechanisms exist for structured interaction between the two governments, and collaboration happens only on an ad hoc basis (Holgate, 2007). It must be recognized at the outset, however, that in many larger urban areas, which comprise two or more local and even state authorities, each authority can act only within its boundaries, so that the overall impact may be limited unless there is horizontal collaboration among neighboring authorities or an overarching strategic metropolitan authority exists to ensure citywide action. Dakar in Senegal provides an extreme example of this problem since the city region comprises forty-three autonomous municipalities (of which nineteen comprise the Dakar department), and there is no metropolitan council (Guèye et al., 2007).

The increased prominence of non-state actors in urban climate change governance has led to growing calls for partnerships across the public–private divide (Osofsky and Koven Levit, 2007; Bontenbal and Van Lindert, 2008; Andonova, 2010). These partnerships play an important role in overcoming gaps in capacity, translating the climate change impacts and response options into language that is meaningful to different groups and individuals, and accelerating the development of solutions. Follow-up analysis of the 2014 MIT-ICLEI Climate survey shows that these partnerships have an important impact on the scope of concrete emissions reductions (Aylett, 2014). Cities that report high levels of partnerships among public, private, and civil-society actors are significantly more able to achieve measurable emissions reductions in areas outside of direct municipal government control (such as residential energy use, emissions from local businesses, or reduced use of private vehicles). For example, in early 2010, the region of Metro Vancouver on the west coast of Canada launched a partnership with seven municipalities and a small sample of small and medium-sized enterprises (SMEs) to conduct a program of GHG management training for SMEs. The three parties agreed to work together to carry out GHG management training, employ a GHGs management tool, and provide technical assistance for SMEs, the costs of which are shared equally among the three partners (Burch et al., 2013).

Networks have emerged that connect diverse stakeholders to create more coordinated, global approaches to climate change mitigation and adaptation (Betsill and Bulkeley, 2004; Burch et al., 2013; Krellenberg et al., 2014). An example of increasingly important global networks that influence climate change responses is ICLEI's Partners for Climate Protection program (Andonova, 2010). Policy action in the Netherlands provides another example of what can be achieved through international partnerships. Even so, the wide variation in jurisdictional power, organizational culture, organizational structure, and political context has constrained the ways that these stakeholders interact with one another and, hence, the effectiveness of mitigation and adaptation policies. The Netherlands has developed a clear methodology of "learning-oriented" environmental policy (National Environmental Policy Plan NEPP 1-IV 1989–2002), and this is being applied to its cities' agendas, specifically using transition experiments to accelerate movement toward sustainability. The development of a Climate Adaptation Partnership has encouraged the sharing of delta-city knowledge and technology and policy transfers between the port cities of Rotterdam and Ho Chi Minh City. This work has the potential to cross cultures and geographies for innovative outcomes.

16.3.3 Justice in Urban Climate Change Governance

Actors vary in the extent to which they have influence over the governance of climate change, legitimacy in the eyes of decision-makers, and the resources to take action. For example, those communities that are most vulnerable to climate change are often not those who are responsible for the bulk of GHG emissions. Climate change also has the potential to exacerbate existing societal inequalities in terms of income distributions and access to resources and options. A growing body of research reveals that climate change governance strategies can produce or reproduce (un)just decision-making processes and outcomes or result in an (in)equitable distribution of climate change risks and resources (see Chapter 4, Mitigation and Adaptation).

Justice in urban climate change governance requires that vulnerable groups are represented in mitigation and adaptation planning processes, that priority setting and framing recognize the adaptation needs of vulnerable groups, and that the impacts of adaptation enhance the freedoms and assets of vulnerable groups in the city (Hughes, 2013). Urban climate change governance systems have multiple entry points at which justice and injustice can be experienced, including decision-making processes, criteria for taking and facilitating actions, and the mechanisms that manage the relationships between climate change impacts and other policy areas (Thomas and Twyman, 2005).

There are scalar dimensions to experiences of justice in urban climate change governance, from individuals through neighborhoods to countries. From a global perspective, cities in low- and middle-income countries have limited adaptive capacity and relatively high reliance on natural resources; as a result, they are considered more vulnerable than cities in high-income countries.

Within cities, there are sociospatial differences that help determine vulnerability and influence on decision-making and thus on differentiated options and assets to respond to risk (Simon, 2011; Kuhlicke et al., 2012; Welz et al., 2014). City authorities may reach out to vulnerable populations but do not necessarily assume responsibility for climate impacts resulting from previous government policies and practices (Bulkeley et al., 2013). Emissions, vulnerability, and risk are linked to equity and poverty in complex ways that require sophisticated policy responses (Hardoy and Pandiella, 2009; Romero-Lankao, Qin, and Borbor-Cordova, 2013). One example of such a strategy is "pro poor adaptation" that uses investments in the assets (both physical and intellectual) of vulnerable and poor communities to reduce vulnerability and improve capacity (Moser and Satterthwaite, 2010).

16.3.4 The Challenge of Fragmentation and Coordinated Action

Although city governments are at the forefront of acting on climate change, it is well documented that the existence of a variety of practical barriers to developing coordinated and cross-sectoral climate change actions is hampering implementation. The importance of the comprehensiveness of these plans is thus also the challenge in regard to horizontal and vertical coordination between actors fragmented across different agencies, utilities, and city administrative departments (Betsill and Bulkeley, 2004; Kern et al., 2008; Bulkeley, 2010; Betsill and Bulkeley 2007).

It has also been observed that, within city governments, climate change expertise often remains concentrated in environmental departments, which makes cross-sectoral coordination within the organizational hierarchy of city government even more challenging because of a limited capacity to implement actions (Kern et al., 2008). The participation of different municipal agencies in climate change planning and implementation is highly uneven. Agencies responsible for environmental planning, land-use planning, and solid waste management tend to be important contributors to local climate action. Those responsible for transportation, water, and building codes occupy a middle ground, whereas other city government agencies responsible for sectors such as health, economic development, and the local electrical utility (where these exist) remain largely on the sidelines (Aylett, 2014). Those agencies currently less engaged represent potential sources of new partnerships, ideas, and resources that could enable even more effective urban responses to climate change.

Fragmentation in governance systems occurs not only as a function of the physical separation of actors. The implementation of climate change activities is also hampered by a multitude of formal and informal institutional constraints and barriers and by the varied visions, interests, and decision-making power of involved actors (Næss et al., 2005; Agrawal et al., 2012; Romero-Lankao and Gnatz, 2013). For example, coordination is more difficult across sectors than within sectors because different sets of institutional rules tend to evolve in departmental divisions. Addressing fragmentation as cross-sectoral climate change planning is essential if undesirable tradeoffs are to be avoided and potential synergies exploited (Biesbroek et al., 2009; Larsen et al., 2012).

The Danish municipality of Aarhus provides an example of how to overcome fragmentation. In 2007, the city set a goal of carbon neutrality by 2030. Networking both within the municipal government and between the government and private actors has helped to produce more coordinated action. The city's Climate Secretariat is driven by visionary leaders willing to circumvent organizational norms and develop informal relationships among the lower levels of the administrative hierarchy. Municipal marketing and the promotion of climate change action with development potential is focused on attracting private companies. Indeed, the Climate Secretariat has brought thirty-two businesses together in a formal network involving housing, clean technology, Aarhus University, and the engineering sector where linkages promote benefits to all parties (Cashmore and Weis, 2014). This networking approach has created demands for urban government to act, which in turn bypasses some intraorganizational constraints. In spite of fragmentation and conflicting ambitions and values, the Climate Secretariat leveraged around 50 million DKR (approximately US$9.2 million) in climate change investments in 2012 (Cashmore and Weis, 2014).

16.4 Other Determinants of Effective Urban Climate Change Governance

In this section, we explore other drivers of effective urban climate change governance to provide explanatory insight and identify possible opportunities for effective actions. The role of each driver is likely to be different in different contexts, and an important area for future research is to further explore the conditions under which the insufficiencies of different drivers, or combination of drivers, act as barriers to effective urban climate change governance. Furthermore, pursuing effective governance measures in and of themselves can mask underlying political tensions and conflicts. For example, while a governance system may reflect or achieve common metrics of good governance,[3] these do not necessarily remove underlying political conflicts (Jessop, 2001).

16.4.1 Legal Frameworks and Mandates

The legal context in which urban climate governance takes place plays a key role in determining the extent to which climate change actions, regulations, and programmatic priorities are legitimized, incentivized, and prioritized. Absent or inadequate laws dealing with mitigation and adaptation can be an obstacle to developing and implementing actions. However, changing legal frameworks is time-consuming and entails complex processes at different political levels. The presence of appropriate legal frameworks can facilitate the development and implementation of mitigation and adaptation action and set the basis for further action. Legal frameworks can also mediate the relationship between the public and decision-makers, providing political structures for participatory planning and decision-making (or not) according to prevailing democratic norms and political cultures.

Legal frameworks determine if mitigation and adaptation action can be undertaken autonomously at the city level. Whereas in centralized systems regional and local actors are significantly conditioned by national decision-making, laws, funding, and distribution of competencies, in decentralized systems municipal governments are more likely to have jurisdiction over climate change–related policy areas such as energy supply, transportation, water supply, and land use. Legal reforms fostering decentralization can sometimes provide opportunities for effective climate change action when there is a coherence between powers and responsibilities on the one hand and available resources and revenue sources on the other, especially when led by one or more champions within the local authorities (Finan and Nelson, 2009; Brockhaus and Kambiré, 2009) (see Section 16.4.4).

Local reforms can also initiate changes to broader legal frameworks. Mexico City's Federal District government has invested heavily in the institutionalization of climate change

governance (Hughes and Romero-Lankao, 2014). The Federal District has established a Climate Change Division within the Ministry of Environment, responsible for tracking and modeling GHG emissions and conducting citywide vulnerability assessments. It has assumed full responsibility for developing and implementing the Mexico City Climate Action Plan. The Division has resisted, or been immune to, changes in administration. Its existence formalizes the organizational structures necessary for mainstreaming climate change expertise, planning, and funding. In 2010, the Federal District passed the Law for Mitigation and Adaptation to Climate Change (Romero-Lankao et al., 2015). Decision-makers in the Climate Change Division fostered a conducive environment for passing the law, and, in doing so, legitimized and shielded climate change policy from the political cycle and administrative and policy shifts. Mexico City's leadership has helped to inform the 2012 General Law of Climate Change passed by the Mexican federal government, which sets GHG reduction targets. Mexico City, however, is faced with challenges, such as fragmented governance arrangements, asymmetries in access to information, and top-down and technocratic decision-making (Romero-Lankao et al., 2013).

Legal frameworks play additional roles. They can influence the adaptiveness of climate change governance (Birkmann et al., 2010). Top-down, inflexible governing mechanisms subject to electoral cycle discontinuities, poor coordination, underfunding, and a lack of local specificity fail to meet the needs of cities. Adaptive legal frameworks, on the other hand, authorize and support an integrated agenda necessary to deal with multiple levels and sectors in the face of uncertainty. Finally, legal frameworks are important because of their ability to develop and channel resources for urban climate change governance. Drawing again on the example of Mexico City, the government of Mexico City passed the Climate Change Act, which established climate change as a line item in the city's budget, thus helping to solidify and institutionalize climate change as a part of the city's normal business. Legal frameworks can also determine the number of staff dedicated to climate change and the information resources that can be marshaled for supporting decision-making. Legal frameworks give mandates and missions that often translate into resources for effective governance.

16.4.2 Generation of and Access to Information

Urban climate change governance requires access to new, context-specific, and complex sources of information, such as future climate projections, GHG inventories, and climate vulnerability assessments. The ways in which this relevant information is generated and conveyed among scientists, practitioners, and decision-making communities will help to determine the effectiveness of climate governance for two reasons. First, the availability and accessibility of information can enhance

---

3 The "good governance" agenda represents an ever-broadening set of concepts, but commonly includes checks and balances in government, decentralization, efficient/equitable/independent judiciary, a free press, and a sound regulatory system (Grindle, 2004). Added to this are cross-cutting principles such as participation/inclusion, nondiscrimination/equality, and rule of law/accountability (UNDP, 2011).

decision-making capacities by helping decision-makers to evaluate and prioritize climate change (Yohe and Tol, 2002; Engle et al., 2011; Romero-Lankao et al., 2013). Second, the process of co-producing and disseminating climate change information can engage stakeholders, raise awareness, and improve the specificity and usability of the information (Dodman and Carmin, 2011; Healey, 2013). Both the process and the outcomes of incorporating science in decision-making, therefore, influence the effectiveness of climate change governance (see Case Study 16.1).

The availability, transmission, and use of information are essential components of the capacity for effectively governing carbon and climate in cities. However, rather than being a technical exercise in information collection and insertion into the policy process, climate-relevant information is politically determined and can reflect the broader priorities of decision-makers (Hughes and Romero-Lankao, 2014). For example, in Latin American cities, information availability, transmission, and use are problematic because there are opaque and limited relationships between the relevant decision-making agencies, between levels of government, and between government and communities (Romero-Lankao et al., 2013). The characterization and communication of uncertainty continues to challenge the relationship of city governments to climate change information. A better understanding of the ways in which governmental and non-governmental actors access climate change information and use it for routine decision-making is needed in order to better incorporate science in urban climate change governance (see Case Study 16.1).

Problems of access to information are particularly important in the area of adaptation policy. Forty percent of cities surveyed by Aylett (2014) report that a lack of information on the local impacts of climate change poses a significant challenge to climate change planning and implementation (compared to the 27% who report being challenged by a lack of information on GHG emissions).

City governments draw in information and guidance from a broad array of sources (see Figure 16.6). Horizontal and vertical links to other governmental agencies are among the most important. Looking across the top-ranking groups, it is clear that professional contacts within government agencies and the networks that facilitate them are critically important. Cities learn from other cities, and government agencies learn (in large part) from other government agencies. In addition, civil-society organizations, educational institutions, and research institutes also play a key role in this space. Private-sector actors are seen to be less important. These global averages also mask the important role of the UN, development agencies, and multilateral development banks in Asia, Latin America, and Africa (Aylett, 2014).

Scientists also have a responsibility to provide carbon and climate information in a way that is easily accessible and usable by decision-makers (Krellenberg and Barth 2014) (see Case Study 16.1). This needs to be a two-way communication that considers the very valuable knowledge of stakeholders and also opens up a platform for ongoing exchange (Ansell and Gash, 2007; Barton et al., 2014).

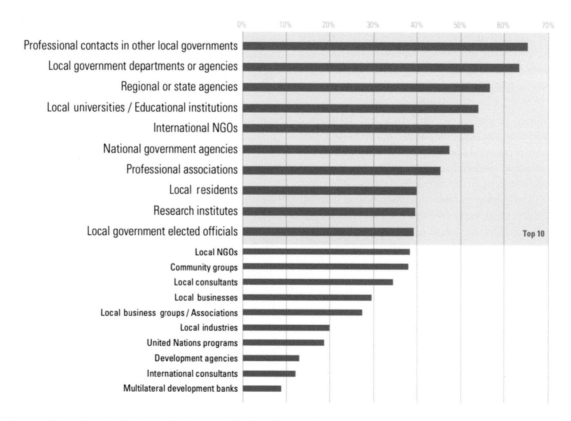

Figure 16.6 *Sources of information and guidance for climate planning. This figure illustrates the percentage of cities that report that they rely significantly on specific groups and organizations for information and guidance related to their climate change planning activities in terms of both mitigation and adaptation.*

Source: Aylett, 2014

Case Study 16.1   Science-Policy Interface in Santiago de Chile: Opportunities and Challenges to Effective Action

Kerstin Krellenberg

Helmholtz Centre for Environmental Research – UFZ, Leipzig

Keywords	Science-policy, impact assessment, adaptation measures
Population (Metropolitan Region)	6,883,563 (Instituto Nacional de Estadísticas [INE], 2010)
Area (Metropolitan Region)	15,403.2 km² (INE, 2010)
Income per capita	US$13,530 (World Bank, 2017)
Climate zone	Csb – Temperate, dry summer, warm summer (Peel et al., 2007)

An inter- and transdisciplinary (ITT) approach was elaborated to develop a Climate Change Adaptation Plan for the Metropolitan Region of Santiago de Chile (MRS). This approach functioned as an interface between 30 social, natural, and engineering scientists, and another 40–50 governmental and non-governmental actors (Krellenberg and Barth, 2014, Barton et al., 2015).

The presence of international and national organizations has been instrumental here, as in other cities, in establishing climate change on the policy agenda (see Section 16.2). Thus, it was only possible given the very strong, long-lasting, and trusting collaboration with all Chilean partners, scientists, and other actors. Participating in transnational networks has opened possibilities for urban authorities in MRS to obtain resources and learn from other cities. Notwithstanding this, the constraints to institutional response capacity embedded in the social and political fabric of the city have thus far prevented effective climate change actions (Romero-Lankao et al., 2013). In

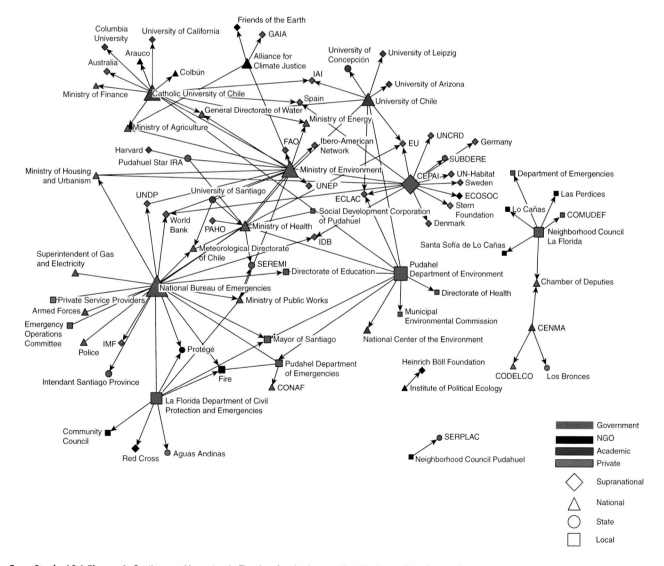

Case Study 16.1 Figure 1 *Santiago working network. The size of nodes is proportional to the number of respondents reporting to work with that actor.*

Source: Romero-Lankao et al., 2013

the city's fragmented governance structure, vertical and horizontal coordination among sectors and tiers of government is still a challenge. For many reasons often related to authoritarian culture or jurisdictional boundaries, environmental authorities seldom interact with development authorities, and tiers of government rarely collaborate (see Case Study 16.1 Figure 1). Priorities in urban planning are dominated by economic concerns. Information is poorly transmitted between levels of government, and the information that is transmitted to the public is for largely informational purposes (e.g., evacuation notices or flood warnings) rather than for learning and engagement (Zunino, 2006). It depends now on a strong administrative leadership if the Plan is to be realized, the commitment of the various implementing organizations guaranteed, and long-term decisions made.

16.4.3 Culture and Constituency

Evaluating the effectiveness of urban climate change governance is itself a contested notion: effectiveness will be determined by the cultures and constituencies affected by, and engaged with, a city's climate change responses. Notions of "the good life," the appropriate role of government, or the demographic make-up of a city can all shape support for climate change responses and their effectiveness. The extent to which urban residents perceive climate change as a risk also influences their support for different policy responses (Zahran et al., 2006) and can produce socioeconomic and institutional barriers to mitigation and adaptation. Furthermore, the existing geographic, economic, and political challenges that cities and urban residents face will determine what mitigation and adaptation responses are feasible and appropriate. In this sense, effectiveness is again a matter of fit – and engenders tradeoffs – between the demands of affected individuals, businesses, and governments and the climate change challenge itself.

One way to further align urban climate change governance with culture and constituency is through ongoing conversations and negotiations with and among governmental and non-governmental urban actors. This means that cities should find ways to develop "an honest and creative deliberative approach that can be more democratic and can yield genuine benefits" for mitigation and adaptation (Few et al., 2007). For example, community risk assessment methods are existing tools cities can use to develop adaptation priorities and strategies jointly with community members (van Aalst et al., 2008. Engaging stakeholders in the climate change decision-making process can also help to build community resilience and social capital (Ebi and Semenza, 2008). Different approaches have been developed in various cities, and ongoing networking can help to work on the transferability of useful approaches to other cities while acknowledging the specific contexts of each city. Urban cultures and constituencies therefore not only shape the options for and barriers to urban climate change response but should also influence our evaluation of the effectiveness of these responses.

16.4.4 Championship and Leadership

Behind the efforts of many cities that are taking steps to address climate change lies the work of one or more leaders, often termed *policy champions* or *institutional entrepreneurs*, who promote climate change as a policy issue and get climate change onto the political agenda (Betsill and Bulkeley, 2007).

The strengths of these actors include "motivation, will, intentionality, interest, choice, autonomy, and freedom," and involve actors' ability to operate somewhat independently of institutional constraints (Battilana and D'aunno, 2009). Effective champions manage to leverage resources to create windows of opportunity and initiate new practices (Maguire et al., 2004). Effectiveness is not only ascribed to a champion's individual qualities, but also to their position within a set of social relationships and ability to navigate within a broader institutional environment (Campbell, 2004; Ziervogel et al., 2016).

Successful leadership strategies include the ability to leverage resources, use and create the right narratives, generate collective consensus, and establish a shared understanding about the city government's direction for the climate work (Cashmore and Wejs, 2014). Narratives able to initiate action may include promoting adaptation for future cost savings and using mitigation to foster green growth (Juhola et al., 2011; Cashmore and Wejs, 2014; Wejs, 2014). Engagement in international networks has been found to allow champions to gather information and mobilize actors and resources (Urwin and Jordan, 2008). Champions have also established local networks to initiate exploratory projects with local businesses, to leverage funding, and to show quick results and kick-start partnerships. Overall, leadership from a mayor, senior elected officials, or senior management has been identified as the most important enabler of successful climate mitigation strategies (Aylett, 2014). However, climate change work within a city government cannot rely on individuals in the long term because these people might not stay in their positions or may lose their legitimacy for action. The work of policy champions must be complemented with legal and regulatory changes. Furthermore, in many cities, the influence of governmental and non-governmental champions in shaping climate agendas and facilitating a learning process has not been enough to push real and effective policy responses. Fragmented governance arrangements, asymmetries in access to information, and top-down and technocratic decision-making pose challenges to effective governance, as does the fact that climate change is still secondary when compared to growth priorities (Romero-Lankao, 2007; Aylett, 2013).

One example of a successful climate change leader comes from Energy Island in Sweden. In 1997, Samsø Municipality, a small island community of 4,300 people, won support from the Danish Ministry of Environment to transform itself through green growth and business development to 100% renewable energy in

10 years. The effort was highly successful: today Samsø isn't just carbon-neutral, it actually produces 10% more renewable energy than it uses. The surplus feeds into the Danish electricity grid, providing revenue to Samsø residents. The 4,300 islanders have made approximately US$80 million of investments in renewables and have reduced their carbon emissions by 140%, which means that each inhabitant emits −3.7 tonnes CO_2 per year. Furthermore, the project has created jobs and businesses. A contributing factor to the island's success was the leadership of Søren Hermansen – environmentalist, teacher, and local community member. Since 2007, he has been CEO of Samsø Energy Academy, a center for renewable energy studies. In 2008, Søren Hermansen was named "Hero of the Environment" by *Time* magazine and received the Göteborg Award in 2009, sometimes called "the Nobel Prize in Environment" (Jakobsen, 2008).

16.5 Conclusions

This chapter analyzed some of the dimensions of the capacity to govern carbon and climate change in cities. It identified an array of urban climate change strategies, going from incremental actions that target mitigation and adaptation in an ad-hoc and isolated way to potentially more transformative strategies. It examined the dynamic mix of actors and networks at multiple scales involved and some of the mechanisms by which issues of interplay such as different mandates, timescales, and ways of knowing create challenges to effective climate governance. It also explored other institutional determinants of the gaps between the policy discourse and the real-world challenges local climate action needs to address.

Although many institutional supports and frameworks for climate change governance reside at the nation-state level, in many jurisdictions, innovative climate change governance has been led by relevant municipal, state, or provincial levels of government, often in spite of the lack of comprehensive or ambitious climate change policy at the national level. Even so, fruitful negotiations at the international level (such as those that took place as part of COP21 in Paris, 2015) may spur domestic policy-making and ratchet up levels of ambition.

For many years, these urban climate governance strategies focused solely on mitigation and were initiated predominantly by cities in high-income countries. More recently, there has been an increase in climate responses by cities in middle- and high-income countries that attempt to integrate mitigation and adaptation actions; however, the bulk of responses in all cities have tended to be incremental and fragmented, with very few cities moving toward transformative urban development pathways that can lead to climate resilience. Due to this lack of transformative pathways, a gap continues to exist between the commitment of cities to respond to carbon and climate change and the effectiveness of their responses.

Part of the reason for this gap lies in a tension that exists between the predictability and stability of institutions and the

flexibility of more informal and unplanned strategies. A contradiction arises because the coordination across sectors and jurisdictions is necessary to create transformative policies, yet the added complication of bringing all the forces and interests at play into alignment often leads to smaller scale actions controlled by local jurisdictions or single institutions and private and community actors.

Another reason for this gap is that while scientific information is necessary for effective governance, scientific information is insufficient to trigger action on its own and often does not mesh with realties on the ground. At play here is a lack or limitation of a connection between the production of science and the production of policy. The science that urban actors look for to support actions is often not produced at the spatial and temporal scale at which it is required nor is it incorporated into decision-making in participatory and iterative ways. Decisions, furthermore, are often based on values, political expediency, and habit, rather than on a rational assessment of scientific information. In some cases, the use of an iterative science policy interface has shown potential to help decision-makers discover the co-benefits of policy actions as a way to foster climate change mitigation and adaptation policies and programs. A more proactive collaboration between science and policy is needed to better address the needs of politicians and practitioners and to better communicate scientific information. Deeper engagement with the social sciences, including social psychology, sociology, and political science, will also serve to reveal powerful drivers of changes in practices and behaviors both at the collective and individual levels.

Another issue hindering effective action is the difficulty many cities have in identifying and realizing co-benefits between their climate change work and other key local development priorities and aspirations. Here, engagement with different ways of knowing can also play a major role in identifying the co-benefits more explicitly.

By understanding these challenges, creative, locally relevant, and equitable strategies may be uncovered and promoted by governmental and non-governmental actors to coordinate efforts in meaningful and effective ways. Urban actors able to maintain these partnerships can significantly increase the scope of their climate change actions. However, overlapping or conflicting jurisdictions or vested interests can prevent such coordination, leading to fragmented actions and policies. Many key urban actors, both inside and outside of city government, remain on the sidelines and represent pools of untapped resources that could be brought to bear as cities continue to address the climate challenge.

Improvements in the metrics used to evaluate mitigation and adaptation efforts will be critical for effective planning and comparison between cities. In mitigation efforts, recent attempts to standardize emissions reporting procedures have been useful in making meaningful comparisons, but gaps still exist at many levels of emission inventorying and in the evaluation of the real mitigation potential of existing actions. Until common frameworks and metrics are also developed for adaptation and longer time horizons are employed, it will be difficult to evaluate and compare the effectiveness of policy responses across cities.

Although they face many obstacles, city governments possess a variety of tools, incentives, and policy options for climate change governance (e.g., land-use plans, transit systems, building codes, and closer ties to constituents than might be found at higher levels of government). These tools can help reinforce and catalyze action by other levels of government and non-state actors. While the level of autonomy and capacity to govern carbon and climate varies across cities, there are still many potential and often untapped synergies available to urban actors to create effective climate actions. Cities and urban actors vary in their levels of leadership, access to information, legal mandates, and financial resources. The most fruitful approaches will, therefore, necessarily include both bottom-up and top-down strategies that can help foster successful responses and achieve effective and fair urban climate change governance.

Annex 16.1 Stakeholder Engagement

The study and practice of urban climate change governance requires the engagement of a variety of stakeholders. As such, this author team endeavored to represent the values of participatory processes and the importance of multiple sources of knowledge in both the process and content of this text. Throughout the writing process, authors discussed the key findings of the chapter with colleagues in the communities of research and practice, considering alternative framings of key governance issues and more inclusive ways of capturing the challenges of justice, collaboration, and fragmentation. The author team itself consisted of individuals who play roles in government and civil society, as well as in academia, thus creating opportunities to improve the central conclusions of the chapter and ground them in the reality of climate change governance. Furthermore, much of the empirical work that informs the core of the chapter involved a variety of stakeholders both in research design and dissemination of findings. The process of stakeholder engagement does not end with the publication of this assessment, but rather continues as its findings are applied and explored in urban contexts around the world.

Chapter 16 Governance and Policy

References

Adger, W. N., Lorenzoni, I., and O'Brien, K. L. (2009). *Adapting to Climate Change: Thresholds, Values, Governance*. Cambridge University Press.

Aggarwal, R. M. (2013). Strategic bundling of development policies with adaptation: An examination of Delhi's Climate Change Action Plan. *International Journal of Urban and Regional Research* 37(6), 1902–1915. doi:10.1111/1468-2427.12032.

Agrawal, A., Lemos, M. C., Orlove, B., and Ribot, J. (2012). Cool heads for a hot world–Social sciences under a changing sky. *Global Environmental Change* 22(2), 329–331.

Andonova, L. B. (2010). Public-private partnerships for the Earth: Politics and patterns of hybrid authority in the multilateral system. *Global Environmental Politics* 10(2), 25–53.

Ansell, C., and Gash, A. (2008). Collaborative governance in theory and practice. *Journal of Public Administration Research and Theory* 18(4), 543–571.

Arup RPA. (2014). Resilience–sustainable cities–Siemens. WCMS3-PortletPage. Accessed October 16, 2015: http://w3.siemens.com/topics/global/en/sustainable-cities/resilience/Pages/home.aspx?stc=wwzcc120526.

Aylett, A. (2013). Networked urban climate governance: Neighborhood-scale residential solar energy systems and the example of solarize Portland. *Environment and Planning C: Government and Policy* 31(5), 858–875.

Aylett, A. (2014). *Progress and Challenges in the Urban Governance of Climate Change: Results of a Global Survey*. MIT Press.

Bai, X. (2007). Integrating global environmental concerns into urban management: The Scale and readiness arguments. *Journal of Industrial Ecology* 11(2), 15–29. doi:10.1162/jie.2007.1202.

Battilana, J., and D'aunno, T. (2009). Institutional work and the paradox of embedded agency. *Institutional Work: Actors and Agency in Institutional Studies of Organizations*, 31–58.

Beg, N., Morlot, J. C., Davidson, O., Afrane, Y.-Okesse, Tyani, L., Denton, F., Sokona, Y., Thomas, J. P., La Rovere, E. L., and Parikh, J. K. (2002). Linkages between climate change and sustainable development. *Climate Policy* 2(2), 129–144.

Betsill, M., and Bulkeley, H. (2007). Looking back and thinking ahead: A decade of cities and climate change research. *Local Environment* 12(5), 447–456.

Betsill, M. M., and Bulkeley, H. (2004). Transnational networks and global environmental governance: The Cities for Climate Protection Program. *International Studies Quarterly* 48(2), 471–493.

Biermann, F., Betsill, M. M., Gupta, J., Kanie, N., Lebel, L., Liverman, D., Schroeder, H., and Siebenhüne, B. (2009). *Earth System Governance: People, Places, and the Planet: Science and Implementation Plan of the Earth System Governance Project*. IDEP, The Earth System Governance Project.

Biesbroek, G. R., Swart, R. J., and Van der Knaap, W. G. M. (2009). The mitigation–adaptation dichotomy and the role of spatial planning. *Habitat International* 33(3), 230–237.

Birkmann, J., Garschagen, M., Kraas, F., and Quang, N. (2010). Adaptive urban governance: New challenges for the second generation of urban adaptation strategies to climate change. *Sustainability Science* 5(2), 185–206.

Bontenbal, M., and Van Lindert, P. (2008). Bridging local institutions and civil society in Latin America: Can city-to-city cooperation make a difference? *Environment and Urbanization* 20(2), 465–481.

Bracken, I. (2014). *Urban Planning Methods: Research and Policy Analysis*. Routledge.

Brockhaus, M., and Kambiré, H. (2009). Decentralization: A window of opportunity for successful adaptation to climate change. *Adapting to Climate Change: Thresholds, Values, Governance*, 399–416.

Broto, V., C. and Bulkeley, H. (2013). A survey of urban climate change experiments in 100 cities. *Global Environmental Change* 23(1), 92–102.

Brown, A., Dayal, A., and Rumbaitis Del Rio, C. (2012). From practice to theory: Emerging lessons from Asia for building urban climate change resilience. *Environment and Urbanization* 24(2), 531–556.

Bulkeley, H. (2010). Cities and the governing of climate change. *Annual Review of Environment and Resources* 35, 229–253.

Bulkeley, H., Carmin, J., Castán Broto, V., Edwards, G. A. S., and Fuller, S. (2013). Climate justice and global cities: Mapping the emerging discourses. *Global Environmental Change* 23(5), 914–925.

Burch, S. (2010). Transforming barriers into enablers of action on climate change: Insights from three municipal case studies in British Columbia, Canada. *Global Environmental* **20**(2), 287–297.

Burch, S., and Robinson, J. (2007). A framework for explaining the links between capacity and action in response to global climate change. *Climate Policy* **7** (4), 304–316. doi:10.1080/14693062.2007.9685658

Burch, S., Schroeder, H., Rayner, S., and Wilson, J. (2013). Novel multisector networks and entrepreneurship: The role of small businesses in the multilevel governance of climate change. *Environment and Planning C: Government and Policy* **31**(5), 822–840. doi:10.1068/c1206.

Burch, S., Shaw, A., Dale, A., and Robinson, J. (2014). Triggering transformative change: A development path approach to climate change response in communities. *Climate Policy*, epub ahead of print, 1–21.

Carmin, J., Nadkarni, N., and Rhie, C. (2012). *Progress and Challenges in Urban Climate Adaptation Planning: Results of a Global Survey.* MIT Press.

Cashmore, M., and Wejs, A. (2014). Constructing legitimacy for climate change planning: A study of local government in Denmark. *Global Environmental Change* **24**: 203–212.

Dodman, D., and Carmin, J. (2011). Urban adaptation planning: The use and limits of climate science. *IIED Briefing Papers.* July 10, 2012.

Ebi, K. L., and Semenza, J. C. (2008). Community-based adaptation to the health impacts of climate change. *American Journal of Preventive Medicine* **35** (5), 501–507. doi:10.1016/j.amepre.2008.08.018.

Engle, N. L., Johns, O. R., Lemos, M. C., and Nelson, D. R. (2011). Integrated and adaptive management of water resources: Tensions, legacies, and the next best thing. *Ecology and Society* **16**(1), 19.

Few, R., Brown, K., and Tompkins, E. L. (2007). Public participation and climate change adaptation: Avoiding the illusion of inclusion. *Climate Policy* **7** (1), 46–59.

Field, C. B., Barros, V., Stocker, T. F., and Dahe, Q. (2012). *Managing the Risks of Extreme Events and Disasters to Advance Climate Change Adaptation: Special Report of the Intergovernmental Panel on Climate Change.* Cambridge University Press.

Field C. B., van Aalst, M., Adger, N., Arent, D., Barnett, J., Betts, R., Bilir, E., Birkmann, J., Carmin, J., Chadee, D., Challinor, A., Chatterjee, M., Cramer, W., Davidson, D., Estrada, Y., Gattuso, J. –P., Hijioka, Y., Hoegh, O.-Guldberg, Huang, H. –Q., Insarov, G., Jones, R., Kovats, S., Romero-Lankao, P., Nymand, J., Larsen, Losada, I., et al. (2014). Technical Summary. In C. B. Field, V. R. Barros, D. J. Dokken, K. J. Mach, M. D. Mastrandrea, T. E. Bilir, M. Chatterjee, K. L. Ebi, Y. O. Estrada, R. C. Genova, B. Girma, E. S. Kissel, A. N. Levy, S. MacCracken, P. R. Mastrandrea, and L. L. White (eds.), *Climate Change 2014: Impacts, Adaptation, and Vulnerability. Part A: Global and Sectoral Aspects. Contribution of Working Group II to the Fifth Assessment Report of the Intergovernmental Panel on Climate Change.* Cambridge University Press

Finan, T. J., and Nelson, D. R. (2009). Decentralized planning and climate adaptation: Toward transparent governance. *Adapting to Climate Change: Thresholds, Values, Governance.* Cambridge University Press

Fisher, D. R. (2013). Understanding the relationship between subnational and national climate change politics in the United States: Toward a theory of boomerang federalism. *Environment and Planning C: Government and Policy* **31**(5), 769–784. doi:10.1068/c11186.

Gore, C. D. (2010). The limits and opportunities of networks: Municipalities and Canadian climate change policy. *Review of Policy Research* **27**(1), 27–46.

Hamin, E. M., and Gurran, N. (2009). Urban form and climate change: Balancing adaptation and mitigation in the U.S., and Australia. *Habitat International, Climate Change and Human Settlements* **33**(3), 238–245. doi:10.1016/j.habitatint.2008.10.005.

Hardoy, J., and Pandiella, G. (2009). Urban poverty and vulnerability to climate change in Latin America. *Environment and Urbanization* **21**(1), 203–224.

Healey, P. (2013). Circuits of knowledge and techniques: The transnational flow of planning ideas and practices. *International Journal of Urban and Regional Research* **37**(5), 1510–1526. doi:10.1111/1468-2427.12044.

Holgate, C. (2007). Factors and actors in climate change mitigation: A tale of two South African cities. *Local Environment* **12**(5), 471–484. doi:10.1080/13549830701656994.

Hughes, S. (2013). Justice in urban climate change adaptation: Criteria and application to Delhi. *Ecology and Society* **18**(4), 48.

Hughes, S., and Romero-Lankao, P. (2014). Science and institution building in urban climate-change policymaking. *Environmental Politics* **0**(0), 1–20. doi:10.1080/09644016.2014.921459.

Huq, S., Kovats, S., Reid, H., and Satterthwaite, D. (2007). Editorial: Reducing risks to cities from disasters and climate change. *Environment and Urbanization* **19**(1), 3–15.

Hutyra, L. R., Duren, R., Gurney, K. R., Grimm, N., Kort, E. A. Larson, E., and Shrestha, G. (2014). Urbanization and the carbon cycle: Current capabilities and research outlook from the natural sciences perspective. *Earth's Future* **2**(10), 473–495.

Jabeen, H., Johnson, C., and Allen, A. (2010). Built-in resilience: Learning from grassroots coping strategies for climate variability. *Environment and Urbanization* **22**(2), 415–431.

Jakobsen, I. (2008). *The Road to Renewables: A Case Study of Wind Energy, Local Ownership and Social Acceptance at Samsø.* Masters thesis, Samso University. Accessed October 5, 2014: http://urn.nb.no/URN:NBN:no-21552

Jessop, B. (2001). *Good Governance and the Urban Question: On Managing the Contradictions of Neo-liberalism.* Department of Sociology, Lancaster University.

Juhola, S., Carina, E., Keskitalo, H., and Westerhoff, L. (2011). Understanding the framings of climate change adaptation across multiple scales of governance in Europe. *Environmental Politics* **20**(4), 445–463.

Krellenberg, K., and Barth, K. (2014). Inter- and transdisciplinary research for planning climate change adaptation responses: The example of Santiagode Chile. *Interdisciplinary Science Review* **39** (4), Special Issue.

Krellenberg, K., Hansjürgens, B., and Barth, K. (2014). Developing a regional climate change adaptation plan: Learning from Santiago de Chile. In *Climate Adaptation Santiago* (207–216). Springer.

Krellenberg, K., Jordán, R., Rehner, J., Schwarz, A., Infante, B., Barth, K., and Pérez, A. (2014). *Adaptation to Climate Change in Megacities of Latin America: Regional Learning Network of the Research Project Climate Adaptation Santiago (CAS).* United Nations Economic Commission for Latin America and the Caribbean (ECLAC), Santiago de Chile.

Krellenberg, K., Müller, A., Schwarz, A., Höfer, R., Welz, J. (2013). Flood and heat hazards in the Metropolitan Region of Santiago de Chile and the socio-economics of exposure. *Applied Geography* **38**, 86–95.

Krellenberg, K., and Welz, J. (2016). Assessing urban vulnerability in the context of flood and heat hazard: Pathways and challenges for indicator-based analysis. *Social Indices Research* doi 10.1007/s11205-016-1324-3

Krellenberg, K., Welz, J., Link, F., and Barth, K. (2016). Urban vulnerability and the contribution of socio-environmental fragmentation: Theoretical and methodological pathways. *Progress in Human Geography.* doi 10.1177/0309132516645959

Kuhlicke, C., Kabisch, S., Krellenberg, K., and Steinführer, A. (2012). Urban vulnerability under conditions of global environmental change: Conceptual reflections and empirical examples from growing and shrinking cities. *Vulnerability, Risks and Complexity: Impacts of Global Change on Human Habitats.* Göttingen: Hogrefe (Advances in People and Environment Studies **3**, 27–38.

Larsen, S. V., Kørnøv, L., and Wejs, A. (2012). Mind the gap in SEA: An institutional perspective on why assessment of synergies amongst climate change mitigation, adaptation and other policy areas are missing. *Environmental Impact Assessment Review* **33**(1), 32–40.

Locke, D. H., King, K. L., Svendsen, E. S., Campbell, L. K., C. Small, Sonti, N. F., Fisher, D. R., and Lu, J. W. T. (2014). Urban environmental stewardship and changes in vegetative cover and building footprint in New York City neighborhoods (2000–2010). *Journal of Environmental Studies and Sciences* **4**(3), 250–262.

Lööf, H., and Pardis, N. (2013). Increasing returns to smart cities. *Regional Science Policy and Practice* **5**, 255–262.

Lutsey, N., and Sperling, D. (2008). America's bottom-up climate change mitigation policy. *Energy Policy* **36**(2), 673–685.

Macdonald, D. (2009). The failure of Canadian climate change policy: Veto power, absent leadership and institutional weakness. *Canadian Environmental Policy: Prospects for Leadership and Innovation*.

Maguire, S., Hardy, C., and Lawrence, T. B. (2004) Institutional entrepreneurship in emerging fields: HIV/AIDS treatment advocacy in Canada. *Academy of Management Journal* **47**, 657–679.

Measham, T. G., Haslam McKenzie, F., Moffat, K., and Franks, D. (2012). Reflections on the role of the resources sector in Australian economy and society during the recent mining boom. *Rural Society* **22**(2).

Mehdi, B., Mrena, C., and Douglas, A. (2006). *Adapting to Climate Change: An Introduction for Canadian Municipalities*. Canadian Climate Impacts And Adaptation Research Network (C-CIARN)

Moser, C., and Satterthwaite, D. (2010). Toward pro-poor adaptation to climate change in the urban centers of low-and middle-income countries. *Social Dimensions of Climate Change*, 231.

Moser, S. C., and Ekstrom, J. A. (2010). A framework to diagnose barriers to climate change adaptation. *Proceedings of the National Academy of Sciences* **107**(51), 22026–22031. doi:10.1073/pnas.1007887107.

Næss, L. O., Bang, G., Eriksen, S., and Vevatne, J. (2005). Institutional adaptation to climate change: Flood responses at the municipal level in Norway. *Global Environmental Change* **15**(2), 125–138.

O'Brien, K. (2012). Global environmental change II from adaptation to deliberate transformation. *Progress in Human Geography* **36**(5), 667–676.

Osofsky, H. M., and Koven Levit, J. (2007). Scale of networks: The local climate change coalitions. *Chicago Journal of International Law* **8**, 409.

Park, S. E., Marshall, N. A., Jakku, E., Dowd, A. –M., Howden, S. M., Mendham, E., and Fleming, A. (2012). Informing adaptation responses to climate change through theories of transformation. *Global Environmental Change* **22**(1), 115–126.

Parris, T. M., and Kates, R. W. (2003). Characterizing a sustainability transition: Goals, targets, trends, and driving forces. *Proceedings of the National Academy of Sciences* **100**(14), 8068–8073.

Pelling, M., and David, M. -N. (2011). From resilience to transformation: The adaptive cycle in two Mexican urban centers. *Ecology and Society* **16**(2), 11.

Rabe, B. G. (2007). Beyond Kyoto: Climate change policy in multilevel governance systems. *Governance* **20**(3), 423–444.

Ranger, N., Reeder, T., and Lowe, J. (2013). Addressing "deep" uncertainty over long-term climate in major infrastructure projects: Four innovations of the Thames Estuary 2100 Project. *EURO Journal on Decision Processes* **1**(3–4), 233–262.

Reckien, D., Flacke, J., Dawson, R. J., Heidrich, O., Olazabal, M., Foley, A., Hamann, J. J. –P., Orru, H., Salvia, M., and De Gregorio Hurtado. S. (2014). Climate change response in Europe: What's the reality? Analysis of adaptation and mitigation plans from 200 urban areas in 11 countries. *Climatic Change* **122**(1–2), 331–340.

Revi, A., Satterthwaite, D. E., Aragón-Durand, F., Corfee-Morlot, J., Kiunsi, R. B. R., Pelling, M., Roberts, D. C., and Solecki, W. (2014). Urban areas. In Field, C.B., Barros, V. R., Dokken, D. J., Mach, K. J., Mastrandrea, M. D., Bilir, T. E., Chatterjee, M., Ebi, K. L., Estrada, Y. O., Genova, R. C., Girma, B., Kissel, E. S., Levy, A. N., MacCracken, S., Mastrandrea, P. R., and White, L. L. (eds.), *Climate Change 2014: Impacts, Adaptation, and Vulnerability. Part A: Global and Sectoral Aspects. Contribution of Working Group II to the Fifth Assessment Report of the Intergovernmental Panel on Climate Change* (535–612). Cambridge University Press.

Roberts, D., and O'Donoghue, S. (2013). Urban environmental challenges and climate change action in Durban, South Africa. *Environment and Urbanization* **25**(2), 299–319.

Romero Lankao, P. (2007). How do local governments in Mexico City manage global warming? *Local Environment* **12**(5), 519–535.

Romero-Lankao, P., and Gnatz, D. M. (2013). Exploring urban transformations in Latin America. *Current Opinion in Environmental Sustainability* **5**(3–4), 358–367. doi:10.1016/j.cosust.2013.07.008.

Romero-Lankao, P., Gnatz, D.M., and Sperling, J.B. (2016). Examining urban inequality and vulnerability to enhance resilience: insights from Mumbai, India. *Climatic Change* **139**(3–4), pp. 351–365.

Romero-Lankao, P., Hardoy, J., Hughes, S., Rosas-Huerta, A., Bórquez, R., and Gnatz, D. M. (2015). 10 Multilevel Governance and Institutional Capacity for Climate Change Responses in Latin American Cities. *The Urban Climate Challenge: Rethinking the Role of Cities in the Global Climate Regime* **4**, 181.

Romero-Lankao, P., Hughes, S., Qin, H., Hardoy, J., Rosas, A.-Huerta, Borquez, R., and Lampis, A. (2014). Scale, urban risk and adaptation capacity in neighborhoods of Latin American cities. *Habitat International* **42**(0), 224–235. doi:10.1016/j.habitatint.2013.12.008.

Romero-Lankao, P., Hughes, S., Rosas-Huerta, A., Borquez, R., and Gnatz, D. M. (2013). Institutional capacity for climate change responses: An examination of construction and pathways in Mexico City and Santiago. *Environment and Planning C: Government and Policy* **31**(5), 785–805.

Romero-Lankao, P., Qin, H., and Borbor-Cordova, M. (2013). Exploration of health risks related to air pollution and temperature in three Latin American cities. *Social Science and Medicine* **83**(0), 110–118. doi:10.1016/j.socscimed.2013.01.009.

Rothman, D., Romero-Lankao, P., Schweizer, V., and Bee, B. A. (2013). Challenges to adaptation: A fundamental concept for the shared socio-economic pathways and beyond. *Climatic Change*, September, 1–13. doi:10.1007/s10584–013–0907–0.

Satterthwaite, D., Huq, S., Reid, H., Pelling, M., and Romero Lankao, P. (2007). *Adapting to Climate Change in Urban Areas*. IIED.

Seto K. C., Dhakal, S., Bigio, A., Blanco, H., Delgado, G. C., Dewar, D., Huang, L., Inaba, A., Kansal, A., Lwasa, S., McMahon, J. E., Müller, D. B., Murakami, J., Nagendra, H., and Ramaswami, A. (2014). Human settlements, infrastructure and spatial planning. In Edenhofer, O., Pichs-Madruga, R., Sokona, Y., Farahani, E., Kadner, S., Seyboth, K., Adler, A., Baum, I., Brunner, S., Eickemeier, P., Kriemann, B., Savolainen, J., Schlömer, S., von Stechow, C., Zwickel, T., and Minx, J. C. (eds.), *Climate Change 2014: Mitigation of Climate Change. Contribution of Working Group III to the Fifth Assessment Report of the Intergovernmental Panel on Climate Change.* Cambridge University Press. Accessed October 31, 2015: http://www.ipcc.ch/pdf/assessment-report/ar5/wg3/ipcc_wg3_ar5_chapter12.pdf

Shaw, A., Burch, S., Kristensen, F., Robinson, J., and Dale, A. (2014). Accelerating the sustainability transition: Exploring synergies between adaptation and mitigation in British Columbian communities. *Global Environmental Change* **25**: 41–51.

Simon, D. (2011). Reconciling development with the challenges of climate change: Business as usual or a new paradigm? In D. J. Kjosavik and P. Vedeld (eds.), *The Political Economy of Environment and Development in a Globalised World: Exploring the Frontiers; Essays in Honour of Nadarajah Shanmugaratnam* (195–217). T. Tapir and Social Scientists' Association of Sri Lanka.

Simon, D. (2012). Climate and environmental change and the potential for greening African cities. *Local Economy*, 1–15. doi:10.1177/0269094212463674.

Stadelmann, M., Michaelowa, A., Butzengeiger-Geyer, S., and Köhler, M. (2011). Universal metrics to compare the effectiveness of climate change adaptation projects. In *Handbook of Climate Change Adaptation* (part IV, 2143–2160). doi: 10.1007/978-3-642-38670-1_128

Stumpp, E.- M. (2013). New in town? On resilience and "Resilient Cities." *Cities* **32**, 164–166.

Thomas, D. S. G., and Twyman, C. (2005). Equity and justice in climate change Adaptation amongst natural-resource-dependent societies. *Global Environmental Change* **15**(2), 115–124.

Tompkins, E. L., and Neil Adger, W. (2005). Defining response capacity to enhance climate change policy. *Environmental Science and Policy* **8**(6), 562–571.

UNDP. (2011). *Governance Principles, Institutional Capacity and Quality, Towards Human Resilience: Sustaining MDG Progress in an Age of Economic Uncertainty.* United Nations Development Programme.

UN-Habitat. (2011). *Cities and Climate Change: Global Report on Human Settlements.* UN-Habitat.

Urwin, K., and Jordan, A. (2008). Does public policy support or undermine climate change adaptation? Exploring policy interplay across different scales of governance. *Global Environmental Change* **18**(1), 180–191.

Vasi, I. B. (2007). Thinking globally, planning nationally and acting locally: Nested organizational fields and the adoption of environmental practices. *Social Forces* **86**(1), 113–136.

Wejs, A. (2014). Integrating climate change into governance at the municipal scale: An institutional perspective on practices in Denmark. *Environment and Planning C: Government & Policy* **32**(6), 1017–1035. doi:10.1068/c1215

van Aalst, M. K., Cannon, T., and Burton, I. (2008). Community level adaptation to climate change: The potential role of participatory community risk assessment. *Global Environmental Change* **18** (1), 165–179. doi:10.1016/j.gloenvcha.2007.06.002.

Yohe, G., and Tol, R. S. J. (2002). Indicators for social and economic coping capacity—moving toward a working definition of adaptive capacity. *Global Environmental Change* **12**(1), 25–40.

Young, O. R. (2002). *The Institutional Dimensions of Environmental Change: Fit, Interplay, and Scale*. MIT Press.

Zahran, S., Brody, S. D., Grover, H., and Vedlitz, A. (2006). Climate change vulnerability and policy support. *Society and Natural Resources* **19**(9), 771–789.

Ziervogel, G., and Parnell, S. (2014). Tackling barriers to climate change adaptation in South African coastal cities. In Glavovic, B. C., and Smith, G. P. (eds.), *Adapting to Climate Change* (57–73). Springer.

Ziervogel, G., Archer Van Garderen, E., and Price, P. (2016). Strengthening the science-Policy interface by co-producing an adaptation plan: leveraging opportunities in Bergrivier municipality, South Africa. *Environment and Urbanization*, 1–20.

Zimmerman, R., and Faris, C. (2011). Climate change mitigation and adaptation in North American cities. *Current Opinion in Environmental Sustainability* **3**(3), 181–187. doi:10.1016/j.cosust.2010.12.004.

Chapter 16   Case Study References

Case Study 16.1   Science-Policy Interface in Santiago de Chile: Opportunities and Challenges to Effective Action

Barton, J. R., Krellenberg, K., and Harris, J. M. (2015). Collaborative governance and the challenges of participatory climate change adaptation planning in Santiago de Chile. *Climate and Development*. doi:10.1080/17565529.2014.934773

Instituto Nacional de Estadísticas (INE). (2010). *Compendio Estadístico* .INE.

Krellenberg, K., and Barth, K. (2014). Inter- and transdisciplinary research for planning climate change adaptation responses: The example of Santiago de Chile. *Interdisciplinary Sciences Review* **39**(4), 360–375.

Krellenberg, K., and Hansjürgens, B. (2014). *Climate Adaptation Santiago*. Springer.

Peel, M. C., Finlayson, B. L., and McMahon, T. A. (2007). Updated world map of the Köppen-Geiger climate classification. *Hydrology and Earth System Sciences Discussions* **4**(2), 462.

Romero-Lankao, P., Hughes, S., Rosas-Huerta, A., Borquez, R., and Gnatz, D. (2013). Institutional capacity for climate change responses: An examination of construction and pathways in Mexico City and Santiago. *Environment and Planning C: Government and Policy* **31**(5), 785–805.

World Bank. (2017). 2016 GNI per capita, Atlas method (current US$). Accessed August 9, 2017: http://data.worldbank.org/indicator/NY.GNP .PCAP.CD

Zunino, H. M. (2006). Power relations in urban decision-making: neoliberalism, "techno-politicians" and authoritarian redevelopment in Santiago, Chile. *Urban Studies* **43**(10), 1825–1846.

Conclusion: Transforming Cities

Scientists and Stakeholders: Essential Partners in Urban Climate Change Mitigation and Adaptation

This ARC3.2 volume focuses on the pathways required for cities to fulfill their potential as climate change response leaders. Cities have an extraordinary potential for transformational change due to their concentration of economic activity, dense social networks, human resource capacity, high levels of investment in infrastructure and buildings, relatively nimble local governments, close connection to surrounding rural and natural environments, and tradition of innovation. As a result, cities can become active players on the world's stage to respond to the new calls coming for enhanced movement toward sustainability. Indeed, cities are increasingly both the centers of investments and economic activities as well as the source of potential solutions to the global sustainability crisis.

The period 2015–2016 was a watershed moment for global sustainability efforts. During this time, a worldwide consensus has emerged that global climate change is now under way, and the world's populations and ecosystems are experiencing the impacts of more frequent extreme events and gradual shifts in the everyday climate. Coupled with accelerating and globally significant biodiversity loss and unprecedented levels of urbanization, human migration, and commerce, we have entered into what many have described as new geological epoch: the *Anthropocene*.

At the same time, several significant global agreements put forward during the 2015–2016 period provide significant benchmarks and capacity for understanding the current trajectory and motivating forward action. In March 2015, the global community came together to ratify the Sendai Framework for Disaster Risk Reduction that provides protocols for addressing extreme events and threats to sustainability.

On September 25, 2015, countries of the world adopted the Sustainable Development Goals (SDGs) agenda designed to help end poverty, protect the planet, and ensure prosperity for all. Each goal has specific targets to be achieved over the next 15 years. With the adoption of the SDGs in September 2015, especially SDG 11 – 'To make cities inclusive, safe, resilient, and sustainable' – much of the discussion has focused on how cities can move toward greater sustainability.

Less than 3 months later, in December 2015, the UN and representatives from almost all of the world's countries ratified a comprehensive Climate Agreement at COP21 just outside of Paris.

And, in October 2016, the urban community came together for Habitat III, to define ways to promote sustainable urban development, including climate change actions.

The ARC3.2 presents a clear five-fold set of pathways to the urban transformation needed to fulfill the mandates from the international agreements. These pathways provide a foundational framework for the successful development and implementation of climate action. Cities that are making progress in transformative climate change actions are following many or all of these pathways.

The pathways can guide the way for the hundreds of cities – large and small; low-, middle-, and high-income – throughout the world to play a significant role in climate change action and sustainability. Cities that do not follow these pathways may have greater difficulty realizing their potential as centers for climate change solutions. The pathways are:

Pathway 1 – Integrate Mitigation and Adaptation: *Actions that reduce greenhouse gas emissions while integrating increasing resilience are a win-win.* Integrating mitigation and adaptation deserves the highest priority in urban planning, urban design, and urban architecture. A portfolio of approaches is available, including engineering solutions, ecosystem-based adaptation, policies, and social programs. Taking the local context of each city into account is necessary in order to choose actions that result in the greatest benefits.

Pathway 2 – Coordinate Disaster Risk Reduction and Climate Change Adaptation: *Disaster risk reduction and climate change adaptation are the cornerstones of resilient cities.* Integrating these activities into urban development policies requires a new, systems-oriented, multi-timescale approach to risk assessments and planning that accounts for emerging conditions within specific, more vulnerable communities and sectors, as well as across entire metropolitan areas.

Pathway 3 – Co-generate Risk Information: *Risk assessments and climate action plans co-generated with the full range of stakeholders and scientists are most effective.* Processes that are inclusive, transparent, participatory, multisectoral, multijurisdictional, and interdisciplinary are the most robust because they enhance relevance, flexibility, and legitimacy.

Pathway 4 – Focus on Disadvantaged Populations: *Needs of the most disadvantaged and vulnerable citizens should be addressed in climate change planning and action.* The urban poor, the elderly, women, minority, recent immigrants, and otherwise marginal populations most often face the greatest risks due to climate change. Fostering greater equity and justice within climate action increases a city's capacity to respond to climate change and improves human well-being, social capital, and related opportunities for sustainable social and economic development.

Pathway 5 – Advance Governance, Finance, and Knowledge Networks: *Advancing city creditworthiness, developing robust city institutions, and participating in city networks enable climate action.* Access to both municipal and outside financial resources is necessary in order to fund climate change solutions. Sound urban climate governance requires longer planning horizons and effective implementation mechanisms and coordination. Connecting with national and international capacity-building networks helps to advance the strength and success of city-level climate planning and implementation.

A final word on urgency: Cities need to start immediately to develop and implement climate action. The world is entering into the greatest period of urbanization in human history, as well as a period of rapidly changing climate. Getting started now will help avoid locking-in counterproductive long-lived investments and infrastructure systems and will ensure cities' potential for the transformation necessary to lead on climate change.

The connections between the academic researcher and urban practitioner communities on the topic of climate change have developed rapidly in recent years, as has the demand for frameworks that promote opportunities for collaborative, co-generation of new knowledge that in turn leads to improved evidence-based

and implementable climate adaptation and mitigation actions. A multifaceted framework is emerging for promoting practitioner–scientist interaction that accelerates the successful co-generation of new climate risk, adaptation, and mitigation knowledge within individual cities and across multiple cities. The framework has emerged out of the process for science–policy interactions through the Urban Climate Change Research Network (UCCRN) and the creation of the ARC3.2.

This integrated framework represents a new *modus operandi* for urban practitioner–scientist interactions at local, regional, and global scales that enables scientifically rigorous, locally based, demand-driven, collaborative new knowledge-generating processes and actions. The need for this new approach was highlighted recently by the IPCC (IPCC, 2016). The framework includes several nested components: (1) city practitioner–scientist panels, (2) regional knowledge and information transfer hubs, (3) a network of networks, and (4) global urban assessments.

These components have emerged as climate change has grown to be an urban policy issue over the past 15 years (Rosenzweig and Solecki, 2001). Their development has been responsive to different interests and opportunities within specific cities, across sets of cities, and within national and international networks of cities. These processes need to be strengthened and interconnected to leverage these linkages for more rapid and transformative climate action. While a foundational component of the framework is local practitioner–scientist collaborative knowledge co-generative partnerships (here defined as the City Panels on Climate Change), a crucial element in the success of these partnerships will be network connections with other cities locally and globally and a sustained assessment process that ensures benchmarked learning through time.

The City Panels are where new knowledge is generated and implemented, and they represent a foundational unit of the framework. City Panels can be connected to any number of existing city-focused, climate change–related networks including the Large Cities Climate Leadership Group (C40), ICLEI—Local Governments for Sustainability (ICLEI), and the United Cities and Local Governments (UCLG) among others, but, for full effectiveness, should be connected to a network explicitly focused on linking practitioner–scientist panels. This type of network provides a horizontal organizational infrastructure to promote new knowledge and experience collection and transfer. Regional hubs can provide a mechanism to accelerate knowledge transfer, experience acquisition, and climate action among cities across regions. The panel network can be connected to the existing international urban climate change networks and describe how the framework enables the continued growth and expansion of new knowledge and actionable science without which the Paris Climate Change Agreement goal of 2°C and the SDG 11 for sustainable cities will be difficult to achieve.

Contact: www.uccrn.org

Annex 1

UCCRN Regional Hubs

Building on a series of scoping sessions with stakeholders and members, the Urban Climate Change Research Network (UCCRN) is transitioning from a report-focused organization to one that leads an ongoing, sustained, global, city-focused climate change knowledge assessment and solutions program. The program is targeted to early-, mid-, and late-adopter cities through the expansion of the UCCRN to include proactive Regional Hubs, with field directors, program coordinators, and researchers who strengthen ongoing collaborations and knowledge exchange both for and with cities.

The Regional Hubs operate at continental-scale and serve to promote enhanced opportunities for new urban climate change adaptation and mitigation knowledge and information transfer, both within and across cities, by engaging in a real-time monitoring and review process with cities through ongoing dialogue between scholars, experts, urban decision-makers, and stakeholders. These activities are achieved through a combination of cities-based activities and workshops held at international city gatherings and accomplished through regular interactions that include monthly coordination calls, e-newsletters summarizing Hub activities, regional meetings hosted by each Hub every year, and an annual UCCRN meeting rotating through the Regional Hub locations.

In addition to hosting and organizing region-specific, climate change, and cities activities and knowledge sharing, Regional Hubs are also responsible for recruitment and outreach to local urban climate experts to expand the UCCRN network and access an increased diversity of knowledge; stakeholder engagement to connect climate change expertise with city leaders; production of locally focused research and downscaled projections for the regions; fundraising to support research projects, coordination activities, staffing, and operational expenses; hosting regional and topical workshops for local scholars and stakeholders to facilitate the exchange of ideas around climate change and cities; promotion of the Urban Climate Change Research Network Assessment Report on Climate Change and Cities (ARC3) series of reports to targeted stakeholders and translation of reports and publications into regional languages; and liaising between the UCCRN Secretariat in New York and the region.

The first UCCRN Regional Hub was launched in Paris, in July 2015, as the European Hub, followed shortly by the launch of the UCCRN Latin American Hub in Rio de Janeiro in October 2015. The UCCRN also announced at COP21 in December 2015 an Australian-Oceania Hub, co-located in Sydney, Melbourne, and Canberra. The UCCRN African Hub in Durban, South Africa, was launched in May 2016, and the UCCRN East Asian Hub was inaugurated in August 2016 in Shanghai. In November 2016, the

UCCRN launched a Hub in Philadelphia to strengthen a North American network of scholars and stakeholders dedicated to climate change and cities.

In addition to the formal Hubs, a Nordic Node has also been established at Aalborg University to help coordinate Northern European urban climate change efforts, and São Paulo State is the home of the UCCRN Center for Multi-Level Governance, whose overall objective is to discuss implementation of climate policies at the subnational level and their jurisdictional circumstances. The UCCRN is in discussion to launch a Southeast Asian Hub at the Ateneo de Manila University and the Manila Observatory, and it is exploring other potential Asian Hubs in Bangkok and Dhaka.

The *UCCRN European Hub*, hosted in Paris at IEES-Paris (Paris Institute of Ecology and Environmental Sciences), in partnership with the Centre National de la Recherché Scientifique (CNRS), University Pierre et Marie Curie (UPMC), and l'Atelier International du Grand Paris, was launched during the pre-COP21 conference "Our Common Future under Climate Change," in July 2015. The UCCRN European Hub is co-directed by Dr. Chantal Pacteau at CNRS, and Prof. Luc Abbadie at UPMC.

The UCCRN European Hub aims to promote integrated climate change responses based on knowledge-sharing and collaboration between European scholars, institutions, local governments, and industry while acknowledging the richness and diversity of Europe's scientists and practitioners. Local and specific solutions are recognized as foundations for lasting solutions. The European Hub will organize workshops to set a research and coordination agenda of the highest priority climate change issues that European cities face. Additionally, the Hub is looking to translate and publish the ARC3.2 Summary for Leaders in French and in Italian.

The European Hub has been able to solidify a core group of partners that include Paris City Hall, Institute of Ecology and Environmental Sciences of Paris (IEES), Paris Climate Agency (APC), Regional Agency for Nature and Biodiversity (Naturparif), and the Institute for Climate Economics (I4CE).

The *UCCRN Latin America Hub* is hosted by the Oswaldo Cruz Institute/FIOCRUZ and the Center for Integrated Studies on Environment and Climate Change at Universidade Federal do Rio de Janiero (COPPE-UFRJ), under the coordination of Dr. Martha Barata at FIOCRUZ, and Prof. Emilio La Rovere at COPPE-UFRJ. It aims to promote enhanced opportunities for new urban climate change adaptation and mitigation knowledge and information transfer through ongoing dialogue between scholars, experts, urban decision-makers, and stakeholders at the regional level.

In October 2015, the UCCRN Latin American Hub, in partnership with UCCRN and the Columbia Global Center in Latin America, hosted a seminar on "Building Resilient Cities: Climate Change Risk Management for Urban Health." Members of the UCCRN presented results of the *Second Assessment Report on Climate Change in the Cities* (ARC3.2) with a focus on Latin American cities. The initiative had positive outcomes within the national and regional media: it connected scientists, experts, and interested authorities on the subject and facilitated their engagement with the new Hub.

Currently, the UCCRN Latin American Hub is engaged in promoting new research on climate change vulnerabilities, impacts, mitigation, and adaptation of Latin American cities; incorporating new members from different parts of the Latin American region; and hosting, promoting, and participating in workshops and scientific networks in order to build capacity for knowledge partnerships with cities' scientists and stakeholders.

The core partners of the Latin American Hub are the Brazilian Environment Ministry, Brazilian Health Ministry, the Rio Resiliente Program, the Environmental Secretary of the City of Rio de Janeiro, and the Environmental Secretary of the State of Bahia.

Additionally, the Latin American Hub is looking to translate the ARC3.2 in Portuguese and Spanish and adapt it to publish an "Education Kit for Municipal Schools and Museums," in order to provide a toolkit to assist in creating more resilient cities.

The *UCCRN Australian-Oceania Hub* is being shaped as a platform to promote a coordinated climate change response based on knowledge sharing and collaboration between Australian and Oceanian scholars, institutions, city and local governments, and industry. It aims to enable local initiatives in the region to inform local initiatives in other countries, as well as to enable those of other countries to inform initiatives in Australian and Oceanian cities. It is co-directed by Dr. Ken Doust at Southern Cross University in Lismore, Prof. Kate Auty at University of Melbourne and the Office of the Commissioner for Sustainability and the Environment, and Prof. Barbara Norman at the University of Canberra.

The Australian-Oceania Hub the richness and diversity of the region's many researchers and practitioners and seeks to channel the potency of this resource into a systematic response to climate change. Local and specific solutions are recognized as foundational for lasting solutions that respect and enhance the individuality and personality of the landscapes, cities, and people that comprise the Australian continent and the Oceania region.

A number of participants in the Australian-Oceania Hub were in attendance at the UNESCO "Our Common Future" Conference and COP21 in Paris. In December 2015, the Hub launched its website http://www.sustainabilitystep2.org/uccrn/ in time for the COP21 events involving UCCRN. The Australian-Oceania Hub is currently discussing the best way to operationalize the collaborative efforts between the universities and how to best engage with city and regional level governments as the first focus. The Hub is building dialogues with mayors and their agencies to explore where they are with the climate change agenda and to understand the initiatives they are leading or embracing and how collaboration with the Hub participants can help in moving down the path of transformational change.

The UCCRN Australian-Oceania Hub's academic partners include Macquarie University, Monash University, Southern Cross University, the University of Canberra, the University of Melbourne, the University of New Castle, the University of New South Wales, the University of Technology Sydney, and the University of Wollongong.

The *UCCRN African Hub* was launched in May 2016, in Durban, South Africa, in partnership with the Durban Research Action Partnership (D'RAP), the University of KwaZulu-Natal (UKZN), and eThekwini Municipality. It is directed by Dr. Sean O'Donoghue at eThekwini Municipality and Prof. Mathieu Rouget at UKZN.

As an African coastal city with a developing economy, Durban's development trajectory is at risk from climate change impacts. Durban leads regional efforts to coordinate climate change responses and in co-producing knowledge products to guide implementation of adaptation activities. Durban's key knowledge partner is the University of KwaZulu-Natal through the Durban Research Action Partnership (DRAP). Although the partnership has been functional since 2011, the Hub's formal launch was in May 2016.

The UCCRN African Hub will seek to coordinate research activities to guide adaptation implementation within its regional partnership of local and district municipalities, called the Central KwaZulu-Natal Climate Change Compact.

The *UCCRN East Asian Hub* was launched in August 2016 during the first Shanghai Forum on Climate Change and Cities. The Hub is co-led by East China Normal University (ECNU) and the Shanghai Meteorological Service (SMS), and co-directed by Prof. Miu Liu at ECNU and Dr. Xiaotu Lei, Director of the Shanghai Typhoon Institute at the Chinese Meteorological Administration.

East Asia experiences several weather- and climate-related disasters every year, including flooding, high temperatures, infrequent rainfall, and typhoons. The UCCRN East Asian Hub will serve to explore further areas for risk assessment and knowledge generation in the region, especially in terms of evaluating the success of implementing adaptation and mitigation plans in East Asian cities. It will provide science to local stakeholders and facilitate the translation of UCCRN publications into regional languages, including Mandarin, Cantonese, Japanese, and Korean.

The UCCRN is also working with the East Asian Hub to establish a Shanghai Panel on Climate Change based on the framework of the New York City Panel on Climate Change. China released its Action Plan on urban climate change adaptation in February 2016, selecting 30 cities to initiate adaptation pilot projects, beginning in Fall 2016. The UCCRN East Asian Hub is planning to collaborate with a few of these city pilot projects to provide them with relevant climate science knowledge to incorporate into their plans.

The *UCCRN North American Hub* was established in November 2016, at Drexel University, in Philadelphia, under the leadership of Dr. Franco Montalto. Three core goals have been established to guide its activities. The first goal is to promote two-way dialogue between urban decision-makers across North America and UCCRN researchers. Hub researchers will provide knowledge regarding climate change impacts and the effectiveness of various adaptation and mitigation strategies to urban decision-makers, while urban decision-makers will help ensure that the Hub researchers are asking policy-relevant research questions with the potential to produce actionable results. The second goal is to consolidate and mainstream relevant domain knowledge, for example regarding information gaps, known opportunities and constraints to different adaptation and/or mitigation strategies and lessons learned from pilot projects and programs. The third and final goal involves networking and mobilizing activities, specifically by enhancing interactions among researchers, cities, students, and other existing networks.

Organizationally, the UCCRN North American Hub, in partnership with the UCCRN Secretariat in New York, will achieve these goals by developing a Research Project Database, an Interaction Zone, and an Urban Climate Resource Exchange. The Research Project Database will be in the form of a searchable listing of applied research projects identified by the Hub Director and other Hub affiliates during interactions with urban decision-makers throughout North America. Students and researchers from North American Hub institutions will use the Database to identify projects that can be incorporated into master's of science (MS) and doctoral theses, project-based learning opportunities in the classroom, independent study projects, or even sabbatical projects. The online portal would also include an upload option to facilitate the addition of new projects to the database remotely.

The Interaction Zone would also be an online portal including a searchable inventory of various climate action plans from North American cities. This would be publicly accessible and is also expected to build capacity for Hub stakeholders interested in rapidly surveying the state of the art in climate preparedness plans in North America. The Urban Climate Resource Exchange would include links to key international guidance documents, as well as links to other related networks.

Hosted by the São Paulo State Environment Secretariat and closely connected to both the UCCRN Latin American Hub in Rio de Janeiro and the University of São Paulo (USP), the *UCCRN Center for Multi-Level Governance* was launched at the Brazilian Embassy in Paris during COP21, in December 2015. Its overall objective is to discuss implementation of climate policies at the subnational level and their specific jurisdictional circumstances. It is co-directed by Dr. Patricia Iglecias, former São Paulo State Secretary and Lecturer at USP, and Dr. Oswaldo Lucon, Climate Change Advisor to the São Paulo State Environment Secretariat and Lecturer at USP. The Center was created after the experience of writing the ARC3.2 report, and its activities will aim to cross-cut and integrate the UCCRN Regional Hubs on these matters.

The *UCCRN Nordic Node* began its formation during the ARC3.2 International Kickoff Workshop in New York, in September 2013. Through the KlimaLab at Aalborg University, Denmark, it has been working with and supporting the UCCRN Secretariat through the completion of the ARC3.2 volume and beyond.

The UCCRN Nordic Node, led by Dr. Martin Lehmann at Aalborg University, is a close affiliate with the UCCRN European Hub located in Paris. It is intended to bring added value to the UCCRN, in particular by tapping into what is sometimes called "The Nordic Way." This approach embodies a close collaboration in partnerships between cities and regions, academic institutions, industry, and business, and civil society.

The Nordic countries have, individually and collectively, some of the most ambitious climate and energy policy agendas in the world. One of the Nordic Node's aims is to ensure that experiences, results, failures, and successes from these agendas are being collected, scrutinized, analyzed, and shared within the wider UCCRN network and beyond. By doing so, the Node aims also to inspire similar actions in other countries and regions of the world and thereby help to underpin the UCCRN response agenda supported by knowledge and rigorous research.

UCCRN Regional Hubs

Annex 2

Climate Projections for ARC3.2 Cities

Presented here are climate projections for the ARC3.2 cities. ARC3.2 Cities include Case Study Docking Station cities, UCCRN Regional Hub cities, UCCRN project cities, and cities of ARC3.2 Chapter Authors.

The projections are for three future timeslices (30-year periods for temperature and precipitation, 10-year periods for sea level rise) centered on the given decade (2020s, 2050s, 2080s).

For each city, the range presented for each variable is the low-estimate (10th percentile) to high-estimate (90th percentile) across 35 GCMs and two RCPs for temperature and precipitation projections (RCP4.5 and RCP8.5). Sea level rise projections are based on a four-component methodology that uses data from 24 GCMs and two RCPs (RCP4.5 and RCP8.5) and values are presented for only coastal cities[1]. Presented are the low-estimate (10th percentile) to high-estimate (90th percentile).

a. 2020s

City	Temperature (2020s)	Precipitation (2020s)	Sea Level Rise (2020s)
Aalborg	+0.5 to 2.1°C	−1 to +11%	
Abu Dhabi	+0.9 to 1.7°C	−25 to +33%	+4 to 19 cm
Accra	+0.7 to 1.1°C	−8 to +10%	+5 to 19 cm
Addis Ababa	+0.7 to 1.5°C	−7 to +10%	
Ahmedabad	+0.7 to 1.4°C	−6 to +37%	
Almada	+0.4 to 1.2°C	−14 to +1%	
Antofagasta	+0.6 to 1.3°C	−26 to +8%	+4 to 17 cm
Antwerp	+0.5 to 1.8°C	−2 to +8%	+4 to 22 cm
Athens	+0.8 to 1.7°C	−12 to +6%	+4 to 16 cm
Atlanta	+0.8 to 1.6°C	−2 to +9%	
Bangalore	+0.6 to 1.2°C	−7 to +13%	
Bangkok	+0.7 to 1.3°C	−8 to +6%	+4 to 17 cm
Bath	+0.6 to 2.1°C	−2 to +11%	
Berkeley	+0.8 to 1.5°C	−3 to +10%	
Berlin	+0.9 to 1.9°C	0 to +10%	
Bilbao	+0.8 to 1.9°C	−4 to +8%	
Bobo-Dioulasso	+0.6 to 1.1°C	−10 to +8%	
Boston	+0.5 to 1.8°C	−3 to +8%	+4 to 23 cm
Boulder	+0.3 to 1°C	−1 to +13%	
Brisbane	+0.8 to 1.8°C	−22 to +17%	+4 to 18 cm

1   Horton, R., Bader, D.A., Kushner, Y., Little, C., Blake, R. and Rosenzweig, C. (2015). *New York City Panel on Climate Change 2015 Report: Climate observations and projections.* Ann. New York Acad. Sci., 1336, 18–35, doi:10.1111/nyas.12586.

Brussels	+0.9 to 2°C	−1 to +12%	
Buenos Aires	+0.7 to 1.1°C	−5 to +7%	+4 to 16 cm
Cairo	+0.7 to 1.2°C	−7 to +8%	
Calgary	+0.5 to 1°C	−12 to −1%	
Can Tho	+0.7 to 1.5°C	−14 to +12%	+3 to 17 cm
Canberra	+0.6 to 1.1°C	−4 to +11%	
Cape Town	+0.6 to 2°C	−2 to +13%	+5 to 17 cm
Chula Vista	+0.6 to 1.3°C	−5 to +9%	+3 to 18 cm
Colombo	+0.5 to 1.2°C	−4 to +9%	+4 to 19 cm
Copenhagen	+0.6 to 1.1°C	−22 to +14%	+4 to 22 cm
Cubatão	+0.6 to 1.1°C	−11 to +9%	+5 to 18 cm
Curitiba	+0.5 to 1.5°C	−9 to +25%	
Dakar	+0.8 to 1.9°C	−4 to +7%	+5 to 17 cm
Dar es Salaam	+0.5 to 1.3°C	−4 to +16%	+5 to 19 cm
Delhi	+0.2 to 1.4°C	−2 to +8%	
Denver	+0.6 to 1.2°C	−5 to +6%	
Dhaka	+0.7 to 1.7°C	−17 to +28%	+3 to 19 cm
Dublin	+0.6 to 1.2°C	−4 to +9%	+4 to 21 cm
Durban	+0.7 to 1.9°C	−5 to +6%	+4 to 18 cm
Enschede	+0.3 to 1.5°C	−1 to +6%	
Faisalabad	+0.6 to 1.1°C	−11 to +6%	
Fort Lauderdale	+0.5 to 1.6°C	−14 to +27%	+5 to 20 cm
Geneva	+0.9 to 2.5°C	1 to +11%	
Genoa	+0.6 to 1.1°C	−7 to +7%	+4 to 16 cm
Glasgow	+0.6 to 1.3°C	−5 to +16%	
Gold Coast	+0.8 to 2°C	−7 to +7%	+4 to 18 cm
Gorakhpur	+0.7 to 1.2°C	−9 to +7%	
Gothenburg	+0.6 to 2°C	−2 to +10%	+5 to 22 cm
Gunnison	+0.8 to 1.8°C	−18 to +1%	
Hayama	+0.8 to 1.6°C	−8 to +4%	+3 to 19 cm
Helsinki	+0.5 to 1.4°C	−8 to +18%	+4 to 25 cm
Hoboken	+0.6 to 1.4°C	−6 to +12%	
Hong Kong	+1 to 1.9°C	−10 to +12%	+5 to 18 cm
Hyderabad	+0.5 to 1.3°C	−6 to +15%	

Indore	+0.7 to 1.2°C	−11 to +13%	
Istanbul	+0.6 to 1.1°C	−4 to +9%	+4 to 16 cm
Jakarta	+0.5 to 1.9°C	−2 to +8%	+4 to 17 cm
Jena	+0.8 to 1.4°C	−6 to +19%	
Jerusalem	+0.4 to 1.6°C	−3 to +6%	
Johannesburg	+0.7 to 1.5°C	−12 to +13%	
Kampala	+0.6 to 1°C	−8 to +5%	
Kassel	+0.4 to 1.6°C	−2 to +8%	
Kathmandu	+0.6 to 1.1°C	−6 to +8%	
Kedarnath	+0.7 to 1.2°C	−10 to +8%	
Khulna	+0.6 to 1°C	−2 to +11%	+3 to 19 cm
Kingston	+0.7 to 1.3°C	−2 to +10%	+4 to 19 cm
Kunshan	+0.7 to 1.8°C	−15 to +5%	+3 to 19 cm
La Ceiba	+0.6 to 1.1°C	−9 to +4%	+5 to 18 cm
Lagos	+0.8 to 1.5°C	−9 to +6%	+5 to 19 cm
Leipzig	+0.6 to 1.2°C	−4 to +9%	
Leusden	+0.5 to 2°C	−2 to +10%	
Leuven	+1.1 to 2.6°C	0 to +14%	
Lima	+0.8 to 1.6°C	−21 to +32%	+4 to 17 cm
Lismore	+0.6 to 1.1°C	−9 to +5%	
London	+0.7 to 1.2°C	−5 to +26%	+5 to 22 cm
Los Angeles	+0.7 to 1.4°C	−6 to +25%	+4 to 18 cm
Makassar	+0.7 to 1.7°C	−10 to +4%	+4 to 17 cm
Manchester	+0.7 to 1.7°C	−3 to +11%	
Manila	+0.9 to 1.8°C	−1 to +10%	+3 to 19 cm
Maputo	+0.9 to 1.8°C	0 to +11%	+5 to 19 cm
Masdar	+0.7 to 1.7°C	−3 to +9%	+4 to 19 cm
Matara	+0.6 to 1.1°C	−10 to +8%	+4 to 19 cm
Medellín	+0.5 to 1.8°C	−4 to +5%	
Ebro Delta (Mediterranean Spain)	+0.9 to 1.9°C	−16 to +15%	+4 to 17 cm
Melbourne	+0.9 to 1.8°C	0 to +12%	+4 to 17 cm
Mexico City	+0.6 to 1.1°C	−11 to +9%	
Miami	+0.8 to 1.3°C	−4 to +11%	+5 to 20 cm

Middelfart	+0.7 to 1.8°C	−18 to +18%	+4 to 22 cm
Montreal	+0.5 to 1.1°C	−10 to +7%	
Moscow	+0.7 to 1.9°C	−9 to +5%	
Multan	+0.4 to 1.7°C	−1 to +9%	
Mumbai	+0.6 to 1.3°C	−8 to +12%	+4 to 19 cm
Nairobi	+0.4 to 1.1°C	−3 to +11%	
Nanjing	+0.5 to 1.3°C	−20 to +5%	
Naples	+0.7 to 1.2°C	−15 to +6%	+4 to 17 cm
New Brunswick	+0.6 to 1.3°C	−5 to +8%	
Newcastle upon Tyne	+0.8 to 2°C	−4 to +7%	
New Songdo City	+0.7 to 1.2°C	−10 to +5%	+3 to 20 cm
New York	+0.7 to 1.7°C	−3 to +11%	+4 to 23 cm
Newark, DE	+0.7 to 1.6°C	−5 to +7%	
Norfolk, VA	+0.6 to 1.3°C	−9 to +7%	+4 to 24 cm
Nova Friburgo	+0.9 to 1.9°C	−9 to +13%	
Ottawa	+0.6 to 1.1°C	−5 to +5%	
Oujda	+0.5 to 1.5°C	−15 to +1%	
Paris	+0.9 to 1.9°C	0 to +10%	
Paveh	+0.9 to 1.9°C	−2 to +8%	
Peshawar	+0.7 to 1.3°C	−5 to +38%	
Philadelphia	+0.7 to 1.2°C	−7 to +8%	
Phitsanulok	+0.6 to 1°C	−3 to +10%	
Portmore	+0.7 to 1.2°C	−9 to +7%	+4 to 19 cm
Quito	+1 to 1.8°C	−8 to +9%	
Rawalpindi	+0.7 to 1.4°C	−5 to +5%	
Rio de Janeiro	+1 to 2.2°C	−1 to +9%	+4 to 18 cm
Rome	+0.7 to 1.2°C	−16 to +10%	
Rotterdam	+0.8 to 1.8°C	−13 to +8%	+4 to 22 cm
San Diego	+0.7 to 2.1°C	−7 to +7%	+4 to 18 cm
San Francisco	+0.8 to 2.2°C	−5 to +9%	+4 to 18 cm
San Jose, Costa Rica	+0.7 to 2.2°C	−1 to +13%	
Santa Fe, Argentina	+0.8 to 1.7°C	−1 to +11%	
Santiago	+0.9 to 2.1°C	−2 to +10%	

Santo Domingo	+0.7 to 1.6°C	−5 to +5%	+4 to 17 cm
São Paulo	+0.9 to 1.7°C	−31 to +43%	
Seattle	+0.9 to 1.8°C	−1 to +12%	+4 to 18 cm
Semarang	+0.8 to 1.4°C	−5 to +32%	+4 to 17 cm
Seoul	+0.6 to 1.1°C	−11 to +9%	
Shanghai	+0.7 to 1.3°C	−11 to +14%	+3 to 19 cm
Shenzhen	+0.5 to 2.3°C	1 to +12%	+5 to 18 cm
Shimla	+0.6 to 2°C	−2 to +10%	
Singapore	+0.9 to 1.8°C	−13 to +7%	+4 to 17 cm
Sintra	+0.9 to 1.8°C	−2 to +11%	
Somerville	+0.7 to 1.5°C	−5 to +5%	
St. Peters	+0.6 to 1.4°C	−8 to +12%	
Surat	+0.5 to 1.9°C	−2 to +9%	+4 to 19 cm
Sydney	+0.6 to 2.1°C	−2 to +10%	+5 to 20 cm
Tacloban	+0.5 to 1.6°C	−8 to +2%	+3 to 19 cm
Tangerang Seltan	+0.7 to 1.9°C	−9 to +5%	+3 to 19 cm
Tehran	+0.5 to 1.8°C	−1 to +9%	
Tempe	+0.4 to 1.6°C	−1 to +8%	
Tokyo	+0.9 to 1.8°C	−1 to +10%	+3 to 19 cm
Toronto	+0.7 to 1.5°C	−5 to +7%	
Townsville	+0.7 to 1.9°C	−7 to +8%	+4 to 18 cm
Trondheim	+0.7 to 1.3°C	−4 to +5%	+5 to 23 cm
Tsukuba	+1 to 2.3°C	0 to +10%	
Tucson	+0.9 to 1.7°C	−20 to −1%	
Udon Thani	+1 to 1.9°C	−12 to +8%	
Venice	+0.4 to 1.6°C	−3 to +6%	+4 to 16 cm
Victoria	+0.7 to 1.8°C	−2 to +7%	+4 to 18 cm
Vienna	+0.7 to 1.2°C	−8 to +7%	
Warsaw	+0.7 to 1.3°C	−7 to +6%	
Washington, D.C.	+0.5 to 1.5°C	−2 to +7%	
Waterloo	+1 to 2.2°C	0 to +9%	
White Plains	+0.4 to 1.7°C	0 to +9%	
Windsor	+0.6 to 1.3°C	−13 to +12%	

b. 2050s

City	Temperature (2050s)	Precipitation (2050s)	Sea Level Rise (2050s)
Aalborg	+1.1 to 3.7°C	−1 to +14%	
Abu Dhabi	+1.6 to 3.4°C	−32 to +49%	+15 to 60 cm
Accra	+1.2 to 2.4°C	−12 to +11%	+17 to 58 cm
Addis Ababa	+1.4 to 3.3°C	−6 to +16%	
Ahmedabad	+1.6 to 3.1°C	−5 to +53%	
Almada	+0.8 to 2.3°C	−26 to −2%	
Antofagasta	+1.2 to 2.9°C	−35 to +30%	+13 to 55 cm
Antwerp	+1.1 to 3°C	−3 to +12%	+18 to 70 cm
Athens	+1.4 to 3.1°C	−22 to +4%	+15 to 55 cm
Atlanta	+1.6 to 3.1°C	−2 to +13%	
Bangalore	+1.3 to 2.6°C	−5 to +18%	
Bangkok	+1.2 to 2.8°C	−9 to +14%	+15 to 57 cm
Bath	+1.3 to 3.6°C	−2 to +16%	
Berkeley	+1.4 to 3.2°C	−7 to +12%	
Berlin	+1.6 to 3.9°C	0 to +14%	
Bilbao	+1.7 to 3.8°C	−4 to +10%	
Bobo-Dioulasso	+1.2 to 2.2°C	−15 to +7%	
Boston	+1.2 to 3°C	−4 to +12%	+16 to 70 cm
Boulder	+0.8 to 2°C	−1 to +15%	
Brisbane	+1.5 to 3.4°C	−31 to +9%	+15 to 58 cm
Brussels	+1.7 to 3.9°C	−1 to +17%	
Buenos Aires	+1.2 to 2.2°C	−3 to +15%	+14 to 54 cm
Cairo	+1.4 to 2.5°C	−10 to +10%	
Calgary	+1 to 1.9°C	−23 to −5%	
Can Tho	+1.3 to 3.1°C	−20 to +11%	+14 to 58 cm
Canberra	+1.1 to 2.3°C	−2 to +19%	
Cape Town	+1.2 to 3.7°C	−1 to +18%	+16 to 57 cm
Chula Vista	+1.1 to 2.6°C	−6 to +15%	+14 to 56 cm
Colombo	+1 to 2.4°C	−3 to +16%	+14 to 59 cm
Copenhagen	+1.1 to 2.3°C	−31 to +27%	+16 to 70 cm
Cubatão	+1.2 to 2.1°C	−14 to +13%	+14 to 56 cm

Curitiba	+1.5 to 3.3°C	−13 to +28%	
Dakar	+1.8 to 3.8°C	−6 to +8%	+15 to 56 cm
Dar es Salaam	+1.2 to 3°C	−5 to +30%	+15 to 60 cm
Delhi	+0.7 to 2.5°C	−2 to +12%	
Denver	+1.2 to 2.3°C	−9 to +7%	
Dhaka	+1.7 to 3.7°C	−21 to +34%	+14 to 57 cm
Dublin	+1.2 to 2.5°C	−5 to +12%	+15 to 65 cm
Durban	+1.3 to 3.5°C	−8 to +6%	+14 to 58 cm
Enschede	+0.8 to 2.6°C	0 to +10%	
Faisalabad	+1.1 to 2.2°C	−14 to +6%	
Fort Lauderdale	+1.6 to 3.3°C	−19 to +33%	+15 to 62 cm
Geneva	+1.6 to 4.5°C	3 to +16%	
Genoa	+1.1 to 2.5°C	−5 to +13%	+15 to 56 cm
Glasgow	+1.4 to 2.9°C	−3 to +30%	
Gold Coast	+1.4 to 3.4°C	−15 to +5%	+15 to 58 cm
Gorakhpur	+1.2 to 2.5°C	−11 to +12%	
Gothenburg	+1.3 to 3.5°C	−2 to +13%	+17 to 70 cm
Gunnison	+1.6 to 3.4°C	−27 to −1%	
Hayama	+1.6 to 3.2°C	−10 to +7%	+13 to 65 cm
Helsinki	+0.9 to 2.9°C	−9 to +26%	+18 to 75 cm
Hoboken	+1.4 to 3°C	−4 to +24%	
Hong Kong	+1.9 to 4.2°C	−12 to +10%	+15 to 60 cm
Hyderabad	+1 to 2.8°C	−6 to +34%	
Indore	+1.2 to 2.4°C	−20 to +14%	
Istanbul	+1.1 to 2.6°C	−10 to +10%	+15 to 55 cm
Jakarta	+1.1 to 3.1°C	−3 to +12%	+14 to 58 cm
Jena	+1.4 to 2.9°C	−7 to +33%	
Jerusalem	+1 to 2.7°C	−4 to +10%	
Johannesburg	+1.4 to 2.9°C	−20 to +9%	
Kampala	+1.1 to 2.1°C	−10 to +11%	
Kassel	+0.9 to 2.8°C	−3 to +11%	
Kathmandu	+1.1 to 2.1°C	−3 to +14%	
Kedarnath	+1.3 to 2.6°C	−15 to +8%	

619

Khulna	+1.1 to 2.1°C	2 to +20%	+14 to 57 cm
Kingston	+1.4 to 2.8°C	−4 to +14%	+14 to 57 cm
Kunshan	+1.3 to 3.2°C	−16 to +3%	+13 to 59 cm
La Ceiba	+1.3 to 2.2°C	−14 to +5%	+15 to 59 cm
Lagos	+1.5 to 3°C	−14 to +7%	+16 to 58 cm
Leipzig	+1.2 to 2.4°C	−5 to +11%	
Leusden	+1.1 to 3.6°C	−2 to +12%	
Leuven	+1.8 to 4.7°C	−1 to +19%	
Lima	+1.9 to 3.7°C	−18 to +40%	+14 to 55 cm
Lismore	+1.2 to 2.2°C	−13 to +7%	
London	+1.3 to 2.5°C	−2 to +43%	+17 to 70 cm
Los Angeles	+1.3 to 2.6°C	−10 to +41%	+14 to 56 cm
Makassar	+1.3 to 3.3°C	−17 to +1%	+15 to 58 cm
Manchester	+1.6 to 3.4°C	1 to +19%	
Manila	+1.7 to 3.7°C	1 to +13%	+14 to 60 cm
Maputo	+1.8 to 3.5°C	1 to +14%	+15 to 58 cm
Masdar	+1.6 to 3.2°C	−2 to +13%	+15 to 60 cm
Matara	+1 to 2.3°C	−14 to +12%	+14 to 59 cm
Medellín	+1.3 to 3.2°C	−6 to +8%	
Ebro Delta (Mediterranean Spain)	+1.8 to 3.9°C	−20 to +20%	+15 to 57 cm
Melbourne	+1.8 to 3.5°C	1 to +13%	+12 to 56 cm
Mexico City	+1.1 to 2.2°C	−21 to +4%	
Miami	+1.3 to 2.6°C	−4 to +19%	+15 to 62 cm
Middelfart	+1.8 to 3.9°C	−21 to +23%	+16 to 70 cm
Montreal	+1 to 2.1°C	−14 to +10%	
Moscow	+1.5 to 3.5°C	−16 to +3%	
Multan	+1.1 to 2.9°C	−2 to +14%	
Mumbai	+1.2 to 2.8°C	−12 to +14%	+15 to 58 cm
Nairobi	+0.9 to 2.2°C	−4 to +16%	
Nanjing	+1.2 to 2.5°C	−31 to −1%	
Naples	+1.1 to 2.4°C	−26 to +2%	+15 to 56 cm
New Brunswick	+1.2 to 2.6°C	−6 to +13%	
Newcastle upon Tyne	+1.4 to 3.7°C	−3 to +10%	
New Songdo City	+1.1 to 2.4°C	−12 to +12%	+13 to 60 cm

New York	+1.5 to 3.4°C	1 to +19%	+16 to 71 cm
Newark, DE	+1.4 to 3.1°C	−2 to +9%	
Norfolk, VA	+1.2 to 2.8°C	−6 to +12%	+15 to 67 cm
Nova Friburgo	+1.9 to 3.9°C	−13 to +17%	
Ottawa	+1.1 to 2.3°C	−5 to +12%	
Oujda	+0.9 to 2.7°C	−28 to −2%	
Paris	+1.7 to 3.9°C	1 to +14%	
Paveh	+1.8 to 4°C	−3 to +11%	
Peshawar	+1.4 to 2.9°C	−2 to +49%	
Philadelphia	+1.3 to 2.4°C	−9 to +11%	
Phitsanulok	+1.1 to 2.1°C	0 to +15%	
Portmore	+1.2 to 2.5°C	−11 to +12%	+14 to 57 cm
Quito	+1.9 to 3.6°C	−12 to +11%	
Rawalpindi	+1.4 to 3°C	−5 to +11%	
Rio de Janeiro	+2.1 to 4.2°C	2 to +12%	+15 to 56 cm
Rome	+1.1 to 2.3°C	−19 to +16%	
Rotterdam	+1.7 to 3.4°C	−14 to +7%	+18 to 70 cm
San Diego	+1.4 to 3.6°C	−10 to +8%	+14 to 56 cm
San Francisco	+1.4 to 3.9°C	−5 to +11%	+14 to 56 cm
San Jose, Costa Rica	+1.5 to 3.9°C	−3 to +13%	
Santa Fe, Argentina	+1.7 to 3.5°C	1 to +14%	
Santiago	+2 to 3.9°C	0 to +14%	
Santo Domingo	+1.5 to 3.3°C	−3 to +10%	+14 to 56 cm
São Paulo	+1.7 to 3.6°C	−32 to +58%	
Seattle	+1.7 to 3.6°C	1 to +13%	+14 to 56 cm
Semarang	+1.5 to 3.1°C	−7 to +47%	+14 to 58 cm
Seoul	+1.1 to 2.2°C	−23 to +2%	
Shanghai	+1.2 to 2.6°C	−20 to +18%	+13 to 59 cm
Shenzhen	+1.4 to 3.9°C	5 to +19%	+15 to 60 cm
Shimla	+1.3 to 3.5°C	−3 to +13%	
Singapore	+1.7 to 3.5°C	−17 to +6%	+15 to 58 cm
Sintra	+1.7 to 3.6°C	0 to +13%	
Somerville	+1.4 to 3°C	−5 to +9%	
St. Peters	+1.2 to 2.9°C	−13 to +13%	

Surat	+1.2 to 3.2°C	−3 to +13%	+15 to 58 cm
Sydney	+1.3 to 3.5°C	−1 to +14%	+16 to 64 cm
Tacloban	+1.1 to 3°C	−15 to −2%	+14 to 61 cm
Tangerang Seltan	+1.4 to 3.5°C	−14 to +4%	+14 to 57 cm
Tehran	+1.1 to 2.9°C	−2 to +13%	
Tempe	+0.9 to 2.8°C	−1 to +10%	
Tokyo	+1.7 to 3.7°C	1 to +13%	+13 to 65 cm
Toronto	+1.6 to 3.4°C	−3 to +12%	
Townsville	+1.8 to 3.8°C	−8 to +9%	+16 to 58 cm
Trondheim	+1.4 to 3°C	−2 to +11%	+17 to 70 cm
Tsukuba	+2 to 4.4°C	2 to +15%	
Tucson	+1.6 to 3.5°C	−29 to −5%	
Udon Thani	+1.9 to 3.9°C	−22 to +9%	
Venice	+1 to 2.7°C	−4 to +9%	+15 to 56 cm
Victoria	+1.4 to 3.4°C	−1 to +10%	+14 to 56 cm
Vienna	+1.4 to 2.8°C	−7 to +13%	
Warsaw	+1.3 to 2.8°C	−3 to +16%	
Washington, D.C.	+1.2 to 2.9°C	2 to +13%	
Waterloo	+2.1 to 4.2°C	1 to +15%	
White Plains	+0.8 to 2.8°C	0 to +13%	
Windsor	+1.3 to 2.7°C	−20 to +11%	

c. 2080s

City	Temperature (2080s)	Precipitation (2080s)	Sea Level Rise (2080s)
Aalborg	+1.6 to 5.4°C	1 to +22%	
Abu Dhabi	+1.9 to 5.6°C	−39 to +78%	+23 to 124 cm
Accra	+1.4 to 3.9°C	−15 to +15%	+23 to 119 cm
Addis Ababa	+1.8 to 5.3°C	−3 to +28%	
Ahmedabad	+1.9 to 5°C	−9 to +90%	
Almada	+1.3 to 3.4°C	−39 to −5%	
Antofagasta	+1.4 to 4.6°C	−29 to +47%	+20 to 116 cm
Antwerp	+1.5 to 4.9°C	−5 to +14%	+25 to 140 cm
Athens	+1.7 to 4.9°C	−33 to +2%	+22 to 116 cm

Atlanta	+2.1 to 5.1°C	−4 to +16%	
Bangalore	+1.5 to 4.5°C	−1 to +45%	
Bangkok	+1.5 to 4.7°C	−12 to +25%	+21 to 122 cm
Bath	+1.7 to 5.6°C	−4 to +18%	
Berkeley	+1.8 to 5.6°C	−11 to +15%	
Berlin	+2.2 to 6.1°C	3 to +19%	
Bilbao	+2 to 6.1°C	−7 to +12%	
Bobo-Dioulasso	+1.4 to 3.9°C	−21 to +11%	
Boston	+1.5 to 4.9°C	−6 to +14%	+23 to 141 cm
Boulder	+1.2 to 3.5°C	0 to +21%	
Brisbane	+1.8 to 5.4°C	−42 to +10%	+23 to 124 cm
Brussels	+2.1 to 6.2°C	−5 to +18%	
Buenos Aires	+1.4 to 3.7°C	−5 to +24%	+19 to 115 cm
Cairo	+1.6 to 4.3°C	−14 to +9%	
Calgary	+1.2 to 3.4°C	−32 to −7%	
Can Tho	+1.8 to 4.9°C	−25 to +15%	+21 to 122 cm
Canberra	+1.4 to 3.8°C	3 to +30%	
Cape Town	+1.7 to 5.4°C	1 to +23%	+22 to 118 cm
Chula Vista	+1.5 to 4.3°C	−7 to +23%	+21 to 118 cm
Colombo	+1.5 to 3.9°C	−2 to +25%	+22 to 122 cm
Copenhagen	+1.4 to 3.8°C	−41 to +43%	+25 to 140 cm
Cubatão	+1.5 to 3.7°C	−19 to +22%	+22 to 118 cm
Curitiba	+2 to 5.6°C	−11 to +38%	
Dakar	+2 to 6.1°C	−7 to +9%	+22 to 118 cm
Dar es Salaam	+1.4 to 4.9°C	−6 to +41%	+22 to 122 cm
Delhi	+0.9 to 3.9°C	−2 to +14%	
Denver	+1.5 to 3.9°C	−10 to +11%	
Dhaka	+2.3 to 5.9°C	−24 to +36%	+21 to 122 cm
Dublin	+1.4 to 4°C	−11 to +13%	+24 to 131 cm
Durban	+2 to 5.8°C	−11 to +8%	+20 to 118 cm
Enschede	+1.2 to 4.1°C	1 to +12%	
Faisalabad	+1.4 to 3.6°C	−21 to +10%	
Fort Lauderdale	+2 to 5.6°C	−15 to +41%	+21 to 123 cm
Geneva	+2.4 to 6.9°C	4 to +24%	

623

Genoa	+1.5 to 4.1°C	−5 to +21%	+22 to 118 cm
Glasgow	+1.6 to 4.9°C	1 to +48%	
Gold Coast	+1.6 to 5.3°C	−21 to +9%	+23 to 124 cm
Gorakhpur	+1.4 to 4°C	−13 to +21%	
Gothenburg	+1.8 to 5.5°C	−4 to +16%	+24 to 140 cm
Gunnison	+1.9 to 5.4°C	−41 to −5%	
Hayama	+2.1 to 5.7°C	−14 to +6%	+20 to 129 cm
Helsinki	+1.4 to 4.7°C	−8 to +44%	+26 to 148 cm
Hoboken	+1.8 to 5.2°C	−3 to +33%	
Hong Kong	+2.3 to 6.8°C	−13 to +20%	+22 to 123 cm
Hyderabad	+1.4 to 4.8°C	−5 to +38%	
Indore	+1.4 to 4°C	−35 to +23%	
Istanbul	+1.5 to 4.4°C	−12 to +14%	+22 to 116 cm
Jakarta	+1.5 to 5.1°C	−5 to +14%	+21 to 122 cm
Jena	+1.7 to 4.7°C	−6 to +63%	
Jerusalem	+1.4 to 4.5°C	−6 to +11%	
Johannesburg	+1.7 to 4.9°C	−21 to +15%	
Kampala	+1.3 to 3.6°C	−13 to +16%	
Kassel	+1.3 to 4.5°C	−5 to +15%	
Kathmandu	+1.4 to 3.6°C	−3 to +21%	
Kedarnath	+1.6 to 4.3°C	−18 to +9%	
Khulna	+1.4 to 3.7°C	1 to +32%	+21 to 122 cm
Kingston	+1.7 to 4.6°C	−6 to +16%	+20 to 119 cm
Kunshan	+1.8 to 5.1°C	−24 to +3%	+21 to 122 cm
La Ceiba	+1.5 to 3.9°C	−20 to +3%	+21 to 122 cm
Lagos	+1.8 to 5.1°C	−23 to +7%	+23 to 120 cm
Leipzig	+1.4 to 4°C	−11 to +11%	
Leusden	+1.4 to 5.2°C	1 to +20%	
Leuven	+2.6 to 7.1°C	1 to +25%	
Lima	+2.3 to 6.1°C	−24 to +41%	+21 to 117 cm
Lismore	+1.5 to 3.8°C	−17 to +10%	
London	+1.6 to 4.4°C	−2 to +64%	+24 to 139 cm
Los Angeles	+1.6 to 4.4°C	−7 to +66%	+21 to 118 cm
Makassar	+1.8 to 5.3°C	−25 to +2%	+22 to 122 cm

Manchester	+2 to 5.5°C	5 to +25%	
Manila	+2.1 to 5.7°C	2 to +19%	+21 to 124 cm
Maputo	+2.2 to 5.6°C	2 to +19%	+22 to 120 cm
Masdar	+2 to 5.1°C	−2 to +16%	+23 to 124 cm
Matara	+1.4 to 4.1°C	−21 to +13%	+22 to 122 cm
Medellín	+1.6 to 5.2°C	−9 to +7%	
Ebro Delta (Mediterranean Spain)	+2.4 to 6.1°C	−27 to +21%	+22 to 119 cm
Melbourne	+2.3 to 5.6°C	2 to +19%	+18 to 116 cm
Mexico City	+1.3 to 3.6°C	−38 to +6%	
Miami	+1.6 to 4.4°C	−6 to +30%	+21 to 123 cm
Middelfart	+2.3 to 5.9°C	−24 to +20%	+25 to 140 cm
Montreal	+1.3 to 3.7°C	−21 to +16%	
Moscow	+2 to 5.6°C	−22 to +2%	
Multan	+1.4 to 4.6°C	−3 to +16%	
Mumbai	+1.5 to 4.4°C	−11 to +22%	+23 to 122 cm
Nairobi	+1.1 to 3.7°C	−3 to +21%	
Nanjing	+1.4 to 4.2°C	−44 to −3%	
Naples	+1.4 to 4°C	−39 to +7%	+22 to 116 cm
New Brunswick	+1.6 to 4.3°C	−7 to +21%	
Newcastle upon Tyne	+1.9 to 5.7°C	−1 to +12%	
New Songdo City	+1.3 to 4°C	−17 to +18%	+20 to 124 cm
New York	+1.9 to 5.5°C	4 to +25%	+23 to 139 cm
Newark, DE	+1.8 to 4.9°C	−3 to +15%	
Norfolk, VA	+1.6 to 4.6°C	−7 to +18%	+23 to 129 cm
Nova Friburgo	+2.2 to 5.9°C	−11 to +28%	
Ottawa	+1.4 to 3.9°C	−4 to +18%	
Oujda	+1.4 to 4°C	−38 to −5%	
Paris	+2.2 to 6.1°C	3 to +19%	
Paveh	+2.2 to 6°C	−4 to +14%	
Peshawar	+1.7 to 4.9°C	−4 to +100%	
Philadelphia	+1.5 to 4.2°C	−13 to +13%	
Phitsanulok	+1.4 to 3.5°C	−2 to +24%	
Portmore	+1.4 to 4°C	−13 to +21%	+20 to 119 cm

Quito	+2.4 to 6°C	−19 to +14%	
Rawalpindi	+1.7 to 4.8°C	−7 to +13%	
Rio de Janeiro	+2.4 to 6.7°C	0 to +17%	+21 to 118 cm
Rome	+1.4 to 4°C	−28 to +17%	
Rotterdam	+2 to 5.9°C	−22 to +10%	+25 to 140 cm
San Diego	+1.9 to 5.8°C	−12 to +9%	+21 to 118 cm
San Francisco	+1.8 to 5.8°C	−7 to +12%	+21 to 118 cm
San Jose, Costa Rica	+2 to 6°C	−2 to +17%	
Santa Fe, Argentina	+2.2 to 5.4°C	2 to +20%	
Santiago	+2.3 to 6.5°C	0 to +18%	
Santo Domingo	+1.9 to 5.6°C	−2 to +15%	+20 to 117 cm
São Paulo	+2.1 to 6°C	−38 to +81%	
Seattle	+2.2 to 5.6°C	2 to +19%	+21 to 118 cm
Semarang	+1.9 to 5.2°C	−5 to +73%	+21 to 122 cm
Seoul	+1.3 to 3.6°C	−38 to +5%	
Shanghai	+1.4 to 4.6°C	−33 to +22%	+21 to 122 cm
Shenzhen	+1.8 to 5.9°C	7 to +24%	+22 to 123 cm
Shimla	+1.8 to 5.4°C	−4 to +16%	
Singapore	+2.1 to 5.8°C	−19 to +8%	+21 to 122 cm
Sintra	+2.1 to 5.7°C	1 to +19%	
Somerville	+1.7 to 4.8°C	−7 to +12%	
St. Peters	+1.5 to 4.5°C	−13 to +21%	
Surat	+1.5 to 5.2°C	−4 to +17%	+23 to 122 cm
Sydney	+1.8 to 5.5°C	−3 to +16%	+24 to 128 cm
Tacloban	+1.4 to 4.7°C	−22 to −3%	+21 to 125 cm
Tangerang Seltan	+1.9 to 5.6°C	−17 to +4%	+21 to 122 cm
Tehran	+1.4 to 4.7°C	−4 to +16%	
Tempe	+1.3 to 4.3°C	−2 to +15%	
Tokyo	+2.1 to 5.7°C	2 to +19%	+20 to 129 cm
Toronto	+1.9 to 5.7°C	−1 to +18%	
Townsville	+2.1 to 6.2°C	−10 to +10%	+22 to 123 cm
Trondheim	+1.7 to 4.8°C	−4 to +13%	+25 to 141 cm
Tsukuba	+2.4 to 6.9°C	3 to +19%	
Tucson	+2.1 to 5.7°C	−44 to −7%	

Udon Thani	+2.4 to 6.3°C	−31 to +9%	
Venice	+1.3 to 4.4°C	−6 to +11%	+22 to 117 cm
Victoria	+2 to 5.6°C	0 to +13%	+21 to 118 cm
Vienna	+1.6 to 4.8°C	−9 to +21%	
Warsaw	+1.5 to 4.6°C	−4 to +24%	
Washington, D.C.	+1.7 to 4.9°C	5 to +19%	
Waterloo	+2.4 to 6.9°C	4 to +19%	
White Plains	+1.2 to 4.1°C	1 to +20%	
Windsor	+1.7 to 4.3°C	−25 to +15%	

Annex 3

Case Study Docking Station: Overview and Methods

UCCRN Case Study Docking Station Team

Martin Lehmann (Aalborg), David C. Major (New York), Patrick A. Driscoll (Copenhagen/Trondheim), Somayya Ali Ibrahim (New York/Peshawar), Wim Debucquoy (Mechelen/The Hague), Samuel Schlecht (Heidelberg), Megi Zhamo (Hamburg/Durres), Jovana Milić (Copenhagen/Niš), Antonio Bontempi (Barcelona/Bergamo), Megan Helseth (Denver), Victoria Ruiz Rincón (Barcelona/Cuernavaca), Jonah Garnick (New York), Stephen Solecki (New York), Annel Hernandez (New York), Carissa Lim (Singapore), Marta De Los Ríos White (Medellin), Usama Khalid (Islamabad), Ioana Blaj (Brussels/Oradea)

CSDS Annex Writing Team

David C. Major, Somayya Ali Ibrahim, Patrick A. Driscoll, Martin Lehmann, Wim Debucquoy

1   Introduction

The Urban Climate Change Research Network's (UCCRN) *First Assessment Report on Climate Change and Cities* (ARC3.1; Rosenzweig et al., 2011), included within its chapters 46 Case Studies on cities around the world and climate topics more generally, including vulnerability, hazards and impacts, extreme events, mitigation actions, and sector-specific themes such as health, transportation, energy, wastewater and flood management. While the cases presented provided concrete examples of the issues discussed within the particular chapters, the lack of an overarching research design, data collection protocol, and analytic matrices made cross-case comparisons difficult. This Annex describes the UCCRN contribution to addressing these issues and considering next steps. This Annex includes the following sections in addition to this Introduction: Section 2, ARC3.1 Background; Section 3, Guidelines for ARC3.2 Case Studies; Section 4, the ARC3.2 Case Study Docking Station; and Section 5, Conclusions.

One of the key strengths of the case study methodology is the ability to provide highly context-specific insights into contemporary phenomena (Yin, 2009; Gerring, 2010; Keskitalo, 2010). However, if case studies are uncoordinated, this benefit is to some extent a weakness if researchers are interested in understanding a broader view of the deeper causal mechanisms that drive urban climate change planning (Ford et al., 2010). Some of the existing literature on, and meta-analyses of, urban climate change issues documents that the study of climate change at the urban scale is characterized by a fragmented research environment where even many of the basic concepts employed remain indistinct and vague (Dupuis and Biesbroek, 2013). For example, in the field of climate adaptation, there are a variety of interpretations about what is baseline adaptation (normal hazard management) and what is climate-driven adaptation (in response to changes beyond the previous status quo), what is adaptation policy, what are suitable measures of success or failure, and what are some standard indicators to measure progress in adaptation strategies and measures (Doria et al., 2009).

To help address these issues, the UCCRN *Second Assessment Report on Climate Change and Cities* (ARC3.2) includes 117 city Case Studies, some placed within ARC3.2 chapters, some in Annex 5, and all incorporated into an online Case Study Docking Station (CSDS) that is under development, a searchable database designed to allow for further exploration and examination of cases. These Case Studies display empirical evidence on what cities are doing on the ground across a diverse set of urban challenges and opportunities. The aim is to develop a mechanism by which to organize the case studies by a variety of metrics and sectoral and content elements; this also provides for engaging a broader and more diverse set of authors for the ARC3.2 Case Studies than for those in the earlier volume.

The ARC3.2 CSDS is designed to inform both research and practice on climate change and cities by enabling initial scientifically valid cross-case comparisons and analysis across a range of social, biophysical, cultural, economic, and political contexts. It is hoped that this first step will lead to new possibilities and pathways of case study research. The ARC3.2 Framework, within which the CSDS was developed, is shown in Figure 1.4 of this volume.

A common data collection protocol for the assembled cases serves as a guideline to achieving a higher level of consistency across the Case Studies. In the ARC3.2 CSDS, all data entries are sourced and traceable. In addition to the data inputs and reviews by expert practitioners, the Docking Station currently uses a straightforward data management program, Caspio Bridge (Version 8.5.5). The technology and software capability of the CSDS system can periodically be advanced as appropriate.

The fullest version of this information is online at the UCCRN CSDS, www.uccrn.org/casestudies. The online database will be available for data extraction and analysis for a variety of different research and practice needs under a creative commons license.

Case Studies in the CSDS provide references to source(s) for additional reading. These sources include both peer-reviewed and gray literature, including city reports and white papers published by international agencies.

2   ARC3.1 Background

2.1   Case Studies in ARC3.1

The role of the ARC3.1 Case Studies, as contributions to the overall objectives of the report, is stated in the ARC3.1 report as follows:

> The ARC3.1 recognizes that there are both similarities and differences between developed and developing city responses to climate change. For example, there is a great deal of fundamental information on climate change projections, vulnerabilities, and risk assessment methods that has a common base in both types of cities. At the same time, there are great differences in the circumstances in developing country cities. These are discussed throughout the chapters, with key points brought forward as city Case Studies. The city Case Studies, which illustrate challenges, "best practices," and available tools to facilitate actions in developing and developed cities, are presented throughout the text. The Case Studies cover the status and activities related to climate change on a city-by-city basis. There are several types of Case Studies included throughout ARC3: those developed by the chapter authors; those invited from others that apply entirely to the chapter topic; and a third category, "cross-cutting Case Studies" that touch on many different urban climate change topics that a particular city or organization is addressing. The Case Studies have been developed by authors drawn from both the research and practitioner

communities; such teams are helping to build a cadre of knowledge-providers to aid in implementation of climate change actions in cities around the world. (Rosenzweig et al., 2011: 6)

The structure and content of the ARC3.1 report was based in part on feedback from selected city decision-makers from around the world. They were asked to express their information needs with regard to climate change hazards, vulnerability, adaptation, and mitigation in their city, addressing questions such as:

- What climate-related challenges does your city face?
- In what fields do you see potential for strong mitigation efforts in your city?
- What policy mechanisms is your city potentially or actually implementing?
- Where adaptation policies and actions are most urgently needed?
- What other special issues would you like this report to address?

The ARC3.1 report was composed of nine chapters to address these responses and broader considerations, including four sector-based topics – urban energy, water and wastewater, transportation, and health. In addition, as noted earlier, the ARC3.1 co-editors solicited Case Studies from urban climate change academics and practitioners from around the world, with the idea that such cases would cover a wide range of climate change mitigation and adaptation aspects in various sectors and cities so as to build a knowledge base of existing policies and actions to be shared worldwide. Case Study authors for ARC3.1 were encouraged to report current climate change conditions as well as future climate change scenarios (i.e., temperature, precipitation, and sea level changes; key vulnerabilities; and mitigation and adaptation programs and policies in place or planned for implementation in the near future).

2.2 Topics and Results of the Case Studies in ARC3.1

The results of this effort are shown in Rosenzweig et al. (2011, Appendix A), with case study topics in vulnerability, adaptation, and mitigation.

There were 46 Case Studies in the ARC3.1 report, inserted within the nine chapters in the volume, of which 35 can be defined as specifically city Case Studies. The other cases discuss selected topics more generally. Of these 35 cases in the ARC3.1 report, 28 were focused on a single city. These 28 single-city Case Studies were reviewed following the publication of ARC3.1 in order to help develop guidelines for the ARC3.2 Case Study Call for Submissions. These new guidelines, presented here, are designed to enable the consistency – insofar as this is

possible – of the Case Studies published in the ARC3.2 report and in the online CSDS.

Among the findings of this review of the ARC3.1 Case Studies were:

- There were wide variations in the data types, sources, and evidentiary standards among the 28 cases reviewed in detail, making comparative analysis difficult.
- The case cities were heavily weighted toward large cities (46% of the sample of 28), highlighting issues of transferability and learning potential, given that a majority of the coming growth of urbanized areas is likely to be concentrated within medium-sized and large cities in the developing world (Dobbs et al. 2012).[1]
- There was no consistent sampling of the prevailing climatic conditions within the case cities; hence, variability, vulnerabilities, and projected or actual climate impacts could not be compared or evaluated.

The ARC3.2 Case Study guidelines presented in this Annex are an effort to deal with these and other problems of Case Study comparison and to provide a preliminary attempt to systemize the CSDS collection process.

3 Guidelines for ARC3.2 Case Studies

Based on the ARC3.1 experience and the objectives of ARC3.2, a Case Study Call for Submissions was sent to members of the UCCRN network and distributed more widely. In particular, the UCCRN sought Case Studies on climate change risks and vulnerabilities in cities, as well on as mitigation and adaptation planning and implementation, with a special emphasis on lessons learned and innovative approaches.

Following the publication of ARC3.1 and a series of scoping sessions held at international conferences on climate change and cities, the UCCRN noted several additional topics of interest from city decision-makers and stakeholders around the world, including climate disasters and risk; urban planning and design; co-benefits of mitigation and adaptation; equity and environmental justice; economics, finance, and the private sector; urban ecosystems and biodiversity; coastal zones; housing and informal settlements; and urban solid waste. As a result, the ARC3.2 volume has almost doubled in size to 16 chapters, and an effort was made to solicit case studies that include these important new topics during the development of the report.

Case Study authors were asked to submit a case study text of 700–1,000 words, not including references, figures, and tables; an abstract of a maximum of 100 words; a completed case study data collection protocol (see Annex 3 Table 1); and a list of data sources.

---

1 In this analysis, large cities are defined as those with a population of 5–10 million and medium-sized, or intermediate, cities with a population of 500,000–1 million (UN-Habitat, 2008).

Case Study submissions were encouraged to cover the following points:

- Clearly identify geography and topic addressed
- Clearly communicate relevance to climate change adaptation and mitigation in terms of action and strategy
- Include visual and textual content and show clarity in each
- Describe and discuss action/policy drivers (e.g., community as a driver, non-governmental organizations (NGOs) as drivers, local authorities or industry or businesses as a drivers of climate change adaptation)
- Describe impact and scale of potential climate changes
- Demonstrate clear lessons from which other cities and stakeholders can learn

Case Studies were submitted by scholars, city leaders and practitioners, stakeholders, and city organizations (see Appendix C) from around the world. The selection of Case Studies for the ARC3.2 report and the CSDS was based on attempting to build a collection that captures the range of socioeconomic, demographic, and geographical conditions that affect cities' vulnerability and responses to climate change. The distribution of Case Studies across topic, geography, income level, climate zone, and city size can be seen in Annex 4, Tables A–E.

3.1 ARC3.2 Case Study Data Collection Protocol

The ARC3.2 Case Study data collection protocol (Annex 3 Table 1), developed in collaboration with the University of Aalborg, Denmark, consists of a table with selected key data. The data within the data collection protocol were selected to be useful in comparative analysis. It is intended that the inputs to the data table be consistent to the extent possible across all ARC3.2 Case Studies. The purpose of the protocol is to ensure a minimum standard of comparability and reliability of the essential data that informs the Case Studies, thereby increasing the validity of any subsequent comparative analysis.

Future cross-case analyses can be expanded and supplemented by additional information such as the items noted in Annex 3 Table 2, some of which are included in the Case Study texts, or from data derived from widely available datasets such as those listed in Annex 3 Table 4.

3.2 Validation of ARC3.2 Case Studies

The integration, review, and verification of the information within the Case Studies was undertaken at the UCCRN Secretariat at Columbia University. Case Studies submitted to the CSDS underwent review by the ARC3.2 co-editors and at least one additional reviewer. The data collection protocol information was verified and sources checked before being uploaded to the online CSDS.

Annex 3 Table 1 *CSDS Data Collection Protocol Template: required information.*

ARC3.2 Case Study Data Collection Template
Case Study Title
Author(s) + institutions
City
Country
Keywords
Abstract (max. 100 words)
Area of city [km²]
Area of metropolitan region [km²]
Population (city/metropolitan region)
Density (city) [/km²]
Density (metropolitan region) [/km²]
Latitude and Longitude
Climate zone (Köppen-Gieger Climate Zones)
Topography (description)
Gross National Income (GNI) per capita (Atlas method) (national) 2017 (World Bank, 2017)
Human Development Index (national) (UNDP, 2014a,b)
Adaptation strategies (max. 25 words)
Mitigation strategies (max. 25 words)

Note For consistency, the data for Gross National Income per capita and the HDI index were taken by the UCCRN from the following sources: GNI per capita (World Bank, 2017), HDI (UNDP, 2014b). These data are regularly updated, and the figures used were those in effect at the time of Annex preparation. This information was included in all the data tables for the Case Studies and it is merely indicative; the Case Study in hand may be referring to an urban area that may be richer or poorer than the national average.

Annex 3 Table 2 *Possible additional data to expand Case Study Data Collection Protocol.*

Gini co-efficient (national)
Governance typology
Projected population and economic growth
Biome
Environmental indicators
Hazards
Vulnerabilities (projected or actual)
Impacts (projected or actual)
Risk assessment
Vulnerability assessment
Ex-post monitoring and evaluation
Funding sources for mitigation and adaptation activities

4 ARC3.2 Case Study Docking Station

The ARC3.2 CSDS consists of an online, searchable database of the ARC3.2 Case Studies. The current Beta version can be accessed at www.uccrn.org/casestudies. In addition to the online search engine, the CSDS will also allow for the selection of Case Studies on an ArcGIS-generated world map.

The database software used for the CSDS is Caspio Bridge (Version 8.5.5), a global cloud platform that allows users to create various applications without the use of coding. By customizing specific search criteria, the user is presented with a relevant list of Case Studies. Case Studies can be searched by keywords, Case Study title, city, country, population size (based on metropolitan population), Gross National Income (GNI per capita), Human Development Index (HDI), and coastal classification (see Annex 3 Figure 1). The CSDS can be searched with up to three keywords as an "OR" search; all other search categories are based on the "AND" logic. Additional search categories may be added as required.

Keywords assigned to the Case Studies may include (1) the key hazard (e.g., flood, heat wave, drought), (2) the type of adaptation or mitigation (policy, infrastructure, ecosystem-based, community-based, etc.), (3) the ARC3.2 topic/chapter in which the Case Study is found or to which it relates (e.g., Coastal Zones, Urban Health, Governance), and (4) additional keywords selected by the Case Study authors that make the Case Study easily searchable in the database.

Annex 3 Table 3 shows the city classification of the CSDS for population size (adapted from UN-Habitat, 2008), GNI per capita (Atlas method, 2016 US$; World Bank, 2017), and national HDI (UNDP, 2014a,b). For climate zones, the Köppen-Geiger classification is used (Kottek et al., 2006; Peel et al., 2007). (For projections of how the Köppen-Gieger climate zones might change under different IPCC scenarios, see Rubel and Kottek [2010]).

After a search is run in the Docking Station, Case Studies fitting the search criteria are displayed with their abstract and keywords. Details of the Case Study and a full download of the Case Study text and data collection protocol can then be accessed. Search criteria and results are still under revision.

One consistency issue should be noted: definitions of city and metro population levels are not consistent from country to country, so that these rankings used as search criteria may not yield completely equivalent results. This is an example of cross-study data collection that requires more detailed work.

Due to space constraints, the presentation of the Case Studies differs between the ARC3.2 volume and the online ARC3.2 CSDS. First, the online version of each Case Study includes the full data protocol table (Annex 3 Table 1), whereas the versions

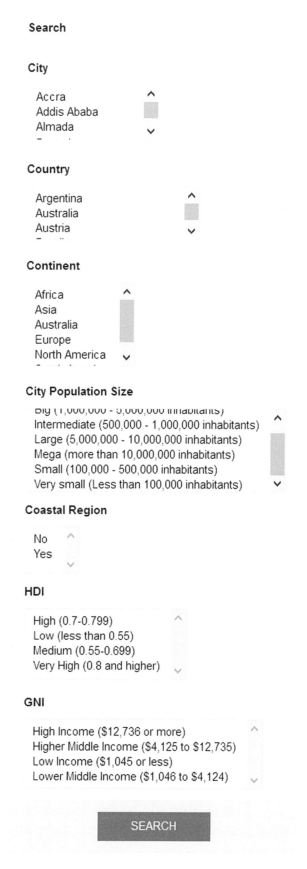

Annex 3 Figure 1 *UCCRN ARC3.2 Case Study Docking Station online search menu.*

Annex 3 Table 3 *City classification by population, GNI, and HDI.*

Criteria	Data Range	Classification
Population of Metropolitan Region	Less than 100,000	Very Small
	100,000–500,000	Small
	500,000–1 million	Intermediate
	1–5 million	Big
	5–10 million	Large
	10 million	Mega
Gross National Income per Capita	US$1,005 or less	Low Income
	US$1,006 to $3,955	Lower Middle Income
	US$3,956 to $12,235	Upper Middle Income
	US$12,236 or more	High Income
Human Development Index	0.550	Low
	0.550–0.699	Medium
	0.700–0.799	High
	0.800	Very High

Source: Adapted from UN-Habitat (2008); Gross National Income per Capita (World Bank, 2017); national Human Development Index (UNDP, 2014a)

published in this volume do not include the full data table, but do include a concise data table box with selected key data (this is also included in the online version for convenience). Second, a brief Case Study executive summary is included in the online version but not the version in printed form.

Case Studies published within the ARC3.2 report itself are distributed within the book in two ways: either placed within a chapter or appended in the back of the report in Annex 5, for reasons of space. Forty-six of the ARC3.2 Case Studies are included in the Case Study Annex of ARC3.2 (Annex 5); these, together with the 71 Case Studies in the chapters, comprise the total of 117 Case Studies in the volume.

Annex 3 Figure 2 shows the geographical distribution of Case Studies within the ARC3.2 report. Case Studies in the CSDS can be selected as described earlier, and it will also be possible to navigate an online ArcGIS world map (beta) and choose a city. A pop-up will appear with a short description of the Case Study, after which the user will be redirected to the full details, including the complete text, the abstract, and the data collection protocol.

Population of Metropolitan Area

- **Mega** (more than 10,000,000 inhabitants)
- **Large** (5,000,000 to 10,000,000 inhabitants)
- **Big** (1,000,000 to 5,000,000 inhabitants)
- **Intermediate** (500,000 to 1,000,000 inhabitants)
- **Small** (100,000 to 500,000 inhabitants)
- **Very Small** (lLess Than100,000 inhabitants)

Annex 3 Figure 2 *Geographical distribution of cities (by population) in the ARC3.2 Case Study Docking Station.*

Annex 3 Table 4 *Selected additional data sources for future research.*

	Data Source	Web Link
1	African Development Bank	http://www.afdb.org/en/knowledge/statistics/open-data-for-africa/
2	Asian Development Bank Open Database	https://sdbs.adb.org/sdbs/index.jsp
3	C40 Cities	http://www.c40cities.org/
4	Carbon Disclosure Project Cities (data is free to extract, but requires a user account)	https://www.cdproject.net/en-U.S./Pages/cities-open-data.html
5	Carbon Disclosure Project data collection	https://www.cdp.net/en-U.S./Pages/cities-open-data.html)
6	CASES [U.S., Adaptation]	http://cses.washington.edu/cig/cases
7	CIRCLE 2 [EU] Climate Adaptation Case Study Database	http://infobase.circle-era.eu/
8	City Forward	http://cityforward.org/wps/wcm/connect/CityForward_en_U.S./City+Forward/Home
9	Climate data	http://www.realclimate.org/index.php/data-sources/
10	Convention on Biodiversity (Adaptation)	http://adaptation.cbd.int/activities.shtml#sec1
11	CSIRO Open Data	https://datanet.csiro.au/dap/home?execution=e1s1
12	DataGov-Global Governance Indicators (IADB/DFID/ADB/UNDP)	http://www.iadb.org/datagob/index.html
13	EU Open Data	http://open-data.europa.eu/en/
14	European Environment Agency Climate Adaptation Database	http://climate-adapt.eea.europa.eu/web/guest/data-and-downloads#
15	Eurostat	http://epp.eurostat.ec.europa.eu/portal/page/portal/eurostat/home/
16	Gapminder [data visualization]	http://www.gapminder.org/
17	Georgetown Climate Center [U.S., Mitigation and Adaptation]	http://www.georgetownclimate.org/
18	Global Cities Indicators Facility	http://www.cityindicators.org/
19	Global City Indicators Facility	www.cityindicators.org
20	IPCC data	http://www.ipcc-data.org/maps/
21	OECD	http://stats.oecd.org/
22	Open Street Map	http://www.openstreetmap.org/#map=5/51.500/-0.100
23	UKCIP [UK] Case Study Database	http://www.ukcip.org.uk/case-studies/
24	UNDP 2014. Human Development Statistical Tables	http://hdr.undp.org/en/data
25	UNFCCC GHG data	http://unfccc.int/ghg_data/items/3800.php
26	UNFCCC Private Sector Initiative	http://unfccc.int/adaptation/workstreams/nairobi_work_programme/items/6547.php
27	UN-HABITAT	http://open.unhabitat.org/
28	United Nations Statistics Division	http://unstats.un.org/unsd/databases.htm
29	U.S. Data.gov	http://www.data.gov/opendatasites
30	weADAPT	http://weadapt.org/
31	World Bank Open Data Initiative	http://data.worldbank.org/

Annex 3 Table 4 *(continued)*

	Data Source	Web Link
32	World Bank GDP (purchasing power parity) (national)	http://data.worldbank.org/indicator/NY.GDP.MKTP.CD
33	World Bank GDP ranking table	http://data.worldbank.org/data-catalog/GDP-ranking-table
34	World Bank GNI per capita, Atlas Method (current US$)	http://data.worldbank.org/indicator/NY.GNP.PCAP.CD
35	World Health Organization	http://www.euro.who.int/en/data-and-evidence/databases

5   Conclusions

The ARC3.2 CSDS is a first step in characterizing the rich diversity in cities' responses to the increasing risks of climate change. The overall goals are to provide a platform for sharing lessons learned, inspire climate action in other cities, and enable new possibilities for research through cross-case comparison and analysis.

The ARC3.2 CSDS will be an ongoing effort to enable city stakeholders and researchers from around the world to share their experience and expertise as response pathways develop over time. The data collection protocol is not static, but represents a learning process that will become more robust as the UCCRN obtains feedback from users.

The Case Studies in the CSDS can be supplemented by additional completed Case Studies drawn from city practitioners, academics, and stakeholders from around the world and submitted to UCCRN. The establishment of the UCCRN Regional Hubs on each continent (see Annex 1) will serve to promote enhanced opportunities for geographically targeted urban climate change knowledge, analysis and information transfer via the solicitation of Case Studies on local climate hazards, vulnerabilities, and adaptation and mitigation projects and programs.

The UCCRN Case Study team recommends that climate change researchers more extensively undertake further systematic reviews and meta-analyses to help establish standard baselines of knowledge within this rapidly advancing field of research. A comprehensive systematic review will be important to better frame the basic questions that continue to arise in relation to urban climate change planning.

Case Study analyses and methodology have limitations and caveats, including questions about overall reliability of data sources, possible biases, replicability/transferability, and other methodological shortfalls. These have begun to be addressed during the course of the ARC3.2 project and constitute important continuing challenges for the future.

Acknowledgments: The UCCRN Case Study Docking Station Team wishes to acknowledge the support of the Joint European Master in Environmental Studies: Cities and Sustainability (JEMES CiSu) program for support of graduate students who worked on the CSDS.

References

Dobbs, R., Remes, J., Manyika, J., Roxburg, C., Smit, S., and Schaer, F. (2012). *Urban World: Cities and the Rise of the Consuming Class*. McKinsey Global Institute.

Doria, M. d. F., Boyd, E., Tompkins, E. L., and Adger, W. N. (2009). Using expert elicitation to define successful adaptation to climate change. *Environmental Science & Policy* 12, 810–819.

Dupuis, J., and Biesbroek, R. (2013). Comparing apples and oranges: The dependent variable problem in comparing and evaluating climate change adaptation policies. *Global Environmental Change 23*, 1476–1487.

Ford, J. D., Keskitalo, E. C., Smith, T., Pearce, T., Berrang-Ford, L., and Duerden, F., et al. (2010). Case study and analogue methodologies in climate change vulnerability research. *WIREs Climate Change 1*, 374–392.

Gerring, J. (2010). Causal mechanisms: Yes, but. *Comparative Political Studies 43*(11), 1499–1526.

Keskitalo, E. C. H. (ed.). (2010). *Developing Adaptation Policy and Practice in Europe: Multi-Level Governance of Climate Change*. Springer.

Kottek, M., Grieser, J., Beck, C., Rudolf, B., and Rubel, F. (2006). World Map of the Köppen-Geiger climate classification updated. *Meteorologische Zeitschrift* 15, 259–263.

Peel, M. C., Finlayson, B. L., and McMahon, T. A. (2007). Updated world map of the Köppen-Geiger climate classification. *Hydrology and Earth System Sciences Discussions* 4(2), 462.

Rosenzweig, C., Solecki, W., Hammer, S., and Mehrotra, S. (eds.). (2011). *Climate Change and Cities: First Assessment Report of the Urban Climate Change Research Network*. Cambridge University Press.

Rubel, F., and Kottek, M. (2010). Observed and projected climate shifts 1901–2100 depicted by world maps of the Köppen-Geiger climate classification. *Meteorologische Zeitschrift* 19, 135–141.

Yin, R. K. (2009). *Case Study Research: Design and Methods* (fourth edition). SAGE Ltd.

UNDP. (2014a). Human development report 2014. Accessed September 12, 2015: http://hdr.undp.org/sites/default/files/hdr14-report-en-1.pdf

UNDP. (2014b). Human Development Index (HDI) 2013. From 2014 Human Development Statistical Tables. Accessed October 25, 2015: http://hdr.undp.org/en/data

UN-Habitat. (2008). *State of the World's Cities 2008/2009: Harmonious Cities*. Earthscan, 12.

World Bank. (2017). 2016 GNI per capita, Atlas method (current US$). Accessed August 9, 2017: http://data.worldbank.org/indicator/NY.GNP.PCAP.CD

Annex 4

Case Studies by Category

The ARC3.2 Case Study Docking Station presents over 100 examples of what cities are doing about climate change on the ground, across a diverse set of urban challenges and opportunities. They are included in the ARC3.2 volume and incorporated into an online website, (www.uccrn.org/casestudies), a searchable database that allows exploration and examination. The ARC3.2 Case Study Docking Station is designed to inform research and practice on climate change and cities by contributing to scientifically valid comparisons across a range of social, biophysical, cultural, economic, and political factors.

Case Studies in this volume can be found either within the Second UCCRN Assessment Report on Climate Change and Cities (ARC3.2) chapters (designated by number, e.g., 2.1) or in the Case Study Docking Station (CSDS) Annex (Annex 3) (designated by a letter, e.g., 2.A). Case Studies have been categorized here by chapter topic (Table A), geography (Table B), income category[1] (Table C), climate zone[2] (Table D), and city size[3] (Table E).

Annex 4 Table A   ARC3.2 Case Studies by Chapter Topic

Note: Some Case Studies allocated to chapters are also relevant to other chapters.

Chapter 2   Urban Climate Science

Chapter 3   Disasters and Risk in Cities

Chapter 4   Integrating Mitigation and Adaptation

---

1 For consistency, the data for Gross National Income per capita and the income categories were taken by the UCCRN from the source for GNI per capita (World Bank, 2017) and country classifications by income level (World Bank, 2017). These data are regularly updated, and the figures and categories used were those in effect at the time of Annex preparation. This information was included in all the data tables for the Case Studies, and it is merely indicative; the Case Study in hand may be referring to an urban area that may be richer or poorer than the national average.
2 For climate zones, the Köppen-Geiger classification is used (Peel et al., 2007).
3 Based on metropolitan population (city sizes adapted from UN-Habitat, 2008).

Chapter 5   Urban Planning and Design

Chapter 6   Equity, Environmental Justice, and Urban Climate Change

Chapter 7   Economics, Finance, and the Private Sector

Chapter 8   Urban Ecosystems and Biodiversity

Chapter 9   Urban Areas in Coastal Zones

Chapter 10   Urban Health

Chapter 11   Housing and Informal Settlements

Chapter 12   Energy Transformation in Cities

Chapter 13   Urban Transportation

Chapter 14 Urban Water Systems

Chapter 15 Urban Solid Waste Management

Chapter 16 Governance and Policy

Annex 4 Table B   ARC3.2 Case Studies by Geography

Africa

Bobo-Dioulasso	Burkina Faso	16.D	Building a Participatory Risks Management Framework in Bobo-Dioulasso	766–768
Cairo	Egypt	6.2	Citizen-Led Mapping of Urban Metabolism in Cairo	190–192
Addis Ababa	Ethiopia	15.2	Challenge of Developing Cities: The Case of Addis Ababa	562–563
Accra	Ghana	15.B	Accra: The Challenge of a Developing a City	753–755
Maputo	Mozambique	6.5	Public-Private-People Partnerships for Climate Compatible Development (4PCCD) in Maputo	203–205
Lagos	Nigeria	13.2	Bus Rapid Transit in Lagos and Johannesburg: Establishing Formal Public Transit in Sub-Saharan Africa	506
Nairobi	Kenya	2.4	Will Climate Change Induce Malaria in Nairobi?	43–44
Dakar	Senegal	11.A	Peri-Urban Vulnerability, Decentralization, and Local-Level Actors: The Case of Flooding in Pikine/Dakar	737–739
Cape Town	South Africa	8.4	Ecosystem Based Climate Change Adaptation in the City of Cape Town	276–278
Durban	South Africa	4.1	Synergies, Conflicts, and Tradeoffs between Mitigation and Adaptation in Durban	111–112

Asia

North America

Latin America

Quito	Ecuador	12.7	Energy and Climate Change in Quito	480–482
Mexico City	Mexico	4.6	Climate Action Program in Mexico City 2008–2012	126
Lima	Perú	11.1	Water-Related Vulnerabilities to Climate Change and Poor Housing Conditions in Lima	407–408

Annex 4 Table C   ARC3.2 Case Studies by Income Category (Gross National Income per Capita – US$)

High Income ($12,236 or more per capita)

Abu Dhabi	5.4	An Emerging Clean-Technology City: Masdar, Abu Dhabi	159–161
Almada	9.C	Storm Surge in Costa da Caparica, Almada, in January 2014	728–732
Antofagasta	16.C	Democratizing Urban Resilience in Antofagasta	763–766
Antwerp	2.A	The Urban Heat Island of Antwerp	659–661
Boulder	3.1	The Boulder Floods: A Study of Decision-Centric Resilience	79–81
Brisbane	9.5	Adaptation Benefits and Costs of Residential Buildings in Greater Brisbane	342–343
Brisbane	10.4	Health Impacts of Extreme Temperatures in Brisbane	376–377
Brussels	2.1	Urban Heat Island in Brussels	32–33
Calgary	10.A	Economic Cost and Mental Health Impact of Climate Change in Calgary	732–733
Canberra	12.4	The Benefits of Large-Scale Renewable Electricity Investment in Canberra	473–474
Cape Town	8.4	Ecosystem-Based Climate Change Adaptation in the City of Cape Town	276–278
Chula Vista	4.4	Sustainable Win-Win: Decreasing Emissions and Vulnerabilities in Chula Vista, California	118–119
Cubatão	8.3	The Serra do Mar Project, Baixada Santista Metropolitan Region (BSMR), São Paulo State	272–274
Denver	14.3	Denver, Seattle, Tucson: How Can Climate Research Be Useful for Urban Water Utility Operations?	537–538
Durban	4.1	Synergies, Conflicts, and Tradeoffs between Mitigation and Adaptation in Durban	111–112
Fort Lauderdale	16.B	Fort Lauderdale: Pioneering the Way toward a Sustainable Future	761–762
Glasgow	5.1	Green Infrastructure as a Climate Change Adaptation Option for Overheating in Glasgow	152–153
Helsinki	9.A	Climate Adaptation in Helsinki	723–725
Hong Kong	5.3	Application of Urban Climatic Map to Urban Planning of High-Density Cities: An Experience from Hong Kong	157–158
Jena	4.3	Jena, Adaptation Strategy as an Essential Supplement to Climate Change Mitigation Efforts	115–118
Jerusalem	8.5	Jerusalem Gazelle Valley Park Conservation Program	282–283
Leuven	4.C	Leuven Climate Neutral 2030 (LKN2030): An Ambitious Plan of a University Town	689–691
London	7.1	The London Climate Change Partnership: Investigating Public and Private Sector Collaboration	238–239
London	13.3	London's Crossrail: Integrating Climate Change Mitigation in Construction and Operations	508–510
London	15.C	Successful Actions of London Municipality	755–757
Los Angeles	2.2	Los Angeles Megacities Carbon Project	36–37

Upper Middle Income ($3,956 to $12,235 per capita)

Lower Middle Income ($1,006 to $3,955 per capita)

Low Income ($1,005 or less per capita)

Annex 4 Table D   ARC3.2 Case Studies by Climate Zone

Equatorial

Equatorial Fully Humid (Af)

Equatorial Monsoonal (Am)

Equatorial Winter Dry (Aw)

Arid

Arid Steppe Hot Arid (Bsh)

Warm Temperate, Fully Humid, Warm Summer (Cfb)

Warm Temperate, Summer Dry, Hot Summer (Csa)

Warm Temperate, Summer Dry, Warm Summer (Csb)

Warm Temperate, Desert, Hot Summer Climate (Cwa)

Snow

Snow, Fully Humid, Hot Summer (Dfa)

Snow, Fully Humid, Warm Summer (Dfb)

Calgary	10.A	Economic Cost and Mental Health Impact of Climate Change in Calgary	732–733
Helsinki	9.A	Climate Adaptation in Helsinki	723–725
Middelfart	14.5	New Citizen Roles in Climate Change Adaptation: The Efforts of the Middle-Sized Danish City of Middelfart	542–543
Moscow	2.C	Temporal and Spatial Variability of Moscow's Urban Heat Island	664–667
Toronto	10.C	City of Toronto Flood: A Tale of Flooding and Preparedness	736–737
Vienna	15.4	Sustainable Waste Management: The Successful Example of Vienna	578–579
Warsaw	16.E	Warsaw and City Sustainability Reporting	769–771

Annex 4 Table E   ARC3.2 Case Studies by City Population

Less than 100,000

Leuven	4.C	Leuven Climate Neutral 2030 (LKN2030): An Ambitious Plan of a University Town	689–691
Ebro Delta	8.1	Coastal Natural Protected Areas in Mediterranean Spain: The Ebro Delta and Empordà Wetlands	262–264
Middelfart	14.5	New Citizen Roles in Climate Change Adaptation: The Efforts of the Middle-Sized Danish City of Middelfart	542–543

Small (100,000–500,000)

Antofagasta	16.C	Democratizing Urban Resilience in Antofagasta	763–766
Boulder	3.1	The Boulder Floods: A Study of Decision-Centric Resilience	79–81
Canberra	12.4	The Benefits of Large-Scale Renewable Electricity Investment in Canberra	473–474
Nova Friburgo	10.2	Health Care and Social Cost: Nova Friburgo, State of Rio de Janeiro	373–374
Santa Fe	3.2	Adaptation to Flooding in the City of Santa Fe, Argentina: Lessons Learned	83–84
Tacloban	3.3	Preparedness, Response and Reconstruction of Tacloban for Haiyan Super-Typhoon (ST) in the Philippines	86–88
Windsor	10.3	Windsor Heat Alert and Response Plan: Reaching Vulnerable Populations	374–375

Intermediate (500,000–1 million)

Gorakhpur	3.A	Integrating Climate Change Concerns in District Disaster Management Plans (DDMP): Case of Gorakhpur	670–672
Jena	4.3	Jena, Adaptation Strategy as an Essential Supplement to Climate Change Mitigation Efforts	115–118
Jerusalem	8.5	Jerusalem Gazelle Valley Park Conservation Program	282–283
Khulna	6.4	Individual, Communal, and Institutional Responses to Climate Change by Low-Income Households in Khulna	200–202
Khulna	9.2	Vulnerabilities and Adaptive Practices in Khulna	327–328

Big (1–5 million)

Large (5–10 million)

Mega (>10 million)

References

Peel, M. C., Finlayson, B. L., and McMahon, T. A. (2007). Updated world map of the Köppen-Geiger climate classification. *Hydrology and Earth System Sciences Discussions* 4(2), 462.

UN-Habitat. (2008). *State of the World's Cities 2008/2009: Harmonious Cities*. Earthscan, 12.

World Bank. (2017). 2016 GNI per capita, Atlas method (current US$). Accessed August 9, 2017: http://data.worldbank.org/indicator/NY.GNP.PCAP.CD

Annex 5

Case Study Annex

Case Study 2.A

The Urban Heat Island of Antwerp

D. Lauwaet, K. De Ridder, B. Maiheu, and H. Hooyberghs

Vlaamse Instelling voor Technologisch Onderzoek (VITO), Antwerp

G. Lambrechts, L. Custers, and I. Gommers

City of Antwerp

Keywords	Heat stress, urban climate model, mitigation, adaptation, climate science
Population (Metropolitan Region)	1,015,000 (Demographia, 2016)
Area (Metropolitan Region)	635 km² (Demographia, 2016)
Income per capita	US$41,860 (World Bank, 2017)
Climate zone	Cfb – Temperate, without dry season, warm summer (Peel et al., 2007)

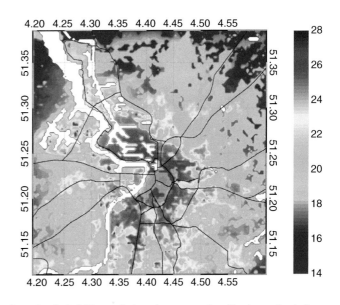

Case Study 2.A Figure 1 *Annual average number of heat wave days in the Antwerp area for the period 2081–2100, under the RCP8.5 climate scenario.*

Cities tend to be warmer than their rural surroundings, a phenomenon called the *urban heat island,* exposing urban residents to much higher levels of heat stress than people living in the nearby rural areas. At the same time, climate projections indicate that the frequency, intensity, and duration of heat waves is very likely to increase, and it is expected that, toward the end of this century, cities will be groaning under the strain of drastically increased levels of heat stress.

Assessing the Urban Heat Island of Antwerp

Commissioned by the city administration of Antwerp (Belgium), the research center VITO started to map the current and future heat stress situation in the city. Since climate projections at the scale of urban agglomerations are lacking, a new urban climate model (UrbClim™) was developed that operates at an unprecedented horizontal resolution of a few hundred meters (De Ridder et al., 2015). First, work was done to evaluate the UrbClim model with meteorological measurements from five automatic weather stations, which were installed both inside and outside the city center to assess the ability of the model to reproduce the urban heat island effect. From this, we

learned that the model achieves accuracy comparable to that of existing traditional models, but at a speed that is more than a hundred times higher. As a result, UrbClim is capable of covering periods long enough (tens of years) to deduce relevant climate statistics.

This fast model was successfully exploited, coupling UrbClim to Global Climate Models (GCMs) contained in the CMIP5 archive of the Intergovernmental Panel on Climate Change (IPCC), and conducting simulations representing present (1986–2005) and future (2081–2100) climate conditions (considering the RCP8.5 scenario). Based on the results for the present period, it was found that the urban area experiences *twice as many heat wave days* than the rural surroundings. Subsequently, when analyzing the climate projections, it was found that, toward the end of the century, *the number of heat wave days is expected to increase by a factor of nearly ten.* Given the higher number of urban heat wave days to start with, city inhabitants will be facing almost one month of heat wave conditions each year (see Case Study 2.A Figure 1).

Mitigation and Adaptation Measures

Motivated by these results, the city of Antwerp decided to implement mitigation and adaptation measures to tackle the problem of heat stress. To achieve optimal results, the measures are implemented simultaneously on three scales: (1) citywide, (2) local, and (3) the individual.

Citywide Scale

The construction of buildings in the city of Antwerp is regulated by a building code, which all inhabitants and developers need to adhere to when renovating or constructing a building for which a city permit is required. In this code, specific instructions are added that will help to reduce the heat stress in the city over time:

- For all roofs with a slope of less than 15% and a surface area of more than 20 square meters, it is obliged to install a green roof on top. This will drastically lower the temperature of the roof and, by retaining and evaporating rain water, the air temperature will be cooled. Additionally, green roofs provide extra isolation for the building.
- All private gardens and open parking lots need to be green and permeable. Only 20 square meters can be paved in gardens of 60 square meters and only one-third in gardens of 60 square meters. All outdoor private parking lots need to have a permeable grassed surface.
- The majority of the buildings in the city center have historical plaster facades. When renovated, these building fronts need to be painted in the original light, preferably white color. White buildings reflect more sunlight and will not warm up as easily as dark buildings, thereby reducing the heat radiation from these buildings.

Local Scale

Regularly, large squares, parks, and neighborhoods in the city are renovated. A new goal is to optimize the thermal

comfort of the people visiting these places. To be able to do this, detailed information is needed on the local microclimate. Therefore, VITO applied a meter-scale computational fluid dynamics (CFD) model (see Case Study 2.A Figure 2) to map the focus areas and assess the effectiveness of potential adaptation measures. It is important to consider not only the air temperatures because human thermal comfort is also affected by radiation, humidity, and wind speed. An internationally recognized indicator for thermal comfort that takes these variables into account is the predicted mean vote (PMV) (Fanger, 1982), which scales between −4 (strong cold stress) and +4 (strong heat stress).

Since simulations with a CFD model are both computationally expensive and time-consuming, it is not possible to do this for every renovation project in the city. Based on several sensitivity experiments and a review of the scientific literature, VITO listed the expected impact of a number of adaptation measures on both the 2 meter air temperature and the PMV value during a typical warm summer afternoon (Table 1). These numbers can be used as first-order estimates for cities in a comparable climate zone as Antwerp. The precise effect will depend on the local situation and should ideally be assessed with new CFD simulations or measurements.

Individual Scale

As mentioned earlier, the Antwerp urban area experiences twice as many heat wave days than the rural surroundings. This is problematic, since so-called heat health action plans are triggered using rural temperature forecasts only (as in most countries). In order to remedy this issue, VITO has set up a short-term (5-day) heat forecast system based on a

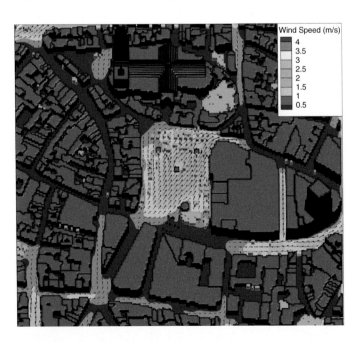

Case Study 2.A Figure 2 *Example of the output of the CFD model for the neighborhood of the cathedral of Antwerp.*

Table 1 *Estimate of the effect of several adaptation measures on mean 2 meter air temperatures and Predicted Mean Vote values during a warm day in summer. The reference situation is a location covered with asphalt and concrete.*

Adaptation measure	Mean 2 m Air Temp.	PMV value
Ref: Asphalt/concrete	=	4
Increase albedo	−0.5°C	3.5
Grass area	−0.5°C	3
Green roof/wall	−0.5°C	3
Large fountain	−1°C	3
Tree with dense crown	−0.5°C	2
Several water sprayers	−1.5°C	2
Row of trees	−1°C	1.5
Water nebulizers	−2°C	1
Large park with water	−1.5°C	0

combination of the regular European forecast model and UrbClim, which delivers specific information for each neighborhood of Antwerp taking into account the urban heat island effect. Since the information is much more detailed, the aid resources, targeting mostly the vulnerable elderly and children, can be devoted more efficiently to those places where they are needed most. Furthermore, heat wave warnings will be displayed on light panels in the streets and on the city's website where citizens can find more information on what to do in case of a heat wave.

More information can be found at the website of the Urban Climate Service Centre (http://www.urban-climate.be).

References

Climate-Data. (2015). Climate-Data.org. Accessed October 9, 2015: http://en.climate-data.org/search/

De Ridder, K., Lauwaet, D., and Maiheu, B. (2015). UrbClim - a fast urban boundary layer climate model. *Urban Climate* **12**, 21–48.

Demographia (2016) Demographia World Urban Areas (Built Up Urban Areas or World Agglomerations). 12th Annual Edition. St. Louis: Demographia. Available from: http://www.demographia.com/db-worldua.pdf.

Fanger, P. O. (1982). *Thermal Comfort*. Robert E. Krieger.

Peel, M. C., Finlayson, B. L., and McMahon, T. A. (2007). Updated world map of the Köppen-Geiger climate classification. *Hydrology and Earth System Sciences Discussions* **4**(2), 462.

World Bank. (2017). 2016 GNI per capita, Atlas method (current US$). Accessed August 9, 2017: http://data.worldbank.org/indicator/NY.GNP.PCAP.CD

Case Study 2.B

Application of Satellite-Based Data for Assessing Vulnerability of Urban Populations to Heat Waves

Stephanie Weber, Natasha Sadoff, and Erica Zell

Battelle Memorial Institute, Columbus

Alex de Sherbinin

Center for International Earth Science Information Network (CIESIN), Columbia University, New York

Keywords	Heat wave, social sensitivity, health impacts, urban heat island, adaptation, vulnerability, remote sensing
Population (Metropolitan Region)	1,567,442 (U.S. Census, 2015)
Area (Metropolitan Region)	347.3 km² (U.S. Census, 2010)
Income per capita	US$56,180 (World Bank, 2017)
Climate zone	Dfa – Cold, without dry season, hot summer (Peel et al., 2007)

Extreme heat, exacerbated by the urban heat island (UHI) effect, is a leading cause of weather-related mortality in the United States and many other countries (Wilhelmi et al., 2012). From 2006 to 2010, an average of 620 U.S. residents died each year owing to heat stroke and/or sun stroke (Barko et al., 2014). The U.S. National Climate Assessment (NCA) and the IPCC Fifth Assessment identify warmer and more extreme temperatures as one of the impacts of anthropogenic climate change (USGCRP, 2014; IPCC, 2014). Compounding the rising air temperatures and increased variability that is occurring with climate change, UHI effects can add 6–8°C to urban air temperatures when compared to surrounding rural areas for many mid-latitude cities (Imhoff et al., 2010). High population densities in urban areas and their social stratification mean that vulnerability to climate change is

also high in certain neighborhoods (Romero Lankao and Qin, 2010). Evidence shows that urban populations with higher levels of sensitivity and lower levels of adaptive capacity generally suffer

Table 2 *Advisory group members.*

	Name	Organization
Academic and Private Sector	Dana Tomlin	Department of Planning, University of Pennsylvania
	Robert Cheetham	Azavea (geospatial analysis firm)
	Shannon Marquez	Drexel University
	Raluca Ellis	Franklin Institute
	Thomas Bonner	Philadelphia Electric Company (PECO)
Policy and Urban Planning	Jeff Moran	City of Philadelphia Department of Public Health
	Palak Raval-Nelson	City of Philadelphia Department of Public Health
	Keith Davis	City of Philadelphia, City Planning Commission
	Mark Wheeler	City of Philadelphia, City Planning Commission
	Scott Schwarz	City of Philadelphia Water Department
	Mami Hara	City of Philadelphia Water Department
	Sarah Wu	City of Philadelphia, Office of Sustainability

greater impacts from a range of climate-related hazards (Reckien et al., 2013; Cutter and Emrich, 2006; Laska, 2006) including heat stress (Ueijo et al., 2011; Johnson and Wilson, 2009).

To help local and regional governments understand the vulnerability of urban populations to heat waves, we developed a set of indicators, listed in Table 1. These indicators map the elements of vulnerability and can be used to target adaptation measures and track their effectiveness.

Guided by an advisory group of local planners and experts in the pilot city of Philadelphia, Pennsylvania (listed in Table 2), we constructed the indicators to leverage the benefits of block group-level socioeconomic data and multi-decadal meteorological station data, along with the broad coverage provided by satellite remote sensing land surface temperatures (LST), land cover, and urban vegetation products.

Our methods of calculating and mapping indicators covered the three components of vulnerability: exposure of urban populations to heat waves (using NASA-derived LST and National Climatic Data Center-derived air temperature data [NCDC, 2014]), social sensitivity to heat wave impacts (tied to age, educational achievement, and race, using U.S. Census Bureau data), and adaptive capacity to cope with urban heat waves (using NASA Normalized Difference Vegetation Index [NDVI] and LST to detect results of adaptive urban "greening" projects). We calculated these indicators for a ten-year period for satellite-based data and a thirty-year period for ground-based temperature data, for Philadelphia.

Since heat-wave health impacts are tied to the duration and intensity of extreme heat, we identified heat wave periods, defined as exceeding the 85th percentile of historical average July and August temperature for Philadelphia (81°F) for three or more consecutive days. We calculated the number of heat wave days per year separately for ground-based weather stations identified as urban (6 locations) versus suburban/rural (13 locations). The monitors show that the total duration (number and length combined) of heat waves per year have been increasing from 4 to 12 days per year in urban areas and staying relatively constant at 5 days per year in suburban/rural areas from 1980 to 2010.

Complete spatial coverage of measures of LST using NASA Moderate Resolution Imaging Spectroradiometer (MODIS) Aqua data confirm elevated temperatures and upward trends in areas flagged as urban, while opposite trends are seen in surrounding suburban and rural areas. Case Study 2.B Figure 1 shows higher LSTs in urban areas, using the example of average July LSTs in 2012. Note that LST, while correlated to ambient air temperature, is not an exact indicator of sidewalk level temperature of interest in this study since the relationship between land and sidewalk-level air temperatures depends on surface type, building height, and other factors (Zhang et al., 2014).

We also mapped areas of high social sensitivity to guide adaptation efforts. Sensitive populations in Philadelphia were identified through U.S. Census American Community Survey (ACS) data and defined using the following population parameters (Ueijo et al., 2011, Johnson and Wilson, 2009):
- % below the poverty line
- % of households with a person over 65 living alone
- % of housing units built before 1960
- % of the population that did not graduate from high school

Table 1 *Indicators of vulnerability of urban populations to heat waves.*

Issue	Indicator name
Exposure	• *Urban Heat Wave Indicator:* An estimate of the intensity and total duration of heat waves for a city • *Urban Heat Island Indicator:* An estimate of the average land surface temperature (LST) difference between urban areas and rural areas for periods of extreme heat
Sensitivity	• *Urban Socioeconomic and Hotspot Indicator:* Classification of sensitivity of census units based on socioeconomic census and urban greenness data
Vulnerability	• *Vulnerability of Urban Populations to Heat Health Impacts Indicator:* Overlap of highly exposed and highly sensitive populations
Adaptive Capacity	• *Urban Adaptation Effectiveness Indicator:* Measured reductions in LST or increases in Normalized Difference Vegetation Index (NDVI) in neighborhoods related to UHI reduction measures

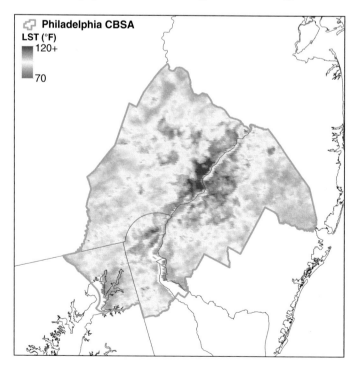

Case Study 2.B Figure 1 *Land Surface Temperature.*

Source: NASA Aqua MODIS LST, 2012

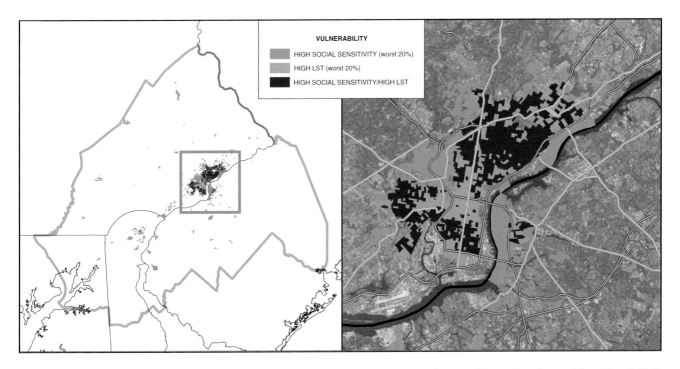

Case Study 2.B Figure 2 *Overlaying the areas with highest LST (2012, shown in orange) and the residential locations of the most sensitive populations (shown in blue) reveals the most vulnerable populations overall to heat wave impacts in Philadelphia (shown in purple).*

An overall Social Sensitivity Index was calculated for each census block group, dividing each sensitivity factor into deciles and averaging the factors. The results in Case Study 2.B Figure 2 show the location of the most sensitive population to heat waves.

Approximately 10% of Philadelphia's population lives within the most vulnerable areas to heat wave health impacts based on our established thresholds for vulnerability, thus facilitating targeting of cooling adaptation measures. One such cooling adaptation measure with a long-term benefit could be increasing urban vegetation, which can provide shade and localized cooling effects (a measure of adaptive capacity of the city to reduce vulnerability). We mapped vegetation in Philadelphia (using NASA MODIS NDVI) to identify which areas currently contain higher and lower levels of vegetation. Tracking changes in NDVI over time could highlight increases in vegetation, particularly resulting from a targeted urban greening or cooling program, or decreases in vegetation, potentially leading to increases in vulnerability. Isolated examples of urban cooling measures were provided by Philadelphia officials, but none is yet at the scale that can be measured by the satellite data used (1 km). A "reverse" example was found of the increase in LST and decrease in NDVI associated with building a large warehouse in a formerly green area, which also provides useful information on potential negative impacts of zoning policies and land-use changes.

The indicators can be used by local decision-makers in Philadelphia to better understand patterns of vulnerability, target adaptation measures, and measure results (LST reduction or NDVI increase) from existing adaptation measures (e.g., tree planting, green/white roofs). Subkilometer-scale data are needed to make these indicators more applicable in mixed decision-making urban landscapes, although the availability of such products is limited, especially for the relatively short time frame (within a few months of data collection) that is most useful to city managers.

The indicator methodology was vetted with stakeholders for different display and visualization options, such as through an interactive tool, and was applied in a second pilot city (New York) as a test of scale-up. That scale-up test showed that modifications to methods (such as the selection of temperature thresholds and designation of urban and suburban locations) may be required based on local contextual factors.

This work was funded under NASA Grant NNX13AN71G, Development and Testing of Potential Indicators for the National Climate Assessment.

References

Berko, J., Ingram, D. D., Saha, S., and Parker, J. D. (2014). Deaths attributed to heat, cold, and other weather events in the United States, 2006–2010. *National Health Statistics Report Number 76*, July 30, 2014. Accessed March 5, 2014: http://www.cdc.gov/nchs/data/nhsr/nhsr076.pdf

Cutter, S. L., and Emrich, C. T. (2006). The long road home: Race, class, and recovery from Hurricane Katrina. *Environment: Science and Policy for Sustainable Development* **48**(2), 8–20.

Imhoff, M. L., Zhanga, P., Wolfe, R. E., and Bounou, L. (2010). Remote sensing of the urban heat island effect across biomes in the continental USA. *Remote Sensing of Environment* **114**(3), 504–513.

Intergovernmental Panel on Climate Change (IPCC). (2014). Climate Change 2014: Impacts, adaptation, and vulnerability. Accessed January 28, 2015: http://www.ipcc.ch/report/ar5/

Johnson, D. P., and Wilson, J. S. (2009). The socio-spatial dynamics of extreme urban heat events: The case of heat-related deaths in Philadelphia. *Applied Geography* **29**, 419–434.

Laska, S. (2006). Social vulnerabilities and Hurricane Katrina: An unnatural disaster in New Orleans. *Marine Technology Society*, **40**(4), 1.

Peel, M. C., Finlayson, B. L., and McMahon, T. A. (2007). Updated world map of the Köppen-Geiger climate classification. *Hydrology and Earth System Sciences Discussions* **4**(2), 462.

National Climatic Data Center (NCDC). (2014). Global Surface Summary of the Day (GSOD), 1980 to 2013 [Data set]. NOAA/NESDIS/National Climatic Data Center, Asheville, North Carolina, Accessed October 15, 2014: https://data.noaa.gov/dataset/global-surface-summary-of-the-day-gsod

Reckien, D., Wildenberg, M., and Bachhofer, M. (2013). Subjective realities of climate change: How mental maps of impacts deliver socially sensible adaptation options. *Sustainability Science* **8**, 159–172.

Romero-Lankao, P., and Qin, H. (2010). Conceptualizing urban vulnerability to global climate and environmental change, *Current Opinion in Environmental Sustainability* **3**(3), 142–149.

Uejio, C. K., Wilhelmi, O. V., Golden, J. S., Mills, D. M., Gulino, S. P., and Samenow, J. P. (2011). Intra-urban societal vulnerability to extreme heat: The role of heat exposure and the built environment, socioeconomics, and neighborhood stability. *Health & Place* **17**, 498–507.

U.S. Census Bureau. (2010). Decennial census, summary file 1. Accessed September 18, 2015: http://www.census.gov/population/metro/files/CBSA%20Report%20Chapter%203%20Data.xls

U.S. Census Bureau. (2016). QuickFacts: Philadelphia City, Pennsylvania. Retrieved November 3, 2016, from http://www.census.gov/quickfacts/table/PST045215/4260000

U.S. Global Change Research Program (USGCRP). (2014). National Climate Assessment. Access January 28, 2015: http://nca2014.globalchange.gov/

Wilhelmi, O., de Sherbinin, A., and Hayden, M. (2012). Exposure to heat stress in urban environments. In Crews, K., and B. King (eds.), *Ecologies and Politics of Health*. Routledge.

World Bank. (2017). 2016 GNI per capita, Atlas method (current US$). Accessed August 9, 2017: http://data.worldbank.org/indicator/NY.GNP.PCAP.CD

Zhang, P., Bounoua, L., and Imhoff, M. 2014. Comparison of MODIS land surface temperature and air temperature over the continental USA meteorological stations. *Canadian Journal of Remote Sensing: Journal Canadien* **40**(2).

Case Study 2.C

Temporal and Spatial Variability of Moscow's Urban Heat Island

Pavel Konstantinov and Eugenia Kukanova

Lomonosov Moscow State University

Mikhail Varentsov

Lomonosov Moscow State University & A. M. Obukhov Institute of Atmosphere Physics RAS

Keywords	Urban heat island, urban climatology
Population (Metropolitan Region)	16,570,000 (Demographia, 2016)
Area (Metropolitan Region)	5,310 km² (Demographia, 2016)
Income per capita	US$9,720 (World Bank, 2017)
Climate zone	Dfb – Cold, without dry season, warm summer (Peel et al., 2007)

Moscow is a very large megalopolis where, according to data from the year 2010 (Russian Federal State Statistics Service, 2010) more than 11 million people live. In July and August 2010, there was a blocking anticyclone over Moscow as well as the central part of European Russia, with anomalously hot weather conditions for a long time. This heat wave lasted from July 4 until August 18. During this period, the warmest day for Moscow in the 143-year period of direct meteorological measurements was recorded on July 29 (+39.0°C in the city center; Konstantinov et al., 2014; Kislov and Konstantinov, 2011). This heat wave led to a rise in mortality rate up to 11,000 (Shaposhnikov et al., 2015), so it was decided to investigate the urban heat island (UHI) phenomenon of Moscow and its climatology due to its propensity to amplify heat wave intensity.

We chose for the basic UHI characterization "UHI intensity." This is defined in Equation (1) as the simple difference between air temperature at an urban station and mean air temperature in a rural area.

$$UHI_{intensity} = T_{urban} - (T_{ruralNord} + T_{ruralSud} + T_{ruralEast} + T_{ruralWest})/4 \qquad (1)$$

The map of the WMO meteorological stations in the Moscow region (Balchug is an classical urban station situated near the Moscow River and the Kremlin) and the elevation can be seen in Case Study 2.C Figure 1. The time period for the investigation was from January 1, 2000 to December 31, 2012. This period was chosen due to the availability of high-quality data for all stations.

Diurnal variation of UHI intensity in different seasons is shown in Case Study 2.C Figure 2. The peak intensity is during the summer and at night. In classical urban climatology (Oke, 1988), the biggest air temperature differences between a city center and its suburbs occur (in the temperate climate zones) in the autumn and at night. Apparently, this phenomenon in Moscow is caused by high-density development in the city center and warm summer conditions.

In autumn and winter, UHI intensity is lower than in the warm seasons. Cities with continental climates display higher UHI than those in environments where sea breezes may have a cooling effect.

The climatology of extreme UHI intensity values (defined as cases with air temperature differences higher than 5°C) is described in Case Study 2.C Figure 3. The frequency and total number of extreme values are higher in the morning and in winter months (especially February). This might be caused by anticyclonic conditions, the probability of which rises at the end of the winter (due to intensification of the Siberian High). The total number of cases with extreme UHI intensity values in February and in the summer months is similar. The month with the lowest

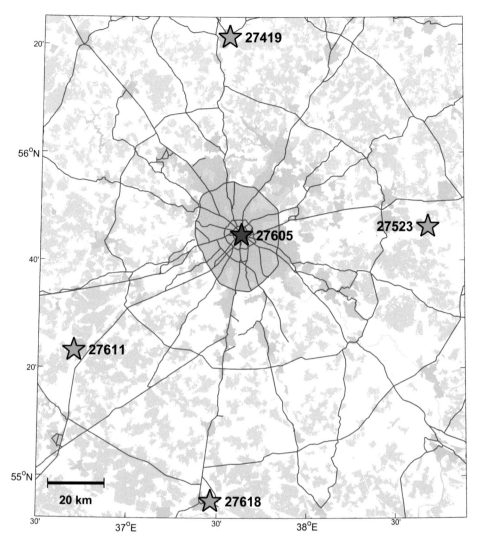

Case Study 2.C Figure 1 *Moscow region's meteorological stations map. Distance between urban and rural stations: 72 km, 92 km, 68 km, 69 km. Data available for period 2000–2012.*

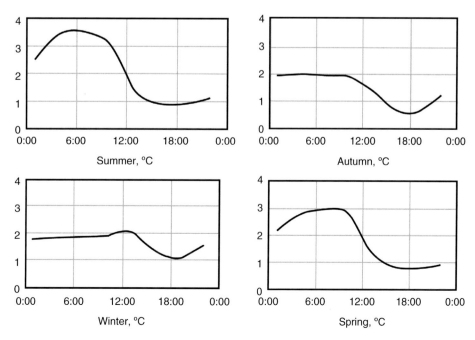

Case Study 2.C Figure 2 *Diurnal urban heat island intensity variations for different seasons in Moscow (2000–2012).*

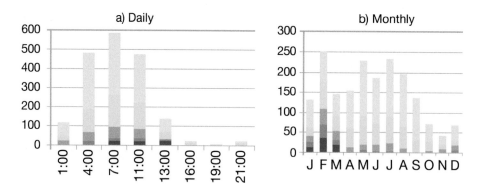

Case Study 2.C Figure 3 *Extreme urban heat island (UHI) intensity values: Number of cases with UHI$_{intensity}$ 5°C daily (a) and monthly (b). Color scale reflects five intervals with intensity increase: [5–6]; [6–7]; [7–8]; [8–9]; [9–∞) – from light yellow to red.*

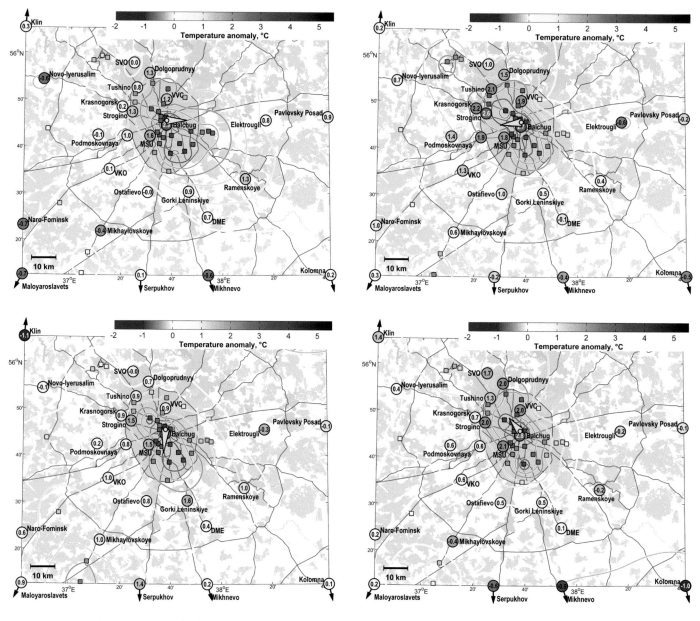

Case Study 2.C Figure 4 *Spatial urban heat island zone displacement due to cases with unified wind direction averaged for the warm season of 2014: (a) western wind, (b) eastern wind, (c) northern wind, and (d) southern wind.*

UHI intensity frequency is November, which also is unusual in comparison with classical theory (Oke, 1988).

Finally, according to the physics of the UHI phenomenon, it can be slightly "blown with the wind," as shown at Case Study 2.C Figure 4. During the warm season of 2014 (May–September), averaged situations with weak (0–1.5 m/s) winds demonstrate that the displacement of the UHI zone is clearly observed, in agreement with theory (Landsberg and Maisel, 1972).

Conclusion

Moscow demonstrated a strong UHI, with a mean annual intensity of 1.8°C in 2000–2012. (Kukanova and Konstantinov, 2014) The strongest UHI conditions occur in the spring and summer months, but frequency of extreme UHI intensity values (5°C) is higher in winter months (especially February). Spatial displacement of the UHI may occur in warm months due to wind influence.

Acknowledgments

This study was supported by the Russian Foundation for Basic Research grants RFBR 13-05-41306-RGO_a, RFBR 15-35-21129-mol_a_ved, RFBR 14105-31384-mol_a, and RFBR 15-55-71004 Arctica_a. The work of Dr. Konstantinov is partially supported by a President of Russia grant for young PhD scientists MK-6037.2015.5.

References

Demographia (2016) Demographia World Urban Areas (Built Up Urban Areas or World Agglomerations). 12th Annual Edition. St. Louis: Demographia. Available from: http://www.demographia.com/db-worldua.pdf.

Kislov, A. V., and Konstantinov, P. I. (2011). Detailed spatial modeling of temperature in Moscow. *Russian Meteorology and Hydrology* **36**(5), 300–306.

Konstantinov P. I., Varentsov M. I., and Malinina E. P. (2014, December). Modeling of thermal comfort conditions inside the urban boundary layer during Moscow's 2010 summer heat wave (case-study). *Urban Climate* **10**(3), 563–572.

Kukanova, E. A., and Konstantinov, P. I. (2014). An urban heat islands climatology in Russia and linkages to the climate change. In *Geophysical Research Abstracts*. EGU General Assembly Vienna, Austria, vol. **15**, p. 10833.

Landsberg, H. E., and Maisel, T. N. (1972). Micrometeorological observations in an area of urban growth. *Boundary-Layer Meteorology* **2**, 365–370.

Oke, T. R. (1988). *Boundary Layer Climates* (second ed.). Methuen.

Peel, M. C., Finlayson, B. L., and McMahon, T. A. (2007). Updated world map of the Köppen-Geiger climate classification. *Hydrology and Earth System Sciences Discussions* **4**(2), 462.

Shaposhnikov D., et al. (2015). Long-term impact of Moscow heat wave and wildfires on mortality. *Epidemiology* **26**(2), e21–22.

World Bank. (2017). 2016 GNI per capita, Atlas method (current US$). Accessed August 9, 2017: http://data.worldbank.org/indicator/NY.GNP.PCAP.CD

Case Study 2.D

Adaptation of the STEVE Tool (Screening Tool for Estate Environmental Evaluation) to Sydney Conditions

M. Bescansa and P. Osmond

Faculty of Built Environment, University of New South Wales, Sydney

S. K. Jusuf

Center for Sustainable Asian Cities, National University of Singapore

N. H. Wong

Department of Building, National University of Singapore

Keywords	Urban heat island, urban planning, STEVE Tool, urban climate science
Population (Metropolitan Region)	4,540,000 (United Nations, 2016)
Area (Metropolitan Region)	2,037 km² (Demographia, 2016)
Income per capita	US$54,420 (World Bank, 2017)
Climate zone	Cfb – Temperate, without dry season, warm summer (Peel et al., 2007)

Australia is particularly vulnerable to the extreme weather events associated with anthropogenic climate change. The City of Sydney and other local government areas that comprise Australia's largest conurbation – the Sydney metropolitan area – are actively collaborating on mitigation and adaptation strategies.

Cities must deal with an additional phenomenon, which exacerbates the effects of global warming. The urban heat island (UHI) effect occurs when a city area experiences significantly warmer temperatures than the surrounding countryside. This localized warming is due to the absorption and trapping of incident solar radiation by buildings and paved surfaces, together with the anthropogenic heat production characteristic of densely urbanized areas. Rising city temperatures can lead to adverse conditions such as risks to public health, poor air quality, and high energy use. Several recent studies have elaborated prediction models in order to quantify and address this phenomenon (e.g., Grimmond et al., 2010, Mirzaei and Hahighat, 2010). A sustainable approach to urban development appears as a key driver to ameliorate UHI effects and reduce building energy consumption.

Research has found that temperatures in Sydney have been rising for the past 40 years above what would be expected through global warming. By 2050, the global warming effect in combination with the UHI phenomenon could increase Sydney

667

temperatures by up to 3.7°C (Argüeso et al., 2013). Along with the many efforts being made to address the UHI effect in Australian cities, the present study aims to investigate how the urban planning process in the city of Sydney can actively incorporate climatology data in order to reach a more sustainable urban development and measure its impact on the environment. This research has been undertaken thanks to a collaborative agreement with researchers from the National University of Singapore who have developed a web-based application (STEVE Tool: The Screening Tool for Estate Environment Evaluation) able to bridge the gap between urban climatology and the urban planning process.

The STEVE Tool is a software tool that can estimate the outdoor thermal comfort (temperature and humidity) performance of developments based on urban form, vegetation, and weather data. It incorporates an empirical model for air temperature prediction with the main objective of evaluating the impact of development on the air temperature and, ultimately, on the UHI phenomenon. It consists of three interfaces: estate existing condition, estate proposed masterplan, and a calculator for air temperature prediction. Based on input of urban morphology and climate data, the STEVE Tool can calculate the air temperature of an existing or planned urban development.

This application has been initialized for Singapore conditions. However, the tool itself appears amenable to initialization for any latitude, provided the data are available. To investigate the tool's adaptability to the Sydney urban context, two scenarios were tested for the same central Sydney location. The first one corresponds to the site before redevelopment, with no vegetation and low building density. The second scenario responds to the site masterplan, incorporating new buildings and landscape features. The hypothesis is that urban morphology and greenery can modify the air temperature and potentially help mitigate the UHI effect.

The project involved the following steps:
1. **Site selection.** A recently redeveloped area in inner Sydney, the Central Park precinct, was selected for the first application of the STEVE Tool. Central Park is a 5.8 hectare urban renewal project 2 kilometers from the Sydney Central Business District (CBD). Formerly dominated by a brewery operation, the new development features medium- and high-rise mixed-use buildings and a large park.
2. **Creation of a 3D model of the site.** 3D models of the pre- and post-development scenarios (see Case Study 2.D Figure 2) were built following the architectural drawings (noting that as the project is staged, some works are yet to be completed).
3. **Insertion into the model of the species indicated in the proposed landscape plans.** Since the plants specified in the plans are mostly Australian and the STEVE Tool has been originally designed for the Singaporean context, new species had to be added to the tool database. This can be done by providing specific data about the species: girth, Leaf Area Index (LAI), and type of vegetation (whether it is a tree or shrub) for each new item. With the species added, the final model was ready to be tested for different scenarios.
4. **Export of both scenarios to the STEVE web application and comparative analysis of air temperature results.** The final stage of the study will evaluate the differences in modeled air temperature and compare these with empirical data collected in previous projects on this site. The software allows the user to insert climate-related values such as the existing air temperature and the solar radiation for any given location. According to these variations, the STEVE Tool will recalculate the air temperature, taking urban morphology parameters (buildings, vegetation and streets) into account.

The results of this project are based on the study of the influence of two major variables relevant to UHI mitigation: the influence of vegetation and the influence of urban morphology on air temperature. Outcomes are expected to be consistent with previous tests of the STEVE Tool in Singapore. Further applications of the tool in Australia will require deeper modification of the program through the calculation of specific air temperature prediction models for each city's conditions based on collection of empirical meteorological data.

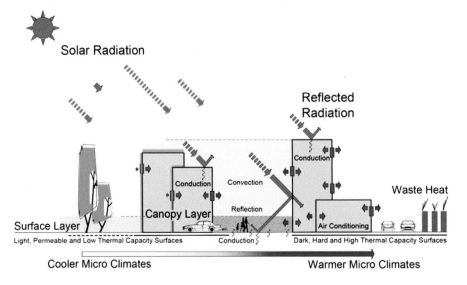

Case Study 2.D Figure 1 *The urban heat island effect in Sydney.*

Source: Sharifi and Lehmann, 2014

(a)

(b)

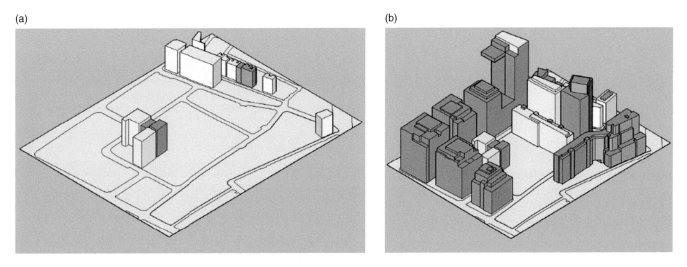

Case Study 2.D Figure 2 *Sydney's Central Park: 3D models of the estate's existing condition and the estate's proposed master plan.*

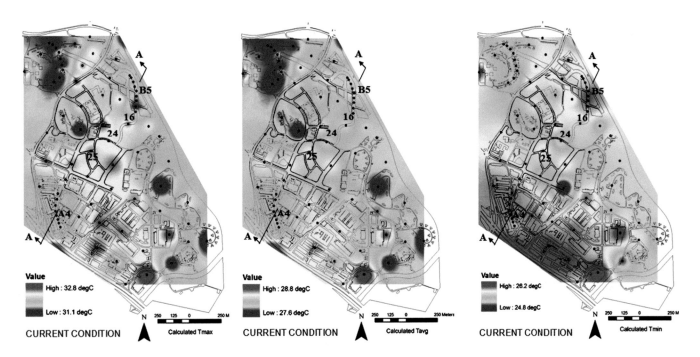

Case Study 2.D Figure 3 *Temperature maps produced by the STEVE Tool for Singapore conditions.*

Source: Jusuf and Wong, 2012

Case Study 2.D Figure 3 shows resulting temperature maps for a similar study undertaken in Singapore, where different climate parameters have been tested for the same urban scenario. The UHI phenomenon can be visualized in the areas colored in red.

Expected Outcomes

The initial conditions scenario shows a low-density urban form with low building heights, which can result in higher daytime air temperatures. As discussed by Jusuf et al. (2011), low building density combined with an absence of vegetation increases solar gain. However, a sparse configuration of buildings (high sky-view factor) facilitates heat release from urban surfaces in the absence of trapping structures.

The post-development scenario responds to the site master plan, incorporating new buildings and landscape features as shown in Case Study 2.D Figure 2. Following the architectural and landscaping plans, the increased building density will be accompanied by the existence of a large park and significant plantings of several tree species. According to Jusuf and Wong (2011), increased building height may not always lead to higher ambient air temperatures when an optimum distribution of greenery, building configuration, and a ratio of paved to unpaved surfaces is met. Higher buildings can provide shade and moderate air temperature, reducing the daytime UHI; conversely, they can increase the nocturnal UHI by trapping heat at night. Vegetation can play a decisive role in developments combining high buildings and substantial open space because it provides shading and

669

evaporative cooling. An additional model of the estate's proposed master plan but excluding greenery will be tested to further evaluate the influence of vegetation on ambient temperature.

Conclusion

The STEVE Tool has the potential to be of major relevance to cities aiming for more sustainable urban form. Through its straightforward implementation, planning professionals can experiment with different design options to help reduce ambient air temperature in densely populated spaces. Together with measures to address anthropogenic heat production, the STEVE Tool can have substantial impact on the mitigation of the UHI effect through facilitating more efficient urban design.

References

Argüeso, D., Evans, J. P., Fita, L., and Bormann, K. J. (2013). Temperature response to future urbanization and climate change. *Climate Dynamics* **42**(7–8), 2183–2199.

Demographia (2016) Demographia World Urban Areas (Built Up Urban Areas or World Agglomerations). 12th Annual Edition. St. Louis: Demographia. Available from: http://www.demographia.com/db-worldua.pdf.

Grimmond, C. S. B., Blackett, M., Best, M. J., Barlow, J., Baik, J. -J., Belcher, S. E., Bohnen-stengel, S. I., Calmet, I., Chen, F., Dandou, A., Fortuniak, K., Gouvea, M. L., Hamdi, R., Hendry, M., Kawai, T.,

Kawamoto, Y., Kondo, H., Krayenhoff, E. S., Lee, S. -H., Loridan, T., Martilli, A., Masson, V., Miao, S., Oleson, K., Pigeon, G., Porson, A., Ryu, Y. -H., Salamanca, F., Shashua-Bar, L., Steeneveld, G. -J., Trombou, M., Voogt, J., Young, D., and Zhang, N. (2010). Initial result for phase 2 of international urban energy balance model comparison. *International Journal of Climatology* doi:10.1002/joc.2227.

Jusuf, S. K., and Wong N. H. (2012). Development of empirical models for an estate level air temperature prediction in Singapore. *Journal of Heat Island Institute International* **7**(2), 111–125.

Jusuf, S. K, Wong, N. H., Tan, C. L., and Tan, A. Y. K. (2011). STEVE Tool, Bridging the gap between urban climatology research and urban planning process. *Proceedings of the International Conference on Sustainable Design and Construction*, Kansas City, Missouri, March 23–25, 2011.

Mirzaei, P. A., and Hahighat, F. (2010). Approaches to study urban heat island—abilities and limitations. *Building and Environment* **45**, 2192–2201.

Peel, M. C., Finlayson, B. L., and McMahon, T. A. (2007). Updated world map of the Köppen-Geiger climate classification. *Hydrology and Earth System Sciences Discussions* **4**(2), 462.

Sharifi, E., and Lehmann, S. (2014). Comparative Analysis of Surface Urban Heat Island Effect in Central Sydney. *Journal of Sustainable Development* **7**(3), 2014.

United Nations, Department of Economic and Social Affairs, Population Division (2016). The World's Cities in 2016 – Data Booklet (ST/ESA/SER.A/392).

World Bank. (2017). 2016 GNI per capita, Atlas method (current US$). Accessed August 9, 2017: http://data.worldbank.org/indicator/NY.GNP.PCAP.CD

Case Study 3.A

Integrating Climate Change Concerns in District Disaster Management Plans (DDMP): Case of Gorakhpur

Shiraz A. Wajih and Nivedita Mani

Gorakhpur Environmental Action Group (GEAG)

Keywords	Flood, adaptation, disaster risk reduction, disaster management plan
Population (Metropolitan Region)	735,000 (Demographia, 2016)
Area (Metropolitan Region)	54 km² (Demographia, 2016).
Income per capita	US$1,680 (World Bank, 2017)
Climate zone	Cwa – Temperate, dry winter, hot summer (Peel et al., 2007)

Gorakhpur is one of the most flood-prone districts in the mid-Gangetic region in eastern India. Although its inhabitants are accustomed to twice yearly flooding during the monsoon seasons, data from the past 100 years show a considerable increase in the intensity and frequency of floods, which are now recurring every 3–4 years and even annually in some

areas. The pilot program was initiated to incorporate climate change considerations into disaster management planning. The aim of the initiative by the District Disaster Management Authority was to respond effectively to more frequent and extreme flooding by planning proactively to minimize the loss of life and property damage. The program, spearheaded by Gorakhpur Environmental Action Group (in collaboration with the National Institute of Disaster Management – Government of India and Institute for Social and Environmental Transition [ISET] International), has improved understanding of how climate change impacts will be manifested at the subnational level. The program team presented relevant scientific analysis of climate change projections in a form that conveyed the urgency, relevance, and implications of climate change to the district's plans and programs.

Large-scale inundation has now become a common feature for the people of Gorakhpur. Historical records date back to 1823, and similar destructive floods have occurred in 1839, 1873, 1889, and 1892. The 1998 flood, in which the Ghaghara and Rapti Rivers along with their many tributaries exceeded their danger levels, caused unprecedented damage. The subsequent embankment failures and drainage congestion disrupted normal life for more than 90 days. With large-scale landscape change from human encroachment, improper infrastructure, urban development and embankment construction, and ineffective flood moderation systems in large dams, medium to high intensity flooding has become more frequent even with average rainfall levels.

India's National Disaster Management Act (2005) provides for constitution of District Disaster Management Authorities

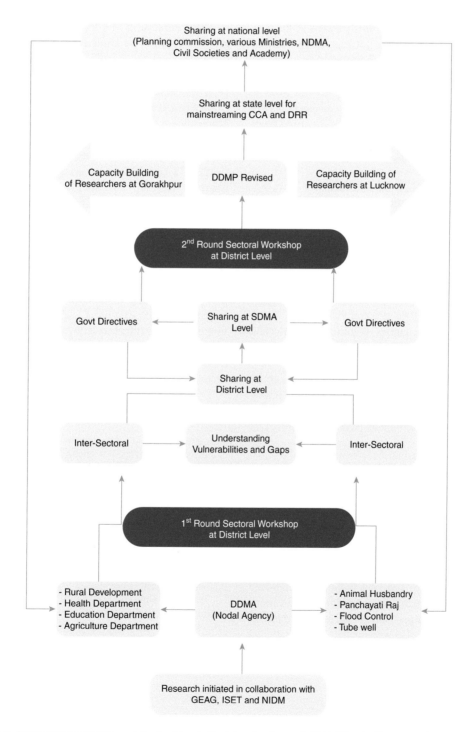

Case Study 3.A Figure 1 *Climate change adaptation and disaster risk reduction Integration Process in District Disaster Management Plan, Gorakhpur.*

Source: Wajih, S. A., and Chopde, S., 2014

(DDMAs), which are entrusted with developing and implementing a District Disaster Management Plan (DDMP) in consultation with all relevant departments. Accordingly, the Gorakhpur DDMA has been constituted and has prepared a DDMP. The Plan used to be focused mainly on disaster response coordination among agencies (i.e., after a flood), with some emphasis on pre-disaster preparedness activities. However, it lacked a systematic approach to hazard risk and vulnerability analysis, and it needed to focus more on pre-disaster risk reduction.

As well, various studies note that flooding patterns in the area are changing and climate projections point to significant changes in patterns of extreme rainfall events in the future. For example, an analysis predicts an increase in intensities of rainfall events of up to 33%, especially for events lasting 12 and 24 hours, and for all return periods (2, 10, and 50 years). This is consistent across all six global climate models used in the analysis for Gorakhpur. To be effective, disaster management planning must include both current and projected climate change impacts. Also, preliminary gap analysis using the Climate Resilience Framework (Tyler and Moench, 2012), which helps assess climate exposure, systems, institutions, and change agents, shows an oversimplified understanding in the DDMP of vulnerability issues and their root causes.

The experience is based on the concept of the Urban Climate Resilience Framework developed under Asian Cities for Climate Change Resilience Network (ACCCRN), supported by the Rockefeller Foundation and the objectives of the CDKN-START program:

- Understanding the systemic factors within the flood-prone Gorakhpur District that contribute to resilience or exacerbate vulnerability
- Understanding specific policy innovations that could help to bridge the vertical gap between the integrated national policy framework and local contexts and the horizontal gap between actions within sectorial development programs to integrate disaster risk reduction and climate change adaptation practice

Developing the relevant capacities of city departments and researchers on climate change adaptation and disaster risk reduction.

The Gorakhpur DDMP has been formulated with the cooperation, collaboration, and participation of different departments and public representatives so that the loss during disasters should be reduced pre-, during- and post-disaster (Case Study 3.A Figure 1). The work plan of the respective departments was also incorporated for reducing losses caused by a disaster. The recommendations and the plan were shared at the state and national level in India for further replication and scaling up. The national government developed a training module on the basis of the Gorakhpur DDMP experience that has been shared with more than 600 districts in India.

As a result of strong buy-in and effective coordination, the program has gone beyond simply making recommendations to publishing a climate-sensitive DDMP. The project demonstrates that a suite of effective initiatives led by credible organizations can result in policy change (Wajih and Chopde, 2014). In Uttar Pradesh, a similar alchemy occurred during the program's rollout. The United Nations Development Programme (UNDP), in close coordination with the state government and the State Disaster Management Authority (SDMA), has supported a capacity-building project for 9,000 *Gram Panchayats* (rural village councils) in the state to develop Village Disaster Management Plans. The networking fostered by the process helped sensitize the SDMA to aspects of integrating disaster risk reduction and climate change adaptation. This in turn led the Gorakhpur DDMA to issue an order for integrating disaster risk reduction into departmental annual development plans. This led to the release of several state government orders to further the disaster risk reduction and climate agenda.

Because of the key enabling factors just highlighted, the program was able to exceed its initial goals. But it also faced some challenges at the district level, such as lack of comprehensive understanding of vulnerability and its contributing factors, as well as a lack of a clear and systematic plan in departments to collect and synthesize relevant data on vulnerability, lack of effective horizontal coordination among departments, and lack of availability of climate projections, downscaled and interpreted in a meaningful way.

References

Demographia (2016) Demographia World Urban Areas (Built Up Urban Areas or World Agglomerations). 12th Annual Edition. St. Louis: Demographia. Available from: http://www.demographia.com/db-worldua.pdf.

Peel, M. C., Finlayson, B. L., and McMahon, T. A. (2007). Updated world map of the Köppen-Geiger climate classification. *Hydrology and Earth System Sciences Discussions* **4**(2), 462.

Tyler, S., and Moench, M. (2012). A framework for urban climate resilience. *Climate and Development* **4**(4), 311–326. doi:10.1080/17565529.2012.745389

Wajih, S. A., and Chopde, S. (2014). Integrating climate change concerns in disaster management planning: The case of Gorakhpur, Uttar Pradesh, India. *Climate and Development Knowledge Network*.

World Bank. (2017). 2016 GNI per capita, Atlas method (current US$). Accessed August 9, 2017: http://data.worldbank.org/indicator/NY.GNP.PCAP.CD

Case Study 3.B

Climate Vulnerability Map of Rome 1.0

A. Filpa, S. Ombuen, F. Benelli, F. Camerata,
L. Barbieri, and V. Pellegrini

*Research Group on Urban Climate Change Adaptation,
Department of Architecture, Roma Tre University, Rome*

F. Borfecchia, E. Caiaffa, M. Pollino,
L. De Cecco, S. Martini, and L. La Porta

UTMEA, ENEA, Rome

Keywords	Rainwater flooding, policy-based adaptation, vulnerability, urban planning
Population (Metropolitan Region)	3,738,000 (United Nations, 2016)
Area (Metropolitan Region)	1,114 km² (Demographia, 2016)
Income per capita	US$31,590 (World Bank, 2017)
Climate zone	Csa – Temperate, hot summer, dry summer (Peel et al., 2007)

Public awareness about climate change adaptation in cities is still limited in Italy: few local authorities have started working toward this aim so far. However, the Italian Ministry of Environment is currently finalizing the National Adaptation Strategy. Moreover, Rome has been included in the first thirty-two cities of the world funded by the Rockefeller Foundation to develop a Resilience Strategy; activities are starting. Therefore, it is likely that, in the near future, the awareness of both citizens and public authorities on adaptation of cities is going to increase.

Within this evolving context, the Department of Architecture of Roma Tre University and the UTMEA department of ENEA (Italian National Agency for New Technologies, Energy and Sustainable Economic Development) have started a joint research project aimed at testing a *quick* yet efficient and *reproducible* procedure that can provide – swiftly and with limited resources – a clear framework of the main climate vulnerability issues of a city.

The purpose of this work is not to suggest adaptation policies or measures, but rather to lay the foundation for future, tailor-made adaptation policies by public authorities. In other words, local policy-makers will be able to use the map as a means to better understand their priorities in terms of adaptation needs. Therefore, the outcome of this research is intended for use by policy-makers, but it will eventually benefit citizens,

enterprises, and the nonprofit sector by increasing awareness of climate vulnerability.

As of May 2015, the first advanced result of this research is the Climate Vulnerability Map of Rome 1.0 (CVMR 1.0): the map is at its first stage and is still open to improvements. Despite the limited available data, the map seeks to show the degree of climate vulnerability of the city. As the map is improved, the city council will be able to use it as a starting point for its future climate strategies. Moreover, a more detailed work is currently under way on a closer focus: a Roman neighborhood that has recently undergone a serious rainwater flooding event is being studied in order to refine the general methodology by using more detailed and comprehensive data, with the aim of proposing more specific urban policy measures and assessing their possible effects on local resilience.

The chosen methodology is similar, though simplified and adapted to the urban scale, to the one used in *Climate Change and Territorial Effect on Regions and Local Economies* developed by the ESPON 2013 Programme (ESPON, 2011). The main concepts – *Exposure, Sensitivity, Impacts, Adaptive Capacity, Vulnerability* – have been incorporated to allow for future integrations.

Within this first phase, the focus has been on the identification and mapping, at an infra-urban scale, of those urban characteristics causing sensitivity to climate phenomena and of those that contribute to relieving its impacts and thus improve the city's resilience. Working on an infra-urban scale requires disaggregated data; because these are not always sufficiently detailed, the research group decided to limit the chosen variables and to use proxies when necessary.

Some stakeholders have been involved in the data collection phase: in particular, the Office for Civil Protection of the City Council of Rome has been interviewed and has provided geo-referenced data on rainwater flooding events of the past years. As regards the next phase of the research – focusing on the critical neighborhood mentioned earlier and its possible adaptation measures – the local citizens' associations, political parties, and submunicipal government level will be involved in collecting all local interests on spatial development and urban policies and analyzing them with the purpose of suggesting improvements and modifications in line with adaptation needs.

To better appreciate the spatial variability of climate vulnerability, the administrative area (Municipality of Rome) has been subdivided into spatial units (SUs). These are the minimum units containing spatial and statistical data, with variable dimensions, corresponding to a neighborhood or part of it, and they identify

673

homogeneous parts of the city in terms of function, urban morphology, and building typology. In this phase, only residential, commercial, and industrial settlements have been taken into account. In the future, the SUs could also be used as areas for the implementation of adaptation policies and actions. Thus, the city council and other local actors will be able to implement climate change adaptation in Rome.

As regards exposure analysis, two phenomena have been considered: the increase in summer temperatures and precipitation intensity. Therefore, the exposure of the city to the intensification of heat wave and flood risks caused by extreme rain events has been assessed.

The predictions regarding such phenomena come from climatic models corresponding to emission scenarios, but the spatial resolution of such models makes them of little use at a local scale unless a substantial downscaling is elaborated. Therefore, the research group sought to render the spatial differentials of

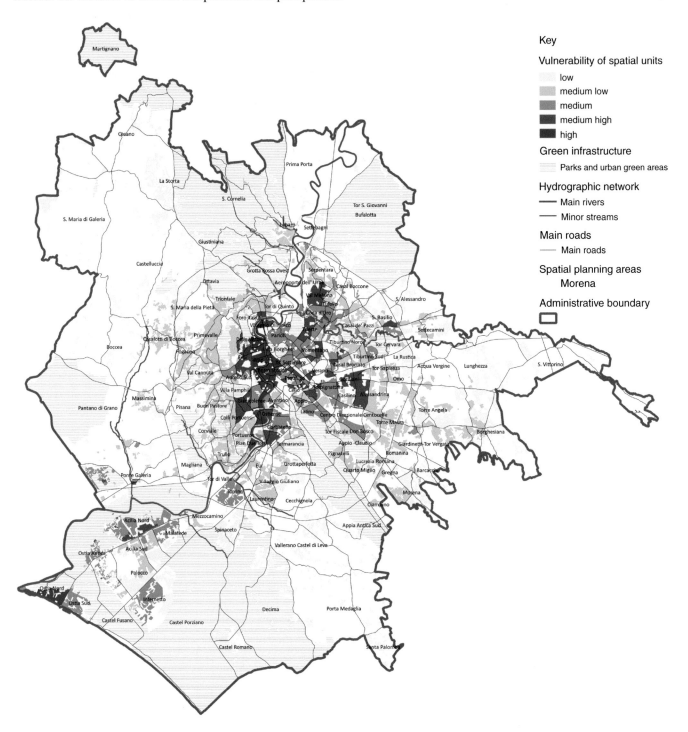

Case Study 3.B Figure 1 *Map of the overall climate vulnerability of Rome, considering the main elements of impact, vulnerability, resilience, and exposure to the threats of summer heat islands, heavy rain floodings, river floodings; 2014 update.*

exposure to chosen phenomena by using observed events and risk maps as proxies:

E1 – exposure to summer night temperatures (retrieved by satellite thermal observations in July 2003; MODIS, 2013)

E2a – exposure to rainwater flooding (maps of rainwater floods observed by the city department for civil protection; Comune di Roma, 2007)

E2b – exposure to river flooding (flood risk maps elaborated by Autorità di Bacino del fiume Tevere (Basin Authority of the Tiber River, 2009a and 2009b)

The next step was an analysis of the degree of sensitivity for each SU. Based on the available literature, three sensitivity factors representing urban and demographic characteristics were selected:

S1 – land cover: use, density and continuity of settlements (Regione Lazio, 2003)

S2 – population density (ISTAT, 2013)

S3 – percentage of elderly population (ISTAT, 2013)

S – aggregate sensitivity (sum of the three above factors)

The *impact analysis* relates exposure and sensitivity elements, aiming to highlight the gravity of possible impacts on each SU. The impact levels for each phenomenon have been calculated as a function of the respective exposure and aggregate sensitivity. Three impact indicators have been elaborated:

I1 – heat wave impact

I2a – rainwater flood impact

I2b – river flood impact

The research team decided to replace the term "adaptive capacity" with "resilience" to emphasize that the research refers to the physical structure of the urban socioeconomic system rather than to resources.

Three *resilience* factors, linked to residual natural elements that characterize the different parts of the city, were considered:

R1 – proximity to green infrastructures (degree of adjacency of each SU to green urban areas and wooded areas; derived from Regione Lazio, 2003)

R2 – presence of vegetation (estimated by using the Normalised Difference Vegetation Index derived from satellite images from Landsat, 2013)

R3 – percentage of permeable soil (inverse value of the soil sealing index elaborated by the EEA at a European level; derived from EEA, 2013)

However, since not all resilience factors influence each impact, three phenomenon-specific aggregate resilience indices have been elaborated for the calculation of vulnerability in the final step:

RS1 – Resilience to summer night temperatures (including all factors)

RS2a – Resilience to rainwater flooding (including the first and third factors)

RS2b – Resilience to river flooding (including the third factor)

Finally, three *vulnerability* indices (V1, V2a, V2b) were calculated, referring to the three phenomena plus an index of aggregate vulnerability (V). The single vulnerability indices were calculated by multiplying the impact indices by the respective phenomenon-specific resilience indices, while the aggregate vulnerability index resulted from the sum of the single vulnerability indices. The CVMR 1.0 is the map of the aggregate vulnerability index and is shown in Case Study 3.B Figure 1.

References

Autorità di Bacino del fiume Tevere. (2009a). Piano stralcio per il tratto metropolitano del Tevere da Castel Giubileo alla foce - PS5 (G.U. n. 114 del 19 Maggio 2009)

Autorità di Bacino del fiume Tevere. (2009b). Primo stralcio funzionale – Aree soggette a rischio di esondazione nel tratto Orte – Castel Giubileo -PS1. (D.P.C.M. del 3 Settembre 1998)

Comune di Roma (2007). Piano Generale di emergenza di Protezione Civile – Scenari di rischio predefiniti e Procedure per l'attivazione e l'intervento dell'Ufficio Extradipartimentale della Protezione Civile e delle Strutture Operative Comunali e di Supporto" – Applicazione Deliberazione G.C. 1099/99 e O.P.C.M. 3606 del 28/08/2007 Accessed May 7, 2015: https://www.comune.roma.it/resources/cms/documents/PIANO_GENERALE_Parte_prima.pdf

Comune di Roma. (2008). Piano Generale di Emergenza di Protezione Civile. Accessed May 7, 2015: https://www.comune.roma.it/resources/cms/documents/PIANO_GENERALE_Parte_prima.pdf

Demographia (2016) Demographia World Urban Areas (Built Up Urban Areas or World Agglomerations). 12th Annual Edition. St. Louis: Demographia. Available from: http://www.demographia.com/db-worldua.pdf.

European Environment Agency (EEA.) (2013). Urban soil sealing 2006. Accessed September 20, 2013: http://www.eea.europa.eu/data-and-maps/data/eea-fast-track-service-precursor-on-land-monitoring-degree-of-soil-sealing

European Spatial Planning Observation Network (ESPON). (2011). Climate change and territorial effects on regions and local economies. Applied Research 2013/1/4, Final Report | Version 31/5/2011, Scientific Report. Accessed May 14, 2014: http://www.espon.eu/export/sites/default/Documents/Projects/AppliedResearch/CLIMATE/ESPON_Climate_Final_Report-Part_C-ScientificReport.pdf

Peel, M. C., Finlayson, B. L., and McMahon, T. A. (2007). Updated world map of the Köppen-Geiger climate classification. *Hydrology and Earth System Sciences Discussions* 4(2), 462.

Regione Lazio. (2003). Carta dell'uso del suolo. Scala 1:25.000. Edizioni Selca, Firenze, pp. 4. https://lpdaac.usgs.gov/data_access/data_pool

Telemart. (2013). TeleatlasMultiNet Digital street network. Accessed September 20, 2015: http://www.tele-mart.com/teleatlas_multinet.php

United Nations, Department of Economic and Social Affairs, Population Division (2016). The World's Cities in 2016 – Data Booklet (ST/ESA/SER.A/392).

World Bank. (2017). 2016 GNI per capita, Atlas method (current US$). Accessed August 9, 2017: http://data.worldbank.org/indicator/NY.GNP.PCAP.CD

The MODIS image from 23 July 2003 was retrieved from the online Data Pool, courtesy of the NASA Land Processes Distributed Active Archive Center (LP DAAC), USGS/Earth Resources Observation and Science (EROS) Center, Sioux Falls, South Dakota, https://lpdaac.usgs.gov/data_access/data_pool.

The Landsat image from 23 July 2003 is available from the U.S. Geological Survey.

Case Study 3.C

Naples, Italy: Adaptive Design for an Integrated Approach to Climate Change and Geophysical Hazards

Mattia Federico Leone

University of Napoli Federico II, Naples

Giulio Zuccaro

PLINIVS-LUPT Study Center, Naples

Keywords	Geophysical hazards, volcanic risk mitigation, building technologies, energy efficiency, adaptive design
Population (Metropolitan Region)	3,114,000 (SVIMEZ, 2016)
Area (Metropolitan Region)	1,023 km² (Demographia, 2016)
Income per capita	US$31,590 (World Bank, 2017)
Climate zone	Csa – Temperate, dry summer, hot summer (Peel et al, 2007)

Complex eruptive scenarios, such as sub-Plinian or Plinian eruptions, produce extremely variable impacts on constructions depending on the specific time history of the event, on the existing building typologies, and on their level of vulnerability. This specific approach has been recently formalized to evaluate the impact of a sub-Plinian eruption in the Vesuvius and Campi Flegrei area (Zuccaro et al., 2008) through the development of the *PLINIVS Volcanic Impact Simulation Model*, a GIS model for the dynamic simulation of impact scenarios, able to evaluate the cumulative damage distribution in time and space from different eruptive phenomena such as earthquakes, pyroclastic flows, and ash fall. Within the main "mitigation scenarios" identified (Zuccaro and Leone, 2012), aimed at the development of local regulations, a priority action concerns the structural reinforcement of roofing systems or the superposition of pitched roofs

to reduce damages from ash fall, especially for areas more distant from the crater not affected by pyroclastic flows, for which, however, overloads between 500 and 1,400 kg/m² are expected (Case Study 3.C Figure 1).

Based on the cost parameters identified in the development of the model for the assessment of economic impacts of eruptive scenarios (Zuccaro et al., 2013) it is possible to compare the costs for the implementation of ash fall mitigation measures with the "avoided cost" of the reconstruction of collapsed buildings. The cost-effectiveness of the mitigation actions is evident in case of ash fall impact since the avoided costs of demolition and reconstruction are in terms of billions of euros against mitigation costs at least an order of magnitude lower. However, the uncertainty related to the wind direction at the time of the eruption does not make such mitigation scenarios feasible since, in the absence of a destructive event, these represent only a cost factor without significant additional benefits.

The integrated approach to disaster risk and climate change frames the mitigation of volcanic risk in the context of a broader strategy aimed at upgrading the energy efficiency of the built environment (contributing to climate change mitigation) and reducing environmental impacts and soil consumption (increasing adaptive capacity). In this context, the need for widespread and cost-effective actions on the built environment requires that the adopted solutions, both with reference to energy retrofitting and risk mitigation objectives, are based on well-established and reliable technologies.

A design simulation has been carried out in the context of preliminary studies related to the development of the new Urban and Building Code for the City of Poggiomarino (Naples, Italy), considering a sample residential building where the risk mitigation from ash fall is connected to an energy retrofitting that provides an increase in gross floor area through the construction

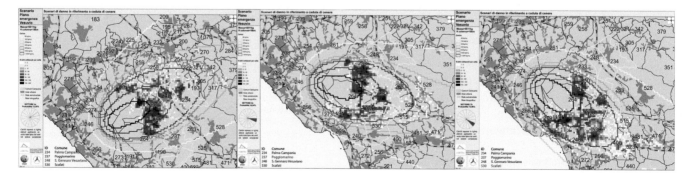

Case Study 3.C Figure 1 *Most probable ash fall scenarios following an eruption of Vesuvius.*

Source: PLINIVS Study Center, University of Napoli Federico II

of an attic space, thus combining the economic benefits deriving from property value increase, the energy savings for insulation of roofs, and the production of energy from photovoltaic panels.

A sloping roof with a cold formed steel structure is overlapped on the existing flat roof thus minimizing the overload on the underlying structure. The insulated and micro-ventilated roof offers high energy performance complementary with the structural retrofitting intervention. Additional economic and energy benefits come from the insertion of high-efficiency photovoltaic modules, with a cost per kWp significantly reduced due to the possibility of integrating the modules in the CFS substructure. The solution does not present any particular complexity in the design and implementation, but it requires the verification of expected loads and connections to the existing structure to provide a further contribution in terms of improving seismic resistance, thus realizing the so-called *box effect*.

Considering a 20-year period, compared to a cost of intervention of €75,000 (US\$88,359), it is possible to estimate significant energy benefits, which in economic terms determine an net present value (NPV) of about €205,000 (US\$241,516), in addition to the increment in the value of the property due to the increased surface area, equal to about €55,000 (US\$64,797). Based on these data, also taking into account the aleatory definition of certain parameters and the need of further details for a complete economic analysis,

it is possible to estimate the return of the initial investment in few years, with particularly relevant cash flows added to the primary benefit of the ability of the building to withstand expected overloads from ash fall (Case Study 3.C Figure 2).

Other relevant co-benefits in terms of climate change adaptation come from the opportunity of integrating rainwater harvesting and recycling systems in the building basement and collecting water from the sloped roof to be reused for multiple purposes (e.g., evaporative coolers; toilet flushing; car washing; indoor plant, pet, and livestock watering; lawn and garden irrigation).

Case Study 3.C Figure 3 shows a possible integrated design approach where new technical systems (solar thermal panels exchange boilers, additional boilers, PV inverters) are located in the attic space, while the lift shaft hosts ducts for connections to the residential units (recycled rainwater for toilet flushes and irrigation; hot/cold water for sanitary uses and radiant floor heating/cooling systems).

This additional retrofitting option is aimed primarily at reducing the amount of water discharged into the sewage system, which aggravates the impacts of flash floods. The increased frequency of extreme precipitation events represents critical risk conditions for municipalities in the Vesuvius area, where the effects of climate change are exacerbated by past and ongoing

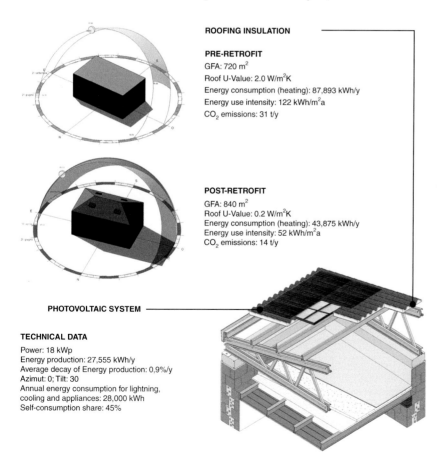

ROOFING INSULATION

PRE-RETROFIT

GFA: 720 m²
Roof U-Value: 2.0 W/m²K
Energy consumption (heating): 87,893 kWh/y
Energy use intensity: 122 kWh/m²a
CO_2 emissions: 31 t/y

POST-RETROFIT

GFA: 840 m²
Roof U-Value: 0.2 W/m²K
Energy consumption (heating): 43,875 kWh/y
Energy use intensity: 52 kWh/m²a
CO_2 emissions: 14 t/y

PHOTOVOLTAIC SYSTEM

TECHNICAL DATA

Power: 18 kWp
Energy production: 27,555 kWh/y
Average decay of Energy production: 0,9%/y
Azimut: 0; Tilt: 30
Annual energy consumption for lightning, cooling and appliances: 28,000 kWh
Self-consumption share: 45%

FINANCIAL AND ENERGY RESULTS

ROOFING INSULATION

Cost of intervention: €40,560
Fiscal incentives: 65%
Potential energy saving: 44,018 kWh/y
Fuel type: natural gas (0.9 €/mc)
Fuel saving: 4,428 mc/y (4,030 €/y)
25 year NPV: €83,092
25 year IRR: 16.2%
Cumulated cash flow (not discounted): €127,728

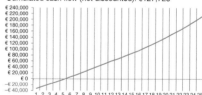

PHOTOVOLTAIC SYSTEM

Cost of intervention: €33,000 (1,833 €/KWP)
Incentives: 50% fiscal deduction+spot exchange rate
Energy production (kWh/y): 27,755
Cost of energy: 0.19 €/kWh
Annual increase in energy costs: 6%
25 year NPV: €122,447
25 year IRR: 25%
Cumulated cash flow (not discounted): €221,473

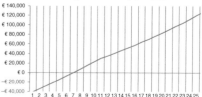

Case Study 3.C Figure 2 *Synthesis of adaptive design technical solutions and cost-benefit analysis for the reference building in located in Poggiomarino Municipality.*

Source: Leone and Zuccaro, 2014

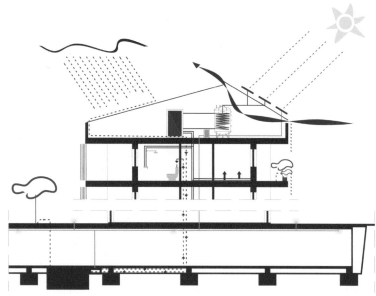

Solar thermal panels sizing	Floor	Stairwell	Building
n. of people	7	42	252
Average daily consumption (l)	350	2,100	12,600
Panel surface (m²)	6	36	**216**
Accumulation volume (l)	400	2,400	
DHW needs average covering	66%		

PV panels sizing	
Average annual solar radiation on 30° surface (az. SSE,SSO)	1,662 kWh/m²a
Calculation value (average system losses: 5%)	1,580 kWh/m²a
Panels surfaces	108 m²
Panels number	67
Single panel power (1,0x1,6 m)	308 Wp
Total power	21.3 kWp
Average annual energy production	33,600 kWh/a

Case Study 3.C Figure 3 *Rainwater harvesting/recycling system and additional solutions for technical systems retrofitting for the reference building in Poggiomarino Municipality.*

Source: M. F. Leone

Case Study 3.C Figure 4 *Pluvial flood event in Scafati and Poggiomarino.*

Source: Local Press. Il Gazzettino Vesuviano, 22.01.2014

local territorial dynamics such as urban sprawl and widespread soil sealing (Case Study 3.C Figure 4).

References

Demographia (2016) Demographia World Urban Areas (Built Up Urban Areas or World Agglomerations). 12th Annual Edition. St. Louis: Demographia. Available from: http://www.demographia.com/db-worldua.pdf.

Leone, M. F., and Zuccaro, G. (2014) The mitigation of volcanic risk as opportunity for an ecological and resilient city. *TECHNE – Journal of Technology for Architecture and Environment* **7**, 101–108.

Peel, M. C., Finlayson, B. L., and McMahon, T. A. (2007). Updated world map of the Köppen-Geiger climate classification. *Hydrology and Earth System Sciences Discussions*, **4**(2), 462.

SVIMEZ – Associazione per lo sviluppo dell'industria nel Mezzogiorno (2016). Rapporto SVIMEZ 2016 sull'economia del Mezzogiorno. Accessed November 2, 2016: http://www.svimez.info/images/RAPPORTO/materiali2016/2016_11_10_linee_appendice_statistica.pdf

World Bank. (2017). 2016 GNI per capita, Atlas method (current US$). Accessed August 9, 2017: http://data.worldbank.org/indicator/NY.GNP.PCAP.CD

Zuccaro, G., Leone, M. F., Del Cogliano, D., and Sgroi, A. (2013). Economic impact of explosive volcanic eruptions: A simulation-based assessment model applied to Campania region volcanoes. *Journal of Volcanology and Geothermal Research* **266**, 1–15.

Zuccaro, G., and Leone, M. F. (2012). Building technologies for the mitigation of volcanic risk: Vesuvius and Campi Flegrei. *Natural Hazards Review (ASCE)* **13**(3), 221–232.

Zuccaro, G., Cacace, F., Spence, R. J. S., and Baxter, P. J. (2008). Impact of explosive eruption scenarios at Vesuvius. *Journal of Volcanology and Geothermal Research* **178**, 416–453.

Case Study 3.D

Surat: The Value of Ad-Hoc Cross-Government Bodies

Manuel Pastor

School of International and Public Affairs (SIPA), Columbia University, New York

Keywords	Floods, early warning systems, sharing knowledge, disaster and risk
Population (Metropolitan Region)	5,685,000 (Demographia, 2016)
Area (Metropolitan Region)	233 km² (Demographia, 2016)
Income per capita	US$1,680 (World Bank, 2017)
Climate zone	Aw – Tropical savanna (Peel et al., 2007)

Flooding is considered the most common stress in Surat (Arup, 2014). With a population of 4.8 million in 2011, the city has experienced one of the highest growth rates in India over the past decade, almost doubling in size between 2001 and 2011. About 20.9% of the population lives in slum areas (Sawhney, 2013), which tend to be along the city waterways and encroach

on about 60% of Surat's public land (Tanner et al., 2009). Surat's main economic activities encompass textile manufacturing and diamond cutting and polishing industries that account for 54% of total employment. Trade and commerce of manufactured goods are the second main source of employment, accounting for a total share of 24% (SMC, 2011). The average annual rainfall in the city is 1,894 millimeters (NDMA, 2010), and climate change scenarios show a projected 200–500 millimeter annual increase in rainfall in the city, thus increasing the risk of flooding (TARU, 2010).

There are two main types of flood events in Surat. The first type, the so-called Khadi floods, takes place along the two streams that go through the city (see Case Study 3.D Figure 1) causing limited levels of damage (SMC, 2011).

The second type, the Ukai Dam floods, tends to cause greater impacts. The dam, which is located about 100 kilometers upstream of the city, is essential for irrigation in the surrounding agricultural area. The increasing demand for water requires maximizing its storage. Therefore, the dam is managed so that it

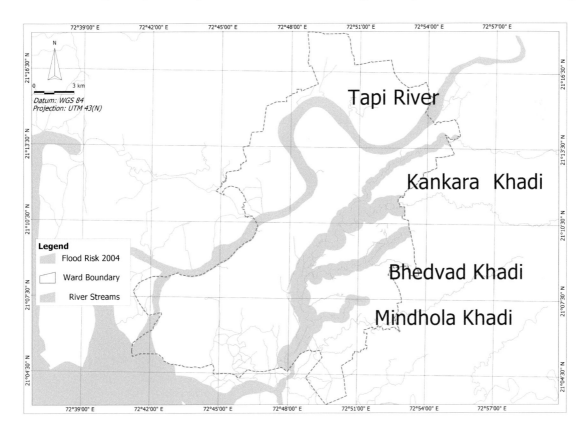

Case Study 3.D Figure 1 *Flood-prone areas in Surat due to Khadi floods.*

Source: SMC, 2011

holds the maximum amount of water by the end of the monsoon season (SMC, 2011). The downside of this policy is that extreme rainfall events can trigger the sudden release of high volumes of water in short periods of time, leading to severe flooding in the city (TARU, 2013). Since the construction of the dam in 1972, there have been four major flooding episodes (1994, 1998, 2002, and 2006) following emergency discharges of the Ukai Dam (525, 700, 325, and 909×10^3 cubic meters/sec, respectively) (Joshi et al., 2012). The flood episode of 2006 was probably one of the worst experienced by Surat in generations. More than 80% of the city was flooded, affecting 2 million people, of which two-thirds lived in low-income areas. People remained without food and drinking water for 4 days (SAARC, 2010). Damages to public infrastructure added up to US$544 million, and industrial losses amounted to US$3.5 billion (Karanth and Archer, 2014).

Case Study 3.D Figure 2 shows the areas within Surat prone to floods following a discharge of the Ukai Dam of 500×10^3 cubic meters/sec (SMC, 2011).

Ukai Dam floods are partly attributed to poor management of the reservoir. City authorities have committed significant infrastructure investments to reduce the risk of floods through its urban drainage systems and flood defenses (da Silva et al., 2012). However, flood control measures rely on Ukai Dam management, which lies beyond Surat's authority. The catchment area spans three states, and further measures require the collaboration between different state administrations (Karanth

and Archer, 2014). Earlier, there were limited platforms to share learning between institutions at different administrative levels, but no integrated actions were taken (TARU, 2013). In addition, there was a lack of information exchange between weather forecasters, dam managers, and Surat's authorities. Ukai Dam discharge was a reactive process based solely on water levels in the reservoir (da Silva et al., 2012).

Recognizing the absence of communication between departments at various administrative levels, Surat authorities established in 2009 the City Advisory Committee (CAC), an ad-hoc cross-government body responsible for coordinating disaster risk reduction initiatives and sharing information among multiple stakeholders (SMC, 2011). In 2010, the CAC led a number of state consultations with key stakeholders to gain a better understanding of the nature of floods in Surat and to discuss effective approaches to flood management (Karanth and Archer, 2014).

As a result of these meetings, the importance of information exchange was acknowledged for a better-informed decision-making process to manage the floods. The need became apparent to both share and generate inputs of rainfall forecasts together with hydraulic analyses of the catchment area to inform future emergency discharges (da Silva et al., 2012). Given the trans-boundary nature of these increasing risks, Surat authorities created the Surat Climate Change Trust (SCCT), a collaborative platform to facilitate data collection and foster dialogue, joint deliberation, and action among the parties involved (Karanth and Archer, 2014).

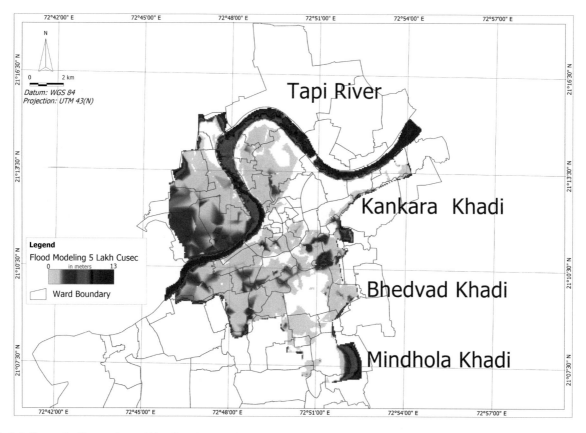

Case Study 3.D Figure 2 *City areas that could be affected by a discharge of the Ukai Dam of 500×10^3 cubic meters/sec.*

Source: SMC, 2011

The SCCT played a crucial role in creating an end-to-end early warning system – one of the short-term strategies proposed by Surat Municipal Corporation – which would span the states of Madhya Pradesh, Maharashtra, and Gujarat (SMC, 2011). A number of data transfer mechanisms were generated among the catchment area, the Ukai Dam, and the city. Reservoir inflow and outflow prediction models were also improved through the installation of several weather stations (TARU, 2013).

The early warning system increased the time available for cities to respond from 1 day to nearly 4 days (Bhat, 2011). In addition, the flow of data among multiple institutions helped to minimize peak water discharges of Ukai Dam, thus reducing flood intensity in Surat (TARU, 2013). In 2013, floods carrying a greater volume of water than those of 2006 had a much lesser impact on the city (Arup, 2014).

The present case study on Surat's end-to-end early warning system shows the importance of the participation of multiple stakeholders when designing climate change adaptation strategies. Furthermore, the case exemplifies how ad-hoc cross-government bodies such Surat's CAC or SCCT can be extremely effective in harmonizing collaborations between different administrations while increasing urban resilience.

References

Arup. (2014). *City Resilience Index: Research Report Volume 2. Fieldwork Data Analysis*. Ove Arup & Partners International Limited.

Bhat, G. K. (2011). Coping to resilience – Indore and Surat, India. Accessed October 7, 2015: http://resilient-cities.iclei.org/fileadmin/sites/resilient-cities/files/Resilient_Cities_2011/Presentations/GKB_ACCCRN_ICLEI_Congress_FIN.pdf

Da Silva, J., Kernaghan, S., and Luque, A. (2012). A systems approach to meeting the challenges of urban climate change. *International Journal of Urban Sustainable Development* 4(2), 125–145.

Demographia (2016) Demographia World Urban Areas (Built Up Urban Areas or World Agglomerations). 12th Annual Edition. St. Louis: Demographia. Available from: http://www.demographia.com/db-worldua.pdf.

Joshi, P., Sherasia, N., and Patel, D. (2012). Urban flood mapping by geospatial technique a case study of Surat City. *IOSR Journal of Engineering* 2, 43–51.

Karanth, A., and Archer, D. (2014). Institutionalising mechanisms for building urban climate resilience: Experiences from India. *Development in Practice* 24(4), 514–526.

NDMA. (2010). *National Disaster Management Guidelines: Management of Urban Flooding*. National Disaster Management Authority, Government of India.

Peel, M. C., Finlayson, B. L., and McMahon, T. A. (2007). Updated world map of the Köppen-Geiger climate classification. *Hydrology and Earth System Sciences Discussions* 4(2), 462.

SAARC. (2010). Urban risk management in South Asia. New Delhi: SAARC Disaster Management Center. Accessed October 7, 2015: http://preventionweb.net/go/14785

Sawhney, U. (2013). Slum population in India: Extent and policy response. *International Journal of Research in Business and Social Science* 2(1), 47–56.

SMC. (2011). *Surat City Resilience Strategy*. Surat Municipal Corporation.

Tanner, T., Mitchell, T., Polack, E., and Guenther, B. (2009). Urban governance for adaptation: Assessing climate change resilience in ten Asian cities. *IDS Working Papers* 2009(315), 01–47.

TARU. (2010). Asian city climate change resilience network: India chapter phase 2: City vulnerability analysis report Indore & Surat. TARU Leading Edge Pvt. Ltd and Surat Municipal Corporation. Accessed October 7, 2015: http://www.imagineindore.org/resource/VA_Report_27Feb/ACCCRN%20II%20Report%20Draft%2025Feb10.pdf

TARU. (2013). *End-to_End Early Warning System for Ukai and Local Floods in Surat City*. TARU Leading Edge Pvt. Ltd and Surat Municipal Corporation.

World Bank. (2017). 2016 GNI per capita, Atlas method (current US$). Accessed August 9, 2017: http://data.worldbank.org/indicator/NY.GNP.PCAP.CD

Case Study 3.E

Digital Resilience: Innovative Climate Change Responses in Rio de Janeiro

Andrés Luque-Ayala and Simon Marvin

Durham University, Durham, UK

Keywords	Resilience monitoring system, emergency response, climate change adaptation
Population (Metropolitan Region)	11,835,708 (IBGE, 2015)
Area (Metropolitan Region)	5,328.8 km² (IBGE, 2015)
Income per capita	US$42,390 (World Bank, 2017)
Climate zone	Am – Tropical monsoon (Peel et al., 2007)

Over the past 15 years, the city of Rio de Janeiro, Brazil, has gained a reputation for being at the forefront in the development of innovative climate change responses. The city's first greenhouse gas (GHG) emissions inventory dates back to 2000, produced shortly after joining ICLEI's Cities for Climate Protection campaign. By 2011, the city had inscribed voluntary emission reduction targets within local law (Law 5,248), aiming for an ambitious 8%, 16%, and 20% reduction by 2012, 2016, and 2020, respectively–against a 2005 baseline. The initiatives directed at achieving this goal were detailed in the 2011 Action Plan for the Reduction of GHG Emissions, collaboratively developed with civil society organizations, academia, and representatives of local industry (Prefeitura do Rio de Janeiro, 2011a). These ideas were further developed with the support of the World Bank and established the foundations for the 2013 Rio de Janeiro Low Carbon City Development Program (World

Bank, 2013). Consequently, it was not a surprise when, in 2014, the city's mayor, Eduardo Paes, assumed the chairmanship of the C40 Cities Climate Leadership Group. But while progress was steadily being made on climate mitigation, a growing concern for the city's climate-related activity was the gradual increase in the frequency and strength of extreme rain events and the resulting vulnerability of the city to flooding and landslides. This case study briefly reviews how the city, in response to these and other threats, embraced narratives around resilience and coupled strategies for responding to climate change with the use of urban digital technologies of control: the mobilization of ICT systems toward both providing urban services and influencing citizen's behavior and the course of key events in the city. It is an example of how climate change narratives and initiatives transcend environmental issues, potentially playing a significant role in emerging forms of ecological and social control.

In early April 2010, the state of Rio de Janeiro experienced a traumatic rain event that resulted in large areas of the city being flooded and hundreds of landslides (New York Times, 2010). While it is common for Rio de Janeiro to experience a high level of rainfall during the month of April, the 288 millimeters of rain that fell between April 5 and 6 represented more than the total rain average for the entire month and was the highest amount of rain ever recorded for a period of 24 hours (Prefeitura do Rio de Janeiro, 2011b). The emergency left a death toll of more than 210 people, with the media reporting more than 15,000 homeless people and a cost of US$12 billion to the nation's economy. The city of Rio de Janeiro, the state's capital, was significantly impacted. The city's main roads were flooded; power, gas, and

water supplies disrupted; and commercial activity paralyzed. Public transport collapsed as hundreds of bus passengers were rescued by fire crews. After 24 hours of continuous rain, the city's Mayor expressed that "The situation is chaos" and ordered all citizens to stay home so that emergency services could focus on helping those in greatest need (BBC, 2010). The death toll affected primarily *favela* inhabitants, low-income residents living in the informal settlements of Rio's hills.

During the event, a major preoccupation of the Mayor was the limited ability of municipal officers to provide and manage emergency response services in a highly disrupted city. As described by a staff member of the municipality, "We didn't have a site from where to manage the city; [a site] with all the required information. Each department functioned in a different locality, and every time we hit a calamity or a crisis it was very difficult to respond" (Interview, 2014).

Once the April 2010 crisis was over, the Mayor enlisted software and ICT companies in the development of an urban operations center capable of increasing the level of integration between different municipal agencies and, in this way, strengthen the city's ability to respond to emergencies. In 2011, the city inaugurated its Center of Operations (also known as COR), a large-scale control room aimed at "interconnecting the information of several municipal systems for visualization, monitoring, analysis and response in real time" (Prefeitura do Rio de Janeiro, 2011b: 14). With more than eighty customizable computer monitors forming a gigantic screen, the room resembles a NASA control room rather than a typical municipal office (Case Study 3.E Figure 1).

Case Study 3.E Figure 1 *Digital resilience: The COR's main control room area.*

Photo: Andrés Luque-Ayala

Its main screens are constantly monitoring the city and its operations through video images captured by more than 800 cameras and maps displaying geo-referenced urban data. These include weather patterns, waste collection functions, public transport movements, and even the location of each of the city's municipal guards. From the COR, city officials make decisions on how to manage the city's everyday infrastructure flows and, when needed, respond to emergencies.

With more than 400 staff members belonging to over forty different organizations, the COR's key abilities are based on its capacity to provide horizontal integration and coordination. In contrast to more traditional control rooms, where a vertical approach emphasizes command and control for a single type of utility or service, the COR operates in a horizontal way by integrating and centralizing dispatch functions and emergency coordination for municipal service providers as well as for other non-municipal agencies. While each agency is still autonomous and maintains its own operative systems and response protocols, the COR provides a digital macro architecture that connects individual systems, as well as a physical location where such integration occurs. It manages to overcome issues of institutional isolation through a digital architecture that facilitates information sharing while maintaining the specialized knowledge and experience that exists within each agency.

As of April 2014, the COR hosted representatives of 32 municipal agencies, 12 private concessions, and a selected number of state level agencies. Among the municipal agencies are the Transport Department, Waste Management Department, Health Department, Social Assistance, the Municipal Guard, Alerta Rio (the city's meteorological monitoring agency), and Civil Defense. The private concessions involved include bus companies as well as Light, the privately owned company in charge of supplying electricity to the city. "We focus on those organizations that are directly linked with citizen's wellbeing on an everyday basis," explains a COR directive (Interview, 2014).

The COR provides a new control capacity to monitor and manage infrastructural flows and ecological conditions on a 24/7 basis with a predictive capacity that enables the mobilization of emergency responses to severe disruption–in particular localized rainfall, flooding, and landslides. It enables infrastructural and ecological disruptions to be more effectively bounded and managed while wherever possible facilitating the maintenance of urban flows and circulations in the rest of the city, from, for example, transport flows and waste collection to trade and other key economic circulations. The city is increasingly managed as a logistical entity in which resilience is the ability to contain disruption and maintain urban circulations even under emergency conditions.

Given its extensive use of urban digital technologies, the COR is globally seen as an exemplar "smart city" initiative (The

Guardian, 2014; New York Times, 2012) based on its ability to integrate and rebundle urban infrastructures and services through digital and communication technologies. But within the world of urban responses to climate change, it has also been hailed as an exemplar urban resilience strategy. Rio's COR features extensively in the literature of the Rockefeller Foundation's 100 Resilient Cities program (Rockefeller Foundation, n.d.). The municipality has designated the COR as the primary site from which the city's resilience strategy is to be designed and implemented, with the director of the COR acting also as the city's director of resilience. Transcending the notion of climate adaptation, the municipality embraces the idea of resilience "because this notion includes, but is not limited to, climate change. … [It] also incorporates social challenges, because a resilient city is one where citizens have access to basic services … [and where] economic, social and financial aspects are capable of faster recovery in case of a national or global crisis" (Prefeitura do Rio de Janeiro, 2015: n.p.).

References

BBC. (2010, April 7). Flooding in Rio de Janeiro state kills scores. Accessed June 11, 2015: http://news.bbc.co.uk/1/hi/8605386.stm

Instituto Brasileiro de Geografia e Estatistica (IBGE). (2015). Table 3.2. Accessed September 18, 2015: http://www.ibge.gov.br/home/estatistica/populacao/censo2010/sinopse/sinopse_tab_rm_zip.shtm

The Guardian. (2014, May 23). World Cup 2014: Inside Rio's Bond-villain mission control. Christopher Frey. Accessed June 11, 2015: http://www.theguardian.com/cities/2014/may/23/world-cup-insiderio-bond-villain-mission-control

New York Times. (2010, April 6). At least 95 are killed as floods paralyze Rio Alexei Barrionuevo. Accessed June 11, 2015: http://www.nytimes.com/2010/04/07/world/americas/07brazil.html?_r=0

New York Times. (2012, March 3). Mission control, built for cities – I.B.M. takes 'smarter cities' concept to Rio de Janeiro. Natasha Singer. Accessed June 11, 2015: http://www.nytimes.com/2012/03/04/business/ibm-takes-smarter-cities-concept-to-rio-dejaneiro.html?_r=0

Peel, M. C., Finlayson, B. L., and McMahon, T. A. (2007). Updated world map of the Köppen-Geiger climate classification. *Hydrology and Earth System Sciences Discussions* 4(2), 462.

Prefeitura do Rio de Janeiro. (2011a). *Plano de ação para redução de emissões de gases de efeito estufa da Cidade do Rio de Janeiro.* Secretaria Extraordinária de Desenvolvimento.

Prefeitura do Rio de Janeiro. (2011b). *Plano de Emergência dara Chuvas Fortes da Cidade do Rio de Janeiro.* Defesa Civil. Accessed June 11, 2015: http://www.rio.rj.gov.br/dlstatic/10112/4402327/4109121/RIODEJANEIRORESILIENTE_2013.pdf

Prefeitura do Rio de Janeiro. (2015). *Rio Resiliente: Diagnóstico e Áreas de Foco.* Secretaria Municipal da Casa Civil. Accessed June 11, 2015: http://www.centrodeoperacoes.rio.gov.br/assets/PEF-0112–14-LVR-210x280-Resiliencia-43.pdf

Rockefeller Foundation (n.d.). Selected cities: Explore the 100 Resilient Cities' second round of cities. Accessed July 20, 2015: http://www.100resilientcities.org/cities#/-_/

World Bank (2013). Rio de Janeiro Low Carbon City Development Program. World Bank Institute/Rio Prefeitura. Accessed June 11, 2015: https://einstitute.worldbank.org/ei/sites/default/files/Upload_Files/RiodeJaneiroLowCarbonCityDe elopmentProgram_PD.pdf

World Bank. (2017). 2016 GNI per capita, Atlas method (current US$). Accessed August 9, 2017: http://data.worldbank.org/indicator/NY.GNP.PCAP.CD

Case Study 4.A

Climate Change Adaptation and Mitigation for Hyderabad City, India

Shalini Sharma

University of Oxford, Oxford

Keywords	Heat, drought, water, mitigation and adaptation, population influx
Population (Metropolitan Region)	9,218,000 (United Nations, 2016)
Area (Metropolitan Region)	7,257 km² (Hyderabad Metropolitan Development Authority, 2016)
Income per capita	US$1,680 (World Bank, 2017)
Climate zone	BSh – Arid, steppe, hot (Peel et al., 2007)

Hyderabad is located in the north of the Deccan plateau, along the banks of the Musi River in South India. The size of the city is 250 square miles (650 km²), with a topography of rocky terrain and predominantly gray granite. Hyderabad has many hills of varying altitudes ranging from 542 to 672 meters, and 140 large and small lakes in and around its neighborhood up until the 1970s. Recently, high population influxes have increased pressures on Hyderabad's infrastructure and resources. A population growth of 87.2% in 10 years has made Hyderabad the fourth most populous city in India. Together, these factors have led to serious climate change issues, which have been experienced mostly over the past 5 years. Major issues pertaining to climate change and probable solutions are detailed in this case study.

Population increases within the city, which add pressure to land and water resources, require strong management measures to be implemented by the state government. The growing population requires various types of infrastructure (facilities, transportation, etc.), which in turn increase the city's greenhouse gas (GHG) emissions.

Power Sector

Data from undivided Andhra Pradesh Central Power Distribution Company Limited (APCPDCL), Government of Andhra Pradesh, for 2012 show that the power requirements of Hyderabad city is approximately 2,000 megawatts (MW) in the summer and 1,400 MW in winter. To meet the demand for projected increases in the city's population by the end of the thirteenth Five-Year Plan period (2021–22), Hyderabad certainly needs to have dedicated power plants with a total installed capacity of 8,000–10,000 MW.

At present, APCPDCL data reveal that demand–supply gap for Hyderabad city remains between 2,000 and 2,350 MW. APCPDCL has implemented new rules to shut down its plants for certain fixed days every week. These steps lead to emissions reduction but also to huge loss for businesses and ultimately impact the economy. This adaptation strategy led to a reduction of 3,000 tonnes CO_2 emission per day (EF 0.76 tCO_2/MWh) from the Southern India grid (CEA, 2011).

The state government is spearheading major retrofits and maintenance of power distribution facilities. As an adaptation strategy, APPDCL is modifying 11 kV substations to 33 kV or higher level substations. This adaptation strategy for the power sector will lead to reductions in load loss, resulting in an efficient power supply and energy efficiency. The government is the driver of this adaptation strategy. Cities in other fast developing economies/least-developed countries can follow this strategy to reap the benefits of energy efficiency.

National Mitigation Strategies

Two national mitigation strategies from the central government are under implementation in all states in India:
- National Solar Mission and its tradable product Renewable Energy Certificate
- National Mission on Energy Efficiency and its tradable product E-certs

Every power distribution licensee has been directed to meet a minimum 5% of energy demand through renewable energy resources (APERC, 2012); the program is initially planned for 2012–2017. This mitigation strategy can lead to a 5% GHG emission reduction that would otherwise have occurred from thermal power plants.

Roof top solar plants can make every house a power house, thus helping villagers to earn their livelihood/establish community business, help prevent crop loss (through solar water pumps), and improve the education of children because many villagers move to cities for access to electricity, jobs, and education. This adaptation strategy of providing power in villages may be helpful in reducing migration to cities. These strategies, if driven by government and communities together, will be helpful in preventing uncontrolled population influx into Hyderabad.

Strategic Urban Development

The improvement of the Hyderabad drainage system is crucial, especially due to the rocky terrain that prevents rainwater drainage and leads to floods in low-lying areas, traffic congestion, and vector borne diseases.

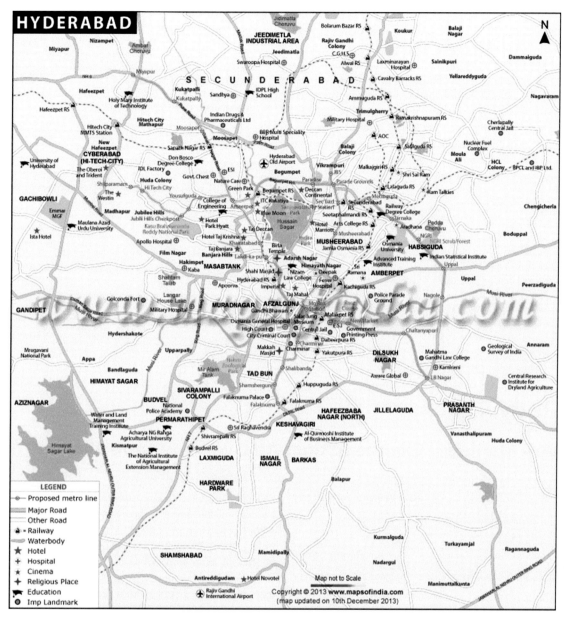

Case Study 4.A Figure 1 *City of Hyderabad with its infrastructure.*

Source: Maps of India, 2014

Hyderabad is undertaking the following adaptation strategies to combat the urban heat island (UHI) effect (Kleerekoper et al., 2012):

1. Kasu Brahmananda Reddy (KBR) national park in middle of the city, a declared reserved bird sanctuary
2. Fountains at cross-roads (wherever possible) to improve ambiance as well as help reduce the temperature.
3. Adjustment of building density in gated communities
4. Heat-reflective coatings are being promoted for use on home/building exteriors to help maintaining cool temperatures in homes/offices
5. The collection of rainwater (through channels under the footpath) into nearby bodies of water, an effective adaptation strategy for increasing/maintaining groundwater level in urban areas. (This is still to be implemented in Hyderabad.)
6. Sustainable water systems to supply trees with enough water to maximize their cooling ability and shallow canals to absorb and discharge heat are strategies being partially used in Hyderabad
7. Adaptation strategies like rainwater harvesting, installation of solar water heaters, and rooftop solar power to meet 3% of the total energy requirements are made mandatory by the state government for implementation in apartments

Discussion and Outcome

In Hyderabad, communities are driving adaptation measures, and mitigation measures are driven by industries at large. Hyderabad no longer has pure water for drinking or domestic use. Groundwater available at few places around the city is not suitable for drinking or domestic use without purification. Therefore, implementation of reverse osmosis plants for water purification or buying water for drinking purpose are basic necessities in any residential area in Hyderabad. Such crises call for the implementation of community-driven actions like rainwater harvesting in apartments, planting vegetation to the maximum extent possible, channelizing rainwater to enable groundwater recharge, and more, even though the rocky terrain of Hyderabad increases the difficulty of taking water-saving measures. Adaptation measures must drive industries to participate in the implementation of such activities outside their boundaries, maybe as a part of corporate social responsibility (CSR) activities.

References

Andhra Pradesh Electricity Regulatory Commission (APERC) (2012). Renewable Power Purchase Obligation Regulation No.1 of 2012. Accessed July 2014: http://www.aperc.gov.in/aperc1/assets/uploads/files/632ea-recregulation1of2012.pdf

Census Population. (2015) Data, Hyderabad (Greater Hyderabad) Urban Region Population 2011 Census. Accessed September 11, 2015: http://www.census2011.co.in/census/metropolitan/342-hyderabad.html

Climate-Data. (2015). Climate-Data.org. Accessed October 9, 2015: http://en.climate-data.org/search/

Central Electricity Authority (CEA), Ministry of Power, Government of India (2011).CO_2 Baseline Database for the Indian Power Sector. Accessed August 20, 2014: http://www.cea.nic.in/reports/planning/cdm_CO_2/user_guide_ver6.pdf

Hyderabad Metropolitan Development Authority. (2016). About Hyderabad Metropolitan Development Authority. Accessed November 3, 2016: http://www.hmda.gov.in/About.aspx

Kleerekoper, L., van Esch, M., and Salcedo, T. B. (2012). How to make a city climate proof, address the urban heat island effect. *Resource Conservation and Recycling* **64**, 30–38.

Peel, M. C., Finlayson, B. L., and McMahon, T. A. (2007). Updated world map of the Köppen-Geiger climate classification. *Hydrology and Earth System Sciences Discussions* **4**(2), 462.

United Nations, Department of Economic and Social Affairs, Population Division (2016). The World's Cities in 2016 – Data Booklet (ST/ESA/SER.A/392).

World Bank. (2017). 2016 GNI per capita, Atlas method (current US$). Accessed August 9, 2017: http://data.worldbank.org/indicator/NY.GNP.PCAP.CD

Case Study 4.B

São Paulo's Municipal Climate Action: An Overview from 2005 to 2014

Laura Valente de Macedo

Environment and Energy Institute at University of São Paulo

Joana Setzer

Grantham Research Institute on Climate Change and the Environment, Department of Geography and Environment at the London School of Economics and Political Science, London

Fernando Rei

Catholic University of Santos, Brazil

Keywords	Subnational climate policy, urban transport, mitigation and adaptation
Population (Metropolitan Region)	19,683,975 (IBGE, 2015)
Area (Metropolitan Region)	7,947 km² (IBGE, 2015)
Income per capita	US$8,840 (World Bank, 2017)
Climate zone	Cfb – Temperate, without dry season, warm summer (Peel et al., 2007)

São Paulo city, located in southeastern Brazil, is São Paulo state's capital and the largest urban conglomeration in South America, with a population of about 11.5 million inhabitants in 590 square miles (IBGE, 2014). In the past decades, the city's economic profile underwent deep transformation, from strongly industrialized to service and technology-oriented. São Paulo has the largest municipal economy in Brazil, with a gross domestic product (GDP) of approximately US$200 billion, 11.5% of the national economy (IBGE, 2011). The scale of these figures compares to those of entire countries such as Bolivia, Portugal, and Denmark.

São Paulo pioneered local engagement in transnational climate networks (ICLEI's Cities for Climate Protection Campaign [CCP] and the C40 Cities Climate Leadership Group). The city's first greenhouse gas (GHG) inventory was undertaken within the CCP. It was published in 2005, considering 2003 as the baseline year. Results demonstrated that primary sources were road transport and waste management (UFRJ, 2005).

The second inventory revised the results of the 2005 report for 2003 and expanded the scope until 2011, focusing on energy and waste emissions. There was no significant variation in sources' contribution to the overall emissions of GHG of the city during that period (Instituto Ekos Brasil et al., 2013).

According to the inventory, emissions totalled 12.9 GgTCO₂e, 84% of which was generated by the energy sector and 15% by solid waste management. Road transport accounted for 88.78% of energy emissions, 68.6% of the total municipal CO_2 emissions, mainly from gasoline-powered passenger vehicles.

São Paulo was the first major city in Brazil to address climate change and adopt municipal regulation and implementation of local policies to reduce GHG emissions. The climate agenda was led by the city's Environment Secretariat, and the Municipal Climate Law (in 2009) stands as its most important milestone. It asserts that mitigation of and adaptation to climate change in the city will contribute toward Brazil's compliance with the United Nations Framework Convention on Climate Change (UNFCCC) objectives. It also establishes a mandatory reduction target of 30% of aggregate municipal emissions in CO_2e, by 2012, relative to the 2003 baseline reported in the municipal inventory published in 2005.

This target, however, proved too ambitious. According to the second inventory, GHG emissions increased by 8.7% in 2011 relative to 2003. National subsidies for car ownership and fossil fuel to address the global economic crisis have played an important role in expanding the car fleet in Brazil, especially after 2008 (INCT, 2013). São Paulo's fleet grew by 41%, from 3.35 million

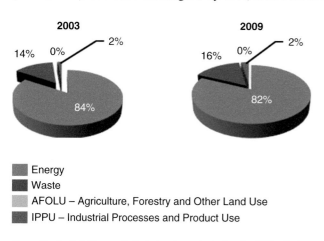

Case Study 4.B Figure 1 *Emissions by sector in the city of* São *Paulo, 2003–2009.*

to 4.73 million, between 2003 and 2011 (Departamento Nacional de Transito, 2015).

Nonetheless, São Paulo developed policies to reduce GHG emissions, control local pollution, and generate revenue beyond the climate agenda (Puppim de Oliveira, 2009). The most successful example is methane gas recovery from landfill sites (United Nations, 2014). The city houses two of the world's largest landfill bio-gas power plants. The first methane-to-energy project began operating in 2004 to recover bio-gas from the Bandeirantes landfill and generate electricity at an on-site power plant (ICLEI, 2009). It was the first such project implemented in Brazil to obtain certified emissions reductions under the Clean Development Mechanism (CDM; UN, 2014; World Bank, 2011, 2012). In the first auction, held in September 2007, the transaction totalled about US$16 million. The other two auctions held in September 2008 and in June 2012 resulted in approximately US$17.5 million and US$2.3 million, respectively (BMF-BOVESPA, 2012).

Energy-efficiency projects begun in 2011 are expected to yield economies of more than 1.5 million KWh monthly by replacing incandescent lamps for LEDs in traffic lights and in tunnels, as well as in schools, hospitals, and other municipal buildings (PMSP, 2012b). Transport policies led to further local emissions reductions despite the growth in car ownership. Transport-related measures implemented between 2003 and 2011 resulted in reductions of 6.3% in local air pollutants, as well as 6.7% of CO_2e emissions, avoiding 7,835 tons of CO_2e per month. The *Ecofrotas* program begun in 2005 expanded the bus fleet using cleaner vehicles, with 1,200 new buses using bio-diesel, 60 buses using ethanol, and 319 buses using sugarcane diesel. Furthermore, the city invested in retrofitting and recovering the tram system of 190 vehicles, 92 of which are new, and it reopened an old tram factory in the city. Implementing bus corridors became a top priority for the new administration, which had added 150 kilometers to the system, with 36 kilometers being built by the end of 2014.

Adaptation measures are also required by the Municipal Climate Law as part of the strategy to reduce climate vulnerability in São Paulo (São Paulo, 2011). In 2010, the city began a program to expand waterfront parks through tree planting,

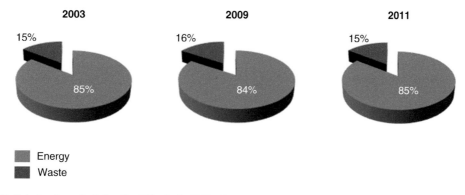

Case Study 4.B Figure 2 *Emissions by sector in the city of* São *Paulo, 2003–2009.*

Source: Instituto Ekos Brasil et al., 2013

687

Table 1 *Energy emissions (GgCO₂e) by subsector. Source: Instituto Ekos Brasil, Geoklock Consultoria e Engenharia Ambiental, 2013*

Sources/Year	2003	2004	2005	2006	2007	2008	2009	2010	2011
Power	865	889	845	781	739	1,264	650	1,399	830
Industry	3,001	3,059	2,555	2,553	2,566	2,652	2,437	2,426	2,477
Transport	8,989	9,117	9,290	179	9,779	9,822	9,239	9,786	10,589
Fugitive[a]	57	0	0	30	29	122	59	31	94
Total	12,912	13,065	12,690	3,543	13,113	13,860	12,385	13,642	13,990

[a] The United States Department of Energy defines Fugitive Emissions as "the release of Green House Gases from pressurized systems" (USDE, 2015)

Table 2 *Emissions (in GtCO₂e) in* São *Paulo city, 2003–2011. Source: Instituto Ekos Brasil, Geoklock Consultoria e Engenharia Ambiental, 2013*

Sector/ Year	2003	2004	2005	2006	2007	2008	2009	2010	2011
Energy	12,911	13,065	12,689	12,544	13,114	13,860	12,384	13,642	13,990
Waste	2,199	2,260	2,335	2,474	2,658	2,307	2,363	2,445	2,440
Total	15,110	15,325	15,024	15,018	15,772	16,167	14,747	16,087	16,430

revitalizing 100 existing parks, and establishing a 1,320,000 hectare park in partnership with the private sector. The municipal housing authority expanded urbanization of shanty towns (*favelas*) and relocation of houses in risk areas (PMSP, 2012a).

References

BMF-BOVESPA. (2012). BM & FBOVESPA. Anuncia resultado do leilão de créditos de carbono realizado nesta terça-feira. Accessed June 22, 2014: http://www.bmfbovespa.com.br/pt-br/noticias/2012/BMFBOVESPAanuncia-resultado-do-leilao-de-creditos-de-carbono-2012-06-12.aspx?tipoNoticia=1&idioma=pt-br

Departamento Nacional de Transito (DENATRAN). (2015). Data on municipal fleet in Brazil. Accessed April 30, 2015: http://www.denatran.gov.br/frota.htm retrieved

Instituto Brasileiro de Geografia e Estatística (IBGE). (2011). Posição ocupada pelos 100 maiores municípios, em relação ao. Produto Interno Bruto a preços correntes e participações percentuais relativa e acumulada. Accessed April 30, 2015: http://biblioteca.ibge.gov.br/visualizacao/periodicos/2/bn_2013_v21.pdf

Instituto Brasileiro de Geografia e Estatística (IBGE). (2014). Cidades. Sao Paulo. Accessed April 9, 2015: http://www.cidades.ibge.gov.br/xtras/perfil.php?lang=&codmun=355030&search=sao-paulo%7Csaopaulo

Instituto Brasileiro de Geografia e Estatística (IBGE). (2015). Instituto Brasileiro de Geografia e Estatistica. Accessed September 18, 2015: http://www.ibge.gov.br/home/estatistica/populacao/censo2010/sinopse/sinopse_tab_rm_zip.shtm

ICLEI. (2009). *Turning Pollution into Profit: The Bandeirantes Landfill Gas to Energy Project.* ICLEI Case Studies.

INCT. (2013). Crescimento da frota de automóveis e motocicletas nas metrópoles brasileiras (2001–2012). In Juciano Martins Rodrigues (ed.), *INCT Report.* Observatório das Metrópoles, Instituto Nacional de Ciência e Tecnologia (INCT), 17.

Instituto Ekos Brasil. (2013). Geoklock Consultoria e Engenharia Ambiental Inventário de emissões e remoções antrópicas de gases de efeito estufa do município de São Paulo de 2003 a 2009 com atualização para 2010 e 2011 nos setores de energia e resíduos. Série Cadernos Técnicos, Volume 12. ANTP.

Peel, M. C., Finlayson, B. L., and McMahon, T. A. (2007). Updated world map of the Köppen-Geiger climate classification. *Hydrology and Earth System Sciences Discussions* **4**(2), 462.

PMSP. (2012a). Ata da 10a Reunião do Grupo de Trabalho Sustentabilidade e Uso do Solo—14/03/2012. Accessed June 22, 2014: http://www.prefeitura.sp.gov.br/cidade/secretarias/upload/chamadas/ata_da_10_reuniao_do_gt_uso_do solo_-_14–03–2012_1331922980.pdf

PMSP. (2012b). Ata da 31ª Reunião do Comitê Municipal de Mudanças Climáticas e Ecoeconomia de São Paulo—22/11/2012. Accessed June 22, 2014: http://www.prefeitura.sp.gov.br/cidade/secretarias/upload/chamadas/ata_31_reuniao_comite_1355854241.pdf

Puppim de Oliveira, J. A. (2009). The implementation of climate change related policies at the subnational level: An analysis of three countries. *Habitat International* **33**(3), 253–259.

São Paulo. (2011). *Guidelines for the Action Plan of the city of São Paulo for mitigation and adaptation to climate change. The Municipal Committee on Climate Change and Eco-economy and the Working Groups for Transportation, Energy, Construction, Land Use, Solid Waste and Health.* São Paulo.

UFRJ. (2005). Inventario de emissões de gases de efeito estufa do município de São Paulo. São Paulo: Prefeitura Municipal de São Paulo and Secretaria do Verde e do Meio Ambiente. Accessed June 22, 2014: http://ww2.prefeitura.sp.gov.br/arquivos/secretarias/meio_ambiente/Sintesedoinventario.pdf

United Nations. (2014). The Bandeirantes Landfill Gas to Energy Project. Accessed June 22, 2016: http://sustainabledevelopment.un.org/index.php?page=view&type=1006&menu=1348&nr=2180

United States Department of Energy. (2015). *Fugitive Emissions.* Accessed June 10, 2015: http://energy.gov/ehss/fugitive-emissions

World Bank (2011). *Guide to climate change adaptation in cities.* Author.

World Bank (2012). Climate change, disaster risk, and the urban poor: Cities building resilience for a changing world. In Judy L. Baker (ed.), *Urban Development Series* (235–267). World Bank.

World Bank. (2017). 2016 GNI per capita, Atlas method (current US$). Accessed August 9, 2017: http://data.worldbank.org/indicator/NY.GNP.PCAP.CD

Case Study 4.C

Leuven Climate Neutral 2030 (LKN2030): An Ambitious Plan of a University Town

Wim Debucquoy

Autonomous University of Barcelona/ Witteveen+Bos

Keywords	Baseline emission inventory, climate action plan, housing and transportation, adaptation and mitigation
Population (Metropolitan Region)	99,075 (IBZ, 2016)
Area (Metropolitan Region)	56.6 km² (IBZ, 2016)
Income per capita[a]	US$41,860 (World Bank, 2017)
Climate zone[b]	Temperate, without dry season, warm summer (Cfb) (Peel et al., 2007)

In 2011, the city of Leuven signed a Declaration of Intent to become "climate neutral" by 2030. It established the ambitious climate goal to cut greenhouse gas (GHG) emissions by 80% by 2030 compared to a 2010 baseline.[1] Although the concept of "climate neutrality" might not be scientifically sound,[2] it has been used successfully in several cities and provinces in Belgium to gather public and political support to implement strong climate action (Vandevyvere, 2014). Leuven Climate Neutral 2030 (LKN2030) has since evolved into a multiactor project and process, engaging a broad spectrum of stakeholders in order to collectively design transition paths to climate neutrality. First, a baseline emission inventory of Leuven was carried out, followed by a citywide project to create a roadmap for LKN2030. In 2013, the nonprofit organization Leuven Climate Neutral 2030 was founded, charged with overseeing the envisioned transition of Leuven.

Baseline Emission Inventory of Greater Leuven

The baseline for the emissions inventory was executed for 2010, the year with the most appropriate available data. For LKN2030, only scope 1 (direct emissions) and scope 2 (indirect emissions from imported energy) emissions were quantified and considered. Scope 3 emissions (indirect emissions from imported goods and services) were qualitatively assessed.[3] In 2010, Greater Leuven emitted 808,000 tons of CO_2eq or 8.5 ton CO_2eq per capita (scope 1+2) (Vandevyvere et al., 2013: 41–58).

Leuven has a strong knowledge-based economy and almost no industry or agriculture. As a result, building-related emissions (household and tertiary sector) amount to nearly 60% of total GHG emissions. Together with transport emissions, they make up 82% of GHG emissions in Leuven. Transition scenarios therefore focus on the renovation of the existing building stock, on greening the energy supply, and on switching to sustainable transport modes (Vandevyvere et al., 2013).

A Scientific Roadmap to Climate Neutrality

Following the Baseline Emission Inventory, a roadmap project was launched, structured along an explicit, combined top-down and bottom-up approach to ensure and maximize public support. The bottom-up process consisted of six thematic groups (energy; built environment; mobility; consumption; agriculture and nature; participation, public support, and transition), each with around fifteen people from civil society, city departments, local businesses, and the university. Simultaneously, there was the local "G20," a transition group with twenty key decision-makers from knowledge institutions, enterprises, local government, and civil society that developed top-down high-level strategies for the city of Leuven. This combined process was supposed to build the necessary local knowledge, create widespread public support for the project, and have a real policy impact (Jones et al., 2012). Both top-down and bottom-up approaches were coordinated by a scientific team of Leuven's University, which simultaneously served as communication link between the two approaches. All the recommendations were combined into a final scientific report (Vandevyvere et al., 2013). In total, some 200 people were active in different sections of the roadmap project. The goal was to shift the process from the city government and the university – the initiators of the process – to a citywide participatory process among civil society, businesses, the university, and the city since participatory processes are vital for building support for the plan and realizing its long-lasting success (Vandevyvere, 2014).

The scientific report made a distinction between immediate and long-term actions and used different scenarios: business-as-usual, Leuven Climate Neutral by 2030, or by 2050. The report focused on five sectors (residential buildings, nonresidential

---

1 These goals were adjusted to GHG reductions of 67% by 2030 and 81% ("climate safe") by 2050, after the scientific report of LKN2030.
2 The concept of "climate neutrality" generally indicates a GHG emission reduction target of 80% or 90%.
3 Scope 3 emissions are estimated at 2,440,000 tons of CO_2, eq

buildings, mobility, nature and agriculture, and energy) and proposed a list of recommendations, concrete measures, and projects for each sector to reduce GHG emissions. For each measure, the reduction in GHG emissions and economic gain/cost was quantified. Some examples of measures proposed are: five refurbishment "waves" to stepwise retrofit the complete building stock of Leuven, starting with the oldest buildings; a modal shift to 33% bike, 33% public transport, 33% car mobility by drastically improving public transport and bike infrastructure and making the city center completely car-free; and the deployment of renewable energy technologies.

The report concludes that, when all sectors are considered simultaneously, aggregated emission cuts of 55% and 67% are possible by 2030 and 2050, respectively. This would lead to a net profit of €39 (US$46) million per year and €34 (US$40) million per year for 2030 and 2050, respectively. These projections include GHG emissions from local industry, a sector that was not considered for intervention in the report. When GHG emissions of local industry are omitted, the LKN2030 and 2050 scenarios lead to a reduction of 67% and 81%, respectively. This means that a climate neutral scenario can be reached by 2050 starting from the proposed LKN-scenarios of the report.

The final recommendation of the scientific report was to formalize LKN2030 into a long-lasting organization. The nonprofit Leuven Climate Neutral 2030 was established by sixty city stakeholders in 2013. The mission of this organization is to inspire, inform, measure, and facilitate and to involve partners and activate them in regards to Leuven's transition to climate neutrality. In 2013, the organization also launched the public

campaign Mission Zero Emission, aimed at increasing the visibility of LKN2030 and putting several initiatives in the spotlight (see Case Study 4.C Figure 1).

In 2015, the Board of Experts – counting ten climate specialists – was charged with translating the scientific report into explicit, yearly operational targets for the nonprofit organization and its partners. At present, some twenty preferential projects have been formulated as transition experiments, and appropriate stakeholders and financing schemes have been identified (Vandevyvere et al., 2015).

Lessons Learned

Several aspects proved vital for establishing a transition project with reasonable chances for success: the scientific support given by the local university, securing engagement from stakeholders and public support by fostering shared ownership in order to gain the required momentum, opening the way for practical implementation through concrete measures and projects, monitoring GHG emissions, and establishing appropriate financing and communication mechanisms. Securing these operational aspects remains a continuous struggle for LKN2030 (Vandevyvere, 2014). Innovation capacity, shared enthusiasm, tangible engagement, and a good balance between bottom-up and top-down actions are promising conditions for the future of LKN2030. Challenges to this optimism are lack of financial input, built-in structural barriers, and the dictates of short-term agendas. Empowerment and leadership now play an important role to secure the success of LKN2030.

Case Study 4.C Figure 1 *Image of public campaign Mission Zero Emission on the Ladeuze square in Leuven (translation: "Insulate with us 500,000 m² of Leuven roofs, 50 times the area of the Ladeuze square, so we save €3 (US$3.5) million each year!").*

Photo: Robbe Maes

References

Jones, P. T., Vandevyvere, H., and Van Acker, K. (2012). Leuven Klimaat-neutraal 2030, project proposal, Stad Leuven/KU Leuven, Leuven. Accessed April 16, 2015: http://www.leuvenklimaatneutraal.be/ontsta-an-en-mijlpalen

IBZ - Ministry of the Interior - General Directorate Institutions and Population (2016). Retrieved November 3, 2016 from http://www.ibz.rrn.fgov.be/fileadmin/user_upload/fr/pop/statistiques/population-bevolking-20160101.pdf

Leuven. (2014). Leuven in numbers. Accessed April 1, 2015: http://www.leuven.be/bestuur/leuven-in-cijfers

Peel, M. C., Finlayson, B. L., and McMahon, T. A. (2007). Updated world map of the Köppen-Geiger climate classification. *Hydrology and Earth System Sciences Discussions* 4(2), 462.

World Bank. (2017). 2016 GNI per capita, Atlas method (current US$). Accessed August 9, 2017: http://data.worldbank.org/indicator/NY.GNP.PCAP.CD

Vandevyvere, H. (2014). Climate neutral city initiatives: Wishful thinking or thoughtful wish? Research paper 7, Steunpunt TRADO, Mol. Accessed April 16, 2015: http://steunpunttrado.be/documenten/papers/trado-research-paper-7-1.pdf

Vandevyvere, H., Jones, P. T., Aerts, J. (2013). De transitie naar Leuven Klimaatneutraal 2030: Wetenschappelijk eindrapport. Februari 2013. Accessed April 16, 2015: https://www.leuven2030.be/sites/default/files/attachments/LKN_Wetenschappelijk%20Eindrapport_1302.pdf

Vandevyvere, H., and Nevens, F. (2015). Lost in transition or geared for the s-curve? An analysis of Flemish transition trajectories with a focus on energy use and buildings. *Sustainability* 7(3), 2415–2436.

Case Study 4.D

The Challenges of Mitigation and Adaptation to Climate Change in Tehran, Iran

Ali Mohamad Nejad

Department of Geography, Shahis Beheshti University, Tehran

Keywords	GHGs emission, land-use management, mitigation and adaptation
Population (Metropolitan Region)	8,516,000 (United Nations, 2016)
Area (Metropolitan Region)	1,632 km² (Demographia, 2016)
Income per capita	US$6,530 (World Bank, 2017)
Climate zone	BSk – Arid, Steppe, cold (Peel et al., 2007)

Climate change is one of the most important challenges of our time. Climate change is mainly caused by the emission of greenhouse gases (GHG), with a significant contribution from fossil fuel consumption in urban areas. In this case study, we investigate the experience and challenges in reduction of GHGs emissions and adaptation to climate change impacts in Tehran Metropolis. Climate change solutions in a centralized planning system such as Iran need multilevel governance and lifestyle changes. These changes are related to sustainability of local communities and many aspects of urban life.

Due to the fossil fuel base of production, distributions, and consumption patterns in Tehran metropolis, as well as the related lifestyle, GHG emissions per capita are very high, making it the largest source of this kind of emissions in Iran. The results

indicate that the building sector (residential, commercial, and administrative buildings) is the most important emitter of carbon dioxide (CO_2) at 21 million tons (Table 1). The consumption of natural gas is the major cause of CO_2 emissions in the building sector.

The second source of CO_2 emissions is urban transportation (14 million tons CO_2). Car-driven land-use planning is the major factor for GHGs emissions, a factor caused by overemphasis on separation of work-residence-leisure activities. This, in turn, has led to 1.5 trips generated for each person and almost 17 million trips a day in Tehran. Bearing in mind the dominance of private

Table 1 *Carbon dioxide emission by sectors in Tehran Metropolis. Source: Tehran Urban Planning and Research Center, 2011*

Sectors	Million tons	Percent	Per capita emission
Residential, commercial and administrative buildings	21	44.3	2.73
Industry	6.9	14.6	0.89
Transportation	14.2	29.9	1.84
Agriculture	0.40	0.9	0.05
Power plants	4.4	9.3	0.57
Other	0.44	1.0	0.06
Total	47.5	100	6.2

cars, the impacts are intensified by the increase in the number of pollutant days and expansion of urban heat island.

Nevertheless, there are successful actions that have reduced air pollution and GHG emissions in Tehran Metropolis. For example, Tehran urban management has expanded public transportation in recent years. Tehran has 160 kilometers in 5 subway lines with 110 stations. Bus rapid transit (BRT) is another example of sustainable action in the public transportation sector. Tehran has 114 kilometers of BRT in 10 lines with 237 bus stations (Table 2).

Table 2 *Climate change mitigation actions in the transportation sector. Source: Tehran Municipality, 2015*

Transportation sectors	Length (kilometer)	Number of lines	Number of stations
Subway	160	5	110
Bus Rapid Transit (BRT)	114	10	237

Table 3 *Area of built environment changes (hectares). Source: Tehran Urban Planning and Research Center, 2011*

Built area (2010)	Built area (2002)	Built area (1988)	Built area
37,411	30,871	22,790	Hectare

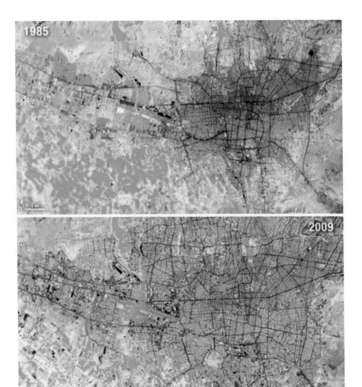

Case Study 4.D Figure 1 *Land use change and green space reduction in Tehran.*

The most important challenge for Tehran urban management in tackling climate change is land-use change and built environment expansion. The area of built environment increased from 22,790 hectares in 1988 to 37,411 hectares in 2010 (Table 3). Rapid urbanization; car-driven land-use planning; and urban management policies such as changes of land-use regulations, expansion of informal settlements, and migration from other parts of the country are the main factors for this change in Tehran Metropolis.

Between 1985 and 2009, the population of Tehran grew from 6 million and to just over 7 million. The city's growth was spurred largely by migration from other parts of the country. In addition to being the hub of government and associated public-sector jobs, Tehran houses more than half of Iran's industry.

Landsat 5 provided false-color images (see Case Study 4.D Figure 1) of Tehran on August 2, 1985 and July 19, 2009. The city is a web of dark purple lines: vegetation is green and bare ground is pink and tan. The images were created using both infrared and visible light to distinguish urban areas from the surrounding desert.

Conclusion

The essential guidelines to reducing GHG and adapting to climate change encompass two categories: (1) The reduction of GHGs by revision of land-use management and zoning by-laws in light of urban sustainability principles and (2) the adaptation to climate change through a reliance on local community assets in the framework of community-based adaptation approach. Tehran urban management strengthens neighborhood management units and subcouncils at the local scale for sustainable community development. The needed radical change in urban lifestyle for coping with climate change impacts cannot be achieved solely by top-down actions. Although considerable efforts have been done by urban management relating to the expansion of public transportation, still extensive measures in all other urban planning dimensions should be taken along with lifestyle changes.

References

Demographia (2016) Demographia World Urban Areas (Built Up Urban Areas or World Agglomerations). 12th Annual Edition. St. Louis: Demographia. Available from: http://www.demographia.com/db-worldua.pdf.

Peel, M. C., Finlayson, B. L., and McMahon, T. A. (2007). Updated world map of the Köppen-Geiger climate classification. *Hydrology and Earth System Sciences Discussions*, 4(2), 462.

Tehran Municipality. (2015). *The Measures and Regulations for the Detailed Plan of Tehran City*. Tehran, Iran, Deputy of Urban Planning and Architecture of Tehran Municipality Publishing, 64. [In Persian]

Tehran Urban Planning & Research Center. (2011). *Tehran City State of Environment: (SoE)*. Tehran, Iran, Tehran Urban Planning & Research Center Publishing, 273. [In Persian]

United Nations, Department of Economic and Social Affairs, Population Division (2016). The World's Cities in 2016 – Data Booklet (ST/ESA/SER.A/392).

World Bank. (2017). 2016 GNI per capita, Atlas method (current US$). Accessed August 9, 2017: http://data.worldbank.org/indicator/NY.GNP.PCAP.CD

Case Study 4.E

Managing Greenhouse Gas Emissions in Rio de Janeiro: The Role of Inventories and Mitigation Actions Planning

Flavia Carolini, Vivien Green, Tomás Bredariol, Carolina Burle S. Dubeux, and Emilio Lèbre La Rovere

Federal University of Rio de Janeiro

Keywords	GHG emissions, local initiatives, mitigation strategies
Population (Metropolitan Region)	11,835,708 (IBGE, 2015)
Area (Metropolitan Region)	5,328.8 km² (IBGE, 2015)
Income per capita	US$8,840 (World Bank, 2017)
Climate zone	Am – Tropical monsoon (Peel et al., 2007)

Rio de Janeiro's Experience

In a pioneer action seen in Brazil and Latin America, Rio prepared in 2000 the first greenhouse gas (GHG) inventory for a city. This study was based on data collected from 1990, 1996, and 1998 (Dubeux and Rovere, 2007). The main methodological challenge was adapting the International Panel on Climate Change (IPCC) Guide to consider emissions resulting exclusively from the socioeconomic activity of the city (Neves and Dopico, 2013; Dubeux, 2007). Another issue was the data collection itself, which was highlighted by the researchers as the most difficult part of the inventory. The problems not only involved being unable to contact a single responsible sector (local, state, or national), but also discrepancies because the data obtained varied at different governmental levels.

In January 2011, Rio enacted Law No. 5248 establishing the Municipal Policy on Climate Change and Sustainable Development (PMMCDS, in Portuguese). Article Number Six states the city's commitment to reduce GHG emissions by 8%, 16%, and 20% in 2012, 2016, and 2020, respectively, compared to the level of emissions from the city in 2005 (Carloni, 2012).

To understand and quantify these goals in terms of volume of GHG emissions, the city conducted a new emission inventory for the year 2005 (Rovere et al., 2012) and updated the previous one. With the purpose of trying to understand how the economic sectors and actions of the city's government could contribute to this reduction, a study of emissions scenarios was also developed for the period 2005–2030 (Rovere et al., 2012). In 2013, the city completed the 2012 inventory and updated the 2005 numbers.

In 2011, Rio, in partnership with the World Bank, developed the Rio de Janeiro Low-Carbon City Development Program with the purpose of tracking the performance of policies and actions with the potential to mitigate emissions (World Bank, 2012). This program allows accomplished reductions to be accounted for in order to monitor, record, and verify achievement of the targets set in PMMCDS and also to certify them for possible commercialization in carbon markets. The International Organization of Standardization (ISO) certifies this program by ISO 14001 and ISO 14064.

The program is structured on two important pillars: (1) Program Roles and (2) Processes for Program Planning and Evaluation. Each new activity that reduces emissions – called an intervention – goes through the same five-step Program Process. This procedure ensures the viability of replicating this initiative in other cities.

Table 1 *The roles of Rio Low Carbon City Development Program. Source: World Bank, 2012*

Fixed Assignments

Coordinating Management Entity (CME): The CME is the central body within the municipality that oversees the coordination and management of the program. Fulfilling this role in Rio is the Mayor's Office (known as "Casa Civil").

Information Management Entity (IME): The IME is the central body that coordinates and manages all information and data related to the Program. The IME must ideally have both coordinating capabilities with all municipal departments and experience in collecting and managing large quantities of data. Fulfilling this role in Rio is the Instituto Pereira Passos (IPP).

Variable, Intervention-linked Assignments

Multisector Municipal Working Group (MWG): The MWG is a working group consisting of members from across the municipality with multiple areas of relevant expertise. It acts as an advisory committee to the CME. The composition and attendance of the MWG may vary from intervention to intervention, but it will always be coordinated by the CME.

Technical Advisory Entity (TAE): The TAE is an entity or consultant with technical expertise in the quantification of emission reductions.

Validation and Verification Entity (VVE): The VVE is an ISO-accredited environmental auditor. It validates and verifies the emission reductions generated by interventions under the Program. For any given intervention, the TAE and VVE must not be the same entity to ensure integrity in the audit process and avoid conflict of interest.

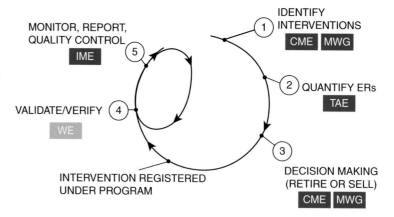

The Program Process has five distinct steps:
1. identify interventions
2. quantify emission reductions (ERs)
3. decision-making (retire or sell)
4. validate/verify
5. monitor, report and quality control

Case Study 4.E Figure 1 *Five steps of the Program process.*

Source: Reproduced with permission from World Bank, 2012

Table 1 explains the roles of groups assigned to carry out the program.

The Program Process advises on the procedures and criteria against which interventions are assessed, as well as on the process of monitoring, reporting, and verifying emission reductions generated by interventions. The Program Process steps are presented in Case Study 4.E Figure 1.

Conclusion

The GHG inventory process is not intended to be an end in itself, but rather a tool for monitoring emissions and a basis for developing strategies to reduce GHG emissions. Quantifying the emission reductions of different mitigation actions can be complex and expensive, but following other cities' examples and using their strategies as base points is a good start.

In this sense, Rio de Janeiro's experience can be helpful. The PMMCDS and Rio's Low Carbon City Development Program may be studied and replicated, and more information will be aggregated if other cities use this pioneering program.

It is important to ensure comparability among different studies from different cities and help promote cooperation among them in mitigation and adaptation to climate change. Therefore, it is necessary to bring all actors together in order to compare experiences and optimize efficiency and tools.

By promoting these associations and the exchange of experiences, cities can be supported by programs and organizations that promote partnership and collaboration. One example is the C40 Cities Climate Leadership Group, a group that connects more than seventy-five of the world's greatest cities, and where the current chair is Rio de Janeiro Mayor. ICLEI—Local Governments for Sustainability is another example of a network connecting cities, towns, and metropolises that are interested in building a more sustainable future. United Cities and Local Governments (UCLG) is another example of how cities are engaged in together and enhancing their experiences.

References

Carloni, F. B. B. A. (2012). *Gestão do inventário e do monitoramento de emissões de gases de efeito estufa em cidades: O caso do Rio de Janeiro.* UFRJ/COPPE.

Dubeux, C. B. S (2007). *Mitigação de Emissões de Gases de Efeito Estufa por Municípios Brasileiros: Metodologias para Elaboração de Inventários Setoriais e Cenários de Emissões como Instrumentos de Planejamento.* Tese de D.Sc, Planejamento Energético, COPPE, Universidade Federal do Rio de Janeiro.

Dubeux, C. B. S., and Rovere, E. L. L (2007). Local perspectives in the control of greenhouse gas emissions – The case of Rio de Janeiro. *Cities* **24**(5), 353–364.

Instituto Brasileiro de Geografia e Estatistica (IBGE). (2015). Table 3.2. Accessed September 18, 2015: http://www.ibge.gov.br/home/estatistica/populacao/censo2010/sinopse/sinopse_tab_rm_zip.shtm

Intergovernmental Panel on Climate Change (IPCC). (2006). *IPCC Guidelines for National Greenhouse Gas Inventories.* National Greenhouse Gas Inventories Programme/IGES.

Peel, M. C., Finlayson, B. L., and McMahon, T. A. (2007). Updated world map of the Köppen-Geiger climate classification. *Hydrology and Earth System Sciences Discussions* **4**(2), 462.

Neves, C. G., and Dopico, Y. B. C. (2013). *Análise de Metodologias de Produção de Inventários de Gases de Efeito Estufa de Cidades – Rio de Janeiro.* UFRJ/ Escola Politécnica

Rovere, E. L. L., Costa, C., Carloni, F. (2012) *Inventário e Cenário de emissões de gases de efeito estufa da Cidade do Rio de Janeiro – 2005–2025.* COPPEUFRJ/Secretaria Municipal de Meio Ambiente.

U.S. Environmental Protection Agency (EPA). (2015). Developing a greenhouse gas inventory. Accessed October 7, 2015: http://www.epa.gov/statelocalclimate/state/activities/ghg-inventory.html

World Bank. (2012, November). *The Rio de Janeiro Low Carbon City Development Program: A Business Model for Green and Climate-Friendly Growth in Cities Directions in Urban Development.* Urban Development and Local Government Unit, 1–6.

World Bank. (2017). 2016 GNI per capita, Atlas method (current US$). Accessed August 9, 2017: http://data.worldbank.org/indicator/NY.GNP.PCAP.CD

Case Study 4.F

Climate Change Adaptation and Mitigation in Sintra, Portugal

Rita Isabel dos Santos Gaspar and Martin Lehmann

Aalborg University

Keywords	Adaptation, mitigation
Population (Metropolitan Region)	2,812,678 (Statistics Portugal, 2015)
Area (Metropolitan Region)	3,015.24 km² (Statistics Portugal, 2015)
Income per capita	US$19,850 (World Bank, 2017)
Climate zone	Csa – Temperate, dry summer, hot summer (Peel et al., 2007)

The municipality of Sintra is located northwest of the capital city of Portugal (see Case Study 4.F Figure 1) and is known for the fairytale scenery of the Natural Park of Sintra. Due to the close connection between nature and culture, Sintra is classified as World Heritage Cultural Landscape (UNESCO, 2016). This makes Sintra an important tourist area in the metro region of Lisbon. Nevertheless, Sintra also has dense urban areas, with more than 375,000 people living in the municipality.

The local government of Sintra expects that the major impacts of climate change in its territory will result from changes in temperature. It is expected that, by 2100, the average daily temperature in Sintra will be 2–3°C higher in the winter season and 5–10°C higher in the summer season in comparison to the current average daily temperatures (Aguiar and Domingos, 2009).

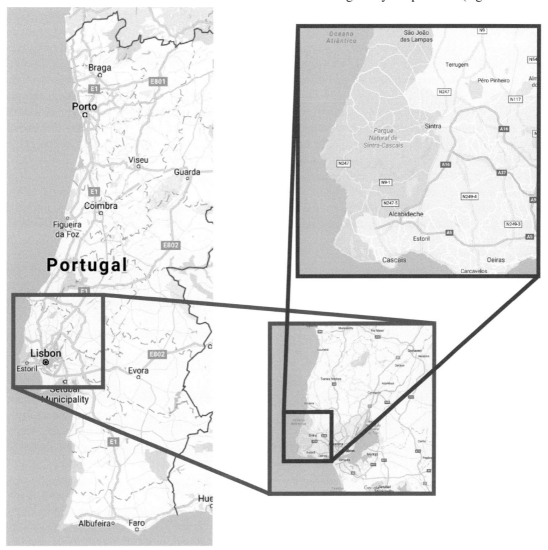

Case Study 4.F Figure 1 *Sintra Municipality location and territory.*

These values may show a spatial variation, being slightly higher in the urban areas (particularly during summer) than in the coastal area of Sintra. Heat waves (with temperature exceeding 35°C) are also expected to be more frequent and longer-lasting in the future (Aguiar and Domingos, 2009).

A consequence of this rise in daily and extreme temperatures is an increased risk of forest fire. In Portugal, the estimated annual cost associated with forest fires is between €60 (US$70.8) million and €140 (US$165.3) million. Apart from the associated costs, forest fires represent a major problem for the municipality because of the great ecological and cultural value, both regionally and nationally, of the Natural Park of Sintra. Case Study 4.F Figure 2 shows the location of the Park in the territory of Sintra. This is a vulnerable zone in matters of forest fires.

Climatic scenarios suggest that, in 2100, the risk of forest fires in Sintra will be 65–100% higher than the present situation (SIAM, n.d. b). The risk of fire is also related to vegetation. Increasing average temperatures have impacted the type of vegetation found in Serra de Sintra. Higher temperatures encourage the proliferation of shrub vegetation, which is prone to combustion (SIAM, n.d. b). Fires in Serra de Sintra are of major concern for the municipality since fire puts at risks not only local populations but also tourists and historic monuments. More than 1 million people visit Sintra's monuments annually. This number is expected to rise with the increased number of warm days (Fundação da FCUL, 2009).

Concerned with these predictions, and in order to reduce the territory's vulnerability to forest fires, the municipality has engaged in climate change mitigation and adaptation planning in its territory. The local government of Sintra has been the key driver in developing climate change strategies for the municipality. Sintra's government considers that it is beneficial to the municipality to have an adaptation and mitigation strategy focused on a regional and local level because, even though climate change is a global issue, its impacts are felt locally (Santos, 2009).

The municipality is collaborating with several partners. The local government collaborates with the CCIAM research group (from the Faculty of Sciences of University of Lisbon) and other institutes, such as the Authority of Civil Protection and the Portuguese Environment Agency, carrying out studies (at a national and regional level) relevant to Sintra's territory. The CCIAM research group is the major partner. The group elaborated a Strategic Plan for the municipality of Sintra to face climate change (PECSAC) by request of the local government of Sintra (SIAM, n.d. a). The plan, adopted in 2009, focuses on the most relevant sectors of the municipality: water resources, coastal areas, biodiversity, forest, agriculture, tourism and leisure, and human health. To further develop PECSAC, climatic and socioeconomic scenarios were used to elaborate an integrated and multisectoral methodology to evaluate impacts and propose adaptation actions to face climate change, following the methodology of the Intergovernmental Panel on Climate Change

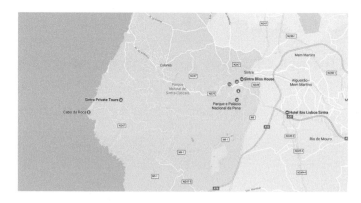

Case Study 4.F Figure 2 *Serra de Sintra.*

(IPCC). The Special Report on Emissions Scenarios (SRES) were, however, not used when planning adaptation actions (Martins, personal communication, February 2015).

After the assessment and identification of vulnerabilities and risks, a list of adaptation measures and processes were set to better prepare the territory and population for the impacts of climate change foreseen in the scenarios. One adaptation measure suggested education of and increased awareness by citizens and tourists in order to prevent behaviors that may increase the risk of fire. The next step in the Sintra adaptation plan is the implementation of the adaptation actions suggested and the involvement of key stakeholders (from the public and private sectors). The final step in Sintra's methodology is the monitoring and evaluation of the actions and adjusting the adaptation strategic plan (Santos, 2009). This is a dynamic process, evolving over time as new data and answers arise.

The mitigation goals of the municipality follow the national goals set by the Kyoto Protocol and the international goals to limit, on a global level, the increase in world temperature to 2°C (Santos F. D., 2009; Aguiar and Ferreira, 2009). Sintra's mitigation strategy focuses on energy efficiency and carbon sequestration actions. The municipality intends to reduce global GHG emissions by implementing local initiatives and by synergies with national policies and measures (SIAM, n.d. c). The mitigation actions proposed by Sintra focus on topics that cross municipal sectorial boundaries, such as buildings, mobility, urban planning, behavior, and renewable energies (biomass and bio-gas) (Santos, 2009).

Sintra's adaptation and mitigation strategic plan is not an isolated project. PECSAC takes into consideration other existing national, regional, and local strategic plans (Santos, 2009), and PECSAC is also integrated into other municipal plans (Martins, personal communication, February 2015).

The municipality is committed to having a strong impact in the adaptation of Portuguese cities to climate change. Sintra collaborates with the ClimAdaPT.Local project (do Vale, 2015), which is a network in which stakeholders share knowledge and experiences with the aim of supporting the development of climate change adaptation strategies at a local level. Sintra, having already adopted a climate change adaptation and mitigation strategic plan, will support other municipalities' work during the project (ClimAdaPT.Local, 2015)

References

Aguiar, R., and Domingos, S. (2009). *Cenários Climáticos de Longo Prazo para o Município de Sintra. INETI e Fundação da FCUL.* Câmara Municipal de Sintra.

Aguiar, R., and Ferreira, A. M. (2009). *Cenários Energéticos e Mitigação no longo prazo para o Município de Sintra.* Project SIAM-Sintra.

ClimAdaPT.Local (2015). Beneficiary Municipalities. Accessed March 10, 2015: http://climadapt-local.pt/en/beneficiary-municipalities/

do Vale, A. Q. (2015). Painel 2 -Experiências Nacionais de Adaptação Local às Alterações Climáticas. Caso PECSAC. Seminário ClimAdaPT. Local - Estratégias Municipais de Adaptação às Alterações Cimáticas em Portugal. Accessed March 10, 2015: http://new.livestream.com/livestreaming-pt/climadaptlocal/videos/73872278

Fundação da FCUL. (2009). Turismo e Lazer sob Alterações Climáticas no Município de Sintra.

Instituto Nacional de Estatistica Portugal (Statistics Portugal) (2015). Statistical Yearbook of Area Metropolitana de Lisboa. Accessed November 3, 2016: https://www.ine.pt/xportal/xmain?xpid=INE&xpgid=ine_

publicacoes&PUBLICACOESpub_boui=279975813&PUBLIC ACOESmodo=2

Peel, M. C., Finlayson, B. L., and McMahon, T. A. (2007). Updated world map of the Köppen-Geiger climate classification. *Hydrology and Earth System Sciences Discussions* **4**(2), 462.

Santos, F. D. (2009). *Plano Estratégico do Concelho de Sintra face as Alterações Climáticas. Introdução.* Projecto SIAM -Sintra. Fundação da FCUL. Câmara Municipal de Sintra.

Santos, F. D., and Miranda, P. (Eds.) (2006). *Alterações Climáticas em Portugal. Cenários, Impactos e Medidas de Adaptação Projecto SIAM II.* Lisbon: Gradiva.

SIAM. (n.d. a). Enquadramento. Alterações Climáticas em Sintra. Proteger o Futuro. Accessed March 10, 2015: http://www.siam.fc.ul.pt/siam-sintra/home.php?id=enquadramento

SIAM. (n.d. b). Florestas e Agricultura: Florestas. Aumento do Risco Meteorológico de Incêndio. Alterações Climáticas em Sintra. Proteger o Futuro. Accessed March 11, 2015: http://www.siam.fc.ul.pt/siam-sintra/home.php?id=florestas

SIAM. (n.d. c). Mitigação: Oportunidade na redução da emissão de gases com efeito de estufa. Alterações Climáticas em Sintra. Proteger o Futuro. Accessed March 12, 2015: http://www.siam.fc.ul.pt/siam-sintra/home.php?id=mitigacao

United Nations Educational, Scientific and Cultural Organisation (UNESCO) (2016). Cultural Landscape of Sintra. Accessed November 11, 2016: http://whc.unesco.org/en/list/723

World Bank. (2017). 2016 GNI per capita, Atlas method (current US$). Accessed August 9, 2017: http://data.worldbank.org/indicator/NY.GNP.PCAP.CD

Case Study 5.A

Urban Regeneration, Sustainable Water Management, and Climate Change Adaptation in East Naples

Mattia Federico Leone

University of Napoli Federico II, Naples

Keywords	Water, hydro-geological risk, flash floods, urban regeneration, Sustainable Drainage Systems (SuDS), adaptation plan
Population (Metropolitan Region)	3,114,000 (SVIMEZ, 2016)
Area (Metropolitan Region)	1,023 km² (Demographia, 2016)
Income per capita	US$31,590 (World Bank, 2017)
Climate zone	Csa – Temperate, dry summer, hot summer (Peel et al, 2007)

East Naples (Napoli), the main expansion area of the Metropolitan City of Naples (about 5 million people, density 2,172 inhabitants per square kilometer (SVIMEZ, 2014), is situated on the alluvial plane of the old Sebeto River (now flowing underground), from the Volla basin (North East) to the Naples port seafront (South). The area is framed on the northwest by the Capodimonte Hill and on the southeast by the Monte Somma-Vesuvio mountain system. Rainwater coming from the hill systems flows almost entirely in the area of the floodplain due to extensive soil sealing following the expansion of the Naples metropolitan area. The reduced depth of the aquifer, combined with high runoff, exposes the entire area to significant hydro-geological risk conditions amplified by increasing extreme precipitation events, thus requiring a more efficient management of flash flood risk.

The shutting down of industrial sites that in the past drained significant amounts of water from the aquifer is "renaturalizing" a traditionally swampy territory, and underground water is increasingly invading foundations, tunnels, and subway areas. Violent tropical-like storms are growing in number and intensity in recent years, and their effects are strongly amplified by urban sprawl. In 2014, two major events struck the area, in June and October, with severe consequences on the built environment and the economy.

Increasing impacts are connected to damage to transport networks and to frequent business interruptions due to flooding of building ground levels hosting productive activities or warehouses, as well as to the reduction in property values due to the risk-proneness of the area. This is leading to significant weaknesses in the implementation of territorial redevelopment plans. Other important impacts have been experienced in the agriculture sector, with large damages to crops and resulting in significant losses in local gross domestic product (GDP).

East Naples is experiencing important territorial and urban redevelopment processes. Large public investments are planned in the area (of about €740 (US$873.8) million), and brownfield

redevelopment is the object of several projects for new mixed-use districts in former industrial areas through the NAPLEST committee (Comitato NaplEST, 2012).

The many issues connected to hydro-geological risk represent a critical factor for public and private investments, and cost-effective adaptation strategies are required to reduce physical and economic impacts in the mid to long term. The complexity of the design and governance processes require a thorough assessment of alternative technological options and design/planning scenarios at building, urban, and territorial scales, supported by multicriteria and cost-benefit analyses, as a key aspect to address public administration policies and to regulate private initiatives.

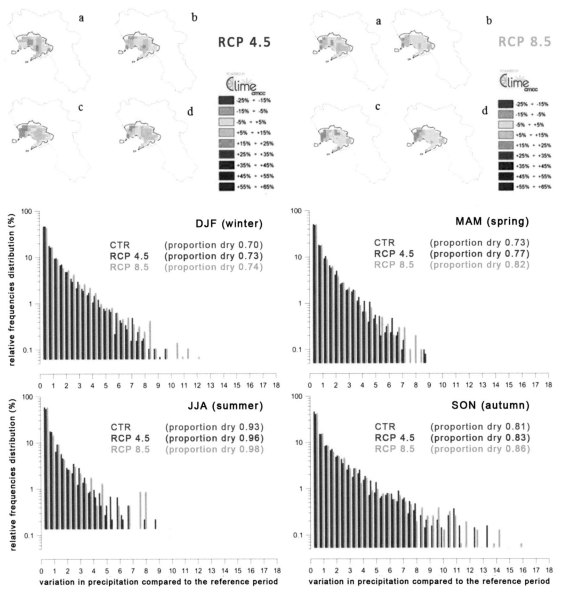

Case Study 5.A Figure 1 *Extreme precipitation scenarios for the period 2021–2050 in Naples Metropolitan Area developed by Euro Mediterranean Centre for Climate Change (CMCC), downscaled according to the RCM COSMO CLM (www. clmassembly.com) through CLIME software. Results, developed for subdaily (a-6h; b-12h; c-24h) and daily (d) scales, show increased extreme events in the autumn-winter season compared to the reference period 1971–2000, of 20% for the RCP 4.5 emission scenario and 40% for RCP 8.5, respectively.*

Source: AdBCC, 2014

Case Study 5.A Figure 2 *Hydrology studies, adaptation plan, and urban regeneration projects in East Naples area.*

Source: UNINA-NAPLEST

The extent of the challenge has created a unique opportunity for a strong multidisciplinary network of different actors at the local level, where local authorities, researchers, and businesses play an active role as action/policy drivers of climate change adaptation. The main stakeholders of this process – Municipality of Naples, Central Campania River Basin Authority (AdBCC), Department of Architecture (DiARC) and PLINIVS-LUPT Study Centre of the University of Napoli Federico II, the Technological District for Safety and Sustainability in Construction (STRESS Scarl), Fintecna (a major public real estate holding company), and Napoli Builders Association (ACEN) – have been committed to identifying adaptive planning and design solutions for the redevelopment of the area, addressing climate change adaptation objectives within the planned urban regeneration investments.

These activities included both hazard and impact modeling based on climate change scenarios for both urban and building design strategies.

Joint efforts by AdBCC and UNINA resulted in the release of the Plan for Hydrogeological Risk-Prone Areas or PSAI (AdBCC, 2014), a unique tool for safeguarding and mitigating risk in the territory through the definition of local guidelines and standards and by updating a previous (2007) version of the plan. The approach is based on the development of supporting tools compatible with models and tools employed by AdBCC, such as the application of CLIME software (developed by Euro-Mediterranean Center for Climate Change [CMCC]) to define 30 years of extreme precipitation simulations in the area based on alternative RCP scenarios. The integration of these simulations within the urban and building design actions allowed a more comprehensive approach to the mitigation of hydro-geological risk and adaptation to extreme precipitation events, thus strengthening the connections among climate science, governance policies, and planning/design solutions.

A more thorough understanding of climate change processes in the area has determined the need to review planned and ongoing urban regeneration actions, thus highlighting weaknesses in conventional approaches (Moccia and Palestino, 2013; Palestino, 2013; Russo, 2012):

- Lack of coordination between sectoral plans and municipal/regional urban and territorial planning (often limited to regulatory advice and not translated into effective local environmental planning and policy actions)
- Persistence of a conflict between municipal and higher level European Union (EU) and Regional planning (the former often based on prescriptive regulations on land, the latter reflecting key environmental issues)
- Experimentation and implementation of good practices for sustainable and adaptive design, such as the redevelopment of the Manifattura Tabacchi area (design by Mario Cucinella Architects), are likely to be undermined if not included in a network-based planning and governance perspective
- "Green infrastructure thinking" needs to be assumed as a guiding principle in municipal and regional planning, with an awareness of the costs and public investment needed

The following measures have been addressed by the NAPLEST Committee and technical advisors from UNINA, focusing on resilience-based design strategies:

- Nearly zero energy standards for new buildings
- Adaptive retrofitting of the existing building stock (buildings and open spaces) through envelope retrofitting (mechanical strengthening, wastewater reduction, etc.) and sanitation systems retrofitting (rainwater harvesting, recycling and reuse of gray water, etc.)
- Sustainable drainage systems (SuDS) implementation, such as permeable paving, stormwater detention, and infiltration solutions
- Evapo-transpiration increase through vegetative surfaces by building green and blue-green infrastructures (bicycle paths and pedestrian areas, green roofs and facades, etc.)

- Technological options and functional-spatial layouts for ground and underground floors of buildings (e.g., pilotis[1], nonresidential uses, water discharge solutions)
- De-sealing of outdoor surfaces and implementation of SuDS for infiltration and rainwater retention, runoff reduction, etc.
- Improvement of wastewater network
- Protection of transport networks and integration of blue and green infrastructures (e.g., greenways, green buffer zones, technical solutions for roads and highways based on surface draining and runoff control)

Lessons learned from the East Naples Case Study show the effectiveness of a networking approach based on knowledge exchange among decision-makers, academics, and practitioners, where assessment tools and cost-effective technological solutions developed through research and decision-support activities allow the integration of climate change adaptation pathways into brownfield and urban regeneration processes.

References

AdBCC (Central Campania River Basin Authority). (2014). *PSAI – Plan for Hydrogeological Risk Prone Areas*.

Comitato NaplEST. (2012). NaplEST: I progetti e le iniziative. Accessed September 21, 2015: http://www.naplest.it/2010/06/10/progetti-naplest-2/

Demographia (2016) Demographia World Urban Areas (Built Up Urban Areas or World Agglomerations). 12th Annual Edition. St. Louis: Demographia. Available from: http://www.demographia.com/db-worldua.pdf.

Grieve, N.F. (2007). *Pilotis*. The Urban Conservation Glossary. Accessed September 20, 2015: http://www.trp.dundee.ac.uk/research/glossary/pilotis.html

Moccia, F. D., and Palestino, M. F. (eds.). (2013). *Planning stormwater resilient urban open spaces*. CLEAN, Napoli.

Palestino, M. F. (ed.) (2013). *Urban open spaces resilient to rainwater under climate change. FARO Project Final Report*. UNINA-DiARC.

Peel, M. C., Finlayson, B. L., and McMahon, T. A. (2007). Updated world map of the Köppen-Geiger climate classification. *Hydrology and Earth System Sciences Discussions* **4**(2), 462.

Russo, M. (2012). Campania. Napoli Est: Prove di sostenibilità. *EcoWebTown. Online Magazine of Sustainable Design*. 5.

SVIMEZ – Associazione per lo sviluppo dell'industria nel Mezzogiorno (2016). Rapporto SVIMEZ 2016 sull'economia del Mezzogiorno. Accessed November 2, 2016: http://www.svimez.info/images/RAPPORTO/materiali2016/2016_11_10_linee_appendice_statistica.pdf

World Bank. (2017). 2016 GNI per capita, Atlas method (current US$). Accessed August 9, 2017: http://data.worldbank.org/indicator/NY.GNP.PCAP.CD

---

1 A piloti is an architectural term referring to posts that support a building, enabling the ground floor to be open for other uses (Grieve, 2007).

Case Study 5.B

Realizing a Green Scenario: Sustainable Urban Design Under a Changing Climate in Manchester, UK

Gina Cavan

Manchester Metropolitan University, Manchester

Aleksandra Kazmierczak

Cardiff University, Cardiff

Keywords	Heat island effect, green space, surface temperature modelling, planning and design
Population (Metropolitan Region)	2,732,854 (Office of National Statistics 2015)
Area (Metropolitan Region)	1,276 km² (Office of National Statistics, 2015)
Income per capita	US$42,390 (World Bank, 2017)
Climate zone	Cfb – Temperate, without dry season, warm summer (Peel et al., 2007)

Executive Summary

Climate change impacts are exacerbated in urban areas by the urban heat island (UHI) effect, which intensifies risks associated with high temperatures to urban populations and infrastructure. Increasing green space, especially in densely built-up areas, is a valuable adaptation response to climate change and addressing the UHI effect. Vegetation influences the microclimate of an urban area, acting to reduce local temperatures through direct shading, evapotranspiration, and storing and reradiating less heat than built surfaces. Urban greening also contributes to creating attractive urban centers essential for economic growth. In urban areas undergoing redevelopment, a collaborative approach with planners, builders, and landowners is required to deliver local action. This case study reports on the modeled impact of redevelopment scenarios based around different amounts of green space on surface temperatures for a knowledge quarter near Manchester City Centre, UK.

Introduction and Case Study Location

Manchester has a pronounced UHI, with temperatures in the core urban areas around 3°C (5°C) warmer on summer days (nights) (Smith et al., 2011). Furthermore, the particular socioeconomic and demographic characteristics of Manchester's communities, in addition to physical environment characteristics, suggests that there are pockets of high social vulnerability to climate change impacts including heat waves and flooding (Kazmierczak and Cavan, 2011). This case study focuses specifically on urban greening as a strategy to cool the urban thermal environment.

This study was conducted in the Oxford Road Corridor area in Manchester, UK. This is a strategically important economic development area and a major transport link, extending south from Manchester City Centre and covering an area of 2.7 square kilometers. The area presents a particular challenge for greening since it is highly built-up, having only two small parks, a few street trees and small areas of green space among buildings, and is largely privately owned.

Methods

A scenario-driven approach was used to investigate the cumulative impact of climate change and land surface cover scenarios on localized maximum surface temperatures (experienced approximately 2 days per summer). Three development scenarios were proposed: Business As Usual (assumes the same 15% green space as the current situation); Deep Green (34% green space); and High Development (4% green space) (Case Study 5.B Figure 1). Future surface temperatures were then modeled under each of the development scenarios using an energy exchange model for the 2050s A1FI emissions scenario – now a freely available online tool – STAR (The Mersey Forest and the University of Manchester, 2011; Cavan et al., 2015). A series of structured interviews were then carried out with a partnership of local stakeholders to (1) transfer the knowledge gained through the scenario modeling exercise, (2) ascertain stakeholder perceptions of the research results, and (3) consider barriers and opportunities associated with implementing the deep green scenario in practice.

Results and Discussion

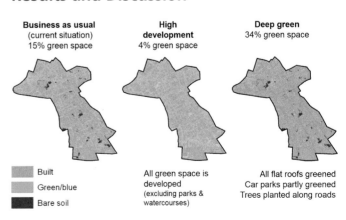

Case Study 5.B Figure 1 *Simulated development scenarios from aerial photograph interpretation.*

Source: Revised from Carter et al. (2015). Base map Crown Copyright and Database Right 2015, Ordnance Survey (Digimap Licence)

Impact of Changing Land Surface Cover on Maximum Surface Temperatures

An increase in green space could help to address rising temperatures associated with climate change (Case Study 5.B Figure 2). If land surface cover ratios remain the same (Business As Usual), climate change will increase maximum surface temperatures by 1.1–3.7°C (2050s A1FI) (UKCP 2009). Under the High Development scenario, projected maximum surface temperatures increase by at least 5°C. In contrast, provision of additional green space under the Deep Green scenario results in around 6°C reduction in projected maximum surface temperatures in relation to the Business As Usual scenario. Around 21% of green space will maintain baseline 1961–1990 temperatures.

Opportunities for Realizing a Green Scenario

Interviews with the local stakeholders revealed that there is a good chance for implementation of elements of the Deep Green scenario. Climate change is perceived to be a significant issue; there are some foundations already in place in the existing local development strategy, and an array of examples of ongoing and planned initiatives of enhancement of green space exist and can stimulate shared learning. Furthermore, positive perceptions of the co-benefits of greening, including an increase in human thermal comfort and quality of life and a reduction in artificial cooling usage (lowering energy bills), could be significant incentives for landowners and developers to invest in green spaces.

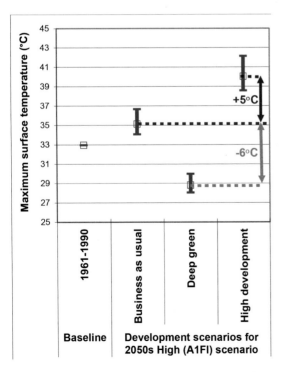

Case Study 5.B Figure 2 *Impact of development scenarios on maximum surface temperatures for the 98th percentile summer day. Bars show the variability of the 100 weather generator runs.*

Source: Revised from Carter et al., 2015. Weather generator data copyright UK Climate Projections 2009.

Financial issues were considered the main challenges to achieving the Deep Green scenario. The stakeholders highlighted that uncertainty about the economic benefits of green space, perceptions that it is not a necessary measure, and long payback times of green space, particularly for green roofs, were all obstacles in justifying investments. Furthermore, maintenance issues, such as the negative impact of trees on utilities, technical difficulties hindering retrofitting green roofs, and uncertainty about maintaining the functionality of green space under changing climate conditions, were highlighted as key threats in realizing the Deep Green scenario.

Transferable Lessons

The high density of cities and restricted opportunities for urban greening present a key barrier affecting local capacity to adapt urban areas. Furthermore, the high private land ownership in cities and low amount of publicly owned land awaiting redevelopment means that planners have limited scope to deliver adaptation actions. A collaborative approach to adaptation, with planners working together with building and land owners, is essential at the local scale. A further key lesson from this case study is the usefulness of quantitative information to clearly illustrate the extent of UHI and climate change to stakeholders, thus helping to build their knowledge of the impact on their specific assets and thus stimulate willingness to take appropriate actions.

This research was conducted as part of the EcoCities project, a joint initiative between the University of Manchester and Bruntwood; further details are available at www.adapting-manchester.co.uk. This contributes to the Greater Manchester Evidence Base for Sustainable Urban Development and Manchester: A Certain Future – the city's shared plan to tackle climate change.

References

Carter, J. G. Cavan, G. Connelly, A. Guy, S. Handley, J., and Kazmierczak, A. (2015). Climate change and the city: Building capacity for Urban Adaptation, *Progress in Planning* **95**, 1–66.

Cavan, G., Butlin, T., Gill, S., Kingston, R., and Lindley, S. (2015). Web-GIS tools for climate change adaptation planning in cities. In Walter L. F. (ed.), *Handbook of Climate Change Adaptation*, pp. 1–27. Springer,

Climate-Data. (2015). Climate-Data.org. Accessed October 9, 2015: http://en.climate-data.org/search/

Kazmierczak, A., and Cavan, G. (2011). Surface water flooding risk to urban communities: Analysis of vulnerability, hazard and exposure, *Landscape and Urban Planning* **103**(2), 185–197.

Manchester City Council (2015). Manchester's population, ethnic groups and ward map. Accessed November 3, 2016: http://www.manchester.gov.uk/info/200088/statistics_and_census/438/public_intelligence/3

The Mersey Forest and the University of Manchester (2011). The STAR Tools. Accessed May 3, 2014: http://maps.merseyforest.org.uk/grabs/

ONS. (2011b). 2011 Census: Digitised boundary data (England and Wales). UK Data Service Census Support. Accessed September 21, 2015: http://edina.ac.uk/census

Office of National Statistics (2015). Population Estimates for UK, England and Wales, Scotland and Northern Ireland: mid-2015. Accessed November 3, 2016: https://www.ons.gov.uk/peoplepopulationandcommunity/populationandmigration/populationestimates/bulletins/annualmidyearpopulationestimates/mid2015

Peel, M. C., Finlayson, B. L., and McMahon, T. A. (2007). Updated world map of the Köppen-Geiger climate classification. *Hydrology and Earth System Sciences Discussions* **4**(2), 462.

Smith, C. L., Webb, A., Levermore, J., Lindley, S.J., and Beswick, K. (2011). Fine-scale spatial temperature patterns across a UK conurbation, *Climatic Change* **109**(3–4), 269–286.

UK Climate Projections (UKCP). (2009). Weather generator data. Accessed September 21, 2015: http://ukclimateprojections.defra.gov.uk

World Bank. (2017). 2016 GNI per capita, Atlas method (current US$). Accessed August 9, 2017: http://data.worldbank.org/indicator/NY.GNP.PCAP.CD

Case Study 5.C

New Songdo City: A Bridge to the Future?

Glen David Kuecker

DePauw University, Greencastle

Keywords	Eco-city, smart city, ubiquitous city, mitigation, planning and design
Population (Metropolitan Region)	45,000 (Gale International, 2016)
Area (Metropolitan Region)	6 km² (Songdo IBD, 2015)
Income per capita	US$27,600 (World Bank, 2017)
Climate zone	Dfa – Cold, without dry season, hot summer (Peel et al., 2007)

The Republic of South Korea's New Songdo City is an urban are occupying 1,500 acres of "reclaimed" land from the Yellow Sea. It is governed by the Incheon Metropolitan City's Free Economic Zone, which is part of the national commitment to develop a green economy. Planners designed New Songdo City to be an international business center that takes advantage of South Korea's geographic location within Northeast Asia. In 2001, the City of Incheon contracted U.S.-based Gale International, a property development firm, to develop the city in a joint partnership with POSCO, the South Korean construction conglomerate(Southerton, 2009). Gale International contracted the architectural firm, Kohn, Pedersen, Fox Associates to design the city. At a cost of US$35 billion, news media sources frequently label it the largest private development project in history. Construction began in 2004 with the landfill, and, by 2014, an estimated 70% of the city was complete. Developers project the city to have a residential population of 65,000 and a workforce of 300,000. While still under development, New Songdo City is a fully functional urban space.

New Songdo City brings together multiple design concepts that potentially make it a bridge to the future of sustainable urbanism. It has the highest concentration of LEED certified buildings in the world and several, such as the Sheraton Hotel,

have won international awards for their sustainable design (Arbes and Bethea, 2014). Green spaces constitute 40% of the landscape (KPF, 2014), and planners designed them for enhancing air circulation, assisting with rain runoff, and promoting pedestrian and bicycle traffic. It has extensive bike trails, and wide, safe sidewalks intended to reduce car dependence. New Songdo City has a cutting-edge waste collection system, one that uses a pneumatic system that takes garbage from offices or residencies to the waste disposal facility. The system does not require garbage trucks (Shwayri, 2013), which reduces carbon emissions and traffic congestion. The city has a network of buses that circulate people within the city but also provides links to Incheon's urban center, the International Airport, and to Seoul. An underground metro line runs around the city's circumference and also links to Incheon and Seoul. New Songdo City also aims to generate efficient energy use through "ubiquitous" Internet technologies designed by Cisco System's "Smart + Connected Cities" program. Ubiquitous technology uses the Internet to link hardware and software to monitoring systems to generate efficient resource consumption (Yingitcanlar and Lee, 2014). Consequently, Songdo consumes 40% less energy per capita than other cities of similar scale.

Gale International, Incheon Metropolitan City, and the Republic of South Korea clearly made strides toward creating a more sustainable urban form in Songdo City. Building from the ground up offers significant advantages for sustainable design because planning does not run into the obstacles of a built environment as well as the customary practices of citizens living in an already existing city. Especially considering that several built-from-scratch eco-cities, such as China's Dongtan eco-city, a zero-carbon project, have failed, New Songdo City appears to be among early success stories, along with Abu Dhabi's Masdar City. Marking its development as a global business hub, the United Nations selected New Songdo City to host the offices of the Green Climate Fund in October 2012.

There are, however, several climate adaptation limitations with the New Songdo City project. First, its creators have not yet established the explicit goal of creating a carbon-neutral city, which was the stated objective for both Dongtan and Masdar City. Second, New Songdo City remains an urban form designed to accommodate the car. City streets are wide boulevards, capable of handling high volumes of car traffic. Intersections are regulated by traffic lights, with timing best suited for the automobile instead of the pedestrian. Planners built a massive underground parking lot, one that encompasses the entire city core and that has the goal of allowing car users quick and easy access to their destination, a convenience that encourages driving over public transit. Third, New Songdo City still has an urban metabolism defined by late capitalism's cradle-to-grave production and consumption system. Despite attempts to close the loop in garbage handling, the city remains an open-loop system (Kuecker, 2013b). Because of its dependence on material supplies from outside the city, New Songdo City contributes to increasing levels of carbon emissions. Fourth, while the city design resulted in an admirable commitment to LEED standards, the jury itself is still out about the carbon footprint of these certified buildings (Ko et al., 2011). Building materials expend carbon in their extraction, production, transportation, and construction. A building may well be climate friendly in its design, but much depends on the habits of its clients, such as their use of air conditioning, lighting, the density of people in work spaces, or the placement and design of furnishing and decorations. Sixth, the city has high electrical demands, especially for running the city's ubiquitous platforms, as well as cooling systems that have a significant energy demand in the South Korean summer. Seventh, the city has eco-unfriendly features, such as the Jack Nicklaus-designed golf course, which is proudly featured in city marketing (Southerton, 2009). Finally, the goal of creating a regional hub locks city users into a jet-set lifestyle, one that burns carbon on each flight to Tokyo, Shanghai, or Hong Kong while fostering a culture of consumption. From this view, New Songdo City lacks the cradle-to-cradle, closed-loop design required for a truly sustainable design needed for successful climate adaptation and mitigation.

Some critics also raise questions about New Songdo City's reputation for sustainability. One of the top concerns is the branding efforts of New Songdo City's developers, who engage in an impressive marketing campaign designed to get tenants for the office towers and residents for the apartments. Gale International, for example, explicitly states that their main goal is to create a city with the highest "quality of life" (Kim, 2010). While their goal may translate into high Human Development Index statistics, close analysis reveals an urban form that replicates features of a gated community built for a transnational class of business elites and wealthy refugees from the inconveniences of a megalopolis like Seoul. In this urban model, the sustainable city becomes a marketable commodity available to the highest bidder, set against the minimal standard of life for billions of urban dwellers. Matched with agendas like the United Nation's Sustainable Development Goals, New Songdo City appears an inappropriate response to the challenges of climate change.

In New Songdo City's short history, it has undergone several re-brandings: business hub, smart city, global university, biotech and hospital hub, and eco-city. The instability reveals uncertainty about how this new urban form fits within 21st-century urbanism. Some see it as a "bridge to the future," a "test-bed" (Kim, 2014) of new technologies and innovative design that serves as a necessary step in climate mitigation and adaptation. Others debunk the bridge metaphor, preferring to view New Songdo City as representing the maladaptive propensities of capitalist markets (Kuecker, 2013a). As investments at unprecedentedly high scales of social and financial capital are made into the building of eco-cities in China, South Asia, and Africa, it is important that this new urban form is critically examined to determine if it is indeed the bridge to the future. These doubts extend beyond New Songdo City and call into question the depth and substance of South Korea's highly publicized green economy agenda.

References

Arbes, R., and Bethea, C. (2014). Songdo, South Korea: City of the Future. Accessed September 21, 2015: http://www.theatlantic.com/international/archive/2014/09/songdo-south-korea-the-city-of-the-future/380849/

Gale International. (2016). Songdo International Business District To Be Featured at Greenbuild 2016 as Exemplar of Sustainable New City Accessed November 3, 2016: http://www.galeintl.com/2016/10/04/songdo-international-business-district-featured-greenbuild-2016-exemplar-sustainable-new-city/

Kim, C. (2010). Place promotion and symbolic characterization of New Songdo City, South Korea. *Cities* 27(1), 13–19.

Kim, J. I. (2014). Making cities global: The new city development of Songdo, Yujiapu and Lingang. *Planning Perspectives* 29(3), 329–356.

Ko, Y., Schubert, D., and Hester, R. (2011). A conflict of greens: Green development versus habitat preservation– The case of Incheon, South Korea. *Environment: Science and Policy for Sustainable Development*. Accessed April 29, 2015: http://www.environmentmagazine.org/Archives/Back%20Issues/2011/May-June%202011/conflict-of-greens-full.html.

Kohn Pedersen Fox Associates. (n.d.). *New Songdo City; Green City*. Accessed April 27, 2015: https://www.kpf.com/projects/new-songdo-city

Kuecker, G. D. (2013a). South Korea's New Songdo City: From neo-liberal globalisation to the twenty-first century green economy. *Papers of the British Association for Korean Studies* 15(2013), 20–36.

Kuecker, G. D. (2013b). Building the bridge to the future: New Songdo city from a critical urbanism perspective. SOAS-AKS Working Papers in Korean Studies II. School of Oriental and African Studies, University of London. Accessed September 21, 2015: http://www.soas.ac.uk/koreanstudies/overseas-leading-university-programmes/soas-aks-working-papers-in-korean-studies-ii/

Peel, M. C., Finlayson, B. L., and McMahon, T. A. (2007). Updated world map of the Köppen-Geiger climate classification. *Hydrology and Earth System Sciences Discussions* 4(2), 462.

Shwayri, S. (2013). *Journal of Urban Technology* 20(1), 39–55.

Songdo IBD (2015). Masterplan. Accessed November 3, 2016: http://songdoibd.com/about/#masterplan

Southerton, D. (2009). *Chemulpo to Songdo IBD: Korea's International Gateway.*

World Bank. (2017). 2016 GNI per capita, Atlas method (current US$). Accessed August 9, 2017: http://data.worldbank.org/indicator/NY.GNP.PCAP.CD

Yigitcanlar, T., and Lee, S.H. (2014). Korean ubiquitous-eco-city: A smart-sustainable urban form or a branding hoax? *Technological Forecasting and Social Change* 89, 100-114

Case Study 5.D

Adaptation in Rotterdam's Stadshavens: Mainstreaming Housing and Education

Tom A. Daamen

TU Delft

Jesse M. Keenan

Harvard University Cambridge, MA

Keywords	Urban flooding, sea level rise, adaptation policy
Population (Metropolitan Region)	1,173,561 (Rotterdam The Hague Metropolitan Area, 2016)
Area (Metropolitan Region)	990 km² (Rotterdam The Hague Metropolitan Area, 2016)
Income per capita	US$51,210 (World Bank, 2017)
Climate zone	Cfb – Temperate, without dry season, warm summer (Peel et al., 2007)

In Rotterdam, the Netherlands, a 16 square kilometer area called Stadshavens or "City Ports" (Case Study 5.D Figure 1) sits between the city center and Europe's leading seaport, the Europoort. The Stadshavens area has become the focal point of the Rotterdam Climate Initiative (RCI). Launched in May 2007 by former Rotterdam Mayor Opstelten during a global summit[1] hosted by former U.S. President Bill Clinton and Mayor Michael Bloomberg in New York, the RCI set forth ambitions to spur projects that aimed to promote the development of a sustainable city that would be at the forefront of research and innovation in climate change adaptation.

National Policy, Local Agenda

The Stadshavens project was initiated in 2002 by the City of Rotterdam as what was then perceived by planning officials as a conventional post-industrial waterfront redevelopment assignment. Port facilities were expected to migrate to other areas in the Port of Rotterdam less encumbered by physical and regulatory

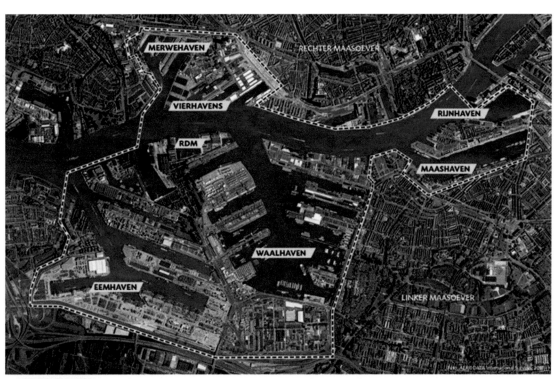

Case Study 5.D Figure 1 *Stadshavens Port Area, Rotterdam, in 2014.*

Source: Stadshavens Rotterdam Programme Office

---

1   C40 Large Cities Climate Summit, May 14–17, 2007, New York.

restrictions. This process was expected to create opportunities to redevelop parts of the Stadshavens area into urban residential waterfront neighborhoods (Daamen, 2010). However, these expectations soon had to be revised because the Rotterdam Port Authority (RPA) – the area's principle landlord – favored an economic development perspective that focused on continued port-industrial uses and combined port-related service functions. The RPA had a strong case because few of the companies in the Stadshavens area were inclined to move their business further downstream to the more modern Europoort and Maasvlakte areas.

While discussions in Rotterdam developed around a revised development trajectory for the Stadshavens area, climate change emerged more emphatically within the Dutch national policy agenda. The National Spatial Strategy established in 2004 led to a period of policy reframing efforts that favored a "working with the water" over the prevailing "fighting against the water" rhetoric common in the economic development discourse. Several national planning programs and decisions followed that focused on broadly preparing the country for the consequences of climate change, such as the Delta Commission Report of 2008. Water-sensitive development and flood risks were dominant topics in these policy briefs, and Rotterdam was soon identified as one of the cities in the country most vulnerable to flood risks within the context of urban development (Lu and Stead, 2013).

Mainstreaming

The Dutch national climate change policies have, in part, been translated by local governments through a process of "mainstreaming," wherein a larger set of social, economic, public health, water management, and other public policy domains are amended to operationalize the broader intent of the national policy to address climate change. By 2008, the Stadshavens area was designated as a potential testing ground for climate change adaptation measures and pilot projects. The area is largely situated outside of the city's protective dike structure, and studies showed that the village of Heijplaat – located in the center of the area – was particularly vulnerable to floods. Likewise, the village housed a particularly sensitive low-income community that was struggling to survive. Given its unique positioning, Heijplaat was selected as an initial testing ground for mainstreaming[2] a climate policy within existing housing and education policy domains. With a clear site and a broad set of objectives, the city and the RPA sought to negotiate a new agreement. In addition, two local educational institutions and a housing association – owner of most of the social housing – soon became major partners in revitalizing the village of Heijplaat and redeveloping the adjacent facilities of the Rotterdam Dry Dock Company (RDM).

Flood protection measures were included into a more specific agreement for the village in 2012, presenting a larger framework of economic, social, and environmental values and objectives under the overall framework of sustainability. Flood protection measures

in the Heijplaat/RDM area are based on the idea of "multilevel security." This concept dictates that protective measures like dams and dikes are complemented by flood risk reduction measures at the district and building scales. For Heijplaat, this has been translated into the development of a small dike (3.6 m above sea level) between the water's edge and the new west side of the village, elevating vulnerable new development locations and designing vital infrastructure, public spaces, and new homes in such a way that flood risks are reduced (Richter, 2014). In addition, new housing prototypes were developed that include passive and autonomous systems that not only reduce risk from flooding but also serve climate mitigation ends through a lower "clean energy" carbon footprint. Although the recent downturn in the Dutch economy has impacted the pace of implementation, partners at Heijplaat seem determined to advance their integrative efforts and turn the village into a showcase of sustainability and resilience.

Aside from housing, economic development was advanced with the RDM shipyard being repositioned as a Research, Design, and Manufacturing campus (Vries, 2014). Together with local educational institutions, the RDM facilities were thus redeveloped to accommodate technical education and research. In addition, space was developed to house start-up companies ranging from an emergency housing material supplier to a manufacturer of electric motorcycles. The organizational and spatial configuration of RDM is designed in such a way that, as students develop their technical competencies, they can laterally transition into providing a well-trained workforce for the emerging companies housed on-site. Likewise, with a capacity to accommodate scientific and engineering research, students at all levels are reciprocally engaged in developing applications, techniques, materials, and designs tested and evaluated on-site. This vertical alignment of education and commerce has served as a model for similar post-industrial facilities around the world.

Advancing Capacities

The next phase under way at RDM/Heijplaat includes a focused expansion of the facilities and corresponding programs to advance capacity for designing and constructing floating structures and water-dependent infrastructure. By building a political coalition around a diverse set of goals including housing and education, Rotterdam has advanced an adaptive capacity that goes beyond physical interventions to the inclusion of social systems that are perhaps the first line of defense in the face of climate change.

References

Daamen, T. (2010). *Strategy as Force: Towards Effective Strategies for Urban Development Projects: The Case of Rotterdam City Ports*. IOS Press.

Lu, P., and Stead, D. (2013). Understanding the notion of resilience in spatial planning: A case study of Rotterdam, The Netherlands. *Cities* **35**, 200–212.

---

2 *Mainstreaming* is a technical concept for the incorporation of climate policies through various horizontal policy domains. See generally, Rauken, T., Mydske, P. K., and Winsvold, M. (2015). Mainstreaming climate change adaptation at the local level. *Local Environment* **20**(4), 408–423.

Peel, M. C., Finlayson, B. L., and McMahon, T. A. (2007). Updated world map of the Köppen-Geiger climate classification. *Hydrology and Earth System Sciences Discussions* **4**(2), 462.

Richter, M. (2014). Private sector participation in flood proof urban development projects: A comparative case study of integrating flood protection measures into urban development projects in the Netherlands. Thesis. Rotterdam/Delft, NL: Master City Developer.

Rotterdam The Hague Metropolitan Area (2016). Municipalities. Accessed on November 3, 2016: http://mrdh.nl/gemeenten

Stadhavens Rotterdam Programme Office (2014). Stadhavens Rotterdam. Accessed on November 3, 2016: http://stadshavensrotterdam.nl/en/

World Bank. (2017). 2016 GNI per capita, Atlas method (current US$). Accessed August 9, 2017: http://data.worldbank.org/indicator/NY.GNP.PCAP.CD

Vries, I. M. J. (2014). From shipyard to brainyard: The redevelopment of RDM as an example of a contemporary port-city relationship. In Alix, Y., Desalle, C., and Comtois, B. (eds.), *Port City Governance* (107–126). EMS.

Case Study 5.E

Climate Change Mitigation in a Tropical City: Santo Domingo, Dominican Republic

Richard Gonzales, AIA

Richard Gonzales Architect, PLLC, New York

Keywords	Natural disasters, urban transportation, forestry conservation, urban planning and design
Population (Metropolitan Region)	3,020,000 (United Nations, 2016)
Area (Metropolitan Region)	427 km² (Demographia, 2016)
Income per capita	US$6,390 (World Bank, 2017)
Climate zone	Am – Tropical monsoonal (Peel et al., 2007)

Introduction

The Dominican Republic needs to justify the environmental impacts of climate change to ensure the country's future growth. As an island nation, threats will be experienced on the three coastal sides of the country as well as on its interior. It will experience increased population shifts as migration gravitates from rural communities to urban environments. Such influx will contribute to urban carbon emissions, and the city's infrastructure can be strained in regard to building operation, transportation, and energy provision. Urban projects will need to address such impacts and seek to reduce greenhouse gas (GHG) emissions.

Santo Domingo: The Oldest City in the New World

The impacts of climate change will affect cities in various ways and disturb current economic, cultural, and social balances. The city of Santo Domingo has been a model for other similar metropolitan areas in the Caribbean context. With its rapid growth and investment in economic sectors, the city has emerged rapidly as an economic center that attracts investment and development opportunities. This model has been considered by other countries (including Haiti) as an example of how smart investment can foster job growth and infrastructure investment within a short time span.

Santo Domingo is in a position to be a leader in the reduction of GHG emissions because internal investment and support is readily available to foster such initiatives. A framework and the necessary tools, such as those used by New York's PlaNYC 2030, can lead sustainability initiatives (New York City Office of Long-Term Planning and Sustainability, 2007).

Island in Context

The Dominican Republic shares the island with the neighboring country of Haiti, occupying two-thirds of the island territory. It has 49,000 square kilometers with 1,288 kilometers of coast lines. Forty percent of the land cover is forested. The Dominican Republic is strategically located between neighboring countries in the Caribbean, Central, South and North America.

Learning from History

The Dominican Republic was the first settlement in the new world, founded by Christopher Columbus during his first travels across the Atlantic Ocean. The City of Santo Domingo was founded in 1496; it is 519 years old and is the oldest functioning city in the Western Hemisphere.

Preserving and Ensuring the Island Economy

With a US$61.6 billion gross domestic product (GDP; World Bank, 2013), the Dominican Republic maintains its economic growth primarily through its service industries that include business and finance investments. Its manufacturing industry includes mining of nickel, gold, and silver; and it hosts a textile industry. Its agricultural sector is a strong exporter of organic fruits and vegetables, tobacco, and sugar. The island maintains a strong tourism industry, with nearly 3 million foreign visitors on a yearly basis.

Population Growth and Forecast

Approximately 10.4 million people live in the Dominican Republic, and currently 70.8% of the population lives in urban areas (United Nations Statistics Division, 2013); 3 million people reside in Santo Domingo alone. It is estimated that the population will increase to 12.1 million by 2030 (Euromonitor International, 2013).

Sustainable Networks of Connectivity through Infrastructure

Since 1986, the country has sustained a resilient construction industry. Capital investment from both public and private sectors has focused on the transportation industry, primarily through road construction. The country has ten international airports with fourteen shipping ports along its costal edges.

Natural Disasters: A Methodology for Resilience Planning

The Dominican Republic is a hot spot for natural disasters. The island has a history of floods, storms, earthquakes, landslides, and tsunamis (the last recorded tsunami occurred in the north province in 1946), tracked since record keeping began in the 1500s. The World Bank indicates the island is at risk to hurricanes, landslides, and earthquakes, in order of priority. About 33.75% of the land area is estimated to be affected, with 66% of the total population projected to be high risk (Dilley, 2005) in the aftermath of a disaster.

Urban Sprawl

Santo Domingo is a dense urban environment. Santo Domingo's growth is directed toward the north and west of the city. These districts are bounded by waterfronts, fault lines, and steep terrain. The majority of commercial and higher end residential building construction is conducted under the rules and regulations required by local building codes, but there are also residential communities expanding through "automated construction," a method whereby individuals utilize construction methods not coherent with local codes or building ordinances

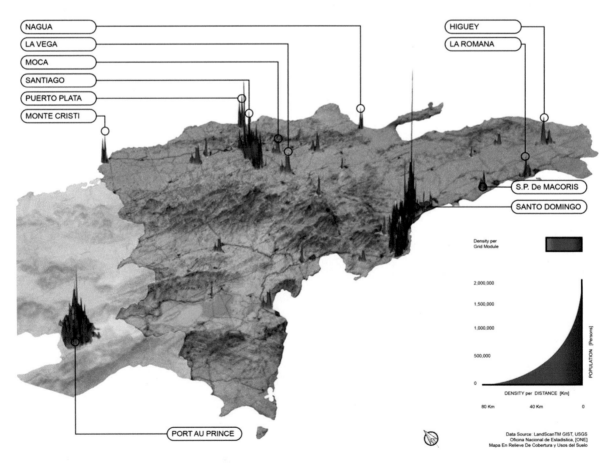

Population density concentration spike mapping on the island of Hispaniola- Dominican Republic.

Image Credit: Richard Gonzalez, AIA

and subjective to individual knowledge. The control and oversight of these building structures remain of concern because many may vulnerable to impacts on life and safety.

Mitigation Strategies: The Shifting of GHG

The Dominican Republic aims to reduce GHG emissions by 25% with a targeted date of 2030 (International Partnership on Mitigation and MRV, 2014). This was announced at a UN Climate Change conference (COP18) hosted in Doha, Qatar. This incentive includes financial support, with the government of Germany as a partner. The three task areas include economic development, reduction of poverty, and securing social inclusion (National Council on Climate Change, 2012). The Dominican Republic has also identified four sectors for the reduction of GHG: transportation, energy, forestry conservation, and water management.

The country is investing largely in its transportation infrastructure. Santo Domingo is expanding its mass transportation system with an underground metro subway. In its second phase of completion, the system will expand to service the east-west districts of the City, with future plans to expand toward the North.

Although the demand for energy outweighs the supply, the Dominican Republic is investing in several alternatives for energy production, including a focus on hydro plants near rivers and solar and wind farms in certain sectors of the country.

Forestry conservation plays a major role in the reduction of carbon. The Dominican Republic plans to conserve existing undisturbed parks and ecological habitats in a nationwide effort. There is also a tree planting initiative for the city of Santo Domingo (Office of the National District, 2010) and a GreenBelt project highlighting eight park areas linked via a series of pathways and water channels (Grupo Tierra Dominicana, 2008).

Water management plays a key issue in regards to catchment, processing, and distribution. The city of Santo Domingo has experienced water overflows and flooding in the aftermath of strong tropical storms. The public works department is planning on expanding the city's sewer lines, which currently services 40% of the city. The system will implement new street catchments,

distribution lines, and treatment plants. This system will also address waste mitigation because small items constantly congest the distribution of water in the underground piping.

References

Demographia (2016) Demographia World Urban Areas (Built Up Urban Areas or World Agglomerations). 12th Annual Edition. St. Louis: Demographia. Available from: http://www.demographia.com/db-worldua.pdf.

Dilley, M., Chen, R. S., Deichmann, U., Lerner-Lam, A. L., and Arnold, M. (2005). *Natural Disaster Hotspots: A Global Risk Analysis.* World Bank.

Euromonitor International. (2013). Dominican Republic in 2030: Future demographic. Accessed May 21, 2015: http://www.euromonitor.com/medialibrary/PDF/Future-Demographic-Dominican-Republic.pdf

Grupo Tierra Dominicana. (2008). Ocho Zonas Ambientales. Accessed May 21, 2015: http://idbdocs.iadb.org/wsdocs/getdocument.aspx?docnum=36483011, p. 11.

International Partnership on Mitigation and MRV. (2014). The Dominican Republic commits to a 25% reduction in greenhouse gas emissions by 2030. Accessed May 21, 2015: http://mitigationpartnership.net/dominican-republic-commits-25-reduction-greenhouse-gas-emissions-2030

National Council on Climate Change. (2012). National Development Strategy. Accessed May 21, 2015: http://www.uncclearn.org/sites/www.uncclearn.org/files/images/estrategia__nacional__para__fortalecer__los_recursos_humanos_republica_dominicana_08_2012.pdf

New York City Office of Long-Term Planning and Sustainability. (2007). PlaNYC 2030: A Greener Greater New York. Accessed May 21, 2015: http://www.nyc.gov/html/planyc/downloads/pdf/publications/full_report_2007.pdf

Office of the National District. (2010). Trees of Santo Domingo. Accessed May 21, 2015: http://www.adn.gob.do/joomlatools-files/docman-files/Arboles%20de%20Santo%20Domingo%20INTEC%20JICA%20ADN%202010%20AR(2).pdf

ONE. (2010). Oficina Nacional de Estadistica. Accessed May 21, 2015: www.one.gob.do

Peel, M. C., Finlayson, B. L., and McMahon, T. A. (2007). Updated world map of the Köppen-Geiger climate classification. *Hydrology and Earth System Sciences Discussions* **4**(2), 462.

United Nations Statistics Division (UNSD). UNData Country Profile 2013. Accessed July 16, 2015: https://data.un.org/CountryProfile.aspx?crName=DOMINICAN%20REPUBLIC

World Bank. (2017). 2016 GNI per capita, Atlas method (current US$). Accessed August 9, 2017: http://data.worldbank.org/indicator/NY.GNP.PCAP.CD

World Bank IBRD IDA. (2013), Accessed May 21, 2015: http://www.worldbank.org/en/country/dominicanrepublic

United Nations, Department of Economic and Social Affairs, Population Division (2016). The World's Cities in 2016 – Data Booklet (ST/ESA/SER.A/392).

Case Study 6.A

The Community-Driven Adaptation Planning: Examining the Ways of Kampungs in North Coastal Jakarta

Hendricus Andy Simarmata

Universitas Indonesia/Indonesian Association of Urban and Regional Planners, Jakarta

Keywords	Flood, community-based adaptation, planning, environmental justice
Population (Metropolitan Region)	10,483,000 (United Nations, 2016)
Area (Metropolitan Region)	3,225 km² (Demographia, 2016)
Income per capita	US$3,400 (World Bank, 2017)
Climate zone	Am – Tropical, monsoon (Peel et al., 2007)

Jakarta, the biggest mega-coastal city in the Southeast Asia region inhabited by more than 10 million people, has experienced climate-related disasters – especially flood events – for decades. Located on land that is sinking at a rate of about 2 centimeters per year over the past three decades, the north coast of Jakarta is potentially at risk of flood from sea level rise, high tides, and extreme rainfall (DKI Jakarta, 2012a). In addition to major infrastructures (e.g., power plants, national harbor, and logistic centers), the northern coastal area is also shaped by dozens of poor informal settlements called named *kampung*. More than 60% of Jakarta's population live in *kampungs* and manage to survive, although their living places are vulnerable to flooding. How they prepare their settlements and deal with flood risks is a major resource that should be incorporated into urban adaptation planning. This form of knowledge needs to be studied in order to increase the effectiveness of urban adaptation.

The Indonesian Association of Urban and Regional Planners (IAP) and the government of DKI Jakarta Province, supported by START (Global Change System for Analysis, Research, and Training), have initiated community-driven adaptation planning in the *kampungs* of Kamal Muara and Kebon Bawang. The initiative aims to know how *kampung* residents plan to adapt their settlements to current and increasing flood risk, how they would provide design guidelines for existing and new developments, and how they control the uses of the land and buildings in their own areas (IAP, 2012). The government can benefit from these types of knowledge in developing effective adaptation strategies, thus complementing the spatial and development planning process. In addition, the collaborative process among *kampung*

residents, planners, and government officers has also bridged different perceptions of the meaning of flood-related vulnerability and adaptation itself.

Both *kampungs* are extremely high-density settlements. For example, Kamal Muara is inhabited by 4,200 people on only 7 hectares. Nearly 90% of the *kampung* area is occupied by buildings, with only a few open spaces left. The residences, small shops, and public facilities, such as mosques and communal bathrooms, are the dominant land uses. Narrow alleys structure the settlement area. The *kampung* residents are mostly low-income groups that work as casual workers, street vendors, and fishermen. They consciously perceive that their living place is a flood-prone area due to extreme rainfall, tidal flood, or overflowing rivers. They know that when a flood occurs, the physical environment not only becomes water-inundated, but also highly polluted with garbage-laden water. By shifting our focus in vulnerability and adaptation assessment from technical experts to a people-centered approach, we can reveal the adaptive practical knowledge of the urban poor.

Kampung residents have locally embedded knowledge of adapting their settlements to floods (Simarmata, 2015). They can intuitively map the inundated area, sense the frequency of flooding, and work together to overcome the impact of floods. They use local resources, such as wood, plywood, bamboo, and sand sacks to raise alley pathways. They have a locally embedded adaptation plan (LEAP) to manage the functionality of houses and follow evacuation strategies when floods come. They can adapt to the flooding but fail to deal with pollution and seawater intrusion. Therefore, the *kampung* residents recognize that they need an adaptation plan, with an aim of not just stopping or avoiding flood events, but of minimizing the pollution and building water protection infrastructure. Case Study 6.A Figure 1 shows how *kampung* resident identified the root causes of their problems through a series of group discussions. They shared the problems that are caused by the floods and prioritized them for development of adaptation strategies.

Public facility improvement was a top priority for private space adaptation. In the group discussion, they agreed to put water infrastructure improvements as top priorities, such as drainage (no. 1) and clean water supply (no. 3). *Kampung* residents also understood the importance of design guidelines for their settlement. They discussed the design principles that were initially introduced by the planners participating in the discussions. They argued that the water infrastructure was ultimately the best way to create adaptive *kampungs*, which should be achieved by applying amphibious building concepts, increasing houses' permeability to water, and accelerating water flow to the estuary. Therefore, they recommended that the stilt house be a preferred building type, that drainage levels should always be

Case Study 6.A Figure 1 *Root causes analysis in Kamal Muara.*

Source: PICAS report, 2012; Photo: Paramitha Yanindaputri

below the road, and that public facilities should be built above the projected flood level and serve as a multipurpose space. Last, alleys should be designed as social spaces as well.

The dynamic conversation also covered how *kampung* residents control land use and building development in their areas. They do not have rules for managing density and did not know that the government has special zoning regulations for floodplain zones. Having had the discussion, they suggested three categories to characterize flood areas in their areas: low-level flood (20 cm), middle-level (20–60 cm), and high-level (60 cm) (DKI Jakarta 2012b). These categorizations resulted from the reflective process on the series of flood events experienced in their areas. They also actively contributed to the discussion on what type of regulations would properly apply for their areas.

The three main parts of the community-driven planning have opened the eyes of government and planners on how local people have the capacity to manage their own settlements to withstand flood impacts. Even though the existing practices did not show aspects of quality and sustainability, government and planners can start the planning process from what they have learned. Knowing how people have adapted for decades is very important because the knowledge has been embedded and institutionalized within the community. Based on this knowledge, the city can develop an inclusive adaptation strategy.

The adaptation knowledge that is embedded in the everyday life of the urban poor has not yet been properly combined with development planning. The tacit form of this knowledge has

created a missing link in establishing a connection to adaptation planning at the city level. We found that by shifting methods from technical experts to a people-centered approach, we can reveal the adaptive practical knowledge of the urban poor. This is a useful and essential data source in assessing vulnerability and providing adaptation options. We suggest that local community adaptation should be incorporated into adaptation planning processes at the city level.

Summary

Flooding is the most typical urban problem, especially for low-lying coastal cities such as Jakarta. The urban poor who have lived in various *kampung*s for many decades have experienced and adapted to floods on their own. However, the adaptation knowledge that is embedded in the everyday life of the urban poor has not yet been properly combined with development planning. The tacit form of this knowledge has created a missing link in establishing a connection to adaptation planning at the city level. By applying a people-centered approach in three *kampung*s in Jakarta, we revealed the adaptive practical knowledge of the urban poor. This is a useful data source in assessing vulnerability and providing adaptation options. We suggest that local community adaptation should be incorporated into adaptation planning process at the city level.

Acknowledgments

Author wishes to thank Dani Muttaqin, Adriadi Dimastanto, Agung Mahesa, and Raka Suryandaru for contributing data and information during and after the Planning for Integrated Coastal

Adaptation Strategies for North Coastal Jakarta (PICAS) project, and to thank the international secretariat of START that provided grants for my field research and the PICAS project in general.

References

Demographia (2016) Demographia World Urban Areas (Built Up Urban Areas or World Agglomerations). 12th Annual Edition. St. Louis: Demographia. Available from: http://www.demographia.com/db-worldua.pdf.

DKI Jakarta. (2012a). *Rencana Pembangunan Jangka Menengah Daerah Provinsi DKI Jakarta 2013–2017*. Draft Final Report. Badan Perencana Pembangunan Daerah DKI Jakarta.

DKI Jakarta. (2012b). *Rencana Tata Ruang Wilayah Provinsi DKI Jakarta*. Draft Final Report. Badan Perencana Pembangunan Daerah DKI Jakarta.

Indonesian Association of Urban and Regional Planners (IAP). (2012). Planning for integrated coastal adaptation strategies for north coastal Jakarta (PICAS). Progress Report in *bahasa*. Unpublished.

Peel, M. C., Finlayson, B. L., and McMahon, T. A. (2007). Updated world map of the Köppen-Geiger climate classification. *Hydrology and Earth System Sciences Discussions* **4**(2), 462.

Simarmata, H. A. (2015). Locally embedded adaptation planning: A trilogy on the adaptation of flood-affected people in Kampung Muara Baru Jakarta. Submitted doctoral dissertation, Universität Bonn. Unpublished.

World Bank. (2017). 2016 GNI per capita, Atlas method (current US$). Accessed August 9, 2017: http://data.worldbank.org/indicator/NY.GNP.PCAP.CD

United Nations, Department of Economic and Social Affairs, Population Division (2016). The World's Cities in 2016 – Data Booklet (ST/ESA/SER.A/392).

Case Study 6.B

Participatory Integrated Assessment of Flood Protection Measures for Climate Adaptation in Dhaka

Anika Nasra Haque

Department of Geography, University of Cambridge

Stelios Grafakos

Institute for Housing and Urban Development Studies, Erasmus University, Rotterdam

Keywords	Climate adaptation, flood protection, multi-criteria analysis, prioritization, participatory approach, environmental justice
Population (Metropolitan Region)	18,237,000 (United Nations, 2016)
Area (Metropolitan Region)	368 km² (Demographia, 2016)
Income per capita	US$1,330 (World Bank, 2017)
Climate zone	Aw – Tropical Savannah (Peel et al., 2007)

Dhaka, the capital of Bangladesh, is one of the largest megacities in the world, and its population is growing rapidly. Due to its location on a deltaic plain, the city is extremely prone to detrimental flooding. Moreover, being located in the active river tidal zone, the low-lying areas are often engulfed by high tides. Risks associated with these are expected to increase further in the coming years due to global climate change impacts as well as the high rate of urbanization the city is facing (Haque et al., 2012). Although the government is planning several adaptive measures to protect the area from floods, it lacks a systematic framework to analyze and assess them.

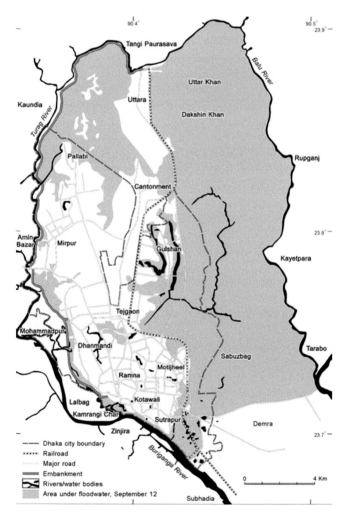

Case Study 6.B Figure 1 *Flood map of Dhaka city during 1998 flood showing inundated study area.*

Source: Nishat et al., 1999

Case Study 6.B Figure 2 *Waterlogging in secondary road of Dhaka; waterlogging in main arterial road of Dhaka.*

Source: Hossain, 2015

Case Study 6.B Figure 3 *Disruption of communication due to flood.*

Source: Habitat International Coalition, 2009

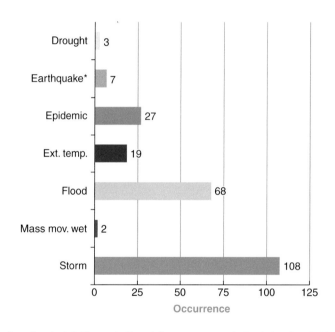

Case Study 6.B Figure 4 *Natural disaster occurrence in Bangladesh from 1980 to 2010.*

Source: UNISDR (2013)

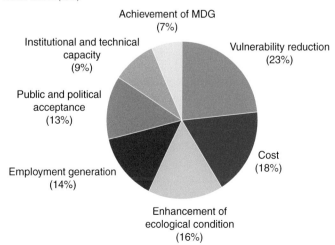

Case Study 6.B Figure 5 *Stakeholders' preferences on evaluation criteria.*

Traditionally, the water-retention areas in Dhaka have efficiently stored the excess water caused by high-intensity rainfall and the canals, which are connected to the rivers, gradually drain the water. The situation is changing. The population of Dhaka is increasing and land scarcity is resulting in the encroachment of these water-retention areas, which mostly lie in Dhaka East. The city drainage system has not improved with the pace of rapid urbanization; consequently, these low-lying areas suffer from inundation. Dhaka East was almost completely inundated in the floods that occurred in 1988, 1998, and 2004, and it was inundated for longer than the other parts of the city. It is not only affected by the river flood, but also by the urban flood caused by excessive runoff due to rainfall. Prolonged inundation caused severe damage to urban agriculture and also to infrastructure (i.e., roads, housing, and piped services). Disruption of communications had severe impact on trade and education, whereas contaminated water led to health hazards.

The study identifies, analyzes, and prioritizes adaptation measures to address flood risks in Dhaka East while applying a multicriteria analysis (MCA) framework. The study emphasizes the importance of the participatory process to an informed decision-making in formulating policy in the context of a least developed country (LDC) at a city level. Furthermore, it provides decision support to both local and national policy-makers regarding the selection of adaptation measures that meet multiple local objectives considering budget, institutional, and technical capacity constraints. The assessment framework involved both stakeholders by eliciting their views and values (normative judgments) and experts (technical expertise). Moreover, the exchange of information from a multitude of stakeholder perspectives makes the outcome of decision-making more legitimate and defensible. Adaptation assessment in Bangladesh is crucial because the level of vulnerability to climate change is very high and resources are very limited. It explores the development of a platform for knowledge generation and sharing that can be used during the MCA process as an important element for enhancing institutional capacity during the decision-making process.

Data have been collected from both primary sources (in-depth interviews, focus group discussion, questionnaire survey, direct observation) and secondary sources (archival records, policy papers, official reports, relevant case studies, relevant literatures in peer-reviewed journals).

The assessment involved the following steps:

Selection of adaptation options: The adaptation options for the study area proposed by the government (flood embankment, pumping stations, regulators, retention basins, construction and upgrading of the road network, flood walls, and canal improvement) were included in the assessment. Additional adaptation options (emergency response mechanism and early warning system) were selected for assessment based on case studies.

Stakeholder criteria selection: In order to assess the adaptation measures, evaluation criteria were identified and selected in a participatory manner (focus group discussions) involving relevant stakeholders (i.e., ward commissioners, community representatives, government and private-sector officials). Thus, the criteria reflect the main objectives and views of different stakeholders, Stakeholders were confronted with the views of other stakeholder group representatives, hence avoiding manipulation of the responses.

Experts' impact judgments: The scoring of each adaptation option against the selected evaluation criteria was conducted by experts on flood protection based on their areas of specialization and experience.

Stakeholders' criteria weighting: Weighting (relative importance) of selected criteria was conducted based on a second round of focus group discussion with the stakeholder group (see Case Study 6.B Figure 5).

Prioritization of options: This step aimed to prioritize the adaptation measures under investigation for the study area based on their weighted scores.

Sensitivity analysis: A sensitivity analysis was conducted to investigate how sensitive the result of the final ranking was to the input variable of criteria weights and to incorporate the uncertainty and range of stakeholders' preferences.

The final outcome of the study is the ranking of adaptation measures (**Case Study 6.B** Table 1), which shows that protection of water-retention areas, enhanced early warning systems, and canal improvements are the highest ranked flood protection measures based on the selected evaluation criteria and stakeholders' preferences. This is an interesting outcome, since the "construction and upgrade of drainage system" is being discussed and considered as a main flood management measure in Dhaka. However, in this study, it has been proved to be a far less prioritized measure. Apparently, if the drainage system is improved, it is expected to reduce flooding, but there are other issues that should also be considered. Construction and upgrading of the drainage system requires a large budget as well as enhanced technical capacity, which are not readily available in a low-income country context. Protection

Case Study 6.B Table 1 *Final ranking and scoring results of adaptation measures.*

Measures	Score	Rank
Protection of water retention areas	0.74	1
Enhancing early warning systems	0.72	2
Canal improvements	0.69	3
Embankments	0.56	4
Construction and upgrading of storm sewer/ drainage systems	0.52	5
Raised roads	0.47	6
Enhancing emergency response mechanisms	0.44	7
Flood walls	0.40	8

of water-retention areas has proved to be the most effective option for reducing the vulnerability of the study area to flooding while simultaneously meeting other important criteria such as cost, enhancement of ecological condition, employment generation, and the like. The sensitivity analysis showed that the results are robust with regard to changes in criteria weights. It is evident from the final ranking results that the highest ranked alternatives performed very well at the most important criteria.

The adaptation assessment undertaken provides significant support for policy design and decision-making for a LDC like Bangladesh, where resources are limited and vulnerability to climate change impacts is high. Although the provision of a structured prioritization of adaptation measures for flood management is a challenging goal, the exchange of information from a large group of stakeholders made the outcome of the decision-making more legitimate and defensible. Moreover, by including both experts and stakeholders, the MCA process contributed to knowledge generation and sharing, which is an important element for enhancing institutional capacity for decision-making.

References

Climate-Data. (2015). Climate-Data.org. Accessed October 9, 2015: http://en.climate-data.org/search/

Demographia (2016) Demographia World Urban Areas (Built Up Urban Areas or World Agglomerations). 12th Annual Edition. St. Louis: Demographia. Available from: http://www.demographia.com/db-worldua.pdf.

Habitat International Coalition. (2009). Accessed February 12, 2010: www.hic-gs.org/news.php?pid=3170

Haque A. N., Grafakos, S., and Huijsman, M. (2012). Participatory integrated assessment of flood protection measures for climate adaptation in Dhaka. *Environment and Urbanization* **24**(1), 197–213.

Nishat, A., Reazuddin, M., Raqibul, A., and Khan, A. R. (eds.) (1999). *The 1998 Flood: Impact on Environment of Dhaka City.* Department of Environment, in conjunction with IUCN Bangladesh.

Peel, M. C., Finlayson, B. L., and McMahon, T. A. (2007). Updated world map of the Köppen-Geiger climate classification. *Hydrology and Earth System Sciences Discussions* **4**(2), 462.

UNISDR (2013). Bangladesh- Disaster statistics. Accessed February 8, 2015: http://www.preventionweb.net/english/countries/statistics/?cid=14

United Nations, Department of Economic and Social Affairs, Population Division (2016). The World's Cities in 2016 – Data Booklet (ST/ESA/SER.A/392).

World Bank. (2017). 2016 GNI per capita, Atlas method (current US$). Accessed August 9, 2017: http://data.worldbank.org/indicator/NY.GNP.PCAP.CD

Case Study 8.A

Parque del Lago, Quito: Reclaiming and Adapting a City Center

Christopher Michael Clark

Office of Research and Development, U.S. Environmental Protection Agency, Washington, D.C.

Michael Flynn

M3PROJECT, New York

Keywords	Urban ecology, reclamation, design, ecosystem services, water supply, adaptation, mitigation
Population (Metropolitan Region)	1,754,000 (United Nations, 2016)
Area (Metropolitan Region)	536km² (Demographia, 2016)
Income per capita	US$5,820 (World Bank, 2017)
Climate zone	Cfb – Temperate, without dry season, warm summer (Peel et al., 2007)

Quito, Ecuador, is the highest official capital in the world, a UNESCO Cultural World Heritage site, and home to over 1.75 million inhabitants (United Nations, 2016.). The city is nestled in the Andes Mountains in the Guayllabamba River basin, on the eastern slopes of the Pichincha volcano. Quito is a central hub of economic and political activity in South America as the headquarters of the Union of South American Nations and a major cultural site with one of the largest and best-preserved historic centers in the Americas.

Quito is already vulnerable to current climate conditions and these risks are expected to intensify with climate change (Magrin et al., 2014; Revi et al., 2014). Nearly 100% of Quito's water supply comes from freshwater streams flowing from the surrounding mountains, glaciers, and protected areas. Climate change in the region is expected to accelerate losses of glacier extent and further increase temperatures, which each could imperil the sustainability of the water supply (Magrin et al., 2014; Rabatel et al., 2013). Anticipated changes in precipitation are more uncertain due to complex feedbacks with the mountainous terrain and regional processes, such as El Niño. However, extreme events such as droughts and storms are expected to become more common (Magrin et al., 2014). These climate risks are exacerbated by other regional pressures such as increasing urbanization, extensification of agriculture and cattle ranching, and demands from hydropower (Revi et al., 2014).

Redevelopment as an Opportunity for Renewal: Climate Adaptation and Mitigation

Within this context, the Metropolitan Municipality of Quito, the Corporation for Environmental Health, and the Architects Association of Pichincha-Ecuador are planning the redevelopment of Quito's former international airport (Mariscal Sucre International Airport) into an ecological urban park, referred to as Parque del Lago, to enhance the sustainability and livability of the urban environs (Case Study 8.A Figure 1; Metropolitan Municipality of Quito, 2008). This redevelopment, similar to projects in other cities globally, offers an opportunity for re-visioning the urban core into a platform that provides a diverse set of social, environmental, and economic benefits while simultaneously addressing the challenges from a changing climate. The efforts to repurpose this urban infrastructure into a vibrant public and ecological site began in 2008, during which time the construction of the city's new international airport outside of central Quito was under way. The new Mariscal Sucre International Airport, located in the Tababela parish about 18 kilometers east of Quito, commenced operations on February 20, 2013. The old Mariscal Sucre International Airport ceased all operations the night before, opening the door for urban revitalization. Currently, the repurposing efforts of the old airport are being assessed within the city's long-term development plan and political agenda. The creation of what would become Quito's largest urban park is envisioned as becoming a vital asset for all demographic groups living in and around Quito's metro area, as well as a guide for regional and international urban planners on promoting urban sustainability in the context of climate change (Metropolitan Municipality of Quito, 2008).

One of the proposed plans for Parque del Lago (Flynn et al., 2012; Figure 2) envisions designing the park's features along three overarching goals: hydrologic sustainability, biodiversity preservation, and ecosystem services. The proposal includes altering the terrain and hydrological networks of the park into an array of interconnected channels and streams. A large water collection and aquifer recharge lake receives the water flows of the site while providing a range of recreational activities for the park's visitors. The former runway is repurposed as a central recreational and commercial promenade, connecting and orienting the park along its dominant north-south axis. A series of paths and trails explore the park's terrain and anchor its thick wooded areas, meadows, grasslands, lagoons, streams, pools, gardens, and fields. The mosaic of wooded areas and clearings provide public space for commercial activities and outdoor concerts, as well as habitat for Quito's rich surrounding biodiversity.

Each of these design features provides a variety of benefits to the urban core with respect to climate change. The hydrologic network functions as a buffer against potential increases in storm severity, absorbing runoff from the surrounding mountains and urban environment while simultaneously increasing groundwater recharge rates below to the city's strained aquifer. A mixture of fast- and slow-growing tree species is selected to provide carbon offsets early in the park's life span and that are sustained over

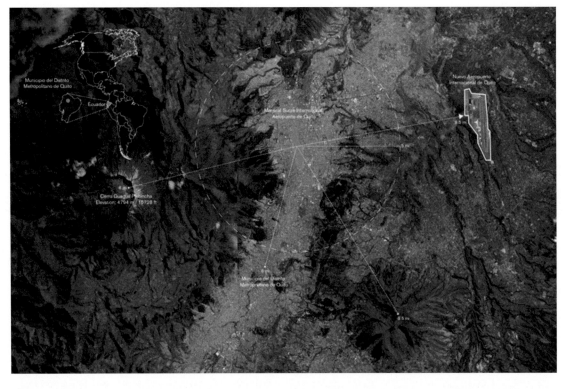

Case Study 8.A Figure 1 *Context map of metropolitan Quito, Ecuador, illustrating the location of the Mariscal Sucre International Airport within the surrounding urbanized landscape.*

Image credits: M3PROJECT, Flynn et al., www.m3project.com

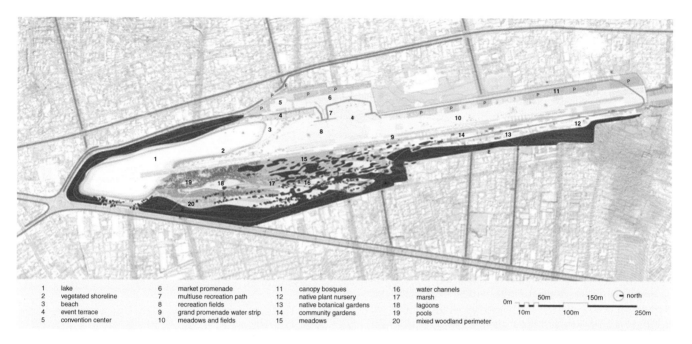

1	lake	6	market promenade	11	canopy bosques	16	water channels	
2	vegetated shoreline	7	multiuse recreation path	12	native plant nursery	17	marsh	
3	beach	8	recreation fields	13	native botanical gardens	18	lagoons	
4	event terrace	9	grand promenade water strip	14	community gardens	19	pools	
5	convention center	10	meadows and fields	15	meadows	20	mixed woodland perimeter	

0m 50m 150m ⊙ north
10m 100m 250m

Case Study 8.A Figure 2 *Illustrative plan to reclaim and adapt Quitos' Mariscal Sucre International Airport into a thriving ecological-urban landscape system designed to increase Quito's preparedness for and resiliency to climate change. The southern lake is fed by channels of grass-lined streams, interspersed with wooded groves to enhance stormwater mitigation, habitat, air quality, and carbon sequestration.*

Image credits: M3PROJECT, Flynn et al., www.m3project.com

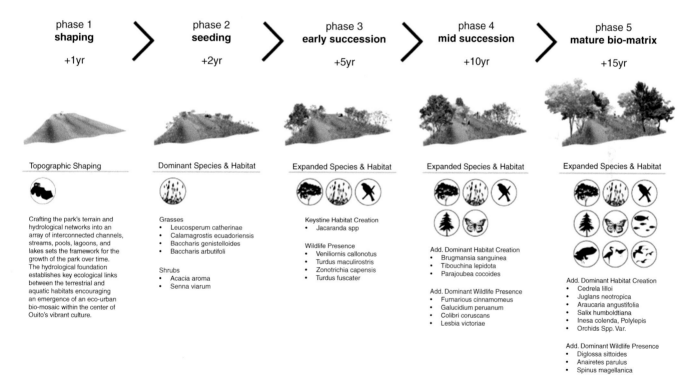

Case Study 8.A Figure 3 *Conceptual diagram illustrating the successional development and diversification of the former Mariscal Sucre International Airport site over the first 15 years. The strategy highlights accelerated species diversity and accumulation of biota over the trajectory of site transformation.*

Image credits: M3PROJECT, Flynn et al., www.m3project.com

time as a climate mitigation measure (Case Study 8.A Figure 3). Evapotranspiration and shade from the plant canopy, as well as evaporation from the large lake, will cool local temperatures and help mitigate the urban heat island effect.

Quito is situated in a global biodiversity hot spot, the Tropical Andes (Meyers et al., 2000), and the proposal selects plant species to represent the rich diversity endemic to the region. The proposal includes the concept of ecological succession intrinsically into the design, where early successional grasses, sedges, and pioneer trees will provide habitat for several endemic bird species long absent in the urban core, including the Pacific horneo and the scarlet-backed woodpecker. Pioneer species such as the leguminous shrub *Aromita* help improve soil quality through nitrogen fixation and will be planted throughout the park early on to remediate the legacy of urban soil degradation. As the park matures, long-lived native species such as the jacaranda, South American cedar, and Ecuadorian walnut will start to dominate the park's landscape. These slow-growing long-lived trees include several species listed as endangered on the International Union for the Conservation of Nature Red List, a global authoritative source for the conservation status of species. Growth of these tree species will promote local biodiversity, provide mixed woodland area for recreational use, and serve as iconic species of the park, embodying habitat resembling that of pre-urban Quito. Finally, the mixture of habitats (e.g., woodland, grassland, wetland) emblematic of the Tropical Andes will provide refuge for many species of bird, butterfly, and others threatened by climate change, while the growth and succession within the park will ensure that many of the ecosystem services are maintained over time.

The park will become a foundational unit in the local urban community and regional ecosystem, benefiting both people and wildlife. Educational placards explaining the identity and function of various species and features of the park will help connect the local population with their regional ecology and the various services these species provide. The FONAG program, a Heritage Trust created by the local Metropolitan Water and Sanitation Department (EPMAPS) with support from the U.S. Agency for International Development and the Nature Conservancy, promotes water conservation and stewardship in Quito. FONAG and EPMAPS can become key allies to the park, supporting its water and biological conservation goals and providing input from their network of local supporters. Financial considerations are also included in design implementation, where a portion of the revenue from FONAG could support park development early on, while, over the longer term, revenue from commercial activities within the park (e.g., paddle boat rental, pavilion rental space, concert ticket sales) could pay back the heritage fund to benefit future generations. Additionally, carbon offsets from the growth of these long-lived tree species can be leveraged in a regional carbon market and tracked to account for the local carbon footprint.

Conclusions from Quito and Lessons for Other Cities

The Parque Del Lago proposal for the city of Quito is an example of the kind of integrative design that can help transform urban areas, which currently cover only 3% of the global land area but appropriate 60% or more of the earth's resources (Grimm et al., 2008), into sustainable systems for human and ecological betterment. The challenges to this city from climate change are great, but the lessons from stakeholder engagement and education, local adaptation planning, and successional design can help propel Quito forward and provide guidance for other cities as they address climate change adaptation and mitigation in the urban context.

References

Climate-Data. (2015). Climate-Data.org. Accessed October 9, 2015: http://en.climate-data.org/search/

Demographia (2016) Demographia World Urban Areas (Built Up Urban Areas or World Agglomerations). 12th Annual Edition. St. Louis: Demographia. Available from: http://www.demographia.com/db-worldua.pdf.

Flynn, M., Clark, C.M., Ryan, T. (2012). Parque del Lago: Keystone Mosaic. A design submission to the Metropolitan Municipality of Quito. Accessed September 4, 2015: www.m3project.com

Grimm, N. B., et al. (2008). Global change and the ecology of cities. *Science* **319**(5864), 756–760.

Magrin, G. O., Marengo, J. A., Boulanger, J. -P., Buckeridge, M. S., Castellanos, E., Poveda, G., Scarano, F. R., and Vicuña, S. (2014). Central and South America. In Barros, V. R., Field, C. B., Dokken, D. J., Mastrandrea, M. D., Mach, K. J., Bilir, T. E., Chatterjee, M., Ebi, K. L., Estrada, Y. O., Genova, R. C., Girma, B., Kissel, E. S., Levy, A. N., MacCracken, S., Mastrandrea, P. R., and White, L. L. (eds.) *Climate Change 2014: Impacts, Adaptation, and Vulnerability. Part B: Regional Aspects. Contribution of Working Group II to the Fifth Assessment Report of the Intergovernmental Panel on Climate Change (1499–1566).* Cambridge University Press.

Metropolitan Municipality of Quito. (2008). Parque del Lago International Idea Competition. Accessed September 4, 2015: (http://akichiatlas.com/en/archives/parque_del_lago.php)

Myers, N., Mittermeier, R. A., Mittermeier, C. G., de Fonseca, G. A. B., and Kent, J. (2000). Biodiversity hotspots for conservation priorities. *Nature* **403**, 853–858.

Peel, M. C., Finlayson, B. L., and McMahon, T. A. (2007). Updated world map of the Köppen-Geiger climate classification. *Hydrology and Earth System Sciences Discussions* **4**(2), 462.

Rabatel, A., Francou, B., Soruco, A., Gomez, J., Cáceres, B., Ceballos, J. L., Basantes, R., Vuille, M., Sicart, J.-E., Huggel, C., Scheel, M., Lejeune, Y., Arnaud, Y., Collet, M., Condom, T., Consoli, G., Favier, V., Jomelli, V., Galarraga, R., Ginot, P., Maisincho, L., Mendoza, J., Ménégoz, M., Ramirez, E., Ribstein, P., Suarez, W., Villacis, M., and Wagnon, P. (2013). Current state of glaciers in the tropical Andes: A multi-century perspective on glacier evolution and climate change. *Cryosphere Discussions* **7**, 81–102.

Revi, A., Satterthwaite, D. E., Aragón-Durand, F., Corfee-Morlot, J., Kiunsi, R. B. R., Pelling, M., Roberts, D. C., and Solecki, W. (2014). Urban areas. In Barros, V. R., Field, C. B., Dokken, D. J., Mastrandrea, M. D., Mach, K. J., Bilir, T. E., Chatterjee, M., Ebi, K. L., Estrada, Y. O., Genova, R. C., Girma, B., Kissel, E. S., Levy, A. N., MacCracken, S., Mastrandrea, P. R., and White, L. L. (eds.) *Climate Change 2014: Impacts, Adaptation, and Vulnerability. Part A: Global and Sectoral Aspects. Contribution of Working Group II to the Fifth Assessment Report of the Intergovernmental Panel on Climate Change* (535–612). Cambridge University Press.

United Nations, Department of Economic and Social Affairs, Population Division (2016). The World's Cities in 2016 – Data Booklet (ST/ESA/SER.A/392).

World Bank. (2017). 2016 GNI per capita, Atlas method (current US$). Accessed August 9, 2017: http://data.worldbank.org/indicator/NY.GNP.PCAP.CD

Case Study 8.B

São Paulo 100 Parks Program

Oswaldo Lucon

São Paulo State Environment Secretariat/University of São Paulo

Keywords	Parks, adaptation, mitigation, ecosystems, landslides, floods, biodiversity
Population (Metropolitan Region)	19,683,975 (IBGE, 2015)
Area (Metropolitan Region)	7,947 km² (IBGE, 2015)
Income per capita	US$8,840 (World Bank, 2017)
Climate zone	Cfb – Temperate, without dry season, warm summer (Peel et al., 2007)

Urban parks (particularly those alongside riverbeds and on steep slopes) are important climate adaptation measures because they prevent risks from landslides and floods, restore microclimates, and improve biodiversity. The recovered vegetation is also an important mitigation measure because plants and soils store carbon. Between 1992 and 2004, only one park was created in São Paulo, a megacity with almost 12 million inhabitants living in 1,522 square kilometers (IBGE, 2014). In 2005, there were 33 parks, totaling 15 million square meters of municipal protected green areas. Still, there were neighborhoods in the city with 700,000 inhabitants without a single local park. From 2005 to August 2008, 17 new parks were established, expanding and distributing more evenly such areas within the macro-regions of the city. The goal for 2010 was to achieve 100 parks and 50 million square meters protected. In 2010, 60 parks (totaling 24 million m²) were implemented, expanded to 133 parks (of which 21 are linear [alongside riverbeds]), totaling 35.66 million square meters or 2.4% of territory by the year 2012, when the project was discontinued after a municipal election (PMSP, 2007).

The 100 Parks Program to São Paulo aimed to identify the highest number of available areas and turn them into parks, thus expanding areas of leisure and contact with nature and, at the same time, more evenly distributing these sites. Among the Program's many objectives can be cited preventing flood hazards and erosions, mitigating heat island effects, providing local leisure options and avoiding trips to distant parks, ensuring soil permeability, sequestering carbon through planting 1.5 million new native trees of the Atlantic Forest, and improving conditions for biodiversity. Beyond urbanism, leisure, and contemplation, parks were considered as a whole system within the city. Several parks at the margins of the Guarapiranga Reservoir were part of a water defense strategy aiming to prevent illegal occupations on areas prone to floods and to prevent supply contamination. In the north, on the edge of the Cantareira Mountains, the implementation of linear parks was proved crucial to establish a barrier to irregular occupation (mostly by slums). Similarly, in the lowland region of the Tietê River, marshes were protected, implementing recreational areas and allowing for better river sanitation and clean-up works.

Linear parks were devised to protect preservation areas, avoid occupation in high-risk areas, help to combat flash floods, create cultural and leisure options for the surrounding population, and recover local stream margins. Natural Parks aim at the preservation of biodiversity in the city according to the commitment that the municipality assumed in Nagoya, Japan – the Aichi Biodiversity Targets at the Tenth Conference to the Parties of the UN Convention on Biodiversity. A Strategic Master Plan introduced the Environmental Riparian Areas Recovery Program as a structural element of urbanization. Linear parks (i.e., recovery of valley riverbeds) were the main line of action for this program, defining utility strips along water courses in order to implement green infrastructure, environmental restoration, and recreation (PMSP, 2010).

The first stage was to identify available areas, their characteristics, and potential uses. More than 200 areas were registered as having potential for parks. If the area was private and there was confirmation of the interest in turning it into park, the municipality started a process of expropriation through Public Utility Decrees. A total of sixty-nine areas were expropriated. The acquisition was often by the Environmental Compensation fund for public and private works, which had carried out tree removals in other places of the city (Ceneviva, 2014). In some cases, parks were created by demand of the population living nearby. The parks were located on compensated lands mainly along the southern branch of the São Paulo Ring Road, allowing for the creation of four natural parks in the region. Where the areas were already publicly owned, the process was faster. In the case of private areas turned into public parks (through expropriation or donations), fencing and surveillance were provided, plus studies of fauna and flora and architectural projects. Before being opened, each park is appointed an administrator with a background in environmental areas. This professional supervises surveillance and management contracts (building services, toilet maintenance, landscaping and gardening). The administrator is also responsible for daily activities, representing the municipality and coordinating – together with Management Centers – the implementation of environmental education programs within the park and in the surrounding community. Once opened, a Steering Council is elected that monitors the life of the park and suggests and proposes uses (PMSP, 2012).

References

Climate-Data. (2015). Climate-Data.org. Accessed October 9, 2015: http://en.climate-data.org/search/

Instituto Brasileiro de Geografia e Estatistica (IBGE). (2014). Cidades. São Paulo. Accessed April 9, 2015: http://www.cidades.ibge.gov.br/xtras/perfil.php?lang=&codmun=355030&search=sao-paulo%7Csao paulo

Instituto Brasileiro de Geografia e Estatistica (IBGE). (2015). Instituto Brasileiro de Geografia e Estatistica. Accessed September 18, 2015: http://www.ibge.gov.br/home/estatistica/populacao/censo2010/sinopse/sinopse_tab_rm_zip.shtm

Peel, M. C., Finlayson, B. L., and McMahon, T. A. (2007). Updated world map of the Köppen-Geiger climate classification. *Hydrology and Earth System Sciences Discussions* 4(2), 462.

PMSP. (2007). Parques urbanos municipais de São Paulo (São Paulo Municipal Urban Parks). Accessed April 9, 2015: http://www.terrabrasilis.org.br/ecotecadigital/pdf/parques-urbanos-municipais-de-sao-paulo-subsidios-para-a-gestao.pdf and http://site-antigo.socioambiental.org/banco_imagens/pdfs/10367.pdf

PMSP. (2010). Guia dos Parques Municipais, 3ª. edição (Municipal Parks Guide, 3rd ed.). Accessed April 9, 2015: http://www.prefeitura.sp.gov.br/cidade/secretarias/upload/meio_ambiente/arquivos/publicacoes/guia_dos_parques_3.pdf

PMSP. (2012). Secretaria do Verde e do Meio Ambiente lança terceira edição do Guia de Parques (Green and Environment Secretariat launches third edition of Parks Guide). Prefeitura do Município de São Paulo. Accessed April 9, 2015: http://www.prefeitura.sp.gov.br/cidade/secretarias/comunicacao/noticias/?p=108899

World Bank. (2017). 2016 GNI per capita, Atlas method (current US$). Accessed August 9, 2017: http://data.worldbank.org/indicator/NY.GNP.PCAP.CD

Case Study 8.C

St. Peters, Missouri, Invests in Nature to Manage Stormwater

Roy Brooke and Stephanie Cairns

Sustainable Prosperity, Ottawa

Keywords	Stormwater management, natural capital, urban climate change adaptation system, ecosystem-based adaptation
Population (Metropolitan Region)	56,971 (U.S. Census, 2016)
Area (Metropolitan Region)	57.93 km² (U.S. Census, 2016)
Income per capita	US$56,180 (World Bank, 2017)
Climate zone	Dfa – Cold, without dry season, hot summer (Peel et al., 2007)

The majority of North American engineered infrastructure faces a backlog of overdue maintenance, pressing needs for modernization and growth-driven expansion, and cost constraints. The 2013 American Society of Civil Engineer report card rates U.S. infrastructure only a D+ overall (American Society of Civil Engineers, 2015). The situation is similar in Canada, where the Federation of Canadian Municipalities has pointed to a chronic infrastructure "funding crunch" (FCM, 2006). This raises vital questions about the extent to which municipalities can meet additional infrastructural and financial burdens imposed by climate change impacts such as increased frequency and intensity in rainfall, increased flood risks, and earlier snowpack melt.

Investment in local nature preservation, restoration, and enhancement can provide a cost-effective approach to increase community resilience to these and other climate impacts. Municipal services such as stormwater management and clean water supply are typically provided through engineered assets such as stormwater systems and water purification facilities or through "green infrastructure" such as engineered bioswales and raingardens.

However, some pioneering municipalities are investing in natural systems such as their local streams, rivers, wetlands, and aquifers to provide these services, in part or whole, and integrating management of these "natural assets" into an overall stormwater management strategy.

The City of St. Peters, in St. Charles County, Missouri, is one municipality that is using this approach. Their initiative is at an early stage but already merits attention as a cost-effective model for municipal infrastructure. Although designed to meet U.S. Environmental Protection Agency (EPA) stormwater quality guidelines, St. Peters' approach is readily transferable to and relevant for urban climate change adaptation strategies. The EPA notes that precipitation in the U.S. Midwest is likely to become more intense (UCS, 2009), which could increase flood risks and reduce summer water availability (EPA, 2015). St. Peters' staff report changes in rainfall patterns, including an increased frequency and intensity of rainfall, and anomalies such the city receiving 50% of its average annual rainfall in May and June of 2015 alone.

St. Peters has a population of nearly 55,000 people and an enviable quality of life, with 25 parks, more than 30 kilometers of hiking trails and 9 out of every 10 residents living within 1.5 kilometers of a park. However, St. Peters has faced challenges in maintaining its built infrastructure. Since the economic downturn in 2007, for example, the city has lost US$2 million in property tax revenue. Meanwhile, acute flooding resulted in stormwater quality that fell short of state and federal standards. This left the city with the daunting challenge of paying for an estimated US$119 million in stormwater upgrades with an annual budget for this purpose of only US$600,000.

St. Peters explored options for a comprehensive stormwater management system to improve water quality and alleviate flooding. The potential to improve and maintain existing natural systems as a key component of this approach emerged for four reasons:

1. St. Peters has a municipal separate storm sewer system (MS4), a system which discharges untreated stormwater runoff into local water bodies. This required a permit from the EPA. The EPA encourages approaches that emulate natural systems by reintegrating rainwater into the natural water cycle (EPA, 2009)– although there is no specific state or federal requirement to protect existing natural systems.

2. St. Peters recognized the capital and operating savings involved in preserving, enhancing, or restoring existing natural systems rather than building new engineered or "green" infrastructure.

3. Many of the natural systems in the St. Peters drainage basins are on private land. Preserving these requires only that the city secure maintenance easements from landowners, a more cost-effective approach than securing easements for capital

projects on private lands or purchasing the land. Public maintenance easements also removed a burden from landowners and community associations, if they were previously maintaining these areas.

4. St. Peters places a high value on its natural spaces and recreational areas. This gave the city a logical starting point for preserving existing natural assets and an approach that was consistent with community values related to parks and nature.

The St. Peters' asset management plan conducted a condition assessment of the city's approximately 75 kilometers of stream corridors, including the streams themselves and engineered detention and retention basins (Smith, 2015). The assessment gave the city its first comprehensive understanding of the role of natural systems in the city's stormwater management and of the interaction between natural systems and the city's engineered assets in the management of stormwater.

The assessment demonstrated that about 1% of the natural systems were in intact condition, about 14% could be easily restored to a similar state, and about 85% were severely degraded and would require major interventions to remove invasive plants, replant native species, change the tree canopy, and stabilize banks. This understanding allowed St. Peters to establish

Case Study 8.C Figure 1 *Spencer Creek in St. Peters. A condition assessment of approximately 75 km of stream corridors gave the city its first comprehensive understanding of the role of natural systems in the city's stormwater management and of the interaction between natural systems and the city's engineered assets in the management of stormwater.*

Source: City of St. Peters

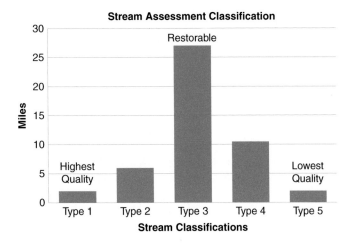

Case Study 8.C Figure 2 *Results of the St. Peters' stream assessment classification. Types 1 and 2 were designated as preservation projects. Types 3 and 4 were designated as restoration projects.*

Source: City of St. Peters

management priorities. For example, the city was able to emphasize the monitoring of intact natural systems to prevent degradation and therefore future costs; intervene quickly in the natural systems that could be easily restored to avoid further deterioration and associated costs; and address any health, safety, and property issues in the most degraded areas. The assessment also gave St. Peters baseline data against which to monitor ecosystem changes and the effectiveness of its efforts.

The assessment also created a foundation for outreach and awareness efforts that resulted in a 72% public approval vote in August 2012 in favor of the "Proposition P" sales tax increase (City of St Peters, 2015). Proposition P funding has enabled St. Peters to undertake capital and maintenance projects worth US$9.4 million to prevent stormwater pollution and erosion of creek banks, maintain detention basins, and improve other natural areas. The goal of the stream stabilization and basin retrofit projects is to mimic nature and use native plantings to filter stormwater and remove pollutants (City of St Peters, 2013) The projects thus far blur the distinction between "engineered" and "natural" systems because many of the latter have been altered or engineered to varying degrees. Nevertheless, they suggest that there are measurable benefits to the City of St. Peters in understanding, managing, and/or restoring *existing* natural systems and allowing them to function, to the greatest extent, as nature intended rather than focusing strictly or primarily on either engineered "gray" assets such as stormwater channels or engineered "green" assets such as bioswales and raingardens.

A comparable example is from the Town of Gibsons, just north of Vancouver, British Columbia. Gibsons has committed to manage and maintain its natural assets (such as green space, forests, topsoil, aquifers and creeks) in the same manner as its storm sewers, roads, and other traditional engineered assets. Their strategy focuses on *identifying* existing natural assets that provide municipal services, including services related to climate change adaptation; *measuring* the value of the municipal

services provided by these assets; and making this information operational by *integrating* it into municipal asset management through governance system changes (Brooke, 2015). The assets under consideration include forested headwaters that convey stormwater, an aquifer, and the ocean foreshore. As with St. Peters, the town of Gibsons faces climate change impacts. These include sea level rise, increased precipitation, and earlier snowpack melt (Vadeboncoeur and Matthews, 2014). Evidence from Gibsons shows that preserving natural assets provides the town with equivalent municipal services to engineered assets at a substantially lower cost and suggests that protecting existing natural systems can increase the town's resilience to climate change impacts (Town of Gibsons, 2014).

In conclusion, while the evidence base is currently limited, it suggests that the assessment and management of existing natural systems as part of an overall asset management strategy can save capital and operating costs and increase municipal resilience to current funding and infrastructural pressures as well as additional ones anticipated as a result of climate change.

References

American Society of Civil Engineers. (2015). Report Card for America's Infrastructure 2013. Accessed June 20, 2015: http://www.infrastructurereportcard.org

Brooke, Roy (2015, May). *Shifting the North American Infrastructure Debate through an Eco-Assets Strategy.* American Public Works Association Report.

City of St. Peters. (2015). Proposition P Information. Accessed October 7, 2015: http://www.stpetersmo.net/proposition-p.aspx

City of St. Peters. (2013) Prop P Progress: City working with residents on stormwater design. My Hometown. Accessed October 7, 2015: http://www.stpetersmo.net/PropositionP/FALL2013MH-PropPProgress.pdf

Environmental Protection Agency (EPA). (2009). Incorporating environmentally sensitive development into municipal stormwater programs. Accessed October 7, 2015: http://water.epa.gov/polwaste/npdes/stormwater/upload/region3_factsheet_lid_esd.pdf

Environmental Protection Agency (EPA). (2015). Climate impacts in the Midwest. Accessed October 7, 2015: http://www3.epa.gov/climatechange/impacts/midwest.html

Federation of Canadian Municipalities (FCM) (2006). Building Prosperity from the Ground Up : Restoring Municipal Fiscal Balance. Ottawa: FCM.

Peel, M. C., Finlayson, B. L., and McMahon, T. A. (2007). Updated world map of the Köppen-Geiger climate classification. *Hydrology and Earth System Sciences Discussions* 4(2), 462.

Smith, A. (2015). Realizing the value of natural systems: An asset management approach. *American Public Works Association Reporter,* February 2015.

Town of Gibsons (2014). *Towards an Eco-asset strategy in the town of Gibsons.*

Union of Concerned Scientists (UCS) (2009). Confronting climate change in the Midwest: Missouri. Accessed October 7, 2015: http://www.ucsusa.org/sites/default/files/legacy/assets/documents/global_warming/climate-change-missouri.pdf

U. S. Census Bureau. (2016). QuickFacts: St. Peter's City, Missouri. Retrieved November 3, 2016, from http://www.census.gov/quickfacts/table/PST045215/2965126,4260000

Vadeboncoeur, N., and Matthews, R. (2014). *Coastal Climate Change in Gibsons, BC.* University of British Columbia Press.

World Bank. (2017). 2016 GNI per capita, Atlas method (current US$). Accessed August 9, 2017: http://data.worldbank.org/indicator/NY.GNP.PCAP.CD

Case Study 9.A

Climate Adaptation in Helsinki, Finland

Simo Haanpaa and Johannes Klein

Department of Built Environment, Aalto University, Espoo

Sirkku Juhola

Department of Environmental Sciences, University of Helsinki

David C. Major

Center for Climate Systems Research, Columbia University, New York

Keywords	Coastal flooding, sea level rise, built environment adaptation
Population	620,715 (City of Helsinki, 2015)
Area	719 km² (City of Helsinki, 2015)
Income per capita	US$44,730 (World Bank, 2017)
Climate zone	Dfb – Cold, without dry season, warm summer (Peel et al., 2007)

The metropolitan region of Helsinki has faced flood events during recent years that have raised flood risk management higher on the city's agenda. Most notable was the record-high coastal flood of January 2005 that in the city of Helsinki saw the establishment of a working group on flood management and subsequent work toward a strategy on flood management (Valkeapää et al., 2010). According to the strategy, the main risks arising from climate change in Helsinki related to flood management are seen to be sea level rise, storm surges, and increased precipitation, with riverine floods causing local challenges.

In Helsinki, exposed bedrock, clay, and moraine and sandy soils are present in about equal shares. In the center of the city, the coastline has been largely modified by landfills over decades (see Case Study 9.A Figure 1). The properties of the soil and the coastline itself have an impact on both construction and adaptation options. Although isostatic rebound (currently at around 4 mm per year) compensates for some of the sea level rise, which is some 20% weaker than the global average, the Helsinki region is facing up to +90 centimeters higher sea level by the year 2100 compared to the year 2000, +33 centimeters being one central estimate and −8 centimeters being the lowest estimate (Kahma et al., 2014). These values do not depict the future flood levels, however, because the projected impact of storm surges have to be added. Currently (in 2011), the mean annual high water level in Helsinki is at +1.21 meters and the estimated 1/250 flood level is +2.08 meters (in N_{2000}, the Finnish national height system) (Kahma et al., 2014). For a comparison, during the flood of January 2005, sea level in Helsinki rose to +1.70 meters (N_{2000}) (Parjanne and Huokuna, 2014).

In 2100, the 1/250 flood level is predicted to rise to +2.73 meters (N_{2000}) (Kahma et al., 2014). Based on this value, the lowest recommended construction level for Helsinki was set at +2.80 meters (N_{2000}), not including the impact of waves, which must be calculated separately for individual coastal stretches.

For new coastal construction sites, following recommended lowest construction heights in detailed planning is a key means of adaptation (Valkeapää et al., 2010). The city of Helsinki has also invested in flood walls and pumping stations at several vulnerable spots that have existing infrastructure. Most of these

Case Study 9.A Figure 1 *Coastline of southern Helsinki.*

Photo: Jari Väätäinen

areas are old bay areas with clay soils reclaimed from the sea due to isostatic lift. The flood walls are typically embankments made of stabilized clay and soil and vary between 30 and 1,300 meters in length (FCG, 2007). In some coastal areas, such as in Arabianranta, a brownfield development located close to the mouth of river Vantaa about 2 km north of the city center, the immediate coastline has also been left undeveloped to serve as floodplains, both because of recreational needs and due to poor soils leaving the sites unsuitable for development (see Case Study 9.A Figure 2). Many largely unprotected flood-prone areas remain, however, most notably in the historic city center around the old harbor and market square, with considerable economic assets invested in the building stock.

Extreme precipitation events can cause increased flooding in the whole metropolitan region, but these events are hard to predict both temporally and spatially (Venäläinen et al., 2009). In order to adapt to both current and foreseen extreme precipitation events, the city of Helsinki has created a strategy for handling stormwater that prioritizes treating rainwater locally by delaying the flow (Nurmi et al., 2008). Examples of this approach are the Kuninkaantammi residential area currently under construction about 10 km north of the city center, where localized rainwater treatment through rain gardens and green roofs will be applied; and Haaganpuro stream, a 12 km long urban stream flowing through the North-Eastern suburbs of Helsinki, where natural flows have been restored to slow the flow to protect areas downstream from flooding. The city has also upgraded sections of the sewage network in the city center to accommodate floodwaters.

Milder and wetter winters as projected for the Helsinki region are also a problem for the long-term durability of man-made

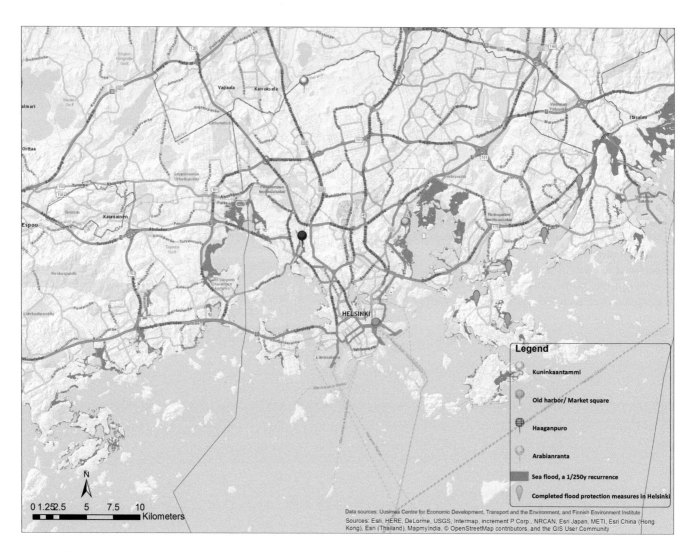

Case Study 9.A Figure 2 *The residential area of Arabianranta with a floodplain in the foreground, shown during a swell of +70 cm in December 2006.*

structures and public infrastructure. These impacts may be mitigated through improving guidance and permit procedures for building and construction and by taking these risks more thoroughly into account in detailed planning (Yrjölä and Viinanen, 2012).

While many of the decisions regarding adaptation can be made at the municipal (city) level, water and flood risk management are also guided by national and European legislation and partly fall to the regional authorities.

Finland was the first European country to develop a National Adaptation Strategy for climate change in 2005 (Swart et al., 2009), further revised in 2015. The strategy identified national causes for concern and helped different sectors to get started on evaluating their own vulnerability to the impacts of climate change. It also served as an impetus for the municipalities to study their vulnerabilities and adaptation needs.

On the municipal and regional level, two organizations have taken a leading role in adaptation: the Helsinki Region Environmental Services Authority (HSY) and the City of Helsinki Environment Center. HSY coordinated the preparation of the Helsinki Metropolitan Area Climate Change Adaptation Strategy (HSY, 2012) and also monitors the progress on adaptation in the metropolitan area. The Metropolitan Area Adaptation Strategy focuses on topics that cross municipal, sectoral, and administrative boundaries, such as land-use planning, traffic management, storm water management, rescue services, and health care. The strategy process was a communication and negotiation exercise that was based on voluntary commitment. Rather than aiming at legal enforcement, it aimed at a broad consensus among the forty-five organizations that contributed actively or provided comments on the strategy. Although the strategy is not legally binding, it has been taken up in the environmental policies and adaptation activities of the municipalities of the metropolitan area.

On the municipal level, the city of Helsinki Environment Center compiled and prioritized all relevant adaptation measures for the city, with a strong focus on the effects of sea level rise and the potential increase in precipitation (Yrjölä and Viinanen, 2012; Haapala and Järvelä, 2014). Furthermore, the Environment Center is active in a network of Finnish municipalities and research institutes that collects best practices and tools for adaptation under a shared webpage (www.ilmastotyokalut .fi). Monitoring and evaluating the implementation of these measures are going to be important activities in coming years. In the long term, risk assessments and understanding the economic implications of the impacts of climate change and adaptation are likely to gain importance.

References

City of Helsinki Urban Facts. (2015). Statistical Yearbook of Helsinki 2015: 13th volume. Accessed November 3, 2016: http://www.hel.fi/hel2/tietokeskus/julkaisut/pdf/16_02_26_Statistical_Yearbook_of_Helsinki_2015_Askelo.pdf

Climate-Data. (2015). Climate-Data.org. Accessed October 9, 2015: http://en.climate-data.org/search/

FCG. (2007). Tulvakohteiden määrittely – Esiselvitys [Mapping of flood-prone sites – Pre-study]. City of Helsinki Public Works Department, Street and Park Division.

Haapala, A., and Järvelä, E. (2014). Helsingin ilmastonmuutokseen sopeutumisen toimenpiteiden priorisointi [Prioritization of Helsinki's climate change adaptation measures]. City of Helsinki Environment Centre, Publications 11/2014.

Helsinki Region Environmental Services Authority (HSY). (2012). Helsinki Metropolitan Area Climate Change Adaptation Strategy (pdf), Helsinki: HSY. Accessed September 21, 2015: https://www.hsy.fi/sites/Esitteet/EsitteetKatalogi/Julkaisusarja/10_2012_paakaupunkiseudun_ilmastonmuutokseen_sopeutumisen_strategia.pdf

Kahma, K., Pellikka, H., Leinonen, K., Leijala, U., and Johansson, M. (2014). Pitkän aikavälin tulvariskit ja alimmat suositeltavat rakentamiskorkeudet Suomen rannikolla [Long-term flooding risks and recommendations for minimum building elevations on the Finnish coast]. *Finnish Meteorological Institute Reports* 2014, 6.

Nurmi, P., Heinonen, T., Jylhänlehto, M., Kilpinen, J., and Nyberg, R. (2008). Helsingin kaupungin hulevesistrategia [Helsinki storm water strategy]. City of Helsinki Public Works Department, Publications 2008:9.

Parjanne, A. and Huokuna, M. (eds.). (2014). Tulviin varautuminen rakentamisessa. Opas alimpien rakentamiskorkeuksien määrittämiseksi ranta-alueilla [Flood preparedness in building – guide for determining the lowest building elevations in shore areas]. Finnish Environment Institute (SYKE), Environment Guide 2014.

Peel, M. C., Finlayson, B. L., and McMahon, T. A. (2007). Updated world map of the Köppen-Geiger climate classification. *Hydrology and Earth System Sciences Discussions* **4**(2), 462.

Swart, R., Biesbroek, R., Binnerup, S., Carter, T. R., Cowan, C., Henrichs, T., Loquen, S., Mela, H., Morecroft, M., Reese, M., and Rey, D. (2009). Europe adapts to climate change: Comparing national adaptation strategies. PEER Report No 1. Helsinki: Partnership for European Environmental Research.

Valkeapää, R., Nyman, T., and Vaittinen, M. (2010). Tulviin varautuminen Helsingin kaupungissa [Adapting to floods in the city of Helsinki]. City of Helsinki, City Planning Department, Reports 2010, 1.

Venäläinen, A., Johansson, M., Kersalo, J., Gregow, H., Jylhä, K., Ruosteenoja, K., Neitiniemi-Upola, L., Tietäväinen, H., and Pimenoff, N. (2009). Pääkaupunkiseudun ilmastotietoa ja -skenaarioita [Climate change data and scenarios for the Helsinki Metropolitan Area]. In *Pääkaupunkiseudun ilmasto muuttuu. Sopeutumisstrategian taustaselvityksiä* [Climate is Changing in the Helsinki Metropolitan Area. Background studies for the Adaptation Strategy]. Helsinki Region Environmental Services Authority 2010. Helsinki Region Environmental Services Authority *Publications* 3(2010), 9–36.

Yrjölä, T., and Viinanen, J. (2012). *Keinoja ilmastonmuutokseen sopeutumiseksi Helsingin kaupungissa* [Climate Change Adaptation Measures in the City of Helsinki]. City of Helsinki Environment Centre Publications 2/2012.

World Bank. (2017). 2016 GNI per capita, Atlas method (current US$). Accessed August 9, 2017: http://data.worldbank.org/indicator/NY.GNP.PCAP.CD

Case Study 9.B

Urban Wetlands for Flood Control and Climate Change Adaptation in Colombo, Sri Lanka

Manishka De Mel

Center for Climate Systems Research, Columbia University, New York

Missaka Hettiarachchi

University of Queensland and World Wildlife Fund, Brisbane

Keywords	Wetlands, flood control, flooding, resilience, adaptation, ecosystem-based adaptation, coastal
Population (Metropolitan Region)	2,251,274 (Sri Lanka Red Cross, 2016)
Area (Metropolitan Region)	676 km² (Sri Lanka Red Cross, 2016)
Income per capita	US$3,780 (World Bank, 2017)
Climate zone	Af – Tropical, rainforest (Peel et al., 2007)

Introduction

The Colombo metropolitan region is situated along the coast, with the Indian Ocean as its western boundary. As a coastal city situated on a floodplain, Colombo is expected to be vulnerable to the effects of climate change, impacting settlements, infrastructure, and other sectors. Vulnerable populations are found in settlements near the beach, in the floodplains of the Kelani River, and along canals and wetland areas. The Colombo Metropolitan region includes Sri Lanka's administrative and financial capitals and adjacent suburban areas. The city of Colombo has a resident population of more than 600,000, while in the larger metropolitan region, the population totals more than 5.8 million people (DCSSL, 2012; SLRCS, 2015).

Climate Change Impacts, Trends, and Projections

An analysis of the historical changes to Colombo's climate is not publicly available and a detailed analysis currently does not exist in literature. Discernible trends in sea level rise have not been identified for Colombo because tide gauge data in the harbor have shown fluctuation over time (Weerakkody, 1997) and changes in sea level rise for Sri Lanka are generally not known (Eriyagama et al., 2010). Furthermore, future projections for temperature, precipitation, and sea level rise are also not available for Colombo or its metropolitan region. The lack of localized climate risk information poses a challenge for effective climate change adaptation planning at the metropolitan region scale.

The main extreme weather events and disasters that occur in Sri Lanka include floods, droughts, landslides, and cyclones, and these are expected to increase in frequency due to climate change (Eriyagama and Smakhtin, 2010). Increases in the frequency and intensity of floods and droughts and in the variability and unpredictability of rainfall patterns are likely to occur in the future in Colombo (MESL and ADB, 2010). Sri Lanka has long had issues with coastal erosion, and this will be exacerbated by storm surges and sea level rise in the future (MESL, 2011). Flooding is identified as a constraint to the city's economic growth. Saline intrusion due to sea level rise is expected to affect aquifers in the coastal area, which could potentially impact the city's drinking water supply. Mosquito-related diseases such as dengue fever are likely to rise if rainfall increases because it influences the spread of the disease (MESL, 2010).

Role of Colombo's Urban Wetlands in Flood Control

Colombo has some existing ecological features that can help build resilience toward climate change impacts and disasters. The Colombo metropolitan region encompasses an interconnected system of natural wetlands that provides flood control. A core area of these urban wetlands has been conserved and designated as the Colombo Flood Detention Area for the primary purpose of mitigating floods (CEA, 1994).

The growth of the city, coupled with ad-hoc expansion, has posed multiple threats to these ecosystems. Recent research (Hettiarachchi et al., 2014) has demonstrated that the wetland system is undergoing transition from a habitat dominated by native grass to one that is comprised of small trees and shrubs (44%), while the peaty soil has changed to a semi-mineral soil. The transition has compromised the water-holding capacity of this wetland system, undermining its flood control function. In-depth research of the largest wetland, the Kolonnawa Marsh in the Flood Detention Area and surrounding rice paddy land, has identified that 13.5% and 60% of these areas, respectively, have been converted to non-wetland uses between 1980 and 2014. While the rate of change is alarming, the protected wetland regions of the Flood Detention Area by Sri Lanka Land Reclamation and Development Corporation (SLLRDC) have remained largely intact from 1999 onward. Conversion of land use together with nutrient pollution have resulted in the wetlands undergoing a major ecological transformation. The removal of certain uses of the broader wetland ecosystem (e.g., rice paddy cultivation) due to urbanization is one of the main drivers of ecosystem transformation, while climate change could impact the current transformation. According to the study, both legal and illegal conversion of wetlands (into non-wetland uses) is continuing, exacerbating the current destruction. At the same time, major or moderate floods in the area are increasing, based on a newspaper content analysis done by researchers. It was reported that a major or moderate flood has been reported every year after between 2005 and 2011 in the Flood Detention Area watershed. The floods of 2011 were recorded as the most severe in the history of Colombo, displacing nearly 15,000 people.

Climate change poses additional risks to these wetlands systems. Changes in rainfall regimes (e.g., intense rainfall events), increase in temperature, intrusion of saline water due to sea level rise, and storm surges are further challenges to Colombo's wetlands. These challenges will be twofold: climate change itself will impact the ecosystems and ecological function. Additionally, extreme rainfall, storms, and floods will require further flood control, highlighting the need for ecosystem-based adaptation. It is of utmost importance that these elements are thoroughly researched to identify and implement appropriate management practices.

Overcoming Challenges and Increasing Resilience

The case study demonstrates that urban wetlands can play an important role in controlling floods in the Colombo metropolitan area, but changes in its ecological regime are compromising its potential of providing ecosystem services. To maximize the potential of ecosystem-based adaptation mechanisms, it is vital to identify climate risks by using downscaled climate projections. This would also allow the identification of direct climate change impacts to ecosystems, while impacts to humans and their coping mechanisms will also have further knock-on effects on ecosystems. These factors could undermine the effectiveness of wetland ecosystems to build resilience. Identifying such risks at the outset will enable appropriate conservation and management mechanisms and will provide the best opportunity to build resilience.

In the broader context, the lack of climate risk information poses a major challenge to identifying impacts and building resilience in the Colombo metropolitan region. Coastal storms, floods, and heat waves in other cities demonstrate that a city's inhabitants, ecosystems, and infrastructure, such as railroads, roads, power stations, water systems, and wastewater treatment plants can be severely affected if resilience measures are not in place. Localized climate risk information, projections, and research are prerequisites for a comprehensive climate risk assessment for the Colombo metropolitan region. Such an assessment is essential to ensure that the correct type of solutions and resilience efforts are implemented. If future climate change projections are not considered, some actions could be maladaptive, worsening the impacts of climate change. Once impacts, projections., and vulnerabilities are identified, developing integrated, locally co-generated solutions will be vital to ensure the resilience of dense human settlements, critical infrastructure, and ecosystems. Ideally, these should begin with small-scale pilot projects that will enable lessons to be learned for scaling up.

Case Study 9.B Figure 1 *Settlements adjacent to Colombo's wetlands.*

References

Central Environmental Authority (CEA). (1994). *Wetland Site Report and Conservation Management Plan – Colombo Flood Detention Areas.* Central Environmental Authority of Sri Lanka, Colombo.

Department of Census and Statistics, Sri Lanka (DCSSL). (2012). Population of municipal councils and urban councils by sex census, 2012. Accessed July 6, 2015: http://www.statistics.gov.lk/Abstract2014/CHAP2/2.4.pdf

Eriyagama, N., and Smakhtin, V. (2010). Observed and projected climatic changes, their impacts and adaptation options for Sri Lanka: A review. In Evans, A., and Jinapala, K. (eds.). *Proceedings of the National Conference on Water, Food Security and Climate Change in Sri Lanka.* BMICH, Colombo, Sri Lanka, June 9–11, 2009. Vol. 2: Water quality, environment and climate change. International Water Management Institute.

Eriyagama, N., Smakhtin, V., Chandrapala, L., Fernando, K. (2010). *Impacts of Climate Change on Water Resources and Agriculture in Sri Lanka: A Review and Preliminary Vulnerability Mapping.* International Water Management Institute.

Hettiarachchi, M., Athukorale, K., Wijekoon, S., and de Alwis, A. (2014). Urban wetlands and disaster resilience of Colombo, Sri Lanka. *International Journal of Disaster Resilience in the Built Environment* 5(1), 79–89.

Ministry of Environment Sri Lanka (MESL). (2010). Sector Vulnerability Profile: Health. Ministry of Environment, Sri Lanka.

Ministry of Environment Sri Lanka (MESL). 2011. *Sri Lanka's Second National Communication on Climate Change.* Climate Change Secretariat, Ministry of Environment.

Ministry of Environment Sri Lanka (MESL) and Asian Development Bank (ADB). (2010). *National Climate Change Adaptation Strategy for Sri Lanka 2011–2016.* Climate Change Secretariat, Ministry of Environment, Sri Lanka.

Peel, M. C., Finlayson, B. L., and McMahon, T. A. (2007). Updated world map of the Köppen-Geiger climate classification. *Hydrology and Earth System Sciences Discussions* 4(2), 462.

Sri Lanka Red Cross Society (SLRCS). (2016). Districts and local authorities. Accessed November 3, 2016: http://www.redcross.lk/sri-lanka-country-profile/districts-and-local-authorities/

Weerakkody, U. (1997). Potential impact of accelerated sea-level rise on beaches of Sri Lanka. *Journal of Coastal Research (Special Issue),* 24, 225–242.

World Bank. (2017). 2016 GNI per capita, Atlas method (current US$). Accessed August 9, 2017: http://data.worldbank.org/indicator/NY.GNP.PCAP.CD

Case Study 9.C

Storm Surge in Costa da Caparica, Almada, in January 2014

Rita Isabel dos Santos Gaspar and Martin Lehmann

Aalborg University

Keywords	Storm surge, adaptation, mitigation
Population (Metropolitan Region)	2,812,678 (Statistics Portugal, 2015)
Area (Metropolitan Region)	3,015.24 km² (Statistics Portugal, 2015)
Income per capita	US$19,850 (World Bank, 2017)
Climate zone	Csa – Temperate, dry summer, hot summer (Peel et al., 2007)

Almada municipality is part of Lisbon Metro Region and it is located on the southern margin of the Tagus River (see Case Study 9.C Figure 1). The municipality is densely populated. It has around 174,000 citizens (more than 2,000 inhabitants/km²). With an area of 70 square kilometers, Almada has 35 kilometers of waterfront: 22 kilometers of riverside and 13 kilometers oceanfront. A wide beach line characterizes the Atlantic coast of Almada (see Case Study 9.C Figure 2).

Almada's oceanfront is extremely popular in the summer season, particularly the beaches of Costa da Caparica, a parish of Almada municipality with 13,400 inhabitants (Câmara Municipal de Almada, Divisão de Estudos e Planeamento, 2014). The beaches of Costa da Caparica are highly visited by tourists and citizens from the nearby municipalities. In 2014, Almada hosted around 130,000 holidaymakers who contributed an estimated €13 (US$15.3) million to the municipality (Cardeira, personal communication, 20 October 2015). Tourism is an important economic sector for the municipality (ICLEI and CEPS, 2013). The maintenance of the beaches is, therefore, of great importance for the municipality.

Unfortunately, Almada's Atlantic coast is also the most vulnerable zone of the municipality. Case Study 9.C Figure 2 shows the location of this zone. The main vulnerabilities to climate change that the territory of Almada faces are related to coastal erosion, storm surges, and coastal flooding (Freitas et al., 2010).

A study carried out nationally has revealed that the coastline of Almada has a medium to very high risk to the impacts of climate change on the Portuguese coast (e.g., rise of sea level, change of maritime agitation patterns, and storm surges) (Santos e Miranda, 2006). In fact, the adaptation strategy of Almada was initiated as a precaution due to its coastal vulnerability. The local government of Almada has decided to carry out studies on a local scale to evaluate potential future scenarios, following the methodology of the Intergovernmental Panel on Climate Change (IPCC) in order to take informed decisions and wisely select measures to effectively adapt Almada's territory to climate change (Freitas, 2015).

In the past century, the sandy beaches of Costa da Caparica have seen, at various periods, a significant volume of their sand

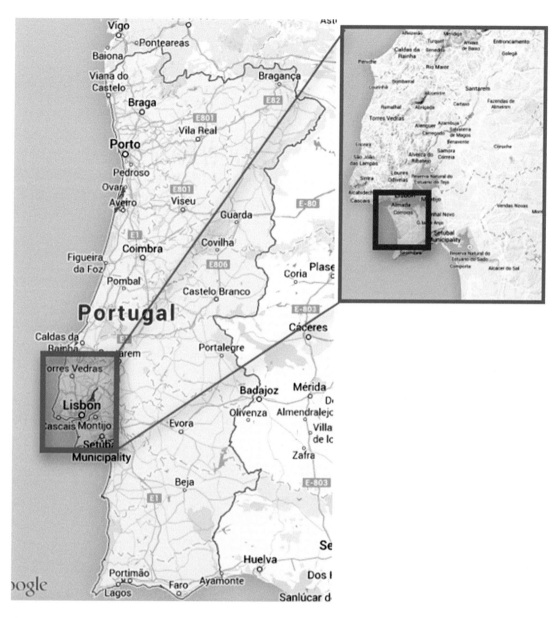

Case Study 9.C Figure 1 *Almada municipality location in Lisbon Metro Region.*

disappear (Veloso-Gomes et al., 2009). Between 1950 and 1972, groins and a seawall were constructed to minimize losses and stabilize the shoreline; these hard measures successfully contributed to the stability of Almada's sandy beaches in the following years. However, since 2000, these beaches have again been showing signs of instability resulting in considerable sand loss (Veloso-Gomes et al., 2009). To oppose this phenomenon, the municipality has implemented soft measure. It artificially nourished Costa da Caparica beaches with more than 2.5 million cubic meters of sand (Veloso-Gomes et al., 2007, 2009). The nourishment of Almada's beaches took place in 2007, 2008, 2009, and more recently, after the winter storm of 2014. According to the Portuguese Environment Agency (APA), the last sand nourishment cost €4 (US$4.7) million.

Costa da Caparica beaches have also experienced storm surges events. The most recent happened in January 2014. That event

was, reports say, a result of a storm–tide interaction (Portuguese Environment Agency, 2014). The combination of high tide level and the occurrence of a storm surge led to an abnormal elevation of the water level in the Atlantic Ocean, which affected the Costa da Caparica beachfront (see Case Study 9.C Figure 3). Storm surge results from a tsunami-like phenomenon of rising water that, in many cases, leads to coastal floods (NOAA, n.d.; Santos e Miranda, 2006). This is of major concern for population and infrastructures located near the coastal frontline. Case Study 9.C Figure 4 shows how close bars, restaurants, camping parks, and the urban area are to the ocean, putting the citizenry and tourists at risk. In 2014, this phenomenon resulted in considerable sand loss from Costa da Caparica beaches and coastal flooding with damage to Almada's oceanfront infrastructure (Portuguese Environment Agency, 2014). According to the APA, the rehabilitation of beach structures, such breakwaters, cost more than €500,000 (US$590,400).

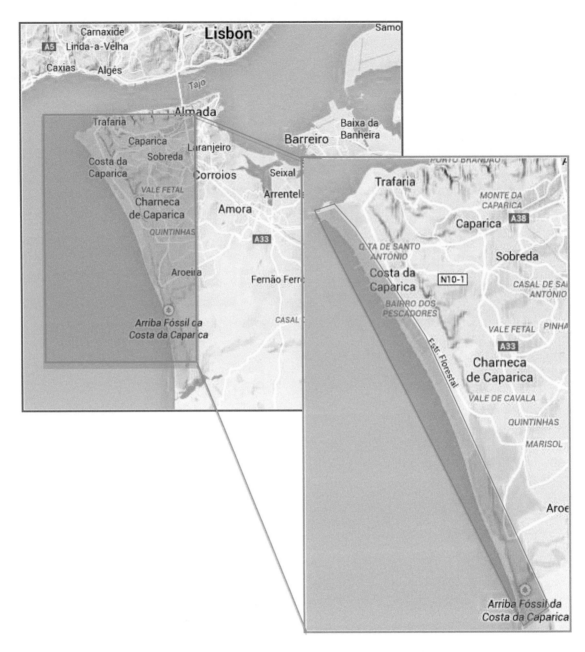

Case Study 9.C Figure 2 *Almada's oceanfront and identification of vulnerable zone.*

In Portugal, adaptive measures have, in most of the cases, been reactive responses to climate change events that have occurred rather than a pro-active response to predicted scenarios (Santos e Miranda, 2006). Apart from the hard measures described, Almada trying to adapt to climate change. In 2007, Almada introduced a strategic mitigation plan to face climate change – ELAC. The strategic adaptation plan is currently being developed.

Even though the adaptation strategic plan is only now being developed, adaptation is included in Almada's local climate change strategy. Adaptation measures have been already introduced in the municipality, such as the increment of flooding quotas in order to prevent the construction of infrastructure in vulnerable zones. Climate change studies are also being integrated into other existing local plans. This integration has the goal of creating synergies between municipal plans (Freitas e Lopes,

2013). Recommendations for climate change adaptation have, for example, been integrated into the recently reviewed local plan of Fonte da Telha that defines the urban development strategy to that area. With this approach, the local government expects to be better prepared for climate change, implementing adaptive measures to minimize its impact on infrastructure and population.

Not only the local government of Almada is apprehensive about coastal erosion. The Portuguese government is concerned with the vulnerability of the Portuguese coast and the potential climate change impacts on the coast. The ministry of Environment and Spatial Planning has requested a vulnerability assessment of the Portuguese coast, and the study revealed that 25% of the coast is exposed to processes of erosion and 66% is at risk (Ministry of Environment, Spatial Planning and Energy, 2015). The Ministry plans to make an investment of

€750 (US$885.6) million until 2050 in adaptation actions such planning and protection measures, elaboration of risk maps, artificial nourishment of beaches, and monitoring the Portuguese coast. These measures are in line with the strategic adaptation actions of the local government of Almada.

Case Study 9.C Figure 3 *Costa da Caparica beachfront and storm surge 2014 detail.*

Apart from the Portuguese government, other stakeholders have been actively collaborating to the climate change adaptation of Almada, including the Department of Environmental Sciences and Engineering (DCEA) and the Center for Environmental and Sustainability Research (CENSE) of the Faculty of Sciences and Technology of University Nova of Lisbon (FCT-UNL), the National Laboratory of Civil Engineering (LNEC), and the Portuguese Environmental Agency (APA).

CENSE and DCEA are collaborating with Almada by using the municipality as a case study on research projects about coastal communities' adaptation and resilience to climate change. They are also supporting Almada in the ClimAdaPT.Local network, of which Almada is a member. DCEA and CENSE are also collaborating on the research group Grupo de Trabalho do Litoral created by the Portuguese government with the goal of introducing solutions to storm surges and coastal erosion on vulnerable coastlines, such as the Atlantic front of Almada (Ferreira, 2015).

LNEC is developing an alarm system to coastal flooding risk. The project (HIDRALERTA) uses the Atlantic coast of Almada as one of its case studies, thus making an important contribution to the municipality's work regarding civil protection (Fortes, 2015). The maps of risks and vulnerability developed by the APA on a national level are very useful for the work being developed in Almada municipality. The findings of the national studies are used to support the local government's own studies.

In relation to the efforts of climate change mitigation, the strategy of Almada is to reduce its greenhouse gas (GHG) emissions by 20% by 2020, as agreed in the Covenant of Mayors (Câmara Municipal de Almada, 2010). The mitigation actions proposed by the local government focus mainly

Case Study 9.C Figure 4 *Costa da Caparica land occupation.*

on the sectors of buildings, public illumination, and industry and transportation, and on the promotion of energy efficiency and the use of renewable sources of energy. By combining its adaptation and mitigation actions with national measures and policies, the local government of Almada expects to achieve the goal set for the municipality and to prevent the occurrence of contradictory impacts (EGENEA, 2007; Câmara Municipal de Almada, 2010).

References

Câmara Municipal de Almada. (2010). *Estratégia Local para as Alterações Climáticas do Município de Almada. Plano de Acção pra a Mtigação.* Câmara Municipal de Almada e AGENEAL.

Câmara Municipal de Almada, Divisão de Estudos e Planeamento. (2014). *Território e População -Retrato de Almada Segundos os Censos 2011.* Almada.

EGENEA. (2007). *Estratégia Local para as Alterações Climáticas no Município de Almada.* Câmara Municipal de Almada, Lisboa.

Freitas, C. (2015). Painel 2 -Experiências Nacionais de Adaptação Local às Alterações Climáticas. Estratégia Local para as Alterações Climáticas, Almada (ELAC). Seminário ClimAdaPT.Local – Estratégias Municipais de Adaptação às Alterações Cimáticas em Portugal. Accessed March 6, 2015: http://new.livestream.com/livestreamingpt/climadaptlocal/videos/73872707

Freitas, C., and Lopes, N. (2013, July 5). Construção da Resiliência territorial em Almada: Projecto EU Cities Adapt. *Conferência Horizontes para uma AML inteligente, sustentável e inclusiva.* Lisboa, Portugal.

Freitas, C., Sousa, C., Lopes, N., and Machado, P. (n.d.). Tourist destination handling climate change: A Mediterranean experience (Almada, Portugal). In van Staden, M., and Musco, F. (eds.), *Local Governments and Climate Change: Sustainable Energy Planning and Implementation in Small and Medium Sized Communities* (243–252). Springer.

ICLEI and CEPS. (2013). Coastal adaptation in Almada, Portugal, a local-urban plan for Fonte da Telha. In *Climate Change Adaptation: Empowerment of Local and Regional Authorities with a Focus on Their Involvement in Monitoring and Policy Design* (18–19). European Union Committee of the Regions

Instituto Nacional de Estatistica Portugal (Statistics Portugal) (2015). Statistical Yearbook of Area Metropolitana de Lisboa. Accessed November 3, 2016: https://www.ine.pt/xportal/xmain?xpid=INE&xpgid=ine_publicacoes&PUBLICACOESpub_boui=279975813&PUBLICACOESmodo=2

Ministry of Environment, Spatial Planning and Energy. (2015). Mantenha-se Atualizado: Portugal tem 25% da sua costa sob erosão e 66% sob risco. Ministro do Ambiente, Ordenamento do Território e Energia. Accessed March 27, 2015: http://www.portugal.gov.pt/pt/os-ministerios/ministerio-do-ambiente-ordenamentodo-territorio-e-energia/mantenha-se-atualizado/20150316-maote-litoral.aspx

NOAA. (n.d.). *Outreach & education: Storm surge resources - Introduction to storm surge guide.* National Weather Service. Accessed February 2015: http://www.nws.noaa.gov/om/hurricane/resources/surge_intro.pdf

Peel, M. C., Finlayson, B. L., and McMahon, T. A. (2007). Updated world map of the Köppen-Geiger climate classification. *Hydrology and Earth System Sciences Discussions* **4**(2), 462.

Portuguese Environment Agency. (2014). *Registo das ocorrências no litoral. Temporal de 3 a 7 de Janeiro 2014.* APA.

Santos, F. D., and Miranda, P. (2006). *Alterações Climáticas em Portugal. Cenários, Impactos e Medidas de Adaptação Projecto SIAM II.* (eds.) Lisbon, Gradiva.

Veloso-Gomes, F., Costa, J., Rodrigues, A., Taveira-Pinto, F., Pais-Barbosa, F., and das Neves, L. (2009). Costa da Caparica Artificial Sand Nourishment and Coastal Dynamics. *Journal of Coastal Research* **56**, 678–682.

Veloso-Gomes, F., Taveira-Pinto, F., Pais-Barbosa, J., Costa, J., and Rodrigues, A. (2007). As Obras de Defesa Costeira Na Costa da Caparica. *2as Jornadas de Hidráulica. Recursos Hídricos e Ambiente*, 23–32.

World Bank. (2017). 2016 GNI per capita, Atlas method (current US$). Accessed August 9, 2017: http://data.worldbank.org/indicator/NY.GNP.PCAP.CD

Case Study 10.A

Economic Cost and Mental Health Impact of Climate Change in Calgary, Canada

Thea Dickinson

University of Toronto

Keywords	Health, extreme events, rainfall, adaptation, flooding
Population (Metropolitan Region)	1,214,839 (Government of Canada, 2015)
Area (Metropolitan Region)	704 km² (Demographia, 2016)
Income per capita	US$43,660 (World Bank, 2017)
Climate zone	Dfb – Snow, fully humid, warm summer (Peel et al., 2007)

The economic costs of climate change to a developed nation city can be illustrated by Calgary, the fourth largest municipality in Canada that sits on the Bow River basin. With a population of 1.2 million, many Calgary residents are either near or on the Bow River and Elbow River floodplains (with some residents living on designated flood ways or in the flood fringe). In June 2013, days of heavy rains in Calgary caused an increase in both rivers. Heavy floodwaters poured into the streets of the downtown core, washing away roadways and bridges. (Munich RE, 2014) With the city under meters of water, 50,000 people had to be evacuated from twenty-six communities.

Historically, both the Bow and Elbow River have flooded (e.g., 1932, 2005); dams were built following these disasters, effectively protecting residence from high-precipitation weather events. In the 1970s, the population in Alberta grew by 30% (Alberta Municipal Affairs, 2015) from the oil boom (Humphries, 2009), with many residents settling in Calgary. This increased the city's density and put pressure on urban infrastructure. Moreover, increasing temperatures in a changing climate have led to increased glacial melt in the Rocky Mountains, increasing the likelihood of high-intensity precipitation events and flooding. The June 19, 2013, flood was the costliest natural catastrophe in Canada's history. More than 30,000 residents were without power and 4,000 businesses were unable to function. In total, the extreme event caused CAD6 billion in total damage with CAD1.7 billion in insured losses. The disaster was also associated with an estimated CAD2 billion future loss to the Canadian economy (and Gross Domestic Product [GDP]) (Menon, 2013).

The flood event triggered the need for change. Calgary promptly moved forward with plans to develop an Expert Management Panel on River Flood Mitigation with six areas of action: managing flood risk; watershed management; event forecasting; storage, diversion, and protection; infrastructure and property resiliency; and changing climate. Their report, *Calgary's Flood Resilient Future*, was released in June 2014. The sixty-two-page paper outlines the panel's major findings under the identified themes with immediate (e.g., urge the Province to regularly review and update official flood hazard maps), mid-term (completed within 2015–2018, e.g., create graduated flood protection level requirements for City infrastructure, long term (initiated within 2015–2018, e.g., develop a time-phased plan to remove buildings from areas with high flood risk, while minimizing the disruption to affected communities), and ongoing actions (e.g., publish up-to-date, graduated flood maps for public information) (Calgary, 2014). Separately, a pilot program, Depave Paradise, was initiated. Increases in impermeable services in dense urban centers vastly contribute to increased flooding. Depave Paradise envisions these spaces being transformed into permeable green spaces. The initiative also creates interim holding tanks for rainfall during extreme events.

Following the flood event, Alberta Health Services recognized the mental health impact of the extreme event. Both Alberta Health Services (AHS) and the Red Cross went door to door to ensure that residents received the emotional help they needed to recover from the flood. AHS immediately provided $25 million in mental health support related to the floods, pledging a total of $50 million in funding to support immediate and future mental health needs. Specifically, the funding provided fifteen experts in mental health to the High River community, with on-site visits

from clinical staff to evacuees housed in hotels. Additionally, they hired child and youth mental health experts and provided training and education for disaster responders and flood victims, including suicide prevention training, loss workshops, and psychological first aid. The province also appointed a Chief Mental Health Officer and distributed information (translated into several local languages) to 85,000 residents on where and how to access counseling services and how to manage stress. The flood highlights the cascading effects of disasters in densely populated urban cities and way the impact of these events extends beyond their initial damages, causing future losses to the economy and unexpected health impacts.

References

Alberta Municipal Affairs. (2015). Alberta's Municipal Population, 1960–2014. Accessed September 14, 2015: http://municipalaffairs.gov.ab.ca/documents/mgb/Alberta_Municipal_Population_History.pdf

Calgary. (2014). Calgary's flood resilient future: Report from the expert management panel on river flood mitigation. Accessed September 14, 2015: http://www.calgary.ca/UEP/Water/Documents/Water-Documents/Flood-Panel-Documents/Expert-Management-Panel-Report-to-Council.PDF

Demographia (2016) Demographia World Urban Areas (Built Up Urban Areas or World Agglomerations). 12th Annual Edition. St. Louis: Demographia. Available from: http://www.demographia.com/db-worldua.pdf.

Government of Canada (2015). Population and dwelling counts, for Canada, census metropolitan areas, census agglomerations and census subdivisions (municipalities), 2011 and 2006 censuses. Accessed September 11, 2015: http://www12.statcan.gc.ca/census-recensement/2011/dp-pd/hlt-fst/pd-pl/Table-Tableau.cfm?LANG=Eng&T=303&CMA=535&S=51&O=A&RPP=25

Humphries, M. (2009). North American oil sands: History of development, prospects for the future. *International Journal of Energy, Environment and Economics* 16(2–3), 291–314.

Menon, N. (2013, June 24). Calgary flooding to cost billions in lost output for Canada. *The Wall Street Journal*. Accessed September 14, 2015: http://blogs.wsj.com/canadarealtime/2013/06/24/calgary-flooding-to-cost-billions-in-lost-output-for-canada/

Munich RE. (2014). Natural catastrophes 2013: Analyses, assessments, positions. *Topics Geo, 2013*. Accessed September 14, 2015: http://www.munichre.com/site/corporate/get/documents_E1043212252/mr/assetpool.shared/Documents/5_Touch/_Publications/302–08121_en.pdf

Peel, M. C., Finlayson, B. L., and McMahon, T. A. (2007). Updated world map of the Köppen-Geiger climate classification. *Hydrology and Earth System Sciences Discussions* 4(2), 462.

World Bank. (2017). 2016 GNI per capita, Atlas method (current US$). Accessed August 9, 2017: http://data.worldbank.org/indicator/NY.GNP.PCAP.CD

733

Case Study 10.B

New York's Million Trees NYC Project

Ashlinn Quinn

Mailman School of Public Health, Columbia University, New York

Keywords	Trees, green infrastructure, public–private partnership, urban forest, health, mitigation
Population (Metropolitan Region)	20,153,634 (U.S. Census, 2016)
Area (Metropolitan Region)	17,319 km² (U.S. Census, 2015)
Income per capita	US$56,180 (World Bank, 2017)
Climate zone	Dfa – Cold, without dry season, hot summer (Peel et al., 2007)

Introduction

MillionTreesNYC is a public–private initiative launched in 2007 in New York with the goal of planting and caring for 1 million new trees across the city's five boroughs by 2017. Involved organizations include NYC's Department of Parks and Recreation, the New York Restoration Project, the Office of the Mayor, and several universities and research institutes. MillionTreesNYC is part of PlaNYC 2030, a comprehensive plan comprising more than 100 initiatives citywide that seek to support the long-term sustainability of New York and to prepare for the challenges of climate change.

A main goal of the MillionTreesNYC program is the reduction of atmospheric CO_2. Trees accomplish this in two ways: first, by carbon sequestration, and second, via avoided CO_2 emissions that result from reductions in heating/cooling energy use. Trees help reduce energy needs via a combination of shading, transpiration-led air cooling, and wind-speed reduction that reduces the movement of outdoor air into interior spaces that results in heat loss. A 2007 study estimated that the shading and climate effects of existing NYC trees resulted in a cost savings for electricity and natural gas of US$27.8 million annually or US$47.63 per tree. The associated net annual CO_2 reduction was 113,016 tons, valued at US$754,947 or US$1.29 per tree (Peper et al., 2007). The cost-effectiveness of trees' ability to reduce carbon, however, differs both by tree species and by planting location. A 2013 study demonstrated that, in the New York region, the London plane tree is the most cost-effective species due to its long life span and large canopy. The most cost-effective tree-planting locations, meanwhile, are those in areas with a higher proportion of low-rise and residential

buildings, where the energy savings from shading will be the greatest (Kovacs et al., 2013).

Expected adaptation benefits of the MillionTreesNYC program include summertime ambient temperature reduction, improvements in air quality, and improved stormwater management. New York like many urbanized areas, is subject to the heat island effect whereby summertime ambient temperatures in the city center are warmer than temperatures in the surrounding suburbs and rural areas. This effect is largely driven by the lack of vegetated surfaces (e.g., grass and trees) in the city center. On average, the daily minimum temperature in New York is 7.2°F (4°C) warmer than the surrounding suburban and rural regions, a discrepancy that becomes more pronounced during heat waves. A 2006 modeling study that focused on the New York area found street trees to be one of the most effective urban heat island reduction strategies, yielding the most cooling per unit area of the heat island mitigation strategies studied (Rosenzweig et al., 2006). Trees' ability to reduce local ambient air temperatures is the result of a combination of shading effects and transpiration.

Reduction in ambient temperatures also results in improvement in air quality, due both to reduced ozone levels and to reduced emissions from power generation. Additional ways that trees improve air quality are via interception and absorption of airborne pollutants. Net annual air pollutants removed, released, and avoided average 1.73 lb per tree and were valued at US$5.27 million (or US$9.02 per tree) in 2007 (Peper et al., 2007).

Finally, the Million Trees program is expected to improve New York's ability to cope with storms and flooding. Plants are able to store water in their tissues and thus can absorb excess rainwater much better than hard surfaces such as asphalt, and they can mitigate the burden placed on stormwater and sewer systems during periods of heavy rainfall. As of 2007, New York's street trees were estimated to reduce stormwater runoff by 890.6 million gallons annually, with an estimated value of US$35.6 million (Peper et al., 2007).

Altogether, the benefits of urban trees are calculated to far exceed the cost of their care. Analyses conducted in conjunction with the project estimate that New York's street trees provide US$5.60 in benefits for every dollar spent on tree planting and care.

Human health and well-being is one of the key areas that MillionTreesNYC aims to improve. Tree-related health benefits include averted heat-related morbidity and mortality from ambient temperature reduction, reduced respiratory illness due

species of tree are important factors to consider in assessing the allergenic potential of urban tree pollen (Lovasi et al., 2013; Kinney et al., 2014).

As of August 2014, MillionTreesNYC was nearing its goal of planting 1 million trees. More than 900,000 trees had been planted, placing the project on a trajectory to completion nearly 2 years ahead of schedule (Case Study 10.B Figure 1).

References

Climate-Data. (2015). Climate-Data.org. Accessed October 9, 2015: http://en.climate-data.org/search/

Kinney, P. L., Sheffield, P. E., and Weinberger, K. R. (2014). *Climate, Air Quality, and Allergy: Emerging Methods for Detecting Linkages.* Humana Press.

Kovacs, K. F., Haight, R. G., Jung, S., Locke, D. H., and O'Neil-Dunne, J. (2013). The marginal cost of carbon abatement from planting street trees in New York City. *Ecology and Economics* **95**, 1–10.

Lovasi, G. S., O'Neil-Dunne J.P., Lu J.W., Sheehan D., Perzanowski M.S., MacFaden S.W., King K.L., Matte T., Miller R.L., Hoepner L.A., Perera F.P., Rundle A. (2013). Urban tree canopy and asthma, wheeze, rhinitis, and allergic sensitization to tree pollen in a New York City birth cohort. *Environmental Health Perspectives* **121**, 494–500.

Peel, M. C., Finlayson, B. L., and McMahon, T. A. (2007). Updated world map of the Köppen-Geiger climate classification. *Hydrology and Earth System Sciences Discussions* **4**(2), 462.

Peper, P. J., McPherson, E. G., Simpson, J. R., Gardner, S. L., Vargas, K. E., and Xiao, Q. (2007). New York City, New York municipal forest resource analysis. [Technical Report]. U.S. Department of Agriculture Forest Service, Pacific Southwest Research Station, Center for Urban Forest Research.

Rosenzweig, C., Solecki, W. D., and Slosberg, R. B. (2006). Mitigating New York City's heat island with urban forestry, living roofs, and light surfaces. New York City Regional Heat Island Initiative, Final Report 06–06. New York State Energy Research and Development Authority.

U.S. Census Bureau. (2010). Decennial census, summary file 1. Accessed October 2, 2014: http://www.census.gov/population/metro/files/CBSA%20Report%20Chapter%203%20Data.xls

U.S Census Bureau. (2016). Table 1. Annual Estimates of the Resident Population: April 1, 2010 to July 1, 2016 – Metropolitan Statistical Area; – 2016 Population Estimates. U.S. Census Bureau. Retrieved March 28, 2017. Accessed March 7, 2017: https://factfinder.census.gov/faces/tableservices/jsf/pages/productview.xhtml?pid=PEP_2016_PEPANNRES&prodType=table

World Bank. (2017). 2016 GNI per capita, Atlas method (current US$). Accessed August 9, 2017: http://data.worldbank.org/indicator/NY.GNP.PCAP.CD

Case Study 10.B Figure 1 *New York street trees.*

Photo: Daniel Avila

to improvements in air quality, and a general increase in physical activity associated with green space. A possible health dis-benefit of tree planting programs, however, is increased allergic disease owing to tree pollen. Research is currently under way to determine the relationship between tree pollen prevalence and the incidence and prevalence of allergic diseases in New York as in the case of the benefits of urban trees, initial studies have determined that both the density of tree canopy and the specific

Case Study 10.C

City of Toronto Flood: A Tale of Flooding and Preparedness

Thea Dickinson

University of Toronto

Keywords	Flood, extreme events, heavy downpour, adaptation
Population (Metropolitan Region)	5,583,064 (Government of Canada, 2015)
Area (Metropolitan Region)	5,906 km² (Government of Canada, 2015)
Income per capita	US$43,660 (World Bank, 2017)
Climate zone	Dfb – Cold, without dry season, warm summer (Peel et al., 2007)

On July 8, 2013, an extreme rain event, only surpassed by Hurricane Hazel (285 mm) in 1954, devastated Canada's largest city with 126 millimeters of rain. Within a few short hours, precipitation changed roads to waterways, causing rivers to rise 3 meters, leading to flooding and erosion in all corners of the urban municipality. The average rainfall in Toronto (the fourth largest city in North America) for the entire month of July is normally is 74 millimeters. That one day saw more than 50 millimeters above the monthly average. The economic costs of the storm were as vast as the devastation: more than CAD1.6 billion in total losses, with CAD920 million in insured losses and property damage in Toronto and surrounding areas and CAD65,325,842 in operating expenses to the City's public sector for the clean-up (City of Toronto, 2013).

In the past decade, Toronto has been impacted by two other extreme rain events. Late in August 2005, 122 millimeters of rain fell causing CAD671 million in damages and triggering the City of Toronto to reflect on its preparedness. Two years after the storm, they developed the Climate Change Action Plan with the objective of exceeding the Kyoto greenhouse gas reduction target (of 80% reduction by 2050) and greening the city's operations. Shortly after, in 2007, the report *Ahead of the Storm* (with multiple images of the 2005 extreme event) outlined a number of short-term actions to start in 2008 that would "improve [Toronto's] ability to cope with climate change, and [included] 29 longer-term actions that [would] result in a comprehensive adaptation strategy." Two years after its publication a 5-day rain event (July 24–28, 2009) left Toronto with a further CAD228 million in losses. In October 2009, the city released the Power to Live Green: Sustainable Energy Strategy that further defined Toronto's mitigation measures.

Despite these strategies and plans, momentum on preparedness and resilience was weakening due to a change in political power. The 2013 event highlighted the city's vulnerability to climate change and the economic damage that a one-day rain event could cause. The July 8, 2013, flood overwhelmed the city's combined sewers, flowed sewage into Lake Ontario, and ushered in a chain of cascading impacts: Toronto beaches, that on average were safe to swim in 90% of the time (Toronto Foundation, 2014), were deemed unsafe for a week during peak tourist season following the storm, and single-family homes and private businesses were inundated with both floodwaters and hazardous sewage, impacting livelihoods and economic activity in the City's core. Those same flooded buildings, if not properly treated, produced mold that had the potential to cause significant chronic respiratory health impacts. The power outage in some communities led to improper refrigeration and citizens being exposed to foodborne illnesses (*Campylobacter enteritis* and salmonellosis). All of these cascading impacts put greater stress on Toronto's health system, increased future economic loss, and created ongoing mental and emotional stress for the city's residents.

The storm underscored the need for a compressive flood prevention strategy and increased investment to the city's aging infrastructure. As a result, Toronto increased investment to reduce the impact of future flood events.

Following the extreme event, the following measures were implemented:

- Investment in Toronto Water Infrastructure, with CAD3.1 billion over 10 years on wastewater and stormwater collection systems
- Basement Flooding Protection Subsidy Program, with CAD962 million provided for the program (an increase from CAD500,000 in 2010 to CAD11 million in 2014)
- "Cause and effect" studies of the water damage to properties, particularly in the west end, to determine the conditions of the sewer system and stormwater storage capabilities
- Financial assistance for damage caused by the storm
- Combined sewer overflows control in Humber River and Black Creek
- Sewersheds and investigation of basement flooding in Study Areas 4 and 5
- Increased public education through partnership with the Toronto and Region Conservation Authority in the form of public meetings, fact sheet distribution to households, and advertisements in local news to help property owners learn how to protect their homes from flooding

The July 8 storm also renewed interest in mitigation as well as adaptation. At the start of 2015, the City Council tabled a motion to establish a Climate Change Mitigation and Adaptation Subcommittee that "makes recommendations to, and reports through, the Parks and Environment Committee" and will "determine and report back on its terms of reference, which are to include a review of City policies, expert advice, and international best practices to mitigate and adapt to climate change." There is optimism that this committee will set in motion policies and measures that will better prepare the city for the next extreme event. The subcommittee is set to meet in March 2015 with a report expected by April 15, 2015 (City of Toronto, 2015).

References

City of Toronto. (2013). Follow–up on the July 8, 2013 storm event. Accessed September 14, 2015: http://www.toronto.ca/legdocs/mmis/2013/ex/bgrd/backgroundfile-62645.pdf

City of Toronto. (2015). Establishing a climate change mitigation and adaptation subcommittee. Accessed September 14, 2015: http://app.toronto.ca/tmmis/viewAgendaItemHistory.do?item=2015.PE1.3

Climate-Data. (2015). Climate-Data.org. Accessed October 9, 2015: http://en.climate-data.org/search/

Government of Canada. (2015). Population and dwelling counts, for Canada, census metropolitan areas, census agglomerations and census subdivisions (municipalities), 2011 and 2006 censuses. Accessed September 11, 2015: http://www12.statcan.gc.ca/census-recensement/2011/dp-pd/hlt-fst/pd-pl/Table-Tableau.cfm?LANG=Eng&T=303&CMA=535&S=51&O=A&RPP=25

Peel, M. C., Finlayson, B. L., and McMahon, T. A. (2007). Updated world map of the Köppen-Geiger climate classification. *Hydrology and Earth System Sciences Discussions* **4**(2), 462.

Toronto Foundation (2014). Toronto's vital signs: 2014 report. Accessed September 14, 2015: http://torontosvitalsigns.ca/full-report.pdf

World Bank. (2017). 2016 GNI per capita, Atlas method (current US$). Accessed August 9, 2017: http://data.worldbank.org/indicator/NY.GNP.PCAP.CD

Case Study 11.A

Peri-Urban Vulnerability, Decentralization, and Local-Level Actors: The Case of Flooding in Pikine/Dakar

Caroline Schaer

UNEP-DTU Partnership (UDP), Copenhagen

Keywords	Flood, community-based adaptation, peri-urban vulnerability, municipal capacity, housing
Population (Metropolitan Region)	3,653,000 (United Nations, 2016)
Area (Metropolitan Region)	194 km² (Demographia, 2016)
Income per capita	US$950 (World Bank, 2017)
Climate zone	BSh – Arid, steppe, hot (Peel et al., 2007)

In Senegal, the effects of climate variability and change are manifested through coastal erosion, the degradation of mangroves, and a decline in overall rainfall coupled with an increase in frequency and intensity of precipitation and sea level rise. In 2012, floods affected 287,400 people in the country and caused the contamination of drinking water sources, nineteen deaths, and the displacement of 5,000 families (OCHA, 2015). In urban areas, the exposure to risk is distributed unevenly because 38.1% of the urban population in Senegal lives in slums (UN Habitat, 2008b). Because of the inadequacy of basic services and infrastructure, poverty, social inequality, and inadequate social security systems, the urban poor are the most vulnerable to adverse weather conditions (Bicknell et al., 2009). In Senegal, a large share

of this population lives in Pikine, a densely populated peri-urban area with a population of 915,300, situated on the periphery of Dakar. The area developed from the early 1950s with uncontrolled urban growth resulting from rural-urban migration and state-ordered evictions from Dakar city center (Fall et al., 2005). Short and more intense outbursts of rain together with a rapid and anarchic urbanization process, nonexistent storm drainage and waste disposal systems, and the rising groundwater table of the Thiaroye aquifer create increased local runoff rates in the area (Faye, 2011). The result is a higher frequency, intensity, and duration of floods reported over the past 13 years (CRED, 2009). In Pikine, floods have had wide-ranging impacts on the food security, health, infrastructure, and housing conditions of the affected communities. Here, the impacts of flooding bring to light the environmental and social inequalities that characterize these areas and aggravate already precarious living conditions by playing the role of multipliers of existing vulnerabilities. Homes and livelihoods are at risk because these people mainly depend on activities that are disrupted by floods, such as small-scale commerce and artisanal trade. In the long term, climate change is likely to exacerbate these conditions due to more extreme weather events and because Senegal is among those countries with the highest proportion of its urban population living in low-elevation coastal zones. In Dakar alone, 61.6% of the population is considered to be at risk from sea level rise and has already been experiencing rising sea levels of 1.5 centimeter per decade over the past 11 years (UN Habitat, 2008b).

The municipalities of Pikine are characterized by high-density shelters, insecure tenure, and primitive shacks built alongside with more modern housing. These hazardous environments

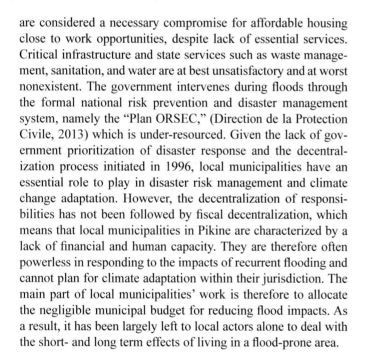

Case Study 11.A Figure 1 *Guinaw Rail Nord/Dalifort (municipalities of Pikine).*

Photo: Caroline Schaer

Case Study 11.A Figure 2 *Guinaw Rail Nord/Dalifort (municipalities of Pikine).*

are considered a necessary compromise for affordable housing close to work opportunities, despite lack of essential services. Critical infrastructure and state services such as waste management, sanitation, and water are at best unsatisfactory and at worst nonexistent. The government intervenes during floods through the formal national risk prevention and disaster management system, namely the "Plan ORSEC," (Direction de la Protection Civile, 2013) which is under-resourced. Given the lack of government prioritization of disaster response and the decentralization process initiated in 1996, local municipalities have an essential role to play in disaster risk management and climate change adaptation. However, the decentralization of responsibilities has not been followed by fiscal decentralization, which means that local municipalities in Pikine are characterized by a lack of financial and human capacity. They are therefore often powerless in responding to the impacts of recurrent flooding and cannot plan for climate adaptation within their jurisdiction. The main part of local municipalities' work is therefore to allocate the negligible municipal budget for reducing flood impacts. As a result, it has been largely left to local actors alone to deal with the short- and long term effects of living in a flood-prone area.

The limited room for maneuvering of the urban poor does not mean that they are powerless. Based on semi-structured interviews, observations, and focus group discussions conducted in these municipalities, it was found that they apply a wide and diverse range of strategies that allow them – in a context of weak state capacity – to react to recurrent flooding with varying degrees of success and sustainability. In Guinaw Rail Nord (GRN) and Dalifort (see Case Study 11.A Figure 1), two municipalities of Pikine, there is an abundance of community-based support networks and associations that seek to provide support to residents. Residents facing flood risks also apply individual disaster risk reduction and adaptation strategies, including improving their houses (i.e., new roof, sanitation), selling off assets, investing in water pumps, and the widespread elevation of the outside and inside floors with garbage, gravel, and sand. The latter strategy

may be considered as an urban alternative to raising houses on stilts. This means that many houses are filled up, to the point where some windows are at ground level and the ground floor becomes uninhabitable. Houses are then heightened to compensate for the lost space. Partly as a result of this modus operandi, 40% of all houses in GRN are abandoned. "If we continue like this we will soon reach the sky" expressed a resident to explain the futility of this strategy. The most vulnerable households often do not have any other options than to continue to live in their flooded homes, by elevating their beds on bricks, sanitizing the flood water to avoid diseases, and keeping children in sight on the beds.

The absence of effective disaster risk management and adaptation planning forces the urban poor in Pikine to cope with disproportionate levels of risk. By acting alone, communities are often forced to cope with recurrent flooding through the application of strategies that are very limited in time and scope, which often has undesirable effects for the most vulnerable. They are left with few possibilities for adaptation, which makes them increasingly vulnerable to future floods. The example from GRN and Dalifort shows that there is a strong need for building the financial and human capacity of local municipalities for disaster risk management and climate change adaptation. A strong support from local government to locally initiated actions by community groups and networks of individuals is a precondition for local action to become a catalyst for adaptation and thereby enable them to tackle the challenges posed by climate change in poor urban areas.

References

Bicknell, J., Dodman, D., and Satterthwaite, D. E. (2009). *Adapting Cities to Climate Change: Understanding and Addressing the Development Challenges*. Earthscan.

CRED (2009). Center for Research on the Epidemiology of Disasters. Accessed September 22, 2015: www.emdat.be

Demographia (2016) Demographia World Urban Areas (Built Up Urban Areas or World Agglomerations). 12th Annual Edition. St. Louis: Demographia. Available from: http://www.demographia.com/db-worldua.pdf.

Direction de la Protection Civile., Ministère de l'intérieur, République du Sénégal. (2013). *Rapport sur l'Etat de la protection civile au Sénégal: Protection civile et participation de la société civile à la prévention des risques et à la gestion des catastrophes.* Direction de la Protection Civile.

Fall, A., Salam, Guèye, C., and Tall, S.M. (2005). Changements climatiques, mutations urbaines et strategies citadines a Dakar. In Fall Salam A., and Guèye, C. (eds.), *Urbain-Rural: l'hybridation en marche.* Dakar: EENDA tiers-monde (191).

Faye, M.M. (2011). *Projet de Gestion des'Eaux Pluviales (PROGEP), Cadre de Gestion environnementale et sociale (CGES), Rapport final.* Dakar: Agence de Développement Municipal

OCHA. (2015). OCHA Annual Report. Accessed September 22, 2015: http://www.unocha.org/about-us/publications/flagship-publications/*/77

Peel, M. C., Finlayson, B. L., and McMahon, T. A. (2007). Updated world map of the Köppen-Geiger climate classification. *Hydrology and Earth System Sciences Discussions* **4**(2), 462.

UN Habitat. (2008a). *Sénégal: Profil urbain de Dakar.* HS/1003/08F. Author.

UN Habitat. (2008b). *State of the World's Cities 2008/2009 Harmonious Cities.* Earthscan.

United Nations, Department of Economic and Social Affairs, Population Division (2016). The World's Cities in 2016 – Data Booklet (ST/ESA/SER.A/392).

World Bank. (2017). 2016 GNI per capita, Atlas method (current US$). Accessed August 9, 2017: http://data.worldbank.org/indicator/NY.GNP.PCAP.CD

Case Study 11.B

Climate Change Adaptation and Resilience Building in Manila: Focus on Informal Settlements

Emma Porio

Department of Sociology and Anthropology, Ateneo de Manila University

Antonia Yulo-Loyzaga and Monica Ortiz

Manila Observatory

Keywords	Adaptation, resilience building, informal settlements, sea level rise, land tenure, housing
Population (Metropolitan Region)	11,855,975 (Philippine Statistics Authority, 2015)
Area (Metropolitan Region)	620 km² (Philippine Statistics Authority, 2015)
Income per capita	US$3,580 (World Bank, 2017)
Climate zone	Aw – Tropical savannah (Peel et al., 2007)

The Philippines is one of the most vulnerable and risk-prone countries in the world (World Risk Report, 2014) to both climate change and disasters. Its capital, Manila, is considered highly vulnerable and has experienced devastating floods in recent years. Manila sits on a semi-alluvial floodplain with a land area of approximately 636 square kilometers (Bankoff, 2003, Yulo-Loyzaga et al., 2014).

Bounded on the east by the Sierra Madre Mountains and the Marikina Watershed, in the southwest by Laguna de Bay, and in the west by Manila Bay (Case Study 11.B Figure 1), the Pasig and Marikina Rivers are Manila's main drainage-ways to Laguna Lake and Manila Bay. The capital's existing drainage system of *esteros*, a network of natural channels constructed during the colonial period, have proved inadequate today (Bankoff, 2003). More than 20% of the historical drainage capacities have been lost to urban developments (Zoleta-Nantes, 2003). Compounding this, eighty-seven chokepoints constrict the flow of many rivers that cross Manila. Thus, the megacity's drainage capacity is significantly compromised not only by elevation, topography, and the morphology of its rivers, but also by industrial, residential, and institutional developments and informal settlements (Siringan, 2013).

Manila has a population of 13.9 million (UN Habitat, 2015) but hosts a daytime population of 16–18 million people. Composed of seventeen political-administrative units (sixteen cities, one municipality), Manila contributes more than a third (37%) of the national gross domestic product (GDP). Drawn by the prospect of income opportunities, its population continues to increase rapidly as a result of high population growth and in-migration (Porio, 2011). Currently, the metropolis has a density of 18,648 persons per square kilometer but is projected to reach about 29,146 in the year 2025 (Porio, 2014). Approximately 4 million persons as of 2010 are living in informal settlements (Ballesteros, 2010).

While Manila's historical rainy season is from June to August, recent years have witnessed an increase in flash floods due to enhanced southwest monsoon events, thunderstorms, and tropical cyclones. A regional climate model (RegCM3) estimates an increase of rainfall associated with the peak of the southwest monsoon season of up to approximately 20% (Narisma et al., 2014). While the typhoon season normally occurs from June to November, the past few years have seen tropical cyclones and extreme rainfall events occurring throughout the year (Porio, 2012).

Areas Vulnerable to 1 Meter Sea Level Rise at Metro Manila, Philippines

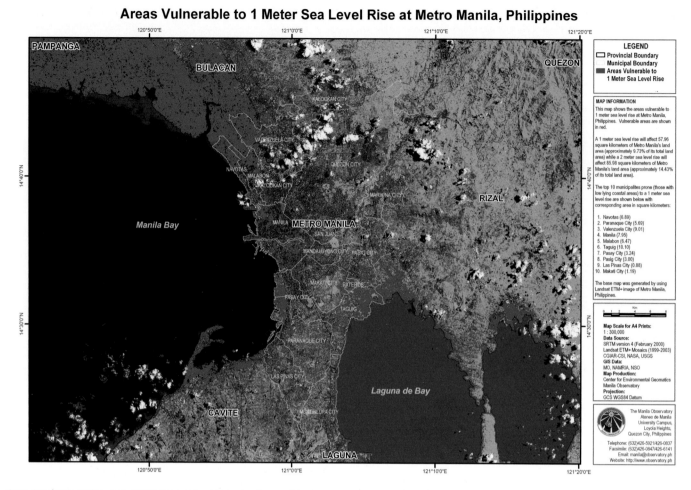

Case Study 11.B Figure 1 *Manila topography and low elevation coastal zones.*

Source: Narisma et al., 2014. Manila Observatory

The challenge of climate change adaptation and resilience in Manila lies in balancing the demands for shelter and basic services for the poor and the restoration and enhancement of drainage capacity through spatial planning and the implementation of risk-sensitive zoning laws.

Informality and Vulnerability

Slum growth is projected to be up to 3.14% per year (Table 1). The highest estimated annual growth in slum population growth is identified at the port area of Manila (Case Study 11.B Figure 2). With an estimated growth of 10% per year, these coastal communities along Manila Bay are natural catchment zones for rural migrants from other Philippine regions and provinces (Yulo-Loyzaga et al., 2014).

Migrants and poor communities often settle in slums or informal settlements in high-risk or hazard-prone areas. Informal settlements typically have little or no infrastructure that helps provide protection from storm events and other natural hazards (Feiden, 2011; Porio, 2012).

Moreover, the 2008 Philippine Asset Reform Card states that a large portion of the population living in Manila does not have

security of tenure in their housing, jobs, and livelihood sources. Only 61% of households in Manila have sufficient access to basic services (Porio, 2011). However, being poor is not the only reason why certain sectors are more vulnerable to floods or other environmental hazards – spatial isolation and lack of participation in decision-making intensify their present and future vulnerability (Zoleta-Nantes, 2003). These patterns of settlement may be connected to the lack of ability of government agencies to impose building and infrastructure standards, which results in unregulated growth and expansion of informal settlements (Porio, 2011).

Adaptive Risk Governance, Equity, and Resilience

Manila was historically governed as one planning unit as recently as 40 years ago. Today, sixteen cities and one municipality independently exercise political and administrative control of the city. These administrative areas are not ecosystem-based and physical plans are not integrated.

Tropical storm Ketsana in 2009 caused severe losses and damages in Manila and in provinces in Luzon. In response, a PhP50 Billion campaign intends to provide resilient housing,

Table 1 *Slum population, urban Philippines. Sources: Family Income and Expenditure Survey; Metro Manila Urban Services for the Poor (MMUSP) Project, HUDCC, 2008; Philippine Institute of Development Studies, 2011*

	Slum Population 2006	% Slum	Slum Annual Growth Rate (%) (2000–2006)	Projected Slum Population		
				2010	2020	2050
Urban Philippines	2,936,011	7.10	3.40	3,819,766	6,572,683	12,967,806
Large towns/cities	978,422	5.57	3.49	1,122,335	1,736,317	10,108,036
Manila	1,351,960	12.17	8.55	1,877,003	4,689,943	6,668,187
Manila[a]	4,035,283	36.33	3.14	4,565,951	6,294,181	8,949,102

[a] Data for Manila based on broader definition of slums, which include squatter or illegal settlements and settlements under informal arrangements (i.e., no formal or legal documentation of arrangement) and blighted area.

Notes: Large towns and cities refer to administrative towns/cities with population as of 2007 above 100,000 to 2,000,000. Slum population growth rate is estimated using exponential growth; r = ln(Pt/Po)/t; for large towns/cities, period covers 2003–06.

Slum population projected is based on estimated slum population growth rate.

Slums are defined as households in illegal settlements (i.e., without consent of owner) or living in makeshift housing.

Case Study 11.B Figure 2 *Samar settlement along Manila Bay, Philippines.*

Photo: Neal Oshima, from Lungsod Iskwater: The Evolution of Informality as a Dominant Pattern in Philippine Cities, Alcazaren et al., editors, co-published by the Luis A. Yulo Foundation for Sustainable Development, Inc., and Anvil Publishing. Inc.

relocation, and income opportunities for Manila's urban poor (Yulo-Loyzaga, et al., 2014). Portions of a new metro-wide flood control masterplan is currently in various stages of implementation.

This effort, however massive, has not yet proved adequate in coping with persistent poverty, income inequality, and population growth as the drivers of vulnerability and exposure in Manila's informal settlements. This plan has also failed to provide coordinated solutions to the removal of river and coastal easements by officially permitted industrial, commercial, and residential developments. It is these developments and the income opportunities that they provide that have drawn informal settlers to hazardous interstices of the city.

The Philippine Supreme Court recently declared the government procedures used to fund the master plan unconstitutional, rendering its future implementation uncertain (Philippine Daily Inquirer, July 20, 2014).

Despite these setbacks, there are initial steps to provide in situ resettlement through construction upgrading and capacity-building in disaster preparedness for informal communities. The institutionalization of green building codes and solid waste management and recycling programs by several local governments in the metropolis is now under way. These plans integrate solid waste management with income opportunities and flood management for the informal settlement communities in the flood-prone areas of the city. Ultimately, local government coordination and scaling up human settlements and ecosystem-based physical planning for Manila is required for developing climate resilience. Efforts to integrate the informal economy by recognizing its value contributions to formal supply and service chains need to be accelerated. The economic integration of informal settlers must be accompanied by the extension of basic social services, including access to health care and housing support options, such as relocation and in situ upgrading. The combination of economic integration and access to basic social services is just the first step in adaptation to the impacts of both rapid and slow-onset extreme weather events. Both Pasig City and Marikina City have relocated informal settlements within the city as well as upgraded the basic services to and drainage systems of marginal communities along the riverlines. Ultimately, a combination of evidence-based government programs and non-governmental organizations in support of livelihood diversification and training in emergency management are essential to community-based climate and disaster resilience.

References

Bankoff, G. (2003). Constructing vulnerability: The historical, natural and social generation of flooding in metropolitan Manila. *Disasters* 27(3), 224–238. Accessed September 21, 2015: http://www.ncbi.nlm.nih.gov/pubmed/14524047

Ballesteros, M.M. (2010). Linking Poverty and Environment: Evidence from Slums in Philippine Cities. Philippine Institute for Development Studies Discussion Paper Series No. 2010-33. Accessed September 21, 2015: http://dirp4.pids.gov.ph/ris/dps/pidsdps1033.pdf

Feiden, P. (2011). Adapting to climate change: Cities and the urban poor. Report by the International Housing Coalition.

Narisma, G. T., Siringan, F., Vicente, M. C., Cruz, F., Perez, J., Dado, J., Gozo, E., Del Castillo, E., and Dayawon, R. (2014). Metro Manila hazards: Extreme weather events and geophysical exposures of a coastal city at risk. Paper presented in the International Conference on Coastal Cities at Risk: Climate Change Vulnerability and Adaptation, Vancouver, Canada, March 2014.

Peel, M. C., Finlayson, B. L., and McMahon, T. A. (2007). Updated world map of the Köppen-Geiger climate classification. *Hydrology and Earth System Sciences Discussions* 4(2), 462.

Philippine Statistics Authority. (2015). Quickstat. On National Capital Region – August 2015. Accessed September 17, 2015: https://psa.gov.ph/sites/default/files/attachments/ird/quickstat/Quickstat_ncr_0.xls

Porio, E. (2011). Vulnerability, adaptation, and resilience to floods and climate change-related risks among marginal, riverine communities in Metro Manila. *Asian Journal of Social Science* 39(4), 425–445. doi:10.1163/156853111X597260

Porio, E. (2012). Decentralization, power and networked governance practices in Metro Manila. *Space and Polity* 16(1), 7–27.

Porio, E. (2014). Climate change vulnerability and adaptation in Metro Manila. *Asian Journal of Social Science* 42(December 2012), 75–102. doi:10.1163/15685314-04201006

Porio, E., Loyzaga, A., Ortiz, M., and See, J. (2014). Climate change mitigation and adaptation in Metro Manila Informal Settlements. Paper presented at the ISA-RC 46 Integrative Session, Addressing Social Inequality Before, During and After Crisis, XVIII World Congress of Sociology, Yokohama, Japan, July 13–19, 2014.[1]

Porio, E., See, J. C., and Dalupang, J. P. (2014). Characterizing social vulnerability and building adaptive capacity and resilience in Metro Manila. Paper presented in the International Conference on Coastal Cities at Risk: Climate Change Vulnerability and Adaptation, Vancouver, Canada, March 2014.

Siringan, F. (2013). "Big River Rising." Christian aid technical report. Accessed September 21, 2015: http://www.christianaid.org.uk/whatwedo/in-focus/big-river-rising

UN Habitat. (2015). State of Asian and Pacific Cities 2015. Bangkok: UN-Habitat.

World Bank. (2017). 2016 GNI per capita, Atlas method (current US$). Accessed August 9, 2017: http://data.worldbank.org/indicator/NY.GNP.PCAP.CD

Yulo-Loyzaga, M. A. (2014). Metro Manila: Coastal city at risk. Paper presented in the International Conference on Coastal Cities at Risk: Climate Change Vulnerability and Adaptation, Vancouver, Canada, March 2014.

Zoleta-Nantes D.B., 2003: Differential impacts of flood hazards among the street children, the urban poor and residents of wealthy neighborhoods in Metro Manila, Philippines. In: Jones J.A.A. and Woo M.-K. (eds.), Coping With Hydrological Extremes, Special Issue, Mitigation and Adaptation Strategies for Global Change, Vol. 7(3). pp. 239–266. Accessed September 21, 2015: http://link.springer.com/article/10.1023/A:1024471412686

---

[1] A large part of this presentation is based on the initial findings of the "Coastal Cities at Risk" (2011–2016), a 5-year research project supported by the International Development Research Center (IDRC) and implemented by Manila Observatory (MO) in partnership with Ateneo de Manila University and other universities in Canada, Thailand, and Nigeria. The overall findings of the project Characterizing Vulnerability in Metro Manila: Coastal City at Risk) was by Yulo-Loyzaga et al., in the Coastal Cities at Risk Mid-Term Conference in Vancouver, Canada, March 16–21, 2014.

Case Study 12.A

Consolidated Edison after Hurricane Sandy: Planning for Energy Resilience

Michael Gerrard

Columbia Law School, New York

Keywords	Heat wave, storm, electricity, utilities, policy regulations
Population (Metropolitan Region)	20,182,305 (U.S. Census, 2015)
Area (Metropolitan Region)	17,319 km² (U.S. Census, 2015)
Income per capita	US$56,180 (World Bank, 2017)
Climate zone	Dfa – Cold, without dry season, hot summer (Peel et al., 2007)

Superstorm Sandy in October 2012 caused a large-scale disruption of electric service in the New York region. Consolidated Edison Company (Con Edison), the principal provider of electricity in New York and Westchester County, suffered major damage to substations, transformers, cables, and other equipment. More than 1,150,000 customers lost electrical service – five times the number of outages caused by the second-largest service disruption, Hurricane Irene of 2011. Some customers were out of service for as long as 2 weeks.

Shortly after the storm, the Sabin Center for Climate Change Law at Columbia Law School prepared a petition (ultimately co-signed by six non-governmental organizations [NGOs], including the Natural Resources Defense Council and Earthjustice) with the New York Public Service Commission (PSC) to require all the utilities it regulates (electricity, natural gas, cable, telephone, and private water) to develop climate change adaptation plans. Shortly thereafter, on January 25, 2013, Con Edison filed with the PSC for its next rate increase for its electric, gas, and steam systems. The request included US$1 billion in storm hardening (i.e., physical improvements to make the system more resilient to extreme weather events). The investments involved four kinds of actions: protecting infrastructure, hardening components, reducing impact, and facilitating restoration.

The Sabin Center, upon examining the filing, determined that it involved efforts to prepare for the next Sandy-like event, such as placement of sea walls around or elevation of sensitive equipment. However, it did not address increasing risks of other climate events that could result from climate change and disrupt service, such as heat waves. The Sabin Center formally intervened in the rate proceeding. It supported the planned expenditures but argued that a broader array of climate-related threats should also be considered.

To inform Con Edison about increasing risks due to climate change, the Sabin Center also brought a scientist from Columbia's Center for Climate Systems Research to a meeting with top Con Edison officials. The scientist, Dr. Radley Horton, gave a presentation about the latest climate projections for the New York region based on the New York City Panel on Climate Change (NPCC, 2013). This presentation helped to persuade Con Edison that climate change does indeed pose a real threat to the company's operations.

The PSC staff convened a collaborative effort involving itself, Con Edison, various representatives of New York State, New York, Westchester County, labor unions, electricity consumers, academic centers (including the Sabin Center), and environmental NGOs. The collaborative met intensively in the summer and fall of 2013 in an effort to resolve a number of issues in the rate proceeding, including climate change preparedness. In December 2013, the parties reached an agreement. One of the elements of that agreement was that Con Edison would retain the services of climate scientists to prepare more fine-grained projections of future climate conditions in the service area than had previously been available, and that, based on these projections, the company would determine whether changes were needed to its capital plans or its operation and maintenance procedures.

Since the ratepayers would pay for this work, the agreement required the approval of the PSC. In February 2014, the PSC issued a decision that not only approved the Con Edison agreement, but also indicated that it expected the other regulated utilities in the state to adopt similar measures.

The study mandated by the agreement is now under way, leading to the preparation of a Climate Change Vulnerability Study. It will examine, among other things, the vulnerability of the electrical utility system to floods, wind, and heat waves, and it will evaluate Con Edison's current design standards for each condition and advise on whether they should be modified. The Sabin Center and the Center for Climate Systems Research are participating in the study.

Con Edison is also proceeding with the physical work that was part of its initial rate proposal, including such activities as hardening the overhead electric distribution system, the gas system, and the steam tunnels; hardening generating stations and substations; and protecting the telecommunications system. Con Edison is also undertaking a quantitative risk assessment in order to identify cost-effective resilience actions.

To ensure that the other utilities in the state follow the PSC's requirements, the Sabin Center has intervened in the next two major rate proceedings to be filed – Central Hudson Gas &

743

Electric and Orange & Rockland Utilities. After negotiations, both companies agreed to review the results of the Con Edison study once it is completed, together with any other climate information submitted by the Sabin Center or others, and then to consider whether further adaptation measures are needed.

The Sabin Center has given a number of presentations around the United States and one in Europe advocating use of this approach outside New York. Environmental advocates in a number of states are now considering this option. Efforts are also now under way to persuade the North American Electric Reliability Corporation (a Congressionally authorized entity that sets reliability standards for the nation's electric grid) to take climate change into account.

This effort was led by one academic center and strongly supported by climate scientists in the same university. Several, NGOs, another university (Pace Law School), and the Law Departments of New York State and New York joined in the action. The scientific evidence persuaded the electric utility company and its state regulators of the wisdom of enhancing the resilience of the system.

This effort can be replicated throughout the United States and potentially in other countries. Each of the fifty states has a public utility regulatory agency charged with the responsibility to ensure the reliability of utility service. Most of these agencies have administrative procedures that would allow citizens to intervene or otherwise participate in their rate cases and other proceedings. Thus, advocates in the other states can utilize the strategy that was successfully utilized with Con Edison.

References

Climate-Data. (2015). Climate-Data.org. Accessed October 9, 2015: http://en.climate-data.org/search/

New York City Panel on Climate Change. (2013): *Climate Risk Information 2013: Observations, Climate Change Projections, and Maps.* C. Rosenzweig and W. Solecki (eds.) NPCC2. Prepared for use by the City of New York Special Initiative on Rebuilding and Resiliency.

Peel, M. C., Finlayson, B. L., and McMahon, T. A. (2007). Updated world map of the Köppen-Geiger climate classification. *Hydrology and Earth System Sciences Discussions* **4**(2), 462.

U.S. Census Bureau. (2010). Decennial census, summary file 1. Accessed September 8, 2015: http://www.census.gov/population/metro/files/CBSA%20Report%20Chapter%203%20Data.xls

U.S. Census Bureau. (2015). Annual Estimates of the Resident Population: April 1, 2010 to July 1, 2015 – Metropolitan Statistical Area; and for Puerto Rico – 2015 Population Estimates. Accessed November 3, 2016: https://factfinder.census.gov/faces/tableservices/jsf/pages/productview.xhtml?src=bkmk

World Bank. (2017). 2016 GNI per capita, Atlas method (current US$). Accessed August 9, 2017: http://data.worldbank.org/indicator/NY.GNP.PCAP.CD

Case Study 14.A

How Can Research Assist Water Sector Adaptation in Makassar City, Indonesia?

Dewi G. C. Kirono

Commonwealth Scientific and Industrial Research Organisation (CSIRO), Canberra

Keywords	Drought, water pollution, infrastructure, community-based adaptation, planning
Population (Metropolitan Region)	1,520,000 (Demographia, 2016)
Area (Metropolitan Region)	179 km² (Demographia, 2016)
Income per capita	US$3,400 (World Bank, 2017)
Climate zone	Am – Tropical, monsoon (Peel et al., 2007)

Makassar, the largest and most urbanized city in eastern Indonesia, is home to 1.4 million inhabitants with a population density of around 8,011 persons per square kilometer (BPS, 2014). Like many cities in Indonesia, Makassar is struggling to provide a clean water supply for its people. Currently, only 62% of the population has access to clean water – the rest rely on groundwater or carted water. The city's Millennium Development Goal target is clean water access for 78% of the population by 2015; however, this might not be achieved because of many existing challenges such as decaying infrastructure, financial and capacity constraints, and the need for sharing surface water resources with other municipalities within the MAMMINASATA metropolitan region (Case Study 14.A Figure 1).

Furthermore, the water resources are highly seasonal and prone to droughts and high sedimentation due to soil erosion and landslides in the upper catchment. The population is projected to increase by 20% by 2020, whereas water demand is expected to increase by more than 120% as people become more affluent and more connected to the water infrastructure. The MAMMINASATA water supply master plan for 2025 recommends the construction of a new dam, infrastructure upgrades of the water treatment plants, and changes to distribution system

coverage in the next 30 years. This plan has not considered the risks of climate change on water resources because an assessment of these impacts had not been conducted for this region at a scale useful for the city's decision-makers.

Research to Understand Problem and Develop Solutions

The first climate adaptation–related initiative in the city was a collaborative effort among Australian and local researchers, local postgraduate students, the Australian aid agency, local governments, and water utilities. It aimed to inform policy to improve access to clean water and sanitation and to manage the impacts of development and climate change (Kirono et al., 2014). First, it developed a better understanding of current and future water services challenges in the region. This included new information on projections of climate and water resources over the coming few decades, as well as the ability of meeting urban clean water demand under a range of plausible scenarios.

The analyses suggested that problems related with water availability and high sedimentation will still occur in the future. Without additional infrastructure upgrades, as outlined in the MAMMINASATA master plan, water shortages will be common from around 2020. Even more, infrastructure upgrades may only provide short-term water security and need further adaptation from around the 2040s. Thus, there is an immediate need to shift reliance on large infrastructure alone to solutions that combine infrastructure and preventive measures, such as demand management and behavioral changes.

Stakeholders then gathered to understand implications, discuss urban water cycle and integrated water management concepts, and identify adaptation strategies to improve the sustainability of water supply. The strategies fall into two categories: (1) managing and protecting existing resources and (2) resource efficiency and exploration of new sources

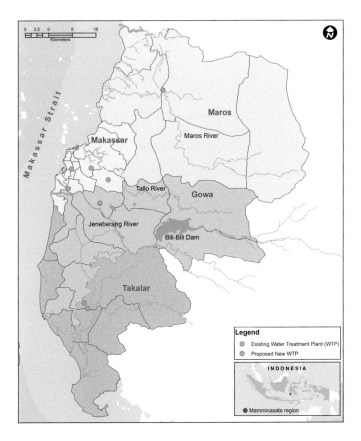

Case Study 14.A Figure 1 *Makassar city as part of the MAMMINASATA metropolitan region, which also encompasses three other municipalities (Maros, Gowa, and Takalar) in South Sulawesi Province, Indonesia. The Jeneberang and Maros rivers feed several water treatment plants in the region. The rivers are also major source for agricultural irrigations.*

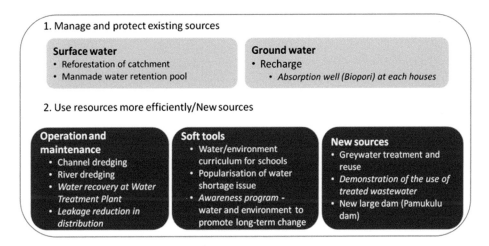

Case Study 14.A Figure 2 *Identified adaptation strategies and options. Italicized options are those on their way for implementation.*

(Case Study 14.A Figure 2). Adaptation options for the former include reforestation of upper catchment and groundwater recharge, for example, through *Biopori* – a locally developed tool designed to aid water infiltration and treatment using organic matter in man-made small-diameter pits. Meanwhile, those for the latter include leakage reduction in the distribution system and an awareness program to promote long-term mindset change.

Some of these options are now finding implementation in the city. For example, a legislative framework for *Biopori* has been discussed with funding provided by the Ecoregion Management Centre for Sulawesi and Maluku (PPE SUMA). However, the performance of the *Biopori* for the soil conditions in Makassar and its long-term infiltration capacity, in view of maintenance and operation practices by the public, still need to be verified, so monitored trials are recommended. Likewise, investment will be required to educate and incentivize residents to maintain the *Biopori* in the long-term.

The feasibility of *water recovery* – that is, reusing the water that was discarded during the treatment process – at the City's water treatment plants (WTPs) was studied by the company operating the WTPs. Early results suggest that it will add a 7% increase to current production at each WTP.

Several awareness programs have been implemented by the PPE SUMA office and the Makassar Public Work Agency (PU). The former promotes water use efficiency and discourages solid waste disposal to waterways for a range of audiences (government, communities, and students) using a variety of tools (e.g., education park, books, and videos). Meanwhile, the PU conducts community education on wastewater management in low-income areas of the city. The PPE office also has implemented a pilot test and demonstration for rainwater harvesting and graywater reuse for an office building.

At the city scale, the PU recently funded an effort to revise its clean water supply master plan to incorporate new information developed by this research initiative.

Lessons Learned

This study was undertaken to address the need for improving the adaptation capability of Makassar's urban water system to multiple drivers including climate change. In doing so, it brought global and regional climate change issues into focus by demonstrating local impacts that people care about (e.g., water supply) and relating them to the local government agenda (e.g., the MAMMINASATA master plan). The approach allows stakeholders to examine when and in what conditions the water supply may or may not meet demand. It also allows them to develop multiple adaptation strategies that are targeted to the local context and to identify when risk management measures will be needed.

The high level of stakeholder involvement within this research played a critical role in fostering multiple modes of communication, participation, and social learning between researchers and stakeholders. It also ensured a match between the knowledge needed and the knowledge produced (Larson et al., 2012). Furthermore, the deliberate stakeholder engagement process is proved to increase the likelihood of research uptake into the city policy formulation domains – as demonstrated by the city's recent effort in revising its water supply master plan to mainstream results from the research.

References

BPS. (2014). *Makassar in Figures*. The Central Board of Statistics (BPS) of Makassar City.

Climate-Data. (2015). Climate-Data.org. Accessed October 9, 2015: http://en.climate-data.org/search/

Demographia (2016) Demographia World Urban Areas (Built Up Urban Areas or World Agglomerations). 12th Annual Edition. St. Louis: Demographia. Available from: http://www.demographia.com/db-worldua.pdf.

Kirono, D. G. C., Larson S., Tjandraatmadja G., Leitch A., Neumann L., Maheepala S., Barkey R., Achmad A., and Selintung M. (2014). Adapting to climate change through urban water management: A participatory case study in Indonesia. *Regional Environmental Change* **14**, 355–367. doi:10.1007/s10113-013-0498-3.

Larson, S., Kirono, D. G. C., Barkey, A. R., and Tjandraatmadja, G. (2012). *Stakeholder Engagement within the Climate Adaptation Through Sustainable Urban Development in Makassar-Indonesia Project. A Report.* CSIRO.

Peel, M. C., Finlayson, B. L., and McMahon, T. A. (2007). Updated world map of the Köppen-Geiger climate classification. *Hydrology and Earth System Sciences Discussions* **4**(2), 462.

World Bank. (2017). 2016 GNI per capita, Atlas method (current US$). Accessed August 9, 2017: http://data.worldbank.org/indicator/NY.GNP.PCAP.CD

Case Study 14.B

Rotterdam's Infrastructure Experiments for Achieving Urban Resilience

Niki Frantzeskaki

DRIFT, Erasmus University Rotterdam

Nico Tilie

Faculty of Architecture, TU Delft

Keywords	Ecosystem-based adaptation (EbA) and disaster risk reduction (DRR), flood water, wetlands, stormwater management
Population (Metropolitan Region)	1,173,561 (Rotterdam The Hague Metropolitan Area, 2016)
Area (Metropolitan Region)	990 km² (Rotterdam The Hague Metropolitan Area, 2016)
Income per capita	US$46,310 (World Bank, 2017)
Climate zone	Temperate, without dry season, warm summer (Cfb) (Peel et al., 2007)

Over the past 150 years, the port city of Rotterdam has experienced a number of changes – both climatic and non-climatic in nature. The population, the economic value of infrastructure, the city's physical features, and its relationship with surrounding peri-urban and rural areas have all increased significantly. The frequency and intensity of extreme weather events and their impact on the conditions of the city's protection dykes are also increasing and are expected to be much greater in the future.

Climate Change Issues

The city is highly vulnerable to coastal flooding. It has had several near-miss types of catastrophic floods, such as the North Sea storm surge flood of 1953 that caused heavy loss of life and property in the Netherlands. The city also was spared severe flooding during the river Meuse floods of 1993, which caused heavy damage in upstream regions. A new precipitation record was set in August 2006 that caused widespread flooding and damage in the city (C40 and CDC, 2015).

Adaptation Strategy

The Municipality of Rotterdam has an adaptive urban planning process that targets citywide interventions at a topical level (district/neighborhood) when considering the national policy on water (Delta Program). Specifically, the City Council took the lead in establishing the Rotterdam Climate Initiative in 2007, which focused on the management of "too much and too little" water as well as increasing the use of green technological solutions. The Regional Program Committee, responsible for implementation of the national water policy at the regional level, cooperates actively with the city office on urban planning. As a result of this collaboration, various multifunctional infrastructures for making the city climate proof were piloted, including submerged parking in the city center (dc. Museum park), water squares that fill with rainwater and act as water storage, and replacing impermeable paved surfaces in the waterfront area with green lawns to increase water infiltration while also improving Boomjes promenade, an amenity of public urban space.

Water Squares

The first water square was built by combining a playground and water Fun Park with water storage during heavy rainfall. Next to the underground Museum parking facility, a 10,000 cubic meter overflow facility was built to hold peak rainfall. The submerged parking lot is inundated in cases of emergency. In addition to the 5,000 square meter vegetation wall at West Blaak parking facility, 50,000 square meters of green roofs have been installed throughout the city, with the ambition to increase this to 160,000 square meters by 2014. Enterprises and citizens are invited to join through awareness projects, showcases, and incentives such as green roof subsidies of €30/m² (US$35.4/m²). During the Green Year of 2008, the City of Rotterdam constructed eight vegetation roofs, the largest of which was on the Sophia Children's hospital. The goal was to show not only the beauty but also the usefulness of green roofs for water storage, insulation, ameliorating air quality, and biodiversity.

Introduction: Floating Urbanization

Rijn-Maashaven showcases delta technology, beginning with the realization of the Floating Pavilion in 2011. Water recreation and events on a floating stage are organized regularly. The floating pavilion is a 12 meter high construction of three half-round transparent hemispheric constellations. It was built in 2010 and is located in the Maashaven on the City Ports area. It serves as a venue for events, hosts an exposition about the area's development and living "on" water concept, is envisaged as location for a National Water Centre (Stadshavens Rotterdam, 2010, and is a new icon in the area, frequently visited by delegations from all over the world. Plans are to transfer it to another part of City Ports in 2015. It is also the first step in constructing 13,000 apartments, 1,200 of which will be on water, a goal outlined in the "Creating on the Edge" vision (Frantzeskaki et al., 2014) and thus freeing land for urban greenery (see Case Study 14.B Figures 1 and 2).

Case Study 14.B Figure 1 *Floating pavilion.*

Green Adaptation Strategy: Boompjes Promenade: Greening the Waterfront

Boompjeskade is a riverbank location where impermeable pavement has been replaced with grass, creating space for water to infiltrate and to be retained as well as space for people to use. It is a pilot site for greening the riverbanks and promoting biodiversity. Attempts to create soft edges between quays and the river failed in past years due to the use of the river as a main artery for shipping with stringent safety regulations that did not allow "greening" to be part of the design. Efforts in this direction for unused harbor inlets remain limited. In strategies where water bodies (including the rivers) and green areas are understood and dealt with as interdependent ecosystems, synergies can point to more robust urban designs and interventions. However, because the greening of riverbanks is also seen as a promising alternative in view of higher risks for flooding (Delta Program, 2012), limiting or restricting "soft edges" or greening is often considered controversial. In places where this is not the case, there is still a lack of soft water-edges. This is also evinced by the number of references to supportive ecosystem services in the visions, plans, and strategies of the city. This approach may be partially the result of limited collaboration between departments of spatial planning and sustainability with the city's ecology office (Frantzeskaki et al., 2014).

Impacts and Lessons Learned

To maintain the city's historical growth and resilience, mainstreaming climate change risk management into urban development planning has been adopted by Rotterdam. Given the uncertain nature of climate change, the city is largely focusing

Case Study 14.B Figure 2 *Well-adapted city parks.*

on no-regret adaptation measures such as strengthening of dykes, water management, and flood disaster risk reduction. The city government is dealing with climate uncertainties by anticipating changes and dealing with them in a flexible manner by making climate resilient urban developments. The City, ranked fourth among European Green Cities in 2014 aims to reduce CO_2 emissions by 50% by 2025 and to be 100% carbon neutral by 2025 (C40 and CDC, 2015). The key lessons for the resilient city are forward planning, good leadership, and scientific management.

References

Delta Program. 2012. *Delta Program Rotterdam, Municipality of Rotterdam, Delta Committee and Ministry of Environment and Infrastructures* (English translation by Frantzeskaki, N and Tilie, N.).

Frantzeskaki, N., and Tilie, N. (2014). The dynamics of urban ecosystem governance in Rotterdam, The Netherlands. *AMBIO* **43**(4), 542–555.

C40 Cities Climate Leadership Group (C40) and Connecting Delta Cities (CDC) (2015). Rotterdam: Climate change Adaptation. Accessed September 28, 2015: http://www.deltacities.com/cities/rotterdam/climate-change-adaptation

Peel, M. C., Finlayson, B. L., and McMahon, T. A. (2007). Updated world map of the Köppen-Geiger climate classification. *Hydrology and Earth System Sciences Discussions* **4**(2), 462.

Rotterdam The Hague Metropolitan Area (2016). Municipalities. Accessed on November 3, 2016: http://mrdh.nl/gemeenten

Stadshavens Rotterdam (2010). Drijvend Paviljoen Geopend. Accessed September 28, 2015: http://stadshavensrotterdam.nl/drijvend-paviljoen-geopend

World Bank. (2017). 2016 GNI per capita, Atlas method (current US$). Accessed August 9, 2017: http://data.worldbank.org/indicator/NY.GNP.PCAP.CD

Case Study 14.C

Environmental Impacts in São Paulo City, Brazil

Eduardo Delgado Assad and Andrea Young

University of Campinas, São Paulo

Flood, landslide risks, drainage system, hydric crisis	Flood, landslide risks, drainage system, hydric crisis
Population (Metropolitan Region)	19,683,975 (IBGE, 2014)
Area (Metropolitan Region)	7,947 km² (IBGE, 2014)
Income per capita	US$8,840 (World Bank, 2017)
Climate zone	Cfb – Temperate, without dry season, warm summer (Peel et al., 2007)

Climate change is expected to increase the risks of flooding and landslides in the São Paulo Metropolitan Area (SPMA) because these are associated with extreme events like storms and heavy rainfall, as well as with the urbanization process, which causes soil compaction and impermeability. Heat waves that have occurred in recent years are also factors that make the city vulnerable. Currently, a major water supply crisis has demonstrated the most important risk factor for the city (http://www2 .sabesp.com.br/mananciais/divulgacaopcj.aspx).

The SPMA is set in the sedimentary basin of the Tiete River centered near 23° 32' S, 46° 38' W, a low-lying region inside the Atlantic Mountain Chain. It occupies approximately 8,000 square kilometers and is surrounded by hills that vary from 650 to 1,200 meters in height above sea level. Its proximity to the ocean influences the atmospheric circulation patterns. The region is encroaching on the remaining portions of the Atlantic forest biome, which, in Brazil, despite the high levels of devastation, is still home to a significant amount of biological diversity. The vegetation of the region is therefore made up of fragments of secondary Atlantic forest (known as the Mata Atlântica). São Paulo is the largest metropolitan area in Brazil with almost 20 million inhabitants. It is located in São Paulo State in southeastern Brazil, approximately 600 kilometers southwest of Rio de Janeiro and 80 kilometers inland from the Atlantic Ocean; it has a population density of about 2,500 inhabitants per square kilometer. The population grew by 0.98% from 2000 to 2010, due mainly to the difference between births and deaths, in spite of an outflow due to migration to other states (Migration rates are –1.62 per thousand inhabitants; amounting to 30,300 people per year) (IBGE, 2010).

According to SEADE (Production Center for the processing, analysis, and dissemination of socioeconomic information about the state of São Paulo) (2010), the SPMA is also home to the largest employment base in the country (9.7 million workers). The gross domestic product (GDP) is about US$347 billion, and it has a GDP per capita of around US$17,666 with a per capita income of about US$470. São Paulo's largest industries are slightly more diverse than those in other metropolitan areas in Brazil. Its largest sector, services, makes up 50% of the economy, and, since 1990, São Paulo has seen the largest growth in its information services (251%) and business services (105%).

With the urban growth scenario developed by Young (2013), it is possible to analyze the projected changes in 2030 in regard to environmental risks. The principal environmental risk factors and threats are droughts, flooding, and landslides

The current situation remains high risk, considering the analysis since 2001. By 2030, the urban area will have spread by approximately 38.7% and will cover 3,254 square kilometers. With this growth, the region will have 807 square kilometers of areas subject to the risk of floods, a 46% increase during the period.

In terms of flooding risks, the areas susceptible to flooding accounted for 23.5% of urban area in 2008, and 22.3% in 2030 (Young, 2013). The areas at risk increased by approximately 254 square kilometers due to urban sprawl of a total area of 1,141 kilometers. Case Study 14.C Figure 1 shows the possible change in flood risks in 2030 (Young, 2013).

In terms of landslide risks, the vulnerable areas in 2030 may be approximately 4.27% of the expansion areas (these usually occur on high slopes and in vulnerable areas). This seemingly small percentage represents a relative increase of upward of 200% in the area, or, in other words, the area currently prone to landslides (0.9%) could almost triple, from 21.21 square kilometers in 2008 to 69.88 square kilometers in 2030 due to projected city expansion.

There are very strong indications that climate change is in progress, demonstrated by the analysis of a time series of climatological and hydrological data (de Carvalho et al., 2014). These have important consequences for water resource management. These climate changes may bring

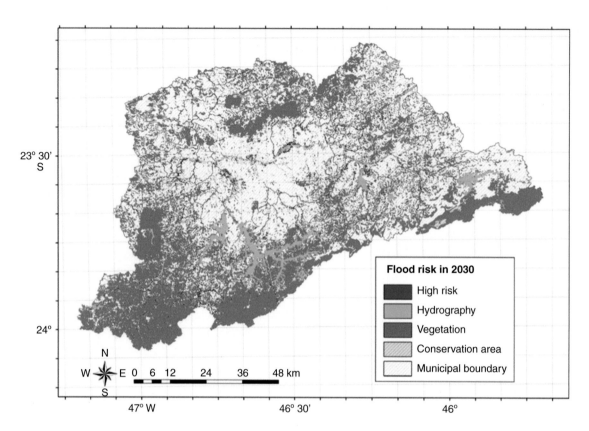

Case Study 14.C Figure 1 *Areas in* São *Paulo Metropolitan Area vulnerable to floods in 2030 considering urban expansion.*

Source: Young, 2013

continuous stress to the long-term water security of people in the São Paulo Metropolitan Area, with implications for the water supply.

Existing information shows that São Paulo's systems do not have sufficient capacity to ensure the flow needed to supply service to the population in the medium and long term, considering that the constant demand for water for SMPA and Campinas region is 20.65 m³/s (average demand in the first half of October 2014).

The prospect of recurring extreme events – such as prolonged droughts alternating with floods – requires a far-reaching vision for planning in São Paulo. To address these conditions, it is urgent to build long-distance channels to transport water, drill wells in the Guarani Aquifer and the transposition basins, or find water in the basins near the Cantareira System.

The preservation of ecosystem services and the protection, conservation, and restoration of biodiversity in the Tiete River Basin is fundamental in the case of São Paulo. The Cantareira Water System that supplies São Paulo is composed of twelve municipalities. A study of Assad et al. (2015) shows the deficit of vegetation in each municipality, especially along rivers that supply the system. Table 1 quantifies the area that must be revegetated to ensure both adequate water supply and biodiversity protection.

General Conclusions

The water crisis in the São Paulo Metropolitan Area, along with the direct influence of climate and hydrological changes, is aggravated by changes in land use, deforestation in the areas containing water sources, and a lack of basic sanitation and sewage treatment. Excess pollution prevents the use of water, even though the causes are relatively well known by managers and organizations that control and monitor the quality of water, air, and soil. On an emergency basis, some actions are suggested to deal with the current drought crisis:

- Immediate changes to the governance system of water resources
- Modernization and streamlining of the management systems. Changes must be implemented in sectorial management response and locally integrated into the ecosystem level (watershed), taking into account ecological, economical, and social impacts
- Implementation of contingency plans
- Reduction of losses in water distribution systems
- A drastic reduction in water consumption
- Major campaigns for rational use for water, reducing waste, and expanding the reuse of water

Table 1 *Revegetation needed the in medium (5 years) and long term (10 years) in Tiete River Basin for the Cantareira Water System.*

Municipality	State	Area (ha)	Natural vegetation (ha)	Natural vegetation (%)	Drainage network length (km)	Area to be revegetated (ha)
Bragança Paulista	SP	51,253.44	5,640.12	11.00	828.52	**5,429.30**
Caieiras	SP	9,620.59	4,797.09	49.86	213.27	**652.14**
Camanducaia	MG	52,821.32	26,410.74	50.00	1,389.92	**2,985.75**
Extrema	MG	24,448.53	7,417.43	30.34	569.75	**2,264.67**
Franco da rocha	SP	13,415.17	4,108.59	30.63	349.91	**1,391.37**
Itapeva	MG	17,727.30	5,295.83	29.87	500.58	**1,518.61**
Joanópolis	SP	37,412.07	17,743.50	47.43	861.00	**3,530.50**
Mairiporã	SP	32,063.92	17,046.96	53.17	848.81	**2,662.75**
Nazaré paulista	SP	32,618.48	15,475.46	47.44	751.07	**4,006.75**
Piracaia	SP	38,539.42	12,832.66	33.30	902.14	**6,322.72**
Sapucaí-mirim	MG	28,490.09	17,201.58	60.38	684.89	**1,464.46**
Vargem	SP	14,257.02	3,231.56	22.67	271.41	**2,313.78**
Total		**352,667.35**	**137,201.52**		**8.171.24**	**34,542.80**

- Immediate investment in long-term measures, such as building long-distance channels for transporting water, drilling wells in the Guarani Aquifer, and transporting water between basins
- Protection, conservation, and restoration of biodiversity along rivers
- Promotion of revegetation actions over the more than 8,000 kilometers of rivers that are part of the Cantareira System, to protect the source and prevent soil erosion

References

Assad, E. D., Peixoto, M., Campagnoli, L. C., and Gonçalves, R. R. V. (2015). Crise Hidrica: Como recuperar a cobertura vegetal. Agroanalysis. Maio. Accessed September 28, 2015: http://www.agroanalysis.com.br/materia_detalhe.php?idMateria=1931

de Carvalho, J. R.P., Assad, E. D., De Oliveira, A. F., and Pinto, H. S. (2014). Annual maximum daily rainfall trends in the Midwest, southeast and southern Brazil in the last 71 years. *Weather and Climate Extremes* **5**, 1–24.

Instituto Brasileiro de Geografia e Estatistica (IBGE). (2010). Primeiros resultados definitivos do Censo 2010. Accessed April 9, 2015: http://site-antigo.socioambiental.org/prg/man.shtm

Instituto Brasileiro de Geografia e Estatistica (IBGE). (2014). *Cidades. São Paulo.* Accessed April 9, 2015: http://www.cidades.ibge.gov.br/xtras/perfil.php?lang=&codmun=355030&search=sao-paulo%7Csa-opaolo

Peel, M. C., Finlayson, B. L., and McMahon, T. A. (2007). Updated world map of the Köppen-Geiger climate classification. *Hydrology and Earth System Sciences Discussions* **4**(2), 462.

World Bank. (2017). 2016 GNI per capita, Atlas method (current US$). Accessed August 9, 2017: http://data.worldbank.org/indicator/NY.GNP.PCAP.CD

Young, A. F. (2013). Urban expansion and environmental risk in the São Paulo Metropolitan Area. *Climate Research* **57** 73–80. doi:10.3354/cr01161

Case Study 15.A

Closing the Loop in Waste Management in Southern Sri Lanka

Manishka De Mel

Center for Climate Systems Research, Columbia University, New York

Missaka Hettiarachchi

University of Queensland and World Wildlife Fund, Brisbane

Achala Navaratne

American Red Cross, Washington, D.C.

Keywords	Solid waste management, disasters, mitigation
Population (Metropolitan Region)	761,370 (Sri Lanka Red Cross Society, 2015)
Area (Metropolitan Region)	1,270 km² (Sri Lanka Red Cross Society, 2015)
Income per capita	US$3,780 (World Bank, 2017)
Climate zone	Af – Tropical, rainforest (Peel et al., 2007)

Solid Waste Management Activities through the Green Recovery Program

Solid waste management is a pressing issue in many urban, peri-urban, and rural areas of Sri Lanka due to the lack of proper collection and disposal systems. Municipal solid waste is a major contributor to climate change, and improper waste management can lead to greenhouse gas emissions such as carbon dioxide and methane, as well as place impacts on health and the environment. Solid waste management is especially a challenge in smaller cities, towns, and peri-urban areas where the waste management infrastructure is not as extensive as in larger cities and town. This case study demonstrates an effective process to manage solid waste with the participation of a range of stakeholders, including local authorities, private businesses, and local communities. The Green Recovery Program, a partnership between American Red Cross and World Wildlife Fund (WWF-US) was launched in 2006 as a post-disaster recovery and resilience project to assist countries devastated by the Indian Ocean tsunami of 2004 (Environmental Foundation and WWF-US, 2010). The Indian Ocean tsunami remains one of the most devastating disasters of recent times with regard to the loss of human lives, economic cost, and post-disaster recovery efforts. The Green Recovery Program is one of the first examples of a humanitarian–environmental group partnership that aimed to build post-disaster resilience.

In Sri Lanka, the partnership joined forces with the Environmental Foundation to address many post-disaster recovery and resilience issues. Solid waste management was one major challenge, and the Program carried out several activities to address this. Composting, recycling, and landfilling were approaches undertaken by local authorities and stakeholders to minimize and manage waste. However, one of the main challenges has been the disintegration of such approaches. Under the Program, several interventions were initiated to close the loop in waste management through an integrated approach.

Composting and Home Gardening as a "Backdoor" Approach to Waste Minimization

The partnership supported disaster-affected families to adopt environmentally friendly waste management practices at the household level. The first step focused on reducing household waste that required disposal. Composting and home gardening were introduced at the community level in coastal urban areas in the Devinuwara, Matara, and Weligama Divisional Secretariat Divisions in the Matara District of Sri Lanka (Environmental Foundation and WWF-US, 2010). This was considered a "backdoor" approach to composting because waste separation was introduced through the promotion of home gardening within these communities. Training on organic home gardening techniques and composting were provided, along with material inputs of compost, seeds, and plants. As a result households were able to enjoy the benefits of composting through pesticide-free vegetables in their home gardens. This in turn helped improve nutrition and increase food security and household savings. While making compost, households were encouraged to separate recyclables, and recycling business networks were developed to collect the recyclables from these peri-urban villages. Additionally, community campaigns and awareness programs were carried out to discourage use of non-biodegradable bags and packaging ("refuse"), and the standard "three-R" approach – reduce, reuse, and recycle – was emphasized to minimize waste generation at the household level. The Environmental Foundation worked with American Red Cross Water and Sanitation Program and the local Sri Lanka Red Cross Society (SLRCS) staff to train volunteers and community leaders as social mobilizers to reach a wide range of beneficiary families at the village level. More than 1,950 beneficiary families received support from the Program for composting, home gardening, and hygiene promotion.

Local-Level Collection Networks and Recycling

Often the missing link between separation and recycling is the collection network for recyclables, especially polythene and plastic. This is less of an issue in large cities where citywide collection systems are in place; however, it is a challenge in urban areas away from large cities. to close the loop in waste management, the Environmental Foundation obtained the support of a local youth group, Sumithuro Api, who were already engaged in recycling from paper-based waste (Environmental Foundation and WWF-US, 2010). The group, after receiving training, engaged in collecting polythene and plastics from villages, which were then sent to local recycling centers.

The program also supported a recycling facility operated by a local business owner (Environmental Foundation and WWF-US, 2010). The program supported the facility by purchasing machinery and by providing funding to construct a small-scale recycling facility. This facility was supported by local recycling networks and helped revive and establish new polythene collection networks to absorb material to fulfill the capacity of the factory by developing a market-based recyclable collection system.

In larger towns in the region, due to the lack of space, people depend on local authority services for waste management. To complete the waste management loop at the larger urban level, the partnership supported the Weligama Urban Council, a pioneer in the country in composting municipal solid waste, to expand its composting yard and recyclable storage facility (Environmental Foundation and WWF-US, 2010).

Lessons Learned

This case study demonstrates how solid waste management in small towns, cities, and peri-urban areas can be effectively managed to mitigate greenhouse gases. It highlights the importance of addressing every component of the waste management cycle. The disintegration of just one component can result in the breakdown of the management process. It highlights that including the private sector, especially for recycling, is vital to ensure demand for recyclables and to make the process sustainable. Another key to the success of this model was the use of social mobilization approaches and community participation in the process. Awareness creation, sensitization, motivation, and voluntary community actions were key in binding the technical components when closing the loop in waste management.

References

Department of Census and Statistics, Sri Lanka DCSSL. (2012a). Population of Municipal Councils and Urban Councils by Sex Census, 2012. Accessed July 6, 2015: http://www.statistics.gov.lk/Abstract2014/CHAP2/2.4.pdf

Environmental Foundation and WWF-US. (2010). *Green Recovery Program. Reducing Vulnerability to Disasters by Building Sustainable Communities Based on Healthy Ecosystems. Achievements of the Humanitarian-Environmental Partnership in Sri Lanka.* Environmental Foundation, Colombo, Sri Lanka.

Peel, M. C., Finlayson, B. L., and McMahon, T. A. (2007) Updated world map of the Köppen-Geiger climate classification. Hydrology and Earth System Sciences Discussions, 4(2), 462.

Sri Lanka Red Cross Society (SLRCS). (2015). Districts and local authorities. Accessed September 4, 2015: http://www.redcross.lk/sri-lanka-country-profile/districts-and-local-authorities/

World Bank. (2017). 2016 GNI per capita, Atlas method (current US$). Accessed August 9, 2017: http://data.worldbank.org/indicator/NY.GNP.PCAP.CD

Case Study 15.B

Accra, Ghana: The Challenge of a Developing City

Martin Oteng-Ababio

University of Ghana, Accra

Keywords	Municipal solid waste, landfill, sustainable waste management, informal recycling sector
Population (Metropolitan Region)	2,316,000 (United Nations, 2016)
Area (Metropolitan Region)	971 km² (Demographia, 2016)
Income per capita	US$1,380 (World Bank, 2017)
Climate zone	BSh – Arid, steppe, hot arid (Peel et al., 2007)

In 2013, the citizens of Accra generated conservatively 2,200 tons per day of municipal solid waste (MSW) (Oteng-Ababio, 2014). The National Environmental Sanitation Policy (NESP) mandates each district assembly to be responsible for solid waste management (SWM), which normally covers collection and sanitary disposal with little or no incentive for waste minimization, thus compromising the ability of city authorities to provide efficient SWM services (MLGRD, 2010). Source separation, for example, remains alien in most cities, while the lack of an exact inventory of the waste stores in the face of illegal dumpsites and disinterested public officials has exacerbated the situation, creating a MSW industry that is vague, disjointed, and dysfunctional (Oteng-Ababio et al., 2013). Typically, waste collection (about 75% of the city's generated waste) and transport to poorly managed dumpsites consume a disproportionate and unsustainable share of municipal budgets (estimated at US$3.45 million or GHS 6.7 million per year), leaving many communities without basic collection and disposal services and driving them to burn their waste – with deleterious health impacts (Oteng-Ababio, 2015).

Solid waste collection in Accra is mostly privatized. The city contracts with about nine waste collection firms that are responsible for all residential, commercial, and industrial waste generated in their respective collection districts. The firms recover their costs by collecting city-regulated fees from waste generators (Alhassan et al., 2014). Waste is collected using different types of vehicles ranging from tricycles to small trucks equipped with compactors, to large trucks with or without compactors. The major reasons for partial waste collection coverage on the part of the private service providers are inadequate or late payment of operational funds and lack of a system for monitoring the performance of private contractors. Among other attempts to increase collection coverage, there is a current attempt by the local authority, in collaboration with Zoomlion, a private company, to mount a project that will provide free plastic waste containers to individual households (Thompson, 2010; AMA, 2015).

Over the past 10 years, AMA has used at least seven temporary (poorly managed) dumping grounds within the city perimeter to dispose of the city's solid waste, thus impacting public health and the environment. This has led to public outrage and resulted in AMA and Ghana's Environmental Protection Agency (EPA) to work toward the closing of open dumps in the city of Accra. The city is served by the new Tema sanitary landfill, located at Kpone, 30 kilometers from the city center (Ranjith and Themelis, 2013). This facility was constructed and is operated by Zoomlion Ghana, a private Ghanaian company. The landfill was constructed to accept 700 tons per day but currently receives more than double that amount (more than two-thirds comes from Accra). The city recently entered into a contract for the construction of a new landfill, but a site has not been determined and financing is not certain. It is unknown whether the new landfill will include leachate and LFG collection systems. The noncollected waste is openly burned or dumped in stormwater and sewage water drains. Open burning of waste is common even in high-income communities.

Currently, nearly all recycling activities in the city in particular and the country as a whole are carried out by the informal recycling sector, with participants typically coming from impoverished and marginalized groups working in hazardous conditions to help address this growing burden (Oteng-Ababio, 2015). These waste pickers collect recyclables by sorting through mixed waste on the streets or at the dumpsites. Some informal recyclers also collect waste by going home to home and by offering incentives to households. Recently, there appears to be a semblance of some collaboration between the informal waste sector and some private companies. For example, the new Tema landfill built and being operated by Zoomlion provides a special shed for informal waste recyclers to sort out recyclables.

The fact that Accra generates a large amount of organic waste (about 65%) that is high in moisture content but without a proper recycling plant makes a sad commentary. There is no official dedicated organics collection service provided by the city, but there are two innovative models currently in place in Accra: a community-based, small-scale composting project and a large-scale, open-windrow facility with a materials recovery unit. The small-scale composting project involves collecting approximately 2 tons of organic waste per day from sixty companies, mainly hotels and restaurants in the tourist area of Osu. These companies receive a 5–10% collection discount depending on volume. The diverted organic waste is sent to neighborhood composting centers where it is converted to compost. The large-scale Accra Composting and Recycling Plant (ACARP) receives approximately 500 tons of MSW per day (organic and nonorganic). The plant is owned and operated by a private company through a public–private partnership with the city. Since its commissioning in 2012, the facility has processed a total of 16,000 tons (CCAC, n.d.)

Case Study 15.B Figure 2 *E-waste scavengers in Accra burning wires to harvest copper.*

Case Study 15.B Figure 1 *Dumping ground in Accra, to dispose of the city's solid waste.*

Ghana reached the middle-income level in 2010 and, more than half (51.9%) of its population lives in cities (GeoHive, 2014). Recognizing that urbanization is growing and that rising incomes also increase waste generation, addressing the downstream consequences of waste collection and disposal practices is clearly a priority because waste is not only an important climate challenge, but one that affects every aspect of life for people in the city and beyond. The city, through its development partners, has shifted focus from its old management practices to attempts at reducing short–lived climate pollutants (SLCPs) through well-managed waste systems to help mitigate climate change and produce significant local and national health, environmental, and economic co-benefits, including improved quality of life and, importantly, dignity for local communities. Consequently, the city of Accra has made much progress toward sustainable solid waste management. It closed the last waste dump within the city, constructed a new sanitary landfill, started operations at the new material recovery and composting facility, and is encouraging the informal waste recycling sector (Ranjith and Themelis 2013; Oteng-Ababio, 2015). The city is now striving, with much financial difficulty, for 100% collection of MSW, increased public awareness and participation in reducing littering, and stopping the open burning of waste that constitutes a major threat to public health through dioxin and particulate emissions.

References

Alhassan, A., Gabbay, O., Arguello, J., and Boakye-Boaten, A. (2014). *Report to the Accra Metropolitan Assembly on Solid Waste Composition*. Earth Institute, Columbia University and the University of Ghana. Accessed August 31, 2015: www.mci.ei.columbia.edu

Accra Metropolitan Assembly (AMA). (2015). Accra Metropolitan Assembly archives. Accessed August 31, 2015: www.ama.ghanadistricts.gov.gh

Climate and Clean Air Coalition (CCAC) (n.d). Solid Waste Management in Accra. Accessed September 2, 2015: http://waste.ccac-knowledge.net/sites/default/files/CCAC_images/city_fact_sheet/Accra_MSW_Fact_Sheet_0.pdf

Demographia (2016) Demographia World Urban Areas (Built Up Urban Areas or World Agglomerations). 12th Annual Edition. St. Louis: Demographia. Available from: http://www.demographia.com/db-worldua.pdf.

Ministry of Local Government and Rural Development (MLGRD). (2010). *National Environmental Sanitation Policy*. Author.

Oteng-Ababio, M. (2014). Rethinking waste as a resource: Insights from a low-income community in Accra, Ghana. *City, Territory and Architecture* 1(10), 1–14.

Oteng-Ababio, M. (2015). Technology must seek Tradition: Re-engineering Urban Governance for Sustainable Solid Waste Management in Low-income Neighbourhoods, in Mohee, R. and Somelane, T. Future Directions in Municipal Solid Waste Management in Africa, (Eds.), African Institute of South Africa.

Oteng-Ababio, M., Melara, J. E., and Gabbay, O. (2013). Solid waste management in African cities: Sorting the facts from the fads in Accra, Ghana. *Habitat International* 39, 96–104.

Peel, M. C., Finlayson, B. L., and McMahon, T. A. (2007). Updated world map of the Köppen-Geiger climate classification. *Hydrology and Earth System Sciences Discussions* 4(2), 462.

Ranjith, A. and Themelis, N.J. (2013). Analysis of Waste Management in Accra, Ghana and Recommendations for further Improvements. Waste-to-Energy Research and Technology Council. Accessed September 4, 2015: http://wtert-ghana.tiswm.com/files/Analysis_of_Waste_Management_in_Accra_Ghana_and_Recommendations_for_further_Improvements.pdf

Thompson, A. (2010). Domestic waste management strategies in Accra, Ghana and other urban cities in tropical developing nations. Accessed August 31, 2015: www.cwru.edu/med/epidbio/mphp439/Waste_Mgmt_Accra.pdf

United Nations, Department of Economic and Social Affairs, Population Division (2016). The World's Cities in 2016 – Data Booklet (ST/ESA/SER.A/392).

World Bank. (2017). 2016 GNI per capita, Atlas method (current US$). Accessed August 9, 2017: http://data.worldbank.org/indicator/NY.GNP.PCAP.CD

Case Study 15.C

The Successful Actions of London Municipality

A.C. (Thanos) Bourtsalas

Earth and Environmental Engineering, Columbia University, New York

Keywords	Municipal waste management, efficient source separation, landfill tax
Population (Metropolitan Region)	8,673,713 (Office of National Statistics, 2015)
Area (Metropolitan Region)	1,572 km² (Office of National Statistics, 2015)
Income per capita	US$42,390 (World Bank, 2017)
Climate zone	Cfb – Temperate, without dry season, warm summer (Peel et al., 2007)

The city of London, like most cities in the developed world, has generally adopted the philosophy of waste prevention and minimization, recovery, incineration, and landfill as the menu for developing their municipal solid waste management (MSWM) systems. In 2013, Londoners produced 3.6 million tons of municipal solid waste (MSW), a 20% decrease since 2000 (Defra, 2015). The treatment of MSW is dominated by waste-to-energy (WTE), 44.1% of the total, followed by recycling/composting (30.4%) and landfilling (25.5%). The landfill disposal exhibited a decrease of 72% since 2000, whereas WTE increased by 65% and recycling/composting increased by 216% (Defra, 2015). Additionally, the CO_2-eq saved from the use of recycling and reusing was calculated by UK authorities to be 5.9 million tons, from composting 1 million tons and from energy recovery 0.18 million tons. Landfilling emitted 2.8 million tons, thus making a total saving of CO_2-eq from the waste disposal methods applied in London 4.3 million tons (Defra, 2015)

The main driver for the rapid growth of sustainable waste management technologies in London was the increase in the landfill tax, from US$90 (or £58) per tonne in 2011 to US$130 (or £80) per tonne in 2014, which actually raised London's annual bill for sending municipal waste to landfills from about US$430 (or £265) million to roughly US$485 (or £300) million (Greater London Authority, 2011). In addition, the cost of recycling was rapidly decreased by −US$40 (−£26) on average in 2014, thus promoting such methods of waste disposal. The landfill tax has also made the cost of generating energy from waste more comparable to landfill and in some cases more commercially attractive, depending on contractual arrangements (Greater London Authority, 2011)

In addition to the increase in the landfill tax, the Department of Environment Food and Rural Affairs (Defra) has revised its definition of municipal waste to include more commercial waste. This brings it into line with other European Union (EU) countries and ensures that the UK is meeting landfill diversion targets under the European Landfill Directive. Implementing this new measure will put considerable pressure on local authorities, communities, and businesses to better manage more of their waste (Greater London Authority, 2011)

The key characteristics of London's waste management is promoting waste management activities to achieve the greatest possible climate change mitigation and energy saving benefits while managing as much of London's waste within London as possible, aiming toward managing 100% of London's waste within London by 2031 (Greater London Authority, 2011).

There are many local and regional education programs and initiatives ("less in your bins, more in your pocket") "nice save" campaign; community composting and reuse; WM facilities producing electricity and heat for local use; volunteer clean-up campaign and others) organized and sponsored by the municipality that provide producers and consumers with the knowledge, infrastructure, and incentives to change the way they manage municipal waste. The main target of these programs is to reduce the amount of waste generated, encourage the reuse of items that are currently thrown away, and to recycle or compost as much material as possible (Greater London Authority, 2011).

London's household recycling and composting services are active in all boroughs and provide curbside collection services for paper, mixed cans, and plastic bottles. All except two boroughs collect glass at the curbside and all except one collect cardboard. Thirteen boroughs collect mixed plastics from curbside services, and nine boroughs collect food and green garden waste together. Some boroughs provide food and green garden waste collections for flats and estates. (Greater London Authority, 2011). There are

forty-one reuse and recycling centers (RRCs) in London, providing drop-off facilities for a range of household waste materials for reuse, recycling, and disposal. They serve a wide community, from the inner city to the semi-rural fringes of London.

In addition, the CO_2eq emissions performance standard (EPS) was implemented for London's municipal waste management activities to work toward achieving rather than prescribing particular waste management activities or treatment technologies. This approach supports waste activities and services that reduce the amount of municipal waste produced and captures the greatest number and highest quality of materials for reuse, recycling, or composting and low-carbon energy generation. A key characteristic of this approach is that it allows flexibility. Waste authorities can look across the whole waste system to find the greatest CO_2eq savings to make an important contribution to achieving the EPS depending on their specific circumstances. In addition to the EPS, the Mayor has set a minimum CO_2eq emissions performance standard that requires all energy generated from London's municipal waste to be no more polluting in carbon terms than the energy it replaces. With this approach, the municipality advocates the development of low-carbon municipal waste management technologies. This is estimated to be possible for about 40% of London's municipal waste after recycling or composting targets are achieved by 2031 (Greater London Authority, 2011).

The municipality has also produced a non-statutory Business Waste Strategy for London's commercial and industrial waste and for construction demolition and excavation waste – waste that is collected and disposed of by waste operators under private contracts rather than by local waste authorities. Waste produced by businesses, be it from shops, restaurants, and offices; industrial processes; or construction and demolition sites – makes up 80% of London's waste: 16 million tonnes a year (Greater London Authority, 2011)

A recent study funded by the municipality concluded that a "do nothing new" approach would lead to an increase in London's annual municipal waste management bill to about US$1,100 (or £680) million by 2031. The results from the study showed that by changing the way Londoners manage their municipal waste, London could save between US$920 (or £573) million and US$1,350 (or £838) million and save between 20 million and 33 million tonnes of CO_2eq emissions by 2031. (Greater London Authority, 2011) These savings can be achieved predominantly by reducing the amount of household waste produced per household each year by approximately 1%, by a gradual decline in municipal waste sent to landfill, by achieving 45–67% recycling and composting rates (including reuse), and by increasing the amount of non-recycled and organic waste used for energy generation.

In addition to making carbon savings, optimizing the treatment of waste can also contribute significantly to a reduction in London's energy bill. Based on the wholesale cost of electricity and gas, London's municipal waste after maximizing recycling could contribute US$150 (or £92) million of savings to London's US$7.1 (or £4.4) billion electricity bill and take US$38 (or £24) million off London's US$4 (or £2.5) billion gas bill (Greater London Authority, 2011)

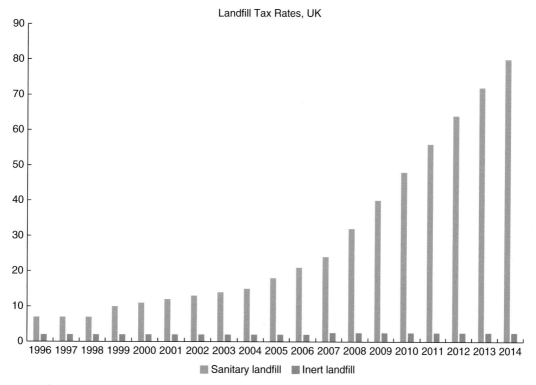

Case Study 15.C Figure 1 *The increase in the UK's landfill tax rates over time. This landfill tax increase helped to incentivize recycling.*

References

Department for Environment, Food and Rural Affairs (Defra) (2015). Digest of Waste and Resource Statistics – 2015 Edition. Accessed September 24, 2015: https://www.gov.uk/government/uploads/system/uploads/attachment_data/file/482255/Digest_of_waste_England_-_finalv3.pdf

Greater London Authority (2011). London's Wasted Resource – The Mayor's Municipal Waste Management Strategy, November 2011. London: Greater London Authority. Accessed September 2, 2015: https://www.london.gov.uk/sites/default/files/municipal_waste_final.pdf

Office of National Statistics (2015). Population Estimates for UK, England and Wales, Scotland and Northern Ireland. Accessed November 3, 2016: https://www.ons.gov.uk/peoplepopulationandcommunity/populationandmigration/populationestimates/datasets/populationestimatesforukenglandandwalesscotlandandnorthernireland

Peel, M. C., Finlayson, B. L., and McMahon, T. A. (2007). Updated world map of the Köppen-Geiger climate classification. *Hydrology and Earth System Sciences Discussions* **4**(2), 462.

World Bank. (2017). 2016 GNI per capita, Atlas method (current US$). Accessed August 9, 2017: http://data.worldbank.org/indicator/NY.GNP.PCAP.CD

Case Study 16.A

Low-Carbon Transition in Shenzhen, China

Weila Gong

Environmental Policy Research Centre (FFU), Free University of Berlin

Keywords	Low-carbon transition, emissions trading system (ETS), China's low-carbon pilots, local carbon policy action, typhoons, extreme rainfalls
Population (Metropolitan Region)	10,828,000 (United Nations, 2016)
Area (Metropolitan Region)	1,748 km² (Demographia, 2016)
Income per capita	US$8,260 (World Bank, 2017)
Climate zone	Cfa – Temperate, without dry season, hot summer (Peel et al., 2007)

China is the world's largest carbon emitter and, at the same time, increasingly determined to address climate change. Because the top leadership announced capping carbon emissions by 2030 (The White House, 2014), this goal will influence Chinese cities in the coming years. Given that 40% of carbon emissions are from dozens of China's largest cities (Dhakal, 2009) and are primarily from their industrial sectors (Wang et al., 2012), these cities are key sites for addressing climate change and are influential in shaping the future direction of China's actions on climate change.

Case Study 16.A Figure 1 *The electronic taxi in Shenzhen. As of November 2014, there were 9,392 new electric vehicles (Shenzhen Municipal Government, 2015); the program is just one of the city's low-carbon successes.*

More than 80% of Chinese cities (at the prefecture level and above) have stated goals to become eco-cities, and more than 40% have set targets to become "low-carbon cities" (Zhou et al., 2012). Shenzhen is one such city, but it is also unique among its peers because it was the first to set targets for peak carbon emissions from 2017 to 2020 (National Development and Reform Commission [NDRC], 2014), and it achieved the lowest energy consumption per gross domestic product (GDP) among major Chinese cities in 2010 (Liu et al., 2012). By continuously expanding green areas to increase carbon sinks and introducing market mechanisms and legislation in the Shenzhen Emission Trading Scheme (Shenzhen ETS), this city enhances its low-carbon transition through reducing carbon emission and is a leader in China's local low-carbon practices.

Shenzhen, the core city of the Pearl River Delta urban cluster, has a population of more than 10 million and is roughly 1,991 square kilometers in size. As one of the first Chinese coastal cities to open up in 1980s, it has achieved rapid economic growth but suffered from a deteriorating environment in the process. In 2005, the Mayor of Shenzhen called for a policy solution to the "four challenges" that threatened Shenzhen's development: limited land and space resources, energy and water resource shortages, a burgeoning population, and weak environmental carrying capacity. Subsequently, the municipal government issued regulations to strengthen urban environmental protection, which laid the foundation for Shenzhen's leading role in local low-carbon policy actions in China.

As the impact and severity of climate change in Shenzhen become increasingly clear, the city's actions to address climate change take on an even greater significance. Shenzhen has encountered climate hazards such as extreme rainfall, typhoons, and more, which as a whole result in the direct economic loss of RMB200 million in 2014. In March 2014, the daily amount of extreme rainfall was more than 150 millimeters, which resulted in flooding occurring in more than 100 municipal places, more than 200 flights delayed, and one person dead (Meteorological Bureau of Shenzhen Municipality, 2014:). To curb these hazards' disastrous impact at the local level, Shenzhen has developed several precautionary measures such as the local hazard monitoring system and the code of practice in times of typhoon and extreme rainfall (Meteorological Bureau of Shenzhen Municipality, 2013). As the first city to issue a code of practice related to extreme weather in mainland China, Shenzhen is increasing its capacity to adapt to climate challenges through educating its public about the extreme weather situation and protecting them from exposure to potential risks under the influence of extreme rainfall and typhoons (Meteorological Bureau of Shenzhen Municipality, 2013). In addition to adapting to climate challenges, the municipality also strengthens the local environmental standards for carbon mitigation in land use and transportation.

Strengthening Local Regulation on Ecological Protection and the Promotion of a Low-Carbon Transition

To address the challenges posed by limited local resources, the local government strengthened environmental regulations to protect land from overexploitation and invested in urban greenway development. In line with Shenzhen's policy on environmental protection, almost half of the land in Shenzhen's jurisdiction has been protected from resource exploitation since 2005. At the same time, the Shenzhen Municipal Government has also invested more than RMB 1 billion in greenway construction. The greenway, which extends over 2,000 kilometers, not only beautifies its urban environment, but also acts as a significant carbon sink to offset the urban heat island effect. Urban projects like the greenway comprehensively strengthen Shenzhen's urban ecological capacity to absorb carbon emissions.

Shenzhen is among the initial leaders of China's local low-carbon transition. The local government prioritizes the development of low-carbon industries, such as the electrical vehicle industry. Since 2010, it has offered the highest level of local subsidies to each electric vehicle consumer, varying from RMB 30,000 to RMB 60,000 depending on the type of electric vehicle – plug-in or purely electric (Qu, 2014). By November 2014, these subsidies, at least partially, had helped to put more than 9,000 new energy-efficient vehicles on Shenzhen's streets. The resulting accumulated carbon emissions reduction reached more than 200 tons. 3.8 billion RMB of the municipal finance, an increase in RMB by 1.4 billion RMB from funds invested in the previous year, will be allocated to the promotion of new-energy vehicles in 2016 (Tang, 2016). In 2012, the local government decoupled energy consumption and economic growth for the first time; this indicates that local economic growth is not wholly dependent on extensive energy consumption and is a milestone for local low-carbon transition practices.

Local Leadership in the Shenzhen Emission Trading System Pilot

Shenzhen's pioneering role in the development of local low-carbon practices was also demonstrated in its legislation on the Shenzhen Emission Trading System (Shenzhen ETS), which guarantees the operation of market mechanisms to reduce carbon emissions. Local leadership, especially that of top local officials, played a major role in putting low-carbon issues on the local policy agenda. As a result of local leadership's seizure of the "window of opportunity" and its organization of research on

Shenzhen's feasibility as an ETS pilot, Shenzhen was selected as one of the initial ETS Pilots by the National Reform and Development Commission. The Shenzhen ETS was launched in June 2013. The corresponding legislation enabled the successful establishment of the Shenzhen ETS because it required participation from relevant industrial sectors. Each industrial participant was tasked with carefully checking its level of carbon emissions and participating in the ETS or otherwise face punishment and penalties for violations. Such legislation on market mechanisms and accountability encouraged the industrial sectors to fully participate in the local low-carbon transition.

With this local legislation on the Shenzhen ETS in place, the system was able to incorporate a large number of entities, including 635 enterprises and 197 public buildings. Participating sectors ranged from electric, industrial, and manufacturing companies to the building sector. As a result of this significant involvement, 3.75 million tons of industrial carbon emissions were mitigated from 2010 to 2013; similarly, carbon intensity – the amount of carbon emitted by a country per unit of GDP – decreased by 33.2% in terms of industrial added value (GDP by industry) from participating entities as compared to the 2010 level (Shenzhen Research Center for Urban Development and China Emission Exchange, 2015). The practices of the Shenzhen ETS demonstrate progress on the low-carbon transition by strengthening low-carbon institutions through legislation, incentivized stakeholder participation, and market mechanisms.

Lessons Learned from Shenzhen's Low-Carbon Transition

Local leadership has played a prominent role in initiating Shenzhen's low-carbon transition. This leadership, and specifically key local leaders, initially raised local awareness of the challenges posed by limited local natural resources and emphasized the importance of balancing economic growth and ecological protection. By fostering the development of less energy-dependent industries and investing in the greening of its urban environment, Shenzhen has steadily transformed its economic growth to include local-carbon features and enhanced its urban capacity to mitigate carbon emissions. After becoming an ETS Pilot region, local leaders strengthened ETS legislation and low-carbon transportation.

Although there has been considerable progress in Shenzhen, there are still many ongoing challenges. There is a need for more and broader public participation in decision-making and implementation. Similarly, more cooperation from enterprises is needed to transform the local economic growth model to become less carbon-dependent. Greater media coverage of the importance of a transition to a low-carbon economy will be important for raising awareness of the big challenges still ahead in Shenzhen. Interviews with officials and expert observers in the city suggest that there needs to be more attention given to the socioeconomic components of Shenzhen's low-carbon energy transition.

References

Chateau, J., Nicholls R.J., Hanson, S., Herweijer, C., Patmore, N., Hallegatte, S., Corfee-Morlot, J. and Muir-Wood, R. (2008). Ranking port cities with high exposure and vulnerability to climate extremes: Exposure estimates. OECD Environment Working Papers, No. 1, OECD Publishing. November 27, 2014. doi:10.1787/011766488208.

China Emissions Exchange (2014) Annual Report on Shenzhen Emissions Trading Scheme First-Year Operation – A Comprehensive Analysis of Shenzhen's Cap-And-Trade Program. Shenzhen, China: China Emissions Exchange. Accessed May 2, 2015: http://www.ieta.org/resources/China/shenzhen-2013etsreport.pdf

Demographia (2016) Demographia World Urban Areas (Built Up Urban Areas or World Agglomerations). 12th Annual Edition. St. Louis: Demographia. Available from: http://www.demographia.com/db-worldua.pdf.

Dhakal, S. (2009). *Urban energy use and carbon emissions from cities in China and policy implications. Energy Policy* **37**, 4208–4219. Accessed November 27, 2014: http://www.sciencedirect.com/science/article/pii/S0301421509003413

Liu, S., Luo, J., Peng, S.H., Liu, P., Zeng, X.T., Jiang, Y., Yu, X.J. and Xie, C.B. (2012). The status quo and counter-measures for challenges of Shenzhen ETS. Accessed May 5, 2015: http://www.szsti.gov.cn/f/services/softscience/14.pdf

Meteorological Bureau of Shenzhen Municipality. (2013). Code of Practice in Times of Typhoon and Extreme Rainfall (for Trial Implementation), Introduction for Content and Approaches. May 8. Accessed October 5, 2016: http://www.szmb.gov.cn/upload/fck/%E6%B7%B1%E5%9C%B3%E5%B8%82%E5%8F%B0%E9%A3%8E%E6%9A%B4%E9%9B%A8%E7%81%BE%E5%AE%B3%E5%85%AC%E4%BC%97%E9%98%B2%E5%BE%A1%E6%8C%87%E5%BC%95%BC%88%E8%AF%95%E8%A1%8C%EF%BC%89.pdf.

Meteorological Bureau of Shenzhen Municipality. (2014). Shenzhen Climate Bulletin. 2014. Accessed October 5, 2016: http://www.szmb.gov.cn/upload/fck/2014%E5%B9%B4%E6%B7%B1%E5%9C%B3%E5%B8%82%E6%B0%94%E5%80%99%E5%85%AC%E6%8A%A5%EF%BC%88%E7%BB%BC%E5%90%88__%E5%85%A8%E5%B9%B4%E5%AE%8C%E6%95%B4%E7%A8%BF%EF%BC%89v3-5-2_2.pdf.

Munning, C., Morgenstern, R.D., Wang Z.M., Liu, X. (2014). Assessing the design of three pilot programs for carbon trading in China. October.

Resources for the Future. Accessed November 27, 2014: http://www.rff.org/RFF/Documents/RFF-DP-14–36.pdf

National Development and Reform Commission. (2014). China's policies and actions on climate change 2014. Accessed May 11, 2015: http://en.ccchina.gov.cn/archiver/ccchinaen/UpFile/Files/Default/20141126133727751798.pdf.

Peel, M. C., Finlayson, B. L., and McMahon, T. A. (2007). Updated world map of the Köppen-Geiger climate classification. *Hydrology and Earth System Sciences Discussions* **4**(2), 462.

Qu, G. (2014). New energy vehicle: National level subsidy rates decrease 5%, Shenzhen's would keep enacted, simultaneously considering charging fees subsidies, parking fees subsidies or purchase tax reduction. Accessed March 9, 2014: http://epaper.nfdaily.cn/html/2014-02/20/content_7274154.htm

Shenzhen Municipal Government. (2015). Work plan for new energy vehicle development in Shenzhen. Accessed May 11, 2015: http://www.sz.gov.cn/zfgb/2015/gb911/201503/t20150304_2822725.htm

Shenzhen Research Center for Urban Development and China Emissions Exchange. (2015). The annual report on first-year operation of Shenzhen ETS. Accessed Oct 4, 2016: http://ets-china.org/wp-content/uploads/2015/05/Annual-report-on-first-Year-SZ-ETS.pdf.

Wang, H., Zhang, R., Liu, M. and Bi, J. (2012). The carbon emissions of Chinese cities. Atmospheric Chemistry and Physics 12, 6179–6202. Accessed May 5, 2015: http://www.atmos-chem-phys.net/12/6197/2012/acp-12-6197-2012.pdf.

Tang, S.K. (2016). Shenzhen Municipality Report on the 2015 Budget Performance and 2016 Budget Plan. Accessed on October 5, 2016: www.mof.gov.cn/zhuantihuigu/2016hb/201602/t20160226_1784270.html&num=1&hl=en&gl=de&strip=0&vwsrc=0.

The White House. (2014). US- China joint announcement on climate change. Accessed May 5, 2015: https://whitehouse.gov/the-press-office/2014/11/11/us-china-joint-announcement-climate-change

United Nations, Department of Economic and Social Affairs, Population Division (2016). The World's Cities in 2016 – Data Booklet (ST/ESA/SER.A/392).

World Bank. (2015a). East Asia's changing urban landscape: Measuring a decade of spatial growth. Urban Development Series. Accessed June 23, 2015: http://www.worldbank.org/content/dam/Worldbank/Publications/Urban%20Development/EAP_Urban_Expansion_full_report_web.pdf

World Bank. (2017). 2016 GNI per capita, Atlas method (current US$). Accessed August 9, 2017: http://data.worldbank.org/indicator/NY.GNP.PCAP.CD

Zhou, N., He, G., and Williams, C. (2012). China's development of low-carbon eco-cities and associated indicator systems. *Ernest Orlando Lawrence Berkeley National Laboratory*, LBNL-5873, 1.

Case Study 16.B

Fort Lauderdale: Pioneering the Way toward a Sustainable Future

Susanne Torriente

City of Miami Beach

Keywords	Sea level rise, community engagement, adaptation, Regional Climate Change Compact
Population (Metropolitan Region)	6,012,331 (U.S. Census, 2015)
Area (Metropolitan Region)	13,150 km² (U.S. Census, 2015)
Income per capita	$56,180 (World Bank, 2017)
Climate zone	Af – Tropical Rainforest (Peel et al., 2007)

The City of Fort Lauderdale is the heart of South Florida on the Atlantic coast of the United States, between Miami and Palm Beach, with a diverse population of more than 170,000. With 7 miles of white sandy beaches, 33 square miles (City of Fort Lauderdale Code of Ordinances, 2015) of an urban metropolis and 300 miles of winding canal coastline it was coined the "Venice of America" many years ago. It is beautiful yet vulnerable to the effects of climate change and sea level rise due to its flat topography, location on a peninsula, dense coastal development, and shallow, porous aquifer.

That vulnerability has become more apparent in recent years. Hurricanes, storms, flooding, and an eroding coastline have affected property, roads, and regional transportation networks and disrupted power and water supplies. Impacts of climate change adversely affect residents and the region's more than 5 million people and 12 million annual visitors. Fortunately, the city and the region have been formally planning and collaborating since 2009 through the leadership of local elected officials and the dedication of staff supporting the Southeast Florida Regional Climate Change Compact, one of the first and largest voluntary, collaborative, and bi-partisan efforts in the country to address climate issues and policy (Southeast Florida Regional Climate Change Compact, 2014).

A seminal moment came in November 2013. Following direct erosional impacts from Hurricane Sandy, additional winds and tides collapsed a portion of Scenic Highway A1A and washed away 2,500 feet of sidewalk, beach showers, a traffic signal, and parking meters along four city blocks. The impact of the storm was a wake-up call that helped the community realize that its future prosperity was in jeopardy and that action was essential to protect its long-term sustainability. The destruction bolstered public support to implement adaptation strategies to reduce the effects of climate change.

The city seized this post-disaster opportunity to rebuild the area for future conditions by creating more sustainable infrastructure with wider sidewalks, protective sand dunes, elevated roadways, and better drainage. This is a model for bouncing forward, not bouncing back – by using the best available data and the public process to build the city of tomorrow.

Fortunately, there was overwhelming political and public support to develop a comprehensive strategy to address growing climate challenges. The City Commission unanimously adopted a community-wide Vision Plan, Fast Forward Fort Lauderdale, and a 5-year strategic plan, Press Play Fort Lauderdale, both of which identify sustainability, climate adaptation, and climate mitigation as the city's top long-term priorities (City of Fort Lauderdale, 2013).

Fort Lauderdale successfully realigned its organizational structure to better integrate and implement these regional and local goals and initiatives. All operations departments have a role in delivering local government services through the lens of climate resilience. Specifically, a new Sustainability Division was created in 2012 and works with subject matter experts in science, environment, planning, fleet management, and waste management across all departments to support this transition and promote the "Greening Our Routine" culture (City of Fort Lauderdale, n.d).

As the sustainability experts shared their knowledge and raised awareness about climate change, employees yearned for additional resources to support ongoing efforts. More than thirty staff, coined "Climate Ambassadors," participated in a 3-day National Oceanic and Atmospheric Administration workshop. They continue to meet quarterly to share experiences regarding planning and implementing sustainable initiatives. This small training program became the catalyst to launch citywide training to educate all 2,600 employees about climate change and their role in mitigation and adaptation efforts. As new employees join the city, their orientation includes a sustainability component (Morejon, 2015). This may be one of the first of few local governments in the nation to institute dedicated climate training organization-wide.

As employees become more mindful of climate change, they are developing new strategies to plan for a sustainable future. A comprehensive Stormwater Master Plan is driving infrastructure investments in drainage, bio-swales, and retention parks to minimize the effects of climate change (Morejon, 2015). Adaptation Action Areas have been incorporated into the comprehensive plan to prioritize funding for infrastructure projects that minimize impacts in vulnerable areas. The city adopted new flood zone maps and instituted tougher regulatory standards to protect future development and minimize the flood risk in low-lying areas.

As momentum builds within the municipal organization, Fort Lauderdale is engaging its community to further adaptation measures. Volunteers serve on the City's Sustainability Advisory Board and participate in events to plant trees and sea oats, collect hazardous waste and recycled materials, and promote walking and biking. A strong network of neighborhood associations serve as sounding boards for proposed initiatives and policy language through workshops, crowdsourcing forums, and Telephone Town Hall Meetings. The annual neighbors survey gauges residents' perception regarding flooding and climate change to identify priorities and allocate future funding (Morejon, 2015).

Fort Lauderdale actively pursues partnerships to explore new ideas, tools, and research. Assistant City Manager Susanne Torriente is a founding member of the Southeast Florida Regional Climate Change Compact. City staff co-authored the Regional Climate Action Plan, a source of strategic climate mitigation and adaptation recommendations. Torriente and staff also serve on several local, state, and national boards dedicated to climate resiliency. The City of Fort Lauderdale was the first municipality on the Compact's Staff Steering Committee. The Compact and Fort Lauderdale model for integration and implementation have opened the door for informational peer-to-peer exchanges with Australia, South Africa, the Philippines, the Canadian consulate, Caribbean Islands, Great Britain, and the Netherlands. Through the ICMA CityLinks program, Fort Lauderdale was chosen by Durban, South Africa, to participate in a peer-to-peer exchange to learn more about the Southeast Florida Regional Climate Compact governance model. This successful relationship led to the creation of several regional compacts in Africa, as well as Broward County and Fort Lauderdale signing the Durban Adaptation Charter. These are the first two U.S. local governments to sign this Charter that promotes local government action to advance climate adaptation (ICMA, n.d.).

Fort Lauderdale partnered with NOAA, the Florida Department of Economic Opportunity, Broward County, and the South Florida Regional Planning Council to serve as a statewide

model to develop an Adaptation Action Area Program (City of Fort Lauderdale, 2015). The Urban Land Institute served as a partner on two Technical Assistance Panels to discuss protection of vulnerable destinations and encouraging growth in more resilient areas. Fort Lauderdale works with partners on tools to understand vulnerability, stormwater modeling, climate adaptation cost-benefit analysis, and sea level rise projections. Through these efforts, Fort Lauderdale is piloting emerging technologies such as tidal valves to reduce flooding at high tides.The City is eager and willing to collaborate, learn, and pilot new ideas and emerging technologies in this ever-changing world of tomorrow.

References

City of Fort Lauderdale (2013). Press Play Fort Lauderdale: Our City, Our Strategic Plan 2018. Accessed October 11, 2015: http://www.fortlauderdale.gov/home/showdocument?id=4642

City of Fort Lauderdale (2015): Fort Lauderdale Adopts Adaptation Action Areas to Protect Against Flooding: Innovative Program Will Serve As Model for Other Communities. Accessed October 12, 2015: http://www.fortlauderdale.gov/Home/Components/News/News/228/16

City of Fort Lauderdale (n.d.). Green Your Routine: Sustainability Division. Accessed October 11, 2015: http://gyr.fortlauderdale.gov/t-government/green-your-routine/sustainability-division

City of Fort Lauderdale Code of Ordinances. (2015). Sec.2.01. Description of corporate limits. Accessed October 15, 2015: https://www.municode.com/library/fl/fort_lauderdale/codes/code_of_ordinances?nodeId=CHFOLAFL_ARTIICOLI_S2.01DECOLI

International City/County Management Association (ICMA) (n.d.). Durban - Southeast Florida Climate Change Partnership (CityLinks). Accessed October 13, 2015: http://icma.org/en/international/projects/directory/Project/1048/Durban__Southeast_Florida_Climate_Change_Partnership_CityLinks

Morejon, J. (2015, September). Green Your Routine: Pioneering A Way Towards A Sustainable Future. Go Riverwalk, 12(8), pp. 26.

Peel, M. C., Finlayson, B. L., and McMahon, T. A. (2007). Updated world map of the Köppen-Geiger climate classification. *Hydrology and Earth System Sciences Discussions* 4(2), 462.

Southeast Florida Regional Climate Change Compact (2014) Who We Are. Accessed October 10, 2015: http://www.southeastfloridaclimatecompact.org/who-we-are/

U.S. Census Bureau. (2015). Annual Estimates of the Resident Population: April 1, 2010 to July 1, 2015 – Metropolitan Statistical Area; and for Puerto Rico – 2015 Population Estimates. Accessed November 3, 2016: https://factfinder.census.gov/faces/tableservices/jsf/pages/productview.xhtml?pid=PEP_2015_PEPANNRES&src=pt

World Bank. (2017). 2016 GNI per capita, Atlas method (current US$). Accessed August 9, 2017: http://data.worldbank.org/indicator/NY.GNP.PCAP.CD

Case Study 16.C

Democratizing Urban Resilience in Antofagasta, Chile

Chiara Camponeschi

University of Guelph

Keywords	Participatory urbanism, governance, innovation, community resilience, co-design
Population (Metropolitan Region)	385,000 (Demographia, 2016)
Area (Metropolitan Region)	28 km² (Demographia, 2016)
Income per capita	US$13,530 (World Bank, 2017)
Climate zone	BWk – Arid, desert, cold (Peel et al., 2007)

The ability to collaboratively plan for resilience is integral to the mandate of empowering the governance of cities. As urban centers continue to grapple with an ever-growing array of social and environmental challenges, the success of their climate strategies is no longer solely determined by how solutions are implemented on the ground but also by who is involved in their design.

Adapting to climate change is necessary to strengthen the resilience of a city's social and economic systems, and public participation is an important goal in formulating adaptive responses (Larsen et al., 2011). Integrating climate adaptation into planning policies can therefore provide new opportunities for decision-makers to work toward greater livability and inclusion at the urban level.

Calls for the inclusion of a broad range of stakeholders are frequently made in major policy documents and are increasingly made by cities in their own adaptation plans. Already in 1992, Article 6 of the United Nations Framework Convention on Climate Change invoked the promotion of "public participation in addressing climate change and its effects and developing adequate responses" (UNFCCC, 1992: 17).

Despite the resurgent interest in inclusive planning, however, climate adaptation responses are typically place-based and therefore context-specific (Adger, 2001), requiring mutual trust between parties and the fair distribution of power and action outcomes in order to be meaningful. Planning for resilience therefore presents an opportunity for cities to move beyond the limitations of consultation (Arnstein, 1969) toward a broader co-design shift (Bason, 2010).

When applied to urban resilience, the practice of co-design no longer restricts adaptation to the domain of "experts" and scientists, but instead provides an opportunity for communities to have a sense of control and ownership over the changes that affect them. Reclaiming and redesigning public space is often the first step toward establishing promising approaches to collaborative adaptation.

The city of Antofagasta, Chile, is an example of just such an approach. A major mining hub in the country's north, during the past decade the city has seen a steady rise in both economic activity as well as in population growth. As a port city, however, Antofagasta is vulnerable to flooding and is equally exposed to Chile's other primary risks: earthquakes and landslides. While Antofagasta boasts the country's highest gross domestic product (GDP) per capita, it also is considered one of the most expensive to live in, and its population is expected to reach 500,000 in the next 10 years (El Mercurio de Antofagasta, 2010) – factors that challenge the city's ecological limits as much as its need for sustainable urban development.

To respond to these growing pressures, Antofagasta works alongside local service providers, schools, businesses, and neighborhood associations to explore solutions to the challenges of socioeconomic resilience. In 2013, the city enlisted Ciudad Emergente – a Chilean urban planning collective – to facilitate a series of participatory initiatives designed to promote an approach to adaptation that takes into account not only ecological concerns but also social equity and livability.

In the spring of 2013, Antofagasta hosted four *Malones Urbanos* – streetwide, open-air neighborhood meals that double as laboratories for collaborative problem-solving (Case Study 16.C Figure 1). The events were the result of the ongoing partnerships developed by the City and Ciudad Emergente with the Universidad Católica del Norte and the Universidad de Antofagasta, as well as with neighborhood associations. As Chile's second largest city, Antofagasta hosted the *malones* to stimulate creative thinking around issues ranging from placemaking to land reclamation, from climate adaptation to waste management – always in a fun, inclusive way that creates accessible entry points for capacity-building and long-term policy change.

Another participatory intervention is the *Okuplaza*, a day-long event format conceived to gather the input of diverse actors while creating an engaging, temporary public space in an otherwise underused area (Case Study 16.C Figure 2). In 2014, Ciudad Emergente and the City of Antofagasta held two *Okuplazas* to develop participatory indicators to be used as diagnostic tools to assess the health of the community. More than 150 participants gathered during the course of the event to develop tools such as socioenvironmental maps and idea trees for tracking the needs of the local population over the short and long term. Citizens also participated in conservation roundtables, public performances,

Case Study 16.C Figure 1 *The Malon Urbano in Antofagasta.*

Source: Ciudad Emergente, 2013

Case Study 16.C Figure 2 *Okuplaza in Plaza de los Colectivos, Antofagasta.*

Source: Ciudad Emergente, 2014

Case Study 16.C Figure 3 *Limpiezas Participativas in Villa Las Condes/La Corvallis, Antofagasta.*

Source: Ciudad Emergente, 2014

and a photographic exhibition to break down barriers to complex information.

Encouraged by the success of these experiences, in March 2014, the municipality launched *Limpiezas Participativas* (in English, Participatory Cleanings), a project developed by Ciudad Emergente in collaboration with several high schools, the local police department, and the neighborhood association for the city's La Corvallis area (Case Study 16.C Figure 3). Together, 300 volunteers joined forces to clear out unauthorized landfills and put an end to illegal waste dumping. Volunteers split into six "eco-teams" to clear these previously degraded areas and turn them into permanent public spaces. A similar initiative was then repeated in the area of La Cantera in May of that same year to create an intergenerational, multipurpose space that could host free neighborhood fairs, workshops on waste reduction and recycling, and programming focused on strengthening both social exchanges and sustainable living skills.

Mayors Adapt (2015), the European Union's Covenant of Mayors Initiative on Adaptation to Climate Change, believes that "strengthening stakeholder participation sets the foundation for fruitful cooperation among citizens and public administration, which may affect further policy areas as well." Meaningful community engagement strategies can contribute to improving policy outcomes by correcting the imbalance of knowledge

that often exists between local government and communities, strengthening local ownership of policy decisions, and enhancing understanding of climate change impacts.

In Chile, rapid urban development has created widespread inequality and continues to put a strain on equitable access to services, infrastructure, and education. In Antofagasta, a political commitment to disaster risk reduction goes hand in hand with a commitment to a better quality of life, increased social cohesion, and support for local responses to climate change. While those presented here are only a handful of examples of the city's approach to co-design, their many positive ripple effects are driving the city's commitment to experimenting with and supporting collaborative adaptation practices that double as best practices for other cities.

Learning from community interventions not only ensures greater relevance on the ground, but also presents an opportunity for policy and action drivers to potentially multiply, going from narrow – often opposing – categories such as "civil society" or "local government" to new and hybrid ones that emerge as more opportunities for input open up in the policy-planning process. Formalizing opportunities for co-design in the adaptation plans of cities like Antofagasta thus encourages social innovation and skill-sharing, building trust and greater social resilience (Christiansen and Bunt, 2012).

For municipalities looking to learn from Antofagasta's experience, the recognition of cities as co-creators of enabling' frameworks will be an important step in the democratization of urban resilience (Camponeschi, 2013). To do so, cities will need to explore new avenues for decision-making, supporting opportunities to embed participation and transparency into the everyday norms that will inform their future responses to climate change and, most importantly, embracing their role as champions of this powerful governance shift.

References

Adger, W. N. (2001). Scales of governance and environmental justice for adaptation and mitigation of climate change. *Journal of International Development* **13**, 921–931.

Arnstein, S. R. (1969). A ladder of citizen participation. *Journal of the American Institute of Planners* **35**, 216–224.

Bason, C. (2010). *Leading Public Sector Innovation: Co-Creating for a Better Society*. Policy Press.

Camponeschi, C. (2013). *Enabling City Volume 2: Enhancing Creative Community Resilience*. Enabling City: Rome.

Christiansen, J., and Bunt, L. (2012). *Innovations in Policy: Allowing for Creativity, Social Complexity and Uncertainty in Public Governance*. NESTA/MindLab.

Climate-Data. (2015). Climate-Data.org. Accessed October 9, 2015: http://en.climate-data.org/search/

Demographia (2016) Demographia World Urban Areas (Built Up Urban Areas or World Agglomerations). 12th Annual Edition. St. Louis: Demographia. Available from: http://www.demographia.com/db-worldua.pdf.

El Mercurio de Antofagasta. (2010). Antofagasta llegará a los 500 mil habitantes. Accessed July 18, 2015: http://www.mercurioantofagasta.cl/prontus4_noticias/site/artic/20101107/pags/20101107000545.html

Larsen, K., Gunnarsson-Ostling, U., and Westholm, E. (2011). Environmental scenarios and local-global level of community engagement: Environmental justice, jams, institutions and innovation. *Futures* **43**, 413–423.

Mayors Adapt. (2015). Mayors Adapt. Climate Change Adaptation in Europe. Accessed July 18, 2015: http://mayors-adapt.eu/about/climate-change-adaptation-in-europe/

Peel, M. C., Finlayson, B. L., and McMahon, T. A. (2007). Updated world map of the Köppen-Geiger climate classification. *Hydrology and Earth System Sciences Discussions* **4**(2), 462.

UNFCCC. (1992). Article 6 – Education, Training and Public Awareness. In *United Nations Framework Convention on Climate Change*. Accessed July 18, 2015: http://unfccc.int/files/essential_background/background_publications_hmtlpdf/application/pdf/conveng.pdf

World Bank. (2017). 2016 GNI per capita, Atlas method (current US$). Accessed August 9, 2017: http://data.worldbank.org/indicator/NY.GNP.PCAP.CD

Case Study 16.D

Building a Participatory Risks Management Framework in Bobo-Dioulasso, Burkina Faso

Liana Ricci

Department of Civil, Building and Environmental Engineering, Sapienza University of Rome, Rome

Basilisa Sanou

UN Habitat, Ouagadougou

Hamidou Baguian

Municipality of Bobo-Dioulasso

Keywords	Heat waves, droughts, risk management, participatory process, adaptation, floods
Population (Metropolitan Region)	735,000 (Demographia, 2016)
Area (Metropolitan Region)	101 km² (Demographia, 2016)
Income per capita	US$640 (World Bank, 2017)
Climate zone	Tropical, savannah (Aw) (Peel et al., 2007)

Bobo-Dioulasso, the second biggest city in Burkina Faso, is situated in the Hauts-Bassins region. Climate change will exacerbate vulnerabilities associated with the combined effects of population growth, land degradation, and reduced rainfall, leading to decreases in agriculture production, proliferation of disease vectors, affecting agro-sylvo-pastoral industries, and undermining local and national development.

Relevance to Climate Change Adaptation and Mitigation in Terms of Action and Strategy

To address the challenges of climate change, the municipality decided to set up a project to strengthen its capabilities. In July 2010, it became a formal partner of the UN Habitat Cities and Climate Change Initiatives (CCCI). CCCI focuses on small- and medium-sized cities in developing countries and aims to increase efforts toward adaptation through four main objectives: (1) promote active climate change collaboration between local governments and their associations, (2) enhance dialogue so that climate change is firmly established on the political agenda, (3) support local governments in making climate sensitive changes, and (4) foster awareness, education, and capacity-building strategies that support the implementation of climate change strategies (UN Habitat, 2011a).

During the first step of the CCCI program (2011–2012), the municipality carried out an initiative called "The City of Bobo-Dioulasso against the Effects of Climate Change: Framework to Improve Dialogue between Local and National Actors." During this initial phase, residents, community actors, territorial officials, and local policy-makers were involved in a study of the challenges and preparatory measures necessary with regard to climate change; an action plan for the management of climate risks and disaster was also developed and approved by local representatives (Ricci et al., 2015) (Case Study 16.D Figure 1).

Actions and Policy Drivers

The implementation of CCCI was promoted and supported by UN Habitat, and the municipality of Bobo-Dioulasso is the key actor for the implementation of the program. However, the multilevel process takes into account existing experiences, knowledge, know-how, and expertise to develop local requirements that are consistent with mitigation and adaptation to climate change (Case Study 16.D Figure 2). The approach then sought to create opportunities to:

- Bring together all inhabitants (men, women, youth), municipal managers, local and national partners, and technical and financial partners
- Encourage an open and constructive discussion to enhance knowledge and awareness of climate change–related issues and build consensus among all stakeholders about the essentials that must underpin the work of the project
- Solicit and obtain the commitment of everyone to get involved in achieving mutually agreed objectives

Subsequent climate interventions formulated by the local population and local authorities focus on the implementation of regulations, decision-making on the control of carbon emissions, construction of energy-efficient buildings, and conservation of urban forests and natural resources. However, further action, such as dissemination of warning systems, emergency services, and support systems for the most vulnerable people and the construction of emergency shelters, still need to be taken by the local authorities (UN Habitat, 2011b, 2013).

Impact and Scale

One of the key outcomes of the initial phase of the CCCI in Bobo-Dioulasso has been the creation of the municipal charter of collaboration and permanent dialogue for a climate-resilient Bobo-Dioulasso ("Charte municipale de collaboration et de dialogue parmanent pour une ville de Bobo-Dioulasso resiliente au climat"). Apart from the efforts of individual stakeholders and groups of stakeholders, the role of administrative districts in participatory risk management is particularly relevant.

Local actors, including municipal and territorial officials, local networks, and experts, were involved in a multistep participatory process (Case Study 16.D Figure 3) for the formulation of a Participatory Climate Risk Management Framework (PCRMF). Under the framework, which was discussed and adopted by

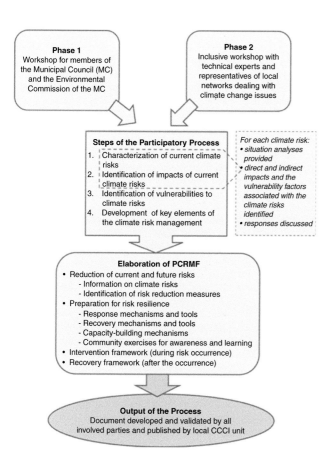

Case Study 16.D Figure 1 *Architecture of the action plan for climate risk and disaster management.*

Source: Ricci et al., 2015

Case Study 16.D Figure 2 *Participants in the fifth consultation workshop with municipal and territorial officials.*

Source: CCCI Team Bobo Dioulass

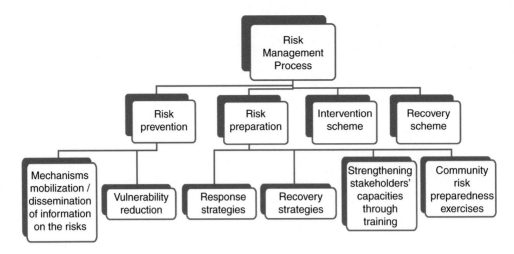

Case Study 16.D Figure 3 *Process for the formulation of the Participatory Climate Risk Management Framework.*

Source: Ricci et al., 2015

municipal representatives and an expert from local climate change network (Institut d'Applications et de Vulgarisation en Sciences [IAVS] – Pôle Bobo-Dioulasso), the local actors agreed that the management of present and future climate risks requires, first, the prevention and reduction of climate change impacts (UN Habitat, 2011b, 2014, Ricci et al., 2015).

Lessons Learned

The PCRMF, formulated in a participatory manner by local actors, is ready for implementation under the institutional leadership of the municipality of Bobo-Dioulasso. A first lesson that can be derived from this participatory work is the detailed collection of observed climate trends and impacts by the population and local actors. As a result, the Municipal Council is ready to lead or support proactively preventive actions for adaptation and mitigation of the negative impacts of climate change. Financing and implementing the set of actions recommended in the framework could mobilize a variety of financial partners as well as stakeholders at different levels: community or local level, national level, and bilateral and multilateral international levels.

Additionally, this case study shows that small- and medium-sized cities have a crucial role in addressing climate change challenges and that adaptation needs to be mainstreamed and implemented at the local level. Risk management is a useful approach to limit vulnerability to current and future hazards. However, different types of constraints need to be considered in the implementation of climate change initiatives. The regulatory capacity of public authorities and the balance of power and resources play major roles in the implementation of adaptation at the local level. Capacity and resources are directly linked to legitimization and information. Therefore, financial resources

and decentralization reforms can be major challenges for climate change adaptation and risk reduction.

References

Demographia (2016) Demographia World Urban Areas (Built Up Urban Areas or World Agglomerations). 12th Annual Edition. St. Louis: Demographia. Available from: http://www.demographia.com/db-worldua.pdf.

Ministere de L'environnement et du Cadre de Vie Pana du Burkina Faso. (2007). Programme D'action National d'adaptation a la Variabilite et aux Changements Climatiques. Accessed September 9, 2015: http://unfccc.int/resource/docs/napa/bfa01f.pdf

Peel, M. C., Finlayson, B. L., and McMahon, T. A. (2007). Updated world map of the Köppen-Geiger climate classification. *Hydrology and Earth System Sciences Discussions* **4**(2), 462.

Ricci, L., Sanou, B., and Baguian, H. (2015, April). Climate risks in West Africa: Bobo-Dioulasso local actors' participatory risks management framework. *Current Opinion in Environmental Sustainability* **13**, 42–48. Accessed September 24, 2015: http://dx.doi.org/10.1016/j.cosust.2015.01.004

UN Habitat. (2011a). *Cities and Climate Change Initiative. Regional Roll-out Strategy for Africa 2012–2021* (vol. **22**). UN-Habitat. Accessed September 9, 2015: http://mirror.unhabitat.org/downloads/docs/10401_1_594147.pdf

UN Habitat. (2011b, December). *Report Technical Workshop on Awareness and Training: Municipal and Territorial Climate in Bobo-Dioulasso. Bobo-Dioulasso Unit Communale.* CICA.

UN Habitat. (2013, March). CCCI pilots Urban and Peri-urban Agriculture in three cities. *Cities and Climate Change Initiative Newsletter.* Available from: http://www.fukuoka.unhabitat.org/programmes/ccci/pdf/9_Cities_and_Climate_Change_Initiative_Newsletter.pdf

UN Habitat. (2014, June). Integrating urban and peri-urban agriculture into city-level climate change strategies. *Cities and Climate Change Initiative Newsletter.* Available from: http://unhabitat.org/integrating-urban-and-peri-urban-agriculture-into-city-level-climate-change-strategies-june-2014/

World Bank. (2017). 2016 GNI per capita, Atlas method (current US$). Accessed August 9, 2017: http://data.worldbank.org/indicator/NY.GNP.PCAP.CD

Case Study 16.E

Warsaw and City Sustainability Reporting

Adam Sulkowski

Babson College, Babson Park

Keywords	Data measurement, reporting, governance, transparency, sustainability, indicators
Population (Metropolitan Region)	1,735,442 (Statistical Office in Warsaw, 2015)
Area (Metropolitan Region)	517.2 km² (Statistical Office in Warsaw, 2015)
Income per capita	US$12,680 (World Bank, 2017)
Climate zone	Dfb - Cold, without dry season, warm summer (Peel et al, 2007)

Sustainability Reporting

Sustainability reporting – publishing data on an entity's environmental, economic, and societal performance and governance – is an essential step in climate change mitigation and adaptation (Sulkowski and Waddock, 2013). As a management tool, it can help city managers evaluate plans, investments, regulations, and incentives to optimize quality of life and climate mitigation and adaptation (Ballantine, 2014). Among other motivations, organizations also engage in sustainability reporting as a way to better communicate with stakeholders, encourage innovation, boost confidence of investors, and attract, retain, and inspire talented people (Hughey and Sulkowski, 2012; KPMG, 2013; Walsh and Sulkowski, 2010; Wu et al., 2011).

Sustainability reporting among cities is at a much earlier phase of adoption than among companies. While a sustainability reporting framework for corporations, the Global Reporting Initiative (GRI), was founded in 1997 (Brown et al., 2009), a similar framework for cities did not appear until late 2012 in North America: the Sustainability Tools for Assessing and Rating Communities (STAR Communities, 2015). The first set of indicators designed for global adoption by cities was not published until 2014 (the International Organization for Standardization's ISO 37120:2014). In 2013, when more than 4,000 companies published sustainability data, including 93% of the largest 250 corporations in the world (KPMG, 2013), it

was still considered innovative for a city to release such data (Ballantine, 2014).

Data Related to Climate Change

All standards suggest that reporting entities publish specific facts related to their organizational profile and governance, plus statistics about environmental, societal, and economic impacts. We present here examples of major metrics that most directly gauge an entity's impacts upon and vulnerability to climate change, plus goals, investments, and progress toward eliminating its negative impacts and adapting to a changed climate:
- Energy consumption within the organization
- Reduction in energy used as a result of efficiency initiatives
- Habitats protected or restored
- Direct greenhouse gas (GHG) emissions
- Indirect GHG emissions
- Reduction of GHG emissions
- Investment in environmental protection expenditures

City of Warsaw Implements Sustainability Reporting

Warsaw, capital of Poland, has an official population of 1.7 million with a total of almost 3 million people in its metropolitan area. It was systematically razed near the end of World War II by occupiers as punishment for resisting and has been rebuilt since 1945. The reconstruction included the largest rebuilding of a Medieval Old Town. At the same time, city planners of the foreign-imposed Communist regime laid out a sprawling new grid of avenues between four- to six-car lanes in width. This context, coupled with the explosive growth of Poland's economy since the country's return to full independence in 1989, has resulted in a mix of opportunities and challenges that, as in other cities, are directly related to climate change.

The 2013 Warsaw Integrated Sustainability Report was the first in the world to use the latest standard (called G4) of the Global Reporting Initiative partly to draw attention to the concept of city sustainability reporting (Cohen, 2013; City of Warsaw, 2013), a distinction that earned it a spot on a list of the top 10 sustainability reports of 2013 (Cohen, 2014). In 2014–15 the city took additional steps of referencing the new ISO standard and engaging local university students under the oversight of professors to complete the reporting exercise (Ballantine, 2014).

Box 1 Highlights of Warsaw's Climate Mitigation and Adaptation

- 25% of Warsaw is green space
- Co-generated heat distribution system from electricity production is third largest (1,700 km) in the world
- Fuel use, because of co-generated power and heat, is 33% more efficient than many power plants elsewhere in the world
- Recently modernized water treatment plant was the largest environmental protection investment in Europe

(totaling €769 (US$908.4) million) and is 33% powered by processed solids from wastewater
- Ranked 16th out of 30 in the European Green City Index
- Biomass-based energy generation to grow from 2% to 10% and waste-to-energy from 1% to 8% from 2010 levels to 2020, with a goal of cutting emissions 20% below 2007 levels by 2020
- Adjacent UNESCO World Biosphere Reservation is home to thousands of plant and animal species

In Warsaw, as elsewhere, sustainability reporting can play a key role in furthering climate change mitigation and adaptation by:
- Optimizing outcomes by helping leaders see positive relationships between climate policy and economic, societal, and local environmental benefits

Encouraging residents (and other stakeholders) with positive-but-underappreciated facts about the city, and inspiring further climate-friendly behavior choices (e.g., related to recycling, lifestyle, and commuting)

Lessons Learned and a Checklist for Implementing Sustainability Reporting

Cities can readily adopt the practice of sustainability reporting. Based on implementing sustainability reporting for Warsaw and municipalities in the United States, sometimes with university student teams, the author offers the following generalized steps to consider:

1. Study the conceptual background of sustainability reporting, standards, and examples.
2. Find a champion of reporting within city government.
3. Explore partnerships with local education institutions and engage university student teams under the guidance of professors, staff, or consultants. This creates many side benefits such as spreading awareness of climate change mitigation and adaptation and other community sustainability issues, inspiring and informing research, and training students in environmental policy, communications, statistical analysis, and public and private sector management.
4. Identify what matters to stakeholders.
5. Choose whether to use or reference a standard.
6. Research and prepare the report, starting with available goals and statistics.
7. Consider a summary of key indicators plus a longer full report.
8. Seek internal or external verification. Corporations often have their sustainability reports audited to make them more credible to investors and other stakeholders.

9. Search for – and incorporate – feedback. Ideally, reporting is a channel for not just sharing data, but also for hearing ideas on improving goals, actions, and outcomes.
10. Before publication, prepare a roll-out and communications strategy. Be aware that, in a politicized climate, even a pragmatic exercise in transparency may be criticized.
11. Consider alternative communication media. Availability of open, real-time city data on mobile devices could help steer behaviors to be more climate neutral (Batty et al., 2012). An example of this is the public transportation navigation application from CityNav in Poland, which updates actual positions of buses, trains, and trams in real time, making mass transit even more predictable and appealing to users.
12. Encourage ambition in goal-setting and the active use of sustainability reporting as a tool for planning and evaluating progress toward, ideally, a net zero emissions future.

References

Ballantine, J. (2014). The value of sustainability reporting among cities. *Cities Today* **15**, 30–35.

Batty, M., Axhausen, K. W., Giannotti, F., Pozdnoukhov, A., Bazzani, A., Wachowicz, M., Ouzounis, G., and Portugali, Y. (2012). Smart cities of the future. *European Physical Journal-Special Topics* **214**(1), 481.

Brown, H. S., De Jong, M., and Lessidrenska, T. (2009). The rise of the Global Reporting Initiative: A case of institutional entrepreneurship. *Environmental Politics* **18**(2), 182–200.

City of Warsaw, Poland. (2013). Warsaw Sustainability Report – English. Accessed November 9, 2015: http://www.um.warszawa.pl/sites/default/files/attach/aktualnosci/warsaw_g4_integrated_sustatinability_report.pdf

City of Warsaw, Poland. (2013). Warsaw Sustainability Report – Polish. Accessed November 9, 2015: http://infrastruktura.um.warszawa.pl/aktualnosci/efektywne-wykorzystanie-energii/zintegrowany-raport-zr-wnowa-onego-rozwoju-warszawy-2013

Climate-Data. (2015). Climate-Data.org. Accessed October 9, 2015: http://en.climate-data.org/search/

Cohen, E. (2013). The first G4 Sustainability Report in the World. CSR-Reporting. Accessed November 9, 2015: http://www.csr-reporting.blogspot.co.il/2013/08/the-first-g4-sustainability-report-in.html

Cohen, E. (2014). The Top Ten Sustainability Reports of 2013. Accessed November 9, 2015: http://www.triplepundit.com/2014/01/top-ten-sustainability-reports-2013

Hughey, C., and Sulkowski, A. J. (2012). More disclosure = better CSR reputation? An examination of CSR reputation leaders and laggards in the global oil & gas industry. *Journal of the Academy of Business and Economics* **12**(2), 24–34.

International Organization for Standardization (ISO). (2014). ISO 37120:2014. Accessed November 9, 2015: http://www.iso.org

KPMG. (2013). The KPMG Survey of Corporate Responsibility Reporting. Accessed November 9, 2015: http://www.kpmg.com/Global/en/IssuesAndInsights/ArticlesPublications/corporate-responsibility/Documents/corporate-responsibility-reporting-survey-2013-v2.pdf

Peel, M. C., Finlayson, B. L., and McMahon, T. A. (2007). Updated world map of the Köppen-Geiger climate classification. *Hydrology and Earth System Sciences Discussions* **4**(2), 462.

Statistical Office in Warsaw (2015). Statistical Yearbook of Warsaw 2015. Accessed on November 3, 2016 from: http://warszawa.gov.pl/en/publications/statistical-yearbooks/statistical-yearbook-of-warsaw-2015,3,14.html

STARS Communities. (2015). Star Communities - Sustainability Tools for Assessing & Rating Communities. Accessed November 9, 2015: http://www.starcommunities.org

Sulkowski, A. J., and Waddock, S. (2013). Beyond sustainability reporting: Integrated reporting is practiced, required & more would be better. *University of St. Thomas Law Review* **10**(4), 1060–1123.

Walsh, C., and Sulkowski, A. J. (2010). A greener company makes for happier employees more so than does a more valuable one: A regression analysis of employee satisfaction, perceived environmental performance and firm financial value. *Interdisciplinary Environmental Review* **11**(4), 274–282.

World Bank. (2017). 2016 GNI per capita, Atlas method (current US$). Accessed August 9, 2017: http://data.worldbank.org/indicator/NY.GNP.PCAP.CD

Wu, J., Liu, L., and Sulkowski, A. J. (2011). Environmental disclosure, firm performance, and firm characteristics: An analysis of S&P 100 firms. *Journal of the Academy of Business and Economics* **10**(4), 73–84.

Case Study 16.F

City of Paris: 10 Years of Climate Comprehensive Strategy

Yann Françoise, Marie Gantois, and Sébastien Emery

City of Paris

Anne Girault and Elsa Meskel

Parisian Climate Agency, Paris

Chantal Pacteau

French National Research Center (CNRS), Paris Research Consortium Climate-Environment-Society, Paris

Keywords	Mitigation, adaptation, policy, heat wave, multisectoral comprehensive strategy
Population (Metropolitan Region)	11,959, 807 (INSEE, 2016)
Area (Metropolitan Region)	12,012 km² (IAU Ile-de-France, 2016)
Income per capita[a]	US$38,950 (World Bank, 201)
Climate zone	Cfb – Temperate, Without dry season, warm summer (Peel et al., 2007)

PARIS Climate Action Plan

In 2007, the City of Paris adopted its First Climate Action Plan. (City of Paris, Urban Ecology Agency, 2012). It was elaborated in an innovative collaborative approach involving all Parisian community: authorities, economic stakeholders, citizens, non-governmental organizations (NGOs), and more. The Plan sets five commitments:

1. Decrease greenhouse gas (GHG) emissions by 75% in 2050 compared with 2004
2. Reduce GHG emissions by 25% in 2020 compared with 2004
3. Reduce energy consumption by 25% in 2020 compared with 2004
4. Get 25% of renewable energy sources (RES) into the energy mix
5. Adapt Paris to climate change

After 5 years and some important successes in mobility, housing retrofitting, and community involvement, the city council decided to update the First Plan in 2012.

The Second Plan reaffirms the objectives of the initial plan. It is a comprehensive strategy trying to mainstream climate change mitigation and adaptation in all sectors. The plan itself is divided into six chapters:

- Urban planning for energy efficiency
- Low-energy and affordable housing
- The service industry in Paris, a new challenge
- Toward transport that improves the climate and air quality
- Toward sustainable food and consumption that generates less waste
- An adaptation strategy

To achieve these goals in all targeted sectors, the City of Paris has undertaken extensive outreach and mobilization action. This includes the Paris Action Climat partnership agreement (City of

Paris, 2013) with community stakeholders who become involved in the momentum of the Paris Climate Action Plan. Other communication tools include the game Clim'Way Paris, available online and free to every person having Internet access: www.climway .paris.fr. Players of the game have a choice of different actions to implement in order to reach the objectives of the Paris Climate Action Plan. The game shows the evolution of Paris from 2004 to 2050. Another information tool the city developed is the solar potential register, available online, which allows each Parisian to check the feasibility of installing solar panels on his or her building in terms of energy production: www.cadastresolaire.paris.fr. Mobilization of citizens is an important action for the Mayor (at that time Bertrand Delanoë, now Anne Hidalgo) and was why he decided to create the Paris climate agency in March 2011.

Paris Climate Agency (APC)

The Paris Climate Agency was set up on the initiative of the City of Paris to help achieve the goals of the Climate Action Plan and to inform Parisians about energy efficiency. The Paris Climate Agency has eighty-five members and partners. It is a key driving force for regional energy transition, seeking to involve the future Metropolis of Greater Paris and assisting with operational projects, particularly the energy retrofit of buildings. By promoting dialogue and debate around these major issues, the APC also helps create a common knowledge basis and initiatives to support urban transformation. It brings stakeholders together in an alliance that aims to achieve a sustainable energy transition in the city and to combat climate change (APC, 2015).

Energy-Efficient Refurbishment of Social and Private Housing and Commercial Stock

Paris, one of the densest cities, occupies more than 130 million square meters of housing and commercial real estate. Most of these buildings were built before the first thermal regulation law in 1975. With the Paris Climate Action Plan, all new buildings must be designed with energy performance better than national standards.

The challenge is to refurbish the stock. To show that it is possible, the City of Paris invested more than €200 (US$236) million in 5 years to retrofit 30,000 housing units. This program improved energy efficiency, decreased GHG emissions, created more than 1,500 local jobs, improved quality of life for tenants, and reduced their energy bills by at least 30%.

The other challenge is to mobilize the private sector to do same. To carry out the strategy in this sector, the City of Paris combines strong financial intervention in selected urban districts together with measures conducted by the APC on the whole territory. For example, a web platform CoachCopro© (www.coach copro.com) helps Parisians conduct their retrofitting projects. This platform will be extended to the professional sector and will facilitate the process of rehabilitation by helping building owners and enterprises find relevant financial tools.

A New Mobility: A Challenge for Paris and Its Metropolis

The City of Paris has one of the densest public transportation networks of both metro and buses, but traffic jams and bad air quality are important issues. In 2001, Mayor Delanoë decided to establish a new mobility policy that would reduce dependence on cars, develop active mobility modes (cycling, walking), and improve public transportation inside and outside Paris. Almost 15 years later, results are 32% less traffic, 25% less air pollution emissions, and 15% less GHG emissions., and now new shared services like Vélib' (24,000 bikes) or Autolib' (2,000 electric cars) are part of the way of life for Parisians.

For the next 20 years, the challenge is to adapt this success model to the wider Metropolis (7 million inhabitants). Greater Paris will see major projects in transportation and urban development to improve quality of life, regional solidarity, and attractiveness with a program worth €32 (US$37.8) billion over 15 years. Significant urban projects will emerge along the Greater Paris Express transport route, such as new neighborhoods with an array of features including housing, economic activities, and university and cultural facilities – especially around the strategic, structural points provided by the stations. The idea is to enable business clusters to take root while rebalancing the Paris region, opening it up to its more remote, isolated areas.

The urban system exemplifies how Paris's economic potential is being harnessed to tackle the new challenge of energy transition, accelerating the move toward a green economy and creating a unique attractiveness factor. The APC motivates stakeholders by means of an alliance for metropolitan energy transition and the fight against climate change

Supported by a complete system of closely networked stakeholders plus highly effective, progressive networks and infrastructure, Paris stands out as a platform for innovation and experimentation driven by a shared, socially responsible political will.

An Innovation Laboratory for the Green Economy

Paris is a skills incubator not only for education but also for research and development, and global economic leaders are investing in the development of clean technologies in Paris. Paris welcomes and actively supports innovative start-ups and arranges local testing in what has become an urban laboratory. For example, Paris Région Lab and the APC have launched two calls for experimentation in energy efficiency (Paris Région Lab, n.d.).

Paris and its metropolis are already moving toward an energy transition through:

- Efficient infrastructure with a focus on new functionalities and practices in the water, energy, and mobility sectors
- Innovative energy systems that are managing workflows as close to producers and consumers as possible; this includes European flagships such as the Paris district cooling network
- Information and communication technologies geared toward the "smart city"

These elements are presented in a dedicated platform (APC, n.d.)

Resilience and Adaptation

Paris not only is taking action on climate change mitigation, it has also assessed its vulnerabilities and opportunities with regard to climate change and resources depletion (water, energy, food, biodiversity) in a comprehensive study in 2012–2013. More than a hundred persons participated in the study (scientific committee, stakeholders, and city representatives) through interviews and workshops that are now summarized into twenty-five sectoral and impact sheets. This initial diagnosis concludes that the main threats the City of Paris will face with climate change are urban heat islands and heat waves, floods and intense rainfall, drought spells and water depletion, and storms and land subsidence. These will affect the energy distribution network and the built environment (especially regarding summer thermal comfort) as well as public health (air quality, heat waves, floods) and the overall economy of the country (Seine flooding).

This diagnosis is now a baseline to co-elaborate the adaptation strategy of Paris as it faces climate change and resource depletion, with Parisian stakeholders involved. The adaptation strategy is completely integrated into the Paris Climate Action Plan. In 2015, the ongoing adaptation actions and presents goals toward the adaptation of Paris will be affirmed. This strategy will be an answer to decrease risk but especially an opportunity to create a new positive urban ecosystem in Greater Paris for improving the quality of life of its citizens in the decades to come.

The main actions implemented in Paris for its resilience and adaptation concern the greening of Paris (new parks, green infrastructures, green roofs and facades, urban agriculture), flood prevention and action plans, a biodiversity preservation plan, an intense rainfall plan, a heat wave warning and assistance plan, and a water savings plan. Sectoral studies and experiments have been instituted to better understand the efficiency of potential adaptation solutions (e.g., quantifying the cooling effects of different green and humid spaces in Paris).

References

Agence Parisienne du Climat (APC). (n.d.). Paris Green. Accessed September 22, 2015: http://www.paris-green.com/

Agence Parisienne du Climat (APC). (2015). Agence Parisienne du Climat. Accessed August 31, 2015: http://www.apc-paris.com

City of Paris. (2013). Paris climate action climate action partnership agreement. Accessed September 22, 2015: http://parisactionclimat.paris.fr/en/

City of Paris, Urban Ecology Agency. (2012). *Paris Climate and Energy Action Plan* (1st ed.).City of Paris.

Climate-Data. (2015). Climate-Data.org. Accessed October 9, 2015: http://en.climate-data.org/search/

French National Institute for Statistics (INSEE). (2011). Séries historiques des résultats du recensement Aire urbaine 2010 de Paris (001). Accessed September 22, 2015: http://www.insee.fr/fr/themes/tableau_local.asp?ref_id=TER&millesime=2012&typgeo=AU2010&typesearch=territoire&codgeo=Paris+%28001%29&territoire=OK

French National Institute for Statistics (INSEE). (2016). Evolution et structure de la population. Accessed November 5, 2016: https://www.insee.fr/fr/statistiques/2522482#titre-bloc-13

IAU Ile-de-France (2016). Paris Region Key Figures 2016. Accessed November 3, 2016: https://www.iau-idf.fr/fileadmin/NewEtudes/Etude_1237/KEY_FIGURES_2016-13_BD_versiondef.pdf

Paris Région Lab. (n.d.). Actualites du lab "efficacité énergétique." Accessed September 22, 2015: http://www.parisregionlab.com/tags/efficacit%C3%A9-%C3%A9nerg%C3%A9tique

Peel, M. C., Finlayson, B. L., and McMahon, T. A. (2007). Updated world map of the Köppen-Geiger climate classification. *Hydrology and Earth System Sciences Discussions*, **4**(2), 462.

World Bank. (2017). 2016 GNI per capita, Atlas method (current US$). Accessed August 9, 2017: http://data.worldbank.org/indicator/NY.GNP.PCAP.CD

Appendix A

ARC3.2 Glossary

Term	Definition[1]
4Rs (reduce, reuse, recycle, resource recovery, and disposal)	*Reduce*: This refers to waste avoidance and materials management, i.e., avoiding or reducing primary/virgin materials for manufacturing and preserving natural resources. This requires reducing financial and environmental resources in the collection, transport, treatment, and disposal of waste. For example, wastage can be minimized through reduced packaging, improved design, and use of durable materials.
	Reuse: This refers to the practice of using materials over and over again for the same purpose for which they were intended. Reusing waste may require collection but relatively little or no processing.
	Recycle: This refers to any activity that involves the collection, sorting, and processing of used or unused items that would otherwise be considered as waste into raw material which is then remanufactured into new products.
	Resource recovery: This encompasses recycling, reprocessing, and energy recovery consistent with the most efficient use of the waste material. Resource recovery includes converting organic matter into useable products (such as compost and digestate) or energy recovery in the form of electricity and/or heat.
	Disposal: If none of the above options is possible, then waste should be disposed of in a controlled manner. This includes using a sanitary landfill or pretreating the waste in other ways in order to prevent harmful impacts on public health or the environment.
Acidification	Acidification refers to a reduction in the pH of the ocean over an extended period time caused primarily by uptake of carbon dioxide (CO_2) from the atmosphere
Actors	Individuals, communities, organizations, and networks that participate in decision-making related to urban adaptation and mitigation.
Adaptation (*or Climate Change Adaptation*)	Adjustment in natural or human systems in response to actual or expected climatic stimuli or their effects, which moderates harm or exploits beneficial opportunities. Various types of adaptation can be distinguished, including anticipatory, autonomous, and planned adaptation.
Adaptation deficit	Failure to adapt adequately to existing climate risks
Adaptive capacity	The degree to which system dynamics can be modified to reduce risk. Traditionally, adaptive capacity focused on human actors and institutions, but, in the context of urban biodiversity and ecosystems, non-human actors, behavior, species interactions, and human–ecological interventions are also important.
	The ability of individuals or communities to utilize their resources and capitals to resist and adapt to present and future hazards. This capacity varies across and within societies, conditioned to the degree to which the geophysical, biological, and socioeconomic systems are susceptible to the adverse impacts of climate change while being dependent on several factors such as economic wealth, technology, information and skills, infrastructure, institutions, and equity.
	The ability of an agent or system to prepare for stresses and changes in advance of the shock or to adjust in the response to the effects of the shock.
	A function of the quality of provision and coverage of infrastructure and services, investment capacity, and land-use management. Urban adaptive capacity during extreme events depends on : (1) proper and simultaneous functioning of lifeline systems including transportation, water, communications, and power; (2) the robustness of critical facilities for public health, public safety, and education; and (3) preparedness programs and response and relief capabilities.
Adaptive management	A structured, iterative process of decision-making in the face of imperfect information. Adaptive management recognizes the uncertainties associated with projecting future outcomes and considers a range of possible future outcomes when formulating interventions. It also explicitly incorporates models of complex systems to support decisions and requires regular updating of models to support institutional learning and iterative decision-making, both of which can facilitate more effective management of complex systems.
Aerobic composting	The degradation of organic waste by micro-organisms in a controlled environment and in the presence of oxygen to produce a stable product: compost. The process, which is ineffective for the management of municipal solid wastes high in plastics, metals, and glass content, can directly emit varying levels of gases including nitrous oxide, depending on how the closed system is managed.

---

1   All key climate change definitions in this Glossary are taken from the Intergovernmental Panel on Climate Change (IPCC) Fifth Report (AR5).

(continued)

Term	Definition
Avoid-Shift-Improve (ASI)	A transport policy and investment paradigm that emphasizes: (1) avoiding unnecessary or low-value travel through smarter urban planning, transport pricing, sound logistics, and telecommunications; (2) shifting travel to lower polluting modes like public transport, walking, cycling, and rail freight; and (3) improving the remaining transport with cleaner vehicles and fuels and more efficient network operations.
Biological diversity	The variability among living organisms from all sources including, inter alia, terrestrial, marine, and other aquatic ecosystems and the ecological complexes of which they are part; this includes diversity within species, between species, and of ecosystems.
Biological technologies	Technologies that require lower temperature than thermal technologies for waste treatment. Examples of these technologies are anaerobic digestion, composting, biodiesel production, and catalytic cracking. They are considered appropriate treatment systems for biodegradable waste. Byproducts of these technologies include electricity, biogas, compost, and chemicals.
Bus rapid transit (BRT)	Bus systems that run on a segregated lane. They currently operate or are being built in more than 300 cities. Cities with established systems, such as Curitiba and Bogota, have seen reduced automobile trips, fuel use, and emissions.
Buyout programs	Government programs reimbursing shorefront landowners for abandoning property in high-risk zones.
Cap-and-trade	A market-based policy tool for protecting human health and the environment by controlling large amounts of emissions from a group of sources. A cap-and-trade program first sets an aggressive cap, or maximum limit, on emissions. Sources covered by the program then receive authorizations to emit in the form of emissions allowances, with the total amount of allowances limited by the cap. Each source can design its own compliance strategy to meet the overall reduction requirement, including the sale or purchase of allowances, installation of pollution controls, and implementation of efficiency measures, among other options. Individual control requirements are not specified under a cap-and-trade program, but each emission source must surrender allowances equal to its actual emissions in order to comply. Sources must also completely and accurately measure and report all emissions in a timely manner to guarantee that the overall cap is achieved.
Carbon sequestration	The process of capturing and storing CO_2 emissions. This may involve removing CO_2 from the atmosphere or directly capturing the emissions at their source. Processes that capture CO_2 emissions from sources like coal-fired power plants, then stores, reuses, and removes the emissions from the atmosphere. Storage is typically provided in geologic formations including oil and gas reservoirs, unmineable coal seams, and deep saline reservoirs.
Catastrophe bond	A financial instrument developed by insurers or governments to pass extreme risks on to private investors who are willing to assume them in exchange for high interest rates.
Cities (see also 'urban systems')	Complex human-dominated social-ecosystems that interrelate dynamically with the economic and technological systems and the built environment.
City Creditworthiness Program	A program designed by the World Bank to help city financial officers conduct thorough reviews of their municipal revenue management systems and take actions to qualify for a rating.
Clean development mechanism (CDM)	One of the "flexibility mechanisms" defined under the Kyoto Protocol. Its objective is to assist developing countries in achieving sustainable development and mitigating greenhouse gas emissions that cause climate change. In addition, the CDM aims to assist industrialized countries in achieving compliance with their quantified emission limitations.
Climate risk factors	The subset of climate hazards that are of most consequence for a given city. They are selected on the basis of interactions between researchers and stakeholders and expert judgment using quantitative and qualitative climate hazard information.
Coastal City Flood Vulnerability Index (CCFVI)	A tool measuring city vulnerability to flooding comprised of a hydro-geological component (sea level rise, river discharge, soil subsidence, cyclones, storm surge), socioeconomic component (exposed populations, vulnerable groups), and a politico-administrative component (institutional organizations, flood risk maps, flood protection measures)
Combined sewer systems	Waste removal systems designed to collect rainwater runoff, domestic sewage, and industrial wastewater in the same pipe and deliver it to treatment plants. During heavy precipitation events, the capacity of combined sewers can be exceeded, leading to discharges of untreated wastewater directly to nearby water bodies.

(*continued*)

Term	Definition
Congestion pricing	Congestion pricing or congestion charge is a system of surcharging users of public goods that are subject to congestion through excess demand such as higher peak charges for use of bus services, electricity, metros, railways, telephones, and road pricing to reduce traffic congestion. This pricing strategy regulates demand, making it possible to manage congestion without increasing supply.
Conservation easement	Transfer of rights to prevent further development while landowner remains on the property
Cost-benefit analysis (CBA)	A cost-benefit analysis is a process by which business decisions are analyzed. The benefits of a given situation or business-related action are summed and then the costs associated with taking that action are subtracted.
Decarbonization	Decarbonization denotes the declining average carbon intensity of primary energy over time.
Decision-centric approach (DCA) (*see also Impact-Centric Approach*)	Urban disaster risk management and climate change adaptation that aims to reduce hazard exposure, nurture resilience, build capacity, and establish governance that enables stakeholders' visions and goals.
Development-accumulated risk	Social construction of risk stemming from government's refusal to address existing vulnerabilities resulting from unplanned or poorly planned development.
Disaster	Severe alterations in the normal functioning of a community or a society due to hazardous physical events interacting with vulnerable social conditions, leading to widespread adverse human, material, economic, or environmental effects that require immediate emergency response to satisfy critical human needs and that may require external support for recovery.
Disaster risk	The likelihood over a specified time period of severe alterations in the normal functioning of a community or a society due to hazardous physical events interacting with vulnerable social conditions, leading to widespread adverse human, material, economic, or environmental effects that require immediate emergency response to satisfy critical human needs and that may require external support for recovery.
Disasters Social Vulnerability Index (DSVI)	A statistical evaluation of a series of indicators related to demographics, living standards, and economic capacity for administrative units or social groups.
Downcycling (*see also Upcycling*)	The process of converting waste materials into products that have less value than the original.
Early warning systems (EWS)	System designed to warn a population of potentially fatal phenomena that might cause casualties and/or structural damage to housing or infrastructure
Ecosystem-based adaptation (EbA)	Climate-resilience techniques that use ecological services for adaptation.
	Adaptation that uses biodiversity and ecosystem services as part of an overall adaptation strategy to help people and communities adapt to the negative effects of climate change at local, national, regional, and global levels.
Ecosystem services (*see also Urban ecosystem services*)	The benefits that people obtain directly or indirectly from ecosystem functions such as protection from storm surges, cooling during heat waves, regulation of air quality, and provision of food, fiber, and freshwater.
El Niño Southern Oscillation	El Niño (La Niña) episodes are characterized by warming (cooling) surface waters of the tropical central and eastern Pacific. The Southern Oscillation is a seesaw pattern of changes in global-scale tropical and subtropical surface pressure, trade winds, circulation, and precipitation.
Environmental gentrification	Low-carbon building and construction standards that undermine socially progressive intentions of housing policy and result in housing designs that are unaffordable for low- and medium-income households.
Equality	A state or quality of correspondence in quantity, degree, value, rank, or ability; for example, with respect to status, rights, or opportunities. In urban areas, crucial components of equality include the right to adequate housing and security of tenure; affordability, accessibility, location, cultural adequacy, and habitability of accommodation; and availability of services, infrastructure, and facilities.
Equity	Fairness or justice in the way that people are treated.
Equity Reference Framework (ERF)	A way to operationalize common but differentiated responsibilities in international climate negotiations.
Evacuation	To remove from a dangerous place.

(*continued*)

Term	Definition
Exposure	The fact or condition of being affected by something or experiencing something : the condition of being exposed to something. The inventory of people, property, or other valued items in areas where hazards may occur.
Fuzzy cognitive mapping	Interview and analysis method that aids in assessing the relative impact of weather events among individuals and groups and supports developing socially feasible adaptation options. A fuzzy cognitive map (FCM) is a weighted, directed and causal network of a situation or system produced by an interviewee. FCMs can be analyzed using network statistics and scenario evaluation.
Gender equality	When women and men have the same conditions for realizing their full human rights and potential to contribute to national, political, economic, social, and cultural development and to benefit from the results; the same valuing by society of both the similarities and differences between women and men and the varying roles that they play.
Gender equity	The process of being fair to women and men. To ensure fairness, measures must often be available to compensate for historical and social disadvantages that prevent women and men from otherwise operating on a level playing field.
Gender mainstreaming	The process of making women's and men's concerns and experiences an integral dimension of the design, implementation, monitoring, and evaluation of policies in political, economic, and social spheres so that women and men benefit equally, and inequality is not perpetuated.
Gender-sensitive	Acknowledging the different rights, roles, and responsibilities of women and men and the relationships between them with a goal of transforming unequal gender relations to promote shared power, control of resources, and decision-making.
Global climate models (GCMs)	Physics-based mathematical representations of the Earth's climate system over time that can be used to estimate how sensitive the climate system is to changes in atmospheric concentrations of greenhouse gases and aerosols.
Green cities	Planning paradigms emphasizing ecological restoration and connected multifunctional green infrastructure in dense, compact cities. These prioritize walkable and mixed land uses and respond to the needs of people and ecosystems.
Green infrastructure	Combined engineering and ecological systems that provide ecological services such as cooling, stormwater management, urban heat island reduction, carbon storage, flood protection, and recreation.
Greenhouse gas (GHG)	Greenhouse gases are those gaseous constituents of the atmosphere, both natural and anthropogenic, that absorb and emit radiation at specific wavelengths within the spectrum of infrared radiation emitted by the Earth's surface, the atmosphere, and clouds. This property causes the greenhouse effect. Water vapor (H_2O), carbon dioxide (CO_2), nitrous oxide (N_2O), methane (CH_4), and ozone (O_3) are the primary greenhouse gases in the Earth's atmosphere. As well as CO_2, N_2O, and CH_4, the Kyoto Protocol deals with the greenhouse gases sulfur hexafluoride (SF_6), hydrofluorocarbons (HFCs), and perfluorocarbons (PFCs).
Hazard (*see also 'Urban climate hazard'*)	The potential occurrence of a natural or human-induced physical event that may cause loss of life, injury, or other health impacts, as well as damage and loss to property, infrastructure, livelihoods, service provision, and environmental resources.
Health	A state of complete physical, mental, and social well-being and not merely the absence of disease or infirmity. In cities, health is associated with social, economic, and environmental determinants, including climate change.
Health dis-benefits	Interventions adopted by different sectors that increase the adverse impacts of climate change on health. For example, urban wetlands designed primarily for flood control may promote mosquito breeding.
Health system	All the organizations, institutions, and resources that are devoted to producing actions principally aimed at improving, maintaining, or restoring health.
Heat wave	Prolonged periods of heat crossing either an absolute or relative threshold above a long-term temperature average that differs by location; typical relative thresholds are two or three standard deviations above mean temperatures.
High occupancy toll (HOT) lanes	Highway lanes with access criterion that encourages carpooling and higher vehicle occupancy, such as cars with two or more passengers. In general, HOT lanes tend to be progressive (impose a tax rate in which people with higher incomes pay proportionally more than people with lower incomes), especially when alleviating congestion on the unpriced lanes.

(*continued*)

Term	Definition
Impact-centric approach (ICA) (*see also Decision-Centric Approach*)	A linear approach in disaster risk management and climate change adaptation decision-making for cities that starts with an examination of the climate hazards to cities and then takes into account vulnerability to hazards by integrating the knowledge of existing city capacities in order to understand the likely impact of disasters and climate change.
Indian ocean dipole (IOD)	A coupled ocean–atmosphere phenomenon in the equatorial Indian Ocean defined as the difference between the sea surface temperature (SST) of the eastern and western parts of the Indian Ocean; it is measured by the Dipole Mode Index (DMI). A negative IOD event is characterized by cooler than normal water in the tropical eastern part of the Indian Ocean and above normal warmer water in the western part of the Indian Ocean basin.
Informality	A state of regulatory flux, where land ownership, land use and purpose, access to livelihood options, job security, and social security cannot be fixed and mapped according to any prearranged sets of laws, planning instruments, and regulations.
Informal settlement/slum	A heterogeneous type of settlement with a certain degree of deprivation. Commonly, slums are considered to lack one or more of five basic housing conditions: (1) durable housing of a permanent nature, 2) sufficient living space, 3) safe access to water, 4) access to adequate sanitation, and 5) security of tenure.
Integrated solid waste management (ISWM)	A comprehensive, multidimensional, and integrated approach to solid waste management in cities that is effective, environmentally sustainable, socially acceptable, and affordable.
Justice	Justice can refer to either social or environmental justice. The social justice movement seeks to establish fair distributions of wealth, opportunity, and privileges by means of fair treatment, proportional distribution, and the meaningful involvement of all people in social decision-making. The goals of the environmental justice movement are healthy environments and protection from environmental hazards for all people, regardless of race, nationality, origin, or income. Environmental justice interacts with environmental risk, exposure, impacts, sensitivity, and adaptive capacity. In that respect, social justice and environmental justice are inextricably linked.
Landscape dynamics	Spatial and temporal scaling of disturbance regimes and its influence on equilibrium/non-equilibrium states in a particular landscape.
Land value capture	Local public sources of finance such as development impact fees, tax incremental financing, public land leasing and development right sales, land readjustment programs, connection fees, joint developments, and cost-benefit-sharing.
Local climate zones (LCZs)	Urban neighborhood classification characterized by typical building heights, street widths, vegetative cover, and paved area.
Low-Carbon City Development Program (LCCDP)	An ISO-compliant environmental management system at the city level helping city governments to plan, implement, monitor, and account for low-carbon investments and climate change mitigation actions across all sectors in the city over time.
Low elevation coastal zone (LECZ)	The area below 10 meters in elevation that is hydrologically connected to the sea. The LECZ is home to approximately 10% of the world's total population and 13% of its urban population.
Madden Julian Oscillation	The dominant mode of tropical intraseasonal climate variability (on the order of 20–100 days). It is a "pulse" of cloudiness and rainfall moving eastward in the equatorial region, in the Indian Ocean, and the western Pacific Ocean, and it can excite atmospheric teleconnections that affect the climate and weather in many regions around the world.
Maladaptation	Practices, even if well intended, that are likely to increase the proportion of a population living in vulnerable conditions. For instance, desalination, while addressing immediate water needs, can be considered maladaptive since it requires an enormous amount of energy and therefore increases GHG emissions.
Managed or strategic relocation/managed realignment (*see also Relocation*)	Process that involves moving the existing built environment near coasts inland. Because people often tend to be unaware of increasing coastal risks due to climate change, setbacks and managed realignment are generally unpopular and politically contentious. Greater stakeholder involvement from the onset may, however, encourage increased public acceptance.
Mega-city	A very large city, typically one with a population of more than 10 million people.
Microgrid	A group of interconnected loads and distributed energy resources within clearly defined electrical boundaries that acts as a single controllable entity with respect to the grid. A microgrid can connect and disconnect from the grid to enable it to operate in both grid-connected or island-mode.

(*continued*)

Term	Definition
Mitigation (or, *Climate Change Mitigation*)	An anthropogenic intervention to reduce the anthropogenic forcing of the climate system; it includes strategies to reduce greenhouse gas sources and emissions and enhancing greenhouse gas sinks.
Multiple criteria analysis (MCA)	A general framework for supporting complex decision-making situations with multiple and often conflicting objectives that stakeholders groups and/or decision-makers value differently.
Municipal separate storm sewer systems (MS4)	Waste system in which stormwater and wastewater are managed separately. Therefore, during intense precipitation events, stormwater can be discharged, but not sewage, thus avoiding combined sewer overflows.
Municipal solid waste (MSW)	Solid waste generated from community activities, e.g., residential, commercial and business establishments. While construction waste and hazardous wastes are excluded as MSW in European countries, they are considered as MSW in most developing countries.
	Commonly called garbage, trash, or refuse, MSW refers to waste generated from the following activities: residential (single and multi-family dwellings); commercial (offices, stores, hotels, restaurants); institutional (schools, prisons, hospitals, airports); industrial (manufacturing, fabrication, etc., when the municipality is responsible for their collection); non-recycled construction and demolition debris; municipal services (street cleaning, landscaping).
National Development Banks	National publicly owned entities that collect funds from the stock market and/or savings and deposits. City governments and municipalities may be able to apply to these domestic public banks for resources under special lines with favorable lender conditions in the field of climate change mitigation and adaptation.
Natural climate variability	Variations in the climate due to natural processes; effects of human activities on the climate system are superimposed on the background of natural variability. Climate varies on daily, monthly, seasonal, and annual timescales; these affect the frequency and intensity of extreme events that can produce natural disasters.
Non-motorized transport (NMT)	Transport options including walking or cycling; this is the lowest greenhouse gas emissions option, accounting for half of all trips in many cities.
North Atlantic Oscillation	An oscillation of atmospheric pressure between subtropical and high latitudes; it is the dominant mode of atmospheric circulation variability in the North Atlantic region. Its influence on climate extends from North America to Europe, Asia, and Africa. The NAO and its influence are stronger in boreal winter, but are present throughout the year. Its index exhibits interannual variability but also a tendency to remain in one phase for long periods, so its impacts may be persistent.
Nuisance flooding	Less extreme tidally related coastal flood events. Two-thirds of U.S. nuisance flood days since 1950 can be attributed to climate change.
Public–private partnerships (PPP)	Coalitions between local authorities and private entities that are commonly used to create critical infrastructure, housing affordability, and urban regeneration. The role of the municipality in PPPs consists of facilitating project development by removing barriers while the private sector assumes part of the risk, provides funding, and manages the project.
Radiative forcing	A measure of the influence a factor has in altering the balance of incoming and outgoing energy in the Earth–atmosphere system and an index of the importance of the factor as a potential climate change mechanism.
Recycling (*see also Upcycling and Downcycling*)	Process that results in products that have the same value as the original or can be used for the same purpose.
	The conversion of waste into reusable material.
Regional climate models (RCMs)	Mathematical code that scientists use to project future climate at the local scale. RCMs simulate climate processes that occur at finer spatial resolution, such as convective precipitation.
Relocation (*see also Managed or strategic relocation/ managed realignment*)	The physical movement of people instigated, supervised, and carried out by either local or national authorities. Relocation practices that are temporary are known as *evacuation*, whereas more permanent efforts to counteract slow-onset effects of climate change imply resettling in another area and usually require planning. *Voluntary relocation* versus *forced relocation* depends on the extent to which communities are driving the process and whether existing frameworks and guidelines are put in place to support them.
Representative concentration pathways (RCPs)	Trajectories of radiative forcing caused by greenhouse gases and other important agents such as aerosols. Each RCP is consistent with a trajectory of greenhouse gas emissions, aerosols, and land-use changes developed for the climate modeling community as a basis for long-term and near-term climate modeling experiments. RCPs serve as inputs to global climate models and are used to project the effects of these climate drivers on future climate.

(*continued*)

Term	Definition
Resilience	The ability of a social or ecological system to absorb disturbances while retaining the same basic structure and ways of functioning, the capacity for self-organization, and the capacity to adapt to stress and change.
Response	Any action taken by governmental and non-governmental actors to manage environmental change, in anticipation of known change or after change has happened. Responses are fashioned through power, through consensus, compromise, or coercion, often by actors who frame mitigation and adaptation in the context of other environmental concerns (e.g., energy and disaster risk management), development pressures and goals (e.g., economic growth and human well-being), and in pursuit of a range of often conflicting values and priorities.
Response capacity	A pool of resources and assets that governmental and non-governmental actors may draw on for climate change mitigation and adaptation while attending to other development needs.
Risk	The potential for consequences where something of human value (including humans themselves) is at stake and where the outcome is uncertain. Risk is often represented as probability of occurrence of a hazardous event(s) multiplied by the consequences if the event(s) occurs.
Salinization	Saltwater intrusion upstream and into coastal aquifers, potentially jeopardizing urban drinking water supplies and contaminating agricultural soils. This is a slow process that can be accelerated by human-induced activities such as excessive extraction from aquifers.
Saving schemes	Strategy designed to encourage savings through small but regular deposits or deductions from salaries. In the context of adaptation, saving schemes commonly serve as coping strategies, allowing households to overcome the impact of a climate event.
Sensitivity	The degree to which an agent or system is affected by a hazard.
Setbacks	Regulations restricting shore construction based on erosion or elevation thresholds.
Shoreline protection	Engineering structures or enhanced natural features designed to withstand current and anticipated sea level rise, storm surge, and shoreline retreat. Protection includes armoring the shoreline with hard infrastructure defenses and soft defenses that mimic natural processes.
Smart growth	A planning framework centered around the conservation of open spaces, enhancement of mixed land use, denser development, and walkable neighborhoods.
Sorting and collection	Waste sorting is the process of separating municipal solid waste into different types. Sorting can occur before or after the waste is collected. Collection involves gathering waste from households, from community and street bins, or from bulk generators into larger containers or vehicles. It extends to activities such as driving between stops, idling, loading, and on-vehicle compaction of waste.
Subnational Technical Assistance (SNTA) program	World Bank program assisting subnational entities to prepare for and obtain credit ratings. The SNTA program provides technical assistance to improve subnational creditworthiness and address weaknesses highlighted by a rating assessment.
Subsidence	Settling of land caused by groundwater overdraft, sediment compaction, long-term geologic processes, enlarging of coastal inlets, dredging of ports and waterways, and upstream trapping of sediments in reservoirs.
Sustainable development goals (SDGs)	Goals passed by the United Nations in September 2015 intended to inspire, guide, and promote sustainable development.
Sustainable drainage systems (SuDS)	Measures aimed at retaining and infiltrating storm water (bio-swales, rain gardens, retention basins, bio-lakes, wetlands, rainwater harvesting systems). This allows for control of water discharge and reduces flood risk.
Thermal technologies	Waste-processing methods that operate at high temperature to produce heat or electricity as a primary byproduct. Gasification and pyrolysis are advanced forms of incineration and suitable for processing dry waste with low moisture content.
Transformation	The conditions under which system-level changes take place in which the risk management regime of a specific site, sector, or institution is fundamentally altered as one management regime is replaced by another regime.
Transit-oriented development (TOD)	Compact, pedestrian-friendly development that incorporates housing, retail, and commercial growth within walking distance of public transportation, including commuter rail, light rail, ferry, and bus terminals.
Transit-synergized development (TSD)	Multisectoral "node-and-network" approach that integrates land use, mass transit, green buildings, green districts, and advanced energy infrastructure to promote healthy, climate-resilient districts and cities.

(continued)

Term	Definition
Turbidity	The measure of relative clarity of a liquid. It is an optical characteristic of water and is an expression of the amount of light that is scattered by material in the water when a light is shined through the water sample. The higher the intensity of scattered light, the higher the turbidity. Excessive turbidity, or cloudiness, in drinking water is aesthetically unappealing and may also represent a health concern.
UCCRN Regional Hubs	The Hubs promote enhanced opportunities for urban climate change adaptation and mitigation knowledge and information transfer, both within and across cities, by engaging in ongoing dialogue between scholars, experts, urban decision-makers, and stakeholders.
Upcycling (*see also Recycling and Downcycling*)	The reprocessing of waste materials into products that have higher value than the original.
Urban design	The arrangement and design of buildings, public spaces, transport systems, services, and amenities; the process of giving form, shape, and character to groups of buildings, to whole neighborhoods, and the city. It blends architecture, landscape architecture, and city planning together to make urban areas functional and attractive.
Urban ecosystems	All vegetation, soil, and water-covered areas that may be found in urban and peri-urban areas at multiple spatial scales (parcel, neighborhood, municipal city, metropolitan region), including parks, cemeteries, lawns and gardens, green roofs, urban allotments, urban forests, single trees, bare soil, abandoned or vacant land, agricultural land, wetlands, streams, rivers, lakes, and ponds.
Urban ecosystem services	Those ecosystem functions that are used, enjoyed, or consumed by humans in urban areas, which can range from material goods (such as water, raw materials, and medicinal plants) to various non-market services (such as climate regulation, water purification, carbon sequestration, and flood control).
Urban heat island	The temperature gradient between dense, built-up environments and rural areas around them. Elevated surface and air temperatures in cities due to the low albedo of urban surfaces (such as rooftops and asphalt roadways), trapping of radiation within the urban canopy, differential heat storage, and greater surface roughness. Reduction in evapotranspiration due to impervious surfaces also contributes to the urban heat island.
Urban planning	The field of practice that uses space as a key resource for development; it takes place at the scale of the city or city-region, whose overall spatial pattern it sets.
Urban water systems	The infrastructure that cities rely on for managing the availability of good-quality water for different users. Key determinants of a well-functioning system include water for health and well-being, economic productivity, recreational and cultural benefits, and environmental services, as well as a design consistent with the available operational capacity.
Urbanization	A set of system-level processes through which population and human activities are concentrated at sufficient densities at which scalar factors become present that in turn can promote further agglomeration effects; conversion of non-urban land to urban land uses.
Vulnerability	A lack of resilience or reduction in adaptive capacity.
	Exposure to a hazard, sensitivity to that exposure, and capacity to adapt to the hazard.
Vulnerability mapping	A tool consisting of a set of indicators for identifying the need for and planning the implementation of transitional adaptation measures in cities, as well as monitoring and evaluating their results.
Waste-to-energy (WTE)	Waste-to-energy or energy-from-waste is the process of generating energy in the form of electricity and/or heat from the incineration of waste
Waste treatment	The process of converting nonrecyclable waste into composts or energy in various forms of useable heat, electricity, or fuel.
Water security	The capacity of a population to safeguard sustainable access to adequate quantities of acceptable quality water for sustaining livelihoods, human well-being, and socioeconomic development; for ensuring protection against water-borne pollution and water-related disasters; and for preserving ecosystems in a climate of peace and political stability. The sustainable availability of water for different uses and the avoidance of water-related disasters.
Water-sensitive urban design (WSUD)	Urban planning and design approach that considers all the elements of the water cycle and their interconnections to achieve an integrated outcome that sustains a healthy natural environment while addressing societal needs and reducing climate-related risks.
Zoning	Land management policy that mandates which areas can be developed and under which conditions.

Appendix B

Acronyms and Abbreviations

4PCCD	Public-Private-People Partnerships for Climate Compatible Development
AAG	American Association of Geographers
ABDI	Brazilian Industrial Development Agency (Agência Brasileira De Desenvolvimento Industrial)
ABS	Australian Bureau of Statistics
ACC	African Center for Cities
ACCA	Asian Coalition for Community Action Program
ACCCRN	Asian Cities Climate Change Resilience Network
ACEEE	American Council for Energy-Efficient Economy
ACHR	Asian Coalition for Housing Rights
ADB	Asian Development Bank
ADB/JFPR	Asian Development Bank/Japan Fund for Poverty Reduction
AIANY	American Institute of Architects New York City chapter
AICP	American Institute of Certified Planners
AIMA	Active Integrated Mitigation and Adaptation
AMO	Atlantic Multi-Decadal Oscillation
AMOC	Atlantic Meridional Ocean Circulation
ANU	Australian National University
APA	American Planning Association
AR5	IPCC Fifth Assessment Report
ARC	Argentine Red Cross
ARC3	First UCCRN Assessment Report on Climate Change and Cities
ARC3.2	Second UCCRN Assessment Report on Climate Change and Cities
ASI	Avoid-Shift-Improve
AURAN	African Urban Risk Analysis Network
BAPPEDA	Planning and Regional Development Agency (Badan Perencanaan dan Pembangunan Daerah)
BCR	Brussels Capital Region
BEST	Integrated Social Economic Development, Tangerang (Bina Ekonomi Sosial Terpadu)
BIS	Bank of International Settlements
BLHD	Regional Environmental Agency (Badan Lingkungan Hidup Daerah)
BMP	Best Management Practices
BNDES	Brazilian Development Bank
BORDA	Bremen Overseas Research and Development Association
BRAC	Bangladesh Rural Advancement Committee
BRT	Building Resiliency Task Force
BRTs	Bus Rapid Transit Systems
BSMR	Baixada Santista Metropolitan Region, São Paulo
BUCEA	Beijing University for Civil Engineering and Architecture
C3	Command, Control, and Communication Centre
C40	C40 Cities Climate Leadership Group
CAP	Climate Action Plan
CARICOM	Caribbean Community Countries
CBA	Cost-Benefit Analysis
CBD	Convention on Biological Diversity
CBO	Collateralized Bond Obligations
CBO	Community-Based Organization
CC	Climate Change
CCA	Climate Change Adaptation
CCAC	Climate and Clean Air Coalition
CCC	Climate Change Commission
CCCI	Cities and Climate Change Initiative, UN Habitat
CCCLs	Construction Control Lines
cCCR	Carbonn Cities Climate Registry
CCD	Climate Compatible Development
CCFVI	Coastal City Flood Vulnerability Index
CCO	Climate Change Office
cCR	Carbon Climate Registry
CCRIF SPC	Caribbean Catastrophe Risk Insurance Facility
CCS	Carbon Capture and Sequestration

CCSR	Center for Climate Systems Research
CCUS	Carbon Capture Utilization and Storage
CDF	Community Development Funds
CDIA	Cities Development Initiative for Asia
CDKN	Climate and Development Knowledge Network
CDM	Clean Development Mechanism
CEBA	Community-Ecosystem-Based Adaptation
CEMC	Community Emergency Management Coordinator
CEO	Center for Economic Opportunity
CERs	Certified Emissions Reductions
CFC	Chlorofluorocarbon
CH_4	Methane
CHP	Combined Heat and Power
CIF	Climate Investment Fund
CIT	Canberra Institute of Technology
CIUP	Community Infrastructural Upgrading Program
CMIM	Center for Mediterranean Integration in Marseilles
CMIP5	Coupled Model Intercomparison Project Phase 5
CNG	Compressed Natural Gas
CNRS	Centre national de la recherche scientifique
CO	Carbon Monoxide
CO_2	Carbon Dioxide
COMLURB	Companhia Municipal de Limpeza Urbana
COP	Conference of the Parties
COPPE	Centre for Integrated Studies on Climate Change and the Environment
CORe	Centralized Organic Recycling Equipment
CPC	Climate Planning Committee
CRED	Center for Research on Environmental Decisions
CRL	Crossrail LTD
CRPP	City Resilience Profiling Program
CSDS	Case Study Docking Station
CSE	Centre for Science and Environment
CSIRO	Commonwealth Scientific and Industrial Research Organization
CSO	Combined Sewer Overflow
CSR	Corporate Social Responsibility
DAC	Durban Adaptation Charter
DCA	Decision-Centric Approach
DCLG	Department for Communities and Local Government
DEP	Department of Environmental Protection
DG	Distributed Forms of Electricity Generation
DHC	District Heating and Cooling
DHW	Domestic Hot Water
DKPP	Department of Cleanliness, Parks, and Cemetery
DMI	Dipole Mode Index
DRM	Disaster Risk Management
D'RAP	Durban Research Action Partnership
DRR	Disaster Risk Reduction
DSM	Demand Side Management
DSNY	New York City Department of Sanitation
DSVI	Disasters Social Vulnerability Index
DTKBP	Dinas Tata Kota Bangunan dan Pemukiman or Department of City Planning Building and Settlement
DUDCP	Department of Urban Development and City Planning
E3	European Environment and Epidemiology
EbA	Ecosystem Based Adaptation
ECDC	European Centre for Disease Control and Prevention
EDF	Environmental Defense Fund
EDGAR	Emissions Database for Global Atmospheric Research
EEA	European Economic Area
EEPCO	Environmental Engineering and Pollution Control Organization
EEQ	Empresa Electrica de Quito
EHCC	Earth Hour City Challenge
EIU	Economist Intelligence Unit

EMDAT	International Disasters Database		IPCC	Intergovernmental Panel on Climate Change
EMSIG	Emission Simulation in Gemeinden		IPP	Intergovernmental Pilot Project
ENCC	National Strategy on Climate Change		IRP	Integrated Resource Plan
ENSO	El Niño Southern Oscillation		ISET	Institute for Social and Environmental Transition
ENSP	Environmental Science and Policy		ISWM	Integrated Solid Waste Management
EPA	Environment Protection Agency		ITCZ	Inter Tropical Convergence Zone
EQCC	Quito Climate Change Strategy (Estrategia Quiteña al Cambio Climática)		ITT	Inter- and Transdisciplinary
EquIA-urban	Urban Equity Impact Assessments		IUWM	Integrated Urban Water Management
ERF	Equity Reference Framework		IWRM	Integrated Water Resource Management
ERP	China's Eastern Route Project			
ES	Ecosystem Services		JELKA	Jena Local Decision Support Tool (Jenaer Entscheidungsunterstützung für lokale Klimawandelanpassung)
ETS	Emissions Trading System			
EU	European Union			
EWMP	Eco-Citizen World Map Project		JWP	Joint Work Program
EWS	Early Warning Systems			
			K4 C	Knowledge Centre on Cites and Climate Change
FAR	Floor Area Ratios		KICAMP	Kinondoni Integrated Coastal Area Management Project
FAS	First Aid Service		KIPRAH	Kita-Pro-Sampah or We-Pro-Waste
FEMA	Federal Emergency Management Agency		KISS	Keck Institute for Space Studies
FERC	Federal Energy Regulatory Commission		KLIMP	Sweden's Climate Investment Program
FFCO$_2$	Fossil Fuel Carbon Dioxide		KMS	Kenya Meteorological Services
FINEP	Funding Authority for Studies and Projects		KSI	Key Sustainability Initiatives
FIOCRUZ	Oswaldo Cruz Foundation		KWASA	Khulna Water Supply and Sewerage Authority
FIRMS	Flood Insurance Rate Maps			
FiT	Feed-in Tariff		LA	Los Angeles
FUNAB	Fundo Nacional do Ambiente (Mozambican Government's Environment Fund)		LAB	Local Action for Biodiversity
			LAC	Latin America and the Caribbean
			LAMATA	Lagos Transit Authority
GAR	Green Area Ratio		LCCDP	Low Carbon City Development
GCF	Green Climate Fund		LCCP	London Climate Change Partnership
GCMs	Global Climate Models		LCZ	Local Climate Zones
GCP	Gross City Product		LECZ	Low Elevation Coastal Zone
GCV	Glasgow Clyde Valley		LED	Light-Emitting Diode
GDP	Gross Domestic Product		LEED	Leadership in Energy and Environmental Design
GEA	Global Energy Assessment		LFGTE	Landfill Gas-to-Energy
GEAG	Gorakhpur Environmental Action Group		LGAs	Local Government Authorities
GEF	Global Environmental Facility		LIDAR	Light Detection and Ranging
GFDRR	Global Facility for Disaster Reduction and Recovery		LNG	Liquefied Natural Gas
GHG	Greenhouse Gases		LPTP	Rural Technology Development Institute, Surakarta (Lembaga Pengembangan Teknologi Pedesaan)
GIS	Geographical Information System			
GLA	Greater London Authority		LST	Land Surface Temperatures
GLAC	Greater Los Angeles County			
GNI	Gross National Income		MA	Millennium Ecosystem Assessment
GPC	Global Protocol for Community-Scale GHG Emissions		MCA	Multiple Criteria Analysis
GPS	Global Positioning System		MCII	Munich Climate Insurance Initiative
GtCO$_2$e	Gigatons of Carbon Dioxide Equivalent		MCUR	Medellin Collaboration on Urban Resilience
GW	Giga Watts		MDGs	Millennium Development Goals
GWP	Global Warming Potentials		MDMQ	Municipality of the Metropolitan District of Quito,
			MENA	Middle East and North Africa
HAP	Heat Action Plan		MEP	Mechanical Electrical and Plumbing
HDB	Housing and Development Board of Singapore		MJO	Madden-Julian Oscillation
HEAT+	Harmonized Emissions Analysis Tool		MODIS	Moderate Resolution Imaging Spectroradiometer
H-E-V	Hazard-Exposure-Vulnerability		MoSE	Experimental Electromechanical Module (Modulo Sperimentale Elettromeccanico)
HFA	Hyogo Framework for Action			
HPFPI	Homeless People Federation Philippines Inc.		MoSSaiC	Management of Slope Stability in Communities
HUDC	Housing and Urban Development Company		MOTF	Modeling Task Force
HVAC	Heating Ventilation and Cooling		MRF	Material Recovery Facilities
			MRS	Metropolitan Region of Santiago de Chile
IADB	Inter-American Development Bank		MRV	Measurement Reporting and Verification method
IBGE	Institute of Geography and Statistics		MS4	Municipal Separate Storm Sewer Systems
IBHS	Institute for Business and Home Safety		MSW	Municipal Solid Waste
ICA	Impact-Centric Approach		MTA	Metropolitan Transit Authority
ICLEI	Local Governments for Sustainability		MtCO$_2$e	Million Metric Tons of Carbon Dioxide Equivalent
ICT	Information and Communications Technology		MW	Megawatts
ICTA	Institute of Environmental Science and Technology			
ICUC9	International Conference on Urban Climate		NAO	North Atlantic Oscillation
IDF	Integrated Development Foundation		NAPA	National Adaptation Program of Action
IEA	International Energy Agency		NASA	National Aeronautics and Space Association
IHS	Housing and Urban Development Studies		NCCAP	National Climate Change Action Plan
IIED	International Institute for Environment and Development		NCE	National Commerce Exchange
IMF	International Monetary Fund		NDRRMC	National Disaster Risk Reduction and Management Council
IMN	National Meteorological Institute			
INDC	Intended Nationally Determined Contributions		NEPP	National Environmental Policy Plan
INDECI	Civil Defense Institute of Peru		NGOs	Non-Governmental Organizations
IOD	Indian Ocean Dipole		NIMBY	Not-in-My-Backyard
			NIST	National Institute of Standards and Technology

NMT	Non-Motorized Transport
NOAA	National Oceanographic and Atmospheric Administration
NPCC	New York City Panel on Climate Change
NPV	Net Present Value
NRC	National Research Center
NYC-EJA	New York City Environmental Justice Alliance
ODI	Open Data Institute
ODU	Old Dominion University
OECD	Organization for Economic Co-operation and Development
OEM	Office of Emergency Management
OR	Odds Ratio
PACCM	Mexico City's Climate Action Program
PAHO	Pan American Health Organizations
PAPD	Participatory Action Plan Development
PAYD	Pay-As-You-Drive
PAYG	Pay-As-You-Go
PCMs	Phase Change Materials
PDO	Pacific Decadal Oscillation
PES	Payment for Ecosystem Services
PIMA	Passive Integrated Mitigation and Adaptation
PM	Particulate Matter
PNA	Pacific North American Pattern
PNRS	Policy on Solid Waste Management
PoA	Program of Activities
PPIAF	Public-Private Infrastructure Advisory Facility
PPPs	Public–Private Partnerships
PSA	Pressure Swing Adsorption
PSE	Puget Sound Energy
PUC	Public Utility Commissions
RAY	Reconstruction Assistance on Yolanda
RB	Recycle-Banks
RCMs	Regional Climate Models
RCPs	Representation Concentration Pathways
RDM	Robust Decision-Making
REMMAQ	Metropolitan Atmospheric Network of Quito
RMI	Royal Meteorological Institute
ROAP	Regional Office for Asia and the Pacific
RRR	Relative Risk Ratios
SACN	South African Cities Network
SACZ	South Atlantic Conversion Zone
SALGA	South African Local Government Association
SBA	Sustainability Benefits Assessment
SCL	Seattle City Light
SCL	ARC3.2 Summary for City Leaders
SDG-F	Sustainable Development Goals Fund
SDGs	Sustainable Development Goals
SDI	Slum/Shack Dwellers International
SDSN	Sustainable Development Solutions Network
SELAVIP	Latin American, Asian, and African Social Housing Service
SERREE	South East Region of Renewable Energy Excellence
SES	Social-Ecological Systems
SFDRR	Sendai Framework for Disaster Risk Reduction
SI	Sustainability Institute
SIDS	Small Island Developing States
SLoCaT	Sustainable Low-Carbon Transport
SLR	Sea Level Rise
SMAC	Municipal Secretariat of Environment, Rio de Janeiro (Secretaria Municipal de Meio Ambiente)
SME	Small and Medium Enterprise

SMIAs	Significant Maritime and Industrial Areas
SNTA	Subnational Technical Assistance
SPNI	Society for the Protection of Nature in Israel
SRA	Sandy Regional Assembly
SSTs	Sea Surface Temperatures
SUDP	Strategic Urban Development Plan
SuDS	Sustainable Urban Drainage Systems
SVI	Social Vulnerability Index
SWM	Solid Waste Management
TMG	Tokyo Metropolitan Government
TOD	Transport/Transit-Oriented Development
TPH	Toronto Public Health
TSB	Technology Strategy Board
TSD	Transit-Synergized Development
UASBs	Upflow Anaerobic Sludge Blankets
UCCRF	Urban Climate Change Resilience Framework
UCCRN	Urban Climate Change Research Network
UCGS	Urban Climate Change Governance Survey
UCLG	United Cities and Local Governments
UHI	Urban Heat Island
UKCIP	United Kingdom Climate Impacts Program
UMIS	Urban Metabolism Information Systems
UN	United Nations
UNDESA	United Nations Department for Economic and Social Affairs
UNDP	United Nations Development Program
UNEP	United Nations Environment Program
UNESCAP	United Nations Economic and Social Commission for Asia and the Pacific
UNESCO	United Nations Educational, Scientific, and Cultural Organization
UNFCCC	United Nations Framework Convention on Climate Change
UNFPA	United Nations Population Fund
UN Habitat	United Nations Human Settlement Program
UNICEF	United Nations Children's Emergency Fund
UNISDR	United Nations International Strategy for Disaster Reduction
UNPF	United Nations Population Fund
UPAF	Urban and Peri-urban Agriculture and Forestry
UPROSE	United Puerto Ricans' Organization of Sunset Park
URBAHT	Urban Heat Tool
URMRCC	Urban Risk Management Response to Climate Change
USAID	United States Agency for International Development
US DOE	United States Department of Energy
UWS	Urban Water Systems
VER	Voluntary Emission Reductions
VIMS	Virginia Institute of Marine Science
VKT	Vehicle Kilometers Travelled
VMT	Vehicle Miles Traveled
VOCs	Volatile Organic Compounds
WASH	Water, Sanitation, and Hygiene
WAVES	Waterfront Vision and Enhancement Strategy
WB/CA	World Bank/Cities Alliance
WEC	World Energy Council
WEDO	Women's Environment and Development Organization
WHO	World Health Organization
WMNY	Waste Management of New York
WSUD	Water-Sensitive Urban Design
WTE	Waste-to-Energy
WWF	World Wildlife Fund/World Wide Fund for Nature
WWTP	Wastewater Treatment Plants

Appendix C

UCCRN Steering Group, ARC3.2 Authors, and Reviewers

Steering Group

ALVERSON, Keith
Nairobi, Kenya
Division on Environmental Policy Implementation,
United Nations Environment Programme (UNEP)

BARATA, Martha M. L.
Rio de Janeiro, Brazil
Institute Oswaldo Cruz (IOC/Fiocruz)

BIGIO, Anthony G.
Washington, D.C., USA
GW Solar Institute, George Washington University

CONNELL, Richenda
Oxford, UK
Acclimatise

DAWSON, Richard
Newcastle upon Tyne, UK
Tyndall Centre, University of Newcastle

DENIG, Stefan
London, UK
Siemens AG

DHAKAL, Shobhakar
Bangkok, Thailand/Kathmandu, Nepal
Department of Energy, Environment & Climate Change,
Asian Institute of Technology

GRIGGS, David
Melbourne, Australia
Monash Sustainability Institute (MSI)

GRIMM, Alice
Curitiba, Brazil
Department of Physics, Federal University of Paraná

HUQ, Saleemul
Dhaka, Bangladesh
International Institute for Environment and Development (IIED)

LEHMANN, Martin
Aalborg, Denmark
Department of Development and Planning, Aalborg University

LIZHONG, Yu
Shanghai, China
New York University, Shanghai

MOLIN VALDÉS, Helena
Paris, France
Climate and Clean Air Coalition to Reduce Short-Lived Climate
Pollutants (CCAC)

NATENZON, Claudia E.
Buenos Aires, Argentina
University of Buenos Aires/FLACSO Argentina

NEILSON, Catherine
Canberra, Australia
Australian Institute of Landscape Architects (AILA)

OMOJOLA, Ademola
Lagos, Nigeria
University of Lagos, Akoka

PACHAURI, R. K.
Delhi, India
The Energy and Resources Institute

REDWOOD, Mark
Ottawa, Canada
Cowater International

ROBERTS, Debra
Durban, South Africa
eThekwini Municipality and University of KwaZulu-Natal

ROMERO-LANKAO, Patricia
Boulder, CO, USA/Mexico City, Mexico
National Center for Atmospheric Research

ROY, Joyashree
Kolkata, India
Department of Economics, Jadavpur University

SANCHEZ-RODRIGUEZ, Roberto
Tijuana, Mexico
Department of Urban and Environmental Studies,
El Colegio de la Frontera Norte

SCHERAGA, Joel
Washington, D.C., USA
Climate Adaptation, U.S. Environmental Protection Agency

TOWERS, Joel
New York, NY, USA
Parsons New School of Design

TUTS, Rafael
Nairobi, Kenya
Urban Planning and Design Branch, United Nations Human Settlements
Programme (UN-Habitat)

WILK, David
Washington, D.C., USA
Climate Change and Sustainability Division (INE/CCS),
Inter-American Development Bank

ZAMBRANO, Carolina
Quito, Ecuador
Avina Foundation

Coordinating Lead Authors

KHAN, M. Shah Alam
Dhaka, Bangladesh
Bangladesh University of Engineering and Technology

BADER, Daniel A.
New York, NY, USA
Center for Climate Systems Research, Columbia University

BARATA, Martha M. L.
Rio de Janeiro, Brazil
Institute Oswaldo Cruz (IOC/Fiocruz)

BLAKE, Reginald
New York, NY, USA/Kingston, Jamaica
New York City College of Technology/CUNY/NOAA-CREST

BURCH, Sarah
Waterloo, Canada
Canada Research Chair in Sustainability Governance and Innovation, University of Waterloo

DAWSON, Richard
Newcastle upon Tyne, UK
Tyndall Centre, University of Newcastle

DEAR, Keith
Kunshan, China
Duke Kunshan University

DELGADO, Martha
Mexico City, Mexico
Fundación Pensar, Planeta, Política, Persona/Secretariat of the Global Cities Covenant on Climate

FOLORUNSHO, Regina
Lagos, Nigeria
Nigerian Institute for Oceanography and Marine Research

GENCER, Ebru
New York, NY, USA/Istanbul, Turkey
Center for Urban Disaster Risk Reduction and Resilience (CUDRR+R)

GORNITZ, Vivien
New York, NY, USA
Center for Climate Systems Research, Columbia University

GRAFAKOS, Stelios
Rotterdam, Netherlands/Athens, Greece
Institute for Housing and Urban Development Studies (IHS), Erasmus University Rotterdam

GRIMM, Alice
Curitiba, Brazil
Department of Physics, Federal University of Paraná

HUGHES, Sara
Toronto, Canada
Department of Political Science, University of Toronto

JEAN-BAPTISTE, Nathalie
Leipzig, Germany/Dar es Salaam, Tanzania
Helmholtz Centre for Environmental Research – UFZ/Ardhi University

KARKI, Madhav
Kathmandu, Nepal
Centre for Green Economy Development, Nepal (CGED-Nepal)/IPBES/UN, Integrated Development Society, Nepal (IDS-Nepal)

KINNEY, Patrick L.
New York, NY, USA
Mailman School of Public Health, Columbia University

LINKIN, Megan
New York, NY, USA
Swiss Reinsurance America Corporation

LWASA, Shuaib
Kampala, Uganda
Department of Geography, Makerere University

MARCOTULLIO, Peter J.
New York, NY, USA
CUNY Institute for Sustainable Cities, Hunter College

MARKANDYA, Anil
Bilbao, Spain/Bath, UK
Basque Centre for Climate Change

McPHEARSON, Timon
New York, NY, USA
Tishman Environment and Design Center, The New School

MEHROTRA, Shagun
New York, NY, USA/Indore, India
Sustainable Development Solutions Center, The New School

MEYER, Peter B.
New Hope, PA, USA
The Energy and Environment Project, The E. P. Systems Group, Inc.

OTENG-ABABIO, Martin
Accra, Ghana
University of Ghana, Legon

PACTEAU, Chantal
Paris, France
French National Research Center (CNRS), Paris Research Consortium Climate-Environment-Society

RAVEN, Jeffrey
New York, NY, USA
Graduate Program in Urban and Regional Design, New York Institute of Technology/Raven A+U

RECKIEN, Diana
Enschede, Netherlands
Faculty of Geo-Information Science and Earth Observation, University of Twente

REDWOOD, Mark
Ottawa, Canada
Cowater International

ROMERO-LANKAO, Patricia
Boulder, CO, USA/Mexico City, Mexico
National Center for Atmospheric Research

SARZYNSKI, Andrea
Newark, DE, USA
School of Public Policy & Administration, University of Delaware

SCHWARZE, Reimund
Leipzig, Germany
Helmholtz Centre for Environmental Research – UFZ

SPERLING, Joshua
Denver, CO, USA
National Renewable Energy Laboratory

VICUÑA, Sebastian
Santiago, Chile
School of Engineering, Center for Global Change, Pontifical Catholic University of Chile

ZUSMAN, Eric
Hayama, Japan
Institute for Global Environmental Strategies (IGES)

Lead Authors

ABBADIE, Luc
Paris, France
University of Pierre and Marie Curie – Institute of Ecology and
Environmental Sciences - Paris

ALI IBRAHIM, Somayya
New York, NY, USA/Peshawar, Pakistan
Center for Climate Systems Research, Columbia University

ANNEPU, Ranjith
New York, NY, USA
Be Waster Wise/Earth Engineering Center, Columbia University

ATKINSON, Larry
Norfolk, VA, USA
Center for Coastal Physical Oceanography, Old Dominion University

AUTY, Kate
Canberra, Australia
Commissioner for Sustainability and Environment, Australian Capital
Territory

AYLETT, Alex
Montreal, Canada
Institut National de la Recherche Scientifique, Centre Urbanisation
Culture Société

BAJPAI, Jitendra N.
New York, NY, USA/Mumbai, India
Earth Institute, Columbia University

BALK, Deborah
New York, NY, USA
Baruch College, CUNY Institute for Demographic Research

BOURTSALAS, A.C. (Thanos)
New York, NY, USA
Earth Engineering Center, Columbia University/Earth and
Environmental Engineering, Columbia University

CASTRO, Ricardo
Buenos Aires, Argentina
Institute of Geography, University of Buenos Aires – CONICET

CHAROENKIT, Sasima
Phitsanulok, Thailand
School of Environment, Resources and Development, Asian Institute of
Technology

CHAVEZ, Abel
Gunnison, CO, USA
Department of Environment and Sustainability, Western State Colorado
University

CLARK, Christopher M.
Washington D.C., USA
Office of Research and Development, U.S. Environmental Protection
Agency

CREUTZIG, Felix
Berlin, Germany
Department of the Economics of Climate Change, Technical University
of Berlin/Mercator Research Institute on Global Commons and Climate
Change (MCC)

DETTINGER, Michael
San Diego, CA, USA
U.S. Geological Survey and Scripps Institution of Oceanography

DHAKAL, Shobhakar
Bangkok, Thailand/Kathmandu, Nepal
Department of Energy, Environment & Climate Change, Asian Institute
of Technology

DICKINSON, Thea
Toronto, Canada
Department of Physical and Environmental Sciences, University of
Toronto

DRISCOLL, Patrick A.
Trondheim, Norway
Norwegian University of Science and Technology

EBI, Kristie L.
Seattle, WA, USA
ClimAdapt, LLC

ESQUIVEL, Maricarmen
Washington, D.C., USA/San José, Costa Rica
Inter-American Development Bank

ESTIRI, Hossein
Seattle, WA, USA
Institute of Translational Health Sciences (ITHS), University of
Washington

GABORIT, Pascaline
Brussels, Belgium
European New Towns & Pilot Cities Platform

GEORGESCU, Matei
Tempe, AZ, USA
School of Geographical Sciences and Urban Planning, Arizona State
University

HAMDI, Rafiq
Brussels, Belgium/Oujda, Morocco
Royal Meteorological Institute of Belgium

HARIRI, Maryam
New York, NY, USA
Environmental Studies Program, New York University

HERZOG, Cecilia
Rio de Janeiro, Brazil
Inverde Institute

HESS, Jeremy
Atlanta, GA, USA
Department of Emergency Medicine, Emory University School of
Medicine

HORTON, Radley
New York, NY, USA
Center for Climate Systems Research, Columbia University

INTHARATHIRAT, Rotchana
Udon Thani, Thailand
School of Environment, Resources and Development, Asian Institute of
Technology

JACK, Darby
New York, NY, USA
Mailman School of Public Health, Columbia University

JACOB, Klaus
New York, NY, USA
Lamont-Doherty Earth Observatory, Columbia University

KATZSCHNER, Lutz
Kassel, Germany
Department of Environmental Meteorology, University of Kassel

KEDIA, Shailly
New Delhi, India
The Energy and Resources Institute (TERI)

KHAN, Iqbal Alam
Toronto, Canada/Dhaka, Bangladesh
Socio-Economic Research and Development Initiative, Bangladesh

KIM, Yeonjoo
Seoul, South Korea
Department of Civil and Environmental Engineering, Yonsei University

KOMBE, Wilbard
Dar es Salaam, Tanzania
Institute of Human Settlements Studies (IHSS), Ardhi University

KRELLENBERG, Kerstin
Leipzig, Germany
Helmholtz Centre for Environmental Research – UFZ

KREMER, Peleg
New York, NY, USA
Tishman Environment and Design Center, The New School

LANDAUER, Mia
Laxenburg, Austria/Espoo, Finland
Risk, and Resilience Program & Arctic Futures Initiative, International Institute for Applied Systems Analysis (IIASA)/Department of Built Environment, Aalto University & Nordic Centre of Excellence for Strategic Adaptation Research (NORD-STAR)

LEMOS, Maria Fernanda
Rio de Janeiro, Brazil
Department of Architecture and Urbanism, Pontifical Catholic University of Rio de Janeiro

LEONE, Mattia Federico
Naples, Italy
Department of Architecture (DiARC), PLINIVS Study Centre (LUPT),
University of Napoli Federico II

LIGETI, Eva
Toronto, Canada
Department of Physical & Environmental Sciences,
University of Toronto - Scarborough

LUCON, Oswaldo
São Paulo, Brazil
São Paulo State Environment Secretariat/University of São Paulo

MALEKI, David
Washington, D.C., USA
Inter-American Development Bank

MANI, Nivedita
Gorakhpur, India
Gorakhpur Environmental Action Group (GEAG)

McEVOY, Darryn
Melbourne, Australia
Department of Civil, Environmental and Chemical Engineering, RMIT

MEHROTRA, Shagun
New York, NY, USA/Indore, India
Sustainable Development Solutions Center, The New School

MILLS, Gerald
Dublin, Ireland
School of Geography, Planning and Environmental Policy, University College Dublin

MONTGOMERY, Mark
New York, NY, USA
Stony Brook University and Population Council, NY

NAKANO, Ryoko
Hayama, Japan
Institute for Global Environmental Strategies

NATENZON, Claudia E.
Buenos Aires, Argentina
University of Buenos Aires/FLACSO Argentina

NOYOLA, Adalberto
Mexico City, Mexico
National Autonomous University of Mexico (UNAM)

OBERMAIER, Martin
Rio de Janeiro, Brazil
Federal University of Rio de Janeiro – UFRJ

OLIVOTTO, Veronica
Rotterdam, Netherlands
Institute for Housing and Urban Development Studies (IHS)

OSORIO, Juan Camilo
Cambridge, MA, USA/New York City, NY, USA
Massachusetts Institute of Technology

PALMER, Matthew I.
New York, NY, USA
Columbia University

PANJWANI, Dilnoor
New York, NY, USA
United Nations Development Program (UNDP)

PATHAK, Minal
Ahmedabad, India
CEPT University

PERINI, Katia
Genoa, Italy
Urban Design Lab (UDL)

PORIO, Emma
Manila, Philippines
Department of Sociology and Anthropology, Ateneo de Manila University

PULLEN, Julie
Hoboken, NJ, USA
Stevens Institute of Technology

QUINN, Ashlinn
New York, NY, USA
Mailman School of Public Health, Columbia University

REPLOGLE, Michael
New York, NY, USA
Deputy Transportation Commissioner for Policy, New York City

ROMÁN DE LARA, María Victoria
Bilbao, Spain
Basque Centre for Climate Change

ROMERO-LANKAO, Patricia
Boulder, CO, USA/Mexico City, Mexico
National Center for Atmospheric Research

ROSENZWEIG, Cynthia
New York, NY, USA
NASA Goddard Institute for Space Studies/Columbia University

SANTIAGO FINK, Helen
Vienna, Austria
University of Natural Resources and Life Sciences/Urban Climate Research Network

SATTERTHWAITE, David
London, UK
IIED/University College London

SCHENSUL, Daniel
New York, NY, USA
United Nations Population Fund (UNFPA)/Brown University, Providence

SILVA SOUSA, Denise
Rio de Janeiro, Brazil
COPPE/Federal University of Rio de Janeiro – UFRJ

SIMON, David
Gothenburg, Sweden/Cape Town, South Africa/London, UK
Royal Holloway University of London

SOLECKI, William
New Brunswick, NJ, USA/New York, NY, USA
Hunter College-CUNY Institute for Sustainable Cities

STONE, Brian
Atlanta, GA, USA
School of City and Regional Planning, Georgia Institute of Technology

SUDO, Tomonori
Tokyo, Japan
Ritsumeikan Asia Pacific University/Japan International Cooperation Agency (JICA)

SURMINSKI, Swenja
London, UK
Grantham Research Institute on Climate Change and the Environment (GRI), London School of Economics and Political Science

TOWERS, Joel
New York, NY, USA
Parsons New School of Design

TSUNEKI, Hori
Washington, D.C., USA
Disaster Risk Management Specialist, IDB

WAJIH, Shiraz
Gorakhpur, India
Gorakhpur Environmental Action Group (GEAG)

WANG, Xiaoming
Clayton, Australia
Commonwealth Scientific and Industrial Research Organization (CSIRO) Land and Water Flagship

YULO-LOYZAGA, Antonia
Manila, Philippines
Manila Observatory

ZIERVOGEL, Gina
Cape Town, South Africa
Department of Environmental and Geographical Science, University of Cape Town

ZIMMERMAN, Rae
New York, NY, USA
Wagner Graduate School of Public Service, New York University

Contributing Authors

ALVERSON, Keith
Nairobi, Kenya
Division on Environmental Policy Implementation, United Nations Environment Programme (UNEP)

ANDERSON, Nancy
New York, NY, USA
Sallan Foundation

BAUTISTA, Eddie
New York, NY, USA
[New York City Environmental Justice Alliance (NYC-EJA)

BROWN, Donald
London, UK
Talisman Sinopec Energy UK Limited

CARVALHO, Mariana
Rio de Janeiro, Brazil
Institute of Psychology, Federal University of Rio de Janeiro – UFRJ

CHEN, Kai
Nanjing, China
School of the Environment, Nanjing University

CLARK, Christopher M.
Washington D.C., USA
Office of Research and Development, U.S. Environmental Protection Agency

CRANE, Stuart
Nairobi, Kenya/London, UK
Climate Change Adaptation Unit, UNEP

DE SHERBININ, Alex
New York, NY, USA
Center for International Earth Science Information Network (CIESIN), Columbia University

DERECZYNSKI, Claudine
Rio de Janeiro, Brazil
Department of Meteorology, Federal University of Rio de Janeiro – UFRJ

DOBARDZIC, Saliha
Washington, D.C., USA
Global Environmental Facility

DOUST, Ken
Lismore, Australia
Southern Cross University

DUBBELING, Marielle
Leusden, Netherlands
RUAF Foundation

FEDIRKO, Lina
New York, NY, USA/San Francisco, CA, USA
Sustainable Development Solutions Center

FERGUSON, Daniel B.
Tucson, AZ, USA
Climate Assessment of the Southwest (CLIMAS), University of Arizona

FERINGA, Wim
Enschede, Netherlands
Faculty of Geo-Information Science and Earth Observation (ITC) of the University of Twente

FERNANDEZ, Blanca
Milan, Italy
The Mercator Research Institute on Global Commons and Climate Change (MCC)

GAFFIN, Stuart
New York, NY, USA
Center for Climate Systems Research, Columbia University

GARNICK, Jonah
New York, NY, USA
City University of New York

GRAFAKOS, Stelios
Rotterdam, Netherlands/Athens, Greece
Institute for Housing and Urban Development Studies (IHS), Erasmus
University Rotterdam

GUERECA, Leonor P.
Mexico City, Mexico
Engineering Institute, National Autonomous University of Mexico
(UNAM)

JAMWAL, Priyanka
Bangalore, India
Ashoka Trust for Research in Ecology and the Environment

KENNARD, Nicole
Atlanta, GA, USA
College of Sciences, Georgia Institute of Technology

KENNEDY, Christopher
Victoria, Canada
Department of Civil Engineering, University of Victoria

LALL, Upmanu
New York, NY, USA
Director, Columbia Water Center

LEE, James
Shanghai, China/Boston, MA, USA
iContinuum Group

LEJAVA, Jeffrey
White Plains, NY, USA
Land Use Law Center, Pace Law School

LEONE, Mattia Federico
Naples, Italy
Department of Architecture (DiARC), PLINIVS Study Centre (LUPT),
University of Napoli Federico II

LIN, Brenda
Aspendale, Australia
Commonwealth Scientific and Industrial Research Organisation (CSIRO)

LINKY, Edward J.
New York, NY, USA
U.S. Environmental Protection Agency (EPA)

LUCON, Oswaldo
São Paulo, Brazil
São Paulo State Environment Secretariat/University of São Paulo

LULHAM, Nicole
Ottawa, Canada
International Development Research Center (IDRC)

MARINHO, Diana P.
Rio de Janiero, Brazil
Fiocruz

MARINHO, Livia
Rio de Janeiro, Brazil
Institute Oswaldo Cruz (IOC/Fiocruz)

McCORMICK, Sabrina
Washington, D.C., USA
Milken Institute School of Public Health, The George Washington
University

NAHON, Claude
Paris, France
EDF France

NAIR, Abhishek
Enschede, Netherlands
Faculty of Geo-Information Science and Earth Observation (ITC),
University of Twente

NATTY, Mussa
Dar es Salaam, Tanzania
Tanzania Electrical, Mechanical, and Electronics Services Agency
(TEMESA)

O'DONOGHUE, Sean
Durban, South Africa
eThekwini Municipality (Durban), University of KwaZulu-Natal

OLAZABAL, Marta
Bilbao, Spain
Basque Centre for Climate Change

OMOJOLA, Ademola
Lagos, Nigeria
University of Lagos, Akoka

OSORIO, Juan Camilo
Cambridge, MA, USA/New York City, NY, USA
Massachusetts Institute of Technology

PANDA, Abhilash
Geneva, Switzerland
UN Office for Disaster Risk Reduction (UNISDR)

PORIO, Emma
Manila, Philippines
Department of Sociology and Anthropology, Ateneo de Manila
University

RASCHID, Liqa
Colombo, Sri Lanka
International Water Management Institute (IWMI–CGIAR)

ROBERTS, Debra
Durban, South Africa
eThekwini Municipality and University of KwaZulu-Natal

RUDD, Andrew
Nairobi, Kenya/New York, NY, USA
Urban Environment and Planning Branch, UN-Habitat

SCHWARZE, Reimund
Leipzig, Germany
Helmholtz Centre for Environmental Research – UFZ

SHARIFI, Ayyoob
Tsukuba, Japan/Paveh, Iran
National Institute for Environmental Studies (NIES), Japan

SVERDLIK, Alice
London, UK
International Institute for Environment and Development (IIED)

TOVAR-RESTREPO, Marcela
New York, NY, USA
Columbia University/Women's Environment and Development
Organization (WEDO)

USHER, Lindsay
Norfolk, VA, USA
Darden College of Education, Old Dominion University

VISCONTI, Cristina
Naples, Italy
Department of Architecture (DiARC), University of Napoli Federico II

VOMMARO, Felipe
Rio de Janeiro, Brazil
Fundação Oswaldo Cruz (Fiocruz)

WEJS, Anja
Aalborg, Denmark
Danish Centre for Environmental Assessment, Aalborg University

WILK, David
Washington, D.C., USA
Climate Change and Sustainability Division (INE/CCS), Inter-American Development Bank

WOUNDY, Matthew
Boston, MA, USA
Sustainable Development Solutions Center/Columbia University

YOON, Susan
New York, NY, USA
Sustainable Development Solutions Center

ZAMBRANO, Carolina
Quito, Ecuador
Avina Foundation

Case Study Authors

AHERN, Jack
Amherst, MA, USA
Department of Landscape Architecture and Regional Planning, UMass Amherst

AL-RASYID, H. H.
Tangerang Selatan, Indonesia
Institute for Economic and Social Development (BEST) Tangerang Selatan

KHAN, M. Shah Alam
Dhaka, Bangladesh
Institute of Water and Flood Management, Bangladesh University of Engineering and Technology

ALI IBRAHIM, Somayya
New York, NY, USA/Peshawar, Pakistan
Center for Climate Systems Research, Columbia University

ALLAN, Chris
Boulder, CO, USA
Pitcher Allan Associates

ALLEN, Charlotte
Los Angeles, CA, USA
Independent Consultant

ANDERSON, Pippin M. L.
Cape Town, South Africa
Department of Environmental and Geographical Science, University of Cape Town

ANGUELOVSKI, Isabelle
Barcelona, Spain
Institute for Environmental Science and Technology
Autonomous University of Barcelona

ASSAD, Eduardo Delgado
São Paulo, Brazil
University of Campinas

ATKINSON, Larry
Norfolk, VA, USA
Center for Coastal Physical Oceanography, Old Dominion University

BAGUIAN, Hamidou
Bobo-Dioulasso, Burkina Faso
Municipality of Bobo-Dioulasso

BARATA, Martha M. L.
Rio de Janeiro, Brazil
Institute Oswaldo Cruz (IOC/Fiocruz)

BARBIERI, L.
Rome, Italy
Roma Tre University

BAUTISTA, Eddie
New York, NY, USA
New York City Environmental Justice Alliance (NYC-EJA)

BENELLI, F.
Rome, Italy
Research Group on Urban Climate Change Adaptation, Department of Architecture, Roma Tre University

BESCANSA, M.
Sydney, Australia
Faculty of Built Environment, University of New South Wales

BORFECCHIA, F.
Rome, Italy
UTMEA, ENEA

BOURTSALAS, A.C. (Thanos)
New York, NY, USA
Earth Engineering Centre, Columbia University/Earth and Environmental Engineering, Columbia University

BRAUMAN, Robert
New York, NY, USA
NYC Department of Environmental Protection

BREDARIOL, Tomas
Rio de Janeiro, Brazil
EDS-PPED, Federal University of Rio de Janeiro – UFRJ

BROOKE, Roy
Ottawa, Canada
Sustainable Prosperity, Canada

CAIAFFA, E.
Rome, Italy
UTMEA, ENEA

CAIRNS, Stephanie
Ottawa, Canada
Sustainable Prosperity, Canada

CAMERATA, F.
Rome, Italy
Roma Tre University

CAMPONESCHI, Chiara
Guelph, Canada
University of Guelph

CARLONI, Flavia
Rio de Janeiro, Brazil
Federal University of Rio de Janeiro – UFRJ

CARRION, Daniel
New York, NY, USA
Mailman School of Public Health, Columbia University

CASTÁN BROTO, Vanesa
London, UK
Bartlett Development Planning Unit, University College London

CAVAN, Gina
Manchester, UK
Manchester Metropolitan University, UK

CHAN, Lena
Singapore, Singapore
National Biodiversity Centre, National Parks Board, Singapore

CHEN, Dong
Canberra, Australia
Commonwealth Scientific and Industrial Research Organisation (CSIRO)

CLARK, Christopher M.
Washington, D.C., USA
U.S. Environmental Protection Agency, Office of Research and Development

COOK, Steven
Canberra, Australia
CSIRO Land & Water Flagship, Australia

CORMIER, Nate
Seattle, WA, USA
SvR Design Company

CUSTERS, L.
Antwerp, Belgium
City of Antwerp

DAAMEN, Tom A.
Delft, Netherlands
TU Delft

DAVISON, Geoffrey W. H.
Singapore, Singapore
National Biodiversity Centre, National Parks Board, Singapore

DE CECCO, L.
Rome, Italy
UTMEA, ENEA

DE GREY, Spencer
London, UK
Foster + Partners

DE LA TORRE, Dennis G.
Manila, Philippines
Climate Change Commission, Philippines

DE MACEDO, Laura Valente
São Paulo, Brazil
Environment and Energy Institute, University of São Paulo

DE MEL, Manishka
New York, NY, USA
Center for Climate Systems Research, Columbia University

DE RIDDER, K.
Antwerp, Belgium
Flemish Institute for Technological Research (VITO)

DE SHERBININ, Alex
New York, NY, USA
Center for International Earth Science Information Network (CIESIN), Columbia University

DEAR, Keith
Kunshan, China
Duke Kunshan University

DEBUCQUOY, Wim
Mechelen, Belgium/The Hague, Netherlands
Witteveen+Bos

DELGADO, Martha
Mexico City, Mexico
Pensar Foundation, Secretariat of the Global Cities Covenant on Climate

DEWI, O. C.
Tangerang Selatan, Indonesia/Institute for Economic and Social Development (BEST) Tangerang Selatan

DICKINSON, Thea
Toronto, Canada
Department of Physical and Environmental Sciences, University of Toronto

DOCZI, Julian
London, UK
Water Policy, Overseas Development Institute

DODMAN, David
London, UK
International Institute for Environment and Development (IIED)

DOS SANTOS GASPAR, Rita Isabel
Aalborg, Denmark
Department of Development and Planning, Aalborg University

DREYFUS, Magali
Paris, France
National Center for Scientific Research (CNRS)

DUBEUX, Carolina Burle S.
Rio de Janeiro, Brazil
Federal University of Rio de Janeiro – UFRJ

DUREN, Riley
Pasadena, CA, USA
NASA Jet Propulsion Laboratory, California Institute of Technology

ECHEVERRI, Leonor
Medellín, Colombia
Administrative Department of Planning, Medellín City Council

ELDRIDGE, Jillian
London, UK
Grantham Research Institute on Climate Change and the Environment (GRI), London School of Economics and Political Science

EMERY, Sébastien
Paris, France
City of Paris

EMMANUEL, Rohinton
Glasgow, UK
Centre for Energy and the Built Environment, Glasgow Caledonian University

ENRIQUEZ, Diego
Quito, Ecuador
Municipality of the Metropolitan District of Quito, Ecuador

ESTIRI, Hossein
Seattle, WA, USA
Department of Biomedical Informatics and Medical Education, University of Washington

EVENDEN, Gerard
London, UK
Foster + Partners

EZER, Tal
Norfolk, VA, USA
Center for Coastal Physical Oceanography, Old Dominion University

FATORIĆ, Sandra
Raleigh, NC, USA / Barcelona, Spain
College of Natural Resources, North Carolina State University &
Department of Geography, Autonomous University of Barcelona

FERGUSON, Daniel
Tucson, AZ, USA
Climate Assessment of the Southwest (CLIMAS), University of Arizona

FILPA, A.
Rome, Italy
Research Group on Urban Climate Change Adaptation, Department of
Architecture, Roma Tre University

FLYNN, Michael
New York, NY, USA
M3PROJECT

FRANÇOISE, Yann
Paris, France
City of Paris

FRANTZESKAKI, Niki
Rotterdam, Netherlands
DRIFT, Erasmus University Rotterdam

GALLOU, Irene
London, UK
Foster + Partners

GANTOIS, Marie
Paris, France
City of Paris

GEBHARDT, Oliver
Leipzig, Germany
Helmholtz Centre for Environmental Research - UFZ

GERRARD, Michael
New York, NY, USA
Columbia Law School

GIANOLI, Alberto
Rotterdam, Netherlands
Institute for Housing and Urban Development Studies (IHS), Erasmus
University Rotterdam

GIRAULT, Anne
Paris, France
Parisian Climate Agency

GOMMERS, I.
Antwerp, Belgium
City of Antwerp

GONG, Weila
Berlin, Germany
Environmental Policy Research Centre (FFU), Free University of Berlin

GONZALEZ, Richard
New York, NY, USA
Richard Gonzalez Architect, PLLC

GORNITZ, Vivien
New York, NY, USA
Center for Climate Systems Research, Columbia University

GRAFAKOS, Stelios
Rotterdam, Netherlands / Athens, Greece
Institute for Housing and Urban Development Studies (IHS), Erasmus
University Rotterdam

GREEN, Vivien
Rio de Janeiro, Brazil
EDS-PPED, Federal University of Rio de Janeiro – UFRJ

GRIMM, Alice
Curitiba, Brazil
Department of Physics, Federal University of Paraná,

GURNEY, Kevin
Tempe, AZ, USA
Global Institute of Sustainability, Arizona State University

HAANPÄÄ, Simo
Espoo, Finland
Department of Real Estate, Planning and Geoinformatics, Land Use
Planning and Urban Studies Group, Aalto University

HAMDI, Rafiq
Brussels, Belgium / Oujda, Morocco
Royal Meteorological Institute of Belgium

HAN, Jun
Canberra, Australia / Dubai, United Arab Emirates
Commonwealth Scientific and Industrial Research Organisation (CSIRO)
/ Heriot-Watt University

HAQUE, Anika Nasra
Cambridge, UK
Department of Geography, University of Cambridge

HERNANDEZ, Annel
New York, NY, USA
New York City Environmental Justice Alliance (NYC-EJA)

HETTIARACHCHI, Missaka
Brisbane, Australia /
University of Queensland, and World Wildlife Fund

HIEU TRUNG, Nguyen
Can Tho, Vietnam
College of Environment and Natural Resources, Can Tho University

HOOYBERGHS, H.
Antwerp, Belgium
Vlaamse Instelling voor Technologisch Onderzoek (VITO)

HOSSAIN, Md. Mohataz
Nottingham UK
University of Nottingham

HUANG, Cunrui
Guangdong, China
School of Public Health, Sun Yat-sen University

JAMWAL, Priyanka
Bangalore, India
Ashoka Trust for Research in Ecology and the Environment

JEAN-BAPTISTE, Nathalie
Leipzig, Germany/Dar es Salaam, Tanzania
Helmholtz Centre for Environmental Research – UFZ/Ardhi University

JUHOLA, Sirkku
Helsinki, Finland
Department of Environmental Sciences, University of Helsinki

JUSUF, S. K.
Singapore, Singapore
Centre for Sustainable Asian Cities, National University of Singapore

KATZSCHNER, Lutz
Kassel, Germany
Department of Environmental Meteorology, University of Kassel

KAZMIERCZAK, Aleksandra
Cardiff, UK
Cardiff University

KEENAN, Jesse M.
Cambridge, MA, USA
Graduate School of Design, Harvard University

KHALIL, Heba Allah Essam E.
Cairo, Egypt
Department of Architectural Engineering, Cairo University

KHAN, Fawad
Islamabad, Pakistan
Institute for Social and Environmental Transition (ISET-Pakistan)

KHOO, Yong Bing
Canberra, Australia
Commonwealth Scientific and Industrial Research Organisation (CSIRO)

KIM, Yeonjoo
Seoul, South Korea
Department of Civil and Environmental Engineering,
Yonsei University

KIRONO, Dewi G. C.
Canberra, Australia
Commonwealth Scientific and Industrial Research Organisation (CSIRO)

KLEIN, Johannes
Helsinki, Finland
Department of Environmental Sciences, University of Helsinki

KNIGHT, Cameron
Canberra, Australia
Environment and Planning Directorate, ACT Government

KOMBE, Wilbard
Dar es Salaam, Tanzania
Institute of Human Settlements Studies (IHSS), Ardhi University

KONSTANTINOV, Pavel
Moscow, Russia
Lomonosov Moscow State University

KRELLENBERG, Kerstin
Leipzig, Germany
Helmholtz Centre for Environmental Research – UFZ

KUECKER, Glen David
Greencastle, IN, USA
Department of History, DePauw University

KUKANOVA, Eugenia
Moscow, Russia
Lomonosov Moscow State University

KWON, Won-Tae
Seoul, South Korea
Yonsei University / National Institute of Meteorological Sciences

LAMBRECHTS, G.
Antwerp, Belgium
City of Antwerp

LA PORTA, L.
Rome, Italy
UTMEA, ENEA

LA ROVERE, Emilio Lèbre
Rio de Janeiro, Brazil
EDS – PPED, Federal University of Rio de Janeiro – UFRJ

LAUWAET, D.
Antwerp, Belgium
Vlaamse Instelling voor Technologisch Onderzoek (VITO)

LECK, Hayley
London, UK
Grantham Research Institute on Climate Change and the Environment
(GRI), London School of Economics and Political Science

LEHMANN, Martin
Aalborg, Denmark
Department of Development and Planning, Aalborg University

LEONE, Mattia Federico
Naples. Italy
Department of Architecture (DiARC), PLINIVS Study Centre
(LUPT), University of Napoli Federico II

LUCON, Oswaldo
São Paulo, Brazil
São Paulo State Environment Secretariat/University of São Paulo

LUQUE-AYALA, Andrés
Durham, UK
Department of Geography, Durham University

MACCHI, Silvia
Rome, Italy
Department of Civil, Building, and Environmental Engineering, Sapienza
University of Rome

MacCLUNE, Karen
Boulder, CO, USA
Institute for Social and Environmental Transition (ISET-International)

MAIHEU, B.
Antwerp, Belgium
Flemish Institute for Technological Research (VITO),

MAJOR, David C.
New York, NY, USA
Center for Climate Systems Research, Columbia University

MALIK, Sharmeen
Islamabad, Pakistan
Institute for Social and Environmental Transition (ISET-Pakistan)

MANI, Nivedita
Gorakhpur, India
Gorakhpur Environmental Action Group (GEAG)

MARANI, Marco
Durham, NC, USA / Padova, Italy
Nicholas School of the Environment and Pratt School of Engineering,
Duke University/ Department of Civil, Environmental, and Architectural
Engineering, University of Padova

MARINHO, Diana P.
Rio de Janeiro, Brazil
ENSP/Fiocruz

MARINHO, Livia
Rio de Janeiro, Brazil
Institute Oswaldo Cruz (IOC/Fiocruz)

MARTINI, S.
Rome, Italy
UTMEA, ENEA

MARVIN, Simon
Durham, UK
Durham University

MESKEL, Elsa
Paris, France
Parisian Climate Agency

MEYER, Manuel
Jena, Germany
Department of Urban Development & City Planning, Jena City Council

MILLS-KNAP, Sarah
Washington, D.C., USA
The World Bank Group

MOGLIA, Magnus
Canberra, Australia
CSIRO Land & Water Flagship

MOHAMAD NEJAD, Ali
Tehran, Iran
Department of Geography, Shahis Beheshti University

MORÉN-ALEGRETA, Ricard
Barcelona, Spain
Department of Geography, Autonomous University of Barcelona

MULINGBAYAN, Mark
Quezon City, Philippines
Manila Water Company, Inc

NAJAFI, Marianne
Paris, France
EDF France

NATENZON, Claudia E.
Buenos Aires, Argentina
University of Buenos Aires/FLACSO Argentina

NATTY, Mussa
Dar es Salaam, Tanzania
Tanzania Electrical, Mechanical, and Electronics Services Agency (TEMESA)

NAVARATNE, Achala
Washington, D.C., USA
American Red Cross

NELSON, David
London, UK
Foster + Partners

NEUMANN, Luis
Canberra, Australia
CSIRO Land & Water Flagship,

NG, Edward
Hong Kong, China
School of Architecture, Chinese University of Hong Kong

NGUYEN, Minh
Canberra, Australia
CSIRO Land & Water Flagship

NORMAN, Barbara
Canberra, Australia
ACT Climate Change Council, Canberra/Urban and Regional Futures, University of Canberra

NUÑEZ, Andrea
San Pedro Sula, Honduras
Catholic University of Honduras

O'DONOGHUE, Sean
Durban, South Africa
eThekwini Municipality (Durban), University of KwaZulu-Natal

OBERMAIER, Martin
Rio de Janeiro, Brazil
Federal University of Rio de Janeiro – UFRJ

OMBUEN, S.
Rome, Italy
Roma Tre University

ORTIZ, Monica
Manila, Philippines
Manila Observatory

OSMOND, P.
Sydney, Australia
Faculty of Built Environment, University of New South Wales

OSORIO, Juan Camilo
Cambridge, MA, USA/New York City, NY, USA
Massachusetts Institute of Technology

OTENG-ABABIO, Martin
Accra, Ghana
University of Ghana, Legon

PACTEAU, Chantal
Paris, France
French National Research Center (CNRS), Paris Research Consortium Climate-Environment-Society

PASTOR, Manuel
New York, NY, USA
School of International and Public Affairs, Columbia University

PELLEGRINI, V.
Rome, Italy
Roma Tre University

PETRETTA, Danielle L.
New York, NY, USA
Graduate School of Architecture, Planning and Preservation, Columbia University

POLLINO, M.
Rome, Italy
UTMEA, ENEA

PORIO, Emma
Manila, Philippines
Department of Sociology and Anthropology, Ateneo de Manila University

PRIYANDANA, Dendi
Tangerang Selatan, Indonesia
Department of City Planning Building and Settlement (DTKBP), Tangerang Selatan

PULHIN, Perlyn M.
Manila, Philippines
The OML Centre

QUINN, Ashlinn
New York, NY, USA
Mailman School of Public Health, Columbia University

RAMSELL, Annie
Nairobi, Kenya
Climate Change Adaptation Unit, UNEP

RANADE, Monali
Washington, D.C., USA
The World Bank Group

REHMAN, Atta
Islamabad, Pakistan
Institute for Social and Environmental Transition (ISET-Pakistan)

REI, Fernando
Santos, Brazil
Professor at the PhD Program in International Environmental Law,
Catholic University of Santos

REN, Chao
Hong Kong, China
School of Architecture, Chinese University of Hong Kong

RICCI, Liana
Rome, Italy
Department of Civil, Building and Environmental Engineering, Sapienza
University of Rome

RICE, Jennifer
Athens, GA, USA
Department of Geography, University of Georgia

ROBERTS, Debra
Durban, South Africa
eThekwini Municipality and University of KwaZulu-Natal

ROHMADI, M. T.
Tangerang Selatan, Indonesia
Department of Cleanliness, Parks and Cemetery (DKPP), Tangerang
Selatan

RON, Dave
Oakland, CA, USA
Ecocity Builders

ROUMANI, Helene
Jerusalem
Jerusalem Bioregion Center for Ecosystem Management.

SADOFF, Natasha
Columbus, OH, USA
Battelle Memorial Institute

SALAM, R.
Tangerang Selatan, Indonesia
Regional Environmental Agency (BLHD), Tangerang Selatan

SANOU, Basilisa
Ouagadougou, Burkina Faso
UN-Habitat

SARA, Liliana Miranda
Lima, Peru
Cities for Life Foro

SCHAER, Caroline
Copenhagen, Denmark
UNEP-DTU Partnership (UDP)

SETZER, Joana
London, UK
Grantham Research Institute on Climate Change and the Environment.
Department of Geography and Environment at the London School of
Economics and Political Science

SHARMA, Somesh
Rotterdam, Netherlands
Institute for Housing and Urban Development Studies (IHS), Erasmus
University Rotterdam

SHARMA, Sonia
Durham, NC, USA
Nicholas School of the Environment and Pratt School of Engineering,
Duke University

SILVESTRI, Sonia
Durham, NC, USA
Nicholas School of the Environment and Pratt School of Engineering,
Duke University

SIMARMATA, Hendricus Andy
Jakarta, Indonesia
Faculty of Economics, University of Indonesia/Indonesian Association of
Urban and Regional Planners

SPERLING, Joshua
Denver, CO, USA
National Renewable Energy Laboratory

SULKOWSKI, Adam
Babson Park, MA, USA
Division of Accounting and Law, Babson College

SURMINSKI, Swenja
London, UK
Grantham Research Institute on Climate Change and the Environment
(GRI), London School of Economics and Political Science

TILIE, Nico
Delft, Netherlands
Faculty of Architecture, TU Delft

TORRIENTE, Susanne M.
Miami Beach, FL, USA
City of Miami Beach

TSUR, Naomi
Jerusalem
Jerusalem Green Fund

VARENTSOV, Mikhail
Moscow, Russia
Lomonosov Moscow State University/A. M. Obukhov Institute of
Atmosphere Physics Russian Academy of Sciences

VENKATESWARAN, Kanmani
Boulder, CO, USA
Institute for Social and Environmental Transition (ISET-International)

VICUÑA, Sebastian
Santiago, Chile
School of Engineering, Center for Global Change, Pontifical Catholic
University of Chile

WAJIH, Shiraz
Gorakhpur, India
Gorakhpur Environmental Action Group (GEAG)

WANG, Chi-Hsiang
Canberra, Australia
Commonwealth Scientific and Industrial Research Organisation

WANG, Xiaoming
Clayton, Australia
Commonwealth Scientific and Industrial Research Organisation (CSIRO)
Land and Water Flagship

WEBER, Stephanie
Columbus, OH, USA
Battelle Memorial Institute

WEJS, Anja
Aalborg, Denmark
Danish Centre for Environmental Assessment, Aalborg University

WONG, N. H.
Singapore, Singapore
Department of Building, National University of Singapore

WOODHOUSE, Connie
Tucson, AZ, USA
School of Geography and Development, University of Arizona

WOUNDY, Matthew
Boston, MA, USA
Sustainable Development Solutions Center/Columbia University

YOON, Susan
New York, NY, USA
Sustainable Development Solutions Center

YOUNG, Andrea
São Paulo, Brazil
Centre for Weather and Climate Research, University of Campinas

YULO-LOYZAGA, Antonia
Manila, Philippines
Manila Observatory

ZAMBRANO, Carolina
Quito, Ecuador
Avina Foundation

ZELL, Erica
Columbus, OH, USA
Battelle Memorial Institute

ZOGRAFOS, Christos
Barcelona, Spain
Institute of Environmental Science and Technology (ICTA), Autonomous University of Barcelona
Department of Environmental Studies, Masaryk University

ZUCCARO, Giulio
Naples, Italy
Department of Structures for Engineering and Architecture (DiST), PLINIVS Study Centre (LUPT) University of Napoli Federico II

Box Authors

ADAMS, Peter
New York, NY, USA
Acclimatise

ALI IBRAHIM, Somayya
New York, NY, USA/Peshawar, Pakistan
Center for Climate Systems Research, Columbia University

ARIKAN, Yunus
Bonn, Germany
ICLEI—Local Governments for Sustainability

BALK, Deborah
New York, NY, USA
Baruch College, CUNY Institute for Demographic Research

BIGIO, Anthony G.
Washington, D.C., USA
GW Solar Institute, George Washington University

BREDARIOL, Tomas
Rio de Janeiro, Brazil
EDS-PPED, Federal University of Rio de Janeiro – UFRJ

CARLONI, Flavia
Rio de Janeiro, Brazil
Federal University of Rio de Janeiro – UFRJ

CARRION, Daniel
New York, NY, USA
Mailman School of Public Health, Columbia University

CASTRO, Ricardo
Buenos Aires, Argentina
Institute of Geography, University of Buenos Aires – CONICET

CHAVEZ, Abel
Gunnison, CO, USA
Department of Environment and Sustainability, Western State Colorado University

DUBBELING, Marielle
Leusden, Netherlands
RUAF Foundation

DUBEUX, Carolina Burle S.
Rio de Janeiro, Brazil
Federal University of Rio de Janeiro - UFRJ

ESTIRI, Hossein
Seattle, WA, USA
Department of Biomedical Informatics and Medical Education, University of Washington

GRADY, Heather
San Francisco, CA, USA
Rockefeller Philanthropy Advisors

GREEN, Vivien
Rio de Janeiro, Brazil
EDS-PPED, Federal University of Rio de Janeiro – UFRJ

GREENWALT, Julie
Brussels, Belgium
Cities Alliance

GUERECA, Leonor P.
Mexico City, Mexico
Engineering Institute, National Autonomous University of Mexico (UNAM)

HOORNWEG, Daniel
Oshawa, Canada
Faculty of Energy Systems and Nuclear Science, University of Ontario Institute of Technology

HOULE, Michaël
Toronto, Canada
ICLEI—Local Governments for Sustainability Canada

KAVANAUGH, Laura
Bonn, Germany
ICLEI—Local Governments for Sustainability

KENNEDY, Christopher
Victoria, Canada
Department of Civil Engineering, University of Victoria

KWON, Won-Tae
Seoul, South Korea
Yonsei University/National Institute of Meteorological Sciences

LA ROVERE, Emilio Lèbre
Rio de Janeiro, Brazil
EDS – PPED, Federal University of Rio de Janeiro – UFRJ

MALEKI, David
Washington, D.C., USA
Inter-American Development Bank

MANI, Nivedita
Gorakhpur, India
Gorakhpur Environmental Action Group (GEAG)

MARX, Sabine
New York, NY, USA
Center for Research on Environmental Decisions, Columbia University

MAYR, Marcus
Nairobi, Kenya
Cities and Climate Change, UN-Habitat

NAJAFI, Marianne
Paris, France
EDF France

NATENZON, Claudia E.
Buenos Aires, Argentina
University of Buenos Aires/FLACSO Argentina

NOYOLA, Adalberto
Mexico City, Mexico
National Autonomous University of Mexico (UNAM)

O'DONOGHUE, Sean
Durban, South Africa
eThekwini Municipality (Durban), University of KwaZulu-Natal

PANDA, Abhilash
Geneva, Switzerland
UN Office for Disaster Risk Reduction (UNISDR)

PODEVIN, Julie
Washington, D.C.
The World Bank Group

RECKIEN, Diana
Enschede, Netherlands
Faculty of Geo-Information Science and Earth Observation, University of Twente

SARZYNSKI, Andrea
Newark, DE, USA
School of Public Policy & Administration, University of Delaware

SCHENSUL, Daniel
New York, NY, USA
United Nations Population Fund (UNFPA)/Brown University, Providence

SHAW, Rajib
Kyoto, Japan
Integrated Research on Disaster Risk Program, Kyoto University

SIMON, David
Gothenburg, Sweden/Cape Town, South Africa/London, UK
Royal Holloway University of London

SOKOLOW, Sharona
Los Angeles, CA, USA
Institute of Environment and Sustainability, UCLA

SOLECKI, William
New Brunswick, NJ, USA/New York, NY, USA
Hunter College-CUNY Institute for Sustainable Cities

SPERLING, Joshua
Denver, CO, USA
National Renewable Energy Laboratory

TIRADO, Cristina
Los Angeles, CA, USA
Institute of Environment and Sustainability, UCLA

TOVAR-RESTREPO, Marcela
New York, NY, USA
Columbia University/Women's Environment and Development Organization (WEDO)

TSUNEKI, Hori
Washington, D.C., USA
Inter-American Development Bank

VINES, Kathryn
Sydney, Australia
C40 Cities Climate Leadership Group

WAJIH, Shiraz
Gorakhpur, India
Gorakhpur Environmental Action Group (GEAG)

Case Study Docking Station Team

ALI IBRAHIM, Somayya
New York, NY, USA/Peshawar, Pakistan

BLAJ, Ioana
Brussels, Belgium/Oradea, Romania

BONTEMPI, Antonio
Barcelona, Spain/Bergamo, Italy

DEBUCQUOY, Wim
Mechelen, Belgium/The Hague, Netherlands

DE LOS RIOS WHITE, Marta
Medellín, Colombia

DRISCOLL, Patrick A.
Trondheim, Norway

GARNICK, Jonah
New York, NY, USA

HELSETH, Megan
Denver, CO, USA

HERNANDEZ, Annel
New York, NY, USA

KHALID, Usama
Islamabad, Pakistan

LEHMANN, Martin
Aalborg, Denmark

LIM, Carissa
Singapore, Singapore

MAJOR, David C.
New York, NY, USA

MILIC, Jovana
Niš, Serbia

RUIZ RINCÓN, Victoria
Barcelona, Spain/Cuernavaca, Mexico

SCHLECHT, Samuel
Heidelberg, Germany

SOLECKI, Stephen
New York, NY, USA

ZHAMO, Megi
Hamburg, Germany/Durres, Albania

Reviewers

The ARC3.2 report was reviewed by a group of climate change and cities experts at the academic, stakeholder, and institutional levels. We thank them for their time and contributions.

ARCHER, Diane
London, UK
International Institute for Environment and Development (IIED)

ARCHFIELD, Stacey
Reston, VA, USA
United States Geological Survey

ARRIGHI, Julie
Kampala, Uganda
American Red Cross in Africa

BALOI, Oana
Nairobi, Kenya
UN-Habitat

BANKS, Nicola
Manchester, UK
School of Environment, Education and Development, University of Manchester

BARONE, Richard
New York, NY, USA
Regional Plan Association

BIGIO, Anthony G.
Washington, D.C., USA
GW Solar Institute, George Washington University

BIRCH, Eugenie
Philadelphia, PA, USA
School of Design, University of Pennsylvania

BIRKMANN, Joern
Bonn, Germany
United Nations University - Institute for Environment and Human Security (UHU-EHS)

BLACK, John
Sydney, Australia
Institute of Environmental Studies, University of New South Wales

BURKETT, Virginia
Reston, VA, USA
United States Geological Survey

CAMPBELL, Mônica
Toronto, Canada
Dalla Lana School of Public Health, University of Toronto

CARBONELL, Armando
Cambridge, MA, USA
Lincoln Institute of Land Policy

CASTÁN BROTO, Vanesa
London, UK
Bartlett Development and Planning Unit, University College London

CHEESEMAN, Chris
London, UK
Department of Civil and Environmental Engineering, Imperial College London

COHEN, Stewart
Vancouver, Canada
Climate Research Division, Environment Canada

COHN, Alan
New York, NY, USA
Climate and Water Policy, New York City Department of Environmental Protection

COLE, Sheila
Halifax, Canada
Nova Scotia Environmental Network

CONFALONIERI, Ulisses
Rio de Janeiro, Brazil
International Human Dimensions Programme (IHDP)

CORVALAN, Carlos
Washington, D.C., USA
Pan American Health Organization/World Health Organization (PAHO/WHO)

COUSINS, Fiona
New York, NY, USA
ARUP

CRAWFORD-BROWN, Douglas
Cambridge, UK
Cambridge Centre for Climate Change Mitigation Research, University of Cambridge

CROWELL, Mark
Washington, D.C., USA
Federal Emergency Management Agency

DAVIDSON, Debra
Edmonton, Canada
Faculty of Agricultural, Life and Environmental Sciences, University of Alberta

DE SHERBININ, Alex
New York, NY, USA
Center for International Earth Science Information Network (CIESIN), Columbia University

DEMUZERE, Matthias
Leuven, Belgium
Department of Earth and Environmental Sciences, KU Leuven

DODMAN, David
London, UK
International Institute for Environment and Development (IIED)

DORA, Carlos
Geneva, Switzerland
World Health Organisation (WHO)

DOUST, Ken
Lismore, Australia
Southern Cross University

DUPUIS, Johann
Lausanne, Switzerland
Institute of Geography and Sustainability, University of Lausanne

DUREN, Riley
Pasadena, CA, USA
NASA Jet Propulsion Laboratory, California Institute of Technology

ELMQVIST, Thomas
Stockholm, Sweden
Stockholm Resilience Center, Stockholm University

EMMANUEL, Rohinton
Glasgow, UK
Centre for Energy and the Built Environment, Glasgow Caledonian
University

FISHER, Dana
College Park, MD, USA
Department of Sociology, University of Maryland

FLEMING, Paul
Seattle, WA, USA
Seattle Public Utilities

FLOATER, Graham
London, UK
LSE Cities, London School of Economics and Political Science

FLYNN, Michael
New York, NY, USA
M3PROJECT

FRANTZESKAKI, Niki
Rotterdam, Netherlands
DRIFT, Erasmus University Rotterdam

FULTON, Lew
Davis, CA, USA
Sustainable Transporation Energy Pathways, University of California,
Davis

FYFE, Angie
Denver, CO, USA
ICLEI—Local Governments for Sustainability USA

GARCIA SANCHEZ, Francisco
Santander, Spain
EDUC (UC Doctoral School), University of Cantabria

GIRONÁS, Jorge
Santiago, Chile
School of Engineering, Center for Global Change, CEDEUS, Pontifical
Catholic University of Chile

GODOY, Alex
Santiago, Chile
Universidad del Desarrollo

GRADY, Heather
San Francisco, CA, USA
Rockefeller Philanthropy Advisors

GRIMMOND, Sue
Reading, UK
Department of Meteorology, University of Reading

GUERREIRO, Sergio
Rio de Janeiro, Brazil
WTERT-Brazil

IGNATIEVA, Maria
Uppsala, Sweden
Department or Urban and Rural Development, The Swedish University of
Agricultural Sciences

JAGANNATHAN, Vijay
Washington, D.C., USA
World Resources Institute

JOHNSON, Craig
Guelph, Canada
Department of Political Science, University of Guelph

KADIHASANOGLU, Aynur
Washington, D.C., USA
American Red Cross

KESKITALO, Carina
Umeå, Sweden
Umeå University

KNOWLTON, Kim
New York, NY, USA
Natural Resources Defense Council

LEE, Marcus
Washington, D.C., USA
The World Bank Group

LEICHENKO, Robin
New Brunswick, NJ, USA
Department of Geography, Rutgers University

LEWIS, Dan
Nairobi, Kenya
Urban Risk Reduction Unit, UN-Habitat

LITTLE, Christopher
Princeton, NJ, USA
Princeton Environmental Institute, Princeton University

LUQUE-AYALA, Andrés
Durham, UK
Department of Geography, Durham University

MADDOX, David
New York, NY, USA
The Nature of Cities

MALEKI, David
Washington, D.C., USA
Inter-American Development Bank

MAYR, Marcus
Nairobi, Kenya
Cities and Climate Change, UN-Habitat

McEVOY, Darryn
Melbourne, Australia
Department of Civil, Environmental and Chemical Engineering, RMIT

MURPHY, Conor
Maynooth, Ireland
Department of Geography, Maynooth University

NORMAN, Barbara
Canberra, Australia
ACT Climate Change Council, Canberra/Urban and Regional Futures,
University of Canberra

OZEKIN, Kenan
Denver, CO, USA
Water Research Foundation

PAULEIT, Stephan
Munich, Germany
Chair for Strategic Landscape Planning and Management, Munich
Technical University

PIERSON, Don
New York, NY, USA
NYC Department of Environmental Protection

PRAKESH, Anjal
Hyderabad, India
South Asia Consortium on Water Resources and Development (SACI-WATERS)

REPLOGLE, Michael
New York, NY, USA
Deputy Transportation Commissioner for Policy, New York City

REVI, Aromar
Bangalore, India
Indian Institute for Human Settlements

RIND, David
New York, NY, USA
Center for Climate Systems Research, Columbia University

ROSA, Rodrigo
Rio de Janeiro, Brazil
Mayor's Office of Rio de Janeiro

RUDD, Andrew
Nairobi, Kenya/New York, NY, USA
Urban Environment and Planning Branch, UN-Habitat

SCHOLZ, Sebastian M.
Washington, D.C., USA
The World Bank Group

SCHROEDER, Heike
Norwich, UK
University of East Anglia

SNEP, Robbert
Wageningen, Netherlands
Wageningen Environmental Research, Wageningen University & Research

SOCCI, Anthony
Washington D.C., USA
U.S. Environmental Protection Agency (EPA)

STONE, Brian
Atlanta, GA, USA
School of City and Regional Planning, Georgia Institute of Technology

THEMELIS, Nickolas
New York, NY, USA
Department of Earth and Environmental Engineering, Columbia University

TITUS, James
Washington D.C., USA
U.S. Environmental Protection Agency (EPA)

TOVAR-RESTREPO, Marcela
New York, NY, USA
Columbia University/Women's Environment and Development Organization (WEDO)

VAN BREDA, Anita
Washington D.C., USA
Disaster Response and Risk Reduction, World Wildlife Fund

VAN HAM, Chantal
Brussels, Belgium
International Union for Conservation of Nature (IUCN)

VIRJI, Hassan
Washington D.C., USA
Global Change System for Analysis, Research and Training (START)

WARNER, Koko
Bonn, Germany
United Nations University - Institute for Environment and Human Security (UHU-EHS)

WHITE-NEWSOME, Jalonne
New York, NY, USA
West Harlem Environmental Action (WE ACT)

WOODRUFFE, Colin D.
Wollongong, Australia
School of Earth and Environmental Sciences, University of Wollongong

ZARRILLI, Daniel
New York, NY, USA
NYC Mayor's Office for Recovery and Resilience

ZIMMERMAN, Rae
New York, NY, USA
Wagner Graduate School of Public Service, New York University

ARC3.2 Partners and Sponsors

The Second UCCRN Assessment Report on Climate Change and Cities (ARC3.2) was supported by:

The views expressed herein do not necessarily represent those of our partners and supporters.

Index

414, 424, 425, 428, 429, 431, 432, 434,
444, 445, 448, 449, 450, 451, 452, 453,
463, 464, 465, 466, 467, 468, 470, 472,
473, 475, 477, 478, 479, 480, 481, 482,
492, 493, 495, 498, 502, 503, 507, 508,
521, 543, 545, 556, 563, 567, 569, 574,
575, 576, 577, 578, 586, 587, 588, 589,
590, 591, 592, 593, 594, 595, 597, 598,
600, 601
monitoring and evaluation, 120, 121, 127, 174,
475
monitoring networks, 28, 29, 30, 34, 35
monsoon, 41, 46, 194, 327, 370, 381, 382, 407,
412, 466, 561
mosquito vectors, 43. *See* maleria
Mozambique, 203
Maputo, 25, 188, 198, 203, 204, 205, 453
municipal
action, 199, 211
governments, 5, 19, 91, 92, 102, 121, 124,
129, 143, 199, 206, 231, 364, 587, 589,
594, 595, 597

National Aeronautics and Space Administration
(NASA), 31, 36, 37
National Oceanographic and Atmospheric
Administration (NOAA), 37, 130, 131,
322, 328, 333, 351
natural systems, 17, 106, 142, 143, 162, 260,
298, 328, 390, 445, 592
natural variability, 29, 39, 40, 42, 44, 45, 446,
524, 525
Nepal, 25, 183, 299, 416, 417, 418
Kathmandu, 3, 25, 257
Netherlands, 30, 248, 291, 299, 334, 339, 340,
348, 349, 353, 467, 565, 570, 595
Rotterdam, 20, 30, 101, 112, 129, 225, 247,
288, 296, 304, 321, 326, 334, 336, 340,
347, 348, 349, 468, 595
Nigeria, 53, 184, 185, 402, 405, 409, 506
Lagos, 5, 13, 53, 61, 178, 184, 185, 188, 198,
404, 453, 454, 468, 493, 506, 510, 519
North America, 31, 40, 42, 45, 50, 51, 273, 322,
402, 419, 432, 502, 589
North Atlantic Oscillation, 28, 39, 40, 42

NAO, 39, 40, 42, 45

ocean acidification, 328

Pacific Decadal Oscillation, 28, 39
PDO, 39
Pacific Ocean, 40, 44, 45, 53, 534
Pakistan, 183, 184, 417, 419, 420, 421, 459, 523
Faisalabad, 51, 419, 420
Multan, 419, 420, 421
Rawalpindi, 419, 420
particulate matter, 38, 378, 452, 505
Pay-As-You-Drive, 132, 510
peri-urban, 5, 19, 108, 111, 258, 259, 260, 269,
272, 274, 279, 284, 285, 288, 289, 290,

291, 292, 299, 301, 303, 304, 345, 347,
402, 462, 520, 521, 529, 530, 531, 532, 536
Peru
Lima, 53, 232, 407, 408, 453, 546
Philippines
Manila, 51, 66, 86, 173, 188, 198, 322, 405,
434, 450, 539, 540, 541
Sorsogon, 25
Tacloban, 86, 87
Poland
Warsaw, 50
population density, 78, 147, 183, 185, 371, 373,
390, 408, 448, 463, 469, 498, 502, 539
poverty, 10, 11, 17, 25, 69, 74, 82, 90, 111, 145,
165, 174, 177, 178, 183, 185, 190, 198,
199, 203, 229, 276, 278, 298, 302, 332,
341, 349, 386, 401, 402, 404, 406, 408,
413, 421, 425, 432, 461, 480, 528, 574,
586, 588, 594, 596
precipitation, 30, 33, 34, 35, 40, 41, 42, 43, 44,
45, 46, 48, 49, 50, 51, 54, 67, 68, 72, 117,
124, 159, 162, 166, 175, 178, 183, 184,
194, 241, 261, 263, 266, 268, 275, 279,
282, 284, 285, 295, 323, 326, 329, 350,
351, 373, 375, 377, 378, 381, 384, 389,
412, 459, 460, 470, 481, 482, 494, 495,
497, 501, 520, 521, 522, 523, 524, 525,
526, 534, 536, 537, 545
extreme, 34, 41, 42, 46, 47, 64, 83, 183, 188,
203, 279, 377, 494, 525
patterns, 33, 45, 275, 329, 377, 459, 481,
495, 545
rainstorm, 41
public-private partnerships, 62, 81, 82, 90, 94,
121, 245, 287, 303, 434, 471, 508, 591

radiative forcing, 49, 108, 284, 292
Regional Hubs, 14, *See* Urban Climate Change
Research Network
relocation, 8, 33, 76, 78, 81, 115, 195, 320, 333,
340, 415, 417, 418, 423, 432, 477, 501
remote sensing, 29, 31, 34, 53, 167, 289
Landsat, 31
MODIS, 31
resilience, 4, 5, 6, 7, 8, 9, 10, 11, 12, 17, 18, 19,
20, 21, 23, 25, 29, 35, 42, 62, 63, 68, 69,
74, 75, 76, 77, 80, 81, 82, 83, 84, 85, 86,
88, 89, 90, 91, 92, 94, 102, 105, 109, 120,
121, 122, 123, 125, 126, 129, 140, 141,
142, 143, 146, 148, 155, 161, 167, 174,
180, 186, 192, 194, 196, 198, 199, 200,
201, 202, 205, 208, 210, 235, 237, 238,
239, 240, 244, 245, 247, 248, 258, 259,
260, 261, 262, 268, 269, 272, 275, 284,
285, 288, 289, 292, 294, 295, 296, 298,
299, 301, 302, 303, 304, 326, 332, 335,
336, 342, 348, 349, 364, 371, 382, 383,
387, 388, 400, 403, 409, 421, 434, 444,
445, 447, 448, 459, 460, 476, 478, 479,
480, 482, 492, 521, 529, 531, 539, 540,
541, 591, 592, 593, 594, 600, 601

building, 10, 63, 68, 75, 76, 85, 86, 89, 91,
299
Rockefeller Foundation, 10, 18, 23, 24, 122,
232
rural-urban migration, 175, 211, 412
Russia, 67, 342, 378, 385
Moscow, 50, 67, 453, 454

saltwater intrusion, 66, 262, 263, 320, 321, 326,
328, 329, 333, 350
Scotland
Glasgow, 152
sea level rise, 25, 28, 45, 51, 52, 53, 63, 65, 72,
73, 120, 124, 184, 185, 190, 203, 241, 262,
275, 277, 279, 294, 295, 320, 321, 322,
323, 324, 325, 326, 328, 329, 330, 332,
333, 334, 336, 337, 338, 339, 340, 342,
348, 349, 350, 351, 352, 401, 406, 407,
412, 455, 479, 525, 575, 588
projections, 51, 320, 351
seasonal forecasts, 44, 386
sea-surface temperature (SST), 43
Senegal
Dakar, 40, 51, 188, 233, 389, 505, 595
Serbia
Belgrade, 40
settlements, 53, 67, 70, 77, 108, 148, 159, 175,
183, 188, 192, 201, 273, 277, 296, 321,
329, 332, 340, 345, 402, 403, 404, 405,
406, 408, 412, 416, 417, 418, 420, 421,
423, 430, 445, 481, 495, 542, 544
human, 10, 11, 24, 104, 105, 122, 278, 329
informal, 16, 69, 70, 71, 86, 175, 183, 188,
192, 198, 199, 200, 349, 370, 381, 389,
400, 401, 402, 404, 405, 406, 408, 409,
412, 413, 414, 418, 419, 423, 424, 428,
431, 432, 462, 466, 467, 481, 498, 520,
542, 561, 589
sewage
combined sewer overflow (CSO), 268
systems, 84, 161, 196, 235, 296
treatment, 80, 540, 544, 545
Siberian high-pressure system, 46
Singapore
Singapore, 40, 132, 150, 288, 294, 295, 296,
299, 475, 476, 489, 510
social-ecological systems, 7, 9, 260, 293
socio-economic factors, 182, 322
solar radiation, 153, 159, 162, 237, 459, 523
South Africa
Cape Town, 104, 184, 202, 276, 277, 429,
430, 431, 447, 450, 468, 541, 542, 585,
589, 595
Durban, 14, 18, 22, 53, 101, 110, 111, 121,
123, 133, 136, 137, 189, 190, 193, 199,
450, 543
Johannesburg, 50, 234, 450, 468, 493, 506
South America, 51, 304, 419, 525, 555
South Atlantic Convergence Zone (SACZ), 41
South Korea
Seoul, 46, 47, 53, 54, 104, 233, 446, 510

flooding, 35, 91, 146, 226, 279, 524, 536

forestry, 17, 102, 114, 261, 288

housing, 400, 401, 406

management, 391, 413

planning, 5, 16, 19, 24, 31, 41, 42, 62, 63, 69, 76, 78, 85, 91, 94, 106, 107, 114, 115, 117, 123, 128, 140, 141, 142, 144, 146, 147, 148, 154, 155, 158, 162, 164, 165, 166, 177, 199, 237, 259, 260, 275, 282, 284, 285, 286, 291, 292, 293, 299, 300, 301, 303, 336, 364, 374, 390, 406, 413, 415, 432, 464, 466, 475, 476, 481, 498, 503, 513, 536, 542, 590, 600

policymakers, 12, 211

pollution, 37, 38, 39

resilience, 23, 83, 85, 285, 304, 403, 593

settlements, 402

sprawl, 33, 114, 141, 248, 275, 277, 345, 373

sustainability, 5, 10, 19, 76, 105, 131, 141, 144, 269

transport, 132, 492, 493, 494, 495, 496, 497, 498, 501, 502, 510, 513

water systems, 72, 520, 521, 522, 523, 526, 527, 530, 543, 544, 546

urban climate change governance, 9

Urban Climate Change Research Network (UCCRN), 5, 11, 14, 167

urban heat island effect, 17, 28, 30, 31, 34, 35, 45, 47, 62, 102, 107, 108, 144, 146, 147, 152, 166, 180, 237, 272, 285, 292

urbanization, 5, 6, 7, 10, 12, 18, 19, 20, 24, 25, 29, 31, 32, 33, 37, 45, 47, 48, 62, 65, 67, 70, 71, 72, 76, 88, 112, 115, 141, 145, 146, 154, 167, 186, 196, 232, 258, 259, 260, 274, 277, 283, 286, 289, 290, 300, 301, 320, 321, 322, 337, 349, 379, 390, 401, 402, 403, 406, 408, 432, 444, 445, 461, 467, 472, 477, 482, 492, 493, 501, 502, 529, 554, 555

urban-rural, 5, 19, 287

variability

temperature and rainfall, 43

vegetation cover, 44, 71, 183

vehicle

electric, 464, 465, 467, 469, 505, 514

hybrid, 115, 243

miles traveled (VMT), 132

Venezuela

Caracas, 40

Vietnam

Can Tho, 529, 531, 532, 550

Ho Chi Minh City, 20, 31, 53, 321, 468, 595

vulnerability, 5, 7, 9, 10, 20, 22, 25, 29, 51, 53, 62, 63, 64, 65, 66, 67, 68, 69, 71, 72, 74, 76, 77, 80, 82, 86, 88, 90, 92, 94, 106, 118, 119, 120, 121, 123, 124, 125, 142, 155, 167, 174, 177, 178, 182, 183, 185, 188, 192, 199, 200, 203, 204, 205, 208, 211, 227, 231, 237, 240, 244, 247, 258, 262, 264, 269, 273, 275, 276, 284, 285, 287, 288, 293, 294, 299, 300, 303, 322, 328, 329, 332, 333, 335, 336, 337, 341, 345, 347, 348, 349, 351, 365, 370, 373, 374, 378, 382, 383, 384, 385, 386, 390, 400, 401, 402, 403, 404, 406, 407, 408, 409, 412, 413, 414, 418, 421, 423, 432, 444, 446, 447, 449, 455, 460, 480, 481, 482, 495, 524, 528, 536, 539, 540, 588, 589, 590, 594, 596, 597

assessments, 74, 121, 125, 231, 240, 300, 455, 597

waste heat, 28, 140, 141, 144, 145, 147, 148, 167, 459, 470

waste management, 11, 102, 106, 108, 114, 122, 123, 124, 129, 131, 175, 178, 197, 204, 364, 371, 389, 405, 467, 471, 554, 556, 558, 559, 561, 562, 563, 565, 566, 567, 568, 569, 570, 575, 577, 578, 579

gas-to-energy, 554, 570

landfills, 108, 471, 481, 509, 510, 554, 557, 558, 559, 562, 563, 570, 572, 577

solid, 198, 344, 412, 413, 466, 471, 554, 556, 563, 565, 567, 575, 577, 596

water

conflicts, 470

demand, 191, 192, 408, 522, 524, 534

ground, 51, 71, 114, 153, 184, 279, 282, 322, 328, 329, 338, 345, 350, 402, 407, 520, 523, 524, 525, 531, 533, 534, 535, 537

policy, 382

recycled, 229, 296, 534

resource management, 277, 537, 541

security, 22, 520, 521, 522, 527, 528, 544, 545, 546, 594

shortages, 5, 23, 114, 183, 184, 277, 279, 375

storm, 143, 148, 241, 266, 268, 297, 323, 333, 521, 522, 523, 525, 526, 530

usage management, 336

watersheds, 71, 279, 282, 521, 522, 537

West Antarctic ice sheet, 28

wetlands, 71, 72, 73, 77, 92, 110, 114, 162, 259, 260, 262, 263, 264, 266, 268, 272, 278, 283, 285, 288, 294, 306, 320, 322, 328, 329, 332, 336, 339, 340, 341, 344, 347, 350, 351, 352, 389, 536

wind farms, 446, 474

World Bank, 32, 36, 41, 43, 46, 72, 79, 83, 86, 111, 112, 115, 118, 122, 126, 130, 152, 154, 157, 159, 177, 184, 186, 190, 194, 200, 203, 208, 227, 230, 231, 232, 233, 234, 235, 236, 238, 241, 246, 248, 249, 262, 266, 272, 276, 282, 283, 286, 292, 294, 297, 299, 323, 327, 333, 335, 337, 342, 345, 348, 351, 370, 373, 374, 376, 381, 402, 403, 407, 412, 413, 414, 416, 417, 419, 429, 445, 452, 453, 461, 466, 473, 475, 477, 480, 499, 504, 506, 507, 508, 510, 521, 529, 533, 537, 539, 542, 555, 556, 558, 559, 561, 562, 565, 566, 578, 579, 591, 599

World Health Organization (WHO), 366, 369, 428, 505

World Resources Institute (WRI), 229, 591

zoning, 25, 71, 77, 78, 105, 107, 120, 146, 163, 186, 188, 190, 193, 261, 297, 401, 498, 510

Printed in the United States
By Bookmasters